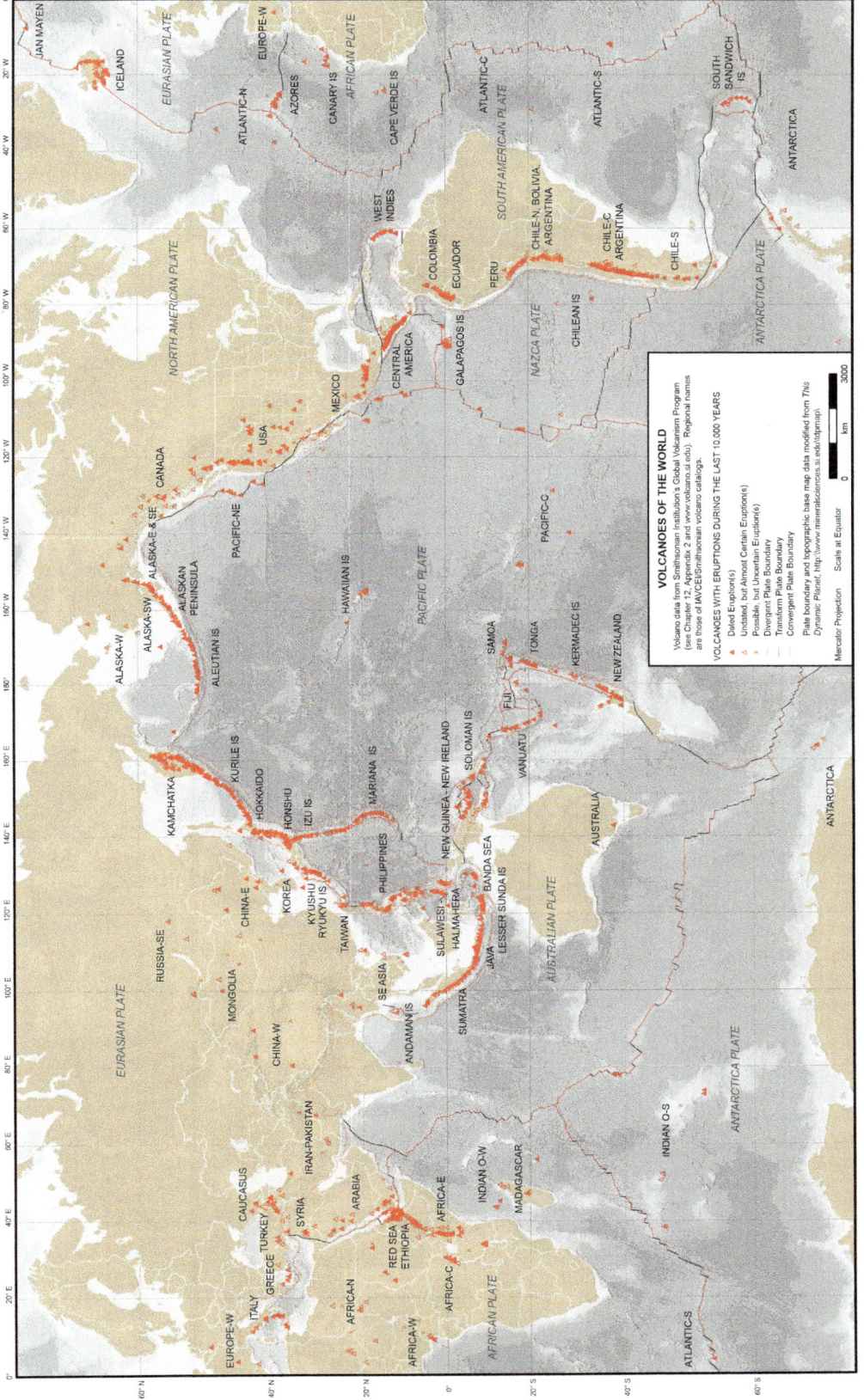

VOLCANOES OF THE WORLD

Volcano data from Smithsonian Institution's Global Volcanism Program
(see Chapter 12, Appendix 2 and www.volcano.si.edu). Regional names
are those of IAVCEI/Smithsonian volcano catalogs.

VOLCANOES WITH ERUPTIONS DURING THE LAST 10,000 YEARS

▲ Dated Eruption(s)
▲ Undated, but Almost Certain Eruption(s)
▲ Possible, but Uncertain Eruption(s)
—— Divergent Plate Boundary
—— Transform Plate Boundary
—— Convergent Plate Boundary

Plate boundary and topographic base map data modified from *This
Dynamic Planet*, http://www.mineralsciences.si.edu/tdpmap/

Mercator Projection   Scale at Equator

0   km   3000

The Encyclopedia of Volcanoes

The Encyclopedia of Volcanoes

Second Edition

Editor-in-Chief

Haraldur Sigurdsson

Graduate School of Oceanography, University of Rhode Island,
Narragansett, RI, USA

Associate Editors

Bruce Houghton

National Disaster Preparedness Training Center, University of Hawai'i,
Honolulu, HI, USA

Stephen R. McNutt

School of Geosciences, University of South Florida, Tampa, FL, USA

Hazel Rymer

Faculty of Science, The Open University, Walton Hall, Milton Keynes, UK

John Stix

Department of Earth & Planetary Sciences, McGill University,
Montreal, Quebec, Canada

ELSEVIER

AMSTERDAM • BOSTON • HEIDELBERG • LONDON • NEW YORK • OXFORD • PARIS
SAN DIEGO • SAN FRANCISCO • SINGAPORE • SYDNEY • TOKYO

Academic Press is an imprint of Elsevier

Academic Press is an imprint of Elsevier
32 Jamestown Road, London NW1 7BY, UK
525 B Street, Suite 1800, San Diego, CA 92101-4495, USA
225 Wyman Street, Waltham, MA 02451, USA
The Boulevard, Langford Lane, Kidlington, Oxford OX5 1GB, UK

First edition 1999

Second edition 2015

British Library Cataloguing in Publication Data
A catalogue record for this book is available from the British Library

Library of Congress Cataloging-in-Publication Data
A catalog record for this book is available from the Library of Congress

Library of Congress Control Number: 20159330593

ISBN: 978-0-12-385938-9

For Information on all Academic Press publications
visit our website at www.store.elsevier.com/

Printed and bound in the USA

Working together
to grow libraries in
developing countries

www.elsevier.com • www.bookaid.org

Publisher: John Fedor
Acquisitions Editor: Louisa Hutchins
Project Manager: Paul Prasad Chandramohan
Cover Designer: Maria Inês Cruz.

Contents

Valerio Acocella Dipartimento di Scienze, Università Roma Tre, Roma, Italy

Graham D.M. Andrews California State University Bakersfield, Bakersfield, USA

Benjamin Andrews Global Volcanism Program, U.S. National Museum of Natural History, Smithsonian Institution, Washington, DC, USA

Silvio De Angelis University of Liverpool, Liverpool, Merseyside, UK

Stefán Arnórsson Institute of Earth Sciences, University of Iceland, Reykjavík, Iceland

Willy Aspinall Aspinall & Associates, Tisbury, and Bristol University, Bristol, England, UK

Jayne C. Aubele Education Department, New Mexico Museum of Natural History and Science, Albuquerque, NM, USA

Jenni Barclay School of Environmental Sciences, University of East Anglia, Norwich, UK

Peter J. Baxter Institute of Public Health, University of Cambridge, Cambridge, UK

Mark Bebbington Statistics Group and Volcanic Risk Solutions, Massey University, Palmerston North, New Zealand

Alexander Belousov Institute of Volcanology and Seismology, Petropavlovsk-Kamchatsky, Russia

Alain Bernard Earth & Environmental Sciences Department, Université Libre de Bruxelles, Brussels, Belgium

Marc Bernstein Department of Geology, University at Buffalo, Buffalo, NY, USA

Jacob Elvin Bleacher Sciences and Exploration Directorate, Code 600, NASA Goddard Space Flight Center, Greenbelt, MD, USA

Russell Blong Aon Benfield Asia-Pacific, Sydney, and Risk Frontiers, Macquarie University, Sydney, Australia

Costanza Bonadonna Department of Earth Sciences, University of Geneva, Geneva, Switzerland

Michael Branney Department of Geology, University of Leicester, Leicester, England, UK

Richard J. Brown Department of Earth Sciences, Durham University, UK

Brandon Browne Department of Geology, Humboldt State University, Arcata, CA, USA

Alain Burgisser CNRS, ISTerre, Le Bourget du Lac, France; Université de Savoie, ISTerre, Le Bourget du Lac, France

Marcus Bursik Department of Geology, University at Buffalo, State University of New York, Buffalo, NY, USA

Ralf Büttner Universität Würzburg, Würzburg, Germany

Eliza S. Calder School of Geosciences, University of Edinburgh, Edinburgh, UK

Steven Carey Graduate School of Oceanography, University of Rhode Island, Kingston, RI, USA

Rebecca J. Carey University of Tasmania, Hobart, TAS, Australia

Simon A. Carn Department of Geological and Mining Engineering and Sciences, Michigan Technological University, Houghton, MI, USA

Ray Cas Department of Geosciences, Monash University, Clayton, Victoria, Australia; School of Physical Sciences, University of Tasmania, Hobart, Tasmania, Australia

Katharine V. Cashman School of Earth Sciences, University of Bristol, Wills Memorial Building, Queens Road, Bristol, UK

Giovanni Chiodini INGV Observatorio Vesuviano, Via Diocleziano, Italy

Raffaello Cioni Dipartimento di Scienze della Terra, via G. La Pira, Firenze, Italy

Amanda Bachtell Clarke School of Earth and Space Exploration, Arizona State University, Tempe, AZ, USA

Bruce D. Clarkson Environmental Research Institute, University of Waikato, Hamilton, New Zealand

Millard F. Coffin Institute for Marine and Antarctic Studies, Hobart, TAS, Australia

Paul D. Cole School of Geography, Earth and Environmental Science, Plymouth University, Plymouth, UK

Chuck Connor School of Geosciences, University of South Florida, Tampa, FL, USA

Charles B. Connor School of Geosciences, University of South Florida, Tampa, FL, USA

Jean-Thomas Cornelis Earth & Life Institute, Environmental Sciences, Université Catholique de Louvain, Louvain-la-Neuve, Belgium

Antonio Costa Istituto Nazionale di Geofisica e Vulcanologia, Bologna, Italy

Elizabeth Cottrell Global Volcanism Program, U.S. National Museum of Natural History, Smithsonian Institution, Washington, DC, USA

Charles M. Crisafulli USDA Forest Service, Pacific Northwest Research Station, Olympia Forestry Sciences Laboratory, Mount St Helens National Volcanic Monument, Amboy, WA, USA

David A. Crown Planetary Science Institute, Tucson, AZ, USA

Larry S. Crumpler Collections and Research Department, New Mexico Museum of Natural History and Science, Albuquerque, NM, USA

Martha J. Daines Department of Earth Sciences, University of Minnesota, Minneapolis, MA, USA

Tim Davies Department of Geological Sciences, University of Canterbury, Christchurch, New Zealand

Simon J. Day Department of Earth Sciences, Institute of Risk and Disaster Reduction, University College London, London, UK

Wim Degruyter School of Earth & Atmospheric Sciences, Georgia Institute of Technology, Atlanta, GA, USA

Jonathan Dehn Geophysical Institute, University of Alaska Fairbanks, Fairbanks, AK, USA

Servando de la Cruz Centro Nacional de Prevencion de Desastres, SEGOB, Mexico, Instituto de Geofisica, UNAM, Mexico, Instituto de Ingeniera, UNAM, Mexico

Natalia Irma Deligne GNS Science, Avalon, Lower Hutt, New Zealand

Pierfrancesco Dellino Università di Bari, Bari, Italy

Pierre Delmelle Earth & Life Institute, Environmental Sciences, Université Catholique de Louvain, Louvain-la-Neuve, Belgium

Cornel E.J. de Ronde Department of Marine Sciences, GNS Science, Lower Hutt, New Zealand

Shan de Silva College of Earth, Ocean and Atmospheric Science, Oregon State University, Corvallis, OR, USA

Josef Dufek School of Earth and Atmospheric Sciences, Georgia Institute of Technology, Atlanta, GA, USA

Marie Edmonds Earth Sciences Department, University of Cambridge, Cambridge, UK

Benjamin R. Edwards Department of Earth Sciences, Dickinson College, Carlisle, Pennsylvania, USA

Patricia Erfurt-Cooper School of Business, James Cook University, Cairns, QLD, Australia

Tomaso Esposti Ongaro Istituto Nazionale di Geofisica e Vulcanologia, Sezione di Pisa, Pisa, Italy

John W. Ewert Cascades Volcano Observatory, US Geological Survey, Vancouver, WA, USA

David Fee University of Alaska Fairbanks, Fairbanks, AK, USA

Tobias P. Fischer Department of Earth and Planetary Sciences, University of New Mexico, Albuquerque, NM, USA

Arnau Folch Barcelona Supercomputing Center, Barcelona, Spain

Jeffrey T. Freymueller University of Alaska Fairbanks, Fairbanks, AK, USA

William Brent Garry Sciences and Exploration Directorate, Code 600, NASA Goddard Space Flight Center, Greenbelt, MD, USA

Paul Geissler US Geological Survey, Center for Astrogeology, Flagstaff, AZ, USA

Mark S. Ghiorso OFM Research, Seattle, WA, USA

Fraser Goff Earth and Environmental Science, New Mexico Institute of Mining and Technology, Socorro, NM, USA

Cathy J. Goff Geothermal Consultant, Los Alamos, NM, USA

Helge Gonnermann Department of Earth Science, Rice University, Houston, Texas, USA

Chris E. Gregg Department of Geosciences, East Tennessee State University, Johnson City, TN, USA

Timothy L. Grove Department of Earth, Atmospheric and Planetary Sciences, Massachusetts Institute of Technology, Cambridge, MA, USA

Guilherme A.R. Gualda Earth and Environmental Sciences, Vanderbilt University, Nashville, TN, USA

Magnús T. Gudmundsson Institute of Earth Sciences, University of Iceland, Reykjavík, Iceland

Jonathan J. Halvorson USDA Agricultural Research Station, Northern Great Plains Research Laboratory Mandan, ND, USA

Andrew J.L. Harris Laboratoire Magmas et Volcans, Université Blaise Pascal, Clermont Ferrand, France

Erik H. Hauri Carnegie Institution of Washington, Washington, DC, USA

Katharine Haynes Risk Frontiers, Department of Environment and Geography, Macquarie University, Sydney, NSW, Australia

James W. Head, III Department of Earth, Environmental and Planetary Sciences, Brown University, Providence, RI, USA

Richard W. Henley Research School of Earth Sciences, Australian National University, Canberra, ACT, Australia

Claire J. Horwell Institute of Hazard, Risk and Resilience, Department of Earth Sciences, Durham University, Durham, UK

Bruce Houghton Department of Geology and Geophysics, National Disaster Preparedness Training Center, University of Hawai'i, Honolulu, HI, USA

C. Ian Schipper School of Geography, Environment and Earth Sciences, Victoria University of Wellington, Wellington, New Zealand

Mikhail A. Ivanov Laboratory of Comparative Planetology, Vernadsky Institute, Russian Academy of Sciences, Moscow, Russia; Department of Earth, Environmental and Planetary Sciences, Brown University, Providence, RI, USA

Richard M. Iverson US Geological Survey, Cascades Volcano Observatory, Vancouver, WA, USA

Michael R. James Lancaster Environment Centre, Lancaster University, Lancaster, UK

Jeffrey Johnson Boise State University, Boise, ID, USA

David Johnston Joint Centre for Disaster Research, GNS Science/Massey University, Lower Hutt, New Zealand

Gill Jolly GNS Science, Wairakei Research Centre, New Zealand

Kazuhiko Kano The Kagoshima University Museum, Kagoshima University, Korimoto 1-chome, Kagoshima, Japan

Jackie E. Kendrick School of Earth, Ocean and Ecological Sciences, University of Liverpool, Liverpool, Merseyside, UK

Christopher R.J. Kilburn Department of Earth Sciences, University College London, London, UK

Anthony A.P. Koppers College of Earth, Ocean and Atmospheric Sciences, Oregon State University, Corvallis, OR, USA

Takehiro Koyaguchi Earthquake Research Institute, University of Tokyo, Tokyo, Japan

Peter C. LaFemina Department of Geosciences, The Pennsylvania State University, University Park, USA

Yan Lavallée School of Earth, Ocean and Ecological Sciences, University of Liverpool, Liverpool, Merseyside, UK

Charles E. Lesher Department of Earth and Planetary Sciences, University of California, Davis, CA, USA, Department of Geoscience, Aarhus University, Aarhus, Denmark

Jan M. Lindsay School of Environment, The University of Auckland, Auckland, New Zealand

Corinne A. Locke University of Auckland, New Zealand

Rosaly M.C. Lopes Earth and Space Sciences Division, Jet Propulsion Laboratory, California Institute of Technology, Pasadena, CA, USA

Bruce D. Marsh Department of Earth & Planetary Sciences, Johns Hopkins University, Baltimore, MD, USA

Warner Marzocchi INGV—Istituto Nazionale di Geofisica e Vulcanologia, Italy

Elena Maters Earth & Life Institute, Environmental Sciences, Université Catholique de Louvain, Louvain-la-Neuve, Belgium

Stephen R. McNutt School of Geosciences, University of South Florida, Tampa, FL, USA

Jocelyn McPhie School of Physical Sciences and CODES, University of Tasmania, Hobart, Tasmania, Australia

John B. Murray Faculty of Science, The Open University, Walton Hall, Milton Keynes, UK

Augusto Neri Istituto Nazionale di Geofisica e Vulcanologia, Sesione Di Pisa, Pisa, Italy

Sophie Opfergelt Earth & Life Institute, Environmental Sciences, Université Catholique de Louvain, Louvain-la-Neuve, Belgium

Clive Oppenheimer Department of Geography, University of Cambridge, Cambridge, United Kingdom; Institut des Sciences de la Terre d'Orléans, Universite d'Orléans, Orléans Cedex, France

John Pallister Volcano Disaster Assistance Program, U.S. Geological Survey and U.S. Agency for International Development, USA

Matej Pec Department of Earth Sciences, University of Minnesota, Minneapolis, MA, USA

Chien-Lu Ping School of Natural Resources and Agricultural Sciences, University of Alaska Fairbanks, Fairbanks, AK, USA

Marco Pistolesi Dipartimento di Scienze della Terra, via G. La Pira, Firenze, Italy

Terry Plank Lamont-Doherty Earth Observatory of Columbia University, Palisades, NY, USA

Fred Prata Nicarnica Aviation AS, Kjeller, Norway

David M. Pyle Department of Earth Sciences, University of Oxford, Oxford, UK

Michael R. Rampino Department of Biology, New York University, New York, NY, USA; Department of Environmental Studies, New York University, New York, NY, USA

Alan Robock Department of Environmental Sciences, Rutgers University, New Brunswick, NJ, USA

Olivier Roche Laboratoire Magmas et Volcans, Université Blaise Pascal-CNRS-IRD, OPGC, Clermont-Ferrand, France

Nick Rogers Department of Environment, Earth and Ecosystems, The Open University, Milton Keynes, UK

Diana C. Roman Department of Terrestrial Magnetism, Carnegie Institution of Washington, DC, USA

Bill Rose Michigan Technological University, Houghton, MI, USA

Mauro Rosi Dipartimento Protezione Civile, via Vitorchiano, Roma, Italy

Scott K. Rowland Department of Geology & Geophysics, School of Ocean and Earth Sciences and Technology, University of Hawai'i at Mānoa, Honolulu, HI, USA

James K. Russell Earth, Ocean and Atmospheric Sciences, University of British Columbia, Vancouver, British Columbia

Hazel Rymer Faculty of Science, The Open University, Walton Hall, Milton Keynes, UK

Bettina Scheu Department of Earth and Environmental Sciences, Ludwig-Maximilians-Universität München, Munich, Germany

Stephen Self Department of Environment, Earth, and Ecosystems, The Open University, Milton Keynes, MK, UK; Department of Earth and Planetary Science, University of California, Berkeley, CA, USA

Payson Sheets Department of Anthropology, University of Colorado, Boulder, CO, USA

Lee Siebert Global Volcanism Program, U.S. National Museum of Natural History, Smithsonian Institution, Washington, DC, USA

Haraldur Sigurdsson Graduate School of Oceanography, University of Rhode Island, Narragansett, RI, USA

S. Adam Soule Department of Geology and Geophysics, Woods Hole Oceanographic Institution, Woods Hole, MA, USA

Frank J. Spera Department of Earth Science, University of California, Santa Barbara, CA, USA

Paul D. Spudis Lunar and Planetary Institute, TX, USA

Hubert Staudigel Scripps Institution of Oceanography, UCSD, La Jolla, CA, USA

Andri Stefánsson Institute of Earth Sciences, University of Iceland, Reykjavík, Iceland

James Stimac Stimac Geothermal Consulting, Santa Rosa, CA, USA

Valerie K. Stucker Department of Marine Sciences, GNS Science, Lower Hutt, New Zealand

Frederick J. Swanson USDA Forest Service, Pacific Northwest Research Station, Corvallis Forestry Sciences Laboratory, Corvallis, OR, USA

Lindsay Szramek Department of Geosciences, Austin Peay State University, Clarksville, TN, USA

Jacopo Taddeucci Istituto Nazionale di Geofisica e Vulcanologia, Rome, Italy

Benoit Taisne Earth Observatory of Singapore, Nanyang Technological University, Singapore

Ronald J. Thomas Department of Electrical Engineering, New Mexico Institute of Mining and Technology, Socorro, NM, USA

Glenn Thompson School of Geosciences, University of South Florida, Tampa, FL, USA

Sverrir Thórhallsson Iceland GeoSurvey, Reykjavík, Iceland

Christy B. Till Volcano Science Center, U.S. Geological Survey, Menlo Park, CA, USA, Now at School of Earth and Space Exploration, Arizona State University, Tempe, AZ, USA

Greg A. Valentine Department of Geology, University at Buffalo, Buffalo, NY, USA

James W. Vallance US Geological Survey, Cascades Volcano Observatory, Vancouver, WA, USA

Alexa R. Van Eaton U.S. Geological Survey, VA, USA

Benjamin van Wyk de Vries Laboratoire Magmas et Volcans, Université Blaise Pascal-CNRS-IRD, OPGC, Clermont Ferrand, France

Edward Venzke Global Volcanism Program, U.S. National Museum of Natural History, Smithsonian Institution, Washington, DC, USA

Sylvie Vergniolle Institut de Physique du Globe, Sorbonne Paris Cité, Université Paris Diderot, UMR CNRS, Paris Cedex 05, France

Paul J. Wallace University of Oregon, Eugene, OR, USA

James D.L. White Geology Department, University of Otago, Dunedin, New Zealand

Glyn Williams-Jones Department of Earth Sciences, Simon Fraser University, Burnaby, BC, Canada

David A. Williams School of Earth and Space Exploration, Arizona State University, Tempe, AZ, USA

Lionel Wilson Lancaster Environment Centre, Lancaster University, Lancaster, UK

Kenneth H. Wohletz Los Alamos National Laboratory, Los Alamos, NM, USA

John A. Wolff School of Earth and Environmental Sciences, Washington State University, Pullman, WA, USA

Bernd Zimanowski Universität Würzburg, Würzburg, Germany

James R. Zimbelman Center for Earth and Planetary Studies, National Air and Space Museum, Smithsonian Institution, Washington, DC, USA

Volcanoes are compelling evidence that the Earth is a dynamic planet characterized by endless change and renewal. Study of volcanoes is thus fundamental to understanding the evolution of the Earth and the surface environments. On an ever more crowded planet the exposure to natural hazards is increasing. An estimated 800 million people live within 100 km of an active volcano in 86 countries worldwide. Volcanoes provide favorable environments for life and bring many benefits to society: eruptions fertilize soils; elevated topography provides good sites for infrastructure (e.g., telecommunications on elevated ground); water resources are commonly plentiful; volcano tourism can be lucrative; and volcanoes can acquire spiritual, aesthetic, or religious significance. Some volcanoes are also associated with geothermal resources, making them a target for exploration and a potential energy resource.

Over the last few decades there has been a dramatic increase in research on active volcanoes and their associated magmatic systems. There are emergent technologies leading to new kinds of information, while advances in computer power lead to much more data and diverse forms of data. The proliferation of scientific information poses an increasing problem on how to access and absorb this information and the latest scientific understanding. There is a critical need to provide mechanisms to enable rapid access to the latest volcano science. The Encyclopedia of Volcanoes provides a solution to this issue by providing articles to cover the complete range of topics on volcanoes. The first edition of the Encyclopedia in 2000 was a seminal publication in volcanology, which became very widely used. It is thus a great pleasure to welcome the updated second addition of the Encyclopedia.

The new edition contains 78 articles covering a great diversity of topics. The authors are of the highest quality and represent many of the leading volcanologists around the world. Some subjects have been updated while there are articles on new topics that reflect emergent issues and new technologies. The Encyclopedia provides an authoritative gateway to knowledge. I congratulate the editors and authors on a great achievement in disseminating the spectacular findings of volcanology in the twenty-first century.

Stephen R.S.J. Sparks
University of Bristol, United Kingdom

This second edition of the Encyclopedia of Volcanoes is a complete reference guide, providing a comprehensive view of volcanism on the Earth and on the other planets of the Solar System that have exhibited volcanic activity. It is the first attempt to gather in one place such a vast store of knowledge on volcanic phenomena. The volume addresses all aspects of volcanism, ranging from the generation of magma, its transport and migration, eruption, and formation of volcanic deposits. It also addresses volcanic hazards, their mitigation, the monitoring of volcanic activity, and economic aspects and, for the first time, analyzes several specific cultural aspects of volcanic activity, including the impact of volcanic activity on archaeology, literature, art, and film. To compose a single volume that is a complete reference for such a far—ranging phenomenon is indeed a daunting task.

ARRANGEMENT OF CONTENT

When the editors developed the fundamental structure of this Encyclopedia, we decided on a thematic approach, where the multitude of volcanic processes are defined, described, and elaborated in a series of chapters. To provide our readers with the best possible treatment of volcanic processes, we have adopted the time-honored custom of Denis Diderot and Jean D'Alembert of recruiting the leading experts in each branch of volcanology and related fields to author the chapters that lie outside the expertise of the editors. When Diderot and D'Alembert began to compile information for their monumental 21-volume Encyclopédie (1751—1765), they went to the carpenters, the masons, the embroiderers, and the other experts for help with the specific terms and concepts, in order to get all the technical details right. Thus they secured an article for the Encyclopédie, penned by the pioneer field volcanologist Nicolas Desmarest (1725—1815), on the volcanic origin of columnar basalt. Similarly, the editors of the Encyclopedia of Volcanoes have recruited the recognized authorities in each field or speciality of volcanology to contribute chapters to this volume. Thus this volume is the product of over 100 volcanologists, petrologists, and other scientists who have specialized knowledge about volcanoes and related processes.

SECTIONAL PLAN

In this volume, the principal aspects of volcanic activity are dealt with in 80 chapters, divided between nine major parts. An introduction to each part written by the editors is provided to give a general perspective of the topics and processes contained in that part. Each chapter covers one fundamental volcanic process in depth and stands alone as a comprehensive overview of the phenomenon in question. Nevertheless, the sequence of major parts and their chapters is intended to be directional and evolutionary, as far as possible, following magma from its place of deep origin to the surface of the Earth. Thus in the first part, the volume commences with 11 chapters on the principal source regions of magmas, the mantle of the Earth, melting processes, magma generation, and transport and geochemical evolution of magmas, as they rise into the Earth's crust. The second part addresses the fundamental aspects of distribution of volcanism in space and time and the range in the scale of volcanic eruptions, in terms of magnitude and intensity. In the three subsequent parts, chapters address the various styles of eruption of magmas, namely, effusive (lavas) versus explosive, and the multitude of types of volcanic deposits that result.

Part V provides comprehensive overviews of volcanic activity on other members of the Solar System, namely, on the Moon, Io, Mars, and Venus and the unusual cryovolcanism on the Galilean satellites of Jupiter. Extraterrestrial volcanology is probably the area of volcanologic research that will experience the greatest growth in the future, as space exploration continues to provide new discoveries in other worlds. In Part VI the interactions of volcanism with hydrosphere and atmosphere are discussed, as well as formation of those mineral deposits that owe their origin to volcanic processes. The impact of volcanoes on society is explored in Part VII, where the various types of volcanic hazards are discussed. Volcano monitoring is treated in Part VIII, along with the mitigating measures that have been developed to counteract volcanic disasters. In the final section of the Encyclopedia, a variety of economic benefits accruing from volcanism are described, and the volume closes with four chapters on the cultural aspects of volcanism and its impact on archaeology, art, literature, and film.

CONVENTIONS OF STYLE

In this volume, we have adopted the convention of using lowercase in the spelling of those adjectives that describe volcanic processes, such as surtseyan and plinian. As this volume is intended for the general reader, such distracting paraphernalia as references or footnotes have been dispensed with as much as possible. Instead, a section of Further Readings, containing a list of half a dozen or more major texts on the subject matter, is provided at the end of each chapter for those interested in the chief sources.

APPENDIXES

Two appendixes are included in the volume. Appendix 1 consists of a series of tables that give common scientific and mathematical units and conversion factors. It also provides a variety of numerical data on the Earth. Appendix 2 is a table of active volcanoes on Earth, as compiled by Lee Siebert of the Smithsonian Institution. A listing containing all the volcanoes on Earth, with data on their eruptions, is beyond the feasibility of this work, as this list alone would equal this volume in length. The editors have therefore opted to list only historically active volcanoes, i.e., the volcanoes whose eruptions have been witnessed and documented, totaling 550 in number. The oldest eruption in this list is therefore the plinian eruption of Vesuvius in Italy on August 24 in AD 79. This selection of the historical period is admittedly somewhat arbitrary, as the "historical period" varies greatly in length from place to place on Earth. In Europe and Japan the historical record of volcanism extends back over 1000 years, whereas in New Zealand and North and South America, for example, it is merely a matter of a few centuries, because written records in these regions begin generally with European settlement.

In Appendix 2 the historically active volcanoes are listed in alphabetical order, and their locations are given in terms of longitude and latitude. The tabulation also includes all of their historic eruptions.

Supplementary Videos and Text files are available in the companion site, http://booksite.elsevier.com/9780123859389/

ACKNOWLEDGMENTS

The number of people who have provided reviews of chapters and assisted the authors and editors in the creation of this volume is literally in the hundreds. A list of those we especially thank is given in the Acknowledgments section at the end of the volume. We single out here, however, the following staff members at Elsevier for their outstanding work: the project's Publisher, John Fedor and the Acquisitions Editor, Louisa Hutchins.

The Editors

The following people have generously provided their time, often at short notice, to review the various chapters of the second edition of the Encyclopedia of Volcanoes and in other ways assisted the editors in making this volume scientifically accurate and complete. The editors gratefully acknowledge their assistance:

Charles Beard
Sonja Behnke
John Benoit
Costanza Bonadonna
Brittany Brand
Jochen Braunmiller
Richard Brown
Alain Burgisser
Rebecca Carey
Steve Carey
Corrado Cimarelli
Raffaello Cioni
Lynn Cole
Jim Cole
Charles Connor
Antonio Costa
Jason Coumans
Martha J. Daines
Joe Dufek
Alexa Van Eaton
Julia Eychenne
Tomaso Esposti Ongaro
Alexandra Farrell
Toby Fischer
Arnau Folch
Cynthia Gardner
Paul Geissler
Dennis Geist
Ophelia George
Guido Giordano
Fraser Goff
Helge Gonnermann
Tim Grove
Leanne Gunn
Jim Head
Grant Heiken
Maria Janebo
Simon Kelley

Peter Kokelaar
Kristine Kosinski
Sarah Kruse
Ryan Libbey
Marc-Antoine Longpré
Rosaly Lopes
Gregor Lucic
Rocco Malservisi
Vernon Manville
Bruce D. Marsh
Heather McFarlin
Maarten de Moor
John Murray
Jean-Luc Le Pennec
Laura Pioli
Marco Pistolesi
Mel Rodgers
Lois Salem
Ian Schipper
Simona Schollo
Steve Self
Cassandra Smith
John Stevenson
Shinji Takareda
Helen Thomas
Glenn Thompson
Thor Thordarson
Graham Tobin
Greg Valentine
James Vallence
Nicholas Voss
Benjamin Van Wyk de Vries
Geoff Wadge
Paul J. Wallace
Samantha Weaver
Peter Webley
James R. Zimbelman
Jeffrey Zurek

Introduction

Haraldur Sigurdsson

Graduate School of Oceanography, University of Rhode Island, Narragansett, RI, USA

Chapter Outline

GLOSSARY

convection The process by which the Earth's mantle loses heat, producing currents of solid but deformable rock that rise toward the surface and contribute to plate motion.

eclogite A rock type common in the Earth's mantle, with the same chemical composition as basalt, but with a mineral assemblage that is stable at high pressure.

intensity The rate of flow of magma out of a volcano during eruption, expressed as mass eruption rate (MER) in kilograms per second (kg/s).

magnitude The size of a volcanic eruption, expressed as the volume of material erupted, usually in cubic kilometers (km^3).

peridotite The principal rock that forms the Earth's upper mantle, consisting mainly of the mineral olivine, with lesser amounts of pyroxene, garnet, and/or spinel. It is the source of basaltic magmas, which are formed from peridotite by partial melting.

photodissociation Chemical reactions that occur in the atmosphere, due to ultraviolet radiation.

proterozoic eon The period in Earth's history that began 2.5 billion years ago and ended 0.57 billion years ago.

pyroclasts Fragmentary material ejected during a volcanic eruption, including pumice, ash, and rock fragments.

radionuclides Chemical elements that undergo spontaneous and time-dependent decay, resulting in the formation of other elements and the release of radioactive energy in the form of heat.

stromatolites The earliest calcium carbonate-secreting organisms on Earth, first appearing during the Proterozoic. Their activity contributed to drawing down carbon dioxide from the atmosphere and fixing it in limestone and other sedimentary rocks, contributing to changing atmospheric chemistry and climate.

subduction The process of sinking of crustal plates into the Earth's mantle. Subduction causes magma generation and results in buildup of island arcs above subduction zones.

volatiles Chemical compounds or elements contained in magmas that are generally released as gases to the atmosphere during a volcanic eruption. They include water, carbon dioxide, and sulfur dioxide. They are generally dissolved in the magma prior to eruption or when the magma is under high pressure, but exsolve during ascent of the magma to the surface.

volcanic aerosol Tiny particles of sulfuric acid droplets, formed by reactions between volcanic sulfur dioxide gas and water vapor in the stratosphere. The resulting aerosol dust veil has important effects on backscattering and absorption of solar radiation and leads to a net surface cooling on the Earth, with possible climate change.

Fifteen years have passed since the publication of the first edition of the Encyclopedia of Volcanoes. In the meantime the field of volcanology has taken enormous strides forward, both in terms of the understanding of generation of magmas and volcanic processes and in applied volcanology. The editors and publisher therefore felt that a second edition was well overdue. The new edition is similar in structure and scope, but the great majority of chapters have been rewritten, with many contributions from a new generation of volcanologists those have now emerged as leaders in their field.

The field of volcanology has benefitted enormously in the past decade from the application of more quantitative methods. This trend was already evident in the study of the geochemistry and petrology of volcanic rocks, as well as in geophysical studies of volcanic processes, such as in seismology and geodesy, but during the past decade the quantitative approach has made possible great advances in physical volcanology, through accurate determination of various properties of volcanic materials and their deposits, their spatial distribution, and numerical treatment of such data to infer and reconstruct volcanic eruption dynamics. Measuring techniques have greatly improved in general, especially in the field of geochemistry. This has resulted for example in much improved knowledge of the concentration of dissolved **volatiles** in magmas. They, including water, sulfur, and carbon dioxide, are not only important as driving forces in explosive eruptions, but can have considerable impact on atmospheric chemistry and environmental quality.

Advances have also been made in our knowledge of volcanic history on Earth. We have, for example, known for a number of years that one of the largest explosive eruptions in the last millennium occurred in 1258, judging from the huge amount of sulfur found in both Greenland and Antarctica ice cores from this time. The source of this great eruption has now finally been identified as Mount Rinjani or Samalas volcano on the island of Lombok in Indonesia.

It is often the case that a major eruption results in advances in the field of volcanology. In the current decade on the other hand, it was a relatively minor eruption that brought volcanology to the forefront in aviation. When the explosive eruption of Eyjafjallajökull volcano in Iceland began in April 2010, it led to the closure of much of European air space and air traffic over the North Atlantic ocean for one week. The figures resulting from this closure are truly staggering: 313 airports closed, 104,000 flights canceled, 10 million passengers stranded at airports in Europe and North America, loss of income by air carriers of $1.7 billion and loss of income by airports estimated in excess of $317 million. It has been estimated that the total economic loss because of the air traffic closures is of the order of $5 billion. The aviation industry is fully aware of the potential danger to aircraft from volcanic ash, but was the closure of air space during the Eyjafjallajökull fully justified? It was a small explosive eruption. The total volume of magma that was erupted is in the range of 50 til 60 million cubic meters or about 0.05 km^3 of magma. As a result of this controversial decision, the study of volcanic ash dispersal and the impact of ash on aircraft have received much renewed interest.

1. THE MELT OF THE EARTH

Volcanic eruptions are the most awesome and powerful display of nature's force. The idea that terra firma may explode under our feet and bombard us with glowing hot ejecta seems almost incomprehensible. Every year about 50 volcanoes throughout the world are active above sea level,

threatening the lives and property of millions of people. A single eruption can claim thousands of lives in an instant. For example, in the 1902 eruption of Mount Pelée on the Caribbean island of Martinique, a flow of hot ash and gases overwhelmed the city of St Pierre, killing all but one of its 28,000 inhabitants. More recently, a mudflow triggered by the 1985 eruption of the volcano Nevado del Ruiz in Colombia killed nearly all of the 25,000 inhabitants of the town of Armero.

The relationship between people and volcanoes is as old as the human race. Our earliest ancestors evolved in the volcanic region of the East African Rift, where their activities and remains are preserved by volcanic deposits, such as the stunning 3.7-million-year-old *Australopithecus* hominid footprints crisscrossing a volcanic ash deposit at Laetoli. In fact, the wealth of information we now have on early hominid evolution has only been made possible because of the rapid burial and excellent preservation of their remains in volcanic deposits. Is it a mere sport of nature that humans evolved in a volcanic region, or was this African volcanic environment especially favorable to human evolution? Was it the abundant game and fertility of the volcanic plains, with their rich soils? We shall probably never know the answer, but humans quickly learned to make use of volcanic rocks in toolmaking and captured fire from volcanoes, to open up the realms of the dark and the cold for further expansion of the race.

When humans first sought explanations for volcanic phenomena, they linked these violent processes to mythology and religion. But with the rise of Western philosophy and learning, Greek scholars in the third century before Christ began the search for the actual physical causes of volcanism. The Greeks speculated that eruptions were the result of the escape of highly compressed air and gases inside the Earth, but later the Romans proposed that volcanoes were natural furnaces in which combustion of sulfur, bitumen, and coal took place. This search for the actual causes of volcanism on Earth and other planets has continued to our day, but we now have many of the answers.

Volcanology is the study of the origin and ascent of magma through the planet's mantle and crust and its eruption at the surface. Volcanology deals with the physical and chemical evolution of magmas, their transport and eruption, and the formation of volcanic deposits at the planetary surface. Some volcanic processes constitute a major natural hazard, whereas others are highly beneficial to society. Thus, the study of volcanism has far-ranging significance for society. For most people, volcanology conjures up a picture of an erupting volcano. Volcanoes and their eruptions, however, are merely the surface manifestation of the magmatic processes operating at depth in the Earth, and thus the study of the volcanism is inevitably highly interdisciplinary, most closely linked to geophysics, petrology, and geochemistry.

In this volume, we examine volcanology in the widest sense, covering not only the traditional aspects of the generation of magmas (traditionally the domain of petrology and geochemistry) and their transport and eruption (the field of traditional volcanology), but also the multitude of effects that volcanism has on the environment and on our society and culture. Volcanism is the best way to probe the interior of Earth. Adopting an anthropomorphic view, we could regard magma as the sweat of the Earth, resulting from the labors of moving the great crustal plates around on the planet's surface and maintaining the Earth's mantle well stirred. The analogy is not that far fetched, because magma and volcanism in general are also a means of heat loss for the Earth. For the Earth scientist, these fluids emerging from the Earth carry valuable clues about the internal constitution of our planet. Just as the medical doctor analyzes the various fluids of the human body, the geochemist samples and analyzes the magmatic liquids that issue from Earth's volcanoes. For the human race, the Earth is a blessed planet, because of its position in the solar system and because of its physical dimensions and vigorous dynamic internal processes, among which volcanism plays a fundamental role. Our Earth is just sufficiently far away from the Sun to benefit from its heat, but not so close as to lose its crucial oceans by rapid evaporation or its precious water by **photodissociation** at the top of the atmosphere and subsequent escape to space. The Earth is cool enough that liquid water stays on its surface—but not so cool that all water freezes. Earth's gravity is sufficiently high to exceed the escape velocity of water and carbon dioxide molecules, allowing it to retain a unique atmospheric composition, with enough carbon dioxide to create a comfortable greenhouse and to provide the building blocks for life, as well as to shield us against harmful solar ultraviolet radiation. Earth's internal heat reservoir is not so hot that it makes life unbearable because of continuous volcanic eruptions, but is sufficient to drive mantle **convection** and plate tectonics. The volcanism resulting from plate tectonics continuously recycles volatile elements such as water, carbon dioxide, and sulfur between the inner Earth and its surface reservoirs: the oceans and atmosphere. It is a very ancient cycle. In the distant history of the primitive Earth, the original atmosphere and oceans probably resulted from the early degassing of the interior of the globe, largely through volcanic activity.

Volcanism is flux of energy and matter. It is an expression of the storehouse of Earth's inner energy, derived in part from cooling of an originally hot planet and in part from heat resulting from the radioactive decay of naturally occurring uranium, potassium, thorium, and other **radionuclides** present deep in the Earth. Thus, volcanic eruptions are the surface expression of these deep Earth processes. When viewed on a geologic timescale, the motions of the inner Earth are veritable storms raging within the planet, with thunderheads rising up through the mantle to form plumes that break the surface as great volcanic hot spots. Other internal storms also lead to convective rollover of the mantle, pulling and pushing along the great crustal plates at the surface, resulting in volcanic activity where plates converge or are pulled apart. When we compare Earth to the other planets in the solar system, we may wonder why our home planet is so prone to volcanic eruptions. The logical question is, however, why does the Earth have such vigorous plate tectonics, the driving force of volcanism?

2. VOLCANISM AND PLATE TECTONICS

Earth may be unique among the planets in the solar system in that its outer rigid skin—the crust and lithosphere—is continuously being destroyed and regenerated. It has active plate tectonics, where the heat and smoke of volcanism rise from the two main battlefields between the plates: the rifts or ocean ridges and the **subduction** zones. The most important consequence of plate tectonics is geologic recycling of materials, turning the Earth into an immense chemical factory where volcanism plays a crucial role. As great crustal plates are pulled apart in the ocean basins, the solid but mobile Earth's mantle below the rift responds to the decrease in overburden and rises upward to fill in the rift. When the rising peridotite mantle experiences a decrease in pressure, it spontaneously undergoes melting, without addition of heat. The reader who does not have a background in petrology or volcanology should pause at this point and ponder the last sentence, for here lies the key to an understanding of the vast majority of volcanic processes: decompression melting. It may be difficult at first to comprehend that rock may melt, without addition of heat, simply because the pressure acting on it decreases, but this is the most common melting process in the Earth. The idea that melting could occur simply as a result of the decrease of pressure (decompression melting) dates back to the beginning of the nineteenth century, but it was not generally accepted by geologists until the latter part of the twentieth century.

Early volcanologists knew that the Earth's interior was hot but solid, and from the basic principles of thermodynamics that were developed in the 1830s, they could theorize that melting would occur if pressure decreased, that is, if the mantle rock were brought upward to a region of shallower depth in the Earth. It was only with the advent of the theory of plate tectonics that a mechanism for vertical motion leading to decompression was discovered. The interstitial melt that forms during decompression, behaving rather like the water that seeps between the sand grains on the beach, forms no more than a few percent of the rising mantle during melting. But the new basaltic magma is less dense and rises faster than the mantle. It collects into pockets that form magma reservoirs, eventually pushing its way upward into the rift in the mid-ocean ridge, to erupt as lava from fissures on the ocean floor. Thus, volcanism is continuously adding new mantle-derived volcanic crust to the plate margins, temporarily welding together the rifted plates, until they break again in another eruption. But why are the plates moving apart? Are they being pushed or are they sliding "downhill" because of the elevated position of the mid-ocean ridge? Are they being dragged along by the convection of the underlying mantle? Or are they pulled along by the subduction of the old, cold, and therefore dense leading edge of the plate into the mantle? To comprehend the fundamental underlying causes of volcanism in island arcs and mid-ocean ridges, we need to answer these important questions. Empirical evidence shows that the plates today are moving primarily because of the force we call slab pull. The plates that are moving fastest are all attached to long and very active subduction zones. Once the cold leading edge of the plate enters the mantle, its fate is sealed: It will subduct. Solid-state phase transitions occur in the minerals making up the plate, converting low-pressure minerals into a much more dense high-pressure assemblage, thus increasing the bulk density of the plate to a point where it exceeds the density of the upper mantle and spontaneously sinks, dragging along with it the attached plate at the surface.

How was this cycle of plate tectonics started, or has it always operated on Earth? We do not yet know the answer, but its operation goes back at least to the **Proterozoic eon** (0.57—2.5 billion years ago). One fascinating idea is that subduction was literally initiated by biological processes on Earth. When the **stromatolites**, the first calcium carbonate-secreting organisms, began to flourish in the Earth's oceans, they mined the abundant CO_2 from the planet's atmosphere and fixed it into dense rock, limestone. In the process, they and other carbonate-fixing organisms gave off the oxygen that created for us an atmosphere that is fit to breathe. When the great masses of dense limestone rock first accumulated around the continental margins late in the Proterozoic, they may have downwarped the underlying oceanic crust to such an extent that the high pressure resulted in conversion of low-pressure basaltic crust to its high-pressure form of **eclogite**. With a density greater than that of the upper mantle, the eclogite crust would sink into the mantle and pull along with it the attached crustal plate, initiating subduction and setting in motion the first plate tectonic cycle (Figures I1.1—I1.5).

3. THE MID-OCEAN RIDGE

A remarkable discovery made only 40 years ago was that the vast majority of volcanism on Earth—perhaps more

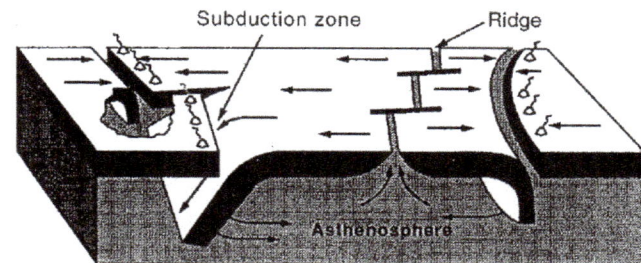

FIGURE I1.1 Most of Earth's volcanism occurs at the junctions of the dozen or so plates that make up the exterior shell of the Earth. Mid-ocean ridge volcanism occurs at the boundaries where plates are pulling apart. Arc volcanism occurs where plates are converging, above the subduction zones. Below the plates resides the asthenosphere, the part of the Earth's mantle from which most magmas originate. *After Allegre (1992).*

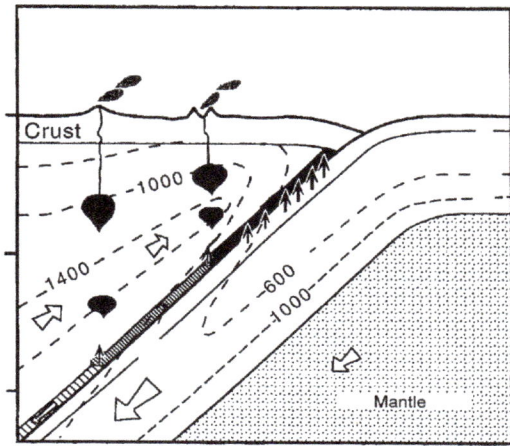

FIGURE I1.2 Arc volcanism above a subduction zone. The descent of the cold subducting plate causes upwelling of hotter mantle below the volcanic arc. Furthermore, volatile components, such as water, are released from the subducting plate to the overlying mantle wedge, promoting melting of the mantle and formation of magmas that feed the arc volcanoes above. The dashed lines are isotherms of temperature distribution in the mantle and subducting plate (in degrees centigrade).

than 80%—occurs at depths beneath the ocean waves. Although we had an inkling of the importance of the global mid-ocean ridge system, its importance was only revealed in the 1960s with the exploration of the ocean floor and establishment of a global seismic network. We cannot generally witness this type of volcanic activity on the ocean floor, but in certain regions, such as Iceland, the mid-ocean ridge literally grows out of the ocean or is exposed on land. In November 1963, as I tossed about in the frigid and turbulent waters south of Iceland, I had the privilege of witnessing the result of the rifting of the Mid-Atlantic Ridge and the growth of the volcanic island of Surtsey.

Normally, the basaltic eruptions on the ridge are effusive and create pillow lava flows on the ocean floor. At these depths in the ocean the pressure is so high that seawater does not boil explosively even in contact with the red-hot lava.

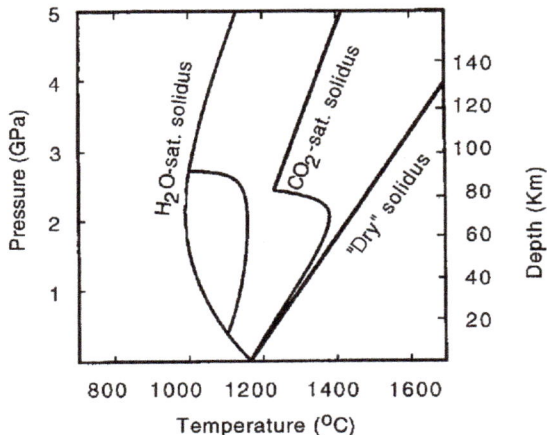

FIGURE 11.3 The solidus defines the location of the beginning of melting in the Earth's mantle. The figure shows the effects of pressure and volatiles (water and carbon dioxide) on the configuration of the solidus of peridotite, the dominant rock type in the upper mantle. Right-hand vertical scale is the depth below the surface; left-hand scale is in units of pressure (GPa). The four solidi shown are the "dry" solidus (neither H_2O nor CO_2 present), the CO_2-saturated solidus, the H_2O-saturated solidus, and (intermediate and unlabeled) the H_2O-undersaturated solidus (water is present, but only in sufficient amount to form hydrous minerals). The presence of either H_2O or CO_2 has a dramatic effect of lowering the beginning of melting of the mantle. Under most Earth conditions, melting is likely to begin in the depth range of several tens of kilometers.

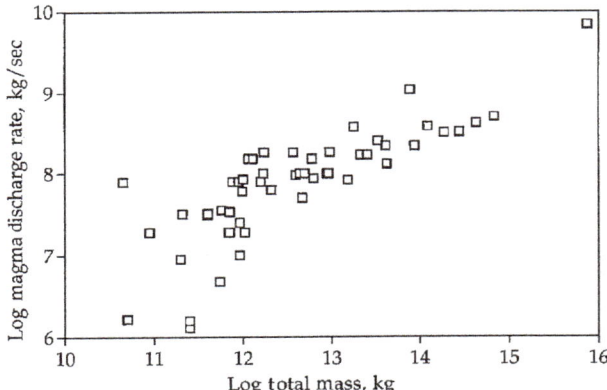

FIGURE 11.4 The figure shows the range in volcanic intensity and magnitude for a number or eruptions on Earth, shown as log units. The magma discharge rate (intensity) during a volcanic eruption can range from 10^6 to 10^{10} kg/s in the most violent events. The total erupted mass (magnitude) can range to more than 10^{15} kg, equivalent to a magma volume of several thousand cubic kilometers.

On the other hand, when the underwater volcano grows and reaches shallow levels, the style of activity changes dramatically, as we observed when Surtsey emerged. Violent explosions sent showers of black ash, lapilli, and muddy steam over our boat, as the hot magma flashed the seawater to superheated steam in shallow water. The steam explosions literally tore apart the molten lava and generated a type of eruption that we now know as surtseyan.

Hydrothermal activity is an important consequence of volcanic activity at the mid-ocean ridge, and this process has made its imprint on the chemical composition of the oceans. The injection of magma into the fractured volcanic crust at the mid-ocean ridge sets in motion a vigorous hydrothermal system. In a way, this system is rather like the radiator on the front of the great magma engine. It circulates cold seawater from the ocean down through the fractured crust, where it encounters hot volcanic or intrusive rocks at depth, in proximity to young dikes or even close to the magma reservoir. Here the water is heated. It exchanges chemicals with the rocks, leaving some chemicals from seawater behind and picking up others from the rock that is hydrothermally altered in the process. The heated seawater transports metal-rich solutions toward the surface. Volcanism on the mid-ocean ridge has a profound impact on the chemistry of the oceans. The hydrothermal circulation process is so forceful at the mid-ocean ridges that the entire mass of the world's ocean is recycled through the hot volcanic crust at a rate of about 1.5×10^{14} kg/year. This rate is sufficient to cycle the entire ocean through the mid-ocean ridge about once in 5 million years. This huge flux of seawater through the oceanic crust has greatly influenced the chemical composition of the oceans through time, resulting in addition and removal of certain chemicals from seawater. This process leads, for example, to the addition of calcium, potassium, and lithium to seawater, from magmas and rocks at the ocean ridges, and the removal of, for example, magnesium from seawater into the oceanic crust. The discovery of mid-ocean ridge hydrothermal activity has explained several puzzles regarding the "sources" and "sinks" of several chemical components in the ocean geochemical cycle. Thus, the mid-ocean ridges are major sinks for magnesium and sulfate and an important source of calcium and manganese. Hydrothermal fluids also transport metals in solution toward the surface. Upon emerging onto the ocean floor, the solutions cool and precipitate metals, leading to the formation of iron—manganese-rich sediments at the mid-ocean ridge. Locally, the hydrothermal solutions emerge through vents on the ocean floor at very high temperature (350 °C). These solutions carry high concentrations of metals and precipitate sulfides, sulfates, and oxides around the vent, to form chimneys up to 10-m high that belch out hot solutions. The solutions precipitate various minerals as they emerge into the ocean, forming black smokers. These solutions are very rich in silica, hydrogen sulfide, manganese, carbon dioxide, hydrogen, and methane, as well as potassium, calcium, lithium, rubidium, and barium. Minerals precipitated on the ocean floor by this process include pyrite (FeS_2), chalcopyrite ($CuFeS_2$), and sphalerite (ZnS). The high concentrations of hydrogen sulfide in these vents support a unique biological assemblage, including sulfide-oxidizing bacteria, which form the basis of a food chain.

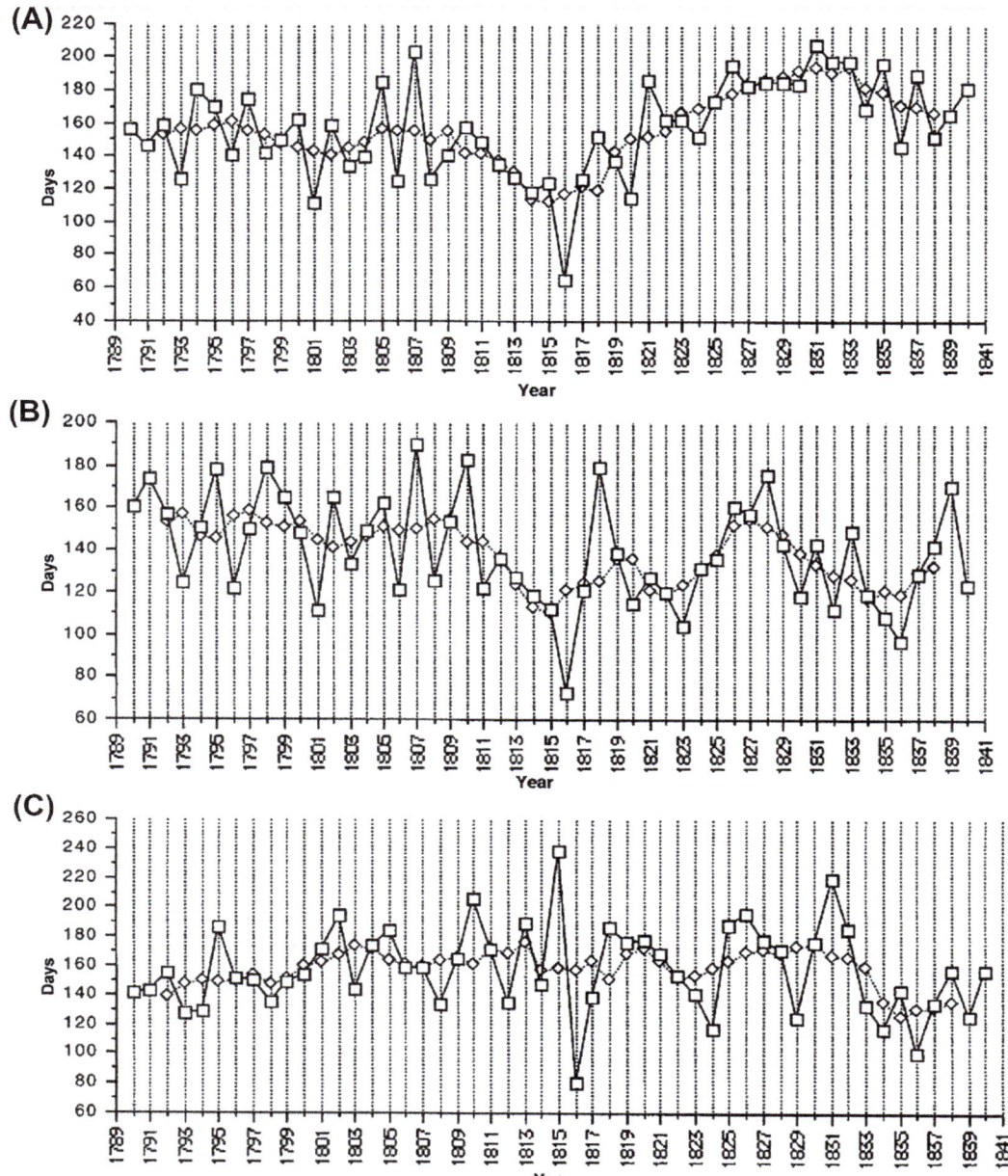

FIGURE 11.5 Volcanic eruptions can change the climate. The great Tambora volcanic eruption in Indonesia in 1815 led to a short-term global climate change that dramatically shortened the growing season in New England the following year. The sulfur dioxide gas emitted by the volcano produced a veil of sulfuric acid aerosol in the stratosphere worldwide, decreasing the amount of solar radiation reaching the Earth's surface, causing global cooling. The figure shows the length of the growing season on an annual basis (dots) from 1789 to 1841 and the 5-year running average (solid line); growing season in the states of (A) Maine, (B) New Hampshire, and (C) Massachusetts. The shortest growing season on record was in the year following the Tambora eruption. *Reproduced from The Year Without a Summer? World Climate in 1816, published by the Canada Museum of Nature, Ottawa, 1992.*

4. THE SUBDUCTION FACTORY

The vast majority of land or island volcanoes on Earth are in volcanic arcs above subduction zones. Although they represent only about 10–20% of the volcanism on Earth, the arc volcanoes are most important in terms of their impact on our society. Because they are subaerial and vent directly into the atmosphere, their eruptions can affect our atmosphere. Furthermore, the regions around arc volcanoes are often densely populated and thus may be regions of high risk. Whatever mechanism may have initiated it, subduction is a dominant component in the great geologic machine that processes and recycles the Earth's oceanic crust and upper mantle. Although the 100-km-thick subducting plate is composed primarily of oceanic crust and upper mantle rocks, it also contains a sliver of sediment at its upper surface and water, carbon dioxide, and other volatile elements bound in clays and other hydrous minerals. In this fashion, the volatile elements are returned to the Earth's mantle from whence they came originally, when they were emitted as volcanic gases in eruptions during the degassing of the early Earth. Water may also facilitate plate motion, lower mantle strength and viscosity, and play a crucial role in lubricating the subduction zone, because the presence of water in the subduction zone environment promotes the formation of structurally weak hydrous silicate minerals that results in rocks of low strength. Similarly, water, even in trace amounts, promotes partial melting by lowering the beginning of melting of mantle rocks. When this occurs in the upper mantle at the base of the lithosphere (the low-velocity layer), plate motion is facilitated. Further, subducted water decreases mantle density in the wedge below the volcanic arc, promoting melting and buoyant upwelling of partially molten mantle material, and thus promoting volcanism at the surface.

The presence of water in the magma further sets the chemical evolution of the magma on an entirely different course from "dry" magmas, leading to the silica enrichment typical of magmas in volcanic arcs above subduction zones. Without the subduction of water, the majority of volcanism on Earth would be submarine, of the type we observe on the mid-ocean ridge: eruption of basaltic magmas of relatively narrow compositional range, releasing dominantly carbon dioxide and sulfur dioxide gases to the ocean, but without emission to the atmosphere. In this subduction-free world, only a small fraction of volcanism would release volcanic gases directly to the atmosphere, that is, the hot spot volcanism of the type we see in Iceland and Hawaii today.

One of the interesting paradoxes of the Earth is that subduction occurs because the leading edge of the plate is cold and dense enough to sink, yet in this apparently frigid environment, volcanism is a characteristic feature. How is it that magma generation occurs in the subduction zone, in response to the descent of the cold plate? The explanation for this paradox is twofold. On one hand, the water introduced into the mantle wedge above the subduction zone lowers the melting point of mantle rock (peridotite), facilitating magma generation and volcanism at the surface. In addition, the descent of the plate stirs up the mantle, bringing upward a current of warmer mantle material from the depth, again facilitating melting by decompression and resulting in magma generation. It has been suggested that the island arcs above subduction zones are the "kitchen" where continental crust is made. The process of magmatic differentiation or geochemical evolution of magmas beneath the volcanic arcs results in the formation of relatively high-silica andesitic or rhyolitic magmas that solidify

as low-density rocks. In the "kitchen," the high-silica magmas may be regarded as the cream that evolves from the primary basaltic magma and rises to the top. The high-silica rocks are too light to subduct, and they accumulate in the arc environment, typically at continental margins. The process by which arc volcanics and associated intrusive rocks accumulate is referred to as continental accretion and may be the principal mechanism for continental growth. In this manner, the subduction zone acts as a giant still, melting the mantle and separating the continent-forming material out of the melts, adding it to the continents and leaving behind a depleted upper mantle. Thus, the great subduction zone still treats chemical elements in two ways. Some are continuously recycled, namely, the volatile components, such as water and carbon dioxide, whereas others are largely retained in the arc crust.

To what extent is the water, which drives explosive volcanism in arcs, derived from the subduction of the oceans? If the recycling is working, then we should find geochemical evidence of subducted material in the magmas erupted from island arc volcanoes. One of the spectacular success stories of geochemistry is its role in providing proof of the recycling of certain trace chemical components. The discovery came with the detection of the beryllium isotope Be^{10} in arc volcanics. This relatively short-lived isotope (half-life about 1.5 million years) is derived solely from reactions in the atmosphere, and it accumulates in marine sediments that may be subducted. The fact that Be^{10} is present in some arc magmas in measurable amounts is proof that relatively young oceanic sediments or fluids must have been subducted and become incorporated in the magma sources underneath the volcanic arcs. Although it is likely that oceanic sediments and related fluids contribute only 1–2% of the magma (the rest is mantle contribution), their chemical imprint is clear and the consequences are evident in the explosive character of these water-rich magmas.

5. VOLCANOES ON HOT SPOTS

The plate tectonic revolution of the 1960s did not explain the origin of a number of better known volcanoes on Earth, particularly the volcanic islands like Hawaii, the Galapagos, and Iceland. In many cases they show no direct relation to plate boundaries and thus do not fit into the paradigm of plate tectonics. They are volcanic, however, and have a high eruption rate, which has earned them the name hot spots. Perhaps the most insidious hot spot is the Yellowstone volcanic field in the United States, a hot spot located in the center of a continental plate. Its prehistoric giant explosive eruptions include some of the largest known on Earth, but because of the long dormant periods, it is a potential continent-wide or even global volcanic threat that we often overlook.

Unusual geochemical signatures in the basaltic hot spot lavas suggest a magma source deeper in the mantle than the mid-ocean ridge source or a source that has not been tapped before by earlier magma extraction. This has led to the idea that hot spot volcanoes are supplied by deep mantle plumes, focused mushroom-like currents of solid but hot rock, rising from either the lower mantle or even from the boundary layer between the mantle and the core of the Earth. This is a radical idea, but if correct, it could mean that volcanoes such as Kilauea in Hawaii bring a signature to the surface from rocks that originally were near the core of the Earth, some 2900 km below the volcanoes.

An alternative model for the origin of hot spot magmas brings us back to subduction. It is likely that great masses of ancient subducted masses of lithosphere reside within the deep mantle. Here the subducted slab of cold lithosphere eventually warms up and its density becomes equal to or even less dense than the surrounding mantle, causing its rise toward the surface as a plume of rock. Nearing the surface, the hot material undergoes decompression melting and magma generation. Again, geochemical evidence from volcanics in some hot spots supports the idea that subducted slabs may be the fuel for mantle plumes and at least some hot spot magmas.

6. VOLCANISM ON OTHER WORLDS

Those who expected that volcanism on other planets of the solar system would be similar to that on Earth were in for a series of big surprises when exploration of the planets began in earnest. Early ideas on extraterrestrial volcanism were largely directed toward the Moon, where craters were seen to be abundant. In 1665 Robert Hooke first proposed the presence of volcanoes on the Moon, but he had observed the lunar craters telescopically. In the eighteenth century, Pierre Laplace proposed that meteorites were volcanic material ejected from the Moon. Volcanism on the Moon did not seem such a bad idea at the time, considering its cratered and pockmarked visage. In 1785, however, Immanuel Kant pointed out that most lunar craters were much larger than any volcanic features known on Earth, and Kant proved to be correct in his assessment that the lunar craters are not of volcanic origin; we now know that most are the products of meteorite bombardment. Exploration of the lunar surface during the Apollo program revealed vast areas or "oceans" of volcanic rocks, known as Maria, but they turned out to be extremely old, mostly on the order of 3.7 billion years. One of the major surprises—and a disappointment to volcanologists—has been that most of the planets are volcanically "dead." That disappointment was, however, compensated by the discovery of spectacular volcanism on Io, where sulfur-rich magma fountains are ejected to heights of hundreds of kilometers. But Io has more surprises. Its volcanism is most likely not the result of internal thermal energy, as on Earth, but rather due to the tidal energy transferred from the gravitational pull of the giant parent planet Jupiter, causing the interior of Io to heat up. We will no doubt have many similar volcanic surprises awaiting us when we explore other strange worlds in the future.

7. VOLCANIC PROCESSES

A bewildering variety of volcanic processes occur at the surface during an eruption, and the list of the volcanic rock types and types of volcanic deposits is seemingly endless. Fortunately, as volcanology has progressed in the latter half of the twentieth century from a descriptive endeavor to a truly interdisciplinary science, we have discovered the major factors that control eruption processes and account for the diversity of the products, as described in parts II, III, and IV of this volume. Although each eruption is unique, and each volcano seems to have its distinct "personality," several fundamental parameters dictate the style of eruption (Table I.1). By far the most important parameter is the mass eruption rate (eruption **intensity**), which depends on the rate of supply of magma during eruption. The observed range in mass eruption rates on Earth is huge, more than at least three orders of **magnitude**, and extending beyond 10^9 kg/s. It is beyond our comprehension to visualize 1 million tons of red-hot molten rock jetting out of the crater of a volcano every second—not for a few seconds, but for hours or days, such as occurs during large ignimbrite-forming eruptions. We are accustomed to referring to such events as "explosions," but that is somewhat of a misnomer, because they are not instantaneous bursts, but rather sustained emissions of magma and gases; only the onset is really explosive. The mass eruption rate is truly a measure of eruption intensity, because it controls the height of the eruption column in explosive events and the length of lava flows in effusive eruptions.

Eruption intensity shows a good correlation with the total mass or volume of magma erupted (magnitude). Although we may intuitively expect that the largest magma bodies within the Earth result in the highest magnitude (largest volume) eruptions, but why they should also be the highest intensity events is not immediately obvious. The flow rate is, of course, a function of the magma pressure and the diameter of the conduit. The explanation may lie in the time it takes for the volcano to widen its conduit by erosion during eruption and allow more magma to pass to the surface. In most major explosive eruptions, the intensity increases with time during the event, and it is most logical to assume that this is simply due to gradual widening of the conduit as it is eroded by the passage of hot magma and the explosive blast of rock, ash, and pumice fragments. During an explosive eruption, mass eruption rate increases, so does the height of the eruption column, attaining more than

TABLE I1.1 Some Factors that Influence Volcanic Processes

Factor	Typical Range	Comments
Magma temperature	800–1250 °C	Influences magma viscosity and affects energy available for eruption plume rise.
Magmatic water	0.1–6 wt% H_2O	Controls magma fragmentation through the vesiculation process.
Magma viscosity	102–1010 P	Determines flow properties of magmas, vesiculation, bubble growth; controls the rise and escape of bubbles out of magmas.
Mass eruption rate	10^6–10^9 kg/s	The rate of flow of magma out of the vent is single most important factor.
External water	0.01–0.5 vol. fract	The fraction of external water that mixes with magma in the conduit or vent.
Crystal content	0–50%	Affects bulk viscosity and thus rheology of magma.

40 km in the Earth's atmosphere for the largest eruptions. Although this correlation is in part related to increase in the height of the jet of magma and **pyroclasts** emerging from the vent, it is really more influenced by the thermal energy from the eruption, available to drive a convective plume. In fact, most of the eruption column above a volcano is a thermal plume, where heat is transferred from the hot pyroclasts to the surrounding air, which in turn convects vigorously, mixes with the ash and other pyroclasts, and drives the plume sky-high.

Paradoxically, however, when the mass eruption rate reaches a critical value, the eruption column ceases to rise; in fact, the column begins to collapse. The process of column collapse generates the most destructive phase of volcanic eruptions and produces pyroclastic flows and surges: hot ground-hugging cloudlike density currents of ash, pumice, volcanic gases, and heated entrained air. It seems at first paradoxical that the column should collapse with increasing mass eruption rate, but the explanation lies in the efficiency of mixing of air with the mixture of pyroclasts erupting from the vent. Under conditions of moderate mass eruption rate, the erupting mixture entrains and mixes with a large volume of air, which is heated, becomes buoyant, and contributes to the rapid rise of the eruption column. As the mass eruption rate increases, however, pyroclasts in the interior of the column become isolated from the atmosphere and have less opportunity to mix with air. Consequently, the entraining of air is no longer contributing to the buoyancy of the column; it becomes denser and collapses as a fountain of hot ash, gases, and pumice, to generate a density current.

The second factor that influences volcanic processes, listed in Table I.1, is the concentration of magmatic water. Water is almost absent in the basaltic magmas erupted on the mid-ocean ridge, but occurs in concentrations of up to 6% in the high-silica magmas of island arcs. The role of water is twofold, both as the agent that causes vesiculation and fragmentation of magmas and as an important

component in accelerating the magma and pyroclasts out of the vent. It is the presence or amount of water in the magma that primarily determines the division between effusive and explosive eruptions, but only a little water goes a long way. With a few percent of water dissolved under high pressure at depth, the magma begins to grow steam bubbles as it rises to the surface. The solubility of water decreases rapidly with decreasing pressure, and the volume fraction of the exsolving steam grows fast. If the bubbles are unable to rise and escape from the magmatic liquid, they will continue to expand until they burst and tear apart the magma into pyroclasts of ash and pumice. As the exsolution and volume expansion of water occur in the conduit of the volcano, this will result in a great volume increase of the erupting magma mixture, propelling it out of the vent.

Magma viscosity is another important factor affecting flow properties of magmas and determining the ability of steam bubbles to rise and escape out of the magma. Viscosity is influenced by temperature, magma composition, and crystal content. In high-temperature and low-silica basaltic magmas, the viscosity is so low that gas bubbles can generally rise in the magma, whereas in high-silica low-temperature magmas, the high viscosity inhibits bubble rise; they move passively with the magma in the conduit, up toward the vent.

All of the factors just mentioned are internal to magma or the volcanic system, but a number of external parameters can also influence the volcanic process. The most important of these factors is external water. Magmas rising through the crust and to the Earth's surface have a high probability of encountering water, either in the form of oceans, lakes, glaciers, or as water-rich rock formations. We saw earlier, in connection with the 1963 Surtsey eruption, that the interaction of external water and magma is relatively passive when it occurs in regions of high hydrostatic pressure, such as in the deep ocean, but violent explosions will, on the other hand, occur in shallow water. Similarly, magma rising through highly porous and water-logged rocks at shallow levels in the crust can result in explosive eruption because of the conversion of water to steam in contact with the magma. Thus even "dry" basaltic magmas, virtually devoid of magmatic water, can erupt with explosive violence if external water is at hand.

8. THE IMPACT OF VOLCANISM

As emphasized earlier, volcanism is the surface manifestation of deep Earth processes, and eruptions are merely the smoke rising from the "subduction factory" or the mid-ocean ridge "kitchen." In the greater scheme of things, volcanic activity is thus a part of a geodynamic system that is essential for maintaining a balanced environment for life on Earth. On a local scale, however, both hazards and benefits are associated with volcanoes. Despite their awe-inspiring, spectacular, and even deadly fireworks, volcanic eruptions have not been nature's most deadly hazard in living memory. The principal volcanic disasters since 1700 have taken a total of 260,000 lives, which pales in comparison with the death toll from earthquakes and tropical storms. The deadliest disaster known was probably the 1556 Huahsien earthquake in China, which killed more than 820,000 people. In 1976 the magnitude 7.8 Tangshan earthquake, also in China, killed more than 240,000 people. The worst hurricane in history occurred in 1970 in the Ganges Delta in Bangladesh, with a loss of 500,000 lives. Yet living with a dormant or active volcano, looming large above village or town, is a visible and permanent threat and qualitatively different from the transient threat of a hurricane or a sudden earthquake.

When we view volcanic hazards, however, we must take the long view and keep in mind that the historical period is too short to display the full range of the magnitude and intensity of volcanic eruptions that the Earth can muster. Even in the 6000-year life span of our civilization, we have been spared a mega-eruption of the type that can occur. We are all familiar with the concept of the storm of the century—an event so large that it occurs only very rarely—but the probability of its recurrence increase with time. Similarly, high-magnitude and high-intensity volcanic eruptions are low-probability events, but they will recur in the future. In our limited human experience, we have the eruption of the type that occurs once in 100 years, such as Krakatau in 1883, or the eruption that occurs once in 1000 years, which must be regarded as the Tambora eruption of 1815. What about the eruption that occurs once in 10,000 or 100,000 or every million years? The Toba eruption on Sumatra, about 75,000 years ago, is probably at the upper limit of eruptions that can occur on the Earth, emitting 2000 km^3 of magma. That there is an upper limit on the magnitude of eruptions is itself an interesting concept, but most likely any larger magma chamber cannot be contained in the Earth's crust and will spontaneously

erupt. Because the Toba event occurred in the far distant past, our information of its environmental impact is still blurred, but a cataclysmic event of this type would have a catastrophic impact on global society today. When we take the geologic perspective of time and volcanism, we are therefore reminded of the chilling words of the American historian Will Durant, "Civilization exists by geologic consent, subject to change without notice."

The Tambora eruption on the island of Sumbawa in Indonesia in 1815 gives an inkling of what may happen in such a volcanic event. The regional impact of the 1815 Tambora eruption in the East Indies has been dimmed by the passage of time, and only incomplete counts exist, but they indicate a natural disaster of enormous gravity for the Indonesian people. The disaster struck both on the island of Sumbawa, where Tambora is located, and on the neighboring islands of Lombok, Flores, South Sulawesi, and Bali. Even Java was not spared. During the climax of the explosive eruption on April 10, 1815, hot pyroclastic flows swept down all flanks of the volcano and wiped out the feudal kingdoms of Tambora on the north side and Sanggar on the east side of the mountain. In the Kingdom of Tambora, the loss of life was so complete that it extinguished the Tambora language, probably the easternmost Austro-Asiatic language at the time, with more than 10,000 people killed outright in the pyroclastic flows. The blanket of ash fallout on Sumbawa also destroyed all the crops, resulting in a famine immediately after the eruption. It was accompanied by a variety of diseases, and clean water was not to be had anywhere. For some reason, the island's climate changed, becoming so hot and dry that crops would not grow—probably because of the destruction of the vegetation and higher runoff. An additional 38,000 people died of starvation and disease, having lost 75% of their livestock, and 36,000 fled the island. It took the island at least one century to recover from this ecological and human disaster.

The effects of the Tambora eruption were not, however, restricted to Sumbawa. The ash fall was principally to the west. In Lombok and Bali, the ash was so thick that many people were killed immediately as the roofs of their homes collapsed under the weight of huge quantities of ash. Even on these distant islands the toll from famine and disease brought about by the eruption was severe. On Lombok, estimates of the death toll ranged between 44,000 and 100,000 and on Bali at least 25,000. The depth of famine and disease was so great that it even claimed lives among the Balinese royalty. A conservative number of dead in the East Indies from the 1815 Tambora eruption is at least 117,000. News of this distant eruption in Indonesia did not reach the Western world for some time, but unusual atmospheric phenomena and climate deterioration set in during 1816—"the year without summer." Tambora-size (c. 100 km^3 of magma) explosive volcanic eruptions emit large quantities of sulfur dioxide gas into the stratosphere, where it is converted into a sulfuric acid aerosol dust veil that encircles the Earth. The aerosol has the effect of reducing solar heat reaching the surface of the planet, leading to global cooling. Strange atmospheric phenomena had been observed in Europe before the summer of 1816. In May 1815 the sunsets in England were exceptionally colorful, and by September the sky seemed to be on fire every evening. But the spectacular skies brought bad weather the following year, and the year 1816 is the worst on record in Europe, with the mean temperature about 3 °C below average across the entire continent from Britain in the north to Tunisia in the south. In England the summer was miserable, with the mean temperature for June the lowest on record, or 12.9 °C. In France it was so cold that the wine harvest was later than at any time since record keeping began—delayed until November in those districts where the grapes were not already frozen on the vineyard. In Germany crops failed, a situation exacerbated by the war with Napoleon, and famine and price inflation were widespread. Conditions in Germany became so intolerable that people in many districts rioted and set off in search of better living conditions, starting the great migration from northern Europe east to Russia and west to America. The severe weather in fact led to crop failure of all the major crops in Europe and a doubling of the price of wheat from 1815 to 1817.

When we consider that in the early part of the nineteenth century agriculture was the mainstay of society, it is clear that the climate change had a strong impact on the economy of the Western world. In North America the year 1816 was also unusually cold and dry, especially in New England and, as mentioned, is known in annals as "the year without summer." In early May, most farmers had planted their corn, the main crop at that time. But by mid-May the weather deteriorated, with severe frost that froze the fields as far south as New Jersey. As late as mid-June frosts also plagued New England, New Jersey, and New York. Then a great drought began, stunting or destroying the few remaining crops. In early July frost again swept over New England, from Maine to Connecticut, and again in late August in Maine, Vermont, and south to Massachusetts. By the end of September a final frost destroyed the few crops that had survived the frigid summer. These crop failures, in both Europe and North America, have rightfully been called "the last great subsistence crisis of the Western world." The growing season, of course, was much shorter than normal. In New England the average growing season is about 120–160 days, but in 1816 it was only 70 days in Maine, 75 in New Hampshire, and 80 in Massachusetts—a reduction of 40–50%. Many farmers suffered such heavy losses that they abandoned their farms and migrated to the western territories.

The abnormal climate conditions in 1816 have frequently been proposed as an important factor in the first worldwide cholera epidemic. The bad weather also caused a series of crop failures in India, and as a consequence a cholera epidemic broke out in Bengal. It was spread further afield by British soldiers, to Afghanistan, Nepal, and Indonesia in 1820. A second epidemic broke out in India in 1826 and spread to Europe in 1831 and North America in 1832. It is likely that the severe climate conditions (abnormal weather, crop failures, subsequent famine, and reduced resistance to disease) brought about by Tambora may have played a role in the origin of a new and aggressive form of cholera that first did its deadly work in Bengal in 1817.

The global climate impact of large eruptions is a major reason that an understanding of volcanic phenomena is important for society today. A Tambora-size eruption in the twenty-first century will have much more profound effects on humans living on an overcrowded Earth, where natural resources are strained to the limit. The effects of a Toba-size eruption (c. 2000 km^3) in the future are incalculable, but it will occur—we hope not in our lifetime.

For the most part, humans not only have learned to live with the slumbering volcanic giant, but have managed to extract a variety of economic benefits from the material and energy resources that lie at the roots of volcanoes. The precious diamond is merely carbon that has been crystallized under the high pressure (equivalent to the weight of the Eiffel tower resting on a 10-cm plate) and the high temperature in the Earth's mantle. Diamonds are brought up to the surface in volcanic pipes, such as the famous diamond pipes of Kimberley in South Africa. Even the carbon in precious diamond may be of mundane origin, reminding us of the steady recycling of material in Earth's kitchen. Thus isotopic analyses show that the carbon in at least some diamonds is probably derived from marine sediments that were subducted into the mantle early in Earth's history.

A wealth of mineral deposits is also associated with the roots of volcanoes, where hydrothermal activity has led to concentration of metals. One of the more promising benefits from volcanic activity is the harnessing of geothermal energy. The Earth's crust and lithosphere are in reality a boundary layer between the hot mantle and the hydrosphere and atmosphere. In volcanic regions the heat flux across this boundary layer is very large, and it can be tapped effectively by drilling. For example, about 85% of all homes in Iceland are heated with geothermal water, making Icelanders the largest users of geothermal energy per capita in the world (24,000 kWh per capita). It is estimated that the use of geothermal energy alone is saving imported oil costs of about $560 per year per person. With the modern technology of deep drilling, the exploitation of this cheap and clean energy resource from the interior of the Earth does not have to be restricted to active volcanic regions such as Iceland, Italy, and New Zealand, but may be one of the major global sources of energy of the future.

FURTHER READING

Allegre, C., 1988. The Behavior of the Earth. Harvard University Press, Cambridge, MA.

Allegre, C., 1992. From Stone to Star. Harvard University Press, Cambridge, MA.

Hawkesworth, C.J., Gallagher, K., Hergt, J.M., McDermott, F., 1993. Mantle and slab contributions in arc magmas. Ann. Rev. Earth Planet. Sci. 21, 175–204.

The History of Volcanology

Haraldur Sigurdsson

Graduate School of Oceanography, University of Rhode Island, Narragansett, RI, USA

Chapter Outline

GLOSSARY

decompression melting The melting that occurs when some portion of the Earth's mantle undergoes a decrease in pressure, such as due to convective rise or upwelling of mantle peridotite beneath mid-ocean ridges and volcanic arcs. The upwelling results in melting, because the melting point (beginning of melting) of peridotite occurs at a lower temperature with decreasing pressure.

eclogite An igneous rock with a mineralogy characterized by high-pressure minerals, but whose chemical composition corresponds to that of basalt.

exothermic reaction A chemical reaction that produces heat. The early chemists proposed that exothermic chemical reactions deep within the Earth, such as reactions between sulfur and iron, or oxidation of the alkali metals could be the cause of heat and volcanic activity.

neptunists The group of eighteenth century geologists who believed that basalt and similar rocks were not of volcanic origin, but had formed by precipitation from a primordial ocean.

obsidian High-silica volcanic glass, generally black or dark gray in color, with the chemical composition of rhyolite.

palagonite A claylike mineral produced by the hydration of basaltic volcanic glass.

peridotite An ultramafic (high-magnesia, low-silica) igneous rock, whose mineralogy is dominated by olivine and pyroxene. Peridotite is the principal rock type of the Earth's upper mantle and the most likely source rock for basalt by partial melting.

plutonists The group of eighteenth century geologists who believed that basalt and related rocks were produced by solidification from magma or silicate melt.

sideromelan Basaltic volcanic glass, which produces palagonite upon hydration.

xenoliths Rock fragments originating in the deep Earth (primarily the mantle), brought to the surface by volcanic eruptions.

1. INTRODUCTION

This chapter is on the history of ideas about the generation of magma within the Earth—the root cause of all volcanic phenomena. It is a process that results in the transport of material and heat from depth to the surface of our planet. It would seem logical that the history of ideas on melting in the Earth would be synonymous with the history of "classical" volcanology. This is not the case, however, because up to the mid-1970s volcanology was essentially a descriptive endeavor, devoted primarily to the geomorphology of volcanic landforms, geography of volcanic regions, and the chronology of eruptions. Until recently, volcanologists have shown little insight as to the physical processes of volcanic action and often ignored the causes of the melting processes that lead to the formation of magma. Volcanology thus became a descriptive field, lacking rigor, on the fringes of science. It is now timely to redefine modern volcanology as the science that deals with the generation of magma, its transport, and the shallow-level or surface processes that result from its intrusion and eruption. Modern volcanology is, however, highly interdisciplinary and draws widely from diverse geoscience subspecialties.

The processes that bring about melting in the Earth had been discovered by physicists in the mid-nineteenth century, but due to the bifurcation of scientific fields, volcanologists were on the whole ignorant of these findings of the early natural philosophers or geophysicists and continued to pursue a descriptive approach to surface features without acquiring an understanding of the fundamental processes at work deep in the Earth. Table I2.1 lists some of the principal historical advances made in the study of volcanism up to the twentieth century.

2. THE BEGINNINGS

In prehistoric times, opportunities to "discover" fire and to observe heat abounded in the natural world. Lightning strikes ignited fires in the primeval forest or on arid, grassy plains. Volcanic eruptions and lava flows threw forth

TABLE I2.1 The Evolution of Ideas About Volcanic Activity

Scientist	Year	Discovery or Hypothesis
Anaxagoras	Fifth-century BC	Eruptions caused by great winds inside the Earth
Seneca	AD 65	Combustion as a source of heat in Earth
Pliny the Elder	AD 79	First scientific expedition to study an eruption
Pliny the Younger	AD 79	First volcanic eyewitness account, Vesuvius eruption
Snorri Godi	AD 1000	Basaltic rock in iceland recognized as of lava flow origin
Liberti Fromondi	1627	Explosive eruptions compared to combustion of gunpowder
Edward Jorden	1632	"Fermentation" chemical reactions of sulfur generate terrestrial heat
René Descartes	1650	Incandescent Earth core as source of volcanic heat
Athanasius Kircher	1665	First global map of the distribution of volcanoes on Earth
Francesco d'Arezzo	1670	First experiments with molten rocks, melts; Mount Etna lava
J.-E. Guettard	1751	Identifies rocks at Volvic in Auvergne, France, as lava
Pierre Grignon	1761	Extracts crystals from glass slags; products of glass furnaces compared to products of volcanoes
Nicholas Desmarest	1763	Prismatic basalt columns in Clermont of volcanic origin
William Hamilton	1776	Pioneer in field of volcanology; recognizes dikes and magmatic veins
Ben Franklin	1783	Volcanic eruptions have atmospheric effects and can cool Earth
James Hutton	1785	Recognizes intrusions of magma into the crust of the Earth
James Hall	1790	Melting and crystallization experiments on basalt
D. Dolomieu	1790	Some sediments are stratified and consolidated volcanic ash
Leopold von Buch	1799	Minerals in lavas are formed by crystallization from magma
Lazzaro Spallanzani	1794	Earliest chemical analysis of volcanic rocks Water vapor is the dominant magmatic gas; causes explosions
John Playfair	1802	Speculates that change in pressure affects melting of rocks
Robert Kennedy	1805	First complete chemical analysis of basalts
Humphry Davy	1808	Exothermic reactions of alkalis provide volcanic heat
Pierre Cordier	1815	Identifies and analyzes the mineral components of lavas
A. von Humboldt	1822	Linear arrangement of volcanoes related to deep-seated tectonics
George P. Scrope	1825	Decompression melting can produce magma in Earth's interior Chemical differentiation produces variety of magma types
Charles Darwin	1835	Crystal settling is the agent of magmatic differentiation
S.-D. Poisson	1835	Recognizes that the high pressure in the deep Earth would lead to solidification of rock at higher temperature

TABLE 12.1 The Evolution of Ideas About Volcanic Activity—cont'd

Scientist	Year	Discovery or Hypothesis
Gustav Bischof	1837	Demonstrates experimentally that the transformation from solid rock to magma results in increase in volume
William Hopkins	1839	Decompression melting as a process in generation of magmas
Robert Bunsen	1851	Magma mixing produces intermediate magma types; recognizes that magmas are solutions
S. Waltershausen	1853	Water vapor expansion causes magma fragmentation and pyroclastic; recognizes importance of submarine volcanic rock formations
Robert Mallet	1858	Map of global distribution of earthquakes and volcanoes
Osmond Fisher	1881	Proposes convection currents within the Earth
John Judd	1881	Volcanic recycling of water from volcanoes to atmosphere, to oceans, and back to magma through solid Earth
Carl Barus	1893	First to determine the melting curve of basalt as a function of pressure
Alfred Lacroix	1902	Documentation of the eruption of Mount Pelée
Frank Perret	1906	Pioneer in application of technology to study volcanic phenomena
Alfred Wegener	1912	Theory of continental drift
Arthur Holmes	1915	Calculates a temperature profile for the Earth based on radioactive generation of heat
Arthur Holmes	1916	Basaltic magma generated by partial melting of peridotite
Arthur Holmes	1928	Decompression melting and magma generation by mantle convection
Norman L. Bowen	1928	Fractional crystallization theory developed to explain magma range

showers of glowing sparks and incandescent heat. Most likely these were the sources our early ancestors borrowed from, carefully nurturing the fires they kindled, for their sources were only fickle and local. With fire at their command, humans could venture into new realms of cold and dark—into polar terrain and mountain caves.

Our knowledge of human precursors and early humans is intimately linked with volcanic eruptions. It is because of the excellent preservation of fossils in volcanic deposits that paleoanthropologists have begun to unravel the mystery of human origins. Most of the oldest remains of early man come from volcanic regions in Africa and Indonesia, and the association of volcanic activity with these fossils is no mere coincidence. The most logical explanation for their relative abundance and good preservation in such environments is that the bones were rapidly covered by volcanic deposits. Just as the Roman cities of Pompeii and Herculaneum were instantly buried by the eruption of Vesuvius in AD 79 and preserved virtually intact to our day, so have earlier volcanic deposits sealed in the remains and fragile artifacts of our more distant ancestors, preserving them for millions of years. The stunning discovery of 3.7-million-year-old *Australopithecus* hominid footprints crisscrossing a volcanic ash deposit at Laetoli in the East African Rift is a graphic testament of the power of preservation by volcanic ash fall.

Volcanic deposits also contain minerals that can conveniently be dated by measuring the decay of radioactive isotopes of potassium and argon, a method that has enabled the dating of fossil remains of the oldest human, *Homo habilis*, at Olduvai as 1.75 million years old. A similar setting in the African rift at Hadar in Ethiopia has yielded fossils of the possible human precursor, *Australopithecus afarensis*, permitting its dating at 3.5 million years. Volcanic stone is the earliest known material used in the creation of tools by genus *Homo*. The earliest stone artifacts are about 2.5 million years old, implements made from lava, found on the shore of Lake Turkana in East Africa. Following their humble beginnings as toolmakers with the fashioning of these crude "cores," early humans gradually improved the techniques of working stone into flakes, choppers, hand axes, cleavers, and, finally, delicate **obsidian** stone blades in the upper Paleolithic.

More mundane materials such as pumice and volcanic ash have been used for nearly 3000 years in making cement. A mixture of volcanic ash and lime produces very durable and water-resistant hydraulic cement. Volcanoes were also a principal source of sulfur. Early on people realized that sulfur burns with a strange and sputtering flame, giving off an evil-smelling and choking vapor. Probably the first uses for sulfur were to kill insects and to bleach wool, feathers, and fur. Homer knew of its fumigation property, praised it the "pest-averting sulfur" and Ulysses called out: "Quickly, bring fire that I may burn sulfur, the cure of all ills." Over the centuries the burning of a sulfur candle was a house-cleaning ritual following a case of contagious disease in the home. Sulfur features in Egyptian prescriptions of the sixteenth century BC, and Pliny the Elder describes its medicinal, industrial, and artisan uses in his *Historia Naturalis* (AD 77). The Romans also found a new application in warfare for the sulfur they mined in Sicily. By mixing it with tar, rosin, and bitumen, they produced the first incendiary weapon.

3. THE LEGENDS AND HELL

Volcanic eruptions (Table I2.2), the most spectacular and awe-inspiring of natural phenomena, have throughout history inspired religious worship and led to the creation of myths (Chapter 80). Even in the twentieth century, these responses can be observed when a volcano erupts. James Hutton, regarded by many as the father of geology, was in

1788 the first to attempt to explode the myth associated with volcanic eruptions. In Hutton's view.

[A] volcano is not made on purpose to frighten superstitious people into fits of piety and devotion; nor to over-whelm devoted cities with destruction; a volcano should be considered as a spiracle to the subterranean furnace, in order to prevent the unnecessary elevation of land, and fatal effects of earthquakes; and we may rest assured that they, in general, wisely answer the end of their intention, without being in themselves an end, for which nature has exerted such amazing power and excellent contrivance.

Primitive people the world over have long believed that volcanoes were inhabited by deities or demons that were highly temperamental, dangerous, and unpredictable. To appease the capricious gods, humans have for centuries made the ultimate sacrifice. Thus the Mayans, Aztecs, and Incas offered humans to their volcanoes. Nicaraguans long believed that their dangerous volcano Coseguina would stay quiet only if a child were thrown into the crater every 25 years. Similarly, young women were thrown into the crater of Masaya volcano in Nicaragua to appease the fire. Until recently, people on Java sacrificed humans on Bromo

TABLE I2.2 Historic Eruptions

Eruption	Significance
Thera (Santorini), 1650 BC	One of the largest explosive eruptions on Earth; may have contributed to the decline of Minoan civilization
Vesuvius, AD 79	First major historical eruption to occur within range of major cities; first eruption to be documented by an eyewitness
Etna, 1669	Major lava flow event with widespread destruction in Catania from lavas
Lakagigar, 1783	Largest lava flow eruption on Earth; catastrophic impact on Iceland population from haze produced by volcanic gas emission
Asama, 1783	Local devastation from pyroclastic flows; country-wide impact in Japan from climate effects
Tambora, 1815	Largest historic volcanic eruption on Earth; highest eruption column (43 km); global sulfuric acid aerosol causes global climate change; largest known death toll of over 92,000 people
Krakatau, 1883	First large eruption of the modern age; severe impact and death toll from tidal waves (tsunami)
Mount Pelée, 1902	The classic example of a small eruption causing severe loss of life; the beginning of the study of pyroclastic flow and surge processes
Paricutin, 1943	The birth and growth of a new volcano observed
Bezymianny, 1956	One of the highest known eruption columns in a historic event (42 km)
Surtsey, 1963	Best evidence of magma—water interaction during explosive eruption
Mount St Helens, 1980	First major eruption to be monitored intensively with modern technology
El Chichon, 1982	Most widespread pyroclastic surges from a historic explosive event
Nevado del Ruiz, 1985	Another example of a small eruption causing severe loss of life; tragic example of lack of hazard mitigation measures
Pinatubo, 1991	Largest eruption of the twentieth century; major societal impact in Philippines; important global atmospheric effects

volcano, and still throw live chickens into the crater once a year. People living near the feared volcanoes of Nyamuragira and Nyiragongo in Central Africa annually sacrificed 10 of their finest warriors to the cruel volcano god Nyudadagora. To those who were skeptical of such rites and pointed out that earlier sacrifices had failed to prevent or stop an eruption, the believers have countered with the argument that things would have been much worse without the sacrifice.

The Aztec people named the volcanoes surrounding the Valley of Mexico after their gods. Popocatepetl and Iztaccihuatl, lying to the east of the valley, were worshiped as deities, linked to a beautiful love story. When Popocatepetl ("smoking mountain") returned victorious from war to claim his beloved, his enemies sent word ahead that he had been killed, and princess Iztaccihuatl ("sleeping woman") died of grief. Popocatepetl then built two great mountains. On one he placed the body of Iztaccihuatl; on the other he stands eternally, holding her funeral torch.

Volcanic eruptions featured high in Greek mythology, which abounds with allusions to volcanoes, associating them with such gods as Pluto, Persephone, Vulcan, and the fearsome Typhon. The idea that volcanic activity represents the stirrings of the Titans, giants imprisoned in the Earth, goes back to the classical time of the Greeks. The mythical association of volcanic eruptions with battles between the Olympian gods and the Titans is very likely to date back to Preclassical antiquity. The Greeks saw the Titans as huge man-creatures, born of the Earth to attack the gods. Confined under various volcanic regions, their appearance on the land or in the air seemed the logical precursor to a volcanic eruption. The most awe-inspiring of all the Greek monsters was Typhon. The firstborn of Gaia (Mother Earth) and Zeus, he was the largest monster that ever lived. His arms, when spread out, reached a hundred leagues, fire flashed from his eyes, and flaming rocks hurtled from his mouth. Even the gods of Olympus fled in terror at his sight. Typhon rebelled against the gods and even opposed Zeus, who then hurled Mount Etna at him, trapping the frightful creature under the mountain. When Typhon was thus imprisoned under Etna, a hundred dragon's heads sprang from his shoulders, with eyes that erupted flames, a black tongue and a terrible voice. Each time Typhon stirs or rolls over in his prison, Etna growls, and the Earth quakes, with eruptions and veils of smoke covering the sky (Figure I2.1).

The Hades of the Greeks and Romans made an easy transformation into the Hell of the early Christians, described in the Bible as a place with an eternal fire that shall never be quenched. St. Augustine spoke of Hell as containing a lake of fire and brimstone. By the Middle Ages, Hell had taken on great importance, with most scholars convinced that it was a real and fiery place. One place most often cited as the gateway to Hell was Mount Etna volcano in Sicily, and "sailing to Sicily" became a euphemism for going to Hell.

ENCELADE SOUS LE MONT-ETNA.

FIGURE I2.1 According to Greek mythology, the giant Typhon is buried under Etna volcano, and whenever the giant stirs, the volcano erupts violently. (Eighteenth-century print, author's collection.)

4. WIND IN THE EARTH

The earliest known ideas on the cause of volcanic eruptions date to the Greek natural philosophers of the fifth-century BC. Anaxagoras proposed that eruptions were caused by great winds stored inside the Earth. When these winds were forced through narrow passages or emerged from openings in the Earth's crust, the friction between the compressed air and the surrounding rocks generated great heat, leading to melting of the rocks and the formation of magma. To anyone who has observed an explosive volcanic eruption, this is a perfectly logical idea, one that in fact was taken up by Aristotle and passed on by scholars until the Middle Ages.

Aristotle (384−322 BC) discussed the origin of earthquakes, attributing the same or similar origin for volcanic eruptions. "The Earth," he wrote in the *Meteorologica*, possesses "its own internal fire," which generates wind inside the Earth by acting on trapped air and moisture, leading to earthquakes and volcanic eruptions. He even makes a comparison with human anatomy in discussing the effect of the "internal wind": "For we must suppose that the wind in the earth has effects similar to those of the wind in our bodies whose force when it is pent up inside us can cause tremors and throbbings." Aristotle thought that the heat associated with volcanic eruptions was generated by friction produced when the wind moved rapidly through restrictions within the Earth. "The fire within the Earth can only be due to the air becoming inflamed by the shock, when it is violently separated into the minutest fragments. What takes place in the Lipari Isles affords an additional proof that the winds circulate underneath the Earth."

The identification of volcanic activity with "wind and fire" by the early Greeks follows logically from actual observations of a volcanic eruption. The most striking phenomena are first the explosive uprush of hot gases or "wind" from the Earth through the crater; then the incandescent red-hot glow of molten rock, giving the appearance of fire. It was on the basis of these phenomena that Aristotle postulated a vast store of pent-up wind within the Earth, which generated friction and high heat and found its escape in volcanoes. After all, the Platonic view was that heat is a kind of motion. Aristotle's concept of volcanism is thus firmly rooted in his ideas about the capacity of motion to generate heat as stated in his *Meteorologica*: "We see that motion can rarefy and inflame air, so that, for example, objects in motion are often found to melt." This theory, which dates in fact to his predecessors, Anaxagoras, Plato, and Democritus, was remarkably long lived and had adherents well into the sixteenth century.

5. INTERNAL COMBUSTION

Initially the Roman philosophers adopted the Greek view of the causes of volcanic eruptions, but eventually they proposed another explanation, one more in keeping with the practical Roman mind. Lucius Annaeus Seneca (2 BC—AD 65) wrote on volcanism in his work on natural philosophy, *Questiones Naturales*. He attributed volcanic eruptions in part to the movement of winds within the Earth, struggling to break out to the surface, thus in part following Aristotle. But his most notable and truly original contribution to theories of volcanism is his proposal that the heat liberated from volcanoes is derived from the combustion of sulfur and other flammable substances within the Earth—an idea that was to have many adherents into and even beyond the Middle Ages. His claim was that there were great stores of sulfur and other combustible substances in cavities within the Earth, and that when the great subterranean wind rushed through these regions, frictional heating would set these fuels on fire. It is in his discussion of hot springs and thermal waters that Seneca first makes reference to sulfur and other combustible materials as a potential heat source, proposing that "water contracts heat by issuing from or passing through ground charged with sulfur." He then extends this process to

explain the blasts from a volcanic eruption in his work *Q. Naturales*:

We must recognise, therefore, that from these subterranean clouds blasts of wind are raised in the dark, what time they have gathered strength sufficient to remove the obstacles presented by the earth, or can seize upon some open path for their exit, and from this cavernous retreat can escape toward the abodes of men. Now it is obvious that underground there are large quantities of sulfur and other substances no less inflammable. When the air in search of path of escape works its tortuous way through ground of this nature, it necessarily kindles fire by the mere friction. By and by, as the flames spread more widely, any sluggish air there may be is also rarefied and set in motion; a way of escape is sought with a great roaring of violence.

Seneca's ideas were long lived and formed the basis of the interpretation of the causes of volcanic eruptions throughout the Middle Ages and even well into the eighteenth century.

Other significant writing on volcanism in the Roman world is found in poetry. The Latin poem *Aetna*, written between AD 63 and 79, is of considerable importance in the history of volcanological thought; its likely author is Lucilius Junior. The poet maintained that the Earth is not solid, but has numerous caverns and passages. Heat, he argues, is more intense and powerful when in action in a confined space within the Earth, with the bellows-like action of winds in subterranean furnaces giving rise to volcanoes. In his view, the flames of Etna were fueled by a combustible substance, such as liquid sulfur, oily bitumen, or alum.

Philostratus the Elder (c. AD 190) also discusses subterranean passages in the Earth, with fires breaking out through volcanoes: "If one wishes to speculate about such matters, the island of Sicily provides natural bitumen and sulfur, and when these are mixed by the sea, the island is fanned into flame by many winds, drawing from the sea that which sets the fuel aflame."

The eruption of Vesuvius in AD 79 marks a watershed in the study of volcanism. When he observed the eruption from Misenum, at a distance of 30 km, Pliny the Elder set out on the first expedition devoted to the study of a volcanic process; he died in the attempt. His nephew Pliny the Younger stayed at home, where he had a spectacular view of the eruption, and wrote the first eyewitness account of the phenomenon, a classic description in the volcanological literature (Figure I2.2).

6. EXOTHERMIC CHEMICAL REACTIONS

Study of the Earth, like many scholarly activities, suffered a setback with the growth of the new Christian religion; and the only role of volcanoes in this new world order was to serve as a reminder of the hellfires burning below. This

FIGURE 12.2 During the AD 79 eruption, Pliny the Younger remained at home in Misenum to do his homework, while his uncle set out on a research and rescue mission toward the volcano, where he perished. *From a painting by Angelica Kauffmann, 1785. Oil on canvas, The Art Museum, Princeton University.*

irrational attitude toward science continued well beyond the Middle Ages. Even by the eighteenth century, most writers on the philosophy of nature still considered that God created the Earth as a habitat for humans, that it had undergone certain stages of evolution since, and that it would ultimately be transformed or destroyed by him. This was still manifest in the writings of John Wesley (1703−1791), the founder of Methodism, who taught that before sin entered the world, there were no earthquakes or volcanoes. These convulsions of the Earth were simply the "effect of that curse which was brought upon the earth by the original transgression."

The Early Middle Ages of Western Europe coincided with a remarkable growth of learning in Asiatic countries, which were in their prime from AD 800 to 1100. Contrary to the Christian philosophy, the Koran encouraged the Islamic scholar to practice the mastery of *taffakur*, or the study of nature. Unlike most learned Europeans, who saw nature as a vivid illustration of the moral purposes of God, the Arabs sought knowledge that would give them power over nature. They were the first alchemists; and one element that figured prominently in their studies was sulfur, which was in part mined from volcanoes. Their discovery that sulfur could also give off heat in chemical reactions was to generate the idea that volcanic action was also fueled by sulfur in the Earth.

The early chemists dealt a death blow to the theory of internal combustion as the source of subterranean fires. In his work on *Discourse of Natural Bathes and Mineral Waters* (1632), Edward Jorden (1632) rejected the notion that the Earth is a hollow and fiery furnace with a universal fire fueled by combustible coal, bitumen, or sulfur.

He pointed out the fundamental problem regarding the internal fire: It would require a tremendous amount of air to keep it burning. Any flame that is confined without access of abundant air would soon be put out, because "fuliginous vapors … choke it if there were no vent for them into the ayre (air)." Such abundance of air could not reach the deeper portions of the Earth to fan the central fire. Instead, Jorden proposed that volcanic regions are underlain at only a shallow depth by "fermenting" material. As a source of the heat, he sought an explanation in the process of chemical reactions as a basis for a new system of the Earth. "Fermentation" or chemical reaction could take place in the presence of water, which was clearly abundant deep in the Earth's crust, whereas combustion could not proceed in the presence of water. Although his reasoning logically should have ended the hypothesis of combustion as the heat source in the Earth, the idea persisted well into the late eighteenth century.

Another pioneer chemist was Robert Hooke (1635−1703), who began his career as Robert Boyle's laboratory assistant in Oxford. Hooke was one of the first to make a clear distinction between heat, fire, and flame, and, together with Boyle, to study the various phenomena of heat. He developed a "nitro-aerial" theory of combustion, in which thunder and lightning were likened to the flashing and explosion of gunpowder, during the combustion of sulfur and niter. These ideas are reminiscent of the work of Liberti Fromondi in the *Metteorologicorum* (1627), in which the explosive power of earthquakes and volcanic eruptions was compared to the effect of gunpowder. Volcanic activity Hooke attributed to "the general congregation of sulfurous, subterraneous vapors." In his work *Lectures and Discourses of Earthquakes and Subterraneous Eruptions* (1668), Hooke also claimed that geologic activity was on the wane, and that subterranean fuel had been more plentiful in the past history of the Earth, when eruptions and earthquakes were apparently more severe: "That the subterraneous fuels do also waste and decay, is as evident from the extinction and ceasing of several vulcans that have heretofore raged." This concept of a finite supply of "subterranean fuels" was widely accepted and was endorsed by Lord Kelvin in 1889.

Isaac Newton (1642−1727) concerned himself with the process of volcanism, but his ideas on the causes of this phenomenon were derived from his contemporary alchemists and early chemists. His ideas on the causes of "burning mountains" are clearly a direct outcome of his secret work on alchemy. In Newton's chemical experiments, he had observed the evolution of heat, or exothermic reactions, when certain substances were mixed together, such as "when aqua fortis, or spirit of vitriol, poured upon filings of iron dissolves the filings with great heat and ebullition." With some other sulfurous mixtures "the liquors grew very hot on mixing as presently to send up a burning flame," and "the *pulvis fulminans*, composed of sulfur, niter, and salt of tartar, goes off with a more sudden and violent explosion than gunpowder." By 1692, Newton had formulated opinions about the origin of heat in the Earth and its intensity, and compared it to the irradiance received by the Earth from the sun: "I consider that our earth is much more heated in its bowels below the upper crust by subterraneous fermentations of mineral bodies than by the sun." In an experiment described in his work *Opticks*, Newton tested his fermentation hypothesis and drew his conclusion on the causes of volcanism:

And even the gross body of sulphur powdered, and with an equal weight of iron filings and a little water made into a paste, acts upon the iron, and in five or six hours grows too hot to be touched and emits a flame. And by these experiments compared with the great quantity of sulphur with which the earth abounds, and the warmth of the interior parts of the earth and hot springs and burning mountains, and with damps, mineral coruscations, earthquakes, hot suffocating exhalations, hurricanes, and spouts, we may learn that sulphureous steams abound in the bowels of the earth and ferment with minerals, and sometimes take fire with a sudden coruscation and explosion, and if pent up in subterraneous caverns burst the caverns with a great shaking of the earth as in springing of a mine.

A new and influential hypothesis on exothermic chemical reactions as a driving force of volcanism was developed with the discovery of the alkali metals by Humphry Davy. The new chemical theory did not call for a fluid interior, a vast storehouse of sulfur, or an internal fire in the Earth, and was consistent with the prevailing idea that volcanic activity in the Earth was progressively decreasing. Davy had a great interest in geology, eventually founding in 1807 the Geological Society of London. That year he was able to isolate by electrolysis for the first time the metals of the alkaline earths—potassium and sodium and later calcium, strontium, magnesium, and barium. The high heat liberated upon the oxidation or "burning" of these metals upon reacting with water, with spectacular flames and explosions, so impressed Davy that it led him to propose a new hypothesis for volcanism in 1808. After his famous chemical discoveries, Davy was convinced that the heat given off during oxidation of the alkalies and the alkaline earths was the source of heat for volcanic action, supposing the existence of large quantities of the metallic alkaline earths within the Earth under volcanic regions.

One of the fiercest critics of Davy's theory about volcanic heat was Gustav Bischof in Germany. The arrangement of volcanoes in great continent-wide lines on the Earth's surface was taken by Bischof as a clear indication of a deep-seated origin of volcanic action, ruling out any superficial source within the crust. Davy had maintained that air could circulate through the Earth via volcanic craters and thus participate in heat-giving oxidation reactions. But

the French chemist L. J. Gay-Lussac had shown by logic that it was impossible for atmospheric air to enter deep into volcanoes, because the pressure of the high-density magmatic liquid is acting outward. Air could not possibly flow into the volcanic system and thus fuel the oxidizing reactions against such a steep pressure gradient.

Charles Lyell in the *Principles of Geology* (1830) considered the possible role of oxidation of the alkalis and alkaline earths as a chemical heat source, following Davy's ideas closely:

Instead of an original central heat, we may, perhaps, refer the heat of the interior to chemical changes constantly going on in the earth's crust; for the general effect of chemical combination is, the evolution of heat and electricity, which, in their turn, become sources of new chemical changes. It has been suggested, that the metals of the earths and alkalis may exist in an unoxidized state in the subterranean regions, and that the occasional contact of water with these metals must produce intense heat. The hydrogen, evolved during the process of saturation, may, on coming afterwards in contact with the heated metallic oxides, reduce them again to metals; and this circle of action may be one of the principal means by which internal heat, and the stability of the volcanic energy, are preserved.

Even near the end of the nineteenth century the chemical theory was still alive.

Despite the many objections, the chemical theory of volcanism was surprisingly long lived and continued to be entertained by prominent scientists well into the middle of the twentieth century. Arthur L. Day was the last of the proponents of the chemical theory, when he proposed in 1925 that chemical reaction between gases plays a leading part in generating volcanic heat. In his view, various gases from different sources in the Earth meet within the crust, and their reactions lead to local fusion and generation of magma. This idea was particularly appealing to Harold Jeffreys, a leading geophysicist in the early twentieth century. He considered volcanism as "local and occasional, not perpetual and worldwide." In accounting for the "local and occasional" eruptions, Jeffreys appealed to Davy's chemical theory of heat generation and melting in the Earth.

7. THE COOLING STAR

During the latter part of the seventeenth century, several philosophers adopted the view that volcanism was due to original or primordial heat in the Earth. The first of these was the French mathematician René Descartes (1596–1650), whose works were to have a profound influence on thinking about the origin of the Earth. The solar system originated, he proposed, as a series of "vortices," with the Earth first appearing as a star, "differing from the sun only in being smaller," and collecting dense and dark matter by gravitational attraction and losing energy as matter falls into place by gaseous condensation. He divided the Earth into three regions: a core consisting of incandescent matter, like that of the Sun; a middle region of opaque solid material (formerly liquid but now cooled); and an outermost region, the solid crust. All of these layers had been arranged in this concentric fashion by virtue of their density. In this scheme, he maintained, there was enough primordial heat remaining to supply any volcano. Descartes's ideas influenced Baron von Leibniz (1646–1716), a German mathematician who proposed that the Earth must have existed originally in a state of fusion, and had thus acquired its spherical form and concentric shell structure, with denser metals concentrated in the center. Lacking an independent source of heat, the planet had cooled by simple conduction over geologic time, forming a stony and irregular crust. The hypothesis of Descartes that the Earth contained a vast store of primordial internal heat was of fundamental significance in evolution of geophysics and volcanology.

8. PLUTONISTS VICTORIOUS OVER THE NEPTUNISTS

Some rocks, which were later found to be of volcanic origin, were thought by the medieval philosophers to have originated by precipitation from water. Thus the Swiss philosopher Konrad Gesner (1516–1565) was convinced that the perfectly symmetrical and hexagonal basalt columns of many lava flows had precipitated as giant crystals from a primordial ocean (Figure I2.3). Two centuries later, Richard Pococke still claimed that the columns of the Giant's Causeway, those huge six-sided pillars of basalt that rise from the sea in Northern Ireland, were formed by precipitation from an aqueous medium. The concept of a crystallization form for these highly ordered rock structures is not surprising, considering their regular shape: The pillars of the Giant's Causeway resemble in many respects giant crystals that might have precipitated out of ocean waters. The debate on volcanic versus aqueous origin of basalt was to become one of the fiercest controversies in the history of Earth science.

The idea of precipitation of rocks from a great ocean had strong theological origins in the legend of the Great Deluge. The leading figure of the Neptunist school was Abraham Gottlob Werner (1749–1817). Among his students were the most famous geologists of his day, including Alexander von Humboldt, Leopold von Buch, Georges Cuvier, Johann Wolfgang Goethe, Jean François d'Aubuisson, and Robert Jameson. Werner considered volcanoes of minor importance, as accidental and relatively recent or postaqueous phenomena on the Earth. They owed their existence, he claimed, to the action of

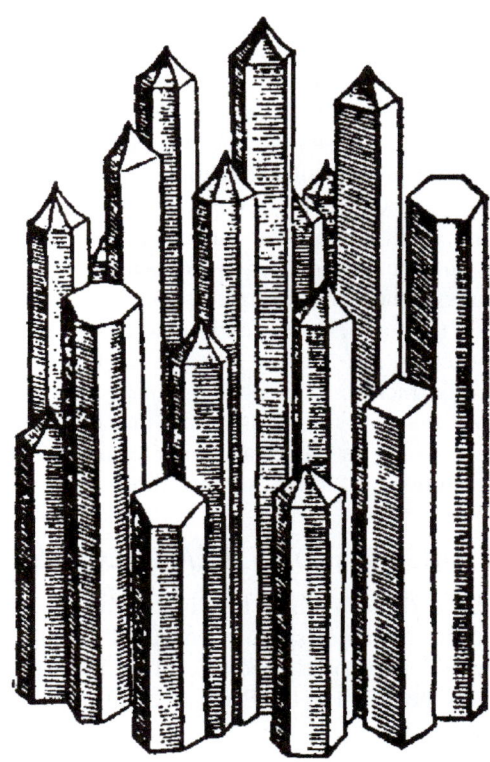

FIGURE I2.3 In the late sixteenth century, Konrad Gesner drew hexagonal basalt columns of lava flows with terminations as crystals, because he believed that they had precipitated as giant crystals from a primordial ocean—in keeping with the Neptunist view of the aqueous origin of basalt.

fire (much as the medieval philosophers had proposed), activated by the ignition of coal deposits or other flammable materials in the Earth: "Most if not all volcanoes arise from the combustion of underground seams of coal." To support this notion, he proposed that deposits of coal or other combustible materials were invariably present in the vicinity of volcanoes.

By the first half of the eighteenth century, chemists were familiar with the formation of crystals in the laboratory as a product of precipitation from an aqueous solution. Because Werner and his followers had recognized that basalt was a crystalline rock, it was perhaps not unnatural that they also deduced its origin as a precipitate from an aqueous solution. It was not until after the middle of the eighteenth century that the idea slowly began to spread that crystallization could also occur as a result of the removal of heat from a silicate melt. In 1761 Pierre Clement Grignon complained that chemists overemphasized aqueous systems, pointing out, after he managed to extract crystals from glass slags, that the products of glass furnaces compared to the products of volcanoes. The idea that magma was a solution that could produce crystals was beginning to emerge, and a group of geologists embraced the view that basalt owes its origin to solidification of magma. The adherents of this theory became known as the **Plutonists**. A profound difference between the Neptunist and Plutonist theories related to the quantity and role of heat in the planet's interior. The **Neptunists** saw a negligible role for heat as a geologic agent and considered volcanoes to be minor phenomena related to shallow-level processes. The Plutonists, on the other hand, saw heat as the fundamental driving force, pointing out the abundance of volcanoes, basalts, and granite as evidence of melting of rocks at high temperature within the Earth. By the early nineteenth century, the Neptunist theory had become severely weakened and was encountering increasing opposition, especially when it was shown that the silicate minerals that compose crystalline rocks such as basalt and granite are insoluble in aqueous solutions at normal temperature.

9. THE FIRST FIELD VOLCANOLOGISTS

Most of the early ideas about heat in the Earth were based on "armchair geology," speculation based on little or no field observation. In the eighteenth century, the approach to the study of the origin of the Earth gradually evolved from pure speculation to a search for answers in rock formations exposed at the surface and in mine shafts. It was the labors of certain men, including those connected with the great mining industry in southern Germany, that brought about the revolution in the study of the Earth and the invention of field geology. A great leap in volcanology was made when geologists were able to identify ancient rocks as of volcanic origin, in regions far removed from active volcanism. This breakthrough occurred in the middle of the eighteenth century in France.

In 1751 Jean-Etienne Guettard (1715—1786) and his friend de Malesherbes made a journey through the Auvergne region, where they noticed some very unusual, black, and porous stones in mile posts, and Guettard immediately suspected that they were lava rock. They then visited Volvic, where Guettard noted dipping layers of the rock with scoriaceous upper and lower surfaces and other unmistakable signs of volcanic activity. He concluded that the stones were indeed from a large lava flow. When they came to the area of Puy de Dome, Guettard recognized that basaltic rock of the hardened lava had flowed from an ancient crater in the cone-shaped hill above (Figure I2.4).

The geologist who first demonstrated the volcanic origin of basalt was Nicholas Desmarest (1725—1815). In 1763 he found prismatic basalt forming hexagonal columns in southwest Clermont. Tracing the rocks to their source, he found similar columns standing vertically in the cliff above, grading upward into the scoriaceous top of the lava flow. His simple observation of the association of columnar basalt as a component of a lava flow stands as a major

FIGURE I2.4 The pioneer geologists Nicholas Desmarest and Jean-Etienne Guettard identified ancient rocks in central France as volcanic in origin, although they were far removed from active volcanoes. This 17th-century print shows hexagonal columnar basalts in the Auvergne region, where Desmarest demonstrated their volcanic origin.

advance in science (Figure I2.5). Although Desmarest had demonstrated unequivocally the volcanic origin of basalt, the debate between the Vulcanists and Neptunists continued to rage into the nineteenth century. His theory of basalt as lava was bitterly opposed by many.

Among the pioneers in field volcanology was William Hamilton, who resided in Naples for several decades and had ample opportunity to study Vesuvius. Among his many observations were precise drawings he made of the changing outline of the volcano during the eruption of 1767 (Figure I2.6). Hamilton was also fortunate to be resident in

FIGURE I2.5 When the volcanic origin of columnar basalt was accepted, the early volcanologists quickly showed that it occurred as lava flows, and was connected to volcanic vents, as shown in this eighteenth-century engraving.

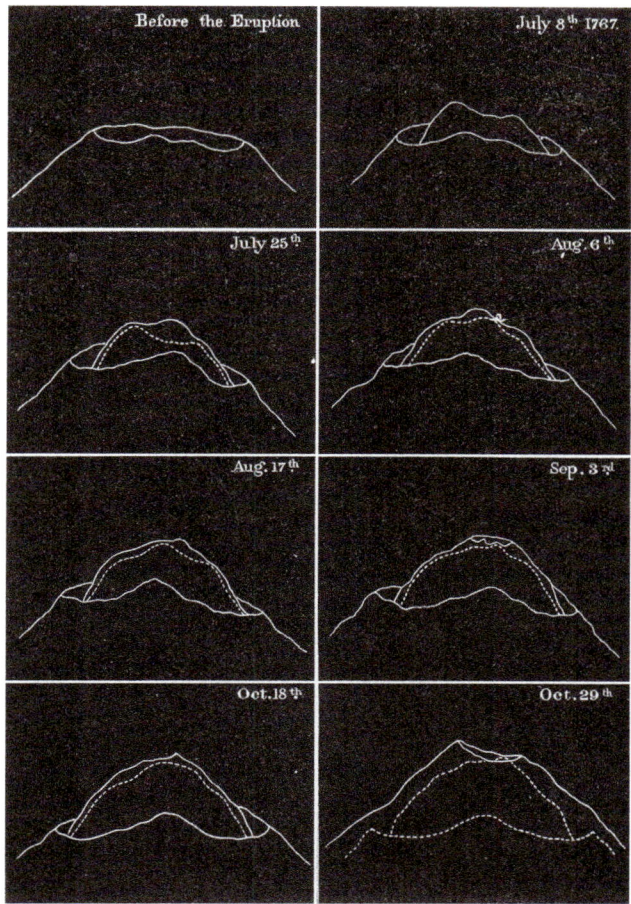

FIGURE 12.6 William Hamilton made some of the first quantitative measurements of volcanic morphology, when he carefully traced the changing outline of the summit of Vesuvius during the 1767 eruption.

FIGURE 12.7 An eighteenth-century engraving showing the excavations of the Roman city of Herculaneum, buried by the AD 79 eruption of Vesuvius. Although the impact of the excavations was primarily on archaeology and art history, it was also a forceful reminder of the total destructive force of a volcanic eruption for visiting scholars like Johann Wolfgang Goethe and William Hamilton.

Naples when the early discoveries and excavations were being made in Herculaneum and Pompeii, findings that brought to light the destructive power of the AD 79 Vesuvius eruption (Figure 12.7).

10. EXPERIMENTS OF THE PLUTONISTS

The undisputed leader of the Plutonists was the Scottish geologist James Hutton (1726−1797). He demonstrated the phenomenon of intrusion of magma into layered strata, and proposed that this molten rock originated in the highly heated interior of the deep Earth. Hutton devoted much energy to field work, examining geologic strata in Scotland, England, and France.

Hutton was an intensely religious man and argued that the Earth's processes followed a divine plan. Thus, volcanoes existed both as safety valves for the release of excess heat from within the Earth, and as a global force, because their expansive and uplifting power was needed to raise the surface of the Earth and contribute to the soils, so that plants and animals might flourish. Hutton thus saw volcanism as a means toward the realization of a higher purpose in the natural order of things.

Hutton and his followers had observed layers in the Earth's crust that, while they resembled basalt in mineralogical and other properties, were sandwiched between layers of sedimentary rock. They recognized that these rocks, which they referred to by the common mining term *whinstone*, were not volcanic but of igneous origin, and had been forcefully intruded between the sedimentary layers. Furthermore, they concluded that these rocks had solidified from magma.

Many of Hutton's opponents had maintained that the melting and solidification of crystalline rocks, such as granite or gabbro, would only yield an amorphous, glassy mass upon cooling, as observed in glass-making furnaces, and therefore the process could not produce basalt, whinstone, or other volcanic rocks. This objection to the Huttonian theory was tackled experimentally by the Scottish geologist and chemist Sir James Hall (1761−1832). To test Hutton's claim that rocks such as whinstone or basalt were derived from magma, Hall set out to do experiments on melting and cooling of basalts. First Hall melted the lava and basalt specimens, and then he converted the melted rock to glass by quenching. Next Hall remelted the crystal-free glass and allowed the melts to cool slowly. In so doing he produced crystalline or partly crystalline rocks, rough and stony in texture, which resembled the original rock samples. These rocks, he showed, had melting and crystallization features exactly the same as those of basalts.

Demonstrating that melts of basaltic rocks precipitate silicate crystals upon cooling was of fundamental importance in establishing the volcanic theory on the origin of basalt. But if basalt rock was merely a product of solidification from a high-temperature silicate melt, then the experimental fusion and recrystallization of the rock should not affect its chemical composition. To test this, Hall submitted specimens of the products of his melting experiments to Robert Kennedy for chemical analysis in 1805, who found that their composition was essentially the same as the original rocks. Hall's demonstration, which was of fundamental importance in establishing the volcanic theory for the origin of basalt, dealt a final blow to the Neptunist theory of an aqueous origin for basalt, granite, and other rocks of igneous origin. Although Hall was a pioneer in experimental work on high-temperature and high-pressure experimental petrology, he was not the first to conduct experiments with molten rocks. The first melting experiments on basalts were done by the Italian scholar Francesco d'Arezzo in 1670, who fused Etna's lavas.

11. A SOLID EARTH

Most geologists studying volcanic activity in the early and mid-19th century continued to envisage a planet with a solid crust and a largely molten interior, influenced by the ideas of Descartes. Heat within the Earth was considered a residue of the central, primitive heat held by the Earth at the time of its formation, a supply that had been diminishing over geologic time. The idea of a central, primitive heat was developed primarily from the observation of increasing temperature with depth in the Earth, based on measurements in mines and drill holes. Such observations led to the theory that the Earth began as a molten sphere, and has been cooling ever since. But if the interior of the globe were molten, it should yield to the gravitational attraction of the Sun and the Moon, and there would be little or no relative movement of the Earth's surface and the ocean's, that is, no sea tide. The response of the Earth to tidal forces, then, became a critical test for the molten Earth hypothesis. The French physicist Andre-Marie Ampere (1833) was the first to point out that observations of tidal action did not support the hypothesis of a fluid interior.

From 1839 to 1842 William Hopkins analyzed the effects of the Moon and the Sun on the rotation axis of the Earth. He concluded that the outer rigid crust must be at least 1500 km in thickness: "We are necessarily led, therefore, to the conclusion that the fluid matter of actual volcanoes exists in subterranean reservoirs of limited extent, forming subterranean *lakes*, and not a subterranean ocean." Hopkins's studies were continued by his student, William Thomson (Lord Kelvin), who showed, on the basis of the effects on the Earth of gravitational attraction of the Sun and the Moon, that the Earth was essentially rigid and that the ideas of James Dwight Dana of "an undercrust fire-sea" were untenable. Geologists were now in a great quandry: There was no evidence for a large body of magma

within the Earth, yet there was a requirement for supplying magma to countless volcanoes throughout the history of the planet. How can you derive magma from the interior of a solid planet?

The Earth's structure at the turn of the century was regarded as a series of concentric, solid shells or layers, with an outermost 30- to 40-km-thick, low-density layer of granitic crust (sial) making up the continents. This was underlain by a very thick layer (sima or the mantle) of high-density, ultrabasic rock with the composition of periodite, and finally by the central core. Those who were searching for a molten region in the Earth as a source of magmas were now faced with a geophysical picture of a largely solid interior. A magma source in the molten core seemed impossible, at a depth of more than 2900 km below the surface.

12. DECOMPRESSION MELTING

The discovery that the Earth beneath the crust was essentially solid completely eliminated the "undercrust fire-sea" that Dana and others had proposed as a source of the magmas that feed Earth's volcanoes. Scholars, now faced with the task of showing how hot magma could be derived from within a solid Earth, found in the science of thermodynamics a partial solution to the problem. In the early years of the eighteenth century, it had been appreciated that pressure influences the temperature at which substances will undergo a change of phase—such as from a liquid to a gas, or from a solid to a liquid. The possible effects of pressure on melting of rocks in the Earth had been qualitatively appreciated very early on in the development of geologic thought. Perhaps the earliest recognition of this fundamental effect was made in 1802 by John Playfair. He pointed out that just as a change in pressure affects the boiling point of water, so would melting in the Earth be influenced by the great pressure exerted by the overlying rocks.

The next to address the importance of pressure in volcanic systems was George Poulett Scrope. By 1825 in his work *Considerations on Volcanos*, Scrope had realized the effect of pressure on the solubility of water in magmas, and was one of the first to point out that decrease of pressure on a water-rich magma could explain volcanic explosions due to the release of dissolved water. He also argued that a change of pressure could either lead to melting or crystallization of magma, without change in temperature:

Having thus far considered the effect of an increase of temperature, or a diminution of pressure, on a mass of lava under such circumstances, let us examine what will follow from the reverse; namely, an increase of pressure, or a diminution of temperature. Upon the solid lava, it is clear, no corresponding change will be produced; but every diminution of temperature, or increase of pressure, on a mass, or a part of the mass, liquified in the manner stated above, must occasion the condensation of a part of the vapour which produces its liquidity, and so far tend to effect its reconsolidation.

Simeon-Denis Poisson had proposed in 1835 that the excessively high pressure in the deep interior of the Earth would lead to solidification of rock material at much higher temperatures than at the low pressures near the surface.

While Playfair and Scrope were struggling with a qualitative approach to the question of the effect of pressure on melting, a more elegant and quantitative approach was being developed in France and Germany, marking the birth of thermodynamics. From the work of Sadi Carnot, Rudolph Clausius, and Benoit-Pierre-Emile Clapeyron came the Clausius-Clapeyron equation, which quantifies the relationship between temperature, volume and pressure of a substance:

$$\frac{dT}{dP} = \frac{T\Delta V}{\Delta H}$$

Consider two phases of the same chemical substance, for example, a liquid and solid (magma and rock), in equilibrium with one another at temperature T and pressure P. By supplying heat slowly to the system, one phase changes reversibly into another to bring about melting, with the system remaining at equilibrium. In the equation, the fraction dT/dP represents the rate of variation of the melting point T with change in pressure P. The value ΔV represents the volume change on melting and is generally positive (i.e., the specific volume of the solid is smaller than the volume of the corresponding liquid), while the value ΔH is the entropy change. The equation expresses the variation of pressure and temperature for a system in equilibrium. The magnitude and sign of the slope of the melting curve, dT/dP reflects the magnitude and sign of the volume change, ΔV, of the substance in question upon solidification or freezing.

It was shown early in the eighteenth century, in melting and crystallization experiments, that the specific volume of a volcanic rock such as basalt is lower than that of the corresponding magma. Many of the geologists who studied columnar basalt correctly interpreted its structure as evidence of contraction of magma upon cooling and solidifying to rock. During cooling and solidification, the material shrunk in volume, forming roughly hexagonal columns separated by a pattern of contraction joints. Thus it was clear that the term ΔV in the Clausius–Clapeyron equation was positive, and consequently, the equation predicts that the pressure–temperature melting curve (dT/dP) of the source rock of basaltic magma has a positive slope in the Earth, that is, the temperature of melting increases with pressure. This interpretation of columnar basalt was consistent with the experimental results of the German chemist Gustav Bischof

in 1837, who carried out one of the first measurements on the volume change of basalt and other volcanic rocks upon fusion and showed that rocks contract upon solidification from magma. He concluded that a melt of granite rock contracted about 25% upon solidification, trachyte about 18%, and a basaltic melt 11%.

Another German chemist, Robert Wilhelm Eberhard Bunsen, was one of the first to experiment on the relation between pressure and melting point of substances. The laboratory facilities available in Bunsen's time (1850) permitted only modest pressures to be achieved, roughly the equivalent of the pressure at a depth of 1 mile in the ocean. He therefore performed his experiments on materials with a relatively low melting point, such as ambergris and paraffin, and extrapolated these results to the high pressures and temperatures prevailing deep in the Earth. His work showed that a pressure increase of only 100 atm increased the melting point of these substances by several degrees centigrade. At about the same time (1851), William Hopkins began to experiment with James Joule, Lord Kelvin, and William Fairbairn, on the effects of pressure on the solidification of the Earth's interior. A large lever apparatus generated pressure up to 5400 atm, equivalent to the pressure at approximately 15-km depth in the Earth. Initially the experiments were on substances with low melting point, such as beeswax and spermaceti, and they essentially confirmed the work of Bunsen.

These pioneers had established that a solid hot substance such as rock at depth in the Earth could spontaneously begin to melt if the presure is decreased, without the addition of heat. They had finally found the secret of the generation of magmas in a solid Earth: decompression melting. But the solution of one problem led to the creation of another, even more daunting problem: What is the mechanism that brings about decompression? Some ingenious proposals were put forth to solve this problem. Thus in 1878 the American geologist Clarence King, realizing that local relief of pressure leads to melting in the Earth, suggested that the pressure release could occur with the erosion and removal of overlying crustal rocks: "So that the isolated lakes of fused matter which seem to be necessary to fulfill the known geological conditions may be the direct result of erosion." This seemed perfectly logical at the time, but if true, the rate of erosion of a given area must occur at a higher rate than that of heat conduction from the rising hot rock. Geologic evidence did not support this theory on two counts. First, many mountainous regions, where erosion is highest, show no signs of volcanism; and second, the rate of pressure relief due to erosion is so slow that heat lost through conduction would prevent melting at depth.

As director of the U.S. Geological Survey, King fostered the fundamental experiments of Carl Barus on rocks at high pressure. Barus determined the volume

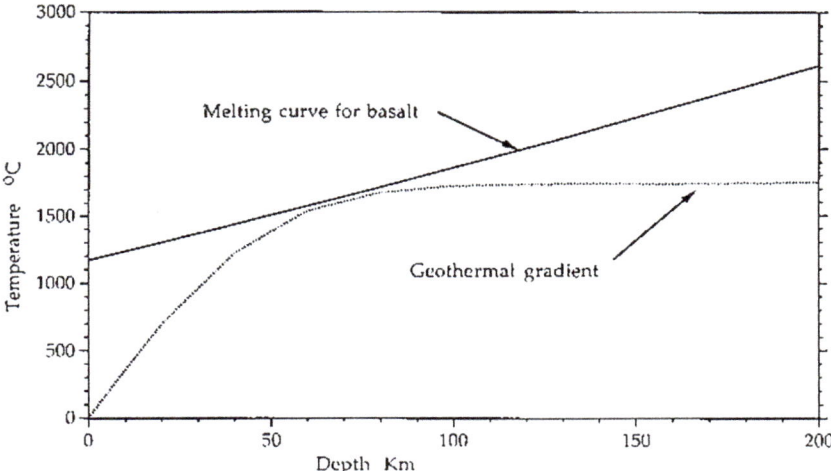

FIGURE I2.8 At the end of the nineteenth century, a geotherm had been calculated for the earth by Clarence King and others. The temperature of melting of basalt as a function of pressure had also been determined experimentally by Carl Barus. It was assumed that melting of the Earth's mantle occurred in the depth range where the geothermal gradient approached the basalt melting curve, that is, in the depth range of about 60–80 km below the surface.

change and latent heat of basalt upon melting. He concluded that the term dT/dP in the Clausius—Clapeyron equation was equivalent to 0.025 for basalt, that is, the melting point increases at a rate of 2.5 °C for every 100-atm increase in pressure. In 1893 Carl Barus was also the first to determine the melting curve of basalt as a function of pressure, which enabled King to propose a geothermal gradient for the Earth (Figure I2.8).

In 1909 the British geologist Alfred Harker summed up his opinion on magma generation in *The Natural History of Igneous Rocks*: "We must seek the immediate cause of igneous action, not in the generation of heat, but chiefly in *relief of pressure* in certain deep-seated parts of the crust where solid and molten rock are approximately in thermal equilibrium. We are thus led by an independent line of reasoning to the principle already enunciated, which connects igneous action primarily with crustal stresses, and so secondarily with crust-movements." Furthermore, "at a sufficient depth, such conditions of temperature prevail that solid and liquid rock are in approximate thermal equilibrium. Any local relief of pressure within that region, connected with a redistribution of stress in the crust, must then give rise to melting." These statements were an incisive expression of the fundamental principle of melting, but what was still lacking was the mechanism that brought about the relief of pressure.

Not all geologists were ready to embrace this as a solution to the origin of magma, because a viable geologic process that could bring about a pressure decrease was not known. The American geologist Reginald Aldworth Daly stated (1933): "Local relief of pressure is hopelessly inadequate and need not be further discussed." Similarly, according to S. James Shand (1949), this pressure effect on

melting of silicates was "a quantity scarcely large enough to have any important petrological consequences."

The progress made in understanding melting and the internal constitution of the Earth by the first part of the twentieth century was truly profound, but it had created a paradox. All the evidence was in favor of a solid interior or an exceedingly thick crust at any rate, yet there was a need to account for the magmas erupted from volcanoes. Experimentalists and geophysicists had discovered a process by which melting could occur simply by decompression, but the geologists were unable to find an acceptable decompression mechanism to produce this melting.

13. STIRRING THE POT: CONVECTION

The discovery of radioactivity caused a great stir among scientists, who quickly applied it to an understanding of the Earth. When Pierre Curie discovered in 1903 that radioactive materials emit heat, the Irish physicist John Joly was the first to point out that the abundance and distribution of radium and other radioactive elements may determine the terrestrial heat. In 1910 the British geologist Arthur Holmes (1890–1965) began a study of the natural radioactivity of rocks, and thus began a career that would make a major contribution to both understanding of heat distribution and melting of the Earth by convective flow of solid rocks toward the surface. By 1915, Holmes had calculated a temperature profile for the Earth based on radioactive generation of heat.

After radioactivity was discovered as an internal heat source, it became even more important to establish the principal means of heat loss from the planet. How is this great flux of heat transferred from the deep Earth toward the surface? Holmes proposed that convection was the most effective mechanism to transport heat from depth to the surface. It was an idea that dates back to Osmond Fisher (1881), who proposed, in the days of an Earth with a totally molten interior, that there were convection currents rising under the ocean and descending beneath the continents. In 1928 Holmes proposed that the excess heat is discharged from the Earth by circulation of material in the solid *substratum*, or the Earth's mantle, forming thermal convection currents. Convection, and in turn continental drift, was seen as a mechanism to rid the Earth of the great heat generated by the radioactive natural reactor. He considered this substratum to be a solid **peridotite** rock, which could flow like a fluid at a rate plausible for a deep Earth process—about 5 cm/year. Holmes's scheme included downward currents below geosynclines, the great sedimentary troughs that today we recognize as subduction zones, and ascending currents below midoceanic "swells," where "a discharge of a great deal of heat" occurs due to upwelling of peridotite mantle, decompression, and melting. Holmes's views in 1928 were amazingly modern and laid the foundations for the concept of plate tectonics, although this has not been generally recognized by Earth scientists.

The concept of plate tectonics had a long gestation period, and it began with the realization in the seventeenth century that volcanoes and earthquakes are distributed in great linear belts over the Earth. This systematic trend of volcanoes and earthquake zones, and the arrangement of the continents, eventually led to the theory known as continental drift in the early twentieth century. The earliest map showing the global distribution of volcanoes was drawn by Athanasius Kircher in 1665. With the spread of geographic knowledge, a more refined picture gradually emerged, one of the systematic, linear or curvilinear arrangement of volcanoes on the surface of the globe. Ideas on the relationship between volcanic activity and major structures and tectonics date to Alexander von Humboldt (1822), who pointed out that the linear arrangement of volcanoes on the Earth was proof that the mechanism that generates volcanism was a deep-seated feature. Robert Mallet (1858) compiled the first map showing the global distribution of both earthquakes and volcanic eruptions in a "vast loop or band round the Pacific," indicating a strong coincidence in their geographic distribution. He also recognized that most of the Earth, especially the interior of the continents, is seemingly free of earthquakes and volcanic activity. He divided the Earth into a series of basins, separated by "girdling ridges," which are present even on the ocean floor: "It is along these girdling ridges, whether mountain-ranges or mere continuous swelling elevations of the solid, which divide these basins beneath the ocean surface one from the other, that all the volcanoes known to exist upon the earth's surface are found, dotted along these ridges or crests in an unequal and uncertain manner." He

regarded these ridges as "sub-oceanic linear volcanic ranges" which mark "the great lines of fracture of the earth's crust." Mallet's recognition of the distribution of these geological structures, which are now basic to the theory of plate tectonics, was probably the earliest.

The recognition of the linearity of volcanoes and earthquake zones coincided with a mobilistic view of the Earth toward the end of the century, a view based both on geophysical reasoning and geologic evidence. Osmond Fisher had argued in 1881 that cooling of the Earth could not be brought about solely by the process of conduction, as had been proposed by Kelvin, but was in part brought about by convection currents in a plastic substratum. In Fisher's view, these convection currents led to lateral movement of the overlying crust, accounting for much of the character of the Earth's surface, including linear arrangement of volcanoes, earthquakes, and mountain chains. The ideas of a mobile Earth were first crystallized into a coherent theory of continental drift in 1912 by Alfred Wegener. One of those who sided with Wegener was Arthur Holmes, who greatly strengthened the theory by proposing in 1928 a more plausible thermal convection mechanism for crustal movement than Wegener had put forward. Holmes's mechanism involved the flow or upwelling of the Earth's mantle, providing a process whereby decompression of the hot mantle would occur, leading to melting and volcanism. Continental drift was an inevitable consequence of this circulation, but the process was much wider in scope: It involved the lateral motion of the oceanic crust as well. By 1931 Holmes had constructed a complete theory of continental drift, sea-floor spreading and subduction, based on this mechanism.

Holmes differentiated between a largely granitic crust in the continents and a basaltic crust in the oceans. He considered the mantle beneath the oceanic crust as most likely of peridotite composition, and argued that although it was probably crystalline and thus highly rigid, it had some of the properties of a fluid in the context of its large-scale dimensions. He proposed that the convection cells have dimensions at the scale of the mantle, some 2900 km in depth, and that their ascending limbs rise beneath the oceans, where they generate oceanic "swells" analogous to our modern concept of midoceanic ridges. Ascending currents he proposed were accompanied by decompression and partial melting of peridotite, to form basaltic magma, giving rise to volcanism. Descending currents were caused in the mantle by sinking of high-density rocks. The process he outlined was essentially the same as the modern concept of subduction: The convergence of crustal plates on the Earth leads to underthrusting of one plate and its descent deep into the mantle. Holmes estimated that the velocities of the convection current were on the order of 5 cm/year, which, as was discovered in the 1960s, is perfectly within the range of typical sea-floor spreading rates. Although widely accepted in Great Britain, Holmes's ideas on mantle convection and a mobile Earth were generally not embraced elsewhere for some time. When the idea of sea-floor spreading was finally accepted by a majority of Earth scientists in the 1960s Holmes's fundamental contribution was frequently ignored. One is reminded of the chilling words of Sir William Osler: "In science the credit goes to the man who convinces the world, not to the man to whom the idea first came." Is it more important in science to convince than to discover?

The discovery of plate tectonics has now provided the long-sought mechanism for decompression in the Earth's mantle—the most important process that brings about melting and generation of magmas. It is clearly the mechanism that generates magmas below the midocean ridges of the Earth and is thus responsible for most of our planet's volcanism.

14. BIRTH OF PETROLOGY

Among the earliest observations on the mineral content or petrography of volcanic rocks were Leopold von Buch's studies in 1799 on volcanic rocks near Rome. Von Buch concluded that the crystals of leucite had formed while the lava was still fluid, and discounted any hypothesis to the effect that they had been precipitated from an aqueous solution and later incorporated in the magma. Basalts and other volcanic rocks, however, are composed of exceedingly small crystals, which are invisible with the naked eye or even through a magnifying glass. The pioneer geologists were therefore unable to investigate the internal structure of these rocks and identify their tiny crystal components. The microscopic study of volcanic rocks was first carried out by Pierre Louis Cordier in 1815, who crushed basalts and other volcanic rocks to a fine powder, separated the various particles by a flotation process, and examined them by microscopic and chemical tests. He concluded that ancient basalts were very much like modern volcanic rocks in texture, mineralogy, and other characteristics—an important step in solving the Neptunist versus Plutonist controversy over the origin of basalts.

The art of making thin sections began with the sectioning of fossil wood, but the method was soon applied to rocks. They could be sliced so thin as to be transparent, and their minerals could be identified and their textures studied in detail. By 1829 the polarizing microscope had been invented by William Nicol and was used widely for the study of crystals. Henry Clifton Sorby (1826—1908) was the first geologist to make thin sections of volcanic and other igneous rocks for microscopic study.

Another crucial step in the study of volcanic rocks was the development of chemical analysis. Studies of the chemical composition of volcanic rocks date back to the work of Abbé Lazzaro Spallanzani (1794), who reported on the chemical composition of volcanic rocks in Italy in terms of weight percent of the five major oxides of silica, alumina, lime, magnesia, and iron. In 1805 the Scottish chemist Robert Kennedy reported chemical analyses of *whinstone* or basaltic rocks. He was able to dissolve basalt by melting the rock powder with caustic potash at high temperature in a silver crucible. He then showed that it contained 48 wt% silica, 16% alumina, 9% lime, 16% iron oxide. Water and other volatile components accounted for about 5%. About 6%, then, was not accounted for. Kennedy suspected this was in part "a saline substance," and, after an elaborate analytical routine, showed that it was about 4% soda. He thus accounted for 99% of the chemical constituents of the rock—a great feat at the time.

When the early geologists began to accumulate data on volcanic rocks, they discovered that the chemical composition of the rocks, even those from the same volcano, varied greatly. Confronted with the problem of how to account for this diversity, George Poulett Scrope proposed in 1825 that all types of igneous rocks were formed from a single parent magma, which gave rise to a variety of types through a process of *differentiation* before crystallization. It was Charles Darwin, however, who laid the foundation for the process of "magmatic differentiation," through his discovery of crystal settling due to density differences between crystals and the magmatic liquid. During his exploration of the volcanic Galapagos archipelago in 1835, Darwin studied a basaltic lava flow on James Island (now San Salvador). He noted that the crystals of feldspar were much more abundant in the base of the lava than in the upper part, and deduced "that the crystals sink from their weight." Von Buch had earlier (1818) noted a similar concentration of crystals in the lower part of obsidian lava flows on Tenerife in the Atlantic, interpreting this as the settling of crystals in the lava during and after flow. In his *Geological Observations on The Volcanic Islands*, Darwin considered this observation as "throwing light on the separation of the high silica versus low silica series of rocks." He realized that once early formed crystals sink, the remaining magma would be chemically different from the initial liquid.

Another process was developed by Robert Bunsen in 1851 to account for the chemical range of volcanic rocks, after his discovery of two magmas in Iceland, with radically different chemical composition. Bunsen was struck by the abundance of two volcanic rock types: yellow and multicolored, silica-rich volcanic rocks (rhyolite), and the dark-colored basaltic rocks. A few volcanic rocks were of intermediate composition, and he proposed that they were the result of mixing of the silicic and basic magmas. Bunsen's contribution was to put forward the first viable hypothesis that could account for the diversity of the chemical composition of volcanic rocks by the mixing of two primary magma: basalt and rhyolite. Traveling with Bunsen in Iceland was Sartorius von Waltershausen. He was the first to discuss the distribution and occurrence of the trace elements, as well as the major oxides in volcanic rocks, and listed 23 trace elements.

In the eighteenth century the volcanic fluid had become known as *magma*, derived from the Greek word which refers to a plastic mass, such as a paste of a solid or liquid matter. The term magma was used early on in pharmacy as in "magnesia magma," or milk of magnesia, and the word passed from pharmacy into chemistry to represent a pasty or semifluid mixture. From chemistry it passed into petrology to replace "subterraneous lava." What was the nature of this fluid inside the Earth? Bunsen, in *Uber die Bildung der Granites*, stressed that magmas are nothing more than a high-temperature solution of different silicates: "No chemist would think of assuming that a solution ceases to be a solution when it is heated to 200, 300 or 400°, or when it reaches a temperature at which it begins to glow, or to be a molten fluid." This was a fundamental breakthrough in thinking about crystallization from magmas. Bunsen further recognized that minerals crystallized from magmas at a much lower temperature than the melting point of the pure minerals. If magmas are true solutions, then they should be able to mix, as indeed Bunsen had proposed on the basis of his Iceland work.

During the latter part of the nineteenth century the chemical analysis of volcanic rocks became commonplace. Leading petrologists considered that most igneous rocks were derived from basaltic magma by a process of differentiation, but no one researcher contributed more to the understanding of the processes that account for the chemical diversity of igneous rocks than the American petrologist Norman L. Bowen. From high-temperature experiments, Bowen discovered that magma could greatly change in composition as it cooled, if the early formed crystals were effectively separated from the liquid, thus providing proof for the fractional crystallization process first proposed by Darwin. Bowen was a strong advocate for fractional crystallization as the primary process responsible for differentiation, and considered that basaltic magma was the primary or parent magma of all volcanic rocks. The new understanding of magmatic differentiation was made possible by Bunsen's recognition that magmas were true solutions, Darwin's concept that settling of crystals resulted in a change in the composition of the liquid, and Bowen's systematic experiments on crystallization of a primary basaltic liquid. But how, then, was this primary basaltic liquid formed?

15. THE SOURCE

Throughout the first half of the twentieth century, two principal hypotheses existed on the nature of the source of magmas. Because the principal type of magma erupted was basalt, it appeared logical to some that the source region was also basaltic in composition, and possibly the high-pressure form of basalt known as **eclogite**. The basaltic magma would then be derived by wholesale melting of the source. The other view was that the magma was derived by partial melting of peridotite. It had long been noted that **xenoliths** of both peridotite and eclogite were ejected from some volcanoes. Because they were brought up in the magma, it seemed logical that these rock types might be an indication of the source region. As early as 1879 it was suggested that peridotite was the source of basaltic magmas, and by 1916 Arthur Holmes proposed that magma was generated by melting of peridotite at a depth of several 100 kilometers.

At the beginning of the twentieth century, geophysical measurements were giving results in favor of a peridotite source. This was done by measuring in the laboratory the physical properties of a number of rock types under high pressure and temperature. By comparing these results with the speed of earthquake waves traveling through the deep Earth, it was shown that properties of the Earth's interior are comparable to those of peridotite and other ultrabasic rocks under great pressure in the laboratory. Geophysicists therefore generally adopted a peridotite composition for the mantle.

Bowen (1928) also proposed that the only rock type that could give rise to basaltic magma by partial melting is peridotite, and suggested that this occurs at depths of 75—100 km. Bowen then addressed the question of whether the partial melting occurred as a result of reheating or due to pressure release. Arthur Holmes's ideas on convective flow in the mantle were published in the same year as Bowen's classic work, but Bowen seemed unaware of them. Having no plausible mechanism to bring about a pressure release, he stated: "It seems necessary to leave open the question whether selective fusion takes place as a result of release of pressure or as a result of reheating." He therefore had considerable difficulty in developing a scenario of pressure-release melting in the peridotite mantle. Only with the development of geodynamics and acceptance of Holmes's theory of mantle convection could it be shown that a peridotite mantle can rise to regions of lower pressure, thus bringing about melting without the addition of heat.

The theory of partial melting of peridotite was still debated and considered on a weak foundation by many in the first half of the twentieth century. Another group of geologists adopted a totally different view, and maintained that basaltic magma was derived from a deep melted layer or from the complete melting of a basaltic layer at depth in the Earth. This view persisted with some until the 1960s. In an influential book on volcanoes, Alfred Rittmann proposed in 1936 that all volcanic rocks were ultimately derived from a global layer of basaltic magma. Rittmann's views were adopted by the petrologist Tom F. W. Barth in 1951: "The fact that wherever the crust is deeply rifted, be it in continents, oceans, or geosynclines, basaltic magma is available and capable of invasion is a proof of the existence of a subcrustal basaltic magma stratum." He was fully aware of the objections of geophysicists, that such a layer could not transmit some seismic waves, contrary to observations, and proposed that the basaltic magma behaved like glass at the high pressures beneath the Earth's crust and was thus able to transmit these waves. In his 1962 work on *Theoretical Petrology* Barth stated: "It is necessary, therefore, that the basaltic substratum is in a molten (noncrystalline) state at its subterranean locale. Otherwise it would not reach the surface with a homogeneous composition as we actually observe."

The American geologists Francis J. Turner and John Verhoogen accepted the seismic evidence for a solid character of the Earth's mantle in 1951, and considered it composed either of peridotite or of eclogite. From this it followed that primary basaltic magma had to be derived from a largely solid mantle, whose composition was other than basalt. The observed temperature increase with depth and heat generation by radioactive decay within the Earth could produce the required heat source. In their classic textbook on *Igneous and Metamorphic Petrology*, Turner and Verhoogen stated: "The problem of the origin of basaltic magma is thus not so much that of finding adequate heat sources as that of explaining how relatively small amounts of heat may become locally concentrated to produce relatively small pockets of liquid in an otherwise crystalline mantle." In what seemed like a fleeting moment, Turner and Verhoogen considered the hypothesis that basaltic magma could be produced by melting of a peridotite mantle due to pressure release, but then quickly rejected it: "This hypothesis of magma generation is possibly the most popular one, although it is unacceptable to the present writers. It is difficult indeed to see how pressure can effectively be reduced at such depths." They realized of course, that melting could occur if convective transport of mantle material occurred to shallower levels, as originally proposed by Arthur Holmes. This was clearly a viable mechanism as long as the deep mantle temperature was greater than the melting point of the mantle at shallower depth. Yet they were not ready to accept the existence of convection in the solid Earth, which led them to conclude: "Whether convection does occur in the mantle is not definitely known, nor is it known whether it could be effective in the upper part of the mantle where magmas are generated. Thus, although convection might lead to melting, it cannot be shown that it does, and the problem of the generation of magma remains as baffling as ever."

This statement is a good reflection of the state of affairs regarding theories of magma generation at the middle of the twentieth century. While great strides had been made in understanding the chemistry and mineralogy of volcanic rocks, the Earth scientists were at a loss to explain melting. This was due almost entirely to the static view taken of the Earth's interior at the time—it seemed inconceivable to most that the rigid and solid mantle could convect. However, the discoveries made on the ocean floor in the 1960s and development of geodynamics were to change all that. Although melting was poorly understood at the time, knowledge of the mantle source region was advancing through studies of xenoliths. A great number of xenoliths are of peridotite composition, giving further credence to the idea that the source region of basalt magma, the Earth's mantle, is dominantly composed of peridotite.

16. THE ROLE OF WATER

Great clouds of gases and steam emitted by volcanoes were long considered to be derived from groundwater, surface waters, such as nearby lakes or streams, or seawater. This was evident, according to many scholars, from the location of volcanoes near the ocean or on islands. The role of water was thought to be crucial in generating explosive eruptions and to have an important effect on the viscosity of magmas. In 1794 Spallanzani recognized that several gases are important in lavas and volcanic regions, including "hydrogenous gas, sulfurated hydrogenous gas, carbonic acid gas, sulfurous acid gas, azotic gas." But another, more powerful agent, he noted, was "water, principally that of the sea," which communicates by passages with the roots of volcanoes. On reaching the subterranean fires it suddenly turns to vapor, and the elastic gas expands rapidly, causing volcanic explosions. Supporting his hypothesis, on a more practical level, Spallanzani cited accidents in glassmaking factories, in which molten glass was poured into molds not completely dry or free of water, causing dreadful steam explosions.

Scrope (1825) attributed the fluidity of magmas to water: "There can be little doubt that the main agent ... consists in the expansive force of elastic fluids struggling to effect their escape from the interior of a subterranean mass of *lava*, or earths in a state of liquification at an intense heat." These "elastic fluids" Scrope considered to be mainly steam and other volcanic gases. It was the expansion of gas in the magma that led to its rise in the Earth's crust, and led to violent and explosive eruption upon reaching the surface.

Scrope discussed in some detail the evolution of steam in magmas at depth, and was perhaps the first to point out that under great pressure, water would be dissolved in the melt, but upon decrease in pressure or increase in temperature the water would be vaporized, leading to explosive eruption. From the increase of temperature with depth in mines, Scrope concluded that at great depth, the Earth was at an intense heat, and that this great accumulation of *caloric* in the deep Earth led to continued flow of caloric toward the surface, in order for the heat to attempt to attain equilibrium. Thus Scrope considered the formation of magma as due to the passage of caloric by conduction from depth to upper levels in the Earth, where the caloric led to melting of rocks. Scrope proposed that the caloric in molten rocks was in large measure due to water: "There is every reason to believe this fluid to be no other than the vapor of water, intimately combined with the mineral constituents of the lava, and volatilized by the intense temperature to which it is exposed when circumstances occur which permit its expansion." The water of the oceans, he theorized, was derived from the interior of the Earth due to degassing through volcanoes. Large quantities of water "remain entangled interstitially in the condensed matter" in the deep-seated rocks, and the escape of steam during volcanic eruptions carries off an immense amount of caloric, leading to rapid cooling and consolidation of magmas at the surface.

By 1865 the French geologist M. Fourque had measured the amount of water in the volcanic products of Etna and estimated the amount of steam discharged from the vent. In the latter part of the nineteenth century geologists began to develop more sophisticated models, proposing that the steam emitted by volcanic eruptions was primordial or of deep-seated origin, rather than just recycled surface waters. Among them was Osmond Fisher (1881), who regarded volcanic gases as original constituents of magma. The potential role of escape of water from volcanoes in the formation of the oceans began to surface as an important hypothesis. The German geologist Eduard Suess proposed that all water in the oceans and atmosphere came from the outgassing of the interior of the Earth. The concept of volcanic recycling of water from the ocean to the atmosphere and back again to the deep Earth was proposed by John Judd (1881), who pointed out that volcanoes are generally located near the ocean, and speculated that fissures may transport seawater from the ocean to the magma at depth. "Volcanic outbursts" could be due to water finding its way down to a highly heated rock mass, lowering the melting temperature and causing melting. Judd considered high temperature and the presence of water and gas as the key factors of volcanic action and argued that magmas can absorb or dissolve large quantities of water, which escapes violently during explosive volcanic eruption, as Spallanzani had first pointed out. The magmas could absorb water or gases either initially, as primordial gases during formation of the globe, or at any stage in geologic history, due to infiltration of water into the Earth's crust.

With the widespread recognition of water in magmas, it was logical to consider its role in explosive eruptions. In the age of steam during the Industrial Revolution of the nineteenth century, John Judd pointed out that "a volcano is a kind of great natural steam-engine." Bonney (1899) also compared volcanoes to a boiler, emphasizing that the steam in magma is the main explosive force in an eruption. He noted that the volume of steam is nearly 1700 times that occupied in the form of water, an enormous expansive

force, adequate to account for volcanic explosions. The origin of volcanic water, according to Bonney, is related to the proximity of volcanoes to the ocean and the percolation of rainwater into the magma. He stressed the importance of addition of water to lower the melting point of rocks, and followed Osmond Fisher and others in attributing eruption to the presence of water. When water is depleted or withdrawn from the system, the eruption ceases.

In the latter part of the eighteenth century, Dolomieu had determined that many deposits in Italy and on Sicily, which at first sight looked like normal stratified sediments, were in fact consolidated volcanic ash deposits, the products of fallout from explosive eruptions. The disruption of magma as it is blown into the air leads to fragmentation and the formation of volcanic ash, tephra, scoria, and other forms of pyroclastic material. We now know that the primary agent of this disruption is the explosive expansion of steam. Sartorius von Waltershausen (1853), a true pioneer in the study of pyroclastic rocks, was the first to attribute their formation to the effects of water on the magma. He proposed that magma rises and erupts due to pressure of water vapor escaping from the magma and that the volume increase of water vapor was also responsible for the fragmenting process—the formation and ejection of pyroclasts such as pumice and ash. He also recognized that peculiar volcanic rock deposits can result when magma enters the ocean or is erupted under water. It was during travels in Sicily in 1835 that he first came across a brown tuff, a homogenous rock, composed largely of a single mineral. He named the rock **palagonite** after the nearby town of Palagonia, and by chemical analysis determined it as one unusually rich in water and iron, with about 12–23 wt% water. He became curious as to the rock's origin, and noting that it was often associated with marine deposits, proposed that it was a volcanic product that forms thick layers in many submarine volcanic formations. In 1846 he and Robert Bunsen studied palagonite mountains in Iceland, and noted that a zone of palagonite tuffs stretches across Iceland. Often in association with palagonite he noted, was a black, water-free and glass-like material, similar to obsidian, which he gave the name **sideromelan**. He was able to show that palagonite is basically sideromelane or basaltic glass that has taken up water, and from its geologic setting, that palagonite is a product of shallow subaqueous or submarine basaltic eruptions. Von Waltershausen was correct in this deduction; we now know that the Icelandic palagonite rocks are formed by volcanic eruptions below the thick ice cap that covered Iceland during the last ice age, and that the Italian palagonites were erupted in the ocean.

Bunsen, on the other hand, disagreed with von Waltershausen, considering the palagonite tuffs as the products of basaltic rocks metamorphosed in the presence of much water and carbonate. He showed by chemical analysis that the tuffs were virtually identical to basalt lava after the high water content had been subtracted. His study was not restricted to Icelandic rocks, because Charles Darwin had given him samples of palagonite from the Cape Verde islands.

Another volcanic rock of basaltic composition is lavalike but composed of rounded or pillow-like forms. The origin of this rock does not have a bearing on water in magmas, but rather on magma in the water. The identification of *pillow lava* dates back at least to the 1870s. Later it was proposed on the basis of observations in Italy that it forms due to submarine eruption. The British geologist Tempest Anderson observed an eruption in Samoa, noting that when the basaltic lava was flowing into the sea, the submarine component of the lava formed bulbous masses and lobes with the shape of pillows. By 1914 a subaqueous origin for pillow lavas had been established. They were already known to be present in the geologic record in association with marine sediments in Scotland, and also as products of subglacial eruptions, such as in Iceland. Oceanographers discovered only in the 1960s that basaltic pillow lava forms the floor of most of the world's oceans and is thus the Earth's most abundant volcanic rock but also the most remote for study.

When high-pressure melting experiments were first carried out at the beginning of the twentieth century, it became evident that magmas residing deep in the Earth can contain much more water in solution, and that this water must be liberated when they are erupted at the surface. In 1903 C. Doelter was one of the first to propose that magmas at great pressure in the Earth's crust may dissolve water and that the magmas may become explosive when they reach the surface of the Earth. In support of this theory, the French geologist Armand Gautier carried out laboratory experiments on volcanic activity in 1906, suggesting that magma rises in the crust as a result of gas expansion, and attributed the violence of volcanic eruptions to the explosive liberation of water from the magmas. By the early twentieth century the fundamental ideas about the causes of explosive volcanism and the importance of water had thus been firmly established.

SEE ALSO THE FOLLOWING ARTICLES

Archaeology and Volcanism • Earth's Volcanoes and Eruptions: An Overview • Mantle of the Earth • Origin of Magmas • Plate Tectonics and Volcanism • Volcanoes in Art • Volcanoes in Literature and Film.

FURTHER READING

Brush, S.G., 1979. Nineteenth-century debates about the inside of the earth: Solid, liquid or gas? Ann. Sci. 36, 225–254.

Kushner, D.S., 1990. The Emergence of Geophysics in Nineteenth Century Britain (Ph.D. thesis). Princeton University, Princeton, NJ.

Laudan, R., 1987. From Mineralogy to Geology. The Foundations of a Science 1650–1830. University of Chicago Press, 278 pp.

Sigurdsson, H., 1999. Melting the Earth; The Evolution of Ideas about Volcanic Eruptions. Oxford University Press, New York, 250 pp.

Origin and Transport of Magma

Haraldur Sigurdsson

Graduate School of Oceanography, University of Rhode Island, Narragansett, RI, USA

Volcanic eruptions are the surface expression of processes that occur deep within the Earth. Many of these processes take place just below the Earth's outer rigid shell, whereas some volcanic eruptions owe their origin to very deep disturbances, even at the boundary between the core and the Earth's mantle at 2890 km below the surface.

The origin and transport of magma are treated in Part I of the second edition of the Encyclopedia of Volcanoes. An understanding of these processes is an essential foundation for an appreciation of the volcanism that is observed at the surface. We set the stage with the Chapter Melting in the mantle of the Earth, describing the chemical and physical properties of the source region of most magmas. Although rarely seen, the mantle is the most voluminous part of the planet, and it is here that the great heat manifest in volcanic eruptions is stored. It is hot, perhaps 1300–1600 °C, but the mantle is for the most part stony and solid. The formation of magma occurs in only certain regions of the mantle by partial melting of the rock peridotite. Melting occurs either by the lowering of pressure in the mantle or by the lowering of the melting temperature of peridotite by introduction of water to the mantle from the Earth's surface during subduction.

When melting begins in the mantle, the hot silicate liquid or magma begins to migrate upward toward the surface, because of its low density compared to the surrounding high-density mantle rock. The migration and transport of magma within the Earth are perplexing problems, treated in the Chapter Migration of Melt. We know that magma begins deep in the mantle, where it forms a small part of the peridotite rock, like water in a sponge, and we know that eventually it pools into large liquid bodies near the surface. The intermediate stage in magma transport, between the source and the shallow reservoir, is one of the major problems in the study of magma evolution, as discussed here. How does the magma percolate upward toward the surface? Does it migrate upward through the action of dikes?

Melting in the mantle determines the location of volcanoes on Earth, but the location of melting is largely determined by plate tectonics, the theory that describes the motion of the great crustal plates that make the rigid exterior shell of the Earth. As described in the Chapter Plate Tectonics and Volcanism, melting occurs beneath plate boundaries where plates are either moving apart, such as below the mid-ocean ridges, or below converging plates in subduction zones. In the former case, melting is due entirely to decompression, as the mantle rises up to fill the void between the diverging plates. In the latter case, however, decompression melting is further aided by the introduction of water and lowering of the solidus. However, a significant fraction of Earth's volcanism occurs totally independent of plate motion, creating hot spots. It is hypothesized—although not yet proven—by many geologists that hot spots are the result of great mantle plumes, which may originate from a region near the boundary of the mantle and the core.

Our most important clues about the formation and origin of magmas come from studies of the chemical and mineralogical composition of volcanic rocks, as described in the Chapter Origin and Composition of Magmas. These silicate liquids, drawn from deep inside the Earth, have much to tell about their peridotite source rock as well as about the history and formation of the planet. The chemical diversity of magmas or volcanic rocks seems at first bewildering, with a plethora of names used by geologists to describe all of these rock types. The reason for this great diversity is explored in this chapter, which explains the great variety of physical and chemical processes leading to the differentiation of magmas.

Rheology and other Physical Properties of Magmas is a chapter that addresses fundamental properties that influence transport and behavior of magma as a liquid. There has been great advance made in quantifying the relationship between the chemical composition and physical properties of magmas, as treated in the new chapter in this volume on Silicate Liquid Phase Relations and Thermodynamics by the acknowledged expert in this field.

Virtually all of the known chemical elements occur in magmas, although most are present only in trace amounts. Many chemical compounds in magmas, including water, carbon dioxide, and sulfur dioxide, are normally dissolved in the magma at high temperatures and pressures, but come out of solution when the magma nears the surface and erupts. These are the volatiles in magmas, whose properties and evolution are dealt with in the Chapter Volatiles in Magmas. The concentration and behavior of the volatiles are particularly important as their abundance in the magma largely determines the explosive nature of volcanic eruptions. The distinction between the quiet effusion of lava flows and the explosive ejection of volcanic ash and pumice is primarily a reflection of the original water content in the magma.

However, the rheological behavior of magmas is also dictated by their physical properties, such as viscosity and temperature, as discussed in the Chapter Physical Properties of Magmas. The importance of the bulk viscosity is, for example, demonstrated by the fact that volatiles can rise and escape freely to the surface as gas bubbles from a low-viscosity basaltic magma, whereas the viscosity of andesitic or rhyolitic magmas is so high that the bubbles cannot rise but are carried with the magma to the surface, with explosive results.

Magmas rising from the mantle may often gather in reservoirs at the base of the crust or within the Earth's crust. Reservoirs may be tens of kilometers in dimension and thus represent huge reserves of magma. The Chapter Magma Chambers explains the behavior of these magma chambers, the geological and geophysical evidence for their size and dimensions, and the processes that occur within them. The Chapter Rates and Timing of Magma Ascent describes how petrologic studies and geochemical research on short-lived isotopes in magma are providing information on the rates of magma ascent.

Most of the magma is transported in dikes from the deep reservoir to the surface, as described in the Chapter Conduits and Dikes. During dike flow the magma experiences extreme gradients in temperature and pressure as it moves through the cold and rigid upper crust of the Earth. It seems at first amazing that magma may flow for several kilometers in a 1-m-wide dike without solidifying, but as explained in this chapter, the magma may retain most of its heat during dike flow and arrive at the surface in a fiery eruption.

During its ascent in a dike or a conduit, magma experiences a very important change, as the volatiles—mainly water, carbon dioxide, and sulfur dioxide—exsolve from the magma to form a separate gas phase. Exsolution occurs when the pressure decreases below the solubility limit of these volatiles in the magma, as described in the Chapter Magma Ascent at Shallow Levels. In the crust above this level, the magma in the dike or conduit behaves as a two-phase flow, consisting of a silicate melt and a gas phase that continues up the ally, expanding as pressure decreases. This is the most dynamic region of magma evolution, and rates of motion are on the order of meters per second, whereas rates of deeper magma flow are centimeters per second. The exsolving gas appears as growing bubbles that expand until they burst, tearing apart the magma in the conduit and expelling it to the surface.

Chapter 1

Melting the Earth's Upper Mantle

Timothy L. Grove

Department of Earth, Atmospheric and Planetary Sciences, Massachusetts Institute of Technology, Cambridge, MA, USA

Christy B. Till*

Volcano Science Center, U.S. Geological Survey, Menlo Park, CA, USA,
*Now at School of Earth and Space Exploration, Arizona State University, Tempe, AZ, USA

Chapter Outline

GLOSSARY

adiabatic The thermodynamic process of expanding or compressing a substance without allowing it to exchange heat with its surroundings.

aggregate primary magma A mixture of near-fractional melts produced during adiabatic decompression melting.

asthenosphere The part of the upper mantle beneath the lithosphere that deforms by plastic flow as a result of high temperatures at a depth in the earth.

batch melting A single stage melting process where the melt is in equilibrium with the solid residue and is removed in a single event.

decompression melting Melting that occurs during adiabatic decompression. Temperature changes very little along an adiabatic gradient, but the solidus of the Earth's mantle is steep allowing the melting point of the mantle to be overstepped.

diapir A less dense crystal + melt mass that ascends buoyantly through denser surrounding material.

flux melting Melting in subduction zones caused by introducing a liquid into a solid matrix. The melt reacts with its surroundings by dissolving crystalline material and bringing the melt + crystal mix into chemical and thermal equilibrium.

fractional melting A melting and extraction process whereby the melt is constantly extracted from the residue and does not react with its subsequent surroundings.

freezing point depression The lowering of the melting point of a compound by adding another compound.

latent heat of fusion (or crystallization) Heat of reaction required to accomplish the phase change from solid to liquid form.

liquidus Temperature above which a magma is completely liquid, or the temperature below which a melt begins to crystallize.

lithosphere The portion the Earth's crust and upper mantle that is rigid, deforms elastically over timescales of thousands of years or longer, and comprises the tectonic plates.

mantle The compositional layer of the Earth that lies below the Mohorovicic discontinuity and above the outer core. The mantle is divided into three parts: the upper mantle that extends to a depth of approximately 410 km, the transition zone that extends to approximately 670 km, and the lower mantle that extends to the core mantle boundary at approximately 2890 km.

mohorovicic discontinuity or moho The seismic velocity discontinuity that separates the Earth's crust and mantle.

peridotite–lherzolite–harzburgite Names indicative of the minerals present in a crystalline rock. Peridotite is a general name for rocks containing olivine, orthopyroxene, and clinopyroxene. Lherzolite is the specific name applied when all these minerals are present. Harzburgite is the name applied when the rock is >90% olivine and orthopyroxene.

picrite A high MgO lava containing >15 wt% MgO.

polybaric near-fractional melting Fractional melting that takes place over a range of decreasing pressures.

primary magmas Liquids produced by partial melting of a source region and unmodified by any postsegregation process.

primitive magmas In a series of genetically related rocks, the sample with the highest MgO content.

reactive porous flow The reaction of a fluid or melt with its surrounding permeable crystalline matrix.

The Encyclopedia of Volcanoes. http://dx.doi.org/10.1016/B978-0-12-385938-9.00001-8

solidus Temperature below which a magma is completely crystallized, or temperature at which a solid begins to melt.

vapor-saturated solidus The solidus of a crystalline material in the presence of a supercritical fluid phase.

1. INTRODUCTION

The magmas that erupt from volcanoes on Earth originate primarily through partial melting processes initiated in the Earth's **mantle**. This chapter will first discuss the structure and physical properties of the Earth's upper mantle that influence magma generation. Second, we will discuss the physical mechanisms that lead to partial melting in the Earth's mantle and the tectonic settings where each of these melting mechanisms occur. The final section will discuss the chemical composition of relatively pure mantle melts prior to their processing in the Earth's crust.

2. STRUCTURE AND PHYSICAL PROPERTIES OF THE EARTH'S MANTLE

2.1. Structure of the Earth's Mantle

The Earth, like the other terrestrial planets, has differentiated into layers that can be described by their chemistry and/or physical behavior (i.e., rheology). The Earth's "mantle" is the compositional layer located between the crust and the outer core, which extends to 2890-km depth and constitutes approximately 84% of the Earth's volume (Figure 1.1). The mantle consists of three regions: the upper mantle, the transition zone, and the lower mantle. The upper mantle begins at the **Mohorovicic discontinuity (or moho)** and

extends to a depth of 410 km. The transition zone extends to 670 km, and the remainder and volumetrically most abundant lower mantle extends down to the core—mantle boundary (Figure 1.1). These different interior regions are defined by changes in the measured velocities of seismic waves and correspond to changes in the chemical composition and/or crystal structure of the materials that make up the mantle. The first boundary, "the moho," marks the profound change in composition that occurs at the transition between the Earth's crust and mantle. The immediately adjacent upper mantle is the part of the vast solid interior of the Earth about which we know the most, and it is within this region that melting processes lead to the generation of magmas. The other seismically defined boundaries in the deeper mantle can be associated with changes in the crystal structure of solid mantle minerals. Below the 410-km discontinuity, the crystal structure of the $(Mg, Fe)_2SiO_4$ mineral in the mantle undergoes a change from the mineral olivine to a phase with the same composition, but with a different crystal structure (a polymorph) called wadsleyite. With increasing depths, wadsleyite becomes unstable and transforms to ringwoodite. Below the 670-km discontinuity $(Mg, Fe)SiO_3$ minerals become unstable and are replaced by a mixture of $(Mg, Fe)SiO_3$ perovskite and $(Mg, Fe)O$ oxide minerals (Figure 1.1).

Although the transition zone and lower mantle have never been sampled directly, their seismic properties are consistent with the experimentally determined pressure—temperature stabilities of the minerals discussed above. One consequence of this is that the pressure—temperature conditions of the Earth's mantle can be estimated (Figure 1.1). An additional consequence is that changes in chemical

FIGURE 1.1 Compositional layers of the Earth. The inset shows olivine polymorphs with increasing temperature and pressure within the mantle.

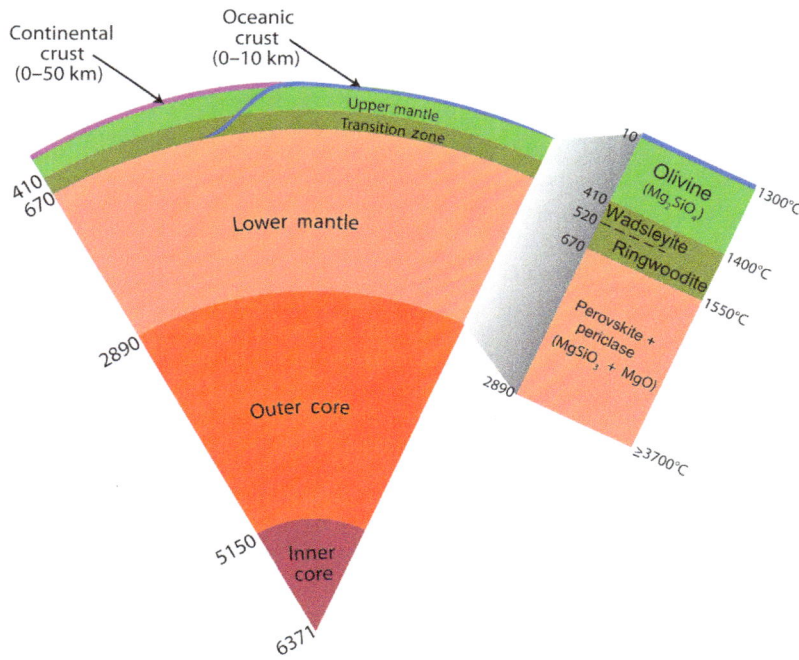

composition are not required to explain the observed seismic discontinuities, and it is generally assumed that the composition of the mantle is relatively constant. The next section discusses the compositional variation in the upper mantle, and there is significant variability. Yet the extent of this variability in the transition zone and lower mantle and/or how it changes with depth in the Earth's mantle is currently unknown, and it is an area of active discussion and debate.

2.2. Dynamic Processes in the Earth's Mantle

Although the mantle is solid, it undergoes solid-state convection and behaves like a fluid over geologic timescales (i.e., millions of years or longer), transporting heat from the hotter, deeper interior to the surface. When the mantle is in a convective state, it is referred to as **asthenosphere**. Where the mantle is convecting, the temperature—pressure path follows an **adiabatic** gradient (or adiabat):

$$[dT/dP]_{s=0} = (\alpha\,T)/(\rho\,Cp),$$

where α is the thermal expansion, ρ is the density, and Cp is the heat capacity. The change in temperature with depth is considered adiabatic or isentropic (constant entropy) because no heat is exchanged with the surroundings through conductive heat loss due to the fact that mantle transport is so fast (1—10 cm/year in plate tectonic processes). As a parcel of mantle ascends, the pressure applied to that parcel decreases, and so it expands in volume. This causes the temperature to fall and the internal energy to decrease. Thus, as the mantle rises along an adiabat, the temperature decreases by 0.3°C/km. Below the moho, convection ceases to operate at shallow depths (with the notable exception of midocean spreading centers where it continues to very shallow crustal depths of ~6 km) and the mantle becomes mechanically rigid and loses heat by conduction. The mantle in this physical state is referred to as **lithosphere**. Strictly speaking, these names are intended to describe the physical state of the mantle, but they are sometimes used by geochemists to refer to the chemical compositional characteristics of different mantle reservoirs. In the following sections, we will see that solid-state convection is an important process that leads to melting.

2.3. Upper Mantle Composition and Structure

What we know of the composition of the upper mantle comes from samples that have been delivered to the Earth's surface. Magmas can carry pieces of the mantle (xenoliths) when they ascend from depth and erupt on the surface. Some of the deepest samples of the upper mantle (~250-km depth) come from a rare magma called kimberlite. The Earth's mantle is exposed at the surface by active tectonic processes in mountain belts (alpine peridotite). The mantle is also exposed on the ocean floor at midocean ridges. Mantle convection causes melting to form midocean ridge basalt (MORB), see Section 3.1, and this mantle is brought to the surface and exposed on the ocean floor where volcanism does not cover it (abyssal peridotite).

The Earth's mantle can rightly be called the most abundant metamorphic rock type on Earth, and this rock type is broadly known as **peridotite**. As we noted in the previous section, pressure—temperature conditions within the mantle change the minerals that are stable. Changes in the chemical composition of Earth's mantle also influence the minerals that are present. These changes can be caused by either the removal of a partial melt or by the addition of or interaction with a melt or fluid. These changes are expressed in the name that is given to the rock. For example, mantle that has not been affected by melt removal contains the minerals olivine $(Mg, Fe)_2SiO_4$, orthopyroxene $(Mg, Fe)SiO_3$, clinopyroxene $Ca(Mg, Fe)Si_2O_6$, and an Al-bearing mineral, namely, plagioclase $(CaAl_2Si_2O_8)$, spinel $([Mg, Fe]Al_2O_4)$, or garnet $([Mg, Fe]_3Al_2Si_3O_{12})$. The name give to this type of mantle sample is **lherzolite** with the prefix of the Al-bearing phase added (e.g., plagioclase lherzolite). Changes in pressure dictate the stable Al-bearing phase with plagioclase stable at low pressure (up to 1 GPa), spinel stable at 1—2 GPa, and garnet stable at higher pressures.

Changes in mantle composition as a consequence of removal of a melt have predictable effects. The melt is always enriched in FeO relative to the minerals that remain behind in the solid mantle (Section 4). A convenient way to keep track of the relative proportions of iron depletion is to measure the proportions of Mg and Fe in a mineral and express it as Mg# (defined as $Mg/[Mg + Fe^{2+}]$ and expressed in molar units). The changes in mantle Mg# are illustrated for two types of peridotite in Figure 1.2. The melt is also enriched in Na_2O and Al_2O_3 relative to the mantle, and the result of melt extraction is to deplete the mantle in these oxide components. This changes the minerals that are present in the rock, decreasing the amounts of clinopyroxene and Al-phase and increasing the proportions of olivine and orthopyroxene. When both clinopyroxene and the Al-bearing phase are lost, the rock name that is applied to this melt-depleted mantle is **harzburgite**. The melt is also enriched in incompatible elements (elements that do not readily fit into the crystallographic sites of the mantle minerals, and they prefer to enter the melt). Most notable among these are K_2O and Na_2O, and these are also depleted as the mantle is progressively melted. Some mantle samples are found to be enriched in these elements, and this enrichment is ascribed to the addition of a melt or fluid phase, a process referred to as mantle metasomatism. Mantle metasomatism often occurs where the mantle is not convecting and is part of the lithosphere. Other rock types can also be recycled into the mantle by plate tectonic processes. Oceanic crust and other fragments of basalt

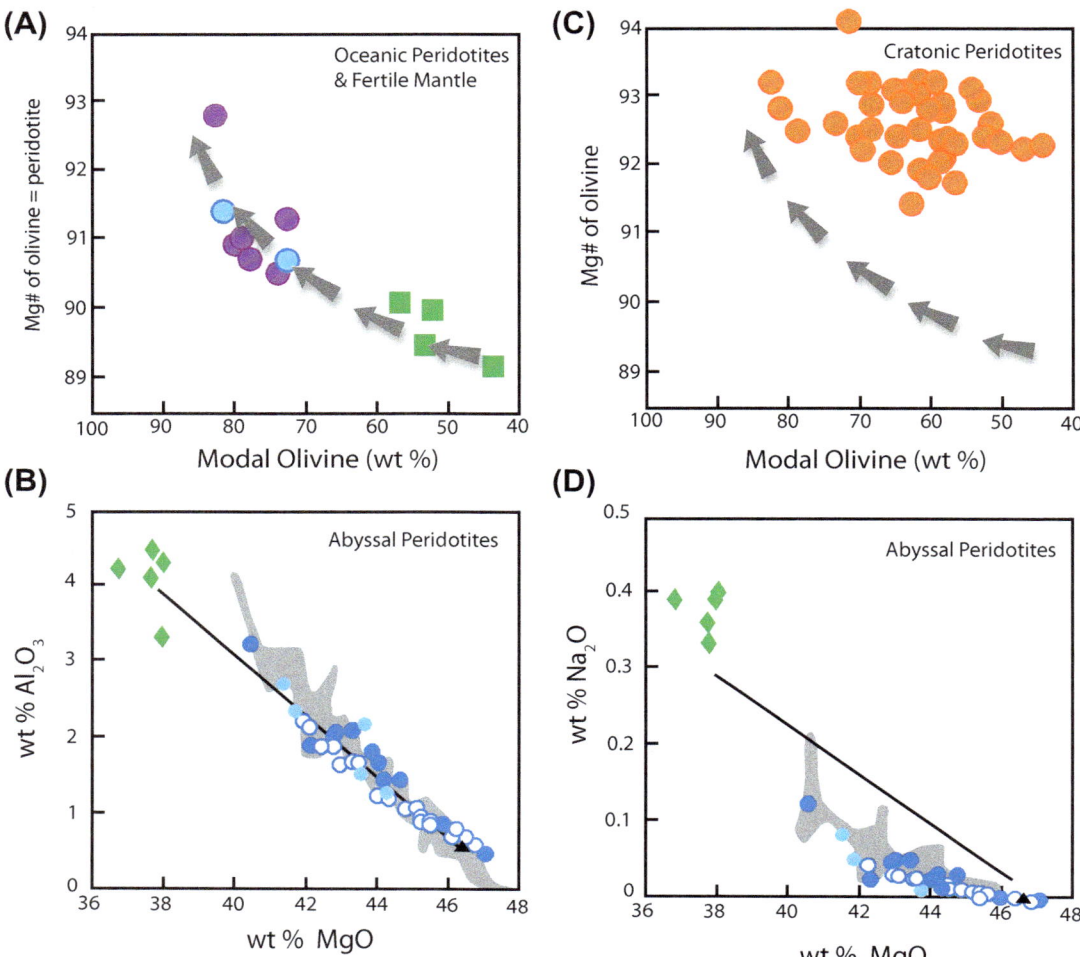

FIGURE 1.2 Upper mantle compositions and melting trends based on mantle samples from different tectonic settings. (A) Green squares are model fertile upper mantle peridotite compositions calculated for the spinel facies; purple circles are harzburgites from ophiolite tectonites and alpine peridotites; and blue circles are abyssal peridotites. Gray arrows indicate a trend from fertile mantle compositions to the upper mantle compositions that represent the residues of mantle melting. (B) Same trend as in (A) plotted with the composition of garnet peridotite xenoliths found in kimberlites from the Kaapvaal craton in South Africa, believed to be the dominant rock type in the lithospheric mantle beneath Archean cratons. These samples may have originated in subcontinental magmatic events. (C) and (D) are primary bulk compositions of site-averaged abyssal peridotites where dark blue circles represent oceanic upper mantle bulk compositions calculated with the knowledge of the relative abundance and average compositions of olivine, orthopyroxene, clino-pyroxene, and spinel, light blue circles where two or three of these mineral compositions are known and open circles where only one is known. Green diamonds represent estimates of fertile upper mantle compositions; the black triangle is the bulk composition of ODP site 895 corrected for alteration and the solid line connects upper mantle compositions. *(The gray field is calculated abyssal peridotite compositions from Niu et al., 1997. (A) and (B) modified from Boyd (1989); and (C) and (D) modified from Baker et al., 1999.)*

erupted from volcanoes are recycled into the mantle and transform at mantle pressures to a garnet- and pyroxene-bearing rock named eclogite. Basaltic melts produced by mantle melting can ascend from their source region into overlying cooler mantle and solidify. These veins are enriched in pyroxene and referred to as pyroxenites.

In addition to the minerals and rock types discussed above, the mantle contains volatile constituents such as H_2O and CO_2. In low abundances, these volatile constituents can be partitioned in vacant sites and defects in the nominally anhydrous minerals discussed above, such as olivine, orthopyroxene, clinopyroxene, and garnet. Low-moderate

amounts of H_2O are likely present in the asthenosphere, perhaps heterogeneously distributed throughout. Estimates of H_2O for the average upper mantle converge on a value of 100 ± 50 ppm or 300−900 ppm H_2O for volatile-rich upper mantle. Or where volatiles are present in sufficient amounts, hydrous minerals that contain hydrogen as an essential structural constituent, such as amphibole, phlogopite, chlorite and serpentine can form in the Earth's upper mantle. Some mantle peridotite samples that erupted in magmas at volcanoes contain hydrous minerals and support the notion that H_2O may be an important minor constituent in the lithospheric upper mantle as well.

3. MANTLE MELTING PROCESSES

3.1. Overview

In this section, we will discuss the conditions that initially cause the Earth's mantle to melt, the mechanisms by which it melts and the tectonic settings where these melting mechanisms occur. Erupted magmas are the best tools we have for understanding and reconstructing the conditions and processes by which the Earth's mantle melts. Melts of the mantle that erupt on the Earth's surface without having been modified in any manner since they were extracted from the mantle are known as **primary magmas**. In reality, all magmas experience some processing en route to the surface. This processing can consist of the growth of crystals from the magma (fractional crystallization) and/or mixing of the magma with new materials (assimilation) or other magmas (magma mixing). We call natural magmas that exhibit the least evidence of processing **primitive magmas**. Near-primary primitive magmas can be found in a variety of tectonic settings on the Earth.

Recall that most of the Earth's mantle (i.e., the asthenosphere) exists as a solid capable of flowing over very long timescales. At a specified pressure, the temperature at which melting first begins is called the **solidus** (Figure 1.3). The **liquidus** is the temperature at which a given rock composition crosses the boundary from solid + liquid to all liquid. The liquidus for a given mantle composition is therefore always greater than or equal to the solidus temperature at a given pressure. The temperature interval over which melting occurs is controlled by pressure, the composition of the mantle, and the presence of volatiles. Different chemical compositions of mantle (e.g., lherzolite and harzburgite) have different solidus and liquidus curves, and there is no one uniform melting curve for the entire mantle. Simulating the melting conditions of the mantle in laboratory studies is one of the main methods scientists have used to understand the melting and crystallization behavior of the mantle. These types of studies are part of a field of the earth sciences known as experimental petrology. The remainder of this chapter will primarily focus on the melting of lherzolite, the main rock type found in the mantle.

There are three ways to melt (or cross the solidus curve of) the mantle: the first is to raise the temperature, the second is to lower the pressure, and the third is to change the chemical composition (Figure 1.3). In the asthenospheric mantle, where mass transport is faster than heat transport and there is presently limited heat production by the decay of radioactive elements, all three mechanisms can lead to melting. The majority of mantle-derived magmas are produced in the asthenosphere or in the mantle that is being transformed from the lithosphere into the asthenosphere.

3.2. Decompression Melting and Melt Transport at Midocean Ridges

Melting the mantle by lowering its pressure or **decompression melting** is the most common and best-understood melting mechanism. Magmas generated at midocean ridges, in the backarc of subduction zones, at ocean islands, and in the interior of many continents are formed by this process. Decompression melting occurs when the adiabat for the convecting mantle crosses the solidus, which has a shallower slope than the adiabat in the upper mantle (Figure 1.4).

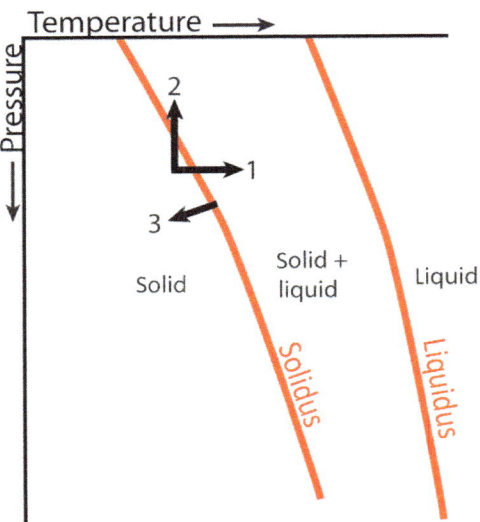

FIGURE 1.3   Schematic diagram of three ways to melt the mantle relative to the solidus (the boundary between solid and solid + liquid) and the liquidus (the boundary between solid + liquid and liquid).

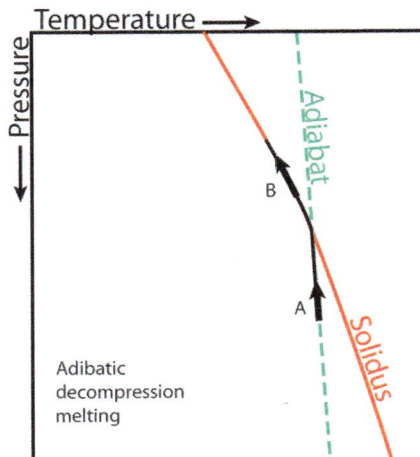

FIGURE 1.4   Schematic illustration of adiabatic decompression melting. Decompression melting occurs when the adiabat for the convecting mantle crosses the solidus, which has a shallower slope than the adiabat in the upper mantle. The thermal energy for melting (or the latent heat of fusion) is obtained from the difference between the temperature that would be achieved if the mantle rose along the adiabat without melting and the temperature that is achieved if the mantle stays on the solidus.

Melting occurs in response to the thermal energy that is obtained from the difference between the temperature that would be achieved if the mantle rose along the adiabat without melting and the temperature that is achieved if the mantle stays on the solidus. This thermal energy supplies the heat of reaction required to accomplish the phase change from solid to solid + liquid at a constant temperature, known as the **latent heat of fusion** (or the "latent heat of crystallization" if the phase boundary is crossed in the opposite direction). Adiabatic decompression melting will continue until a depth is reached at which the temperature of the parcel of mantle is lower than the solidus temperature for the mantle. The solidus is encountered at different depths in the mantle in different tectonic settings, and this location depends on the ambient mantle temperature and composition of the upwelling mantle. The rate of melt production is controlled by the latent heat of fusion for mantle peridotite (~ 130 cal/g), the heat capacity (0.3 cal/g°C), the ambient mantle temperature, and the pressure and compositional dependence of the mantle solidus. For example, if the mantle rises 100 °C above its solidus, cooling it to the solidus would supply 33 cal/g of thermal energy that could be used to melt the mantle by 25%.

Whether or not the melt remains in contact with the crystalline material that is being melted or separates and rises through the melting regime independently influences both the volume and the composition of the melts and the crystalline materials left behind by the melting process. If the melt stays with its solid residue during ascent and is only extracted at the shallowest depth of melting, this is called **batch melting**. In batch melting, the bulk composition of the residue + melt is constant, and the pressure signature of the melt produced will be equivalent to the pressure of segregation. In this case, the volume and composition of the magma produced will be a function of the starting composition of the mantle, the pressure of segregation, and the extent of melting achieved, which is a function of temperature. If the melt is constantly extracted from the residue and rises without reacting with its surroundings, this is called **fractional melting**. When melts segregate from the rising mantle as they are produced, then the bulk composition of the melting system changes continuously. True fractional melting requires that the phases in the solid residue are chemically homogeneous and therefore requires solid-state diffusion to bring the interiors of mineral grains into chemical equilibrium with the outer edges where melt removal has occurred. In fractional melting, the pressure signature associated with the magma formed by aggregating these small fractions of melt at a shallower depth will be an average of the range of pressures over which the melt fractions were produced. This process in natural systems is therefore also known as **polybaric near-fractional melting** and is thought to be the mechanism by which

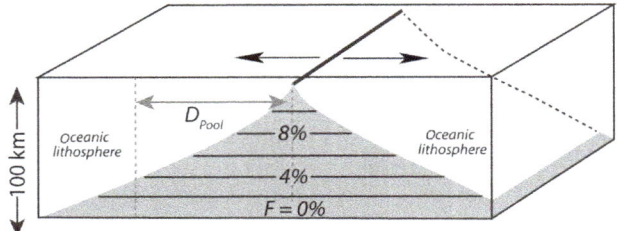

FIGURE 1.5 Three-dimensional melt region beneath the midocean ridge, Melts are generated by adiabatic decompression melting due to passive upwelling in the three-dimensional region shown in gray. D_{pool} is the distance within which melts are focused at the ridge axis to form aggregate primary magmas.

melts are produced beneath midocean spreading ridges at divergent plate boundaries.

The evidence for polybaric near-fractional melting comes from the compositions of MORBs that have been sampled globally at spreading ridges. A schematic of the mantle melting processes that occur in the midocean ridge environment is illustrated in Figure 1.5. At midocean ridges, magma production is dominated by passive upwelling of hotter, deeper mantle. As the mantle ascends and crosses its solidus at a depth determined by the ambient mantle temperature, near-fractional adiabatic decompression melting begins. The ending of melting beneath midocean ridges depends upon the thickness of the overlying oceanic lithosphere. The thinnest lithosphere is right under the midocean ridge, where newly formed oceanic lithosphere is pulled apart and melts rise to form zero-age oceanic crust. With time, the newly formed oceanic lithosphere is pulled in opposite directions away from the ridge and cools and thickens. This leads to the formation of a three-dimensional triangular prism-shaped melting region, where melting ends at greater depths as one moves away from the ridge. Up to a critical distance from the ridge, the melts in this region focus at the ridge by flowing along the sloping asthenosphere—lithosphere boundary. The melts aggregate at shallow depths and erupt as a mixture of melts produced over a range of depths from the mantle that has experienced progressively higher extents of melt depletion as it rises to shallower depths. This effect is illustrated in Figure 1.6 by the progressive depletion of Na_2O in the melts as near-fractional melting continues. A break also occurs in the composition of the melt when the solid mantle transforms from spinel to plagioclase lherzolite. During near-fractional decompression melting, the melt separates from the ascending mantle after only a small extent of melting (0.1—1%). The mixture of these melts produced over the total depth range of melting in the triangular melt regime is referred to as an **aggregate primary magma**.

Decompression melting can also occur at subduction zones. Here, convection set up by the sinking of cool dense oceanic lithosphere into hotter asthenosphere causes the

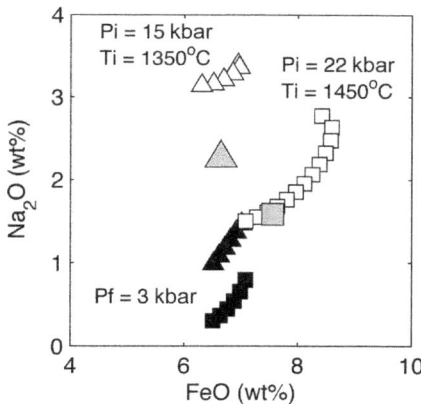

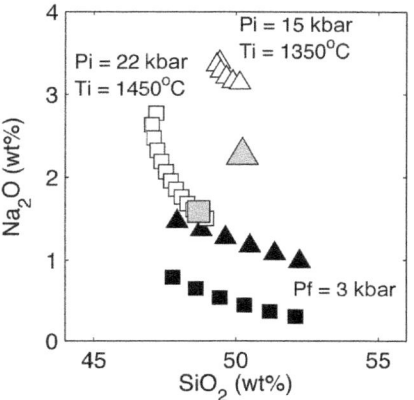

FIGURE 1.6   Composition of MORBs generated with the melting model of Till et al., 2012 for a depleted mantle composition. Closed symbols denote melting in the plagioclase lherzolite stability field and open symbols melting in the spinel lherzolite stability field. Melting in the spinel stability field starts at 22 kbars ■ and 15 kbars ▲ for the plagioclase stability field and all melting finishes at 3 kbars. Each symbol represents 1% melting at increments of 1 kbar. The aggregate melts are denoted by the gray symbols.

flow of hot mantle into the wedge-shaped region of the mantle bounded by the overlying crust and lithosphere and the underlying oceanic plate (or slab) (Figure 1.7). The consequence of the downward slab pull is to draw hot mantle up into the wedge. Adiabatic decompression melting can happen along the upwelling limb of the mantle flowing into the wedge corner. This decompression melting occurs behind the main arc volcanic front and can develop into back-arc spreading centers at oceanic convergent margins like the Marianas or persist as widespread, isolated volcanism in continental convergent margins like the Cascades.

3.3. Flux Melting and Melt Transport in Subduction Zones

A completely different style of melt generation occurs in subduction zones to form the volcanoes of the main

arc. In general, this second way to melt the mantle, melting by changing composition, occurs when the chemical composition of mantle is changed such that the solidus temperature for the new composition is lower than the present temperature. This process can be explained through the concept of **freezing point depression**, where the addition of a solute to a solvent expands the liquid stability field and lowers the freezing point of the solvent. The most common cause of this type of melting in the Earth's mantle is the addition of volatiles to mantle lherzolite at subduction zones (Figure 1.8). The subducting slab contains abundant volatiles stored in H_2O-bearing metamorphic minerals found in the layers of sediment and basalt on the upper surface of the plate, as well as within the oceanic mantle lithosphere where infiltrating volatiles transform mantle lherzolite to serpentine and chlorite (Figure 1.7). As the slab encounters higher temperatures and pressures, the hydrous minerals therein reach the limits of their stability and release their volatiles into the overlying mantle as an H_2O-rich fluid or melt. It is generally agreed that this fluid initiates melting at the base of the mantle wedge when it ascends into hotter overlying mantle and encounters pressure–temperature conditions of the **vapor-saturated solidus** (P1, T1 in Figure 1.9).

These melting conditions can only be achieved when the temperature structure of the subducting slab and the overlying mantle wedge fall within a narrow window (Figure 1.7) that allows the H_2O-bearing minerals in the slab and adjacent mantle wedge to remain stable to sufficient depths so that the overlying mantle wedge is hot enough to melt. These pressure–temperature conditions are defined by the point at which a vertical path through the oceanic plate and mantle wedge just reaches a high enough temperature in the hottest, shallowest part of the mantle wedge to allow vapor-saturated melting (path A in Figure 1.7). Vertical paths that lie farther from the wedge corner (e.g., path B in Figure 1.7) pass through the hotter parts of the mantle and melting can occur to a greater extent.

Melts formed at the vapor-saturated solidus will be extremely H_2O rich (~30 wt% H_2O) and will be positively buoyant. This melt ascends into hotter, shallower overlying mantle and reacts with it, dissolving the mantle minerals into the melt phase and lowering the H_2O content of the melt. Through this process, called **reactive porous flow**, the melt percolates up through the mantle and reacts with the hotter, shallower surrounding mantle as it ascends (P2, T2 in Figure 1.9). The melt will continually reequilibrate with the surrounding silicate mineral matrix by dissolving the solid minerals in the peridotite to approach chemical and thermal equilibrium and decrease in relative volatile content. The ascent velocity of the melt and the permeability of the mantle will control how long the melt remains in contact with the mantle and thus the composition of the melts formed by reactive porous flow. This style of melting

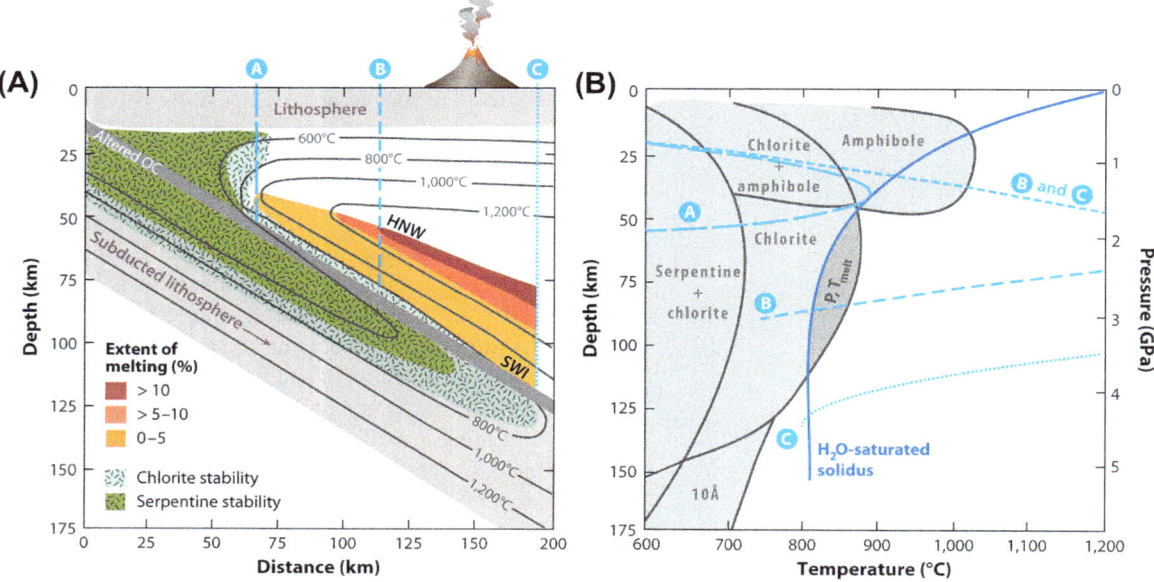

FIGURE 1.7   Two diagrams that schematically illustrate the conditions that limit hydrous flux melting in subduction zones. (A) A typical cross-section through a subduction zone using a thermal model for a slab dip of 30° and a convergence rate of 40 km per million years. Also shown are the stability limits for high-H_2O (>10 wt%) minerals. The melting model is used to predict the extents of hydrous flux melting in the bottom half of the wedge. HNW denotes the hottest, shallowest nose of the mantle wedge where the maximum amount of melting occurs. The part of the mantle wedge directly above the interface between subducted oceanic lithosphere and the mantle is denoted as SWI and extends from path A to path C. (B) Phase diagram with the temperature—depth paths from panel a superimposed. The dark gray region labeled P,Tmelt shows the region of pressure—temperature space where the stability of hydrous phases in the base of the mantle wedge allows H_2O-saturated melting to begin. Abbreviations: HNW, hot nose of the wedge; OC, oceanic crust; SWI, slab—wedge interface. *From Grove et al., 2012.*

is referred to as **flux melting**. It is distinctively different from adiabatic decompression melting, because the melt continually reacts with the hotter shallower overlying mantle. If the melting conditions were to be adiabatic, no melting would occur. Adiabatic decompression paths that leave the vapor-saturated solidus would enter into the stability field of crystals + vapor and no melting could occur and any melt that was present would freeze within the mantle wedge. Therefore, when volatiles are present, melting will continue only if the ascending mass is small enough to be thermally equilibrated with the surrounding hotter mantle and heat can be transferred into it to drive further melting. Some models of flux melting call upon the development of small **diapirs** where melt flow is restricted to channels that contain partially molten mantle that rise into the mantle wedge and undergoes further melting.

3.4. Plume Melting: A Case Study from Hawaii

For many, the giant shield volcanoes of Hawaii are the first examples of volcanoes that come to mind. Such occurrences of voluminous lavas outpouring from a single volcanic center far from the plate margins represent a unique type of mantle melting. As originally hypothesized, ocean islands such as Hawaii are the result of stationary vertical mantle

upwellings called plumes, extending from the core—mantle boundary to the surface, where anomalously high mantle temperatures trigger mantle melting. However, there is currently considerable debate over whether such temperature variations exist in the mantle and their vertical extent, as well as whether variations in chemistry can lead to the observed behavior. Hawaii is an excellent case study for these types of volcanoes because it is so well studied. Many samples of Hawaiian lavas have been collected, both on land and underwater around the flanks of the volcano and from the youngest volcano in the Hawaiian chain, the undersea volcano named Lo'ihi. In addition, geophysical methods have determined the thickness of the lithosphere under Hawaii to be approximately 90 km, and earthquakes associated with magma ascent and ponding suggest the active magma migration zones under Hawaii are at depths of 30—45 km, which is well within the lithospheric mantle.

Like most of the volcanic systems that we have discussed, the near-primary primitive magmas necessary to interrogate mantle melting are very rare at Hawaiian volcanoes. Most of the lavas erupted from Hawaiian shield volcanoes have experienced extensive fractional crystallization in the shallow crust beneath the shield volcano. From the primitive magmas, at least three types of mantle melting can be recognized at Hawaii. The most voluminous lavas in the Hawaiian systems are the lavas that erupt

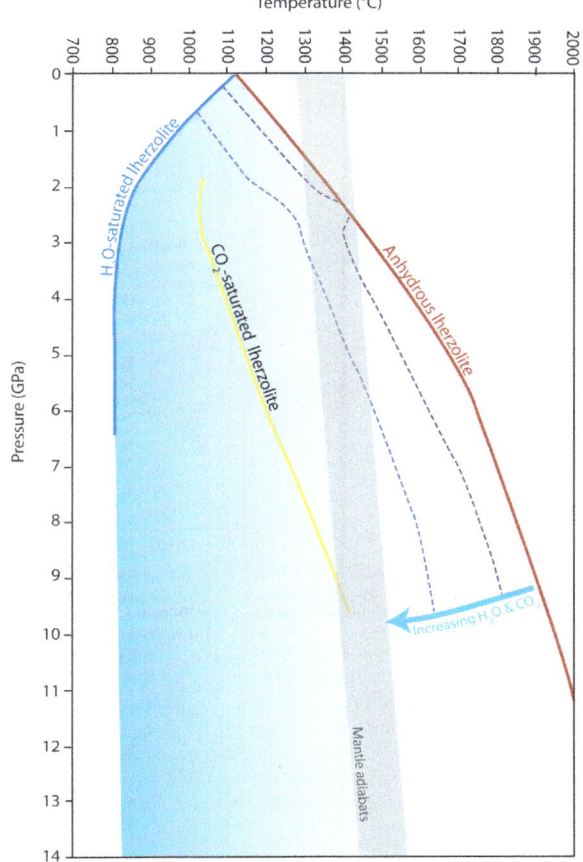

FIGURE 1.8   Volatile depression of the lherzolite solidus. The red curve is the anhydrous lherzolite solidus, the blue curve is the location of the solidus when the mantle is saturated with H_2O, and the yellow curve is the solidus when the mantle is saturated with CO_2. The dashed lines show the progressive lowering of the solidus temperature at a given pressure when small amounts of H_2O or CO_2 are added for the range of volatile concentrations thought to be present in the ambient asthenospheric mantle. The gray bar illustrates the range of adiabatic gradients estimated for the Earth's mantle.

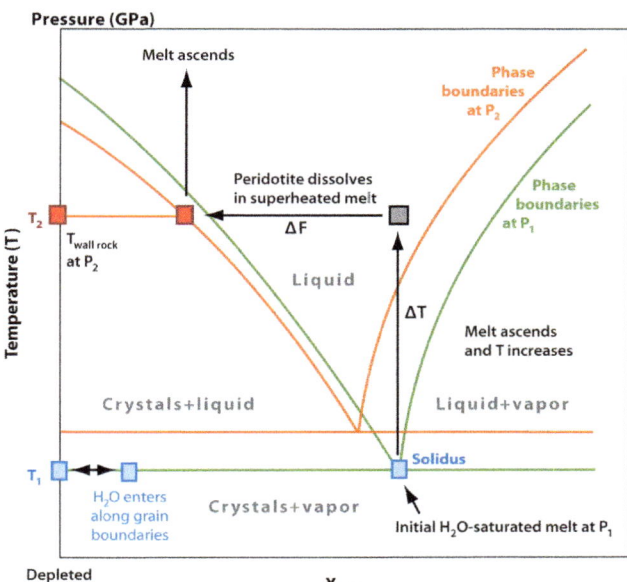

FIGURE 1.9   A polybaric temperature—composition phase diagram that illustrates the process of hydrous flux melting of the mantle wedge by reactive porous flow or by ascent of small, but buoyant, low-viscosity diapirs of mantle plus melt. At P1, T1, melting begins at the vapor-saturated solidus. This fluid-saturated melt ascends to shallower depths (P2, T2) in the inverted temperature gradient of the mantle wedge, where it is out of equilibrium with its surrounding mantle. The melt contains more H_2O than does an equilibrium melt at the same temperature, and the melt dissolves the surrounding mantle, increasing melt fraction and decreasing melt H_2O content. *From Grove et al., 2012.*

on the shield. These voluminous shield lavas are demonstrated to be related to a primary magma that is very high in MgO called **picrite**. Experimental studies have been performed on Hawaiian picrites and indicate that these melts are in equilibrium with a depleted harzburgite source at approximately 30—45 km. Another set of lavas that can be demonstrated to be primary magmas are found to be in equilibrium with garnet lherzolite at depths near the asthenosphere—lithosphere boundary (~90 km), and these are found as eruptive products that form during the early stages of Hawaiian volcano growth (present at Lo'ihi). These lavas are compositionally distinct and are called alkali basalts. A possible origin that links these two compositionally distinct primary magmas at Hawaii is described in Figure 1.10. The picrites form when asthenospheric magmas produced at a 90-km depth (i.e.,

alkali basalts) ascend into the mantle lithosphere and react with the colder solid lithosphere. Here, the alkali basalts heat the lithosphere above its melting point and assimilate orthopyroxene, a high MgO high SiO_2 mineral phase, and crystallize olivine to alter the melt composition to a picrite, a process demonstrated to occur at shallow depths. This process requires thermal energy to melt the lithosphere, and this process can be inferred from the temperature of the picrite when it was present at 40 km, the temperature of the alkali basalt delivered from the lithosphere, and the thermal energy needed to assimilate the lithopshere (Figure 1.10). In order to erupt and sample the primary alkali basalt magma produced in the asthenosphere, an additional process must allow the alkali basalts to escape reaction with the shallow lithosphere and/or picrite genesis may depend on the eruptive stage of the volcanoes in Hawaii. For example, magmatic conduits may be established that lead to heating and assimilation of the lithosphere as magmatic flux increases during the beginning of the shield-building stage at Hawaiian volcanoes.

A third type of primary magma at Hawaii is generated when CO_2 stored in the lithosphere lowers the melting point and produces melts saturated in this volatile component.

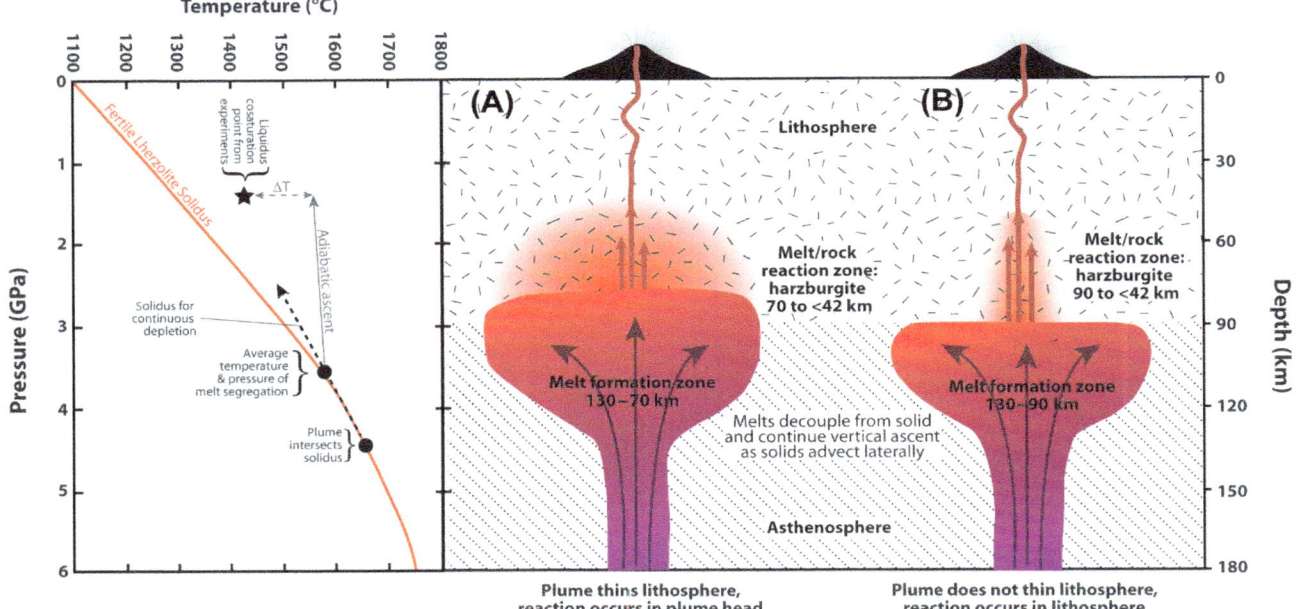

FIGURE 1.10 Schematic diagram of melt generation and melt—rock reactions inferred to form tholeiitic lavas erupted at the Kilauea volcano, Hawaii after Wagner and Grove, 1998. (A) The plume is able to thin the lithosphere and melts generated deep in the plume react with the harzburgite lithosphere at the plume top. The central area of the plume undergoes more ascent than the plume periphery and becomes harzburgite as it experiences >20% melting. As the periphery of the plume advects laterally, garnet lherzolite melts continue to ascend vertically and pass through the harzburgite zone. Melt/rock reaction occurs in the harzburgite to a mean depth of <42 km. (B) The plume does not thin the lithosphere, and the melts generated in the plume react with harzburgite as they ascend through the lithosphere. Melt/rock reaction occurs over a depth range of 90 km to a mean depth of <42 km. The reaction zone must contain 22—44 wt% orthopyroxene based on density constraints and the height corresponds to the vertical distance between the estimated thickness of the lithosphere (90 km) and the pressure of the liquidus cosaturation with olivine and orthopyroxene.

These melts are rare, but ubiquitous at Hawaii and other hot spots and form the third mantle-derived melt associated with intraplate volcanism. The CO_2 is stored in carbonate minerals that are stable as solid phases and melt to form a unique liquid composition enriched in CO_2. After melting in the presence of excess H_2O, such carbonate-saturated melting is the most common volatile-bearing melting process that occurs in the Earth's upper mantle.

4. MANTLE MELT COMPOSITIONS

The compositions of a representative suite of primary mantle melts and their pressure—temperature conditions of melting are tabulated in Table 1.1 and Figure 1.11. The MORB magmas are unique among primary mantle melts because they are an aggregate of melt fractions that are produced continuously during near-fractional adiabatic decompression melting (shown for two different melting paths by the lines in Figure 1.11(A). The two MORB melt compositions shown in Figure 1.11(B)—(E) are aggregated melts produced all along the decompression melting path. The primary controls on melt composition are temperature, pressure, and mantle composition. Mantle melts contain higher MgO and FeO as the depth of melting increases. The temperature range for melt initiation shown in Figure 1.11(A) is

representative of the melting temperature range observed for the global midocean ridge system. In the Hawaiian mantle environment, temperature and pressure of melting both exercise the most striking controls on primary magma composition. The deepest and hottest primary magmas are from Hawaii, and these have the highest amounts of MgO and FeO (Figure 1.11(B)) and the lowest amounts of SiO_2 and Al_2O_3 (Figure 1.11(C) and (D)). The temperature range within the subduction zone environment spans the range of Hawaii and MORB melting environment, and this is reflected in the wide range of MgO contents found in primary subduction zone magmas. At subduction zones, an additional added variable is the role of H_2O in melting. The major influence of H_2O is to lower the melting temperature, and this influences the melting composition by increasing SiO_2 and Al_2O_3 contents and decreasing FeO and MgO contents. The decreased FeO and MgO concentrations are the result of the lower temperatures of melting in subduction zones and the temperature dependence of FeO and MgO partitioning. The variation in Na_2O (Figure 1.11(E)) for primary mantle melts is dominantly controlled by mantle composition and the extent of melting. The low Na_2O primary melt compositions include one MORB and two Hawaiian primary magmas. For these compositions, the MORB source has been depleted by a prior melting event and the

TABLE 1.1 Chemical compositions of mantle melts and their pressure - temperature conditions of melting.

#/location	T	P	H_2O	SiO_2	TiO_2	Al_2O_3	FeO	MnO	MgO	CaO	Na_2O	K_2O	P_2O_5	Mg#
MORB1	1450	2.2	0	48.7	0.76	16.65	7.53		12.18	12.1	1.59	0.04		
MORB2	1350	1.5	0	50.24	0.82	17.26	6.62		10.32	11.46	2.27	0.04		
Subduction Zones														
TGI J	1125	1.2	7.5	59.59	0.44	13.55	6.32	0.12	9.65	0.24	2.66	1.3	0.13	0.73
SD-438 J	1300	1.2	0	49.76	1.02	15.48	8.87	0.16	11.68	8.91	2.6	1.28	0.22	0.7
SH-85-41c C	1180	1	6	57.79	0.6	14.46	5.74	0.11	9.14	8.17	3.11	0.71	0.15	0.74
SH-85-44C	1200	1	4.5	51.68	0.6	16.4	7.93	0.16	10.79	9.67	2.24	0.42	0.11	0.71
ML-82-72f C	1310	1.1	0	47.4	0.59	18.5	8.2	0.15	10.52	12.02	2.16	0.07	0.06	0.7
M.102 M	1300	1.7	6	49.4	0.99	14.84	7.63	0.12	11.63	8.3	3.53	1.4	0.44	0.73
Av-MGB I			0.3	49.1	0.82	15.9	8.9	0.17	10.5	11.3	2.21	0.34	0.1	0.68
BON IB	1480	1.5	2.5	51.88	0.7	8.71	8.94	0.12	20.69	7.1	1.7	0.16		0.8
V35-5 A				56.54	0.61	16.68	4.84	0.12	6.62	7.2	3.89	1.3	0.13	0.75
ID16 A	1320	1.2	0	48.9	0.7	16.01	8.9	0.17	11.42	10.89	2.21	0.52	0.12	0.7
Hawaii														
K98-R-15 L	1553	3.5		45.16	2.05	10.9	12.38	0.13	18.6	8.37	1.94	0.28	0.19	0.73
Kil KSF	1420	1.3	0	48.8	1.77	10.3	11.6	0.17	17.2	8.25	1.52	0.27	0.17	0.73
B19 KSF	1375	3	0.61	41.5	1.99	10.9	12.62	0.15	17.2	9.53	3.86	1.22	0.29	0.71

References for the melt compositions can be found in Grove et al. (2012) with the exception of the MORB compositions in the spinel lherzolite stability field which are from Till et al. (2012).
MORBs are calculated using Till et al. (2012) for spinel lherzolite melting and Kinzler and Grove (1992) for plag lherzolite melting.
Key to location: J = Japan, C = Cascades, M = Mexican, I = Indonesia, IB = Izu-Bonin, A = Aleutians, L = Loihi, KSF = Kilauea South Flank.
See Grove et al. (2012) for references to subduction zone data.
Hawaii compositions from Grove et al. (2013) and Wagner and Grove (1998).
Temperature in °C, Pressure in GPa, and H_2O contents in %wt.

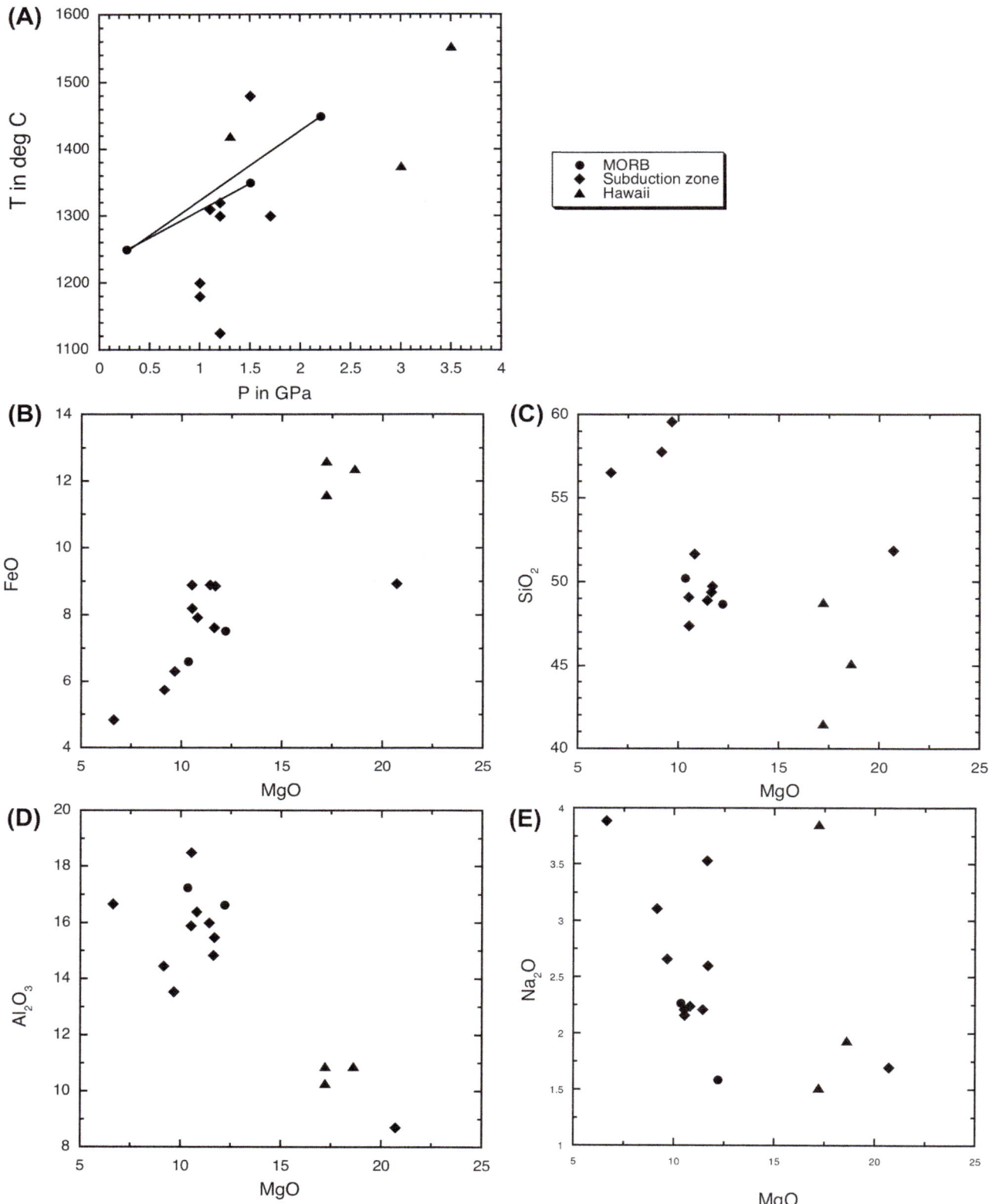

FIGURE 1.11   Primary mantle melt compositions and variations with tectonic setting from Table 1.1. The MORB magmas (circles) represent an aggregate of melt fractions that are produced continuously during near-fractional adiabatic decompression melting, as illustrated by the lines representing two different melting paths in a). Temperature is in degrees centigrade, pressure in gigapascals.

two Hawaiian melts represent high extents of melting or mantle—melt reaction. The highest Na_2O abundances are found in the subduction zone magmas because the mantle has been chemically modified by the addition of a Na_2O-bearing fluid-rich component derived from the subducted slab. The low-extent melts of carbonated mantle also have high Na_2O, and this is a result of compositionally modified mantle present in the lithosphere beneath Hawaii.

FURTHER READING

Anderson, D.L., 2007. New Theory of the Earth. Cambridge University Press, New York, 400 pp.

Boyd, F.R., 1989. Compositional distinction between oceanic and cratonic lithosphere. Earth and Planetary Science Letters 89, 15—26.

Baker, M.B., Beckett, J.R., 1999. The origin of abyssal peridotites: a reinterpretation of constraints based on primary bulk compositions. Earth and Planetary Science Letters 171, 49—61.

Langmuir, C.H., Klein, E.M., Plank, T., 1992. Petrological systematics of mid-ocean ridge basalts: constraints on melt generation beneath ocean ridges. In: Phipps Morgan, J.P., Blackman, D.K., Sinton, J.K. (Eds.), Mantle Flow and Melt Generation at Mid-ocean Ridges, Geophysical Monograph, vol. 71. American Geophysical Union, Washington, DC.

Grove, T.L., Holbig, E.S., Barr, J.A., Till, C.B., Krawczynski, M.J, 2013. Melts of garnet lherzolite: experiments, models and comparison to melts of pyroxenite and carbonated lherzolite. Contributions Mineralogy Petrology 166, 887—910. http://dx.doi.org/10.1007/s00410-013-0899-9.

Grove, T.L., Till, C.B., Krawczynski, M.J., 2012. The role of H_2O in subduction zone magmatism. Annual Review of Earth and Planetary Sciences 40, 413—439. http://dx.doi.org/10.1146/annurev-earth-042711-105310, 183—280.

Niu, Y., Langmuir, C.H., Kinzler, 1997. The origin of abyssal peridotites: a new perspective. Earth and Planetary Science Letters 152, 251—265.

Till, C.B., Grove, T.L., Krawczynski, M.J., 2012. A melting model for variably depleted and enriched lherzolite in the plagioclase and spinel stability fields. Journal of Geophysical Research 177 (B6). http://dx.doi.org/10.1029/2011JB009044.

Wagner, T.P., Grove, T.L., 1998. Melt/harzburgite reaction in the petrogenesis of tholeiitic magma from Kilauea volcano, Hawaii. Contributions to Mineralogy and Petrology 131, 1—12.

Chapter 2

Migration of Melt

Martha J. Daines and Matej Pec

Department of Earth Sciences, University of Minnesota, Minneapolis, MA, USA

Chapter Outline

GLOSSARY

compaction A process in which the solid matrix deforms to expel intergranular melt. Compaction processes could include rearrangement of matrix grains without any distortion of individual grains (granular flow) or changes in grain shape as a result of intracrystalline plasticity or diffusive transfer processes.

compaction length The distance over which the compaction rate in a column of partially molten rock decreases by a factor of e. The compaction length, δ_c, is the characteristic length scale over which gradients in pressure and melt fraction can be sustained in partially molten rocks. This distance is a function of the viscosity and permeability of the matrix and the viscosity of the melt migrating through the matrix.

deviatoric stress Stress at a point can be described by a stress tensor containing nine components, with the principal stresses (σ_{11}, σ_{22}, σ_{33}) on the main diagonal and shear stresses (σ_{12}, σ_{13}, σ_{23}, σ_{21}, σ_{31}, σ_{32}) off the diagonal. Deviatoric stress is obtained by subtracting the mean pressure (i.e., the isotropic part ($\sigma_{11} + \sigma_{22} + \sigma_{33}$)/3) from the stress tensor. Deviatoric stress causes distortion of the stressed body.

dihedral angle Measured at triple junctions where two grains and melt meet; it is the angle at which the two grains intersect. The value of the dihedral angle is a function of the ratio of the solid–solid interfacial energy to the solid–melt interfacial energy. Dihedral angles lower than $60°$ imply that an interconnected melt network will form. The lower the dihedral angle is, the lower the percolation threshold.

melt alignment The alignment of melt at the grain scale. Alignment can be caused by deviatoric stresses or be due to crystalline anisotropy with respect to surface energy resulting in anisotropic wetting properties of crystals.

melt segregation The formation of melt-rich and melt-depleted regions at distances greater than the grain scale but smaller than the compaction length.

permeability A measure of how easily fluid flows through a porous medium. Ease of flow is a function of the size, shape, and connectivity of the porosity network.

viscosity A measure of the resistance of a material to flow in response to stress.

Magma erupted on the Earth's surface is generated by partial melting of rocks at depth. How does this melt separate from the residual solid and migrate to its eruption location? Recent models of physical processes, combined with constraints imposed by the chemical and isotopic compositions of erupted lavas and magmas, suggest that melt migration in the Earth's interior occurs by porous flow along grain-size channels and by flow through larger conduits.

1. POROUS FLOW MODEL OF MELT MIGRATION

Because melting of rocks commences at the grain scale, it results in a distribution of melt throughout the rock. The separation of melt from the residual solid requires that melt flows for at least some distance through a connected network of grain-size channels. This type of flow is referred to as **two-phase** or **porous flow**; the partially molten region is regarded as a porous medium. As such, the flow of melt can be described by Darcy's law:

$$q = \frac{K}{\mu}\frac{dP}{dx} \tag{2.1}$$

where q is the flux of melt (i.e., the volume of melt per unit time passing a unit cross-sectional area perpendicular to the flow direction), K is the **permeability**, μ is the **viscosity** of the fluid, and dP/dx is the pressure gradient that drives melt flow. The driving force for melt migration can be chemical and/or physical in nature. Chemical driving forces can be related to differences in interfacial energies; physical forces can be related to the buoyancy of melt relative to residual solids or to applied **deviatoric stress**. All three of these types of driving forces have been exploited in laboratory studies of melt migration.

The permeability K is a measure of how easily melt flows through the partially molten rock. Ease of flow, as measured by permeability, is a function of the size, shape, and connectivity of the network of porosity or melt channels. Because the permeability of partially molten rocks is difficult to measure directly, permeability is usually calculated using an expression such as

$$K = \frac{d^2\phi^n}{C} \tag{2.2}$$

where d is the grain size, ϕ is the melt fraction and n is the melt fraction exponent dependent on microstructure and its evolution, and C is a geometric constant quantifying the tortuosity of melt flow. Basically, as grain size and melt fraction increase, permeability increases because the melt channels become larger. The larger the melt channel, the easier it is for melt to flow.

Darcy's law is applicable as long as the total amount of melt in the system remains constant. If the melt volume entering the system is different from the volume leaving the system, then Darcy's law must be modified to take this fact into account. Gains in melt volume, perhaps due to increases in melting rate relative to melt extraction rate, result in either increases in rock volume or increases in the melt pressure. Ramifications of high melt or pore pressure will be discussed in a later section. Reduction of the total amount of melt, perhaps due to rapid melt extraction relative to the melt production rate, implies that the solid matrix must deform to accommodate the change in fluid volume. The reduction in pore volume requires that the matrix compact. **Compaction** processes could include rearrangement of matrix grains without any distortion of individual grains (granular flow), or changes in grain shape as a result of intracrystalline plasticity or diffusive transfer processes. Compaction means that the mechanical properties of the matrix must be considered when determining the melt flux.

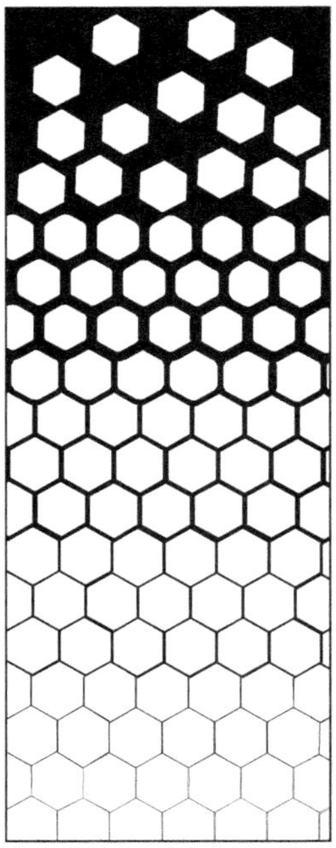

FIGURE 2.1 Illustration showing the stages of compaction that result from gravity-driven melt migration. At the base of the column, melt volume decreases as the matrix compacts. Above the compacting zone is a layer in which melt fraction does not change. At the top of the column, melt accumulates.

Gravity-driven compaction models of melt migration provide useful first-order insights into the problem of melt extraction. In these models a column of partially molten rock with an underlying impermeable base compacts due to density differences between melt and solid. Figure 2.1 shows the stages of this compaction process. At the base of the column, melt volume decreases as the matrix compacts. Above this compacting layer, the upward flux of melt from the compaction zone maintains the melt volume. At the top of the column, melt accumulates, forming a melt-rich layer. Essentially, the movement of melt due to buoyancy is resisted by the strength of the matrix and the viscosity of the melt.

The thickness of the compacting zone depends on the properties of both the matrix and melt; the matrix can compact only if melt is removed and melt will be expelled only if the matrix compacts. Compaction occurs most rapidly at the base of the column; the rate of compaction decreases exponentially with distance from the base. The distance over which the compaction rate decreases by a factor of e is the characteristic length scale of the compaction process. Known as the **compaction length**, δ_c, this length scale is

$$\delta_c = \left[\frac{\zeta + \frac{4}{3}\eta}{\mu}K\right]^{1/2} \tag{2.3}$$

where ζ and η are the effective bulk and shear viscosities, respectively, of the partially molten rock. The expression in the numerator $(\zeta + 4\eta/3)$ describes how the partially molten rock deforms in response to the applied forces. In this model both the melt and the matrix are assumed to be linearly viscous. That is, the flow of matrix and melt depends linearly on the applied stress. A characteristic timescale for this process is given by τ_0, which is the time required to reduce the melt fraction by a factor e at the base of the compacting layer:

$$\tau_0 = \frac{\delta_c}{w_0(1-\phi)} \qquad (2.4)$$

In this equation, w_0 is the relative velocity between the melt and matrix in the interior layer where no compaction is occurring. If melt migration is driven by density contrasts only, then

$$w_0 = \frac{K(1-\phi)(\rho_s - \rho_f)g}{\mu\phi}, \qquad (2.5)$$

where $(\rho_s - \rho_f)$ is the density contrast between solid and melt and g is the gravitational acceleration (~ 9.8 ms^{-2}).

Another way to compare characteristic extraction times is to calculate the time, t_h, to reduce the porosity of a layer of thickness h, by a factor of e:

$$t_h = \tau_0 \left(\frac{h}{\delta_c} + \frac{\delta_c}{h} \right) \qquad (2.6)$$

If the thickness of the layer h is much greater than the compaction length, then

$$t_h = \frac{\tau_0 h}{\delta_c} = \frac{h}{w_0 - (1-\phi)} \qquad (2.7)$$

In essence, this means that if the total thickness of the partially molten region is much greater than the compaction length, the time to reduce the melt volume by a factor of e is independent of the viscosity of the matrix.

These equations suggest that the important physical parameters in the two-phase flow model of melt migration are melt viscosity, μ; permeability, K; and, depending on the ratio of compaction length to layer thickness, the mechanical properties of the matrix, $(\zeta + 4\eta/3)$. Substituting the values of the physical parameters into these equations allows one to calculate characteristic times and distances over which melt migration via porous flow occurs, allowing one to evaluate the likelihood of sufficiently rapid or volumetrically significant melt extraction occurring in a particular geologic environment by this process.

The next three sections of this chapter focus on our current understanding of the physical factors (i.e., melt viscosity, permeability, and matrix mechanical properties) that affect rates of melt migration in both the crust and the mantle. Although it is certain that the behavior of a layer containing an initially constant melt fraction is much simpler than that of any real geologic problem, examining the effects of physical parameters using this one-dimensional model can provide a great deal of insight into the melt extraction process by porous flow. The solutions to more complicated models show that various instabilities can develop during melt migration by porous flow. Similarly, deviatoric stress enhances the driving force for melt migration and modifies the permeability structure of the partially molten rock. These effects may allow **melt segregation** into melt channels that are larger than those present along individual grains, allowing for much more rapid melt migration. These and other flow-focusing mechanisms of melt extraction are discussed in the later portion of this chapter.

2. MELT VISCOSITY

The rate of melt migration depends on the viscosity of the melt that is percolating through the rock. Melt viscosity can vary by more than a factor of 10^{14} depending on the composition of the melt (mainly the silica and volatile content). For example, a dry rhyolitic melt at 700 °C has a calculated viscosity on the order of 5×10^{11} Pa s, whereas a carbonatite melt has a viscosity of less than 5×10^{-3} Pa s. These differences translate into large variations in compaction length, separation velocity, and compaction time. These characteristic lengths, velocities, and timescales were calculated using Eqns (2.3)–(2.7) and are listed in Table 2.1 for melts with different viscosities. These values were calculated assuming the same melt fraction (0.1), grain size (1 mm), matrix viscosity (10^{18} Pa s), permeability ($K = 1 \times 10^{-12}$ m^2), and driving force ($\rho_s - \rho_f = 0.5 \times 10^3$ kg m^{-3}). Hence, the differences listed in Table 2.1 are only the result of variations in melt viscosity.

Mantle melts of basaltic or picritic composition have viscosities of ~ 1 Pa s. These melts separate rather easily from the residual solid matrix. If the average melt fraction produced beneath a mid-ocean ridge is 0.1, then the compaction length is ~ 1 km, assuming that the average grain size of mantle rocks is 1 mm. If the partially molten region extends vertically for roughly 50 km, h/δ_c is ~ 50. From Table 2.1, τ_0 is ~ 800 years, implying that t_h, or the time to reduce the melt fraction by a factor of e, is $\sim 40,000$ years. Since this time is about thousand times less than the cooling time for a lithospheric plate, melt extraction beneath the ridge should be rapid enough to keep the melt fraction well below 0.1 at any given time. Geophysical data support this conclusion. Although shear wave velocities beneath mid-ocean ridges are anomalously slow, a melt fraction of less than 0.03 can account for these anomalies as long as the melt is present as an interconnected network. Reducing the thickness of the partially molten layer to 100 m, however, changes the efficacy of

TABLE 2.1 Characteristic Compaction Length, Separation Velocity, and Compaction Time for Melts with Different Viscosities

Melt Composition	Melt Viscosity (Pa s)	Compaction Length, σ_c	Separation Velocity (m/year)	Compaction Time, τ_0 (year)
Dry rhyolite	5×10^{11}	1.4 mm	2.8×10^{-12}	5.5×10^8
Wet granite	5×10^4	4.5 m	2.8×10^{-5}	1.8×10^5
Basalt	1	1 km	1.42	8×10^2
Carbonatite	5×10^{-3}	14 km	280	56

melt extraction as h/δ_c is reduced to 0.1. The time for significant melt extraction t_h then becomes ~ 8000 years; comparison of this time with the time required for cooling a 100-m-thick partially molten region suggests that not much compaction or melt segregation would occur before the melt crystallized due to cooling.

Crustal melts with characteristically high viscosities take much longer to separate from their solid residuum than more fluid mantle-type melts. Consider a 1-km-thick layer of partially molten crustal rocks with a grain size of 1 mm. If 10% of the layer is melted to form a dry rhyolitic melt, how long will it take for most of the melt to separate by gravity-driven grain-scale flow? From Table 2.1, the compaction length is ~ 1 mm, so h/δ_c is $\sim 1 \times 10^6$. The value of τ_0 is $\sim 10^8$ years, implying that t_h, or the time to reduce the melt fraction by a factor of e, is $\sim 10^{14}$ years. In other words, insignificant volumes of melt would segregate over the whole of geologic time.

Reducing the viscosity of the melt by adding water, however, does make gravity-driven segregation in crustal environments somewhat more feasible. If the same 1-km-thick layer of partially molten crust now contains 10% of wet granitic melt, the compaction length is increased to about 5 m with h/δ_c decreased to $\sim 10^2$; τ_0 is also reduced

to $\sim 10^5$ years, implying that t_h, or the time to reduce the melt fraction by a factor of e, is $\sim 10^7$ years.

Even this timescale is problematic for efficient segregation of melt in crustal environments by gravity-driven compaction alone. Separation of melt will be more rapid than that described here if the permeability of the residual solid is higher than that assumed above. Separation of high-silica melts may also require that the driving force for melt extraction be related to pressure gradients in excess of those induced by density contrasts between melt and solid. Deviatoric stresses resulting from tectonic activity or due to contrasts in matrix strength with composition are discussed in a later section.

3. PERMEABILITY

Permeability is a measure of how easily fluid flows through a porous media. From Darcy's law it is clear that melt flux scales directly with permeability. So, if permeability increases by a factor of 10, melt flux will increase by a factor of 10. Table 2.2 shows the values of the characteristic length, velocity, and compaction time as a function of permeability. These values were calculated using Eqns (2.3)–(2.7), assuming the same melt fraction (0.1), matrix viscosity

TABLE 2.2 Characteristic Compaction Length, Separation Velocity, and Compaction Time for Different Matrix Permeability

Permeability, K (m^2)	Compaction Length, δ_c (m)	Separation Velocity, ω_0 (m/year)	Compaction Time, τ_0 (year)	Time for Significant Melt Extraction in a 50-km-Thick Layer, t_h (year)
10^{-10}	10^4	6	2×10^3	10^4
10^{-12}	10^3	6×10^{-2}	2×10^4	10^6
10^{-14}	10^2	6×10^{-4}	2×10^5	10^8
10^{-16}	10	6×10^{-6}	2×10^6	10^{10}
10^{-18}	1	6×10^{-8}	2×10^7	10^{12}

(10^{18} Pa s), melt viscosity (1 Pa s), and driving force $(\rho_s - \rho_f) = 0.5 \times 10^3$ kg m^{-3}. Hence, the differences listed in Table 2.2 are only the result of variations in permeability. Clearly, the permeability of the partially molten rock can have a huge effect on the rates of melt extraction.

At the Earth's surface the permeability of rocks to the flow of water and other low-temperature fluids can be directly measured. At depth or in rocks at their melting temperature, these direct measurements are difficult to impossible to make. As a result, the permeability of partially molten rocks is constrained using observations about the geometry of melt networks present in the rock or through laboratory experiments.

Textural observations of melt networks can be made both on rock analogues synthesized in a laboratory or on natural rocks. Observations made on natural samples are often difficult to interpret because the original geometry of the melt phase may be obscured by the effects of crystallization or other subsequent processes.

3.1. Models of Permeability

Permeability models can be used to estimate the permeability of partially molten rock. In these models, melt flow is generally assumed to be steady and nonturbulent, with a parabolic velocity structure across the melt channel. This assumption of Poiseuille flow means that simple relationships can be derived between permeability and melt fraction or porosity if the melt channel geometry is not very complex. For example, the permeability relationships used in most two-phase flow models assume that melt flows either through parallel cracks ($K \propto \phi^3$) or through cylindrical tubules ($K \propto \phi^2$). Although these relationships can be useful as first-order approximations and have been used very extensively, observations of partially molten rocks

indicate that melt networks are more complex than these simple relationships suggest.

More accurate approximations of permeability are derived using one of two approaches. In the first approach, the rock is approximated by an equivalent porous medium or an idealized representation of the melt network in order to derive analytical solutions for permeability as a function of measurable physical characteristics of the rock. Examples of these measurable characteristics include melt fraction, pore shape and size, and the tortuosity of the melt network. Tortuosity is the ratio of the length of the actual path to the length of the apparent path (Figure 2.2). Essentially, the partially molten rock is presumed to be equivalent to a channel with a highly complex cross section of constant area. Permeability in these models is often defined by the Kozeny–Carman equation:

$$K = \frac{\phi^3}{bC^2S^2(1 - \phi)^2}, \qquad (2.8)$$

where b is a constant, C is the tortuosity, and S is the surface area of pores per unit volume of solid. Although this approach has been applied with some success to the flow of low-temperature fluids in sandstone, it has at least two drawbacks. First, both S, the melt–solid interfacial area, and C, the tortuosity, are difficult to measure accurately in partially molten rocks. This difficulty is partly due to the fact that both require detailed knowledge of the three-dimensional structure of the melt network. Second, the model assumes that the flow network is homogeneous, implying that the network has no dead ends or preferred flow paths. Because melt channels in partially molten rocks are typically quite heterogeneous in terms of shape and size, the equivalent channel or medium approach may not accurately predict permeability.

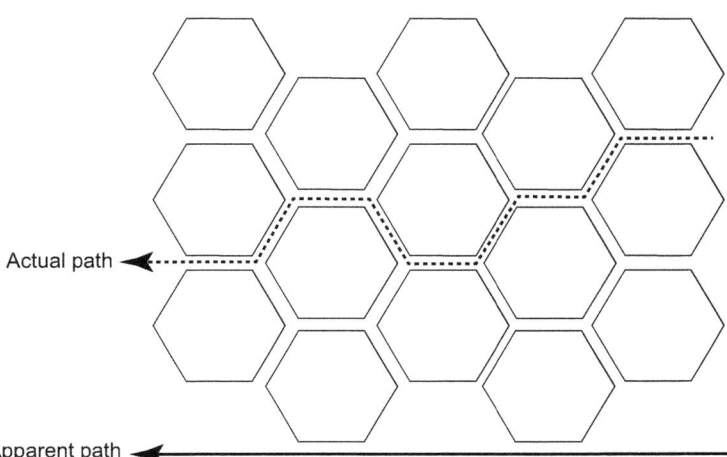

FIGURE 2.2 Tortuosity is the ratio of the actual flow path (the zigzag line) to the shortest flow distance (straight line). Very complex flow paths have high tortuosity.

A second approach to estimating or measuring permeability uses numerical simulations of flow in networks that approximate those found in natural materials. The melt network in some of these models consists of a collection of tubes, penny-shaped or sheetlike channels, and spherical pores of different sizes. These individual porosity elements are distributed on a grid; equant pores are connected by tubes or sheetlike channels. The permeability is controlled by the size and connectivity of the channels; the pores contribute to the storage capacity of network. By changing the relative proportion, distribution, and connectivity of these elements, the permeability of different melt networks and melt fractions can be determined. For example, a melt network that has spherical pores connected by penny-shaped or sheetlike channels will have lower permeability than a network with the same pores connected by cylindrical tubes.

3.2. Geometry of Melt Networks in Partially Molten Rocks

Application of these permeability models to partially molten rocks requires a detailed knowledge of the melt network geometry. The initial melting of a rock occurs at the junctions of grains taking part in the reaction. Once melt is present, the melt–rock system adjusts to minimize its total energy. The energy of the system is reduced by minimizing the area of high-energy interfaces and by changing the shapes of interfaces to reduce chemical potential gradients. This energy reduction results in the redistribution of melt, with melt wetting grains if melt–crystal interfaces are of lower energy than crystal–crystal boundaries or melt pooling if the opposite is true. This equilibrium grain-scale distribution of melt in partially molten rock can be characterized by the **dihedral angle**. The dihedral angle is measured at triple junctions where two grains and the melt meet; it is the angle at which the

two grains intersect (Figure 2.3). In an ideal rock, composed of identical grains with no variation of surface energies with crystal orientation, dihedral angles lower than $60°$ imply that an interconnected melt network will form; the lower the dihedral angle, the lower is the percolation threshold. Dihedral angles less than $60°$ imply that the melt network is composed of triangular-shaped tubules along three-grain junctions, whereas dihedral angles greater than $60°$ imply that at low melt fractions melt is isolated in more spherically shaped pockets at four-grain corners (Figure 2.4). In rocks with dihedral angles in the range of $1-40°$, permeability should follow the relationship for flow in tubes, that is, $K \propto \phi^2$. In this ideal rock, boundaries between two grains are free of melt unless the dihedral angle is zero or the melt fraction exceeds a critical value.

Most rocks contain more than one mineral. Because surface energies vary among minerals, these additional phases will alter the connectivity of the melt network. The addition of a mineral with a high dihedral angle will decrease the connectivity of the melt network; the reduction in connectivity and change in shape of melt channels will depend on the concentration and distribution of additional minerals.

Dihedral angles in a wide range of rock types have been measured and interpreted using this ideal model. Almost all measured angles for melt in contact with crystals are $\leq 60°$. As a first-order approximation of melt network geometry,

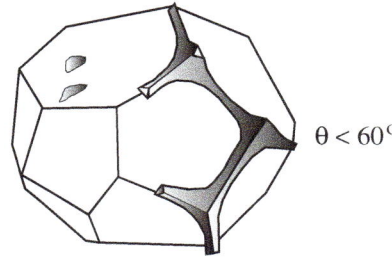

$$\theta < 60°$$

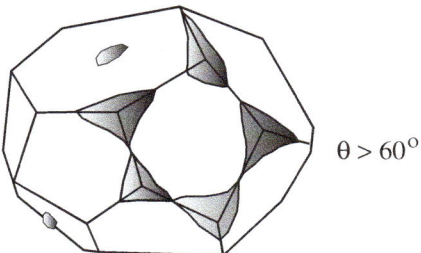

$$\theta > 60°$$

FIGURE 2.4 Idealized melt networks in partially molten rocks with dihedral angles (A) <$60°$ and (B) >$60°$. For dihedral angles <$60°$ melt forms an interconnected network along three- and four-grain junctions or along some grain boundaries. (Adapted from Wark and Watson, 1998, with permission from Elsevier Science).

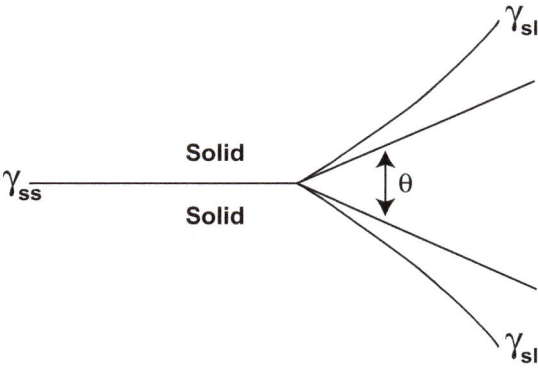

FIGURE 2.3 The dihedral angle is the angle at which the two grains intersect with the melt. The value of the dihedral angle is a function of the ratio of the solid-solid interfacial energy (γ_{ss}) to the solid-melt interfacial energy (γ_{sl}).

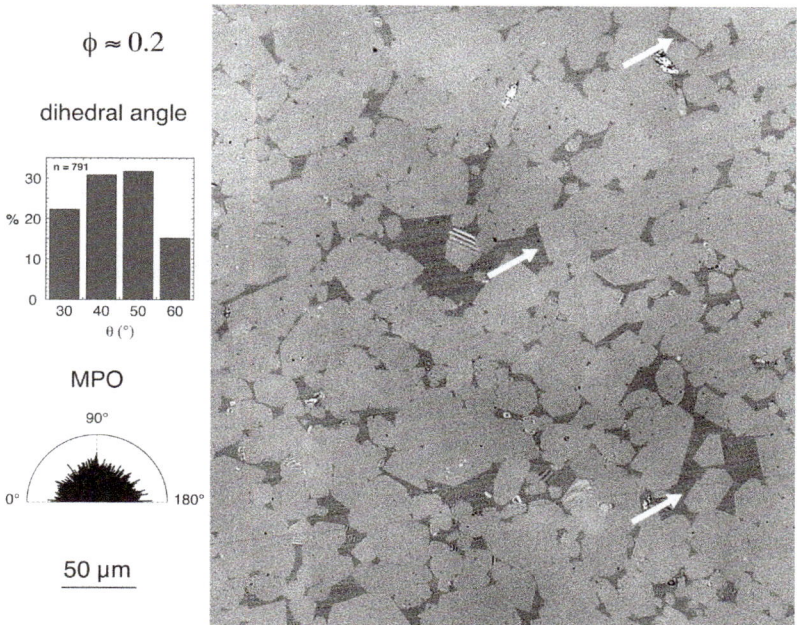

$\phi \approx 0.2$

dihedral angle

MPO

50 μm

FIGURE 2.5 Reflected light image of an undeformed olivine-basalt aggregate. Olivine is light; quenched melt is dark. Arrows point to planar mineral-melt interfaces illustrating the effect of crystalline anisotropy with respect to interfacial energy on the melt distribution. Histogram shows the measured distribution of dihedral angles; rose diagram shows the surface orientation distribution of melt pockets.

these measurements suggest that partially molten rocks in both the crust and the mantle will have an interconnected melt network at very low melt fractions. Work on the melt distribution in olivine-rich mantle rocks, however, points to a more complicated melt geometry in these and other rock types due to the effects of crystalline anisotropy with respect to surface energy. Olivine has been demonstrated experimentally to be anisotropic with respect to surface energy; the contact angle between basaltic melt and (100) olivine crystal surfaces are a factor of 3—10 times greater than between other surfaces. Flat crystal—melt interfaces, rather than the smoothly curved boundaries predicted for isotropic minerals, are prevalent and preferentially developed perpendicular to *b* axes, parallel to (010) (Figure 2.5). Qualitatively, more melt is present in these planar channels between two grains than is found in tubules as would be predicted from simply examining the characteristic dihedral angle. As melt fraction increases, this effect is amplified, because more penny-shaped channels develop as two-grain junctions become melt-filled. Increasing the melt fraction in these rocks essentially causes the growth of penny-shaped or sheetlike channels rather than simply increasing the size of tubular channels. These penny-shaped channels contribute very little to the overall melt flow until they reach a critical thickness. Hence, permeability may increase slowly with increasing melt fraction until a critical melt fraction is attained.

Because all rock-forming minerals show some degree of anisotropy with respect to surface energy, the dihedral angle-based tubule model in Figure 2.4 may not be the best melt network geometry to assume when estimating permeability. This conclusion may be especially applicable to rocks containing minerals such as amphibole, feldspar, and biotite, which are highly anisotropic with respect to surface energy. Biotite, for example, because of its layered structure and high degree of anisotropy, develops very prevalent planar rational crystal faces in contact with melt. Melt topology is also profoundly influenced by deviatoric stress and deformation. The ideal model is derived assuming isotropic stress conditions and, thus, no inelastic deformation of the rock. The distribution of melt in rocks that have been deformed while partially molten, however, can be significantly different from that predicted by the isotropic model depending on the mechanism of deformation and boundary conditions. The effects of deformation are treated in more detail in later sections.

3.3. Laboratory Constraints on Permeability

Constraints on melt extraction rates have been given by experiments where a partially molten rock is placed next to a porous reservoir at high pressures and temperatures. The pressure in the reservoir can be controlled and the compaction rate recorded. These experiments indicate that permeability depends on melt fraction to the cube ($K \propto \phi^3$) in the range of $0.05 < \phi < 0.35$ in olivine—basalt aggregates. However, other experiments designed to look at migration rates of basalt in olivine-rich rocks at lower melt fractions indicate that $K \propto \phi$, a relationship not predicted by any theoretical considerations of flow in any shaped network. Melt migration in these experiments was driven by differences in initial melt content; capillary forces resulting from interfacial energy considerations led to the movement of melt from a melt-rich disk to a melt-poor disk. In some experiments the melt-poor disk initially contained no melt, whereas in other experiments the melt-poor disk initially contained a melt fraction of ~ 0.01. For both starting conditions, permeability in these experiments was determined through comparison of numerical simulations of two-phase flow with the actual melt migration profiles of the experiment. From these comparisons permeability was found to range from 5.2×10^{-15} to $1 \times 10^{-18} \phi \, m^2$, depending on the initial quantity of melt in the melt-poor disk and on the composition of the melt. Only in experiments in which the melt-poor disk was melt-free was permeability found to be linearly dependent on melt fraction. The unexpected linear dependence of permeability on melt fraction yields relatively high permeability at low melt fractions, implying that melt extraction by porous flow through grain-size channels is efficient enough to keep melt fractions beneath mid-ocean ridges below 0.001.

Direct permeability measurements on monomineralic aggregates of quartz and calcite synthesized at high pressure and temperature in the presence of water suggest that, for melt–fluid fractions greater than 0.01, the functional dependence of permeability on melt or fluid fraction is cubic ($K \propto \phi^3$), the same as that predicted by some theoretical models. In these experiments, the annealed sample contains a network of pores filled with a water-rich fluid during the synthesis process. To measure the permeability of this now-empty network, a pressurized air reservoir is placed against one end of the sample. As air moves through the pore network in rock, the decrease in the pressure of the air reservoir is recorded. From the decay in pressure, permeability is calculated. For melt–fluid fractions above about 0.01, permeability in quartz aggregates is

$$K = \frac{d^2 \phi^3}{1800(1 - \phi)^2} \qquad (2.9)$$

At melt–fluid fractions below 0.01, however, permeability levels off or increases if the dihedral angle is $<60°$. This permeability relationship should apply to other natural fluid- or melt-bearing rocks provided no major differences exist in the degree of crystalline anisotropy or the extent of development of flat faces.

Clearly, the network of melt channels present in partially molten rocks is much more complicated than the simple relationship $K \propto d^2 \phi^n$ would imply. Significant uncertainties in permeability estimates exist especially at very low melt fractions, which—unfortunately—are likely prevalent in nature.

4. MECHANICAL PROPERTIES OF THE MATRIX

Partially molten material will deform if subjected to stress. In gravity-driven compaction models, the stress arises from the density differences between the solid and the interstitial melt. The deformation or compaction of the matrix results in melt pressure gradients; melt flows in response to these gradients. The importance of matrix deformation during melt migration by porous flow depends on the relative values of the compaction length and the height of the partially molten column. Table 2.3 shows the values of the characteristic length, velocity, and compaction time as a function of the mechanical properties of the matrix. These values were calculated from

TABLE 2.3 Characteristic Compaction Length, Separation Velocity, and Compaction Times Calculated Assuming Different Matrix Viscosities

Matrix Viscosity (Pa s)	Compaction Length, δ_c (m)	Separation Velocity, ω_0 (m/year)	Compaction Time τ_0 (year)	Time for Significant Melt Extraction in a 50-m-Thick Layer, t_h (year)	Time for Significant Melt Extraction in a 50-km-Thick Layer, t_h (year)
10^{14}	10	1.5	7.4	38.6	3.7×10^4
10^{16}	10^2	1.5	7.4×10	1.9×10^2	3.7×10^4
10^{18}	10^3	1.5	7.4×10^2	1.5×10^4	3.7×10^4
10^{20}	10^4	1.5	7.4×10^3	1.5×10^6	3.85×10^4
10^{22}	10^5	1.5	7.4×10^4	1.5×10^8	1.85×10^5

Eqns (2.3)–(2.7), assuming the same melt fraction (0.1), melt viscosity (1 Pa s), and driving force $(\rho_s - \rho_f) = 0.5 \times 10^3$ kg m^{-3}. Hence, the differences listed in Table 2.3 are only the result of variations in matrix viscosity. The times for significant extraction of melt from two layers of different thickness (50 km and 50 m) are also listed. Clearly, compaction of the matrix has little effect on melt migration rates unless the thickness of the partially molten region is less than or about the same as the compaction length. When the matrix viscosity is relatively high, the compaction length is relatively large, so that the deformation of the matrix is more likely to control the rate of melt expulsion.

Calculation of the compaction length requires that one know the values of ζ, the bulk viscosity, and η, the shear viscosity of the matrix. These values are difficult to determine independently from typical deformation experiments. Deformation experiments can determine flow laws that govern the partially molten rock's response to stress. These laws can be used to calculate an effective viscosity (η_{pm}) that can be substituted into the equation for compaction length. For example,

$$\eta_{pm} = \sigma/\dot{\varepsilon} = A\sigma^{1-n}d^{-p}\exp(Q/RT), \qquad (2.10)$$

where σ is differential stress ($\sigma_{11} - \sigma_{33}$), ε is the axial strain rate, d is grain size, p is the grain size exponent, n is the stress exponent, A is a materials parameter, Q is the activation energy for creep, R is the ideal gas constant, and T is absolute temperature. The effective viscosity of a partially molten rock is dependent on the amount of melt present in the rock. Increasing the melt fraction decreases the viscosity of the rock. The degree to which melt affects the rheology of the rock depends on the mechanism by which the rock is deforming. A range of deformation mechanisms has been observed in partially molten rocks exposed to stress. These mechanisms are dependent, in part, on the amount of melt in the rock. One of several deformation mechanisms may be dominant: diffusion creep, dislocation creep, or granular flow with or without fracturing.

At relatively low melt fractions, diffusion and dislocation creep mechanisms may dominate. In diffusion creep-dominated deformation, deformation may be limited by the rate of diffusion through the matrix grains or grain boundaries or by the rate of interface reaction. The presence of melt along grain boundaries can accelerate creep by providing high-diffusivity paths, and by enhancing the stress concentration at melt-free boundaries. For a melt fraction of about 0.03 and a dihedral angle of 35°, the presence of melt enhances the rate of diffusion creep by a factor of 2. Because higher melt fractions lead to more two-grain boundaries containing melt, this enhancement factor also increases as melt fraction increases. In partially molten fine-grained olivine-rich aggregates deformed in the laboratory, a marked change in diffusion creep rate or effective viscosity occurs at melt fractions above about 5%, the melt fraction at which high percentages of grain boundaries contain melt. However, based on model results and geophysical evidence described earlier, this melt fraction is not likely to be present, at least in mid-ocean ridge environments, because melt extraction processes are efficient enough to maintain melt fractions below this value. Experiments on nominally melt-free olivine as well as on theoretical models for solid–liquid composites point out the importance of very low melt fractions on viscosity in the diffusion creep regime. As the melt fraction increases from zero, there exists a critical melt fraction for shear viscosity, η, below which the rate-limiting process changes from diffusion through solid grain boundaries to diffusion through the liquid, and for bulk viscosity, ζ, there exists a critical melt fraction below which the rate-limiting process changes from diffusion through grain boundary to reaction at the pore surface. In other words, different processes may limit the kinetics of bulk and shear viscosities at small melt fractions. Extrapolations to millimeter mantle grain sizes suggest that a marked drop in bulk and shear viscosities can be caused by melt fractions as low as $10^{-4} - 10^{-8}$ in nature.

An enhancement of deformation rates with increasing melt fraction has also been observed in partially molten fine-grained mantle rocks deformed in the dislocation creep regime. Theoretically, this enhancement should be limited to that associated with the increased concentration of stress at melt-free grain boundaries, and so should be quite small, depending on the geometry of the melt network. Models based on the dihedral angle melt network geometry predict a creep rate enhancement of less than 15% for a melt fraction of 0.03. Measured enhancements in mantle rocks, however, can be much greater. For example, the presence of melt fractions of 0.03 and 0.1 increased the creep rate by factors of 1.5 and 25, respectively, in olivine-rich rocks deformed in the dislocation creep regime. This increased enhancement probably results from significant contributions of grain boundary deformation mechanisms to the creep process. These processes include grain boundary sliding as well as grain boundary diffusion. The involvement of grain boundary processes implies that this increased creep rate enhancement in the dislocation creep regime is in part a grain-size dependent phenomenon; estimating its applicability in partially molten regions of the mantle requires grain-size constraints. Furthermore, the nonlinear dependence of stress on strain rate in the dislocation creep regime (stress exponent, $n = 3-5$) significantly complicates the theoretical analysis of such systems. Nevertheless, for nonextreme melt fractions, the diffusion and dislocation creep data for the viscosity of partially molten mantle rocks can be fit by the following empirical relationship:

$$\eta_{pm} = \eta_s \exp(-\lambda \phi), \qquad (2.11)$$

where η_{pm} and η_s are the viscosity of the solid with and without melt, respectively, and λ is the material-dependent parameter determined as $\lambda \approx 20-30$ in diffusion creep regime and $\lambda \approx 30-45$ in dislocation creep.

From Eqn (2.11) it is clear that the viscosity of partially molten rock strongly depends on the melt fraction and hence any perturbation in melt fraction will result in large variations in viscosity. Further, deviatoric stress causes the distribution of contact area between solid grains to become anisotropic. In turn, this change causes anisotropy of the matrix viscosity and permeability at the continuum scale. The consequences of this viscous anisotropy are discussed in Section 5.

In crustal rocks, because of low dihedral angles and crystalline anisotropy, the penetration of melt into grain boundaries may be even more prevalent than in mantle lithologies, suggesting that enhancements of diffusion or dislocation creep rates could be even greater in these rocks. Unfortunately, it is challenging to deform partially molten granites in the diffusion or dislocation creep regime in the laboratory. Instead, these materials often deform by a combination of cataclastic deformation of grains and granular flow. Experimental results suggest that the relationship of viscosity and melt fraction is nonlinear. At melt fractions $\phi < 0.07$ the dependence of viscosity on ϕ is significantly greater than at melt fractions $\phi > 0.07$. The first significant strength drop at low melt fractions is interpreted as a result of melt interconnectivity. The second strength drop at higher melt fractions is interpreted as the result of breakdown of the solid framework.

5. MELT LOCALIZATION AND FLOW FOCUSING

Although the initial segregation of melt from the solid residuum must occur by the two-phase model for percolation described earlier, calculation of melt migration rates and length scales, combined with field evidence and geochemical constraints, suggests that at some stage in the extraction process melt must be focused into larger conduits for more rapid melt extraction.

In mantle rocks this conclusion is based in part on chemical and isotopic signatures of both melt and residuum. Abyssal peridotites, assumed to be the residues of partial melting that formed mid-ocean ridge basalts (MORB), are strongly depleted in light rare earth elements and so are not in trace element equilibrium with MORB. MORB is also not saturated in orthopyroxene, one of the major constituents of these residual peridotites. This disequilibrium requires the efficient extraction of very small amounts of melt (<0.01) throughout the melting region, and little or no reequilibration between melt and solid as melt is transported from depth to the surface. Essentially, the melt extraction is efficient enough to generate near-fractional melting. Pervasive two-phase flow of large volumes of melt through grain-scale channels would result in extensive chemical interaction, wiping out these disequilibrium features.

$^{238}U-^{230}Th-^{226}Ra$ abundances in MORB and other tholeiites also indicate a fractional mode of melting. Field relationships in ophiolites also indicate that melt flow is focused into larger conduits at some stage of melt extraction. Elongate dunite bodies and ductile shear zones in ophiolites and ocean floor drill core have both been interpreted as former channels for focused melt flow in the shallow mantle.

In the crust, calculations of compaction times and segregation velocities indicate that gravity-driven flow alone cannot efficiently segregate significant volumes of melt over geologic timescales unless melt viscosity is less than $\sim 10^4$ Pa s. Partially molten crustal regions must be subjected to an applied stress if significant melt segregation is to occur in most of these rocks. Or high viscosity crustal melts must collect and migrate through a network of veins and dikes, minimizing the distance over which melt must migrate through intragranular channels. Field evidence abounds, both for the inefficiency of gravity-driven compaction and for the focusing of flow into veins and larger conduits. For example, layers, pods, or veins of granitic material (leucosomes) found in plastically deformed metamorphic rocks could represent melt crystallized in a drainage network in the end stages of melt extraction or melt crystallized before any appreciable melt migration has even occurred. Shear zones, however, undoubtedly represent relatively high-permeability channels for melt extraction.

The next sections examine several mechanisms for localizing melt and focusing melt flow into larger conduits. These larger conduits serve to increase the permeability of the partially molten region and to chemically isolate the melt from the residual solid as the melt moves to the surface.

5.1. Effects of Deformation

Melting occurs predominantly at divergent and convergent plate boundaries and plumes, which are also zones of intense deformation. Therefore, it is important to understand the influence of deviatoric stresses and deformation on melt migration. As melt migrates toward the surface, it crosses a wide range of pressure and temperature conditions, which control whether melt transport occurs by viscous or brittle processes.

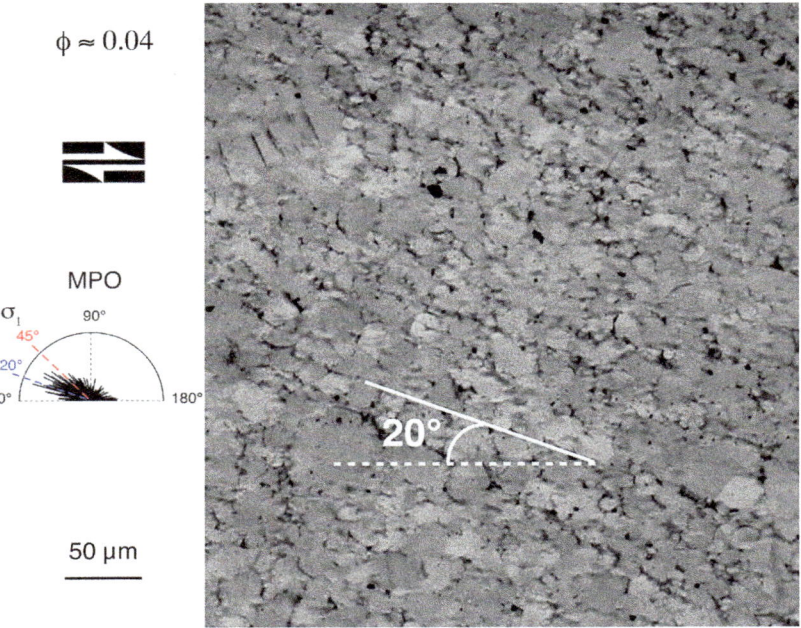

$\phi \approx 0.04$

MPO

50 μm

FIGURE 2.6 Reflected light image of a sheared olivine-basalt aggregate. Olivine is light; quenched melt is dark. Rose diagram shows surface orientation distribution of melt pockets. Note the alignment at 20-30° antithetic with respect to shear plane.

5.1.1. Stress-Driven Melt Alignment

The distribution of melt in rocks that have been deformed while partially molten can be significantly different from that predicted by the isotropic model based on the dihedral angle (see Section 3.2). At low differential stresses, the melt geometry appears unaffected and is indistinguishable from that present in undeformed rocks. As differential stress increases, the melt geometry is modified and a melt-preferred orientation (MPO) develops at the grain scale (Figure 2.6). In coaxial compression of olivine−basalt aggregates, the melt aligns at 15°−30° with respect to the principal stress (σ_1). In shear experiments on olivine−basalt aggregates, melt aligns at $\sim 20°$ antithetic to the shear direction, i.e., $\sim 25°$ with respect to σ_1 (Figure 2.6). Deviatoric stress applied to a partially molten rock leads to variations in the stress field associated with the melt network. Depending on the orientation of melt pockets or segments of the network, local pressures may be high or low. These pressure gradients lead to the growth of melt pockets as melt migrates into those pockets preferentially oriented at some angle to the principal stress. The MPO then develops due to the concentration of stress that develops at the edges or ends of melt pockets. If the melt network is treated as an interconnected network of elliptical melt inclusions, then in a compressive stress field, melt inclusions inclined at an angle ψ to σ_1 will yield the greatest tensile stress concentration. The value of ψ is given by

$$\cos 2\psi = 0.5\left(\sigma_1' - \sigma_3'/\sigma_1' + \sigma_3'\right) \qquad (2.12)$$

where σ_3' is the least compressive stress. The primes in this equation indicate effective stress components, meaning that each has been reduced by a value equal to the melt pressure (i.e., $\sigma_{ij}' = \sigma_{ij} - \alpha\delta_{ij}P_m$ where σ_{ij} is the applied stress, δ_{ij} is the Kronecker delta, P_m is the melt pressure, and $\alpha = 1$). In this formulation, the melt inclusions act as defects in the rock; these defects produce local stress concentrations and gradients in mean pressure, which depend on the orientation of the defect in the stress field. In response to these gradients, melt flows into regions of the melt network that have the lowest mean pressure, essentially extending the melt inclusions in the most favorable orientation with respect to stress (Figure 2.7(A)). In the case of partially molten mantle rocks deformed in the laboratory, substituting appropriate experimental conditions into Eqn (2.12) gives $20° < \psi < 30°$. These values are in excellent agreement with the preferred orientation of melt measured in laboratory experiments.

For this mechanism of **melt alignment** to operate effectively in the mantle, the pressure gradients due to stress in the mantle must be greater than the pressure gradients produced by surface tension, which are generally trying to redistribute melt homogeneously. The driving force for melt redistribution due to applied stress scales approximately with the square root of the grain size (d); the driving force for melt redistribution due to surface tension, based on the relationship between pressure and radius of curvature ($P \propto 1/r$), scales approximately with the reciprocal of grain size. Comparison of these relationships between grain size and driving force indicates that as grain

(A)

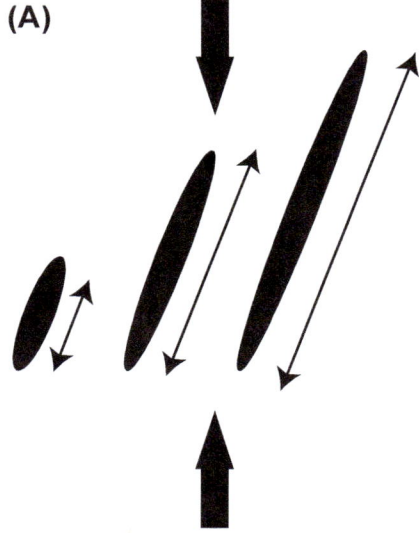

(B)

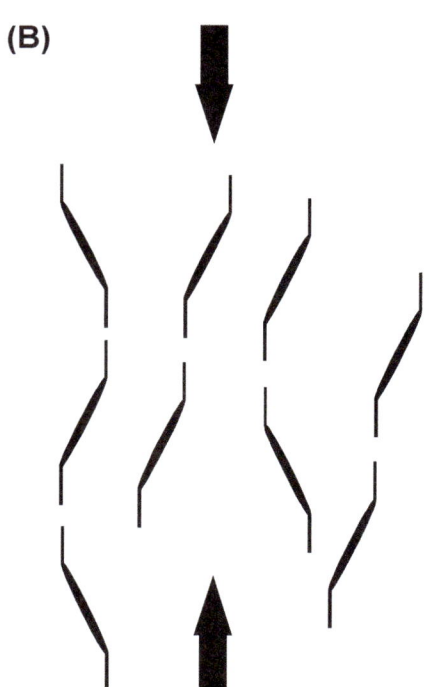

FIGURE 2.7 Anisotropic permeability structures can develop in deforming partially molten rocks either by (A) extension of melt pockets favorably oriented relative the maximum compressive stress (σ_1) or by (B) microcracking. Propagation and coalescence of cracks results in the development of veins parallel to σ_1 whereas extension of the melt pockets results in the development of veins oriented at an angle of $\sim 30°$ to σ_1.

size increases the driving force for development of a stress-induced melt distribution will increase by a factor of $a^{1.5}$ relative to the driving force for melt redistribution due to surface tension forces. This observation, combined with

laboratory results, suggests that the stress-induced melt distribution should be observed at differential stresses of a few kilopascals for rocks with grain sizes of ~ 1 mm. These stresses are considerably lower than those expected in the upper mantle, implying that stress-induced anisotropic melt distributions will dominate in deforming regions of the mantle.

Furthermore, the MPO introduces strong anisotropy in viscosity and permeability of the partially molten aggregate. Since anisotropy causes a direct coupling between shear and isotropic components of the stress tensor, the role of shear deformation in melt migration is significantly increased. Hence, viscous anisotropy at the grain scale drives further melt redistribution over distances greater than the grain scale and leads to melt segregation in partially molten rocks described below.

A similar MPO as in peridotites has also been observed in deformed partially molten crustal rocks, where after deformation melt occupies grain boundary segments and transgranular cracks parallel to compression leaving grain boundaries normal to compression essentially melt free. This MPO, however, is attributed to melt redistribution in response to cracking of the rock. This cracking or brittle failure of the rock occurs if local pressure gradients near a melt pocket or inclusion are high enough to exceed the strength of the rock. Cracks are melt-filled and oriented with respect to the applied stress; cracks initially open at orientations similar to those described earlier for partially molten peridotites, but eventually grow or extend parallel to the axis of maximum compression (Figure 2.7(B)).

5.1.2. Stress-Driven Melt Segregation

Partially molten rocks deformed with a large simple shear component in the laboratory can develop planar, anastomosing melt-rich bands surrounded by melt-depleted lenses (Figure 2.8). These bands develop around a shear strain, γ, of ~ 1 and persist at an angle of $15°-20°$ antithetic to the shear direction up to very high shear strains ($\gamma > 10$). The solid grains rotate with increasing strain toward the infinitesimal stretching direction and define a foliation. However, since the band angle is independent of the finite strain, melt must continuously reorganize to maintain its constant orientation. In other words, melt flows from bands rotated to a steeper angle into bands oriented at a lower angle and a steady-state melt network develops.

Because the viscosity of partially molten rocks is strongly dependent on melt content (recall Eqn (2.11)), any perturbation in melt fraction on a length scale larger than the grain size will lead to a spatial variation in viscosity and thus in shear stress, provided the partially molten rock is deforming at the same rate throughout. The perturbations in shear stress translate into perturbations in mean pressure that drive melt flow toward the

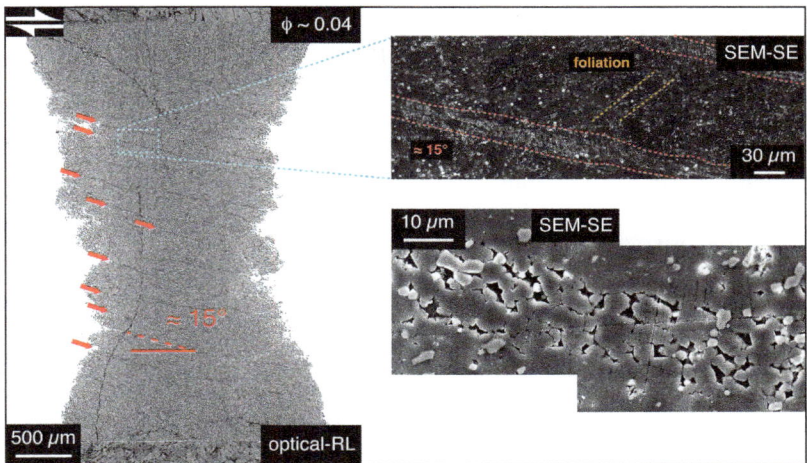

FIGURE 2.8 Melt-rich bands produced during stress-driven melt segregation. Reflected light (RL) and secondary electron scanning electron microscope (SEM-SE) images of a sample deformed in torsion. Olivine is medium gray, chromite is light and quenched melt is dark. Note the alignment of bands at ~15° antithetic to the shear plane (red arrows).

low-pressure regions and therefore a positive feedback develops—eventually, the melt will self-organize into weak, melt-rich shear zones surrounded by strong, melt-depleted lenses at length scales larger than the grain size and smaller than the compaction length. Once this melt segregation occurs, the compaction length varies over a length scale smaller than the original value for the homogeneous melt distribution and thus loses meaning as an average value of the system. Laboratory studies and theoretical considerations suggest that once melt-rich shear zones form, surface tension will not homogenize bands thicker than ~1 cm for a grain size of ~1 mm and differential stress of 10 MPa, suggesting that melt-rich channels could be preserved in natural rocks.

Considerable progress was achieved over the last 5 years in the theoretical understanding of deforming solid–liquid composites. Numerical models based on viscous constitutive equations cast in terms of grain boundary contiguity successfully predict the segregation of melt in various flow geometries in agreement with laboratory observations and provide a necessary step for confident extrapolation from the laboratory scales to the mantle.

5.1.3. Brittle Failure

Perhaps the most obvious mechanism for melt localization and focusing of melt flow has been brittle failure or magma fracturing. On the laboratory scale, this process may begin with microcracking along grain boundaries and through grains. In partially molten rocks these cracks are filled with melt. With continued deformation, these cracks link, eventually forming networks of veins and dikes at scales larger than the grain size. Failure can be

distributed throughout the partially molten region or localized in shear zones.

In a rock containing melt, the melt pressure acts to reduce the effective stresses. If the effective confining pressure is low enough and the differential stress is high enough, microcracking will occur. As stated earlier, the effective stresses that govern failure in rocks containing melt or fluid are

$$\sigma'_{ij} = \sigma_{ij} - \sigma \delta_{ij} P_{\mathrm{m}}. \qquad (2.13)$$

If melt does not flow easily out of the partially molten rock so that deformation occurs at constant volume, the conditions are described as **undrained**. Most deformation experiments are carried out under undrained conditions. The melt pressure when volume is constant is

$$\sigma_3 \leq P_{\mathrm{m}} \leq (\sigma_1 + \sigma_2 + \sigma_3/3). \qquad (2.14)$$

where σ_3 is the confining pressure and the quantity on the right-hand side is the mean pressure. In rocks in which the fluid can move freely in response to stress, the pore pressure equals the confining pressure, whereas in rocks in which the fluid cannot move freely, the pore pressure is more likely to equal the mean pressure.

If the melt network is again treated as an interconnected network of elliptical melt inclusions, then in a compressive stress field, the same analysis used earlier for stress-induced melt distribution holds true. That is, melt inclusions inclined at an angle ψ to σ_3 will yield the greatest tensile stress concentration. Melt migrates into those most favorable oriented cracks as above, but in this case, the maximum local tensile stress around the melt inclusion exceeds the tensile strength of the rock and cracking of the solid matrix occurs. The stress condition for failure is

$$\sigma_3 - P_m = T, \qquad (2.15)$$

where T is the tensile strength of the rock. Cracks will be oriented initially at the same angle to σ_1 as described for stress-induced melt redistribution, but will propagate more nearly parallel to σ_1 (Figure 2.7(B)). On failure, melt pressure decreases within the crack; this decrease results in additional pressure gradients that drive melt into the crack.

If melt is distributed throughout the rock, these cracks will develop throughout the rock. As cataclastic failure continues, these cracks grow and coalesce, leading to the formation of a network of veins and fractures in the rock. Under certain conditions (e.g., low temperature and pressure), cracks may coalesce to form localized shear zones. The melt distribution observed in many partially molten granites deformed in the laboratory is due primarily to this process.

The development of fractures or veins depends on the ability to generate melt pressures that exceed the tensile strength of the solid matrix as well as on the temperature, pressure, and stress conditions.

Conditions in the crust are generally favorable for the formation of veins and fractures due to the relatively low pressures and temperatures at which deformation occurs. High melt pressures can also develop in partially molten crustal rocks as a consequence of the high viscosity of the melt and positive volume changes associated with melting. As described in Section 2, high melt viscosity considerably reduces the ability of melt to migrate in grain scale-size channels. If melt cannot migrate quickly enough to keep up with the imposed rate of deformation, melt pressure will increase locally, possibly leading to failure. If slow melt migration within grain-scale channels is combined with melting in which a positive change in volume occurs, this effect is exacerbated. A positive change in volume occurs on melting in relatively dry granitic systems. Melt pressure is increased not only by the positive volume change on melting but also by the stepwise manner in which melt is produced in these systems. Laboratory experiments of melting in dry crustal rocks indicate that melt fraction increases rapidly over very small temperature intervals due to the breakdown of hydrous minerals such as biotite, muscovite, or hornblende. These rapid increases in melt fraction combined with the positive volume changes associated with melting favor the development of high melt pressures, cataclastic failure, and the development of vein or fracture networks.

In the mantle, brittle failure seems unlikely, primarily because, at typical pressures and temperatures, mantle rocks can deform viscously at differential stresses much lower than the tensile strength of the rocks. However, as discussed earlier, brittle failure can occur if melt pressure is high enough to exceed the confining pressure and the tensile strength of the rock (Eqn (2.16)). The mechanisms described for inducing high melt pressures in crustal rocks generally do not apply to mantle rocks, primarily because melt movement in response to deformation and imposed

pressure gradients is relatively rapid. The maximum possible difference between the melt pressure and the confining pressure that can be generated due to melt buoyancy in the compacting mantle is

$$\sigma_3 - P_m = (\rho_s - \rho_f)g\delta_c. \qquad (2.16)$$

Fracture will occur if the right-hand side of this equation is $\geq T$, the tensile strength of the rock. Estimates of T, based on laboratory experiments, suggest that T is at least ~ 50 MPa. This minimum estimate means that, for appropriate density contrasts, the compaction length δ_c must be greater than 10^4 for fracture to occur. In contrast, most estimates of compaction length in the mantle beneath mid-ocean ridges are less than 10^3. In other mantle environments, it may be possible that δ_c is $\sim 10^4$ but only if mantle viscosities are high or if melt is already focused into high-porosity channels with coarse grain sizes. Focusing melt into coarse-grained channels increases the permeability; since $\delta_c \propto K^{1/2}$, δ_c also increases. Brittle failure in the mantle, then, may only occur after melt has been localized to form veins and may not be an effective mechanism for vein formation.

5.2. Effect of Reaction

Melt−rock reactions in the mantle not only may account for the compositions of erupted magmas and rocks through which melt has moved but also may provide a mechanism for the focusing of melt flow into larger conduits. Geochemical data suggest that many dunites found in ophiolites and other mantle sections are features formed by reaction between ascending basaltic melt and mantle peridotite. These dunites occur in a variety of shapes and sizes and have sharp boundaries in which pyroxene content changes from <2% to >15% over distances of millimeters. In many mantle peridotites, these dunites form an interconnected network of anastomosing channels. These dunite channels are interpreted as former conduits for focused melt flow through the mantle. Focusing of flow into dunite channels essentially isolates the melt from further interaction with the surrounding mantle, leading to the geochemical signatures for near-fractional melting described above.

These dunite channels are thought to develop because mineral solubilities are dependent on pressure. For example, melt produced at depth in the mantle may become saturated in olivine, but undersaturated in pyroxene as it ascends through the overlying rock. At some critical depth or degree of undersaturation, melt will begin to dissolve the pyroxene in the rock through which it is moving. This reaction increases the melt fraction, which in turn enhances the permeability of the rock. The increased melt flow that results in turn enhances the rate of reaction, so that a positive feedback loop exists between reaction and melt flow. Although initially the reaction occurs throughout the rock, small differences in pyroxene content may cause an instability to develop. For example, an area with an initially larger proportion of pyroxene will have a higher melt fraction after all the pyroxene is dissolved. This higher melt fraction implies greater permeability; this effect focuses melt into the region. With continued reaction and focusing of flow, olivine-rich channels containing high melt fractions form and grow. This process is called the reaction infiltration instability (RII, Figure 2.9). Experiments at high pressures and temperatures where reactive basalt melt was placed next to a harzburgitic or lherzolitic rock reproduce the principal features of the mineralogical layering observed on natural ophiolites. The infiltration of melt causes the pyroxene(s) to dissolve and a dunite—harzburgite—lherzolite layering develops. The growth of the olivine-rich reaction layer scales with the square root of time and seems to be limited by SiO_2 diffusion in the melt phase. In most of these dissolution couple experiments, however, reaction infiltration instabilities have not developed, probably due to small sample sizes used in some experiments or to other sources of experimental uncertainty.

Nevertheless, numerical and some laboratory experiments indicate that this reaction instability could operate in the mantle. Figure 2.9 shows a channel developed during

laboratory experiments in which melt undersaturated in pyroxene migrated into a fine-grained rock containing equal amounts of pyroxene and olivine. The initial melt migration—reaction front was planar; after several hours at high temperatures and pressures, several channel features developed. These channels had sharp boundaries defined by changes in melt and pyroxene content. Within the channel no pyroxene remained; outside the channel the initial pyroxene content was unchanged. Experiments in which water flowed through a mixture of glass balls and salt show very similar results. Although these reaction channels seem likely to develop in the mantle, questions still remain about the time and length scales for this channeling process. For example, channels that develop by reaction during melt migration may have only a limited ability to focus flow in the mantle if the length of channel that develops is limited or if extensive branching of channels occurs. However, even with limited development, this process could provide a starting point for the melt-focusing effects of deformation described earlier. For example, the high melt fractions developed in the reaction zones will cause these regions to be significantly weaker than the surrounding region. As discussed earlier, strength contrasts can cause melt to flow from strong to weak areas during deformation. Deformation would then draw more melt into the existing channels, increasing the length of these features beyond that obtained by reaction alone. If the channels become long enough, fracturing could occur as melt pressure at the top of the column of melt exceeds the rocks strength.

Although the relationship between aqueous fluid flow and reaction in crustal rocks has been examined, the effect of reaction on the focusing of melt flow in crustal rocks has not been considered in any detail.

6. FUTURE DIRECTIONS

Melt migration in both the crust and the mantle is a complicated process that requires an understanding of microscopic and macroscopic properties of partially molten rocks. Considerable progress has been made in the theoretical understanding of the extraction of melt, particularly when this separation is driven by density differences between melt and solid and by shear. Numerous models of large-scale melt migration have been developed based on the two-phase flow theory; many of these apply to melt migration beneath mid-ocean ridges. Examination of these models, however, suggests that our understanding is still limited by our incomplete knowledge of the basic physical properties of partially molten systems and the complications of applying our current knowledge. For example, extrapolations from experimentally derived flow laws for olivine to mantle conditions suggest that olivine will deform by dislocation creep or dislocation-accommodated grain boundary sliding, indicating that

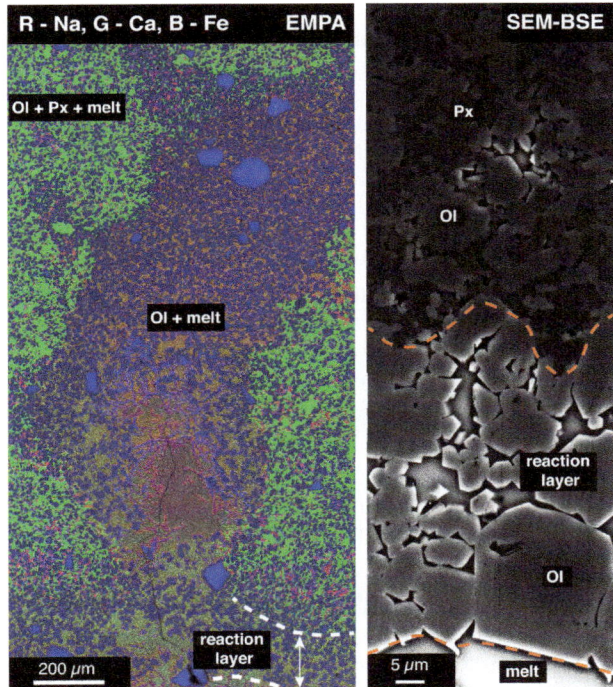

FIGURE 2.9 Reactive melt migration. Electron microprobe (EMPA) and SEM backscattered electron (SEM-BSE) image with Ol — olivine, Px — pyroxene. On the left, the EMPA color composite image is made by combining RGB channels where R — sodium, G — calcium and B — iron. Note the infiltration instability formed of only Ol + melt. On the right is a typical dunite (ol + melt) reaction layer adjacent to a harzburgitic rock (Ol + Opx). These features form by dissolution of pyroxene and precipitation of olivine form the melt.

$n > 1$ in the upper mantle, rather than $n = 1$ as assumed in the early sections of this chapter. Similarly, partially molten rocks undergoing stress-driven melt segregation in the laboratory exhibit stress exponents, $n > 1$. Significant computational effort is required for modeling materials deforming in power-law creep with stress exponents, $n > 1$. In addition, the effect of melt and stress-driven melt segregations on lattice-preferred orientations (LPO) was described, where with segregation of melt the flow-parallel alignment of olivine a-axes is weakened and ultimately becomes flow-orthogonal. This change in alignment has important implications for the interpretation of seismic data; however, the mechanistic understanding of melt influence on LPO is yet to be developed. Overall our understanding of melt migration mechanisms, especially at high pressures and temperatures is incomplete. Numerical models suggest that melt might migrate in forms of "porosity waves" or shear stress-driven "vug waves" focusing and enhancing flow. RII is still a poorly constrained phenomenon in the laboratory and much work remains to be carried out to assess the influence of various physical parameters on the topology of reactive infiltration instabilities. At last, we need to know more about the permeability and viscosity of both mantle and crustal rocks especially at very low melt fractions, which are very hard to study in the laboratory.

Advances in synchrotron microtomography allowed imaging of melt networks in three dimensions constraining the melt topology to melt fractions as low as 0.02 and confirmed observations performed on 2D sections. Nevertheless, despite considerable work on the distribution of melt in partially molten rocks, permeability estimates remain qualitative at best.

In the last 30 years, numerous laboratory studies have examined the mechanical properties of deforming partially molten rocks. In peridotites and other mantle analogues, detailed flow laws obtained through these experiments provide very good constraints on the effective viscosities of these rocks during melt migration. Despite the development of these flow laws, a fundamental understanding of the ongoing microscopic processes that result in these flow behaviors is still incomplete.

Much work remains to be completed to constrain mechanical properties of partially molten crustal rocks. Experiments need to be performed at lower stresses and strain rates to examine deformation in the diffusion and dislocation creep regimes; current quantitative results apply only to deformation by cataclastic flow. Technological advances that allow for resolution of lower stresses and strain rates make these experiments possible. Not only will these experiments provide data to develop flow laws, but also they will provide insight into the role of deformation in the localization of melt and the focusing of melt flow in crustal rocks.

FURTHER READING

Bagdassarov, N., Dorfman, A., 1998. Granite rheology: magma flow and melt migration. J. Geol. Soc. London 155, 863—872.

Daines, M.J., Kohlstedt, D.L., 1997. Influence of deformation on melt topology in peridotites. J. Geophys. Res. 102, 10257—10271.

Hirth, G., Kohlstedt, D.L., 1995. Experimental constraints on the dynamics of the partially molten upper mantle: deformation in the diffusion creep regime. J. Geophys. Res. 100 (B2), 2156—2202.

Holtzman, B.K., Kohlstedt, D.L., 2007. Stress-driven melt segregation and strain partitioning in partially molten rocks: effects of stress and strain. J. Petrol. 48 (12), 2379—2406.

Jackson, I., Faul, U.H., Fitz Gerald, J.D., Morris, S.J.S., 2006. Contrasting viscoelastic behavior of melt-free and melt-bearing olivine: implications for the nature of grain-boundary sliding. Mater. Sci. Eng. A 442 (1—2), 170—174.

Keleman, P.B., Hirth, G., Shiumzu, N., Spiegelman, M., Dick, H.J.B., 1997. A review of melt migration processes in the adiabatically upwelling mantle beneath oceanic spreading ridges. Philos. Trans. R. Soc. Lond. A 355, 283—318.

King, D.S.H., Hier-Majumder, S., Kohlstedt, D.L., 2011. An experimental study of the effects of surface tension in homogenizing perturbations in melt fraction. Earth Planet. Sci. Lett. 307 (3—4), 349—360.

Kohlstedt, D.L., Holtzman, B.K., 2009. Shearing melt out of the Earth: an experimentalists perspective on the influence of deformation on melt extraction. Annu. Rev. Earth Planet. Sci. 37, 561—593.

LaPorte, D., Watson, E.B., 1995. Experimental and theoretical constraints on melt distribution in crustal sources: the effect of crystalline anisotropy on melt interconnectivity. Chem. Geol. 124, 161—184.

McKenzie, D., 1984. The generation and compaction of partially molten rock. J. Petrol. 25, 713—765.

Morgan, Z., Liang, Y., 2005. An experimental study of the kinetics of lherzolite reactive dissolution with applications to melt channel formation. Contrib. Mineral. Petrol. 150 (4), 369—385.

Nicolas, A., 1986. A melt extraction model based on structural studies in mantle peridotites. J. Petrol. 27, 999—1022.

Riley Jr., G.N., Kohlstedt, D.L., 1991. Kinetics of melt migration in upper mantle-type rocks. Earth Planet. Sci. Lett. 105, 500—521.

Rosenberg, C., Handy, M.R., 2005. Experimental deformation of partially melted granite revisited: implications for the continental crust. J. Metamorph. Geol. 23 (1), 19—28.

Rutter, E.H., Brodie, K.H., Irving, D.H., 2006. Flow of synthetic, wet, partially molten "granite" under undrained conditions: an experimental study. J. Geophys. Res. 111, B06407.

Spiegelman, M., Kelemen, P., Aharonov, E., 2001. Causes and consequences of flow organization during melt transport: the reaction infiltration instability in compactible media. J. Geophys. Res. 106 (B2), 2156—2202.

Takei, Y., Holtzman, B.K., 2009. Viscous constitutive relations of solid-liquid composites in terms of grain boundary contiguity: 1. Grain boundary diffusion control model. J. Geophys. Res. 114 (B6), 2156—2202.

Takei, Y., Katz, R., 2013. Consequences of viscous anisotropy on a deforming, two-phase aggregate. Part 1. Governing equations and linearized analysis. J. Fluid Mech. 734, 424—455.

Wark, D.A., Watson, E.B., 1998. Grain-scale permeabilities of texturally-equilibrated, monomineralite rocks. Earth Planet. Sci. Lett. 164, 591—605.

Zimmerman, M.E., Kohlstedt, D.L., 2004. Rheological properties of partially molten lherzolite. J. Petrol. 45 (2), 275—298.

Chapter 3

Plate Tectonics and Volcanism

Peter C. LaFemina

Department of Geosciences, The Pennsylvania State University, University Park, USA

Chapter Outline

GLOSSARY

asthenosphere The part of the upper mantle between the lithosphere and mesosphere that deforms plastically. It is the source of mid-ocean ridge basaltic (MORB) magmas. This part of the mantle is typically >350 km thick if the base is taken as the 660 km seismic discontinuity.

Coulomb failure stress The stress required to drive a fault toward failure. This stress is equal to the change in shear stress on a fault plane plus the product of the coefficient of friction, accounting for pore fluid pressure, and the change in normal stress. A positive change in the Coulomb failure stress promotes fault slip and a positive normal stress change "unclamps" a fault or dike, leading to failure or formation, respectively.

crust The rigid outer shell of the Earth and the upper layer of the lithosphere. This layer is defined by lower seismic velocities above the Mohorivicic discontinuity (Moho), the boundary between the crust and upper mantle. The composition and thickness of the crust is variable between oceanic (basaltic, 7 km) and continental (granodiorite, 40–75 km) regions.

dike A vertical or near-vertical tabular magmatic intrusion within the brittle upper crust. They are responsible for the accommodation of relative plate motions at divergent plate boundaries, the growth and formation of oceanic crust and lithosphere, and feed fissure eruptions.

dynamic stresses The instantaneous stress change in the crust caused by the passage of seismic surface waves. These stresses can trigger instantaneous failure of fault planes, the opening of pore spaces and fractures, and the migration of upper crustal fluids.

These stresses decrease by $1/r^{1.66}$, where r is the distance from the earthquake or intrusion.

lithosphere The rheoligially defined rigid outer shell of Earth, comprised of the crust and upper mantle, and broken into tectonic plates that move relative to each other at active plate boundaries and the underlying asthenosphere. It has variable thickness between oceanic (2–140 km) and continental (40–280 km) regions. Transition to the asthenosphere (i.e., the lithosphere-asthenosphere boundary) occurs at a temperature of ∼1300°C.

magmatism Processes related to the formation, evolution, and migration of magmas in the lithosphere and crust. Magmatism includes the intrusion and eruption of magmas and lavas, respectively.

mantle The layer of the Earth between the crust and the outer core. The mantle is subdivided based on changes in seismic velocity into the upper mantle located between the Moho and the 440 km seismic discontinuity, the low-velocity zone between 440 and 660 km, and the lower mantle below the 660 km seismic discontinuity. The lithosphere-asthenosphere boundary separates the upper mantle into rigid upper (lithosphere) and plastic lower (asthenosphere) sections.

neovolcanic zone The active zone of magmatism and volcanism along divergent plate boundaries.

ophiolite Fragments of oceanic crust and lithosphere that have been accreted onto the edge of continents and allow geologists to explore the formation of oceanic crust and lithosphere. The geology of some indicates that they are mid-ocean ridge segments.

static stress The instantaneous stress change in the crust related to an earthquake or magmatic intrusion. These stresses decrease by $1/r^3$, where r is the distance from the earthquake or intrusion.

The Encyclopedia of Volcanoes. http://dx.doi.org/10.1016/B978-0-12-385938-9.00003-1

supercontinent A continent that is composed of most or all of the continental lithosphere on the Earth. The formation and breakup of supercontinents occurs on ~350 Myr cycles termed the supercontinent cycle. There have been at least six supercontinents during Earth history.

the Mohorovicic discontinuity The boundary between the Earth's crust and upper mantle. The boundary was first defined by Andrija Mohorovicic as a sharp increase in seismic velocity due to the compositional difference between the crust and mantle. It is commonly referred to as the Moho.

volcanism The end result of magmatism; the formation, evolution, and migration of magmas to the surface of the Earth. At a first order, volcanic processes are compositionally dependent, with composition of magma controlled by the tectonic location of formation.

In this instance, however, at the same hour when the whole country around Concepcion was permanently elevated, a train of volcanoes situated in the Andes, in front of Chiloe, instantaneously spouted out a dark column of smoke, and during the subsequent year continued in uncommon activity. It is, moreover, a very interesting circumstance, that, in the immediate neighbourhood, these eruptions entirely relieved the trembling ground…

Charles Darwin, *Voyages of the Beagle* (1839).

1. INTRODUCTION

Plate tectonics is the paradigm in Earth system science that connects solid Earth processes leading to the formation and destruction of oceanic and continental **crust** and **lithosphere**, with processes leading to the formation of the atmosphere, hydrosphere, and cryosphere. Plate tectonics on Earth has been ongoing for greater than 3.5 Ga, and there has been a clear spatial and geochemical correlation between it and **magmatism** and **volcanism** since it's initiation. Plate tectonics is driven by thermal processes acting within and on the surface of the earth, including planetary cooling related to primordial heat associated with initial planetary formation and heat generated by radioactive decay. Magmatism is the end result of these same thermal processes acting within the earth and the interaction of the lithospheric plates. Magmatism and volcanism play a crucial role in planetary cooling, formation of the crust, lithosphere and atmosphere, and recycling of material, especially the cycling of H_2O, from Earth's **mantle** to the atmosphere and back to the mantle.

The *lithosphere* formed during initial cooling and differentiation of Earth, and is comprised of the *crust* and upper *mantle*. The boundary between the crust and mantle is defined by changes in seismic velocities and is called **the Mohorovicic Discontinuity** (Moho). There are two types of lithosphere on Earth, oceanic and continental, defined by their bulk geochemical compositions. Oceanic lithosphere is formed at mid-ocean ridges by the repeated intrusion of **dikes** into the crust, and eruption of lavas along the ridge axis. Partial melting of upper mantle peridotite by decompression melting forms mid-ocean

ridge basaltic (MORB) magmas. Oceanic lithosphere is approximately 2 km thick at the mid-ocean ridge axis (i.e., the lithosphere is defined solely by the thickness of the crust), but increases to 50−140 km thick as the lithosphere moves away from the ridge axis and cools. Continental crust and lithosphere are formed through accretion of continental terranes and magmatism along convergent margins, and the crust has a bulk composition of granodiorite. Continental lithosphere is typically between 40 and 280 km thick. There are currently seven first order and eight second-order lithospheric plates (Figure 3.1). The number of plates has varied throughout the history of the earth, and will continue to vary as lithosphere is created and destroyed along active plate boundaries and plates amalgamate to form supercontinents.

Lithospheric plates move relative to or with the underlying **asthenosphere** driven by body forces or traction forces, respectively. Body forces acting within the lithosphere are generated by gravity and density contrasts between the plates and the mantle. The two main body forces are ridge push and slab pull. Traction forces are imparted by convecting asthenosphere on the base of the lithosphere. The relative motion between lithospheric plates occurs at rates on the order of cm a^{-1} and is accommodated along plate boundaries or plate boundary zones (Figure 3.1). In plate tectonic theory it is assumed that the interior of lithospheric plates are rigid; however, this is being redefined by new geodetic observations. It is also important to note that the regions of deformation associated with the boundaries can vary dramatically in width due to variations in crustal rheology, with a clear correlation between plate boundaries within oceanic (narrow) versus continental (wide) lithosphere.

There are three main plate boundary types, designated based on the relative motion between the plates (Figure 3.2). These plate boundary types are:

1. Divergent plate boundaries: Lithospheric plates move away from each other. Divergence occurs across mid-ocean ridges (e.g., the East Pacific Rise or the Mid-Atlantic Ridge) and continental rifts (e.g., the East African Rift). Passive upwelling of mantle results in decompression melting and the formation of magmas. In the case of mid-ocean ridges, and some continental rifts, magma accumulates in axial magma reservoirs or central volcanic systems where they reside until they are injected as sheeted dikes and erupted on the seafloor or rift valley floor. The rate of divergence and flux of magma to the ridge axis are fundamental to the morphology of the plate boundary.

2. Convergent plate boundaries: Lithospheric plates move toward each other. Oceanic lithosphere of higher density subducts into the mantle beneath younger, less dense oceanic or continental lithosphere. The subduction process leads to the formation of primary basaltic magmas

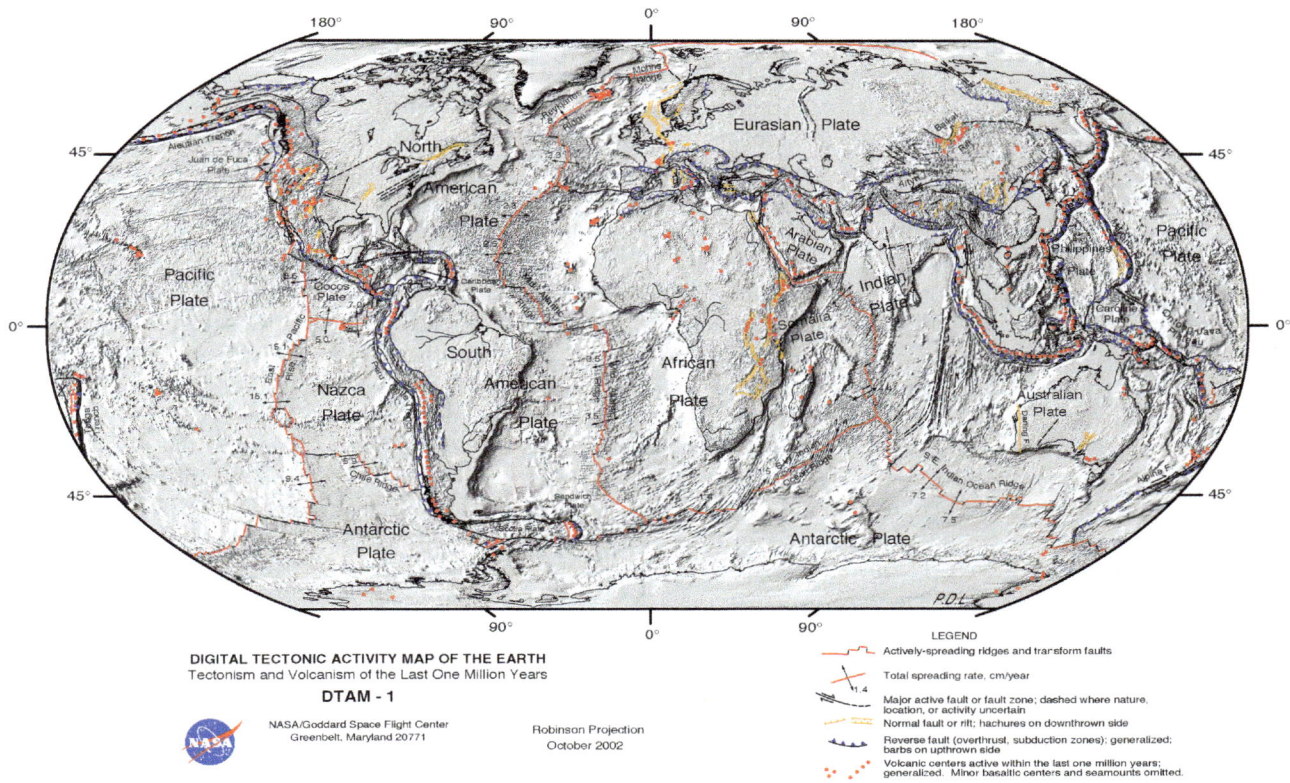

FIGURE 3.1   Plate tectonic map of Earth showing the major tectonic plates, plate boundaries, Holocene volcanic centers, and relative plate motion rates. *Lowman, P., Yates, J., Masuoka, P., Montgomery, B., O'Leary, J., Salisbury, D., 1999. A digital tectonic activity map of the earth. J. Geosci. Ed. 47 (5), 428–437.*

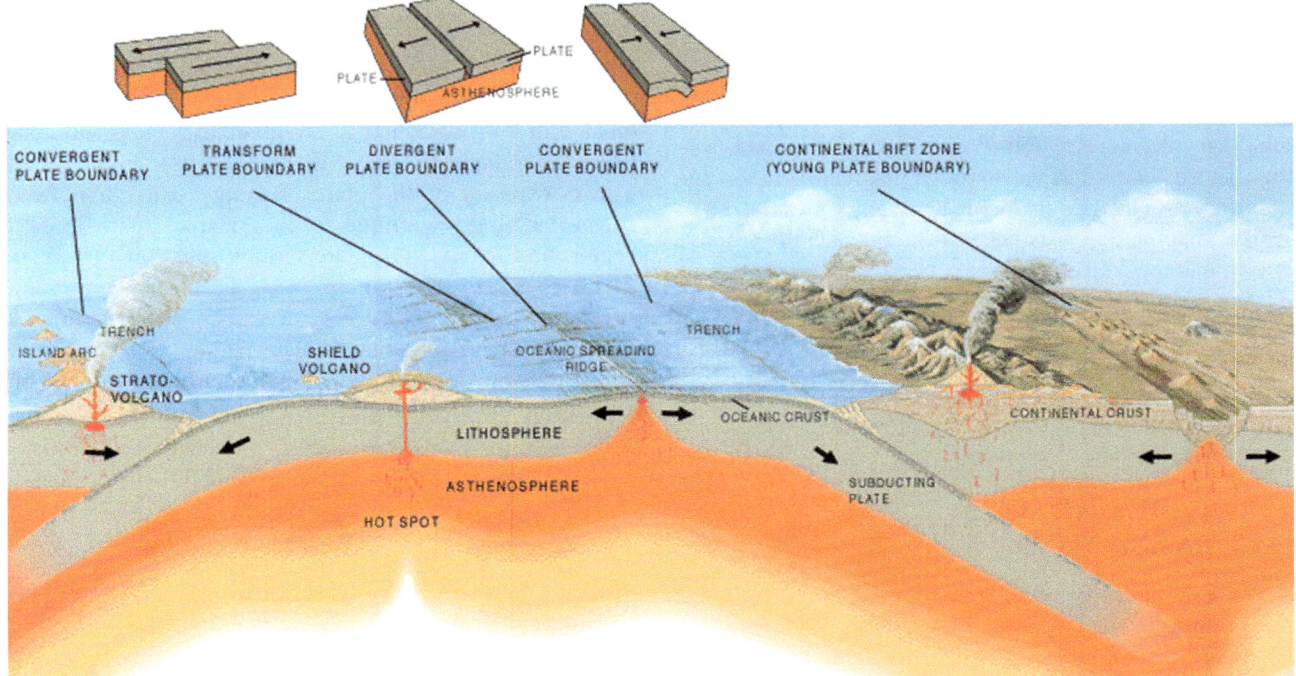

FIGURE 3.2   Cross-section of the earth and block diagrams highlighting the main plate boundary types. USGS. *URL: http://pubs.usgs.gov/publications/ text/Vigil.html.*

and ultimately volcanic arcs that are synonymous with subduction zones (e.g., the circum-Pacific Ring of Fire). In the case of convergence between two continental plates, continental collision takes place leading to thickening of the continental lithosphere and orogenesis (i.e., mountain building). The Indo-Eurasian collision, which began >35 Ma, and formation of the Himalayan Mountains and Tibetan Plateau are the best example of continental collision and orogenesis on the Earth today.

3. Transform plate boundaries: Plates move laterally past each other. Oceanic transform faults accommodate motion between active mid-ocean ridge segments. Continental transform faults accommodate motion between mid-ocean ridge segments, continental terranes, and in collisional tectonic regions. Magmatism and volcanism are not typically associated with this plate boundary type. However, in regions of transtension (i.e., regions where relative plate motions are oblique to the transform system allowing for extension of the crust) volcanism may occur.

Magma-tectonic interactions in plate boundary zones encompass a range of processes and phenomena linking magmatic and tectonic systems across vast spatial and temporal scales. On geologic time and plate boundary distance scales, these interactions are the association of magmatism with plate tectonics (e.g., subduction zone arc volcanism) and the accommodation of relative plate motions by magmatism (e.g., divergent plate boundaries). These interactions are responsible for formation of oceanic and continental crust and lithosphere. On shorter (i.e., human) time and distance scales, these interactions are reflected in the temporal and spatial correlations between intermediate to large magnitude earthquakes and volcanic unrest. There is a growing set of observations that indicate intermediate to large magnitude earthquakes can trigger or promote magmatic processes, including volcanic eruptions. Additionally, magmatic systems can trigger earthquakes through magma migration and induced stress changes, including earthquake swarms, intermediate magnitude earthquakes that may herald eruptive activity, and tectonic earthquakes. This chapter will provide an introduction to the topic of plate tectonics and associated magmatism, and will focus on the interaction between plate tectonics and magmatism at both long and short timescales. Readers will be provided with references to other chapters that provide more in depth coverage of related topics.

2. PLATE BOUNDARIES AND MAGMATISM

Lithospheric plates move relative to each other and to the underlying asthenospheric mantle. The relative motion between the plates is accommodated at dynamic plate boundaries by permanent deformation of the lithosphere on

plate boundary faults and fault systems, and by magmatism and volcanism. At these boundaries new oceanic and continental lithosphere is formed, continental lithosphere grows through accretion, continental lithosphere is broken into smaller fragments, and oceanic and continental lithosphere is recycled into the mantle. Whether continental lithospheric growth is increasing, decreasing, or keeping pace relative to its destruction is an area of active investigation, and relevant to our discussion here, as continental growth occurs partially through magmatism and volcanism at active plate boundaries. The amalgamation of continents into a single continent or several large continents, called **supercontinents**, and later break up of the continental lithosphere is termed the supercontinent cycle. The formation and closing of oceanic basins is termed the Wilson cycle, after J. Tuzo Wilson, an early pioneer in plate tectonic theory. Magmatism and volcanism that take place as part of these cycles and along active plate boundary zones have distinct geochemical signatures that can be used to decipher the tectonics of a region long after the plate boundaries have shifted. We describe here the three main types of plate boundaries, with a focus on the interaction of magmatism and tectonics in the accommodation of plate boundary zone deformation and relative plate motions. We do not describe magma genesis in detail here, as the topics are covered thoroughly in Chapters 1—*Melting in the Mantle of the Earth* and 4—*Origin and Composition of Magmas*. Hotspots and mantle plumes are not described in a separate section here; however, their influence and association with the three main plate boundary types is described where relevant.

2.1. Divergent Plate Boundaries

Divergent plate boundaries are defined by the relative motion of tectonic plates away from each other and can be located in both oceanic (i.e., mid-ocean ridges) and continental lithosphere (i.e., continental rifts) (Figure 3.2). The orientation of the plate boundary can range from normal to highly oblique to the relative plate motion vector, and the relative plate motion rate can vary from <1 to >12 cm a^{-1} (Figure 3.3). There are important similarities between mid-ocean ridges and continental rifts. The active magmatic zones within divergent plate boundaries are called **neovolcanic zones**. Magmatism within the neovolcanic zone can be segmented along strike and represented by central volcanic systems (i.e., distinct magmatic systems spaced kilometers apart). Lateral injection of magma from central volcanoes into associated fissure swarms as *dikes*, accommodates plate motion at these boundaries and forms new crust. The rate of magma supply to individual ridge or rift segments can be highly variable, and is related to the relative plate motion rate and proximity of the boundary to a mantle plume or hot spot. Divergence is accommodated by a

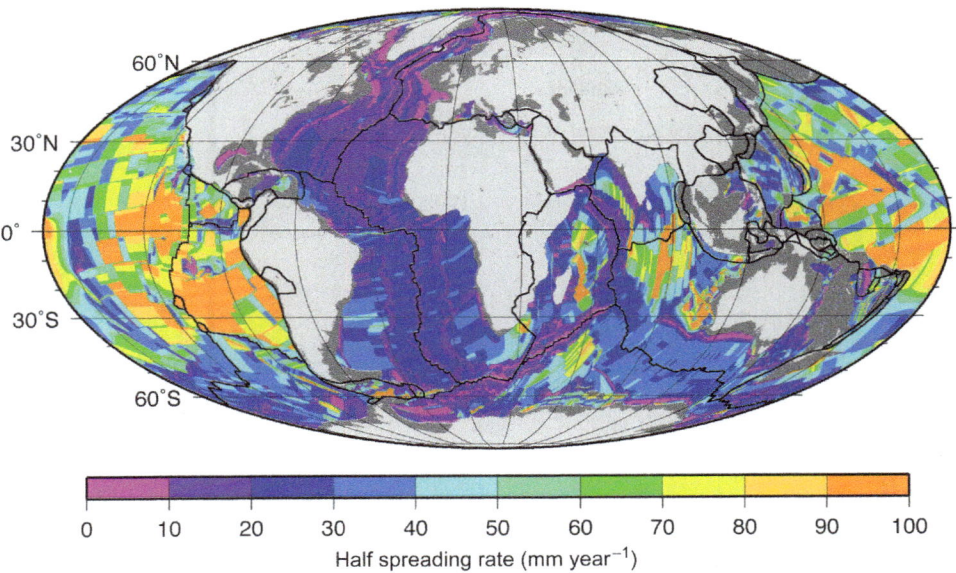

FIGURE 3.3 Map of half-spreading rate for Earth's mid-ocean ridges. *From Muller, R.D., Sdrolias, M., Gaina, C., Roest, W.R., 2008. Age, spreading rates and spreading symmetry of the world's ocean crust. Geochem. Geophys. Geosyst. 9, Q04006, http://dx.doi.org/10.1029/2007GC001743.*

combination of tectonic and magmatic processes that change as the plate boundary evolves, and the above factors influence the style of tectonic deformation and magmatism, and therefore the morphology along and across divergent boundaries. Here, we discuss divergent plate boundaries located in oceanic and continental lithosphere, and the role that magmatism plays in the formation of crust and lithosphere and the accommodation of relative plate motions at these boundaries.

2.1.1. Mid-Ocean Ridges

Mid-ocean ridges are the most prominent plate boundaries and the most active volcanic features on Earth (Figure 3.1). In Earth's current lithospheric plate configuration, the mid-ocean ridge system extends connected for over 60,000 km making them also the longest mountain chain and plate boundary type. Mid-ocean ridges are related to heat generation in the mantle by radioactive decay and other thermal sources, which result in mantle convection and upwelling. Decompression melting of upwelling mantle along mid-ocean ridges forms MORB magmas (see Chapter 1—*Melting in the Mantle of the Earth* and Chapter 4—*Origin and Composition of Magmas*), which are intruded into the crust and erupted at the surface forming oceanic crust and lithosphere.

Oceanic lithosphere is comprised of four main layers. The upper three layers comprise the crust and are up to 6–7 km thick (Figure 3.4). The bottom layer (layer 4) comprises the upper mantle. Layer 1 is typified by several hundred meters of pelagic sediment. As lithosphere moves away from the ridge axis, and therefore increases in age, the

sediment cover thickens. This sediment is deposited on top of layer 2a, a layer of up to 0.5 km of basaltic lava flows erupted as pillow or sheet lava flows within the neovolcanic zone along the axis of the mid-ocean ridge (see Chapters 19—*Submarine Lavas and Hyaloclastite* and 21—*Mid-Ocean Ridge Volcanism*). Layer 2b is approximately 1.5 km thick and is comprised of vertically oriented, sheeted diabase dikes. These dikes accommodate the <1 to >12 cm a^{-1} of relative plate motion at mid-ocean ridges, when they are periodically intruded into the crust. Studies of **ophiolite** complexes (i.e., fragments of mid-ocean ridges or oceanic crust and lithosphere that have been accreted onto the edge of a continent), rare exposures along oceanic rifts (e.g., the Hess Deep at the propagating tip of the Galapagos Rise), and Tertiary dike swarms in Iceland indicate dike thicknesses ranging from <0.01 to >13.0 m with a mean thickness of ~1.0 m. Dikes are formed by the injection of MORB magmas from central volcanic crustal or subcrustal magma chambers along the ridge axis. Along one segment of mid-ocean ridge there may be one or more axial magma chamber where magma is stored before it is intruded vertically and laterally along the ridge axis (see Chapter 8—*Magma Chambers*). The number of magma chambers is related to the relative plate motion rate, length of the ridge segment, and magma supply to the ridge. These magmatic systems have been well imaged geophysically (e.g., Sinton and Detrick, 1992) and geologically mapped in ophiolites and the Tertiary lava pile of Iceland (Figure 3.5). The gabbroic lower crust, accounting for up to 5 km of crustal thickness and layer 3 (Figure 3.4), forms by lateral flow and cooling at the edges of these magmatic bodies.

Axial summit graben
Melt lens
Basalts 200–800 m thick
Sediments
Layer 1
Layer 2a
Layer 2b
Layer 3
Basalts
Dikes
Gabbros
Partially molten low-velocity zone
~4 km

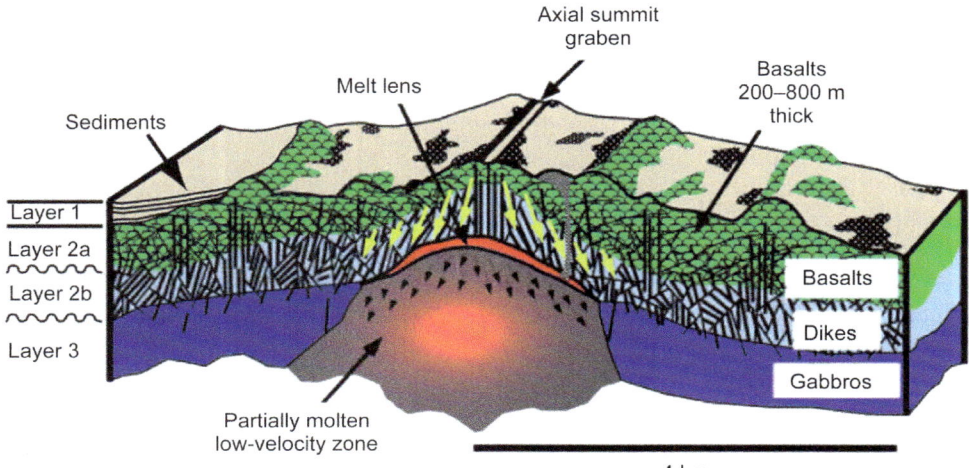

FIGURE 3.4 Geologic cross-section through oceanic crust at a fast-spreading, mid-ocean ridge. *Modified from Karson, J.A., Klein, E.M., Hurst, S.D., Lee, C.E., Rivizzigno, P.A., Curewitz, D., Morris, A.R., Hess Deep '99 Scientific Party, 2002. Structure of uppermost fast-spread oceanic crust exposed at the Hess Deep Rift: implications for subaxial processes at the East Pacific Rise. Geochem. Geophys. Geosyst. 3, http://dx.doi.org/10.1029/2001GC000155 and retrieved from http://www.odp.tamu.edu/publications/prosp/206_prs/prosp15.html on 02/01/2015.*

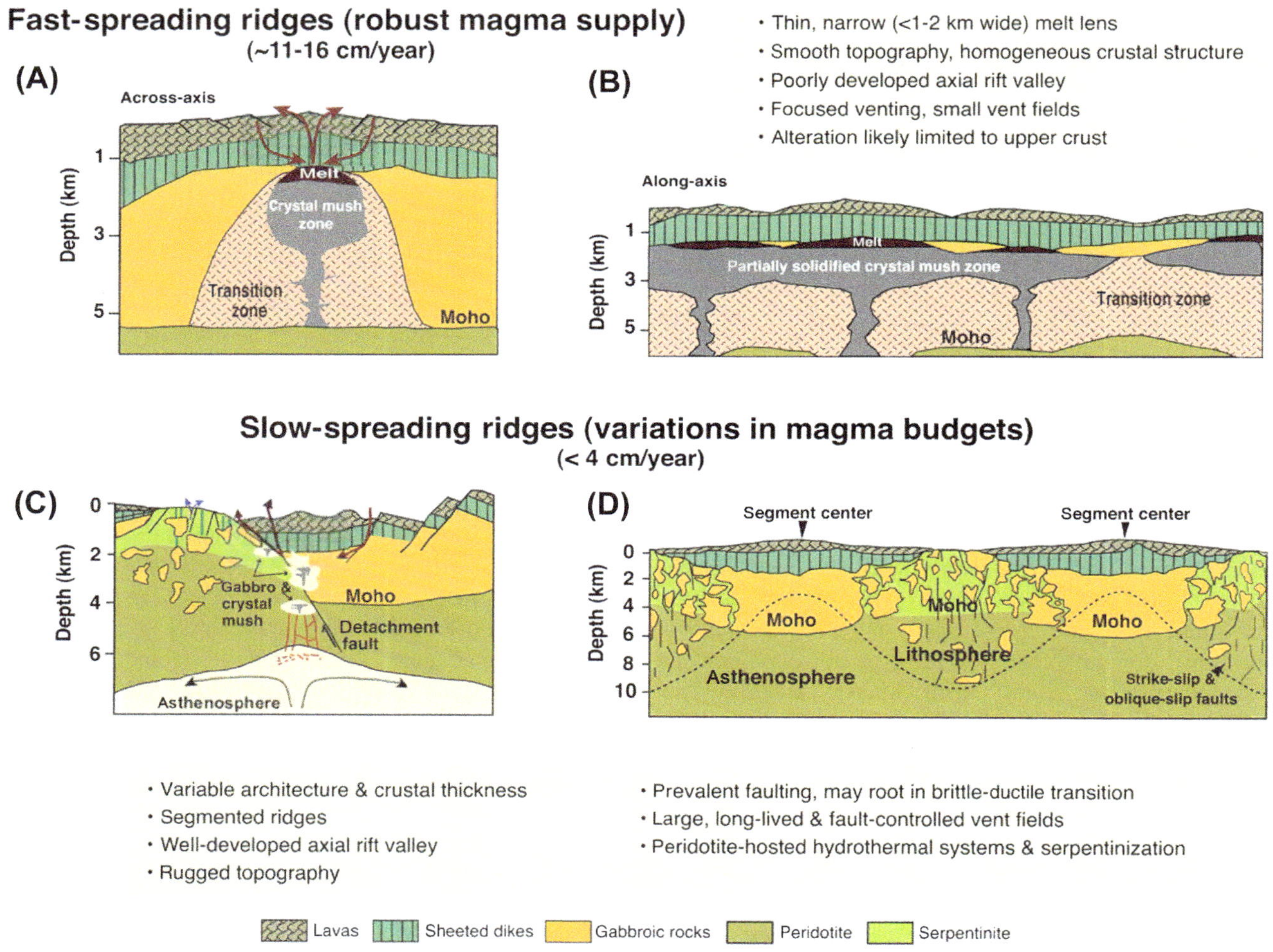

Fast-spreading ridges (robust magma supply)
(~11-16 cm/year)
(A)
Across-axis
Depth (km)
1
3
5
Melt
Crystal mush zone
Transition zone
Moho
(B)
• Thin, narrow (<1-2 km wide) melt lens
• Smooth topography, homogeneous crustal structure
• Poorly developed axial rift valley
• Focused venting, small vent fields
• Alteration likely limited to upper crust
Along-axis
Melt
Partially solidified crystal mush zone
Transition zone
Moho
Slow-spreading ridges (variations in magma budgets)
(< 4 cm/year)
(C)
0
2
4
6
Gabbro & crystal mush
Moho
Detachment fault
Asthenosphere
(D)
Segment center
Segment center
0
2
4
6
8
10
Moho
Moho
Moho
Lithosphere
Asthenosphere
Strike-slip & oblique-slip faults
• Variable architecture & crustal thickness
• Segmented ridges
• Well-developed axial rift valley
• Rugged topography
• Prevalent faulting, may root in brittle-ductile transition
• Large, long-lived & fault-controlled vent fields
• Peridotite-hosted hydrothermal systems & serpentinization
Lavas
Sheeted dikes
Gabbroic rocks
Peridotite
Serpentinite

FIGURE 3.5 Schematic representation of fast and slow-spreading mid-ocean ridges and descriptions of the dominant features and process found at each. *From Bach, W., Früh-Green, G.L., 2010. Alteration of the oceanic lithosphere and implications for seafloor processes. Elements, vol. 6, p. 173–178, http://dx.doi.org/10.2113/gselements.6.3.173.*

The upper mantle is composed of the ultramafic rock peridotite. The boundary between layers 3 and 4, that is the Moho, has been defined in oceanic lithosphere both petrologically and seismically. The thickness of oceanic lithosphere is a function of its age. At the ridge axis, where active accretion of new crust is taking place, lithospheric thickness is approximately 2 km (i.e., the thickness of layer 2). As oceanic crust moves away from the ridge axis it cools by conductive cooling. Oceanic lithosphere forms by cooling of the crust and upper mantle and thickens to between 50 and 140 km thick, following the simplified relation:

$$h \cong 2\sqrt{kt} \qquad (3.1)$$

where, h is the thickness of the lithosphere, k is the thermal diffusivity of the lithosphere, and t is the age of the lithosphere. The age (t) can be approximated by the quotient of the distance from the ridge axis (l) and the relative rate of plate motion (v). In addition to becoming thicker with distance from the ridge axis, the lithospheric mantle becomes denser. This is an important factor in global tectonics, because after ~ 10 Myr oceanic lithosphere is denser than the asthenospheric mantle and can therefore subduct into the mantle. The oldest oceanic lithosphere on Earth is <280 Ma, because as new oceanic crust and lithosphere are created at mid-ocean ridges, they are consumed at subduction zones (Figure 3.6).

Mid-ocean ridges are defined by the relative rate of plate motion between the diverging plates and are subdivided into ultrafast (>12 cm a^{-1}), fast (<12 to >8 cm a^{-1}), intermediate (<8 to >5 cm a^{-1}), slow (<5 to >2 cm a^{-1}), and ultraslow (<2 cm a^{-1}) spreading ridges (Figure 3.3). These differences in spreading rate give rise to distinctive mid-ocean ridge morphologies (Figure 3.7). At fast-spreading ridges the bathymetry gently increases from the abyssal plain to the ridge or rise with an elevation increase of ~ 500 m. The neovolcanic zone or region of active crustal accretion is confined to a roughly 100 m wide by 10 m deep axial depression. This region is bound by a several 100 m wide zone of fissuring and up to 10 km of active normal faulting. The bathymetry is subdued by the volume of lava flows erupted long the ridge axis (Figure 3.8). The classic example of a fast-spreading ridge is the East-Pacific Rise (EPR) at $\sim 9°$N, where the spreading rate is ~ 10 cm a^{-1} (Figure 3.9(A)). The rate of spreading along the entire length of the EPR ranges from 7 to 17 cm a^{-1}. Along the EPR there is a shallow axial trough where lavas are erupted, but topography here is subdued due to the larger volume of magmas injected as dikes and lavas erupted along the ridge axis. The bathymetry clearly shows that eruptive centers also occur off axis. Common features of fast to ultrafast-spreading ridges are overlapping or propagating ridge segments (Figure 3.9(A)). Overlapping ridges occur at a number of scales, but could eventually lead to the formation

of microplates; for example, the Easter and Juan Fernandez microplates along the southern EPR.

The morphology of intermediate to slow-spreading ridges is dramatically different to that of fast-spreading ridges (Figures 3.7 and 3.10). The neovolcanic zone is defined by a >10 km wide by 0.5–2.5 km deep axial graben, bound by inward dipping normal faults. The region of active faulting is up to 50 km wide. The Mid-Atlantic Ridge is a classic example of a slow-spreading ridge with rates in the North Atlantic of ~ 2 cm a^{-1} (Figure 3.9(B)). Lavas are erupted within the neovolcanic zone, and in some cases, erupted outside of the axial trough. Normal faulting at the ridge axis is important for several reasons. First, normal faulting accommodates a fraction of relative plate motion across the ridge axis. Relative plate motion along the ridge axis, however, is mostly accommodated by the intrusion of sheeted dikes. Second, normal faulting provides pathways for ocean water to penetrate the crust where it is heated by the high thermal gradient near the ridge axis. Hydrothermal circulation of ocean water through the oceanic crust causes alteration (i.e., serpentinization) and cooling of the crust. Hydrothermal circulation is dramatically displayed at hydrothermal vents or "black smokers" observed within the axial trough (see Chapter 47—*Deep Ocean Hydrothermal Vents*). The hydration of oceanic crust and lithosphere at mid-ocean ridges is also important for the cycling of fluids from the hydrosphere into the mantle at convergent margins. In addition to the steeply dipping normal faults found along ridge axes, low-angle normal faults and oceanic core complexes are found along slow- to ultra-slow spreading ridge segments and at segment ends adjacent to oceanic transform faults and nontransform accommodation zones (Figure 3.10). Oceanic core complexes or megamullions are formed by low-angle detachment faulting, and expose peridotite upper mantle and serpentinized shear zones and crust.

Ultraslow mid-ocean ridges accommodate relative plate motions by magmatic and amagmatic accretion processes. The spreading ridges are well defined in bathymetric data (Figure 3.9(C)), sharing some characteristics with slow-spreading ridges; however, the mode of accommodation is different from either fast- or slow-spreading ridges. Magmatic centers occur along the spreading axis, punctuated by regions of amagmatic accretion, where mantle peridotites are emplaced directly to the ridge axis through low-angle normal faulting and exhumation (i.e., formation of oceanic core complexes) (Figure 3.10). Amagmatic accretion is especially prevalent along highly oblique, ultraslow-spreading ridge segments. The Southwest Indian Ridge (Figure 3.9(C)) and Gakkel Ridge are the type localities for ultraslow-spreading ridges. The Gakkel Ridge has one of the slowest spreading rates for a mid-ocean ridge on Earth at 1.1 cm a^{-1}. The spreading rate across the Western Volcanic Zone, Iceland, a dying ridge segment in a

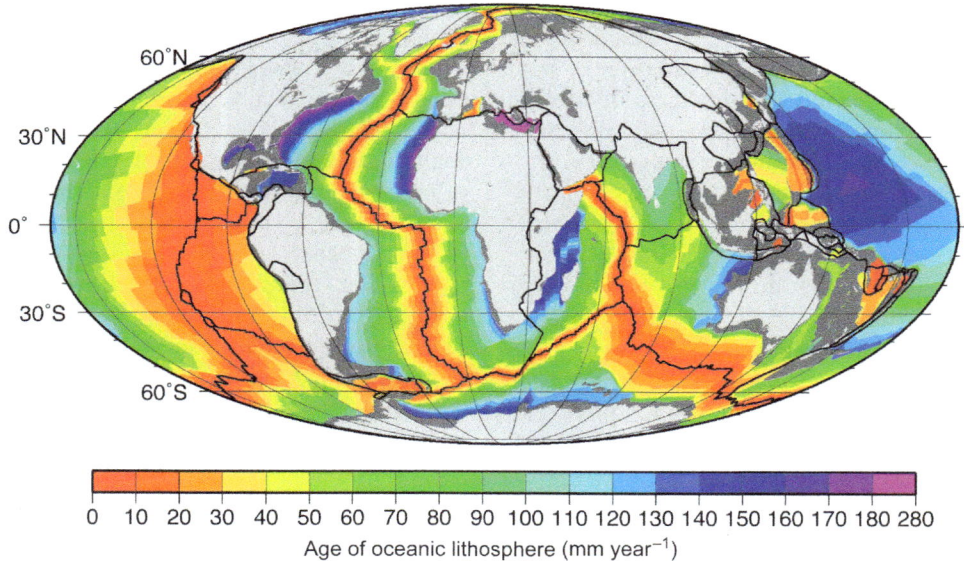

FIGURE 3.6   Map of the age of oceanic lithosphere. *From Muller, R.D., Sdrolias, M., Gaina, C., Roest, W.R., 2008. Age, spreading rates and spreading symmetry of the world's ocean crust. Geochem. Geophys. Geosyst. 9, Q04006, http://dx.doi.org/10.1029/2007GC001743.*

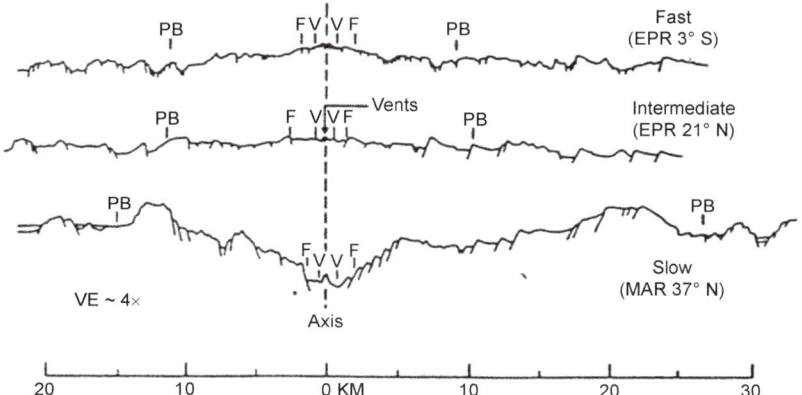

FIGURE 3.7   Bathymetric profiles across fast-, intermediate-, and slow-spreading ridges showing the change in morphology and width of the plate boundary (PB). The extent of the neovolcanic zones (V's) and fissuring (F's) is shown. *From Macdonald, K.C., 1982. Mid-Ocean Ridges: fine scale tectonic, volcanic and hydrothermal processes within the plate boundary zone. Ann. Rev. Earth Planet. Sci. 10, 155.*

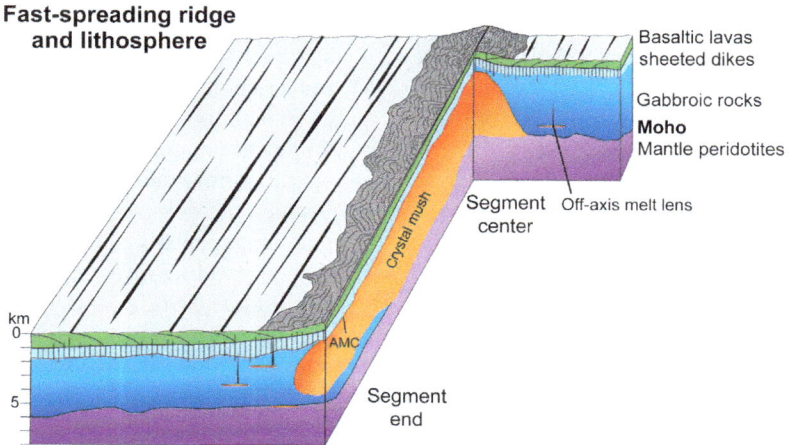

FIGURE 3.8   Block diagram of a fast-spreading mid-ocean ridge segment. Note subdued topography. Lava flows (dark grey) are erupted along the neovolcanic zone, fed by an axial magma chamber (AMC). *From Karson, J.A., Fornari, D.J., Kelley D.S., Perfit, M.J., Shank, T.M., 2015. Discovering the Deep: A Photographic Atlas of the Seafloor and Oceanic Crust, Cambridge University Press.*

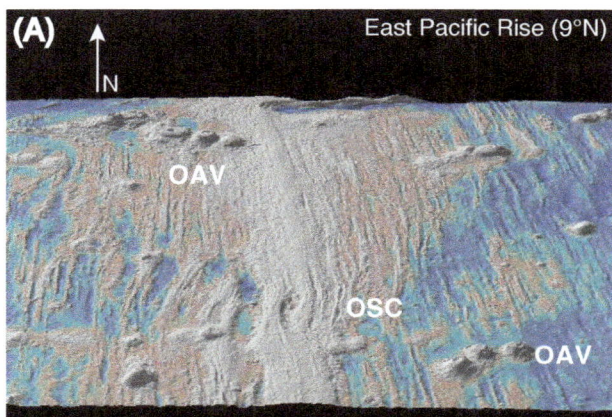

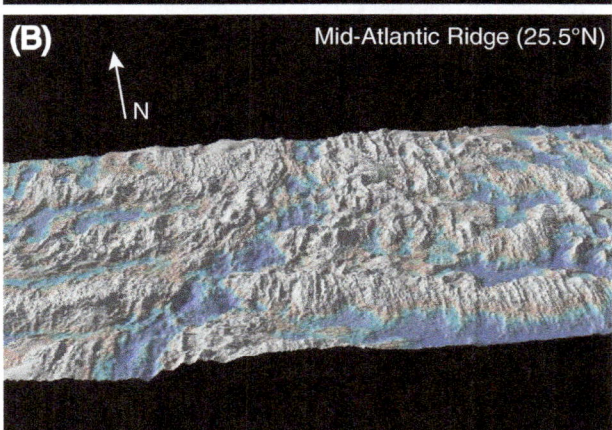

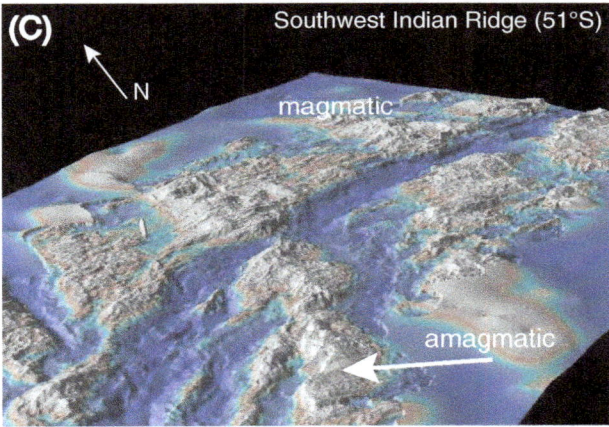

FIGURE 3.9 Bathymetry of fast, slow and ultra-slow spreading ridges. (A) East Pacific Rise near 9°N. Note the off-axis volcanism (OAV) and the overlapping spreading centers (OSC) at the lower center of the image. (B) Mid-Atlantic Ridge at 25.5°N. (C) Southwest Indian Ridge at 51°S. The magmatic segment is in the upper right of the image and amagmatic segment is in the lower left of the image.

propagating ridge system, is <0.8 cm a^{-1}, and the spreading rate across the Red Sea is as low as 0.7 cm a^{-1}.

Magmatic and tectonic processes at mid-ocean ridges form oceanic crust and accommodate relative plate motions. Oceanic lithosphere forms and becomes denser as the plate moves away from the mid-ocean ridge and cools. As we continue to explore and make new geophysical and geochemical observations of the vast ocean basins that comprise approximately 70% of the surface of the Earth, we will develop new insights into the magmatic and tectonic processes that lead to the formation of oceanic crust and lithosphere, and accommodation of relative plate motions at mid-ocean ridges.

2.1.2. Continental Rifts

Continental rifts are the representation of divergent plate boundaries in continental lithosphere. Continental rifting causes the fragmentation of continental lithosphere into smaller lithospheric plates or microplates, and is the initial stage in the formation of ocean basins (i.e., the Wilson cycle). Extensional stresses within the lithosphere are imparted by either relative plate motions or by uplifting and doming of the lithosphere by a mantle plume. Continental rifts can be prominent topographic features on the Earth, with relief exceeding 2 km (e.g., the East African Rift (EAR)). Rift topography is generated by displacement on inward dipping normal faults and by doming of the lithosphere by upwelling asthenospheric mantle. The roles of tectonic versus magmatic processes in the accommodation of divergence are time dependent in continental rift systems, and the mechanisms by which divergence is accommodated change as the systems evolve. In the early stages of continental rifting, tectonic processes accommodate relative plate motions (Figure 3.11); where as magmatic processes dominate later stages. Here we will describe the tectonic and magmatic processes and features associated with continental rifting, and the link between tectonics and magmatism in the accommodation of relative plate motions.

At the initiation of rifting, lithospheric extension and thinning are accommodated in the crust by normal faulting and the formation of horst and graben structures (Figure 3.11). As the lithosphere is thinned, warm asthenosphere flows toward the thinned region. The flow of mantle toward the thinned region results in partial melting of the mantle by decompression melting. This scenario is highly variable and is dependent on several factors, including the rate of extension and existence or proximity of a mantle plume. With the formation and migration of melts, magmatism initiates and the plate boundary system begins to switch from one that is dominated by tectonic processes to one where magmatism accommodates divergence. The system will eventually evolve to full seafloor spreading and generation of oceanic crust and lithosphere. It is important to note here that not all continental rifts evolve to seafloor spreading and the formation of oceanic lithosphere. Continental rifts often meet at triple junctions (i.e., a point in space and time where three plates and their plate boundaries meet).

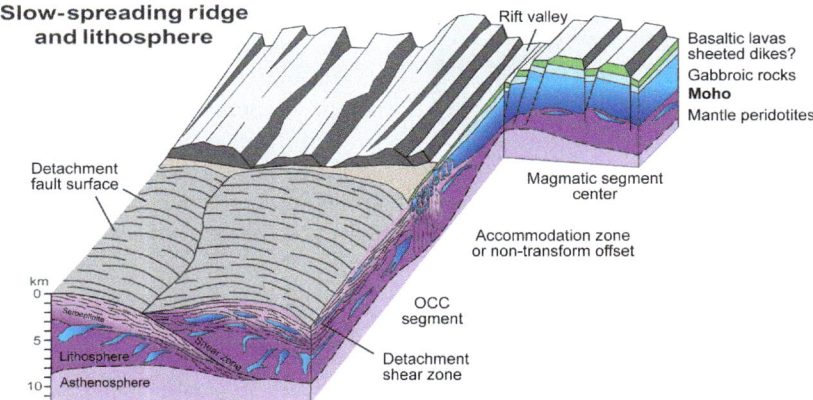

FIGURE 3.10   Block diagram of a slow-spreading mid-ocean ridge segment, including the oceanic crust and lithosphere. Note the high-angle normal faulting and rifted nature of the ridge axis. At mid-ocean ridge segment ends and accommodation zones, extension is accommodated amagmatically on low-angle normal faults and oceanic core complexes (OCC) are formed. *From Karson et al. (2015).*

A modern example of an active ridge-ridge-ridge triple junction is the Afar Triple Junction, where the Nubian, Arabian, and Somalian plates meet. In some instances, one continental rift arm may fail to evolve. Failed rifts are called aulacogens.

Geologic and geophysical observations of rifted margins, as well as recent, well-observed rifting events indicate that topography is also generated when magmas are intruded, but not necessarily erupted at Earth's surface (Figure 3.12). These observations follow from simple elastic dislocation theory for a pressurized vertical crack (i.e., a magmatic dike in the case of rifting) in a homogeneous elastic medium, where tensile stresses are generated above and in front of the crack as it propagates through the medium (Rubin and Pollard, 1988; Figure 3.13). As a dike approaches the Earth's surface, normal faulting and the formation of a structural graben, and vertical fissures, accommodate the extensional stresses. The faults dip steeply (i.e., $\geq 45°$) toward the dike and can extend to the top of the dike. Faults will not form in the crust adjacent to an intruding dike, as the dike imparts compressive stresses in this region, inhibiting faulting (Figure 3.13). Vertical fissures are often found at the surface above an intruded dike and their formation is enhanced in lava piles through the opening of vertically oriented cooling joints. In the case of eruption, the structural depression may be partially or completely filled with lavas and tephra. Once formed, normal faults and fissures can be reactivated during future rifting events and continue to grow, creating rift topography. Although tectonic structures (i.e., faults and fissures) are formed at continental rifts and play an important role in the early stage of rift growth and as pathways for magma to migrate in the upper crust, they are incidental with regards to the accommodation of plate motion once magmatism is initiated, as it is the repeated intrusion of dikes that accommodates plate motion at depth (Figures 3.12 and 3.14).

The initiation of continental rifting is often aided by a mantle plume or hot spot, which provides the geodynamic forces to drive extension within the lithosphere and the magma to accommodate the extension. Because of this association with mantle plumes, the initiation of continental rifting is marked by the eruption of large volumes of lavas as flood basalts or large igneous provinces (LIPs) (Figure 3.15) (see Chapter 24—*Flood Basalt Provinces—LIPS*). For example, the ∼135 Ma Parana—Etendeka LIP in eastern South America and Western Africa is associated with the rifting and break up of the Gondwana supercontinent and the formation of the South Atlantic Ocean basin. As continental rifts evolve, the source, geochemistry, and petrology of erupted lavas evolves and magmatism plays a more dominant role in the accommodation of relative plate motion. Eventually, the lithosphere transitions from that of continental composition to oceanic lithosphere. Lavas erupted during the initial stages of rifting are typically tholeiitic basalts formed through decompression melting of the mantle. However, initial magma chemistry and petrology can be highly variable depending on the mantle source (i.e., were the magmas formed by partial melting of lithospheric mantle or from an asthenospheric plume source), and the amount of crustal contamination (see Chapter 4—*Origin and Composition of Magmas*).

As a rift system evolves and magmatic systems associated with rifting evolve through time, the entire petrologic spectrum of lava compositions can be formed and erupted. The magmatic systems within more mature rifted margins are analogous to those found along intermediate- to slow-spreading mid-ocean ridges; discrete central volcano magmatic centers are found within the neovolcanic zones of individual rift segments (Figures 3.14 and 3.16). These central volcanoes can host a range of magmatic compositions from basaltic to dacitic. Petrologic and

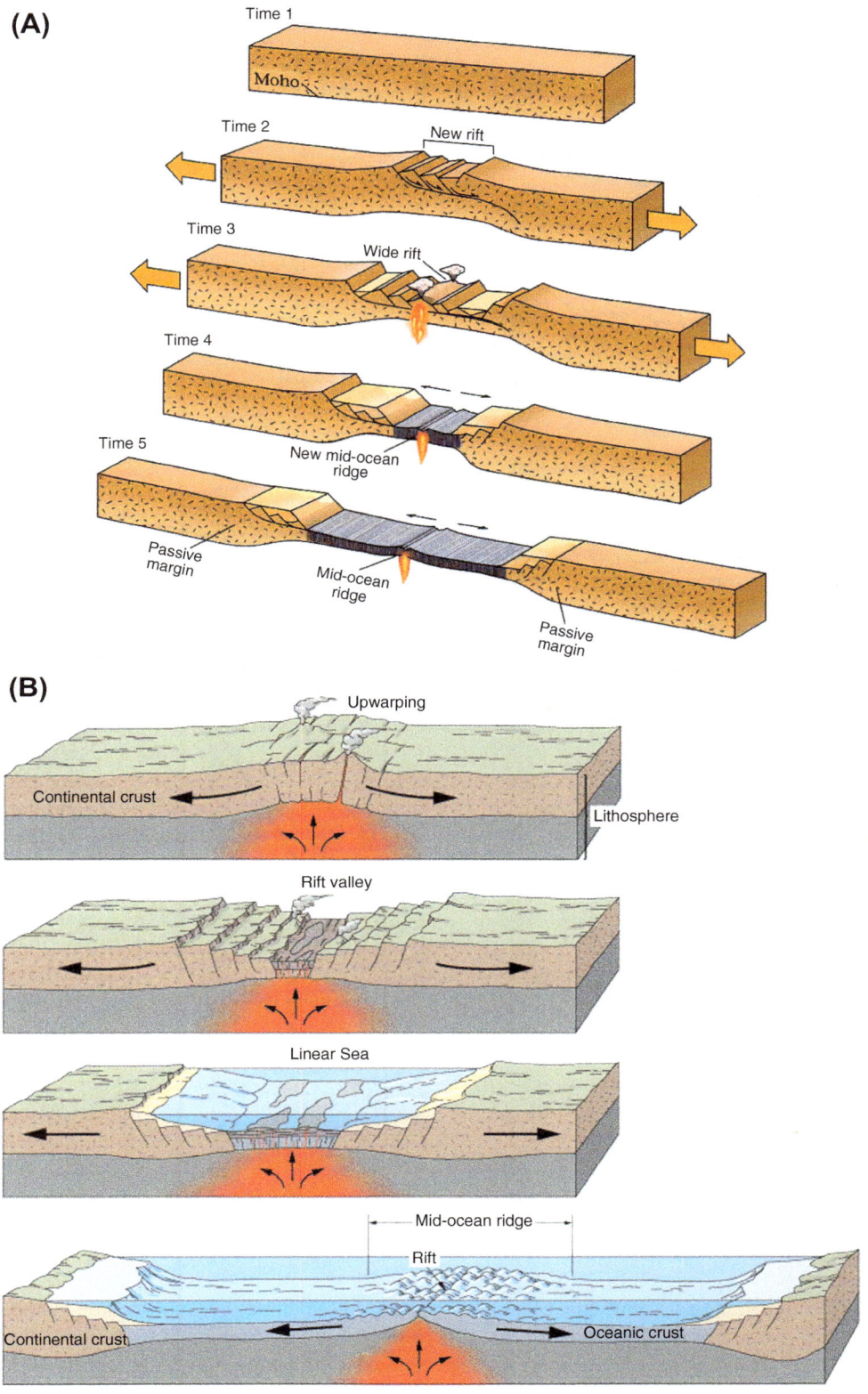

FIGURE 3.11 Continental rifting. (A) Continental rifting driven by far-field plate motions (*Marshak, S., Earth Portrait of a Planet, 4th Edition, October 2011, ISBN 978-0-393-93518-9, W. W. Norton & Company.*) (B) Continental rifting initiated by a mantle plume (*Tarbuck, E.J., Lutgens, F.K., & Tasa, D.G., Earth: An Introduction to Physical Geology, 11th Edition, 2014, ISBN-10: 0321813936, Prentice Hall.*)

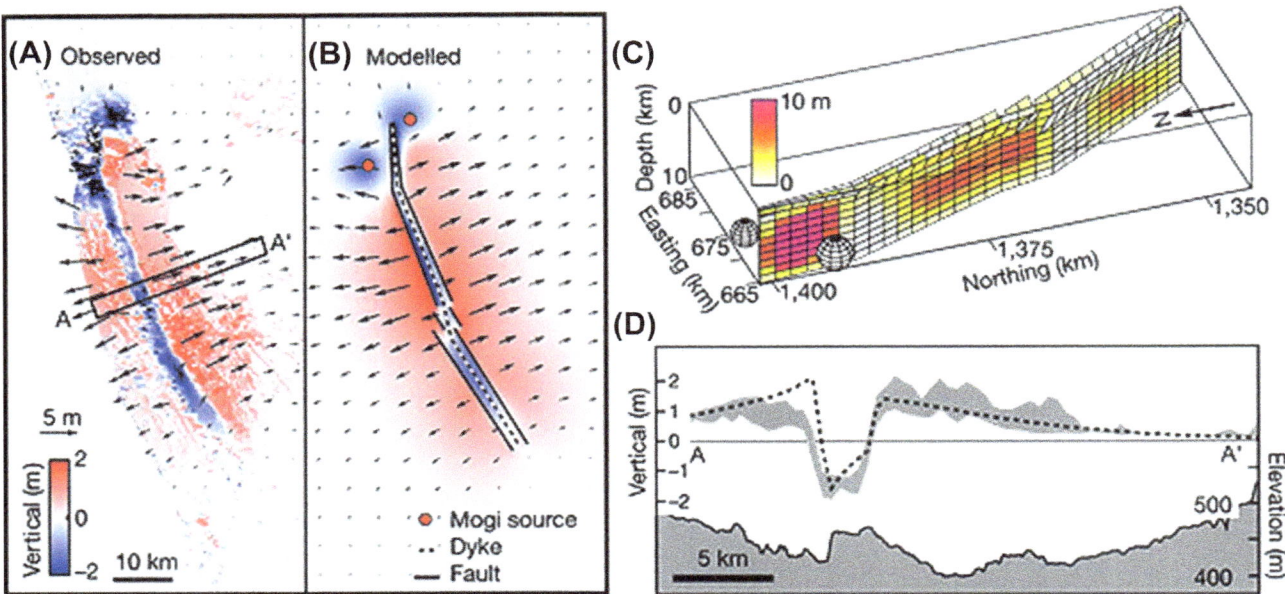

FIGURE 3.12   Geodetic observations and models of the 2004 Dabbahu, Ethiopia rifting event. (A) Radar interferogram showing deformation due to rifting episode, (B) Model of rifting deformation, (C) Three-dimensional model of rifting and magmatic deformation, and (D) Observed (gray) and modeled (dashed) topography generated during the rifting event. *From Wright, T.J., Ebinger, C., Biggs, J., Ayele, A., Yirgu, G., Keir, D., Stork, A., 2006. Magma-Maintained Rift Segmentation at Continental Rupture in the 2005 Afar Dyking Episode, 442, http://dx.doi.org/10.1038/nature04978.*

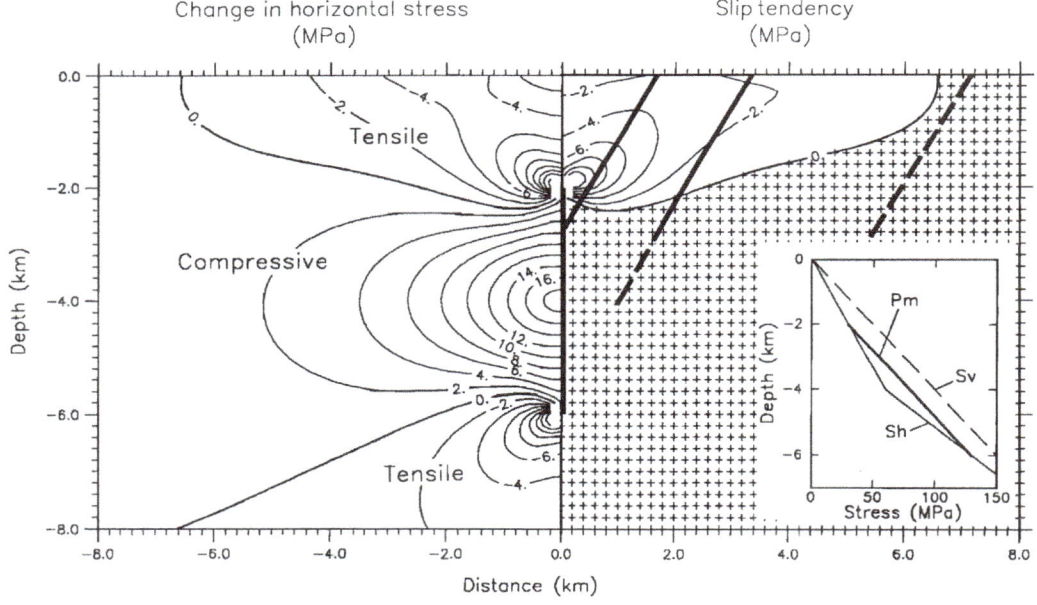

FIGURE 3.13   Change in horizontal stress and slip tendency for normal faults around an intruding magmatic dike. *From Rubin and Pollard (1988).*

geochemical observations indicate that the central volcanoes feed fissure swarms with less evolved basaltic magmas, even if the central volcano erupts highly evolved magmas. These magmatic systems aid in the accommodation of divergence through the repeated injection of magma from the central volcano into the fissure swarm. Crustal growth also takes place by the eruption of lava and tephra,

and formation of volcanoes within the neovolcanic zone of the rift. To a lesser extent volumetrically, volcanic centers also occur on the flanks and up to 100 km from the neovolcanic zone of continental rifts.

Geologic and geophysical observations from continental rift settings indicate a wide range of dike widths. Mesozoic dikes associated with continental rifting and the opening of

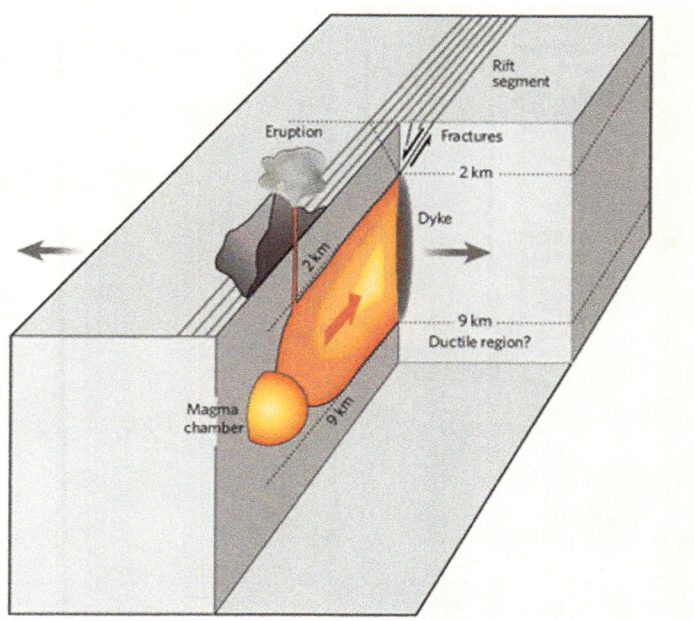

FIGURE 3.14 Block diagram of central volcano—fissure swarm magmatism. *From Sigmundsson, F., 2006. Magma Does the splits. Nature 442, 251—252, http://dx.doi.org/10.1038/442251a.*

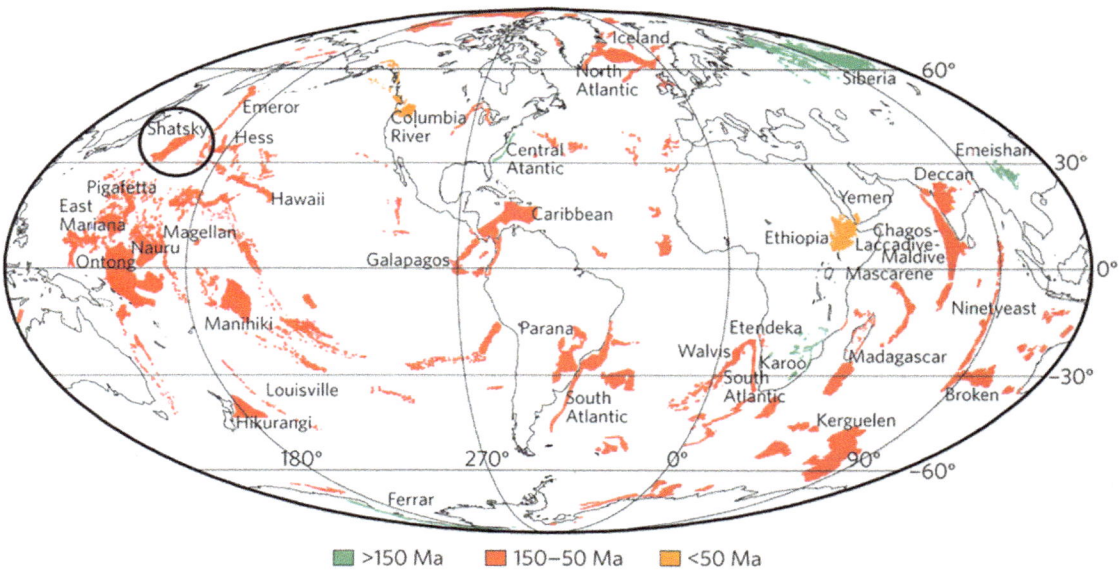

FIGURE 3.15 Map of large igneous provinces on Earth. *From Uenzelmann-Neben, G., 2013. Volcanology: magma giant. Nat. Geosci. 6, 902—903, http://dx.doi.org/10.1038/ngeo1958.*

the Atlantic Ocean basin can exceed 100 m in width. Geodetic observations of recent rifting events in Iceland (i.e., the 1975—1984 Krafla rifting episode) and Afar (i.e., the 2005—2009 Dabbahu—Manda Hararo rifting event) indicate the formation of topography above intruded dikes and dike widths ranging from <1.0 to ≤10.0 m (Figure 3.12). It is important to note here that these dike widths are estimates based on inversions of terrestrial (e.g., leveling and distance measurements) and space geodetic (Global Positioning System and Interferometric Synthetic Aperture Radar) observations using elastic dislocation models. They are not direct measurements of dike widths, as has been discussed for mid-ocean ridges and the Mesozoic dikes above. Nevertheless, these observations indicate that during rifting events, relative plate motions are accommodated by the intrusion of magma from central volcanoes into the neovolcanic zones of the rifts and that topography (i.e., normal faulting and grabens) can be formed above intruded dikes.

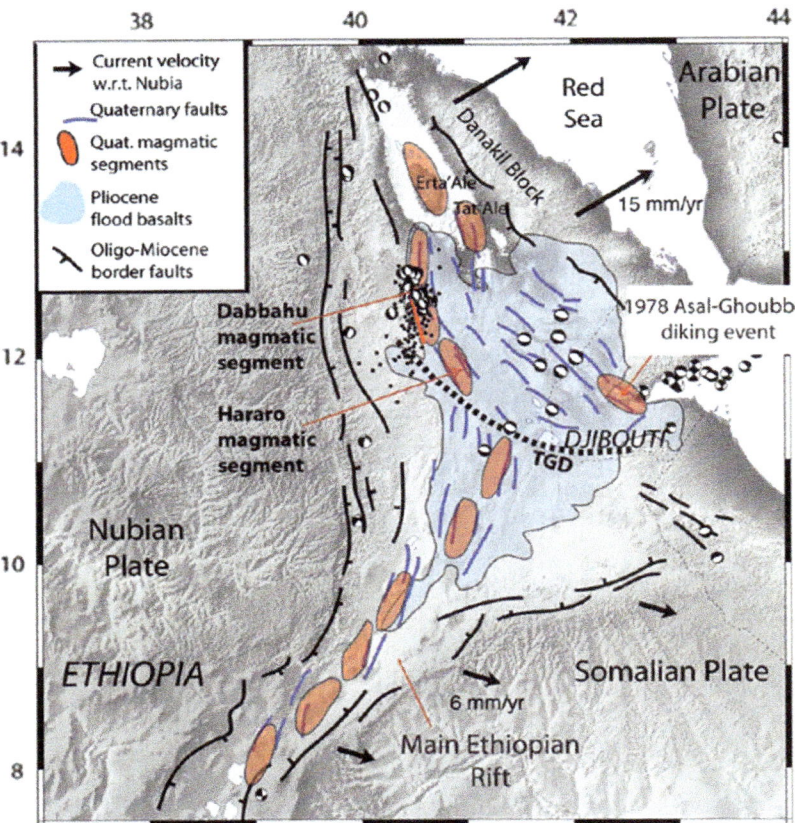

FIGURE 3.16   Magmatic segmentation within the East African Rift. *From Egger, A., Rooney, T., Shillington, D., (Eds.), Report from the Magmatic Rifting and Active Volcanism Conference, Afar Rift Consortium (Addis Ababa, Ethiopia). GeoPRISMS Newsletter, 28, Spring 2012. Retrieved from: http:// geoprisms.org.*

Rifting of the Nubian, Somalian, and Arabian plates is the best example of current continental rifting on Earth and has allowed geoscientists to observe the different stages of continental rifting along the three main rift segments. Rifting here began >40 Ma with the eruption of the Amaro and Gamo flood basalts (Figure 3.17), hypothesized to be driven by the Afar plume. There have been several additional discrete plume pulses at ~30 and ~22 Ma, and the magmatic history is likely more complicated. In the cases of the Nubia—Arabia and Somalia—Arabia plate boundaries, continental rifting has proceeded through to seafloor spreading in the Red Sea and Gulf of Aden, respectively. The relative plate motion rates across the Red Sea, ranging from ~0.7 cm a^{-1} (northwest) to ~1.7 cm a^{-1} (southeast), and across the Gulf of Aden ranging from ~1.6 cm a^{-1} (southwest) to ~1.9 cm a^{-1} (northeast), put these spreading centers in the ultraslow spreading rate regime. These ridge segments come onshore at the Afar Triple Junction, meeting with the northern extent of the EAR system. In the EAR, continental rifting between the Nubia and Somalia plates is ongoing at ~0.6 cm a^{-1} (Figure 3.16). Historical rifting events in the EAR and Afar Triple Junction; for example,

the Dabbahu rifting event (Figure 3.12), have provided geoscientists the ability to study the processes associated with continental breakup, including the dynamics of segmented rift margins, neovolcanic zones and central volcano - fissure swarm interactions.

Continental rifting breaks apart larger continents into smaller tectonic plates. Magmatic and tectonic processes at continental rifts modify continental lithosphere and accommodate relative plate motions. Continental rifts evolve from tectonically dominated in the early stages to full seafloor spreading and the formation of oceanic crust and ocean basins (i.e., the Wilson cycle in the later stages). Crustal growth occurs through dike injection and volcanism within the neovolcanic zones of continental rifts.

2.2. Convergent Plate Boundaries

Convergent plate boundaries are defined by the relative motion of tectonic plates toward each other and can be located in both oceanic and continental lithosphere (Figure 3.2). Convergent plate boundaries, specifically subduction zones, are just as extensive as mid-ocean ridges, with a cumulative length of ~55,000 km

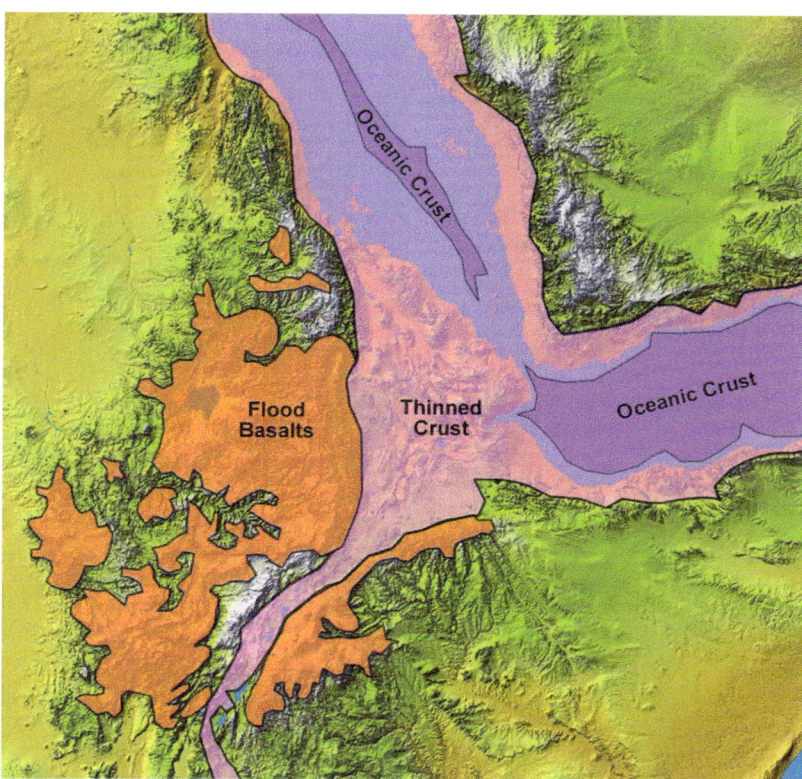

FIGURE 3.17 Geology of the Afar Triple Junction showing regions of magmatism associated with plume (orange, flood basalts) and continental rifting and seafloor spreading processes (pink and purple, thinned continental and oceanic crust). *From Wood, J., Guth, A., East Africa's Great Rift Valley: A Complex Rift System. Geology.com (accessed 27.11.14).*

(Figure 3.1). The plate boundary orientation can range from normal to highly oblique to the relative plate motion vector, and the convergence rate can vary from <1 to <12 cm a^{-1}. As compared to divergent margins where relative plate motions are accommodated predominantly by magmatism, relative motion between converging plates is almost entirely accommodated by permanent deformation along plate boundary fault systems (i.e., subduction zone megathrust faults). The subduction process leads to the formation of parental basaltic magmas and subsequently volcanic arcs, which aid in the formation of continental crust and lithosphere through volcanic eruptions and the emplacement of large volume intrusive complexes. At convergent margins where convergence is highly oblique, plate boundary zone strain is partitioned between the plate boundary megathrust and the overriding continental lithosphere, resulting in the formation of a continental transform fault within or adjacent to the volcanic arc. The proximity of the transform fault to the active volcanic arc leads to interesting magma—tectonic interactions discussed in Section 2.3. We describe below the process of subduction, the formation of primary magmas and volcanic arcs, and therefore continental lithosphere, and the interaction between the plate boundary and the volcanic arc.

Convergence between oceanic—oceanic and oceanic—continental lithosphere will lead to subduction, the result of thermal process acting on the surface of the earth. Recall, that as oceanic lithosphere moves away from the mid-ocean ridge it cools, and the thickness and density increase. Driven by the density difference between converging plates, oceanic lithosphere being of higher density compared to continental lithosphere will underthrust the continental lithosphere and subduct into the mantle. Once oceanic lithosphere begins to underthrust the overriding lithosphere and penetrates the upper mantle, slab pull forces (i.e., body forces driven by the density contrast between the oceanic lithosphere and mantle) take over. The subduction of oceanic lithosphere into the mantle will lead to the partial melting of the mantle wedge and the formation of primary magmas (Figure 3.18).

Formation of primary melts and magmas is a thermal and hydrogeological problem. Recall again the formation of oceanic lithosphere at the mid-ocean ridge and the hydrothermal circulation and alteration within the oceanic crust that takes place there. In addition to hydrothermal circulation at the ridge, bending of the oceanic plate during subduction leads to the formation of normal faults and horst and graben structures. The normal faults and grabens allow for additional hydrothermal circulation and deposition of

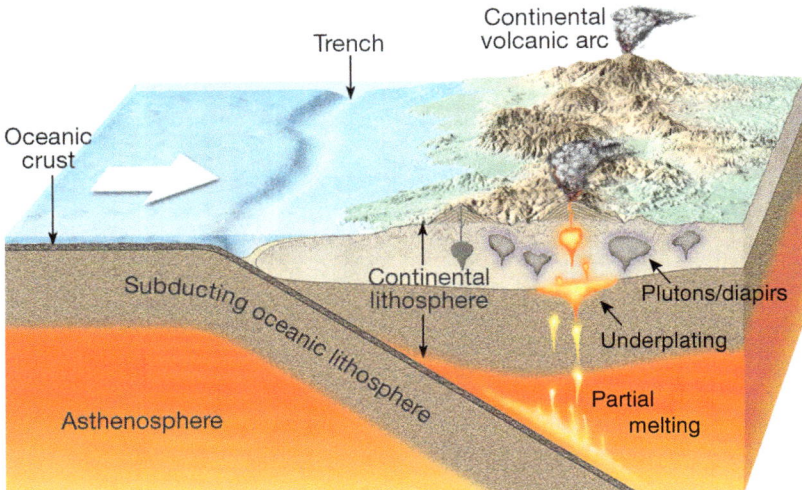

FIGURE 3.18   Subduction of oceanic lithosphere beneath continental lithosphere. Partial melting occurs in the mantle wedge. Note magma migration as dikes and diapers, underplating at the Moho, formation of magma chambers, cooled plutons, and the volcanic arc. *TasaGraphics, 2009.*

sediment, respectively. Hydrothermal circulation of seawater through the oceanic crust and lithosphere leads to metamorphism of the subducting plate. Seismic tomography and electrical resistivity studies across several subduction zones indicate serpentinization of the subducting oceanic lithosphere. Serpentinization is the result of hydrothermal circulation and low-temperature metamorphism, and produces a suite of hydrated mineral phases. The hydration of the oceanic crust and the existence of pelagic sediment with interstitial pore water play an important role in the formation of magmas and potentially in the seismogenic behavior of plate boundary faults. As the subducting oceanic lithosphere is pulled into the mantle, the increasing pressure and temperature drive metamorphic reactions that release fluids from the crust into the overlying mantle wedge. This addition of fluids (i.e., hydration) reduces the mantle solidus (i.e., lowers the melting temperature) resulting in partial melting of the mantle wedge and formation of basaltic primary melts (see Chapters 1—*Melting in the Mantle of the Earth* and 4—*Origin and Composition of Magmas*) (Figure 3.19). This process has been termed volatile flux melting and typically occurs between depths of 80 and 120 km.

Once primary melts are formed in the mantle wedge, they will rise through the wedge and to the overriding lithospheric plate due to buoyancy forces. Here, several processes occur separately or in concert with each other (see Chapters 2—*Migration of Melt* and 9—*Rates and Timing of Magma Ascent and Storage*). The magmas may underplate or accumulate at the Moho of the overlying lithosphere (Figure 3.18). This addition of heat to the base of the crust will cause partial melting of the continental lithosphere and the production of more chemically evolved magmas (e.g., rhyolitic magmas). The primary

basaltic magmas may also rise through the lithosphere. In the lithospheric mantle and ductile lower crust, these magmatic bodies are likely diapers or inverted teardrop shaped bodies (Figure 3.18), rising due to buoyancy forces. In the upper brittle crust (i.e., at depths less than 20 km), magmas migrate as dikes or are intruded as sills (see Chapter 10—*Conduits and Dikes*). In general, magmatic bodies may rise directly through the crust or stall along the way. For example, magmas may accumulate at the Moho or within the upper crust, forming magma chambers (see Chapter 8—*Magma Chambers*). As the magmas rise, migrate, and accumulate in the crust they may undergo a number of processes that lead to the differentiation and evolution toward more evolved magmas (e.g., andesitic to rhyolitic compositions). These processes include assimilation and melting of crust, fractional crystallization, and magma mixing of primary or less evolved basaltic magmas with more evolved rhyolitic magmas in crustal (mid to upper crust) magma chambers. Recent studies have suggested that the mixing of basaltic and rhyolitic magmas is the main process in the formation of andesitic magmas. The geochemistry and petrology of erupted lavas or intruded plutonic rocks may give insight to the processes that took place between the source and final crystallization.

Volcanic arcs are synonymous with subduction zones and form above the region of partial melting (Figure 3.18). Volcanic arcs can form in oceanic or continental lithosphere. For example, the Izu-Bonin-Mariana Island Arc is the result of oceanic lithosphere of the Pacific plate subducting beneath oceanic lithosphere of the Philippine Sea plate. And the Andean volcanic arc is the result of the subduction of oceanic lithosphere of the Nazca plate, beneath continental lithosphere of the South

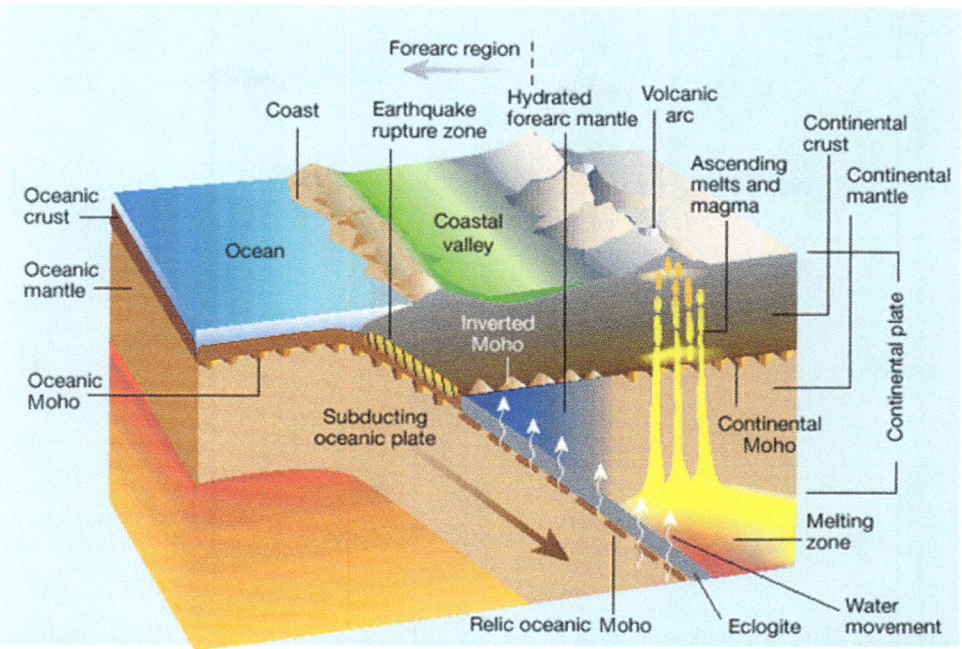

FIGURE 3.19 Cross-section through a subduction zone showing volatile release from the oceanic crust, partial melting, magma migration (dikes and diapers), magma chambers, and volcanoes. *From Zandt, G., 2002. Earth science: the slippery slope. Nature 417, 497—498, http://dx.doi.org/10.1038/417497a.*

American plate (Figure 3.1). It is the formation of these volcanic arcs that grow continental lithosphere at convergent plate boundaries by the accumulation and cooling of magmas in plutonic or batholith complexes, the addition of magmas in the upper crust as dikes and sills, the eruption of lavas and tephra, and construction of volcanoes (Figure 3.18). Global magmatic arc productivity rates have been estimated to be on average $90 \text{ km}^3 \text{ km}^{-1} \text{ Myr}^{-1}$ (Clift and Vannuchi, 2004). This estimate includes intrusive and extrusive components, and represents an addition of less than 1% of total crustal volume per Myr. The continental arc volcanoes grow with an average extrusive flux of $4.5 \times 10^{-3} \text{ km}^3 \text{ yr}^{-1}$ (White et al., 2006) (see Chapter 14—*Volcanic Episodes and Rates of Volcanism*). It has been suggested that crustal growth in some volcanic arcs is balanced by destruction of continental lithosphere through subduction erosion, the process of physically eroding the overriding plate during subduction.

The location of volcanic arcs is ultimately constrained by the location of the melt production zone in the mantle wedge. However, emplacement of magmas and the construction and morphology of volcanic arcs are controlled by the tectonic stress field in the upper plate (Acocella, 2014 and Chapter 10—*Conduits and Dikes*). The stress field in the overriding lithosphere can be extensional (e.g., Taupo volcanic zone, New Zealand), compressional (e.g., the Central Andes or northern Japan), shear (e.g.,

Central America or Sumatra), or a combination thereof (e.g., the transtensional arc in Nicaragua) (Figure 3.20). The morphology of volcanic centers and alignment of monogenetic vents have been used as indicators of the upper plate stress field. At a first order this is a good assumption; however, individual volcanoes or volcanic complexes modify the regional stress field locally, which provides interesting feedbacks between magmatism and tectonics. This feedback between the stress field and magmatic systems is important for the discussion of triggering of volcanic unrest, as the orientation (i.e., strike and dip) and geometry of magmatic plumbing systems relative to the plate boundary are intrinsic factors in the stress and strain changes imparted onto the system. This will be discussed in Section 3.

Convergent plate boundaries and subduction zones are dynamic tectonic environments, with the rate and azimuth of convergence, as well as the age and morphology of the subducting oceanic lithosphere, changing over time. The convergence and subduction of aseismic and seismic ridges can greatly affect the subduction process and dynamics of the overriding lithospheric plate. Aseismic ridges are ridges formed by magmatism and volcanism associated with passage of oceanic lithosphere over a mantle plume or hot spot. For example, the Cocos and Carnegie Ridges formed by passage of the Cocos and Nazca plates, respectively over the Galapagos hot spot. The subduction of these ridges has led to the formation of forearc terranes and in the case of

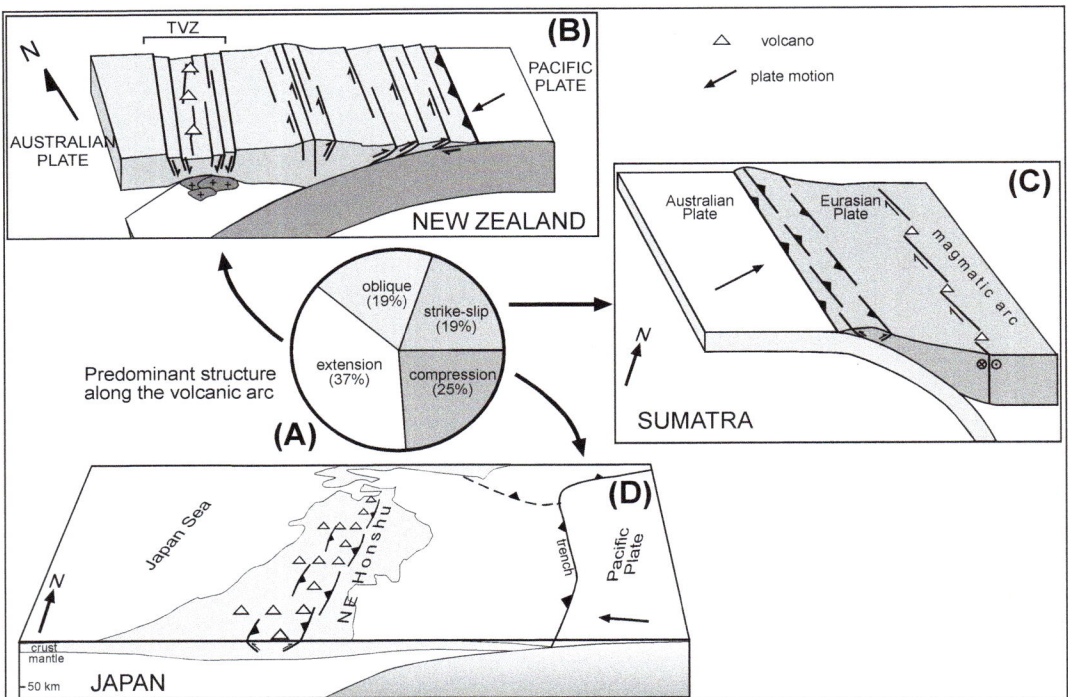

FIGURE 3.20 Tectonics within the overriding lithospheric plate at convergent plate boundaries. *From Acocella, V., Funiciello, F., 2010. Kinematic setting and structural control of arc volcanism. Earth Planet. Sci. Lett., http://dx.doi.org/10.1016/j.epsl.2009.10.027.*

the Cocos Ridge, the displacement and extinction of the volcanic arc and exhumation of young plutonic rocks in southern Costa Rica. The subduction of active spreading ridges can also have profound effects on subduction and magmatism. Subduction of active spreading centers will lead to slab break off and formation of a slab window along the volcanic arc. These processes and phenomena result in a change in the location of magmatism and volcanism. Magmatism and volcanism may be reduced or cease entirely in the arc and initiate in extensional regions in the forearc. The opening of a slab window allows warm asthenospheric mantle to flow into the region, inducing melting of the adjacent slab and partial melting of mantle. This may reinitiate volcanism in the arc or form a new arc with distinct geochemical and petrologic signatures. There is also warm asthenosphere associated with the subducting ridge segment. The subduction of the South Chile Rise in southern Chile is the best modern example of this process (Figure 3.21), while geologic examples are found at recently active convergent margins; for example, western Panama and southern Costa Rica.

At convergent plate boundaries, tectonic processes predominantly accommodate relative plate motions, mainly through the permanent deformation across plate boundary megathrusts and forearc regions. Primary basaltic melts and magmas are formed by partial melting in the mantle wedge, enhanced by the flux of volatiles from the subducting oceanic lithosphere. Crustal and

lithospheric growth takes place through the formation of intrusive magmatic complexes (i.e., plutonic and batholith complexes), and volcanic arcs in overriding lithosphere. However, these processes account for a small fraction of continental lithospheric growth, as compared to accretion of continental fragments and island arcs.

2.3. Transform Plate Boundaries

Transform plate boundaries are defined by the relative motion of tectonic plates moving laterally past each other and can be located in both oceanic and continental lithosphere, although the former dominate (Figure 3.2). Oceanic transforms connect the neovolcanic zones of mid-ocean ridge segments and are pervasive along the mid-ocean ridge system (Figure 3.1). Continental transforms can connect ridge segments on adjoining oceanic plates (e.g., the San Andreas fault), can accommodate forearc sliver transport or terrane migration in obliquely convergent margins (e.g., the Sumatran fault system) or can accommodate extrusion of continental lithosphere in collisional zones (e.g., the Anatolian fault). The plate boundary orientation can range from parallel to highly oblique to the relative plate motion vector, and the relative plate rates across continental transforms are typically <10 cm a^{-1}. Continental transforms that are oriented oblique to the plate motion direction or have stepovers along the strike of the fault system will result in

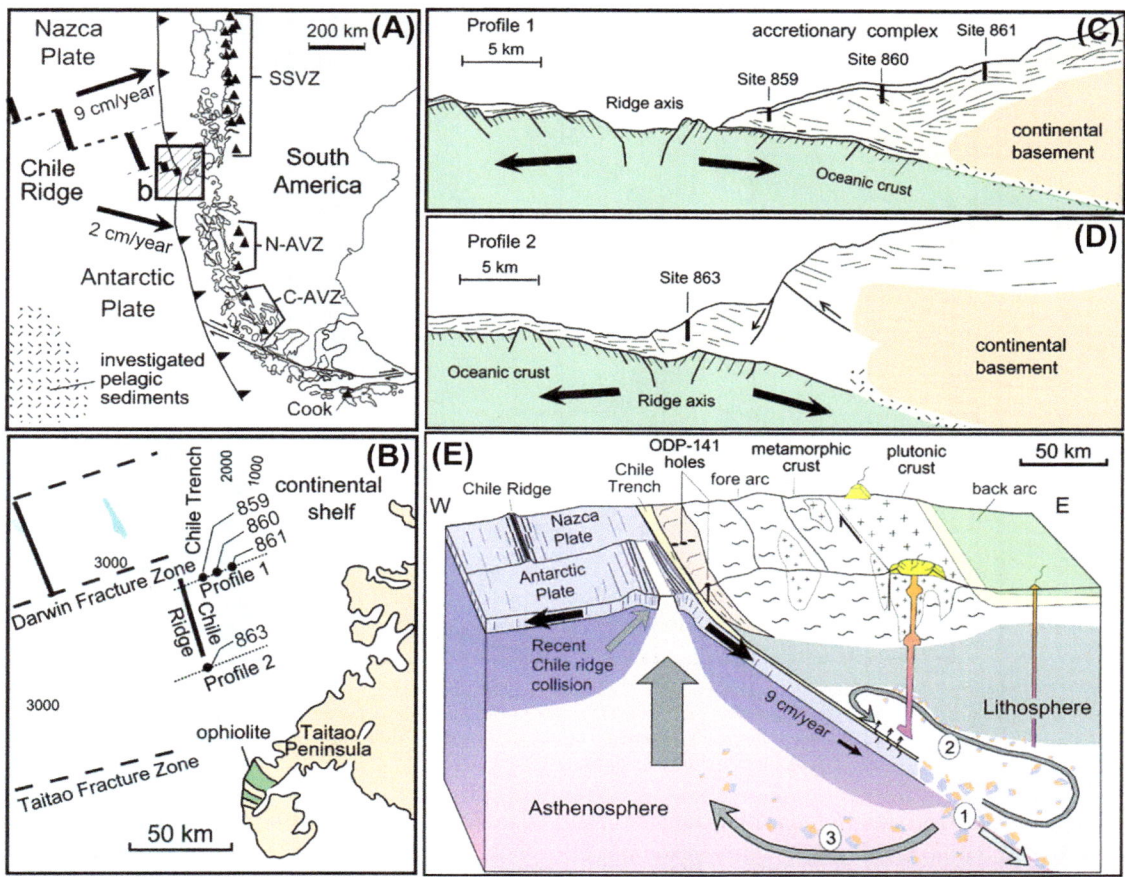

FIGURE 3.21 Subduction of the South Chile Rise and magmatism in the overriding plate. *From Killian, R., Behrmann, J.H., 2003. Geochemical constraints on the sources of Southern Chile Trench sediments and their recycling in arc magmas of the Southern Andes. J. Geol. Soc. Lond., 160, 57—70.*

either transtensional or transpressional stress regimes. Transtensional stress regimes lead to extension of the crust and lithosphere, and can result in transtensional volcanism. Volcanism in transtensional settings can vary from monogenetic basaltic volcanism to the formation of more magmatically evolved stratovolcanoes. Compared to divergent and convergent boundaries, the flux of magmatic and volcanic material added to the crust and lithosphere is low at transform plate boundaries. We describe the tectonics and magmatism of several different transtensional plate boundary settings below.

The Basin and Range province in the western United States was a focus of volcanism throughout the Tertiary. Crustal extension dominated through this period and still dominates at its eastern boundary (e.g., the Wasatch fault zone). Since Pliocene time the western extent of this system has been dominated by transtension accommodating >1.0 cm a^{-1} of Pacific—North America relative plate motion. Volcanism in this region is dominated by basaltic volcanic fields that are intimately linked with tectonic structures (see Chapter 23—*Basaltic Volcanic Fields*). Extension and transtension within the region have resulted

in crustal and lithospheric thinning. As is the case for continental rifting, this thinning leads to decompression melting of the underlying mantle and the formation of magmas. Although, the magmas in this region are also associated with hydration of the mantle during subduction of the Farallon plate. Because of the interaction of the magmatic and volcanic system with tectonic structures, crustal growth in this region is mainly through dike injection and volcanism.

Oblique convergence and strong mechanical coupling along the plate boundary megathrust at convergent boundaries, results in strain partitioning and the development of transform faults in the overriding lithosphere oriented roughly parallel to the plate boundary. These transform faults are located within or in close proximity to active volcanic arcs (e.g., the Sumatran fault; Figure 3.20(C)), influence magmatism and volcanism within the arc, and accommodate motion of a distinct forearc terrane relative to the overriding lithosphere. Shear across a volcanic arc will cause deformation within volcanic complexes. Additionally, transtensional tectonics can develop along these transform faults, leading to crustal extension

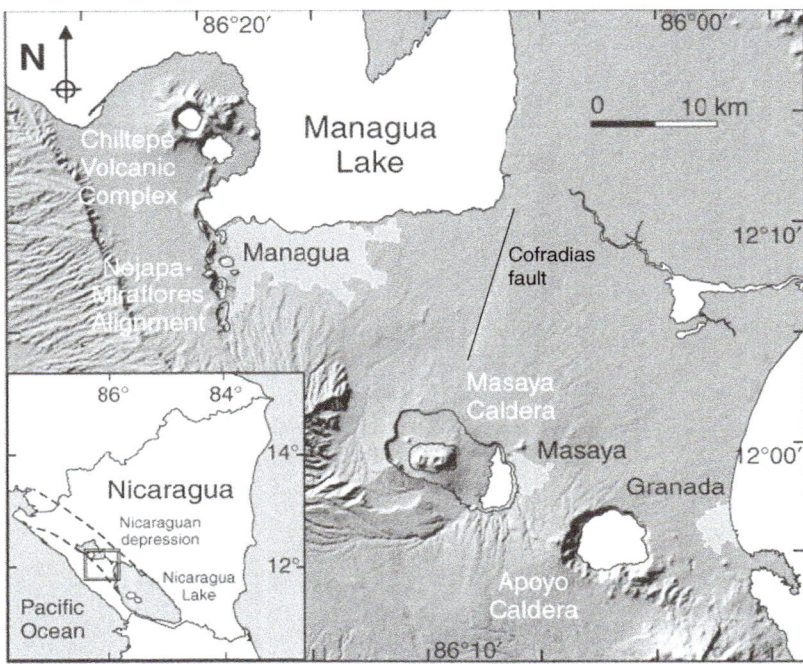

FIGURE 3.22   The Managua Graben, Nicaragua. The graben is bound by the Nejapa-Mira Flores volcanic alignment and Mateare fault (not-shown) (west) and the Cofradias fault (east). *Modified from: Perez, W., Freund, A., 2006. The youngest highly explosive basaltic eruptions from Masaya Caldera (Nicaragua): Stratigraphy and hazard assessment, in Volcanic Hazards in Central America, GSA Special Paper, v. 412, p. 189–207, Rose, W.I. Editor, Geological Society of America; http://dx.doi.org/10.1130/2006.2412(10).*

(i.e., pull-apart basin formation) and magmatism. The Holocene volcanic arc in Nicaragua is a good example of these phenomena. The Nicaraguan forearc migrates to the northwest relative to the Caribbean plate at ~ 1.0 cm a^{-1}. Dextral shear across the volcanic arc manifests in volcanic complexes as roughly north-trending normal faults, resulting in lateral growth and spreading of the complexes. North-trending volcanic alignments of maars and monogenetic volcanoes (e.g., the Cerro Negro-Cerro La Mula, Nejapa-Miraflores, and Granada) are associated with right steps in the volcanic arc and dextral transform fault system. Magmas erupted along these alignments are less evolved than in the arc and indicate almost direct passage through the crust and lithosphere from the mantle. The Managua graben is a pull-apart basin at a right step in the volcanic arc and is a center of volcanism, with the Masaya caldera complex at its southern boundary and monogenetic volcanism along the edges of the graben, and associated with strike-slip faults within the graben (Figure 3.22). Dike injection along basin bounding faults partially accommodates extension across the western boundary of the graben and forms the Nejapa-Miraflores volcanic alignment.

In regions of continental collision, continental transform faults develop to accommodate the motion of lithosphere away from the collision zone (i.e., lithospheric extrusion). These regions are very complex zones of deformation and active research in these areas is ongoing. Like the transform systems described above,

the Anatolian fault system, with relative rates across the fault of ~ 2 cm a^{-1}, is transtensional along several of its segments. There have been multiple phases of transtensional volcanism along the Anatolian fault starting in Miocene time (Figure 3.23). Like the western Basin and Range province, the magmas erupted along the Anatolian fault are likely associated with the hydration of the mantle during an earlier phase of subduction.

Transform fault plate boundaries accommodate the relative motion between mid-ocean ridges and continental terranes. Transform fault systems are not typically associated with magmatism. However, in transform systems that have components of extension (i.e., transtensional systems), magmatism does occur. Extension in these regions is accommodated by both tectonic and magmatic process, with magmatic accommodation in the form of dike injection. Magmas in transtensional systems are formed by decompression melting that could be enhanced in regions of hydrated mantle from earlier periods of subduction. Volcanism in transtensional regions can vary from monogenetic basaltic volcanic fields to the formation of stratovolcanoes.

3. MAGMA–TECTONIC INTERACTIONS

Magma–tectonic interactions encompass a range of processes and phenomena linking magmatic and tectonic systems across vast spatial and temporal scales. In Section

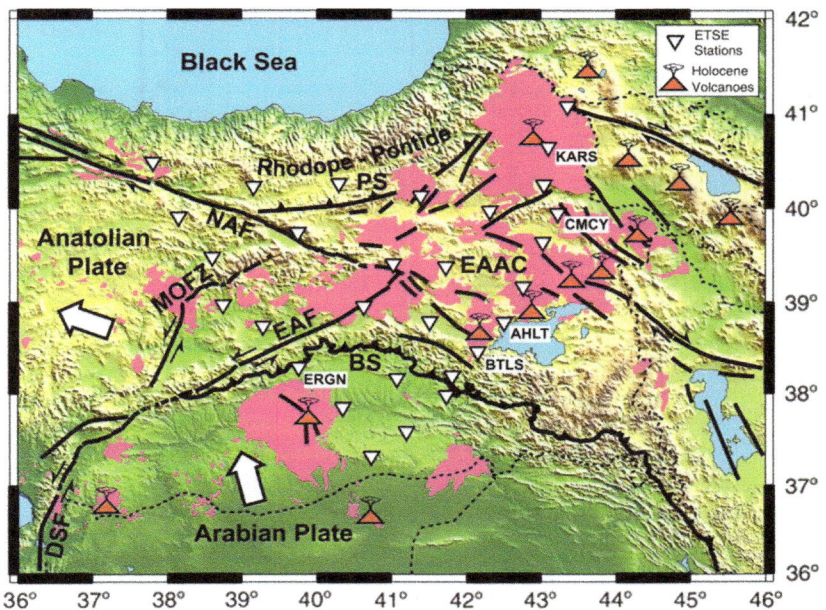

FIGURE 3.23 Volcanism along the eastern North Anatolian fault. Neogene volcanic rocks (pink). *From Zacar, A.A.O., Zandt, G., Gilbert, H., Beck, S.L., 2010. Seismic images of crustal variations beneath the East Anatolian Plateau (Turkey) from teleseismic receiver functions. In: Sosson, M., Kaymakci, N., Stephenson, R.A., Bergerat, F., Starostenko, V. (Eds.), Sedimentary Basin Tectonics from the Black Sea and Caucasus to the Arabian Platform, Geological Society, London, Special Publications, 340, 485–496, http://dx.doi.org/10.1144/Sp340.210305-8719/10, The Geological Society of London.*

2, we described magma—tectonic interactions at geologic time and plate boundary distance scales, including the association of magmatism with plate tectonics (e.g., subduction zone arc volcanism) and the accommodation of relative plate motions by magmatism (e.g., divergent plate boundaries). Furthermore, we demonstrated how these interactions are responsible for formation of oceanic and continental crust and lithosphere. On shorter spatial and temporal scales, there is a growing set of observations and statistical studies that suggest intermediate to great earthquakes can promote or trigger magmatic processes including the migration and eruption of magma (Eggert and Walter, 2009). Furthermore, magmatic processes can trigger tectonic earthquakes and the eruption of other volcanoes. Our understanding of these correlations and potential processes is far from complete. However, improved geophysical and geochemical monitoring of tectonic and magmatic systems (see Chapter 66—*Integration of Volcano Monitoring*), laboratory experimentation and numerical modeling studies continue to improve our understanding of this correlation and the processes that may lead to promotion and triggering of volcanic eruptions.

The temporal and spatial correlation between earthquakes and eruptions makes a strong case for causality between the phenomena. Linde and Sacks (1998) first presented data demonstrating a temporal and spatial link between earthquake and volcanic eruptions. Clearly, at short temporal and distance scales, there is an increase in volcanic eruptions following magnitude greater than seven earthquakes (Figure 3.24). But what are the mechanical, chemical, and/or petrologic processes that actually trigger magma migration and eruptions? The short-term interaction between tectonic and magmatic systems is likely caused by coseismic and postseismic changes in stress and strain on magmatic systems and resulting physicochemical processes within the magmatic system. Large magnitude earthquakes change the state of stress in the Earth's crust. In the near-field, within several fault lengths of the hypocenter, coseismic permanent *strain* and **static stress** changes take place. In the near- and far-fields, propagating seismic waves cause near-instantaneous *dynamic strain* and *stress* changes. In addition, larger magnitude earthquakes will induce viscous flow in the lower crust and upper mantle, allowing for translation of stress and strain changes through the crust over longer temporal and spatial scales. Large magnitude earthquakes also induce static and dynamic stress and strain changes affecting an entire arc, not just individual magmatic systems. Finally, in the case of magma-triggered tectonic earthquakes, changes in the **Coulomb failure stress** within the crust surrounding a magma body will promote failure along optimally oriented faults.

In addition to these tectonically induced stresses, external forcing and periodic stress changes at short and long temporal and spatial scales affect the lithosphere. The Earth and ocean tides induce stress changes of ∼0.001 and ∼0.01 MPa, respectively and with a period of 12 h.

Number of eruptions as a function of time

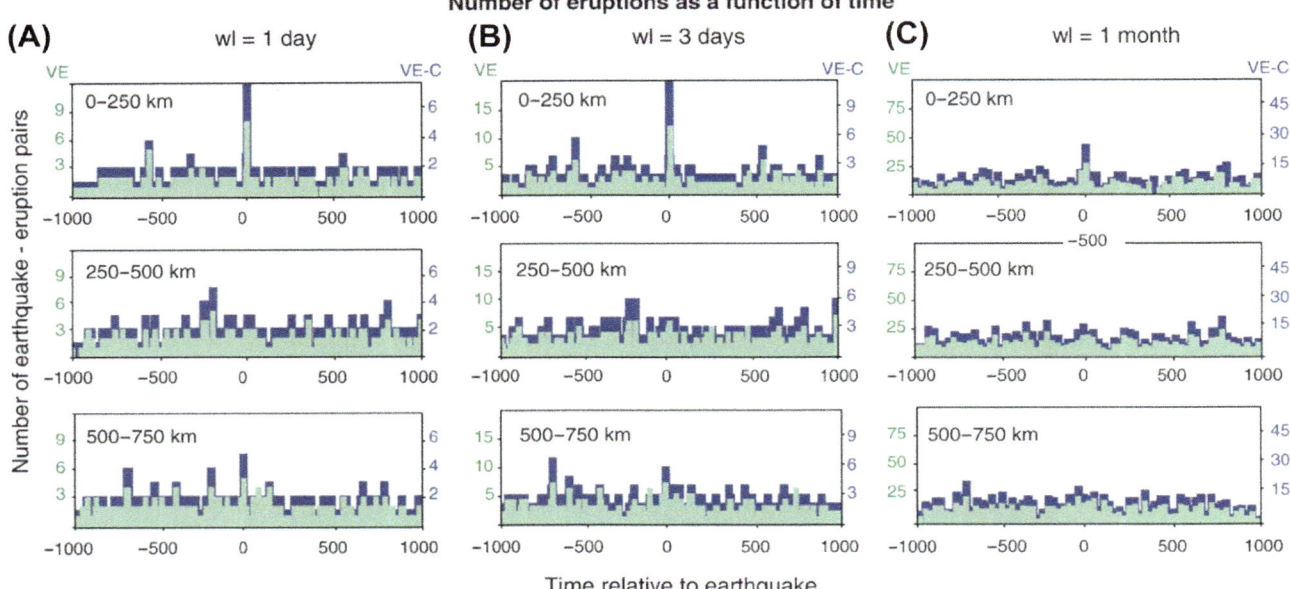

FIGURE 3.24 Histograms showing the number of volcanic eruptions (VE) following M ≥ 7 earthquakes binned in 1 day, 3 day, and 1 month increments and at distances of 0−250 km, 250−500 km, and 500−750 km from the earthquake. Statistics for volcanic eruptions minus ongoing eruptions (VE-C) are also shown on the histograms (right y-axis). *From Eggert and Walter (2009).*

The magnitude of tidal induced stress changes is often used as the minimum stress required to promote or trigger an earthquake or eruption. At timescales of days to years, hydrologic loading can induce stress changes on the order of 0.1−0.001 MPa, and on millennial timescales glacial loading can range from 10 to 100 MPa. All of these phenomena have been linked to triggered seismicity and magmatic activity. In comparison to these phenomena, earthquakes produce similar stress changes, but have higher stressing rates. The magnitudes of earthquake induced stress changes (i.e., static and dynamic stress changes) are dependent on the magnitude of the earthquake, distance between the source and receiver fault of magmatic system, and geometry and orientation of the receiver fault or magmatic system. In this section, we will provide an introduction to the topic of magma−tectonic interactions at short spatial and temporal scales with a focus on triggering of volcanic unrest by static and dynamic stress changes, strain changes, and associated magmatic processes. In addition, we discuss and give several examples of magma-triggered tectonic earthquakes. For a discussion of magma-induced volcanic seismicity, the reader is referred to Chapter 59—*Volcano Seismicity.*

3.1. Static Stress Changes

Coseismic static stress changes on faults and magmatic systems are of significant magnitude in the near-field. That is, they are the result of elastic strain released during an

earthquake and decay by $1/r^3$, where r is the distance from the hypocenter. For example, the static stress change for an M8 earthquake at 100 km is 0.1 MPa, but 0.0001 MPa at 1000 km (Manga and Brodsky, 2007). Therefore, the magnitude of an earthquake and the proximity of the magmatic system to the hypocenter of the earthquake are important factors in the magnitude of static stress changes. We use the Coulomb failure criteria, the stress required to drive a fault toward failure, to describe the stress changes on a fault. The change in Coulomb failure stress:

$$\Delta\sigma_f = \Delta\tau_s + \mu'\Delta\sigma_n \qquad (3.2)$$

is equal to the change in shear stress (τ_s) on a fault plane plus the product of the coefficient of friction accounting for pore fluid pressure (μ'), and the change in normal stress (σ_n). Therefore, an equally important factor in the magnitude of stress changes induced by an earthquake is the orientation of the receiver fault relative to the source fault or magmatic system, as stress changes are related to the strike, dip, and rake of a fault. A positive change in the Coulomb failure stress promotes fault failure. In the case of normal stress changes, the stress change is independent of the rake, and a positive normal stress change promotes fault failure.

In the discussion of earthquake-triggered eruptions in the near-field, changes in normal stress ($\Delta\sigma_n$) on a magmatic system are often invoked. A positive change in normal stress can cause "unclamping" and formation or opening of a feeder dike, allowing for magma migration and eruption

Static Stress & Strain Triggering Mechanisms

Thrust faulting

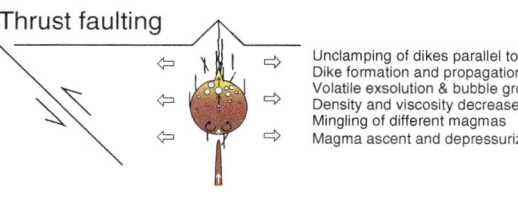

Unclamping of dikes parallel to margin
Dike formation and propagation
Volatile exsolution & bubble growth
Density and viscosity decrease
Mingling of different magmas
Magma ascent and depressurization

Normal faulting

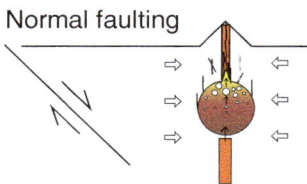

Unclamping of dikes normal to margin
Dike formation and propagation
Volatile exsolution & bubble growth
Compression of magma chamber
Density and viscosity decrease
Mingling of different magmas
Magma ascent and depressurization

FIGURE 3.25 Mechanisms for the triggering of volcanic eruptions by earthquakes. Magmatic and tectonic systems not to scale. *Modified from Walter, T.R., Amelung, F., 2007. Volcanic eruptions following M ≥ 9 megathrust earthquakes: implications for the Sumatra-Andaman volcanoes. Geology 35, 539–542, http://dx.doi.org/10.1130/G23429A.1.*

(Figure 3.25). The regional tectonic stress field controls the orientation (i.e., strike and dip) and geometry (i.e., dike, sill, and/or magma chamber) of magmatic systems in the upper crust (Figure 3.20) (see Section 2 and Chapters 8—*Magma Chambers* and 10—*Conduits and Dikes*). The magnitude of unclamping will be a function of source strength and geometry, distance between the source and receiver (r), and receiver geometry. Therefore, magmatic systems (i.e., dikes) oriented parallel to extensional and compressional plate boundaries and normal to transform plate boundaries will be favored for positive normal stress changes (Figure 3.26). Normal stress changes between >0.0001 and <10.0 MPa for earthquake-triggered eruptions have been estimated and reported in the literature.

The static stress triggering of volcanic eruptions is typically not instantaneous. The timing between source earthquakes and receiving magmatic systems is highly variable and can be on the order of hours to months to years. This temporal range likely reflects the state of the magmatic system prior to the stress change (i.e., how close was the magma to eruption) and processes related to magma migration and eruption. Unclamping or opening of a conduit will promote formation of a dike and magma ascent. If there are nested magma chambers throughout the crust, magmas migrating from depth may be injected into shallower magma chambers, inducing magma mingling and mixing, vesiculation, and eruption. There is a preponderance of petrologic evidence that explosive silicic eruptions are triggered by the injection of more mafic magmas. Therefore, the time between an earthquake and triggered volcanic eruption likely reflects the rate and distance of magma ascent (see Chapter 9—*Rates and Timing of Magma Ascent and Storage* and Chapter 11—*Magma Ascent and Vesiculation at Shallow Levels*), and the timing

of mingling and mixing between distinct magmas prior to eruption. The ascent rate of magmas is fairly well constrained by geochemical and geophysical studies, with average rates between 0.001 and 0.015 m s^{-1}. Recent studies of magma mixing suggest full mixing of different magmas at the timescales of days to weeks.

Several of the greatest historical earthquakes, as well as lower magnitude earthquakes, have been temporally and spatially correlated with large, volcanic explosivity index (VEI) >3 volcanic eruptions (see Chapter 13—*Sizes of Volcanic Eruptions*), and have observed petrology consistent with magma mixing prior to eruption. For example, the 1707 VEI 5 Hoei eruption of Mt Fuji took place 49 days after the M_w > 8.7 Hoei earthquake along the Nankai subduction zone, Japan. This was the largest historical eruption of Mt Fuji and the largest historical earthquake in Japan prior to the M_w 9.1 2011 Tohoku-Oki earthquake. Here, the earthquake changed the normal stress on the magmatic plumbing system by ~0.1 MPa beneath Mt Fuji, promoting formation of a deep dike and rise of basaltic magma from ~20 km depth (Figure 3.27). Mingling and mixing of distinct andesitic and dacitic magmatic bodies, possibly triggered by the basaltic magma, occurred prior to their eruption and the basaltic plinian eruptive phase. These observations suggest an average magma ascent rate of 0.005 m s^{-1}, although it is likely that the ascent rate was variable between the different magmas. Continued investigation of large magnitude earthquake—explosive eruption pairs will provide further insights into the processes leading to the triggering of volcanic eruptions.

3.2. Dynamic Stress Triggering

Our ability to monitor and image magmatic systems and related phenomena has expanded rapidly over the last several decades. New geochemical and geophysical monitoring techniques and more extensive, multiparameter networks provide better synoptic views of magmatic and volcanic activity, and more importantly changes in activity, and allow for a holistic approach to investigating these systems (see Chapter 66—*Integration of Volcano Monitoring*). Increased seismic monitoring of volcanoes has led to the development of new techniques to image the internal structure of volcanic systems and changes in crustal parameters (e.g., velocity structure) within and around volcanoes, and allows for observing the passage of seismic waves from tectonic earthquakes through the volcano. Seismic waves generated by large magnitude earthquakes cause near-instantaneous changes in crustal stress and strain at great distances from the earthquake hypocenter. These *dynamic stress* changes decay by $1/r^{1.66}$, where r again is the distance from the earthquake hypocenter. The magnitude of these stresses can be quite high, on the order

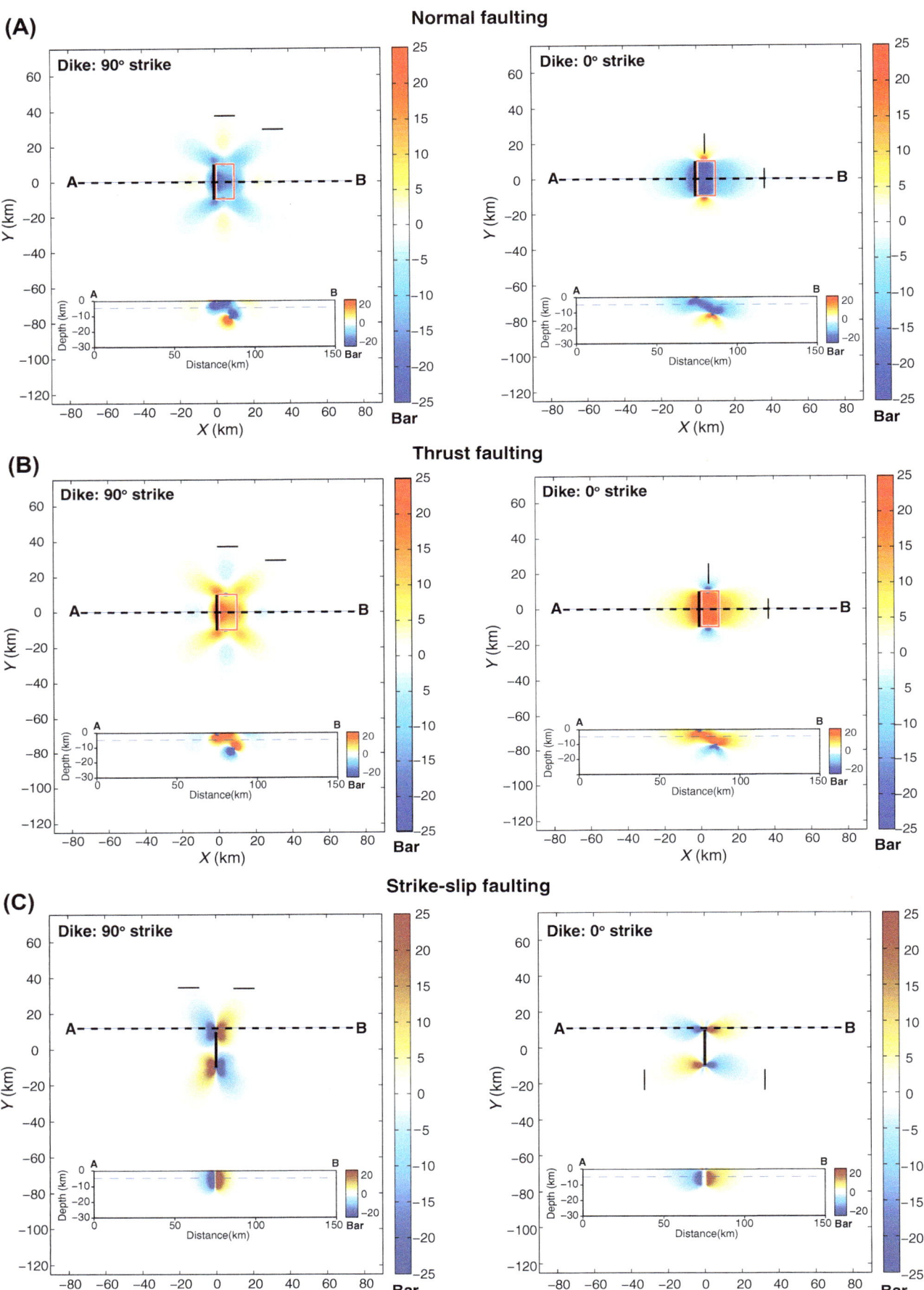

FIGURE 3.26 Normal stress change for a M7 earthquake calculated at 5 km depth on vertical planes (i.e., dikes or faults) striking parallel and normal (thin black lines) to the fault that slipped during the earthquake. Surface projection of faults shown as thick black lines and the faults are shown by red boxes. Cross-sections (A-B) through the static stress change field are shown for each case. The dashed line is at 5 km depth. (A) Normal faulting on a 45° east-dipping fault. (B) Thrust faulting on a 45° east-dipping fault. (C) Strike-slip faulting. Note the change in scale of the normal stress changes. 1 bar equals 0.1 MPa.

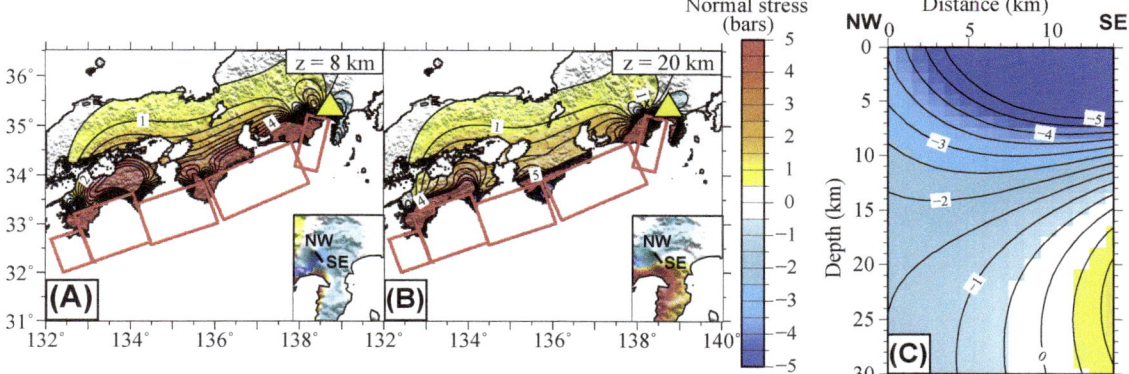

FIGURE 3.27 Normal stress change on the Mt. Fuji magmatic system following the 1707 M > 8.7 Hoei earthquake. *Chesley, C., LaFemina, P.C., Puskas, C.M., Kobayashi, D., 2012. The 1707 M8.7 Hoei Earthquake triggered the largest historical eruption of Mt. Fuji, Japan. Geophys. Res. Lett. 39, L24309, http://dx.doi.org/10.1029/2012GL053868.*

of 1 MPa in the near-field (100 km) to 0.01 MPa in the far-field (1000 km) (Manga and Brodsky, 2007).

Dynamic stresses are larger in magnitude and more far-reaching than static stresses, and there is a clear correlation between the passage of seismic waves and triggering of seismicity in hydrothermal systems (Figure 3.28). Dynamic stresses and strains associated with the passage of seismic waves may stress preexisting fractures in the shallow crust to failure or cause dilation of fractures and subsequent fluid flow that triggers seismicity. However, there are few studies indicating the direct triggering of volcanic eruptions by dynamic stress changes. Nevertheless, dynamic stresses have been invoked for the triggering of mud volcano eruptions, changes in volcanic activity and are linked mechanically to several processes that have been postulated to increase magmatic overpressure (Manga and Brodsky, 2007 provide a thorough discussion).

Processes within magma chambers that enhance the formation of bubbles (i.e., vesiculation) will increase the magmatic overpressure and potentially trigger magma migration and eruption. Several processes related to dynamic stresses have been called upon to increase bubble formation; these include rectified diffusion, advective overpressure, direct bubble nucleation by the passage of seismic waves, and crystal settling or collapse of a crystal mush roof. Rectified diffusion is the increase in growth of existing bubbles caused by the passage of seismic waves. This mechanism produces small pressure variations $(10^{-2}-10^{-4}$ MPa) and is likely not significant enough to promote eruption. Advective overpressure is the increase in magmatic overpressure caused by the rise of bubbles from the bottom to the top of a magma chamber or conduit. Because magmas are compressible, volatiles diffuse in and out of bubbles, and magmatic systems deform in response to increases in pressure, the effects of advective overpressure are also not large. The passage of seismic waves can also trigger bubble nucleation by changing the pressure

regime in the magma chamber. In this case, the magma must be supersaturated and already near the critical pressure for bubble nucleation. Once bubbles are nucleated they may grow rapidly, increasing magmatic overpressure. Finally, dynamic strain and stress changes associated with passage of seismic waves may trigger the settling of crystals or crystal mushes formed on the roof of a magma chamber. The settling of the crystals can cause bubble nucleation and convection within a magma chamber resulting in an increase in magmatic overpressure. This process can lead to magmatic overpressures of high enough magnitude (>1 MPa assuming basaltic magmatic systems) to trigger a volcanic eruption.

3.3. Strain Changes on Magmatic Systems

Static and dynamic stress changes are inherently linked to co-seismic strain changes in the crust via Hooke's Law. These strain changes have also been invoked with regards to the triggering of volcanic unrest and eruptions (Figure 3.25). Specifically, compressional and dilatational stresses have been discussed with regards to eruption triggering, and again, the geometries of the source and receiver systems play a very important role in the triggering process. Coseismic strain changes that increase compression on a magmatic system (i.e., magma chamber) are thought to trigger magma migration and eruptive activity by "squeezing" the magma. When combined with unclamping of a feeder dike system, the two processes could lead to eruption. Compression is typically associated with divergent plate boundaries and normal faulting earthquakes. Although, compression of magmatic systems could occur in convergent margins, dependent on the location and orientation of the magmatic systems relative to the plate boundary (e.g., see the case of the 1707 eruption of Mt Fuji above). One issue with compression as an eruption trigger is the compressibility of magmas. Magma compressibility is a function of magma chemistry,

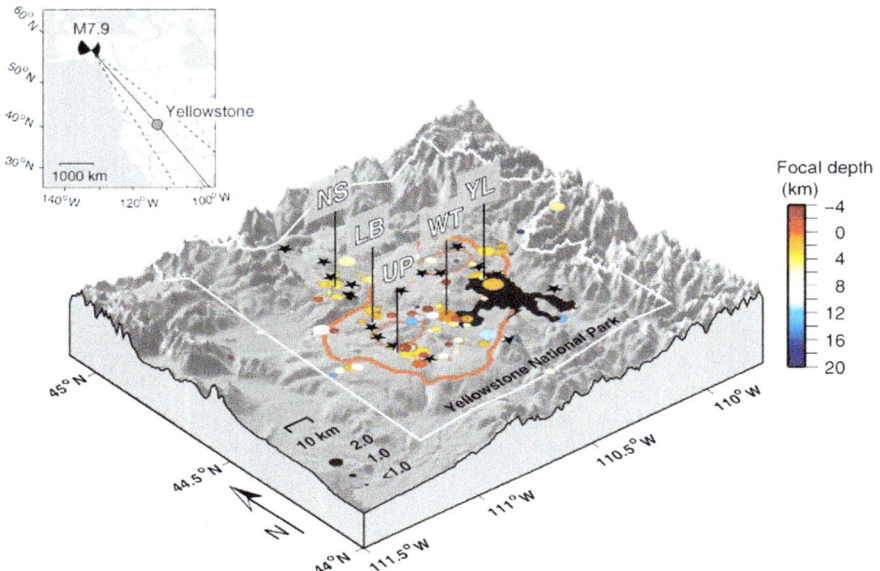

FIGURE 3.28 Dynamic triggering of seismicity (colored circles) in the Yellowstone caldera following the 2002 Denali earthquake. *From Husen, S., Taylor, R., Smith R.B., Healser, H., 2004. Changes in geyser eruption behavior and remotely triggered seismicity in Yellowstone National Park produced by the 2002 M 7.9 Denali fault earthquake, Alaska. Geology 32 (6), 537–540, http://dx.doi.org/10.1130/G20381.1.*

specifically the concentration and solubility of volatiles in the magmas, and therefore the depth of the magma chamber. Earthquake induced strain changes imparted on the magmatic system could be absorbed by compression of the magma.

Coseismic displacements associated with large magnitude earthquakes at convergent margins cause extension (i.e., positive dilatation) across volcanic arcs and could promote the opening of margin parallel magmatic systems (i.e., magmatic dikes and conduits). Reducing the pressure on a magma chamber could induce vesiculation, magmatic overpressure, and eruption (Figure 3.25). This mechanism has been discussed for the triggering of volcanic eruptions at convergent margins following large magnitude earthquakes. The large magnitudes of dilatation associated with large magnitude earthquakes may not lead to eruption, but could still trigger changes in the magmatic systems. Recent observations of the Holocene volcanic arcs in Chile and Japan following M > 8 earthquakes, indicate ∼15 cm of relative subsidence of the arcs and magmatic systems along the arcs. These large subsidence signals were hypothesized to be caused by fluid flow in the upper crust within the magmatic—hydrothermal systems and by deformation of magma bodies beneath the arc, respectively. In the case of Japan, the subsidence signal was also associated with instantaneous (i.e., dynamically triggered) changes in seismic velocity, which were attributed to increases in pressure within the shallow (<5 km) crust, again suggesting the affect of fluids in triggering changes in crustal parameters and magmatic—hydrothermal activity.

Although there are clear temporal and spatial correlations between earthquakes and volcanic eruptions, our understanding of these complex interactions is far from complete. Future studies must investigate both the mechanics of earthquake rupture and the mechanical, physicochemical, and petrologic processes that can lead to triggering of volcanic eruptions.

3.4. Magmatic Triggering of Tectonic Earthquakes

Magmatism and volcanism have been linked to seismicity at a range of temporal and spatial scales since Pliny the Younger's observations of the 79 AD eruption of Mt. Vesuvius. Magma chamber pressurization, magma migration, and eruption all induce seismicity (see Chapter 59—*Volcano Seismicity*). Our interest here is in the triggering of regional tectonic earthquakes by magmatic processes. The pressurization of magma chambers and the intrusion of magmas into the crust can induce stress changes on fault systems in the surrounding crust, promoting failure. The pressurization or change in volume of a magma chamber will induce changes in the radial (σ_{rr}) and tangential ($\sigma_{\theta\theta}$) components of the stress field and will enhance or hinder failure on vertical strike-slip faults (Figure 3.29). Where as a change in $\sigma_{\theta r}$ is important in the promotion of failure on dipping faults (i.e., normal or thrust faults). Following Coulomb failure criteria, shear stress changes will be greatest for vertical faults oriented 45° to the source and normal stress changes will be maximum for faults oriented transverse to the source, clamping the faults and hindering failure. Therefore, vertical faults oriented

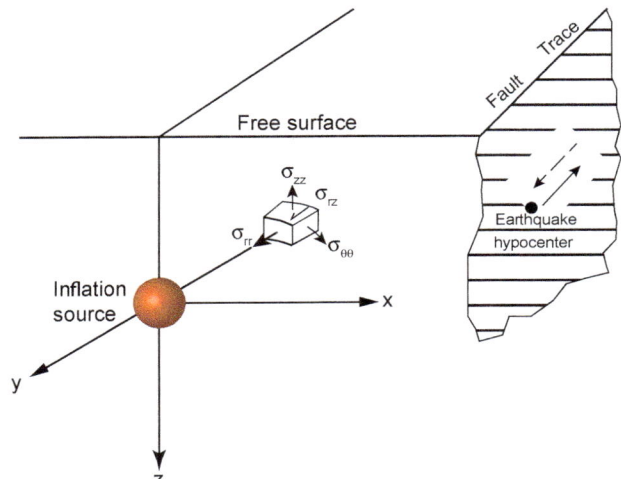

FIGURE 3.29 Stress field geometry for an inflating magmatic source and vertical strike-slip fault promoted by changes in the shear and normal stresses. *Modified from Thatcher, W., Savage, J.C., 1982. Triggering of large earthquakes by magma chamber inflation, Izu Peninsula, Japan. Geology 10, 637—640.*

radial to the source will be promoted by normal stress changes. For example, pressurization and inflation episodes of the Long Valley Caldera, CA in the 1970s and 1980s triggered several M6 earthquakes by increasing the shear stress on north-trending faults south of the caldera. Rifting events have also been correlated temporally and spatially with tectonic earthquakes (see Figure 3.13 and earlier discussion on stress changes associated with dike intrusion in Section 2.1.2). Following the initial dike intrusion event for the 1975—1984 Krafla, Iceland rifting episode, there was an M6.2 earthquake on the Tjornes fracture zone, a transform fault zone connecting the Northern Volcanic Zone (i.e., the on-land expression of the Mid-Atlantic Ridge in northern Iceland to which Krafla belongs) to the offshore Kolbeinsey Ridge. In some regions, it has also been shown that tectonic and magmatic systems are synchronized, promoting failure and triggering between each other. For example, intermediate magnitude earthquakes on normal faults in the Apennine Mountains and inflation and eruption of Mt. Vesuvius can promote failure in each other. And there are observed interactions between rift zone activity on Mauna Loa and earthquakes on basal decóllement faults beneath the volcano. In total, these observations indicate interesting interactions between magmatic and tectonic systems that are both acting to accommodate relative pate motions.

4. SUMMARY

Earth is a dynamic planet driven by internal heat loss and resulting plate tectonic and magmatic processes. Plate tectonics is the paradigm in Earth system science that best explains the processes acting within and on the surface of the Earth that result from this internal heat and planetary cooling. Magmatism and volcanism are processes resulting from heat generation in the core and mantle, and plate tectonics. Relative lithospheric plate motions are accommodated by tectonic (i.e., elastic and plastic deformation of the lithosphere) and magmatic (i.e., intrusive and extrusive) processes. Magma—tectonic interactions occur across vast temporal and spatial scales. At geologic timescales, magma—tectonic interactions are the formation of oceanic and continental crust and lithosphere, and the accommodation of plate motions. Repeated lateral and vertical dike injections along divergent plate boundaries accommodate the roughly 1 to <12 cm a^{-1} of relative motion and form new oceanic crust and lithosphere. The intrusion of large volume magmatic bodies (i.e., batholiths and plutons) and formation of volcanic arcs at convergent margins, aid in the growth of continental lithosphere, albeit at a small percentage of total continental lithospheric volume. Volcanism occurs along transform plate boundaries where transtension takes place.

On human timescales, there is a growing set of observations and studies indicating that earthquakes trigger explosive volcanic eruptions and magmatic processes trigger tectonic seismicity. Large magnitude subduction zone earthquakes induce arc-wide static stress and strain changes that can relieve stresses on magma chambers and allow for the opening of magmatic conduits. Dynamic stress and strain changes associated with the passage of seismic waves can also trigger vesiculation processes in magma chambers that increase the magmatic overpressure, therefore triggering eruption. The development and expansion of multiparameter volcano monitoring networks, combined with geochemical and petrologic analyses of eruptive products, will lead to new understanding of the complex physicochemical processes that lead to explosive volcanic eruptions.

FURTHER READING

Acocella, V., 2014. Structural control on magmatism along divergent and convergent plate boundaries: overview, model, problems. Earth Sci. Rev. 136, 226—288. http://dx.doi.org/10.1016/j.earscirev.2014.05.006.

Clift, P., Vannucchi, P., 2004. Controls on tectonic accretion versus erosion in subduction zones: implications for the origin and recycling of the continental crust. Rev. Geophys. 42, RG2001. http://dx.doi.org/10.1029/2003RG000127.

Darwin, Charles, 1839. Journal of Researches into the Natural History and Geology of the Countries Visited during the Voyage of H.M.S. Beagle Round the World, Under the Command of Capt. Fitz Roy, R.N., from 1832—1836. Henry Colburn, London.

Dick, H.J., Lin, J., Schouten, H., 2003. An ultraslow-spreading class of ocean ridge. Nature 426, 405—412.

Eggert, S., Walter, T., 2009. Volcanic activity before and after large tectonic earthquakes: observations and statistical significance. Tectonophysics 471, 14–26. http://dx.doi.org/10.1016/j.tecto.2008.10.003.

Linde, A.T., Sacks, I.S., 1998. Triggering of volcanic eruptions. Nature 395, 888–890.

Manga, M., Brodsky, E., 2006. Seismic triggering of eruptions in the far field: volcanoes and geysers. Annu. Rev. Earth Planet. Sci. 34, 263–291. http://dx.doi.org/10.1146/annurev.earth.34.031405.125125.

Michael, P.J., Langmuir, C.H., Dick, H.J.B., Snow, J.E., Goldsteink, S.L., Graham, D.W., Lehnertk, K., Kurras, G., Jokatq, W., Mühe, R.,

Edmonds, H.N., 2003. Magmatic and amagmatic seafloor generation at the ultraslow-spreading Gakkel ridge, Arctic Ocean. Nature 423, 956–961.

Rubin, A.M., Pollard, D.D., 1988. Dike-induced faulting in rift zones of Iceland and Afar. Geology 16, 413–417.

Sinton, J.M., Detrick, R.S., 1992. Mid-Ocean Ridge magma chambers. J. Geophys. Res. 97 (B1), 197–216.

White, S.M., Crisp, J.A., Spera, F.J., 2006. Long-term volumetric eruption rates and magma budgets. Geochem. Geophys. Geosyst. 7, Q03010. http://dx.doi.org/10.1029/2005GC001002.

Chapter 4

The Composition and Origin of Magmas

Nick Rogers

Department of Environment, Earth and Ecosystems, The Open University, Milton Keynes, UK

Chapter Outline

GLOSSARY

accessory mineral Accessory minerals are not an essential component of a rock and are often rich in a trace element(s). For example, sphene, a calcium titanate, is found in many granites and diorites but is not ubiquitous and is not an essential component of all granites and diorites. Other common accessory minerals include zircon (Zr), monazite (Th), rutile (Ti), allanite (REE), apatite (P), and garnet. They are usually present in small quantities ($<1\%$).

adiabatic The thermodynamic process of expanding or compressing a substance without allowing it to exchange heat with its surroundings.

basalt A volcanic rock dominated by a mineralogy of clinopyroxene, plagioclase, olivine, and iron ore. Basaltic magma is the major product of mantle melting.

compatible element A compatible element has a high partition co-efficient (D-value) between a mineral and a coexisting melt.

continental flood basalt (CFB) Continental flood basalt provinces are large volumes of dominantly tholeiitic basalt, in the form of sheet flows with few intervening sedimentary or other beds. They are thought to have been erupted in geologically short periods of time at high extrusion rates.

fractionation index (index of fractionation) A compositional parameter, such as MgO, SiO_2 or a trace element, that varies in a systematic (and predictable) way as a magma evolves. Plotted as the abscissa (x-axis) of a variation diagram.

geotherm The geotherm is defined by the variation of temperature with depth within the earth.

granite A plutonic (intrusive) igneous rock dominated by alkali feldspar quartz and mica. Granite magma is the major product of crustal melting.

HFSE High field strength elements are those with a high ionic charge ($+3$, $+4$, $+5$) and a small ionic radius. Examples include Y and the heavy REE, Yb and Lu ($+3$); Zr, Hf, Ti, Th ($+4$); Nb, Ta ($+5$).

incompatible element An incompatible element has a low partition coefficient (D-value) between a mineral and a coexisting melt.

LILE Large ion lithophile elements are those trace elements with low ionic charge ($+1$, $+2$) and relatively large ionic radii. Examples include the alkali elements K, Rb, Cs ($+1$), and the alkaline earth elements, Sr, Ba ($+2$).

lithosphere The outer, rigid part of the Earth in which heat is transported by conduction. Comprises the crust and the uppermost part of the mantle.

mafic Mafic rocks have high proportions of minerals rich in MgO, FeO, and CaO, that is, olivine, pyroxene, amphibole and mica. Basalt is a mafic rock. Mafic is the opposite of felsic.

magma Molten rock that comprises three phases—solid, liquid, and vapor—in the form of crystals and bubbles suspended in a medium of silicate melt. Surface volcanic rocks and intrusive igneous rocks are derived from solidified magma that has either erupted on to the Earth's surface or cooled more slowly at depth.

major element An element that comprises more than 1 wt.% of the composition of an igneous rock. Usually expressed as weight% oxide (e.g., SiO_2, 52.13 wt.%).

mantle plume A body of rock that rises as a consequence of its thermal buoyancy through the mantle. Mantle plumes can produce basaltic magma if they reach shallow depths in the mantle.

metaluminous Metaluminous rocks are poor in aluminum (Al_2O_3) relative to the total of calcium, sodium, and potassium ($Al_2O_3 < CaO + Na_2O + K_2O$ expressed as molecular %). They are also rich in feldspars and aluminous mafic minerals such as biotite, hornblende, and melilite.

The Encyclopedia of Volcanoes. http://dx.doi.org/10.1016/B978-0-12-385938-9.00004-3

minor element An element that comprises between 0.1 wt.% and 1.00 wt.% of a rock composition. Usually expressed as weight% oxide (e.g., TiO_2, 0.78 wt.%).

mid-ocean ridge basalt (MORB) Mid-ocean ridge basalts are produced at a constructive plate margin and are tholeiitic.

ocean island basalt (OIB) Ocean island basalts are generated away from plate boundaries and are commonly alkaline but also tholeiitic. Generally produced from a mantle plume.

partition coefficient (D-value) This is the ratio of the concentration of a trace element in a mineral to its concentration in a coexisting liquid or melt.

peralkaline Peralkaline rocks are depleted in aluminum (Al_2O_3) relative to the alkali elements Na_2O and K_2O ($Al_2O_3 < Na_2O + K_2O$, expressed as molecular %). Peralkaline rocks are rich in feldspars and feldspathoids (if SiO_2 is also low) and contain Na-rich pyroxenes and amphiboles such as aegerine, riebeckite, and richterite.

peraluminous Peraluminous rocks are rich in aluminum (Al_2O_3) relative to the total of calcium, sodium, and potassium ($Al_2O_3 > CaO + Na_2O + K_2O$ expressed as molecular %). They are rich in feldspars and the excess aluminum is accommodated by minerals such as muscovite, topaz, tourmaline, garnet, and corundum.

polybaric Describes processes that occur over a range of pressures (literally "many pressures"). e.g., polybaric melting implies melting that takes place over a range of pressures (depths).

primary magma A magma produced by melting of its source region that has not been modified by subsequent igneous processes.

REE The rare earth elements are the 15 4f series elements between La (atomic number 57) and Lu (atomic number 71). They generally have +3 ionic charge and their ionic radii decrease systematically from La–Lu.

solidus The solidus is the melting temperature of a rock and it varies with pressure and volatile content.

trace element An element that comprises less than 0.1 wt% of a rock composition, usually expressed as a part per million (ppm) or as $\mu g\ g^{-1}$ (e.g., Sr 353 ppm).

variation (Harker) diagram A binary plot of a compositional parameter (major, minor or trace element) against a fractionation index.

1. INTRODUCTION

The generation of melts and the movement and crystallization of those melts is the primary mechanism whereby planet Earth has evolved into a core, mantle, and crust. At the present time melting is limited to the uppermost 200 km of the Earth, within the crust and upper mantle, but early in Earth history it is now considered likely that most of the Earth was molten and that the Earth is still cooling from that initial hot state. Such primordial heat and the heat derived from long-lived radioactive isotopes keep the interior of the Earth in a dynamic state, the surface expression of which is plate tectonics and volcanic activity.

Because melts are generally less dense than the rocks from which they are derived, they tend to rise toward the surface and while some may crystallize at depth, many reach the surface at active volcanoes. Molten rocks are rarely erupted at the surface as pure liquids. Rather they are complex mixtures of silicate liquid and crystals that may have crystallized from the liquid or have been picked up by the melts as they move toward the Earth's surface. **Magma** is the term used to describe these mobile and largely molten mixtures of liquid and solid material that also include an inventory of dissolved gases, such as water, carbon dioxide, and other more exotic compounds. When it approaches the surface, magma generates volcanic activity during which mobile silicate lava may erupt and solidify to form extrusive or volcanic igneous rock; alternatively it may stall at depth and crystallize slowly to form intrusive igneous rocks that are eventually revealed by erosion long after the surface volcano has become extinct. Thus it is tempting to equate the compositions of magmas with igneous rocks, but because the processes of crystallization, emplacement, and eruption allow volatiles to be released to the surface and liquids to be expelled from compacting crystal mushes, igneous rocks typically have compositions that are distinct from the magmas from which they crystallized.

The most extensive volcanic activity on Earth is located along constructive plate margins, submerged beneath the oceans and only rarely observed. Here, between 10 and 20 km^3 of basaltic magma is added to the oceanic crust each year. Much of this solidifies beneath the surface in the form of dykes and larger intrusions of gabbro, but a significant amount is erupted as lava flows on the ocean floor. Other more obvious manifestations of volcanic activity occur at destructive plate margins where arcs of active volcanoes give rise to the most explosive and destructive eruptions. Yet other volcanoes, such as those of Hawaii, are located remotely from plate margins and their origins appear to be independent of processes related to plate tectonics. Thus volcanoes are located above zones of melting in the deep Earth, but their location in different tectonic regimes implies a variety of mechanisms whereby melt can be produced and leads to a range of compositions.

2. ROCK NAMES AND COMPOSITIONS

Igneous rocks are increasingly classified according to their bulk composition, conventionally expressed in terms of major, minor, and **trace elements**. Major and **minor elements** are expressed as oxides: SiO_2, Al_2O_3, Fe_2O_3, FeO, CaO, MgO, and Na_2O (**major elements**) and TiO_2, MnO, K_2O, and P_2O_5 (minor elements). Major elements are by definition those with oxide abundances above 1 wt.%, minor elements those between 0.1 and 1 wt.%. Some elements, such as potassium and titanium are present in minor element quantities in some rocks but can attain major element proportions in others. Elements with concentrations below 0.1 wt.% are defined as trace elements, the concentrations of which are conventionally expressed in terms of ppm (parts per million or $\mu g\ g^{-1}$) of the element alone. Thus conventional analyses of igneous rocks appear as in Table 4.1,

TABLE 4.1 Major and Trace Element Compositions of Representative Igneous Rocks

	Peridotite	Tholeiitic Basalt MORB	Alkali Basalt OIB	Andesite	Metaluminous Granite	Peraluminous Granite	Peralkaline Granite
Major and Minor Elements (wt.%)							
SiO_2	44.2	48.77	47.52	59.89	67.89	69.08	70.87
TiO_2	0.13	1.15	3.29	0.95	0.45	0.55	0.1
Al_2O_3	2.05	15.9	15.95	17.07	14.49	14.3	14.78
Fe_2O_3	0.75	1.33	3.16	3.31	1.27	0.73	2.64
FeO	7.54	8.62	8.91	3	2.57	3.23	
MnO	0.13	0.17	0.19	0.12	0.08	0.06	0.06
MgO	42.21	9.67	5.18	3.25	1.75	1.82	0.1
CaO	1.92	11.16	8.96	5.67	3.78	2.49	0.34
Na_2O	0.27	2.43	3.56	3.95	2.95	2.2	6.47
K_2O	0.06	0.08	1.29	2.47	3.05	3.63	4.19
P_2O_5	0.03	0.09	0.64	0.31	0.11	0.13	0.02
H_2O^+		0.3	1.16				0.33
F (ppm)		210	1150				
Total	99.29	99.67	99.81	99.99	98.39	98.22	99.9
Trace Elements (ppm)							
V	70	262	350	125	74	72	9
Cr	3200	528	421	484	27	46	5
Ni	2300	214	153	38.6	9	17	3
Rb	0.64	0.56	31	75.4	132	180	148
Sr	21	90	660	886	253	139	1.8
Y	5	28	29	12	27	32	188
Zr	11	74	280	240	143	170	772
Nb	0.7	2.3	48	15	9	11	100
Ba	7	6.3	350	886	520	480	
La	0.7	2.5	37	38	29	31	59.6
Ce	1.8	7.5	80	67	62.9	69	152
Nd	1.4	7.3	39	32	23.4	25	93.7
Sm	0.44	2.63	10	5.82	4.25	4.89	25.4
Eu	0.17	1.02	3	1.57	2.23	1.74	1.99
Gd	0.6	3.68	7.62	4.73	4.16	4.35	27.8
Tb	0.11	0.67	1.05	0.66	0.59	0.64	4.8
Yb	0.49	3.05	2.16	1.64	1.81	2.25	16.3
Lu	0.07	0.46	0.3	0.25	0.27	0.34	2.46
Ta	0.04	0.13	2.7	0.88	0.42	0.56	6.68
Hf	0.31	2.05	7.8	5	4.1	5.5	26.3
Th	0.08	0.12	4	6.5	16	19	17.8
U	0.02	0.05	1.02	1.89	3	3	4.59

MORB, mid-ocean ridge basalt; OIB, ocean island basalt.

which lists a variety of igneous rocks from different locations. Volatile elements are less frequently reported but can be added to the list as H_2O, CO_2, SO_2, F, Cl, etc.

Two points are clear from Table 4.1. First, igneous rock compositions vary considerably, from those with low SiO_2, for example, to others dominated by SiO_2 almost to the exclusion of other elements. The second point is that it is very difficult to assimilate such information from a list of numbers alone and igneous petrologists and geochemists have devised a number of graphical methods for displaying and summarizing such data that render them easier to interpret. Indeed, the so-called **variation diagram** (sometimes referred to as a Harker diagram after one of the first petrologists to propose its use) is now a standard tool of modern igneous petrology. One in particular, a plot of total alkalis ($Na_2O + K_2O$) against SiO_2, is used to define common rock names (Figure 4.1).

Rock names also depend on mineralogy and grain size and, to a lesser extent, on the presence or absence of specific minerals, resulting in an even greater variety of names than those given in Figure 4.1. In Figure 4.2 the mineralogical variations of some of the more common rock types are illustrated graphically as a function of bulk composition. Thus **basalt** is the name given to a rock with olivine plus abundant pyroxene and plagioclase that has relatively high Mg, Fe, and Ca, but lower Si, Na, and K. If basaltic magma cools more slowly beneath the surface then a coarse grained rock called gabbro is produced. Similarly rhyolite is a volcanic rock with high Si, Na, and K, low Mg, Ca, and Fe, and is dominated by quartz and alkali feldspars with small amounts of mica and amphibole. Should such a rhyolitic magma cool

slowly it would produce its coarse-grained equivalent, **granite**.

Despite the plethora of specific names igneous rocks are dominated by two major compositional types: basalts and granites. Rocks of basaltic composition dominate the crust beneath the world's oceans whereas granitic rocks are the most abundant igneous rock type of the continental crust. Indeed it is the lower density of granites, resulting from their low iron and high silica composition, which causes the continents to have on average a 5 km higher elevation than the dense, iron-rich, and silica-poor basaltic oceanic crust.

3. BASALTIC MAGMAS

3.1. Mid-ocean Ridge Basalts

Basalts dominate the floor of the oceans, forming the igneous oceanic crust which is generated at constructive plate margins. Magma is produced in response to the separation of two plates, allowing the mantle to rise up to occupy the space made available by the divergent plate motion. In so doing, the mantle melts as a result of **adiabatic** decompression.

Adiabatic decompression melting is one of the major mechanisms of melt generation within the Earth and occurs when hot mantle (peridotite) rises from depth (high pressure) to shallower levels (lower pressures). As the mantle rises adiabatically it cools at a rate of 0.5 °C for every km of upward movement. Given that the gradient of the peridotite **solidus** is greater than the adiabatic gradient, a parcel of mantle originally below the solidus at high pressure can rise through the mantle into a region where the ambient

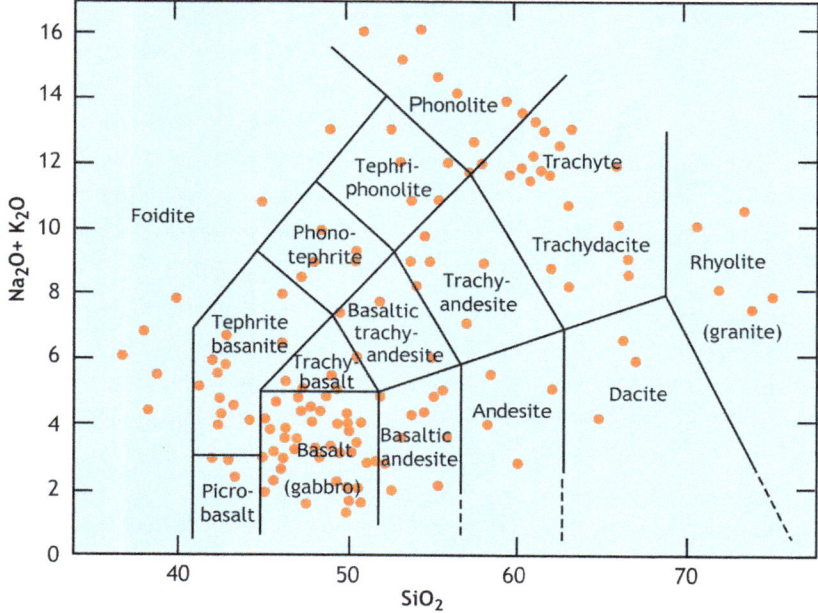

FIGURE 4.1   Total alkalis versus silica diagram with ~100 volcanic rock analyses illustrating the enormous variation in rock compositions. Superimposed are the boundaries that delineate different rock types according to the International Union of Geological Sciences (IUGS) (Le Bas and Streckheisen, 1991) with the names of volcanic and selected plutonic rocks types superimposed.

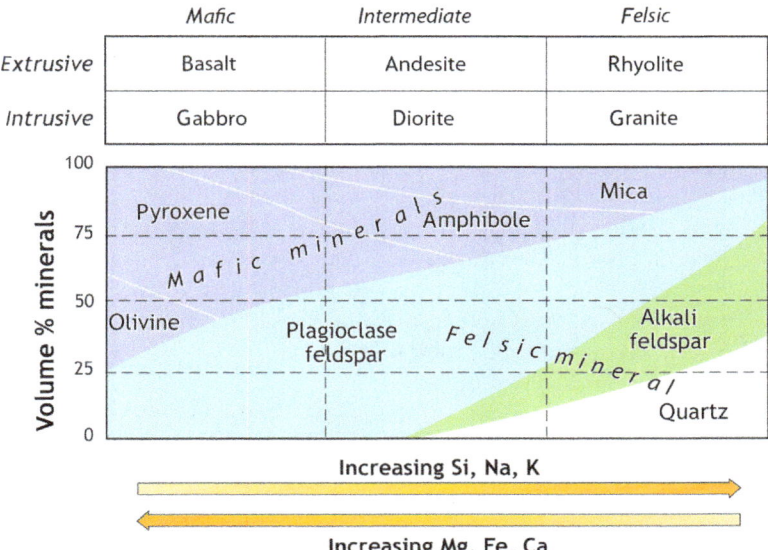

FIGURE 4.2   A chart summarizing the mineralogy and compositional characteristics of the most common igneous rock types.

conditions are above the solidus, and it consequently melts. This is illustrated in Figure 4.3 (See also Chapter 3).

The mantle is made predominantly of peridotite, a rock type which is in turn made up of olivine, *ortho-* and clinopyroxene and an aluminous phase, either spinel at low pressures or garnet at high pressures. As a result, the mantle melts incongruently, some phases, clinopyroxene and spinel or garnet contributing to the melt more than orthopyroxene and olivine. Thus melts of peridotite are richer in the components of clinopyroxene, spinel, and garnet than their peridotite source and so are richer in CaO and Al_2O_3 than peridotite and also have higher SiO_2 and lower MgO contents—these melts are known as **basalts** (Figure 4.1). Furthermore the rate at which different phases are consumed or produced during melting change with pressure, leading to changes in the melt composition. So the composition of a basalt can give useful information about the depth and temperature at which it was produced, and hence on conditions in the interior of the Earth.

Melting beneath ocean ridges produces a basalt that contains ~50% SiO_2 but is poor in Na_2O and K_2O (Figure 4.4) known as a tholeiitic basalt (after the type locality, Tholey, in Germany). Once emplaced within the crust, the basalt cools and starts to crystallize. However, melts differ in composition from their source region, so the material that first crystallizes from a basalt also has a different composition from the basaltic liquid—it is the MgO-rich mineral olivine. Olivine is denser than basaltic melt and the two are easily separated by crystal settling due to gravity. This separation leaves the basalt liquid poorer in MgO and slightly enriched in CaO, Al_2O_3, Na_2O, and K_2O, and eventually a second mineral crystallizes that includes these elements. That mineral is plagioclase feldspar and it too separates from the melt, depleting the latter in CaO and

Al_2O_3. If the minerals once formed are separated rapidly then this process is described as *fractional crystallization* and is illustrated in Figure 4.5. It is one of the most important processes in modifying the composition of magmas and is the cause of much of the diversity in the compositions of igneous rocks.

The effects of fractional crystallization on ocean floor basalts (mid-ocean ridge basalts or MORB, as they are frequently known) are illustrated in Figure 4.6, using the MgO content as an index of fractionation (see Box 4.1). The diagram shows

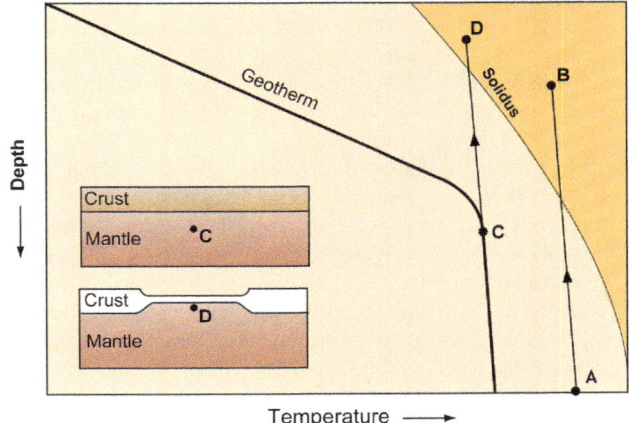

FIGURE 4.3   The principles of decompression melting. A parcel of mantle lies on the geotherm at point C. If the overlying crust is stretched and thinned, then the parcel of mantle rises along an adiabatic thermal gradient to point D, at a lower pressure but with a minimal drop in temperature (0.3 °C km^{-1}). Point D is above the solidus at that pressure and so the mantle melts. The vector A-B shows the trajectory of a similar parcel of mantle at a higher initial temperature, for example as in a rising mantle plume. Decompression melting of this parcel starts at a greater depth and has the potential to produce a greater volume of melt.

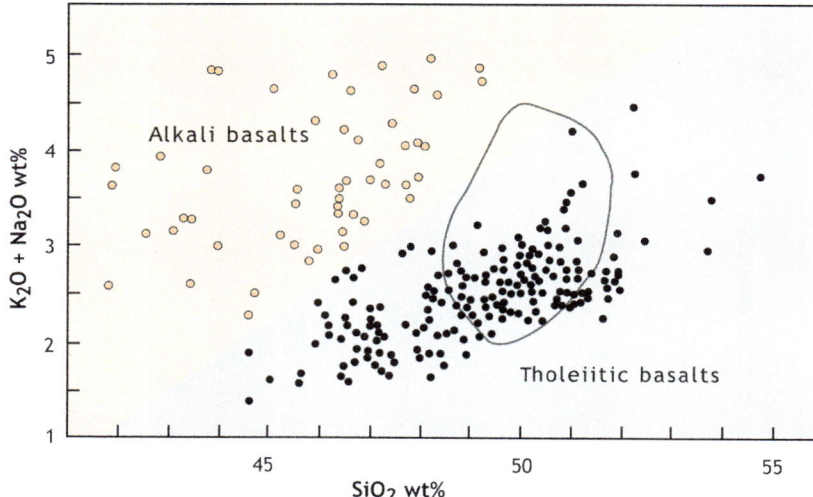

FIGURE 4.4 A diagram showing the variation in total alkalis and silica in Hawaiian basalts and the division between alkali and tholeiitic types (after Macdonald and Katsura, 1964). The outlined field represents the region in which most MORB plot.

BOX 4.1 Fractionation Indices, Primitive and Parental Magmas

The representation of the compositions of associated lavas from distinct volcanoes or volcanic provinces allows the assessment of the role of fractional crystallization in developing those compositions. Figures 4.1 and 4.2 use SiO_2 as an **index of fractionation** with basaltic compositions with low SiO_2 representing magmas that are parental by fractional crystallization to more evolved lavas with higher SiO_2. However, SiO_2 is not always the best index of fractionation. Consider, for example a tholeiitic basalt with 50 wt.% SiO_2 crystallizing a mixture of olivine (42 wt.% SiO_2) and plagioclase (55 wt.% SiO_2). The mean concentration of SiO_2 in this mixture will be close to 50 wt.% and so while separation of this mixture will lead to changes in some elements, such as CaO and K_2O, SiO_2 will remain virtually unchanged. An alternative **fractionation index** for basaltic compositions is MgO. This is a major constituent of olivine, the first mineral to crystallize from basaltic magma, and has a high MgO content (30–45 wt.%) compared with basalt (10–15 wt.%). Separation of olivine or an olivine-rich mixture of minerals therefore reduces the MgO content of the basalt fairly rapidly. An example of the use of MgO as a fractionation index is shown in Figure 4.6 where the Na_2O contents of three groups of **mid-ocean ridge basalt** (MORB) lavas from three different locations are plotted against MgO.

The three groups in Figure 4.6 plot along distinct trends implying that they were derived from distinct high-MgO parent magmas. Because MgO decreases with fractionation, basaltic rocks with the highest MgO contents are considered representative of the parental magma to an evolving suite of lavas; thus the Kolbeinsey suite has a parent with lower Na_2O than the EPR suite. But what was the composition of the magma that was generated within the mantle but before it crystallized any olivine—the so-called **primary magma**?

Magmas generated by melting peridotite are all produced with a residue that is rich in olivine. It is known from analyses of mantle samples that the olivine in the mantle has a limited compositional range represented by the Mg#. This value is the molar proportion of MgO relative to FeO in a mineral or melt:

$$Mg\# = 100 \times \left[Mg^{2+}\right] / \left(\left[Mg^{2+} + Fe^{2+}\right]\right)$$

$$= 100 \times (wt.\% \ MgO/40.311) / [(wt.\% \ MgO/40.311) + (wt.\% \ FeO/71.846)]$$

where Fe^{2+} and Mg^{2+} are the molar concentrations of Fe and Mg in the magma or olivine (or any other mineral containing MgO and FeO). Olivine in mantle peridotite usually falls in the limited range of 87 < Mg# < 91.

Exchange reactions control the distribution of FeO and MgO between olivine and melt and can be represented by an equilibrium constant defined as:

$$K_D = \left[Fe^{2+}/Mg^{2+}\right]_{Ol} / \left[Fe^{2+}/Mg^{2+}\right]_{Liq}$$

Remarkably, over a wide range of compositions the exchange coefficient, K_D, has a constant value of 0.3 and so it is possible to use the limited range of mantle olivine Mg# to determine the Mg# of a primary melt, which according to the above equations lies in the range 70 < Mg# < 75.

Using this criterion, a plot of MgO versus Mg# can be used to define the MgO content of the primitive magma which for MORB varies from 10 wt.% to 15 wt.% (McKenzie and Bickle, 1988).

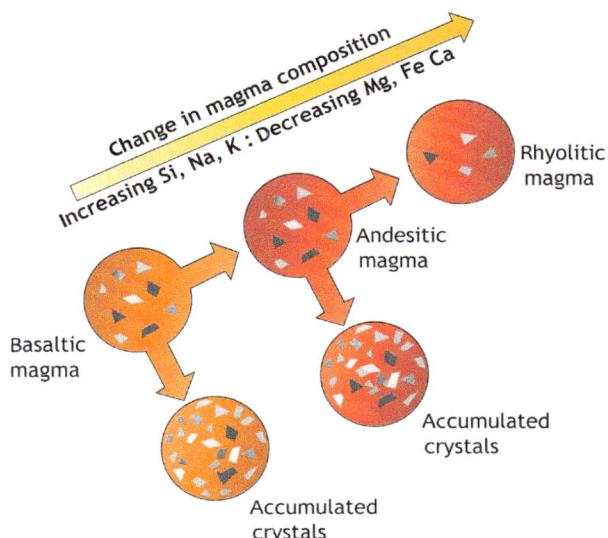

FIGURE 4.5 Cartoon illustrating the evolution of a magma by fractional crystallization as a result of cooling and the removal of crystals from the gradually solidifying melt. This process may be continuous or occur in a number of stages.

how Na_2O increases as fractionation proceeds, consistent with the removal of Mg-rich and Na-poor phases such as olivine. All of the basalts in Figure 4.6 are tholeiitic and the trends labeled Australia-Antarctic discordance (AAD) shows an upward inflection at about 8 wt.% MgO marking the start of plagioclase crystallization in addition to olivine.

A second important feature of Figure 4.6 is that the basalts form three separate trends and do not converge on a single Na_2O content at a high MgO abundance. This discrepancy strongly suggests that each area has a distinct primitive magma composition which in turn relates to

differences in the pressure and temperature or the composition of the underlying mantle source or a combination of the two. These regional differences in MORB composition have been the subject of detailed analysis over the past 25 years and a broad consensus has emerged that they can be related to the extent and degree of melting in the MORB source which in turn relates to the prevailing temperature gradient in the upper mantle (McKenzie and Bickle, 1988).

Figure 4.3 shows how two parcels of mantle lying on different **geotherms** intersect the mantle liquidus at different points. The hotter mantle, A intersects the liquidus at a higher pressure (greater depth) than cooler mantle represented by C. As both mantle parcels continue to rise toward the surface they produce more melt, however, the cooler mantle has less potential to produce melt than the hotter mantle. Therefore, oceanic crust overlying cool mantle will be thinner than oceanic crust overlying hot mantle (see Chapter 1).

An added complication is that once melt forms it moves along grain boundaries toward the surface much more rapidly than the mantle rises by convection. Thus melts produced from all parts of the melting column accumulate in magma chambers beneath the spreading center. Because the rate at which different minerals are consumed varies with pressure and temperature, the composition of the melts produced under different conditions varies and the aggregate melts that accumulate at the surface also differ from any particular primary melt produced directly from the mantle. While such aggregate fractional melting processes can be modeled, the details of the calculations are complex and beyond the scope of this chapter. However, some broad conclusions can be drawn from experimental studies that reveal that melts derived from greater depths tend to have lower SiO_2, and higher FeO; those formed at higher

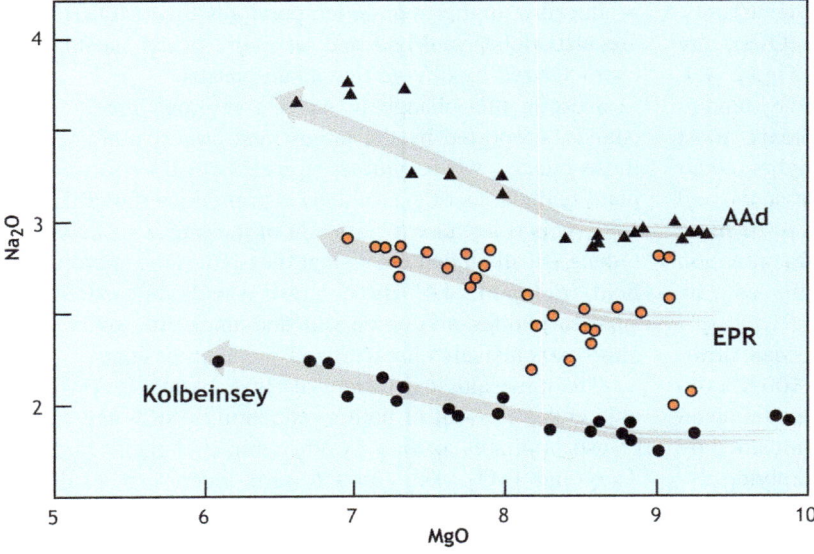

FIGURE 4.6 The variation of sodium with magnesium (both expressed as wt.% oxides) in basalts from three segments of the mid-ocean ridge system. The Kolbeinsey Ridge is the extension of the mid-Atlantic ridge, north of Iceland, the EPR is the East Pacific Rise, and the AAD lavas are from the Australia—Antarctic discordance, a segment of ridge south of Australia in the Southern Ocean. Each region shows an increase in sodium as magnesium decreases, but the trends start at different sodium contents at high magnesium abundances. *After Klein and Langmuir (1987).*

temperatures will have greater MgO; and that the Na_2O content of a melt decreases as the amount of melt increases. Thus magmas with high Na_2O and SiO_2 and lower MgO and FeO produce thin oceanic crust overlying cool mantle and the converse (lower Na_2O and SiO_2, higher MgO and FeO) produce thick crust overlying hotter mantle. This prediction is borne out by observation that basalts from the AAD are from one of the deepest parts of the ocean ridge system and are underlain by cool mantle. The crust is thin and the basalts, as can be seen from Figure 4.6, have high Na_2O. Conversely the Kolbeinsey lavas are from the mid-Atlantic ridge, north of Iceland where the ocean crust is thick and much shallower and they have much lower Na_2O contents. The East Pacific Rise (EPR) samples represent basalt compositions from a more typical intermediate depth ridge and they have compositions intermediate between the other two.

In summary the composition and origins of basaltic magma generated beneath MORB are the result of **polybaric**, aggregate, fractional, decompression melting of a mantle peridotite source region that produces a range of parental magmas. When these magmas cool they crystallize first olivine and subsequently plagioclase to produce discrete series of MORB with Na_2O and FeO contents that reflect the temperature of the mantle beneath the ridge. Indeed variations in MORB composition have been used to infer temperature variation of up to 150 °C between regions of thick shallow crust (e.g., Kolbeinsey $\sim$1480 °C) and thin deep crust (AAD $\sim$1320 °C) (Chapter 1).

3.2. Ocean Island Basalts

A second important location where basalts are found is ocean islands, such as Hawaii, Tahiti, Galapagos, Reunion, the Azores, and the Canary Islands, to name but a few. These active volcanoes are remote from plate boundaries, located in an intraplate or within-plate tectonic environment. The compositions of basalts in these islands, commonly known as **ocean island basalt** (OIB), are distinct from MORB. As you can see from Figure 4.4, which shows analyses of basalts from Hawaii, they tend to have a wider range of SiO_2 and alkali contents, most significantly they vary to lower SiO_2 and higher alkali contents than MORB, suggesting higher pressures and lower degrees of melting. In addition they tend to be more magnesium-rich than MORB and their fractionation sequence is dominated by olivine followed by clinopyroxene.

Many ocean islands are located on thick ocean **lithosphere**. Hawaii for example is located on 80–100 Ma old lithosphere that is about 80 km thick. They are also situated on uplifted areas of ocean crust which are accompanied by gravity anomalies on the same length scale, implying dynamic uplift supported from sublithospheric depths. The presence of basaltic volcanism at the surface further

implies the presence of hot mantle and these so-called hot spots appear stationary with respect to plate movements. In the Hawaiian chain the currently active volcanoes on Big Island will within a few million years move to the NW and become inactive, their place being taken by volcanic activity forming the Loihi Seamount that lies to the SE of the currently active island. The Hawaiian chain of islands and seamounts stretches all the way to Midway Island and represents a more or less continuous record of 50 Ma of hot spot basaltic volcanism.

The combination of these observations has led to the development of the **mantle plume** hypothesis in which areas of intraplate magmatism, such as Hawaii, are thought to be underlain by a jet of mantle that is hotter than the ambient upper mantle through which it is rising. Mantle plumes can occur beneath continents as well as the ocean basins and are an integral part of the convective overturn of the Earth's mantle. They have their origins at great depth, possibly at the core—mantle boundary and if the plume is hot enough and the overlying lithosphere thin enough, then as the plume approaches the base of the lithosphere it can undergo decompression melting (Watson and McKenzie, 1991).

Two aspects of this melting regime are, however, different from ocean ridges. First, while the high pressure initiation of melting is determined by the plume temperature, melting is cut off when the plume impinges on the base of the lithosphere. Thus the melt regime does not extend to shallow depths and the mean pressure of melting is greater than for MORB. Second, the pressure at which melting begins can be in the region of P-T space where the dominant Al-bearing phase in peridotite is garnet. The presence of garnet has a profound effect on the composition of the melts produced at the liquidus. These two effects tend to generate melts that are richer in alkalis and iron than MORB, but poorer in SiO_2. Moreover because OIB are produced at higher mantle temperatures than MORB they are also richer in MgO and so many ocean islands are characterized by olivine rich alkali basalts.

Despite this change in melting regime, larger ocean islands, supported by the largest and hottest plumes, and those created where plumes upwell beneath constructive plate boundaries (e.g., Iceland) also produce tholeiitic basalts. This is because the amount of melting is sufficient to reduce the alkali content so that they fall in the appropriate field in Figure 4.4. By contrast islands located above smaller plumes and/or on old and thick lithosphere are almost exclusively characterized by alkali basalts.

The mineralogical variety of alkali basalts is considerable and is a result of both a reduction in SiO_2 and an increase in alkalis as well as other minor elements such as TiO_2 and P_2O_5. The trend toward increasing alkalinity produces magmas with compositions outside the field of alkali basalts which are generally called basanites and foidites (Figure 4.1). These tend to be the products of even smaller degrees of melting than are responsible for alkali basalts and in extreme cases reflect the effect of CO_2 on mantle melting at great depths where carbonate can be a stable subsolidus phase. Rock types produced in this way included nephelinites, leucitites, and melilitites and are found on smaller ocean islands or during the final phases of volcanic activity on larger islands, such as Hawaii.

As a result of the contrasting phase relations of silica saturated and undersaturated basalts (Box 4.2), the evolutionary paths followed by tholeiitic and alkali basalts are quite distinct. Tholeiitic basalts crystallize olivine first followed by plagioclase, resulting in iron enrichment with minimal increase in SiO_2. Eventually the iron content builds up to such an extent that magnetite appears on the liquidus causing a dramatic fall in FeO in the magma and a rapid evolution to silica-rich compositions. This evolutionary pattern was first recognized in Iceland in lava compositions associated with large central volcanoes. By contrast, alkali basalts crystallize olivine followed by clinopyroxene. This largely prevents the increase in iron and silica while pushing the liquid composition to become more alkali-rich. Thus evolved lavas on ocean islands tend to vary with the nature of the parental basalt and also with the magma flux into the island from the mantle. Large, active islands such as Hawaii and Reunion are dominated by basalts and basic intermediate rocks often described as

Hawaiites. They are olivine-poor and rich in clinopyroxene and plagioclase feldspar. By contrast on small ocean islands such as the Canary Islands and the Azores the more modest magma fluxes allow high level crustal magma chambers to develop beneath large central volcanoes within which magmas can evolve undisturbed for much longer periods. These islands frequently have silica-undersaturated lavas that are rich in alkalis for their silica contents and are dominated by trachytes and occasionally phonolites (Figure 4.1). Islands dominated by tholeiitic lavas, such as Iceland tend to have more silica-rich evolved lavas, the rock type known as Icelandite, being the silica-saturated equivalent of Hawaiite, and the most evolved lavas that develop in areas where magma fluxes are lowest are alkali-poor rhyo-dacites or rhyolites.

3.3. Subduction-Related Basalts

3.3.1. Island Arc Basalts

The third location where basalts are produced is above active subduction zones where cold oceanic lithosphere descends back into the mantle. Subduction zones are consequently regions of low heat flow and at first glance it may appear surprising that melt should be produced in regions where the mantle is cold. However, the subducting oceanic lithosphere carries with it an inventory of deep sea

---

BOX 4.2 Fractional Crystallization of Basalt and the Thermal Divide

Alkali and tholeiitic basalts both evolve by fractional crystallization of phenocrysts phases, initially olivine and subsequently plagioclase and clinopyroxene and an oxide, usually magnetite. However, the sequence in which these minerals crystallize is different for the two basaltic magma types and produces different evolved magmas. More significantly tholeiitic magmas do not evolve into alkali magmas or vice versa.

The reason for this is related to reactions between minerals and melt that consume silica. For example there is a reaction relationship between nepheline and quartz that produces albite:

$$NaAlSi_2O_6 + SiO_2 = NaAlSi_3O_8$$

$$Ne + q = Ab$$

Most magmas either have a deficiency of quartz, so have the potential to crystallize albite + nepheline, or an excess of quartz. In those with an excess of quartz, there is a further reaction relationship between olivine and quartz to produce orthopyroxene (enstatite):

$$Mg_2SiO_4 + SiO_2 = 2MgSiO_3$$

$$Fo + q = En$$

Consequently enstatite and nepheline can never occur in the same rock and be in equilibrium. These reactions form the basis of the concept of silica saturation: those rocks with the potential to crystallize orthopyroxene are *silica saturated* whereas those with potential to crystallize nepheline, *undersaturated*.

This difference is best explained with the use of a phase diagram and one of the more simple examples is the system Di−Ne−Q (Figure 4.7). This particular diagram is chosen because it includes two of the essential end member minerals that make up a basalt, specifically, Di for clinopyroxene, and Ab (intermediate between Q and Ne) for plagioclase. Of course some phases are missing, most notably those containing iron, but it illustrates the effect of the nepheline−albite−quartz reaction on basalt fractionation at low pressures. In Figure 4.7, the effect of these reactions is such that basalts that fall either side of the Ne−Di line follow markedly different crystallization paths toward low temperature points that are either silica-saturated or silica-undersaturated, labeled respectively M_q and M_{Ne}. The join albite−diopside acts as a barrier to fractionation. It is commonly described as a thermal divide and it exists in any low pressure system that includes the Ne−q join, although it breaks down at high pressures.

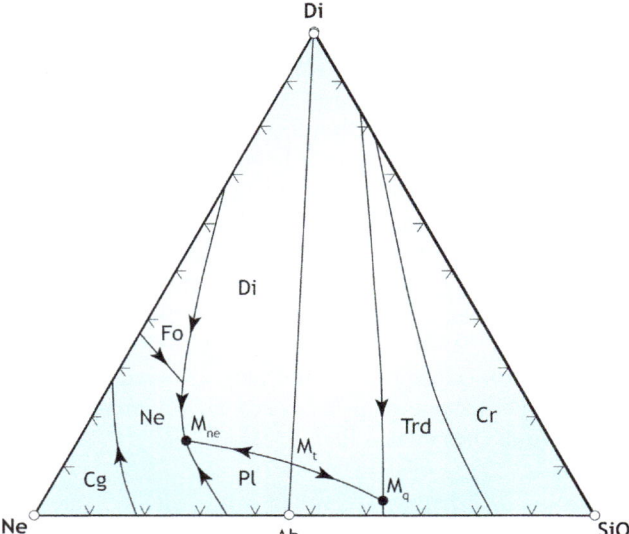

FIGURE 4.7 The system diopside—nepheline—quartz (Schairer and Yoder, 1960). The reaction between nepheline and quartz produces a thermal divide corresponding to the diopside—albite join. Compositions to the left of this line evolve to a silica-undersaturated composition, represented by the point M_{ne}, while compositions to the right of the divide evolve toward a silica-saturated composition, M_q. If the composition lies on the Di—Ab join then the final liquid composition will correspond to the temperature minimum at M_t.

sediments and altered oceanic crust (oceanic crust modified by its interaction with ocean water by hydrothermal processes shortly after its creation at an ocean spreading center). These hydrated rocks readily release their water as the slab descends into the mantle and warms up. Water has a dramatic effect on the liquidus temperature of the mantle, lowering it by as much as 300 °C (Chapter 1). Thus not only is the mantle capable of generating melts at significantly lower temperatures in the presence of H_2O, the magmas so produced will be very water-rich and this affects their physical and chemical characteristics.

Basaltic magma produced in island arc settings tend to be tholeiitic in character and the name, *island-arc basalt (IAB)* is used to distinguish them from MORB on account of their higher H_2O, K_2O and Al_2O_3 contents. More significantly their high water contents have a dramatic effect on the crystallization sequence. First, water expands the phase field of olivine at the expense of clinopyroxene and plagioclase. Second, the reduction in the liquidus temperature of plagioclase means that when it does appear on the liquidus it has a much more calcic composition—more anorthite-rich—and so has a lower silica content than plagioclase that crystallizes from MORB. Finally while water lowers the temperature at which silicate minerals crystallize, it has a much reduced effect on the temperature of oxide crystallization. Thus a typical crystallization sequence in an IAB is olivine, calcic plagioclase, and magnetite, all of which have low SiO_2, resulting in a much more rapid increase in SiO_2 in the

fractionating magma. Thus island arc lavas tend to evolve from basalts through low-K andesites to dacites and rhyolites.

3.3.2. Andean Margin Magmatism

In regions where subduction takes place beneath a continental margin, such as beneath the Andes, Indonesia, Japan, and western N. America, the basaltic magma may become trapped in the overlying continental crust. In these regions the continental crust can be many tens of kilometers thick and provides ample opportunity for the magma to accumulate in magma chambers and undergo fractional crystallization and interaction with the silica-rich granitic crust. Basaltic magma is denser than continental crust and so can only rise to a level of neutral buoyancy, often at the junction of layers with contrasting density, such as the Moho or at tectonic boundaries within the crust. Here it accumulates and undergoes fractional crystallization in the same way as arc tholeiites until the magma density is such that it becomes buoyant once more and continues its passage toward the surface. This critical density is reached after a significant amount of olivine and magnetite fractionation has reduced the magnesium and iron and increased the alkali and silica content of the magma, such that it has become an andesite. It is the density filtration of the continental crust that increases the proportion of andesites over basalts erupted in Andean margins.

The intrusion of significant amounts of basaltic magma into the continental crust can lead to melting and remobilization of the crust. Melts of crustal rocks tend to have granitic (or rhyolitic) compositions (see Section 4) and these can mix with the basaltic magma to make a hybrid rock that also has a broadly andesitic composition. Thus andesites may originate through magma mixing and contamination of basaltic magma as well as closed system fractional crystallization.

The general fractionation trend observed in Andean and arc basalts systems is described as calc-alkaline, the reasons for which are historic and the subject of continual debate (Arculus, 2003). However, the term hides some important and fundamental differences between MORB, IAB and andesitic magmatism, in particular the contrast between the lack of iron enrichment in IAB compared with MORB. Indeed this contrast is sometimes seen within arc volcanoes as at Also in Japan (Figure 4.8) in which the Fe/Mg ratio in the tholeiitic magmas increases much more significantly that in the so-called calc-alkaline series. The reasons for this difference could be different parent magmas, different P—T conditions for fractionation or more H_2O in the calc-alkaline group.

3.4. Continental Basalts

Basaltic magmas are not restricted to the oceans and plate margins—they are also found widely in the continents.

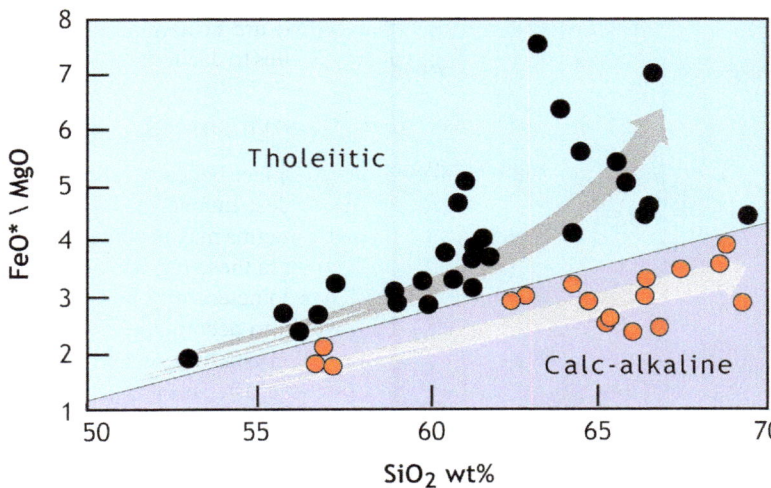

FIGURE 4.8 Contrasting FeO/MgO ratios in tholeiitic and calc-alkaline rocks from Aso, a volcano in Japan. Fractionation of olivine, and plagioclase in the tholeiitic suite produces iron enrichment and hence high FeO/MgO ratios as silica increases, whereas early fractionation of calcic plagioclase and oxide (magnetite) in the calc-alkaline suite produces less iron enrichment for a given increase in silica. *Data from Hunter (1998), calc-alkaline—tholeiite division after Miyashiro (1974).*

Indeed the continents, because continental crust is not destroyed by subduction, contain a record of basaltic activity back to earliest Archaean times. Contemporary and Phanerozoic continental basaltic volcanism is invariably associated with tectonic extension, because the continental lithosphere (crust and mantle) is frequently more than 100 km thick and beneath Archaean cratons can be as much as 200 km thick. This effectively puts a lid on any mantle plumes that impinge on the base of the continental lithosphere preventing much of the potential melt production within them. However, when the continents are stretched over the site of a mantle plume, as in some rift zones, and particularly during continental break-up, basaltic magmas are generated and erupted at the surface.

Modern day rift zones include the east African Rift system, the Baikal Rift in Siberia, the Rhine graben in Europe and the Rio Grande rift in the USA. All are characterized by alkali basalts, basanites, and nephelinites and their evolved derivatives (trachyte and phonolite). This is because the amount of extension across such rift zones is small and so only small melt fraction alkali-rich magmas are produced from the mantle plume, similar to OIB.

By contrast, **continental flood basalts** (CFB) provinces (or large igneous provinces) contain large volumes (perhaps $>10^6$ km^3) of tholeiitic basalts that appear to have erupted geologically rapidly (1–2 Myr). Examples include the Deccan traps of India (trap is a German word that describes the step-like profile of the topography of some CFB provinces), the Karroo lavas of S. Africa and the Parana basalts of Brazil. They represent large scale magmatic and volcanic events and in many cases it is speculated that they were linked with mass extinctions, such as the eruption of the Siberian traps and the Emeishan lavas in China that are coincident with the end-Permian mass extinction. Many are also associated with continental break-up and the

development of a new ocean, and some occur at the end of hot-spot traces in the oceans and so can be linked to the presence of a mantle plume. Such large volumes of magma clearly require the presence of anomalously hot mantle as in a plume and an association between CFB and the initiation of a mantle plume is often invoked.

As more data have become available it appears there are differences in major and trace element compositions (e.g., Figure 4.9) and also in the average eruption rates of different CFB. The differences may reflect differences in the source regions and in the causes of partial melting in the different CFB provinces, and an alternative view to the plume initiation hypothesis relates CFB to the conductive heating and melting of sources within the mantle section of the continental lithosphere. As with the oceans, the continental mantle lithosphere is thought to be largely residual after partial melting and therefore relatively incapable of generating further melts. However, if it contains a small amount of water locked up in hydrous phases, such as mica or amphibole, then these phases reduce the mantle solidus so that melt can be generated if the lithosphere is heated. A mantle plume impinging on the base of the lithosphere could supply sufficient heat by conduction while the release of water from the hydrous phases would ensure the generation of tholeiitic basalt. However, distinguishing between such processes is fraught with difficulty because most CFB have undergone significant amounts of fractionation. Few examples have more than 8 wt.% MgO or any olivine phenocrysts, the majority are multiply saturated in plagioclase pyroxene and oxides and CFB with the compositional characteristics of primary magmas are rare. Tracking the evolutionary history of such evolved rocks to yield unambiguous indications of their origins remains one of igneous petrology's more challenging problems.

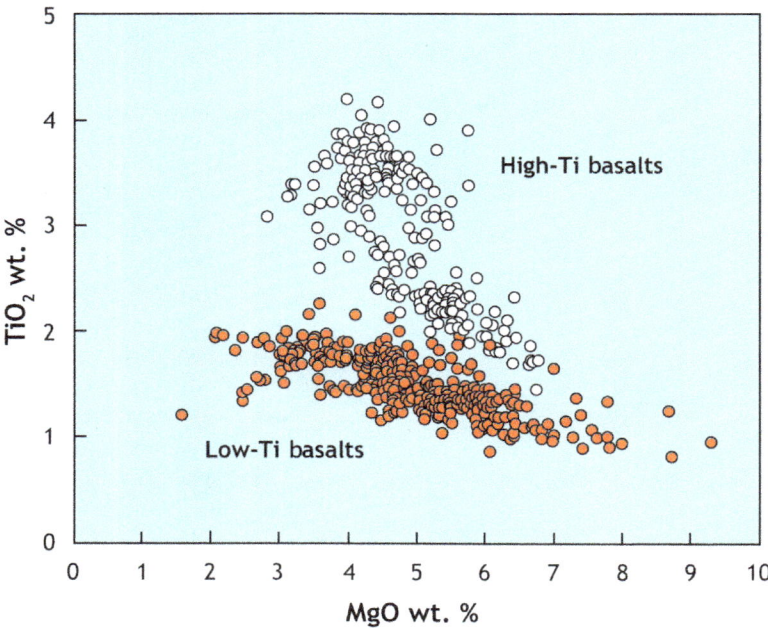

FIGURE 4.9   Plot of wt.% TiO_2 against MgO in basaltic lavas from the Parana CFB province, Brazil. Two distinct magma groups are recognized on the basis of their TiO_2 contents and are thought to be derived from distinct parent and primary magmas. *Modified after Peate (1997).*

4. SILICIC MAGMAS: GRANITES AND RHYOLITES

Silicic magmas have a more restricted range in major element composition than basalts, a feature that results from the processes that lead to their production. As their name implies silicic magmas have high silica, alumina, and alkali contents and low MgO and FeO (Figures 4.1 and 4.2), and their mineralogy is dominated by quartz and feldspars (orthoclase and albite-rich plagioclase). As with basaltic magmas, silicic magmas can crystallize at depth to form coarse-grained intrusive rocks called granites and this is their most common fate. Sometimes they erupt at the surface to form fine-grained or glassy rocks known as dacite and rhyolite or obsidian if the rock is almost completely glassy.

All granites contain alkali feldspar, plagioclase feldspar and at least 20% quartz, and classification schemes for granitic rocks depend on a combination of mineralogy and bulk composition. The detailed nomenclature of plutonic rocks, summarized in Figure 4.10, is based on the relative proportions of alkali and plagioclase feldspar. Of these different rock types, granite and granodiorites are most abundant in the continents, followed by diorites and gabbros, whereas gabbros dominate the deeper parts of the oceanic crust. Syenites and monzonites are less common and tend to occur in association with alkali basalts and continental rifts.

Chemical classifications add further complexity to this system but reveal significant differences between granites possibly related to their modes of formation. The primary classification depends on the molecular ratio $Al_2O_3/(Na_2O + K_2O + CaO)$. If this ratio is <1, then the granite is defined as **metaluminous**, whereas if it is >1 it is **peraluminous**. If the molecular ratio $(K_2O + Na_2O)/Al_2O_3 > 1$, then the granite is said to be **peralkaline**. Metaluminous granites tend to be sodium-rich and are often associated with a broad range of other igneous rocks. Peraluminous granites have lower sodium contents, are richer in potassium and restricted in their association to other high silica magmas. Peralkaline granites are more likely to be associated with a range of other silicic rocks, such as syenites and monzonites and as with alkali basalts are frequently found in continental rift zones.

It has long been recognized that the compositions of granites lie close to that of the lowest temperature melt fraction of crustal rocks; those dominated by quartz and feldspars. The ternary system quartz—albite—orthoclase (q—ab—or) is part of a larger system, quartz—nepheline—leucite, known as petrogeny's residua system (Figure 4.11). This system contains two temperature minima separated by a thermal divide, as with the Di—q—Ne system. One of these minima corresponds to silica-saturated granitic or rhyolite compositions, whereas the other corresponds to silica undersaturated phonolite or the intrusive equivalent, nepheline syenite. Consequently if, for example, a basaltic rock is itself melted then the first melt to form will correspond either to the granite minimum or to the phonolite minimum, depending on whether the basalt is silica saturated or undersaturated. Likewise a fractionating basalt will theoretically produce a

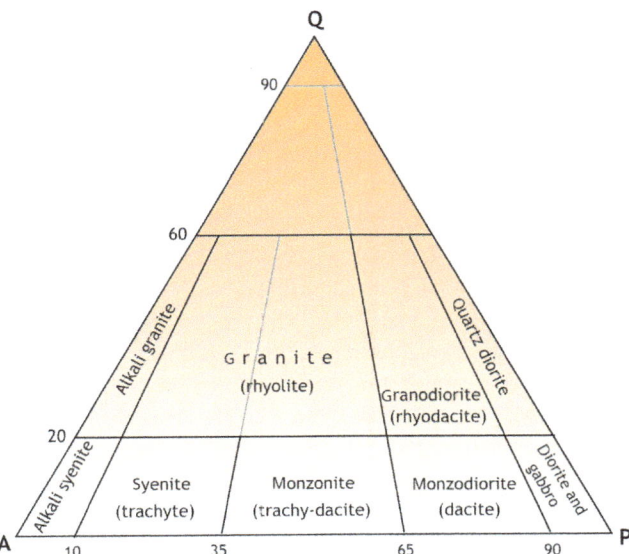

FIGURE 4.10 The mineralogical classification of plutonic rocks. Q refers to the amount of quartz in the rock, P the amount of plagioclase and A alkali feldspar. Granites (*sensu lato*) include all those rock types with >20% Q. Note that on this diagram rocks with a basaltic or andesitic composition (gabbroic and dioritic) are restricted to a small region close to the P apex, because they are dominated by plagioclase feldspar, pyroxene, and olivine.

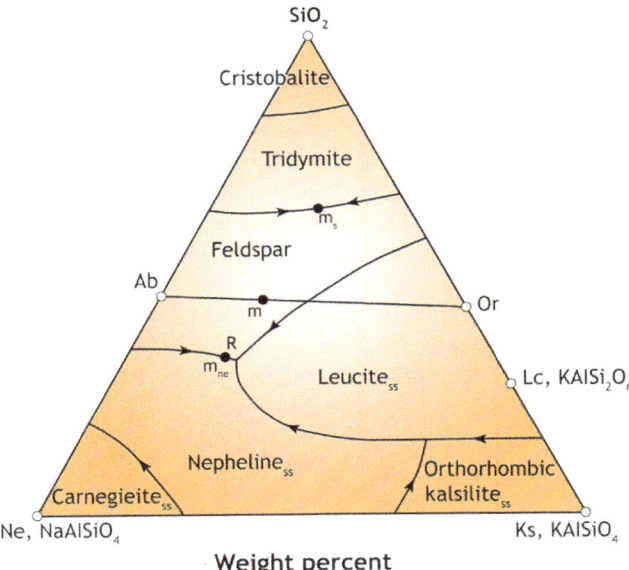

Weight percent

FIGURE 4.11 The system quartz–nepheline–leucite, known as petrogeny's residua system after Schairer (1950). The reactions between nepheline kalsilite and quartz produce a thermal divide between silica-saturated and silica-undersaturated compositions. There are however, three minima in this system, one labeled M_s, which corresponds to granitic compositions, a second at M_{ne}, which corresponds to nepheline syenites, and a third, M_t on the albite–orthoclase join, which corresponds to the compositions of trachytes.

small amount of granitic or phonolitic liquid, once all the **mafic** and calcic minerals have been exhausted from the melt. Thus, there are two dominant processes whereby granitic magmas can be generated, melting of pre-existing crustal rocks and fractional crystallization of mafic magmas, as discussed above.

Granites are dominantly found in association with subduction or in areas of continental collision—Andean margins and mountain belts. Volatile species, particularly water are critical in the evolution of granitic magmas and this may well explain their relative paucity in constructive plate margins and within-plate environments. Water decreases the melting temperature of crustal rocks, in much the same way it does in the mantle, and it is the dehydration of hydrous minerals, such as biotite, muscovite, and amphibole that allows the production of large volumes of granitic magma at moderate temperatures within the crust. Furthermore magmas rich in SiO_2 are viscous because the Si-O bonds link to form chains and complex three-dimensional structures within the melts. In general the higher the silica content of a liquid, the more highly structured the melt is and the higher its viscosity. Water breaks the Si—O bonds and so reduces the viscosity, enabling the melt to move more readily through the crust.

The water introduced into the magma source by subduction clearly plays an important role in the generation of the range of subduction-related magmas and that effect has a most dramatic impact on the nature of volcanic eruptions associated with subduction zones. While many regions of active subduction erupt basalts and andesites from small volcanic centers, others build up considerable volcanic structures from repeated eruptions of increasingly silica-rich lavas and pyroclastic rocks, producing the classic steep-sided strato-volcano with a central vent, such as Mt Fuji in Japan. As these large volcanoes grow and magma continues to evolve beneath them by assimilation of crustal materials and fractional crystallization, the volatile content builds up with the silica content and the magma body rises further into the crust as a result of its buoyancy, producing a potentially explosive system. The magma, charged with dissolved water, remains stable only if it is kept under a suitable confining pressure (Chapter 1). If the magma chamber is subsequently breached by tectonic activity, or as in the case of Mt St Helens gravitational instability of the volcanic cone, the water is rapidly exsolved and the magma is erupted with explosive violence. Examples of such eruptions include Karakatau and Tambora in Indonesia and more recently Mt St Helens in the USA, Pinatubo in the Philippines and Montserrat in the Caribbean, all of which incurred a societal toll in displacing local communities and causing significant casualties.

While recent historic eruptions, such as Tambora, Krakatoa, and Mt St Helens gave captured the public imagination, these are dwarfed by supereruptions in which magma volumes are measured in thousands of cubic kilometers that are a feature of the geological record. Such supereruptions include Yellowstone and the Long Valley in the USA, Lake Toba in Indonesia and Lake Taupo in New Zealand, all of which have erupted within the last million years. In prehistory, the eruption of Thera (now Santorini) in the Mediterranean devastated the Minoan civilization approximately 3600 years b.p. while the eruption of Toba in Indonesia is thought to have led to sudden and dramatic global climate change that reduced the human population significantly and consequently changed the course of human evolution. The eruption of a comparable volume of magma today could have a catastrophic effect on modern society, both in the short and longer term as a consequence of hemispheric or even global environmental change.

The ability of a granite magma to dissolve water increases with pressure, in much the same way that basaltic magma can dissolve more water at high pressures. Hydrous minerals are, however stable to higher pressures than the water-saturated granite solidus, and if the temperature of a crustal rock containing a hydrous mineral is increased and the mineral breaks down, the water so released can induce melting. This mechanism is illustrated in Figure 4.12, which shows the relationship between the stability curve for muscovite, a geotherm of 25 °C km^{-1} and the water-saturated granite solidus. Muscovite is stable as long as the geotherm lies above its stability curve. When the two cross, as at point A, muscovite breakdown liberates a large amount of water that in turn catalyzes the production of large amounts of granitic melt. What is equally important is that because the magma is now at a temperature well above its liquidus it can rise to higher levels in the crust (lower

pressures) without freezing immediately. Eventually its upward progress is halted when conditions match those of the water-saturated solidus (point B), but the relationship between the solidus and muscovite stability does mean that granitic magmas can be generated in large volumes and be mobile within the continental crust. Furthermore, it is the upward limit on the mobility of granitic magmas imposed by the shape of the water-saturated solidus that means they seldom erupt at the surface, solidifying at depth to form granites, rather than reaching the surface to erupt as rhyolites.

On those occasions when large volumes of silica-rich magma do approach the surface, they create the potential for the most voluminous, violent and hazardous of volcanic eruptions—supereruptions.

5. TRACE ELEMENTS

By definition the major elements dominate the bulk composition of igneous rocks, and their abundances and relative variations reflect the depths at which melting took place and the amount of magmatic evolution that occurred after the melt segregated from its source. Further information on these processes can be extracted from the behavior of the trace and minor elements which are widely reported in contemporary analyses of igneous rocks. Their use lies in their sensitivity to melting and crystallization which can have much more dramatic effects on their concentrations than major elements, the abundances of which are frequently buffered by the presence of particular minerals.

For example, the major element composition of a melt changes only gradually as the melt fraction increases, indeed the variation of MgO in a melt is sensitive to the temperature of melting but insensitive to the amount of melting. By contrast the concentration of some trace elements can change by two orders of magnitude between 0.05% and 5% melting, in essence because the element partitions strongly into the melt.

Similarly the minor and trace element concentrations in mantle peridotite vary much more than the major elements and these can be reflected in the trace element concentrations in any melt. Consequently trace element variations in basaltic rocks have proven critical in assessing the magnitude and origins of chemical variations in the Earth's mantle as well as placing important constraints on the operation of specific magmatic processes during the evolution of suites of basaltic rocks.

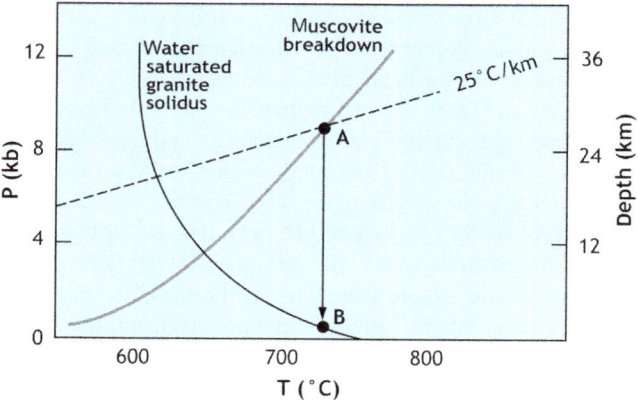

FIGURE 4.12 A graph of pressure versus temperature showing the muscovite breakdown curve and the water-saturated solidus of granite, i.e., the temperature at which granite melts in the presence of water. While muscovite is stable no melt forms even above the granite solidus, because there is no free water in the assemblage. When the ambient temperature (as shown by the geotherm) crosses the stability curve of muscovite, the latter breaks down, releasing water and producing a large quantity of crustal melt. This melt can then rise through the crust because of it buoyancy until it reaches point B, where it solidifies and can move no further.

5.1. Trace Elements in Oceanic Basalts: MORB and OIB

The analyses shown in Figure 4.13 include an average MORB and an average OIB. The differences between the two are strikingly clear, the OIB being more enriched in the

BOX 4.3 Partition coefficients, compatible and incompatible elements

The distribution of trace elements between a mineral and a coexisting liquid is often described in terms of a **partition coefficient**. These are simply defined as the concentration of an element in a mineral divided by the concentration in the coexisting melt:

$$K_d = K = D = X^i_{mineral}/X^i_{Melt}$$

The expressions K_d, K, and D are used variably and often interchangeably.

Partition coefficients of a trace element vary enormously and depend on its ionic radius, its ionic charge and the size of the crystallographic sites in the mineral into which the element substitutes. In general those elements with D values >1 are described as *compatible*, those with D values <1 are *incompatible*.

Examples of **compatible elements** during mantle melting include Ni in olivine, Cr in pyroxene and Yb (one of the rare earth elements or **REE**) in garnet. Others such as Nb and Zr, both second row transition elements, and the light REE (La, Ce, Nd) are incompatible because they do not fit easily into the structure of the common rock-forming minerals of lherzolite.

Incompatible trace elements are further subdivided into a number of groups according to their specific geochemical properties. The lanthanides or *rare earth elements* (REE) are a coherent group of transition elements that all exhibit +3 ionic charge. Because of the way in which their electronic structure develops as atomic number increases from La to Lu, their ionic radii decrease in a systematic way. This results in coherent behavior during melting and fractional crystallization. The minor and trace alkali and alkaline earth elements, K, Rb, Cs, Sr, and Ba, have low ionic charge and large ionic radii and so are

known as the *large ion lithophile elements* or **LILE**. By contrast transition elements with high ionic charge and small ionic radii including Zr, Nb, Hf, Ta, and Ti, are known as the *high field strength elements* (**HFSE**).

Within the crust a different set of elements are compatible because it has a more granitic composition and feldspars are the dominant minerals. Thus Sr and Eu (another REE) are both incompatible in the mantle, but are more easily accommodated by feldspar and so behave more compatibly during low pressure fractional crystallization involving feldspars or during crustal melting (see later section). The relative abundances of such elements can therefore be used to identify the minerals that were present either during melting or during fractional crystallization. Moreover, because certain minerals are stable over particular temperature and pressure ranges, trace elements can be used in conjunction with major elements to infer the depths at which melting and fractional crystallization took place.

A convenient way for displaying and comparing trace element abundances in basaltic rocks is via the use of mantle or chondrite (meteorite) normalized diagrams, such as those in Figure 4.13. The elements are ordered according to their incompatibility in the mantle, the most **incompatible elements** to the left. Each point is then the concentration of the element in the rock sample divided by its abundance in the normalizing composition. Note that the elements also include the minor elements, K, P, and Ti as these all behave incompatibly in the mantle and are not essential components in the widespread silicate minerals of the mantle (olivine, pyroxene, spinel, and garnet).

---

most incompatible elements than MORB which show a progressive depletion in the same elements—a feature that is typical of MORB worldwide. These variations can be modeled from a knowledge of the mineralogy of the mantle, experimental and theoretically determined partition coefficients of the various elements between minerals and melt (Box 4.3), and the melting reaction determined from phase equilibria. Assuming fractional melting equations the trace element distribution patterns of varying melt fractions can be calculated and compared with analyses of real rocks. What these calculations reveal is that once the melt fraction exceeds more than a few percent, the shape of the trace element pattern in the melt will be closely similar to that of the mantle source, albeit at higher abundances. Thus the convex upwards pattern in MORB reflects a similar pattern in the MORB source.

In comparison to MORB, OIB are significantly more enriched in the most incompatible elements and have slightly lower abundances of the least incompatible, notably Y, Yb, and Lu, which lie to the far right of the

diagram. These differences reflect two aspects of the generation of OIB. The higher concentrations of the most incompatible elements (Cs through to La) are due to smaller overall melt fractions than MORB. This is consistent with the models of decompression melting developed in Section 3.2. Second, the lower abundances of Yb and also the way in which elements such as La and Dy are fractionated from Yb—high La/Yb and Dy/Yb ratios—implies that the Heavy Rare Earth Elements (HREE) are behaving more compatibly during the generation of OIB melts than during MORB genesis. This behavior reflects their compatibility in the mineral garnet, the stable aluminous phase in the mantle at depths >75 km. Once again this is consistent with the model of deeper, hotter decompression melting developed in Section 3.2.

These differences can be modeled more quantitatively using trace element ratios. A particularly useful diagram in the context of mantle-derived magmas is a plot of Dy/Yb versus La/Yb (Figure 4.14). On this diagram melts of garnet-bearing mantle and garnet-free mantle define very

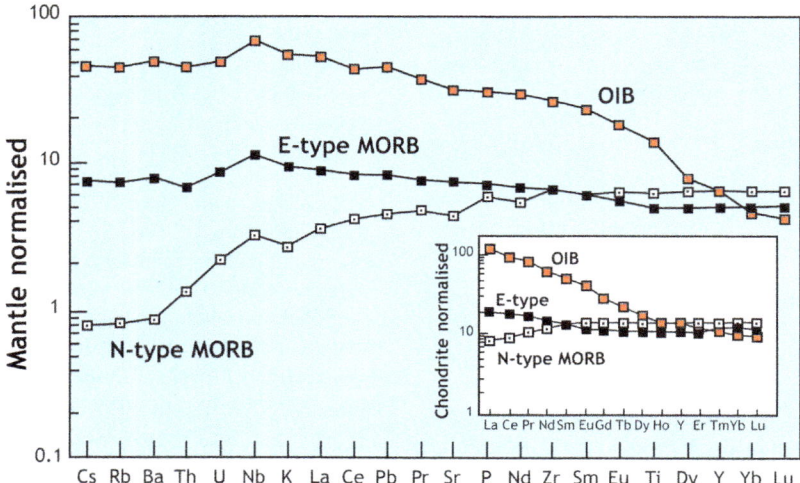

FIGURE 4.13   The variation of incompatible elements in typical MORB (both normal and enriched) and OIB. Trace elements are normalized to concentrations in the upper mantle and arranged in order of their incompatibility, the most incompatible elements to the left of the diagram. The inset shows the REE for the same exemplars normalized to their abundances in chondritic meteorites. Note the depletion of the most incompatible elements in N-MORB and the low abundance of the most compatible elements (Y, Yb, and Lu) in OIB. E-MORB are found sporadically along the mid-ocean ridge system, sometimes close to ocean islands, and reflect a more trace element-enriched source region, emphasizing the compositional heterogeneity of the earth's shallow mantle. *After Sun and McDonough (1989).*

different trajectories. The garnet-free mantle has a modest influence over the REE elements, only La being significantly more incompatible during melting; thus melts from low pressure melting in the mantle plot at low values of Dy/Yb while La/Yb shows some variation. By contrast, garnet-bearing mantle lithologies retain Yb in preference to either Dy or La and the compositions of melts produced in the

presence of garnet extend to higher values of Dy/Yb. For comparison, MORB analyses and data from Hawaii are plotted on the same diagram and it reveals that MORB have little or no garnet influence whereas there is a clear garnet signature in the Hawaii basalts. It is noteworthy, however, that the Hawaiian basalts do not plot along the garnet trajectory but occupy the space between the garnet-free and

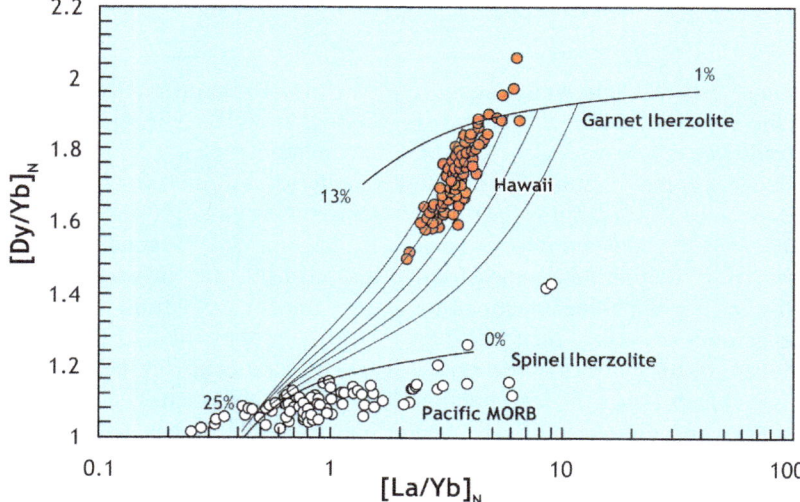

FIGURE 4.14   A plot of the variation of La/Yb and Dy/Yb ratios, both normalized to chondritic abundances, during melting of spinel- and garnet-bearing mantle lherzolite with MORB-source trace element abundances. The curves show how the La/Yb ratio is sensitive to the degree of melting whereas the Dy/Yb ratio is sensitive to the presence or absence of garnet. The steep curves are models for decompression melting (accumulated fractional melts) of a mantle source that starts in the garnet stability field but moves into the spinel field as it rises and melts. Analyses of Pacific MORB define a shallow gradient at low values of Dy/Yb, indicative of shallow melting of garnet-free lherzolite. By contrast the steep gradient shown by the Hawaii lavas reflects the influence of garnet and decompression melting that extends to depths where garnet is unstable. *MORB data from Niu and Batiza (1997), Hawaiian data from, Huang and Frey (2003), trace element model courtesy of Ian Parkinson.*

garnet-bearing melting curves. This implies that the melt regime crosses the boundary where garnet becomes stable/ unstable which is again consistent with the detail of the decompression melting model of a plume impinging on the base of ocean lithosphere, if the lithosphere is no more than 70 km thick.

The abundances of trace elements in some OIB exceed the concentrations that can be readily generated from a lherzolite mantle source region with primitive mantle incompatible element concentrations. This is well illustrated by the element Ti which has a well constrained mantle composition from analyses of peridotite massifs and xenoliths in comparison with abundances in meteorites. Titanium abundances in some OIB, by contrast are highly variable and range to values >3 wt.%. This is more than can be supplied from a fertile mantle source region assuming realistic melt fractions and partition coefficients (Prytulak and Elliott, 2007). This observation has encouraged some authors to invoke the presence of pyroxenitic components with higher complements of Ti and other incompatible elements in magma source regions. Such pyroxenites may be related to recycled oceanic lithosphere or patches of mantle that have been subject to melt infiltration during previous magmatic episodes. Abundances of first row transition elements in high Mg# olivines in OIB and some CFB have also been used to infer the presence of pyroxenitic components in their source regions (Sobolev at al. 2007). Thus the origins of OIB remain the subject of considerable debate and the role of pyroxenitic components in the mantle are as yet unconstrained.

5.2. Trace Elements in Subduction-Related Basalts

Figure 4.15 illustrates the trace element analyses of four basalts (IAB) from the Marianas arc in the western Pacific. In this case the trace elements are normalized to abundances in N-MORB so that on this diagram a typical MORB sample would plot with a value of one for all elements. The differences between these profiles and those of MORB (Figure 4.13) are quite striking, in that the IAB have very high abundances of elements such as K, Rb, Ba, and Sr (the so-called LILE), and low abundances of Nb and Ta. These latter elements, along with Zr, Hf, and Ti are referred to as high field strength elements or HFSE and their apparent depletion in IAB and other subduction-related basalts is often referred to as the niobium anomaly. The magnitude of the Nb anomaly relative to other elements can be expressed in trace element ratios such as Ba/Nb or La/Nb and high values of these ratios have long been regarded as distinctive characteristics of subduction-related igneous rocks.

In detail, basalts from different arcs show a wide variety of trace element abundances with varying amounts of enrichment in the light REE and LILE and varying depletions in the HFSE. These variable concentrations are considered to be a feature of the magma source region because it is difficult to see how such dramatic variations could be developed in response to the fractional crystallization of phenocryst phases or contamination/mixing with materials derived from the oceanic lithosphere. The unique aspect of arc-related magmas is that they are produced in

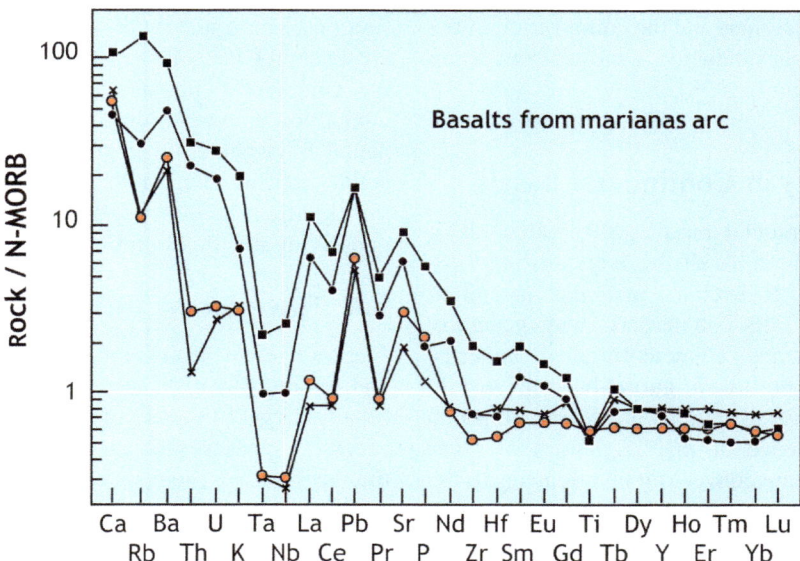

FIGURE 4.15 Incompatible element abundances in selected basalts from the Marianas arc, western Pacific, normalized to trace element abundances in N-MORB (cf. Figure 4.9). Note the low abundances of Nb and Ta, the so-called Nb anomaly, and the relatively high concentrations of the LILE (Cs, Rb, Ba, K, Pb, and Sr). *Data from Elliot et al. (1997).*

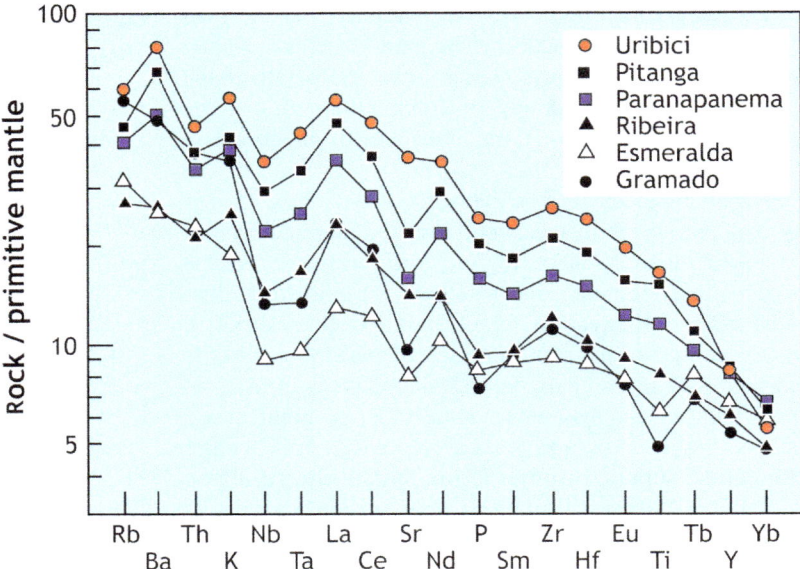

response to the introduction of water-rich fluids from subducting oceanic lithosphere. In addition subducting lithosphere also includes a veneer of sediments and these can also be mobilized as hydrous melts during subduction. Thus the most widely accepted models for arc magmatism invoke variations in the proportions of hydrous fluids and sediment melts to account for the variability in trace element ratios and abundances. There is strong evidence in support of the presence of materials recycled from the surface, including the presence of measurable cosmogenic [10]Be in some arc basalts (Morris and Tera, 1989) that can only have been incorporated into the magma source region by sediment subduction, and correlations between the trace element contents of arc magmas and the composition of the sedimentary package being subducted in the adjacent ocean trench (Plank and Langmuir, 1998).

5.3. Trace Elements in Continental Basalts

The two types of continental basalts, rift-related alkali basalts and continental flood basalts have contrasting trace element characteristics. Rift-related magmas are often closely similar to their OIB counterparts with generally smooth incompatible trace element-enriched patterns, indicating limited melting of a garnet-bearing mantle source region. Once again deviations are found from this generality but the influence of mantle plumes or other thermal perturbations in the convecting mantle beneath the continents is clear in the trace element contents of rift basalts. Continental flood basalts, by contrast, show much greater variability, including the presence of Nb anomalies and LILE enrichment over the REE, similar characteristics to those seen in subduction-related basalts.

Figure 4.16 illustrates the trace element abundances in selected lavas representing different magma groups from the Parana CFB province in Brazil. Some have fairly smooth profiles, not dissimilar from OIB, implying a plume-related source regions. These are generally the high-titanium magma types as described above. By contrast the low-titanium lavas tend to show uneven profiles strikingly different from MORB and OIB, but broadly comparable with subduction-related magmas. Much of the continental crust is thought to have been generated above destructive plate margins and the crust certainly contains a long record of repeated subduction-related magmatic and tectonic events. Thus it may be that the continental lithosphere becomes imprinted with these characteristics and their presence in CFB is testament to magmatic interaction with the continental lithosphere. This may be the result of interaction between plume-derived magmas and the continental lithosphere, either the crust or the mantle parts, or could indicate that melts are wholly derived from the mantle lithosphere as a result of conductive heating by an underlying mantle plume.

5.4. Trace Elements in Granitic Magmas

The trace element contents of granites and their volcanic equivalents, rhyolites are highly variable, more so than basalts, largely because of the effect of small amounts of accessory phases such as zircon, monazite, allanite and titanite both in their source regions and during granite crystallization. However, the abundances of critical trace elements can be used to discriminate granites formed in different tectonic environments, in a similar way to basalts. An example is shown in Figure 4.17, which illustrates that granites associated with subduction, either in volcanic arc

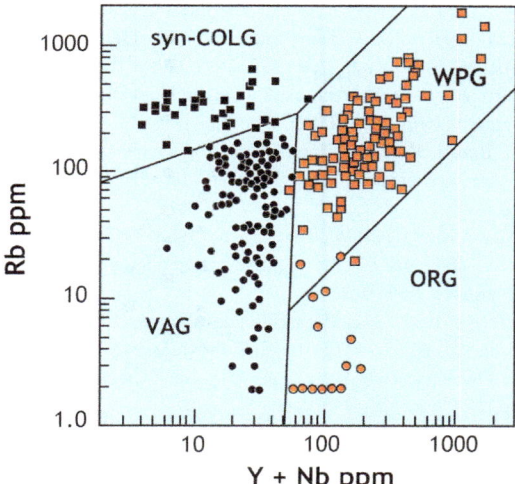

FIGURE 4.17 Trace element discriminant diagram for granites from different tectonic regimes. VAG; volcanic arc granites, ORG; ocean ridge granites, WPG; within-plate granites, COLG; collision zone granites. *From Pearce et al. (1984).*

granitic in composition. The huge compositional variety of other magmas result from processes that modify and alter the primary composition, mainly fractional crystallization and hybridization (mixing and contamination). There are subtle differences in primary magma composition that relate to tectonic environment, and the role of volatiles, especially water, is critical in both the generation of magma in partic-ular environments and how that magma subsequently evolves.

Basaltic magmas are generated by partial melting of the mantle by adiabatic decompression beneath mid-ocean ridges (tholeiitic MORB), by adiabatic decompression melting of mantle plumes (alkali and locally tholeiitic ba-salts, OIB), and by depression of the liquidus consequent on addition of water to the mantle above subduction zones (IAB, tholeiite, and calc-alkaline basalts). In continental regions melting also occurs by adiabatic decompression within mantle plumes (rift and some CFB) although conductive heating of the mantle lithosphere is another mechanism for generating magma. These different mech-anisms are summarized in Figure 4.18.

These processes are reflected in the major, minor and trace element compositions of igneous rocks, the interpre-tation of which give insights not only into the formation of magmas but also the evolution of the Earth as a whole. Partial melting is the dominant process responsible for the evolution of the Earth into mantle core and crust. The continental crust has been derived from the mantle over the age of the Earth and there is evidence for basaltic magmas in the most ancient of continental rocks. The modern day flux out of the mantle is basaltic, yet the average compo-sition of the continental crust is considerably more evolved, probably andesitic or dioritic. Given that andesites and diorites currently develop thorough a process of fractional

or continental collision zones, have variable Rb abun-dances, but in general low $(Y + Nb)$—features which compare with basalts from the same tectonic regimes. Within-plate granites, by contrast have high Rb and high $(Y + Nb)$, again similar to associated basaltic rocks.

6. SUMMARY

Despite the huge range of magma compositions, there are in general only two types of primary magma—those that derive from the mantle and are broadly basaltic in composition, and those that are derived from the crust and are dominantly

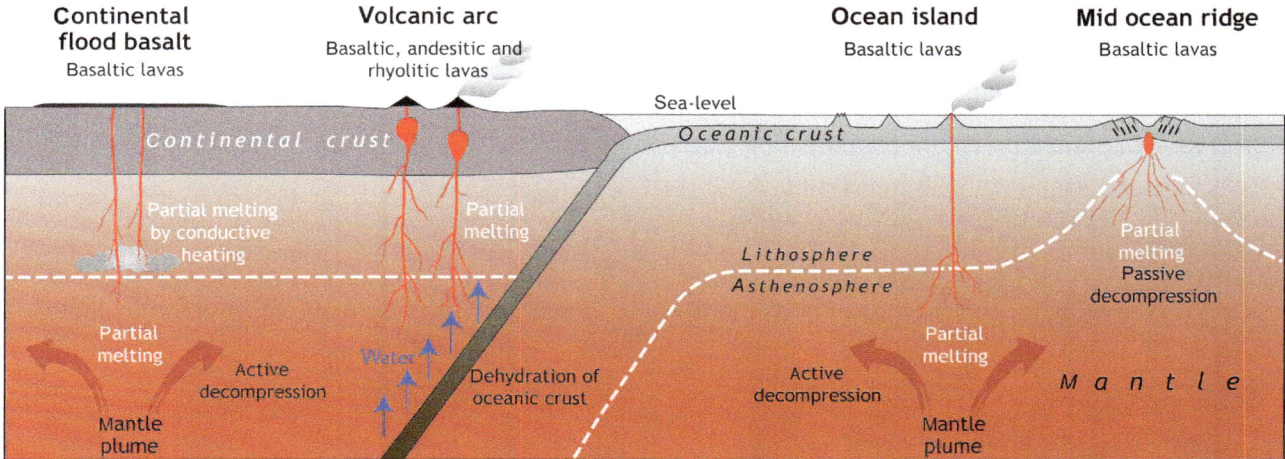

FIGURE 4.18 A schematic section through the outer parts of the Earth showing the crust, mantle, and lithosphere and melting mechanisms that give rise to MORB, OIB, IAB, and continental flood basalts. Granite formation in this diagram would be largely restricted to the continental crust above the subduction zone.

crystallization and the removal of mafic minerals (mainly olivine and clinopyroxene) from basaltic magma, the inescapable conclusion is that mafic cumulate material once in the crust has been recycled into the mantle over geological time. Thus, igneous processes are at one time a major consequences of the plate tectonic cycle but only one part of the continual cycle of material from the interior of the Earth to its surface and back again.

FURTHER READING

Arculus, R.J., 2003. Use and abuse of the terms calcalkaline and calcalkalic. J. Petrol. 44, 929−935. http://dx.doi.org/10.1093/petrology/44.5.929.

Elliott, T., Plank, T., Zindler, A., White, W., Bourdon, B., 1997. Element transport from slab to volcanic front at the Mariana arc. J. Geophys. Res. 102 (B7), 14991−15019.

Huang, S., Frey, F.A., 2003. Trace element abundances of Mauna Kea basalt from Phase 2 of the Hawaii Scientific Drilling Project: Petrogenetic implications of correlations with major element content and isotopic ratios. Geochem. Geophy. Geosy. 4, 2002GC000322.

Hunter, A.G., 1998. Intracrustal controls on the coexistence of tholeiitic and calc-alkaline magma series at Aso Volcano, SW Japan. J. Petrol. 39, 1255−1284.

Klein, E.M., Langmuir, C.H., 1987. Global correlations in ocean ridge basalt chemistry with axial depth and crustal thickness. J. Geophys. Res. 92, 8089−8115.

Kushiro, I., Yoder, H.S., 1969. Melting of forsterite and enstatite at high pressures under hydroius conditions. Carnegie Inst. Washington Yearbook 67, 153−158.

Le Bas, M.J., Streckheisen, A.L., 1991. The IUGS systematics of igneous rocks. J. Geol. Soc. 148, 825−833. London.

Macdonald, G.A., Katsura, T., 1964. Chemical composition of Hawaiian lavas. J. Petrol. 5, 82−133.

McKenzie, D.P., Bickle, M.J., 1988. The volume and composition of melt generated by extension of the lithosphere. J. Petrol. 29, 625−679.

Morris, J., Tera, F., 1989. 10Be and 9Be in mineral separates and whole rocks from volcanic arcs: Implications for sediment subduction. Geochim. Cosmochim. Acta 53, 3197−3206.

Miyashiro, A., 1974. Volcanic rock series in island arcs and active continental margins. Am. J. Sci. 274, 321−355.

Niu, Y., Batiza, R., 1997. Trace element evidence from seamounts for recycled oceanic crust in the Eastern Pacific mantle. Earth Planet. Sci. Lett. 148, 471−483.

Pearce, J.A., Harris, N.B.W., Tindle, A.G., 1984. Trace element discrimination diagrams for the tectonic interpretation of granitic rocks. J. Petrol. 25, 956−983.

Peate, D.W., 1997. The Paraná-Etendeka province. In: Mahoney, J., Coffin, M. (Eds.), Large Igneous Provinces: Continental, Oceanic, and Planetary Flood Volcanism, AGU Geophysical Monograph, 100, pp. 217−245.

Plank, T., Langmuir, C.H., 1998. The chemical composition of subducting sediment and its consequences for the crust and mantle. Chem. Geol. 145, 325−394.

Prytulak, J., Elliott, T., 2007. TiO_2 enrichment in ocean island basalts. Earth Planet. Sci. Lett. 263, 388−403.

Schairer, J.F., 1950. The alkali feldspar join in the system $NaAlSiO_4$-$KAlSiO_4$-$SiO2$. J. Geol. 58, 512−517.

Schairer, J.F., Yoder, H.S., 1960. The nature of residual liquids from crystallisation, with data on the system nepheline-diopside-silica. Am. J. Sci. 258A, 273−283.

Sobolev, A.V., Hofmann, A.W., Kuzmin, D.V., Yaxley, G.M., Arndt, N.T., Chung, S.-L., Danyushevsky, L.V., Elliott, T., Frey, F.A., Garcia, M.O., et al., 2007. The amount of recycled crust in sources of mantle-derived melts. Science 316, 412−417.

Sun, S.-S., McDonough, W.F., 1989. Chemical and isotopic systematics of oceanic basalts: implications for mantle composition and processes in Magmatism in the Ocean Basins. In: Saunders, A.D., Norry, M.J. (Eds.), Geological Society Special Publication No. 42, pp. 313−345.

Watson, S., McKenzie, D., 1991. Melt generation by plumes: A study of Hawaiian volcanism. J. Petrol. 32, 501−537.

Chapter 5

Thermodynamic and Transport Properties of Silicate Melts and Magma

Charles E. Lesher

Department of Earth and Planetary Sciences, University of California, Davis, CA, USA,
Department of Geoscience, Aarhus University, Aarhus, Denmark

Frank J. Spera

Department of Earth Science, University of California, Santa Barbara, CA, USA

Still, however, the highest value is set on glass that is nearly colorless or transparent, as nearly as possible resembling crystal. For drinking vessels glass has quite superseded the use of silver and gold

Pliny the Elder (c. AD 23–79)

GLOSSARY

basaltic magma Most typical type of magma erupted on Earth and other terrestrial planets. Basaltic magma is erupted at temperatures 1000–1300 °C, and is made up mainly of SiO_2 (50 wt%), Al_2O_3 (15%), CaO (12%), and roughly equal amounts of FeO and MgO (10%). The alkalis ($Na_2O + K_2O$) make up most of the difference although trace amounts of every naturally occurring element are invariably present. About 25 km^3 of basaltic magma is generated and erupted or emplaced each year on Earth, primarily along the 75,000 km of diverging plate boundaries. The density, specific heat, viscosity, and thermal conductivity of a typical basaltic magma at 1200 °C at low pressure are 2600 kg/m^3, 1450 J/kg K, 100 Pa s, and 0.6 W/m K, respectively. The heat

needed to completely fuse (melt) gabbro, the plutonic (crystalline) equivalent of basaltic magma, is about 500 kJ/kg.

magma High-temperature multiphase mixture of solids (cognate crystals and exotic lithic fragments), silicate or carbonatitic liquid, and H–O–C–S–Cl-rich gas or supercritical fluid formed by partial or total melting of parental source material.

magma rheology The rheological properties of magma depend on temperature, bulk composition, pressure, phase assemblage (melt ± crystals ± vapor), particle size and shape distribution, spatial arrangement of particles (structure), and shear rate in a complex and intertwined manner characterized by multiple feedback loops.

magma transport phenomena This term encompasses all the dynamical processes responsible for the generation, ascent, eruption or emplacement of magma, its subsequent quenching or solidification, and interaction in hydrothermal magmatic systems. These processes involve simultaneous consideration of heat, mass, and momentum transport between relevant magmatic subsystems. Rates of momentum, heat, and mass transfer are widely varied ranging from creeping (percolative) flow of small degree partial melts through an otherwise solid parental source rock in response to pressure, buoyancy, and viscous forces to the rapid (several hundred meters per second) eruption of low density highly expanded magmatic mixtures characteristic of high-silica, volatile-rich rhyolitic magma that erupts to form pyroclastic flows of great natural hazard. The flow of magma in fractures, buoyant rise of magma diapirs and evolution of magma within crustal magma bodies that undergo simultaneous assimilation of wall rock, recharge of fresh parental magma and fractional crystallization and convective mixing (or unmixing) are all examples of magma transport phenomena.

metasomatism Metasomatism refers to the process whereby a preexisting igneous, sedimentary, or metamorphic rock undergoes compositional and mineralogical transformations associated with chemical reactions triggered by the reaction of fluids (so-called metasomatic agents), which invade the protolith. A large-scale example is provided in subduction environments. Dehydration of hydrothermally altered oceanic crust and attached upper mantle generates fluids that migrate upwards and metasomatize the ultramafic mantle wedge lying above the subducting slab. Because the peridotite solidus temperature is lowered by the presence of water, metasomatism of the mantle wedge can trigger partial melting there. A smaller scale environment in which metasomatism is important occurs in contact metamorphic aureoles surrounding granitic plutons when emplaced into cooler preexisting crust. The magma body acts as a heat engine and drives hydrothermal circulation in the surrounding country rock. Economically important mineral deposits form in contact metamorphic environments.

rhyolitic magma Rhyolitic magma is erupted at temperatures 750−1000 °C, and is made up mainly of SiO_2 (75 wt%), Al_2O_3 (13%), and the Na_2O and K_2O in roughly equal amounts (3−5%). Only a small fraction of a cubic kilometer of rhyolitic magma is erupted each year mainly in continental environments although a significant fraction of rhyolitic and intermediate-composition magmas may stagnate at depth within the crust crystallizing there to form granitic plutons. Granitic batholiths are generally associated with subduction zone magmatism and can extend for hundreds to thousands of kilometers roughly parallel to the strike of oceanic trenches and along active continental margins. The density, specific heat, viscosity, and thermal conductivity of typical rhyolitic magma bearing 2 wt% dissolved H_2O at 900 °C at low pressure, are 2260 kg/m^3, 1450 J/kg K, 500,000 Pa s, and 1 W/m K, respectively. A water-free rhyolitic melt of otherwise identical composition has a density and viscosity of 2350 kg/m^3 and 1.2×10^{10} Pa s, respectively. The heat needed to completely fuse (melt) granite, the plutonic (crystalline) equivalent of rhyolitic magma, is ∼300 kJ/kg. Pristine volatile contents of typical rhyolitic magmas lie in the range 1−6 wt% and are mainly H_2O although finite amounts of CO_2 and sulfur-rich gasses (H_2S, SO_2, etc.) are also present. Volcanic gasses are often corrosive and contain HCl and HF as minor constituents.

thermodynamic properties Magma thermodynamic properties may be separated into two families: thermal functions and the equation of state (EOS). Thermal functions relate the temperature of a substance to its internal energy. These properties include the enthalpy, entropy, and heat capacity. The EOS relates the density of a substance to its composition, pressure, and temperature, and important properties include the isothermal compressibility and isobaric expansivity. The reactivity of a material depends on its chemical potential that involves both thermal functions and EOS data.

transport properties In typical dynamic (nonequilibrium) magmatic systems, gradients in pressure, stress or velocity, temperature, and chemical potential give rise to transport of momentum, heat, and mass, respectively. Momentum, heat, and mass can be transported by advective or diffusive flow. The transport properties that govern diffusive flow of momentum, heat, and mass are the viscosity, thermal conductivity and self, tracer and chemical diffusivities, respectively.

NOMENCLATURE

T Temperature (K)
P Pressure (Pa)
ρ Density (kg/m^3)
C$_p$ Molar isobaric heat capacity (J/mol K)
c$_p$ Specific isobaric heat capacity (J/kg K)
C$_{pi}$, c$_{pi}$ Partial molar, specific isobaric heat capacity (J/mol K), (J/kg K)
S Entropy (J/K mol)
H Enthalpy (J/mol)
σ Interfacial surface energy (N/m)
β$_T$ Isothermal compressibility,$(\partial \ln\rho/\partial P)_T$ (Pa^{-1})
α$_P$ Isobaric expansivity, $-(\partial \ln\rho/\partial T)_P$ (K^{-1})
X$_i$ Mole fraction of ith component
V$_i$ Partial molar volume of ith component (m^3/mol)
K$_T$ Isothermal bulk modulus (Pa)
K' Pressure derivative of bulk modulus
η Viscosity (kg/m s ≡ Pa s)
E$_A$ Activation energy for viscous flow (J/mol)
V$_A$ Activation volume for viscous flow (m^3/mol)
η$_r$ Relative viscosity, ratio of mixture viscosity to melt viscosity
φ Volume fraction dispersed phase (solid or vapor)
k$_R$ Radiative (photon) conductivity (J/m K s)
k Thermal conductivity (J/m K s)
κ Thermal diffusivity, k/ρc$_p$ (m^2/s)
D Self, tracer or chemical diffusivity (m^2/s)
E$_a$ Activation energy for diffusion (J/mol)
V$_a$ Activation volume for diffusion (J/mol)

Subscripts

fus fusion
form formation
conf configuration
mix mixing
r relative

1. INTRODUCTION

A central goal of studies in igneous petrology and volcanology is to understand the factors that lead to the compositional diversity of magmatic rocks and the related issue of the origin of the Earth's crust and mantle. In the crust are found essentially all of the material and energy resources accessible to the world population of 7 billion. This understanding is most powerful when framed in terms of petrogenesis—the origin of **magma** and the rocks formed from cooling and solidification of magma—within the context of the coupling between the thermal evolution and chemical differentiation of Earth. The growth of oceanic and continental crust throughout the ~ 4560 million years of earth history, the extent of recycling of crust and lithosphere by subduction, the relationship of mantle plumes to the compositional differentiation of Earth, and the role of subduction in island arc magmatism and growth of continental crust are problems that studies of magmas and magmatic processes shed light upon. In addition, a close connection exists between magmatic diversity and practical problems, such as volcano forecasting, the mitigation of volcanic hazards, and the discovery of material and energy resources such as ore deposits and geothermal heat. The **thermodynamic** and **transport properties** of magma are central to all of these considerations.

There has been a sustained and accelerating effort in the last century to determine the properties of magma both in the laboratory and by application of models based on the chemistry and physics of materials. It is no exaggeration to claim that without these foundational measurements and theoretical models, petrology could not have evolved far beyond the purely descriptive stage. In this chapter, we provide an overview of current knowledge of the subject and future directions.

The composition of lava emitted from a volcanic center reflects both the composition of its source as well as myriad dynamical phenomena operating during its generation, segregation, ascent, residence time in crust (storage) and eruption. For the purposes of this chapter, magma is defined as a high temperature (generally >900 K), multiphase mixture of crystals, liquid, and vapor. The vapor can be either a gas or a supercritical fluid. Herein, the term liquid is used interchangeably with melt (i.e., they are synonyms) and fluid generally restricted to a phase that is rich in H, O, C, N, Cl, and S. The solid fraction of magma is primarily made up of oxide and silicate crystals, the relative abundance of which depends strongly on composition, temperature, pressure, and additional nonequilibrium or kinetic factors, such as cooling rate, rate of decompression, and rates of mass transfer by molecular diffusion, mechanical dispersion, and advective transport. Bits of local wall rock (lithic inclusions), cognate crystal cumulates, or exotic xenoliths are also commonly found in plutonic and volcanic rocks and provide clues to understanding magma genesis and dynamics as well as a providing constraints on the bulk composition of the source. The liquid or melt portion of magma is generally a multicomponent (O—Si—Al—Ca—Mg—Fe—K—Na—H—C) silicate liquid; crustal melts (silicic or rhyolitic to intermediate or andesitic) are rich in O—Si—Al—Na—K—H whereas melts generated within the Earth's mantle are richer in O—Si—Al—Ca—Mg—Fe (mafic

or basaltic). Carbonatitic melts containing >50 wt% carbonate are also generated within the mantle. Although the volumetric rate of eruption of carbonatitic magma is very small compared to the roughly 25 km^3/year of mid-ocean ridge basaltic magma produced, rare carbonatitic magmas may be important agents for mantle **metasomatism**. Carbonatites also have an affinity with economically important diamond-bearing kimberlitic magmas, which are rapidly erupted from depths of several hundred kilometers, if not more.

Common magmas vary nearly continuously in composition from basaltic (50 wt% SiO_2) to rhyolitic (75 wt% SiO_2). These magmas usually contain small amounts (on the order of a few weight percent) of dissolved H_2O, CO_2, and other volatile species such as H_2S, N_2, HCl, HF, COS, and SO_2. H_2O dissolved in melts occurs in two forms: molecular H_2O and as the hydroxyl polyanion OH^-. The ratio of molecular water to hydroxyl depends on the composition (devolatilized) of the melt. When sufficient volatile components are present, melt becomes supersaturated and a discrete vapor or supercritical fluid forms. This fluid has a low viscosity and density compared to silicate melts and is particularly rich in the components O—H—C—S—N—Cl—F. The speciation of this fluid depends on bulk composition, temperature, and pressure. At crustal pressures and temperatures, supercritical fluids, especially those rich in molecular H_2O, are quite corrosive and can dissolve a few percent or more by mass of other oxide components. In part, this is due to H_2O's dipolar nature and large dielectric constant making it a good solvent. At very high pressure and temperature even larger quantities of solid may dissolve in these fluids, while at conditions of subduction zone magmatism, i.e., depths of ~ 100 km (3 GPa) and temperatures of 700—1400 °C, silicate melt and hydrous fluids can be completely miscible. The concentration of dissolved volatiles in a melt is strongly dependent on pressure because the partial molar volume of a volatile species in the dissolved state is much smaller than its molar volume in the gaseous or supercritical fluid state. It is this difference in volume that relates the Gibbs free energy, and hence solubility, to pressure at fixed temperature. The huge volumetric expansion of a magmatic mixture that accompanies exsolution of volatiles is one of the primary causes of explosive volcanism as pressure—volume (PV) expansion work is converted to kinetic energy. Although at depths of a few kilometers magma ascent rates may be only a fraction of a meter per second, once volatile exsolution commences and PV work is converted into kinetic energy, magma eruption speeds of order 100—300 m/s can easily develop. Other factors, such as magma viscosity and the rates of volatile component diffusion, and their variations with temperature and pressure are also important in assessing the dynamics of explosive volcanism.

TABLE 5.1 Estimated Properties of Common Natural Melts at 1 bar (10^{-4} GPa) at Their Respective Liquidus Temperatures. All Compositions Are Anhydrous Except Where Indicated. See Tables S5.1−S5.7 for More Exhaustive Compilations of These Properties.

Composition	Liquidus Temperature (°C)	Specific Heat of Fusion (kJ/kg) Δh_{fus}	Density (kg/m^3) ρ	Specific Isobaric Heat Capacity (J/kg K) c_p	Melt Viscosity (Pa s) η
Granite (dry)	900	220	2349	1375	1.2×10^{10}
Granite (2 wt% H_2O)	900	250	2262	1604	5×10^{5}
Granodiorite	1100	354	2344	1388	1.3×10^{6}
Gabbro	1200	396	2591	1484	30.0
Eclogite	1200	570	2591	1484	−
Komatiite	1500	540	2748	1658	0.15
Peridotite	1600	580	2689	1793	0.25

Volcanic rocks, in particular, play a unique role in efforts to understand magmatic transport phenomena because, unlike plutonic rocks, they are quenched relatively rapidly and provide more or less direct information regarding the composition of natural melts. In order to understand the significance of the chemical composition of volcanic rocks, the dynamical and physical aspects of its parental magma evolution must first be unraveled. This is not an easy task. The range of transport phenomena of potential relevance to petrogenesis is quite varied and covers large spatiotemporal scales. Although the dynamics can be complex, it is clear that the thermodynamic and transport properties of magma are absolutely pivotal to the success of any quantitative dynamical theory of magma genesis and transport. If twentieth-century research in petrology can be summarized in a few words, it was the century of quantification—quantification of the energetics of magma. Although far more needs to be learned, it is reasonable to claim that geologists now have a first-order understanding of the basic properties of the common magma types and are beginning to understand how such properties relate to igneous petrogenesis—the origin of igneous rocks. In the twenty-first century, new instrumental techniques are providing a treasure trove of information regarding the composition of crystals, glass inclusions, and grain boundaries at the micron- to nanoscale. This information must ultimately be connected to dynamics of magmatic systems through the macroscopic petrogenetic prism. This will require an even better understanding of the properties of melts and magmas. In this chapter, the most critical properties will be briefly reviewed in the context of petrogenesis. Representative data are presented in the figures, Tables 5.1 and 5.2, and 11 data compilations archived as Supplementary Material Tables S5.1− S5.11. These compilations are not exhaustive and particular research applications may require a return to the original sources for additional details. Some of these sources are listed in Further Reading and many more are provided in the Supplementary Material section. The Nomenclature includes symbols used in the text.

2. MAGMATIC SYSTEMS: TIME AND LENGTH SCALES

Lifetimes of magmatic systems and processes vary widely—from the rapid radiative quenching of a millimeter-sized melt droplet (pyroclast) during its high-speed ejection from a vent (several seconds), to the slow cooling of a single lava flow (weeks or months), to the hundred thousand to million-year timescale of a large pluton cooling mainly by phonon (heat) conduction and hydrothermal convection. Individual magmatic hydrothermal systems, such as at Yellowstone National Park, USA, can remain active for a few millions of years because heat conduction is intrinsically a slow process (rocks are good insulators) and because many magmatic systems are open systems replenished by entry of magma rising from deeper in the crust or mantle below. This replenishment or recharge magma is often hotter than resident magma because it comes from greater depths where higher temperatures are usually found. In terms of dynamical process, creeping percolation flow in a partial melt region is of the order of several meters or less per year and may be contrasted with the explosive eruption of a volatile-rich magma at several hundred meters per second. Although rates of heat, mass, and momentum transfer are wildly different in these systems or subsystems, knowledge regarding the

TABLE 5.2 Representative Phonon Thermal Conductivity of Geosilicate Liquids and Glasses, and Olivine. Data Sources and More Exhaustive Data Compilations Are Provided in Tables S5.8 and S5.9. Experimental Measurements Are Given in Roman Type. Values in Italic Type Are Determined From Recent MD Simulations (Tikunoff and Spera, 2014).

Composition	Temperature (°C), Pressure (GPa)[1]	Thermal Conductivity (W/m K)
Olivine (Fo_{90})	800	2.69
	1400	2.18
Olivine (Fo_{100})	1400	1.59
Basalt (glass)	300	1.48
Obsidian (rhyolite glass)	300	1.67
Rhyolite (liquid)	800–1100	1.5
$NaAlSi_3O_8$ (glass)	800	1.56
$NaAlSi_3O_8$ (liquid)	1200	1.59
	1800	*1.45*
	1800, 10	*2.16*
$CaMgSi_2O_6$ (glass)	800	1.46
$CaMgSi_2O_6$ (liquid)	1100	1.21
	1800	*1.14*
	1800, 10	*2.02*
Mg_2SiO_4 (liquid)	3000	*1.06*
	3300	*0.90*
	3300, 10	*2.11*

[1]Pressure is 1 bar (10^{-4} GPa) unless noted.

hundreds to thousands of kilometers deep may have existed. Modeling this system demands knowledge of the properties of magma and phase equilibria for temperatures in the range 1000–5000 K and pressures at the Earth's surface (10^{-4} GPa = 1 bar) to the core—mantle boundary (135 GPa). Thus, knowledge of the physical properties of magmas is essential to virtually all facets of magmatism throughout Earth history at a wide variety of scales in space and time, including the environmental impacts and hazards of volcanism on virtually all inhabitants of our planet. This is the realm of Earth System Petrology. Because the terrestrial planets and minor bodies (asteroids) of the solar system have a similar origin, study of magmatism is also a key element in understanding solar system origin and early evolution. It has been estimated that tens of billions of terrestrial planets, that is, planets composed of metal and rock, exist in the Milky Way galaxy, one of about 300 billion galaxies in the observable Universe. The properties of high-temperature silicate melts and magmas are therefore relevant to phenomena at cosmological scales throughout the past 13.8 billion years.

3. MAGMA THERMODYNAMIC PROPERTIES

Magma properties can be separated into two groups—equilibrium thermodynamic quantities and transport properties. Important equilibrium thermodynamic quantities include density (ρ), heat capacity (C_P), third law and configurational entropies (S), enthalpies of formation (ΔH_{form}), fusion (ΔH_{fus}), mixing (H_{mix}), and interfacial surface energy (σ). Application of equilibrium thermodynamic data are most useful in addressing the influence of source bulk composition, pressure, and temperature on the composition and amount of melt generated during partial fusion, as well as the composition of solids and residual melts during solidification of magma. Although chemical and thermal equilibrium are not perfectly attained in nature, the concept of local equilibrium is useful because equilibrium is often closely approximated at some scale and because it represents a reference state from which deviations from equilibrium can be assessed. When chemical equilibrium is assumed, all the classical theory of chemical thermodynamics can be applied and many times this greatly simplifies a problem and allows one to test and discriminate between competing hypotheses. For example, at constant temperature and pressure, minimization of the Gibbs free energy for a fixed bulk composition allows one to compute the composition and abundance of each phase in the equilibrium assemblage. To compute the energetics of melting, heat capacities and enthalpies of fusion of appropriate phases are essential. Although not strictly an equilibrium process, calculation of magma heat transport

thermodynamic and transport properties of magma is critical in order to meaningfully analyze and ultimately predict the relevant dynamics regardless of the particular scale.

Length scales relevant to magma transport and genesis also vary over many orders of magnitude. At the smallest scale, submicrons to millimeters, nucleation and growth of crystals, or vapor bubbles depend on local fluctuations in melt structure and the rates of mass diffusion. At larger scale, differentiation involves the physical separation of crystals, melt, and vapor in conduits and chambers over meters to kilometers, while the generation, transport, and ponding of magmas can extend over hundreds of kilometers. Batholithic terrains, the product of multiple emplacement of many individual plutons have aerial extents on the order of thousands of square kilometers. In the early part of Earth history, a globe-encircling magma ocean requires heat capacity and fusion enthalpy data, in addition to thermal diffusivities. The variation of melt density with temperature, pressure, and composition is needed for analysis of momentum transport since buoyancy is a significant factor driving the segregation, ascent, and eruption of magma. The differentiation of an emplaced magma body by gravity-driven crystal fractionation critically depends on the density difference between melt and newly formed crystals. Like an iceberg in the ocean, there are conditions in temperature and pressure space where crystals grown in a melt float rather than sink. The final solidified state of such a system is obviously quite dependent upon the equation of state (EOS)—the relationship between density, temperature, and pressure—for all relevant phases. For example, the compressibility of magma, defined according to $\beta_T = (\partial \ln \rho / \partial P)_T$, informs us about the variation of magma density with pressure and is therefore important for constraining buoyancy force(s) driving magma ascent and the depths of ponding where the density contrast between magma and host rock vanishes. Magma compressibility also is an important factor governing the explosivity of magma at or near the earth's surface, especially for degassing, volatile-rich magmas. These examples show that there is a deep connection between the equilibrium thermodynamic properties of magma and its petrogenetic transport history. Without fundamental property data, it is impossible to exploit the equations governing conservation of energy, mass, species, and momentum to solve quantitatively magma transport problems. The equilibrium chemical thermodynamic model provides a logical starting point for the analysis of magma production.

Natural silicate melts contain SiO_2, Al_2O_3, MgO, CaO, iron oxides, the alkalis (Na_2O and K_2O), P_2O_5, transition metals, and rare earth elements, and minor but important amounts of volatile compounds made up of $H-O-C-S-Cl-F-N$. A complete thermodynamic description of such a complex multicomponent system over the relevant range in temperature–pressure–composition space is beyond the scope of this chapter (see Chapter 6—Chemical Thermodynamics and the Study of Magmas by M. Ghiorso and G. Gualda—for details). Many excellent reviews have also been published on the thermodynamic properties of silicate liquids including the relationship between structure and properties listed. A few of these are listed in Further Reading. We summarize the salient features here.

3.1. Density and EOS

The density of a silicate melt may be determined by evaluating the quotient

$$\rho = \Sigma X_i M_i / \Sigma X_i V_i \qquad (5.1)$$

where X_i is the mole fraction, M_i is the molar mass, and V_i is the partial molar volume of the ith oxide component in

the melt. Although V_i depends weakly on composition, to a good approximation, it may be taken independent of composition and as a function of temperature and pressure for most petrological calculations. Partial molar isothermal compressibilities for the major oxide components have been determined from ultrasonic sound speed laboratory experiments. These parameters can be used to compute melt density as a function of temperature, composition, and pressure up to several GPa using a simple empirical EOS

$$V(T, P, X) = \sum X_i \left[V_{i, T_r} + \left(\frac{\partial \overline{V}_i}{\partial T} \right)_P (T - T_r) \right.$$
$$\left. + \left(\frac{\partial \overline{V}_i}{\partial P} \right)_T (P - P_r) \right] \qquad (5.2)$$

where T_r and P_r are reference conditions (generally $1400\,°C$ and 10^{-4} GPa (1 bar), respectively) and V_i is the partial molar volume of the ith component. Parameters for computing melt density as a function of pressure (up to several GPa) and temperature is presented in Table S5.1. Densities of some common natural melts at their respective approximate liquid temperatures are given in Table 5.1 (and Table S5.2) and shown graphically as a function of temperature and volatile content in Figures 5.1 and 5.2. As a rule of thumb, adding FeO, Fe_2O_3, MgO, TiO_2, and CaO to a melt increases its density, whereas adding alkalis (Li_2O, Na_2O, and K_2O) and volatiles (H_2O, CO_2) have the opposite effect. The temperature derivative of the partial molar volumes of SiO_2 and Al_2O_3, are both effectively zero. As a

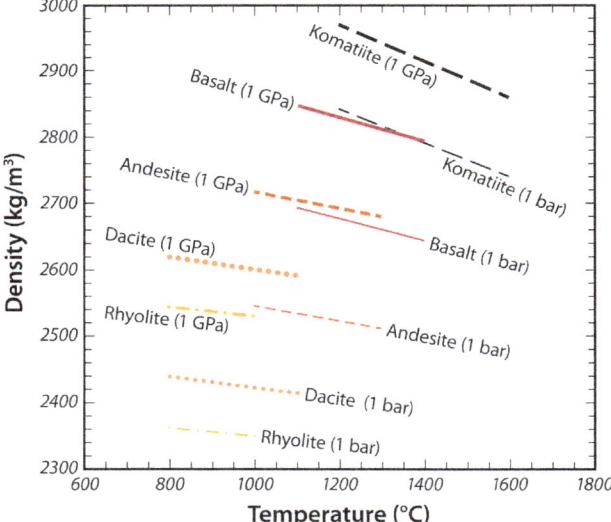

FIGURE 5.1 Melt density as a function of temperature for natural melts spanning the compositional range rhyolite to komatiite at 1 bar (10^{-4} GPa) and 1 GPa pressure. All compositions are volatile-free.

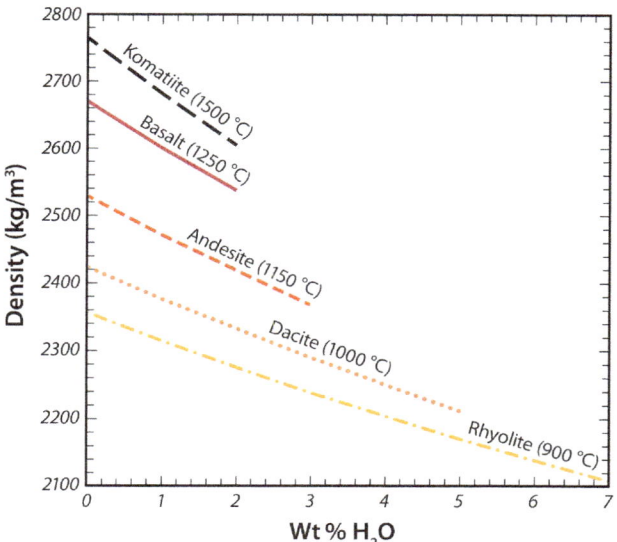

FIGURE 5.2 Melt density as a function of dissolved water content for natural melts spanning the range rhyolite to komatiite. Temperatures for each composition are characteristic eruption temperatures. Densities are calculated at a pressure of 1 bar (10^{-4} GPa).

consequence, the density of a melt rich in silica and alumina is quite insensitive to temperature. Densities of melts vary from 2500 to 2900 kg/m^3 at magmatic temperatures (~ 1100 °C) for silicic through intermediate to mafic and ultramafic compositions at low pressure. Density increases as pressure rises, although increasing the concentration of the alkaline earth metals, Ca, and Mg, serves to decrease the pressure dependence of melt density. The alkali metals and water are relatively compressible and contribute significantly to the pressure dependence of density. It is important to note that small differences in composition can compensate for relatively large differences in temperature because of the strong dependence of melt density on composition. Whereas a typical value for the isobaric expansivity is $\alpha_P = -(\partial \ln \rho / \partial T)_P \approx 5 \times 10^{-5}$/K, the analogous quantity describing the variation of density with composition is of order 0.1 to 3 depending on the component. The volatile components H_2O and CO_2 have an especially large effect on melt density. The partial specific density (M_i/V_i) for CO_2 and H_2O in natural melts are ~ 1400 and 900 kg/m^3, respectively, for typical crustal magmas. For comparison, M_i/V_i for K_2O, MgO, FeO, and SiO_2 are ~ 2000, 3500, 5300, and 2200 kg/m^3, respectively. Thus, a small amount of dissolved water dramatically lowers melt density and significantly affects transport and thermodynamic melt properties such as viscosity, mass diffusivities, thermal conductivity, and isobaric heat capacity. The temperature at which rocks begin to melt (the solidus temperature) is substantially lowered when volatile constituents such as H_2O are present. Because oxygen is volumetrically by far the most abundant anion in the silicate Earth, the major incorporation mechanism of H_2O into nominally anhydrous silicates and oxides, including melts, is as the hydroxyl anion, OH^-. Almost all of the nominally anhydrous minerals that compose the earth's crust and mantle can incorporate measurable amounts of "water." The amount of water (generally as OH^-) incorporated into nominally anhydrous minerals generally increases with pressure and sometimes with temperature, and is typically in the range 50–300 ppm in the dominant minerals of the earth's upper most mantle (olivine, pyroxenes, and garnet). The solubility of water in the $(Mg,Fe)_2SiO_4$ polymorphs wadsleyite and ringwoodite stable in the transition zone (~ 410 to ~ 660-km-depth) is markedly higher, and in total it has been estimated that several ocean's worth of water may be present in Earth's mantle. The recent discovery of upward of 1 wt% water in ringwoodite encapsulated in a natural diamond from a Brazilian kimberlite bolsters the claim that much of the water recycled into the deep mantle is trapped by the transition zone. This has important implications for the phase relations, dynamics, and magmas formed by partial melting of the mantle.

The change in volume of a crystalline silicate upon fusion under isobaric conditions is important because it influences the slope of the melting curve in pressure–temperature space and therefore the generation of magma by partial fusion. For small degrees of partial melting under nearly isochoric conditions, melt pressure can rise above lithostatic values and the small melt overpressure can drive magma fracture and vertical transport. Additionally, melt density influences the separation rate of melt from its source. Volume changes accompanying melting of some silicates are given in Table S5.2. Typically, the molar volume of a crystalline phase increases by about 10% upon fusion although silica (like ice) may shrink (i.e., become more dense) when fused at low pressure. Less is known regarding density–temperature–pressure relations or EOS for natural carbonatite melts. Some data pertaining to the density and viscosity of simple carbonate liquids as well as supercritical H_2O are given in Table S5.3. Relative to common silicate melts, carbonatite liquids are both less dense (by about 30%) and less viscous by several orders of magnitude compared to mafic melts and up to 10 orders of magnitude compared to silicic ones.

At pressures corresponding to depths greater than about 100 km (>3 GPa), the density of melts at superliquidus temperatures can be determined from high-pressure static (sink–float) and shock wave experiments. More recently, a vast array of spectroscopic tools, including X-ray absorption, tomography and diffraction/scattering, and ultrasonic measurements in conjunction with static high-pressure conditions are being employed to augment data obtained by more traditional approaches—in some cases accessing subliquidus conditions more directly applicable to magma genesis and differentiation. First-principles or classical molecular dynamics (MD) simulations are increasingly being utilized as a complement to experimental measurements, and enable predictions beyond experimental reach.

A useful quasi-theoretical relationship is the Birch–Murnaghan EOS that relates liquid (of fixed composition) density to pressure and temperature according to:

$$P = \frac{3}{2}K_T\left[\left(\rho/\rho_{T,0}\right)^{7/3} - \left(\rho/\rho_{T,0}\right)^{5/3}\right]$$
$$\left[1 - \frac{3}{4}(4 - K')\left(\rho/\rho_{T,0}\right)^{2/3} - 1\right] \tag{5.3}$$

where $\rho_{T,0}$ is the density of melt at 10^{-4} GPa and temperature T, K_T is the isothermal bulk modulus ($K_T = 1/\beta_T$) at 1 bar (10^{-4} GPa), K' is the pressure derivative of K_T, and ρ is the melt density at pressure P and temperature T. Many other equations of state have been proposed and find application in geoliquids. Compressibility data for silicate melts are presented in Table S5.4. Normalized density–pressure relations for silicate liquids ranging from felsic to ultramafic compositions are shown in Figure 5.3. Polymerized rhyolite liquid possessing an open network structure is vastly more compressibility than basalt or andesite, while highly depolymerized and compact melts such as komatiite and peridotite are the least compressible. An interesting application of high-pressure EOS studies is that

MgO-rich komatiitic melts, probably the most common magma type erupted in the first billion or so years of earth history, become denser than olivine crystals at pressure corresponding to depths of roughly 250–400 km. This has important implications for differentiation of the silicate portion of Earth including internal mineralogical layering as well as possible chemical stratification. Moreover, the compressibility of silicate liquids also bears on the dynamic stability of melts in the mantle, and the interpretation of low seismic velocity regions there. Models calling on the presence of (neutrally buoyant) melt immediately above the core–mantle boundary (ultra low-velocity zone), at the top of the transition zone (~410 km), and at a depth of about 220 ± 30 km (Lehmann discontinuity) depend critically on the details of how silicate melts densify as pressure increases. This is an area of very active research.

3.2. Enthalpy, Entropy, and Heat Capacity

In addition to volumetric or EOS data, the calorimetric properties of high temperature and molten silicates are needed to analyze magma transport problems. These properties inform us regarding the internal energy of melts and crystals and how internal energy and other closely related thermodynamic functions change with temperature. These properties include the enthalpy, entropy, and heat capacity and are inextricably bound to the thermal evolution of magma and relevant to processes such as partial melting, solidification, and the advective transport of heat. Melting temperatures, enthalpies of fusion, entropies of fusion, and specific heats of fusion for some common phases are listed in Table S5.5. The specific enthalpy of fusion is the heat per unit mass needed at a reference pressure (generally 10^{-4} GPa or 1 bar) to transform a crystal or crystalline assemblage to the liquid state. There is a wide variation in the specific enthalpy of fusion from 100 to 300 kJ/kg for silica polymorphs, albite, and sanidine, all important components in crustal-derived melts, to ~1000 kJ/kg for refractory phases, such as forsterite, pyrope, enstatite, and transition metal oxides relevant to mafic and ultramafic compositions. Table 5.1 provides estimates of fusion enthalpies for common compositions computed using data from Table S5.5 and typical modal abundances. There is a factor of three difference in the specific heat of fusion for a model granite (~200 kJ/kg) compared to an ultramafic composition (~600 kJ/kg). The more refractory nature of the ultramafic composition is a reflection of the higher fusion enthalpy of olivine and the pyroxenes relative to quartz, alkali feldspar, and plagioclase. Garnet also has a high specific fusion enthalpy. Melting eclogite (pyroxene plus garnet) at high pressure requires more heat (570 kJ/kg) than a corresponding low-pressure plagioclase-pyroxene gabbro (~400 kJ/kg). The relatively low fusion enthalpy for mineral phases comprising continental crust implies

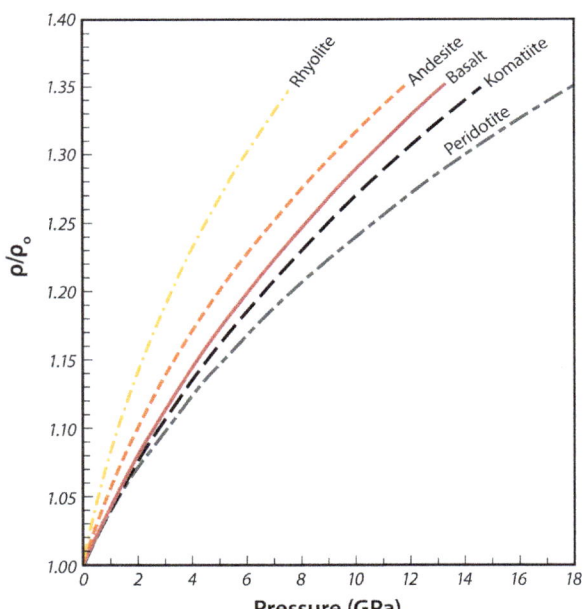

FIGURE 5.3   Normalized density (ρ/ρ_o) at 2000 K as a function of pressure for natural silicate liquids ranging in composition from rhyolite to peridotite computed using Eqn (5.3) and parameters from Table S5.4, assuming an average $dK_T/dT = -0.022$.

that anatexis of crust by heat exchange between mafic magma and its surroundings is thermally efficient. It is worth noting that the heat required to completely melt Earth's mantle $\sim 3 \times 10^{30}$ J, is less than 10% of the kinetic energy delivered to Earth by impact of a Mars-sized body (15% mass of Earth) with an impact velocity equal to the Earth's escape velocity of 11.2 km/s. If, in fact, the earth did possess a magma ocean in the Hadean, then its lifetime would have been controlled by the rate at which $\sim 10^{30}$ J of heat could be dissipated by conduction, radiation, and convection to space.

In addition to heat effects associated with solid to liquid phase changes, it is important to consider sensible heat effects or the variation of the enthalpy of the melt with temperature. This is measured by the molar isobaric heat capacity, C_p. For most aluminosilicate glasses and melts of petrologic significance, C_p can be approximated as an additive function of oxide component partial molar isobaric heat capacities. As temperature increases for glasses, an increased number of vibrational modes become excited and so the isochoric heat capacity $C_V \equiv (\frac{\partial E}{\partial T})_V$ increases toward the high temperature Dulong–Petit limit as vibrational modes become saturated. The isobaric heat capacity is related to the isochoric one by the expression $C_P = C_V + \frac{T\alpha_T^2}{\rho\beta_P}$. Parameters for the common oxide components to estimate the isobaric heat capacity of a glass as a function of composition and temperature at low pressure are listed in Table S5.6. At the glass transition temperature (T_g), there is generally a discontinuous jump in isobaric heat capacity. The magnitude of this jump depends on the atomic structure of the liquid. Highly polymerized melts—so-called, strong liquids—exhibit little or no discontinuity in heat capacity at T_g, while depolymerized, i.e., fragile, melts do. For petrologic purposes, it is sufficient to note that the jump in C_p at the glass transition temperature is small for silica-rich liquids (e.g., rhyolitic melts) and it becomes larger for more mafic liquids (andesitic and basaltic melts).

For silicate melts (unlike silicate glasses), the temperature dependence of C_p can generally be neglected unless great accuracy is required. Parameters for estimating C_p for melts as a function of composition at magmatic temperatures are presented in Table S5.7. Table 5.1 gives specific isobaric heat capacities (c_p) for some common naturally occurring anhydrous silicate melts. Silicic anhydrous melts have c_p around 1300–1400 J/kg K, whereas melts with higher MgO contents have specific heats around 1600–1700 J/kg K. Because the advective transport of heat per unit volume is given by $\rho c_p \Delta T$ magmas are good transporters of mantle heat to crustal levels since ΔT is of order 100–1000 K. Magmatic water content can also have a large effect on the heat capacity of magmas. The partial isobaric heat capacity of water dissolved as hydroxyl

$(OH)^{-1}$ is $\sim 13,000$ J/kg K, while water dissolved in its molecular form (H_2O) has a partial isobaric heat capacity of ~ 4700 J/kg K. For comparison, c_p for ambient tap water is 4184 J/kg K. Likewise, the isobaric heat capacity of silica glass increases by $\sim 30\%$ with the addition of 1200 ppm OH, while adding 2 wt% H_2O to an otherwise anhydrous granitic melt will increase c_p by 20%. Thus, the addition or loss of magmatic volatiles (namely water) can have significant enthalpic effects that, in turn, will influence ascent rates, eruptibility, solidification timescales, and countless other magmatic processes.

4. MAGMA TRANSPORT PROPERTIES

Transport properties of magma play a crucial role in development of petrogenetic theory. Simple consideration of the various stages of magma transport from source to surface makes this abundantly clear. Some examples include the generation of magma by partial melting, melt segregation by porous or channel flow, and magma ascent by viscous flow in magma-filled propagating cracks. Other examples include differentiation of magma by crystal fractionation and the processes of magma mixing and crustal anatexis. In all these phenomena, significant amounts of momentum, heat, and chemical component transport occur. Transport of momentum, heat, and chemical species by magmas involves the material properties of viscosity, thermal conductivity, and mass diffusivity, respectively. Although data and models for the composition and temperature dependence of melt viscosity are available, less is known regarding the quantitative rheological properties of magmatic mixtures. There are relatively few experimental data bearing on the thermal conductivity of melts at high temperatures; this is perhaps the most uncertain of all magma transport properties but one with significance with respect to the thermal history of the Earth. Chemical, self, and tracer diffusivity data exist and correlations have been developed for estimating diffusivities as a function of species charge and size for melts of various temperature and composition, and treating more complicated multicomponent diffusion and associated isotopic effects. The effects of pressure are not well understood, but systematics are emerging for network former and network-modifying cations exhibiting different pressure dependencies. A brief review of these properties is presented below. The original references should be consulted for more detail.

4.1. Melt Viscosity

The dynamic viscosity (η) of magma is the material property that connects the shear stress in the fluid to the rate of strain. The viscosity determines the rate of diffusion of momentum. More precisely, the kinematic viscosity, defined $v = \frac{\eta}{\rho}$, measures the tendency of a melt or magma to diffuse velocity gradients. Critical issues include the effect of composition, pressure, and temperature on melt viscosity, as well as the rheological properties of magmatic suspensions. Although to a good approximation silicate liquids well above the glass transition temperature behave as Newtonian fluids (that is, they show a linear relationship between shear rate and shear stress at fixed temperature and pressure), this is not the case for magmatic multiphase suspensions that generally exhibit complex rheological properties or for natural glasses that are described as viscoelastic materials.

The temperature dependence of melt viscosity can be described by several models. The simplest is the Arrhenian model for a melt of fixed composition

$$\eta(p, T) = \eta_o \exp[(E_A + PV_A)/RT] \qquad (5.4)$$

In expression (5.4), η_o is the asymptotic viscosity as $T \rightarrow \infty$, E_A is the activation energy for viscous flow, V_A is the activation volume for viscous flow, and R is the universal gas constant. The activation energy and volume are constants for melt of fixed composition. The preexponential term is approximately a "universal" constant that varies quite weakly with composition and is in the range 10^{-4} to 10^{-5} Pa s. An optimal value of $\eta \approx 10^{-4.6}$ Pa is found empirically for melts in the range of natural compositions. Activation energies vary systematically with composition from values around 200 kJ/mol for mafic and ultramafic melts to 400 kJ/mol for more silicic compositions. Increasing the concentration of nonframework components such as the alkalis, the alkaline earths, and especially H_2O results in creation of oxygens with only a single nearest neighbor of Si or Al. The presence of these nonbridging oxygens (NBOs) destroys the network structure created by the linkage of tetrahedra to form 3 to 6+ membered rings at intermediate atomic scales of order $0.5-1$ nm. For the ideal network tetrahedral melt, each oxygen is bound to two Si, two Al or one Si, and one Al simultaneously. The archetypal example is molten silica in which each oxygen has two nearest neighbors of silicon and each silicon is surrounded by four nearest neighbors of oxygen. The increase in the concentration of NBO by addition of network modifiers (e.g., H_2O) lowers E_A. Small (several percent by mass) concentrations of water have a dramatic effect in lowering the viscosity of silicate liquids because water is 89% oxygen (by mass) unlike any other common oxide component. This is illustrated in Table 5.1 for the granitic composition where the addition of 2 wt% H_2O to a granitic melt lowers melt viscosity by a factor of 10^5. H_2O also tends to lower the viscosity of more mafic compositions, although the effect is less dramatic because fewer bridging oxygen (BO) exist in anhydrous mafic melts.

The relationship between liquid structure at the atomic level and macroscopic properties (both transport and thermodynamic) is an essential keystone of **magma transport phenomena**. The activation volume, a measure of the pressure dependence of viscosity, is usually a small fraction of the molar volume of the melt and changes sign, from negative to positive, as the fraction of NBO increases and in the case of polymerized melts pressure increases. A typical value for V_A for a network tetrahedral melt such as $NaAlSi_3O_8$ at low pressure is around -6×10^{-6} m^3/mol (-6 cm^3/mol), whereas for more depolymerized melt (e.g., $CaMgSi_2O_6$) V_A is around 3×10^{-6} m^3/mol ($+3$ cm^3/mol). Melts with a negative value of V_A show a decrease in viscosity as pressure increases—the opposite is true for melts with positive V_A.

Fully polymerized network melts, for which essentially all the oxygen is BO typically possess intermediate range order defined by the formation of n-membered rings ($n = 4-10$) of tetrahedra at low pressure. As pressure increases, the ring structure collapses and the "anomalous" effect of viscosity ($-V_A$) diminishes to a point that free-volume effects dominate and viscosity increases with pressure ($+V_A$). Elevated pressure also drives oxygen into higher (five and sixfold) coordination with Si and Al and such changes in the network structure also have been linked the change from anomalous to normal pressure dependence of viscosity.

Not all silicate melts follow the Arrhenian temperature—viscosity relationship. This is especially true for fragile liquids, e.g., melts containing a high proportion of NBO. An empirical relationship for predicting the temperature dependence of viscosity data in fragile non-Arrhenian melts at 1-bar pressure is the so-called Tammann—Vogel—Fulcher (TVF) expression

$$\eta = \eta_O \exp[B/(T - T_K)] \qquad (5.5)$$

where B and T_K are functions of composition but not temperature. T_K is a constant closely related to both the glass transition temperature and the Kauzmann catastrophe temperature. At $T < T_K$, the entropy of supercooled liquid computed by extrapolation of high-temperature data is *less than* that of its corresponding crystal (the Kauzmann Paradox). The computed TVF viscosity (Eqn (5.5)) explodes toward an infinite singularity as $T \rightarrow T_K$. The relationship between the TVF and the Arrhenian model is found by examining the temperature derivative of the viscosity. In the Arrhenian model, $\partial \ln \eta_{Arr}/\partial(1/T) = E_A/R$ whereas for a TVF fluid $\partial \ln \eta_{TVF}/\partial(1/T) = BT^2/(T - T_K)^2$. Thus, in the limit $T_K \rightarrow 0$, the TVF and the Arrhenian models are identical with $B = E_A/R$. In general, T_K is close to but usually somewhat less than T_g, the glass transition temperature. Typically, T_g for geosilicate compositions is ~ 1000 K. T_g for pure silica is somewhat higher (1500 K).

The TVF model therefore shows that the variation in melt viscosity with reciprocal temperature decreases as temperature increases and is not a constant (E_A) as in the Arrhenian model. The TVF empirical model draws some theoretical foundation from the Adams–Gibbs–DiMarzio (AGD) configurational entropy model. In AGD, viscosity depends on temperature according to

$$\eta = \eta_o \exp(D/TS_{conf}(T)) \qquad (5.6a)$$

where S_{conf} is the configurational entropy of the melt. S_{conf} includes a contribution computed from liquid and glass isobaric heat capacity data, as well as the residual entropy of glass (frozen liquid) at the glass transition temperature, T_g. The value of $S_{conf}(T)$ is computed from

$$S_{conf}(T) = \int_{T_K}^{T} \frac{\Delta C_p(T)}{T'} dT' \qquad (5.6b)$$

where the heat capacity term in Eqn (5.6b) is the experimentally observed difference in liquid and solid heat capacities. The parameter D is temperature independent but varies with composition. An important aspect of the AGD theory is that it connects thermodynamic and transport properties. It is based on a theory of cooperative behavior during dynamic rearrangement of atomic configurations in the liquid state. The AGD theory collapses to the empirical TVF expression provided the heat capacity difference is given by the hyperbolic expression $\Delta C_P(T) = \frac{CT_K}{T}$ in Eqn (5.6b). Although this identification is one of several possibilities, the collapse of the AGD expression to the TVF provides some theoretical support for the validity of the otherwise empirical TVF formulation.

An empirical model that gives melt viscosity as a function of composition and temperature based on the TVF formulation has been calibrated and in wide use (Giordano et al., 2008). This empirical model gives the added benefit of providing estimates of the glass transition temperature and melt fragility as a function of melt composition at low pressure. Pressure effects are not considered in the present form of this model, however. The Arrhenian pressure correction can be used to roughly account for pressure by simple extension of the TVF. For less precise work, the Arrhenian model still finds application because it is so simple to implement, although extrapolating beyond the bounds of the experimental measurements is risky at best. The viscosity of naturally occurring silicate melts can span over 10 orders of magnitude due to variations in composition and temperature (Table 5.1 and Figure 5.4). Dissolved water is especially effective in lowering the viscosity of polymerized melts (e.g., three orders of magnitude with addition of 3 wt H_2O for rhyolite melt), while it has comparative small effects on the viscosity of depolymerized (mafic and ultramafic) liquids (see Figure 5.5).

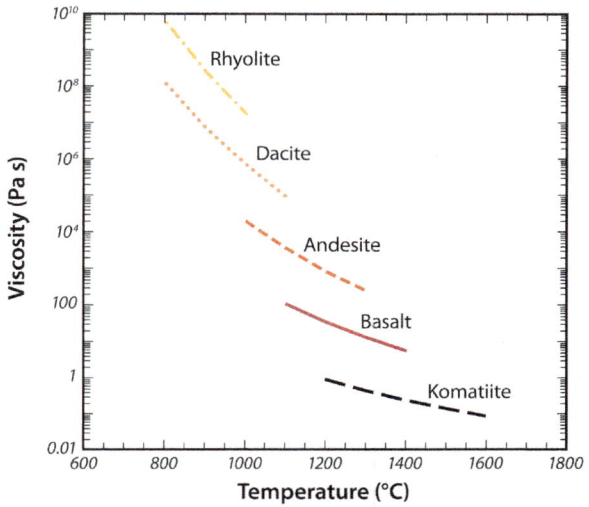

FIGURE 5.4 Viscosity as a function of temperature at 1 bar (10^{-4} GPa) for natural melts spanning the compositional range rhyolite to komatiite. All compositions are volatile-free. The temperature range is illustrative of typical eruption temperatures for each composition.

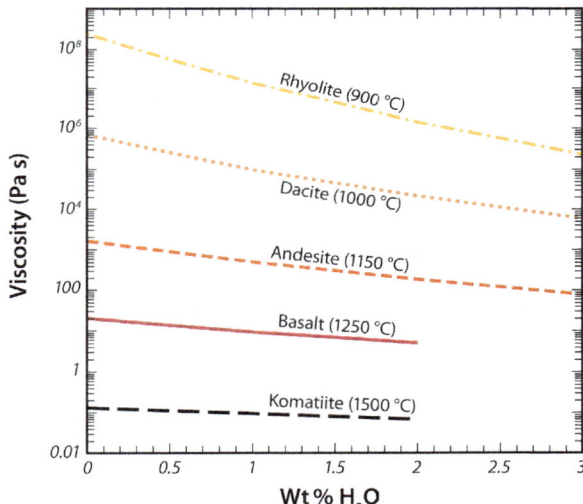

FIGURE 5.5 Melt viscosity as a function of dissolved water content for natural melts spanning the range rhyolite to komatiite. Temperatures for each composition are characteristic eruption temperatures. Viscosities are calculated for pressure equal to 1 bar (10^{-4} GPa). For the rhyolite composition, two different models are shown. Note the very dramatic effect of dissolved water on the viscosity of natural melts. Viscosity models are from Shaw (1972) and Giordano et al. (2008).

4.2. Magma Viscosity

Far fewer data exist on the rheological properties of magmatic suspensions compared to silicate melts. Magma, a mixture of suspended crystals and vapor bubbles in a melt matrix, may be expected to exhibit complex rheological behavior when significant amounts of suspended crystals and bubbles ("particles") are present. Silicate foams, defined as concentrated emulsions with porosity greater than about 60 volume percent ($\phi > 0.6$) are found in small quantities at virtually all volcanic centers because magma typically contains volatiles (0.1−10 wt%) and the common volatiles (H_2O and CO_2) are practically insoluble in melts at low pressure. Magma decompression during magma ascent commonly leads to volatile saturation. The important variables governing the viscosity of multiphase mixtures include the shear rate, the volume fraction of solid and bubbles, the size and shape distributions of these "particles" (solids and bubbles), temperature, pressure, and melt viscosity. There are no experiments exploring the rheological properties that account for these variables collectively and comprehensively although some effects, such as the effects of variable solid loading fractions, have been explored. A complication is that in natural systems the flow itself influences the distribution of particles via mechanical (sorting) and thermal effects. For example, viscous forces acting on particles can cause their movement relative to melt and each other, e.g., flow differentiation. A heterogeneous distribution of particles can evolve from an initially homogeneous distribution under the influence of particle-melt viscous drag and pressure forces. Similarly, thermal gradients can effect the local concentration of phenocrysts thereby modifying velocity gradients. The rheology of multiphase fluids is complex and currently the subject of intensive study. Fortunately, there has been some progress on two-phase mixtures (both crystal-melt suspensions and melt-vapor emulsions) applicable to magmatic systems. These studies enable one to approximate the viscosity of magmatic suspensions when suspensions are composed of monosized "particles" in the creeping flow regime at very low rates of shear. These models are valid from the dilute limit to the random packing volume fraction limit of $\phi \approx 0.64$.

To first order, the effect of increasing the volume fraction of crystals, ϕ, in a melt is to increase the relative viscosity, η_r, according to

$$\eta_r = \eta_{mix}/\eta_{melt} = (1 - \phi/\phi_o)^{-5/2} \qquad (5.7)$$

where η_{mix} is the viscosity of the crystal−liquid suspension or mixture, η_{melt} is the viscosity of aphyric melt, and ϕ_o is the maximum packing fraction of the solid. Equation (5.7) shows that the loading level of crystals is limited by ϕ_o, the geometric limit for contact of solids in the most efficient packing arrangement. For rhombohedral packing of spherical particles, $\phi_o = 0.74$ whereas for simple cubic packing and body-centered cubic packing $\phi_o = 0.52$ and 0.60, respectively. In practice, random packing of spheres generally gives ϕ_o in the range 0.60−0.65. Equation (5.7) shows that the volume fraction and the maximum packing fraction ϕ_o of solids are the two important parameters governing the viscous response of magma. Polydispersity, the presence of a nonunimodal distribution of particles,

tends to reduce the viscosity of a magma at a fixed ϕ because smaller particles can fit in between the larger ones. The viscosity of a unimodal crystal-bearing magma with $\phi = 0.3$ is about 10 times higher than its value for the melt phase alone based on Eqn (5.7). In more detail, it is both the size and shape variation of individual crystals, as well as the overall structure of the mixture, that strongly controls rheological properties. An alternative model that purportedly accounts for the presence of melt entrapped between particles gives smaller relative viscosities compared to Eqn (5.7). The expression is

$$\eta_r = \frac{\eta_{mix}}{\eta_{melt}} = \left(1 - \frac{\phi}{\phi_o}\right)^{-C_1 \phi_o} \qquad (5.8)$$

where the C_1 equals 2.5 and ϕ_o represents the random close packing limit. The parameters C_1 and ϕ_o vary depending on the particle size and shape distributions. For nonspherical particles, C_1 is larger than the Einstein value of 5/2, used in Eqn (5.7). For high shear rates, some particle arrangements can form where spherical particles tend to form clusters and the mixture is capable of flow at volume fraction that exceeds the zero shear rate limit $\phi_o \approx 0.64$. Presently, the most accurate model for a unimodal, solid + melt suspension at zero shear rate is

$$\eta_r = \frac{\eta_{mix}}{\eta_{melt}} = \left(\frac{1 - \phi}{1 - \frac{\phi}{\phi_o}}\right)^{\frac{C_1 \phi_o}{1 - \phi_o}} \qquad (5.9)$$

Values for C_1 and ϕ_o suitable for **magma rheology** calculations accounting for nonspherical crystals and the close pack limit are 3 and 0.64, respectively. Some typical relative suspension viscosities as a function of solid volume fraction are shown in Figure 5.6 based on Eqns (5.7)−(5.9) for creeping flow (low shear rate) of a suspension of rigid spherical monodisperse particles with the packing limit $\phi_o = 0.64$. Also shown is an approximation to the more realistic case of inequant particles based on Eqn (5.9) with $C_1 = 3$. At high-volume fraction solids, these various models exhibit order of magnitude differences in the relative viscosity—small variations in C_1 and ϕ_o amplify these differences.

In noncreeping flows, a nonlinear relation exists between the shear stress sustained by the mixture and the shear rate. Such suspensions exhibit power-law behavior. On a plot of shear stress versus shear rate, the tangent at a given shear rate is the effective viscosity of the suspension at that shear rate. Because the effective viscosity is a monotonically decreasing function of shear rate at fixed temperature and solid fraction, a value for the yield stress can sometimes be approximated by extrapolation of the tangent to the locus of points in the shear stress−shear rate plane to the stress axis at zero shear rate. More experimental work is required to

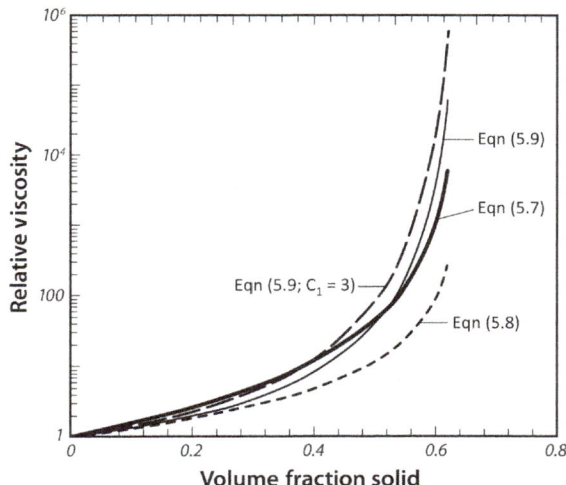

FIGURE 5.6 Relative viscosity of magmatic suspensions containing solids as a function of volume fraction of solids. The models refer to Eqns (5.7)–(5.9). Note the congruity of models for low crystal loading volume fraction and the divergence as the close packing limit of ~0.64 is approached. Relative viscosity models from Brouwers (2010). The coefficient $C_1 = 3$ allows for particle nonsphericity, approximately based on Eqn (5.9).

develop quantitative expressions that relate volume fraction solid to the apparent yield stress. In the meantime, Eqn (5.9) may be used as an approximation keeping in mind its inherent limitation to low shear rate flows.

In bubble-melt mixtures, λ viscosity can be either an increasing or decreasing function of the volume fraction of the low-viscosity phase (ϕ) depending on the rate at which the mixture is sheared. At low rates of shear ($\dot{\gamma}$), bubbles act as nondeformable inclusions and η_r increases with increasing ϕ similar to when solids (e.g., crystals) are added to melt. In distinction, at high shear rate, low-viscosity bubbles readily deform and the mixture viscosity decreases with increasing bubble fraction. A dimensionless parameter termed the capillary number (Ca) is useful in determining the appropriate rheodynamic regime. The capillary number is defined $Ca = \frac{\eta \dot{\gamma} r_b}{\sigma}$ where η, $\dot{\gamma}$, r_b, and σ represent the melt viscosity, shear rate, bubble radius, and melt-vapor interfacial tension, respectively. The capillary number can be thought of as the ratio of viscous tractions acting on the boundaries of a bubble that distort it from a spherical shape relative to interfacial surface forces that tends to preserve its spherical shape. Small values of Ca correspond to conditions where surface tension dominates and bubbles retain their spherical shapes. This is favored in finely dispersed (small bubble size), low-viscosity emulsions subjected to low rates of deformation. One expects a relative viscosity relation of the form $\eta_r = f(\phi, Ca)$ such that for small capillary number (roughly $Ca < 1$), η_r is an increasing function of ϕ. In contrast, for regimes with $Ca > 1$, the relative viscosity decreases with

increasing bubble content. Natural systems span the range from $Ca < 1$ to $Ca \gg 1$ so it is important to consider the dynamic regime when considering the effect of bubbles on the viscosity of magma. Relative viscosity can drop by a factor of 10 as the bubble content increases from 0 to ~50 volume percent in the high Ca regime.

Self-organization of dispersed particles (solids or bubbles) to form structured mixtures can give rise to large spontaneous changes in viscosity during flow. The formation of structure has been observed in laboratory models and probably quite relevant during flow of crystal slurries or vesiculated magma in chambers and conduits. Further experimental, analytical, and numerical work is needed to quantify the rheology of magmatic crystal-melt-vapor suspensions.

4.3. Thermal Conductivity

Transport of magmatic heat occurs by several mechanisms including convection (heat transported by bulk flow), phonon conduction (heat transported by atomic vibration), and radiation (an electromagnetic phenomenon involving photon transfer). Radiation travels through a vacuum at the speed of light; most gases are transparent to radiation. Radiative transport of heat may be important in transition metal poor melts due to their relative transparency. Since most gases are transparent, radiative heat transfer may also be important across bubbles in magmatic emulsions and foams. In contrast, radiative heat transport can generally be ignored in mafic melts containing significant amounts of transition metals (e.g., Fe, Ti, Ni) because such melts are relatively opaque to thermal radiation, i.e., the mean free photon path is relatively short—on the order of millimeters. The photon (radiative) conductivity (k_R) can be approximated in this case as

$$k_{rad} = \frac{16}{3}\sigma n^2 T^3 \Lambda \qquad (5.10)$$

where σ is the Stefan–Boltzmann constant (5.67×10^{-8} J/ m^2 K^4 s), n is the index of refraction, and Λ is the mean free path for photons. The thermal transfer by photons depends on Λ as well as the emissivities of the surfaces across which heat is being transferred. When $\Lambda \sim 0$ (as in an opaque melt) or when one is concerned with transport of heat across distances much larger than Λ, radiative transfer is negligible. It is only when Λ is relatively large, as in transparent high-silica melts or in transparent vapor-melt emulsions that photon conduction may become appreciable. Because geometric factors are often relevant, the importance of radiative heat transfer in geological processes (as opposed to laboratory small-scale experiments) should be carefully considered.

The phonon thermal conductivity (k) provides a quantitative measure of the importance of phonon heat conduction in solids and melts. The thermal diffusivity (κ) defined $\kappa = k/\rho C_p$ is the relevant parameter in transient heat conduction problems. The thermal diffusivity involves a combination of thermodynamic properties and the phonon thermal conductivity. Quantized thermal waves called phonons carry heat in silicate crystals, glasses, and melts. Thermal resistivity (W), which is inversely proportional to thermal conductivity (i.e., $k \propto W^{-1}$), arises due to both phonon—phonon interaction and structural disorder. Because disorder is an outstanding characteristic of the glassy and molten state, one may expect the mean free path for the dominant thermal phonons to approximately equal the scale of structural disorder, roughly 0.3—0.6 nm in a typical silicate glass or melt. The thermal conductivity of a solid or melt may be estimated according to the relation

$$k = \frac{1}{3}\rho C_V c \Lambda \qquad (5.11)$$

where Λ is the phonon mean free path length, c is the sonic velocity (several kilometers per second), ρ is the melt density and C_V is the isochoric specific heat capacity. Because the structure of a melt or glass is sensitive to pressure, one might anticipate the mean free phonon length and hence k (other factors constant) to increase with pressure. In general, for a silicate liquid k decreases as temperature increases but increases as pressure increases, consistent with expectations (see Table 5.2).

Because radiative transfer tends to dominate at high temperatures, it can be difficult to cleanly separate the effects of phonon conduction from radiative transfer in laboratory experiments. There are surprisingly few reliable measurements of the thermal conductivity of liquid silicates, fewer for geochemically important compositions, and virtually no laboratory data regarding the effects of pressure on phonon conduction. Fortunately, within the past few years, there have been some important advances. Values for the phonon thermal conductivity of some geosilicate liquids, glasses, and crystals for a range of temperatures and pressures are given in Table 5.2 (Tables S5.8 and S5.9 provide more exhaustive compilations). These data show that as temperature increases, the thermal conductivity typically decreases for both solids and liquids. Values if k for silicate liquids lie in the range 0.8—1.6 W/m K with k generally higher from more polymerized liquids. Finally, very recently, phonon conductivities for molten $NaAlSi_3O_8$, $CaMgSi_2O_6$, and Mg_2SiO_4 have been determined from MD simulations in the range 2000—5000 K and 0—30 GPa (Tables 5.2 and S5.9). Values for molten albite and diopside at 0 GPa agree well with laboratory measurements at 1 bar, and confirm a weak negative temperature dependence of k at constant pressure. They further show that at constant temperature, k increases rather significantly as pressure increases. These

relationships have important implications for the efficiency of heat transfer associated with molten regions of the mantle.

4.4. Diffusion

Mass is transported in magmatic systems by bulk flow (advection) and by diffusion. Their operation separately and collectively is responsible for convection in magma bodies. There is an extensive body of information pertaining to diffusion in silicate melts and glasses as a function of temperature, composition, and, to a lesser extent, pressure. This is due in large measure to myriad industrial and materials processing applications of diffusion as well as the importance of diffusion in petrologic processes such as crystal nucleation and growth, the growth of vapor bubbles, the homogenization of melt inclusions, the resetting of geochronological clocks, and the mixing of magmas. There is a deep connection between the atomic structure of a melt or glass and the mobility of its constituents. Hence study of the systematics of diffusion bridges the gap between the microscopic and macroscopic realms relevant to magma transport phenomena. Several excellent reviews of diffusion for both industrial and geological materials are available and listed in Further Reading. In this section, the types of diffusion important for magmatic systems are reviewed and typical values are given to indicate the main trends and orders of magnitude. We review theory and highlight applications of these data. Since we are mainly concerned with magmatic systems discussion is limited to properties for the liquid state, i.e., temperatures above the glass transition ($\eta < 10^{12}$ Pa s).

Diffusion involves the motion of different constituents at the molecular or atomic scale leading to local (inter) mixing or net transport of mass. From an atomistic standpoint, diffusion is the consequence of the random walk of particles, often thermally activated, that depends on temperature, the size and charge density of the diffusing species, and the viscosity of the surrounding medium. Motion by random walk determines the intrinsic mobility of the diffusing species and is described by the self-diffusion coefficient (D*). The pathways for random walk (self-diffusion) cannot be measured directly in the laboratory; however, approximate values can be obtained by tracking the rate of transfer of readily identifiable components that are either chemically indistinguishable, i.e., different isotopes of a given element, or elements that are in such dilute concentration that their presence has no resolvable effect on the concentration of other components. The term intradiffusion is used to describe the bidirectional exchange of isotopes of major, minor, or trace elements in an otherwise homogeneous system, while the term tracer diffusion is reserved for cases where a trace element (or one of its isotopes) diffuses solely down its own concentration gradient. Both are commonly assumed to be equal measures of self-diffusion provided (1) a constant activity coefficient—which in the case of trace diffusion means the tracer obeys Henry's law (the concentration limits for Henrian behavior must be determined empirically, but is commonly <1000 part per million), (2) pathways for diffusion are the same as for random walk, and (3) there are no significant mass-dependent (isotope) effects. There are instances where these essential requirements are not met and here diffusion coefficients provide at best semi-quantitative constraints on intrinsic mobility in natural systems only in so far as they mimic laboratory conditions.

More commonly in natural systems diffusion arises from and occurs in the presence of chemical (potential) gradients due to non-Henrian behavior of the tracer or concentration gradients of other components at some location and time. This is referred to as chemical diffusion and is distinguished by simultaneous fluxes of components. From a phenomenological perspective, the chemical flux of a given component can be described by Fick's first law relating the net flux of that component to a negative gradient in its chemical potential or activity. Where chemical diffusion involves just two components of a multicomponent system a simple binary solution to Fick's law is sufficient; however, if there are net displacements of more than two components a full description of chemical diffusion requires a $(n - 1)$ by $(n - 1)$ diffusion matrix with elements D_{ij}. That is, for an n-component system, the flux of the ith component depends on the chemical potential gradient of any independent set of $n - 1$ components and not simply the ith component. Because the chemical diffusion rate of the ith component depends on $n - 1$ chemical potential gradients whereas isotopic equilibration of the ith component depends on the self-diffusivity of that species, decoupling between elemental and isotopic concentrations is possible. Furthermore, due to nonideality and/or diffusive coupling effects, even for ideal solutions, chemical diffusion can involve a net flux of elements up, rather than down, their concentration gradients. This is referred to as uphill diffusion and is well documented in experiments. In nature, enrichment in K_2O along the granite/gabbro interface and biotite-rich rinds often surrounding mafic enclaves in granitic or granodioritic magma has been attributed to uphill diffusion.

In recent years, there have been a number of experimental studies of chemical diffusion involving three and four component systems (e.g., $CaO-MgO-Al_2O_3-SiO_2$) that have succeeded in constraining the full diffusion coefficient matrix. These studies have shown that the on-diagonal D_{ii} correspond closely to self-diffusivities, while the off-diagonal terms are typically smaller, although not always the case, and commonly negative, which can lead to uphill diffusion. Given the challenges constraining the diffusion coefficient matrix for even up to four components

it is not likely that this approach will have much practical utility for modeling natural magmatic systems $(n \gg 4)$ anytime soon.

An alternative and much simplified approach is to treat chemical diffusion in magmas as an effective binary process. This is possible when chiefly the component of interest varies in concentration relative to all other components in the system. In the limit of Henry's law this is tracer diffusion. However, once the concentration exceeds the Henrian limit, the flux is no longer controlled solely by the magnitude of the concentration gradient, but in how the activity of the species changes along that gradient—this is the domain for chemical diffusion. The effective binary diffusion approach has also been used to quantify interdiffusion between vastly different magmas (such as basalt and rhyolite) where often the diffusion of SiO_2 (Si and O being the slowest diffusing elements) is considered relative to all the other components or where all the network formers (Si, Al, Fe^{+3}, etc.) are lumped together as one component and all the network modifiers (Fe^{+2}, Mg, Ca, Na, and K in excess of that needed to charge balance tetrahedral Al, Fe^{+3}, etc.) are combined as the other. While again the diffusivities determined by this approach are only applicable to natural systems in so far as the laboratory conditions simulate nature, effective binary diffusivities especially for SiO_2 or network formers do place limits on the length and time-scales for chemical homogenization during magma commingling. This enables us to evaluate how species with high intrinsic mobility are locally redistributed by chemical diffusion within the more sustained major element gradient due to sluggish diffusion of network-forming cations and oxygen (Figure 5.7). Consider the diffusive length during the 10^5 year solidification time of a 10^3 km^3 gabbroic pluton (characteristic length ~ 10 km). For O and Si, diffusion can effect changes in concentration over ~ 3 m, while in the case of Li and H_2O having diffusivities two to three orders of magnitude larger diffusive exchange can occur over ~ 60 m. Possible ramifications for isotopic composition of fast and slow diffusion species such as Sr and Nd, respectively, are illustrated in Figure 5.8.

Nearly 65 years ago L.S. Darken proposed that the chemical diffusivity D_i^c can be expressed as

$$D_i^c = kTM_i \left[1 + \frac{d \ln \gamma_i}{d \ln x_i} \right] \qquad (5.12)$$

where k is the Boltzmann constant, T is temperature, γ_i and x_i are the activity coefficient and mole fraction of species i, respectively, and M_i is the intrinsic mobility (velocity of species i per unit force). Equation (5.12) provides an explicit expression relating the chemical diffusivity to both species i intrinsic mobility (random walk velocity) and its activity gradient causing a directional flux. Consider first the case of uniform composition. Since there is no chemical

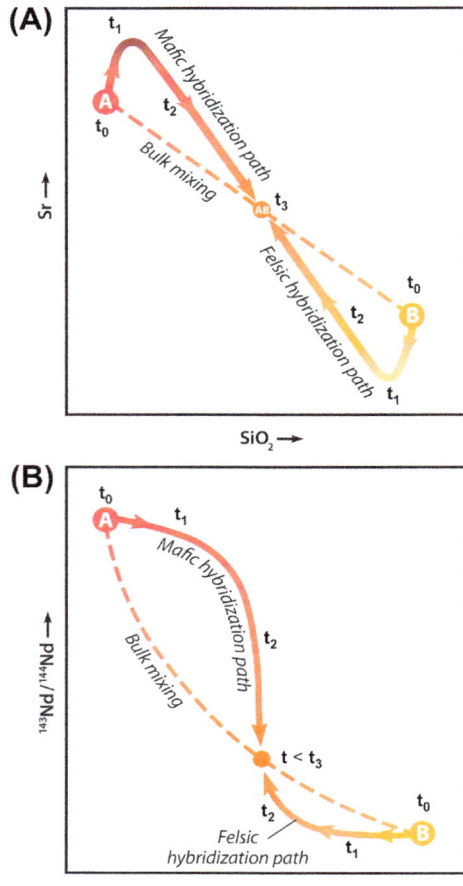

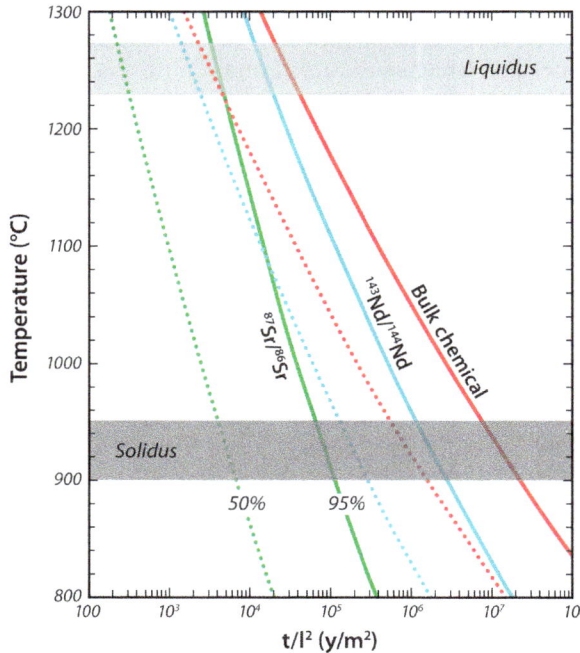

FIGURE 5.8 Effects of temperature, time, and diffusion length scale for Sr and Nd isotopic homogenization and bulk chemical homogenization of basalt with entrained blobs of rhyolite. Diffusive impedance is the time (t) in years over the length scale of the compositional heterogeneity (l) in meters squared, t/l^2, is the diffusive impedance. Contours for 50 and 95% of the way to complete homogenization of isotopic composition and bulk chemical composition are represented by the dotted and solid curves, respectively. Approximate liquidus and solidus temperatures indicated by stippled horizontal bars delimiting the temperature interval where hybridization of basalt is likely to occur. *Modified after Figure 12 in Lesher, C.E., 1994. Kinetics of Sr and Nd exchange in silicate liquids — theory, experiments, and applications to uphill diffusion, isotopic equilibration, and irreversible mixing of magmas. Journal of Geophysical Research 99, 9585–9604 with permission from John Wiley & Sons Publications.*

FIGURE 5.7 Schematic diagram showing progressive hybridization of mafic (A, red) and felsic (B, yellow) magmas when mixing is limited by a large viscosity contrast and diffusion. (A) Solid curves with arrows show the paths for the mafic magma (upper trajectories) and complementary felsic magma (lower trajectories) as a function of time leading eventually to bulk homogenization in time t_3. These trends deviate markedly from those predicted for simple binary mixing shown by the dashed curves. Uphill diffusion of Sr is depicted when the gradient in activity is opposite to the gradient in concentration, negative Darken diffusivity (see text). In (B) the covariation of $^{87}Sr/^{86}Sr$ and $^{143}Nd/^{144}Nd$ ratios reflects the fact that the intrinsic mobility of Sr is greater than of Nd (see Tables S5.10 and S5.11). Isotopic equilibrium is practically achieved in time t, which may be less than that required for the system to achieve complete chemical homogenization (t_3). *Modified after Figure 2 in Lesher, C.E., 1990. Decoupling of chemical and isotopic exchange during magma mixing. Nature 344, 235–237 with permission Nature Publishing Group.*

flux, $\frac{d \ln \gamma_i}{d \ln x_i} = 0$ and $D_i^c = kTM_i$. This is tantamount to our definition of self-diffusivity, D^*, so that

$$D_i^c = D_i^* \left[1 + \frac{d \ln \gamma_i}{d \ln x_i} \right] \qquad (5.13)$$

Now consider species i present in dilute concentration and obeying Henry's law, i.e., $\gamma_i = $ constant. In this case, while $d\ln x_i$ is nonzero, $d\ln\gamma_i$ will be zero and thus $\frac{d \ln \gamma_i}{d \ln x_i} = 0$.

Herein lies the common assumption that tracer diffusion and self-diffusion are equivalent measures of intrinsic mobility—but again only in the limit of Henry's law.

The equivalency of the self-diffusion and intradiffusion as might be derived from an isotope diffusion couple experiment is not so straightforward, since isotopes of the same element have mass differences and thus their self-diffusivities may differ. Drawing again on Darken's work, the diffusivity for exchange of two isotopes of the same element (isotope 1 with mass m_1 and isotope 2 with mass m_2) in an otherwise homogeneous solution depends on their relative concentrations and self-diffusivities

$$\overline{D} = x_1 D_2^* + x_2 D_1^* \qquad (5.14a)$$

where

$$D_1^* = D_2^* \left(\frac{m_2}{m_1} \right)^\beta \qquad (5.14b)$$

β is typically 0.5 for gases but markedly smaller for condensed phases. In the case of silicate melts, ^{6}Li has been shown to diffusive $\sim$12% faster than ^{7}Li giving $\beta = 0.2$, while the differences in diffusivities of isotopes of Fe, Mg, and Ca are well below 2%, constraining $\beta < 0.06$ for these elements. β for Si is expected to be markedly smaller ($\leq$0.025) meaning that differences in self-diffusivities for Si isotopes are likely far too smaller to be resolved by current mass spectrometry methods. Furthermore, from Eqn (5.14b) it can be appreciated that as $\frac{m_2}{m_1} \to 1$, $D_1^* \to D_2^*$ so that for most petrologic applications involving traditional radiogenic systems (Sr, Nd, Hf, and Pb) the differences in isotope self-diffusivities will be so small that diffusive fractionation effects will be negligible. On the other hand, this also means that, in practice, the self-diffusivity can be determined directly by measuring the flux of one isotope relative to others of that element in an otherwise homogeneous medium—in other words, from an intradiffusion couple experiment.

Returning to Eqn (5.13), in most situations encountered in magmatic systems the ratio $\frac{d \ln \gamma_i}{d \ln x_i}$ will have a value other than zero, while its magnitude is expected to vary in space and time. Thus, the Darken diffusivity has the peculiarly feature of being either positive, negative, or zero depending on whether the activity of species i increases, decreases, or is constant with concentration, respectively. While the former case is usually encountered and certainly prevails in the absence of gradients in other components, the latter situations arise due to strong nonideality accompanying gradients in other components, particularly silica content. The occurrence of uphill diffusion in this context can then be appreciated as a direct consequence of a negative Darken diffusivity for that element, while for rapidly diffusing species concentration differences can be established essentially under quasiequilibrium conditions where the activity gradient tends to zero. These concentration differences can persist as long as there remain differences in, for example, silica content. The main point regarding magma mixing is that ultimately mass transfer depends on chemical diffusion operating at the molecular scale. The process is generally not explicable by linear mixing, and due to differences in self-diffusivities and the complexities of the activity–composition relations significant decoupling among chemical components and isotopic systems is possible (Figures 5.7 and 5.8). Armed with this appreciation can help in deciphering the often complex compositional relationship among hybrid rocks formed by magma mingling.

A great deal is known about the effects of temperature and composition on self/tracer and effective binary chemical diffusivities at atmospheric pressure, while our understanding of the effects pressure is meager. Following from Eqn (5.4), the Arrhenius relation can also describe the dependence of diffusivity on pressure and temperature

$$D = D_0 \exp[-(E_a + PV_a)/RT] \qquad (5.15)$$

where D_0 is constant for a fixed composition, and E_a and V_a represent the activation energy and activation volume for diffusive transport, respectively. Equations (5.15) is a useful starting point for any analysis of the temperature and pressure dependence of diffusivities; however, certainly in the case of silicate melts the assumption of constant E_a and V_a may only be valid for restricted temperature and pressure intervals—for many of the same reasons that Eqn (5.4) fails for fragile liquids. However, for diffusion the preexponential term (D_0) depends strongly on melt composition and the size and charge of the species. As such, Eqn (5.15) often does not provide sensible results if the diffusivities being considered are chemical diffusivities that depend on the details of the compositional gradients used to determine D_{ij}. It is best to limit consideration to experimental measurements that serve as good proxies for self-diffusivities, i.e., tracer and intradiffusivities (see Tables S5.10 and S5.11). E_a is found to be of order 300–400 kJ/mol for network-forming atoms like Si and Al and in the range 250–180 kJ/mol for transition metals and the alkaline earth metals. Alkali metals have activation energies in the range 120–220 kJ/mol.

Less is known about V_a for silicate melts. V_a is typically positive for network modifiers such as Ca and Mg (+2 to +12 cm^3/mol) and can be positive or negative for network-forming cations (Si) and oxygen, depending on composition and pressure. Positive V_a is consistent with more conventional free-volume models of ionic diffusion, while to explain negative V_a one often appeals to cooperative modes of diffusion. For basalt to rhyolite compositions at low to moderate pressures, V_a for Si and O range from −2 to −15 cm^3/mol with more negative values associated with more silicic (polymerized) melts. Both experiments and MD simulations show that V_a for Si and O can change from negative to positive at high pressure. The maximum in the diffusivity has been attributed to progressive stabilization of high coordinated network species with pressure, but there is also mounting evidence that the anomalous pressure dependence of network formers up to 5–6 GPa is intimately connected to the compressible nature of polymerized silicate melts over this pressure range. In contrast, V_a for Si and O is positive even at low pressure in depolymerized nature liquids such as diopsidic melt suggesting that given their more compact structure diffusion is restricted by available free-volume and more readily accomplished by motion of individual ions rather than through cooperative motion. Granted, these expectations are based on limited data for

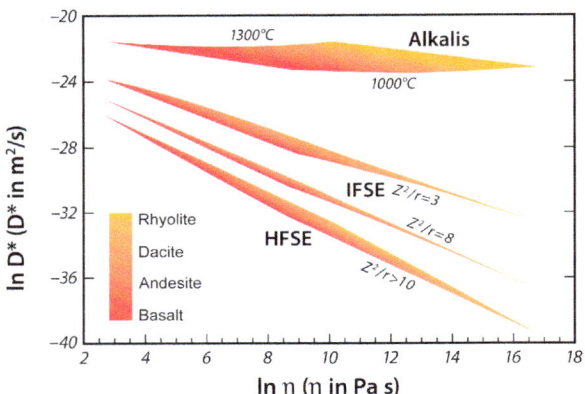

FIGURE 5.9 Tracer diffusivities for the alkalis, intermediate field strength elements (IFSE bounded by $Z^2/r = 3$ and 8) and high field strength elements (HFSE; $Z^2/r > 10$) from 1300 °C (upper bound) to 1000 °C (lower bound) and spanning composition of basalt (red, left hand bound) to rhyolite (yellow, right hand bound) computed from equations presented by Mungall (2002). Diffusivities are plotted against log viscosity from Giordano et al. (2008).

high-pressure conditions and further laboratory measurements and first-principle calculations are sorely needed to reliably constraint processes involving diffusion in the deep interior of our own and other planetary bodies. Studies of diffusion in conjugation with viscous flow are also critical for understanding the connections between these transport processes on a fundamental physical/ structural level.

Roughly, E_a and V_a for oxygen self-diffusion in a melt are equivalent to E_A and V_A for viscous flow because viscous flow in a silicate melt is ultimately related to the mobility of oxygen, the predominant constituent. This expectation derives from the classic expressions proposed by Einstein, Glasstone, Eyring, among others, that assume $D^\alpha \eta^{-1}$. This simple proportionality has been very useful in understanding relative magnitudes of self-diffusivities and their dependence on magma composition (Figure 5.9). As a rule, oxygen together with Si and other high field strength cations tend to have the smallest self-diffusivities for a given temperature and melt composition, while diffusivities for the transition, alkaline earth, and alkali metals are uniformly larger. In general, network-modifying ions of high charge and large size have smaller self-diffusivities than small ions of low charge, other factors remaining the same. For example, in a basaltic melt at 1300 °C, oxygen and the network cations Si and Al have $D^* \sim 3 \times 10^{-12}$ m^2/s. This compares with rare earth and transition metals (e.g., Fe, Ti), and alkaline earths (e.g., Mg, Ca, Sr, Ba) for which $D^* \sim 1 \times 10^{-11}$ m^2/s and $\sim 2 \times 10^{-11}$ m^2/s, respectively. A typical ranking for basaltic melts is $D^{Ba} \geq D^{Sr} > D^{Ca} > D^{Mg} > D^{Fe} > D^{REE} > D^{Ti}$. The alkalis are significantly more mobile with $D^* \sim 3 \times 10^{-10}$ m^2/s and exhibit a strong dependence on ionic radius with

$D^{Li} \geq D^{Na} > D^{K} > D^{Rb} > D^{Cs}$. The diffusivity of the carbonate group (CO_3) is $\sim 1.2 \times 10^{-10}$ m^2/s whereas for H_2O, $D = \sim 1 \times 10^{-9}$ m^2/s, and depends on the concentration of dissolved water. At fixed temperature, D^*'s for oxygen and most high field strength cations, including Si, are larger in basalt melt compared to rhyolite, by a factor of 5–500; but interestingly the alkalis shown little to no dependence on magma composition. These relationships are illustrated in Figure 5.9 using an empirical model based on the assumption that $D^\alpha \eta^{-1}$ and fit to experimental data for self-diffusion and independent constraints on viscosity. While such empirical models do have limitations, notably missing is a comprehensive picture of the effects of volatile species (H_2O, CO_2, etc.) and pressure, they do permit order of magnitude estimates of self-diffusivities for ions of different size and charge over a broad range of natural magma compositions and temperature. This information is essential for modeling the dynamics and timescales of magmatic processes fundamentally controlled by diffusion at the molecular to atomic length scales.

5. CONCLUSIONS

The important thermodynamic and transport properties of magma have been reviewed in this chapter. The complete description of multicomponent and multiphase magma at temperatures from 800 to 5000 K and pressures from 10^{-4} GPa (surface) to 135 GPa (base of mantle) is an ambitious program that will take many years to accomplish. Although an enormous task, there are reasons to be optimistic. Within the last quarter century, knowledge of the thermodynamic properties of high-temperature crystalline, molten, and glassy silicates including melts of natural composition has grown dramatically. Although still far from complete, thermodynamic properties for many crystalline phases relevant to magmatic systems are known at the high temperature and pressure conditions extending at least to the base of the lithosphere and within the asthenosphere where primary liquids are formed beneath spreading ridges, within upwelling mantle plumes and at subduction zones. An approximate model for the Gibbs free energy of natural silicate liquids has also been developed in the past 15 years and is undergoing continuous revision and expansion. These advances have improved efforts to describe and predict equilibrium phase relations involved in the melting and solidification of magma. Similarly, studies of the temperature, pressure, and composition dependence of melt viscosity, the rheology of magmatic (solid–melt–vapor) mixtures, chemical, tracer, and self-diffusion and thermal conductivity have been made for a variety of natural and synthetic melts and glasses. Correlations have been developed for some properties valid in specific regions of temperature–pressure–composition space that enable one to obtain order of magnitude or better estimates of many critical properties. These have been greatly aided by significant advances in experimental techniques, especially for high pressures and utilizing a variety of radiation sources and spectroscopic tools, and analytical methods that are probing at ever smaller length scales with hitherto unfathomable precision and accuracy. Development of models based on condensed matter physics has come to supersede earlier purely empirical models and provide for rational interpolations and a basis for careful extrapolation. In the computational realm, MD simulations are increasingly being applied to investigate the structure and properties of molten silicates at very high temperatures and pressures. This method, which involves use of a potential energy expression describing the electrostatic interactions between the various atoms in a fluid or melt, enables one to compute thermodynamic, spectroscopic, and transport properties under a variety of conditions. Although properties so

computed are estimates, they do provide a means for understanding, at the atomic level, the effect of temperature, pressure, and composition on physical properties and for identifying critical laboratory experiments that should be performed.

Study of the properties of molten, glassy, and high-temperature crystalline silicates and oxides of geological relevance is a burgeoning field for last research. The rate of acquisition of new information is accelerating as new techniques are applied to old problems. There are hundreds of volcanic eruptions each year and many of these present dangers involving the loss of life and property. The first step in taming magma is to understand its nature from a fundamental and broad perspective. At the same time a deeper understanding of the evolution of planet Earth and, by analogy, other terrestrial planets both within our solar system and beyond depends on knowledge regarding the physical properties of magma.

SUPPLEMENTARY MATERIAL FOR THERMODYNAMIC AND TRANSPORT PROPERTIES OF SILICATE MELTS AND MAGMA

TABLE S5.1 Partial Molar Volumes, Thermal Expansions and Compressibilities of Oxide Components. Note That $(\partial V_i/\partial P)_T \equiv -\beta_T V_i$, Where β_T Is the Isothermal Partial Molar Compressibility of the ith Component and $(\partial V_i/\partial T)_p \equiv \alpha_P V_i$, Where α_P Is the Isobaric Expansivity.

$$V_{liq}(T, P, X) = \Sigma X_i [V_{i,1673K} + (\partial V_i/\partial T)_p (T - 1673) + (\partial V_i/\partial P)_T P]$$

	$V_{i,1673K}$ $(10^{-6}\,m^3/mol)$	$(\partial V_i/\partial T)_p$ $(10^{-9}\,m^3/mol \cdot K)$	$(\partial V_i/\partial P)_T$ $(10^{-6}\,m^3/mol \cdot GPa)$
SiO_2	26.86 ± 0.03	0.0	-1.89 ± 0.02
TiO_2	23.16 ± 0.26	7.24 ± 0.46	-2.31 ± 0.06
Al_2O_3	37.42 ± 0.09	0.0	-2.26 ± 0.09
Fe_2O_3	42.13 ± 0.28	9.09 ± 3.49	-2.53 ± 0.09
FeO	13.65 ± 0.15	2.92 ± 1.62	-0.45 ± 0.03
MgO	$11.69 \pm .08$	3.27 ± 0.17	0.27 ± 0.07
CaO	16.53 ± 0.06	3.74 ± 0.12	0.34 ± 0.05
Na_2O	28.88 ± 0.06	7.68 ± 0.10	-2.40 ± 0.05
K_2O	45.07 ± 0.09	12.08 ± 0.20	-6.75 ± 0.14
Li_2O	16.85 ± 0.15	5.25 ± 0.81	-1.02 ± 0.06
H_2O	26.27 ± 0.5	9.46 ± 0.83	-3.15 ± 0.61
CO_2	25.4 ± 1.0	10.86 ± 1	-3.82 ± 0.8

Modified after Lange and Carmichael (1990), Lange (1997), Ochs and Lange (1997), and Ghiorso and Gualda (2014, this volume). Uncertainties are 1σ.

TABLE S5.2 Molar Volume of Typical Silicate Crystals and Melts at High Temperature and 10^{-4} GPa. Depolymerized Compositions Typically Decrease in Density 10–20% upon Fusion. The Volume Change Is Less for Tetrahedral Fluids Such as SiO_2. Molar Volumes at 1400 °C, Except for H_2O at 273 K.

Mineral	Melting Temperature (K)	V_{melt} (10^{-6} m^3/mol)	$V_{crystal}$ (10^{-6} m^3/mol)	V_m/V_c	ΔV_{fusion} (10^{-6} m^3/mol)
SiO_2 (cristobalite)	1999	26.91	27.44	0.98	−0.53
$NaAlSi_3O_8$	1393	112.83	104.13	1.08	8.64
$KAlSi_3O_8$	1473	121.24	111.72	1.09	9.75
$CaAl_2Si_2O_8$	1830	108.41	103.42	1.05	5.48
$CaSiO_3$	1817	43.89	41.87	1.05	2.31
$MgSiO_3$	1830	38.85	33.07	1.18	5.99
$CaMgSi_2O_6$	1665	81.80	69.74	1.18	12.62
$Ca_2MgSi_2O_7$	1727	99.70	97.10	1.03	2.60
$Ca_3MgSi_2O_8$	1848	118.09	105.92	1.12	13.07
Fe_2SiO_4	1490	53.15	48.29	1.11	5.49
Mg_2SiO_4	2174	52.37	47.41	1.12	5.53
$CaTiSiO_5$	1670	67.61	57.57	1.17	10.04
H_2O (ice)	273	18.01	19.64	0.92	1.63

Modified from Lange and Carmichael (1990).

TABLE S5.3 Density and Viscosity of Molten Carbonate (Jones et al., 1995 and Sykes et al., 1992) and Supercritical H_2O (Haar et al., 1984).

Composition	Pressure (GPa)	Temperature (°C)	Density (kg/m^3)	Viscosity (Pa s)
$K_2Ca(CO_3)_2$	10^{-4}	975	2014	—
	2.5	950	2750	0.032
	2.5	1150	2580	0.018
	4.0	1050	2800	0.023
K_2CO_3	4	1500	3100	0.023
$K_8Ca_3Mg(CO_3)_8$	2	1250	—	0.065
$K_2Mg(CO_3)_2$	3	800	—	0.036
	3	900	—	0.022
	5.5	1200	—	0.006
H_2O	0.3	500	—	10^{-4}
		800	—	7×10^{-5}

TABLE S5.4 Approximate Values of the Isothermal Bulk Modulus (K_T) and Its Pressure derivative ($K' = dK_T/dP$) for Use in the Birch–Murnaghan Equation of State for Some Simple and Naturally Occurring Silicate Melts at High Temperature (T_o). The Temperature Derivative, dK_T/dT, has Been Constrained Only for a Few Compositions.

Composition	K_T (GPa)	K'	dK_T/dT	T_o (K)	Data Source*
Simple Systems					
$CaMgSi_2O_6$	22.4	6.9	—	1773	(1)
$CaAl_2Si_2O_8$	17.9	5.3	—	1923	(1)
Fe_2SiO_4	19.4	5.3	—	1773	(2)
Naturally Occurring					
Rhyolite (0–8 wt% H_2O)	11.5	6.5	−0.0016	1273	(3)
Andesite (0– 9 wt% H_2O)	17.2	6.3	—	1273	(4)
Basalt (anhydrous)	19.3	4.4	—	1673	(5)
Basalt (anhydrous)	20.8	4.6	—	1673	(6)
Basalt (2 wt% H_2O)	18.2	4	—	2573	(6)
Basalt (8 wt% H_2O)	14.5	3.2		2473	(6)
Komatiite (anhydrous)	26.0	4.3	—	2073	(5)
Komatiite (anhydrous)	23.1	4.9	—	2073	(6)
Peridotite (anhydrous)	23.4	6.2	—	2273	(6)
Peridotite (anhydrous)	24	7.3	−0.0027	2100	(7)
Peridotite (5 wt% H_2O)	8.8	9.9	−0.0022	1773	(8)
Peridotite (~2 wt% CO_2)	23	8.5	−0.01	1800	(9)

*(1) Rigden et al., (1989); (2) Thomas et al., 2012; (3) Malfait et al., 2014a; (4) Malfait et al., 2014b; (5) Agee, 1998; (6) refits by Jing and Karato, 2008, of published sink/float constraints, see additional references therein; (7) Sakamaki et al., 2010; (8) Sakamaki et al., 2009; (9) Sakamaki et al., 2011.

TABLE S5.5 Melting Temperature, Molar and Specific Enthalpy of Fusion and Entropy of Fusion at 10^{-4} GPa for Some Silicate and Oxide Phases.

Formula	Melting Temperature (K)	ΔH_{fus} (kJ/mole)	ΔS_{fus} (J/mole-K)	Δh_{fus} (kJ/kg)
H_2O	273	7.401	27.11	116
NH_3	195	5.657	29.01	333
CH_4	90.6	0.937	10.34	58
ZrO_2 (baddeleyite)	3123	87.03	27.87 ± 0.02	706
$Fe_{0.947}O$	1652	31.3 ± 0.2	19.0 ± 0.1	454
$FeTiO_3$	1640	21.7 ± 0.1	13.2 ± 0.1	143
TiO_2 (rutile)	1870	67.0 ± 0.1	35.8 ± 0.1	838
Fe_2O_3 (hematite)	1895	114.5 ± 0.2	60.4 ± 0.2	717
PbO (Massicot)	1170	25.52 ± 0.01	21.81	114

(Continued)

TABLE S5.5 Melting Temperature, Molar and Specific Enthalpy of Fusion and Entropy of Fusion at 10^{-4} GPa for Some Silicate and Oxide Phases.—Cont'd

Formula	Melting Temperature (K)	ΔH_{fus} (kJ/mole)	ΔS_{fus} (J/mole-K)	Δh_{fus} (kJ/kg)
Fe_3O_4 (Magnetite)	1870	138.07 ± 0.05	73.83	596
$LiAlO_2$	1883	87.9 ± 0.3	46.7 ± 0.5	1333
$MgAlO_2$	2408	107 ± 11	44 ± 4	752
$Ca_2Fe_2O_5$	1750	151.0 ± 0.5	86.3 ± 0.1	556
Al_2O_3	2323	107.5 ± 54	46.3 ± 23	1054
SiO_2 (quartz)[a]	1700	9.40 ± 1.0	5.53 ± 0.56	157
SiO_2 (cristobalite)	1999	8.92 ± 1.0	4.46 ± 0.50	149
$MgSiO_3$[a]	1834	73.2 ± 6.0	39.9 ± 3.3	729
$CaSiO_3$ (wollastonite)[a]	1770	61.7 ± 4.0	34.9 ± 2.3	531
$CaSiO_3$ (pseudowoll)	1817	57.3 ± 2.9	31.5 ± 1.6	493
$CaMgSi_2O_6$	1665[b]	137.7 ± 2.4	82.7 ± 1.4	636
$Ca_2MgSi_2O_7$	1727	123.9 ± 3.2	71.7 ± 1.9	454
$Ca_3MgSi_3O_8$[b]	1848	125 ± 15	67.5 ± 8.1	350
Fe_2SiO_4	1490[a]	89.3 ± 1.1	59.9 ± 0.7	438
Mn_2SiO_4	1620	89.0 ± 0.5	55.2 ± 0.3	633
Mg_2SiO_4	2174	142 ± 14	65.3 ± 6	1010
$CaTiSiO_5$	1670	123.8 ± 0.4	74.1 ± 0.2	755
$NaAlSi_3O_8$	1393	64.5 ± 3.0	46.3 ± 2.2	246
$NaAlSi_2O_6$[a]	1100	59.3 ± 3.0	53.9 ± 2.7	293
$NaAlSiO_4$[a]	1750	49.0 ± 2.1	28.0 ± 1.2	345
$Na_2Si_2O_5$	1147	35.6 ± 4.1	31.0 ± 3.6	196
$KAlSi_3O_8$[a]	1473	57.7 ± 4.2	39.2 ± 2.8	207
$CaAl_2SiO_8$	1830	133.0 ± 4.0	72.7 ± 2.2	478
$Mg_3Al_2Si_3O_{12}$[a]	1500	243 ± 8	162 ± 5	603
$Mg_2Al_4Si_5O_{18}$[a]	1740	346 ± 10	199 ± 6	591
$KMg_3AlSi_3O_{10}F_2$	1670	308.8 ± 1.3	185 ± 1	733

[a]Metastable congruent melting.
[b]Incongruent melting.
Data compiled from Kelly (1936), Robie and Hemingway (1995), Ghiorso (2004) and references therein.

TABLE S5.6 Coefficients for Estimating the Isobaric Heat Capacity of Silicate Glasses Valid for Temperatures in the Range 400 K < T < 1000 K at 10^{-4} GPa. X_i Is the Mole Fraction of the ith Oxide Component. The gfw Is Defined According to gfw $= \sum X_i M_i$, Where M_i is the Molar Mass of the ith Component.

$$C_p(T) = \sum a_i X_i + \sum b_i X_i T + \sum c_i X_i T^{-2} \ \left(J \ K^{-1} \ gfw^{-1}\right)$$

	a_i	$b_i \times 10^2$	$c_i \times 10^{-5}$
SiO_2	66.354	0.7797	−28.003
TiO_2	33.851	6.4572	4.470
Al_2O_3	91.404	4.4940	−21.465
Fe_2O_3	58.714	11.3841	19.915
FeO	40.949	2.9349	−7.6986
MgO	32.244	2.7288	1.7549
CaO	46.677	0.3565	−1.9322
Na_2O	69.067	1.8603	2.9101
K_2O	107.194	−3.2194	−28.929

Modified after Stebbins et al. (1984).

TABLE S5.7 Partial Molar Isobaric Heat Capacity for Molten Oxide Components Applicable to Silicate Melts at 10^{-4} GPa. C_p is Approximately Independent of Temperature at T ≥ 1400 K.

$$C_p = \sum C_{pi} X_i \ \left(J \ K^{-1} \ gfw^{-1}\right)$$

	Molar C_{pi} (J/gfw K)	Specific c_{pi} (J/kg K)
SiO_2	80.0 ± 0.9	1331
TiO_2	111.8 ± 5.1	1392
Al_2O_3	157.6 ± 3.4	1545
Fe_2O_3	229.0 ± 18.4	1434
FeO	78.9 ± 4.9	1100
MgO	99.7 ± 7.3	2424
CaO	99.9 ± 7.2	1781
SrO	88.7 ± 8	855
BaO	83.4 ± 6.0	544
H_2O molecular	41 ± 14.0	2278
$(OH)^-$ hydroxyl	153 ± 18	8500
Li_2O	104.8 ± 3.2	3507
Na_2O	102.3 ± 1.9	1651
K_2O	97.0 ± 5.1	1030
Rb_2O	97.9 ± 3.6	524

Modified from Stebbins et al. (1984). Water values from Bouhifd et al. (2006).

TABLE S5.8 Thermal Conductivity of Some Geosilicate Crystals, Glasses and Melts at 1 bar.

Composition and state	Temperature (°C)	Thermal Conductivity (W/m K)	Data Source*
Olivine (Fo$_{90}$)	800	2.69	(1)
	1400	2.18	(1)
Forsterite	400	2.47	(1)
	800	1.84	(1)
	1400	1.59	(1)
Orthopyroxene (bronzite)	0	4.62	(1)
	300	3.05	(1)
Diopside	20	4.27	(1)
Diabase	300	2.09	(1)
Obsidian (rhyolite glass)	0	1.34	(1)
	300	1.67	(1)
	500	1.89	(1)
Basalt glass	0	1.15	(1)
	300	1.48	(1)
SiO$_2$ (glass)	600	1.76	(1)
	1230	1.87	(1)
NaAlSi$_3$O$_8$ (glass)	25	1.37	(2)
	800	1.56	(2)
CaAl$_2$Si$_2$O$_8$ (glass)	25	1.13	(2)
	800	1.43	(2)
CaMgSi$_2$O$_6$ (glass)	25	1.24	(2)
	800	1.46	(2)
KAlSi$_3$O$_8$ (glass)	25	1.12	(2)
	800	1.44	(2)
SiO$_2$ (liquid)	1400	1.32	(2)
	1700	1.41	(2)
CaMgSi$_2$O$_6$ (liquid)	1100	1.21	(3)
CaAl$_2$SiO$_8$ (liquid)	1200	1.45	(3)
NaAlSi$_3$O$_8$ (liquid)	1200	1.59	(3)
KAlSi$_3$O$_8$ (liquid)	1200	1.45	(3)
Rhyolite (liquid)	800−1100	∼1.5	(3)

*(1) Clark (1966); (2) Hofmeister et al. (2009); (3) Pertermann et al. (2008).

TABLE S5.9 Thermal Conductivity for Geosilicate Liquids at Elevated Pressure Determined by Molecular Dynamics Simulation.

Temperature (°C)	Pressure (GPa)	Thermal Conductivity (W/m K)
$CaMgSi_2O_6$ liquid		
1800	0	1.14
1800	10	2.02
2800	0	1.02
2800	10	1.94
4000	0	0.77
4000	10	1.85
$NaAlSi_3O_8$ liquid		
1800	0	1.45
1800	10	2.16
2800	0	0.99
2800	10	2.55
3750	0	0.89
3750	10	1.89
Mg_2SiO_4 liquid		
3000	0	1.06
3300	0	0.90
3300	6.3	1.83
3300	10	2.11
3800	0	0.89
4300	0	0.68

Data from Tikunoff and Spera (2014).

TABLE S5.10 Activation Energies (E_a) and Preexponential Factors (D_o) for Self/Tracer Diffusivities in Anhydrous Basaltic (46–49.7 wt% SiO_2) Melts. Extreme Caution Should Be Exercised Extrapolating Beyond the Experimental T and P Range.

Species	P (GPa)*	T Range (K)	E_a (kJ/mol)	$\ln D_o$ (D_o in m^2/s)	Data Source[#]
Li	10^{-4} (a)	1569–1673	116	−11.8	(1)
Na	10^{-4} (a)	1577–1692	164	−9.3	(2)
Cs	10^{-4} (a)	1567–1675	274	−4.4	(2)
Ca	10^{-4} (a)	1538–1723	184	−9.8	(3)
Sr	10^{-4} (a)	1538–1723	182	−10.5	(3)
Sr	10^{-4} (a)	1567–1675	191	−9.7	(2)
Sr	1 (gr)	1528–1738	136	−14.4	(4)

(Continued)

TABLE S5.10 Activation Energies (E_a) and Preexponential Factors (D_o) for Self/Tracer Diffusivities in Anhydrous Basaltic (46–49.7 wt% SiO_2) Melts. Extreme Caution Should Be Exercised Extrapolating Beyond the Experimental T and P Range.—Cont'd

Species	P (GPa)*	T Range (K)	E_a (kJ/mol)	$\ln D_o$ (D_o in m^2/s)	Data Source[#]
Ba	10^{-4} (a)	1538–1723	165	−12.0	(3)
Ba	10^{-4} (a)	1572–1692	170	−11.7	(2)
Sc	10^{-4} (a)	1572–1675	198	−10.3	(2)
Mn	10^{-4} (a)	1567–1671	166	−11.2	(2)
Fe^{+3}	10^{-4} (a)	1570–1675	265	−4.8	(2)
Co	10^{-4} (a)	1567–1671	197	−9.0	(2)
Co	10^{-4} (a)	1538–1723	152	−12.1	(3)
Si	1 (gr)	1593–1873	167	−12.8	(6)
O	10^{-4} (v)	1593–1773	251	−6.6	(7)
O	1 (gr)	1593–1873	172	−12.4	(6)
Eu^{+3}, Gd	10^{-4} (a)	1593–1713	170	−12.1	(5)
Eu^{+3}	10^{-4} (a)	1567–1673	268	−5.0	(8)
Eu^{+3}	10^{-4} (a)	1577–1672	248	−9.3	(8)

*(a), run in air; (v), run at variable log fO_2 between 0 and −3.3; (gr), enclosed in graphite capsule.
[#](1) Lowry et al., 1981; (2) Lowry et al., 1982; (3) Hofmann and Magaritz, 1977; (4) Lesher, 1994; (5) Magaritz and Hofmann, 1978; (6) Lesher et al., 1996; (7) Canil, 1990; (8) Hendersen et al., 1985.

TABLE S5.11 Activation Energies (E_a) and Preexponential Factors (D_o) for Self/Tracer diffusivities in Anhydrous Rhyolitic (69–76 wt% SiO_2) Melts. Extreme Caution Should Be Exercised Extrapolating Beyond the Experimental T and P Range.

Species	P (GPa)*	T Range (K)	E_a (kJ/mol)	$\ln D_o$ (D_o in m^2/s)	Data Source*
Li	10^{-4} (a)	1573–1673	84	−13.8	(1)
Na	10^{-4} (a)	1573–1673	152	−9.6	(2)
Cs	10^{-4} (a)	1573–1673	219	−10.7	(2)
Cs	10^{-4} (a)	1063–1573	201	−13.8	(3)
Ca	10^{-4} (a)	903–1203	284	−1.6	(3)
Ba	10^{-4} (a)	1573–1673	130	−16.6	(2)
Ba	10^{-4} (a)	1433–1588	199	−11.2	(2)
Sr	1 (gr)	1528–1738	136	−16.3	(4)
Mn	10^{-4} (a)	1573–1673	210	−10.6	(2)
Fe^{+3}	10^{-4} (a)	1573–1673	177	−14.4	(2)
Co	10^{-4} (a)	1573–1673	167	−13.9	(2)
B	1 (gr)	1573–1773	400	−1.7	(5)
Ga	1 (gr)	1573–1773	347	−4.3	(5)
Si	1 (gr)	1573–1773	139	−23.0	

TABLE S5.11 Activation Energies (E_a) and Preexponential Factors (D_o) for Self/Tracer diffusivities in Anhydrous Rhyolitic (69–76 wt% SiO_2) Melts. Extreme Caution Should Be Exercised Extrapolating Beyond the Experimental T and P Range.—Cont'd

Species	P (GPa)*	T Range (K)	E_a (kJ/mol)	$\ln D_o$ (D_o in m^2/s)	Data Source*
Si	1 (gr)	1733–1935	380	−3.8	(6)
Si	2 (gr)	1733–1935	305	−8.3	(6)
Si	4 (gr)	1733–1935	163	−15.9	(6)
O	1 (gr)	1733–1935	293	−8.9	(6)
O	2 (gr)	1733–1935	264	−10.2	(6)
O	4 (gr)	1733–1935	155	−15.9	(6)
Eu^{+3}	10^{-4} (a)	1473–1673	253	−9.0	(2)
Eu^{+3}	10^{-4} (a)	973–1323	289	−7.2	(3)
Ce	10^{-4} (a)	1148–1373	490	6.3	(3)
Nd	1 (gr)	1528–1738	198	−13.7	(4)
U^{+6}	10^{-4} (a)	1250–1923	364	−6.7	(7)
Th	10^{-4} (a)	1250–1923	369	−7.0	(7)

*(a), run in air; (gr), enclosed in graphite capsule.
*(1) Cunningham et al., 1983; (2) Hendersen et al., 1985; (3) Jambon, 1982; (4) Lesher, 1994; (5) Baker, 1992; (6) Tinker et al., 2001; (7) Mungall, 1997.

ACKNOWLEDGMENTS

This material is based upon work supported by the National Science Foundation under grants EAR-1019887 and EAR-1215714, and the U.S. Department of Energy under contract DE-FG-03-91ER-14211. This research also used resources of the National Energy Research Scientific Computing Center (NERSC) supported by U.S. Department of Energy contract DE-AC02-05CH11231. CEL acknowledges support from the Danish National Research Foundation for the Niels Bohr Professorship at Aarhus University during production of this chapter. Any opinions, findings, and conclusions or recommendations expressed in this chapter are those of the authors and do not necessarily reflect the views of agencies funding this work.

FURTHER READING

Agee, C.B., 1998. Crystal-liquid density inversions in terrestrial and lunar magmas. Phys. Earth Planet. Int 107, 63–74.

Baker, D.R., 1992. Tracer diffusion of network formers and multicomponent diffusion in dacitic and rhyolitic melts. Geochim Cosmochim Acta 56 (2), 617–631.

Bouhifd, M., Whittington, A., Roux, J., Richet, P., 2006. Effect of water on the heat capacity of polymerized aluminosilicate glasses and melts. Geochimica et Cosmochimica Acta 70, 711–722.

Brouwers, H.J.H., 2010. Viscosity of a concentrated suspension of rigid monosized particles. Physical Review E 81, 051402.

Canil, D., Muehlenbachs, K., 1990. Oxygen diffusion in a Fe-rich basalt melt. Geochim Cosmochim Acta 54, 2947–2951.

Clark Jr, S.P., 1966. Thermal conductivity. GSA Memoirs 97, 459–482.

Cunningham, G.J., Henderson, P., Lowry, R.K., Nolan, J., Reed, S.J.B., Long, J.V.P., 1983. Lithium diffusion in silicate melts. Earth Planet Sci Lett 65, 203–205.

Dingwell, D.B., Webb, S.L., 1989. Structural relaxation in silicate melts and non-Newtonian melt rheology in geologic processes. Physics and Chemistry of Minerals 16, 508–516.

Dobson, D.P., Jones, A.P., Rabe, R., Sekine, T., Kurita, K., Taniguchi, T., Kondo, T., Kato, T., Shimomura, O., Urakawa, S., 1996. In-situ measurement of viscosity and density of carbonate melts at high pressure. Earth Planet. Sci. Lett 143, 207–215.

Ghiorso, M.S., 2004. An equation of state for silicate melts. III. Analysis of shock compression data and mineral fusion curves. Am. J. Sci 304, 752–810.

Ghiorso, M., Gualda, G., 2015. Chemical Thermodynamics and the Study of Magmas. In: Sigurdsson, H., Houghton, B.F., McNutt, S.R., Rymer, H., Stix, J. (Eds.), Encyclopedia of Volcanoes. Elsevier, pp. 143–161.

Giordano, D., Russell, J.K., Dingwell, D.B., 2008. Viscosity of magmatic liquids: a model. Earth and Planetary Science Letters 271, 123–134.

Haar, L., Gallagher, J., Kell, G., 1984. NBS/NRC Steam Tables: Thermodynamic and Transport Properties and Computer Programs for Vapor and Liquid States of Water in SI Units. Hemisphere Publishing Corporation, Washington, 320 pp.

Henderson, P., Nolan, J., Cunningham, G.C., Lowry, R.K., 1985. Structural controls and mechanisms of diffusion in natural silicate melts. Contrib Mineral Petrol 89, 263–272.

Hofmann, A.W., Magaritz, M., 1977. Diffusion of Ca, Sr, Ba, and Co in a basalt melt - implications for geochemistry of mantle. J Geophy Res 82, 5432−5440.

Hofmeister, A.M., Whittington, A.G., Pertermann, M., 2009. Transport properties of high albite crystals and near-endmember feldspar and pyroxene glasses and melts to high temperature. Contrib. Mineral. Petrol 158, 381−400.

Jambon, A., 1982. Tracer diffusion in granitic melts - Experimental results for Na, K, Rb, Cs, Ca, Sr, Ba, Ce, Eu to 1300 Degrees C and a model of calculation. J Geophy Res 87, 797−810.

Jing, Z., Karato, S., 2008. Compositional effect on the pressure derivatives of bulk modulus of silicate melts. Earth Planet Sci Lett 272, 429−436.

Jones, A.P., Dobson, D., Genge, M., 1995. Comment on physical properties of carbonatite magmas inferred from molten salt data, and application to extraction patterns from carbonatite-silicate magma chambers. Geol. Mag 132, 121−121.

Kelley, K.K., 1936. Contributions to data on theoretical metallurgy: V. Heats of fusion of inorganic compounds. U.S. Bur. Mines Bull 393, 166.

Lange, R.A., Carmichael, I.S.E., 1990. Thermodynamic properties of silicate liquids with emphasis on density, thermal expansion and compressibility. Reviews in Mineral 24, 25−59.

Lange, R.A., 1997. A revised model for the density and thermal expansivity of K_2O-Na_2O-CaO-MgO-Al_2O_3-SiO_2 liquids from 700 to 1900 K: extension to crustal magmatic temperatures. Contrib. Mineral. Petrol 130, 1−11.

Lesher, C.E., 2010. Self-diffusion in silicate melts: theory, observations and applications to magmatic systems. Reviews in Mineralogy and Geochemistry 72, 269−309.

Lesher, C.E., Hervig, R.L., Tinker, D., 1996. Self diffusion of network formers (silicon and oxygen) in naturally occurring basaltic liquid. Geochim Cosmochim Acta 60, 405−413.

Lowry, R.K., Henderson, P., Nolan, J., 1982. Tracer diffusion of some alkali, alkaline-earth and transition element ions in a basaltic and an andesitic melt, and the implications concerning melt structure. Contrib Mineral Petrol 80, 254−261.

Lowry, R.K., Reed, S.J.B., Nolan, J., Henderson, P., Long, J.V.P., 1981. Lithium tracer-diffusion in an alkali-basaltic melt - an ion-microprobe determination. Earth Planet Sci Lett 53, 36−40.

Magaritz, M., Hofmann, A.W., 1978. Diffusion of Sr, Ba and Na in obsidian. Geochim. Cosmochim. Acta 42, 595−605.

Malfait, W.J., Seifert, R., Petitgirard, S., Mezouar, M., Sanchez-Valle, C., 2014b. The density of andesitic melts and the compressibility of dissolved water in silicate melts at crustal and upper mantle conditions. Earth Planet Sci Lett 393, 31−38.

Malfait, W.J., Seifert, R., Petitgirard, S., Perrillat, J.-P., Mezouar, M., Ota, T., Nakamura, E., Lerch, P., Sanchez-Valle, C., 2014a. Supervolcano eruptions driven by melt buoyancy in large silicic magma chambers. Nature Geoscience 7, 122−125.

Mungall, J.E., Dingwell, D.B., 1997. Actinide diffusion in a haplogranitic melt: Effects of temperature, water content and pressure. Geochim Cosmochim Acta 61, 2237−2246.

Mungall, J.E., 2002. Empirical models relating viscosity and tracer diffusion in magmatic silicate melts. Geochimica et Cosmochimica Acta 66, 125−143.

Mysen, B.O., Richet, P., 2005. Silicate Glasses and Melts. Elsevier, New York.

Navrotsky, A., 1995. Energetics of silicate melts. Reviews in Mineral 32, 121−149.

Nevins, D., Spera, F.J., 1998. Molecular dynamics simulations of molten $CaAl_2Si_2O_8$: Dependence of structure and properties on pressure. Am Mineral 83, 1220−1230.

Ochs III, F.A., Lange, R.A., 1997. The partial molar volume, thermal expansivity, and compressibility of H_2O in $NaAlSi_3O_8$ liquid: new measurements and an internally consistent model. Contrib Mineral Petrol 129, 155−165.

Pertermann, M., Whittington, A.G., Hofmeister, A.M., Spera, F., Zayak, J., 2008. Transport properties of low-sanidine single-crystals, glasses and melts at high temperature. Contrib Mineral Petrol 155, 689−702.

Richet, P., 1983. Viscosity and configurational entropy of silicate melts. Geochim Cosmochim Acta 48, 471−483.

Richet, P., Bottinga, Y., 1995. Rheology and configurational entropy of silicate melts. Rev.. in Mineral 32, 67−89.

Rigden, S.M., Ahrens, T.J., Stolper, E.M., 1989. High-pressure equation of state of molten anorthite and diopside. J. Geophys. Res 94, 9508−9522.

Rivers, M.L., Carmichael, L.S.E., 1987. Ultrasonic studies of silicate melts. Jour. Geophys. Res 92, 9247−9270.

Robie, R.A., Hemingway, B.S., 1995. Thermodynamic properties of minerals and related substances at 298.15 K and 1 bar (10^5 pascals) pressure and at higher temperatures. USGS Bulletin 2131.

Sakamaki, T., Ohtani, E., Urakawa, S., Hidenori Terasaki, H., Katayama, Y., 2011. Density of carbonated peridotite magma at high pressure using an X-ray absorption method. Am. Mineral 96, 553−557.

Sakamaki, T., Ohtani, E., Urakawa, S., Suzuki, A., Katayama, Y., 2009. Measurement of hydrous peridotite magma density at high pressure using the X-ray absorption method. Earth Planet Sci Lett 287, 293−297.

Sakamaki, T., Ohtani, E., Urakawa, S., Suzuki, A., Katayama, Y., 2010. Density of dry peridotite magma at high pressure using an X-ray absorption method. Am. Mineral 95, 144−147.

Scarfe, C.M., Mysen, B.O., Virgo, D., 1987. Pressure dependence of the viscosity of silicate melts. In: Mysen, B.O. (Ed.), Magmatic Processes: Physicochemical Principles, Geochem. Sec, 1, pp. 59−67. Special Publication.

Shaw, H.R., 1972. Viscosities of magmatic silicate liquids: an empirical method of prediction. American Journal of Science 272, 870−893.

Spera, F.J., Ghiorso, M., Nevins, D., 2011. Structure, thermodynamic and transport properties of liquid $MgSiO_3$: comparison of molecular models and laboratory results. Contributions to Mineralogy and Petrology 75, 1272−1296.

Stebbins, J.F., Carmichael, I.S.E., Moret, L.K., 1984. Heat capacities and entropies of silicate liquids and glasses. Contrib Mineral Petrol 86, 131−148.

Stixrude, L., Lithgow-Bertelloni, C., 2010. Thermodynamics of the earth's mantle. In: Wentzcovitch, R., Stixrude, L. (Eds.), Theoretical and Computational Methods in Mineral Physics, Reviews in Mineral and Geochemistry, 71, pp. 465−484.

Sykes, D., Baker, M.B., Wyllie, P.J., 1992. Viscous properties of carbonate melts at high pressure, EOS. Trans. Am. Geophy. Union 73, 372.

Thomas, C.W., Liu, Q., Agee, C.B., Asimow, P.D., Lange, R.A., 2012. Multi-technique equation of state for Fe_2SiO_4 melt and the density of Fe-bearing silicate melts from 0 to 161 GPa. J. Geophys. Res 117, B10206.

Tikunoff, D., Spera, F.J. Thermal conductivity of molten and glassy NaAlSi$_3$O$_8$, CaMgSi$_2$O$_6$ and Mg$_2$SiO$_4$ by non equilibrium molecular dynamics at elevated temperature and pressure: part 1 — methods and results. American Mineralogist, 99(11—12), pp. 2328—2336.

Tinker, D., Lesher, C.E., 2001. Self diffusion of Si and O in dacitic liquid at high pressures. Am Mineral 86, 1—13.

Urbain, G., Bottinga, Y., Richet, P., 1982. Viscosity of liquid silica, silicates and alumino-silicates. Geochimica et Cosmochimica Acta 46, 1061—1072.

Zarzycki, J., 1991. In: Chan, R.W. (Ed.), Glasses and the Vitreous State. Cambridge University Press, Cambridge, pp. 505.

Zhang, Y., Ni, H., Chen, Y., 2010. Diffusion data in silicate melts. Reviews in Mineralogy and Geochemistry 72, 311—408.

Chapter 6

Chemical Thermodynamics and the Study of Magmas

Mark S. Ghiorso

OFM Research, Seattle, WA, USA

Guilherme A.R. Gualda

Earth and Environmental Sciences, Vanderbilt University, Nashville, TN, USA

Chapter Outline

GLOSSARY

activity A thermodynamic measure of the energetically effective concentration of a component in a solution.

chemical potential A thermodynamic quantity that describes how the Gibbs free energy of the system changes by addition of one mole of a system component; formally the partial derivative of the Gibbs free energy with respect to the number of moles of a particular component, evaluated at constant temperature, pressure, and the molar abundances of all other components in the system.

(thermodynamic) component An independent variable that describes the abundance of a chemical constituent in a thermodynamic system.

enthalpy A thermodynamic potential that characterizes the equilibrium state under conditions of fixed entropy content, pressure, and bulk composition; the "heat content" of a thermodynamic system at fixed pressure; this thermodynamic potential is minimized when entropy, pressure and bulk composition are set, with other thermodynamic quantities (e.g., temperature, volume, etc.) being derived quantities.

entropy In an equilibrium (reversible) state, a measure of the total heat content of a thermodynamic system; under more general (irreversible) conditions, a measure related to the energetic drive (the chemical affinity) that brings a system to an equilibrium state.

equilibrium state The state under which a thermodynamic system can no longer evolve; a state of rest characterized by a global minimum in the thermodynamic potentials (e.g., Gibbs free energy, enthalpy, Helmholtz free energy) of the system.

fugacity The energetically effective partial pressure of a component in a thermodynamic system.

Gibbs free energy A thermodynamic potential that characterizes the equilibrium state under conditions of fixed temperature, pressure, and bulk composition; this thermodynamic potential is minimized when temperature, pressure and bulk composition are set, with other thermodynamic quantities (e.g., entropy, volume, etc.) being derived quantities.

Gibbs—Duhem relation A differential equation that relates variation in the chemical potentials of system components to changes in temperature or pressure; a more generalized statement of the petrologic phase rule.

Helmholtz free energy A thermodynamic potential that characterizes equilibrium under conditions of fixed temperature, volume, and bulk composition; this thermodynamic potential is minimized when temperature, volume and bulk composition are set, with other thermodynamic quantities (e.g., entropy, pressure, etc.) being derived quantities.

The Encyclopedia of Volcanoes. http://dx.doi.org/10.1016/B978-0-12-385938-9.00006-7

internal energy A thermodynamic potential that characterizes equilibrium under conditions of fixed entropy content, volume, and bulk composition; this thermodynamic potential is minimized when entropy, volume and bulk composition are set, with other thermodynamic quantities (e.g., temperature, pressure, etc.) being derived quantities.

isenthalpic process A process in a thermodynamic system at fixed enthalpy content; a reversible adiabatic process at fixed pressure.

isentropic process A process in a thermodynamic system at fixed entropy content; any reversible adiabatic process.

isochoric process A process in a thermodynamic system at fixed volume.

Korzhinskii potential Any thermodynamic potential applicable to a system that is open to mass transfer of a perfectly mobile component across its boundaries; the potential is defined in such a way that the energetic properties of that mobile component are fixed externally and imposed upon the thermodynamic system.

latent heat Heat generated by a system as a result of a change in the thermodynamic state of a phase (e.g., crystallization of a magma).

(thermodynamic) phase Any compositionally homogeneous macroscopic region of a thermodynamic system.

(thermodynamic) potential An energetic measure of a thermodynamic system whose value is uniquely determined by a set of independent variables (i.e., the state of the system); a function that depends only on the current state of the system, and not on the manner in which the system acquired that state.

sensible heat Heat generated by a system as a result of changing the temperature in the absence of a change in the thermodynamic state of a phase (see latent heat).

(thermodynamic) system A macroscopic region that contains a collection of one or more phases; the system is closed if the boundary is impermeable to mass transfer.

1. INTRODUCTION

Chemical thermodynamics provides a theoretical framework that relates the composition of a **system** to its energy. A system is something that you are interested in studying. It can be broadly defined, as for example a magma body residing in the shallow crust, or narrowly focused, as a phenocryst or a rock. Composition in the system need not be homogeneously distributed, nor even contiguous. In chemical thermodynamics, we speak of the system being partitioned into compositionally homogeneous regions and refer to these as **phases**. We express the composition of the system in terms of a linearly independent set of variables, called **components**. The chemical thermodynamics relates how the energy of the system varies as components are partitioned between the phases and as the proportions of phases vary. This relationship is often nonintuitive; gravity is intuitive, Newton's laws are intuitive, but how the structure of a material and its composition determine its **potential** energy is not. The practical challenge of chemical thermodynamics when applied to magmatic systems is to illuminate as clearly and simply as

possible how the energy of the system relates to its phase assemblage.

It is important to appreciate that chemical thermodynamics is applicable to a dynamically evolving system. The common perception is that thermodynamics is the study of the **equilibrium state**, that is, the state of the system for which the energy is at a minimum and from which there can be no further spontaneous evolution of phase compositions and proportions. This view stems from the fact that the principles of thermodynamics uniquely characterize the system at equilibrium, which is determined by minimizing some energetic measure of the system. Yet, thermodynamics allows us to assess the energy of a system in all states, not just the equilibrium one; it thus allows calculation of differences in energy between diverse disequilibrium states and the unique equilibrium state of the system. In this way, chemical thermodynamics gives us a measure of the energetic drive to achieve an equilibrium state, which through the application of kinetic theory, can inform the rates of chemical reactions and the timescales of system evolution.

Despite this direct connection to kinetics, in the application of chemical thermodynamics to magmatic systems we are usually concerned with the equilibrium state. The estimation of the temperature and pressure of coexistence between phases (geothermometry and geobarometry), the analysis of trace element partitioning between phases, the modeling of volatile phase saturation in silicate liquids, and the estimation of phase stabilities and elemental partitioning between phases, all assume that the system of interest is at equilibrium. Even computational tools that do not conform to the theory of thermodynamics still assume an underlying state of equilibrium. While tools may or may not be based upon computational thermodynamics, those rooted in the foundations of thermodynamics generate the most comprehensive view of elemental partitioning in the system. There is an intrinsic advantage of using a long-established (and never refuted!) theory for the dependence of composition, temperature, and pressure on energy; thermodynamics provides constraints on the functional form of the underlying mathematical expressions, which in turn reveals the best method of extrapolation of experimental data sets.

In this chapter we cannot make an encompassing review of the application of chemical thermodynamics to magmatic systems, but instead we will focus on an overview of the application of thermodynamics to the estimation of phase relations in magmatic systems, utilizing a broad spectrum of boundary constraints. Additionally, we will provide a summary of available tools in computational thermodynamics for the calculation of phase relations, volatile solubilities, and geothermometry and geobarometry. For comprehensive treatments of the chemical thermodynamics, presented with an orientation

TABLE 6.1 Thermodynamic Potentials and Their Differential Forms

Name	Independent Variables	Definition	Differential	
Gibbs free energy (G)	$T, P, n_1, n_2,..., n_c$	$G = E - TS + PV$	$dG = -SdT + VdP + \sum_i^c \mu_i dn_i$	Isobaric crystallization
Korzhinskii oxygen (L)	$T, P, n_1, n_2,..., n_c$ for all components except O_2, μ_{O_2}	$L = G - n_{O_2}\mu_{O_2}$	$dG = -SdT + VdP + \sum_{i \neq O_2}^c \mu_i dn_i + n_{O_2} d\mu_{O_2}$	Isobaric, fixed f_{O_2} crystallization
Enthalpy (H)	$S, P, n_1, n_2,..., n_c$	$H = G + TS$	$dH = TdS + VdP + \sum_i^c \mu_i dn_i$	Magma mixing, wall rock assimilation
Helmholtz free energy (A)	$T, V, n_1, n_2,..., n_c$	$A = G - PV$	$dA = -SdT - PdV + \sum_i^c \mu_i dn_i$	Isochoric (constant volume) crystallization
Internal energy (E)	$S, V, n_1, n_2,..., n_c$	$E = G + TS - PV$	$DE = TdS - PdV + \sum_i^c \mu_i dn_i$	Crystallization by heat loss under variable volume constraints

In the above table S denotes the entropy of the system.

directed to the earth sciences, the reader should consult the excellent texts by Ganguly (2008) and Spear (1993). The classic texts by Pitzer and Brewer (1961) and Prigogine and Defay (1954) are timeless and definitive reference works.

2. THERMODYNAMIC POTENTIALS AND MODELING THE EQUILIBRIUM STATE

Solution of most problems in chemical thermodynamics rests on the exercise of calculating an equilibrium assemblage and its properties. This is done through the minimization of a thermodynamic potential (or state function), a measure of the energy content of the system as a function of a set of independent variables (see Table 6.1). The practical question is then which thermodynamic potential is appropriate in which scenario.

The Gibbs[1] free energy (G) is the thermodynamic potential that is minimal in a chemical system at equilibrium for the *necessary* conditions of fixed bulk composition, temperature (T), and pressure (P). The omnipresence of the **Gibbs free energy** in discussions of chemical thermodynamics in earth sciences suggests some higher standing for this potential. This is not so. The apparent obsession for the Gibbs free energy is simply due to (1) the fact that it is relatively easy to perform experiments at fixed T and P in

the laboratory and (2) the fact that it is in many cases reasonable to assume that geologic processes take place at fixed T and P.

The necessary conditions associated with any thermodynamic potential are essential, and should be understood and applied with care. An open system, e.g., one for which oxygen or hydrogen transfer occurs across the boundaries, or for which a phase or part of a phase is sequestered from chemical communication with the other phases in the system, does not have fixed bulk composition. Such a system can still be in equilibrium, but the necessary conditions must be altered to account for mass transfer across the permeable boundary. In this case, the Gibbs free energy is not the thermodynamic potential of choice. Thermodynamic potentials that are applicable to equilibrium calculations in open systems are called **Korzhinskii potentials**[2] (Ghiorso and Kelemen, 1987) and may be constructed via suitable mathematical transformations of the Gibbs free energy. The reader is referred to the discussion in Ganguly (2008, p. 53) or Callen (1985, Chapter 5) on the method of Legendre transforms for more details. An open system is at equilibrium when the **chemical potential** (μ, definition below) of the mobile component is specified and fixed. A typical example that is applicable to magmatic systems involves equilibrium calculations at fixed redox state. The redox state of a

---

1 Josiah Willard Gibbs, 1839—1903, see the preface of Bumstead and Van Name (1906).

2 After Dmitri Sergeyevich Korzhinskii, 1899—1985, a petrologist who first formulated and applied the phase rule to rock systems open to volatile transfer.

magma is generally specified by constraining the oxygen **fugacity** (f_{O_2}), which is simply the "effective" partial pressure of oxygen (p_{O_2}); the difference between f_{O_2} and p_{O_2} is due to energetic interaction between oxygen and other system components. The fugacity can be greater than or less than the partial pressure, depending on whether the system energy is raised or lowered by the energetic interaction; the specific relation requires experimental measurement or first-principles computational assessment. The oxygen fugacity is related to the chemical potential of oxygen via the definition

$$\mu_{O_2} = \mu_{O_2}^0 + RT \ln \frac{f_{O_2}}{f_{O_2}^0} \tag{6.1}$$

where R is the universal gas constant and the superscript zero denotes the *standard state*, which generally refers to the properties of the pure substance[3]. Formally, the chemical potential is defined as the partial derivative of the Gibbs free energy with respect to the molar concentration of a component in the system:

$$\mu_{O_2} = \left(\frac{\partial G}{\partial n_{O_2}}\right)_{T,P,n_{i \neq O_2}} \tag{6.2}$$

The derivative is taken while holding T, P, and the concentrations of all other components in the system constant. Physically, the chemical potential assesses the infinitesimal change in the Gibbs free energy of the system associated with an infinitesimal increase in the concentration of the specified component. The chemical potential is the essence of chemical thermodynamics. All we desire to know about the relation between composition and energy is embodied in the system chemical potentials. Their definition is simple, and the great triumph of Gibbs. Their practical calculation is complex and often elusive, but nevertheless represents the holy grail of chemical thermodynamic modeling and the focus of attention of practitioners of chemical thermodynamics for over a century.

Gibbs (1878; eq. 88, p. 87 in Bumstead and van Name, 1906) derived a remarkable equation (which we now know as the **Gibbs–Duhem relation**) that establishes that in a system at equilibrium, the chemical potential of a given component must be the same in all phases. This condition is often taken as the definition of an equilibrium state, and is equivalent to the previously stated criterion that the Gibbs free energy of the system is minimal at equilibrium.

As stated above, equilibrium in an open system can be calculated by constructing a Korzhinskii potential for the case where the chemical potential of the mobile component is specified. In our example case of a fixed oxygen redox state, this potential (L, see Table 6.1 and Ghiorso and

Kelemen, 1987) is minimal at equilibrium when T, P, μ_{O_2}, and the concentrations of all other system components are held fixed. Oxygen is released or absorbed across the system boundary in order to maintain the fixed chemical potential constraint. It is worth recalling that a large number of experimental determinations of phase equilibria in magmatic systems are performed at fixed oxygen redox

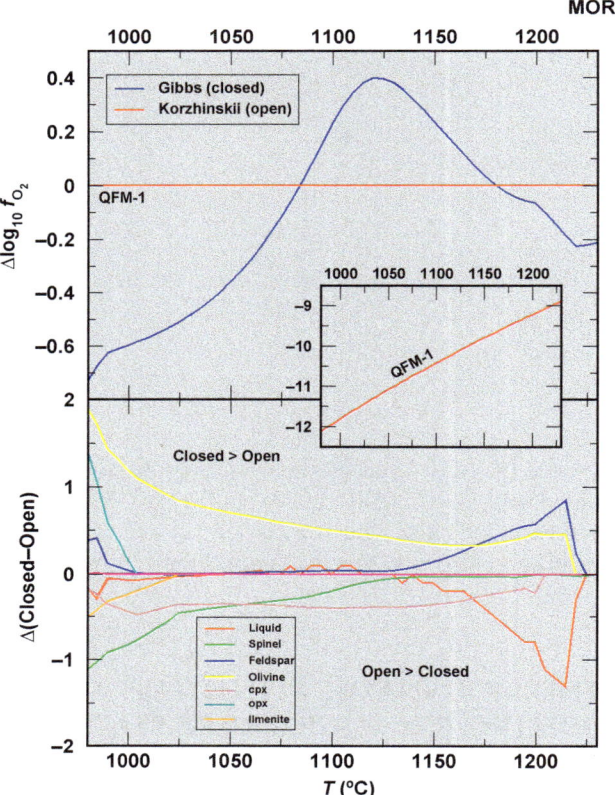

FIGURE 6.1 Equilibrium crystallization of MORB under closed-system conditions (minimization of the Gibbs free energy) and open-system conditions (boundary open to oxygen transfer; minimization of the Korzhinskii potential). Top panel: redox state of the system, expressed as the base 10 logarithm ($\log_{10}$) of the fugacity of O_2 (f_{O_2}) relative to the quartz–fayalite–magnetite buffer minus one $\log_{10}$ unit (QFM-1), plotted as a function of T. Red line shows the open system, blue curve the closed system. Note that the phase assemblage for the closed system is plotted as a function of T in Figure 6.2. The inset shows the variation of absolute $\log_{10} f_{O_2}$ as a function of T for the open system; the oxygen buffer itself is a strong function of temperature over the entire crystallization interval. Lower panel: difference in phase abundances between closed- and open-system evolutions plotted as a function of T. Calculations performed using MELTS (Ghiorso and Sack, 1995) with a 5 °C resolution. In the open-system scenario the redox state is buffered. In the closed-system evolution the system initially oxidizes because precipitating solid phases are enriched in ferrous iron relative to ferric iron, thereby lowering the ferrous to total iron ratio of the liquid with crystallization. The oxidation trend is reversed when a solid phase (here magnetite) appears on the liquidus that has a ferrous to total iron ratio lower than that of the liquid. cpx: clinopyroxene, opx: orthopyroxene. *(See Ghiorso and Carmichael (1985) and Ghiorso (1997) for further discussion.)*

---

3 In this case the chemical potential or the fugacity of pure oxygen gas at the temperature of interest and the specific pressure of 0.1 MPa.

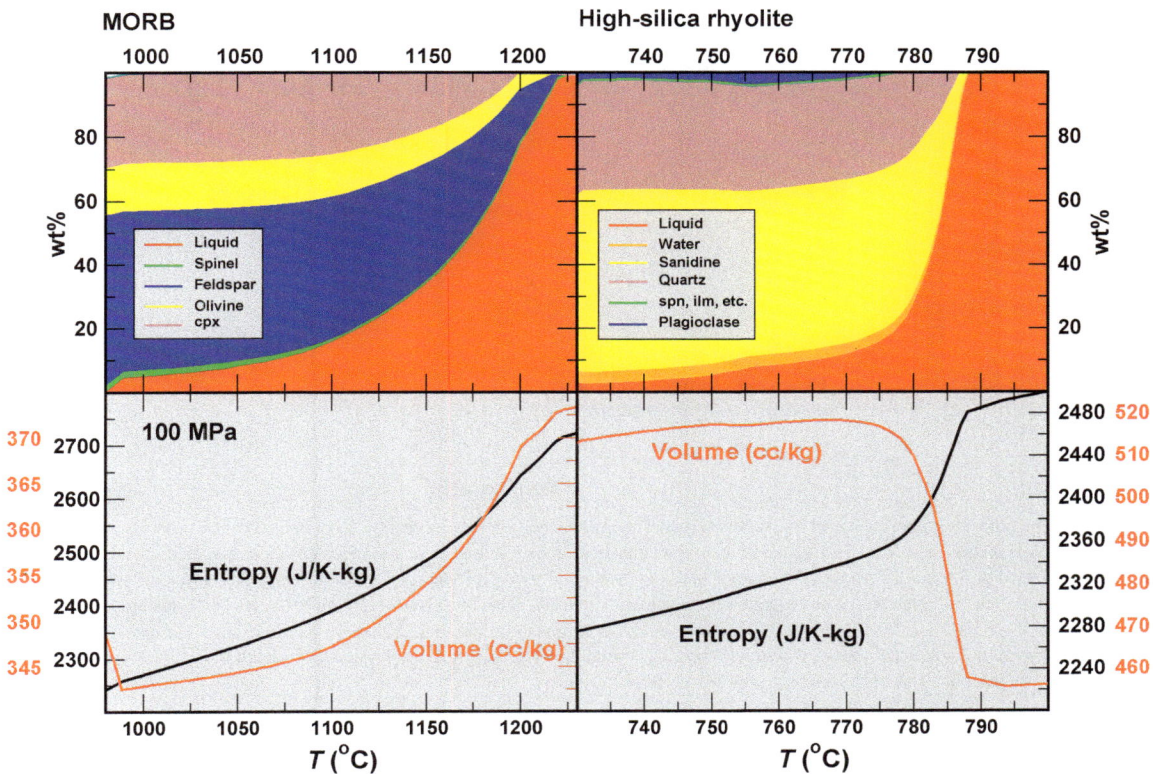

FIGURE 6.2 Closed system equilibrium crystallization of MORB (left) and high-silica rhyolite (right) as a function of T at fixed pressure (100 MPa). Upper panels show phase proportions in wt%. Note that crystallization for the MORB extends over 250 °C, while that for the rhyolite is substantially completed over a 20 °C interval, highlighting the fact that rhyolite bulk composition is near-eutectic in nature. The lower panels show variation in entropy and volume with T. The contribution of latent heat to the entropy production is clearly seen for the rhyolite around the "eutectic," which is not apparent for the MORB composition. The decrease in system volume with crystallization for the MORB case is typical, and contrasts with the rapid increase in volume accompanying volatile exsolution demonstrated by the rhyolite composition. Compare the volume and entropy trends for the high-silica rhyolite with the contour diagrams plotted in Figure 6.4. Calculations are performed using rhyolite-MELTS (Gualda et al., 2012) with a 1 °C resolution. MORB composition (grams): SiO_2, 48.68; TiO_2, 1.01; Al_2O_3, 17.64; Fe_2O_3, 0.89; Cr_2O_3, 0.042; FeO, 7.59; MgO, 9.1; CaO, 12.45; Na_2O, 2.65; K_2O, 0.03; P_2O_5, 0.08; H_2O, 0.2. High-silica rhyolite composition (grams): SiO_2, 77.8; TiO_2, 0.09; Al_2O_3, 12.0; Fe_2O_3, 0.196; FeO, 0.474; MgO, 0.04; CaO, 0.45; Na_2O, 3.7; K_2O, 5.36; H_2O, 3.74. cpx: clinopyroxene, ilm: ilmenite, spn: spinel.

state, e.g., with an imposed oxygen fugacity buffer. As such, these experiments seek to minimize the Korzhinskii potential in their quest to describe the equilibrium state.

To illustrate the consequences of Gibbs or Korzhinskii potential minimization on magmatic phase relations, closed- and open-system equilibrium crystallization of Mid-ocean ridge basalt (MORB) magma is illustrated in Figures 6.1 and 6.2. The results are obtained by application of the MELTS (Ghiorso and Sack, 1995; Gualda et al., 2012) thermodynamic modeling software (see below) and illustrate the very different evolution paths experienced by the magma cooling under these alternate constraints. Note that in the closed system scenario the redox state, relative to a fixed oxygen buffer, varies in response to the solid phase assemblage that is crystallizing, with the redox state effectively set by the ratio of ferric to total iron in the liquid phase (Kress and Carmichael, 1991); this ratio varies as a consequence of the differential partitioning of ferrous and

ferric iron in the coexisting crystals whose identity and compositions vary as a function of crystallization. Some details are presented in the figure legend. For a more thorough discussion of the redox evolution accompanying magmatic crystallization see Osborn (1959, 1962), Ghiorso and Carmichael (1985), Carmichael and Ghiorso (1990), and Ghiorso (1997).

Not all near-equilibrium magmatic processes and evolution scenarios can best be described as Gibbs or Korzhinskii potential minimization problems, despite the fact that intuitively, we are comfortable with using T and P as constraint variables in thinking about the state and evolution of magma bodies. This intuition comes naturally from the realization that both variables are common controls on experimental phase relations, both are accessible to our everyday experience, and both are independent of the mass of the system being considered. Our intuition to choose these constraint variables, however, does not always lead us

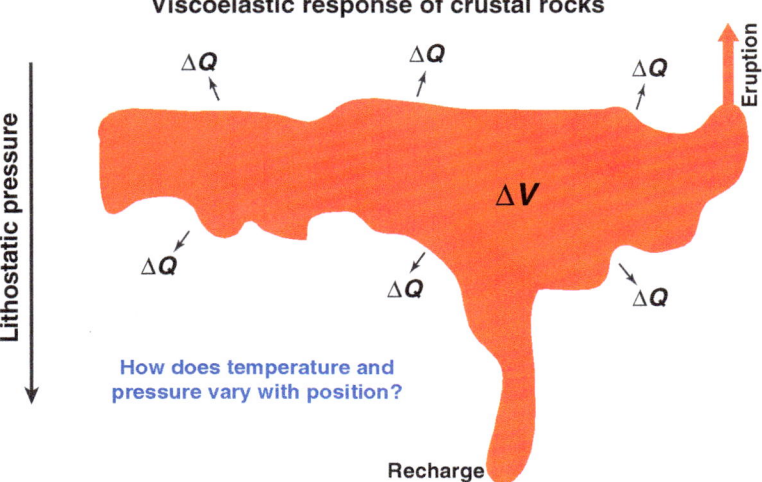

FIGURE 6.3   Magma body residing in the mid- to shallow-crust. The magma body is surrounded by a viscoelastic container of country rocks. Lithostatic pressure is vertical. Heat is lost (ΔQ) from the magma body to the country rock by cooling. Heat sources within the magma body are derived from sensible and latent contributions. The latter arise from phase change. Cooling also induces volume change (ΔV) of the magma body, similarly arising from sensible and latent contributions. The volume change associated with magmatic crystallization is generally negative, but it can be strongly positive in the case of volatile exsolution at low pressures (see Figure 6.2). If the magma ΔV accompanying magmatic heat loss is not exactly compensated by the viscoelastic response of the crustal rocks, an internal pressure field will develop within the magma reservoir that does not match the external lithostatic pressure. This pressure differential has the potential of driving either recharge or eruption.

down the best path of understanding how magma bodies respond to the dynamically changing mantle and crustal environments in which they reside. Consider, for example, a magma body in a mid- to shallow-crustal setting. How does this body evolve? Discounting for the moment the consequences of eruption, recharge, and assimilation, the body evolves principally by withdrawal of heat to the country rock (Figure 6.3). The heat withdrawal (ΔQ) takes place by conductive cooling and perhaps some degree of convective removal supported by fluid circulation through the medium of the country rocks. The magma body—regardless of the nature or style of internal convection—partitions this withdrawn heat between lowering the temperature of the body (the portion of heat known as **sensible heat**), crystallizing the liquid phase (the portion known as **latent heat**), and perhaps converting one solid phase to another (generating a negative or positive heat of reaction). The natural operative variable in this scenario is the heat content of the body and *not the temperature*. The temperature is a derived value, determined by the nature of the phase diagram that describes crystallization behavior of the bulk composition of interest. To see how the nature of the phase diagram controls the thermal evolution, compare the evolution of a MORB with that of a high-silica rhyolite, in which crystallization takes place over a very narrow temperature interval, leading to a very nonuniform partitioning between sensible and latent heat (Figure 6.2). Additionally, temperature is critically controlled by the size of the magma reservoir, because heat content is an *extensive* thermodynamic quantity; it depends on mass and

hence, the size of the body. It would be more natural, therefore, to consider modeling the evolution of this crustal magma reservoir by specifying the heat content and the pressure of the body, and determining phase compositions, phase abundances, and T as modeled outcomes.

The equilibrium crystallization scenario described in the previous paragraph can easily be modeled. The second law of thermodynamics tells us that for a *reversible* process—a process modeled as a series of equilibrium steps—the change in heat content of a system is equivalent to a change in **entropy** (S),

$$\Delta S = \frac{\Delta Q_{rev}}{T} \qquad (6.3)$$

The thermodynamic potential that is minimal in a system at equilibrium subject to fixed bulk composition, fixed pressure, and fixed entropy (i.e., heat content) is the **enthalpy** (H, see Table 6.1). Consequently, by minimizing the enthalpy of the magma for specified S, P, and bulk composition, we can calculate the phase compositions, phase proportions, and the temperature of the magma body; the entropy content can be linked to the heat flow out of the body dictated by heat transfer rates through the surrounding country rocks. An example of a calculation of this nature is shown in Figures 6.4 and 6.5. For this calculation the software PhasePlot (www.phaseplot.org) has been utilized, which implements the underlying thermodynamic models of the rhyolite-MELTS package described in Gualda et al. (2012). The bulk composition chosen for this example is a high-silica rhyolite (same composition as in Figure 6.2). For reference, Figure 6.4 shows the equilibrium phase assemblage over a $T-P$ grid, determined by minimizing the Gibbs free energy at each grid point. Phase relations are displayed (see legend), and the entropy of the magma is contoured (inset, upper right); in this case, entropy is a derived quantity. Compare this reference case with the enthalpy minimization case, displayed in Figure 6.5, which shows the equilibrium assemblages on an $S-P$ grid. Again, phase relations are displayed, but in this case the temperature of the magma is a derived quantity and is shown contoured. The observation that derives immediately from examination of these figures is that the release of latent heat of crystallization is evident at the quartz + plagioclase + sanidine + fluid saturated "minimum" (compare with Figure 6.2, which is for a given pressure of 100 MPa—a horizontal section through Figures 6.4 and 6.5). In Figure 6.4, this latent heat effect is shown by the density of entropy contours over a narrow temperature span of 770–730 °C, and in Figure 6.5, by the rotation of temperature contours from the near vertical to horizontal across the entropy span of 2250–2450 J/K-kg. The latent heat is not uniformly distributed across the crystallization interval of this magma type and this has a profound effect on the cooling history. Indeed, if the flux of heat withdrawal was approximately constant across the boundaries of the magma chamber, then entropy would tend to decrease as a linear function of time, and more time would be spent in the relatively uniform temperature regime about the "minimum" than any other interval of crystallization. In other words, in our attempt to understand the time-dependent evolution of magma bodies, entropy is a much better proxy for time than temperature, given that the surrounding rocks control the flux of heat, not the change in temperature. For this bulk composition, the view from an entropy (heat content)-centric perspective is quite different than that derived from a traditional temperature–pressure phase diagram. Nature views magma bodies as reservoirs of dissipating entropy, with temperature going along for the ride.

The inset contour diagram at the lower right in Figure 6.4 shows the variation of magma volume as a function of temperature and pressure. Notice that the magma volume is a strongly nonuniform function of T and P, and the reasons for this stem from the fact that the specific volume of phases formed during solidification of the magma are not the same as the specific volumes of the parental liquid phase. Another way of saying this is that the densities of phases produced by crystallization of the magma are not the same as the density of the liquid phase, and the rate of crystallization is not constant. The molar

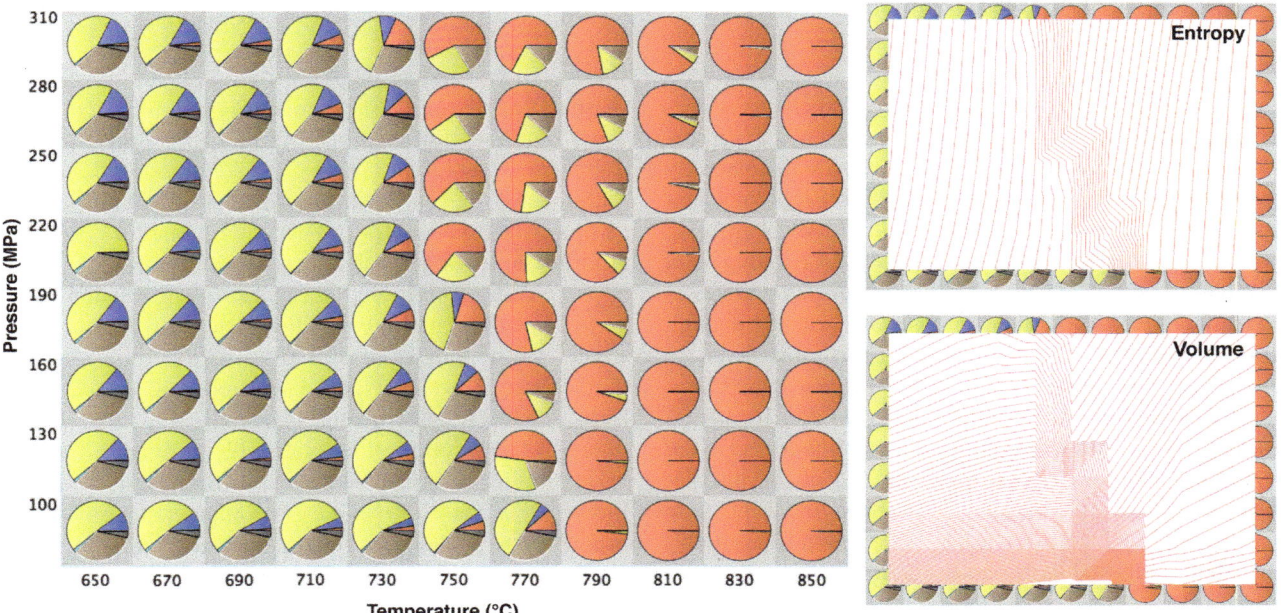

FIGURE 6.4 Phase relations computed by Gibbs free energy minimization over a temperature–pressure grid of a bulk composition corresponding to a high-silica rhyolite (see legend for Figure 6.2). Phase relations are depicted as pie diagrams centered on each grid node. Colors are: red = melt, yellow = sanidine, brown = quartz, blue = plagioclase, other phases include water, magnetite, pyroxenes, and ilmenite. The phase diagram visualization software PhasePlot (www.phaseplot.org) is used to generate the diagram. The inset on the upper right contours an overlay of the entropy of the system, with 10 J/K-kg contours displayed across the range 2125 J/K-kg (left)–2570 J/K-kg (right). The inset on the lower right contours an overlay of the system volume, with 1 cm³/kg contours displayed across the range 420 cm³/kg (left top)–526 cm³/kg (left bottom). Note that the most rapid change in system entropy and volume is associated with the compact interval of solidification across the narrow band of temperature corresponding to the quartz + two-feldspar + fluid-saturated cotectic. The large and positive volume change across the cotectic is driven entirely by volatile exsolution.

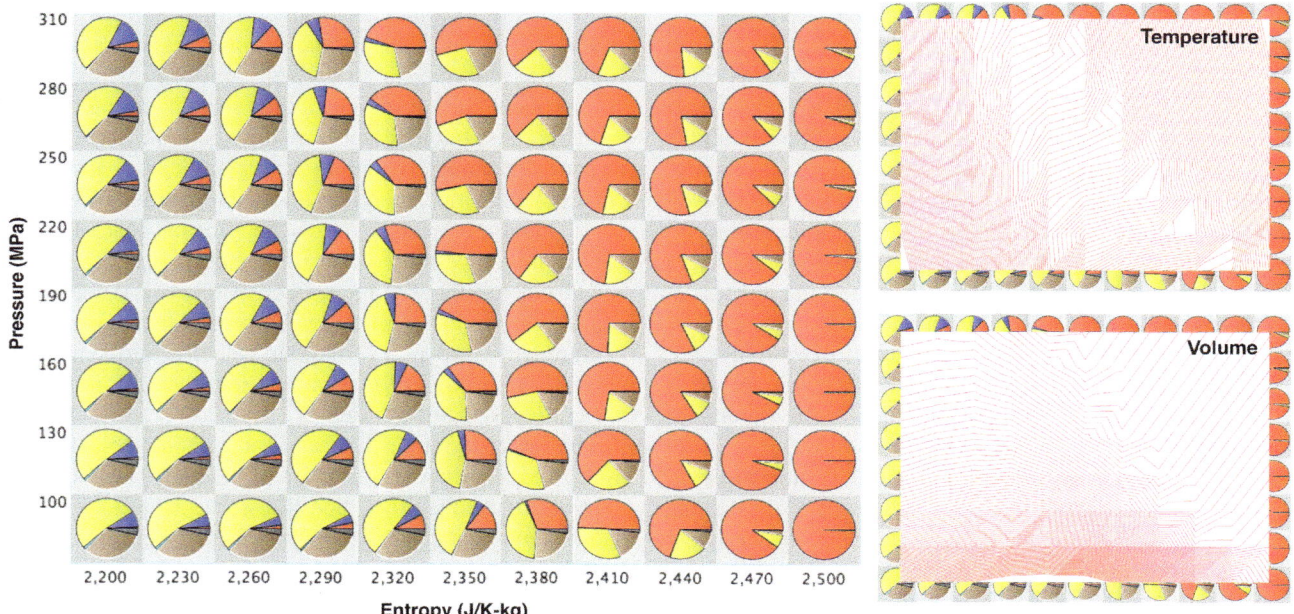

FIGURE 6.5 Phase relations computed by enthalpy minimization over an entropy–pressure grid of a bulk composition corresponding to a high-silica rhyolite. Phase relations are depicted, labeled, and computed as in Figure 6.4. The inset on the upper right contours an overlay of the temperature of the system, with 1 °C contours displayed across the range 676 °C (left)–818 °C (right). The inset on the lower right contours an overlay of the system volume, with 1 cm^3/kg contours displayed across the range 419 cm^3/kg (top)–525 cm^3/kg (left bottom). Note that the temperature contours change orientation from nearly vertical to nearly horizontal as the quartz + two-feldspar + volatile-saturated cotectic is traversed. Compare the volume contour diagram with that shown in Figure 6.4 and note that the compact volume change in temperature space is extended over a much broader interval of entropy space. As the entropy (or heat content) of the magma is changed, the temperature response is dominated by sensible heat, where the temperature contours are vertical, or latent heat, where the temperature contours approach the horizontal.

volume of a volatile phase (principally H_2O–CO_2 fluid solutions) is, at shallow crustal pressures, much larger than the partial molar volumes of dissolved H_2O or CO_2 in the melt, and the specific volumes of solid phases are generally smaller than the corresponding compositionally equivalent mass of liquid. These generalizations are intriguing because if one views the magma body as residing in a deformable pile of crustal rocks, then the question arises as to whether the crustal container can deform quickly enough to compensate the volume change associated with phase transformations in the solidifying magma and maintain a constant pressure within the magma body. If crustal deformation is not fast enough, a pressure drop or rise in the magma will follow, leading to decoupling of the internal pressure field of the magma body from the lithostatic pressure field of the crustal rocks. The critical thermodynamic variables that govern the internal pressure of the magma are consequently the volume of the magma container, and the bulk compressibility (β; or its inverse the bulk modulus, K) of the magmatic liquid + solid + fluid assemblage. The latter quantifies how the volume changes due to a change in applied pressure.

In order to access the consequences of magmatic crystallization on the condition of chamber volume dictated by crustal deformation rates, we must find a

thermodynamic potential that is minimal under conditions of specified volume (V), temperature, and bulk composition. That potential is the **Helmholtz free energy** (A, see Table 6.1). Minimizing A at fixed V, T, and bulk composition yields the equilibrium compositions and proportions of phases and values for the dependent quantity pressure. An illustration of this calculation, again for the high-silica rhyolite composition used in Figures 6.2, 6.4, and 6.5, can be performed with PhasePlot and is presented in Figure 6.6. One sees from the pressure contour diagram that overlays the V–T phase relation grid (inset, lower right), that pressure changes dramatically as a consequence of exsolution of a volatile phase, a not altogether surprising result given the drastic difference in density of the volatile phase compared with that of the solids and liquids. The pressure contours are subhorizontal in the V–T subgrid where volume change is dominated by thermal deflation; pressure contours become vertical where the ΔV of phase change induced by a temperature drop dominates (see figure legend for additional discussion). A detailed application of Helmholtz minimization to phase equilibria and evolution of the Campanian magma body is presented by Fowler et al. (2007) for the specific end-member case of **isochoric** (constant volume; infinitely rigid crust) crystallization.

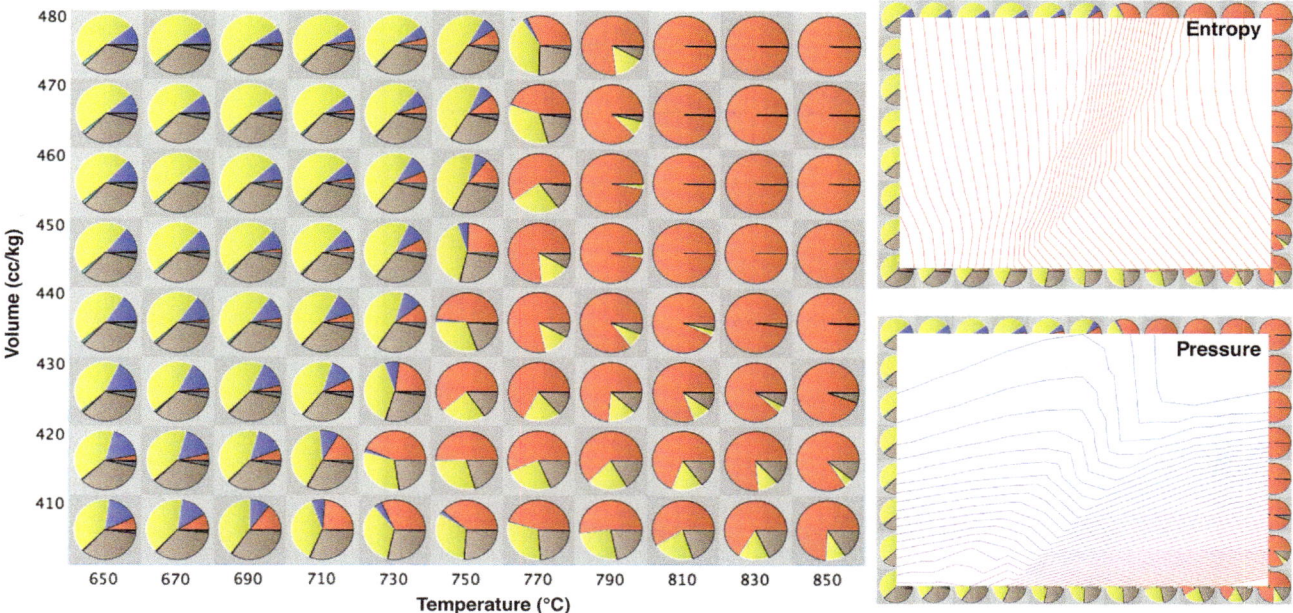

FIGURE 6.6 Phase relations computed by Helmholtz free minimization over a temperature—volume grid of a bulk composition corresponding to a high-silica rhyolite. Phase relations are depicted, labeled, and computed as in Figure 6.4. The inset on the upper right contours an overlay of the entropy of the system, with 10 J/K-kg contours displayed across the range 2100 J/K-kg (left)—2570 J/K-kg (right). The inset on the lower right contours an overlay of the system pressure, with 20 MPa contours displayed across the range 76 MPa (right top)—966 MPa (right bottom). The abrupt change in the slope of the pressure contours corresponds to the onset of volatile saturation.

None of the results presented in Figures 6.4, 6.5, or 6.6 bear the greatest fidelity to the more realistic boundary constraints on upper crustal magma chamber evolution that are dictated by simultaneous and competing rates of heat withdrawal and crustal deformation. The operative thermodynamic variables in this general case are the total heat content of the magma (the entropy in the equilibrium approximation) and the chamber volume. The thermodynamic potential that is minimal for specified values of entropy, volume, and bulk composition is the **internal energy** (E, see Table 6.1). Minimizing E at fixed S, V, and bulk composition yields the equilibrium proportions and compositions of phases in the system as well as values for the dependent variables, T and P. We illustrate such a calculation with results from PhasePlot in Figure 6.7, again using the same high-silica rhyolite composition as before. Note that the temperature—pressure field changes in a nonlinear fashion over the $S-V$ grid. The key thermodynamic quantities that translate heat and volume to temperature and pressure are the isobaric heat capacity (C_P) and, as noted above, the isothermal compressibility (β). Both quantities have been measured experimentally (for compilations see Lange and Carmichael, 1990; Ghiorso and Kress, 2004). S and V are the "natural" thermodynamic variables for understanding the thermal and mechanical coupling between the state of a magma body and surrounding environment. They are not the intuitive variables that we are used to

thinking and working with. But, viewing the evolution of a magma body as a pool of dissipating entropy residing in a deforming container is the proper mind-set for understanding the thermodynamic constraints under which the system evolves.

How the pressure—temperature field of a magma body evolves in detail will be mandated—at both a local and global scale—by heat content and volume constraints, which are in turn a product of a combination of internal and external factors. Phase transformations, or more generally the phase diagram of the magma, dictate the internal contributions to S and V, while the mechanics of deformation and the vagaries of heat flow dictate the external factors. Predictions of specific outcomes are difficult to generalize and must be computed by numerical models that couple thermodynamic and dynamic (and kinetic) contributions. Such models are state of the art in computational petrology and beyond the realm of this brief chapter, but several important features of these coupled models should be appreciated. Even the simplest coupled models of magma-body crustal evolution must admit the compressibility of the magmatic fluid, which is commonly ignored by imposing the Boussinesq approximation in standard dynamical modeling. Volume change on phase transformation must appear in a realistic coupled model as must the dependence of the latent heat production on volume fraction of melt. These are first-order physical phenomena

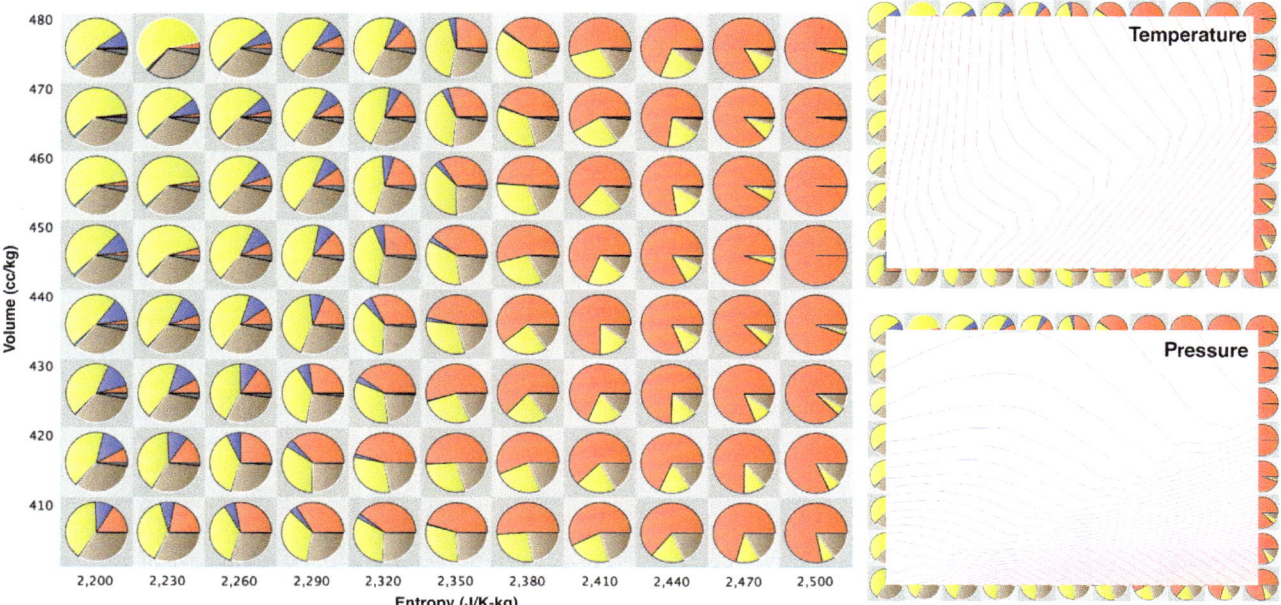

FIGURE 6.7   Phase relations computed by internal energy minimization over an entropy—volume grid of a bulk composition corresponding to a high-silica rhyolite. Phase relations are depicted, labeled, and computed as in Figure 6.4. The inset on the upper right contours an overlay of the temperature of the system, with 5 °C contours displayed across the range 680 °C (left)—860 °C (right). The inset on the lower right contours an overlay of the system pressure, with 20 MPa contours displayed across the range 76 MPa (top)—992 MPa (right bottom).

that prescribe the evolution of the $T-P$ field in the magma reservoir, and all these phenomena are dictated by the phase diagram of the magma. Phase diagrams can be determined by experiment under controlled T and P conditions. We make thermodynamic models of these data—meaning that we generate expressions for Gibbs free energies of the phases that contribute to the phase diagram. We require the theory of chemical thermodynamics, however, to utilize these data in the broadest possible context, which allows us to calculate outcomes that are not directly obtainable from experiment, but that can be computed in ways that are a direct and internally consistent consequence of the primary observations.

3. THE MINIMIZATION OF A THERMODYNAMIC POTENTIAL

Often the statement that the equilibrium state of a system is given by the minimum in an appropriately constrained thermodynamic potential leaves the student in a quandary. How is that minimization actually done? It turns out that the method is not really all that complicated, but in most cases the process is tedious, and inevitably for real systems, must be accomplished by iterative approximation.

Any thermodynamic potential that is minimal in an equilibrium state can be thought of as defining a geometrical or topographical landscape with hills and valleys

(Figure 6.8). The height of the landscape is the value of the system energy. The coordinates of the landscape (and in general, there are more than two) constitute the temperature or entropy, pressure or volume, and independent composition variables of the system. In thermodynamic modeling we seek the lowest point on this energy landscape (Figure 6.8). All other points represent states that are not in

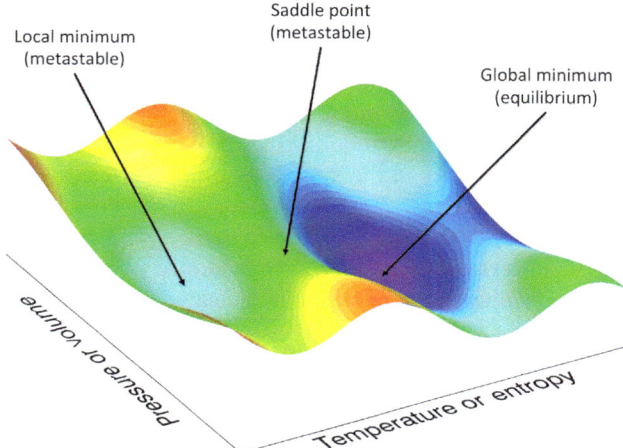

FIGURE 6.8   Hypothetical thermodynamic potential surface for a one-component system, illustrating an energy landscape with independent variables temperature and pressure (G), temperature and volume (A), entropy and pressure (H), and entropy and volume (E). See Table 6.1. Various topological features are displayed. The global minimum is the deepest valley in the energy landscape.

equilibrium and correspond to metastable phase assemblages. The system might reside at an apparently stable local minimum and not the global minimum, but under such circumstances it would be metastable (Figure 6.8); the phases may be in exchange equilibrium with respect to one or more components, or some critical phase may be suppressed from the assemblage due to kinetic barriers to its formation.

To determine the equilibrium state of a system the task in computational thermodynamics is twofold: (1) the identity of the equilibrium phase assemblage must be determined and (2) the system components must be partitioned between those phases in order to reduce the energy of the system as much as possible. The first task is the key to success. A global minimum will not be achieved unless the correct phase assemblage is specified. Simply choosing an arbitrary point on an energy landscape and riding downhill to the nearest minimum does not guarantee that you will end up at the global minimum and consequently does not guarantee that the correct equilibrium phase assemblage will be deduced. The procedure that must be employed to assure attainment of an equilibrium assemblage involves a periodic interrogation of the stability or instability of all potential phases that *might contribute* to the assemblage. Identifying the stability condition of a phase *vis-a-vis* a potential assemblage of phases that are

candidates for an equilibrium assemblage, can be formulated and solved as an iterative but algebraic problem with a guaranteed solution (Ghiorso, 2013; illustrated also in Figure 6.9). Accordingly, the most complex task of the equilibrium algorithm is the one of finding the minimum energy of the system once a hypothetical phase assemblage has been identified. There are many ways to solve this computational problem. The method employed in Phase-Plot or the MELTS family of phase calculators (Ghiorso, 1985; Ghiorso and Sack, 1995) is illustrated in Figure 6.10.

No matter what the order of complexity of the energy landscape for a thermodynamic system, the topology can always be locally approximated about any point of interest by a tangential paraboloid (the multidimensional equivalent of a parabola). This approximation is done by constructing the paraboloid using the first and second derivatives of the true energy surface, evaluating both at the point of interest (see Figure 6.10). The first derivative of the energy surface in practice is a vector and the second derivative is a symmetric matrix of second partial derivatives, but regardless of these details, this quadratic geometrical approximation is always possible. The point of interest, of course, is an initial guess to the phase compositions and proportions as determined utilizing the phase stability algorithms alluded to previously. A minimum of the locally tangential paraboloid can always be located by construction or numerical

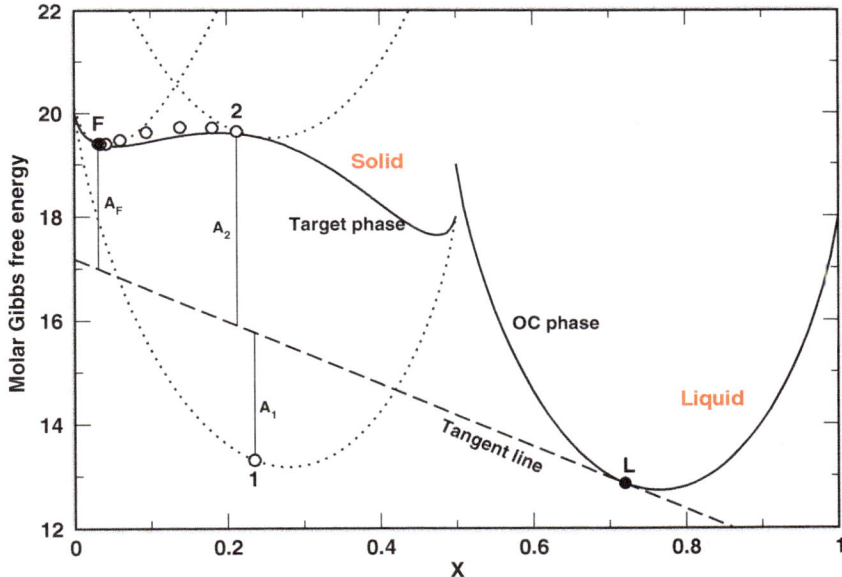

FIGURE 6.9 Illustration of a saturation state detection algorithm for a solid phase referenced to a liquid phase of composition X_L. The Gibbs free energies of the two phases are illustrated by the heavy solid curves. The dashed line labeled "tangent line" is tangent to the liquid phase at "composition" X_L. This line has the same slope as the tangent to the target phase at X_F. Points 1, 2, etc. refer to the sequence of intermediate solutions that converge to F. The chords labeled A_i are geometrical representations of the chemical affinity, which is a direct measure of the degree to which the solid phase is undersaturated with respect to the liquid phase. When the affinity is zero (i.e., when the tangent line touches both the liquid and solid Gibbs free energy surface) the solid is in equilibrium with the liquid phase. The dotted curves are "ideal mixing" approximations of the solid phase. The algorithm for locating X_F and a value for A_F is taken from Ghiorso (2013). The procedure is to approximate the solid phase as a succession of pseudoideal solutions, until that approximation converges on the properties of the real solution. For implementation details, consult the reference. OC phase: omnicomponent phase.

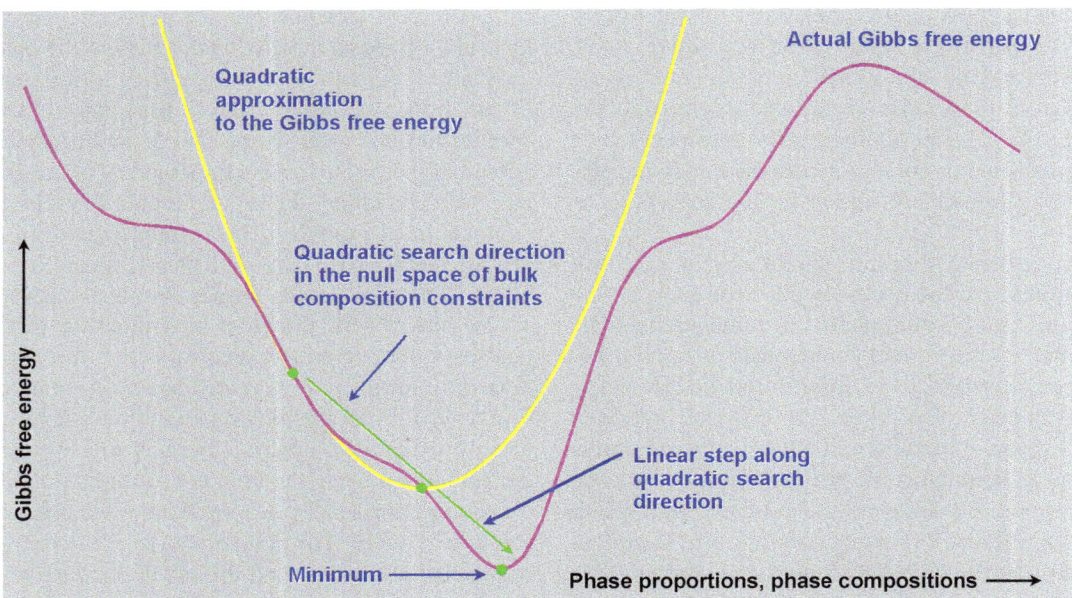

FIGURE 6.10   The minimum of a thermodynamic potential, like that of the system Gibbs free energy illustrated, is obtained by choosing an initial guess to the energy minimum, illustrated here as the left most filled green circle, and constructing at that point a parabolic approximation to the energy, which has the same first and second derivatives as the energy surface at the point of interest. In this figure that parabolic approximation is shown as a yellow curve. The initial guess must be independently estimated, perhaps using some form of the saturation state algorithm illustrated in Figure 6.9; it consists of approximate phase proportions and compositions that might characterize a good guess of the equilibrium assemblage. A search direction is poised starting from the initial point and projecting through the minimum of the parabolic approximation. A "linear search" is extended along that direction (the illustrated green vector) until a lower point on the actual energy surface is reached; this procedure derives new and better values of phase compositions and proportions. This new "low point" becomes the next guess, and the procedure is repeated until the true minimum is reached. Typically this takes four or five parabolic or "quadratic" iterations. The illustrated algorithm is given in detail in Ghiorso (1985). Mathematically, the method is a concrete example of a projected second-order Newton method.

evaluation. The only complication to this exercise is that the location of the parabolic minimum must satisfy, as did the initial feasible point, all of the imposed constraints that qualify the equilibrium condition. For the Gibbs free energy, these would be fixed T, P, and bulk composition, and for the Helmholtz energy, fixed T, V, and bulk composition, etc. The mechanics of imposing these constraints is straightforward and discussed elsewhere (Ghiorso, 1985), but the important thing to understand is that the parabolic minimum is chosen in such a way that the constraints are enforced and the solution is physically viable.

As shown schematically in Figure 6.10, the vector from the initial point of interest to the parabolic minimum is taken as a feasible direction of extrapolation along which a "linear" search is made to locate the lowest position on the energy landscape. Once this location is found, a new paraboloid is constructed tangential to this position and the whole process is repeated until the true minimum is located. Generally, four to five parabolic approximations suffice to locate the true minimum. Once found, phase-stability checks either discredit the minimum by the addition of more stable phases to the assemblage, or verify the minimum as global if no additional phases are found to be more stable than the current assemblage. This procedure

can be programed for very rapid execution, particularly if the thermodynamic derivatives of all of the phases involved in the calculation are coded analytically. As an example, execution time for a typical Gibbs free energy minimization in rhyolite-MELTS (Gualda et al., 2012) for a five-phase system with ~ 25 optimal variables and 10 active constraints requires about 100 ms on a 3-GHz processor.

4. COMPUTATIONAL THERMODYNAMICS TOOLS AVAILABLE TO THE VOLCANOLOGIST/IGNEOUS PETROLOGIST

We will end this brief chapter with a summary and notes regarding freely available tools for computational chemical thermodynamics applications in volcanology and igneous petrology. These applications fall into four general categories: (1) geothermometers, geobarometers, and geohygrometers, (2) mineral and melt thermodynamic property calculators, (3) solubility calculators for volatiles in magmatic systems, and (4) phase equilibrium calculators. Our summary is by no means exhaustive, and emphasis has been placed on calculators that are web-based or run on a variety of computer systems. Commercial software products are not included.

4.1. Geothermometers, Geobarometers, and Geohygrometers for Magmatic Systems

Geothermometers are tools that utilize compositional partitioning of elements between coexisting phases to infer something quantitative about the temperature under which the assemblage formed. Geobarometers do the same for pressure. Geohygrometers describe the effect of dissolved water in one phase (generally the liquid) on elemental partitioning between the liquid and another coexisting solid phase, allowing estimation of water content in the liquid in equilibrium with the solid. These tools are always calibrated from underlying experimental data sets and generally their formulation is thermodynamically "inspired," but incomplete; the formalism is based on thermodynamic theory but is generally not developed into a complete internally consistent description of the Gibbs free energy of all phases that participate in the assemblage. All geothermometers, geobarometers, and geohygrometers rely on the establishment of exchange equilibrium that governs predictable partitioning of an element or elements of interest between the phases of interest. It is important to bear in mind that two phases can be in exchange equilibrium and still constitute a metastable phase assemblage (e.g., exchange equilibrium could correspond to a local minimum or a saddle point, see Figure 6.8); as noted above, only at the global minimum of the energy landscape is the true equilibrium phase assemblage manifest.

An excellent summary of geothermometers and geobarometers applicable to volcanic rocks is provided in Putirka (2008); see Table 6.2. Also in the table, we list additional sources of information and direct links to other commonly utilized formulations. Of particular note are the Ti-in-quartz (Wark and Watson, 2006) and the Ti-in-zircon (Watson et al., 2006) geothermobarometers. These tools rely on knowing the partitioning behavior of Ti between melt and mineral phases and have wide application in the study of silicic rocks. Both require estimation of the **activity** of TiO_2 in the coexisting liquid phase; for this purpose see the method of Ghiorso and Gualda (2012). A number of thermometric calculators derived from solid−solid elemental exchange are also listed in Table 6.2, most notably the elegant method of Andersen et al. (1993; QUILF), which is an example of a generalized geothermobarometric approach that examines elemental partitioning between multiple phases, thereby generating a more reliable estimate of T and P than methods based on a single exchange. The widely utilized Fe−Ti oxide geothermometer and oxygen barometer is also listed. This method relies on the calibration of Fe−Ti

TABLE 6.2 Computational Thermodynamic Tools

Source	Type Notes/URLs
Putirka (2008)	Volcanic geothermometers and geobarometers Review and summary. Links to Excel workbook implementations: Clinopyroxene P-T Feldspar-liquid P-T-H2O Mantle Potential Temperatures Olivine and glass thermometers Opx thermobarometers Silica activity barometers Two-feldspar thermometers Two-pyroxene thermobarometers
Anderson et al. (2008)	Plutonic geothermometers and geobarometers
Wark and Watson (2006), Thomas et al. (2010), Huang and Audetat (2012)	Ti-in-quartz geothermometer and geobarometer TitaniQ
Hayden et al. (2008)	Sphene (titanite) geobarometer and geothermometer
Watson et al. (2006), Ferry and Watson (2007)	Ti-in-zircon, Zr-in-rutile geothermometer
Berman (2007)	Thermobarometric calculations using an extended Berman (1988) database http://serc.carleton.edu/research_education/equilibria/twq.html
Tim Holland	Mineral activity models, geothermometers, and geobarometers https://www.esc.cam.ac.uk/research/research-groups/holland (AX: activity models, HBPL: hornblende-plagioclase geothermometer)
Ghiorso and Evans (2008)	Fe−Ti oxide geothermometer oxygen barometer http://ctserver.ofm-research.org/OxideGeothrm/OxideGeothrm.php
Sack and Ghiorso (1991)	Olivine−spinel geothermometer http://ctserver.ofm-research.org/Olv_Spn_Opx/index.php
Lange et al. (2009)	Plagioclase−liquid hygrometer and thermometer http://www.lsa.umich.edu/earth/people/faculty/ci.langerebecca_ci.detail
Gualda and Ghiorso (2014)	Quartz + feldspar + melt barometry
Andersen et al. (1993)	Multivariate equilibrium assessment QUILF. Program may be downloaded from the journal archives

(Continued)

TABLE 6.2 Computational Thermodynamic Tools—cont'd

Source	Type Notes/URLs
Berman (1988)	Thermodynamic database: end-member properties, reaction properties, univariant curves http://ctserver.ofm-research.org/ThermoDataSets/Berman.php
OFM Research	Sack and Ghiorso mineral solution models http://ctserver.ofm-research.org/phaseProp.html http://melts.ofm-research.org/CalcForms/index.html
Kress (1997, 2000, 2007), Kress et al. (2008)	Solution models of sulfide liquids The system O−S−Fe−Ni−Cu
OFM Research	Web services: REST and SOAP based http://ctserver.ofm-research.org/webservices.html
Newman and Lowenstern (2002)	Melt−H_2O−CO_2 fluid solubility model http://volcanoes.usgs.gov/observatories/yvo/jlowenstern/other/software_jbl.html
Papale et al. (2006)	Melt−H_2O−CO_2 fluid solubility model http://ctserver.ofm-research.org/Papale/Papale.php
Moore (2008)	H_2O, CO_2, and mixed fluid volatile solubility in magmas Review and critical summary; his Table 3—General models for mixed fluids
Wallace and Carmichael (1992), Moretti and Ottonello (2005), Moretti and Baker (2008)	Sulfur solubility and speciation in magmas
Kress et al. (2004)	C−O−H−S−Cl−F gas speciation Excel workbook available as journal supplement
Connolly (2005)	Phase equilibrium calculations PerpleX: http://serc.carleton.edu/research_education/equilibria/perplex.html
Powell et al. (1998)	Phase equilibrium calculations THERMOCALC: http://serc.carleton.edu/research_education/equilibria/thermocalc.html
de Capitani and Petrakakis (2010)	Phase equilibrium calculations Theriak/Domino: http://serc.carleton.edu/research_education/equilibria/theriak-domino.html

TABLE 6.2 Computational Thermodynamic Tools—cont'd

Source	Type Notes/URLs
Frank Spear	Phase equilibrium calculations Program Gibbs: http://ees2.geo.rpi.edu/MetaPetaRen/Software/GibbsWeb/Gibbs.html
Ariskin (1999)	Phase equilibria in basaltic magmas COMAGMAT: http://geo.web.ru/~kbs/
Ghiorso and Sack (1995), Asimow and Ghiorso (1998), Ghiorso et al. (2002), Gualda et al. (2012)	Magmatic phase equilibria MELTS—low pressure, broad composition range, pMELTS—1−3 GPa, mantle-like bulk compositions, rhyolite-MELTS—like MELTS, with improved modeling of silicic two-feldspar + quartz + fluid-saturated systems http://melts.ofm-research.org/
Asimow et al. (2004), Smith and Asimow (2005), Thompson et al. (2007), Antoshechkina et al. (2010)	Mantle phase equilibria alphaMELTS—pMELTS plus trace elements and water in nominally anhydrous mantle minerals. Numerous option enhancements; batch processing http://magmasource.caltech.edu/alphamelts/
OFM Research	PhasePlot Gridded phase equilibrium calculations for magmatic systems, utilizing the MELTS and pMELTS thermochemical data/model collections http://phaseplot.org/ Macintosh App Store
Bohrson and Spera (2001, 2003, 2007), Spera and Bohrson (2001, 2002, 2004), Bohrson et al. (in press)	Energy-constrained open-system magmatic processes http://magma.geol.ucsb.edu/papers/ECAFC.html
Kress and Ghiorso (2004)	Postentrapment crystallization of melt inclusions http://brimstone.ess.washington.edu/

exchange between coexisting spinel and ilmenite. It should be noted that alternative calibrations of the Fe−Ti oxide geothermometer oxygen barometer exist and can be accessed via QUILF. Finally, the plagioclase hygrometer/thermometer of Lange et al. (2009) is an extraordinarily useful tool for understanding the mutual effect of temperature and melt water content on the stability relations of plagioclase in intermediate to silicic magmas.

4.2. Mineral and Melt Thermodynamic Property Calculators for Magmatic Phases

Table 6.2 lists several sources of information (all online) for thermodynamic property calculators (pure end-member properties and mixing properties, e.g., activities). When using any of these tools it is very important to understand that pure end-member properties and solution or mineral activities are related through the definition of the chemical potential, as

$$\mu_i = \mu_i^0 + RT \ln a_i \qquad (6.4)$$

The only bit of reality in Eqn (6.4) is the end-member chemical potential, μ_i. That quantity and its temperature and pressure derivatives are calibrated against experimental measurement. The chemical potential is partitioned by convenience into a pure end-member contribution (μ_i^0) and an end-member activity (a_i), and unless there is independent experimental evidence to fix one or the other, the two are correlated quantities and the model that derives one of these quantities must be internally consistent with the calibration of the other. So, the reader is advised to be very careful in mixing and matching mineral/melt activities with arbitrary pure end-member properties. The two are not always consistent unless care is taken to make them so. Failure to heed this warning can lead to incorrect estimates of solution properties and perhaps erroneous predictions of phase stability, elemental partition, temperature, or pressure.

4.3. Solubility Calculators for Volatiles in Magmatic Systems

There are a number of thermodynamically based methods for estimating volatile solubilities in magmatic systems. The majority of work has focused on pure H_2O or pure CO_2 fluids; Moore (2008) provides a comprehensive review. For mixed-fluid (H_2O-CO_2) saturation applications a number of compositionally restricted formulations exist (e.g., Iacono-Marziano et al., 2012), but two general methods are in common use and are listed in Table 6.2. The calibration of Papale et al. (2006) covers the broadest compositional range, and is most generally applied to the interpretation of melt inclusions. Moore (2008) assesses the accuracy of this calibration against experimental measurements.

Almost all the models of H_2O solubility in magmatic liquids make the assumption that the solubility of silicate material in the fluid phase is either negligible or has little effect on the chemical potential of H_2O in the fluid. This assumption is thermodynamically inconsistent, but experimental data are limited to assess the consequences quantitatively. The main exceptions are the SiO_2-H_2O and

$NaAlSi_3O_8-H_2O$ systems, for which experimental data are available to define the extent of solution on both sides of the binary. Hunt and Manning (2012) have calibrated an elegant thermodynamic model for the SiO_2-H_2O binary that accounts for the mutual solubility of end-members and the closure of the miscibility gap to form a continuous solution above the second critical endpoint of the system. As interest focuses on the nature of volatile saturation at mantle temperatures and pressures, it will become necessary to extend all of the melt–volatile thermodynamic solubility treatments to incorporate the concept of mutual solubility and complete solution. This is an exciting challenge for the future.

Thermodynamic models describing the solubility of reduced and oxidized sulfur gas species in silicate melts are also listed in Table 6.2. These formulations have yet to be integrated with ferric–ferrous redox calibrations (Kress and Carmichael, 1991) for the melt phase. Coupling the two models and merging the sulfur saturation state models with those for water and carbon dioxide remains another challenge for future thermodynamic models of volatile saturation.

We are not aware of any thermodynamic models of volatile saturation for other components such as Cl, F, or, for that matter, reduced carbon species. If the focus is on the gas state, however, the 2004 model of Kress et al. (2004) may be utilized to estimate species abundances, activities, and fugacities in the system $C-O-H-S-Cl-F$, on the assumption of ideal mixing.

4.4. Computation of Phase Equilibria

We list at the bottom of Table 6.2, descriptions and links to software that perform calculations of phase equilibria (generalized phase diagrams) for both subsolidus crustal- and mantle-like bulk compositions, and for magmatic systems. The latter category includes COMAGMAT, which is optimized for low-pressure, basaltic magmas; MELTS (and rhyolite-MELTS), which is applicable over a broader compositional range, but still restricted to pressures below about 1 GPa; and pMELTS, which is a calibration optimized for mantle-like bulk compositions in the range 1–3 GPa. MELTS and pMELTS consist of collections of thermodynamic models for silicate liquid, mineral, and fluid phases, with the liquid-phase properties derived in part from experimental liquid–solid phase equilibrium data. These models suffer from the serious limitations of incomplete data coverage, inconsistent data, and inadequacies of the underlying thermodynamic solution theory. One particular hindrance is a lack of high-quality experimental data characterizing elemental partitioning between amphibole and liquid and between biotite and liquid. This data deficiency precludes the development of appropriate solution models for these solid phases and prevents modeling magmatic evolution scenarios where either phase plays a dominant role. PhasePlot is an implementation of the MELTS and pMELTS computational infrastructure with a user interface best suited for visualization of phase relations on grids, and for the computation of phase diagrams.

A derivative of the pMELTS thermochemical data/model collection is alphaMELTS (previously phMELTS and Adiabat_1ph) with addition of trace element partitioning and H_2O partitioning into nominally anhydrous phases. It is optimized for a number of computing platforms and can function without a graphical user interface. It is the tool of choice for modeling melt production in the mantle by fractional or batch melting along adiabatic ascent paths.

While MELTS and pMELTS can be utilized to perform modeling in closed or open (to oxygen) systems in either equilibrium crystallization or fractionation modes, they can also be used for isochoric (constant volume, Helmholtz free energy minimization), **isentropic** (constant entropy, enthalpy minimization), and **isenthalpic** (constant enthalpy, entropy maximization) thermodynamic paths. The last is especially useful for exploring energy-constrained assimilation scenarios at constant pressure (Bowen, 1922; Ghiorso and Kelemen, 1987). The EC-AFC models of Bohrson and Spera (Table 6.2) explore this concept and develop a modeling infrastructure for tracking major- and trace-element evolution in a fractionating magma body subjected to energy-constrained assimilation, magma recharge, and periodic eruption. The latest incarnation of this modeling software is termed the Magma Chamber Simulator (Bohrson et al., 2014) that utilizes rhyolite-MELTS as the computational phase equilibrium engine.

The final entry in Table 6.2 is a modeling package aimed at simulating (and reverse calculating) the effects of postentrapment crystallization within melt inclusions trapped within host crystals. The scheme accounts for volume change on crystallization and the constraints imposed upon the inclusion by the thermal expansion and compressibility of the enclosing crystal. The phase equilibrium engine utilized for these calculations is MELTS.

5. FUTURE DIRECTIONS OF CHEMICAL THERMODYNAMIC APPLICATIONS IN MAGMATIC SYSTEMS

The application of chemical thermodynamics to the study of magmas has come a long way since the pioneering work of Verhoogen (1949) and Carmichael et al. (1970). These prescient applications of chemical thermodynamics to the

study of the origin and evolution of magmas laid the foundation for all subsequent work. Although much has been accomplished over the last half century, much still needs to be done. The future is bright for applications of chemical thermodynamics to magmatic systems. One could say that the whole field is just beginning to mature to facilitate the creation of a working infrastructure for general applications.

One of the main research objectives driving the field is the integration of thermodynamics-based phase property engines into computational fluid dynamics (CFD) simulation of magma chamber processes and melt generation. Until very recently, the computer cycles required to compute phase properties and proportions at each grid node in a CFD computation rendered this level of integration impractical. With the advent of cheap, cluster-based computing and the improvement in numerical algorithms for nonlinear multidimensional optimization, this obstacle is rapidly disappearing. Couple this evolution with the realization that thermodynamic calculations are intrinsically local and can be largely decoupled from neighboring computations in a spatial grid, and it becomes clear that the rate-limiting step in CFD simulations is no longer the time consumed in evaluating the constitutive relations. Casting the CFD simulation in terms of a heat content field rather than a temperature field (e.g., Figure 6.5) serves to nearly eliminate Stefan boundary issues that would otherwise require high-density grid spacing around phase transformation fronts.

Refinement and broadening of chemical thermodynamic models for magmatic systems awaits acquisition of more and better data on the physical properties of earth materials as well as the chemical partitioning of elements between magmatic phases. These data can be acquired by either physical or computational experiment, the latter arising from first-principles and molecular dynamics simulations. Increasingly, the accuracy of first-principles-potential and empirical-potential molecular dynamics simulations of high-temperature silicate liquids and minerals has shown that these tools are capable of generating fundamental data for the purpose of calibration of thermodynamic models (Karki, 2010; Ghiorso and Spera, 2010; Stixrude and Lithgow-Bertelloni, 2010), especially at elevated pressures where traditional physical experimentation is not yet achievable, too costly, or too difficult for routine systematic studies. But, the emergence and success of theoretical computation methods does not diminish the need for high-quality physical experiments. What is desperately needed in this regard are experimental studies aimed specifically at generating data for the calibration of thermodynamic models. This objective requires fully characterized samples through careful chemical analysis and textural documentation, and studies that focus on illuminating the relations between the concentration of an element in a phase and its effective energetic concentration, that is, its activity. That perspective is seldom pursued in phase equilibrium experimentation, but when it is, the results can be spectacular and rewarding beyond the mere documentation of phase relations. Additionally, experimental data that illuminate the fundamental properties (entropy, enthalpy of formation, heat of solution, heat capacity, and equation of state) are by no means complete for magmatic phases, even under crustal conditions. Many of the fundamental data that are used in models that calculate temperatures, pressures, volatile solubilities, or phase relations in magmatic systems are based, in part, on assumed values of thermodynamic constants and internally consistent interpolations or extrapolations of primary data. These models are only as good as their underlying data, and improvement in data quality is both welcome and necessary.

The objective of this chapter on application of chemical thermodynamic to magmatic processes is to highlight the role chemical thermodynamics can play in the day-to-day assessment of petrologic and volcanologic hypotheses. Thermodynamics is the platform upon which kinetic theory is built. It is the framework for our understanding of the systematics of phase equilibria, which provides us with a means of understanding why minerals appear in rocks. It is the underlying theory behind the validity of all geothermometers, geobarometers, and any methods used for the estimation of other intensive variables in magmatic systems. Chemical thermodynamics will continue to play a vital role in making the connection between the chemistry of a magma and its energy, thereby permitting the dynamical evolution of magmatic systems to be coupled directly and quantitatively to the chemical signatures observed in the rock record.

ACKNOWLEDGMENTS

Material support from the National Science Foundation (EAR 11-19297 to Ghiorso; EAR 1151337 and EAR 0948528 to Gualda) is gratefully acknowledged. The first author is indebted to Jean Verhoogen, Hal Helgeson, and Ian Carmichael, whose astounding contributions to chemical thermodynamics of Earth materials continue to inspire.

FURTHER READING

Anderson, J.L., Barth, A.P., Wooden, J.L., Mazdab, F., 2008. Thermometers and thermobarometers in granitic systems. In: Putirka, K.D., Tepley III, F.J. (Eds.), Minerals, Inclusions and Volcanic Processes. Reviews in Mineralogy and Geochemistry, vol. 69, pp. 121–142.

Andersen, D.J., Lindsley, D.H., Davidson, P.M., 1993. QUILF: a Pascal program to acces equilibria among Fe-Mg-Ti oxides, pyroxenes, olivine, and quartz. Computers in Geosciences 19, 1333–1350.

Antoshechkina, P.M., Asimow, P.D., Hauri, E.H., Luffi, P.I., 2010. Effect of water on mantle melting and magma differentiation, as modeled using Adiabat_1ph 3.0. In: American Geophysical Union, Fall Meeting 2010 abstract #V53C-2264.

Ariskin, A.A., 1999. Phase equilibria modeling in igneous petrology: use of COMAGMAT model for simulating fractionation of ferro-basaltic magmas and the genesis of high-alumina basalt. Journal of Volcanology and Geothermal Research 90, 115–162.

Asimow, P.D., Dixon, J.E., Langmuir, C.H., 2004. A hydrous melting and fractionation model for mid-ocean ridge basalts: application to the Mid-Atlantic Ridge near the Azores. Geochemistry Geophysics Geosystems 5. http://dx.doi.org/10.1029/2003GC000568.

Asimow, P.D., Ghiorso, M.S., 1998. Algorithmic modifications extending MELTS to calculate subsolidus phase relations. American Mineralogist 83, 1127–1132.

Berman, R.G., 1988. Internally-consistent thermodynamic data for minerals in the system Na_2O-K_2O-CaO-MgO-FeO-Fe_2O_3-Al_2O_3-SiO_2-TiO_2-H_2O-CO_2. Journal of Petrology 29, 445–522.

Berman, R.G., 2007. winTWQ (version 2.3): a software package for performing internally-consistent thermobarometric calculations. Geological Survey of Canada. Open File 5462, (ed. 2.32), 41 pp.

Bohrson, W.A., Spera, F.J., 2001. Energy-constrained open-system magmatic processes. II: application of energy-constrained assimilation-fractional crystallization (EC-AFC) model to magmatic systems. Journal of Petrology 42, 1019–1041.

Bohrson, W.A., Spera, F.J., 2003. Energy-constrained open-system magmatic processes. IV: geochemical, thermal and mass consequences of energy-constrained recharge, assimilation and fractional crystallization (EC-RAFC). Geochemistry, Geophysics, Geosystems 4. http://dx.doi.org/10.1029/2002GC00316.

Bohrson, W.A., Spera, F.J., 2007. Energy-constrained recharge, assimilation, and fractional crystallization (EC-RA/É'/FC): a Visual Basic computer code for calculating trace element and isotope variations of open-system magmatic systems. Geochemistry, Geophysics, Geosystems 8. http://dx.doi.org/10.1029/2007GC001781.

Bohrson, W.A., Spera, F.J., Ghiorso, M.S., Creamer, J. Thermodynamic model for energy-constrained open system evolution of crustal magma bodies undergoing simultaneous assimilation, recharge and crystallization: the magma chamber simulator. Journal of Petrology, in press.

Bowen, N.L., 1922. The behavior of inclusions in igneous magmas. Journal of Geology 30, 513–570.

Bumstead, H.A., Van Name, R.G. (Eds.), 1906. The Scientific Papers of J. Willard Gibbs. Thermodynamics, vol. 1. Longmans Green and Co., 464 pp.

Callen, H.B., 1985. Thermodynamics and an Introduction to Thermostatics, second ed. John Wiley and Sons, New York. 493 pp.

Carmichael, I.S.E., Ghiorso, M.S., 1990. The effect of oxygen fugacity on the redox state of natural liquids and their crystallizing phases. In: Nicholls, J., Russell, J.K. (Eds.), Modern Methods of Igneous Petrology: Understanding Magmatic Processes. Reviews in Mineralogy and Geochemistry, vol. 24, pp. 191–212.

Carmichael, I.S.E., Nicholls, J., Smith, A.L., 1970. Silica activity in igneous rocks. American Mineralogist 55, 246–263.

Connolly, J.A.D., 2005. Computation of phase equilibria by linear programming: a tool for geodynamic modeling and its application to subduction zone decarbonation. Earth and Planetary Science Letters 236, 524–541.

de Capitani, C., Petrakakis, K., 2010. The computation of equilibrium assemblage diagrams with Theriak/Domino software. American Mineralogist 95, 1006–1016.

Ferry, J.M., Watson, E.B., 2007. New thermodynamic models and revised calibrations for the Ti-in-zircon and Zr-in-rutile thermometers. Contributions to Mineralogy and Petrology 154, 429–437.

Fowler, S.J., Spera, F.J., Bohrson, W.A., Belkin, H.E., de Vivo, B., 2007. Phase equilibria constraints on the chemical and physical evolution of the Campanian Ignimbrite. Journal of Petrology 48, 459–493.

Ganguly, J., 2008. Thermodynamics in Earth and Planetary Sciences. Springer-Verlag, Berlin, 501 pp.

Ghiorso, M.S., 1985. Chemical mass transfer in magmatic processes. I. Thermodynamic relations and numerical algorithms. Contributions to Mineralogy and Petrology 90, 107–120.

Ghiorso, M.S., 1997. Thermodynamic models of igneous processes. Annual Reviews of Earth and Planetary Sciences 25, 221–241.

Ghiorso, M.S., 2013. A globally convergent saturation state algorithm applicable to thermodynamic systems with a stable or metastable omni-component phase. Geochimica et Cosmochimica Acta 103, 295–300.

Ghiorso, M.S., Carmichael, I.S.E., 1985. Chemical mass transfer in magmatic processes. II. Applications in equilibrium crystallization, fractionation and assimilation. Contributions to Mineralogy and Petrology 90, 121–141.

Ghiorso, M.S., Evans, B.W.E., 2008. Thermodynamics of rhombohedral oxide solid solutions and a revision of the Fe-Ti two-oxide geothermometer and oxygen-barometer. American Journal of Science 308, 957–1039.

Ghiorso, M.S., Gualda, G.A.R., 2012. A method for estimating the activity of titania in magmatic liquids from the compositions of coexisting rhombohedral and cubic iron–titanium oxides. Contributions to Mineralogy and Petrology 165, 73–81.

Ghiorso, M.S., Hirschmann, M.M., Reiners, P.W., Kress III, V.C., 2002. The pMELTS: A revision of MELTS for improved calculation of phase relations and major element partitioning related to partial melting of the mantle to 3 GPa. Geochemistry, Geophysics, Geosystems 3. http://dx.doi.org/10.1029/2001GC000217.

Ghiorso, M.S., Kelemen, P.B., 1987. Evaluating reaction stoichiometry in magmatic systems evolving under generalized thermodynamic constraints: examples comparing isothermal and isenthalpic assimilation. In: Mysen, B. (Ed.), Magmatic Processes: Physiochemical Principles, Geochemical Society Special Publication, vol. 1, pp. 319–336.

Ghiorso, M.S., Kress, V.C., 2004. An equation of state for silicate melts. II. Calibration of volumetric properties at 10^5 Pa. American Journal of Science 204, 679–751.

Ghiorso, M.S., Sack, R.O., 1995. Chemical mass transfer in magmatic processes. IV. A revised and internally consistent thermodynamic model for the interpolation and extrapolation of solid-liquid equilibria in magmatic systems at elevated temperatures and pressures. Contributions to Mineralogy and Petrology 119, 197–212.

Ghiorso, M.S., Spera, F.J., 2010. Large scale simulations. In: Wentzcovitch, R., Stixrude, L. (Eds.), Theoretical and Computational Methods in Mineral Physics. Reviews in Mineralogy and Geochemistry, vol. 71, pp. 437–462.

Gibbs, J.W., 1878. On the equilibrium of heterogeneous substances. Transactions of the Connecticut Academy III, 108–248 and 343–524.

Gualda, G.A.R., Ghiorso, M.S., Lemons, R.V., Carley, T.L., 2012. Rhyolite-MELTS: a modified calibration of MELTS optimized for silica-rich, fluid-bearing magmatic systems. Journal of Petrology 53, 875–890.

Gualda, G.A.R., Ghiorso, M.S., 2014. Phase-equilibrium geobarometers for silicic rocks based on rhyolite-MELTS. Contributions to Mineralogy and Petrology 168, 1033.

Hayden, L.A., Watson, E.B., Wark, D.A., 2008. A thermobarometer for sphene (titanite). Contributions to Mineralogy and Petrology 155, 529–540.

Huang, R., Audetat, A., 2012. The titanium-in-quartz (TitaniQ) thermobarometer: a critical examination and re-calibration. Geochimica et Cosmochimica Acta 84, 75–89.

Hunt, J.D., Manning, C.E., 2012. A thermodynamic model for the system SiO_2–H_2O near the upper critical end point based on quartz solubility experiments at 500–1100 °C and 5–20 kbar. Geochimica et Copsmochimica Acta 86, 196–213.

Iacono-Marziano, G., Morizet, Y., le Trong, E., Gaillard, F., 2012. New experimental data and semi-empirical parameterization of H_2O–CO_2 solubility in mafic melts. Geochimica et Cosmochimica Acta 97, 1–23.

Karki, B.B., 2010. First-principles molecular dynamics simulations of silicate melts: structural and dynamical properties. In: Wentzcovitch, R., Stixrude, L. (Eds.), Theoretical and Computational Methods in Mineral Physics. Reviews in Mineralogy and Geochemistry, vol. 71, pp. 355–386.

Kress, V.C., 1997. Thermochemistry of sulfide liquids I. The system O-S-Fe at 1 bar. Contributions to Mineralogy and Petrology 127, 176–186.

Kress, V.C., 2000. Thermochemistry of sulfide liquids II. Associated solution model for sulfide liquids in the system O-S-Fe. Contributions to Mineralogy and Petrology 139, 316–325.

Kress, V.C., 2007. Thermochemistry of sulfide liquids III. Ni-bearing liquids at 1-bar. Contributions to Mineralogy and Petrology 154, 191–204.

Kress, V.C., Carmichael, I.S.E., 1991. The compressibility of silicate liquids containing Fe_2O_3 and the effect of composition, temperature, oxygen fugacity and pressure on their redox states. Contributions to Mineralogy and Petrology 108, 82–92.

Kress, V.C., Ghiorso, M.S., 2004. Thermodynamic modeling of post-entrapment crystallization in igneous phases. Journal of Volcanology and Geothermal Research 137, 247–260.

Kress, V.C., Ghiorso, M.S., Lastuka, C., 2004. Microsoft EXCEL spreadsheet-based program for calculating equilibrium gas speciation in the C-O-H-S-Cl-F system. Computers and Geosciences 30, 211–214.

Kress, V.C., Greene, L.E., Ortiz, M.D., Mioduszewski, L., 2008. Thermochemistry of sulfide liquids IV: density measurements and the thermodynamics of O–S–Fe–Ni–Cu liquids at low to moderate pressures. Contributions to Mineralogy and Petrology 156, 785–797.

Lange, R.A., Carmichael, I.S.E., 1990. Thermodynamic properties of silicate liquids with emphasis on density, thermal expansion and compressibility. In: Nicholls, J., Russell, J.K. (Eds.), Modern Methods of Igneous Petrology: Understanding Magmatic Processes. Reviews in Mineralogy and Geochemistry, vol. 24, pp. 25–64.

Lange, R.A., Frey, H.M., Hector, J., 2009. A thermodynamic model for the plagioclase-liquid hygrometer/thermometer. American Mineralogist 94, 494–506.

Moore, G., 2008. Interpreting H_2O and CO_2 contents in melt inclusions: constraints from solubility experiments and modeling. In:

Putirka, K.D., Tepley III, F.J. (Eds.), Minerals, Inclusions and Volcanic Processes. Reviews in Mineralogy and Geochemistry, vol. 69, pp. 333–361.

Moretti, R., Baker, D.R., 2008. Modeling the interplay of fO_2 and fS_2 along the FeS-silicate melt equilibrium. Chemical Geology 256, 286–298.

Moretti, R., Ottonello, G., 2005. Solubility and speciation of sulfur in silicate melts, the Conjugated Toop–Samis–Flood–Grjotheim (CTSFG) model. Geochimica Cosmochimica Acta 69, 801–823.

Newman, S., Lowenstern, J.B., 2002. VolatileCalc: a silicate melt–H_2O–CO_2 solution model written in Visual Basic for excel. Computers and Geosciences 28, 597–604.

Osborn, E.F., 1959. Role of oxygen pressure in the crystallization and differentiation of basaltic magma. American Journal of Science 257, 609–647.

Osborn, E.F., 1962. Reaction series for sub-alkaline igneous rocks based on different oxygen pressure conditions. American Mineralogist 47, 211–226.

Papale, P., Moretti, R., Barbato, D., 2006. The compositional dependence of the saturation surface of H_2O + CO_2 fluids in silicate melts. Chemical Geology 229, 78–95.

Pitzer, K.S., Brewer, L., 1961. Thermodynamics. McGraw-Hill, New York, 723 pp.

Powell, R., Holland, T.J.B., Worley, B., 1998. Calculating phase diagrams involving solid solutions via non-linear equations, with examples using THERMOCALC. Journal of Metamorphic Geology 16, 577–588.

Prigogine, I., Defay, R., 1954. Chemical Thermodynamics. Longmans Green and Co., London, 543 pp.

Putirka, K.D., 2008. Thermometers and barometers for volcanic systems. In: Putirka, K.D., Tepley III, F.J. (Eds.), Minerals, Inclusions and Volcanic Processes. Reviews in Mineralogy and Geochemistry, vol. 69, pp. 61–120.

Sack, R.O., Ghiorso, M.S., 1991. Chromium spinels as petrogenetic indicators: thermodynamics and petrological applications. American Mineralogist 87, 79–98.

Smith, P.M., Asimow, P.D., 2005. Adiabat_1ph: a new public front-end to the MELTS, pMELTS, and pHMELTS models. Geochemistry, Geophysics, Geosystems 6. http://dx.doi.org/10.1029/2004GC000816.

Spear, F.S., 1993. Metamorphic Phase Equilibria and Pressure-Temperature-Time Paths. Mineralogical Society of America Monograph, 799 pp.

Spera, F.J., Bohrson, W.A., 2001. Energy-constrained open-system magmatic processes. I: general model and energy-constrained assimilation and fractional crystallization (EC-AFC) formulation. Journal of Petrology 42, 999–1018.

Spera, F.J., Bohrson, W.A., 2002. Energy-constrained open-system magmatic processes. III: energy-constrained recharge, assimilation and fractional crystallization (EC-RAFC). Geochemistry, Geophysics, Geosystems 3. http://dx.doi.org/10.1029/2002GC00315.

Spera, F.J., Bohrson, W.A., 2004. Open-system magma chamber evolution: an energy-constrained geochemical model incorporating the effects of concurrent eruption, recharge, variable assimilation and fractional crystallization (EC-E'RAFC). Journal of Petrology 45, 2459–2480.

Stixrude, L., Lithgow-Bertelloni, C., 2010. Thermodynamics of the Earth's mantle. In: Wentzcovitch, R., Stixrude, L. (Eds.), Theoretical and Computational Methods in Mineral Physics. Reviews in Mineralogy and Geochemistry, vol. 71, pp. 465–484.

Thomas, J.B., Watson, E.B., Spear, F.S., 2010. TitaniQ under pressure: the effect of pressure and temperature on the solubility of Ti in quartz. Contributions to Mineralogy and Petrology 160, 743–759.

Thompson, R.N., Riches, A.J.V., Antoshechkina, P.M., Pearson, D.G., Nowell, G.M., Ottley, C.J., Dickin, A.P., Hards, V.L., Nguno, A.-K., Niku-Paavola, V., 2007. Origin of CFB magmatism: multi-tiered intracrustal picrite–rhyolite magmatic plumbing at Spitzkoppe, Western Namibia, during Early Cretaceous Etendeka magmatism. Journal of Petrology 48, 1119–1154.

Verhoogen, J., 1949. Thermodynamics of a magmatic gas phase, vol. 28. University of California Publications. Bulletin of the Department of Geological Sciences, 91–135.

Wallace, P., Carmichael, I.S.E., 1992. Sulfur in basaltic magma. Geochimica et Cosmochimica Acta 56, 1863–1874.

Wark, D.A., Watson, E.B., 2006. TitaniQ: a titanium-in-quartz geothermometer. Contributions to Mineralogy and Petrology 152, 743–754.

Watson, E.B., Wark, D.A., Thomas, J.B., 2006. Crystallization thermometers for zircon and rutile. Contributions to Mineralogy and Petrology 151, 413–433.

Chapter 7

Volatiles in Magmas

Paul J. Wallace
University of Oregon, Eugene, OR, USA

Terry Plank
Lamont-Doherty Earth Observatory of Columbia University, Palisades, NY, USA

Marie Edmonds
Earth Sciences Department, University of Cambridge, Cambridge, UK

Erik H. Hauri
Carnegie Institution of Washington, Washington, DC, USA

Chapter Outline

GLOSSARY

exsolved gas Gas that has been released from solution in a magma. Gas exsolution occurs when pressure is decreased, such as when the magma moves toward the Earth's surface.

fugacity A gas is considered to be perfect if its behavior can be described by the ideal gas law: $PV = nRT$, where P is the pressure, V is the volume, n is the number of moles, R is the ideal gas constant, and T is the temperature. Most gases are not perfect, especially at high pressures or low temperatures. The fugacity is a kind of equivalent pressure that accounts for the deviation of a real gas from ideal gas behavior.

glass If a liquid is cooled very rapidly, it may harden to a glass and not crystallize. Magma is a silicate liquid which, if cooled rapidly enough, forms glass.

immiscible Some types of liquids, such as molten iron sulfide, will not dissolve into a silicate melt to form a homogeneous solution. Such coexisting liquids are described as immiscible.

magma Natural silicate melt with or without suspended crystals and bubbles.

melt (glass) inclusion During crystallization of magma, many crystals grow imperfectly, trapping small blebs of silicate melt inside the crystals. If the magma is erupted and cooled rapidly, then these trapped inclusions of silicate melt become quenched to glass.

The Encyclopedia of Volcanoes. http://dx.doi.org/10.1016/B978-0-12-385938-9.00007-9

saturation A melt can be saturated with respect to certain chemical elements in three ways: it may contain bubbles of a gas or gas mixture (e.g., H_2O-CO_2), it may contain crystals rich in the element such as anhydrite ($CaSO_4$, rich in S), or it may contain droplets of immiscible liquid rich in the element such as drops of sulfide melt (rich in S).

solubility The maximum amount of a volatile species or component that can be dissolved in a silicate melt under a given set of conditions such as pressure, temperature, and melt composition.

volatile An element or compound, such as H_2O or CO_2, that forms a gas at a relatively low pressure and magmatic temperature. Volatiles can be dissolved in silicate melts, can occur as bubbles of exsolved gas, and can crystallize in minerals such as biotite and amphibole. The gas phase is also commonly described as vapor. Because its density increases with pressure, making it more liquid-like at crustal and mantle depths, it is also sometimes referred to as a fluid phase.

1. DEFINITION AND SIGNIFICANCE OF MAGMATIC VOLATILES

The observatory worker who has lived a quarter of a century with Hawaiian lavas frothing in action, cannot fail to realize that gas chemistry is the heart of the volcano magma problem.

T.A. Jaggar, 1940

The above quotation from T.A. Jaggar, founder of the Hawaiian Volcano Observatory and an influential volcanologist in the first half of the twentieth century, attests to the importance of gases in governing volcanic eruptive phenomena. As early as the middle of the nineteenth century, it was recognized that gas plays a fundamental role in forcing **magma** to the Earth's surface and generating explosive eruptions. Many different species of gas can dissolve in a molten silicate liquid in the same way that carbon dioxide can dissolve in water. Such gas species are referred to as **volatile** components or magmatic volatiles because of their tendency to form gas bubbles at a relatively low pressure. The amount of a volatile component that can dissolve in silicate melt increases with pressure. As magma ascends toward the Earth's surface, pressure decreases, thereby decreasing the **solubility** of volatiles, and causing them to come out of solution to form bubbles. In addition, the bubbles expand with a further decrease in pressure. At one atmosphere pressure and typical magma temperatures, an amount of exsolved H_2O equivalent to 0.1% of total magma mass is sufficient to produce a magmatic foam that has >90% bubbles by volume. Tremendous expansion occurs near the surface as a result of such volatile exsolution, and this provides the driving force for volcanic eruptions, similar to the fountaining of champagne from an uncorked bottle.

Water and carbon dioxide are the most important volatile components in natural magmas. At higher pressures, deeper within the Earth, most of the water and carbon dioxide are dissolved in the silicate melt portion of the magma, and they have important effects on magma crystallization, temperature, density, and viscosity. In the upper mantle, where basaltic magmas are generated by partial melting, volatiles affect the composition of melts and their physical segregation from the residual mantle minerals. After H_2O and CO_2, the most abundant volatiles are sulfur, chlorine, and fluorine; sulfur is particularly important for understanding the composition and fluxes of volcanic gas, and all three can be important in the formation of ore deposits. Many other volatiles, such as He, N, Ar, Br, and I, can dissolve in natural silicate melts (magmas), but they are generally much less abundant. The noble gases, for example, are usually present at concentration levels <1 ppm (0.0001% by weight) in natural magmas. Despite such low concentrations, the isotopic compositions of noble gases have been studied extensively because of their value in understanding large-scale degassing of the Earth and the formation of the atmosphere.

In this chapter, we focus on the most abundant volatiles: H_2O, CO_2, S, Cl, and F. We review (1) experimental data on the solubility of volatiles in natural silicate melts, (2) analytical and sampling techniques used for measuring the abundances of volatiles in magmas, (3) volatile abundances in basaltic and silicic magmas typical of various tectonic environments, and (4) Earth degassing and volatile recycling by subduction. We also discuss the evidence for preeruptive vapor **saturation**, as this is important for understanding the explosive eruptive behavior of volcanoes, fluxes of volcanic gases, and the formation of ore deposits.

2. SOLUBILITY OF VOLATILES IN SILICATE MELTS

2.1. Water and Carbon Dioxide

Solubility refers to the maximum amount of a volatile species or component that can be dissolved under a given set of conditions, such as pressure, temperature, and melt composition. Laboratory measurements of the solubilities of H_2O, CO_2, S, and many other volatile components have been made for a number of different melt compositions. In such laboratory experiments, it is possible to control the parameters that affect solubility. As an example, a silicate melt can be equilibrated with H_2O vapor at high temperatures and a variety of pressures in order to determine the solubility of water (Figure 7.1). Such melt is described as "H_2O saturated" because in the experiment, the melt coexists with water vapor, though it should be noted that the vapor phase is not pure because it contains at least a minor amount of other components from the silicate melt. As seen in the experimental data, the solubility of H_2O in silicate melts is strongly dependent on pressure. This pattern, in

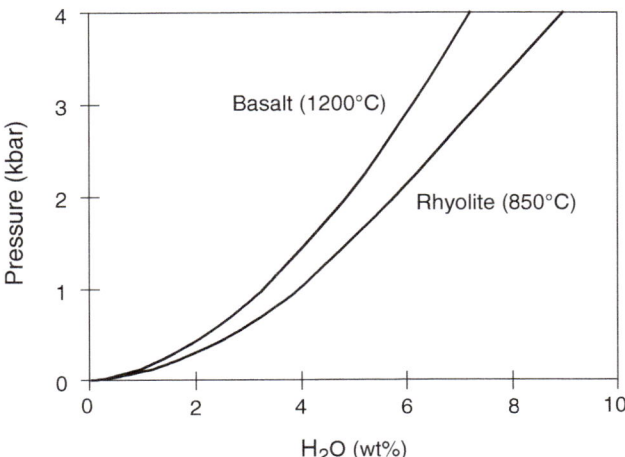

FIGURE 7.1 H₂O solubility in basaltic and rhyolitic melts at typical magmatic temperatures based on experimental data. *(See review of experimental solubility data in Baker and Alletti, 2012.)*

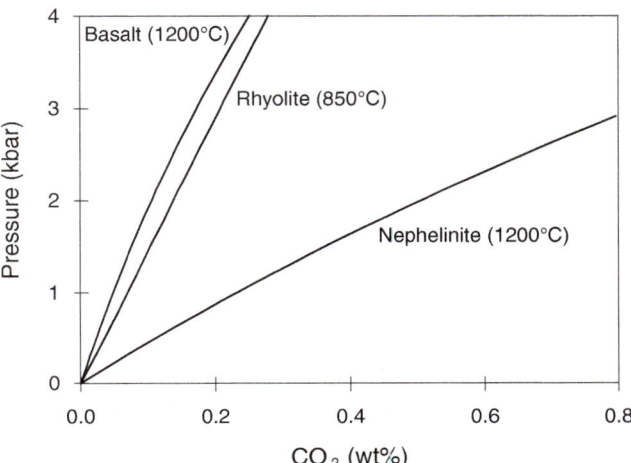

FIGURE 7.2 Carbon dioxide solubility in basaltic and rhyolitic melts at typical magmatic temperatures. Also shown for comparison is the solubility in a nephelenitic melt to show the strong increase in solubility in alkali-rich and silica-poor melts. *(Solubilities calculated using Iacono-Marziano et al., 2012, for mafic melts and Newman and Lowenstern, 2002, for rhyolitic melt.)*

which the solubility increases with increasing pressure, is observed for many volatile components, and occurs because the gaseous form of a volatile component has a larger molar volume than does the same component when it is dissolved in a silicate melt. The solubility of H₂O in silicate melts is also dependent on both melt composition and temperature. The H₂O solubility is greater in rhyolitic melts than in basaltic melts, but much of this difference is actually due to temperature differences rather than to compositional effects (Figure 7.1).

Early studies on the solubility of water in silicate melts recognized that there is a linear relationship between the partial pressure (or **fugacity**) of H₂O and the square of the mole fraction of dissolved water in the melt. Such a relationship is consistent with a solution mechanism in which water dissolves by forming OH⁻ groups that are structurally bound to the aluminosilicate network of the melt. More recent studies using infrared spectroscopy have shown that dissolved water in silicate melts occurs as two different species: OH⁻ groups and H₂O molecules. The relative proportions of the two species vary systematically with total water concentration. At low total dissolved water concentrations, virtually all the water is present as OH⁻ groups. As total dissolved water concentration increases, the relative proportion of molecular H₂O increases as well. In rhyolitic melts, the proportions of OH⁻ groups and H₂O molecules are equal at about 3 wt% total dissolved water, whereas in basaltic melt, the proportions of the two species are equal at about 3.5 wt% total water. The speciation is also dependent on temperature and quench rate, and this forms the basis of a geospeedometer, from which the cooling rates of hydrous rhyolitic **glasses** can be calculated using measurements of OH⁻ groups and H₂O molecules.

As with H₂O, the solubility of CO₂ in silicate melts is strongly dependent on pressure (Figure 7.2). However, CO₂

solubilities are much lower than for H₂O. The amount of CO₂ that dissolves into both basaltic and rhyolitic melts is 50−100 times less, by weight, than the solubility of H₂O at a comparable pressure and temperature. Carbon dioxide dissolves in silicate melts in two different species, but in contrast to water, the speciation is controlled by the bulk silicate melt composition. Thus, in silica-poor glasses (basalts, basanites, and nephelinites), CO₂ is present as structurally bound carbonate (CO_3^{2-}), whereas in silica-rich glasses (rhyolite), it occurs as CO₂ molecules. Glasses with intermediate silica contents (andesitic) contain both species. Experimental studies show that the solubility of CO₂ is similar in basaltic, andesitic, and rhyolitic melts, despite the differences in speciation across this compositional range. However, increases in alkali content strongly increase CO₂ solubility such that nephelinite and phonolite magmas have much higher solubilities (Figure 7.2). In addition, in basaltic melts, the melt CaO content has a strong positive effect on solubility.

In discussing solubility, it is important to distinguish between saturation and undersaturation with a particular volatile component. Solubility refers to the maximum amount of a volatile species or component that can be dissolved in silicate melt at a given set of conditions. In both natural and experimental silicate melts, however, the melt can have a concentration of a dissolved volatile component that is less than the maximum possible. Such a melt is described as undersaturated with respect to the volatile component of interest. As an example, laboratory experiments show that a rhyolitic melt can contain about 6 wt% dissolved H₂O at 2 kbar, but this does not mean that all natural rhyolitic melts that are at a 2-kbar pressure in the

Earth's crust contain this much water. Instead, they could have any amount from 0 wt% to 6 wt%. A rhyolitic melt containing less than the dissolved H_2O needed to saturate the melt at a given temperature and pressure is thus "H_2O undersaturated."

It is possible in laboratory solubility experiments, as described above, to equilibrate a melt with a single gas component such as H_2O. However, in magmas, many volatile components are present together. It is therefore important to distinguish between the dissolved concentrations of individual volatile components and their concentrations in the separate **exsolved gas** phase. If the sum of the partial pressures of all dissolved volatile components in a magma is equal to the local confining pressure, then the magma is saturated with bubbles of a multicomponent vapor phase containing H_2O, CO_2, S, and other minor gases such as Cl, F, and noble gases. Because volatile species have different solubilities, their relative abundances dissolved in the melt will differ from their relative abundances in the gas phase.

For vapor-saturated silicate melts, the solubilities of individual volatile components are dependent in part on the abundances of other volatiles that are present in the system. This is because the amount of a given component that is dissolved in the melt is proportional to the partial pressure of the volatile component in the coexisting equilibrium vapor. As an example, consider a silicate melt that is saturated with vapor consisting only of H_2O and CO_2. The sum of the partial pressures (P_{H_2O} and P_{CO_2}) of these components in the vapor phase must equal the total pressure in a vapor-saturated melt. Thus, at a constant total pressure, an increase in P_{H_2O}, and hence dissolved H_2O in the melt, must result in a decrease in P_{CO_2} and dissolved CO_2. On a diagram of dissolved H_2O versus dissolved CO_2 in silicate melts, vapor saturation isobars, which show the maximum dissolved H_2O and CO_2 at a given total pressure, have negative slopes (Figure 7.3). However, there are some indications that at pressures greater than approximately 2–3 kbar, low to moderate amounts of dissolved H_2O may actually increase CO_2 solubility slightly. In many common igneous systems, the partial pressures of other volatile components (e.g., Cl, F, and S) are probably relatively small compared with those of H_2O and CO_2, but this may not always be the case. A knowledge of the preeruptive concentrations of H_2O and CO_2 in a magma as measured in melt inclusions can thus potentially constrain the pressure at which the magma was stored before eruption, provided the magma is vapor saturated.

2.2. Sulfur

The behavior of S in silicate melts is complex because the solubility of S is controlled by the stabilities of sulfide and

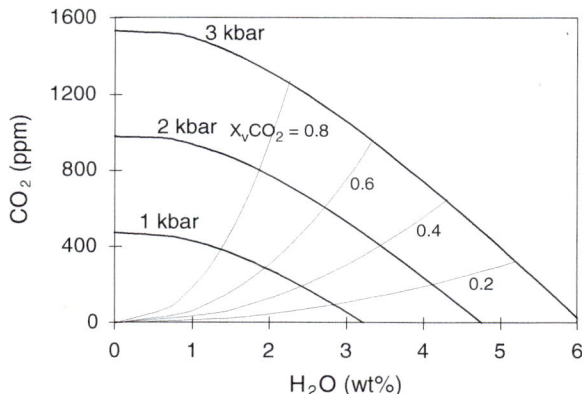

FIGURE 7.3 H_2O and CO_2 solubility (heavy solid curves) in vapor-saturated basaltic melts at 1200 °C and 1–3 kbar pressure based on experimental data and thermodynamic models. The curvature of the saturation curve for a given pressure results from the nonlinear dependence of H_2O solubility on pressure (Figure 7.1). The light curves show the composition (in mole fraction CO_2) of the vapor phase in equilibrium with the melt. (*Diagram calculated using Newman and Lowenstern, 2002.*)

sulfate-bearing phases and because dissolved S can occur in multiple valence (oxidation) states. The speciation of dissolved S in silicate melts is related to the relative magmatic oxygen fugacity (Figure 7.4). At relatively low oxygen fugacities, S is present in solution mainly in the reduced form sulfide (S^{2-}), whereas under more oxidizing conditions, sulfate (S^{6+}) appears to be the dominant species. For basaltic to andesitic magmas in the range of

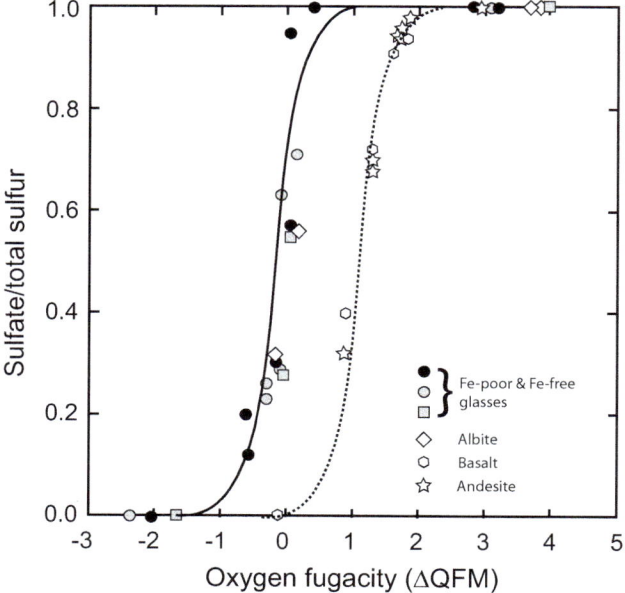

FIGURE 7.4   Fraction of total sulfur that is present as sulfate (sulfate/total sulfur) as a function of oxygen fugacity, relative to the quartz–fayalite–magnetite buffer curve (ΔQFM). The curves show the experimental relationship for basaltic and andesitic melts compared to Fe-poor silicic melts. (*Figure is modified from Klimm et al., 2012.*)

oxygen fugacities from slightly above the fayalite–magnetite–quartz (FMQ) buffer to approximately 1.5 $\log_{10}$ units above FMQ, both dissolved sulfide and sulfate are present in significant quantities (Figure 7.4). In Fe-poor silicic melts, the region where both S species are present is shifted to lower oxygen fugacities by about 1.5 $\log_{10}$ units (Figure 7.4). The maximum S that can be dissolved in silicate melt is controlled by saturation of the melt with an S-bearing phase. At relatively low oxygen fugacities, this phase is either an **immiscible** iron-rich sulfide liquid (with as much as 10% oxygen), which occurs in high-temperature basaltic melts, or crystalline pyrrhotite ($Fe_{1-x}S$) in intermediate (andesitic) and silicic (rhyolitic) melts. A Cu–Fe sulfide mineral may also crystallize from andesitic to rhyolitic melts. At higher oxygen fugacities, the sulfate mineral anhydrite ($CaSO_4$) crystallizes in intermediate to silicic melt compositions. In high-pressure laboratory experiments, mafic melts at high temperatures can be saturated with an immiscible sulfate melt, but evidence for this in natural samples has only rarely been found. Silicate melts at intermediate oxygen fugacities can crystallize both pyrrhotite and anhydrite, an assemblage that was found, for example, in trachyandesite tephra from the 1982 eruption of El Chichón in Mexico. Experimental solubility studies show that S is more soluble at high oxygen fugacities, under which conditions the dissolved S is present as sulfate, and anhydrite is the stable S-bearing mineral (Figure 7.5). The solubility of reduced S (S^{2-}) increases with the Fe concentration in the silicate melt

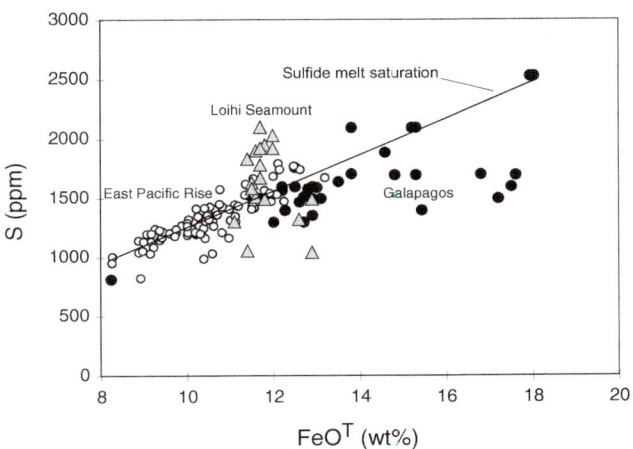

FIGURE 7.6 Covariation of sulfur and total iron (FeO^T) in submarine MORB glasses from the Galapagos spreading center (Perfit et al., 1983) and the East Pacific Rise (le Roux et al., 2006) compared to submarine basaltic glasses from an oceanic island volcano (Loihi seamount). The line shows the S saturation limit in MORB melts. The saturation limit for Fe-poor Galapagos basalt is calculated at 1170 °C and an oxygen fugacity that is 1.5 log units more reduced than the nickel–nickel oxide (NNO) buffer. For Fe-rich Galapagos basalt, the S concentration is calculated at 1070 °C and an oxygen fugacity that is 0.5 log units more reduced than NNO. *(Modified from Wallace and Edmonds, 2010, using additional data from Dixon and Clague, 2001.)*

(Figure 7.6), which indicates that S^{2-} dissolves largely by complexing with Fe^{2+}. Sulfur solubility under both oxidizing and reducing conditions increases with temperature (Figure 7.5). Pressure has contrasting effects, depending on the oxidation state. For oxidizing conditions, the solubility of S increases with increasing pressure, but for reducing conditions, where S^{2-} is the dominant species, solubility decreased with increasing pressure. As a result of the latter effect, a basaltic magma with low oxygen fugacity formed in equilibrium with residual sulfide in the mantle becomes undersaturated in sulfide during adiabatic ascent.

2.3. Halogens (Cl and F)

As with S, the solubility of Cl is complex because silicate melt can be saturated with an immiscible alkali chloride melt (molten salt). If water is present, then the immiscible alkali chloride melt will also contain dissolved water and is referred to as a hydrosaline melt. Chlorine solubility in silicate melts is strongly dependent on silicate melt composition, and the solubility generally increases as the ratio $(Na + K)/Al$ in the silicate melt increases. A silicate melt saturated with molten salt has the maximum dissolved Cl. The saturation concentration of Cl varies with pressure, temperature, dissolved H_2O concentration, and silicate melt composition. For silicate melts that are saturated with an H_2O- and CO_2-rich vapor phase at middle to upper crustal pressures, maximum Cl

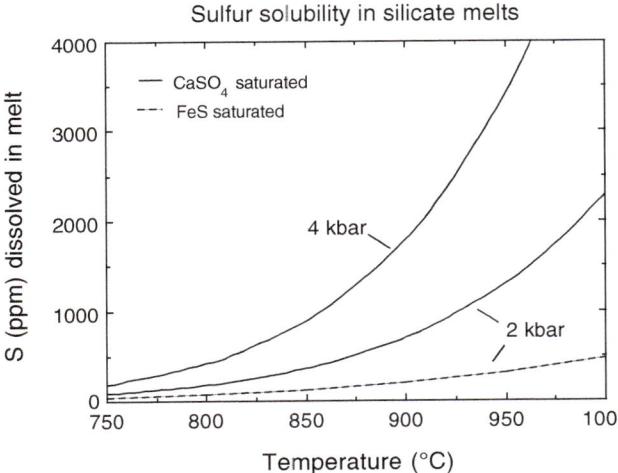

FIGURE 7.5 Experimental determinations of sulfur solubility versus temperature for hydrous silica-rich melts at different pressures and oxygen fugacities. Curves are shown for both sulfide-saturated melts (oxygen fugacity at the NNO buffer) and sulfate-saturated melts (oxygen fugacity at the MnO-Mn$_3$O$_4$ [MNO] buffer). At 2-kbar pressure, S is more soluble under oxidizing conditions (MNO) than under reducing conditions (NNO). Solubility curve for sulfate-saturated melts (MNO buffer) at 4 kbar shows that sulfate solubility increases with increasing pressure. *(Figure is based on experimental data of Luhr, 1990.)*

solubilities range from nearly 3000 ppm in rhyolitic melts to nearly 3 wt% in basaltic melts. Thus, Cl solubility in silicate melts is somewhat intermediate between that of H_2O and that of CO_2. In addition, at low pressures (≤ 1.2 kbar at $\sim 800\,°C$), a silicate melt can be saturated with both hydrosaline melt and an H_2O- and CO_2-rich vapor phase.

A natural silicate melt may contain less dissolved Cl than the amount needed to saturate the melt with either molten salt or hydrosaline melt. Many common magma types do not contain Cl-rich mineral phases or evidence of hydrosaline melt. By comparison with experimental Cl solubility studies, most magmas probably have insufficient dissolved Cl to be saturated with hydrosaline melt. In some Cl-rich magmas, however, formation and separation of a hydrosaline melt phase or alkali-chloride-rich vapor may play an important role in the formation of magmatic-hydrothermal ore deposits. For magmas that are saturated with an H_2O- and CO_2-rich vapor phase, some Cl partitions into the vapor. Experimental studies suggest that the concentration of Cl in the vapor phase coexisting with melt compositions ranging from basalt to rhyolite is about $1-8$ times greater than the concentration dissolved in the melt. This results in fairly minimal loss of Cl from melt by degassing until very low (near-surface) pressures are reached, at which point large mass fractions of H_2O are exsolved and some Cl is lost.

The solubility and speciation of dissolved F in silicate melts are complex and still not well understood, but available data clearly show that solubility is strongly dependent on silicate melt composition. In general, F is highly soluble in silicate melts, similar to H_2O. In granitic composition melts, as much as 10 wt% F can be dissolved. Basaltic and andesitic magmas typically have low dissolved F, on the order of a few hundred parts per million or less. Rare melts, such as those in topaz-bearing rhyolitic magmas, may contain as much as 5 wt% F, and ultra-potassic mantle-derived magmas can contain as much as 2 wt% F. Many granitic magmas have sufficient F to become saturated with fluorite. Because of its high solubility, F is much more strongly partitioned into the liquid phase than in the vapor phase, and therefore experiences minimal degassing even upon eruption, in contrast to the other volatiles described above.

3. APPROACHES FOR ASSESSING MAGMATIC VOLATILE CONTENTS

All terrestrial magmas contain dissolved volatiles, the most abundant being H_2O, CO_2, S, Cl, and F. If the dissolved concentrations of volatiles are sufficiently high, a magma may be vapor saturated such that the magma also contains an exsolved vapor (gas) phase. Assessing the total volatile

content of magma requires that both dissolved and exsolved volatile abundances are known. Because the solubility of most volatile components is strongly pressure dependent, magmas exsolve gas during ascent to the surface. In order to understand the crystallization and physical properties of magma, as well as to understand eruptive behavior, it is important to know the total volatile content of magma during its preeruptive storage within the Earth's crust. Dissolved volatiles may be estimated in many ways, but exsolved volatiles are difficult to assess. The following discussion will focus primarily on dissolved volatiles; we will discuss the relationship to exsolved volatiles in a subsequent section, and the reader is also referred to the Chapter 45 on *Volcanic, Magmatic and Hydrothermal Gases* for additional information.

3.1. Experimental Phase Equilibria

One of the most valuable methods for estimating dissolved volatile concentrations in magmas, particularly H_2O and CO_2, is through the use of phase equilibrium experiments. The assemblage of minerals that crystallizes from a melt of a given bulk composition and the sequence and temperature of appearance of those minerals during cooling vary depending on pressure and dissolved H_2O concentration (Figure 7.7). By equilibrating a given rock composition at

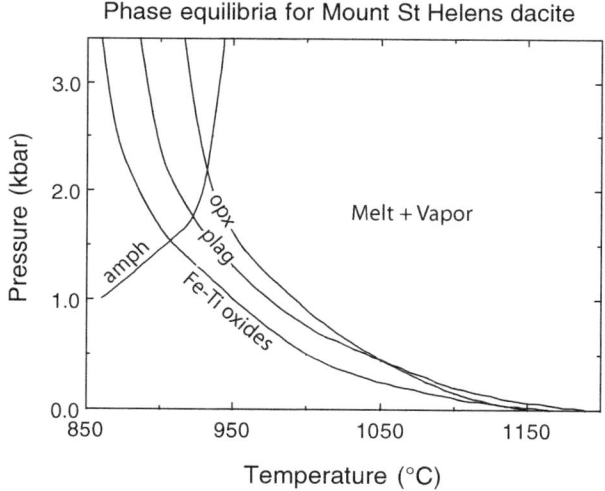

Phase equilibria for Mount St Helens dacite

FIGURE 7.7 Experimental phase equilibria for 1980 Mount St Helens dacite as a function of temperature and pressure for water-saturated melts. Because dissolved H_2O increases with pressure (Figure 7.1), the water contents of the experimental melts increase along the vertical axis of the diagram. With increasing pressure (and dissolved H_2O), the temperatures at which anhydrous minerals (pyroxene, plagioclase, and Fe—Ti oxides) begin to crystallize are significantly diminished. In contrast, increasing pressure and dissolved H_2O increases the stability of amphibole, which therefore crystallizes at higher temperatures as pressure increases. Comparison with the actual observed mineral assemblage in the 1980 Mount St Helens dacite indicates preeruptive temperature and pressure of 920 °C and approximately 2.2 kbar. *(Modified from Rutherford et al., 1985.)*

various temperatures, pressures, and H_2O concentrations, it is frequently possible to find a relatively narrow range of conditions under which the observed phenocryst assemblage in the rock is in stable equilibrium. This technique is particularly suited to constraining the dissolved concentrations of H_2O during preeruptive crystallization and storage of magma within the Earth's crust. Experimental studies have been applied with great success to a number of volcanic eruptions, most notably, the eruptions of Katmai (1912), Parícutin (1943—1951), Mount St Helens (1980), El Chichón (1982), Mt Pinatubo (1991), Soufriere Hills (1995—2011), and Chaitén (2008).

In addition to its effect on phase stability, dissolved H_2O also affects the composition of minerals that crystallize from silicate melts. The most useful effect related to magmatic H_2O concentrations is the change in plagioclase composition that occurs with variations in dissolved H_2O. As the dissolved H_2O concentration in a melt is increased, the equilibrium plagioclase composition becomes more calcic (anorthite-rich) and also crystallizes at a lower temperature than in dry magmas. As will be discussed below, anorthite-rich plagioclase phenocrysts are common in subduction-related basaltic magmas, providing evidence of relatively high concentrations of H_2O compared to basaltic magmas from other tectonic environments.

3.2. Analysis of Quenched Glass (Submarine Glasses, Melt Inclusions, Matrix Glass)

Subaerially cooled volcanic rocks and tephra have lost nearly all their original magmatic volatiles because of the low pressures of Earth's atmosphere. In contrast, during deep submarine eruptions, the confining pressure exerted by the overlying water may prevent exsolution and loss of many volatile species. Such pressures can be as great as 500—600 bars. Submarine-erupted magmas typically form glassy rinds where hot magma is rapidly quenched against cold seawater. Volatiles may be retained dissolved within such glassy basalt, which can be analyzed for volatiles directly. H_2O is relatively soluble in basaltic melt and as much as 2 wt% may be retained in deep sea basaltic glass without eruptive loss. In contrast, for relatively insoluble CO_2, the pressures on the seafloor are commonly insufficient to prevent exsolution. This results in small vesicles in quenched submarine glasses, usually <1% by volume, except in submarine island arc volcanoes and some backarc regions, where initial melts are more H_2O rich and hence exsolve substantial gas at seafloor pressures. For typical, vesicle-poor glasses, the small size of the vesicles and their relatively low volumetric abundance result in isolated (noninterconnected) bubbles trapped in the quenched glass. Bulk analyses of these glassy rocks can therefore be used to

assess total volatile concentration in the erupted magma, but the magma may have gained or lost bubbles before eruption. Bulk analyses of low solubility volatiles such as CO_2 in samples of glass with trapped bubbles typically give greater concentrations than do direct microanalyses of the glass.

Another important source of information on the dissolved volatile concentrations of magmas is through the analysis of **melt (glass) inclusions** trapped inside crystals. When melts cool they crystallize, and the crystals often grow imperfectly, causing small amounts of melt to be trapped inside the crystals. If the magma ascends, erupts, and cools rapidly, then these trapped melt inclusions quench to glass (Figure 7.8). Because the crystalline host for the inclusions is relatively rigid, they act as tiny

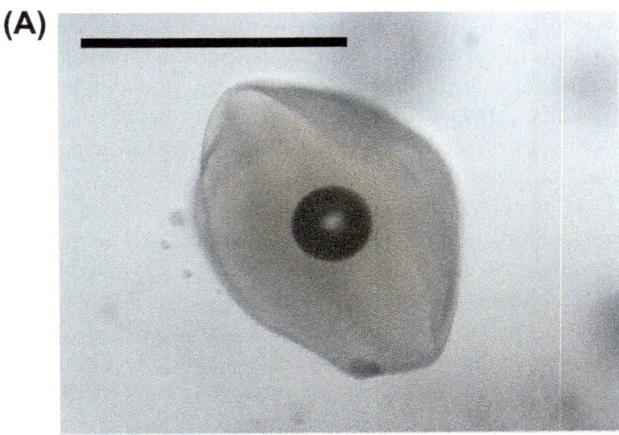

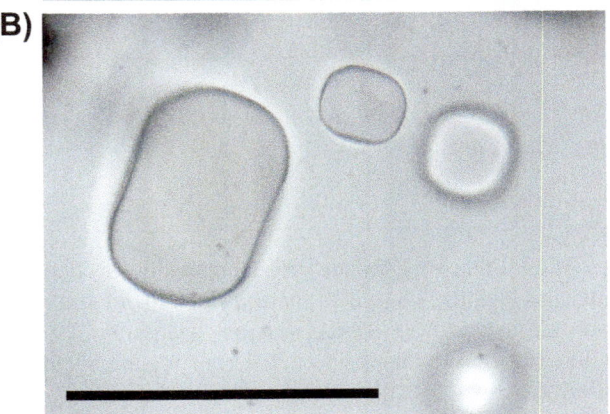

FIGURE 7.8 Glass (melt) inclusions in crystals from volcanic rocks. (A) Basaltic glass inclusion with a vapor bubble in an olivine phenocryst from Mauna Loa, Hawaii. The bubble likely formed during cooling after the inclusion was trapped because the melt in the inclusion contracts more on cooling than does the surrounding olivine host crystal and because the crystallization of olivine along the inclusion—host interface causes additional volume loss from the inclusion. (B) Rhyolitic glass inclusions in quartz phenocryst from the Huckleberry Ridge Tuff, Yellowstone. A 100-μm scale bar (black) is shown in both photographs.

pressure vessels and keep the trapped melt at high pressure, even though the bulk magma decompresses to surface pressure during eruption. For this reason, trapped melt (glass) inclusions can retain a record of dissolved volatiles at the time of trapping. However, caution must be exercised when interpreting melt inclusion data because the trapped volatile concentrations in the melt can be modified after entrapment by several processes. One such process is the formation of a shrinkage bubble during postentrapment cooling, which causes the loss of low solubility CO_2 from the trapped melt into the vapor bubble. Another effect is the potential loss or gain of H_2O in an inclusion due to diffusion of H through the mineral host. Laboratory experiments have found that diffusive reequilibration of H_2O in melt inclusions can occur through mineral hosts on the order of hours to days (olivine) and days to weeks (quartz) at magmatic temperatures. Finally, melt inclusions may in some instances leak volatiles during explosive eruption due to cracks in the host mineral, in which case textural evidence such as cracks or a large vapor bubble or multiple bubbles are typically present. This is especially problematic for inclusions in minerals with relatively good cleavages, such as hornblende and plagioclase.

Melt inclusions in phenocrysts are trapped during crystal growth and thus sample the liquid from which the crystals grew at the time of entrapment. Inclusions that are trapped at different times during crystallization may preserve a record of changing magma compositions due to processes such as crystal fractionation, magma mixing, and degassing. Glassy reentrants and hourglass inclusions, which are connected to the outside of the host crystal by a melt-filled neck, reveal the composition of melt just before

CO_2 values, reflecting trapping at different pressures from melts that have variably degassed before entrapment. Basaltic melt inclusions from arc volcanoes also commonly show a range of H_2O values that can also be caused by variable preentrapment degassing as well as postentrapment diffusive H loss. Sulfur concentrations can also be variable, reflecting partial degassing before trapping or loss of S to sulfides. Various methods have been used to assess the original H_2O content of magmas from such data sets, but the simplest approach is to select the maximum H_2O concentration or H_2O/K_2O ratio recorded within a melt inclusion population as representative of the parental magma.

Quenched glassy scoria or pumice that formed during subaerial eruptions can also be analyzed for volatiles. Generally, such material has extensively degassed during eruption and quenching and can also be affected by post-emplacement, low temperature hydration, so measured volatile contents do not reflect those before eruption. Analyses of relatively fresh (nonhydrated) scoria glass or pumice can, nevertheless, be useful for interpreting degassing and eruptive processes.

3.3. Thermodynamic Calculation Based on Mineral Compositions

One method that has commonly been used to estimate volatile concentrations in magmas is the application of thermodynamic calculations. For example, reactions involving hydrous minerals and their anhydrous breakdown products can be used along with thermodynamic properties of the phases to estimate water concentrations. One such reaction is the equilibrium involving

$$KFe_3AlSi_3O_{10}(OH)_2 \;+\; \tfrac{1}{2}O_2 \;=\; KAlSi_3O_8 \;+\; H_2O \;+\; Fe_3O_4$$

$$\text{biotite} \qquad\quad +\; \text{gas} \;=\; \text{feldspar} \;+\; \text{gas} \;+\; \text{magnetite}$$

eruption. Pressure-dependent concentrations of volatiles in melt inclusions, hourglass inclusions, and reentrants can provide a record of decompression, degassing, and crystallization during magma ascent. These processes can make it difficult to assess the original volatile content of basaltic magmas before their ascent into the crust. Suites of melt inclusions from a single eruption often have a wide range of

This equation can be used to calculate the water fugacity at the time of crystallization for any volcanic rock containing biotite, feldspar, and magnetite, provided the oxygen fugacity can also be estimated.

The most commonly used thermodynamic method is a hygrometer based on the equilibrium exchange of anorthite and albite components between plagioclase and silicate melt:

$$CaAl_2Si_2O_8 \;+\; NaAlSi_3O_8 \;=\; CaAl_2Si_2O_8 \;+\; NaAlSi_3O_8$$

$$\text{plagioclase} \qquad \text{melt} \qquad\quad \text{melt} \qquad\quad \text{plagioclase}$$

Increases in dissolved H_2O in silicate melt shift the equilibrium to the left, favoring an increase in anorthite content in plagioclase. Calibration of this relationship using experimental data for a wide range of melt compositions from basalt to rhyolite yields a relationship from which one can calculate magmatic H_2O concentrations with an uncertainty of about ± 0.25 wt% if the temperature and composition of plagioclase and coexisting melt are known. Hygrometers based on amphibole compositions have also been experimentally calibrated, but have larger uncertainties than those of the plagioclase-melt hygrometer.

A relatively new approach to estimating preeruptive H_2O contents of magmas is to study the H_2O contents of phenocrysts that grew in the magma during ascent. Small amounts of H_2O can be contained in the structure of silicate minerals (olivine, clinopyroxene, orthopyroxene, and feldspar), and the equilibrium partitioning of H_2O between melts and minerals has been experimentally studied for basaltic compositions over a wide range of pressures. These studies have determined the mineral−melt partition coefficient for H_2O:

$$D_{H_2O}^{mineral/melt} = \frac{H_2O(mineral)}{H_2O(melt)}.$$

The H_2O partition coefficients are dependent on pressure and mineral composition, particularly for pyroxenes, where higher Al contents result in higher partition coefficients. The use of these partition coefficients permits an estimate of the magmatic H_2O content if the H_2O concentrations of coexisting minerals can be accurately determined.

3.4. Fluid Inclusions

The presence of primary fluid inclusions in minerals from volcanic or plutonic rocks provides evidence that crystallizing melts were vapor and/or brine saturated. However, it can be difficult to distinguish between primary fluid inclusions, trapped during igneous crystallization, and secondary inclusions formed by lower temperature hydrothermal processes. Some volcanic phenocrysts from basaltic rocks contain dense CO_2-rich, primary fluid inclusions, indicating that the basaltic magma was CO_2-saturated during crystallization. Such inclusions can be used to estimate the pressure of inclusion entrapment. A knowledge of this, coupled with experimentally determined CO_2 solubilities in basaltic melt, can be used to estimate the dissolved CO_2 concentration of the basaltic magma at depths. Phenocrysts in more evolved volcanic rocks (dacites, rhyolites) may contain inclusions containing one or more of the following phases: fluid, vapor, hydrosaline melt, and daughter crystals. The presence of these inclusions also can be taken as evidence for vapor saturation during crystallization, but the multicomponent nature of such inclusions (H_2O-CO_2-NaCl) makes it difficult to infer the original dissolved volatile concentrations in the coexisting silicate melt.

3.5. Saturation with Volatile-Rich Minerals and Immiscible Liquids

The most common volatile-rich minerals and immiscible liquids that occur in magmas are sulfide and sulfate minerals and immiscible sulfide (Fe−S−O) liquid. The presence of one or more of these phases in a quenched volcanic rock indicates preeruptive saturation with an S-rich phase. By comparison with experimental solubility data, this makes it possible to infer a preeruptive dissolved S concentration. There are both thermodynamic and empirical solution models available that can be used to calculate the dissolved S content for either sulfide- or sulfate-saturated melts for a wide range of bulk compositions and oxygen fugacity conditions.

For volatile species whose solubility in a melt is controlled by the presence of a volatile-rich mineral or melt, such as a sulfide phase, complex changes in solubility can occur during magma cooling and differentiation. As an example, it has been observed in quenched glassy submarine pillow rinds from midocean ridge basalts (MORBs) that only a small amount (<1.5% by mass) of the total S in the magma is present as quenched sulfide globules. This observation may seem somewhat surprising when one considers that the strong temperature dependence of S solubility should cause considerable precipitation of immiscible sulfide liquid during cooling. However, the observed low quenched sulfide globule abundance is consistent with thermodynamic and experimental assessment of the effects of differentiation on S solubility. During the cooling and crystallization of midocean ridge basaltic magmas, residual liquids become increasingly Fe-enriched due to fractional crystallization of silicate phases. The increase in Fe content increases the S solubility in the melt by an amount that is greater than the decrease in S solubility brought about by cooling. This results in differentiated Fe−Ti basalts that have a higher dissolved S than do the parental (unfractionated) magmas. This effect, combined with the increase in residual melt S brought about by crystallization, keeps midocean ridge basaltic magmas finely balanced just at the S saturation point, causing the total mass of S in immiscible sulfide liquid to be small throughout the course of differentiation. In contrast, Kilauean tholeiitic magmas do not show an Fe-enrichment trend because plagioclase crystallization occurs later than in MORBs. In the absence of Fe-enrichment, if the melt is saturated with immiscible sulfide liquid, then the solubility of S will decrease as the temperature decreases.

4. ANALYTICAL TECHNIQUES FOR MEASURING VOLATILE ABUNDANCES IN SILICATE GLASSES

A large number of analytical techniques have been used to measure volatile concentrations in silicate glasses. These techniques fall into two main categories—bulk extraction techniques, in which a bulk rock or glass sample is crushed and analyzed, and microanalytical techniques, in which a tiny portion of glass is analyzed. Bulk extraction techniques can make it difficult to distinguish between the volatiles that are actually dissolved in quenched magmatic liquid (e.g., glassy submarine pillow rinds) and the volatiles that may be present in unopened vesicles or networks of cracks formed during magma solidification. Furthermore, it is critical to select material for analysis that is free of alteration. Microanalytical techniques offer a high spatial resolution, making it possible to analyze a tiny portion of alteration-free glass. The most commonly used microanalytical techniques are secondary ion mass spectrometry (SIMS; H_2O, CO_2, S, Cl, and F), Fourier transform infrared spectroscopy (FTIR; H_2O, CO_2), Raman spectroscopy (H_2O), and electron microprobe analysis (S, Cl, and F). Technical advances, improved calibrations, and increasing availability of FTIR, SIMS, and Raman in the last 15 years have led to a dramatic increase in the number of studies of volatiles in melt inclusions and submarine glasses as well as experimental studies of volatile solubilities and solution mechanisms in silicate melts and nominally anhydrous minerals.

5. VOLATILE CONCENTRATIONS IN BASALTIC MAGMAS

As an introduction to discussing volatiles in basaltic magmas, we first briefly review the abundance and storage of volatiles in the Earth's mantle, from which basaltic magmas are generated by partial melting. Volatiles in the Earth's mantle can reside in a number of potential minerals. Recent experimental studies have demonstrated that significant quantities of water can be present as structurally bound OH^- in the nominally anhydrous minerals olivine, pyroxene, and garnet. Chlorine solubility in these nominally anhydrous minerals is quite low, but F solubility is relatively high, sufficient for F to be primarily stored in olivine and orthopyroxene in the upper mantle. Water, Cl, and F could also be stored in hydrous minerals such as amphibole, phlogopite, and apatite. Carbon is present primarily in magnesian carbonate minerals, low degree carbonatitic melts, or in the reduced form as diamond, graphite, or carbide. Sulfur is bound in sulfide minerals. Estimates of the water content of the depleted upper mantle that is the source region for normal MORBs (N-MORB) are

approximately 70—160 ppm H_2O, based on studies of water in MORB glasses. If these concentrations occur throughout the entire mantle, then an amount of water equivalent to approximately 10—50% of the mass of water in the ocean could be contained in the upper mantle. The mantle source regions for enriched MORBs (E-MORB) probably contain 300—500 ppm H_2O.

During partial melting, water and other volatiles generally behave as incompatible elements—elements that are enriched in the melt phase—because they are not stoichiometric constituents of the major mantle silicates. Even S, which is contained in sulfides, behaves as an incompatible element because of the low modal abundance of sulfide. Thus, basaltic magmas act as agents for outgassing the Earth's mantle by carrying volatiles to the surface.

5.1. Midocean Ridge Basalts

Midocean ridge basaltic magmas (N-MORB) have relatively low H_2O contents, typically <0.4 wt%, but frequently as low as 0.1—0.2 wt% (Figure 7.9). Enriched MORB magmas (E-MORB), which form a volumetrically less abundant component of the oceanic crust, have systematically higher concentrations, as great as 1.5 wt%. Water concentrations from comagmatic suites of MORB glasses commonly show positive covariation with nonvolatile, incompatible trace elements such as K_2O and Ce. In fact, the H_2O/Ce ratio is remarkably uniform in MORB glasses (and also ocean island basalt glasses), generally falling within values of 100—300. This uniformity indicates that H_2O behaves as an incompatible element during partial

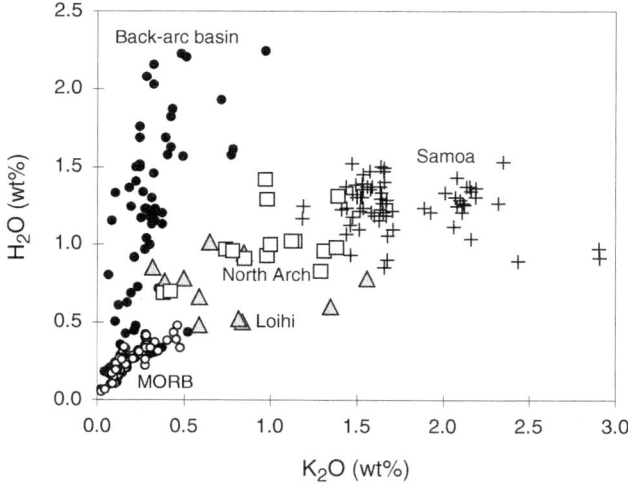

FIGURE 7.9 Variations in H_2O and K_2O in submarine basaltic glasses from different tectonic environments. Data are shown for MORBs, backarc basin basalts, and ocean island glasses from Hawaii (Loihi and North Arch) and Samoa. *(Data are from le Roux et al. (2006), Taylor and Martinez (2003), Dixon and Clague (2001), Dixon et al. (1997), and Workman et al. (2006).)*

melting of the mantle and fractional crystallization in crustal magma reservoirs and that significant H_2O is neither lost by degassing nor gained by assimilation in magma reservoirs. Rare andesites and dacites along midocean ridges can contain 1.0—2.5 wt% H_2O, and their geochemical characteristics indicate that assimilation of altered oceanic crust is important along with fractional crystallization.

Dissolved CO_2 concentrations in quenched MORB glasses vary from 50 to nearly 400 ppm. Most MORB glasses contain small vapor-filled bubbles as well, indicating vapor saturation of the melt at the time of eruption and quenching. Equilibrium thermodynamic calculations, using measured H_2O and CO_2 contents and experimental solubility data, suggest that the coexisting vapor phase would be 93—100 mol% CO_2. Actual measurements of the vapor composition in MORB vesicles are limited, but most have >90 mol% CO_2. Because MORB basaltic magmas are saturated with a nearly pure CO_2 vapor, the dissolved CO_2 measurements can be used to estimate a pressure of quenching. Such calculations indicate that many MORB melts were supersaturated with CO_2, that is, they contain more CO_2 than should be dissolved given the pressure equivalent to their depth of eruption on the seafloor. This supersaturation is caused by slow diffusion of CO_2 in the basaltic melt compared to the timescale of ascent, eruption, and quenching.

Sulfur concentrations in MORB glasses are typically between about 800 and 1400 ppm, but may reach as high as 2500 ppm in Fe—Ti-rich basalts. Sulfur contents show a strong positive correlation with the FeO content of the melt, consistent with experimental solubility relations (Figure 7.6). Most MORB glasses contain small amounts of quenched immiscible Fe—S—O liquid, indicating that the melts were S-saturated at the time of eruption. Primitive, uncontaminated MORB magmas contain 10—50 ppm Cl, but Cl is particularly susceptible to the effects of assimilation of wall rocks that have been hydrothermally altered by seawater, and values up to several hundred parts per million Cl are commonly found. For example, highly differentiated Fe—Ti rich basalts contain as much as 1100 ppm Cl, an amount that greatly exceeds what would be expected during closed-system fractional crystallization of basaltic magma. Fluorine contents of MORB glasses are low, generally about 100—600 ppm. Fluorine contents in suites of comagmatic glasses show F increasing with increasing fractionation, in accordance with enrichment in residual liquids during closed-system fractionation.

5.2. Ocean Island Basalts

Dissolved H_2O in melt inclusions and submarine basaltic glasses from ocean island volcanoes range from about 0.2 to 1.5 wt%, a range that overlaps with normal midocean ridge basaltic glasses (MORB; see Figure 7.9) but extends to higher values. Typical values for tholeiitic basalts are 0.3—0.7 wt% H_2O (Hawaii), 0.6—1.0 wt% (Galapagos), and 0.5—0.7 wt% (Iceland). Alkalic compositions generally have higher values, and some well-studied examples include alkali basalts from Samoa (0.6—1.5 wt% H_2O), mildly alkalic basalts from the 2010 Eyjafjallajökull eruption in Iceland (0.8—1.0 wt%), basanites recovered from Loihi Seamount, Hawaii (0.9—1.0 wt%), and basanites and nephelinites recovered from the submarine North Arch volcanic field, Hawaii (0.8—1.4 wt%). The systematics between incompatible trace elements and H_2O contents in some of the tholeiites and alkalic compositions indicate higher H_2O concentrations in their mantle source regions compared to MORB. Sulfur concentrations in ocean island basalts are commonly similar to those in MORB, suggesting they are saturated with immiscible Fe—S—O liquid before eruption (Fig. 7.6). Some ocean island basaltic magmas contain as much as 3000 ppm dissolved S but are not highly enriched in FeO, in contrast to S-rich MORB magmas, which are FeO-rich. This contrast is consistent with higher oxygen fugacities for the S-rich ocean island basaltic magmas. The higher oxygen fugacities decrease the stability of immiscible Fe—S—O liquid, and a higher dissolved S concentration is required before the basaltic melt is saturated. Chlorine concentrations in ocean island basaltic magmas are quite variable, from low values similar to uncontaminated MORB to much higher values. Some higher Cl values have been interpreted as the result of the presence of deeply recycled crustal components, which are enriched in Cl by hydrothermal alteration, in the mantle plume sources of hotspots. However, Cl/K ratios in some ocean island basalts are similar to those in MORB, suggesting that recycled ocean crust in ocean island basalt mantle sources may not be uniformly Cl-enriched. Fluorine concentrations in ocean island basaltic magmas are mostly between 100 and 1000 ppm and show considerable overlap with the values found in MORB glasses. In general, F concentrations increase with differentiation, as indicated by K_2O contents, and higher values occur in more alkalic compositions. The F/Nd ratio (Nd has a similar incompatibility as F during melting and crystallization) is nearly constant in MORB and ocean island basalts.

The best understood oceanic island volcano, and probably the most thoroughly studied volcano on Earth, is Kilauea on the island of Hawaii. The dissolved H_2O concentrations in a variety of basaltic melt inclusions and submarine glasses from Kilauea are shown in Figure 7.10. The variations in H_2O are controlled by five main factors: (1) heterogeneity in mantle-derived magma compositions, (2) fractional crystallization, (3) variable degassing before entrapment of melt inclusions, (4) mixing of surface degassed and undegassed magmas at depths as a result of

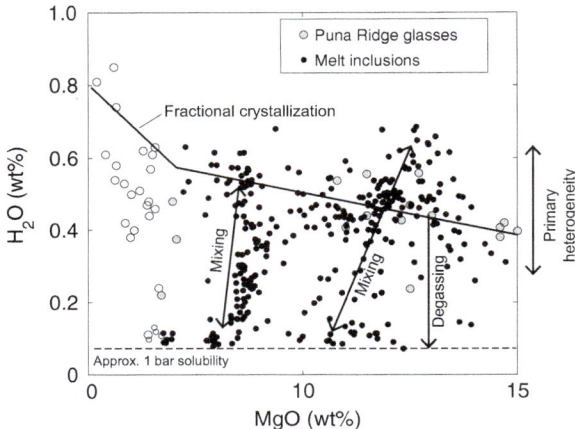

FIGURE 7.10 H_2O contents of olivine-hosted melt inclusions in tephra erupted from the summit of Kilauea and submarine glasses from the Puna Ridge. Lines show the effects of fractional crystallization, magma mixing (typically between more evolved degassed and primitive undegassed melts), and degassing or diffusive loss of H from the melt inclusions. Magma mixing may be due to drain back after fountaining episodes or vertical convection in the magma storage system beneath the summit. It is likely that a degree of compositional heterogeneity exists in primary melts (with MgO >15 wt%). The melt inclusion compositions have been corrected for the effects of postentrapment crystallization. *(Figure is modified from Edmonds et al., in press.)*

drainback, and (5) H loss from melt inclusions by diffusion through the olivine host.

The interpretations of the H_2O data fit with our general understanding of degassing at Kilauea, which is inferred from gas monitoring studies to occur mainly in two stages that are related to subsurface storage and transport of magma. New batches of mantle-derived magma that ascend beneath Kilauea are vapor saturated with a gas rich in CO_2. The solubility of CO_2 at the pressures of the summit magma reservoir is low, and the first stage of degassing occurs as most of the CO_2 is lost through noneruptive degassing, generating a persistent CO_2-rich plume at the summit. The amount of H_2O that is lost with the CO_2 in the first stage depends on the amount of CO_2 and the pressure, but is generally small, because H_2O is highly soluble in the melt. The second stage of degassing occurs when CO_2-depleted magma from the summit reservoir is transported laterally through the rift system in dikes, at depths of generally >1−2 km, and then erupts, degassing the more soluble volatile components and generating a gas plume rich in H_2O and SO_2.

As mentioned above, the volatile contents of magmas at depths may be modified by drainback of surface-degassed magma. Drainback was well documented for the 1959 Kilauea Iki eruption: magma that degassed during eruption collected into a surface lava lake and then drained back into the vent. Magma that has degassed in a lava lake and then drained back down into the conduit can mix with unerupted magma stored at depths to yield partially degassed hybrid magmas. Degassed hybrid magmas, with variable and in

part low concentrations of H_2O (0.1−0.85 wt%) and S (0.02−0.15 wt%), have been erupted on the deep submarine flanks of Kilauea. The submarine basalts erupted and were sampled from depths at which hydrostatic pressure would have prevented significant coeruptive degassing of H_2O and therefore the low concentrations of H_2O have been interpreted as reflecting low-pressure degassing prior to submarine eruption at higher pressures. Such low pressure degassing could have occurred by drainback and mixing prior to magma being transported down the rift. The shallow mixing interpretation is borne out by recent observations from the 2008−2013 summit eruptions at Halemaumau, which show this process at work: melt inclusion data from this period show that degassed magma sank down the conduit and mixed with undegassed magma at a depth of approximately 1 km.

The compositions of primitive melts at Kilauea influence the style of volcanic activity at the surface. The magmas that feed the high fountain eruptions that typically occur just outside the caldera have higher H_2O and CO_2 and are more primitive. In contrast, the magmas that erupt from the summit magma reservoir and feed effusive and lava lake eruptions have lower volatile contents and are more evolved, having mixed extensively with summit-stored magmas. There are broad changes in parental melt compositions at Kilauea that occur over 50- to 100-year timescales and are thought to be caused by the changes in the degree of partial melting coupled to small-scale mantle source heterogeneities.

5.3. Island Arc and Continental Margin (Subduction-Related) Basalts

It was first suggested in the early 1960s, before the advent of the modern concept of plate tectonics, that subduction of altered oceanic rocks beneath the Aleutian arc was responsible for recycling water into the mantle source region of arc magmas. Early work on melt inclusions in arc basalts during the 1970s used the deficit from 100% on the sum of the oxides measured by electron microprobe to estimate H_2O concentrations. Such sum deficits were not very precise, but they turned out to provide an accurate view of the water-rich nature of arc magmas. FTIR and SIMS ion microprobe techniques now provide a much higher precision on H_2O analyses. As a result, there has been an explosion of data, both from melt inclusions and experiments, on the volatile contents of basaltic magmas parental to arc volcanoes because of interest in geochemical recycling at subduction zones. The results of these studies show that mafic arc magmas typically contain 2−6 wt% H_2O (Figure 7.11). These values are much higher than the H_2O contents of MORBs and ocean island basalts, supporting the interpretation that water is recycled into the mantle by subduction, where it fluxes the overlying

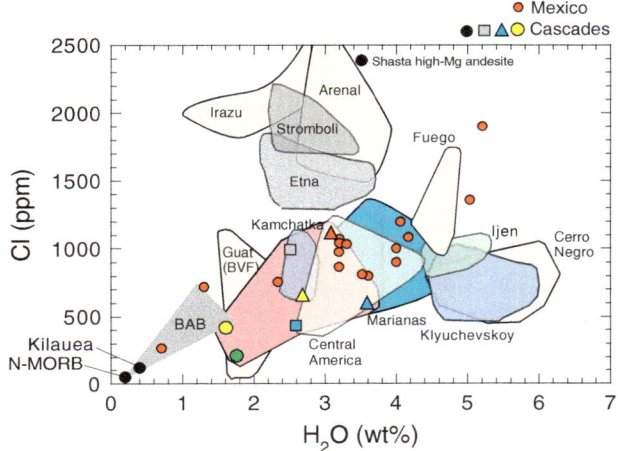

FIGURE 7.11 Variations in H_2O and Cl in basaltic melt inclusions from various subduction-related arcs and specific arc volcanoes. Also shown for comparison is a shaded field for the Lau and Marianas back-arc basins (BAB). Guat (BVF) is for cinder cones behind the volcanic front in Guatamala. (*Figure is modified from Metrich and Wallace, 2008.*)

asthenospheric mantle wedge to generate hydrous basaltic melts.

Given the large number of volcanoes that have now been studied, average H_2O values can be calculated for individual arcs, and the results show a surprisingly narrow range, from 3.2 wt% (for the Cascades) to 4.5 wt% (for the Marianas), with a global average of 3.9 ± 0.4 wt% H_2O. The narrow range and common average value for H_2O in each arc are unexpected because the concentrations of most other incompatible elements in arc magmas, such as Nb or Ba, vary by orders of magnitude, and H_2O might be expected to show similar large variations. The narrow range of H_2O concentrations could reflect a crustal control, in which magmas commonly stall at depths of 2–10 km, allowing more deeply formed melt inclusions in olivine to diffusively reequilibrate via H loss. Alternatively, the narrow range might reflect a mantle control, in which the addition of H_2O to the mantle causes larger degrees of melting, thus diluting the concentration of H_2O in partial melts. Such a negative feedback could cause a narrow range of H_2O contents for mantle melts compared to the ranges observed for incompatible trace elements.

Despite the similarities in averages and ranges in H_2O contents from different arcs, when data for individual volcanic centers within a given arc are examined, there are important variations that in some cases correlate with other geochemical parameters. For example, in the Aleutians, Cascades, Central America, and Trans-Mexican Volcanic Belt, there are correlations between magmatic H_2O concentrations and the abundances of many incompatible trace elements, indicating that transfer of a hydrous component from the subducted plate upward into the mantle wedge also controls many trace element characteristics of arc magmas.

Although arc magmatism is dominantly calc-alkaline in composition, some arcs contain a spectrum of basaltic magma types that extends to tholeiitic magmas, which show an increase in FeO^T content during differentiation. Detailed melt inclusion studies of Aleutian volcanoes show that the calc-alkaline types have >2 wt% H_2O, whereas the tholeiitic types have <2 wt%. A combination of higher H_2O, which suppresses plagioclase crystallization, and higher oxygen fugacity, which enhances Fe–Ti–oxide crystallization, is responsible for the calc-alkaline fractionation trend of decreasing FeO^T. Both the Cascades and Trans-Mexican Volcanic Belt also contain some basaltic magmas with lower H_2O ($\leq$1.5 wt%), and they likely form by decompression melting of the mantle rather than by fluxing with a hydrous component from the subducted slab.

Arc basaltic magmas also contain significant amounts of dissolved CO_2 (present as carbonate), with reported concentrations as high as 2500 ppm. Isotopic data for C in volcanic gases suggest that the CO_2 is partly mantle derived and partly recycled from subduction of carbonate-rich marine sediments and organic carbon. Because melt inclusions are trapped at crustal pressures after at least some CO_2 has been degassed, it is likely that the initial CO_2 contents of arc basaltic magmas are higher than the melt inclusion values, possibly as high as approximately 1 wt%.

Olivine-hosted melt inclusions from basaltic arc magmas typically have 900–2500 ppm S, with some magmas ranging up to 5000 ppm. These values are higher than for MORB magmas of the same FeO^T but are comparable to some ocean island basalts. The high S contents of many arc basaltic magmas require that they have a higher oxygen fugacity than do MORB magmas and also require enrichment of S in the mantle wedge above subducting slabs. Sulfur isotope data for fumarolic gases, submarine basaltic glasses, and subaerial whole rock samples have been interpreted to indicate recycling of seawater sulfate. The Cl concentrations of basaltic arc magmas (mostly between 250 and 2500 ppm) and Cl/K ratios are also much higher than those of MORBs and ocean island basalts, consistent with recycling of seawater-derived Cl from subducted sediments or altered oceanic crust.

5.4. Backarc Basin Basalts

Basaltic magmas from backarc basins show wide variations in H_2O concentration, ranging from values as low as those in the most H_2O-poor N-MORB (0.1 wt%) to values as high as 2.3 wt% (Figure 7.9). Such variations reflect the range of bulk chemical composition of backarc basin basalts, which can vary from MORB-like to showing many similarities to island arc volcanics, reflecting their formation in spreading centers that overlie subduction zones. Dissolved CO_2 in submarine basaltic glass samples from the Mariana Trough are between 100 and 200 ppm. Negative covariation between H_2O and CO_2 in these glasses and pressure estimates based on experimental solubility data suggest that the samples were vapor saturated at their depth of eruption and quenching, consistent with the presence of vesicles in the glassy pillow rims. Therefore, the bulk magmatic concentrations of H_2O and CO_2 may be significantly greater than the dissolved concentrations. Chlorine concentrations vary from 80 ppm to nearly 1 wt% and show a positive correlation with H_2O in comagmatic suites of glasses. Glasses with the higher values (>2000 ppm) as well as many with values in the range from 200 to 2000 ppm also have high Cl/K_2O and therefore appear to have been affected by assimilation of chlorine-rich altered oceanic crust at shallow levels. By contrast, other glasses in the range from 80 to 2000 ppm with a very low Cl/K_2O likely reflect variably amounts of subduction-related enrichment from the underlying slab into the mantle wedge.

Some H_2O-rich basaltic magmas erupted in backarc basins or the submarine portions of island arcs have a high-enough water concentration that they vesiculate and degas dissolved H_2O even during eruption in very deep water. These glasses are distinguished by their relatively high vesicularities (as much as 50% vesicles by volume). As expected, they also have anomalously low concentrations of volatile components, such as S, that exsolve from magma together with H_2O.

Measurements of H_2O content in backarc basin glasses generally show negative correlations with TiO_2, which, as an incompatible element during mantle melting, can be used as a proxy for the extent of melting. The correlation therefore shows that the degree of mantle melting is controlled in part by how much H_2O is added to the mantle from the underlying subducted plate. The H_2O concentrations estimated to be in the mantle beneath backarc basins, based on these correlations, suggest that mantle H_2O contents increase toward the trench, as would be predicted from the presence of a subducted slab that progressively dehydrates with depth. Backarc spreading segments that lie above the mantle with the highest estimated H_2O concentrations are at anomalously shallow water depths, probably because the high H_2O increases mantle melt productivity and therefore results in a greater thickness of backarc basin crust. These interpretations have been somewhat controversial, however, and an alternative possibility is that the negative correlations between H_2O and TiO_2 are the result of mixing of magmas from H_2O--poor mantle, similar to that beneath midocean ridges, and subduction-modified mantle, like that which is beneath island arcs.

5.5. Flood Basalts and Large Igneous Provinces

Huge outpourings of basaltic lava to form continental flood basalts and oceanic plateaus (known as large igneous provinces, LIPs) have occurred throughout the history of the Earth and are likely linked to plumes of hot, upwelling mantle. Data on the volatile contents of LIP basaltic magmas, which are dominantly tholeiitic in composition but include alkalic and more silicic compositions in some cases, are still relatively limited. Water concentrations in submarine basaltic pillow rims from the Kerguelen and Ontong Java plateaus are relatively low (0.15−0.5 wt%), similar to values for midocean ridge and some ocean island basalts. Chlorine concentrations vary from low values similar to MORB (<50 ppm) to values as high as 1000−3000 ppm at Ontong Java, where basalts were likely affected by the assimilation of hydrothermally altered wall rocks. Sulfur concentrations are similar to or even lower than MORB values. Melt inclusions from several continental flood basalts have S concentrations of about 800−2000 ppm and Cl concentrations of 300−500 ppm. Much higher concentrations (as much as 0.5 wt% S and 0.9 wt% Cl) are found in some alkalic compositions or in magmas that have been extensively contaminated.

The formation of many LIPs coincides in time with significant perturbations to the global carbon cycle and mass extinctions, including the end-Permian Siberian Traps, the Central Atlantic Magmatic Province at the Triassic−Jurassic boundary, and the Deccan Traps, erupted at the Cretaceous−Tertiary boundary. It has been recognized recently that the LIPs associated with the greatest impact on the environment, such as the Siberian Traps, were those that erupted through volatile-rich sediments, raising the possibility that the volatile budget of such eruptions might be augmented by the thermal decomposition of oil shales and evaporites. This would have enhanced the emission of methane and organohalogens that could warm the climate over long timescales and destroy the ozone layer, respectively.

In addition to carbon-rich gases, the other climate-forcing gas of interest with regard to LIPs is sulfur dioxide (SO_2). The magmatic sulfur budget of LIP eruptions is not well understood, owing to the fact that there are few data on preeruptive magmatic volatile abundances in these types of magmas. Some of the eruptions, however, have yielded estimates of sulfur emissions. The eruption of the Siberian Traps is estimated to have released 6.8×10^7 Tg SO_2 over less than 1 million years, and S isotope compositions of melt inclusions indicate that most of this sulfur was mantle derived. Sulfur species in volcanic gases (SO_2 and H_2S) oxidize in the atmosphere to form sulfate aerosol, which scatters, absorbs, and reflects incoming solar radiation, causing short lived, but perhaps catastrophic, tropospheric cooling. The climate impact of these kinds of gas injections over such timescales are not clear, but if the effects of recent, smaller eruptions can be scaled up, it is likely that decades or longer of low temperatures occurred, which could have had a catastrophic impact on life on Earth.

6. VOLATILE ABUNDANCES IN SILICIC MAGMAS

6.1. Andesites

Andesitic magmas typify convergent plate margins, and their volatiles play major roles in planetary degassing and differentiation. Most andesite magmas are complex hybrids arising from mid to upper crustal differentiation of mafic magma, long-timescale residence of crystal-rich mush in upper crustal magma reservoirs, and repeated injections of mafic magma into such reservoirs. Most andesites have 30–45 vol% phenocrysts coexisting with a rhyolitic interstitial melt. The volatile budget is similarly complex: the magma reservoirs likely coexist with a significant fraction of vapor, rich in H_2O and S. It is likely that the recharging mafic magmas supply volatiles during mixing and mingling with the resident andesites, either by second boiling during quenching and crystallization at the mingling interface, or by changes in volatile solubility arising from differences in temperature and oxygen fugacity between the two magmas.

Standard methods for assessing volatile budgets are problematic for andesites: melt inclusions in plagioclase or pyroxene are often subject to the loss of volatiles through leakage and, if they are pristine, it remains unclear how they relate to the bulk magma and to the recharging magmas. All the observations thus far, however, suggest that the melts are vapor saturated and H_2O-rich at mid to upper crustal depths. Rhyolitic melt inclusions in plagioclase in andesite from Soufriere Hills contain up to approximately 6 wt% H_2O, suggesting that the melts are H_2O-rich and that the magma resides in the midcrust (perhaps at depths >8 km, once the CO_2 content of the melt is taken into account) prior to eruption. Andesitic to rhyolitic melt inclusions hosted by plagioclase and pyroxene contain up to 3–4 wt% H_2O in andesites from Colima and Popocatepetl volcanoes, Mexico. Rhyolitic melt inclusions from Mt Hood in the Cascades contain up to 5.5 wt% H_2O.

Experimental volatile studies are difficult to design for a system characterized by disequilibrium on all scales. Some andesites contain volatile-bearing mineral phases, such as hornblende and apatite, which lend further constraints on the volatile budget and magma pathways through the crust, and some mineral–melt systems may be used as hygrometers. From phase equilibrium studies, a minimum of approximately 3 wt% dissolved H_2O is necessary to stabilize hornblende, a common phenocryst mineral in andesites, although hornblende stability is also strongly affected by other factors, such as the Na_2O content of the coexisting silicate melt.

Sulfur concentrations measured in melt inclusions in andesites from subduction zone magmas are normally ≤ 1000 ppm, with typical values of <200 ppm in more silicic melts. Sulfur has low solubility in Fe-poor oxidized rhyolitic melts, and as a consequence, most of the sulfur partitions into hydrous vapor in the magma reservoir. Sulfur solubility is strongly dependent on oxygen fugacity; for typical andesitic magmas, this varies from slightly below the Ni-NiO (NNO) buffer reaction to about 2 log units more oxidized than NNO. Over this range, sulfur speciation changes from dominantly sulfide to dominantly sulfate (Figure 7.4).

There remains limited data on CO_2 abundances in andesites, although it seems likely that there is a pervasive flux of deeply derived CO_2-rich vapor released from the volatile-rich mafic melts that pond and fractionate in the lower arc crust. Melt inclusions in andesites are typically poor in CO_2, with abundances of <500 ppm. Volcanic gas measurements at arc andesitic volcanoes, however, suggest that CO_2 fluxes are several times the flux of SO_2 by mass, which lends support to the idea that these gases are sourced in the lower to middle arc crust.

Melts in andesites are typically rich in halogens, which are derived ultimately from the subducting slab. Andesitic melt inclusions from the Izu volcanic arc contain >800 ppm fluorine and up to 0.9 wt% chlorine and are similarly rich in other elements that behave in a volatile way during subduction such as lithium and boron. Rhyolitic melt inclusions in andesites from Augustine and Soufriere Hills volcanoes contain up to 0.6 wt% Cl, which is close to the solubility limit for this melt composition. At Colima Volcano, Mexico, melt inclusions contain up to 0.24 wt% Cl. For particularly chlorine-rich melts, an immiscible hydrosaline melt (brine) phase may coexist with silicate melt and vapor, which can sometimes be recognized as an additional phase in melt inclusions. Chlorine is relatively soluble and remains largely dissolved at magma reservoir pressures, but it partitions into hydrous vapor at low pressure and is thus partially lost by degassing during magma ascent and eruption.

6.2. Dacites and Rhyolites

Dacitic and rhyolitic eruptions span a wide range of volume, volcano type (e.g., stratovolcanoes, dome complexes, and large calderas), and tectonic association, including both arc and intraplate settings. Preeruptive H_2O concentrations are often relatively high providing the energy for the powerful explosive eruptions that typify silicic magmas. Glassy melt inclusions are common in rapidly quenched phenocrysts from explosive silicic eruptions, particularly in quartz and sanidine, and analyses of the inclusions commonly yield values of 4–6 wt% H_2O (Figure 7.12). In this way, the range of H_2O contents in dacites and rhyolites is remarkably similar to that in basalts and andesites.

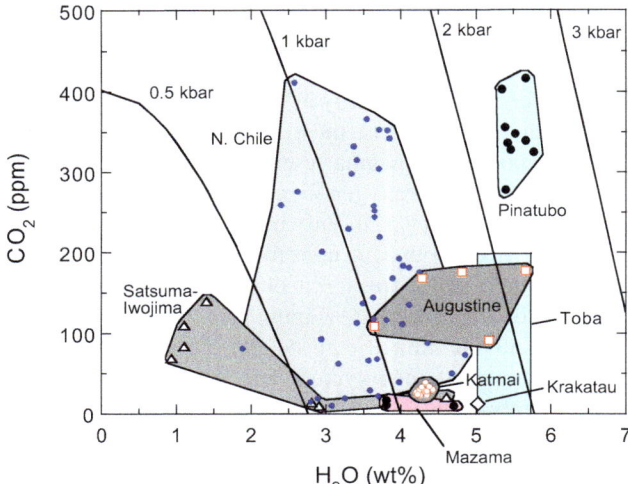

FIGURE 7.12   Dissolved H_2O and CO_2 in melt inclusions from dacites and rhyolites from subduction-related arcs. Diagonal lines show the vapor saturation pressures for rhyolitic melt from 0.5 to 3 kbar pressure. *(From Wallace, 2005).*

Complementary to the melt inclusion data for silicic magmas are experimental phase equilibrium studies that provide information about temperatures and depths of equilibration before eruption and dissolved H_2O concentrations. The analytical (melt inclusion) and experimental results give broadly consistent H_2O values.

Water-rich melt inclusions are typical of low temperature (700–800 °C) rhyolites containing hydrous phenocrysts of biotite and amphibole from both arc and intracontinental settings. Phase equilibrium data for hotter (≥900 °C) rhyolites on the Snake River Plain, which are associated with the Yellowstone hotspot, require much lower values (≤1.5 wt%), consistent with their lack of hydrous phenocrysts. Dacitic and rhyolitic magmas from the catastrophic 1883 eruption of Krakatau, Indonesia, demonstrate, however, a higher amount of H_2O in the melt (4 wt%) at similar temperatures. Interestingly, melt inclusions in fayalitic olivines from Hekla volcano in Iceland contain relatively high H_2O (5–6 wt%), despite their occurrence in a hotspot setting. Peralkaline rhyolites (pantellerites) also contain high H_2O (4–5 wt%), which as with Hekla, are likely to be caused by the large amounts of fractional crystallization needed to produce rhyolite from parental basaltic magma.

Wide variations in dissolved CO_2 are found in rhyolitic melt inclusions, from <30 ppm to about 1200 ppm, but most values are <200 ppm. Based on the pressure-dependent solubilities of H_2O and CO_2 in rhyolitic melts, the melt inclusion data can be used to compute the pressure at which the melt would be saturated with a vapor phase (the vapor-saturation pressure) and thereby constrain pressures of crystallization (inclusion formation). For vapor-saturated magmas, the calculated vapor-saturation

pressure equals the pressure of crystallization. For magmas lacking vapor (vapor undersaturated), the actual pressure of crystallization is greater than the vapor-saturation pressure indicated by melt inclusions. Vapor-saturation pressures based on melt inclusion H_2O and CO_2 for rhyolites vary from about 1 to 4 kbar, implying a range of upper crustal depths through which silicic volcanic magmas differentiate and reside before eruption.

Detailed experimental phase equilibrium studies have been performed on the 1912 rhyolite from Katmai, Alaska, the 1980 Mount St Helens dacite (Figure 7.7), the 1991 Mt Pinatubo dacite, and the 2008 and 2011 rhyolitic eruptions of Chaitén and Cordón Caulle, respectively, in Chile. The results indicate crystallization pressures of 0.5–1.0 kbar (Katmai, Cordón Caulle), 1.2–2.0 kbar (Chaitén), and approximately 2.2 kbar (St Helens, Pinatubo) at temperatures ranging from 780 to 920 °C. The experimental data require dissolved H_2O contents for Pinatubo, Katmai and St Helens that agree with melt inclusion studies. For the Mount St Helens dacite, the experimentally synthesized mineral assemblage and mineral compositions match the natural assemblage when there is about 4.6 wt% H_2O dissolved in the melt. The experimental data for Mt Pinatubo dacite indicates a higher H_2O concentration of approximately 6 wt%. The data for Katmai indicate lower H_2O contents (approximately 2.5–4.5 wt%) given their lower equilibration pressures. The preeruptive magmas at Katmai, Mount St Helens, and Pinatubo all appear to have contained a vapor phase during crystallization, because the vapor-saturation pressures calculated from melt inclusion data and pressures based on experimental phase equilibria are equal.

One of the best studied rhyolitic eruptions in terms of magmatic volatile abundances is the 0.76 Ma explosive, caldera-forming eruption that formed the >600-km³ Bishop Tuff. Melt inclusions in Bishop quartz phenocrysts reveal systematic, preeruptive gradients in dissolved H_2O and CO_2 during crystallization. Melt inclusions from early erupted pumice fall deposits contain 5.3 ± 0.4 wt% H_2O and 60 ± 40 ppm CO_2, whereas those from the middle of the eruption contain higher H_2O (5.7 ± 0.2 wt%) and CO_2 (120 ± 60 ppm). Thus, early erupted magma contained less dissolved H_2O and CO_2 than did magma from the middle of the eruption. Compared with melt inclusions from early and middle Bishop Tuff samples, inclusions from late-erupted magma have much lower H_2O (4.1 ± 0.3 wt%), and higher and variable CO_2 (150–1100 ppm), and the most CO_2-rich inclusions were trapped in the rims of crystals shortly before eruption, following an episode of magma mixing and partial quartz dissolution. The estimated pressures of crystallization (vapor-saturation pressures) are lower for inclusions from the earliest erupted samples than in the samples from the middle of the eruption and highest for the rims of late erupted quartz. This pattern is consistent with sequential tapping of a large zoned body of vapor-saturated magma, because the first material erupted would likely come from the shallowest (lowest pressure) portion of the magma body.

Dissolved S concentrations in silicic magmas tend to be relatively low (≤ 200 ppm), and are frequently below the minimum detection limit of common analytical techniques. These low concentrations occur despite the fact that most silicic magmas are saturated with a sulfide (pyrrhotite) and/or sulfate phase (anhydrite). The low S solubility is caused by a combination of low melt FeO content and the low temperature of silicic magmas. The former is an important factor at a relatively low magmatic oxygen fugacity where dissolved S is dominantly sulfide, whereas decreasing temperature significantly decreases both sulfide and sulfate solubilities (Figure 7.5).

Fluorine and chlorine abundances are quite variable in metaluminous dacites and rhyolites (F, 200−1500 ppm; Cl, 600−2700 ppm). In the Bishop Tuff melt inclusions, F varies from 160 to 460 ppm and Cl varies from 550 to approximately 800 ppm. Both elements are more abundant in the most highly differentiated, early erupted Bishop Tuff, whereas lower concentrations occur in the least differentiated magma that erupted later. Melt inclusions from the 1991 Mt Pinatubo dacite contain approximately 1000 ppm Cl. In contrast, Cl concentrations in melt inclusions from Taupo Volcanic Field rhyolites (New Zealand) and Augustine volcano dacites (Alaska) range as high as 2700 ppm, sufficient to cause the preeruptive magma to be saturated with an immiscible hydrosaline melt in addition to a vapor phase. Compared to halogen abundances in metaluminous magmas, values for peralkaline rhyolites (pantellerites) are much higher, with as much as 1.3 wt% Cl and 1.5 wt% F. These high concentrations reflect their derivation from parental mafic magmas created by small degrees of mantle melting, the large amount of fractional crystallization necessary to form the rhyolitic magmas (which causes Cl and F to increase like other incompatible trace elements), and to the effects of alkalis in increasing Cl and F solubilities in silicate melts. Peraluminous tin and topaz rhyolites of the western United States and central Mexico also contain high halogen concentrations, with Cl reaching 0.5 wt% and F as great as 5 wt%.

6.3. Gradients in Volatile Concentration and Vapor Saturation in Magmas

Gradients in dissolved volatiles, especially H_2O, within individual silicic magma bodies have been inferred from (1) variations in phenocryst assemblages and compositions; (2) variations in whole-rock abundances of F and Cl; and (3) the observation that many volcanic eruptions evolve with time from high-energy Plinian columns to pyroclastic flows, or culminate with quiescent dome extrusion. Gradients in dissolved H_2O in silicic magma bodies have more recently been confirmed using ion probe and infrared spectroscopic techniques on melt inclusions in phenocrysts. In some cases, as at Mount St Helens, changes in eruptive style from Plinian to subplinian and Vulcanian are reflected in decreases in melt inclusion H_2O concentrations. In others, like the Bishop Tuff, Plinian and pyroclastic flow deposits have similar melt inclusion H_2O values, probably because the different types of deposits were coeval rather than sequential. Even effusive (nonexplosive) eruptions of silicic magma appear to involve magma with initially high dissolved H_2O concentrations. The fact that such magmas are erupted nonexplosively requires that water be lost prior to extrusion. This may occur when magma decompression induces sufficient vesicularity to form a magmatic foam that is highly permeable to gas loss or when magma rises so slowly that bubbles can segregate efficiently.

It has commonly been thought that magmas of intermediate to silicic composition only become vapor saturated during shallow ascent and emplacement, during eruptive decompression, or during advanced (pegmatitic) stages of plutonic crystallization. However, studies of eruptive dynamics, CO_2 and SO_2 emissions, and melt compositional features all suggest that many magmas in subvolcanic reservoirs are saturated with a multicomponent vapor phase. Volcanoes could be viewed primarily as gas vents. In particular, the small but finite solubility of CO_2 in silicic magmas at crustal pressures and the relatively high CO_2 contents of mafic magmas that provide heat and mass to create crustal magmatic systems could result in saturation with H_2O-CO_2 vapor during silicic magma formation as well as during ascent, emplacement, and crystallization. High-temperature fluid inclusions in phenocrysts from some silicic volcanic rocks provide direct evidence of preeruptive vapor saturation. Gradients in alkali and trace element abundances might result in part from upward flux of exsolved vapor in a large magma body.

Correlations between CO_2 and trace elements in melt inclusions from the Bishop tuff indicate that the magma was vapor saturated during preeruptive crystallization. Quantitative models of vapor exsolution during crystallization of the Bishop magma are consistent with exsolved gas contents varying from about 1 wt% in the deeper regions of the magma body to nearly 6 wt% near the top. Thus, the data and models suggest that there was a gradient in exsolved gas, with mass fractions of gas increasing upward toward the roof of the magma body. Because the exsolved gas phase would be H_2O-rich, such large exsolved gas contents imply that the early erupted Bishop magma had a bulk (dissolved + exsolved) H_2O content of approximately 10 wt%. However, these interpretations assume that gas remained in the magma during most of its evolution. The presence of significant mass fractions of exsolved gas in Bishop magma is consistent with H_2O and CO_2 dissolved in variably degassed rhyolitic obsidians at Mono Craters, California. Similar mass fractions of preeruptive exsolved gas have been inferred for the magma bodies of the 1883 Krakatau, 1982 El Chichón and 1991 Mount Pinatubo eruptions, and many other smaller silicic eruptions, by assuming that all SO_2 released (measured by remote sensing techniques) was stored in the erupted volume of magma.

6.4. Volcanic Gases: Total Volatile Budgets and Climate Impacts of Eruptions

An important constraint on the volatile budgets of volcanic eruptions comes from measurements of gas fluxes vented directly to the atmosphere, which represent the integrated sum of degassing over the magma column beneath the volcano. Gas flux measurements are also important for evaluating the climate and environmental impacts of eruptions.

For magmas that are vapor saturated at or below the depths at which crystallization occurs, melt inclusions in phenocrysts record only a fraction of the volatiles that were originally present in the melt before vapor saturation. Volcanic gas flux measurements therefore provide important additional information about the total volatile budgets of volcanic eruptions that is complementary to data from melt inclusion studies. For intermediate and silicic magmas, and for relatively oxidized basaltic magmas in arc settings, the concentrations of S in melt inclusions are far too low to account for the total amount of S degassed during eruptions, implying that much of the preeruptive S budget is partitioned into a magmatic vapor phase rather than dissolved in silicate melt. In order to characterize S in volcanic systems, measurements of the fluxes of SO_2 and H_2S gases at the surface are therefore required. For volatile components with a very low solubility, such as CO_2, the problem of evaluating the degassing budget is even more severe: magmas may become saturated with a CO_2-rich vapor phase at depths of tens of kilometers in the crust, and perhaps even in the upper mantle in some cases.

Gas flux measurements from volcanoes are made using spectroscopic techniques that collect ultraviolet (UV) radiation, utilizing the strong absorption signature of SO_2 in this region of the electromagnetic spectrum. For larger eruptions, the gas emissions might be very large and injected to high altitudes in the stratosphere. In this case, satellite-based spectrometers are more appropriate for capturing the full mass of SO_2 in the volcanic gas cloud, again utilizing absorption in the UV (e.g., the Ozone Mapping Instrument), and also in the infrared (e.g., the Infrared Atmospheric Sounder Interferometer).

Determination of carbon dioxide (CO_2) fluxes at persistently active volcanoes has become more commonplace, and this is of importance for quantifying the carbon output from volcanoes. The method involves measuring the ratio between CO_2 and SO_2 in the volcanic gas, typically using infrared spectroscopy and electrochemical sensors, and combining it with the SO_2 flux measured using UV spectroscopy as described above, to yield an estimate of CO_2 flux. It has been shown that the sustained flux of CO_2 from volcanoes such as Kilauea and Mount Etna is very large, at times >20 kilotons per day (kt/day), but more generally <7 kt/day. The explosive eruption of Eyjafjallajökull in Iceland in 2010 was associated with the emission of approximately 150 kt/day CO_2, sustained for several days. These amounts of CO_2 are still very small when compared to the time-averaged CO_2 emissions from anthropogenic sources (which is around 100–200 times larger). For example, the integrated daily output of CO_2 from automobiles in a single metropolitan area —Washington DC—is approximately equal to the daily global CO_2 output from all volcanoes worldwide.

The magnitude of these CO_2 emissions, when the magma supply rate is known, places some constraints on the CO_2 contents of primary melts in these settings, which probably lie within the range 0.5–1.0 wt%. For andesitic arc volcanoes, where volatiles are sourced from complex hybrid magmas that are likely continually recharged at depths, the volcanic CO_2 flux at the surface cannot be easily related back to a primary melt, but the overall magnitude of CO_2 emission suggests that parental arc magmas may also have CO_2 contents in the 0.5–1.0 wt% range.

As described in the section above for eruptions involved in the formation of LIPs, volcanic gases have environmental and climate impacts. Sulfur-rich gases are converted to sulfate aerosol rapidly in the atmosphere, where they act to scatter, absorb, and reflect incoming solar radiation. Satellite-based spectroscopic measurements are the primary means of observing this conversion and quantifying it. Large aerosol clouds arising from eruptions in the tropics can circumnavigate the Earth very rapidly, within days, causing global climatic effects. Eruptions that occur at high latitudes, however, have only local or regional effects. This is mainly due to the circulation pattern of air in the atmosphere, in which air is drawn from the equator and then flows out to high latitudes. Additionally, there is relatively low solar flux at high latitudes, which means that the effect of scattering by aerosol is limited.

7. ORIGIN OF VOLATILES, EARTH DEGASSING, AND VOLATILE RECYCLING BY SUBDUCTION

Studies of inert gases such as helium, argon, and neon are consistent with the theory that early in its history, the Earth was covered by a deep magma ocean, caused by conversion into heat of some of the gravitational potential energy of accreting material. Inert volatiles would have become predominantly degassed and concentrated in the atmosphere during the Earth's formation. Such an early atmosphere could have been largely lost from the Earth as a result of solar activity or violent impacts. Significant amounts of reactive volatiles such as H_2O could have dissolved in the magma ocean. Subsequent crystallization of the magma ocean and formation of minerals poor in H_2O would promote degassing by concentrating the reactive volatiles in a diminishing residue of liquid, thereby leading to the formation, rise, and escape of bubbles of gas.

It is conceptually important that present degassing rates of primordial, nonradiogenic inert gas from midocean ridge magmatism (based on the ^3He rate and ratios of other inert gases to ^3He), when multiplied by the age of the earth, yield much less total nonradiogenic neon and argon than is present in the earth's atmosphere. Thus, it is inferred that the rate of degassing from the earth either was much greater in the past or that significant inert gas has been separately added to the outermost earth by meteoritic infall. The isotopic composition of nonradiogenic inert gases that is lost from the earth is noticeably different from that in air, and intermediate between that of inert gases from the sun and from meteorites. The discrepancy can therefore be reconciled with either a mixture of sources for the earth's inert gases or preferential loss from the earth's atmosphere of less massive isotopes.

Subduction provides a mechanism by which atmospheric volatiles, contained in sediments and altered oceanic crust, are recycled back into the mantle. The sediments and altered crust are relatively rich in the two major reactive volatiles: carbon dioxide stored as carbonates and organic carbon, and water, both as pore water and as water bound in hydrous minerals such as clays. Seawater-derived chlorine is also added to the oceanic crust during its alteration prior to subduction. The pore water is buoyant, and increasing pressure during subduction collapses pores and forces water out of deeply buried sediments and rocks. Most of the hydrous minerals break down to anhydrous minerals and release water vapor, which is also buoyant, as the subducting rock heats up. Some deep storage of water is possible, however, in view of experimental studies that show that some dense hydrous magnesian silicates are stable to very high pressures and temperatures. Also, small concentrations of water occur as structurally bound OH in nominally water-free minerals such as olivine and pyroxene. Subduction-related basaltic magmas are notably more H_2O-rich than midocean ridge and ocean island basaltic magmas, and such magmas may return most or all the subducted water to the surface.

The isotopic composition of water is unaffected by radioactive decay (except for bomb tritium) and has probably been negligibly affected by preferential escape from Earth of hydrogen compared to more massive deuterium. Ratios of deuterium to hydrogen (D/H) in various water

reservoirs mainly reflect the geological processes that partition water into these reservoirs. Water initially released from magmas at igneous temperatures is eventually redistributed by formation of new low-temperature minerals in the reservoirs of altered rocks, seawater, fresh water, and ice. Isotopic fractionation during this redistribution causes the oxygen and hydrogen isotopic composition of seawater to differ from that dissolved in parental basaltic magmas; the balance is made up mainly by the isotopically light water bound in altered rocks. There is no isotopic evidence for a significant source of surface water beyond that which degasses from basaltic magmas. However, the D/H ratio of seawater entering subduction zones is considerably higher, by approximately 7%, compared with upper mantle water. If the H_2O content and D/H ratio of the mantle are both at steady state, as suggested by an approximately constant ocean mass through geologic time, this requires that H_2O delivered to the mantle sources of arc magmas must be preferentially enriched in deuterium compared with MORB. This is generally supported by the D/H ratios measured in arc melt inclusions.

The isotopic composition of carbon in parental basaltic magmas and various terrestrial reservoirs is not certain. There may be a significant isotopic shift upon degassing, and this makes it difficult to assess the isotopic composition of carbon in parental basaltic magmas. The predominant recognized reservoir of terrestrial carbon is in sedimentary limestones and dolomites, and it differs isotopically from that which degasses from basaltic magmas at igneous temperatures. Photosynthesis yields reduced carbon in biological tissues that is relatively poor in ^{13}C, and this affects the isotopic composition of carbon in surficial reservoirs. Existing knowledge permits the amount and isotopic composition of buried organic matter (mainly reduced C in shales and coals) to balance the amount and isotopic composition of carbon in carbonate rocks yielding a sedimentary average that is consistent with igneous gas. However, significant additions of isotopically distinctive carbon to the Earth's early atmosphere during the final stages of accretion, as well as storage of carbon in Earth's core, are possible. Furthermore, it is possible that the isotopic composition of carbon now released from magmas reflects a recycling steady state, and it may have differed in the remote past, reflecting a smaller proportion of recycled carbon. The sum of these processes has led to an H/C ratio in the Earth's exosphere that is higher than that in its interior, and significantly higher than that of chondritic meteorites.

The downgoing view of carbon dioxide is illuminated by the fact that at fairly shallow mantle depths the carbonate mineral dolomite becomes stable in a typical assemblage of mantle silicates. Thus, a deep residence of mineralogically bound CO_2 is feasible. Under the conditions of pressure and temperature where dolomite is stable in the mantle, CO_2 is quite soluble in silicate melts and can flux partial melting at comparatively low temperatures. The upgoing view of ascending carbonate-rich magmas is marked by the expected release of CO_2 gas at shallow upper mantle pressures. Plausibly this pressure-sensitive release of CO_2 gas is partially responsible for the explosive nature of rare magmas bringing diamonds and fragments of the mantle to the surface.

Despite the measured differences in the isotopic composition of H_2O and CO_2 in various terrestrial reservoirs, the isotopic composition of hydrogen, carbon, and chlorine—as well as nitrogen—are all similar to the range of isotopic compositions measured in carbonaceous chondrites. This striking isotopic similarity is convincing evidence that the terrestrial budgets of these volatile elements were sourced from the late addition, after magma ocean solidification, of a mass of carbonaceous chondrite equivalent to approximately 2% of the present mass of the Earth.

Yearly amounts of reactive volatiles H_2O and Cl that are subducted are in the range of rough estimates of what is returned to the surface by midocean ridge and arc magmatism. If there were no surficial return of subducted H_2O and Cl, then the formation, alteration, and subduction of oceanic crust would comprise a net drain on the water and salt in the oceans. The geological record and understanding of tectonism reveal that continental crust and ocean water have existed for most of earth's history; subduction has not caused the oceans to dry up. Thus, water lost to altered oceanic crust is probably largely returned to the oceans via dewatering of subducted crust and subduction-related magmatism.

8. FUTURE DIRECTIONS

The study of magmatic volatiles lies at the interface of igneous petrology and volcanology. Much of our present state of knowledge has been derived relatively recently due to major advances in techniques of microanalysis. Increasingly, it is possible to measure the isotopic composition of volatile and nonvolatile species in silicate melt inclusions, as well as trace metals in trapped fluid or vapor + melt inclusions in volcanic phenocrysts. Combined analyses of volatile species, trace elements, and isotope ratios (e.g., H, O, S, C, He, and Sr) will contribute to a better understanding of volatile recycling in the Earth, melting processes, magmatic differentiation, and magma degassing.

The pressure-dependent solubilities of the major volatiles H_2O and CO_2 in silicate melts make it possible to use melt inclusions as geobarometers. Combined with major, trace, and isotopic data, and a detailed knowledge of time—volume—compositional variations in eruptive deposits, melt inclusions can be used to place constraints on magma chamber configurations, and magmatic and eruption processes. In this context, petrologic techniques for estimating the bulk (dissolved + exsolved) volatiles in magma bodies need to be further developed and refined to better understand volcanic gas fluxes, low pressure degassing, and the relationship between volatiles and eruption dynamics. The extent to which volatile species leak from partially enclosed inclusions or diffusively reequilibrate through crystals is potentially among the fastest chronometers in geology, clocking the minutes of magma ascent prior to eruption. Although the interpretation of melt inclusion data is frequently complex due to effects of magma mixing, crystallization, and volatile leakage, the ability to sample melt inclusions from different stratigraphic levels of an eruptive deposit makes it possible to get an unprecedented look at the workings of magma bodies.

It is increasingly being recognized that magmatic volatiles, including volatile trace metals, make important contributions to magmatic-hydrothermal ore deposits. Future analytical developments that make it possible to measure the stable and radiogenic isotope ratios of melt inclusions will significantly add to our knowledge of magmatic inputs to ore-forming systems. Another important area of future research will be the relationship between volatile abundances in volcanic rocks and their plutonic equivalents. The goal of such work will be to gain a better understanding of the relationship between plutonic and volcanic rocks, and in particular, whether volcanic eruptions tap the volatile-enriched upper portions of larger crustal magma reservoirs, leaving behind less volatile-rich, uneruptable material that eventually solidifies as a pluton.

FURTHER READING

Baker, D.R., Alletti, M., 2012. Fluid saturation and volatile partitioning between melts and hydrous fluids in crustal magmatic systems: the contribution of experimental measurements and solubility models. Earth-Science Reviews 114, 298–324.

Baker, D.R., Moretti, R., 2011. Modeling the solubility of sulfur in magmas: a 50-year old geochemical challenge. Reviews in Mineralogy and Geochemistry, Mineralogical Society of America 73, 167–213.

Dixon, J.E., Clague, D.A., Wallace, P., Poreda, R., 1997. Volatiles in alkali basalts from the North Arch Volcanic field, Hawaii: extensive degassing of deep submarine-erupted alkalic series lavas. Journal of Petrology 38, 911–939.

Dixon, J.E., Clague, D.A., 2001. Volatiles in basaltic glasses from Loihi Seamount, Hawaii: evidence for a relatively dry plume component. Journal of Petrology 42, 627–654.

Edmonds M, Sides I, Maclennan J., 2015. The role of volatiles in magma transport and eruption at Kilauea Volcano, Hawaii. In: Hawaiian Volcanoes: From Source to Surface, Rebecca Carey, Valerie Cayol, Michael Poland, Dominique Weis (Eds.), AGU Monograph, 323–350.

Klimm, K., Kohn, S.C., Botcharnikov, R.E., 2012. The dissolution mechanism of sulphur in hydrous silicate melts. II: solubility and speciation of sulphur in hydrous silicate melts as a function of fo_2. Chemical Geology 322–323, 250–267.

Le Roux, P.J., Shirey, S.B., Hauri, E.H., Perfit, M.R., Bender, J.F., 2006. The effects of variable sources, processes and contaminants on the composition of northern EPR MORB (8−10°N and 12−14°N): evidence for volatiles (H_2O, CO_2, S) and halogens (F, Cl). Earth and Planetary Science Letters 251, 209−231.

Luhr, J.F., 1990. Experimental phase relations of water- and sulfur-saturated arc magmas and the 1982 eruptions of El Chichón Volcano. Journal of Petrology 31, 1071−1114.

Iacono-Marziano, G., Morizet, Y., Le Trong, E., Gaillard, F., 2012. New experimental data and semi-empirical parameterization of H_2O−CO_2 solubility in mafic melts. Geochimica et Cosmochimica Acta 97, 1−23.

Newman, S., Lowenstern, J.B., 2002. VolatileCalc: a silicate melt−H_2O−CO_2 solution model written in Visual Basic for excel. Computers and Geosciences 28, 597−604.

Perfit, M.R., Fornari, D.J., Malahoff, A., Embley, R., 1983. Geochemical studies of abyssal lavas recovered by DSRV Alvin from eastern Galapagos Rift, Inca Transform, and Ecuador Rift 3. Trace element abundances and petrogenesis. Journal of Geophysical Research 88, 10551−10572.

Plank, T., Kelley, K., Zimmer, M., Hauri, E.H., Wallace, P.J., 2013. Why do mafic arc magmas contain ∼4 wt% water on average? Earth and Planetary Science Letters 364, 168−179.

Metrich, N., Wallace, P., 2008. Volatile abundances in basaltic magmas and their degassing paths tracked by melt inclusions. Minerals, Inclusions, and Volcanic Processes Reviews in Mineralogy and Geochemistry, Mineralogical Society of America 69, 363−402.

Rutherford, M.J., Sigurdsson, H., Carey, S., Davis, A., 1985. The May 18, 1980, eruption of Mount St Helens: melt composition and experimental phase equilibria. Journal of Geophysical Research 90, 2929−2947.

Taylor, B., Martinez, F., 2003. Back-arc basin basalt systematics. Earth and Planetary Science Letters 210, 481−497.

Wallace, P., 2005. Volatiles in subduction zone magmas: concentrations and fluxes based on melt inclusion and volcanic gas data. Journal of Volcanology and Geothermal Research 140, 217−240.

Wallace, P.J., Edmonds, M., 2011. The sulfur budget in magmas: evidence from melt inclusions, submarine glasses, and volcanic gas emissions. Reviews in Mineralogy and Geochemistry, Mineralogical Society of America 73, 215−246.

Workman, R.K., Hauri, E., Hart, S.R., Wang, J., Blusztajn, J., 2006. Volatile and trace elements in basaltic glasses from Samoa: implications for water distribution in the mantle. Earth and Planetary Science Letters 241, 932−951.

Chapter 8

Magma Chambers

Bruce D. Marsh

Department of Earth & Planetary Sciences, Johns Hopkins University, Baltimore, MD, USA

Chapter Outline

GLOSSARY

batholith A vast collection of solidified magmatic bodies (plutons) commonly forming a mountain range like the Sierra Nevada or Andes.

capture front The position in a solidification front denoting a region inward of which crystals are able to move freely relative to one another and outward of which crystals move in concert with their neighbors.

critical crystallinity The point in a solidification front where the packing of crystals is close enough to allow formation of a crystalline network.

dike A discordant sheetlike body of magma, commonly near vertical, cutting the country rock.

dilatancy The tendency of an assemblage of particles to expand or dilate during shear deformation.

mush zone The region in a solidification front where crystals and solids tend to move altogether even though the solids are not all physically connected.

neck A stalklike magmatic feeder connecting a volcano to a deeper supply of magma.

plug The solidified lump or mass of magmatic rock filling a volcanic vent or crater.

pluton An isolated solidified parcel of magma found in the Earth's crust and thought to represent an extinct magma chamber.

Rayleigh number A pure, dimension-free number the magnitude of which indicates the tendency of a layer to undergo sustained thermal convection. The larger this number, the more vigorous is convection.

rigid crust The region marking the trailing half of a solidification front where all the crystals are interconnected to form a matrix possessing strength.

sill A sheetlike concordant magmatic body usually emplaced horizontal in the Earth's upper crust.

solidification front The marginal zone encompassing all active magmatic and volcanic bodies within which most crystallization takes place.

stock Any cylindrical subterranean solidified magmatic body with no clear connection to a volcanic vent.

suspension zone The inward leading region of the solidification front in which small, newly formed crystals are able to move with ease relative to one another.

1. INTRODUCTION

Magma chambers are the virtual and, sometimes, actual classical homes of the physical and chemical processes bringing about differentiation, which is the process to which planet Earth owes its gross structure and great

The Encyclopedia of Volcanoes. http://dx.doi.org/10.1016/B978-0-12-385938-9.00008-0

diversity of its igneous rocks. Occurring during the crystallization and transport of magma, the outcome of differentiation is in many ways abundantly obvious. Yet, what exactly goes on in magma to produce these end results is still not yet entirely clear. As early as the voyage of the Beagle (1828–1833), Charles Darwin noted in examining Galapagos lavas that magma can evolve chemically simply through the sinking of crystals. Because crystals are always of a different composition than their parent magma, sorting of crystals from melt chemically distills or fractionates magma. The ensuing trail of chemical compositions defines a differentiation sequence that was greatly enunciated experimentally by the brilliant petrologist Norman Bowen (1887–1956).

Earth's outer core is undoubtedly the solar system's largest magma chamber, consisting of mainly molten iron that consists about 30% of Earth's mass. The slow crystallization of the outer core produces crystals that sink to form the inner core, which perhaps from this progressive transfer of momentum then rotates slightly faster than Earth itself. Even buried at the great depth of about 3000 km, the general spherical shape and size ($\sim$2500 km thick) of this magma chamber is well known from seismology. Oddly enough, although the physical nature of this active magma chamber is known in detail, its chemical composition is not nearly as well known. This is quite the reverse of magma chambers associated with volcanism, where chemical composition is known in excruciating detail, but the shapes, sizes, and cooling rates of the active reservoirs are not well known.

2. MAGMA BODIES: TYPES AND SIZES

Magma itself, as also defined by Darwin, is any molten mass of Earth material containing any combination of crystals, liquid, and gas. Postmortem studies of old, now solid, magma reservoirs or chambers (generally called **plutons**) exhumed by erosion are commonplace. This is where the most direct information comes from on the nature of near-surface chambers. The message from these studies is clear. Magma bodies can be almost of any imaginable shape and size, but some types are much more common than others.

2.1. Sills and Dikes

Planar sheets, which when conformable with local rock strata are called **sills** and when cutting strata are called **dikes**, are by far the most common igneous bodies. In general, dikes transport magma and sills store magma. They each can be of almost any size from a few centimeters to a kilometer thick and with aspect ratios from 10 to well over 1000. The Basement Sill in the Dry Valleys region of Antarctica, for example, is generally 300–400 m thick and

can be traced for 100 km. Single dikes a few meters thick and traceable over tens of kilometers are commonplace in many terrains. Their sheetlike form reflects the fact that these are propagating elastic cracks filled with buoyant magma, and the overall shapes of these bodies are spreading disks and pods with very thin edges. Over a distance of 10 km, the Basement Sill, for example, thickens from 1 cm at its edge to 400 m near the center.

2.2. Necks, Plugs, and Stocks

Other once molten magmas were clearly not sheets, but vertical cylindrical intrusions. Called **necks**, **plugs**, pipes, and **stocks**, these were conduits linking deeper reservoirs to volcanoes. Eroded remnants of these feeders, such as Devils Tower in Wyoming and the Hopi Buttes diatremes—necks that apparently penetrated upward almost as locomotives—in northern Arizona, are particularly distinctive. Although these forms could occur at depth, they are probably only common near the surface where the original crustal rock can be punched out or excavated to the surface by the rising magma. The diamond-bearing pipes of South Africa, for example, connect downward over several kilometers into more dikelike features. The diameter of most necks is in the range of 100 m–1.5 km.

2.3. Plutons

Plutons are bulbous masses that commonly develop beneath strings of volcanoes associated with plate subduction. **Batholiths** may contain vast nests of hundreds of plutons intimately crowded against or penetrating one another. The Sierra Nevada range of California and the Andes literally define the notion of batholiths. Yet, the individual plutons within these batholiths are horizontally flattened and, to some degree, sheetlike, although with much smaller aspect ratios (2–20) than sills and dikes. Rather than being injected like sills and dikes, plutons rise diapirically in the fashion of a sluggish thunderhead, eventually losing buoyancy at the leading edge, slowing, and spreading as the lower reaches continue to ascend. The body inflates in place just as do other bodies that begin as necks and locally balloon into a pluton. Although the area of plutons can be enormous (hundreds of squares of kilometers), many plutons are equivalent to spheres of diameters of 2–10 km.

2.4. Volumes

To put these numbers in perspective, the volumes of composite and stratocone volcanoes are well represented by a cone of a height equal to its radius, which is equivalent to a sphere of diameter 1.25 times that of the volcano's height. With volcano relief commonly in the range of 1—2 km, although the largest are 3—4 km, equivalent magma sphere diameters are in the range of 1—9 km. And considering that most stratocones are made principally of pyroclastic materials, which is of high porosity, this range should be reduced to 0.5—5 km. This result should not be construed to mean that volcanoes are simply the outpouring of a pluton of magma, but that volumes of magma of this magnitude (i.e., 0.1—60 km^3) are not unusual to the Earth. Individual volcanoes are the accumulation of often a long series of periodic eruptions, each of which may be a fraction of large bodies arriving in the subsurface that go on to build a pluton. The Dufek (Antarctica) and Bushveld (South Africa) intrusions, each on the order of 10^5 km^3, are the largest known basaltic plutons; both are also strongly sheetlike. Close behind in volume (~35,000 km^3) is the 1.85-Ga-old Sudbury impact feature, which was originally a multi-ringed crater perhaps 250 km wide and several kilometers deep filled with extremely high temperature (~1700 °C) crystal-free melt. Single lava flows, either basaltic or silicic ash flows, reach volumes of 2000—3000 km^3, and these are often part of much larger complexes of either flood basalts or silicic calderas. Some cogenetic igneous complexes, such as oceanic plateaus and rift-related diabase or dolerite terrains, attain volumes of more than 10^6 km^3, but these are clearly often the result of volcanism over millions of years. For even the largest plutons it is not clear how much of the system was simultaneously molten and dynamically connected. The dynamics of molten systems are determined by the behavior of **solidification fronts** (SFs), which link the smallest-scale processes of crystal nucleation and growth to the largest-scale dynamics and also determine the nature of the final rock record.

3. MAGMA CRYSTALLIZATION AND SFs

A central key to appreciating the dynamic evolution of magma is understanding the crystallization of silicate melts. Magma crystallizes over a wide range of temperature (~200 °C) and, unlike aqueous solutions and molten metals, silicate melts generally do not form large dendritic crystals (see Figure 8.1). Moreover, regardless of the size of the magmatic body, crystal sizes commonly range from 1 to 50 mm. This tiny-length scale is crucial to the chemical and physical evolution of the magma as a whole. Silicates in magma generally grow in tiny, essentially symbiotic, clusters. Individual crystals grow by diffusion of chemical components through the melt at similar rates. The effective rates are similar because coupled multicomponent diffusion depends critically on the slowest diffusing component. Because diffusion is so slow (e.g., ~100 years/cm), components rejected by one mineral locally build up and the melt saturates in a new mineral; eventually, enough minerals appear of just the right kinds as dictated by the composition and abundance of the elements on hand to use all the local melt. In essence, these minerals systematically consume the melt.

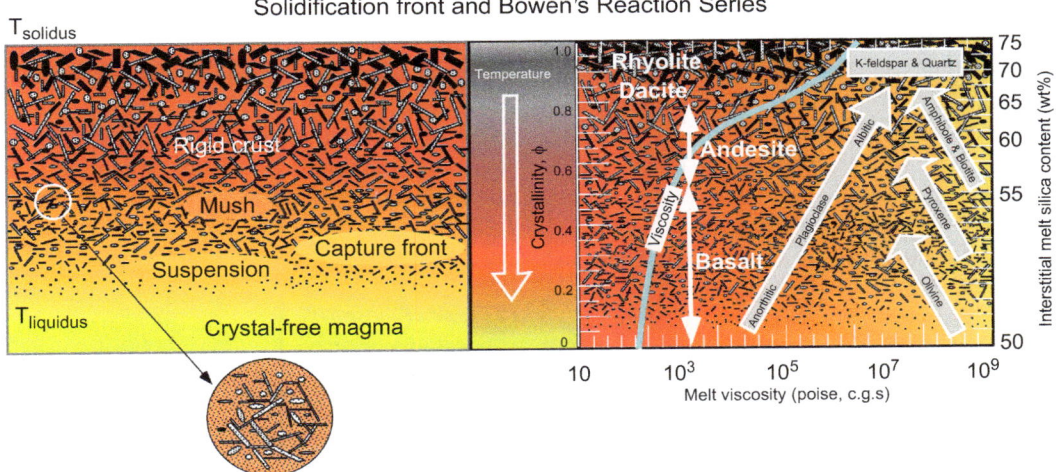

FIGURE 8.1 Left side: Upper solidification front (SF) of a sheetlike magma chamber of basaltic composition. The base (top here) of the front is defined by the solidus and the leading upper edge is the liquidus. The overall thickness of the front depends on the thermal regime and especially the age of the front or how long the cooling event has been running and the temperature of the roof-rock. The inset depicts the framework-like structure of the crystals at about 30% crystals when the crystals begin to attach to one another and anneal into increasing larger crystals. The right side shows how Bowen's Reaction Series operates within a SF. As crystallization proceeds, the crystallinity increases (upward at cooler temperatures) and the interstitial residual melt becomes increasingly enriched in silica (see axis on right-hand side) such that the melt has the compositions as labeled from basalt to rhyolite. The difficulty at high crystallinities in extracting these melts is clear not only due to the strongly decreasing permeabilities but also due to the increasing viscosity of the melts (lower axis).

This is in contrast to dendritic crystallization, where large crystals with large buoyant boundary layers of low-viscosity ($<\sim 1$ P) rejected melt allow continual circulation of melt among the growing crystals. This is in striking contrast to silicate melts of much larger viscosity containing compact clusters of tiny crystals inhibiting melt motion and, instead, favor local equilibration with stagnant melt.

In most common magmas several minerals appear after relatively little cooling ($\sim 25\ ^\circ$C) near the liquidus (see below) and these minerals dominate the chemical evolution of the melt over most of the cooling history. Other minerals appear as needed to consume melt components rejected by the existing minerals. This crystallizing assemblage forms a dynamic, growing feature of all magmas: the SF.

3.1. Solidification Fronts

The spatial relation between the liquidus (all melt) and solidus (all solid) defines a SF about any magmatic body (see Figure 8.1). Across this front, magma goes from being fully liquid at the leading or innermost and hottest edge to fully solid at the outer or trailing and coolest edge. The bundle of isotherms between these two extremes envelops the body and collectively propagates inward, meeting in the interior at the point of final solidification. Although thin and sharp just after emplacement, with time the SF progressively thickens in proportion to the square root of time, as is well known from the drilling of Hawaiian lava lakes.

Within any SF, at least four dynamic and rheological divisions can be identified (see Figure 8.1).

1. **Suspension zone**: Beginning at the leading edge or liquidus, the volumetric percentage of crystals, N, equals zero; nucleation commences and crystals grow to an overall concentration of about $N = 25\%$ at the trailing edge. The small, sparse crystals can move freely relative to one another and perhaps even escape downward from the front itself. But because they are so small, motion is limited relative to the overall rate of advance of the SF itself.
2. **Capture front**: The boundary marking the trailing edge of the suspension zone ($N = 25\%$), where the bulk viscosity has increased by a factor of 10 over that at the liquidus and behind which the increasing crystal concentrations inhibit settling of crystals, thus prohibiting escape from the advancing SF.
3. **Mush zone**: Bounded by the capture front ($N = 25\%$) at the leading edge and the point of **critical crystallinity** at the maximum packing of the solids at the trailing edge, where $N = 50-55\%$; individual crystal migration is now highly unlikely due to touching and mutual hindrance.
4. **Rigid crust**: Outward of and adjacent to the mush zone and bounded by the solidus ($N = 100\%$) and the point of critical crystallinity ($N = 50-55\%$). This is the drillable portion of the SF commonly referred to as the "crust" in studies of Hawaiian lava lakes.

Two aspects of this subdivided SF are important to emphasize. First, these are generalized definitions and are closely tied to the specific phase equilibria of the magma under study; the defining crystallinities will naturally vary somewhat with magma type. In plagioclase-rich basalts, for example, an interlocking network of significant strength may form at crystallinities as low as about 30%. Second, the entire SF, which is always present, is a dynamic feature of solidification; it propagates inward at an ever-decreasing rate in response to cooling and thickens with time. All physical and chemical processes occurring as a result of crystallization do so within the SF; the rates and duration of these processes are limited by the rate of motion of the front itself. Thus, the entire evolution of the magma, whether being transported or held in a chamber, is reflected by the dynamic behavior of the SF, even when the chamber experiences additional injections.

In an ideal magma emplaced free of crystals, nucleation and growth begin at the inwardly advancing liquidus; the viscosity there is essentially that of the crystal-free magma and only when the crystallinity increases to about 25% does the viscosity increase by a factor of about 10 at the capture front. At the other extreme at the trailing edge of the SF, near the solidus, the viscosity is enormous, as the rock is nearly solid. Inward from this point crystallinity decreases, but the crystals form a strong interlocking network and viscosity remains enormous until the crystallinity decreases to about 50-55%. Here the viscosity decreases dramatically as the magma makes the transition from a partially molten solid to a mushy liquid. This is the point of **critical crystallinity** marking the region of maximum packing of the solids, where at all higher crystallinities the crystals form an interlocking network of great strength. Moreover, beyond this point, at all higher crystallinities, the magma is a dilatant solid, where under shear it expands and chokes any conduit. Thus it is uneruptible. High-alumina basalts in some island arcs erupt carrying 40-50% large crystals, and because they are so near to **dilatancy**, they can only erupt explosively. At smaller crystallinities, the network of crystals is not fully interlocking and the material behaves as a crystal-laden fluid possessing a large effective viscosity. The crystals are, nevertheless, unable to move freely relative to one another and the region behaves as a **mush**. This mush-like behavior continues inward with decreasing crystallinity (i.e., increasing temperature) until reaching the **suspension zone**, which connects to the crystal-free deep interior of the body.

3.2. Bowen's Reaction Series and SFs

As mentioned earlier, Norman Bowen demonstrated in detail experimentally many of the fundamental phase relations relating the chemical evolution of silicate magmas during crystallization. He showed how with progressive cooling and crystallization the composition of the solids and residual melt evolves to generally become progressively more silicic or granitic. Beginning with a basaltic magma, after prolonged cooling and separation or **fractionation** of the crystals and melt, continental-like compositions could be produced. And he was able to generalize this process into a sequence of crystallization that has become known as Bowen's Reaction Series. Not only do new and sometimes different crystals nucleate and grow as needed to use up the available melt, the so-called **discontinuous series**, but in the **continuous series** existing crystals progressively change their composition through ongoing diffusive exchange with the local residual melt to allow prolonged chemical evolution of the melt. Although in a laboratory experiment this happens in a small crucible where the temperature is uniform throughout the charge, which represents a specific, unchanging, spatial position within an actual magma, in the Earth the temperature field always varies systematically with space and time, which is reflected in the concept of SFs. In nature, therefore, Bowen's Reaction Series exists within SFs; in fact, SFs are the spatial representation of Bowen's series, which is depicted also in Figure 8.1. And herein emerges the dilemma in the process of fractional crystallization: Early in the crystallization sequence, where crystals could potentially easily move around and be separated from the melt, the net effect on melt composition is tiny, and where the melt compositions are greatly evolved, far back in the SF, the crystals are all welded to each other and separation from melt, by the usual means of sinking, is impossible. This enigma lies at the center of understanding the behavior of magma and magma chambers, which is investigated henceforth below.

4. PHENOCRYSTS AND SFs

4.1. Phenocryst-Bearing Magma

Magmas can be put into two convenient categories: those that carry phenocrysts upon emplacement, and those that do not. Phenocrysts, being relatively large cognate crystals, are easily recognized in lavas, but much less so in plutons where once embedded with other large crystals they can be mistaken for crystals grown after emplacement. Unusually large crystals in the chilled margins of the body are one indication of initial phenocryst presence. Another is the internal distribution of phenocrysts. Because SFs initially grow exceedingly fast upon emplacement, they capture all

phenocrysts in the magma near the contacts. But as the SFs move inward they advance ever more slowly and, at some point, all but the smallest phenocrysts can escape capture and settle to the floor, which is formed by a commensurate upward-moving basal SF. The net result is a loss of phenocrysts in the upper half of the body and an accumulation of phenocrysts in the lower half. The final distribution of phenocrysts from top to bottom is S-shaped (see Figure 8.2). The efficiency of this process depends on phenocryst size and density and also on magma viscosity, however, many basaltic bodies show this classic distribution; Shonkin Sag laccolith in Montana and Kilauea Iki lava lake are good examples (Figure 8.2).

A cumulate layer of phenocrysts within a sheetlike body is also an indication of phenocryst emplacement. Many large diabase sills have orthopyroxene or, more rarely, olivine cumulate layers in the central or lower half of the sill, but these phenocrysts are not found elsewhere in the sill. Phenocrysts are most commonly cognate and thus represent an earlier (i.e., prior to emplacement) crystallization event. This distribution reflects not only that the emplacement of the phenocryst-rich magma was late in the injection sequence, but also that the magma was **flow differentiated**. That is, the phenocrysts were segregated in the ascending magma into a tongue-like central concentration trailing the leading, phenocryst-free magma (see Figure 8.3). This is the **Simkin Sequence**. The leading magma, which forms the margins of the body, therefore, is fractionated, and the central tongue rock is more primitive. Neither rock is a true measure of the original magma composition. Bodies with more evolved compositions at the margins are thus due to flow differentiation, and the spectrum of compositions is not something produced in situ but is, instead, a reflection of the dynamics of transport. Although seen in many sills, this process is particularly well exemplified in the Basement Sill, Antarctica where an immense slurry or tongue of large orthopyroxene crystals (Opx Tongue) is present throughout the sill reflecting in its thickness the history of dynamics of emplacement (Figure 8.3). (A photomicrograph of a thin section of a typical assemblage of these large orthopyroxenes is also given as one of the color plates at the end of this section.)

4.2. Phenocryst-Free Solidification

Magma emplaced free of phenocrysts is, of course, also dominated by SFs, but now all crystals are formed after emplacement; nevertheless, few magmas are probably ever free of nuclei and very small crystals. As the SFs move inward, crystal ensembles are nucleated and continue growing until the solidus isotherm arrives when all melt has been consumed. So crystallization can be viewed as the passage of a packet of isotherms through a particular local volume of melt. The rate of cooling determines the rate of

FIGURE 8.2 The settling and capture of phenocrysts during solidification of a sheet of magma. The intensity of the colors depicts the prevailing temperature at each location. The overall temperature field is shown at the upper left, and the upper right diagram depicts the settling and escape of large crystals from the upper solidification front (SF) and accumulating in the lower, upward moving, SF. The three lower diagrams show examples of the final S-shaped profile distribution of crystals after complete solidification of three sheet-like bodies.

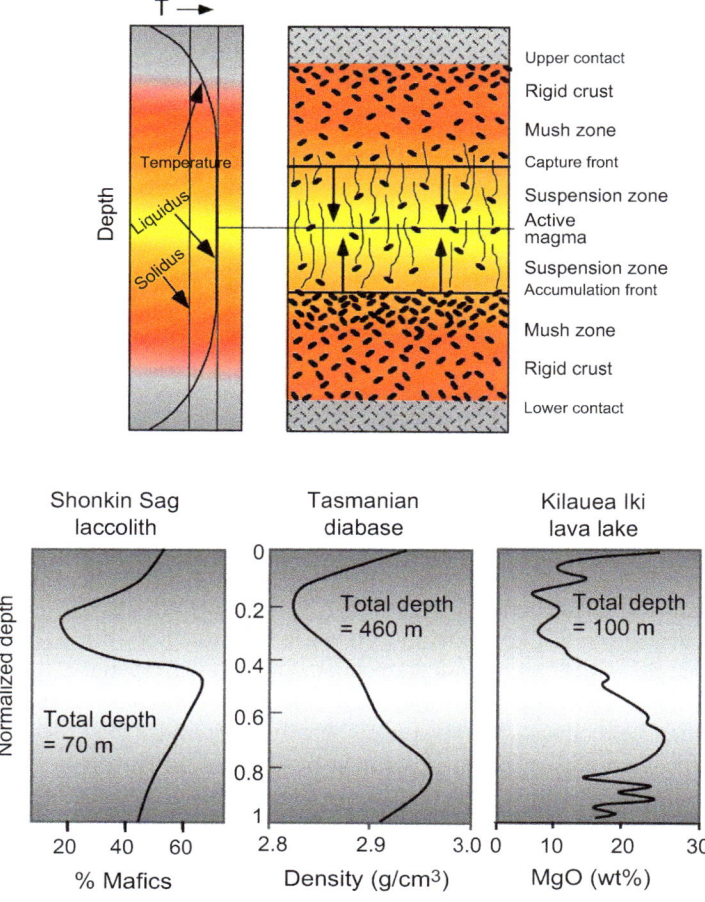

SF advancement and the melt eventually becomes solid regardless of the time given for solidification. The melt does this through a combination of crystal nucleation and growth. Because growth is diffusion controlled, it is slow and relatively insensitive to changes in cooling rate. Nucleation, on the other hand, is highly sensitive to cooling rate. A balance between growth and nucleation thus sustains solidification. Fine-grained chilled margins and coarse interiors reflect this balance.

Differentiation by crystal fractionation is difficult in phenocryst-free magmas because the crystals are confined to the SFs from which they must escape to fractionate. Once the magma reaches a crystallinity of about 25% (vol.) it is within the capture front and escape of individual crystals is unlikely. At lower crystallinities, in the suspension zone at the leading edge of the upper SF, plumes of slightly cooled magma containing small crystals may drop from the SF, traverse the core of the body containing the hottest magma, and, if the intrusion is not too thick, reach the leading edge of the lower, upward-advancing SF. But since there is no net separation of crystals from the initial melt, chemical differentiation is negligible. The plumes now enter a position

in the lower SF commensurate with the position vacated in the upper SF with no net chemical effect. Moreover, if the body is sufficiently thick, the plumes will become heated during descent, losing their negative buoyancy, destroying any resident crystals, and become incorporated into the interior melt before reaching the floor. This is why phenocryst-free bodies, especially sheetlike intrusions, are frequently uniform in composition from top to bottom.

In striking contrast to the Basement Sill already mentioned as containing a major slurry of entrained phenocrysts (Opx Tongue), the 350-m-thick Peneplain Sill of the Dry Valleys region of Antarctica is a good example of a phenocryst-free intrusion (see Figure 8.4). Except for the horizon of silicic segregations near the roof, which will presently be considered, the composition is essentially uniform from top to bottom. Although it might be imagined that this could only occur in relatively small bodies like sills, on a grander scale the Sudbury Igneous Complex, which is about 3 km thick, shows the same feature. Sudbury is an exceptional example in this respect because it was produced in a matter of minutes by meteorite impact. Because the initial melt was heated to at least 1700 °C, it

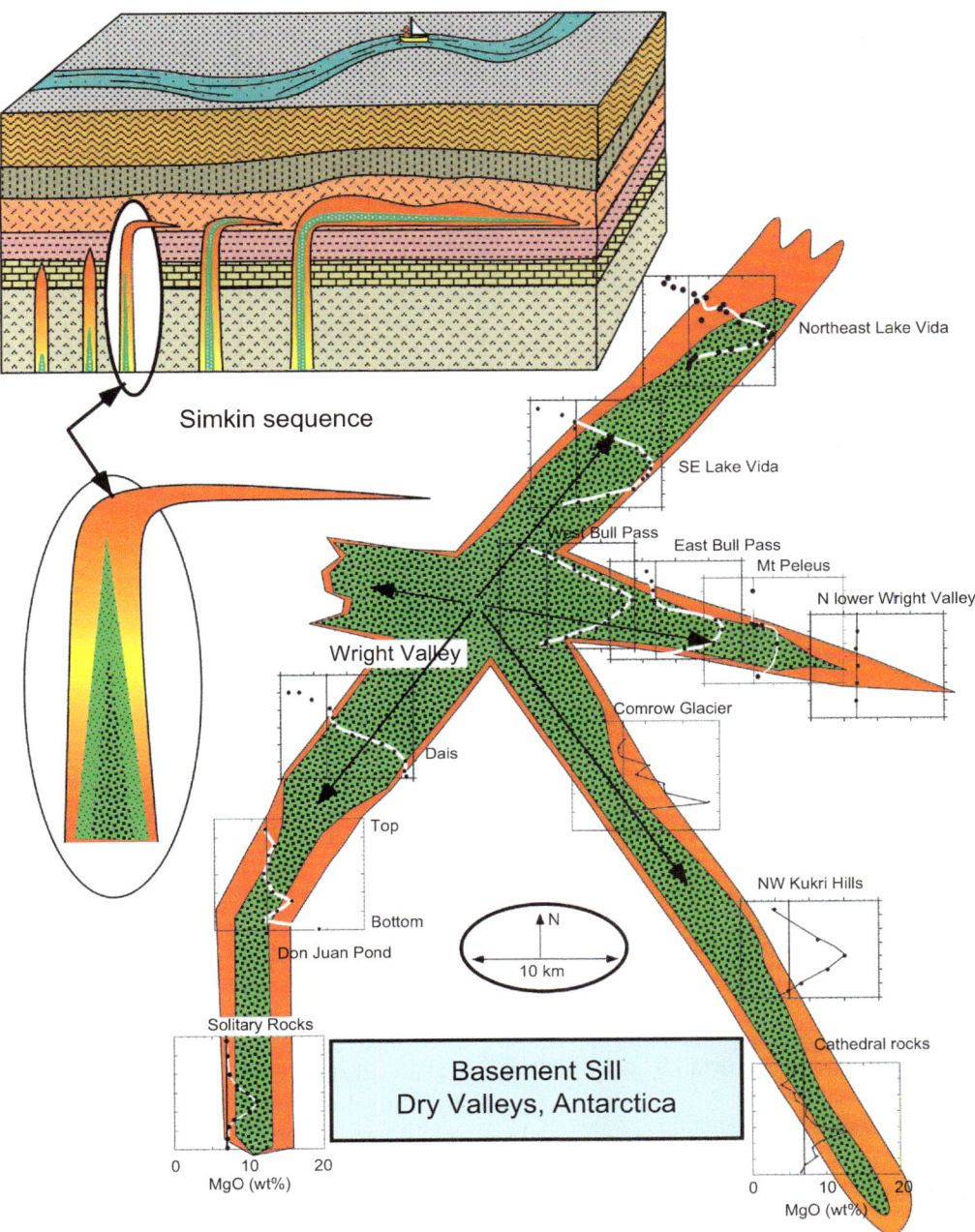

FIGURE 8.3 Upper left: Schematic sequential depiction of the Simkin Sequence of flow differentiation or sorting of crystals according to size and density during vertical transport and emplacement of crystal-laden magma as a sill. The adjoining octopus-shaped diagram shows the actual distribution of orthopyroxene phenocrysts in the Basement Sill, Antarctica. This feature, the so-called Opx Tongue, thins outward in all directions from the main filling point in central Bull Pass.

was essentially initially free of all crystals. In more detail, Sudbury consists of a ∼2 km layer of granitic rock overlying ∼1 km of dioritic (actually, norite) rock. The granitic layer is much too thick to have been produced from differentiation of the diorite and is clearly the result of impact melting of granitic continental crust. The diorite may simply be melted lower crustal rock and doleritic sills; the

transient cavity reached depths of 25–30 km in 1–2 min. But what is interesting in the present context is that both layers, granitic and basaltic, are essentially uniform in composition. From these and similar bodies, it is becoming clear that differentiation is critically dependent on the presence of crystals in the initial magma. How, then, is the wide range of diversity of igneous rocks produced?

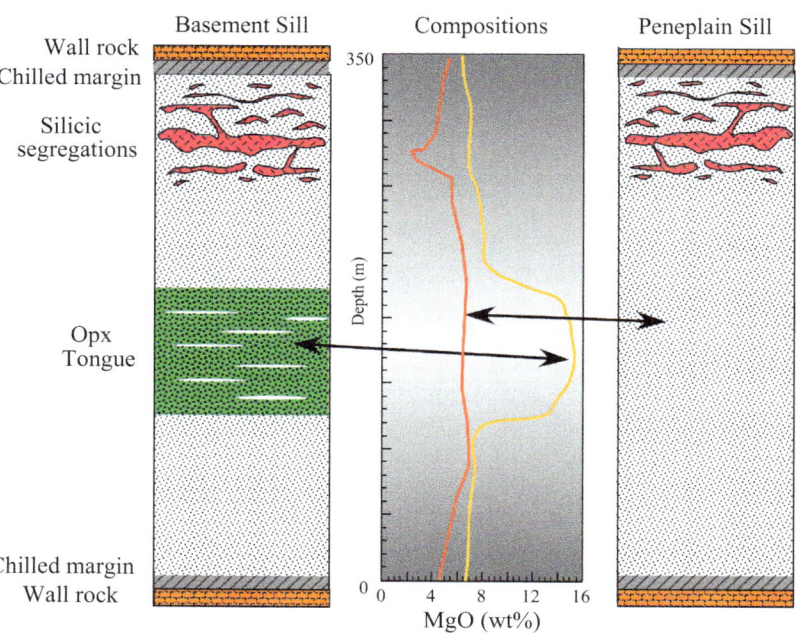

5. DIFFERENTIATION PROCESSES IN CHAMBERS

There is no doubt whatsoever that chemical differentiation in igneous systems results from the physical separation of crystals and melt. Crystal compositions are always distinct from any natural melt, and progressive crystallization produces residual melts that match the observed compositions of naturally occurring rocks. Thus, it would seem simple to produce compositional diversity by removing crystals (by sinking or floating) from the melt. As mentioned already, this process of differentiation by crystal fractionation was elucidated in great detail by N.L. Bowen in the first half of the last century.

Numerous mass balance calculations involving reasonable solid phases and derived melts show the great feasibility of crystal fractionation as a means of efficient chemical differentiation. But as reasonable as this process is chemically, the details of the physical process of extensive crystal removal have remained obscure. Many massive basaltic magmatic systems, like Hawaii, produce no silicic magmas whereas other small ($\sim$1 km^3) island arc volcanoes erupt lavas of a wide range of compositions. With the recognition of the central role of SFs in crystallization, the controls on the processes of separating crystals and melt have become clear. Refined or fractionated melt is buried deep within the SF, and this makes the separation of crystals from melt difficult. Over the first 50% of crystallization, there is relatively little chemical evolution of the melt. As an extreme example, if 50 vol% olivine is fractionated from basalt, the silica content increases by only 5 wt%. It is

over the last 50% of crystallization that the melt becomes highly fractionated. This overall sequence describes Bowen's classic differentiation sequence (see Figure 8.2), but it is contained within the SF and does not normally occur deep in the hottest part of the chamber where there are no crystals and thus herein lies the enigma mentioned briefly earlier: Where the melt and crystals are easiest to separate (i.e., at the lowest crystal contents), the melt is hardly fractionated. And where chemical fractionation is strongest, the melt is held in a rigid network and virtually impossible to separate from the crystals. Unless this interstitial melt can be collected into an eruptible mass, the system will show little diversity. The SF must somehow be physically disrupted in order to be purged of the interstitial melt. There are three fundamental ways to do this.

5.1. Solidification Front Instability and Silicic Segregations

Because SFs are relatively cool and are partially solid, they are denser than the underlying uncrystallized melt. Once a SF becomes sufficiently thick, it becomes gravitationally unstable. Anywhere the internal strength is locally exceeded by the weight of the leading half of the front, gashes or tears develop. As the tears open, local interstitial melt is drawn into them, producing silicic pods and lenses. Interdigitating lenses of silicic rock, 2–3 m thick and 40–50 m long, are common in the upper parts of tholeiitic sills regardless of the initial phenocryst content. Plagiogranite segregations of the oceanic crust are undoubtedly products of solidification front instability (SFI). The silica contents of silicic segregations are in the range of 60–70 wt% and the bulk composition of the sill is near 50%. The more siliceous the sill, the more siliceous are the segregations. Mass balance calculations show that the segregations are produced from local melt and that the tears open when the local crystallinity is about 65%. This is spatially just behind the point of critical crystallinity when the mush becomes locked into a rigid, but still somewhat weak, network.

If the intrusion is thick enough, the instability may mature to the extent that a large portion of the front detaches, sinking into the core of the chamber. This frees the silicic segregations to collect in the roof area into eruptible masses of magma; repeated buildup and failure of the SF may thus produce sizable volumes of silicic magma. Overall, however, a characteristic of suites produced by SFI is that they are distinctly bimodal. There are both basaltic and silicic compositions, but little of intermediate composition. This is a characteristic of many volcanic and plutonic suites, especially in oceanic islands such as Rapa Nui (Easter Island), Galapagos, and Iceland, which will be considered shortly.

5.2. Mush Flow Fractionation

When the center of a magmatic body reaches an advanced stage of crystallization, it is ripe for differentiation by flow fractionation. This occurs when new magma invades the system, forcing its way through the established crystalline meshwork, pushing out the fractionated interstitial melt. The original body is thus purged of fractionated melt by replacing it with new magma. Not only is fractionated melt freed to erupt, but also the new magma is contaminated by older crystals of the meshwork, which are thermally and chemically out of equilibrium. The SF meshwork will be partly dissolved, perhaps even disrupted, and the escaping fractionated melt may also carry small telltale pieces of the original SF in the form of clots or clusters of crystals.

The ideal location for this process is in reactivated fissures associated with long-standing flank eruptive systems of large active volcanoes, like the east rift of Kilauea. During periods of repose, magma remaining in the dominant fissures can solidify to the point of critical crystallinity (~55%). New activity displaces the residual melt, yielding an eruption of fractionated melt on the flanks of the volcano. This is seen at Hawaii (e.g., the 1955 eruption), where only the fractionated lava issues from fissures on the far flanks; and this melt carries small clots of the earlier SF. Deep in these same fissures thick accumulations of phenocrysts form (at Hawaii these are olivine cumulates) and may also be reentrained and similarly purged by the invading magma, yielding unusually primitive (i.e., olivine-rich) basalt compositions.

The amount of melt available for purging is proportional to the size of the existing SFs. Because magma will always follow the path of least resistance, it will not normally invade an existing SF unless that is the only course available. Moreover, as the residual melt in the SF becomes enriched in silica, its viscosity rises dramatically, making displacement increasingly difficult. It is thus unlikely that highly silicic melt could be freed in this fashion unless enough volatiles built up in the SF to offset the effect of silica on viscosity. In extensive, geometrically diverse, and long-lived magmatic systems, such as mush columns (see below), this form of melt fractionation is most effective.

5.3. Wholesale Reprocessing

The formation of silicic segregations is a normal consequence of solidification. Some alkalic basalts become siliceous only at the latest stages of crystallization when the SF is too strong to be torn, and some bodies are too small to undergo SFI. But in general most basaltic bodies will produce silicic segregations, and once formed they remain as silicic noise in otherwise basaltic systems. Even if the entire body were remelted, these silicic pods, being so much more viscous than the parent basalt, are physically (but not chemically) immiscible in basaltic magmas. Stirring is ineffective in reblending these silicic blobs. Wholesale remelting of basaltic crust frees swarms of segregations to collect into eruptible volumes of rhyolitic magmas. This is very likely the means of producing the strongly bimodal association of rhyolite and basalt in Iceland.

Unlike almost every other oceanic volcanic center, about 15% of the surface rocks of Iceland are rhyolitic. There is a preponderance of basalts and a paucity of lavas of intermediate composition. The suite is strikingly bimodal. Iceland is a prime location for reprocessing of the basaltic crust because it straddles the Mid-Atlantic Ridge. Ongoing spreading of the oceanic crust promotes periodic jumps of the ridge axis, refocusing from time to time magmatism at new locations in older crust. Prolonged use of new fissure swarms steadily heats the older, still warm crust, forming outward propagating—progressive instead of regressive—solidification (i.e., melting) fronts. These fronts systematically break down and reprocess the crust, freeing silicic segregations to collect as rhyolitic magma (see Figure 8.5). The ensuing rhyolites, such as those at the large Torfajokull silicic center, are nearly crystal-free, but carry ubiquitous, diagnostic crystalline debris from the original SF, including pieces of granophyric rock in various stages of melting. The massive 1875 pumice eruption of Askja volcano in northern Iceland, for example, is laced with chunks of silicic segregations quenched in all degrees of fusion. The oxygen isotopes of the rhyolites show the effects of extensive prior hydrothermal alteration when the

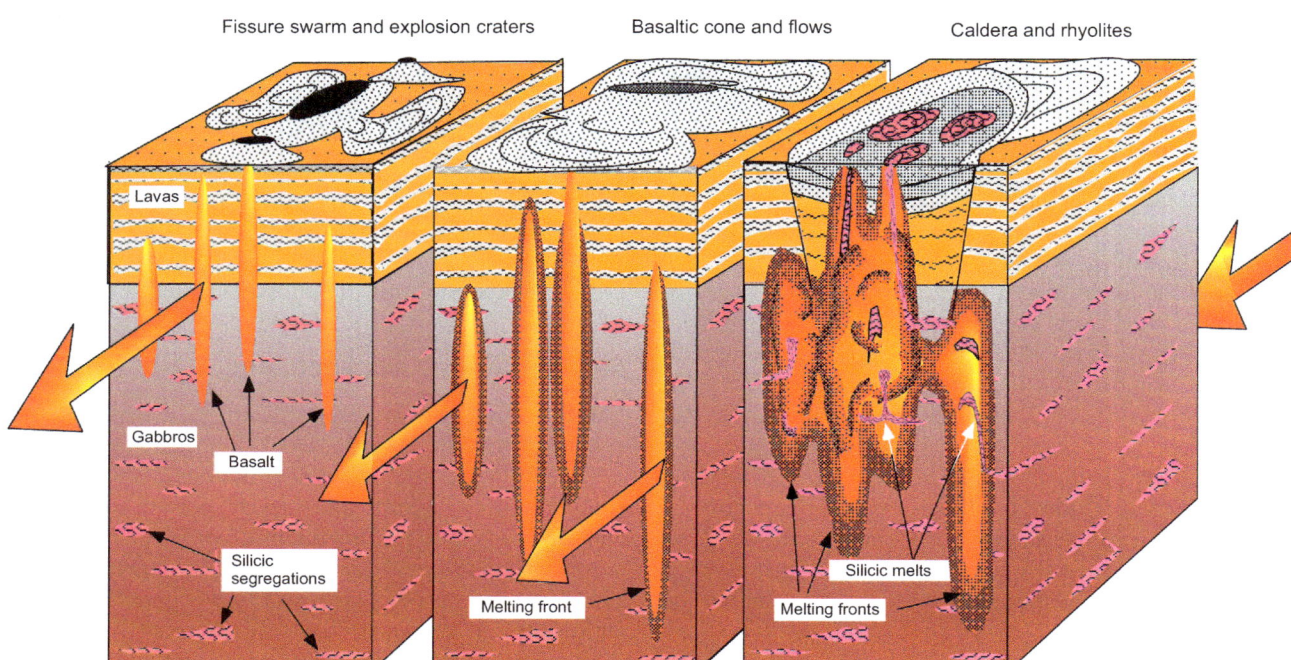

FIGURE 8.5   Generation of rhyolitic magma through reprocessing of older heterogeneous basaltic crust. Left: As fissures propagate into older crust they heat the crust, which through prolonged use leads to extensive melting in outward propagating solidification fronts (center) and the eventual breakdown of large areas of the crust (right). This process is enhanced by the spatial variations in melting characteristics of the country rock due to variations in composition and water content. As regions of silicic material are freed during melting it collects and coalesces into eruptible masses of rhyolitic magma.

silicic segregations were a part of the crust. This is also even true for comagmatic basalts. Thus the Iceland crust undergoes wholesale differentiation through remelting and reprocessing.

5.4. Compaction

SFs growing upward on the floors of magma chambers can become so thick or distended that their weight is enough to deform the crystalline matrix and squeeze out the interstitial melt. This is a slow process that can only occur when the growth of the front itself is slow, which occurs either at advanced stages of crystallization in large bodies or deeper in the crust where the wall/floor rock is suitably hot, making inward propagation of the solidus sluggish. Depending on the melt density, the evicted melt may accumulate within the upper reaches of the front or continue upward through the core of the body and lodge within the upper SF. In thick sheetlike chambers the changeover from solidification control to compaction may be sudden enough to produce a distinct compositional horizon higher in the system, and these segregations may also be collected during reprocessing. This process is

particularly effective when dense loads of crystals are literally dumped on the chamber floor as crystal-laden slurries replenish the chamber.

5.5. Chemical Fractionation Effects

The chemical aftermath of SF fractionation is virtually indistinguishable from massive crystal fractionation however it is enacted. All processes involving separation of an ensemble of slowly grown indigenous solids produce essentially indistinguishable chemical records. There are, strictly speaking, small systematic chemical differences between extracting a melt from an equilibrium assemblage of solids fairly late in the crystallization sequence and separating crystals, crystal by crystal, from the melt as they appear. But there is no physical process that can systematically cleanse a melt of tiny crystals over a wide crystallization interval. Moreover, it is doubtful whether the resulting rocks contain enough discriminatory evidence to reveal the details of the fractionation process. Nevertheless, certain constraints due to mass balance considerations are generally consistent with SF processes. It is the physical evidence coupled with the chemical record that reveals process. This also is evident in the deciphering of the dynamics of large magmatic systems.

6. CHAMBER PROCESSES REVEALED IN THE PLUTON ROCK RECORD

Exotically layered and compositionally varied plutons would seem to hold the key to understanding magmatic processes. The stratigraphic records left on the floors of the classic magma chambers of Skaergaard, Stillwater, Rum, Bushveld, the Great Dike, and Dufek, among others, have been carefully studied with the hope of discovering the fundamental dynamics, chemical and physical, of magma chambers. All other bodies would then simply be subsets of these more dynamically rich systems. An easy reading of this rock record has proven elusive. It has become clear, however, that the chemistry of the rock record is not enough, in and of itself, to reconstruct the solidification history.

Unlayered plutons have been viewed as containing definitive evidence of processes distinct from those of layered plutons. Moreover, lavas and ash flows, lacking any information on pre-eruption stratigraphy, have been examined strictly in terms of chemistry. There thus exists three separate sets of principles garnered from the study of layered, unlayered, and erupted bodies, and they fail to produce a consistent dynamic depiction of magma chambers.

6.1. Layered Plutons and Initial Conditions

The historical approach to interpreting layered plutons has been to assume an instantaneous emplacement of a more or less homogeneous magma at or near its liquidus temperature. All subsequent layering and differentiation come from processes occurring in the chamber. Progressive crystallization throughout the magma either rains crystals to the floor as in a snowfall, producing more or less homogeneous gabbroic cumulates, which evolve upward into more fractionated compositions, or makes sedimentary layers on the floor through periodic overturns of the crystal-laden melt. Some crystallization at the roof is unavoidable and an upper border group also develops, but it is deemed to be much thinner than the floor sequence because most of the new crystals fall to the floor. Where the floor and roof sequences meet, a sandwich horizon occurs, and this is where the most evolved rocks should reside. So the expected sequence from these initial conditions (instantaneous and crystal-free) is a thick basal sequence of cumulate crystals, layered or unlayered, a thin upper border sequence, and a sandwich horizon.

These irrefutable facts came from the earliest field observations at Shonkin Sag laccolith by L. V. Pirsson (1860−1919) in 1905. This thin (~70 m) disklike (~ 3 km) intrusion in Montana has all the fundamental features of the large classic intrusions, namely, a thick basal sequence of large crystals, a relatively thin upper margin, and in between a sandwich horizon of differentiated melt. From the very beginning, these have been interpreted to reflect strong cooling from the roof, promoting strong crystallization and convective circulation throughout the interior and major deposition of crystals on the floor. The inversion of this rock record would seem to be straightforward. But, oddly enough, in most intrusions showing these characteristics, other than the usual sporadic silicic or granophyric segregations, significant volumes of silicic or otherwise highly differentiated rock are rarely found at the inferred sandwich horizon. They could have been erupted away, but perhaps there were none there in the first place.

Even more perplexing has been the realization that thick floor sequences are absent in phenocryst-free diabase sills, regardless of thickness and lateral extent. The style of cooling and the expected position and nature of the true sandwich horizon can be precisely determined in many sills and lava lakes from the mismatch in roof and floor joints, which are shrinkage or cooling cracks that track the inward propagation of the upper and lower SFs. They generally meet approximately in the middle, reflecting more or less equal cooling from top and bottom. Thermal modeling, both with and without convection, bears this out; cooling is not significantly more intense from the roof. Thus, the absence of basal cumulates and of differentiated sandwich horizons at the point of thermal symmetry points to a fundamental problem with the historical concept of magma solidification in chambers. The field evidence has another interpretation.

Thick accumulations of crystals on the floors of bodies such as Shonkin Sag are from the deposition of crystals (phenocrysts) carried in by the initial magma. The key observation is that the crystals are much too large and numerous to have been generated in place. They are essentially snowfalls of crystals upon emplacement; this is sudden and massive cumulate formation. Phenocryst-bearing magmas, which are slurries, injected into chambers of any sort—sills, dikes, plutons—form cumulates; it is unavoidable as long as the body is not so small as to cool instantaneously and prohibit the process.

Initially, crystal-free magma, on the other hand, such as the Sudbury igneous complex, crystallizes to almost featureless rock. The instantaneous emplacement of such magma of any significant size is a rare occurrence. Yet, it does happen. Individual eruptions of continental flood basalt reaching volumes of 1000−2000 km^3 often contain very few phenocrysts (2−5%) and are chemically homogeneous. And 100-m-thick ponds of these basalts solidify into featureless rock with some silicic segregations. Were these magmas to cool, instead, in sheetlike chambers at depth, as some must, they would be homogeneous gabbros with signs of SFI.

The central problem in interpreting plutonic rock sequences is that the initial conditions of the system are unknown, and these conditions are absolutely vital to know. Was the body formed of one massive injection? Or did it form by many periodic emplacements, as in long-term volcanic eruptions at Hawaii? What was the phenocryst content of these inputs? Although general impressions are sometimes attainable from nearby dikes and sills, more generally the answers to these and many similar questions have been inaccessible. Some answers may come from layered sequences.

6.2. Layering

Phenocryst-charged magma cannot solidify without crystal sorting into layers, however cryptic or well formed; it is unavoidable. The separation of solids from fluid slurries is a vast subject rich in process and dynamics. Almost any conceivable final sequence of solids of arbitrary shape, size, and density is attainable through reasonable sedimentation scenarios. Moreover, the ensemble of possible scenarios increases greatly if the shape of the magmatic container is also varied. In a relatively narrow chamber with outward-slanting walls, for example, the well-known Boycott effect controls sedimentation by greatly enhancing the rate of deposition. Crystals near the walls settle quickly to the wall and avalanche **en masse**, sorting crystals according to size, shape, and density, which dictate angles of repose, in a delicate and rapid fashion (see Figure 8.6). Moreover, repeated injections of varying intensity supply inputs to the chamber of varying phenocryst concentration and size. The net effect of crystal sorting, chamber form, and input history is to establish a vast dynamic regime that can give rise to intricate layered sequences.

Knowing the dynamic regimes of layering it would seem straightforward to read the history of the injection process, but annealing clouds this inversion, making it highly nonunique. That is, once any mechanical segregation of crystals takes place, local thermodynamic equilibrium controls further maturation of the texture. Certain minerals may be unstable and dissolve to refine cryptic layering into nearly monomineralic layers. This form of annealing or textural ripening is known from metallurgy and ceramics and has been carefully enunciated for magmatic systems by A. Boudreau. Annealing can only be avoided by rapid cooling, which itself is possible only in small bodies.

Large bodies, then, commonly show good layering for two principal reasons. First, they were formed by multiple injections; hence, the probability is high that crystal-laden magma (slurries) was present. Second, unusually long cooling times provide maximum opportunity for textural ripening. In addition, large bodies may form extensive mush regions, extending at some point throughout the entire body. This allows interstitial melt to flow and supply nutrients to layer-forming crystals and to destabilize other crystals. This late-stage flow through the porous mush blurs the initial chemical spatial state of the system, making inversion to an active magma chamber using the final rock record nonlinear and difficult.

6.3. Thermal Convection

If magma did not crystallize and form SFs, the convective regime in chambers could be easily predicted from the shape, thickness, temperature, and viscosity of the magma. From these and a few other physical properties, the magnitude of a pure number (i.e., free of dimensions) called the **Rayleigh number** (Ra) can be determined. Everything dynamic about noncrystallizing fluid can be learned from Ra; the temperature distribution, rate of convection, and rate of cooling are each set by Ra. The enormous literature on thermal convection in fluids of almost every conceivable kind tells this in great detail, but it tells relatively little about fluids undergoing solidification. Although the general concepts still hold, the details of the

FIGURE 8.6 The enhanced sedimentation leading to layering due to inclined walls or solidification fronts. The particles settle to the nearby wall and accumulate enough to initiate an avalanche, which gives rise to a general circulation in the slurry suspension and provides a steady supply of crystals to the wall area. Because small and large particles support different angles of repose, avalanching promotes layering as depicted in the inset figures.

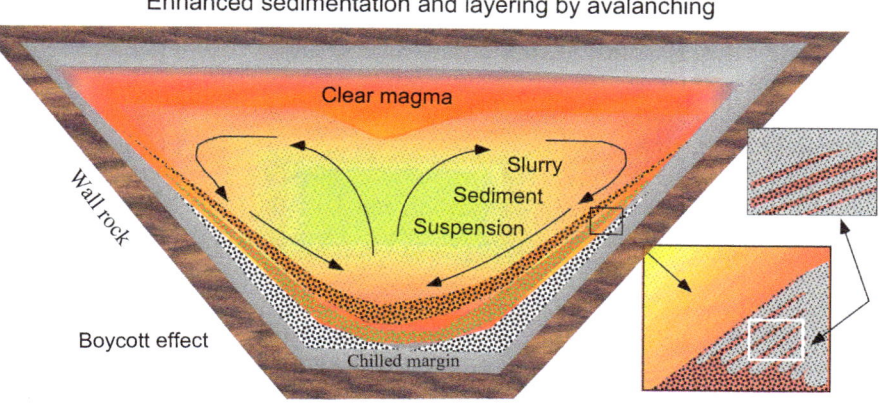

Enhanced sedimentation and layering by avalanching

dynamic regime cannot be safely predicted from the conventional consideration of Ra. It is particularly unclear, for example, exactly how to estimate the temperature difference driving convection.

The main reason for this is that the cool, denser magma along the roof, which would normally form descending plumes to be replaced by ascending hotter, lighter magma going on to form a systematic field of large cells, is mainly locked in SFs. This minimizes the amount of cool fluid available to drive convection. The SFs also propagate inward with a velocity that tends to override incipient plumes forming at the leading edge of the SF, trapping them forever within the SF. And the incipient onset of nucleation throughout the interior may produce enough heat from latent heat of crystallization to offset the heat potentially transferable by convection. Although the detailed physics is only now emerging, it is clear from a growing set of experiments that solidifying fluids do not convect much, if at all, once they are below their liquidus temperature. If the fluid is superheated (i.e., above the liquidus and point of initial nucleation), however, vigorous convection occurs in the conventional sense. The absence of erupting superliquidus magma may well reflect the efficiency of convection in dispatching superheat. The absence of superheat indicates a diminished role for thermal convection as a dominant process in chambers, but this does not mean that chambered magma is motionless. Along the walls in tall chambers, for example, crystal-laden flows may cascade down and sweep across the floor, possibly forming troughs in the basal front and at the same time depositing crystals.

6.4. Convection in Sedimentation

Swarms of crystals in freshly injected magma slurries can never be uniformly distributed and so the uneven density along horizontal (i.e., equipotential) surfaces will spawn convection. The motion can take on many forms, but descending plumes are the most common. If the crystals are small ($< \sim 0.1$ mm), have a low-density contrast, and their concentration is low ($< \sim 5\%$), the crystals form a dusty suspension that may slowly circulate in cells as the dust is deposited on the floor. The flow will be relatively gentle, as it will also be for highly concentrated ($> \sim 40\%$) mushes, which are too viscous and sluggish to move vigorously. In between these extremes there is a wide range of dynamic behavior where crystal sedimentation can strongly stir the magma. The more basic the magma, the lower its viscosity and the more pronounced the effects of sedimentation. The aftermath is a sequence of layering clearly due to convection, but not thermal convection. Considering the large size of magmatic systems, the physics of crystallization, and the possible shapes of chambers, some motion, however small, is probably always present.

7. CHAMBER EVOLUTION AND PLUTON FORMATION

The age-old dilemma in interpreting plutonic rocks is whether to read the sequences as due to purely in situ processes or as a result of a series of injection events. This issue, which is historically almost a philosophical matter in petrology, still plagues interpretations; for it is always easy to interpret a complex rock sequence as due to a specific string of injection events, which moves the problem to another venue. Given the geochemical tools and physical insight now available on crystallizing systems, however, this question can be approached with far more objectivity and certainty than ever before. Thick continuous sequences of cumulates on the floor reflect a chamber formed of phenocryst-laden magma (e.g., Shonkin Sag). Thick expanses of uniform texture gabbro or diabase, norite or gabbro reflect crystallization of mainly phenocryst-free magma (e.g., Sudbury). These are perhaps end-members (see Figure 8.7), and the larger the body the higher the probability that it was formed over a significant period of time (say, hundreds to thousands of years) involving many injections of magma with varying amounts of phenocrysts. The net result can be exotically layered sequences sandwiched within monotonous nonlayered sequences, or even a thick, coarsely layered sequence followed by a finely layered sequence, marking perhaps two distinct injection episodes, much as seen in volcanism itself. The detailed isotopic record through such sequences generally shows the number of injections to increase with the size of the body. Bodies as large as Stillwater, Bushveld, and Dufek, for example, which are each over 5 km thick, may have experienced hundreds of injections during formation. In spite of this, the thickness of highly fluid magma at any time may have been only a few hundred meters or less, as has been recognized by R. Hunter at Rum in piles of pieces of the lower SF dislodged by the injections. The lower SF, including any cumulates, was locally disrupted time and again by these injections. Crystals were reworked and redeposited and crystal growth was interrupted many times. Individual crystals may carry zoning, telling of a long and dynamic history.

Further variations on this basic theme come from the geometry of the wall rock container holding the magma. Injection of phenocryst-laden magma along steeply inclined walls, as mentioned already, promotes rapid sedimentation, avalanching, and sorting, leaving a telling sequence reflecting an intricate dynamic record that continues to evolve via diffusion until dropping temperature finally freezes in the record.

The net result is a pluton whose size, container geometry, and overall composition greatly influence the constituent rock record. The larger the body, the more injections are necessary to build it; the more the injections,

FIGURE 8.7 Depictions of the final rock records from three idealized magma chambers. The top depicts a classically conceived chamber with a thick basal sequence of layered rocks and a thin upper sequence stemming from the initial instantaneous injection of crystal-free magma. The middle depiction shows what can actually be expected under the initial conditions of emplacement of the classical chamber, which is similar to the Peneplain Sill in being nearly of a uniform composition. The lower depiction shows the result the serial emplacement of crystal-laden slurries, which, depending on the magnitude and frequency of delivery, can produce a highly varied final result.

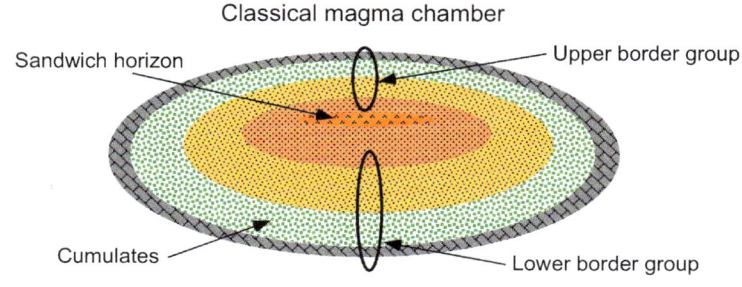

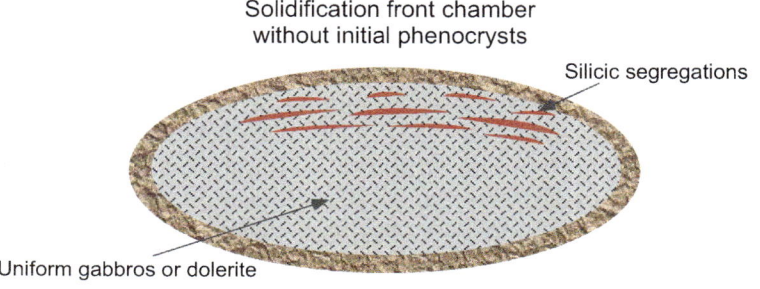

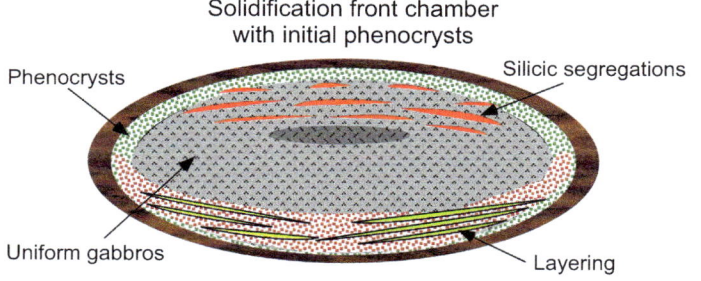

the higher is the probability of injecting a wide variety of phenocryst concentrations and sizes. The wider the variety of phenocryst assemblages, the greater is the crystal sorting and layering. Inclined walls enhance this variety, but increasing silica content stifles sorting and variety.

Small, single injection sills and dikes are most often unremarkable in variety because they are small and of simple geometry. Yet, their story is still important. They tell what systems become when they are not steadily reinjected and cool quickly. At another extreme are lavas. Volcanic sequences are clearly the result of long periods of subaerial reinjection or eruption, which means they are actually magma chambers. But the small size of individual injections (i.e., flow thickness) promotes rapid cooling, thereby prohibiting crystal sorting, annealing, and extensive self-organization. At yet another extreme are large, long-lived, and heavily reinjected chambers.

The lesson learned is that the three dynamic regimes of magma, characterized by large chambers, sills, and lava flows, each shows different aspects of the same spectrum of physical and chemical processes common to magma. Magma chambers are thus integrated systems behaving in response to system size, nature of holding regime, and state of crystallinity of the magma itself. An integrated magmatic system is more aptly characterized as a mush column.

8. MAGMATIC SYSTEMS AS MUSH COLUMNS

Mush columns are vertically extensive active magmatic systems in the crust and lithosphere (Figure 8.8); in essence, a diverse collection of related magma chambers. At depth, they are a blend of sills, stocks, and original wall rock, where with time all boundaries, both physical and chemical, become blurred. Thick expansive sills cool slowly and, on average, always contain a core of magma containing few crystals. Narrow interconnecting passages, on the other hand, cool much faster and are choked with

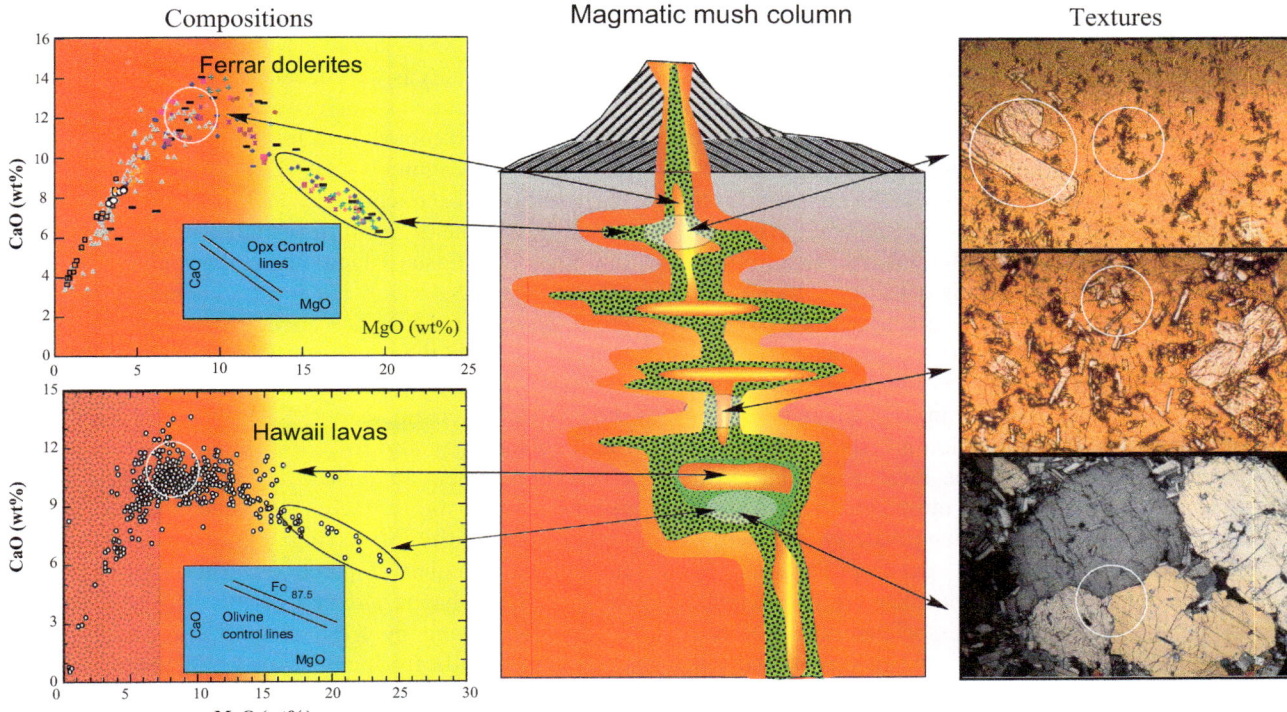

FIGURE 8.8 An idealized magmatic mush column (center) with associated chemical compositions (CaO vs MgO) of lavas from Hawaii (left lower) and plutons (Ferrar dolerite sills) from the McMurdo Dry Valleys, Antarctica (left upper). Representative textures of actual magmas that might reside at these locations within the mush column are shown on the right in thin sections from Hawaiian lava lakes.

crystals in a relatively short time. Yet, as long as the magnitude and frequency of magmatism sustains super-solidus temperatures, a distinct, however tortuous, core region is always available to transport magma upward. In the narrow, mushy necks, crystals are entrained and flushed away. The local mush state thus depends on the local magmatic flux or the magnitude and frequency of magma transmission.

8.1. Geophysical Detection

Direct evidence of this kind of system comes from studies of deeply eroded subvolcanic terrains combined with geophysical inferences (e.g., earthquake distributions) on large active systems like Hawaii and the ocean ridges.

Beneath volcanoes, contrary to petrologic expectations, seismic studies find little sign of large magma-filled caverns or chambers. Teleseismic studies generally do find a general slowing of seismic waves (i.e., travel-time residuals) around large calderas, like at Yellowstone and Long Valley, but more detailed local surveys do not find anything easily ascribable to a large concentration of highly fluid magma. Near-source studies at Katmai on the Alaska Peninsula showed strong S-wave shadowing presumably due to magma, and the distribution of earthquakes both

prior to and post-eruption at Mt Pinatubo, Philippines as well as at Kilauea, Hawaii shows a diffuse magmatic region, but nowhere is the picture very clear on in situ magmas beneath volcanoes.

Recognizing the dominant role of SFs in magmatic systems, these results are not surprising. Mature SFs, as already mentioned, are made mostly of an interconnected network of crystals that give the SF elastic strength and the ability to transmit seismic waves at velocities not much slower than the solid rock itself. Moreover, large active magmatic systems are generally mature (i.e., size is age) and posses extensive mush regions and relatively small volumes of low-crystallinity magma. And individual bodies themselves are small relative to the wavelengths of seismic waves (3–5 km), making detection difficult. Magmatic regions may thus appear seismically solid, and were it not for the manifestations such as surface inflation and eruption itself, the presence of the magma would not be suspected. Detailed seismic reflection studies at ocean ridges find a thin (50–150 m) lens or sill of magma residing just below the sheeted dike complex and perched at the top of an extensive mush column. At fast-spreading ridges like the East Pacific Rise the lens is clear and near the surface, but at slow ridges like the Mid-Atlantic Ridge the lens presence is less clear, perhaps ephemeral, and possibly deeper due to the more subdued thermal regime associated with slower plate spreading.

Gravity and magnetic surveys also do not reveal any simple magma chamber system. Near-surface magmas being almost neutrally buoyant have small density contrasts, resulting in gravity anomalies that are subtle to detect among background anomalies. The high temperature of magmas makes them nonmagnetic and without magnetic contrast to most country rock. But volcanic areas are often magnetically noisy again making detection difficult.

The arrival of magma into the near surface swells the reservoir and inflates the surface, which is directly measured through leveling. This is commonly seen at large-shield volcanoes like Kilauea in Hawaii and Krafla in Iceland. Modeling of this inflation gives some idea of the general position and size of the local reservoir, and this inference is also consistent with mush column systems (see Figure 8.8). In island arc-style magmatic systems, characterized by small-volume central vent eruptions as at Mt St Helens, inflation is much more localized to the immediate area of the volcano itself. The associated mush column is similar in basic features, but is weaker, cooler, and much more spatially restricted.

Thus, large swimming pools of magma immediately beneath volcanoes and ocean ridges, a model held for nearly a century, have not proven out. Instead, the actual magma chambers are more similar to thin sills and the overall magmatic system is more like a vertically distended mush column. And it is the individual dynamic and chemical characteristics of mush columns that are responsible for the distinctive petrologic character of these systems.

8.2. Igneous Diversity and Bimodality

It may seem odd that neither petrologic diversity nor bimodality (i.e., a preponderance of basaltic and rhyolitic rocks with a paucity of rocks of intermediate composition) is proportional to system size. This is a direct reflection of mush column dynamics.

In relatively small, sluggish, and tectonically immobile magmatic systems, the magnitudes and frequencies of eruption are small, and magma repeatedly traverses similar pathways and encounters earlier intrusions at various stages of crystallization. Tectonic immobility of the magmatic center along with intermittent eruption over a 1−5 million-year period, for example, allows most intrusions within the underlying magmatic column, regardless of size, to become highly mushy or even locally to solidify and yet remain in place to be reprocessed through reheating by later magmas. Immobility is central to reprocessing, and reprocessing is central to bimodality. This is the primary difference between Hawaii and Iceland. Hawaii is highly mobile relative to its deep-seated source and escapes reprocessing, whereas Iceland straddles the Mid-Atlantic Ridge, which periodically relocates and extensively reprocesses the crust of Iceland. Bimodality is well developed only at Iceland. The silicic lavas contain pervasive restitic debris and mingled lavas are not uncommon; both features reflect reprocessing.

If the mush column is relatively cool and sluggish, only fairly evolved or fractionated basalts reach the surface, which is the rule at small, intermittent magmatic centers like Rapa Nui (Easter Island) and perhaps many island arc centers. The basalts will not form a chemically cohesive group easily related to one another through simple fractionation, instead the compositions of each eruptive batch, in detail, will each be distinct, and the suite as a whole will show a characteristic scatter. Reactivating a cool, weak mush column may also force earlier stagnated magmas to ascend and sequentially invade one another as new magma climbs toward the surface. Stagnated magma has the highest probability of having compositionally evolved through any number of processes. The erupting magma may thus be a haphazard mix of less and more evolved magma. Reheating by fresh magma also locally remobilizes mushy sills, freeing SF-bound silicic segregations to collect and eventually erupt as silicic magma.

If, on the other hand, a resumed flux of magma through the mush column is strong, recently formed cumulate beds can be disrupted and the crystals entrained, partially resorbed, and redeposited higher in the column. They can also be carried in part to the surface. In this fashion a strong vigorous system can erupt primitive, olivine-laden picritic lavas. At Hawaii, for example, as shown by K. Murata and D. Richter both olivine abundance and size are proportional to the eruptive flux of Kilauea. The compositions of these Hawaiian lavas are characterized by strong olivine control, and olivine fractionation produces a basalt with ∼7 wt% MgO and multiply saturated, at low pressure, in olivine, clinopyroxene, and plagioclase. The compositions of successive lavas are easily related to one another through simple crystal fractionation, and the suite as a whole will show a certain chemical cohesiveness reflective of this process.

The net chemical result of the mush column, thus, depends on the magmatic strength of the system. Weak, sluggish columns produce a series of disjoint magma compositions due to a series of distinct local differentiation histories, each related to the others through the basic chemistry of the ultimate source of the entire magmatic event itself. Strong systems, especially if immobile, show good chemical cohesiveness due to continual crystal entrainment, transport, and deposition. A general feeling for the evolution of the entire system can be gained by considering the system as a whole, tied together by crystal fractionation in a virtual magma chamber, which at some level must always be true. It is the details of the rock chemistry, petrography, and temporal occurrence that record the details of the local magmatic processes. It is these processes that define magma chambers. Physical processes, buttressed by chemistry, thus govern magmatic processes.

FURTHER READING

Arndt, N., 2008. Komatiite. Cambridge University Press, 467 p.

Cawthorn, R.G. (Ed.), 1996. Layered Intrusions. Elsevier, p. 531.

Davidson, J.P., Charlier, B., Morgan, D.J., Harlou, R., Hora, J.M., 2007. Microsampling and isotopic analysis of igneous rocks: implications for the study of magmatic systems. Annu. Rev. Earth Planet. Sci., 274–310.

Garcia, M.O., Pietruszka, A.J., Rhodes, J.M., 2003. A petrologic perspective of Kilauea volcano's summit magma reservoir. J. Petrol. 44, 2313–2339.

Gibb, F.G.F., Henderson, C.M.B., 1989. Discontinuities between picritic and crinanitic units in the Shiant Isles sill: evidence of multiple intrusion. Mineral. Mag. 2, 127–137.

Gunnarsson, Bjorn, Marsh, B.D., Taylor Jr., H.P., 1998. Generation of Icelandic rhyolites: silicic lavas from the Torfajokull central volcano. J. Volcanol. Geotherm. Res. 83, 1–45.

Irvine, T.N., Andersen, J.C.O., Brooks, C.K., 1998. Included blocks (and blocks within blocks) in the Skaergaard intrusion: geologic relations and the origins of rhythmic modally graded layers. Geol. Soc. Am. Bull. 110, 1398–1447.

Jaeger, H.M., Nagel, S.R., Behringer, R.P., 1996. The physics of granular materials. Phys. Today, 32–38.

Mahoney, J.J., Coffin, M.F. (Eds.), 1997. Large Igneous Provinces: Continental, Oceanic, and Planetary Flood Volcanism. Geophysical Monograph 100, American Geophysical Union, Washington D.C., p. 438.

Marsh, B.D., 1996. Solidification fronts and magmatic evolution. Mineral. Mag. 60, 5–40.

Marsh, B.D., 2013. On some fundamentals of igneous petrology. Contrib. Mineral. Petrol. 166, 665–690.

McBirney, A.R., 1993. Igneous Petrology, second ed. Jones and Bartlett Pubs., Boston, MA. 508 p.

Murata, K.J., Richter, D.H., 1966. The settling of olivine in Kilauean magma as shown by lavas of the 1959 eruption. Am. J. Sci. 264, 194–203.

Simkin, T., 1967. Flow differentiation in the prictic sills of North Skye. In: Wyllie, P.J. (Ed.), Ultramafic and Related Rocks. John Wiley and Sons, New York, pp. 64–69.

Upton, B.G.J., Wadsworth, W.J., 1967. A complex basalt-mugearite sill in Piton des Neiges Volcano, Reunion. Am. Mineral. 52, 1475–1492.

Wright, T.L., Klein, F.W., 2014. Two Hundred Years of Magma Transport and Storage at Kilauea Volcano, Hawaii, 1790–2008. U.S. Geological Survey Professional Paper, Washington, D.C, in press.

Chapter 9

Rates of Magma Ascent and Storage

Brandon Browne

Department of Geology, Humboldt State University, Arcata, CA, USA

Lindsay Szramek

Department of Geosciences, Austin Peay State University, Clarksville, TN, USA

Chapter Outline

GLOSSARY

conduit A pathway used by magma as it ascends to the surface from a magma reservoir. For many volcanoes, these pathways are thought to be pipelike in shape and several tens of meters in diameter.

groundmass A term used to describe the portion of the erupted magma that solidified quickly, resulting in a fine-grained and glassy texture that surrounds and encases larger phenocrysts that crystallized at depth.

hornblende A member of the amphibole family of silicate minerals. It commonly occurs in igneous rocks and is recognized by its black, brown, or dark green color and 60°/120° cleavage angles. Hornblende contains up to 4% water in its chemical structure.

hypocenter The origination point of rupture during an earthquake in the Earth's lithosphere.

lithosphere The rigid outermost layer of the Earth, which includes the crust and uppermost mantle.

magma extrusion The effusive, or nonexplosive, eruption of magma at the surface, which typically produces lava flows and lava domes.

magma reservoir A region of the Earth's lithosphere characterized by the presence of molten—or partially molten—rock or magma. Many actively erupting volcanoes are thought to be connected to a magma reservoir located 3—15 km beneath the surface via a conduit.

microlite Tiny crystals comprising the groundmass portion of volcanic rocks that are typically needlelike in shape and form quickly during magma ascent and eruption.

phenocryst Large crystals surrounded by fine-grained groundmass in volcanic rocks that are typically prismatic and/or tabular in shape and form slowly during magma storage at depth.

reaction rim A layer of crystals that forms around a crystal of different composition at the crystal—melt contact in response to disequilibrium induced through decompression, heating, or a change in the composition of the melt.

undercooling Equating the change from equilibrium in pressure to the same change from equilibrium in temperature.

volatile exsolution The processes whereby dissolved volatiles in magma separate from the magma and form bubbles.

Vulcanian A style of volcanic eruption characterized by short-lived and violent explosions that eject mostly solidified lava and ash.

1. INTRODUCTION

Volcanic eruptions take on many forms, from violent explosive eruptions that produce rapidly expanding volcanic clouds and ground-hugging pyroclastic density currents to effusive eruptions where viscid lava domes amass around a vent or fluid lava flows pour down a volcano's flanks. These different styles of volcanic eruptions represent radically different types of volcanic hazards. Our capacity to decrease the threat of casualties and property damage by volcanic eruptions depends on our ability to understand what controls eruption style, which may change during the course of a single eruption.

To a large degree, the style in which a volcano erupts is a product of the rate of syneruptive magma ascent and the geometry of the volcanic **conduit** connecting the volcanic vent at the surface to the **magma reservoir** region in the shallow crust. Syneruptive magma ascent rate also affects the morphology of resulting lava flows and domes, magma degassing, vesiculation, fragmentation, magma supply and withdrawal, earthquake type and occurrence, magma mixing, and mineralogical reactions in the rising magma. Factors controlling the rate in which magma ascends to the surface during volcanic eruptions include the physical and chemical properties of the magma, such as its temperature, composition, volatile budget, and crystallinity, all of which affect its viscosity and density, as well as the permeability of the conduit walls. The degree to which ascending magma is obstructed by or confined in volcanic conduits, as well as the magmatic pressure within the reservoir region also influence magma ascent rates.

Several techniques are utilized to examine and constrain syneruptive magma ascent rates (Table 9.1). A particularly direct approach involves careful measurements of **magma extrusion** rate during effusive volcanic eruptions.

Once limited by inclement weather and/or surface degassing, both of which obscured the observers view of the volcanic vent during eruptions, recent studies that combine digital photogrammetric analysis of oblique aerial photographs with digital elevation models and precise Global Positioning System control network have been able to overcome challenging visibility and accurately track magma extrusion rate. These measurements provide valuable insight in terms of how syneruptive magma ascent rate evolves during an eruption, which greatly advances our ability to predict potential changes in the style of eruption. For example, whereas an increasing extrusion rate may precede a shift toward more explosive and dangerous eruption style, a prolonged decreasing extrusion rate is often observed as volcanic eruptions expire. Magma extrusion rate measurements during explosive volcanic eruptions have

also been utilized to constrain syneruptive ascent rates at depths greater than ~ 1 km of the surface, although this technique requires an understanding of variables such as magma mass flux as well as the density and viscosity of the erupting magma, which typically are not determined until after the eruption has occurred. Thus, while this approach has increased our appreciation for magma ascent rates during explosive eruptions, it is not utilized in defining syneruptive magma ascent rate in real time, which precludes it from being used as a monitoring tool or as a means to predict possible changes in eruptive style.

Other techniques employed to examine and constrain syneruptive magma ascent rates rely on the chemical and mineralogical reactions that occur as magma rises toward the surface prior to and during volcanic eruptions, some of which can be used to examine and constrain syneruptive magma ascent rates petrologically. One such example is **volatile exsolution**, where volatile species once dissolved in the melt exsolve into bubbles as magma rises toward the surface and is subjected to decreasing pressure. Because different volatiles possess different solubilities and exsolve at different rates as a function of magma composition and pressure, studies examining the bubble textures in the erupted products as well as the degassing behavior of different volatile species have enhanced our understanding of magma ascent rates. Magma ascent rates may also be constrained by examining **groundmass** textures in the erupted material, where the size distribution, morphology, and abundance of **microlites** all have been shown to record the style and rate of syneruptive magma ascent (usually within 1000 m of the surface) due to the relation between microlite nucleation rate and **undercooling** driven by

TABLE 9.1 Summary of Syneruptive Magma Ascent Rates

Volcano	Observation	Explosive Ascent Rate (m/s)	Extrusive Ascent Rate (m/s)
Mt St Helens, Washington	Groundmass crystallization	1–3	0.01–0.02
Mt St Helens, Washington	Hornblende reaction rims	>0.18	0.004–0.015
Mt St Helens, Washington	Extrusion rate	1–2	0.005–0.0001
Mt St Helens, Washington	Seismicity	0.6	0.007–0.01
Pinatubo, Philippines	Groundmass crystallization	>0.2	0.002–0.05
Soufrière Hills, Montserrat	Hornblende reaction rims	>0.2	0.001–0.012
Soufrière Hills, Montserrat	Extrusion rate	>0.2	0.0001–0.02
Black Butte, California	Hornblende reaction rims and plagioclase growth	Not present	>0.1
Arenal Volcano, Costa Rica	Groundmass crystallization	0.05–0.9	<0.05
Chaitén, Chile	Extrusion rate	>0.5	0.02–0.5

Modified from Rutherford (2008).

decompression. Reactions between **phenocrysts** and melt also can be used to constrain magma ascent rate, particularly in the case of hydrous phenocrysts. **Hornblende**, for example, often crystallizes from a water-saturated melt in midcrustal magma storage reservoirs where it incorporates up to 2 wt% water in its crystal structure. During ascent to the surface, hornblende becomes unstable and reacts with the surrounding melt as the concentration of dissolved water in the melt decreases at lower pressures. If magma ascends at a sufficiently slow rate, a **reaction rim** of anhydrous minerals such as pyroxene, plagioclase, and iron and titanium oxides may develop around a crystal where it is in contact with melt. Experimental studies have successfully replicated this reaction, the results of which can further constrain the rate of syneruptive magma ascent.

If a sufficiently dense monitoring network exists on an active volcano, seismological observations may also function as a means to broadly define syneruptive magma ascent rate. One seismological technique that has produced insight into pre- and syneruptive magma ascent rates utilizes realtime seismic amplitude measurements, where ascending magma has been approximately correlated with seismic signals at progressively shallower depths. Readers are directed to the Volcanic Seismicity chapter in this book for a discussion of this technique.

2. MAGMA STORAGE REGIONS

Results from seismological, geodetic, and petrological studies indicate that magma ascent during volcanic eruptions typically initiates from magma storage regions in midcrustal depths (3–15 km) beneath volcanoes. The 2006 eruption of Augustine Volcano, located on Augustine Island, Alaska, is one example where seismological, geodetic, and petrological data sets collected during the 3 months of eruption implicate the existence of a shallow crustal magma storage zone (Figure 9.1).

Seismological data acquired prior to, during, and after the 2006 eruption of Augustine Volcano indicated the existence of a magma storage region located at a depth of 3.5–5 km below mean sea level. In particular, characteristics of volcano-tectonic earthquakes located at depths of 2–5 km below mean sea level **following** the 2006 eruption were interpreted to reflect a stress response caused by the removal of magma from this region. Interferometric synthetic aperture radar, which measures ground deformation at millimeter-scale accuracy, detected both a contracting magma storage region at 2–4 km depth and an inflating magma storage region at 7–12 km depth beneath Augustine Volcano through from 1992 to 2005. Whereas the shallow magma storage region was interpreted to represent a short-lived magma reservoir tapped during the eruption, the deeper storage region was interpreted as a longer-term storage zone. Geodetic observations from the same eruption were used to define a magma storage region based on deflationary pressures at 3.5–5 km beneath the Augustine Volcano summit. Finally, attempts to constrain the location of a preeruptive magma storage region through petrological means generally agreed with geophysical results, where volatile concentrations in trapped melt inclusions record depths beneath Augustine Volcano of 4–8 km, and microlite and matrix glass compositions of natural andesite erupted by Augustine Volcano in 2006 most closely resemble andesite samples experimentally reproduced at pressures of 140–150 MPa, which corresponds to depths of 5.3–5.7 km assuming a crustal density of 2700 kg/m^3.

The 2006 eruption of Augustine Volcano is only one example of where a magma storage region was identified beneath an actively erupting volcano, as magma storage regions have similarly been identified during many other well-studied eruptions at Mt St Helens, USA; Soufriere Hills, Montserrat; Unzen, Japan; and Mt Pinatubo, Philippines. In contrast, locating magma storage regions beneath volcanoes that are not erupting has been less successful. Results from an emerging suite of studies aimed at constraining the rates of magmatic processes within magma storage regions continue to shed light on why subvolcanic magma storage regions can be detected geophysically during eruptions but not during periods of inactivity. For example, studies utilizing major- and trace-element compositional zoning patterns in phenocrysts and diffusion modeling are appropriate methods to constrain relatively short-lived processes ($10^0–10^2$ days), like the duration of time associated with the intrusion of mafic magma into a storage region followed by eruption. On the other hand, studies aimed at determining the rates of more prolonged process operating in magma storage regions may utilize radioactive isotopes from whole-rock, mineral separates, or through multiage determination of single crystals. Results from these studies suggest that longer-term processes such as magma assimilation and crystal–melt fractionation occur over timescales of $10^3–10^6$ years, which requires equivalent timescales for the accumulation and storage of magma in the shallow crust, particularly in the case of silicic magmas emplaced as a result of voluminous caldera-forming eruptions. However, an increasing number of studies contend that magma stored in shallow crustal storage regions spends the majority of its time as crystalrich and highly viscous "mush" existing at temperatures only slightly above the solidus punctuated by brief periods of magma remobilization and eruption triggered by intrusion of hotter and more fluid magma.

3. MAGMA EXTRUSION RATES

Geological observations of magma extrusion provide valuable insight into syneruptive magma ascent rates—at

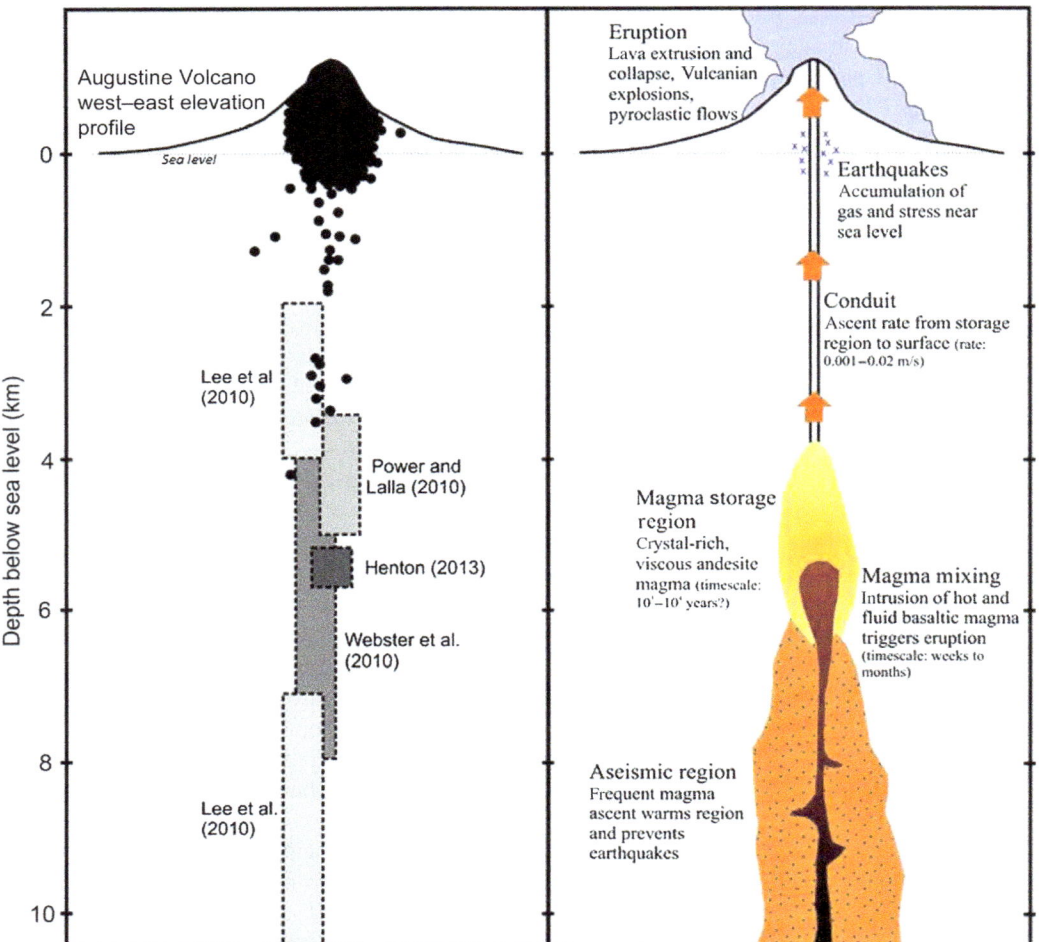

FIGURE 9.1　Left: Schematic west–east cross section through Augustine Volcano, Alaska. Filled circles are earthquake **hypocenters** between 1993 and 2007 with standard horizontal and vertical errors of less than 5 km (from Power and Lalla, 2010), where hypocenters 2–4 km deep occurred in the 5 months following the end of the 2006 eruption. Blacked-out summit region represents densely overlapping hypocenters. Elevation profile of Augustine Volcano and depth (in km) below sea level (thin dash line) shown on map. Boxes represent locations of magma storage regions during the 2006 eruption (Power and Lalla, 2010; Lee et al., 2010; Henton, 2013) and during older eruptions (Webster et al., 2010) based on seismic, geodetic, petrologic, and experimental phase equilibria studies. Right: Schematic diagram modified from Larsen et al. (2010) showing interpretation of processes occurring within magma storage region and conduit.

least in the upper few hundred meters of the conduit—and have been particularly useful in advancing our understanding of how, and over what timescales, volcanic eruptions transition from effusive behavior to explosive eruptions takes place. And, because careful observations of magma extrusion rate can be done continuously, these observations can be closely linked with real-time measurements from seismicity, gas emissions, and ground deformation, which allows for a more comprehensive appreciation for volcanic processes. Recent eruptions at Mt St Helens, Washington; Mt Redoubt, Alaska; Chaitén, Chile; Santiaguito, Guatemala; Soufrière Hills, Montserrat; and Mt Unzen, Japan have provided magma extrusion rate data sets that can be used to evaluate the evolution of syneruptive ascent rate over the course of eruptions.

An important drawback to syneruptive magma ascent rates calculated from observations of magma extrusion is that they likely only represent ascent conditions occurring within a few hundred meters of the surface as magmas undergo extensive microlite crystallization caused by decompression, and to a lesser extent, cooling. Consequently, magmas in the shallowest stretch of the conduit are dramatically more viscous than they were when ascent initiated from the magma reservoir several kilometers beneath the surface. This explains why ascent rates calculated from extrusion measurements are often inconsistent with those produced through decompression experiments, where experimentally grown mineral textures used to calibrate ascent rates preferentially form at greater depths.

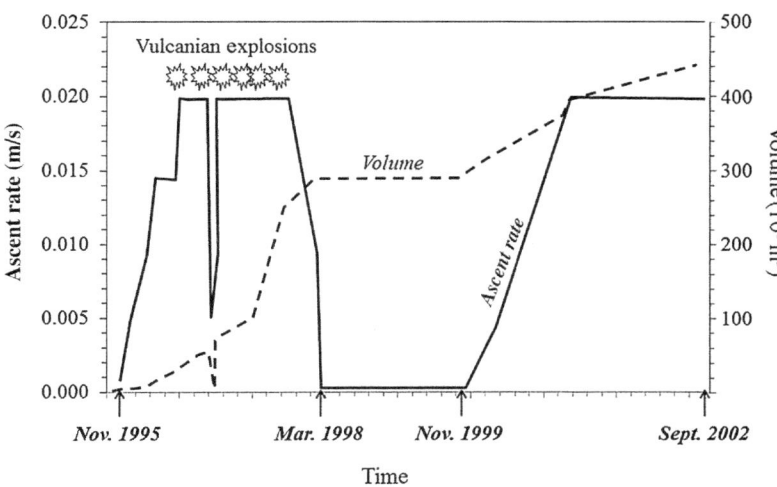

FIGURE 9.2 Combined plot showing ascent rate (in m/s) and cumulative erupted volume (in 10^6 m^3) over time during the 1995–2005 eruptions of Soufrière Hills, Montserrat (modified from Rutherford and Devine (2003)). Ascent rate curve is based on experimentally grown hornblende reaction rims. Note that **Vulcanian** explosions coincide with the fastest ascent rates.

Direct observations coupled with digital aerophotogrammetry and thermal imagery of the 2004–2006 eruption of crystal-rich dacite from Mt St Helens in Washington, USA provide useful constraints on syneruptive magma ascent rates in the uppermost conduit if reasonable approximations of conduit geometry are made, as several studies have shown that the erupting magma had mostly solidified within 1 km of the surface. Assuming that the 2004–2006 dacite magma ascended through a conduit ranging in diameter from 50 to 100 m, ascent rates in the uppermost conduit ranging from 0.001 to 0.005 m/s correspond with direct observations of initial dome growth rate in October 2004. Near-surface ascent rates slowed from 0.003 to 0.0008 m/s in November 2004 to 0.002–0.0004 m/s by the beginning of January 2005 to 0.0001–0.0005 m/s by December 2005.

Observations of andesite lava dome extrusion during the 1995–2002 eruption of Soufrière Hills Volcano on the island of Montserrat in the British West Indies can also constrain near-surface syneruptive magma ascent rates and clarify the role of magma ascent rate in the threshold between effusive and explosive eruption behavior (Figure 9.2).

Previous studies approximate a 30-m-diameter conduit for Soufrière Hills Volcano, which yields magma ascent rates that correspond to dome growth observations steadily increasing from 0.0001 m/s in November 1995 to 0.0007 m/s in February 1996. Ascent rates in the uppermost conduit associated with dome extrusion peaked at 0.02 m/s during the second half of 1997, coinciding with major dome collapse events and frequent Vulcanian explosions between August and October. This was interpreted to represent that ~0.02 m/s was the maximum velocity that magma could ascend and still produce effusive eruptions. If magma ascended faster than 0.02 m/s, explosive eruptions occurred, which evacuated the conduit to depths of 0.5 to

>2 km as magma accelerated through the upper several hundred meters of the conduit at velocities of 40–140 m/s. Magma ascent rate declining to zero between March 1998 and mid-November 1999, when no dome extrusion was observed before resuming to between 0.015 and 0.02 m/s by 2002, as shown in Figure 9.2.

The 2008–2009 eruption of rhyolite from Chaitén Volcano in Chile is a particularly interesting example because of the unusually fast lava extrusion rate, which is attributed to an abnormally low magma viscosity due to its high temperature (~800 °C). Geologic field mapping coupled with photogrammetric analysis of oblique aerial photographs and digital elevation models was used to calculate lava extrusion rates, which averaged 66 m^3/s during the first 2 weeks of the eruption followed by 45 m^3/s for the next 4 months. Decompression experiments on the Chaitén rhyolite suggest ascent rates of >0.5 m/s from a magma reservoir located at a depth of at least 5 km during the 4 h preceding the initial eruption. Little information is known about the conduit geometry of Chaitén Volcano, but assuming a diameter somewhere between 20 and 50 m, near-surface magma ascent rates corresponding to observed extrusion rates during the first 2 weeks would have been 0.2–0.03 m/s compared to 0.1–0.02 m/s over the next 4 months. Remarkably, these rates are 10–100 times faster than that observed for the 2004–2006 eruption of dacite magma from Mt St Helens and 1–10 times faster than that observed for the 1995–2002 eruption of andesite magma from Soufrière Hills Volcano.

4. BUBBLE FORMATION AND EVOLUTION DURING MAGMA ASCENT

Bubbles form when magma ascends from depth and certain components are no longer soluble and exsolve as gas. This gas forms vesicles, which are preserved within pumice deposited from volcanic eruptions. Often, these volcanic deposits must be used to infer information about the dynamics of past eruptions. This leads to the question if ascent rate can be estimated from the preserved bubbles within deposits.

While volcanic volatiles include H_2O, CO_2, SO_2, Cl, and F, only H_2O and CO_2 are typically considered in the formation of vesicles because of their relativity high concentrations. During the ascent of magma, the majority of all volatiles are exsolved from the melt because of lower solubility at lower pressures. This gas can either remain as bubbles within the melt or can escape the melt as gas through a permeable network. Either way volcanic samples at the surface, from explosive or effusive eruptions, do not contain dissolved volatiles and are considered dry. Therefore, magma ascent, no matter the rate, will allow for volatiles to exsolve.

Models that explore the eruption dynamics of explosive eruptions often model degassing as in equilibrium with the melt. For this to be true, it means that some bubble must be present prior to ascent, for equilibrium degassing never permits the melt to become supersaturated in water and nucleate new bubbles. The rate of degassing will directly control the rate at which magma is allowed to ascend, which has led to experiments to examine this problem (Figure 9.3). A water-rich high-silica rhyolite melt with an initial population of bubbles was hydrated at temperatures between 750 and 800 °C and a pressure of 150 MPa. The hydrated samples were then decompressed, at 875 or 750 °C, to final pressures of 41.6−100 MPa at varying rates of decompression. It was seen that high-temperature samples were able to degas in equilibrium at rates of 0.1−1.2 MPa/s, but an increased rate of 3.7 MPa/s was unable to maintain equilibrium. Low-temperature samples showed similar results but at slower decompressions. Rates of 0.015−0.025 MPa/s maintained equilibrium whereas higher rates of 0.05 MPa/s did not. This shows that a temperature increase can drastically alter the rate at which magma can degas in equilibrium, hotter temperatures have faster ascents possible. However, most conduit models are overestimating the rates of ascent by an order of magnitude. Current models cannot address magma that has ascent rates too fast for equilibrium melting and also has bubbles that nucleate during ascent.

The majority of information on bubble growth rates comes from models based on experiments in rhyolitic melts at temperatures varying from 460 to 925 °C. Current models are able to estimate the growth rates with acceptable precision for temperatures higher than 500 °C, pressures higher than 0.1 MPa, and water contents lesser than 4 wt% in silicic melts. For compositions that are varied, information on viscosity and H_2O diffusivity is not well understood and therefore not well modeled. While most work in this topic centers on experiments, natural samples have also been worked on. At Mt St Helens, physical parameters were used to estimate that bubbles had between 960−4120 s to grow. This time range was used to determine that bubbles grew at rates of $5.2 \times 10^{-7}-1.2 \times 10^{-7}$ cm/s, an order of magnitude slower than rates modeled with decompressive and diffusion growth. A detailed study of basaltic rocks that form scoria or reticulate determined an estimated growth rate of 9×10^{-4} cm/s for the forming bubble, three orders of magnitude faster than effusive basalt bubble growth rates. More work is needed to better understand the bubble growth rates of melts at various temperatures and compositions for better determination of magma ascent rates.

Numerical models have also been used to model how volatiles in magma will behave upon ascent. While most models simply address water in the melt, some work has been done that also considers the role of CO_2, which is typically at concentrations of 0.1−0.01 times that of the H_2O present. This small percent of CO_2 is enough to increase gas saturation pressures by several orders of magnitude and to alter the distribution of volatile species in the melt versus gas phase. Therefore, models that do not use CO_2 are likely using incorrect parameters to model eruption dynamics such as magma ascent. Numerical models demonstrate that CO_2 in the melt will counteract the role of H_2O in the melt. For example, it is possible to decrease the mass flow rate by as much as 30% by the addition of CO_2 in the melt. Examining the other volatiles present in the system may lead to similar findings what would enable better constraints of magma ascent based on bubble formation.

One of the important aspects of the findings on the formation of bubbles is that water is able to diffuse out of

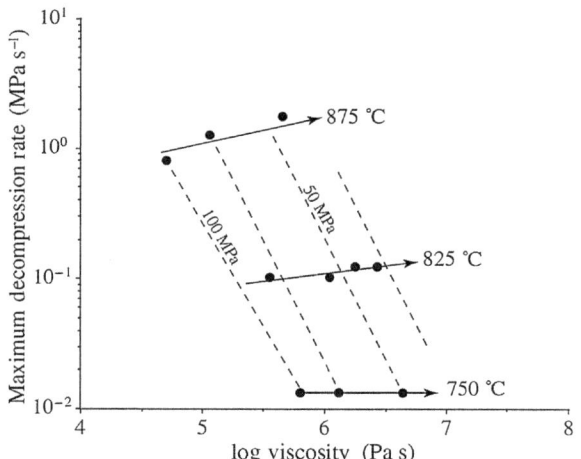

FIGURE 9.3 Maximum decompression rate for range samples from Gardner (2009). The maximum decompression rate is shown for viscosities and water contents based on various models. Values calculated for equilibrium degassing and arrows show trends of degassing for increasing viscosity. *Figure from Gardner (2009).*

silicate melts at a rapid rate during magma ascent. This rate of water diffusion means that other processes occurring in the melt will not be limited by water, but rather by kinetics as determined from the specific processes such as groundmass crystallization.

5. GROUNDMASS CRYSTALLIZATION DURING MAGMA ASCENT

As magma ascends, decompression forces water to exsolve from the melt, thus resulting in crystallization of the groundmass. Microlites, as these crystals are known, are typically less than 100 μm in size, with varying morphologies. The rate at which the magma is decompressed will cause various disequilibrium textures to form in the microlites. In the early 1970s, extensive experiments were carried out that linked microlite textures to cooling rate in mafic magmas. Since then it has been observed that similar changes also occur when magma, both mafic and felsic, is decompressed experimentally. This observation linking microlite textures with decompression rate has allowed scientists to compare experimentally decompressed magmas with naturally decompressed magmas to obtain estimates of ascent rates (Figure 9.4).

Early work on microlite texture and ascent rate was done on the eruptive products of the 1980–1986 Mt St Helens eruptions. Having determined that the storage depth

(8–11 km), temperature, and water content remained constant during the eruptions, it was assumed that the variations in microlite textures were a result of difference in ascent rates. Using experimental and natural data, it was determined that the Plinian phase of the eruption had an ascent rate of 1–3 m/s whereas the dome-building phase was an order of magnitude slower, at ~0.3 m/s.

Building on this work, detailed experiments on the kinetics that control microlite growth in felsic systems have been conducted. One such study focused on the kinetics that govern microlite crystallization in felsic magmas by conducting experiments on Pinatubo dacite that where decompressed to final pressures of 175-5 MPa and held for 0.3–931 h. These experiments imposed effective undercoolings of $\Delta T_{eff} = 34$–$266\,°C$. It was found that for the range of $\Delta T_{eff} = 125$–$241\,°C$, or final pressures of 10–50 MPa, the dominant activity was microlite nucleation whereas slower decompressions grew normally zoned and euhedral feldspar crystals. Chemical equilibrium was not achieved with the experiments, but it was noted that slowly decompressed samples are further from chemical equilibrium than rapidly decompressed samples. This is caused by the relative rates of feldspar crystallization and nucleation experienced during decompression.

Determining that ascent rates can indeed be recorded by microlite textures, a study of the andesite of Soufriére Hills Volcano that explores the differences between dome lavas

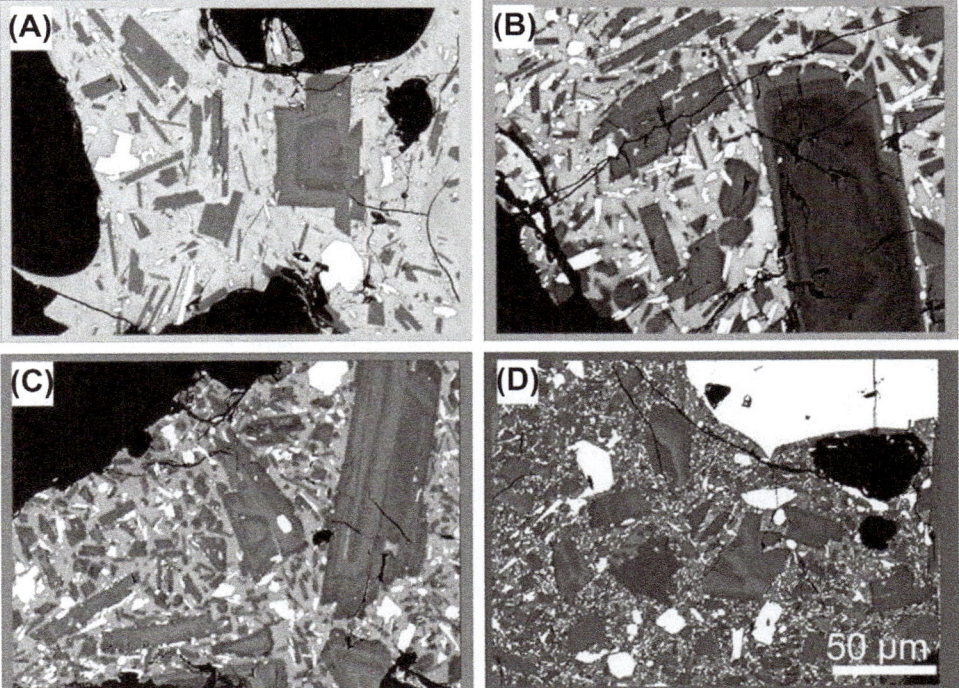

FIGURE 9.4 Back-scattered electron images showing microlite textures found in ash samples from Tungurahua volcano, Ecuador. (A)–(D) show increasing complexity of morphology and phases found in ash samples. Light gray is glass, dark gray is plagioclase, and white is mafic and Fe–Ti oxides. *Used with permission from Heather Wright, USGS.*

and pumice samples was conducted. Experiments conducted at $\Delta T_{eff} = 40-150\,°C$ showed that with increasing ΔT_{eff}, plagioclase morphology changed from tabular, hopper, swallowtail, dendritic to chain forms. A nucleation lag was also identified of 1–4 h, depending on the ΔT_{eff}. The experiments determined that the dome samples ascended at a variety of rates, all longer than 4 h whereas the pumice samples ascended at a rate of less than 4 h. The amount of time taken for ascent translates into ~0.2 m/s for the pumice and ~0.05–0.002 m/s for the dome-building eruptions. The important takeaway from this study is that there is a nucleation lag in felsic magma where no new microlites form and microlite textures are relatable to not only ascent rate but also explosivity of the magma. The link between explosivity and ascent rate was also seen at Mt St Helens between the dome and Plinian eruption.

Up to this point, only felsic systems have been considered. There are a number of reasons for this. The first and most important is that felsic systems tend to produce eruptions that are more explosive than mafic systems. Therefore knowing how the magma behaves during ascent, which is linked to explosivity, is vital. The other issue is experimental equipment. Lower temperature experiments, such as on dacite, are run in cold-seal pressure vessels whereas higher temperature experiments, such as on basalt, are run in TZM (titanium-zirconium molybdenum) pressure vessels. The higher temperature that the TZM pressure vessels allow requires a more complex experimental setup. The more complex a setup becomes, the more difficult it is to conduct experiments. The last issue addressed is the rate of kinetics in a mafic system is faster than that in a felsic system and better control is needed to look at what happens during ascent. However, this has not limited us from trying to understand the ascent of magmas in more mafic systems.

Arenal Volcano, in Costa Rica, has been erupting basaltic andesite since 1968. The eruptive sequence began with a Plinian eruption, which transitioned into Strombolian eruptions, lava flows, and pyroclastic flows. Over the course of the eruption the composition of the magma has remained relatively constant, making it an ideal location to study ascent rate and explosivity in a more mafic system. Using samples from all the eruptive types seen at Arenal Volcano, experimental data suggest that the Plinian eruption had magma that ascended at 0.05–0.9 m/s whereas all the non-Plinian eruptions examined had magma that ascended at less than 0.05 m/s. The microlites in the non-Plinian eruptions were too complicated to determine more definite ascent rates.

The kinetics of mafic magmas add another layer of complexity to determining the ascent rate of natural samples. In mafic systems, crystallization occurs at a rapid rate, which means that if the samples ejected from the volcano during the eruption are large enough, they will continue to crystallize after ascent and during cooling. It has been seen

in samples where the core takes as little as 500 s to cool that plagioclase and magnetite can undergo extensive crystallization caused by cooling. One method to deal with this issue is to only use the rim of natural pumice as an analogue for comparing decompression experiments or to only use small particles such as ash.

Tungurahua volcano, Ecuador, has produced Strombolian to Vulcanian eruption over the years of 1999–2006. During this time, ash was collected at sample locations on a continual basis, allowing for microlite textures to be examined from known eruptions with a known magma supply rate (MSR). MSR is determined from the size of the ash plume generated, with more explosive eruptions producing a larger MSR. While magma ascent rate has not been determined, the MSR has been shown to correlate with the microlite textures formed. Further study of magmas with different compositions will hopefully lead to a refinement of the understanding of magma ascent.

6. PHENOCRYST–MELT REACTIONS DURING MAGMA ASCENT

Volatile-bearing minerals such as hornblende are particularly useful tools to constrain syneruptive magma ascent rate because they become unstable as the concentration of dissolved water in the melt decreases during decompression. If magma ascends at a sufficiently slow rate, a reaction rim of anhydrous minerals, such as pyroxene, plagioclase, and iron and titanium oxides, may develop around a hornblende crystal where it is in contact with melt (Figure 9.5).

The kinetics of reactions between melt and volatile-bearing phenocrysts such as hornblende, biotite, and anhydrite in an ascending magma have yet to be resolved fully due to the complex nature of the reactants, which involve viscous silicate melt and volatile-bearing minerals with elaborate chemical compositions. However, experimental studies have successfully replicated reactions between volatile-bearing phenocrysts and surrounding melt during decompression, and results from these experiments have been used to further constrain the rate of syneruptive magma ascent.

In a groundbreaking study by Rutherford and Hill (1993), rates of hornblende breakdown in response to magma ascent were reproduced in a suite of experiments using dacite erupted from Mt St Helens between 1980 and 1986. The authors of this study observed no reaction rims surrounding hornblende in pumice samples erupted during the Plinian eruption on May 18, 1980, even though these samples had lost all but about 0.3 wt% of the ~4.6 wt% dissolved water they contained before eruption. In contrast, hornblende phenocrysts were enclosed by reaction rims in

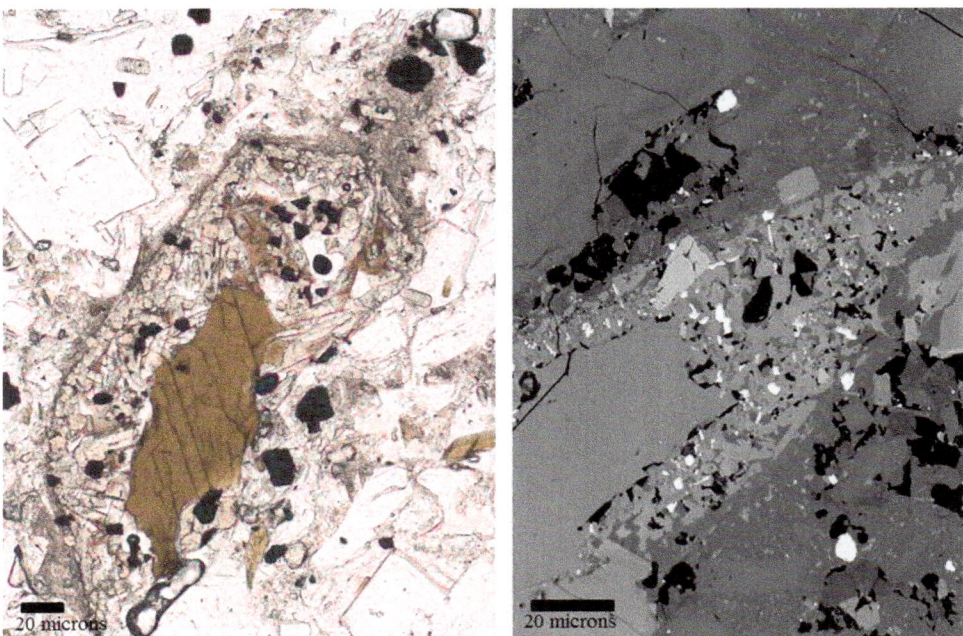

FIGURE 9.5   Photomicrograph (left) and back-scattered electron image (right) of hornblende phenocrysts erupted in January 1990 from Mt Redoubt, Alaska showing reaction rims at the contact between the phenocryst and surrounding melt composed of plagioclase, orthopyroxene, and Fe–Ti oxides.

virtually all subsequently erupted dome material. Most reaction rim widths ranged in thickness from 0 to 50 μm, although another population of thicker reaction rims was also observed. Hornblende with thinner reaction rims was interpreted to have originated from the main pulse of ascending magma, whereas the population of hornblende characterized by thicker rims was interpreted to have originated from batches of leftover or stalled magma in conduit. Rutherford and Hill (1993) experimentally calibrated the rate of hornblende reaction rim growth at 900 °C during constant-rate decompression from ~210 MPa, or approximately 8 km depth. Experiments simulating ascent from 8 km depth in less than 4 days produced no reaction rims, whereas experiments simulating ascent from 8 km depth in 7 and 20 days resulted in reaction rim widths of 2 and 32 μm, respectively (Figure 9.6).

Their results showed that hornblende reaction rim formation initiates when the melt no longer contains enough water to stabilize hornblende and that slower decompression rates result in progressively thicker reaction rims surrounding hornblende phenocrysts. Comparing the experimental findings to the observations made of natural samples erupted from Mt St Helens during the 1980–1986 eruptions, ascent rates from 0.004 to 0.015 m/s were determined for dome-building eruptions compared to 2–3 m/s for the Plinian eruption. These ascent rates closely resemble those experimentally derived for the 2004–2008 eruption of Mt St Helens (~0.005–0.015 m/s) despite the lower temperature (860 ± 15 °C) and higher crystallinity

(~40%) of the 2004–2008 magma. Hornblende reaction rims in the 2004–2008 deposits are also 50–75% thinner than those observed in 1980–1986 lavas due to the delayed onset of rim formation during ascent resulting from high concentrations of fluorine in their structure (Rutherford and Devine, 2008), which extends their stability to lower

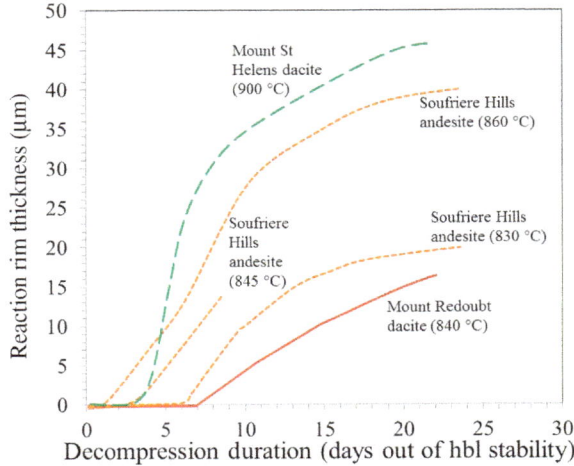

FIGURE 9.6   Experimentally derived relationship between decompression duration (time spent at pressures below hornblende stability) and thickness of reaction rims enclosing hornblende in samples from Mt St Helens (Rutherford and Hill, 1993), Soufrière Hills (Rutherford and Devine, 2003), and Mt Redoubt (Browne and Gardner, 2006). Reaction rim thickness decreases as the magma temperature decreases and the silica content increases. hbl, hornblende stability.

pressures ($\sim$50 MPa) unlike the typical stability limit of 100 MPa for fluorine-free hornblende.

A similar series of decompression experiments were conducted on lavas erupted during 1995–2002 from Soufrière Hills, Montserrat to determine syneruptive magma ascent rates. Although the phenocryst assemblage remained constant throughout the several-year long eruption and recorded crystallization conditions of 5–6 km depth (130 MPa) and 830 °C, the widths of hornblende reaction rims changed significantly. Hornblende phenocrysts in magma produced at the start of the eruption in December 1995 were enclosed by 120-μm-wide reaction rims, but over time the widths of reaction rims decreased. For example, hornblende phenocrysts in samples collected from lava domes emplaced in February and April 1996 were enclosed by 18- and 10-μm-thick reaction rims, respectively. Magma erupted between July and September 1996 contains hornblende virtually free of reaction rims. Decompression experiments on natural samples suggest that magma ascent rates increased from 0.001 to 0.012 m/s during the first 300 days of the eruption, which corresponds to observations of increased extrusion rate over the same period.

A series of decompression experiments using dacite pumice samples erupted in December 1989 from Redoubt volcano, Alaska shed more light on the rate of hornblende reaction rim growth, as well as the utility of this reaction as a "geospeedometer" for syneruptive magma ascent rate. These experiments were similar to previous experimental decompression studies in that they examined the influence of magma ascent rate on the rate of hornblende reaction rim formation through isothermal (840 °C) decompression experiments. For example, experiments simulating ascent

from 5 km depth to the surface in less than 7 days produced no reaction rims, whereas experiments simulating ascent from 5 km depth in 11 and 21 days resulted in reaction rim widths of 3–7 μm and 13–19 μm, respectively. However, the design of these experiments differed from previous studies in that they involved decompressed to pressures ranging from 100 to 2 MPa ($\sim$4 km–70 m depth) for 1–30 days before being quenched. This approach allowed for incremental examination of hornblende reaction rim development at progressively lower pressure as magma ascended toward the surface during volcanic eruptions (Figure 9.7).

In this regard, one key finding from these experiments was that reaction rims did not grow around hornblende when decompressed to pressures immediately below the hornblende stability field ($\sim$100 MPa). Rather, hornblende phenocrysts dissolved, which was marked by both a decrease in modal hornblende and the development of rounded hornblende crystal faces. This process was also observed in experiments conducted using basaltic magmas, suggesting that hornblende dissolution may play an important role in the early development of reaction rims. Another finding was that hornblende reaction rims formed preferentially at pressures from 10 to 40 MPa ($\sim$300–1300 m depth), and that this favorable pressure range narrowed and decreased as the duration of decompression increased. Finally, the authors observed no hornblende reaction rim growth at pressures below 10 MPa (<300 m depth), regardless of decompression style or experiment duration, which the authors interpreted as a result of limited hornblende dissolution combined with extremely high viscosity of the near-solidus interstitial melt.

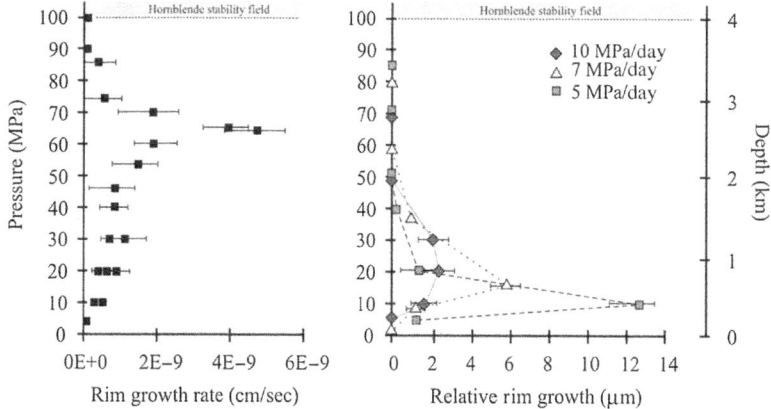

FIGURE 9.7   Reaction rim growth plotted against pressure and corresponding depths (in km) for single-step (left) and multistep (right) decompression experiments of dacite pumice erupted in 1989 from Mt Redoubt, Alaska (from Browne and Gardner, 2006). Reaction rim growth rate is calculated by dividing the measured rim thickness by the total time held below the hornblende stability field (shaded). Relative rim growth is the difference in rim thickness from sequentially decompression experiments quenched at lower and lower pressures. For both decompression styles, growth rate is extremely slow near the hornblende stability field and at pressures below 10 MPa. Growth occurs at the fastest rate at pressures between 60 and 70 MPa or 40–30 MPa below the hornblende stability field for single-step runs, whereas reaction rims form preferentially at lower pressures (10–50 MPa) for multistep runs with increasing decompression duration.

7. SUMMARY

Resolving rates of magmatic processes while stored in shallow crustal reservoirs regions requires a combination of several approaches, including major and trace element compositional zoning patterns in crystals coupled with diffusion modeling as well as the use of radioactive isotopes from whole-rock samples and mineral separates. Whereas short-lived processes like magma mixing and eruption triggering through the intrusion of mafic magma into crystal-rich magma "mush" are thought to occur over days, weeks, or months, magma assimilation and crystal—melt fractionation are thought to occur over timescales of tens to thousands of years. Future work will continue to address the role of phase equilibria experiments, geothermobarometric models, and geophysical records in accurately identifying long- and short-lived magma storage systems.

Determining the rate at which magma ascends during volcanic eruptions is challenging for several reasons because we cannot directly observe it and ascent rate is influenced by a number of factors, including the physical and chemical properties of the magma, such as its temperature, composition, volatile budget, and crystallinity, all of which change during ascent and necessarily affect magma viscosity and density. In addition, ascending magma may be obstructed by or confined in volcanic conduits, which will also influence magma ascent rates. Thus, although magma ascent rates appear to be relatively consistent over a wide range of magma compositions, as shown in Table 9.1, it is important to remember that these rates should be considered as averages. Moreover, every technique that is used to constrain magma ascent rates has limitations, whereas extrusion rate may be a preferred approach to resolve magma ascent rates in the upper few hundred meters of the conduit, groundmass crystallization and experimentally reproduced phenocryst—melt reactions are more appropriate for understanding magma ascent at depths between ∼4 km and a few hundred meters.

Constraining magma ascent rates for explosive eruptions continues to present even more challenges than extrusive eruptions, in part because of the way volatiles are exsolved and expand in rising magmas that ascend at rates in excess of ∼0.1 m/s. The radical transformation that ascending magma undergoes as it initially ascends at velocities of a few millimeters per second at depths of several kilometers to velocities in excess of hundreds of meters per second during violent eruptions at the vent argues that more attention focused on advancing our understanding of volatile diffusion and degassing may eventually allow for better constraints on magma ascent rates during explosive eruptions.

SEE ALSO THE FOLLOWING ARTICLES

Plate Tectonics and Volcanism ● Thermodynamic and Transport Properties of Silicate Melts and Magma ● Volatiles in Magmas ● Magma Chambers ● Magma Transport in Dikes ● Magma Ascent and Degassing at Shallow Levels ● Calderas ● Lava Flows and Rheology.

FURTHER READING

Browne, B.L., Gardner, J.E., 2006. The influence of magma ascent path on the texture, mineralogy, and formation of hornblende reaction rims. Earth Planet. Sci. Lett. 246, 161—176.

Carey, S., Sigurdsson, H., 1985. The May 18, 1980 eruption of Mount St. Helens: 2. Modeling of dynamics of the Plinian phase. J. Geophys. Res. 90, 2948—2958.

Castro, J.M., Dingwell, D.B., 2009. Rapid ascent of rhyolitic magma at Chaitén Volcano, Chile. Nature 461, 780—783.

Cooper, K.M., Kent, A.J., 2014. Rapid remobilization of magmatic crystals kept in cold storage. Nature 506, 480—483.

Couch, S., Sparks, R.S.J., Carroll, M.L., 2003. The kinetics of degassing induced crystallization at Soufrière Hills Volcano, Montserrat. J. Petrol. 44, 1477—1502.

Druitt, T.H., Young, S.R., Baptie, B., Bonadonna, C., Calder, E.S., Clarke, A.B., Cole, P.D., Harford, C.L., Herd, R.A., Luckett, R., Ryan, G., Voight, B., 2002. Episodes of cyclic Vulcanian explosive activity with fountain collapse at Soufrière Hills Volcano, Montserrat. Mem. Geol. Soc. Lond. 21, 281—306.

Gardner, J.E., 2009. The impact of pre-existing gas on the ascent of explosively erupted magma. Bull. Volcanol. 71, 835—844.

Geschwind, C.H., Rutherford, M.J., 1995. Crystallization of microlites during magma ascent: the fluid mechanics of 1980—1986 eruptions of Mount St Helens. Bull. Volcanol. 57, 356—370.

Hammer, J.E., Rutherford, M.J., 2002. An experimental study of the kinetics of decompression-induced crystallization in silicic melt. J. Geophys. Res. 107, 8-1—8-24.

Henton, S.M., 2013. Experiment vs nature: Using amphiboles to test models of magma storage and pre-eruptive magma dynamics preceding the 2006 eruption of Augustine Volcano. Ph.D. Alaska, University of Alaska Fairbanks, 147 pages.

Larsen, J.F., Nye, C.J., Coombs, M.L., Tilman, M., Izbekov, P., Cameron, C., 2010. Petrology and Geochemistry of the 2006 Eruption of Augustine Volcano. In: Power, J.A., Coombs, M.L., Freymueller, J.T. (Eds.), The 2006 eruption of Augustine Volcano. Alaska: U.S. Geological Survey Professional Paper 1769.

Lee, C., Lu, Z., Jung, H., Won, J., Dzurisin, D., 2010. Surface deformation of Augustine Volcano, 1992—2005, from multiple-interferogram processing using a refined small baseline subset (SBAS) interferometric synthetic aperture radar (InSAR) approach. In: Power, J.A., Coombs, M.L., Freymueller, J.T. (Eds.), The 2006 eruption of Augustine Volcano. Alaska: U.S. Geological Survey Professional Paper 1769.

Liu, Y., Zhang, Y., 2000. Bubble growth in rhyolitic melt. Earth Planet. Sci. Lett. 181, 251—264.

Pallister, J.S., Diefenbach, A.K., Burton, W.C., Muñoz, J., Griswold, J.P., Lara, L.E., Lowenstern, J.B., Valenzuela, C.E., 2013. The Chaitén rhyolite lava dome: eruption sequence, lava dome volumes, rapid effusion rates and source of the rhyolite magma. Andean Geol. 40, 277–294.

Power, J.A., Coombs, M.L., Freymueller, J.T., 2010. The 2006 Eruption of Augustine Volcano. U.S. Geological Survey Professional Paper 1769, Alaska.

Power, J.A., Lalla, D.J., 2010. Seismic observations of Augustine Volcano, 1970–2007. In: Power, J.A., Coombs, M.L., Freymueller, J.T. (Eds.), The 2006 eruption of Augustine Volcano. Alaska: U.S. Geological Survey Professional Paper 1769, pp. 3–40.

Rutherford, M.J., Hill, P.M., 1993. Magma ascent rates from amphibole breakdown: an experimental study applied to the 1980–1986 Mount St. Helens eruptions. J. Geophys. Res. 98, 19667–19685.

Rutherford, M.J., Devine, J.D., 2003. Magmatic conditions and magma ascent as indicated by hornblende phase equilibria and reactions in the 1995–2002 Soufrière Hills magma. J. Petrol. 44, 1433–1453.

Rutherford, M.J., 2008. Magma ascent rates (in Minerals, inclusions and volcanic processes) Reviews in Mineralogy and Geochemistry 69 (1), 241–271.

Scandone, R.S., Malone, S., 1985. Magma supply, magma discharge and readjustment of the feeding system of Mount St. Helens during 1980. J. Volcanol. Geotherm. Res. 23, 239–262.

Sherrod, D.R., Scott, W.E., Stauffer, P.H., 2008. A Volcano Rekindled: The Renewed Eruption of Mount St. Helens, 2004–2006. U.S. Geological Survey Professional Paper 1750.

Szramek, L.A., Gardner, J.E., Hort, M., 2010. Cooling-induced crystallization of microlite crystals in two basaltic pumice clasts. Am. Mineral. 95, 503–509.

Szramek, L., Gardner, J.E., Larsen, J., 2006. Degassing and microlite crystallization of basaltic andesite magma erupting at Arenal Volcano, Costa Rica. J. Volcanol. Geotherm. Res. 157, 182–201.

Turner, S., Costa, F., 2007. Measuring timescales of magmatic evolution. Elements 3, 267–272.

Webster, J.D., Mandeville, C.W., Goldoff, B., Coombs, M.L., Tappen, C., 2010. Augustine Volcano; the influence of volatile components in magmas erupted A.D. 2006 to 2,100 years before present. In: Power, J.A., Coombs, M.L., Freymueller, J.T. (Eds.), The 2006 eruption of Augustine Volcano. Alaska: U.S. Geological Survey Professional Paper 1769.

Wright, H., Cashman, K.V., Mothes, P.A., Hall, M.L., Ruiz, A.G., Le Pennec, J., 2012. Estimating rates of decompression from textures of erupted ash particles produced by 1999–2006 eruptions of Tungurahua volcano, Ecuador. Geology 40, 619–622.

Chapter 10

Magma Transport in Dikes

Helge Gonnermann
Department of Earth Science, Rice University, Houston, Texas, USA

Benoit Taisne
Earth Observatory of Singapore, Nanyang Technological University, Singapore

Chapter Outline

GLOSSARY

brittle A material that has little capacity to deform plastically and is not able to relax stresses that become localized at crack-like defects during an applied deformation. In such a material, such defects will propagate rapidly and spatially localized deformation that is brittle fracture. Such brittle deformation involves the breaking of chemical bonds, whereas plastic and viscous deformation involves microscopic motion of atoms or molecules past one another.

buoyancy Buoyancy is a force induced by the density difference between two bodies. In the case of dikes, the buoyancy force is proportional to the density of the host rock minus the density of the magma. Therefore, this force can be either positive, and drive the magma toward the surface, or negative, and prevent the ascent of the magma. The level at which the buoyancy changes sign is called the level of neutral buoyancy.

buoyancy length The length of a dike at which its buoyancy force is sufficient to result in its upward propagation without the need of additional magma supply from a pressurized reservoir.

constitutive equation A material-specific relation between two physical quantities, such as the response of a material to an externally applied force.

crack A crack is a defect around which atomic bonding is completely broken. This results in a large volume expansion and stress concentration and, hence, a long-range stress field surrounding cracks.

cubic law The volumetric flow rate of a Newtonian viscous fluid through a planar channel of high aspect ratio and that is bounded on both sides by solid walls and scales with the cube of channel thickness. The functional relationship between flow rate and viscosity, pressure gradient, and channel thickness cubed is called the cubic law.

elastic Upon an applied stress a material will deform. If the deformation is instantaneous and recoverable it is called elastic.

feeder dike Dikes that supply magma to the surface during volcanic eruptions.

fracture Interaction among cracks during an applied stress leads to localized deformation that is discernible macroscopically, that is brittle fracture. In other words, brittle fracture starts with the formation of microcracks, which interact to form a macroscopically discernible fracture.

fracture toughness The fracture toughness, specific to the material and opening mode (see mode I definition), is the capacity of the material to resist fracturing when stressed.

The Encyclopedia of Volcanoes. http://dx.doi.org/10.1016/B978-0-12-385938-9.00010-9

Griffith's theory According to Griffith's theory, an existing crack will propagate if the strain energy released is greater than the surface energy created by the formation of the two new material surfaces.

Hooke's law The constitutive equation for a material that undergoes a recoverable deformation where the strain is linearly proportional to the applied stress.

level of neutral buoyancy (LNB) Theoretically, the level at which buoyantly ascending magma has the same density as the surrounding rock.

linear elastic fracture mechanics (LEFM) Fracture mechanics is the study of the propagation of cracks in elastic materials where the relationship between applied stress and strain is linear. The basic assumptions of LEFM are (1) the presence of crack-like defects; (2) a crack is a free, internal, planar surface with stresses near the crack tip that can be calculated from linear elasticity; and (3) the growth of such cracks leads to the failure, that is, brittle fracture, of the material.

lithosphere The lithosphere can be defined based on seismic, mechanical, or thermal considerations. The seismic lithosphere of the Earth was initially defined as a static unit lying above a low velocity zone for shear waves in the upper mantle, presumably associated with partial melting and weak mechanical behavior, that is, the asthenosphere. The mechanical lithosphere comprises those rocks that remain a coherent part of the plates on geological timescales and deform in an elastic manner, even on long timescales, whereas the underlying asthenosphere undergoes permanent (plastic) deformation. Studies of flexure and deformation due to topographic loads, such as volcanic edifices and mountain belts, have led to a definition of lithosphere as the layer that sustains elastic stresses to support these loads. The thermal lithosphere is taken to be the upper thermal boundary layer of a planet, where heat is transported by conduction, as opposed to the underlying asthenosphere where convection is the dominant heat transfer mechanism. Because the mechanical properties of rocks are strongly temperature-dependent, the mechanically and thermally defined lithospheres are essentially equivalent.

Maxwell model A model for the deformational behavior of viscoelastic materials, which is based on the timescale at which deformation changes from recoverable to nonrecoverable.

mode I There are three ways of applying a force to enable crack propagation. One is a tensile stress normal to the plane of the crack and the two others are shear stresses parallel to the plane of the crack. The mode in which a crack opens under tensile stress normal to the plane of the crack is called mode I, whereas the other two are called mode II and mode III.

Newtonian viscosity If for a viscous material the relationship between applied shear stress and resultant shear rate is linear, the constant of proportionality is called the Newtonian viscosity.

plastic If a solid material undergoes a time-dependent and nonrecoverable deformation under an applied stress it is called plastic. The deformation may be distributed homogeneously throughout the material or it may be localized in a narrow region.

stress intensity factor The growth of a crack under an applied stress is related to the tensile stress acting at the crack tip. For a given opening mode, the relationship between applied stress and crack length is characterized by a constant, called the stress intensity factor.

thermal boundary layer It corresponds to the layer of magma closest to the dike walls where heat transfer is dominated by conduction toward the dike wall, as opposed to along-dike advection by magma flow.

viscoelastic A material that undergoes recoverable deformation on short timescales and nonrecoverable deformation on long timescales is called viscoelastic. The time at which deformation changes from recoverable to nonrecoverable is often referred to as the Maxwell time.

viscous Under an applied stress a viscous fluid undergoes a time-dependent, nonrecoverable deformation that is homogeneous throughout the fluid volume.

yield stress Upon an applied stress a material will deform. If the material will undergo recoverable deformation below some threshold stress value and permanent deformation above this value, this threshold stress value is called the yield stress. After yielding, nonrecoverable deformation is usually referred to as plastic deformation. However, sometimes the terms viscous and plastic deformation are used interchangeably.

1. INTRODUCTION

Dikes are the principal mode of magma transport within and through planetary **lithospheres**. Dikes represent the critical link between regions of melt production and regions of melt accumulation, such as magma chambers. In some cases, dikes propagate to a planet's surface, resulting in fissure eruptions. In other cases, dikes may give way to more spatially localized conduits at shallow depths, supplying magma to individual volcanic vents. This chapter provides a summary of the mechanics and

FIGURE 10.1   Aerial view of a 8-km-long dike that radiates from Ship Rock, New Mexico, USA, a volcanic plug that rises 550 m above the valley floor. Three prominent dikes radiate out from Ship Rock (Nature/Universal Images Group/Getty Images).

dynamics of dike propagation and magma transport through dikes.

2. OVERVIEW

Dikes and sills are sheet-like bodies of igneous rock that were emplaced as magma within preexisting rock. They are typically of the order of $1-10$ m in thickness and of the order of $1-100$ km in lateral extent (Figure 10.1). Dikes are formed by fracturing the host rock, due to magma injection (Figure 10.2). Their propagation is oriented preferentially in the direction perpendicular to the minimum total compressive stress of the surrounding rock, but may also follow preexisting planes of weakness, such as bedding planes or preexisting **fractures**.

The formation and propagation of dikes may allow the rapid transport of magma without extensive solidification due to cooling. Estimates of dike propagation speeds, deduced from seismic observations, are of the order of meters per second. In order for dikes to propagate, magma needs to exert sufficient stress at the dike tip to overcome the fracture strength of the rock. Furthermore, magma pressure within the dike needs to exceed the stresses exerted by the surrounding rock on the dike walls, in order to keep the dike from closing. Pressure gradients within the magma also need to balance **viscous** resistance, in order to allow magma to flow into the dike as it grows in length. Processes that may result in the required magma pressures may include the change in volume associated with melting, compaction of partially molten rock, magma **buoyancy** relative to the surrounding rock, and in the case of dikes emanating from inflated magma reservoirs, the **elastic** energy stored in the reservoir's wall rock, commonly known as reservoir overpressure.

Theoretical treatments of dike propagation therefore have to consider the fracture mechanics at the dike tip, magma flow within the dike, deformation of the surrounding wall rock, as well as cooling of the dike (Rubin, 1995). We will define several characteristic pressure scales, in order to facilitate the distinction of dominant stress balances for dike propagation. End-member cases for propagating dikes are (1) dikes of finite volume that are "self-propagating" due to magma buoyancy; (2) dikes that propagate within partially molten rock and are supplied with melt from the surrounding porous wall rock; and (3) dikes that are connected to a magma reservoir and are supplied by magma from within the reservoir.

3. DEFORMATIONAL BEHAVIOR OF MATERIALS

Upon an applied stress a continuous material will deform. For a fluid of **Newtonian viscosity,** the resultant deformation is nonrecoverable. For a linearly elastic solid the deformation is instantaneous and recoverable. If the solid deforms elastically below some threshold stress, but yields to nonrecoverable deformation above this **yield stress** it is called **plastic**. Nonrecoverable deformation can also be **brittle** and localized, which involves the breaking of chemical bonds and is called fracture.

3.1. Elastic Deformation

The **constitutive equation** for linear elastic deformation is

$$\sigma = M\epsilon, \quad (10.1)$$

where σ is the applied stress, M is the elastic constant, and ϵ is strain. This relation is also called **Hooke's law**. There are five elastic constants. For example, Young's modulus, E, applies to the case of uniaxial stress

$$\sigma = E\epsilon \quad (10.2)$$

and the shear modulus, μ, in the case of shear stress

$$\sigma = \mu\epsilon. \quad (10.3)$$

The relationship between E and μ is given by

$$E = 2\mu(1 + \nu), \quad (10.4)$$

where ν is the Poisson's ratio.

3.2. Viscous or Plastic Deformation

The constitutive equation for linear viscous deformation is

$$\sigma = \eta\dot{\epsilon}, \quad (10.5)$$

where η is the Newtonian viscosity and $\dot{\epsilon}$ is the strain rate. For many natural materials, the relation of stress to strain rate is not linear and constitutive models for such non-Newtonian rheology often relate stress to strain rate via a power law.

3.3. Viscoelastic Deformation

In addition many materials behave like an elastic solid on short timescales, but viscously over longer times. The most well-known constitutive model for such **viscoelastic** deformation is the **Maxwell model**, where the total strain rate is the sum of an elastic strain rate and a viscous strain rate, with the latter becoming increasingly dominant at times greater than or equal to the Maxwell time.

3.4. Brittle Fracture

Under application of a stress, the fragmentation of a solid volume into two or more parts is referred to as fracture. Although plastically deformable solids can undergo

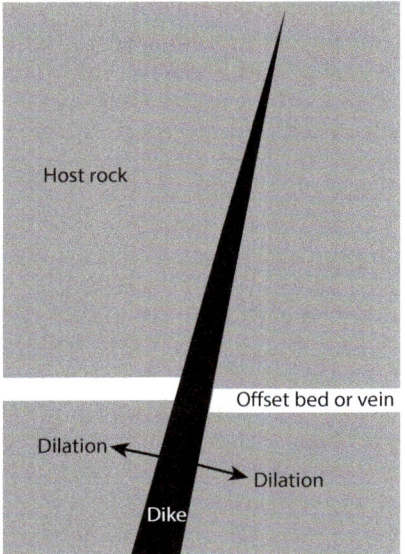

FIGURE 10.2 Schematic illustration of a dike with wall displacements that correspond to dilation by magma injection, as indicated by the offset of preexisting planar features. This type of fracture wall displacement is called opening mode or mode I displacement.

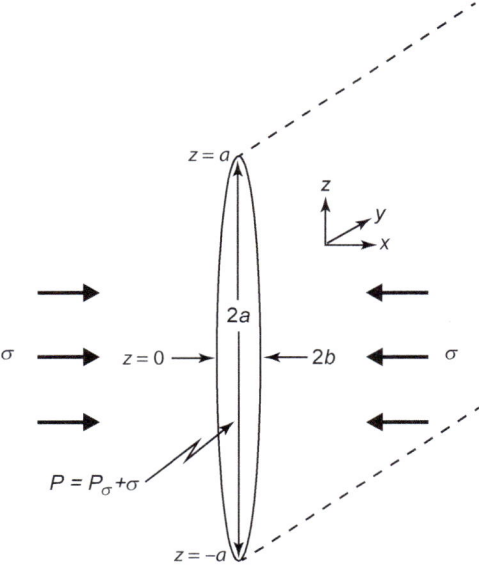

FIGURE 10.3 Geometry of an elastically walled dike of infinite extent in the *y* direction. The dike is of height $2a$, width $2b$, and has an internal pressure of P that exceeds the stress in the surrounding rock is σ by P_σ.

ductile fracture, we will focus, herein, on brittle fracture. The brittle fracture of a solid body typically involves the accumulation of damage, resulting in the nucleation of **cracks** or voids and finally, their growth. On a macroscopic level, the mechanical properties of solids tend to be limited by their imperfections, also known as defects. Linear elastic fracture mechanics (LEFM) provides a conceptual framework that is based on crack propagation from such defects, in order to asses a material's **fracture toughness**. Because the in situ stresses at depth are never tensile, dike formation and propagation requires that magma pressure exceeds the least principal stress by some critical amount in order for a tensile fracture to grow.

4. LINEAR ELASTIC FRACTURE MECHANICS

LEFM provides a framework for quantifying the stress required to propagate a fracture, through the use of a parameter known as the fracture toughness.

4.1. Elastic Pressure

Consider a two-dimensional crack (Figure 10.3) of length $2a$ and subject to a uniform excess internal pressure P_σ, relative to a uniform remote stress σ. This crack will have a half thickness of

$$b(z) = \frac{(1-\nu)P_\sigma}{\mu}\sqrt{a^2 - z^2} \quad |z| < a, \qquad (10.6)$$

where z is the vertical coordinate. From the pressure required to open such an idealized crack by $2b$ at $z=0$, one can define a characteristic elastic pressure as

$$P_e \sim \frac{\mu}{(1-\nu)}\frac{b}{a}. \qquad (10.7)$$

4.2. Fracture Pressure

According to Griffith, an existing crack will propagate if the strain energy released is greater than the surface energy created by the formation of the two new crack surfaces. The practical application of **Griffith's theory** is difficult, in part because of the challenges posed by measuring quantities such as surface energy. Instead material fracture is conventionally treated within the framework of LEFM, which postulates the existence of microcracks within any real solid volume. Because stress becomes concentrated at the tips of these microcracks they can grow within a process zone located at the fracture tip, resulting in the opening and propagation of a macroscopic fracture.

As a consequence of these material imperfections, actual fracture strengths are considerably less—by about one to two orders of magnitude—than the theoretical cohesive material strength, also called cleavage stress. Moreover, the fracture of a material can be predicted in terms of the tensile stress acting in the crack tip region.

Under pure tensile failure (**mode I**) this stress is proportional to the **stress intensity factor**, defined as

$$K = Y \Delta P_\sigma \sqrt{\pi a}, \qquad (10.8)$$

where Y is a constant that depends on the crack opening mode and geometry, and for the geometry under consideration has a value of 1. The minimum value of K at which fracture occurs or a crack propagates can be determined experimentally and is known as the fracture toughness, K_c. Laboratory measurements at atmospheric conditions indicate that $K_c \sim 1$ MPa m$^{1/2}$ and that it increases with pressure, although the actual fracture toughness under field conditions remains poorly resolved and the range of plausible values is from about 1 MPa m$^{1/2}$ to 100 MPa m$^{1/2}$. The characteristic fracture pressure scale is thus defined as

$$P_f \sim \frac{K_c}{\sqrt{a}}. \qquad (10.9)$$

5. MAGMA FLOW IN DIKES

Most treatments of magma flow within dikes approximate the flow as one-dimensional, laminar, and incompressible. In the case of a vertically oriented dike, the volumetric flow rate is

$$Q = 2bw\bar{u}_z = -\frac{w}{3\eta}b^3\frac{dP'}{dz}, \qquad (10.10)$$

where $\bar{u}_z$ is the average magma velocity in the z-direction, η is the magma viscosity, P' the nonhydrostatic magma pressure, and w is the dike breadth, which is the dimension perpendicular to the direction of propagation (z) and to the width (x) of the dike. Equation (10.10) yields the characteristic viscous pressure scale

$$P_\eta \sim \frac{\eta a Q}{b^3 w} \sim \frac{\eta a \bar{u}}{b^2}. \qquad (10.11)$$

Furthermore, using the average magma velocity, $\bar{u}_z$, together with mass balance results in an equation for the change in dike width

$$\frac{db}{dt} + \frac{d}{dz}(b\bar{u}_z) = 0. \qquad (10.12)$$

6. DIKE FORMATION AND PROPAGATION

6.1. Dikes within Melting Zones

The isotopic disequilibria found in mid-ocean ridge, Ocean Island and island arc basalts require average melt ascent rates of 1–50 m yr^{-1} (Kelemen et al., 1997; Turner and Bourdon, 2011). It is thought that these high ascent rates

and isotopic disequilibria are most likely a consequence of channelized porous flow by reactive melt transport within the asthenosphere. However, there are other processes that have the potential to enhance rates of melt transport in melting zones, such as the formation of melt-filled veins during deformation of the partially molten rock (Kohlstedt and Holtzman, 2009).

To what extent dikes contribute to melt transport within melting zones, however, remains unclear (Kelemen et al., 1997; Weinberg, 1999). In the case of highly viscous felsic melts, the flow of melt into the dike from the surrounding partially molten rock is very slow and it has been suggested that a viable alternative could be that upward heat transfer may facilitate pervasive melt migration through ductile deformation, rather than elastic fracture (Weinberg, 1999).

Dike growth in partially molten peridotite could be a consequence of excess pore pressure due to the volume increase associated with melting and compaction (McKenzie, 1984). Melt could thus be forced into preexisting melt pockets and exceed the rock fracture toughness, thus facilitating fracture propagation (Rubin, 1998). If dike growth in partially molten peridotite is treated as a purely elastic process, it is found that the upward propagation and width of dikes are primarily controlled by the balance between the rate of porous inflow from the surrounding rock and upward flow of melt within the dike. Because the melt flux into the dike is, to first order, dependent on the ratio of permeability to melt viscosity the porous influx for felsic dikes is rate limiting, making dike propagation difficult. For asthenospheric dikes the balance between porous inflow and upward propagation results in theoretical dike widths of the order of 10^{-2} m (Rubin, 1998). Such dikes, if they exist, would be too thin to propagate within the lithosphere without freezing. Consequently, lithospheric transport of asthenospheric melts requires that dikes are fed by large accumulations of melt, perhaps at the top of the melting column or from a channelized tributary network (Weinberg, 1999).

6.2. Buoyancy-Driven Dike of Finite Volume

Vertically oriented dikes of constant volume can "self-propagate," once they reach a critical length, called the **buoyancy length**, denoted as a_b. At this length, the balance between the body force exerted on the magma within the dike and the lithostatic force against the dike walls causes the dike tip to propagate upward. At the same time the tail is being squeezed shut, thereby displacing the magma upward. The stress due to buoyancy scales as the characteristic buoyancy pressure

$$P_\rho \sim \Delta \rho g a, \qquad (10.13)$$

where $\Delta\rho = \rho_r - \rho$ is the difference between host-rock density, ρ_r, and magma density, ρ. It is balanced by the fracture pressure, P_f (Eqn (10.9)). Equating these two pressure scales one obtains a scaling relation for the buoyancy length, that is, the half length required for a "self-propagating" dike, given by

$$a_b \sim \left(\frac{K_c}{\Delta\rho g}\right)^{2/3}. \qquad (10.14)$$

An important implication of this relation is that for reasonable geological conditions, that is, $\Delta\rho \sim 100$ kg m^{-3} and $K_c \sim 1$ MPa m$^{1/2}$, once a magma-filled fracture has formed and grown to a length of the order of 100 m, the fracture resistance of the rock to further propagation is negligible.

For a dike of volume, V, Taisne et al. (2011) found that during the early stages of dike propagation the dike increases in width. Once the dike has reached a width of

$$w_f \sim \left(\frac{\mu}{1-\nu}\frac{V}{\Delta\rho g}\right)^{1/4}, \qquad (10.15)$$

obtained from a balance between P_e and P_ρ, the increase in dike width becomes negligible, due to the fracture toughness of the solid. Instead, the dike will predominantly increase in length, at a rate governed by a balance between P_η and P_ρ, and at a velocity of $\overline{u} = a/t$. Its length thus scales as

$$a \sim \left(\frac{\Delta\rho g V^2}{\eta w_f^2} t\right)^{1/3}. \qquad (10.16)$$

Because the velocity of melt flow depends on the square of dike width (Eqn (10.10)), it becomes virtually impossible for dike tail to completely close and some fraction of melt will become "trapped" within the tail. Consequently, the net volume of melt within the dike above the tail decreases with time (Figure 10.4) and the dike decreases in thickness until it eventually stalls.

6.3. Dike Connected to a Magma Source

For a dike that remains connected to regions of subsurface magma accumulation, that is, magma reservoirs, the volume of melt within the dike will increase over time. Magma flow into the dike may be driven by a combination of buoyancy and reservoir pressure, with the latter due to elastic deformation of the reservoir's host rock. Assuming that on the timescale of dike propagation the rock can be treated as elastic and that the magma is incompressible, the relationship between the change in magma pressure, relative to the ambient stress within the surrounding wall rock, is linearly proportional to the change in volume of stored magma, ΔV_{ch} within a reservoir of volume, V. The resultant pressure scale is

$$P_{ch} \sim \mu \frac{\Delta V}{V}. \qquad (10.17)$$

Changes in magma storage arise due to an imbalance between volumetric rate of magma supply to the chamber, Q_s, and magma volumetric flow rate into the dike, Q. The resultant rate of change in chamber pressure is

$$\frac{dP_{ch}}{dt} \sim \frac{\mu}{V}[Q_s(t) - Q(t)]. \qquad (10.18)$$

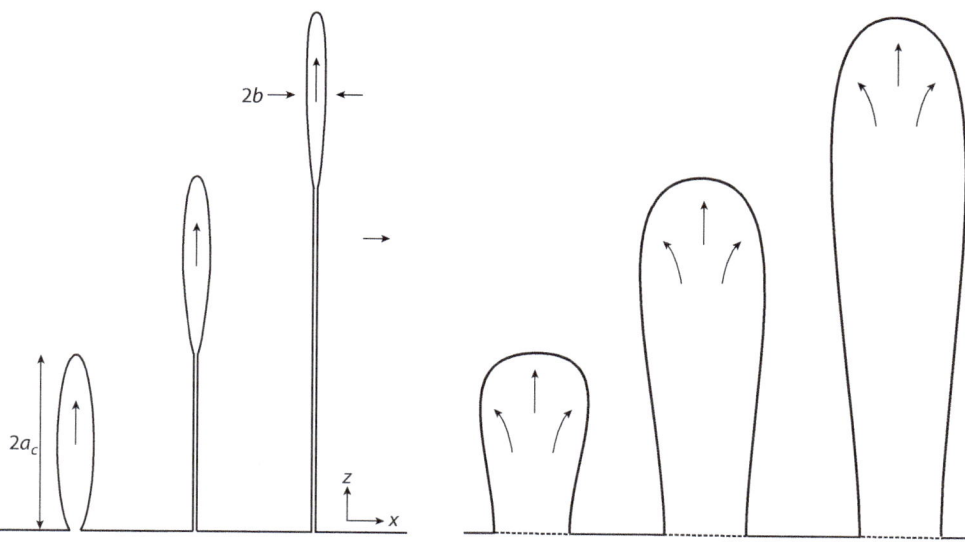

FIGURE 10.4 Schematic illustration showing the shape of a propagating dike of finite volume. As the dike ascends a small volume of magma remains trapped in the tail, which does not close completely. Consequently, the head of the dike decreases in thickness as it rises (Taisne et al., 2011).

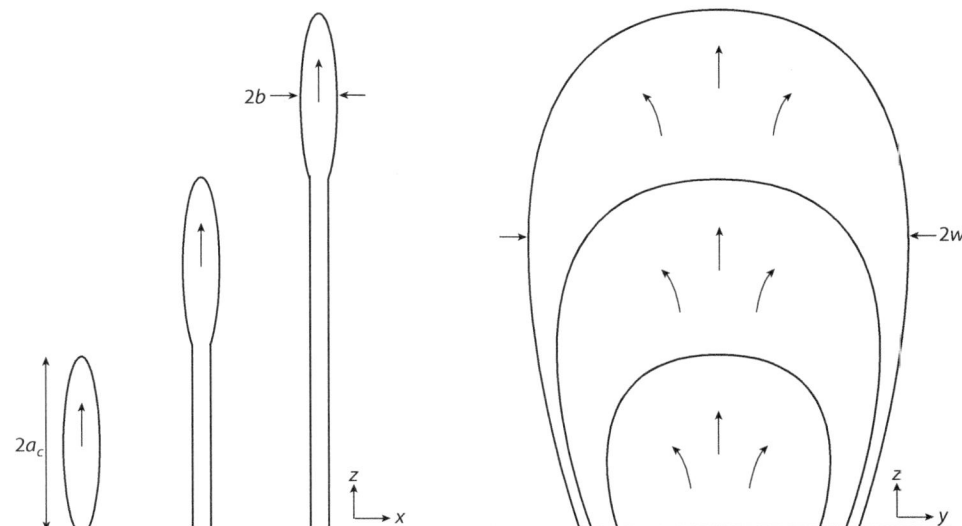

FIGURE 10.5 Schematic illustration of a dike fed by a pressurized magma reservoir, propagation in the steady-state regime (modified after Menand and Tait (2002)).

Typically, the tensile strength of the surrounding wall rock is of the order of $1-10$ MPa, implying that excess magma chamber pressures will be of that magnitude (Gudmundsson, 2012). The relative importance of P_{ch} and P_ρ depends on dike length, dike orientation, and magma density, relative to the surrounding rock. Menand and Tait (2002) showed that for a constant P_{ch}, dikes will initially propagate in a regime where P_ρ is negligible and the dominant balance is between P_{ch} and P_f. In this initial transient regime dike propagation is radial with both a and w increasing (Figure 10.5). For vertically oriented dikes, also called **feeder dikes** because they typically feed surface eruptions from magma reservoirs, the flow eventually transitions to a balance between P_ρ and P_f in the bulbous dike tip and between P_ρ and P_η in the thinner dike tail (Figure 10.5). This transition occurs at a critical length, a_c, which is defined by this balance and the thickness of the dike tail can be estimated from the balance between P_ρ and P_η as (Lister and Kerr, 1991)

$$b_t \sim \left(\frac{\eta Q}{w \Delta \rho g} \right)^{1/3}. \tag{10.19}$$

The volumetric flow rate of magma into an incipient dike is inversely proportional to magma viscosity and proportional to the dike thickness cubed (Eqn (10.10)). Consequently, there is an expected time delay of the order of hours to days for basalts and years for rhyolites before an incipient dike tip becomes sufficiently pressurized to propagate (McLeod and Tait, 1999). Once a dike has exceeded a length of a_c, an important implication of the above results is that for vertically oriented dikes connected to a source reservoir, the negligible resistance of rock to fracture will allow them

to propagate until they either reach a level of neutral buoyancy (LNB) or until the magma within the dike freezes. The rate of propagation will be controlled by the viscous resistance to flow of magma into the dike tip and Eqn (10.19) implies that dikes of highly viscous felsic magma will be considerably thicker than mafic dikes and/or propagate a lower speeds.

7. THE FATE OF DIKES

7.1. The Level of Neutral Buoyancy

For buoyantly ascending feeder dikes, the flow adjusts so that the local viscous pressure gradient, dP_η/dz is balanced by local hydrostatic pressure gradient, dP_ρ/dz, in order to transfer the volume flux, Q, from below. As a result, the dike thickness varies with depth according to the balance between elastic stress, P_e, and excess internal pressure, P_σ. The balance between P_η and P_ρ in theory requires that magma will not ascend beyond the regional **LNB**, the depth above which dP_ρ/dz changes from values of less than zero to greater than zero (Figure 10.6). In the bulbous region near the tip, however, P_η can be neglected and the pressure balance is between P_e and P_ρ. In order to accommodate the volume flux, Q, the head of the dike will inflate thereby increasing P_e and allowing an overshoot beyond the LNB by slow propagation of the dike tip into the upper layer where magma is negatively buoyant (Taisne and Jaupart, 2009).

Eventually, because of magma accumulation and local stress buildup, magma will spread laterally along the LNB, either as vertically oriented dikes or horizontally oriented sills, depending on the stress field. Repeated intrusions may result in the formation of crustal hot zones, wherein

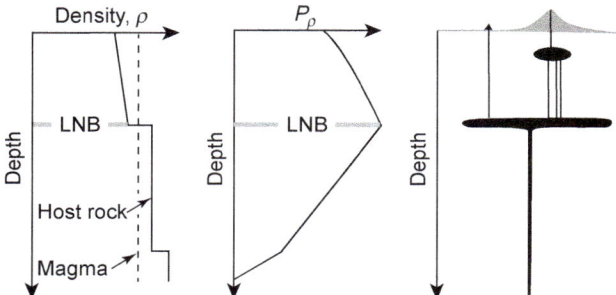

FIGURE 10.6 Schematic illustration of the density ρ (left) and buoyancy pressure P_ρ (center) for a feeder dike at the scale of a density stratified lithosphere (adapted from Lister and Kerr, 1991). The level of neutral buoyancy (LNB) is defined by the change in sign of dP_ρ/dz. At the LNB magma may accumulate and differentiate in "hot zones," from where shallow crustal reservoirs are supplied.

fractional crystallization and assimilation allow magma to evolve toward more felsic compositions and sufficient buoyancy to ascend to shallower levels.

7.2. Ambient Stress Field

Dikes preferentially intrude in the plane perpendicular to the direction of minimum compressive stress. This behavior can, however, be modulated in the presence of significant variations in fracture toughness of the surrounding rock, for example, due to stratification (Figure 10.7; Maccaferri et al., 2010). It is thus frequently observed that dikes invade such planes of weakness or preexisting fractures, perhaps becoming sills or incipient magma reservoirs in the process. However, if rock fracture is negligible in the overall force balance, intrusion along preexisting structures may offer negligible mechanical advantage and dike orientations will be predominantly determined by the regional and/or local stress fields, with

the latter potentially changing abruptly if there is strong stratification. Dikes may thus become arrested, if the local stress field is unfavorable to dike propagation, and magma supply from depth is insufficient for magma accumulation and breakthrough (Gudmundsson, 2006).

The upper crustal stress field gets further modified due to the presence of volcanic edifices at the Earth's surface. As the volcanic edifice grows in size, it generates an increasingly significant compressive stress field that may impede vertical dike propagation, especially for less evolved, more dense magmas (Pinel and Jaupart, 2000). The result may be predominant magma storage at depth. By the same token, for large and shallow reservoirs the edifice load may produce tensile stresses surrounding the magma chamber, which reach a maximum at the top of the chamber. Consequently, dike formation directly beneath the summit may be favored, resulting in the development of a central vent system. In detail, the distribution of compressive and tensile stresses surrounding a magma chamber at depth beneath a volcanic edifice depends on edifice size (height and diameter), as well as chamber size and depth.

7.3. Thermal and Viscous Effects

Thermal effects of magma ascent are perhaps most important during basaltic eruptions, which often are associated with dikes that form linear fissures at the Earth's surface. Whereas small eruptions typically cease within less than a day, larger eruptions tend to become localized into individual vents after sometime. These vents may perhaps become reactivated during subsequent eruptions. This localization is a consequence of the strong dependence of flow rate on dike thickness (Eqn (10.10)) and ensuing thermal and viscous feedbacks (Figure 10.8). Because most of the flow becomes channeled to the thicker sections of the dike, the rate of magma cooling as the magma ascends will

FIGURE 10.7 Schematic illustration of stress field (contours are schematic and not drawn at equal stress increments) associated with an idealized inflated magma reservoir and overlying rock strata (adapted from Gudmundsson, 2006; Karlstrom et al., 2009). Note that dikes within some zone will be "captured" by the magma chamber, as well as the effect of stratification on the stress field.

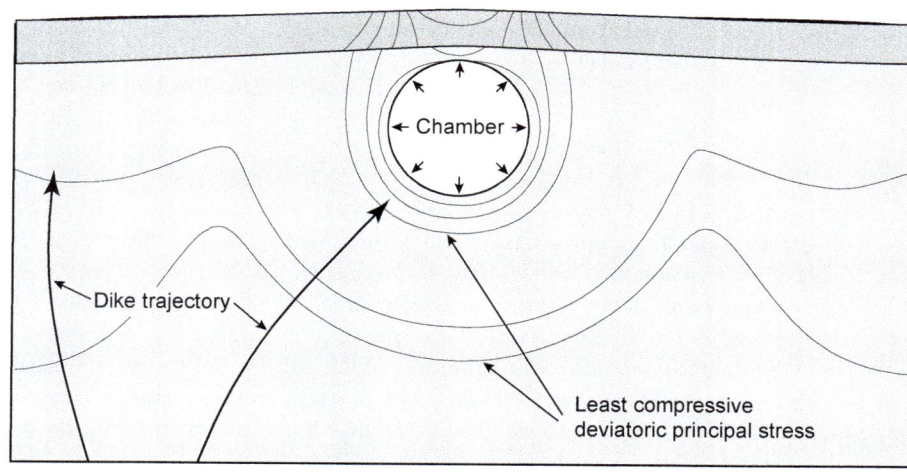

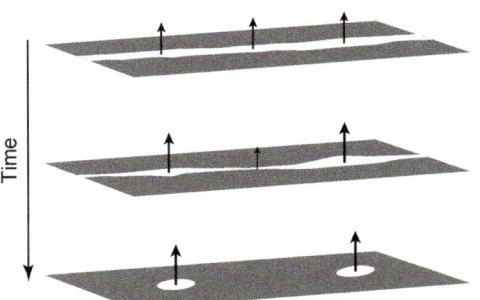

FIGURE 10.8 Schematic diagram illustrating flow localization due to feedbacks between dike thickness, magma flow rate, magma cooling, and magma viscosity (adapted from Bruce and Huppert, 1989). Because magma flow rate is slower in narrower sections of the dike, the ratio of advective heat transport to conductive magma cooling is smaller, resulting in cooler magma, higher viscosity, and, hence, a further decrease in flow rate, as well as magma solidification against dike walls. Ultimately, this may result in the localization of flow into conduit-like pathways, or in the case of eruptive fissures into the localization of the activity into individual vents.

be lowest there and perhaps even result in the melt-back of dike walls. In contrast, sections that are of small thickness will have lower flow rates and faster rates of magma cooling. This will result in increased magma viscosity, which is strongly dependent on temperature, and perhaps magma solidification against the dike walls (Figure 10.9).

Magma cooling and/or solidification and melt-back are a consequence of the conduction of heat from the flowing magma into the dike walls. The degree to which magma cools, perhaps resulting in solidification, depends on the balance between along-dike advective heat transport by the

flowing magma (which depends on the magma velocity) and the across-dike diffusion of heat to the conduit walls (which depends on the gradient in temperature across the magma within the **thermal boundary layer** close to the dike walls). The ratio of these two competing heat transfer mechanisms leads to the following dimensionless parameter (Bruce and Huppert, 1989)

$$\Pi \sim \frac{b^4 \Delta P}{\kappa \eta a^2},\qquad(10.20)$$

where $\Delta P/a$ is the average pressure gradient driving magma flow within the dike and κ is the thermal diffusivity. For small Π magma cooling will be dominant and the dike will close by the positive feedback between viscous resistance to magma flow and solidification (Wylie et al., 1999). Cooler magma will have a higher viscosity, which will reduce magma velocity, all else being equal (Figure 10.10). Hence, Π will decrease further, implying a further decrease in advective heat transfer through the dike, relative to conductive heat loss to the dike walls. The result will be flow localization, that is, sections of the dike with small Π tend to close, whereas sections with large Π remain open, perhaps even increasing in width and magma flux due to melt-back of dike walls. The latter occurs when Π is sufficiently large and the magma ascends fast enough for the advective heat flux to maintain a high magma temperature and steep thermal gradients in the magma closest to the dike walls.

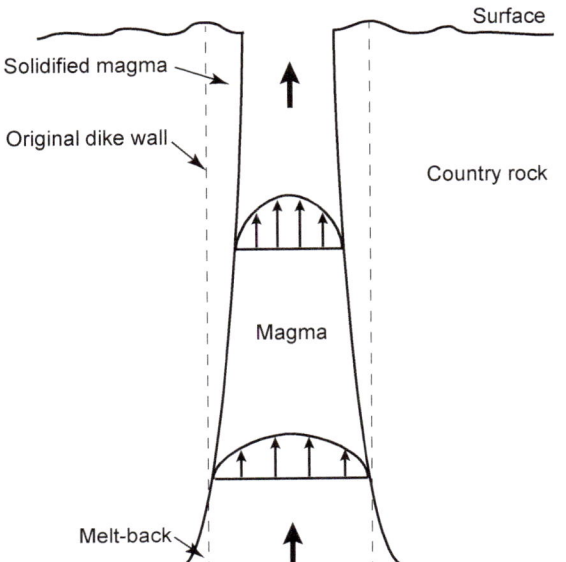

FIGURE 10.9 Schematic diagram illustrating the evolution of a two-dimensional dike undergoing widening due to melt-back near its bottom and narrowing due to solidification of magma against conduit walls (adapted from Bruce and Huppert, 1989).

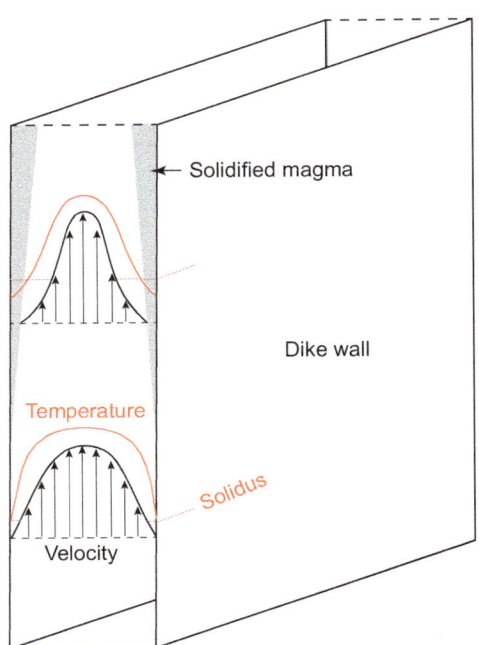

FIGURE 10.10 Schematic diagram illustrating the evolution of the velocity and temperature profiles for a magma with a temperature-dependent viscosity. As the magma cools against the conduit walls, viscosity increases. Consequently, velocity decreases, which reduces the advective heat transport, resulting in a positive feedback between cooling, increasing viscosity, and decreasing velocity (e.g., Wylie and Lister, 1995).

FURTHER READING

Bruce, P.M., Huppert, H.E., 1989. Thermal control of basaltic fissure eruptions. Nature 342, 665–667.

Gudmundsson, A., 2006. How local stresses control magma-chamber ruptures, dyke injections, and eruptions in composite volcanoes. Earth Sci. Rev. 79, 1–31.

Gudmundsson, A., 2012. Magma chambers: formation, local stresses, excess pressures, and compartments. J. Volcanol. Geotherm. Res. 237-238, 19–41.

Karlstrom, L., Dufek, J., Manga, M., 2009. Organization of volcanic plumbing through magmatic lensing by magma chambers and volcanic loads. J. Geophys. Res. 114 (B10204).

Kelemen, P.B., Hirth, G., Shimizu, N., Spiegelman, M., Dick, H.J.B., 1997. A review of melt migration processes in the adiabatically upwelling mantle beneath oceanic spreading ridges. Phil. Trans. Roy. Soc. A 355 (1723), 283–318.

Kohlstedt, D.L., Holtzman, B.K., 2009. Shearing melt out of the earth: an Experimentalist's perspective on the influence of deformation on melt extraction. Annu. Rev. Earth Planet. Sci. 37, 561–593.

Lister, J.R., Kerr, R.C., 1991. Fluid mechanical models of crack propagation and their application to magma transport in dykes. J. Geophys. Res. 96, 10049–10077.

Maccaferri, F., Bonafede, M., Rivalta, E., 2010. A numerical model of dyke propagation in layered elastic media. Geophys. J. Int. 180, 1107–1123.

McKenzie, D., 1984. The generation and compaction of partially molten rock. J. Petrol. 25, 713–765.

McLeod, P., Tait, S., 1999. The growth of dykes from magma chambers. J. Volcanol. Geotherm. Res. 92, 231–246.

Menand, T., Tait, S.R., 2002. The propagation of a buoyant liquid-filled fissure from a source under constant pressure: an experimental approach. J. Geophys. Res. 107, 2306.

Pinel, V., Jaupart, C., 2000. The effect of edifice load on magma ascent beneath a volcano. Phil. Trans. Roy. Soc. A 358, 1515–1532.

Rubin, A.M., 1995. Propagation of magma-filled cracks. Annu. Rev. Earth Planet. Sci. 23, 287–336.

Rubin, A.M., 1998. Dike ascent in partially molten rock. J. Geophys. Res. 103, 20901–20919.

Taisne, B., Jaupart, C., 2009. Dike propagation through layered rocks. J. Geophys. Res. 114 (B09203).

Taisne, B., Tait, S., Jaupart, C., 2011. Conditions for the arrest of a vertical propagating dyke. Bull. Volcanol. 73, 191–204.

Turner, S.P., Bourdon, B., 2011 (pp. 102–115). In: Dosseto, A., Turner, S.P., Van Orman, J.A. (Eds.), Melt Transport from the Mantle to the Crust — Uranium-series Isotopes. John Wiley & Sons Ltd., Chichester, UK.

Weinberg, R.F., 1999. Mesoscale pervasive felsic magma migration: alternatives to dyking. Lithos 46, 393–410.

Wylie, J.J., Helfrich, K.R., Dade, B., Lister, J.R., Salzig, J.F., 1999. Flow localization in fissure eruptions. Bull. Volcanol. 60, 432–440.

Wylie, J.J., Lister, J.R., 1995. The effects of temperature-dependent viscosity on flow in a cooled channel with application to basaltic fissure eruptions. J. Fluid Mech. 305, 239–261.

Chapter 11

Magma Ascent and Degassing at Shallow Levels

Alain Burgisser

CNRS, ISTerre, Le Bourget du Lac, France, Université de Savoie, ISTerre, Le Bourget du Lac, France

Wim Degruyter

School of Earth & Atmospheric Sciences, Georgia Institute of Technology, Atlanta, GA, USA

GLOSSARY

bubble Void space in a silicate melt that is filled by a gas phase. Bubbles are spherical unless deformed by stress or coalescence.

coalescence Process by which two or more bubbles merge into each other, forming a network through which gas can flow.

crystal Solid phase of a magma.

degassing General process by which magma loses its volatile elements. Degassing can occur by exsolution and/or by outgassing.

exsolution Process by which dissolved volatiles come out of solution to join a coexisting gas phase (generally present as bubbles).

melt Liquid phase of a magma.

microlite Small crystals (tens of micrometers across) occurring in magmas. They are generally formed during magma ascent as a consequence of volatile loss from the melt.

phenocrysts Large crystals (hundreds of micrometers across) occurring in magmas. They are generally inherited from the magmatic chamber.

outgassing Loss of the gas phase from a magma.

permeability (magmatic) A material property quantifying the resistance encountered by gas flowing through an interconnected network of bubbles.

rheology Physical description of the way magma deforms when under stress. It generally involves the property of viscosity.

saturation Maximum quantity of volatile element that can be dissolved in a silicate melt.

vesicle Void space left in a solidified magma. It could correspond to a bubble having lost its gas, or to stress-generated fractures.

vesicularity The ratio of void space to total magma volume. Also named porosity or gas volume fraction.

viscosity Material property linking the amount of deformation of a fluid to the amount of applied stress.

volatiles Chemical elements or species that preferentially partition into the gas phase of a magma.

1. GENERAL PRINCIPLES

The most dramatic change that magma undergoes while ascending to the surface is volatile loss because of pressure drop. The process by which the silicate melt loses its dissolved volatile is called degassing. It is the main mechanism by which magma can reach the volcanic vent either as a mostly degassed lava that oozes quietly or as a fragmented mixture of gases and particles that explosively rushes out into the atmosphere. Degassing causes gases to accumulate into the magma in the form of **bubbles** that transform the magma into foam. It causes some **crystals** to grow and the melt **viscosity** to increase. The appearance of bubbles and new crystals transforms the magma into an unstable mixture of liquid, crystals, and gas bubbles, which flows and reacts to stress in complex ways. Some of these

The Encyclopedia of Volcanoes. http://dx.doi.org/10.1016/B978-0-12-385938-9.00011-0

changes occur at the submillimeter scale while others affect the entire plumbing system. The outcome of the impending volcanic eruption is most often controlled by the interactions between small- and large-scale processes rather than by one dominant length scale.

Small-scale magmatic processes that are important at shallow levels concern the three phases that make magma. The first phase, melt, contains dissolved **volatiles** that have evolving solubilities and that are regulated by chemical reactions. The second phase is composed of gas bubbles that successively nucleate, grow, and coalesce, possibly creating pathways for the gas to escape from the magma. The third phase is composed of crystals that also nucleate, grow, and may change shape during ascent. Each of these small-scale processes has been studied experimentally or theoretically in isolation of the others so as to yield a detailed vision of how degassing affects the three phases composing magma.

There are four main large-scale magmatic processes. One process is how magma as a whole (melt, bubbles, and crystals) flows under stress, which drove scientists to study its rheological properties. Another is how the geometry of the volcanic conduit influences magma ascent, which involves the characterization of the various plumbing systems feeding volcanoes. The third process is how the gas can escape out of the foamy magma, which brought forth the concept of **outgassing**. The last process involves the interactions between the magma-filled conduit and the rest of the volcanic edifice.

Both small- and large-scale processes interact in complex ways. Natural observations, whether of the actual eruption or its products, show an integrated result of all ascent-related processes. To understand the interactions between these processes, the most-often used tools are numerical models issued from fluid dynamics. Conduit flow models in particular are amenable to select a subset of small- and large-scale magmatic processes and to generate a general picture of the ascent regimes they sustain.

2. SMALL-SCALE PROCESSES

2.1. Chemistry of Volatile Loss

Silicate melts contain up to a few weight percent of dissolved volatiles. These volatiles, which involve elements, such as O, H, C, S, and halogens, such as F and Cl, are named so because they preferentially partition into the gas phase as opposed to the crystalline phase. Measures of volcanic gases emitted at the vent indicate that H_2O is the dominant gas, followed by CO_2, SO_2, and minor amounts of H_2S, CO, HCl, and HF. Figure 11.1 shows representative gas compositions for a basaltic melt at Etna and a rhyolitic melt at Mt St Helens. These high temperature gases were measured at the vent at atmospheric pressure. At depths, when the gas phase was

Etna gas composition (basaltic melt)

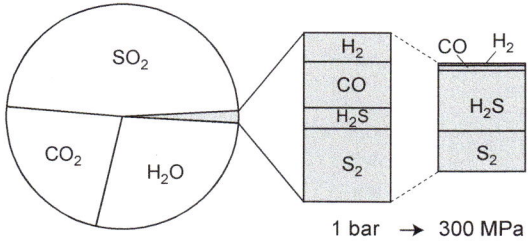

Mt St Helens gas composition (rhyolitic melt)

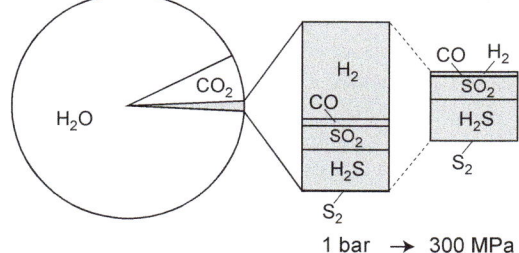

FIGURE 11.1 Representative compositions of volcanic gases emitted during eruptions involving basaltic and rhyolitic melts, respectively. Pie charts represent chemical species proportion with respect to 100% in molar fraction. Bar columns in the middle represent the molar proportions of minor species as measured at the vent (0.1 MPa). Bar columns on the right represent the molar proportions of minor species when the gas is compressed back to magma chamber levels (300 MPa) without interacting with the host melt. Recompression does not affect major species proportions in a visible manner, so pie charts at 300 MPa were omitted. *Data from Symonds et al. (1994) in Carroll and Holloway (1994).*

trapped into bubbles in the volcanic conduit or the magma chamber, these gases had different compositions for two reasons: first, because of the high temperature, there are fast chemical reactions within the gas itself that dictate the proportions of all present species. These reactions are sensitive to pressure, and so at a depth gas composition changes. These changes mostly affect minor species such as H_2S or CO (Figure 11.1). Second, another set of chemical reactions occur between the gaseous and the dissolved species. The maximum amount of volatile species that remains in solution is referred to as solubility and varies from a few parts per million to a few percents depending on the species considered. The solubility of most gas species diminishes drastically with pressure (Figure 11.2). This also modifies the gas composition at depths, although there are complications when one wants to quantify such modification. One complication is linked to the difficulty in establishing solubility laws for the relevant species because they depend on melt composition, temperature, and oxidation state (Figure 11.2). Another complexity is linked to kinetics. Magma ascent and the associated pressure drop cause volatiles to come out of solution and concentrate in the gas phase. The amount of gas thus increases, turning the magma into bubbly foam. The

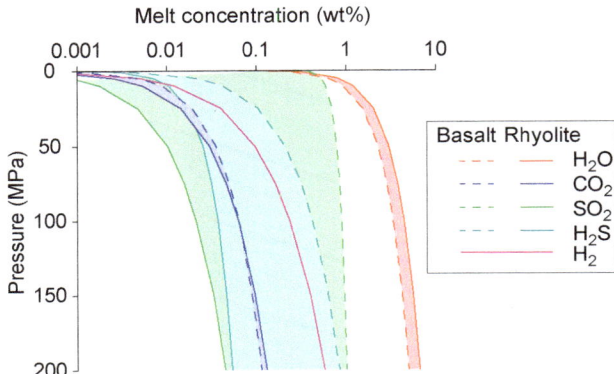

FIGURE 11.2 Maximum dissolved concentrations of five chemical species present in volcanic gases. Curves define the maximum values (solubilities) of each species taken individually. When several species are present, as in natural magmas, the melt can accommodate less of each respective species than the solubilities calculated here. Note that H_2O, H_2, and CO_2 are much less sensitive to melt composition than are sulfur-bearing species. Basaltic melt is at 1000 °C and rhyolitic melt is at 850 °C.

reactions involved, however, are much more sluggish than those occurring into the gas phase because the dissolved volatiles have to migrate (or diffuse) through the melt to reach the bubbles. Depending on the eruption style, some or all volatiles may not be able to keep up with the decompression rate and remain concentrated into the melt to values above that of solubility. These complexities make the prediction of the gas composition during ascent difficult. One method used on basaltic magmas to track the loss of volatile species during ascent is to measure their abundances in small melt pockets trapped in crystals called melt inclusions. Another method that is also valid on a restricted range of magma compositions is thermodynamical modeling.

Water being the dominant gaseous and dissolved species, it is often considered that the physics of degassing is controlled by how water degases during ascent. One exception is the degassing of CO_2-rich basalts, which is a little explored case because of experimental difficulties. Water also changes melt density much more than do other volatiles. A change in 1 wt% dissolved water in a basalt has the same effect as a temperature increase of 400 °C or a pressure decrease of 500 MPa at a constant melt water content.

2.2. Gas Bubbles

As seen in the section above, water is the dominant volcanic gas. Bubbles are thus mainly filled with water molecules and we ignore the presence of other volatiles in describing the physics of bubble nucleation, growth, **coalescence**, and collapse.

Bubbles do not appear as soon as the pressure drops below the **saturation** value. Creating a small nucleus of gas molecules takes energy, and thus, new bubbles nucleate only with a certain amount of oversaturation, which is

generally measured as a pressure difference. In the simplest case, the magma is devoid of crystals, and thus, bubbles nucleate in a pure melt. Such nucleation is referred to as homogeneous and is well understood. Nucleation rate depends on water diffusivity, temperature, surface tension, and amount of oversaturation. Typically, a rhyolitic melt at saturation at 200 MPa needs to be decompressed by 120−150 MPa to nucleate bubbles homogeneously. In an ascending magma, these variables can be combined such that the bubble number density (i.e., the number of bubbles per cubic meter) can be related to decompression rate. The faster the decompression is, the more bubbles nucleate in a given volume. When the magma contains crystals, bubbles may be able to nucleate preferentially on the sides of some crystals and heterogeneous nucleation occurs. This tendency is measured by surface tension and depends on mineral species. Some minerals such as plagioclase do not modify nucleation dynamics while others, such as pyroxenes, or Fe−Ti oxides bring the necessary oversaturation from >100 MPa to values as low as 1−5 MPa. The link between bubble number density and decompression rate is maintained only if enough crystals are present. In other words, if there are not as many crystals present as the number of bubbles that would like to nucleate, one cannot predict the resulting bubble number density.

As in the case of nucleation, bubble growth is chiefly caused by the loss of pressure linked to magma ascent. Growth occurs for two distinct reasons: First, the gas expands following, to first order, the ideal gas law, and second, the amount of gas contained into the bubbles increases as pressure drops because the volatiles dissolved into the melt come out of solution (Figure 11.3). Bubbles, however, cannot grow freely because they have to push away from

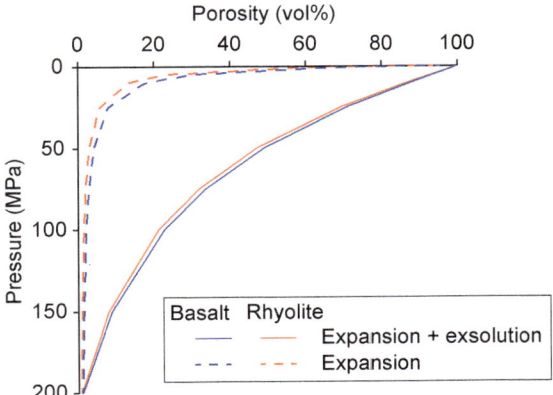

FIGURE 11.3 Evolution of magma porosity as a function of pressure. All curves start with 1 vol.% porosity at 200 MPa and bubbles are assumed to remain with the melt at all pressures (i.e., the system is closed with respect to gas). Dashed curves only take into account gas expansion, and solid curves take into account both gas expansion and exsolution. The basaltic melt is at 1000 °C and rhyolitic melt is at 850 °C. The difference between basalt and rhyolite is essentially due to temperature.

their centers the surrounding melt, crystals, and other bubbles. Leaving aside crystals and other bubbles, the simplest case is one single bubble growing in a silicate melt. At rest, the bubble internal pressure is slightly higher than the pressure of the surrounding melt because maintaining the bubble wall spherical takes some force. This force is a function of surface tension and is proportional to the bubble radius. When pressure drops because of magma ascent, the bubble walls have to push the melt away, which takes a force proportional to melt viscosity. Simultaneously, the dissolved water migrates through the melt into the bubble at a rate proportional to water diffusivity and bubble surface. The faster the pressure drops, the harder it is to work against melt viscosity and the less time water has to diffuse into the bubbles. Growth is thus a function of pressure, decompression rate, bubble size, water diffusivity, and melt viscosity. Each of the processes involved can be described by a representative timescale: τ_{dec} that quantifies decompression rate, τ_{visc} that quantifies melt flow induced by bubble growth, and τ_{diff} that quantifies diffusion of volatiles into the bubble. The relative importance of the processes is assessed by comparing these timescales. This is done by rearranging them into two dimensionless numbers that assess how important viscosity ($\Theta_V = \tau_{visc}/\tau_{dec}$) and diffusion ($\Theta_D = \tau_{diff}/\tau_{dec}$) are compared to decompression rate. If decompression dominates, bubbles grow in equilibrium, that is, bubbles stop growing as soon as decompression stops. If either viscosity or diffusion dominates, growth occurs in disequilibrium, that is, bubbles continue growing after decompression stops because their internal pressure is still much higher than the melt pressure. Disequilibrium growth thus generates overpressurized bubbles compared to the melt. This is important because this is one way to fragment the magma and generate an explosive eruption. Typically, a bubble growing from small to large has a diagonal path in a Θ_V vs Θ_D diagram (Figure 11.4). Overpressurized bubbles are thus most likely to occur when bubbles are large and decompression rates are fast.

As bubbles grow, their crowding increases the likelihood that they interact with each other and merge into one another. Depending on melt viscosity, merging, or coalescence, can occur for different reasons. In low-viscosity melts, such as basalts, bubbles can rise faster than the surrounding melt. Coalescence is then driven by the differential rise speed of bubbles or, if bubbles are trapped and close packed, by gravity-driven drainage of the melt film confined between two neighboring bubbles. The new bubble quickly recovers a spherical shape. If this process involves enough parent bubbles, the resulting bubble can reach several meters in diameter at shallow depths. Its rise is then fully decoupled from that of the magma, and it can rapidly reach the surface before bursting. In higher viscosity melts, such as rhyolites, bubbles do not move relatively to each other, and coalescence occurs because of the rupture of the melt

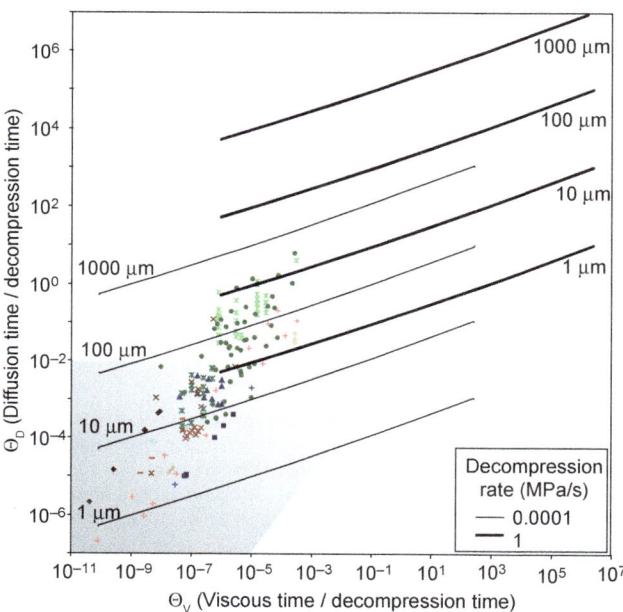

FIGURE 11.4 Bubble growth regimes as characterized by two dimensionless parameters Θ_D and Θ_V. The solid lines indicate trajectories of bubbles of a size given by the labels when they decompress from 200 MPa (lower left) to 0.1 MPa (upper right). Two decompression rates are represented, which correspond to typical effusive rates (0.0001 MPa/s) and extreme explosive rates (1 MPa/s). The melt is a rhyolite at 850 °C with a viscosity evolution with pressure shown in Figure 11.6. The gray area covers the equilibrium growth regime, whereas the area left blank characterizes disequilibrium growth. The limit between these regimes is model dependent and is symbolized by the gradient from gray to white. Symbols represent data from 14 experimental decompression studies on natural melts carried out from 1994 to 2011.

films. The exact mechanisms are still under investigation but involve shear, differential inner bubble pressure, and growth rate. Coalescence is promoted when bubbles are so crowded that they deform and create a flat melt film between each other. The coalesced bubbles keep the shape of the original bubbles for a sufficient time that bubble chains are created as more bubbles join the coalesced cluster. This process creates an interconnected network of bubbles through which the gas can flow freely; the magma is then permeable to gas.

If the gas is able to escape from the interconnected bubbles, the foamy magma collapses. As a result, the gas leaks gently out of the magma without fragmenting. Studies carried out on the mechanisms ruling bubble collapse are recent. They show that the rate at which gas is lost is controlled by **permeability**, melt viscosity, and the differential stress applied to the foam. If the magma is subjected to a sufficient pressure increase, the gas dissolves back into the melt and bubbles disappear, leaving the melt fully healed and homogeneous.

Natural observations regarding bubbles have been obtained through textural analyses of eruptive products and thanks to geophysical methods such as infrasound

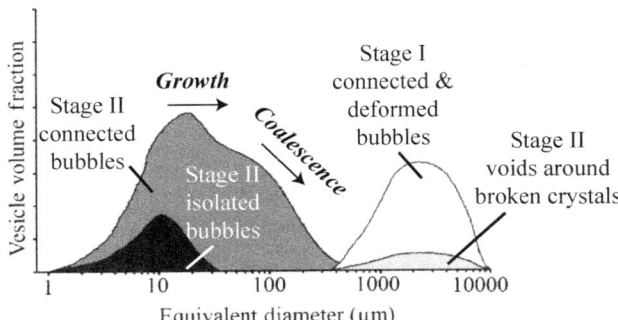

FIGURE 11.5 Schematic vesicle size distribution from explosive products of the 1997 Vulcanian explosions at Soufrière Hills volcano, Montserrat. Four vesicle populations can be identified based on size and shape. One population was generated during the first stage of degassing linked to the magma ascent from chamber to conduit (Stage I). The other populations were generated during the sudden evacuation of the conduit that fed the Vulcanian explosions (Stage II). The arrows indicate the respective effects of growth and coalescence. *Modified from Giachetti et al. (2010).*

surveying of bubble bursting during Strombolian explosions. Textural analyses give quantities such as **vesicularity**, bubble number density, and bubble size and shape, which can be linked to the various stages of bubble evolution. Briefly, nucleation controls bubble number density, growth changes bubble size and magma vesicularity, coalescence modifies bubble shape and reduces bubble number density, and collapse reduces bubble size and vesicularity. Figure 11.5 shows a schematic bubble size distribution from Vulcanian eruptive products where a few of these stages have been identified.

2.3. Crystallization

Decompression-driven water loss affects the phase equilibrium of the ascending magma. The less water is dissolved in the melt, the larger is the stability field of most anhydrous crystalline phases. In other words, water loss promotes crystallization of phases such as plagioclase, pyroxenes, and Fe−Ti oxides while destabilizing phases such as hornblende. Crystallization occurs by enlarging existing crystals and nucleating new ones. Existing crystals are often large (hundreds of micrometers across) and inherited from the magmatic chamber. Decompression-driven crystallization adds a growth rim on such crystals, which are named phenocrysts because of their size. New crystals are named microlites. They are numerous and small (tens of micrometers across) because growth time is restricted between ascent and quench at the surface.

Travel time between the chamber and the surface is too short to dissolve entire phenocrysts. Instead, they undergo fast phase transformation starting at their rims. Crystal rims break down into other water-poor phases that are stable at a low melt water content. The rate of such a reaction has been quantified experimentally and has first been used to link amphibole breakdown rate to ascent rate during the 1980 eruptions of the Mt St Helens dacite. Travel times also affect the way microlites grow. As a result, explosive products contain fewer microlites than do effusive ones.

Numerous experiments have reproduced microlite crystallization in high silica melts. General relationships can be drawn between water content and microlite volume fraction. Number densities and microlite shape, however, are controlled by kinetics, the details of which are still under investigation. Order of magnitude growth rate has been measured experimentally $(10^{-11}-10^{-7}\,\text{mm/s})$. As in cooling-driven crystallization, fast decompression triggers the fast growth of dendritic microlites while slow decompression generates more blocky shapes. The microlites found in natural products are, however, difficult to reproduce experimentally. It seems that this is due to multiple cycles of decompression and stalling in the conduit. At low pressure and low water content, which are conditions prevailing in lava domes, silica precipitation can occur. Its density is lower than that of the melt, which enables it to fill preexisting **vesicles**, possibly sealing gas pathways and stopping outgassing.

Because of experimental difficulties, microlite crystallization in basaltic melt is not well constrained. Field samples of Plinian basaltic eruptions show that up to 30 vol % microlite may be created during fast ascent and degassing, while recent experiments indicate rates on the order of $10^{-4}-10^{-5}$ mm/s.

3. LARGE-SCALE PROCESSES

3.1. Magma Rheology

Rheology studies address the way magma deforms when under stress. It is perhaps the domain where the unstable nature of the mixture of liquid, crystals, and gas bubbles takes its fullest meaning. As seen below, the three phases composing magma have each a distinct influence on rheology.

Melt viscosity is a function of melt composition, which includes the amount of dissolved volatiles. Water and fluorine are the two volatiles that influence most melt viscosity, the effect of water being dominant. As the amount of dissolved water decreases because of magma ascent in the volcanic conduit, melt viscosity increases by several orders of magnitude. This increase is so large that it dominates rheological changes during decompression. At low shear rates, melt viscosity is Newtonian, that is, it deforms linearly with the applied stress. As shear rate increases, however, the melt becomes viscoelastic and finally breaks. This implies that pure silicate melt fragments under sufficient shear rate. Experiments have shown that these rates too depend on melt composition through a linear dependence on viscosity; at magmatic temperatures, silica-rich melts such as rhyolites break under shear rates on the order of $10^{-2}-10^{4}\,\text{s}^{-1}$, whereas silica-poor melts such as basalt break under rates on the order of $10^4 - 10^8$ s^{-1}. As explained below, conduit flow models predict that the lower end of these shear rates are achievable either at the conduit margin or during the decompression-driven acceleration that occurs at shallow levels. Strain rates deduced from mass flux and conduit diameter from dome-building eruption at Montserrat and Unzen are on the order of $10^{-6} - 10^{-2}$ s^{-1}. Despite being in the brittle range of shear rates, the melt involved into these two effusive eruptions reaches the surface unfragmented. It is thus possible for the melt to break during ascent without explosive fragmentation; after having accommodated the strain and lowered the strain rate it is subjected to, the fracture heals and the melt flows again as a Newtonian fluid.

When crystals are present in the melt, they influence the rheology in two ways: first, the larger the crystal volume fraction is, the harder it is to deform magma. In other words, the rheology of magma is a strong function of crystal volume fraction. Second, the more elongated the crystals are, the more the suspension resists deformation. Since fast-growing, ascent-related microlites are highly elongated, they influence magma rheology even when a small amount crystallizes. The details of the relationship between crystal shape and rheology is a complex issue—if microlite is platy and of the same size, it may align under deformation and instead ease motion. At high crystal volume fraction, transient shear localization occurs, that is, preferential planes of motion are constantly created and abandoned within the deformed suspension. At a still higher volume fraction, a rheological threshold is reached, and the mixture resists nonbrittle deformation. One popular way to approach these complexities is to treat melt and crystals as a perfectly homogeneous mixture with a bulk rheology. Figure 11.6 presents the simplest case of such an approach, where the bulk behavior of the mixture is assumed to be Newtonian. This approach has recently been extended to non-Newtonian behavior. A number of rheological models link together the non-Newtonian equivalent of viscosity to crystal volume fraction, crystal shape, and shear rate.

Bubbles can play a role similar to that of crystals if they remain spherical at all times, which tends to occur at low shear rates. Such rigid behavior increases bulk viscosity to a lesser degree than in the case of crystals because of the internal motion of the gas within the bubbles. At high shear rates, however, bubbles deform and ease flowage. The situation becomes quickly complex as high shear and deformation promote bubble coalescence and gas channeling. The common simplification is to calculate two extreme bulk viscosities, one for fully deformed bubbles and one for nondeforming spheres (Figure 11.6).

When the three phases are together, experiments have shown that shear localization and complex rheological response occur. Field studies have brought corroborating observations. Pyroclasts and dissected lava domes feature

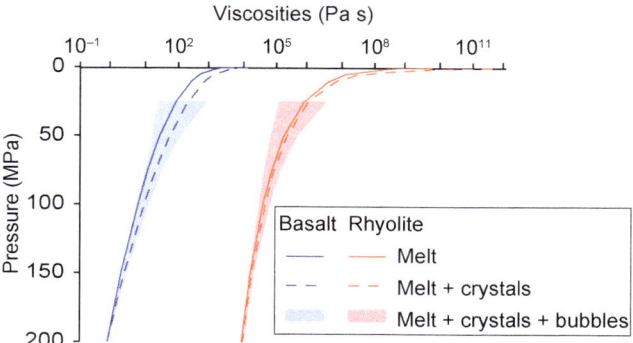

FIGURE 11.6 Evolution of magma bulk viscosities as a function of pressure. Solid curves consider melt viscosity alone and show the effect of water loss. Dashed curves add the effect of microlites, which are assumed to grow linearly from 0 to 30 vol.% for basalt and exponentially from 3 to 60 vol.% for rhyolite. Colored areas cover the two limit cases when taking gas bubbles into account, which correspond to deformable bubbles and rigid bubbles, respectively. Bubble calculations are stopped when porosity exceeds approximately 65 vol.% because foam rheology changes fundamentally above that value. Basaltic melt is at 1000 °C and rhyolitic melt is at 850 °C. Bubble volume fractions are calculated as in Figure 11.3.

shear bands of various types (alternating microlite content, melt water content, and crystal orientations) that document such phase reorganization. Figure 11.6 takes a broad stance to show that, overall, water loss and crystallization during ascent both conspire to increase magma viscosity, yielding a highly viscous lava to emerge at the surface.

3.2. Geometry of Conduit

Studies of conduit geometry brought forth a puzzling fact. Dissected volcanoes show that at a depth, magma moves preferentially along dikes and sills, that is, vertical or horizontal fracture planes that enlarge to yield a way to magma. Most volcanoes, however, erupt through vents that are shaped like vertical cylinders. There is thus a transition between shallow cylindrical conduits and deeper penny-shaped dikes.

Generally, plumbing systems of basaltic volcanoes (i.e., edifices mostly erupting fluid lavas) are composed of a series of interconnected dikes and sills feeding a central vent and several adventive vents. In such a geometrical context, experiments have shown that horizontal plumbing elements such as sills may act as bubble traps. Sills may accumulate bubbles until the foam formed collapses into a large bubble that rises rapidly to the upper reaches of the plumbing system. This mechanism drives intermittent explosions of the Strombolian type. Such conduits can be filled with magma and open to the surface for decades to millennia, as exemplified by long-lived lava lakes such as Erebus, Antarctica, and by continuously erupting volcanoes such as Stromboli, Italy. Field studies of regions with older, partially eroded, central volcanoes have shown two types of sheet-like conduits: series of regional, subvertical, and thick dikes that may span an area much larger that a single volcanic edifice, and series of local, inclined, and thin sheets. A few of these magma pathways lead to the surface.

Conduits feeding more silicic volcanoes seem to involve fewer interconnected pathways. While dikes have also been observed to dominate at a depth, the development of shallow cylindrical pipes is promoted by mechanical erosion during explosive phases. This seems to be also valid for lava dome eruptions, which commonly start by phreatic and phreatomagmatic explosions that excavate near-surface, axisymmetric vents. Not all conduits feeding silicic eruptions, however, are vertical. Localization of seismic signals at Unzen volcano, Japan, has mapped out a several-kilometer-long magmatic pathway that is inclined at 30–40° with respect to the vertical. Conduit geometry is much more complex for large-scale eruptions involving caldera formation. Caldera collapse often creates a ring fracture trough which magma is expelled. This shape breaks down if the caldera floor subsides chaotically.

The horizontal area of the conduit greatly influences how much magma can erupt. For a given decompression rate, the simplest approximation of ascent dynamics, Poiseuille's law, states that the flux of magma rising (in mass or volume per second) is proportional to the square of conduit horizontal area. In the case of a cylindrical conduit, this relationship means that the ascent rate is proportional to the square of the conduit radius. One consequence is that changes in conduit shape during eruption by erosion, transition from fissure-fed to conduit-fed, or caldera collapse, have major effects on ascent speed and thus eruptive dynamics.

An open conduit is an unstable structure that eventually collapses if empty. During an explosive eruption, there is a sharp decompression near the fragmentation level that brings a part of the conduit to pressures inferior to those of the surrounding rocks. This underpressure might induce collapse of the conduit walls and eruption shutdown. When the eruption is over, a significant part of the conduit might be left empty, which also induces an underpressure and wall collapse.

3.3. Outgassing and Permeability

During magma ascent, segregation of the gas phase from the magma, also called outgassing, occurs. This can happen through (1) buoyant bubble rise, (2) the development of a permeable bubble and fracture network, and (3) magma fragmentation.

In the case of low-viscosity magma, bubbles are able to rise buoyantly through the magma. When gas volume fraction and bubble sizes are small enough, bubbles rise together with the magma to form either a lava lake where they will follow the convective flow of the magma or a lava flow where they ooze out at low flow rates together with the magma. Larger bubbles can make their way up in the conduit individually and burst at the surface where they release gas to the atmosphere. If gas volume fraction is high enough, individual bubbles can coalesce to form large gas pockets called slugs. Slugs can fill up almost the entire width of the conduit. The bursting of these gas slugs near the surface is characteristic of Strombolian type eruptions. Even further increase in gas volume fraction might lead to annular flow, i.e. gas with particles in suspension, which has been proposed to cause Hawaiian-style fire fountain eruptions.

Bubbles in high-viscosity magmas have low mobility and remain coupled to the magma at all times. The magma–bubble mixture will therefore move as a single fluid. Unlike the case of low-viscosity magma, when bubbles coalesce, they form a permeable network, and the magma can develop a foamy structure. If the bubbles are able to connect to the surface or to the conduit walls, the gas is able to move at a different velocity compared to that of the magma, which causes low ascent rates. Such permeable pathways through which gas escapes might also be formed by fractures that develop due to friction. This friction, or shear rate, is the highest near the conduit walls and can lead to melt fracturing, which aids the gas escape to the surface. Such behavior is associated to dome-forming eruptions and coulees.

If gas and magma remain coupled, rapid ascent occurs, regardless of melt viscosity. As the gas cannot separate from the flow, the gas phase will continue to expand, pressurizing the magma and accelerating the magma–bubble mixture. This positive feedback continues until the stresses within the magma become so high that it fragments into parcels of magma called pyroclasts, which are then carried upward by the gas to the Earth's surface. The reader is referred to the chapter 25 (Volume 1) on magma fragmentation for a presentation of the different fragmentation mechanisms. Depending on the efficiency of fragmentation, this type of outgassing can result in behavior ranging from fire fountaining (usually associated with low viscous magmas) to the most violent type of explosive eruption, so-called Plinian eruptions (usually associated with more viscous magmas). The pyroclasts produced by these explosive eruptions can partially preserve the state of the magma at fragmentation and have thus been used to gain an insight into outgassing behavior. In particular, analysis of the bubble textures contained within pyroclasts has been linked to the development of permeability during magma ascent. The dominant control on permeability is vesicularity and the radius of the permeable pathways, which is related to the size and number density of bubbles. Another effect is the shape of these pathways, which is controlled by the deformation the bubbles have been subjected too. Highly deformed, large bubbles have the most favorable shape to develop high permeability.

3.4. Interactions between Conduit and Edifice

When the magma is highly viscous, it does not flow easily along conduit walls. This interaction has several consequences. First, slipping can occur at the interface. Such slipping is intermittent because it suddenly releases stress and some time is needed for new stress buildup. This is the mechanism proposed to explain sudden lava dome motions such as at Santiaguito volcano, Guatemala (Figure 11.7). The lava dome moves up several times a day in a stick-slip fashion, generating explosive events with accompanying ejecta-laden plumes. Second, the walls deform elastically under the stress imposed by the ascending magma. The elastic energy stored temporarily by the wall rock is released as soon as the dynamic pressure of the magma diminishes. This feedback mechanism influences ascent dynamics, and the resulting edifice deformations can be captured by geophysical tools such as tiltmeters or differential global positioning system measurements.

In some cases, changes of the stress field in the edifice can influence ascent dynamics. Arenal volcano, Costa Rica, has been in continuous eruption since 1968. Its conduit is open to the surface and gives way to lava extrusions and ash emissions. The conduit at Arenal is extremely sensitive to changes in stress field, as its eruptive activity can be linked to Earth's tides. Such effects, however, seem to be confined to small changes in eruptive activity rather than acting as major controls of ascent dynamics.

Shallow degassing is not confined to the magmatic column. In many instances, the volcanic gas can flow out of the column through the conduit walls. The ascending magma becomes permeable to gas when bubble networks are formed. The resulting outgassing may leak through the wall rock. Such a gas leak has been calculated to be efficient enough to reduce magma porosity and turn an explosive eruption into an effusive one. While gas leakage through conduit walls is difficult to measure, there are geological evidences of such interactions. Studies of fossil conduits have shown the presence of tuffisite veins along the magma/wallrock interface (Figure 11.8). These veins are filled with magma fragments that are less than a millimeter across and that form sedimentary structures including planar and crossbedding. Tuffisite veins are interpreted as shear fractures formed by the brittle response of highly viscous magma to flow-related shear in the conduit. The magma fragments produced during shear were redeposited through the fracture system by a gas—particle mixture, the gas phase of which was probably derived from magma outgassing.

4. PROCESS INTERACTIONS

4.1. Conduit Flow Models

Direct observations of volcanic conduits during eruptions are not possible, and therefore, numerical models are used to understand the wide variety of eruption styles at volcanoes. Modeling of magma ascent requires integrating the small- and large-scale processes that were discussed above. Each of these presents interesting and sometimes complicated physics. Conduit modeling looks at the feedbacks that occur between these processes, which present additional difficulties. We therefore rely on assumptions that reduce the complexity of the problem in order to develop such models.

The simplest approach is to present a basic conduit model that can be used to study, to a first order, the conduit dynamics during an explosive eruption. This reduces the complexities of magma ascent to a one-dimensional steady flow through a cylindrical and vertical conduit with a constant radius. The temperature of the magma does not change during ascent, the volatile phase in the magma is only made out of water, and bubbles grow at their equilibrium rate. Near the conduit inlet, at depths, the magma is considered as a homogeneous and compressible fluid made

FIGURE 11.7 Rapid uplift of Santiaguito lava dome, Guatemala with accompanying degassing and ash emission. Such explosions last tens of seconds and repeat several tens of times a day. They are inferred to be triggered by a sudden slip at the magma/wallrock interface. *Reproduced from Johnson et al. (2008).*

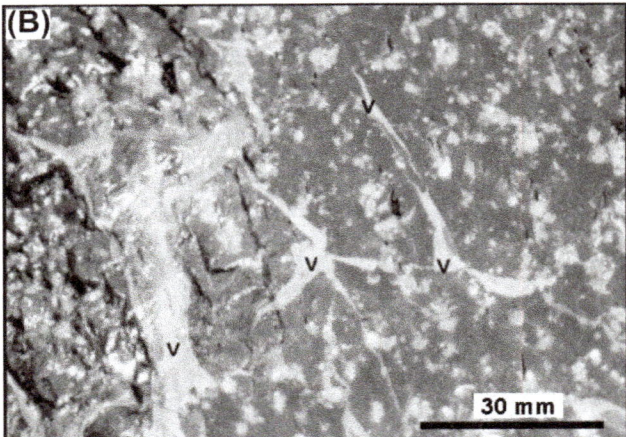

FIGURE 11.8 Eroded volcanic conduit in southeast Rauðufossafjöll, Iceland. (A) Overview of the conduit, showing dark gray obsidian walls (o) and near-vertical flow banding in the devitrified interior (arrow). (B) Network of angular, branching, pale gray tuffisite veins (v) in conduit−-wall obsidian. Pale blobs are feldspar phenocrysts. *Modified from Tuffen et al. (2003).*

of a perfect mixture of melt and bubbles. The viscosity of this mixture is assumed to depend solely on melt properties, which includes melt water content. The presence of crystals is thus ignored. Fragmentation is postulated to occur when the gas volume fraction reaches a critical value. After fragmentation, the ascending magma becomes a fluid made up of a perfect mixture of gas and pyroclasts in suspension.

These assumptions can be used to write down the governing equations of the flow. The conservation of mass is

$$\frac{d(\rho u)}{dz} = 0,$$

where z is the vertical coordinate, ρ is the mixture density, and u is the ascent velocity. This equation stipulates that there is no mass gained or lost during ascent and thus that

mass flow rate $q = \rho \times u$ (kg/s) during ascent remains constant. The conservation of momentum is

$$\rho u \frac{du}{dz} = -\frac{dP}{dz} - \rho g - F,$$

where P is the pressure of the magma—bubble mixture, g is the gravitational acceleration, and F is the friction force. The term on the left-hand side quantifies how the momentum changes during ascent. The three terms on the right-hand side state that momentum changes are due to (1) decompression of the magma, (2) gravitational forces, and (3) friction due to interaction with the wall, respectively.

These two equations need to be completed by defining the magma density ρ and the friction term F. The mixture density is defined as the volume average of the melt and the gas phase:

$$\rho = (1 - \varphi)\rho_m + \varphi \rho_g,$$

where φ is the gas volume fraction, ρ_m is the melt density, and ρ_g is the gas density. The melt is incompressible, and thus, its density will not change during ascent. The gas phase is compressible, and its volume expands as the pressure drops. The simplest approach is to consider that the gas phase behaves as an ideal gas.

The gas volume fraction is controlled by how much gas is exsolved at each point during ascent (Figure 11.3). Equilibrium bubble growth means that the amount of gas follows the solubility curve of water (Figure 11.2). As a first approximation, the solubility curve is described by Henry's law:

$$m_{H_2O} = s\sqrt{P},$$

where m_{H_2O} is the mass fraction of the dissolved water in the magma and s is an experimental saturation constant. This in turn can be related to the mass fraction of water that is exsolved in the mixture through the equation of mass conservation of water

$$m_{H_2O,0} = m_{H_2O,m}(1 - m_g) + m_g$$

with $m_{H_2O,0}$ being the initial mass fraction of water before exsolution and m_g the gas mass fraction within the mixture. Note that m_g is considered to be zero when $m_{H_2O,0} < m_{H_2O}$. These last three equations can be combined to calculate the gas volume fraction:

$$\frac{1}{\rho} = \frac{1 - m_g}{\rho_m} + \frac{m_g}{\rho_g},$$

$$\varphi = \frac{\rho m_g}{\rho_g}.$$

The friction term F depends on whether the magma is fragmented or not. The simplest fragmentation criterion is a critical gas volume fraction, φ_c. When the gas volume fraction is below this point, the friction with the conduit wall is governed by the magma and is thus that of a viscous flow in a pipe. Above this critical gas volume fraction, the friction with the wall is governed by the gas phase and is that of a turbulent flow in a pipe. The wall friction term can thus be written using classical works of fluid dynamics addressing viscous and turbulent flows in a circular pipe:

$$F = \frac{8\mu u}{r^2} \quad \text{if } \phi \leq \phi_c,$$

$$F = \frac{0.01 u^2}{4r} \quad \text{if } \phi > \phi_c,$$

where μ is the magma–bubble mixture viscosity and r is the conduit radius. For silicic magmas, μ is calculated thanks to an experimentally derived expression that relates viscosity to the temperature and volatile content (Figure 11.6). The transition from viscous to turbulent flow is a defining feature of an explosive eruption. It causes the friction term F to become very small, which allows the velocity to increase tremendously once fragmentation occurs.

The unknowns in the above set of equations are the velocity u and the pressure P. These quantities can be solved using these equations with inflow and outflow conditions at the conduit extremities. At the inlet of the conduit, the pressure can be assumed to be equal to the pressure of the magma chamber P_0. At the outlet of the conduit, one might expect the pressure to reach the atmospheric pressure. However, when fragmentation occurs, the gas–pyroclast mixture can accelerate to high velocities. It can reach but not exceed the sound velocity of the mixture at the vent. This is called the choking condition and it fixes the value of u at the vent, regardless of conduit dimensions and magma viscosity. This condition implies that the exit pressure at the vent is larger than the atmospheric pressure, which results in a dramatic expansion at the conduit exit that generates shock waves. The solutions of these equations depend on the values of the conduit geometry (length z, radius r) and magma properties (volatile content, temperature). Parameters s and g are constant.

The dynamics of magma ascent in the conduit during an explosive eruption change dramatically. Figure 11.9 presents model solutions involving a rhyolitic melt. Initially, the gas volume fraction is zero, and the magma consists only of melt. Once the pressure is sufficiently low, the gas starts to exsolve. The feedback between the exsolution of volatiles and pressure drop creates a run-away effect by which the magma continues to accelerate until it reaches fragmentation. After fragmentation, the friction drops by many orders of magnitude and the gas–pyroclast mixture continues to accelerate until it reaches the speed of sound of the mixture at the top of the conduit. One of the solutions in Figure 11.9 has a constant melt viscosity, whereas the other

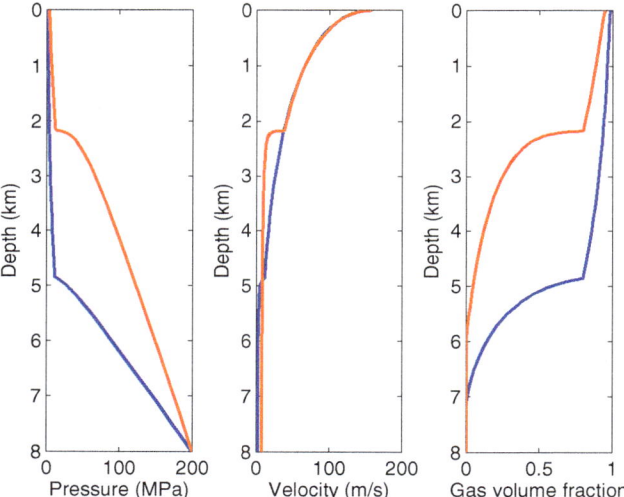

FIGURE 11.9 Results of the heuristic conduit model. Respective changes of pressure, velocity, and gas volume fraction as a function of depth in the conduit. The kinks in the profiles are generated by fragmentation and the ensuing drastic change in friction force. Blue curves represent solutions of the model using a constant melt viscosity of 10^6 Pa s. Red curves represent solutions of the model using the same empirical equation for rhyolite melt viscosity as in Figure 11.6, which depends on the magma temperature and dissolved water content. Both runs have a constant temperature of 850 °C, 5 wt.% total (and initial) water content, an inlet pressure of 200 MPa, and a conduit 8 km in length and a radius of 25 m. Following Koyaguchi (2005), a critical gas volume fraction of 0.8 is used as a fragmentation criterion.

has a viscosity that depends on melt water content. These two solutions illustrate the strong impact of melt viscosity on the depth of fragmentation.

It is clear from comparing the described model to the processes discussed in the previous sections that strong simplifications were made to arrive at a solution. In order to improve our understanding of magma ascent, the basic conduit model has been improved upon in several ways:

1. Outgassing in the basic model could only occur through magma fragmentation. In order to account for buoyant bubble rise or development of a permeable network, a two-phase model is needed such that the gas phase can move relatively to the magma. Under some conditions, such models predict solutions in which fragmentation is never reached because gas permeability allows the gas volume fraction to decrease by outgassing during ascent. These models have been predominantly used to explore the transition from effusive to explosive regimes. They have established that this transition is mostly controlled by the development of permeability relative to the wall friction.

2. Variations of magma properties across the conduit, especially rheological changes, change the velocity profile across the conduit. Such variations have been taken into account by two-dimensional models.

Permeable outgassing does not occur homogeneously within the conduit. At shallow depths, it leads to the formation of a 30- to 100-m-thick degassed magma cover at the top of the conduit and a thin blanket of degassed magma near the conduit walls.

3. Although explosive eruptions can last for hours to days, their intensity varies over time due to changes in properties that were considered constant in the simplified model (e.g., conduit radius or inlet pressure). Adding time dependence to the equations has shed light on such transient behavior. Together with outgassing and crystallization, time dependence leads to complex oscillation of the exit velocity. The strong feedback mechanisms existing between these processes greatly amplify the effect on extrusion rates of small changes of chamber pressure, conduit dimensions, or magma viscosity. When the ascent rate is such that microlite has time to crystallize, there can be multiple steady solutions for fixed conditions. Such nonlinear dynamics can cause large changes in dome extrusion rate and cyclic patterns of dome growth.

4. The shape and dimensions of the conduit are most probably not constant from chamber to vent. Some models exploring the effects of conduit shape found that the fragmentation level is typically deeper in a dyke than in a cylinder. For flows in wide dykes, the pressure at the fragmentation depth can be lower than the surrounding lithostatic pressure by several tens of megapascals, possibly leading to the collapse of the walls, which implies that this conduit geometry has a natural upper limit for dyke width. Models including a transition from dyke shape at a depth to cylindrical near the surface have shown that the fragmentation level moves into the cylindrical region, which causes the flow pressure to approach the lithostatic value and impedes wall collapse.

5. The isothermal assumption is adequate for rapid ascent. At slow ascent, however, heat loss to the conduit walls and heating due to wall friction become important mechanisms. Magma may either cool or heat up during its journey from a shallow crustal level to the volcanic vent. Cooling can be induced by heat loss at the conduit wall, or by the expansion of the gas bubbles that magma carries. Heating can be caused by latent heat release during crystallization or viscous deformation at conduit walls. Viscous heating and cooling due to gas expansion have each been calculated theoretically. Latent heat, on the other hand, has been quantified using crystal contents and mineral geothermometry of samples from explosive eruptions. It was found that these three processes can change magma temperature by up to 100 °C. To which extent these effects coexist and counteract each other in a given eruption is, however, yet unclear because of the differing methodologies employed.

When these effects are accounted for, they concentrate shear in narrow zones along the conduit margin. The resulting reduction in friction drastically reduces the zone of low pressure predicted by isothermal models and moves the fragmentation level closer to the surface.

6. By assuming equilibrium exsolution, we avoided having to model bubble nucleation and growth. Delaying nucleation until oversaturation pressures of 150 MPa restricts degassing to within approximately 1500 m of the surface, which brings the fragmentation level to much shallower depths. The nucleation delay leads to higher pressures at equivalent depths in the conduit, and higher mass flux and exit pressures. The introduction of disequilibrium degassing reduces the deviation from lithostatic pressure, the flow acceleration before fragmentation, and the associated decompression rate. The presence of carbon dioxide is important in steering these dynamics as well. An increase in the proportion of carbon dioxide produces a decrease in the mass flow rate and an increase in the exit gas volume fraction and depth of the fragmentation level.

7. Unlike assumed in the simple model of Figure 11.7, the rheology of the magma depends on the presence of bubbles and crystals. Models using bulk expressions that take their effect on viscosity in account have shown that bubbles cause a decrease in calculated fragmentation depth and an increase in calculated eruption rate. Models based on non-Newtonian formulations of magma rheology have shown that shear bands are most likely to initiate at the junction of the conduit and base of the dome, where the shear stress experienced between new lava entering the dome and existing lava is the greatest.

8. The fragmentation criterion of a critical gas volume fraction is simplistic. Other criteria, such as that of a critical stress, or that of critical bubble overpressure, do not change the outputs of 1D models much because of the narrowness of the region where large decompression rates are attained. This is not true for 2D models, as different criteria lead to fragmentation occurring in different regions; critical stresses, for instance, are reached at conduit walls before being reached in the center.

4.2. Eruptive Dynamics at the Surface

Shallow magma ascent gives rise to a rich diversity of processes that cause volatile to come out of solution from a silicate melt to produce a highly expansible gas phase. The surficial expression of these processes has been broadly classified into effusive eruptions that produce lava flows and lava domes and explosive eruptions that range from short-lived explosions of the Hawaiian, Strombolian, or Vulcanian type to long-lived (sub-) Plinian eruptive columns.

There are several key parameters that control this diversity. Melt viscosity and thus magma type control bubble mobility. In low-viscosity magmas, highly mobile bubbles have time to migrate, coalesce, and escape from the magma, yielding eruptive styles combining lava flows and Hawaiian or Strombolian bubble bursts. If fast enough, ascent rate can counteract bubble mobility and yield more explosive behavior such as basaltic Plinian eruptions. In high-viscosity magmas, less mobile bubbles are constrained either to coalesce into a permeable network that allows for outgassing and lava dome formation or to accumulate until fragmentation liberates the gas phase and allows for Vulcanian or (sub-) Plinian gas and ash column to form.

The first and most complete paradigm of magma ascent and degassing is that of the Plinian eruptive regime illustrated in Figure 11.9. Similar frameworks have been then proposed for effusive regimes with outgassing and for regime where individual bubble motion predominates. Although most of these templates capture the essentials of shallow degassing processes, many mechanism descriptions still need to be refined. One example is phreatic eruptions because the links between small-scale processes involved in magma degassing in the presence of surficial water and large-scale processes involved in steam and ash generation remain unclear. These modeling efforts where fed by experimental data and natural observations and build canonical cases of eruptive regimes. In parallel, the past decade has seen the emergence of numerous observations and quantifications of lower intensity eruptive regimes. A significant fact is that the boundaries between these canonical eruptive regimes blur as more and more observations are collected. One of the many examples of such refinement is the current research effort to distinguish the subtleties between "violent Strombolian" and "Vulcanian" explosions.

FURTHER READING

Burgisser, A., Scaillet, B., 2007. Redox evolution of a degassing magma rising to the surface. Nature 445, 194—197.

Carroll, M.R., Holloway, J.R., 1994. Volatiles in magmas. In: Reviews in Mineralogy, vol. 30. Virginia.

Giachetti, T., Druitt, T.H., Burgisser, A., Arbaret, L., Galven, C., 2010. Bubble nucleation, growth and coalescence during the 1997 Vulcanian explosions of Soufrière Hills Volcano, Montserrat. Journal of Volcanology and Geothermal Research 193, 215—231.

Giordano, D., Russell, J.K., Dingwell, D.B., 2008. Viscosity of magmatic liquids: a model. Earth and Planetary Science Letters 271, 123—134.

Gonnermann, H.M., Manga, M., 2007. The fluid mechanics inside a volcano. Annual Review of Fluid Mechanics 39, 321—356.

Gonnermann, H.M., Manga, M., 2013. Dynamics of magma ascent in the volcanic conduit. In: Fagents, S.A., Gregg, T.K., Lopes, R.M.C. (Eds.), Modeling Volcanic Processes: The Physics and Mathematics of Volcanism. Cambridge University Press, Cambridge, pp. 55—84.

Hammer, J., 2008. Experimental studies of the kinetics and energetics of magma crystallization. Reviews in Mineralogy and Geochemistry 69, 9—59.

Johnson, J.B., Lees, J.M., Gerst, A., Sahagian, D., Varley, N., 2008. Long-period earthquakes and co-eruptive dome inflation seen with particle image velocimetry. Nature 456, 377—381. http://dx.doi.org/10.1038/nature07429.

Keating, G.N., Valentine, G.A., Krier, D.J., Perry, F.V., 2007. Shallow plumbing systems for small-volume basaltic volcanoes. Bulletin of Volcanology 70, 563—582.

Koyaguchi, T., 2005. An analytical study for 1-dimensional steady flow in volcanic conduits. Journal of Volcanology and Geothermal Research 143, 29—52.

Llewellin, E.W., Manga, M., 2005. Bubble suspension rheology and implications for conduit flow. Journal of Volcanology and Geothermal Research 143, 205—217.

Martel, C., 2012. Eruption dynamics inferred from microlite crystallization experiments: application to Plinian and dome-forming eruptions of Mt. Pelée (Martinique, Lesser Antilles). Journal of Petrology 53, 699—725.

Métrich, N., Wallace, P., 2008. Volatile abundances in basaltic magmas and their degassing paths tracked by melt inclusions. Reviews in Mineralogy and Geochemistry 69, 363—402.

Navon, O., Lyakhovsky, V., 1998. Vesiculation processes in silicic magmas. In: Gilbert, J.S., Sparks, R.S.J. (Eds.), The Physics of Explosive Volcanic Eruptions. Geological Society Special Publication No 145, London, pp. 27—50.

Pistone, M., Caricchi, L., Ulmer, Burlini, L., Ardia, Reusser, E., Marone, F., Arbaret, L., 2012. Deformation experiments of bubble- and crystal-bearing magmas: rheological and microstructural analysis. Journal of Geophysics Research 117, B05208. http://dx.doi.org/10.1029/2011JB008986.

Stasiuk, M.V., Barclay, J., Carroll, M.R., Jaupart, C., Ratte, J.C., Sparks, R.S.J., Tait, S.R., 1996. Degassing during magma ascent in the Mule Creek vent (USA). Bulletin of Volcanology 58, 117—130.

Tuffen, H., Dingwell, D.B., Pinkerton, H., 2003. Repeated fracture and healing of silicic magma generate flow banding and earthquakes? Geology 31, 1089—1092.

Zhang, Y., Xu, Z., Zhu, M., Wang, H., 2007. Silicate melt properties and volcanic eruptions. Reviews of Geophysics 45, RG4004. http://dx.doi.org/10.1029/2006RG000216.

Part II

Eruptions

Hazel Rymer

Faculty of Science, The Open University, Walton Hall, Milton Keynes, UK

Eruptions are exciting. It is almost impossible to think of a volcano without contemplating the awesome power displayed during an eruption. But what exactly is an eruption? How long do eruptions last? How long does a volcano remain dormant between eruptions? Have volcanoes always erupted as they do now and where they do now? These are just a few of the fundamental questions addressed in this section.

We begin with an overview Chapter, Earth's Volcanoes and their Eruptions, in which Lee Siebert and colleagues provide an introduction to the various types of eruptions that occur at the Earth's surface and how they vary with geographic location. The theory of plate tectonics is used to explain the present-day distribution of active volcanoes, which are defined as those historically active or active within the last 10,000 years.

It is sobering to remember that our own history is pitifully short compared with the lifetime of an average volcano. It is quite possible for a volcano not to have erupted at all since humans began documenting their surroundings, but to be simply dormant right now, between periods of activity. The longer ago that an eruption took place, the harder it is to gather evidence of what exactly took place; the products of the eruption are subject to the same erosion processes as any other rocks. Therefore, the further back in geological time that we look, the larger the eruptions need to have been for us to identify them. On the other hand, large eruptions are less common than small eruptions, and this has probably always been true. Products from huge caldera-forming eruptions therefore will be found not in the most recent volcanic deposits, but buried deep in the geological record.

Detailed analysis of all the available evidence from the last 10,000 years has led the authors to deduce the approximate frequency at which we can expect eruptions of various sizes. For example, small eruptions, such as that at Nevada del Ruiz in Colombia in 1985 killing 23,000 people, occur typically several times a year, although fortunately usually without such tragic consequences. The Mount St. Helens event of 1980 occurs on average once each decade. It is unclear, however, how frequent much larger eruptions are.

Almost as important as predicting the onset of volcanic activity, and certainly more difficult, is predicting when an eruption will cease. With a few notable exceptions, such as Stromboli in the Aeolian Sea offshore Italy, which has been erupting steadily for at least 2500 years, the average eruption duration is about 7 weeks. Some last for only a day and the longer the dormant period, the larger the final eruption. The median interval between successive eruptions is about 12 years, but 20% of eruptions follow a period of more than 50 years of quiescence. Just as earthquakes can be defined using scales of magnitude and intensity, so can volcanic eruptions. Estimates of the magnitude and intensity of eruptions can be made for very large events millions of years ago using a comparison of the distribution and volume of erupted products (ash and lava) with those of smaller, recent events. In the Chapter Sizes of Volcanic Eruptions, David Pyle defines eruption magnitude and intensity and illustrates the use of these properties for understanding the mechanisms driving the colossal eruptions of the past. There is a very interesting comparison too of the power output of an erupting volcano and its similarity to the time-averaged rate of heat loss from Earth's interior ($\sim 5 \times 10^{13}$ W). In Chapter 14, Natalia Deligne and Haraldur Sigurdsson address the question of how the rate of volcanism has varied through geological time. Because there have been no really large caldera-forming eruptions since humans have been on Earth, it might be inferred that such things will not occur in the future. However, we know that large eruptions occur less frequently than smaller eruptions. In the Chapter Global Rates of Volcanism and Volcanic Episodes, they consider the last 65 million years of Earth history and put into geological and geographical perspective the fluctuations in the magnitude and intensity of volcanism over this time. Evidence shows that there have been several episodes of increased volcanism worldwide in the past and that these link with global climate change. Whether climate change is a cause or an effect of variations in the rate of volcanism remains an intriguing question.

Chapters 15 and 16 describe the landscape features produced by volcanoes. First Shan de Silva and Jan Lindsay look at the role composition of magma plays in the landforms produced. These range from the iconic snow peaked mountain to an indistinct broad depression. In the final chapter in this section, Mike Branney and Valerio Acocella describe in detail the range of features and activity found at calderas. These huge craters can be found in all tectonic settings and are usually the product of a series of eruptions. Calderas have often evolved through many cycles of eruptive activity and untangling their history can be an exciting detective job.

Chapter 12

Earth's Volcanoes and Their Eruptions: An Overview

Lee Siebert, Elizabeth Cottrell, Edward Venzke and Benjamin Andrews

Global Volcanism Program, U.S. National Museum of Natural History, Smithsonian Institution, Washington, DC, USA

Chapter Outline

GLOSSARY

eruption The explosive ejection of fragmented new magma or older solidified material and/or the effusion of liquid lava.

tephra Greek term for ash, used for explosively ejected material of all size classes.

volcanic explosivity index (VEI) A measure of explosive magnitude of an eruption incorporating both subjective and objective criteria, influenced by factors including volume of ejected pyroclastic material and eruption column height.

volcano An accumulation of explosively or effusively erupted materials originating from single or multiple vents or fissures at the surface of the Earth or other planets.

1. INTRODUCTION

Volcanoes and their **eruptions** are among the most awe-inspiring expressions of the natural world and have played a major role in the evolution of the Earth. **Volcano** lifetimes are long, compared with ours, and most of those lifetimes are spent in repose. However, this quiescence can be broken suddenly by violent eruptions that too often take us by surprise. In recent decades, major advances in understanding how volcanoes work have helped mitigate their impact. In part, these have been spurred by dramatic natural events, such as the 1980 eruption and debris avalanche at Mt St Helens, USA, or the climatic impact of eruptions like Mexico's El Chichón (1982) and Mt Pinatubo in the Philippines (1991). However, enhanced monitoring efforts by scientists at volcano observatories or other institutions, as well as improvements in global communication and interdisciplinary cooperation, have equally advanced volcano science. In light of these advances, this chapter will attempt to answer the following questions: Where are the world's volcanoes?, What do they do?, and How often do they do it?

Understanding the impact of and hazard posed by volcanic activity requires information on the frequency, duration, magnitude, and rates of volcanism in both space and time. We will review what has been learned, with primary emphasis on the Holocene, the past 10,000 years. This record is short, by geological standards, but is a rich and important resource that provides our best statistical basis for extrapolating the observations and measurements of contemporary eruptions to better understand those of both the past and the future. Our review draws primarily on two resources of the Smithsonian's Global Volcanism Program (GVP), both available online at www.volcano.si. edu: (1) our reporting of current volcanic activity around

The Encyclopedia of Volcanoes. http://dx.doi.org/10.1016/B978-0-12-385938-9.00012-2

FIGURE 12.1 Volcano types. Volcano shapes reflect the wide range in lava chemistry, vent orientation, gas content, and eruptive style. (A) Strato-volcanoes include roughly conical constructs such as Mexico's Popocatepetl volcano (right) and irregular profiles produced by eruptions from multiple vents, such as at adjacent Iztaccihuatl volcano (left). *Photo by Lee Siebert (Smithsonian Institution).* (B) Accumulation of ejecta from moderate explosive eruptions produces pyroclastic cones, such as Eve Cone on Edziza volcano in Canada, that are the Earth's most common volcanic feature. *Photo by Ben Edwards (Dickinson College).* (C) A tephra blanket surrounds a low-rimmed maar resulting from magma interaction with shallow groundwater during a 1977 eruption at Ukinrek Maars in Alaska. *Photo by Chris Nye (Alaska Division of Geology and Geophysics/USGS).* (D) Low-angle shield volcanoes, such as Galápagos Islands volcanoes Darwin, cut by a summit caldera in the foreground, and Wolf in the background, grow through repeated effusion of

the world since 1968 and (2) our database of historical and other age-dated volcanism over the past 10,000 years (Holocene time). Today GVP publishes online both the monthly *Bulletin of the Global Volcanism Network* and the Smithsonian/USGS *Weekly Volcanic Activity Report*. The database began in 1971, expanding on the regional *Catalog of Active Volcanoes of the World* (1951−1975) series published by the International Association of Volcanology and Chemistry of the Earth's Interior. It has grown steadily, through careful compilation of information from a multitude of sources, to cover more than 1500 volcanoes and 10,000 dated eruptions, regionally listed in Appendix 2 at the end of this volume.

2. VOLCANOES—TYPES, DISTRIBUTION, NUMBER, AND SETTING

Most volcanoes are not the stereotypical towering conical, snow-capped giants. They in fact range from small cinder cones (by far the most common type of volcano) to huge piles of lava, standing 10 km above the ocean floor. Some, like Yellowstone or New Zealand's Lake Taupo, do not even identify themselves by rising above the surrounding landscape. Volcanoes reflect the processes that shape them, exhibiting widely varying plumbing systems, magma types and supply rates, and resulting eruption styles. It is thus not surprising that a wide range of features results. Furthermore, volcanic processes include many that destroy and modify earlier constructions, adding to the variety of forms that fall under the name "volcano."

2.1. Types

Volcanoes are born when subsurface magma, normally less dense than surrounding rock, rises buoyantly toward the surface. In its path upward, magma understandably exploits local fracture systems and, where long linear fractures are present, a "curtain of fire" fissure eruption may result. Alternatively, magma may follow the line of weakness formed where two vertical fracture planes intersect, resulting in a pipelike vertical conduit building a symmetrical cone (Figure 12.1(A), right). Many gradations and complexities exist, even changing during the course of a single eruption. The supply of magma to the surface is commonly sporadic, and if the supply ceases for too long, the next batch will be forced to find another route to the surface because the old conduit will have been blocked with cooled, solidified magma. For example, a lateral shift in the plumbing system of a simple stratovolcano turns it into an asymmetric compound volcano (Figure 12.1(A), left), or a large volcanic edifice dotted by smaller cones.

Perhaps the most influential factor in shaping volcanic landforms, though, is the manner in which gas exits the magma (Oppenheimer, 2005). As magma nears the surface, the attendant decrease in pressure permits exsolution of dissolved gases that then drive the eruption vertically (the only direction in which it is free to expand). Both gas content and viscosity vary widely in magmas, and the more gas-rich, viscous, and lower-temperature compositions common at continental margins tend to explode violently, fragmenting the liquid into countless small particles that cool swiftly as volcanic **tephra**. This process can form small cones of ash (Figure 12.1(B) and (C)), layer upon layer, or (depending on the violence and longevity of the explosions) thin, widely dispersed ash layers. In contrast, gas separates more easily from the less viscous, hotter, and less gas-rich magmas of both spreading centers and oceanic islands, where lava flows passively from vents to form gentle slopes as on Hawaii and the Galápagos Islands (Figure 12.1(D)). Depending on the gas content, magma chemistry, and eruptive style, products range from gentle flows, through near-vent spatter, to scoria, cinder, ash, and pumice, with resulting widely varying volcanic landforms. Often the most violent explosive eruptions are followed by slow, "toothpaste tube"-like extrusions of viscous, degassed magma to form steep-sided domes above the vent (Figure 12.1(E)). Extrusion of lava in meltwater cavities in thick glacial ice caps produces distinctive landforms known as tuyas, or table mountains (Figure 12.1(F)).

Destructive processes can significantly modify the wide variety of initial constructional shapes during a volcano's lifetime. Water and glacial ice are relentless long-term erosional modifiers in many parts of the world. However, slow constructional growth can be suddenly interrupted by catastrophic destructive processes such as caldera collapse (Figure 12.1(G)) (as at Crater Lake in the Oregon Cascade Range, 7700 years ago) or, even more commonly, major slope failure (Figure 12.1(H)) (as at Mt St Helens in 1980)

thin lava flows and form some of the Earth's largest volcanic edifices. *Photo by Patricio Ramon (Escuela Politecnica, Quito)*. (E) Extrusion of viscous lava forms lava domes, such as those that grew at Mt St Helens from 1980 to 1986 (lower left) and from 2004 to 2008 (right-center). *Photo by Steve Schilling (USGS)*. (F) Extrusion of lava beneath glacial ice forms flat-topped tuyas, or table mountains, as seen in aptly named The Table (lower right) in British Columbia, with Garibaldi volcano in the background. *Photo by Lee Siebert (Smithsonian Institution)*. (G) Calderas, such as this one cutting Nemrut volcano in Turkey, are formed by inward collapse following eruption of large volumes of magma. Note postcaldera lake, cones, and lava flows. *(NASA Space Shuttle image.)* (H) Another destructive process, resulting from large volcanic landslides, forms horseshoe-shaped depressions like this one created at Bezymianny volcano in Kamchatka in 1956 and later largely filled by a growing lava dome. *Photo by Yuri Doubik (Institute of Volcanology, Kamchatka)*.

that dramatically change a volcano's shape. Such varied processes lead inevitably to the enormous diversity of volcanic forms on the planet.

2.2. Distribution

Volcanoes are not randomly distributed on the Earth's surface. The global volcano map (inside book covers) shows the distribution of about 1500 terrestrial volcanoes with known or possible activity in the last 10,000 years. This broader time interval provides a more uniform coverage of active volcanism than the regionally variable historical record, and includes most volcanoes likely to erupt again. Tectonic setting, discussed in more detail below, clearly plays a major role in volcano distribution. Although some volcanoes are scattered, most are concentrated in linear belts along tectonic plate boundaries that have accounted for >94% of known historical eruptions. Volcanic belts mostly located above sea level total 32,000 km in length and (assuming a generous 100-km belt width) cover less than 0.6% of the Earth's surface. Most subaerial volcanoes are adjacent to deep oceanic trenches, and it is clear that subduction of one plate beneath another (where oceanic crust is being recycled into the deep Earth at converging plate margins) causes most of the volcanism that we see. The complementary tectonic process of rifting (the creation of new crust at diverging plate margins) takes place largely on the deep ocean floor, unwitnessed by humans, but contributes another ∼70,000 km of volcanic ridges to the global tally. The left half of Figure 12.2 shows the apparent small proportion of the world's documented Holocene eruptions that have taken place in the oceanic or continental rift environment (and in the intraplate environment, where "hot spots" deep in the Earth's mantle feed molten rock to the surface through any plate passing

overhead). It is possible, however, to use relative plate motions to calculate the budget of new lava reaching the Earth's surface every year and these estimates are shown in the right half of Figure 12.2 (Crisp, 1984). Mid-oceanic ridge volcanism overwhelmingly dominates the planet and, although most of it poses no threat to humans, this dominant but mostly unseen volcanism must be remembered in a global perspective.

The geometry of the Earth's tectonic plates means that the distribution of volcanoes by nations is also quite uneven. This national aspect has gained significance as attention to volcanic hazards—with attendant governmental expenditures for monitoring and other studies—has grown. Fully 80% of the world's population live in, and presumably pay taxes to, nations with responsibility for at least one Holocene volcano. For some, these volcanoes are in a distant overseas territory, while for others they loom over the nation's most populous city. Indonesia leads the world in the number of historically active volcanoes, but many would be surprised by the prominent position of the United States (including the Marianas and American Samoa, as well as Alaska and Hawaii), which leads all countries in the number of Holocene volcanoes. The resources for dealing with volcanic hazards are also unevenly distributed, with more than half of the world's volcanoes located in nations with a per capita GDP less than one-fifth that of the United States. But volcanoes disregard national boundaries. Ashfall can be devastating to a downwind nation (e.g., Argentina during the 1991 eruption of Chile's Cerro Hudson); ash clouds from major explosive eruptions threaten airplanes along well-traveled flight paths (e.g., great circle routes to Japan over Alaska and trans-Atlantic flights, such as those impacted by the 2010 eruptions of Eyjafjallajökull in Iceland). Climatic effects of unusually large eruptions can be serious on the opposite side of the globe (e.g., June snowfalls—and crop failures—in New England following the 1815 eruption of Indonesia's Tambora). Clearly volcanic hazards are an international problem, calling for multinational collaboration.

2.3. Number

How many active volcanoes are there in the world? The answer to this common question depends upon use of the word "active." At least 20 volcanoes will probably be erupting as you read these words; roughly 70 erupt each year; 575 have had historically documented eruptions; at least 1250 have erupted in the Holocene (past 10,000 years); and some estimates of young seafloor volcanoes exceed a million. Because dormant intervals between major eruptions at a single volcano may last hundreds to many thousands of years, dwarfing the relatively short historical record in many regions, it is

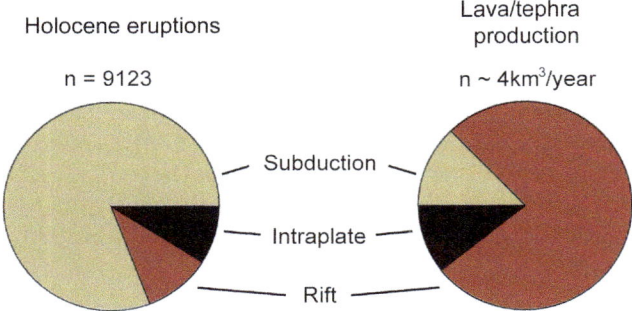

FIGURE 12.2 Volcanism distributed by tectonic setting (see text and inside covers map of the world). Left diagram from Smithsonian data shows proportion of confirmed Holocene eruptions from subduction zones, mid-ocean and continental rifts, and hot spot and intraplate settings. Both Holocene and historical eruptions are strongly biased by dominance of field studies and observations of subaerial volcanism. Right diagram (Crisp, 1984) shows proportion of estimated annual magma budget in the same settings, contrasting the volcanism that we see with that we do not.

misleading to restrict usage of "active volcano" to recorded human memories: we prefer to add modifiers such as "historically active" or "Holocene."

The definition of "volcano" is as important in answering the number question as the definition of "active." Usage has varied widely, with "volcano" applied to individual vents, measured in meters, through volcanic edifices measured in tens of kilometers, to volcanic fields measured in hundreds of kilometers. We have tended toward the broader definition in our compilations, allowing the record of a single large plumbing system to be viewed as a whole, but this approach often requires careful work in the field and laboratory to establish the common magmatic link. The problem is particularly difficult in Iceland, where eruptions separated by many tens of kilometers along a single rift may share the same magmatic system, and be immediately adjacent to parallel rifts from a different volcano. A "volcanic field," such as Michoacán-Guanajuato in Mexico (comprising nearly 1000 cinder cones, derived from a single magmatic system, dotting a 200×200 km area), may be counted the same as a single volcanic system. Perhaps the most honest answer to the number question is that we do not really have an accurate count of the world's volcanoes, but that there are at least 1000 identified magma systems on land, in addition to those along oceanic rifts, likely to erupt in the future.

2.4. Tectonic Setting

Tectonic setting governs the distribution, behavior, and chemistry of volcanoes on the Earth because the divergence and convergence of tectonic plates leads to volcanism in fundamentally different ways. As such, tectonic setting (Figure 12.3) provides a valuable framework in which to categorize and understand volcanism.

2.4.1. Convergent Margins (Subduction Zones)

The most common eruptions observed by humans, and by far the most dangerous to human populations, are those from volcanoes overlying the world's subduction zones. Because subduction zone volcanism can propel volcanic gases and particulates into the stratosphere, these are also the volcanoes most likely to affect short-term climate.

As cold water-laden oceanic plates descend into the hot interior at oceanic arcs and continental edges, they dehydrate. The efflux of water lowers the temperature at which the overlying rocks of the solid mantle melt, leading to magma generation where it would not otherwise occur. The unique mix of sediments and fluids bourn by these subducting plates to depth, in combination with a longer path to the surface wherein magmas can stall, assimilate continental crust, crystallize, and degas, results in a great diversity of magmatic chemistries (from basalt to rhyolite,

encompassing 45—80% SiO_2) and behaviors expressed at the surface. Indeed, it is the eventual exsolution of subducted water and carbon that leads to the explosive character of subduction zone volcanoes.

Volcanic arcs vary enormously in length from tens to nearly 10,000 km. Active volcanism is typically concentrated in a narrow band 20—60-km wide located just over 100 km above the surface of the descending slab, though both arc length and width can vary substantially due to the subducting plate geometry, rate of descent, and other factors. In the Andes, for example, volcanism is weak or absent above downgoing slabs that are inclined at only a gentle angle to the horizontal. We have found that relatively weak volcanism also correlates with (1) thin overriding plates, (2) young downgoing slabs, (3) a decreased angle between the direction of the downgoing slab and the bearing of the arc itself, and (4) low aseismic slip rates (where great earthquakes indicate strong coupling between converging plates).

2.4.2. Divergent Margins (Mid-oceanic Ridges, Back-Arc Basins, and Intercontinental Rifts)

Although the majority of Earth's eruptions occur in this setting, they are not normally seen because they erupt under 1—4 km of water. Here, 70,000 km of volcanic mountain ranges demarcate where tectonic plates are diverging and Earth's convecting solid mantle is melting and erupting in response to decompression. New magma completely resurfaces the ocean floor on 100-million-year timescales. This magma is dominantly basaltic and, because it does not transit thick continental crust, represents the most direct means we have of sampling Earth's mantle. Its extrusion is generally in a narrow 1- to 2-km-wide zone and ranges from linear fissures, through elongate shields, to discontinuous chains of central volcanoes constructed of fresh pillow lavas. Central volcanoes, or seamounts, are also found outside the rift itself, particularly near large offsets in the ridge system and near topographic highs of the faster spreading ridges; they outnumber subduction zone and hot spot volcanoes by several orders of magnitude.

While effusive and constant relative to the explosive and punctuated volcanism at arcs, mid-ocean ridge volcanism is now known to be highly variable, capable of generating large-scale pyroclastic activity (Sohn et al., 2008) as well as driving hydrothermal circulation that regulates ocean and atmospheric chemistry and provides energy to thriving deep-sea ecologies (Martin et al., 2008).

Iceland is one of the few places where the world's rift system emerges above sea level. Here the historical record suggests that, although crustal spreading is probably steady, the rift zone's response to it is episodic: rift eruptions and intrusions lasting several years cause a few meters of separation every 100 years or so, rather than a constant few

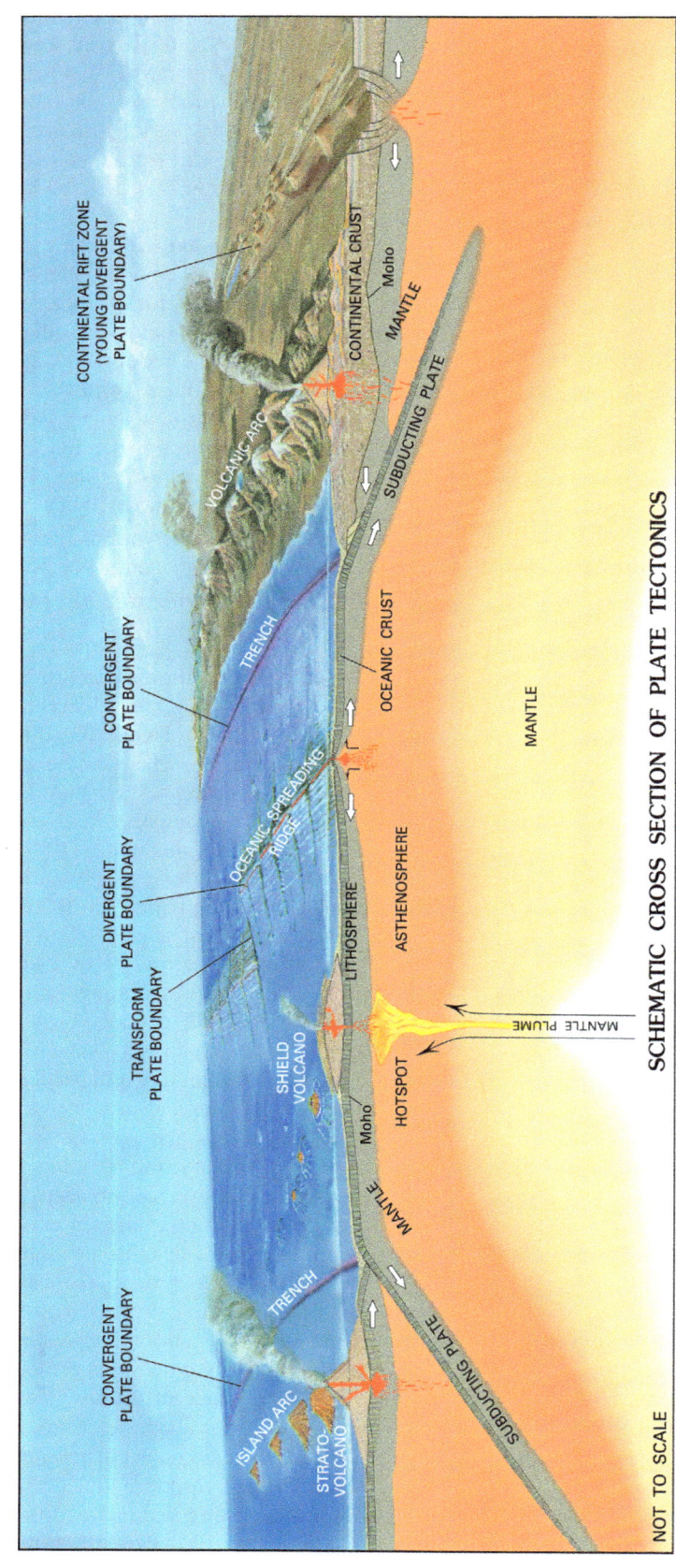

FIGURE 12.3   Plate Tectonics: schematic cross-section illustrating processes. *(modified from Simkin et al., 2006.) Artists José F. Vigil and Robert I. Tilling.*

centimeters of separation every year. Areas of plate divergence are not limited to the oceans but can propagate into continental interiors where they manifest in the thinning of continental crust, the rise and decompression of the Earth's mantle, active volcanism, and ultimately new ocean basins. The volcanoes of the East African Rift, Russia's Baikal Rift, and the Rio Grande Rift of the American Southwest are examples of intercontinental rifts.

2.4.3. Hot Spots (and Flood Basalts)

Hot spot volcanoes are believed to be formed by long-lived plumes of hot upwelling mantle that originate from relatively stationary anomalies at thermal boundary layers such as the mantle transition zone at a 400- to 600-km-depth or the core–mantle boundary at a 2900-km-depth (Montelli et al., 2006). Mantle plumes often bring ancient and enigmatic material to the surface, such that study of hot spot volcanism critically informs our understanding of the Earth's history and the underlying structure of the planet.

As tectonic plates move over these stationary upwellings, or "hot spots," magma production from decompression melting fuels active volcanism at the surface. This geometry results in age-progressive chains of dead and dying volcanoes emanating from the loci of active volcanism centered at the present plume location, which can be on land or at sea. Hawaii is the best-known example of the dominantly basaltic oceanic hot spot volcanism, and the variable volumes of the progressively older volcanoes to the northwest testify to the direction in which the Pacific Plate has moved during the last 70 million years.

The volumes of plume magmatism are enormously variable, ranging from relatively tiny erupted volumes at isolated oceanic islands like Reunion to midsized oceanic plateaus as in the Azores, to the massive thermal anomalies and volcanism of the African "Super Plume." Hot spot volcanism is believed by many to begin with the first batch of magma rising buoyantly through the mantle to reach the surface as a giant outpouring of flood basalts. Flood basalt provinces, such as India's Deccan Plateau (erupted 60–65 Ma) and the Columbia Plateau in the northwestern United States (around 15 Ma), have produced hundreds of thousands of cubic kilometers of lava and individual flows that covered thousands of square kilometers in just a few months or even days. When plume heads impinge on continental crust, they can be more silicic and produce the most violent and least well-understood forms of volcanism on the planet. Such eruptions can produce huge volumes of ash distributed over thousands of kilometers, such as the Cenozoic volcanoes of Colorado's San Juan Mountains, or the calderas of Yellowstone and the Snake River Plain that trace the movements of the North American Continent over the Yellowstone Hot Spot. Studies of nonsubduction volcanism in the western United States emphasize the

relatively long life of these centers (perhaps 5 million years, or 10 times the active life of many subduction zone volcanoes). Such catastrophic episodes are fortunately rare in the planet's history.

3. ERUPTIONS

3.1. Types

Volcanic eruptions vary widely in both magnitude and duration and display a broad spectrum of eruptive styles and processes. Volcanologists traditionally describe eruptions in terms derived from activity types characteristic of particular volcanoes or volcanic regions.

Two terms originate from volcanoes in the cradle of volcanology, Italy's Aeolian Islands. *Strombolian* eruptions (Figure 12.4(A)) derive their name from the long-term eruptions at Stromboli volcano. These discrete, sometimes rhythmic explosions common at low-viscosity, basaltic to basaltic–andesite volcanoes eject incandescent lava fragments that can form spectacular nighttime displays. The nearby island of Vulcano lent its name to more energetic *Vulcanian* eruptions (Figure 12.4(B)), common at stratovolcanoes and lava domes. These events are typically very short duration blasts or explosions that eject blocks, bombs, and tephra accompanied by atmospheric shock waves.

Eruptions where water has access to the vent are named *Surtseyan* (Figure 12.4(C)) after the 1963–1967 submarine and emergent eruptions that formed the island of Surtsey off the south coast of Iceland. These explosive bursts eject a column of ash, mud, water, and steam, with individual-ejected blocks characteristically trailing cockscomb sprays of ash and steam. Base surges often also produce a ring-shaped cloud that travels horizontally above the water surface following collapse of the vertical column. This type of magma–water interaction is common during shallow submarine eruptions and at crater lakes, such as at New Zealand's Ruapehu volcano.

Pelean eruptions are named for the renowned 1902 eruption of Pelée volcano on the West Indies island of Martinique (Figure 12.4(D)), when explosive eruptions accompanying collapse of summit lava domes (Figure 12.4(E)) produced pyroclastic flows that devastated the town of St. Pierre and its 28,000 inhabitants. These dome-collapse pyroclastic flows are often referred to as Merapi-type pyroclastic flows, which are common at that very active Indonesian volcano on the island of Java.

The most powerful explosive eruptions are named not for a volcano, but from the observations of Pliny the Younger during the renowned AD 79 eruption of Italy's Vesuvius volcano. *Plinian* eruptions (Figure 12.4(F)) form vent-sourced convective plumes that propel large volumes

FIGURE 12.4 Eruption types. (A) Incandescent ejecta from *Strombolian* eruptions such as these at Parícutin volcano in Mexico in 1944 produce spectacular nighttime displays. *Photo by Carl Fries (USGS).* (B) *Vulcanian* eruption column rises above Karymsky volcano in Kamchatka in 2004. *Photo by Alexander Belousov (Institute of Volcanology, Petropavlovsk-Kamchatsky).* (C) Water−magma interaction produces *Surtseyan* eruption column at Ruapehu volcano in New Zealand in 1971. *Photo by Peter Otway (New Zealand Geological Survey).* (D) Pelée volcano in the West Indies overlooks the city of St. Pierre in 2002, a century after pyroclastic flows from *Pelean* eruptions destroyed the city. *Photo by Lee Siebert (Smithsonian Institution).* (E) Incandescent blocks spill down the flanks of a growing lava dome in the crater of Kelut volcano in Indonesia in 2007. *Photo by Tom Pfeiffer (Volcano Discovery).* (F) A large explosive column from a 1991 *Plinian* eruption at Pinatubo volcano in the Philippines towers above Clark Air Force base. *Photo by Dave Harlow (USGS).* (G) A pyroclastic flow sweeps down the flanks of Soufrière Hills volcano in the West Indies in 1997. *Photo by Richard Heard (Montserrat Volcano Observatory).* (H) A lava fountain feeds lava flows during an eruption of Hawaii's Kilauea volcano in 1983. *Photo by Norm Banks (USGS).*

of ash to high altitudes where it is dispersed by stratospheric winds over huge areas. Pyroclastic flows are commonly generated in these eruptions when the magma eruption rate exceeds the capacity of the plume to rise convectively, resulting in partial or complete collapse of the eruption column. Although pyroclastic flows can also occur during smaller eruptions (Figure 12.4(G)), they often are an integral component of particularly large eruptions that can form calderas when so much magma erupts that the shallow crust founders into the emptying magma reservoir. Quantification of grain-size characteristics has led to subdivision of Plinian activity into sub-Plinian, Plinian, phreatoplinian (major interaction with external water), and ultra-Plinian eruptions.

Dominantly effusive eruptions are named after two volcano-rich regions, Hawaii and Iceland. *Hawaiian* eruptions can produce long-term activity feeding summit-crater lava lakes and the effusion of fluid, low-viscosity lava flows fed by sustained lava fountains along radial fissures (Figure 12.4(H)), cumulatively building massive shield volcanoes. *Icelandic* eruptions display similar activity, but originate from regional fissures such as those that produced the voluminous lava flows of the 1783 "Laki fires" eruption from Grímsvötn volcano. Icelandic-style shield volcanoes are typically much smaller than their Hawaiian counterparts.

Looking at the distribution of volcanism through time again raises questions of definition: Is an individual explosion an "eruption" or should the word be used (as we tend to) for clearly linked events that may be separated by hours, days, or even months of surface quiet? Arrival of solid volcanic products or liquid lava at the Earth's surface is normally termed an eruption, but the historical record is often unclear. Steam venting is commonly only a by-product of subsurface heat, and is rarely dignified by the word "eruption," but steam may be all that was seen by a sea captain reporting an "erupting" volcano 200 years ago. Volcanic heat is also the prime source of "phreatic" (or "steam-blast") explosions in which heat and water combine to fragment preexisting volcanic materials and hurl the resulting tephra into the sky. These events can be quite violent, however, and are rightfully called eruptions, even without the direct involvement of new magma. Common accounts of "smoking" volcanoes in the early historical records, however, leave uncertainty as to whether the plumes observed were restricted to vigorous steam emission or included ash ejection.

Gas is an unequivocal volcanic product, but the gentle, continuous emission of fumarolic gases—common to most volcanoes—is not considered an "eruption." Occasionally, though, large quantities of gas are suddenly vented, as in the fatal eruption of the Dieng volcano complex on Java in 1979. In 1984 and 1986, the African nation of Cameroon suffered catastrophic expulsions of CO_2 from a crater lake, resulting in many fatalities by suffocation. The origin of the African events is controversial, but there is evidence suggesting that they resulted not from an eruption, but from the sudden overturn of crater lake water, with catastrophic exsolution of volcanic gas that had gradually accumulated in the bottom waters over the course of many years.

Eruptions may release quite different amounts and types of gas, with comparably different effects on climate. Although larger eruptions inevitably receive the greatest share of scientific (and media) attention, the steady pulse of smaller events adds up to significant contributions. In assessments of annual volcanic SO_2 production, for example, the steady, unspectacular contributions of the world's fumaroles and small, ongoing eruptions are found to outweigh those of the less frequent larger eruptions.

3.2. Durations and Culminations

Clearly some eruptions last for a very long time, from decades to centuries, and currently more than a dozen volcanoes have been in eruption for a quarter century or more. However, other eruptions end swiftly: 10% of those for which we have accurate durations lasted no longer than a single day (Figure 12.5(A)), most end in less than 3 months, and few last longer than 3 years. The median duration is about 7 weeks.

Of particular importance to volcanic hazard mitigation is the time interval between an eruption's onset (the initial ejection of tephra or effusion of lava) and its culminating, paroxysmal phase. Several famous eruptions (e.g., Krakatau, 1883; Mt St Helens, 1980; Pinatubo 1991) have culminated months after the start of low-level eruptive activity, but a look at the 288 most explosive eruptions for which we have adequate data (Figure 12.5(B)) emphasizes the sobering fact that many catastrophic explosions provide little advance warning. More than two-fifths of the eruptions reached their climax within the eruption's first day (42%) and more than half (52%) within the first week. Where data are available, half of the first-day paroxysms (including Tarawera, 1886; Bandai-san 1888, Hekla, 1947; and Shiveluch, 1964) came within only an hour of the eruption's start.

The range of eruptive behavior is wide, however, and history's largest explosive eruption, Tambora in 1815, provides sobering lessons about eruption duration. Mild eruptive activity over 3 full years was followed by a dramatic eruption, with cloud height estimated at 33 km, but even that was not its climax. After a lull of 5 days, the culminating eruptive cloud reached heights estimated at 44 km and caused 3 days of total darkness 500 km from the volcano. Over 60,000 people died as a result of this eruption. It is understandably difficult to sustain public awareness of hazards during long-term, low-level

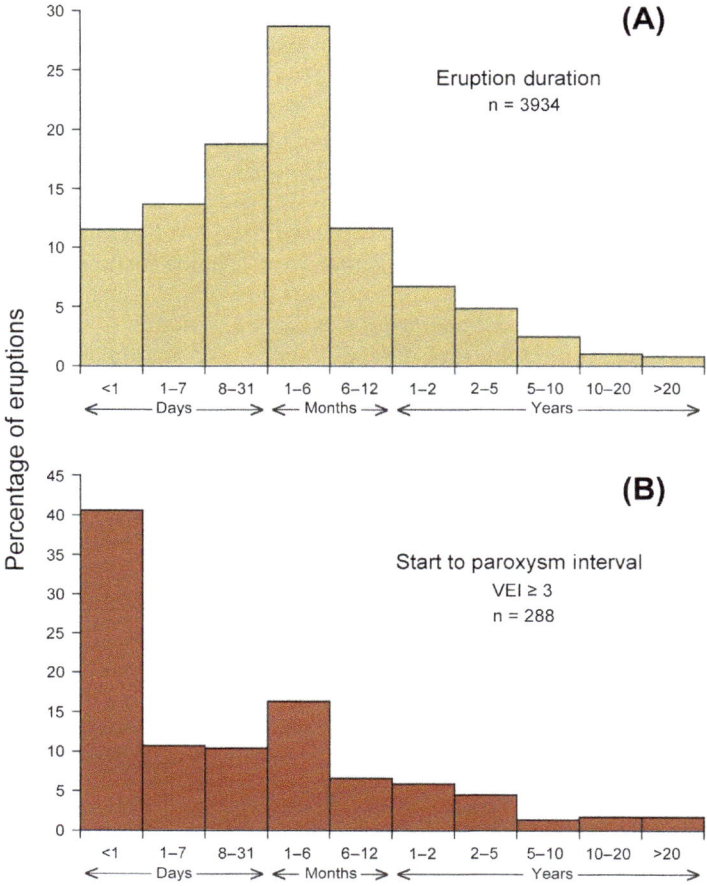

FIGURE 12.5 Eruption duration and development. (A) 3934 eruptions with known durations. (B) Interval between the onset of an eruption and its paroxysmal phase for 288 highly explosive eruptions (VEI $\geq$ 3, producing $>10^7$ m^3 of tephra, or, in the absence of volume estimates, described as "severe" or "violent"). VEI, volcano explosivity index.

eruptive activity, but it is dangerous to assume that the worst is over after the initial explosive phase. Predicting an eruption's end is even more difficult than predicting its beginning.

3.3. Patterns through Time

Has global volcanism changed through historical time? The answer is likely "no." A look at the number of volcanoes active per year over the last 600 years (Figure 12.6) shows a dramatic increase, but one that is closely related to increases in the world's human population and enhanced communication methodologies. We believe that this represents an increased reporting of eruptions, rather than increased frequency of global volcanism: more observers, in wider geographic distribution, with better communication, and broader publication. The past 200 years (Figure 12.7) show this generally increasing trend along with some major "peaks and valleys," which suggest global pulsations. A closer look at the two largest valleys, however, shows that they coincide with the two world wars, when people

(including editors) were preoccupied with other things. Many more eruptions were probably witnessed during those times, but reports do not survive in the scientific literature.

If these apparent drops in global volcanism are caused by decreased human attention to volcanoes, then it is reasonable to expect that increased attention after major, newsworthy eruptions should result in higher-than-average numbers of volcanoes being documented in the historical literature. The 1902 disasters at Mt Pelée, St. Vincent, and Santa María (Figure 12.7) were highly newsworthy events. They represent a genuine pulse in Caribbean volcanism, but we believe that the higher numbers in following years (and following Krakatau in 1883) result from increased human interest in volcanism. People reported events that they might not otherwise have reported and editors were more likely to print those reports.

Further evidence that the historical increase in global volcanism is more apparent than real comes from the lower plot of Figure 12.7. Here only the larger eruptions (generating at least 0.1 km^3 of tephra, the fragmental

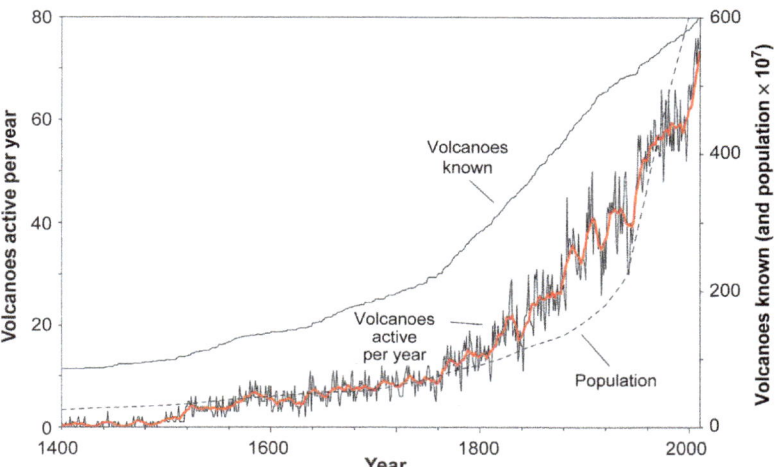

FIGURE 12.6 Six hundred years of volcano reporting. Total number of volcanoes erupting per year (black line) and 10-year running mean of the same data (red line). Uncertain eruptions and those with dating uncertainties larger than 1 year are not shown. "Volcanoes known" is the total number with historically recorded eruptions by any given year. Dashed line shows world's estimated human population data.

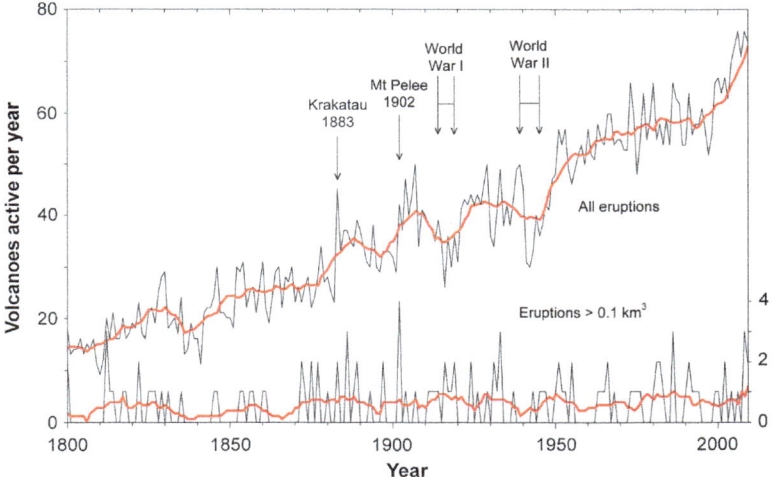

FIGURE 12.7 Volcano reporting since AD 1800. Total number of volcanoes erupting per year (upper black line) and 10-year running mean of the same data (upper red line). Lower plot shows the annual number of volcanoes producing large eruptions (≥ 0.1 km^3 of tephra), and scale is enlarged (on right axis) with lower red line showing a 10-year running mean.

products of explosive eruptions) are plotted. The effects of these larger events are often regional, and therefore less likely to escape documentation even in remote areas. The frequency of these events has remained impressively constant for more than a century, and contrasts strongly with the apparent increase in the number of smaller eruptions with time.

While short-term phenomena, such as Earth tides and large-magnitude tectonic earthquakes (Linde and Sacks, 1998), have been proposed as potential eruption triggers, we conclude that volcanism within the past few centuries appears to have been relatively constant. On a longer scale, however, factors such as increased eruption rates due to glacial rebound-induced decompression melting following Pleistocene and early Holocene deglaciation

have been noted by Huybers and Langmuir (2009) and Kutterolf et al. (2013).

3.4. Magnitude and Frequency (Global)

How much do eruptions vary in size? Like earthquakes, the frequency of eruptions decreases with increasing size. We have borrowed from the seismologists' magnitude–frequency plots, using as our best measure of magnitude the **volcanic explosivity index** (VEI; Newhall and Self, 1982). This closely parallels the volume of erupted tephra, which increases by an order of magnitude with each larger VEI integer. More rigorously quantitative criteria are available for eruptions from well-monitored and well-studied volcanoes, but are not readily extrapolated to less well-documented eruptions of the past. The VEI therefore offers a relative magnitude scale that has been applied to more than three-fourths of reported Holocene eruptions.

The explosive magnitude as expressed by the VEI has a strong peak at VEI 2, with small eruptions of VEI 0–2 accounting for nearly four-fifths of all eruptions (Figure 12.8(A)). Although VEI 2 is used as a default assignment for explosive eruptions in the absence of other criteria, it remains the dominant VEI for eruptions during the past 50 years (when more quantitative data are available), with VEI 0–2 eruptions increasing to more than 80%. Only 5% of documented Holocene eruptions are VEI 4 (the size of the devastating 1902 eruption of Pelée in the West Indies), and only 2.4% of explosive

eruptions are VEI 5 or larger, the size of the 1980 eruption of Mt St Helens.

Nearly 90% of volcanoes have eruptions with average VEIs of less than 3 (Figure 12.8(B)). This figure increases to more than 97% when only historical eruptions are considered due to the more comprehensive documentation of smaller-volume eruptions. This implies that the vast majority of eruptions are fairly modest affairs, although proximal effects can still be considerable.

The Holocene record is clearly biased by the significant underdocumentation of the small-magnitude volcanic eruptions that are more likely to be missed in the stratigraphic and historical record (Deligne et al., 2010). Extrapolation of better-known rates of recent historical eruptions suggests that less than 2% of Holocene eruptions have currently been documented (Siebert et al., 2010).

This recent record thus provides a better window into eruption frequencies. Data for the past half century (Figure 12.7) show that 50–70 eruptions occur annually, with a recent increase attributable to enhanced remote sensing observations. Subaerial eruptions producing at least 10^6 m^3 of tephra (VEI 2) take place at an average rate of once every few weeks somewhere on the Earth. Those producing $\geq 10^7$ m^3 of tephra, such as the VEI 3 Ruiz eruption that generated devastating mudflows in 1985, take place several times a year. Eruptions the size of Mt St Helens in 1980 (10^9 m^3, or 1 km^3 and VEI 5) occur perhaps once a decade, and those such as Krakatau in 1883 (≥ 10 km^3 and VEI 6) on the order of once a century. The known Holocene record implies that VEI 7 eruptions, such as at Tambora in 1815, occur less than once a millennium, although this record is also incomplete.

3.5. Quiescent Interval and Subsequent Magnitude

While Figure 12.7 illustrates the frequency of eruptions on a global basis, most people are more interested in how much longer a particular volcano will remain quiet and what kind of eruption will break that quiet. Such information can come only from detailed study and monitoring, as described elsewhere in this volume, but the volcanological record provides some guidance that is both useful and sobering. One of the certain contributors to large volcanic death tolls is the fact that the repose interval between eruptions is very much longer than the historical records in many parts of the world. Figure 12.9(A) shows the relationship between eruption quiescent interval and the proportion of eruptions in each VEI class. Eruptions of low explosivity follow variable but often shorter repose intervals of a decade or less, but with higher explosive magnitudes the proportion of longer quiescent intervals

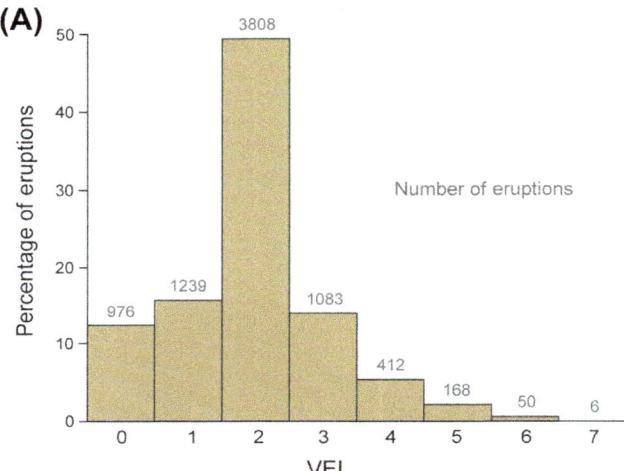

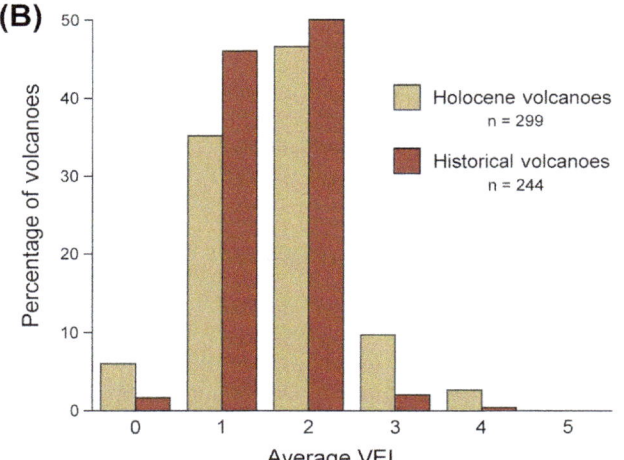

FIGURE 12.8 (A) Percentage and number of volcanic explosivity index (VEI) assignments for all Holocene eruptions. (B) Average VEI of volcanoes with five or more Holocene VEI assignments compared with volcanoes with five or more historical VEI assignments.

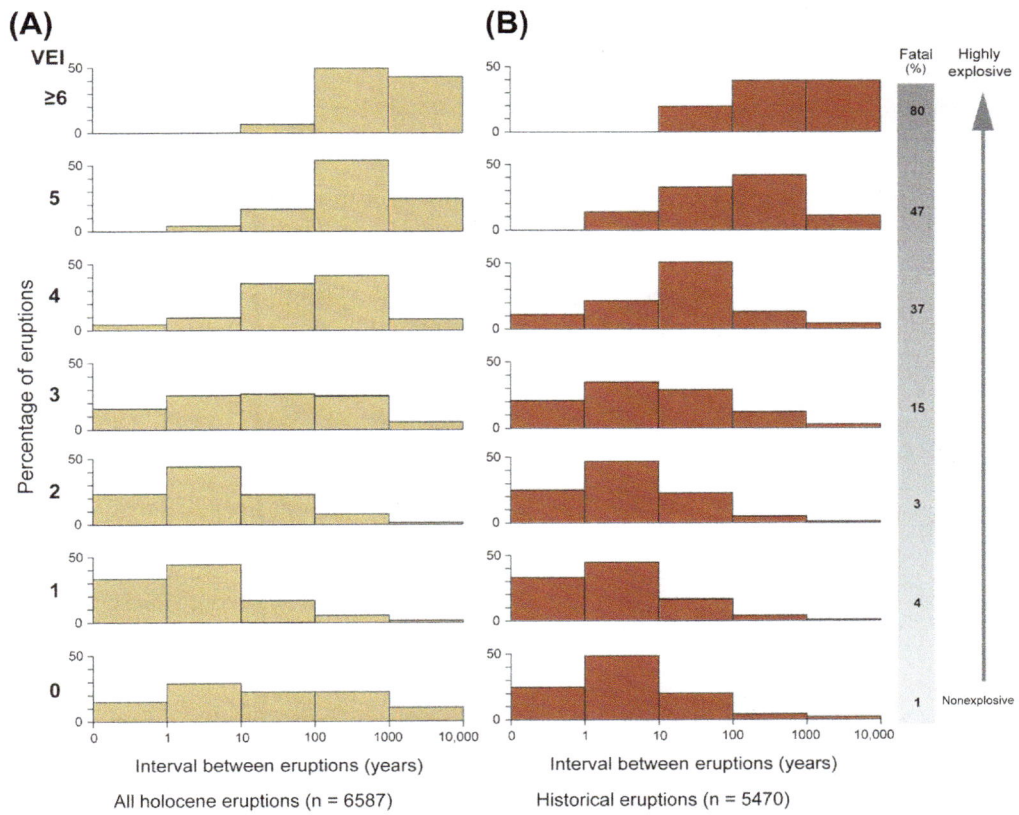

FIGURE 12.9 Explosivity and quiescent intervals preceding eruptions. For each volcano explosivity index (VEI) unit, percentages of eruptions are grouped by time interval from the start of previous eruption. (A) Intervals prior to Holocene eruptions in VEI groups 0 to ≥6 are 635, 1031, 3383, 970, 376, 137, and 30, respectively. (B) Intervals prior to historical eruptions are likewise 369, 999, 3173, 718, 166, 36, and 5, respectively. For each VEI group, the percentage of historical eruptions that have been fatal is also shown to emphasize the danger of large explosive eruptions from volcanoes that have appeared to be quiet for hundreds to thousands of years.

increases. These data can be biased by the underrecording of smaller eruptions between older larger eruptions often dated by radiometric techniques, but a similar pattern (Figure 12.9(B)) is seen when looking at intervals prior to historical eruptions.

The median interval between successive eruptions of a volcano is only about a dozen years (Figure 12.10(A)). About two-fifths of eruptions follow quiet intervals of more than a half century, often placing the last eruption outside the direct memory of those living near the volcano. More than 60% of volcanoes with dated Holocene eruptions have currently been in repose for more than a half century, and the median period of current inactivity is slightly more than a century (Figure 12.10(B)). Many volcanoes have thus currently been quiescent for significantly longer periods of time than is typical. The frequency data discussed above imply that we are not currently in a period of reduced volcanism, but the current long quiescence at many volcanoes (which can precede larger eruptions) is not necessarily a reason for reassurance.

It would be a mistake to overemphasize the historical record in selecting volcanoes at the highest risk to erupt in the near term. For at least two centuries we have had an average of one or two eruptions per year from volcanoes with no previous historical activity; these include some of history's worst natural disasters, such as at the Indonesian volcanoes of Tambora in 1815 and Krakatau in 1883. In regions with short historical records, the human population is commonly unprepared. The results are often tragic.

4. VOLCANOES AND HUMANS

Volcanoes have from time immemorial been the object of both fear and fascination. Volcanoes are not only well-known agents of destruction, but also benefit mankind by providing rich agricultural soils, geothermal heat, the land on which to grow crops and locate cities, and economic benefit ranging from volcanic ore deposits to tourism and recreation at some of our planet's most spectacular terrains.

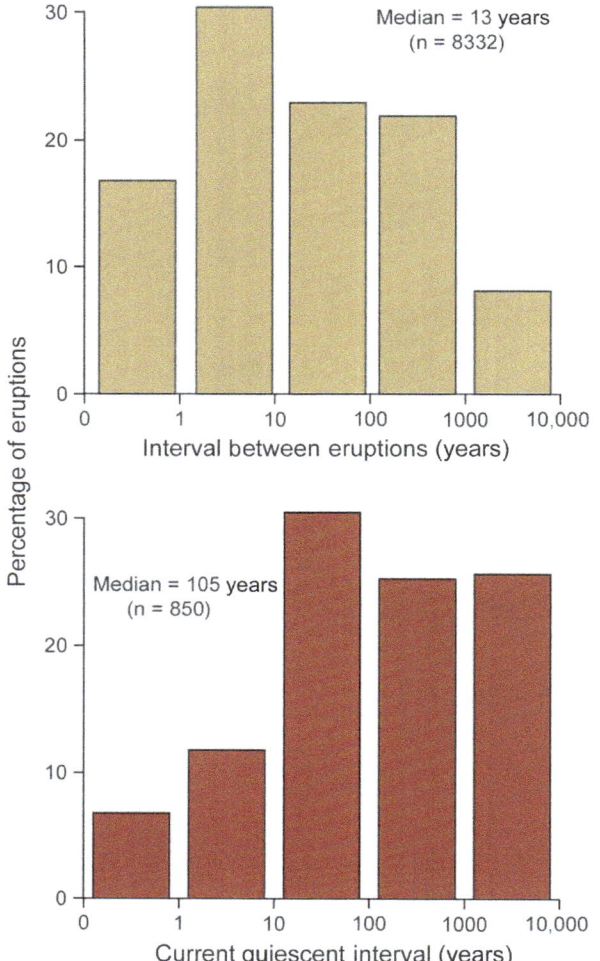

FIGURE 12.10 Global distribution of intervals between successive eruptions of Holocene volcanoes (top) and length of current quiescence (bottom; as of 2010).

Volcanoes have caused the destruction and abandonment of entire cities, but have also been an inspiration to artists, photographers, and all who are drawn to observe their rugged beauty and to witness some of the Earth's most dramatic geological processes. Coexistence with volcanoes requires the full efforts and attention of scientists, public officials, and those living in harm's way.

4.1. Proximity to Volcanoes

The numbers of persons living in proximity to volcanoes have risen commensurately with increases in global population. Population data sets now allow regional and global assessment of persons living near volcanoes. Preliminary analysis of data from LandScan population algorithms generated by Oak Ridge National Laboratory suggests that about 58 million people reside within a

10 km radius of ~1300 Holocene volcanoes (or volcanic-field vents), where evacuation may be needed for small to moderately large eruptions. More than 200 million live within a 30 km radius, and more than 11% of the world's 6.5 billion people live within 100 km of a Holocene volcano. Examination of population densities near volcanoes with respect to the degree of activity of those volcanoes (Figure 12.11) shows a somewhat disturbing trend from a volcanic hazards perspective, with the highest populations at risk from volcanoes living adjacent to the most active volcanoes.

Hazards from volcanic eruptions are not restricted to people living near volcanoes, but also include those in transit over volcanic or even nonvolcanic areas. The near-catastrophic jumbo jet—ash encounter at Galunggung volcano in Indonesia in 1982 first raised concerns about aircraft—ash interactions, prompting major response efforts by the volcanological and aviation communities. The 2010 eruption of Eyafjallajökull volcano in Iceland underscored these concerns, when ash plumes that drifted over trans-Atlantic flight paths caused more than 100,000 flight cancellations, with severe economic ramifications.

4.2. Fatalities and Evacuations

Although fatalities from volcanic eruptions are substantially fewer than those from other natural disasters such as earthquakes, floods, and extreme weather events, more than 500 fatality-producing events have been documented globally during the Holocene (Siebert et al., 2010). Although the historical record is likewise incomplete, more than 275,000 volcanic fatalities have been recorded since AD 1600 (Auker et al., 2013). Pyroclastic flows have claimed the most lives. However, the highest number of fatal events has resulted from tephra, which includes all fragmental material thrown from volcanoes. The most tephra deaths have come from collapse of ash-covered roofs, but projectile impact has caused by far the largest number of fatal tephra accidents (Simkin et al., 2001).

The impact of volcanic events has been offset by increased numbers of evacuations due to monitoring and mitigation efforts by volcanologists and public officials. The latter include educational campaigns to heighten awareness of hazards. Evacuations were noted at nearly 300 eruptions from 1976 to 2009 (Siebert et al., 2010). The number of evacuees has been tallied for less than half of these events and yields a minimum of 1,900,000 persons who left for safer ground. Although data are approximate and uncertain, Auker et al. (2013) noted a dramatic drop during the twentieth century in the volcanic fatality index that factors in population growth and the likely underreporting of both eruptions and low-fatality events (Figure 12.12). This suggests that investments in

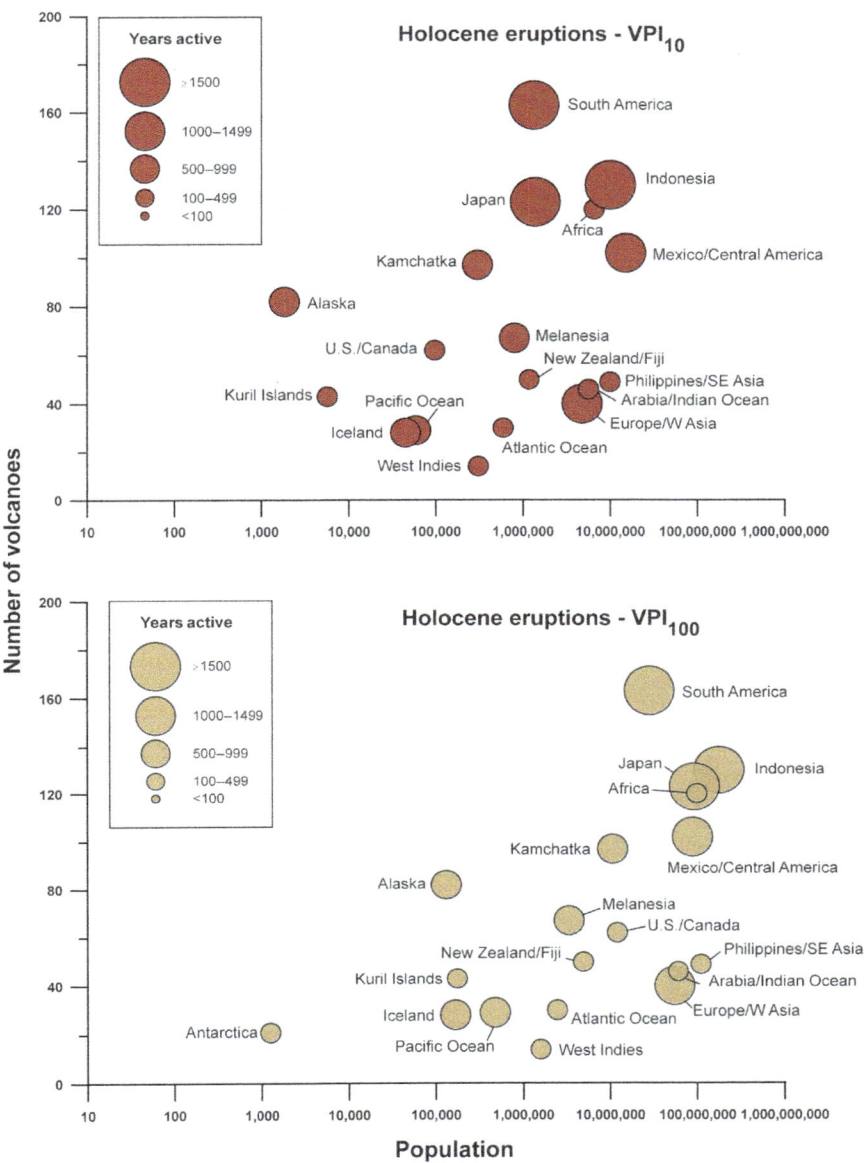

FIGURE 12.11 Regional population data for Holocene volcanoes at 10 km radius (top) and 100 km radius (bottom). Circle size indicates the number of cumulative Holocene years active within each region.

mitigation, civil protection, and preparedness have had the desired effect of reducing the human impact of volcanic eruptions.

Humans have not been successful in controlling volcanoes in a substantive way and are not likely to be so in the future. We can, however, develop a more successful coexistence with volcanoes by improving our understanding of the huge eruptions of the geologic past as well as the smaller, more frequent and more familiar events of our own lifetimes.

Although analysis of historical eruption frequencies suggests that documentation of a uniform global eruption rate is near at hand, even in the twenty-first century, eruptions can occur unnoticed in sparsely populated and poorly instrumented areas. Ongoing efforts to improve documentation of eruptive activity at both ends of the time continuum are essential to understanding our planet's volcanism and anticipating what future volcanic events might be in store. Recognizing that volcanoes can resume activity after very long periods of quiescence, even exceeding the full duration of the Holocene, underscores the importance of extending the record back into the Pleistocene to include all volcanoes that could erupt next year. That effort has begun, including generation of a database of large-volume Pleistocene eruptions (Crosweller et al., 2012). These efforts, along with expanded devotion internationally of resources to volcano monitoring and hazards mitigation,

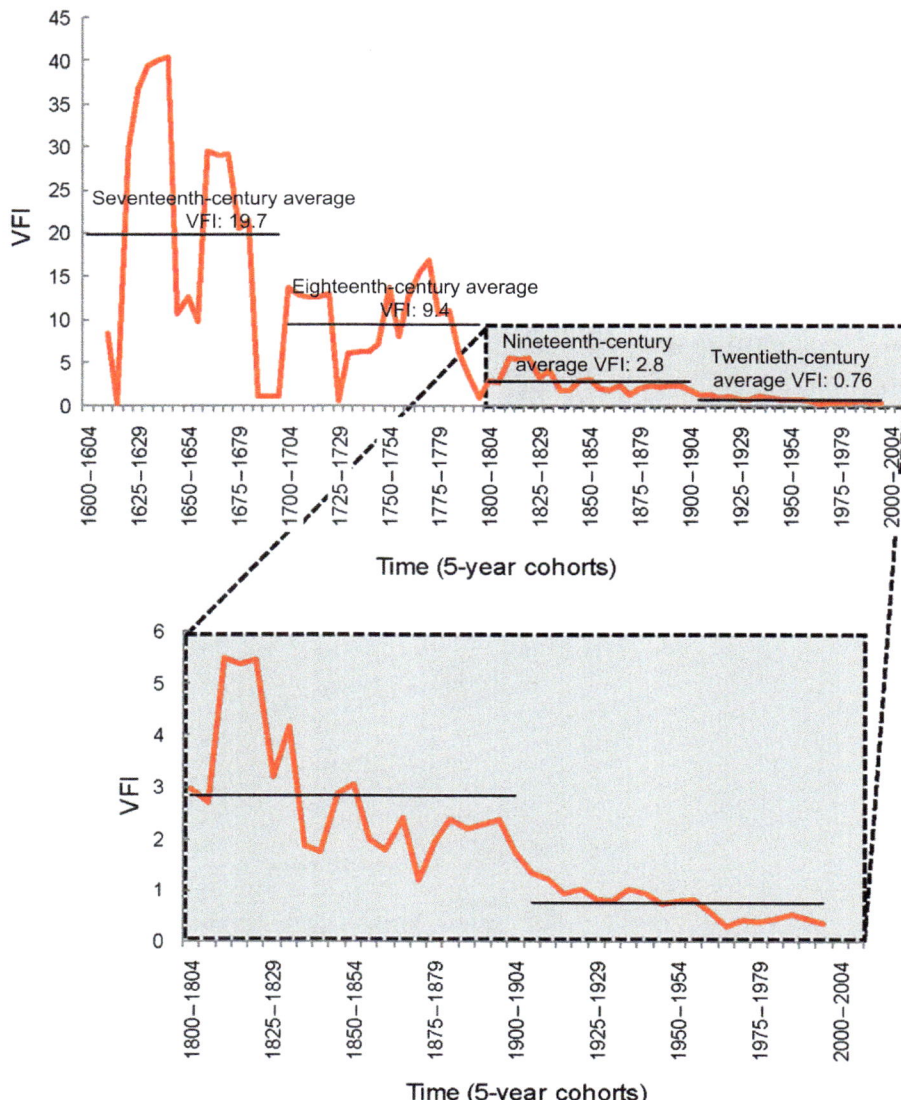

FIGURE 12.12 Twenty-five-year running average of volcano fatality index (VFI). The numbers of fatalities divided by the product of the number of recorded volcanic eruptions and population at fixed intervals (Auker et al., 2013). Century-wide average VFI figures in black. Inset plot shows 25-year running average VFI over the period 1800–2010. The largest 10 disasters have been removed.

will help to mitigate the impact of volcanoes in the years to come.

FURTHER READING

Auker, M.R., Sparks, R.S.J., Siebert, L., Crosweller, H.S., Ewert, J., 2013. A statistical analysis of the global historical volcanic fatalities record. J. Appl. Volcanol. 2 (2). http://www.appliedvolc.com/content/2/1/2.

Catalog of Active Volcanoes of the World, vol. 1–22, 1951–1975. International Association of Volcanology and Chemistry of Earth's Interior, Rome.

Crisp, J., 1984. Rates of magma emplacement and volcanic output. J. Volcanol. Geotherm. Res. 20, 177–211.

Crosweller, H.S., Arora, B., Brown, S.K., Cottrell, E., Deligne, N.I., Ortiz Guerrero, N., Hobbs, L., Kiyosugi, K., Loughlin, S.C., Lowndes, J., Nyembil, M., Siebert, L., Sparks, R.S.J., Takarada, S., Venzke, E., 2012. Global database on large magnitude explosive volcanic eruptions (LaMEVE). J. Appl. Volcanol. 1 (4). http://www.appliedvolc.com/content/1/1/4.

Deligne, N.I., Coles, S.G., Sparks, R.S.J., 2010. Recurrence rates of large explosive volcanic eruptions. J. Geophys. Res. 115 (B06203). http://dx.doi.org/10.1029/2009JB006554.

Huybers, P., Langmuir, C., 2009. Feedback between deglaciation, volcanism, and atmospheric CO_2. Earth Planet. Sci. Lett. 286, 479–491.

Kutterolf, S., Jegen, M., Mitrovica, J.X., Kwasnitschka, T., Freundt, A., Huybers, P.J., 2013. A detection of Milankovitch frequencies in global volcanic activity. Geology 41, 227–230.

Linde, A.T., Sacks, I.S., 1998. Triggering of volcanic eruptions. Nature 395, 888–890.

Martin, M., Baross, J., Kelley, D., Russell, M.J., 2008. Hydrothermal vents and the origin of life. Nat. Rev. Microbiol. 6, 805–814.

Montelli, R., Nolet, G., Dahlen, F.A., Masters, G., 2006. A catalogue of deep mantle plumes: new results from finite-frequency tomography. Geochem. Geophys. Geosyst. 7 (11) http://dx.doi.org/10.1029/2006GC001248.

Newhall, C.G., Self, S., 1982. The volcanic explosivity index (VEI): an estimate of explosive magnitude for historical volcanism. J. Geophys. Res. 87, 1231–1238.

Oppenheimer, C., 2005. Volcanic degassing. In: Rudnick, R.L. (Ed.), The Crust, vol. 3. Elsevier, Amsterdam, pp. 123–166.

Siebert, L., Simkin, T., Kimberly, P., 2010. Volcanoes of the World, third ed. University of California Press, Berkeley.

Simkin, T., Siebert, L., Blong, R., 2001. Volcano fatalities: lessons from the historical record. Science 291, 255.

Simkin, T., Tilling, R.I., Vogt, P.R., Kirby, S.H., Kimberly, P., Stewart, D.B., 2006. This Dynamic Planet — World Map of Volcanoes, Earthquakes, Impact Craters, and Plate Tectonics. U.S. Geol. Surv. Geol. Inv. Ser. Map I-2800 scale 1:30,000,000.

Sohn, R.A., Willis, C., Humphris, S., Shank, T.M., Singh, H., Edmonds, H.N., Kunz, C., Hedman, U., Helmke, E., Jakuba, M., Liljebladh, B., Linder, J., Murphy, C., Nakamura, K., Sato, T., Schlindwein, V., Stranne, C., Tausenfreund, M., Upchurch, L., Winsor, P., Jakobsson, M., Soule, A., 2008. Explosive volcanism on the ultraslow-spreading Gakkel ridge, Arctic Ocean. Nature 453, 1236–1238.

Chapter 13

Sizes of Volcanic Eruptions

David M. Pyle

Department of Earth Sciences, University of Oxford, Oxford, UK

Chapter Outline

GLOSSARY

dense rock equivalent (DRE) volume The calculated volume of erupted materials, corrected to an assumed (constant) density, equivalent to the density of bubble-free magma.

intensity A logarithmic scale used to quantify mass eruption rate (kg/s).

magnitude A logarithmic scale used to quantify the mass of material ejected during a volcanic eruption (kg).

plinian eruption A typical large explosive eruption, characterized by the formation of a sustained eruptive plume that rises tens of kilometers into the atmosphere.

volcanic explosivity index (VEI) A widely used metric to describe the size of explosive volcanic eruptions, based primarily on the erupted volume of a deposit or on the peak eruption column height.

1. INTRODUCTION

Subaerial volcanic eruptions span a wide range of erupted volumes of lava, pyroclastic rock, and gas across a spectrum of eruptive styles. They also vary widely in their violence, destructiveness, and wider impacts. Consequently, there are many different ways that eruption "size" could be measured and quantified. The two principal quantities that are most widely used to define the scale of an eruption are magnitude, the mass of material erupted; and intensity, the mass eruption rate. These two quantities can be determined more or less precisely for both modern and ancient eruptions, and for both effusive and explosive styles of eruption. Studies of the sizes of past and present volcanic eruptions allow volcanologists to compare recent "smaller" events with the rare, but colossal, eruptions of past millennia. By so doing, we may learn more about the processes that control eruptions of all scales.

2. SCALES AND SIZES OF VOLCANIC ERUPTIONS

Volcanic eruptions, like many geophysical phenomena, span a vast range of scales of size, rate, and duration. The mass of magma ejected during an eruption, the quantities of gas released, and the eruptive flux all vary by many orders of magnitude. Consequently, logarithmic scales are needed to categorize the sizes of volcanic eruptions, in much the same way as the Gutenberg–Richter magnitude scale is used for earthquakes. The first attempts to devise such a scale for volcanoes used a single integer to describe the size of explosive eruptions. This scale, proposed by Tsuya (1955), divided eruptions up into different magnitude classes (I–IX), based on the logarithm of the bulk volume of pyroclastic ejecta.

A revised scale called the volcanic explosivity index (VEI) was proposed by Newhall and Self (1982), who developed this concept into a practical and widely used metric for categorizing the scale of explosive eruptions (Table 13.1). The VEI, as originally defined, uses an integer scale from 0 to 8 to broadly describe the erupted volume and the eruption plume height. This index is therefore defined by both eruption magnitude (volume) and intensity (eruption column height). As an example, a VEI 4

The Encyclopedia of Volcanoes. http://dx.doi.org/10.1016/B978-0-12-385938-9.00013-4

TABLE 13.1 Categories of the Volcanic Explosivity Index

Index	0	1	2	3	4	5	6	7	8
Bulk tephra volume (m^3)	$<10^4$	$<10^6$	$<10^7$	$<10^8$	$<10^9$	$<10^{10}$	10^{11}	$<10^{12}$	$>10^{12}$
Eruption plume column height (km)	<0.1	$0.1-1$	$1-5$	$3-15$	$10-25$	>25			
Qualitative description	Gentle	Effusive	Explosive		Cataclysmic, paroxysmal				
Stratospheric injection	None	None	None	Possible	Certain				
Percentage of known eruptions in the past 10,000 years	13	16	49	14	5	2	<1	<0.1	0
Typical recurrence interval	Days to weeks			0.3 years	3 years	20 years	80 years	500 years	7×10^5 years

Based on Newhall and Self (1982), Mason et al. (2004), Siebert et al. (2010), and Brown et al. (2014).

eruption is defined to have a bulk volume of $0.1-1$ km^3 of tephra and an eruptive column height of between 10 and 25 km.

The VEI was originally intended mainly to be a semiquantitative tool for comparing the sizes of both ancient and modern explosive eruptions, and it has been very successful. In practical use, though, the VEI of an eruption is based mainly on the volume of ancient deposits and the column height of observed eruptions. This index has been adopted by the Smithsonian Institution's Global Volcanism Program (www.volcano.si.edu) for use in their catalogs of volcanic eruptions of the past 10,000 years (Siebert et al., 2010) and, for this reason, is very widely used. The scale is, however, not useful for effusive eruptions, which are predominantly nonexplosive and therefore receive a default classification of 0 or 1. Nor can the VEI scale be readily applied to the very small eruptions that can now be routinely detected and analyzed using modern monitoring techniques, including infrasound and high-speed video. At the top end of the scale, there is also no VEI "9" category, which becomes relevant when considering the scale of the largest known volcanic eruptions on the Earth. Finally, the VEI scale is also not easily applied to long-lasting or intermittent eruptions, where the classification of an event based on the total erupted volume might differ from that based on the explosivity of any individual eruptive phase. For these reasons, there have been a number of attempts to broaden the scope of the VEI in recent years, to extend its potential range (Pyle, 1995; Houghton et al., 2013).

One assumption that is implicit in the VEI is that the magnitude and intensity of eruptions are related in some way so that a single integer can fully describe the different elements of the size of an eruption. Instead, if the full spectrum of effusive and explosive volcanic activity is considered, it is clear that there cannot be a single simple relationship between eruption intensity and magnitude. To cover the spectrum of eruption styles and yet allow different eruptions to be compared, two separate scales are needed: one for magnitude, the other for intensity. As we shall see, magnitude and intensity are not independent variables (both are a function of the erupted mass), but using two measures of eruption size rather than one makes it possible to describe and compare both explosive and effusive eruptions (at least, those for which both parameters are known) and to compare both short-lived and long-lived eruptions.

The magnitude scale is based on a logarithmic index of that is defined as follows:

$$\text{magnitude} = \log_{10}(\text{erupted mass, kg}) - 7$$

For most eruptions, magnitudes defined in this way will be numerically similar to their VEI, but the magnitude scale is continuous, while the VEI scale is discrete. For example, the 1991 eruption of Pinatubo had a VEI of 6 and a magnitude 6.1, while the Tambora eruption of 1815, the largest known eruption of the past 1000 years, had a VEI of 7 and a magnitude of 6.9.

The intensity scale is based on a logarithmic index of eruption rate and is defined as follows:

$$\text{Intensity} = \log_{10}(\text{mass eruption rate, kg/s}) + 3.$$

On this scale, a very vigorous eruption will have an intensity of $10-12$, while a small, or gentle, eruption might have an intensity of $4-6$. One useful property of these scales is that both can be used to describe historic, prehistoric, and geological events. Another useful property is that since the scales are logarithmic, large errors in estimates of erupted mass or mass eruption rate do not translate into large errors in magnitude or intensity: an error of a factor of 2 in mass or mass eruption rate leads to an uncertainty of only 0.3 in magnitude or intensity. One weakness of the intensity scale is that it is still quite hard to measure, particularly for the largest eruptions for which there are no modern observations of events of anything like a comparable size.

3. TYPICAL ERUPTIONS AND RECURRENCE RATES

As with earthquakes, the recurrence rates of eruptions of a certain size become much longer as the eruptions become larger. So, for example, a magnitude 4 eruption will occur somewhere around the globe approximately every 2–3 years, while a magnitude 7 eruption may occur about every 1000 years, and magnitude 8 eruptions may occur every million years or so (Table 13.1). Table 13.2 lists the largest known eruptions of the past 100, 1000, and 10,000 years based on a recent compilation of Large Magnitude Explosive Volcanic Eruptions (LaMEVE database, http://www.bgs.ac.uk/vogripa). This list has changed in recent years and could still change in the future as volcanologists continue to make new discoveries and piece together the fragmentary records of eruptions from the geological past.

4. MASS AND VOLUME OF VOLCANIC DEPOSITS

Determining the mass of a volcanic deposit is not a trivial matter. Lava flows can have variable thicknesses and a complex distribution pattern, and deposits may extend both on land and offshore. Poorly consolidated tephra deposits have a low preservation potential and can rarely be reconstructed in full without detailed work and good fortune. Large proportions of distal tephra deposits may be emplaced in terrestrial environments, where they will never leave a trace as a result of reworking or erosion, or in marine environments, where their extent will never be determined without ambitious seafloor coring expeditions. Finally, both lava flows and tephra deposits have variable densities, which need to be known before mapped volumes can be converted to an equivalent mass. If all magmas and all pyroclastic deposits had the same physical properties, this would not be a problem; however, this is not the case. Molten rock that is free of gas bubbles typically has a

density between 2300 and 2700 kg/m^3, depending on composition. Tephra may have a bulk density as low as 400–600 kg/m^3 in the case of ash fall deposits and 1600–2000 kg/m^3 for block and ash flow deposits. Consequently, there is no single correction factor that can be universally applied to convert from volume to mass. The same is true for lavas, which can erupt with a spectrum of vesicle contents and bulk densities in the range 1800–2700 kg/m^3.

To make comparisons between different deposits, volumes are often corrected to a dense rock equivalent (DRE) volume. This is an estimate of the volume of dense, non-vesicular magma that erupted to form the pyroclastic deposit. The common usage of both bulk and DRE volumes remains a source of confusion in the literature because it is not always clear which volume is being referred to. DRE volumes are also of little value unless the assumptions about densities of both bulk and dense deposits are stated. There is no such confusion with erupted mass, which is the reason why mass should be used as the basis for magnitude and intensity scales.

5. MAGNITUDES OF VOLCANIC ERUPTIONS

The mass of ejecta thrown out during explosive eruptions spans at least 10 orders of magnitude (Figure 13.1). The smallest explosive eruptions are rarely documented, and their deposits are rarely anything other than ephemeral. Discrete ash-venting episodes during an eruption, such as those at the Soufrière Hills volcano, Montserrat, between 1995 and 1998, or discrete explosions during Hawaiian or Strombolian activity may last for only seconds to tens of seconds, release very small quantities of material, and have magnitudes <0 (Figure 13.1). At the other end of the scale, catastrophic and caldera-forming events like the historical eruptions of Tambora in 1815 and Rinjani in 1257, or the prehistoric eruption of Toba

TABLE 13.2 Magnitudes and Intensities of Some Large Eruptions

Timescale	Largest Known Eruption	Volcanic Explosivity Index	Magnitude	Intensity
Last 100 years	Pinatubo, Philippines, 1991	6	6.1	12.0
Last 1000 years	Tambora, Indonesia, 1815	7	6.9	12.0
Last 10,000 years	Akahoya Tephra, Kikai Caldera, Japan, 7.4 kyr BP	7	8.1	11.8
Last 1,000,000 years	Younger Toba Tuff, Toba, Indonesia, 74 kyr BP	8	8.8	—

Adapted from the LaMEVE database, Crosweller et al. 2012 and Brown et al. (2014).

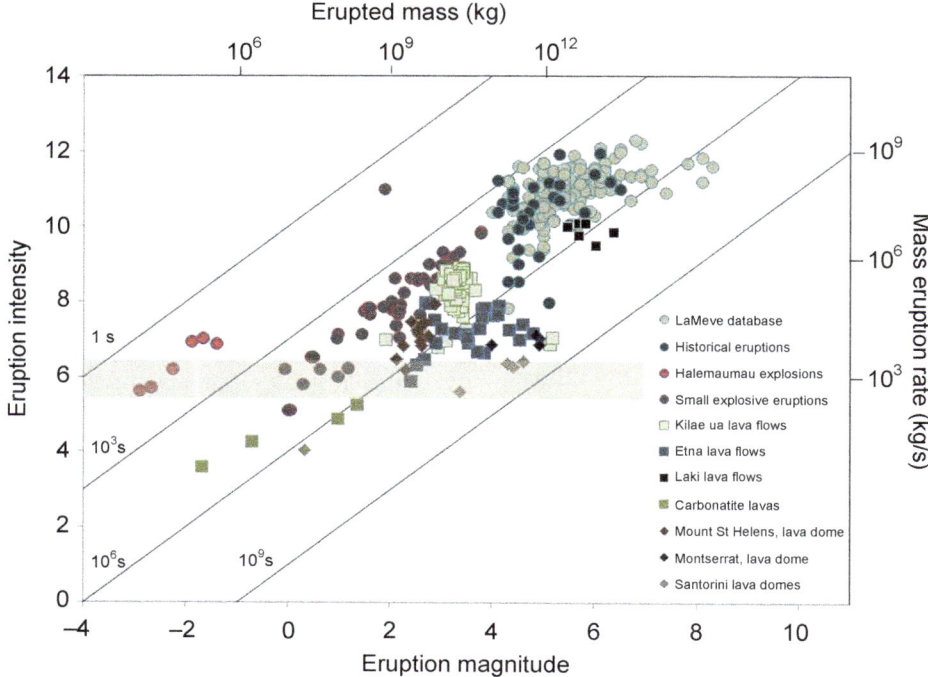

FIGURE 13.1 Plot of the magnitudes and intensities of selected recent and historical eruptions, as well as examples from the geological record. Explosive eruptions are plotted as circles, eruptions dominated by lava flows as squares, and dome-forming eruptions as diamonds. For the explosive eruptions, the intensity is usually the peak intensity derived from the eruption plume height. For large magnitude explosive eruptions, which are usually dominated by ignimbrites, the actual peak intensity during eruption may well have been one or two log units higher. For most of the effusive eruptions, the intensity is the time-averaged intensity, which may be one or two log units lower than the peak eruption intensity (Table 13.3). The diagonal lines are contours of eruption timescale, in seconds (from 1 to 10^9 s). The shortest-lived events are Strombolian, Hawaiian, or Vulcanian explosions; the longest-lived events are sustained lava flow and lava-dome eruptions. The shaded bar shows the typical rate of volcanic growth estimated from studies of individual volcanoes or volcanic edifices in a range of plate tectonic settings (White et al., 2006). All eruptions with intensities >6−7 (10^3−10^4 kg/s) will invariably be intermittent, since they exceed the typical long-term rates of magma supply to volcanic systems. Eruptions with intensities <5−6 could be plausibly sustained for considerable periods of time by magma supply from depth.

(74 kyr BP) may have magnitudes of 7−9 and erupt 10^{14}−10^{16} kg of magma.

Eruptions of lava also span at least 8 orders of magnitude (Figure 13.1). At the smallest end, individual carbonatite lava flows at Oldoinyo Lengai volcano (Tanzania) range up from just a few cubic meters or a few thousand kilograms. In contrast, 15 km³ or over 10^{11} kg of lava was erupted during the largest effusive eruption of the last 300 years at Laki, Iceland, in 1783−1784. Some effusive eruptions attain a large mass simply because the eruption lasts for a long time. For example, the predominantly lava-dome-forming eruption of Soufrière Hills volcano, Montserrat has extruded at least 3×10^{12} kg of magma in five major pulses of activity since the eruption began in 1995; while at Kilauea, Hawaii, at least 10^{13} kg of lava has been erupted over the course of a 30-year long phase of activity (Table 13.3). Examples from the geological record extend to even more colossal scales. Individual lava flows from large igneous provinces, such as those from Miocene-age Columbia River Basalts or the Cretaceous Deccan Traps of India, have erupted DRE volumes of many

thousands of cubic kilometers, erupted masses exceeding 10^{16} kg (magnitude 9), and are parts of a flood basalt flow field provinces that are 100 times larger (Table 13.3).

6. INTENSITIES OF VOLCANIC ERUPTIONS

In an explosive eruption, the intensity of the eruption, or the rate at which volcanic ejecta leave the vent, is the main factor that controls the height of an eruptive plume. For a sustained eruption, the height reached by the eruption column is approximately proportional to the fourth root of the intensity: $H \sim I^{0.25}$. The plume height in turn controls the dispersal of the pyroclasts, so it is possible to infer the intensities of ancient eruptions from studies of the deposits. For recent eruptions, direct visual and satellite observations of the heights reached by eruption columns and rates of plume spreading mean that the intensities of a wide range of eruptions can be determined. The intensities of selected explosive eruptions are summarized in Table 13.2. The

TABLE 13.3 Volumes and Magnitudes of Some Very Large Eruptions

Unit Name	Location	Age	Volume (Dense Rock Equivalent, km³)	Magnitude (M)	Sources
Three Largest Known Subaerial Basaltic Lava Flow Fields Associated with Large Igneous Provinces					
Mahabaleshwar–Rajahmundry Traps	Deccan Traps, India	64.8 Ma	9300	9.4	Bryan et al. (2010)
McCoy Canyon flow	Columbia River, USA	15.6 Ma	4278	9.1	Bryan et al. (2010)
Sand Hollow flow	Columbia River, USA	15.3 Ma	~2750	8.9	Bryan et al. (2010)
Three Largest Known Silicic Ignimbrites from Large Igneous Provinces					
Guarapuava–Tamarana–Sarusas	Paraná–Etendeka province	132 Ma	8587	9.3	Bryan et al. (2010)
Santa Maria	Paraná–Etendeka province	132 Ma	7808	9.3	Bryan et al. (2010)
Guarapuava Ventura	Paraná–Etendeka province	132 Ma	7571	9.3	Bryan et al. (2010)
Three Largest Known Silicic Eruptions, Not from Large Igneous Provinces					
Fish Canyon Tuff	Colorado, USA	27.8 Ma	4500	9.2	Mason et al. (2004)
Toba, Indonesia	Sumatra, Indonesia	75 ka	2800	8.8	Mason et al. (2004)
Lund Tuff	Utah/Nevada, USA	29 Ma	2500	8.8	Mason et al. (2004)

most powerful Plinian eruptions have mass eruption rates that exceed 10^8 kg/s (intensity >11) and may send buoyant plumes of ash and gas to between 25 and 55 km above the Earth's surface. The intensities of small ash eruptions can be as low as 5 or 6, with mass fluxes of as little as 100–1000 kg/s, supporting very weak plumes.

Explosive eruptions that undergo a transition from a Plinian phase to an ignimbrite-forming phase may do so as a result of the increasing intensity of the eruption. In many cases, these changes in intensity may be linked to the increasing size of the eruptive vent or the opening of new vents as ring fractures open up. The intensity of ignimbrite-forming eruptions may be as much as 1 or 2 orders of magnitude larger than during the Plinian phase (Carey and Sigurdsson, 1989). While there are a number of models available that can be used to derive mass eruption rates for pyroclastic density current deposits, there are still relatively few examples of sufficiently well-studied eruptions that have included estimates of the intensity of the ignimbrite-forming phase of the eruptions.

The intensities of effusive eruptions are even more difficult to constrain, unless they happen to have been observed. Many effusive eruptions may show dramatic changes in mass eruption rate as the eruption proceeds, but quantitative data are almost nonexistent for prehistoric eruptions. For this reason, intensity data for lava flows are likely to be time-averaged estimates and based on the ratio of known erupted mass to the known eruption duration. Some examples are given in Table 13.4.

For comparison, the average rate of magma supply to a volcanic center over its lifetime is in the range from 4×10^{-3} to 2×10^{-2} kg/s for many volcanic systems across all settings, other than large igneous provinces. This corresponds to an intensity of 5.5–6.3 (Figure 13.1; White et al., 2006). At a very simple level, this explains why high-intensity eruptions are expected to be intermittent: eruptions with intensities >6–7 (10^3–10^4 kg/s) exceed the typical long-term rates of magma supply to volcanic systems, so will tend to be discrete events separated by periods of repose. Eruptions with intensities <5–6 could be plausibly sustained for considerable periods of time by magma supply from depth.

7. VERY LARGE ERUPTIONS

At the top end of the magnitude scale, explosive eruptions with magnitudes 8 and 9 have caught the popular imagination and are often called "supereruptions." Although this term suggests that there may be something particular about these sorts of eruptions that makes them stand out, it is still not known what particular combination of physical conditions (crustal thickness and rheology, magma flux and heat flow, and so on) allows volcanoes to supersize. In fact, the majority of eruptions of the very largest scale (9 and above) that are now known are all associated with large igneous provinces—which contain both the deposits of very large flows of basalt lava and the welded deposits of very large volume silicic ignimbrites (Table 13.3). The real challenge

TABLE 13.4 Size and Intensity of Some Sustained Volcanic Eruptions

Eruption	Eruption Style	Time Span[1]	Total Erupted Mass (kg)	Time-Averaged Magma Flux (kg/s)	Peak Eruption Rate (kg/s)	Average Intensity	Peak Intensity
Kilauea, Hawaii	Lava flows	1983−2012	10^{13}	10^4	9×10^5	7	9
Soufrière Hills volcano, Montserrat	Lava domes	1995−2013	3×10^{12}	5×10^3	10^8	6.7	11
Santiaguito, Guatemala	Lava domes	1922−2013	5×10^{12}	2×10^3	10^5	6.3	8
Nyiragongo, Congo	Lava lake	1901−1977	—	1000		6	
Stromboli, Italy	Lava pond	Historical	—	0.3	4×10^5	2.5	8.6

[1]All of these volcanoes are still in eruption.

with eruptions of this scale is that there are no modern analogues that are even close in terms of size, which means that there is a lot of uncertainty about the energetics of the eruptions—how long they might have lasted and the nature of the eruptive vents or fissures that they might have erupted from. Most known deposits from explosive eruptions of magnitude 8 and above do not have pyroclastic fall deposits such as those formed during Plinian eruptions; instead, they are dominated by ignimbrites and associated ash fallout deposits. So, for these eruptions, there are no good estimates of eruption intensity, and the apparent upper limit of 12 on the intensity scale ($\sim 10^9$ kg/s) may well be artificially low as a result.

8. THE RELATIONSHIP BETWEEN THE INTENSITY AND MAGNITUDE OF A VOLCANIC ERUPTION

There are a growing number of eruptions for which there are sufficient data to determine both eruption magnitude and intensity. For example, the LaMEVE data set (http://www.bgs.ac.uk/vogripa; Crosweller et al., 2012; Brown et al., 2014) contains information for many Quaternary eruptions of magnitude 4 and larger; a large number of which also contain estimates of eruption intensity. These data, together with published data from a variety of other types of eruption (lava flows, lava domes, and smaller explosive eruptions), are plotted in Figure 13.1. In each case, the magnitude is based on the total erupted mass; whereas the intensity is based on an estimate of either the *peak* eruption intensity, as inferred from the eruption plume height (in the case of explosive eruptions), or the *mean* intensity, determined by dividing the total erupted mass by the eruption duration. These eruption data span 10 orders of magnitude in erupted mass (from $<10^5$ to $>10^{15}$ kg), 8 orders of magnitude in eruption intensity (from <10 to

$\sim 10^9$ kg/s), and 7 orders of magnitude in eruption duration (from <10 s to >1 year). The scatter of the data shows that there is a significant range in eruption duration, both for eruptions of similar intensities and for eruptions of similar magnitude. In terms of historical eruptions, the largest explosive eruptions are larger in both magnitude and intensity than the largest known effusive eruptions, but this is not the case for the eruptions known from the geological record (Table 13.3). There is also a significant overlap in both eruption magnitude and intensity between eruptions of very different character such as small to moderate scale silicic lava domes, mafic lava flows, and small explosive eruptions of intermediate composition. In these examples, parameters such as the vent shape and size, the dissolved volatile content of the magma, and magma rheology will be the main factors influencing eruption style.

9. ENERGY OF VOLCANIC ERUPTIONS

During an eruption, most volcanic energy is dissipated as heat. The thermal energy release will depend on both the composition and the temperature of the erupted materials, and this ranges from c. 1 MJ/kg for rhyolite erupted at 850 °C to c. 1.5 MJ/kg for basalt erupted at 1150 °C. The thermal power output during an eruption will range from 10^{13} to 10^{15} W for typical explosive Plinian eruptions and from 10^{10} to 10^{13} W for large basaltic fissure eruptions. The total thermal energy release for eruptions of magnitude 5−8 will be of the order of $10^{18}−10^{22}$ J. In comparison, the time-averaged rate of heat loss from the Earth's interior is $\sim 5 \times 10^{13}$ W, while the solar flux to the Earth's surface is 1.5×10^{17} W.

The kinetic energy of particles ejected during volcanic eruptions is comparatively small: even for a violently explosive eruption with ejecta emitted at the local speed of sound, the kinetic energy release will only be of the order of 50 kJ/kg or less than 5% of the thermal energy release.

TABLE 13.5 Estimates of Eruption Magnitude and Peak Intensity for Selected Recent Eruptions

Eruption	Total Erupted Mass (kg)	Peak Eruption Plume Height (km)	Peak Eruption Rate (kg/s)	Magnitude (M)	Peak Intensity	Total SO₂ Release (kg)
Kasatochi, Aleutians, August 2008	$3-6 \times 10^{11}$	13.7	1.6×10^{7}	4.5–4.7	10.2	1.5×10^{9}
Chaitén, Chile, May 2008	$5-10 \times 10^{11}$	19	4×10^{7}	4.7–5	10.6	10×10^{6}
Etna, Italy, 2002–2003	$1-2 \times 10^{12}$	6	4×10^{4}	5.1–5.3	7.6	10^{9}
Pinatubo, Philippines, June 1991	1.3×10^{13}	40	9×10^{8}	6.1	12.0	20×10^{9}
Mount St Helens, USA, May 1980	1.3×10^{12}	19	2×10^{7}	5.1	10.3	1×10^{9}

Seismic energy release during eruptions can be very important in terms of the local consequences of the eruptions, but is usually a small fraction of the thermal energy release ($<0.1-0.001\%$).

10. OTHER MEASURES OF ERUPTION SIZE

There are few other metrics to describe eruption size that have been adopted by the volcanological community. One parameter that volcanologists are increasingly able to measure with some precision for many eruptions is the volatile release, in particular the quantity of sulfur dioxide released during an eruption. This parameter is potentially of great importance because the quantity of gas emitted in an eruption and the way in which it is released are factors that help to determine the potential local, regional, or global environmental consequences of eruptions. This is an area where the challenge of working out what happened during past eruptions is substantial, and the data set of modern observations currently only spans about the past 30 years during which time satellite remote sensing of volcanic emissions has become routine (Table 13.5; Carn et al., 2003). Even though this data set is small, one thing that is clear from modern observations, analysis, and experiment is that the amount of sulfur released during an eruption is not related simply to the mass of magma erupted or to the intensity of the eruption. Instead, the potential for a volcano to release large quantities of sulfur and the potential for global environmental consequences depend critically on the magma composition and the conditions under which the magma is stored (pressure, temperature and oxidation state, or oxygen fugacity), and may also be influenced significantly by the compositions of the rocks through which the magma flows prior to eruption. As yet, there are no measures or indices that have been widely used to compare the environmental or climatic consequences, or the destructiveness or destructive potential of past or future eruptions.

Full references to the data sets used to create Tables 13.2-13.5 and Figure 13.1 are available in the companion site (http://booksite.elsevier.com/9780123859389)

FURTHER READING

Brown, S.K., Crosweller, H.S., Sparks, R.S.J., Cottrell, E., Deligne, N.I., Guerrero, N.O., Hobbs, L., Kiyosugi, K., Loughlin, S.C., Siebert, L., Takarada, S., 2014. Characterisation of the quaternary eruption record: analysis of the large magnitude explosive volcanic eruptions (LaMEVE) database. J. Appl. Volcanol. 3, 5.

Bryan, S.E., Peate, I.U., Peate, D.W., Self, S., Jerram, D.A., Mawby, M.R., Marsh, J.S., Miller, J.A., 2010. The largest volcanic eruptions on Earth. Earth-Sci. Rev. 102, 207–229.

Carey, S.N., Sigurdsson, H., 1989. The intensity of Plinian eruptions. Bull. Volcanol. 51, 28–40.

Carn, S.A., Krueger, A.J., Bluth, G.J.S., Schaefer, S.J., Krotkov, N.A., Watson, I.M., Datta, S., 2003. Volcanic eruption detection by the total ozone mapping spectrometer (TOMS) instruments: a 22-year record of sulphur dioxide and ash emissions. In: Oppenheimer, C., Pyle, D.M., Barclay, J. (Eds.), Volcanic Degassing, vol. 213. Geol. Soc., London, Spec. Pub., pp. 177–202.

Crosweller, H.S., Arora, B., Brown, S.K., Cottrell, E., Deligne, N.I., Guerrero, N.O., Hobbs, L., Kiyosugi, K., Loughlin, S.C., Lowndes, J., Nayembil, M., Siebert, L., Sparks, R.S.J., Takarada, S., Venzke, E., 2012. Global database on large magnitude explosive volcanic eruptions (LaMEVE). J. Appl. Volcanol. 1, 4.

Houghton, B.F., Swanson, D.A., Rausch, J., Carey, R.J., Fagents, S.A., Orr, T.A., 2013. Pushing the volcanic explosivity index to its limit and beyond: constraints from exceptionally weak explosive eruptions at Kilauea in 2008. Geology 41, 627–630.

Mason, B.G., Pyle, D.M., Oppenheimer, C., 2004. The size and frequency of the largest eruptions on Earth. Bull. Volcanol. 66, 735–748.

Newhall, C.G., Self, S., 1982. The volcanic explosivity index (VEI) — an estimate of explosive magnitude for historical volcanism. J. Geophys. Res. 87, 1231–1238.

Pyle, D.M., 1995. Mass and energy budgets of explosive volcanic eruptions. Geophys. Res. Lett. 22, 563−566.

Siebert, L., Simkin, T., Kimberley, P., 2010. Volcanoes of the World, third ed. University of California Press.

Tsuya, H., 1955. Geological and petrological studies of volcano Fuji, V. Bull. Earthquake Res. Inst. Tokyo 33, 341−383.

White, S.M., Crisp, J.A., Spera, F.J., 2006. Long-term volumetric eruption rates and magma budgets. G-Cubed 7, Q03010.

Yokoyama, I., 1957. Energetics in active volcanoes. Bull. Earthquake Res. Inst. Tokyo 35, 75−97.

Chapter 14

Global Rates of Volcanism and Volcanic Episodes

Natalia Irma Deligne
GNS Science, Avalon, Lower Hutt, New Zealand

Haraldur Sigurdsson
Graduate School of Oceanology, University of Rhode Island, Narragansett, RI, USA

Chapter Outline

GLOSSARY

continental breakup Breakup of a continental plate.

divergent plate volcanism Volcanism associated with the separation of plates.

episodicity Nonuniform behavior; when used to describe volcanism indicates that there is some temporal clustering of volcanic activity.

hot spot Region, typically not on a plate boundary, where there is excess magma production relative to what one would expect given plate tectonic framework.

ignimbrite The deposit of large scale pyroclastic density currents.

large igneous province (LIP) Thick and extensive crustal region composed of volcanic and intrusive rocks that have formed by processes other than normal seafloor spreading or subduction, with aerial extent >0.1 million km^2, volume >0.1 million km^3, and emplacement time <50 Myrs, although typically $>75\%$ of volume is emplaced in <5 Myrs. LIPs can occur in a variety of settings and can be predominately mafic or silicic.

large scale silicic eruption Silicic (>65 wt% SiO_2) eruption producing $\geq 10^{15}$ kg magma (e.g., magnitude ≥ 8), corresponding to roughly ≥ 400 km^3 dense rock equivalent magma.

mass extinction event Time period when most species become extinct as recorded in the fossil record.

ocean ridge The localization of divergent plate volcanism.

plate tectonics Theory that Earth consists of semirigid plates which separate, converge, or slide past one another; volcanism is associated with the separation (divergent plate volcanism) and convergence (subduction zone volcanism).

production rate The amount of magma erupted in given time, often expressed in units of km^3 per year.

subduction zone Zone in which one plate is going underneath another.

tectonic setting Setting in plate tectonic framework, e.g., subduction zone, continental interior, ocean ridge.

volcanic arc Localization of subduction zone volcanism.

volcanic ash Volcanic material fragmented via conduit processes with diameter <2 mm.

1. INTRODUCTION

In the eighteenth and nineteenth centuries, the early days of geology, geologists clashed over two fundamental views of the rates of geologic processes. On one side were the uniformitarians, whose mantra was "the present is the key to the past"; they argued that the rates of volcanism and other Earth processes were in steady state and had not changed noticeably over geologic time. On the other side were the catastrophists, who maintained that the steady progression of geologic time was punctured by brief cataclysms or periodic catastrophes, when volcanism, mountain building, and major upheavals occurred, often resulting in species extinctions. We will explore whether on a global scale volcanism is more aligned with uniformitarian or catastrophic viewpoints; in the end we will find examples of both depending on **tectonic setting**.

The Encyclopedia of Volcanoes. http://dx.doi.org/10.1016/B978-0-12-385938-9.00014-6

While the current tectonic and volcanic state of Earth is not necessarily typical and/or representative of all past regimes, we can gain insight into the long-term geochemical and tectonic evolution of Earth by studying accessible volcanic rocks and deposits. This chapter examines the geologic record for the evidence of episodes in volcanism. Due to issues with preservation of geologic evidence, we focus primarily on the Cenozoic Era (past 65 million years). Three categories of terrestrial volcanism are considered: (1) **ocean ridge** volcanism, where the magma **production rates** have been largely determined by geophysical techniques, primarily by paleomagnetic methods, (2) arc volcanism above **subduction zones**, where the magma production rates are primarily inferred from tephra layers found in adjacent regions, and (3) **large igneous province** volcanism and some volcanic island chains, where magma production rates have mostly been determined through geologic mapping.

2. OCEAN RIDGES (DIVERGENT PLATE VOLCANISM)

The floor of the ocean basins represents about 70% of Earth's surface and the crust which constitutes it is volumetrically the most important product of volcanism on the planet. Ocean basins are a consequence of plate motion: subducting slabs pull on their plates, leading to spreading at divergent plate boundaries. In areas of **continental breakup**, divergent plates are expressed as rift systems. Within ocean basins, where the vast majority of divergent plate boundaries are located (and hence the focus of this section), submarine ridges form at divergent plate boundaries. The volcanism that builds up these ridges is a direct result of the mantle upwelling where plates are diverging; here, basaltic magma forms via the mechanism of decompression melting of the mantle source. Within ocean basins, the combined oceanic crust and lithosphere (c. 100 km thick) form the outermost layers of the solid Earth, separating the mantle from the overlying hydrosphere. While this boundary layer has been forming throughout most of Earth's history, most of the older oceanic crust has been subducted back into the mantle and as such has been destroyed; the oldest preserved oceanic crust is only Late Permian (270 Ma) in age (Figure 14.1). The rate of volcanic (magmatic) activity at the ocean ridges is measured by the rate of plate generation, which in turn is calculated from the plate tectonic framework, primarily developed from paleomagnetic studies.

Ocean ridges currently generate a surface area of 3 km^2 per year of new plate (Figure 14.2(A)), formed by the solidification of magmas, in part as volcanic rocks (basalts), and in part as plutonic rocks (gabbros, ultramafic rocks). As the average thickness of the oceanic crust is 6−7 km, the flux of basaltic magma from the mantle to the surface crustal layer at ocean ridges is 18−21 km^3 per year. Assuming a density of 3.3 g/cm^3 (3.3×10^{12} kg/km^3), this corresponds to roughly 6×10^{13} kg per year, or about one order of magnitude greater than the rate of volcanism generated at island arcs and **hot spots**.

Any given plate's growth rate depends on the length of its bounding ridges and their spreading rates. The global length of spreading ridges is 6.8×10^4 km with a mean half-spreading rate of 20 mm per year (Figure 14.2(B)). Due to the large range in spreading rate from ridge to ridge (from half spreading rates of below 10 mm per year in the North Atlantic and the northern Weddell Sea to values approaching 100 mm per year at the East Pacific Rise; Figure 14.3), and variable ridge length, the rate of ocean ridge volcanism varies from region to region. Presently, the Antarctic plate, which is entirely bound by divergent plate boundaries (i.e., no portion of the Antarctic plate is subducting), is the fastest growing plate, growing at a rate of 0.561 km^2 per year. The aggregate global crustal production rate has not varied greatly in the Cenozoic (Figure 14.2(A)) and while crustal production rates were demonstrably higher in the Cretaceous, all known geologic rates have been within a factor of 2 of modern rates.

The modern (i.e., unsubducted) ocean floor formed by spreading has a mean age of 64 million years and is up to 270 million years old (Late Permian); the oldest in situ oceanic crust is located in the Ionian Sea and eastern Mediterranean (Figure 14.1). We do not know the history of oceanic plates prior to 270 Ma, because all of the older crust has been subducted. However, there is broad agreement that **plate tectonics** have been operating for at least the past 1.9 Ga, so if we assume that plate generation and subduction has been a steady-state process over that period, the modern ocean floor is only 5% of the oceanic crust created in Earth's history. In this steady-state scenario, 5.4×10^9 km^2 of oceanic crust has already been subducted throughout Earth's history. Assuming 100 km as the crust and lithosphere thickness and no raw material subducted twice, this implies that 80% of the mantle has been through the "ridge and subduction factory" system, albeit only 5% of it as oceanic crust (lithosphere is the dominant component of subducting plates). As mass is not being added to Earth, it is assumed that subduction has occurred at roughly the same rate, although the occasional suggestion of a cooling contracting Earth would involve slightly greater rates of subduction compared to spreading.

These rates of ocean crust generation reflect modern steady-state production, yet plate reconstruction studies have demonstrated that while Cenozoic crustal production rates have been relatively constant, production rates fluctuated considerably in the Cretaceous Era (Figure 14.2(B)). Cretaceous ocean crust production rates were up to 30−70% greater than modern rates (the value depends on

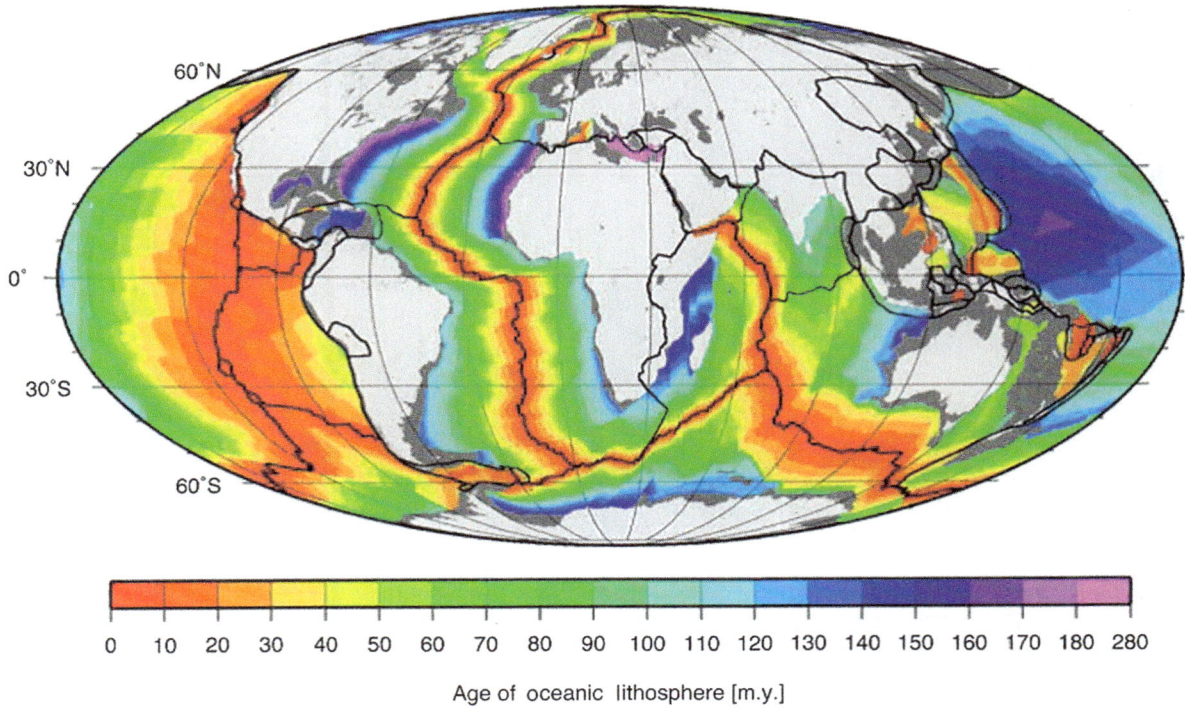

FIGURE 14.1 Age of the oceanic crust. Continental margins are gray, continents are light gray, and modern plate boundaries are shown with a black line. *Figure from Müller et al. (2008).*

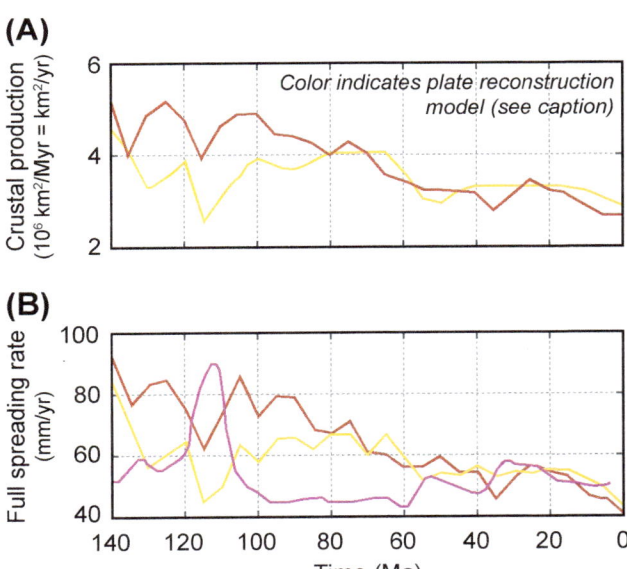

FIGURE 14.2 (A) Crustal production rates and (B) full cumulative mid-ocean spreading rate from the Cretaceous to the present based on various plate reconstruction. Red and orange lines are based on CK95G94 and GTS2004 plate reconstruction models applied in Seton et al. (2009), respectively, and purple line (B only) is from Conrad and Lithgow-Bertelloni (2007), based on plate reconstruction models CLB07 and CK95G94. For details on reconstruction models, see Seton et al. (2009) and references therein. *Figure after Seton et al. (2009).*

the duration of the Cretaceous Normal Superchron in time model used to reconstruct production rates), with production peaks in the Early to Mid-Cretaceous. This may be due to the breakup of Pangea and the formation of the Indian and North Atlantic ridge systems. Thus, major tectonic reorganization events can alter global production oceanic crustal production rates. Research in the coming years will provide insight on whether Cenozoic oceanic crustal production rates reflect typical production values or are abnormally constant in the context of Earth's history.

3. VOLCANIC ARCS (SUBDUCTION ZONE VOLCANISM)

Volcanism above subduction zones plays a fundamental role in the exchange of mass and energy between the solid Earth and its oceans and atmosphere. Subduction zone volcanism is often explosive and has had the greatest societal impact, providing added motivation to understand rates of volcanism in these settings. Furthermore, episodes in volcanic arcs (i.e., above subduction zones) will be likely to have a greater immediate environmental impact than variations in ocean ridge volcanism rates, as the former are dominantly subaerial and thus discharge climate-modifying gases and aerosols directly into Earth's atmosphere. Just as variations in ocean ridge volcanism have been the focus of

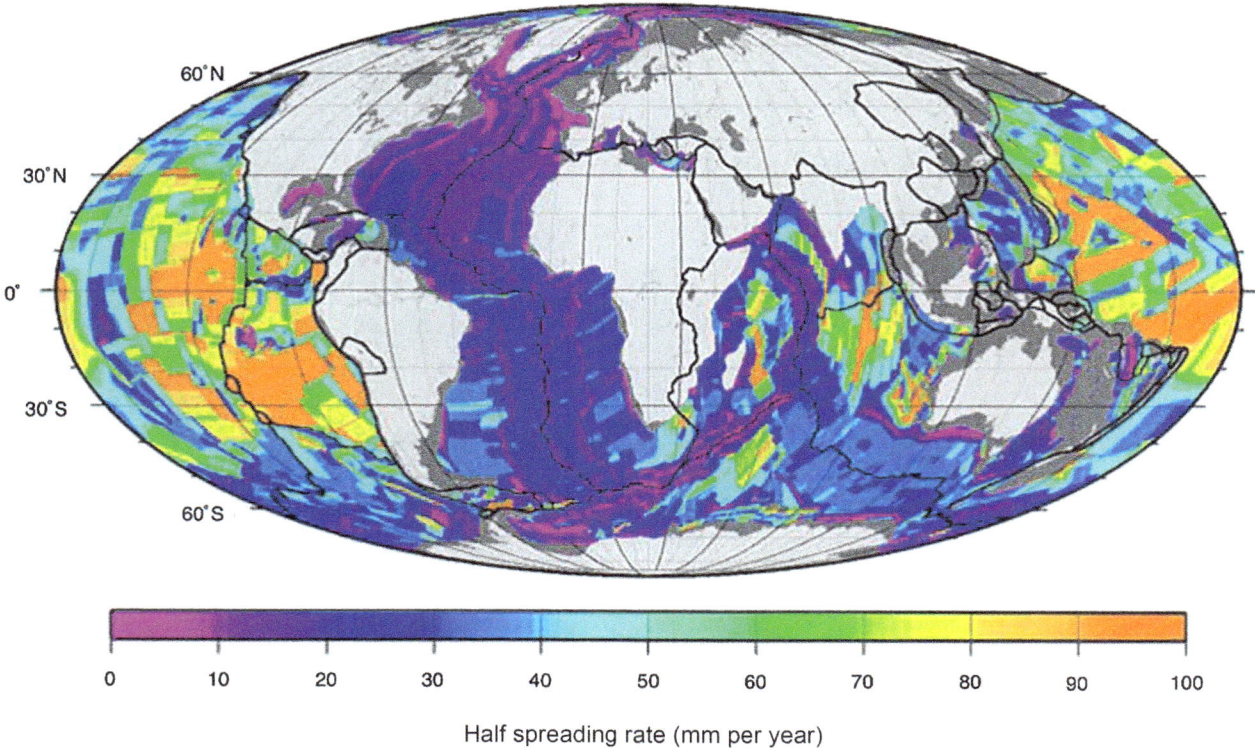

FIGURE 14.3   Half spreading rate of ocean ridges. Modern plate boundaries are shown with a black line. *Figure from Müller et al. (2008).*

considerable study, we will summarize long-term variations in volcanism associated with subduction zones.

Unlike rates of seafloor spreading and oceanic crustal formation, the rates of island arc magma generation and volcanism are difficult to establish quantitatively due to preservation and accessibility issues. Products of arc volcanism include edifices, **volcanic ash**, and pyroclastic density current deposits (referred to here as **ignimbrites**). Volcanic edifices are typically eroded away within several hundred thousand years of activity cessation from glaciations and general erosional processes. Volcanic ash and ignimbrite records generally extend further back. On land, ignimbrites form massive coherent deposits most often preserved in arid settings and volcanic ash layers can be preserved in long-standing lakes and bogs. However, unraveling the land-based record of arc volcanism is hampered by the paucity and poor precision of many radiometric age dates, lack of mappable stratigraphic levels, and poor preservation of deposits. The best preservation is often in adjacent oceanic basins, where marine tephra layers can be interbedded with deep-sea deposits. Marine tephra layers are millimeters to tens of centimeters thick and result from the fallout of airborne volcanic ash hundreds or even thousands of kilometers from the source. As such, the marine tephra record predominantly records only the biggest eruptions whose deposits cover great distances. While the marine tephra record is condensed in thickness compared with the terrestrial one, it is more likely to be complete for large, environmentally impacting eruptions and dateable by biostratigraphic methods and radiometric dating of phenocrysts found in the tephra layers. However, is difficult to use the marine record to establish eruption volumes, so the absolute values of magma production rate remains poorly determined.

The Holocene record of effusive volcanism in volcanic arcs indicates that magma production is positively correlated with the rate of plate convergence, with lower production in areas with oblique plate convergence. There is a poor record of effusive arc volcanism further back, so at present we can only assume the observed positive correlation between convergence rate and effusive volcanic output is representative of typical behavior. Likewise, one might assume that explosive volcanism is similarly correlated with the rate of plate convergence, although this has not been definitively demonstrated.

Considerable work this past decade has focused on cataloging and analyzing records of explosive volcanism; while not all such eruptions are associated with volcanic arcs, the vast majority are. The Holocene record indicates that global eruption frequency decreases with eruption size (larger eruptions are less frequent), although the majority of mass erupted comes from larger eruptions. Under-reporting of volcanic eruptions is a major issue in the Holocene and earlier record, which makes it difficult to estimate production rates, but a conservative estimate is that 0.5 km^3 per year is produced at volcanic arcs from explosive eruptions.

Work in the late twentieth century on tephra layers in deep-sea cores indicated that explosive arc volcanism is remarkably episodic in time, although little subsequent work has focused on exploring possible causes to this. The tephra layers found in Deep Sea Drilling Project and the Ocean Drilling Program cores suggest a peak in explosive volcanism in the Quaternary (0−5 Ma), the middle Miocene (c. 13−17 Ma), and potentially the Eocene (c. 37−42 Ma). These could reflect multimillion year peaks in volcanic output at arcs. Recent work suggests a possibility of shorter term localized variability, as it appears that there is a peak in explosive activity following deglaciation, potentially due to the mass unloading causing depressurization of magmatic systems; this effect lasts at most a few millennia. There is still considerable uncertainty in the overall scale of variability of volcanic arc output and how it compares with other volcanic settings. However, it is important to note that the observed convergent volcanism rate variations are almost entirely derived from the tephra records derived from high-silica magmatic processes, whereas in other volcanic settings these rates are inferred from records left by low-silica magmatic processes.

4. LARGE IGNEOUS PROVINCES

Large igneous provinces (LIPs) are thick and extensive crustal regions composed of volcanic and intrusive rocks that have formed by processes other than normal seafloor spreading or subduction (see Chapter 24). LIPs have worldwide distribution (Figure 14.4) and occur in a variety of structural settings. By definition they have aerial extents >0.1 million km^2, volumes >0.1 million km^3, and emplacement times <50 Ma, although typically >75% of volume is emplaced in <5 Ma. Generally "LIPs" refer to mafic volcanism with silicic equivalent referred to as silicic LIPs; we follow this convention here although note that the aerial, volumetric, and temporal definitions of silicic LIPs are the same as "regular" LIPs. Refer to http://www.largeigneousprovinces.org for a compilation of published research and a reference list for known events.

4.1. (Mafic) Large Igneous Provinces

The volcanic components of LIPs are principally subaerial basaltic (tholeiitic) lava flows, but felsic and intermediate volcanic rocks may also be produced. In contrast to other types of volcanism, LIPs are strongly episodic in nature. The long-term (Phanerozoic) average is one LIP event every 10−20 Ma, although LIPs tend to be temporally clustered (Figure 14.5) and are thus likely associated with particular tectonic and/or mantle conditions. LIPs are linked to continent and especially supercontinent breakup, and often, although not always, precede plate (continental and oceanic) breakup, and are conspicuously infrequent during periods of supercontinent assembly. LIPs initially were thought to be caused by new mantle plumes rising from the core−mantle boundary, although this view has been repeatedly challenged in the past decade as predictions of the mantle plume model are shown to be unfounded. In the coming years the origin and role of LIPs will continue to be the subject of much debate; the website www.mantleplumes.org is recommended for the following discussion and new findings.

The largest LIPs are oceanic plateaus: broad, flat-topped volcanic regions that rise about 2 km above the surrounding seafloor. The largest is the Ontong Java plateau in the Pacific. This LIP likely also comprises the Manihiki and Hikurangi plateaus, which are chemically and temporally equivalent to the Ontong Java plateau but presently geographically separated by ocean basins that have formed since plateau emplacement. The composite Ontong Java−Manihiki−Hikurangi plateau covers 3.5 million km^2 with an estimated volume 59−77 million km^3. This massive mid-Cretaceous LIP was erupted over a relatively short time span, from 125 to 119 Ma; this implies emplacement rates of tens of km^3 per year. The Kerguelen plateau in the southern Indian Ocean is the second largest LIP, covering 2.3 million km^2 with a volume of 15 million km^3 emplaced primarily 110−114 Ma. Not all LIPs are oceanic plateaus; for example, the North Atlantic volcanic province, derived from a hot spot presently centered beneath Iceland, has an area of 1.3 million km^2 and was emplaced between 55 and 58 Ma. Also notable are the Deccan traps in India (1.8 million km^2/9.3 million km^3), emplaced between 64 and 67 Ma and the Columbia River Basalts of the western United States (0.164 million km^2/0.234 million km^3), erupted between 15 and 17 Ma; these latter two are continental LIPs (as opposed to oceanic plateau LIPs). All LIPs appear to have relatively fast emplacement period given their huge volumes, and can be considered transient magmatic episodes that greatly increase global magma production rates during a short period.

In the context of Earth history, LIPs are truly catastrophic events. The big five **mass extinction events** (end-Ordovician, mid-Devonian, end-Permian, end-Triassic, and end-Cretaceous) all coincide with LIPs, although not all LIPs or LIP clusters are associated with mass extinction events (conspicuously, Ontong Java−Manihiki−Hikurangi plateau emplacement does not coincide with a mass extinction event). Given that not all LIPs have coincident mass extinction events, there is no consensus on a "kill mechanism", although some speculate that LIPs serve as tipping points for fragile or stressed global ecosystem

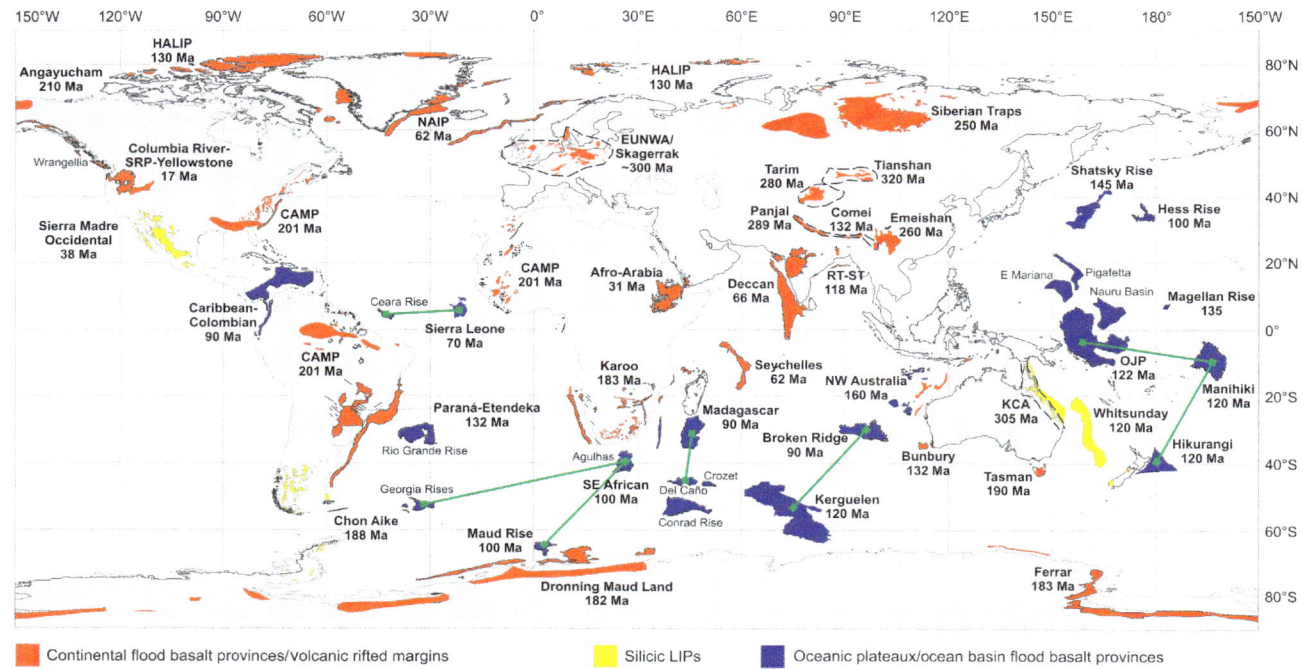

FIGURE 14.4 Know LIPs (mafic and silicic) erupted after assembly of Pangea c. 320 Ma. Ages are onset of main pulse of magmatism, and green lines tie provinces that were geographically continuous at the time of emplacement but have since been rifted apart. Dashed lines show inferred extent. *Figure from Bryan and Ferrari (2013).*

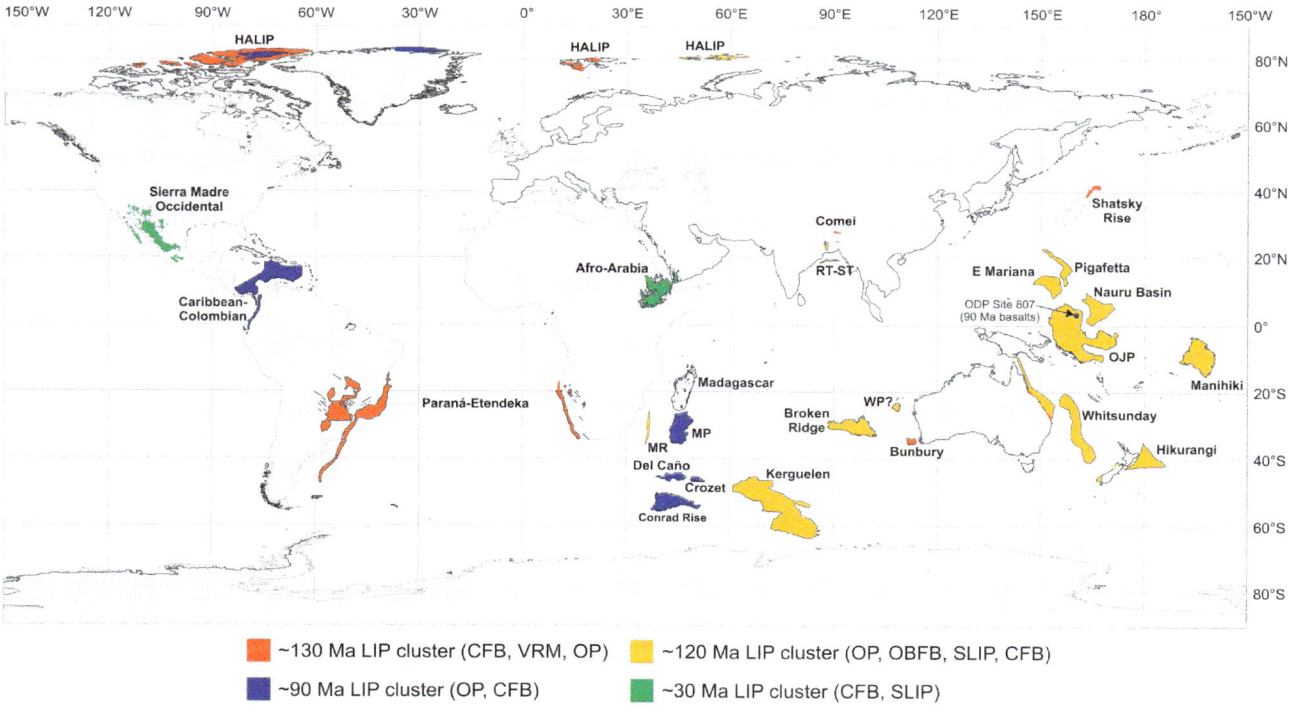

FIGURE 14.5 Known clusters of LIP events at 130, 120, 90, and 30 Ma along with province type, as follows: CFB, continental flood basalt; VRM, volcanic rifted margin, OP, oceanic plateau; OBFB, ocean basin flood basalt; SLIP, silicic large igneous province. Geographic abbreviations are same as for Figure 4 with the following additional abbreviations: WP, Wallaby Plateau; OPB, Ocean Drilling Program. *Figure from Bryan and Ferrari (2013).*

assemblages. Interestingly, each of the big five mass extinction events are associated with a single continental LIPs rather than a clusters of LIPs; it is unclear why a single LIP is evidently more lethal than multiple LIPs occurring geologically contemporaneously. LIPs likely release vast volumes of volcanic gases and aerosols, which can lead to global warming from greenhouse gases or global cooling from injection of SO_2 into the stratosphere, ocean acidification, anoxia (oxygen depletion) and/or euxinia (sulfidic conditions), and numerous other challenging conditions for life. However, recent studies suggest that prolonged LIP-induced ozone-layer depletion and destruction may be the most damaging LIP environmental impact; research in the coming years will provide greater understanding of severity of and mechanisms driving environmental consequences of LIP events.

4.2. Silicic Large Igneous Provinces

Silicic LIPs, like mafic LIPs, have aerial extents >0.1 million km^2, volumes >0.1 million km^3, and emplacement times <50 Myrs, but are predominately >65 wt% SiO_2 compositionally. Unlike mafic LIPs, they are only found in continental settings. LIPs in continental settings are often compositionally bimodal with a mafic and a silicic component; whether they are deemed mafic or silicic reflects the dominant composition. The same processes drive continental mafic and silicic LIPs, but differences in crustal setting dictate which kind of LIP will form, with silicic LIPs thought to occur on continental margins where subduction processes have created hydrated fertile crust. Silicic LIPs follow similar clustering behavior to mafic LIPs.

A brief note on **large-scale silicic eruptions** is warranted. These are the largest type of explosive volcanic event known on Earth, and are defined as erupting $\geq 10^{15}$ kg of magma, i.e., magnitude (M) ≥ 8. Assuming a magma density of 2500 kg/m^3, M ≥ 8 eruptions have volumes ≥ 400 km^3 dense rock equivalent (DRE) magma; we assume this magma density throughout this chapter. Such eruptions are of great interest due to their potential to generate global environmental change and far-reaching catastrophic devastation. However, unlike the large-scale mafic eruptions which form mafic LIPs, not all large-scale silicic eruptions are associated with silicic LIPs—large scale silicic eruptions can occur in other settings. Indeed, none of the 23 quaternary M ≥ 8 (≥ 400 km^3 DRE) eruptions (average one every ~ 0.1 Myr) are considered LIP eruptions.

Silicic LIPs producing large scale eruptions produce several such eruptions and as such generate a large scale silicic eruptions "pulse" in the overall record. For example, the Paraná-Etendeka silicic LIP produced nine large scale silicic eruptions around 132 Ma with magnitudes ranging from 8.7 to 9.3 (2000—8000 km^3 DRE), six of which were

M ≥ 9 (≥ 4000 km^3 DRE). This illustrates another point: the geologic record suggests silicic LIP eruptions are larger than "typical" large scale silicic eruptions, often with M ≥ 9 (≥ 4000 km^3 DRE). It appears different mechanisms drive "regular" and "silicic LIP" large-scale silicic eruptions: the former generally follows a typical eruption sequence (ash fall followed by large-scale ignimbrites with post-eruption small scale rhyolite dome emplacement) whereas large-scale silicic LIP eruptions have few or no associated ash fall (Plinian) deposits and no post-eruption small scale domes. The lack of Plinian deposits suggests a mass eruption rate exceeding that required for a stable buoyant Plinian eruption column.

4.3. "Mantle Plumes"

As previously discussed, plate tectonics appear to play a critical role in the occurrence of LIPs. LIPs are temporally clustered, and appear to precede supercontinent breakup. Given the huge volumetric contribution of LIPs to magma productions, the supercontinent cycle is a driver for global magma production and periodicity in output. In the past, a plume hypothesis has been invoked to explain LIPs, but as previous mentioned, in the past decade key predictions made by the plume hypothesis had been contradicted and so this hypothesis is on shaky grounds. However, there are several areas of intraplate volcanism with volcanic activity often explained by a mantle plume; while we will not comment on driving processes for this volcanism, we will briefly consider the rates of this intraplate volcanism.

Hawaiian hot spot volcanic output has increased steadily in the last 35 million years to a modern value of about 0.25 km^3 per year for the past 1 Ma and 0.095 km^3 per year over the past 6 Ma (Figure 14.6), up from 0.010 km^3 per year when the hot spot formed the Emperor Seamount chain. Other hot spots appear to be small remnants of former LIPs (e.g., reunion hot spot and the Deccan Traps LIP). In all, hot spot volcanism presently contributes ~ 2 km^3 per year to global magma production.

5. CONCLUSIONS

In this chapter, we have surveyed eruptive volumes and production rates in different volcanic settings. We find that volcanic production rates fluctuate over geologic time, and appear to be intrinsically linked to the geologic context of the time. In particular, circumstances around supercontinent assembly and breakup profoundly impact magma production rates. Below we briefly summarize today's understanding of volcanic production rates and episodity, which will change with future discoveries and data synthesis and interpretation.

Divergent plate (ocean ridge) volcanic output has been within a factor of two of modern rates over geologic time

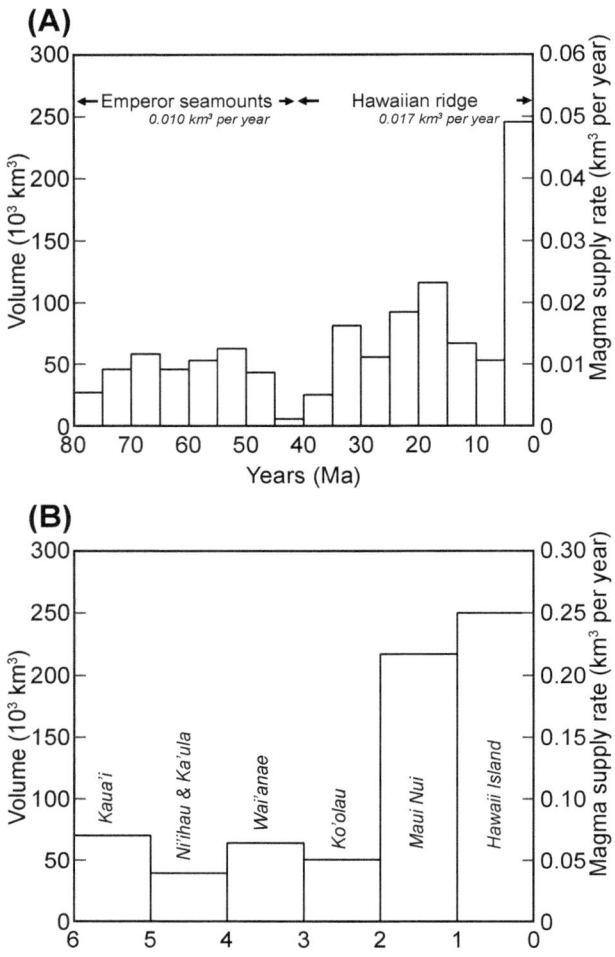

FIGURE 14.6 Emperor—Hawaiian hot spot volcanic output since the (A) 80 Ma and (B) 6 Ma; labeled are the main volcanic island complex for the million year period. Note the increase in production toward the present. *Figure after Robinson and Eakins (2006).*

for which the rate is discernable. Presently ~20 km³ per year are produced. Rates have been quite steady over the Cenozoic Era and were higher in the Cretaceous Era. Compared to other settings divergent plate volcanism is steady in output.

Convergent plate (subduction zone) volcanic output is difficult to quantify, but a modern conservative estimate is 0.5 km³ per year. There is evidence that output has varied in the past, but the scale remains challenging to assess.

LIP volcanic output is extremely variable. LIPs are often temporally clustered and appear linked to the super-continent cycle. Individual LIPs can produce over 5 km³ per year for a few million years surplus to "regular" volcanism for a total output in the 10s of millions km³.

Hot spot volcanism presently contributes ~2 km³, although it is unknown how and if it will change during an LIP flare up.

FURTHER READING

Brown, S.K., Crosweller, H.S., Sparks, R.S.J., Cottrell, E., Deligne, N.I., Guerrero, N.O., Hobbs, L., Kiyosugi, K., Loughlin, S.C., Siebert, L., Takarada, S., 2014. Characterisation of the Quaternary eruption record: analysis of the large magnitude explosive volcanic eruptions (LaMEVE) database. J. Appl. Volcanol. 3 http://dx.doi.org/10.1186/2191-5040-3-5.

Bryan, S.E., Ferrari, L., 2013. Large igneous provinces and silicic large igneous provinces: progress in our understanding over the last 25 years. GSA Bull. 125, 1053−1078. http://dx.doi.org/10.1130/B30820.

Müller, R.D., Sdrolias, M., Gaina, C., Roest, W.R., 2008. Age, spreading rates, and spreading asymmetry of the world's ocean crust. Geochem. Geophys. Geosyst. 9, Q04006. http://dx.doi.org/10.1029/2007GC001743.

Seton, M., Gaina, C., Müller, R.D., Heine, C., 2009. Mid-Cretaceous seafloor spreading pulse: Fact or fiction? Geology 37, 687−690. http://dx.doi.org/10.1130/G25624A.1.

Primary Volcanic Landforms

College of Earth, Ocean and Atmospheric Science, Oregon State University, Corvallis OR, USA

Jan M. Lindsay

School of Environment, The University of Auckland, Auckland, New Zealand

GLOSSARY

caldera volcano A volcanic landform that is a large collapse depression resulting from collapse of the roof of a magma reservoir during catastrophic eruption of tens to thousands of cubic kilometers of magma. A "negative" volcano surrounded by a partial or complete apron of pyroclastic deposits, dominantly ignimbrite. May contain a central peak in the form of a resurgent dome or block. The largest caldera volcanoes are now commonly known as "supervolcanoes."

cinder or scoria cone A steep conical hill of tephra (volcanic debris) that accumulates around and downwind from a volcanic vent. They are dominantly composed of fragments of scoria, a highly vesicular, usually lightweight, volcanic rock of basaltic to andesitic composition.

composite volcanoes All conical or broadly conical polygenetic volcanoes constructed of accumulations of lava and pyroclastic deposits, sometimes alternating, erupted from vent(s) located at the summit (eruptions can also on occasion occur from the flanks) of the volcano.

diatreme A typically funnel-shaped, steep-sided subvolcanic structure filled with a mixture of fragmented juvenile and country rock material and with a feeder dike at the base that forms due to shallow phreatomagmatic processes that cut deeply into the country rock. Diatremes underlie maar craters, tuff rings, and tuff cones, and are usually an order of magnitude larger than the surface volcanic edifices.

flood basalt plateau A plateau consisting of regional-scale flows of basalt. Often called **traps**, because of a characteristic stair-step morphology.

hot spot volcano A volcanic center, 100–200 km across and persistent for at least a few tens of millions of years, that is thought to be the surface expression of a persistent rising plume of hot mantle material.

ignimbrite plateau A plateau consisting of large ignimbrite sheets, the depositional product of extensive pyroclastic flows, that share a petrogenetic and volcanic history. Comparable in size to flood basalt plateaux. The largest are referred to as silicic large igneous provinces.

kimberlite A rare, blue-tinged, coarse-grained intrusive igneous rock sometimes containing diamonds, found in South Africa and Siberia. Many kimberlites are emplaced as carrot-shaped, vertical intrusions termed "pipes" resulting from explosive maar–diatreme volcanism.

large igneous provinces (LIPs) A massive composite landform of $>10^6$ km^3 of mafic volcanic and associated plutonic rocks that did not form by seafloor spreading or subduction but are commonly associated with hot spots. Thought to be constructed rapidly (<5 Myr) at a high rate (0.1 to >1 km^3/yr). LIPs occur within continents and oceans, and include continental flood basalts in the former, and volcanic passive margins, oceanic plateaux, submarine ridges, and ocean-basin flood basalts in the latter.

maar volcano A flat-bottomed roughly circular crater, often filled with water, that formed during shallow explosive interaction between magma and groundwater. Maar crater floors lie below the preeruptive surface and are surrounded by low rims of ejected debris.

monogenetic volcano Discrete minor volcanic landform that forms during one eruptive cycle. Can be mafic (cinder or scoria cones, maars, diatremes, tuff cones, tuff rings, and scutulum shields) or silicic (lava domes and coulées) in composition. Typical lifetimes of years to decades for mafic centers, decades to centuries for silicic ones.

pit crater Also called a subsidence crater or collapse crater, this is a depression formed by a sinking or collapse of the surface above a void or empty magma chamber.

polygenetic volcano Large discrete volcanic landform that forms during several episodic eruptive cycles. Typical lifetimes of 10^4-10^5 years. Can be mafic or silicic in composition. Most common types are composite volcanoes and shield volcanoes.

pyroclastic or ignimbrite shield A shield-shaped volcano dominated by pyroclastic deposits on its flanks. Some may show a collapse caldera on top, others lava flows and domes capping pyroclastic deposits with no evidence of collapse.

scutulum shield Small shield volcanoes found in monogenetic volcanic fields or flood basalt terrains like the Snake River Plain of Idaho, USA. From the Latin, scutulus means small shield.

shield volcanoes All shield-shaped polygenetic volcanoes constructed predominantly of basaltic lava flows erupted from summit vent(s) and from the flanks of the volcano. Some composite volcanoes with bimodal compositions and pyroclastic shields are included.

silicic large igneous provinces (SLIPs) A large igneous province dominated by silicic compositions and are most common on continental crust.

supervolcano A large (>25-km diameter), often resurgent, caldera volcano from which at least one supereruption (>500 km^3 of magma or 1000 km^3 of erupted material) has vented.

tuff ring Broad low-profile aprons of tephra with slopes $2-10°$, typically found surrounding maar—diatreme volcanoes. The tephra is often unaltered and thinly bedded, and generally comprises a mixture of fall and proximal pyroclastic density current deposits.

tuff cone Steep-sloped and cone-shaped volcanic features with wide craters that comprise typically highly altered, thickly bedded tephra resulting from phreatomagmatic eruptions through shallow surface water. They are considered to be a taller variant of a tuff ring, formed by less powerful eruptions.

volcano-tectonic depression A caldera volcano where development, eruption, and final form are strongly linked to local tectonic structures and history.

1. INTRODUCTION

What is a volcano? If you ask this question, generally people conjure up the image of Mt Fuji, a composite volcano in Japan. Many are familiar with the shield shape of Mauna Loa on the big island of Hawaii or stacks of basaltic lava that form flood basalt provinces (Figure 15.1). These are of course the iconic images and exemplify that broadest and simplest definition of a volcano as *the site of emission of volcanic products*, or more elaborately, *the site on the surface of the Earth where material from the interior of the Earth is vented onto the surface as lava (molten rock on the Earth's surface), pyroclastic rock (fragmented magma ejected explosively), and/or hot vapor or gas.* Most commonly these sites are marked by positive landforms

FIGURE 15.1 What is a volcanic landform? The classic pyramidal cone of Mt Fuji, Japan (top) comes to most peoples' minds when this question is asked. Some may think of the low shield shape of Mauna Loa seen in the center. But what about extensive lava plateaux like the Columbia River Plateau (bottom), seen here at Lake Billy Chinook in Oregon, that dwarfs Mt Hood (in the distance)?

where erupted material piles up around the actual opening or rupture (a central or fissure vent) from which the eruption issues and forms a volcanic edifice or landform (cone, ring, dome, shield) that most people refer to as a "volcano."

However, in some cases individual volcanoes are "negative" features, taking the form of craters that can be as large as 80 km in maximum dimension. Thus, rather than all volcanoes being like Fuji or Mauna Loa, there are actually many different landforms. However, if we are talking about extraterrestrial volcanic landforms, our definitions and perspective needs to be extended to include the exotic "magma" types and unique landforms that erupt on some extraterrestrial bodies. Here we attempt to describe the main volcanic landforms found on the Earth. While recognizing that the greatest volume of Earth's volcanism happens beneath the ocean (see Chapters 19 and 21), and that significant volcanism occurs under ice cover (see Chapter 20) as well as on other bodies in the solar system (see Chapters 39—43), we restrict ourselves to subaerial (on land) volcanic landforms on the Earth as organized in Figure 15.2. We do not attempt to deal with volcanic landforms produced by magma—ice interactions (see Chapter 20).

Volcanoes are the surface expression of thermal activity and magmatism in the interior of a planet. Thus, the general characteristics of the main volcanic landforms reflect fundamental controls of magma composition and other intrinsic properties that control eruption character. Since magmatism on the Earth (and other terrestrial planetary bodies for that matter) is fundamentally basaltic, originating in the mantle, the character of volcanoes is the result of many factors, in particular tectonic environment, melt production and transport, and the nature of the lithosphere

the magma must traverse to from source to surface. In addition, if magma accumulates and ponds in a shallow crustal environment prior to eruption, although not all do, processes of differentiation and magma movement from the reservoir to the surface and the nature of the conduit environment are vital to the final volcanic landform that is formed.

What is described as a volcanic landform is a matter of scale and here we define them as landforms that have a distinct temporal and spatial identity at the scale being considered. A single edifice is of course the simplest landform, but a prolonged history often complicates this simple identity. An individual long-lived volcanic landform represents the interplay between eruptive and erosional histories and as it evolves, it may change in form from one landform type to another, and may build an areally diffuse massif. Some volcanic landforms are collections of different types of other volcanoes that may define spatially and temporally related clusters, and fields of volcanoes that collectively have an individual identity. We focus here on describing the typical features of primary landforms that result from eruptive processes and present key quantitative morphometric data where relevant (Tables 15.1 and 15.2). The characteristic eruptive activity and tectonic associations are also presented. This is a summary treatment with reference to details in the relevant thematic chapters.

A useful initial framework is to consider the main volcanic landforms in terms of magma composition and the volume of erupted material (Figure 15.2). Magma composition is important because it is the main control on the rheology (resistance to flow) that ultimately controls eruptive style (see Chapter 17). Simply put, mafic magma (basalt) that is poor in network-forming molecules like SiO_2, and is formed and erupted at high temperatures, will be more fluid and produce lava-dominated volcanic landforms with low-angle slopes (**shield volcanoes**), or form extensive flat plateaux (*Flood Basalts*; Chapter 24). On the other hand, intermediate to silicic magmas, by virtue of being rich in SiO_2 and having a lower temperature, are more viscous and tend to form more steep-sided volcanic landforms known collectively as **composite volcanoes** (stratovolcanoes, composite cones, compound volcanoes) that produce effusive (lava-dominated) eruptions as well as *explosive* (pyroclastic-dominated) eruptions. However, there are important departures from these general characteristics. Small mafic volcanoes can be steep-sided **cinder or scoria cones** and may be dominated by explosive activity. If external water is involved in the eruptions, highly explosive "fuel—coolant interactions" (see Chapter 26) may result, producing negative volcanoes like the eponymous "Hole-in-the-ground" **maar volcano**, an explosive crater in Oregon. On the other hand, the product of the most catastrophic eruptions on the Earth are from silicic magma,

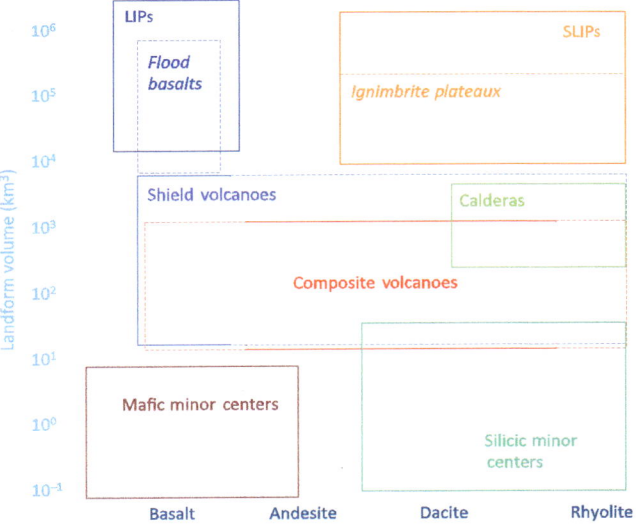

FIGURE 15.2 The primary volcanic landforms can be conveniently placed in composition—volume space. Solid lines represent the characteristic range of composition—volume. Dashed lines indicate subordinate ranges. Composition can be broadly considered a proxy for viscosity and eruption style, while landform volume insinuates source productivity and longevity. SLIPs, silicic large igneous provinces; LIPs, large igneous provinces.

TABLE 15.1 Average Characteristics of Primary Individual Landforms (i.e., Unmodified by Posteruptive Processes excluding small mafic centres)

Landform	Height of Cone or Rim (H_{co})	Average Basal Diameter (W_{co} or D_{co})	Crater or Rim Diameter (W_{cr} or D_{cr})	Primary Slope of Construct (α)
Composite volcano	2–3 km	Simple cones, 10–20 km	Vents $W_{cr} < 1$ km $D_{cr} < 300$ m Calderas $W_{cr} < 10$ km $D_{cr} < 1$ km	30°
Oceanic shield volcano	Small, 100–500 m Galapagos [1], ~1300 m Hawaiian, >10,000 m	Small, up to 10 km Galapagos, 25–48 km Hawaiian, >50 km	Small, <100 m Galapagos, 3.5–9 km Hawaiian, 3–4 km	Small, <<10° Galapagos, 4° to >25° Hawaii, <<10°
Other shield-shaped volcanoes	Bimodal basalt– rhyolite shields, 1.5 km	50–60 km	7–12 km Medicine Lake 7 km Newberry	1–5° flanks >30° central complex (rhyolitic)
	Ignimbrite shields, 300–1300 m	45–50 km [4] Purico, Panizos	Summit lava complex 25 km	2–10° [5]
Caldera volcanoes	1–2 km [7]	100–300 km [4]	30–80 km	1–5° 10–15° [8]

[1] Depth to basal platform included.
[2] Subaerial basal diameter.
[3] Area of Newberry: 3100 km² as per http://volcanoes.usgs.gov/volcanoes/newberry/ and Area of Medicine Lake: 2200 km² as per http://volcanoes.usgs.gov/volcanoes/medicine_lake/.
[4] Width of the ignimbrite outcrop defining the landform.
[5] Normalized for basal topography.
[6] A_{basal} is the areal footprint of the ignimbrite shield, $A_{lava\ complex}$ is the areal extent of the summit lava complex.
[7] Normalized average elevation above local base level.
[8] Average slope of the domal topography associated with large calderas.
[9] Areal footprint of landform (outflow ignimbrite plateau + collapse).
[10] Total duration of activity.
Data sources in references cited (see online version for complete list) also unless stated.

TABLE 15.1 Continued

Ratios of Note (H, Height; W, Width; D, Depth; A, Area)	Composed Primarily of	Volume of Erupted Material (km³)	Eruption Style	Typical Time for Formation (Years)
H/W, 0.15—0.33	*Oceanic—basalt to basaltic andesite* *Continental: andesite—dacite* Lava and pyroclastic rocks Volcaniclastics	Intraplate ≤ 5000 Arcs — 10 to 300	Strombolian Vulcanian Plinian Phreatoplinian	Simple and subcones, 10^4–10^6 Complexes and massifs, 10^7
Hawaiian H/W ~1/20 $W_{basal}{}^2/W_{caldera}$ Galapagos, 2.3—4 Hawaii >10 $A_{basal}/A_{caldera}$ Galapagos, 28—51 Hawaii, 156—277 $V_{volcano}/V_{caldera}$ Galapagos, 16—103 Hawaii, 830—1030	Basaltic lava Pyroclastic deposits <1%	Small, 1—15 Galapagos, 1000 Hawaiian, >10,000	Hawaiian *Strombolian Phreatomagmatic*	~10^6
$A_{basal}{}^3/A_{caldera}$ Newberry, 62 Medicine Lake, 36	Flanks—Basalt Central—Rhyolite		Flanks—Strombolian, Hawaiian Central—Dome building Vulcanian Plinian Phreatoplinian	10^5–10^6
$A_{shield}/A_{lava\ complex}{}^6$ 3.24	Dacite to rhyolite	100—600 km³	Plinian Effusive	Ignimbrite— catastrophic Lava complex 10^5–10^6 [10]
$A_{landfrom}{}^9/A_{collapse}$ 30 (Taupo, Campanian)—4 (Toba, Galan)	Dacite to rhyolite	100—1000 (maximum 6000 km³)	Plinian, Ultraplinian	Caldera —catastrophic 10^4–10^6 [10]

TABLE 15.2 Average Characteristics of Primary Landforms (i.e., Unmodified by Posteruptive Processes) Associated with Small Mafic Volcanoes

Landform	Height of Cone or Rim (H_{co})	Average Basal Diameter (W_{co} or D_{co})	Crater or Rim Diameter (W_{cr} or D_{cr})	Primary Slope of Construct (α)
Cinder (scoria) cone	≤300 m high [1] 50–200 m but Paricutin and Cerro Negro are ~500 m high [2]	[3]900 m [2]Range: 0.25–2.5 km, average = 0.8 km	0.4 W_{co} [2]50–600 m	25–38° [4] ~30° [5] Angle of repose = 33°
Spatter cone		0.08 km [7] Can be as low as 0.2 km [5]		Sometimes >33° because fragments are welded together [5]
Tuff cone	Often >100 m, but <300 m [8] 50–330 m [5]		<0.1–1.5 km [8]	20–30° [1] >25° [2] 10–30° [4]
Tuff ring	<50 m [8] But Capelinhos tuff ring is ~200 m thick [9]	1.6 km [8]	0.2–3.0 km [8] [5]Average = 0.7–0.8 km	2–10° [1] Subhorizontal to 20° near rim [4] <25° [5]
Maar crater	Height of rim <30 m [8]; but Joya Honda maar rim sequence >100 m [5] **Depth:** ≤300 m deep [13] 10 to >500 m deep [5]	[3]1 km [8]1.4 km	0.2–3.0 km [8] [5]Average = 0.7–0.8 km Typically 0.2–1.5 km [13]	Subhorizontal to 20° near rim [4] <25° [5]

[1]Francis and Oppenheimer, 2004
[2]Schminke, 2004
[3]Wood, 1980
[4]White and Ross, 2011
[5]Vespermann and Schminke, 2000
[6]Inbar and Risso, 2001
[7]Wood, 1980
[8]Head et al., 1981
[9]Waters and Fisher, 1971
[10]Lorenz, 1986
[11]Houghton and Schmincke, 1989
[12]Kereszturi and Németh, 2013
[13]Ross et al., 2011

TABLE 15.2 Continued

Ratios of Note	Composed Primarily of	Volume of Erupted Material	% Juvenile Material in Deposits	Time for Formation
$^3H_{co}:W_{co}=0.18$ $^3D_{cr}:W_{co}=0.4$ $^6H:W_{co}=0.26$	Scoria lapilli, blocks and bombs deposited through fallout or grain avalanches	^{2}Average $=$ $4\times10^7\,m^3$ (Schminke, 2004) $10^5-10^9\,m^3$ (cones + flows)[4]	Close to 100%	Usually <1 year[3] Few weeks to months, with eruptive rates usually highest on the first day[2] 50% take <30 days, 95% take <1 year[5]
$W_{cr}/W_{co}=0.36$ km $H_{co}/W_{co}=0.22$ km[7]	Agglutinated and welded spatter deposited through fallout		Very close to 100%	
$H_{co}:D_{cr}=$ $0.5-0.2$ km[8]	Ash and lapilli deposited from fallout and base surges	$10^5-10^9\,m^{3}$[4]	Over 95%[4]	
$H_{co}:D_{cr}=$ $0.13-0.05$ km[8]	Ash and lapilli deposited from fallout and base surges	$10^5-10^9\,m^{3}$[4]	Variable, may be 95–99%[10] or as little as 45%[11]	Days to a few weeks[12]
$H_{co}:D_{cr}=$ $0.5-0.2$ km[8] Freshest maar craters have D_{cr}:Depth $=$ 3:1–7:1[13]	Ash and lapilli deposited from fallout and base surges	Volume esti-mates difficult	As little as 10%[4]	

where the entire crust above a massive magma chamber collapses after the magma is withdrawn during a catastrophic eruption to form a *caldera* (see Chapter 16). Some silicic volcanoes dominated by viscous lava domes are called cumulovolcanoes, some shield volcanoes can have central portions dominated by silicic magma, and classic intermediate composition composite cones may start life as lava shields.

In this chapter we group volcanic landforms based on the complexity of their eruptive history into **polygenetic volcanoes** and **monogenetic volcanoes**. Polygenetic volcanoes have experienced several eruptive episodes in their history. These are here considered to be large volcanoes that have been built over at least tens of thousands, but usually hundreds of thousands, or even millions of years to form a major landform that defines a location above a long-lived stable thermal anomaly in the crust. In this group we include composite volcanoes such as Mt Fuji (Japan) and Mt St Helens (Washington, USA), shield volcanoes such as Mauna Loa (Hawaii, USA) and Fernandina (Galapagos), and silicic calderas such as Toba (Indonesia) and Yellowstone (Wyoming, USA).

In contrast, *monogenetic volcanoes* are defined as those where eruptive activity ceases after only one episode of activity. An episode could consist of one eruption of few weeks to months, or quasi-continuous activity over a few years to decades. The landforms included in this category are mafic minor centers like the iconic scoria cones of Parícutin (Mexico) or Lava Butte (Oregon, USA), maar volcanoes like the Ubehebe crater (California, USA) or Crater Elegante (Mexico), and **tuff rings** like Fort Rock (Oregon, USA). Also discussed are silicic lava domes, like the impressive Chao lava (Chile), the most common form of silicic minor center.

Finally, we discuss areally distributed but spatially, temporally, and magmatically connected volcanic landforms that are not normally discussed as part of the pantheon of "volcanic landforms." These clusters, fields, and complexes find their largest expression in **large igneous provinces** (LIPs) that represent the most inclusive expression of volcanism and the most obvious volcanic landform at the planetary scale. These often include the entire range of individual volcanic landforms.

2. POLYGENETIC VOLCANOES

2.1. Composite Volcanoes

Often used synonymously, the terms stratovolcanoes, lava cones, composite cones, volcanic centers, or compound volcanoes are all used to describe polygenetic volcanic landforms formed by repeated eruptions from a single vent or migrating vents related to a common magmatic system. The prefix "strato" (from the Latin stratum,

meaning "a covering") is used in reference to the observation that some composite volcanoes comprise layer upon layer (or strata) of lava and pyroclastic deposits. Since many volcanoes known as "stratovolcanoes" do not display such layering, and are actually composites of many different stages of evolution, some confocal, some not, it is more appropriate to use the term "composite" to describe this group of landforms. Composite volcanoes thus include all the conical or broadly conical edifices constructed of stacked lava and pyroclastic deposits erupted from a central vent(s) located at the summit (eruptions can also on occasion occur from the flanks) of the volcano. It is common for the center of volcanism to shift around during the lifetime of a volcano, allowing for the development of multiple, overlapping edifices. Many composite volcanoes have a cyclic growth history that includes long periods of buildup punctuated by rapid partial collapse of the edifice. Three main morphologies are found (Figure 15.3):

1. Large steep-sided cones such as Mt Fuji (Japan), Mt Hood (Oregon, USA), El Misti (Peru), and Mt Mayon (Luzon, Philippines);
2. Asymmetric, broader to ridge-form edifices or subcones such as Lascar (Chile) and Ruapehu (New Zealand); and
3. Compound edifices or massifs constructed from overlapping edifices forming a distinct massif separated from other large volcanoes. Examples include Aucanquilcha (Chile), Coropuna (Peru), Tongariro (New Zealand). The term "cumulovolcano" is also used for a dome-shaped volcano constructed of multiple lava domes and flows such as Mammoth Mountain (California, USA). This morphotype would also include volcanic centers such as Lassen Volcanic Center (California, USA), which is a long-lived (>1 Ma), large-volume, composite edifice.

Composite volcanoes occur in all tectonic environments, but most typically at subduction zones, particularly around the Pacific Ring of Fire. They are less commonly found in an intraplate setting such as the East African Rift zone (e.g., Mount Kilimanjaro) and San Francisco Mountain in Arizona, USA; these volcanic zones are thought to be associated with continental rifting. In Iceland, which is situated on a divergent plate boundary, large central volcanoes such as Askja and Hekla, and Torfajökull are composite volcanoes. Typically these are constructed from viscous intermediate to silicic magma of andesitic to dacitic composition that erupt explosively (Strombolian, Vulcanian, sub-Plinian to Plinian eruption styles; see Chapters 27—29) to produce pyroclastic cones, density currents, and falls, and effusively to produce lava flows, coulées, and domes (see Chapters 17 and 18). The magmatic arcs of the Andes (the type location for andesites), the Cascades,

FIGURE 15.3 The three main forms of composite volcanoes. Left—Mt Hood looming above the city of Portland, Oregon, USA, is a classic steep-sided cone approximately 2 km in height and with circular basal footprint approximately 15 km in diameter. Top right—The asymmetric, *ridge-form edifices* or *subcone* of Volcan Lascar in northern Chile has maximum basal diameter of ~10 km. The summit region (5650 m above sea level) consists of an overlapping and nested group of craters aligned approximately east to west. Bottom right—The compound massif of Volcan Coropuna in Peru (6300 m above sea level) has a maximum basal diameter of almost 30 km. The massif consists of at least five overlapping cones, the peaks of which can be distinguished on this view.

Indonesia, and New Zealand host some of the most iconic andesite to dacite composite volcanoes on the planet. Large Plinian eruptions may punctuate the more typical activity and result in the formation of a large caldera in the summit. This happened at Mt Mazama/Crater Lake 7700 years BP (Oregon, USA; Figure 15.4), Santorini (Aegean Sea, Greece), Mt Tambora, 1815 (Sumbawa, Indonesia), and Krakatau, 1883 (Java, Indonesia). These Plinian eruptions are usually associated with dacitic to rhyolitic composition magma (see Chapter 29).

Composite volcanoes are not always intermediate and silicic. Lassen Volcanic Center erupted a diverse assemblage of rock types from basalt to rhyolite. Kamchatka hosts some of the largest composite cones, some of which are basaltic, and island arcs such as the Marianas, Aleutians, and Japan tend to produce composite volcanoes with more mafic compositions. Relatively rare mafic-dominated composite volcanoes are also found on oceanic islands like the Azores, the mid-ocean ridge of Iceland, and the continental rift environment of the East African rift.

The size of a volcano is largely a function of long-term magma supply and rate. If supply is at a rate greater than the cooling timescale of the system, the volcano will continue to grow. Also important is the hydraulic equilibrium between the growing volcano and the driving pressure from the magma reservoir, as well as the ease with which magma can move up the conduit. Typical composite volcanoes have edifice heights (base to summit) of 1–3 km, and volumes of 10–100 km^3. The largest examples at subduction zones include Mt Shasta (California) with a volume of about 300 km^3, and Kliuchevskoi (Kamchatka), about 250 km^3. The largest composite volcanoes on the Earth are found in intraplate settings. Kilimanjaro (Tanzania), with an edifice height of over 5 km (summit elevation is 5895 m) and a volume of almost 5000 km^3, is one of the largest composite volcanoes on the Earth.

The basic conical shape reflects the dominance of eruptive activity from the central summit vent. However, the complex interplay between growing volcano and driving pressure—as the volcano grows it becomes more

FIGURE 15.4 Key volcanomorphic features of composite volcanoes. (A) Steep-sided conical edifice of Volcan Misti, Southern Peru. The view of the northern flank reveals the composite nature of the edifice with intercalated lava and a few pyroclastic beds interfingering with more massive lava bodies. (B) Many composite cones have parasitic vents like the 200-m-high La Poruna scoria cone on the flanks of the majestic ~2.5-km-high edifice of Volcan San Pedro in Chile. The 8-km-long lava flow erupted from the site La Poruna is in the foreground. The inset shows detail of the San Pedro summit area where thick stubby dacite lava flows result in steepening of the uppermost flanks of the volcano. (C) Collapse of composite volcanoes is a common stage in their evolution. The classic amphitheater of Mt St Helens, Washington, USA, formed during the May 18, 1980 flank failure is a common feature of composite volcanoes. The scar is now being healed by a series of lava domes. (D) Mt Washington, Oregon is the eroded remnant core of a previously majestic cone (dashed lines). (E) The iconic Crater Lake caldera, Oregon (8 km in diameter) formed when the upper parts of the 13,000-ft-tall Mt Mazama collapsed into the magma chamber during the 7350 BP climactic eruption. Inset shows aerial view courtesy of Google Earth (See also Figure 15.10D). Explosive caldera-forming eruptions are a rare but probable event in the history of large composite volcanoes.

difficult to push magma out and the common trend toward more silicic (more viscous) compositions with time both also play a role in the steepness. The classic concave profile of mature composite volcanoes results from gently dipping flanks composed mainly of pyroclastic and volcaniclastic deposits reflecting the interplay between effusive and explosive activity, and aggradation and degradation (glaciation, collapse, and erosion) most completely recorded on the flanks of the volcano (Figure 15.5).

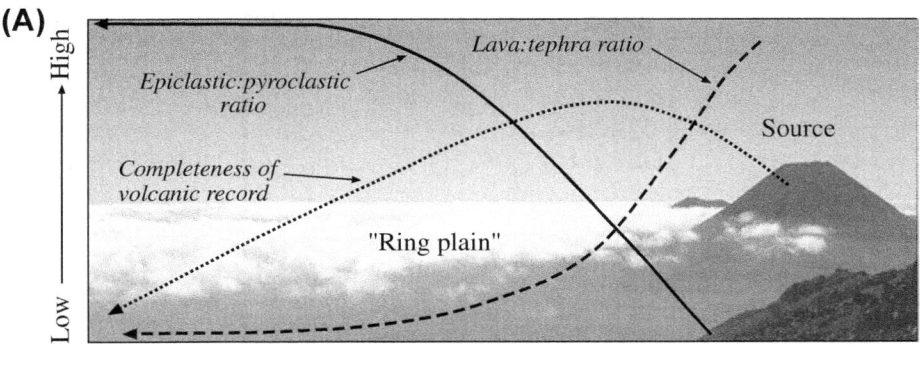

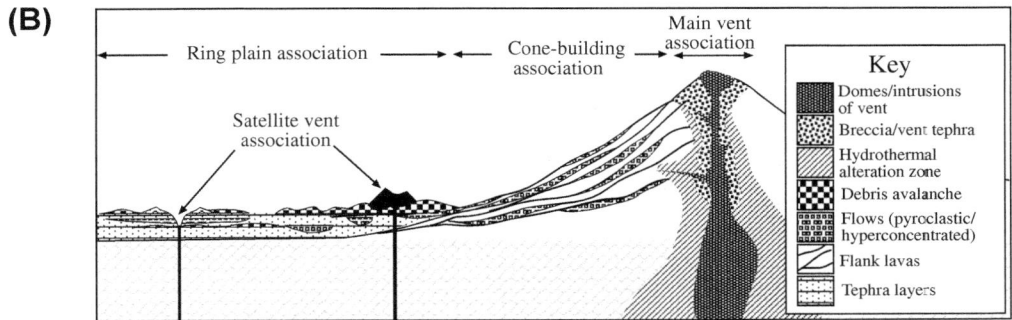

FIGURE 15.5 (A) Schematic illustration of the distribution of deposits relative to the vent for a composite volcano. Completeness of the volcanic record refers to the Proportion of events recorded in a succession at a given location. The most complete record is preserved in tephra that are deposited over a wide area. On the slopes tephra is rapidly reworked. Thus a more complete, albeit more condensed section is found in the ring plain—the volcaniclastic apron surrounding the volcano. (B) Schematic illustration of lithofacies associations for a typical composite cone, showing the *types* of deposits. Compare with *distribution* of deposits in (A). *(From Davidson and de Silva (2000). first edition of the Encyclopedia of Volcanoes.)*

Detailed studies of composite volcanoes are now revealing that common associations of cones, subcones, and massifs in volcanic regions that share a common tectonic and magmatic history may reflect an evolutionary sequence. Morphology may be a clue to the complexity of the history and evolutionary stage, but is not always. Many simple cones might be thought to have a simple constructive evolution, however many may have undergone a major edifice collapse that has been "healed" like at Shasta (California), or Socompa and Parinacota (Chile). General sequences where simple cones grow into large cones, or where they widen into subcones and massifs once they grow to a critical size, have been proposed where large cones can undergo sector collapse and/or gravitational spreading, without significant final morphometry change as the volcano "heals" itself. Simple cones may also evolve by vent migration to elliptical subcones and massifs before reaching the critical height. Such general trends in morphology representing different growth stages can be related to magma flux, edifice strength, structure, and tectonics. However, there are many exceptions. Many composite volcanoes in the Cascades start their life as lava-shield volcanoes; some are simply large lava-armored pyroclastic cones. Thus, any general scheme is necessarily incomplete, and understanding the fundamental controls on edifice evolution and realization of a more inclusive framework remains a challenge.

2.2. Shield Volcanoes

Lava-shield volcanoes, such as those that comprise the Hawaiian Islands, USA, are the classic basaltic volcanic landform (Figures 15.1 and 15.6A,B). Other well-known examples are those of the Galapagos Islands (Figure 15.6C and D). Shield volcanoes are also the most recognizable volcanic landforms in the solar system, with Olympus Mons on Mars being the most iconic of all. Oceanic basaltic shield volcanoes are constructed primarily by successive lava flows and minor pyroclastic layers, and are commonly characterized by relatively low slopes of 4—8°. In Hawaiian shields steep-walled calderas and smaller **pit craters** are found in the summit. In addition, two, sometimes three well-defined linear rift zones extend up to 250 km from the summit along which most eruptions are concentrated. Eruptions are dominated by fluid lava flows often fed by spectacular fire fountains that merge directly from the dike-fed fissure systems. For this reason,

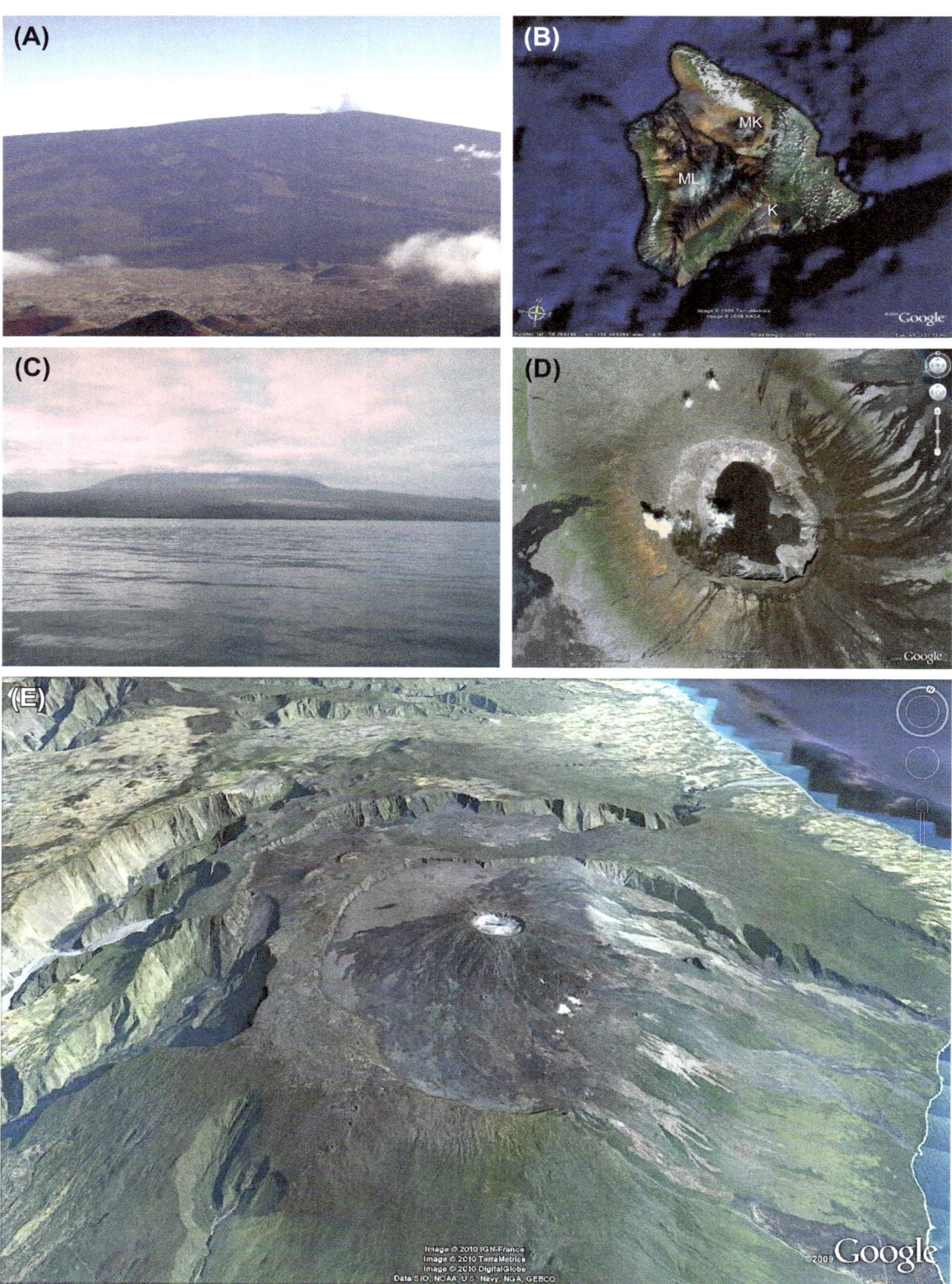

FIGURE 15.6 (A) The 4169-m summit of Mauna Loa from the summit of neighboring Mauna Kea. View to the south at the northeast rift zone. Cinder cones of Mauna Kea in middle ground. *(Image credit: Scott Rowland.)* (B) Google Earth Image of the Big Island of Hawaii (120 km across) which is made of overlapping shield volcanoes. Mauna Loa (ML), the largest occupies about two-thirds of the island. Summits of Mauna Kea (MK) and Kilauea (K) are also indicated. (C) Overturned soup plate profile of Volcan Wolf, Galapagos Islands, Ecuador. Summit elevation is 1707 m. *(Image credit: Scott Rowland.)* (D) Google Earth image extract of the summit crater and flanks of Volcan Wolf. Although only about 5 km in diameter, the summit crater is large relative to the size of the volcano compared with Hawaiian shields. (E) Perspective Google Earth image of Piton de la Fournaise on the eastern end of Reunion island. The 10-km-diameter shield volcano built within an older collapse has its summit crater at an elevation of 2632 m above sea level.

Hawaiian shield volcanoes tend to be elongated in the direction of the fissure. The low-angle slopes reflect the fluidity of the basaltic lava, high effusion rates that drive lava fast and far, the dominance of flank eruptions, and the spreading and widening of the volcano in the summit and rifts. Small cinder and scoria cones and spatter ramparts may build at eruption foci along these rifts and lines of pit craters often develop. The caldera and rift zones are underlain by dike swarms consisting of several thousand dikes representing several kilometers of added width to the edifice. This spreading along the rift zones results in seaward movement of the flanks of the volcano that weakens the edifice, eventually leading to catastrophic rock avalanches that carry debris hundreds of kilometers along the ocean floor. Mauna Loa, which has an elevation 4169 m above sea level, is only the upper 25% of an edifice that extends beneath the sea to the ocean floor about 5 km below. With an areal footprint of 5300 km^2 and a volume of $\sim$80,000 km^3 it is the largest individual volcano on the Earth. Its immense mass causes the base of the volcano to sag a further 8 km, making the summit 17 km above its base.

Active oceanic shield volcanoes are thought to mark hot spots in the lithosphere produced by plumes in the mantle (see Chapter 3). Hawaiian volcanoes go through "life stages" during their development and these can be related to their relative relationships to the Hawaiian hot spot (Figure 15.7). The three most important constructive stages are as follows:

1. Early alkalic stage. As a "new" portion of lithosphere (it is actually $\sim$90 million years old in the vicinity of Hawai'i) moves over the hot spot, the effects are not instantaneous; the degree and amount of partial melting and magma production are small. Small percentages of partial melting produce magma that is rich in alkali elements relative to silica (called alkali basalt). This low-efficiency melting produces only small amounts of magma, meaning that the eruption rate of a young **hot spot volcano** is not high.

2. Main tholeiite shield stage. As heating of the lithosphere continues, the degree of partial melting increases and the absolute volume of magma produced really increases. A higher degree of partial melting produces tholeiite basalt, which has a slightly higher percentage of silica than alkalic basalt. Catastrophic edifice collapses can occur during this time.

3. Postshield alkalic stage. As the volcano moves off the hot spot, the amount and degree of partial melting both become lower. The lower degree of partial melting leads to a return to alkali lava production, characterized by a greatly reduced eruption rate manifested in small fields of monogenetic volcanoes. A result of this reduced eruption rate is a much longer period of time between resurfacing at any one place on the volcano, so that erosion and weathering can be extensive.

Once growth has completed the islands undergo erosion and subsidence that eventually reduces them to sea level and with continued subsidence, the islands become *coral atolls*. These coral reefs may die if conditions are not favorable. After the reef dies, the volcano continues to subside. Once below sea level, these flat-topped, coral-capped volcanoes are called "guyots."

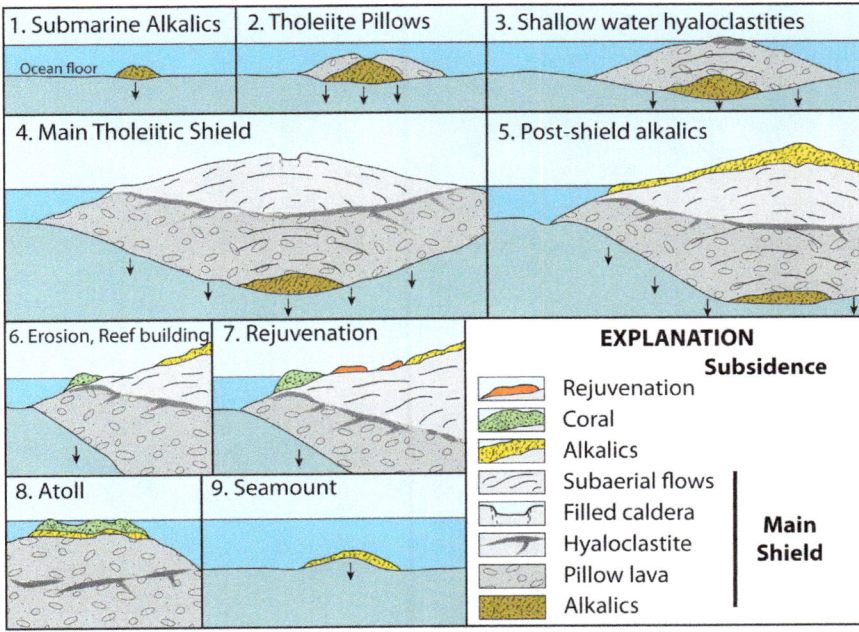

FIGURE 15.7 Life stages of a Hawaiian volcano. Note that because of subsidence, about half of the volume of the volcano is below the level of the ocean floor. Within the main southeast Hawaiian chain, Lo'ihi is in stage 2. Kilauea is a young stage 4 and Mauna Loa is a mature stage 4. Hualalai and Mauna Kea are both in stage 5. Kohala is in stage 6. Moving to Maui, East Maui is in stage 7 whereas West Maui is between stages 7 and 8. East Moloka'i is also between stages 7 and 8. Lana'i (which skipped stage 5), Kaho'olawe, and West Moloka'i are in stage 6, perhaps never to go through stage 7. On O'ahu, Ko'olau is in (still in?) stage 7 (but also skipped stage 5), and Wai'anae is probably between stages 7 and 8. Kaua'i and Ni'ihau are between stages 7 and 8. As you can see, even though a lot of work has gone into figuring out this sequence, the volcanoes themselves have a good deal of variation and have not perfectly "followed" the sequence. *Used with permission from http://volcano. oregonstate.edu/book/export/html/113.*

Shield volcanoes elsewhere may deviate from the Hawaiian model. The Galapagos shields differ from Hawaiian shields by having radial dike swarms leading to a more radially symmetric shape in plan view. They have a distinct "inverted soup-plate" profile with a clear break in slope from lower slopes up to 4° to steep upper slopes and a broad flat top (the summit platform) with a deep summit crater (Figure 15.5C, D). Two morphotypes have been suggested,

1. Deep calderas (depth 40−60% of subaerial height of the volcano) and >20° maximum slopes for more than 60% of the volcano height (Azul, Fernandina, and Wolf) and
2. Shallow calderas (<20% of subaerial height) with steep slopes only in the uppermost 10% (Alcedo, Darwin, and Sierra Negra).

A circumferential (annular rift) zone around the caldera rim is a feature of Galapagos shields. Piton de la Nieges (Reunion Island, Indian Ocean) is an ocean lava shield cut by swarms of sills rather than dikes. Small flat shields of only a few hundred meters elevation and slopes as low as 1° and diameters of about 10 km are a distinct landform in Iceland, the Snake River Plain, Idaho, and elsewhere. They are known as **scutulum shields**. These have volumes of only a few tens of cubic kilometers, a fraction of the giant shields of Hawaii or the smaller ones of the Galapagos.

Some shield volcanoes have persistent active lava lakes in their summit calderas. Halemaumau pit crater in Kilauea

was active for more than a century until 1924. Erta Ale (Afar, Ethiopia), Mount Erebus (Antarctica), and Nyiragongo (Congo) have active lava lakes at the time of writing. Like Kilauea, these latter three volcanoes are also intraplate, and over hot spots, although they are built on continental crust.

Nonbasaltic volcanoes with a shieldlike profile are also referred to as shield volcanoes (Figure 15.8). These are not related to hot spots but occur in a variety of tectonic associations. Two examples are the bimodal volcanoes of Newberry (Figure 15.8(B), 15.10(A)) and Medicine Lake in the subduction-related setting of the Cascades arc of western North America. Referred to sometimes as "central" volcanoes, these are centered on a mafic shield-form edifice with a central part made up of rhyolitic rocks, and the flanks consisting of flood basalts and minor basaltic vents. At Newberry, the basaltic lower flanks extend out as far as 65 km, sloping at 1−3° outward with over 400 cinder cones arranged along fissures that run broadly north-northwest to south-southeast through the volcano. The central core is made up of multiple rhyolitic vents, domes, and flows within which one or more calderas may develop during large explosive eruptions that punctuate the evolution of the system (Figure 15.10A).

Also deviating from the classic basaltic lava shields are **pyroclastic shields**. These are broad shallow volcanic constructs that are dominated by pyroclastic rocks. These may be basaltic or silicic. Basaltic pyroclastic-dominated

FIGURE 15.8 Nonbasaltic shield-shaped volcanic landforms. (A) The Cerro Purico ignimbrite shield, Chile has gently dipping (<10°) flanks of dacite ignimbrite extending out from a flat summit where a broad ∼15-km diameter complex of dacite to andesite domes overlie the ignimbrite. Total relief seen here is ∼1300 m in a distance of 15 km. (B) The Newberry Volcano in Oregon, USA, has low flanks of basalt lava flows and minor centers (fore to middle ground), and a high central region (distant horizon) of silicic lava flows, domes, and calderas. The view is to the south and shows the ∼45-km east to west width of the shield-like landform. Vertical relief is ∼1000 m in a distance of 30 km.

shield volcanoes include the continental hot spot-related Jebel Marra volcano (Sudan) and Emi Koussi volcano (Tibetsi, Chad). Both have shieldlike morphological similarities to those of the Galápagos, in particular, with gradients upward of 13° at the central crater rim and a relatively abrupt leveling off of approximately 3° toward the midflank and base regions. Among the silicic varieties are **ignimbrite shields** described from the subduction-related volcanic environment of the Central Andes. These consist of a large-volume ignimbrite apron dipping gently (8–10°) away from a central lava dome complex that occupies more than half the areal footprint of the shield. Unlike calderas where clear evidence for collapse is found, ignimbrite shields have a "downsag" at their summits in which ignimbrite dips inward with only weakly developed caldera collapse, if present at all. The lavas also clearly overlie the ignimbrite (Figure 15.8(A)). These observations have led to the suggestion that these are a class of explosive silicic volcanoes, separate from calderas, that may owe their form to deeper or more prolate magma chambers. In general, both the basaltic and silicic pyroclastic shields remain poorly studied.

3. CALDERA VOLCANOES (INCLUDING SUPERVOLCANOES)

The largest, most catastrophic eruptions result when hundreds to thousands of cubic kilometers of magma is erupted in a matter of days during a single event. Such eruptions, known as supereruptions if they produce more than

1000 km^3 of pyroclastic material (500 km^3 of magma) (see Chapter 13), occur relatively rarely, approximately every 100,000 years. They result in inverse volcanoes, or central collapse depressions, surrounded by flanks of pyroclastic rocks that are dominated by large-volume pyroclastic flow deposits or ignimbrites. The formation of these calderas, or **supervolcanoes** as they have become known, is attributed to subsidence related to rapid withdrawal of magma (see Chapter 16). However, unlike their smaller (<10 km) counterparts, where Plinian eruptions and related withdrawal of magma truncate just the upper parts of preexisting composite volcanoes or pyroclastic shield volcanoes, these large calderas represent collapse of the entire crust with maximum dimensions as large as 80 km, covering areas of 2500–3000 km^2 (Figure 15.9). Some smaller eruptions have been remarkably violent. The 1800 yr B.P. 30-km^3 Taupo ignimbrite eruption covers more than 20,000 km^2 of the North Island of New Zealand. The ~37,000 BP, 200-km^3 Campanian ignimbrite eruption covers an even more impressive 30,000 km^2. Both these erupted from relatively small caldera collapses of the 30-km Taupo volcano (New Zealand) and 13-km Phlegrean Fields (Italy), respectively. A significant range of violence is therefore associated with caldera-forming eruptions.

The central depressions of the largest calderas often host thick sequences of caldera-fill ignimbrites, that after caldera formation can be uplifted to form resurgent domes by renewed magmatic activity. Examples are shown in Figure 15.10(B, C). They are found in all tectonic environments on continental crust, the latter required for production of the batholith-scale bodies of silicic magmas that are

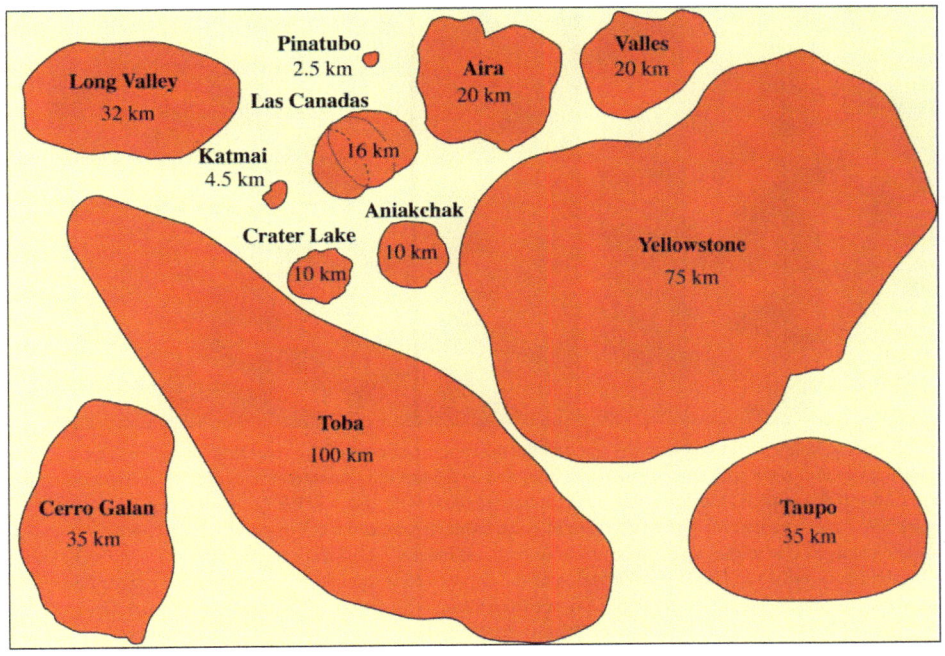

FIGURE 15.9 Relative sizes of calderas. Smaller sizes less than 20 km are calderas that form in the summit of preexisting composite volcanoes or shield volcanoes. Calderas of sizes greater than 20 km are "negative" volcanoes that are defined by the collapse depression. The largest calderas of sizes greater than 30 km are commonly called supervolcanoes and commonly have a structural resurgence after the climactic eruption. *Image credit: Adrian Pittari.*

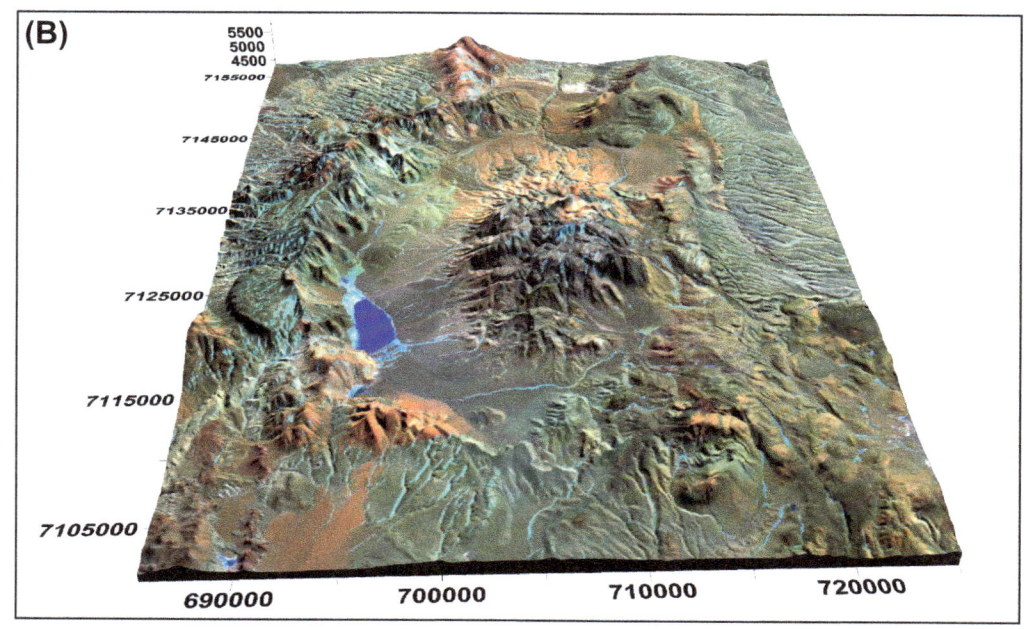

FIGURE 15.10 (A) The summit calderas of Newberry Volcano in Oregon are formed in the silicic central core of the shieldlike complex. The singular collapse was the result of an explosive pyroclastic eruption; resurgent volcanism has divided the ~6.5 km maximum diameter central collapse into two basin now filled with lakes. The ridged surface of the 1000-year-old Obsidian flow is seen on the bottom left. (B) ASTER false color perspective view of the 35 × 20 km 2.0 Ma Cerro Galan caldera in NW Argentina demonstrating the classic large **caldera volcano** form of an elliptical depression with a central dome formed when the floor of the caldera resurged due to magmatic pressure from remnant magma. (C) Toba caldera, Sumatra ~80 × 30 km is the site of one of the largest eruptions in the Earth history ~74 ka. The N—S elongation of the depression is thought to represent the strong influence of regional structures making this a type example of a volcanic-tectonic depression. *(Extract from Google Earth.)* (D) The classic scalloped quasicircular Crater Lake caldera formed when the upper 6000 ft of Mt Mazama collapsed into the magma chamber during the climactic 7700 yr eruption. The caldera has a diameter of 8 km. *(Extract from Google Earth.)*

erupted. The development of a large long-lived caldera system is thought to reflect a magmatic flare-up, where elevated thermal fluxes from the mantle into the crust result in prodigious magma production rates, much higher than "normal" magmatism that produces composite volcanoes (see Chapters 4 and 14).

The term **volcano-tectonic depression** is used to describe large calderas or nested calderas where collapse and structural development is strongly linked to regional faults, graben, or rifts, although cause and effect are debated. Examples of such large calderas include the 1 Ma–74 ka Toba caldera (Sumatra, Indonesia) (Figure 15.10(C)), the 2 Ma–640 ka Yellowstone caldera (Wyoming, USA), and the 5.6–4 Ma La Pacana caldera (Chile). Each of these measures >65 km in maximum dimension and has had at least two major eruptions attesting to the long-lived cyclic nature of these systems that are clearly associated with regional structures. Importantly, each of these has a central resurgent dome showing that postcollapse magmatic activity is focused beneath the caldera. Other examples include the classic ~2–1 Ma, 25-km-diameter, Valles caldera (New Mexico, USA) and the 30-km-diameter Taupo volcano (New Zealand). No resurgent dome is found at Taupo or other calderas in the Taupo Volcanic Zone of New Zealand suggesting that the extensional stress field there may hinder the processes of resurgence.

The key features of youthful, well-preserved large calderas are as follows:

1. A depression bounded by steep walls defining an escarpment, the topographic rim of which marks the overall areal extent of subsidence. The shape of the depression can range from circular/equant to elliptical/subequant to polygonal, depending on relative influence of regional stress fields. A lake may occupy the depression, as at Toba, or extensive lake deposits may record the former existence of a lake, as at Aso (Japan). At Yellowstone, the caldera has been glaciated and infilled with 1000 km^3 of lava, masking the depression. At La Pacana and Cerro Galan (Argentina), the arid high-desert environment has led to superb preservation of the main features of very complex calderas.

2. A central uplift within the depression made up of dipping welded intracaldera ignimbrite representing the uplifted floor. Sometimes the dips define an antiformal structure resulting from arching associated with uplift, as at La Pacana and Valles. At La Pacana and Toba, elongate resurgent domes indicate strong regional structural influence mimicking the general elongation of the depression, whereas at Yellowstone, Valles, and Long Valley (California), the resurgent domes are more equant. Dipping lake sediments capping the ignimbrite on the resurgent dome are found at Toba.

3. A broadly radially distributed apron of ignimbrite flanking the depression. Depending on preeruption topography and volume of erupted material, the ignimbrites could be valley confined or could inundate the topography forming a plateau. All the large calderas mentioned here have **ignimbrite plateaux** surrounding them in various stages of dissection.

Diffusion of these features with time due to erosion and mass wasting will "soften" many of these characteristics making identification of calderas in older or rapidly eroded or vegetated areas a much more difficult task.

4. MONOGENETIC VOLCANOES

4.1. Mafic Monogenetic Volcanoes

Small mafic volcanoes are found in their thousands all over the world in all different tectonic environments, although they seem to particularly favor extensional regimes. They typically occur either as part of a monogenetic volcanic field formed due to distributed volcanism (e.g., East and West Eifel, Germany; Chaine des Puy in the Auvergne, France; and the Auckland Volcanic Field, New Zealand; see Chapter 23) or as vents on the flanks of larger volcanoes (e.g., Rotomahana maar on the flank of Mt Tarawera rhyolite dome, New Zealand; satellite scoria cones on the flanks of Mt Etna composite volcano, Italy). Mafic monogenetic volcanoes are well studied, in part because hazard and risk posed by monogenetic eruptions is of concern for ever-growing nearby population centers such as Auckland (New Zealand), Mexico City (Mexico), Al-Madinah (Saudi Arabia), and Portland and Bend (Oregon, USA). Moreover, they usually erupt the most primitive magmas in a region yielding important insight into the origin and production of magma, and economically important deposits (diamonds) are hosted by some maar–**diatreme** structures like **kimberlites**.

The main forms of mafic minor volcanic landforms (Figure 15.11) are as follows:

1. *Scoria* or *cinder* cones where morphology depends on the interplay of many factors, including total eruptive volume, eruptive style, and posteruption modification. Typically, the cones are ≤300-m high and cone shaped, although they are often asymmetrical due to a dominant wind direction during eruption, a breached crater, a migrating vent, or rafting away of parts of the cone by late-stage lava flows. They often have large bowl-shaped craters that are proportionately large with respect to the size of the edifice as a whole (the ratio of crater diameter to basal diameter is on average 0.4 (Table 15.2)). Young, fresh cones typically have a simple geometrical profile, with their slopes defined by

FIGURE 15.11 Examples of small mafic volcanic landforms. (A) The late Quaternary Motukorea volcano (also known as Browns Island) in the Auckland Volcanic Field, New Zealand, experienced a range of eruption styles during the course of a single eruption resulting in a hybrid landform, with a scoria cone and rafted cone remnants within an eroded tuff ring surrounded by late-stage lava flows. *(Photograph courtesy of B. Hayward.)* (B) The late Quaternary Meke Gölü maar volcanic complex of the Karapinar Volcanic Field, Turkey, comprises a large, elongated maar volcano with a well-preserved tuff ring, and a late-stage scoria cone in the crater. (C) The late Quaternary Narköy maar, near Nevsehir in Central Anatolia, Turkey, is a deep maar volcano with a steep crater wall and surrounding tuff ring, with pre-maar lava flows exposed in the crater walls. (D) A quarried section of a Quaternary scoria cone in western Mexico, near Volcán Ceboruco, exposes typical features of a scoria cone dominated by coarse ash to fine lapilli pyroclastic successions. (E) An older scoria cone from the Central Anatolian Volcanic Province, Turkey. The gentle slopes reflect advanced erosion of a coarse ash to fine lapilli-dominated scoria cone. *(Photographs (B–E) courtesy of K. Németh.)*

the angle of repose for loose scoria (33°), although gentler slopes may occur due to factors such as crater breaching and interaction of erupting magma with preexisting topography (e.g., 22—30° in Holocene scoria cones on Tenerife). The crater area of scoria cones is commonly red or purplish red due to oxidation by hot gases streaming through the central part of the edifice; the outer walls are commonly black and non-oxidized. These cones are commonly associated with late-stage lava flows that are typically fed by lava fountains or erupted from the base of the cone. Lava flow volumes are generally higher than the associated cone volumes.

2. *Maar—diatremes* are circular depressions that have crater floors that lie below the preeruptive surface and are surrounded by low rims of ejected debris (unless the debris has been eroded). These rims may be asymmetrical if the eruption occurred during strong winds (e.g., Cerro Colorado maar, Pinacate Volcanic Field, Mexico). Maar craters are underlain by a ≤2-km-deep diatreme, a typically cone-shaped, steep-sided structure filled with a mixture of fragmented juvenile and country rock material and with a feeder dike at the base. *Diatremes* are usually an order of magnitude larger than the

surface volcanic edifices. *Maars* are, by definition, holes that have been excavated into the preeruptive surface rather than structures built up above the ground, and for this reason often fill with water and manifest as lakes. The immediate posteruptive walls of maar craters can be very steep, with slopes up to 70° inferred for some *kimberlite* maar craters or "pipes."

3. *Tuff rings* and **tuff cones**, unlike maars that are excavated into the substrate, are built on the substrate. Morphologically, tuff rings are broad flat features with slopes 2—10°. Tuff cones are smaller, with thick near-vent deposits that thin rapidly outward, resulting in steeper (20—30°) slopes. The diameters of craters in tuff cones are comparable to tuff rings and maars, but the elevation of the crater rims are higher, reaching up to 300 m. Tuff rings are underlain by a shallow diatreme; diatremes are typically absent beneath tuff cones.

A progression of small basaltic landforms from scoria or cinder cones in totally dry environments, through tuff rings where groundwater is present, to tuff cones formed in shallow surface water is seen (Figure 15.12). This progression can produce a range in eruption styles from Hawaiian-style passive emission of lava producing small spatter cones and lava flows, through Strombolian-style fire-fountaining

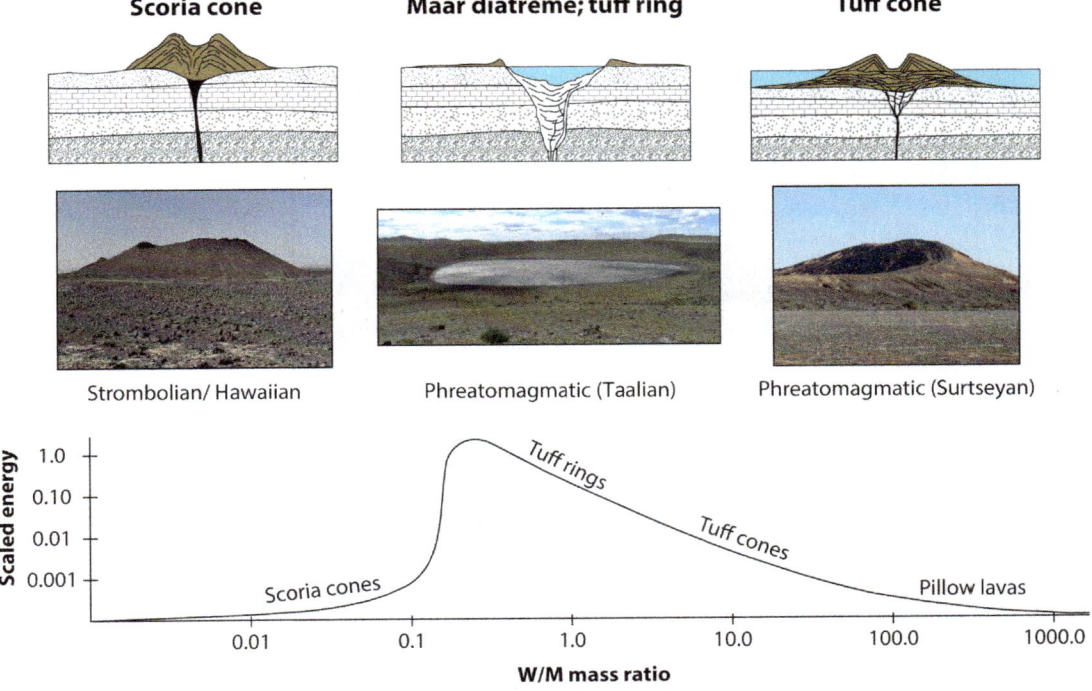

FIGURE 15.12 Diagram illustrating the effect of water/magma (W/M) ratios on eruption style and volcanic landform. Scoria cones result from dry Hawaiian or Strombolian magmatic eruptions (low W/M ratios), tuff rings from Taalian phreatomagmatic eruptions (relatively low W/M ratios), and tuff cones from Surtseyan phreatomagmatic eruptions (relatively high W/M ratios). *(Figure modified from Kereszturi and Németh (2013) and Vespermann and Schmincke (2000).)* Photographs are courtesy of K. Németh and are of a Pleistocene scoria cone in Harrat Rahat, Western Saudi Arabia (left), a Pleistocene maar volcano from the Pali Aike Volcanic Field in Southern Patagonia, Argentina (center) and Cerro Colorado, a Quaternary tuff cone in the Pinacate Volcanic Field in Sonora, Mexico (right).

producing scoria cones, to violent phreatomagmatic explosions producing maar—diatremes and tuff rings including kimberlites (see Chapters 27 and 30). It is also possible to have a range of these eruption styles during the course of a single eruption, resulting in hybrid landforms. Motukorea volcano in the Auckland Volcanic Field, for example, comprises an intra-tuff ring scoria cone, with the tuff ring breached by late-stage lava flows (Figure 15.11(A)); all of these are thought to have formed during a single eruptive event.

Various morphological features and ratios have been used to distinguish between the different landforms (such as tuff rings and scoria cones) of small mafic volcanoes (Table 15.2). These morphometric parameters are used in landform recognition in both terrestrial and extraterrestrial environments.

4.2. Silicic Monogenetic Volcanoes

Small silicic minor centers are relatively rare, possibly reflecting the propensity for explosive eruption in these magmas. The most common minor silicic volcanic landforms are discrete extrusions collectively called lava domes (see Chapter 18). These are thick bodies of limited extent representing sluggish flows owing to the high viscosities of the dacitic to rhyolitic lava composition (Figure 15.13(A)). Often these lavas are crystal-rich, adding to their viscosity, although glassy obsidian bodies are also found. These lava domes are most typically parasitic extrusions related to a larger composite cone or caldera where they may be located along faults. They are most commonly associated with explosive eruptions as the final eruptive events of the cycle, when degassed remnant magma is rejuvenated or induced to erupt by new intrusions of more mafic magma. While most commonly dacite, many of these domes are andesitic in composition, maybe representing intimate mixing between invading mafic and resident silicic magmas that produces spectacularly banded lavas.

Four different forms of silicic lava body morphologies can be recognized:

1. Low domes or tortas are the most common type, forming flat-topped, roughly symmetrical steep-sided cakes or tortas (although talus often masks any internal structure). They have rubbly tops and large-scale

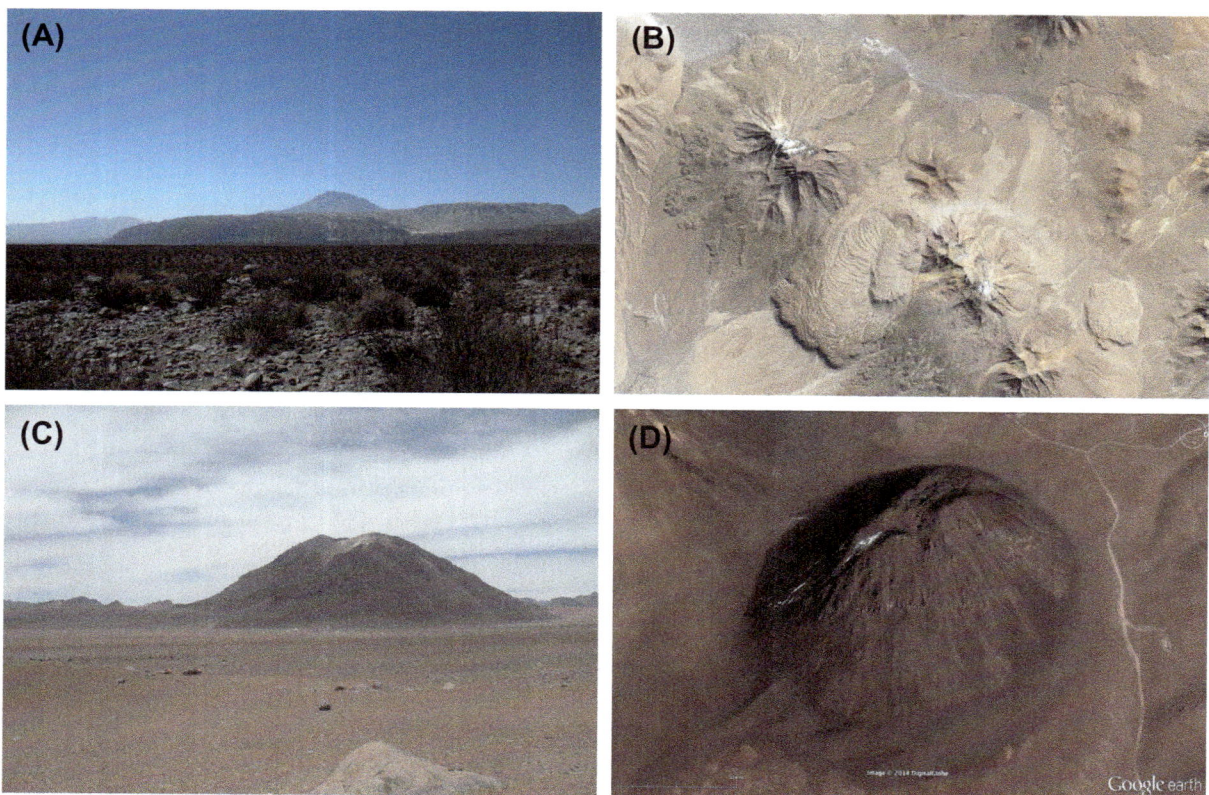

FIGURE 15.13 (A) Chao dacite coulée, North Chile. The 400-m-thick flow front attests to the viscous nature of the lava. View from 25 km to the southeast. (B) Landsat 8 view of the Chao dacite showing the scale of this monogenetic compound lava body—it is 15-km long. The ridged surface is produced by ramp structures resulting from bulldozing of the front of the lava mass from behind. Note the relative size of large composite volcanoes flanking Chao. *(Image courtesy of NASA.)* (C) Cerro Aspero in North Chile is an example of a Pelean dome. The height is 500 m and the circular plain has a diameter of about 2 km. (D) Google Earth view of Aspero, showing the distinct summit fissure and thick stubby flank lava flows.

pseudocolumnar jointing, often revealed in the steep walls. Some have a concentric, onion-like structure indicating endogenous or internal growth as successive extrusions push earlier lava outward. Others clearly show a composite plain-form made up of several abutting lobes attesting to exogenous growth. The steep sides attest to very high yield strength.

2. Coulées form where tortas are erupted onto slopes resulting in elongation of the lava body as it flows under the influence of gravity and internal shear overcomes the resistive force of the yield strength. As a consequence of the flow, large flow ridges or ogives produce a pleated "elephant-skin" upper surface. Although superficially like giant pahoehoe "ropes," these are actually pressure ridges that are the surface expression of ramp structures that form as successive lava extrusions bulldoze preexisting lava forward. The 15-km-long, 400-m-high Chao dacite of Chile is one of the most spectacular examples of a coulée (Figure 15.13(A) and (B)).

3. Peléean domes are more conical in shape, and characterized by craggy topography and lava spines at the crest. Collapse of spines and associated avalanching produces a steep collar of debris that encircles the domes, enhancing the steep conical profile of these domes.

4. Upheaved plugs are rock masses pushed upward as magma rises from depth. These may have country rock and lake sediments on top.

Lava domes are among the best documented of volcanic bodies. They accumulate over many tens of years to centuries and their eruption progress can often be documented relatively safely. The most recently documented examples are those of Soufríere Hills, Montserrat, which has been erupting since 1995, and the Santiaguito dome, since 1922. The Showa Shinzan plug in Japan grew completely during World War II.

5. VOLCANIC CLUSTERS, FIELDS, OR PROVINCES

Imagine looking down on the Earth from space. Zoom out, and one can see that individual volcanic landforms described above do not occur in isolation, but rather are often part of a volcanic cluster. Good examples are the Auckland Volcanic Field, a cluster of c. 55 small basaltic volcanoes spread over 350 km^2, or the Lassen Volcanic National Park, a coherent locus of volcanism that is spatially (430 km^2), temporally (3.5 Ma of activity), and volumetrically (>500 km^3) more extensive than the largest individual volcanoes in the region. Zoom out some more from the Lassen region and you see that the cluster itself is part of a regionally coherent volcanic terrain, the Cascades

arc. Arcs like the Cascades or the Andes are rarely discussed as "volcanic landforms" per se; however, on a planetary scale the most obvious volcanic landforms are going to be such spatially extensive but coherent collective volcanic landforms that occupy the large-volume (10^3 to >10^4 km^3) sector of Figure 15.1. These are aggregates of many individual volcanoes and their products.

The largest of these, the highest order grouping of volcanoes on the Earth, are known as LIPs that are dominated by basaltic volcanic features (see Chapter 24). The Ontong Java Plateau LIP in the South Pacific is 2 million square kilometers in area and 30-km thick, making it one of the largest of any type of landform on the Earth; it rivals the Tibetan Plateau in area but is at least five times thicker. Iceland as a whole, with an area of over 1 million square kilometers, could be classified as an LIP, and represents a distinct landform. The silicic-rock dominated siblings, **silicic large igneous provinces** (SLIPs) like the Sierra Madre Occidental in Mexico, with an areal extent of ~350,000 km^2 and a volume of 200,000 km^3, are smaller, but massive nonetheless on the scale of volcanic landforms. In many ways these are the primary manifestation of a planet's magmatism defining a spatially and temporally related group of volcanic landforms linked to some distinct thermotectonic event in the planet's history. The most common type of LIP and SLIP are **flood basalt plateaux** and ignimbrite plateaux, respectively. Approximate coincidences of some LIPs and mass extinctions have been noted (see Chapter 61).

5.1. Flood Basalt Plateaux

Flood basalt plateaux consist of voluminous and extensive sheets of lava flows erupted from scattered monogenetic fissure vents. They tend to flood the landscape and generate a new landscape of subdued relief. They occur above hot spots where significant crustal spreading has occurred. For example, the Columbia River Plateau of Oregon and Washington, the High Lava Plains of Oregon, and the Snake River Plain of Idaho in the northwest of the United States, are all thought to be related to the interaction of the Yellowstone plume with the North American lithosphere above it. The even more massive Deccan Flood Basalt Province (Figure 15.14(A)) is the result of the Indian lithosphere interacting with the Reunion hot spot (now under Reunion Island). Even larger is the giant Siberian Traps flood basalt field in Russia.

Many individual lava flows have volumes exceeding 10 km^3, but individual lava flows exceeding 1000 km^3 have been identified in the Columbia River Plateau and the Deccan province. Some have traveled hundreds of kilometers. Individual flow units are thickest on shallow slopes, where they commonly exceed tens of meters thick and are spectacularly columnar jointed. Evidence that many lava

(A)

(B)

FIGURE 15.14   (A) The Deccan Traps flood basalt plateau consists of more than 6500 feet (>2000 m) of flat-lying basalt lava flows and covers an area of nearly 200,000 square miles (500,000 km^2) similar to the Columbia River basalts of the northwestern United States. This photo shows a thick stack of basalt lava flows north of Mahabaleshwar. *(Photograph by Laszlo Keszthelyi, January 28, 1996.)* (B) Part of the dissected plateaux of the Sierra Madre Occidental silicic large igneous province. The photo shows an approximately 1-km-thick ignimbrite pile exposed in the Copper Canyon, northern Sierra Madre Occidental, reaching an elevation of 2240 m above sea level, with the base of the canyon at 1320 m above sea level. The exposed section includes six major ignimbrite units emplaced from 29 to 30 Ma. *(Photo taken by Luca Ferrari at 27°30′40″ N, 107°50′09″ W looking to the northeast.)*

flows are compound and have structures indicating endogenous growth has lead to the recognition that many of these flows were inflated as fresh lava was injected beneath a surface crust that was lifted up. A complex network of transient tubes or tunnels is seen as responsible for the long travel distances as interior flowing lava was well insulated during travel. Smaller flood basalt fields like those in Iceland and the Snake River Plain consist of many overlapping low-angle shields of the scutulum type. These shields in turn are made up of accumulations of pahoehoe flow units.

Eroded flood basalt fields are normally seen to have inwardly directed dip, most pronounced where the lava pile is deepest, and thought to indicate loading-related subsidence. Linear dike swarms accounting for 5–20% of the total volume of rock are common features in proximal areas, consistent with linear vents that are marked by crater rows of spatter, agglutinate, and cinders. Commonly fissure vents evolved to point sources as vent widening occurred, mimicking the process, albeit at a much larger scale, seen in Hawaiian fissure eruptions.

5.2. Ignimbrite Plateaux

Like flood basalt plateaux, *ignimbrite plateaux* are the composite of overlapping and superimposed products of clusters of large caldera complexes. As the name implies, the dominant features are the extensive ignimbrite sheets that define an areally extensive but spatially and temporally coherent volcanic history. Other volcanic intermediate to silicic volcanic features are often present but the volcanic episode is dominated by large catastrophic caldera-forming eruptions. Examples include the caldera clusters and associated ignimbrites of the San Juan

Mountains of Colorado, the Indian Peak and Central Nevada caldera clusters of the Great Basin of North America, and the Altiplano–Puna Volcanic Complex of the Central Andes of Bolivia, Chile, and Argentina. The largest well-exposed plateau is the Sierra Madre Occidental of Mexico (Figure 15.14(B)), although few calderas or source vents have been unequivocally identified there. Another is the Snake River Plain–Yellowstone province, although this is a more obviously bimodal province than the others.

These and many others identified through space and time on the Earth suggest that such plateaux occur in a variety of tectonic environments where continental crust is present, whether at subduction zones or at intraplate hot spots resulting in continental rifting. The requirement of continental crust points to a dominant mechanism for producing the large volumes of silicic magma involving crustal melting and assimilation by mantle-derived basalt. The coherence of volcanic style, magmatic composition, and temporal transience of these episodes has been suggested to be a "flare-up" in magma production resulting from an anomalous thermal input from the mantle. The impact of a mantle plume on continental lithosphere (continental hot spot) or a change in the mantle–crust interface through loss of insulating continental–mantle lithosphere (delamination) allowing hot upper mantle to rise and fuel melt production are among the most commonly suggested mechanisms.

The main distinctive features are stacks of ignimbrite sheets defining a "plateau" several tens to hundreds of thousands of square kilometers in extent. At the core of the plateau are the calderas and other source vents for the ignimbrites. The ignimbrite plateau is thickest where the calderas are located; ignimbrite stacks of several kilometers may be present within the collapse structures. The ignimbrite stratigraphy on the flanks of the calderas is considerably thinner but can overlap and combine to produce an aggregate thickness of over 1 km or greater if preexisting topography is infilled.

6. CONCLUDING REMARKS

What we call a volcanic landform is a matter of scale, and we have presented volcanic landforms as having a distinct temporal and spatial identity at the scale being considered. On the local scale this may mean that a single lava flow or small cone is the volcanic landform; on the planetary scale, the Big Island of Hawaii, Iceland, and LIPs like the Ontong Java Plateau, should be considered as volcanic landforms. Such planetary-scale volcanic landforms may contain all of the familiar volcanic landforms that we typically classify into polygenetic and monogenetic volcanoes regardless of tectonic environment. This suggests some primary controls on the character of the volcanic landforms. Magmatism on the Earth is fundamentally basaltic, but the variety of volcanic landforms reflects a variety of compositions and resulting eruptive styles. Thus, at the most primal level it is how the basaltic magma from the mantle is differentiated by the lithosphere that is responsible for the variety of compositions. The final volcanic landform that is formed reflects this but is ultimately the result of final modifications in the uppermost lithosphere and the conduit environment.

FURTHER READING

General

Bishop, M.A., 2009. A generic classification for the morphological and spatial complexity of volcanic (and other) landforms. Geomorphology 111 (1−2), 104−109.

Borgia, A., Aubert, M., Merle, O., et al., 2010. What is a volcano? Geological Society of America Special Paper 470. http://dx.doi.org/10.1130/2010.2470(01).

Cashman, K.V., Sparks, R.S.J., 2013. How volcanoes work: a 25 year perspective. Geological Society of America Bulletin 125 (5-6), 664−690. http://dx.doi.org/10.1130/B30720.1.

Francis, P.W., Oppenheimer, C., 2004. Volcanoes. Oxford University Press, Oxford, UK, 521 pp.

Fisher, R.V., Heiken, G., Hulen, J.B., 1997. Volcanoes. Crucibles of Change. Princeton University Press, Princeton, NJ.

Hildreth, E.W., 2007. Quaternary Magmatism in the Cascades − Geologic Perspectives. USGS. Professional Paper 1744.

Schminke, H.-.U., 2004. Volcanism. Springer-Verlag, Berlin Heidelberg, p. 324.

de Silva, S.L., Francis, P.W., 1991. Volcanoes of the Central Andes. Springer-Verlag, Heidelberg, 216 pp.

Thordarsson, T., Larsen, G., 2007. Volcanism in Iceland in historical time: volcano types, eruption styles and eruptive history. Journal of Geodynamics 43, 118−152.

Thouret, J.-C., 1999. Volcano geomorphology − an overview. Earth Science Reviews 47, 95−131.

Composite Volcanoes

Cotton, C.A., 1944. Volcanoes as Landscape Forms. Whitcombe and Tombs, Christchurch, New Zealand.

Clynne, M.A., Muffler, L.J.P., 2010. Geologic Map of Lassen Volcanic National Park and Vicinity, California. U.S. Geological Survey. Scientific Investigations Map 2899, scale 1:50,000.

Davidson, J.P., de Silva, S.L., 2000. Composite volcanoes. In: Sigurdsson, H., Houghton, B.F., McNutt, S.R., Rymer, H., Stix, J. (Eds.), Encyclopedia of Volcanoes. Academic Press, San Diego, pp. 663−680.

Grosse, P., van Wyk de Vries, B., Petrinovic, I.A., Euillades, P.A., Alvarado, G.E., 2009. Morphometry and evolution of arc volcanoes. Geology 37 (7), 651−654. http://dx.doi.org/10.1130/G25734A.1.

Hackett, W.R., Houghton, B.F., 1989. A facies model for a quaternary andesitic composite volcano: Ruapehu, New Zealand. Bulletin of Volcanology 51, 51−68.

Halsor, S., Rose, W.I., 1988. Common characteristics of paired volcanoes in northern Central America. Journal of Geophysical Research 93, 4467−4476.

Hildreth, E.W., Lanphere, M., 1994. Potassium-argon geochronology of a basalt-andesite-dacite arc system: the Mount Adams volcanic field, Cascade Range of southern Washington. Geological Society of America Bulletin 106, 1413−1429.

Hobden, B.L., Houghton, B.F., Davidson, J.P., Weaver, S.D., 1999. Small and short-lived magma batches at composite volcanoes: time windows at Tongariro Volcano, New Zealand. Journal of the Geological Society of London 156, 865−867.

Pike, R.J., Clow, G.D., 1981. Revised Classification of Terrestrial Volcanoes and a Catalog of Topographic Dimensions with New Results on Edifice Volume. US Geol. Surv. Open File Rep. OF 81−1038.

Singer, B.S., Thompson, R.A., Dungan, M.A., Feeley, T.C., Nelson, S.T., Pickens, J.C., Brown, L., Wulff, A., Davidson, J.P., Metzger, J., 1997. Volcanism and erosion during the past 930 k.y. at the Tatara−San Pedro complex, Chilean Andes. Geological Society of America Bulletin 109, 127−142.

Wood, C.A., 1978. Morphometric evolution of composite volcanoes. Geophysical Research Letters 5 (6), 437−439.

Shield Volcanoes

Moore, J.G., Clague, D.A., 1992. Volcano growth and evolution of the island of Hawai'i. Geological Society of America Bulletin 104, 1471−1484.

Moore, J.G., Clague, D.A., Holcomb, R.T., Lipman, P.W., Normark, W.R., Torresan, M.E., 1989. Prodigious submarine landslides on the Hawaiian ridge. Journal of Geophysical Research 94, 17,465−17,484.

Mouginis-Mark, P.J., Rowland, S.K., Garbeil, H., 1996. Slopes of western Galapagos volcanoes from airborne interferometric radar. Geophysical Research Letters 23 (25), 3767−3770.

Munro, D.C., Rowland, S.K., 1996. Caldera morphology in the western Galápagos and implications for volcano eruptive behavior and mechanisms of caldera formation. Journal of Volcanology and Geothermal Research 72, 85−100. http://dx.doi.org/10.1016/0377-0273(95) 00076−3.

Naumann, T., Geist, D., 2000. Physical volcanology and structural development of Cerro Azul Volcano, Isabela Island, Galápagos: implications for the development of Galápagos-type shield volcanoes. Bulletin of Volcanology 61, 497−514.

Permenter, J., Oppenheimer, C., 2007. Volcanoes of the Tibesti massif (Chad, northern Africa). Bulletin of Volcanology 69, 609–626. http://dx.doi.org/10.1007/s00445-006-0098-x.

Peterson, D.W., Moore, R.B., 1987. Geologic History and Evolution of Geologic Concepts, Island of Hawaii. U.S. Geol. Surv. Prof. Pap. 1350: pp. 149–189.

Simkin, T., 1972. Origin of some flat-topped volcanoes and guyots. Geological Society of America Memoir 132.

Volcano World http://volcano.oregonstate.edu/hawaiian.

Walker, G.P.L., 1990. Geology and volcanology of the Hawaiian Islands. Pacific Science 44, 315–347.

Walker, G.P.L., 2000. Basaltic volcanoes and volcanic systems. In: Sigurdsson, H., Houghton, B.F., McNutt, S.R., Rymer, H., Stix, J. (Eds.), Encyclopedia of Volcanoes, first ed. Academic Press, San Diego, pp. 283–290.

Calderas

Acocella, V., 2007. Understanding caldera structure and development: an overview of analogue models compared to natural calderas. Earth-Science Reviews 85 (3–4), 125–160. http://dx.doi.org/10.1016/j.earscirev.2007.08.004.

Bacon, C.R., Lanphere, M.A., 2006. Eruptive history and geochronology of Mount Mazama and the Crater Lake region, Oregon. Geological Society of America Bulletin 118 (11–12), 1331–1359. http://dx.doi.org/10.1130/B25906.1.

Bailey, R.A., Dalrymple, G.B., Lanphere, M.A., 1976. Volcanism, structure, and geochronology of Long Valley Caldera, Mono County, California. Journal of Geophysical Research 81 (5), 725–744.

Chesner, C.A., Rose, W.I., 1991. Stratigraphy of the Toba tuffs and the evolution of the Toba caldera complex, Sumatra, Indonesia. Bulletin of Volcanology 53 (5), 343–356.

Christiansen, R.L., 2001. The Quaternary and Pliocene Yellowstone Plateau Volcanic Field of Wyoming, Idaho, and Montana. United States Geological Survey. Professional Paper 729-G, 120 pp.

Cole, J.W., Milner, D.M., Spinks, K.D., 2005. Calderas and caldera structures: a review. Earth-Science Reviews 69 (1–2), 1–26.

Druitt, T.H., Francaviglia, V., 1992. Caldera formation on Santorini and the physiography of the islands in the late Bronze Age. Bulletin of Volcanology 54, 484–493.

Francis, P., 1983. Giant volcanic calderas. Scientific American 248, 60–70.

Folkes, C.B., Wright, H.M., Cas, R.A.F., de Silva, S.L., Lesti, C., Viramonte, J.G., 2011. A re-appraisal of the stratigraphy and volcanology of the Cerro Galán volcanic system, NW Argentina. Bulletin of Volcanology 73, 1–28.

Geyer, A., Martí, J., 2008. The new worldwide collapse caldera database (CCDB): a tool for studying and understanding caldera processes. Journal of Volcanology and Geothermal Research 175 (3), 334–354. http://dx.doi.org/10.1016/j.jvolgeores.2008.03.017.

Gregg, P.M., de Silva, S.L., Grosfils, E.B., Parmigiani, J.P., 2012. Catastrophic caldera-forming eruptions: thermomechanics and implications for eruption triggering and maximum caldera dimensions on Earth. Journal of Volcanology and Geothermal Research 241–242, 1–12. http://dx.doi.org/10.1016/j.jvolgeores.2012.06.009.

Kennedy, B., Wilcock, J., Stix, J., 2012. Caldera resurgence during magma replenishment and rejuvenation at Valles and Lake City calderas. Bulletin of Volcanology 74 (8), 1833–1847. http://dx.doi.org/10.1007/s00445-012-0641-x.

Lavallée, Y., Silva, S.L., Salas, G., Byrnes, J.M., 2005. Explosive volcanism (VEI 6) without caldera formation: insight from Huaynaputina volcano, southern Peru. Bulletin of Volcanology 68 (4), 333–348. http://dx.doi.org/10.1007/s00445-005-0010-0.

Lindsay, J.M., de Silva, S., Trumbull, R., Emmermann, R., Wemmer, K., 2001. La Pacana caldera, N. Chile: a re-evaluation of the stratigraphy and volcanology of one of the world's largest resurgent calderas. Journal of Volcanology and Geothermal Research 106 (1), 145–173.

Lipman, P.W., 2000. Calderas. In: Sigurdsson, H., Houghton, B.F., McNutt, S.R., Rymer, H., Stix, J. (Eds.), Encyclopedia of Volcanoes. Academic Press, San Diego.

Rowe, M.C., Wolff, J.A., Gardner, J.N., Ramos, F.C., Teasdale, R., Heikoop, C.E., 2007. Development of a continental volcanic field: petrogenesis of pre-caldera intermediate and silicic rocks and origin of the Bandelier magmas, Jemez Mountains (New Mexico, USA). Journal of Petrology 48 (11), 2063–2091. http://dx.doi.org/10.1093/petrology/egm050.

de Silva, S., 2008. Arc magmatism, calderas, and supervolcanoes. Geology 36 (8), 671–672.

de Silva, S., Zandt, G., Trumbull, R., Viramonte, J.G., Salas, G., Jiménez, N., 2006. Large ignimbrite eruptions and volcano-tectonic depressions in the Central Andes: a thermomechanical perspective. Geological Society, London, Special Publications 269 (1), 47–63.

Smith, R.L., Bailey, R.A., 1968. Resurgent cauldrons. In: Coats, R.R., Hay, R.L., Anderson, C.A. (Eds.), Studies in Volcanology, vol. 116. Geological Society of America Memoir, pp. 153–210.

Walker, G.P., 1984. Downsag calderas, ring faults, caldera sizes, and incremental caldera growth. Journal of Geophysical Research 89 (B10), 8407–8416.

Wilson, C.J.N., Houghton, B.F., McWilliams, M.O., Lanphere, M.A., Weaver, S.D., Briggs, R.M., 1995. Volcanic and structural evolution of Taupo Volcanic Zone, New Zealand: a review. Journal of Volcanology and Geothermal Research 68, 1–28.

Mafic Monogenetic Volcanoes

Bemis, K., Walker, J., Borgia, A., Turrin, B., Neri, M., Swisher III, C., 2011. The growth and erosion of cinder cones in Guatemala and El Salvador: models and statistics. Journal of Volcanology and Geothermal Research 201 (1–4), 39–52.

Hasenaka, T., 1994. Size, distribution, and magma output rate for shield volcanoes of the Michoacán-Guanajuato volcanic field, Central Mexico. Journal of Volcanology and Geothermal Research. ISSN: 0377-0273 63 (1–2), 13–31. http://dx.doi.org/10.1016/0377-0273(94)90016-7.

Head, J.W., Sparks, R.S., Bryan, W.B., Walker, G.P.L., Greely, R., Whitford-Stark, J.L., Guest, J.E., Wood, C.A., Shultz, P.H., Carr, M.H., 1981. Distribution and Morphology of Basalt Deposits on Planets. Basaltic Volcanism on the Terrestrial Planets. Pergamon Press Inc, New York, 701–800.

Hooper, D.M., Sheridan, M.F., 1998. Computer-simulation models of scoria cone degradation. Journal of Volcanology and Geothermal Research 83 (3–4), 241–267.

Houghton, B.F., Schmincke, H.-U., 1989. Rothenberg scoria cone, East Eifel: a complex Strombolian and phreatomagmatic volcano. Bulletin of Volcanology 52, 28–48.

Inbar, M., Risso, C., 2001. A morphological and morphometric analysis of a high density cinder cone volcanic field — Payun Matru, south-central Andes, Argentina. Zeitschrift für Geomorphologie 45 (3), 321–343.

Kereszturi, G., Németh, K., Cronin, S.J., Agustín-Flores, J., Smith, I.E.M., Lindsay, J., 2013. A model for calculating eruptive volumes for monogenetic volcanoes — implication for the Quaternary Auckland Volcanic Field, New Zealand. Journal of Volcanology and Geothermal Research 266, 16—33. http://dx.doi.org/10.1016/j.jvolgeores.2013.09.003.

Kereszturi, G., Németh, K., 2013. Monogenetic basaltic volcanoes: genetic classification, growth, geomorphology and degradation. In: Németh, K. (Ed.), Updates in Volcanology—New Advances in Understanding Volcanic Systems. InTech, ISBN 978-953-51-0915-0, pp. 3—88. http://dx.doi.org/10.5772/51387.

Lorenz, V., 1986. On the growth of maars and diatremes and its relevance to the formation of tuff rings. Bulletin of Volcanology 48, 265—274.

Luhr, J.F., Simkin, T., 1993. Paricutin: The Volcano Born in a Mexican Cornfield. Geoscience Press, Phoenix.

Ross, P.-S., Delpit, S., Haller, M.J., Németh, K., Corbella, H., 2011. Influence of the substrate on maar—diatreme volcanoes — an example of a mixed setting from the Pali Aike volcanic field, Argentina. Journal of Volcanology and Geothermal Research 201 (1—4), 253—271.

Valentine, G.A., Perry, F.V., Krier, D., Keating, G.N., Kelley, R.E., Cogbill, A.H., 2006. Small-volume basaltic volcanoes: eruptive products and processes, and posteruptive geomorphic evolution in Crater Flat (Pleistocene), southern Nevada. Geological Society of America Bulletin 118 (11—12), 1313—1330. http://dx.doi.org/10.1130/B25956.1.

Vespermann, D., Schminke, H.-U., 2000. Scoria cones and tuff rings. In: Sigurdsson, H., Houghton, B.F., McNutt, S.R., Rymer, H., Stix, J. (Eds.), Encyclopedia of Volcanoes. Academic Press, San Diego, pp. 683—694.

Waters, A.C., Fisher, R.V., 1971. Base surges and their deposits: Capelinhos and Taal volcanoes. Journal of Geophysical Research 76, 5596—5614.

White, J.D.L., Ross, P.-S., 2011. Maar-diatreme volcanoes: a review. Journal of Volcanology and Geothermal Research 201 (1—4), 1—29.

Wohletz, K.H., Sheridan, M.F., 1983. Hydrovolcanic explosions II. Evolution of basaltic tuff rings and tuff cones. American Journal of Science 283, 385—413.

Wood, C.A., 1979. Monogenetic volcanoes in terrestrial planets. In: Lunar and Planetary Science Conference, 10th, Houston, Tex., March 19—23, 1979, Proceedings. Volume 3 (A80-23677 08-91). Pergamon Press, Inc., New York, pp. 2815—2840.

Wood, C.A., 1980. Morphometric evolution of scoria cones. Journal of Volcanology and Geothermal Research 7 (3—4), 387—413.

Silicic Monogenetic Volcanoes

Blake, S., 1989. Viscoplastic models of lava domes. In: IAVCEI Proceedings in Volcanology. Lava flows and domes, vol. 2. Springer Verlag, Heidelberg, pp. 88—126.

Eichelberger, J., 1995. Silicic volcanism: ascent of viscous magmas from crustal reservoirs. Annual Review of Earth and Planetary Sciences 23, 41—64.

Eichelberger, J.C., Carrigan, C., Westrich, H.R., Price, R.H., 1986. Nonexplosive silicic volcanism. Nature 323, 598—602.

Griffiths, R.W., Fink, J.H., 1993. Effect of surface cooling on the spreading of lava flows and domes. Journal of Fluid Mechanics 252, 667—702.

Fink, J.H., Anderson, S.W., 2000. Lava domes and Coulees. In: Sigurdsson, H., Houghton, B.F., McNutt, S.R., Rymer, H., Stix, J. (Eds.), Encyclopedia of Volcanoes, first ed. Academic Press, San Diego, pp. 307—319.

Fink, J.H., Anderson, S.W., Manley, C.R., 1992. Textural constraints on effusive silicic volcanism: beyond the permeable foam model. Journal of Geophysical Research 97, 9073—9083.

Nakada, S., Miyake, Y., Sato, H., Oshima, O., Fujinawa, A., 1995. Endogenous growth of dacite dome at Unzen volcano (Japan), 1993—1994. Geology 23, 157—160.

de Silva, S.L., Self, S., Francis, P.W., Drake, R.E., Ramirez, C., 1994. Effusive silicic volcanism in the Central Andes: the Chao dacite and other young lavas of the Altiplano-Puna Volcanic Complex. Journal of Geophysical Research 99, 17,805—17,825.

Watts, R.B., de Silva, S.L., Jimenez de Rios, G., Croudace, I., 1999. Effusive eruption of viscous silicic magma triggered and driven by recharge: a case study of the Cerro Chascon-Runtu Jarita Dome Complex in Southwest Bolivia. Bulletin of Volcanology 61 (4), 241—264.

Watts, R.B., Herd, R.A., Sparks, R.S.J., Young, S.R., 2002. Growth patterns and emplacement of the andesitic lava dome at Soufriere Hills Volcano, Montserrat. Geological Society, London, Memoirs 21 (1), 115—152. http://dx.doi.org/10.1144/GSL.MEM.2002.021.01.06.

Flood Basalts and LIPs

Coffin, M.F., Eldholm, O., 1993. Large igneous provinces. Scientific American. ISSN: 0036-8733 269 (4), 42—49.

Duncan, R.A., Richards, M.A., 1991. Hotspots, mantle plumes, flood basalts, and true polar wander. Reviews of Geophysics 29 (1), 31—50.

Jerram, D.A., 2002. Volcanology and facies architecture of flood basalts. In: Volcanic Rifted Margins, pp. 121—135.

Self, S., Thordarson, T., Keszthelyi, L., 1997. Emplacement of continental flood basalt lava flows. In: Large Igneous Provinces: Continental, Oceanic, and Planetary Flood Volcanism, pp. 381—410.

Self, S., Thordarson, T., Keszthelyi, L., Walker, G.P.L., Hon, K., Murphy, M.T., Finnemore, S., 1996. A new model for the emplacement of Columbia River basalts as large, inflated pahoehoe lava flow fields. Geophysical Research Letters 23 (19), 2689—2692.

White, R.S., McKenzie, D.P., 1995. Mantle plumes and flood basalts. Journal of Geophysical Research 100, 17,543—17,586.

Ignimbrite Plateaux and SLIPs

Best, M.G., Gromme, S., Deino, A.L., Christiansen, E.H., Hart, G.L., Tingey, D.G., 2013. The 36—18 Ma Central Nevada ignimbrite field and calderas, Great Basin, USA: multicyclic super-eruptions. Geosphere 9, 1562—1636. http://dx.doi.org/10.1130/GES00945.S4.

Bryan, S.E., Riley, T.R., Jerram, D.A., Stephens, C.J., Leat, P.T., 2002. Silicic volcanism: an undervalued component of large igneous provinces and volcanic rifted margins. Special Papers-Geological Society of America, 97—118.

Ferrari, L., López-Martínez, M., Rosas-Elguera, J., 2002. Ignimbrite flareup and deformation in the southern Sierra Madre Occidental, western Mexico: implications for the late subduction history of the Farallon plate. Tectonics 21 (4), 17-1.

Lipman, P.W., 2007. Incremental assembly and prolonged consolidation of Cordilleran magma chambers: evidence from the Southern Rocky Mountain volcanic field. Geosphere 3 (1), 42. http://dx.doi.org/10.1130/GES00061.1.

de Silva, S.L., 1989. Altiplano-Puna volcanic complex of the central Andes. Geology 17 (12), 1102—1106.

de Silva, S.L., Gosnold, W.D., 2007. Episodic construction of batholiths: insights from the spatiotemporal development of an ignimbrite flareup. Journal of Volcanology and Geothermal Research 167 (1—4), 320—335. http://dx.doi.org/10.1016/j.jvolgeores.2007.07.015.

Chapter 16

Calderas

Michael Branney

Department of Geology, University of Leicester, Leicester, England, UK

Valerio Acocella

Dipartimento di Scienze, Università Roma Tre, Roma, Italy

Chapter Outline

GLOSSARY

caldera A wide topographic depression formed by subvertical subsidence into a partly drained magma reservoir, typically during a large volcanic eruption.

caldera fill Material deposited within a caldera during and after the caldera-forming eruption: typically thick ignimbrite, megabreccia, breccias, lake sediments, and lavas.

caldera floor The foundered precaldera strata beneath the caldera fill and above the magma reservoir.

caldera-forming eruption sequence (CFE-sequence) A characteristic vertical succession of deposits that records a caldera-forming eruption, its precursors, and aftermath.

caldera volcano A volcano with a prominent caldera.

cauldron A subsidence structure of width >1 km with a central collapsed mass surrounded by linked peripheral faults and/or ring dykes. Some cauldrons are exhumed deep levels of a former caldera but with no preserved caldera; others (e.g., bell-jar intrusions) may not have propagated to the original land surface.

downsag Subsidence involving inward (centripetal) tilting of strata. Commonly peripheral and associated with arcuate extensional faulting.

ignimbrite Deposit of a pumiceous pyroclastic density current. Typically largely composed of lapilli-tuff that contains ash, pumice lapilli and accidental rock fragments.

magma reservoir Hot melt plus crystals assembled beneath an active caldera volcano. It may be homogeneous body, a compositionally zoned body, or several interconnected lenses of crystal mush and variously segregated melt. Only parts of the reservoir may erupt when a caldera collapses.

megabreccia Deposit of rock fragments, each megablock too large to be seen at a field exposure without mapping. It forms by rapid mass wasting into a subsiding caldera.

mesobreccia Fragmental deposit of blocks up to 1 m in size, typically introduced to a caldera by rock-fall avalanches during or shortly after the subsidence.

outflow sheet The part of an ignimbrite that is deposited outside the caldera.

pyroclastic density current A ground-hugging current of hot ash and gases that spreads from a volcano because it is denser than the enclosing atmosphere.

resurgence Magmatic-induced uplift of part of a caldera floor and fill, commonly forming a central structural dome with an apical graben.

The Encyclopedia of Volcanoes. http://dx.doi.org/10.1016/B978-0-12-385938-9.00016-X

ring-fault system An annular, elliptical, or polygonal system of intersecting faults around a caldera or cauldron enclosing a central subsided area.

ring dyke Vertical to steeply outward-dipping circumferential magmatic or pyroclastic intrusion commonly along a caldera ring-fault system.

supereruption A large caldera-forming explosive eruption of magnitude 8 or larger, VEI 8 or larger, or with a volume exceeding 450 km³.

vent The top opening of a volcanic conduit from which material is ejected.

volcanic crater A bowl- or funnel-shaped depression around a vent, typically smaller than a caldera, formed by the explosive fragmentation and ejection of country rock.

volcanotectonic fault A fault that moved due to subsurface movement or eruption of magma.

1. INTRODUCTION

A **caldera volcano** is a volcano of any composition that includes a large area of ground that has subsided subvertically toward a partly drained **magma reservoir**. Caldera volcanoes range from broad, rhyolitic **ignimbrite** shields to steep andesitic–dacitic composite volcanoes to basaltic lava shield volcanoes. The subsidence is typically in response to the removal of several cubic kilometers of subvolcanic magma during a large explosive or effusive eruption, or due to subsurface magma flow along dykes or sills. A **caldera** is the wide topographic basin that results from this collapse and normally has inward-facing caldera walls or scarps surrounded by an elevated topographic rim (Figure 16.1).

Caldera eruptions vary widely in composition and volume. Some basaltic calderas form during moderate-volume (<1 km³) effusive eruptions that persist for days to months, whereas most silicic calderas are related to large-volume explosive eruptions (tens to thousands of cubic kilometers) lasting just hours to days. Fernandina (Galápagos; Figure 16.1) is an example of a basaltic caldera volcano, where a 5-km-wide **caldera floor** subsided 300 m during a 0.2-km³ lava eruption in June 1968. In contrast, the 10-km-wide caldera at Crater Lake (Oregon, USA), formed spectacularly during the

FIGURE 16.1 (A) The flooded 8 × 12 km explosive caldera of Santorini volcano in Greece, with steep arcuate caldera-wall scarps topped by the pyroclastic deposits (white layer) of the most recent (Minoan; 3.6 ka) caldera-forming eruption. *Photo: Mike Branney*. The island (left) is an emergent central postcollapse lava shield. (B) The basaltic Fernandina caldera, Galápagos, showing the caldera wall, rim (skyline), and terraces of subhorizontal lavas that partly filled the caldera prior to further caldera collapses. Peripheral crevasses (foreground) indicate radial extension around the caldera. *Photo: Dennis Geist*.

FIGURE 16.2   Crater Lake caldera, Oregon, collapsed during a 7.7-ka, ∼50-km^3 explosive eruption in the Cascades Arc, USA. Radial pyroclastic density currents deposited the compositionally zoned (rhyodacite to andesite) ignimbrite outflow sheet (right). The deep caldera lake is 8 × 10 km. Wizard Island is a postcollapse emergent andesite volcano. *Photos: Mike Branney.*

∼50 km^3 explosive eruption of a compositionally zoned rhyodacitic ignimbrite (Mazama tuff) ∼7700 years ago (Figure 16.2). Caldera diameters generally increase with the mass of the associated eruption, and may reach 80 km. In plan view, they range from polygonal (e.g., Ishizuchi, Japan; Glencoe, Scotland; Ossippee, New Hampshire), subcircular (e.g., Hwasan, Korea), to elongate (e.g., Toba, Sumatra) and embayed (e.g., Aso, Japan). They are occupied by a thick **caldera fill**, typically of ponded ignimbrite deposited during the **caldera-forming eruption** (CFE) (or lava in some basaltic calderas), commonly overlain by postcollapse lacustrine to subaerial volcaniclastic sediments, tephras, lavas, or domes (Figure 16.3). Buried beneath the caldera fill is the foundered caldera floor made of precollapse strata that may be variously tilted and faulted, intruded, hydrothermally altered, and locally contact-metamorphosed adjacent to a central subvolcanic magma reservoir or intrusion. Magma reservoirs beneath active calderas may have the form of tabular regions filled with crystal mushes and variously segregated melt lenses and interconnected sills. Their form will influence both the erupted products and the subsidence geometry.

Calderas are generally larger than **volcanic craters** and form by subsidence, whereas craters form by explosive excavation and ejection, and buildup of ejecta around a **vent**. Calderas may be symmetrical or asymmetrical, but they are distinct from open-sided sector-collapse scarps (horseshoe-shaped valleys) that form by outward-directed sector-collapse landslides down flanks of unstable volcanoes. That said, some calderas are open sided where one or more segments of the caldera rim have been removed by erosion or landsliding (e.g., at Santorini, Greece). Calderas on the terrestrial planets may resemble large impact craters, and are distinguished from them by the presence of an associated volcanic edifice, tephras, lavas, and domes, and by their imperfect circular shape.

Caldera volcanoes are typically polygenetic and undergo several minor eruptions from flanks, rift zones, or centrally, both before and after the main caldera-forming eruption. Some undergo a single major caldera-forming event (e.g., Crater Lake, Oregon, USA), but many undergo two or several, successive caldera-forming eruptions (e.g., Valles, New Mexico, USA; Santorini, Greece; Aso, Japan), sometimes over a protracted period (e.g., 2 Ma in the case of Las Cañadas, Tenerife). With each successive caldera collapse, such multiple calderas become increasingly complex, with the development of variously overlapping, nested, reactivated, and overprinted structures. The resulting complexity is evident at some exhumed caldera volcanoes, such as at Mull and Glencoe (Scotland), but is nearly impossible to resolve

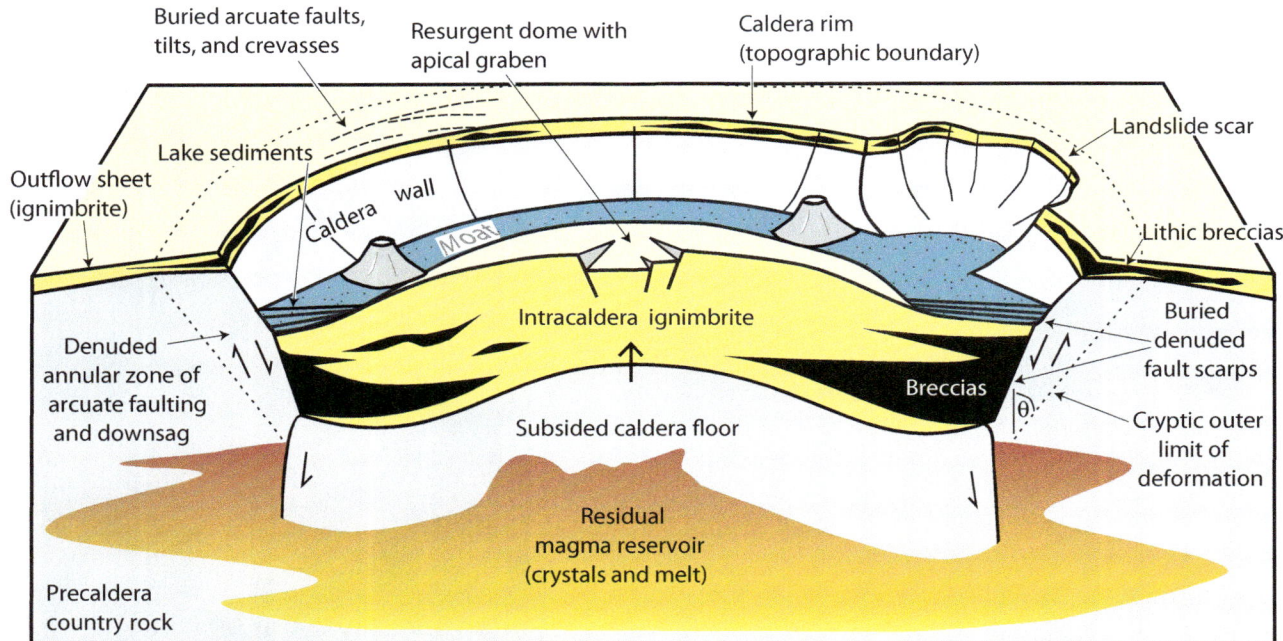

FIGURE 16.3 General features of a resurgent explosive caldera (not to scale and structural detail omitted), showing the caldera rim, a partly buried caldera wall with scalloped landslide scarps, a subsided caldera floor overlain by a fill of thick ignimbrite, landslide breccias, caldera-lake sediments, and angle of draw, θ. Note that the cryptic structural boundary lies outside the caldera rim. A thin outflow ignimbrite sheet contains proximal lithic breccias, and may partly conceal a conical-shaped zone of peripheral extension and slight inward tilting. The ring-fracture zone is largely denuded and buried. A central resurgent dome is flanked by postcollapse eruptive vents. *Partly adapted from Lipman (1984) and Cole et al. (2005).*

at recent volcanoes (despite gravity studies and drilling) where much of the key evidence is concealed.

Basaltic calderas, such as in the Galápagos and Hawaiian archipelagos, have some similarities to silicic explosive calderas. They may undergo repeated subsidence and infill events, and may be the site of hydrothermal activity. However, many basaltic calderas are occupied predominantly by ponded lavas, sometimes with stepped terraces (Figure 16.1) and multiple arcuate faults. They subside during explosive or effusive eruptions, or due to subsurface lateral flow of magma from a subvolcanic reservoir. Explosive basaltic eruptions can result when rising hot magma interacts violently with groundwater in basalt aquifers, generating widespread layers of fine-grained basaltic ash, such as seen on Hawai'i.

1.1. Significance of Calderas

Calderas are a fundamental manifestation of planetary differentiation and outgassing. They occur on all continents in a variety of geological settings, from within-plate hot spots to continental and oceanic rifts and volcanic arcs. Their presence informs us about the location and evolution of major tectonomagmatic events on the Earth, and their formation relates to the emplacement and evolution of long-lived crustal magma reservoirs, which

solidify to form sills and plutons. Understanding the magmatic and structural development of calderas may help us understand how magma is emplaced into the shallow crust, differentiated, and erupted. For example, caldera-forming eruptions provide us with chilled early taps from magma reservoirs that continued to evolve and crystallize into sills, plutons, granitic batholiths, and alkalic- and mafic-ring complexes. Eroded calderas inform our understanding of how subvolcanic faults and intrusions are formed, including mechanisms of magmatic inflation, stoping, and lateral injection.

Caldera-forming explosive eruptions are, along with large meteorite impacts, the most catastrophic events to affect the Earth's surface. They present widespread and complex hazards, with devastating impact upon the regional environment, topography, and drainage, and they are known to perturb global climate. For example, the $5-8 \text{ km}^3$ 1991 eruption that produced a 2.5-km-diameter caldera at Mt Pinatubo, Philippines, affected global climate for $2-3$ years. All explosive eruptions greater than about 5 km^3 probably produce a caldera, and this volume threshold may be lower where the magma reservoir is shallow. Where a caldera from an eruption greater than 5 km^3 has not been detected, it may simply be buried beneath volcanic products. **Supereruptions** (explosive eruptions of magnitude $8-9$, roughly equivalent to eruptions of $\geq 450 \text{ km}^3$) are invariably caldera forming. The largest volcanoes on the Earth (e.g., Hawai'i, Tenerife, Toba, La Garita, Yellowstone) all have calderas, as do many large volcanoes on the other terrestrial planets (e.g., Olympus Mons on Mars).

Caldera volcanoes commonly host long-lived hydrothermal systems and these can form valuable geothermal resources for low-carbon generation of electricity (e.g. Los Humeros, Mexico). Several host economic metal mineralization resulting from the circulation of fluids within large hydrothermal systems such as at calderas of the Archean Abitibi Greenstone Belt in Canada. Flooded calderas form excellent natural harbors (e.g., Aira, Kyushu, Japan; Santorini, Greece; Rabaul, New Guinea) and large caldera volcanoes across the world form outstanding scenic landscapes of natural beauty; many have been designated as nature reserves, national parks, and sites of tourism.

1.2. Calderas Research

Field observations by Lyell, Forque, and Verbeek in the nineteenth century led to the notion that calderas form by subsidence during large volcanic eruptions. In the early twentieth century, further evidence was derived from meticulous geological studies of exhumed calderas and associated ring complexes, such as the Silurian Glencoe caldera (Scotland), and the Paleogene central igneous complexes of northwest Scotland. The subsidence origin of many calderas was reviewed in 1941 in the well-known *Calderas and Their Origin* by Howel Williams.

The second half of the twentieth century saw detailed field and petrological studies that significantly advanced understanding of the extent to which caldera volcanoes vary, and how they variously subside during major eruptions. Seminal studies of the resurgent Valles caldera (New Mexico) by Smith and Bailey in 1968, and of the calderas of the San Juan Mountains (Colorado) by Peter Lipman were particularly influential. Accounts of modern caldera-forming eruptions focused mainly on interpreting the eruption products, for example, at Taupo (NZ), Santorini (Greece), Askja (Iceland), Kilauea (Hawai'i), Krakatau (Indonesia), Fernandina (Galápagos), Katmai (Alaska), and Pinatubo (Philippines). These were complemented by investigations at exhumed, older calderas, where deeper elements, including the caldera floor, faults, vents, and fills are accessible: this work revealed the complex structure of caldera margins, with **ring-fault systems** of multiple arcuate faults, peripheral graben, and inward-tilted strata. The studies analyzed the geometry, formation, and evolution of caldera faults, pseudotachylite, extensional crevasses, and fault-block tilts, central and ring vents, filled conduits, intracaldera landslides, avalanches, and lacustrine sedimentation. Detailed field studies in the western USA, southern Japan, and Korea, and in the Lake District

(England), Snowdonia (Wales), and Scottish Highlands (UK), have significantly improved the understanding. Petrological studies of homogeneous and zoned ignimbrites and their crystal assemblages are leading to improved understanding of the contrasting types and behaviors of subvolcanic magma reservoirs. Calderas research has extended to remote investigations of calderas on Mars, Venus, and Io.

Analogue modeling has clarified some key parameters that control the shape and structural development of calderas. Recent numerical models have explored the dynamics of partially erupting magma reservoirs and conduits. Modeling enables specific parameters to be varied and tested individually in a way that can rarely be isolated at natural examples where, for example, strain rates and fracture propagation can be difficult to quantify. Models require robust validation, and further geological studies involving detailed structural mapping of well-exposed exhumed caldera volcanoes are needed.

1.3. How to Identify a Caldera

Many calderas have yet to be discovered. They can be identified by the juxtaposition of a topographic depression with proximal volcanic products and peripheral extensive **outflow sheets** or fans. Elongate and polygonal calderas may be underrepresented in the literature simply because circular calderas are easier to spot. Reconnaissance using remote imaging is best followed by the field identification of thick, ponded caldera-fill deposits in a down-dropped area surrounded by a structural margin of the same age. Recognizing a "CFE-sequence" within outflow sheets (see below) will indicate that a caldera exists and its location may then be inferred using isopleth maps of the associated fallout layers, and palaeoflow indicators in the outflow ignimbrite(s). Exhumed calderas are commonly identified by the juxtaposition of hypabyssal intrusions and very thick ignimbrite with dramatic thickness changes across **volcanotectonic faults**.

Characteristic features of silicic calderas are described below. Basaltic calderas are best identified by their topography and particularly the subsidence margins.

2. THE ANATOMY OF A CALDERA VOLCANO

The morphology and internal architecture of a caldera varies with the depth, dimensions, and morphology of the subvolcanic magma reservoir; the state and extent of crystallization of the reservoir; the amount of subsidence; the strain rate; the strength of the rocks; the pattern of preexisting fractures; and the stress field. The latter may be influenced by both regional tectonics and local factors, such as intrusion and gravitational spreading of the volcanic edifice. The following description divides a caldera volcano into three contrasting regions: (1) the "extracaldera" region around the caldera, (2) the interior, or "intracaldera" region, and (3) the more complex caldera margin, which is the site of the most intense deformation and separates the other two regions.

2.1. Around the Caldera

During an explosive caldera-forming eruption, the region surrounding the caldera is covered by an extensive ejecta blanket. In vertical section, this pyroclastic layer exhibits a distinctive *CFE-sequence* (Figure 16.4) that records the onset, climax, and aftermath of the caldera-forming eruption. The CFE-sequence outside the caldera is thinner, more stratigraphically condensed, and less complete than the intracaldera CFE-sequence (Figure 16.3); however, it is often more accessible and provides useful information about the onset and development of the CFE (Figure 16.4). Proximally, for example, thin ashfall layers may record minor precursory explosive phreatic or phreatomagmatic eruptions. A Plinian pumice fall layer may record the opening of the main eruption, with inverse-grading

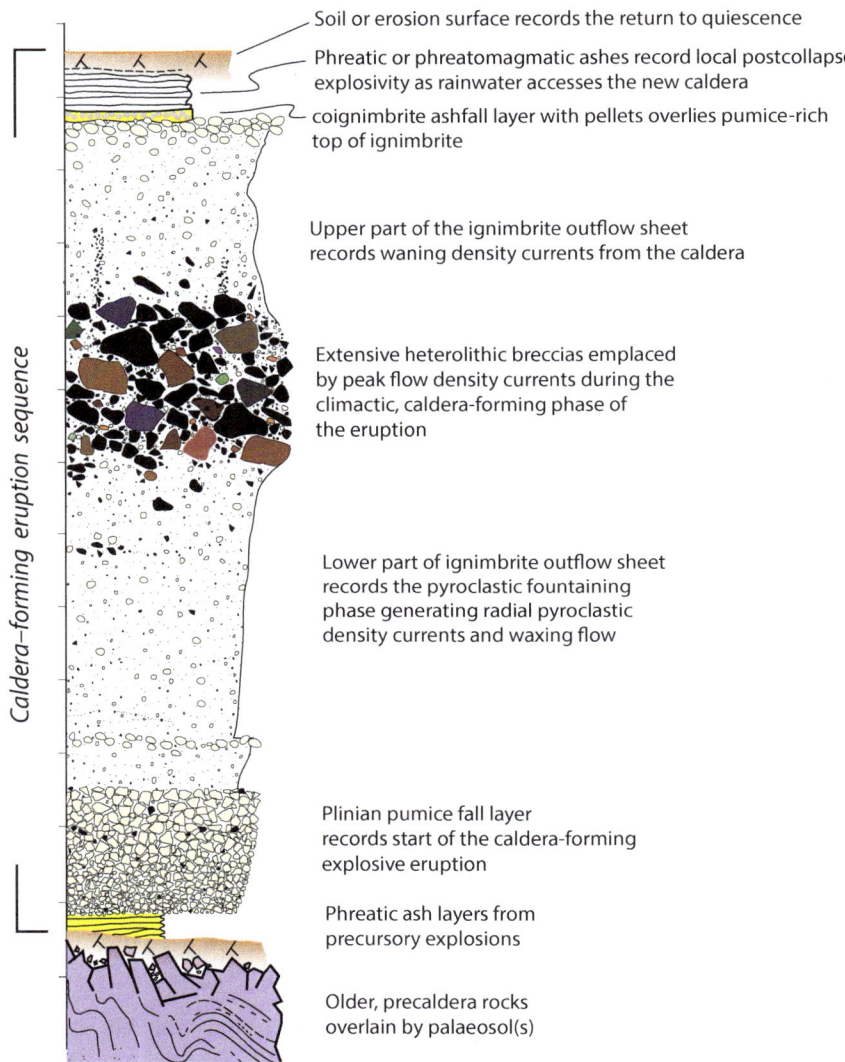

Caldera–forming eruption sequence

Soil or erosion surface records the return to quiescence

Phreatic or phreatomagmatic ashes record local postcollapse explosivity as rainwater accesses the new caldera

coignimbrite ashfall layer with pellets overlies pumice-rich top of ignimbrite

Upper part of the ignimbrite outflow sheet records waning density currents from the caldera

Extensive heterolithic breccias emplaced by peak flow density currents during the climactic, caldera-forming phase of the eruption

Lower part of ignimbrite outflow sheet records the pyroclastic fountaining phase generating radial pyroclastic density currents and waxing flow

Plinian pumice fall layer records start of the caldera-forming explosive eruption

Phreatic ash layers from precursory explosions

Older, precaldera rocks overlain by palaeosol(s)

FIGURE 16.4 A common "caldera-forming eruption (CFE) sequence" from outside an explosive caldera (not to scale). A paleosol on precaldera rocks is overlain by ashfall deposits from precursory phreatic eruptions. A pumice fall layer (not always present) may record a Plinian early phase of the CFE. The ignimbrite outflow sheet records catastrophic radial density currents and may be compositionally zoned. It coarsens upward to proximal heterolithic breccias that record the climactic, caldera-forming phase of the eruption. Ignimbrite above this records waning stages of the eruption and is overlain by an ashfall deposit from the residual dilute atmospheric plume. Laminated ash layers above this record, ash postcollapse phreatomagmatic or phreatic explosivity. The top of the CFE-sequence is marked by a paleosol that records the return to eruptive repose. CFE-sequences differ widely from caldera to caldera, and the top may be removed by erosion. A much thicker CFE-sequence, topped by caldera lake deposits, occurs within the caldera.

recording how the eruption column increased in height as the mass flux of the eruption waxed prior to phase of the pyroclastic fountaining (column collapse). An overlying ignimbrite outflow sheet, normally ≤100-m thick, (Figure 16.2B) contains pumice lapilli and lithic fragments supported in a poorly sorted ash or tuff matrix, and records the radial emplacement of ground-hugging **pyroclastic density currents** during the eruption climax. The ignimbrite outflow sheet may be monotonous and crystal rich, or crystal poor with some form of chemical or mineralogical zoning. Zoned ignimbrites record the progressive evacuation of a heterogeneous magma reservoir, which may be density stratified and/or composed of partly connected lenses of melt and crystal mush, into which new magma may have intruded.

Lithic fragments in the outflow ignimbrite typically increase in size and abundance with height to a peak, where extensive layers and lenses of lithic breccia (Figures 16.4 and 16.5) record the climactic phase of the caldera-forming eruption—a time when abundant blocks were entrained into the pyroclastic currents as the caldera subsided with attendant conduit flaring, erosion, faulting, and rock avalanching. Lithic breccias associated with caldera collapse thicken proximally and into valleys, and may locally form low-angle antidune-like bedforms, as at Santorini (Greece). Their bases commonly show scouring and loading into the subjacent ignimbrite. Blocks show imbrication, and curviplanar fracturing around their margins caused by thermal spalling (e.g., at Cañadas volcano, Tenerife) and they commonly include cognate intrusive lithologies, and variously fresh or hydrothermally altered volcano-basement lithologies accidentally incorporated.

At calderas that were flooded during the eruption, the lithic breccias are commonly associated with thick scoria agglomerates: clast-supported accumulations of imbricated, rather dense juvenile magma rags and bombs with fluidal, chilled, or bread-crusted surfaces. These occur at Santorini, Taal, and Scafell calderas. Fines-poor elutriation pipes ramify upward through the coarse-grained ignimbrite facies and into the overlying ignimbrite, where an upward decrease in the size of lithic fragments, coupled with concentrations of abraded pumice blocks toward the top of the ignimbrite records the waning stages of the pyroclastic density current after the caldera has collapsed. Some outflow ignimbrite sheets exhibit columnar cooling joints and some are welded (e.g., eutaxitic) as a result of hot emplacement, but less so than the corresponding caldera fill. They may extend beyond 75 km from the caldera, and commonly thicken into basins.

Thin, pellet-bearing coignimbrite ashfall layers on top of the outflow ignimbrite (Figure 16.4) derive from lofted dilute phoenix plumes that persist in the atmosphere after the catastrophic ground-hugging currents have dissipated. Upper parts of a CFE-sequence may include thin, phreato-magmatic ashes that record postcaldera, rootless phreatic explosions within the ignimbrite sheet, or derive from minor explosions in the caldera as water began to accumulate there (e.g., Pinatubo, 1991, Philippines). The top of the CFE-sequence may record wind ablation, erosional scour, and fluvial incision. Distal extremities may be overlain by volcaniclastic alluvial aprons, derived from erosion and sedimentary reworking in the aftermath of the eruption.

Deformation of extracaldera rocks is far less marked than within the caldera. However, proximal pre- and syncaldera strata may undergo peripheral sagging and extension (Figure 16.3). Concentric sets of peripheral extensional arcuate faults, for example, are commonly seen on upper flanks of many planetary caldera volcanoes. Precaldera strata around some calderas have developed arcuate peripheral folds (e.g., Mull, Scotland) and swarms of cone sheets (e.g., Tejeda, Gran Canaria), probably due to magmatic pressure from the central subvolcanic reservoir. The intensity of intrusion and hydrothermal alteration falls off rapidly with distance from the caldera.

FIGURE 16.5 Proximal lithic breccias in outflow ignimbrites record the climactic, caldera-forming phase of a large explosive eruption. (A) Near the top of the 1991 ignimbrite of Mt Pinatubo, Philippines (person for scale). (B) Heterolithic breccia in the extensive "Aso 4" ignimbrite of Kyushu, Japan. *Photos: Mike Branney.*

2.2. Within the Caldera

A thicker ("intracaldera") CFE-sequence occurs within the caldera (Figure 16.3). It comprises the foundered caldera-floor strata buried by a thick caldera fill of massive **ignimbrite, mesobreccias, and megabreccia**, and overlying caldera-lake sediments.

2.2.1. Caldera-Fill Ignimbrite and Breccias

Intracaldera ignimbrites are thicker, more intensely welded, contain larger blocks, and show more hydrothermal alteration than the equivalent outflow ignimbrite. Thicknesses range from a few hundred meters to over 2 km. The thickest ones, such as in the San Juan Mountains (Colorado, USA), accumulated rapidly during individual supereruptions and represent the thickest single-event deposits on the Earth. As with outflow sheets, they may be monotonous and crystal rich, crystal poor, or variously zoned. Intense welding is common and promoted by rapid accumulation close to source. However, polygonal columnar cooling joints are less common than in the thinner outflow sheet. Some caldera-fill ignimbrites are so intensely welded that their clastic nature is obscured; these can resemble dense silicic lava, a resemblance sometimes enhanced by the development of hot-state, rheomorphic flow folding and flow brecciation. Angular, framework-supported, and locally jigsaw-fit blocks of welded tuff may be partly annealed back together (e.g., at Scafell caldera, England, and Bachelor caldera, Colorado). Such deformation of the caldera-fill tuff may be widespread or localized such as near a caldera-floor fault.

Lenses of **mesobreccia** are abundant in most caldera fills and record frequent rock-fall avalanches from steep and growing caldera wall scarps as the caldera subsides. Megablocks of country rock, hundreds of meters to a couple of kilometers in size and sometimes arcuate in shape, reside within many calderas, such as in the Late Cretaceous Tucson Mountains, Arizona, USA. These record segments of unstable caldera wall material and fault blocks that detached and moved into the caldera as it rapidly subsided and filled. Some megablocks are internally brecciated and so large that they can be confused with parts of the caldera floor.

2.2.2. Caldera Lakes and Sediments

Volcaniclastic sediments mass wasted or washed in from the caldera walls commonly overlie the caldera-fill ignimbrite. Some calderas (Campi Flegrei, Italy; Snowdon, Wales; Santorini, Greece) are partly flooded with seawater (Figure 16.1). Others have caldera lakes (Figures 16.2 and 16.6), which, because they have no exit, gradually deepen to hundreds of meters and accumulate sedimentary

FIGURE 16.6 The dacitic 1991 caldera of Mt Pinatubo, Philippines, 3 years after its formation, showing talus cones and prograding alluvial fans below sheer caldera walls surrounding a deepening, acidic caldera lake with dacite dome (center) and hydrothermal plume (lower center).

successions that in some cases exceed 500 m in thickness (e.g., Scafell, England). The volcaniclastic sediments include a large component reworked from the recently erupted pyroclastic deposits. An overall fining-up sedimentary sequence is typical and reflects the gradual stabilization of the initially steep and tephra-covered caldera walls. Early rock-fall avalanches and pumiceous lahars form alluvial cones and fans, and the succession becomes lacustrine as standing water accumulates (Figure 16.6). Initial lake waters are highly acidic and host fumaroles, hot springs, and silicic lava domes, which may become emergent and decrepitated, shedding aprons of silicic lava and perlite. Pumiceous parts of domes may float off and dock at lake margins as giant pumice blocks (<10 m in size), which eventually lodge and become buried by laminated silts (e.g., at La Primavera, Mexico). With time, caldera lakes become less acidic as the fumarolic activity wanes. Sedimentation in deeper, central parts of the lake is below wave base and dominated by turbidity currents and suspension sedimentation to form massive and laminated, commonly diatomaceous silts and muds. Marginal areas are dominated by coarser, prograding emergent fans of sediment and littoral pumiceous sands and gravels (e.g., at Lake Taupo, NZ). Further explosive eruptions may give rise to catastrophic incursions of coarse pyroclastic material into the caldera, and spectacular soft-state deformation results from the rapid loading of water-saturated sediment together with caldera-related tilting and seismicity. Resurgent doming of the central caldera fill may produce an island with a surrounding lacustrine moat (Figure 16.3) such as at Creede and Valles (USA). Elsewhere, postcollapse volcanoes grow and emerge from the caldera lake forming islands (Figures 16.1 and 16.2); these may be tuff rings, scoria cones, lava fields, or composite volcanoes, and they develop their own fringing volcaniclastic aprons.

2.2.3. Caldera Floor: Deformation, Intrusions, and Hydrothermal Activity

The floor of the caldera beneath the fill is commonly cut by hypabyssal intrusions and faults during and after the collapse event. Where caldera subsidence was uniform, large sectors of the floor may be relatively coherent and platelike. Structurally coherent, exhumed caldera floors are exposed at several sites in the western USA such as at the Stillwater Range, Questa, and Tucson Mountains. In contrast, the 16-km-diameter caldera floor of Scafell caldera (England) fractured during the collapse into numerous horsts and graben. What controls how much the caldera floor breaks up as it subsides is likely to be influenced by the morphology and depth of the magma reservoir, which in some systems may comprise several connected lenses, and the rigidity (plus any preexisting fractures) of the subsiding floor. Where subsidence was

greater in one sector than in another, the caldera floor tilts like a trapdoor. The side with the deepest subsidence may correspond with where the magma reservoir was shallowest, thickest, or more eruptible. Caldera floors can develop inward, centripetal dips (**downsag**). Arcuate zones of compression inboard of such inward-tilted strata occur in nature (Miyakejima, Japan) and in analogue models (Figure 16.7). However, because the caldera fill accumulates during the subsidence, any deformation structures (faults, tilting, compressional, or extensional structures) in the caldera floor that develop during the subsidence have a growth geometry within the fill. This means that whereas lower levels of the caldera-fill ignimbrite are significantly disrupted and tilted, higher levels of the fill increasingly bury and conceal the deformation. However, caldera faults, once formed, are prone to reactivation (unrest) during subsequent magmatic activity (e.g., Glencoe, Scotland) and may go on to influence caldera-lake sedimentation and its architecture, e.g., at Scafell (England).

Below the caldera floor lies the magma reservoir. Large bodies of liquid magma are not indicated beneath modern calderas by geophysical studies, and may be transient. The floor of a caldera and, to a lesser extent, the fill are typically cut by dykes, sills, intrusive domes, and bosses, particularly in calderas that have undergone a protracted eruption history. Where exhumed, their crosscutting relations can be deciphered given sufficient exposure and dedication such as in centres of the North Atlantic Igneous Province. A large postcollapse hydrothermal system is characteristic, with fumaroles, geysers, and intense alteration of the caldera fill and underlying lavas. It may persist long after caldera subsidence and typically causes alkali and silica mobility, and development of alteration assemblages such as sericite, epidote, and carbonate.

2.3. Caldera Margins

The caldera margin is the most complex part of a caldera (Figures 16.3, 16.6 and 16.7). It is the site of intense deformation, intrusion, and topographic instability, susceptible to landslides and erosion. It is the site of eruption vents, proximal deposits, and intense hydrothermal alteration. An abrupt increase in thickness of the ignimbrite occurs from the external thin outflow to the thick, ponded caldera fill.

The *topographic* margin of a caldera is a circumferential caldera rim, just outboard of an inward-facing caldera wall, which is commonly an arcuate fault scarp (Figure 16.3). In contrast, the *structural* boundary of a caldera is the outermost limit of deformation (Figures 16.3 and 16.7); this differs from the topographic boundary because the first-formed prominent caldera wall scarps are unstable and retreat outward by as much as several kilometers as the result of erosion, avalanches, and landslides

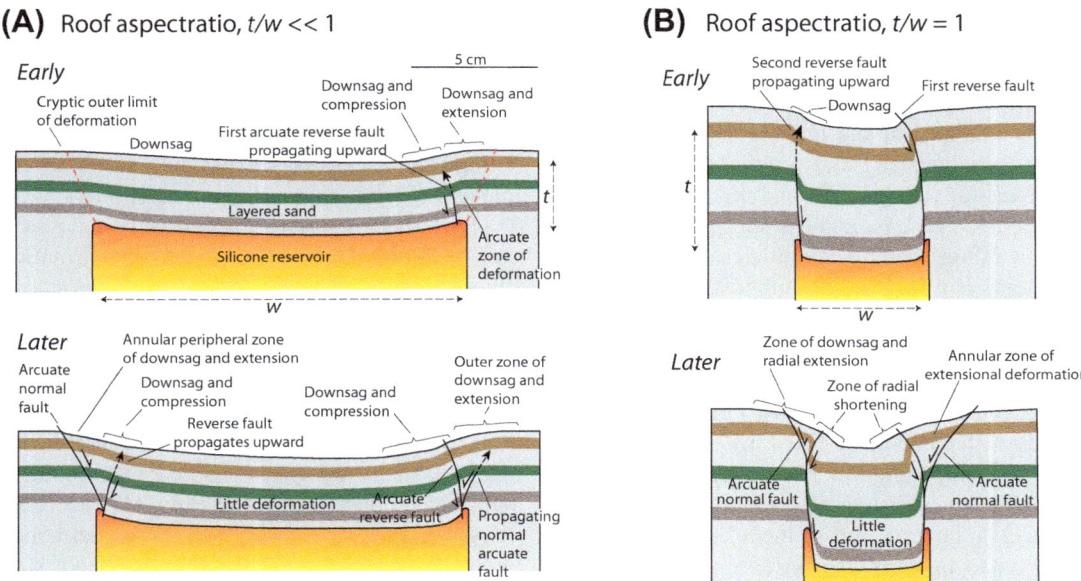

(A) Roof aspectratio, $t/w \ll 1$

(B) Roof aspectratio, $t/w = 1$

FIGURE 16.7 Progressive structural development of calderas as inferred from two- and three-dimensional analogue models. In both (A) wide and (B) narrow calderas, initial downsag is followed by propagation of vertical to outward-dipping reverse faults that gradually link around the caldera. Slightly later, normal arcuate faults and tilting form outboard of this, with formation of an annular zone of intense deformation, which in real calderas involves multiple arcuate normal faulting, crevasses dilation, brecciation, rapid denudation, and burial by caldera fill. In sandbox models, subsidence is caused by withdrawal from a silicone reservoir. *Adapted from Roche et al. (2000).*

that may leave embayment-shaped scarps as at Creede and La Garita (USA) and at Aso (Japan). The distance of retreat of the wall varies with height of the wall, its erodibility, and time elapsed since the subsidence. But the structural margin of the caldera was probably not at a prominent ring fault even prior to the erosion and landsliding. This is because deformation, in the form of extension and slight inward tilting (downsag), commonly also occurs outside the caldera walls (Figure 16.3). The outer boundary, or limit of deformation during subsidence in geological materials, as revealed in analogue models and mining subsidence, typically projects upward and outward from the outer margin of the magma reservoir to the land surface at the "angle of draw" ($\sim 35°$ from vertical). This limit of deformation is cryptic (unfaulted). Thus, arcuate extensional faults and crevasses occur outboard of the topographic caldera rim, typically accompanied by slight inward rotation (downsag) of flanking strata. These structural features are reported at several silicic calderas such as at Bolsena (Italy) and Chegem (Russia), and around basaltic shield volcanoes such as Fernandina (Figure 16.1B), Kilauea (Hawai'i), and the 110-km-wide caldera of Arsia Mons (Mars). However, they are readily masked by surficial burial and reworking on depositional slopes. Their development depends upon the tensile and flexural strength of the outlying, precaldera strata.

The structure of a caldera margin is influenced by the thickness and rigidity of the subsiding block and the depth of subsidence relative to the caldera diameter (Figure 16.7).

Small amounts of subsidence and subsidence of less-rigid strata may involve significant inward tilt or peripheral downsag. The width of the zone affected by tilting may be several kilometers wide and extends from inside the caldera to well outside the rim, although some of this is only of a few degrees. Given the high strain rates, the hinge zones of inward tilting are brittle and accompanied by arcuate extensional fracturing parallel to the caldera margin, such as the north side of Campi Flegrei and the south part of Bolsena caldera in Italy. Steeper tilts within the caldera are associated with the dilation of spectacular arcuate crevasses around the caldera, some 50- to 100-m wide and >300-m deep, filled with ignimbrite and breccia. Because the tilting accompanies infill of the caldera, the dips tend to decrease with height through the succession. Thus, little-dissected calderas typically show inward tilts of only a few degrees, whereas exhumed calderas reveal inward dips steeper than 50° in precaldera strata, as recorded at Glencoe (Scotland), Scafell (England), and Kumseongsan (Korea). Dips of strata around some **cauldrons** may approach vertical, and account for more than half the total subsidence in some cases, such as at Ossippee Mountains (New Hampshire). Such steep tilting has yet to be explored in subsidence models.

Calderas tend to collapse along a linked system of interconnected arcuate faults. With increasing subsidence, the deformation may increasingly focus at a single fault. The fault systems of young calderas are seldom observed because they are eroded and buried. A rare example is at

Miyakejima in Japan, where an outward-dipping fault formed during subsidence in 2000, and the later movement was accompanied by, and then followed by, inward-dipping peripheral, concentric faulting outboard of it. This is consistent with analogue models of caldera collapse in which steep outward-dipping arcuate faults propagate upward near the caldera margins in response to the sudden stress reorientation caused by the decreased support from the magma reservoir (Figure 16.7). With continued subsidence these connect to form an annular ring-fault system, leaving a crustal wedge in the hanging wall unsupported so it collapses along a new set of concentric, arcuate normal faults outboard of the initial outward-dipping fault(s). This pattern of subsidence was first described in other natural subsidence structures (e.g., due to melting of buried ice) and mining subsidence, and is evidently quite scale independent. Strata between the reverse and normal arcuate fault-sets are commonly tilted inward and cut by minor faults, pervasive extensional fractures, and dilated crevasses, as observed at Miyakejima in 2000. Similar, larger sets of intersecting arcuate faults are well exposed around the margin of the 16-km-diameter Scafell caldera (England), with variously tilted arcuate graben, and spectacular tuff-filled crevasses, some exceeding a depth of 300 m.

The presence of reverse arcuate faulting is indicated by seismic clustering along outward-dipping zones below the rims of several calderas, as recorded at Rabaul (New Guinea) in the 1980s, and during the 1991 eruption of Pinatubo (Philippines). The normal faults outboard of the reverse fault are more commonly observed in the field because they tend not to be so deeply buried.

The depth of subsidence, for a given caldera diameter, is a factor in shaping the structural margin. Where one side of a caldera subsides more than the other (e.g., trapdoor-like subsidence), that margin may develop a pair of concentric arcuate faults, whereas the opposite, shallower side may be dominated by tilting (downsag). At Bolsena caldera (Italy), an outer zone subject to normal faulting is narrower where subsidence is greatest, compared to the broader marginal zone of inward tilting developed at the side where subsidence was less.

Regional faults commonly affect the structure of a caldera. Suitably oriented fault segments can be reactivated to form a polygonal caldera shape, as recently inferred at Deception Island (Antarctica). At Glencoe (Scotland), successive caldera subsidence events during regional transtension had the form of straight-sided graben. Elongate calderas along the East Pacific Rise and the collapsed rifts in Central Afar (Ethiopia) are influenced by tectonic extension. They formed in rapidly extending crust, and some of their bounding structures are reactivated normal tectonic faults. In the Iceland rift zone, elongate graben subsidence associated with basaltic fissures has been interpreted to be linear calderas in a strongly extensional setting.

Seismic and ground tilt data from calderas, and analogue models suggest that the rate of subsidence during caldera formation may be irregular and discontinuous. Once a reverse ring-fault system has propagated to the surface, the collapse becomes discontinuous and incremental as a consequence of friction along the faults. The outward-dipping reverse faults have a dilatational component during subsidence, and this may facilitate the rise of pyroclastic material or magma to form ring dikes, which may act as eruption conduits. Recent eruptions at Rabaul were from the ring-fracture zone.

2.4. Caldera Vents and Conduits

Caldera-forming eruptions may start at a point source, either at a central location or at some point within the marginal fracture system of a forming caldera. Stress patterns change as the caldera subsides and as successive parts of the magma reservoir are depressurized, and new fractures propagate and dilate, causing vents to shift and multiply. At Santorini (Greece) and Long Valley (USA), initial Plinian explosivity from a central location shifted to caldera margins during the climactic collapse phase, as indicated by changing lithologies of lithic blocks within the ignimbrites (e.g., Druitt and Sparks, 1984).

Vents and conduits are exhumed at some older calderas. Subcylindrical pyroclastic eruption conduits are exposed at Weolseong and Kumseongsan (Korea). They cut the caldera floor, and are filled with intensely welded, vertically lineated rheomorphic ignimbrite. In contrast, arcuate dyke-like eruption conduits, ≤100-m wide and filled with steep eutaxitic ignimbrite, have been described crosscutting the exhumed caldera floor at Sabaloka (Sudan), and spectacular ignimbrite-filled eruption conduits ≤400-m wide encircle some exhumed calderas, such as Hwasan (Korea) and Ishizuchi (Japan). At Hwasan, compositional changes in the pyroclastic **ring dyke** match the compositional zoning of the caldera fill and reveal that an arcuate fissure conduit unzipped progressively during the caldera subsidence to form a complete ring. Elsewhere, such as at Long Valley (USA) rapid unzipping of eruptive fissures has been inferred from petrological variations within the ignimbrites. Exhumed ring-shaped pyroclastic eruption conduits are also seen at Loch Bá (Isle of Mull), Slieve Gullion (Ireland), and Sultepec-Goleta (Mexico).

3. ACTIVITY AT CALDERA VOLCANOES

3.1. Precaldera Events

Long before a silicic caldera and its magma reservoir forms, local centers typically produce small- to medium-scale eruptions of mafic and intermediate compositions, forming scattered lava fields and cones.

3.2. Precursory Events

A hot silicic magma reservoir accumulating within the upper crust prior to a caldera-forming eruption may uplift of the land surface by tens to hundreds of meters. Such uplift is known as precaldera tumescence, and with elastic deformation it causes structural doming, although faulting is usually involved. The area affected depends upon the size and architecture of the magma reservoir, and may be broader than that of the future caldera. Precaldera tumescence is magmatic in origin and distinct from regional uplift, for example, of a volcanic arc, or due to a mantle plume. How common it occurs is not well known, as it can be difficult to discern from the geological record.

A good example of tumescence is at Solitario (Texas, USA), where the formation of a caldera at the site of structural doming was linked to the intrusion of an asymmetric laccolith, ~1.5-km deep. At Grizzly Peak (Colorado, USA), there is evidence for a topographic high and associated faulting prior to caldera collapse. Some basaltic caldera volcanoes, such as at Isabela and Fernandina (Galápagos), undergo precollapse tumescence associated with radial fracturing. In northeast Honshu (Japan), Miocene uplift along the Backbone Range was associated with the formation of several calderas. The collapse of Kakeya caldera, in southwest Japan, was preceded by more than 350 m of uplift on faults. In central Italy, several hundreds of meters of uplift of marine deposits in the Quaternary volcanic belt may be tumescence related to subvolcanic magma beneath Bolsena, Latera, Vico, and Bracciano calderas, although this may include a regional component.

The geometry of tumescence relates to the morphology and location of the developing subvolcanic magma reservoirs, and to the mechanical and thermal effects of intrusion and inflation at shallow levels in the crust. For example, a deeper magma reservoir may cause less uplift but affect a wider region. Magma ascent by stoping may cause little tumescence, whereas shallow, sill-like magma reservoirs may readily raise the overlying crust as they inflate. Swarms of radial fractures, radial dykes, or cone sheets, related to magmatic uplift may form prior to, or following, caldera eruptions. Spectacular examples occur on Gran Canaria and Tenerife (Canary Islands), and Mull and Ardnamurchan (Scotland).

Other events likely to precede a caldera-forming eruption include changes in the flux and species of gases released as magma rises to shallow levels, increases in hydrothermal activity, phreatic eruptions, and intense subvolcanic seismicity such as that occurred prior to the 1991 eruption of Pinatubo (Philippines). However, precursory activity to a caldera-forming supereruption has not been witnessed, and the geological evidence commonly lies concealed beneath caldera fills.

Once a shallow, evolved subvolcanic magma reservoir has been established, a perturbation may trigger a caldera-forming eruption. The evolution of the reservoir may take several tens of thousands of years, but the triggering occurs in a much shorter time frame (e.g., at Toba, Sumatra). The trigger may be an injection of new magma, a gravitational instability, a melt segregation, or an external structural event. Earthquakes and faulting may enhance vesiculation, magma mingling, and dike propagation. Injection of hot mafic magma from depth may heat, vesiculate, and mobilize accumulated magma. Broad sill-like magma bodies may be substantially wider than calderas formed above, and part of such a magmatic body may be an uneruptible, locked crystal mush. How much of this becomes eruptible may depend upon the time-averaged magma flux and associated heat flux: systems with substantial and relatively frequent input may remain largely eruptible across their width, whereas lower inputs may engender pluton formation, sometimes with no eruption.

3.3. Caldera-Forming Events

A caldera-forming eruption may begin with sustained Plinian explosivity producing a tall convective eruption column. As the vent widens, this changes to sustained pyroclastic fountaining with the catastrophic emplacement of radial, more-or-less sustained pyroclastic density currents that deposit ignimbrite widely across surrounding slopes. After a few cubic kilometers of magma is erupted, the caldera starts to subside: initially sagging, then with arcuate fault propagation and proliferation. Steep and overhanging fault scarps grow suddenly, spall, and crumble, generating rock avalanches that may increase as the caldera deepens. Secondary, normal faults develop in the hanging wall; vents widen, migrate, and lengthen into arcuate fissures that unzip; and as the eruption waxes to a climax the runout distance of the density currents increases. Successive melt lenses and/or mush zones beneath the volcano decompress, exsolve, and erupt. Around the caldera, peripheral arcuate fault-blocks tilt and founder, crevasses open to hundreds of meters wide and rapidly fill with ash and breccia. At the peak mass flux, the erupting dispersion and proximal currents entrain myriad rock fragments, and carry them tens of kilometers to form heterolithic breccias within the aggrading outflow sheet (Figure 16.5). As the subsiding caldera fills with hot ignimbrite, arcuate megablocks calve into it, are shed from unstable scarps in the fracturing peripheral zone, and are rapidly buried by ignimbrite (Figure 16.3).

A caldera-forming eruption develops its own weather system. Cold air is sucked inward and entrained into upper parts of the hot density currents, where it expands and lofts to form a vast coignimbrite or phoenix cloud that, together with the Plinian plume, disperses fine vitric ash and aerosols around the Earth, affecting global climate, as in the 1991 eruption of Pinatubo (Philippines). After several hours or days, the caldera-forming eruption wanes and the distance that the density currents reach from the volcano decreases, leaving retreating strandlines of pumice cobbles across the top of the ignimbrite sheet. When the eruption ceases, the landscape has profoundly changed: hills are deforested and ash covered, and valleys and lower hills are buried under hot ignimbrite. At the site of the new caldera, the landscape is unrecognizable, with vigorous fumaroles and minor explosivity. The depth of the extant caldera basin at this time depends on the extent that it was partially filled or completely buried by ignimbrite. This may in part reflect the timing of the subsidence: caldera collapse early in the eruption may result in more complete filling, whereas subsidence later in the eruption may leave a less-filled depression.

3.4. Postcollapse Sedimentation, Eruptions, and Resurgence

A newly formed caldera is one of the most dynamic types of sedimentary basin, with steep and fractured, sometimes hydrothermally altered, unvegetated scarps several hundred meters high, and with abundant unconsolidated ash and debris (Figure 16.6). Landslides, avalanches, rock falls, talus cones and continuing small explosive eruptions are characteristic of early stages. Thick ignimbrite can remain hot for several years, and rootless phreatic explosions are common as the caldera starts to fill with water. Volcaniclastic sediment fans prograde across the deepening caldera lake, and further eruptions produce lavas, domes, volcanic cones, and tuff rings, which emerge to form islands such as Volcano Island in Taal caldera lake (Philippines) and Wizard Island within Crater Lake, Oregon (USA; Figure 16.2A). Rising caldera-lake levels may eventually breach the caldera rim, with spectacular lake-water outbursts. The floods and lahars generated can pose a major post-eruption hazard to downstream regions unless remedial engineering (e.g., tunneling) is undertaken promptly. Natural breaches through the caldera rim may develop into deep permanent drainages, and lead to the erosional incision of the caldera-fill sediments (e.g., Aso, Japan).

Some larger calderas (>15-km diameter) undergo **resurgence**, which is magma-induced uplift of part of a caldera after collapse (Figure 16.3). Resurgence is rare at basaltic calderas, possibly in part reflecting the more fluidal magma which can degas and escape via fractures more readily. Even so, uplift of several tens of meters has occurred at the basaltic Sierra Negra caldera (Galápagos). Resurgence is generally recognized by unusual elevations and tilts of the caldera-fill ignimbrite or caldera-lake sediments.

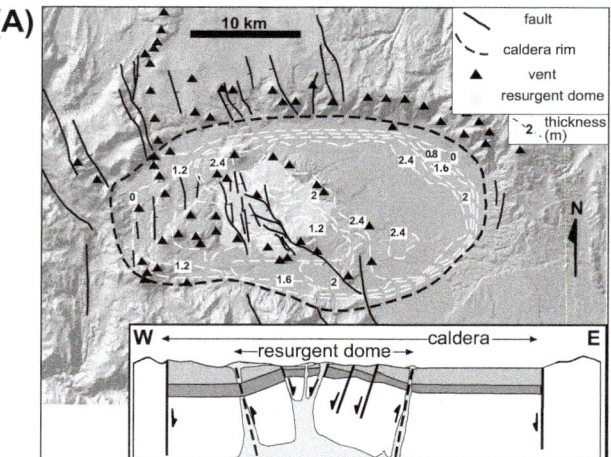

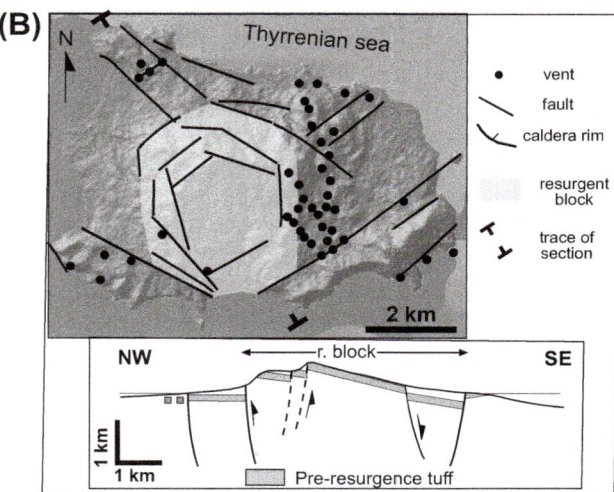

FIGURE 16.8 Examples of resurgent structures. (A) Map and cross-section through Long Valley caldera in the USA, showing the thickness of the caldera-fill ignimbrite (white isopachs) and the resurgent dome. (B) Map and cross-section of the asymmetrical resurgent block in Ischia caldera, Italy.

The uplifted area is commonly subcircular, with a diameter between 5 and 50 km; some are polygonal or elongate (Figure 16.8). Uplift varies from a few tens to over a thousand meters, and can result in radial erosional incision. The time between collapse and the onset of resurgence varies from 10 to 100,000 years. Resurgence is commonly intermittent or episodic, and may continue for up to a few tens of thousands of years with rates from a few centimeters per year to a few decimeters per year—an order of magnitude faster than tectonic uplift. The record of alternating resurgence and subsidence of the flooded Snowdon caldera (Wales) relative to constant sea level is well documented.

Resurgent domes occur at Valles, Lake City, and Long Valley calderas (USA). They typically have apical graben elongated parallel to dominant regional structures or at a

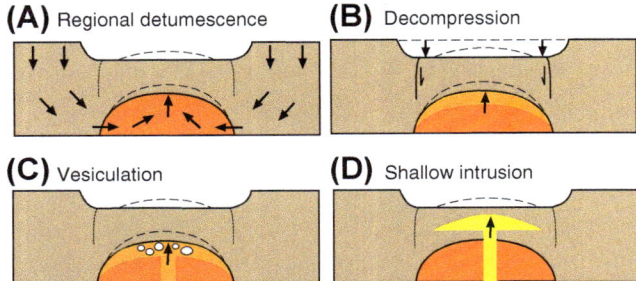

FIGURE 16.9 Four possible causes of resurgence (postcollapse magmatic uplift) at calderas. (A) regional relaxation of originally inflated crust squeezes residual magma inward and upward; (B) decompression associated with collapse may induce injection of new magma within the system; (C) magmatic vesiculation due to decompression (B) or caused by new injection of hot magma into the magma reservoir; and (D) new shallow intrusion of magma.

high angle to the local extension direction (Figures 16.3 and 16.8A). During resurgence, eruptions may occur from vents on the dome, or around it, near the ring-fault system. In some cases, resurgence uplifts a fault block, commonly asymmetrically, as at La Primavera (Mexico) and Ischia (Figure 16.8B) and Pantelleria (Italy). The structural form of the resurgence may relate to the aspect ratio (thickness/width) of the caldera floor (i.e., the magma chamber roof). Aspect ratios ~1 tend to develop resurgent blocks, whereas aspect ratios ~0.4 develop resurgent domes. Boundaries of many resurgent structures are sharp and may be faulted. Analogue models generate bounding, subvertical to steeply inward-dipping, reverse faults; cone sheets may form at this time. Peripheral normal faults may develop as a result of gravitational instability at the margins of the uplifted area.

Resurgence results from a volume increase or pressure buildup in the magma reservoir. Several models have been proposed to explain how this occurs (Figures 16.9A–D). Of these, regional detumescence and decompression are closely related to the caldera-forming event, whereas shallow intrusion does not require involvement of residual magma in the reservoir and can occur at any time.

3.5. Hydrothermal Activity

Calderas host a hydrothermal system in which fluids from shallow or deep aquifers are heated, pressurized, and contaminated by magmatic volatiles. Vapor and liquid accumulate and convect in fracture networks within the caldera, above the crystallizing magma reservoir, in some cases partly confined by impermeable layers, such as lacustrine clays or welded tuffs. Spectacular surface manifestations include hot springs, geysers, fumaroles, and mud pools, famously seen at Campi Flegrei (Italy), Yellowstone (USA), and Rotorua (New Zealand).

Convecting hydrothermal fluids may cause chemical and mineralogical changes to rocks locally or pervasively in the caldera floor and fill, and particularly in fractured marginal zones. These may involve leaching, alkali exchange, silicification, and the formation of minerals such as sericite, carbonate, chlorite, sulfates, sulfides, epidote, and clays. The alteration is commonly arranged into argillic, propylitic, and phyllic zones with increasing depth and temperature. Ore mineralization may derive directly from the magma, magmatic fluid, or hot groundwater, and mostly occurs along permeable fracture networks that tend to self-seal with mineral growth. Mineralization may begin soon after caldera collapse, or millions of years later. Examples include the Colorado-type epithermal silver base metal veins and stockworks (e.g., Creede, USA) and porphyry and epithermal mineralization (e.g., Tavua, Fiji).

3.6. Caldera Unrest

Unrest is a deviation (from hours to decades) from the normal, nonerupting state of a volcano toward a state that may lead to an eruption. It occurs at nearly 20 calderas in a typical year, and reflects interaction between tectonic, magmatic, and hydrologic processes. Most eruptions are preceded by unrest, but unrest does not always result in an eruption. More than a half of 225 large Quaternary caldera volcanoes have experienced historical unrest.

Unrest is identified by monitoring changes in seismicity, microgravity, surface deformation, and gas emissions (Figure 16.10). Most commonly reported features are local earthquake swarms; uplift; subsidence; tilt; ground fissuring; changes in the temperature of soil, water, or gas; changes in fumarolic activity; and minor eruptions.

Most seismic events in calderas are discrete earthquakes of magnitude <3, shallower than 15 km. Some result from brittle failure of country rock in response to magma intrusion; others reflect release of tectonic stress, shear of viscous magma along conduit walls, magma explosions, or collapse following subsurface magma flow. Ground deformation on all scales can occur. It can be sufficiently dramatic to be witnessed directly, but normally rates are just millimeters or centimeters per year and can be detected only through leveling or gravity surveys, GPS or InSAR. Subtle years-to-decades-long uplift has occurred within several large calderas, including Campi Flegrei, Rabaul, Aira, Iwo-Jima, Long Valley, Yellowstone, Kilauea, Mauna Loa, Askja, and Krafla. A common pattern is dome-like inflation, with maximum uplift centered within the caldera.

Whereas deformation may occur throughout the caldera, most seismic activity is restricted, for example, to caldera margins. However, seismicity, uplift, increased thermal activity, and eruptions at a caldera do not always share a common center, and the centers may shift

(A)

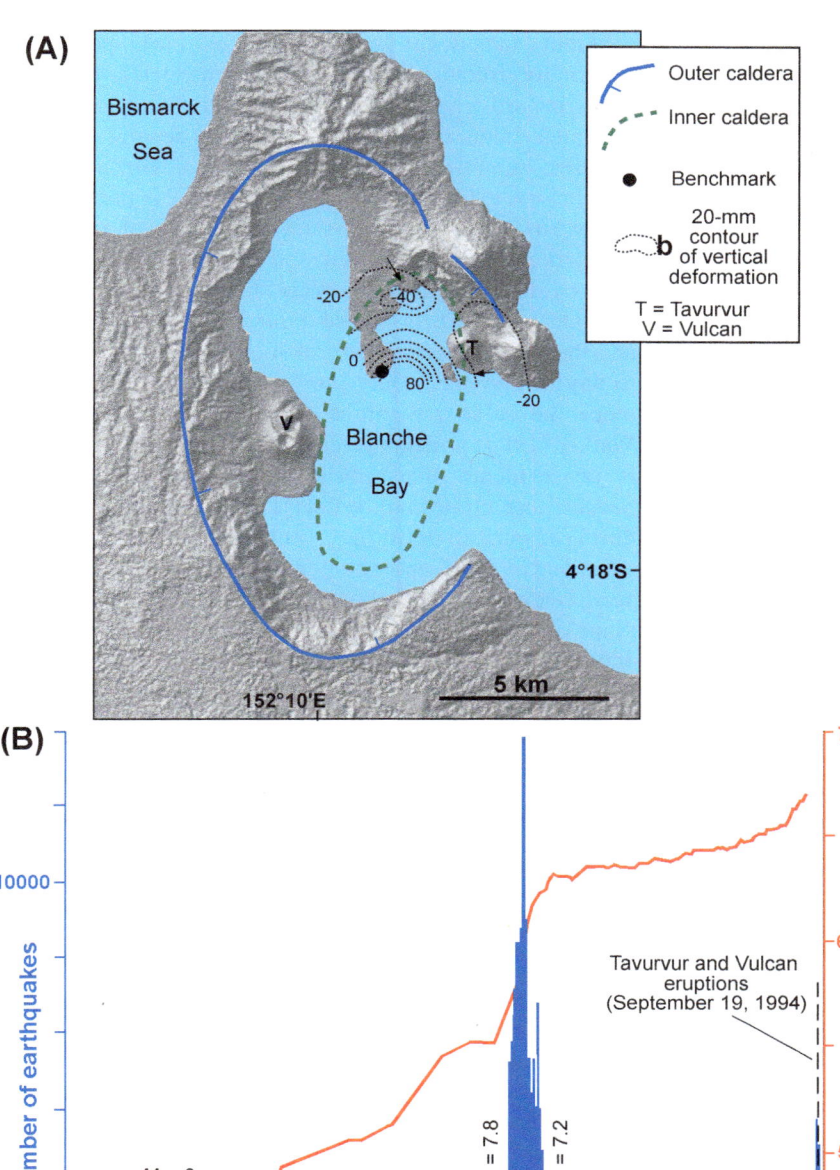

FIGURE 16.10 (A) Simplified structure of Rabaul caldera volcano, showing an inner caldera nested within an outer caldera, the two recently active vents, Tavurvur and Vulcan, and the deformation between July 1985 and September 1989, highlighted by the uplift contours. (B) Monthly total earthquakes (blue histogram) and compounded uplift (red line) at Rabaul between 1968 and 1994; the occurrence of major regional earthquakes in the area is also indicated.

during unrest. At calderas with thermal areas (e.g., Uzon, Kamchatka; Yellowstone, USA), changes may occur in the composition or flux of fumaroles or hot springs, or in the level and composition of groundwater or caldera-lake water. Common changes are increases in total discharge, in the discharge and proportion of acid gases (SO_2, H_2S, HCl, and HF), and in CO_2 emission. Temperature changes of caldera lakes have been reported and the temperature of fumaroles may change by several hundred degrees.

Unrest can be intermittent, posing a challenge for eruption forecasting. It may persist for weeks or centuries at large calderas, such as Aira (Japan), Campi Flegrei (Italy), Rabaul (Papua New Guinea), and Iwo-Jima (Philippine Sea). The activity may wax or wane several times before culminating in an eruption, a shallow intrusion, or returning to quiescence. Seismicity commonly occurs in repeated swarms, and uplift may alternate with subsidence. For example, subsidence at Yellowstone caldera (USA) during 1985–1987 followed decades of net uplift. Uplift at Campi Flegrei caldera in Italy has alternated with decades to centuries of subsidence. Ground deformation may be episodic, as indicated by stepped terraces at Toba and Iwo-Jima volcanoes. Monitoring unrest has a proven record of short-term (hours to months) forecasting, but uncertainties remain. Precursory unrest typically begins a few hours to decades before an eruption, but dramatically increased uplift rates to several meters per day and intense and shallowing seismicity are the only changes that reliably immediately precede an eruption. A common problem in forecasting is that different processes can produce similar symptoms of unrest. An example of this is the still-debated 1982–1984 unrest at Campi Flegrei in Italy, which was not followed by an eruption.

3.7. Further Caldera-Collapse Events

Many caldera volcanoes undergo more than one caldera-forming eruption. For example, during the last 360 ka, Santorini volcano in Greece has undergone 12 substantial explosive eruptions, at least 4 of which were caldera forming. The Santorini pyroclastic succession records two types of magmatic cycle: 20000- to 40000-year cycles and 180000-year cycles. The shorter cycles involved early construction of basaltic–andesite lava cones and shields, followed by a major explosive eruption of more evolved magma, and are thought to relate to the slow growth, partial discharge, and then cooling and crystallization of a shallow magma body, whereas the longer-duration cycles probably record deeper, larger-scale thermal maturation of the crust. Other calderas (e.g., Las Cañadas and Tejeda, Canary Islands) have undergone more protracted histories spanning millions of years, with temporal clusters of caldera-forming eruptions separated by periods of limited explosivity. The resultant calderas are compound, that is, formed of several nested or shingled (overlapping) smaller structures, which can be challenging to resolve.

4. DISCUSSION

4.1. Caldera Types

Calderas have been classified in various ways such as according to size, shape, single versus multiple collapse, and tectonic setting. They can usefully be divided into the following types: (1) calderas at basaltic shield volcanoes, (2) calderas at summits of intermediate-composition composite volcanoes at arcs, and (3) rhyolitic ignimbrite calderas, which characteristically form during large-volume explosive eruptions. Some of the last type are resurgent.

To emphasize the possible wide variety of calderas, George Walker proposed the existence of several contrasting "structural types," including pistonlike (with coherent subsidence on a ring fracture), piecemeal (internally fractured), trapdoor (asymmetric), downsag (with inward tilts), and funnel-shaped. Field studies of exhumed calderas have since revealed several of these features, sometimes at a single caldera. However, any simple structural classification of calderas is problematic for two reasons. (1) The internal structure of most calderas is poorly constrained and challenging to resolve even by drilling and geophysical studies. The structure of a modern caldera is buried beneath eruption products, and the surface morphology of such a caldera, even prior to any resurgence, rarely provides sufficient evidence to place the caldera within a structural classification. As Walker (1984) noted, a circle of postcollapse lava domes is not evidence that subsidence was on a ring fault, because the circle may reflect a fissure that propagated during magmatic uplift (e.g., a cone sheet) not during the subsidence. Even an exhumed caldera exposed in cross section only exposes those structures that intersect the plane of section. Thus, most modern calderas, and several ancient ones yield insufficient structural data to ascribe them to particular structural type. (2) At rare examples where the caldera floor and structural margins are sufficiently exhumed and well exposed to gain a decent overall picture, it seems that several of the proposed "end-member" structures coexist in the same example, just as they do in models. For example, the nonresurgent Scafell caldera in England widely exhibits downsag of the caldera floor (with $\geq 45°$ inward tilts), arcuate caldera-bounding faults with downthrows exceeding a kilometer, an overall asymmetric trapdoor-like shape with minor downthrow along one margin, and a piecemeal caldera floor cut by numerous spaced high-angle faults. The caldera-floor faults, spaced 100–500 m apart, demonstrably moved as much as 500 m, forming horsts and graben during the main caldera-forming eruption. Such a caldera is not readily categorized as either trapdoor, piecemeal, or downsag type, as it possess all these characteristics at the same time. Snowdon (Wales) is piecemeal in one section, yet is trapdoor-like with downsag when viewed perpendicular to this. Calderas elsewhere such as Rotorua (NZ) and Bolsena (Italy) also may have pistonlike portions, downsagged strata, trapdoor-like asymmetry, and multiple faults (piecemeal). It seems that downsag, trapdoor asymmetry, caldera-floor faults, and ring-fault systems are best viewed as common structural *attributes* of calderas that commonly develop in tandem, just as in analogue models (Figure 16.7). They are not mutually exclusive.

Caldera models show progressive structural evolution with increasing depth of subsidence, expressed as the diameter/subsidence ratio, d/s. Limited subsidence ($d/s > 40$) is mostly accommodated by downsag. Then, as subsidence progresses (d/s 18–40), a steep outward-dipping ring fault propagates. With further subsidence (d/s 14—18), downsag develops outboard of this, and with more subsidence ($d/s < 14$) the outer zone develops arcuate normal faults and crevasses. Thus, various structural features develop successively at the same caldera as it subsides. Also influencing caldera structure is the ratio between the thickness of the floor (= roof of magma reservoir), t, and the extraction width, w. This is the "roof aspect ratio," t/w (Figure 16.7). Mining subsidence and models show that the pattern of surface deformation varies: in subcritical extraction, $t/w > 0.7$, a single point of compression develops in the center of the subsided area, surrounded by a zone of extension, whereas in supercritical extraction, $t/w < 0.7$, a platelike, nondeformed central zone forms, bounded by a narrow zone of compression that passes outward into a marginal zone of extension as seen at calderas (Figure 16.7). An overall asymmetric, trapdoor-like subsidence geometry then, may reflect initial variations in the thickness of the caldera floor above an asymmetric magma reservoir.

Physical subsidence models have yet to assess the possible influence of subvolcanic crystal mushes with multiple melt lenses. The depth of subsidence, for example, may be limited by the thickness and crystal content of a mush-filled sill-like reservoir, and the subsidence may migrate laterally during an eruption when partly interconnected melt lenses are erupted in rapid succession. At Scafell caldera (UK), for example, the subsidence shifted markedly southeast as the caldera-forming explosive eruption changed from andesitic to rhyodacitic, reflecting the evacuation of different parts of a complex reservoir: this may have contributed to its piecemeal subsidence behavior. Subsidence (strain) rate, and the existence of preexisting faults or weak (e.g., hydrothermal) zones also may influence the pattern of subsidence deformation.

4.2. Further Research

Understanding of calderas has made large advances, but intriguing questions remain to excite future research. What is the global impact of a large caldera-forming eruption? How can they be predicted and mitigated? How frequent are they globally? How do the magma reservoirs assemble, evolve, and erupt? How common is pretumescence, and what controls whether a caldera undergoes resurgence? Does tectonic setting influence caldera magmatic and structural development, and what features of a caldera might indicate its tectonic setting? Does the caldera structure affect the frequency and style of eruptions? Are most calderas cyclic, or do some simply undergo repeated caldera-forming eruptions, say following recharge events? How might better structural and facies models of calderas be used to improve mineral and geothermal prospecting?

FURTHER READING

Acocella, V., 2007. Understanding caldera structure and development: an overview of analogue models compared to natural calderas. Earth Science Reviews 125, 125—160.

Acocella, V., Palladino, D.M., Cioni, R., Russo, P., Simei, S., 2012. Caldera structure, amount of collapse and erupted volumes: the case of Bolsena Caldera, Italy. Geological Society of America Bulletin 124, 1562—1576.

Branney, M.J., 1995. Downsag and extension at calderas: new perspectives on collapse geometries from ice-melt, mining, and volcanic subsidence. Bulletin of Volcanology 57, 3030—3318.

Branney, M.J., Kokelaar, P., 1994. Volcanotectonic faulting, soft-state deformation and rheomorphism of tuffs during the development of a piecemeal caldera, English Lake District. Geological Society of America Bulletin 106, 507—530.

Cole, J.W., Milner, D.M., Spinks, K.D., 2005. Calderas and caldera structures: a review. Earth Science Reviews 69, 1—96.

Druitt, T.H., Sparks, R.S.J., 1984. On the formation of calderas during ignimbrite eruptions. Nature 310, 679—681.

Druitt, T.H., Edwards, L., Mellors, R.M., Pyle, D.M., Sparks, R.S.J., Lanphere, M., Davies, M., Barriero, B., 1991. Santorini volcano. Geological Society of London, Memoirs 19, 1—165.

Geshi, N., Shimano, T., Chiba, T., Nakada, S., 2002. Caldera collapse during the 2000 eruption of Miyakejima volcano, Japan. Bulletin of Volcanology 64, 55—68.

Gudmunsson, A., Nilsen, K., 2006. Ring faults in composite volcanoes: structures, models, and stress fields associated with their formation. Journal of the Geological Society, London 269, 83—108.

Lipman, P.W., 1984. Subsidence of ash-flow calderas in Western North America: windows into the tops of granitic batholiths. Journal of Geophysical Research 89, 8801—8841.

Lipman, P.W., 1997. Subsidence of ash-flow calderas: relation to caldera size and magma-chamber geometry. Bulletin of Volcanology 59, 198—218.

Newhall, C.G., Dzurisin, D., 1988. Historical unrest at large calderas of the world. U.S. Geological Survey 1, 1109.

Marsh, B.D., 1984. On the mechanics of caldera resurgence. Journal of Geophysical Research 89, 8245—8251.

Michon, L., Massin, F., Famin, V., Ferrazzini, V., Roult, G., 2011. Basaltic calderas: collapse dynamics, edifice deformation, and variations of magma withdrawal. Journal of Geophysical Research 116, B03209. http://dx.doi.org/10.1029/2010JB007636.

Mori, J., Mckee, C., 1987. Outward-dipping ring-fault structure at Rabaul Caldera as shown by earthquake locations. Science 235, 193—195.

Moore, I., Kokelaar, P., 1998. Tectonically controlled piecemeal caldera collapse: a case study of Glencoe caldera volcano, Scotland. Geological Society of America Bulletin 110, 1448—1466.

Roche, O., Druitt, T.H., Merle, O., 2000. Experimental study of caldera formation. Journal of Geophysical Research 105, 395—416.

Smith, R.L., Bailey, R.A., 1968. Resurgent cauldrons. Geological Society of America Memoirs 116, 613—662.

Stix, J., Kennedy, B., Hannington, M., Gibson, H., Fiske, R., Mueller, W., Franklin, L., 2003. Caldera-forming processes and the origin of submarine massive sulfide deposits. Geology 31, 375—378.

Walker, G.P.L., 1984. Downsag calderas, ring faults, calderas sizes, and incremental caldera growth. Journal of Geophysical Research 89, B10, 8407—8416.

Part III

Effusive Volcanism

John Stix

Department of Earth & Planetary Sciences, McGill University, Montreal, Quebec, Canada

This section of the Encyclopedia is concerned with volcanoes and volcanism dominated by comparatively quiet outpourings of lava in all its guises. This style of volcanic activity is found on both land and in the oceans, and the sizes of eruptions range from very small to very large. Although lava compositions are dominated by basalt, the most common volcanic rock on Earth, lavas do exhibit a large compositional range, with attendant differences in rheology hence eruptive behavior.

Chapter 17 describes the myriad types of aa and pahoehoe lavas which are produced by basaltic volcanoes. These are then contrasted with blocky lavas which are generated by magmas of more felsic composition. These lavas resemble aa in their overall structure but they are generally much thicker; nevertheless, large scale surface and internal folding can also be observed in felsic lava flows. The internal structure of aa lava is characterized by a brittle crust, adjacent visco-elastic layers, and a hot core zone. Lavas commonly undergo inflation where the front of the lava stalls while the interior of the lava fills with fluid lava. In terms of transport, lava is transported in open channels and in enclosed tube systems, both of which range from simple to complex distribution systems. As lava is transported downslope, it undergoes changes in its temperature and rheology. Cooling is enhanced in open channels compared to tubes. As the cooling progresses, the lava crystallizes and becomes stiffer rheologically.

Chapter 18 addresses the extrusion of highly viscous magma of more felsic composition. Lava domes are notable, as they are commonly associated with explosive activity, attesting to the complex interplay of processes which control effusive versus explosive behavior. Another remarkable aspect of lava domes is that sometimes they can grow very fast, within days to weeks; other times they can be slow-growing or even static. Many lava dome complexes show highly episodic behavior, with periods of fast and slow growth. A few lava domes can be active on timescales of tens of years. Similar to lava flows, domes have highly variable rheology including a deformable core and a carapace which is brittle generating significant talus. The chapter examines two historic lava dome eruptions that are probably the best studied in the world. Mt St Helens experienced a series of lava domes during two time periods in 1980−1986 and 2004−2008; the contrast between them is interesting, as the first period occurred after the May 18, 1980 Plinian eruption, while the 2004−2008 activity was characterized nearly exclusively by dome growth. Soufrière Hills volcano on the island of Montserrat began experiencing lava dome activity in 1995, and this activity continues to today. Soufrière Hills taught volcanologists much about lava domes, including the delicate transitions between effusive and explosive activity, and how eruptions can escalate during different time frames.

Chapter 19 discusses submarine lavas, highlighting the fact that these eruptions are dominated by basaltic eruptions, which due to their generally low volatile contents mean that the eruptions are either not explosive or only weakly so, usually in their initial phases. The lavas are characterized by pillow, lobate and sheet forms, which can be found together. Because these different lava flow morphologies are largely a function of effusion rate, their presence (or absence) can be profitably used to make such inferences. In some situations, basaltic lava can break. Commonly this happens when the lava is rapidly chilled by seawater, or when lava flows down steep slopes; the result is a fragmental glassy material termed hyaloclastite. In other situations, basaltic magma interacts with unconsolidated water-saturated sediments in the shallow subsurface; the product is peperite, a complex association of the quenched magma and the sediment. Submarine felsic lavas, while much less abundant than basaltic compositions, nevertheless produce some unusual features such as columnar joints of various sizes which can be used to infer the subaqueous environment. The combination of high magma viscosity and rapid quenching by water means that fragmented lava facies are both common and commonly associated with the coherent lava facies. Where sedimentation rates are high, synvolcanic intrusions may be present, particularly when the density of the felsic magma is higher than that of the water-saturated sediment.

Chapter 20 looks at glaciovolcanism in its myriad manifestations. Tuyas or mobergs are volcanoes produced beneath ice. Typically basaltic magma erupts into ice, melting it and creating a subsurface cavern. Eruptive products include pillow lavas and hyaloclastite, and these are typically found together. As pillows accumulate to form a subglacial edifice, slopes steepen, favoring the production of hyaloclastite, which is also enhanced by decreasing lithostatic pressures and greater degrees of expansion and explosivity. If the lava reaches the surface of the ice, a capping of coherent lava can form which protects the friable hyaloclastite facies beneath from erosion. Volcanoes of more felsic composition, such as andesitic arc volcanoes, can also have substantial volumes of ice resting on their edifices, so glaciovolcanism in these environments is significant. However, recognizing such products can be difficult, as the features are more subtle. Again, jointing patterns can be helpful to deducing their origin. Glaciovolcanism can be used for paleoclimate studies, including spatiotemporal distributions in a given area; under favorable circumstances, ice thicknesses can be estimated as well. The amount of heat transferred from magma during subglacial eruptions can be calculated and quantified. The chapter ends with an examination of three recent subglacial eruptions: Gjálp in 1996 and Eyjafjallajökull in 2010 (both in Iceland) and Redoubt in 2009 (Alaska, USA).

Chapter 21 explores volcanoes associated with mid-ocean ridges, the largest magma system on Earth. An important point to note is that most of the magma generated by mantle melting beneath mid-ocean ridges ends up as intrusive rock forming the new oceanic crust, with a relatively minor amount of magma erupting as lava. Shallow magma chambers appear to be common along fast-spreading mid-ocean ridges but much less common along slow-spreading ridges, attesting to the more discontinuous magma supply in these slower-spreading environments. Where imaged seismically, the tops of magma chambers at fast-spreading ridges also appear to be spatially focused and shallower than those at slow-spreading ridges. In terms of eruptive products, pillowed, lobate and sheet flows are observed, with a greater abundance of sheet flows along fast-spreading ridges and more pillow lavas along slow-spreading ridges. This difference may be ascribed to the different magma supply and effusion regimes.

Chapter 22 looks at the development of seamounts on the ocean floor. Seamounts form in a range of tectonic settings, including on or near mid-ocean ridges, within intraplate settings, and associated with subduction zones. Seamounts near mid-ocean ridges may develop from mantle processes similar to those responsible for magmatism at the ridge itself. Seamounts also result from plume—ridge interactions, and these seamounts tend to be larger in size. Seamounts have a variety of morphologies ranging from isolated cones through complex clusters to flat-topped structures, which are formed either by erosion at wave base or by lava infillings of summit calderas. Interestingly, since the seafloor is still not very well known or studied, we do not have an accurate estimate of the total number of seamounts; estimates for those with heights of at least 1 km range from 45,000 to 350,000, a range of nearly an order of magnitude. The model for seamount formation starts with an early stage followed by a shield-building stage, and terminating with more alkalic, smaller volume magmatism. These changes are generally correlated with mantle melting events, with the largest melting associated with the shield-building stage. Much of our knowledge regarding seamount and ocean island volcanism is based on the well-developed Hawaiian chain of islands and seamounts. Seamount lithologies include pillow lavas, volcaniclastic rocks, and intrusive rocks, such as dikes and sills, and many seamounts exhibit hydrothermal activity as well during and after the time they were magmatically active.

Chapter 23 focuses upon basaltic volcanic fields, which are quite common in a range of tectonic settings. These terrains are interesting due to their rather unusual characteristics. First, a field may comprise hundreds if not thousands of individual volcanic centers which are generally of basaltic composition. Second, the volcanoes are generally monogenetic, meaning that they erupt only once, then die. Third, the volume of an individual volcano is generally small and sometimes very small. Overall, the magmatic output from these fields can extend in time for up to a few million years and covers a large area. Despite this fact, individual volcanic centers within a given field commonly show vent clustering and vent alignments, attesting to control by the underlying structure including faults and the tectonic stress regime. Within such fields the character of the volcanism typically reflects both magmatic and phreatomagmatic styles of activity. Magmatic eruptions include scoria cones while phreatomagmatic activity is characterized by maars, tuff rings, and tuff cones. Statistics can be applied to estimate eruptive recurrence rates, while geochemistry can characterize magma source characteristics and magmatic processes, which can be quite diverse and variable. The main hazards from such eruptions are tephra falls, pyroclastic density currents, and lava flows.

Chapter 24 studies the remarkable magmatism and volcanism associated with flood basalts and large igneous provinces. Dwarfing volumes of currently active volcanoes by many orders of magnitude, these truly monumental eruptions are one type of supervolcano which have occurred repeatedly during the evolution of the earth (the other type of supervolcano are calderas, see Chapter 16). Large igneous provinces are found both on the continents and in ocean basins. The volume of individual eruptions may extend up to 10^3-10^4 km^3, while the total volume of an individual province can range from 10^4 to 10^6 km^3. These are truly enormous numbers, and furthermore, the eruptions appear to occur over a very short time geologically speaking. For example, the output from a large igneous province may extend over only $1-3$ million years, and it is possible that the bulk of this output may occur in an even shorter time frame. While flood basalts and large igneous provinces appear in some cases to be associated with divergent plate boundaries and mantle plumes, many large igneous provinces are surprisingly not well characterized in terms of their tectonic environment. Nevertheless, the occurrence of mantle plumes provides an attractive means to transport large volumes of material from regions deep in the mantle to the surface. Geochemical studies indicate that the sources of flood basalts have characteristics akin to mid-ocean ridge basalts and ocean island basalts, with a significant lithospheric component as well. Flood basalt eruptions are thought to occur through fissure systems, based upon smaller-scale examples and modeling studies.

Chapter 17

Lava Flows and Rheology

Andrew J.L. Harris

Laboratoire Magmas et Volcans, Université Blaise Pascal, Clermont Ferrand, France

Scott K. Rowland

Department of Geology & Geophysics, School of Ocean and Earth Sciences and Technology, University of Hawai'i at Mānoa, Honolulu, HI, USA

Chapter Outline

GLOSSARY

'a'ā Type of lava with auto-brecciated surface and base of rough-surfaced, *spinose,* clasts, and a coherent, typically non-vesicular interior. Name is Hawaiian in origin as assigned by Dutton (1883).

block lava Type of lava, typically intermediate or silicic, with surface crust of metric, typically angular, blocks, and a coherent non-vesicular interior. Name originally assigned by Finch (1933).

facies Means of describing an outcrop in terms of the unit (i) geometry (thickness, width, length), (ii) lithology and texture (composition, colours, crystals, bubbles), (iii) structures (bedding, layering, folding, fractures) and their character (granular, coherent), (iv) movement patterns (ramping, shearing, tilting, deformation, cooling structures), and (v) fossils/alien objects (tree moulds, entrainment, and type of entrained material).

inflation Process by which a stationary lava unit, with a still-fluid interior and solid crust, is supplied by new lava so that its surface crust is lifted upwards.

pāhoehoe Type of lava with smooth, non-brecciated, billowy (sometimes ropey) surface. Name is Hawaiian in origin as assigned by Dutton (1883).

lava channel Stream of fluid lava flowing between two stationary banks (called levees).

lava flow field Lava flow comprising many lava flow units that may be spread laterally, stacked vertically, or both.

lava flow unit A body of coherent lava with a cooling surface at its top and base, these being the surface and basal crusts.

lava rise A general term to describe an inflation-induced feature.

lava rise pit Low point or pit in an inflated surface which has not inflated, but around which all other lava has been up-lifted.

lava tube Conduit with static, stationary roof, walls and floor through which lava flows, thus sometimes termed a *"tunnel"* or *"pyroduct"*.

levee Term used for the static banks of a lava channel.

pipe-vesicle bearing (p-type) pāhoehoe Type of pāhoehoe that is relatively poor in bubbles, with thick coherant glassy crust, often with pipe vesicles in its lower half.

rate of flow This is a loose term which requires clarification as to whether it is a rate expressed in terms of velocity (m/s), mass flux (kg/s) or volume flux (m³/s). For lava volume flux, three terms can be defined:

  (i) **Effusion rate** is the instantaneous flux of lava from the vent at any given moment in time;

The Encyclopedia of Volcanoes. http://dx.doi.org/10.1016/B978-0-12-385938-9.00017-1

(ii) **Time-averaged discharge rate** is the flux of lava from the vent averaged over a known period of time;

(iii) **Mean output rate** is the flux of lava from the vent averaged over the entire eruption: it is total volume erupted divided by the duration of the eruption.

sheet lobe Broad plateau of lava, often with cleft running around plateau perimeter, up-lifted by inflation; termed a "lava rise" by Walker (1991).

simple lava flow A long, thin lava flow erupted in a short duration event and comprising a small number of flow units.

spongy (S-type) pāhoehoe Type of pāhoehoe that is rich in rounded vesicles, which increase in size, but decrease in number, towards the unit interior.

transitional lava Rough surfaced pāhoehoe that is transitional to 'a'ā, and often with isolated swirls and clasts of 'a'ā firmly fixed to the surface.

tumulus Dome-like mound of lava up-lifted by inflation with prominent cleft(s) running along the crest.

viscosity The internal resistance of a fluid to flow.

yield strength The amount of stress required before a fluid will begin to deform, i.e., move.

1. INTRODUCTION

Lava flows dominate the surfaces of many of the world's basaltic centers, such as Hawai'i, the Galápagos, Iceland, Vesuvius and Etna (Italy), the Afar, Nyiragongo and Nyiramuriga (Democratic Republic of Congo) and Piton de la Fournaise (Réunion), to name but a few, as well as the natrocarbonitite center of Ol Doinyo Lengai (Tanzania). Lava is also commonly encountered at monogenetic basaltic fields, such as Paricutín (México) and the Chaine des Puys (France), as well as across ancient flood basalt provinces such as the Deccan (India) and Columbia River (USA). Felsic centers, such as Unzen (Japan), Chao (Chile), Newberry (USA), Lipari (Italy), San Pietro (Sardinia, Italy) and Karismibi (Rwanda) also host thick, stubby silicic lava flows, often referred to as "coulees". Although coulees are typically viewed as lava domes which deform down-slope so that the dome becomes elongated in the down-slope direction, such silicic flows can also have morphologies similar to basaltic lava flows, as at Santiaguito (Guatemala). Lava flows of almost exclusively basaltic composition comprise the great majority of the ocean floors and are known as Mid-Ocean-Ridge Basalts (MORBs). Lava-related morphologies are also common across many planetary landscapes shaped by effusive volcanism, including those of Mercury, Venus, Earth's Moon, Mars and Io. Understanding the structure, architecture and morphology of lava flows, as well as the processes that produce the various lava-flow **facies**, is thus a fundamental starting point in volcanic studies. Defining the characteristic facies of various lava flow types is thus the main objective of this Chapter.

2. TYPES OF LAVA

Active lava is a mixture of molten rock (liquid), crystals (solids), gas (bubbles) and other voids. In addition, the mixture may contain xenoliths of basement rock carried up with the magma during eruption.

Compositions of molten rock that comprise the mixture range from ultramafic (e.g., komatiite) through felsic (e.g., rhyolite). The mixture arriving at the vent usually contains crystals that have grown either in the magma chamber or during ascent. These are phenocrysts and/or microphenocrysts (if less than 30–100 μm). The size, type, and number of crystals depends on the lava chemistry and pre-eruptive history. Whereas lavas at Kīlauea, for example, may have less than 1 vol% olivine phenocrysts upon eruption, it is not uncommon for Etna lavas to contain up to 35 vol% phenocrysts of pyroxene and plagioclase. The void component of the mixture comprises bubbles of exsolved gas, intercrystal void spaces created by water exsolution during crystallisation (diktytaxitic voids that are usually ragged and typically <0.2 mm in size) and other, typically elongate-to-oval, spaces that can open up within and around shear zones. Those voids due to bubble formation are called vesicles once lava has solidified. Lavas can be completely non-vesicular or extremely bubbly (Table 17.1). Vesicularities of up to 94% have been measured in near-vent lavas emplaced during 1997 at Kīlauea's Pu'u 'Ō'ō vent. Furthermore, the number, size and shape of vesicles can vary dramatically over only a few centimetres both vertically and horizontally.

Macdonald (1953) distinguished three types of lava: **pāhoehoe**, **'a'ā** and **block lava** as follows:

Pāhoehoe is characterized by a smooth, billowy, or ropy surface, and spheroidal vesicles. 'A'ā is characterized by a fragmental and spinose surface, and irregularly shaped vesicles. Block lava differs from 'a'ā in greater regularity of the shapes of the fragments in the breccia phase, and less spinose surfaces. Beneath the fragmental tops of 'a'ā and block lava flows is a nearly continuous non-fragmental layer

(Macdonald, 1953, p. 169).

Within each of these three lava types, are several sub-types, which we describe in the following sections.

2.1. 'A'ā

'A'ā is characterised by auto-brecciated surface and basal crusts, and a coherent interior which is usually termed the core (Figure 17.1). It is the core that is fluid and deformable while the flow is active (Figure 17.2(A)). 'A'ā flows are typically 0.5–20 m thick and, in outcrop, are characterised by a coherent core sandwiched between two layers of breccia (Figure 17.2(B)).

TABLE 17.1 Some Published Vesicularities for Lavas at Etna and Kilauea

| Location | Lava Type | Vesicularity (%) | | | | Source |
		Min	Max	Mean	St. Dev	
(1) Etna Samples						
Etna	a'a	14	35	19	6	Gaonac'h et al. (1996)
Etna	Etnean pāhoehoe	17	33	25	6	Gaonac'h et al. (1996)
Etna	a'a, pāhoehoe & channel	4	42	20	10	Gaonac'h et al. (1996)
Etna	a'a and pāhoehoe	9	71	22	12	Herd and Pinkerton (1997)
All Etna:		4	71	22	8	
(2) Kilauea Samples						
Kilauea	P-type pāhoehoe	2	28	17	8	Wilmoth and Walker (1993)
Kilauea	S-type pāhoehoe	33	70	47	8	Cashman et al. (1994)
Kilauea	P- and S-type	2	70	37	16	Cashman et al. (1994)
Kilauea	Episode 48	10	73	35	15	Cashman et al. (1994)
Kilauea	Episode 53 (tube-samples)	18	45	32	8	Cashman et al. (1994)
Kilauea	Episodes 48 & 53	10	73	34	14	Cashman et al. (1994)
Kilauea	Lava channel overflows	41	74	57	9	Robert et al. (2014)
All Kilauea:		2	74	37	11	

2.1.1. 'A'ā Breccia

Clasts comprising the auto-breccia are termed *"clinker"*. By virtue of being derived from the flow core, they have the same geochemical and petrographic character as the core, although they are often more oxidized. Individual clinkers range in size from a few centimetres to metres across. They are sub-rounded, with rough, spiny, jagged surfaces with vesicular tendencies, so that it is often referred to as being *"scoriaceous"*. However, unlike scoria, 'a'ā clinkers are usually quite dense, especially in their interiors because they are derived from the vesicle-poor flow core. Moreover, they are not pyroclastic in origin, as the term *"scoria"* implies. The favoured model for clinker formation invokes a velocity profile which increases from zero at the flow edge, to highest at the flow centre (Figure 17.3). Facing down flow, and considering the right half of the flow, the increase in velocity towards the flow centre means that the left hand—flow-centre—side of a molten lava blob will

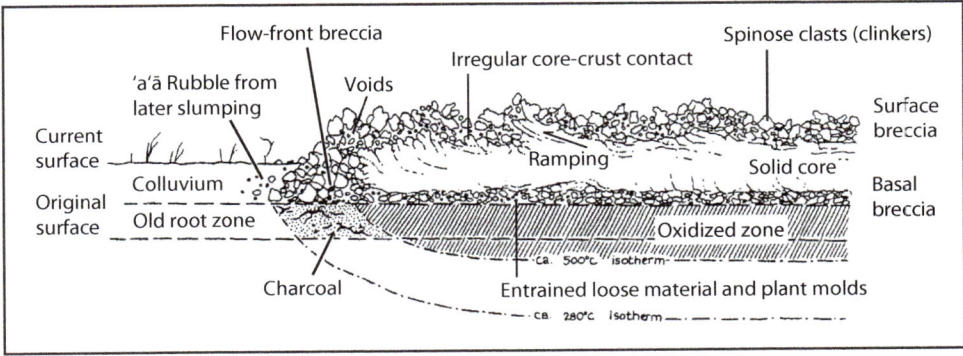

FIGURE 17.1 Sketch of Lockwood and Lipman (1980) showing an 'a'ā flow, in section, that has moved through (and burnt) a vegetated zone. Note heating and oxidation of underlying soil, ramp structures that intrude the surface clinker, and the uneven form of the clinker-interior contact.

FIGURE 17.2   (A) Active 'a'ā flow front photographed in 1992, about 1 km N of Pu'u 'Ō'ō (Kilauea). The flow front is 3—4 m high and the cooler surface crust and hot, fluid interior are clearly visible. (B) 'A'ā unit cropping out in a quarry about 100—200 m from the front of Etna's 1981 flow. Outcrop is 8—10 m high and shows a solid interior sandwiched between two layers of breccia—these being the basal and surface crusts. (C) Lobe of active pāhoehoe oozing from beneath the rafted up, stalled crust of a ropey pāhoehoe unit. Active lobe is about 1 m across — note the thin, dark, glassy crust already beginning to form on flow front. (D) Pāhoehoe in outcrop photographed at Makapuu Point, Oahu. Unit is about 40 cm high and consists of a main, lower unit onto which an upper breakout has been extruded, the two being separated by a "hanging" contact. The basal breccia of overlying 'a'ā can be seen above the pāhoehoe.

move faster than the right hand—flow-margin—side. This will cause the blob to rotate in a clockwise direction. Left of the centre the situation is the contrary, so that motion is anti-clockwise. These rotations tear the lava surface, continually exposing hotter core lava and causing individual clinkers to rotate relative to each other. Each time they stick to a neighbour they are then caused to rotate relative to that neighbour and pull apart, either producing spines, if still partially molten, or fractured surfaces, if mostly solidified.

The same velocity profile acts in the vertical direction, increasing from zero at the flow base to maximum at the top. At the flow front this causes an over-steepening, forward-toppling and rolling, "*caterpillar tread*", motion, as well as near-continuous flow front collapse. Surface clinker arriving at the flow front tumbles down the front to be overridden by the advancing core, thus creating the basal clinker layer. Rotational structures found at the flow base, with their roots in the flow core, show that some basal clinkers also form in situ via rotation created by the cross-flow velocity gradient apparent at the flow base, just as at the surface.

The surface of an 'a'ā flow is often a mix of clinkers, accretionary lava balls, slabs of pāhoehoe from farther up the lava-flow system, scoria that has fallen from fountaining or strombolian activity at the source vents and, occasionally, large chunks of cone material that collapsed onto the flow surface, such as at Pīmoe on East Maui (Hawai'i). Charred tree trunks also may litter the flow surface, as well as various anthropogenic artefacts if the flow has entered an urban area. Photos of Etna's 1928 flow front moving through the town of Mascali taken by Gaetano Ponte (1876—1955) show the flow front to be a mixture of masonry, metal bars, mangled balcony frames and clinkers. The basal clinker can likewise be a mix of 'a'ā clasts with natural and anthropogenic foreign material that has been mixed into the loose crust. This may include stones and pebbles of underlying material including clinker

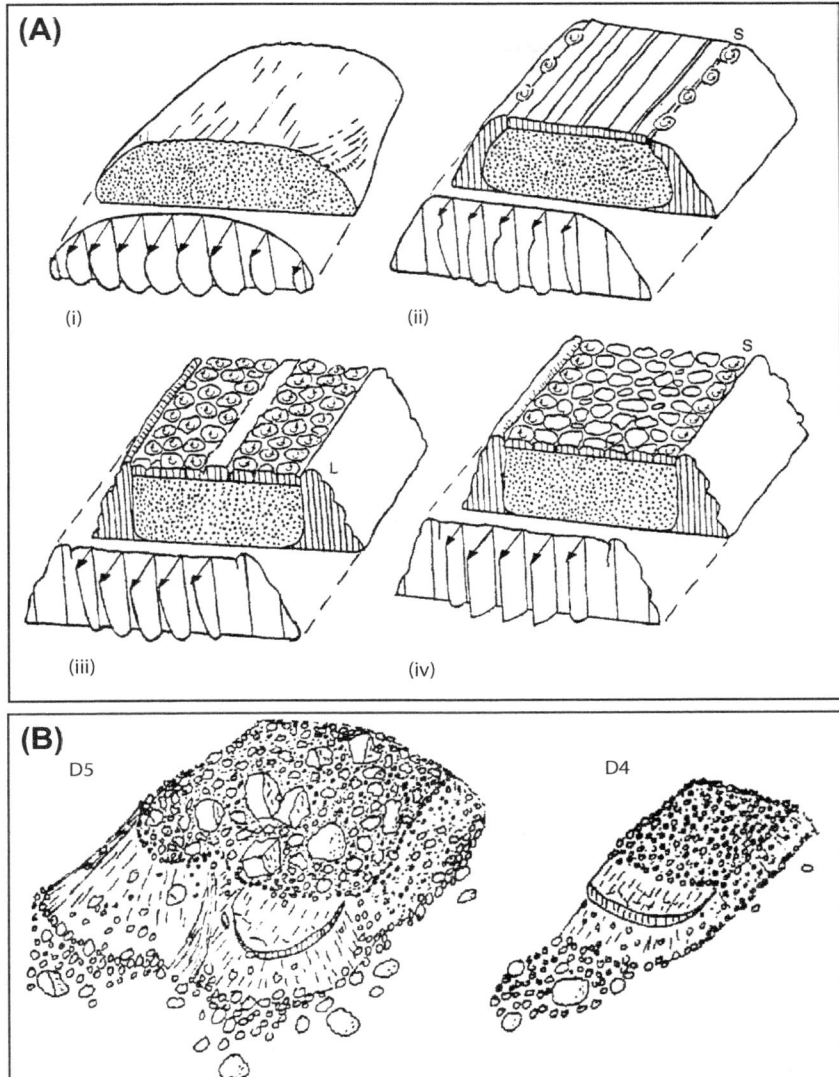

FIGURE 17.3 (A) Block diagrams (not to same scale) of Rowland and Walker (1987) to show clinker formation on four types of Hawaiian lava (arrows are proportional in length to velocity). (i) Pāhoehoe: The gentle variation in velocity across the flow means that shearing is minimal and is accommodated by stretching of the surface skin. (ii) Toothpaste lava: Shearing (S) produces clinker at either side of a central slab, which is essentially solid with no cross-plug variation in velocity. (iii) Proximal-type 'a'ā: The variation in velocity across the flow causes tearing and rotation of portions of the skin. This, due to the form of the velocity profile (which decreases bank-wards either side of a central plug) rotate. (iv) Distal-type 'a'ā: A wide zone of plug flow carries loose rubble forward, with marginal zones of shearing (S) and rotated rubble. (B) Sketches of two 'a'ā flow front types by Kilburn and Guest (1993). In D4 proximal-type 'a'ā forms a flow-front scree. In D5 distal-type (rubbly) 'a'ā, with a larger clinkers than D4, forms the scree.

and broken slabs of pāhoehoe from the surface of underlying lava flows, scoria, fumarolically altered blocks, pebbles of basement rock, logs, human detritus, etc.

2.1.2. 'A'ā Flow Core

The flow core is occasionally quite vesicular, particularly in thinner 'a'ā flows. However, high degrees of shearing, associated with the strong velocity gradients within the flow, causes vesicles to elongate in the flow direction. More commonly the cores of 'a'ā flows lack vesicles visible to the naked eye, particularly in thicker flows.

Fractures and flow-parallel shear structures are often apparent in the interior. Fractures often run in and out of bands of small vesicles, formed by the depressurisation effect of the fracture opening. Ramping also can occur, whereby movement (along curved, concave-up, shear planes) causes core lava to slide upwards towards the flow surface. This causes blocks of solid core lava, with striated faces, to ramp upwards to sometimes extrude from the flow surface or front (Figure 17.1). 'A'ā cores sometimes include entrained pieces of surface and basal clinker. Clinker that was entrained for a short time before the flow came to rest will have sharp contacts with the host lava and retain its original morphology. However, smaller clasts with longer residence times may take on more ghost-like forms, sometimes being distinguished merely as a rounded pocket of vesicluar material.

2.2. Pāhoehoe (Figure 17.2(B,C))

Pāhoehoe is characterised by a smooth, glassy, coherent surface — sometimes with surface folds, termed "*ropes*". Classically an individual pāhoehoe unit has a lobate form. A pāhoehoe flow field invariably consists of hundreds to thousands, even tens of thousands of individual lobes, sometimes referred to as toes. The surface topography is undulating, termed "*hummocky*" by Swanson (1973), with "*billowy*" being another frequently used descriptor. Pāhoehoe flows are commonly associated with very low **effusion rates** and advance rates. Mauna Loa's 1880–1881 pāhoehoe flow field advanced 48 km in 9 months, to give a time-averaged flow front velocity of around 180 m per day. However, pāhoehoe can also form at high effusion rates, as during emplacement of Laki's 1783 eruption in Iceland, and can be associated with high velocities as witnessed, for example, during the creation of pāhoehoe at the centres of fast moving channels and sheet flow.

An active pāhoehoe lobe consists of a fluid core with an outer crust that thickens with time. New flow units emerge from fractures in this crust. These fractures may be the clefts where two flow lobes are joined or may be created at the base of the flow front where the lobe front is pushed upwards or where the flow front fractures due to pressurization of the brittle surface crust (Figures 17.2(C–D)). Upon emplacement, lobes are typically <0.5 m in dimension; thicknesses measured for 3000 primary pāhoehoe flow units in Hawaii gave a minimum of 3 cm and median of 40 cm (Rowland and Walker, 1987).

2.2.1. Pāheohoe Surfaces and Outcrops

Pāhoehoe surfaces are typically glassy due to quenching of the surface in the first few seconds of exposure. Entrapment of phenocrysts and microphenocrysts in the glass causes little bumps and mounds, each marking the presence of a glass coated crystal. If vesicular, the glassy rind can be extremely fragile, and often spalls off within a few hours of emplacement or even during emplacement itself. This leaves a fragmented detritus of thin, slightly curved, glassy flakes on the surface, typically a few cm in size. These result in the characteristic crunching sound made when walking on pāhoehoe. Spalling is due to cooling-induced contraction of the thin surface glass at the same time as

expansion of the fluid interior, causing sections of the glassy crust to pop off. This makes a popping/tinkling sound, and projectiles can fly a meter or more.

A pāheohoe lobe in outcrop is typically oval in section perpendicular to flow direction (Figure 17.2(D)). A glassy rind a few mm to a few cm thick may be found running around the lobe. The interior is typically vesicular, and vesicles increase in size, but decrease in number, from the surface to the interior (Figure 17.4). In thicker lobes, bubbles may coalesce in the centre to form a large void. If sufficient gas accumulates so that the upper and lower parts of the lobe are completely separated this void becomes a gas blister. Because pāhoehoe comes to rest while it is still fluid, bubbles that have been deformed by shearing are able to recover their spherical shape before the lava solidifies. Vesicles are thus largely spherical, except in and near the glassy crust where they are stretched across the expanding surface, and then quickly frozen into the solidifying fluid.

2.2.2. S-Type and P-Type Pāhoehoe

Following Wilmoth and Walker (1993), pāhoehoe can be split into two main classes: **spongy (S-Type)** and **pipe vesicle-bearing (P-type)**:

S-Type pāhoehoe (Figure 17.4(A)) contains "very abundant, small and approximately spherical vesicles. Vesicles commonly comprise >40 vol.% of the rock and most of them are <4 mm in diameter" (Wilmoth and Walker, 1993, pp. 130–131). The great quantity of bubbles and sponge-like appearance of this pāhoehoe type gives it its name. The large quantity of vesicles also means that some are stretched across the surface to form a fabric of tiny fibres separated by hollows, these being the bubble walls and interiors, respectively. Light can reflect off of the glassy selvage giving fresh S-Type pāhoehoe a silvery sheen that lasts just a few days. However, as the glass spalls off, the rougher and less reflective vesicular layer immediately beneath the selvage becomes exposed.

P-Type pāhoehoe (Figure 17.4(B)) is less common than S-Type and is typically emplaced when underlying surface slopes are <4°. It is less vesicular than S-Type pāhoehoe and has larger vesicles. The upper glassy zone may contain large vesicles that form a band just below the surface. Due to the lack of vesicles, the surface glassy layer is coherent and usually few centimetres thick, lacking the stretched vesicle fabric of the S-Type surface. The lack of vesicles also means that P-Type pāhoehoe tends not to spall its surface glass. Instead, P-type pāhoehoe usually contains pipe vesicles near the lobe base. Pipe vesicles have their roots in the basal glassy rind, which is commonly 2–3 cm thick. One model for formation of pipe vesicles is that they formed as gas bubbles which expanded by scavenging gas molecules while the lobe solidified. The idea is that the

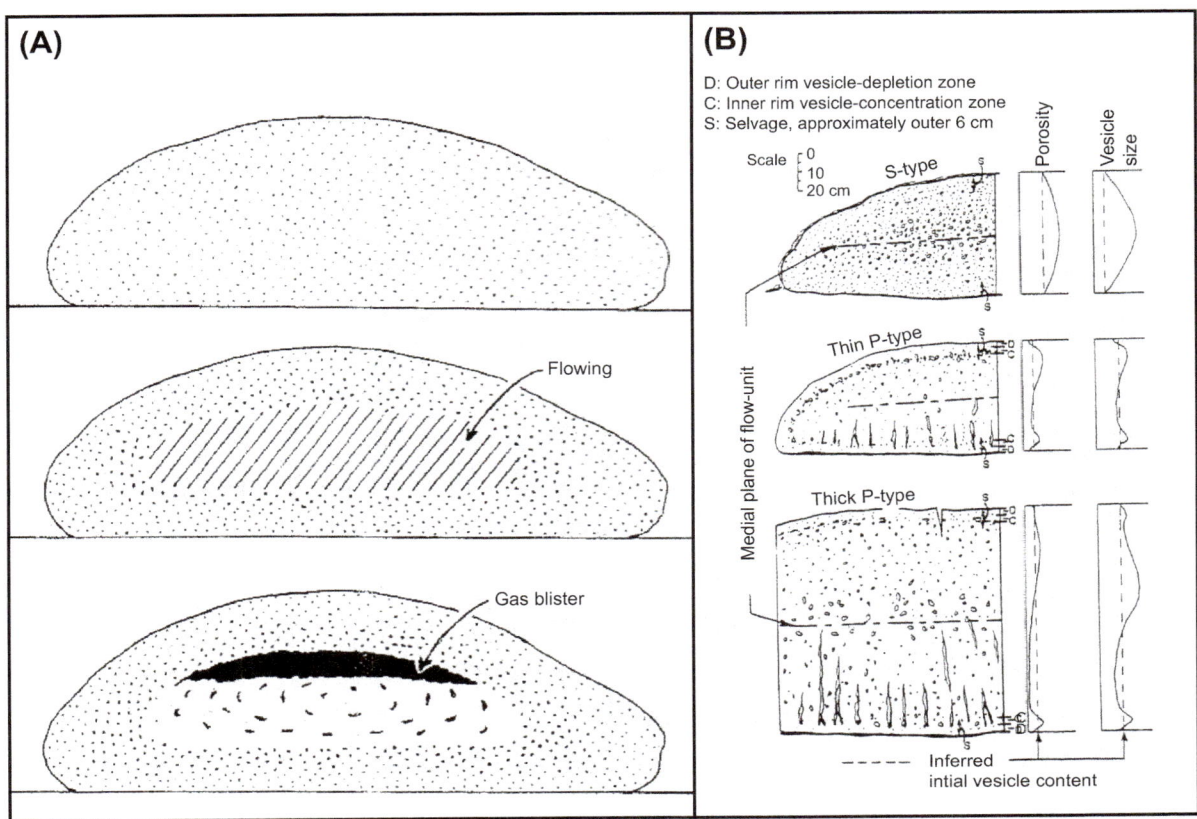

FIGURE 17.4 (A) Schematic of Walker (1989) showing "sequence of events leading to formation of broad central zone of late flowage and associated gas blister in … … … spongy pāhoehoe." (Walker, 1989, p. 206). (B) Cross-sections of S- and P-type pāhoehoe from Wilmoth and Walker (1993).

vesicles can only expand into fluid lava, i.e., toward the lobe interior, while the cooling front that also migrates toward the lobe interior prevents the trailing part of the vesicle from closing. It is the pipe vesicles that give this lava type its name. When fresh, the surface is blue and glassy, so that another name is "*blue glassy*", although the blue sheen usually disappears in a few months, so the name is less useful. P-Type pāhoehoe results from eruption of lava that has had some residence time in the flow system, and has thus undergone prolonged degassing, but minimal cooling. In contrast, S-Type pāhoehoe tends to be erupted without storage and degassing so that the full bubble population is retained.

2.2.3. Pāhoehoe Sub-Types

There are a large number of other variants on the pāhoehoe-lava type. Seven of the most frequently encountered are described here. As we descend the list, we move from gas-rich pāhoehoe to lavas that are increasingly degassed and/or emplaced at lower effusion rates.

Pāhoehoe sheet flows are extensive, relatively flat, sheets of pāhoehoe lacking the bulbous surface morphology of the classic, low effusion rate, pāhoehoe expanse. As their name suggests, they flow as sheets fed at high effusion rates, typically hundreds of m³/s, and flow fronts can advance at several kilometres per hour. Sheets can be hundreds of meters across and very thin. A pāhoehoe sheet flow erupted from Pu'u 'Ō'ō during August 1997 was 20 cm thick, but about 215 m wide. Sheet flows tend to characterise near-vent areas, but they also can form from particularly vigorous lava-tube breakouts several km from the vent. Most sheet flows are S-type pāhoehoe, but occasionally they can be P-type. Sheet flows can be very vesicular. The vesicularity of the Pu'u 'Ō'ō lava mentioned above was 82−94%, and the vesicles were up to 1 cm across, oval and interconnected in the interior, but highly deformed where frozen into the thin layer of surface glass where they were stretched in the down flow direction. Sheet flow surfaces are characterised by parallel, low relief (cm-high) ridges. These are separated by several meters and orientated down flow and extend tens-to-hundreds, even thousands, of meters. Those on Krafla's (Iceland) 1984 flow surface are separated by 200−400 m and can be traced for 1.4 km down flow. They appear to be narrow shear zones and may mark the location of embryonic **levees**.

Shelly pāhoehoe is a gas-rich near-vent pāheohoe type with frothy surfaces and large voids inside. Usually the

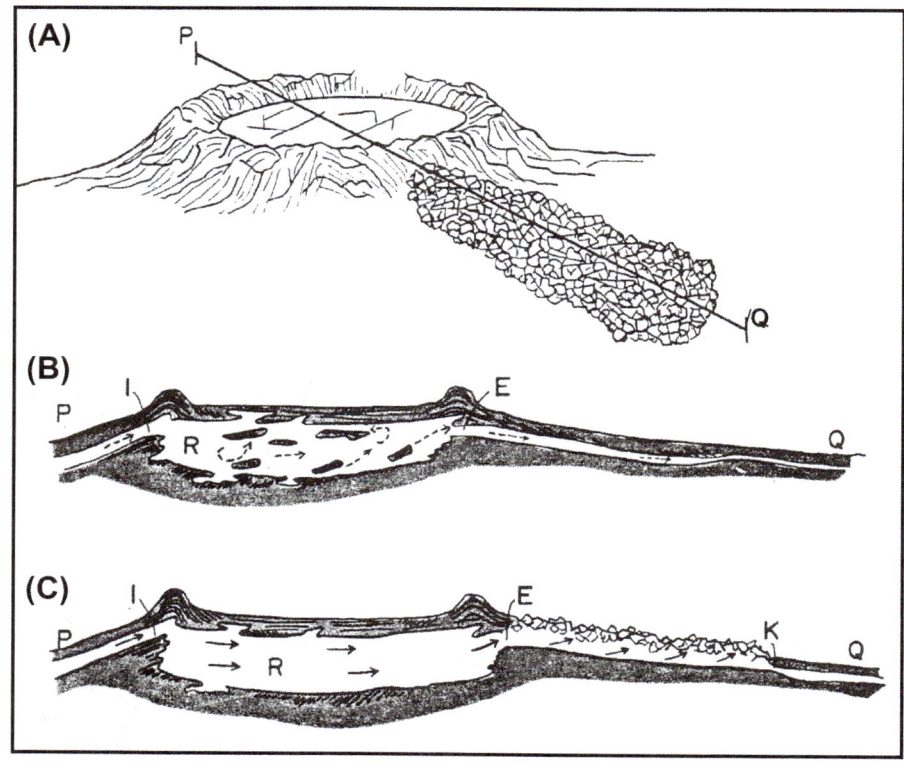

FIGURE 17.5   Schematic diagram and profiles (through P—Q) of Peterson and Tilling (1980) showing one mechanism for the formation of a slabby pāhoehoe flow. Lava stored in reservoir (R) cools and degasses, while being fed by an inlet tube (I) and feeding an outlet tube (E). Increase in supply from I increases lava pressure in R and **rate of flow** through E. Constriction K causes pressure to rise upstream, rupturing the overlying crust between E and K. Sluggish lava engulfs the fractured crust and the entire mass moves off as a flow.

upper crust is 1—2 cm thick and overlies an the internal void that is 0.1—1.5 m high. Hence shelly pāhoehoe units collapse easily when walked upon, the walker crashing chest deep into an empty space surrounded by the broken jagged crust. Swanson (1973) ascribes this form of lava to segregation of gas and lava within a still-fluid pāhoehoe flow, and the inability of the gas to escape the quenched upper skin. Upon solidification, the gas eventually escapes to be replaced by air which then fills the void.

Slabby, rough or rubbly pāhoehoe can form if a flow stalls for a period of time. Lava builds up beneath the crust and when the crust fails it breaks into slabs which are carried along on the once-again flowing lava (Figure 17.5). Slabby pāhoehoe was described by Peterson and Tilling (1980) as being "*composed of a jumble of jagged slabs (each) several centimetres thick and as much as 1 m in maximum dimension*". The slabs form a chaotic and loose jumble atop the flow, with the slabs being intact roughly planar fragments of the once-solid pāhoehoe surface. They can be tilted upward, sometimes standing near-vertical to be imbricately stacked. Slabs can also form by break up of a central slab of pāhoehoe in an active channel, or sheet flow, with the slabs then being transported down channel to be found littering the surface of the 'a'ā flow that the channel feeds. Rough pāhoehoe is also composed of broken fragments of once-solid pāhoehoe, but the debris is less slab-like. Fragments can range from angular to rounded and can be roughly equant. Also termed "*rubbly pāhoehoe*" by Keszthelyi et al. (2004) it is common in both pāhoehoe and 'a'ā **lava flow fields**.

Finger and entrail pāhoehoe is formed by an individual lobe extruded at an extremely low local effusion rate. If the crust fails locally, a small finger of pāhoehoe will be squeezed out. This will slowly form a stubby, finger-like projection typically around 10 cm across and extending 15—20 cm, sometimes farther. These can stand near-vertically and on emplacement are black with glowing tips. Localized fields of multiple short but narrow pāhoehoe lobes, also fed at extremely low local effusion rates, have sometimes been termed "*entrail lava*". In this case, hundreds of small units will stack vertically and horizontally to form intricate drapery's typically on extremely steep slopes, such as cliffs and fault scarps.

Toothpaste lava (Figure 17.6) was described by Rowland and Walker (1987) as lava that has the rheology of 'a'ā, but because it flows very slowly due to near-horizontal underlying slopes, for example, it does not break into clinkers. A secondary formation mechanism is when late

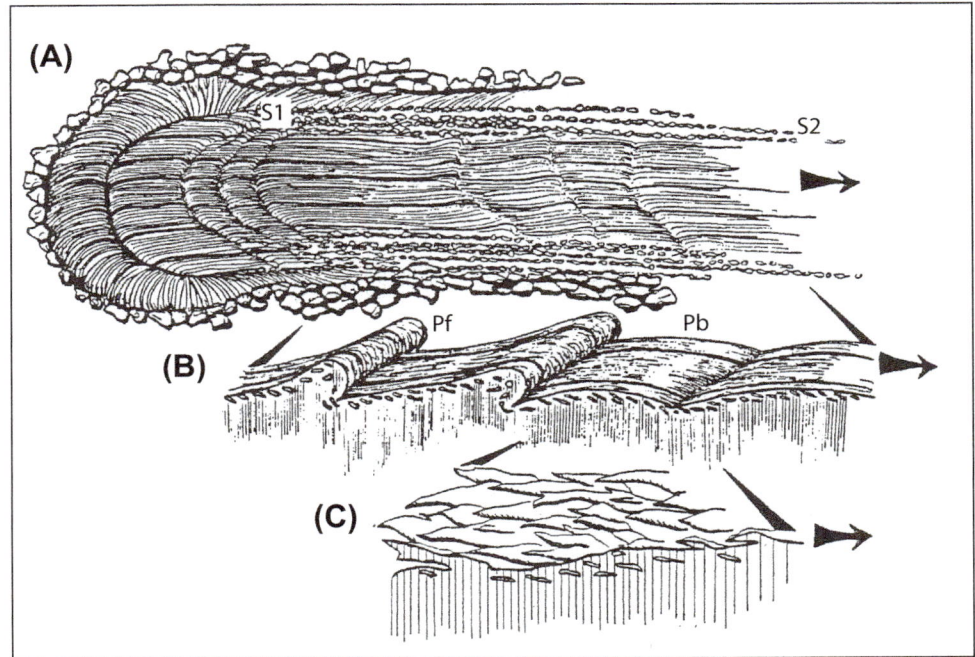

FIGURE 17.6 Sketch of features characteristic of toothpaste pāhoehoe by Rowland and Walker (1987). Flow direction is left to right; arrows approximately 1, 0.5 and 0.01 m long in a, b and c, respectively. (A) "Toothpaste tongue issuing from curved bocca. S1-S2, lateral shear zone evidenced by imbricate shearing and clinker. Note how longitudinal lineations maintain same spacing along entire length of tongue. (B) Cut away view showing pulse buckles (Pb) and pulse flaps (Pf), features of discontinuous extrusion. Note how vesicles deformed because crust was retarded relative to underlying lava. (C) Close up showing spines pointing back toward bocca."

stage lava oozes from within the primary pāhoehoe field where it has been stored (Peterson and Tilling, 1980). Both processes form tongues and lobes of degassed lava, containing a few, large, vesicles. Stretching of the large vesicles over the surface of the sticky extrusion results in the formation of spines, typically 1−5 cm long and 1 cm wide. These are remnants of the stretched vesicle walls. Spines can also be the remnants of stretched fibers of pasty lava that adhered to the ceiling of the bocca that fed the flow. These then became stretched down flow, before breaking off and being carried along on the flow surface. Thus, this lava type may also sometimes be called "*spiny pāhoehoe*". Just below the surface a band containing a few large vesicles stretched in the flow direction can usually be found. The surface may be characterized by grooves and ridges, orientated parallel to the flow direction, that are inherited from irregularities in the bocca openings which gouge the extruding lava surface. Commonly there are also transverse undulations with wavelengths of tens of cm, which likely result from pulsing of the flow. Flow units are significantly thicker than for primary pāhoehoe, typically being 60 cm, and up to 150 cm. Termed "*slab-crusted flow*" on Etna by Guest and Stofan (2005), a toothpaste lava stream often consists of a central slab of spiny pāhoehoe bounded by zones of clinker, with the central slab often breaking up down flow.

Transitional pāhoehoe is likely the result of emplacement of relatively degassed and crystallized lava at relatively high local effusion rates and advance velocities. The surface is rough and dull, and the interior relatively dense with large, sheared vesicles. Frozen into the surface are occasional swirls of 'a'ā clinker that show signs of rotation in their fabrics, as well as coils of pāhoehoe. However, the surface remains intact, and clinkers are firmly rooted to the core, so that this is not loose-surfaced, rough pāhoehoe.

2.3. Block Lava

Blocky surfaces are typical of dacitic and rhyolitic lava flows, i.e., silicic lava flows, as well as some andesitic and basaltic-andesite lava flows. Block lava is distinguished from 'a'ā by the fact that clasts are "*relatively smooth polyhedral blocks bounded by dihedral angles, lacking the exceedingly rough and spinose character of typical 'a'ā*" (Macdonald, 1953, p. 182). Some silicic lava flows lack blocks and instead have a surface that comprises a coarse grained rubble. The 1999−2003 silicic lava flow of Santiaguito had such a crust. This was comprised of subrounded boulders up to several meters across.

The internal structure of a block lava flow is much like that of an 'a'ā flow, comprising a basal and surface breccia,

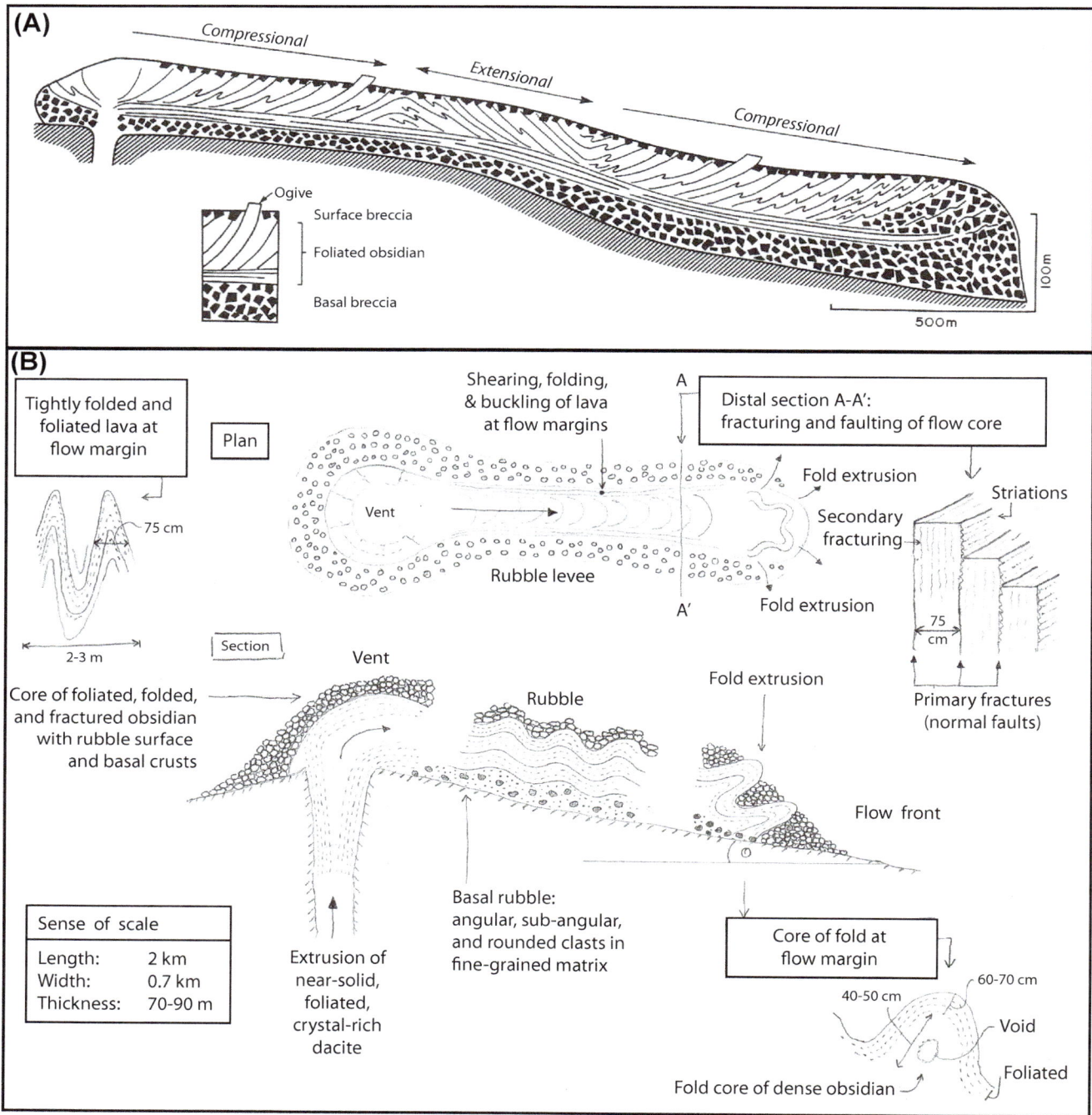

FIGURE 17.7 (A) Cross-section through the length of the Rocche Rosse coulee with generalised foliation patterns. This is Figure 4.28 of Cas and Wright (1987), which in turn is modified from Hall, S.H. (1978), *The stratigraphy of northern Lipari and the structure of the Rocche Rosse rhyolite flow and its implications* (PhD Thesis, University of Leeds, UK). (B) Schematic plan, section and detail of key structures in the San Pietro silicic flow units.

and a coherent core (Figure 17.7(A)). However, block lava flows are much thicker than their 'a'ā counterparts, being tens or even hundreds of meters thick. The cores of block lava flows are dense, vesicle-free to vesicle-poor and are often crystal-rich. The Chao dacite in Chile, for example, has a phenocryst content of up to 40–60 vol% (de Silva

et al., 1994). Some interiors are obsidian, such as Lipari's Rocche Rosse flow in the Aeolian Islands of Italy. At Vulcano's Pietre Cotte flow, also in the Aeolian Islands, the interior is composed of hundreds of cm-thick alternating bands of obsidian and pumice. Pumice layers have grooves in them, typically 10–20 cm long, suggesting movement by short slip events during which the foam layer became gouged by irregularities in the underlying glass layer. Indeed, although block lavas may move as a true viscous flow close to their source vents. Macdonald (1953, p. 183) suggested that "*examination of old flows suggests that block lava flows more commonly advance by sliding or gliding over the surface of the underlying material . . . . in much the same manner as some glaciers*".

Ramping and fold structures are common (Figure 17.7). The silicic lavas of Sardinia's San Pietro Island show multiple generations of folds, foliations, thrusts and fractures, with folding effecting the whole flow thickness (Cioni and Funedda, 2005). The main surface expression of internal folding on thick silicic lava flows are called ogives. These are cross-flow ridges that are convex down flow with limbs that run close and parallel to the flow margins (Figure 17.7(B)). Ogives on the Chao dacite are up to 30 m high and have wavelengths of 50–100 m (de Silva et al., 1994). Flow front advance velocities are extremely low. Those of Santiaguito's silicic lava flow between 1999 and 2002 were between 2 and 13 m per day. Emplacement durations also can be very long, where de Silva et al. (1994) estimated that the 22.5 km^3 Chao dacite was erupted over a period of 100–150 years.

2.4. Thermal and Rheological Structure of a Lava Interior

Rare measurements of the internal thermal structure of active lava were made by Hon et al. (1994) on active pāhoehoe at Kīlauea. Their results allow us to define five thermally and rheologically different layers in an active lava, as sketched for an active basaltic pāhoehoe and 'a'ā flow in Figure 17.8. The depth of each layer will depend on lava composition and the following description is for Hawaiian, basaltic, pāhoehoe.

Layer 1: Surface Crust. This is a relatively low temperature solid zone of brittle lava, which begins life as an extremely thin glassy selvage. Rapid heat loss by radiation results in reduction of the surface temperature by 50–300 °C almost immediately after emerging from the lava source. Measurements on active pāhoehoe at Kīlauea revealed surface temperatures of 850–1000 °C after 1 s, compared with an interior temperature of 1150 °C. This means that the surface is rapidly air quenched to form the selvage. According to Hon et al. (1994) the base of surface crust is defined by the 800 °C isotherm, whose depth (D_{800}, in m) increases with time (t in hours) following.

$$D_{800} = 0.0473\sqrt{t} - 0.0233.$$

Layer 2: Upper Visco-Elastic Layer. This is a partially molten elastic layer whose top is defined by the 800 °C isotherm and whose base is defined by the 1070 °C

isotherm. The base of this layer was also encountered during drilling at Hawaiian lava lakes where the drill bit always began to fall under its own weight when it arrived at the 1070 ± 5 °C isotherm (Wright and Okamura, 1977). At this temperature, crystals and melt are present in roughly equal proportions. According to Hon et al. (1994) the depth of the 1070 °C isotherm (D_{1070}, in m) increases following.

$$D_{1070} = 0.0779\sqrt{t},$$

so that the variation in thickness of the upper visco-elastic layer (H_{VE}) with time can be obtained from

$$H_{VE} = \left(0.0779\sqrt{t}\right) - \left(0.0473\sqrt{t} - 0.0233\right).$$

Layer 3: Core. This is a low **viscosity** and high temperature zone at the interior of the flow. Although fluid, the lava core is a mixture of melt, crystals and bubbles. For a simple pāhoehoe unit, core thickness will decrease with time as the upper and lower crusts thicken. However, for an inflating flow (see below), the core thickness will increase as long as supply is maintained. It will only begin to thin once supply, and **inflation**, ends.

Layer 4: Lower Visco-Elastic Layer. This is a second, lower, visco-elastic layer between the flow core and the basal crust. Because it is insulated by the overlying hot core, its thickness ($H_{VE\text{-}low}$, in cm) will increase more slowly than the upper visco-elastic layer, increasing with

$$H_{VE-low} = 5.41\sqrt{t}.$$

Layer 5: Basal Crust. This is a solid glassy layer in pāhoehoe, or zone of clinker in 'a'ā, between the lower visco-elastic layer and the cold underlying surface.

2.5. Inflation

Both pāhoehoe and 'a'ā can undergo inflation. Inflation is the process whereby fluid lava is injected into a unit that has already established basal and surface crusts, so that the unit balloons outwards with surfaces expanding in all directions. Most of the resisting strength to pressurization of the unit is in the visco-elastic crust. This expands to contain and accommodate the new fluid volume flowing into the core (Figure 17.9). Instead, the brittle crust will fracture to accommodate the expansion, with cracks widening and deepening with time as inflation progresses.

Typical pāhoehoe lobes are usually a few tens of cm thick upon emplacement, but subsequent inflation over a period of hours to days can produce a final unit that is many meters thick. The measurements of Hon et al. (1994) showed that the rate of surface uplift due to inflation decayed following a power law, so that the flow thickness

Pāhoehoe   'a'ā lava flow

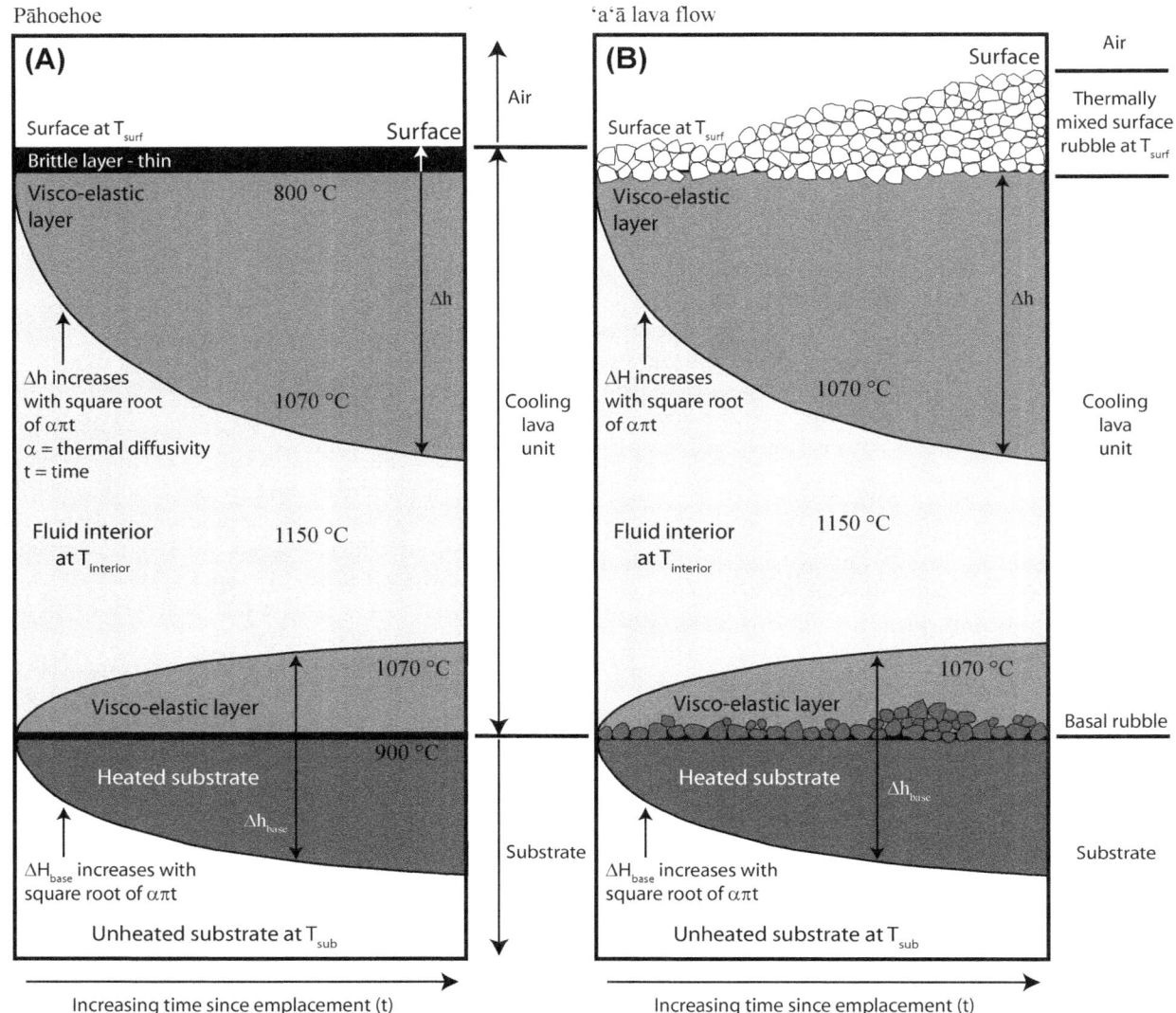

FIGURE 17.8   Thermally-defined layers in (A) pāhoehoe and (B) 'a'ā, and the variation in depth of each boundary with time. *Modified from Figure 4.17 of Harris (2013).*

(H in m) at any point in time (t, in hours) could be approximated from

$$H_{VE-low} = 0.677t^{0.278}.$$

Inflation results in three main features: tumuli, sheet lobes and **lava rise pits**.

2.5.1. Tumuli

Walker (1991) described tumuli as *"positive topographic features that are common on Hawaiian pāhoehoe lava flow fields, particularly on shallower slopes."* Such features have been observed and described since the nineteenth century. The inflating fluid mass causes the brittle crust to crack into plates which are then up-tilted. The resulting

tumulus is typically dome-like in section, rounded to oval in plan, with a main axial cleft running parallel to the long axis, from which several lesser clefts branch (Figure 17.10).

2.5.2. Sheet Lobes and Lava Rises

Termed *"lava rises"* by Walker (1991), these features have roughly horizontal surfaces, are tens to hundreds of meters across, and are bounded by escarpments of up-tilted crustal slabs (Figure 17.11). They can be thought of as very large, flat-topped, tumuli. They were named *"sheet lobes"* by Self et al. (1998), with lava rise being a term better used to describe inflation-induced features in general (Walker, 2009). Formed by coalescence, inflation and thickening of individual pāhoehoe units, sheet lobes can be 10s of meters thick and several kilometres across in flood basalts. In

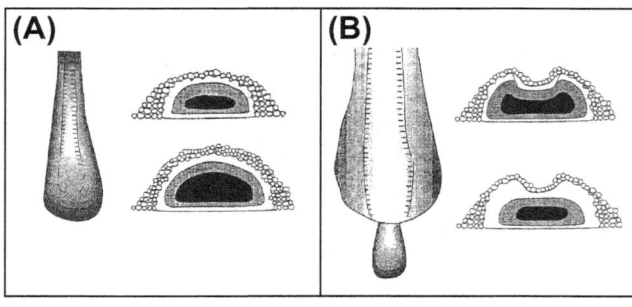

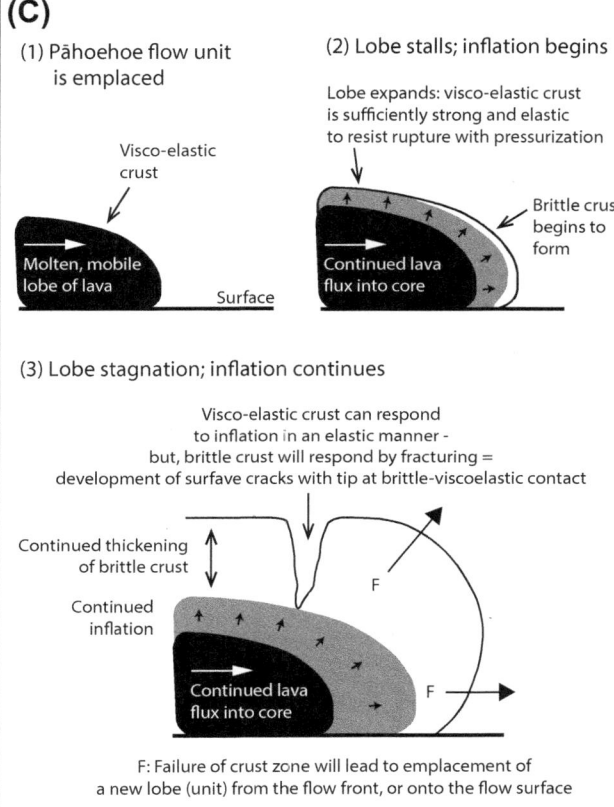

FIGURE 17.9 Sketches of (A) inflation and (B) deflation of an 'a'ā flow front by Calvari and Pinkerton (1998). The flow front stalls and the fluid interior (black), contained within the viscoelastic crust (grey), expands; ballooning the flow front zone vertically and laterally. When the internal pressure exceeds the strength of the viscoelastic crust, the crust fails and lava flows out from an opening in the flow front. Withdrawal of volume due to the breakout causes deflation. This repeated process results in a tube characterised by caverns (marking zones of inflation) linked by narrow passages (marking zones of breakout). (C) Sketch summarizing the inflation process in pāhoehoe.

Kilauea's Kalapana flow field sheet lobes of the order of 10 m thick and 100–150 m across can be found. According to Hon et al. (1994), "*increased hydrostatic pressure is distributed evenly through the liquid lava core of the flow, producing essentially uniform uplift of the otherwise stationary crust*".

2.5.3. Lava-Rise Pits

Pits are commonly found in lava rises. They generally widen downwards, with concave walls and overhanging rims of thin pāhoehoe slabs (Figure 17.11). These form where the lava failed to inflate. They are not formed by collapse, hence the name assigned to them by Walker (1991). Occasionally the pre-flow surface is exposed in the bottom of a lava-rise pit.

3. DISTRIBUTION SYSTEMS

Lava is distributed from the vent to the flow front by open channels and **lava tubes**. The flows that these feed may consist of a single unit, several units, or thousands to hundreds-of-thousands of units. Walker (1972) made the following definitions:

1. **Flow Unit.** This is lava which has "*a top which cooled significantly and solidified before another flow-unit was superimposed on it. Each flow-unit is a separate cooling unit … … …*".
2. **Simple Lava Flow.** Such a flow may not be divisible into flow-units, or is composed of just a few units.
3. **Compound Lava Flow or Lava Flow Field.** This is "*a lava which is divisible into flow-units*". Such lava flow fields may be compound either laterally, vertically or both. That is, multiple simple flows erupted during the same eruption may be emplaced next to one another, on top of one another, or both.

Simple lava flows tend to be emplaced during short duration eruptions. Such flows are long and narrow, so that the ratio of length to width is high. Even a simple lava flow will likely contain more than one unit. We may, for example, have a principle 'a'ā unit within which a channel develops. Overflows from the channel then emplace 'a'ā and pāhoehoe units on top of the channel levees, and late stage-flow in the channel emplaces further units within the channel (Harris et al., 2009).

In contrast, lava flow fields tend to be emplaced during long duration eruptions. In these cases, once the first simple flow has been emplaced, subsequent simple flows will build next to, or on top of, the previously emplaced simple flows, so that the flow field builds by widening and/or thickening. Thus flow fields can be as broad as they are long, so that the ratio of length to width is low; flow fields can be extremely thick.

Unit is probably a better term for lava in outcrop. Here the units can be divided in vertical sequence according to the presence of cooling surfaces, i.e., the presence of an 'a'ā and pāhoehoe crust. At the outcrop-scale, it may not be known whether the unit is part of a simple or compound flow.

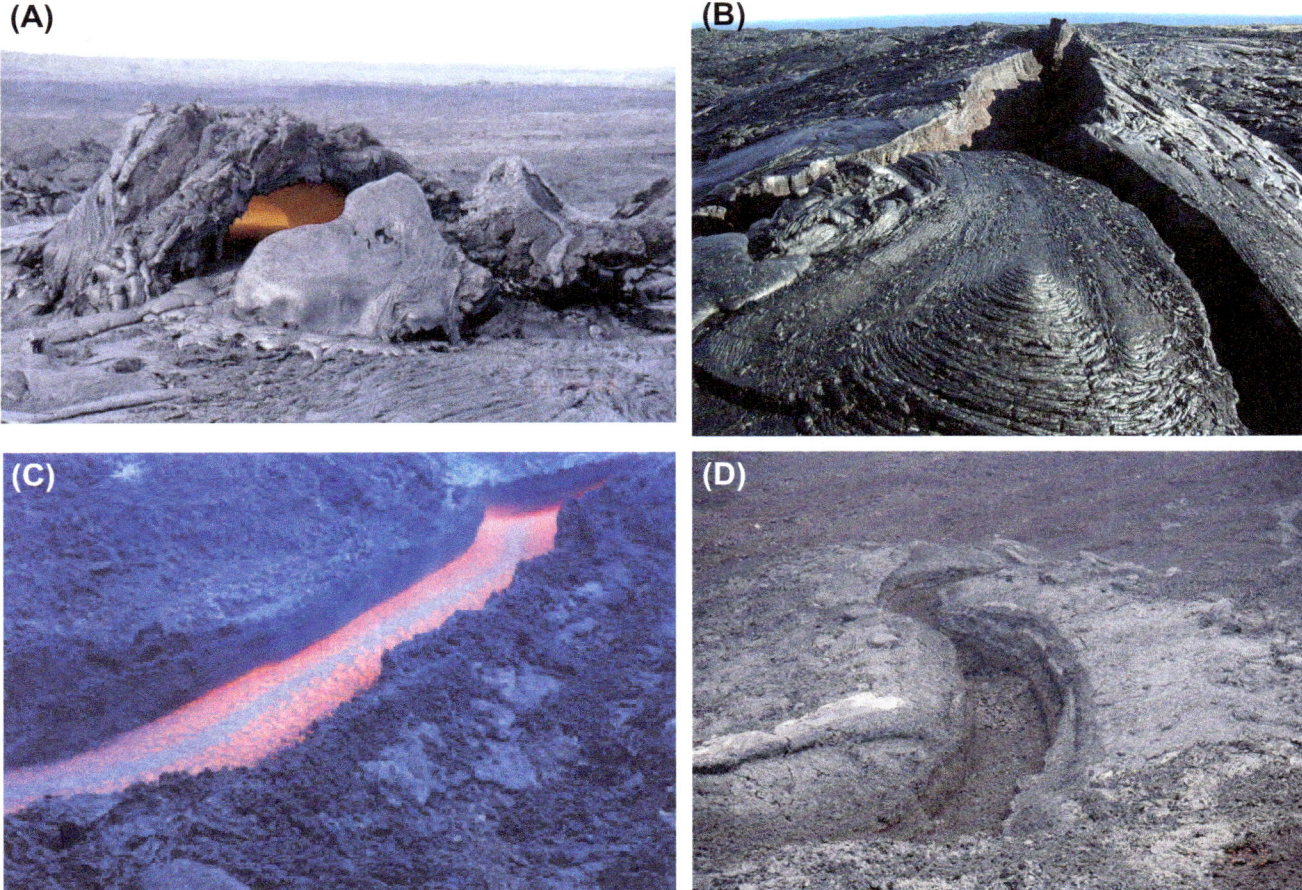

FIGURE 17.10 (A) Active tumulus at Kilauea. Tumulus has failed, where two blocks cut by the axial and secondary fractures have rolled forward, releasing the fluid interior, causing the tumulus to empty and expose the—normally full of fluid lava—tumulus interior. Lava is still flowing into the tumulus from a source at the back, right. Tumulus is about 2 m high. (B) View along the axial cleft of a tumulus at Kilauea. Cleft is 1.5 m wide, and slabs each side are up-tilted at an angle of 30°. (C) Lava channel active at Mt. Etna in May 2001. Channel is 3 m wide, with pāhoehoe and 'a'ā overflow levees. In the channel we see a central slab of cooler pāhoehoe, surrounded by hotter zones of 'a'ā forming in the lateral shear zones. Note overhanging rims due to repeated overflow and that the flow level is below bank. (D) Drained lava channel in Kilauea's Mauna Ulu flow field. Channel is 6 m wide, surrounded by pāhoehoe overflow levees and floored by a late-stage 'a'ā flow.

3.1. Channel Systems

Channels are molten lava streams whose surface is exposed to the sky. The stream is contained between two banks of stationary lava termed "*levees*" (Figure 17.10(C)). Channels develop because the lava margins cool faster and experience more drag than the centre. These differences generate a positive feedback to the point that the margins stop flowing altogether while the centre continues to flow rapidly, meaning that the cross-channel velocity gradient becomes increasingly steep.

3.1.1. Levees

Levees often form early in the life of a lava flow and usually extend almost the entire length of the flow system. For example, by length, 97% of Mt. Etna's 2001 LFS-1 flow

was channelized. Sparks et al. (1976) devised a four-fold classification of levee types:

1. **Initial levees** are formed by stationary lava that each side of, and just behind, an advancing flow front.
2. **Rubble levees** are a form of initial levee comprising flow front breccia that has been pushed aside by the advancing axial stream.
3. **Overflow levees** are formed by lava that overflows the channel to build units on top of the existing levees. Repeated overflows and drain backs can build overhanging levees, giving the channel a triangular shape (Figure 17.10(C)).
4. **Accretionary levees** form during low, below-bank, flow. Slow-moving pasty blebs of lava at the channel margin will roll against the levee wall. If they stick, other blebs may become stuck behind them and/or stick to their outer

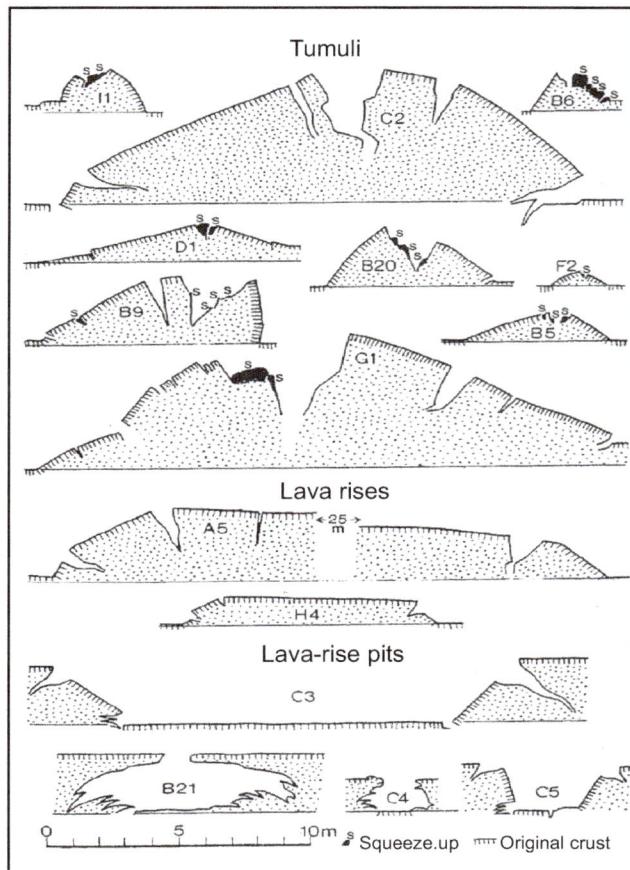

FIGURE 17.11 (A) Profiles, on the same scale, across typical tumuli, lava rises and lava-rise pits of Walker (1991). Note the flat, plateau-like form of lava rises, as opposed to the dome-like form of the tumulus. The term "lava rise" has since been replaced by "sheet lobe", lava rise now being a generic term for any feature resulting from inflation.

edge, thereby building an accretionary levee that grows toward the centre of the flow from the inner wall of the levee. These form bench-like structures, where the individual blebs that comprise the levee are apparent as a rough, bumpy surface of welded, cm-scale, clasts, often elongated in the down flow direction.

Lipman and Banks (1987) also defined "*deformation levees*". These are the result of lateral pressure exerted by the flowing lava on the levees. This pressure can push stationary, but still plastic lava, on the inner edge of the levee upwards into ridges parallel to the flow direction.

3.1.2. Channel Zones

Based on observations of Mauna Loa's 1984 channel system, Lipman and Banks (1987) defined four channel zones (Figure 17.12). This system can be applied to channels of all lengths and types:

1. **Flow toe.** This is the flow front where there is little velocity differential between the flow centre and margins so that the front moves forward as a single mass.
2. **Zone of dispersed flow.** Within this zone, some parts of the lava start to slow relative to other parts, although the differences are not great. Shear zones, usually marked either by flow-parallel depressions and/or by flow-parallel ridges, develop between parts of the lava moving at different velocities.
3. **Transitional channel zone.** This is marked by a distinct faster-flowing central stream, bounded by lava still capable of deformation and slow forward motion. Outboard of the central stream lava is beginning to stagnate and the initial levees are beginning to form.
4. **Stable channel zone.** In this zone, the levees have stagnated completely and all motion is confined to the central channel. Levees become ever more complex as the overflow, accretionary and deformation processes described above occur. Thus, older and/or proximal levees are more complex than young or distal levees.

3.2. Tube Systems

Also termed "*pyroducts*", Greeley (1987) defines a lava tube as "*a conduit beneath the surface of solidified lava through which molten lava flows*". According to Greeley (1987), the key distinction between a **lava channel** and a **lava tube** is:

in regard to the roof crust, so long as the crust remains mobile and free-floating on the active flow, the structure is regarded as a channel; sections in which the crust is continuous (and stationary) across the active flow and fixed to the immobile parts of the flow are considered lava tubes.

(Greeley, 1987, p. 1590)

Although being much more common in pāhoehoe, tubes can also form in 'a'ā (Calvari and Pinkerton, 1998). They can vary in length from a few meters to 100 kilometres or more. Tubes in the Undara volcanics of North Queensland (Australia) are more than 100 km long. They can branch and braid, undergo dramatic changes in height and width over extremely short distances, and are subject to collapse both during eruptions and after (Figure 17.13). They can also have multiple tiers and, when active, may be marked by a line of gas, a warmer surface zone, a linear zone of inflation or a line of skylights, these being holes in the tube roof through which the flowing lava can be observed.

3.2.1. Tube Formation

Formation of tubes by surface cooling was recognised by Rev. William Ellis during his 1823 tour of Hawaii.

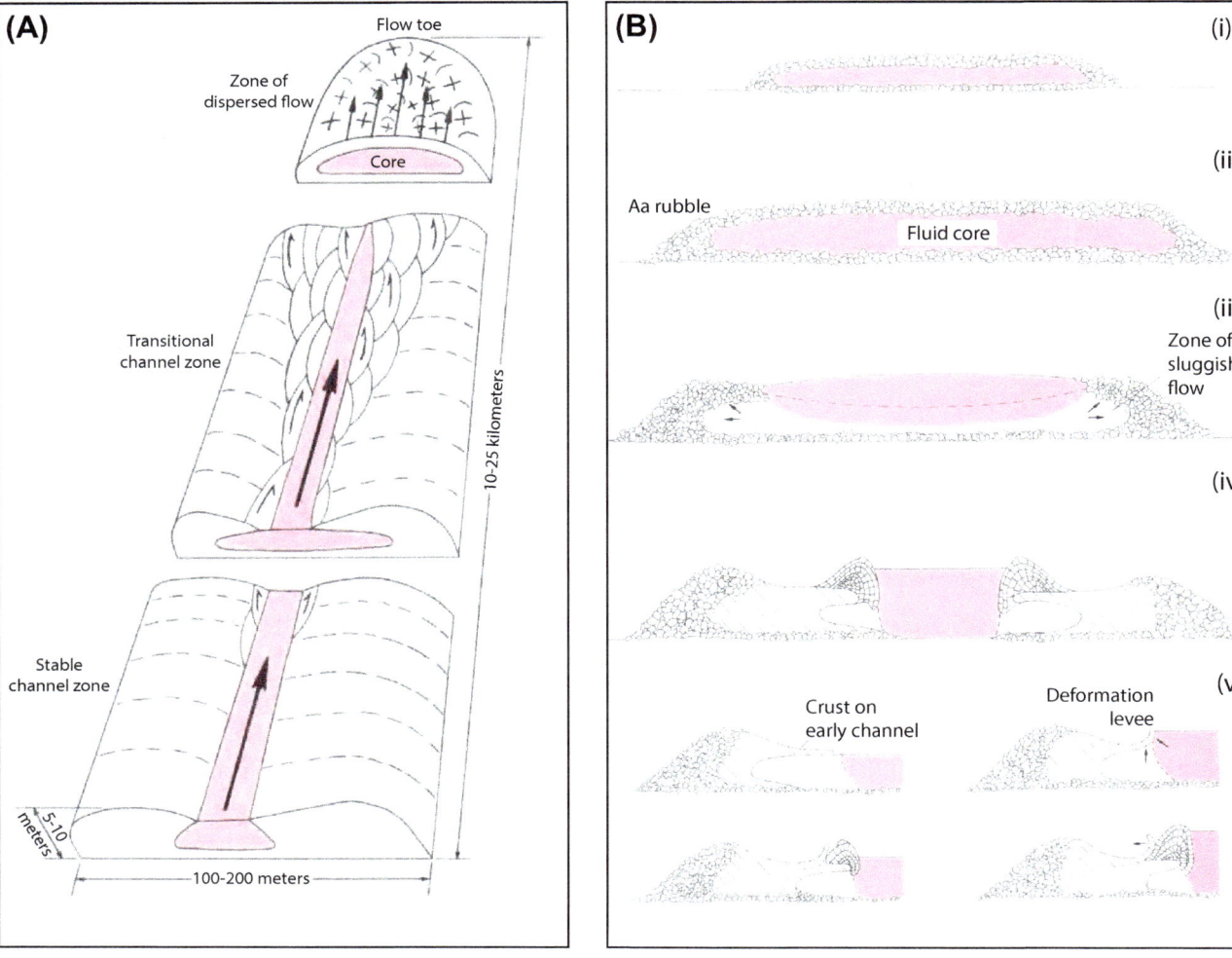

FIGURE 17.12 The definitive lava channel: Mauna Loa 1984. (A) Channel zone scheme of Lipman and Banks (1987, Figure 57.5, p. 1533). (B) Development of 'a'ā channel and levees from emplacement of (i) initial dispersed flow, through (ii—iii) formation of transitional (rubble) levees, finishing with formation of (iv) overflow levees, and finally (v) deformation levees. Each zone is thus characterised by an increasingly complex levee morphology. *Modified from Lipman and Banks, 1987, Figure. 57.9, p. 1536—1537.*

However, although surface cooling is one primary way to create a tube, there are several other ways a tube can form. Peterson et al. (1994) listed and sketched five formation mechanisms:

1. **Crustal growth.** Accretion of pasty lava blebs to a channel margin can build an accretionary levee. If accretion continues across the channel surface from both sides, a stationary, hanging—that is, lava continues to flow beneath it—levee can extend entirely across the channel hence roofing it over to cause a tube with a semi-circular, square, or rectangular form depending on the shape of the original channel.

2. **Repeated channel overflow.** Repeated overflows form a channel with increasingly overhanging levees, with each overflow plastering a new layer of lava over the levee rim. When the rims join, an arched roof is created

and the tube is formed which typically has a light-bulb or triangular section.

3. **Jamming.** Plates of solidified crust floating down stream will, if large enough, become jammed at channel constrictions. Lava that oozes up between the fragments of crust will weld them into a solid roof. Other plates arriving behind the blockage will build the roof upstream.

4. **Lobe advance.** Progressive lobe-by-lobe extension of pāhoehoe with a solid upper crust will build a pathway through which lava is fed to the flow front. The new break out may then stall, inflate, and breakout again. The crust of the stalled flows behind the breakouts will form a tube with a bulbous plan, each bulge being a zone of stalling and inflation; bulges being linked by narrower tube sections that mark the zone of breakout. Tubes in 'a'ā can from in a similar manner (Calvari and Pinkerton, 1998).

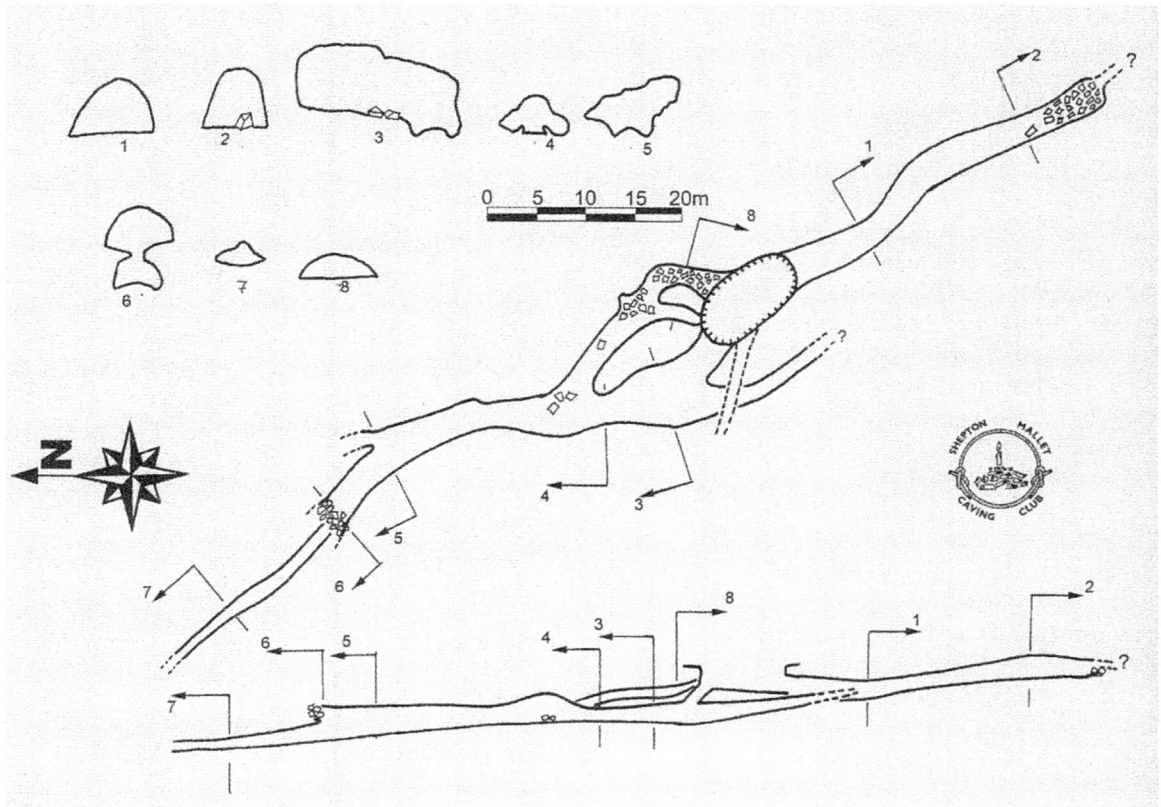

FIGURE 17.13 Map (and cross-sections) of a section of lava tube in Iceland's Syðri-Lautarhellir flow field showing skylights, dimension changes, multiple levels and collapse rubble. *Map drafted by the Shepton Mallet Caving Club, published in Hrarsson (2008)* Hellahandbkin, Leiðsögn um 77 islenska hruanhella *(Mal og Menning, Reykjavik), p. 78 (in Icelandic).*

5. **Littoral.** When lava enters the ocean, waves can abruptly chill the crust across the entire surface of a molten stream crossing through the surf zone, thus forming a tube across that zone.

4. FLOW DYNAMICS, COOLING AND RHEOLOGY

In 1925 Harold Jeffreys introduced an equation that allows calculation of the mean velocity ($\bar{v}$) of a Newtonian fluid flowing in a channel that is (i) semi-circular or square, and (ii) much wider than it is deep:

$$(i) \quad \bar{v} = \frac{d^2 \rho g \sin\left(\theta\right)}{8\eta} \quad (ii) \quad \bar{v} = \frac{d^2 \rho g \sin\left(\theta\right)}{3\eta}$$

Here, d is the depth of the lava in the channel, ρ is lava density, g is acceleration due to gravity, θ is the angle of the slope down which the lava is flowing and η is the lava viscosity. Nichols (1939) used a re-arranged version of this equation to calculate the viscosity of Mauna Loa's Alika (1919) flow. Using the mean flow velocity (4.9 m/s), depth (6.1 m), density (1400 kg/m^3) and slope (7.2°) he obtained

a viscosity of 4.3×10^3 Pa s. If we compare this with silicic lavas active at Santiaguito during 1999–2002 with a typical flow velocity of 7.5 m per day, flow depths of 30 m, a lava density of 2500 kg/m^3 and slope of 11°, we obtain a viscosity of 1.6×10^{10} Pa s.

However, Jeffreys formula is for Newtonian fluids, specifically water. Such fluids will flow when an infinitesimal amount of force or shear stress is applied. For a Bingham fluid, however, the shear stress must exceed a critical value before flow will occur. The shear stress at the base of a flow (τ) can be estimated from.

$$\tau = d\rho g \sin\left(\theta\right)$$

so that for the 1919 Mauna Loa and 1999–2000 Santiaguito flows we have basal shear stresses of 10^4 and 10^5 Pa, respectively. The critical value of shear stress that needs to be reached before flow will occur is determined by the **yield strength** of the fluid (τ_0). This, in turn, is related to a critical flow thickness (h_0) below which the shear stresses are insufficient to force the fluid to move down a given slope:

$$\tau_0 = h_0 \rho g \sin\left(\theta\right)$$

Lavas tend to be Bingham fluids. Basaltic mixtures of fluid, crystals and bubbles typically have a yield strength of around 10^3 Pa, that of silicic lavas being around 10^5 Pa. Measurements on active lava at Etna, for example, have indicated yield strengths of 370 ± 30 Pa, with estimates based on the thickness of the Chao dacite being 8×10^5 Pa.

Thus, for our two cases, the critical thicknesses the lava must attain before flow will occur are 0.6 m and 21.4 m, respectively. Because the Jeffreys equation does not have a term that takes into account the yield strength effect, it will over-estimate velocities and under-estimate viscosities if applied to a Bingham fluid. Moore (1987) thus provided revised versions of this equation which take into account the non-Newtonian character of lava. For a channel that is wider than it is deep we now have:

$$\bar{v} = \left(\frac{d^2 \rho g \sin(\theta)}{3\eta}\right)\left(1 - \frac{3}{2}\frac{\tau_0}{\tau} + \frac{1}{2}\left(\frac{\tau_0}{\tau}\right)^3\right).$$

Whereas for a semi-circular or square channel.

$$\bar{v} = \left(\frac{r^2 \rho g \sin(\theta)}{8\eta}\right)\left(1 - \frac{4}{3}\frac{\tau_0}{\tau} + \frac{1}{3}\left(\frac{\tau_0}{\tau}\right)^4\right)$$

Effectively, the term on the right is a damping factor that takes into account the degree to which the basal shear stress exceeds the yield strength. For lava at Kīlauea, the dampening factor has a value of 0.9; for Santiaguito, where yield strengths are higher, it is 0.1. The recalculated Bingham viscosities are 5040 Pa s and 1.4×10^{11} Pa s, respectively. To understand lava dynamics, we therefore need to define the controls on viscosity and yield strength.

4.1. Viscosity Relationships

The dynamic viscosity (η) of lava will increase as temperature (T) decreases. For lavas below the liquidus, and in which crystallization has begun, the Vogel-Tammann-Fulcher (VTF) equation has been adopted as an empirical means of providing a best fit to available temperature and viscosity data for a given melt:

$$Log\ \eta(T) = D + \frac{E}{T(K) - F}.$$

where D, E and F are adjustable best–fit parameters that depend on melt composition. Thus, as the lava cools the viscosity will increase logarithmically.

As the lava cools, crystals will grow adding a population of solid particles to the mixture. To take into account the effect of crystals on the mixture viscosity, the Einstein-Roscoe equation is typically applied.

$$\eta(T, \phi) = \eta(T)(1 - R\phi)^{-2.5}$$

in which ϕ is the volume fraction of crystals and R is equal to one over the maximum crystal packing content, this being the crystal content that makes additional movement rheologically impossible. This value is generally thought to be in the range 50–70% crystals by volume.

4.1.1. Viscosities of Lavas

Table 17.2 summarizes some viscosity measurements made at lava flow systems. For basaltic flows, we see that viscosities are typically in the range 10^2–10^3 Pa s, increasing towards 10^5 and 10^6 Pa s for cooler and more crystallized basalts. Viscosities increase with silica content, so that the interior of an andesite may have a viscosity of 10^5–10^7 Pa s, increasing to 10^7–10^{10} Pa s for rhyolites.

4.2. Down-Flow Cooling

As lava moves away from the vent, heat losses due to radiation and convection will cause the lava to cool. Radiative heat losses per meter advanced (W m^{-1}) can be estimated from

$$Q_{rad} = \varepsilon\sigma(T_{Lava}^4 - T_{amb}^4)w$$

in which ε is the surface emissivity, σ is the Stefan-Boltzman constant (5.67×10^{-8} W m^{-2} K^{-4}), T_{Lava} is the lava surface temperature (in K), T_{amb} is the temperature of the environment into which the surface is radiating (in K), and w is the flow width (in m). Heat loss due to convection can be calculated using the convective heat transfer coefficient (h$_c$),

$$Q_{conv} = h_c(T_{Lava} - T_{amb})w$$

Over active lavas, the convective heat transfer coefficient has generally been found to have a value in the range 10–125 W m^{-2} K^{-1}. These two heat losses can be used to approximate cooling per unit distance, dT/dx, of a lava core.

$$\frac{dT}{dx} = \frac{Q_{rad} + Q_{conv}}{E_r \rho \left(c_p + L\frac{d\phi}{dT}\right)}$$

in which E_r is the lava volume flux, c_p is the lava specific heat capacity, L is the latent heat of crystallization and $d\phi/dT$ is the amount of crystallization per degree cooled. It turns out that for surface temperatures $>500\,°C$ heat loss is radiation-dominated and all other factors are negligible, so that for these temperatures cooling rate will increase with the fourth power of the surface temperature. That is,

$$dT/dx \approx aT_{Lava}^4$$

For a channel 10 m wide fed at 100 m^3/s (assuming $\varepsilon = 0.95$, $\rho = 2100$ kg/m^3, $c_p = 880$ J/kg K, $L = 3.5 \times 10^5$ J K^{-1} and $d\phi/dT = 0.0018$ K^{-1}), coefficient a has a value of 1.73×10^{-15} K^{-4} m^{-1}. For this case, a surface

TABLE 17.2 Some Viscosity Measurements on Lavas with a Range of Compositions, and with a Range of Temperatures and Crystallinities

Location	Viscosity (Pa s)	Approach	References
Basalt			
0–30% crystals, T ≈ 1130–1150 °C			
Kilauea (Makaopuhi)	650–750	Viscometer	Shaw et al. (1968)
Kilauea (Pu'u 'O'o)	60–2000	Jeffreys	Fink and Zimbleman (1990)
Kilauea (pāhoehoe)	600–6000 (12 000 toothpaste)	Crystal settling	Rowland and Walker (1988)
Mauna Loa (1984)	100–2000	Modified Jeffreys	Moore (1987)
30–45% crystals, T ≈ 1070–1090 °C			
Etna (1966)	$3–38 \times 10^3$	Jeffreys	Walker (1967)
Etna (1966)	$5–74 \times 10^3$	Jeffreys	Tanguy (1973)
Etna (1971)	$10^3–10^5$	Viscometer	Gauthier (1973)
Etna (1975)	9400 ± 1500	Viscometer	Pinkerton and Sparks (1978)
Etna (1983)	1385–1630	Viscometer	Pinkerton and Norton (1995)
Etna (1991–93)	$8–19 \times 10^3$	Jeffreys	Calvari et al. (1994)
25–50% crystals, T ≈ 1000 °C			
Stromboli (2003)	$10^5–10^6$	Jeffreys	Harris et al. (2005b)
Andesite			
Mount Hood (60–61 wt% SiO_2)	$10^5–10^7$	Lab-based (900–1000 °C)	Murase and McBirney (1973)
Colima (59 wt% SiO_2)	$10^9–10^{10}$	Jeffreys	Navarro-Ochoa et al. (2002)
Trachyte & dacite			
Karisimbi (61 wt% SiO_2)	$7.5 \times 10^{10}–5.2 \times 10^{11}$	Surface folds	McKay et al. (1998)
Santiaguito (62.5 wt% SiO_2)	$4 \times 10^9–6.9 \times 10^{10}$	Jeffreys	Harris et al. (2004)
Chao (67–69 wt% SiO_2)	4.9×10^9 (interior) $10^{15}–10^{24}$ (surface)	Surface folds	Fink (1980)
Medicine Lake (68.2 wt% SiO_2)	$10^7–10^9$ (interior) $10^9–10^{11}$ (crust)	Surface folds	Lescinsky et al. (2007)
Rhyolite			
Newberry (72–74 wt% SiO_2)	$10^7–10^{13}$	Lab-based (600–1000 °C)	Murase and McBirney (1973)
Badlands (75–77 wt% SiO_2)	$3.5 \times 10^9–1.2 \times 10^{10}$	Lab-based	Manley (1996)

temperature of 1000 °C gives cooling of 4.5 °C per kilometer, decreasing to 0.13 °C per kilometer once the surface temperature decreases to 250 °C. The temperature of the lava will thus decline with distance from the vent. However, a number of studies have shown that this decline is not linear down flow. Cooling rates tend to be quite low down the channelized and tubed portions of the flow system. They then increase rapidly near the flow front. With cooling, crystallization will also progress, so that the crystal content should also increase with distance from the vent.

4.2.1. Cooling and Crystallization in Channels

The hot exposed surface of the lava in a channel means that heat losses, and thus cooling and crystallization rates, should be high. Temperature data for the interior of Mauna Loa's 1984 channel-fed flow, obtained by a thermocouple by Lipman and Banks (1987), showed cooling by around 0.67 °C per km over the first 15 km of flow, increasing to 4 °C per km over the next 10 km (Figure 17.14(A)). This channel had a typical width of 23 m, a depth of 2−7 m, and was fed, initially, at effusion rates of 800 m³/s. Measurements made by Cashman et al. (1999) at a 3 m-deep and 5-10 *m*-wide channel on Kīlauea, fed at 20−40 m³/s, showed internal temperatures of 1142 °C at 1.7 km, 1140 °C at 1.8 km and 1099 °C at 1.9 km. Given a vent temperature of 1150 °C, this yields cooling of 4.7 °C/km between the vent and 1.7 km, increasing to 20 °C/km at 1.8 km, and 410 °C/km at 1.9 km. Cooling per unit distance therefore appears to increase gradually along most of the flow with a rapid increase near the flow toe (Figure 17.14(B)).

For the Kīlauea case, crystallinity was also measured from water-quenched samples scooped from the channel at each measurement station. These showed typical crystallization rates of 0.007% per meter, or 1.2% per degree cooling, until 1.8 km, and then 0.33% per meter between 1.8 and 1.9 km (Figure 17.14(B)). At Etna, cooling rates calculated for a somewhat smaller channel (2 m wide and deep, fed at a volume flux of 0.16 m³/s) were much higher, being 110 °C/km (Harris et al., 2005a).

4.2.2. Cooling in Tubes

The presence of a tube roof greatly reduces heat loss from flowing lava. As a result, tube-fed lavas can extend great distances even at quite low volume fluxes because the cooling rates are so low. If we take our cooling equation and consider a flow that is 2 m wide and fed at 2.5 m³/s. If the surface temperature of the tube roof is at 60 °C, then the cooling rate is only 0.7 °C/km. Indeed, measurements and calculations in mature tubes reveal cooling rates typically of the order of 1 °C/km, and as low as 0.35 °C/km (Keszthelyi, 1995; Clague et al., 1999).

4.3. Down-Flow Variations in Rheology

Decreasing temperature and increasing crystallinity with distance should mean that the rheology changes down flow. That is, viscosity and yield strength will increase with distance. Unfortunately, only a few down-flow rheological measurements have been made at an active channel, such as those made by Moore (1987) down Mauna Loa's 1984 channel. Summarized here in Table 17.3, Moore's measurements show an increase in both viscosity and yield strength with distance. This agrees with the results of Fink and Zimbelman (1990) who made a post-emplacement analysis of 'a'ā units emplaced during the first five episodes of the Kilauea's Pu'u 'Ō'ō eruption in 1983. From topographic profiles perpendicular to the flows, Fink and Zimbelman (1990) calculated that when active, the flow

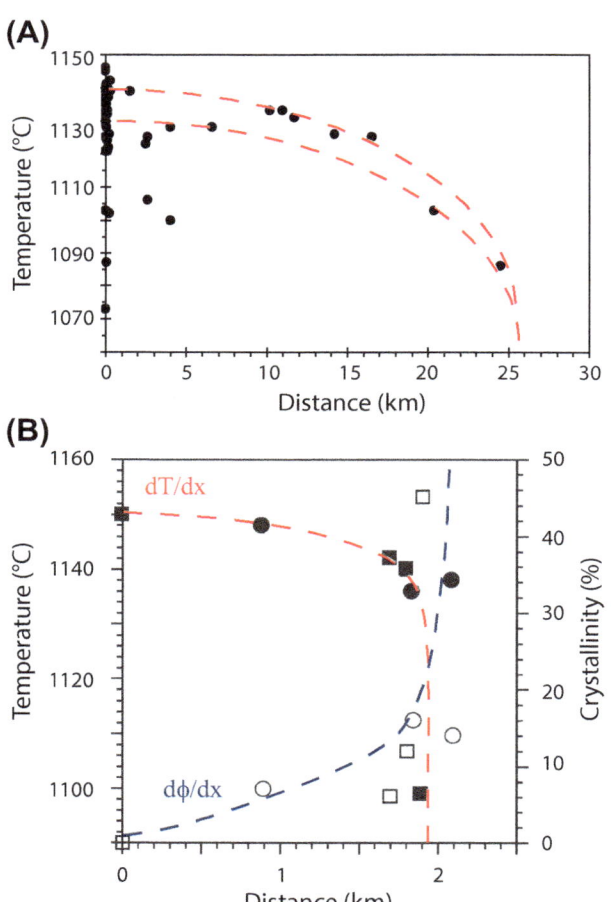

FIGURE 17.14 (A) Down flow temperature data for Mauna Loa's 1984 lava channel as given by Lipman and Banks (1987). (B) Down flow temperature and crystallinity data for a lava channel active on Kilauea in May 1997 as given by Cashman et al. (1999).

TABLE 17.3 Summary of Temperature and Crystallinity Measurements Down Mauna Loa's 1984 Channel, With Moore's (1987) Results for Yield Strength and Viscosity

Down-Channel Distance:	0 km	3 km	9 km	15 km	26 km
Temperature (°C)	1140	1130	1130	1125	1086
Crystallinity (%)			6	19	22
Yield strength (Pa)	66−220				10^3−10^4
Viscosity (Pa s)	140	1150	6000	1×10^5	6×10^6

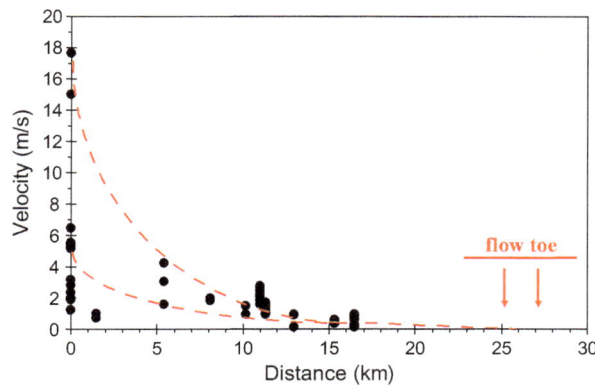

FIGURE 17.15 Down flow velocity data for Mauna Loa's 1984 lava channel as given by Lipman and Banks (1987).

viscosities were 60–2000 Pa s near the vent and increased to $1.2–32 \times 10^4$ Pa s at the flow fronts. Likewise, yield strengths increased from 300 to 2500 Pa near the vent to $3.6–36 \times 10^3$ Pa at the flow fronts.

If we look at the velocity equation of Moore (1987), as given above, these increases in viscosity and yield strength with distance should mean that the flow velocity will decline with distance. An examination of the down-flow velocity data given by Lipman and Banks (1987) for lava flowing in the active channel during Mauna Loa's 1984 eruption shows that this is indeed the case (Figure 17.15).

5. SUMMARY

The terms pāhoehoe and 'a'ā were introduced to western science by Dutton in 1888, and block lava was defined by Finch in 1947. This tripartite scheme was formalized by Macdonald in 1953 and has thus now been in place for more than 60 years. The initial descriptions and illustrations given in these early works were excellent. However, modern contributions such as those by Kilburn and Guest (1993), Swanson (1973) and Cioni and Funedda (2005) have done much to define and detail the sub-types, as well as the variety of structures within each of the pāhoehoe, 'a'ā and block lava groupings. At the same time, studies such as Lipman and Banks (1987), Calvari and Pinkerton (1998) and Walker (2009) have defined the components of, as well as structures associated with, the feeder systems — channels and tubes — and inflation. We need to continue to fill out these classifications, and descriptions of their attendant structures, while fixing on an agreed and consistent nomenclature. We also need to define the emplacement, textural and rheological conditions responsible for formation of each lava facies. In spite of decades of work, this latter step remains elusive, mainly due to the difficulty of making direct observation and measurement at the interior and base of an active lava unit.

FURTHER READING

Cas, R.A.F., Wright, J.V., 1987. Chapter 4: Lava flows. In: Volcanic Successions. Chapman & Hall, London, pp. 59–92.

Cashman, K.V., Thornber, C., Kauahikaua, J.P., 1999. Cooling and crystallization of lava in open channels, and the transition of pāhoehoe lava to 'a'a. Bull. Volcanol. 61, 306–323.

Cashman, K., Mangan, M., Newman, S., 1994. Surface degassing and modifications to vesicle size distributions in active basalt flows. J. Volcanol. Geotherm. Res. 61 (1–2), 45–68.

Calvari, S., Pinkerton, H., 1998. Formation of lava tubes and extensive flow field during the 1991–1993 eruption of Mount Etna. J. Geophys. Res. 103 (B11), 27291–27301.

Calvari, S., Coltelli, M., Neri, M., Pompilio, M., Scribano, V., 1994. The 1991–93 Etna eruption: chronology and lava flow-field evolution. Acta Vulcanol. 4, 1–4.

Cioni, R., Funedda, A., 2005. Structural geology of crystal-rich, silicic flows: a case study from San Pietro Island (Sardinia, Italy). Geol. Soc. Am. Spec. Pap. 396, 1–14.

Clague, D.A., Hagstrum, J.T., Champion, D.E., Beeson, M.H., 1999. Kilauea summit overflows: their ages and distribution in the Puna district, Hawaii. Bull. Volcanol. 61, 363–381.

Dutton, C.E., 1883. Hawaiian Volcanoes, 2005 edition. University of Hawaii Press, Honolulu. 235 pp.

Finch, R.H., 1933. Block lava. J. Geol. 41 (7). 769–770.

Fink, J.H., Zimbelman, J.R., 1986. Rheology of the 1983 Royal Gardens basalt flows, Kilauea volcano, Hawaii. Bull. Volcanol. 48, 87–96.

Fink, J.H., Zimbelman, J., 1990. Longitudinal variations in rheological properties of lavas: Puu Oo basalt flows, Kilauea Volcano, Hawaii. In: Fink, J. (Ed.), Lava Flows and Domes. Springer-Verlag, Berlin, pp. 159–173.

Fink, J., 1980. Surface folding and viscosity of rhyolite flows. Geology 8, 250–254.

Gaonac'h, H., Stix, J., Lovejoy, S., 1996. Scaling effects on vesicle shape, size and heterogeneity of lavas from Mount Etna. J. Volcanol. Geotherm. Res. 74, 131–153.

Gauthier, F., 1973. Field and laboratory studies of the rheology of Mount Etna lava. Phil. Trans. R. Soc. Lond. 274, 83–98.

Greeley, R., 1987. The role of lava tubes in Hawaiian volcanoes. U. S. Geol. Surv. Prof. Pap. 1350, 1589–1601.

Guest, J.E., Stofan, E.R., 2005. The significance of slab-crusted lava flows for understanding controls on flow emplacement at Mount Etna, Sicily. J. Volcanol. Geotherm. Res. 142, 193–205.

Harris, A.J.L., 2013. Thermal Remote Sensing of Active Volcanoes: A User's Manual. Cambridge University Press, Cambridge.

Harris, A., Bailey, J., Calvari, S., Dehn, J., 2005a. Heat loss measured at a lava channel and its implications for down-channel cooling and rheology. Geol. Soc. Am. Spec. Pap. 396, 125–146.

Harris, A.J.L., Dehn, J., Patrick, M., Calvari, S., Ripepe, M., Lodato, L., 2005b. Lava effusion rates from hand-held thermal infrared imagery: an example from the June 2003 effusive activity at Stromboli. Bull. Volcanol. 68, 107–117.

Harris, A.J.L., Favalli, M., Mazzarini, F., Hamilton, C.W., 2009. Construction dynamics of a lava channel. Bull. Volcanol. 71, 459–474.

Harris, A.J.L., Flynn, L.P., Matias, O., Rose, W.I., Cornejo, J., 2004. The evolution of an active silicic lava flow field: an ETM+ perspective. J. Volcanol. Geotherm. Res. 135, 147–168.

Hon, K., Kauahikaua, J., Denlinger, R., McKay, K., 1994. Emplacement and inflation of pāhoehoe sheet flows: observations and measurements of active lava flows on Kilauea volcano, Hawaii. Geol. Soc. Am. Bull. 106, 351–370.

Herd, R.A., Pinkerton, H., 1997. Bubble coalesence in basaltic lava: its impact on the evolution of bubble populations. J. Volcanol. Geotherm. Res. 75, 137–157.

Keszthelyi, L., 1995. A preliminary thermal budget for lava tubes on the earth and planets. J. Geophys. Res. 100, 20411–20420.

Keszthelyi, L., Thordarson, T., McEwen, A., Haack, H., Guilbaud, M.-N., Self, S., Rossi, M.J., 2004. Icelandic analogs to Martian flood lavas. Geochem. Geophys. Geosyst. 5 (11), Q11014. http://dx.doi.org/10.1029/2004GC000758.

Kilburn, C.R.J., Guest, J.E., 1993. 'A'ā lavas of Mount Etna, Sicily. In: Active Lava. UCL Press, London, pp. 73–106.

Lescinsky, D.T., Skoblenick, S.V., Mansinha, L., 2007. Automated identification of lava flow structures using local Fourier spectrum of digital elevation data. J. Geophys. Res. 112, B05212. http://dx.doi.org/10.1029/2006JB004263.

Lipman, P.W., Banks, N.G., 1987. 'A'ā flow dynamics, Mauna Loa. U. S. Geol. Surv. Prof. Pap. 1350, 1527–1567.

Lockwood, J.P., Lipman, P.W., 1980. Recovery of datable charcoal beneath young lavas: lessons from Hawaii. Bull. Volcanol. 43, 609–615.

Manley, C.R., 1996. Physical volcanology of a voluminous rhyolite lava flow: the Badlands lava, Owyhee Plateau, southwestern Idaho. J. Volcanol. Geotherm. Res. 71, 129–153.

MacDonald, G.A., 1953. pāhoehoe, 'a'ā and block lava. Am. J. Sci. 251, 169–191.

McKay, M.E., Rowland, S.K., Mouginis-Mark, P.J., Garbeil, H., 1998. Thick lava flows of Karisimbi Volcano, Rwanda: insights from SIR-C interferometric topography. Bull. Volcanol. 60, 239–251.

Moore, H.J., 1987. Preliminary estimates of the rheological properties of 1984 Mauna Loa lava. USGS Prof. Pap. 1350, 1569–1588.

Murase, T., McBirney, A., 1973. Properties of some common igneous rocks and their melts at high temperatures. Geol. Soc. Am. Bull. 84, 3563–3592.

Navarro-Ochoa, C., Gavilanes-Ruiz, J.C., Cortes-Cortes, A., 2002. Movement and emplacement of lava flows at Volcan de Colima, Mexico: November 1998–February 1999. J. Volcanol. Geotherm. Res. 117, 155–167.

Nichols, R.L., 1939. Viscosity of lava. J. Geol. 47 (3), 290–302.

Peterson, D.W., Tilling, R.I., 1980. Transition of basaltic lava from pāhoehoe to 'a'ā, Kilauea, Hawaii: field observations and key factors. J. Volcanol. Geotherm. Res. 7, 271–293.

Peterson, D.W., Holcomb, R.T., Tilling, R.I., Christiansen, R.L., 1994. Development of lava tubes in the light of observations at Mauna Ulu, Kilauea volcano, Hawaii. Bull. Volcanol. 56, 343–360.

Pinkerton, H., Sparks, R.S.J., 1978. Field measurements of the rheology of lava. Nature 276, 383–385.

Pinkerton, H., Norton, G., 1995. Rheological properties of basaltic lavas at sub-liquidus temperatures: laboratory and field measurements on lavas from Mount Etna. J. Volcanol. Geotherm. Res. 68, 307–323.

Robert, B., Harris, A., Gurioli, G., Médard, E., Sehlke, A., Whittington, A., 2014. Textural and rheological evolution of basalt flowing down a lava channel. Bull. Volcanol. 76 http://dx.doi.org/10.1007/s00445-014-0824-8.

Rowland, S.K., Walker, G.P.L., 1987. Toothpaste lava: characteristics and origin of a lava structural type transition between pāhoehoe and 'a'ā. Bull. Volcanol. 49, 631–641.

Rowland, S.K., Walker, G.P.L., 1988. Mafic-crystal distributions, viscosities, and lava structures of some Hawaiian lava flows. J. Volcanol. Geotherm. Res. 35, 55–66.

de Silva, S.L., Self, S., Francis, P.W., Drake, R.E., Carlos Ramirez, R., 1994. Effusive silicic volcanism in the Central Andes: the Chao dacite and other younger lavas of the Altiplano–Puna Volcanic complex. J. Geophys. Res. 99 (B9), 17,805–17,825.

Self, S., Keszthelyi, L., Thordarson, T., 1998. The importance of pāhoehoe. Ann. Rev. Earth Plan. Sci. 26, 81–110.

Shaw, H.R., Wright, T.L., Peck, D.L., Okmaura, R., 1968. The viscosity of basaltic magma: an analysis of field measurements in Makaopuhi lava lake, Hawaii. Am. J. Sci. 266, 225–264.

Sparks, R.S.J., Pinkerton, H., Hulme, G., 1976. Classification and formation of lava levees on Mount Etna, Sicily. Geology 4, 269–271.

Swanson, D.A., 1973. Pāhoehoe flows from the 1969–1971 Mauna Ulu eruption, Kilauea volcano, Hawaii. Geol. Soc. Am. Bull. 84, 615–626.

Tanguy, J.C., 1973. The 1971 Etna eruption: petrology of the lavas. Phil. Trans. R. Soc. Lond. 274, 45–53.

Walker, G.P.L., 1967. Thickness and viscosity of Etnean lavas. Nature 213, 484–485.

Walker, G.P.L., 1972. Compound and simple lava flows and flood basalts. Bull. Volcanol. 35, 579–590.

Walker, G.P.L., 1989. Spongy pāhoehoe in Hawaii: a study of vesicle-distribution patterns in basalt and their significance. Bull. Volcanol. 51, 199–209.

Walker, G.P.L., 1991. Structure, and origin by injection of lava under surface crust, of tumuli, "lava rises", "lava-rise pits", and "lava-inflation clefts" in Hawaii. Bull. Volcanol. 53, 546–558.

Walker, G.P.L., 2009. The endogenous growth of pāhoehoe lava lobes and morphology of lava-rise edges. In: Studies in Volcanology: The Legacy of George Walker. Geological Society, London, pp. 17–32.

Wilmoth, R.A., Walker, G.P.L., 1993. P-type and S-type pāhoehoe: a study of vesicle distribution patterns in Hawaiian lava flows. J. Volcanol. Geotherm. Res. 55, 129–142.

Wright, T.L., Okamura, R.T., 1977. Cooling and crystallization of tholeiitic basalt, 1965 Makaopuhi lava lake, Hawaii. USGS Prof. Pap. 1004, 78.

Chapter 18

Lava Dome Eruptions

Eliza S. Calder
School of Geosciences, University of Edinburgh, Edinburgh, UK

Yan Lavallée and Jackie E. Kendrick
School of Earth, Ocean and Ecological Sciences, University of Liverpool, Liverpool, Merseyside, UK

Marc Bernstein
Department of Geology, University at Buffalo, Buffalo, NY, USA

Chapter Outline

GLOSSARY

block-and-ash flow deposits The deposits of pyroclastic density currents generated by lava dome collapse.

brittle behavior A deformation style, macroscopically resulting in complete material failure. Also the process of brittle fracturing.

cataclasite Granulated rocks formed along a slip zone during faulting.

coulée A thick flow or dome of lava, which has flowed away from the vent, transitional between a typical lava dome and lava flow.

cryptodome An accumulation of magma at a shallow level within a volcanic edifice or just below the surface.

ductile behavior Substantial strain deformation without the tendency to localize the flow into faults; in other words, flow does not macroscopically induce complete failure, although microscopic fracturing may take place.

endogenous growth Enlargement of a lava dome due to influx of magma into the dome interior.

exogenous growth Enlargement of a lava dome due to discrete lobes of lava being emplaced at the surface or above each other.

lava spine A coherent slab or block of solid lava extruded, usually vertically, from a vent or exogenously through the carapace of a dome.

magmatic degassing The exsolution of volatiles from the magma resulting in gas bubbles.

outgassing Gas loss, largely occurring through faults and fractures in the conduit and dome, to the atmosphere, or to surrounding country rocks.

plug A body of near-solid magma within the conduit, the upward movement of which is characterized by strain localized along or near the conduit margin.

pseudotachylite Fault rock that has undergone frictional melting.

rheology The study of flowing materials or the flow behavior of materials.

shear lobe A large, often arcuate shaped, body of solid lava extruded through a vent or exogenously through the carapace of a dome.

The Encyclopedia of Volcanoes. http://dx.doi.org/10.1016/B978-0-12-385938-9.00018-3

The margins of a shear lobe with the adjacent rock are often fault or shear zone bound.

shear thinning Characteristic of non-Newtonian rheological behavior whereby the viscosity of the fluid decreases with increasing applied stress and strain rate.

viscoelastic Property of a liquid that exhibits both viscous and elastic deformation responses to an applied stress.

1. INTRODUCTION

Lava dome eruptions are among the most unpredictable and hazardous volcanic phenomena and their common occurrence makes an improved understanding of their mechanisms important. This chapter addresses the topic of lava dome eruptions, providing a definition of lava dome types, structures, and general behavior. These descriptions are complemented by a review of eruptive activity and related hazards during three recent, well-monitored dome eruptions, framed in terms of the dynamics of magma ascent and extrusion. Also highlighted is the balance of forces and magma response that drives the common transitions in activity between effusive and explosive eruptions, and the potential for predicting this behavior. A closing section details the main hazards associated with lava dome eruptions.

2. DEFINITION

Lava domes are mounds of viscous lava and rocks that pile up and accumulate around a volcanic vent. They form as magma cools and degasses relatively quickly after erupting onto the Earth's surface. They are also observed on the Moon, Mars, and Venus. Dome-building eruptions are a common style of volcanism, especially in convergent margin settings and the resultant domes may stand alone, form in the crater of a volcanic cone, or form part of a cluster or dome complex (Fink, 1990). Around 6% of eruptions worldwide are lava dome forming eruptions.

Lava domes can form relatively rapidly, over days to weeks, but eruptive episodes can extend for years to decades. Lava dome eruptions vary from those that are characterized by relatively passive and unthreatening effusion, to others that can exhibit varied, highly unpredictable, and menacing activity. Hazards associated with lava domes include the collapse of portions of a lava dome, pyroclastic density currents, moderate to major explosive activity, and when inundated by intense rainfall, lahars. The propensity for lava domes to collapse with little or no warning, the impulsive transitions from benign effusive to dangerous explosive behavior, and their protracted episodes of eruptive activity make lava dome eruptions a uniquely challenging type of volcanic crisis to manage. Because of the complexity of the eruptive processes involved, and as volcanic provinces become ever more populated, the study

and improved understanding of lava dome eruptions are vital in order to mitigate the risks posed by such eruptions.

Dome lavas cover a wide compositional range from basaltic (e.g., Semeru, 1946 eruption; and at mid-ocean ridges) through to rhyolitic (e.g., Chaitén, 2010 eruption), although the majority are of intermediate composition (andesitic and dacitic). Dome morphology and eruptive behavior, however, owe more to the magmas' characteristic high viscosity and poor ability to flow from the vent than to their chemical composition. Petrographically speaking, there are two types of lava domes: crystal-poor, obsidian domes, which tend to be rhyolitic or rhyodacitic in composition and crystal-rich domes, which tend to be rhyolitic to basalt andesite in composition. Active lava domes are typically made of an inner, somewhat ductile lava core and a cooler, brittle outer rock carapace (Fink, 1990). At steep dome margins, or at the fronts of active lobes, loose or unsupported portions of the carapace commonly collapse to form a talus apron. Domes range in diameter from a few tens of meters to a few kilometers, and their height can reach up to 1 km. Their morphology can vary from steep-sided to tabular in cross-section and circular or elliptical to irregular in plan view. The morphology of lava domes also depends on the underlying topography; domes are commonly restricted to crater areas and in these instances their shape is nearly circular, when erupted on the flanks of volcanic edifices, the shape tends to elongate downhill. Lava domes may also develop as complexes along linear or arcuate chains up to 30 km long.

Protracted lava dome eruptions generally last for years to decades, although the growth of a dome may be reset by destruction during explosive phases. Extensive monitoring of active lava domes in recent decades has unraveled episodic growth dynamics with emplacement timescales spanning hours to decades (Barmin et al., 2002; Voight et al., 1999). Effusion rates vary widely from as little as $0.01 \, \mathrm{m^3 \, s^{-1}}$ to over $100 \, \mathrm{m^3 \, s^{-1}}$, with important implications for the structural stability of the dome (Voight, 2000). Sudden disruption or collapse of lava domes may generate pyroclastic density currents, which have resulted in substantial fatalities in the last century (Table 18.1).

3. TYPES OF LAVA DOMES

The shapes and habits of lava domes are diverse (Figure 18.1). Their morphology is controlled by a combination of magma **rheology**, substrate topography, ascent dynamics, and the mechanism of dome growth (i.e., endogenous or exogenous). **Endogenous growth** refers to the enlargement of a lava dome due to expansion caused by intrusion of new magma. **Exogenous growth** refers to dome enlargement as a result of magma forcing its way through a preexisting carapace to the surface or

TABLE 18.1 Important Historical Lava Dome Eruptions Including Emplacement and Collapse, and Associated Fatalities

Volcano	Year	Estimated Fatalities	Interesting Facts
Mount Sinabung (Indonesia)	2014 (2013–2015)	16	—
Merapi (Indonesia)	2010	353	350,000+ people evacuated
Chaitén (Chile)	2008 (2008–2009)	1	Rhyolite dome
Augustine (United States)	2006 (2005–2006)	—	—
Mount St. Helens (United States)	2004—2008	—	Extrusion of magma spines
Guagua Pichincha (Ecuador)	2000	2	Only 8 km from Quito city center
Volcán de Colima (Mexico)	1998–2013	—	One of the most active volcanoes in North America
Soufrière Hills (Montserrat)	1997 (1995–2015)	19	Caused 2/3 of the population to leave the island
Popocatépetl (Mexico)	1996 (1994–2015)	5	Persistently active since 1300s
Lascar (Chile)	1984–1993	—	Cyclic dome building and degassing
Galeras (Colombia)	1993 (1988–1993)	9	—
Pinatubo (Philippines)	1991	847	Successful prediction of eruption and evacuation
Unzen (Japan)	1991 (1990–1995)	43	Conduit magma sampled by Unzen Scientific Drilling Project
El Chichón (Mexico)	1982	3500	Believed dormant prior to eruption
Mount St. Helens (United States)	1980 (1980–1986)	57	Large sector collapse
Bezymianny (Russia)	1956 (1955–1956) and ongoing	—	Active after >1000 years repose
Usu (Japan)	1943–1945	1	First detailed record of growth of a lava dome
Santiaguito volcano (Guatemala)	1929 (1922–2015)	∼5000	Continuously active
Kelut (Indonesia)	1919	5100	—
Novarupta (United States)	1912	1	Most voluminous volcanic eruption of the twentieth century
La Soufrière (Saint Vincent)	1902	1680	-
Mount Pelée (Martinique)	1902	30,000	The worst volcanic disaster of the twentieth century
Unzen (Japan)	1792	15,000	Accompanied by large earthquake, sector collapse, and tsunami
Kelut (Indonesia)	1586	10,000	Produced devastating lahar

flowing directly from the vent and forming discrete lobes of lava that pile on top of, or adjacent to, each other. This latter growth style sometimes results in the extrusion of **lava spines** (Sherrod et al., 2008). Peléan-type lava domes (Figure 18.1(A) and (B)), typically of andesitic to dacitic composition, characteristically build up a significant blocky mound surrounded by a talus apron. Regular dome

collapses from Peléan domes feed extensive fans of block-and-ash flow deposits (Figure 18.1(A)). Eruptions of this type typically result in much greater volumes of material residing in the pyroclastic deposit fan surrounding the dome than in the dome itself (Wadge et al., 2009). The eruption of Soufrière Hills volcano (SHV), Montserrat, has produced over 1 km³ of magma, but only between 100

FIGURE 18.1 Different styles and aspects of dome building volcanoes: (A) View of the Soufrière Hills Volcano (SHV), Montserrat, a lava dome complex surrounded by low-angle pyroclastic fans. The recent lava dome, the high parts in the background with fumaroles, was built in a crater at the summit of a cluster of preexisting domes (partially green flanks). (B) The summit of the active andesitic dome at SHV. (C) The dome at Volcán de Colima in 2011 prior to the most recent explosive activity starting in 2013. (D) A view of the Santiaguito dome complex (a line of four domes) from the summit of Santa Maria volcano, Guatemala. (E) Lava coulées at Mount Sinabung, Indonesia, during a period of unrest in 2014, photo courtesy of H. Wright. (F) The rhyolite lava dome at Chaitén volcano, Chile published with permission of Andean Geology. (G) Chillahuita dome, in Northern Chile courtesy of S. de Silva. A crystal-rich dacitic coulée, with margins 200−400 m high, and a volume of ∼5 km^3. This is a classic dome type, prevalent in the Central Andes and often referred to as a "torta" dome.

and 200 × 10^6 m^3 of that volume currently sits in the dome. This is important when assessing the extent of historical activity, as it is sometimes hard to constrain the vast erupted volume represented by the pyroclastic component (Wadge et al., 2009).

Peléan type domes can also extrude spines, whereby a **plug** of dense, mostly undeformable magma is thrust above the vent. The 1902 eruption of Pelée, Martinique, exemplifies this eruption style. Such structures are typically very unstable and end up collapsing, fragmenting, and generating block-and-ash flows. Their formation is related to brittle processes and tends to occur at the beginning of a dome growth phase, or at the end of an extrusion phase when the magma viscosity is too high to allow for flow. Such spine extrusion dynamics occurred at Unzen volcano

in 1991 and 1995, i.e., at the beginning and end of that 4-year-eruption.

Coulées represent types of low-lying lava domes that have flowed downhill and are essentially transitional to lava flows (Figure 18.1(E) and (G)). In the Andes, such domes are often referred to as tortas (pies). They can be extremely thick (up to 400 m), and are usually of dacitic to rhyolitic composition. The volume of these dome types can be immense—the Chao dacite coulée in Chile is around 15 km^3, but they do not tend to produce the pyroclastic fans associated with Peléan lava domes. Unlike their Peléan counterparts, the volume of these lava domes (minus any associated tephra component) essentially represents the total erupted volume. Coulées can display the morphological features typical of lava flows, including levees and ogives (large flow-transverse ridges that are pushed up as the carapace deforms), but in plan view their typically oval to circular shape and domed upper surface legitimize their classification as a type of lava dome. Obsidian domes, such as the Inyo domes in California and Big Obsidian Flow in Newberry Crater, Oregon, are of this category, as are many of the large volume dacite domes typical of the Central Andean Altiplano (Figure 18.1(G)).

Some lava domes do not extrude large volumes of lava but instead form dome-shaped plugs or caps at the top of magma conduits within a crater. In this regard, these lava domes are simply the manifestation of the magma's free surface at the top of the conduit, which has undergone some lateral spread at the surface. These lava domes can display inflation and deflation episodes, concentric fracture opening and closing, and explosion pits as magma properties in the upper conduit vary. These types of domes have formed intermittently within the crater of Lascar, Chile, Galeras, Colombia, and Popocatépetl, Mexico. They are often associated with systems that have relatively frequent explosive eruptions, so they can be short-lived. After an explosion removes the dome and conduit magma, the magma's free surface rises up again and often reforms a domed cap at the summit.

4. LAVA DOME STRUCTURES AND TEXTURES

At first glance, a lava dome resembles a pile of blocks (Figure 18.1(B)). A closer look, however, reveals that this pile is actually characterized by a wide range of structures and textures (Table 18.2, Figures 18.2 and 18.3). The architecture of each lava dome reflects a unique interplay between internal conditions of magma forcing its way through the upper crust and the local emplacement conditions at the point of extrusion (Hale et al., 2009). It is the spectrum of structures and textures (Figure 18.2) distinct to lava domes that control their permeability and thus their ability to lose gas—their buoyant driving force—and erupt passively or explode catastrophically (Druitt and Kokelaar, 2002; Edmonds and Herd, 2007).

A lava lobe consists of an inner lava core and an outer rock carapace, further flanked by a brecciated talus (Figure 18.3(B)). The distinction between the core, the carapace, and the talus rests in their rheological properties. The core is characterized by a ductile mode of deformation, whereas the carapace is subject to brittle fracture, and the talus is composed of fragmental blocks resulting from brittle failure. Low-viscosity lava extrusion may lead to a very smooth dome surface, though it is usually not preserved for long. Eruptive stress conditions, degassing, and thermal contraction enhance the brittle response of the

lava (Sparks, 1997), inducing brecciation and occasionally polygonal (columnar-like) jointing. Throughout a lava dome eruption, the clastic or brecciated component of total erupted products increases to >50 vol% and may constitute nearly the total volume by the conclusion of cooling.

Externally, lava domes display a field of non-sintered? to poorly sintered blocks ranging in size from centimeters to tens of meters. Near the extrusive vents of low-viscosity lava domes, crease structures may form where a tensile fracture opens progressively as lava diverges laterally (Figure 18.3(E) and (F)). Away from the vent area, the carapace sometimes exhibits compressional ridges, generated by folding of the dome surface due to internal shear traction from a spreading lava core. Lava dome surfaces and surroundings are also characterized by small pit craters reamed out by gas-and-ash explosions, a common phenomenon occurring hourly to daily during some dome eruptions (Lavallée et al., 2012).

Internally, lava domes reveal a certain degree of strain in material subjected to shear localization along the conduit and dome margins (Figure 18.4). A lava dome may occasionally develop an "onion-skin" foliation when it is intruded by magma, pushing the first extruded lava to the outermost dome margin; as the dome inflates, the process repeats itself and additional concentric layers can be emplaced. Macrofractures are also common within the core of the lava dome; they can form subparallel and also perpendicular to the extrusion direction. Those that are horizontal may reflect unloading conditions during ascent. Close examination of the fractures reveals the accumulation of variably sintered fragmental material, solidified as tuffisite veins (Castro et al., 2012); the degree of sintering increases with emplacement depth (Figures 18.4(D) and (G)). Alternatively, secondary minerals (salts, clays, zeolites, and silica polymorphs such as cristobalite and tridymite) can also precipitate and grow in fractures, reducing permeability in the process. Veins hosting tuffisite and precipitated minerals show a tendency to be altered, weathered, and sometimes completely eroded.

Spine extrusion reflects internal brittle processes, where the core of the spine is bordered by defined shear zones (e.g., Sherrod et al., 2008) showing signs of both plastic and brittle strain in the crystal phase (Figure 18.4(A)), further mantled by **cataclasite**, breccia, and occasionally **pseudotachylite** (Figure 18.4(C)).

Texturally, lava domes are extremely diverse. Crystal-rich domes tend to be uniformly crystalline (Figure 18.2); locally, however, the crystal-size distributions range from 10 to ~40 vol% of phenocrysts in a groundmass containing 10 to ~50 vol% of microlites and up to ~50 vol% interstitial melt, solidified as glass. The porous network, and in turn the permeability of lava domes varies significantly,

TABLE 18.2 Lava Dome Textures and Features

Structure	Location in Dome	Extent	Interpretation	Role in Eruptive Dynamics
Shear lobe	Anywhere, but often extruded at or near the summit region	Few tens to hundreds of meters	Usually a fault or shear zone-bound unit of lava that extruded from a vent or through the carapace	Shear lobes represent the basic building blocks, or structural unit, of many domes
Spine	Commonly above the vent, though further magma extrusion pushes it aside	Few tens to hundreds of meters	A coherent block of solid lava extruded, usually vertically, from a vent or exogenously through the carapace of a dome	Indicative of magma that has essentially solidified before the final stages of extrusion
Whaleback lobe	Direct continuation of magma extruding from the vent pushing up into the air	Few tens to hundreds of meters	A whaleback lobe is a type of shear lobe, distinctive because of its smooth arcuate back. It can also be a recumbent spine	Represent the basic building block of many domes
Talus	A diamict comprised of lava blocks in an ash matrix. Forms around the dome margin	Tens of meters	Fragmented blocks, some of which have been comminuted to ash, which have collapsed from the carapace	Can inhibit spread of the dome, and increase stability by forming a substantial buttress
Carapace	Outer surface of lava lobes	Up to tens of meters	Cooler, pervasively fractured, rind of a lava dome. Often covered with loose blocks	Will insulate dome interior, but can also form a relatively permeable layer through which outgassing will occur
Blocks	In the carapace and talus	Centimeters to several meters	Stress conditions for brittle failure have been met	A fragmented carapace can promote outgassing
Vesicular textures	Anywhere	Nanometers to meters	Vestige of high exsolved gas content in bubbles. Often associated with rapid extrusion rates	Increase possibility of gas storage, decreases the strength of the dome
Explosion pits	Anywhere on or around a dome surface, but especially prevalent around dome margins	Few meters to tens of meters	Indicate zone of weakness allowing for sudden outgassing of magma and rocks at depth	Bleeds pore pressure in the rock and magmas

	Location	Thickness	Description	Effect on permeability
Polygonal joints	Across the carapace	Centimeters to tens of meters	Formed by contraction during cooling	Permanently increases the permeability and weakens a dome
Compressional ridges	Across the carapace	Tens of meters	Formed by viscous flow in the dome core, which induces compressional shear and folding of the high viscosity carapace. Used to estimate the viscosity of a dome	Damage in the carapace may increase the bulk permeability of the dome and weaken the structure
Crease structure	Above the extrusion point	Meters to few hundreds of meters	Tensile fracture where magma diverges laterally above the vent	Locally increases the permeability
Shear zones	Near the margin of conduits, spines, shear lobes, or lava sheets	Up to a few meters thick	Regions in which magma strains during ascent	Controls magma ascent (up to the point of failure or fragmentation); may act as a permeable pathway
Fault zones	Near the margin of conduits, spines, or lava sheets	Up to 1–2 m thick	Regions in which magma sheared beyond its elastic limit and failed	Fault-control magma ascent dynamics; may induce stick-slip motion; may act as permeable pathway
Pseudotachylite	Near the margin of conduits, spines, or lava sheets	Centimeters to decimeters meters	Regions in which magma sheared beyond its elastic limit, failed, slip, and frictionally melted	Fault-control magma ascent dynamics; may induce stick-slip motion; may act as permeable pathway
Gouge	Near the conduit margin and mantling spines	Centimeters to several meters thick	Regions in which magma sheared beyond its elastic limit, failed, and induced cataclasis	Fault-control magma ascent dynamics; may induce stick-slip motion; may act as permeable pathway and a source for gas-and-ash explosion
Tuffisite	Throughout the domes	Millimeters to decimeters thick	Pyroclasts from a fragmentation event, got trapped in fracture network	Momentarily or permanently increase the permeability and weaken a dome
Veins	Throughout the domes	Millimeters to centimeters thick	Mineral precipitation during fluid flow	Decrease the permeability

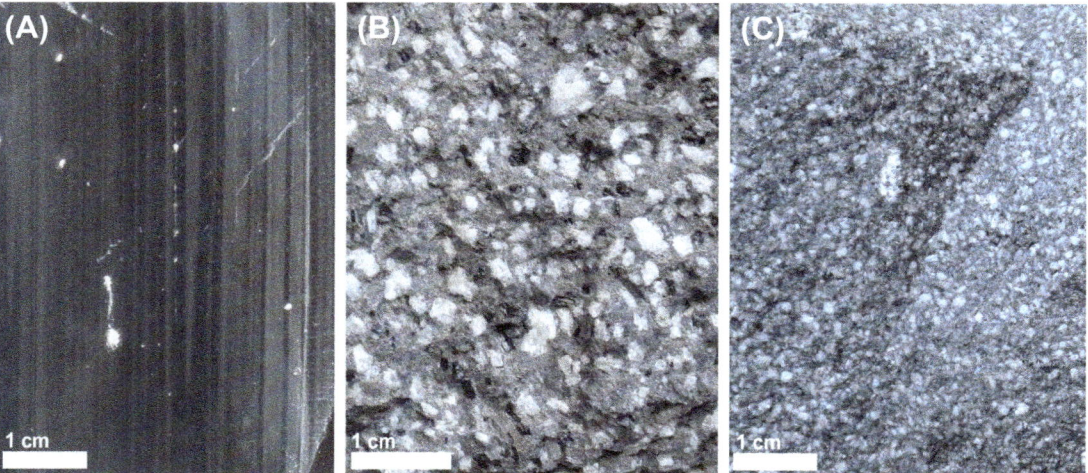

FIGURE 18.2   Lava dome rock textures in hand specimen: (A) Crystal-poor obsidian rhyolite from Newberry caldera showing parallel flow bands. (B) Dacite from the last eruption of Unzen, with large phenocrysts set in glass. (C) Finely crystalline andesite from Volcán de Colima.

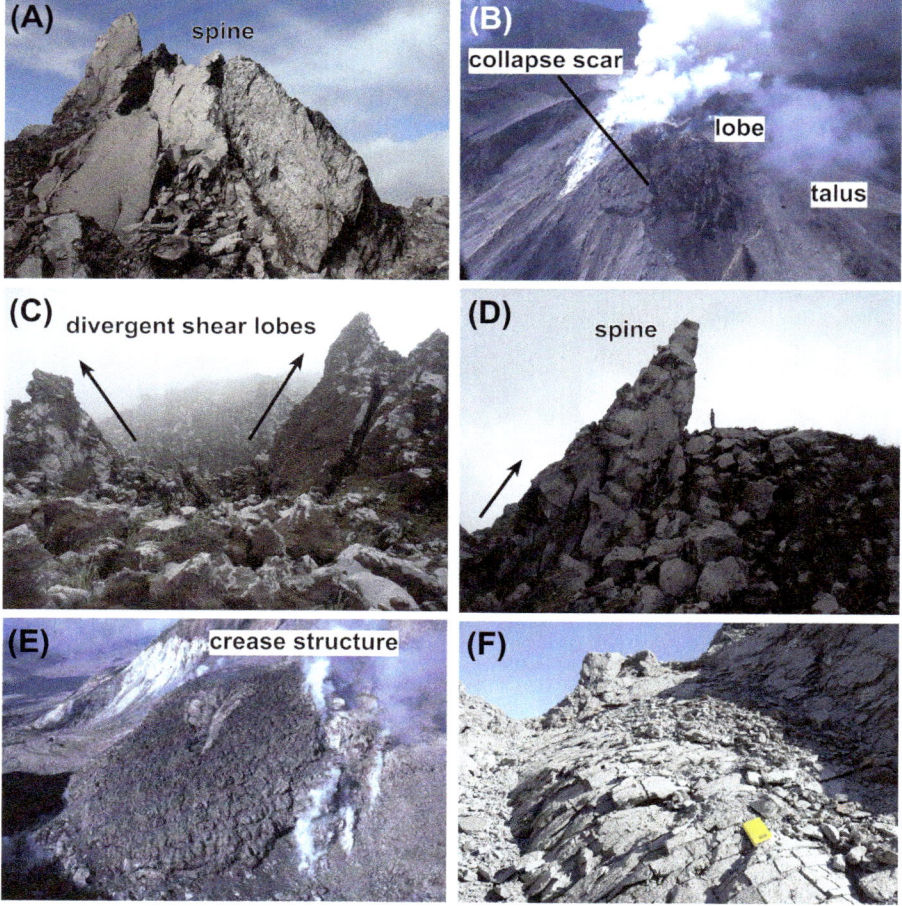

FIGURE 18.3   Large-scale lava dome structures: (A) Spine at Unzen showing intact core to the left, a shear zone running at 60° from vertical in the center and bounded by breccia on the right. (B) The lava dome at Soufrière Hills volcano, Montserrat, showing a lobe of younger lava growing into and filling the scar from a previous dome collapse. An extensive talus apron is developed around the dome. (C) A pair of divergent shear lobes at the summit of one of the inactive Santiaguito domes, Guatemala. (D) A spine extruded at the summit of one of the inactive domes at Santiaguito, with a person for scale. (E) A crease structure in a young, post 1980, dome at Mount St. Helens. Credit: US Geological Survey. (F) The surface of a crease structure on the 1980–1986 lava dome at Mount St. Helens.

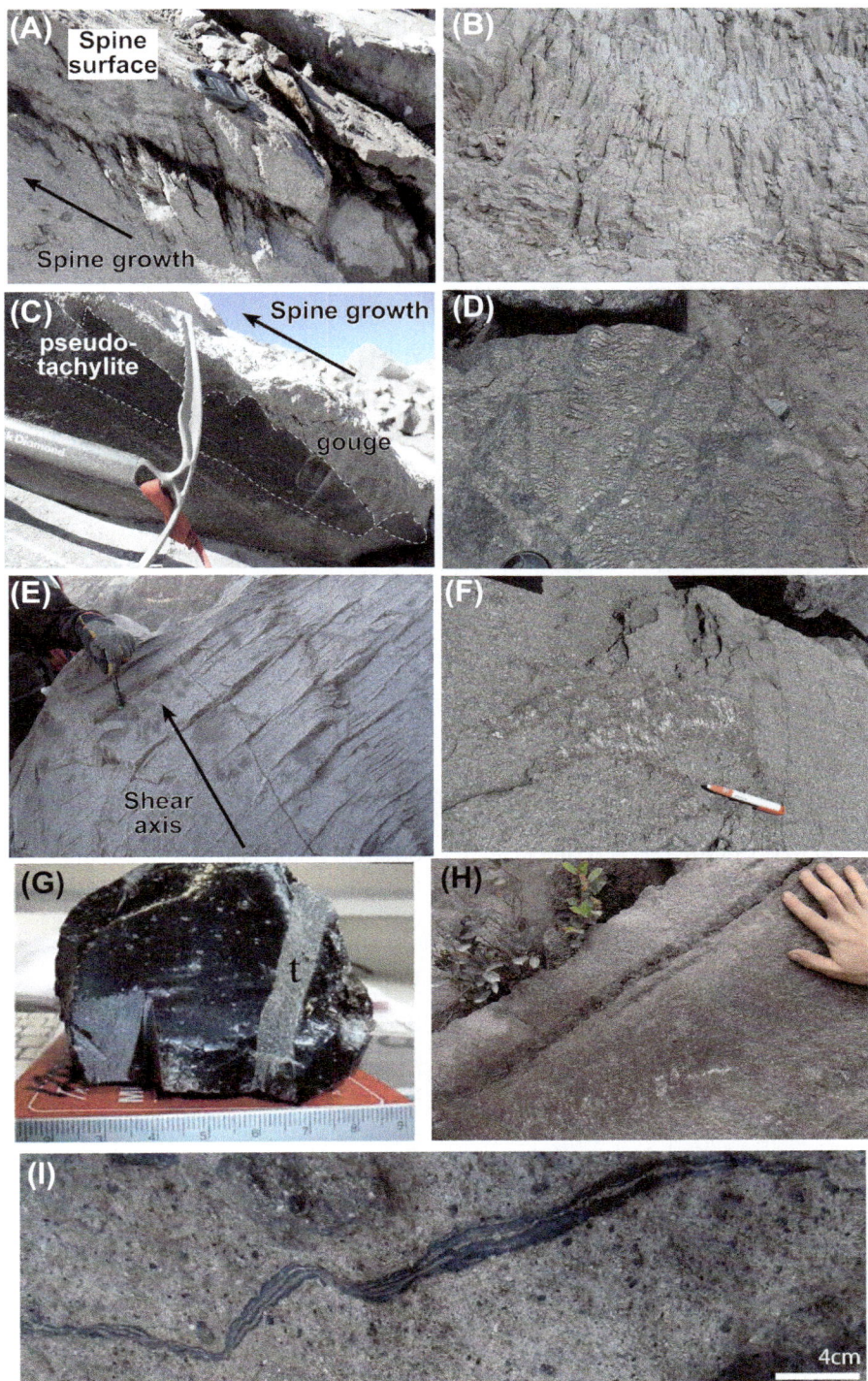

FIGURE 18.4 Closer inspection of lava dome structures: (A) A shear zone at the surface of a spine on the 2004–2008 Mount St. Helens lava dome. (B) Polygonal joints in the 1980–1986 Mount St. Helens lava dome. (C) Pseudotachylite and gouge at the spine surface from the 2004–2008 eruption at Mount St. Helens. (D) Tuffisite on the surface of a block on the lava dome at Volcán de Colima. (E) A ductile shear zone inside the 1980–1986 lava dome at Mount St. Helens. (F) Friction marks on the surface of a block from Unzen. (G) A tuffisite vein in a block of obsidian from the 2008 Chaitén lava dome. (H) Dilational band formed by shearing. (I) A spectacular aphanitic pseudotachylite on a dome block from Soufrière Hills volcano, Montserrat.

especially laterally, throughout a lava dome. Lava domes are generally composed of dense, microvesicular lavas, and field-based density measurements have demonstrated that their porosity is generally lower than 35 vol%, although during a few rapid dome growth episodes more porous material has been erupted. Generally, a lava dome carapace is more porous than the core but both can be locally heterogeneous. Pores are often branching or reentrant; providing evidence for gas percolation in the crystal-rich liquids and in some cases, low confining pore pressures. Where strain is localized, the porous network develops strong anisotropy, preferentially directing **outgassing** from the conduit center to the marginal shear zone, where gas migrates upward to the surface. Despite their low porosity, dome rocks are also characterized by high permeability due to pore connectivity in the form of microfractures. Vertical dilational lenses of porous material also develop during magma ascent in conduits (Figure 18.4(H)). These features demonstrate that lava domes tend to outgas via localized permeable structures. Explosive events, which occasionally disrupt lava domes, expose the underlying magma that feeds lava domes, and reveal higher porosity; more volatile-rich magmas are located at depth.

Obsidian lava domes are structurally similar to crystal-rich lava domes, but important distinctions exist. They too divide into core, carapace, and talus, but examination of the latter two reveals common block sintering, owing to relatively high healing efficiency. In the near absence of crystals during its ascent to the Earth's surface, the magma does not develop thick shear zones; instead, the development of thin flow bands is favored, with fracturing and healing cycles also common near the conduit margin (Castro et al., 2012). Some fractures are subsequently infiltrated by fine-grained fragmental materials and develop into tuffisites (Figure 18.4(G)).

Texturally, obsidian domes display stark contrasts to crystal-rich domes. For instance, the carapace generally comprises highly porous, or vesicular bands, whereas the core is dense. Obsidian reveals mostly rounded pores ranging from nanometers to meters in diameter, with a tendency for the largest pores to exhibit an increased degree of flattening with depth in the dome. Classically, this observation has been interpreted in two ways. One model suggests that dense lava emerges and foams during flow, forming distinct finely and coarsely vesicular zones in the carapace. An alternative explanation, known as the permeable foam model, suggests that decompression induces volatile exsolution, foaming, and bubble coalescence; upon eruption, outgassing occurs and the pores subsequently viscously collapse. Microanalysis of flow bands shows that the evidence favors a pore collapse origin. Further overprinting by microlite and spherulite nucleation takes place at the liquid/gas interface on the pore boundaries. Upon cooling, the obsidian glass is subjected to

intense alteration, which takes the form of spherulites formed by devitrification and crystallization of anhydrous minerals. Obsidian dome eruptions sometimes end with the extrusion of crystal-rich lavas, probably representing residual reservoir magma mushes.

5. CLASSIC LAVA DOME ERUPTIONS

Over the last three decades, dome-building eruptions have occupied an important place in volcanology. This focus is rooted in the very rich contribution that dome-building eruptions have made to our understanding of magma transport and eruptive styles, as well as the obvious attention that the eruptions themselves draw. In particular, the eruptions of Mount St. Helens, USA (1980–1986 and 2004–2008), Soufrière Hills Volcano (SHV), Montserrat (1995–ongoing as of 2015), and Unzen Volcano, Japan (1990–1995) have provided us with comprehensive data sets on which to build models of crystal-bearing magma ascent and lava dome eruption dynamics. During the 1980–1986 eruption of Mount St. Helens, advances were made in understanding the morphological evolution and kinetics of lava domes, while during the later dome eruptions, the breadth of eruptive styles and improvements in monitoring techniques have resulted in improved understanding and refining of the models. These models have also recently been expanded to include obsidian by complementary observations of the eruption of a crystal-free obsidian dome at Chaitén and the extensive obsidian lava flows at Puyehue-Cordón Caulle volcanoes, Chile. Here we review activity at three of these volcanoes.

5.1. Mount St. Helens 1980–1986 and 2004–2008

Mount St. Helens is a stratovolcano consisting of dacite lava domes, layered pyroclastic products, and basaltic to andesitic lavas. It is also the most active volcano in the Cascade Volcanic Arc. Mount St. Helens has experienced two widely contrasting eruptive phases, in 1980–1986 and 2004–2008, which represent two of the best-monitored volcanic eruptions to date (e.g., Pallister et al., 1992; Sherrod et al., 2008).

The 1980 activity began in March with the intrusion of a **cryptodome** causing bulging of the North flank at up to 2 m per day (Figure 18.5(A)), and reducing the stability of the edifice. By May 18th, the cryptodome had grown to approximately 0.13 km^3 when a magnitude 5.1 earthquake triggered a catastrophic flank failure (or sector collapse). The collapse removed approximately 2.9 km^3 of material, resulting in a debris avalanche that traveled at speeds up to 200 km h^{-1}. The sudden unloading of the cryptodome induced the decompression of the underlying magma, triggering a lateral blast that traveled at up to 1000 km h^{-1} and

FIGURE 18.5 Recent eruptive activity at Mount St. Helens: (A) Mount St. Helens cryptodome, pictured on April 27th 1980 (US Geological Survey). (B) Photograph of the crater and lava dome taken on September 13, 1984 (by Lyn Topinka, US Geological Survey). (C) Panoramic photograph from the southern rim of Mount St. Helens' crater rim looking north on July 27, 2006 (US Geological Survey).

flowed up to 30 km from the volcano. An eruptive column reaching >20 km high followed, depositing 540×10^6 tons of ash in the first 9 h, equivalent to approximately 7% of the volume of the debris avalanche. During the May 18th eruption column collapse pyroclastic flows and lahars, formed when the ice cap and glaciers melted, caused widespread destruction in surrounding drainages.

The initial eruption had a volcanic explosivity index (VEI) of 5, and formed an amphitheater in the edifice, exposing the top of the conduit. In the following months, several episodes of lava dome growth and intermittent explosive activity took place. The initial domes were destroyed by explosions and subsequently reformed. From October 1980 through the end of the eruption in 1986, many episodes of endogenous, exogenous, and mixed lava dome growth took place, eventually building a 1 km

diameter, 350 m high lava dome with a volume of 74×10^6 m^3. Monitoring of seismicity and crater floor deformation allowed several of these eruptions to be predicted to within a few hours.

Throughout the eruption, an increase in degassing efficiency resulted in a gradual decline in average extrusion rate, reflected by a decrease in water content and changes in rock textures from predominantly scoriaceous to predominantly smooth crease structures. Studies of the in situ magma, revealed a porphyritic dacite with around 15−25 vol% porosity and average 63 wt% SiO$_2$ ($\sim$2% less than early ash) with $\sim$45 vol% phenocrysts, $\sim$30 vol% microlites, and <25 vol% rhyolitic interstitial glass and few lithic fragments. Crystallinity was found to increase slightly through the eruption.

The end of the eruption in October 1986 marked the beginning of 18 years of near-quiescence, interrupted by two short periods of unrest in 1989−1991 and 1995. This ended on September 23rd 2004 when the volcano reawakened. Localized ground uplift and shallow seismicity (<1 km) was followed by small phreatic explosions. Seismic signals centered under the 1980−1986 dome increased in intensity and frequency. From October 1st to 5th, a number of explosions created ash columns and opened a vent. Then, in contrast to the explosive onset of the 1980−1986 eruption, on October 11th a solid magma plug began to emerge in the crater between the older dome and the crater wall. The magma, having resided in the conduit for many years, was degassed and largely devoid of interstitial melt. This enabled it to rise, intact, to many tens of meters above the crater floor in a series of seven spines or **shear lobes**, over 4 years. The dome grew to 460 m above the crater floor, and during this time there was almost no explosive activity. High initial extrusion rates of 5.9 m^3 s^{-1} declined to 0.7 m^3 s^{-1}. The early spines extended south from a fixed vent atop an inclined conduit under the 1980−1986 dome, forming whaleback morphologies (see Table 18.2). Later, the vent shifted west and the last spine emerged more slowly, forming a dome-shaped rubbly mass. The eruption ended in 2008, having built a 92×10^6 m^3-lava dome (Figure 18.5(C)).

The 2004−2008 dome consisted of porphyritic dacite ($\sim$64.5 wt% SiO$_2$), which was holocrystalline, low porosity (<5 vol%), and nearly volatile-free. By the time magma reached 1 km below the surface, it was nearly completely degassed and crystallized. Interstitial glass content decreased from $\sim$30 vol% to <2 vol% in the first 2 months, and the remainder of erupted products were glass-poor. By the time it reached the surface, thermal imaging recorded temperatures of up to 600 °C at the surface, and 730 °C in fractures and gashes. Carapace surface temperatures rapidly cooled below 100 °C within 50 m of the vent with higher temperatures in fractures (Sherrod et al., 2008).

Each of the seven spines was mantled by ash-gouge, bordered by a zone of highly localized deformation that included tensile and shear fractures, cataclasite, and occasionally pseudotachylite (see Table 18.2). The surface character evolved as extrusion rate decreased and the magma became denser and stronger. Most of the deformation textures were restricted to the outer surface of the spine, decreasing from >3 to <1 m during the course of the eruption, while the bulk of the spine remained largely intact, indicating that the outer surface character had an important control on magma ascent and extrusion (Kendrick et al., 2014 and references therein). Spine extrusion was accompanied by small (generally <M1.5) long period or hybrid earthquakes that occurred at approximately 1 km depth on average every 40–80 s throughout the eruption. This regular occurrence led to the term "drumbeats," which have also been noted during other dome-building eruptions at Soufrière Hills and Unzen. It has been proposed that stick-slip motion was responsible for the seismicity, linking the events to slip events along the conduit margin, although their precise source mechanism remains elusive (Sherrod et al., 2008).

These two eruptions, so narrowly separated in time, show how a dome-building volcano can have significantly different eruption styles. By studying lava domes and linking structures to monitored processes, we can begin to understand the mechanisms driving these eruptions.

5.2. Soufrière Hills Volcano, Montserrat, 1995–Ongoing as of 2015

The SHV, on the small island of Montserrat, British West Indies, has been erupting since July 1995. Unique aspects of the eruption include the diverse repertoire of activity associated with the periodic extrusion of the lava dome, the unusual increase in activity throughout the eruption, the breadth and intensity of the monitoring program, advances in the understanding of andesite eruptions that have arisen as a result of the intense scientific interest, and the effectiveness of the disaster management program with respect to communities that are situated in relatively close proximity to the volcano.

The eruption has consisted of the growth and collapse of a sequence of andesite lava domes reaching volumes of $200–300 \times 10^6$ m^3, at an average extrusion rate of $3–5$ m^3 s^{-1} and peak extrusion rates >15 m^3 s^{-1}, punctuated by five distinct pauses in extrusion (Wadge et al., 2014) (Figure 18.6). The extrusion pauses have extended for between 1.5 and 4.5 years (as of January 2015), and

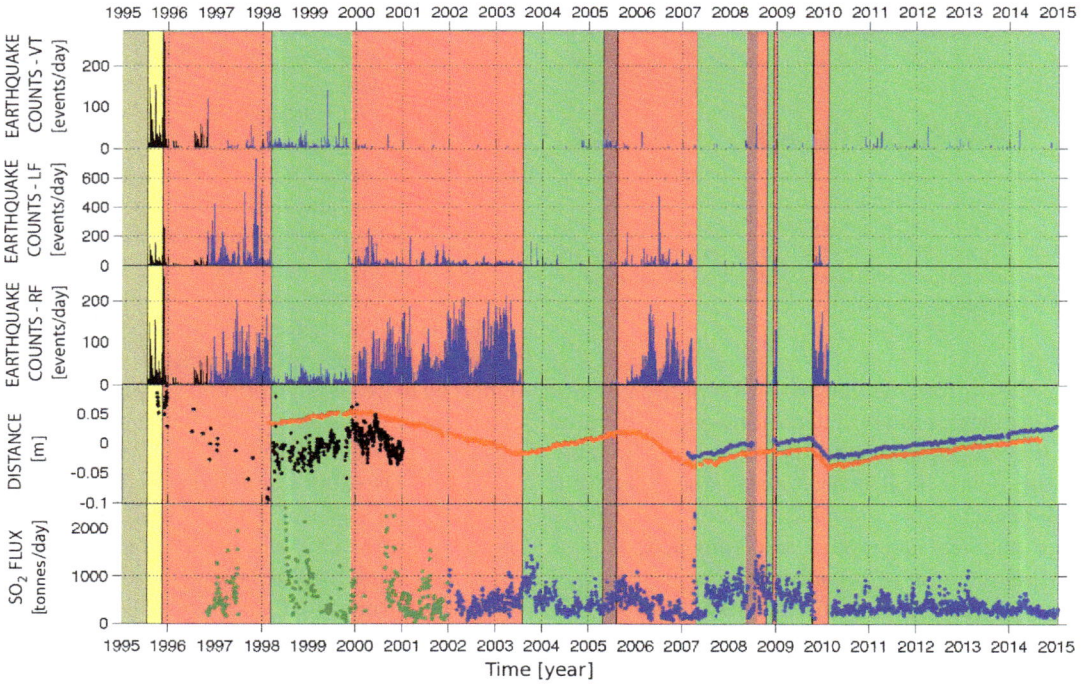

FIGURE 18.6 Multiparametric monitoring data from the Soufrière Hills volcano from 1995 until 2015. The colors in the plot are: yellow (initial phreatic phase), pink (lava dome extrusion periods), dark-pink (transitional/precursory) and green (pauses in lava dome extrusion). The five plots from top to bottom display daily frequency of volcano tectonic (VT) earthquakes, low-frequency (LF) earthquakes (both long-period and hybrid earthquakes), and earthquake counts related to rockfalls (RF), which are all closely tied to magma extrusion phases, as well as ground deformation (distance) and SO$_2$ degassing measurements. The ground deformation data show the radial displacement of stations MVO1 (red) and GERD (blue), with a 7-day median filter, and the elevation of station HARR is shown in black. The bottom plot shows the daily SO$_2$ flux filtered with a 7-day median filter (Green: COSPEC, Blue: DOAS). Although currently (January 2015), there has been no active lava dome growth since 2010, continued ground deformation and SO$_2$ signals provide indication of continued magmatic unrest at depth. Figure courtesy of MVO.

fall within the duration range of pauses recorded at other dacitic to andesitic lava dome eruptions. The Soufrière Hills eruption, however, is now one of the longest-lived dome forming eruptions in recorded history (Wadge et al., 2014).

The activity catalog through this eruption has been tremendously diverse. Accompanying lava dome growth throughout the eruption have been rockfalls and pyroclastic density currents generated by dome and shear lobe collapse (Calder et al., 1999). Many of the pyroclastic flows have entered the sea, at least one of which is known to have formed a significant hydromagmatic explosion, and most of which have contributed to coastal deltas as well as submarine fans extending tens of kilometers offshore. There have also been phreatic activity and cold base surges before the magma reached the surface, a 9-h-long subplinian eruption, three periods characterized by closely spaced sequences of vulcanian explosions many of which generated fountain collapse pyroclastic flows, frequent and vigorous ash venting episodes, a debris avalanche, and many mudflows (Druitt and Kokelaar, 2002). What is perhaps so fortunate is that throughout the eruption, activity escalated sufficiently slowly that monitoring capabilities and understanding of the continuing eruption managed to keep pace with the situation. Only one event, a dome collapse on 25th June 1997, which occurred during a period of heightened activity, caught people by surprise and resulted in the tragic loss of 19 lives (Wadge et al., 2014).

Frequent collapses of the Soufrière Hills lava dome occur as a natural consequence of growth by extrusion of lava, and range from conventional rockfalls, through small pyroclastic flows, to major dome collapse events involving up to $210 \times 10^6 \, \text{m}^3$ of material. Since the lava dome first emerged in November 1995, it has generated tens of thousands of individual rockfalls, and several hundred pyroclastic flows with runouts >2 km. On around 50 occasions, pyroclastic flows were formed during large dome collapse events which removed $>1 \times 10^6 \, \text{m}^3$ of the dome, after which the domes commonly resumed growth, refilling the resulting collapse scars. For a number of these individual events, precursory activity, event chronology, inundation area, and flow parameters such as duration, volume, mass fluxes have been well documented (Druitt and Kokelaar, 2002). However, despite intense efforts toward understanding the physical mechanisms behind dome collapse at Montserrat, as well as elsewhere, the propensity for lava domes to collapse with little or no apparent warning remains a serious issue that still jeopardizes surrounding communities.

Although the last magma extrusion occurred in 2010, seismic swarms and ash venting continued until 2012, and degassing and ground deformation still continue to date (January 2015). The contributions of the data sets from this extensively monitored eruption continue to provide rich material for research (Figure 18.6). Furthermore, geophysical monitoring data sets such as those of the 6−12 h seismic and ground deformation cycles have enabled some degree of event forecasting to be achieved at times. Understanding the processes involved during the repetitive cycles of explosive activity largely through the associated seismicity (e.g., Neuberg et al., 2006), detected by a sophisticated array of broadband seismometers, has significantly advanced the understanding of magma failure during ascent in volcanic conduit.

5.3. Chaitén, Chile, 2008−2009

On the evening of May 1st, 2008 Chaitén Volcano, located in the Southern Volcanic Zone of the Chilean Andes, erupted suddenly and violently producing a 21-km-high eruption column. The eruption was noteworthy for several reasons: for giving such little precursory warning; for being one of only three rhyolitic composition lava eruptions observed in historic time (the others being the 1912 Katmai-Novarupta, Alaska and the obsidian lava flow at Puyehue-Cordón Caulle, Chile in 2011); and for the very rapid growth of a large compound lava dome, associated with synchronous explosive activity.

Due to the relative remoteness of the area, impacts were mostly limited to the nearby town of Chaitén, whose 5000 residents were evacuated before the town was destroyed by lahars. Prior to the 2008 eruption, Chaitén had two known Plinian eruptions about 9400 and 5000 years ago, as well as a probable lava dome eruption in the seventeenth century. The structure comprised a $>500 \times 10^6 \, \text{m}^3$-rhyolite lava dome surrounded by a moat within a 3-km-diameter caldera that probably formed during the event 9400 years ago.

While the Chaitén eruption was much better tracked than its 1912 Katmai-Novarupta cousin in Alaska, monitoring was intermittent due to the location's logistical and weather limitations. The result is that the eruption record is somewhat incomplete, particularly with regards to the timing of some of the intrusive and extrusive events. Increasing seismicity, which was only identified 36 h before the start of the eruption, tracked magma rising rapidly along an inclined dike from a depth of 5−9 km in about 4 h. The initial explosive phase of the eruption lasted from May 1st to May 12th and produced three VEI 4 to 5 Plinian events on May 2nd, 6th, and 8th with column heights up to 22 km, as well as several smaller events (Pallister et al., 2013). These resulted in extensive ash deposition downwind to the East, across the border in Argentina impacting transportation, utilities, and agriculture.

The Plinian events generated modest pyroclastic density currents on the North, Northeast, and East flanks of the edifice resulting in tree blowdown. Ashfall was remobilized by rain-generated lahars in the Chaitén River drainage on May 11th and 12th. This was followed by a transitional explosive-effusive phase lasting from May 12th to 31st, during which continued ash emissions were accompanied by initial extrusion of lava. The exogenous lava dome phase lasted from June through September of 2008, and produced several large flow lobes that nearly covered the preexisting lava dome. In October 2008, a spine extrusion phase commenced, accompanied by continuing endogenous growth. This phase ended on February 19th, 2009 when a large dome collapse removed about 10% of the dome volume and generated block-and-ash flows, which drained through a low point in the caldera wall and entered the Chaitén River valley. Dome growth soon resumed within the collapse scar forming a large endogenous lobe on the west side of the lava dome complex, which continued to grow until late 2009/early 2010.

The composition of the lava at Chaitén has been remarkably constant, not only throughout the 2008–2009 eruption, but also during the older dome-forming events. The lava is a phenocryst-poor rhyolite with 75% SiO_2. Phase equilibria indicate the preeruptive magma was stored at a temperature of about 800 °C and a depth of 5–10 km. Rising through the conduit, the magma was apparently sufficiently fluid to allow rapid H_2O loss, foam formation, and fragmentation (Castro et al., 2012). The low viscosity also explains the very high extrusion rates that peaked at 66 $m^3 s^{-1}$ during the first two weeks of the lava extrusion and averaged 45 $m^3 s^{-1}$ over the first 4 months. The volume of lava added to the lava dome complex is about 800 × 10^6 m^3, while that emitted during the explosive phase was about 300 × 10^6 m^3, for a total output of approximately 1.1 km^3 (Pallister et al., 2013). Thermal infrared mapping showed variations in flow morphology as a function of extrusion rate, and also indicated that multiple flow paths within the lava dome were required to sustain simultaneous endogenous and exogenous growth, as well as spine extrusion.

In concert, these dome eruptions as well as others at Merapi Volcano, Indonesia, Volcán de Colima, Mexico, Santiaguito dome complex, Guatemala, Galeras, Colombia, Unzen, Japan, etc. have provided us with an increasingly better-defined portrayal of lava dome eruption dynamics.

6. DYNAMICS AND MECHANICS OF LAVA DOME EXTRUSION

The wealth of collected data that have emerged from recent lava dome eruptions has spurred efforts to replicate the observed petrological and rheological processes with experiments (e.g., Kendrick et al., 2014) and models (e.g., Hale et al., 2009) to constrain eruptive conditions. These model and experimental inputs used in tandem with multiparametric monitoring have provided us with an increasingly complex description of magma ascent dynamics (e.g., Lavallée et al., 2012).

6.1. Cyclic Magma Ascent

The discharge rate associated with lava dome eruptions is generally low when compared to lava flow activity and exclusively explosive eruptions, and commonly demonstrates cyclicity on multiple timescales (Voight et al., 1999; Chouet and Matoza, 2013).

On timescales of months to years, lava domes often demonstrate relatively long episodes of steady growth, alternating with periods of differing discharge rates or even quiescence. Effusion is commonly accompanied by explosive activity occurring at a frequency and intensity that increases with extrusion rate. Volcán de Colima illustrates this relationship well, where two phases of low discharge rates and explosive activity (1997–2001 and 2007–2013) were interspersed with a period of high discharge rate and heightened explosive activity (2001–2005). In May 2011, Volcán de Colima entered a 20-month-period of quiescence, which ended in January 2013 with a renewal of explosive activity. Similarly, the 1980–1986 dome eruption at Mount St. Helens was followed by 18 years of relative quiescence before the extrusion of a sequence of seven magma spines in 2004–2008.

On the timescale of several days to weeks, growth pulses are generally attributed to episodes of dyke propagation, feeding magma into the dome (Barmin et al., 2002). For instance, in 1997 at Soufrière Hills Volcano, the sharp onset of tilt cycles and seismic swarms have been modeled to reflect a 300–500 m long, by 3–6-m-wide magmatic intrusion.

On the timescale of hours to a few days, pulsations in lava dome growth are associated with gas, and sometimes ash, venting activity (e.g., Lavallée et al., 2012; Wadge et al., 2014). The May–June 1997 record of dome eruption at Soufrière Hills typifies this behavior where ~10-h-cycles of edifice inflation/deflation and seismicity were triggered by pulses of magma ascent in the shallow conduit as well as magma failure and slip (Neuberg et al., 2006). It has also been proposed that pulses in gas fluxing through the conduit may drive such cyclic magma ascent (Edmonds and Herd, 2007).

On the timescale of seconds to minutes, magma ascent tends to happen in rhythmic pulses of millimeters up to a few meters of ascent at rates of about 1 mm s^{-1} to 1 m s^{-1}. This activity tends to be defined by repetitive drumbeat seismicity, which was interpreted as a result of magma failure and stick-slip motion near the conduit margin as described for the 2004–2008 spine extrusions at Mount St. Helens (Kendrick et al., 2014; Sherrod et al., 2008) as well as at Soufrière Hills and Unzen. Recent research on lava domes has shown that longer cycles reflect the periodic dynamics of the magmatic system, while shorter cycles and changes in eruptive behavior are seen to be the result of a competition between chemical, petrological, rheological, and mechanical processes taking place during magma ascent within the shallower conduit.

6.2. Petrology

The relatively low ascent rate driving lava dome eruptions provides sufficient time for magma to react to local changes in pressure and temperature conditions. Petrological studies point to efficient volatile exsolution and variable degrees of crystallization from a magma that is relatively depleted in gas and crystal-poor at its outset in dome building eruptions (Cashman and Blundy, 2000). The decompression associated with an eruption reduces the solubility of volatiles which exsolve, forming gas bubbles. This in turn triggers the crystallization of microlites at a rate inversely proportional to the viscosity of the interstitial melt. In exceptional circumstances, silica-rich melts can forgo crystallization and quench to form crystal-poor obsidian. Hydrous phases of the mineralogical assemblage become unstable upon decompression and react to recrystallize as anhydrous minerals, liberating a gas phase in the process. It has been demonstrated that an increase in decompression rate results in (1) bubbles and crystals with smaller sizes, (2) a lower crystallinity and thus higher glass fraction, and (3) a higher abundance of unstable hydrous phases. The microtextural information delivered by eruptive products throughout the course of protracted dome eruptions has become an important indicator of physicochemical changes, which dictate magma ascent dynamics.

6.3. Rheology

Lava dome eruptions sometimes evolve from relatively benign effusions to catastrophic explosions. The cause of this sudden change rests in the rheological behavior of ascending magma. Silicate melts are **viscoelastic** liquids, which share the properties of a viscous body and that of an elastic solid, with a divide between the two defined in terms of strain rate and temperature. This divide, the glass transition, rheologically contributes to the ductile–brittle transition of magma, which marks either the onset of magma failure, or conversely the healing of fragments (Castro et al., 2012). Due to these rheological attributes, lava dome behavior can shift on short timescales in response to changes in strain rate.

The physicochemical changes experienced by magmas feeding active lava domes have important rheological consequences, resulting in magma properties increasingly resembling those of rock as it nears the surface. Experiments on magma at realistic pressure/temperature conditions have shown that viscosity has a first order control on deformation mechanisms and hence on magma ascent and extrusion (e.g., Lavallée et al., 2012). In addition, analogue experiments have helped unravel broader structural characteristics of lava domes (Fink and Griffiths, 1998). The advantage of analogue experiments is that parameters can be varied independently, demonstrating how changes in viscosity or extrusion rate affect the morphology and stability of lava domes. For instance, it has been shown through analogue modeling that the morphology of a dome can evolve from spine growth, to lobate, platy, and axisymmetric morphologies, with an increase in effusion rate and reduction in cooling rate.

The chemical evolution associated with degassing causes a nonlinear increase in the viscosity of the liquid, which lowers the strain rate required for fragmentation. The relatively low water content dissolved in most dome lavas has an important control on their morphology. This is especially the case for obsidian, which has been argued to quench due to devolatilization (Castro et al., 2012); the rheological contribution of volatiles on the viscosity exceeds that of temperature within the eruptive temperature range. Crystallization induces opposing chemical effects as the volatile phase remains in the melt. Crystallization also releases latent heat, which contributes to enhanced thermal conditions in the shallow magmatic column. Physically, the crystals are strong and generally remain intact; however, they may deform plastically and fail when subject to strain localization near the conduit margin.

The presence of bubbles and crystals mechanically alters ascending magma and extruded dome lavas. Magmas behave with a non-Newtonian rheology, nearing that of a Bingham fluid; i.e., their viscosity is relatively high and strain rate dependent, yet they do not have a yield strength (although strain rates are very low at low-applied stresses). Bubbly magma may compact by foam collapse, and the resultant rheology may exhibit strain hardening; that is, apparent viscosity increases with strain (Lavallée et al., 2007). During eruptions, dome lavas bearing high crystallinities (>50%) tend to have low porosities (<35%), and their viscosity empirically adheres to a singular rheological law:

$$\log \eta = -0.993 + 8974/T - 0.543^* \log \gamma$$

where η is the apparent viscosity (Pa s), T the temperature (°C), and γ the strain rate (s^{-1}). Although more precise laws have been defined for specific lava dome scenarios, this law provides a first order constraint on the shear-thinning behavior of crystalline dome-building magmas (Lavallée et al., 2007). Such a rheology characteristically results in strain localization in regions of high stresses (e.g., the conduit margin) and promotes plug flow dynamics. Upon strain, the viscosity may seek to decrease (i.e., strain weakening), as the crystals and bubbles organize themselves into flow bands, and crystals plastically deform or fail. Crystals and bubbles in magmas modify the ductile—brittle transition, shifting its onset to lower stress values for a given temperature. The strength of magma has been described to be inversely proportional to the pore fraction and size distribution of vesicles and arguably even to the crystal fraction. It is possible that the strain frozen into vesicles and crystals may be used as a strain marker for the ductile—brittle transition of magmas.

6.4. Seismogenic Magma Failure and Eruptive Style

Magma failure results from the nucleation of microscopic fractures, which propagate and coalesce macroscopically. This increasing scale of damage results in a characteristic acceleration in released seismicity, which may be used as a proxy to forecast the timing of failure and, in certain instances (often retrospectively), of an explosive eruption.

During lava dome eruptions, magma failure has two manifestations; failure may be followed by slip along a fault plane or may trigger fragmentation. These failure mechanisms carry important consequences for the evolution of lava extrusion. Magma failure and slip have been increasingly recognized as common processes, taking place in regions of high strain rates near the conduit margins (Edmonds and Herd, 2007). The repetitive character of the associated seismic signals has been associated with the failure and slip of magma at a level in the conduit where the criterion for failure is met (Neuberg et al., 2006), or to the resonance of fluids inside a fracture. In each scenario, the conduit margin fault zone may reach the surface and act as a pathway for outgassing. Magma faulting is increasingly accepted as associated with the second-to-minute stick-slip cycles observed during magma ascent and spine extrusion (Chouet and Matoza, 2013). Failure and slip relieves some of the stress accumulated in the conduits and upon repose, magma may heal along the fault plane. The geological record has demonstrated that extreme slip events may occasionally induce enough heat to melt magma or rocks along a fault plane, producing pseudotachylites (Kendrick et al., 2014). Recurring slip events and comminution along the fault may also generate ash gouge in the fault plane, which may be entrained and erupted during gas-and-ash venting events (e.g., Sherrod et al., 2008).

Magma failure may also take the form of fragmentation. In this important scenario, the energy trapped in overpressurized gas bubbles may momentarily exceed the stress that the melt structure can accommodate by viscous relaxation, causing catastrophic failure of magma (Lavallée et al., 2012). If sufficient energy remains, it will be converted to kinetic energy to drive an explosive eruption, ejecting gas, ash, and ballistics. On the other hand, if little energy remains after fragmentation, the densest and largest fragments produced will remain in the conduit as breccia, and only gas and ash will be expelled, with a fraction remaining at depth to form tuffisites, while the other fraction reaches the surface, causing ash venting (Castro et al., 2012). In recent years, tuffisites have been extensively investigated, as it is becoming increasingly apparent that they are important indicators of fragmentation, and changes in the permeability and strength of domes.

6.5. Permeable Porous Networks

During lava dome building eruptions, development of a permeable, porous network is crucial as it regulates outgassing and dome pressurization (Sparks, 1997). To a first order, dome permeability (κ) is nonlinearly proportional with the porosity (ϕ) of the rock according to the relationship:

$$\kappa = 10^{-17}\phi^{3.4}$$

Although not always applicable, this empirical relationship illustrates that dome rocks are highly permeable, due to the abundance of pervasive fractures connecting the pores. At depth, however, fractures are shut due to high normal stresses, and permeability is closed off. Recent work illustrates how the presence of localized defects (e.g., tuffisites, shear fractures, cataclasite, pseudotachylite) superimposed on the primary dome rock structure can affect permeability. Strain localization near the conduit margin constructs an anisotropic permeability network, which favors lateral outgassing toward the conduit margin, followed by vertical gas fluxing; a scenario supported by the occurrence of outgassing and gas-and-ash explosions near dome margins (as is common at Santiaguito, Figure 18.1(D)).

While the presence of crystals in magma determines dome morphology and growth characteristics, it is the presence of gas that creates the threat. It is the potential energy resting in buoyant ascending magma which dictates the eruption style, whether effusive or explosive, and ultimately, the intensity of a fragmentation event. Numerical models have been used to illustrate how gas fluxing though the magma column and the dome occurs in pulses or waves; that is, the gas is decoupled from the magma due to its high permeability (Sparks, 1997; Michaut et al., 2013). Yet, lava domes are particularly dangerous as they are unstable. When domes collapse, rapid decompression of the dome core and the underlying magma column occur, which may trigger explosive activity. The 1980 explosive eruption at Mount St. Helens typifies such an unloading event as large-scale sector collapse exposed the shallow cryptodome to atmospheric pressure, triggering fragmentation (Pallister et al., 1992). Recent efforts to constrain the energy budget in active lava dome systems have made use of ballistics scattered around domes to estimate magma overpressure at the source of explosive eruptions to a few tens of MPa (Lavallée et al., 2012).

The competition between these rheological and mechanical processes provides some understanding of the cyclicity associated with lava extrusion and the transition in eruptive behavior so characteristic of lava domes.

7. HAZARDS ASSOCIATED WITH LAVA DOMES

Lava dome eruptions can be dangerous. They can subject surrounding populations to prolonged periods of hazards ranging from frequent ash falls to the acute hazards posed by violent lava dome collapses (Druitt and Kokelaar, 2002). The primary hazards associated with dome collapse events include pyroclastic flows and surges, and sometimes lateral blasts. Lava domes are also often associated with explosive eruptions, which produce ash falls, ballistic showers, and sometimes column collapse pyroclastic flows. Secondary hazards may include lahars; in tropical regions, these comprise seasonal bouts of sediment transport by debris flows sourced from unconsolidated block-and-ash flow deposits and talus fans. Old, inactive, and hydrothermally altered lava domes and lava dome complexes also have the propensity to undergo sector collapse forming debris avalanches. These events are often unassociated with volcanic activity, but external forces such as rainfall and/or regional earthquakes therefore represent a hazard type that is difficult to forecast.

7.1. Explosive Activity

Lava dome growth is associated with a diverse repertoire of explosive volcanic activity ranging from frequent but mild gas exhalations, moderate to violent Vulcanian explosive activity, through to major sustained Plinian eruptions.

Repetitive gas-and-ash exhalations are common at the Santiaguito, Colima, and Galeras lava domes, among others, and this activity is sometimes referred to as ash venting, or gas puffing. In these cases, mild explosive activity, usually lasting only a few tens of seconds, involves the release of gas and ash-laden plumes that rise 2−6 km. These explosions often emanate from concentric fissures or vents near the margins of the dome, and are characteristically nondestructive in nature, exploiting zones of high permeability. In other words, the lava dome or plug remains for the most part intact, and the outgassing occurs synchronous with continued magma extrusion. Lava domes that display this activity can continue doing so for periods of months to years, and as such this activity is characteristic of degassing processes and fluxes through the permeable fracture system commonly constructed in the shallow plumbing of these highly viscous magmas. Only rarely are these explosion types associated with collapsing plumes, and when this does occur, very thin, ephemeral pyroclastic deposits dominated by fine ash are generated (Druitt and Kokelaar, 2002).

Vulcanian explosions at lava domes can occur as discrete individual events, dispersed clusters, or in sequences of rapid succession which are often cyclical in nature. These events commonly involve significant removal of dome material, and in many cases tap hotter more gas-rich (and sometimes only moderately crystal-bearing) magmas lower in the conduit. These explosions result in substantial, lithic, and pumice-laden eruption plumes reaching a maximum height of about 15 km which can lead to column or fountain collapse pyroclastic flows extending for several kilometers radially around the volcano (typically VEI ≤ 3). The deposits from these flows are generally rich in pumice and are often referred to as pumice flow deposits, but can have a significant component of lithics derived from the preexisting dome and thus also be referred to as **block-and-ash flow deposits**. Tephra dispersal from these events can have regional effects. In many cases, these explosions are triggered by rapid decompression of deep-seated, gas-charged dome magma after significant removal of material during more superficial lava dome collapse events. The Soufrière Hills eruption is a classic case where after an initial collapse-driven decompression event in 1997, 88 cyclical Vulcanian explosions then occurred at intervals of between 6 and 12 h for around 4 weeks. Enduring questions remain about how such systems work, in particular, how the system is replenished and repressurized in order to generate the repeating sequence of events. Various models exist, but central to all of them is the understanding that magma viscosity, driven by degassing induced crystallization, increases by several orders of magnitude as the magma approaches the surface. Pressure buildup combined with reduction and segregation of permeability results in the explosive events that remove the plug. In terms of hazard mitigation, this style of eruptive activity poses serious challenges. The transition from benign dome growth to explosive activity can come with few precursory indicators. In some cases, however, clear indicators for a transition to explosive behavior have become evident, such as long-period earthquakes (for example, tornillo events at Galeras), and cyclic swarms of hybrid earthquakes and associated cyclic ground deformation (e.g., Soufrière Hills).

Major explosive Plinian eruptions are also associated with some lava dome eruptions. In some cases (e.g., Lascar, 1993 eruption), these major eruptions are the culmination of a rapid succession of Vulcanian explosions where rapid decompression of the dome and material in the shallow conduit results in a runaway decompression process drawing up magma from deeper in the plumbing system. Around 95% of historical lava dome-forming eruptions were associated with explosive activity. Most of these explosions were Vulcanian-type events, but large Plinian explosions with a VEI ≥ 4 are also common occurrences, and some of the most significant eruptions of the twentieth century have been in this category (e.g., May 8, 1902, Mt. Pelée, Martinique; May 18, 1980 Mount St. Helens; and the June 15, 1991 eruption of Mt. Pinatubo, Luzon, Philippines). More recently, the Chaitén 2008–2009 eruption involved 10 days of major explosive activity followed by lava dome growth. The fatal October–November 2010 VEI-4 eruption at Merapi, Central Java, Indonesia, occurred during a period of rapid dome growth. A recent study of the temporal relationship of major explosive lava activity and dome growth found that the majority of large explosive eruptions (>VEI 4) associated with lava dome growth occurred at andesitic volcanoes. However, a greater percentage of dacitic and rhyolitic dome-growth episodes were associated with such explosions.

7.2. Dome Instability

Rockfalls comprise the small, end-member of collapse phenomena from lava domes, and occur frequently, with anywhere up to 200 rockfalls occurring daily during periods of active dome growth (see earthquake counts related to rockfall activity; Figure 18.6). Rockfalls occur along high-angle failure planes on the outer, largely degassed, carapace of lava domes and may involve discrete blocks that roll, bounce, or slide downhill, to significant avalanches. Rockfalls rarely travel far beyond the base of the talus apron from the dome summit and are commonly only recorded seismically. At both Soufrière Hills, Montserrat and Volcán de Colima, Mexico, they have been found to be a proxy for magma extrusion rate (e.g., Calder et al., 2005, Hale et al., 2009).

Pyroclastic flows are distinguished from rockfalls by their larger size, longer runouts ($\gg 0.5$ km), greater component of fine material, fluidization, and appreciable buoyant ash clouds. In practice, there is a continuum of collapse phenomena between rockfalls and pyroclastic flows, and their frequency has also shown a correlation with extrusion rate at Unzen and Soufrière Hills. The deposits of pyroclastic flows generated by lava dome collapse are often referred to as block-and-ash flow deposits, because they typically comprise dense blocks of fragmented lava in an ash matrix. Pyroclastic flow deposit volumes are typically on the order of $10^4 - 10^5$ m^3 for small to medium-sized (1–3 km runout) flows, while large dome collapse events are loosely defined as those which generate pyroclastic

flows and have deposit volumes in excess of 1×10^6 m^3 and up to 210×10^6 m^3.

Major lava dome collapse episodes commonly occur as retrogressive failures over minutes to hours and excavate spoon-shaped cavities as deep as 200 m. Upon collapse, each large slice of dome material generates discrete, energetic flow pulses producing discrete peaks in seismic signal energy. The dynamics of this class of pyroclastic density current are notorious for their ability to form bipartite currents with a dense basal avalanche and dilute turbulent ash cloud, which often separate during flow. A large number of the deaths associated with inundation by pyroclastic density currents are directly due to this separation, the ash cloud surge component travels over and out of the drainages in unexpected directions. Topographic effects, such as changing retaining capacity of the drainage channel, bends in the flow path, and abrupt slope changes can all help induce separation of the dense and dilute part of the flows.

The challenge now is to better understand the causative mechanisms and specifically temporal controls including the nonstationarity of lava dome mass-wasting patterns as well as monitoring signal precursors to lava dome collapses, in order to directly tackle the issue of event forecasting.

7.3. Collapse Mechanisms

Understanding causal mechanisms of lava dome instability and subsequent pyroclastic flow formation are key objectives in volcanology (Voight, 2000). Several potential mechanisms for these failures have been proposed. Some are directly related to magma extrusion characteristics: thrust forces associated with an active lava lobe intrusion or extrusion can literally push adjacent dome portions; gas overpressurization, especially in association with rapidly ascended gas-rich magmas, makes domes prone to autobrecciation, or explosive collapse; and slope oversteepening and propagation of thermal fractures become important when lava lobes are advancing over steep slopes. One model for collapse mechanisms, which views the outer carapace as variably coherent and bound by frictional stresses, and the cohesive core, proposes that rapid influx of low-viscosity magma into the core favors the development of deeply transecting failures over those of a spalling nature occurring on the chilled and oversteepened carapace. External forces also affect dome stability, such as destabilization by seismic acceleration during earthquake swarms. There is also evidence that intense bouts of rainfall can initiate the remobilization of talus and induce dome failure by undermining the lava dome and/or promoting the development of cooling fractures that facilitate structural weakening. Intense rainfall has been associated in some cases with overpressurization of the dome interior, although explanations for the causal mechanisms for this are diverse. For the most part, these mechanisms are not mutually exclusive and it is clear from the study of individual dome collapse events that the relative importance of different mechanisms is highly variable.

7.4. Debris Avalanches

A significant hazard at lava dome-forming volcanoes is sector collapse that generates debris avalanches. Debris avalanches can be induced at inactive volcanoes due to external factors such as tectonic earthquakes, seasonal weather changes, and long-term hydrothermal alteration, or as a result of renewed activity or during ongoing eruptive phases. It is noteworthy that a large proportion of debris avalanche deposits around the world involve altered, collapsed portions of lava domes, or lava dome complexes, indicating that they represent inherently unstable structures over the long term (hundreds to tens of thousands of years postemplacement).

A long-term result of water percolating through hot lava domes is the development of hydrothermal systems. Within these systems, dome lavas can be altered to clay mineral assemblages (e.g., smectite clays, kaolinite, and alunite), and glass can devitrify to spherulite or cristobalite, which weakens the edifice, reduces slope stability, and can ultimately result in slope failure. Additionally, clay-rich alteration materials not only offer a low strength and high permeability but have the potential to absorb and channel groundwater while locally increasing pore fluid pressure within an edifice, which may promote the expansion and/or formation of low-strength zones and exacerbate the risk of slope failure. Upon collapse, clay-hosted pore water may lubricate the moving masses and result in the generation of highly mobile cohesive debris flows, which can extend the effects of a sector collapse far beyond the potential run-out of dry material. Two examples of collapses of hydrothermally altered lava domes are the 1997 debris avalanche at Soufrière Hills, Montserrat, and the 1998 debris flow at Casita in Nicaragua. In both cases, these low-strength, low-permeability, fines-rich alteration materials were thought to have concentrated water and lubricated structural discontinuities, reducing the shear strength of the rocks which ultimately led to catastrophic destabilization of the edifices.

8. SUMMARY

Lava domes, although geologically defined as a pile of rocks and lava plugging volcanic conduits, are increasingly recognized as a specific class of eruption, where the gas phase competes with a relatively slowly ascending magma whose rheology progressively evolves from a viscous liquid to that of a brittle rock. Lava dome

emplacement is concordant with the development of a wide spectrum of structures and textures, which reflect locally contrasting deformation mechanisms that construct a variably efficient permeable network. Magma ascent and eruptions generate an indicative set of geophysical and geochemical signals, which serve to facilitate the development of increasingly accurate and complex models. The rich data sets offered by extensive monitoring in the last several decades suggest that dome eruptions are characterized by pulsatory magma ascent at multiple timescales. The cyclicity involved provides information about the state of the magma that feeds a lava dome, and to its structural stability, which provides information about potential threats, whether caused by fragmentation and explosive eruptions or collapse of the lava dome and generation of pyroclastic density currents.

The complexity of lava dome eruptions, from the dynamics of magma ascent in conduits, through to extrusion, emplacement, and destruction-related hazards, make their study an important challenge in volcanology today. As our view of their behavior evolves, it becomes increasingly clear that the challenge of lava dome monitoring and related hazard mitigation requires an integrated multidisciplinary effort.

FURTHER READING

Barmin, A., Melnik, O., Sparks, R.S.J., 2002. Periodic behavior in lava dome eruptions. Earth Planet. Sci. Lett. 199 (1–2), 173–184.

Calder, E.S., Cole, P.D., Dade, W.B., Druitt, T.H., Hoblitt, R.P., Huppert, H.E., Ritchie, L., Sparks, R.S.J., Young, S.R., 1999. Mobility of pyroclastic flows and surges at the Soufrière Hills Volcano, Montserrat. Geophys. Res. Lett. 26 (5), 537–540.

Calder, E.S., Cortes, J.A., Palma, J.L., Luckett, R., 2005. Probabilistic analysis of rockfall frequencies during an andesite lava dome eruption: The Soufrière Hills Volcano, Montserrat. Geophys. Res. Lett. 32, L16309. http://dx.doi.org/10.1029/2005GL023594.

Cashman, K., Blundy, J., 2000. Degassing and crystallization of ascending andesite and dacite. Philos. Trans. R. Soc. London 358, 1487–1513.

Castro, J.M., Cordonnier, B., Tuffen, H., Tobin, M.J., Puskar, L., Martin, M.C., Bechtel, H.A., 2012. The role of melt-fracture degassing in defusing explosive rhyolite eruptions at Volcán Chaitén. Earth Planet. Sci. Lett. 333–334, 63–69.

Chouet, B.A., Matoza, R.S., 2013. A multi-decadal view of seismic methods for detecting precursors of magma movement and eruption. J. Volcanol. Geotherm. Res. 252, 108–175.

Druitt, T.H., Kokelaar, B.P. (Eds.), 2002. The Eruption of Soufrière Hills Volcano, Montserrat, from 1995 to 1999, vol. 21. The Geological Society of London Memoir, London.

Edmonds, M., Herd, R.A., 2007. A volcanic degassing event at the explosive-effusive transition. Geophys. Res. Lett. 34 (21), L21310.

Fink, J.H., 1990. Lava Flows and Domes. Springer, Berlin Heidelberg, New York.

Fink, J.H., Griffiths, R.W., 1998. Morphology, eruption rates, and rheology of lava domes: insights from laboratory models. J. Geophys. Res. 103 (B1), 527–545.

Hale, A.J., Calder, E.S., Loughlin, S.C., Wadge, G., Ryan, G.A., 2009. Modelling the lava dome extruded at Soufrière Hills Volcano, Montserrat, August 2005—May 2006: Part II: rockfall activity and talus deformation. J. Volcanol. Geotherm. Res. 187 (1—2), 69—84.

Kendrick, J.E., Lavallée, Y., Hirose, T., Di Toro, G., Hornby, A.J., De Angelis, S., Dingwell, D.B., 2014. Volcanic drumbeat seismicity caused by stick-slip motion and magmatic frictional melting. Nat. Geosci. 7, 438—442.

Lavallée, Y., Hess, K.U., Cordonnier, B., Dingwell, D.B., 2007. Non-newtonian rheological law for highly crystalline dome lavas. Geology 35 (9), 843—846.

Lavallée, Y., Varley, N., Alatorre-Ibargüengoitia, M., Hess, K.U., Kueppers, U., Mueller, S., Richard, D., Scheu, B., Spieler, O., Dingwell, D., 2012. Magmatic architecture of dome-building eruptions at Volcán de Colima, Mexico. Bull. Volcanol. 74 (1), 249—260.

Michaut, C., Ricard, Y., Bercovici, D., Sparks, R.S.J., 2013. Eruption cyclicity at silicic volcanoes potentially caused by magmatic gas waves. Nat. Geosci. 6, 856—860. http://dx.doi.org/10.1038/ngeo1928.

Neuberg, J., Tuffen, H., Collier, L., Green, D., Powell, T., Dingwell, D., 2006. The trigger mechanism of low-frequency earthquakes on Montserrat. J. Volcanol. Geotherm. Res. 153 (1—2), 37—50.

Pallister, J.S., Diefenbach, A.K., Burton, W.C., Muñoz, J., Griswold, J.P., Lara, L.E., Lowenstern, J.B., Valenzuela, C.E., 2013. The Chaitén rhyolite lava dome: eruption sequence, lava dome volumes, rapid effusion rates and source of the rhyolite magma. Andean Geol. 40, 277—294.

Pallister, J.S., Hoblitt, R.P., Crandell, D.R., Mullineaux, D.R., 1992. Mount St. Helens a decade after the 1980 eruptions: magmatic models, chemical cycles, and a revised hazards assessment. Bull. Volcanol. 54 (2), 126—146.

Sherrod, D.R., Scott, W.E., Stauffer, P.H. (Eds.), 2008. A Volcano Rekindled: The Renewed Eruption of Mount St. Helens, 2004—2006. U.S. Geological Survey. Professional Paper 1750.

Sparks, R.S.J., 1997. Causes and consequences of pressurisation in lava dome eruptions. Earth Planet. Sci. Lett. 150 (3—4), 177—189.

Voight, B., Sparks, R.S.J., Miller, A.D., Stewart, R.C., Hoblitt, R.P., Clarke, A., Ewart, J., Aspinall, W.P., Baptie, B., Calder, E.S., Cole, P., Druitt, T.H., Hartford, C., Herd, R.A., Jackson, P., Lejeune, A.M., Lockhart, A.B., Loughlin, S.C., Luckett, R., Lynch, L., Norton, G.E., Robertson, R., Watson, I.M., Watts, R., Young, S.R., 1999. Magma flow instability and cyclic activity at Soufrière Hills Volcano, Montserrat, British West Indies. Science. 283, 1138—1142.

Voight, B., 2000. Structural stability of andesite volcanoes and lava domes. Philos. Trans. R. Soc. London 358, 1663—1703.

Wadge, G., Ryan, G., Calder, E.S., June 2009. Clastic and core lava components of a silicic lava dome. Geology 37, 551—554.

Wadge, G., Robertson, R.E.A., Voight, B. (Eds.), 2014. The Eruption of Soufrière Hills Volcano, Montserrat from 2000 to 2010, vol. 39. The Geological Society Memoir, London.

Chapter 19

Submarine Lavas and Hyaloclastite

James D.L. White

Geology Department, University of Otago, Dunedin, New Zealand

Jocelyn McPhie

School of Physical Sciences and CODES, University of Tasmania, Hobart, Tasmania, Australia

S. Adam Soule

Department of Geology and Geophysics, Woods Hole Oceanographic Institution, Woods Hole, MA, USA

Chapter Outline

GLOSSARY

autoclastic facies Clastic facies generated by nonexplosive fragmentation accompanying lava effusion and flowage. Autobreccia and hyaloclastite are the two most common kinds of autoclastic facies.

coherent facies Facies generated by the solidification of molten lava or magma.

effusive eruption An eruption style involving passive discharge of magma from a volcanic vent, producing lava flows and/or lava domes.

felsic Relatively high SiO_2 (silica) composition, including dacite (64–69 wt% SiO_2) and rhyolite (>69 wt% SiO_2).

hyaloclastite Clastic aggregate generated by the quench fragmentation of molten lava in contact with water and/or ice.

mafic Relatively low SiO_2 (silica) composition, rich in magnesium and iron, most notably basalt (45–52 wt% SiO_2).

peperite A rock formed essentially in situ by the disintegration of magma intruding and mingling with an unconsolidated or poorly consolidated, typically wet, clastic host.

perlite Volcanic glass dissected by abundant arcuate overlapping fractures ("perlitic fractures") around generally intact cores or kernels; the cores or kernels typically have diameters in the range 0.5 mm–5 cm (macroperlite).

pumice Highly vesicular felsic or intermediate volcanic glass.

viscosity The ratio between shear stress and strain rate of a substance, or the ease with which a substance deforms and flows in response to applied stress. High viscosity substances resist flowage (strain) unless a high shear stress is applied.

volcanic breccia Clastic aggregate composed predominantly of angular volcanic clasts.

volcanic glass noncrystalline rock produced by the quenching of molten lava.

1. INTRODUCTION

Submarine lavas are important because they are the most widespread surficial igneous rocks on Earth. They are erupted at a variety of ocean water depths, display a wide spectrum of sizes, shapes, and compositions, and comprise up to 2 km of the uppermost ocean crust created at mid-ocean ridges or back-arc spreading centers (Chapter 3). Volcanoes at intra-oceanic convergent margins begin as small submarine volcanoes in deep water consisting of submarine lavas, **hyaloclastite**, intrusions, and in some cases submarine pyroclastic facies (Chapter 31). In submarine settings, rising magma commonly encounters thick successions of water-saturated, unconsolidated sediments

The Encyclopedia of Volcanoes. http://dx.doi.org/10.1016/B978-0-12-385938-9.00019-5

and may fail to erupt, instead spreading laterally into the soft sediments to form shallow intrusions. Volcanoes also form on oceanic crust away from plate boundaries, and may grow large enough to form islands (e.g., Hawaii-Emperor chain). This setting also includes huge submarine lava-dominated volcanic plateaus such as the Ontong-Java Plateau. Volcanic rocks from ancient submarine settings may be exposed on land, uplifted as a result of plate collisions. In this chapter, we summarize eruption styles and modern versus ancient products, present results from new studies of the modern seafloor, address the formation of hyaloclastite and **peperite** associated with seafloor lavas and intrusions, and review submarine **felsic** lava complexes.

1.1. What Controls Eruption Style?

Volcanic eruptions can be divided into effusive styles that produce lava flows versus explosive styles that produce pyroclastic deposits. One fundamental control on this behavior is the silica (SiO_2) content of the magma that erupts. Basaltic magma with <52 wt% SiO_2 is hot and very fluid (low **viscosity**) so it commonly erupts as a spray of droplets (fountains) or quietly as lava flows. In contrast, magmas that are richer in SiO_2, such as andesite (52–64 wt% SiO_2), dacite, and rhyolite (>69 wt% SiO_2), generally erupt at lower temperatures and have higher viscosities, which commonly result in eruptions being explosive. Most submarine eruptions in deep water are basaltic in composition and so are generally not, or only weakly, explosive. In most cases, basaltic magmas have lost volatiles during ascent and reach the seafloor with only minor volatiles which, at high hydrostatic pressures, remain mostly in solution. Exsolved volatiles form bubbles. As a generalization, vesicles are less voluminous in basaltic lavas generated by deep-water eruptions.

Explosive eruptions can result when magma reaches the seafloor without losing its original volatiles, or where shallow magma is fluxed by volatiles from deeper magma (Chapter 31). Explosions can also be caused by rapid transfer of heat from magma to water (Chapter 26). At surface pressures, the change from liquid water to steam results in a volume expansion of about a 1000 times, and the interaction of magma and water can be highly explosive. At confining pressures greater than the critical pressure of water (at about 3-km water depths), heating water significantly increases its volume, but there is no sharp phase boundary across which explosive "flashing" can occur, for either water or the other important volatile in seafloor magmas, carbon dioxide (critical depth ~700 m in seawater).

The nature of the eruptive site, particularly whether covered with thick loose sediment or formed of rigid rock, provides another control on eruptive behavior. Magma approaches the surface in dikes that advance by crack propagation. Unconsolidated sediment is insufficiently rigid to crack, and the dike tip stalls. Simultaneously, the heat of the magma heats and expands water in the sediment; the expansion pushes grains apart and allows the mixture to behave as a soft plastic or viscous fluid. These effects allow the magma to spread laterally into the sediment at shallow depths below the seafloor, commonly resulting in intrusions of complex form, and mingled combinations of magma and sediment termed peperite.

1.2. Modern versus Ancient Examples

We discuss submarine lavas and shallow intrusions from modern deep-sea environments and their ancient equivalents on land. Modern and ancient deposits provide complementary information. For example, volcanic rocks from modern deep-sea settings can be fresh and unaltered, in contrast to ancient examples that have been subject to hydrothermal fluids or long-term diagenesis, and on the modern seafloor it is possible to map the areal distribution and surface features of lavas and related rocks over many tens of square kilometers, which is not possible in most ancient successions. Also, modern seafloor examples have the advantage that their eruption depth is known. In the case of ancient examples, the depth of eruption can only be approximated but, critically for interpreting eruption processes, exposures of ancient successions reveal stratigraphy, which allows the investigation of successive events at a single site, as well as the geometry, dimensions, and relationships of eruption products. In contrast, our view of the modern seafloor from cameras, manned submersibles, and remotely operated vehicles (ROV) is restricted to largely two-dimensional surface exposures, providing a snapshot of the present that rarely extends even through the duration of a single eruption. Drilling into the deep seafloor by the Deep Sea Drilling Project (DSDP) and its successors has provided invaluable information about site histories, but drill holes cannot substitute for good three-dimensional exposure. Geologic work on the sea floor is also inherently more difficult (and more expensive) than comparable work on land, particularly at the outcrop (50- to 100-m) scale. Shallow intrusions and peperite form below the surface, and are understood almost exclusively on the basis of ancient examples.

2. MAFIC LAVAS

2.1. Basaltic Lavas

By far the most voluminous submarine volcanic facies are of basaltic composition. **Effusive eruption**s are most common, but explosive eruptions also occur (Chapter 31). Lava of this type is dominant along the global mid-ocean ridges, in back-arc basins, at intraplate seamounts, ocean volcanic islands, and oceanic plateaus. Due to their remote locations and great depths, direct observations of seafloor basaltic eruptions are extremely rare, but recent high-resolution seafloor acoustic mapping and ROV surveys have greatly improved our understanding of the physical characteristics of these lavas and the eruptions that produce them.

Submarine basaltic lavas differ morphologically from subaerial lavas due primarily to efficient crust formation (i.e., quenching) that results from eruption into cold seawater rather than air. The quench rate in submarine basaltic glass is up to several orders of magnitude greater than in air, a consequence primarily of the high thermal capacity of seawater. Thick quenched crusts coupled with emplacement via tubes and roofed channels keep submarine basaltic lavas well insulated, so they typically do not cool or crystallize extensively during emplacement (Gregg and Fornari, 1998). The primary mineral phases (plagioclase feldspar, olivine, pyroxene) are present in small volumes (<5%) as are vesicles (<<5%). The three primary types of submarine lava based on morphology are—pillow, lobate, and sheet—and lie along a spectrum (Figure 19.1), distinguished from one another by the size and

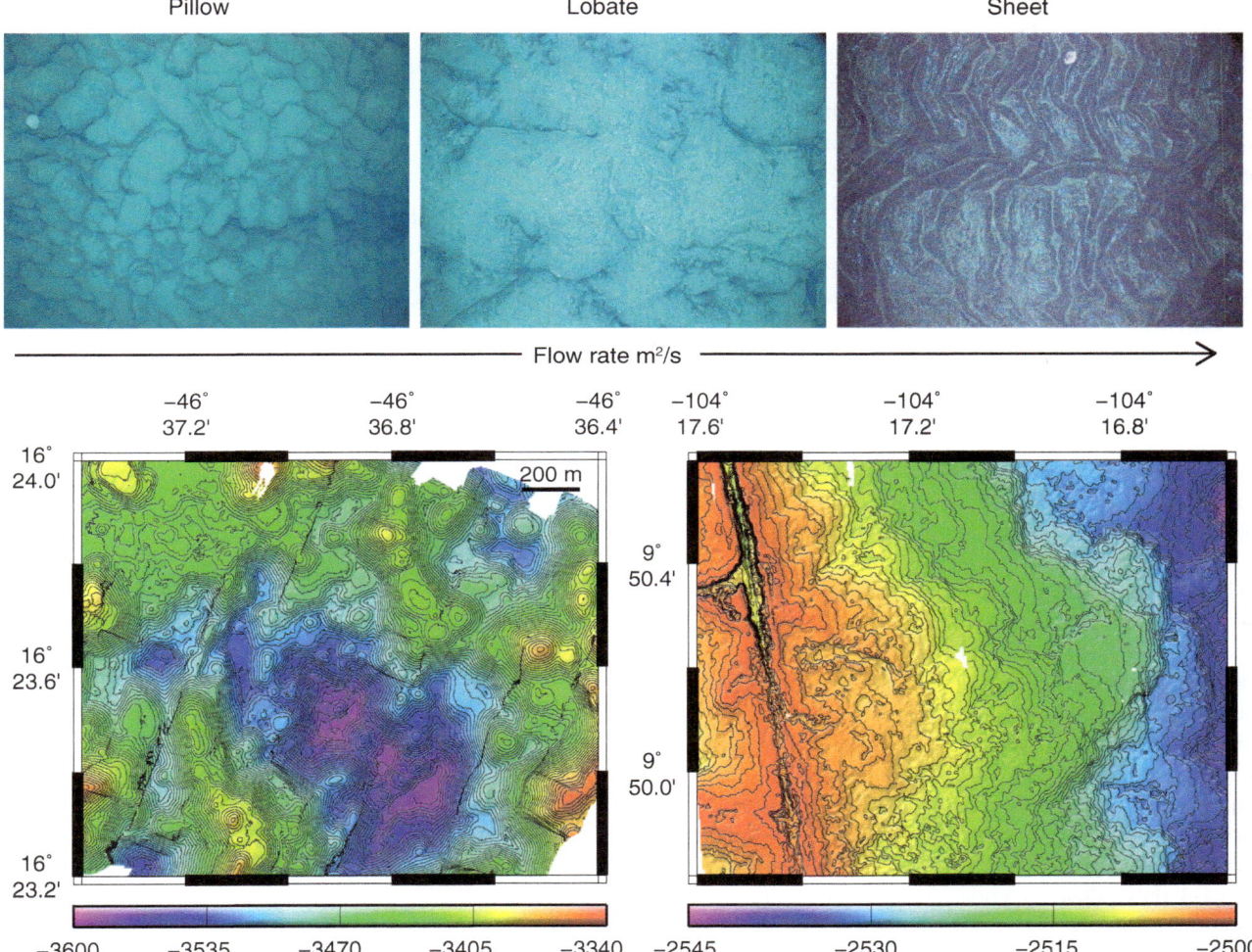

FIGURE 19.1 (*Top*) Examples of pillow, lobate, and sheet morphology from submarine basaltic lavas. Photographs were collected by a deep-sea digital camera system at ∼3 m above seafloor with a field of view of 4—5 m across. Pillow image is from the Mid-Atlantic Ridge at 37.3°N; lobate and sheet images are from the East Pacific Rise near 9.8°N. The images are arranged in order of increasing flow rate as well as increasing length scales of flow subunit and lava connectivity and decreasing flow lengths (see text). (*Bottom*) High-resolution bathymetric maps collected by autonomous underwater vehicles Autonomous Benthic Explorer (ABE) and Sentry are shown at the same scale and compare seafloor relief of sites dominated by pillow lavas (*left*) and lobate and sheet lavas (*right*). The pillow lava area is from the axis of the slow-spreading Mid-Atlantic Ridge and is characterized by hummocky relief of overlapping pillow mounds 500 m and less in diameter and 50—100 m in height. Contour intervals of 5 m are shown over shaded relief. The lobate and sheet lava area is from the fast-spreading East Pacific rise and is characterized by low relief lavas that produce overlapping lobes extending 1—4 km from the eruptive vents (narrow, NNW-striking graben at the left of the map). Contour intervals of 1 m are shown over shaded relief, with 5-m contours in bold for comparison to the Mid-Atlantic Ridge (MAR) map. *Photo copyrights Woods Hole Oceanographic Institute (WHOI), bathymetry data courtesy of D. Smith and D. Fornari.*

interconnectedness of the unit cells that make up the lava. Pillow lavas are composed of meter-scale, poorly connected bulbs of lava; lobate lavas have larger ($\sim$5–10 m wide), moderately connected pads of lava; and sheet lavas comprise fully connected sheets of lava. Laboratory experiments using wax extruded into cold water to simulate the efficient formation of crusts on deep sea lavas indicate that these types reflect different flow rates and lava viscosities. In these experiments, morphology can be predicted from a dimensionless parameter Ψ that measures the ratio of the timescales of cooling (crust formation) and lava advance (crust disruption). In this framework, submarine lavas that advance slowly relative to the rate of crust formation produce pillow lavas, while lavas that advance quickly relative to crust formation rates produce sheet lavas.

The thermal gradient between molten lava and seawater for a reasonable range of eruption temperatures (1000–1200 °C) and water temperatures (2–10 °C) is large regardless of eruption depth and tectonic setting. Further, the major-element compositional range and hence potential for differences in melt viscosity is small for most basaltic eruptions along mid-ocean ridges, ocean islands, seamounts, and oceanic plateaus. As a result, lava morphology is commonly used as a proxy for lava flow rate and eruption rate for submarine basalts.

2.2. Pillow Lava

Pillow lava is commonly cited as the most abundant geologic landform on Earth's surface. As described above, pillows form at low, though imprecisely known, flow rates. Pillow lavas have been observed while forming on the shallow submarine slopes of Hawaii, where subaerial lava flows enter the sea, and on W. Mata Volcano, a seamount in the NE Lau back-arc basin (Resing et al., 2011). These direct observations along with the studies of seafloor deposits, subglacial deposits, and ophiolites have led to a good understanding of the mechanics of pillow lava formation. Single pillows are commonly equant, but may also form elongate pillow tubes on steep slopes. The surfaces of pillows typically have a variety of "decorations", including parallel grooves-and-ridges that form as the pillow crust fails and the pillow expands along circumferential fractures, and pillow buds that are similar to squeeze-outs on subaerial pahoehoe lavas. Single pillows stop growing when crusts become sufficiently thick to impede inflation. Once pillow growth ceases, failure of the pillow crust leads to the formation of a new pillow at the leading edge of the flow. Alternatively, flow may be redirected within the feeder systems to produce new pillows elsewhere. In this way, pillow lava tends to produce accumulations that do not extend significant distances from their source vent, and commonly produce high-relief pillow mounds, hummocks, and ridges (Figure 19.1).

FIGURE 19.2 Frame grab from high-definition video collected by Alvin at the East Pacific Rise showing low-relief surface of lobate lava flows suggestive of high degrees of connectivity of the molten flow interior and inflation to a common depth across the flow. Large collapse pit at the right of the image shows the hollow flow interior and the lava selvages or bathtub rings that form as the flow interior drains away. *Photo copyright WHOI and courtesy of T. Shank, B. Lange.*

2.3. Lobate Lava

Lobate lava is the dominant morphology on many fast and intermediate spreading-rate ridges, and reflects higher eruption rates than produce pillow lavas. Single lobes are many times wider than they are thick. Lobate lava is similar to pillow lava in that individual flow units (lobes) are easily recognized on the flow surfaces, but boundaries between lobes are commonly destroyed within the lava interior. Lobe surfaces are smooth, lacking the "decorations" of pillow lavas, but may have ropy folds on the surface of the largest lobes. Very low relief across lobate lava surfaces suggest that lobes may be fully interconnected in the flow interior, and inflated to a common level (Gregg and Fornari, 1998). In addition, near-vent lobate lavas are commonly hollow, as the molten interior has drained downslope or back into the eruptive fissures. The large cavities beneath the solid crust, visible through collapsed regions, can be much larger in extent than the single lobe boundaries observed at the surface. It is common for the upper crust in a hollow lobate lava to be supported by lava pillars, which are hollow columns of lava that extend from the base to the upper surface of the lavas (Figure 19.2). Within collapsed regions, regular sequences of subhorizontal glass selvages or "bathtub rings" mark pauses in lava withdrawal (Chadwick, 2003).

2.4. Sheet Lava

At the highest eruption/flow rates, sheet lavas are produced. Their interiors are fully interconnected and the surfaces lack any indication of separate units. Instead, the surfaces may be smooth, ropey, lineated in the direction of flow, or jumbled. Jumbled surfaces reflect continued deformation of a folded or lineated surface causing the crust to brecciate. Sheet lavas are typically restricted to near-vent regions and areas where flow is constricted, e.g., by topographic barriers or flow channelization (Soule et al., 2005). Sheet lavas are commonly deflated and drained, consistent with high flow rates and little solidification before magma supply ceases (Figure 19.1).

2.5. Flow Scale Features

Only a handful of submarine basaltic lavas have been mapped in enough detail to determine their extent and volume (Rubin et al., 2012). Examples at mid-ocean ridges include the Aldo-Kihi lobate lava of the south East Pacific Rise (14 km^2, 25—140 × 10^6 m^3), the North Cleft pillow mounds (8 total, 2—3 km^2, 50 × 10^6 m^3), the 2005—2006 lava of the north East Pacific Rise (14.6 km^2, 22 × 10^6 m^3), and the 2011 lava at Axial Volcano on the Juan de Fuca Ridge (7.8 × 10^6 m^2, 27 × 10^6 m^3). Areally larger and more voluminous mid-ocean ridge lavas have been proposed based on acoustic mapping, but have not been confirmed by other means. At slow-spreading ridges, longer repose intervals, greater tectonic disruption, and generally rougher terrain make mapping single lavas difficult, but recent studies on the Mid-Atlantic Ridge at 45°N suggest individual eruptions may produce single pillow mounds that average 3 × 10^5 m^3 (Searle et al., 2010). At ocean islands, single pillow mound eruptions, similar in size to those on mid-ocean ridges, have been identified on submarine rift zones such as the Puna Ridge, and larger, flat-lying sheet/lobate lavas, >1 km^3, have been identified on the abyssal plains around Hawaii and the Galapagos. Flat-topped volcanic "terraces", akin to Venusian pancake domes, are common on the submarine flanks of ocean islands and thought to be constructed by lavas where emplacement extent was limited by cooling, rather than by the amount of lava erupted.

More commonly, submarine lava lengths are limited by the volume erupted. At mid-ocean ridges, lavas typically erupt within a narrow neo-volcanic zone from ridge-parallel fissures and traverse the ridge crest or axial valley depending on the spreading rate. Lava flows advance downslope to distances dictated by a cutoff of lava supply from the vent or where they are blocked by topography. The resulting lavas may acquire thicknesses of a few to more than 10 m. Thicker lavas generally result from flooding a pre-existing depression. These massive lavas are easily recognized by their fine-grained, low-porosity interiors in vertical sections in comparison to thinner lavas and pillowed lavas that contain abundant glass and zones that are highly porous to breccia-like (e.g., Tominaga and Umino, 2010).

2.6. Basaltic Hyaloclastite and Peperite

As for basaltic lavas, studies of the modern seafloor contribute to our knowledge of hyaloclastite. Hyaloclastite is a glassy clastic deposit (Figure 19.3), formed by nonexplosive thermal granulation ± spalling of solidified lava rinds as the result of deformation. Sheetform deposits of hyaloclastite are a distinctive type exemplified by thin (<1-m-thick) layers of mm-size dense glass fragments on Seamount 6, an off-axis volcano along the East Pacific Rise, formed by thermal shock granulation and spalling of thin lava flows. Most of the glass fragments in hyaloclastite are compositionally identical to underlying glassy lavas, and the lavas have very irregular tops. Thin lava tendrils broke up to shed contorted glass fragments into the overlying contemporaneous hyaloclastite. A small proportion of thin, folded sheet-form particles in the hyaloclastite resembles "limu of Pele" produced when lava from Kiluaea volcano enters the sea and interacts with seawater. Some authors consider such "limu" indicative of explosive subaqueous magmatic eruptions (Chapter 31). However, simple seawater entrapment can explain limu formation even in very deep water where neither water nor carbon dioxide exist as vapor (Schipper and White, 2010).

Another common type of hyaloclastite is associated with the effusion of basaltic lavas on slopes such as seafloor pillow cones (Figure 19.4). Fragmentation is by both quenching processes and deformation of brittle lava crusts. Additional clast breakage occurs during transport down steep pillow cone slopes. Some facies have recognizable fragments of lava pillows (pillow breccia), whereas others comprise centimetric clasts of microcrystalline lava dispersed in a matrix of glassy particles generated both at the vent and by shattering of glassy margins during downslope transport. It is common for hyaloclastite to form as lavas flow over steep slopes, either pre-existing slopes or those built by the lavas themselves. Chilling and fragmentation are facilitated by increased flow velocities that enhance brittle fracturing of lava crusts, as well as stretching and pulling apart of the lava, and these processes operate in varying proportions across a range of magma compositions. Where erupting magma is so vesicular that broken-off pieces float rather than being carried downslope (see Section III), however, these slope-driven processes do not take place. There are systematic variations at larger scales in basaltic hyaloclastite formed in other settings. Work in Iceland on basaltic hyaloclastite on slopes showed that prograding deposits of coarse hyaloclastite reduce initial 60° slopes to about 35° (Figure 19.5), whereas for initial slope angles of 30—40°, somewhat finer-grained hyaloclastite forms a delta capped by lava. Shallow seafloor intrusions, also called hypabyssal intrusive complexes, and associated peperite have been characterized primarily from ancient successions, but new insights into their origins have come from experiments, coring and dredging, and submersible dives. At large scales, seismic imaging for hydrocarbon exploration in the North Sea revealed the relationships among shallow-subsurface and deeper intrusions and peperite formed in unconsolidated

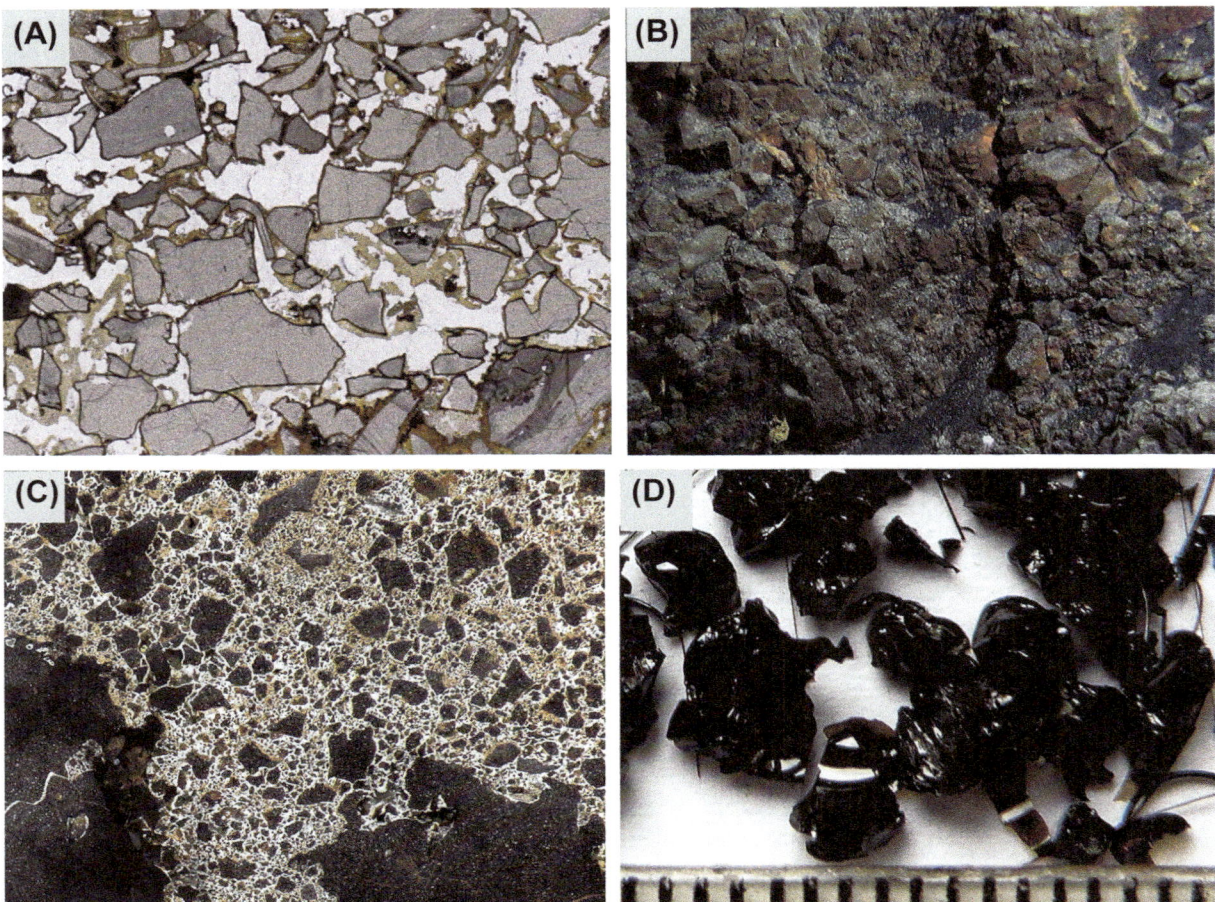

FIGURE 19.3 Basaltic hyaloclastite. (A) Photomicrograph of blocky sideromelane glass fragments from Seamount Six deposits from about 2 km depth, off-axis to the East Pacific Rise. Largest fragment is 4 mm across. (B) Photo from submersible of jigsaw-fit hyaloclastite on Loihi seamount at 1.2 km depth. (C) Calcite-cemented (white) hyaloclastite from Oligocene continental shelf volcano, Oamaru, NZ. Fragment at lower right is 15 cm across. (D) Glassy hyaloclastite granulate formed by pouring molten basalt into water. Scale at base is in mm.

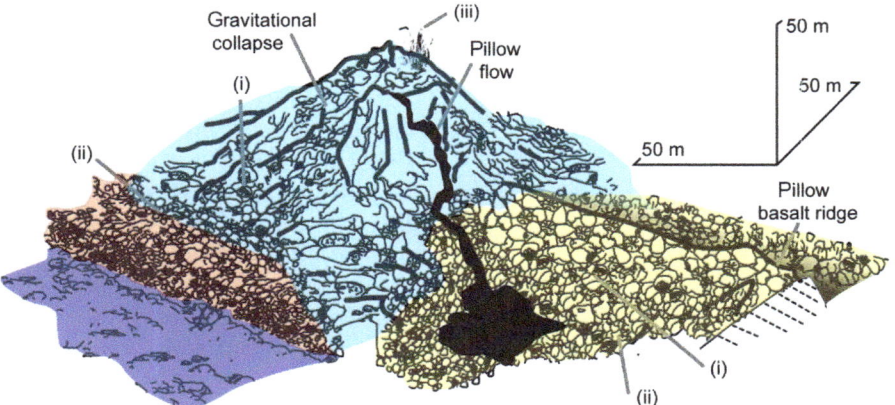

FIGURE 19.4 Facies model for the production of glassy fragmental rocks at mid-oceanic ridge seafloor (dark blue) on the slopes or in a basin close to a pillow cone (cyan) with a lateral pillow ridge (yellow). Processes include: (i) in situ and/or talus accumulation of pillow-fragment breccia; (ii) grain flow deposits derived from unconsolidated pillow-fragment breccia and pillow basalt above; and (iii) minor fountain deposits. Gravity collapse from destabilization of the erupting vent generated thick hyaloclastite and pillow-fragment breccia aprons (red) along pillow cone slopes. Advancing pillow lava tubes (labeled pillow flow) may push accumulated pillow-fragment breccia debris down slope. The scale of features is schematic. Thin black lines show cross-sectional relief. *Modified from Dickinson et al. (2009).*

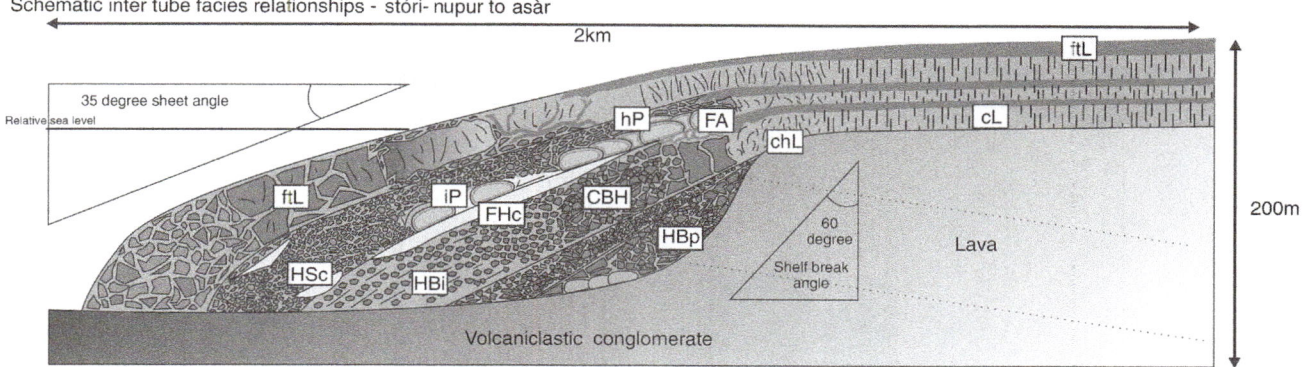

FIGURE 19.5 Hyaloclastite associated with tube-fed basaltic lavas occurs in a variety of lithofacies (those with "H" in facies code below), mostly breccias, that reflect the role of high depositional slopes in fragmentation, producing large-volume deposits. In this diagram (from Watton et al., 2013), ftL is flow breccia; HBp, CBH, and hP are primary deposits from quenching and autobrecciation of lava; HSc, HBi, and FHc are redeposited hyaloclastite.

basinal sediments. These observations are being combined, along with outcrop observations of ancient sequence and experiments in which fluids are intruded into "sand boxes," to reveal the fluid-mechanical controls on the transition from shallow intrusion to submarine eruptions, as well as the styles and rates of magma passage as it approaches the surface. Among the most impressive products of such intrusive magma—sediment interaction formed where the Gulf of California spreading ridge lies beneath the thick sediment shed from adjacent rift shoulders. Shallow intrusion of basalt, rather than eruption to the seafloor, takes place because the density of rising basaltic magma is greater than that of wet sediment. In the Gulf of California, magma erupting from the spreading center encounters such uncompacted, waterlogged sediments, and cannot rise upward through them, instead spreading laterally within the sediments to form sill-sediment complexes (Einsele, 1985). The margins of such sills are expected to display peperite or intrusive pillows, but coring during the 1979 Deep Sea Drilling Project (DSDP) cruise failed to recover the sill-sediment contacts. Much smaller-volume magma-sediment interaction is revealed by peperite formed at or near the seafloor. Coring during International Ocean Discovery Program (IODP) leg 330 recovered multiple intervals of peperite associated with lava emplaced over sediment on guyots of the Louisville Seamount Trail, for example, and dredge hauls from "Petit Spot" volcanoes near the Japan Trench also contain peperite from interaction of erupted basalts with seafloor sediment. Despite these recent examples on the seafloor, the great majority of what we know about peperite comes from ancient examples, simply because most peperite is associated with subsurface intrusions. Studies of ancient peperite show differences in scale and texture that can be related both to properties of the magma and to the nature of intruded sediment (Skilling et al., 2002).

Experiments have provided additional information on how peperite forms. Some researchers have used analog fluids together with thermal modeling to assess how fluid with the rheology and thermal properties of basalt will mingle, and under what conditions magma—water explosions (Chapter 26) might take place. Others have directly introduced molten basalt into both water and fluid sediment, revealing that the cooling of magma in a fluid sediment is orders of magnitude slower than in water alone. Fast quenching in water produced hyaloclastite, while peperitic mingling yielded fluidal glass fragments and even welding of still-hot blobs that formed when melt entered the sediment and then came to rest against one another within the sediment at the bottom of the apparatus (Figure 19.6).

3. FELSIC LAVAS AND ASSOCIATED ROCKS

To an even greater degree than for basaltic magmas, submarine felsic lavas and domes are poorly known compared to their subaerial counterparts. Felsic magmas are far subordinate to basaltic magmas in ocean basins and much less frequently encountered by ocean floor surveys. Very few examples on the modern seafloor have been mapped and sampled (e.g., Allen et al., 2010) and uplifted examples are incompletely exposed and/or preserved. Nevertheless, felsic lavas and domes are important in modern submarine arc and back arc settings, and locally present at mid-ocean ridges where basalt has fractionated sufficiently to produce rhyolite. They are an integral component of numerous ancient submarine volcanic successions, including those that host massive sulfide ore deposits (e.g., Paulick and McPhie 1999).

Regardless of setting, felsic magmas have high viscosities and produce lavas that are thicker and less extensive than basaltic lavas. Submarine felsic effusive eruptions generate tabular lavas, equant domes and dome complexes, and lobe-hyaloclastite lavas. All three types consist of combinations of coherent and **autoclastic facies**. The **coherent facies** forms from the solidification of molten

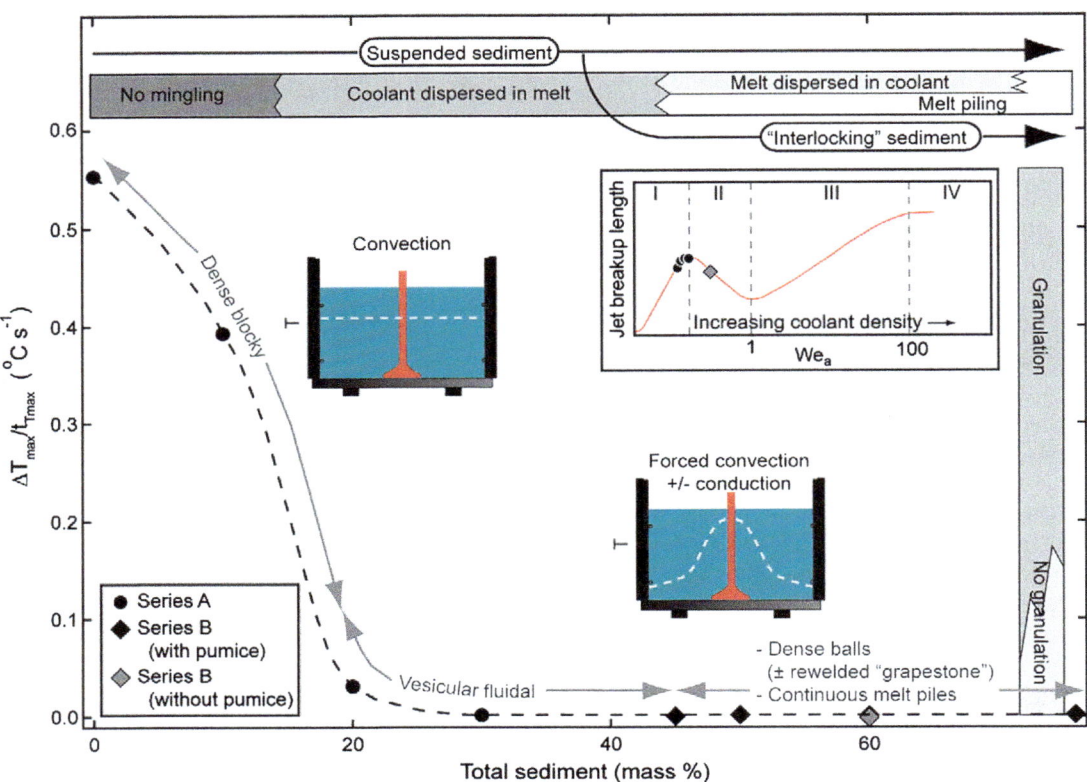

FIGURE 19.6 Experiments simulating magma-sediment mingling address peperite and hyaloclastite formation. In this summary diagram (from Schipper et al., 2011), fragmentation processes are related to coolant properties of water vs. a host sediment. Heat transfer, expressed as the ratio $\Delta T_{max}/T_{max}$, is the primary control on whether hyaloclastite forms by thermal granulation, shown qualitatively in vertical gray-scale bar. Total sediment plotted ranges from 0%, pure water, to 80%, i.e., sediment with 20% porosity; it is taken as a proxy for coolant viscosity, and is the primary control on hydrodynamic mingling, shown qualitatively in horizontal gray-scale bar.

lava. The autoclastic facies include hyaloclastite (resulting from quench fragmentation of molten lava) and autobreccia (generated by dynamic stressing of molten lava) though hyaloclastite is typically dominant in submarine settings.

3.1. Submarine Felsic Coherent Facies

Coherent facies represents molten lava that escaped autoclastic brecciation during eruption and emplacement. Felsic coherent facies are characterized by porphyritic or aphyric texture; the most common phenocrysts are feldspar (plagioclase and sanidine) and quartz. Groundmass textures show wide ranges in crystallinity and vesicularity that mainly relate to gradients in cooling rate and volatile content, respectively. Glassy groundmass is very common in submarine felsic coherent facies, reflecting the high viscosity of the high SiO_2 melt and efficient quenching underwater. Except in freshly erupted units, the glassy groundmass domains show perlitic fractures due to hydration of the glass. Groundmasses of coherent facies that cool relatively slowly are partly glassy and partly crystalline. The crystalline domains include spherulites (radial aggregates of fine, needle-like crystals) and lithophysae

(spherulites with a central vug). Coherent facies that cool very slowly are completely crystalline, though the crystals are very small (<0.5 mm). Micropoikilitic texture (interlocking equant quartz crystals that contain abundant feldspar microlites) may be present in crystalline domains. The crystals in crystalline groundmass domains are feldspar and quartz, reflecting the felsic composition of the melt.

Vesicles may be abundant in the groundmass of coherent facies, so much so that the texture is pumiceous. Pumiceous texture is restricted to zones where the volatile content in residual melt was relatively high, the confining pressure was relatively low, and the cooling rate allowed bubble nucleation and growth before quenching to solid glass. These conditions are most commonly met in the outer zones of lavas and domes. Large domains of felsic coherent facies may be columnar jointed. The columns may be regular in size, shape and orientation, reflecting the simple geometry of isotherms during solidification. Highly variable sizes, shapes and orientations over short distances indicate that isothermal surfaces had very complex shapes. Flow bands are also found in felsic coherent facies; they are defined by variations in groundmass crystallinity and vesicle content. Flow bands are produced by laminar shear during flowage; because the melt viscosity is high, the bands remain after laminar shear stops.

3.2. Submarine Felsic Autoclastic Facies

Autoclastic facies are generated at the expense of coherent facies and therefore, have the same composition and mineralogy as the associated coherent facies. The combination of the high viscosity of felsic lava and the underwater setting means that autoclastic facies are invariably present, and the most abundant autoclastic facies is hyaloclastite. Hyaloclastite is generated by cooling contraction of molten lava quenched on contact with water, and forms where curviplanar first-order quench fractures intersect, creating coarse, equant, polyhedral glassy or partly glassy clasts (Figure 19.7(A)). Closely spaced (cm),

FIGURE 19.7   A) In situ rhyolitic hyaloclastite; Ponza, Italy. (B) Dacite clast generated by quenching, showing the typical polyhedral shape, curviplanar surfaces and "tiny normal joints" (arrow); Cabo de Gata, Spain. (C) Submarine dacite dome, showing the transition from the coherent core into in situ hyaloclastite; Hariba, Japan. The in situ hyaloclastite changes outward from having no matrix to having abundant apparent matrix. (D) Resedimented dacitic hyaloclastite; Milos, Greece. Hammer inside circle. (E) Submarine dacitic autobreccia; Cabo de Gata, Spain. (F) Giant pumice clast in thinly bedded mudstone; Milos Greece. Hammer inside circle.

short (cm), second-order fractures perpendicular to the first-order fractures may be present along clast margins ("tiny normal joints," Yamagishi 1987; Figure 19.7(B)). If there is no disturbance of the fractured glass, the pattern of fractures defines jigsaw-fit texture that is characteristic of in situ hyaloclastite (Figure 19.6(A)). In situ hyaloclastite forms adjacent to, and grades into, coherent facies. In situ hyaloclastite close to coherent facies generally lacks matrix, but there may be a transition to apparent matrix-rich hyaloclastite farther away (Figure 19.7(C)). The apparent matrix comprises relatively fine (mm) glassy clasts that are identical to the coarser clasts and show jigsaw-fit texture. This apparent matrix occupies the more intensely fragmented domains along first-order quench fractures; the less intensely fractured domains in between appear as coarse clasts. If in situ hyaloclastite is disturbed, for example, by the movement of adjacent molten lava, then the clasts are progressively rotated and the jigsaw-fit texture is lost, producing clast-rotated hyaloclastite. Further disturbance may dislodge the clasts so they move freely and independently downslope, forming resedimented hyaloclastite (Figure 19.7(D)). Transport may be sufficient for sorting, clast shape modification, and mixing of clasts from different sources. In situ hyaloclastite, clast-rotated hyaloclastite and resedimented hyaloclastite are all **volcanic breccias**, i.e., clastic aggregates dominated by angular volcanic clasts. In situ hyaloclastite and clast-rotated hyaloclastite are strictly monomictic facies; resedimented hyaloclastite may be monomictic or weakly polymictic.

Subaqueous autobreccia is essentially the same as subaerial autobreccia, and results from the break-up of the more viscous parts of molten lava propagating by laminar shear. This process requires lava to be both molten and moving, a combination that is not commonly met by felsic lavas in submarine settings where quenching dominates. Autobreccia is composed of coarse slabby clasts that are internally flow banded, and occurs in close spatial association with coherent facies of the same composition (Figure 19.7(E)).

3.3. Submarine Felsic Tabular Lavas and Domes

Felsic tabular lavas and domes erupted under water are similar in size and shape to their subaerial counterparts. They are typically thick (tens to hundreds of meters) but short (for tabular lavas, hundreds of meters to ~ 10 km; for domes, tens to a couple of 100 meters). Felsic tabular lavas are mainly composed of coherent facies that may be crystalline, or partly crystalline and partly glassy. In situ hyaloclastite occurs at the base, top, and margins; clast-rotated hyaloclastite and resedimented hyaloclasite may be present at the margins, adjacent to in situ hyaloclastite. In submarine felsic domes, in situ hyaloclastite surrounds a coherent core, though the proportion of hyaloclastite versus coherent facies varies widely. In situ hyaloclastite may grade outward to clast-rotated and resedimented hyaloclastite. Some felsic submarine domes are entirely glassy, and some have outer zones of pumiceous hyaloclastite and pumiceous coherent facies. Giant (up to ~ 10 m), pumiceous clasts may spall from the upper margins of active submarine felsic lavas and domes (Allen et al., 2010). Providing the vesicles remain filled by hot gas, the spalled clasts are less dense than seawater and rise through the water column, in some cases, reaching the sea surface (e.g., Kano 2003). Once waterlogged, the clasts settle from suspension to the seafloor (Figure 19.7(F)).

3.4. Submarine Felsic Lobe-Hyaloclastite Lavas

Lobe-hyaloclastite lavas have no subaerial equivalent, but share at a much larger scale some characteristics of basaltic pillow breccias. Lobe-hyaloclastite lavas consist of multiple coherent domains or lobes (m to tens of m across) separated by domains of hyaloclastite (Figure 19.8). The lobes form where viscous molten lava emerges from a feeder dyke that intrudes cogenetic hyaloclastite (Yamagishi 1987). The lobes detach from the feeder dyke and are pushed aside as more lobes and more hyaloclastite forms at the dyke. This process can create lava units tens to hundreds of m thick with an overall tabular or dome morphology. Most of the hyaloclastite is in situ hyaloclastite and clast-rotated hyaloclastite although resedimented hyaloclastite may be present at the margins.

3.5. Shallow Synvolcanic Intrusions

Ocean basins are depocenters for sediment derived from the continents as well as from intrabasinal sources. The rate of aggradation of sediment can be very high, especially in and near active submarine volcanoes. Magma rising in a conduit beneath the seafloor may meet a layer of unconsolidated wet sediment and in most cases, the magma will be more dense than the wet sediment. As a result, the magma will intrude the unconsolidated sediment rather than erupt, forming a shallow, syn-volcanic intrusion. Synvolcanic intrusions are commonly conformable or partly conformable because their emplacement relates to the level of neutral buoyancy within the sediment pile. They can be any composition but their shapes reflect the magma viscosity which to a large extent is controlled by composition, especially silica content. High-viscosity felsic magmas form thick tabular intrusions or equant cryptodomes (Figure 19.9(A); e.g., Stewart and McPhie 2003). Low-viscosity **mafic** magmas form thin sills and sill complexes (e.g., Einsele 1985). Intermediate magmas may behave like either the low-viscosity or the high-viscosity end-members.

FIGURE 19.8 A) Submarine dacitic lobe-hyaloclastite lava; Cabo de Gata, Spain. The lava is about 50-m thick (top and base outlined by dashed line); conspicuous lobes of coherent facies outlined by thin white line. (B) Detail of coherent lobe in the dacitic lobe-hyaloclastite lava in A; Cabo de Gata, Spain. The coherent lobe is columnar jointed and passes outward into in situ hyaloclastite. Notebook inside circle.

Most syn-volcanic intrusions are composed entirely of coherent facies that has the same range of characteristics as the coherent facies of lavas, although glassy groundmasses and vesicular domains are less common than in lavas. Some felsic syn-volcanic intrusions display large-scale flow bands (e.g., Goto and McPhie 1998; Figure 19.9(B)) parallel to the intrusion margins. The flow bands result from laminar shear affecting the intrusion immediately inside the intrusive contact, implying that these intrusions increased in size by inflation (addition of new magma into the interior; Goto and McPhie 1998). Well-developed columnar joints are common in syn-volcanic intrusions of all compositions (e.g., Stewart and McPhie 2003; Goto and McPhie 1998; Figure 19.8(B)). Column axes are perpendicular to the intrusive contacts, reflecting the close coincidence of isothermal surfaces and contact surfaces for syn-volcanic intrusions. Column diameters increase systematically from the margins to the interior, reflecting the cooling rate control on size. Pillows may be present in basaltic syn-volcanic intrusions. In thick (>5 m) basaltic syn-volcanic intrusions, the groundmass is typically crystalline, except for a narrow (mm) glassy zone at the contacts, and increases in grain size inwards; the innermost portions display a doleritic texture.

Because the level of intrusion is very shallow, and the host is unconsolidated wet sediment, syn-volcanic intrusions have internal textures and structures that are closely similar to lavas of the same composition. Their recognition depends critically on the nature of the contacts and the correct identification of peperite. Peperite is a facies generated by the mingling of molten lava or magma with unconsolidated sediment (Skilling et al., 2002). It is characterized by the presence of intricately intermingled clastic or "sedimentary" domains (derived from the unconsolidated sediment) and domains with igneous textures (derived from the lava or magma) (Figure 19.9(C)). The igneous domains may be dominant and contain clasts of sediment, or the sedimentary domains may be dominant and contain igneous clasts. In blocky peperite, the igneous clasts are angular and equant, while in fluidal peperite, the igneous clasts are fluidally shaped (Busby-Spera and White 1987). Peperite is most commonly found along the contacts of syn-volcanic intrusions and the bases of lavas that have overridden unconsolidated sediment. The lower contacts of these intrusions and lavas can be very similar, and hence are not useful for distinguishing intrusions from lavas. The upper contacts are far more informative and the presence of peperite is taken as a reliable basis for distinguishing

FIGURE 19.9   A) Kalogeros dacitic cryptodome, Milos Greece. The cryptodome has intruded a submarine felsic volcanic succession. (B) Large-scale flow bands and columnar joints in the marginal zone of the Momo-Iwa dacitic cryptodome, Japan. (C) Small basaltic intrusion and basalt-mudstone peperite; Merimbula, Australia.

syn-volcanic intrusions from lavas. However, upper contacts may also be sharp and marked only by the local induration of the adjacent host sediment. Cryptodomes and other felsic intrusions in some cases include marginal domains of glassy monomictic breccia that shows jigsaw-fit texture. This breccia passes outward into monomictic breccia in which clasts are separated by the narrow seams of the host sediment. This texture implies that the magma in contact with the wet sediment was fragmented by quenching during intrusion, forming intrusive hyaloclastite. Intrusive hyaloclastite can be considered a variety of blocky peperite.

FURTHER READING

Allen, S.R., Fiske, R.S., Tamura, Y., 2010. Effects of water depth on pumice formation in submarine domes at Sumisu, Izu-Bonin arc, Western Pacific. Geology 38, 391–394.

Busby-Spera, C.J., White, J.D.L., 1987. Variation in peperite textures associated with differing host-sediment properties. Bull. Volcanol. 49, 765–775.

Chadwick, W.W., 2003. Quantitative constraints on the growth of submarine lava pillars from a monitoring instrument that was caught in a lava flow. J. Geophys. Res. 108, 2534. http://dx.doi.org/10.1029/2003jb002422.

Einsele, G., 1985. Basaltic sill-sediment complexes in young spreading centres – genesis and significance. Geology 13, 249–252.

Kano, K., 2003. Subaqueous pumice eruptions and their products: a review. AGU Geoph. Monog. 140, 213–229.

Goto, Y., McPhie, J., 1998. Endogenous growth of a Miocene submarine dacite cryptodome, Rebun Island, Hokkaido, Japan. J. Volcanol. Geotherm. Res. 84, 273–286.

Gregg, T.K.P., Fornari, D.J., 1998. Long submarine lava flows: observations and results from numerical modeling. J. Geophys. Res. 103, 27517–27531.

Paulick, H., McPhie, J., 1999. Facies architecture of the felsic lava-dominated host sequence to the Thalanga massive sulfide deposit, Lower Ordovician, Northern Queensland. Austral. J. Earth Sci. 46, 391–405.

Resing, J.A., Rubin, K.H., Embley, R.W., Lupton, J.E., Baker, E.T., Dziak, R.P., Baumberger, T., Lilley, M.D., Huber, J.A., Shank, T.M., Butterfield, D.A., Clague, D.A., Keller, N.S., Merle, S.G., Buck, N.J., Michael, P.J., Soule, A., Caress, D.W., Walker, S.L., Davis, R., Cowen, J.P., Reysenbach, A.-L., Thomas, H., 2011. Active submarine eruption of boninite in the Northeastern Lau Basin. Nature Geosci. 4, 799–806.

Reynolds, M.A., Best, J.G., Johnson, R.W., 1980. 1953–1957 eruption of Tuluman volcano: Rhyolitic volcanic activity in the northern Bismarck Sea. Papua New Guinea Geol. Surv. Mem. 7, 44.

Rubin, K., Soule, S.A., Chadwick, W., Fornari, D., Clague, D., Embley, R., Baker, E., Perfit, M., Caress, D., Dziak, R., 2012. Volcanic eruptions in the deep sea. Oceanography 25, 142–157.

Schipper, C., White, J., 2010. No depth limit to hydrovolcanic limu o Pele: analysis of limu from Lō'ihi Seamount, Hawai'i. Bull. Volcanol. 72, 149–164.

Searle, R.C., Murton, B.J., Achenbach, K., LeBas, T., Tivey, M., Yeo, I., Cormier, M.H., Carlut, J., Ferreira, P., Mallows, C., Morris, K., Schroth, N., van Calsteren, P., Waters, C., 2010. Structure and development of an axial volcanic ridge: Mid-Atlantic Ridge, 45°N. Earth Planet. Sci. Lett. 299, 228–241.

Skilling, I., White, J.D.L., McPhie, J., 2002. Peperites: processes and products of magma-sediment mingling. Elsevier, Amsterdam, 289 pp.

Soule, S.A., Fornari, D.J., Perfit, M.R., Tivey, M.A., Ridley, W.I., Schouten, H., 2005. Channelized lava flows at the east pacific rise crest 9–10 N: the importance of off-axis lava transport in developing the architecture of young oceanic crust. Geochem. Geophys. Geosyst. 6, Q08005.

Stewart, A.L., McPhie, J., 2003. Internal structure and emplacement of an upper Pliocene dacite cryptodome, Milos Island, Greece. J. Volcanol. Geotherm. Res. 124, 129–148.

Tominaga, M., Umino, S., 2010. Lava deposition history in ODP Hole 1256D: insights from log-based volcanostratigraphy - Tominaga - 2010. Geochem. Geophys. Geosyst. 11, Q05003.

Waters, J.C., Wallace, D., 1992. Volcanology and sedimentology of the host succession to the Hellyer and Que River VHMS deposits, Northwestern Tasmania. Econ. Geol. 87, 650–666.

Yamagishi, H., 1987. Studies on the Neogene subaqueous lavas and hyaloclastites in Southwest Hokkaido. Geol. Surv. Hokkaido Rep. 59, 55–101.

Chapter 20

Glaciovolcanism

Benjamin R. Edwards

Department of Earth Sciences, Dickinson College, Carlisle, Pennsylvania, USA

Magnús T. Gudmundsson

Institute of Earth Sciences, University of Iceland, Reykjavík, Iceland

James K. Russell

Earth, Ocean and Atmospheric Sciences, University of British Columbia, Vancouver, British Columbia

Chapter Outline

GLOSSARY

glaciostatic pressure The pressure at the base of a glacier generated by gravity acting on the load of the overlying ice.

glaciovolcanism (Syn. volcano—ice interactions, ice-contact volcanism, subglacial volcanism, englacial volcanism) encompassing any interactions either direct or indirect between magma/lava and solid H_2O (ice, firn, or snow), or water derived directly from melting of solid H_2O.

hyaloclastite Vitric ash to block-size clasts derived by quench fragmentation of magma.

hydrostatic pressure The pressure at a specified depth of water generated by gravity acting on the load of the overlying water.

ice cauldron A depression in the surface of a glacier, resulting from subglacial melting induced by geothermal or volcanic activity, which can be (1) bounded by vertical ice walls or (2) saucer-shaped with crevasses at the margins.

jökulhlaup A sudden burst of water released from a glacier, which is frequently mixed with substantial volume fractions of ice and sediment. Jökulhlaups caused by volcanic eruptions commonly feature very high discharge rates (10,000—100,000 m^3/s) that can be sustained for several hours.

lava delta An aggrading deposit of dominantly volcaniclastic material produced by the flow of lava into water; common diagnostic components include pillow lava, hyaloclastite (sensu stricto), and volcanic breccia with clasts derived from gravitational collapse of pillow lava.

móberg A Pleistocene/Quaternary-aged stratigraphic formation in central Iceland that encompasses a number of different, dominantly fragmental lithofacies including hyaloclastite (sensu stricto), volcanic breccia, lapilli tuff, and hyalotuff. The deposits have a distinctive brown color and are usually ascribed to Quaternary "subglacial" eruptions.

palagonite An amalgamation of clay minerals, zeolites, unreacted glass, and mineraloids resulting from hydration and alteration of volcanic glass (frequently sideromelane, a common form of basaltic glass). It generally has a distinctive orange-red color in the field, which is also apparent in thin section (palagonitized, palagonitization).

passage zone A depositional surface separating subaqueous from subaerial volcanic deposits. Originally and mainly recognized in effusive sequences as the boundary between subaerial lava and underlying lava delta; also identified in explosive volcanic sequences.

The Encyclopedia of Volcanoes. http://dx.doi.org/10.1016/B978-0-12-385938-9.00020-1

surtseyan eruption A style of eruption that frequently occurs where water floods active vents, resulting in large-scale magma—water interaction forming highly fragmented and fine-grained tephra (e.g., ash to fine ash).

tindar A glaciovolcanic edifice with a length to width ratio of 2:1 or greater. Tindars most commonly comprise pillow lavas, tuff-breccia, and lapilli tuff. In rare instances they may feature a minor subaerial lava cap.

tuya A glaciovolcanic edifice with a length to width ratio less than 2:1. A tuya commonly appears to have a pillow lava base overlain by a volcaniclastic cone flanked by lava deltas underlying a flat lava cap. However, it should not be applied to isolated glacio-volcanic deposits that cannot be directly tied to known or inferred vents.

1. INTRODUCTION

Volcanic eruptions beneath continental ice sheets have occurred frequently during the past 2.5 million years on Earth, producing distinctive landforms that dominate the landscapes of Iceland and parts of northern British Columbia (BC), Canada (Figure 20.1). This type of volcanism is known as **glaciovolcanism** and is formally defined as encompassing volcanic interactions with ice in all its forms (including snow and firn) and, by implication, any meltwater created by volcanic heating of that ice. Our understanding of glaciovolcanism derives from documentation of contemporary eruptions associated with ice and snow cover and the forensic study of ancient volcanic deposits formed at times when ice cover was more widespread than at present. Glaciovolcanism includes volcanic eruptions occurring (1) beneath glaciers that are up to several hundred meters thick such as observed in Iceland, (2) within even thicker continental-scale ice sheets (e.g., Antarctica, Canada, and Iceland during glacial periods), (3) within ice-filled calderas (e.g., Iceland and Alaska), (4) on the ice (thin)-covered flanks of large stratovolcanoes (e.g., Kamchatka, Alaska, and Chile), or (5) on top of or beneath snow (e.g., Etna, Iceland, and Kamchatka). Most of the well-documented glaciovolcanic deposits on Earth formed during the past 10 million years, although glaciovolcanism has likely occurred since the earliest Precambrian glaciations. Glaciovolcanism is also thought to have occurred on Mars.

Research on glaciovolcanism has expanded exponentially in recent years for several reasons (Russell et al., 2014). Firstly, the 2010 eruption of Eyjafjallajökull highlighted at least two of the hazards that are unique to glaciovolcanism: local massive flooding and enhanced tephra production leading to greater disruption of airline travel. Secondly, the Neogene to Recent geological history of many parts of the planet, such as Iceland, BC, and Antarctica, cannot be fully understood without taking into account the role glacier—volcano interactions play in controlling landscape evolution. Thirdly, research on the

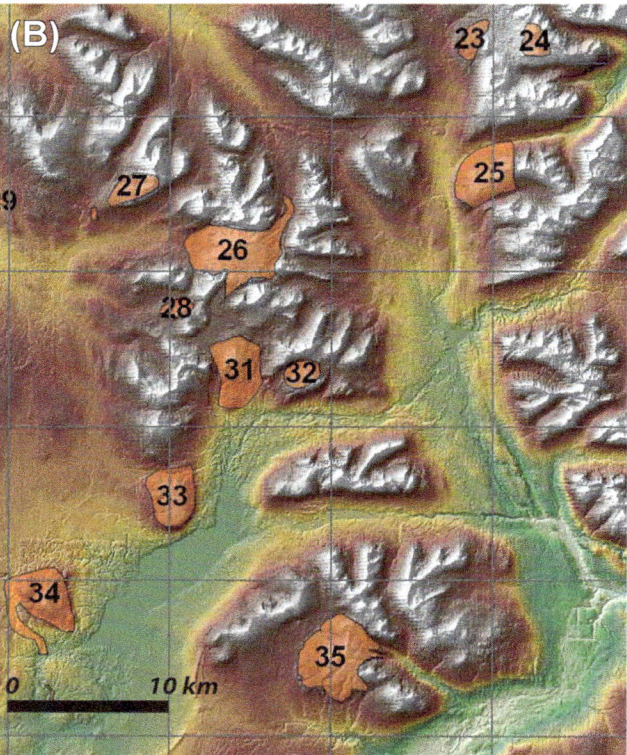

FIGURE 20.1 Characteristic glaciovolcanic landscapes. (A) Aerial view looking to the south of north-central Iceland during the winter, showing large-volume tuyas and linear tindars distributed across low-relief landscape. Tindar in the foreground is approximately 2 km in length. (B) False-colored hillshade map derived from a 30-m Digital Elevation Model (DEM) for Tuya-Teslin region of British Columbia showing distribution and plan-view shapes of tuyas (orange) situated in the northern Canadian cordillera.

geology of Mars based on various remote sensing techniques has called for improved understanding of terrestrial analogs, including glaciovolcanic eruptions and deposits. Lastly, and perhaps most importantly, glaciovolcanic deposits are now recognized as important proxies for determining the distributions of ice sheets in space and time and, thus, provide to critical paleoclimate reconstructions. One of the consequences of this increase in interest in glaciovolcanism is that the breadth of the community of researchers participating in the field has expanded greatly to include, but not be limited to, volcanology in its various forms, glaciology, geomorphology, climatology, and planetary science.

The purpose of this review is to provide an updated introduction to the products and processes ascribed to glaciovolcanism. We begin with a brief overview of nomenclature and key deposit characteristics, and then focus on the substantial advances in our understanding of glaciovolcanism over the last 15 years. These advances have been driven by new studies of intermediate to felsic glaciovolcanic edifices, new methods for recovery of paleoclimate information, more sophisticated understanding of explosive eruption processes, and multidisciplinary real-time studies of active glaciovolcanic eruptions. For more in-depth treatment the reader is referred to recent reviews on various aspects of glaciovolcanism and references therein (e.g., Edwards et al., 2009; Jakobsson and Gudmundsson, 2008; Smellie, 2006).

2. THE BASALTIC TUYA PARADIGM

Deposits produced by basaltic glaciovolcanism have been studied since the early 1900s in Iceland and the 1940s in BC. In Iceland, the **Móberg** Formation is a formal stratigraphic designation that encompasses most of the basaltic volcanic rocks that are of Pleistocene age in the central part of the country, and the term "móberg" is frequently used as a general term for brown-colored volcaniclastic rocks inferred to be of glaciovolcanic origin. Similarly, the Tuya Formation in northern BC comprises Pleistocene volcanic deposits, and the term "tuya" has become the standard designation for all volcanoes formed in glaciovolcanic environments. Basaltic tuyas are the most common expression of glaciovolcanism on Earth, as expressed by dozens of individual edifices in BC and Antarctica, and hundreds in Iceland.

2.1. The "Classic" Model

The archetypal model for a glaciovolcanic eruption has been developed over the course of 70 years of field studies of extinct, mainly basaltic, glaciovolcanic edifices. The generalized eruption model has four stages (Figure 20.2) and assumes a basaltic eruption through a substantial ice sheet where the temperature of the ice at the base is at the melting point for water. The model serves as a useful starting point for discussing the general processes extant during a glaciovolcanic eruption (Figure 20.2), although we emphasize that many aspects of this general model (e.g., effusive versus explosive initiation, recognition of magmatic versus phreatomagmatic components, tuyas as monogenetic versus polygenetic edifices) are presently

being actively debated. During **stage 1** (Figure 20.2(A)), propagation of a dike to the bedrock—ice interface initiates an effusive eruption into the base of the ice; theoretical studies have suggested that it is possible, if not likely, that the dike can overshoot this interface and extend for some distance into the overlying ice (e.g., Wilson and Head, 2007). Ice thicknesses are assumed to be sufficient (>500 m) to suppress volatile exsolution preventing magmatic fragmentation and to restrict volcanic explosivity to phreatomagmatic processes. The initial pressure on the magma/lava is the pressure at the base of the glacier, which is generated by the combined load of the overlying ice (glaciostatic) and any accumulated meltwater (hydrostatic). Observations from recent eruptions, field evidence, and theoretical models all indicate that the volume of melted ice exceeds the volume of erupted material. Thus, a meltwater lens/englacial lake rapidly forms at the base of the ice. The main effusive product is pillow lava, with or without minor volumes of **hyaloclastite** (Figure 20.2(A)). Meltwater will flow away from the eruption site at the glacier base toward the edge of the glacier. The demarcation between the end of **stage 1** and the beginning of **stage 2** is a switch in the character of the volcanic deposits from dominantly coherent (e.g., pillow lava) to dominantly fragmental (Figure 20.2(B)). This change in lithofacies is broadly linked to the growth in edifice height within the englacial lake in at least two ways. Firstly, as the growing pillow mound enlarges and steepens, gravitational collapse will increasingly generate deposits of volcaniclastic material (e.g., pillow breccia, hyaloclastite). Secondly, as the eruption proceeds, the effective water pressure on the erupting magma decreases. This is due to the combined effects of two processes: (1) the increasing height of the edifice within the englacial lake and (2) the deformation and subsidence of the ice above it. As the overlying ice thins due to melting, its surface subsides creating an **ice cauldron** above the growing volcanic edifice. Subsidence allows for part of the load of the central ice block to be supported by the surrounding ice (Figure 20.3). The reduced pressure supports greater magmatic and phreatomagmatic explosivity by allowing for higher amounts and rates of vesiculation and greater expansion of the gas phase (magmatic or external H_2O). **Stage 3** occurs when the bulk of the edifice has breached the surface of the englacial lake and the largely degassed lava can move laterally away from the vent until it reaches the shoreline of the surrounding englacial lake, where it builds a delta formed by pillow lavas and hyaloclastite (Figure 20.2(C)). As the delta continues to prograde, the subaerial top of the edifice is resurfaced by subaerial lava flows. This boundary between the subaqueous and subaerial eruption products is referred to as a **passage zone**, and is one of the critical paleoclimate proxies produced by glaciovolcanism (Jones, 1969; Smellie, 2006; Russell

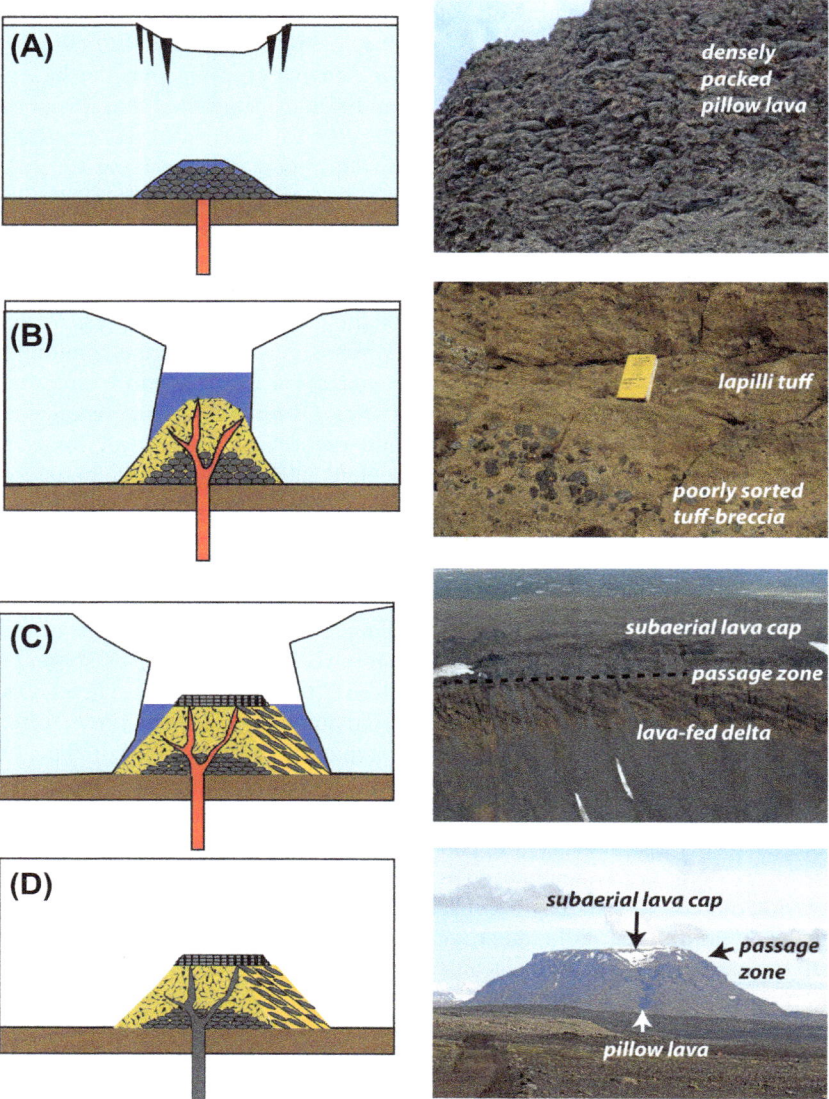

FIGURE 20.2   Standard model for formation of lava-capped tuyas and characteristic lithofacies. (A) Initial eruption under thick ice is effusive, creating a sub-ice cavern partially filled by meltwater, and forming a pillow lava mound. Photograph to right shows stacked pillow lavas from Pillow Ridge, BC. (B) Decreased vent pressures due to reduced ice thickness and/or communication with the atmosphere supports volatile exsolution and magmatic fragmentation. Simultaneously, melt water flooding the vent drives increased and sustained phreatomagmatic explosivity producing heterogeneous fragmental lithofacies. Photograph to the right shows poorly sorted, palagonitized lapilli tuff from Kima' Kho tuya, BC. (C) Volcanic edifice breaches surface of englacial lake, allows drying out of vent, and results in return to effusive volcanism producing stacks of horizontally bedded lavas and an associated lava-fed delta. The passage zone is the surface separating subaerial from subaqueous lithofacies. Photograph to right shows a passage zone at Kima' Kho tuya, BC, that separates subaqueous pillow lava and tuff-breccia delta (orange) from overlying flat-lying lava sheets (gray). (D) Longer term climatic change leads to removal of enclosing ice sheet, leaving the "classic" flat-topped tuya landform and a stratigraphic succession that includes pillow lavas and hyaloclastite, mixed polygenetic tephra, and subaerial lavas and associated lava-fed pillow breccia deltas. Photograph to the right shows Hlöðufell tuya, Iceland.

et al., 2014). Variations in passage zone elevations can record changes to the lake level that are controlled by the glacial hydrology, or subglacial and/or supraglacial drainage. After the eruption, the enclosing ice sheet decays over glacial timescales (**stage 4**) to reveal a unique, flat-topped, near-vertical-walled, basaltic volcano that is commonly perched well above the surrounding landscape

(Figure 20.2(D)). As tuyas collapse due to the loss of the bounding ice (buttressing effect), they expose a substantial portion, if not all, of the volcanic stratigraphy allowing for reconstruction of the glaciovolcanic eruption history. The capping lava flows shield the edifice allowing much of the volcanic stratigraphy to survive and be preserved through multiple subsequent glaciations.

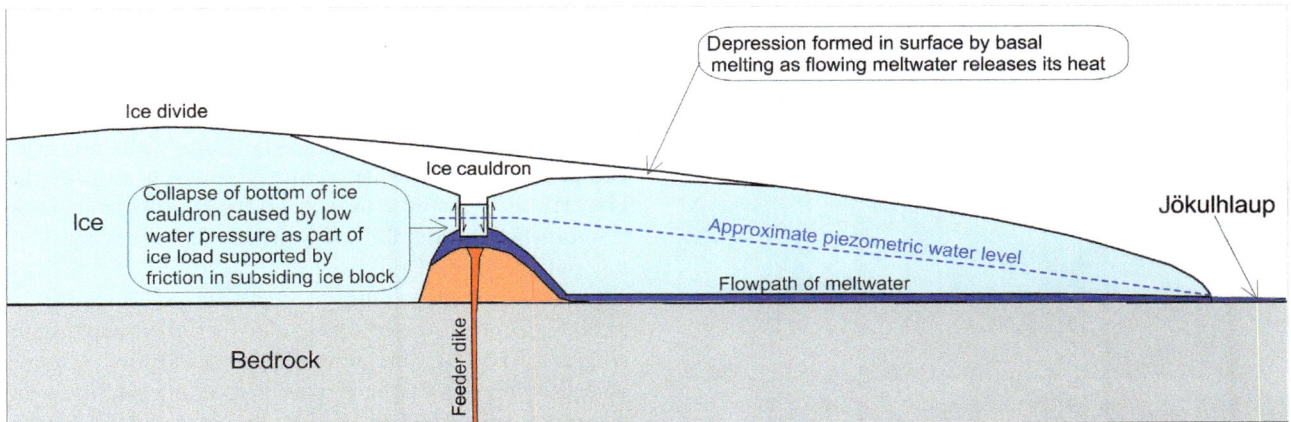

FIGURE 20.3 Effects of subglacial hydrology and ice mechanics on water pressure and glacier response to a subglacial eruption. Subglacial water drainage can lower water pressure at the vent, leading to collapse of the central ice block.

2.2. Landforms

Early workers were struck by the unique morphologies of distinctly flat-topped volcanoes, which they deduced to have formed within an ice-confined environment (e.g., Mathews, 1947). Numerous subsequent studies have identified a range in edifice and deposit morphologies that are frequently accepted as forming mainly in glaciovolcanic environments. While the term "tuya" can be applied to any edifice constructed predominantly of glaciovolcanic deposits, the three most common edifice scale morphologies for glaciovolcanoes are flat-topped tuyas, elongate tuyas or tindars, and conical tuyas (Russell et al., 2014). Flat-topped tuyas have distinctive morphologies in plan view and profile (Figure 20.4(A)) and generally have a length-to-width ratio of 2:1 or less. The flat top generally comprises subhorizontal lava flows that were erupted above the level of the enclosing englacial lake. The tuya morphology can form edifices produced during one eruption, or volcanoes built over longer timescales. Tindars are elongate, steep-sided ridges that generally have a length-to-width ratio of >2:1, which can range in overall length from ~2 km to more than 20 km (Figure 20.4(B)). They are the most common morphology in Iceland, where their overall direction of elongation (approximately north–south) broadly records the tectonic-scale stress field of their extensional setting (Jakobsson and Gudmundsson, 2008; Russell et al., 2014). Eruptions that fail to produce lava caps can produce conical landforms (Figure 20.4(C)). While the vast majority of glaciovolcanic landforms fit one of these types, other types of deposits have been documented, including extensive sheets, often predominantly composed of pillow lavas, but sometimes comprising an ensemble of columnar jointed lava at the base overlain by pillow lavas and hyaloclastites. Individual units can be 100–200 m thick and have been interpreted as being subglacially erupted lava flows. Such stratigraphic packages are understood to cover broad areas in southeast Iceland.

2.3. Lithofacies

Two elements are key to identifying glaciovolcanic lithofacies: (1) evidence for confinement and (2) evidence for the presence of water (Table 20.1). The link between these two elements is ice. Ice sheets are ephemeral relative to the timescales of erosion generally required to completely remove a volcanic edifice and its deposits. Thus, glaciovolcanic deposits have a longevity that far outlasts the duration of the ice that was critical to their formation (e.g., timescales of glacial-interglacial climate cycles). For example, the presence of thickly stacked sheets of subhorizontal lavas on a plateau where there is no obvious physical barrier can suggest confinement by ice. In a similar manner, the presence of glaciovolcanic meltwater can be deduced from sequences of waterlain volcaniclastic deposits (or other subaqueous lithofacies) found at elevations above recorded sea level or in areas known to be devoid of terrestrial paleolakes.

Coherent volcanic lithofacies important for recognition of glaciovolcanism include pillow lava, massive lava, and intrusions (Table 20.1). As pillow lavas only form in water-dominated environments, their presence documents an abundance of water or water-saturated sediments. Again, the presence of preserved pillow lavas in areas with no obvious means of impounding water can be a record of glaciovolcanism (Figure 20.5(A)). Dikes are common at glaciovolcanic edifices, and the presence of dikes with pillowed margins is a common indicator that the host volcaniclastic deposits were water saturated (Figure 20.5(B)). Lava morphologies can also sometimes suggest an

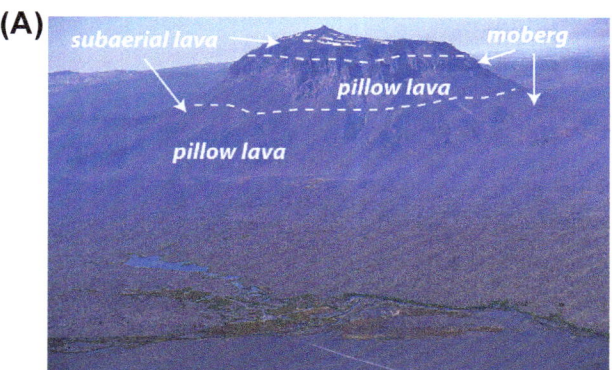

FIGURE 20.4   Field photographs showing large-scale forms of glacio-volcanic landforms. (A) Aerial photograph of the eastern face of Herðu-breið tuya, central Iceland, illustrating the iconic tuya profile with a fragmental base (sloping lower sides) capped by subaerial lava flows (upper cliffs) and a summit cone. (B) Aerial view of Pillow Ridge tindar, north-central British Columbia, showing the elongated nature of a typical pillow-dominated glaciovolcanic ridge. (C) View of conical summit of South tuya, northern British Columbia, which is constructed of palago-nitized tuff-breccia and lapilli tuff lying on top of a basal platform of pillow lava.

ice-dominated eruption environment. Overthickened indi-vidual lava flows or stacks of lava flows featuring vertical cooling surfaces provide evidence for ephemeral confine-ment of lava or water or both (Figure 20.5(C)). Radial or convoluted cooling fractures may also record anomalously rapid cooling due to the presence of water (Figure 20.5(D)).

Diagnostic fragmental lithofacies for basaltic glacio-volcanic deposits are ones whose clasts preserve evidence for rapid cooling, multiple fragmentation mechanisms, and confined/unconfined deposition in water (Table 20.1). Grain size distributions can vary from well-sorted tuff comprising hyaloclastite (Figure 20.6(A); also hydro-clastite or hyalotuff) to coarse lapilli tuff comprising sideromelane ash/lapilli and lithic basalt formed by gravi-tational collapse of pillow lava flows (Figure 20.6(B)) to sequences of well- to poorly sorted volcaniclastics varying in componentry from blocks to ash over short lengthscales (Figure 20.6(C)). Frequently glaciovolcanic eruptions generate deposits that have clasts recording multiple frag-mentation processes because the water-rich environment promotes rapid heat extraction from the melt ("quenching") leading to high thermal stresses as well as phreatomagmatic explosions as trapped water is converted to steam. In many cases, evidence for magmatic fragmentation is also present as partly to fully vesiculated clasts. Bedding may record confinement via anomalously shallow dips along edifice flanks that have significantly steeper slopes. In general, the volume fractions of nonscoriaceous fragmental deposits found at many basaltic tuyas is significantly higher than would be expected for subaerial basaltic volcanism, as is their very common high degree of palagonitization (Figure 20.6(B)).

3. INTERMEDIATE TO FELSIC GLACIOVOLCANISM

Significant recent advances in our knowledge of glacio-volcanic processes come from field-based studies of non-basaltic glaciovolcanoes and associated deposits (Table 20.2). A substantial number of studies have now been published on intermediate to silicic glaciovolcanism in Iceland, BC, the western United States, the Andes in South America, and even Antarctica. Detailed field studies of andesitic to dacitic deposits in Iceland, BC, and the western United States show that intermediate composition lavas appear to produce much less fragmental material than basaltic lavas (Table 20.2). For example, a summary of rhyolitic glaciovolcanism in Iceland (McGarvie, 2009) re-ported that effusive deposits were dominant over explosive deposits, and that deposit morphology and orientations of cooling joints were key to recognition of glaciovolcanic origins (Figure 20.7(A)). Explanations for this field-based observation are still being developed. For example, in BC many of these deposits are situated in mountainous land-scapes having extreme relief, where gravity and drainage systems reduce preservation of fragmental units. Likewise, in the western United States many of these deposits are on the flanks of steep stratovolcanoes, where preservation may be unlikely. The number of published studies of

TABLE 20.1 Criteria Used to Ascribe a Glaciovolcanic Origin to Volcanoes or Volcanic Deposits (Modified from Smellie, 2000). Volcanism Occurring in a Region with an Established History of Glaciation(s) Should Be Evaluated for Characteristics Listed Below. Volcanic Sequences Erupted in Ice-Dominated Environments are Characterized by Deposits that May Record the Effects of Confinement (by Ice), Enhanced Cooling (by Ice or Water), Multiple Fragmentation Processes (Thermal Shock, Magmatic and Phreatomagmatic Fragmentation), and Subaqueous/Subaerial Transitions

Features Indicating Confinement by Ice
1. volcanic edifice has a higher aspect ratio than conventional volcanoes
2. primary cooling surfaces of lavas are near vertical as indicated by orientations of thermal fractures
3. volcaniclastic deposits have bedding attitudes that are too steep
4. anomalously thickened lava flows (thickness will depend on composition as well)
5. pervasive **palagonite** replacement of glass suggesting retention of heat in water-rich environment
6. volcaniclastic deposits contain accidental faceted or striated lithic clasts, or associated deposits of glacial diamict or glaciolacustrine sediments

Features Indicating Enhanced Cooling
7. thermal Fractures (e.g., columnar joints, cooling columns) are pervasive, have anomalously small widths, and anomalous orientations (horizontal or fanning to radial)
8. pillow lava interbedded with hyaloclastite, pillow lava breccias
9. volcaniclastic Deposits are dominated by vitric particles or palagonitized vitric particles

Features Indicating Multiple Fragmentation Processes
10. blocky ash-size fragments, which may or may not be vesicular (thermal shock, phreatomagmatic)
11. abundance of fine to very fine ash (phreatomagmatic)
12. shard-shaped ash bounded by cuspate margins fractured across vesicles (magmatic)
13. relatively poor sorting and all three fragment types potentially present together at thin section scale

Features Indicating Subaqueous/Subaerial Transitions
14. lava-fed delta directly overlain by subaerial lava flows (bounding surface is effusive passage zone)
15. subaqueous volcaniclastic cone overlain by subaerial pyroclastic deposits (bounding surface is explosive passage zone)
16. elevations of all bounding surfaces may show sudden changes in elevation up to several meters

Discrimination from Terrestrial (Subaerial) Environments
Terrestrial (subaerial) environments will likely not have many of the elements listed above, especially 1−10; 11 may be present at maar eruptions, and fragmental deposits will be dominated by 12.

Discrimination from Fluvial Environments
Fluvial environments will likely not have 1, 3, 4, and will have lower abundances of 10−12; they will also likely show rounding of individual clasts and may show unidirectional current structures (e.g., cross-bedding) and preserve evidence for paleochannels. Fine-ash fraction characteristic of phreatomagmatic eruptions will likely not be readily deposited due to small grain size, and palagonitization will be absent or greatly reduced as constant flow of water will more efficiently remove heat.

Discrimination from Submarine/Lacustrine Environments
Submarine/lacustrine environments can be nonunique, but will likely be missing 1,3,4, and will only form in locations accessible by sea level or topographic basins. Volcaniclastic deposits may preserve fossils (freshwater or marine), and compositions of zeolite minerals associated with palagonite may have a marine chemical signature. Submarine/lacustrine environments can contain 10−12; however, they are likely to be less abundant in vent proximal setting.

intermediate to felsic tuyas is also substantially fewer than the number of corresponding studies of basaltic deposits, which could bias perceptions. Studies of Hoodoo Mountain, a major peralkaline volcano in northern BC, which erupted repeatedly over the past 100 k.y., show that relatively fluid intermediate peralkaline lava flows can be impounded by ice to become overthickened (Figure 20.7(B)). Rhyolitic glaciovolcanic lavas are frequently extensively perlitized (Figure 20.7(C)). Lava flows on stratovolcanoes in the Cascades and Andes (Lescinsky and Fink, 2000; Mee et al., 2006) show more subtle features preserved as cooling

fractures that are used to infer glaciovolcanic origins (Table 20.2).

The edifices and deposits produced by glaciovolcanic eruptions of more evolved (intermediate to felsic) magma compositions can be more difficult to identify. This is because the glaciovolcanic signature is more subtle than that recorded in basaltic glaciovolcanic deposits, reflecting the substantial differences in magma properties. For example, the relatively low viscosity of basaltic magmas means that in subaerial environments mafic lava flows have low aspect ratios; confinement by ice generates anomalous,

FIGURE 20.5 Field characteristics of diagnostic coherent glaciovolcanic lithofacies. (A) A thick stack of lava pillows (~25 m) in a glaciovolcanic ridge, Undirliðar quarry, Iceland. (B) Outer "pillowed" margin at tip of shallow-level dike that intrudes palagonitized lapilli tuff, northern British Columbia. (C) Ice-confined flow of pillow lavas, northern British Columbia. (D) Radially fanning columnar jointing, south-central Iceland.

but diagnostic, lava morphologies having high aspect ratios. However, evolved lavas will naturally produce thick, high-aspect-ratio lavas due to their higher viscosities. Thus, it can be more difficult to recognize evolved lavas that may have been confined by ice. Similarly, because the glass transition temperatures for basaltic magma are substantially lower than their eruption temperatures, sideromelane is not a main constituent of subaerial basaltic lavas. Thus, glass-rich basaltic deposits can be a clear indicator of "water-enhanced" cooling. However, for rhyolitic magmas glass transition temperatures are much closer to eruption temperatures, so even subaerial rhyolite lava flows can have significant volumes of obsidian and "enhanced" cooling rates are not as evident.

Intermediate to felsic glaciovolcanoes have a significant advantage over their basaltic counterparts for paleoclimate studies. Their compositions (relative enrichment in K) make them better targets for high-precision geochronometric studies (e.g., $^{40/39}$Ar geochronometry) than basaltic rocks. Thus, while this compositional range of subglacial volcanoes appears, at present, to be volumetrically minor relative to basaltic glaciovolcanoes, they are important as paleoclimate proxies.

4. GLACIOVOLCANISM AS A PALEOCLIMATE PROXY

Our understanding of global climate variability over the past 5 Ma on Earth is largely based on records from ice cores and from paleo-ocean temperatures derived from oxygen isotopic analyses of benthic and planktonic foraminifera. Terrestrial records for climate derive mainly from scattered glacial deposits, which are poorly preserved and difficult to study with the conventional geochronometric techniques employed on igneous and metamorphic rock samples. While a growing list of promising techniques (e.g., exposure and burial geochronometers) is adding better constraints on the timing of recent glacial deposits, these techniques still have limited applicability for deposits older than the Last Glacial Maximum (ca. 20 k.y.a.). The identification of volcanic rocks that record magma—ice interactions provides critical paleoclimate proxy information because these rocks can be directly dated. These rocks record the presence of ice and, thus, their ages give us the distribution of terrestrial ice in space and time. Glaciovolcanic deposits are also commonly more resistant to erosion than other types of glaciogenic deposits and, thus, should become increasingly relied upon

FIGURE 20.6 Field characteristics of diagnostic fragmental glaciovolcanic lithofacies. (A) Cross-laminated finely bedded hyaloclastite deposits underlying a sequence of pillow lavas, Pillow Ridge, BC. (B) Componentry of poorly sorted lapilli tuff, with fresh sideromelane and angular clasts of lithic basalt in a palagonitized matrix. (C) Sequence of glaciovolcanic lithofacies ranging from poorly sorted tuff-breccia, to discontinuous lenses of lapilli tuff, to well-sorted and more laterally continuous tuff, southwestern Iceland.

stratigraphic relationships that allow for estimation of the depth and elevation of the englacial lakes from which the minimum thicknesses of the enclosing ice can be indirectly calculated (see below).

4.1. Constraints on Spatial Distribution of Ice

While glaciovolcanic deposits can be used as paleoclimate proxies in a number of ways, simply recognizing an edifice/deposit as having a glaciovolcanic origin immediately constrains the spatial distribution of a past ice body. In areas where ice is still present (e.g., many active volcanoes in Iceland, Antarctica, the western United States, or South America), flank deposits provide evidence for expansion or contraction of existing ice. However, in areas like northern BC, which are now ice-free, glaciovolcanic deposits provide the only evidence for the existence and minimum extent of former glaciers. With increasingly high-resolution images from Mars, it will soon be possible to use morphometric studies of Martian edifices in the same way to confidently constrain extents of past ice sheet there.

4.2. Constraints on Temporal Distribution of Ice

Glaciovolcanism produces deposits of coherent igneous rocks whose eruption ages can be confidently constrained by an array of geochronometers, hence documenting the timing of ice presence. However, the overwhelming majority of glaciovolcanism is basaltic (low K concentrations), which limits the precision of age determinations, even using the most advanced geochronologic techniques. These uncertainties limit our capacity to correlate terrestrial-based glaciovolcanic events to Marine Isotopic Stages; these stages currently represent the most detailed records of changes in global ice volumes over the last 3 Ma. For example, in BC, basaltic glaciovolcanism at Mathews Tuya records glaciation at 730 ka +/− 40 k.y., which closely correlates with Marine Isotope Stage (MIS) 18 (e.g., Edwards et al., 2010). However, the uncertainties in the age preclude determining whether the eruption corresponded to waxing or waning of the ice sheet. McGarvie et al. (2007) have used Ar—Ar geochronology on felsic rocks to more closely tie a rhyolitic eruption in Iceland to MIS 5. Recently, Licciardi et al. (2007) have used exposure age dating to attempt to document ages of ice retreat based on tuyas formed at the end of the last glacial maximum. The most exhaustive study to demonstrate the importance of glaciovolcanic deposits comes from James Ross Island, in the Antarctic Peninsula, where Smellie et al. (2008) have developed a detailed record of changing ice thicknesses over the past 6 Ma.

as a tool for filling gaps in our terrestrial paleoclimate databases. This concept is also being applied to help constrain changes in Martian paleoclimates; the recognition of volcanoes with morphologies very similar to tuyas has been the basis for estimating thicknesses of past Martian ice extents (up to several hundreds of meters; cf. Edwards et al., 2009). In addition, some glaciovolcanic successions host

TABLE 20.2 Some Examples and Characteristic Properties of Nonbasaltic Glaciovolcanism

Composition	Examples[1]	Characteristic Properties
Rhyolite	Torfajokull (TR),[2] Kerlingarfell, (KR)[3]	Confined lava flows, radial jointing, fine-scale jointing, phreato-magmatic tephra (KR), peperite (TR)
Dacite	Garibaldi (GR)[4]	Confined lava domes (GR,HD)
Trachyte/phonolite	Hoodoo (HD),[5] Takahe (TK)[6]	Confined lava flows, radial jointing, fine-scale jointing, hydro-thermally altered tephra deposits (HD); pillow lava delta (TK)
Andesite	Garibaldi (GR),[4] Kerlingarfell (KR),[3] Cascades (CS)[7]	Stacked lava flows (GR); pillow lava (KR); ice-bounded flows with pseudo-pillow fractures (CS)

[1]Torfajokull and Kerlingarfell from Iceland; Garibaldi and Hoodoo from British Columbia, Canada; Takahe from Antarctica; Cascades from Washington, Oregon, and California, U.S.A.
[2]Tuffen, H., Gilbert, J., and McGarvie, D. (2001). Products of an effusive subglacial rhyolite eruption: Blahnukur, Torfajokull, Iceland. Bull. Volcanol. **63**, 179–190.
[3]Stevenson, J.S., Smellie, JS, McGarvie, D, Gilbert, JS, and Cameron, B (2009). Subglacial intermediate volcanism at Kerlingarfjöll, Iceland: magma–water interactions beneath thick ice. J. Volcanol. Geotherm. Res. **185**, 337–351.
[4]Kelman, M.C., Russell, J.K., and Hickson, C.J. (2002). Effusive intermediate glaciovolcanism in the Garibaldi Volcanic Belt, SW BC. In: Smellie, J., and Chapman, M. (eds.) Volcano-Ice Interaction on Earth and Mars. Geol. Soc. Lond., Sp. Pub. **202**, 115–148.
[5]Edwards, B.R., Russell, J.K., and Anderson, R.G. (2002), Subglacial, phonolitic volcanism at Hoodoo Mountain volcano, northwestern Canadian Cordillera. Bull. Volcanol. **64**, 254–272.
[6]Le Masurier, W.S. (2002). Architecture and evolution of hydrovolcanic deltas in Marie Byrd Land, Antarctica. In: Smellie, J., and Chapman, M. (eds.) Volcano-Ice Interaction on Earth and Mars. Geol. Soc. Lond. Sp. Pub. **202**, 195–211.
[7]Lodge, R.W.D., and Lescinsky, D.T. (2009). Fracture patterns at lava-ice contacts on Kokostick Butte, Oregon and Mazama Ridge, Mount Rainier, Washington. J. Volcanol. Geotherm. Res. **185**, 298–310.

FIGURE 20.7 Field characteristics of diagnostic intermediate-to-felsic glacio-volcanic lithofacies. (A) Radially oriented columnar jointed subglacial lobe of obsidian exposed at Bláhnúkur, Iceland. (B) Ice-confined lava flows (>200 m thick) at Hoodoo Mountain, BC. (C) Pervasive development of perlite in obsidian, Bláhnúkur, Iceland. (D) Jointed rhyolitic lava lobes surrounded by tuff-breccia, Bláhnúkur, Iceland.

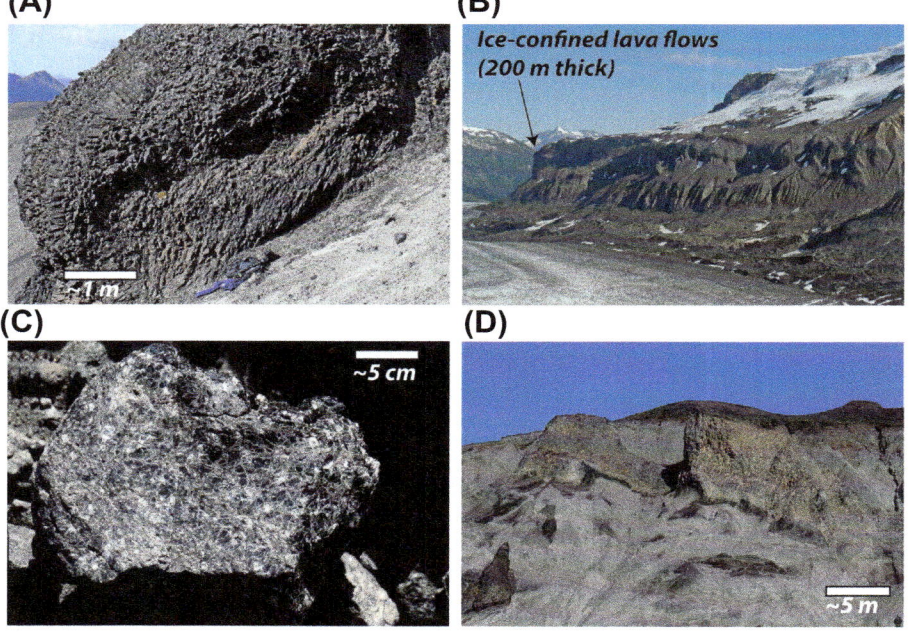

4.3. Constraints on Ice Thickness

Two main methodologies are presently used to constrain syn-eruptive ice thicknesses: elevations of passage zones and the presence of undegassed volatiles from the vitric rims of pillow lavas. Passage zones, which are thought to record the height of an enclosing englacial lake surrounding the building volcanic edifice, have been recognized, in principle, since at least 1947 (Mathews, 1947). Jones (1970) coined the term "passage zone" and Smellie (2006) has described some of the implications that the geometry of a passage zone has for reconstructing glaciovolcanic processes and history. Recently, Russell et al. (2013) presented evidence for pyroclastic passage zones as well. The importance of passage zones is that, by recording the depth of an englacial lake, they also indirectly constrain the minimum thickness of the enclosing ice. As passage zones are sharp lithofacies boundaries, their elevations can be measured very accurately. However, at present considerable uncertainty exists regarding the relationship between englacial lake level and initial ice thickness in an ice sheet. Clearly, the depth of the englacial lake (D) alone defines the minimum thickness of the surrounding ice (H). Furthermore, an equality of glaciostatic (ice thickness) and hydrostatic (water depth) pressures can be argued to estimate an alternate and greater minimum ice thickness (D ~0.9*H). The only observational evidence comes from Icelandic eruptions at Gjálp (1996) and Grímsvötn (2004) where the lake levels were about 150 m lower than the preeruption ice surface. At Gjálp, ice thickness before the eruption was 600−750 m, in which case the lake level elevation corresponded to 75−80% of the original ice thickness. The 2004 Grímsvötn eruption occurred under thinner ice and, there, the transient lake levels were also well below the levels required to float the glacier (i.e., 0.9*H). These few observations may suggest that passage zone elevations are always substantially lower than the maximum ice thicknesses at the time of eruption. For example, given constraints on potential ice volumes melted during glaciovolcanic eruptions, a tuya whose total thickness is 500 m could supply enough thermal energy to melt through 1000 m of ice. Thus, even if a passage zone developed at the uppermost few meters of the tuya, the calculated minimum ice thickness would be substantially lower than the actual ice thickness.

The second method that has been used by many workers to infer **glaciostatic/hydrostatic pressure** during glaciovolcanic eruptions is using measurements of the concentrations of volatile species (mainly H_2O, CO_2, SO_2) in the glassy rims of pillow lavas. The formation of a separate volatile phase in magma and the subsequent distribution of volatiles between the vapor and melt phases is highly pressure dependent. So, if lava is emplaced in water or under an ice load, the quickly cooled melt in contact with the coolant can lock in the concentrations of volatile species dissolved in the melt at that point, which can then be analyzed and used to estimate the pressure extant at the time of quenching (e.g., Schopka et al., 2006).

5. PHYSICAL CONSTRAINTS ON GLACIOVOLCANIC PROCESSES

The distinctive landforms and lithofacies produced by glaciovolcanic eruptions (e.g., relative to terrestrial eruptions in ice-free environments) are consequences of

interaction between meltwater derived from ice, the magma (i.e., rapid cooling, quenching, and fragmentation), and the enclosing ice, which serves to confine the deposits. Thus, quantitative analysis of heat transfer from the magma to the surrounding environment and of fragmentation processes is essential to understanding modern glaciovolcanic eruptions and for forensic interpretations of ancient glaciovolcanoes.

5.1. Calorimetry of Subglacial Eruptions

During the early stages, a glaciovolcanic eruption is likely to be fully subglacial, so the heat released by rapid cooling of the erupted material is used almost exclusively for melting ice and heating of meltwater. During these early stages the eruption rate is likely to be the highest. Under these circumstances, energy transfer between magma and ice can be modeled effectively by simple calorimetric relations. The thermal energy carried by magma (E_m) is not all released instantaneously and the temperature of the deposited tephra/lava is significantly higher than that of its surroundings. The energy used for ice melting is E_i, while the energy carried to the glacier base is E_m. The ratio of these two energies is the melting efficiency, f:

$$E_i = fE_m \qquad (20.1)$$

For a fully realistic description of the physical problem, the efficiency should be evaluated as the ratio of the energy fluxes, meaning that E_i and E_m should denote energy/unit time. In practical terms, changes in ice volume melted in an eruption are evaluated over discrete time periods. The value of f, which only evaluates heat transfer during the early stages of the eruption, does not explicitly take into account any energy that is used to heat meltwater or produce water vapor, so it evaluates the maximum amount of ice that can be melted for a given volume of erupted magma. The ice melting energy is obtained from

$$E_i = V_i \rho_i [C_i(T_0 - T_i) + L_i] \qquad (20.2)$$

where V_i and ρ_i are, respectively, the volume and density of ice melted, T_i is the initial temperature of the ice, T_0 is its melting point ($\sim 0\,°C$), C_i the specific heat capacity and L_i is the latent heat of fusion for ice. The total thermal energy of the magma, E_m, is given by

$$E_m = V_m \rho_m [L_m + C_m(T_m - T_0)] \qquad (20.3)$$

Here the subscript m refers to magma/lava/tephra properties, with V_m being the volume of volcanic material erupted and ρ_m being its density. In some eruptions, crystals may be a significant fraction of the erupted volume, whereas fragmentation facilitates cooling and suppresses crystallization by pushing the melt to its glass transition temperature faster than rates of crystal nucleation and growth. In the former case, the latent heat must be multiplied by the volume fraction of crystals. In the latter case, no crystallization occurs and thus no latent heat is released ($L_m \sim 0\ °C$). The efficiency value f is not generally easy to determine, but for eruptions where fragmentation (magmatic, phreatomagmatic, or quench) is dominant, $f = 0.5-0.8$ is appropriate (Figure 20.8). For effusive eruptions (e.g., pillow lavas), the value is much less, and over short timescales the order of magnitude of f is considered to be $0.1-0.3$ (Figure 20.8). Due to higher magma temperatures and heat capacities, basaltic magmas have higher heat contents than evolved magmas such as rhyolites.

The density of pillow lavas is considerably higher than that of a pile of fragmented material. If this is taken into account, efficiencies of $0.1-0.3$ for basalts lead to melting of ice volumes of $1.2-5$ times the lava volume. Corresponding values for fragmentation, having efficiencies of $0.5-0.8$, allow for melting of ice volumes $4-7$ times that of the tephra volume. Thus, the volume of ice melted can be generally expected to be larger than the volume taken up by the eruption products.

5.2. Controls on Eruption Initiation and Fragmentation Processes

While the majority of modern glaciovolcanic eruptions have been dominated by highly explosive activity,

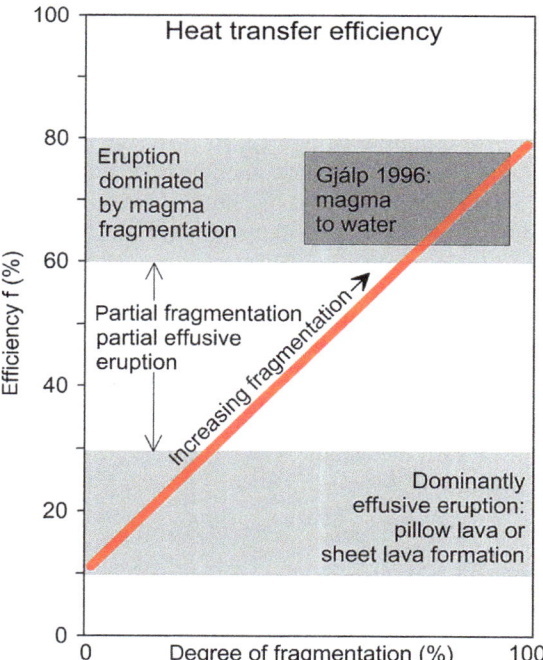

FIGURE 20.8 Schematic representation of heat transfer efficiency (f) mapped as a function of degree of fragmentation for effusive glaciovolcanic eruptions where quench fragmentation dominates ($f < 30$), and for explosive glaciovolcanic eruptions ($f > 60$) driven by magmatic fragmentation (modified from Gudmundsson, 2003).

abundant evidence from the geologic record shows that many glaciovolcanic eruptions also have significant effusive episodes. As for all volcanic eruptions, one of the main controls on explosivity during glaciovolcanic eruptions is the relationship between external pressure and the solubility of volatile phases for a given magma (Figure 20.9). In general, high glaciostatic/hydrostatic pressures and "dry" magmas will favor effusive eruptions, while low pressure and "wet" magmas will favor explosive eruptions (Figure 20.9). Thus, even some

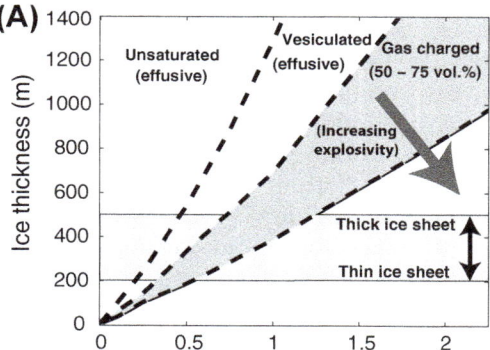

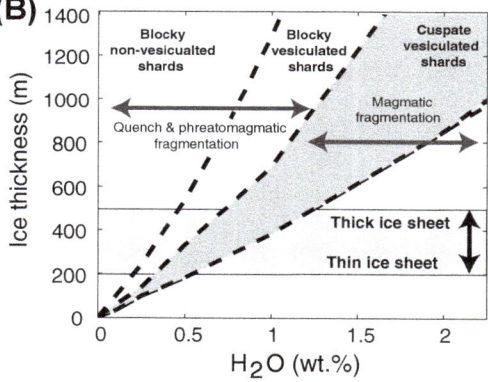

FIGURE 20.9 Schematic diagram showing state of volatile saturation, style of eruption, and pyroclast properties as a function of water content and ice thickness (i.e., load pressure) space for the initial eruption of a basaltic magma. (A) Low water contents and thick ice cause effusive eruption of undersaturated magma; vitric tephra results from quench fragmentation and/or phreatomagmatic fragmentation. Higher water contents or thinner ice allow for effusive eruption of highly vesiculated magmas, and tephra largely results from phreatomagmatic explosive processes and quench fragmentation. At water contents in excess of 0.5 wt % basaltic magma erupting under ice sheets as thick as 500 m will be highly gas charged (>50 vol.% bubbles). An explosive onset to the glaciovolcanic eruption is a distinct possibility at these gas fractions (50–75 vol.%); exsolved volatile pressures overcome the glaciostat allowing for gas exsolution at pressures below the H_2O critical point allowing for explosive eruption driven by rapid expansion of exsolved fluid phase. (B) Regions in water content–ice thickness (i.e., load pressure) space where fragmentation will be driven dominantly by (i) thermal quenching (i.e., water or ice), (ii) phreatomagmatic explosivity, or (iii) exsolution and expansion of magmatic volatiles.

"thick" ice eruptions may initially be explosive, which favors high heat transfer efficiencies and rapid melting. The pressure extant during glaciovolcanic eruptions is complicated by the properties of the ice "overburden," which can melt and fracture during the course of an eruption.

The presence of external water adds an additional complication to fragmentation at any stage of eruption, as it can cause high thermal stresses in the magma leading to "quench" fragmentation, as well as phreatomagmatic explosions driven by shockwave breakup of the magma followed by rapid expansion of trapped liquid water being converted to steam (Figure 20.9). The relative importance of quench, phreatomagmatic, and magmatic fragmentation will ultimately depend on the unique magmatic and glacial conditions for any given eruption.

6. OBSERVATIONS ON MODERN ERUPTIONS

While observations from glaciovolcanic eruptions have long been an important source of information (cf. Jakobsson and Gudmundsson, 2008), opportunities for close monitoring of eruptions have only been abundant since the 1990s. In the intervening 20 years, well over a dozen eruptions have been studied where lava interacted with snow/ice. These eruptions take many forms (Figure 20.10). Truly subglacial eruptions, where the vents are covered by ice throughout, are at the present time not common, but geological evidence indicates they occurred frequently in volcanic regions during glaciations. Modern glaciovolcanic eruptions may be crudely divided into three main classes. The first includes those eruptions that occur under/within glaciers where eruption style and behavior are dominated by the presence of the ice and meltwater. For this class two end-member styles exist: eruptions through thick ice (>500 m) and eruptions through thin ice (<200 m) (Figure 20.10). The first class of eruption is common in Iceland, including the Gjálp eruption in 1996, several Grímsvötn eruptions, and the first part of the Eyjafjallajökull eruption in 2010. It is likely that some eruptions of this type have occurred in Antarctica in the recent past, but due to the size, remoteness, and short observation periods for Antarctica, the scale and frequency is not known. The majority of Pleistocene tuyas found globally are of this type.

The second class included glaciovolcanic eruptions where the style of vent activity is not significantly affected by the ice presence, but the surrounding ice exerts considerable influence on properties and distributions of the eruption products. These eruptions are typical of the largest number of snow- and ice-covered volcanoes,

(A) Class 1A : Thick ice

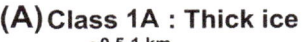
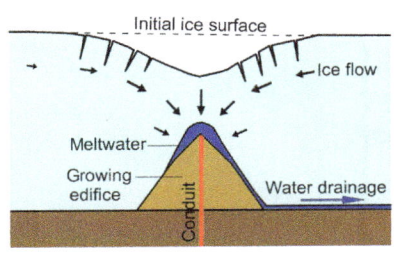

(B) Class 1B : Thin ice

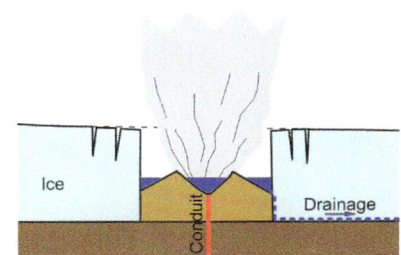

(C) Class 2: Pyroclastic flow /dome collapse over ice

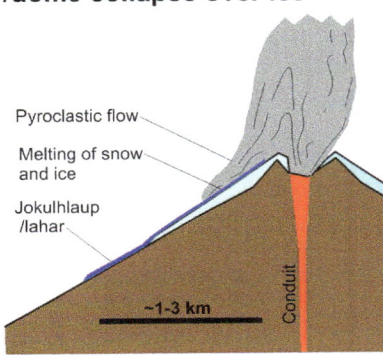

(D) Class 3 example: Lava-ice contact

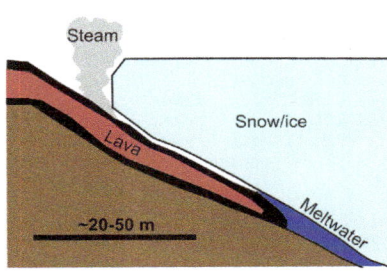

FIGURE 20.10 Schematic illustrations summarizing the three main classes of historic glaciovolcanic eruptions. (A) Class 1A: Subglacial eruption beneath a thick (>500 m) ice sheet showing simultaneous melting and deformation of ice and growth of volcanic edifice (see text). (B) Class 1B: Eruption from beneath thin ice; melting of ice cover produces a cauldron with steep walls and little ice flow. (C) Class 2: Supraglacial eruption common to stratovolcanoes where the resulting pyroclastic flows can produce local hazards. (D) Class 3: Small-volume lavas originate from above the ice surface but flow over and/or beneath the snow/ice surface.

including many in South America (e.g., Villarica), Alaska, and Kamchatka (e.g., Klyuchevskoy). This second class of glaciovolcanic activity includes recent eruptions in Redoubt (1989–90 and 2009), where ice melting generated a significant flood that damaged local infrastructure, as well as the effusive portion of the Eyjafjallajökull eruption (Figure 20.10).

The third class involves minor volcano–ice interactions that still produce identifiable glaciovolcanic deposits. While this class does not construct a large edifice and only generates local hazards (e.g., Belousov et al., 2011), it is likely by far the most widespread type of glaciovolcanism, and has the potential to generate deposits that can record the presence of snow/ice across the spectrum of earth latitudes (e.g., many low-latitude stratovolcanoes are snow covered for at least part of the year now, and presumably would have had much more significant cover during glacial epochs) (Figure 20.10).

An indication of the variety in style and settings encountered is given by the following accounts of three recent eruptions (Gjálp 1996; Eyjafjallajökull 2010; Redoubt 2009; Table 20.3).

6.1. Gjálp 1996

The Gjálp eruption in October 1996 is the only recorded example of an eruption under a thick glacier (Table 20.3; Figure 20.11(A)). It lasted for 13 days and started as a 4-km-long fissure eruption under 600 m thick ice in the northwest part of the 8100-km^2 Vatnajökull ice cap, Iceland. No preceding thermal activity was observed but the eruption quickly formed two ice cauldrons above the main vents, each about 2 km wide and 100 m deep. These cauldrons had crevassed margins but in cross-section the depressions were V-shaped indicating focused melting above the vents. On average 6000 m^3/s of meltwater was generated during the first 4 days (melting rate of ~0.5 km^3/day). This meltwater was continuously draining away from the eruption site, 15 km to the south into the Grímsvötn caldera where it accumulated in a subglacial lake. This lake is semipermanent but with a varying lake level, sustained by a geothermal area within the caldera. The eruption melted through the 600 m of ice in 31 h, opening a vent that was active for the duration of the eruption. The volcanic fissure grew toward the north, reaching a length of 6 km on the second day, and forming an ice cauldron where the ice

TABLE 20.3 Observed, Measured and Calculated Properties of Four Modern Glaciovolcanic Eruptions

	Gjálp 1996[1]	Grímsvötn 2004[2]	Eyjafjallajökull 2010[3]	Redoubt 2009[4,5]
Magma Composition	Basaltic andesite	Basaltic	Trachyandesite	Dacite
Eruptions Style(s)	Subglacial/Explosive	Explosive	Explosive/Effusive	Dome/Explosive
Subglacial Phase (hours)	31	0.5	4	0
Volume Erupted (km^3 DRE)	0.45	0.02	0.17	0.10
Volume % Erupted Subglacially	~95%	~10%	~5%	Minor
Duration (days)	13	6	39	13
Volume, Airborne tephra (km^3 DRE)	~0.02	0.02	0.14	0.02
Volume, Lava (km^3)	0	0	0.03	0.08
Initial Ice Thickness (m)	600–750	150–200	200	100
Maximum Cauldron Width (km)	7	0.5-1.0	0.6	1.5
Volume Ice Melted (km^3)	4.2	0.07	0.22	0.10–0.25
Jökulhlaup Discharge (Max. m^3/s)	50,000	3,000	~5,000	~100,000
Jökulhlaup Duration	2 days	7 days	hours	<1 hour

[1] Gudmundsson, M.T., Sigmundsson, F., Björnsson, H., and Högnadóttir, þ. (2004). The 1996 eruption at Gjálp, Vatnajökull ice cap, Iceland: efficiency of heat transfer, ice deformation and subglacial water pressure. Bull. Volcanol. **66**, 46–65.
[2] Jude-Eton, T.C., Thordarson, T., Gudmundsson, M.T., and Oddsson, B. (2012). Dynamics, stratigraphy and proximal dispersal of supraglacial tephra during the ice-confined 2004 eruption at Grímsvötn volcano, Iceland. Bull. Volcanol. **74**, 1057–1082.
[3] Gudmundsson, M.T., Thordarson, T., Höskuldsson, Á., Larsen G., Björnsson, H., Prata, A.J., Oddsson, B., Magnússon, E., Högnadóttir, T., Pedersen, G.N., Hayward, C.L., Stevenson, J.A., and Jónsdóttir, I. (2012). Ash generation and distribution from the April-May 2010 eruption of Eyjafjallajökull, Iceland. Scientific Reports **2**, 572. http://dx.doi.org/10.1038/srep00572.
[4] Bull, K., and Buurman, H. (2013). An overview of the 2009 eruption of the Redoubt Volcano, Alaska. J. Volcanol. Geotherm. Res. **259**, 2–15.
[5] Waythomas, C.F., Pierson, T.C. Major, J.J., and Scott, W.E. (2013). Voluminous ice-rich water-rich lahars generated during the 2009 eruption of Redoubt Volcano, Alaska. J. Volcanol. Geotherm. Res., **259**, 389–413

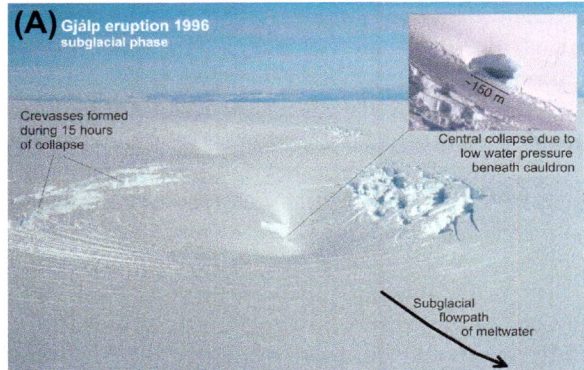

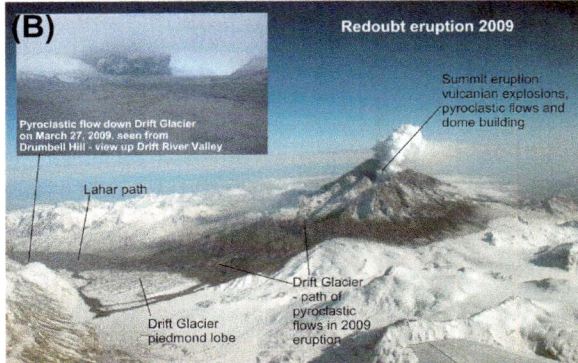

FIGURE 20.11 Field images from recent glaciovolcanic eruptions highlighting characteristic behavior. (A) The 1996 eruption at Gjálp, central Iceland, occurred within relatively thick ice. (B) The 2010 eruption at Eyjafjallajökull, southern Iceland, occurred within relatively thin ice. (C) The 2009 eruption at Redoubt produced a lava dome that periodically collapsed onto surrounding alpine glaciers.

thickness was 750 m prior to the eruption. After day four the rate of ice melting dropped substantially, but remained at 500–1000 m³/s so that by the end of the eruption (13 days after start) about 3 km³ of ice had melted. A further 1 km³ was melted in the 3 months that followed and about 0.5 km³ in the next 5 years (Jarosch et al., 2008). About 80% of the melting happened at the eruption site, with the remaining 20% occurring above the path of meltwater flow into Grímsvötn, within Grímsvötn, and on the first 6 km of the path of the **jökulhlaup** out of

Grímsvötn. The Gjálp eruption produced a 6-km-long, 0.5- to 2-km-wide, and up to 450-m-high hyaloclastite ridge or **tindar** (bulk volume 0.7 km³) of basaltic andesite. A further 0.1 km³ of pyroclasts was transported with the meltwater into the Grímsvötn caldera. Meltwater drainage was subglacial except during the latter part of the eruption, when a 3.5-km-long supraglacial ice canyon transported a part of the meltwater before it cascaded back to the glacier bottom. Most of the Gjálp meltwater was released 3 weeks after the eruption, when the Grímsvötn lake was drained in a massive jökulhlaup with a maximum discharge of ∼50,000 m³/s onto the outwash plains south of the glacier (Table 20.3). Measurements indicate that the permeability of the edifice was reduced significantly in the first year after the eruption, possibly because of consolidation and palagonitization of the edifice.

6.2. Eyjafjallajökull 2010

The Eyjafjallajökull eruptions occurred in Iceland in March–May 2010 (Table 20.3; Figure 20.11(B)). The main, explosive eruption is best known for the disruption it caused to air traffic in Europe and across the Atlantic Ocean. But these eruptions also displayed a large variety of interactions between magma, snow, and ice. Eyjafjallajökull is a 25-km-long, 15-km-wide, and 1600-m-high gently sloping stratovolcano with its upper slopes covered by an 80 km² and 50–200 m thick ice cap. As with Gjálp, no increase in geothermal activity was detected before the eruptions. The first phase was effusive and occurred in an ice-free area on the southeastern flank of the volcano where the seasonal snow pack was 1–5 m thick. The basaltic lava mainly flowed on top of the snow, since the rate of melting by lava–snow contact was much too slow to remove the snow ahead of the advancing lava (Edwards et al., 2012). Gradual melting of the snow followed, taking weeks to months. The main explosive eruption (bulk tephra volume of 0.27 km³) of mostly very-fine-grained trachyandesite tephra started in the shallow, ice-filled summit caldera on 14 April. It melted through the 200-m-thick ice in about 4 h, leading to the development of the plume that carried ash to Europe. In contrast to Gjálp, this eruption occurred through ice cauldrons with vertical walls (Magnússon et al., 2012). Due to limited ice thickness, ice deformation was relatively minor outside the cauldrons. Ice melting was rapid in the first 24 h (500–1000 m³/s), but declined somewhat (∼200 m³/s) in the second day. By day three melting around the vent was largely finished as the ice cauldron had reached a width of about 500 m and tephra within the cauldron largely insulated the surrounding ice. In these first days, meltwater was drained northward down the steep outlet glacier Gígjökull in several swift jökuhlaups with discharge of the order of 1000–10,000 m³/s. The meltwater carried with it large amounts of tephra, filling a proglacial lagoon in the first 24 h. These jökuhlaups drained subglacially for the first 1−1.5 km but then emerged on both sides of the glacier and flowed mainly supraglacially down the less than 100 m thick Gígjökull. On the eighth day of the eruption, lava started to flow subglacially toward the north, progressing very slowly for about a week, while it gradually melted through Gígjökull, developing about a 100-m-thick lava flow in the process. The lava advanced about 2 km in a few days as it progressed through 50- to 100-m-thick ice, forming a ∼ 20-m-thick lava flow by melting the ice above the advancing lava front. The lava stopped flowing early in May, as the eruption intensity increased and became fully explosive again. Magma—ice interaction and melting was minor after the lava stopped flowing. Although tephra fallout amounted to several meters on top of the glacier in the vicinity of the craters and minor base surges occurred, this did not cause surface melting.

6.3. Redoubt 2009

Redoubt is a 3110-m-high, steep stratovolcano in Alaska (Figure 20.11(C)). It has an ice-filled summit crater and glaciers on the slopes, and typifies dozens of glacierized stratovolcanoes in Alaska, Kamchatka, and South America. The 2009 eruption was preceded by 8 months of unrest and escalating thermal activity causing some melting before the outbreak of the eruption (Table 20.3). About 0.1 km^3 dense rock equivalent of andesite was erupted, partly as lava domes. Although subsidence and crevassing occurred on the glacier surface, melting was not confined to ice cauldrons. The lava dome repeatedly collapsed during the eruption producing hot debris avalanches that descended the Drift Glacier and melted large volumes of ice. These events resulted in short-duration lahars with peak discharges near the source of the order of 100,000 m^3/s (over 10 times the peak discharge of the Eyjafjallajökull floods) that flowed down the Drift Valley toward the ocean (Table 20.3). Some lahars were essentially slurries of meltwater and snow and ice from the surface of the glacier. Others were hyperconcentrated flows of meltwater and ample juvenile material from the debris flows mixed with ice and snow. Importantly, in contrast with the Gjálp and Eyjafjallajökull eruptions, the meltwater was generated mainly by hot pyroclasts flowing down the slopes and interacting with the surface of the glacier, including the seasonal snow and firn.

6.4. Summary

The three eruptions outlined above had very different characteristics. To a considerable extent, these differences arise from differences in ice thickness and the presence or absence of water above the vents. In Gjálp, the great thickness of the ice ensured water saturation of the volcanic edifice and the crater throughout the eruption, leading to piling up of the volcanic material over the vents. Thus, the edifice shape and distribution of products were predominantly controlled by the glacier—volcano interaction. At Eyjafjallajökull, where the ice thickness was intermediate to small, the glacier apparently had major effects on the activity at times and modified the morphology and extent of the lava formed during the effusive phase. At other times, the presence of the ice had apparently little impact on the activity. At Redoubt, the products (the lava dome and the explosively formed pyroclasts) were only marginally, if at all, affected by the ice. However, the presence of the ice resulted in large-scale ice melting and hazardous lahars that damaged local infrastructure.

7. FUTURE RESEARCH

We suspect that the interest in glaciovoclanism will continue to grow in the future. The emerging field of large-scale, high-temperature experimentation will allow for better constraints on melting rates, heat transfer mechanisms, and morphologies of ice-confined deposits. New advances in the accuracy of geochronometry for basaltic systems will allow for more accurate testing of hypotheses linking glaciovolcanism to global climate changes recorded in the marine record. Advances in analytical precision and in interpretation of preserved H_2O and CO_2 contents of volcanic glasses will allow more accurate reconstructions of syn-eruption glaciostats/hydrostats, and provide better information about ice sheet geometries. Further research is also needed combining studies of edifice morphology, glacial hydrology, and ice mechanics with passage zone heights and ice sheet thickness in past eruptions. For example, it has been suggested that the stable water levels of englacial lakes as recorded by extensive horizontal passage zones can only be explained by supraglacial drainage of the meltwater, as opposed to subglacial drainage. The latter is fundamentally unstable and should lead to cycles of gradual accumulation and catastrophic drainage. Moreover, field-based and theoretical studies of eruptions within cold-based glaciers are needed. The deposits associated with eruptions within cold-based glaciers have been identified in Antarctica and are expected to be of major importance on Mars. Finally, advances in remote sensing techniques will improve real-time monitoring of future glaciovolcanic eruptions, enabling testing of hypotheses regarding the dominance of specific fragmentation mechanisms and how this dominance might shift during the course of the eruption. Recent identification of volcanic tremor beneath the East Antarctic ice sheet is a reminder that the potential for eruptions through major ice sheets still is very real, and future Antarctic eruptions will be invaluable for our understanding of the effects of Pleistocene glaciovolcanism on former ice sheets.

ACKNOWLEDGMENT

We thank many colleagues and students for fruitful and provocative discussions over the past two decades, especially John Smellie, Ian Skilling, Dave McGarvie, Hugh Tuffen, Sarah Fagents, Thorvaldur Thordarsson, Bjorn Oddsson, John Stevenson, Alex Lloyd, Meagen Pollock, Lucy Porritt, Steinunn Hauksdottir, Melanie Kelman, and Cathie Hickson.

FURTHER READING

Belousov, A., Behncke, B., Belousova, M., 2011. Generation of pyroclastic flows by explosive interaction of lava flows and ice/water-saturated substrate. J. Volcanol. Geotherm. Res. 202, 60—72.

Edwards, B.R., Tuffen, H., Skilling, I.P., Wilson, L., 2009. Introduction to special issue on volcano-ice interactions on Earth and Mars: the state of the science. J. Volcanol. Geotherm. Res. 185, 247—250.

Edwards, B.R., Russell, J.K., 2002. Glacial influence on the morphology and eruption products of Hoodoo Mountain volcano, northwestern British Columbia. In: Smellie, J., Chapman, M. (Eds.), Volcano-Ice Interaction on Earth and Mars, 202. Geol. Soc. London Spec. Pub, pp. 179—194.

Edwards, B.R., Russell, J.K., Simpson, K.A., 2010. Volcanology and petrology of mathews tuya, northwestern british Columbia, Canada: glaciovolcanic constraints on 0.730 Ma cordilleran paleoclimate. Bull. Volcanol. 73 (5), 479—496. http://dx.doi.org/10.1007/s00445-010-0418-z.

Edwards, B., Magnússon, E., Thordarson, T., Gudmundsson, M.T., Höskuldsson, A., Oddsson, B., Haklar, J., 2012. Interactions between snow/firn/ice and lava/tephra during the 2010 Fimmvörðuháls eruption, south-central Iceland. J. Geophys. Res. 117, B04302. http://dx.doi.org/10.1029/2011JB008985.

Furnes, H., Fridleifsson, I.B., Atkins, F.B., 1980. Subglacial volcanics - on the formation of acid hyaloclastites. J. Volcanol. Geotherm. Res. 8, 95—110.

Gudmundsson, M.T., 2003. Melting of ice by magma-ice-water interactions during subglacial eruptions as an indicator of heat transfer in subaqueous eruptions. In: White, J.D.L., Smellie, J.L., Clague, D. (Eds.), Explosive Subaqueous Volcanism, 140. Am. Geophys. Union Geophys. Monograph, pp. 61—72.

Gudmundsson, M.T., Sigmundsson, F., Björnsson, H., Högnadóttir, Þ., 2004. The 1996 eruption at Gjálp, Vatnajökull ice cap, Iceland: efficiency of heat transfer, ice deformation and subglacial water pressure. Bull. Volcanol. 66, 46—65.

Jakobsson, S.P., Gudmundsson, M.T., 2008. Subglacial and intraglacial volcanic formations in Iceland. Jökull 58, 179—197.

Jones, J.G., 1969. Intraglacial volcanoes of the laugarvatn region, southwest Iceland. Quart. J. Geol. Soc. London 124, 197—211.

Licciardi, J.M., Kurz, M.D., Curtice, J.M., 2007. Glacial and volcanic history of icelandic table mountains from cosmic [3]He exposure ages. Quat. Sci. Rev. 26, 1529—1546.

Lescinsky, D.T., Fink, J.H., 2000. Lava and ice interaction at stratovolcanoes: use of characteristic features to determine past glacial extents and future volcanic hazards. J. Geophys. Res. 105, 23711—23726.

Magnússon, E., Gudmundsson, M.T., Sigurdsson, G., Roberts, M., Höskuldsson, F., Oddsson, B., 2012. Ice-volcano interactions during the 2010 eyjafjallajökull eruption, as revealed by airborne radar. J. Geophys. Res. 117, B07405. http://dx.doi.org/10.1029/2012JB009250.

Mathews, W.H., 1947. "Tuyas," flat-topped volcanoes in northern BC. Am. J. Sci. 245, 560—570.

McGarvie, D.W., 2009. Rhyolitic volcano-ice interactions in Iceland. J. Volcanol. Geotherm. Res. 185, pp. 367—389. http://dx.doi.org/10.1016/j.jvolgeores.2009.06.003.

McGarvie, D.W., Stephenson, J.A., Burgess, R., Tuffen, H., Tindal, A.G., 2007. Volcano-ice interactions at prestahnukur, iceland: rhyolite eruption during the last glacial-interglacial transition. Ann. Glaciol. 45, 38—47.

Russell, J.K., Edwards, B.R., Porritt, L.A., 2013. Pyroclastic passages zones in glaciovolcanic sequences. Nat. Comm. 4, pp. 1788—1795. http://dx.doi.org/10.1038/ncomms2829.

Russell, J.K., Edwards, B.R., Porritt, L.A., Ryane, C., 2014. Tuyas: a descriptive genetic classification. Quat. Sci. Rev. 87, 70—81.

Schopka, H., Gudmundsson, M.T., Tuffen, H., 2006. The formation of helgafell, southwest iceland, a monogenetic subglacial hyaloclastite ridge: sedimentology, hydrology and volcano-ice interaction. J. Volcanol. Geotherm. Res. 152, 359—377. http://dx.doi.org/10.1016/j.jvolgeores.2005.11.010.

Smellie, J.L., 2000. Subglacial eruptions. In: Sigurdsson, H. (Ed.), Encyclopedia of Volcanoes. Elsevier, New York, pp. 403—419.

Smellie, J.L., 2006. The relative importance of supraglacial versus subglacial meltwater escape in basaltic subglacial tuya eruptions: an important unresolved conundrum. Earth Sci. Rev. 74, 241—268.

Smellie, J.L., Johnson, J.S., McIntosh, W.C., Esser, R., Gudmundsson, M.T., Hambrey, M.J., Wyk de Vries, B., 2008. Six millions years of glacial history recorded in volcanic lithofacies of the James Ross Island volcanic group, antarctic peninsula. Paleogeog. Paleoclim. Paleoeco. 260, 122—148.

Tuffen, H., 2006. Models of ice melting and edifice growth at the onset of subglacial basaltic eruptions. J. Geophys. Res. 112. http://dx.doi.org/10.1029/2006JB004523.

Tuffen, H., McGarvie, D.W., Gilbert, J.S., Pinkerton, H., 2002. Physical volcanology of a subglacial-to-emergent rhyolitic tuya at rauðufossafjöll, torfajokull, iceland. In: Smellie, J.L., Chapman, M.G. (Eds.), Volcano-Ice Interaction on Earth and Mars, 202. Geol. Soc. London Spec. Pub, pp. 213—236.

Wilson, L., Head, J.W., 2002. Heat transfer and melting in subglacial basaltic volcanic eruptions: implications for volcanic deposit morphology and meltwater volumes. In: Smellie, J.L., Chapman, M.G. (Eds.), Volcano-Ice Interaction on Earth and Mars, 202. Geol. Soc. London Spec. Pub, pp. 5—26.

Chapter 21

Mid-Ocean Ridge Volcanism

S. Adam Soule

Department of Geology and Geophysics, Woods Hole Oceanographic Institution, Woods Hole, MA, USA

Chapter Outline

GLOSSARY

axial magma chamber Also called an axial magma lens, an **axial magma chamber** is a shallow crustal melt body 1–4 km deep where magma is stored at the plate boundary.

axial valley A graben that forms at a mid-ocean ridge axis from faulting related to tensional stresses due to plate separation.

central seamount A volcanic edifice formed by multiple eruptions within the axial valley of an intermediate- or slow-spreading ridge segment.

extrusive layer A layer in the igneous oceanic crust of extrusive basalts that is typically recognized by its low-seismic velocity (high porosity), commonly referred to as seismic layer 2A.

lobate flow A submarine lava flow morphology characterized by large (1–5 m) pads of lava that form sequentially to produce a lava flow. It is most similar to pahoehoe flows in terrestrial basaltic systems.

mid-ocean ridge A divergent plate boundary, typically located in the ocean basins, where ascending mantle melts produce new igneous oceanic crust through magmatic and volcanic accretion.

neovolcanic zone The neovolcanic zone is a band along the plate boundary in which new eruptive fissures are expected to form. It can be described as the region of active volcanism.

overlapping spreading center A location where two adjacent ridge segments are both active.

pillow flow A submarine lava flow morphology characterized by small (∼1 m) bulbs of lava that form sequentially to produce a volcanic mound.

ridge axis The ridge axis is a fictional line that marks the location at which plates diverge.

sheeted dikes A layer in the igneous oceanic crust of solidified, magma-filled cracks that represent repeated pathways for magma from shallow lenses to the seafloor, commonly referred to as seismic layer 2B.

sheet flow A submarine lava flow morphology characterized by smooth, ropey, or jumbled surfaces on low relief, areally expansive lava flows.

spreading rate The full spreading describes the rate at which an observer on one oceanic plate would see a point recede that sits on the other oceanic plate when looking across the spreading center. The half-spreading rate describes the rate at which a point would recede from an observer standing directly on the plate boundary.

1. INTRODUCTION

The global **mid-ocean ridge** (MOR) system is considered the largest magmatic system on Earth. This linear feature extends over 56,000 km through the ocean basins at depths up to 5 km—and above sea level at Iceland—and produces more than two-thirds of the Earth's annual volcanic output (Figure 21.1). This chapter describes the current understanding of MOR magmatic and volcanic processes from magma generation, through ascent and storage, and eruption onto the seafloor. I describe the variations in magmatic and volcanic processes in the context of ridge **spreading rate** and furthermore how these variations are manifested in the generation of igneous oceanic crust.

2. MID-OCEAN RIDGES AND MAGMA GENERATION

At divergent plate boundaries, separation of the tectonic plates is matched by upwelling of the underlying mantle resulting in the partial melting of mantle peridotites and the generation of magma. The major element chemistry of MOR magmas is dominantly basaltic although rare high-silica lavas

The Encyclopedia of Volcanoes. http://dx.doi.org/10.1016/B978-0-12-385938-9.00021-3

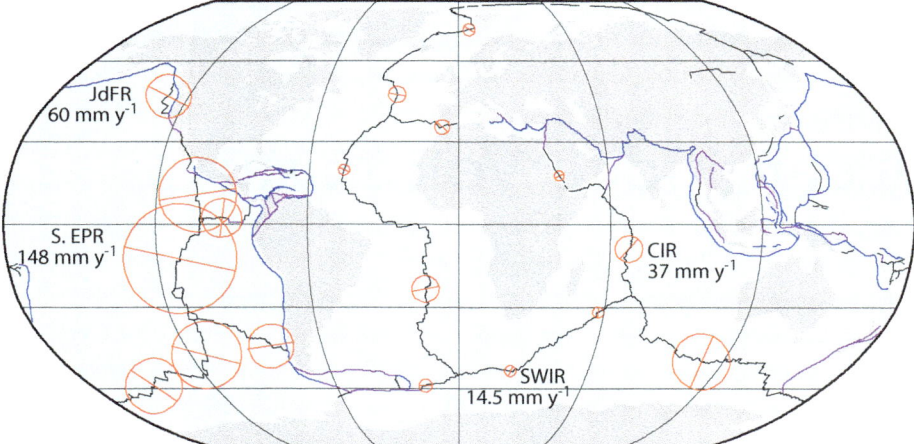

FIGURE 21.1 Global map of plate boundaries with spreading centers shown in black, transforms shown in purple, and convergent margins in blue. Spreading rates are shown by red circles, which are scaled according to the divergence rates calculated from NUVEL01. Spreading direction is shown by the lines bisecting the circles. Full spreading rates for the superfast Southern East Pacific Rise (S. EPR), intermediate Juan de Fuca Ridge (JdFR), slow Central Indian Ridge (CIR), and ultraslow Southwest Indian Ridge (SWIR) are listed.

generated by extensive fractionation and crustal assimilation have been recovered. Within the MOR basaltic (or MORB) system, two main groups are identified as normal (N-) and enriched (E-MORB). E-MORB is distinguished from N-MORB by enrichment in compatible trace elements (Figure 21.2) and is correspondingly identified by increased trace element ratios (e.g., K/Ti, Zr/Y, La/Sm). These geochemical differences are attributed to heterogeneous enrichment in the MORB mantle source. N-MORB is generated through partial melting of a mantle depleted by melt extraction (e.g., during formation of the continental crust

early in Earth's history). E-MORB cannot be generated by partial melting of this depleted mantle and instead requires an enriched source that is commonly attributed to metasomatism of the mantle by fertile melts (e.g., from nearby hot spots) or inclusions of enriched material (e.g., subducted oceanic crust or subcontinental lithosphere).

Buoyant, MOR magmas ascend through the upper mantle and lower crust to pool beneath the **ridge axis**. During ascent, magma may stall in lower- and mid-crustal melt lenses that solidify in place, or may ascend to shallow melt lenses where a balance of replenishment from below and solidification from above may produce stable, long-lived magma bodies. The majority of the magma cools in the crust, producing gabbros. The remaining magma is injected into dikes, some of which penetrate the crust and deposit lava on the seafloor. Collectively these processes produce the oceanic crust that is commonly represented by a layered structure defined in part from the study of ophiolite exposures that from top to bottom includes the follows: extrusive lavas, **sheeted dikes**, gabbros, and mantle rocks. This structure is also evident in vertical variations of seismic velocity that display a progressive increase in velocity with depth due to decreases in permeability and changes in lithology (Figure 21.3).

Dike injection into the upper crust and tectonic deformation along ridge-parallel faults episodically accommodate plate separation, which over time produces steady divergence. Many properties of MORs are correlated with rates of plate divergence including geochemical composition of the magma, frequency of volcanic eruptions, and ridge depth. Perhaps the clearest difference between ridges of different spreading rates is the morphology of their ridge axes (Figure 21.4). At fast-spreading rates, ridges are shallow and display an axial high at the plate boundary. As

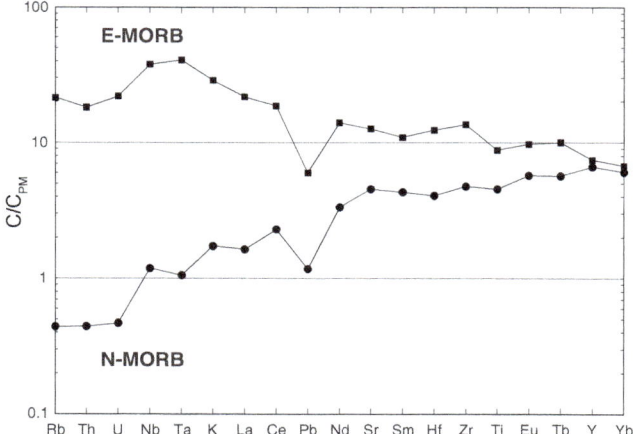

FIGURE 21.2 Trace element concentrations (C) normalized to primitive mantle concentrations (C_{PM}) of a representative normal (N) and enriched (E) mid-ocean ridge basaltic (MORB) from the Mid-Atlantic Ridge. N-MORB is depleted in compatible rare earth elements, which is thought to reflect its origin from a mantle source that has experienced melt extraction in the past. E-MORB is enriched in compatible rare-earth elements suggesting additions from a more primitive source, possibly from the deeper mantle.

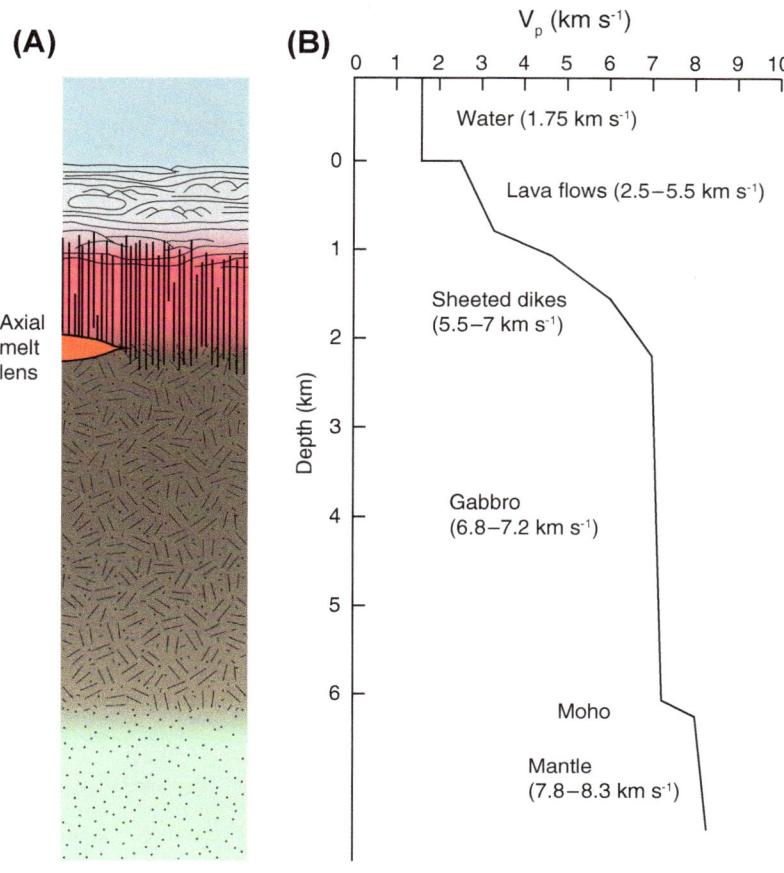

(A)

Axial
melt
lens

(B)

V_p (km s^{-1})

0   1   2   3   4   5   6   7   8   9   10

Depth (km)

Water (1.75 km s^{-1})

Lava flows (2.5–5.5 km s^{-1})

Sheeted dikes
(5.5–7 km s^{-1})

Gabbro
(6.8–7.2 km s^{-1})

Moho

Mantle
(7.8–8.3 km s^{-1})

FIGURE 21.3 (A) Vertical section through idealized oceanic crust comprised by extrusive lava flows, sheeted dikes, gabbro, and mantle rocks. An axial melt lens (white) is shown at the transition from sheeted dikes to gabbro. (B) Approximate P-wave velocities are shown for young oceanic crust (i.e., without overlying sediments). Velocity variations reflect changes in porosity between extrusive and intrusive rocks and with increasing lithostatic pressure. Gradients in velocity between units reflect the diffuse nature of the lithologic boundaries. The relative thickness of crustal layers will vary between ridges depending on spreading rate and may evolve as the crust ages due to burial and compaction, fracturing, and mineralization infilling pore space. Alternatives to this idealized crustal structure have been suggested for slower spreading systems.

spreading rates decrease, ridges deepen and the plate boundary is marked by deep and wide **axial valleys**. These morphologic contrasts reflect differences in the thermal structure of the ridge and the relative proportion of magmatic- and tectonic-accommodated spreading, which are largely driven by the rate of plate separation.

3. SEGMENTATION

MORs are not continuous features, but are segmented at a variety of length scales. Segment boundaries are classified by the magnitude of the offset between successive spreading centers. The largest offsets, where ridge segments are separated by more than 50 km, are accommodated by oceanic transform faults oriented roughly 90° to the strike of the ridge (Figure 21.5). These first-order segments are commonly hundreds of kilometers in length and persist for millions of years. Smaller offsets take the form of **overlapping spreading centers** and ridge offsets that displace the ridge axis by >1 km. These second- and third-order segments can range in length from 50 km to 200 km and nominally define chemically distinct or isolated magmatic systems along the ridge, fitting the classical definition of discrete volcanoes. Finer scale

segmentation that offsets the ridge <1 km persists for much shorter timescales (e.g., one eruption cycle) (Table 21.1).

4. MAGMA STORAGE

Magma generation beneath MORs occurs through decompression melting of upwelling mantle in response to plate separation. The melt-producing region is believed to be on the order of 100 km wide at its base and narrows at shallower depths as mantle flow lines become horizontal (and thus cease melting) producing a triangular melting region. Ultimately, melt accumulates beneath the ridge in shallow crustal magma reservoirs that are elongated parallel to the ridge and generally located beneath the axis of spreading. The processes by which melts focus from a 100 km wide region to narrow melt lenses less than a few kilometers in width are debated. Among the proposed mechanisms are coalescence of ascending melt channels driven by instabilities and positive feedbacks of a reactive melt flowing through a porous media and lateral flow in high-porosity decompaction layers located along the sloped base of the oceanic lithosphere.

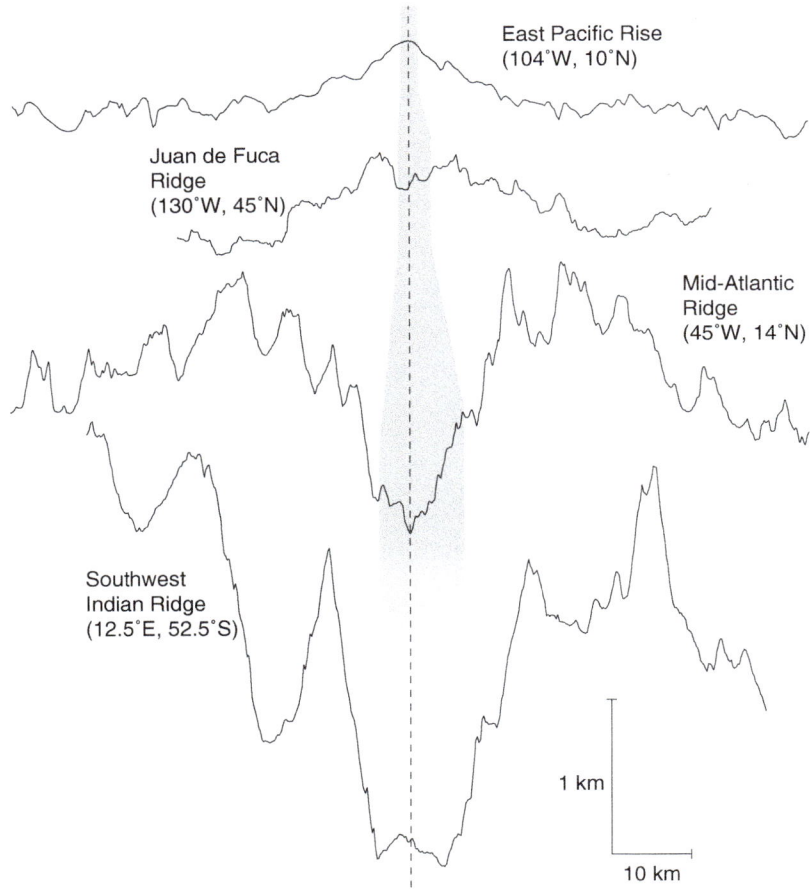

FIGURE 21.4   Bathymetric profiles across fast-, intermediate-, slow-, and ultraslow-spreading mid-ocean ridges. All profiles are shown at the same scale (vertical exaggeration = 20×). The ridge axis is shown by the dashed line and the approximate width of the neovolcanic zone is shown by the gray region. The Southwest Indian Ridge is nominally avolcanic. Note the increasing width and depth of the axial valley with decreasing spreading rate. The ridges shown (and their full spreading rate) are the East Pacific Rise (10.6 cm per year), Juan de Fuca Ridge (6 cm per year), Mid-Atlantic Ridge (1.5 cm per year), and the Southwest Indian (0.14 cm per year).

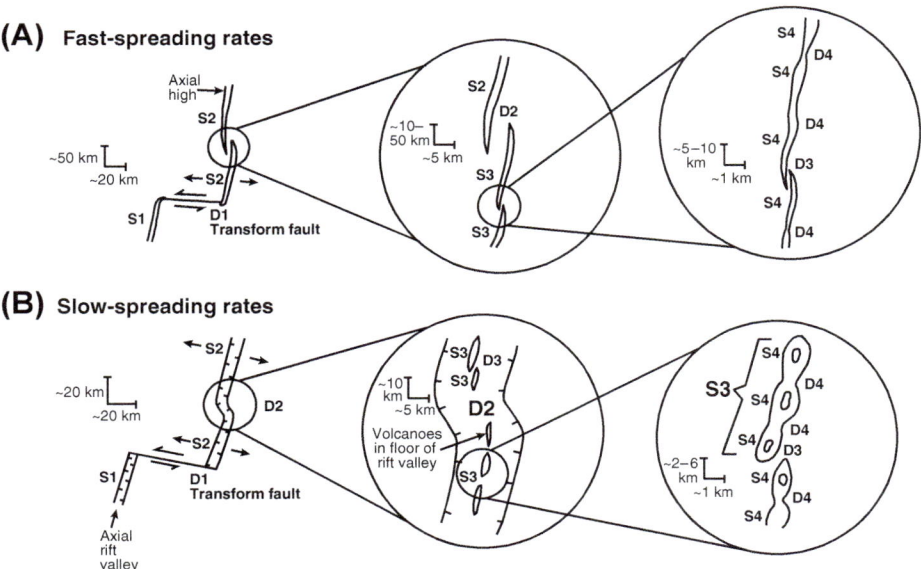

FIGURE 21.5   Mid-ocean ridges are segmented at a variety of length scales. The order of segmentation is classified by the type, offset, and longevity of the ridge discontinuity. In this cartoon, segments (S) and discontinuities (D) and their order (e.g., 1, 2, 3, 4) are indicated. The manifestation of these discontinuities can differ between fast- and slow-spreading ridges. The characteristics of segment boundaries can differ between fast (A) and slow (B) spreading ridges. In particular, a second-order discontinuity at a fast-spreading ridge is manifested as an overlapping spreading center, whereas at a slow-spreading ridge, it is a nontransform bend in the ridge axis. *Figure after Macdonald (1988).*

TABLE 21.1 Characteristics of Segment Boundaries. Where Values Differ for Slow-Spreading Ridges, They Are Shown in Parentheses

Segment Order	First	Second	Third	Fourth
Segment length (km)	600 ± 300 (400 ± 200)	140 ± 90 (50 ± 30)	50 ± 30 (<25)	<20 (<15)
Discontinuity type	Transform fault	Overlapping spreading center (OSC)	Small OSC (intervolcano gaps)	Gaps, changes in ridge axis orientation
Offset length (km)	>30 km	2–30 km	0.5–2 km	<1 km
Segment longevity (ka)	$>5 \times 10^6$	0.5–5×10^6 $(0.5$–$30 \times 10^6)$	10^4–10^5	10^2–10^4

Shallow crustal melt lenses have been seismically imaged along many portions of the global MOR system. In general, melt lens depths at fast-spreading ridges (>10 mm per year) are typically 1–2 km, at intermediate-spreading ridges (5–10 mm per year) depths are 2–3 km, and at slow-spreading rates (<5 mm per year) depths are greater than 3 km. The depths of imaged melt lenses are correlated with spreading rate and correspond with predictions of the thermal gradients within the crust that account for heat input from ascending magma and heat loss by conduction and by advection from hydrothermal fluids (Figure 21.6).

Melt lens depths can vary over short length scales due to variations in magma supply from below and/or heat extraction from above. At fast and intermediate-spreading ridges, melt lenses are thought to persist over 10^5 years. Most slow-spreading ridges although volcanically active,

do not display seismically imaged melt lenses. Only two melt lenses have been seismically imaged at slow-spreading ridge, the Lucky Strike Segment (37°N) and Reykjanes Ridge (57°N) along the Mid-Atlantic Ridge, both of which have enhanced melt supply due to local hot spots. These slow-spreading melt-lenses are laterally constrained along axis on shorter length scales than the observed ridge segmentation. This implies that shallow melt bodies are ephemeral at slow-spreading rates, consistent with cooler thermal structure and decreased melt supply.

Melt lenses that are replenished or recently emplaced have the potential to erupt. Eruptions at MORs are fed by planar, ridge-parallel magma conduits (i.e., dikes) that originate from shallow melt lenses. Observations from exposed sections of upper oceanic crust at ophiolites and in submarine fractures zones indicate typical dike widths of ~1 m, and lateral extents of as much as several tens of kilometers. The repeated injection of dikes, only a fraction of which reach the surface to erupt lava, represents the magmatic accommodation of plate spreading, with faulting accounting for the rest, and results in the development of a semicontinuous layer of dikes within the crust (i.e., sheeted dike complex) (Figure 21.3).

The region of diking is largely constrained to a narrow band along the plate boundary called the **neovolcanic zone** (Figure 21.4). The band varies in width and continuity at fast and slow-spreading ridges. At fast-spreading ridges, the neovolcanic zone is narrow, <500 m, and is fairly continuous along the ridge axis with breaks at larger segmentation boundaries. It is commonly marked by a shallow and narrow graben that forms in response to deformation above dikes that do not reach the surface and is made wider and deeper by tectonic stretching and shallower and narrower by volcanic overprinting. At slow-spreading ridges, the neovolcanic zone is wider, up to 10 km, and less continuous along axis. Volcanism is largely contained within an axial valley, which forms due to the relative increase in tectonic versus magmatic extension that allows

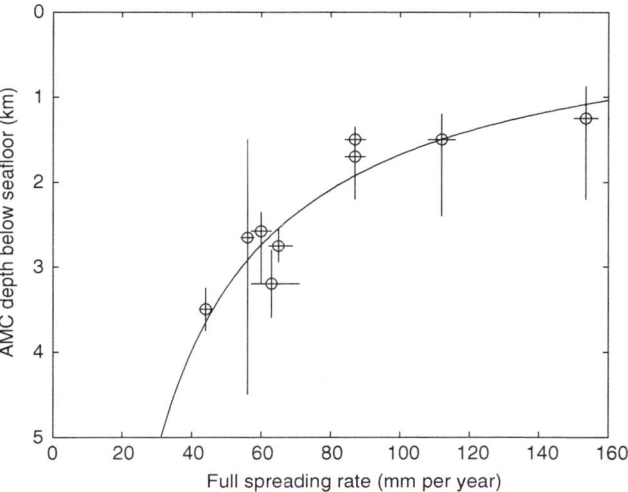

FIGURE 21.6 Seismically imaged axial melt lenses (AMCs) vary in depth with spreading rate. The variation is consistent with the predicted thermal structure beneath the ridge, in particular, the depth of the 700 °C isotherm (approximated by the black line) that is thought to represent the brittle–ductile transition in oceanic crust.

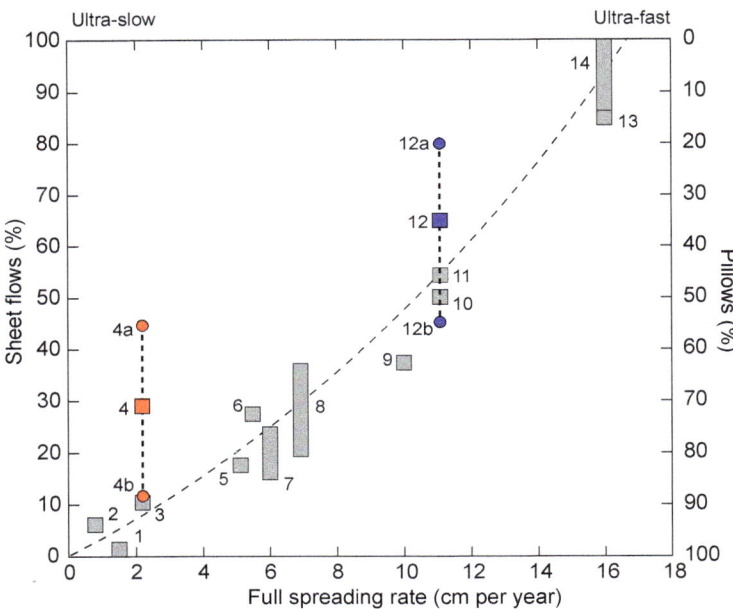

FIGURE 21.7 Compilation of morphology distributions from mapped portions of the global mid-ocean ridge system as a function of spreading rate. The relative proportions of pillows to sheets decrease exponentially with increasing spreading rate. Where **lobate flows** are mapped as a separate morphology, they are divided equally between pillows and sheets. Two sites: Mid-Atlantic Ridge—Lucky Strike (red) and East Pacific Rise—2005–2006 Eruption (blue) are shown in as an average (●) bracketed by minima and maxima (■). For the Mid-Atlantic Ridge (MAR), the bracketing values represent mapping off and on the **central seamount** (and the underlying melt lens). For the East Pacific Rise (EPR), the bracketing values represent mapping at the distal ends and central portion of the flow as shown in Figure 21.6. The MOR sites are indicated by numbers as follows: (1) Red Sea, 18°N; (2) Gakkel Ridge, 86°E; (3) MAR, 36°N; (4) MAR, 37°N (4a, on seamount; 4b off seamount); (5) Galápagos Spreading Center—GSC, 92°W; (6) GSC, 95°W; (7) EPR, 21°N; (8) GSC, 86°W; (9) EPR, 12.5°N; (10) EPR 9°50′N; (11) EPR, 9°–10°N; (12) EPR 2005–2006 eruption (12a, eruption center; 12b eruption ends); (13) EPR, 20°S; (14) EPR 17.5°–21.5°S.

the development of large graben-bounding faults. Intermediate-spreading rate ridges show characteristics of both fast- and slow-spreading neovolcanic zones, depending on the relative magma supply.

Not all eruptions occur within the narrowly defined neovolcanic zone. A number of eruptions much younger than the spreading age of the crust on which they are emplaced have been identified through radiogenic isotope dating. These eruptions have been found at fast- to ultraslow-spreading ridges and are assumed to have been emplaced in situ at distances of 4–20 km from the plate boundary, fed by off-axis melt bodies. The extent of "off-axis" volcanism is not well constrained due to limited exploration beyond the neovolcanic zone of most ridges. However, magma should be available to off-axis regions as it traverses the base of the lithosphere toward the plate boundary, and recent 3D seismic reflection and refraction surveys have identified off-axis melt bodies at both the East Pacific Rise and Juan de Fuca spreading centers.

5. MID-OCEAN RIDGE ERUPTIONS

Dikes that reach the surface produce lava flows with unique morphological characteristics relative to their subaerial counterparts due to efficient quenching by cold seawater.

The resulting lava flows display morphologies characterized as pillow, lobate, and sheet, each representing an increase in local volume flux (i.e., eruption rate), intraflow hydraulic connectivity, and decrease in relief (i.e., flow thickness vs flow length) (cf. Chapter 19). Globally, the relative proportion of sheet versus **pillow flows**, which are commonly thought to reflect high- and low-volumetric effusion rates respectively, is correlated with the spreading rate of ridges. Slower spreading rate ridges produce a greater proportion of pillow flows and faster spreading ridges a greater proportion of **sheet flows** (Figure 21.7). This variation in flow morphology is also present at the scale of individual ridge segments (i.e., at constant spreading rate) where a greater abundance of pillow flows is observed toward the segment ends and a greater abundance of sheets toward the segment center. This can be seen in Figure 21.7 for the Lucky Strike segment of the Mid-Atlantic Ridge at 37°N, where the mean proportion of sheets to pillow flows of ~0.3 is bracketed by observations for on (~0.45) and off (~0.1) the **central seamount**. In this case, the off-seamount value is closer to the globally defined trend and perhaps more representative of slow-spreading systems. The on-seamount value may reflect increased magma supply as it sits above one of the few melt lenses that have been identified on a

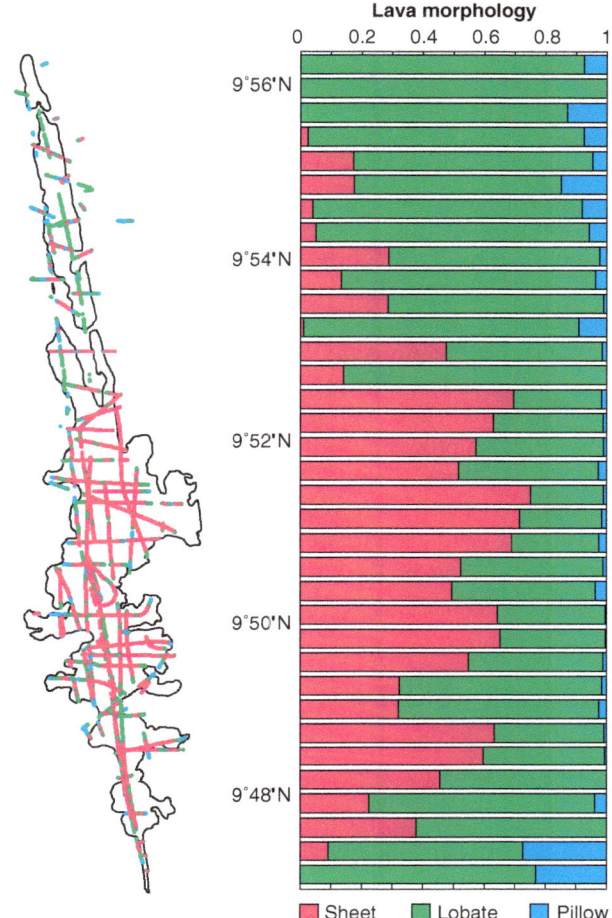

Lava morphology

■ Sheet   ■ Lobate   ■ Pillow

FIGURE 21.8 The 2005—2006 eruption of the East Pacific Rise was mapped in detail by seafloor photography, submersible dives, and acoustic seafloor imaging. Flow morphology was characterized by over 10,000 seafloor photographs along towed camera tracks. The morphology, binned in 500 m intervals along the eruptive fissure, shows a high proportion of sheet flows (red) toward the center of the eruption and an increase in the proportion of lobate (green) and pillow flows (blue) toward the ends of the eruption. *Figure after Fundis et al. (2010).*

slow-spreading ridge. Similar variations are also apparent within individual eruptions. In the 2005—2006 eruption of the East Pacific Rise near 9°50'N (Figure 21.8), pillow morphologies are more abundant at the distal ends of eruptive fissures and sheet flows in the central region of the eruptive fissure. Lower eruption rates at the ends of the fissures are consistent with decreased magma driving pressures that would have accompanied a cessation of fissure opening.

The covariation of spreading rate and flow morphology has been ascribed to variations in lava viscosity and variations in tectonic stress accumulation across ridges of differing spreading rates. The former, lava viscosity variations, appears unlikely as global compilations of MOR basalt chemistry suggest relatively uniform melt

compositions and higher eruption temperatures (thus lower crystallinity) at slower spreading rates than at faster spreading rates. This difference is attributed to the shallower depth and more efficient cooling of melt lenses at faster spreading rates. The latter, tectonic stress variations, implies that stress release resulting from frequent eruptions at faster spreading rates limits the accumulation of tectonic stresses and thus greater magmatic overpressure is required to initiate a dike. This model, however, may not explain variations in flow morphology that occur at the ends of magmatic segments and eruptive fissures as described above. The clear correlation between flow morphology and spreading rate reveals a fundamental underlying physical process common to MOR magmatic systems. However, our capacity to develop or validate a model of controls on morphology and eruption rate is limited by the paucity of direct measurements of magma ascent and eruption rate.

6. VOLCANIC DEPOSITION

The accumulation of lava flows on the seafloor results in the development of a layer of extrusive basalts that is, tens to hundreds of meters thick. The mechanisms and modes of accumulation can vary along and between ridge systems. The frequency of eruptions is presumed to correlate with spreading rate and magma supply. Based on mass balance calculations, rough repose timescales of 10, 1000, and 10,000 years are suggested for fast-, intermediate-, and slow-spreading ridges. Two repeat eruptions (i.e., at the same location) have been documented on the fast-spreading East Pacific Rise at 9°20'N—9°50'N (1991—2005), consistent with the proposed repose interval. The increasing repose interval with decreasing spreading rate may be counterbalanced by increases in eruption volume leading to a relative global uniformity in the extrusive volcanic thickness layer. At present, however, too few MOR lava flows have been documented in sufficient detail to accurately estimate their volume.

At fast-spreading ridges, volcanic deposition is distributed semicontinuously along the ridge axis, although it may be locally diminished at discontinuities in the underlying melt lenses (White et al., 2000). The dikes feeding the eruptions are thought to ascend vertically with little lateral transport in the subsurface. Eruptions emanate from ridge-parallel fissures within the neovolcanic zone and produce lava flows that move across the ridge crest (i.e., normal to the strike of the ridge) to distances of 1—4 km. At fast-spreading ridges, the dominance of lobate and sheet lava flows results in lava flows with large areal extents, but limited thickness (<5 m). The succession of lava flows produces an extrusive volcanic layer that ranges in thickness from tens of meters at the ridge axis to several hundred meters within a few kilometers of the ridge axis. This thickening of the extrusive volcanic pile can be modeled by imposing a narrow zone of intrusion and wide zone of deposition, which is consistent with observations of recent lava deposition at the ridge crest.

At slow-spreading ridges, volcanic deposition is typically focused at the center (along-axis) of spreading segments. The resulting morphology is the development of a central volcano within the axial valley that reaches shallow depths relative to basins at the segment ends. Geophysical measurements of seismic velocity reductions and gravity lows at the segment center are thought to reflect thicker crust and greater accumulations of melt in these regions. Despite the greater depth and thinner crust, volcanism does still occur away from the center of slow-spreading ridge segments. In these settings, it is inferred that melt is transported along the rift axis by lateral diking over tens of kilometers. This behavior is consistent with similar focused magmatic centers along subaerial rifted margins in Iceland and the Afar in Ethiopia. At ultraslow-spreading rates (<10 mm per year), the degree of magmatic and volcanic focusing is even more extreme. In these settings, large volcanic centers are spaced along the ridge, separated by hundreds of kilometers of spreading with minimal volcanic deposition where mantle rocks are tectonically exhumed to the seafloor.

In the absence of central seamounts, volcanic accretion at slow-spreading ridges is dominated by pillow lava flows that are emplaced as ridges comprised of mounds or hummocks. The ridges are typically longer (along-axis) than they are wide reflecting the orientation of fissures controlled by the regional tensional stress field, and can rise hundreds of meters above the axial valley floor. Eruptive fissures are more widely distributed across the axial valley in slow-spreading systems, although volcanic activity will commonly focus in one part of the axial valley at any given time before shifting elsewhere. The relief produced by volcanic ridges and the highly faulted seafloor within the axial valley together produce topographic obstacles that limit the potential width of a lava flow and often increase thickness as flows pond in topographic lows. The narrow zone of volcanic deposition coupled with the wider zone of magmatic intrusions contrasts with fast-spreading ridges and results in extrusive volcanic layer that does not show systematic thickening with increasing distance from the ridge axis.

7. SUMMARY

Volcanism at MORs varies in character with the rate of plate divergence. The same is true for many other ridge properties including axial depth, lava geochemistry, melt lens depth and temporal stability, and the relative proportions of magmatic and tectonic spreading. These relationships are obvious in global-scale compilations, but important exceptions exist. One emblematic example is the Galápagos

Spreading Center between 98°W and 91°W (a distance of ~750 km), which shoals from −4500 m to −1500 m and displays an increasing proportion of sheet flows from west to east despite very minor changes in spreading rate. This change occurs as the ridge approaches the Galápagos hot spot, where the addition of hot spot material increases the magma supply to the ridge. At a smaller scale, the relative abundance of sheet and pillow flows on and off the central seamount of the Lucky Strike segment of the Mid-Atlantic Ridge (Figure 21.7) further exemplifies this process. At Lucky Strike, magmatic focusing in the lower crust and mantle redistributes magma supply such that it is higher at the segment center than at the ends. Exceptions to these trends illustrate how the volcanic properties of MORs directly reflect processes of melt generation, transport, and aggregation in spreading systems. At a global scale, rates of magma production scale with plate divergence rate, thus providing a central framework in which to understand the volcanic behavior of the largest magmatic system on our planet.

FURTHER READING

Ballard, R.D., van Andel, T.H., 1977. Morphology and tectonics of the inner rift valley at lat 36 50′ N on the Mid-Atlantic Ridge. Geol. Soc. Am. Bull. 88, 507−530.

Carbotte, S.M., Marjanović, M., Carton, H., Mutter, J.C., Canales, J.P., Nedimović, M.R., Han, S., Perfit, M.R., 2013. Fine-scale segmentation of the crustal magma reservoir beneath the East Pacific Rise. Nat. Geosci. 6, 866−870.

Chadwick Jr, W.W., Embley, R.W., 1998. Graben formation associated with recent dike intrusions and volcanic eruptions on the mid-ocean ridge. J. Geophys. Res. 103, 9807−9824.

Dick, H., Lin, J., Schouten, H., 2003. An ultraslow-spreading class of ocean ridge. Nature 426, 405−412.

Fundis, A.T., Soule, S.A., Fornari, D.J., Perfit, M.R., 2010. Paving the seafloor: Volcanic emplacement processes during the 2005−2006 eruptions at the fast spreading East Pacific Rise, 9°50′N. Geochem. Geophys. Geosyst. 11, Q08024.

Haymon, R.M., Fornari, D.J., Damm, Von, K.L., Lilley, M.D., Perfit, M.R., Edmond, J.M., Shanks III, W.C., Lutz, R.A., Grebmeier, J.M., Carbotte, S., 1993. Volcanic eruption of the mid-ocean ridge along the East Pacific Rise crest at 9 45−52′ N: direct submersible observations of seafloor phenomena associated with an eruption event in April, 1991. Earth Planet. Sci. Lett. 119, 85−101.

Hooft, E.E., Schouten, H., Detrick, R.S., 1996. Constraining crustal emplacement processes from the variation in seismic layer 2A thickness at the East Pacific Rise. Earth Planet. Sci. Lett. 142, 289−309.

Lin, J., Purdy, G.M., Schouten, H., Sempere, J.C., Zervas, C., 1990. Evidence from gravity-data for focused magmatic accretion along the Mid-Atlantic Ridge. Nature 344, 627−632.

Macdonald, K.C., Fox, P.J., Perram, L.J., Eisen, M.F., Haymon, R.M., Miller, S.P., Carbotte, S.M., Cormier, M.H., Shor, A.N., 1988. A new view of the mid-ocean ridge from the behaviour of ridge-axis discontinuities. Nature 335, 217−225.

Moores, E.M., Vine, F.J., 1971. The Troodos Massif, Cyprus and other ophiolites as oceanic crust: evaluation and implications. Philos. Trans. R. Soc. A: Math. Phys. Eng. Sci. 268, 443−467.

Perfit, M.R., Chadwick, W.W., 1998. Magmatism at mid-ocean ridges: constraints from volcanological and geochemical investigations. In: Buck, W.R., Delaney, P.T., Karson, J.A., Lagabrielle, Y. (Eds.), Faulting and Magmatism at Mid Ocean Ridges. American Geophysical Union, Washington DC, pp. 59−116.

Phipps Morgan, J., Chen, Y.J., 1993. The genesis of oceanic crust: magma injection, hydrothermal circulation, and crustal flow. J. Geophys. Res. 98, 6283−6297.

Plank, T., Langmuir, C.H., 1992. Effects of the melting regime on the composition of the Oceanic-crust. J. Geophys. Res. 97, 19749−19770.

Rubin, K.H., Sinton, J.M., 2007. Inferences on mid-ocean ridge thermal and magmatic structure from MORB compositions. Earth Planet. Sci. Lett. 260, 257−276.

Smith, D.K., Cann, J.R., 1999. Constructing the upper crust of the Mid-Atlantic Ridge: a reinterpretation based on the Puna Ridge, Kilauea Volcano. J. Geophys. Res. 104, 25379−25399.

Soule, S.A., Fornari, D.J., Perfit, M.R., Rubin, K.H., 2007. New insights into mid-ocean ridge volcanic processes from the 2005−2006 eruption of the East Pacific Rise, 9°46′N−9°56′N. Geology 35, 1079−1082.

White, S.M., Macdonald, K.C., Haymon, R.M., 2000. Basaltic lava domes, lava lakes, and volcanic segmentation on the southern East Pacific Rise. J. Geophys. Res. 105, 23519−23536.

Chapter 22

Seamounts and Island Building

Hubert Staudigel

Scripps Institution of Oceanography, UCSD, La Jolla, CA, USA

Anthony A.P. Koppers

College of Earth, Ocean and Atmospheric Sciences, Oregon State University, Corvallis, OR, USA

Chapter Outline

GLOSSARY

Abyssal hill Any topographic feature on the seafloor smaller than 100 m formed by volcanic or tectonic processes defining the roughness of the seafloor. The upper size of an abyssal hill may be as tall as 1000 m, depending on overall seafloor roughness.

Alkali basalt Basalts with relatively high Na and K contents, typically found in intraplate volcanic settings

Atoll A truncated, commonly volcanic island capped by a coral reef composed of closely spaced small coral islands that encircle a shallow lagoon. The complex is surrounded by deep water and atolls are restricted to tropical waters.

Calc-Alcaline Basalt Basalt relatively rich in Na and K as well as Ca, typically found at volcanic arcs above subduction zones

Feeder Dike A near-vertical planar conduit through which magma rises to the earth's surface.

Hyaloclastite A deposit consisting largely of glassy fragments, commonly bedded and associated with the eruption and or quenching of lava under water.

Pillow lava Interconnected, elongated lava tubes formed in a subaqueous environment. Cross-sections of tubes resemble pillows with a convex upper and flat convex lower surface and display radial fractures. Pillow lavas are surrounded by glassy margins.

Sill Intrusive sheet of magma emplaced horizontally or parallel to the bedding of its host rock.

Tholeiitic basalt The most common basaltic rock type on earth, low in Na and K, typically found at mid-ocean ridges but also in the shields of some very large ocean islands.

1. INTRODUCTION

Isolated volcanoes on the seafloor forming seamounts are the most common physiographic features on Earth (Figure 22.1) with hundreds of thousands to millions of them found widely distributed over all ocean basins. Just counting the ones taller than 1 km defines a seafloor area the size of Europe that remains largely unexplored with only a few 100 seamounts visited by scientists so far. Nevertheless, studies of this very small fraction of seamounts gives a good basic understanding of their geology

The Encyclopedia of Volcanoes. http://dx.doi.org/10.1016/B978-0-12-385938-9.00022-5

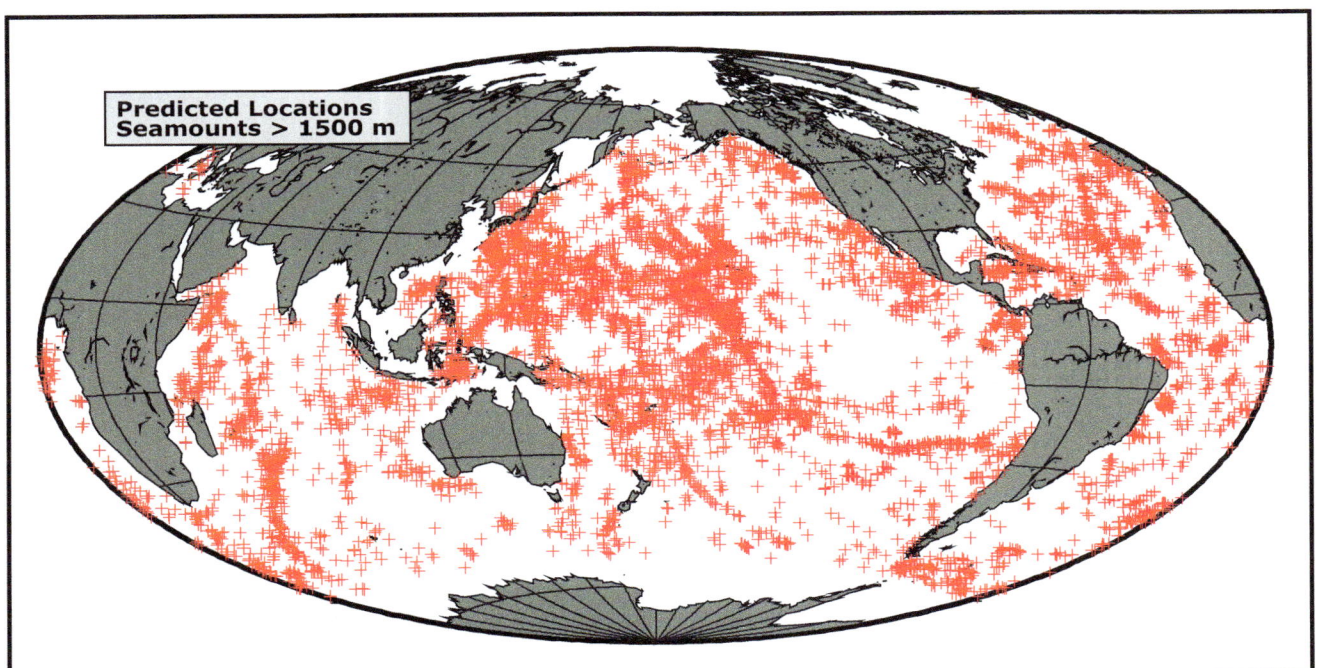

FIGURE 22.1   Global distribution of seamounts taller than 1.5 km based on echosoundings from seagoing expeditions.

and biology and helped explore important related issues, such as the motion of tectonic plates, the composition and heterogeneity of the earth's mantle, earthquakes at subduction zones or the mixing of the oceans. Furthermore, seamounts are of substantial societal interest, in particular due to their rich fishing grounds, many of which have been depleted by overfishing and are now subject to marine conservation measures. Mining Mn-encrustations or massive sulfides at seamounts may soon provide a much-needed supply for a range of key industrial metals.

With very few exceptions, seamounts are volcanic features built by extrusive, intrusive, and volcano-tectonic processes such as crater and caldera formation or sector collapse. Very large seamounts may temporarily emerge as volcanic islands, be subject to surface erosion, provide the foundation for coral reefs and ultimately subside below sea level again. Seamounts can be found in nearly all seafloor tectonic settings, including mid-ocean ridges, subduction zones, and intraplate settings (Figures 22.2, 22.3 and 22.4). They may also be associated with major nonspreading ridges such as the Walvis Ridge or the Ninety-East Ridge. In some occasions seamount rocks and sequences may be found in the on-land geological record, in particular in accreted terrains and in the uplifted core of ocean islands.

In this chapter, we will explore the definitions of the term seamount in the literature, their geology, geophysics, and geochemistry, as well as their relevance in the physical, life, and social sciences.

2. DEFINITION OF THE TERM SEAMOUNT

The term seamount is not consistently defined across all subdisciplines of seamount research. Probably the most important inconsistency relates to their minimum size. The International Hydrographic Organization (IHO) defines a seamount, as "An isolated or comparatively isolated elevation rising 1000 m or more from the sea floor" whereby smaller features are referred to as **abyssal hills** or, more rarely, knolls. This size limitation is at odds with the common practice to using the term seamount for structures with an elevation exceeding 100 m. In practice, the 100 m threshold is useful for seamounts built on smooth abyssal terrain, while 1000 m is more appropriate for seamounts in rough terrain.

In addition to size, there are some other criteria that play a role when defining the term seamount:

- **Shape:** Seamounts may be defined as conical or elongated features with a pointy summit, while many flat-topped features may be referred to as tablemounts or guyots (IHO). Fishermen commonly use terms such as bank, or shoals, in particular for shallow features close to shore.
- **Plate tectonic setting:** Many marine geologists and geochemists originally restricted the term seamount to volcanic features in an intraplate setting, specifically

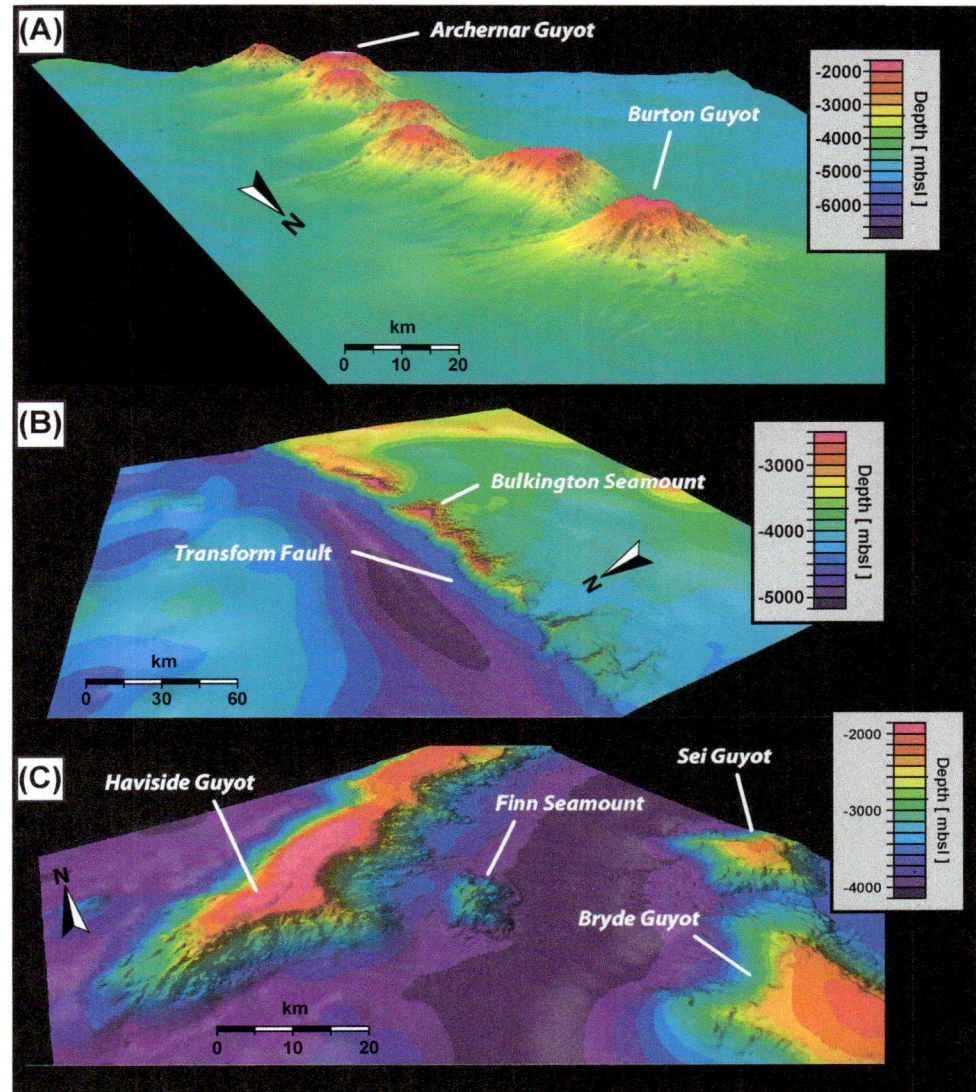

FIGURE 22.2   Three-dimensional views of seamounts. (A) Seven small guyots and seamounts within the Louisville seamount chain in the South Pacific of which two seamounts recently were drilled by the International Ocean Drilling Program (IODP). (B) Seamounts formed along and intersected by a large-scale transform fault system at the Walvis Ridge in the Southeast Atlantic. (C) Group of large guyots and a single small volcano at the Walvis Ridge in the Southeast Atlantic.

excluding features at converging plate boundaries or mid-ocean ridges.

- **Submergence:** Biologists tend to use the term seamount exclusively for features that are in the deep ocean, well below the photic zone to exclude any photosynthetic life.
- **Island inclusion:** Ocean islands are geologically very large seamounts that emerged above sea level, before they subside again below sea level.

The above examples show why particular scientific communities define seamounts differently, but it also shows the importance of disclosing the definition used before the results of one study may be compared with another. For example, earlier seamount counts that restricted the definition of seamounts to intraplate volcanoes resulted in too low an estimate when determining the size of the entire seamount biome.

We adopt here a broad definition of a seamount as any geographically isolated topographic feature on the seafloor taller than 100 m including ones whose summit regions may temporarily emerge above sea level (Staudigel et al., 2010b). In this definition, seamounts may form in any plate tectonic setting, they may currently be buried beneath sediments on the seafloor, or they may now be exposed on land by tectonic and erosion processes.

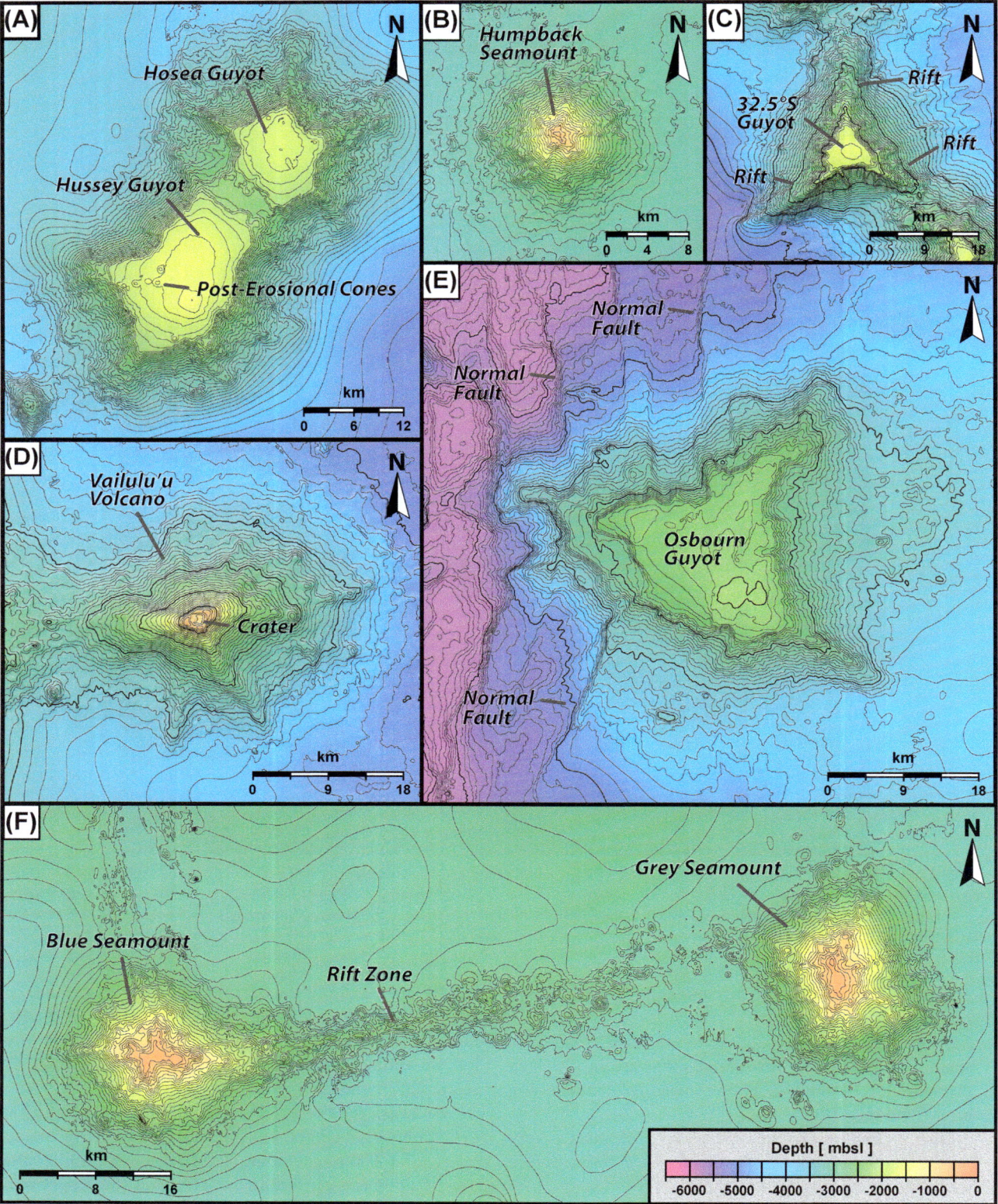

FIGURE 22.3   Different seamount shapes and configurations. (A) Hussey and Hosea guyots forming a coalesced seamount couple at the Walvis Ridge. (B) Conical Humpback seamount at the Walvis Ridge. (C) Starfish-shaped 32.5°S guyot with three well-developed rift zones in the Louisville seamount chain. (D) Volcanically active Vailulu'u volcano with a summit crater at the leading edge of the Samoan seamount chain. (E) 79 million years old Osbourn guyot in the Louisville seamount chain that is subducted into the Kermadac-Tonga Trench causing normal faulting on its western flank. (F) Blue and gray seamounts at the Walvis Ridge that are connected via a 30-km-long volcanic rift zone that runs west to east and starts at blue seamount.

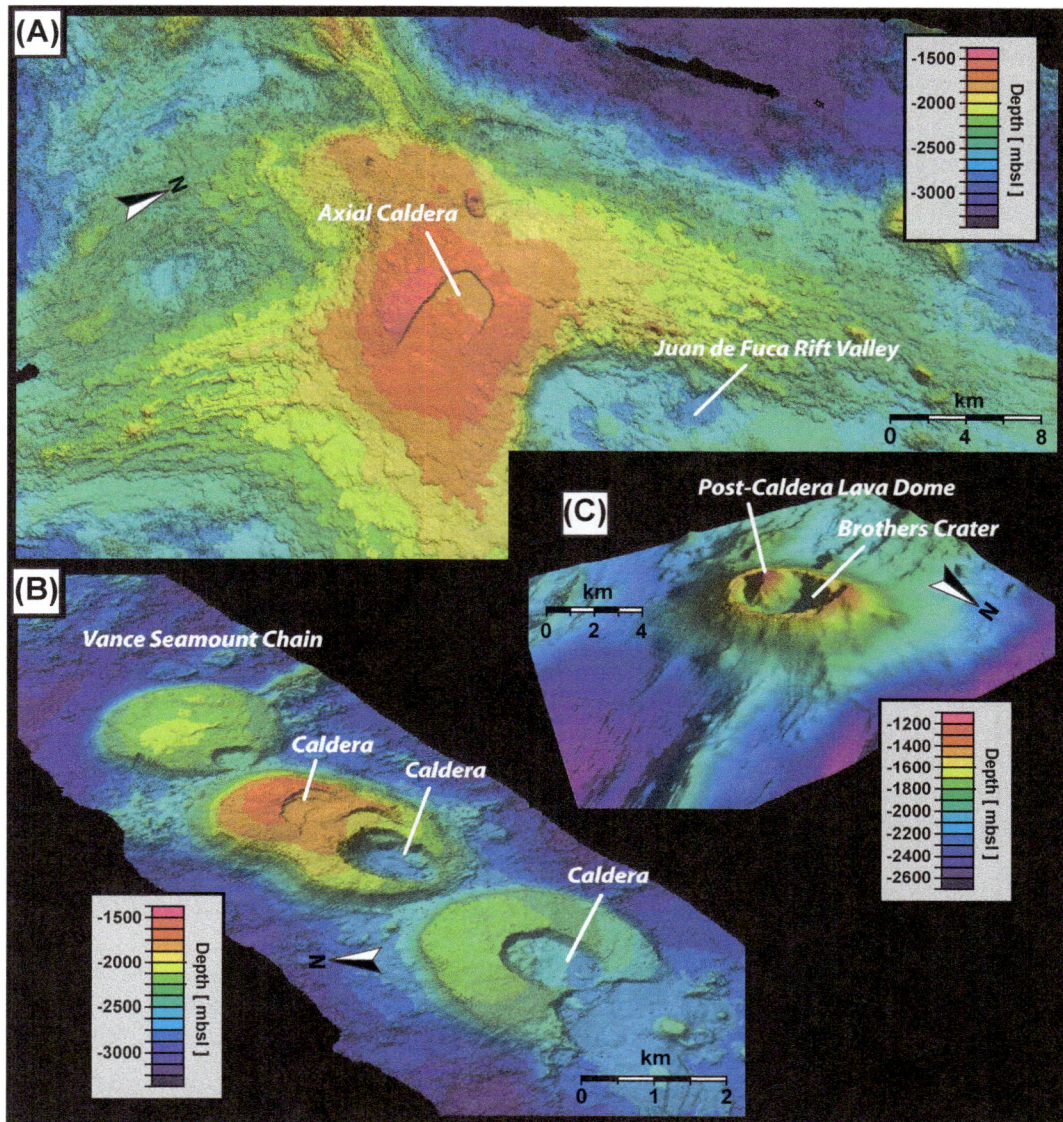

FIGURE 22.4 Caldera and crater formation on active seamount volcanoes. (A) Caldera formation on axial seamount that formed on top of the Juan de Fuca mid-ocean ridge spreading center. (B) Pancake-shaped Vance seamounts northwest of Axial Seamount showing multiple caldera formation stages. (C) Crater of Brothers Seamount forming in the Kermadac island arc showing the formation of a postcaldera lava dome. Credits: Multibeam data by Dr S. Merle at National Oceanic and Atmospheric Administration (NOAA).

3. TYPES OF SEAMOUNTS

3.1. By Origin

The bulk of all seamounts are of volcanic origin formed by igneous processes that are initiated by partial melting of the underlying mantle producing mostly basaltic rocks in a range of tectonic settings. These include tholeiitic and alkalic seamounts at mid-ocean ridges or in intraplate settings, and **calc-alkaline** seamounts from arc settings. A very small fraction of seamounts, however, is not formed by igneous processes. Some of those are made up of tectonically emplaced materials near or at plate boundaries or they

may be shaped by erosion before they submerged. For example, *Eratosthenes* seamount south of Cyprus is a carbonate platform that likely was raised above the surrounding seafloor by tectonic forces.

There are also seamounts that are formed by the expulsion of fluids from the underlying sediments or mantle, including mud, asphalt and serpentinite volcanoes. Most mud volcanoes are formed by water-rich fluids expelled from underlying sediments during compaction and diagenesis. Asphalt volcanoes are rooted in similar processes, but fluids are dominated by hydrocarbons in seafloor petroleum seeps. Serpentinite volcanoes are a particularly interesting type of mud volcano found in forearcs above subduction zones. They form as fluids are released from the downgoing sediments into the shallow overlying mantle. Because these mantle regions are too cold for partial melting, but warm enough for serpentinization, they produce a serpentine-rich mud that rises (through **feeder dikes**) to the surface to form serpentine mud volcanoes. Mud, asphalt, and serpentinite volcanoes tend to be relatively small when compared to other seamounts but they have attracted much attention as they also create very exciting microbial biomes.

3.2. Tectonic Setting

Seamounts can be found in all major tectonic settings on the seafloor, at divergent and convergent plate boundaries, and in the plate interiors. Magma generation processes may be quite different in these settings, whereby seamounts in intraplate settings are typically attributed to the activity of mantle plumes or hot spots, and in some cases to mantle upwelling due to plate extension. Arc-related seamounts such as Brothers Seamount (Figure 22.4(C)) are formed by processes that involve the devolatilization of the downgoing slab in a subduction zone and fluid-flux melting in the overlying mantle wedge. Small seamounts at mid-ocean ridges are likely to be formed by melting processes closely related to the extensional regime of mid-ocean ridges, but larger volcanoes in the same regions can also be formed when a plume or hot spot closely interacts with the mid-ocean ridge. Such plume—ridge interactions are inferred for major volcanic features such as Iceland, Galapagos, or the Walvis Ridge and the Rio Grande Rise. Special tectonic situations may include the intersection of fracture zones with a mid-ocean ridge, such as the Azores at the intersection of the Gibraltar Fracture Zone and the Mid-Atlantic Ridge or Axial Seamount (Figure 22.4(A)) close to the intersection of the Blanco Fracture Zone and the Juan de Fuca Ridge. Seamounts and recently active volcanic islands may also be found on relatively old lithosphere as it bends during subduction in the Japan Trench or on top of older oceanic plateaus such as the Canary Platform.

3.3. Shape and Arrangement

Seamounts may show a range of shapes and arrangements (Figures 22.2, 22.3 and 22.4). They are commonly of conical shape, in particular small to moderate-sized structures. As seamounts grow, they may develop craters or a caldera (Figure 22.4) and often display rift zones, mostly two in opposing directions, giving them an elongate shape, but occasionally three or four, resulting in a star-pattern, defining a radial geometry (Figure 22.3). However, seamounts at hot spots with rather prolific magma supply systems often form on the flanks of a previous seamount in the same seamount trail, buttressing

against or overlapping with the preexisting volcanic edifice (Figure 22.3(A)). This results in complex composite seamounts with multiple eruption centers. The largest seamounts also can show the results of erosion during their emergent phase of development. This in particular includes guyots, which have a well-developed summit platform typically caused by continued wave erosion (Figures 22.2(A) and (C), 22.3(A), (C) and (E)). These flat-topped tablemounts also may show characteristic wave-cut terraces due to changes in sea level or drowned valleys formed subaerially by erosion of rivers.

Slope angles of seamounts tend to be steeper than those of subaerial volcanoes, partly due to the more rapid cooling of submarine lavas, and partly due to the common presence of steep, sub-vertical fault scarps. There is obviously a great variance between volcano shapes and slope angles, because of differences in lava flow characteristics that are a function of diverse chemical compositions and viscosities. For example, the overall slope angles of Pacific seamounts are about $18 \pm 6°$ whereby larger seamounts may have somewhat shallower mean slope angles than intermediate-sized volcanoes.

Seamounts may form conspicuous seamount chains such as the Hawaiian-Emperor, Louisville (Figure 22.2(A)), Ninety-East Ridge, and Walvis Ridge chains, extending over thousands of nautical miles across a significant fraction of an ocean basin. There are however many shorter seamount chains in all of the major ocean basins, but none of them are as voluminous, and most of them tend to be discontinuous. Discontinuous chains might have extended sections missing or have sections that still are to be identified due to their small size or complex interference with other chains. Alternatively, seamounts may be solitary features or form groups or clusters, in near-ridge settings or in intraplate settings, such as the Azores.

4. ABUNDANCE OF SEAMOUNTS

The bulk of the ocean floor remains unmapped by modern multibeam acoustic mapping, preventing a direct global seamount count based on reliable bathymetric maps. Instead, single-track echo soundings may be used, but they also do not cover all of the ocean basins and require extrapolations for global seamount numbers. By contrast, satellite altimetry coverage is complete for all oceans, but it is problematic in detecting small seamounts at great water depth, providing reliable data only for seamounts taller than 2 km.

Single-track echo soundings and satellite-based data may be used to estimate global seamount abundances through extrapolations in a graph of seamount height versus their cumulative global count (Figure 22.5). These two methods yield two independent estimates with lower estimates for single-track bathymetric profiles (white triangles,

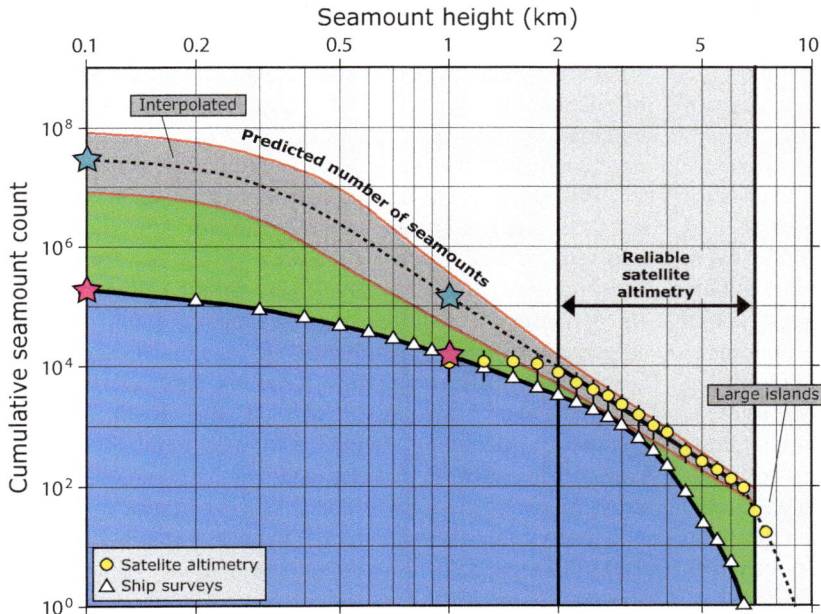

FIGURE 22.5 Global seamount counts versus seamount size as derived from satellite altimetry and single-beam ship track data. *Adapted from Wessel et al. (2010).*

red stars) than for satellite-based estimates (yellow dots, blue stars). The latter are considered more reliable for seamounts 2–7 km in height as they do not require an extrapolation for global coverage. The linear correlation of satellite-based data may be used for an extrapolation of the abundance of seamounts with a minimum height of 1 km, yielding an average abundance of approximately 125,000, with a probable range of 45,000–350,000 (in gray). Even the low end of this satellite-based estimate is substantially higher than the 15,000 estimate from single-track data. Estimates of seamount numbers >100 m are even more uncertain. They require a flattening of the satellite-based curve due to the finite size of the ocean basins with a likely range of 8–80 millions while single-track data suggest only a total amount of about 200,000. All of these estimates for seamounts >100 m are highly uncertain, but all estimates are very high, orders of magnitude larger than the seamounts that have been visited to date.

5. GEOCHEMISTRY

5.1. Introduction

Seamount geochemical studies have revealed much about the specific geochemical evolution of particular volcanoes and have provided key insights into broader issues of mantle melting and the geochemical evolution of the silicate earth. Much work has been focused on intraplate oceanic volcanoes, because they are considered to be a reliable window into the Earth's mantle, without the crustal contamination processes that may occur at volcanoes in an arc or continental crust setting. Seamounts in arc settings are relevant as they reflect the processes that ultimately lead to the growth of the continents by accretion.

5.2. Major and Trace Element Composition of Intraplate Volcanoes

While seamounts near mid-ocean ridges and some very large hot spot volcanoes have volcanic shields that are predominantly constructed of **tholeiitic basalts**, the shields of the majority of seamounts and intraplate oceanic volcanoes are made of mildly alkalic to alkalic lavas. Nearly all well-studied intraplate volcanoes appear to progress through a characteristic geochemical evolution that is commonly divided into four stages, patterned after the "standard model" developed for Hawaiian volcanoes:

1. Initiation of the volcano as a small seamount that is built by relatively small volumes of diverse lavas ranging from tholeiitic to alkalic;
2. The shield building phase, consisting entirely of tholeiitic or slightly alkalic basalts forming the bulk of the volcano;
3. The alkalic cap, a small volume of alkalic lavas capping the shield structure; and
4. Highly alkalic, rejuvenated volcanism that may occur after an extended period of volcanic quiescence and erosion, millions of years after the shield has been completed.

The first stage is represented by Loihi, a small submarine volcano located east of the Big Island of Hawaii. A good example of the last stage is Diamond Head on Ohau Island and these late stage eruptions may contain some highly alkalic lava, such as basanites and nephelinites.

This sequence can be interpreted in terms of a series of distinct mantle melting processes. The beginning includes diffuse and less intense mantle melting with an immature magma delivery system that delivers small batches of chemically distinct magmas to a juvenile seamount such as Loihi. This stage transitions to a massive melting event with large volumes of magma produced from a relatively large portion of the mantle beneath. Naturally, such a melting process establishes a more mature and voluminous plumbing system that will allow for fractional crystallization as well as magma mixing and homogenization as melts rise from the mantle into and through the plumbing system. The shield phase ends with some late stage "capping" melts that are formed by smaller degrees of partial melting and higher degrees of fractionation in magma chambers and delivered in relatively small volumes to the volcanoes' surface. The latest, rejuvenated eruptions are caused by extremely small volumes of magma and very small degrees of melting and are produced and delivered after the shield building magmatic plumbing system has entirely frozen out and the volcano has moved away from the hot spot.

This "standard model" is mostly derived from data on very large intraplate volcanoes such as Hawaii, the Canaries or Samoa, but it also applies to many other ocean islands and moderately sized seamounts such as Jasper seamount. Key similarities amongst all of these oceanic intraplate volcanoes include the development of distinct volcanic stages with systematic increases in alkalinity from their early voluminous shields to late stages of less voluminous melts often separated by periods of volcanic quiescence lasting up to several millions of years. However, truly tholeiitic shields are rare for most ocean intraplate volcanoes and Loihi Seamount (Hawaii) remains the singular example for the first stage of intraplate geochemical evolution.

5.3. Mantle Source Regions

The geochemistry of intraplate seamounts and ocean islands has critically shaped our understanding of how the silicate Earth evolved into a heterogeneous mantle and continental crust. Radiogenic isotope studies (Rb–Sr, Sm–Nd, U/Th–Pb) have shown that the oceanic mantle is depleted in elements that are enriched in continental crust, witnessing a long-term fractionation of the mantle–crust system into a depleted mantle and enriched continental crust. However, the same research also demonstrated that the oceanic mantle is not uniformly depleted. It evidently also contains components that are enriched in elements that are normally found only in sediments or in the continental crust or altered ocean crust. The presence of these mantle components

is considered proof for the return of some sediment or continent-derived materials into the oceanic mantle by subduction. These findings effectively created the discipline of chemical geodynamics, the study of how plate tectonics and mantle convection cause the silicate earth to fractionate chemically into the present-day mantle–crust system.

This global set of isotope data from seamounts not only has been critical to our understanding of how the Earth evolved as a large scale geochemical system but also offers an opportunity to chemically fingerprint the source regions of particular tectonic settings or mantle plumes. This fingerprinting allows distinction of seamounts formed in association with subduction zones as well as different intraplate hot spots. Especially in the South Pacific, intraplate seamounts show extreme enrichments in components such as recycled sediment, lower continental crust or recycled oceanic crust. Many of these hot spots appear to retain these distinct signatures for possibly over 100 million years helping them to be uniquely identified in the maze of interfering hot spot chains in the Western Pacific Seamount Province and the Equatorial Pacific.

6. GEOCHRONOLOGY

6.1. Introduction

Seamount geochronology has been crucial to our understanding of absolute plate motion and hot spot motion. In addition, it taught us much about the life cycle of oceanic intraplate volcanoes and the dynamics of their mantle magma generation processes. However, the geochronology of seamounts and the interpretation of seamount age data have been plagued by a number of problems. The most vexing of them relates to the fact that old seamount samples in particular are notoriously altered from prolonged exposure to seawater. This makes it difficult, often even impossible, to determine their eruption age with confidence. A second problem relates to the fact that seamounts, in particular large ones, can remain volcanically active for long periods of time, well past the time they arrived at and moved over a hot spot. Hence, the ages of lavas from late stage, rejuvenated volcanism will not yield a reliable age for the time when the bulk of a seamount was formed above the hot spot. It is thus important to understand the growth cycle of each dated seamount before the along-chain age progressions may be used for absolute plate motion models.

6.2. Obtaining Reliable Ages for Seamount Samples

Alteration with seawater or superheated hydrothermal fluids profoundly alters the chemical and isotopic composition of seamount rocks. This issue severely affects all isotopic systems that are currently used for seamount geochronology. This is particularly true for rocks that have relatively low concentrations of parent isotopes, such as K and Rb, as those are typical for tholeiitic to moderately alkalic basalts from the shield building stage. More alkalic lavas have substantially more K and Rb and therefore are intrinsically more suitable to dating using K/Ar and Rb/Sr techniques, but they would tend to reflect later stages of volcanism. In addition, a rock has to remain closed with respect to the abundance of both the parent and daughter isotopes, so the radiogenic ingrowth since crystallization and eruption can be reliably determined. Unfortunately, submarine alteration influences elements such as K, Rb, and Sr in particular. It also can limit the retention of the gaseous ^{40}Ar daughter isotope in the rocks and minerals.

While these issues severely impact our capabilities to determine reliable ages from seamounts, there are a number of techniques that allow determination of reliable ages in a large number of cases. These techniques include analytical approaches to work around alteration issues as well as sample selection techniques to pick the most useful samples for dating. Groundmass $^{40}Ar/^{39}Ar$ dating is probably the single most-used technique that is capable of overcoming alteration-related problems in seamount geochronology. Key to the success of this technique is the fact that groundmass tends to concentrate a relatively large fraction of K during the cooling and crystallization process, yielding the largest radiogenic ingrowth of daughter isotopes within a rock. In addition, well-crystallized groundmass has a dense network of microcrystals with very little interstitial glass minimizing the potential for seawater alteration related K-uptake. After a thorough acid-cleaning to remove any superficial alteration in veins or vesicles and along grain boundaries, groundmasses often allow for a coherent release of radiogenic ^{40}Ar during $^{40}Ar/^{39}Ar$ isotopic analysis. Selection of samples for groundmass dating therefore has to be carefully centered on the appropriate grain size and crystallinity of the groundmass and the ability to clean these samples from clays and other alteration phases. While groundmass dating has been very successful, it is also important to build in redundancies by analyzing multiple groundmass samples and, when possible, also separates of mineral phases, such as plagioclase and hornblende, to explore the concordance between multiple dates from the same sample.

6.3. The Volcanic History of Seamounts

Most of what we know about seamount evolution comes from geochronological data from volcanic islands. The Hawaiian Islands again serve here as the "standard model" but in this case for the temporal evolution of oceanic intraplate volcanoes. Our understanding of (recent) absolute plate motion from satellite geodesy and the presence of particular evolutionary stages on the Island of Hawaii allows us to calibrate the initiation and duration of volcanism as it passes over the hot spot. This time period is bracketed by the concurrent volcanic activity at Loihi, Kilauea and Mauna Loa that suggests that these three volcanoes are located above a 100 km diameter hot spot. Taking a 10 cm/year absolute plate motion and the distance between Loihi for the initiation of shield volcanism and Mauna Loa for its termination, provides constraints on its overall duration of about one million years. Mauna Kea offers another constraint, in this case for the **alkali basalt** capping stage that occurs within a few 100 thousands years of the shield building stage. Rejuvenated volcanism on both West Maui and Koolau suggests that posterosional volcanism may occur sporadically over millions of years, but typically after a period of several millions of years in volcanic quiescence and erosion.

While Hawaii remains the most intensely dated hot spot, some other island chains are also well documented. From studies on Samoa and the Canary Islands it appears that these volcanic islands experience a rapid onset of volcanism with a massive shield that is less alkalic in composition, followed by smaller volumes of more alkalic late stage volcanism for relatively long periods of time, well over 10 million years. Furthermore, on these islands late stage volcanism may comprise a much larger relative fraction of a given volcano when compared to Hawaii. For example, the original shield-building phase of Savai'i (Samoa) could be recovered only on the deepest submarine flanks of the volcano while its entire subaerial top is blanketed by late stage volcanic products. Jasper seamount offshore of Baja California is the smallest ocean intraplate volcano with a well-documented Hawaiian sequence. It has three distinct volcanic phases: the shield building "Flank Transitional Series," the "Flank Alkalic Series," and the "Summit Alkalic Series" that together span an evolution of 7 million years, again also separated by periods of volcanic inactivity.

6.4. Plate Motion and Hot Spot Motion

Not surprisingly the Hawaiian-Emperor volcanic lineament also has been the standard-bearer in our understanding of how age progressive seamount and volcanic island chains may be used to constrain absolute plate motion (Figure 22.6(A)). The age progression along the Hawaii-Emperor seamount trail validates the hot spot concept and is corroborated by GPS measurements of Pacific plate motion. However, there is considerable discussion about the interpretation of age progressions along the Emperor seamounts. Paleomagnetic evidence suggests that the Emperor hot spot showed relatively rapid hot spot motion towards the South between roughly 80 and 50 million years

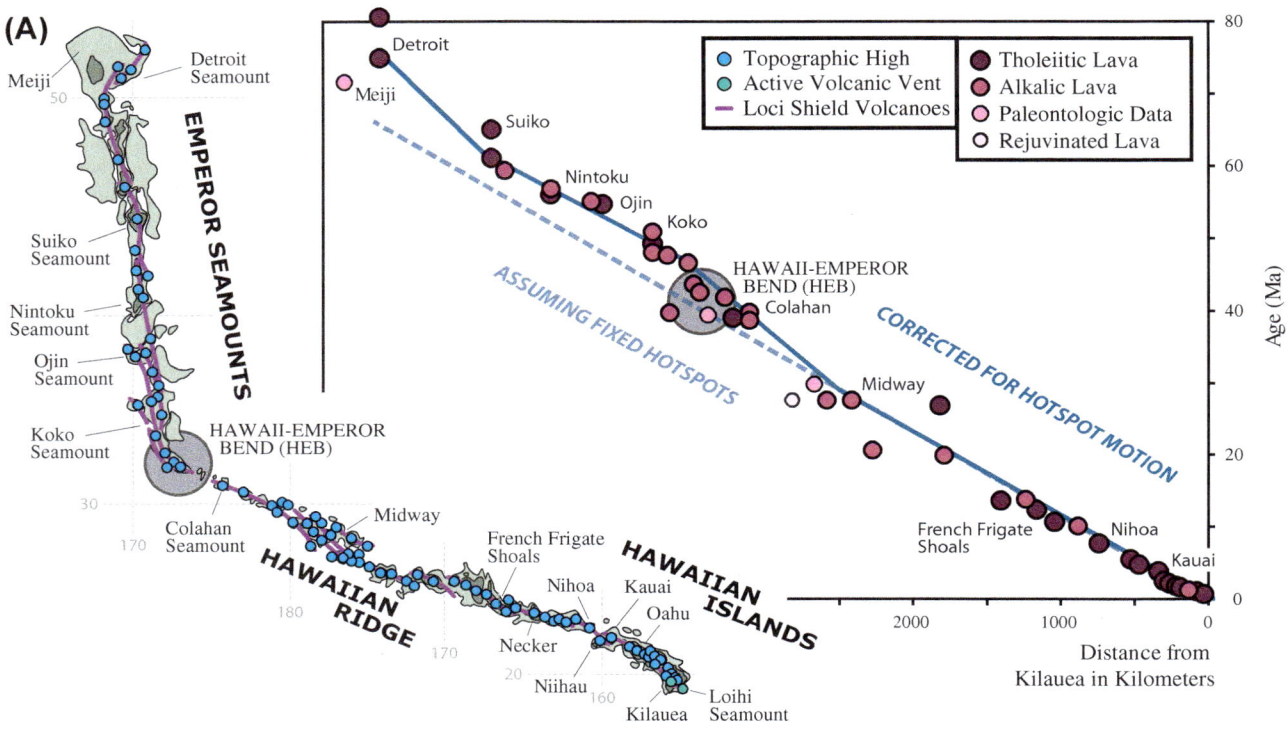

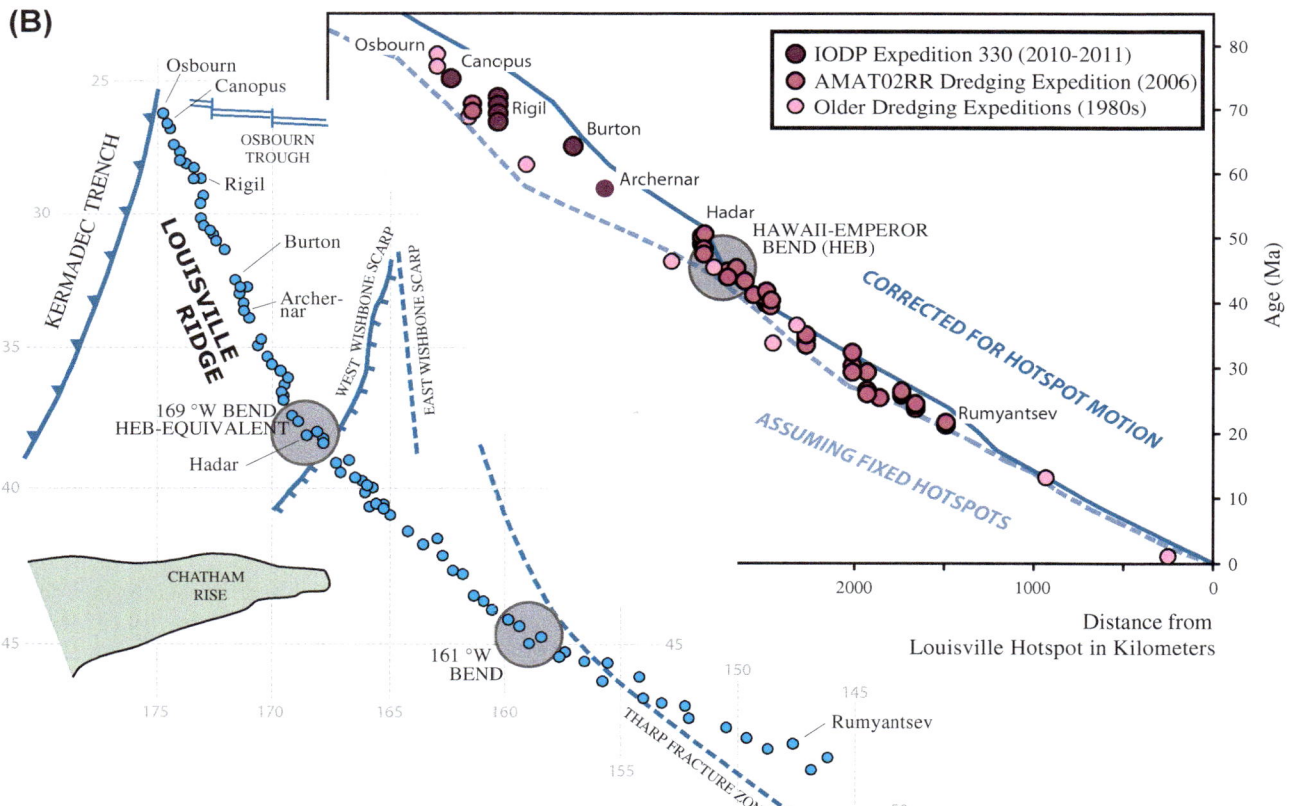

FIGURE 22.6 Age-distance relationships and locations of dated samples for the Hawaiian-Emperor (A) and Louisville seamount chains (B). Hawaiian-Emperor seamount chain samples are plotted in different colors for shield, post-shield and post-erosional stages. Particular evolutionary stages cannot be assigned to the alkali basaltic Louisville seamount samples so colors indicate specific sample expeditions.

ago. While the motion of hot spots can be viewed as a necessary product of a convecting mantle, it invalidates the assumption of hot spot "fixity" and makes the interpretation of plate motion nonunique. This problem can be eliminated by studying multiple age progressive seamount chains on the same tectonic plate, adding for example the Louisville (Figure 22.6(B)) and Rurutu seamount chains in plate motion modeling efforts. Adding to the complexity are recent results from scientific ocean drilling of the Louisville seamounts, which provided evidence for each mantle plume having a different motion history, with the southern-hemisphere counterpart of the Hawaiian hot spot in fact not moving as much. It therefore is more likely that plume motions are controlled by the regional presence of subduction zones and upwelling "Super-Plumes."

Another problem arises when a tectonic plate is passing over multiple hot spots throughout its geological history. Clearly this will result in a complex morphology of interfering and crossing seamount chains, which only can be untangled using modern $^{40}Ar/^{39}Ar$ dating techniques and using Sr, Nd, Pb, and Hf isotope fingerprinting to detect their diverse mantle sources. The Rurutu, Rarotonga, Macdonald, Samoa, and Tuvalu seamount chains in the Southwest Pacific are an example where multiple seamount chains are overprinting each other in what has been called the "hotspot highway" of intraplate volcanism.

7. STRUCTURAL ROCK TYPES AT SEAMOUNTS

7.1. Introduction

Structural rock types at seamounts may be observed on the seafloor, in drill cores, and in some cases in land exposures, with each category of observations having its own biases, weaknesses and strengths. Surface observations on volcanically extinct or dormant seamounts obviously are limited to the outermost layer consisting of lava flows or volcaniclastics belonging to the latest eruptions. It is rare to catch a particular rock type as it is formed, with a few exceptions such as the video documentations of **pillow lava** formation and Strombolian eruptions at seamounts (e.g., NW Rota; http://www.youtube.com/watch?v=QwTr0dw2FbQ). Drill cores give a time sequence through a continuous eruptive-intrusive sequence, but they are obviously limited to a one-dimensional view and they currently do not reach deeper than ~ 500 m into the igneous basement of a seamount. The most detailed information on seamount lithologies, however, comes from subaerial exposure, in particular in the >2 km thick seamount sequence on the Island of La Palma (Canary Islands) that is made up of pillow lavas, volcaniclastics and intrusives. Other land exposures, especially those of seamounts in accreted terrains, are more abundant but also

more fragmented due to severe deformation occurring during emplacement.

7.2. Pillow Lavas

Pillows are the most common lava flow type at seamounts as well as in other submarine settings (Figure 22.7(A) and (B)). They are of tubular shape, often with circular or oval cross-sections. They commonly have a distinct top-to-bottom asymmetry, whereby the bottoms of these lava tubes tend to be molded around the preexisting topography into "V-shapes" while the tops are round and often flattened as pillows grow larger. When pillows are formed they occasionally bifurcate, expanding at their fronts as they flow downslope. Pillow lavas are well known for their thick quenched glassy margins, their radial cooling cracks and their concentric vesicle banding.

Pillow lavas at seamounts tend to have smaller diameters than pillow lavas at mid-ocean ridges, and they show a characteristic variation in tube diameters. Pillows at the base of a pillow sequence are larger than the ones on the top, and a decrease in pillow size also can be observed from the center of an eruption to the periphery of a flow. Near an eruption center, large pillow lavas may buttress each other and possibly combine to form massive, relatively coarsely crystallized flows with very thin quenched margins. Pillow lavas have very gently sloping flow surfaces when they are extruding at relatively high mass eruption rates, but they may also form "pillow-walls" along-strike fissures with a relatively small supply of magma. Often large pillow flows terminate in steep fault scarps that are formed by the syn-eruptive collapse of their flow fronts.

While massive lava flows are common in mid-ocean ridge settings, they are rare or nonexistent at seamounts. In particular near eruption sites, pillows may grow to be several meters in thickness, but they appear to not form the tabular lava bodies observed at mid-ocean ridge settings, where such flows are supported by very high lava eruption rates and along-ridge-axis depressions where they can pond and cool coherently.

7.3. Volcaniclastics

Seamount volcaniclastic rocks include a range of rock types formed by explosive outgassing, by thermal tension-induced fragmentation, and by the collapse of lava flows or entire sectors of a seamount. Volcaniclastics are particularly common in the summit regions of shoaling seamounts and in volcaniclastic aprons surrounding them. We describe here the most common types of volcaniclastics at seamounts, which include fine-grained **hyaloclastites**, autochthonous pillow breccias and allochthonous pillow-fragment breccias.

FIGURE 22.7 Images of different structural rock types from Vailulu'u seamount (A) and the seamount series of La Palma (Canary Islands, B—H): (A) Pillow on the surface of Vailulu'u seamount lavas at a steep slope where collapse truncated some of the pillow lava tubes. (B) Pillow lavas three dimensional exposure. (C) Pillow lavas with interpillow hyaloclastite. (D) Well bedded hyaloclastite. (E) Pillow fragment breccia. (F) Layered pillow fragment breccia with interbedded dike. (G) Amoeboidal pillow breccia. (H) Sheeted sills underlying and interbedded with the layered seamount series of La Palma.

Fine-grained hyaloclastites are mostly made of sand-sized glass fragments that may be formed in submarine lava fountains or by spallation of volcanic glass during the eruption and cooling of pillow lavas (Figure 22.7(C) and (D)). They may fill pillow interstices or define continuous units that are relatively well sorted and bedded or cross-bedded on scales of centimeters; they also may form bedded units up to several meters in thickness that are deposited on the seafloor by enhanced bottom currents around seamounts and possibly by thermal convection in the bottom water during eruptions. The presence of rare pillow fragments may indicate that some fine-grained hyaloclastites may be the distant products of collapsed pillow flows.

Autochthonous pillow breccias can form deposits of tens to hundreds of meters in thickness (Figure 22.7(G)). They typically are made up of coarse blocks of highly vesicular glassy material, intermingled with amoeboidal-shaped pillow lava stringers, which appear to have intruded into the poorly sorted glassy matrix. Their fragile nature suggests that they were deposited in situ because any transport would have destroyed their delicate lava stringers.

Allochthonous pillow-fragment breccias (Figure 22.7(E) and (F)) may be well-bedded and consist of individual layers of volcanic rock fragments often in the form of "pie-slice" pillow cross-sections that are recognizable by their telltale curved glassy margins and that formed by breakage along the radial cracks of pillows. Pillow fragment breccias can form individual layers from 10 to 50 cm that may be reversely graded to thick (>10 m) units that overall are normally graded.

Possibly the most widespread and volumetrically the most important materials are volcaniclastics deposited in the aprons of seamounts. Explosive eruptions, collapse of steep flow fronts or entire sections of the seamount flanks, all contribute to these volcaniclastic sediments. Volcaniclastic aprons of seamounts may be more than a kilometer thick, in particular if they are deposited in the deep moat formed by the volcano itself while it is loading and bending down the underlying oceanic lithosphere. Seamount volcaniclastics may also be deposited in distal environments as fine grained sediments, but nearby deposits may contain individual boulders or blocks that can be extremely large. One such block, Tuscaloosa seamount offshore Koolau, Oahu (Hawaii), is several kilometers in size and likely resulted from a mass wasting event that collapsed a large portion the north-western side of this volcanic island into the ocean. Volcaniclastic apron deposits are functionally turbidites, and their internal structure reveals their rapid turbulent downslope transport.

7.4. Intrusives

Dikes and **sills** are the most important intrusive rock types in the upper sections of seamounts while plutonic rocks play a larger role in their deep structure. The most conspicuous dike intrusions include subvertical feeder dikes (i.e., conduits feeding lava eruptions at the surface) that tend to form radial arrays in conical seamounts or which may cluster along the rift zones of more elongated seamounts. Continued intrusion of dikes in rift zones can cause "volcanic spreading" due to the added volume as was observed in the slowly expanding eastern rift zone of Kilauea Volcano. Such rift zones can include a significant fraction of a volcano and hence accommodate massive, sometimes more distal, magma delivery systems. For example, Kilauea's East Rift and its submarine Puna Ridge extend about 100 km from its central caldera. At greater depth inside a seamount, however, horizontal sill intrusions may comprise a significant fraction of the seamount volume. In the case of the La Palma Seamount Series, sills have a total thickness of 1.8 km. Horizontal intrusion of those sills implies an equivalent uplift of the overlying rock units, functionally vertically inflating the seamount by 1.8 km (Figure 22.7(H)). The deeper portions of seamounts accommodate the bulk of their magma plumbing system with magma reservoirs that mediate the ascent of magma from the mantle to the volcanic surface. Upon cooling, these magma reservoirs are likely to form plutonic rocks, often including ultramafic cumulates and felsic fractionated rocks, as also demonstrated by the chemical composition of lavas found in the extrusive sections of seamounts.

8. HYDROTHERMAL PROCESSES

Seamount hydrothermal systems show interesting and unique features that make them exciting research targets with significant impacts on global geochemical fluxes and the microbial ecology of the deep biosphere.

As a seamount grows from an abyssal hill to a massive undersea or subaerial volcano, it develops a hydrothermal system that involves upwelling of heated water in the summit region and influx of cold seawater from its flanks. Such a hydrothermal system is recorded in the La Palma Seamount Series where geothermal gradients in the summit region are high at 200–300 °C/km while these are substantially lower in the flanks. Active hydrothermal systems have been explored, in particular at Loihi in Hawaii and Vailulu'u in Samoa, and a range of seamounts in the Marianas and Tonga volcanic arcs. Deposits from seamount hydrothermal systems include a range of rare sulfidic black smokers (as observed at Axial Seamount) and Fe-hydroxide deposits that typically are expressed in fragile chimney-type structures or more massive deposits. Such volcanogenic massive sulfide deposits may form in association with calderas in seamounts at convergent margins, for example at Brothers Seamount.

During the growth of a seamount, the confining pressures of the hydrothermal systems systematically change, potentially causing the boiling of hydrothermal fluids in shallow systems. It may also impact the physical state of volatile phases such as CO_2 that is venting as buoyant liquid droplets at Northwest Eifuku and Vailulu'u seamount at water depths of 1600 m and 1000 m, respectively.

Seamount hydrothermal systems may remain active for a very long time, well past their period of magmatic activity, albeit at much lower geothermal gradients. This long duration relates to the fact that bare-rock exposure at seamounts offers opportunities for the underlying oceanic crust to vent and recharge fluids, because they are not sealed off by pelagic sediment deposition. This makes seamount "hydrothermal siphons" that are driven by very small thermal gradients in the oceanic crust. This long duration of fluid flow is likely responsible for the highly altered nature of seamount rocks found in seamount drill cores (e.g., DSDP Hole 417A and ODP Site 1205).

The exact duration of seamount mediated exchange of fluids between the oceanic crust and seawater is not known. It may last for extremely long periods of time, possibly throughout their lifetime, as long as there are nearby seamounts with bare rock exposure that allow for circulation of seawater. A study of Henry Seamount has shown evidence for fluid flow, even >120 Ma after its birth (e.g. Henry Seamount). This extremely long duration of hydrothermal exchange may have important implications for geochemical mass balances as well as the microbial ecology of the deep biosphere in the oceanic crust.

Throughout their geological history, seamount rocks offer substrates for Mn-oxy-hydroxide deposition, certainly during their period of hydrothermal activity, but also after they have turned dormant and purely hydrogenous deposition continues. This deposition is potentially providing a major nucleation site for elements that are relatively insoluble in seawater, such as REE, Th, Co, Tb, and Te. Therefore, Mn crusts may provide commercially relevant deposits of Co, Tb, and other elements that are important for the high-tech industry.

9. GROWTH AND COLLAPSE OF SEAMOUNTS

The life cycle of seamounts may be divided into six stages (Figure 22.8) whereby their growth may end at the first construction stage or cycle through all six of them. Their life cycle typically ends by subduction, but only rarely they may be preserved in the geological record by accretion during plate convergence.

Stage 1. Small seamounts (100−1000 m in height) are structurally and geochemically distinct, and by far the most abundant type of seamount. This makes them an important class of seafloor features and a potentially relevant geochemical reservoir in seafloor weathering budgets. Small seamounts are made up mostly of pillows, less abundant massive flows, and approximately 20% volcaniclastics. The abundance of volcaniclastics is higher than in other ocean floor assemblages and consistent for the range of small seamounts, including the 150 m tall seamount drilled at DSDP Site 417A and the >660 m thick deep water portion of the La Palma Seamount Series. Intrusive processes play only a minor role in small seamounts unless they are integrated into the mid-ocean ridge volcanism such as at Axial Seamount (Figure 22.4(A)). Rock samples from seamount drill cores display extreme enrichments in seawater-derived chemical components, indicating substantial chemical fluxes between seawater and the seamount strata, which potentially support key habitats for lithosphere-based microbial communities.

Stage 2. Mid-size seamounts (>1000 m in size) are tall enough to begin establishing an internal magma plumbing system inside the volcano itself and above the oceanic crust. Mid-sized seamounts have proportions of eruptive rock types similar to small seamounts, with only a moderate amount of explosive/pyroclastic activity. Mid-to large sized seamounts begin to develop their own characteristic internal distribution of stress, largely imposed by their intrusive geometries. This results in the formation of rift systems and development of summit craters and calderas from syn-eruptive magma chamber collapse. Intrusive systems and their volcano-tectonic activity control the location and formation of hydrothermal systems (e.g., within the crater at Vailulu'u Seamount) and the inflation/deflation of a seamount with potential sector collapse.

Stage 3. The transition to **shallow seamounts** is mostly characterized by an increase in the production of submarine volcaniclastic rocks. The depth of transition from the deep water dominance of effusive pillow lavas to shallow water explosive/clastic rock types is likely not to occur at a single depth for all volcanoes. Explosive volcanic outgassing may occur at depths well below 2000 m, but the effects of hydromagmatic processes become dominant at much shallower water depths, probably closer to 700 m. At the La Palma Seamount Series, the transition from effusive volcanism to volcaniclastics is found at approximately 1000 m water depth. Here the abundance of volcaniclastics increases to well over 60% of the exposed section. Shallow seamount eruptions, in particular, those in island arcs settings above subduction zones, may be of sufficient magnitude to involve caldera collapse (e.g., Figure 22.4(C)) and produce sizeable pumice rafts

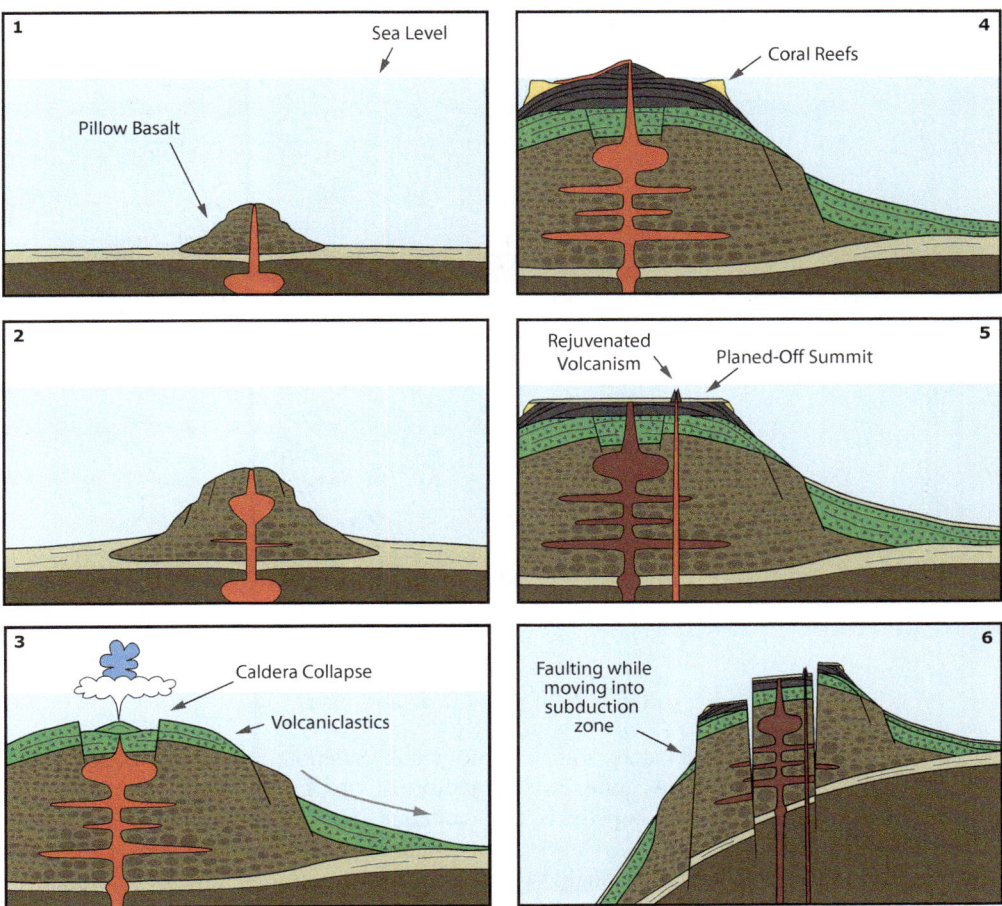

FIGURE 22.8 Cycle of the birth, rise, caldera-formation, emergence, subsidence, and the collapse of a seamount, ending with fragmentation of the seamounts in subduction zones. *Modified from Staudigel and Clague (2010).*

that may drift substantial distances away from their source.

Stage 4. Islands and reefs are both formed when the summits of very large seamounts emerge above sea level and shift into subaerial volcanic activity (Figure 22.8). Emergence is a major step in the evolution of a seamount that typically is successful only under special circumstances. In fact, most newly born volcanic islands known today disappeared shortly after their emergence (e.g., Metis Shoal in Tonga, Myojinsho/Izu in Ogosoawara, Graham Island in the Mediterranean Sea) as the largely volcaniclastic tops of shallow water seamounts are vulnerable to erosion by wave action. A newly born volcanic island persists only if its emergence is associated with the outpouring of voluminous lava flows that outpaces wave erosion over prolonged periods of time.

While islands are less likely to produce clastic eruptions than shoaling seamounts, they do produce substantial quantities of volcaniclastics as well. Major Plinian eruptions are common in arc volcanoes in particular, often dispersing ash over large distances. Lava flows entering

the sea are also known for the production of volcaniclastics. Explosive activity at near-sea-level islands and seamounts has led to some of the largest human tragedies in history, including the catastrophic eruptions and tsunamis associated with the 5200 BCE Santorini cataclysm that likely caused the collapse of the Minoan culture, and the 1883 Krakatoa eruption and tsunami that destroyed many coastal communities all around the Indian Ocean. Sector collapse at islands and very large seamounts/guyots have led to some of the largest landslides observable on the seafloor, mobilizing up to 3000 km^3 producing a debris flow covering 10,000–15,000 km^2. Such events leave prominent scars on the side of the volcano, and almost always results in a significant tsunami that may have local run-ups of up to 400 m above sea level, constituting a major ocean basin-wide natural hazard.

Coral reefs, developing on (or around) tropical islands, display very characteristic landforms (Figure 22.9) as these islands subside together with the cooling lithosphere on which they were formed and the growth of the coral reefs need to compensate for the subsidence. Usually islands are

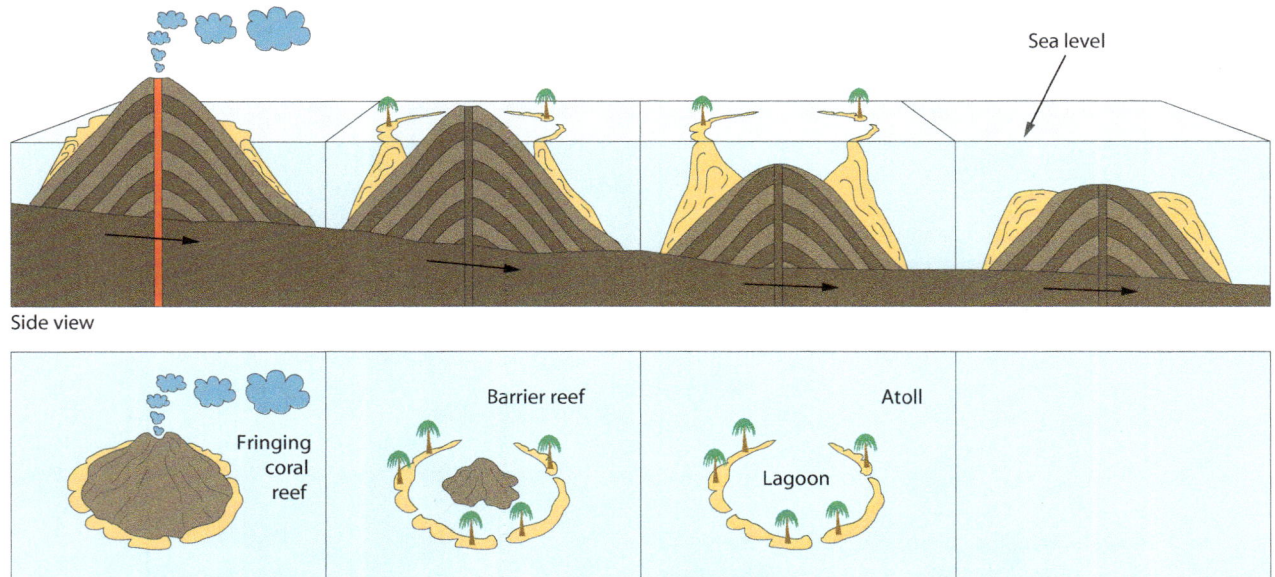

Side view

Top view

FIGURE 22.9   Morphological evolution of ocean islands rimmed by a fringing reef to a barrier reef, atoll and guyot as it subsides due to lithosphere cooling.

first surrounded by a fringing reef, that develops into a barrier reef, and ultimately into an **atoll**. A coral reef eventually drowns, as coral reef growth is outcompeted by the subsidence rate of the island.

As soon as the volcanic islands are formed they begin to weather and erode, producing characteristic landforms that are unlikely to be formed in a submarine setting. Such landforms include erosional canyons, valleys or wave-cut platforms or terraces. They are associated with indicative "terrestrial" sedimentary deposits and soil formation, such as the beach cobble conglomerates recorded in scientific drilling cores from the Louisville seamounts and "red" soil horizons encountered in cores from the Emperor seamounts. When mapped and explored in their current settings, these landforms may be used to reconstruct the subaerial history of a seamount. In addition, even submerged and apparently dormant seamounts, such as Vlinder Seamount may experience some late stage volcanic activity well after drowning, as evidenced by a small volcanic cone on top of an otherwise entirely flat summit of a volcano (see stage five in Figure 22.7).

Stage 5. Volcanically **extinct seamounts** do not grow any further from intrusive/extrusive processes, they don't have an internal heat supply that drives hydrothermal convection, and they are not likely to collapse any further. They may be covered with coral reefs that may keep the seamount summit emerged for prolonged periods despite lithosphere subsidence (Figure 22.9). Once completely drowned, seamounts do not erode any more, making some of the older seamounts, for example, the 145 million years

old Look Seamount in the Western Pacific the oldest landforms on Earth. The shapes of some of these "mountains in the sea" have not changed for over 100 million years. The main activities occurring at these seamounts include potential hydrothermal exchange with the underlying oceanic crust, microbial activity, and the slow accumulation of Mn crusts and pelagic sediments on their surface rocks.

Stage 6. Subducting seamounts: The transport of seamounts to subduction zones typically ends with their descent into the trench. Steep subduction possibly involves the fragmentation of a seamount, while subduction into a shallower sloping accretionary wedge typically causes them to "plow" into the sediment while deforming the smooth and soft front of this wedge. Some seamounts can accrete onto a continent, as has happened along the northern California coastal range. These processes are likely to result in significant seismic activity, in particular when a subducted seamount offers a point of resistance to subduction causing earthquakes with magnitudes reaching more than eight on the Richter Scale.

10. IMPORTANCE OF SEAMOUNTS

Seamounts are the most abundant type of volcano and least explored physiographic feature on Earth. Much less than one percent of them have been studied in any detail, yet these limited studies have nevertheless contributed substantially to our understanding of seamounts and a number of key broader and societally critical issues. As active volcanoes, seamounts have major societal impacts, ranging from navigational obstacles to explosive volcanism, tsunamogenic landslides, and seamount flank collapses. Seamount subduction is associated with some of the largest earthquakes. Seamounts are also recognized for being a crucial factor in ocean mixing by acting as stirring rods that help to mix the highly stratified oceans. In life sciences, seamounts are used as microbial observatories for Fe-oxidizing microbes and their benthic communities are studied for processes such as the dispersal, isolation and evolution of benthic organisms. Over the last decades seamounts provided important fishing grounds, providing a valuable source of protein for a rapidly growing world population. Yet, overfishing has led to habitat destruction and justifies global conservation efforts aimed at sustainable harvesting of fish. In the future, seamounts may offer a much needed resource for high-tech metals, such as Te, which are crucial in the semiconductor industry, the production of photovoltaics and telecommunication devices. Most seamounts with economic significance are located outside territorial waters, making them subject to international agreements within the Law of the Sea.

Above all, seamounts are largely of volcanic origin and volcanologists will continue to play a crucial role in their study and understanding these broad societally relevant but scientifically complex issues at the interface between lithosphere, hydrosphere, and biosphere.

FURTHER READING

Embley, R.W., DeRonde, C.E.J., Merle, S.G., Davey, B., Caratori, F., 2012. Detailed morphology and structure of an active submarine arc Caldera: Brothers volcano, Kermadec arc. Econ. Geol. 107, 1557–1570.

Pitcher, T.J., Morato, T., Hart, P.J.B., Clark, M.R., Haggan, N., Santos, R.S. (Eds.), 2007. Seamounts: Ecology, Fisheries & Conservation. Blackwell Publishing, Oxford, p. 527.

Staudigel, H., Clague, D.A., 2010. The geological history of deep sea volcanoes. Oceanography 23, 58.

Staudigel, H., Koppers, A.A.P., Lavelle, J.W., Pitcher, T.J., Shank, T.M., 2010a. Mountains in the Sea. A Special Issue of Oceanography 23 (1), 16–213.

Staudigel, H., Koppers, A.A.P., Lavelle, J.W., Pitcher, T.J., Shank, T.M., 2010b. Defining the word seamount. Oceanography 23 (1), 20.

Wessel, P., Sandwell, D.T., Kim, S.-S., et al., 2010. The global seamount census. Oceanography 23, 24.

Internet Resources

Seamount Biogeoscience Network (http://earthref.org/SBN/) a multidisciplinary network for seamount sciences.

Seamount Catalogue (http://earthref.org/SC/) for seamount maps and bathymetry data.

Chapter 23

Basaltic Volcanic Fields

Greg A. Valentine

Department of Geology, University at Buffalo, Buffalo, NY, USA

Charles B. Connor

School of Geosciences, University of South Florida, Tampa, FL, USA

Chapter Outline

GLOSSARY

agglomerate rampart Elongate accumulation of bombs and lapilli associated with an eruptive fissure.

maar-diatreme A volcanic crater cut below the previous ground surface, with proximal ejecta forming a low tephra (ejecta) ring; the crater is underlain by a funnel-shaped body filled with volcaniclastic material, country rock breccia, and hypabyssal intrusions, which extends downward in a funnellike shape.

magmatic eruptions Effusive and explosive activity driven primarily by exsolution and expansion of volatiles that are derived from the juvenile magma.

phreatomagmatic eruptions Explosive activity dominated by violent interaction between hot magma and surface water or groundwater.

scoria cone A conical accumulation of lapilli and bombs, typically with a summit crater or a crater that opens toward one side.

spatial vent density The number of vents per unit area in a given region on the surface.

temporal recurrence rate The number of volcanic eruptions per unit time, within some defined volcanic system (such as a volcanic field).

tuff cone Conical accumulation of ash, lapilli, blocks, and bombs, typically poorly sorted, and with a summit crater.

volcanic field One or more volcanoes within an area defined by elevated spatial vent density, and typically comprising a single structural/tectonic setting.

1. INTRODUCTION

Most subaerial volcanoes on Earth and on other terrestrial planets have the following characteristics: (1) basaltic in composition (note that we use the term basalt in a broad sense to include a range of compositions generally with less than about 52 wt% SiO_2); (2) monogenetic in their eruptive history, meaning that each volcano has a single eruptive episode before becoming extinct (eruptive episodes may have durations from weeks to decades); and (3) occur in **volcanic fields**, where one or more volcanoes within a defined age range occur within an area that defines the extent of the field and that is separate from other volcanic areas (the exact application of these definitions is subjective and varies somewhat from case to case). Most of these volcanoes are small in total erupted volume (typically <1 km^3). They include a variety of landforms and there is commonly a small proportion of landforms such as lava domes that are related to more evolved magma compositions, among the predominant basaltic volcanoes. Basaltic volcanic fields occur in subduction, rift, and intraplate tectonic settings. They can also be associated with major calderas, stratovolcanoes, and shield volcanoes, forming on the flanks or in the areas immediately surrounding these longer lived systems. While individual volcanoes within the

The Encyclopedia of Volcanoes. http://dx.doi.org/10.1016/B978-0-12-385938-9.00023-7

fields have geologically short life spans, the fields themselves can be active for several million years. Basaltic volcanic fields provide important insights into the nature of basaltic magmatism and its interaction with tectonism, into eruptive styles, and are an important source of volcanic hazards that involves many challenges because of the complex spatial-temporal controls on volcano location and timing.

2. GENERAL CHARACTERISTICS OF BASALTIC VOLCANIC FIELDS

The number of volcanoes in basaltic volcanic fields can range from one to more than a thousand (Table 23.1; Figure 23.1). Volumes of individual volcanoes, including pyroclastic products and lavas, tend to fall in the range of $\sim 0.1-5$ km^3. Monogenetic basaltic landforms (Figure 23.2) are directly related to eruption processes and erupted volumes and the rates at which the volumes are erupted. An important controlling factor of eruption processes is whether or not ascending magma interacts explosively with groundwater or surface water (phreatomagmatic styles), or if its ascent and eruption is driven primarily by its own internal volatiles (magmatic styles). The most abundant landforms are **scoria cones**. Agglomerate (or spatter) ramparts are similar to scoria cones but build up along elongate vents. Both of these, along with their attendant lava fields, are produced by eruptions that are dominated by effusive magmatic styles with a component of explosive activity. Larger volume **magmatic eruptions** dominated by lava effusion, rather than by explosive activity, can form small shield volcanoes. Maars, tuff rings, and **tuff cones** that are formed by **phreatomagmatic eruptions** typically comprise $\sim 10\%$ of the volcanoes in a given volcanic field, although in some cases they can exceed >50% (Table 23.1; Németh, 2010; Brown and Valentine, 2013). Many monogenetic volcanoes have phases during their lifetimes where different magmatic and phreatomagmatic styles dominate and where the eruptive vents migrate or change shape (e.g., from elongate fissures to central vents). The resulting landform can be complex as a result of the interplay of these factors.

Historical monogenetic activity has been quite limited. Examples in the twentieth century are limited to Parícutin (scoria cone, tephra deposit, and lava field; 1943–1952), Eldefell (scoria cone, tephra, and lava field on island of Heimey, Iceland, 1973), Tolbachik (four scoria cones formed in Kamchatka, Russia, 1975–1976), Navidad (scoria cone and lava field in Chile, 1990), Nilahue (maar in Chile, 1955), Taal (maar in Philippines, 1965), and Unkinrek (two small maars in Alaska, 1977) and submarine eruptions. Until another of these volcanoes erupts, most of what we can learn about them must come from analyzing the landforms and deposits of these and older eruptions.

This chapter uses the term "basalt" very loosely, and in detail the rocks in basaltic fields can include alkaline compositions (basanite, tephrite, trachybasalt) to tholeiitic and calc-alkaline, and can range to the mafic side of intermediate compositions (such as basaltic andesite or andesite; Table 23.1). Fields that occur in intraplate settings tend to be alkali basaltic, while arc settings are dominated by calc-alkaline magmas. Eruptive flux is an important parameter for describing and ultimately understanding the mechanisms of volcanic fields. On the scale of an entire field and its lifetime, estimated eruptive fluxes range from ~ 0.5 to 800 km^3/Ma (Southwest Nevada volcanic field, Michoacán-Guanajuato volcanic field, respectively). For comparison with other volcanic systems, the low value is similar to a few years of output for Kilauea (Hawaii) but spread over a million years. The high end of the spectrum is similar to an arc-related stratovolcano, but spread out over a large area rather than from a single location. At the scale of individual volcanoes and their eruptions, limited information suggest that most basaltic monogenetic eruptions involve lava discharge rates of <10 m^3/s, and probably most commonly on the order of ~ 1 m^3/s. This can be compared with a typical Plinian eruption from a more silicic volcano, with discharge rates of >10^4 m^3/s.

3. ERUPTIVE PROCESSES AND PRODUCTS

Magmatic eruption styles are driven primarily by the expansion of volatiles that were originally dissolved in the magma at depth, and similar processes occur at monogenetic volcanoes as at polygenetic volcanoes with mafic compositions (Valentine and Gregg, 2008). As magmas ascend, those volatiles (dominantly H_2O, but CO_2 can play an important role in some cases) exsolve and form bubbles that can grow through a combination of decompression and coalescence. Magmatic eruptions produce scoria cones when the ascent of liquid magma is relatively slow and bubbles are able to coalesce and form gas-rich slugs that drive Strombolian bursts, ejecting relatively coarse pyroclasts, which accumulate in a cone-shaped pile of vesicular lapilli and fluidal bombs around a central vent. These deposits can be locally welded if their parent Strombolian burst produced a fountain that rapidly deposited still-hot clasts, but most of the deposits are not welded (Figure 23.3(A); Table 23.2). Scoria lapilli and bomb beds in the cones typically are lenticular and often reversely graded, recording final emplacement by avalanching down cone slopes (Figure 23.3(B)). Relatively rapid ascent of magma that precludes appreciable bubble coalescence can produce sustained jets or eruption columns dominated by lapilli and ash; these violent Strombolian (also sometimes referred to as

TABLE 23.1 Physical Characteristics of Selected Volcanic Fields

Volcanic Field	Total Volcanoes (Main Magma Compositions)	% of Maars, Tuff Cones	Evolved Volcanoes	Age Range (Ma)	References
West Eifel, Germany	~240	30%	2	0.7–0.01	Schmincke, H.-U., 2007. In: Ritter, R., Christensen, U. (Eds.), Mantle Plumes—A Multidisciplinary Approach. Springer Heidelberg. pp. 241–322.
East Eifel, Germany	~100	"Rare"	3	0.46–0.01	Schmincke, H.-U., 2007. In: Ritter, R., Christensen, U. (Eds.), Mantle Plumes—A Multidisciplinary Approach. Springer Heidelberg. pp. 241–322.
Lamongan, Indonesia	~90 (Basalt, basaltic andesite)	32%	1 composite volcano		Carn, S.A., 2000. J. Volcanol. Geotherm. Res. 95, 81–108.
Michoacán-Guanajuato, Mexico	1040	2%	43 domes	~3–0	Hasenaka, T., Carmichael, I.S.E., 1985. J. Volcanol. Geotherm. Res. 25, 105–124.
Springerville, USA	409 (Tholeiite, alkali basalt)	1%	None	5.3–0.3	Condit, C.D., Connor, C.B., 1996. Geol. Soc. Am. Bull. 108, 1225–1241.
Newer Volcanic Province, Australia	~400 (Basalt, trachybasalt, basanite)	10%	None	4.6–0.005	Hare, A.G., Cas, R.A.F., 2005. Aust. J. Earth Sci. 52, 59–78. Blaikie, T.N., et al., 2012. J. Volcanol. Geotherm. Res. 235–236, 70–83.
Pali Aike, Argentina–Chile	139 (Basanite, alkali basalt)	24%	None	3.8–0.17	Mazzarini, F., D'Orazio, M., 2003. J. Volcanol. Geotherm. Res. 125, 291–305.
Pinacate, Mexico	400 (Alkali basalt)	2%	None	2–0.01	Gutmann, J.T., 2002. J. Volcanol. Geotherm. Res. 113, 345–356.
Auckland, New Zealand	49 (Basanite, alkali basalt)	69%	None		Houghton, B.F., et al., 1999. J. Volcanol. Geotherm. Res. 91, 97–120. McGee, L.E., et al., 2012. Lithos 155, 360–374.
Hurricane, USA	10 (Basalt, basanite)	None	None	0.35–0.26	Smith, E.I., et al., 1999. J. Geol. 107, 433–448.
Southwest Nevada, USA	17 (Basalt, trachybasalt)	None	None		Valentine, G.A., Perry, F.V. 2007. Earth Planet. Sci. Lett. 261, 201–216.
Lunar Crater, USA	>100 (Trachybasalt, tephrite)	~4%	3 domes	6–0.04	Yogodzinski, G.M., et al., 1996. J. Geophys. Res. 101, 17425–17445. Valentine, G.A., et al., 2011. Bull. Volcanol. 73, 753–756.
Sabatini, Italy	45 (Wide range)	31%	2 calderas	0.3–<0.09	Sottili, G., et al., 2012. Bull. Volcanol. 74, 163–186.
Camargo, Mexico	>300 (Alkali basalt)	<10%	None	4.7–0.09	Aranda-Gomez, J.J., et al., 2003. Geol. Soc. Am. Bull. 115, 298–313.
San Francisco, USA	606 (Alkali basalt)	<5%	8 dome Complexes	5.6–0	Tanaka, K.L., et al., 1986. Geol. Soc. Am. Bull. 97, 129–141. Conway, F.M., et al., 1997. J. Geophys. Res. 102, 815–824.
Coso, USA	54 (Alkali basalt, rhyolite)	None reported	38 domes	2.0–0	Duffield, W.A., et al., 1980. J. Geophys. Res. 85, 2381–2404. Bacon, C.R., 1982. Geology 10, 65–69.
Big Pine, USA	24 (Alkali basalt)	None	1 dome	1.2–0.03	Blondes, M.S., et al., 2008. Earth Planet. Sci. Lett. 269, 140–154. Gazel, E., et al., 2012. Geochem. Geophys. Geosyst. 13. Q0AK06.
Cima, USA	70 (Alkali basalt)	1	None	7.6–0	Farmer, G.L., et al., 1995. J. Geophys. Res. 100, 8399–8415.

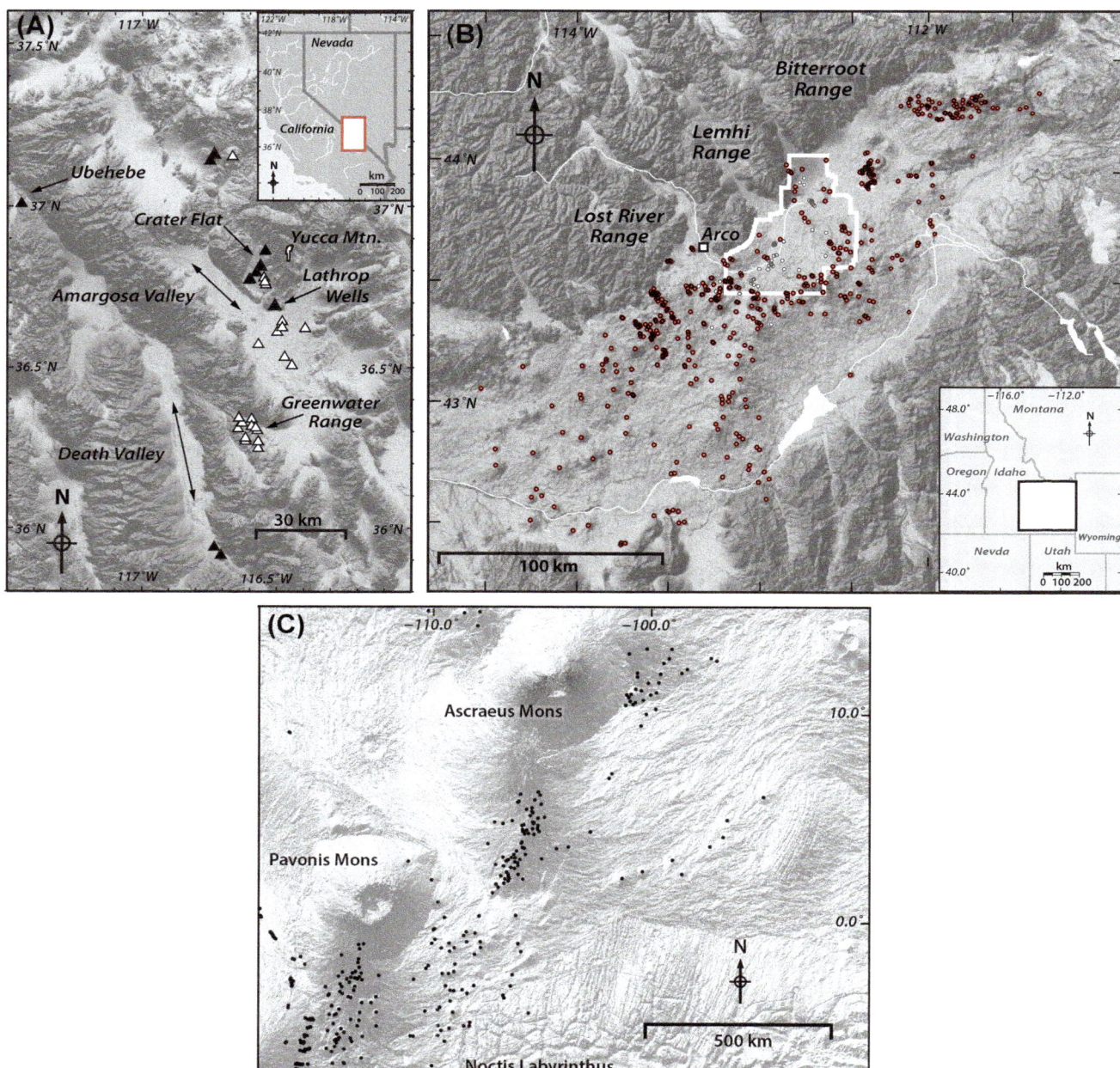

FIGURE 23.1 Maps of volcanic fields on Earth and Mars showing range of sizes. (A) Monogenetic volcanoes (mainly scoria cones and lavas) in southern Nevada and southeastern California (USA). Black triangles are volcanoes of Pleistocene Southwest Nevada Volcanic Field. White triangles are Pliocene volcanoes. (B) Monogenetic volcanoes mapped on the Eastern Snake River Plain (open circles) and inferred buried vents from borehole data (white circles). Idaho, USA. (C) Monogenetic shield volcanoes on Mars (solid circles) near Pavonis Mons in the Tharsis region, superimposed on a digital elevation model showing larger shield volcanoes and related landforms and based on Mars Orbiter Laser Altimeter data (see Richardson et al., 2013). Each group or cluster of monogenetic volcanoes can be referred to as a volcanic field.

microplinian) eruptions can also produce a scoria cone but with an attendant tephra fall deposit that reach tens of kilometers downwind from the vent. Their fall deposits can extend into the cones themselves as well-sorted, parallel-bedded horizons that differ from the avalanche beds described above (Table 23.2). Scoria cones are commonly accompanied by lava fields that extend hundreds of meters to tens of kilometers. In some cases these lava fields were fed directly from the summit vents of scoria cones, but more often they are fed by smaller vents, referred to as bocas

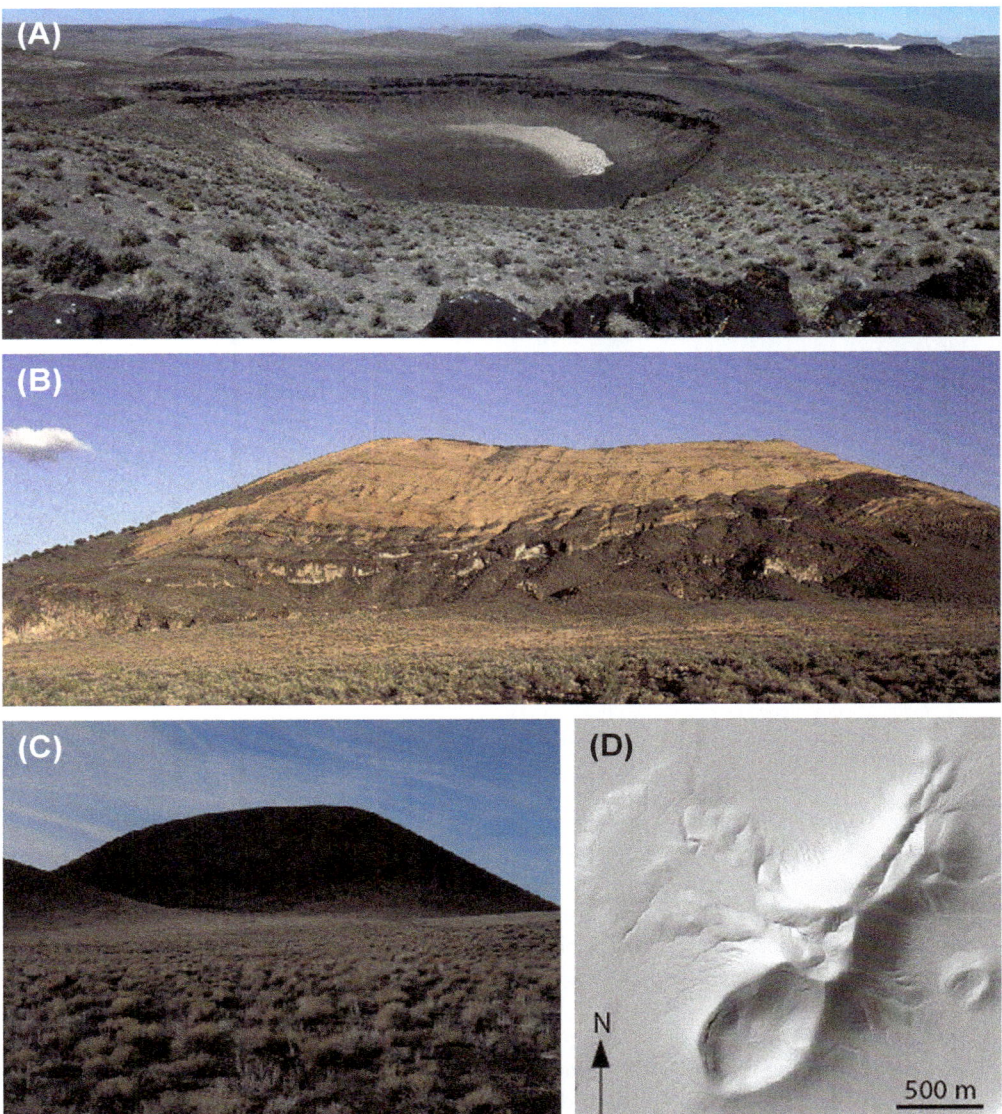

FIGURE 23.2   Examples of monogenetic basaltic volcanic landforms. (A) Lunar Crater maar (Nevada, USA). Crater diameter ~1 km. Note numerous scoria cones in the background. (B) Pahvant tuff cone (Utah, USA, photo by J.D.L. White). Tan colored deposits are altered (palagonotized) tuff. Tuff cone is ~200 m high. (C) About ~160 m high Marcath scoria cone (Nevada, USA). (D) Shaded relief map of a complex landform comprising agglomerate ramparts, central scoria cones, lava field, and maar, from a single eruptive episode along a ~2.5-km-long fissure vent system (Easy Chair volcano, Nevada, USA).

(Italian for mouth), which occur on the flanks and/or at the bases of the cones. Sustained discharge of magma with large gas bubbles causes Hawaiian eruptions that produce cones and lava fields, which are often fed by rapid deposition and coalescence of hot, still-fluid clasts around a lava fountain. Pyroclastic deposits from Hawaiian eruptions tend to be moderately to densely welded over much of their extents due to the high accumulation rates of coarse, hot fluidal clasts (Table 23.2). If the vent is an elongate fissure, the resulting landform is a spatter or **agglomerate rampart**—essentially an elongate scoria cone—that might

also be accompanied by a lava field (the terms spatter and agglomerate are used when deposits are highly welded or variably welded, respectively). Monogenetic basaltic volcanoes with volumes toward the larger end of the spectrum, which are dominated by effusive eruption of lava, can form low-profile shield volcanoes with basal diameters of several kilometers.

Older, eroded scoria cones and ramparts allow direct measurement of their feeding systems. Basaltic magma ascends through dikes that range from one to a few meters in width. Flared conduits form in the upper tens of meters

FIGURE 23.3 Examples of proximal deposits from monogenetic basaltic eruptions. (A) Agglomerate with fluidal bombs (Tower Hill, Australia). Note that bomb bottoms mold around underlying lapilli. (B) Scoria cone deposits (near Pasto, Colombia) showing lenticular bedding (enclosed by dashed lines) and reverse grading characteristic of emplacement by grain avalanches down cone slopes. (C) Ballistic blocks in lapilli tuff deposits, tephra ring (Easy Chair volcano, USA). (D) Bedded tuff from a tephra ring around a maar (Alban Hills, Italy).

of Earth's crust (Figure 23.4(A)). This flaring results from erosion of the conduit walls as magma accelerates as well as by collapse of walls as magma pressure fluctuates and the magma column rises and falls during the eruptive episode. At any given moment during an eruptive episode this flared structure is likely filled mostly with debris from the walls and from slumped cone material, with only a small part being occupied by active magma. Magmatic eruption styles thus only disrupt the very uppermost parts of their feeder systems, which merge quickly downward into the dikes that delivered magma from its deep source or reservoir (Valentine and Gregg, 2008; Brown and Valentine, 2013).

If there is a strong interaction between rising magma and externally derived water, eruptions tend to be dominated by numerous phreatomagmatic explosions. Maars are "negative" volcanic landforms, forming craters that cut into the preeruption landscapes and are surrounded by low-profile rings of ejecta (tephra ring), that form when phreatomagmatic explosions occur mainly in the subsurface (magma–groundwater interaction; White and Ross, 2011). Maars range in diameter from 100 to 8000 m with most falling between 400 and 1200 m (diameters larger than this are often associated with multiple craters that coalesced to form a single landform). Crater depths range from tens of meters to ~350 m; the fact that maars are closed basins means that their original depths are quickly modified by accumulation of sediment fill. The ejecta

deposits around maars (referred to as tephra rings) commonly include lapilli tuffs with variably abundant blocks and bombs up to a few meters in size, bomb sag structures, cross-bedding and cross-lamination, dune forms, and poorly sorted massive tuff units, all of which suggest powerful explosions with ballistic ejection of coarse clasts and formation of pyroclastic density currents with a range of properties (Figure 23.3(C) and (D); Table 23.2). Well-sorted, parallel-bedded deposits indicative of deposition by fallout also occur, and some of these are likely related to phases when magmatic activity (e.g., Strombolian, violent Strombolian) were important even though phreatomagmatic processes may have dominated most of the volcano's eruptive episode.

Tuff cones are positive landforms with a low conical shape and a summit crater that form when phreatomagmatic explosions occur at the Earth's surface with abundant water, such as when magma ascends into a lake or sea. The term tuff cone relates to the fact that the cones are composed mainly of ash-sized particles, which are created when magma is rapidly quenched and fragmented during its interaction with water. Data on tuff cone dimensions are sparse, but the features can reach ~200 m height with summit craters whose floors can be nearly at the level of the preeruptive ground surface. Tuff rings are similar to tuff cones but have lower profiles and are thought to result from phreatomagmatic activity where surface water is less abundant, such as a shallow, ephemeral lake in a desert basin. Tuff rings are constructive landforms similar to tuff cones and their craters do not cut into the preeruptive landscape, but tuff rings have low profiles as do the tephra rings around maars. The deposits within tuff cones and rings tend to be dominated by bedded, massive lapilli tuffs that record emplacement by fallout or density currents of wet, poorly sorted ejecta (Table 23.2). Typically, all of the basaltic glass in these deposits is altered to palagonite, and this alteration can mask original sedimentary structures to make deposits appear massive.

While external (ground or surface) water is an essential ingredient for phreatomagmatic activity, overly abundant water and/or high hydrostatic pressure can suppress explosion power. Thus, tuff cones record weak phreatomagmatic explosive activity where erupted material was not ejected very far, while tuff rings record more powerful explosions that spread ejecta over a larger distance (hence the lower profile of the landform). Tuff rings represent phreatomagmatic activity that is intermediate between the lower energy tuff cone-forming explosions and maar-forming eruptions that are the most energetic of the spectrum because in their eruptions water supply is limited by aquifer properties. In maar-forming eruptions, all (or nearly all) of the available water is rapidly converted to steam, and pyroclastic density currents involve

TABLE 23.2 Characteristics of Deposits from Different Eruptive Styles

Eruptive Style	Bedding	Texture and Grading	Clast Size	Clast Shape	Clast Vesicularity	Welding, Induration
Proximal (Cone, Tephra Ring, Tuff Ring, Tuff Cone) Deposits						
Strombolian (magmatic)	Mainly lenticular over several meters to ~10 m, decimeters to ~1 m thick	Reverse graded to massive	Coarse lapilli, blocks and bombs	Aerodynamic shapes (e.g., ribbon, spindle) of coarsest clasts, angular to slightly rounded smaller clasts	Moderate vesicularity with a wide range of sizes	Local moderate to dense welding very close to vent, nonwelded elsewhere
Hawaiian (magmatic)	Lenticular to continuous over more than tens of meters	Massive to reverse graded	Coarse lapilli and bombs	Aerodynamic and fluidal (e.g., wrapping around underlying clasts), ragged margins	Moderate vesicularity with a wide range of sizes	Densely to partially welded over much of cone extent
Violent strombolian (magmatic)	Mainly planar, continuous over more than tens of meters, locally lenticular. Localized, thin, ash-rich cross-bedded horizons	Massive to graded, internal planar stratification, local reverse-graded lenses	Coarse ash to coarse lapilli, sparse blocks and bombs	Mainly blocky and angular to slightly rounded. Sparse aerodynamic shapes	Moderately to highly vesicular with abundant small vesicles	Nonwelded
Maar forming (phreatomagmatic)	Thin to thick, planar to low-angle cross-bedding, scour and fill structures, draping topography	Massive to internally graded, laminated, and cross-laminated, bomb/block sags	Blocks and bombs to fine ash	Coarse clasts dominantly blocky, may contain proportion of fluidal juvenile clasts	Poorly to highly moderately vesicular	Nonwelded, but may be variably indurated by glass alteration products (palagonite)
Tuff cone-, tuff ring-forming (phreatomagmatic)	Thick beds, high-angle cross-bedding with soft-sediment deformation structures	Massive to internally graded, cross-bedded, block/bomb sags	Blocks and bombs to fine ash	Coarse clasts dominantly blocky, may contain proportion of fluidal juvenile clasts	Poorly to highly moderately vesicular	Nonwelded but typically indurated by alteration to palagonite, which can mask bedding structures
Medial to Distal (Fallout) Deposits						
Strombolian (magmatic)	Planar, localized within hundreds of meters of cone base	Massive to reverse or normally graded	Fine to coarse ash, sparse fine lapilli	Blocky to bubble-wall shapes	Moderate to high	Nonwelded
Hawaiian (magmatic)	Planar, localized within hundreds of meters to a few kilometers from cone base	Massive to reverse or normally graded	Fine to coarse ash, lapilli close to cone	Small ribbons, aerodynamic shapes in lapilli, bubble wall shapes in ash, and reticulite	Moderate to extremely high (e.g., reticulite), large vesicles common in lapilli	Possibly partly welded close to cone
Violent strombolian (magmatic)	Planar, extending about a kilometer to a few tens of kilometers from vent. Localized ash-rich, cross-bedded horizons	Massive to reverse or normally graded	Fine to coarse ash and lapilli sizes	Small ribbon fragments and blocky to slightly rounded lapilli, bubble wall to blocky shapes in ash	Moderate to high with abundant small vesicles in lapilli (possible pumice texture)	Nonwelded
Maar forming (phreatomagmatic)	Planar, extending several kilometers from vent	Massive, may contain accretionary lapilli	Fine ash	Blocky ash particles, pellets and accretionary lapilli	Nonvesicular to moderately vesicular ash	May be indurated by palagonitic alteration
Tuff cone-, tuff ring-forming (phreatomagmatic)	Planar, extending several kilometers from vent	Massive, may contain accretionary lapilli	Fine ash	Blocky ash particles, pellets and accretionary lapilli	Nonvesicular to moderately vesicular ash	Typically indurated by palagonitic alteration

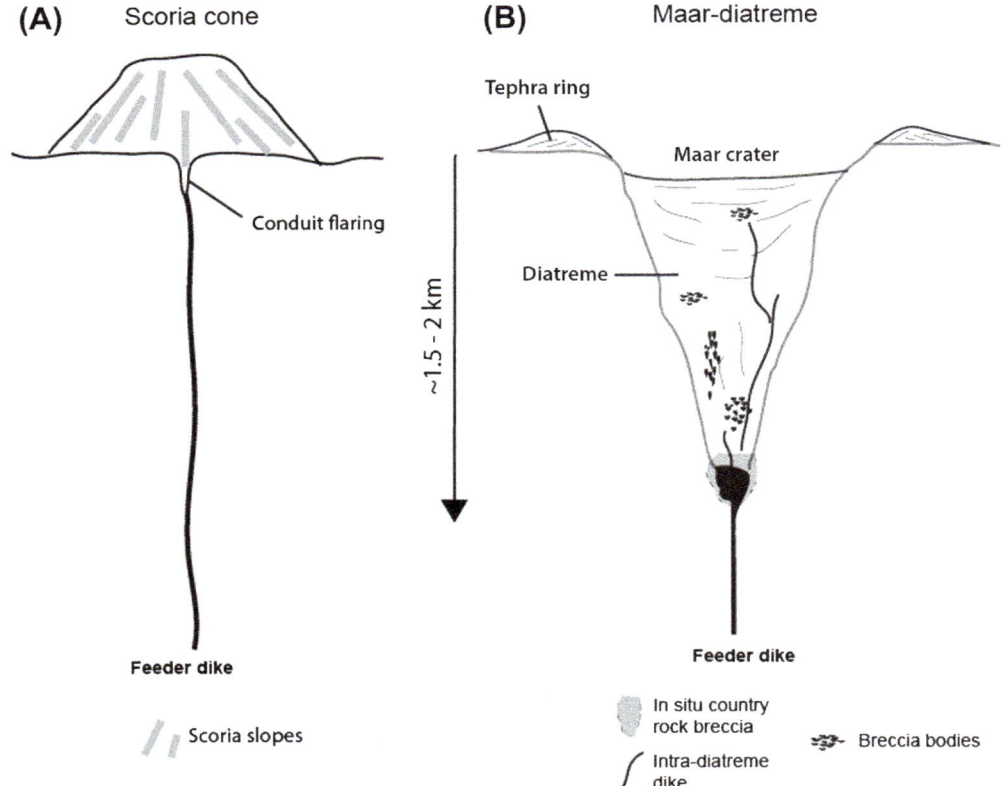

FIGURE 23.4   Comparative diagram of surface expressions and plumbing of (A) scoria cones and (B) maar-diatremes, based upon field observations. Most of the explosive activity associated with a scoria cone is near or at the surface, while in a maar-diatreme it is mostly underground.

deposition with saltation of noncohesive, individual grains, resulting in relatively low-angle bed forms. In tuff cone- and ring-forming eruptions, only part of the water is rapidly converted to steam; thus, ejecta is wet, poorly sorted, and cohesive when deposited. However, these are rules of thumb, and in any given eruption there might be phases that are not typical for that particular landform. Many questions remain about the underlying mechanisms and the details of how magma interacts with water.

Eroded volcanic fields expose the shallow feeding systems for phreatomagmatic volcanoes, but in contrast with the relatively shallow flaring of conduits for volcanoes dominated by magmatic activity, these can form much more extensive subsurface structures (Figure 23.4). In particular, maar volcanoes are typically underlain by crudely funnel-shaped bodies of fragmented country rock, erupted material, contorted dikes and sills, and variable proportions of juvenile material (Figure 23.4(B); White and Ross, 2011). These bodies, referred to as diatremes, can extend up to 2 km downward before merging with their feeder dike(s), and can have volumes as large, possibly larger, as the volume of ejecta preserved in the tephra ring. Diatremes demonstrate the importance of subsurface activity, with phreatomagmatic explosions occurring at depth where aquifers can supply water to interact with magma.

As magma supply and water supply vary during an eruptive episode, subsurface activity might range from quite explosive to passive intrusion of magma into the diatreme. Material from different depths within a diatreme will gradually mix due to subsurface explosions such that deep-seated wall rock fragments might eventually end up near the surface. Most explosions that actually erupt at the surface probably are relatively shallow (upper ∼200 m), but they may eject some of these originally deeper seated wall rock clasts. Thus, tephra rings can contain fragments of wall rock from a range of depths, but typically they are dominated by shallow-seated clasts. Subsidence during and after eruption also plays an important role in diatreme growth and maar crater widening, such that tephra ring deposits can subside into a diatreme. Tuff cones and tuff rings, on the other hand, only have very shallow diatremes because most of their explosive activity occurs at the surface rather than underground.

4. SPATIAL DISTRIBUTION OF VOLCANOES

The episodic and varied nature of eruptions in volcanic fields is further revealed by spatial patterns in vent distribution that appear to be common in nearly all volcanic fields. Spatial distribution of volcanism indicate that (1) vents are distributed in patterns that reflect the regional tectonic setting of volcanism, implying that the setting influences the location of mantle partial melting and generation of small-volume magma batches, and the ascent of these magmas through the mantle and crust; (2) vents cluster within volcanic fields, often on several scales, suggesting that rates of activity vary substantially within volcanic fields and may shift through time; and (3) vent alignments are ubiquitous. The latter include short alignments (generally <10 km in length) consisting of several vents often formed during the same episode of volcanic activity, and regional alignments (often >20 km in length) consisting of numerous vents that may have formed during several distinct episodes of activity with significant gaps between these episodes.

4.1. Distribution of Volcanic Fields

Regional tectonic setting plays a critical role in the distribution of volcanic fields. Consider the Chugoku region of southwest Honshu, Japan (Figure 23.5). Distinct

Quaternary volcano clusters (e.g., the Abu monogenetic volcano group) and three stratovolcanoes (e.g., Diasen) are located in northern Chugoku, above the leading edge of the subducted Philippine Sea plate. In contrast, most Pliocene volcanic fields are located in central Chugoku. The change in position of active volcanic fields from Pliocene to Quaternary reflects a change in geometry of the subduction zone during this time period. It is an open question why in arcs such as Chugoku, monogenetic volcanic fields form in some areas and central vents form in others, but this transition is certainly reflected in comparatively low rates of volcanism in the volcanic fields compared to individual stratovolcanoes. Similarly, in Baja California, Mexico (Figure 23.6), postsubduction volcanism formed a series of distributed monogenetic volcanic fields, including approximately 900 vents within an area ∼700 km long and 70−150 km wide (Germa et al., 2013). Statistical analysis of the distribution of Baja vents indicates that overall these volcanic fields are elongate parallel to the paleosubduction zone, indicating that partial melt zones reflected the geometry of the subducted slab. On a more local scale, alignments of vents within these fields are parallel and

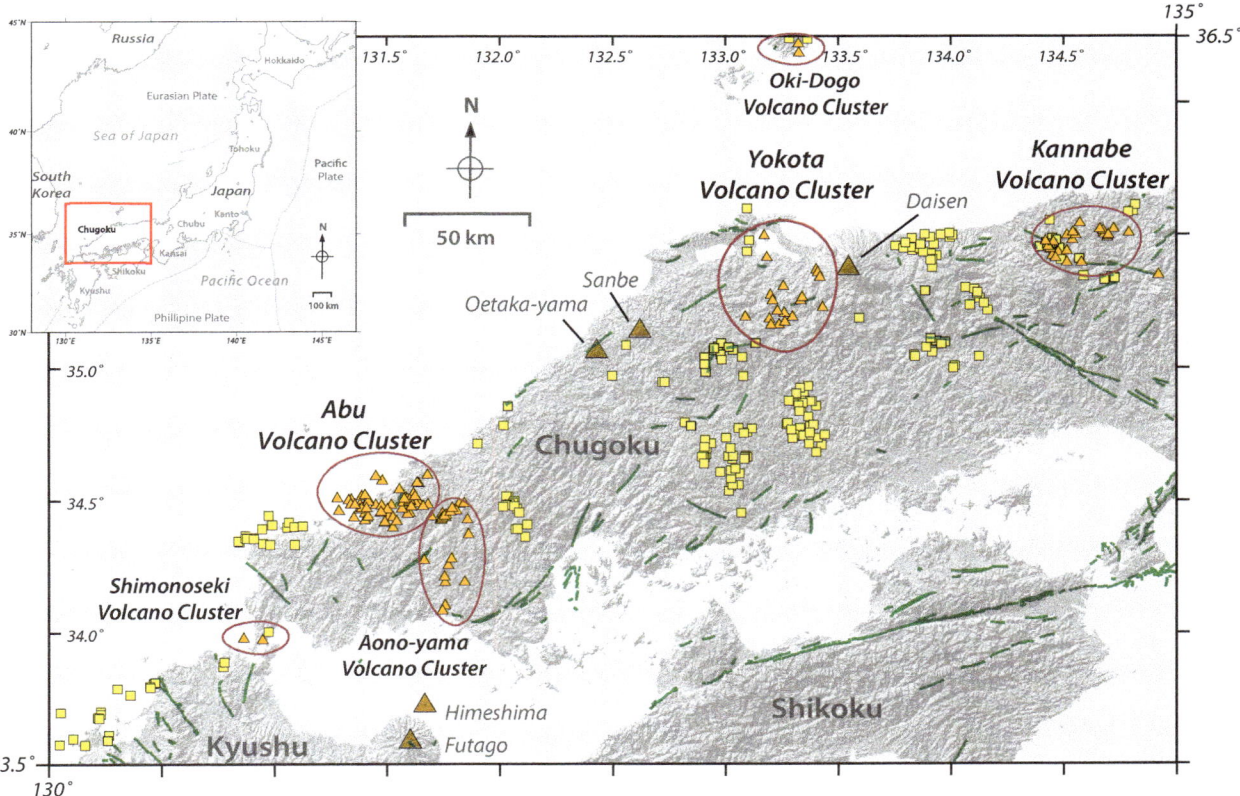

FIGURE 23.5 The regional pattern of distributed basaltic volcanic fields at convergent margins is illustrated by volcano clusters in Chugoku, southwest Honshu, Japan. The Quaternary Shimonseki, Abu, Yogata, Yono-yama, and Kannabe clusters (volcanoes shown as dark yellow triangles) are located near the leading margin of the subducted Philippine Sea Plate, or at a significant bend in this subducted slab. Older (Pliocene) basaltic volcanic fields (light yellow squares) are mostly located south of these active volcanic fields, indicating a change in the location of clusters with changing tectonic setting. Stratovolcanoes (large triangles) and faults (green lines) are also shown.

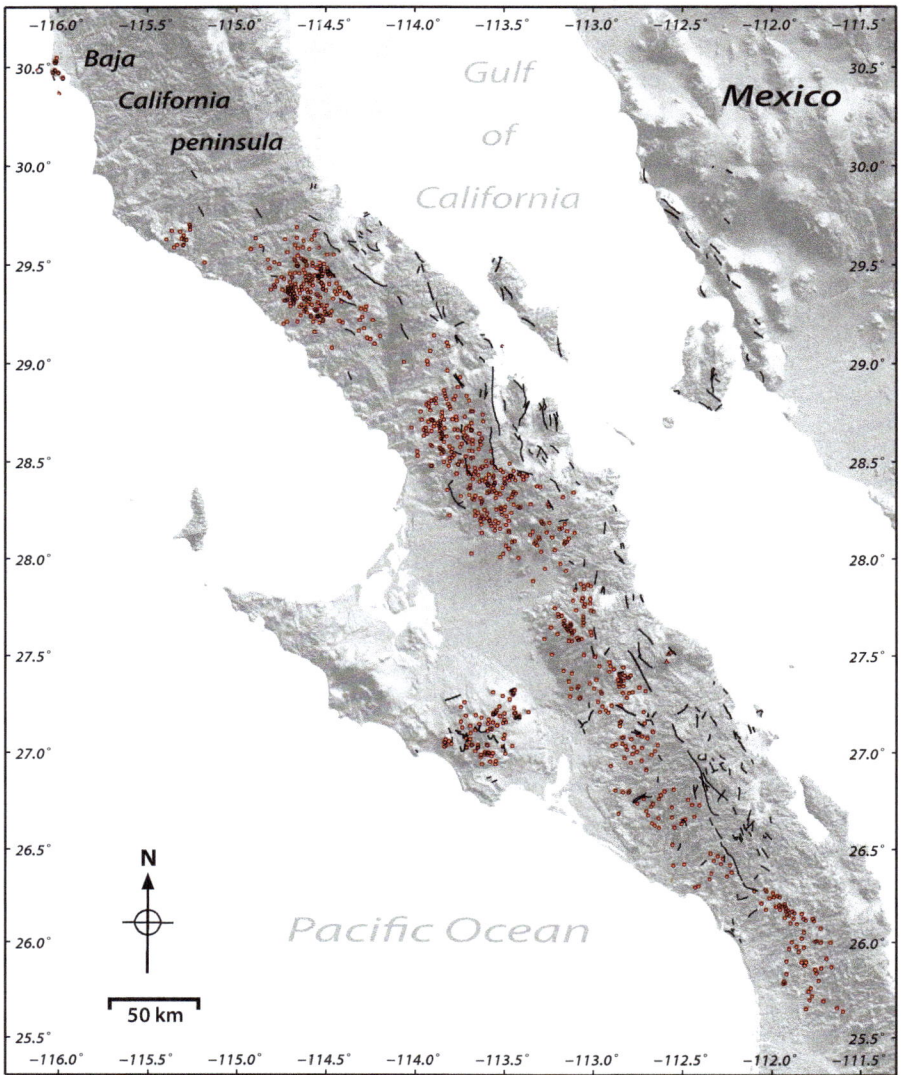

FIGURE 23.6   More than 900 scoria cones and related volcanoes (solid black circles) are distributed in volcanic fields along Baja California (Mexico). The overall NW trend of these volcanic fields reflects the orientation of paleosubduction in the region. Clusters of volcanoes are indicative of nonuniform melt production, and rates of volcanic activity, along the peninsula. Vent alignments in this region tend to parallel the arc, and parallel predominant fault orientations.

subparallel to individual faults. As illustrated by these two examples, regional patterns of distributed monogenetic volcanism are consistent with tectonic setting, shifting patterns of activity, on a regional scale, reflect changes in regional plate setting.

4.2. Vent Clusters

Vent clusters occur on a more local scale, within individual volcanic fields. For example, in the Michoacan-Guanajato (Mexico), Springerville (USA), and San Francisco (USA) volcanic fields, individual vent clusters consist of 10−100 individual vents and have diameters up to 50 km. In smaller volcanic fields, clusters of 10−20 vents are

common. In some of these volcanic fields, such as Springerville, detailed mapping and radiometric age determinations have shown that the recurrence rate of new vent formation within individual clusters waxes and wanes on short timescales compared to overall activity in the field. Elsewhere, such as in the Michoacán-Guanajuato volcanic field, adjacent clusters are active simultaneously, at similar rates over long periods of time. Parícutin and Jorullo volcanoes are both part of the Michoacán-Guanajuato volcanic field, but are in spatially distinct clusters. Thus, these clusters are both active today and the clusters themselves represent heterogeneity in spatial distribution without necessarily representing change in the recurrence rate of volcanism.

A simple explanation for vent clustering within volcanic fields is that magma supply varies across volcanic fields. Mantle rocks beneath vent clusters are close to their solidus and produce partial melts more readily in response to small changes in temperature, pressure, or composition. Hence, magmas in these areas are generated repeatedly as long as the fraction of partial melting is small so that the source region within the mantle is not depleted as a result of repeated melt generation. The scale and longevity of clusters reflect the scale of temperature, pressure, and compositional variations within the mantle. An alternative explanation for vent clustering is that lithosphere discontinuities (e.g., the Moho) and the development of shallow magmatic plumbing systems can focus magmatism and create vent clusters. For example, if sills form in response to repeated magma injection in the lower or mid crust, the changed geothermal gradient may make it more likely that subsequent magmas would reach the surface in that location rather than in another, forming clusters through time. Generally, the distribution of vent clusters within volcanic fields does not reflect changes in the abundance or orientation of mapped crustal structures, such as faults, and the formation of clusters is not readily attributable to shallow crustal processes.

4.3. Vent Alignments

Alignments of volcanic vents are common in volcanic fields, on shield and stratovolcanoes, and within calderas. These alignments and their related structures are sometimes used to infer the presence and orientation of subsurface dikes and dike swarms, if eruptions along an alignment can be shown to be coeval. Geologic mapping of deeply eroded volcanic fields can be used to better understand the relationship between volcanic vents, alignments of vents, and intrusions such as dikes and sills. One such field is the San Rafael (east-central Utah, USA; Kiyosugi et al., 2012), where approximately 800 m of erosion has revealed the relationship among dikes, sills, and conduits that transported magmas to the surface and fed pyroclastic eruptions and lava flows during the Pliocene. Geologic mapping in the San Rafael region shows that dikes may be 10–100 times more abundant than volcanic vents in distributed volcanic fields, that vent clusters are often coincident with sills in the subsurface, and that dikes are often obliquely oriented in the subsurface with respect to alignments of vents, reflecting rotation of the stress field in the shallow crust. Nevertheless, a strong correlation exists between structural trends (faults, joints, fractures) and vent alignments consisting of contemporaneous scoria cones. Examples include historically observed alignment formation at Jorullo and Parícutin, Mexico, and the 1975 Tolbachik scoria cones, Russia. In the geologic record, paleomagnetic data, radiometric age determinations, and cone morphology are used to indicate that short cone alignments sometimes form over very geologically brief periods of time.

One explanation for the occurrence of volcanic vents along faults is that during ascent a dike may deviate from its orientation perpendicular to the least principle stress in favor of a more energy-efficient path along a preexisting joint or fault plane. The orientation of the preexisting structure relative to the least principle stress plays an essential role in determining whether a fault is more or less likely to dilate in response to dike injection and redirect the ascending dike. Faults oriented roughly perpendicular to the least principle stress are termed high-dilation-tendency faults, and are more likely to provide a low-energy path to the surface compared to faults of other orientations. Similarly, in the shallowest part of the system, stress effects near the Earth's surface may cause the ascending dike to break out of a fault zone and ascend vertically to the surface, leading to an offset between the mapped fault and the resulting vent alignment. Numerical modeling indicates that such near-surface effects are important to a depth of approximately 500 times the dike width.

Faults and dikes are further related because faulting and dike injection play the same role in accommodating crustal stress, by slip in the case of a fault and by increasing total crustal volume in the case of a dike. In some volcanic fields, such as in the eastern Snake River Plain (USA) and the Michoacán-Guanajuato volcanic field (Mexico), topography due to fault slip is suppressed near vent clusters because stress is accommodated by dilation of the crust during dike injection, rather than by fault slip. In other volcanic fields, such as the Lassen-Caribou region (USA), rates of magmatism are not sufficiently high to accommodate all deviatoric stress, and fault-related topography is not suppressed (Figure 23.7). This relationship can explain much of the variation observed in the structure of volcanic fields. Vent alignments are common in low-volume, low-density volcanic fields, such as the Lassen-Caribou. In larger volume volcanic fields, mapped faults are rare and vent alignments, though present, are less pervasive. In the former case, dike injection rates are not sufficient to fully accommodate crustal stress. As a result, fault systems persist and dikes and vent alignments tend to parallel or inject into these fault systems. In the latter case, rates of dike injection are sufficient to completely accommodate regional tectonic stresses within the field. Dike injection equalizes, or nearly equalizes the magnitudes of principle horizontal stresses and their orientations may vary substantially across the volcanic field and over time. Thus, vent alignments are less common in volcanic fields with comparatively high rates of volcanic activity and relatively low deviatoric stress.

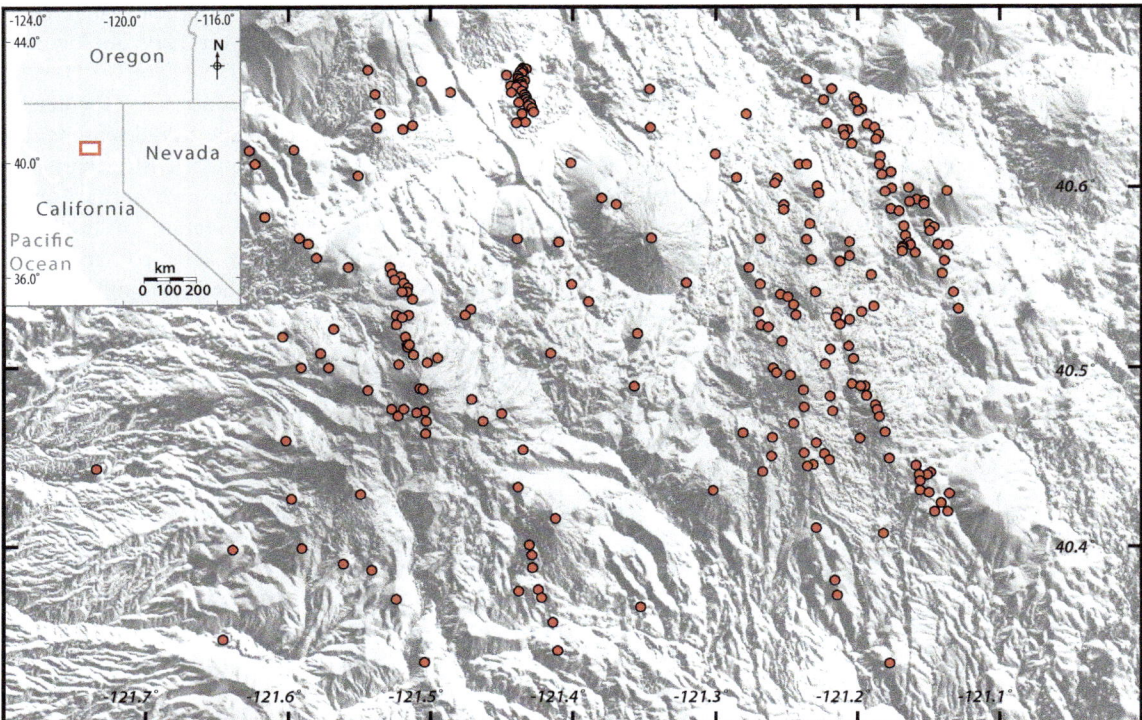

FIGURE 23.7   Volcanoes in the Lassen-Caribou volcanic field (USA) (solid gray circles) are frequently distributed in vent alignments that parallel NNW-trending Basin and Range structures. The topography of the region, here represented by a shaded-relief digital elevation model, reflects Basin and Range faulting. The rate of volcanism and dike injection is not sufficient to suppress topography in this region that is produced by faults to accommodate deviatoric stress in the upper crust.

5. TEMPORAL RECURRENCE RATE OF ACTIVITY WITHIN BASALTIC FIELDS

The timing of volcanism is a central issue in the study of volcanic fields. Age determinations for individual volcanic vents and their lava flows are used to estimate recurrence rate of vent formation and to correlate rates of activity with other factors, such as plate motion, rates of crustal extension, and composition. Age dating techniques applied to volcanic fields include relative dating of lava flows using stratigraphic relationships, paleomagnetic age dating, and geomorphology of scoria cones. Radiometric age determinations are made primarily using ^{40}Ar-^{39}Ar, K-Ar, and cosmogenic isotopic methods. Normally, in any volcanic field a variety of age determination methods are used in order to best characterize the chronology of volcanic events.

Even today, most volcanic fields have relatively few radiometric age determinations and details of the change in recurrence rate with time are unknown. In many cases, especially where relatively few age determinations are available, a straightforward method for estimating average recurrence rate (λ) is given by

$$\lambda = \frac{N-1}{t_o - t_y}$$

where N is the total number of volcanic events (e.g., mapped vents), t_o is the oldest known volcanic vent and t_y is the youngest known volcanic event. Long-term average recurrence rates are typically on the order of 10^{-4} to 10^{-5} volcanic events per year (v/yr). These average recurrence rates are low compared to the frequencies of eruptions at individual stratovolcanoes. Characterization of rates of volcanic activity using a long-term average may smooth out significant short-term (<100 kyr) variations that occur in some volcanic fields. For example, the Michoacán-Guanajuato volcanic field has a recurrence rate of approximately 2×10^{-3} v/yr over the last 40 ka, significantly greater than the long-term average recurrence rate in the Quaternary. In contrast, average recurrence rates of vent formation in the Southwest Nevada volcanic field are about 8×10^{-6} v/yr during the last 1 Ma. Thus, although average recurrence rates are low compared to individual stratovolcanoes, they span orders of magnitude between volcanic fields.

In detail, the number of new vents formed through time or the volume erupted through time may vary substantially in volcanic fields. In the Abu volcanic field (Japan), very low rates of activity occurred prior to approximately 400 kyr. After 400 kyr, the number of eruptions and volume erupted increased dramatically.

This change in rate was accompanied by a shift from predominately basaltic volcanism to a mixture of basalts and andesites (Figure 23.8(A)). Since 400 ka, the volume of erupted magma over time in the Abu field appears to be relatively constant, and successive eruptions occur at intervals that depend on the volume of the previous eruptions. A similar pattern of activity has been observed in other volcanic fields where volumes are well constrained, such as the Coso and Springerville volcanic fields.

Linear behavior of recurrence rate or volume erupted over time may indicate constant rates of partial melt generation, and/or a correlation with rates of extension or plate motion, which are relatively constant over the life spans of individual volcanic fields. Consider recurrence rate in the distributed Nejapa-Miraflores volcano alignment (Nicaragua) (Figure 23.8(B)). Here, detailed stratigraphic studies have enabled volcanologists to construct a record of Holocene volcanic events, each of which is likely associated with the formation of a new volcanic vent. The cumulative frequency of these events is stationary, with random variations around the average rate of volcanic

events through time. Unfortunately, as is often the case, the volumes of individual eruptions are poorly known due to erosion and burial. Nevertheless, it is clear that the recurrence rate of new vent formation in this Nejapa-Miraflores volcano alignment is about one volcano every 800 years, and that there is no indication of a significant deviation from this rate. The last eruption along the alignment was approximately 1000 years ago.

6. MAGMA SOURCES AND ASCENT

Monogenetic volcanoes are fed by magmas generated in the upper mantle, at depths of $\sim 50-150$ km, depending on the mantle's composition and the geothermal gradient in a given region. The Southwest Nevada volcanic field (USA) is thought to be an example where magmas are generated in ancient, enriched lithospheric mantle that extends to depths of ~ 100 km. Lithospheric mantle by definition does not convect or upwell, therefore melting processes may be different than for asthenospheric mantle. This source is reflected in radiogenic isotope compositions such as $^{87}Sr/^{86}Sr$ and $^{143}Nd/^{144}Nd$ and in trace element

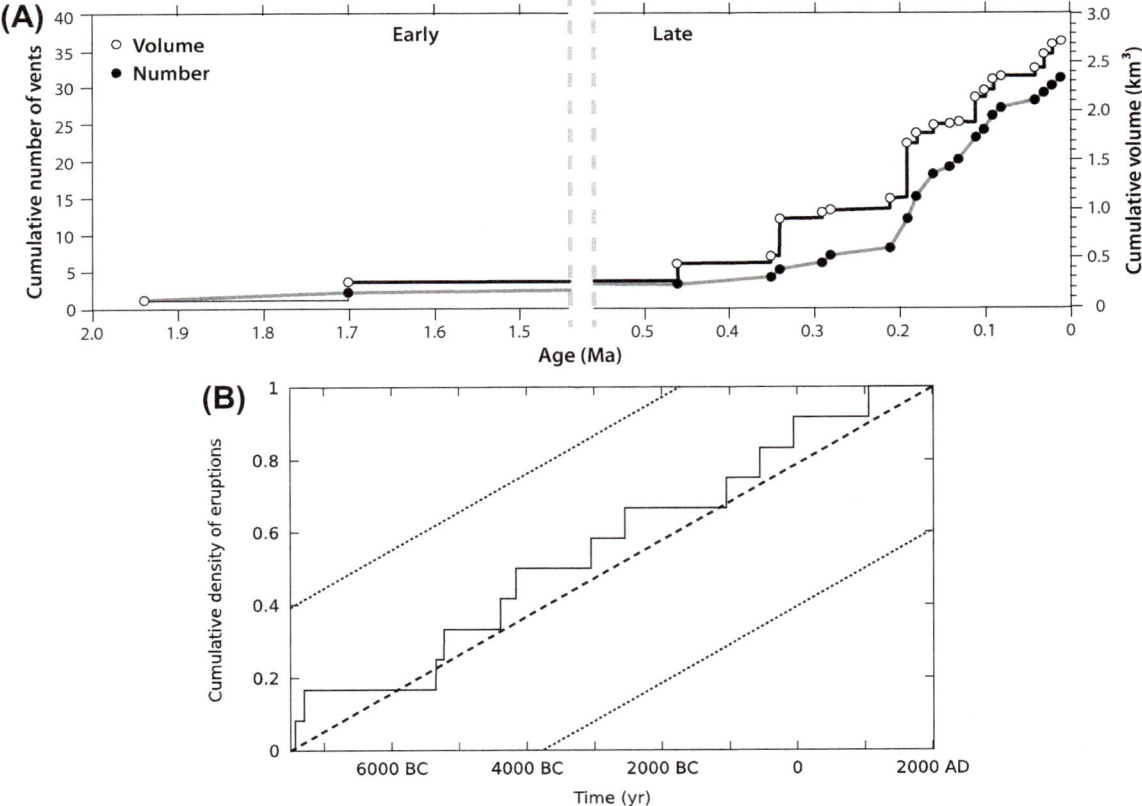

FIGURE 23.8 Recurrence rates of new volcano formation in basaltic volcanic fields. (A) The Abu volcanic field (Japan) underwent a change in the rate of new volcano formation approximately 400 ka. This increase in rate of activity was concomitant with a change from predominantly basaltic magmas to a combination of basalt and andesite magmas. Such orders of magnitude change in recurrence rate is common in volcanic fields during their overall lifetimes. (B) Cumulative rate of known Holocene eruptions in the Nejapa-Miraflores volcanic field (Nicaragua) (solid stair-step) can be modeled as a stationary process with constant recurrence rate with a high degree of confidence (straight lines show 95% confidence bounds on the stationary model).

compositions. Geochemical data indicate that the mantle source rocks are enriched in hydrous phases such as phlogopite, which can suppress solidus temperatures at a given pressure. Physical volcanological data (e.g., eruption volumes, feeder dike lengths) and geochemical data support an integrated conceptual model wherein the lithospheric mantle is heterogeneous, with domains with length scales of kilometers that contain slightly higher partial melt fraction, and slow deformation of the lithosphere causes mechanical focusing of the melts until they are able to trigger buoyancy-driven dikes that propagate upward. Magma ascends rapidly through these dikes and erupts onto the surface, forming a volcano whose size and eruption style are related to the size and composition of its melt source domain.

In other fields, magmas are sourced in asthenospheric mantle that may undergo decompression by upwelling. Recently studied examples include the Auckland volcanic field (New Zealand) and Jeju Island volcanic field (South Korea). Geochemical data from these volcanic fields also suggest that individual monogenetic volcanoes tap individual partial melt domains that have length scales on the order of kilometers (Brenna et al., 2010; McGee et al., 2013). In some cases, magmas for individual volcanoes are from multiple source domains, such as when a deeply sourced magma batch ascends through a shallower partial melt domain and takes on some of its geochemical characteristics. The degree of crustal interaction may be proportional to the size of a magma batch; in other words, small magma batches may not have sufficient thermal energy or residence time to partly melt the rocks through which they ascend, compared to larger batches. Even in asthenosphere-sourced volcanic fields, ongoing detailed work is revealing an important role of source heterogeneity. At the Big Pine volcanic field (USA), it appears that mechanical focusing of melts due to deformation of mantle beneath crustal faults can play an important role in triggering magma ascent, similar to the conceptual model for Southwest Nevada volcanic field, even though the mantle sources are very different (Gazel et al., 2012; Valentine and Perry, 2007).

New research is also focusing on estimating volatile contents, and melting temperatures and pressures. Volatile contents can be deduced relatively directly by studies of melt inclusions in phenocrysts, although not all volcanic fields contain these. Indirect methods include experiments that aim to reproduce mineral assemblages and textures of erupted rocks, and calculations of intrinsic parameters associated with mineral or mineral—glass equilibria. In contrast to the more voluminous midocean ridge or ocean island basalts that typically have low volatile contents, those that form monogenetic volcanic fields often have several weight percent H_2O and hundreds of parts per million CO_2. Recent work that combines data from melt inclusions with physical volcanological parameters such as mass eruption rates indicates that monogenetic basaltic volcanoes such as Jorullo (active 1759—1774) and Parícutin, both in Mexico, can emit gases such as SO_2 at rates comparable to emissions at Mt Etna (Italy) in recent years (Johnson et al., 2010). Volatile contents and degassing processes affect the minerals and textures that form as the magmas ascend, and, importantly, the eruptive styles in that more explosive activity such as violent Strombolian styles can result. New methods of estimating melting pressures and temperatures are focusing on whole-rock chemistry, but more work is required in order for these to be applicable to many volcanic fields, especially those with alkaline affinities.

In some cases, erupted magmas appear to have ascended from source depths nearly directly as stated above, but in many others the magmas stall in the middle to upper crust. There they may undergo fractionation and, if the magma batch is sufficiently large, assimilation of wall rocks (e.g., Erlund et al., 2009). Volatile concentrations can increase through these processes, which again can affect eruption style once the magmas continue their ascent to the surface. As with all magma types, at shallow levels basaltic magmas and eruptive styles can be strongly affected by interaction with the environment, for example, by losing gas to the host rocks, causing degassing-induced crystallization, or by interaction with externally derived water (phreatomagmatism). All of these factors complicate our ability to predict likely processes should a new monogenetic eruption occur.

7. HAZARD ANALYSIS

Hazards associated with basaltic volcanic fields stem from their distributed nature. Unlike central vents, volcanic fields cover areas of hundreds to thousands of square kilometers. Renewed activity in these volcanic fields involves the ascent of a dike or dike swarm along a new pathway and eruption of a volcano in a new location. These new vents may produce a wide range of eruptive phenomena, including tephra fallout, pyroclastic surges, lava flows, and volcano ballistic projectiles with potential impacts on nearby populations and infrastructure. Furthermore, although scoria cone eruptions are of short duration geologically, their potential duration of years to decades is very long compared to many hazardous phenomena that communities face (Sandri et al., 2012).

Assessment of long-term hazards in basaltic volcanic fields is a two-step process. First, statistical models must be prepared for the spatial distribution and recurrence rate of volcanism in order to forecast the probability of future eruptions (Jaquet and Carniel, 2006). Spatial distribution models generally use information about the location of existing vents, based on the premise that past patterns of activity will reflect the most likely locations of future activity (e.g., El Difrawy et al., 2013). Additional information is often incorporated into these models of spatial distribution, such as seismic tomographic anomalies (e.g., that may indicate areas of potential partial melting in the mantle), gravity anomalies (e.g., indicating zones of high extension rate and basin formation), and the locations of faults (e.g., high-dilation-tendency faults that may indicate preferred magma ascent pathways in the shallow crust). Such information can be used to weight the probable locations of future eruptions. These spatial density models must be capable of modeling clusters (multimodal distributions), they must be capable of modeling anisotropy in vent distribution due to tectonic stress or shape of the magma source region, and they must be sensitive to additional geological and geophysical data that may inform the analysis of vent distribution. Statistical models for spatial distribution of vents include nonparametric kernel methods (Figure 23.9), Cox process models, and near-neighbor models. These models yield an estimate of the spatial density of volcanism, $\lambda(x,y)$, or the probability of volcanic eruptions within a given small area, Δx, Δy, of the volcanic field, given an eruption somewhere in the field. Similarly, the recurrence rate of volcanism, $\lambda(t)$, must be modeled. Recurrence rate models include long-term averages, as described previously, or more complex models of clustering of eruptions through time. These models estimate the probability of future volcanic eruptions within some small area of the volcanic field, Δx, Δy, and during some time interval ΔT:

$$P[\text{volcanic eruptions within } \Delta x, \Delta y, \Delta T]$$
$$= 1 - exp[-\lambda(x,y)\Delta x\Delta y \cdot \lambda(t)\Delta T].$$

This equation can be integrated over some larger area than Δx, Δy, to estimate probabilities of volcanic eruptions over all or part of the volcanic field, while still capturing small-scale variations in the spatial density.

Once the probability of a volcanic eruption is estimated, the area affected by potential eruptions and the severity of these affects is estimated. One approach to estimating hazard impacts is to estimate the maximum distance from the potential new volcanic vent that might be impacted by specific phenomena. For example, a screening distance of 5 km might be chosen as a reasonable limit for ballistic impacts associated with the formation of a new vent. Or, the screening distance might be defined by a hazard attenuation function, in which the probability of impacts diminishes with increasing distance from the vent. Such screening distances are estimated from volcanological studies of specific eruptive phenomena in volcanic fields and by accounting for current features of the volcanic field (such as the distribution of surface water and shallow groundwater that will promote the potential for phreatomagmatic eruptions). If such screening distances are defined, convolving the probability of new vent formation with the screening

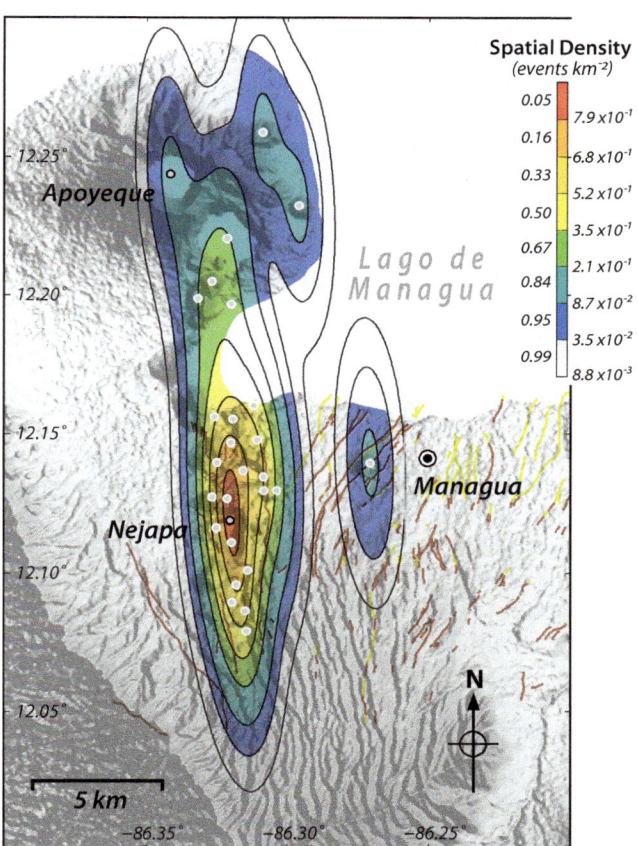

FIGURE 23.9 A spatial density model of volcanic hazards associated with new vent formation in the elongate Nejapa-Miraflores volcanic field (Nicaragua). The kernel density model is estimated from the distribution of mapped eruptive vents, and contoured to show the spatial density (probability of new vent formation) per unit area. For example, the model forecasts there is a 95% chance that future eruptions, when they occur, will be located in areas $>3.5 \times 10^{-2}$ vents/km^2 (blue), and a 50% chance future eruptions will be located in areas $>3.5 \times 10^{-1}$ vents/km^2 (yellow). Such probability maps are coupled with models of tephra fallout, lava flow, surge and related phenomena in order to construct a comprehensive forecast of hazards in volcanic fields.

distance yields an estimate of the hazard for specific locations, Δx, Δy, within the volcanic field.

Specific eruptive products, such as tephra fallout or lava flows, may also be simulated using numerical models of these phenomena in order to refine hazard assessments. Tephra fallout or lava flow inundation can be simulated using a Monte Carlo approach by sampling the statistical model of vent distribution for potential vent location, and then sampling distribution functions of eruption characteristics to simulate areas affected by eruptions (e.g., Connor et al, 2012). Although computationally expensive, such simulations have the advantages of allowing alternative scenarios (e.g., relatively explosive eruptions, effusive eruptions) to be tested and accounting for specific characteristics of the volcanic field, such as terrain. Eruption simulations depend on input data, such as the expected lava flow volume or explosive energy of future eruptions, and these data are derived from geologic studies of the volcanic field and analogous volcanic fields. The outcome of these simulations is an assessment of the types of hazards that exist across the volcanic field, the probabilities and the uncertainties associated with these hazards.

8. FUTURE DIRECTIONS

Much remains to be learned about basaltic volcanic fields, including the links between magma sources in the upper mantle and eruptive processes on the surface, shallow plumbing systems and their controls on eruptive styles, and the dynamics of phreatomagmatic eruptions. New advances in these topics at monogenetic volcanoes also provide insights into polygenetic basaltic volcanoes that have similar eruption styles. Although basaltic volcanic fields represent a sizeable flux of magma from the Earth's interior to its surface, little is known about the nature of long-term degassing and hydrothermal processes that couple volcanism with the atmosphere and hydrosphere; this is an important topic for future research. Although this chapter emphasizes subaerial volcanic fields, new studies are demonstrating the existence of monogenetic, small-volume, basaltic volcanic fields in intraplate sea floor settings. Understanding the similarities and differences between subaerial and submarine basaltic fields will provide new insights into petrogenesis (Valentine and Hirano, 2010). Hazard assessment methods for volcanic fields are ripe for meaningful progress on several fronts. Hazard models can be improved by better resolution of the geological and geophysical setting of volcanic fields, better age resolution of eruptions, and improvement of statistical and numerical models of activity. Currently, little attention is paid to monitoring of volcanic fields because of the long quiescence between eruptions and the fact that activity within a volcanic field is spread over a large area, rather than a focused area such as at a stratovolcano. A critical issue involves the allocation of monitoring resources to ensure that precursors to eruptions in volcanic fields might be detected.

Perhaps most importantly, significant progress is required to better understand the impacts of eruptions within volcanic fields on nearby communities and infrastructure, and to raise awareness of the potential consequences of these eruptions. A well-known anecdote in geology relates the total surprise expressed by people in the area near Parícutin volcano, and in fact by people worldwide, that a new volcano could simply grow in a cornfield (Luhr and Simkin, 1993). We will face such a scenario again, perhaps next time within a major metropolitan area. Studies of volcanic fields will better prepare us to anticipate the duration and impacts of such eruptions.

ACKNOWLEDGMENT

We thank Laura Connor (University of South Florida) for her work in preparing figures for this chapter.

FURTHER READING

Brenna, M., Cronin, S.J., Smith, I.E., Sohn, Y.K., Németh, K., 2010. Mechanisms driving polymagmatic activity at a monogenetic volcano, Udo, Jeju Island, South Korea. Contrib. Mineral. Petrol. 160, 931–950.

Brown, R.J., Valentine, G.A., 2013. Physical characteristics of kimberlite and basaltic intraplate volcanism and implications for a biased kimberlite record. Geol. Soc. Am. Bull. 125, 1224–1238.

Connor, L.J., Connor, C.B., Meliksetian, K., Savov, I., 2012. Probabilistic approach to modeling lava flow inundation: a lava flow hazard assessment for a nuclear facility in armenia. J. Appl. Volcanol. 1, 1–19.

Carn, S.A., 2000. J. Volcanol. Geotherm. Res. 95, 81–108.

Erlund, E.J., Cashman, K.V., Wallace, P.J., Pioli, L., Rosi, M., Johnson, E., Delgado-Granados, H., 2009. Compositional evolution of magma from Paricutín volcano, México. J. Volcanol. Geotherm. Res. 197, 167–187.

El Difrawy, M.A., Runge, M.G., Moufti, M.R., Cronin, S.J., Bebbington, M., 2013. A first hazard analysis of the Quaternary Harrat Al-Madinah volcanic field, Saudi Arabia. J. Volcanol. Geotherm. Res. 267, 39–46.

Germa, A., Connor, L.J., Cañon-Tapia, E., Le Corvec, N., 2013. Tectonic and magmatic controls on the location of post-subduction monogenetic volcanoes in Baja California, Mexico, revealed through spatial analysis of eruptive vents. Bull. Volcanol. 75, 1–14.

Gazel, E., Plank, T., Forsyth, D.W., Bendersky, C., Lee, C.-T.A., Hauri, E.H., 2012. Lithospheric versus asthenospheric mantle sources at the Big Pine Volcanic Field, California. Geochem. Geophys. Geosyst. 13. http://dx.doi.org/10.1029/2012GC004060. Q0AK06.

Gazel, E., et al., 2012. Geochem. Geophys. Geosyst. 13. Q0AK06.

Granados, H., 2009. Compositional evolution of magma from Paricutín volcano, México. J. Volcanol. Geotherm. Res. 197, 167–187.

Jaquet, O., Carniel, R., 2006. Estimation of volcanic hazards using geostatistical models. In: Mader, H.M., Coles, S.G., Connor, C.B., Connor, L.J. (Eds.), Statistics in Volcanology. Sp. Pub. IAVCEI 1, The Geological Society, London, pp. 89–104.

Johnson, R.R., Wallace, P.J., Cashman, K.V., Delgado Granados, H., 2010. Degassing of volatiles (H_2O, CO_2, S, Cl) during ascent, crystallization, and eruption at mafic monogenetic volcanoes in central Mexico. J. Volcanol. Geotherm. Res. 197, 225–238.

Kiyosugi, K., Connor, C.B., Wetmore, P.H., Ferwerda, B.P., Germa, A.M., Connor, L.J., Hintz, A.R., 2012. Relationship between dike and volcanic conduit distribution in a highly eroded monogenetic volcanic field: San Rafael, Utah, USA. Geology 40, 695–698.

Luhr, J.F., Simkin, T., 1993. Parícutin: The Volcano Born in a Mexican Cornfield. Geoscience Press, Inc., Phoenix, Arizona.

McGee, L.E., et al., 2012. Lithos 155, 360–374.

McGee, L.E., Smith, I.E.M., Millet, M.-A., Handley, H.K., Lindsay, J.M., 2013. Asthenospheric control on melting processes in a monogenetic basaltic system: a case study of the Auckland Volcanic Field, New Zealand. J. Petrol. 54, 2125–3153. http://dx.doi.org/10.1093/petrology/egt043.

Németh, K., 2010. Monogenetic volcanic fields: origin, sedimentary record, and relationship with polygenetic volcanism. In: Cañon-Tapia, E., Szakács, A. (Eds.), What is a Volcano? Geol. Soc. Am., pp. 43–66. Sp. Pap. 470.

Richardson, J.A., Bleacher, J.E., Glaze, L.S., 2013. The volcanic history of Syria Planum, Mars. J. Volcanol. Geotherm. Res. 252, 1–13.

Sandri, L., Jolly, G., Lindsay, J., Howe, T., Marzocchi, W., 2012. Combining long-and short-term probabilistic volcanic hazard assessment with cost-benefit analysis to support decision making in a volcanic crisis from the Auckland Volcanic Field, New Zealand. Bull. Volcanol. 74, 705–723.

Valentine, G.A., Gregg, T.K.P., 2008. Continental basaltic volcanoes—processes and problems. J. Volcanol. Geotherm. Res. 177, 857–873.

Valentine, G.A., Hirano, N., 2010. Mechanisms of low-flux intraplate volcanic fields—basin and range (North America) and Northwest Pacific Ocean. Geology 38, 55–58.

Valentine, G.A., Perry, F.V., 2007. Tectonically controlled, time-predictable basaltic volcanism from a lithospheric mantle source (central basin and range province, USA). Earth Planet. Sci. Lett. 261, 201–216.

Valentine, G.A., et al., 2011. Bull. Volcanol. 73, 753–756.

White, J.D.L., Ross, P.-S., 2011. Maar-diatreme volcanoes: a review. J. Volcanol. Geotherm. Res. 201, 1–29.

Chapter 24

Large Igneous Provinces and Flood Basalt Volcanism

Stephen Self

Department of Environment, Earth, and Ecosystems, The Open University, Milton Keynes, MK, UK; Department of Earth and Planetary Science, University of California, Berkeley, CA, USA

Millard F. Coffin

Institute for Marine and Antarctic Studies, Hobart, TAS, Australia

Michael R. Rampino

Department of Biology, New York University, New York, NY, USA; Department of Environmental Studies, New York University, New York, NY, USA

John A. Wolff

School of Earth and Environmental Sciences, Washington State University, Pullman, WA, USA

Chapter Outline

GLOSSARY

asthenosphere The relatively plastic low (seismic) velocity zone in the Earth's upper mantle that underlies the more rigid lithosphere that forms tectonic plates.

basalt (basaltic) A volcanic rock with high iron and magnesium and low silica content and composed mainly of small crystals of the minerals plagioclase and clinopyroxene, with $Ti-Fe-Mg$ oxides and often olivine or orthopyroxene.

eclogite A rock of basaltic chemical composition, but composed primarily of the minerals clinopyroxene and garnet, which are stable under mantle conditions of temperature and pressure.

fissure A line of volcanic vents.

hot spot A volcanic center, 100–200 km across and persistent for at least a few tens of millions of years that is thought to be the surface expression of a persistent rising plume of hot mantle material.

lithosphere The brittle outer layer of the Earth, made up of the crust and uppermost part of the mantle that "floats" on the asthenosphere and constitutes the tectonic plates.

magma Mobile, hot rock material, created by the partial melting of rocks at high temperature.

mantle The largest part of the planet Earth, between the core and thin outer skin or crust.

mantle plume The hypothetical cause of a hot spot, originating in the deep mantle, with a large plume head and a much narrower plume tail. A thermochemical feature.

MORB Mid-ocean ridge basalts, formed by the decompressional melting of mantle peridotite previously depleted by partial melting. MORB forms new oceanic crust.

nephelenite Strongly alkalic volcanic rock with low-silica content.

pāhoehoe (lava) (Hawaiian, pronounced "pa·ho·e·ho·e"): smooth and ropey-surfaced flows that are common in Hawaii.

The Encyclopedia of Volcanoes. http://dx.doi.org/10.1016/B978-0-12-385938-9.00024-9

picrite A volcanic rock with a large proportion of the magnesium-rich mineral olivine.

supereruption An eruption of any magma composition that releases $>1 \times 10^{15}$ kg of material, which is equivalent to ~360 km^3 of basaltic lava, or ~450 km^3 of rhyolitic lava, or ~1000 km^3 of rhyolitic volcanic ash.

silicic A magma or volcanic product type (pyroclastic or lava) that is rich in silica.

syenite An alkalic intrusive rock.

tholeiite A relatively silicic and iron-rich basalt.

1. INTRODUCTION

The creation of large igneous provinces (LIPs, Coffin and Eldholm, 1994 [R]) are exceptional volcanic events in Earth history because of the large total volume of dominantly mafic **magma** released (up to millions of km^3), erupted over a brief period of time (usually less than 1 million years). (Reviews of LIPs and LIP-forming eruptions are useful sources of reference material; those cited here are designated by [R] after the first citation.) Furthermore, the volume of magma emitted during each of the individual eruptions that make up an LIP (frequently 10^3-10^4 km^3) is also exceptional (Bryan et al., 2010 [R]). The combination of large erupted volumes and relatively frequent eruptions, either on land or in the ocean basins, led to the geologically rapid construction of extensive lava plateaus, up to several kilometers thick. LIPs come in two broad compositional varieties: **basaltic** LIPs, which are the most numerous and the only type forming oceanic plateaus, and **silicic** LIPs, which are few in number (Bryan and Ferrari, 2013 [R], see Figure 24.1). Silicic LIPs form some of Earth's major ignimbrite provinces (e.g., Sierra Madre Occidental, Mexico), see Bryan et al. (2010) and Bryan and Ferrari (2013).

The volcanic and plutonic products of LIPs collectively may cover areas well in excess of a million km^2 and attain thicknesses (intrusive and extrusive rocks) up to 40 km. Oceanic plateaus define the upper limits of the areal and volumetric dimensions of terrestrial LIPs, with the suggested reconstruction of the Ontong Java, Hikurangi, and Manihiki plateaus as originally one LIP having a prerift areal extent of 2–3 million km^2 (larger than the Indian subcontinent), a maximum crustal thickness of 35–40 km, and a total possible maximum igneous volume of $6-7.7 \times 10^6$ km^3, both including intruded material.

LIPs are composed of predominantly iron- and magnesium-rich mafic lavas that form by processes other than normal seafloor spreading. They are the dominant form of near-surface magmatism on the other terrestrial planets and moons of our solar system. Research between the late 1980s and the turn of the millennium (summarized by Courtillot and Renne 2003) showed that LIPs usually form within a geologically brief period of time, especially the climax, or peak, of volcanic output, which may be less

than one million years in duration (Rampino and Stothers, 1988). LIPs have formed throughout much of Earth's history and are a manifestation of Earth's normal, but in this case episodic, internal processes (see Mahoney and Coffin (1997) [R] for a monograph on LIP characteristics). The most common hypothesis to explain the cause of LIPs is that they are generated when a rising **mantle plume** impacts Earth's **lithosphere** (e.g., Campbell, 2005; this paper is part of a useful set on LIPs, see Saunders, 2005 [R]). Alternative hypotheses for LIP occurrences also exist (see Saunders (2005) and later in this chapter). This chapter will concentrate on basaltic LIPs on Earth, and the terms LIP (volcanism) and flood basalt province (volcanism), pertaining to subaerial basaltic LIPs, are used interchangeably.

2. COMPOSITIONAL RANGE AND DIVERSITY OF PROVINCES, GLOBAL DISTRIBUTION, AND TECTONIC SETTINGS

LIPs occur on continental crust (flood basalt provinces) and in ocean basins (oceanic plateaus), and at the transition between continents and oceans ("volcanic-"divergent margins) (Figure 24.1). In detail, some basaltic LIPs, which are dominated by vast piles of basalt lava flows (Figure 24.2), have silicic igneous rocks among their sequences, and vice versa for silicic LIPs. Several important LIPs were formed just prior to major continental breakup events on Earth, and the remnant rock sequences of the same LIP are found on separate continents. Examples are the Central Atlantic Magmatic Province, the Etendeka-Paraná provinces in Africa and South America, and the North Atlantic Volcanic Province (Bryan and Ferrari, 2013; Figure 24.1). While oceanic plateaus define the upper size limits for LIPs, the largest known subaerial, or continental, LIP in volume and area is the Siberian Traps (see Box 24.1), a vast flood lava province, which is largely basaltic and formed about 252 million years ago (Reichow et al.,

---

BOX 24.1 Trap(s)

The term *trap* or *traps* is often used in the name of continental flood **basalt** provinces, such as the Siberian Traps or the Deccan Traps of India. The term is of disputed origin, derived either from a Sanskrit word for south or the Swedish/Dutch word for a set of steps (stairs; *trappa*), and was first used by geologists in Victorian times (Geological Survey of India, 2012). The stepped topography of most flood basalt lava sequences (see Figure 24.2) is thought to be the reason for the use of this term.

Geological Survey of India, Deccan Basalt volcanism. Accessed May 23, 2012. www.portal.gsi.gov.in/pls/portal/url/page/.../GSI_STAT_DECCAN.

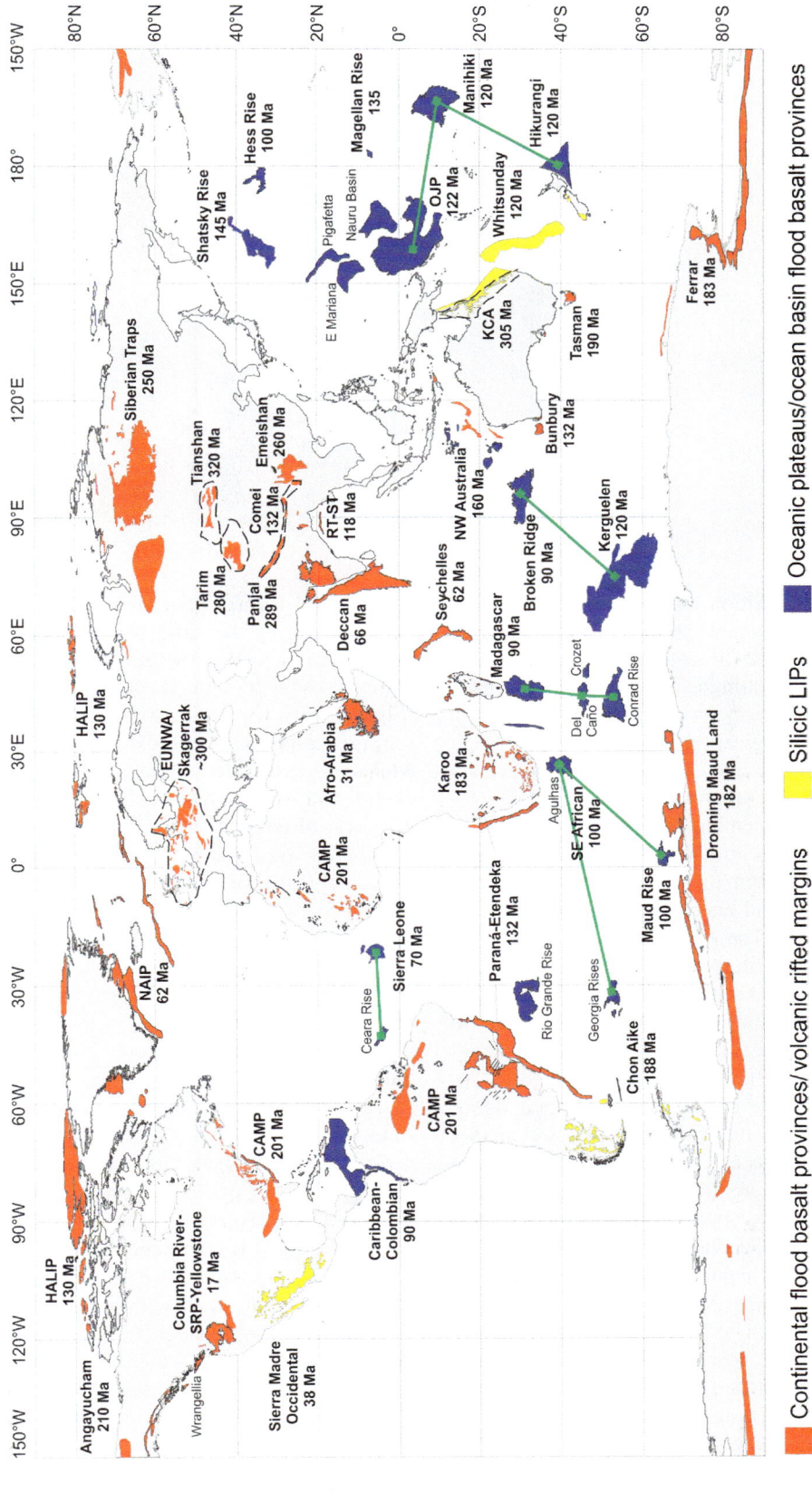

FIGURE 24.1 Large igneous provinces (LIPs) of the world, including continental flood basalt provinces, oceanic plateaus, and silicic LIPs. (*After Bryan and Ferrari (2013).*) Mentioned in this chapter are CAMP, Central Atlantic Magmatic Province; NAIP, North Atlantic Igneous Province; OJP, Ongtong Java Plateau. Ages are those of main phase or pulse of volcanism, according to above-mentioned authors. Inferred extent of some provinces shown by dashed lines; tie-lines join parts of same original province.

Continental flood basalt provinces/volcanic rifted margins

Silicic LIPs

Oceanic plateaus/ocean basin flood basalt provinces

FIGURE 24.2   Stack of lava flows, typical of the "trap" or stepped topography shown by flood basalts, exposed along the Grande Ronde River in Washington State of the United States. The >600 m-thick pile of lavas seen is part of Columbia River Basalt Group, Earth's youngest large igneous province. *Photo, Stephen Self.*

2005). It totals about 5 million km^2 in area and contains about 4 million km^3 of lava and associated volcanic rocks, which erupted over a period of about 1 Ma.

LIP rocks are readily distinguishable from the products of other types of magmatism on the basis of petrologic, geochemical, geochronological, geophysical, and physical volcanological data. LIPs and **hot spot** volcanoes, are commonly attributed to decompression melting of hot low-density **mantle** material ascending from the Earth's interior in mantle plumes, and thus provide a window onto mantle processes. This type of magmatism currently accounts for about 10% of the mass and energy flux from the Earth's deep interior to its crust. The episodic nature reveals dynamic nonsteady state circulation within the Earth's mantle, perhaps extending far back into Earth history, and suggests a strong potential for LIP emplacements to contribute to, if not initiate, major environmental changes (Eldholm and Coffin, 2000).

LIPs are defined by the characteristics of their dominantly iron- and magnesium-rich (basaltic or mafic) extrusive rocks; these typically consist of subhorizontal subaerial basalt flows (Figure 24.2). Silica-rich rocks also occur as lavas and intrusive rocks (see section 5 of this chapter on Petrology and Geochemistry). Relative to mid-ocean ridge basalts, LIPs include higher MgO lavas, basalts with more diverse major-element compositions, rocks with more common fractionated components, both alkalic and tholeiitic differentiates, basalts with predominantly flat light rare-earth element patterns, and lavas erupted in both subaerial and submarine settings.

Worldwide, LIPs occur in both continental and oceanic crust, in purely intraplate settings, and along present and former plate boundaries, although the tectonic setting at the time of formation is unknown for many examples. If a LIP forms at a plate boundary, the entire crustal section is composed of LIP-type crust (Figure 24.3). Conversely, if a LIP forms in an intraplate setting, the preexisting crust must be intruded and sandwiched by extrusive and intrusive LIP rocks, albeit to an extent that is not resolvable by current geological or geophysical techniques.

Continental flood basalts (CFBs), which are the most intensively studied LIPs owing to their surface exposure, were probably mainly erupted from **fissures** in the continental crust. Most CFBs are associated with sedimentary basins that formed via extension, but it is not always clear what happened first, the magmatism or the extension. Volcanic divergent margin LIPs form as a result of excessive magmatism during continental breakup along the trailing rifted edges of continents.

As the extrusive component of LIPs is the most accessible for study, nearly all of our knowledge of LIPs is derived from the lavas forming the uppermost lava sequences i.e., sections through CFB provinces. The extrusive layer may exceed 10 km in thickness in the Ontong Java Plateau and is commonly 3-km thick. On the basis of geophysical, predominantly seismic, data from LIPs and from comparisons with normal oceanic crust, LIP crust beneath the extrusive layer is believed to consist of a dense intrusive layer and a lower crustal body, characterized by P-wave velocities of 7.0−7.6 km s^{-1}, at and below the base of the crust (Figure 24.4). Beneath continental or oceanic crust this body may be considered as a magmatically underplated layer (e.g., Ridley and Richards, 2010).

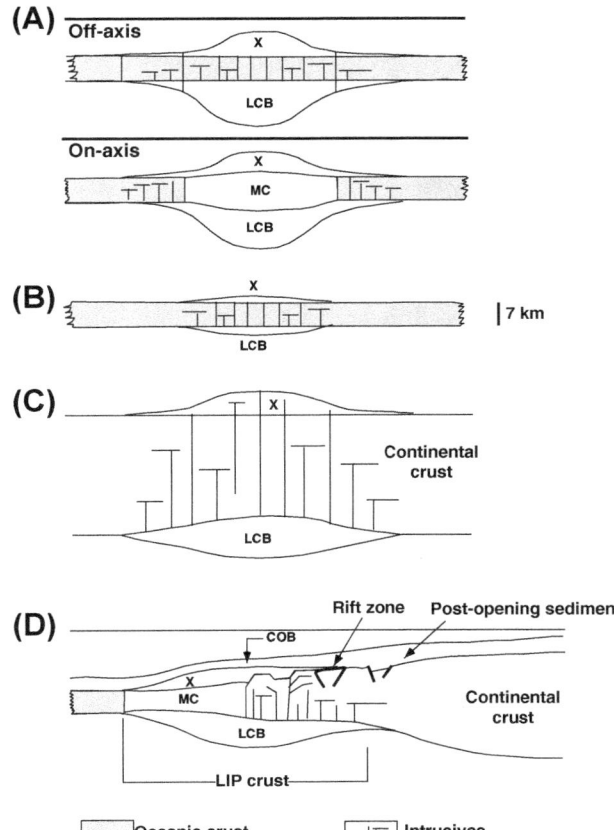

FIGURE 24.3 Schematic of large igneous province (LIP) plate-tectonic settings and gross crustal structure. LIPs are emplaced in various plate-tectonic settings, yet are characterized by a common three-layer crustal structure, although crustal thickness varies considerably. LIP crustal components are as follows: extrusive upper crust (X), middle crust (MC), and lower crustal body (LCB). Normal oceanic crust, 7 km thick, is shown in gray, and intrusives are denoted by vertical lines. Continent-ocean boundary (COB) is indicated for volcanic margin. Horizontal scale varies from a few hundred to more than thousand kilometers. (A) Oceanic plateau, (B) ocean basin flood basalt, (C) continental flood basalt, and (D) volcanic margin. *Modified after Eldholm and Coffin (2000).*

Seismic-wave velocities suggest a gabbroic intrusive layer and a lower crust that is ultramafic. If the LIP forms on preexisting continental or oceanic crust or along a divergent plate boundary, dikes and sills are probably common in the middle and upper crust and can be a voluminous component of LIPs.

3. PHYSICAL PARAMETERS: SIZE, AGE, PERIODICITY/TIMING

A feature of LIPs is an overall high-magma emplacement rate in which aggregate magma volumes of ∼1 million km^3 or more from a focused source were emplaced over periods of 1−3 million years. For most of the duration of province formation, there was probably no volcanic activity

(but there must have been intrusive activity); the total time interval when lavas were being erupted was probably less than 100,000 years (based on a theoretical 1000 eruptions, each lasting a century), and featured high-magma output during eruptions. Constraining the duration of CFB eruptions, or even a whole eruptive pulse, is difficult because of the precision limits of most geochronological determinations.

Consequently, it is the volume of magma emitted during individual LIP eruptions, the frequency of such large-volume eruptions, and the total volume of magma intruded and released during the main igneous pulses that make LIP events so exceptional in Earth's history. Without LIP-forming igneous events, basaltic **supereruptions** would not have occurred during Earth's history, nor indeed many of Earth's largest volcanic deposits (Bryan et al., 2010). By comparison, approximately 30 years of almost constant basaltic magma eruption at Kilauea volcano, Hawaii, from 1983 to present has produced about 4 km^3 of lava (USGS, 2013). At that rate it would take around 110 years to produce a Laki-sized eruption (15 km^3), let alone a CFB-sized lava field (∼1000 km^3), which would take ∼7500 years.

The largest CFB province, the Siberian Traps at 252 Ma ago, might have originally totaled 4−5 million km^3 of lava (not including the volume of intrusive sills below). It is the combination of large erupted volumes and the high frequency of eruptions that led to the rapid construction of extensive lava plateaus, which internally show few signs of major time breaks such as erosion surfaces and regional unconformities. Interbedded terrestrial and lake deposits are common, and lateritic soils, sometimes called boles, also exist on some flow tops and are evidence of a hiatus in activity during province growth.

LIPs have not been distributed uniformly over Earth's history (Figure 24.4). Such episodicity probably reflects variations in rates of mantle circulation. The $^{40}Ar/^{39}Ar$ dating technique (combined with uranium−lead (U−Pb) and rhenium−osmium (Re−Os) age determinations) has had a particularly strong impact on studies of LIP volcanism. Studies using these geochronological methods have shown that many LIPs erupted and grew over quite brief periods of geologic time (Courtillot and Renne, 2003; see Kelley, 2007 [R]), especially the major, more intense pulses of lava production within an LIP event. Some LIPs show evidence of two or more peaks in output. The total extent, and thus importance, of older LIPs is difficult to establish because the continuous subduction of ocean crust means that as much as four times the present area of ocean floors has been lost in the last 400 million years; older oceanic plateau LIPs will never be detected. By considering only the last 270 million years, it is possible to capture most of the recent oceanic LIPs and almost certainly all the continental LIPs. Courtillot and Renne

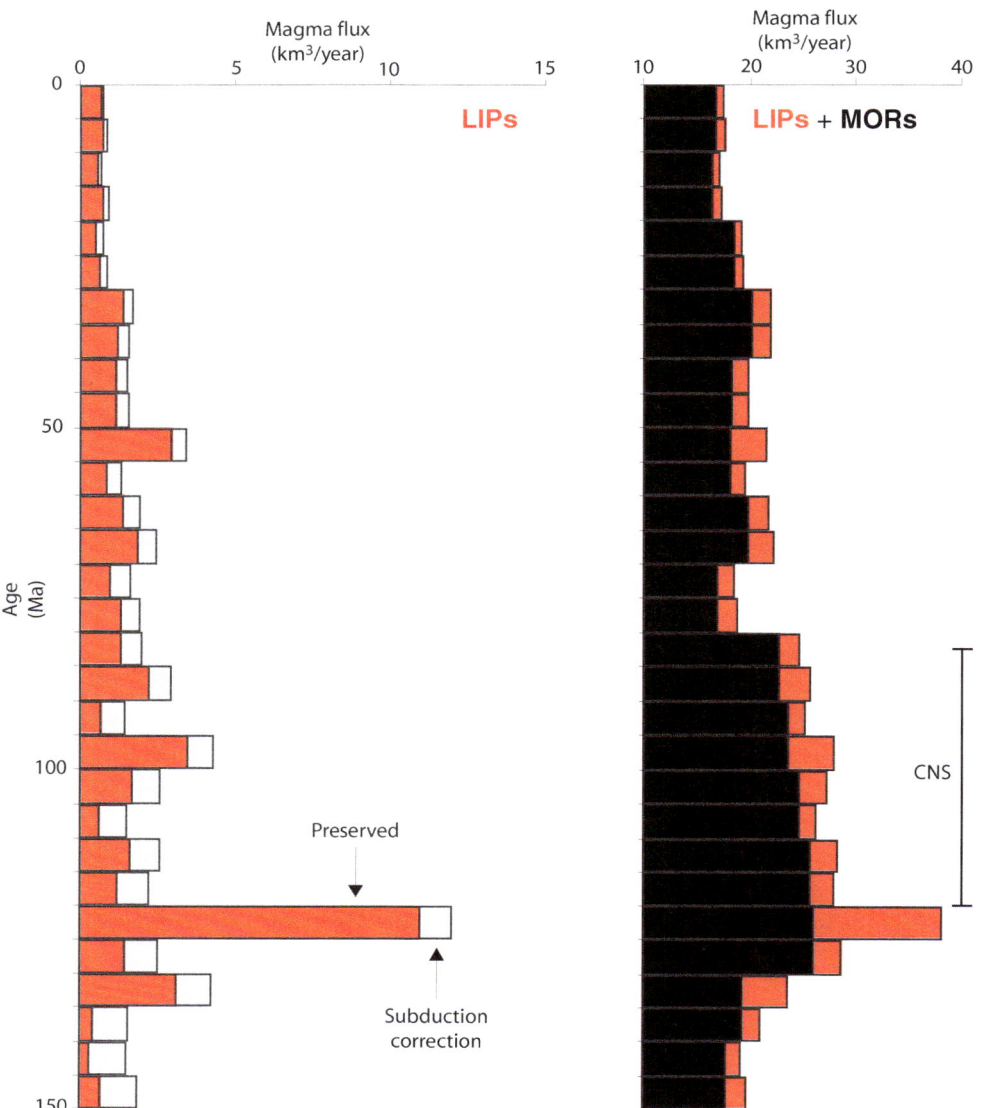

FIGURE 24.4 Large igneous province (LIP) magma production, corrected for subduction (left), and summed LIP and mid-ocean ridge (MOR) magma production (right) since 150 Ma. Red is LIP magma and black is MOR basaltic magma. Overall, mafic magma flux from both LIPs and MOR was highest in mid-Cretaceous time. Subduction correction assumes 1) that all oceanic crust is recirculated in mantle over 200 million years, and 2) a constant average LIP 5-million-year production volume over the same period. Note difference in x-axis scales; CNS is Cretaceous Normal Superchron. *Modified after Eldholm and Coffin (2000).*

(2003) reviewed the geochronology of LIPs and produced a database of the peak lava eruption times (see Table 61.1 in Chapter 61). Statistical studies of LIP episodes suggest periodicities in LIP production (and mantle plumes) at about 30−35 and 60 million years (Rampino and Prokoph, 2013).

4. SOURCE OF MAGMAS AND MANTLE DYNAMICS

Geochronological studies of CFBs, and other evidence, suggest that most LIPs result from mantle plumes, which initially transfer huge volumes (c. 10^5-10^7 km^3) of mafic rock into localized regions of the crust over short intervals (c. 10^5-10^6 years). The large-volume magmatism during LIP formation is commonly attributed to mantle-plume "heads" reaching the crust following transit through all or part of the Earth's mantle (Figure 24.5), whereas persistent (hot spot) magmatism is considered to result from mantle-plume "tails" penetrating the lithosphere, which may be moving relative to the plume. However, not all LIPs have obvious connections with mantle plumes or even hot spot tracks, suggesting that more than one source model may be required to explain some LIPs.

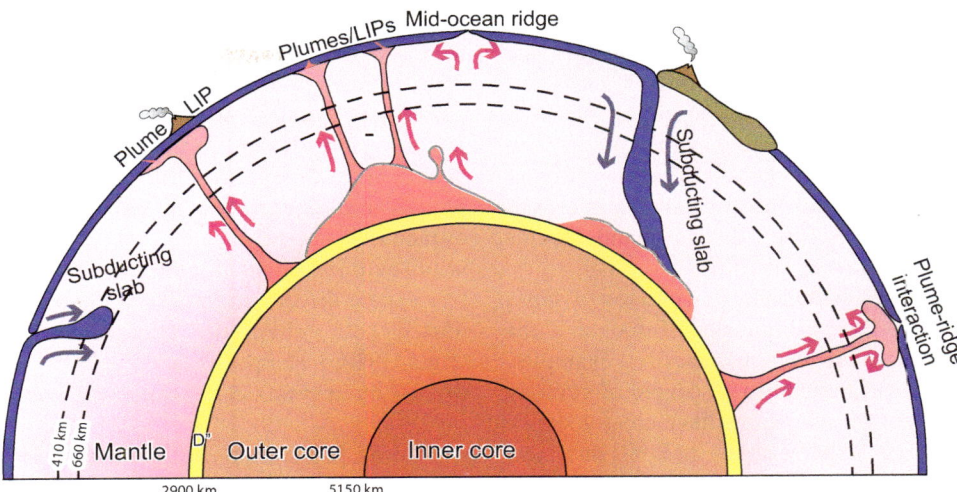

FIGURE 24.5 Schematic crosssection of Earth depicting sources of LIPs and hot spots, and fate of subducting lithosphere. LIPs are transient and believed to originate either from a thermal boundary layer near the core-mantle boundary (D″) or from thermal or mechanical boundary layers associated with large low-velocity zones. *Modified after Coffin and Eldholm (1994).*

The formation of various sizes of LIPs in diverse tectonic settings on both continental and oceanic lithosphere suggests that a variety of thermal anomalies in the mantle (e.g., higher temperatures than normal) give rise to LIPs and that the lithosphere strongly controls their formation. Equivalent thermal anomalies beneath continental and oceanic lithosphere should produce more magmatism in the latter setting, as oceanic lithosphere is thinner, allowing more decompression melting. Similarly, equivalent thermal anomalies beneath an intraplate region (e.g., Hawaii) and a divergent plate boundary (e.g., Iceland) should produce more magmatism in the latter setting, again because decompression melting is enhanced. Seismic tomographic images of mantle-velocity (a proxy for temperature) structure beneath Iceland and Hawaii show significant differences between the two, supporting this contention (Eldholm and Coffin, 2000).

Modeling of the impingement of a plume head on the underside of the lithosphere shows that the large head spreads out to form a thermal dome with the potential to form basaltic eruptions by decompressional melting over an area that may be as much as 2000 km in diameter (Campbell, 2005) The unusually large melt fractions formed in the mantle by this process cause the high-extrusion rates that typify flood basalt provinces. As lithospheric plates move across a plume, the plume head disperses, leaving only the plume tail erupting basaltic magma at a much lower rate. Such plume tails create chains of volcanic islands that terminate in active volcanoes marking the current position of the mantle plume, such as at Hawaii or Reunion.

Recently, seismic tomography has revealed that slabs of subducting lithosphere can penetrate the entire mantle to the D″ layer just above the boundary between the mantle and the core, at a depth of approximately 2900 km. If we assume that the volume of the Earth's mantle has remained roughly constant throughout geological time, then the mass of crustal material fluxing into the mantle must be balanced by an equivalent mass of material fluxing from the mantle to the crust. Most, if not all, of the magmatism associated with the plate-tectonic processes of seafloor spreading and subduction is believed to be derived from the upper mantle (above c. 660 km depth), on the basis of geochemistry and seismic tomography. It is reasonable to assume that lithospheric material entering the lower mantle is eventually recycled, in some part contributing to mantle plumes that rise to the Earth's surface.

5. PETROLOGICAL AND GEOCHEMICAL CHARACTER OF BASALTIC LIP MAGMAS

Flood basalt provinces are dominated by tholeiitic mafic lavas consisting of plagioclase, clinopyroxene (augite and often pigeonite) and iron—titanium oxides, usually with olivine and (less frequently) orthopyroxene. Plagioclase is the dominant phenocryst and may form very large crystals up to several cm in length, for example, in the Giant Plagioclase Basalts of the Deccan Volcanic Province (India: henceforth, Deccan) and the Steens and Imnaha formations of the Columbia River Basalt Group (CRBG). Volume-averaged compositions of **tholeiite** differ significantly from province to province with some provinces, most notably the CRBG, dominated by basaltic andesites; however, true basalts with <53% SiO_2 are more common globally. Some provinces (e.g., North Atlantic) erupted very primitive picritic magmas in their early stages. The tholeiitic character and the presence of **picrites** are consistent with high-degree partial melting of mantle at shallow depths. Nonetheless, in many provinces the onset of flood volcanism is preceded or accompanied by eruption of alkalic, strongly silica-undersaturated magmas (Karoo, Deccan, Siberia) that represent much smaller proportions of partial melt and, in some cases, distinct mantle source compositions (e.g., Reichow et al., 2005).

Many CFB provinces include large volumes of high-temperature rhyolites; Parana-Etendeka, Karoo, and the Proterozoic Keweenawan are well-characterized examples. Formerly, the CRBG was thought to be devoid of strongly silicic rocks, but recent investigations have clearly shown that it is part of the larger "Yellowstone Hot Spot" province, incorporating voluminous contemporaneous and later rhyolites. Rhyolites associated with CFBs typically have distinctive field characteristics; the ignimbrites are intensely welded to the point of resembling lavas, while true lavas may be voluminous and extensive. Emplacement of both rock types is favored by high-magmatic temperatures and low-magmatic water contents. Petrologic evidence is fully consistent with physical characteristics; the rhyolites usually lack hydrous phenocrysts, mineral geothermometers return high-magmatic temperatures (850–1050 °C), and both thermodynamic calculations based on mineral compositions and melt inclusion analyses indicate magmatic water contents of 2–3.5 wt% H_2O, significantly lower than common orogenic rhyolites. All these features contrast with more familiar dacites and rhyolites associated with active continental margins and define a distinct class of silicic volcanic rocks. The petrologic significance is that they demonstrate the importance of crustal melting associated with the generation and emplacement of CFBs.

Radiogenic isotope and incompatible trace element abundance ratios are thought not to be modified by mantle partial melting and early stages of magmatic evolution (dominated by fractionation of olivine) of basalts, and therefore their values reflect the compositions of the mantle sources. The dominant tholeiites of basaltic LIPs have isotope and trace element ratios that span almost the entire range of common basalts on Earth, with the exception of the most strongly depleted mid-ocean ridge basalts (**MORB**). It seems reasonable, therefore, that they are derived from a similar range of mantle sources as are invoked for common "nonflood basalts."

Broadly speaking, these sources are depleted mantle, enriched mantle similar to that for ocean island basalt (OIB), subduction-modified mantle, continental lithospheric mantle, and, perhaps, primitive mantle. Continental basalt products derived from any or all of these sources may be additionally modified by incorporation of continental crust during transport to the surface. Over the past 40 years,

a substantial literature with a database of many thousands of isotopic analyses has grown up around the relative contributions of these source components to basalts. A closely allied question is the ultimate cause of LIP and intraplate magmatism. At the time of writing, a majority of geoscientists favor a role for deep-seated (lower) mantle plumes in the generation of at least some LIPs, OIB, and intracontinental basalts, but a consensus does not exist, with some investigators preferring a purely upper-mantle, lithosphere-driven origin. The argument has been conducted both within the context of a single province (e.g., CRBG: Hooper et al., 2007 [R], and the ensuing discussion-in-print) and globally (e.g., www.mantleplumes.org).

The Sr, Nd, and Pb characteristics of fresh basalts from oceanic plateaus closely resemble those of the MORB–OIB spectrum, which is dominated by source region mixtures of depleted mantle (the chief source of MORB) and "FOZO," a mantle reservoir with distinct Pb isotope ratios that is prominent in most OIB (see Stracke et al., 2005), variably modified by ancient subduction of sediment to form, with time and isotopic ingrowth, additional distinct enriched mantle reservoirs. The rare "HIMU" OIB mantle reservoir, with very high Pb isotope ratios, appears to be absent from the major oceanic plateaus. Therefore, the mantle sources for oceanic LIPs appear to be chemically identical to those for ocean basin basalts generally.

All well-studied CFB provinces also have values of Sr, Nd, and Pb isotope ratios that are "rooted" within the MORB–OIB field (Figure 24.6) but radiate out to broad arrays of typically high $^{87}Sr/^{86}Sr$, low $^{143}Nd/^{144}Nd$, and very diverse Pb isotope ratios, features that characterize continental lithosphere. It is reasonable to infer common mantle sources for oceanic and continental LIP magmas, with the latter prone to modification via interaction with continental lithosphere, the only major Earth reservoir for high $^{87}Sr/^{86}Sr$ and low $^{143}Nd/^{144}Nd$. If so, then the involvement of lithosphere must be near-ubiquitous, because most CFB do not fully overlap with OIB in isotopic space (Figure 24.6). The lithosphere, of course, consists of both crust and mantle, and explanations for the origins of this signature in CFB tend to favor either continental crust or subcontinental lithospheric mantle (SCLM) as the source of the distinct isotopic signatures in the basalts, with different geodynamic implications. In the SCLM source model, the mantle lithosphere melts to supply a fraction of the basalt. This may be achieved via heating by the subjacent plume head, heating by intrusion of hot, primitive basalt derived from a plume into relatively low-solidus enriched SCLM, and plume-induced uplift and extension of the lithosphere. In the crustal contamination model, mantle-derived magma incorporates crust during transport through, and transient

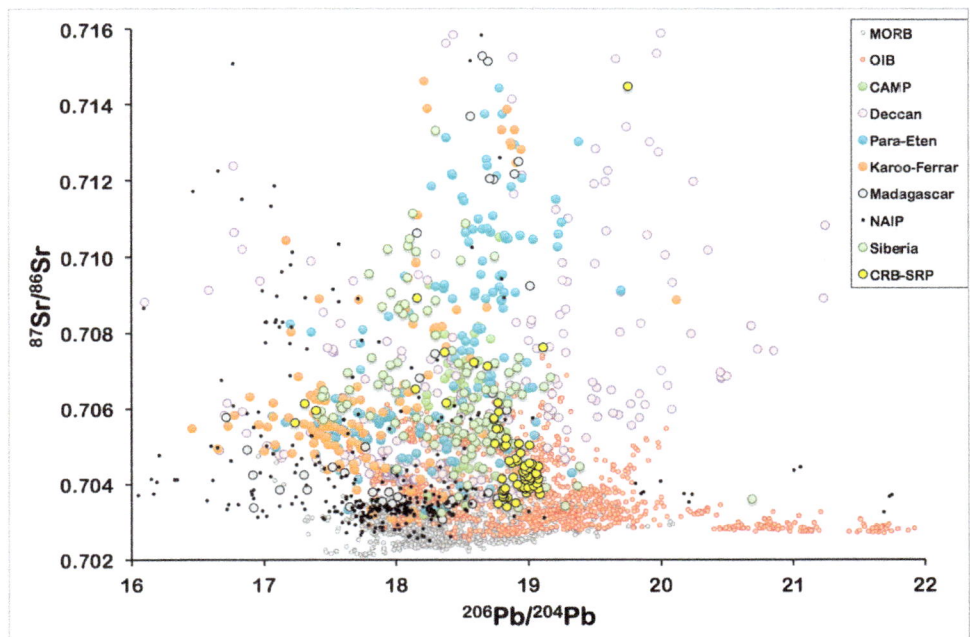

FIGURE 24.6 Compilation of CFB Sr and Pb isotope data from the GEOROC database; MORB-OIB data from Stracke et al. (2003). Some outlying points plot beyond the limits of the diagram and are excluded. MORB, Mid-ocean ridge basalt; OIB, ocean island basalt; CAMP, Central Atlantic magmatic province; Deccan, Deccan Traps (India); Para-Eten, Parana-Etendeka province, Brazil and Namibia; Karoo-Farrar, South Africa and Antarctica; NAIP, North Atlantic igneous province, British Isles, Faroes, Iceland and offshore Norway; Siberia, Siberian Traps, Russia; CRB−SRP, Columbia River basalt−Snake River plain province (a.k.a. the Yellowstone hot spot track). Note that several provinces are dispersed across more than one continent today due to plate motions subsequent to their emplacement.

storage within, the continental crust, with few implications for the dynamic behavior of the mantle source. In some provinces there is strong circumstantial evidence, such as correlations between isotope ratios and indices of magmatic differentiation, as well as isotopic disequilibrium among phenocrysts grown at crustal depths, in favor of crustal contamination (Wolff and Ramos, 2013 [R]; Borges et al., 2014). The common association between CFB and high-temperature rhyolites, noted earlier, also constitutes strong evidence for interaction between basaltic magma and continental crust. A case can also be made for an SCLM contribution, if some reasonable assumptions are made about the major element, trace element, and isotopic compositions of crust and SCLM. The arguments are frequently complex but may, in sum, be compelling on a case-by-case basis.

Much effort has been devoted to the question of SCLM versus crustal contribution to CFB genesis. The fundamental difficulty is that isotope ratios of the elements Sr, Nd, and Pb do not allow clear distinction between SCLM and continental crust, so an alternative approach is to seek a diagnostic isotopic signature in the CFB lavas themselves. Osmium isotopes provide such a fingerprint. Unlike other commonly used radiogenic isotope tracer systems, in which both parent and daughter are incompatible elements, osmium is highly compatible (effective olivine/melt

partition coefficients are $>>10$ and maybe >100), hence relatively abundant in refractory upper mantle, while the parent ^{187}Re is moderately incompatible. Ancient melting events have therefore resulted in low $^{187}Os/^{188}Os$ ratios being "locked in." Evidence from SCLM xenoliths shows these low ratios are resistant to later enrichment events that dominate the incompatible element SCLM budget, probably because the enriching melts (or fluids) contribute insignificant Os, and insufficient Re to perturb the Os reservoir through ingrowth of new ^{187}Os. SCLM therefore has lower ratios than any other major mantle reservoir, in contrast to continental crust, which has high Re/Os and hence extremely radiogenic $^{187}Os/^{188}Os$. Low $^{187}Os/^{188}Os$ has been found in some CFB provinces, notably the Karoo and the Siberian Traps, but not the CRBG. Much of the low $^{187}Os/^{188}Os$ data come from early erupted picrites which, because of their high-Os contents, are resistant to perturbation of the ratio through crustal contamination. But because Os is so compatible, even a modest amount of olivine fractionation results in drastic reduction of melt Os concentrations, and typical basalts (5−8% MgO) are thus highly sensitive to Os contamination by crustal additions. The Grande Ronde basalts and basaltic andesites, which volumetrically dominate the CRBG, have very high $^{187}Os/^{188}Os$, consistent with other evidence for a significant crustal component in these lavas.

Of course, roles for SCLM and crust as contributors of incompatible element enrichment in CFB are not mutually exclusive, but SCLM components, if present, should in principle be detectable among the most primitive basalts ($\geq$8% MgO) of any one province, whereas a crustal contribution is expected to increase in proportion among more differentiated lavas. The CRBG is a case in point. The dominant Grande Ronde lavas (52–58% SiO_2, 3–6% MgO) are clearly contaminated by crust (e.g., Wolff and Ramos, 2013) and have $^{87}Sr/^{86}Sr$ up to 0.706. Later, more magnesian, Saddle Mountains basalts with $^{87}Sr/^{86}Sr$ up to ~0.712 are interpreted by most workers to contain a significant SCLM component, introduced as the North American craton overrode the upwelling mantle that is ultimately responsible for the CRBG.

The broad compositional correspondence of LIPS basalts and OIB complements the "head–tail" relationship of LIPs and oceanic hot spot tracks, suggesting a common mantle source. The FOZO component may reside in the lower mantle, whereas DM, which is more refractory and therefore less dense, is probably mostly located in the upper mantle. The trace element signature of FOZO points to an origin by recycling and melting of ancient slabs subducted into the lower mantle, and is consistent with, but does not require, a deep-seated plume origin. The clearest evidence for deep-seated plumes comes from He isotopes. As a noble gas, He is efficiently lost from the mantle during melting. Radioactive decay (chiefly of U, Th and Sm) continually replenishes ^4He, but ^3He is primordial, hence high $^3He/^4He$ ratios signify incompletely degassed mantle, which is thought not to survive above the 660 km discontinuity. Depleted mantle (determined by analysis of MORB) has RA ~ 8, where RA is the atmospheric value of $^3He/^4He$. Helium isotope ratios in OIB vary from low values relative to MORB, consistent with slab recycling, to very high values (RA > 30) indicating the presence of primitive, incompletely degassed and presumably therefore lower mantle material in the source. Helium isotope data is available for some LIPS and associated OIB along connected hot spot tracks. In summary, CRBG-Yellowstone track, Deccan-Reunion track, Kergulelen-Heard Island, and the Siberian Traps all have RA > 8, while Tristan (Parana-Etendeka hot spot) has RA < 8. Geochemical data therefore support a deep-seated plume origin for many LIPs, including the relatively small-volume CRBG-Yellowstone province.

6. OCEANIC PLATEAUS

Submarine plateaus are broad, typically flat-topped features generally lying 2000 m or more above the surrounding seafloor. They can be oceanic, continental, or hybrid in origin; only those of dominantly mafic composition or with a significant mafic carapace are considered to be LIPs.

Among the best studied oceanic plateaus are the Ontong Java Plateau in the western equatorial Pacific, the Kerguelen Plateau in the southern Indian Ocean, and Shatsky Rise in the northwest Pacific. Although morphologically not a plateau, the Caribbean ocean basin flood basalt province is analogous to oceanic plateaus and is relatively well studied. A hybrid plateau with a significant basaltic carapace is the Naturaliste Plateau in the southeast Indian Ocean. The crustal nature of some submarine plateaus is unknown, e.g., the Madagascar Ridge and Conrad Rise.

Oceanic plateaus can form at divergent plate boundaries (e.g., the Shatsky Rise), or in intraplate settings (e.g., Bermuda Rise), the latter analogous to CFB provinces. However, it appears that most oceanic plateaus form proximal to mid-ocean ridges, suggesting that mid-ocean ridges are pinned or tied where mantle plumes and mid-ocean ridges interact.

Relative to CFB provinces and volcanic divergent margins (see below), oceanic plateaus offer better opportunities to examine the formation and petrogenesis of LIPs, because they erupt at seafloor spreading centers or through ~7 km-thick mafic oceanic crust and are uncontaminated by felsic crust. Furthermore, some oceanic plateaus are partially obducted, enabling observational opportunities and sampling densities similar to those that CFB provinces offer.

Significant results to date from oceanic plateaus indicate subsidence behavior both analogous (Kerguelen) and anomalous (e.g., Ontong Java) relative to normal oceanic lithosphere. Results also show (1) petrological and geochemical homogeneity (Ontong Java) and heterogeneity (Kerguelen/Broken Ridge) extending over 10^6 km^2; (2) an uppermost crust consisting of alternating massive and thin lava flows; (3) enormous individual volcanoes may form a significant part of an oceanic plateau; and (4) temporal correlations between oceanic anoxic events and oceanic plateau emplacements.

Outstanding questions include how oceanic plateaus are constructed and hence their internal magma plumbing systems; the spatiotemporal history of individual LIP emplacements; the nature of the ubiquitous high-velocity (7.0–7.6 km s^{-1}) lower crustal bodies; the nature of low-velocity upper mantle roots beneath some oceanic plateaus; the origin and nature of thermochemical anomalies in the mantle that give rise to LIPs; and the physical, chemical, and biological mechanisms associated with LIPs that were synchronous with major environmental changes.

7. VOLCANIC DIVERGENT MARGINS

Continental rifting and breakup are accompanied in many instances by voluminous magmatism. In fact, more than 50% of rifted passive margins worldwide are magma dominated. In most occurrences, magma-dominated divergent continental margins are offshore equivalents to CFB provinces (e.g., North Atlantic Volcanic Province; Central Atlantic Magmatic Province; Parana-Etendeka and South Atlantic volcanic margins). Because many magma-dominated margins have thick sediment sections, much of our knowledge comes from seismic data and drilling younger margins where sediment overlying volcanic rock is relatively thin.

Drilling of seaward dipping reflector (SDR) wedges of the North Atlantic Volcanic Province confirmed them to be a thick series of subaerial lava flows covering large areas. On the landward side of the SDRs, geochemistry of the lavas indicates contamination by continental crust, implying ascent through continental crust during early rifting. On the seaward side of the SDRs, lavas appear to have formed at a seafloor spreading center resembling Iceland. Drilling results from the British, Norwegian, and SE Greenland margins document extreme magmatic productivity over a minimum distance of 2000 km during continental rifting and breakup, with temporal and spatial influence of the Iceland plume clearly visible during rifting, breakup, and early seafloor spreading. In addition to the SDRs, igneous intrusions (high-seismic velocity bodies) at the base of the crust are observed.

Magmatism along divergent continental margins is a function of the degree of lithospheric thinning, the degree of asthenospheric upwelling, and the strain distribution across the margins. The timing, volume, chemistry, and style of magmatism are the key parameters to investigate. Three primary competing mechanisms for excessive magmatism are: a mantle plume with elevated temperatures; small-scale convection at the base of the mantle; and heterogeneities in mantle source composition.

Key questions remaining to be addressed include: melt sources and melting conditions; timing of magmatism; spatial and temporal variations of volcanism; eruption environment and vertical movements; along-axis variations in melt production; and consequences of excessive magmatism for environmental change (e.g., the Paleocene-Eocene thermal maximum).

8. CONTINENTAL FLOOD BASALTS

8.1. Eruption and Lava Flow-Field Sizes

Despite the huge total cumulative eruptive volumes of lava, and the fact that the overall timing of LIP events is reasonably well-constrained, the scientific understanding of the size, duration, and frequency of individual flood basalt eruptions, which are the best exposed and studied products of LIP volcanism, is very limited. Almost all information on the size of individual flood basaltic eruptions comes from the many studies undertaken on the Columbia River flood basalt province in the US Pacific Northwest (e.g., the

monograph edited by Reidel et al., 2013 [R]). This is the smallest (roughly 250,000 km^3 of lava) and youngest example of a CFB province; most eruptions occurred between 16 and 15 million years ago, in the mid-Miocene geologic epoch. Larger provinces, such as the Siberian Traps (Reichow et al., 2005), originally contained in total at least 10 times as much lava. During the growth of flood basalt provinces, hundreds to possibly thousands of eruptions produce immense lava flows, each the product of a vent or a group of vents along a fissure (Self et al., 1998). It is only since 2005 that some understanding has been gained of the magnitude of flood basalt eruptions from other flood basalt provinces, mainly the Deccan Traps (e.g., Self et al., 2008).

The basaltic lava flow fields in LIPs in many cases are so extensive, and the provinces so widespread and fragmented or eroded, that it has taken many years of study to determine the products of a "typical" flood basalt eruption (White et al., 2009 [R]). The Journal of Geophysical Research issue on long lava flows (edited by Cashman et al., 1998 [R]) contains many useful papers. Recently, proposed chemically correlated flow types such as the McCoy Canyon or Cohassett of the Grande Ronde Basalt Formation of the CRBG, indicates much larger volume flows may have been emplaced during the interval when ~70% of the volume of the Columbia River province was erupted (Reidel et al., 2013), see Figure 24.7. Also, studies on the Ambenali and Mahabaleshwar Formations of the Deccan province indicate that single Deccan formations have volumes similar to the entire CRBG, ~230,000 km^3, with volumes of individual flow fields ranging from ~2000 to >8000 km^3 (Self et al., 2008). This general upper magnitude for flood basalt lavas is similar to the dimensions of associated sills in flood basalt provinces, such as the enormous Peneplain Sill in the Dry Valleys, Antarctica, with an estimated volume of 4750 km^3.

8.2. Character of Lavas

Dominantly, **pāhoehoe** lava flow fields form the flood basalt lavas in LIPs; 'a'ā flows are in the minority (Brown et al., 2011). All flood basalt provinces from the very ancient geologic eras to the youngest on Earth consist almost entirely of pāhoehoe (Self et al., 1998). Some flood basalt lavas occur as rubbly pāhoehoe flows (e.g., Kezsthelyi et al., 2006) where short-lived pulses at higher effusion rates, potentially up to 10^6 m^3 s^{-1}, disrupted the thick, insulating crust and prevented the more commonplace pāhoehoe mode of emplacement. One explanation is that these surges of higher effusion rate may reflect enhanced periods of magma output rate over greater lengths of the fissure system.

Each flow field is interpreted to be the product of one, but potentially sustained, eruptive event producing several

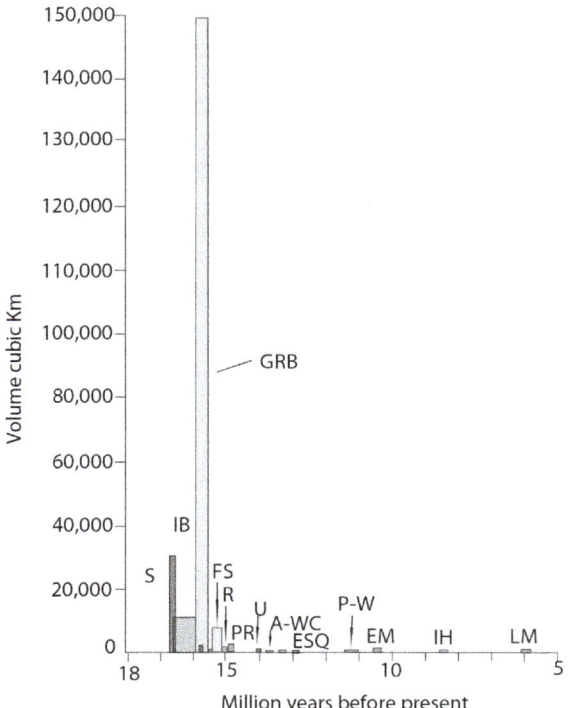

FIGURE 24.7 Volume of lava plotted against time (based on latest $^{40}Ar/^{39}Ar$ age dates for Columbia River Basalt Group, showing large "spike" of main pulse of volcanism, the GRB (Grande Ronde Basalt Formation). Other groups of lava identified by letters are not mentioned in this chapter, except Roza (R).

major compound lava flows that in turn consist of multiple sheetlike flow lobes. The sheet lobes contain the majority of the lava volume, and in the Columbia River and Deccan LIPs are commonly 20−30 m thick and several kilometers wide. In the Karoo and Etendeka flood basalt provinces, however, studied sections generally reveal only a few lava flows >20 m thick, suggesting that the majority of the flood lava volume is not always expressed in thick lava flow units. Nevertheless, the flood basalts from all these provinces show features consistent with in situ flow thickening by endogenous growth and inflation (Hon et al., 1994). These extensive lobes, with aspect ratios (length/thickness) ranging from ∼50 to 500, are the basic building-blocks of CFB provinces and give the provinces their "layer-cake," or steplike appearance (White et al., 2009).

Sheet lobes show a relatively simple internal structure of crust and lava core facies that are ubiquitous in the lavas regardless of thickness (Figures 24.8 and 24.9), persisting over the extent of the entire flow field (10^2−10^3 km^2). Parts of some CFB provinces, e.g., Deccan and Faroes (North Atlantic province), are formed of flow fields dominated by thinner, smaller lava units stacked up to form compound lavas. Also, complexities such as pillowed facies are relatively common in the Columbia River Basalt lava flow fields, and occur when lava entered lakes on the flat-surfaced, developing lava pile. It should be noted that the regional slopes in many flood-basalt provinces once the first few pioneer flow fields have been erupted is remarkably low, mostly much less than 1°. This implies that the surface of the previous flow field is very important in

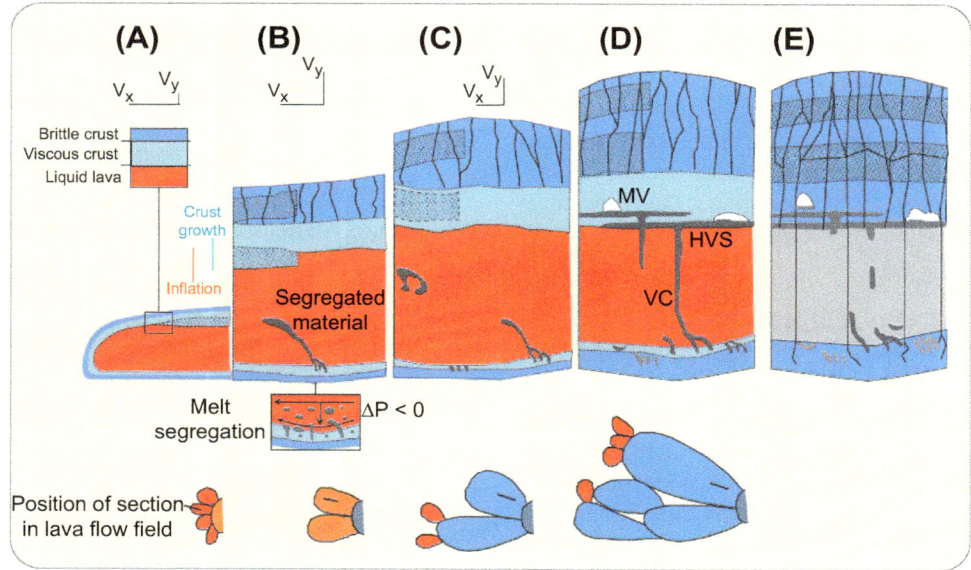

FIGURE 24.8 Cartoon showing development of a pāhoehoe basalt lava-field over time as lava sheet-lobes grow outward and downslope by advance; also coalescence of smaller lobes and thickening by inflation occurs. *(After Thordarson and Self (1998).)* This scheme applies to flood basalt lava flow-fields. Red shows liquid lava; blue and gray show solidified lava. MV, megavesicles; HVS, horizontal vesicular sheets; VC, vesicle cylinders. Relative lateral and vertical expansion of lobes shown by Vx and Vy; bottom sketches show plan view of growing lava lobes. (A) Hours, (B) days, (C) weeks, (D) months, and (E) solidified sheet lobe.

FIGURE 24.9   Part of Sand Hollow flood basalt flow-field showing internal morphology of a single large-magnitude, ~60 m thick, pāhoehoe sheet-lobe. Upper and lower colonnade jointed regions are separated by more closely spaced jointed central entablature region. Location is Palouse Falls, Washington, the United States (46° 39.730′ N 118° 13.554′W), in Columbia River Basalt province. People standing on eroded sheet lobe top provide scale; behind them is a gorge and continuation of same lobe. Lava is light colored due to surface coating of mud-spray from adjacent waterfall. *Photo, Stephen Self.*

controlling the pathways and character of the succeeding lava lobes in the next eruption.

8.3. Eruption Styles

Studies have shown that: (1) at least parts of flood basalt eruptions were "Hawaiian-like" in nature at the vent (Reidel and Tolan, 1992), but sometimes the vent edifices are on a larger scale; (2) the flow fields most likely originated from prolonged eruptions lasting years to decades, with the larger flow fields fed by very long fissures; (3) the length of flow fields was supply-limited rather than landscape- or cooling-limited (Keszthelyi et al., 2006), with several CRBG flows reaching the sea >400 km from vent; (4) at any one time during a CFB eruption, activity was more likely confined to distinct fissure segments on the vent system (Thordarson and Self, 1998), resulting in

incremental flow advancement and flow field growth; and (5) postulated eruption rates of ~4000 m^3 s^{-1} (Self et al., 1998) are similar to the maximum sustained eruption rates of the largest historic flood lava eruptions (e.g., the Laki (Iceland) event in AD 1783–1784; Thordarson and Self, 1993). Given the inferred duration and sporadic and restricted activity along the vent fissure system of flood basalt eruptions, defining the product of one eruption remains a difficult task in flood basalt provinces. It also appears that the putative longest flows on Earth (>1000 km long) from the Deccan province (Self et al., 2008), were also limited in length by flowing into the ocean.

The eruptive style of flood basalt lavas is important in controlling the atmospheric and climatic effects during these events. In fissure-fed basaltic lava flow eruptions, the volumetric eruption rate is a critical parameter in determining the height of fire fountains over active vents, the convective rise of the volcanic gas plumes in the atmosphere, and hence the climatic impact of an eruption. Flood basalt lavas were originally envisioned to flow as turbulent sheets 10–100 m thick that covered large areas in a matter of days. By contrast, it is believed now that flood basalts are erupted mainly as fissure-fed pāhoehoe flows that inflate to their great thicknesses during the time that they reach their great extent. This suggests a more gradual emplacement of the flows over longer periods of time, most likely years to decades. Yet, the average eruption rates must still have been higher than in any historic eruption except perhaps the extremes of some Icelandic lava outpourings. For example, the time required to emplace a large 1000–2000 km^3 lava flow at the range of peak output rates of the largest historical lava flow eruption, Laki, would be approximately 10–20 years.

8.4. Vent Systems

The vent system for only one moderate-sized flood basalt eruption (Roza) has been studied in detail (Brown et al., 2014), and the nature and extent of activity along other fissure vent systems and dykes for other flood basalt flow fields remain poorly known. However, given the ∼180 km length of the total fissure system for the Roza flow, at the above mentioned (1000s $m^3 s^{-1}$) effusion rates, only part of the fissure system must have been active at any one time. The eruption rate averaged over the whole fissure length would be exceedingly low (∼0.03 $m^3 s^{-1} m^{-1}$ length of fissure) and could possibly result in magma freezing in the dykes/fissures in transit to the surface. This may be one mechanism by which effusion becomes concentrated or localized along the fissure system. Voluminous sheet lava flow-fields, therefore, do not require rapid extrusion of mafic magma at rates much higher than in historic young flood volcanism (e.g., Laki, 1783), nor do long fissures imply high-eruption rates, as only single segments of the fissure were likely active at any one time.

8.5. Lava Feeders and Shallow Plumbing

Determining the discharge rate of CFB lavas has implications in terms of requirements for vast plumbing systems and magma reservoirs, such that the magma is delivered quickly to the surface, perhaps via lateral magma flow in dikes (Wolff and Ramos, 2013 [R].)

The potential similarity in terms of discharge rate and therefore eruption style between small-volume historic (≤20 km^3; e.g., Laki) and flood basalt (≥1000 km^3) eruptions raises the possibility that there is a physical limitation to how much, and how quickly, basaltic magmas can be erupted. Despite the apparent significant length of fissure vents, the effusion rates appear supply rate-constrained, resulting in eruptions that must last years to decades in order to empty stored magma reservoirs in the

crust. Insight into why flood basalt eruptions do not have extraordinarily high-discharge rates given the huge volumes of magma erupted may come from temporal compositional trends observed in intraplate basalt eruptions. In these cases, rates of melt extraction from the mantle and source-to-surface melt velocities of ∼10^0–10^1 km year^{-1} are indicated, and may provide the limiting constraints on effusion rates at the vent. Some flood basalt lava eruptions in the Columbia River flood basalt province have also been demonstrated to preserve temporal compositional variations (Reidel, 1998) suggesting that a similar process may operate for larger volume LIP eruptions.

9. SUMMARY

The creation of LIPs are exceptional volcanic events in Earth history because of the large total volume of dominantly mafic magma released (up to millions of km^3), erupted over a brief period of geologic time. The volume of magma emitted during each individual eruption that makes up an LIP (frequently 10^3–10^4 km^3) is also exceptional. LIPs fall into two broad types: basaltic LIPs, which this chapter describes, and silicic LIPs. Submarine LIPs, or oceanic plateaus, are always composed of mafic magmas.

LIPs occur on continental crust (flood basalt provinces) and in ocean basins (oceanic plateaus), and at the transition between continents and oceans ("volcanic" divergent margins). Basaltic LIPs usually form within a geologically brief period of time, especially the climax, or peak, of volcanic output, which may be less than 1 million years in duration. The eruptions that form flood basalt provinces have been proposed as a cause of major environmental perturbations throughout much of Earth history, including mass extinctions. Some of these eruptions were possibly up to 10^4 km^3 of magma in size, and the provinces may consist of up to 5 million km^3 of lava.

Relative to mid-ocean-ridge basalts, basaltic LIPs include higher MgO lavas, basalts with more diverse major-element compositions, and rocks with more common fractionated components. Flood basalt provinces are dominated by tholeiitic mafic lavas and contamination by either continental crust or SCLM have been proposed to be the source of the distinct isotopic signatures in the basalts. Further, associated silicic volcanic rock in many provinces demonstrate the importance of crustal melting associated with the generation and emplacement of CFBs. Lavas may have been fed from a source zone by lateral transport in dikes in some provinces.

Pāhoehoe-type lavas dominate in flood basalt provinces, and eruptions are thought to have durations of decades to perhaps centuries. The longest and largest eruptive units on Earth are confined to LIPs. The huge flow fields of pāhoehoe lavas are dominated by sheet lobes in many provinces, as well as alternations with flow fields dominated by smaller inflated lava lobes. Vents are poorly known, but some, at least, were long fissure-type systems.

ACKNOWLEDGMENTS

We dedicate this chapter to the memory of Peter Hooper, author of the entry on Flood Basalts in the First Edition, who died before this Second Edition of EOV was undertaken.

FURTHER READING

Borges, M.R., Sen, G., Hart, G.L., Wolff, J.A., Chandrasekharam, D., 2014. Plagioclase as a recorder of magma chamber processes in the Deccan Traps: Sr-isotope zoning and implications for Deccan eruptive event. J. Asian Earth Sci. 84, 95–101. http://dx.doi.org/10.1016/j.jseaes.2013.10.034.

Brown, R.J., Blake, S., Thordarson, T., Self, S., 2014. Pyroclastic edifices record vigorous lava fountains during the emplacement of a flood basalt flow field, Roza Member, Columbia River Basalt Province, USA. Geol. Soc. Am. Bull. 126, 875–891.

Bryan, S.E., Ferrari, L., 2013. Large igneous provinces and silicic large igneous provinces: Progress in our understanding over the past 25 years. Geol. Soc. Am. Bull. 125, 1053–1078, 10.1130/B30820.1.

Bryan, S.E., Ukstins Peate, I.A., Self, S., Peate, D., Jerram, D.A., Mawby, M.R., Miller, J., Marsh, J.S., 2010. The largest volcanic eruptions on earth. Earth-Science Rev. 102, 207–229.

Campbell, I.H., 2005. Large igneous provinces and the mantle plume hypothesis. Elements 1, 265–268.

Cashman, K., Pinkerton, H., Stephenson, J., 1998. Introduction to special section: long lava flows. J. Geophys. Res. 103, 27,281–27,289.

Coffin, M.F., Eldholm, O., 1994. Large igneous provinces: crustal structure, dimensions, and external consequences. Rev. Geophys. 32, 1–36.

Courtillot, V.E., Renne, P.R., 2003. On the ages of flood basalt events. C. R. Geosci. 335, 113–140.

Eldholm, O., Coffin, M.F., 2000. Large igneous provinces and plate tectonics. In: Richards, M.A., Gordon, R.G., van der Hilst, R.D. (Eds.), The History and Dynamics of Global Plate Motions, Geophysical Monograph, vol. 121. American Geophysical Union, Washington, DC, pp. 309–326.

Hon, K., Kauahikaua, J., Denlinger, R., Mackay, K., 1994. Emplacement and inflation of pahoehoe sheet flows: observations and measurements of active lava flows on Kilauea Volcano, Hawaii. Geol. Soc. Am. Bull. 106, 351–370.

Hooper, P.R., Camp, V.E., Reidel, S.P., Ross, M.E., 2007. The Columbia River Basalts and their relationship to the Yellowstone hotspot and Basin and Range extension. In: Foulger, G.R., Jurdy, J.M. (Eds.), Plumes, Plates, and Planetary Processes, Geol. Soc. Am. Bull. Special Papers, vol. 430, pp. 635–668.

Kelley, S.P., 2007. The geochronology of large igneous provinces, terrestrial impact craters, and their relationship to mass extinctions on earth. J. Geol. Soc. 164, 923–936.

Keszthelyi, L., Self, S., Thordarson, T., 2006. Flood lavas on Earth, Io and Mars. J. Geol. Soc. 163, 253–264.

Mahoney, J.J., Coffin, M.F. (Eds.), 1997. Large Igneous Provinces: Continental, Oceanic, and Planetary Flood Volcanism. Geophys. Monograph., vol. 100. Amer. Geophys. Union, Washington, DC, p. 438.

Rampino, M.R., Prokoph, A., 2013. Are mantle plumes periodic? Eos, Trans. Am. Geophys. Union 94, 113–114.

Rampino, M.R., Stothers, R.B., 1988. Flood basalt volcanism during the past 250 million years. Science 241, 663–668.

Reichow, M.K., Saunders, A.D., White, R.V., Al'Mukhamedov, A.I., Medvedev, A.Y., 2005. Geochemistry and petrogenesis of basalts from the West Siberian Basin: an extension of the Permo–Triassic Siberian Traps, Russia. Lithos 9, 425–452.

Reidel, S.P., 1998. Emplacement of Columbia River Basalt. J. Geophys. Res. 103, 27,393–27,410.

Reidel, S.P., Camp, V.C., Ross, M.E., Wolff, J.A., Martin, B.S., Tolan, T.L., Wells, R.E. (Eds.), 2013. The Columbia River Basalt Province. Geol. Soc. Amer. Bull. Special Papers, vol. 497, p. 440.

Reidel, S.P., Tolan, T.L., 1992. Eruption and emplacement of flood basalt; an example from the large-volume Teepee Butte Member, Columbia River Basalt Group. Geol. Soc. Am. Bull. 104, 1650–1671.

Ridley, V.A., Richards, M.R., 2010. Deep crustal structure beneath large igneous provinces and the petrologic evolution of flood basalts. Geochem. Geophys. Geosyst. 11. http://dx.doi.org/10.1029/2009GC002935.

Saunders, A.D., 2005. Large igneous provinces: original and environmental consequences. Elements 1, 259–263.

Self, S., Jay, A.E., Widdowson, M., Kesthelyi, L.P., 2008. Correlation of the Deccan and Rajahmundry Trap lavas: are these the longest and largest lava flows on Earth? J. Volcanol. Geotherm. Res. 172, 3–19.

Self, S., Keszthelyi, L., Thordarson, T., 1998. The importance of pahoehoe. Ann. Rev. Earth Planet. Sci. 26, 81–110.

Stracke, A., Bizimis, M., Salters, V.J.M., 2003. Recycling oceanic crust: quantitative constraints. Geochem. Geophys. Geosyst. 4 http://dx.doi.org/10.1029/2001GC000223.

Stracke, A., Hofmann, A.W., Hart, S.R., 2005. FOZO, HIMU, and the rest of the mantle zoo. Geochem. Geophys. Geosyst. 6 http://dx.doi.org/10.1029/2004GC000824.

Thordarson, T., Self, S., 1993. The Laki (Skaftar Fires) and Grimsvotn eruptions in 1783–1785. Bull. Volcanol. 55, 233–263.

Thordarson, T., Self, S., 1998. The Roza Member, Columbia River Basalt Group: A gigantic pahoehoe lava flow field formed by endogenous processes? J. Geophys. Res. 103, 27411–27445.

Website: USGS, 2013 Website: USGS, 2013. http://hvo.wr.usgs.gov/kilauea/summary/ (accessed 30.09.13).

White, J.D.L., Bryan, S.B., Ross, P.-S., Self, S., Thordarson, T., 2009. Physical Volcanology of Large Igneous Provinces: update and review. In: Thordarson, T., Self, S., Larsen, G., Rowland, S.K., Hoskuldsson, A. (Eds.), Advances in Volcanology (The Legacy of George Walker), Special Publications of IAVCEI, vol. 2. Geological Society of London, London, pp. 291–321.

Wolff, J.A., Ramos, F.C., 2013. Source materials for the main phase of the Columbia River Basalt Group: Geochemical evidence and implications for magma storage and transport. In: Reidel, S.P., Camp, V.C., Ross, M.E., Wolff, J.A., Martin, B.S., Tolan, T.L., Wells, R.E. (Eds.), The Columbia River Basalt Province, vol. 497. Geological Society of America, Special Paper, pp. 273–292.

Part IV

Explosive Volcanism

Bruce Houghton

Department of Geology and Geophysics, National Disaster Preparedness Training Center, University of Hawai'i, Honolulu, HI, USA

Explosive volcanic eruptions are among the most spectacular displays of nature. The quality and frequency of observations of explosive processes and products have undergone a complete transformation since the 2000 edition of the encyclopedia. There is now a wealth of high resolution data and sophisticated numerical and analogue modeling of eruptive process. This trend is also reflected in the inclusion of 23 new authors in section 4 of the 2015 edition.

The nature and products of explosive volcanism are treated in 14 chapters as Part IV of the Encyclopedia of Volcanoes. These chapters follow a pattern that parallels the evolutionary structure of the entire encyclopedia. We discuss first, the mechanisms for volcanic explosions (Chapters 25—26) and the styles of eruption that result (Chapters 27—31), then processes of transport and deposition and their products (Chapters 32—38).

MECHANISMS AND STYLES OF EXPLOSIVE ERUPTIONS

Chapters 25 and 26 set the stage, defining and modeling the two mechanisms for producing volcanic explosions. Kathy Cashman and Bettina Scheu (Chapter 25) describe how the rapid, often disequilibrium, release of dissolved gases from swiftly ascending magma "drives" most volcanic explosions. Magma is transformed from a foam of gas bubbles in silicate liquid into a rapidly accelerating stream of gas carrying "pyroclasts" (liquid and solid particles). The chapter cuts across four disciplines to meld exciting new insights from experimental studies of degassing, numerical and analogue models for magma ascent and fragmentation, and studies of pyroclasts, which carry the fingerprints of these processes.

The violent interaction of magma and surface (lake or sea) or groundwater is the focus of Chapter 26 by Bernd Zimanowski, Ralf Büttner, Pierfrancesco Dellino, and James White. The spectrum of activity ranging from passive quenching of the magma to explosive ejection of pyroclasts reflects the hydrology of the environment in which magma comes into contact with water. The authors blend field observations with laboratory simulations and analogies with industrial explosions to understand this violent, unpredictable volcanic process.

In the block of five chapters that follow we describe five classical styles of explosive eruptions. The spectacular pyrotechnics of Strombolian explosions and Hawaiian fountaining are described and modeled by Jacopo Taddeucci, Marie Edmonds, Bruce Houghton, Michael James, and Sylvie Vergniolle in Chapter 27. Hawaiian and Strombolian eruptions are the least violent yet best understood and most majestic form of volcanism. Such volcanoes are frequently the locations of large and growing volcano-tourism operations and their eruptions pose constant issues for management agencies because the eruption sites are highly accessible. New data using high resolution and exceptionally rapid acquisition rates have transformed our understanding of the relative roles of magma and magmatic volatiles in these eruptive styles.

Vulcanian eruptions, as described in Chapter 28 by Amanda Clarke, Tomaso Esposti Ongaro, and Sasha Belousov, are transient (seconds to minutes) discrete explosions, capable of ejecting incandescent lava bombs and blocks of fragmented wall rock to distances less than 5 km. Our understanding of the causes and timing of Vulcanian eruptions has improved dramatically due to an increase in the number of volcanoes under observation, improvements in seismic and acoustic monitoring techniques, and sophisticated numerical models that consider both subsurface and surface processes. Detailed study of explosions at numerous volcanoes has been possible in the last 15 years. New approaches within laboratory experiments and numerical models have elucidated complex nonlinear processes at these volcanoes and identified their scaling properties.

Plinian and subplinian eruptions (Chapter 29) are characterized by the formation of sustained, often steady high eruption plumes resulting in atmospheric particle injection, and dispersal of pyroclasts by winds over huge areas. They produce blankets of coarse-grained, well-sorted pumice in the proximal area, grading downwind to widely dispersed ash beds in the far field. The name is forever linked to the AD 79 eruption of Vesuvius that provoked the death of Pliny the Elder. The dynamics of these awesome events are modeled by Raffaello Cioni, Marco Pistolesi, and Mauro Rosi, drawing from the classical eruptions of Vesuvius and at Mount St Helens in 1980.

Bruce Houghton, James White, and Alexa Van Eaton examine examples of magma mingling with external water in Chapter 30. This chapter replaces two chapters in the first edition that dealt separately with Surtseyan and phreato-plinian eruptions. They emphasize the complexity of processes and heterogeneity of products of transient, mostly basaltic phreatomagmatic eruptions. In sustained phreato-plinian eruptions, magma interacts violently with abundant water often residing in caldera lakes. Chapter 30 reexamines the deposits of classical phreatoplinian eruptions in Iceland and New Zealand and the role of the water in the formation of clusters and aggregates of ash particles in the eruption plumes (Chapters 32, 33).

In a new addition to the encyclopedia, James White, Ian Schipper, and Kazuhiko Kano characterize the least observed, most elusive style of explosive volcanism, eruptions occurring in deep submarine settings (Chapter 31). Study of such submarine explosive volcanism is in its infancy and presents a series of exciting challenges to marine volcanologists.

PYROCLAST TRANSPORT AND DEPOSITION

Volcanic plumes are produced by a variety of explosive eruptions. Plumes form from jets of gas and pyroclasts, which entrain air and atmospheric moisture and continue to rise buoyantly, to as high as at least 40 km. Large volcanic plumes that penetrate the tropopause may trigger global environmental effects and produce regional short-term changes of climate. Impact to aircraft in flight that inadvertently encounters volcanic plumes is particularly significant, for example, the 2010 eruption of Eyjafjallajökull in Iceland that virtually shuts down commercial air traffic in Europe at an estimated cost of some $1.7 billion (US). Elegant models for eruption plumes are presented in Chapter 32 by Steve Carey and Marcus Bursik.

Chapter 33 by Costanza Bonadonna, Antonio Costa, Arnau Folch, and Takehiro Koyaguchi is a new addition to the second edition reflecting the emergence of computation models for the dispersal and sedimentation of tephra.

Tephra dispersal and sedimentation represents a critical aspect of volcanic explosive eruptions on many temporal and spatial scales, including damage to vegetation, infrastructures and lifelines, contamination of land and water ecosystems, impacts on several economic sectors, disruption of air and ground transportation networks, threat for human and animal health, and accumulation of loose deposits that can generate secondary hazards, such as ash resuspension and lahars. The authors describe and critique the many empirical, analytical, and numerical models that have been developed to investigate tephra transport and assess ground deposition and atmospheric dispersal in both strong and weak plumes.

Important clues about eruption processes come from the study of the textures and fabrics of pyroclastic fall deposits, as described in Chapter 34 by Bruce Houghton and Rebecca Carey. Pyroclastic fall deposits, produced by the rainout of clasts through the atmosphere, are the simplest of pyroclastic products, and their value is in the ease with which their properties can be used to infer eruption parameters.

The perception of pyroclastic density currents (PDCs) as a highly dangerous and devastating type of explosive volcanic activity was brought about dramatically by the 1902 eruptions of Mount Pelee, Martinique, which killed 30,000 people. Since then several destructive pyroclastic-flow eruptions have occurred every decade. They are described here in the form of two linked chapters. Processes and models for PDCs (pyroclastic, surges, and flows) are described in Chapter 35 by Joe Dufek, Tomaso Esposti Ongaro, and Olivier Roche, while the architecture and characteristics of the resultant deposits are described by Richard Brown and Graham Andrews in Chapter 36.

Lahars (Chapter 37) occur when large masses of volcanic sediment, and water, sweep down and off volcano slopes incorporating additional sediment. James Vallance and Richard Iverson describe how both liquid and solid interactions determine the unique behavior of lahars and distinguish them from debris avalanches and floods. The rock fragments carried by the flows make them especially destructive, and people in distal areas commonly neither expect the danger nor anticipate the destructive power of lahars.

A debris avalanche is the product of a large-scale collapse of a sector of a volcanic edifice often triggered either by gravitational deformation, intrusion of new magma, or a phreatic explosion or an earthquake. The AD 1792 debris avalanche at Unzen volcano killed 15,190 people, including 11,000 taken by an avalanche-triggered tsunami. In Chapter 38, Ben van Wyk de Vries and Tim Davies describe the architecture and internal fabric of deposits of volcanic debris avalanches, and landslides and the models used to describe their emplacement.

Chapter 25

Magmatic Fragmentation

Katharine V. Cashman

School of Earth Sciences, University of Bristol, Wills Memorial Building, Queens Road, Bristol, UK

Bettina Scheu

Department of Earth and Environmental Sciences, Ludwig-Maximilians-Universität München, Munich, Germany

GLOSSARY

ash: Volcanic particles that are ≤ 2 mm in diameter.

bomb: Volcanic particles that are ≥ 64 mm in diameter.

coalescence: Process by which two or more bubbles join.

cryptodome: Magma that stagnates at very shallow levels before crystallizing (solidifying).

degassing: Process by which melt loses its dissolved volatile components; degassing can occur by exsolution or diffusion.

dome: A viscous flow of lava that ponds over, rather than flowing away from, a volcanic vent.

exsolution: Process by which dissolved volatiles come out of solution to form a gas phase.

fragmentation: Process by which a continuous liquid or solid phase is transformed to a continuous gas phase with suspended particles (liquid and/or solid).

glass (volcanic): Solid phase of melt; it has a noncrystalline (amorphous) structure and crosses a glass transition when heated toward the liquid state.

glass transition: The reversible transition in melts between a brittle (solid-like) and viscous (fluid-like) response to deformation; depends on the structural (Maxwell) relaxation time of the melt.

lapilli: Volcanic particles that are 2–64 mm in diameter.

magma: A suspension of up to three phases: melt (liquid), bubbles (gas), crystals (solid).

maxwell relaxation time: The characteristic timescale of stress relaxation after sudden deformation of a viscoelastic material.

melt: Liquid phase of magma.

outgassing: Physical process by which gas escapes from magma to the atmosphere or host rock by permeable flow or bubble rise and rupture at the magma surface.

overpressure: Pressure in excess of lithostatic/magmastatic pressure in the magma. If overburden is removed, overpressure refers to pressure in excess of either hydrostatic or atmospheric pressure.

permeability (magmatic): Measure of the ability of a network of connected bubbles/fractures within the magma to transmit gas in response to a pressure gradient.

porosity: Measure of the void space in solidified magma (in %); includes both bubbles and fractures.

pyroclasts: Juvenile particles formed by fragmentation of magma/lava; may also include accessory lithic clasts and broken crystals.

Reynolds number: A dimensionless number that describes the relative importance of inertia and viscosity in a flowing liquid.

rheology: The flow of matter in response to an applied stress. In Newtonian fluids, the rheology is described by a single (temperature-dependent) coefficient of viscosity.

suspension: A heterogeneous mixture of solid particles in a fluid (liquid or gas).

tephra: Fragmental material (pyroclasts) produced by a volcanic eruption; the word is used for material of all compositions, sizes, and emplacement mechanisms.

vesicle: A "frozen" bubble within tephra or lava.

vesicularity: Percentage of vesicles in a rock (often assumed to represent the percentage of bubbles in the melt phase prior to solidification).

viscosity: Material property of a fluid linking the amount of deformation to the amount of applied shear or tensile stress.

volatiles: Chemical elements or species that preferentially partition into the magmatic gas phase at low pressure.

volcanic plumes: Columns of hot gas and tephra produced by explosive volcanic eruptions.

The Encyclopedia of Volcanoes. http://dx.doi.org/10.1016/B978-0-12-385938-9.00025-0

1. INTRODUCTION

Explosive eruptions are the most powerful and destructive type of volcanic activity; they can produce large quantities of **tephra** that are carried upward in **volcanic plumes** or transported laterally in energetic pyroclastic flows. Recent eruptions of volcanoes, such as Chaiten (Chile, 2009), Eyjafjallajökull (Iceland, 2010), Mount Pinatubo (Philippines, 1991), Mount St Helens (United States, 1980), Puyehue-Córdon Caulle (Chile, 2011), and Soufriere Hills (Montserrat, 1995—present) serve as vivid reminders that pyroclastic material can have devastating effects on nearby human populations. For example, pyroclastic density currents completely destroy areas close to a volcano, falling tephra disrupts daily life in downwind communities, volcanic **ash** injected into the atmosphere can severely disrupt air traffic, and injection of ash and aerosols into the stratosphere during very large eruptions may modify the global climate for years (as seen after the 1815 of Tambora, Indonesia).

Two different mechanisms cause explosive eruptions. Magmatic, or "dry," eruptions are driven solely by gases originally dissolved in the **magma**, while phreato-magmatic, or "wet," eruptions occur when magma interacts with external water. In this chapter we focus on magmatic eruptions, and specifically on the process of **fragmentation**, which is central to our understanding of both how volcanoes work and the hazards they pose. Fragmentation transforms magma from a continuous liquid phase with dispersed gas bubbles (+/− crystals) to a gas phase with dispersed magma fragments (**pyroclasts**). The explosive force of magmatic eruptions is generated during fragmentation, where the potential energy of expanding magma is converted to the kinetic energy of the gas phase and individual pyroclasts and then enhanced by thermal expansion of the gas-particle mixture within volcanic plumes. Primary fragmentation occurs (1) when magma ascends rapidly during explosive eruptions, (2) after rapid decompression caused by collapse of volcanic edifices or lava **domes**, (3) by shearing of magma at the conduit walls, or (4) by impact-induced explosion of hot solid blocks during dome collapse. A single eruptive episode may involve all of these primary fragmentation mechanisms if the eruption style changes from inception to termination. Fragmentation of individual particles can also be sequential in time; secondary fragmentation of primary pyroclasts reduces the particle size during transport within volcanic conduits and after pyroclast ejection from the vent, particularly within pyroclastic density currents.

Direct observation of magmatic fragmentation during volcanic eruptions is not possible; for this reason, our understanding of fragmentation derives from studies of volcanic deposits, theoretical analysis, and experiments on natural and analogue materials. Drivers for explosive fragmentation can be viewed as either "bottom-up"—that is, driven by upward acceleration of bubbly magma—or "top-down," where fragmentation occurs in response to a downward-propagating decompression event (Figure 25.1). These two models are not mutually exclusive, as bottom-up processes that initiate eruptions may induce subsequent top-down decompression. Primary fragmentation can occur in the liquid or solid state, and reflect either the local or bulk properties of the magma (e.g., **rheology**, **porosity**, permeability). Secondary fragmentation can occur by shearing and extension of fluids, or collision and abrasion of solids. This range in fragmentation processes is reflected in the eruptive products, which vary from large blocks and fluidal **bombs** to very fine ash. Below we use data derived from studies of both experimental and natural samples to relate magma properties (composition, rheology) and conditions of ascent and emplacement to the spectrum of fragmentation styles and products observed in volcanic eruptions.

FIGURE 25.1 Simple illustration of two primary fragmentation models discussed in the chapter. Fragmentation by rapid acceleration operates from the bottom-up, when bubbles grow rapidly during decompression; fragmentation may be caused by inertial processes or by brittle breakage, when high strain rates cause the melt to cross the glass transition. Fragmentation by rapid decompression operates from the top-down, when bubble-bearing magma is suddenly exposed to a lower pressure; fragmentation is typically brittle. Arrows indicate directionality of process; photographs are labeled by illustrative eruption style. Note: During Plinian eruptions, both fragmentation models may act in concert.

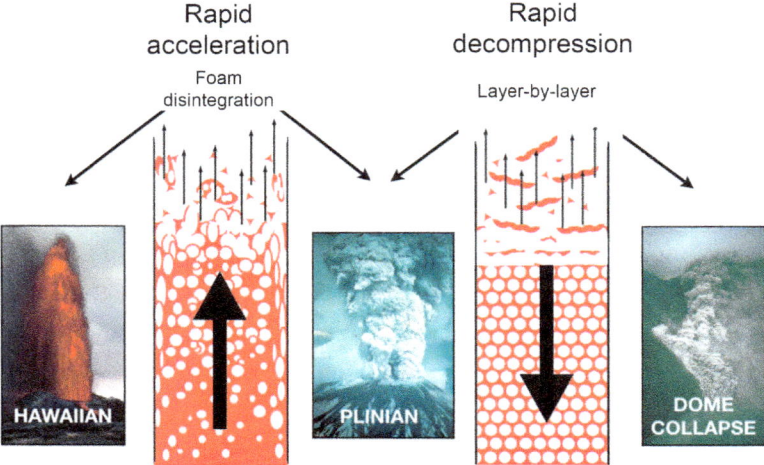

2. FROM SOURCE TO SURFACE

The processes that cause magma to fragment start deep within the Earth, where magma is stored at pressures of 100—1000s of atmospheres (typically 100—200 MPa) and volatile elements, particularly water, are dissolved within silicate **melts**. At lower pressures, water exsolves from the melt to form bubbles; loss of dissolved water from the melt can also cause magma to crystallize. Once bubbles form, the behavior of the gas phase determines, in large part, whether and how magma reaches the surface. If gas escapes from magma storage regions, then the magma is likely to solidify in place to form an intrusive rock. If, on the other hand, the gas phase cannot escape and expands within rising magma, bubble growth can drive magma toward the Earth's surface. If bubbles grow faster than they either rise individually or form permeable networks, the magma will continue to accelerate, and ultimately, fragment. If bubbles rise through the ascending magma, or if bubble networks are sufficiently interconnected to allow permeable gas escape, the magma may reach the surface without fragmenting to form lava. In this way, the relative rates of bubble growth, magma transport, and gas loss control the conditions of eruption. Details of gas behavior are provided in the Chapter 11 "Magma Ascent and **Degassing** at Shallow Levels." Here, we briefly summarize key points that relate directly to magmatic fragmentation, including rates of bubble nucleation, mechanisms of bubble growth, degassing-induced crystallization, and conditions of gas escape by either bubble rise or flow through permeable networks.

Bubbles form (nucleate) at a rate determined by the relevant chemical driving force (supersaturation) and the activation energy required to overcome the excess energy of the bubble surface. Experiments on silicic (high SiO_2, moderate-to-high **viscosity**) melts show that large supersaturations (large **overpressure**, ΔP) are required for spontaneous bubble formation (*homogeneous nucleation*; Figure 25.2(A), where ΔP is illustrated by the pressure difference between the two large asterisks; Mangan et al., 2004). Under these conditions, the volatile concentration in the melt remains high during decompression to low pressures, bubble nucleation is rapid, bubble nuclei are numerous, and most bubbles are small (Figure 25.2(B)). Activation energies are substantially reduced, however, if there are crystals within the melt that can act as nucleation sites (*heterogeneous nucleation*). Under these conditions, degassing approaches equilibrium, bubble nuclei are less numerous, and bubble size distributions are Gaussian (Figure 25.2(C)). The conditions of bubble formation within mafic (low SiO_2, low viscosity, high temperature) melts are more poorly constrained because high-temperature experiments are difficult to conduct and bubbles form rapidly. Recent results suggest, however, that when the primary volatile phase is water, degassing occurs under equilibrium conditions; carbon dioxide degassing, in contrast, may occur under disequilibrium conditions that are similar to those of silicic melts (Pichavant et al., 2013).

Once formed, bubbles grow by both continued volatile **exsolution** and gas expansion in response to decreasing pressure. When bubbles expand within the melt (*closed system degassing*), the magma volume increases with increasing bubble volume and the mixture accelerates toward the Earth's surface. Importantly, even small initial amounts of dissolved water will expand to very large volume fractions (>99%) at atmospheric pressure. For this reason, the amount of gas retained (or lost) during magma transit from the storage region to the surface contributes

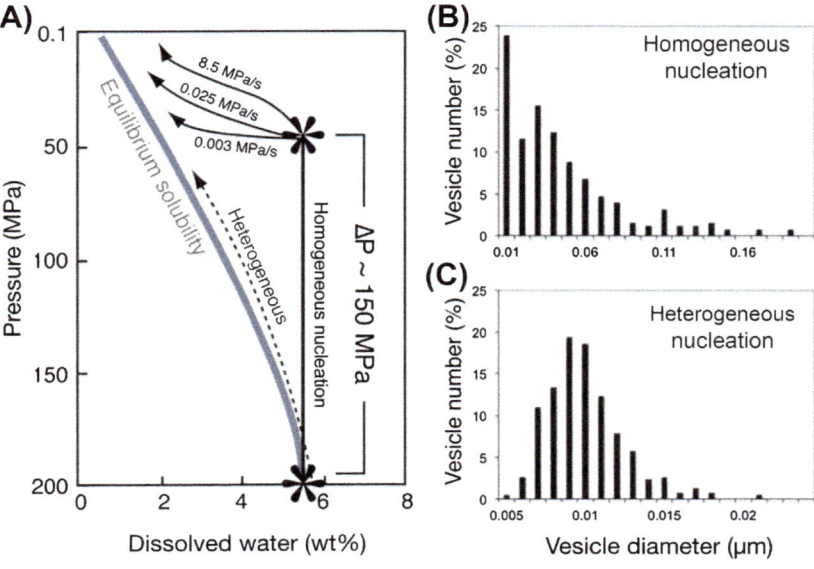

FIGURE 25.2 Experimental vesiculation of hydrous rhyolites. (A) Changes in dissolved water content as a function of pressure for different rates of decompression and starting conditions. Experiments start at $P = 200$ MPa and 5.5 wt% H_2O (water saturation), as shown by the lower asterisk. When the starting melt has no crystals, the pressure must be reduced to ~50 MPa to produce bubbles, regardless of the rate of decompression (upper asterisk). At this point, bubbles form by homogeneous nucleation, and continued decompression produces different P-H_2O paths for different decompression rates. When small crystals are present, they can act as nuclei for H_2O bubbles (heterogeneous nucleation) and degassing occurs under equilibrium conditions except when decompression rates are very high. (B) Typical vesicle size distributions resulting from homogeneous nucleation. (C) Typical vesicle size distributions resulting from heterogeneous nucleation. *All figure components redrafted from Mangan et al. (2004).*

substantially to the explosivity of the resulting eruption. Also important is the depth of vesiculation, which depends on the volatile species (H_2O, CO_2, etc.), the amount of dissolved **volatiles**, the melt composition, and the kinetics of degassing. The depth and extent of vesiculation, in turn, together with the conduit geometry, determine the magnitude of magma acceleration as it approaches the surface. Importantly, downward propagation of the fragmentation surface during an eruption may (1) cause vesiculation intensity to increase with time because of increasing depth of fragmentation (i.e., reduction of lithostatic pressure) and (2) mobilize magma from increasingly deeper storage regions.

The most abundant (and therefore most important) volatile phase in most magmatic systems is water. Not only does water degassing provide the driving force for magma ascent, but it also plays a key role in controlling the physical properties of magma within volcanic conduits (Cashman, 2004). First, melt viscosity is sensitive to the dissolved water content and increases dramatically as the dissolved water content drops below $\sim 1-1.5$ wt%. Second, the stability of crystalline phases, particularly plagioclase feldspar, is strongly affected by the melt water content. For this reason, it is possible for water-rich magma ascending at a constant temperature to crystallize completely en route to the surface, if the rate of crystallization is sufficiently high relative to the rate of magma ascent. Crystallization affects the viscosity of the melt phase by changing its composition, and the rheology of the magma by adding solid particles. As fragmentation mechanisms are strongly dependent on magma rheology, the relative timing, and extent, of both vesiculation and crystallization strongly affect fragmentation conditions (Figure 25.3).

Conditions that allow gas to escape from rising magma before it reaches the surface include (1) bubble rise through the melt (*separated flow*) or (2) formation of connected pore networks (*permeable flow*; Rust and Cashman, 2011). Bubble rise rates are strongly controlled by melt viscosity, such that separated flow is most efficient in low-viscosity (mafic) melts, and when magma ascent rates are slower than ~ 1 m/s. Creation of permeable gas pathways requires first gas expansion to sufficient bubble volume fractions that neighboring bubbles are close to each other, and then thinning and rupture of the bounding melt films; the latter process will be enhanced by shear. Numerical (percolation) models predict that bubble networks should form when bubble volume fractions approach 30 volume%. Experiments on viscous fluids, however, suggest that gas pathways are not likely to form until bubble volume fractions reach $\sim 65-75\%$. The difference between theoretical and measured values can be explained by the added requirement of melt film thinning, which is not considered in percolation models. As soon as permeable networks form,

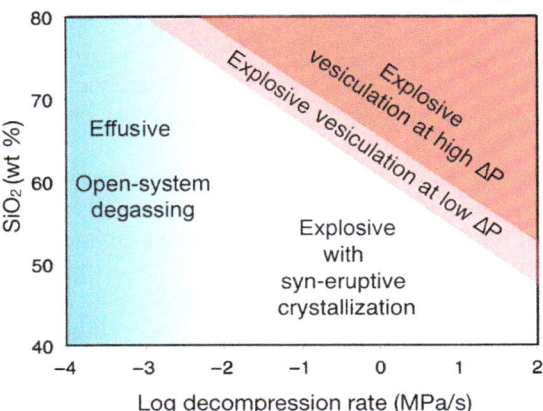

FIGURE 25.3 Schematic representation of experimentally constrained degassing and crystallization regimes, plotted as a function of bulk composition (shown as weight% SiO_2) and decompression rate (*dP/dt*). Explosive eruptions occur at high decompression rates; under these conditions, degassing occurs under disequilibrium conditions when bubble nucleation is homogeneous, and under equilibrium conditions when bubble nucleation is heterogeneous. When the melt is intermediate to mafic in composition, degassing may be accompanied by some amount of crystallization on eruptive timescales. Under most conditions, eruptive behavior changes from explosive to effusive when the degassing behavior changes from closed to open. *Approximate fields drawn using data from Larsen and Gardner (2004), Mangan et al. (2004), Szramek et al. (2006), Castro and Dingwell (2009), Shea et al. (2010), Cichy et al. (2011), Rust and Cashman (2011), Pichavant et al. (2013).*

gas can escape either laterally through conduit walls or vertically through the magma. The directionality of gas escape is determined by both the anisotropy of the permeable network and the pressure differential driving gas flow. **Outgassing** through permeable networks is often accompanied by partial collapse (densification) of the magmatic foam; under these conditions permeability—porosity curves are hysteretic as gas pathways are maintained during densification (Figure 25.4). Variations in conditions of gas loss control the steadiness, and cyclicity, of eruptive behavior.

3. FRAGMENTATION PROCESS

Prior to an eruption, magma is stored at high pressures beneath a volcano, where pressures may exceed the lithostatic load because of the confining strength of the surrounding rocks. The pressure difference that drives an eruption acts when the magma source is connected to a region of lower pressure, such as the atmosphere, across the fragmentation surface. By Newton's laws, the force imbalance, or thrust, causes the magma to accelerate. Two processes have been proposed to power explosive eruptions. In the first, vesiculation and bubble growth provide the driving force for expansion. Fragmentation may be controlled by inertia in low-viscosity melts or by brittle fracture when the tensile strength of the melt is exceeded;

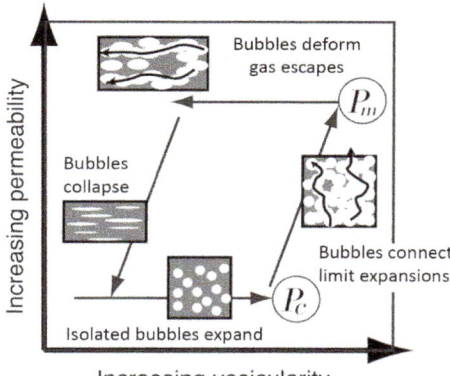

FIGURE 25.4 Illustration of hysteresis in permeability—porosity data. Initial expansion occurs in isolated bubbles until percolation threshold (P_c) is reached at ~60% vesicularity. Permeability then increases rapidly to a permeability threshold (P_m) that limits further expansion. Deformation of vesicular magma permits gas loss (porosity reduction) without loss of permeability as bubbles deform, until bubble collapse reduces permeability (bubbles again become isolated). *Modified from Cashman and Sparks (2013).*

in either case fragmentation is caused by high strain rates in the expanding two- (or three-) phase mixture. In the second, fragmentation is accomplished by the failure of suddenly unloaded, already vesicular, magma. This model describes explosive eruptions from volcanic domes, such as lateral blasts, explosive dome collapse, and Vulcanian eruptions. The fragmentation mechanisms may act in concert, with varying impact, to cause explosive disintegration of magma. Fragmentation can also occur after eruption, within volcanic conduits and eruptive plumes, during transport in pyroclastic density currents, accompanying advance of lava flows fronts, and by disintegration or comminution of viscous lava domes and flows. Below we discuss these mechanisms in more detail and then use textural analysis of eruptive products to illustrate both conditions of magma ascent and degassing and the fragmentation mechanisms operating during different types of explosive eruptions.

3.1. Fragmentation Resulting from Rapid Acceleration

Fragmentation by rapid acceleration is initiated at depth by either intrusion of new hotter magma or volatile overpressure caused by crystallization of anhydrous phases. In either case, sufficient pressure is required within a magma reservoir to initiate upward movement of magma to start the process. In the simplest case, bubble growth and expansion in response to decompression establish a feedback, such that each increment of bubble growth (expansion) drives the magma upward, which decreases the pressure, which in turn triggers more bubble growth by

volatile exsolution and expansion. The feedback can cause runaway acceleration and, ultimately, fragmentation (Figure 25.1). Laboratory experiments on low-viscosity liquids show that as bubbles grow, they merge; eventually the bubble walls get so thin that the liquid "foam" becomes unstable and breaks up. In low-viscosity (mafic) liquids, buildup of overpressure during rapid acceleration toward the Earth's surface is inhibited by the short structural relaxation time of these liquids. Under these conditions, fragmentation is controlled by *inertia*. In high-viscosity (silicic) liquids, conditions of fragmentation are determined by both the magnitude and rate of decompression. Fragmentation occurs when vesiculation is sufficiently rapid to prevent bubble **coalescence** and permeable gas escape (Rust and Cashman, 2011). Critical timescales for fragmentation include both the relaxation timescale characteristic of the **glass transition** (Figure 25.5; Papale, 1999), and the timescale that controls viscous bubble expansion (Kurzon et al., 2011; see section below on Fragmentation Models and Criteria). Under these conditions, fragmentation is controlled by *brittle breakage*.

3.2. Fragmentation Resulting from Rapid Decompression

In contrast to fragmentation by rapid acceleration, where an initially slow process accelerates, *fragmentation by rapid decompression* starts with the (near-instantaneous) application of a large overpressure (e.g., by dome collapse), and the underlying magma experiences decompression on a timescale where other processes, such as bubble nucleation and growth, are of minor importance. Under these conditions, the dominant behavior of magma is brittle (Figure 25.6). This type of explosive behavior has been applied primarily to vesicular (near) solids, such as disruption of viscous lava following plug removal during Vulcanian eruptions and collapse of lava domes. Rapid downward propagation of decompression (unloading) may also be important in other eruption styles, however, and enhance fragmentation of shallow (and degassed) bubbly magma within conduits.

When shallow degassed magma is decompressed, bubble expansion, rise, and coalescence are all largely counteracted by high melt viscosity. For this reason, bubbles contained within viscous melts can achieve high gas overpressures, particularly in either near-static magma beneath (dense) low-permeability lava plugs or flow surfaces. Plug removal or dome collapse creates a new surface and exposes bubbly magma to rapid decompression, resulting in an unloading wave that propagates into the magma body at the compressional seismic velocity of the magma. The bubbles close to the new surface thus experience a near instantaneous buildup of overpressure controlled by the pressure difference across the decompression front in addition to possibly gas over-pressure stored in the bubbles. When the pressure difference exceeds the strength of the magma, the magma fragments brittlely create new surfaces. If permeable gas loss from the underlying magma is slow relative to the speed of one fragmentation event, the buildup of over-pressure and consecutive fragmenchen repeats in a layer-by-layer process as long as the overpressure is sufficient to overcome the magma's strength (Figure 25.1, Fowler et al., 2010).

Experimental studies of fragmentation by rapid decompression use both natural and analogue materials (Figure 25.6). High-speed video recordings of natural samples exposed to a rapid decompression event elucidate the process of fragmentation itself, and show that the fragmentation front propagates at 10−150 m/s, depending on magma porosity and overpressure (video 1). These experiments constrain minimum pressures for the onset of stable fragmentation (the *fragmentation threshold*), which is inversely proportional to sample porosity but appears independent of sample composition (Figure 25.7). Importantly, an overpressure of 5 MPa is sufficient to fragment volcanic rocks with only 25−30% porosity, as long as the permeability is sufficiently low ($<10^{-12}$ m^2). These values are typical for dome rocks and show that, under rapid decompression, fragmentation may be initiated at significantly lower vesicularities than previously anticipated. Porosity also affects the velocity of the fragmentation front, which increases with increasing porosity (Scheu et al., 2008).

Supplementary video related to this article can be found online at http://booksite.elsevier.com/9780123859389/.

3.3. Fragmentation Models and Criteria

Early models of magma ascent and eruption assumed that fragmentation occurred when expanding magma exceeded a critical **vesicularity**. For silicic Plinian eruptions, the vesicularity criterion has variously been placed between 60 and 80 vol% based on the observed range of vesicularity preserved in (crystal-poor) pyroclasts; the lower values assume that higher vesicularities record postfragmentation bubble expansion prior to quenching. However, more recent experimental and theoretical investigations have shown that (1) fragmentation can occur well below these critical vesicularities and (2) high-vesicularity magma may be erupted effusively. These observations required more refined models that are based on physical aspects of the different types of fragmentation described above. Physically based fragmentation criteria include magma/melt failure when (1) bubble overpressure exceeds the strength of the surrounding magma/melt (Koyaguchi et al., 2008; Fowler et al., 2010; Kurzon et al., 2011) or when (2) the

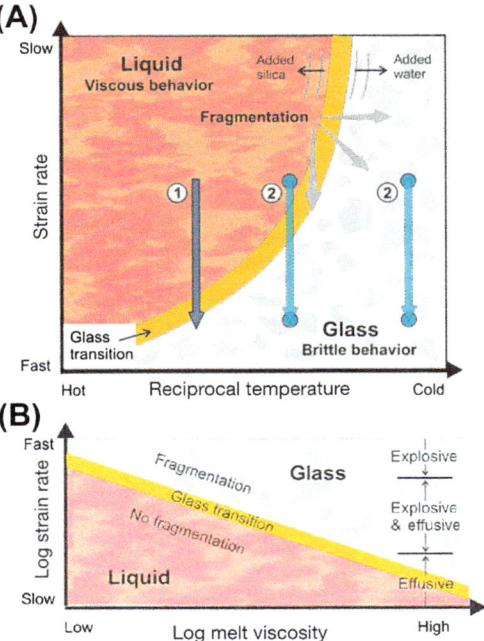

FIGURE 25.5 Illustration of the glass transition. (A) Strain rate−temperature space. The glass transition is a kinetic barrier dividing the behavior of silicate melts into two states: viscous liquid and brittle glass. The liquid field refers to the relaxed state of the melt, resulting in viscous response to (slow) deformation. The glassy field refers to the solid material, where the deformation is faster (high strain rate) than structural relaxation time of the melt, resulting in elastic response and brittle failure. The glass transition is very sensitive to even small variations in H$_2$O and SiO$_2$ content of the melt (top right corner), highlighting the important role of crystallization and degassing during magma ascent and eruption. The glass transition may be crossed several times during the formation of volcanic glasses. Fragmentation of melt occurs by crossing the glass transition, either by increasing the rate of deformation (strain rate), cooling of the melt, or a combination of both (gray arrows). Dark blue arrow (1) indicates the response of a hot (low viscous) melt to rapid acceleration, the major part of the melt's response is viscous, only the final stage might share a brittle component. The light blue arrows (2) show the response of a cooler (high viscous) melt to rapid decompression (note: actually only the beginning and the end state indicated by the arrow might be realized), here the melts react either purely or predominantly brittle, however, a viscous deformation prior to the initialization of fragmentation is possible. Figure modified from Dingwell (1996). (B) Glass transition in strain rate−viscosity space, plotted as log−log distribution. This is an equivalent way of depicting the glass transition to (A). For low strain rates the melt reacts as a liquid and no fragmentation occurs. High strain rates cause brittle response and thus fragmentation. The transition depends on melt viscosity. At the right side eruptions styles are indicated based on typical strain rates. *Figure modified from Gonnermann and Manga (2003).*

expanding melt exceeds either (a) a critical velocity producing instabilities in the fluid phase (inertial criterion; Namiki and Manga, 2008) or (b) a critical strain rate that crosses the glass transition (brittle criterion; Dingwell, 1996; Papale, 1999).

Laboratory experiments on low-viscosity liquids suggest that a criterion for inertia-driven fragmentation can be

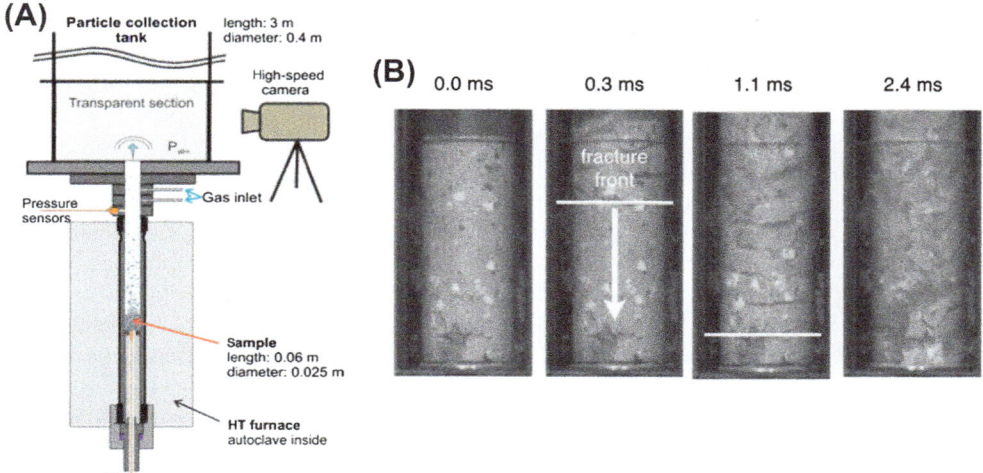

FIGURE 25.6 (A) Experimental setup to investigate magma fragmentation by rapid decompression, based on a shock tube apparatus at LMU Munich. The main parts of this setup are similar in different labs and comprise a high-pressure section (autoclave) with the sample (natural rock, magma analogue) inside, which is separated from a low-pressure section, into which the magma may be ejected after fragmentation. The autoclave is pressurized by argon gas (up to 40 MPa) and can be heated up to 900 °C. A set of diaphragms separates the autoclave from a large low-pressure tank from which the generated clasts can be fully recovered and analyzed. For high-speed video recording a transparent extension of the tank is inserted, or a transparent autoclave is used (ambient temperature only). Figure modified from Mueller et al. (2008). (B) Photographic sequence showing fragmentation due to rapid decompression of Unzen dacite (connected porosity: 48%, sample length: 0.06 m). Frames taken from a high-speed movie of a fragmentation experiment at 6 MPa applied pressure and ambient temperature (Video 1, supplementary material). Time of video frame shown above; location of the fragmentation front is indicated by a horizontal white line. Sample fragments in a "layer-by-layer" fashion; after the sample is fully fragmented, some clasts continue to disintegrate by repeated collisions. Analysis of pressure sensors and high-speed video showed that the fragmentation front propagated at 55 m/s. *Modified from Fowler et al. (2010).*

defined using the balance of inertial forces, that drive, and viscous forces, that limit, expansion. This balance is represented by the **Reynolds number** $Re = \frac{\rho(1-\phi_i)vL}{\eta_e}$, where ρ is the melt density, ϕ_i is the initial vesicularity, v is the expansion velocity, L is the vertical length scale of the bubbly fluid, and η_e is the elongation viscosity of the melt; if $Re > \sim 1$ the bubbly liquid will fragment (Namiki and Manga, 2008). This fragmentation condition is applicable to low-viscosity (crystal-poor) mafic magmas, where fragmentation is most likely to be controlled by inertia. For melts of higher viscosity, or for conditions of very high strain rate, rapid acceleration may cause the melt to cross the glass transition and therefore fail brittlely. This *strain rate* condition for failure was derived initially from fiber elongation experiments that showed a strong correlation between the **Maxwell relaxation time** (τ_m) and brittle failure of silicic melts: $d\varepsilon/dt = A/\tau_m$, where ε is strain and $A = 0.01$ is an experimentally constrained coefficient. This criterion has been extended by substituting A with the critical strain $\varepsilon_{cr}(\mu)$ required for fragmentation as a function of the shear modulus μ (Kurzon et al., 2011).

Brittle failure is often modeled with reference to the gas overpressure (ΔP) developed within individual bubbles (*overpressure strength or stress criterion*). In these models, overpressure (ΔP) is defined as the difference between the gas pressure (p_g) and the mean pressure of the liquid–gas mixture (p_{mix}): $\Delta P = p_g - p_{mix} = (1 - \phi)(p_g - p_l)$, where ϕ is the vesicularity and p_l is the pressure of the

liquid (melt). When permeable gas flow is neglected, the fragmentation criterion is derived from the stress distribution in the melt surrounding isolated spherical **vesicles** and fragmentation is determined by conditions at the bubble wall (Koyaguchi et al., 2008). An alternative (empirical) fragmentation criterion has been derived from experiments. Here, rapid decompression of variably porous volcanic rocks suggests that a critical threshold pressure ΔP_{FR} required to initiate fragmentation can be described as $\Delta P_{FR} = (a\,k^{1/2} + \sigma_m)/\phi$, where σ_m is the effective tensile strength of the magma, k is the permeability, and a is an experimentally constrained constant (Mueller et al., 2008) (Figure 25.7). This empirical criterion also accounts for permeable gas flow that causes a shift of threshold pressures for highly permeable samples ($k > 10^{-12}$ m^2), whereas bubbles in samples with a low permeability can be considered isolated (no significant gas loss on the timescale of fragmentation). Finally, when permeable gas flow is included explicitly in theoretical models, the failure criterion (effective stress > yield stress of the magma: $p + \sigma > \sigma_y$) is reached by coupling slow gas flow out of the pressurized vesicles to fast decompression of the solid phase (i.e. melt + crystals; Fowler et al., 2010). Key in the formulations of the various fragmentation criteria is the balance between structural relaxation time and decompression rate (e.g., Kameda et al., 2008).

The process of magmatic fragmentation has important implications for eruption dynamics. In sustained (Plinian)

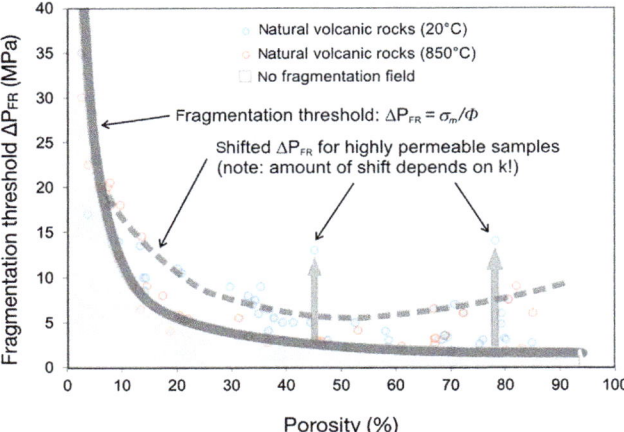

FIGURE 25.7 The pressure needed to fully fragment a sample as function of sample porosity. The gray line depicts the threshold curve, separating the response of a sample to rapid decompression in two fields: in the light gray field below the curve no fragmentation occurs, whereas in the white field above the energy supplied as gas in the vesicles is sufficient to reach the fragmentation criterion and the sample fragments; here the empirically determined criterion is shown relating threshold pressure ΔP_{FR} to sample porosity Φ and effective tensile strength of the magma σ_m. ΔP_{FR} was investigated for a vast amount of samples from basaltic to rhyolitic and found to be independent from chemical composition. Also the experimental temperature has only minor effect on the threshold behavior owned by the very fast deformation time (high strain rate) resulting in a brittle response (see Figure 25.5). The fragmentation threshold is very sensitive to high permeabilities ($k > 10^{-12}$ m^2); highly permeable sample looses the pressurized gas phase in the vesicles (the "fuel for fragmentation") too efficiently, resulting in a shift to higher ΔP_{FR} compared to minor permeable samples (gray dashed line). The shift of ΔP_{FR} strongly depends on the permeability (gray arrows). *Figure modified from Mueller et al. (2008).*

eruptions, the position of the fragmentation surface is a balance of the (upward) speed of magma ascent and the (downward) speed of magma decompression and fragmentation. Conversely, in transient (Vulcanian) explosions the fragmentation front propagates downward within the conduit. The creation of new surfaces during fragmentation consumes a significant amount of energy. An energy balance for magmatic fragmentation based on evaluation of fragmentation and ejection behavior shows that fragmentation reduces the ejection velocity, density, and the mass discharge rate per unit area of the gas–particle mixture compared to models not considering fragmentation (Alatorre-Ibargüengoitia et al., 2010). Finally, fragmentation mechanisms are not exclusive to a single eruption style. Plinian and subplinian eruptions may include initially fast ascent (rapid acceleration) of volatile-rich melt and, subsequently, dehydration, increased melt viscosity and brittle fragmentation in response to pressure differences across the decompression surface. From this perspective, the steadiness of volcanic eruptions may reflect the relative

rates of magma ascent and downward propagation of the decompression/fragmentation.

3.4. Secondary Fragmentation

Particles created by primary fragmentation, by either rapid acceleration or rapid decompression, are often subject to further (secondary) fragmentation by continued decompression, thermal stresses, mechanical collisions, and abrasion. Here, we use the term *secondary* simply to denote sequential fragmentation (which may include several different individual breakage events) either before or after the pyroclasts exit the volcanic vent.

When primary fragmentation is accompanied by rapid cooling, the resulting glass may be thermally stressed and subject to further spontaneous rupture. This process is well known in the glass industry, which is why glass objects are annealed (cooled slowly) or tempered (by putting the outer surface under compressive stress) to improve durability. Although thermal stresses are acknowledged to be important in phreatomagmatic fragmentation, the importance of thermal stresses in magmatic fragmentation has not been extensively studied. A well-known problem in field studies of pyroclastic rocks, however, is the fragility of pumice samples caused by residual stresses.

Particles fragmented within a volcanic conduit can continue to break because of turbulence within the expanding mixture of gas and solid particles. The extent of secondary fragmentation under these conditions is a function of particle fragility (controlled by vesicularity, thermal stresses, and the presence of crystals), the transport distance within the conduit (the fragmentation depth), and the particle size (large particles are more likely to break than small particles). Secondary fragmentation will reduce the average particle size exiting the vent, and thereby increase the efficiency of heat transfer to the volcanic plume, and hence its height; secondary fragmentation also increases the abundance of widely dispersed fine ash that can affect both civil aviation and human health.

Large particles that survive transport through the conduit may experience further fragmentation by breakup (fracture) and abrasion within pyroclastic density currents (PDCs). The relative importance of breakup and abrasion can be assessed from particle shape, which will be angular for collisional breakage and round for abrasion. The overwhelming predominance of rounded clasts in PDC deposits provides clear evidence of extensive abrasion. The abundant fine ash produced by abrasion affects both the mobility of the density currents and the amount of co-PDC ash that is contributed to the eruption column (Dufek et al., 2012). In fact, most of the widely dispersed ash in very large eruptions is probably secondary (co-PDC) in origin.

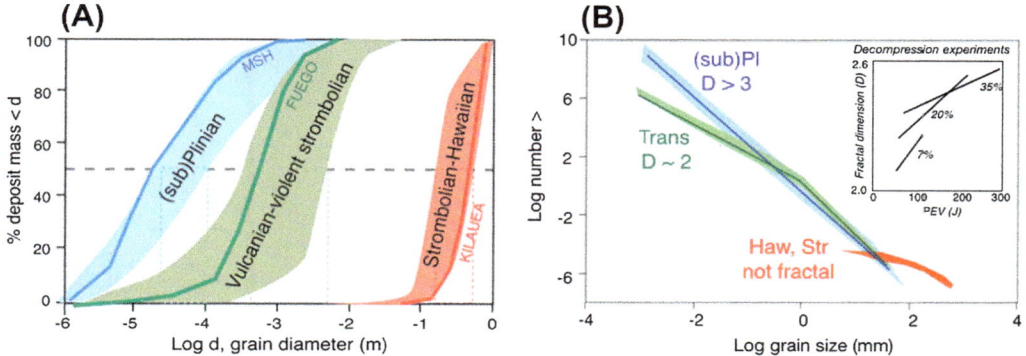

FIGURE 25.8 (A) Examples of total grain size distributions (TGSDs) for eruptions of different styles. TGSDs for specific eruptions mentioned in the text are highlighted in each field and labeled MSH (Mount St Helens, USA, 1980), Fuego (Guatemala, 1974), and Kilauea (USA, 1959). Dashed line at 50% shows the median values of the grain size distributions; intersections of the median with the boundaries of each field are projected at dashed colored lines to the size axis. (B) Power law plots of TGSDs, plotted as number-based rather than mass-based distributions. Colors and data are the same as in (A). Cumulative TGSDs of (sub) Plinian eruptions have power law distributions with slopes (D values) > 3, that small particles comprise most of the sample. Transitional (violent Strombolian and Vulcanian) eruptions have TGSDs with power-law slopes ~2 for the smaller grain sizes; steeper power law trends in the large grain sizes are probably a function of sampling, as larger clasts are likely to be emplaced as ballistics or density flows, which are not included in the TGSD calculations. TGSDs from Strombolian and Hawaiian eruptions do not follow power law trends, consistent with fragmentation by a different mechanism than in the other eruption styles. Inset shows similar data for rapid decompression experiments, which show D values that increase from ~2.0 to 2.6 with increasing potential energy for fragmentation and initial sample porosity. *Figures modified from Kueppers et al. (2006), Rust and Cashman (2011).*

4. PRODUCTS OF FRAGMENTATION

Any model of fragmentation must explain both the size distributions and textural characteristics of the resulting particles. Pyroclast characteristics, in turn, provide important insight into the fragmentation process (Rust and Cashman, 2011). One measure of fragmentation efficiency is the total grain size distribution (TGSD), which also provides a measure of eruption energy (smaller particles have larger surface areas, which requires higher energies for formation). TGSD data (Figure 25.8(A)) show that large explosive silicic eruptions have very small median particle sizes (<1 mm). When converted to number-based (rather than mass-based) distributions, TGSDs from silicic eruptions form power law trends with slopes (D values) > 3 (Figure 25.8(B)). Eruptions of intermediate magnitude (that is, eruptions classified as violent Strombolian and Vulcanian) have intermediate median particle sizes (1−10 mm) and D ~ 2, while the few published TGSDs from basaltic Hawaiian and Strombolian eruptions have large median grain sizes (>10 mm) and show log-linear rather than power law TGSDs; the latter suggest that these deposits form via a fundamentally different fragmentation mechanism than evidenced by silicic deposits.

The relationship between fragmentation efficiency and eruption energy has also been explored by rapidly decompressing natural volcanic samples at eruptive temperatures. At energies just above the fragmentation threshold, the full spectrum of grain sizes is already produced (including fine ash). With increasing energy, the fragmentation efficiency increases, as measured by a shift in the resulting grain size toward smaller particles (illustrated by the increase in fractal

dimension with increasing potential energy for fragmentation (PEV), a product of ΔP, sample volume and porosity; inset Figure 25.8(B)). Interestingly, fragmentation efficiency decreases (the overall grain size is larger under the same conditions) for samples with higher permeabilities ($> \sim 10^{-12}$ m^2). This observation underpins the significance of permeability in fragmentation models.

Details of processes important in both fragmentation and postfragmentation evolution of erupted pyroclasts are preserved in the physical characteristics of the eruptive products. Most striking in the field are large pyroclasts that continue to expand after fragmentation. For example, mafic *bombs* often have highly inflated interiors and dense or poorly vesicular outer rinds that indicate rapid exterior quenching at the same time as interior expansion. In mafic eruptions, where expansion is rapid and melt viscosities are low, the outer surface may be smooth and very thin (Figure 25.9(A)); in silicic eruptions, where expansion is slower, the outer rind is commonly thick and fractured to form "breadcrust" patterns (Figure 25.9(B)). Perhaps, the most spectacular example of postfragmentation expansion is *reticulite*, highly expanded (~98%) polyhedral basaltic foam that forms in high Hawaiian fire fountains (Figure 25.9(C)). The same Hawaiian eruptions also produce relatively dense (30% vesicles) droplets that show limited expansion (Figure 25.9(D)). Their droplet morphology provides strong evidence of inertia-driven fragmentation followed by fluidal deformation of fragmented clasts. That a range of clast vesicularity from <30% to >90% can be generated by a single eruption also provides a vivid reminder that simple pyroclast vesicularity thresholds for fragmentation cannot be applied to natural

FIGURE 25.9 Volcanic pyroclasts that have experience postfragmentation expansion. (A) Expanded bombs erupted from Stromboli volcano, Italy, during a paroxysmal eruption in April 2003. (B) "Breadcrust" bomb erupted from Pichincha volcano, Ecuador, in 1999. (C) Tomographic rendering of a reticulite clast erupted from Kilauea volcano, Hawaii, USA; the gas cells form a polyhedral foam of ∼98% vesicularity. (D) Variably expanded Pele's tears from Kilauea fire fountain, modified from Porritt et al. (2012).

samples without considering the postfragmentation history of individual particles.

More important for understanding fragmentation, however, is the wide range of pyroclasts that comprise the bulk of pyroclastic deposits from explosive eruptions. Silicic eruptions, in particular, are characterized primarily by abundant fine ash and variable amounts and sizes of larger pumice clasts. Ash particles include platelike glass shards, broken crystals, and micropumice (e.g., Figure 25.10(A)); all provide evidence of brittle fragmentation in the form of bounding fracture surfaces. Similar fracture surfaces are seen on ash particles produced by experimental decompression of natural samples (Figure 25.10(B)). Larger pumice clasts are typically highly vesicular (Figure 25.10(C)), with preserved vesicle populations that have power law size distributions that are similar to the power law grain size distributions shown in Figure 25.8(B). This correlation shows that the size distribution of small bubbles controls the size distribution of ash particles; it also suggests that the larger clasts were preserved during fragmentation because they had sufficient permeability at the time of fragmentation to allow expanding gas to escape (Rust and Cashman, 2011). The external morphology of pumice clasts is typically angular (unless abraded in pyroclastic flows), reflecting brittle breakage. When internal vesicles are highly elongated ("tube" pumice), individual pumice clasts are typically elongated in the direction of vesicle elongation. This suggests that bubble morphology may also play a role in fragmentation, either by anisotropic thinning of melt films or anisotropic development of permeability.

The role of permeability in controlling magma fragmentation can be further assessed through measurements of individual pyroclasts. Crystal-poor silicic pumice clasts have both high vesicularities and high permeabilities

($>10^{-13}$ m^2; Figure 25.11). Moreover, there does not appear to be a strong correlation between eruption intensity and pumice characteristics. These data are consistent with experimental evidence for rapid bubble nucleation at high ΔP (rapid nucleation; Figure 25.2, Figure 25.3), which may be controlled primarily by a downward-propagating decompression wave, although in sustained eruptions this must be balanced by magma rise from below. As described above, pyroclasts formed in eruptions of crystal-poor mafic magmas have much more variable vesicle properties (Figure 25.10(C) and (D)) and suggest that low-viscosity magmas experience fragmentation by fluid instabilities as well as extensive postfragmentation expansion of larger clasts. A related observation is that both the size and internal textures of mafic pyroclasts vary with eruption intensity. Moreover, individual pyroclasts from high-intensity eruptions commonly preserve evidence of both brittle and viscous deformation mechanisms, which suggests that during mafic Plinian eruptions, strain rates may approach those of the glass transition (Figure 25.5). Finally, mafic pyroclasts typically show a correlation between measured vesicle number densities and eruption intensity that appears to reflect the dependence of bubble nucleation rate on decompression rate (Poland et al., 2014).

An additional control on fragmentation derived from the study of natural samples relates to groundmass crystallinity, which may be extensive. Pyroclasts formed during eruptions of hydrous mafic to intermediate magma, in particular, commonly show groundmass textures that are dominated by acicular crystals of plagioclase that grew in response to degassing during magma ascent (Figure 25.3). The groundmass crystals limit the bulk vesicularity (bubbles form only in the melt phase), although vesicles deformed around groundmass crystals (Figure 25.10(D)) provide evidence for limited postcrystallization bubble

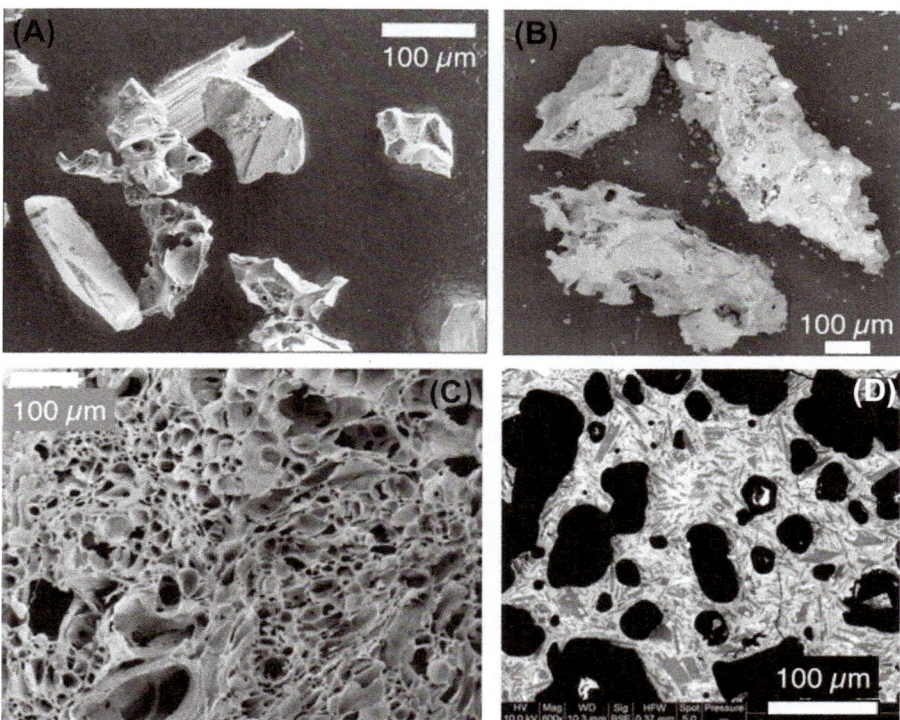

FIGURE 25.10   Electron microscope images of pyroclast textures; images in ((A)–(C)) are created using secondary electrons; image in (D) created using backscattered electrons. (A) Crystals (dense, with cleavage) and micropumice in ash fall from Mount St Helens, 1980; (B) experimentally generated pyroclasts from a sample of Unzen dacite with 36% connected porosity that was fragmented at 850 °C. The clasts are elongated and angular, showing clear evidence of brittle fragmentation (Photo courtesy of U. Kueppers, LMU); (C) "spherical" pumice clast interior from Mount St Helens, 1980; (D) scoria from 1974 eruption of Fuego, Guatemala; note low vesicularity and irregular shape of vesicles (black) and numerous plagioclase crystals (dark gray) in quenched glass (light gray).

expansion. Vesicles in crystal-rich pyroclasts also tend to be smaller, and show less evidence for coalescence, than in comparable crystal-poor pumice clasts, although clast permeabilities are often similar, because gas is forced into convoluted, but interconnected, three-dimensional pathways. In fact, crystal-rich mafic pyroclasts are often poorly vesicular because of extensive gas loss through permeable pathways during magma ascent. The same clasts also tend to be larger than either individual bubbles or crystals, as indicated by both the median size and low variance of typical TGSDs (Figure 25.8). This suggests that crystals may not only facilitate development of connected gas pathways within the melt phase, but also that these connected pathways may determine the spatial scales of fragmentation.

Finally, another quite spectacular consequence of magmatic fragmentation is volcanic lightning, which is commonly observed in plumes from ash-rich eruptions. Two main electrification mechanisms of tephra are invoked: *triboelectrification* (electrification of solids through friction) and *fractoemission* (emission of electrons and ions from newly created surfaces resulting in a residual

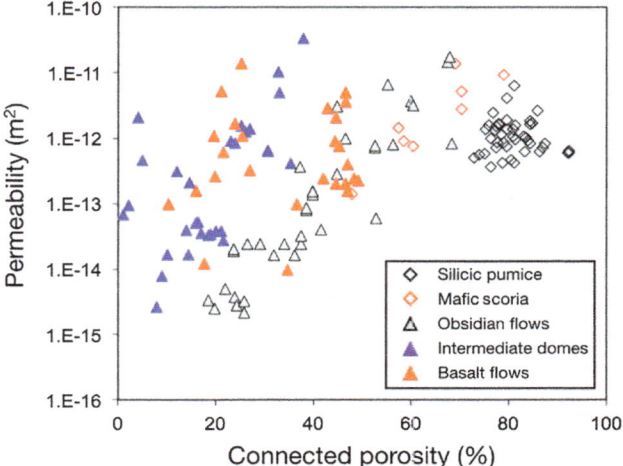

FIGURE 25.11   Illustration of porosity–permeability relations in volcanic samples. Critical porosity for the permeable networks varies from <10% (in lava flows; see Figure 25.4) to ∼30% (in crystal-rich pumice samples) to >70% (in crystal-poor pumice samples). Once the critical threshold is reached, permeability is high (generally >10^{-13} m²). *Simplified from Wright et al. (2009).*

FIGURE 25.12 (A) Decompression experiment using 250 µm-size Popocatépetl ash at 10 MPa. Still from high-speed video, showing a lightning occurring in the particle-laden jet 3.16 ms after decompression. Figure from Cimarelli et al. (2014) *Geology* **12**, 79-82. Right: Volcanic lightning occurring during mild Vulcanian explosions at Showa crater, Sakurajima (Japan) in July, 2013 (photo by B. Scheu, LMU). Note the different dimension of vent or cater: 0.03 m for the vent in the laboratory, and ~200 m for Showa crater.

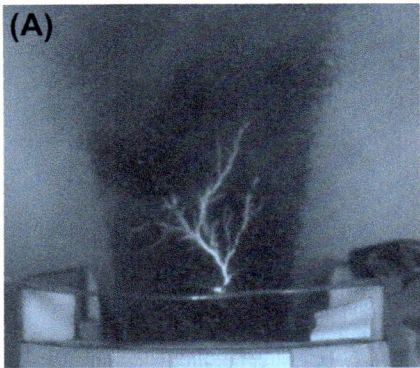 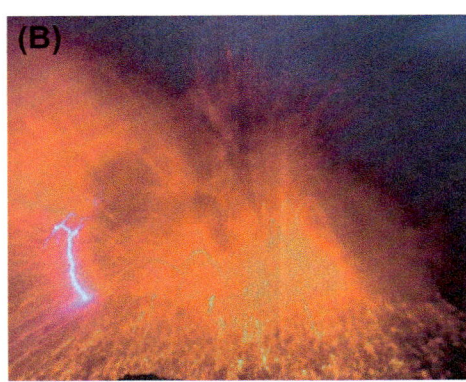

charge). Volcanic lightning reproduced in rapid decompression experiments shows a direct relation between the number of electrical discharges and the abundance of finer ejected ash. This correlation suggests that efficient fragmentation and particle clustering in the plume provide favorable conditions for charge generation, as well as lightning discharge, within volcanic plumes (Figure 25.12).

5. SUMMARY

The past few decades have seen substantial advances in our understanding of magma fragmentation by combining laboratory experiments on natural and analogue materials with quantitative studies of the products of explosive volcanic eruptions and using these together to develop and test fragmentation models. A challenge for the next decade is to merge the different views of fragmentation presented above, and to account for both viscous and brittle behavior in response to magma ascent and decompression. From this perspective, inertia-driven fragmentation provides the viscous end member (analogous to *ductile failure* in material sciences), whereas fragmentation by rapid decompression presents the brittle end member (*brittle failure*). More detailed studies of fragmentation in expanding (decompressing) three-phase (liquid + gas + crystal) systems of both natural and analogue material are required to understand the complex interplay between degassing and crystallization, and the resulting competition between rheological changes and evolution of gas permeable networks. A quantitative description of these relations would allow allocation of specific grain characteristics (size, shape, porosity) to different fragmentation conditions, information that is required as input to ash dispersion models. Also critical is an improved understanding of secondary fragmentation, particularly as it contributes to the formation of fine ash that may be widely dispersed by volcanic plumes, and is easily remobilized by the wind.

FURTHER READING

Alatorre-Ibargüengoitia, M.A., Scheu, B., Dingwell, D.B., Delgado-Granados, H., Taddeucci, J., 2010. Energy consumption by magmatic fragmentation and pyroclast ejection during vulcanian eruptions. Earth Planet. Sci. Lett. 291 (1–4), 60–69. http://dx.doi.org/10.1016/j.epsl.2009.12.051.

Cashman, K.V., 2004. Volatile controls on magma ascent and degassing. State Planet: Front. Challenges Geophys. Am. Geophys. Union Monogr. 150, 109–124.

Cashman, K.V., Sparks, R.S.J., 2013. How volcanoes work: a 25 year perspective. Geol. Soc. Am. Bull. 125, 664–690. http://dx.doi.org/10.1130/B30720.1.

Dingwell, D.B., 1996. Volcanic dilemma: flow or blow? Science 273, 1054–1055.

Dufek, J., Manga, M., Patel, A., 2012. Granular disruption during explosive volcanic eruptions. Nat. Geosci. 5, 561–564.

Fowler, A.C., Scheu, B., Lee, W.T., McGuinness, M.J., 2010. A theoretical model of the explosive fragmentation of vesicular magma. Proc. R. Soc. A 466, 731–752. http://dx.doi.org/10.1098/rspa.2009.0382.

Kameda, M., Kuribara, H., Ichihara, M., 2008. Dominant time scale for brittle fragmentation of vesicular magma by decompression. Geophys. Res. Lett. 35, L14302. http://dx.doi.org/10.1029/2008GL034530.

Koyaguchi, T., Scheu, B., Mitani, N.K., Melnik, O., 2008. A fragmentation criterion for highly viscous bubbly magmas estimated from shock tube experiments. J. Volcanol. Geotherm. 178 (1), 58–71. http://dx.doi.org/10.1016/j.jvolgeores.2008.02.008.

Kurzon, I., Lyakhovsky, V., Navon, O., 2011. Bubble growth in viscoelastic magma: implications to magma fragmentation and bubble nucleation. Bull. Volcanol. 73, 39–54. http://dx.doi.org/10.1007/s00445-010-0402-7.

Mangan, M., Mastin, L., Sisson, T., 2004. Gas evolution in eruptive conduits: combining insights from high temperature and pressure decompression experiments with steady-state flow modeling. J. Volcanol. Geotherm. Res. 129, 23–36.

Mueller, S., Scheu, B., Spieler, O., Dingwell, D.B., 2008. Permeability control on magma fragmentation. Geology 36 (5), 339–402. http://dx.doi.org/10.1130/G24605A.1.

Namiki, A., Manga, M., 2008. Transition between fragmentation and permeable outgassing of low viscosity magmas. J. Volcanol. Geotherm. Res. 169, 48–60.

Papale, P., 1999. Strain-induced magma fragmentation in explosive eruptions. Nature 397, 425–428.

Pichavant, M., Di Carlo, I., Rotolo, S.G., Scaillet, B., Burgisser, A., Le Gall, N., Martel, C., 2013. Generation of CO_2-rich melts during basalt magma ascent and degassing. Contrib. Mineral Petrol. 166, 545–561.

Poland, M.P., Takahashi, T.J., Landowski, C.M., (Eds.), 2014. Characteristics of Hawaiian volcanoes: U.S. Geological Survey Professional Paper 1801, Chapter 8. http://dx.doi.org/10.3133/pp1801.

Rust, A.C., Cashman, K.V., 2011. Permeability controls on expansion and size distributions of pyroclasts. J. Geophys. Res. 116 http://dx.doi.org/10.1029/2011JB008494.

Scheu, B., Kueppers, U., Mueller, S., Spieler, O., Dingwell, D.B., 2008. Experimental volcanology on eruptive products of Unzen volcano. J. Volcanol. Geotherm. Res. 175 (1–2), 110–119. http://dx.doi.org/10.1016/j.jvolgeores.2008.03.023.

Taddeucci, J., Spieler, O., Kennedy, B., Dingwell, D.B., Pompilio, M., Scarlato, P., 2004. Experimental and analytical modeling of basaltic ash explosions at Mt. Etna, Italy, 2001. J. Geophys. Res. 109, B08203. http://dx.doi.org/10.1029/2003JB002952.

Wright, H.M., Cashman, K.V., Gottesfeld, E.H., Roberts, J.J., 2009. Pore structure of volcanic clasts: measurements of permeability and electrical conductivity. Earth Planet. Sci. Lett. 280, 93–104.

Chapter 26

Magma–Water Interaction and Phreatomagmatic Fragmentation

Bernd Zimanowski and Ralf Büttner
Universität Würzburg, Würzburg, Germany

Pierfrancesco Dellino
Università di Bari, Bari, Italy

James D.L. White
Geology Department, University of Otago, Dunedin, New Zealand

Kenneth H. Wohletz
Los Alamos National Laboratory, Los Alamos, NM, U.S.A

Chapter Outline

GLOSSARY

ATP Ambient temperature and pressure conditions. SAPT refers to standard atmospheric pressure (0.1 MPa) and temperature of 293 K (20 °C).

ductile–brittle transformation and fragmentation Transition of magma from a plastic state to an elastic state by rapid cooling (quenching) or by exceeding a critical strain rate—beyond this transition, strain accumulates as elastic deformation.

hydrodynamic mingling Fragmentation process caused by relative movement of two immiscible liquids.

hydrovolcanism A term comprising all kinds of volcanism involving magma–water interaction, explosive or nonexplosive, surface or subsurface, subaerial or submarine.

interactive particles During explosive magma–water interaction, only parts of the melt and the water are directly involved in the escalating feedback of heat transfer and fragmentation. Interactive particles are the product of this process and are characterized by peculiar surface textures and typical grain sizes.

Molten Fuel–Coolant Interaction (MFCI) A term used in safety engineering, sometimes FCI (fuel–coolant interaction) is also used. Industrial melts like iron, aluminum, etc., also corium (forming by nuclear reactor core meltdown), are the fuels that interact explosively with industrial coolants (mostly water) after they get accidentally mixed.

phreatomagmatic explosion Most intensive type of magma–water interaction, occurring when a mixture of magma and water evolves into a thermohydraulic explosion.

shock waves are nonlinear, impulse-like perturbations that travel at supersonic speed and always damage the structural integrity of a material, leading to cavitation in fluids. In solids, low-energy shock waves lead to micro cracks and material fatigue, and shock waves of moderate energy lead to brittle-type fragmentation.

spall strength If a solid material is exposed to strain exceeding this, it starts breaking.

superheated liquid A liquid with a temperature above its boiling temperature at ambient pressure; fast heating of liquids can result in this metastable state. The upper limit of superheat, as described by classic (linear) thermodynamics, is the homogeneous nucleation temperature (HNT), which for water at SAPT is about 595 K.

The Encyclopedia of Volcanoes. http://dx.doi.org/10.1016/B978-0-12-385938-9.00026-2

supersonic crack propagation Crack velocities in solids, as observed in static bending tests, cannot exceed the shear-wave velocity (V_s). Crack propagation becomes supersonic if the crack growth is driven by shock waves or supersonic penetration.

thermal granulation Fragmentation process driven by rapid cooling-contraction during nonexplosive magma–water interaction.

thermohydraulic explosion Most intensive type of MFCI and magma–water interaction. If, in a premix of hot melt and coolant under incompressible conditions (hydraulic coupling), the initial interface is large enough to cause a critical thermal flux, the expanding coolant pressurizes the melt beyond its mechanical strength and leads into a positive feedback mechanism of heat transfer, pressurization, and fragmentation.

vapor films—film boiling At ambient pressures well below the critical pressure of water, the contact of water with a surface of a temperature well above HNT provokes the "Leidenfrost phenomenon." In this state, a stable film of vapor forms, which strongly reduces the heat flux.

1. INTRODUCTION

Geoscientists are familiar with geological timescales. Every geologist is well aware of the difference between one year (a tectonic plate has moved 30 mm) and 1 million years (the plate has moved 30 km). Processes of explosive volcanism, and especially of explosive magma–water interaction, however, act on the timescales of short-term physics. It is essential to consider, that between 1 s and 1 μs, there is also a factor of one million and, as elsewhere in nature, this makes a huge difference.

More than 70% of the earth's surface is covered by water. The majority of volcanic activity is oceanic, linked to the mid-oceanic ridges, and must be influenced by magma–water interaction. However, a hydrosphere also exists in the subaerial crust as a consequence of groundwater and hydrous fluids, stored in porous sedimentary rocks and circulating in joints and faults. Therefore, ascending magma will normally encounter some type of water, either groundwater in aquifers and fractures or surface water in ocean basins, rivers, and lakes.

Water is an excellent thermodynamic working fluid. Practically all thermal power plants in the world use water for converting thermal energy into mechanical energy and finally into electricity. Where rising magma contacts water (in contrast to dry rocks or atmospheric gases) the major effect is an increase of thermal energy flux from the magma and, by analogy to a thermal power plant, of power generation. Thermal energy represents by far the greatest usable energy contained in magma. The intensity of volcanic eruptions scales with the amount of energy released per unit time. A good proxy for the energy release during an eruption is the mass eruption rate (Sparks 1986). The violence of volcanic eruptions depends on the coupling of this thermal energy to the surroundings, i.e., the heat flux.

As magma is a very poor heat conductor, the heat flux is determined by the contact area or the interfacial area. The size of this interfacial area depends on the intensity of magma fragmentation. Therefore, fragmentation of magma plays the key role in explosive volcanism and consequently the most explosive volcanic eruptions are linked to the highest production rate of new surface area and produce the highest amounts of small particles, especially fine volcanic ash (grain size <64 μm).

Fragmentation of magma comes at a cost, and the mechanical energy needed to drive fragmentation and to produce new surface area must be considered as an essential part of the total energy budget of an explosive eruption (Zimanowski et al. 2003). Magma is a complex system consisting of melt containing variable amounts of juvenile crystals (phenocrysts) and crystals and rock fragments entrained during the ascent from the magmatic source region (xenocrysts and xenoliths). Once magma ascends to regions of lower magmastatic (hydrostatic) pressure in the crust, exsolution of supersaturated volatiles can occur and form vesicles. Magma under low stress and strain can be described as a complex liquid, and fragmentation is dominated by fluid processes. Under high stress and strain, typical for the regime of explosive fragmentation, magma reacts like a complex solid, and fragmentation is dominated by brittle mechanics (Wohletz et al. 2013). Fine-ash fractions in deposits produced by explosive eruptions consist almost entirely of products from brittle fragmentation. Energy considerations of possible fragmentation processes show that volcanic eruptions in the gravitational field of the earth can only produce significant amounts of volcanic ash (particularly fine ash) by brittle-type fragmentation.

The consequences of magma–water interaction strongly depend on the material and environmental parameters and on contact dynamics. They range from underwater effusion of lavas to explosive pulses so strong that up to 90% of the resulting deposits consist of powdered country rock that formerly hosted the volcanic conduit in a hard rock environment (Figure 26.1).

2. HISTORY OF OBSERVATIONS AND RESEARCH

The relationship between magma and water at volcanoes has been noted for at least two centuries. By the early 1800s, volcanic craters called *Maars* (including tuff rings and cones) in the Eifel district of Germany were recognized for their filling by lake water and distinctive ash-rich ejecta; these landforms have since been associated with explosive magma–water interactions. In the early 1830s during his voyage on the *Beagle*, Charles Darwin documented the formation of tuff rings along shorelines by the interaction of basaltic magmas with seawater. In North America,

FIGURE 26.1 (A) Photo of lava from Kilauea volcano entering the coast near Kalapana in 1988, taken by J. D. Griggs (USGS). (B) Photo of the May 24th, 2011 eruption column at Grimsvotn volcano taken by Magnús Tumi Gudmundsson. (C) One of many eruptions at Mt. Ruapehu in 1971 displaying a Surtseyan explosive style, taken by Peter Otway (New Zealand Geological Survey). (D) Photo of the two Ukinrek Maars, taken about 2 years after eruption by Volker Lorenz in 1979.

geologists described the detailed structures of eruption craters that formed within or near lakes and areas of abundant groundwater but observed that not all magma—water interactions were explosive. Jaggar and Finch (e.g., Jaggar 1949) gave the first eyewitness accounts of magma—water interaction in Hawaii in 1924. Where groundwater and magma are involved in eruption, the term *phreatomagmatic* became the common descriptive term; but magma—water interaction is so broadly recognized that the terms *hydrovolcanic* or *hydromagmatic* are used to include a spectrum of nonexplosive interactions.

The 1950s and 1960s brought new eyewitness accounts of phreatomagmatic eruptions, including the formation of Nilahue Maar in Chile in 1955, of Capelinhos Volcano on Faial (Azores) in 1957—1958, of Surtsey (Iceland) in 1963—1965, and of base-surge eruptions within Lake Taal (Philippines) in 1965. Some important observations included:

- The magma—water explosions were commonly episodic or pulsating.
- Many explosions occurred directly after water was observed to enter into the vent.
- The amount of water entering the vent and the apparent depth of explosions both greatly affected the manner of pyroclast ejection.
- Pyroclastic density currents were produced at the base of the eruptive jets ("base surges").

One branch of research in recent years has focused on distinguishing the physical properties of tephra produced by phreatomagmatic fragmentation from tephra produced by other, wholly magmatic fragmentation (see Section 4).

Phreatomagmatic explosions result from complex interactions of four components which are independent and often internally heterogeneous (magma, magmatic volatiles, external water, and wall rock). Experimental studies of more simple systems offer powerful analogues to understand the eruption dynamics, and these experiments developed from adaptation of research on industrial accidents in which molten fluids (metals, salts) explosively interacted with water, generally termed FCI. Theoretical and experimental studies at Sandia National Laboratory (New Mexico, USA) and at Institut für Kernenergetik und Energiesysteme (Stuttgart, Germany) in the 1970s and 1980s provided a strong basis for understanding FCIs and a framework for the study of hydrovolcanic activity. These experiments were designed to investigate steam explosions during core meltdown at nuclear reactors and other industrial environments such as metal smelters. The primary goal of the experiments was to constrain the conditions for a steam explosion when a coolant comes into contact with a fuel. Fuel (melt) used in the experiments ranged from molten salts to thermite-generated melts.

Experiments conducted at Los Alamos National Laboratory (LANL) were designed to simulate phreatomagmatic activity using an Fe-Al thermite melt and water (Wohletz

FIGURE 26.2 Schematic drawings of the five thermite—water interaction experiments conducted at Los Alamos National Laboratory. (A) Emplacement of 10 kg of thermite in an iron pipe above a second pipe filled with water; (B) immersion of a Plexiglas tube in a large Plexiglas box filled with >1 cubic meter of water; (C) emplacement of 90 kg of thermite held above water by an aluminum disk in a sealed steel cylinder with a vent at the top that burst open when the internal pressure exceeded 7 MPa; and (D) same as design (C) except the vent pipe extended down through the thermite into the water compartment. (E) The lift-off experiment designed similarly to (D) except the burst valve is located at the base of the vessel.

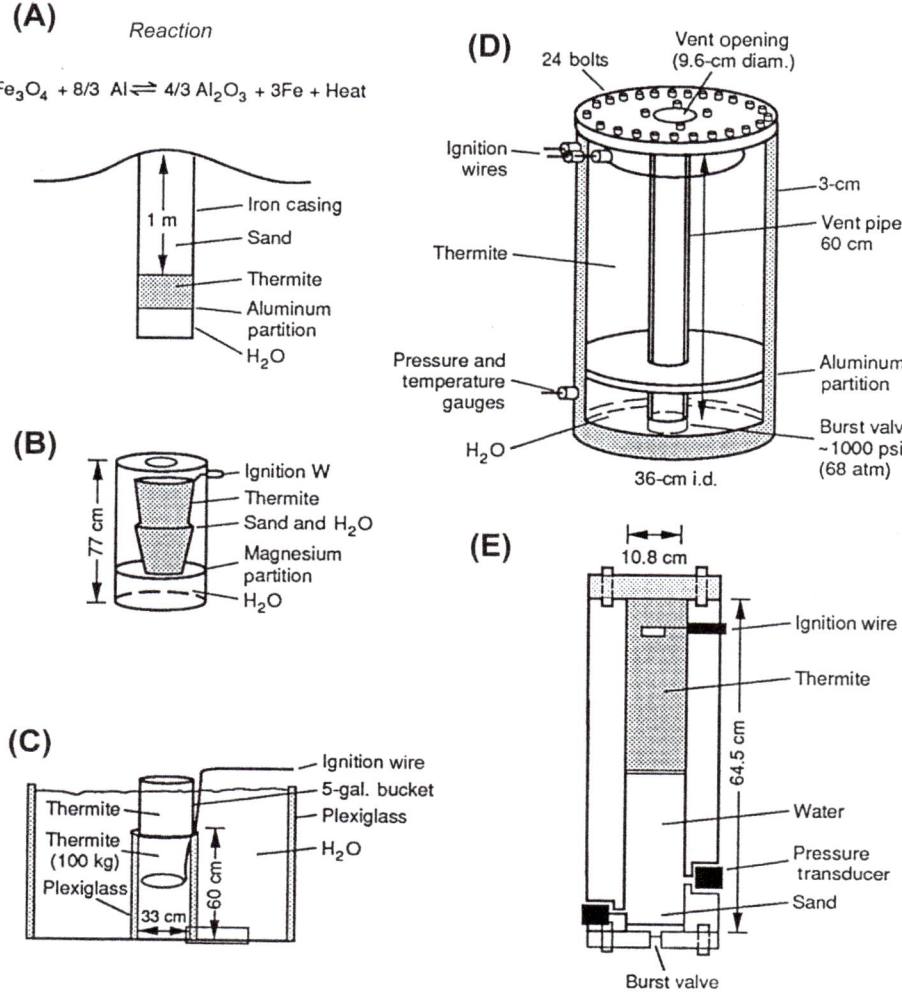

(A) Reaction

$$Fe_3O_4 + 8/3\ Al \rightleftharpoons 4/3\ Al_2O_3 + 3Fe + Heat$$

1983, 1986). The objectives were to monitor dynamic interactions and to determine what controls the rapid conversion of the melt's thermal energy into mechanical energy when it interacts with water. Four designs (Figure 26.2) were employed to evaluate the effects of contact geometry, the mass ratio of water to melt, and the confining strength (or pressure) on the explosivity of the experimental phreatomagmatic eruption. High-speed cameras were used in all experiments and samples of ejected thermite were collected from some of the experiments.

More than one style of melt ejection, each resembling a type of volcanic activity, resulted from each of the experimental designs:

1. Continuous fountaining of centimeter-sized liquid fragments of melt, which resembled a Hawaiian eruption.
2. Pulsating ballistic ejections of partially quenched and vesicular ("scoriaceous") fragments, which resembled a Strombolian eruption.
3. Dry vapor-rich explosions ejecting micron-sized fragments in expanding jets of superheated steam, which resembled a Surtseyan eruption.
4. Wet vapor explosions ejecting millimeter-sized fragments in plumes of condensing steam.
5. Passive chilling of the thermitic melt into centimeter-sized fragments with quenched surfaces, which was analogous to the formation of pillow lavas or peperites.

A fifth experimental design employed an interaction vessel that functioned like a rocket engine (Figure 26.2(E)) in order to quantify the propulsion (mechanical energy) as a function of water-to-melt mass ratio and confining pressure. In this design, ejection of high pressure steam and fragmented thermite passing through the vent caused the vessel to lift off the ground. Measurements of the internal pressure history and ejecta characteristics were recorded as was the maximum height reached by the vessel allowing to measure the kinetic energy release and (by using the thermal energy content of the thermite reaction) the calculation of a mechanical conversion ratio (MCR). The results from these experiments include three distinct pressurization histories inside the vessel which were found to correspond to three specific ranges of MCR and produced particles with distinct morphologies. Group 1 experiments produced predominantly mosslike grains; Group 2 produced mainly spherical or droplike and aggregate grains; Group 3 experiments produced mainly blocky and the irregularly shaped grains (see also Section 4). The three corresponding pressurization histories are characterized by (1) exponential, (2) approximately linear, and (3) parabolic, pressure rise with time.

The pressure—time histories reflect the degree to which water—thermite melt interaction approached thermal equilibrium. Thus, Group 1 water—melt interactions were the most efficient and explosive, and came closest to thermal equilibrium. These equate to "dry" phreatomagmatic events. Group 3 interactions were the least efficient and explosive, and did not approach thermal equilibrium. These experiments were considered analoguous to submarine phreatomagmatic events. Group 2 experiments fell between Groups 1 and 3 and were considered analogous to "wet" subaerial phreatomagmatic events.

Experiments on phreatomagmatic fragmentation have been carried out since the 1990s at the Physikalisch Vulkanologisches Labor at Würzburg University (Germany) using remelted volcanic rock samples at magmatic temperatures with compositions ranging from ultramafic to high-silica rhyolite (Austin-Erickson et al. 2008, Büttner et al. 2006). Experiments have evaluated the effect of the mode of contact between water and melt, the mixing conditions, the fragmentation processes during a **thermo-hydraulic explosion**, and the expansion phase. All experiments involve high resolution measurements of physical processes and properties and evaluation of fragmented melt. These experiments include various geometrical configurations on an experimental length scale of 5—50 cm including: (1) water entrapment by injection of water into melt, (2) melt entrapment achieved by the stratification of water and melt, and (3) melt jets intruding into water and water—sediment mixes (e.g., Schipper et al. 2011).

The results indicate that heat loss from the melt during initial contact and premixing is critical for the generation of an explosion. When the melt is dispersed in excess water, the heat transfer, even under stable conditions of vapor film boiling, is very effective, such that melt particles with diameters <1 cm completely solidify and a thick crust forms around larger melt particles within about 1 s. In this case explosive behavior cannot occur. In the case of water entrapment in excess melt, explosive behavior becomes highly probable. Here, the intensity of resulting phreatomagmatic explosion was found to be proportional to the

initial direct contact area, indicating that the mixing stage was the primary control on the intensity of explosive magma—water interaction (see Section 3). In experimental phreatomagmatic explosions three fragmentation regimes can be distinguished:

1. Inside the crucible without measurable expansion of the premix during the phase of direct contact (thermohydraulic phase).
2. During acceleration of the melt by expansion of superheated steam inside the crucible.
3. During high-speed transport outside the crucible.

Particles produced during the thermohydraulic phase 1, resembled the ash-sized particles generated by the LANL experiments. This phase was found to be the crucial phase for a phreatomagmatic explosion, during which the most kinetic energy (explosion energy) is released (see Section 3). This phase produced the peculiar shaped angular particles with diameters between 30 and 130 μm and a high surface/volume ratio (see Section 4). The total new (i.e., fresh) surface area of these particles was found to correlate linearly with the explosion intensity recorded during their formation. Fragmentation during melt acceleration in a confined geometry produced coarser fragments with distinctively smaller surface/volume ratios. Fragmentation during decelerated movement of melt on trajectories in free air produced elongated to hairlike shapes known as Pelé's hair, depending on the ejection velocity (see Section 4).

3. PROCESSES AND PHYSICAL MODELS

Our present understanding of physical processes is based on studies of explosive energy release and fine fragmentation observed in experiments and phreatomagmatic eruptions and on physics of detonation wave phenomena and their experimental verification (Wohletz et al. 2013). The interaction between magma and water (or water—salt—sediment mixes) can range from very mild with slow heat exchange that can last hours and emit very little mechanical energy (Schipper et al. 2011) to extremely violent detonation-like thermohydraulic explosions, where more than 30% of the available thermal system energy (the thermal energy of the batch of melt that interacts with water) is converted into mechanical energy within a fraction of a millisecond, and about 60% of this mechanical energy is emitted as **shock waves**, that have the potency to pulverize even solid rocks within the conduit and vent region of a volcano (Büttner et al. 2005). The efficiency and intensity of these processes depend on internal factors (i.e., the thermal and hydro-mechanical properties of magma and coolant) and on external factors (i.e., interaction geometry and dynamics).

The history of magma—water interaction begins with a **premix-phase**, where liquids meet and the stage is set for the following evolution. Premixes leading to a phreatomagmatic explosion, similar to critical mixtures of components for chemical explosions, require a critical initial contact area (i.e., interfacial area) that needs to be established fast enough to prevent significant temperature loss of the magma. Experiments have shown that about 1 m^2 interfacial area needs to form per second in a premix volume of 1 m^3. In the case of basaltic volcanism **hydrodynamic mingling** is plausible (Büttner and Zimanowski 1998). High-silica magma, because of its much higher viscosity, would require unrealistic hydrodynamic mingling energy. In experimental studies, it is shown that explosive premixing conditions can be generated by mechanical deformation of the magma that promotes brittle-type fragmentation at the melt—water interface (Austin-Erickson et al. 2008). The results from this series of experiments indicate that, under natural conditions, stress-induced magmatic fracturing can lead to a critical magma—water interface growths and trigger phreatomagmatic explosions of high-silica magma.

Hydrodynamic mingling occurs exclusively between two immiscible liquids and generally depends on material parameters, mingling geometry, **ATP**, the available mechanical energy of mingling, and, in the case of magma—water interaction, the initial thermal coupling (Zimanowski and Büttner 2002, White et al. 2003). The size distribution of the dispersed phase therefore reflects the physical parameters of both liquids and the mingling intensity (shear rate or differential flow speed). Under otherwise constant physical parameters the mean size of the dispersed liquid domains will decrease with increasing differential flow speed. It is important to understand that this mingling intensity does not represent the mingling energy, because the energy needed to establish a shear flow strongly depends on the viscosities of both liquids and the ratio of their viscosities. Depending on thermal coupling, mingling time, and mingling intensity, nonexplosive behavior results in formation of lava sheets, pillows, or hyaloclastites by **thermal granulation** of magma (Figure 26.3). Within the shaded regime critical premixes are produced that can lead to phreatomagmatic explosion. In nonexplosive premixes of magma and coolant (including water, aqueous solutions, suspensions, and all kinds of wet sediments) the magmatic domains are subjected to enhanced cooling, compared to cooling of subaerial pyroclasts. Experiments (Figure 26.4) have shown that once advection of the coolant is induced, the heat flux from the magmatic domains causes intensive magma fragmentation by cooling contraction of magma (Schipper et al., 2011), a process referred to as **thermal**

granulation. The typical grain size of this product is in the coarse ash range (about 0.1 mm < granules < 2 mm) and the typical shape is angular and platy, or blocky with smooth fracture planes (see Section 4).

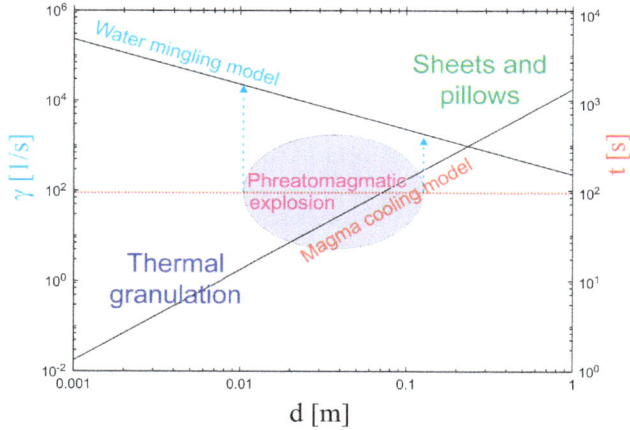

FIGURE 26.3 Effects of premixing of basaltic magma and pure water. Water-mingling model gives the size d of water domains that form in a specific basaltic magma host as a consequence of the shear rate γ, (i.e., the differential flow speed of both liquids). Magma cooling model gives the time t within which domains of size d cool from magmatic temperature to the glass transition temperature at stable film boiling conditions. The shaded area marks the explosive premix conditions as observed in experiments. The dotted red line gives the mingling time found for the formation of the optimum premix that would drive the strongest phreatomagmatic explosion possible. The dotted blue lines mark the shear rate on the water-mingling model needed to produce the critical domain sizes.

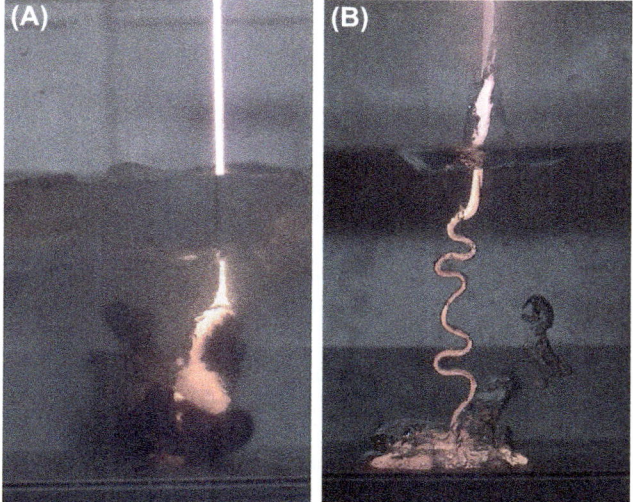

FIGURE 26.4 Images of thermal granulation experiments showing centimeter-thick jets of remelted basaltic rock (1400 K) entering water at temperatures of (A) 275 K (near freezing point) and (B) 370 K (near boiling point). The pictures were shot at two kfps using a high-speed camera. The insulating vapor films get thicker and more stable at higher water temperatures and thus more insulating, as can be identified in the pictures: In (A) the melt is solid when it reaches the ground while in (B) it is still liquid.

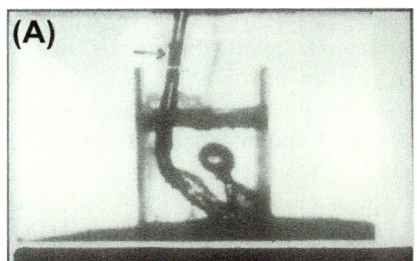 

FIGURE 26.5 Frames from a high-speed recording at 10 kfps. A water jet at 293 K, 4.5 mm in diameter, has been injected into a transparent salt melt at 1173 K by means of an injection tube (arrow). The crucible has an internal diameter of 50 mm. Frame (A) shows the state of the jet before arrival of the trigger signal and in frame (B) after its passage. Note the complete collapse of the insulation vapor film in the system.

Even when a premix of magma and water is in the critical field (Figure 26.3) additional conditions must be met to enter the explosive regime. The ambient pressure must be well below the critical pressure of water to allow stable vapor film boiling. This keeps the thermal flux low so the premix can form and persist long enough for a detonation-like process to evolve. Without **vapor films** the magmatic domains are quickly quenched, preventing a coherent reaction of a sufficient volume of premix. Next, all vapor films need to be quasi-simultaneously destroyed during the **trigger phase**, by a pressure pulse of significant strength (Büttner and Zimanowski 1998) traveling through the premix (Figure 26.5). The sources of such **trigger signals** can be internal, by local collapses of vapor films within the premix, or external, by thermal cracking of host rocks or seismic events in the surroundings. The maximum premix volume of a single explosion is limited by the speed of sound within the premix and by the time window between the onset of vapor film destruction and the onset of significant vaporization of the superheated water. The volume of premix through which a trigger signal passes in this time window is the volume that can be coherently triggered and contribute to the energy release. This "ignitable" volume ranges from practically 0 to about 100,000 m^3, depending on the temperatures of magma and water (the higher the contrast, the wider the time window), on the vesicularity of magma (the higher the vesicularity, the lower the speed of sound), on the nature of the trigger (sonic or supersonic), and on the amount of incondensable gases in the premix (the higher the amount, the stronger the absorption of trigger energy). If several premix volumes have formed separately within an extended magmatic feeder, it is very likely, that one explosion will trigger others. Unfortunately no experimental or observational data exist about whether such a scenario can lead to additional escalations. We hope that large scale experimentation can be performed in the near future.

After the passage of the trigger signal the vapor films are destroyed within about 1 ms and the **phase of direct contact** between magma and water is established. Now two important changes in the state of the premix occur: (1) the

heat flux increases by 2 orders of magnitude, and (2) the system changes from a compressible to an incompressible state. The increasingly heated water, because of its much higher thermal expansion rate, needs to expand, but the cooling melt cannot provide nearly enough volume. Thus, the water is pressurized at high rates and the magma is stressed until it cracks (Figure 26.6). Due to its high pressure the water intrudes into the cracks and undergoes more superheating from the new hot surface. This again increases the pressure and drives growth of the cracks at rates that exceed the speed of sound within the melt, resulting in **supersonic crack propagation**. A positive feedback mechanism is started, a detonation-like coupling of heat transfer and fragmentation, leading to the emission of destructive shock waves into the surroundings: the **thermohydraulic explosion**. Thermodynamic modeling of this highly nonlinear process requires nonequilibrium thermodynamics (Büttner et al. 2005, Wohletz et al. 2013). The timescale of this process is in the range of microseconds and it is dominated by solid state properties, because liquid relaxation cannot take place. The energy release is

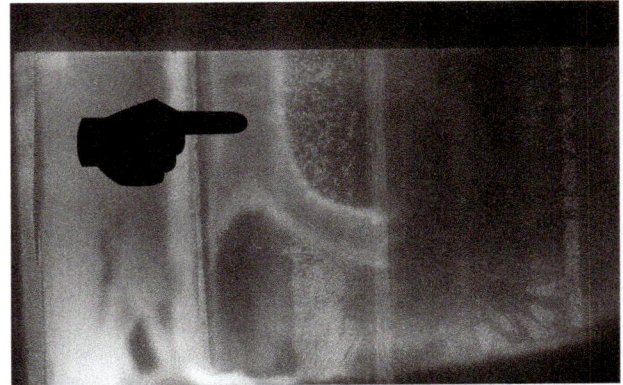

FIGURE 26.6 Frame from a high-speed recording at 2.5 Mfps (shutter time 100 ns). A water jet at 293 K of 4.5 mm in diameter has been injected into a transparent salt melt at 1173 K by means of an injection tube (finger). The frame shows the state of the system 20 μs after establishment of direct contact conditions. Note the onset of melt fragmentation and the formation of cracks in the melt.

determined by the elastic properties of the magma, mainly the **spall strength**. Gas bubbles in the magma (vesicularity) will have a strong influence, for example an exponential decrease of the spall strength has been measured in porous ceramic materials (Wagh et al. 1993). It is important to understand that a porous (and thus weak) magma needs much less deformation energy to produce volcanic ash. Therefore volume and grain size of volcanic ash do not alone represent the energetics of explosive eruptions, because the weaker a magma, the less energy can be stored during deformation, and the less energy can finally be released at and after fragmentation.

Once ATP forces a phase transition of the superheated water into superheated vapor, thermal and hydraulic decoupling of the system takes place and terminates the thermohydraulic explosion. More than 80% of the total explosion energy has already been released at this stage. Cause of this termination normally is a distinct pressure drop by the explosive opening of the system. Now the **expansion phase** starts and the superheated steam plus the residual elastic energy of the particles create a two-phase system, flowing up the conduit and expanding in the surroundings: a phreatomagmatic eruption takes place.

4. PHREATOMAGMATIC DEPOSITS AND DIAGNOSTIC TOOLS

Phreatomagmatism occurs over orders of magnitude of magma volume, mass flow rate, and energy release, and we have to add that explosions do not occur as single events, but as a complex series of pulsations. If each pulse is to be related to a discrete magma fragmentation event and if pulses can be linked to layers in the volcano stratigraphy, particles features from deposits can be used to reconstruct the fragmentation mechanism.

Since pioneering work (e.g., Heiken and Wohletz 1985, Lorenz 1986, Sheridan and Wohletz 1983), proposed the first definitions, classifications, and interpretations of phreatomagmatic eruptions and associated products, experimentally generated particles have been systematically compared to the natural counterparts to interpret and quantify the fragmentation processes. It is now widely recognized that phreatomagmatic events are not only peculiar to the small eruptions of tuff rings, tuff cones, and maar volcanoes, but occur across the full range of mass eruption rates and conditions (e.g., phreatoplinian eruptions). They occur not only with basaltic melts, but also with more evolved magmas, even with rhyolites. Finally, it is common that in an eruption phreatomagmatic fragmentation overlaps or alternates with vesiculation and mechanisms of magmatic fragmentation. Since it is important to recognize the fragmentation process, one question readily comes to mind: how can we identify phreatomagmatism in

the products of past eruptions? Or, how can we distinguish phreatomagmatic fragmentation from other magma fragmentation processes?

An important tool to answer this is the detailed study of the surface features of glassy fine ash (e.g., Büttner et al. 1999). It is not only a significant amount of fine ash that is typical of phreatomagmatic deposits, because this just indicates intensive brittle fragmentation, which also can occur in stress-induced explosive magmatic fragmentation (Dellino et al. 2012, Dürig et al. 2012). The most revealing particles for the interpretation and discrimination of fragmentation processes, however, are fine ash–particles (diameter <100 µm), because they show the crucial diagnostic features on their surface.

Only part of a magma batch involved into explosive magma–water interaction has actually experienced direct contact with water and has been finely fragmented (active particles). Other particles (passive particles) are fragmented by the stresses resulting from the thermohydraulic explosion. It has been demonstrated that active particles can represent less than 10% of the total fragmented mass. They preserve a record of the direct fast energy transfer between magma and water, which generates enough energy to also fragment the remainder of particles and expel the gas-particle mixture from the volcanic vent.

During nonexplosive magma–water interaction, fragmentation can occur by passive cooling and contraction, e.g., thermal granulation. This process is thought to be typical of oceanic deepwater eruptions, but it is likely also in any other water-rich environment. It is even possible that some of the particles produced during explosive phreatomagmatic events are fragmented in this way. Also they are to be regarded as passive particles. Finally, the prolonged contact of newly formed particles with hot fluids (both in the eruptive mixture and during transportation in density currents), may lead to the formation of surface features superimposed on those directly acquired upon fragmentation.

4.1. Active particles as Fingerprints of Explosive Magma-Water Interaction

The most diagnostic features characterizing active particles are quenching cracks within the nonhydrated glass, stepped features, and mosslike patterns. These can be found both on the surface of fine-ash glassy particles from explosive **MFCI** experiments and in natural deposits, both of low and high silica content. Quenching cracks are represented by a branching network of fractures. These cracks form immediately after fragmentation due to the sudden quenching and consequent contraction of still-hot particles. The occurrence of quenching cracks is not accompanied by any differences in the chemical composition compared to fresh glass; thus the formation of these patterns on the surfaces of

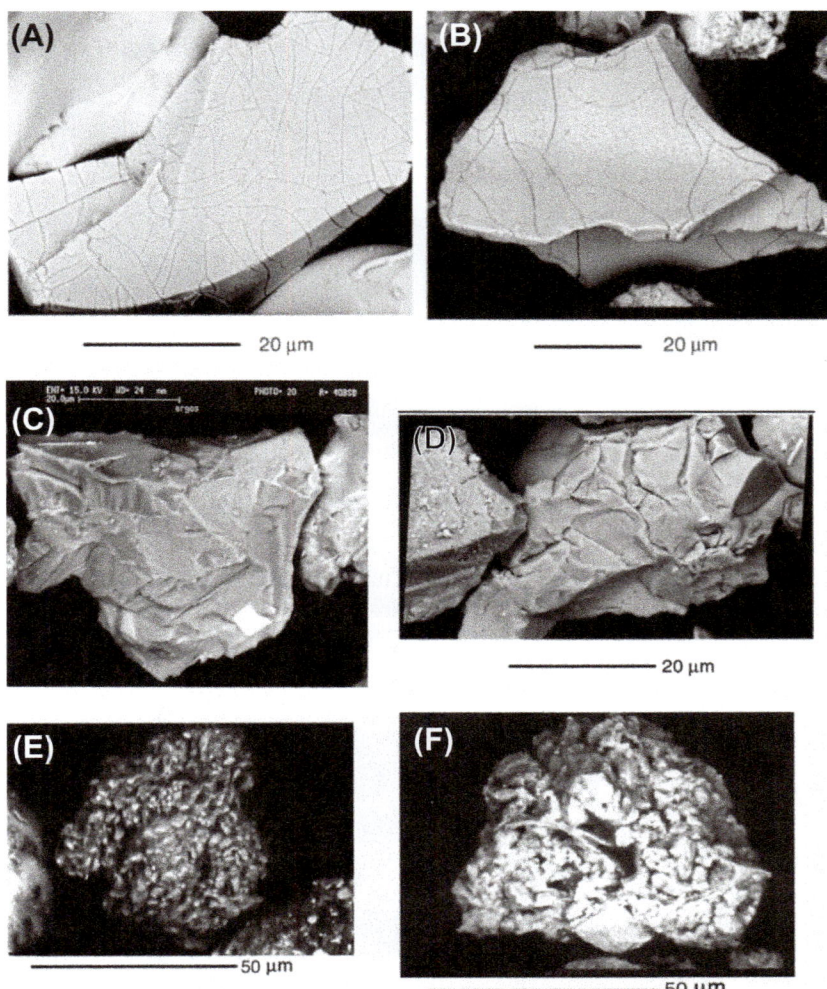

FIGURE 26.7   Scanning electron microscope (SEM) images of active particles. (A) A particle with quenching cracks from MFCI experiments. (B) A particle with quenching cracks from natural deposits. (C) A basaltic particle with stepped features. (D) A rhyolitic particle with stepped features. (E) A particle with mosslike pattern from MFCI experiments. (F) A natural particle with mosslike pattern.

experimentally produced particles (Figure 26.7(A)) by hydration processes can therefore be excluded, because of the extremely short contact time between the hot particles and the water (<1 ms). Quench patterns of the same type are also found on corresponding particles from natural phreatomagmatic deposits (Figure 26.7(B)). Such patterns can thus be related to the fast passage (a few ms) of newly fragmented particles through a domain of liquid water. This mechanism results in a high cooling rate and effective chilling. Stepped features refer to particles with an uneven surface made up of three-dimensional polyhedral elements that, when highly developed, confer an irregular outline to the particle. They are a consequence of the extreme and intensive brittle fracturing of the melt occurring in basalts (Figure 26.7(C)) as well in rhyolites (Figure 26.7(D)). Mosslike patterns consist of angular elements bonded together to form complex irregularly

shaped grains. They are interpreted in terms of annealing that occurs immediately after brittle fragmentation of very fine particles before effective cooling could act. They occur both in experimental (Figure 26.7(E)) and natural particles (Figure 26.7(F)).

4.2. Passive Particles

Passive particles in basaltic magmas, where the melt has low viscosity, are fragmented by a hydrodynamic mechanism after effective MFCI, which leads to two types of particles: rounded and elongated as Pele's hairs and Pele's tears. Rounded (Figure 26.8(A)) or subrounded (Figure 26.8(B)) shapes with smooth surfaces result from fragmentation in a ductile regime under confined conditions that occur during the early expansion stage of MFCI. Coarser particles typically show elongated smooth shapes

FIGURE 26.8 Scanning electron microscope (SEM) images of passive particles. (A) A rounded particle. (B) A subrounded particle. (C) A particle of the type of Pele's hairs. (D) A particle with blocky shape. (E) A platy particle. (F) A cuspate particle.

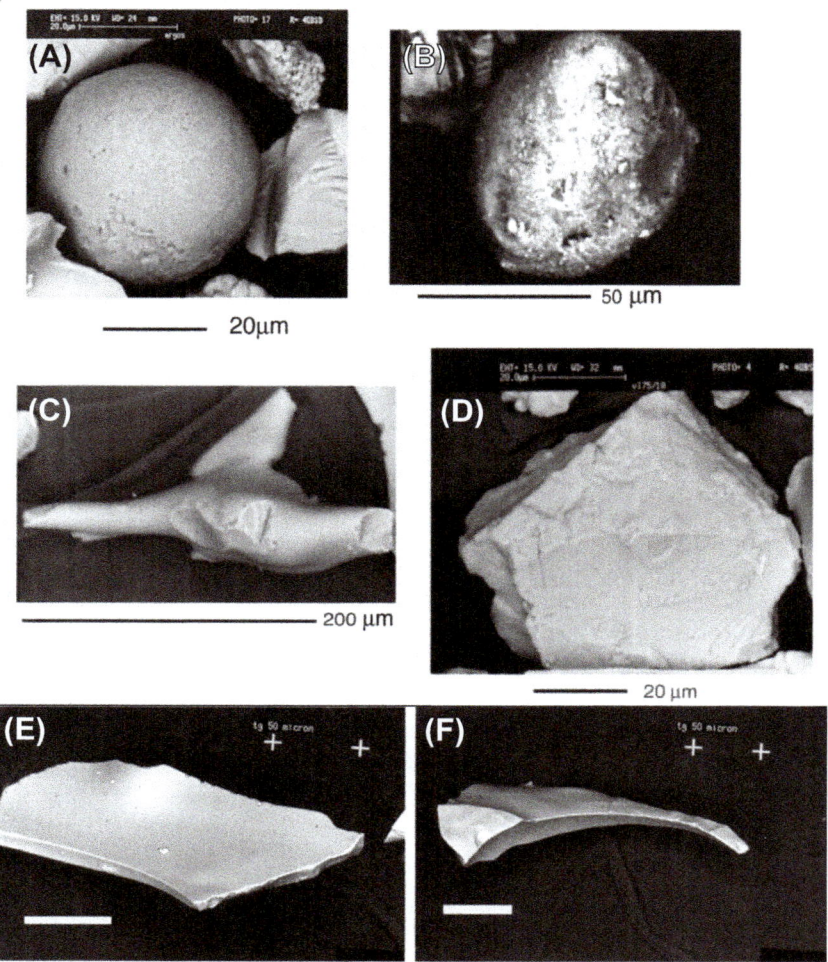

(Figure 26.8(C)), similar to Pele's hairs. They relate to free air fragmentation in a ductile regime during the late expansion phase of MFCI, likely caused by fast ejection of the melt out of the vent. Passive particles generated from rhyolite magmas, due to their higher viscosities are fragmented after the decompression wave produced by MFCI with a brittle-type fragmentation mechanisms, they generally show blocky shapes and are devoid of quenching cracks, stepped features, and pitting (Figure 26.8(D)). The fragments produced by nonexplosive thermal granulation typically show angular, elongated shapes, and planar glass surfaces with platy (Figure 26.8(E)) and cuspate morphologies (Figure 26.8(F)).

4.3. Particle Surface Features Acquired by Contact with Hydrothermal Fluids after Fragmentation

These features are acquired by the glass surface during contact with hydrothermal fluids in the eruptive column and during transportation in density currents. They consist of pitted surfaces, hydration skins (overgrowth film), and adhering particles. Pitting of particle surfaces is evident from the occurrence of scattered spherical holes, which are typically a few μm in diameter (Figure 26.9(A)). The chemical composition of pitted surfaces is slightly different from that of fresh glass, being depleted in the highly mobile alkaline elements and enriched in magnesium and iron. Pitting probably results from the interaction of glass with corrosive fluids. Hydration "skin" is a few μm-thick overgrowth film that envelopes glass fragments (Figure 26.9(B)) and is significantly depleted in alkaline elements and enriched of magnesium and iron. A "skin" probably represents the end product of the alteration that generates pitting. Some particles show a complete range of variation between the two features (Figure 26.9(C)). Adhering particles (Figure 26.9(D)) are μm to sub-μm glass particles stuck to the surfaces of larger glass fragments by electrostatic forces during steam condensation in the eruptive cloud or

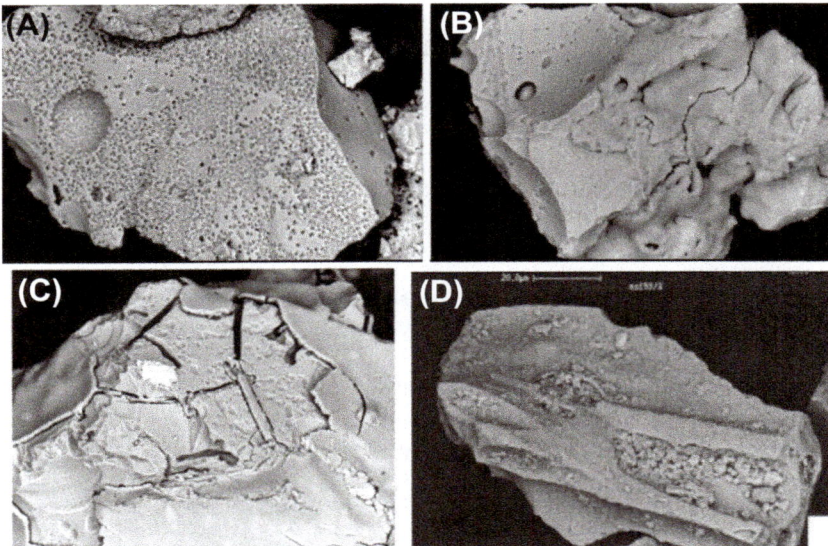

FIGURE 26.9 Scanning electron microscope (SEM) images of particle surface features due to alteration by hydrothermal fluids. (A) Pitting. (B) Overgrowth film. (C) Hydration skin grown over pitting. (D) Adhering particles.

associated density current. Adhering particles and the larger glass fragments both have the composition of fresh glass.

4.4. Mixed Magmatic-Phreatomagmatic Eruptions

This can occur with melts with a variable degree of vesicularity. Active particles tend to show the same features as those illustrated above, but can also display moderate vesicularity. A good example is the phreatomagmatic/magmatic eruption of the Eyjafjallajökull volcano (Iceland) in 2010 (Dellino et al. 2012).

4.5. Quantitative Discrimination of Fragmentation Processes as Based on Shape Parameters

Because each fragmentation process required a characteristic energy for particle surface generation, the quantification of particles to assign to specific processes or the amount of particles to be attributed to magmatic vs. phreatomagmatic fragmentation in mixed eruptions is useful for the calculation of the energy budget of eruptions. Linear combination of simple shape parameters as compactness, elongation, circularity, and rectangularity (Büttner et al., 2002) or the fractal and multifractal dimension of particles, calculated by the Mandelbrot's diagram on the particle contour, have been used. This quantitative approach has corroborated the hypothesis that active particles represent only a relatively small fraction of grains in phreatomagmatic deposits, most being passive particles. This is in full agreement with the experimental results, and calculations based on the energy required to

produce new surfaces and mechanical energy released by the process show that enough residual energy is available for transportation in the conduit and to drive a typical phreatomagmatic eruption.

FURTHER READING

Austin-Erickson, A., Büttner, R., Dellino, P., Ort, M.H., Zimanowski, B., 2008. Phreatomagmatic explosions of rhyolitic magma: experimental and field evidence. J. Geophys. Res. 113, B11201.

Büttner, R., Dellino, P., Zimanowski, B., 1999. Identifying modes of magma/water interaction from the surface features of ash particles. Nature 401, 688–690.

Büttner, R., Dellino, P., LaVolpe, L., Lorenz, V., Zimanowski, B., 2002. Thermohydraulic explosions in phreatomagmatic eruptions as evidenced by the comparison between pyroclasts and products from molten fuel coolant interaction experiments. J. Geophys. Res. 107, B11–B2277.

Büttner, R., Zimanowski, B., Mohrholz, C.-O., Kümmel, R., 2005. Analysis of thermohydraulic explosion energetics. J. Appl. Phys. 98, 043524.

Büttner, R., Dellino, P., Raue, H., Sonder, I., Zimanowski, B., 2006. Stress induced brittle fragmentation of magmatic melts: Theory and experiments. J. Geophys. Res. 111, B08204.

Büttner, R., Zimanowski, B., 1998. Physics of thermohydraulic explosions. Phys. Rev. E 57, 5726–5729.

Dellino, P., Gudmundsson, M., Larssen, G., Mele, D., Stevenson, J., Thordarson, T., Zimanowski, B., 2012. Ash from the Eyjafjallajökull eruption (Iceland): fragmentation processes and aerodynamic behaviour. J. Geophys. Res. 117, B00C04.

Dürig, T., Sonder, I., Zimanowski, B., Beyrichen, H., Büttner, R., 2012. Generation of volcanic ash by basaltic volcanism. J. Geophys. Res. 117, B01204.

Heiken, G., Wohletz, K., 1985. Volcanic Ash. University of California Press, Berkeley, California.

Jaggar, T.A., 1949. Steam blast eruptions. Hawaiian Volcano Observatory 137, 4' Sp. Rpt.

Lorenz, V., 1986. On the growth of maars and diatremes and its relevance to the formation of tuff rings. Bull. Volcanol. 48, 265–274.

Schipper, C.I., White, J.D.L., Zimanowski, B., Büttner, R., Sonder, I., Schmid, A., 2011. Experimental interaction of magma and "dirty" coolants. Earth Planet. Sci. Lett 303, 323–336.

Sheridan, M.F., Wohletz, K.H., 1983. Hydrovolcanism: basic considerations and review. J. Volcanol. Geotherm. Res 17, 1–29.

Sparks, R.S.J., 1986. The dimensions and dynamics of volcanic eruption columns. Bull. Volcanol. 48, 3–15.

Wagh, A.S., Singh, J.P., Poeppel, R.B., 1993. Dependence of ceramic fracture properties on porosity. J.Mater. Sci. 28, 3589–3593.

White, J.D.L., Smellie, J.L., Clague, D., 2003. Explosive Subaqueous Volcanism. In: AGU Monograph, Vol. 140. Washington.

Wohletz, K.H., 1983. Mechanisms of hydrovolcanic pyroclast formation: grain-size, scanning electron microscopy, and experimental results. J. Volcanol. Geotherm. Res. 17, 31–63.

Wohletz, K., 1986. Explosive magma-water interactions: thermodynamics, explosion mechanisms, and field studies. Bull.Volcanol. 48, 245–264.

Wohletz, K., Zimanowski, B., Büttner, R., 2013. Magma-water interactions. In: Fagents, S.A., Gregg, T.K.P., Lopes, R.M.C. (Eds.), Modeling Volcanic Processes. Cambridge University Press.

Zimanowski, B., Wohletz, K.H., Büttner, R., Dellino, P., 2003. The volcanic ash problem. J. Volcanol. Geotherm. Res. 122, 1–5.

Zimanowski, B., Büttner, R., 2002. Dynamic mingling of magma and liquefied sediments. J. Volcanol. Geotherm. Res. 114, 37–44.

Chapter 27

Hawaiian and Strombolian Eruptions

Jacopo Taddeucci
Istituto Nazionale di Geofisica e Vulcanologia, Rome, Italy

Marie Edmonds
Earth Sciences Department, University of Cambridge, Cambridge, UK

Bruce Houghton
Department of Geology and Geophysics, National Disaster Preparedness Training Center, University of Hawai'i, Honolulu, HI, USA

Michael R. James
Lancaster Environment Centre, Lancaster University, Lancaster, UK

Sylvie Vergniolle
Institut de Physique du Globe, Sorbonne Paris Cité, Université Paris Diderot, UMR CNRS, Paris Cedex 05, France

Chapter Outline

GLOSSARY

scoria A vesicle-bearing fragment of quenched magma from an explosive eruption, usually millimeters to decimeters in size, made of volcanic glass and crystals. With respect to pumice, typical scoriae have larger and more interconnected vesicles, are characterized by a lower silica content, and display a darker, brown, reddish, or black color.

rampart Linear embankments of coarse pyroclasts that accumulate parallel to the sides of an eruption fissure.

melt inclusion A micron-scale portion of glass enclosed in a crystal. Melt inclusions represent relics of melt entrapped in growing crystals and can be analyzed to provide information on melt composition at early, preeruption stages of its evolution.

infrasound Acoustic waves at frequencies below those of audible sound (<20 Hz). With respect to audible sound, infrasound carries more energy and propagates further through the atmosphere, and is thus more suitable for eruption studies and monitoring.

slug or Taylor bubble A particular type of gas bubble flowing through a liquid-filled pipe. A slug, also called Taylor bubble, is characterized by having a width comparable to that of the pipe and a round- or bullet-shaped head and a cylindrical body. The motion of a slug in a pipe differs from that of smaller bubbles, and is described by a different set of parameters.

1. OVERVIEW

Hawaiian and Strombolian eruption styles are characteristic of silica-poor magmas, often of basaltic to basaltic andesite composition, that erupt at temperatures in the 1000–1200 °C range. The low viscosity of these magmas (10–10^4 Pa s) allows for efficient segregation of gas bubbles and ductile melt deformation over timescales that are short relative to the ascent of the magma. As a result, the eruptions are characterized by the violent release of relatively large volumes of gas that fragment the melt and eject coarse, still molten pyroclasts. Strombolian eruptions are often best described as the bursting of buoyant gas bubbles at the free surface of a vent or lava lake while, for

Hawaiian eruptions, the term fountaining is often the most appropriate way to describe the sustained jetting of gas and fluidal pyroclasts. These eruption styles represent the most frequent manifestations of subaerial explosive volcanism on Earth, occurring in all geodynamic and volcanological settings, e.g., at monogenetic centers (**scoria** cones) distributed over volcanic fields (e.g., Jorullo and Paricutin, Mexico; Croscat, Spain; Lathrop Wells, USA) or on the summit or flanks of large shields or stratovolcanoes (e.g., Mt Etna and Stromboli, Italy; Kīlauea, USA; Villarrica, Chile; Ruapehu, New Zealand), but also at caldera complexes (e.g., Latera caldera, Italy).

2. TERMINOLOGY

Eruptions are classified either on the basis of observed characteristics during the events or on the properties of the resulting deposits. However, deposit-based classifications are not necessarily compatible with observed Hawaiian and Strombolian activity styles. For example, all historical explosions at Stromboli would plot in Walker's Hawaiian field, and the one well-documented Hawaiian deposit from Kīlauea plots in his Strombolian field.

Hereon, we define hierarchical classes of eruptive phenomena based on their characteristic time and mass scales (Figure 27.1). In order of increasing time and mass scale, we define: (1) *pulses*, often identifiable at second to subsecond scales, in which magma ejection is the result of a single, discrete pressure release event; (2) *explosions*, on timescales of tens of seconds to minutes including from one to tens of closely spaced—but still identifiable—ejection pulses; (3) *episodes*, in which magma is ejected continuously for hours to days, although with fluctuations that may represent near-continuous pulses; and finally, (4) *eruptions*, during which multiple Strombolian explosions and/or Hawaiian episodes, often associated with lava effusion, invariably concur and recur multiple times. Following this definition, Stromboli volcano (Italy), hosting persistent

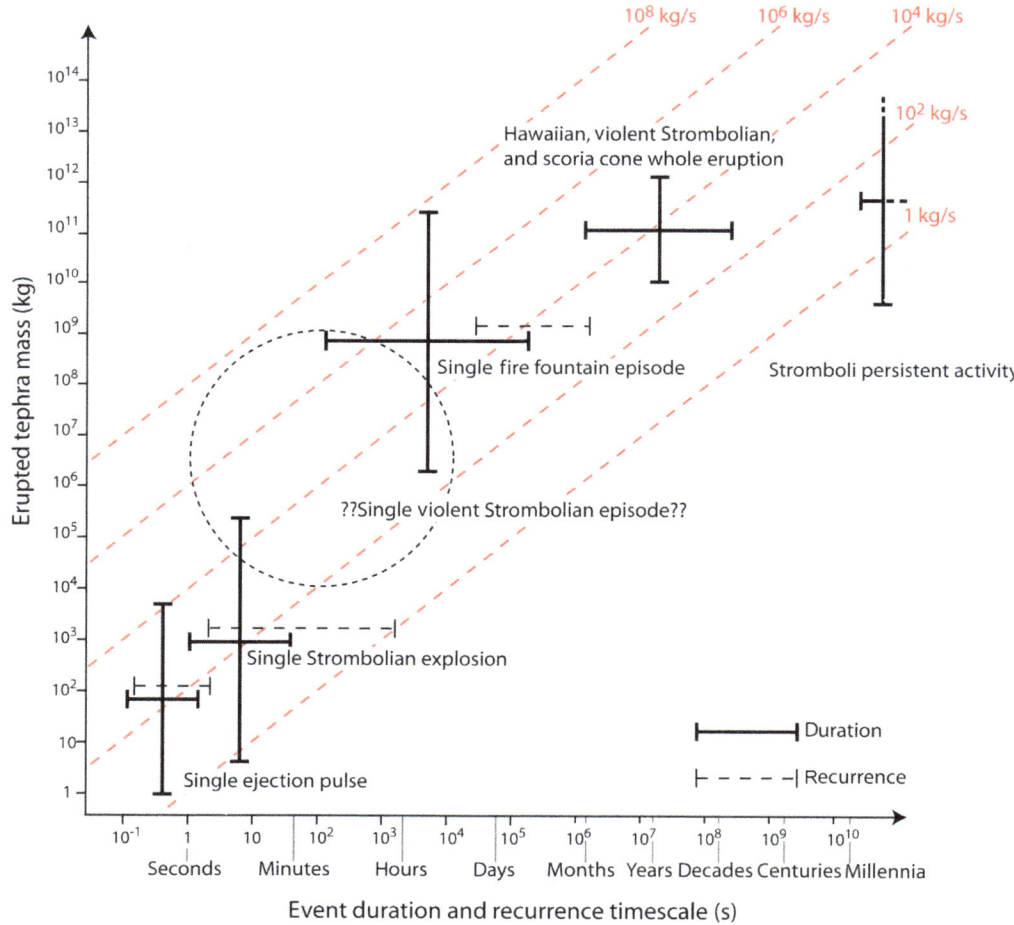

FIGURE 27.1   Ranges of duration, recurrence time, and erupted mass (dense rock equivalent) defining Hawaiian and Strombolian eruptive phenomena. One or more ejection pulses occur during one Strombolian explosion. Hawaiian fountain episodes and violent Strombolian episodes are sustained over longer times, while combinations of the above events recur in one eruption. Dashed circle is the poorly defined area of violent Strombolian episodes. Red dashed lines identify equal mass eruption rates.

Strombolian explosions and occasional lava flows since the eighth century AD, is currently undergoing a single eruption more than 1400 years long (Rosi et al., 2013, where, as in all other references, the reader is redirected to the section "Annex 1: Further Readings") and Kīlauea has been in eruption for more than 30 years.

3. HAWAIIAN AND STROMBOLIAN ERUPTIVE STYLES

The rare combination of low-intensity and long-lasting activity at several accessible and well-monitored volcanoes allows Strombolian activity to be one of the best-documented styles of explosive volcanism. Strombolian activity is characterized by weak, discrete, short-lived explosions that are the surface manifestation of the impulsive release of pressurized pockets of gas. The rapid expansion of this gas fragments and ejects adjacent molten magma clots and ash, plus minor lithic debris occasionally present in the vent (Figure 27.2, see companion site for a high-speed video of a Strombolian explosion at Stromboli volcano).

Pyroclasts ejected from Strombolian explosions typically reach only a few hundreds of meters in height if coarse (up to meter sized), but may include a fraction of ash-sized particles that form a minor buoyant plume reaching less than a kilometer in height (Patrick et al., 2007). Pyroclast ejection often occurs in short-lived jets with velocities exceeding 400 m/s, thus being supersonic with respect to the surrounding atmosphere, but averages of 50–100 m/s are more typical for bomb velocities (Taddeucci et al., 2012). Typical erupted masses at Stromboli are of order 10^3–10^4 kg per explosion (Gaudin et al., 2014). While no one-to-one relationship links the return time (order of minutes) and intensity of individual explosions, at Stromboli, the two parameters correlate positively when time-averaged over hours to days, so that

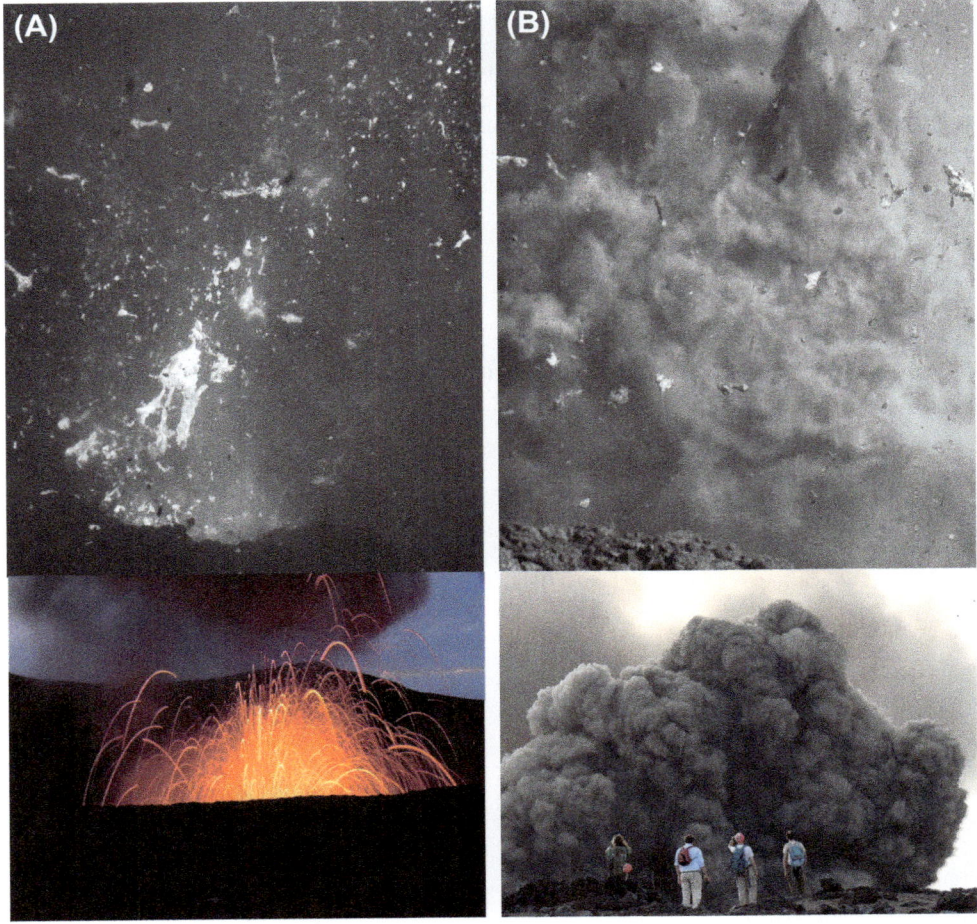

FIGURE 27.2 Still frames from high-speed video of Strombolian explosions at Yasur volcano (Vanuatu). Rising red-hot lapilli and bombs appear with very bright and darker gray tones, in contrast with darker, colder, settling ones. (A) An ash-poor explosion, showing a car-sized bomb plastically deforming and stretching in flight. Field of view is 15 m wide. Below, a long exposure photo of similar activity by night. (B) An ash-poor explosion. Note the fast-rising ash jets (top right) piercing through the slower-ascending plume. Field of view is 15 m. Below, the resulting ash plume some tens of second after the explosion. A high-speed video of a Strombolian explosion at Stromboli is available in the companion site (http://booksite.elsevier.com/9780123859389/).

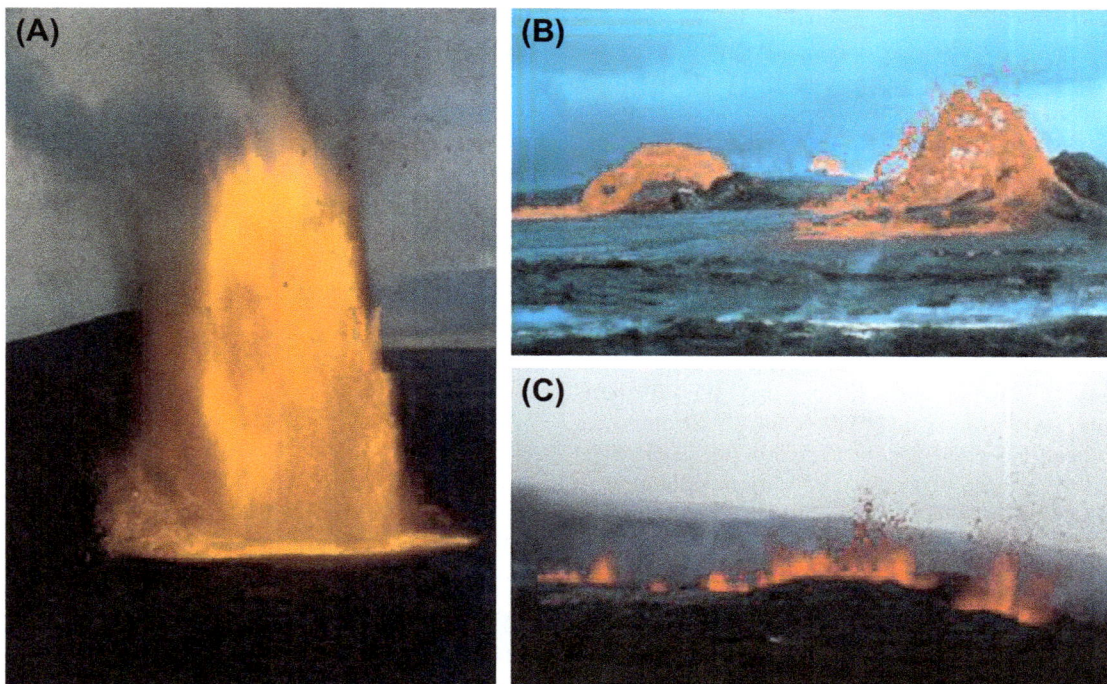

FIGURE 27.3 Images of Hawaiian explosive eruptions. (A) High fountain at Kīlauea Iki at 7:00 am on December 11, 1959 with downwind drift of ejecta to the south. Photograph Jerry Eason. (B) Mauna Ulu "dome" fountain on June 29, 1970. *Photograph Don Swanson.* (C) Weak fissure-fed activity, Kamoamoa March 6, 2011. *Photograph Matthew Patrick.*

periods with more explosions per day are also characterized by stronger explosions on average.

Hawaiian activity style is characterized by explosive episodes that are sustained on timescales of hours to days, producing fountaining of molten magma clots and little ash, in the absence of wall-rock lithic debris, typified by the well-documented activity at Kīlauea between 1959 and 1986 (Figure 27.3). High fountains can reach elevations of

TABLE 27.1 Statistics for the Three Highest Fountaining Eruptions Recorded at Kīlauea Volcano

Eruption date	Kīlauea Iki 1959	Pu'u O'o 1983—1986	Mauna Ulu 1969
Number of high-fountaining episodes	16	45	12
Average episode duration (h)	20^1	43	18
Maximum episode duration (h)	167	384	34
Average of maximum fountain height (m)	314	256	229
Maximum fountain height	579	467	540
Average discharge rate (kg/s)	7.8×10^5	1.5×10^5	4.7×10^6
Highest average discharge rate (kg/s)	1.7×10^6	9.7×10^5	1.2×10^6
Duration of high-fountaining phase (days)	36	1271	221
Time spent high-fountaining eruption (days)	13	47	10
Total duration of eruption (days)	36	$9114+^2$	874
Volume for high-fountaining episodes (m³)	1.5×10^8	5.0×10^8	6.5×10^7
Total volume (m³)	1.5×10^8	1.2×10^9	1.9×10^8

[1]10 h excluding episode 1.
[2]The eruption is still ongoing at the time of writing.

hundreds of meters and are typically episodic, with the duration of episodes and the length of the repose periods being highly variable, especially between eruptions (Table 27.1). Typical erupted masses at Kīlauea are of order 10^8-10^{11} kg per fountaining episode. Time-averaged mass eruption rates for high fountains range from 10^4 to 10^6 kg/s. The start and end of high-fountaining episodes usually show behaviors transitional toward Strombolian activity. At these times, magma is generally ponded in and above the vent and activity consists of explosions that are spaced seconds to tens of seconds apart and associated with pockets of gas bursting through the free surface. Often multiple overlapping "jets" form from adjacent point sources.

Many other eruptions at Kīlauea, which could be termed "low fountains," differ significantly from the type examples described above. Low fountains typically eject pyroclasts to less than 50 m above the vent and are also generally associated with fissure sources. Typically, magma is ponded to, or above, the level of the fissure/vent and the eruptions consist of extended series of bursting gas pockets, through a static free surface, at intervals only a few seconds apart. The location of the gas bursts remains remarkably constant, from sites generally a few meters apart along the fissure. Near-identical styles of activity have been recorded at some vents during recent eruptions at Etna. Each burst ejects only a few tens to hundreds of kilograms of magma but the time-averaged eruption rates along fissures up to several kilometers in length can be equivalent to that of a high fountain from a single vent.

Pulsations during Strombolian and Hawaiian Eruptions

Within a single Strombolian explosion, the ejection of pyroclasts occur in pulses that are characterized by a well-defined, nonlinear trend of decaying ejection velocity over time (Figure 27.4(A)) which reflects the decay in the pressure differential between the expanding gas and the atmosphere. The occurrence of one or more ejection pulses in all high-speed videos of Strombolian explosions and their tight link with explosion energetics (Gaudin et al., 2014), suggest that such pulses may be an important component of Strombolian activity.

Hawaiian episodes at Kīlauea and Etna are sustained on timescales of hours but there is evidence that the behavior is unsteady and pulsatory on timescales of seconds to minutes (Figure 27.4(B) and (C)). These patterns are strongly suggestive that Hawaiian high fountaining is also marked by pulsed discharge of the largest bubble population present in the conduit, as seen clearly in low-fountaining eruptions.

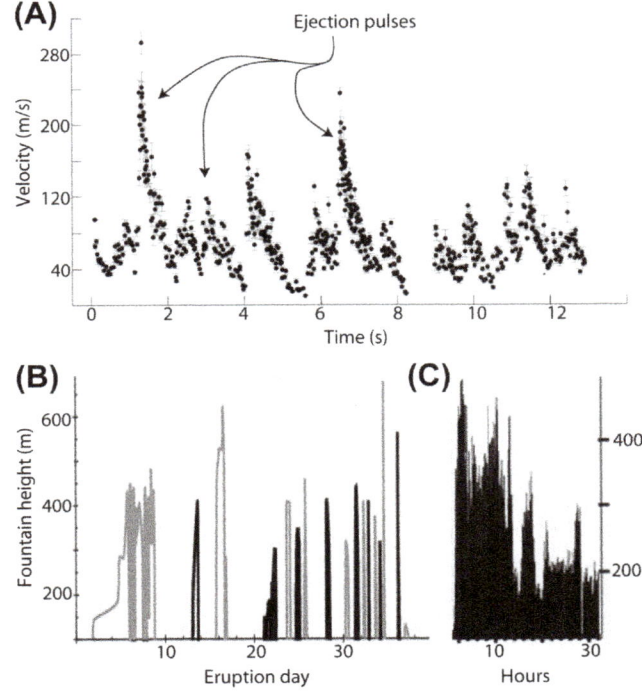

FIGURE 27.4 Pulsatory behavior during Strombolian and Hawaiian activity. (A) Ejection velocity of centimeter-sized pyroclasts progressively erupted during one single Strombolian explosion at Yasur, measured from high-speed imaging. Although the explosion visually appeared as a single, jet-like event, individual ejection pulses are clearly visible, characterized by a nonlinear trend of decaying ejection velocity over time. (B) Plot of fountain height with time for Kīlauea Iki 1959 eruption. Alternate episodes are outlined in gray and black for clarity. *Data from Richter et al. (1970) and unpublished Hawaii Volcanoes National Park records.* (C) Fountain height with time for Pu'u O'o episode 16. *Data from Wolfe et al. (1988).* After a relatively sharp onset fountain height decays exponentially with time over approximately 30 h, but with substantial fluctuations on a timescale of minutes.

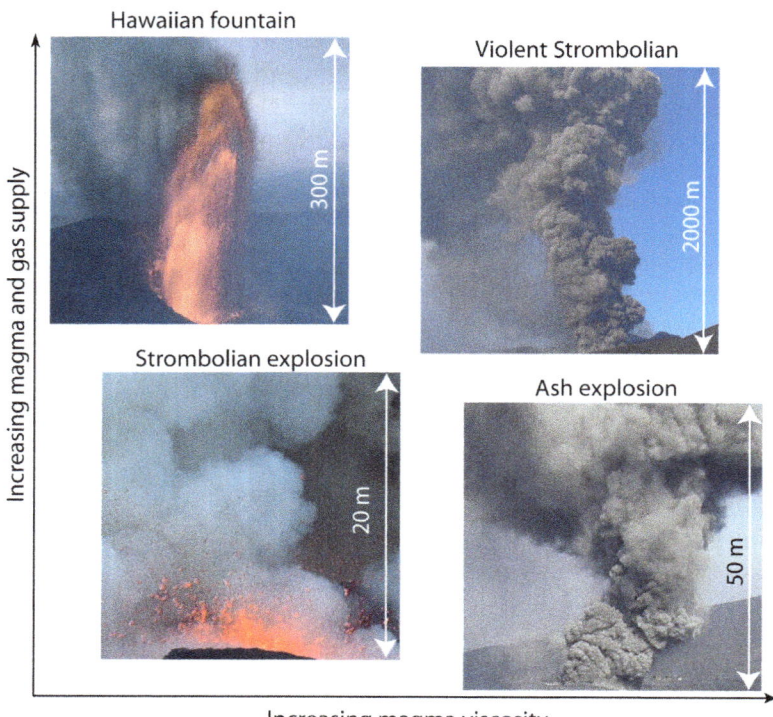

FIGURE 27.5   Representative pictures of Hawaiian— Strombolian activity and related variations, including ash explosions and violent Strombolian activity, roughly ordered in increasing magma and gas supply rates and increasing bulk magma viscosity. *Lower row and upper row, right, are from Etna activity in 2001 and 2002, courtesy of INGV-Catania, Osservatorio Etneo; upper left is from Kīlauea Iki in 1959 courtesy of Jerry Eaton.*

Several other styles of explosive activity are akin to, and interspersed with, Strombolian and Hawaiian styles (Figure 27.5). The term *spattering* is used to describe the repetition, every few seconds, of small weak magma ejection pulses likely from meter-sized or smaller gas bubbles, while *gas pistoning* indicates cycles of upwelling and draining of a standing column of lava in the vent (e.g., Edmonds and Gerlach, 2007). *Ash-rich explosions* are characterized by small, impulsive explosions at intervals from minutes to hours, simultaneously ejecting limited volumes of ash-sized particles and blocky bombs. *Violent Strombolian* defines a style of activity characterized by the fast (seconds to minutes) repetition of ash- and lapilli-rich explosions that form kilometer-high, pulsating eruption columns that last for hours to days. Shorter-lived (seconds to minutes) but intense explosions, ejecting up to meter-sized lithic blocks and molten bombs up to a few kilometers away, and producing transient convective plumes and occasional small pyroclastic density currents, punctuate every few decades of Strombolian activity at Stromboli volcano. Such events are classically termed *paroxysms*, although this term is somewhat ambiguous and local, being used to describe different styles of activity at different volcanoes or the climactic phases of an eruption.

4. DEPOSIT GEOMETRY

A key characteristic of Strombolian and Hawaiian deposits is that they are relatively coarse grained, well sorted, and locally dispersed relative to all other pyroclastic deposits. Coarse pyroclasts do not enter into a convective plume but accumulate rapidly around the vent. Eruption observations show effective partitioning of gas and coarse-grained pyroclasts such that the latter (mainly coarse lapilli and bombs) settle proximally along ballistic trajectories from elevations of generally no more than few hundred meters whereas the former (with coupled fine lapilli to ash particles) rise buoyantly up to a few kilometers as relatively small, wind-advected thermal plumes. As a result, the bomb population is fractionated into cones (point sources) or elongate **ramparts** (fissure events), both of which pass distally and abruptly into lapilli-dominant tephra sheets, resulting from the progressive deposition of pyroclasts from one Hawaiian episode or from a large number of superimposed Strombolian explosions, each of which deposits only a few scattered bombs and lapilli. A plot of deposit thickness (or mass/area) versus distance (or √area) (Figure 27.6), a common tephra dispersal metric, highlights the much more local dispersal of Hawaiian and Strombolian paroxysmal tephra sheets with respect to those of, for

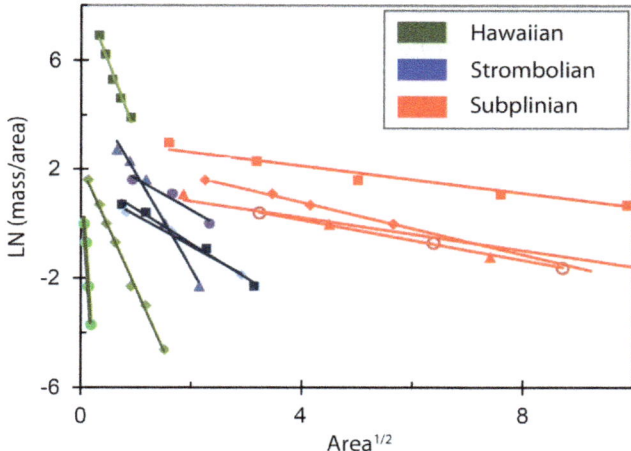

FIGURE 27.6 Thinning relationships for well-constrained fall deposits at Stromboli, Arenal, Kīlauea, and Etna. Green are Hawaiian, Blue Strombolian paroxysms and red are subplinian products from Etna.

example, basaltic subplinian falls from Etna (Houghton and Gonnermann, 2008).

5. PYROCLAST TEXTURES

Pyroclasts from Hawaiian and Strombolian activity are typically 10^{-2} to 10^{-1} m in size (but range from 10^{-6} to 10^{0} m) and are blocky to fluidal in shape, reflecting variable conduit and eruptive conditions and magma rheology. *Scoria*, i.e., moderately vesicular, mostly glassy pyroclasts dominated by millimeter-to centimeter-sized vesicles in hand specimen, is the most common product of Strombolian explosions and Hawaiian fountains. *Bombs and spatter* are commonly found in proximal settings, often displaying fluidal shapes and a variable degree of agglutination (i.e., postdepositional welding of multiple bombs), up to complete rehomogenization in spatter-fed lava flows. *Pele's hair and tears* are smaller fluidal-shaped ash-sized fragments. Rarer, blocky *breadcrust bombs* are associated with ash-rich explosions.

Vesicles in Hawaiian/Strombolian products span a broad range of number densities (number of vesicles per unit volume) and size distributions. The most vesicular (c. 95–99% vol of voids) of all types of pyroclasts, *reticulite*, is formed during Hawaiian high-fountain episodes. With respect to other types of explosive eruption styles, bubble coalescence plays an important role in Hawaiian–Strombolian eruptions, illustrated by the generally smaller number and larger size of vesicles in scoria with respect to, e.g., pumice (Figure 27.7(A) and (B)).

Wall-rock lithic clasts are sparse (<5 wt%) in most Strombolian deposits and absent in Hawaiian deposits. Crystals in pyroclasts from Hawaiian to Strombolian activity range from centimeter-sized phenocrysts to micron-sized microlites (and to sub-micron nanolites) and

from scarce to >90% by volume. In particular, microlite abundance is one of the defining parameters of the two end members of pyroclast groundmass: *sideromelane*, characterized by a brownish to golden glass with few or no microlites; and *tachylite*, dominated by the presence of abundant microlites. In ash, *sideromelane* is usually associated with abundant, spherical vesicles, and fluidal overall particle morphology, while *tachylite* is accompanied by larger, scarce, irregular vesicles, and blocky morphologies. Sideromelane and tachylite textures reflect hotter, gas-rich and cooler, more outgassed and crystallized magma end members respectively, with sideromelane dominating Strombolian and Hawaiian products and tachylite dominating the products of ash-rich explosions at Etna (Figure 27.7(C) and (D)). Recycled ash particles are also common in Strombolian eruptions. These clasts are erupted again after having fallen back into the vent a first time and acquire textures akin to tachylite, although bearing a different volcanological significance.

Complex vesicle and microlite textures highlight frequent mingling of magmas with different thermal and/or degassing histories in the vent and shallow conduit of Strombolian to Hawaiian eruptions. Mingling textures have been used to assess magma residence times and flow dynamics in Strombolian conduits and the transition between Strombolian and violent Strombolian activity (e.g., Lautze and Houghton, 2005).

6. STYLE TRANSITIONS

Transitions between different styles are frequent during Strombolian and Hawaiian activity, and occur both temporally and spatially (e.g., Valentine and Gregg, 2008). Temporally, activity at the same vent may shift from Strombolian to Hawaiian and vice versa. For instance,

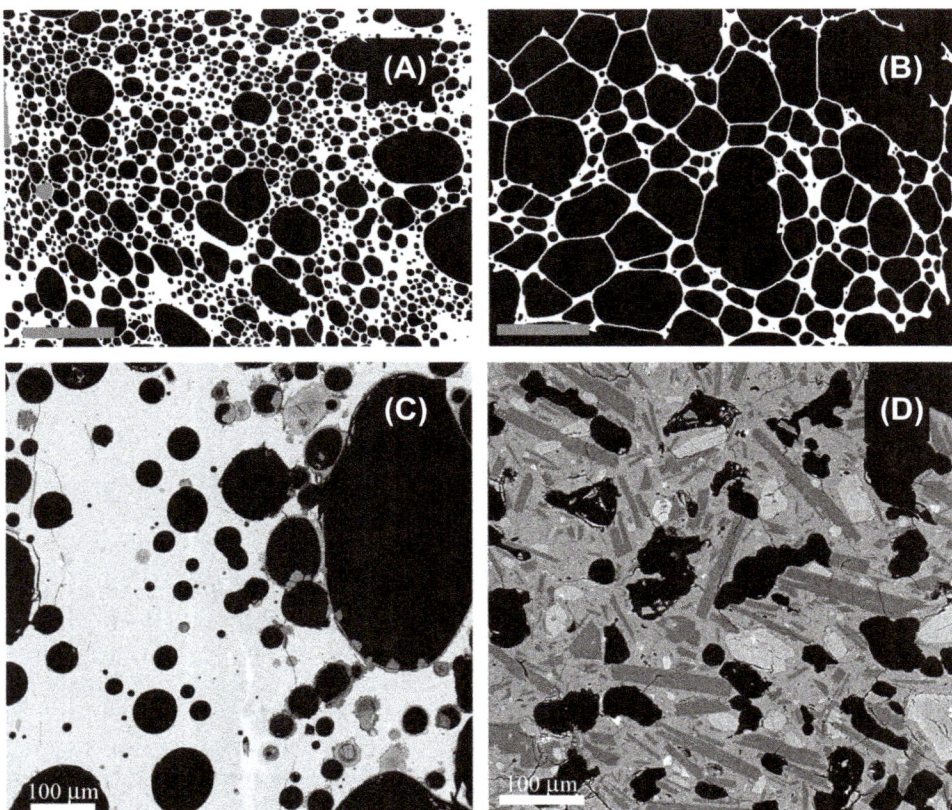

FIGURE 27.7   25x images showing the vesicle sizes of scoria produced by postfragmentation expansion of bubbles (A) and achneliths which have undergone rapid quenching (B) Scale 1 mm. Scanning electron microscope images (backscattered electrons) of glassy, well-vesiculated sideromelane (C) scoria from the Strombolian activity phase of the Croscat scoria cone, Spain) and less vesicular, microlite-rich tachylite (D) scoria from the 2002 Etna eruption). Black: vesicles; dark gray: plagioclase; gray: glass; light gray: pyroxene.

a gradual increase in the number and intensity of Strombolian explosions often precedes the onset of fountaining episodes at Etna, Strombolian explosions also occur between fountaining episodes at Kīlauea. Ash-rich explosions often characterize the final phases of activity at Strombolian vents. Spatially, eruptions fed by basaltic magmas often begin along eruptive fissures, when magma reaches the surface along feeder dykes. As the eruption proceeds, activity becomes localized at one or more aligned vents, tens to hundreds of meters apart, which may erupt simultaneously but with different styles.

Multiple processes govern the transitions between eruptive styles. The eruption of slightly more gas-rich magmas sometimes modulates the transition from Strombolian to violent Strombolian activity. For a given magma chemical composition and volatile content, magma—gas segregation processes deeper in the volcanic plumbing system are expected to control Strombolian to Hawaiian transitions or a transition from normal Strombolian activity to paroxysm, as described later in this chapter.

Vent and shallow conduit processes play a role in controlling the eruption. The rheological properties of magma at the shallowest levels are key in this regard, as an increase in magma viscosity usually promotes efficient fragmentation and the formation of ash. The combination of cooling, degassing, and crystallization of magma, exacerbated by the reentrainment of chilled pyroclasts falling back into the vent, increases magma viscosity. This increase hinders bubble mobility and coalescence, and leads to segregation of crystal-rich, semisolid magma. Conversely, the frequent arrival of large gas pockets may efficiently stir and homogenize the magma, favoring open vent conditions. Local rates of gas supply and temporal and spatial shifts in vent and conduit geometries, well documented both in eruption observation and the geological record, control these competing processes.

7. THE ROLE OF VOLATILES

Strombolian/Hawaiian magmas contain up to a few weight percent of volatile species, dominated by water (H_2O) and carbon dioxide (CO_2). CO_2 exsolves from basaltic magmas at high pressures as a dense, supercritical fluid. Consequently, CO_2 exsolution does not affect magma buoyancy

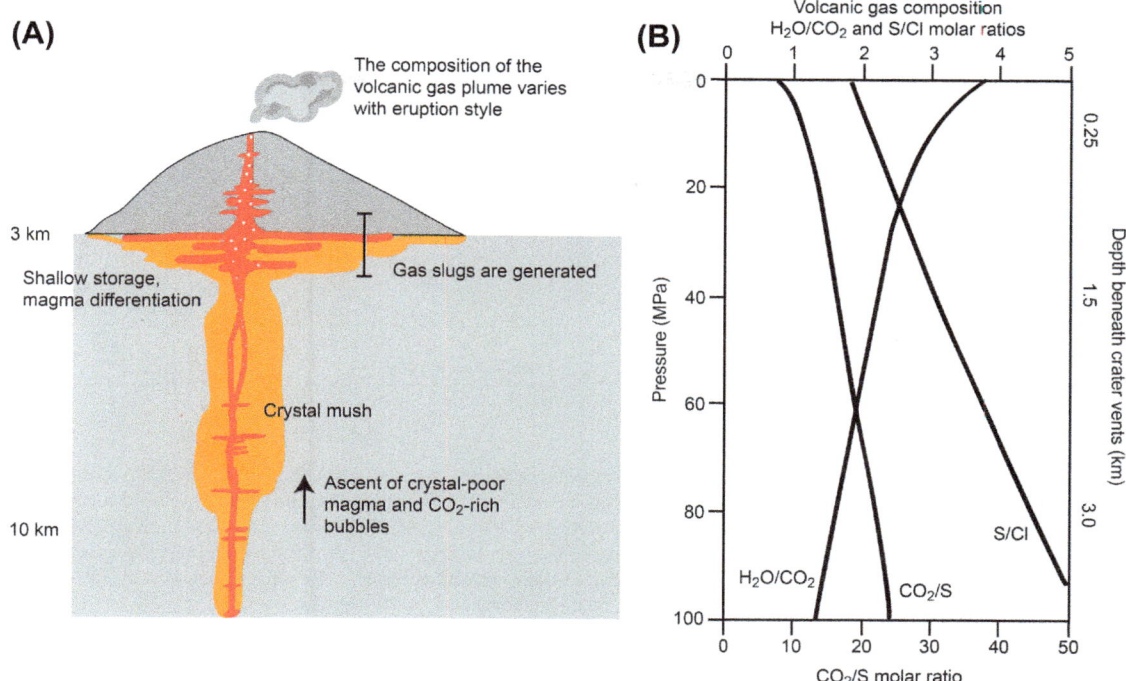

FIGURE 27.8 (A) Schematic cross-section of Stromboli, deduced from petrological and gas geochemical studies. (B) The evolution of gas composition with pressure, for melt and gas in equilibrium, from thermodynamic models. The gas plume emitted from Stromboli displays wide variability in gas composition which has been proposed to be due to mixing between deep gas segregated at the base of the edifice, which forms gas pockets which rise up the conduit, and more shallow gases derived from magmas ascending and degassing at lower pressures.

significantly until the magmas reach shallow depths, when the density of the exsolved volatile phase becomes low. At lower pressures, water exsolution becomes increasingly significant and the exsolved phase evolves from a CO_2-rich to a H_2O-rich gas through the shallow plumbing system. It is likely that both species are critical for the volatile-related processes driving Strombolian and Hawaiian activity, namely: (1) gas-melt segregation, whereby gas bubbles rise relative to melt (Jaupart and Vergniolle, 1988); (2) rapid magma ascent and "runaway" vesiculation, which may lead to fragmentation; (3) influx of primitive, volatile-rich melts into a shallow reservoir, which might cause rapid crystallization and vesiculation (Sides et al., 2014); and (4) changes in the rheological properties of magmas as they ascend and exsolve H_2O.

The evidence for the importance of these processes comes from measurements of volcanic gases (see also Chapter 7) and **melt inclusion** geochemistry (see also Chapter 8). Volcanic gases at the surface represent the depth-averaged composition of the volatile phases that exsolve from magma en route to the surface, and may be monitored continuously (Figure 27.8). Melt inclusions (i.e., melt trapped in the interiors of rapidly growing phenocrysts) may preserve the preeruptive chemical evolution of melt and have been of great value in the forensic analysis

and reconstruction of magma pathways prior to explosive eruption at the surface.

Geochemical data can provide evidence for the fundamental trigger mechanism of Hawaiian–Strombolian activity. Some controls on these styles of activity are undoubtedly shallow seated, with changes in magma ascent rate and segregation efficiency of the gas phase controlling dynamics and transitions (Jaupart and Vergniolle, 1988), as discussed earlier in the chapter. An interesting question though, is whether the melts that drive the more explosive styles of activity are fundamentally more enriched in volatiles. At Kīlauea, eruptions that produce higher fountains are subtly, yet statistically more primitive and more enriched in incompatible elements, suggesting that the enriched primary melts saturate deeper and may be slightly more buoyant as they approach the surface, which might be enough to promote one style of activity over another (Sides et al., 2014). On the other hand, melt inclusions in pyroclasts erupted during high-fountaining episodes show that their volatile contents are not dramatically different to those measured in the products of effusive activity. This indicates that the driving forces for Hawaiian activity are linked to exsolved H_2O in the shallow plumbing system and that gas-melt segregation processes remain the dominant control on style of activity and the transitions between

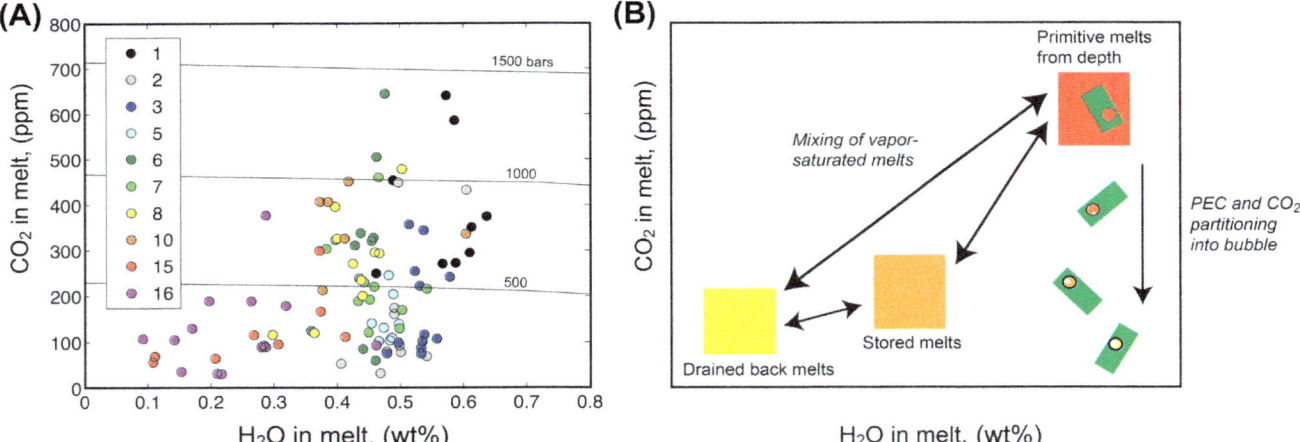

FIGURE 27.9 Olivine-hosted melt inclusion volatile data for the 1959 Kīlauea Iki eruption. (A) H_2O plotted against CO_2, color-coded for episode number. There were 17 episodes of high fountaining (Figure 27.4). Isobars are marked. (B) Schematic diagram to show the main controls on the volatile data. The array is dominated by magma mixing. The H_2O concentrations may be explained by mixing between incoming primitive melts, stored melts, and drained back melts, the proportions of each of which have been estimated from geophysical data. The CO_2 concentrations are complicated by post-entrapment crystallization (PEC), which has promoted the sequestration of CO_2 into a shrinkage bubble. This is of most importance in those olivines that have undergone the most postentrapment cooling, caused by mixing with drained back and stored magmas.

them. At Kīlauea, some high-fountaining eruptions occur as far as 80 km from the zone of magma ascent and CO_2 outgassing. If magmas lose most of their exsolved CO_2 before lateral transport to the eruption zones, as has been deduced from volcanic gas studies, then the high fountains are likely to be driven dominantly by near-surface H_2O degassing as magmas make their final ascent.

Exsolved vapor may be quantified by the direct measurement of volcanic gas flux and composition. Gas emissions during regular Strombolian explosions at Stromboli are richer in CO_2 than those during quiescent degassing between eruptions (Burton et al., 2007). A dramatic increase in CO_2 gas flux has been observed beginning 2 weeks prior to paroxysmal activity at Stromboli. These increases in gas emissions are interpreted as due to "leakage" of a CO_2-rich foam that had accumulated at depth. The accumulation of exsolved volatiles at depth creates the instability required for the rapid ascent of gas-rich magma and the paroxysmal explosion at the surface. Melt inclusion studies have shown that the feeder melts are geochemically diverse and the paroxysms are triggered by the influx of hot, volatile-rich melts from depth.

At Etna, observations of volcanic gas composition during transitions between effusive and Hawaiian styles of activity showed that the fountaining episodes were associated with more CO_2-rich gases (Allard et al., 2005), further reinforcing the role of exsolved CO_2-rich vapor in driving magma ascent and fragmentation during these styles of activity. Melt inclusion studies have shown that the early stages of the 2002 flank eruption, which involve high fountains, preserve H_2O-CO_2 concentrations consistent with rapid, closed system degassing, similar to melt inclusions in

basaltic tephra from Irazu and Arenal. Later phases of the 2002 Etna eruption showed that melts had been flushed by a CO_2-rich vapor, consistent with the results from volcanic gas studies, which shows that Etna outgasses a large flux of CO_2 gas at the surface.

Magma mixing is an important control on magma rheology and vesiculation and hence on eruption style and transitions. At Etna, mixing between gas-rich magmas and cooler, stored melts gives rise to a marked change in the rheology of the magma and has been linked to transitions between Hawaiian and Strombolian styles of activity. Mixing between hot, primitive magmas and stored magmas may generate bursts of crystallization and vesiculation, thereby stiffening the magma and prompting violent Strombolian activity. Also at Kīlauea repetitive high-fountaining episodes were preceded by mixing between hot incoming magmas carrying exsolved vapor and cool, degassed, drained back lavas (Figure 27.9) (Sides et al., 2014). The episodes with the highest fountains occurred after the highest degree preeruptive mixing and cooling, indicating a critical link between the mixing process and Hawaiian—Strombolian dynamics (Figure 27.10).

8. INSIGHTS FROM INFRASONIC MEASUREMENTS

Infrasound is the oscillation of atmospheric pressure below audible sound. Measurement and modeling of eruption-related infrasound is a powerful tool to investigate Hawaiian and Strombolian activity (e.g., Vergniolle et al., 1996; Lane et al., 2013). For instance, infrasound, in combination with other gas measurements, can be used to deduce the

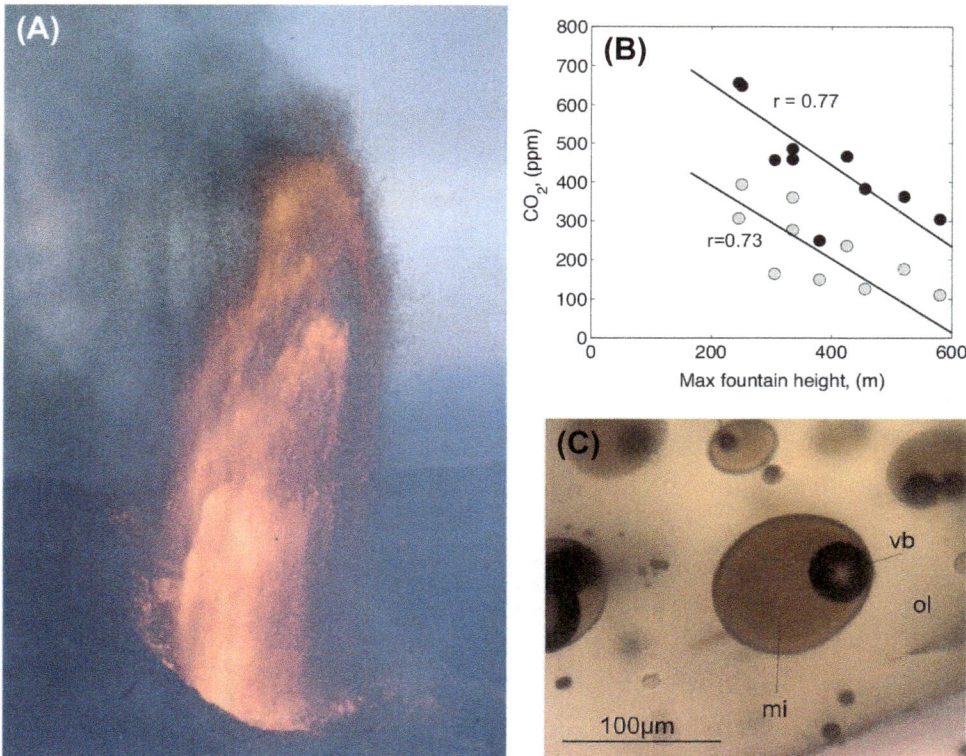

FIGURE 27.10 (A) Representative picture of Hawaiian fountain at Kīlauea Iki in 1959 courtesy of Jerry Eaton, USGS. (B) The mean and maximum CO_2 concentration in the melt inclusion for each episode shows a correlation with the maximum fountain height for that episode (see Figure 27.5). This correlation arises owing to the highest fountains being generated as a result of the largest degree of cooling of incoming melts due to mixing with stored magmas, which created a large burst of crystallization and degassing. The cooling caused postentrapment crystallization on the walls of the melt inclusion (mi) in (C) and the resulting pressure drop promoted the loss of CO_2 from the melt into the vapor bubble (vb).

minimum volume of overpressurized gas associated with individual Strombolian explosions (Figure 27.11) (e.g., Vergniolle and Ripepe, 2008). This technique can be performed continuously to produce long time series of data or to monitor ongoing activity routinely.

Calculating acoustic power provides an estimate of gas velocity and hence the gas volume if the radius of the source is known, which, if activity is confined to the top of the conduit, can be estimated precisely. The gas volume can also be determined from inversion of acoustic waveforms along with the initial gas overpressure (Figure 27.11) (e.g., Vergniolle and Brandeis, 1996). Although the results depend on the model used, deriving these parameters independently is one of the strengths of acoustic waveform inversion.

Inversion of acoustic data was used to deduce the temporal evolution of gas pocket length and overpressure during Strombolian activity at Shishaldin volcano (Alaska) (Figure 27.12). A similar inversion performed at Etna during a transition from Strombolian to Hawaiian activity (Figure 27.12) showed that the transition was accompanied by a simultaneous increase in the number of explosions and

the length of the gas pockets (Vergniolle and Ripepe, 2008). Furthermore, the gas pocket oscillates in length prior to its arrival at the surface, thereby pushing the overlying magma column up and down while simultaneously producing gravity waves at the surface (Figure 27.11). Both of these modes are sometimes seen as precursory signals to the main pulse.

Continuous acoustic measurements performed at Stromboli have also shown the breaking of weakly overpressurized gas pockets at intervals of seconds, each of these small pressure transients being termed a *puff* (e.g., Ripepe et al., 2002).

Performing acoustic or seismic measurements is the main way to estimate the volume of overpressurized gas that comes from depth. However, the inversion of acoustic waveforms, similarly to that of seismic waveforms, leads to calculated average degassing rates that are about one order of magnitude smaller than those obtained by other techniques. This suggests that the overpressurized gas, although playing a crucial role in the eruptive behavior due to its deep origin, corresponds to a much smaller volume than the outgassing that occurs at pressure equilibrium.

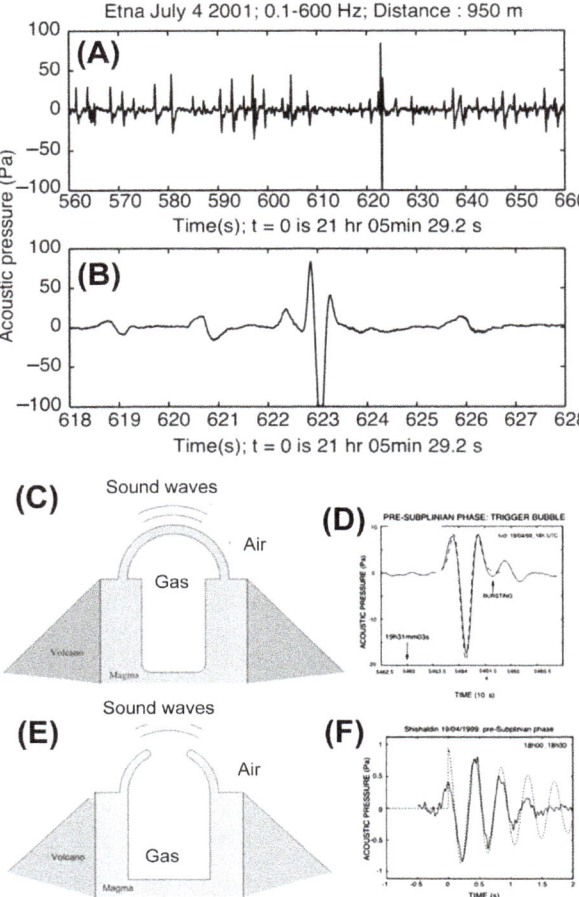

FIGURE 27.11　(A) Acoustic pressure measured from the Southeast crater at Etna during a typical Strombolian activity. (B) Each main pulse in acoustic pressure results from breaking a large bubble at the top of the magma column. Two possible models to explain the sound produced by large gas pocket breaking at the top of the magma column. (C) Oscillations in the pocket volume can radiate a strong acoustic pressure prior to its breaking. (D) Best fit between synthetic (dashed line) and measured (solid line) acoustic pressure radiated by a Strombolian explosion at Shishaldin volcano. (E) Sketch of a weakly overpressurized pocket radiating acoustic pressure by letting its inner gas escape quickly through a narrow hole. (F) Best fit between synthetic (dashed line) and measured (solid line) acoustic pressure radiated by a weakly overpressurized large pocket, measured during the pre-Subplinian phase of the 1999 Shishaldin eruption.

Multiparametric Studies of Eruptive Activity

Relatively accessible and persistent activity renders some volcanoes ideal "natural laboratories" for the study of Strombolian and Hawaiian activity, e.g., Stromboli, Kīlauea, Etna. At these volcanoes, permanent surveillance networks and temporary expeditions record the activity with a broad range of sensors, simultaneously detecting, for example, visible, infrared, ultraviolet, seismic, acoustic, geodetic, magnetic, and gravimetric eruptive signals. The combined interpretation of this variety of signals has provided for decades—and still provides—powerful understanding of the eruptions that cannot be achieved by individual approaches.

9. CONDUIT FLUID DYNAMICS

In low-viscosity magmas like basalt, the dynamics of very small bubbles are strongly influenced by surface tension, while large gas pockets are controlled by a combination of inertial and viscous forces. In vertical, cylindrical conduits this gives the largest gas pockets a stereotypical "slug" geometry, with a hemispherical nose leading the cylindrical body of the slug, which is surrounded by a falling liquid film (Figure 27.13). Characteristic gas pocket ascent velocities in basalts are of order a meter per second, with pockets in larger conduits ascending more rapidly. At depth, and for constant magma properties and conduit size, gas pocket velocities are independent of their volume. As a gas pocket ascends, it decompresses and expands. Constrained within a conduit, this expansion is dominantly lengthwise, inducing more magma to be accommodated in the falling film around the pocket, further reducing the overburden pressure, and driving greater expansion.

Gas pocket formation during Strombolian activity has been attributed to bubble coalescence, either linked to (1)

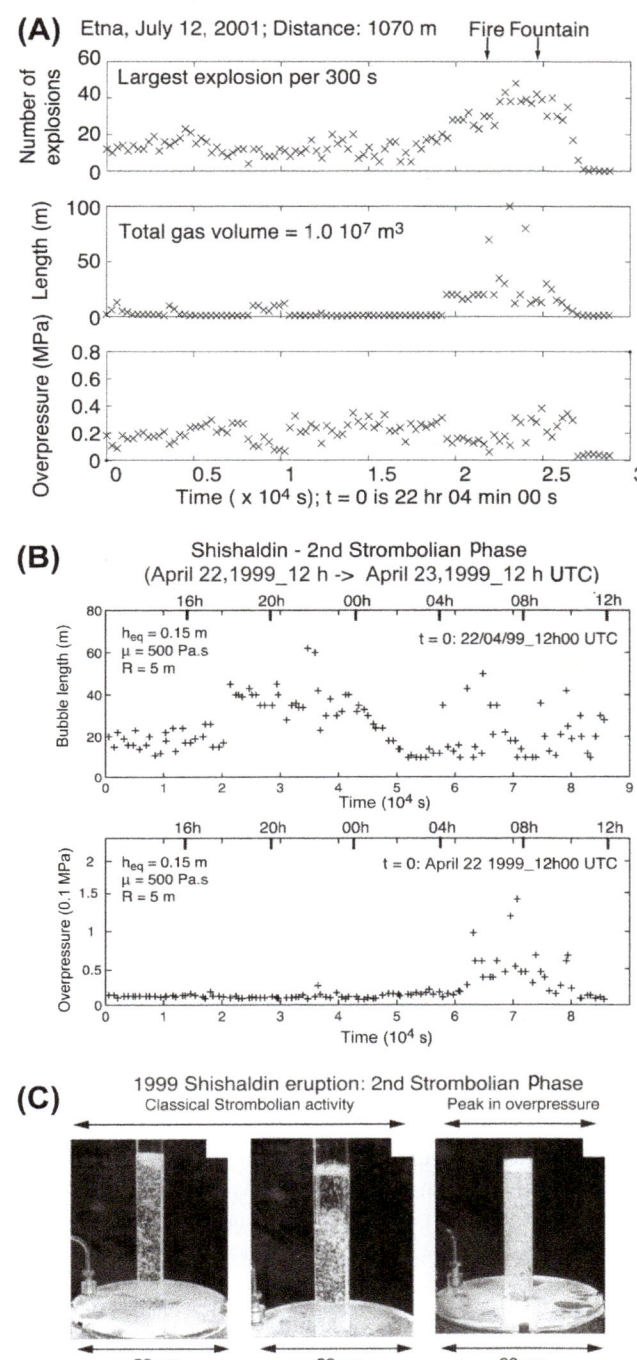

FIGURE 27.12 (A) Temporal evolution of the number of explosions per 10 min, bubble length, and bubble overpressure at vent during the activity of Etna on July 12, 2001. The transition from a Strombolian to fire fountain activity is marked by an increase in both the explosions number and gas pocket length. (B) Temporal evolution of the pocket length and overpressure at vent during the second Strombolian phase of the 1999 Shishaldin eruption. (C) Laboratory experiments mimicking how a classical Strombolian explosion is produced in the conduit (left and middle photos) and how the rise of a large gas pocket within a magma enriched in small bubbles can significantly increase the initial bubble overpressure, by increasing the magma viscosity (right photo).

static bubble accumulation and foam development at structural discontinuities or (2) dynamic coalescence in the conduit of bubbles with different rise speeds (e.g., Jaupart and Vergniolle, 1988; Parfitt, 2004). At a certain critical shallow depth, the ascent of large pockets evolves from a relatively steady rise and expansion, into a runaway burst process in which the nose can accelerate to many tens of meters per second. Gas pockets bursting at the surface at Stromboli during Strombolian explosions are rich in CO_2, consistent with the last equilibration with melt at pressures as high as 100 MPa, which may correspond to the zone of shallow storage and magma differentiation at 3 km depth (the base of the edifice). During the expansion of the gas pocket the magma above the pocket must also be accelerated upward to accommodate the extra gas volume. It is the force required to do this that results in a gas overpressure (i.e., a pressure greater than from a stationary overlying column of fluid alone) being retained in the pocket as it approaches the surface. For smaller gas pockets, or at volcanoes where the near-surface conduit flares (e.g., Mt Erebus) or where a lava lake exists, gas pocket expansion can also be accommodated laterally which, if all other parameters remaining constant, reduces burst overpressures at the surface (James et al., 2009). Lateral expansion and the lack of shear around the sides of the gas pocket(s) also probably promote the life of the membrane formed at the surface prior to burst. The bowing upward and bursting of such membranes has been observed at the surfaces of lava lakes, forming a primary observation of the large gas pockets inferred to be responsible for Strombolian activity. The strong variability in the rate of occurrence of ejection pulses during each explosion at Stromboli points toward the occurrence of complex fluid dynamics during the release of the gas pockets (Taddeucci et al., 2012). Relatively long-lasting, isolated pulses seem to reflect the bursting of conduit-filling slugs, while shorter, closely spaced pulses represent near-continuous bursts of multiple, transient gas pockets (Figure 27.13). It appears that fluid flow very close to the free surface is strongly heterogenous, perhaps turbulent and churn-like, and that gas pockets may take on very different forms and dimensions in this region. Interpretations of conduit flow leading to Hawaiian eruptions differ more widely than for Strombolian explosions, ranging from annular flow following foam collapse in a storage region (Jaupart and Vergniolle, 1988) to sustained bubbly flow (Parfitt, 2004). Spectroscopic data for the exsolved volatiles outgassed during high fountaining at Etna suggest that they are related to decoupled two-phase flow involving a CO_2-rich (and hence deep-derived) gas phase (Allard et al., 2005), while at Kīlauea magma has much lower volatile contents than Etna, and there seems to be little potential for a role for deep-decoupled CO_2-rich fluids in Hawaiian activity. At Mt Etna, bubble number densities correlate with eruption intensity, suggesting that syneruptive degassing of

(A)   **(B)**   **(C)**   **(D)**

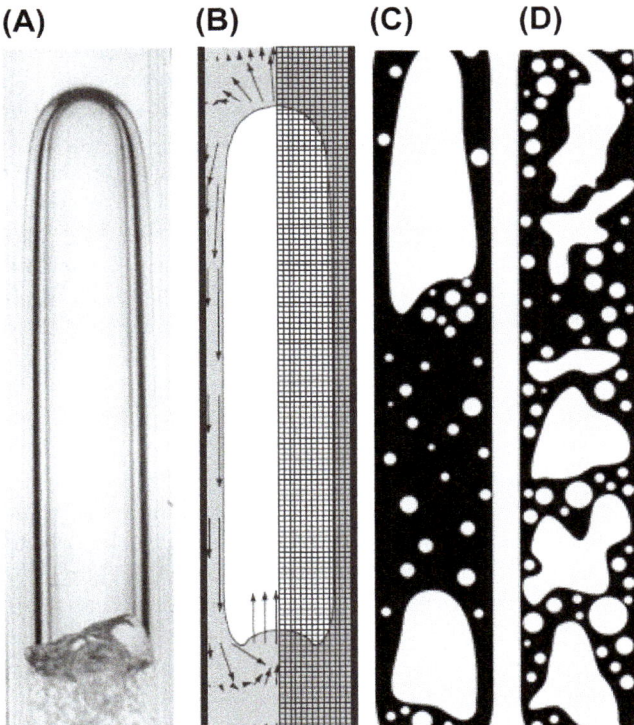

FIGURE 27.13 Gas pockets ascending vertical tubes of stagnant (stationary) liquid. (A) A slug in water in a ∼7.5-cm diameter laboratory pipe. With the low viscosity of water, the dynamics are controlled by the inertia of the liquid, resulting in a relatively flat but oscillating base to the gas pocket and a turbulent, bubble-rich wake. (B) The results of a numerical model of a gas pocket ascending a more viscous fluid. Although the main shape of the gas pocket remains similar, the increased influence of viscosity suppresses turbulence in the tail and gives a domed base to the bottom of the gas pocket. Arrows give relative velocities for the liquid. (C) During a single Strombolian explosion, an initial pocket may cause an individual ejection pulse, followed at regular intervals by similar ones. (D) Long pockets may fragment before or during bursting, causing transient blockages of the vent and multiple, closely repeated pulses.

H_2O at fast magma rise speeds could be important in driving the largest Hawaiian episodes. In addition, melt inclusions data suggest that, both at Etna and Kīlauea, some Hawaiian episodes involved magma rising rapidly from depth and vesiculating without time for bubble segregation and melt outgassing.

Here, we have considered only relatively straightforward conduit or plumbing geometries. In reality, dynamics will also be influenced by complexities such as wall irregularities and branching. In any long-lived magmatic system (e.g., such as at Yasur, Mt Erebus or Stromboli), heat flow must be sufficient to keep the magma pathways open, implying that magma circulation (driven by the degassing and persistent activity) must be pervasive. Consequently, not only does conduit geometry influence the flow dynamics (James et al., 2006), but the flow dynamics also exerts control on the evolution of the geometry.

10. NEW DIRECTIONS AND CONCLUDING REMARKS

Strombolian and Hawaiian eruptions occur at very active volcanoes frequently associated with volcano tourism. Their eruptions pose constant issues for management agencies because the eruption sites are highly accessible and there is a need to balance the strong popular interest in viewing the eruptive activity against the issue of public safety. There is both a scientific desire for better knowledge of the volcano's behavior and a need for improved forecasting of the likely course and footprint of all future and ongoing eruptions. For this, we need improved understanding of the two-phase flow and outgassing in the uppermost part of the conduits that modulate such explosive eruptions.

A key research objective is thus to obtain high-resolution and in situ parameterization of the range of Hawaiian, Strombolian, and related eruption styles at basaltic volcanoes via synchronized geophysical (e.g., infrasonic, seismic) and geochemical (e.g., gas chemistry) data coupled with high-speed imagery. Merging these data to combine properties of both pyroclasts, including size, mass, temperature, and ejection velocity and angle (high-speed and thermal cameras), and gases, including temperature (thermal cameras), amounts (SO_2 camera, fixed SO_2 sensors), and composition (gas spectroscopy and in situ gas sensor) will provide rigorous constraints for new flow models.

ACKNOWLEDGMENTS

Strong, constructive reviews by Marco Pistolesi and Laura Pioli greatly helped us improve this chapter. We acklowedge funding support from NSF, the USGS Volcano Hazards Program, CNRS-INSU, Université Paris Diderot 7 (BQR), IPGP (BQR and PPF) and ANR (Undervolc, Arc-Vanuatu, Risk-Volc-An, Domoscan) for their support.

FURTHER READING

Allard, P., Burton, M., Muré, F., 2005. Spectroscopic evidence for a lava fountain driven by previously accumulated magmatic gas. Nature 433, 407—410.

Burton, Mike, Allard, Patrick, Muré, Filippo, Spina, Alessandro La, 2007. Magmatic gas composition reveals the source depth of slug-driven strombolian explosive activity. Science 317 (5835), 227—230.

Edmonds, M., Gerlach, T.M., 2007. Vapor segregation and loss in basaltic melts. Geology 35, 751—754.

Gaudin, D., Taddeucci, J., Scarlato, P., Moroni, M., Freda, C., Gaeta, M., Palladino, D.M., 2014. Pyroclast tracking velocimetry illuminates bomb ejection and explosion dynamics at Stromboli (Italy) and Yasur (Vanuatu) volcanoes. J. Geophys. Res. 119, 5384—5397. http://dx.doi.org/10.1002/2014JB011096.

Houghton, B.F., Gonnermann, H.M., 2008. Basaltic explosive volcanism: constraints from deposits and models. Chem. Erde-Geochem. 68, 117—140. http://dx.doi.org/10.1016/j.chemer.2008.04.002.

James, M.R., Lane, S.J., Chouet, B.A., 2006. Gas slug ascent through changes in conduit diameter: laboratory insights into a volcano-seismic source process in low-viscosity magmas. J. Geophys. Res. 111, B05201. http://dx.doi.org/10.1029/20 05JB0 03718.

James, M.R., Lane, S.J., Wilson, L., Corder, S.B., 2009. Degassing at low magma-viscosity volcanoes: quantifying the transition between passive bubble-burst and Strombolian eruption. J. Volcanol. Geotherm. Res. 180, 81—88. http://dx.doi.org/10.1016/j.jvolgeores.2008.09.002.

Jaupart, C., Vergniolle, S., 1988. Laboratory models of Hawaiian and Strombolian eruptions. Nature 331, 58—60.

Lane, S.J., James, M.R., Corder, S.B., 2013. Volcano infrasonic signals and magma degassing: first-order experimental insights and application to Stromboli. Earth Planet. Sci. Lett. 377—378, 169—179. http://dx.doi.org/10.1016/j.epsl.2013.06.048.

Lautze, N.C., Houghton, B.F., 2005. Physical mingling of magma and complex eruption dynamics in the shallow conduit at Stromboli volcano, Italy. Geology 33, 425—428. http://dx.doi.org/10.1130/G21325.1.

Parfitt, E.A., 2004. A discussion of the mechanisms of explosive basaltic eruptions. J. Volcanol. Geotherm. Res. 134, 77—107. http://dx.doi.org/10.1016/j.jvolgeores.2004.01.002.

Patrick, M.R., Harris, A., Ripepe, M., Dehn, J., Rothery, D.A., Calvari, S., 2007. Strombolian explosive styles and source conditions: insights from thermal (FLIR) video. Bull. Volcanol. 69, 769—784. http://dx.doi.org/10.1007/s00445-006-0107-0.

Ripepe, M., Harris, A.J.L., Carniel, R., 2002. Thermal, seismic and infrasonic evidence of variable degassing rates at Stromboli volcano. J. Volcanol. Geotherm. Res. 118, 285—297.

Rosi, M., Pistolesi, M., Bertagnini, A., Landi, P., Pompilio, M., Di Roberto, A., 2013. Stromboli Volcano, Aeolian Islands (Italy): Present Eruptive Activity and Hazards, 37. Geological Society, London, Memoirs, 473—490. http://dx.doi.org/10.1144/M37.14.

Sides, I.R., Edmonds, M., Maclennan, J., Swanson, D.A., Houghton, B.F., 2014. Eruption style at Kīlauea volcano in Hawaii linked to primary melt composition. Nat. Geosci. 7 (6), 464—469.

Taddeucci, J., Scarlato, P., Capponi, A., Del Bello, E., Cimarelli, C., Palladino, D.M., Kueppers, U., 2012. High-speed imaging of Strombolian explosions: the ejection velocity of pyroclasts. Geophys. Res. Lett. 39. L02301. http://dx.doi.org/10.1029/2011GL050404.

Valentine, G.A., Gregg, T.K.P., 2008. Continental basaltic volcanoes — processes and problems. J. Volcanol. Geotherm. Res. 177, 857—873.

Vergniolle, S., Brandeis, G., 1996. Strombolian explosions: 1. A large bubble breaking at the surface of the lava column as a source of sound. J. Geophys. Res. 101, 20433—20448.

Vergniolle, S., Brandeis, G., Mareschal, J.C., 1996. Strombolian explosions. 2. Eruption dynamics determined from acoustic measurements. J. Geophys. Res. 101, 20449—20466.

Vergniolle, S., Ripepe, M., 2008. From Strombolian explosions to fire fountains at Etna volcano (Italy): what do we learn from acoustic measurements?. In: Lane, S.J., Gilbert, J.S. (Eds.), Fluid Motions in Volcanic Conduits: A Source of Seismic and Acoustic Signals, vol. 307. Geological Society, London, Special Publications, pp. 103—124.

Annex 1: Further Reading

Hawaiian and Strombolian eruptive styles: observations and dynamics

Blackburn, E.A., Wilson, L., Sparks, R.S.J., 1976. Mechanisms and dynamics of Strombolian activity. J. Geol. Soc. Lond. 132, 429—440. http://dx.doi.org/10.1144/gsjgs.132.4.0429.

Chouet, B., Hamisevicz, N., McGetchin, T.R., 1974. Photoballistics of volcanic jet activity at stromboli, Italy. J. Geophys. Res. 79, 4961—4976. http://dx.doi.org/10.1029/JB079i032p04961.

Gaudin, D., Moroni, M., Taddeucci, J., Scarlato, P., Shildler, L., 2014. Pyroclast tracking velocimetry: a particle tracking velocimetry-based tool for the study of Strombolian explosive eruptions. J. Geophys. Res. 119, 5369—5383. http://dx.doi.org/10.1002/2014JB011095.

Decker, R.W., 1987. Dynamics of Hawaiian volcanoes: an overview. In: Decker, R.W., Wright, T.L., Stauffer, P.H. (Eds.), Volcanism in Hawai'i, vol. 1350, pp. 997—1018. U.S. Geological Survey Professional Paper.

Fedotov, S.A., Chirkov, A.M., Gusev, N.A., Kovalev, G.N., Slezin, Y.B., 1980. The large fissure eruption in the region of Plosky Tolbachik volcano in Kamchatka, 1975—1976. Bull. Volcanol. 43, 47—60.

Harris, A., Ripepe, M., 2007. Temperature and dynamics of degassing at Stromboli. J. Geophys. Res. 112, B03205. http://dx.doi.org/10.1029/2006JB004393.

Harris, A.J.L., Delle Donne, D., Dehn, J., Ripepe, M., Worden, A.K., 2013. Volcanic plume and bomb field masses from thermal infrared camera imagery. Earth Planet. Sci. Lett. 365, 77—85. http://dx.doi.org/10.1016/j.epsl.2013.01.004.

Head, J.W., Wilson, L., 1987. Lava fountain heights at Pu'u'O'o, Kilauea, Hawaii: indicators of amount and variations of exsolved magma volatiles. J. Geophys. Res. 92, 13715—13719.

Heliker, C., Mattox, T.N., 2003. The first two decades of the Pu'u 'Ō'ō-Kupaianaha eruption: chronology and selected bibliography, the Pu'u 'Ō'ō-Kupaianaha eruption of Kīlauea volcano, Hawai'i: the first 20 years. US Geol. Surv. Prof. Pap. 1676, 1—27.

Luhr, J.F., Simkin, T., 1993. Paricutín: The Volcano Born in a Mexican Cornfield. Geoscience Press Inc, Phoenix, Arizona, 427 p.

Patrick, M.R., Harris, A., Ripepe, M., Dehn, J., Rothery, D.A., Calvari, S., 2007. Strombolian explosive styles and source conditions: insights from thermal (FLIR) video. Bull. Volcanol. 69, 769–784. http://dx.doi.org/10.1007/s00445-006-0107-0.

Richter, D.H., Eaton, J.P., Murata, K.J., Ault, W.U., Krivoy, H.L., 1970. Chronological narrative of the 1959–60 eruption of Kilauea volcano, Hawaii. U. S. Geol. Surv. Prof. Pap. 537-E, 73.

Ripepe, M., Rosi, M., Saccorotti, G., 1993. Image processing of explosive activity at Stromboli. J. Volcanol. Geotherm. Res. 54, 335–351. http://dx.doi.org/10.1016/0377-0273(93)90071-X.

Rosi, M., Bertagnini, A., Harris, A.J.L., Pioli, L., Pistolesi, M., Ripepe, M., 2006. A case history of paroxysmal explosion at Stromboli: timing and dynamics of the April 5, 2003 event. Earth Planet. Sci. Lett. 243, 594–606.

Rosi, M., Pistolesi, M., Bertagnini, A., Landi, P., Pompilio, M., Di Roberto, A., 2013. Stromboli volcano, Aeolian Islands (Italy): present eruptive activity and hazards, 37. Geological Society, London, Memoirs, pp. 473–490. http://dx.doi.org/10.1144/M37.14.

Swanson, D.A., Duffield, W.A., Jackson, D.B., Peterson, D.W., 1979. Chronological narrative of the 1969–71 Mauna Ulu eruption of Kīlauea volcano, Hawaii. U. S. Geol. Surv. Prof. Pap. 1056, 55.

Taddeucci, J., Pompilio, M., Scarlato, P., 2002. Monitoring the explosive activity of the July–August 2001 eruption of Mt. Etna (Italy) by ash characterization. Geophys. Res. Lett. 29, X1–X4. http://dx.doi.org/10.1029/2001GL014372.

Taddeucci, J., Spieler, O., Kennedy, B., Pompilio, M., Dingwell, D.B., Scarlato, P., 2004b. Experimental and analytical modeling of basaltic ash explosions at Mount Etna, Italy, 2001. J. Geophys. Res. 109, B08203. http://dx.doi.org/10.1029/2003JB002952.

Taddeucci, J., Alatorre-Ibargüengoitia, M.A., Moroni, M., Tornetta, L., Capponi, A., Scarlato, P., Dingwell, D.B., De Rita, D., 2012. Physical parameterization of Strombolian eruptions via experimentally-validated modeling of high-speed observations. Geophys. Res. Lett. 39, L16306. http://dx.doi.org/10.1029/201 2GL052772.

Taddeucci, J., Palladino, D.M., Sottili, G., Bernini, D., Andronico, D., Cristaldi, A., 2013. Linked frequency and intensity of persistent volcanic activity at Stromboli (Italy). Geophys. Res. Lett. 40, 3384–3388. http://dx.doi.org/10.1002/grl.50652.

Wolfe, E.W., Neal, C.A., Banks, N.G., Duggan, T.J., 1988. Geologic observations and chronology of eruptive events, in the Pu'u 'O'o eruption of Kilauea volcano, episodes 1–20, January 3, 1983 to June 8, 1984. In: Wolfe, E. (Ed.), US Geological Survey Professional Papers, vol. 1463, pp. 471–508.

Deposit geometry and pyroclast textures

Andronico, D., Cristaldi, A., Scollo, S., 2008. The 4–5 September 2007 lava fountain at South-East crater of Mt Etna, Italy. J. Volcanol. Geotherm. Res. 173, 325–328.

D'Oriano, C., Bertagnini, A., Cioni, R., Pompilio, M., 2014. Identifying recycled ash in basaltic eruptions. Sci. Rep. 4, 5851. http://dx.doi.org/10.1038/srep05851.

Gurioli, L., Harris, A.J.L., Houghton, B.F., Polacci, M., Ripepe, M., 2008. Textural and geophysical characterization of explosive basaltic activity at Villarrica volcano. J. Geophys. Res. 113, B08206. http://dx.doi.org/10.1029/2007JB005328.

Gurioli, L., Colò, L., Bollasina, A.J., Harris, A.J.L., Whittington, A., Ripepe, M., 2014a. Dynamics of Strombolian explosions: Inferences from field and laboratory studies of erupted bombs from Stromboli volcano. J. Geophys. Res. 119, 319–345. http://dx.doi.org/10.1002/2013JB 010355.

Gurioli, L., Harris, A.J.L., Colò, L., Bernard, J., Favalli, M., Ripepe, M., Andronico, D., 2014b. Classification, landing distribution, and associated flight parameters for a bomb field emplaced during a single major explosion at Stromboli, Italy. Geology 41, 559–562. http://dx.doi.org/10.1130/G33967.1.

Houghton, B.F., Gonnermann, H.M., 2008. Basaltic explosive volcanism: constraints from deposits and models. Chem. Erde-Geochem. 68, 117–140. http://dx.doi.org/10.1016/j.chemer.2008.04.002.

Keating, G.N., Valentine, G.V., Krier, D.J., Perry, F.V., 2008. Shallow plumbing systems for small-volume basaltic volcanoes. Bull. Volcanol. 70, 563–582. http://dx.doi.org/10.1007/s00445-007-0154-1.

Lautze, N.C., Houghton, B.F., 2005. Physical mingling of magma and complex eruption dynamics in the shallow conduit at Stromboli volcano, Italy. Geology 33, 425–428. http://dx.doi.org/10.1130/G21325.1.

Mangan, M.T., Cashman, K.V., 1996. The structure of basaltic scoria and reticulite and inferences for vesiculation, foam formation, and fragmentation in lava fountains. J. Volcanol. Geotherm. Res. 73, 1–18. http://dx.doi.org/10.1016/0377-0273(96)00018-2.

Parcheta, C.E., Houghton, B.F., Swanson, D.A., 2013. Contrasting patterns of vesiculation in low, intermediate, and high Hawaiian fountains: a case study of the 1969 Mauna Ulu eruption. J. Volcanol. Geotherm. Res. 255, 79–89.

Parfitt, E.A., 1998. A study of clast size distribution, ash deposition and fragmentation in a Hawaiian-style volcanic eruption. J. Volcanol. Geotherm. Res. 84, 197–208.

Pioli, L., Erlund, E., Johnson, E., Cashman, K., Wallace, P., Rosi, M., Delgado Granados, H., 2008. Explosive dynamics of violent Strombolian eruptions: the eruption of Parícutin volcano 1943–1952 (Mexico). Earth Planet. Sci. Lett. 271, 359–368. http://dx.doi.org/10.1016/j.epsl.2008.04.026.

Polacci, M., Corsaro, R.A., Andronico, D., 2006. Coupled textural and compositional characterization of basaltic scoria: insights into the transition from Strombolian to fire fountain activity at Mount Etna, Italy. Geology 34, 201–204.

Polacci, M., Baker, D.R., Bai, L., Mancini, L., 2008. Large vesicles record pathways of degassing at basaltic volcanoes. Bull. Volcanol. 70, 1023–1029. http://dx.doi.org/10.1007/s00445-007-0184-8.

Polacci, M., Baker, D.R., Mancini, L., Favretto, S., Hill, R.J., 2009. Vesiculation in magmas from Stromboli and implications for normal Strombolian activity and paroxysmal explosions in basaltic systems. J. Geophys. Res. 114, B01206. http://dx.doi.org/10.1029/2008JB005672.

Stovall, W.,K., Houghton, B.F., Gonnermann, H., Fagents, S.A., Swanson, D.A., 2011. Eruption dynamics of Hawaiian-style fountains: the case study of episode 1 of the Kīlauea Iki 1959 eruption. Bull. Volcanol. 73, 511–529. http://dx.doi.org/10.1007/s00445-010-0426-z.

Taddeucci, J., Pompilio, M., Scarlato, P., 2004a. Conduit processes during the July–August 2001 explosive activity of Mt. Etna (Italy): Inferences from glass chemistry and crystal size distribution of ash particles. J. Volcanol. Geotherm. Res. 137, 33–54. http://dx.doi.org/10.1016/j.jvolgeores.2004.05.011.

Valentine, G.A., Krier, D., Perry, F.V., Heiken, G., 2005. Scoria cone construction mechanisms, Lathrop Wells volcano, southern Nevada, USA. Geology 33, 629–632. http://dx.doi.org/10.1130/G21459.1.

Valentine, G.A., Krier, D.J., Perry, F.V., Heiken, G., 2007. Eruptive and geomorphic processes at Lathrop Wells scoria cone volcano. J. Volcanol. Geotherm. Res. 161, 57–80. http://dx.doi.org/10.1016/j.jvolgeores.2006.11.003.

Walker, G.P.L., Croasdale, R., 1972. Characteristics of some basaltic pyroclastics. Bull. Volcanol. 35, 303–317.

Style variations and transitions

Allard, P., Behncke, B., D'Amico, S., Neri, M., Gambino, S., 2006. Mount Etna 1993—2005: anatomy of an evolving eruptive cycle. Earth-Sci. Rev. 78, 85—114. http://dx.doi.org/10.1016/j.earscirev.2006.04.002.

Andronico, D., Cristaldi, A., Del Carlo, P., Taddeucci, J., 2009. Shifting styles of basaltic explosive activity during the 2002—03 eruption of Mt. Etna, Italy. J. Volcanol. Geotherm. Res. 180, 110—122. http://dx.doi.org/10.1016/j.jvolgeores.2008.07.026.

Arrighi, S., Principe, C., Rosi, M., 2001. Violent Strombolian and Subplinian eruptions at Vesuvius during post-1631 activity. Bull. Volcanol. 63, 126—150. http://dx.doi.org/10.1007/s004450100130.

Cimarelli, C., Di Traglia, F., Taddeucci, J., 2010. Basaltic scoria textures from a zoned conduit as precursors to violent Strombolian activity. Geology 38, 439—442. http://dx.doi.org/10.1130/G30720.1.

Houghton, B.F., Wilson, C.J.N., Smith, I.E.M., 2000. Shallow-seated controls on styles of explosive basaltic volcanism: a case study from New Zealand. J. Volcanol. Geotherm. Res. 91, 97—120.

Lautze, N.C., Houghton, B.F., 2007. Linking variable explosion style and magma textures during 2002 at Stromboli volcano, Italy. Bull. Volcanol. 69, 445—460. http://dx.doi.org/10.1007/s00445-006-0086-1.

Orr, T.R., Thelen, W.A., Patrick, M.R., Swanson, D.A., Wilson, D.C., 2013. Explosive eruptions triggered by rockfalls at Kīlauea volcano, Hawaiʻi. Geology 41, 207—210. http://dx.doi.org/10.1130/G33564.1.

Parfitt, E.A., 2004. A discussion of the mechanisms of explosive basaltic eruptions. J. Volcanol. Geotherm. Res. 134, 77—107. http://dx.doi.org/10.1016/j.jvolgeores.2004.01.002.

Parfitt, E.A., Wilson, L., 1995. Explosive volcanic-eruptions, 9: the transition between Hawaiian-style lava fountaining and Strombolian explosive activity. Geophys. J. Int. 121, 226—232.

Walker, G.P.L., 1973. Explosive volcanic eruptions - a new classification scheme. Geol. Rundsch. 62, 431—446.

Wilson, L., 1980. Relationships between pressure, volatile content and ejecta velocity. J. Volcanol. Geotherm. Res. 8, 297—313.

Wilson, L., Head III, J.W., 1981. Ascent and eruption of basaltic magma on the Earth and Moon. J. Geophys. Res. 86, 2971—3001.

The role of volatiles

Aiuppa, A., Burton, M., Caltabiano, T., Giudice, G., Guerrieri, S., Liuzzo, M., Murè, F., Salerno, G., 2010. Unusually large magmatic CO_2 gas emissions prior to a basaltic paroxysm. Geophys. Res. Lett. 37, L17303. http://dx.doi.org/10.1029/2010GL043837.

Aiuppa, A., Moretti, R., Federico, C., Giudice, G., Gurrieri, S., Liuzzo, M., Papale, P., Shinohara, H., Valenza, M., 2007. Forecasting Etna eruptions by real-time observation of volcanic gas composition. Geology 35, 1115—1118. http://dx.doi.org/10.1130/G24149A.1.

Allard, P., 2010. A CO_2-rich gas trigger of explosive paroxysms at Stromboli basaltic volcano, Italy. J. Volcanol. Geotherm. Res. 189, 363—374.

Allard, P., Burton, M., Muré, F., 2005. Spectroscopic evidence for a lava fountain driven by previously accumulated magmatic gas. Nature 433, 407—410.

Benjamin, E.R., Plank, T., Wade, J.A., Kelley, K.A., Hauri, E.H., Alvarado, G.E., 2007. High water contents in basaltic magmas from Irazù volcano, Costa Rica. J. Volcanol. Geotherm. Res. 168, 68—92.

Bertagnini, A., Métrich, N., Landi, P., Rosi, M., 2003. Stromboli volcano (Aeolian Archipelago, Italy): an open window on the deep-feeding system of a steady state basaltic volcano. J. Geophys. Res. 108, 2336. http://dx.doi.org/10.1029/2002JB002146.

Edmonds, M., Gerlach, T.M., 2007. Vapor segregation and loss in basaltic melts. Geology 35, 751—754.

Ferlito, C., Viccaro, M., Cristofolini, R., 2009. Volatile-rich magma injection into the feeding system during the 2001 eruption of Mt. Etna (Italy): its role on explosive activity and change in rheology of lavas. Bull. Volcanol. 71, 1149—1158.

Gerlach, T.M., McGee, K.A., Elias, T., Sutton, A.J., Doukas, M.P., 2002. Carbon dioxide emission rate of Kīlauea Volcano: Implications for primary magma and the summit reservoir. J. Geophys. Res. 107, 1—15. http://dx.doi.org/10.1029/2001JB000407.

Hauri, E., 2002. SIMS analysis of volatiles in silicate glasses. 2: Isotopes and abundances in Hawaiian melt inclusions. Chem. Geol. 183, 115—141.

Métrich, N., Wallace, P.J., 2008. Volatile abundances in basaltic magmas and their degassing paths tracked by melt inclusions. Rev. Mineral. Geochem. 69, 363—402.

Métrich, N., Bertagnini, A., Di Muro, A., 2010. Conditions of magma storage, degassing and ascent at Stromboli: new insights into the volcano plumbing system with inferences on the eruptive dynamics. J. Petrol. 51, 603—626.

Sides, I., Edmonds, M., Maclennan, J., Houghton, B.F., Swanson, D.A., Steele-MacInnis, M.J., 2014. Magma mixing and high fountaining during the 1959 Kīlauea Iki eruption, Hawaiʻi. Earth Planet. Sci. Lett. 400, 102—112.

Spilliaert, N., Allard, P., Métrich, N., Sobolev, A., 2006. Melt inclusion record of the conditions of ascent, degassing and extrusion of volatile-rich alkali basalt during the powerful 2002 flank eruption of Mount Etna (Italy). J. Geophys. Res. 111, B04203. http://dx.doi.org/10.1029/2005JB003934.

Wade, J.A., Plank, T., Melson, W.G., Soto, G.J., Hauri, A.H., 2006. Volatile content of magmas from Arenal volcano, Costa Rica. J. Volcanol. Geotherm. Res. 157, 94—120.

Wallace, P.J., Anderson Jr., A.T., 1998. Effects of eruption and lava drainback on the H_2O contents of basaltic magmas at Kilauea Volcano. Bull. Volcanol. 59, 327—344.

Insights from infrasonic and other geophysical measurements

Cannata, A., Montalto, P., Privitera, E., Russo, G., Gresta, S., 2009a. Characterization and location of infrasonic sources in active volcanoes: Mt. Etna, September—November 2007. J. Geophys. Res. 114, 15. http://dx.doi.org/10.1029/2008JB006007.

Cannata, A., Montalto, P., Privitera, E., Russo, G., Gresta, S., 2009b. Tracking eruptive phenomena by infrasound: May 13, 2008 eruption at Mt. Etna. Geophys. Res. Lett. 36, L05304.

Chouet, B., Dawson, P., Ohminato, T., Martini, M., Saccorotti, G., Giudicepietro, F., De Luca, G., Milana, G., Scarpa, R., 2003. Source mechanisms of explosions at Stromboli Volcano, Italy, determined from moment-tensor inversions of very-long-period data. J. Geophys. Res. 108, 219. http://dx.doi.org/10.1029/2002JB001919.

Colò, L., Ripepe, M., Baker, D.R., Polacci, M., 2010. Magma vesiculation and infrasonic activity at Stromboli open conduit volcano. Earth Planet. Sci. Lett. 292, 274—280. http://dx.doi.org/10.1016/j.epsl.2010.01.018.

Fee, D., Matoza, R.S., 2013. An overview of volcano infrasound: from hawaiian to plinian, local to global. J. Volcanol. Geotherm. Res. 249, 123—139. http://dx.doi.org/10.1016/j.jvolgeores.2012.09.002.

Gerst, A., Hort, M., Kyle, P.R., Voge, M., 2008. 4D velocity of Strombolian eruptions and man-made explosions derived from multiple Doppler radar instruments. J. Volcanol. Geotherm. Res. 177, 648—660.

Harris, A., Ripepe, M., 2007. Synergy of multiple geophysical approaches to unravel explosive eruption conduit and source dynamics - a case study from Stromboli. Chem. Erde-Geochem. 67, 1–35. http://dx.doi.org/10.1016/j.chemer.2007.01.003.

Johnson, J.B., Lees, J.M., 2000. Plugs and chugs — seismic and acoustic observations of degassing explosions at Karymsky, Russia and Sangay, Ecuador. J. Volcanol. Geotherm. Res. 101, 67–82.

Johnson, J.B., Aster, R.C., 2005. Relative partitioning of acoustic and seismic energy during Strombolian eruptions. J. Volcanol. Geotherm. Res. 148, 334–354.

Johnson, J.B., Ripepe, M., 2011. Volcano infrasound: a review. J. Volcanol. Geotherm. Res. 206, 61–69. http://dx.doi.org/10.1016/j.jvolgeores.2011.06.006.

Johnson, J.B., Aster, R., Jones, K., Kyle, P., McIntosh, W., 2008. Acoustic source characterization of impulsive Strombolian eruptions from the Mount Erebus lava lake. J. Volcanol. Geotherm. Res. 177, 673–686.

Jones, K.R., Johnson, J.B., Aster, R., Kyle, P.R., McIntosh, W.C., 2008. Infrasonic tracking of large bubble bursts and ash venting at Erebus Volcano, Antarctica. J. Volcanol. Geotherm. Res. 177, 661–672. http://dx.doi.org/10.1016/j.jvolgeores.2008.02.001.

Lane, S.J., James, M.R., Corder, S.B., 2013. Volcano infrasonic signals and magma degassing: first-order experimental insights and application to Stromboli. Earth Planet. Sci. Lett. 377–378, 169–179. http://dx.doi.org/10.1016/j.epsl.2013.06.048.

Lopez, T., Fee, D., Prata, F., Dehn, J., 2013. Characterization and interpretation of volcanic activity at Karymsky Volcano, Kamchatka, Russia, using observations of infrasound, volcanic emissions, and thermal imagery. Geochem. Geophys. Geosys. 14, 5106–5127. http://dx.doi.org/10.1002/2013GC004817.

Lyons, J.J., Waite, G.P., Rose, W.I., Chigna, G., 2010. Patterns in open vent, Strombolian behavior at Fuego volcano, Guatemala, 2005–2007. Bull. Volcanol. 72, 1–15.

Marchetti, E., Ripepe, M., Harris, A.J.L., Delle Donne, D., 2009. Tracing the differences between Vulcanian and Strombolian explosions using infrasonic and thermal radiation energy. Earth Planet. Sci. Lett. 279, 273–281. http://dx.doi.org/10.1016/j.ep sl.2009.01.004.

Marchetti, E., Ripepe, M., Delle Donne, D., Genco, R., Finizola, A., Garaebiti, E., 2013. Blast waves from violent explosive activity at Yasur volcano, Vanuatu. Geophys. Res. Lett. 40, 5838–5843. http://dx.doi.org/10.1002/2 013GL057900.

Ripepe, M., Harris, A.J.L., Carniel, R., 2002. Thermal, seismic and infrasonic evidence of variable degassing rates at Stromboli volcano. J. Volcanol. Geotherm. Res. 118, 285–297.

Ripepe, M., Marchetti, E., Ulivieri, G., 2007. Infrasonic monitoring at Stromboli volcano during the 2003 effusive eruption: Insights on the explosive and de-gassing process of an open conduit system. J. Geophys. Res. 112, B09207. http://dx.doi.org/10.1029/2006JB004613.

Ripepe, M., Delle Donne, D., Harris, A., Marchetti, E., 2008. Dynamics of Strombolian activity. In: Calvari, S., et al (Eds.), The Stromboli Volcano: An Integrate Study of the 2002–2003 Eruption. AGU, Washington D. C, pp. 39–48. Geophysical Monograph Series 182.

Ripepe, M., Marchetti, E., Bonadonna, C., Harris, A.J.L., Pioli, L., Ulivieri, G., 2010. Monochromatic infrasonic tremor driven by persistent degassing and convection at Villarrica volcano, Chile. Geophys. Res. Lett. 37, L15303. http://dx.doi.org/10.1029/2010GL043516.

Taddeucci, J., Sesterhenn, J., Scarlato, P., Stampka, K., Del Bello, E., Pena Fernandez, J.J., Gaudin, D., 2014. High-speed imaging, acoustic features, and aeroacoustic computations of jet noise from Strombolian (and Vulcanian) explosions. Geophys. Res. Lett. 41, 3096–3102. http://dx.doi.org/10.1002/2 014GL05992 5.

Ulivieri, G., Ripepe, M., Marchetti, E., 2013. Infrasound reveals transition to oscillatory discharge regime during lava fountaining: Implication for early warning. Geophys. Res. Lett. 40, 3008–3013. http://dx.doi.org/10.1002/grl.50592, 2013.

Vergniolle, S., Brandeis, G., 1996a. Strombolian explosions: 1. A large bubble breaking at the surface of the lava column as a source of sound. J. Geophys. Res. 101, 20433–20448.

Vergniolle, S., Brandeis, G., Mareschal, J.C., 1996b. Strombolian explosions. 2. Eruption dynamics determined from acoustic measurements. J. Geophys. Res. 101, 20449–20466.

Vergniolle, S., Boichu, M., Caplan-Auerbach, J., 2004a. Acoustic measurements of the 1999 basaltic eruption of Shishaldin volcano, Alaska — 1. Origin of Strombolian activity. J. Volcanol. Geotherm. Res. 137, 109–134. http://dx.doi.org/10.1016/j.jvolgeores.2004.05.003.

Vergniolle, S., Caplan-Auerbach, J., 2004b. Acoustic measurements of the 1999 basaltic eruption of Shishaldin volcano, Alaska: 2) precursor to the Subplinian activity. J. Volcanol. Geotherm. Res. 137, 135–151.

Vergniolle, S., Ripepe, M., 2008. From Strombolian explosions to fire fountains at Etna Volcano (Italy): what do we learn from acoustic measurements?. In: Lane, S.J., Gilbert, J.S. (Eds.), Fluid Motions in Volcanic Conduits: A Source of Seismic and Acoustic Signals, vol. 307. Geological Society, London, Special Publications, pp. 103–124.

Vidal, V., Ripepe, M., Divoux, T., Legrand, D., Geminard, J.C., Melo, F., 2010. Dynamics of soap bubble bursting and its implications to volcano acoustics. Geophys. Res. Lett. 37, L07302. http://dx.doi.org/10.1029/2009GL042360.

Woulff, G., McGetchin, T.R., 1976. Acoustic noise from volcanoes: theory and experiment. Geophys. J. R. Astron. Soc. 45, 601–616.

Conduit fluid dynamics

Brown, R.A.S., 1965. The mechanics of large gas bubbles in tubes I. Bubble velocities in stagnant liquids. Can. J. Chem. Eng. 43, 217–223. http://dx.doi.org/10.1002/cjce.5450430501.

Del Bello, E., Llewellin, E.W., Taddeucci, J., Scarlato, P., Lane, S.J., 2012. An analytical model for gas overpressure in slug-driven explosions: insights into Strombolian volcanic eruptions. J. Geophys. Res. 117, B02206. http://dx.doi.org/10.1029/2011JB008747.

James, M.R., Lane, S.J., Chouet, B., 2004. Pressure changes associated with the ascent and bursting of gas slugs in liquid-filled vertical and inclined conduits. J. Volcanol. Geotherm. Res. 129, 61–82. http://dx.doi.org/10.1016/S0377-0273(03)00232-4.

James, M.R., Lane, S.J., Chouet, B.A., 2006. Gas slug ascent through changes in conduit diameter: laboratory insights into a volcano-seismic source process in low-viscosity magmas. J. Geophys. Res. 111, B05201. http://dx.doi.org/10.1029/20 05JB0 03718.

James, M.R., Lane, S.J., Corder, S.B., 2008. Modelling the rapid near-surface expansion of gas slugs in low viscosity magmas. In: Lane, S.J., Gilbert, J.S. (Eds.), Fluid Motion in Volcanic Conduits: A Source of Seismic and Acoustic Signals. Geological Society Special Publications, London, pp. 147–167.

James, M.R., Lane, S.J., Wilson, L., Corder, S.B., 2009. Degassing at low magma-viscosity volcanoes: quantifying the transition between passive bubble-burst and Strombolian eruption. J. Volcanol. Geotherm. Res. 180, 81–88. http://dx.doi.org/10.1016/j.jvolgeores.2008.09.002.

James, M.R., Llewellin, E.W., Lane, S.J., 2011. Comment on "It takes three to tango: 2. Bubble dynamics in basaltic volcanoes and ramification for modelling normal Strombolian activity". In: Suckale, J., et al. (Eds.), Journal of Geophysical Research, vol. 113, p. B06207. http://dx.doi.org/10.1029/2010JB008167.

Jaupart, C., Vergniolle, S., 1989. The generation and collapse of a foam layer at the roof of a basaltic magma chamber. J. Fluid Mech. 203, 347–380.

Llewellin, E.W., Del Bello, E., Taddeucci, J., Scarlato, P., Lane, S.J., 2012. The thickness of the falling film of liquid around a Taylor bubble. Proc. R. Soc. Lond. A 468, 1041–1064. http://dx.doi.org/10.1098/rspa.2011.0476.

Manga, M., 1996. Waves of bubbles in basaltic magmas and lavas. J. Geophys. Res. 101, 17457–17465. http://dx.doi.org/10.1029/96JB01504.

Menand, T., Phillips, J.C., 2007. Gas segregation in dykes and sills. J. Volcanol. Geotherm. Res. 159, 393–408. http://dx.doi.org/10.1016/j.jvolgeores.2006.08.003.

Namiki, A., Manga, M., 2008. Transition between fragmentation and permeable outgassing of low viscosity magmas. J. Volcanol. Geotherm. Res. 169, 48–60.

Nogueira, S., Rietmuler, M.L., Campos, J.B.L.M., Pinto, A.M.F.R., 2006. Flow in the nose region and annular film around a Taylor bubble rising through vertical columns of stagnant and flowing Newtonian liquids. Chem. Eng. Sci. 61, 845–857. http://dx.doi.org/10.1016/j.ces.2005.07.038.

Pioli, L., Azzopardi, B.J., Cashman, K.V., 2009. Controls on the explosivity of scoria cone eruptions: magma segregation at conduit junctions. J. Volcanol. Geotherm. Res. 186, 407–415. http://dx.doi.org/10.1016/j.jvolgeores.2009.07.014.

Pioli, L., Bonadonna, C., Azzopardi, B.J., Phillips, J.C., Ripepe, M., 2012. Experimental constraints on the outgassing dynamics of basaltic magmas. J. Geophys. Res. 117, B03204. http://dx.doi.org/10.1029/2011JB008392.

Saar, M.O., Manga, M., 1999. Permeability-porosity relationships in vesiculating basalts. Geophys. Res. Lett. 26, 111–114. http://dx.doi.org/10.1029/1998GL900256.

Seyfried, R., Freundt, A., 2000. Experiments on conduit flow and eruption behavior of basaltic volcanic eruptions. J. Geophys. Res. 105, 23727–23740.

Vergniolle, S., Gaudemer, Y. From reservoirs and conduits to surface: a review on the role of bubbles in driving basaltic eruptions. AGU monography, in Hawaiian Volcanism: From Source to Surface, Chap 14. (in press).

Vergniolle, S., Jaupart, C., 1986. Separated two-phase flow and basaltic eruptions. J. Geophys. Res. 91, 12842–12860.

Vergniolle, S., Jaupart, C., 1990. Dynamics of degassing at Kilauea volcano, Hawaii. J. Geophys. Res. 95, 2793–2809.

Viana, F., Pardo, R., Yanez, R., Trallero, J., Joseph, D., 2003. Universal correlation for the rise velocity of long gas bubbles in round pipes. J. Fluid Mech. 494, 379–398.

Chapter 28

Vulcanian Eruptions

Amanda Bachtell Clarke

School of Earth and Space Exploration, Arizona State University, Tempe, AZ, USA

Tomaso Esposti Ongaro

Istituto Nazionale di Geofisica e Vulcanologia, Sezione di Pisa, Pisa, Italy

Alexander Belousov

Institute of Volcanology and Seismology, Petropavlovsk-Kamchatsky, Russia

Chapter Outline

GLOSSARY

Adiabatic Flow An idealized flow condition in which heat does not enter or escape the fluid system due to perfect insulating boundaries.

Andesite An extrusive igneous rock type or magma of intermediate silica content falling between that of basalt and dacite; typically contains phenocrysts of plagioclase feldspar and pyroxene, and may contain hornblende depending on the water content of the magma.

Specific buoyancy flux The rate at which buoyancy per unit mass is injected into the atmosphere, expressed in terms of the volume flux and the density contrast between the injected fluid and the ambient fluid, such as the atmosphere. Buoyancy represents the force acting on an object or a parcel of fluid when it is immersed in a fluid with a different density and in a gravity field. Buoyancy is positive when the surrounding fluid has a higher density.

Country rock The rocks through which the magma travels on its way to the surface during eruption.

Dacite An extrusive igneous rock type or magma of intermediate silica content falling between that of andesite and rhyolite; typically contains phenocrysts of potassium feldspar and plagioclase feldspar, and may contain quartz, biotite, and hornblende.

Decompression wave This is a also rarefaction wave. A continuous perturbation of a flow field generated by a pressure discontinuity, traveling at the local speed of sound throughout the region at higher pressure.

Inviscid flow An idealized flow condition that assumes that the fluid involved has zero viscosity, and therefore loses no energy to viscous dissipation. The inviscid assumption is typically made when inertial forces far outweigh viscous forces.

Isothermal A fluid system is **isothermal** when its temperature (and the temperature of all its components) remains constant during motion or throughout the course of a thermodynamic transformation.

Juvenile A fragmented and solidified clast of erupted magma.

Lithic A dense pyroclast of solidified juvenile magma or an erupted fragment of country rock.

Specific momentum flux The rate at which momentum per unit mass is injected into the atmosphere, expressed in terms of the volume flux and the injection velocity.

Pseudofluid A fluid consisting of a mixture of at least two phases (solid, liquid, gas) that can be treated as a single fluid in which all phases have the same temperature and velocity.

Shock waves Waves characterized by a discontinuous change in pressure, density, temperature, and velocity, and by **supersonic** propagation velocities.

Self-similar solutions A solution of the equations of a dynamical system that has the same form (i.e., it is represented by the same function) at any point. Self-similarity usually allows a simple

The Encyclopedia of Volcanoes. http://dx.doi.org/10.1016/B978-0-12-385938-9.00028-6

parameterization of the dynamic variables in terms of constant scaling factors. For example, for jets and plumes, the scaling factor is the vertical distance from the source.

Speed of sound of a fluid The speed at which a pressure disturbance travels in a fluid.

Subsonic flow Flow of a fluid whose velocity always and everywhere remains less than that of the **speed of sound** in the fluid.

Supersonic flow Flow of a fluid whose velocity exceeds that of the speed of sound in the fluid.

Thermal Flow resulting from an instantaneous release of a buoyant fluid.

Vesicularity The volume fraction of bubbles in a magma or volcanic rock.

Volcanic bomb A fragment of magma or wall rock ejected by a volcanic explosion that is sufficiently large (formally >64 mm) to be transported in a ballistic trajectory. A **bread-crust bomb** is a volcanic bomb that in the process of transportation developed a specific surface texture resembling the cracked surface of a loaf of bread. This surface indicates that the interior of the magma fragment was still partially molten and continued to expand (due to the growth of gas bubbles) while its outer surface solidified and became brittle (chilled by ambient air).

1. INTRODUCTION

Since the early days of volcanology, classification of explosive activity has been based on the visual characteristics of eruption clouds. Modern investigations have demonstrated that, in most cases, visual qualitative differences between eruption clouds are closely related to fundamental differences in the eruption mechanisms, which, in turn, produce specific types of pyroclastic deposits recognizable in the field. Such links are rather straightforward for Hawaiian, Strombolian, Plinian, and Surtseyan types of explosive activity.

Vulcanian eruptions, however, are more complex to classify. Vulcanian eruptions were first distinguished by Mercalli and Silvestri (1891) who noticed that the 1888–1890 eruption of Vulcano in the Aeolian Islands was somewhat different from eruptions of nearby Stromboli volcano. Both volcanoes produced small to moderate scale, short-lived intermittent explosions, but explosions of Vulcano were louder, perhaps due to shock waves, eruption clouds were darker in color (almost black due to the presence of abundant ash), and ejected material had lower temperatures (few or no glowing ejecta were visible during daytime). The morphology of the juvenile products indicated higher viscosity and lower vesicularity magma at Vulcano; ballistics ranged from "bread-crust bombs" to dense, angular, glassy blocks. Mercalli thus suggested that Vulcanian activity is typical for magmas of intermediate composition.

Later investigations of other volcanoes have shown that visually similar "Vulcanian" eruptions can deposit very different pyroclastic products, and thus may have different eruptive mechanisms. Two major eruptive mechanisms were suggested: phreatomagmatic and magmatic. Phreatomagmatic "Vulcanian" explosions commonly occur during initial "throat-clearing" and/or final stages of volcanic eruptions of other types, when rising/receding magma explosively interacts with groundwater or hydrothermal fluids surrounding the upper part of the conduit. Such explosions can be produced by magma of any composition (e.g., the 1924 eruption of Halemaumau crater in Hawaii was basaltic; Jaggar, 1947) and the resulting pyroclastic material contains a significant percentage of fragmented country rock.

However, in most cases (like the 1888–1890 classic eruption of Vulcano) there is no evidence of contact between magma and groundwater or hydrothermal fluids. Deposits of these eruptions contain few to no country rock fragments, and the participating magma is almost always of intermediate composition. Intermittent Vulcanian explosions of such eruptions commonly continue for months to years, sometimes with little variation in frequency and intensity. The mechanisms associated with such eruptions emphasize the critical role of a gas-impermeable plug composed of degassed, partly solidified magma that periodically forms in the upper part of a slowly ascending magma column. Volatiles gradually exsolve from crystallizing magma and accumulate as pressurized gas bubbles under the plug. Eventually the plug mechanically fails, initiating the eruption, and the released gas expands, fragmenting and ejecting magma of variable vesicularity.

Thus, currently the term "Vulcanian eruption" can be used in a broad sense, when dark intermittent "Vulcanian" clouds are observed irrespective of their eruptive mechanism (purely magmatic or phreatomagmatic), or in narrow sense, when such explosions are produced by the magmatic eruption mechanism. Here we consider only the purely magmatic mechanism of eruption initiation (see Chapter 30 for an account of phreatomagmatic mechanisms).

2. HAZARDS

Vulcanian explosions can precede large Plinian eruptions, as at Mt Pinatubo (Philippines; June 12–14, 1991); produce dangerous pyroclastic flows, as at Mount St Helens (USA; summer of 1980), and at Soufrière Hills volcano (Montserrat, BWI; July, August, and September of 1997); and present a significant hazard to aircraft, as at Galunggung (Indonesia) in 1982 and Redoubt (Alaska, USA) in 1989. At several volcanoes around the world, including Semeru (Indonesia), Sakurajima (Japan), and Karymsky (Russia), Vulcanian eruptions occur daily and can persist for years, potentially representing a large cumulative erupted mass. Ash and gases produced by Vulcanian eruptions are not typically ejected into the stratosphere where they would have a global effect; nevertheless explosion products can have devastating effects on local crops and nearby populations. Vulcanian eruptions occur much more frequently around the world than Plinian eruptions offering excellent opportunity for detailed field observation. To date, however, such observations are rare in the literature. Continued observation of Vulcanian eruptions will play a critical role in advancing general theoretical understanding of explosive eruption dynamics.

3. FIELD EXAMPLE: KARYMSKY VOLCANO

Karymsky volcano is located in the eastern volcanic belt of Kamchatka Peninsula, Russian Far East. The cone-shaped stratovolcano (1553 m asl in 2010, 900 m above the base) of andesitic to dacitic composition occupies the central part of the 4.5-km-wide Karymsky caldera that formed 7700−7800BP (Braitseva and Melekestsev, 1991). Karymsky volcano is among the most active volcanoes in the world; it has had over 20 historical eruptive periods, 9 of which occurred in the twentieth century (Siebert and Simkin, 2002). Small to moderate scale short-lived Vulcanian

explosions with eruptive clouds 0.3−3-km high and a frequency of one every several minutes to every several days are characteristic of Karymsky's activity (Figure 28.1(A) and (B)). Concentrations of pyroclastic material in the eruptive clouds notably fluctuate; the weakest explosions produce light gray clouds containing few ash and ballistics, whereas the dark grey to black clouds of the strongest explosions are heavily ash laden and commonly accompanied by small-volume hot avalanches originating from abundant ballistic fallout (Figure 28.1(D)). Relatively uniform explosions with regular repose intervals that commonly span over periods of days to months gradually change their average intensity and frequency (Figure 28.1(A) and (B)). Some periods of explosive activity are combined with the extrusion of small intra-crater domes and/or viscous blocky lava flows up to 20−30-m thick, with the explosions completely or partially destroying these domes/flows. Products of the explosions are bread-crust bombs (Figure 28.1(E)), as well as lapilli and ash composed of poorly vesiculated blocky particles (Figure 28.1(F)).

Investigations of paleosol sections have shown that periods of long-lasting Vulcanian activity were common

FIGURE 28.1 A) 1996 and (B) 2005 eruptions of Karymsky volcano. These consisted of frequent moderately strong explosions with few ballistics. Eruption cloud is approximately 1 km above the crater. Eruption clouds of previous explosions that occurred minutes before are drifting downwind; (C) Inner slope of the crater of Vulcano, Aeolian Islands, Italy. Multiple layers of bombs and lapilli were deposited by numerous transient explosions of the 1888−1890 eruptions; (D) Strongest Vulcanian-type explosions of Karymsky volcano, July 2004. Quickly rising eruption cloud is more than 1 km above the vent. Final height of the eruption cloud will be approximately 3 km. Massive ballistic fallout forms multiple hot avalanches on the volcano slope. A curtain of ash fallout is visible on the lee side of the eruption cloud; (E) Bread-crust bomb of Karymsky volcano, 1999. Lens cap is 6 cm across; (F) Scanning electron microscope image of ash particles of Karymsky volcano eruptions in 2003. Blocky sharp-edged particles with no/few gas bubbles indicate fragmentation of degassed highly viscous magma. *All photos by A. Belousov.*

throughout the 5000 years of history of the stratovolcano; such periods are indicated in paleosols by high concentrations of dispersed fine-grained ash. The last eruption cycle of Vulcanian activity started at Karymsky in 1996 and continues today. Intensity of the eruptions has fluctuated through time: periods were characterized by either pure explosive activity of varying intensity or by explosions associated with lava extrusions; also there were several episodes of complete inactivity that lasted up to several months.

As an example we present a description of the volcano's activity in July 2008. The explosions were associated with the slow extrusion of a small andesitic lava dome in the summit crater, and were separated by periods of vigorous gas venting (in the form of a continuous vertical white plume with an audible "jet engine" sound) from the dome apex. The gas venting ceased abruptly several tens of minutes before each explosion indicating plugging and pressurization of the magma conduit. Each explosion started with a sudden violent ash-and-ballistic-rich loud outburst, commonly consisting of a rapid succession of several inclined jets. Minutes-long pulsatory expulsions of ash clouds of declining intensity and ash concentration followed this initial stage, generating ash plumes up to 3 km above the crater. All of the observed explosions produced large volumes of ballistic materials that landed on the upper flanks of the volcano, although the specific volume varied from event to event (up to several tens of thousands of cubic meters per explosion). Massive ballistic fallout from the largest explosions produced small hot avalanches on the volcano slopes (Figure 28.1(D)). No incandescence was visible during daylight hours and dull red ballistics were observed during the night. Ejected pyroclasts indicate that the fragmenting andesitic magma was represented by two end members: vesicular with large irregular interconnected gas bubbles (vesicularity 40%), and dense with small isolated gas bubbles (vesicularity 5%).

4. GENERAL PHENOMENOLOGICAL FEATURES OF VULCANIAN ERUPTIONS

Vulcanian eruptions may have a wide range of dispersal areas and degrees of fragmentation, making them difficult to classify from deposit characteristics alone. For example, Walker (1973) proposed in a preliminary sense that Vulcanian eruptions may have dispersal areas similar to those of subplinian eruptions, but should contain a larger proportion of fine-grained pyroclasts (Figure 28.2). This increased "explosivity" of Vulcanian eruptions reflects dynamic factors (i.e., it is not related solely to rheological or compositional differences), and was attributed in particular to plugging of the vent. Additional synthesis of field data by Cas and Wright (1987) demonstrates that

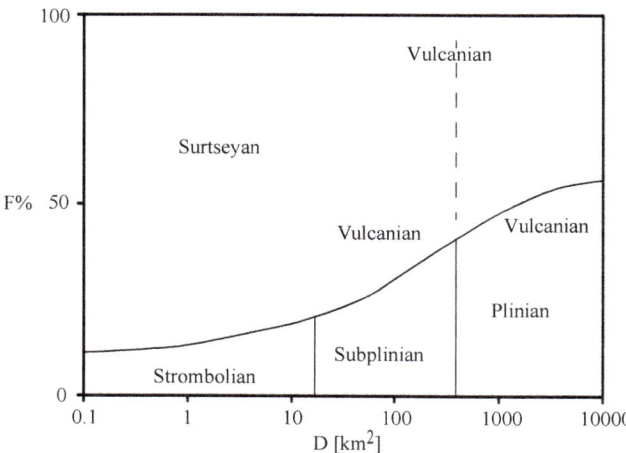

FIGURE 28.2 Classification of Vulcanian eruptions in terms of the percentage of (vertical axis) clasts <1 mm in diameter on the main dispersal axis where it crosses the isopach of thickness equal to 10% of the maximum thickness of the deposit (F%), and (horizontal axis) the area enclosed by the same isopach (D). Data defining the Vulcanian fields are from Wright et al. (1980), based on eruptions of Cerro Negro (Guatemala, 1968 and 1971), Fuego (Guatemala, 1971), Ngauruhoe (New Zealand, 1974 and 1975), Mt Edgemont (New Zealand, 1665), and Irazu (Costa Rica, 1963). *Modified from Walker (1973) and Cas and Wright (1986).*

Vulcanian eruptions equal subplinian to Plinian in terms of dispersal areas, but almost always contain a larger proportion of fine-grained pyroclasts, sometimes significantly more (Figure 28.2).

However, Vulcanian eruptions share some common observable phenomenologies that make them identifiable.

- Distinctive deposit characteristics
  - *Relatively small magnitude.* The juvenile mass erupted in a single Vulcanian event does not usually exceed that stored in the conduit, and is not typically greater than 10^{11} kg (VEI 3–4). However, cyclic activity may result in much larger cumulative erupted volumes, producing thick coarsely layered deposits proximal to the vent (Figure 28.1(C)).
  - *Relatively fine ejecta.* With respect to other explosive eruptions with similar volumes and dispersal areas, the average mean particle size of Vulcanian eruptions is relatively fine (Figure 28.2). This phenomenon is explained by the higher specific energy related to the initial rapid decompression stage.
  - *Low-vesicularity pyroclasts.* The average vesicularity of pyroclasts is lower (and bubble walls are thicker) than those typical of other types of magmatic eruptions (Figure 28.1(F)).
  - *Variable clast vesicularity.* It testifies to the progressive sampling of variably vesiculated conduit magma by the deepening fragmentation wave.
  - *Blocky shape of ash particles.* It indicates brittle fragmentation of highly viscous magma (Figure 28.1(F)).

- *Strong ballistic ejection.* Large numbers of decimeter-to meter-sized clasts are generated by explosive plug disruption. Typically, ballistics are bread-crust bombs composed mostly of poorly vesiculated and degassed magma ejected in an almost solid state, representing the plug material (Figure 28.1(E)). Ballistics are common for other eruption types as well, but they are different in notable ways; they are either more vesicular in Plinian and Strombolian eruptions, consisting of large pumice or scoria clasts, or less vesicular in phreatic or phreatomagmatic eruptions, dominated by lithic country rocks. Unusually large ballistic ranges in Vulcanian explosions can be caused by the cumulative effect of a strong initial pressure gradient and the accelerating drag of the ejected pyroclastic mixture.

- *Transient behavior.* Vulcanian eruption duration is limited, such that the injection lasts seconds to minutes (Figure 28.1(A) and (B)). Vulcanian eruptions often serve as the opening stage in some subplinian or Plinian eruptions.

- *Supersonic regimes.* Vulcanian explosions are distinguishable by a leading shock wave and a subsequent formation of an underexpanded jet. This stage can be recognized by infrasonic and acoustic measurements. The transient injection and the underexpanded jet lead to the typical "mushroom" shape of the jet immediately above the vent (Figure 28.1(A), (B) and (D)).

- *Short-lived atmospheric plume.* The evolution of the volcanic plume is controlled by the large vortex structure at its head, which controls atmospheric air entrainment and also partly explains the typical "mushroom" shape (Figure 28.1(A), (B) and (D)). As a consequence, ascent velocity of the plume follows a different temporal evolution than a steady, sustained plume.

Despite these common features, the wide range of variability of eruptive conditions makes a simply phenomenological definition of Vulcanian eruptions not fully satisfactory. We therefore proceed by defining Vulcanian eruptions from a dynamical point of view.

5. ERUPTION MECHANISM

Vulcanian eruptions result from the sudden decompression of a volcanic conduit that contains highly pressurized crystallized bubbly magma of intermediate composition (Figure 28.3). These eruptions initiate when a conduit plug or dome is disrupted due to a sufficiently high pressure gradient in the underlying magma. Upon plug disruption, a decompression wave, followed by a fragmentation front, travels down the conduit, while a compression shock propagates into the atmosphere. At the fragmentation front, vesicular magma is disrupted into a gas–pyroclast mixture,

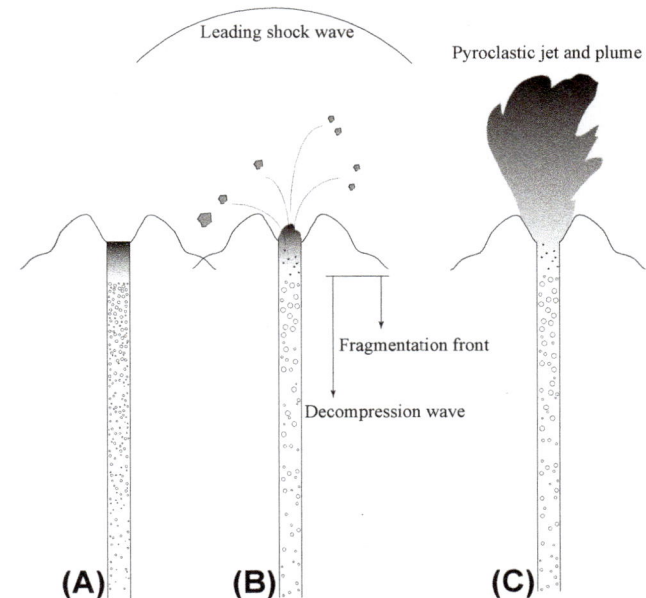

FIGURE 28.3 Vulcanian Eruption Schematic. (A) A dense plug seals a conduit containing bubble-bearing magma. (B) Eruptions initiate when the plug is disrupted, evidenced by the launching of ballistic clasts. A decompression wave, followed by a fragmentation wave, travels into the conduit, while shock waves propagate through the atmosphere. (C) An overpressured mixture of fragmented magma and expanding gas is ejected into the atmosphere.

propelled upward, and ejected from the vent into the atmosphere as an underexpanded jet at sonic to supersonic velocities. This ejection is characteristically impulsive and unsteady. Transition to the subsonic regime can occur very rapidly above the vent, so that the jet may evolve into a buoyant plume, collapse gravitationally to form pyroclastic density currents, or both may occur simultaneously.

Typically, only a portion of the magma in the conduit is fragmented and evacuated, such that Vulcanian eruptions characteristically are of relatively small volume and last only seconds to minutes. They may occur as single events or as a sequence of explosions spaced sufficiently far apart in time to produce distinguishable, discrete, unsteady events.

The short duration and unsteady vent conditions of Vulcanian eruptions define them and make them distinct from quasi-steady Plinian, subplinian, or Hawaiian eruptions, for which it is generally assumed that bubbly magma rises to meet the fragmentation front and thus steadily feeds vent flux over hours to days. Strombolian activity can also be characterized as short-lived and impulsive, however, in Vulcanian events bubbles cannot ascend quickly through the highly viscous magma as they do at Stromboli, making the dynamics significantly different. Column heights typically exceed heights associated with Strombolian eruptions (see Chapter 27) and are less than those associated with Plinian and subplinian eruptions (see Chapter 29).

6. ERUPTION INITIATION

Vulcanian eruptions initiate when a coherent magma plug or dome that sealed the conduit is suddenly disrupted. Observational evidence of this initiation process consists of the formation of crystallized and degassed domes, ballistic clasts launched at the very beginning of an eruption, and dense ballistic clasts in deposits. Plug disruption occurs when pressure in the underlying conduit rises sufficiently (up to 10–15 MPa, but usually <5 MPa, based on the strength of typical magmas) to exceed the mechanical strength of the lava or when the overlying dome collapses. Two key processes are involved in creating conduit plugs: magma outgassing and subsequent microlite crystallization in the volcanic conduit. As magma rises, water exsolves from the melt in response to decreasing pressure, and a corresponding shift in the liquidus causes anhydrous phases to crystallize, especially plagioclase feldspar. Crystallization and degassing increase magma viscosity and, by concentrating volatiles in the remaining melt, force further degassing. Bubble connections develop at some threshold vesicularity. Simultaneous crystallization and degassing tends to concentrate vesicles in the interstices between crystals, and may enhance permeable gas loss via connected bubble networks. The consequent outgassing may cause vesicle collapse and formation of a dense viscous plug that leads to the stagnation of underlying magma (Figure 28.4).

Many of these same processes lead to increased pressure below the viscous plug. For example, magma pressure may increase due to rheological stiffening of the ascending magma (Figure 28.4) and volatile pressure may increase because bubble growth is inhibited by high viscosity. Eventually pressures reach values sufficient to disrupt the overlying plug resulting in fragmentation of both the plug and underlying magma (Figure 28.3(B)).

Ballistic blocks or bombs mainly represent the disrupted sealing plug. Although ballistic block fields have been documented carefully for several types of eruptions (including the 1977 phreatomagmatic eruptions of Ukinrek Maars, Alaska, the 1992 subplinian eruptions of Crater Peak Vent, Mount Spurr volcano, Alaska, the 1997 Vulcanian eruptions of Soufrière Hills volcano, and the 1999 Vulcanian eruptions of Guagua Pichincha volcano, Ecuador, among others) ballistic launch in Vulcanian eruptions (or in the Vulcanian stage of complex eruptive sequences) is particularly intense and characterized by unusually large ranges. Blocks on the order of a half meter in diameter can be launched to 3 km from the vent, and in some cases smaller blocks can reach >6 km from the vent.

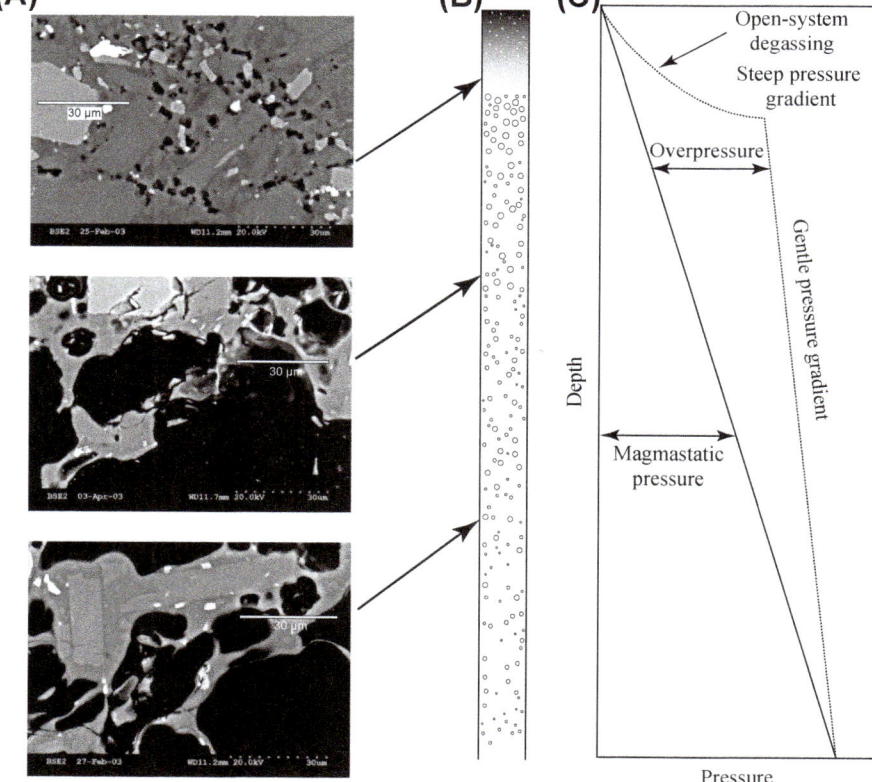

FIGURE 28.4 Schematic representation of conduit conditions prior to Vulcanian eruptions. (A) SEM images of pyroclasts from Soufrière Hills volcano, in which bubbles (black) and crystals (white and light gray) can be distinguished from rapidly quenched melt (glass, darker gray). (B) Relative vertical source positions in the conduit for each sample are indicated by arrows pointed at the conduit schematic. (C) The pressure profile for the bubbly magma in an unplugged state (magmastatic) and a preeruption over-pressured state. *Modified from Clarke et al. (2007).*

They may exhibit a bread-crusted texture indicating that interior gas expanded after the surface was quenched. In general, ballistic size decreases with distance from the vent. However, in some cases the opposite is true because complex drag interactions between the blocks and the expanding pyroclastic mixture and the surrounding air lead to very high drag-to-weight ratios for some small blocks.

Observations and numerical models resolving the full Lagrangian equations for individual particles have shown that, in Vulcanian eruptions, clast trajectories quite often deviate from parabolic trajectories. In particular, accelerating drag exerted by the starting jet and the added effect of a strong pressure gradient lead to launch distances up to 70% greater than those predicted by a simple ballistic analysis (Figure 28.5).

7. DECOMPRESSION AND FRAGMENTATION

Upon plug disruption, a decompression wave travels at the local sound speed into the conduit (Figure 28.3(B)). The decompression wave is followed by a fragmentation wave that travels more slowly than the pressure wave through the bubbly magma (Figure 28.3(B)). The mathematical formulation of such a model is analogous to that for flow in a one-dimensional shock tube. In this configuration, a one-dimensional tube is subdivided into a high-pressure region at the base of the tube and a low-pressure region in the upper portion, separated by a diaphragm. Upon rupture of the separating diaphragm, a compression wave (shock) propagates upward, while a rarefaction wave decompresses the underlying fluid.

Behind the fragmentation front, a mixture of expanding gases and freshly produced pyroclasts is projected upward and expelled from the volcanic vent. The fragmentation wave is generally thought to fragment and quench the magma faster than dissolved gases can exsolve in response to the decompression. Therefore, to first order, exsolution of magmatic volatiles is assumed to be insignificant during the decompression and fragmentation process, and thus only previously exsolved volatiles participate in the eruption. Eruptive products therefore preserve to some extent the preexplosion state of magma vesiculation. However, up to several percent by volume bubble expansion can occur during eruption, due to bubble growth, and thus erupted clasts may be more vesicular than the preeruptive magma. The velocity of the fragmentation wave is thought to

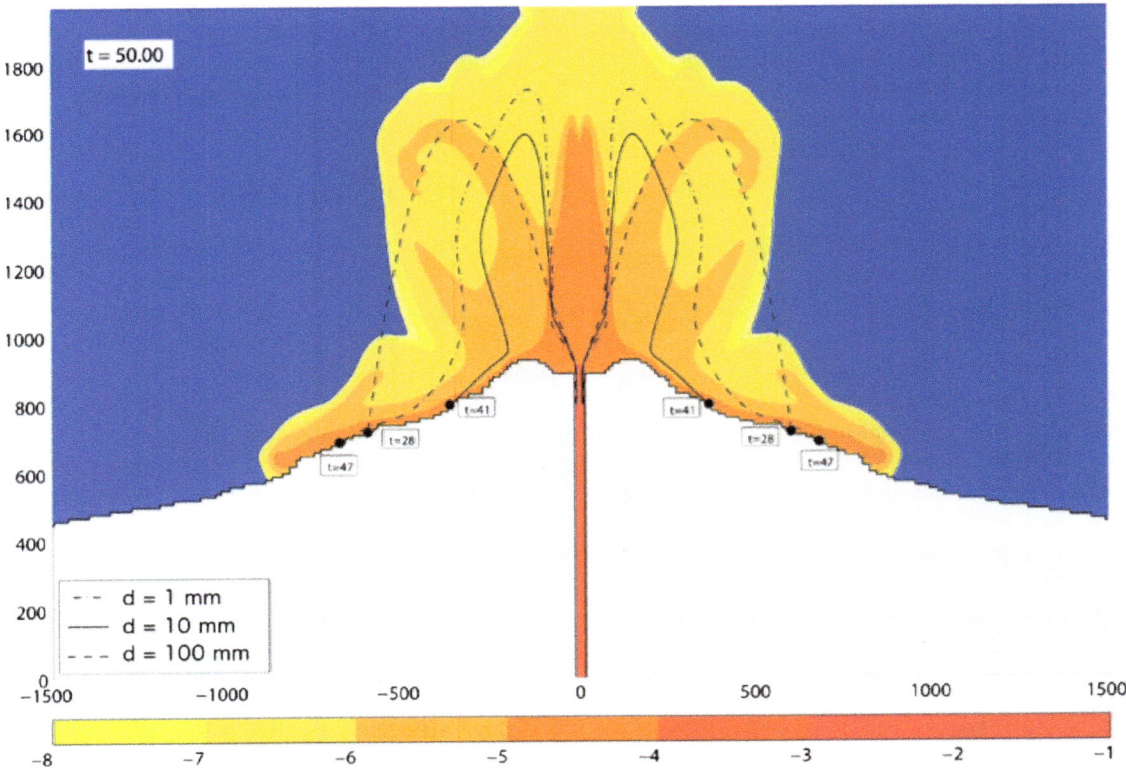

FIGURE 28.5 Computed trajectories and flight times of three different sets of particles based on the 1997 eruption of Soufrière Hills volcano, Montserrat. Vertical and horizontal scales in are in meters; the color scale represents particle volume fraction in log$_{10}$ scale (unitless); travel time labels are given in seconds. Lines indicate trajectories for different particle diameters (d). Note that the ballistics have been influenced to varying degrees by the dynamics of the modeled pyroclastic jet. *Modified from de'Michieli Vitturi et al. (2010).*

greatly exceed magma ascent velocity, and the magma is assumed to have stagnated prior to fragmentation. Therefore the fragmentation front meets magma with varying degrees of crystallinity, vesicularity, and viscosity. The observed variability of pyroclast densities and crystallinities in Vulcanian deposits supports this concept of a progressive downward-migrating fragmentation front.

The details of the fragmentation process remain a subject of active research in volcanology, but several models have been presented. Fragmentation is thought to occur due to high shear or elongational strain rates, high tensile stresses at bubble walls due to overpressured bubbles, or a combination of the two. When magma is subject to very high strain rates its characteristic relaxation time may exceed the time over which stresses are applied, preventing the magma from stretching or flowing, leading to brittle behavior and fragmentation. Magma relaxation time increases with magma viscosity, meaning that fragmentation generally occurs more readily in higher viscosity magmas. The tensile stress threshold may be exceeded in bubbly magma when the stress on bubble walls exceeds the tensile strength of the magma. Fragmentation due to this condition is thought to occur by disruption of bubbles near the free surface of a vesicular magma, where there can be a significant pressure gradient between the ambient pressure and the bubble gas pressure. According to a series of numerical solutions and experiments on natural melts, bubbles should stop growing long before explosive fragmentation, primarily because of increasing melt viscosity. These solutions led to the concept that bubble volume fraction never exceeds 66–83% vesicularity. This range of vesicularity is often used as a fragmentation criterion in numerical models of magma ascent.

It is likely that each of these mechanisms contributes to fragmentation in Vulcanian eruptions. Upon plug disruption, and decompression of the underlying bubbly magma, one or both of the strain rate thresholds may be exceeded due to magma acceleration in response to the sudden decompression. The decompression also leads to rapid bubble expansion, which should produce very high strain rates within bubble walls such that bubble walls behave as a brittle solid and interbubble partitions are ruptured instead of stretched. The tensile stress criterion may also apply because plug disruption suddenly exposes the underlying magma to a very low ambient pressure, which rapidly increases bubble overpressure and thus the magma exceeds the threshold overpressure criterion. In some cases, upon plug disruption, the volume fraction criterion may be reached as bubbles expand in response to the decompression.

Laboratory experiments have been used to test fragmentation theories regarding the nearly instantaneous decompression of vesiculated magma under conditions appropriate for Vulcanian eruptions, as described above. The magnitude of the pressure drop and the vesicularity were varied in order to define a fragmentation threshold. According to results, the minimum required pressure drop varies linearly with the effective tensile strength of the magma (~ 1 MPa) and is inversely proportional to the vesicularity.

This relationship holds for a wide range of magma compositions, crystallinities, and porosities. An interesting point to note is that for magmas with >20% vesicularity, a sudden decompression of magnitude 5 MPa results in fragmentation, whereas low vesicularity magmas ($\leq 10\%$ vesicularity) may require a drop in excess of 15–30 MPa.

Experiments have also shown that onset of permeability may relieve bubble pressure during propagation of a fragmentation wave (syn-fragmentation) by allowing high-pressure volatiles to escape via connected bubble pathways in magma below the fragmentation front, thus increasing the fragmentation threshold.

The corresponding propagation speed of the fragmentation front, as measured experimentally, generally falls between 2 and 70 m s^{-1} for magmas with 20–60% vesicularity and increases roughly linearly with the magnitude of the sudden decompression. Vulcanian explosions stop when the decompression front reaches unfragmentable magma, which may occur when the front reaches a depth in the conduit where: (1) the magma has insufficient vesicularity; (2) the bubbles are insufficiently overpressured; (3) the magma has sufficiently low viscosity allowing it to respond quickly to high strain rates; or (4) the front has weakened such that the pressure gradient is below the fragmentation threshold.

8. VENT CONDITIONS

The vent flux associated with Vulcanian eruptions is highly impulsive because the eruptions represent the rapid discharge of magma from a pressurized conduit of finite volume. Initially, the flux rapidly accelerates, then may become steady for a relatively short period, and then quickly wanes. When the fragmentation wave reaches unfragmentable magma the conduit is fully depressurized, and the vent flux decays to near zero, although sustained gas exhalations following the main pulse have been documented for several eruptions.

The motion of the fragmented pyroclast/gas mixture at the vent can be calculated by assuming that, upon decompression, available gas in the underlying magma expands to atmospheric pressure as an ideal gas and accelerates the pyroclasts to the same velocity as the gas itself (pseudofluid approximation). The mixture of pyroclasts and gas is assumed to be isothermal when heat is transferred from the clasts to the gas on timescales much shorter than the duration of the explosion. This assumption applies when the average diameter of pyroclasts is $\ll 1$ mm. Solutions for a wide range of initial gas mass fractions (0.01–0.1), initial temperatures (600−1400 K), and initial pressure ratios across the plug (0−100) reveal several trends (some shown in Figure 28.6): vent velocity increases nonlinearly with increasing pressure ratio and increasing volatile mass fraction; mass flux normalized by conduit/vent size increases with increasing pressure ratio due to a rise in vent velocity and decreases with increasing volatile mass fraction due to a decline in mixture density.

The isothermal assumption is not appropriate when a significant portion of the pyroclasts are $> \sim 1$ mm. Mixture velocity decreases and mixture density increases with increasing levels of thermal disequilibrium associated with increasing proportions of large clasts (Figure 28.6). These lower vent velocities may push the system toward gravitational collapse and formation of pyroclastic density currents.

A more complete description of the flow conditions leading to the development of Vulcanian eruptions comes from the multidimensional solution of the nonequilibrium, multiphase flow equations for a mixture of expanding gas and pyroclasts in the atmosphere. On exiting from the vent, the expansion fan radiates as a three-dimensional wave. Analogous to a flow exiting a supersonic nozzle, velocity at the vent cannot exceed the speed of sound in the multiphase mixture and the subsequent expansion in the atmosphere leads to a complex three-dimensional shock wave pattern. Applications to the eruptive conditions at Soufrière Hills volcano have shown that flow at the conduit exit was likely characterized by a stage of (unsteady) sonic to supersonic flow conditions lasting up to about 30 s, with vertical velocities above the vent slightly exceeding the sound speed in an equivalent pseudofluid mixture and higher-than-atmospheric pressure (with pressure ratio as high as 40, Figure 28.7). The initial overpressured jet stage was then followed by a later stage of subsonic flow and decreased mass flow rate, corresponding in time to column collapse and the formation of pyroclastic density currents.

9. ATMOSPHERIC DYNAMICS: SHOCK WAVES

Shock waves propagating ahead of the pyroclastic mixture have been documented during many Vulcanian eruptions, and are referred to as leading shock waves. They have been observed at, for instance, Sakurajima volcano, Japan (Ishihara 1985), and Mt Ngauruhoe in New Zealand (Self et al. 1979). These shock waves result from the initial pressure difference between the high-pressure, gas-rich magma in the conduit and the atmosphere, which, up until the moment of eruption initiation, are separated by the conduit plug. These waves represent a pressure, temperature, and density discontinuity and travel at sonic to supersonic speeds ahead of the pyroclastic mixture. Leading shock waves are sometimes visible because they condense atmospheric water vapor, allowing their velocities to be estimated. The passing of one of these shock waves is marked by an N-shaped pressure variation in time, in which a sharp increase in atmospheric pressure is followed by a dip to a pressure that is less than atmospheric, followed by a return to atmospheric pressure. These N-shaped waves have been documented by stationary pressure sensors at several volcanoes.

For the simple case of adiabatic and inviscid flow of an ideal gas, shock characteristics have been derived in terms of the preexplosion pressure ratio across the plug by solving the conservation of mass, momentum, and energy equations over a control volume that encompasses and travels along with the shock wave. Note that this control volume considers only two pressures that represent a single pressure above and a single pressure below the plug, and does not consider vertical or horizontal variations in pressure nor pressure differences between phases that may occur in a volcanic conduit. Solutions show that, at a given distance, the velocity and amplitude of the wave, or the strength of the shock defined as the ratio of pressure before,

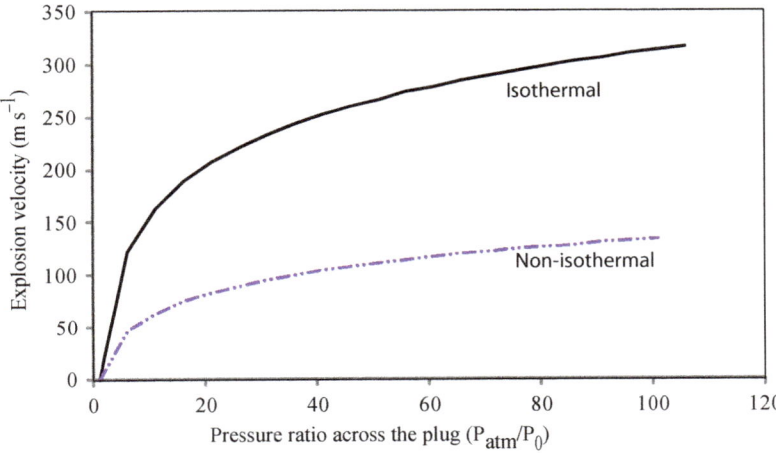

FIGURE 28.6 Calculated vent velocities for a range of initial pressure ratios, using a form of the shock tube equations for isothermal and nonisothermal cases. The nonisothermal solutions assume 0.01 mass fraction volatiles at 1000 K (dashed−dotted line) and account for 3% of particles >1 mm in diameter. The isothermal solutions (solid black line, also for 0.01 mass fraction volatiles and initial temperature of 1000 K) significantly exceed the nonisothermal solutions.

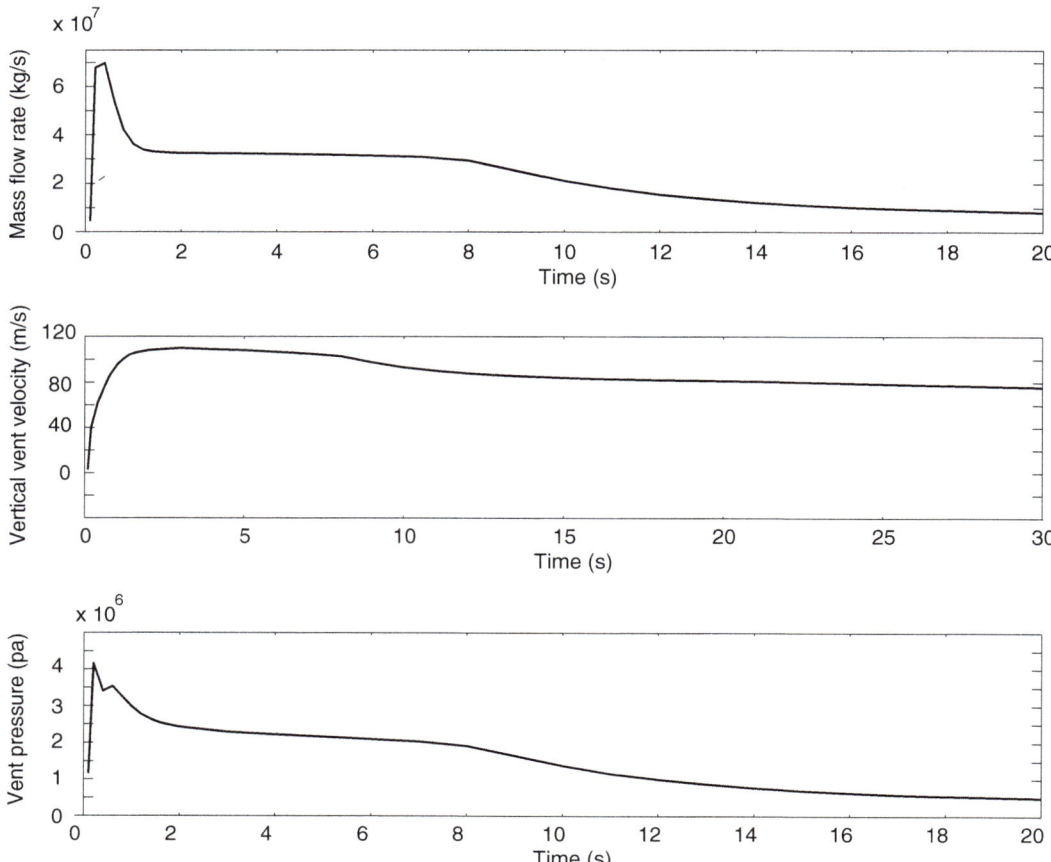

FIGURE 28.7   Vent conditions calculated by multiphase simulations of the 1997 Vulcanian eruptions of Soufrière Hills volcano, Montserrat. Note that the mass flow rate reaches nearly 10^8 kg s^{-1} for less than 2 s, and then declines to a nearly constant rate of 3×10^7 kg s^{-1} for 10 s before tapering to very low values. Mixture velocity above the vent is supersonic for 10 s (using a calculated sound speed for the pyroclastic mixture of 100 m s^{-1}) and remains nearly constant for 10 s of seconds. The vent pressure is in excess of 20 times atmospheric pressure for the first 8 s of the event and remains overpressured for the first 20 s of the event. *Modified from Clarke et al. (2002b).*

to pressure after the wave, decay nonlinearly with decreasing initial pressure ratio across the plug.

Recent experimental and numerical work shows that, in Vulcanian eruptions, the presence of the particulate phase tends to reduce the velocity and strength of shock waves relative to those formed by single gas phases. This trend is in part due to the fact that, generally speaking, the shock wave is formed by the expanding gas alone, and particles hinder gas expansion by reducing the gas volume and hinder gas motion via an interphase drag force. The magnitude of this reduction is independent of particle size, although particle volume fraction is important; shock speed and strength decrease with decreasing gas volume fraction.

Shock waves generated at Sakurajima volcano moved with velocities of 440–500 m s^{-1} according to high-speed video observations; experiments suggest corresponding preexplosion conduit pressures of 1.5–10 MPa, which compares reasonably well with values of 0.2–5 MPa calculated using a numerical approach using recorded shock wave strength.

10. UNDEREXPANDED JET STAGE

Because Vulcanian eruptions initiate via a sudden decompression of a high-pressure bubbly magma, the resulting pyroclastic mixture may enter the atmosphere at supersonic velocities and in an overpressured state. Evidence of higher-than-atmospheric pressure at the conduit exit and of sudden-onset impulsive source is provided by empirical and theoretical analysis of the shape of volcanic jets (stalks with nearly vertical edges, rather than conical stems, and a large vortex front—"mushroom shape"), as well as from infrasonic and acoustic measurements of shock waves. As a consequence, the erupting mixture must equilibrate to the ambient pressure through gas decompression and expansion. The decompression of the gas–particle mixture is a complex nonlinear process, which is affected by the conduit flow and fragmentation dynamics and by the shape of the volcanic crater, among other factors.

To understand the decompression process further, we can apply scaling relationships derived from experiments on unsteady underexpanded jets of gases having thermodynamic properties similar to those of a gas–particle mixture (Orescanin et al., 2010). For initial conduit pressures of 15 MPa (pressure ratio of ~150), fluid properties of dilute gas–particle mixtures, and conduit diameters ~30 m, it is found that Vulcanian jets should equilibrate to atmospheric pressure at a distance of about 240 m from the source. At lower pressure ratios (K ~ 5) the jet decompresses within the first 50 m above the vent. Mixture density decreases during the decompression and expansion stage, however, its value at atmospheric pressure may still exceed that of the ambient fluid. The subsequent transition to a buoyant plume is then controlled by the rate at which air is mixed into the jet.

The combination of field observations from Soufrière Hills volcano, and multiphase flow models for gas–particle mixtures shows that pressure inside the volcanic jets might have adjusted to atmospheric pressure within about 200 m from the vent, but transient vent conditions significantly affected the stability properties of the jet, which collapsed generating radially spreading pyroclastic density currents.

11. PLUME STAGE

Above the jet expansion region, at subsonic velocity, compressibility can largely be neglected and the dynamics of Vulcanian eruptions are dominated by the balance between specific momentum flux $\dot{M} = \frac{\beta}{\alpha}R^2U^2$ and specific buoyancy flux $\dot{B} = g\left(\frac{\beta - \alpha}{\alpha}\right)R^2U$. Here β indicates the eruptive mixture density, α the atmospheric density, g the gravitational acceleration, R the plume radius, and U its average vertical velocity. For stationary turbulent plumes, the Morton length scale $L = \frac{\dot{M}^{3/4}}{\dot{B}^{1/2}}$ characterizes the transition between momentum- (*jet*) and buoyancy-dominated (*plume*) regimes. Experimental and theoretical results predict a buoyancy-dominated regime above a height of about $5L$ from the vent.

For several different unsteady vent conditions, similarity solutions of the fluid dynamics equations are possible for Vulcanian jets and plumes. Depending on the duration of the injection with respect to the plume ascent time, Vulcanian eruptions can either be described as thermals (when the timescale of release is much less than the timescale of flow propagation) or short-lived releases of buoyancy and momentum. Vertical motion of a thermal is controlled entirely by the total buoyancy injected (*B*). Thermals commonly have spherical morphology and resemble a spherical vortex in which flow is nonuniform, with upflow in the center and downflow at the edges. The vertical velocity of a thermal decreases linearly as the height above the source increases, and decreases with the square root of time after the release ($U \sim t^{-1/2}$). Thermals

are low-energy, very small volume, end members of Vulcanian eruptions and have been observed at Fuego in Guatemala, as well as during the 1982 eruption of Sakurajima volcano, Japan (Figure 28.8).

Short-lived jets characterize more energetic Vulcanian eruptions. In such cases, throughout most of the flow, the vertical velocity of the flow front decreases with time raised to the $^{3}/_{4}$ power ($U \sim t^{-3/4}$). This scaling breaks down far from source, where atmospheric stratification plays a dominant role in the ascent dynamics. The well-documented 1975 eruptions of Ngauruhoe appear to have been dominated by momentum forces (Figure 28.8). Still other eruptions are best explained by a short injection of both momentum and buoyancy. These eruptions exhibit more rapid deceleration than either a purely buoyant thermal or a purely momentum-driven unsteady jet ($U \sim t^{-1}$). Examples of this third type include a February 1990 eruption of Lascar volcano in Chile; a July 1980 eruption of Mount St Helens in the Western US; and two eruptions of Soufrière Hills volcano (Figure 28.8).

For steady plume-forming eruptions, collapse and pyroclastic density current formation is favored for large vent radii, low vent velocities, and high solid mass fractions at the vent and for large particles. Examples of Vulcanian plumes that have collapsed and produced pyroclastic density currents include Mount St Helens during the summer of 1980 and Soufrière Hills volcano, Montserrat, in July, August, and September of 1997. The conditions that favor collapse are similar to, although more complex than, those for steady plumes. In particular, the rapidly changing vent conditions and the varied ways in which Vulcanian jets entrain atmospheric air play a significant role, and are not definitively characterized. Unsteady jets and plumes are

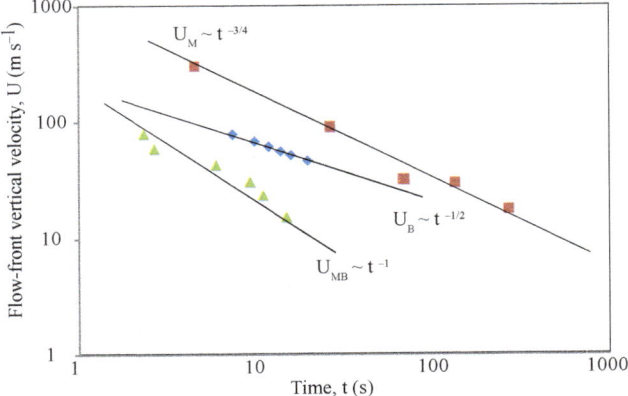

FIGURE 28.8 Vertical flow front velocity vs. time for three different Vulcanian eruptions. The 1975 eruption of Ngauruhoe (New Zealand) was dominated by momentum M as shown by the relationship $U \sim t^{-3/4}$ (squares). The 1982 eruption of Sakurajima (Japan) was dominated by buoyancy B (thermal) as shown by the relationship $U \sim t^{-1/2}$ (diamonds). The August 7, 1997 eruption of Soufrière Hills volcano (Montserrat, British West Indies), was controlled by both momentum M and buoyancy B, as shown by the relationship $U \sim t^{-1}$ triangles. *Modified from Clarke et al. (2009).*

thought to entrain atmospheric air at different rates with respect to their well-documented steady equivalents; entrainment in Vulcanian jets and plumes depends on the temporal evolution of large-scale eddies and may be a function of local flow conditions, which rapidly change in time and space. These complexities are important because column collapse is in part controlled by entrainment, especially in the near-vent region.

Additional complexities exist, causing natural Vulcanian events to vary to some extent from idealized models. Detailed analysis of the 1997 Vulcanian eruptions of Soufrière Hills volcano reveals that explosion initiation involves multiple, individual finger jets that have distinct characteristics and progressively increasing velocities. These observations indicate that there were preexplosive gradients in volatiles in the shallow conduit, and that fragmentation, especially in the initial stages, was heterogeneous in time and space. Corresponding rate of

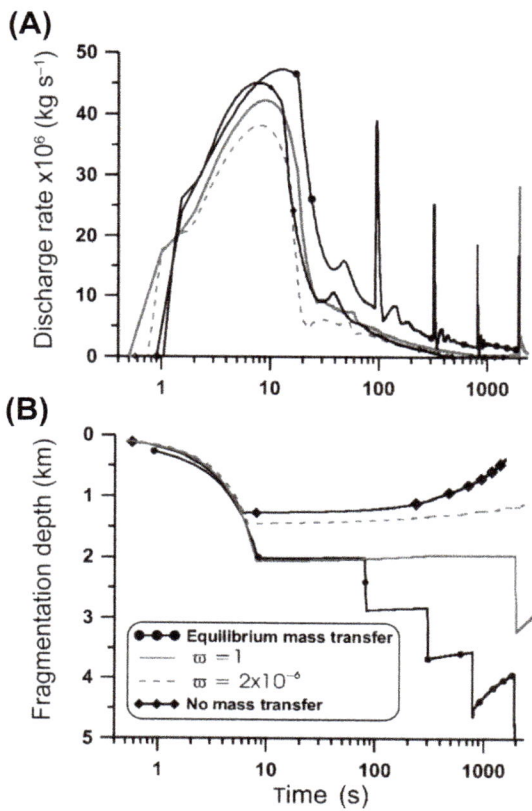

FIGURE 28.10 Calculated variation in discharge rate (A) and fragmentation front position (B) with time using the overpressured bubble fragmentation criterion. The system behaves differently depending on the effective rate of mass transfer of gas from the dissolved phase to the gas phase (rate of gas exsolution). As end members, the cases of no volatile mass transfer (Peclet number much greater than 1) and equilibrium mass transfer (Peclet number much less than 1) are shown, along with two intermediate cases expressed in terms of a gas diffusion factor which scales with gas diffusion rate and bubble number density. Note that the equilibrium volatile mass transfer case favors repeated pulses. *Modified from Melnik and Sparks (2002).*

FIGURE 28.9 Images of the typical morphology of Vulcanian eruptions, as seen at Sakurajima volcano, Japan, (top left, Smithsonian Institution image), Semeru volcano, Indonesia (top right image, photo by Jean-Francois Smekens), and Soufrière Hills volcano, Montserrat (bottom). The image of Soufrière Hills eruption is a hybrid of a photograph by B. Voight and multiphase numerical simulations of the event, the same simulations as represented in Figure 28.7.

entrainment of atmospheric air is very low for individual jets, and this inefficient entrainment is thought to have contributed to collapse and pyroclastic flow formation.

Two-dimensional (axisymmetric) modeling and numerical simulations of Vulcanian explosions at Soufrière Hills volcano (Figure 28.10) have shown that, for the most intense events, initial rapid expansion and the subsequent sudden decrease of mass flow rate generate complex collapse patterns qualitatively and quantitatively consistent with observed plume and PDC morphology.

12. CYCLIC ACTIVITY, TRANSITIONS IN ERUPTION STYLE

Vulcanian eruptions may transform into sustained, quasi-steady explosive eruptions or may end suddenly to be replaced by effusive dome building. Due to their evolved magma compositions, it is reasonable to assume that gas does not diffuse from the melt into bubbles during propagation of the fragmentation front into the conduit. However, when volatile diffusion is fast and efficient, such as for high bubble number densities and high volatile diffusion rates, this assumption becomes invalid. The role of syn-fragmentation gas exsolution can be determined by examining the Peclet number (Pe), which is the ratio of the timescale for gas exsolution to the timescale of fragmentation.

The timescale of exsolution is determined by the rate of diffusion of a volatile in the melt and the length scale in question. This timescale is a function of bubble number density and the diffusion coefficient of gases (mainly water) in the melt. End members are zero gas exsolution and the equilibrium case in which gas exsolution instantaneously responds to decompression. The timescale of fragmentation is a function of the distance between the decompression wave and the fragmentation front and the velocity of the fragmentation front relative to the (possibly) ascending unfragmented magma; this relative velocity is simply the difference between the two velocity vectors. When volatile diffusion is slow or the fragmentation front closely follows the decompression wave (fast fragmentation), the Peclet number is much greater than 1, and syn-fragmentation gas exsolution can be ignored. When volatile diffusion is fast or the fragmentation front lags the decompression wave (slow fragmentation), the Peclet number is much less than 1, and syn-fragmentation gas exsolution must be considered. For example, fragmentation velocity tends to increase when the initial pressure ratio and gas fraction in the conduit increase, while diffusion rate tends to increase when melt viscosity decreases (e.g., diffusion is faster in andesites than in dacites) and when volatile concentrations are higher.

However, because of complex, competing relationships, the expected outcome for a given system cannot be easily predicted. In order to explore the relationships more thoroughly, the system has been represented by conservation equations for a multicomponent magma rising in a one-dimensional conduit, accounting for the difference in pressure between the bubbles and the magma, gas exsolution during fragmentation depending on Peclet number, and crystallization. For these cases, numerical solutions have been used to explore the effects of syn-explosion gas diffusion on eruption characteristics. Model solutions show that no gas exsolution during fragmentation results in a single short-lived explosive pulse, which does not repeat until magma slowly ascends in response to conduit evacuation and prepares for another explosion. The equilibrium case results in repeated pulsatory eruptions (Figure 28.9); greater total volatile content for a given set of conditions pushes the system toward multiple pulses or quasi-steady behavior. The addition of crystals tends to increase the depth to which the fragmentation front reaches for a single pulse because of increased viscosity. However, magmas with low crystal fractions and relatively low viscosities, are more likely to stabilize into a steady-state eruption because the magma will more easily ascend to meet and feed the fragmentation wave.

The choice of fragmentation criterion in magma ascent models does not significantly affect numerical results at the vent. However, strain rate and overpressure criteria are able to produce pulsatory behavior, whereas the volume fraction criterion tends to produce a single pulse. Moreover, maximum calculated fragmentation front velocities exceed 200 m s^{-1} for the first two criteria and are less than 50 m s^{-1} for the volume fraction criterion. Finally, transition from periodic explosions to effusive activity may occur when permeability, rather than fragmentation, develops either throughout the magma or along conduit walls, and therefore suppresses explosivity. A sequence of periodic explosive eruptions at Soufrière Hills volcano in 2003 is thought to have ended in this way via permeable escape of gas along conduit walls.

13. SUMMARY AND FUTURE PERSPECTIVES

A large variety of explosive volcanic phenomena fall under the category of Vulcanian eruptions. Their unpredictable (and hazardous) nature is related to the wide variability in the magma composition, volatile content, degassing and crystallization patterns (in space and time), and different preeruptive distribution of vesicularity and pressure, all of which control magma decompression in the feeding conduit or dike, leading to different textural features in the resulting deposits. However, Vulcanian eruptions are well defined by their eruption mechanism, by means of which we can explain most of the features of their deposits and those commonly observed during eruption. Laboratory experiments and numerical models represent valuable tools for elucidating complex nonlinear processes and identifying their scaling properties. Future research work will need to address, in particular, the problem of nonequilibrium processes related to syn-eruptive degassing and crystallization and the complex rheology of the mixture of melt, crystals, and bubbles involved in the process.

Nonetheless, the conditions sufficient to produce a transition from quiescent or effusive states to Vulcanian events are still extremely difficult to establish and the forecasting of new explosive events, even during periods of persistent activity, is challenging.

New promising multiparameter-monitoring techniques are being developed on active volcanoes. Their development in the next years will likely provide further constraints for theoretical and experimental investigations and for the forecasting of Vulcanian explosions and volcanic risk mitigation.

ACKNOWLEDGMENTS

Much of our understanding of Vulcanian eruptions was gained in the field and through interactions with collaborators. In particular we acknowledge the contributions by members of the Montserrat Volcano Observatory, as well as Tim Druitt, Barry Voight, Augusto Neri, Marina Belousova, and Jeffrey Johnson. This compilation was possible via generous support from the National Science Foundation (USA), the INGV-Pisa (Italy), and the Institute of Volcanology and Seismology (Russia).

FURTHER READING

Alatorre-Ibargüengoitia, M.A., Scheu, B., Dingwell, D.B., 2011. Influence of the fragmentation process on the dynamics of Vulcanian eruptions: An experimental approach. Earth Planet Sci. Lett. 302 (1), 51–59.

Alidibirov, M., Dingwell, D.B., 1996. Magma fragmentation by rapid decompression. Nature 380 (6570), 146–148.

Alidibirov, M.A., 1994. A model for viscous magma fragmentation during volcanic blasts. Bull. Volcanol. 56 (6–7), 459–465.

Belousov, A., Belousova, M., 2001. Eruptive process, effects and deposits of the 1996 and ancient basaltic phreatomagmatic eruptions in Karymskoye lake, Kamchatka, Russia. In: White, J.D., Riggs, N.R. (Eds.), Lacustrine Volcanoclastic Sedimentation, IAS Special Volume, 30, pp. 235–260.

Braitseva, O.A., Melekestsev, I.V., 1991. Eruptive history of Karymsky volcano, Kamchatka, USSR, based on tephra stratigraphy and C14 dating. Bull. Volcanol 53, 195–206.

Cas, R.A.F., Wright, J.V., 1987. Volcanic Successions: Modern and Ancient. Allen & Unwin, London, Boston, Sydney, Wellington xviii + 528.

Chojnicki, K.N., Clarke, A.B., Phillips, J.C., 2006. A shock-tube investigation of the dynamics of gas-particle mixtures: Implications for explosive volcanic eruptions. Geophys. Res. Lett. 33 (15).

Chojnicki, K.N., Clarke, A.B., Adrian, R.J., Phillips, J.C., 2014. The flow structure of jets from transient sources and implications for modeling short-duration explosive volcanic eruptions. Geochem. Geophys. Geosyst 15, 4831–4845. http://dx.doi.org/10.1002/2014GC005471.

Chojnicki, K.N., Clarke, A.B., Phillips, J.C., Adrian, R.J., 2014. Rise dynamics of unsteady laboratory jets with implications for volcanic plumes. Earth and Planet. Sci. Lett. 412, 186–196.

Clarke, A.B., Phillips, J.C., Chojnicki, K.N., 2009. An investigation of Vulcanian eruption dynamics using laboratory analogue experiments and scaling analysis. Stud. Volcanol. 2, 155–166.

Clarke, A.B., Voight, B., Neri, A., Macedonio, G., 2002. Transient dynamics of vulcanian explosions and column collapse. Nature 415 (6874), 897–901.

Clarke, A.B., Neri, A., Macedonio, G., Voight, B., Druitt, T.H., 2002b. Computational modelling of the transient dynamics of the August 1997 Vulcanian explosions at Soufrière Hills volcano, Montserrat: influence of initial conduit conditions on near-vent pyroclastic dispersal. In: Druitt, T.H., Kokelaar, B.P. (Eds.), The eruption of Soufrière Hills Volcano, Montserrat, from 1995 to 1999, Geological Society, 21. Memoir, London, pp. 319–348.

Clarke, A.B., Stephens, S., Teasdale, R., Sparks, R.S.J., Diller, K., 2007. Petrological constraints on the decompression history of magma prior to Vulcanian explosions at the Soufrière Hills volcano. Montserrat. J. Volcanol. Geotherm. Res. 161, 261–274.

de' Michieli Vitturi, M., Neri, A., Esposti Ongaro, T., Lo Savio, S., Boschi, E., 2010. Lagrangian modeling of large volcanic particles: Application to Vulcanian explosions. J. Geophys. Res. 115, B08206. http://dx.doi.org/10.1029/2009JB007111.

Druitt, T.H., Young, S.R., Baptie, B., Bonadonna, C., Calder, E.S., Clarke, A.B., Voight, B., 2002. Episodes of cyclic Vulcanian explosive activity with fountain collapse at Soufrière Hills Volcano, Montserrat. Memoirs-Geol. Soc. London 21, 281–306.

Edmonds, M., Herd, R.A., 2007. A volcanic degassing event at the explosive-effusive transition. Geophys. Res. Lett. 34 (21).

Formenti, Y., Druitt, T.H., Kelfoun, K., 2003. Characterisation of the 1997 Vulcanian explosions of Soufrière Hills Volcano, Montserrat, by video analysis. Bull. Volcanol. 65 (8), 587–605.

Ishihara, K., 1985. Dynamical analysis of volcanic explosion. J. Geodyn. 3 (3), 327–349.

Johnson, J.B., Lees, J.M., 2000. Plugs and chugs—seismic and acoustic observations of degassing explosions at Karymsky, Russia and Sangay, Ecuador. J. Volcanol. Geotherm. Res. 101, 67–82.

Lopez, T., Fee, D., Prata, F., Dehn, J., 2013. Characterization and interpretation of volcanic activity at Karymsky Volcano, Kamchatka, Russia, using observations of infrasound, volcanic emissions, and thermal imagery. Geochem. Geophys. Geosyst. 14, 5106–5127.

Mason, R.M., Starostin, A.B., Melnik, O.E., Sparks, R.S.J., 2006. From Vulcanian explosions to sustained explosive eruptions: the role of diffusive mass transfer in conduit flow dynamics. J. volcanol. geotherm. res. 153 (1), 148–165.

Melnik, O., Sparks, R.S.J., 2002. Modelling of conduit flow dynamics during explosive activity at Soufrière Hills Volcano, Montserrat. In: Druitt, T.H., Kokelaar, B.P. (Eds.), The Eruption of Soufrière Hills Volcano, Montserrat, from 1995 to 1999, Geol. Soc. vol. 21. Mem, London, pp. 307–317.

Morrissey, M., Garces, M., Ishihara, K., Iguchi, M., 2008. Analysis of infrasonic and seismic events related to the 1998 Vulcanian eruption at Sakurajima. J. Volcanol. Geotherm. Res. 175 (3), 315–324.

Morrissey, Mastin, 1999. Vulcanian eruptions. In: Sigurdsson, H., Houghton, B., Rymer, H., Stix, J., McNutt, S. (Eds.), Encyclopedia of Volcanoes. Academic Press.

Mueller, S., Scheu, B., Spieler, O., Dingwell, D.B., 2008. Permeability control on magma fragmentation. Geology 36 (5), 399–402.

Orescanin, M.M., Austin, J.M., Kieffer, S.W., 2010. Unsteady high-pressure flow experiments with applications to explosive volcanic eruptions. J. Geophys. Res. 115 (B6), B06206.

Self, S., Wilson, L., Nairn, I.A., 1979. Vulcanian eruption mechanisms. Nature 277, 440–443.

Siebert, L., Simkin, T., 2002. Volcanoes of the World: An Illustrated Catalog of Holocene Volcanoes and their Eruptions. Global Volcanism Program Digital Information Series GVP-3. Smithsonian Institution.

Spieler, O., Kennedy, B., Kueppers, U., Dingwell, D.B., Scheu, B., Taddeucci, J., 2004b. The fragmentation threshold of pyroclastic rocks. Earth Planet. Sci. Lett. 226 (1–2), 139–148.

Walker, G.P.L., 1973. Explosive volcanic eruptions — a new classification scheme. Geol. Rundsch. 62 (2), 431–446.

Woods, A.W., Sparks, R.S.J., Ritchie, L.J., Batey, J., Gladstone, C., Bursik, M.I., 2002. The explosive decompression of a pressurized volcanic dome: the 26 December 1997 collapse and explosion of Soufrière Hills Volcano, Montserrat. Geol. Soc., London, Memoirs 21 (1), 457–465.

Woods, A.W., 1995. A model of Vulcanian eruptions. Nucl. Eng. Des. 155, 345–357.

Wright, J.V., Smith, A.L., Self, S., 1980. A working terminology of pyroclastic deposits. J. Volcanol. Geotherm. Res. 8, 316–336.

Chapter 29

Plinian and Subplinian Eruptions

Raffaello Cioni and Marco Pistolesi

Dipartimento di Scienze della Terra, via G. La Pira, Firenze, Italy

Mauro Rosi

Dipartimento Protezione Civile, via Vitorchiano, Roma, Italy

Chapter Outline

GLOSSARY

ash Pyroclastic fragments finer than 2 mm. A further distinction can be made between coarse (2 mm–63 μm) and fine (finer than 63 μm) ash.

ballistic clasts Decimetric to metric clasts ejected from the vent, whose size was too large to be entrained into the vertical jet of the eruptive plume. **Ballistic clasts** are ejected with an important horizontal component of velocity.

co-PDC ash Fine ash deposit resulting from the delayed sedimentation of an ash-rich, convective plume formed by the buoyant detachment of a gas–ash mixture from the top of a pyroclastic density current (PDC). In presence of wind, co-PDC ash settles far away from the PDC deposit.

co-plinian ash The coarse and fine ash related to delayed sedimentation from the umbrella region. Well identifiable as a normally graded bed at the top of the proximal and medial plinian fallout deposits from eruptions under no-wind conditions.

eruption intensity (I) An index related to mass discharge rate (MDR, in kg/s) and eruption column height. Calculated as $I = \log_{10} (MDR) - 3$.

eruptive plume A general term indicating the convective region of the eruptive column.

isopach Line connecting points of equal deposit thickness.

isopleth Line connecting points of equal value for particle size, e.g. diameter of maximum lithic clasts.

juvenile clasts Fragments derived from the explosive disruption of the erupting magma.

magma chamber A shallow reservoir in which magma coming from the source region may accumulate and evolve before erupting. Magma chamber shape can be largely variable, from tabular bodies (dyke, sills) to more sub-equant shapes (laccoliths, plutons).

magnitude An index related to the mass (kg) erupted during an eruption. Magnitude (M) scale is calculated as $M = \log_{10}$ (erupted mass) $- 7$.

mass discharge rate Mass eruption rate (kg/s). Mass discharge rate (MDR) scales directly (approximately as the fourth root) to the total height reached by the eruptive column.

phreatomagmatic activity Activity related to the involvement of external water during an eruption, by direct contact with the ascending magma. The more general term hydromagmatic activity is used here to describe the interaction with either groundwater or surface water bodies (sea, lake).

pyroclastic density current (PDC) Gas–pyroclast mixture flowing along the ground and driven by the density contrast (negative buoyancy) with the surrounding fluid (atmosphere, water). PDCs can be generated by various eruptive processes: eruption column collapse, lateral explosions or landslides of hot material, such as a lava dome.

volcanic explosivity index (VEI) A scale for the explosive phases of an eruption, mainly based on the column height or on the mass

The Encyclopedia of Volcanoes. http://dx.doi.org/10.1016/B978-0-12-385938-9.00029-8

ejected as pyroclasts. It relies on the assumption of a direct proportionality between magnitude and intensity of an eruption, and, as first introduced, varied from 0 to 8. The use of a negative VEI has been proposed to describe small-scale, mainly basaltic, eruptions.

wall-rock lithic clasts Clasts derived from the fragmentation of rocks from the nonvolcanic basement (accidental lithics), or comagmatic rocks emplaced during preceding eruptions (accessory or cognate lithics).

1. INTRODUCTION

In modern volcanology, the term "plinian" encompasses explosive eruptions characterized by the quasi-steady, hours-long, high-speed discharge into the atmosphere of a high-temperature, multiphase mixture (gas, solid, and liquid particles), forming a buoyant vertical column that reaches heights of tens of kilometers (Figure 29.1(A)). After having attained its maximum height, the column eventually spreads laterally into an "umbrella" cloud (Figure 29.1(A) and (C)), which maintains its identity for hundreds of kilometers. Conversely, when buoyancy of the erupting mixture is not achieved, the basal part of the column collapses and forms a sustained, ground-hugging cloud of hot gases and pyroclasts, which disperse around the volcano (Figures 29.1(B) and (D)).

The volcanic phenomenon was superbly described for the first time in two letters written by Pliny the Younger to Tacitus to report on his uncle's death during the eruption of Vesuvius (Italy) in AD 79. Pliny's description of the eruptive cloud as a pine tree with a high vertical trunk enlarging into several branches is fully evocative of the actual phenomenon.

I cannot give you a more exact description of its appearance than by comparing to a pine tree; for it shot up to a great height in the form of a tall trunk, which spread out at the top as though into branches. ... Occasionally it was brighter, occasionally darker and spotted, as it was either more or less filled with earth and cinders.

Plinian columns disperse large quantities of highly vesicular pyroclastic material, which settles to the ground over vast areas as a continuous shower. Resulting deposits consist of blankets of coarse-grained, well-sorted pumice in the proximal area, grading downwind to widely dispersed **ash** beds.

Although plinian eruptions by definition include a sustained phase, they typically consist of a complex succession of volcanic pulses. These may include sustained, quasi-steady convective plumes that alternate and overlap with pulsatory explosions of different style, intensity, and dynamics (from vulcanian explosions, to phases of prolonged ash emission, to **phreatomagmatic activity**, to the emission of lava flows or domes). As a result, simple classification schemes merely based on the dynamics and the products of the sustained phase are in some cases inadequate, as they ignore the complex time-evolution of

FIGURE 29.1 Eruption columns of plinian and suplinian eruptions. Mount St. Helens (USA) during (A) and immediately after (B) the 1980 eruption; (C) and (D) 2011 eruption of Puyehue-Córdon Caulle (Chile). Note in (D) contemporaneous convective and collapsing regimes with pyroclastic flow generation. *Photos (A) and (B) are by Richard G. Bowen, courtesy of the Bowen family. Picture (C) is courtesy of NASA.*

the eruption dynamics. From this perspective, a plinian eruption is more realistically described as a succession of different eruptive phases. In the following text, the term *plinian pulse* will be used to specifically address the sustained plinian part of more complex eruptive scenarios. For example, the dominant phase of a *plinian eruption* can consist of a combination of *plinian pulses*, whereas the onset can be represented by *vulcanian pulses*. The term *plinian regime* will be instead used as a more general term, to address the sustained high flow rate discharge of fragmented magma, which may either result in the formation of a convective column or of a collapsing cloud generating a **pyroclastic density current**.

The term *subplinian* was first introduced to describe eruptions of lower scale but dynamics similar to plinian events. The study of subplinian eruptions has revealed that an important distinguishing feature is the occurrence of high-frequency fluctuations or of temporary breaks in the discharge, with the repeated generation of short-lived convective plumes often alternated with phases of quiescence or of lower intensity, explosive or effusive activity.

All the eruptive events sharing a plinian regime have been often grouped under the general term of "plinian eruptions," which encompasses subplinian, plinian, and ultraplinian styles.

2. CLASSIFICATION ISSUES AND GENERAL CHARACTERISTICS

Explosive eruptions can be classified according to different criteria. Recent events at monitored volcanoes are classified by using ground-based geophysical data and satellite remote-sensing observations, as well as features of the tephra deposits. In contrast, past eruptions are classified solely on the basis of deposit features. These two different classification approaches have been analyzed and discussed by many authors with the aim of establishing a coherent classificatory grid (Walker, 1973; Pyle, 1989). However, the results of the two classification approaches are not fully tested for high intensity eruptions (with eruptive columns higher than 30 km), mainly due to the scarcity of observed cases. For this reason, it is still common practice to use deposit characteristics to classify high-intensity eruptions.

Volcanic Explosivity Index (VEI) values in the range of 4−6 characterize subplinian and plinian eruptions. Recurrence rates of about 10 plinian (VEI 5−6) and 30−40 subplinian (VEI 4) events per century strongly contrast with the very low recurrence rate of ultraplinian eruptions (VEI 7, less than 1 per 1000 years).

2.1. Plinian Eruptions

Plinian and ultraplinian eruptions share common eruption dynamics, but different values of **eruption intensity** and

magnitude (10−12, and 4−8, respectively, according the scale of Pyle, 2000).

During plinian pulses, a mixture of gas and particles (juvenile, cognate, and lithic material) is discharged from a vent at high speed (typically 150−600 m/s). Variations in the discharge or in parameters controlling the eruption dynamics (e.g. magma composition, volatile content, vent and conduit geometry) generally occur over a time scale longer than the characteristic times of the different processes, which dominate magma ascent, magma fragmentation, and plume development. This results in a sustained, quasi-steady eruption column.

In the gas-thrust region, the eruptive jet incorporates air and eventually gains a positive buoyancy with respect to the ambient atmosphere (Figures 29.1(A)−(C)), shifting to a convective regime. Above the gas-thrust, the buoyant column rises in the atmosphere and reaches neutral buoyancy, then starts to propagate laterally as a radially spreading density current, advected by winds. Powerful eruption columns typically penetrate the local tropopause and spread into the stratosphere (Figure 29.2).

The volume of ejected material produced by individual plinian pulses ranges typically between 0.1 and 10 km^3, with peak **Mass Discharge Rates** (MDR) between 10^6-10^8 kg/s (Table 29.1). Due to the very high discharge rate, the duration of plinian pulses generally ranges between a few hours (3 h for the 1815 Tambora eruption, Sunda arc, Indonesia) to days (about 48 h for the 1991 Cerro Hudson eruption, Chile). Many plinian eruptions consist of repeated plinian pulses (at least three for the 1982 El Chichón eruption, Mexico, and for the 1912 Katmai eruption, Alaska), each separated by hiatuses of hours. Although considered quasi-steady, many plinian pulses show a progressive increase of MDR, which is recorded as inverse grading (coarsening upward) of the resulting fallout deposits.

In plinian and ultraplinian eruptions, interplay of the source parameters (exit velocity, MDR, magma gas content, total grain-size distribution, conduit geometry) has a fundamental role in determining the fate of the eruption column (Figure 29.3). Within the typical range of MDRs for the plinian regime, sustained, buoyant plumes may rapidly shift to collapsing behavior and generation of PDCs (Figure 29.1(D)). Consequently, fall and PDC deposits are commonly interbedded in plinian sequences. PDC deposits may represent only a small fraction of total erupted volume (<20% for the AD 79 Vesuvius eruption) or, conversely, the larger part (40% for 1912 Katmai eruption, to 60% for the Minoan eruption of Santorini, to 95% estimated for the 1815 Tambora eruption). The transition from sustained to collapsing conditions is generally preceded by a phase of column unsteadiness (Figure 29.3), during which partial collapse of the column produces pulsating, dilute PDCs (transitional regime; Neri et al., 2002).

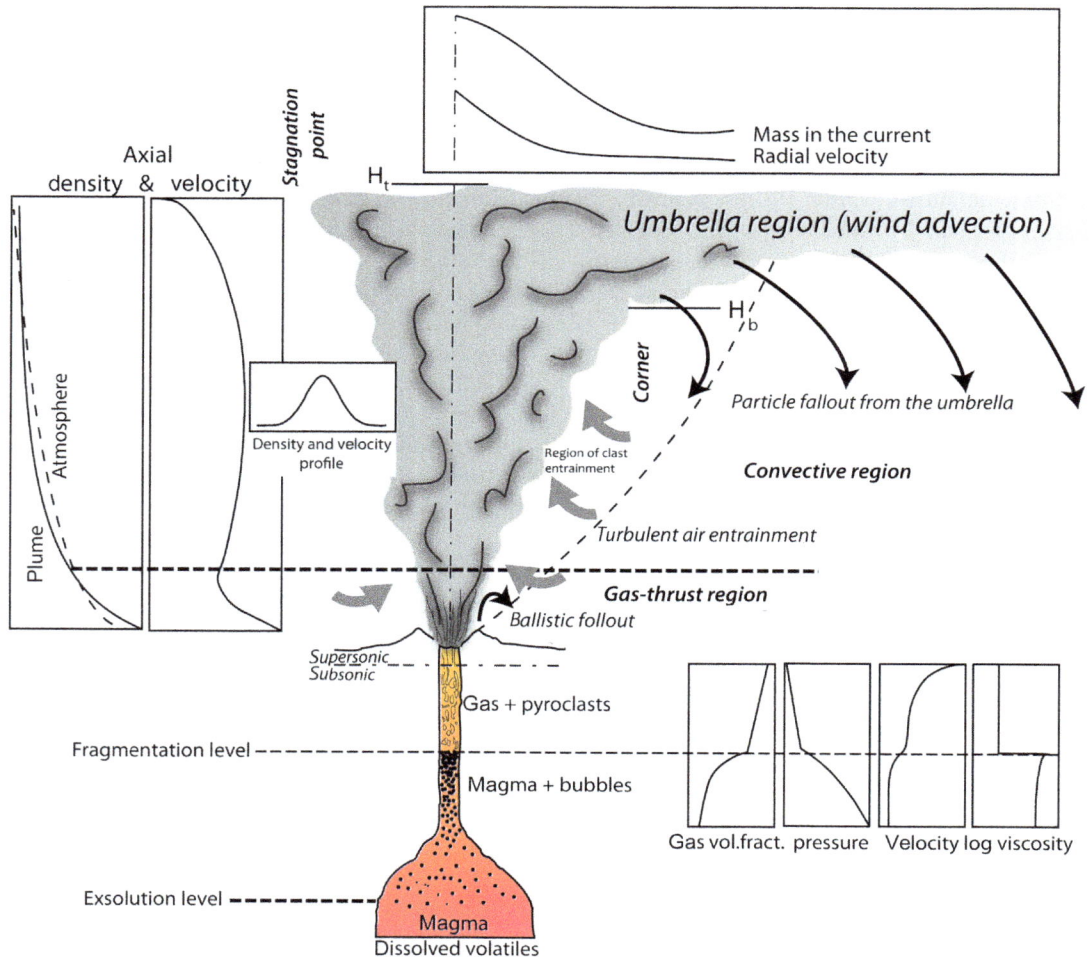

FIGURE 29.2 General scheme of eruptive regimes for a strong plume, and variation of physical parameters during plinian eruptions.

In a number of well-studied eruptions, generally described as *ignimbrite eruptions* (AD 1270 Quilotoa, Ecuador; Bronze Age Minoan, Santorini, Greece), the passage from sustained convective to collapsing column appears to be driven by progressive increase of the MDR (Figure 29.3), and marked by a gradual increase of the flow/fall ratio in the deposit sequence.

Ultraplinian eruptions differ from classical plinian events only by their higher mass flow rate, which reflects in higher columns and larger dispersal (dispersive power). The exceptionally high intensity of the ultraplinian regime, and the very large amount of fine ash generated during magma fragmentation, results in convective columns up to 55 km high (the maximum theoretical height for a convective column maintaining plume stability), which generally evolve into a collapsing column phase with the generation of high-mobility PDCs. Walker (1980) first introduced the term ultraplinian eruption, describing the fallout deposits of the AD 186 Taupo eruption

(New Zealand). At present, the initial fallout phase of the 39 ka Campanian Ignimbrite eruption (Italy) is the only other described example of an ultraplinian event, suggesting that the extremely high intensity that characterizes this eruptive style could be related to the emplacement of low-aspect-ratio ignimbrites. In addition, a recent detailed study of the fallout deposits of the Taupo eruption suggests that they are better described as the superposition of the products of several powerful plinian pulses (Houghton et al., 2014).

2.2. Subplinian Eruptions

Subplinian eruptions have lower values of magnitude ($M = 4$) and intensity ($I = 10$) with respect to plinian events. Subplinian eruptions generally consist of unsteady events characterized by phases of short-period oscillations (minutes) with time breaks that can repeat several times over longer periods (days, weeks). These dynamics result in

TABLE 29.1 Main Eruptive Parameters for Plinian and Subplinian Eruptions (H_T = Column Height; MDR = Mass Discharge Rate)

Eruption	Country	Date	Composition	H_T (km)	MDR (kg/s)	Mass (kg)
El Chichon A[1]	Mexico	1982	Trachyand.	27	8.0E+07	7.5E+11
El Chichon B[1]	Mexico	1982	Trachyand.	32	1.5E+08	1.3E+12
El Chichon C[1]	Mexico	1982	Trachyand.	29	8.5E+07	1.0E+12
Santa Maria[2]	Guatemala	1902	Dacite	34	1.7E+08	2.2E+13
Mt St Helens[3]	United States	1980	Dacite	19	1.9E+07	7.1E+11
Katmai[4]	United States	1912	Rhy./Dac.	32	1.7E+08	2.5E+13
Askja[5]	Iceland	1975	Rhyolite	26	7.9E+07	8.9E+11
Vesuvius[6]	Italy	AD79	Phonolite	32	1.5E+08	6.1E+12
Santorini[7]	Greece	BC1470	Rhyolite	36	2.5E+08	3.3E+13
Tarawera[8]	New Zealand	1886	Basalt	34	1.8E+08	3.3E+13
Hatepe[9]	New Zealand	BP1820	Rhyolite	33	1.8E+08	3.8E+12
Taupo[10]	New Zealand	BP1820	Rhyolite	51	1.1E+09	7.7E+13
Tambora[11]	Indonesia	1815	Trachyand.	43	2.0E+09	2.6E+13
Etna[12]	Italy	BC122	Basalt	26	8.5E+07	9.0E+11
Hudson[13]	Chile	1991	Andesite	18	7.0E+07	2.7E+12
Krakatoa[14]	Indonesia	1883	Dacite	≈40	5.5E+08	2.9E+13
Redoubt[15]	Alaska	1989	And./Dacite	9	7.0E+06	7.5E+10
Hekla[16]	Iceland	1947	Andesite	28	4.6E+06	1.5E+12
Chaitén[17]	Chile	2008	Rhyolite	20	4.2E+07	9.6E+11
Cordon Caulle[18]	Chile	2011	Rhyodacite	14	1.5E+07	1.1E+12

[1]Carey and Sigurdsson, 1986.
[2]Williams and Self, 1983.
[3]Carey and Sigurdsson, 1985.
[4]Fierstein and Hildreth, 1986.
[5]Sparks et al., 1981.
[6]Sigurdsson et al., 1985.
[7]Watkins et al., 1978
[8]Walker et al., 1984.
[9]Walker, 1981.
[10]Walker, 1980.
[11]Kandlbauera and Sparks, 2014.
[12]Coltelli et al., 1998.
[13]Scasso et al., 1994; Naranjo et al., 1993.
[14]Self, 1992.
[15]Miller and Chouet, 1994; Scott and McGimsey, 1994.
[16]Thorarinsson, 1949; Thorarinsson and Sigvaldason, 1972.
[17]Alfano et al., 2011.
[18]Bonadonna et al. (2014).
References in supplementary material. Please see companion site, http://booksite.elsevier.com/9780123859389/

the formation of fairly short-lived convective pulses, frequently followed by the generation of small-volume PDCs. In some cases, during time breaks between subplinian pulses, episodes of lava effusion can occur. Unsteadiness (destabilization of the convective column and occurrence of time breaks) is possibly related to the decline of ejection velocity and decoupling between magma supply at depth and magma discharge at the surface (Scandone and Malone, 1985), more evident at the lower discharge rates (10^6–10^7 kg/s), typical of these eruptions. Duration of individual subplinian pulses generally varies from minutes to hours, and higher frequency fluctuations in the discharge are recorded in the resulting products as thin, alternating beds of coarser to finer-grained fallout deposits. Subplinian convective plumes generally do not cross the tropopause; in many cases the vertical velocity of the convective column is

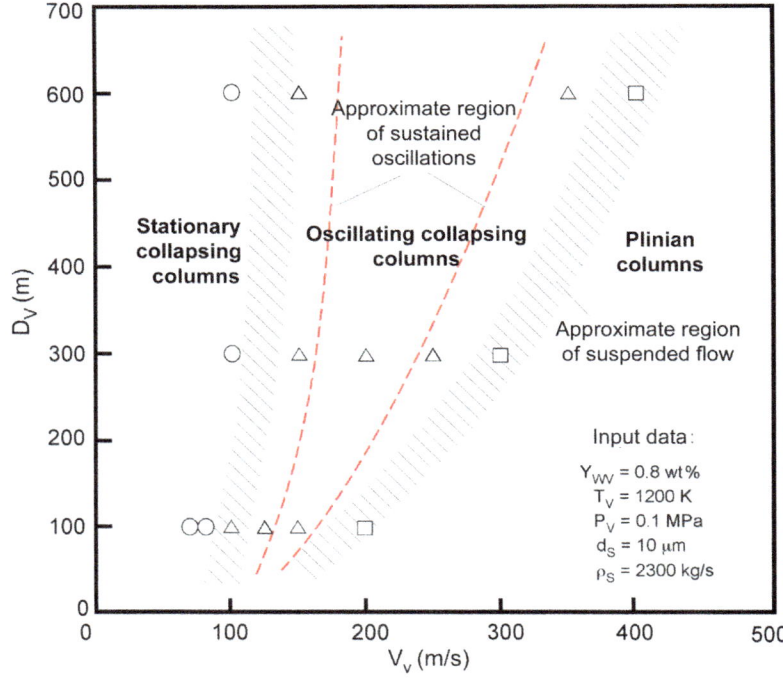

FIGURE 29.3 Relationship between the vent diameter and vent exit velocity showing different regions of volcanic columns. The cross-hatched regions identify the transition among stationary collapsing columns, oscillating columns, and plinian columns. Y_{WV} = water content, T_V = temperature, P_V = pressure, d_S = particle size, ρ_S = density. *Redrawn after Neri and Dobran (1994). References in supplementary material. Please see companion site, http://booksite.elsevier.com/9780123859389/*

moderate (weak plumes), and the plume can be noticeably bent-over in the presence of strong winds. At least five subplinian eruptions occurred at Mt St Helens in the six months after the 18 May 1980 climactic eruption; several subplinian events occurred in a period of months at Redoubt (1989–1990) and at Spurr (1992) volcanoes (Alaska). This activity was characterized by generation of sustained (although in some cases short-lived) plumes and PDCs by (partial) column collapse. In a review of past activity of Somma-Vesuvius (Italy), Cioni et al. (2008) discriminated between two different types of subplinian eruptions on the basis of the presence or absence of associated PDC deposits. In many cases, subplinian activity is gradational to violent strombolian activity in terms of characteristics of the eruptive cloud and related fallout deposits. Vulcanian explosions generating short-lived, convective ash plumes that rapidly detach from the eruptive source and are laterally advected by winds have been sometimes erroneously described as subplinian pulses.

3. THE PLUMBING SYSTEM OF PLINIAN AND SUBPLINIAN ERUPTIONS

The composition of magma driving most of plinian and subplinian eruptions is intermediate to silicic. From a compilation of 45 Pleistocene and Holocene plinian eruptions (Carey and Sigurdsson, 1989), about 80% have evolved composition (SiO_2 higher than 60 wt%). Only few

basaltic plinian eruptions are known (Tarawera 1886, New Zealand; Etna 122 BC, Italy; Fontana Lapilli eruption, Masaya, Nicaragua; Figure 29.4(A)), while trachyandesitic to andesitic eruptions are more frequent (e.g. Cotopaxi, Ecuador, or Ruapehu, New Zealand). Conversely, subplinian eruptions are generally linked to intermediate magma compositions.

Magma chamber features (size and depth), as well as magma composition, may have strong influence on the dynamics of plinian eruptions. Uniform composition or vertical zoning of plinian deposits suggest that magma feeding the eruption may be tapped from homogeneous, possibly large, regional magma reservoirs (plutons), or from volumetrically smaller, compositionally stratified magma chambers. Compositionally zoned plinian deposits generally record a reversal of the magma chamber stratification, with the topmost, most evolved region of the reservoir being erupted first (Figure 29.4(B)). Despite this, the observed compositional gradient is often the result of a combination of factors such as the sequence of magma extraction from the reservoir and the occurrence of syn-eruptive mixing of different parcels of magma within the magma chamber and along the conduit. The dynamics of magma tapping and ascent, the internal gradients of the physical properties of the magma, as well as the shape of the magma chamber, control the syn-eruptive magma mixing process.

Efficient syn-eruptive magma mixing can also occur inside the magma chamber, in the course of the eruption,

FIGURE 29.4 (A) Deposit of the basaltic plinian Fontana lapilli eruption, Masaya (Nicaragua). (B) Compositionally zoned deposit of the Bronze Age eruption of Avellino Pumice (Vesuvius, Italy). *(A) is courtesy of C. Bonadonna.*

triggered by partial emptying and local decompression of the reservoir, or by the catastrophic collapse of the magma chamber walls during caldera collapse. Physically mingled, often banded pumice clasts carrying portions of contrasting magma compositions record magma mingling that takes place shortly before the eruption. Compositional magma hybridization requires more time, and can eventually be observed in the final products of the eruption. The sudden change of magma composition during the course of an eruption can result in important changes of eruption dynamics, due to the link between magma composition, type/amount of dissolved volatiles, and the rheological and physical properties of the erupting magma. During the AD 79 Pompeii eruption of Vesuvius, the shift from phonolitic to tephriphonolitic magma composition (marked in the deposits by the passage from white to grey pumice) triggered the first partial collapse of the convective column, and favored the progressive increase of MDR mainly due to the important decrease of magma viscosity.

At several volcanoes (Novarupta, Mt St Helens, Pinatubo), the presence of deep volcano-tectonic seismicity at the end of a plinian eruption has been related to important changes of the local stress field due to the rapid removal of some cubic kilometers of magma from beneath the volcano. Such enhanced, deep fracturing possibly triggers the ascent of more mafic magma from depth, which may contribute to the final phases of degassing and eventually mixes with the unerupted magma in the reservoir, determining the ejection of a more mafic magma during the late stages of the eruption.

4. MAGMA ASCENT AND FRAGMENTATION

The depth of the magma reservoir, as estimated by petrological or geophysical methods, is used to infer the length of the conduit. If the local stratigraphy is known, then the type of accessory lithic fragments in the deposits represents

an effective way to infer the lateral dimension and length of the conduit. Shea et al. (2011) used the mass of total ejected lithics to evaluate the diameter of an assumed cylindrical conduit during the AD 79 Vesuvius eruption, finding values from few meters at the beginning of the eruption to 65 m by the end of the plinian phase (corresponding to an average enlargement of the conduit radius of 1.5—2 cm/min over a conduit length of 6 km). Carey and Sigurdsson (1989) suggested a slightly higher rate of enlargement of conduit radius (around 5 cm/min) through a general analysis of 45 past plinian eruptions, in the assumption that the commonly observed increase of MDR with time is related to erosion of conduit walls. In the case of subplinian eruptions, available numerical models suggest that smaller conduit diameters (around 10—15 m) are needed.

Magma ascent during plinian eruptions can be approximated to a quasi-steady process, sustained by the pressure difference between the magma chamber and the surface, and mainly dominated by decompressional degassing. Magma degassing generally occurs under closed conditions, as a consequence of the mechanical coupling between growing vesicles and the high-viscosity magma (bubbly flow regime). However, degassing may pass from closed to open conditions if magma reaches an internal high permeability before fragmentation, and if the characteristic transit time in the conduit is long enough to allow gas release through the magma column. This process is likely to occur on a large scale in the final stages of those plinian eruptions characterized by a decrease in MDR, allowing a massive gas release, which can be responsible for long-lasting phases of ash emission and/or extrusion of viscous lava flows or lava domes. During ascent, magma density (Figure 29.2) steadily decreases as a result of decompression-driven gas exsolution and vesicle expansion; such a process contributes to the acceleration of the magmatic foam in the conduit. In contrast, magma bulk viscosity increases several orders of magnitude until the fragmentation level is reached (Figure 29.2). Available numerical models of conduit flow provide, under certain assumptions, estimates of the continuous variation of the main parameters governing magma ascent.

High magma ascent velocities are needed to maintain the high MDR of plinian eruptions. Lateral gradients of the physical and rheological parameters in the magma likely accompany the rapid acceleration of the vesiculating magma in the conduit. Given the generally high viscosity of plinian magmas, the development of boundary layers along the conduit walls is important in reducing drag effects during magma ascent. In crystal-poor magmas, shearing along the walls results in stretching and alignment of vesicles, locally reducing the viscosity of the magma. This effect is clearly recorded in a large proportion of tube (woody) pumice in the deposits of these eruptions. In crystal-rich magmas, the high friction along the conduit walls results in crystal milling and temperature increases by viscous heating, with a consequent local decrease of the viscosity. This may promote partial crystal resorption and heterogeneities in glass composition, sometimes observed in mingled pumice clasts of large plinian eruptions (Pinatubo 1991, Philippines; Quilotoa AD 1220, Ecuador).

The decompression rate strongly controls final magma vesicularity (in terms of size and number of vesicles), and hence the style of magmatic foam fragmentation. For a conduit length of 5−8 km (the common depth range of crustal magma chambers), and MDR typical of a plinian regime, ascent times of the orders of minutes are estimated using the available numerical models, resulting in average decompression rates in the order of 1 MPa/s, and in peak decompression rates up to 10−100 MPa/s (Shea et al., 2011). Mechanisms of magma fragmentation in plinian eruptions are classically related to the reaching of a critical concentration of bubbles in the magma (about 75 vol%), or to processes of strain-rate dependent brittle fragmentation. Both processes are possible and not mutually exclusive. At high vesicle concentrations, decompressional and diffusional growth of bubbles is in part hampered by the increase of viscosity of the residual, volatile-poor liquid. Rupture of the liquid films separating gas bubbles can also occur following a progressive increase of the differential pressure, which may develop during final expansion of bubbles until the failure strength of the surrounding melt (*stress criterion*). Analog experiments have shown that explosive disintegration of the erupting foam can occur in coincidence with the highest acceleration rates in the conduit. If the timescale for deformation is shorter than the timescale for viscous relaxation of the liquid (the time needed for the liquid to keep pace with an applied strain), the bubble−liquid mixture crosses the glass transition, behaving in a brittle fashion and disintegrating (*strain criterion*). The mechanisms of fragmentation and the magma permeability play an important role in the final grain-size distribution of pyroclasts. Permeability (through bubble coalescence) will favor gas escape and consequent

formation of lapilli versus ash. Delayed bubble nucleation and expansion related to disequilibrium degassing may result conversely in a very fine-grained disintegration of the magma, due to the explosion of a highly vesicular melt, characterized by a high number density of unconnected small vesicles. Comminution of fragmented magma clasts may continue during ascent in the conduit by two additional processes: (1) high-velocity particle−particle collisions, as clearly demonstrated by the experimental and theoretical analysis of Dufek et al. (2012); (2) disruption of the vesicles trapped in the glass due to the rapid decompressional gas expansion. As a consequence of these two processes, the depth of magma fragmentation may also exert an important control on the final size of the clasts.

Fragmentation acting during subplinian events reflects the unsteady character of these eruptions. The alternation of sustained columns with periods of quiescence has been explained in terms of syn-eruptive degassing and viscosity increase of the magma column during the slow ascent in the conduit. The partially degassed, high-viscosity portion of the magma crystallizes rapidly, causing magma stagnation in the upper portions of the conduit (in some cases culminating in a dome extrusion), and a progressive pressure buildup that eventually reopens the system. This process repeatedly occurred, for example at Mt St Helens, following the climactic phase of the 18 May 1980 eruption. The discontinuous nature of several subplinian eruptions is likely to be regulated by differences between the rate of magma discharge at the surface (MDR) and the rate of magma supply from the magma chamber (Magma Supply Rate, MSR). If MDR is greater than MSR, the fragmentation surface migrates downward in the conduit, until fragmentation ceases and eruption stops (Scandone and Malone, 1985). Conditions that may favor a decrease of MSR are high viscosity, increase of magma during syn-eruptive degassing, and groundmass crystallization, as well as decrease of the conduit diameter by lining of crystallizing magma. During plinian eruptions, a fully connected conduit between the reservoir and the surface may develop, allowing the sustained discharge of magma for a prolonged time, whereas during subplinian events this possibility could be hampered. Scandone et al. (2007) suggested that mid-intensity, discontinuous eruptions like subplinian events could be related to the periodic release of distinct pulses of evolved magma ("magma quanta") from a reservoir through a network of opening−closing fractures, implying an average low MSR and resulting in distinct, short phases of high discharge.

Vent geometry is fundamental in modulating the exit conditions of the eruptive mixture during high-intensity explosive eruptions. Acceleration to ultrasonic velocities and equilibration to the ambient pressure of the erupting mixture occur inside the crater and are preconditions to the formation of a vertically directed eruptive jet.

The modalities of decompression contribute to shape the crater, by erosion and land sliding of the steep, often incoherent walls, creating a feedback process that favors early pressure equilibration and sonic/supersonic transitions of the erupting mixture. Phases of vent widening are often recorded in the plinian fallout deposits as lithic-enriched beds dominated by shallow-seated wall rock fragments.

The tapping of large amounts of gas-rich magma during plinian eruptions may eventually trigger the collapse of the roof of the magma reservoir and the formation of a caldera. When caldera collapse occurs during the eruption, it modifies eruption dynamics, recompressing the residual magma in the reservoir, and causing the massive input of the host rock blocks and the contact with the magma of external fluids, favoring the onset of phreatomagmatic activity. A concurrent transition from central to fissural (ring-fault) activity may also occur. The massive incorporation of lithic fragments in the eruptive mixture during caldera collapse may in turn promote the collapse of the **eruptive plume** and the formation of PDCs.

5. ATMOSPHERIC DYNAMICS

The general physics of plinian column behavior is captured in buoyant plume theory; in no-wind or moderate wind conditions the column is roughly axisymmetric and parameters such as density and velocity of the erupting mixture are maximum along the central axis; lateral decay of these parameters can be approximated by a Gaussian function (Figure 29.2). Density and velocity also show a general upward decrease. Velocity decrease may or may not be monotonous; plumes generated by high-MDR eruptions, for example, may show significant acceleration at the passage between the gas thrust and the convective region (*superbuoyant columns*), due to efficient incorporation and heating of ambient air, or from release of latent heat from water condensation. Average vertical velocity in the ascending column controls the transport capacity of the plume as well as the distribution of the transported pyroclasts (in terms of size and density). Carey and Sparks (1986) proposed a model of the velocity field in the eruptive column in which the maximum size of the clasts that can be transported at different elevations in the convective column or in the umbrella region is related to the mass flow rate of the eruption. For this reason, the maximum size of pyroclasts found in the mid-distal fallout deposits decreases from plinian to subplinian eruptions. The diameter of the maximum lithic clasts dispersed by the umbrella region, and hence expected to be found in the deposits beyond the column corner distance (Figure 29.2) is in the order of 4–5 cm for plinian phases (for values of MDR higher on average than 10^7 kg/s), and not larger than 1–2 cm for large subplinian events (MDR between 10^6–10^7 kg/s).

The physics of the umbrella region in plinian eruptions is dominated by a centrifugal lateral intrusion into the atmosphere. Wind plays an important role by producing an elongated plume shape (and resulting fall deposits) in the downwind direction. The occurrence of westerly, large-scale meandering jet streams immediately under the tropopause at latitudes between 30° and 60° in both the hemispheres (at an altitude variable of 6–10 km and 10–16 km for the polar and the subtropical jet streams, respectively) often results in a strong lateral advection of the topmost part of the ascending plumes. In some cases, these strong winds shear the top of volcanic plumes, producing elongate tephra deposits. Due to their lower flux rates, subplinian plumes are more influenced by lateral winds in the upper troposphere, as shown by long, narrow, often asymmetric deposit dispersal.

6. PRODUCTS—DENSITY AND COMPONENTRY

During plinian eruptions, a mixture of finely fragmented multiphase magma, solid fragments of preexisting rocks and gas is injected in the atmosphere. Magma fragments (juveniles; Figure 29.5) mainly consist of highly vesicular pumice clasts and ash fragments (pumice and glass shards), along with minor amounts of variably vesicular to dense, comagmatic clasts (dense magma fragments and magma chamber-derived intrusive rocks).

Juvenile fragments forming plinian deposits generally show a narrow range in vesicularity and small textural variability (clast morphology, vesicle shape, crystal content), which contrast with the wider variability often observed in subplinian products. Bulk vesicularity of pumice clasts from plinian deposits ranges on average between 65 and 85 vol%, and it may vary along the eruption sequence due to variations in MDR and eruption dynamics. Conversely, bulk vesicularity of subplinian juvenile fragments may range from 10 up to 80 vol% within a single eruption, with multiple frequency modes between 40 and 75 vol%. The coexistence of variably vesicular fragments within the same deposit can be related to the presence of large physical gradients in the ascending magma as compared with plinian eruptions. For example, the lower MDR of subplinian eruptions likely involves narrower conduits, implicating a major role of boundary effects at the margins of the conduit, leading to lateral vesicularity gradients in magma. The simultaneous expulsion of variably vesicular magma fragments can also be related to the pulsatory nature of subplinian eruptions; rapid decompression of discrete parcels of volatile-rich magma can trigger the fragmentation of the degassed magma still residing in the conduit.

Wall-rock lithics are more abundant in plinian than in subplinian eruptions, reflecting a more effective erosion of

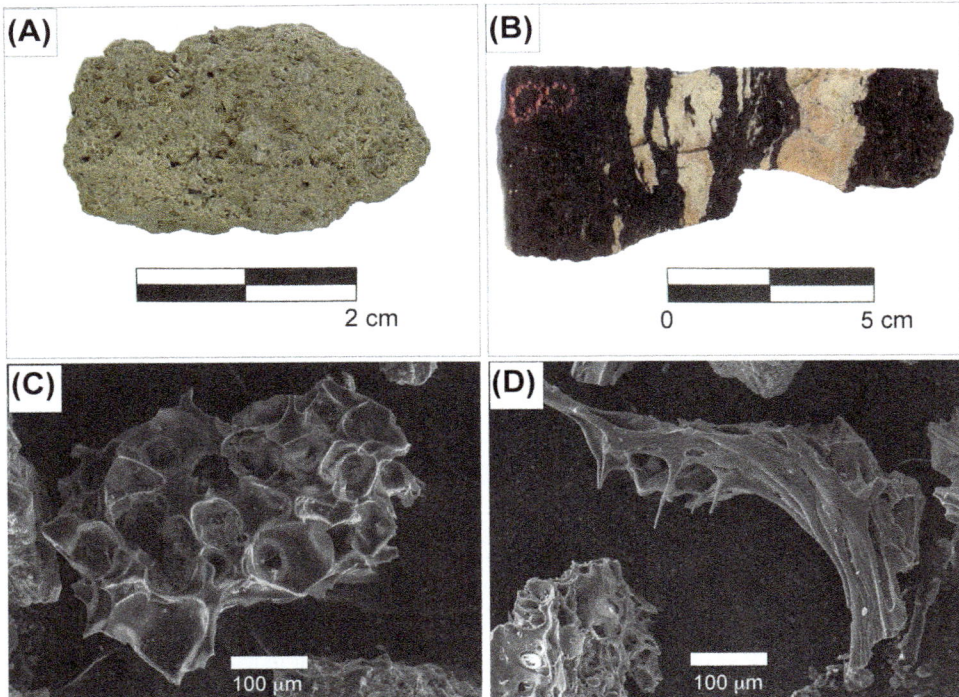

FIGURE 29.5   Variability of the juvenile component of plinian products. (A) Pumice clast. (B) Rhyolite-andesite banded pumice (1912 Novarupta-Katmai eruption, Alaska). (C) Micropumice fragment. (D) Glass shard.

the conduit/crater system. Variations in the composition and relative abundance of lithic fragments during the eruption may also provide information on the evolution, collapse, and clearing of the conduit/crater. The study of magma chamber-derived lithic fragments (such as comagmatic, salic-intrusive, mafic cumulites, thermometamorphic, or deep-hydrothermally altered rocks) may also provide clues on the position of magma chamber and hydrothermal systems. Strong increases of fragments of rocks hosting regional aquifers can also suggest phreatomagmatic activity.

7. DEPOSITS

Due to the complexity of the eruptive regimes and occurrence of multiple eruptive pulses, plinian eruption deposits typically exhibit variability in terms of types, products, dispersal and sedimentological features. A first main division for the description of the deposits can be conveniently done between fallout and flow deposits, although they can often be interlayered.

7.1. Fall Deposits

Fall deposits result from the settling on the ground of pyroclasts from the eruptive plume or the volcanic cloud, forming sheet-like layers dispersed over large areas. Due to their wide dispersal, plinian deposits represent excellent,

isochronous marker beds for stratigraphic correlations; in addition, their study can be successfully used for the reconstruction of the eruptive dynamics and assessment of eruptive parameters like volume, MDR, column height, etc. **Isopach** and **isopleth** maps (Figure 29.6), as well as the study of the erupted products in terms of their grain-size, componentry, density, compositional and textural features of the **juvenile clasts**, represent fundamental steps in order to describe and quantify the eruption dynamics (Figure 29.7).

Plume dynamics and wind field at the time of eruption control tephra dispersal during plinian events (Figure 29.6(A) and (B)). Asymmetric dispersal of fallout deposits (or multilobed deposits) can be the result of variable wind directions at different altitudes during a prolonged sustained eruption characterized by a progressive change in MDR. Asymmetry in the dispersal fan can also be related to the juxtaposition of different lobes with variable dispersal, as a result of a change in wind directions during different plinian pulses. These lobes generally grade one into the other at their margins, and are often difficult to separate.

Pumice fragments are likely to break at impact especially at proximal sites, where they are coarser and often fractured by cooling contraction before landing. In contrast, lithic clasts do not break upon impact and their size in the deposit truly reflects dispersal and sedimentation processes. In distal,

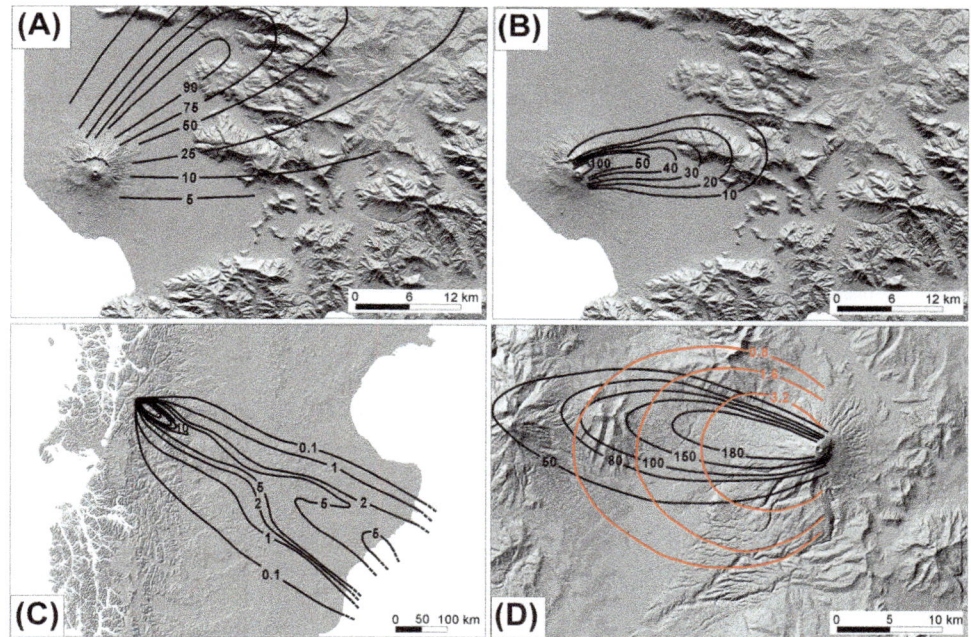

FIGURE 29.6 Isopach maps of (A) Avellino Pumice (Sulpizio et al., 2010) and (B) 1631 (Rosi et al., 1993) eruptions of Vesuvius, Italy. (C) Isopach map of the 1991 Hudson eruption, Chile (Scasso et al., 1994). Note secondary maxima in the fallout deposit. (D) Isopach (black lines; kg/m^2) and isopleth data (red lines; cm) for layer 5 at Cotopaxi volcano, Ecuador (Barberi et al., 1995). *References in supplementary material. Please see companion site, http://booksite.elsevier.com/9780123859389/*

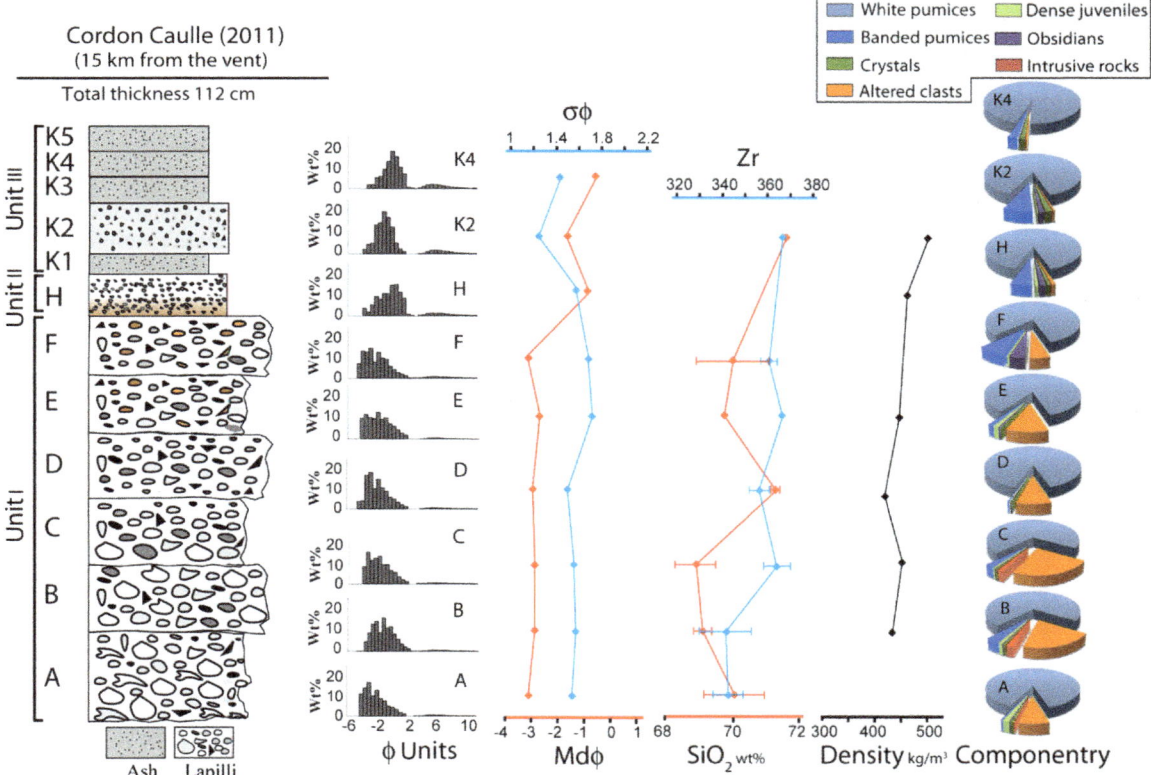

FIGURE 29.7 Stratigraphic column of the 2011 Puyehue-Córdon Caulle eruption (Chile) at 15 km east of the vent. Variations of grain-size parameters, SiO$_2$ and Zr content, density of pumice clasts, and componentry are also shown on the right.

ash-dominated regions of the volcanic cloud, aggregation of the airborne fine ash particles (<250 μm) frequently occurs due to electrostatic attraction or wet cohesion. This phenomenon increases the apparent terminal fall velocity of the particles. Because ash aggregation becomes effective at a certain distance (tens to hundreds km) from the vent, it may result in the formation of a secondary thickness maximum in the deposits (Figure 29.6(C)).

Grain-size distribution of samples of plinian deposits is generally polymodal (Figures 29.7 and 29.8). This is, in part, related to the fallout of clasts with different density, but roughly equivalent fall velocity. For example, large, low-density pumice clasts fall synchronously with smaller, denser lithics or crystals.

In the rare case of weak or no-wind conditions (more commonly corresponding to eruptions in the tropics, with weaker, high-level jet streams), the deposit has symmetrical dispersal around the eruptive vent. The lack of a strong asymmetry is accompanied by poor to moderate sorting of the coarse-grained deposits, and these are topped by a normally graded (fining upward) ash bed. This **co-plinian ash** is the result of the slow settling of fine particles after the end of the eruption, and is modulated only by the fall velocity of the particles (Figure 29.9(A) and (B)).

The rate of decrease of deposit thickness away from source reveals features of the eruption dynamics. These patterns can be described in a variety of ways on log thickness—area$^{1/2}$ diagrams, by multiple exponential segments, or by power laws or Weibull distributions (Pyle, 1989; Bonadonna and Houghton, 2005; Bonadonna and Costa, 2012). When tephra deposits are completely preserved on land and can be traced over long distances in a

continental setting (e.g. Mt St Helens 18 May 1980; Quizapu 1932; Hudson 1992), plinian fall deposits formed in presence of wind typically show increase of thickness up to a maximum value moving away from the source vent and along the dispersal axis, followed by a slow decrease into mid-distal regions, and a secondary maximum at distal sites. The general fine-grained total grain-size distribution associated with plinian eruptions favors the efficiency of ash aggregation. Once compaction is completed, the bulk density of the deposit shows a general increase with distance, due to the combined effect of increased density with fining of the grain-size and the lower porosity of fine-grained ash beds.

Fine-grained particles injected in the stratosphere by plinian columns travel for weeks to months before settling to the ground, circumnavigating the globe several times. Most of the total ejected volume consists of fine ash dispersed over huge areas and is rarely preserved as discrete beds in continental settings, unless rapid burial protect them from erosion and remobilization. Many recent plinian events have occurred on volcanic islands or close to the coast, where most of the pyroclasts settled over the ocean. The difficulty in mapping distal deposits in oceanic settings results in major uncertainty in assessing the total erupted volumes and total grain-size distribution.

7.2. Proximal Deposits

Deposits within 5—10 km from the vent result from the rapid accumulation of coarse material from the gas-thrust region and from the margins of the convective column, up to the corner of the umbrella region. Due to the direct dependence of the corner position on column height, this region may range up to 20 km from the vent (Sparks, 1986). Ballistic ejecta are generally dispersed within 4—5 km of the crater; an accurate study of dispersal, shape, and size of ballistics may provide important information on exit velocity and vent position (Wilson, 1976).

Additional complexities are at work in the proximal area. For example, clasts falling from the umbrella cloud may be recycled back into the rising column by advective inflow circulation (Figure 29.2); PDCs may erode (or overthicken) proximal fall deposits by scouring (or depositing). Simultaneous sedimentation of material from different regions of the eruption column generally results in coarse-grained, thick, moderately to poorly-sorted, often lithic- and bomb-rich deposits. Preservation of these deposits varies with distance from the vent, often being very poor in the first kilometers due to the steepness of volcanic edifices, occurrence of caldera collapse, and erosion by PDCs. Very-proximal (<100—200 m from the vent) deposits are rarely preserved: Fierstein et al. (1997) interpreted an ejecta ring

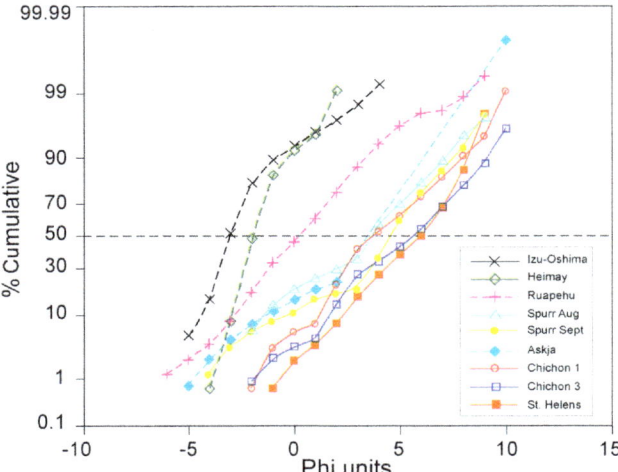

FIGURE 29.8 Total grain-size distribution of plinian and subplinian deposits, with phi = $-\log_2 D$, where D is the particle diameter in millimeters.

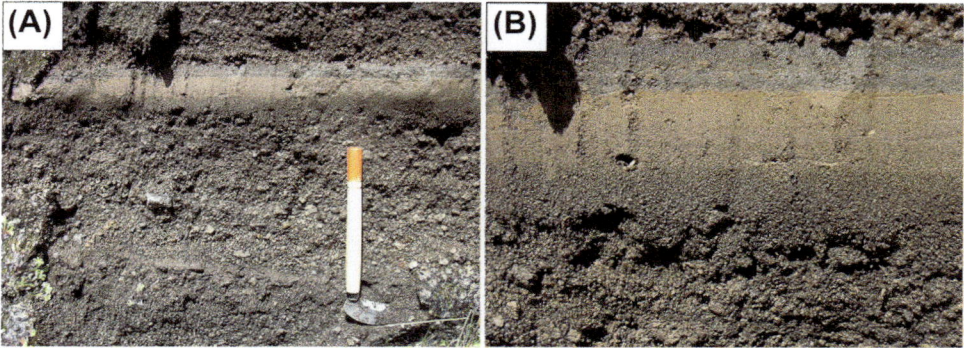

FIGURE 29.9   (A) Deposit of Layer 9, no-wind eruption of Cotopaxi volcano, Ecuador (Barberi et al., 1995). (B) Details of the Layer 9, normally graded, co-plinian ash deposit.

for the 1912 Novarupta eruption (Alaska) as the product of an irregular collar of low-fountaining ejecta encircling the main plinian plume. Welding of proximal fallout deposits related to plinian activity has been described for the Askja 1875 eruption, as well as at Cotopaxi volcano (Figure 29.10(A)).

7.3. Medial Deposits—Classification Issues

Medial deposits typically extend between 5 and 10 km and up to 40–50 km, depending on the size of the eruption and wind strength. They are generally the best preserved products and represent the classical pumice-rich, well-sorted, lapilli beds.

Medial plinian fallout deposits can be conveniently classified according to their origin process into three main categories labeled as *Simple*, *Simple-Stratified*, and *Multiple* (Rosi, 1996). This is especially useful during field surveys of proximal and medial deposits in the attempt to carry out and order all critical observations.

Simple plinian deposits result from the sedimentation of tephra from a steady eruption column. Related deposits are massive (nonstratified) or reversely graded, reflecting an increase in MDR (Figure 29.10(B)). Plinian deposits are often multiple-graded as the result of the gradual variation of the column height. Although thickness and mean grain-size decrease as a function of distance, other sedimentological features (e.g. vertical changes in the pumice/lithic ratio) do not vary with distance from vent and help to correlate among different outcrops. Due to aeolian fractionation, thinning and fining of tephra deposits away from vent can also be accompanied by changes in the relative proportions of components mainly because of the variable density of the material. Simple plinian deposits are among the most common; examples are those of Santorini Minoan (Greece), Unit five Taupo lapilli (New Zealand), and 1902 Santa Maria (Guatemala).

Simple-stratified plinian deposits commonly contain internal bedding or interlayering of PDC deposits within the fall sequence. Despite their internal stratification, these deposits are related to single eruptive events, which undergo column instability and transitions between convective and collapsing phases. Partial column collapses, particularly common during subplinian events, are frequent in this type of eruption. Reverse grading often characterizes these fall deposits; alternation of distinct tephra beds eventually grades with distance into a massive plinian fall deposit (Figure 29.10(C)). Due to their directionality, partial collapses of the eruptive column may not leave PDC deposits interbedded with the plinian fallout; in this case, they are recorded by **co-PDC ash** beds or by beds of reduced grain-size due to the temporary lowering of the eruption column. Simple-stratified deposits formed during the May 1980 eruption of Mount St. Helens and the AD 79 eruption of Vesuvius.

Multiple plinian deposits are related to separated eruptive pulses that occur in a short time interval (days to months). The rapid succession of events does not allow significant erosion or interbedding of reworked material between successive fallout beds. Individual plinian packages actually result from the juxtaposition of different tephra layers deposited under different atmospheric and eruptive conditions (Figure 29.10(D)). "Multiple" deposits can be distinguished from "simple stratified" deposits only after careful field mapping of the constituent beds, generally on the basis of dispersal axis, granulometric changes, presence of co-ignimbrite ash beds, or changes in the proportions of components. Assessment of eruptive parameters (volume and column height) demands that each single bed is studied separately in order to avoid overestimation of eruptive parameters. In very distal areas, the separation of different beds eventually becomes problematic. Eruptions with multiple plinian deposits are 1815 Tambora, 1982 El Chichón, or 2008 Chaitén. The 15 June 1991 eruption of Pinatubo (Philippine) can be also classified as a multiple plinian, as it was preceded by at least one subplinian pulse on 12 June.

FIGURE 29.10 (A) Welded fallout deposit on the eastern sector of Cotopaxi Volcano (Ecuador). (B) Simple plinian deposit: Layer 3 eruption of Cotopaxi volcano. (C) Simple-stratified plinian deposit: the grey pumice of the AD 79 Pompeii eruption (Vesuvius, Italy). (D) Multiple plinian deposit: the layer 1 and 2 eruptions of Cotopaxi volcano.

7.4. Distal Deposits

Distal deposits of plinian eruptions (from 50 km up to several hundreds of kilometers from the source) are represented by fine lapilli to coarse ash layers that slowly thin out to millimeter-thick, fine-ash beds. Complete data sets fully describing these deposits exist only from recent eruptions. Their preservation in the geological record is, in fact, hampered by their easy reworking by wind, erosion by running water, or incorporation into the soil. Dry, fine-ash deposits are, in particular, prone to substantial wind-reworking immediately after the eruption. Fallout and sedimentation of distal ash is variably controlled by aggregation; it is common to find local thickness variations (Figure 29.6) related to differences in the aggregation efficiency in the ash cloud. Locally, preferential accumulation may be also related to the effect of low-level winds on the slowly falling material, to rapid remobilization after settling, or to differential post-depositional compaction. Due to the very low inertia when settling, ash tends in fact to form soft, uncompacted beds, which progressively compact from rain or snow. Data collected on the 1932 eruption of Quizapu (Chile) demonstrated that distal thickness nearly halved

compared to data collected one year after the eruption, while this effect was insignificant for coarser, medial deposits.

7.5. Subplinian Deposits

Due to lower intensity, and hence column height, subplinian eruption clouds tend to inject less material into the stratosphere, if at all. Tephra sedimentation is thus mainly controlled by low-level, (typically) weaker winds. In addition, subplinian events also exhibit lower vertical velocities than their plinian counterparts, so that both the umbrella and the vertical regions of the eruption column are more prone to become bent-over in the presence of strong winds. All these factors are reflected in the dispersal and sedimentological features of the fall deposits, which often form strongly elongate fans that can be asymmetrical (both for thickness and grain-size) with respect to the dispersal axis, as shown by the 1980 post-climactic subplinian eruptions at Mt St Helens (Waitt et al., 1981). Subplinian eruptions often consist of multiple pulses. Tephra deposits consist of several fallout beds each characterized by its own dispersal axis, leading to

FIGURE 29.11 (A) Subplinian deposit of the Greenish Pumice eruption (Vesuvius, Italy). (B) Close view of the thinly stratified deposit of the AD 512 eruption of Vesuvius. (C) The complex proximal sequence (total thickness about 15 m) of PDC and fallout deposits of the AD 79 eruption (Vesuvius). (D) Tephra deposits of the 1.2 ka Quilotoa eruption (Ecuador); the fallout deposit of the plinian phase is overlain by a strongly stratified sequence of fallout and PDC deposits of the transitional phase of the eruption.

complex deposit architectures. These are generally thinly stratified due to grain-size alternations (Figure 29.11(A) and (B)) as seen in the recent example of the 2011 Puyehue-Córdon Caulle eruption (Chile). Compared to plinian events, subplinian deposits tend to be thinner and finer-grained at medial sites and show greater variability in the juvenile components (Figures 29.6 and 29.7).

7.6. Co-Plinian vs co-PDC Ash

Ash deposits associated with convective plinian columns are not only present at distal sites. Co-plinian, fining-upward, ash beds topping the main pumice bed have been described also at proximal-medial sites for eruptions occurring under no-wind conditions (Pululahua and Cotopaxi volcanoes; Figure 29.9). They are the result of the delayed settling of the finer-grained portion of the erupted material, which remained suspended into the umbrella region and was not widely dispersed due to the windless conditions. Conversely, co-PDC ash beds are the result of the settling of fine ash elutriated from PDCs; they may also occur interbedded in coarser, pumice-bearing tephra beds, and are generally characterized by better sorted, glass-shards-enriched, finer-grained grain-size populations. Co-PDC clouds detach from above the top of flowing PDCs

with a very low vertical velocity component, allowing the elutriation of the finest particles suspended in the flow. Co-PDC fine ash beds produced by large scale pyroclastic flows can blanket vast areas.

At distal sites, co-plinian ash deposits are difficult to separate from co-PDC ash beds due to the similar size range of the two deposits; in these cases, the erroneous attribution of a co-ignimbrite ash to the plinian deposit may result in an overestimation of the total fallout volume.

7.7. PDC Activity during Plinian Eruptions

PDCs produced by the collapse of plinian columns are heterogeneous mixtures of volcanic particles and gas at high temperature, which flow across the volcano flanks driven by gravity. Although many plinian fall deposits are the result of steady- or quasi-steady state, hours-lng convective columns, steady-state conditions may rapidly shift to phases in which convection is no longer reached, leading to repeated cycles of partial or total collapses of the erupting mixture and consequent PDC generation. This dynamic is recorded in the proximal deposits as an alternation of fallout and PDC deposits (Figure 29.11(C)). When this occurs, often in response to modifications of conduit/crater geometry, magma properties, MDR and magma fragmentation style, the erupted mass collapses back to the ground under the action of gravity, spreading as a PDC. Each time a fraction of the erupted mass is subtracted from the convective system, this results in a temporary decrease of the height of the convective column and of the dispersal of the fall deposits. Plinian eruptions commonly record repeated shifts from convective to collapsing regime, resulting in an interbedding of thin fallout and PDC beds, with the latter representing a variable, sometimes important percentage of the total erupted volume (Figure 29.11(D)).

Beyond the run-out limit attained by PDC deposits, column collapse events are recorded within the plinian fallout tephra as finer-grained beds related to the sedimentation from the lower-level, coexisting convective column, and/or by sedimentation of co-PDC ash. The presence of hybrid fallout beds consisting of centimeter-sized lapilli mixed with fine co-PDC ash accumulated synchronously also attests to the passage of the eruption through the transitional regime in which column partial collapses and sustained, buoyant phases coexist (e.g. the 1991 Pinatubo and the 1280 AD Quilotoa eruptions). In these cases, despite the increase of the total mass flow rate, the partial mass partitioning into PDCs actually results in a decrease of the height of the convective column. The coexistence of buoyant and collapsing regimes in the same eruption column may result in atypical sedimentological features of the fall deposits. If sedimentation rates of fall and PDCs are similar, poorly-sorted deposits form in which coarse and fine material coexists, sometimes showing hybrid features of fall/flow type. If sedimentation of the fallout material prevails, a poorly sorted deposit is formed, varying from a coarse bed containing an anomalous amount of ash, up to a coarse bed in which the coarse fragments are completely covered by a very fine ash patina. Conversely, if sedimentation from the PDC cloud dominates, a poorly-sorted, ash-rich flow deposit forms, bearing sparse, coarse angular fragments settled by fallout.

8. CONCLUDING REMARKS

Plinian and subplinian eruptions are very complex events consisting of different pulses of variable style, magnitude, and intensity. Although the study and modeling of the convective regime of plinian eruptions have been for decades a field of fundamental development in modern physical volcanology, much still remains to understand of the nature of these events, of the processes that control the shifts between different eruptive phases, as well as of the causes of the instabilities, which modulate subplinian dynamics. All these topics can be of utmost importance for assessing and reducing the volcanic hazards related to this type of events that, although infrequent, yet have a tremendous impact on human beings and environment.

ACKNOWLEDGMENTS

We would like to thank the many colleagues and students who, in the years, have shared with us stimulating discussions and pleasant periods in the field, and in particular Daniele Andronico, Antonella Bertagnini, Costanza Bonadonna, Kathy Cashman, Bruce Houghton, and Roberto Santacroce. The comments by Alexa Van Eaton and Simona Scollo greatly helped to improve the manuscript.

FURTHER READING

Bonadonna, C., Houghton, B.F., 2005. Total grain-size distribution and volume of tephra-fall deposits. Bull. Volcanol. 67, 441–456.

Bonadonna, C., Costa, A., 2012. Estimating the volume of tephra deposits: a new simple strategy. Geology 40, 415–418.

Carey, S., Sigurdsson, H., 1989. The intensity of plinian eruptions. Bull. Volcanol. 51, 28–40.

Carey, S., Sparks, R., 1986. Quantitative models of the fallout and dispersal of tephra from volcanic eruption columns. Bull. Volcanol. 48, 109–125.

Cioni, R., Bertagnini, A., Santacroce, R., Andronico, D., 2008. Explosive activity and eruption scenarios at Somma-Vesuvius (Italy): towards a new classification scheme. J. Volcanol. Geotherm. Res. 178, 331–346.

Dufek, J., Manga, M., Patel, A., 2012. Granular disruption during explosive volcanic eruptions. Nat. Geosci. 5, 561–564.

Fierstein, J., Houghton, B.F., Wilson, C., Hildreth, W., 1997. Complexities of plinian fall deposition at vent: an example from the 1912 Novarupta eruption (Alaska). J. Volcanol. Geotherm. Res. 76, 215–227.

Houghton, B.F., Carey, R.J., Rosenberg, M.D., 2014. The 1800a Taupo eruption: 'III wind' blows the ultraplinian type event down to Plinian. Geology 42, 459–461.

Neri, A., Di Muro, A., Rosi, M., 2002. Mass partition during collapsing and transitional columns by using numerical simulations. J. Volcanol. Geotherm. Res. 115, 1–18.

Pyle, D.M., 1989. The thickness, volume and grainsize of tephra fall deposits. Bull. Volcanol. 51, 1–15.

Pyle, D.M., 2000. Sizes of volcanic eruptions. In: Sigurdsson, H., et al. (Eds.), Encyclopedia of Volcanoes. Academic Press, pp. 263–269.

Rosi, M., 1996. Quantitative reconstruction of recent volcanic activity: a contribution to forecasting of future eruptions. In: Monitoring and Mitigation of Volcano Hazards. Springer, Berlin Heidelberg, pp. 631–674.

Scandone, R., Malone, S.D., 1985. Magma supply, magma discharge and readjustment of the feeding system of Mount St. Helens during 1980. J. Volcanol. Geotherm. Res. 23, 239–262.

Scandone, R., Cashman, K.V., Malone, S.D., 2007. Magma supply, magma ascent and the style of volcanic eruptions. Earth Planet. Sci. Lett. 253, 513–529.

Shea, T., Gurioli, L., Houghton, B.F., Cioni, R., Cashman, K.V., 2011. Column collapse and generation of pyroclastic density currents during the A.D. 79 eruption of Vesuvius: the role of pyroclast density. Geology 39, 695–698.

Sparks, R., 1986. The dimensions and dynamics of volcanic eruption columns. Bull. Volcanol. 48, 3–15.

Waitt Jr., R.B., Hansen, V.L., Sarna-Wojcicki, A.M., Wood, S.H., 1981. Proximal air-fall deposits of eruptions between May 24 and August 7, 1980. Stratigraphy and field sedimentology. The 1980 eruptions of Mount St Helens, Washington. In: Lipman, P.W., Mullineaux, D.R. (Eds.), Geological Survey Professional Paper 1250, pp. 601–616.

Walker, G., 1981. Plinian eruptions and their products. Bull. Volcanol. 44, 223–240.

Walker, G.P., 1973. Explosive volcanic eruptions—a new classification scheme. Geol. Rund. 62, 431–446.

Walker, G.P.L., 1980. The Taupo pumice: product of the most powerful known (ultraplinian) eruption? J. Volcanol. Geotherm. Res. 8 (1), 69–94.

Wilson, L., 1976. Explosive volcanic eruptions—III. Plinian eruption columns. Geophys. J. R. Astron. Soc. 1, 543–556.

Chapter 30

Phreatomagmatic and Related Eruption Styles

Bruce Houghton

Department of Geology and Geophysics, National Disaster Preparedness Training Center, University of Hawai'i, Honolulu, HI, USA

James D.L. White

Geology Department, University of Otago, Dunedin, New Zealand

Alexa R. Van Eaton

U.S. Geological Survey, VA, USA

GLOSSARY

accretionary lapilli Spherical aggregates of volcanic ash, commonly having a concentric structure and formed by clumping of moist ash. Often used as a general term for all ash aggregates including structureless "ash lumps/pellets."

armored lapilli Solid fragments coated with one or more layers of volcanic ash; essentially a variant of "accretionary lapilli."

ash aggregation Formation of clusters of ash particles during transport and sedimentation.

ash lump/ash pellet Ash aggregates of irregular shape and with no discernible, concentric internal structure.

cock's tail jets A common type of tephra jet with an arching, multifingered form likened to the shape of a rooster's tail. Also called cypressoid jet.

continuous uprush Sustained phreatomagmatic eruption column that is fed by ongoing or closely spaced explosions.

external water Any water phase involved in explosive volcanism that was not originally dissolved in the magma, including surface, ground, and atmospheric water.

mud rain Ash-bearing droplets of water from an eruption plume.

phreatoplinian deposits Widespread phreatomagmatic fall deposits, defined by a dispersal index of ≥ 50 km^2, and a fragmentation index generally $>80\%$.

spall dome Subhemispherical, gas-driven shock zone expanding outward from a subaqueous explosion.

tachylite/sideromelane Varieties of basaltic glass. Sideromelane is translucent; tachylite is opaque as a result of abundant nanolites and microlites, inferred to result from a slightly slower cooling.

The Encyclopedia of Volcanoes. http://dx.doi.org/10.1016/B978-0-12-385938-9.00030-4

tephra jets Discrete ejections containing tephra and water vapor, driven by steam expansion and momentum of large clasts.

vesiculated tuff Ash deposit containing spherical to irregularly shaped cavities, reflecting entrapment of air or water droplets in wet, fine-grained ash.

1. INTRODUCTION

Here, we describe the spectrum of "wet" explosive eruptions that result from the combination of magmatic heat and "external water," such as groundwater, lakes, or oceans (Figure 30.1). These eruptions have been grouped in the past under descriptors such as *phreatomagmatic*, *phreatic*, and *hydrothermal*. Although definitions vary, most authors use *phreatomagmatic* to characterize eruptions where magma interacts directly with external water, *phreatic* where magma heats surrounding fluids without actually erupting at the surface, and *hydrothermal* to describe eruptions of geothermal fluids. However, water-influenced eruptions in nature occur well outside the boundaries of these simplified definitions. The case studies given below illustrate some of the defining characteristics of "wet" eruptions, and how they can be recognized through direct observation and in the geological record. In revisiting these examples, we illustrate how water can modify or, in some cases, govern the dynamics of volcanic explosions, and address the following open questions:

1. How does water involvement modulate eruption style, transport processes, and the resulting deposits?
2. How do eruption dynamics reflect the source of water (magma, surface water, groundwater, and atmosphere) and its physical state (ice, liquid, and vapor)?

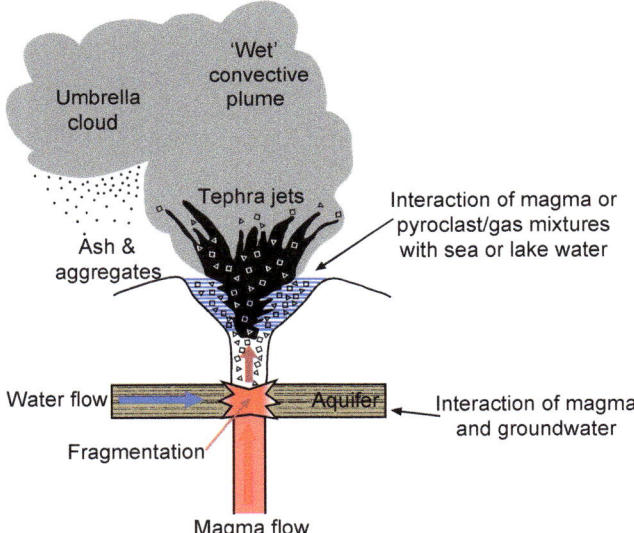

FIGURE 30.1 Conceptual cartoon for subaerial phreatomagmatic eruptions showing two alternative external sources for the water that is flashed into steam. These are either surface water in lakes or streams, or groundwater held in pores and cracks within the shallow rocks.

3. Are phreatomagmatic products necessarily finer grained than their dry counterparts?
4. Does water involvement at the volcanic source necessarily leave behind "wet" signatures in the deposits?

2. TRANSIENT EXPLOSIONS

2.1. Drain-Back, Heated Wall Rock, and Groundwater: Halema'uma'u 1924, Hawaii

A 20-day sequence of multiple discrete explosions at Halema'uma'u crater on Kilauea volcano occurred when magma withdrawal from the lava lake back down the conduit led to multiple collapses of the conduit walls, and significant crater widening (Figure 30.2). This event has been called an example of phreatic activity, that is, explosions culminating from the indirect transfer of magmatic heat to external water via wall rock, but the presence of small amounts of juvenile ejecta suggests that it was phreatomagmatic.

Explosions began on May 10, 1924, reaching a peak intensity on May 18. Several "large" explosions and numerous smaller events occurred each day, sometimes <1 min apart. The last explosions were recorded on May 29. The locus of the activity was not fixed, and each explosion produced a discrete apron of ejecta, but the northeastern wall of the crater was the main source. The activity generated transient plumes to heights of 6.5 km (Figure 30.2(D)), depositing ash, lapilli, "mud rain," and ash aggregates, and ejecting ballistic blocks (up to 10 tonnes) to 1 km from vent. The blocks ranged greatly in temperature, from ambient to approximately 700 °C, and were in part breadcrusted. Ash fall was recorded >30 km downwind, consisting almost entirely of wall rock, principally preexisting lava flows.

Before 1924, Halema'uma'u had contained a long-lived, active lava lake. The lava lake subsided and disappeared in February 1924 as part of a more widespread, 1- to 4-m subsidence of the volcano summit, accompanying magma removal from the shallow summit reservoir. Geodetic data suggest the draining of approximately 0.4 km^3 of magma into the East Rift Zone. In February, the floor of Halema'uma'u sank to a depth of 115 m, leaving the crater 520 m wide. By the end of the eruption, the depth of the crater was 417 m, and its diameter was 915 × 1065 m. Jagger estimated the collapse volume to be approximately 2×10^8 m^3, and the ejecta volume was 7×10^5 m^3.

A model for these explosions requires there to have been thermal quasiequilibrium established during the extended period when lava was ponded in the Halema'uma'u crater (Figure 30.2(A)). Adjacent to the shallow, magma-filled

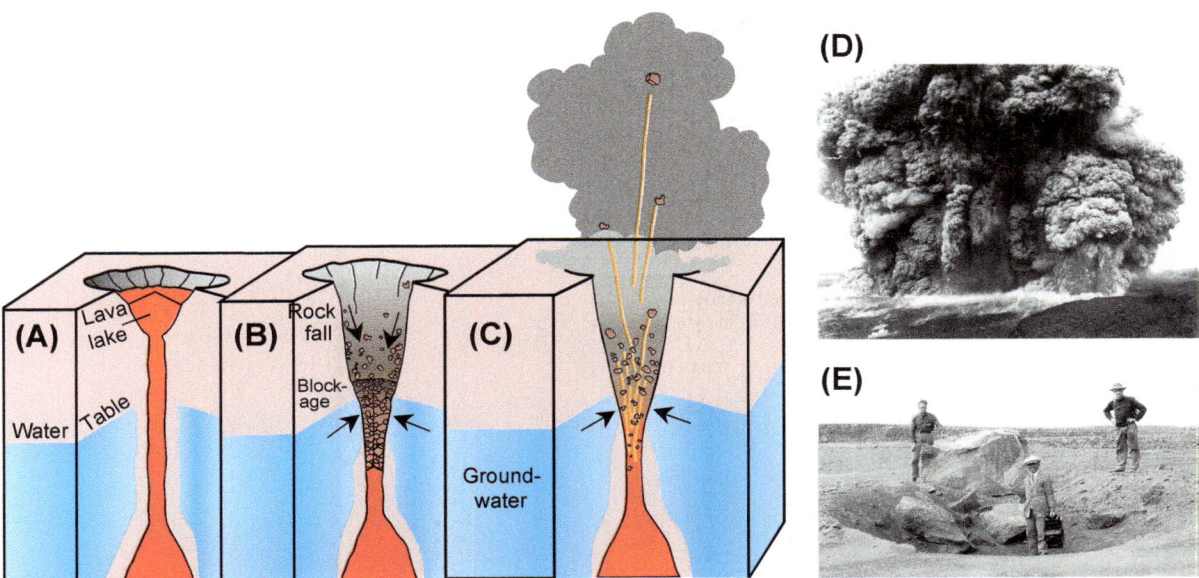

FIGURE 30.2 (A) Cartoon, *modified after Swanson et al. (2011)* showing the model for 1924 explosions. Drainage of the long-lived lava lake (A) dropped the free surface below the water table and withdrew support from the steep, hot walls of the conduit. (B) Wall collapse and vent blockages triggered (C) a series of transient explosions involving hot wall rock and a small proportion of largely outgassed magma. (D) Photograph of the plume formed by explosion at 11:15 (Hawaii Standard Time) on May 18 at the peak of the 1924 explosive activity. Observers noted fallout of ballistic blocks, hot ash, and accretionary lapilli. Photographer K. Maehara, HVO archive. (E) Impact crater and 8- to 10-tonne block thrown 1 km (SE) by the explosion in (D). Photographer H.T. Stearns, Hawaiian Volcano Observatory archives.

conduit was extremely hot wall rock (probably at least 700 °C), which formed the incandescent blocks seen during explosions. Outboard from this region, colder wall rock contained groundwater in cracks, fractures, joints, and pore space. The water phase was necessarily in a liquid state to enable significant expansion on heating. This state was maintained until support was removed from the conduit walls by withdrawal of magma from the conduit (Figure 30.2(B)). The collapses of the conduit walls that followed brought these thermally heterogeneous wall-rock materials into rapid, intimate contact. It is inferred that pockets of water were trapped during the collapses and heated rapidly, flashing to steam, expanding perhaps 5000x in volume, and powering impulsive, short-lived explosions (Figure 30.2(C) and (D)). The observed thermal heterogeneity of the ejected blocks is explained by the short time scales of the mixing events. The explosions thus involved magmatic heat interacting with an external source of liquid water, but only after temporary storage in conduit wall rock. By this mechanism, explosive activity was initiated without direct involvement of magma.

2.2. Wall-Rock Collapse and Decoupled Magmatic Volatiles: Halema'uma'u March 19, 2008, Hawaii

The March 19, 2008, eruption at Halema'uma'u, Kīlauea is included as an example of an eruption where neither

external water nor magma was involved (Figure 30.3). Instead, a rock fall blocked the release of magmatic volatiles from the conduit, leading to a transient buildup of pressure. The eruption was thus powered by a sudden expansion of the decoupled, but trapped, magmatic volatiles.

Enhanced seismic tremor at Halema'uma'u was first observed in November 2007, followed by increasing SO_2 emissions in December. Microearthquakes increased in frequency on March 12, 2008, when the gas emissions became centered upon a 150-m-wide area at the base of the southern wall of Halema'uma'u, which was to become the future vent (Figure 30.3(B)). Incandescence (rock temperatures >500 °C) was first seen on March 14. On March 17, there were at least 16 incandescent gas vents in an area of 15 × 30 m. The final measurement of SO_2 flux prior to the explosion was 1630 ± 130 tons/day on March 18, 2008, up from a background of approximately 150 tons/day.

The explosion on March 19, 2008, was preceded by three small, high-frequency earthquakes, inferred to have been caused by rock falls (Figure 30.3(C)). It was associated with a long-period seismic event and a decompressive infrasound signal, attributed to a pressure release at source (Fee et al., 2010). Its total duration, inferred from geophysical techniques, was only 53 s. The March 19, 2008, explosion produced a steep-walled 35-m-wide crater at least 30 m deep (Figure 30.3(D)), in the center of the degassing area, and ejected only wall-rock particles (Swanson et al.,

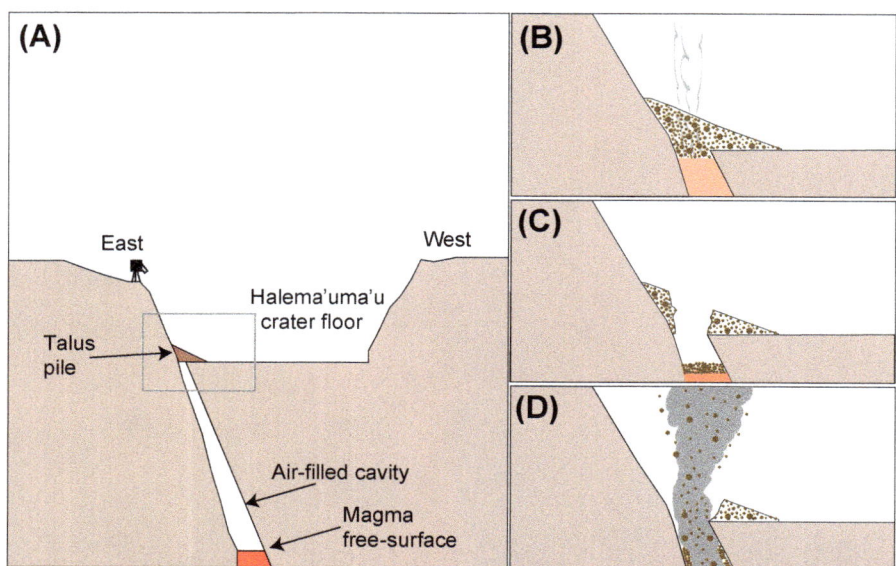

FIGURE 30.3 Three-stage model for the March 19, 2008, at Halema'uma'u crater, Kilauea. The explosion occurred well above the approximately 600-m-deep water table and was powered by the confinement and subsequent expansion of decoupled magmatic volatiles exsolved at depths of several hundred meters below the floor of the crater. (A) Sketch of vent geometry. The new vent formed close to the foot of the wall of the preexisting Halema'uma'u crater, through a loose talus scree of fallen blocks. The subsequent disparity between the volume of the vent and that of the ejecta suggested that there was considerable void space beneath the talus, represented schematically here, in an oversimplified way. (B) Rise of basaltic magma from the shallow storage region at depths >1 km was accompanied by the release of magmatic volatiles which, a week prior to the eruption, were being discharged from numerous sources across a 150-m-wide region on the surface of the talus fan. By 5 days prior to the explosion, the discharge was incandescent and concentrated on perhaps 10–15 points immediately above the future vent. (C) Rock falls accompanied by three small impulsive high-frequency seismic events at and just after 2:55 a.m. on March 19, 2008, temporarily blocked the system, inhibited gas expansion, and release. (Note for simplicity, the blockage is shown just below at the base of the talus fan but could have been significantly deeper.) (D) Three minutes later, the vent is explosively cleared by the expanding trapped volatiles ejecting only wall-rock particles in a short-lived plume.

2009; Patrick et al., 2011). The volume of ejecta was, at most, 1% of the volume of the collapse crater, producing a total mass of 4.1×10^5 kg (wall-rock particles only), which equates to a time-averaged eruption rate of 8×10^3 kg s^{-1}. The major volume discrepancy between ejecta and the collapse crater indicates that a significant preeruptive cavity existed before eruption, which was only partially blocked at the surface by the talus derived from the Halema'uma'u wall. This cavity was a pathway for the free escape of magmatic gas, which before March 19, was discharged through a network of pore space (Figure 30.3(B)). The magma, however, was significantly deeper in the conduit and was not involved in this explosion. Juvenile ejecta first appeared four days later (Swanson et al., 2009) and has been followed by more than six years of magmatic activity.

It can be reasonably inferred that the gas phase involved on March 19, 2008, had the same origin as the decoupled magmatic volatiles that were being discharged freely prior to and after the eruption. There is also no evidence for involvement of external water; the water table in the Kīlauea caldera is approximately 600 m below the floor of the caldera. The flux of magmatic volatiles immediately after the explosion (1620 tons/d) was essentially the same as on March 18. The erupted volatiles represent a pocket of

magmatic gas that was temporarily trapped between the precursory rock falls and the explosive eruption 3 min later.

No external water or magma was involved in this eruption, yet the products look identical to phreatic deposits. It cannot be accommodated at all in existing classifications of eruption styles.

2.3. Lava and Surface Water: Kalapana 1988–2013, Hawaii

When pahoehoe lava enters the ocean, it forms a broad volcaniclastic delta traversed by lava tubes. Interaction with seawater becomes most explosive when lava fluxes are high and focused in relatively few tubes. Major littoral explosions are often linked to external mechanical triggers, such as cracking or collapse of lava tubes or bench collapses. Activity in these cases takes three dominant forms: episodic "tephra jets" (Figure 30.4 (C)) lithic-rich directed explosions, and bubble bursts (Figure 30.4 (D)).

Episodic tephra jets: This activity is linked to the severing of lava tube by bench collapse (Figure 30.4(E) and (F)). Breaking waves then disrupt the stream of melt, producing closely spaced successions of tephra jets up to tens of meters high (Figure 30.4(B) and (C)). The explosions are

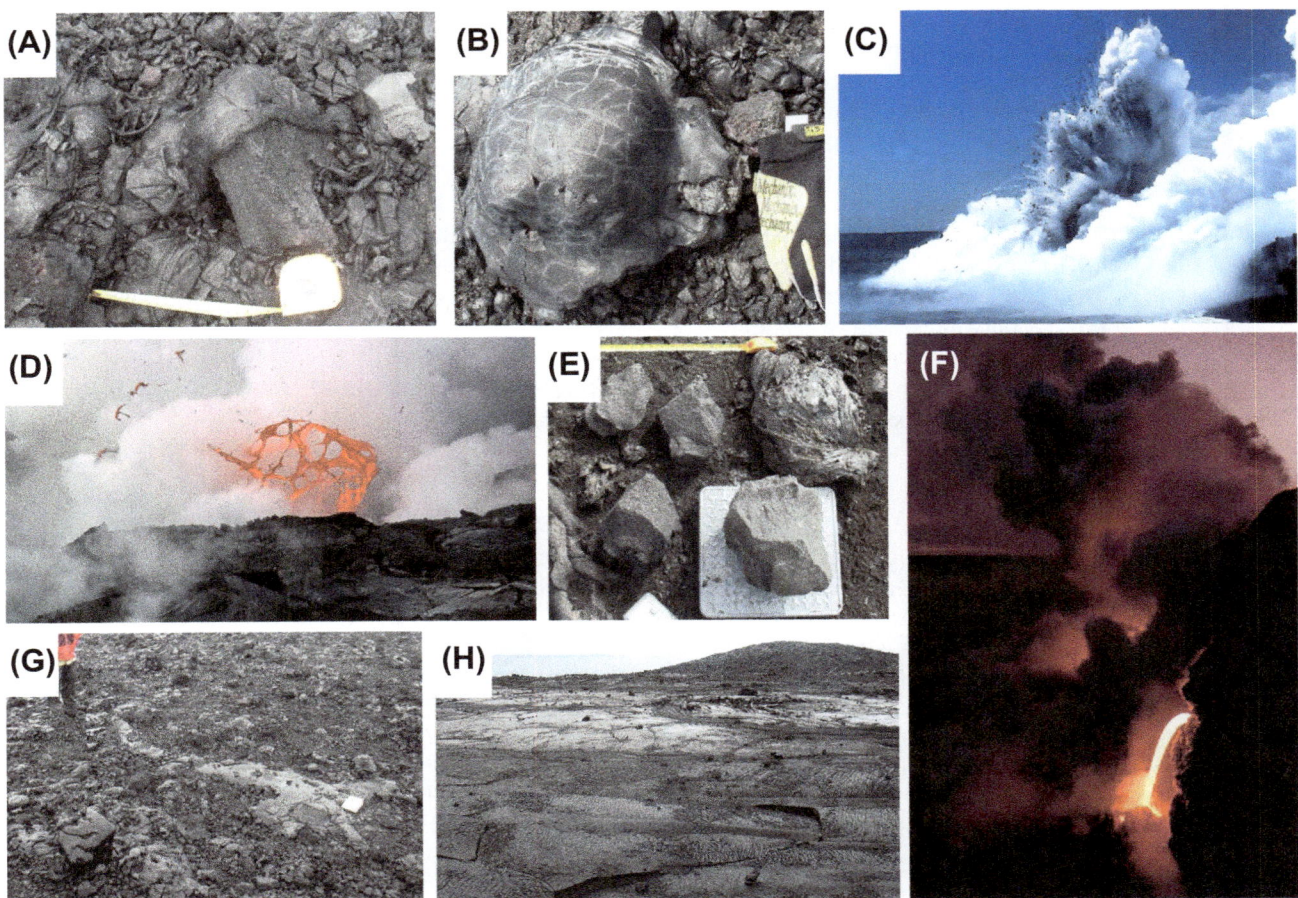

FIGURE 30.4 Photographs of explosions and products of Kalapana littoral explosions. (A) Pele's locks and (B) fluidal, outgassed clast enclosing an angular wall-rock block, formed during episodic tephra jets in 2008 of the type shown in (C). (C) Tephra jetting at Kalapana ocean entry of lava. *J.D. Griggs, USGS, February 3, 1988.* (D) Burst of steam bubble formed under fluid lava, an excellent example of hydrodynamic fragmentation. *J.D. Griggs, USGS, October 5, 1988.* (E) Angular, blocky fragments torn from the walls of lava tubes during major explosions accompanying collapse of the lava delta. (F) Outpouring of a lava stream from a severed tube following one such collapse. *J.D. Griggs, USGS, November 27, 1989.* (G) Exceptionally fluidal, meter-long clast (under yellow notebook) formed during events as seen in (D). (H) Scattered blocks from the collapse-triggered explosions in the foreground overlying pahoehoe lava. In the background, the littoral cone formed during more sustained periods of tephra jet activity in 2008.

exceptionally thermally inefficient, with incandescent temperatures visible in large clasts for tens of seconds after deposition. This activity results in two types of eruption products: (1) proximal littoral cones, which can grow rapidly at rates up to meters per day, and (2) a downwind tephra blanket. The cone is constructed mostly from largely outgassed but exceptionally fluidal bombs, which spatter on landing and often form thin sheets. Other ejecta take the form of centimeter wide lava ropes, known informally as "Pele's locks" (Figure 30.4(A)), Pele's hairs, fluidal ash, fine lapilli, and glass flakes. The distal blanket contains a mix of Pele's hair, fluidal ash, and glass flakes.

Directed explosions: During collapse of unstable lava deltas (Figure 30.4(F)), incandescent bench faces are exposed, and brought into contact with seawater, which triggers explosions, ejecting blocks from the hot "wall rock" of the lava delta (Figure 30.4(E)) and lesser amounts

of fluid lava. Ballistic clasts are produced and dispersed significantly more widely than tephra jets, and the ejecta forms discontinuous and asymmetric block fields that can extend up to 200 m inland (Figure 30.4 (H)). On April 19, 1993, a collapse killed one visitor and the resulting block fall injured several others (Mattox and Mangan, 1997).

Bubble bursts: Any process that traps seawater, for example, fracturing of the roof of lava tubes close to sea level, may produce 1- to 10-m diameter steam bubbles. Explosive ruptures of bubbles (Figure 30.4(D)) eject a mixture of fluidal pyroclasts (Figure 30.4(G)), limu-o-Pele and Pele's hair. On rare occasions, tube fracture instead produces continuous fountaining lasting tens of minutes.

Mattox and Mangan (1997) summarize the activity as representing either open or confined mixing of seawater with largely outgassed lava. Tephra jets and lithic blasts are the result of "open" or unconfined mingling with an extended contact area caused by tube or bench collapse. The formation of tephra jets mostly involves hot fluid lava; blasts incorporate very hot glassy material from the head of the tube. Bubble bursts (and fountains) are the result of confined mingling, whereby seawater trapped in tubes or elsewhere develops small confining pressures, permitting vapor expansion to disrupt overlying melt.

Littoral cones have an inherently small preservation potential when formed on lava deltas or eroding coasts, due to the dynamic environment in which they form. Pahoehoe-fed cones are generally only preserved when they are welded or armored by subsequent lava. Larger cones form when ʻaʻā lava flows into the ocean, generally at much higher lava fluxes. Tephra jets are the dominant transport mechanism, and the ejecta consist of ragged fluidal bombs and lapilli.

2.4. Lava and Groundwater or Ice: Laki, Iceland

The processes giving rise to rootless cones are similar in many respects to those producing littoral cones in coastal areas. Both are driven by explosive entrapment and mingling of lava flows and external water rather than a direct magmatic source. In the case of rootless cones, the water comes from a saturated or icy substrate, and debris from the substrate is often included in the ejecta. It is clear from the accumulation of rootless cones on top of lava flows fed by the same magma that the cones formed after initial advance of the lava (Figure 30.5).

After pahoehoe overrides a wet substrate, the lava interior remains molten. The flow thickens by inflation, with lava tubes feeding the advancing flow front. As the flow thickens, it begins to sink into the underlying deforming substrate, most strongly at sites where the lava is the thickest, where there is also sustained lava supply. Cracking of the base of the lava during this subsidence provides direct access to underlying water-saturated sediment, producing blasts

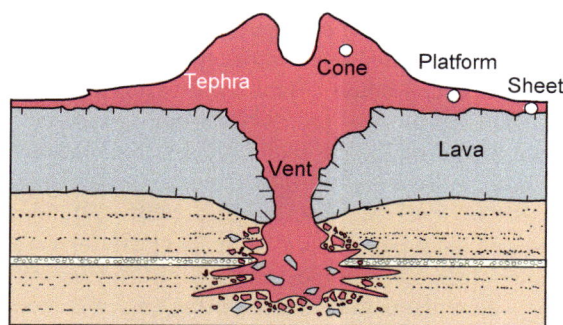

FIGURE 30.5 Conceptual model of a rootless cone at Laki, after *Hamilton et al. (2010)*. The deposit has typically three facies: a distal sheet of material from multiple source vents, an intermediate tephra platform, and a proximal cone centered on a crater above the rootless vent.

and tephra jets like those of littoral cones. In some cases, this also results in localized density currents. The ejected tephra contains both lava fragments and sediment, and many show a progressive coarsening or "drying-out" with time.

A prominent feature of rootless cones is their distribution along lava "pathways" in cone fields. In other words, they tend to be localized along paths of sustained lava passage, in channels or tubes. During the eruption of Laki (1783—1784), hundreds of rootless cones, grouped into clusters, formed where lava advanced over swampy lowlands. Channel-fed rootless cones often have a U-shape in plan view, the result of tephra being carried away by flowing lava in the channel. In contrast, rootless cones formed above lava tubes are more equant.

2.5. Vesiculating Magma and Surface Water: Surtsey, Iceland

Once Surtsey built above the sea, Thorarinsson (1967) noted " ... *two* clearly *discernible types of activity.* When *the sea had easy access to one or more of the vents ... the activity was the typical one for submarine eruptions... After each explosion a tephra-laden mass rushes up, and out of it shoot numerous lumps of liquid or plastic lava, called bombs, each with a black tail of tephra.*

Within a few seconds these black tails turn grayish white and furry as the superheated vapor ... cools and condenses" (pp 17—18).

Such activity from a flooded vent is termed tephra jetting, or cock's tail plumes (as seen in Figure 30.4(C)). In contrast, "continuous uprush" activity was linked with tephra accumulation that blocked seawater from pouring into the vent. "Instead of intermittent explosions the uprush of vapor and tephra was then continuous... at the base of the eruption column the speed of the uprush was about 400 feet per second. The black columns of tephra frequently reached half a mile in height... accompanied by a heavy rumbling noise. The tephra production from such a continuous uprush was much greater than from the intermittent explosive activity... uprush of this kind might last for a few hours at a time." (Thorarinsson, 1967; pp. 18—19).

Tephra jets carried tephra and steam along modified ballistic paths outward from the vent. Several jets per minute were commonly produced, and jets also occurred in isolation at widely spaced and irregular intervals, or in increasingly close succession at the onset of continuous uprush episodes. Strong, pulsatory vapor plumes commonly rose from the vent while jetting took place and were associated with lightning, whirlwinds, and tephra-hail showers, in which each hailstone was cored by a sand-sized tephra grain. As the inner parts of larger jets at Surtsey descended, they formed steam and tephra density currents that flowed down the cone and out over the sea. Continuous uprush

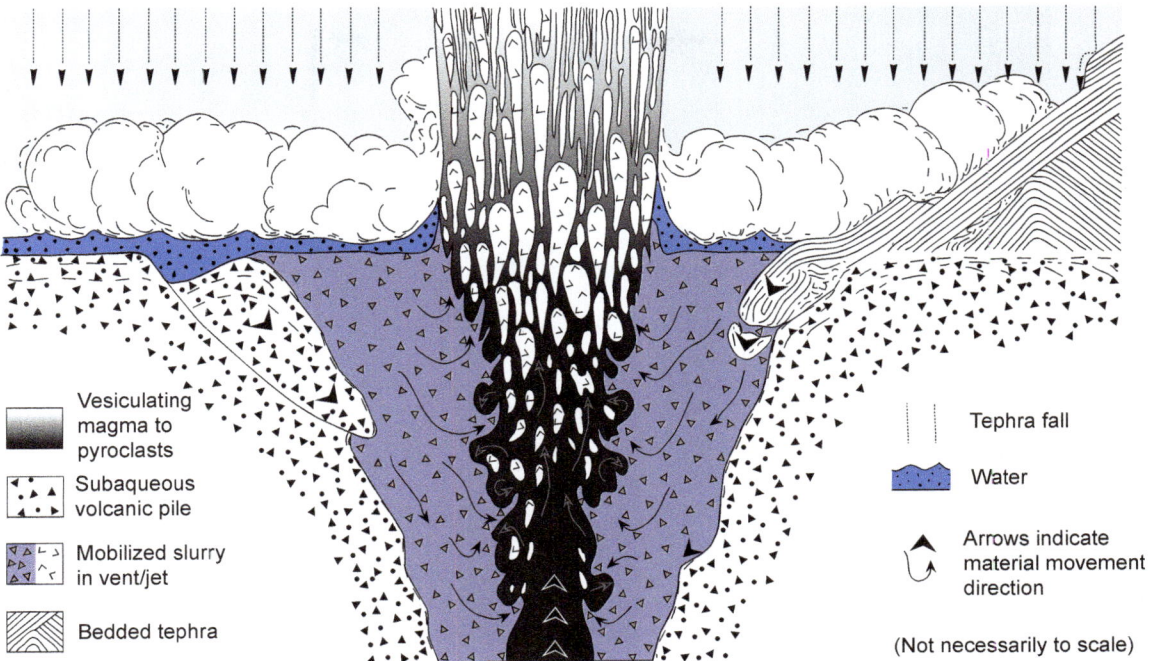

FIGURE 30.6   Conceptual model for processes in the Surtseyan vent in which shear-instability entrainment and mixing of a slurry of recycled tephra and water with rising magma drives discrete or continuous uprush activity. Depending on the setting, the wall rock may be loose to consolidated sediment in addition to buried volcanics. *Modified after Kokelaar (1983).*

activity formed relatively stable, dark-colored eruption columns up to a couple of kilometers in height, carrying water vapor and heavily charged with tephra (Figure 30.6). Incandescent clasts were abundant in such columns and fell back onto the cone. Buildup of tephra around the vent (tephra dams) was observed to prevent flooding of the vent during uprush activity.

Subaerially deposited tephra at Surtsey is poorly sorted and, in many places, includes "vesiculated tuff" rich in "accretionary lapilli" and sideromelane. Deposits contain widely scattered pebbles and local blocks of indurated marine sediments from the preeruption seafloor, and many fragments of thin vesicle-layered dikes. Vesiculated tuff beds on steep cone slopes ($>20°$) are broken by channels formed by small tephra debris flows. On even steeper slopes ($>35°$), these tuff beds failed en masse, breaking into plastic slabs. Quenched bombs, with ubiquitous incorporated wall rock, and a rugged, glassy exterior are present in late-stage beds; they indicate heterogeneous interaction of water and recycled tephra with erupting fluidal magma.

2.6. Vesiculating Magma and Groundwater: Ukinrek Maars, Alaska

A 10-day eruption at Ukinrek, Alaska, resulted from rising and ponded magma interacting with small amounts of shallow, perched, initially frozen groundwater within glacial till and underlying silicic pyroclastic deposits

(Kienle et al., 1980). It formed two maar craters (volume 4×10^6 m^3) and ejected 2×10^7 m^3 of olivine basalt and wall-rock lithic clasts. West Maar formed over the first three days of eruption and was 170 m wide and 35 m deep. The larger East Maar formed over the following 7 days, and is 300 m wide and 70 m deep (Figure 30.7(A)). Crater formation was gradual over the course of the eruption. Magma formed an incandescent pond in the vent late in the eruption sequence (Kienle et al., 1980).

Eruption plumes occasionally reached 6 km high (Figure 30.7(B)), and distal ash fall was recorded over an area of 20,000 km^2. The eruptions were phreatomagmatic throughout, but activity at both vents showed a progressive trend toward drier activity (transitional to Strombolian eruptions), marked by a lower content of wall-rock clasts and an increased abundance of ragged scoria lapilli in the deposits. This change reflects depletion of the water supply, delivered by flow through and collapse of the shallow aquifers. From day 5 after the start of eruption, contrasting eruptive styles were observed from two vents within East Maar. One vent within the lava pond experienced relatively "dry explosions" coeval with dark ash-rich plumes from the second vent (Kienle et al., 1980).

The deposits, which reach a maximum thickness of 26 m on the rim of the East Maar, are of four types: (i) lithic-rich ash and lapilli fall deposits, (ii) scoria-rich lapilli fall deposits, (iii) late-stage lithic-block aprons extending to only 800 m from the rims of the maars, and

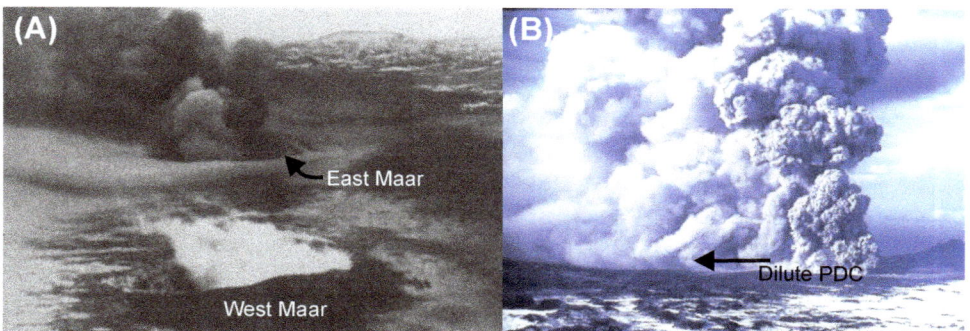

FIGURE 30.7 (A) Easterly view of Ukinrek Maars, Alaska, on April 3, 1977, showing black ejecta surrounding the steaming West Maar, and darker, ash-rich phreatomagmatic plume from East Marr. Photograph by Larry Conyers. (B) Photograph looking north at East Ukinrek Maar in vigorous phreatomagmatic eruption April 5, 1977, 16:30 AST. Note the chevron-shaped, dilute PDC. *Photograph by Jim Faro. Modified from Kienle et al. (1980).*

(iv) minor deposits of dilute pyroclastic density currents (PDCs) (Figure 30.7(B)). The lithic-rich lapilli fall deposits contain up to 75 weight% wall-rock clasts and subordinate rounded dense cauliflower juvenile lapilli, yet are well sorted but, on grain size criteria alone, the phreatomagmatic units would not be distinguished from Strombolian ones.

3. SUSTAINED ERUPTIONS

3.1. Characteristics of Sustained "Wet" Plumes

Large-scale wet plumes are distinctive from their dry counterparts in several ways. Classic Plinian ("dry") eruption plumes produce well-sorted lapilli beds that become progressively finer grained with distance. Coarser, lapilli-sized particles settle more or less individually, allowing them to fractionate by size and density during atmospheric transport. In contrast, wet plumes form poorly sorted and fine-grained deposits even close to source. The disparity arises from two key processes: (1) additional fragmentation during magma–water interaction leading to increased ash production, and (2) wet aggregation of ash, which rapidly scavenges fine-grained particles out of the atmosphere. However, a number of complexities exist within this simplified framework for "wet" versus "dry" volcanism. Even the type examples of wet volcanism on the largest scale, known as *phreatoplinian* eruptions, exhibit a spectrum of eruptive styles and behaviors. Here, we use these type examples of phreatoplinian volcanism to explore the characteristics of sustained, water-rich plumes.

3.2. Outgassed Magma and Surface Water: Rotongaio, New Zealand

The 1.8-ka Rotongaio Ash, Taupo, New Zealand, has a similar dispersal and similar whole-deposit grain size characteristics to the preceding Hatepe Ash, formed during the same eruption, but is different in three key ways. Its distinguishing characteristics are exceedingly fine grain size with close to 100 wt% of the deposit finer than 1 mm, a finely laminated character even in the most proximal exposures, and a lack of lateral continuity for laminae between proximal sites.

There is a marked contrast in the dispersal pattern of relatively coarse (lapilli-bearing) and relatively fine-grained subunits. Coarse-grained packages were significantly more wind advected, and hence have more restricted crosswind dispersals similar to the Plinian phases of the eruption, ash-rich subunits have much more symmetrical patterns of dispersal around the vent. Concentrically zoned accretionary lapilli are rare in the Rotongaio Ash, but many fall laminae consist of partially flattened, structureless, <1- to 2-mm diameter ash "pellets." The products of PDCs are, subordinate in volume to the fall deposits, confined to proximal areas and are interpreted as the products of relatively dilute suspension currents that temporarily interrupted fall deposition at any particular locality.

The Rotongaio Ash is distinctive as a superb and relatively rare example of syneruptive reworking due to local rainstorms precipitated by the volcanic event itself. A delicate balance apparently existed between accumulation, reworking, and erosion during deposition. Locally reworked beds are juxtaposed with, or transitional into, primary beds. Individual beds show a complete range of textures from primary fall through slope-affected material to water-reworked ash on meter scales. Plastic deformation and thickness variations along outcrop reflect the local topography. In some cases, ash has clearly flowed off high ground as a viscous liquid to produce flat-topped gully fills (Figure 30.8). Plastic deformation and thickness variations along outcrop reflect the local topography. These field relationships clearly show that the water reworking was syneruptive, and a complex and intimate relationship existed between primary pyroclast deposition and secondary reworking from downbursts generated by the wet volcanic activity.

(A)   **(B)**

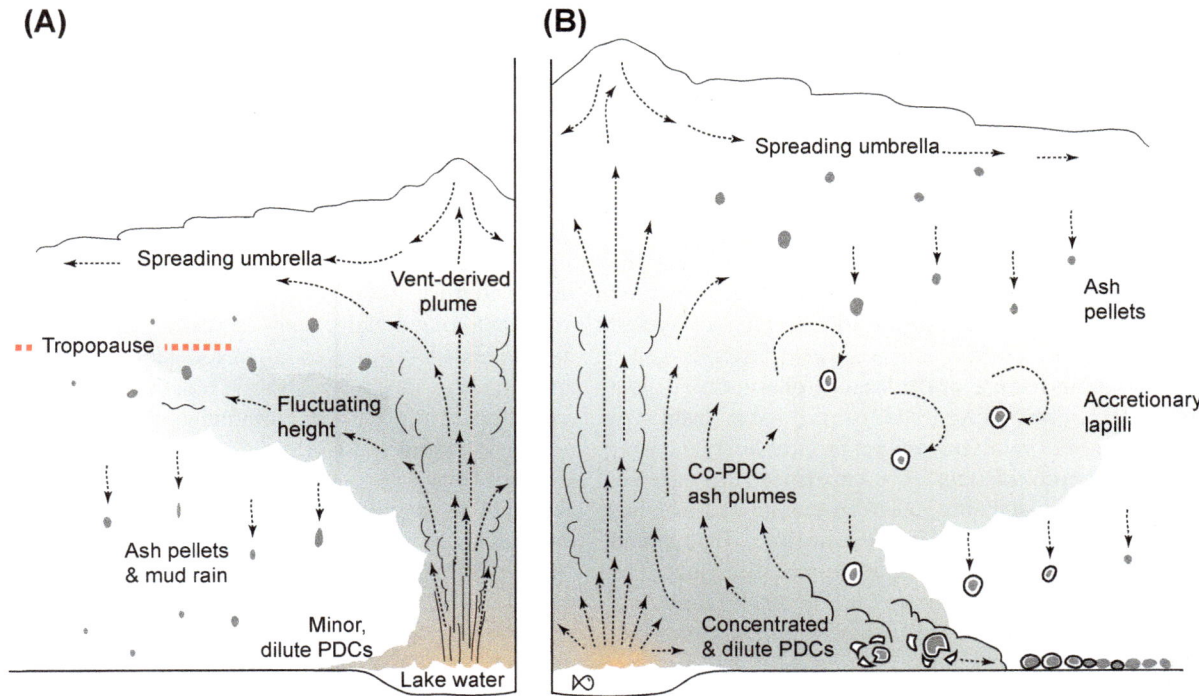

FIGURE 30.8 Cartoon showing two styles of sustained, wet plume development and associated deposits during phreatoplinian volcanism. (A) During an event like the 1.8-ka Hatepe Ash at Taupo, New Zealand, vent-derived plumes are characterized by fluctuating heights, and dilute PDCs play a relatively minor role. Ash aggregates fall mainly as structureless pellets or coalesce into vesiculated tuff. In an Oruanui-like scenario (B), voluminous PDCs (dilute or concentrated) significantly affect transport processes in the proximal area, leading to a hybrid eruption column made up of vent-derived and co-PDC plumes. Weakly structured ash pellets that are reentrained (especially into co-PDC plumes) develop a distinctively finer-grained outer rim. This leads to a general decrease in aggregate size and structure with distance away from proximal PDCs. Note that wet plumes can significantly propagate upwind due to their high density.

This clearly shows that such secondary processes were syneruptive, and a complex and intimate relationship existed between primary pyroclast deposition and secondary remobilization.

The Rotongaio Ash shows conspicuous evidence for prolonged episodicity. Like the Ikedako ash from Ikeda caldera in Japan, and Unit E of the Terra Blanca Jovan tephra from Ilopango caldera in El Salvador, it contains numerous sharply defined millimeter- to centimeter-thick beds that were the products of numerous powerful but discrete explosions that gave rise to individual short-lived plumes and numerous short-lived episodes of deposition. The model for the eruption involves abundant lake water coming into contact with magma well after the peak of degassing, transient powerful explosions, and the development of unstable, short-lived eruption plumes with wide fluctuations of the water–pyroclast ratio. The Rotongaio Ash has an extremely low wall-rock lithic content, and a predominance of dense glassy juvenile clasts. Fragmentation appears to have taken place after outgassing, and explosivity driven by rapid vesiculation appears to have been unimportant.

3.3. Vesiculating Magma and Surface Water: Oruanui, New Zealand

The Oruanui eruption of Taupo volcano, New Zealand (ca. 25.4 ^{14}C-ka) is one of the largest phreatomagmatic eruptions documented worldwide. Over a period of weeks to months, the eruption produced approximately 530 km^3 of high-silica rhyolite (dense-rock equivalent) in 10 major phases of activity (Wilson, 2001). Each phase produced interlayered fall and PDC deposits, and is inferred to have undergone some level of interaction with water. The following observations are consistent with eruption through a large, long-lived lake:

- Extremely fine-grained volcanic deposits, rich in ash aggregates;
- Entirely nonwelded ignimbrite even where >200 m thick, suggesting relatively low-temperature emplacement; and
- Widespread lacustrine sedimentation leading up to the eruption, with ubiquitous incorporation of freshwater diatoms (algae skeletons) in the volcanic deposits.

Ejecta are dominated by vesicular pumice and bubble wall shards, consistent with an actively vesiculating magmatic foam meeting lake water during rapid ascent to the surface. However, the eruption appears to have alternated between "wet" and "dry" phases of deposition throughout, as seen in changes in the abundance of fine ash and aggregates between units. For example, deposits with a clear phreato-magmatic signature (e.g., units 3, 6, and 8) are dominated by PDC deposits and associated co-PDC fall layers. The ash aggregates in these units span a wide range of internal microstructures, from mud rain to complexly layered accretionary lapilli.

An intriguing feature of the wettest phase, unit 3, is a reversal of the usual fining trend; overall mean grain size is finest near the vent, becoming progressively coarser outward to approximately 110 km from the source. Similar outward coarsening has been recognized in other deposits, including the 1.8-ka Hatepe ash, Taupo, and 1991 co-PDC ash from Unzen, Japan, and is attributed to early fallout of fine ash particles (<250 μm) by vigorous, wet aggregation. Despite their water-rich nature, the wet Oruanui units lack the kind of syn-depositional remobilization observed in deposits from the Rotongaio Ash, suggesting somewhat more modest liquid water contents overall (less than ~ 30 wt.%, the slurrying threshold for fine ash).

In contrast, units 2 and 5 are similar to "dry" Plinian deposits, comprising relatively well-sorted lapilli fall layers that become progressively finer away from source. The subordinate amounts of ash aggregates are mainly weakly bound particle clusters and small, massive ash pellets, reflecting lower water contents in the erupted mixture overall. These drier units also contain smaller proportions of PDC deposits, indicating a dominantly stable, vent-derived plume (Figure 30.9(A)).

It is interesting that the most pronounced eruption column instability appears to have occurred during the wettest phases of eruption. The complexly interlayered fall and PDC deposits point to a hybrid plume system, arising from both vent-derived and co-PDC clouds for sustained periods (Figure 30.9(B)). Given the difficulty of distinguishing emplacement mechanisms on the basis of grain size alone (due to their fine-grained, poorly sorted nature overall), the microstructures of ash aggregates become a particularly useful tool for these kinds of deposits. In this regard, the wettest Oruanui units show a distinct lateral transition. Where high-level plumes interacted with vigorous updrafts from PDCs in the proximal region, deposits contain accretionary lapilli with ultrafine outer rims (ash dominantly <10 μm, see Figure 30.10(B)). Beyond the runout distances of PDCs ($>40-80$ km from source, depending on the unit), ash aggregates become progressively smaller, less internally complex, and land purely as fall deposits (Figure 30.9(B)). Generation of voluminous PDCs is thus inferred to be an important factor in the growth of large and complex aggregates during this style of volcanism.

3.4. Vesiculating Magma and Groundwater: Phase C Askja 1875 Eruption, Iceland

The 17-h-long rhyolitic eruption of Askja volcano, Iceland in 1875 was a complex but well-documented silicic explosive eruption characterized by abrupt and reversible shifts in eruption style, for example, from "wet" to "dry" eruption conditions, and transitions between fall to PDC transport and sedimentation (Figure 30.11(A)). We include it here because the phreatoplinian eruption phase C interacted with a very small pond of water fed by groundwater

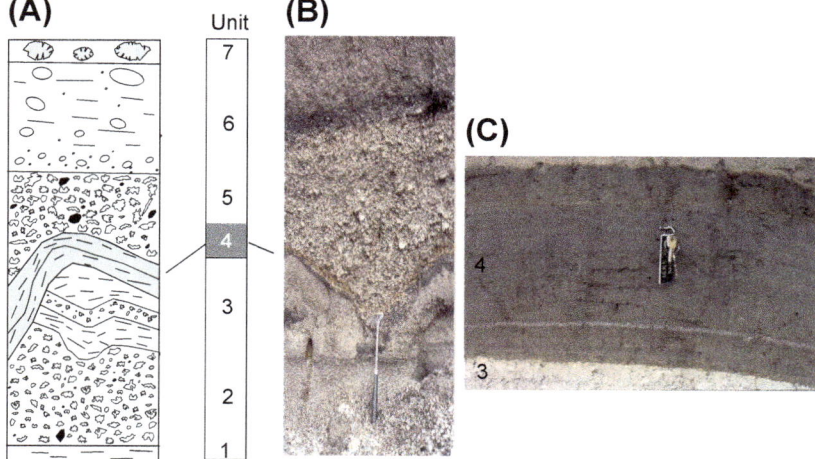

FIGURE 30.9 Stratigraphic log (A) and photographs (B,C) of the products of the 1.8-ka eruption of Taupo volcano. Unit 4, the Rotongaio Ash, is noticeably fine grained in even the most proximal outcrops (C) and distinctively dark in color (B) due to the abundance of poorly to moderately vesicular pyroclasts derived from outgassed magma.

FIGURE 30.10 (A) Deposits of the 25.4 ka Oruanui eruption, Taupo, New Zealand, 85 km from the source, showing ignimbrite scouring the top of the fall layers. Units 1—8 are labeled on image. Note the scalloped base of units 7 and 8 at this site, caused by ash aggregates impacting into underlying layers. Dry hand samples show the matrix-supported ash aggregates characteristic of Oruanui ignimbrite (B) and clast-supported ash aggregate structure of fall deposits (C).

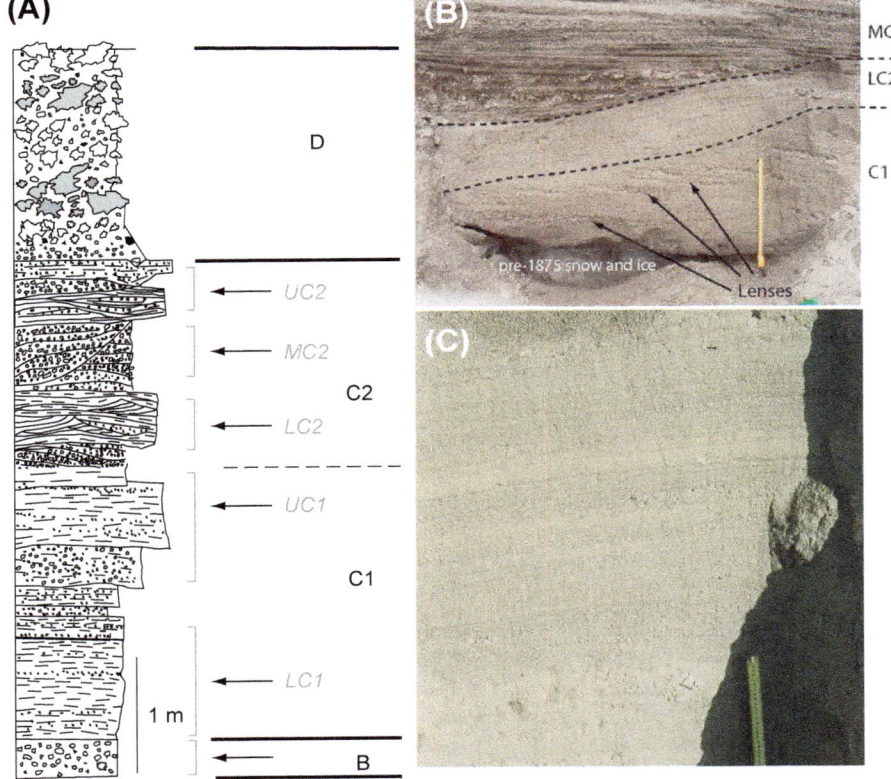

FIGURE 30.11 Proximal deposits of the 1875 Askja eruption, Iceland. (A) Stratigraphic log showing the contrast of fine-grained phreatoplinian unit C and coarse-grained subplinian unit B and Plinian unit D, *after Carey et al. (2009)*. (B) Image showing a proximal section for Unit C. The lower fall ash has discontinuous partings of fine-medium lapilli. The overlying PDC deposits show indications of a progressive decrease in the water/magma ratio. (C) Image showing the close-up view of finely bedded nature of the C1 fall deposit at a proximal site.

and ice melt runoff rather than a large caldera lake. The phreatoplinian C phase reflects the interaction of rapidly vesiculating magma and shallow groundwater, which was contained within joints and fractures in basaltic lavas infilling an older Holocene caldera.

The main eruption began with a subplinian dry phase (B), followed by a shift in vent location and the onset of the phreatoplinian fall phase (unit C1). Dilute density currents (unit C2) were emplaced after the phreatoplinian fall (Figure 11(B)) but were largely confined to the larger Askja caldera region. A shift in vent position was then associated with an abrupt change in eruptive style producing Plinian Unit D. The phreatoplinian phase lasted for approximately 1 h, depositing 0.45 km^3 of "wet, sticky gray ash." This C1 fall deposit consists of very fine, pale gray, uniformly massive ash up to 2 m thick (Figure 11(C)), which contains vesicles in the matrix, but only rare ash aggregates. Lack of fine bedding suggests steady, sustained eruption conditions and the time-averaged mass discharge rate was 7×10^7 kg s^{-1}, and is very similar to those inferred for phreatoplinian phases at Taupo volcano. Grain size analyses conducted by Sparks et al. (1981) suggest that 99 wt.% of this deposit is finer than 1 mm. The fall tephra fall deposits are dominated by vesicle wall fragments, suggesting that the magma was actively vesiculating and a foam at fragmentation. Vesicle size and volume distributions suggest that the processes of nucleation at depth, decompression, and ascent rates were similar between wet (C) and dry (B,D) phases of the 1875 eruption, and therefore, water source and availability were primary factors influencing the eruption style and plume behavior. Unit C is widely dispersed to the east–northeast, extending to Scandinavia where it cannot be distinguished from the Plinian fall. In the proximal area, unit C1 is overlain by C2 dilute density current deposits. Three major intervals of PDC deposition occurred, of successively "drier" character.

Transitions between "dry" and "wet" styles were driven by shifts in vent position, permitting movement of the vent in and out of groundwater. The shift between C1 and C2 is inferred to be a function of the reduced availability of water, and vent widening, decreasing the exit velocity of the jet.

4. DISCUSSION

4.1. "Ash Aggregation" in "Wet" Plumes

One of the key ways in which external water affects the airborne transport of volcanic products is by promoting aggregation of ash. Aggregation presents a challenge to traditional deposit characterization technique of grain size and componentry. Grain size analysis typically ignores any aggregated clasts that were originally present on deposition, due to unintentional breakup during sieving—a practice that obscures the sizes and types of the particles as they originally landed.

Investigating how and when aggregation occurs is crucial to our understanding of plume dynamics, particularly from "wet" eruptions. In the simplest sense, aggregation takes place whenever attractive forces between particles overcome the dispersive forces breaking them apart. However, there are factors at a range of length scales that influence aggregate growth and structure: particle binding and cementation processes (at the scale of microns), volcanic plume dynamics (meters to kilometers), and interaction with large-scale weather patterns (hundreds of kilometers).

On the particle scale, binding forces operate in two distinct categories: hydrostatic and electrostatic. Weak, but long-range electrostatic attraction is thought to be omnipresent in volcanic plumes, forming fragile, porous clusters akin to dry, snowflakes. This mechanism mainly affects particles <63 μm in the absence of a liquid phase. In contrast, wet aggregation dominates when liquid is abundant (more than ∼5% by weight in the case of water). The particles adhere by hydrostatic forces due to cohesion of water molecules, forming liquid bridges between grains. In general, increasing the amount of liquid water involvement leads to larger, denser aggregates that trap larger grains, although fine ash <250 μm tends to be most strongly affected. Once individual particles have been brought into close range by these binding forces, more permanent cementation occurs by precipitation of salts out of solution, ice formation, and perhaps the products of glass alteration.

On the scale of dynamic plume processes (meters to many kilometers), the evolving style of aggregation responds to two factors: (1) availability of liquid water and (2) presence of turbulent updrafts. There is some indication that in water-rich eruptions, wet aggregation in the eruption column and near-source umbrella cloud can occur extremely rapidly, within minutes to tens of minutes. In these cases, cyclic collection, freezing and remelting of ashy water droplets contribute to aggregate growth during moist convection. There is also a complex interplay between rapid, wet growth in warm regions of the plume, and weaker, electrostatic aggregation elsewhere. This is reflected in the detail of aggregate structures; the most complexly layered types represent multiple stages of growth by recycling through warm, moist updrafts—especially those rising from ground-hugging PDCs (Figure 30.9(B)). Growth of the outermost ultrafine rim (rich in <10-μm ash) seen on many accretionary lapilli is thought to occur during final passage through a low-level cloud of elutriated, co-PDC ash, which may or may not still be connected to a laterally moving density current (Brown et al., 2010; Van Eaton and Wilson, 2013).

On the largest scale of continental or global transport, aggregation becomes primarily a meteorological phenomenon—in other words, controlled by background weather patterns. In these distal regions, well downwind of the eruption column, volcanic particles serve as nucleation sites for precipitation processes, and are eventually scavenged out of the atmosphere by rain, snow, or as delicate, electrostatically bound clusters. These processes generate distal fall deposits with a low-density, porous texture.

4.2. Models for "Wet" Plumes

Volcanic plumes are remarkably sensitive to the addition of external water. Glacial melting or water flashing to steam fundamentally redistributes the energy in a rising column of particles and gases. Although an influx of external water cools the erupting mixture, it does not necessarily produce a sluggish, low-level plume. In fact, "phreatoplinian deposits" are among the most powerfully dispersed deposits (Self and Sparks, 1978), leading to the apparent paradox of widespread dispersal despite cool, water-rich, and often unstable plume behavior.

The key factor here is that water vaporization transfers an enormous amount of energy to the gas phase, where it is stored as latent heat. As soon as the erupted mixture expands outward and entrains enough cool, ambient air for condensation (and eventually freezing) to occur, that stored energy is released, leading to rapid expansion of the cloud (Koyaguchi and Woods, 1996). Other phase changes of water provide sources and sinks of energy and buoyancy during the lifetime of a volcanic plume. This process, known as moist convection, is characteristic of conventional thunderstorms, with the key difference being that volcanic clouds usually start with more thermal energy and momentum, allowing them to puncture the tropopause and ascend into the stratosphere (Van Eaton et al., 2012).

Early one-dimensional (1D) models of wet plumes concluded that the most intense water-rich eruptions (e.g., Oruanui) were likely to collapse before condensation could occur, along with its boost in buoyancy, thereby producing low-temperature PDCs confined to lower levels of the atmosphere (Koyaguchi and Woods, 1996). However, more recent 3D multiphase models accounting for microphysical processes show that ash plumes can rise vigorously from these ground-hugging flows (Van Eaton et al., 2012). Convective columns of fine ash feed into, and merge with, the main jet issuing from vent, producing a hybrid system that expands outward as a powerfully spreading umbrella cloud (Figure 30.9(B)). Such "dirty thunderstorm" dynamics, powered by moist convection, provide an overall picture that is more consistent with the observational evidence for powerful dispersal in the most intense wet eruptions. Interestingly, this redistribution of energy helps delineate wet and dry behavior in another

way. While dry, plinian eruption columns tend to have the highest vertical velocities near the vent, where energy and momentum are the greatest, unstable, water-rich plumes experience a delayed onset of the maximum vertical velocities, once rapid growth during moist convection occurs some distance vertically (and laterally) away from the vent.

Although wet plumes containing water contents of >15% are more unstable and prone to collapse than are dry ones, the reality is that water involvement does not have a simple, predictable effect on eruption style. Three-dimensional numerical models suggest that changing the external water content of the erupted mixture by even a few weight percent produces nonlinear effects in the dynamic system that cannot be categorized in a simple way (Koyaguchi and Woods, 1996; Van Eaton et al., 2012). However, in a broad sense, increasing water involvement does seem to increase the likelihood of rapid oscillations between Plinian-style (buoyant) ascent and partial to wholesale column collapse (Figure 30.9(B)). This leads to overprinting of multiple transport levels; coarse ash and lapilli are transported farther during stable phases of eruption, yet deposited closer to source during periods of fluctuating or lower plume heights (Figure 30.10(A)). Further, in cases where moist convection dominates the plume behavior, a substantial percentage of erupted mass stays in the troposphere (like a thunderstorm), where a temperature inversion inhibits the rise of weaker portions of the eruption column. Here, the plume spreads outward radially as a gravity current. This combination of fluctuating, multilevel transport and upwind/crosswind expansion at the tropopause (Figure 30.10(A) and (B)) may contribute to the more radially symmetric distribution of wet eruption deposits around their volcanic source areas.

4.3. Role of the Vent and Shallow Conduit

Vent/conduit wall properties: The walls of the vent allow, in one or more ways, external water to encounter magma and contribute to initial and/or secondary fragmentation and eruption. The simplest way in which this happens, exemplified by activity at Surtsey, Iceland, or Taupo, New Zealand, is for surface water to occupy or flow directly into the vent during eruption. In such a situation, the first phreatomagmatic explosions take place by interaction of magma with some small part of an effectively unlimited supply of clear ocean, river, or lake water, and the ejecta consist of juvenile pyroclasts and water. Under these circumstances, the form of the vent and shallow conduit strongly influences the geometry of the interaction.

Many subaerial phreatomagmatic eruptions begin under different circumstances than those of subaqueous vents. In particular, there is no body or sizeable domain of water with which magma can interact. Instead, magma encounters interstitial water in fractures of the surrounding wall rocks, or contained within a porous and permeable framework in these deposits. Two end-member modes of interaction can be described for this situation: (1) water flows through pores or fractures in the wall rock to interact with magma; (2) the wall rock and contained interstitial water are disrupted, and together interact with magma as wet blocks or an impure fluid.

In many explosions, the yield strength of the shallow wall rock is negligible, and the primary grain size of vent-wall material has a strong effect on the grain size of the wall-rock lithic clast population of the resulting pyroclastic deposit. In particular, if there is abundant ash-sized wall-rock material, its inclusion (often without any modification to its size distribution by the explosions) can lead to a misleadingly high estimate of the efficiency of fragmentation. Surtsey exemplifies another type of wall instability, in that the fragmental deposits formed early in the eruption form the walls of the vent later in the eruption. In this situation, the vent is occupied by a slurry of seawater and tephra that sloughed and slid back into the vent (Figure 30.6), which strongly affects eruption style.

Vent geometry: The geometry of the vent will change as an eruption proceeds, and this may feed back to modify the style of the eruption. Cone growth will influence the surface hydrology of the vent, and in general act to restrict the ingress of water. Growth of the cone or tuff ring may also permit magma to "pond" in the growing vent at levels above the water table, bringing an end to "wet" phreatomagmatic fragmentation. Vent-wall collapse is more important in phreatomagmatic than in dry magmatic eruptions. Changes of vent geometry, such as collapse of the inner walls of the growing tuff cone or ring, may promote ongoing magma–water interaction or may temporarily block the vent leading to episodic vent-clearing explosions. Alternatively, vent widening as a result of shallow collapse may promote a change to wetter explosions if surface water can pond in the crater.

Recycling of clasts is a process strongly influenced by vent geometry. Most pyroclastic deposits contain both first-cycle juvenile clasts, derived from magma at the instant of eruption, and recycled juvenile clasts that were fragmented in earlier explosions but then fell or collapsed back into the vent. The recycled clasts are similar to wall-rock lithic clasts in that they contribute no heat to further magma/water interaction, but they can be difficult to separate from "first-cycle" juvenile clasts. Textural features, such as a combination of uneven mud coating and a mixture of fluidal and angular faces on many clasts, indicate that they were fragmented at least twice—once while the magma was fluid and subsequently in a brittle state.

4.4. Role of the Magma in Wet Volcanism

As in all volcanic eruptions, magma is the source of energy for phreatomagmatic explosions. Many phreatomagmatic events are modifications of explosive eruptions that would have occurred even in the absence of external water, essentially the phreatomagmatic equivalents of Hawaiian, Strombolian, subplinian, and plinian eruptions. Exceptions are the equivalent of extrusive events, such as the Rotongaio Ash, which in the absence of water would have erupted only lava, or Kilauea 1924, which prior to the explosive eruption was an actively receding lava lake. In interpreting the style and products of phreatomagmatic eruptions, it is important to consider the influences of physical properties of the magma and the "background" magmatic processes (cooling, vesiculation, crystallization, etc.).

Intrinsic properties of the magma include temperature, viscosity, dissolved volatile content, vesicularity, and crystallinity. During fragmentation, thermal energy is transferred to the external water phase (latent and sensible heat) and/or is converted to seismic and acoustic energy, fragmentation energy, and kinetic energy. This energy partitioning has major influences of the form of the volcanism. Preeruptive magma (and water) temperatures influence the heat transfer rates and the temperatures in the eruptive jet and/or plume. The absolute amount of heat available is clearly also a factor.

Magma viscosity is strongly dependent on chemical composition, volatile content, temperature, and crystal content. An increase in viscosity will retard the mixing of magma and water, yet, by slowing the ascent of magma, it may prolong the opportunity for contact between the two phases. The magmas described in this chapter show a range of viscosity at liquidus temperatures from few tens of Pascal seconds, in the case of basalts, to 10^7 to 10^9 Pa-s for silicic magmas. A 200 °C decrease in temperature will produce a 10−100 times decrease in viscosity. Viscosity of the magmatic phase at the time of fragmentation may reflect both lowered temperatures as a result of cooling during ascent and early interaction with the conduit walls and groundwater, and the effects of degassing and microlite crystallization that accompanied ascent. Vesiculation also increases melt surface area while reducing the density of the bulk melt phase and may therefore promote better geometry of mixing.

Dynamic properties of the magma include ascent rate, discharge velocity and steadiness, and extent of vesiculation and outgassing. Processes of magma ascent close to the fragmentation zone have a major influence on the ensuing style of phreatomagmatism. The extent of degassing and microlite crystallization is strongly influenced by the rates of ascent in the deeper conduit, and degassing and crystallization and, in turn, influences magma rheology and rise rate at shallower levels.

A magma that rises rapidly and continuously to the surface is likely to undergo most of its vesiculation at shallow levels and be relatively fluid, hot, and homogeneous at the point of contact with external water. Relatively simple phreatomagmatic eruption styles result, as typified by most phreatoplinian eruptions. If magma ascends in a staged fashion and/or subsequently stagnates in the shallow conduit, then at least some portion of this melt will have undergone quenching, and outgassing, often becoming physically and thermally heterogeneous. Complex magma−water interaction and phreatomagmatism often result.

4.5. Role of the Preeruptive State of Water

The availability of water or other external fluid has a first-order influence on the style of phreatomagmatism. Relative proportions of water and magma, and the efficiency with which they mix, determine the thermodynamic states of water phases in phreatomagmatic eruptions. If very little water interacts, it can be heated to high temperatures, and high pressures if confinement is available, but its expansion to vapor can fragment only a relatively small volume of magma (Sheridan and Wohletz, 1981). Most of the work of fragmentation is done during the volume increase from liquid water to superheated fluid, rather than by subsequent expansion in the vapor phase. Alternatively, if the external phase is already steam (see below), then its ability to expand on further heating, and thus drive fragmentation, is highly limited. At the other extreme, water is present beyond the mixing domains, and is ejected or slowly boiled during postfragmentation expansion; this water contributes nothing to magma fragmentation. At intermediate ratios of interacting magma and water, the available heat is capable of vaporizing all the water, thus resulting in efficient magma fragmentation.

A variety of relatively shallow influences can disturb the supply of external water or the effectiveness of mixing and interaction, so most wet eruptions are characterized by abrupt changes in the proportions of water and magma and hence eruptive behavior. The simplest example is the formation of a deposit that excludes influx of surface water; thus, many phreatomagmatic eruptions end with "dry" phases. In models and some field situations, the fluid involved in magma−water interactions is a liquid initially at ambient temperatures. However, phreatomagmatic eruptions commonly also include examples of magma contacting one or more of

1. vapor-dominant geothermal fluid at temperatures up to 800 °C,
2. superheated geothermal liquids at temperatures of 200−250 °C,
3. snow or ice,
4. impure fluids in which sediment particles are suspended in water,
5. vent slurries with significant yield strengths.

Where magma intrudes a preexisting hydrothermal system, the water phase prior to interaction may be vapor, two-phase vapor−water mixtures, or hot water. Vapor (dry steam) is a minor contributor to explosivity because the major volume change is that of vaporization, and subsequent (super)heating causes only a minor volume increase. The state of water most effective in forming phreatomagmatic explosions is superheated liquid water; it will undergo large volume changes upon vaporization, which can be accomplished by small additions of heat from interaction with magma and by subsequent disruption of the hydrothermal system.

Water involved in the interaction typically contains suspended particles (sediment or tephra) or dissolved species (e.g., salt in seawater, or silica and chlorine in hydrothermal fluids). These impurities change the physical properties of the water phase and introduce nonvaporizable components. Depending on the concentration of impurities in the water, its viscosity may increase up to an order of magnitude, the density may double and the heat capacity may decrease by 25%. Precipitation of dissolved ions during vaporization absorbs energy.

There is a subtle distinction that could be made between "wet" and phreatomagmatic eruptions. In the latter, it is assumed that fragmentation is induced or enhanced by the flashing of water to steam during the magma's ascent to the surface. However, water can also become involved after fragmentation. A challenge therefore exists in relating clear evidence of wet deposition to a case for magma−water interaction in the vent. Larger eruptions have an opportunity to interact with external water well beyond the vent area, including large-scale weather systems. While there are plenty of wet eruptions for which the source of external water is known or reasonably inferred, others are not so clear cut, including the preclimactic eruptions of Pinatubo 1991. The source of external water remains cryptic in these cases, despite the noted role of moist convection and precipitation processes. Might some unconventional pathways, such as the incorporation of wall rock derived from a water-saturated volcanic edifice, or vaporization of rivers, streams, and moist vegetation by hot PDCs, deliver sufficient water to induce particle aggregation? Further work is required to learn how and when these mechanisms play a role in blurring the line between wet and dry plume behavior.

4.6. Heterogeneity of Magma−Water Mixing and Thermal Efficiency

The extent/efficiency with which magmatic heat can be used during fragmentation and transport is important; this is strongly influenced by a number of processes described above. The unifying characteristics of phreatomagmatic eruptions are the involvement of external water and heterogeneity of eruption processes and products. In particular, many of the case studies above illustrate how heterogeneous heat transfer can be.

Dry magmatic eruptions (Strombolian, Hawaiian, etc., see Chapter 27) involve two phases, the silicate liquid and a gas/vapor phase, which we can often treat as thermally, compositionally, and physically homogeneous. In wet systems, the starting point is magma at high temperature in contact with a source of external water and any material that contains or confines the water, both of which are at much lower, commonly ambient temperatures. The end product is a mixture of juvenile particles, magmatic gas, steam, wall-rock particles, and often liquid water (Figure 30.1). Both the starting ingredients and the end products are heterogeneous (Figure 30.11).

Total mixing is rarely achieved, and a wide temperature range often characterizes the products of a single small explosion (i.e., it is not uncommon for both liquid water droplets and steam to be present in the eruption and for cold and hot clasts to be ejected simultaneously). The inescapable conclusion from this observation is that, in many explosions, some water is ejected that is never in significant contact with magmatic heat and some magma never interacts with external water. For larger sustained eruptions, thermal and physical mixing is often more complete. but rapid and reversible temporal shifts in the magma—water ratio are recorded, and usually ascribed to changing vent locations, construction of tephra barriers, or ground deformation (inflation/deflation).

4.7. Concluding Thoughts

We have a clear need for a more comprehensive set of criteria to define phreatomagmatic eruption styles. This is one of two great challenges remaining; the other is comprehensive modeling of the thermodynamics of eruptions of the types described in the case studies above.

FURTHER READING

Brown, R.J., Branney, M.J., Maher, C., Dávila-Harris, P., 2010. Origin of accretionary lapilli within ground-hugging density currents: evidence from pyroclastic couplets on Tenerife. Geol. Soc. Am. Bull. 122, 305–320.

Carey, R.J., Houghton, B.F., Thordarson, T., 2009. Abrupt shifts between wet and dry phases of the 1875 eruption of Askja Volcano: microscopic evidence for macroscopic dynamics. J. Volcanol. Geotherm. Res. 184, 256–270.

Fee, D., Garcés, M., Patrick, M., Chouet, B., Dawson, P., Swanson, D., 2010. Infrasonic harmonic tremor and degassing bursts from

Halema'uma'u Crater, Kilauea Volcano, Hawaii. J. Geophys. Res. 115, B11316,. http://dx.doi.org/10.1029/2010JB007642.

Hamilton, C.W., Thordarson, T., Fagents, S.A., 2010. Explosive lava—water interactions I: architecture and emplacement chronology of volcanic rootless cone groups in the 1783–1784 Laki lava flow, Iceland. Bull. Volcanol. 72, 449–467.

Houghton, B.F., Smith, R.T., 1993. Recycling of magmatic clasts during explosive eruptions: estimating the true juvenile content of phreatomagmatic volcanic deposits. Bull. Volcanol. 55, 414–420.

Kienle, J., Kyle, P.R., Self, S., Motyka, R.J., Lorenz, V., 1980. Ukinrek Maars, Alaska, I. April 1977 eruption sequency, petrology and tectonic setting. J. Volcanol. Geotherm. Res. 7, 11–37.

Kokelaar, P., 1983. The mechanism of Surtseyan volcanism. J. Geolo. Soc. London 140, 939–944.

Kokelaar, B.P., 1986. Magmaewater interactions in subaqueous and emergent basaltic volcanism. Bull. Volcanol. 48, 275–289.

Koyaguchi, T., Woods, A.W., 1996. On the formation of eruption columns following explosive mixing of magma and surface water. J. Geophys. Res. 101, 5561–5574.

Mattox, T.N., Mangan, M.T., 1997. Littoral hydrovolcanic explosions: a case study of lava—seawater interaction at Kilauea Volcano. J. Volcanol. Geotherm. Res. 75, 1–17.

Patrick, M., Wilson, D., Fee, D., Orr, T., Swanson, D., 2011. Shallow degassing events as a trigger for very-long-period seismicity at Kīlauea Volcano, Hawai'i. Bull. Volcanol. 73, 1179–1186.

Self, S., 1983. Large-scale silicic phreatomagmatic volcanism: a case study from New Zealand. J. Volcanol. Geotherm. Res. 17, 433–469.

Self, S., Sparks, R.S.J., 1978. Characteristics of widespread pyroclastic deposits formed by the interaction of silicic magma and water. Bull. Volcanol. 41, 196–212.

Sheridan, M.F., Wohletz, K.H., 1981. Hydrovolcanic explosions: the systematics of water—pyroclast equilibration. Science 212, 1387–1389.

Sparks, R.S.J., Wilson, L., Sigurdsson, H., 1981. The pyroclastic deposits of the 1875 eruption of Askja, Iceland. Phil. Trans. R. Soc. London Ser. A 299, 241–273.

Swanson, D., Wooten, K., Orr, T., 2009. Buckets of ash track tephra flux from Halema'uma'u Crater, Hawai'i. Eos Trans. Am. Geophys. Union. 46, 427–428.

Thorarinsson, S., 1967. Surtsey. The New Island in the North Atlantic. The Viking Press, New York, pp. 47.

Van Eaton, A.R., Wilson, C.J.N., 2013. The nature, origins and distribution of ash aggregates in a large-scale wet eruption deposit: Oruanui, New Zealand. J. Volcanol. Geotherm. Res. 250, 129–154.

Van Eaton, A.R., Herzog, M., Wilson, C.J.N., McGregor, J., 2012. Ascent dynamics of large phreatomagmatic eruption clouds: the role of microphysics. J. Geophys. Res. 117, B03203. http://dx.doi.org/10.1029/2011JB008892.

Walker, G.P.L., 1981. Characteristics of two phreatomagmatic ashes and their water-flushed origins. J. Volcanol. Geotherm. Res. 9, 395–407.

White, J.D.L., 1991. Maar-diatreme phreatomagmatism at Hopi Buttes, Navajo Nation (Arizona), USA. Bull. Volcanol. 53, 239–258.

Wilson, C.J.N., 2001. The 26.5 ka Oruanui eruption, New Zealand: an introduction and overview. J. Volcanol. Geotherm. Res. 112, 133–174.

Chapter 31

Submarine Explosive Eruptions

James D.L. White

Geology Department, University of Otago, Dunedin, New Zealand

C. Ian Schipper

School of Geography, Environment and Earth Sciences, Victoria University of Wellington, Wellington, New Zealand

Kazuhiko Kano

The Kagoshima University Museum, Kagoshima University, Korimoto 1-chome, Kagoshima, Japan

GLOSSARY

deep submarine eruptions occur below storm wave base, typically beyond the continental shelf. In rock successions, the distinction between above versus below wave base is commonly the only depth indicator.

eruption-fed density currents are negatively buoyant mixtures of fluid, generally water ± vapor and volcanic gases, with particles that were delivered directly into the water column by an eruption.

explosive subaqueous eruptions are those that produce volcanic jets, plumes, or density currents from vents beneath water.

hydrostatic pressure is the pressure exerted by the overlying column of water at a given water depth.

lava balloons are large floating clasts with glass exteriors and frothy or hollow interiors.

pyroclastic deposits are the primary deposits of explosive eruptions, formed from eruption-fed currents, plumes, or jets. Particles comprising these deposits are pyroclasts.

shallow submarine eruptions occur above storm wave base.

surtseyan eruptions are explosive basaltic eruptions, like those of Surtsey, that build toward the water surface and typically emerge (see Chapter 30).

supercritical fluid is the physical state of a substance at pressures and temperatures exceeding its critical point, where neither liquid nor vapor exists separately. For seawater the critical point is 30 MPa and 405 °C; for carbon dioxide it is 7.4 MPa and 31 °C.

volatile-coupled system also called closed-system degassed, describes magma in which volatiles exsolved during its ascent and decompression remain in vesicles within the melt that exsolved them.

1. INTRODUCTION

The submarine realm has been called the last frontier on Earth, and understanding of submarine eruptions naturally lags behind that of their subaerial counterparts. Submarine volcanoes build oceanic crust along spreading ridges, form abundant intraplate seamounts, and build island arcs (Table 31.1). Though almost none of these volcanoes are observed in eruption under water, their abundance makes it probable that they produce more eruptions than our familiar volcanoes on land. Some submarine eruptions are explosive and this chapter summarizes current knowledge and understanding of these elusive events. Submarine eruptions that produce primary volcaniclastic deposits through nonexplosive processes are addressed in the context of submarine effusive eruptions in Chapter 19; subaerial and

The Encyclopedia of Volcanoes. http://dx.doi.org/10.1016/B978-0-12-385938-9.00031-6

TABLE 31.1 Abundance and Significance of Explosive Submarine Eruptions. Subaqueous eruptions, quantitatively dominated by spreading ridge eruptions, are more commonly basaltic than are subaerial ones, and make up a much larger share of total global volcanism. Evidence summarized in this chapter demonstrates that some subaqueous eruptions are explosive. Given total submarine eruptive volumes, even a small proportion of such eruptions is enough to be important in the global transfer of heat and volatiles (sulfur provided as representative of nonwater volatiles) to the oceans. Volumes are derived from Crisp (1984), with a sixth of subduction volcanism considered subaqueous; basalt versus nonbasalt proportions are estimates. Sulfur contents are estimated after Palais and Sigurdsson (1989). Estimated proportions of eruptions that are explosive are illustrative only

	Subaerial	Subaqueous
Estimated % global volcanism	15% (.6 km^3/yr)	85% (3.5 km^3/yr)
• basaltic volcanism	0.19 km^3/yr (30%)	2.8 km^3/yr (80%)
• nonbasaltic volcanism	0.43 km^3/yr (70%)	0.697 km^3/yr (20%)
Sulfur	∼2.4 Mt/yr	∼14.6 Mt/yr
• basalt ∼5 x 10^6 tons/km^3	∼0.93 Mt/yr	∼13.9 Mt/yr
• other ∼10^6 tons/km^3	1.4 Mt	∼0.68 Mt/yr
Volcanic heat released as percentage of global total	14% (30% at 1200C + 70% at 800C)	86% (80% at 1200C + 20% at 800C)
Estimated proportion explosive	80% (generous) of 15% → 12% all eruptions are explosive subaerial ones	5% (pessimistic) of 85% → ∼4% of all eruptions are explosive subaqueous ones

emergent eruptions involving interactions with water are treated in Chapter 30. Fundamental processes of magma–water interaction are described in Chapter 26.

1.1. Volcanoes in the Ocean

All submarine eruptions interact with the ambient water into which they erupt, including groundwater in the permeable seafloor. **Explosive subaqueous eruptions** inject gas + particles into an environment (Table 31.2) that affects shallow magma ascent, fragmentation, and particle cooling and dispersal.

Ambient water influences subaqueous eruptions *indirectly* through the elevated **hydrostatic pressure** present under a water column and *directly* by thermal and mechanical exchanges between magma and water.

1.1.1. Indirect Effect of Water: Higher Confining Pressure

The preeruption volatile budget of a magma is determined by its origin and differentiation history, while the volatile solubility is controlled by temperature, composition, and, most strongly, pressure (see Chapter 26). Confining pressure, whether hydrostatic from an overlying water column (∼1 MPa per 100 m of water depth) or increased atmospheric pressure on other planets (e.g., Venus—9.2 MPa), reduces the exsolution and expansion of magmatic volatiles

(mainly $H_2O + CO_2$), including at shallow levels in the conduit. Eruption under high hydrostatic pressures in Earth's oceans has been inferred in the past to preclude extensive volatile exsolution and expansion, effecting a "volatile fragmentation depth" (VFD), below which magmatic fragmentation (see Chapter 25) and explosive volcanism are not possible. Seafloor deposits suggest otherwise.

For many subaerially erupted magmas and for submarine silicic magmas, it is acceptable to consider H_2O as the only magmatic volatile influencing explosive volcanism. Primitive mid-ocean ridge basalt (MORB) and ocean island basalt (OIB) have low parental H_2O, no more than 1.2 wt%, making these most common submarine magmas undersaturated in pure H_2O until a pressure of ∼10 MPa, the hydrostatic pressure at ∼1 km below sea level. For vent depths much greater than this, H_2O alone will not exsolve from these magmas, let alone to play a significant role in eruption dynamics. Intraplate, and especially arc-related, basalts have the potential for higher H_2O contents, thus somewhat deeper VFDs. More differentiated silicic magmas generally have much higher preeruptive H_2O contents, and therefore deeper VFDs.

As vent depth increases, so too does the potential importance of CO_2 in driving eruptions. Submarine lavas called "popping rocks" show that at ocean ridge depths, CO_2 can significantly vesiculate magma, and CO_2 is a major component of fluids discharged from active

TABLE 31.2 Some Important Properties of Water, Air, and Water Vapor/Steam. The key information is that water vapor, or steam, is more like air than water in most properties. Because water in contact with magma invariably boils to produce steam, microscale air-like environments are formed; the size and stability of these vapor zones varies greatly with interaction dynamics, and can give rise to subaerial-like features or behavior at different scales

	Air	Water	Steam (vapor)
Density	1.207 kg/m^3 at 20 C 0.101 MPa (1 atm) 83.98 kg/m^3 at 20 C, 6.90 MPa (1000 psi) [A]	998.22 kg/m^3 at 20 C 0.101 MPa	0.5897 kg/m^3 at 100 C, 0.101 MPa [B] 225.2 kg/m^3 at 375 C, 22.2 MPa [C]
Viscosity	17.9 μPa s at 20 C, 0.101 MPa	1000 μPa s at 20C, 0.101 MPa	23.41 μPa s at 375 C, 0.1 MPa 31.7 μPa s at 375 C, 22.2 MPa [C]
Specific Heat Capacity	1158 J/kg K at 27 C , 0.101 MPa	4181.8 J/kg K at 20 C, 0.101 MPa	4039.2 J/kg K at 100 C, 0.101 MPa
Thermal Conductivity	0.025 W/m K at 0.101 MPa,	.6072 W/m K at 25 C, 0.1 MPa [D]	.0251 W/m K at 100 C 0.1 MPa [D] 0.1132 W/m K at 600 C 10 MPa [D]
Thermal Diffusivity	2.17 x 10^{-5} m^2/sec =21.7 mm^2/sec at 20 C, 0.101 MPa [E]	0.1456 mm^2/sec at 25 C, 0.1 MPa [D]	20.83 mm^2/sec at 100 C, 0.1 MPa [D] 2.240 mm^2/sec at 800 C, 10 MPa [D]

Data from The Physics Hypertextbook, physics.info, accessed 15 November 2013, unless otherwise indicated.
[A]http://www.engineeringtoolbox.com/air-temperature-pressure-density-d_771.html, accessed 15 Nov 2013. Converted from imperial units.
[B]http://www.efunda.com/Materials/water/steamtable_general.cfm, accessed 15 Nov 2013.
[C]Watson JTR, Basu RS, Sengers JV (1980) An improved representative equation for the dynamic viscosity of water substance. Journal of Physical and Chemical Reference Data 9(4):1255−1290.
[D]Sengers JV, Watson JTR, Basu RS, Kamgar-Parsi B, Hendricks RC (1984) Representative equations for the thermal conductivity of water substance. Journal of Physical and Chemical Reference Data 13(3):893−933.
[E]ftp://ftp1.esrl.noaa.gov/users/cfairall/wcrp_wgsf/flux_handbook/andreas_Handbook_BF2.doc, accessed 15 Nov 2013. U. S. Army Cold Regions Research and Engineering Laboratory, Physical constants and functions for use in Marine Meteorology, Appendix A, 2005.

submarine volcanoes. CO_2 begins to exsolve at very high pressures (e.g., ~2200 MPa: the lithostatic pressure at a depth of ~70 km, for a saturated tholeiite containing 0.5% CO_2), deep below the vent, where it affects subaerial and submarine eruptions identically. At a deep submarine vent, however, magma is released to hydrostatic pressures of 10−20 MPa, rather than atmospheric pressure, resulting in significantly different proportions of exsolved CO_2 versus magmatic H_2O.

Bubbles of exsolved CO_2 accumulated below submarine conduits can coalesce and rise as decoupled slugs that burst as at Stromboli (Head and Wilson, 2003; Clague et al., 2009). Calculations of vesicle volumes based on exsolution from initial magma compositions cannot address such volatile-decoupled (also called open-system degassing) systems in which CO_2 is exsolved and accumulated from large volumes of unerupted magma. These same accumulation processes that can allow CO_2 to drive abyssal eruptions make it exceedingly difficult to establish the parental CO_2 contents of magmas. Melt inclusion studies of MORB from recent eruptions at the Juan de Fuca Ridge indicate preeruption CO_2 in excess of 9000 ppm for some samples, the highest yet measured for any magma, and greatly exceeding even those of ocean island basalt inclusions. Another important effect of exsolved CO_2, even in small amounts in a volatile-coupled system, is that

CO_2-rich fluids can cause H_2O to exsolve at higher pressures than it would on its own (see Chapter 25). For eruptions with vents up to 2 km deep, such excess H_2O exsolution could increase vesiculation and explosivity more than the addition of CO_2 alone.

1.1.2. Direct Effects: Thermomechanical Interactions of Magma with Water

Water can assist or drive fragmentation of magma in several ways (see Chapter 26), depending particularly on eruption depth, prefragmentation vesiculation, mass discharge rate, and vent geometry (Kokelaar, 1986). Water can cause explosive fragmentation in a submarine eruption that would, without water, be effusive or fail to erupt at all (i.e., cause explosions involving shallow intrusions); it can fragment lava in nonexplosive ways, and it can enhance some types of magmatic fragmentation.

Heat transfer to water causes submarine melts to quench quickly. Rapid quenching under hydrostatic pressures means that submarine pyroclasts preserve better the syn-fragmentation microtextures (bubbles + crystals) of the erupting magma, and at high hydrostatic pressures, quenching freezes residual dissolved volatiles into matrix glasses. These glasses can provide information about the depths of fragmentation and the extent of prior degassing.

In quenched clasts, phenocryst-hosted "melt" inclusions, captured during crystal growth prior to eruption, also remain as glass.

After fragmentation, the water column greatly influences how pyroclasts are dispersed and deposited. Compared with subaerial eruptions, the high density, viscosity, and heat capacity of ambient water (Table 31.2) limit the height of submarine eruption plumes by preventing them from achieving convective buoyancy (Head and Wilson, 2003). Even if buoyancy is achieved, the plume is firmly capped at the water–atmosphere boundary, only a few kilometers at most above explosive submarine vents. Hydrodynamic sorting by water is more effective for most grains, except that water cannot sort clasts that are buoyant within it, such as dry or vapor-saturated pumice. These float and clasts of many different sizes and masses can float together in pumice rafts. Very small particles with high specific surface areas are easily transported by ocean currents and in weak thermal plumes.

1.2. Known and Inferred Explosive Eruptions on the Modern Seafloor

Explosive submarine eruptions take place in intraplate settings and along all types of plate margin, concentrated along spreading centers and the Pacific Ring of Fire. Most submarine eruptions of "historical" age are certain to have gone undetected. Monitoring networks exist only at a few persistently active submarine volcanic centers. Two eruptions have been observed at depth by remotely operated vehicles (ROVs; see below), but most submarine eruptions are known from some surface expression; they have formed islands (e.g., Surtsey, 1963–1967), subaerial eruption plumes (Havre Seamount 2012, backtracked by satellite weeks after eruption), pumice rafts (Havre, first seen 2 weeks after eruption), or have simply roiled the ocean surface (El Hierro volcano, Canary Islands 2012). Remotely detected thermal, seismic, or acoustic signals allow inference of a submarine eruption's occurrence, but not certainty. Some submarine eruptions have been inferred entirely from their still-submerged deposits, examined by submersible or ROV, or sampled from ships using dredges and sediment cores. The first observed wholly submarine explosive eruption was of NW Rota-1 in 2008.

2. EXPLOSIVE BASALTIC ERUPTIONS

Hydrostatic pressure is a formidable impediment to purely magmatic explosive submarine eruption of basalts, which typically have low-preeruption volatile contents (Table 31.2). Nevertheless, for the right combinations of depth, volatile exsolution and coupling, and magma–water

interaction dynamics, submarine explosive eruptions of basalt are possible to depths of a few kilometers (Head and Wilson, 2003; Clague et al., 2009; Schipper et al., 2010; Helo et al., 2011). The specific mechanisms of deep basalt explosivity remain subjects of debate, with a few key eruptions addressed in different ways: by syn-eruptive observations at depth, by syn-eruptive observations at the water's surface, and by posteruptive interpretation of submarine deposits and particles. Ancient successions rarely preserve the pyroclast glass needed to assess fragmentation/quench depth, but outcrops provide stratigraphic control and information on depositional process.

2.1. Mafic Eruptions Observed at Depth

To date, only two small submarine explosive eruptions have been observed at their vents, providing information about submarine explosive eruption dynamics that can only be gained by direct observation.

2.1.1. NW Rota-1 Volcano, Marianas Arc

Repeated ROV dives at NW Rota-1 Volcano in the Marianas Arc yielded the first observations of a submarine explosive eruption, at ~560 m below sea level (Chadwick Jr. et al., 1008; Deardorff et al., 2011). In 2004–2005, pulsating plumes of molten sulfur droplets in water were visible, with proximal dispersal of ash and lapilli around the small (<15 m diameter) vent (Figure 31.1). Dives in 2006, fortuitously began just after a flank collapse, had exposed the vent region to close observation, and molten material was observed erupting along with identifiable bubbles of CO_2, hot water plumes bearing molten sulfur droplets, and quenched pyroclasts.

The NW Rota-1 eruption lasted from at least 2004 through 2010. The observed cyclicity, with 2- to 6-min eruptive bursts separated by 10–100 s pauses, inspired observers to draw parallels with Strombolian eruptions (see Chapter 27 (Chadwick Jr. et al., 2008)). Clear CO_2 bubbles were discharged between episodes of pyroclast ejection, whereas whitish plumes of sulfur globules accompanied pyroclast discharge, indicating that segregation of different magmatic volatiles according to their relative solubilities gives them differing roles during the eruption. NW Rota-1 pyroclasts have a range of textures showing effects of magmatic and hydromagmatic fragmentation, particle recycling, and sulfur agglutination (Deardorff et al., 2011). Despite very low mass eruption rates (<0.03 m³/s; Stromboli typically ~0.3 m³/s; see Chapter 27) and limited particle dispersal, the eruption persisted through many cycles of tephra accumulation and collapse that have controlled the long-term morphological evolution of the edifice (Chadwick Jr. et al., 2008).

NW Rota 1, Brimstone Pit

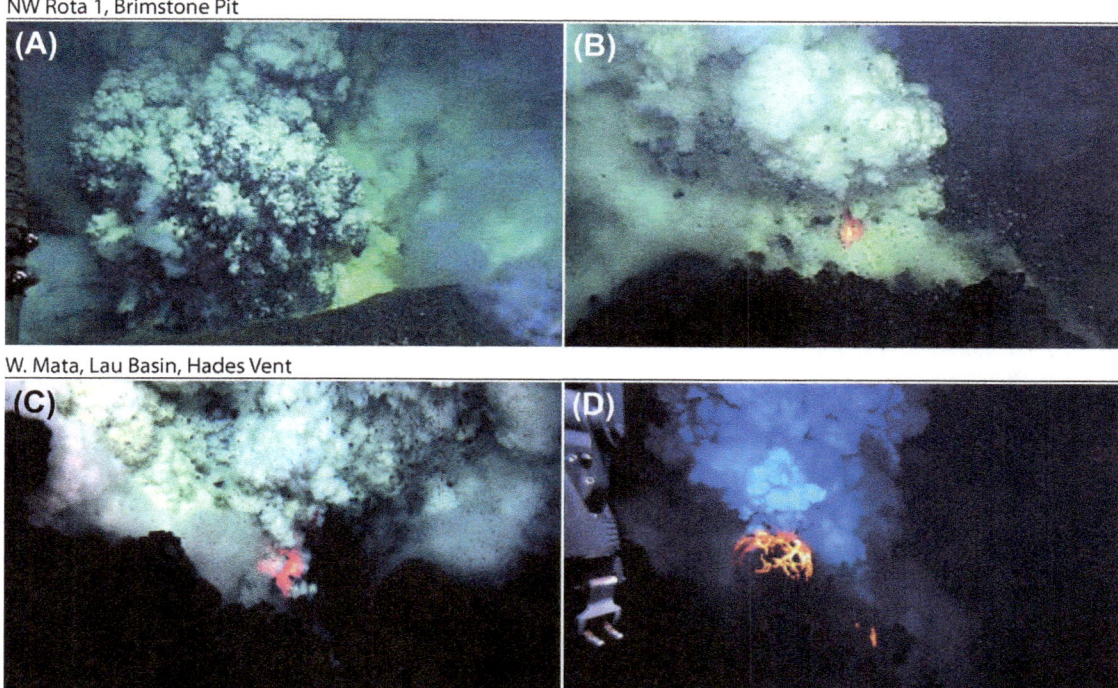

W. Mata, Lau Basin, Hades Vent

FIGURE 31.1 Video frames from the two observed submarine explosive mafic eruptions. (A,B) Basaltic andesite eruption from Brimstone pit at NW Rota-1 Volcano, Mariana Arc, 550 mbsl. *(From Chadwick W.W. Jr. et al., 2008)* (C,D) Boninitic eruption from the Prometheus (A) and Hades (B) vents at West Mata volcano, NE Lau Basin, 1200 mbsl. *(From Resing et al., Active submarine eruption of boninite in the northeastern Lau Basin, Nature Geoscience 4:799–806, 2011).*

2.1.2. West Mata Volcano, Lau Basin

Explosive and effusive activity was observed by ROV at West Mata volcano in the Lau Basin in 2008 from two intermittently active vents ~1200 m below sea level. Effusion produced pillow lavas, and individual lava bubbles expanded up to a meter in diameter before bursting to disperse clasts near the vent. There was weak ejection of lapilli and plumes of elemental sulfur (Figure 31.1). The lava bubbles have been interpreted as submarine equivalents to those that drive eruptions at Stromboli and indicating that weak submarine explosive activity is a viable mechanism for producing deep-sea limu o Pele "bubble wall" fragments (see below).

Vigorous sulfur plumes accompanied the activity and while no CO_2 bubbles were observed, there was dissolved CO_2 in the hydrothermal plume above the eruption site. Primary pyroclasts included vesicular Pele's hair, limu o Pele, spatter, and scoria ejected during higher-intensity explosive phases, and altered, denser particles that were possibly recycled during waning activity.

Both the NW Rota-1 and West Mata eruptions were from small vents with low mass discharge rates, but persisted long enough to modify local volcano morphology and create significant hydrothermal anomalies. Observation periods were brief, so it is uncertain whether the observed activity represents well the long-term behavior. Year-to-year changes to vent structure at NW Rota-1, and acoustic signals at West Mata that were subdued during observation periods compared to the long-term background, suggest that both volcanoes have seen other, more-intense, activity. NW Rota-1 erupted basaltic andesite typical of the volatile-rich Mariana Arc subduction zone, and West Mata erupted very hot and volatile-rich boninite magma in a nascent arc setting. Submarine explosive eruptions of the MORB or OIB magmas that volumetrically dominate in the world's oceans thus remain unobserved.

2.2. Mafic Eruptions Observed at the Surface

2.2.1. Lava Balloons

Mafic submarine eruptions can form large floating "balloon" clasts along with seawater-discoloring fines. Balloon diameters range from several tens of centimeters to a few meters, and are buoyant because they have large interior cavities, or cores of frothy silicic glass from melting and vesiculation of marine sediments. Hollow basaltic **lava balloons** have been observed: west of Hawaii

(1877); near Pantelleria in the Straits of Sicily (1891); off Socorro Island, Mexico (1994); and at Serreta, NW of Terceira Island, Azores (1998–2001), and balloons with frothed sediment cores at Teishi Knoll, Japan (1989), and El Hierro, Canary Islands (2011–2012).

Hollow lava balloons have interior temperatures up to 900 °C, and usually break upon reaching the ocean surface. Lava balloons have not been observed erupting from their vents, but they have been collected floating at the surface, and posteruptively from the seafloor. The vent offshore of Pantelleria was identified during ROV surveys and is surrounded by proximal spatter deposits. Hollow balloons form when vesicular clasts are erupted at depth then rise rapidly upward before their outer rind fully solidifies (Kelly et al., 2014). Strong decompression during ascent allows rapid expansion of gas inside the balloon so that bubbles coalesce and the balloon inflates. Cored balloons form when melt entraps, melts, and vesiculates marine sediments that froth to form pumice-like domains within magma that is then fragmented; expansion of the froth during rise and strong decompression inflates the balloons. Eruption depths for balloon-forming eruption are estimated at up to 1 km, with the mounting number of known lava balloon eruptions indicating they are a common type of seafloor eruption.

2.2.2. Surtseyan Volcanism

From when jets first pierced the ocean surface (Figure 31.2), Surtsey's eruption was extensively observed, and Sigurdur Thorarinsson gave an integrated account in the book *Surtsey. The New Island in the North Atlantic* (1967, New York, The Viking Press, 47 p). The eruption lasted about 3.5 years, during which is greater than 10^9 m^3 of magma was erupted from five sites to form three islands, but knowledge of the submarine parts of the eruptions is limited, and as new vents became active their edifices apparently grew rapidly from seafloor to emergence. Here we focus on the earliest surface observations made during the eruptions, before islands were formed; Chapter 30 addresses the subsequent subaerial eruptive activity. For the whole eruption, the total period of fully subaqueous eruption probably totaled only a few weeks. Surtsey's eruption began 130 m below sea level on the seafloor south of Iceland, proceeding unnoticed until the volcano had built to perhaps 10 m below the surface. When first observed on

FIGURE 31.2 Surface observations of underwater eruptions, prior to edifice emergence (A) Surtla vent, Surtsey eruption, 29 December 1963 *(Thorarinsson S (1967) Surtsey. The New Island in the North Atlantic. The Viking Press, New York, p 47)*. (B) Tephra jet from Kavachi volcano, Solomon Islands, from vent 2–5 m below water on 14 May 2000; photograph courtesy Pamela Brodie. (C) Inferred processes accompanying tephra jets breaching from slightly submerged vents as at start of Surtsey eruption. Vent occupies one part of broader edifice building to emergence. "cw" = concentric waves; dune-bedded tephra ~10 m underwater results from density currents interacting with surface waves. *(Redrawn after White, J.D.L. (1996) Pre-emergent construction of a lacustrine basaltic volcano, Pahvant Butte, Utah (USA). Bull. Volcanol. 58, 249–262.).*

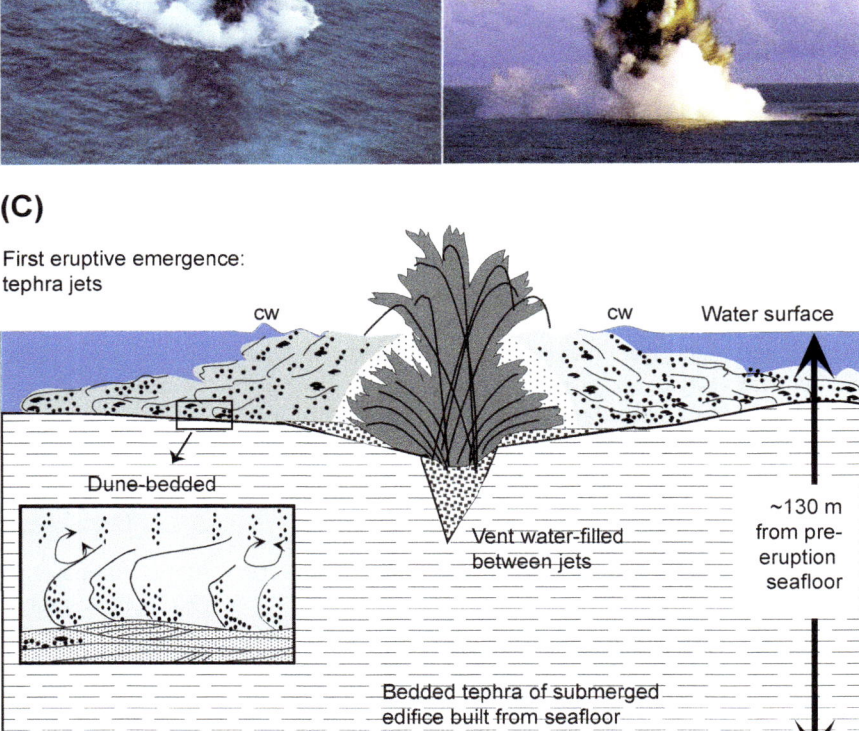

14 November 1963, a boat captain saw black eruption columns that soon reached ~70 m rising out of the sea in two or three places; the eruption was silent. A large circle of whitened water surrounded the vent, representing a zone of agitation containing vapor bubbles, and long-wavelength concentric waves traveled away from the vent. By the next morning Surtsey emerged from the sea as a small island and was in fairly constant eruption. Three satellite vents to Surtsey (Surtla, Syrtlingur, and Jolnir) were later observed as they shoaled. Prior to emergence, broad, low-relief tephra mounds were visible below the surface. The mound surfaces accreted tephra from varied eruptive activity and were disturbed both by normal sea waves and concentric eruption-generated waves.

The surface waters at the Surtsey vents were agitated by hot, convecting water, with rising steam from vapor bubbles bursting at the water surface. The whitened circles of water surrounding vents during explosive eruption were likely "spall domes" intersecting the surface; the circles thus represented the zones within the primary pressure waves generated by the explosions. Slightly later, subaerial tephra jets joined with the turbulence and steam, fed by subaqueous tephra jetting that resulted from explosive hydrovolcanic disruption of magma at shallow water depths. Vigorous tephra jets from vents slightly below the surface had sufficient momentum to clear the water surface. Less vigorous ones did not breach the surface but generated sufficient steam to feed a vapor column rising through the surface; the weakest agitated the water and sent heated pockets of water to roil the surface, but steam was recondensed in the cold ambient seawater. The subaqueous flashes resulted from explosive interaction of incandescent magma with water, and outward-traveling concentric surface waves formed during shoaling. Explosion-induced shaking would have caused local failure of the growing edifice. Wide craters, as observed at Syrtlingur and perhaps Jolnir, would have accumulated a tephra slurry by fallback from nonbreaching jets. This "almost emergent" phase of edifice construction, with tephra ejected into the air from an underwater vent with no island formed, lasted for more than 2 weeks at Syrtlingur vent. Once Surtsey's edifices emerged, the eruptive activity and resulting deposits took on characteristics shared with other subaerial phreatomagmatic volcanoes (Chapter 30).

A unique resource from Surtsey is a core that transects almost the full thickness of the volcano. It has been interpreted as penetrating only vent-filling deposits remobilized downward from the surface (Moore, 1985), but work now underway suggests a richer record. The lowest retrieved deposits consist of loose fragments of unaltered translucent basalt glass of moderate vesicularity with simple subspherical vesicles suggesting quenching soon after vesicle nucleation (Chapter 25). Multiple thin and irregular steep-dipping dikes penetrate lapilli tuff in the core, suggesting an abundance of small steep intrusions into the edifice. This is confirmed by common dike fragments found in Surtsey's surface deposits. Both in the core and, especially, in surface ejecta are composite clasts containing recycled particles.

Information complementing that from Surtsey comes from similar volcanoes formed in deep glacial-age lakes of the western USA. Pahvant Butte has a form and scale like Surtsey's (Figure 31.3), and erupted into ~100 m of water, while Black Point formed largely under water, with only small tephra rings above the water's surface (Figure 31.3). Both of these volcanoes, like Surtsey, began with explosive eruptions; there is no evidence at any of the three for initial pillow lavas. Density currents produced the great majority of subaqueous deposits exposed in outcrop. Bedding dips in the subaqueous deposits are low; they accumulated as stacks of shallow-dipping beds, not as steep-sided subaqueous cones. The pyroclasts of Surtsey are similar to those of Pahvant Butte and Black Point, which also have simple subspherical vesicles in translucent glassy grains (Figure 31.3) as well as recycled clasts within them. Subaqueously emplaced fine ash at Black Point (Figure 31.4) shows evidence for molten fuel—coolant interaction fragmentation (Chapter 26) as well as fluidal particles, while detailed study of its lapilli show a broad range of vesicle populations (Murtagh and White, 2013). Its lapilli have vesicle number densities that span those known from lapilli of almost all other basaltic explosive eruptions, from Stromboli to basaltic Plinian ones (Figure 31.4).

2.3. Basaltic Eruptions Inferred from Modern Seafloor Deposits and Particles

For Surtsey, surface observations help us infer at least the shallow-subsurface eruption processes. To infer eruption styles from submarine deposits alone requires interpretation of deposit stratigraphy and particle characteristics. A tool used for subaerial eruptions, the dispersal index, could not be directly applied for submarine eruptions even if the required thickness/distribution data were available (they are not), because dispersal from submarine eruptions is governed by water, which has very different properties from air.

Many geochemical studies of seafloor volcanism infer that because of the low volatile budgets of MORB and OIB magmas, volatiles must be exsolved from volumes of magma larger than those that erupt. This is evident at Axial Seamount, where melt inclusions reveal very high preeruption concentrations of CO_2 inferred to drive submarine explosivity (Helo et al., 2011). Similarly, melt inclusions from an OIB pyroclastic sequence at Loihi Seamount, Hawaii (Figure 31.5) indicate decoupled (open-system) discharge of CO_2. This ~1100-m-deep Loihi site has

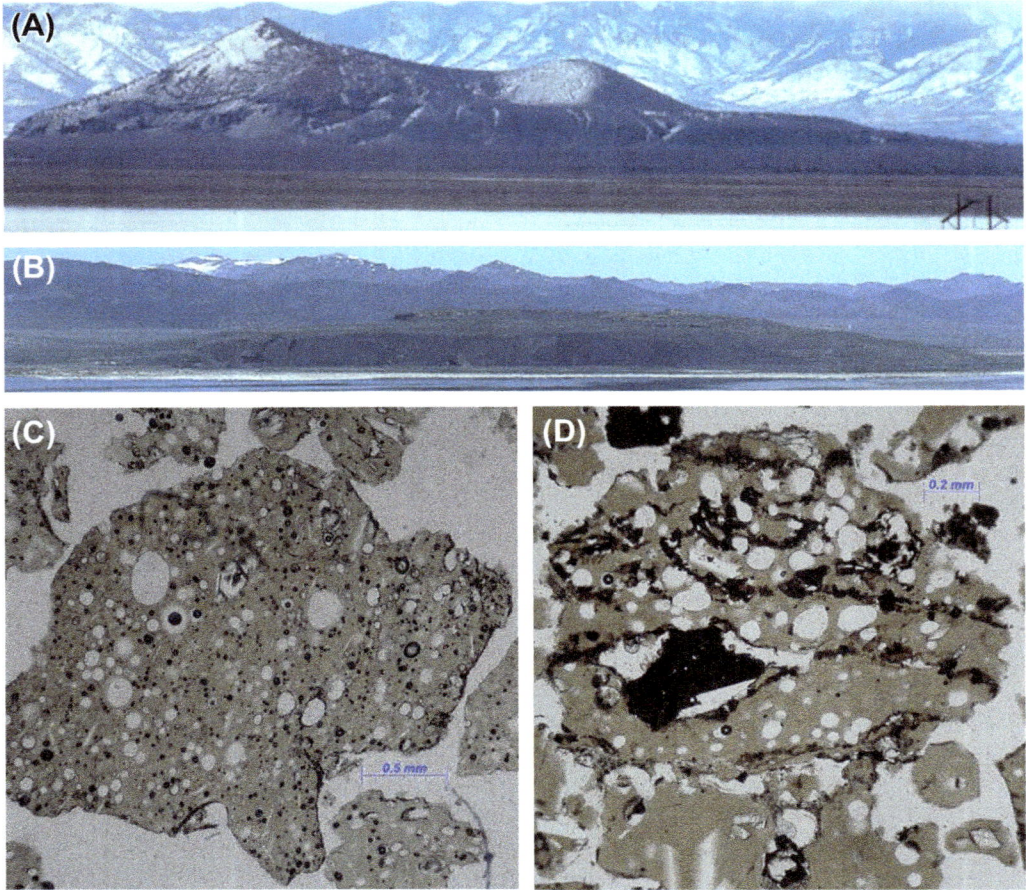

FIGURE 31.3 Basaltic volcanoes erupted subaqueously in Surtseyan style, (A) Pahvant Butte, Utah, and (B) Black Point, California. Typical translucent basaltic glass pyroclasts from Black Point (C), and Pahvant Butte (D), with subspherical vesicles and in (D) a previously erupted particle recycled into the new pyroclast.

deposits more than 30 m thick, pyroclasts with a wide range of vesicularity and groundmass crystallinity, and some limu o Pele particles. Nearby and at similar depth, however, other **pyroclastic deposits** at Loihi were produced by volatile-coupled exsolution ("closed-system degassing") of CO_2 and H_2O during magma ascent (Schipper et al., 2010). These deposits are relatively coarse-grained cones with larger clasts having higher and more restricted vesicularity, no limu o Pele, and a greater proportion of quenched sideromelane particles plus fine particles indicative of explosive fuel—coolant interactions (see Chapter 26).

There is one type of particle particularly influential in the discussion of basaltic explosive volcanism; the tiny folded sheets of sideromelane called "limu o Pele" (Figure 31.6), which have been found at nearly all investigated sites of mafic deep submarine volcanism. Many papers infer their formation from CO_2-driven, strombolian-like eruptions, in which the fragmented melt films are inferred to represent the skins of isolated bubbles that

burst to eject limu from the vent(s). Activity at West Mata volcano (Figure 31.1), 1.2 km under water produced limu during the growth and bursting of large volatile-filled bubbles as lava was extruded from the vent. Based on limu deposits explosive eruptions have been inferred from submarine volcanoes as deep as 4 km on the Gakkel Ridge in the Arctic Ocean. Some field and experimental work, as reviewed by Schipper and White (2010), has been used to caution against using the presence of limu o Pele as diagnostic of eruption style because limu can also form by nonexplosive magma—seawater interaction, as where it was first identified as a product of littoral interactions where Kilauea's lava flows enter the sea.

3. EXPLOSIVE SILICIC ERUPTIONS

3.1. Calderas and Composite Volcanoes

Unlike mafic magmas such as basalt, silicic magmas are largely restricted to volcanoes rooted in continental or

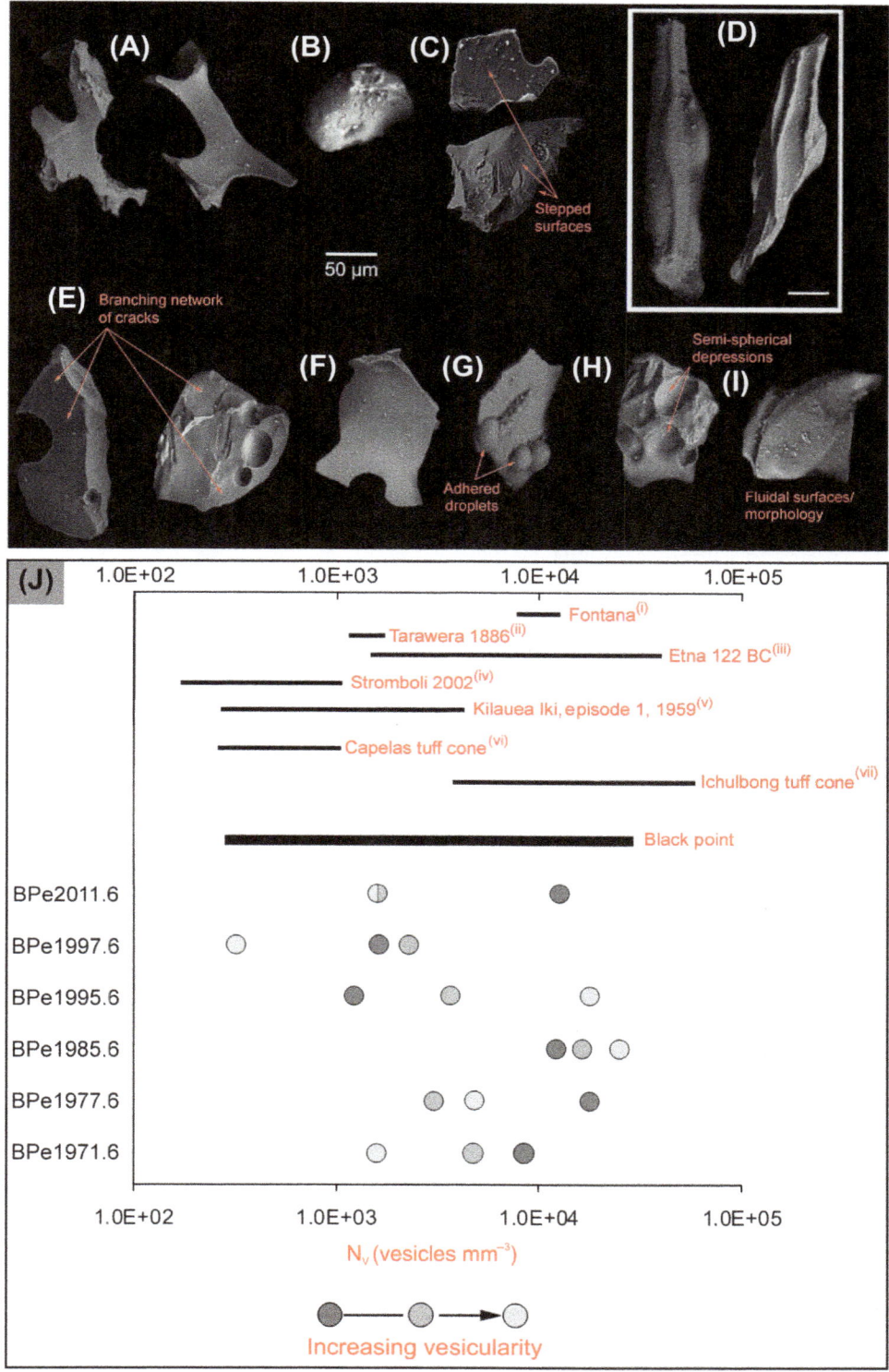

FIGURE 31.4   Fine ash grains from Black Point (A—I) show both fluidal forms and stepped fracture indicative of molten fuel—coolant interaction (see Chapter 26). (J) Vesicle number density of pyroclasts from Black Point shows a broad range, overlapping almost all types of subaerial basaltic eruption. *From Murtagh and White (2013).*

FIGURE 31.5   Example of deposits and particles from unseen explosive submarine eruptions. (A) 30 plus meter thick massive, unconsolidated deposit of lapilli ash at ∼1100 mbsl on Loihi Seamount, Hawaii. Enlargements of (i) broken pillow near base of deposit; (ii) thick bed of lapilli ash; (iii) bulbous exposure of unconsolidated lapilli; (iv) weakly bedded lapilli ash; (v) bulbous exposure of unconsolidated lapilli. (B,C) Glassy vesicular lapilli and (D) sideromelane ash from succession. *Modified from Schipper, C.I., White, J.D.L. and Houghton, B.F. (2011)* Textural, geochemical, and volatile evidence for a Strombolian-like eruption sequence at Lōʻihi Seamount, Hawaiʻi. *J. Volcanol. Geoth. Res.* 207, 16−32.

transitional crust, and are volatile-rich; subaerially they produce composite cones and calderas that erupt explosively. Intraoceanic arcs develop with many submarine volcanoes and can produce large volumes of more-evolved magmas as well as basalts and andesites, for example, in the Kermadec arc north of New Zealand (Figure 31.7), or intraoceanic arcs south of Japan (Kano, 2003). The submarine composite volcanoes studied so far erupt basalt

Natural	Experimental

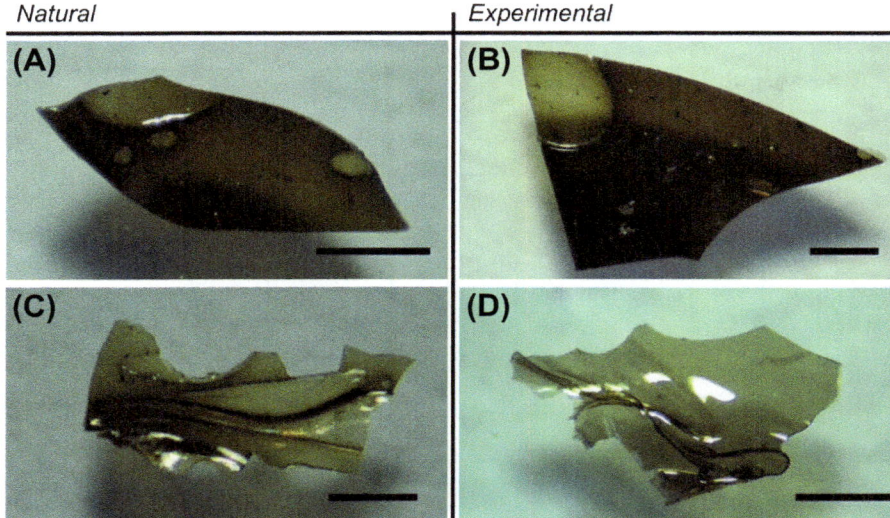

FIGURE 31.6 Limu o Pele fragments from Loihi (A,B)—these are known to form by expansion of isolated magma-walled bubbles either during *effusion* of magma/lava that entraps and expands seawater (Chapter 19); they are also widely considered to form from strombolian-like submarine explosive eruptions. Experimental fragments (C,D) formed in laboratory experiments when basalt melt is poured into water. *Modified from* Schipper, C.I., Sonder, I., Schmid, A., White, J.D.L., Dürig, T., Zimanowski, B., and Büttner, R. (2013) Vapour dynamics during magma-water interaction experiments: Hydrodynamic origins of submarine volcaniclastic particles (limu o Pele) Geophys. J. Int. 192: 1109—1115.

or/and andesite, but other submarine eruptions, of more-evolved magmas, have formed lava domes, pyroclastic cones, and calderas. Many submarine calderas result from collapse following large-volume eruptions. Submarine calderas are also represented in ancient deep-marine successions. Modern submarine calderas are circular to elliptical and 2—10 km across, with flat or slightly concave floors 1—7 km wide underlain by pyroclastic deposits; their walls rise at shallow slopes (14°—30°) up to 1 km above the caldera floor. Submarine calderas are of similar size to subaerial calderas and shaped by similarly large and explosive eruption processes (Yuasa and Kano, 2003).

Pyroclastic deposits have been dredged from many submarine cones and calderas. The Macauley deposits contain dacitic pumice clasts with two clast-density modes; one mode is weakly peaked at ∼600 kg/m^3, while the other is sharply peaked at 200 kg/m^3 indicating extremely high vesicularity compared to equivalent subaerial pumice from Macauley with 400 kg/m^3 modal density (Barker et al., 2012). Expansion of vesicles during rise through the water column is inferred, and may be likened to inflation of basaltic lava balloons. Pumice clasts dredged from the rim of West Rota caldera, Mariana arc, are ball shaped with prismatic joints (Stern et al., 2008) and may represent weakly explosive eruptions of fluid rhyolite clasts that were reshaped by surface tension before deposition. Both Macauley and West Rota caldera pyroclasts suggest a less intense and explosive style of eruption than inferred for the 2012 Havre eruption.

Dredged seafloor pumice is difficult to interpret, because one dredge haul may contain both primary and reworked clasts, even clasts from multiple eruptions. The best information comes from submersible study and sampling, e.g., of rhyolite pyroclastic deposits in the walls of

Myojin Knoll caldera (Izu-Ogasawara arc; Fiske et al., 2001) and West Rota caldera (Mariana arc). Such in situ sampling is crucial for detailed description of bedding and grain-size characteristics of caldera deposits. Drilling into the 1000-m-deep seafloor of Sumisu Rift, Izu-Ogasawara arc sampled a 250-m-thick Pliocene-to-Pleistocene succession of rhyolite pumice lapilli and ash deposits linked to nearby submarine calderas (Fiske et al., 2001). Such thick successions of volcaniclastic deposits are very common along island arcs and associated basins.

Subaerially exposed ancient successions provide additional information critical for interpreting submarine eruptive processes. Products of a submarine caldera-forming eruption 15,000 years ago are exposed on Shinjima Island in Kagoshima Bay, Japan. The deposit was originally part of the submarine caldera floor, but uplift in 1780 created the island and exposed the rhyolitic Shinjima Pumice originally deposited as much as 140 m below sea level, sandwiched within marine sediments. Acoustic profiles suggest the deposit extends across the entire floor of the still-submarine part of the caldera.

The Shinjima Pumice consists mainly of inferred density current deposits (Figure 31.8(A)) made of pumice and ash with a very small proportion of lithic fragments. Units are 1—10 m thick, diffusely stratified and poorly sorted with upward coarsening of pumice. Parallel to wavy diffuse stratification is defined by alignment of pumice clasts, especially in the lower half of each unit (Figure 31.8(A)). The uppermost part of the Shinjima Pumice has extremely well sorted, parallel laminated ash beds 2—3 m thick (Figure 31.8(B)). Large pyroclasts are highly vesicular, commonly with fracture surfaces, and mostly indistinguishable from those erupted subaerially, except that mm- to cm-wide polyhedral or prismatic joints extend inward

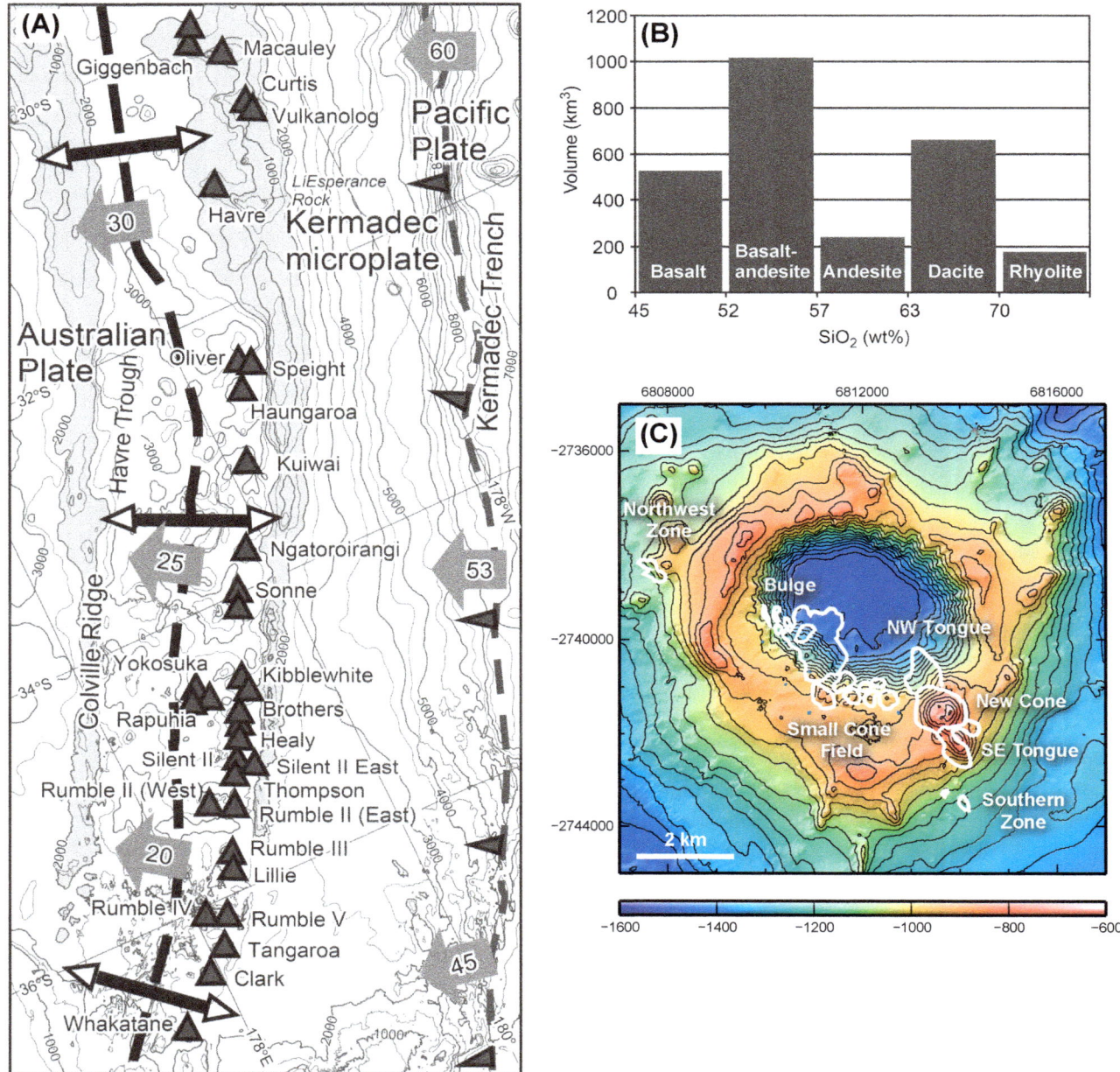

FIGURE 31.7   The southern Kermadec arc (A,B) from Wright et al. (2006). Triangles show 33 submarine volcanoes, with ∼800 km³ of silicic dacite and rhyolite ones that include calderas like (C), Havre volcano *(from Carey et al., 2014)*, which erupted explosively in 2012 to form an eruption plume captured on satellite, and a pumice raft spreading from an initial 400 km² to cover ∼20,000 km² within days.

from pyroclast surfaces. Ash-sized pumice particles are commonly blocky, and many dense glass shards are platy, both morphological features widely considered diagnostic of explosive magma–water interaction (Figure 31.8(C)). These features suggest that the Shinjima Pumice was emplaced from aqueous density currents directly fed by a submarine explosive eruption.

Eruption columns in the ocean are expected to ingest seawater because of turbulence and Rayleigh–Taylor and/or

Kelvin–Helmholtz instabilities at the head and walls of the rising plume, with the volume of water drawn into the eruption column determining the column's fate (Figure 31.9). If enough water is ingested to condense most magmatic water vapor, the eruption column collapses. If the volume of incorporated water is small enough to be vaporized by the internal heat of pyroclasts and volcanic gas, the eruption column expands and there is additional time for quench fragmentation of juvenile materials. In both cases,

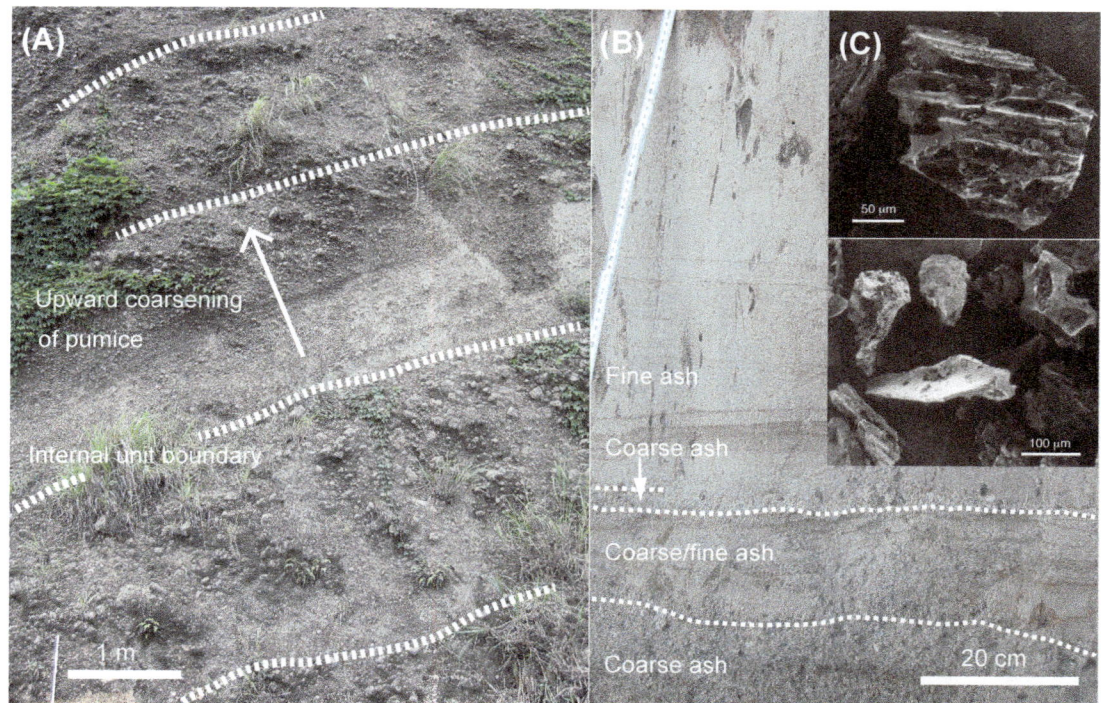

FIGURE 31.8 (A) The main Shinjima Pumice. Each internal unit shows reverse grading of pumice and stratification in its lower interval. (B) Uppermost low-angle cross- to parallel-laminated ash bed, transitional from the pumice-rich main part through upward-thinning and upward-fining interlayers of coarse and fine ash. (C) Scanning electron photomicrographs of blocky pumice with long-tube vesicles and curviplanar fracture surfaces cutting across the vesicles and platy or curved angular glass shards in the Shinjima Pumice. *Modified from Kano, K., Yamamoto, T. and Ono, K. (1996)* Subaqueous eruption and emplacement of the Shinjima Pumice, Shinjima (Moeshima) Island, Kagoshima Bay, SW Japan. *J. Volcanol. Geotherm. Res.* 71, 187–206.

the column collapse produces density currents, such as the one inferred to have deposited the Shinjima Pumice. A gas-supported hot density current could be produced from a low "boiling over" eruption fountain dominated by dense ash, allowing deposition of submarine deposits that are welded at high temperature. Even such low fountains and their currents will lose much material to overriding aqueous currents from interaction with water, and are not expected to travel far. In contrast, if an eruption produces an intense gas thrust that reaches the water surface, pyroclasts in the core of the gas thrust region will be erupted to the atmosphere. This may have happened in the 2012 eruption of Havre volcano, Kermadec Arc, New Zealand, at 0.7- to 1.5-km water depth in the Kermadec arc, New Zealand (Figure 31.7). It was the first-ever documented major explosive submarine eruption, estimated to have erupted twice the volume of the 1980 subaerial eruption of Mt St Helens. Havre produced a large and quickly dispersed raft of floating silicic pumice, ~ 1.5 km^3 of seafloor deposits, and a subaerial eruption plume imaged by satellite.

Seafloor-deposited pumice clasts emitted in such major submarine eruptions range in density from 200 to 800 kg/m^3 with a mean ~ 500 kg/m^3, indistinguishable from subaerial equivalents (Kano, 2003). In primary deposits, they are closely associated with glassy ash shards inferred

to represent intense fragmentation of viscous magma comprising a foam of gas bubbles, beyond which bubble size cannot increase without magma fragmentation. With a typical dissolved water content of 3–6 wt.% at 850 °C, water exsolved from these magmas can produce 74% vesicularity at depths of 500 to 1300 m, respectively. On the other hand, a rhyolite magma at 850 °C with 1 wt.% H$_2$O cannot erupt explosively even in water as shallow as 200 m. This estimate is based on Henry's law and water solubility in rhyolite magma, and consistent with reported and inferred depths of 1000 m or less for explosive silicic eruptions.

Ash and pumice clasts from smaller-volume eruptions can be carried to the water surface by thermal convection and drift with surface water currents, while coerupted large dense clasts settle quickly to the seafloor. Clast-by-clast suspension settling from a water—clast mixture ejected into the water column produces normally graded (by settling velocity) deposits, like fallout from a subaerial Plinian eruption. In contrast, deposits of submarine **eruption-fed density currents** may show upward coarsening of water-quenched pumice clasts and reduced grain-size sorting. As proposed for the upward-fining submarine beds in Dogashima, Izu Peninsula, Japan, water-saturated pumice clasts will settle from aqueous suspension clouds together with dense clasts

FIGURE 31.9 A conceptual model of subaqueous flow-generating eruption. A = gas-jet. B = mixing zone between eruption jet and water. C = buoyancy-driven convective plume carrying pumice clasts and ash. D = suspension cloud of pumice clasts and ash. E = fallout of waterlogged pumice clasts and other dense materials. F = gas- and water-supported density current. G = gas-supported hot pyroclastic flow. H = pumice deposit. Hot pumice clasts are buoyant as the vesicles are filled with steam. At high temperatures, steam films protect pumices from rapid cooling and water invasion. *Modified from Kano, K., Yamamoto, T. and Ono, K. (1996) subaqueous eruption and emplacement of the Shinjima Pumice, Shinjima (Moeshima) Island, Kagoshima Bay, SW Japan. J. Volcanol. Geotherm. Res. 71, 187–206.*

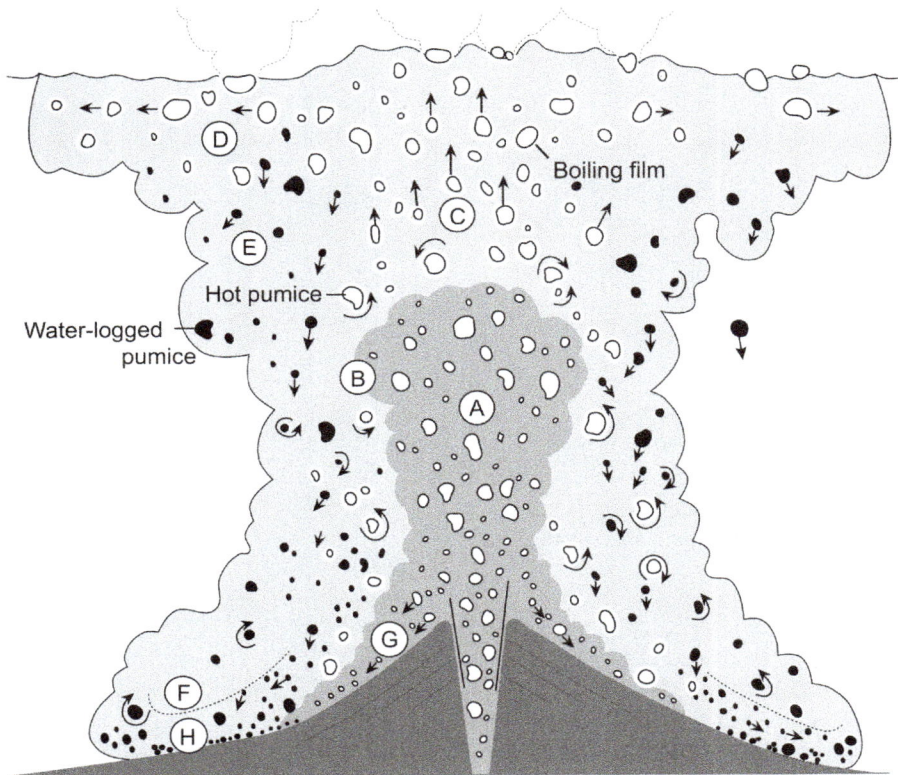

of equivalent settling velocity. Such sorting by settling velocity also takes place in the body and tail of dilute turbulent density currents such as turbidity currents.

3.2. Lava Domes and Related Deposits

As in subaerial settings, many submarine calderas host postcaldera lava domes, and lava domes also can grow atop on composite volcanoes. Some domes are associated with tephra cones and rings (Kano, 2003). When silicic magma ascends slowly, a lava dome may grow over the vent, or the conduit may be plugged by solidified lava in direct contact with water. In either case, a gas pocket can be produced beneath the lava crust by accumulation of gas exsolved from the magma. If the pocket becomes strongly overpressured or is breached or unloaded by sudden gravitational collapse, it can burst and eject blocks and finer clasts of the lava crust that may accumulate as a tephra cone or ring. No modern example has yet been described, but this type of eruption is envisaged for Miocene Tayu volcaniclastic beds in the Shimane Peninsula, SW Japan (Kano, 1996). The Tayu deposits include pumice clasts, notably large blocks, with abundant nonvesicular to poorly vesicular lava crust clasts at the base of the deposit, suggesting that the ejecta originated mainly from a pumiceous zone developed beneath a lava crust.

Similar explosions with a different origin can arise from lava domes when water enters the dome interiors through cracks, producing localized phreatomagmatic explosions. Repetition of such subaqueous phreatomagmatic explosions also can produce tephra cones or rings enclosing lava domes or conduits. Examples include the 1934–1935 eruption of Shin-Iwojima, Kikai caldera 100 km south of Kyushu, Japan; 1952–1953 eruption of Myojinsho 400 km south of Tokyo; and 1986 eruption of Fukutoku-Oka-no-Ba 1300 km south of Tokyo (Kano, 2003).

The 1934–1935 eruption of Shin-Iwojima, Kikai caldera 100 km south of Kyushu, Japan started on the caldera floor about 300 m below sea level (Kano, 2003). Initially meter-sized rhyolite pumice blocks rose in swarms to the sea surface, with the sea surface doming to a height of 1–2 m (Figure 31.10(A)). The floating pumice blocks boiled surrounding seawater to form steam that rose in white plumes to 800–1000 m. Floating pumice blocks up to 30 m³ sank abruptly when water invaded their hot interiors through cracks opened by water-cooling contraction and expansion of internal gas. Other pumice blocks and lapilli remained afloat and drifted downcurrent. The eruption subsequently changed to shallow phreatomagmatic eruptions with growth of an emergent tephra cone, followed by lava effusion onto the new island of Shin-Iwojima.

In very deep water, high hydrostatic pressures can inhibit explosions even in domes formed from silicic magmas. Slabby giant rhyolite pumiceous clasts observed on the seafloor in the Sumisu Rift are inferred to have

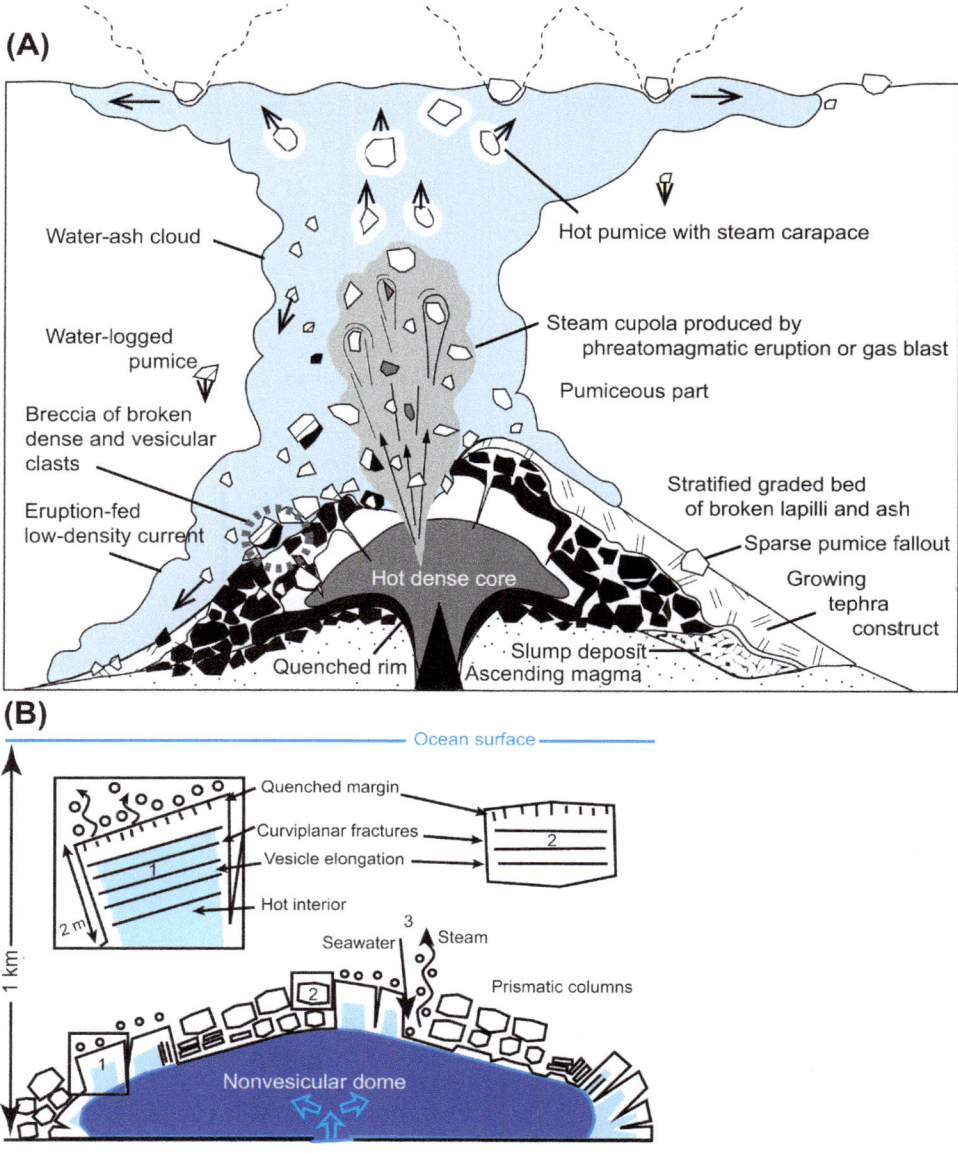

FIGURE 31.10 (A) Eruption from a submarine pumiceous dome with phreatomagmatic or gas explosion and generation of a density current. Fragments from quench-mechanical fragmentation and/or explosion accumulate as breccias around the dome. Hot pumice blocks may ascend to the water surface, drift, then settle separately when saturated with water. *(Conceptual model proposed by Kano (1996, 2003))*. (B). Emplacement of deep-water silicic domes involves frothing of magma at the seafloor, chilling-in of vesicles in an outer carapace, and degassing of the lava core through interconnected vesicles that collapse. Numbers 1−3; 1 = in situ pumice carapace, 2 = giant pumice slab, 3 = possible magma−water explosions. *Redrawn from Allen et al. (2010)*.

broken from a dome, in the hot core of which connected vesicles allowed escape of volatiles and then vesicle collapse to produce a nonvesicular inner dome (Figure 31.10(B)). Despite high vesicularity, the dome fragments probably did not reach the surface before saturating with water and sinking to the seafloor near the dome (Allen et al., 2010).

The Yali Pumice Breccia is a deposit from a submarine dome eruption (Allen and McPhie, 2000). It comprises moderately to well-sorted beds, 3 cm−3 m thick, of framework-supported, rhyolite pumice with less than 10 wt.% ash and less than 1 vol.% dense juvenile and basement lithic clasts. Pumice blocks (up to 3 m) are prismatic with quenched margins and internal polyhedral joints, formed by spalling and fragmentation of submarine pumiceous lava and probably deposited by settling. Smaller pumice clasts are polyhedral, angular to subrounded, and blocky with sharp curviplanar surfaces that cut across vesicle boundaries. They form massive or diffusely laminated, wedge-shaped beds up to 2 m thick and formed by (1) passive and explosive disintegration of the larger pumice clasts, plus (2) phreatomagmatic explosions.

4. SUMMARY

Though there can be no "dry" eruptions under water, the range and influence of water's effects is varied, and many surprisingly "subaerial-like" features have been identified in submarine eruption products, including fluidal spatter-like fragments, local welding, and highly vesicular pumice. These observations must be reconciled with the certainty that submarine-erupted pyroclasts are either formed in contact with water or encounter it soon afterward and while still hot. Determining the precise role(s) of water in submarine fragmentation is challenging because pyroclast "signatures" of magma—water interaction are limited to thermomechanical surface textures on *some* fragments, and to strong quenching of *some* fragments. Fresh young deposits, with unaltered fine ash, are the only ones likely to allow researchers to address this question, and, even here, additional experimentation and observation may be needed to assess the range of textures resulting from direct magma—water interaction across magmas of different chemistry, vesicularity, and crystallinity. One tool commonly used to assess the state of magma at the time of fragmentation, the volatile content of matrix glass, is made problematic because water absorption into silicic glass may happen before pyroclasts are cooled in subaqueous eruptions. Studies of basaltic seafloor deposits at Loihi demonstrate a range of different behaviors, from fast rise of magma where exsolved volatiles remain fully coupled as in Hawaiian fountaining eruptions, to slower rise with volatile decoupling and probable discrete bursts of activity like those seen subaerially at Stromboli. The deposits of these two eruptive styles on Loihi are broadly similar low-relief cones of glassy coarse lapilli. This similarity in deposits of volatile-coupled versus decoupled basalt contrasts with the differing deposits of fire fountains versus discrete showers of incandescent clasts subaerially. It is a reminder that not only is pyroclast formation different subaqueously, dispersal of pyroclasts is even more strongly, fundamentally, different subaqueously. Ballistic transport is ineffective, hot particles with gas in vesicles can float, and even water-saturated particles have low excess densities and sink slowly or/and drive only slow-moving density currents from which fine particles are strongly partitioned into suspension and deposited elsewhere. Cooling during particle transport increases fragmentation and limits accumulation of hot clasts that agglutinate or weld together.

Study of submarine explosive eruptions is in its infancy, and there is much still to learn. The thousands of explosive eruptions that have been observed subaerially can be contrasted with the two submarine eruptions with video observation to date, both of mafic arc volcanoes, both with extremely low magma flux. Moreover, the volcano tourism and ubiquitous cell phone video that are rapidly increasing observations available for study subaerially have no equivalent for submarine eruptions—they are likely to remain underobserved for the foreseeable future.

FURTHER READING

Allen, S.R., McPhie, J., 2000. Water-settling and resedimentation of submarine rhyolitic pumice at Yali, eastern Aegean, Greece. J. Volcanol. Geoth. Res. 95, 285—307.

Allen, S.R., Fiske, R.S., Tamura, Y., 2010. Effects of water depth on pumice formation in submarine domes at Sumisu, Izu-Bonin arc, western Pacific. Geology 38, 391—394.

Barker, S.J., Rotella, M.D., Wilson, C.J.N., Wright, I.C., Wysoczanski, R.J., 2012. Contrasting pyroclast density spectra from subaerial and submarine silicic eruptions in the Kermadec arc: implications for eruption processes and dredge sampling. Bull. Volcanol. 74, 1425—1443.

Carey, R.J., Wysoczanski, R., Wunderman, R., Jutzeler, M., 2014. Discovery of the largest historic silicic submarine eruption. Eos, Trans. Am. Geophys. Union 95, 157—159.

Chadwick Jr, W.W., Cashman, K.V., Embley, R.W., Matsumoto, H., Dziak, R.P., de Ronde, C.E.J., Lau, T.K., Deardorff, N.D., Merle, S.G., 2008. Direct video and hydrophone observations of submarine explosive eruptions at NW Rota-1 volcano, Mariana arc. J. Geophys. Res. 113, B08S10. http://dx.doi.org/10.1029/2007jb005215.

Clague, D.A., Paduan, J.B., Davis, A.S., 2009. Widespread strombolian eruptions of mid-ocean ridge basalt. J. Volcanol. Geotherm. Res. 180, 171—188.

Deardorff, N.D., Cashman, K.V., Chadwick Jr, W.W., 2011. Observations of eruptive plume dynamics and pyroclastic deposits from submarine explosive eruptions at NW Rota-1, Mariana arc. J. Volcanol. Geotherm. Res. 202, 47—59.

Fiske, R.S., Naka, J., Iizasa, K., Yuasa, M., Klaus, A., 2001. Submarine silicic caldera at the front of the Izu-Bonin Arc, Japan; voluminous seafloor eruptions of rhyolite pumice. Geol. Soc. Am. Bull. 113, 813—824.

Head III, J.W., Wilson, L., 2003. Deep submarine pyroclastic eruptions: theory and predicted landforms and deposits. J. Volcanol. Geotherm. Res. 121, 155—193.

Helo, C., Longpré, M.-A., Shimizu, N., Clague, D.A., Stix, J., 2011. Explosive eruptions at mid-ocean ridges driven by CO_2-rich magmas. Nat. Geosci. 4, 260—263.

Kano, K., 1996. A Miocene coarse volcaniclastic mass-flow deposit in the Shimane Peninsula, SW Japan: product of a deep marine eruption? Bull. Volcanol. 58, 131—143.

Kano, K., 2003. Subaqueous pumice eruptions and their products: a review. In: White, J., et al. (Eds.), Explosive Subaqueous Volcanism, Am. Geophys. Union Geophys. Monograph, 140, pp. 213—230.

Kelly, J., Carey, S., Pistolesi, M., Rosi, M., Croff-Bell, K., Roman, C., Marani, M., 2014. Exploration of the 1891 Foerstner submarine vent site (Pantelleria, Italy): insights into the formation of basaltic balloons. Bull. Volcanol. 76, 1—18.

Kokelaar, B.P., 1986. Magma-water interactions in subaqueous and emergent basaltic volcanism. Bull. Volcanol. 48, 275—289.

Moore, J.G., 1985. Structure and eruptive mechanism at Surtsey volcano, Iceland. Geol. Mag. 122, 649—661.

Murtagh, R.M., White, J.D.L., 2013. Pyroclast characteristics of a subaqueous to emergent Surtseyan eruption, Black Point volcano, California. J. Volcanol. Geotherm. Res. 267, 75—91.

Schipper, C.I., White, J.D.L., 2010. No depth limit to hydrovolcanic limu o Pele: analysis of limu from Loihi Seamount, Hawaii. Bull. Volcanol. 72, 149–164.

Schipper, C.I., White, J.D.L., Houghton, B.F., Shimizu, N., Stewart, R.B., 2010. "Poseidic" explosive eruptions at loihi seamount, Hawaii. Geology 38, 291–294.

Stern, R.J., Tamura, Y., Embley, R.W., Ishizuka, O., Merle, S.G., Basu, N.K., Kawabata, H., Bloomer, S.H., 2008. Evolution of West Rota Volcano, an extinct submarine volcano in the southern Mariana Arc: evidence from sea floor morphology, remotely operated vehicle observations and 40Ar-39Ar geochronological studies. Isl. Arc. 17, 70–89.

White, J.D.L., Smellie, J.L., Clague, D.A., 2003. Introduction: a deductive outline and topical overview of subaqueous explosive volcanism. Geophys. Monograph, 140, 1–20.

Wright, I.C., Worthington, T.J., Gamble, J.A., 2006. New multibeam mapping and geochemistry of the 30°–35° S sector, and overview, of southern Kermadec arc volcanism. J. Volcanol. Geotherm. Res. 149, 263–296.

Yuasa, M., Kano, K., 2003. Submarine silicic calderas on the northern Schito-Iwojima Ridge, Izu-Ogasawara (Bonin) Arc, western Pacific. Geophys. Monograph, 140, 231–243.

Chapter 32

Volcanic Plumes

Steven Carey

Graduate School of Oceanography, University of Rhode Island, Kingston, RI, USA

Marcus Bursik

Department of Geology, University at Buffalo, State University of New York, Buffalo, NY, USA

Chapter Outline

GLOSSARY

advection horizontal transport of material

buoyancy force exerted by the contrast in density between objects immersed in fluid

chaotic advection the stretching of air parcels into complex shapes by wind

coignimbrite plume plume produced from a pyroclastic density current

elutriation loss of small particles by the upward flow of gas through a pyroclastic flow

entrainment the process of picking up and carrying along

hemispheric flow the largest scale of the atmospheric flow, generally in distinct latitudinal bands; also called zonal flow

jet high-velocity stream of gas and particles usually flowing under laminar conditions

latent heat of condensation heat released upon conversion of vapor to liquid

neutral buoyancy level height at which the plume density is equal to that of the surrounding atmosphere

plume mixture of gas and particles released during explosive volcanic activity and dispersed in the atmosphere

stratosphere the upper portion of the atmosphere from an altitude of about 10–50 km

thermal a discrete mass of air or particles and gases that rise in the atmosphere by buoyancy

troposphere the lower portion of the atmosphere from ground level to about 10 km

tropopause boundary between the troposphere and stratosphere

turbulent diffusion mixing by turbulent flow

1. INTRODUCTION

Volcanic **plumes** are mixtures of volcanic particles, gases, and entrained air that are produced by a variety of explosive eruptions. The material is injected into the atmosphere and can be dispersed on a global scale. Plumes vary tremendously in scale ranging from small discharges that resemble industrial smokestacks to continental-scale plumes generated during supereruptions. **Buoyancy** plays a fundamental role in the motion of plumes and determines how they interact with the atmosphere. Energetic plumes can rise as high as 40 km above the Earth's surface. Their structure and behavior are controlled by factors such as magmatic composition, the amount and nature of volatile components, rate of magma discharge, and geometry of the source vent. Plumes are generally able to disperse volcanic particles and gases over large areas by transport and fallout through the atmosphere. This type of activity poses a relatively low risk to downwind populations except in areas very close to the source. An exception, however, is the hazard and economic impact to aircraft in flight that inadvertently encounter volcanic plumes. This was highlighted during the recent 2010 eruption of Eyjafjallajokull in

Iceland that virtually shut down commercial air traffic in Europe at an estimated cost of some $1.7 billion (US) to the airline industry. Volcanic plumes have the potential for impacting the Earth's climate by the injection of volcanic aerosols into the **stratosphere**, and subsequent modification of the solar energy flux that reaches the Earth's surface. In contrast, collapsing plumes generate hot mixtures of gases and particles (pyroclastic density currents) that move down the slopes of a volcano at high speeds and can result in a large number of fatalities. Significant progress has been made in the modeling of the source structure and evolution of plumes as they interact with the atmosphere, and in the tracking of plumes by remote sensing techniques once they disperse downwind from the source. The models and plume tracking have led to an improved capability to forecast the potential impact of volcanic plumes on the environment and human populations.

2. GENERATION OF VOLCANIC PLUMES

Volcanic plumes are generated by a wide variety of explosive volcanism styles. In all cases, magma is fragmented into small pieces and a mixture of solids, gases and, in some cases liquids, which is discharged from a vent. Rapid expansion of entrained air, and gases that come out of solution in the magma or are generated by heating of external water, results in buoyancy generation and the rapid acceleration of the material into the atmosphere.

Magmas typically contain small amounts of dissolved volatile components such as water, carbon dioxide, and sulfur dioxide prior to eruption when they are stored beneath the Earth's surface. As magmas rise to the surface in a sustained eruption (Figure 32.1(A), Plinian style), these volatile components begin to come out of solution and form bubbles. High internal pressures develop within the bubbles, and the bubbles are sheared, and eventually the bubble walls rupture. At this point, the magma is transformed into a mixture of broken bubble fragments and gas that accelerates rapidly upward through a conduit. Alternatively, cooling and crystallization of magma emplaced at shallow levels can lead to buildup of pressure from gas exsolution. If the conduit is blocked, catastrophic failure of the blockage can lead to fragmentation and transient acceleration of a volume of magma out of the vent (Vulcanian style).

Volcanic eruptions often involve the interaction of hot magma with external water at, or near, the Earth's surface (Figure 32.1(B)). This type of activity is referred to as phreatomagmatic (Chapter 30). Factors that influence whether magma−water interaction will be explosive and potentially plume generating include the mass ratio of water to magma, the degree to which the interaction is confined, and the rate at which the interaction takes place. Volcanic plumes generated by this mechanism may contain larger quantities of water and be cooler than those produced solely by degassing of magmatic volatiles.

Another important class of plumes can be generated from material rising off the surface of pyroclastic density currents. These differ from central vent plumes in that they originate from a laterally extensive area. As PDCs travel away from the source, sedimentation of particles from their base and heating of entrained air lead to a decrease in bulk density. Secondary plumes, or **coignimbrite plumes**, are then generated from the tops of flows by buoyant rise (Figure 32.1(C)). During the initial phase of the Mount St Helens eruption in 1980, a giant plume was generated from the top of a pyroclastic blast that traveled up to 29 km from the volcano.

Particle-poor volcanic plumes may also be generated during eruptions that discharge large volumes of basaltic magma. In general, basaltic magmas have a lower gas content compared to that of more silicic magmas. In addition, their low viscosity allows for the differential movement and coalescence of bubbles to take place during magma ascent. Consequently, the degree of fragmentation is poor, and the acceleration of material out of the vent is not as great as during the explosive eruption of viscous magma (Chapter 24). The ejected material commonly forms fountains that rise to only a few hundred meters above the vent. Plumes can be generated, however, by heat from the fountain itself and the limited number of small particles, as well as high temperature gases exsolved from the magma (Figure 32.1(D)).

3. STRUCTURE AND BEHAVIOR OF VOLCANIC PLUMES

3.1. Sustained Plumes from a Single Vent

Large-scale volcanic plumes are generated during the continuous discharge of fragmented silicic magma and gases from a single vent (Plinian eruptions, Chapter 29). Plinian plumes are fed at their base by the discharge of hot gas and particles rising at velocities up to several hundreds of meters per second. As the material exits the vent, its bulk density is usually greater than that of the surrounding atmosphere, and the rise of the material as a **jet** is due to the momentum imparted by gas depressurization. Rapid deceleration occurs under gravity and by drag from the surrounding atmosphere. This jet phase typically extends up to several kilometers above the vent and is characterized by highly turbulent flow (Figure 32.2(A)). Air is drawn into the gas thrust region by the development of turbulent eddies along the plume margins. The air is heated, and expansion results in a decrease in plume density with height.

A critical transition occurs when the bulk density of the gas thrust material becomes less than that of the surrounding atmosphere. The forces driving the motion of the

FIGURE 32.1 Volcanic plumes produced by (A) Plinian-style eruption (Cordon Caulle, Chile 2011), (B) phreatomagmatic eruptions (Surtsey, Iceland 1963), (C) PDCs (Mt Mayon, Indonesia 1984), (D) fissure eruption (Kafla, Iceland 1984).

plume are now dominated by rapid expansion of air and buoyancy, and continued rise occurs. This convective phase represents a large part of the plume "trunk" and may extend for tens of kilometers into the atmosphere (Figure 32.2(A)). The width of the plume increases progressively with height as **entrainment** of air increases and continuously adds to the mass flux. Vertical velocities are of the order of tens to hundreds of meters per second. Integral plume models predict that vent exit velocity plays a fundamental role in the resulting plume structure (Figure 32.3). At a given vent size and low exit velocities (50 m/s), limited entrainment of air does not allow for the mixture to become less dense than the atmosphere and collapse of the jet occurs to form PDCs. At higher exit velocities (~ 75 m/s), initial deceleration is followed by rapid expansion and accelerating vertical ascent as the mixture becomes buoyant. This condition is referred to as superbuoyancy. At the highest exit velocities (~ 200 m/s), rapid deceleration is followed by buoyant rise that slows progressively until the maximum plume height is attained.

Entrainment of air into the eruption plume clearly plays a major role in determining the dynamics of the resulting column. Recent numerical simulations suggest that the efficiency of entrainment varies significantly with height in the plume. In the area just above the vent, entrainment is controlled by the jet structure of the gas thrust region. Far from the vent, in the upper portions of the plume, the entrainment efficiency is the highest in areas where large turbulent eddies are developed when the plume is positively buoyant. In between, the entrainment efficiency drops to its

minimum value and is likely related to a transition in the dimensions of vortical structures with height in the plume.

The Earth's atmosphere decreases in density with height, and eventually, the convective part of a volcanic plume will reach a level where the densities of the plume and surrounding atmosphere are the same (Figure 32.2(A)). At this point, buoyancy is no longer the main driving force, and material will begin to move laterally at a level of neutral buoyancy, H_b. The ultimate top of the plume, H_t, will, however, be higher than this level because of the inertia that the plume has when it reaches the **neutral buoyancy level**. Between H_t and H_b, the material spreads out laterally as a gravity current to form a large mushroom-shaped intrusion into the atmosphere known as the umbrella region (Figure 32.2(A)). The dynamics and dimensions of umbrella regions can be remarkable. A good example is the umbrella cloud formed during the 1991 eruption of Mount Pinatubo in the Philippines. Lateral expansion of material produced an umbrella region that had reached a diameter of 400 km after about 5 h and 1000 km after about 14 h. At least, 200 km of horizontal expansion took place against the prevailing northeasterly winds, attesting to high mass flux. Eventually, the influence of the gravitational spreading will diminish, and the plume will be subject to **advection** by atmospheric winds. The transition point, in terms of distance from the source, is strongly dependent on the mass eruption rate of the event.

An important aspect of plumes from maintained sources is the height to which they rise in the atmosphere. The maximum height, H_t, is a function largely of the **thermal**

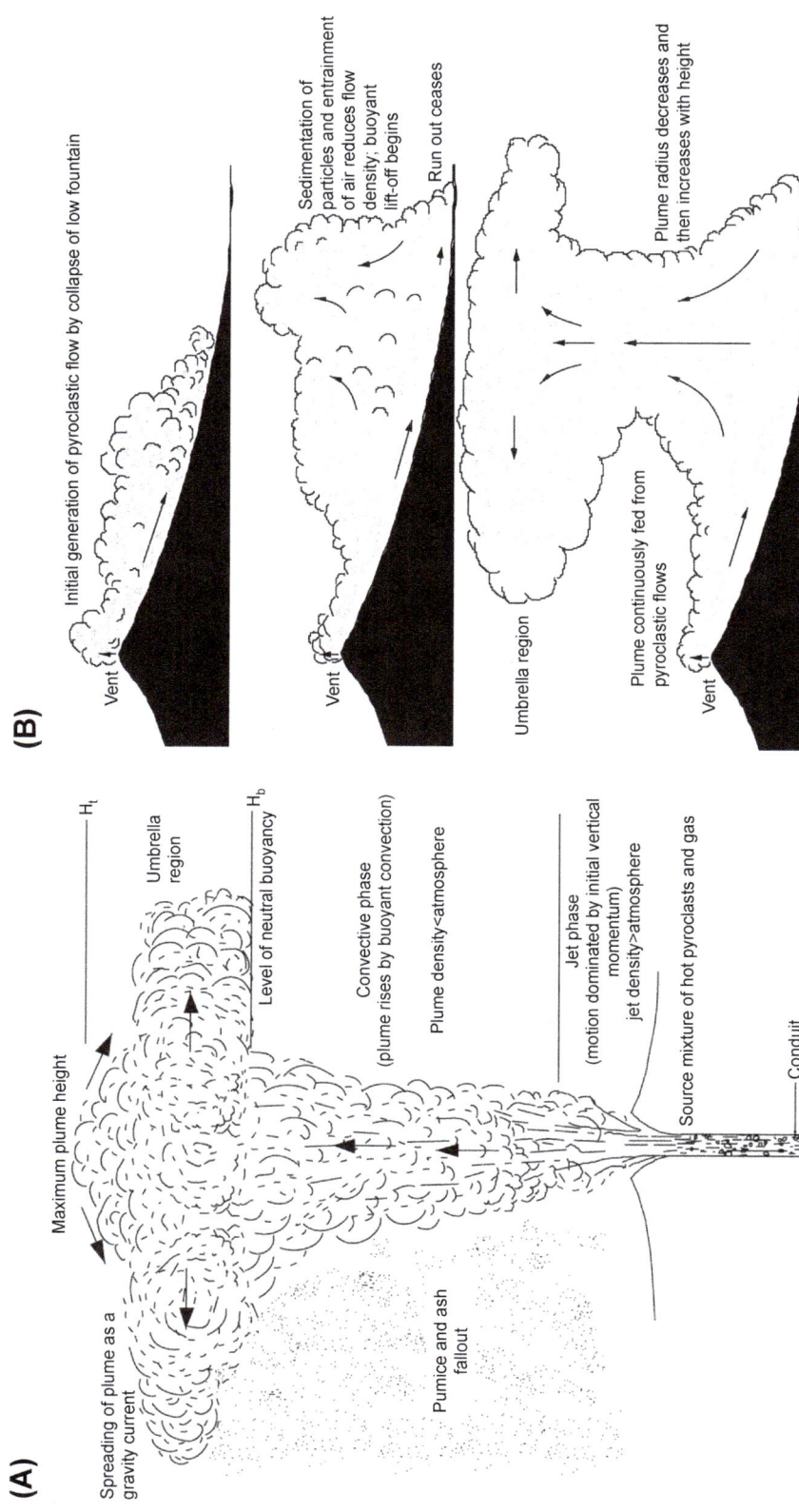

FIGURE 32.2   (A) Structure of a volcanic plume generated by a sustained Plinian-style eruption. (B) Generation and form of a volcanic plume generated from the top of a pyroclastic flow.

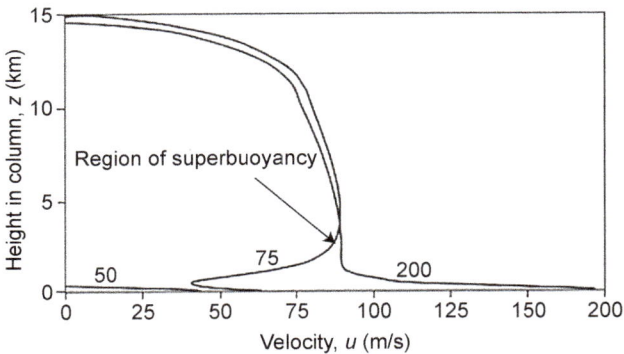

FIGURE 32.3 Modeled variation of ascent velocity in a sustained volcanic plume as a function of height above the source vent. Curves are shown for initial exit velocities of 50, 75, and 200 m/s. *Modified from Woods (2013).*

flux at the vent, the stratification and moisture content of the atmosphere, and the volatile content of the magma. Thermal flux is the most important factor and is related to the mass discharge rate of magma and its heat content. A simple approximation for column height can be found from the following equation:

$$H_t = 1.67Q^{0.259}, \qquad (32.1)$$

where H_t is the maximum plume height (in kilometers), Q is the volume discharge rate of magma in cubic meters per second, and 1.67 is a constant related to the stratification of the atmosphere. Observations of historic eruptions are in relatively good agreement with this simple relationship (Figure 32.4), and integral models of sustained plumes also predict a similar relationship, although for weak plumes, Eqn (32.1) does not provide reliable predictions due to the influences of the pyroclasts, wind, and external water. Given sufficient observations of plume characteristics and using numerical models that are more sophisticated than Eqn (32.1), it is in fact possible to "invert" the data to derive discharge rate, eruption speed, vent size, and local atmospheric conditions. For eruptions where direct observations of plume heights were not possible, for example, ancient events, estimates of plume heights can be derived through inversion methods that utilize quantitative information from the resulting tephra fall deposit. The technique attempts to optimize the fit between a predicted areal distribution/thickness of a fall deposit with the observed one by automated repetition of a forward distribution model with variable input source parameters such as plume height and wind speed. The technique allows for rapid testing of a large number of potential combinations of source parameters to achieve the best fit with the observed fall deposit.

Because of the importance of volcanic source conditions to atmospheric dispersal, numerous methods have been developed to monitor plume height and other source conditions. Generally, ground-based methods are used for near-vent observations. Among the ground-based methods, Doppler weather radar has been used to characterize plume height and rise speed above the vent, and particle size and concentration. Using techniques from meteorology, the reflectivity of plume material in a given weather radar band is converted to plume characteristics. Infrasound and thermal imaging yield information about discharge rate, exit speed from the vent, temperature, pyroclast size, and concentration. Infrasound signals are analyzed by using an array of sensors to track acoustic signals back to the source. The acoustic pressure for a given distance is then related to discharge rate and speed of the moving plume source. Thermal techniques can directly image hot particles, yielding their temperature as well as size and speed. Seismic techniques can even be used to estimate discharge rate, since an eruption generates an unusual, scaled seismic signal.

Two factors that contribute to variability in the establishment of a convectively rising plume are (1) thermal disequilibrium between pyroclasts and gas, and (2) reentrainment of pyroclasts along the plume margins. Thermal disequilibrium is related to the timescale for heat exchange between pyroclasts and surrounding gas compared to the residence time of the pyroclasts in the rising plume. Larger pyroclasts may not be able to reach thermal equilibrium during their rise in the column compared with much smaller particles. The net result is a decrease in the available thermal energy to drive buoyant convection and a reduction in the effective column height. Similarly, reentrainment of cold particles into the plume during fallout in the atmosphere results in a slower decrease in the net vertical mass flux, and a reduction in the vertical rise velocity.

Sustained eruptions may also include interactions of magma with external water. These interactions typically result in an increase in the degree of fragmentation and the production of large volumes of fine ash (phreatomagmatic eruptions). The presence of additional water has important effects on plume behavior. One effect is that as liquid water is converted to steam the bulk density of the erupting mixture will decrease, and the formation of buoyant plumes is enhanced. However, an offsetting effect is that incorporation of relatively low temperature water results in an overall decrease in the temperature of the erupting mixture. A lower temperature reduces the potential buoyancy of the plume and decreases its ultimate height. Under conditions of no external water, buoyant plumes are generated for mass eruptions rates up to 10^8 kg/s. With the addition of 10–30% external water buoyant plumes can be generated at higher mass eruption rates ($>10^9$ kg/s) because of the higher initial buoyancy flux of the mixture. However, when the amount of external water is $\geq$40%, the plume becomes unstable at lower mass eruption rates owing to the lower initial temperature of the mixture and its higher density.

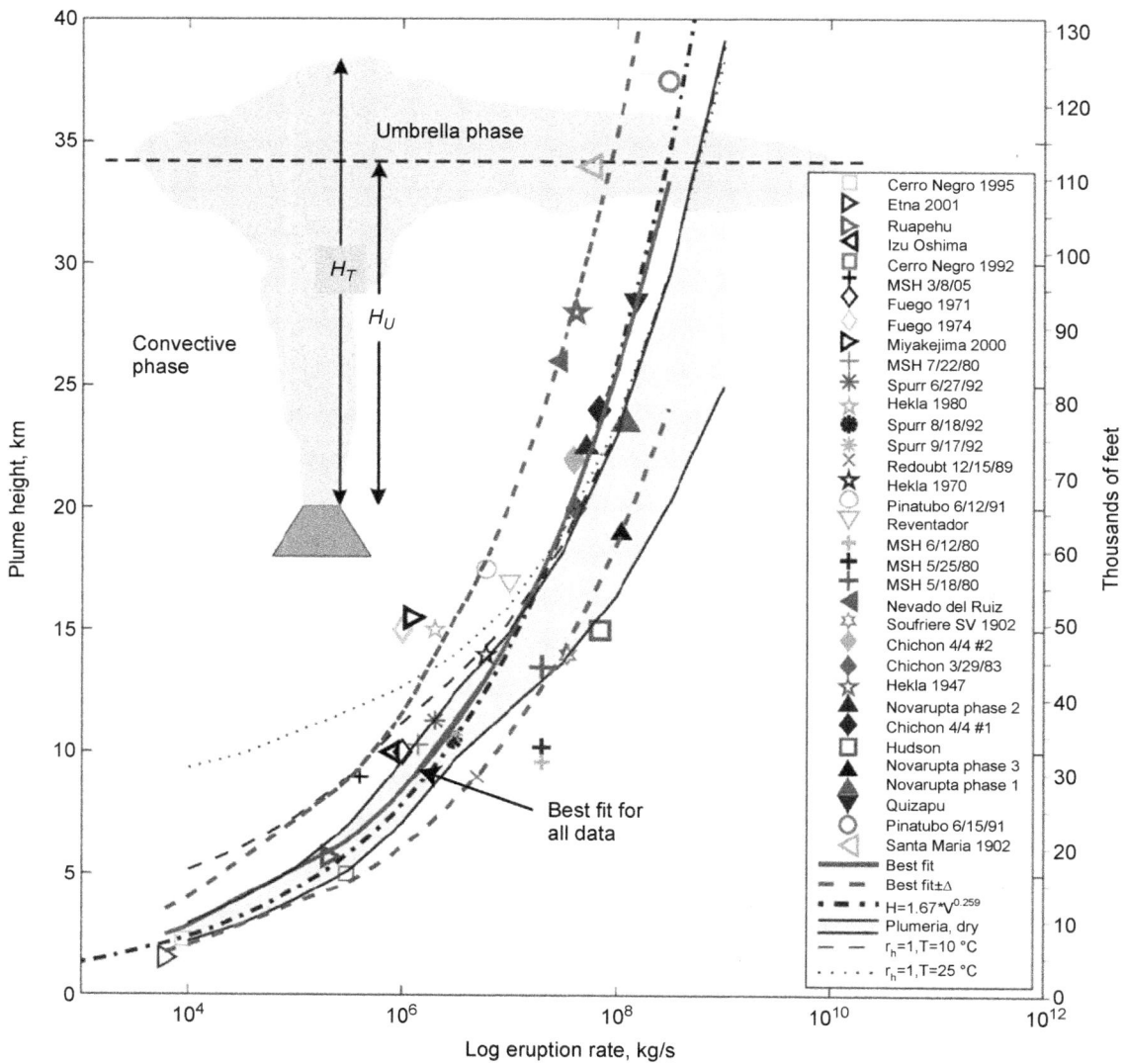

FIGURE 32.4 Plume height versus eruption rate for numerous historic explosive eruptions. The bold solid line is the best fit to eruption data. The bold dashed line represent the 50% confidence interval based on uncertainties in estimating the two parameters. The upper light solid line represent results from a steady-state plume model (Plumeria), and the lower solid line indicates the estimates of H_u based on the model results. *Modified from Mastin et al. (2009).*

3.2. Discrete Explosions from a Single Vent

Vulcanian eruptions differ from Plinian events in that the discharge of gases and particles occur as a series of discrete explosions separated by a few minutes to several hours (Chapter 28). Observations and measurements of the frequent Vulcanian eruptions at the Soufriere volcano on Montserrat have led to significant improvements in the understanding of plume-generating processes. A common evolutionary sequence for the Montserrat Vulcanian eruptions involves an initial rapid acceleration of gas and particles out of the vent and development of a collapsing veil of material that produces a distinctive overhanging plume shape. After a short time, PDCs emerge from the veil and

move downslope. As the flows spread out, a vertically directed plume rises over the central vent and is joined by convective plumes of fine ash derived from the laterally moving flows. Multiphase numerical simulations of transient magma discharge at Montserrat have successfully captured many of the essential elements of plume generation behavior (Figure 32.5). The models are tightly constrained by vent geometry characteristics, petrologic estimates of volatile content, and representative pyroclast grain size distributions. An important variable is the extent to which volatiles are lost by outgassing during magma storage prior to eruption. If the rate of volatile loss is low, simulations correctly reproduce the overhanging plume,

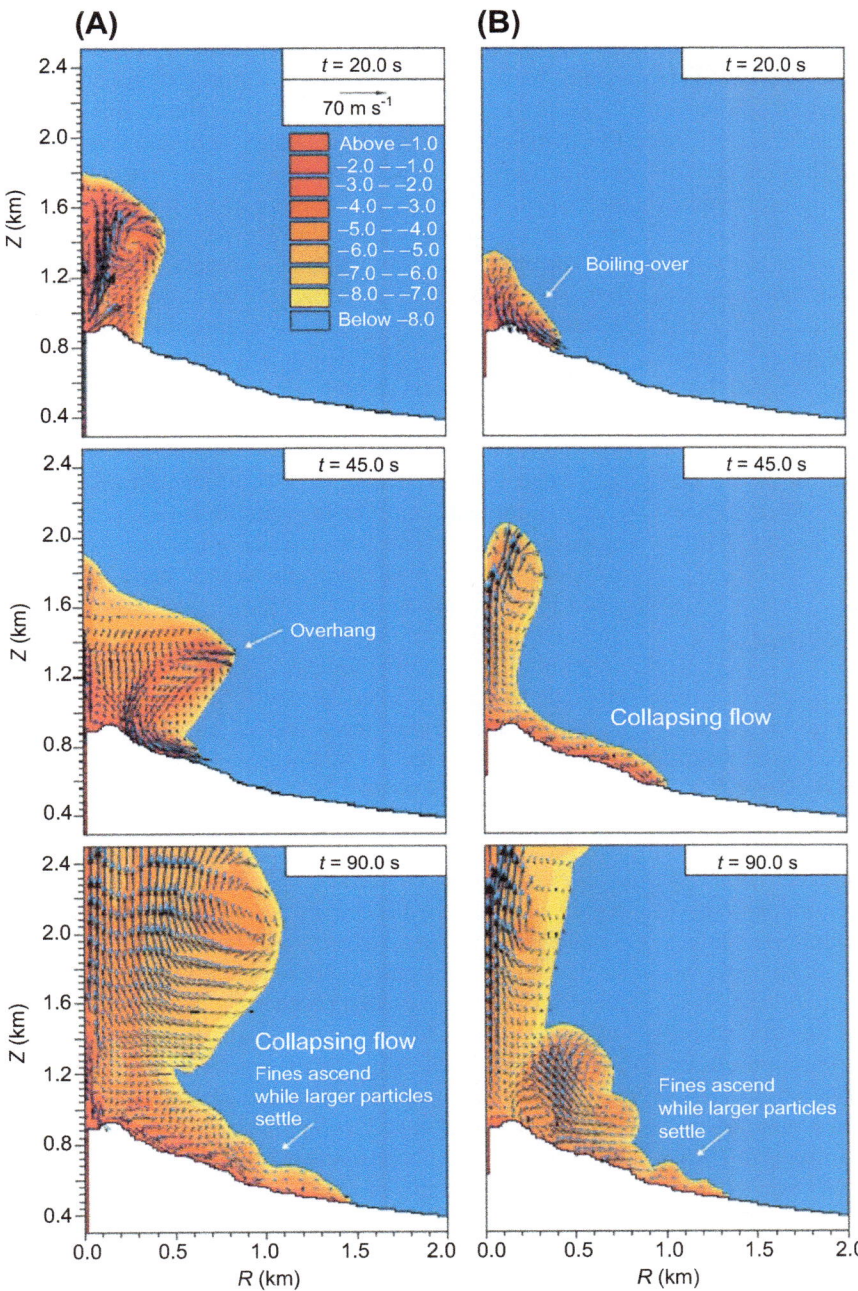

FIGURE 32.5 Model simulations of Soufriere Hills (Montserrat) Vulcanian eruptions. (A) Initial parameters set with no gas leakage and (B) initial parameters set to allow for some gas leakage prior to eruption. The color scale indicates relative particle concentration in the plume. Simulation time is given in seconds (s). *Modified from Clarke et al. (2002).*

development of pyroclastic flows through a collapsing veil, and rise of a vigorous central plume (Figure 32.5(A)). A very good agreement is also found between simulated and observed plume rise velocities that define three principal periods of plume evolution: jet-dominated, stagnation phase, and buoyancy dominated. On the other hand, if the rate of volatile loss is high, then the simulations produce a rapid "boiling-over" of material at the vent that feeds laterally moving PDCs with no overhanging plume structure (Figure 32.5(B)). A distinct rising plume centered

over the main vent develops shortly thereafter and is joined by rising plumes shed from the tops of the PDCs. The good correspondence between the multiphase simulation results and observations of actual vulcanian eruptions at Montserrat demonstrates that consideration of the transient nature of this type of activity and knowledge of pre-eruption volatile content are critical to developing models that can be used effectively to understand eruption dynamics and their potential hazards in the proximal environment.

Vigorous steam-rich plumes are often generated when magma is erupted in shallow water. A Surtseyan eruption sends steam-rich cocks-tail jets along parabolic trajectories (Chapter 25). The tephra-laden jets are produced as a series of discrete events that may take place on a variety of timescales ranging from minutes to hours. The vent area of a Surtseyan eruption is a slurry of hot water and cold tephra mixed with new magma. Upward migration of the mixture (pressure reduction) and heating cause the water to flash to steam and be rapidly accelerated out of the vent as a jet. Observations of the 1963 Surtsey eruption in Iceland indicated that during ejection of material there is rapid decoupling and fallout of large poorly fragmented clasts. The remaining steam-rich parts of the plume rise as a series of thermals or, if the frequency of explosions is high, they may coalesce to form a steam-rich convective plume that carries small highly fragmented particles. Such plumes rarely rise to high altitudes because the thermal flux is generally low.

3.3. Maintained Plumes from Distributed Sources

An important mechanism for the generation of volcanic plumes is the buoyant rise of gas and particles from the top of laterally moving PDCs (Figure 32.2(B)). These PDCs are concentrated mixtures of hot particles with gases and entrained air that form by a variety of mechanisms during explosive volcanism (Chapter 35). Evidence from the geologic record indicates that large volume explosive eruptions have generated enormous coignimbrite plumes that are likely to have caused significant global environmental change. For example, an eruption of the Toba caldera, Sumatra, Indonesia, 75,000 years ago spread ash from a coignimbrite plume over much of the northern Indian Ocean and may have triggered a posteruption global cooling of 4 °C with regional reductions in precipitation.

Even though PDCs begin as mixtures that are denser than the atmosphere, entrainment, heating and expansion of air, and sedimentation occur during flow, both of which allow for plume development. The lateral movement of PDCs is driven primarily by gravity, but fluxing of gas through the flow can fluidize the particles, reducing drag on the bed, and enabling the current to move downslope at high velocities. Explosive eruptions produce a large range of particle sizes, and some of the smaller particles are lost from the flow by **elutriation** with the flux of gases into a more dilute, overlying portion of the PDC (Figure 32.2(B)). Heating of entrained air is a particularly important mechanism in the generation of coignimbrite plumes. Air enters a moving current at its head and along its upper surface where a turbulent mixing zone is established. A second process for the reduction of pyroclastic current density and the generation of buoyancy is sedimentation of particles.

Eventually, the current will reach a point where the bulk density becomes less than the atmosphere and convective rise at least of its upper part will occur (Figure 32.1(C)). Depending on the initial flow conditions, this buoyant liftoff may occur over an extensive area and can generate plumes from a source area with a width of tens to hundreds of kilometers. This is probably the main plume-generating mechanism during large-scale PDC-forming eruptions.

The structure of a maintained coignimbrite plume differs substantially from a large-scale convective plume generated from a central vent. First, the coignimbrite plume does not have the high-velocity jet phase associated with a single vent eruption (Figure 32.2(A)). It begins its ascent with very low velocity and little initial momentum. Second, the source area, and thus radius, of the plume base can be very large. Entrainment of air is therefore quite different for a coignimbrite plume. The large initial radius means that it entrains a smaller fraction of dense lower atmospheric air compared to a central vent plume that starts out with a small initial radius and then expands upward. Thus, during the early stages of development, the buoyancy of the coignimbrite plume is reduced compared to a single vent eruption having a similar thermal flux.

Three-dimensional modeling of coignimbrite plumes highlights the complex entrainment domain as the plume develops. Multiple convection cells, with both updrafts and downdrafts, are developed within the internal structure of the plume (Figure 32.6(A)). As with a plume from a single vent, a coignimbrite plume will also form a well-defined umbrella region where the plume density is similar to the surrounding atmosphere, and lateral advection of material takes place as a large-scale gravity current. However, compared to plinian plumes, 3D models predict relatively lower levels of neutral buoyancy as shown by using SO_2 as a tracer component. This has important implications for the potential climate impact of very large-scale explosive eruptions (supereruptions) that likely produce enormous coignimbrite plumes. The models suggest that such plumes may not be that efficient in transferring large amounts of climate-modifying aerosols to the upper stratosphere. Even large-radius PDCs (70 km) result in maximum SO_2 concentrations only just above the **tropopause** (Figure 32.6(B)). The ultimate height of the plume is primarily related to the thermal flux. However, during PDC-forming eruptions, only a fraction (ca. 30−40% maximum) of the total thermal energy can be utilized for coignimbrite plume generation. The remainder is retained in the hot PDC deposits formed from the basal part of the moving flow. Consequently, for a given mass eruption rate, a coignimbrite plume will not rise as high as a plume from a central vent. However, mass eruption rates are often significantly higher during the PDC-forming stages of explosive eruptions.

Large-scale basaltic eruptions often take place from fissures that may extend several to tens of kilometers in

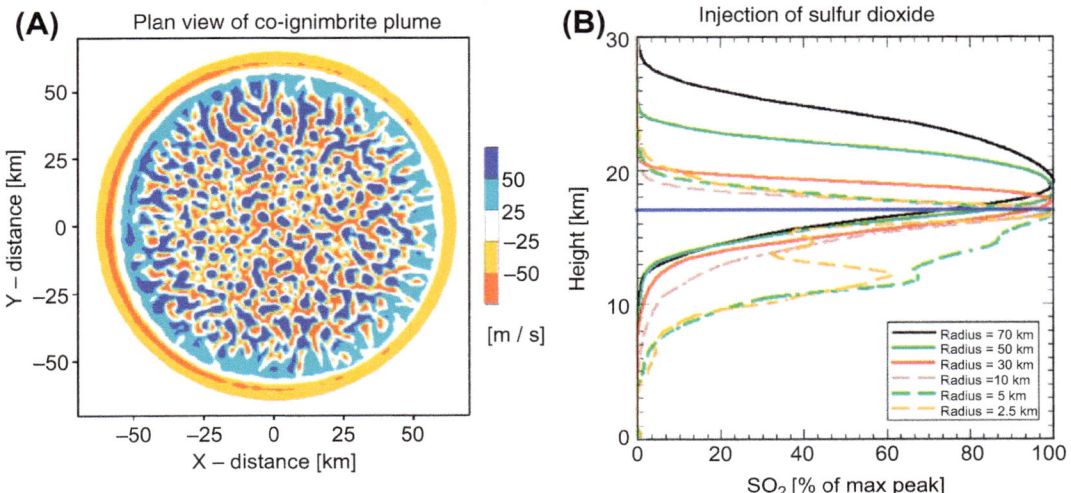

FIGURE 32.6 (A) Vertical velocities at a height of 5 km over a coignimbrite plume with an initial 50-km radius based on the 3D Active Tracer High Resolution Atmospheric Model (ATHAM). Note the complex upward and downward flow patterns. (B) Predicted SO_2 loading from coignimbrite plume of different starting radii (2.5–70 km) as a function of the height based on 3D ATHAM modeling. *Modified from Herzog et al. (2010).*

length (Chapter 24). Plumes are generated above the fissures if fragmentation can produce a source of sufficient small particles and gas. The generally low fragmentation of basaltic magmas means that only a small fraction of the total thermal energy can be utilized to drive a convective plume above the vent. As with ignimbrites, a large fraction of the heat is retained in the clasts that fall back to the surface near the vent at high temperatures. There are three main sources of thermal energy for the production of plumes above a large fissure eruption. The first are small fragmented particles (millimeters or less in size) that can be carried aloft by the gas. This fraction is generally <20% of the total erupted magma. A second source is the hot gas, a mixture of water, carbon dioxide, and sulfur dioxide that is decoupled from the magma upon eruption. A third source is heat exchange between the fire fountain and surrounding air. Temperature decreases of a few to tens of degrees have been inferred for the main volume of magma in the fountain. These three sources can contribute to the development of a gas and particle plume that ascends above the top of the fountain by convective rise (Figure 32.1(D)).

Plumes developed from large basaltic fissure eruptions are important because of their potential for climatic impact. Basalt carries higher amounts of sulfur per unit mass than does more evolved magmas, and sulfur is an especially important agent for altering the heat budget of the atmosphere (Chapter 53). However, for sulfur injection to be important for climate change, it must reach altitudes >10–15 km in the atmosphere. The height to which a basaltic plume will rise in the atmosphere is a function of the mass eruption rate of magma, the volatile content, the fraction of ash produced by fragmentation, and the amount of cooling experienced by the eruptive jet. For a given mass eruption rate, the eruption of basaltic magma generates lower plumes than eruptions of more evolved magma. This is mainly because of the inefficient use of thermal energy to drive the buoyant plume. The 1783 basaltic fissure eruption of Laki in Iceland generated a plume that rose to a height of at least 12 km. This height implies an eruption rate of about 1 m^3/s/m of vent length based on a theoretical modeling of basaltic plumes. Although such rates are unusual for historic eruptions, there is evidence in the geologic past for very large fissure eruptions that exceeded discharge rates of 10 m^3/s/m. For example, the great Roza basaltic flow in eastern Washington, USA, is inferred to have been fed by a discharge of 10 m^3/s/m. It is likely that major atmospheric plumes are generated during such events, with the potential for significant climatic impact.

4. ATMOSPHERIC DISPERSAL OF PLUMES

4.1. Crosswind Interactions

Once a volcanic plume is generated, the dominantly horizontal flow in the atmosphere will begin to distort its structure and transport material downwind. The nature of the plume–atmosphere interaction depends on many factors, but at least initially, the changes will be governed by the vertical and horizontal scales of the plume relative to atmospheric motions. It is important to emphasize that explosive eruptions are capable of creating plumes that vary over several orders of magnitude in scale. At the low end of the spectrum are small explosive eruptions with magma discharge rates of $\leq 10^4$ kg/s (Figure 32.4) that produce plumes rising only several kilometers in height and forming thin elongated downwind dispersal patterns.

The difference between relatively weak and strong plumes can be expressed in terms of the ratio of the horizontal wind speed to the upward plume speed. A weak plume, for which this ratio is large (>1), is greatly affected by the wind, taking on a characteristic "bent-over" shape like an industrial smokestack plume without a distinctive umbrella region. In the case of weak plumes, a vertical eruption column is unable to develop. Unlike strong plumes that form umbrella clouds, weak plumes maintain a cloudlike structure as they reach their maximum rise height, since their flow characteristics are not reorganized in the upper atmosphere by gravity current motion. The plume is not only bent over by the wind but it also mixes with the turbulent atmosphere as it is carried along. This enhanced entrainment results in lower total plume heights for a given mass eruption rate as the wind speed increases because of the more rapid dilution of the plume's thermal energy by the entrained air. In plan view, such plumes often consist of downwind-elongated lenses of gas and ash that become detached from the volcano when the eruption ceases. Sometimes, however, smaller plumes can have rather irregular outlines because of their susceptibility to local atmospheric motions. They can even have several lobes at different heights due to wind shear effects. The plume–wind interaction in weak plumes can also result in spectacular organized motions. The combination of an upward-directed plume meeting a horizontal wind can create vorticity in the downwind side of the plume. This vorticity has occasionally manifested itself as a line of tornadoes shed from the plume margin, as occurred during the eruption of Surtsey in 1963.

Intermediate scale eruptions with magma discharge rates of 10^4-10^8 kg/s have ratios of wind speed to vertical

plume velocity less than approximately 1 and typically form well-developed umbrella regions (Figure 32.4). In general, for plumes with a total rise height >20 km, no typical values of wind speed are able to affect plume rise substantially. Most tropospheric plumes in weak winds are strong plumes, as well as most plumes that reach the stratosphere. Spreading of the plume within the umbrella region behaves as a gravity current intrusion into the atmosphere and will encounter resistance in the upwind direction until a stagnation point is reached. At this point, the wind speed and the gravity current speed are balanced, and further upwind expansion of the current is impeded. For very powerful eruptions, the stagnation point can occur at substantial distances upwind of the source. For example, the umbrella cloud from the May 18, 1980, Mount St Helens blast plume spread upwind 15 km before it was balanced by stratospheric winds. In the crosswind direction, the plume may continue to spread as a gravity current to hundreds of kilometers from the vent, as was the case with the Pinatubo eruption of June 15, 1991 (Figure 32.7). In the downwind direction, the plume is carried along in the wind field. As the plume propagates progressively further downwind, its speed approaches the wind speed while the plume loses pyroclasts and the gravity-driven spreading weakens. Further downwind from the vent, the wind completely dominates plume motion. Shearing stresses between the upper and lower interfaces of the downwind plume and the atmosphere and direct atmospheric entrainment result in downwind transport at the wind speed. The plume loses its density contrast with the atmosphere and can be advected as a lens of aerosols and gas with a nearly constant width. Only slowly might it thin, spread, and disperse as shearing and small-scale atmospheric

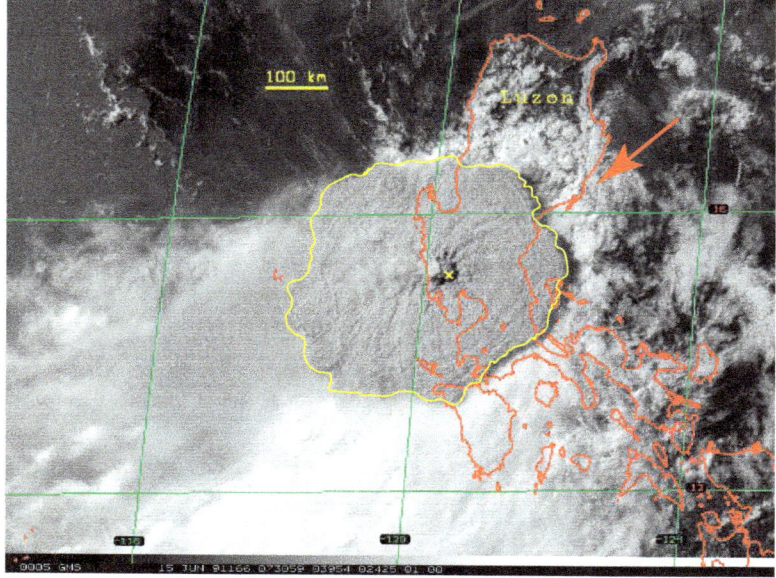

FIGURE 32.7 Satellite image of the umbrella region (yellow outline) of the 1991 eruption plume from Pinatubo volcano in the Philippines. Local wind direction at the level of the umbrella region is shown by the red arrow. Location of Pinatubo volcano is indicated by the yellow x and island coastlines in red. *Reprinted courtesy of the USGS.*

turbulence act at its margins and as very fine ash settles out. The elongated shape of such plumes is illustrated by the aerosol band generated during the El Chichon eruptions of 1982. This band maintained a nearly constant width and became gradually more dilute due to shearing at its margins as it completely encircled the globe. Further information about the dispersal of volcanic plumes can be found in Chapter 33.

Of particular interest to the concern with hazards to jet aircraft that might encounter volcanic plumes are the role of upper tropospheric—lower stratospheric jets and their interactions with source plumes. Jets are focused regions of high wind speed that occur dominantly between the latitudes 30° and 60°. Located at altitudes of about 10 km, average speeds range from 10 to 40 m/s but with some high intensity core regions reaching up to 130 m/s. Moderate-sized plumes that interact with high-speed jets are more susceptible to modification due to the higher ratio of horizontal wind speed to upward plume velocity. Modeling of plume behavior that takes into account large vertical variations in horizontal wind speed (jet patterns) has been able to accurately reproduce the observed shapes of actual tropospheric plumes that were injected into regions of jet flow (Figure 32.8). An important result is that interactions of plumes with strong jets has the effect of limiting the plume rise height for a given magma discharge rate. Moderate- and large-sized eruptions are thus able to inject larger quantities of pyroclasts and gases into jet levels because of the net reduced plume height and strong horizontal jet velocities that rapidly transport particles downwind. This can lead to considerable uncertainties in the mass loading of pyroclasts and gases at tropospheric altitudes especially if a particular eruption experiences variations in magma discharge rate during an eruptive episode.

These uncertainties carry significant implications for the interaction of jet aircraft and eruption plumes because many commercial airlines plan flight paths to either take advantage of, or avoid the jet stream. Eastbound flights use the high-velocity jets to reduce fuel consumption, whereas westbound flights strive to avoid the excess drag and maximize efficiency. Flights that utilize jet streams for enhanced transport are likely to be more susceptible to encountering ash loaded levels in the atmosphere.

At the highest end of the plume-producing spectrum are eruptions that discharge magma at rates $>10^9$ kg/s. Eruptions of this scale, often referred to as supereruptions are yet to be observed during historical times, and much about their eruptive conditions has been inferred through the study of their resulting deposits and theoretical modeling. These events are typically characterized by enormous discharges of magma as PDCs and consequently a principal plume-generating mechanism is coignimbrite plumes (Figure 32.2(B)). It is suggested that magma discharge rates of this magnitude would generate umbrella clouds with initial diameters from 600 to 6000 km. Because of their exceptional scale, the motion of the clouds would be controlled by a balance between gravity and the Coriolis force, leading to the development of a huge spinning cloud that is relatively unaffected by the local atmospheric wind system. Internal expansion and rotational velocities are predicted to be tens of meters per second even at distances of several thousand kilometers from source. Such plumes would be able to spread pyroclasts and gas over continental-scale areas before being significantly dispersed by the prevailing winds. This is in accord with observations of widespread volcanic ash layers, such as the Toba ash layer from a 75K Ybp eruption in Indonesia, whose distribution pattern cannot be easily explained by wind dominated transported from a single point source.

4.2. Effects of Latitudinal Variations in Temperature and Moisture

The rise height and dispersal of plumes can be affected by differences in the atmospheric structure in the vicinity of the source volcano. Variations in the atmospheric stratification result from vertical temperature variations caused by the differential heating of the solid earth and air. In particular, the height of the tropopause is strongly dependent on surface heating, and hence on latitude. In general, the **troposphere** is only weakly stratified, while the stratosphere is more strongly stratified. In a typical mid-latitude atmosphere, the temperature decreases steadily to the tropopause, which is located at about 11 km. The temperature then remains nearly constant up to a height of about 20 km where it begins to increase, and the atmosphere becomes strongly stratified. In the tropical

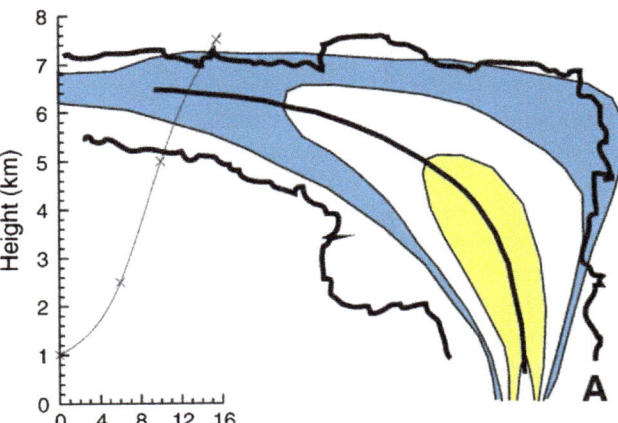

FIGURE 32.8 A. Comparison of observed (heavy black irregular outline) and model-predicted (colored patterns) plume shape of the 1994 eruption of Kliuchevskoi volcano, Kamchatka. Model results produced using the BENT integral model of steady plume motion in a high-velocity jet wind field (shown to the left of the figure). *Modified from Bursik et al. (2009).*

atmosphere, the tropopause is located at about 16—18 km. The temperature increases steeply above this height, producing a very stably stratified atmosphere. In polar latitudes, the tropopause may occur as low as 8 km, and the stratosphere above is much less strongly stratified than it is in the tropics or midlatitudes.

As a result of such variations in the ambient stratification, the maximum rise height of an eruption column at a fixed eruption rate can vary by a few kilometers. In a given atmosphere, the rate of increase of column height as a function of the mass eruption rate decreases for columns that penetrate the tropopause. In conjunction with this effect, as the latitude of a hypothetical eruption changes from tropical to polar, a given column will penetrate the tropopause at ever-decreasing heights. Therefore, the plume ascends to a lower height for a given mass eruption rate and eruption temperature in a polar atmosphere (lower tropopause) because the plume propagates through a deeper region in which the temperature increases with height.

During plume rise, large quantities of air are typically entrained from the surrounding atmosphere. If the air is moist, significant amounts of water vapor are added to the plume. This air will be cooled as it rises and become saturated. Once saturated, the vapor condenses into liquid water, releasing its **latent heat of condensation**. The latent heat increases the temperature of the gases in the eruption column, thereby increasing the buoyancy and ultimately the eruption column height. For low mass eruption rates, the entrainment and convection of atmospheric moisture upward in the plume can provide the dominant energy source driving the motion of the eruption column. An analogy is provided by forest and brush fires, which heat the air near the ground. As this heated air rises, the water vapor within it becomes saturated and condenses. The latent heat released provides thermal energy, which enables the hot smoke to continue to rise. Such a fire plume thus ascends higher when the relative humidity is higher than it does when the relative humidity is low.

In humid air, the height to which a small volcanic plume rises is likewise considerably greater than it is in dry air. This is because the thermal energy released by the condensing atmospheric vapor is comparable to or even greater than is the thermal energy associated with the erupted material. For small eruptions in relatively dry air, the column remains unsaturated. In larger eruptions or in moister air, the column becomes saturated, and therefore, the height of rise of the column increases significantly. However, as the erupted mass flux increases above about 10^6 kg/s, the erupted thermal energy begins to significantly exceed the energy that can be extracted from condensing water vapor even at 100% relative humidity. This is because the ambient moisture is confined to the lower atmosphere. Only smaller columns can entrain a sufficient mass of water vapor relative to the erupted thermal energy so that the

latent heat produced on condensation is comparable to or exceeds the thermal energy erupted in the pyroclasts. Larger columns ascend far above the region in which most of the atmospheric water vapor is found, and so, only a small fraction of the entrained air is moist. As a simple criterion, eruption columns whose total ascent heights exceed the tropopause height are not significantly affected by atmospheric moisture.

4.3. Downwind Evolution

As an eruption comes to a close, the accumulated volcanic gas and ash ejected into the atmosphere assume one of two typical forms. Smaller plumes are often markedly elongated in the downwind direction, with a prominent taper toward the upstream end (Figure 32.9(A)). The taper is the result of both the gravity current spreading and atmospheric **turbulent diffusion** having affected the downstream end of the plume for a longer time than the upstream end. Larger umbrella clouds are more often nearly circular in plan view, with an elongation in the downstream direction that makes them occasionally ovoid. Because they are often the result of a complex sequence of events, larger, developed umbrella clouds may be compounded from several separate clouds. They can therefore display a complex digitate geometry. In addition, the most massive umbrella clouds, such as that from the 1991 eruption of Pinatubo, display another type of digitate margin that indicates they have reached the Rossby radius, the size at which the Earth's rotation begins to affect cloud motion (Figure 32.7).

When any volcanic plume has persisted for a sufficiently long time (more than several hours), it becomes fully subjected to atmospheric motions. These motions fall into a spectrum of characteristic size scales, from microscopic molecular agitation, to **hemispheric flow**. To gain an understanding of the effects of the atmosphere on the long-range dispersal of eruption clouds, horizontal and vertical motions can be thought of separately. Horizontal dispersal is controlled by the nearly two-dimensional nature of the largest atmospheric motions that transport ash on regional and global scales. The vertical dispersal of the particles is controlled by the smallest scale turbulence and gravitational settling, which act on much smaller length scales. In addition, aggregation of small ash and aerosol particles in the months and years following an eruption results in the gradual increase of particle size. Eventually, particles become sufficiently large that their settling speed causes the particles to be deposited.

The large-scale, quasi-two-dimensional atmospheric motions responsible for the distal transport of ash clouds can be described by a "turbulence energy spectrum." The effect of the different scales of motion in this spectrum on plumes of different characteristic sizes is vastly different despite the fact that all atmospheric agitation has its origin

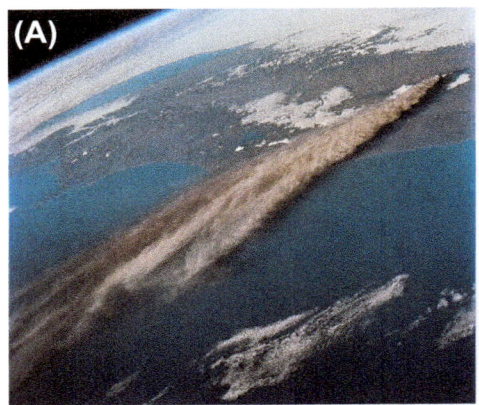 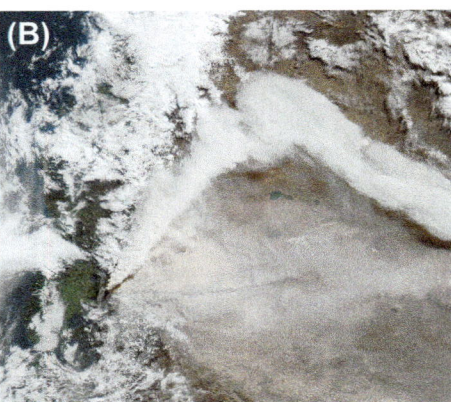

FIGURE 32.9 Satellite images of the downwind dispersal of volcanic plumes from A) 1994 eruption of Kliuchevskoi volcano, Kamchatka, and B) 2011 eruption of Puyehue-Cordon Caulle volcano, Chile. *Images courtesy of NASA.*

in the heating of the Earth and air by the Sun. The largest scale motions ($> \sim 10^3$ km) primarily advect virtually all plumes, only slowly deforming them, and causing little growth or turbulent spreading. Observations of plumes carried in these zonal winds suggest that distortion primarily occurs as simple stretching and very moderate bending of plume outlines. Intermediate scale turbulence structures ($10^2 - 10^3$ km) are primarily responsible for complex bending and relative diffusion (spreading and dilution) of small- to moderate-sized ash clouds. The complex bending or contortion is known as **chaotic advection**, as plume-fluid elements are distorted in a chaotic fashion, despite the completely deterministic nature of the velocity field itself. The plume from the Puyehue—Cordon Caulle eruption of 2011 exhibited this phenomenon as it drifted at the edge of the polar vortex (Figure 32.9(B)). The motion of the largest volcanic ash clouds, such as that generated by the eruption of Pinatubo, seem to show almost exclusively relative diffusion by the intermediate scale motion, with little apparent stretching or bending.

Vertically, a downwind volcanic plume does disperse, but at rates very different from the horizontal dispersion rates discussed above. This vertical dispersion occurs because of the coupled effects of the settling properties of the particles and of the turbulence structure of the stratosphere. The stratosphere consists of alternating layers of a 1- to 2-km thickness. Every other layer seems to display turbulent motions with vertical wind speeds between −0.5 and 0.5 m/s. The intervening layers are relatively quiescent. These observations are consistent with the layered temperature structure observed in stratospheric soundings. The turbulent layers will be able to retain particles longer than will the quiescent layers because the turbulence will hinder particles from settling from the convecting layer. Particles will fall out relatively rapidly from the quiescent layers that lack a counteracting agitation. Because of this alternating turbulent and quiescent structure in the stratosphere, volcanic ash and aerosol clouds will tend to separate into layers over long time periods, lending to them a distinct banded appearance when viewed from the side. Over a period of months to years however, virtually all volcanic particles settle completely from the atmosphere.

4.4. Tracking of Volcanic Plume Dispersal in the Atmosphere

An essential part of hazard mitigation associated with volcanic plumes is the ability to track the movement, size, and composition of plumes as they travel away from source. Satellite sensors have proved to be one of the most effective tools in long-range monitoring of volcanic plumes and play a fundamental role in providing critical data for models used to forecast areas that are likely to be impacted downwind. Efforts to track the progress of plumes have focused largely on trying to detect suspended ash particles and sulfur dioxide gas. A significant problem with the monitoring of volcanic plumes from space is that, in many cases, it is difficult if not impossible to distinguish distal volcanic plumes from regular meteorological clouds, especially in the visual light spectrum. The use of infrared imagery has proved to be a much more effective tool in isolating volcanic plumes from weather clouds. A method has been developed that utilizes the difference in two images collected at different infrared wavelengths (10 and 12.5 μm). Ash plumes contain abundant small solid particles (pyroclasts), unlike weather clouds that are dominantly water and ice. Each type of cloud absorbs and reflects infrared radiation in a slightly different way as a result of this contrast in composition. This difference can be captured by looking at the differences in the images collected at the two wavelengths. Using this so-called "split-window" technique and false color assignments, an image can be created that shows areas of volcanic ash that can be discriminated from normal water clouds (Figure 32.10). It is also possible to extract quantitative information about ash particle size and mass loading from infrared imagery. Such information is particularly

FIGURE 32.10 Satellite image of volcanic plume from explosive eruption of Karthala volcano in the Indian Ocean. Occurrence of volcanic ash (color scale) is based on Advanced Very High Resolution Radiometer (AVHRR) split-window technique. Portions of the cloud that are ash-free are shown as white/gray. *Image courtesy of NASA.*

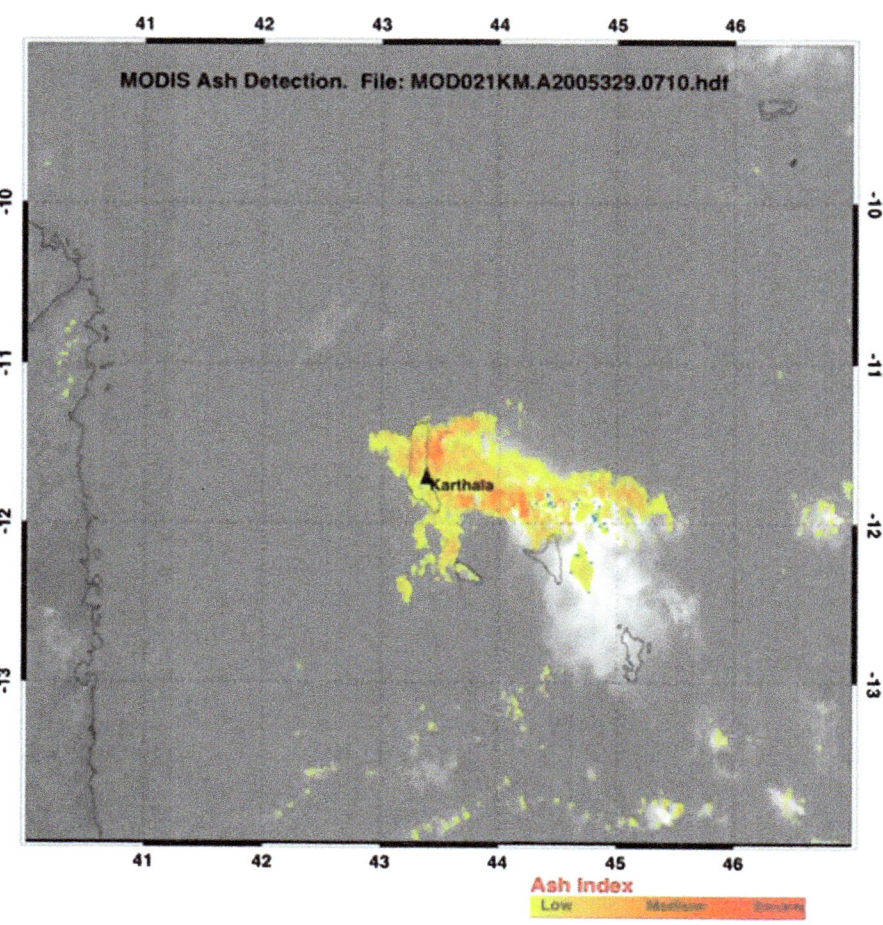

important because damage to aircraft that encounter volcanic plumes in flight is critically dependent on the concentration of suspended ash. Some satellite instruments with a higher spectral resolution and a greater number of channels may allow for specific identification of the solid particle composition, that is, volcanic or nonvolcanic. If this technique is refined, there is the potential to further discriminate a volcanic plume from other types of particle-laden cloud such as dust storms. This is significant because the melting temperature of nonvolcanic dust is considerable highly than volcanic glass and will have a less serious impact on jet engine functioning if ingested during flight.

Satellite imagery has also been used to identify excess sulfur dioxide gas in the atmosphere, an important component of volcanic ash plumes. Detection of volcanic SO_2 was first accomplished using the Total Ozone Mapping System on several polar-orbiting satellites. More recently, the Ozone Mapping Instrument on National Aeronautics and Space Administration's (NASA's) Aqua platform provides near real-time SO_2 data. The presence of SO_2 can be sensed using scattered solar ultraviolet radiation at wavelengths of 0.28 and 0.32 μm. Infrared imagery has also been

used to detect excess SO_2 but with less precision and accuracy. The use of the SO_2 signal to identify and track volcanic plumes has proved to be very useful, but from a hazards perspective must be used with discretion. Despite the fact that both volcanic ash and SO_2 are discharged together at the source volcano, there are indications that physical separation of SO_2 from the solid particles often occurs after eruption. Following the 1991 eruption of Pinatubo, Philippines, there was clear evidence that the SO_2-rich cloud was present at a higher level (25 km) and transported in a slightly different direction than the ash-rich cloud (18 km). This decoupling could lead to unnecessary alerts for areas impacted by SO_2-rich clouds that contain relatively little volcanic ash.

5. FUTURE RESEARCH

Volcanic plumes have the potential for significant impacts on both the global economy and environment. The 2010 disruptions of air traffic during the eruptions of Eyjafjallajokull volcano in Iceland highlighted the toll that even a relatively small eruption can have in an area transected by numerous flight paths. The need for accurate forecasts of volcanic plume trajectories and concentrations is of utmost importance to mitigate the economic consequences of potential aircraft–plume interactions. Improvements are being made continually to existing models of volcanic ash transport and sedimentation, a key area of research being focused on modifications of ash settling by processes of aggregation and gravity currents. The ability to predict the downwind concentrations of ash in a volcanic plume hinges on a fundamental understanding of how the solid particles are lost during transport. Potential areas of new research on ash concentration should be directed toward technological developments for sensing and sampling of plumes. The use of unmanned aerial vehicles to explore downwind plumes for ash concentration would appear to be productive area of investigation.

Geological evidence points to the occurrence of volcanic plumes on a scale that has not yet been observed during historic times. Initial theoretical studies of the enormous plumes produced by supereruptions suggest dynamics of an entirely different scale from those for which direct observations exist. A concentrated effort to model and better understand these unique and potentially catastrophic events is needed to assess the complete dynamic range of environmental impacts associated with explosive volcanism. Improved models of such plumes could be coupled with global climate models to predict the magnitude and duration of large-scale cooling events that would likely be triggered by the resulting atmospheric ash loading. At the same time, the model results could provide key constraints to a detailed evaluation of the socioeconomic toll to human populations and the development of strategies to cope in the aftermath of a supereruption.

Sophisticated supercomputer models coupled with analog laboratory experiments are two of the principal tools that volcanologists can develop further to better understanding the dynamic processes of volcanic plumes, including those affecting aviation safety as well as those generated by supereruptions.

ACKNOWLEDGMENTS

We thank Tomaso Esposti Ongaro, Arnau Folch, and an anonymous reviewer for providing many excellent suggestions and comments that were used in revising the chapter.

FURTHER READING

Baines, P., Sparks, R.S.J., 2005. Dynamics of giant volcanic ash clouds from supervolcanic eruptions. Geophys. Res. Lett. 32 (L24808), 1–4.

Bursik, M., Kobs, S., Burns, A., Braitseva, O., Bazanova, L., Melekestev, I., Kurbatov, A., Pieri, D., 2009. Volcanic plumes and wind: Jetstream interaction examples and implications for air traffic. J. Volcanol. Geotherm. Res. 186, 60–67.

Carey, S., 2005. Understanding the physical behavior of volcanoes. In: Marti, J., Ernst, G. (Eds.), Volcanoes and the Environment. Cambridge University Press, pp. 1–54.

Charpentier, I., 2007. Adjoint modelling experiments on eruptive columns. Geophysical J. Int. 169, 1356–1365.

Clarke, A., Voight, B., Macedonio, G., 2002. Transient dynamics of vulcanian explosions and column collapse. Nature 415, 897–901.

Fero, J., Carey, S., Merrill, J., 2009. Simulating the dispersal of tephra from the 1991 Pinatubo eruption: implications for the formation of widespread ash layers. J. Volcanol. Geotherm. Res. 186, 120–131.

Herzog, M., Graf, H.F., 2010. Applying the three-dimensional model ATHAM to volcanic plumes: dynamics of large co-ignimbrite eruptions and associated injection heights for volcanic gases. Geophys. Res. Lett. 37 (L19807), 1–5.

Lacasse, C., Karlsdottir, S., Larsen, G., Soosalu, H., Rose, W.I., Ernst, G.G.J., 2004. Weather radar observations of the Hekla 2000 eruption cloud, Iceland. Bull. Volcanol. 66, 457–473.

Mastin, L., Guffanti, M., Servranckx, R., Webley, P., Barsotti, S., Dean, K., Durant, A., Ewert, J., Neri, A., Rose, W., Schneider, D., Siebert, L., Stunder, B., Swanson, G., Tupper, A., Volentik, A., Waythomas, C., 2009. A multidisciplinary effort to assign realistic source parameters to models of volcanic ash-cloud transport and dispersion during eruptions. J. Volcanol. Geotherm. Res. 186, 10–21.

Oberhuber, J., Herzog, M., Graf, H., Schwanke, K., 1998. Volcanic plume simulation on large scales. Jour. Volc. Geotherm. Res. 87, 29–53.

Prata, A., 2009. Satellite detection of hazardous volcanic clouds and the risk to global air traffic. Nat. Hazards 51, 303–324.

Ripepe, M., Harris, A.J.L., Carniel, R., 2002. Thermal, seismic and infrasonic evidences of variable degassing rates at Stromboli volcano. J. Volcanol. Geotherm. Res. 118, 285–297.

Self, S., Walker, G.P.L., 1994. Ash clouds: characteristics of eruption columns. In: Casadavall, T.J. (Ed.), Volcanic Ash and Aviation Safety: Proceeding of the First International Symposium on Volcanic Ash and Aviation Safety. U.S. Geological Survey Bulletin 2047, pp. 65–74.

Suzuki, Y.J., Koyaguchi, T., 2010. Numerical determination of the efficiency of entrainment in volcanic eruption columns. Geophys. Res. Lett. 37, L05302. http://dx.doi.org/10.1029/2009GL042159.

Sparks, R.S.J., Bursik, M.I., Carey, S.N., Gilbert, J.G., Glaze, L.S., Sigurdsson, H., Woods, A.W., 1997. Volcanic Plumes. John Wiley and Sons, Chichester.

Stohl, A., Prata, A.J., Eckhardt, S., Clarisse, L., Durant, A., Henne, S., Kristiansen, N.I., Minikin, A., Schumann, U., Seibert, P., Stebel, K., Thomas, H.E., Thorsteinsson, T., Tørseth, K., Weinzierl, B., 2011. Determination of time- and height-resolved volcanic ash emissions and their use for quantitative ash dispersion modeling: the 2010 Eyjafjallajokull eruption. Atmos. Chem. Phys. 11, 4333–4351.

Woods, A.W., 2013. Sustained explosive activity: volcanic eruption columns and Hawaiin fountains. In: Fagents, S., Gregg, T., Lopes, R. (Eds.), Modeling Volcanic Processes: The Physics and Mathematics of Volcanism. Cambridge University Press, Cambridge, pp. 153–172.

Chapter 33

Tephra Dispersal and Sedimentation

Costanza Bonadonna
Department of Earth Sciences, University of Geneva, Geneva, Switzerland

Antonio Costa
Istituto Nazionale di Geofisica e Vulcanologia, Bologna, Italy

Arnau Folch
Barcelona Supercomputing Center, Barcelona, Spain

Takehiro Koyaguchi
Earthquake Research Institute, University of Tokyo, Tokyo, Japan

Chapter Outline

GLOSSARY

ash Tephra particles smaller than 2 mm in diameter; fine ash is smaller than 63 μm

blocks and bombs Tephra particles larger than 64 mm in diameter.

lapilli Tephra particles between 2 and 64 mm in diameter

strong plume Subvertical volcanic plume for which the upward velocity is significantly larger than the horizontal wind velocity

tephra Particles injected into the atmosphere during an explosive volcanic eruption regardless of the composition, shape, and size

volcanic cloud Cloud of volcanic gas, ash particles, and aerosols forming during an explosive volcanic eruption that spreads laterally at the neutral buoyancy level, also called umbrella cloud

volcanic plume A convective mixture of volcanic gas, ash particles, and aerosols rising above a continuous and constant source of buoyancy, typically a single vent, a system of vents, or an extended source, such as Pyroclastic Density Currents (PDCs)

weak plume Bent-over volcanic plume for which the upward velocity is significantly lower than the horizontal wind velocity

1. INTRODUCTION

Powerful explosive eruptions are typically associated with subvertical or bent-over **volcanic plumes** that can inject large quantities of gases and particles of various sizes and shapes into the atmosphere, altogether known as **tephra**. Tephra consists of different components with variable density of both juvenile (fresh magma) and lithic (wall rock) particle and can vary from meter-sized **blocks and bombs**, which are ejected from the vent as ballistics falling within a few kilometers from the source, to micron-sized particles, which can be transported by atmospheric winds at continental or global scales. Tephra particles may sediment individually, may be clustered in various types of aggregates, or may be entrapped within sedimentation instabilities and, depending on their sizes, represent different threats. In particular, the impact of ballistic blocks and bombs ($\geq$64 mm) can significantly damage infrastructure close to the vent; accumulation of **lapilli** (2–64 mm) and **ash** (<2 mm) can cause a wide range of damage to communities and ecosystems, while fine ash (<63 μm) can jeopardize civil aviation and the finest micrometric particles (e.g., PM_{10} and $PM_{2.5}$, i.e., particulate matter smaller than 10 and 2.5 μm, respectively) can also threaten human health.

Depending on the ratio of horizontal wind velocity to plume rise velocity, volcanic plumes can develop as strong (subvertical) or weak (bent-over) plumes and eventually reach the neutral buoyancy level (NBL), where their density equals that of the surrounding atmosphere, and start spreading laterally around this level (Figure 33.1). When the rising plume velocity is significantly larger than the horizontal wind velocity, the plume rises beyond the NBL because of momentum at the top of the plume and, from there, collapses toward the NBL spreading as a gravity current to form an umbrella cloud. In contrast, when the horizontal wind velocity dominates, the plume bends over, generating horizontal turbulent currents that spread around the NBL. More accurately, the influence of wind advection on plume dynamics can be characterized by the dimensionless number Π defined as:

$$\Pi = \frac{\overline{N}H}{1.8\overline{v}}\left(\frac{\alpha}{\beta}\right)^2 \qquad (33.1)$$

which determines which of the two fundamental terms controlling plume dynamics is dominant (i.e., radial expansion versus wind entrainment), where H is the plume height above the vent (m); $\overline{N}$ and $\overline{v}$ are the average buoyancy frequency (1/s) and the average wind velocity (m/s) across the plume height, respectively; α is the radial entrainment coefficient; and β is the wind entrainment coefficient (Degruyter and Bonadonna, 2012). Very large Π implies that the radial entrainment term is most important and the associated plume would mostly develop as subvertical.

As it spreads far from source, **volcanic clouds** are progressively diluted by entrained air and particle sedimentation and a transition occurs to a passive transport regime, which is dominated by wind advection and atmospheric turbulence. The distance at which this transition occurs will depend on the volumetric flow rate at the NBL (e.g., Costa et al., 2013).

Grain-size distribution of tephra inside a volcanic plume is related to sorting processes acting on the initial total grain-size distribution (TGSD) of particles mostly resulting from magma fragmentation. Depending on the ratio between particle terminal velocity and plume vertical velocity, the mean grain size of particles in the volcanic plume tends to decrease with height. In particular, large blocks and bombs separate from the gas thrust region of the column and deposit near the vent, while sedimentation from plume margins and umbrella clouds are dominated by lapilli and ash, respectively. **Strong plumes** developing in a moderate-to-strong wind field are associated with elongate but wide deposits with a certain degree of upwind

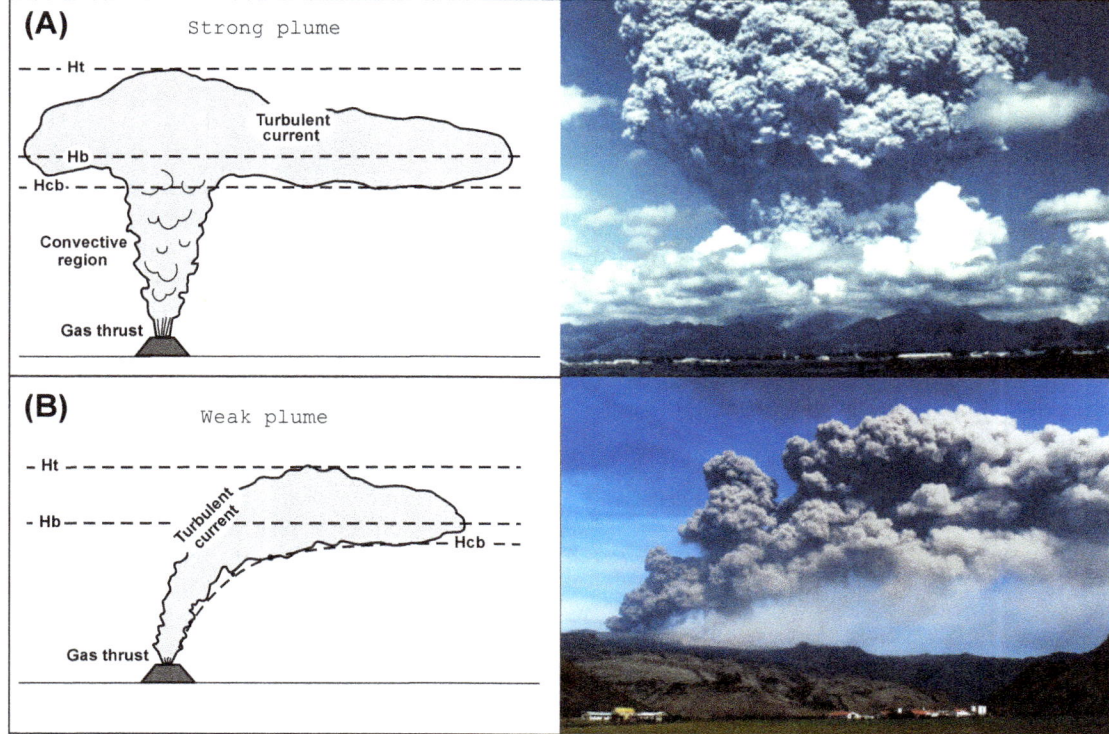

FIGURE 33.1   Sketch showing the main characteristics of (A) a strong volcanic plume and (B) a weak volcanic plume ($\Pi \gg 1$ and $\ll 1$, respectively, in Degruyter and Bonadonna 2012). Examples of a strong plume (18 km-high volcanic plume from one of a series of explosive eruptions of Mount Pinatubo beginning on June 12, 1991; photograph by David H. Harlow, USGS) and a weak plume (Eyjafjallajökull plume spreading toward the southeast of Iceland on May 4, 2010) are also shown. Ht, Hb, and Hcb indicate maximum plume height, height of NBL, and height of based current, respectively.

sedimentation due to gravity current spreading at NBL. In contrast, **weak plumes** generate narrower deposits substantially elongated in the downwind direction with negligible upwind sedimentation. Blocks, lapilli, and most coarse ash are lost from the bent-over section of the plume that can typically extend a few tens of kilometers from the vent, where the fine-ash-rich turbulent current starts spreading horizontally around the NBL.

Volcanic plumes can develop from a single vent, a system of vents, or from extended sources, i.e., large basaltic fissures and pyroclastic density currents (PDC) plumes (see Chapter 32 on "Volcanic Plume" for more details). For example, in eruptions like those of Kilauea (Hawaii) multiple vents along a fissure may form discrete jets that eventually converge in a single plume. In contrast, co-PDC plumes develop above moving PDCs by particle elutriation as particle sedimentation and air entrainment reduce PDC density.

The generation, structure, and dynamics of volcanic plumes are addressed in detail in the Chapter 32 "Volcanic Plumes", while tephra deposits are discussed in the Chapter 34 "Pyroclastic Fall Deposits." This chapter presents the principal processes that characterize tephra sedimentation (i.e., particle settling, aggregation, and cloud processes) and overviews the main approaches to modeling atmospheric tephra dispersal and sedimentation. These include the characterization of eruption source parameters (ESPs) (principally height of eruption column, erupted mass, mass eruption rate (MER), and TGSD) that, together with the wind field, represent the main inputs for tephra dispersal models.

2. TEPHRA SEDIMENTATION FROM VOLCANIC PLUMES AND CLOUDS

2.1. Particle Settling Velocity

Particle settling velocity is a complex function of particle size, density, and shape and has a primary control on the residence time of tephra in the atmosphere and, hence, on tephra deposition. For simplicity, it is commonly assumed that tephra particles in plumes and clouds settle at their terminal velocity. The terminal velocity v_j (m/s) of particles of characteristic size d_j (m) can be derived from the balance between gravity, buoyancy, and drag forces:

$$\frac{\pi}{6}d_j^3 g(\rho_p - \rho_a) = \frac{C_D}{2}\rho_p v_j^2 A \qquad (33.2)$$

where g (m/s^2) denotes the acceleration due to gravity; ρ_p (kg/m^3) and ρ_a (kg/m^3) are the particle and air densities, respectively; A (m^2) is the characteristic cross-sectional area of the particle; and C_D is the drag coefficient (dimensionless), which is a function of the particle shape and the Reynolds number $\text{Re} = d_j\rho_a v_j/\mu_a$, where μ_a is the air dynamic viscosity (Pa s). Tephra particles are typically

nonspherical and irregular, and the value of C_D depends not only on their shape but also on their orientation, which varies due to particle rotation and tumbling (Figure 33.2). A large amount of experimental aerodynamic data exists for simple regular shapes but only a few studies have directly measured the terminal velocities of irregular volcanic particles (see Alfano et al. (2011) for a review). As a first approximation, terminal velocity of tephra particles can be computed analytically under the assumption of spherical shape with the mean diameter defined by the average of the three main axes or the diameter of the enclosing sphere. For particles falling at high Reynolds numbers (Re > 1000), inertial forces are dominant and the drag coefficient only weakly depends on the Reynolds number. Dedicated experimental studies have suggested that drag coefficients for volcanic particles at Re > 1000 are more similar to those of cylinders than spheres, with drag coefficients around ∼1. In the case of intermediate Reynolds numbers, the drag coefficient is typically described using analytical expressions between the settling laws for low and high Reynolds number regimes or parameterizations of experimental data. One of the most commonly used methods for estimating the drag coefficient of nonspherical particles is that of Ganser (1993), which uses the equal volume sphere diameter and the sphericity of particles. Results suggest that, irrespective of altitude, the variations in terminal velocity due to the effect of density (between 500 and 2500 kg/m^3) and shape (between sphericity 0.5 and 1) are between 50% and 80% and 20% and 70%, respectively (Figure 33.3). The slope change occurring within the field of coarse ash is interpreted as due to the variation in sedimentation regime

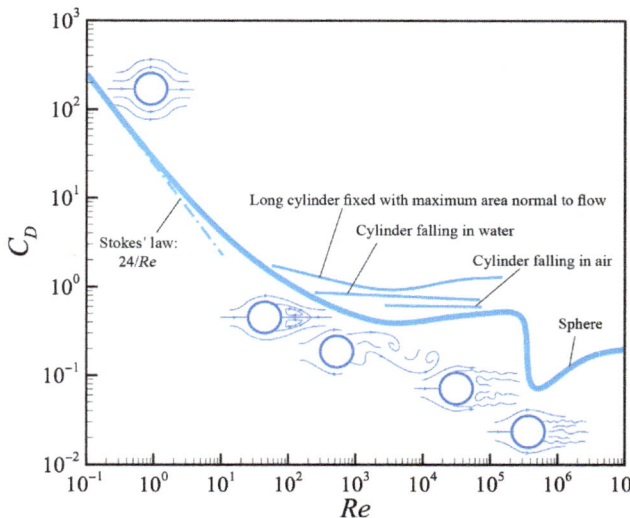

FIGURE 33.2 Drag coefficient for fixed or freely falling sphere and cylinder versus particle Reynolds number in various fluids. Spheres freely falling or fixed in air or water have similar values of drag coefficients. *Courtesy of G. Bagheri.*

(i.e., particle Reynolds number). Nonetheless, large uncertainty is associated with the calculation of critical morphological parameters of irregular particles, such as the particle surface area, necessary to determine sphericity.

Complexities of constraining the terminal velocity of tephra particles, and, therefore, of modeling tephra sedimentation accurately, are related not only to the characterization of particle shape. but also to the description of processes that modulate tephra deposition. As an example, instead of considering single values of terminal velocity, particles would be described better by a range of terminal velocities associated with various orientations during fall, and, therefore, various projected areas. The range of terminal velocities should also account for the fact that ϕ categories typically used to describe TGSD do not correspond to individual particle diameters, but enclose a range of particle sizes (e.g., ϕ class 0 and class -1 are equivalent to particle diameters between 1 and 2 mm and 2 and 4 mm, respectively, with $\phi = -\log_2$(particle diameter in millimeters)). In addition, fine ash typically settles as aggregates and/or through various types of sedimentation instabilities (see Sections 2.2 and 2.3). As a result, terminal velocity of fine ash cannot be calculated simply based on the analytical or experimental models described above, but should account for both processes of particle aggregation and gravitational instabilities.

2.2. Particle Aggregation

Particle aggregation is a fundamental process that controls the dispersal and sedimentation of ash with diameter <100 μm and reduces its atmospheric residence time, making it fall closer to the vent than expected. Considering that at least 30% of the total erupted mass of most silicic eruptions consists of fine ash (diameter <63 μm) (Rose and Durant, 2009), particle aggregation also affects depositional patterns generating secondary maxima of accumulation, locally increasing deposition rate, and altering patterns of tephra thinning with distance from the vent.

The timescale of particle aggregation is short and particles collide and cluster mainly because of complex particle–particle processes in the presence of surface liquid layers, electrostatic forces, mechanical interlocking, secondary minerals, turbulence, and/or differences in settling velocities. Depending on the water content, particle aggregation results in the formation of particle clusters (including ash clusters and coated particles) or accretionary pellets (including poorly structured pellets, pellets with concentric structures, and liquid pellets) (Brown et al., 2012 and references therein) (Figure 33.4). Water in volcanic clouds mainly originates from volatiles in the magma (up to about 6 wt%), by entrainment of moist lower

FIGURE 33.3 Terminal velocity (meters per second) versus particle diameter (millimeters) (A) at sea level and (B) at 20 km a.s.l. (above sea level) for three densities (500−2500 kg/m^3) and two sphericity values (0.5 and 1).

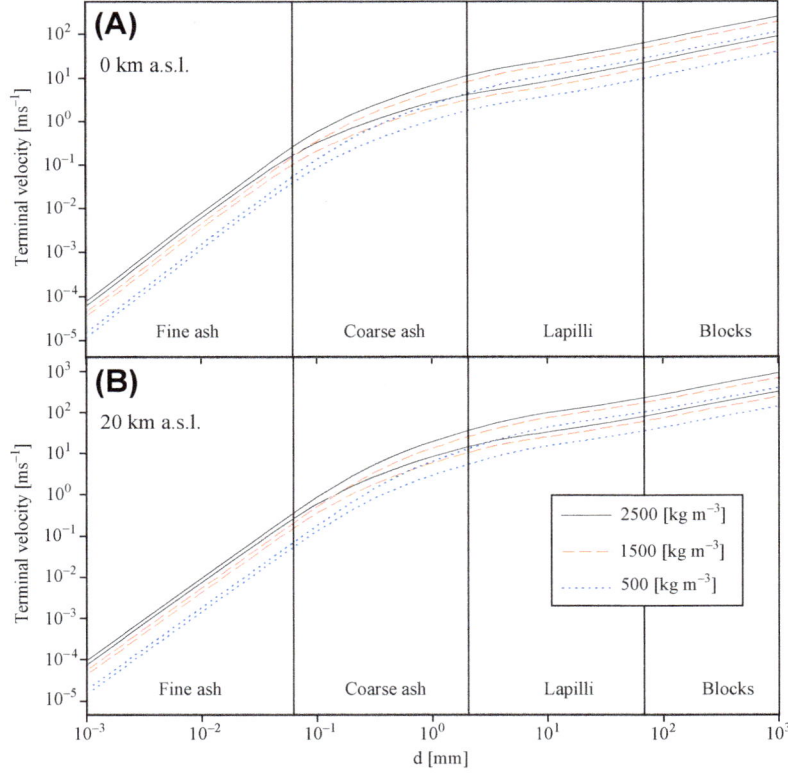

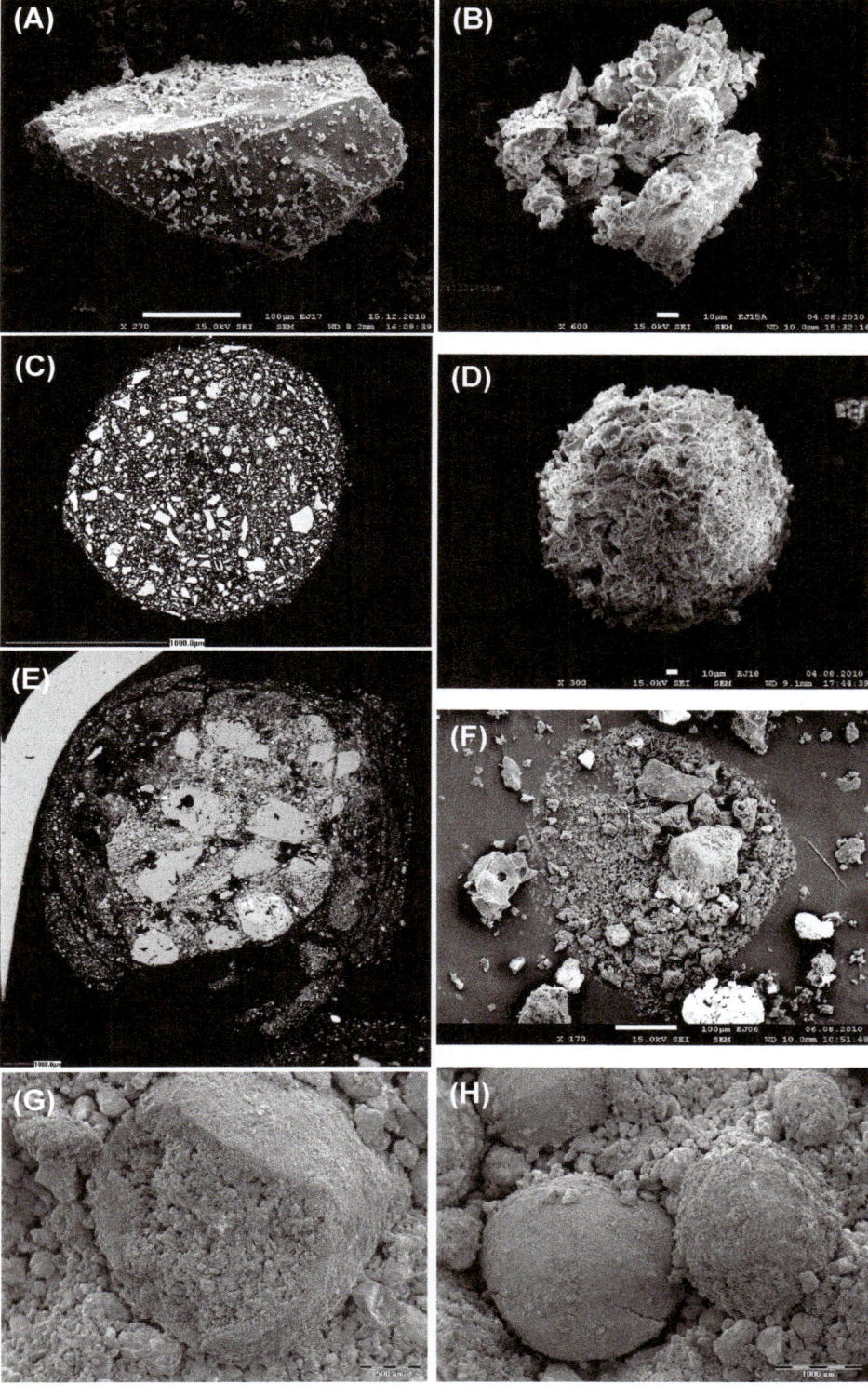

FIGURE 33.4 Examples of particle aggregates: (A) coated particle (May 6, 2010; 20 km from vent; Eyjafjallajökull); (B) ash cluster (May 5, 2010; 10 km from vent; Eyjafjallajökull); (C) poorly structured pellet (September 28, 1997, Vulcanian explosion; 7 km from vent; Montserrat); (D) poorly structured pellet (May 6, 2010; 56 km from vent; Eyjafjallajökull); (E) structured pellet (December 26, 1997, dome collapse; 4 km from vent; Montserrat); (F) liquid pellet (May 4, 2010; 20 km from vent; Eyjafjallajökull); (G) structured pellet (Keanakak'oi ash, lower footprint layer, Kilauea; 7 km from vent; *(Courtesy of A. Durant.)*); (H) structured pellets (Keanakakoi ash, lower footprint layer, Kilauea; 7 km from vent; *(Courtesy of A. Durant.)*).

tropospheric air, and through interaction with sources of external water (e.g., ocean, lakes, groundwater, glaciers). Magma/water interaction also plays an important role in plume dynamics and in the formation of PDCs, leading to

overprinting of eruptive styles and transport processes that can enhance particle aggregation.

Aggregation processes have been observed associated with sedimentation from ash-rich volcanic plumes and clouds from Vulcanian to plinian eruptions, but they are even more striking for co-PDC plumes associated with both large pumiceous flows and smaller block-and-ash flows, as they mostly consist of particles <1 mm. Even though ash aggregates are difficult to document and analyze due to their low preservation potential, drag coefficients, aggregation coefficients, and particle size distribution have been parameterized through laboratory experiments that show good agreement with field data (Brown et al., 2012 and references therein). More than 60% of ash aggregates observed in laboratory experiments are characterized by fall velocities between 0.2 and 0.4 m/s, in agreement with numerical results. Densities of ash aggregates are typically in the range of 100−1000 kg/m^3 as shown by both field and experimental observations. Aggregate size and typology may vary with distance from the vent in a nonsystematic manner, indicating variability in eruptive style and a complex interaction between the various sources of liquid water and ice. However, the mode of aggregating particles is typically around 4−5 phi. Most aggregates, with the exception of pellets with concentric structures (also defined as accretionary lapilli; Figure 33.4(G)−(H)), are not preserved in deposits as they are too fragile to survive impact. Their presence can be often inferred from poorly sorted deposits and polymodal grain-size distributions.

Both dry and wet aggregation can occur during the same volcanic event with electrostatic ash aggregates forming extremely rapidly and, in the presence of water, accretionary pellets more efficiently scavenging large particles (up to a few hundred micrometers) due to the associated stronger binding forces. Dry conditions promote size-selective aggregation for the sub-63 μm fraction, while wet conditions also allow for scavenging of millimetric particles. Gilbert and Lane (1994) and James et al. (2003) gave some quantitative estimation of wet and dry aggregation efficiency based on experimental analysis, showing a dependence on the grain size. In fact, not all particles that collide will adhere, and the sticking probability is shown to decrease with particle size.

Costa et al. (2010) demonstrated how wet aggregation depends on the ratio between the residence time of aggregating particles within the cloud region where liquid water exists and the time required for aggregates to form. Wet aggregation is likely to be enhanced within the plume region characterized by temperatures between the melting and freezing points of water, as aggregation due to the presence of ice seems to be less effective. Bulk sedimentation would then occur due to the onset of gravitational instability (analogous to mammatus clouds). Nonetheless, wet aggregation in the distal dilute regions of volcanic clouds can also happen, due to subsidence of the plume, which results in the formation of liquid water as frozen aggregates pass through the melting isotherm.

2.3. Gravitational Instabilities

Other processes that increase sedimentation rate and enhance deposition of fine ash include gravitational instabilities (Figure 33.5). In fact, tephra fallout can be characterized by descending gravity currents, i.e., series of discrete and highly concentrated protrusions that reach the ground at higher speed than individual particles and seem to generate preferential paths for fine ash to settle (e.g., Carazzo and Jellinek, 2013). These "fingers" produced by gravitational instabilities forming in density-stratified fluids have been widely observed in volcanic eruptions over a range of atmospheric conditions (e.g., Soufrière Hills volcano 1997, Montserrat; Ruapehu 1996, New Zealand; Eyjafjallajökull 2010, Iceland). Based on analog experiments and observations of the Eyjafjallajökull 2010 plume and deposit, Manzella et al. (submitted for publication) have shown how there is no single explanation for fine ash deposition close to source and that a range of distinct origins is possible. Due to their high particle concentration, fingers not only enhance sedimentation of fine ash but also can promote particle aggregation. Gravitational instabilities associated with the Eyjafjallajökull 2010 plume initially took the form of turbulent downward propagating fingers that formed continuously at the base of the cloud and were advected passively at the wind speed and eventually

FIGURE 33.5 Gravitational instabilities observed at the base of the umbrella cloud associated with the eruptions of (A) Mount Saint Helens on May 18, 1980 (*mammatus*) (*Photograph by D. Miller*) and (B) Eyjafjallajökull 2010 (*fingers*) (*Photograph by J. Elíasson*).

merged into a diffuse and well-mixed layer below the parental ash cloud (Figures 33.1(B) and 33.5(B); Manzella et al., submitted for publication). As a possible triggering mechanism of fingers, Durant et al. (2009) proposed a conceptual model to explain the role of mammatus clouds in transporting very fine ash from the May 18, 1980, Mount St Helens eruption. Following their model, ash particles initiate ice hydrometeor formation high in the troposphere. As mammatus clouds develop from increased particle loading, the volcanic cloud rapidly subsides. At this stage rapid sedimentation occurs as the cloud passes through the melting level in a process analogous to snowflake aggregation, forming distal secondary maxima in the deposit thickness, as observed in many recent volcanic tephra deposits.

3. MODELING TEPHRA DISPERSAL AND SEDIMENTATION

Most existing models that quantify tephra dispersal and sedimentation from volcanic plumes and clouds are based on the solution of the advection—diffusion—sedimentation (ADS) equation either analytically or numerically (e.g., Bonadonna and Costa (2013), Folch (2012) and references therein). During the past decades, several models have been developed for different purposes. The main approaches consist of: (1) one-dimensional (1D) analytical models that consider tephra to be dispersed mainly as a gravity current around the NBL and are commonly used to investigate plume sedimentation in the near-source region; (2) two-dimensional (2D) analytical models that describe tephra dispersion as passive transport due to atmospheric wind advection, turbulent diffusion, and particle sedimentation; they are commonly used to compile long-term probabilistic assessment of hazards related to tephra sedimentation and to solve the inverse problem to reproduce distal tephra distribution; (3) three-dimensional (3D) numerical models that compute mass concentration in the atmosphere and loading on the ground solving mass conservation equations using Eulerian, Lagrangian, or hybrid formulations. Given their higher computational cost, 3D models are typically used for particle tracking, short-term forecasting, and, with the use of computer clusters, also for solving inverse problems to reconstruct past events and for long-term probabilistic hazard assessments.

3.1. Eruption Source Parameters

Irrespective of the approach used, modeling tephra dispersal requires the characterization of the ESPs, principally total erupted mass, plume height, MER, eruption duration, TGSD, and vertical distribution of mass along the column. Strategies for quantifying these physical parameters might differ whether they are estimated in real time or a posteriori (mainly from data analysis and field observations) and might, therefore, be characterized by different levels of

uncertainty. An accurate real-time definition of ESP can be achieved mainly through a synergistic application of geophysical and modeling strategies with different application limits and resolution, and is necessary for real-time dispersal forecasting. The a posteriori ESP definition relies on the detailed study of tephra deposits and provides probability density functions for selected activity scenarios both to compile long-term hazard assessments and to help with preliminary real-time simulations of dispersal forecasting when no real-time data are available.

Erupted mass: Erupted mass of past eruptions is determined based either on the integration of various empirical fits of deposit thickness versus square root of area of corresponding isopach contours or on the analytical inversion of mass/area data (see Chapter 34 "Pyroclastic Fall Deposits"). Associated uncertainties can be large depending on the deposit exposure. Real-time determination of erupted mass relies on the determination of MER and eruption duration.

Plume height: Even though plume height is usually the easiest parameter to constrain in real time, measurements are always characterized by a relatively large degree of uncertainty. For example, determinations of plume heights of recent Icelandic eruptions using radar (Grímsvötn 2004; Eyjafjallajökull 2010) were affected by 10—40% uncertainty for values of 8—12 km above sea level due to the step resolution of the weather radars used. Plume heights of past eruptions have been typically determined based on semi-empirical models and analytical inversion strategies (see Chapter 34 "Pyroclastic Fall Deposits"). In particular, the semiempirical models are based on the distribution of the largest clasts around the vent, and, therefore, are affected by the uncertainty associated with the selection and averaging of clasts. Analytical inversions consider full granulometry data but depend on good characterization of physical processes and of meteorological data. Either way, plume height of past eruptions is typically constrained within about 20% of uncertainty.

MER: MER is difficult to measure directly and is commonly derived from plume height using semiempirical relationships based on buoyant plume theory (e.g., Mastin et al., 2009; Sparks et al., 1997). Such estimations can have up to a factor 10 of uncertainty. Recent studies have shown the importance of accounting for wind effects in the determination of MER using either numerical models or analytical relationships (Bursik, 2001; Degruyter and Bonadonna, 2012; Woodhouse et al., 2013). It is important to assess the propagation of error both for real-time and a posteriori determination of MER. In fact, given the approximately fourth power relation between MER and plume height, a relatively small error in the plume height generates a large uncertainty in MER estimates.

TGSD: the determination of TGSD commonly relies on the detailed characterization of tephra deposits, and, in particular, on various strategies applied to average all grain-size data determined at individual locations or on both analytical and numerical inversion strategies (see Chapter 34 "Pyroclastic Fall Deposits"). It is important to highlight how all current averaging techniques used for the determination of TGSD cannot account for poor deposit exposure, and, therefore, resulting distributions could be depleted in either the fine or coarse tail (or both) if large parts of the deposit are missing. A comprehensive real-time technique that can provide the erupted mass associated with the whole particle size spectrum does not yet exist.

3.2. 1D Models

Generally, the spreading of a volcanic cloud is governed by the complex interplay between a gravity current intruding at the NBL and wind advection. Its dynamics is described by a gravity current in the near-source region and by wind advection and ambient turbulence (passive dispersion) in the more distal downwind region (e.g., Sparks et al 1997). The size of the transition region depends upon the characteristics of the eruption (mainly intensity) and atmosphere. In particular, defining the Richardson number, Ri, as the ratio of the square of the umbrella cloud front velocity, U_b, to the squared mean wind velocity, U_b, i.e., $Ri = U_b^2/U_w^2$, when $Ri > 1$, the transport is mainly density driven, whereas for $Ri < 0.25$, transport is substantially passive, and for $1 < Ri < 0.25$ both transport mechanisms are relevant (Figure 33.6; (Costa et al., 2013)). Numerical simulations have shown how, for the Mt St Helens eruption, the first 200 km were dominated by a gravitational spreading proportional to the cloud thickness, beyond which the umbrella cloud was dispersed passively. For the Pinatubo 1991 Plinian eruption, consistent with satellite observations, simulations indicate that the first 5 h of the climatic phase of the eruption, corresponding to a radius of ~450 km, was mainly density driven, whereas passive dispersion approximation was fully valid only for times larger than 40 h (corresponding to a radius of ~1800 km).

The 1D models assume that ejected pyroclasts are conveyed laterally within the umbrella cloud, are mixed due to turbulence, and eventually fall out to the ground (e.g., Figure 33.7). Considering a specific volume in a horizontally expanding cloud, as particles are removed from the base of the specific volume at their terminal fall velocities, v_j, the mass of the particles in the specific volume, $M(t,v_j)$, will decrease with time. In this model, it is assumed that (1) particles are homogeneously distributed inside the umbrella cloud because of turbulent mixing and (2) they fall at their terminal velocities from the base of the volcanic cloud where turbulence diminishes. Under these assumptions $M(t,v_j)$ decreases as:

$$M(t, v_j) = M(0, v_j) exp\left(\frac{-v_j t}{h}\right) \qquad (33.3)$$

where h is the thickness of the volcanic cloud. Under the assumptions of steady conditions, and that the rate of change in thickness of the volcanic cloud is sufficiently small compared with the radial velocity of the specific volume, Eqn (33.3) can be written expressing M as a function of the position (the distance from the vent, r), and, since the mass of particles with terminal velocity v_j per unit area of the deposit, $s(r,v_j)$, is proportional to the sedimentation rate at r:

$$s(r, v_j) = s(0, v_j) exp\left(\frac{-\pi v_j r^2}{\dot{V}}\right) \qquad (33.4)$$

where $\dot{V}$ is the volumetric flow rate at NBL. Therefore, Eqn (33.4) can be used to estimate the dynamics of the umbrella cloud (i.e., $\dot{V}$) and the mass of the deposit at the source for each grain size with terminal velocity v_j (i.e., $s(0,v_j)$) from the data describing the deposit load $s(r,v_j)$. This method is a powerful tool to reconstruct the dynamics of volcanic clouds from grain-size data of Plinian deposits particularly for intense eruptions where the expansion velocity of the umbrella cloud due to the gravity current is much greater than the wind velocity. For example, the expansion rate of the umbrella cloud estimated from this method has shown a good agreement with that observed in satellite images during the climactic phase of the Pinatubo 1991 eruption (Koyaguchi and Ohno (2001) and references therein).

The 1D model is also applicable to relatively weak eruptions with some modifications. In fact, Eqns (33.3) and (33.4) can also be adapted to describe sedimentation from weak plumes with the assumption that $\dot{V}$ varies with distance from vent due to the effect of turbulent mixing with the atmosphere and that the volcanic cloud width is described by Fickian diffusion (Bonadonna and Costa (2013) and references therein).

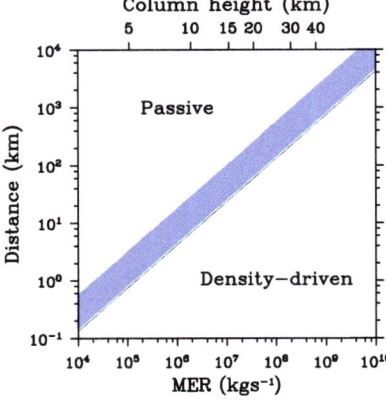

FIGURE 33.6 Tephra transport regimes in the umbrella region as a function of MER and distance from the source. The colored area shows the transition zone, delimited by Ri > 1 (density driven regime) and Ri < 0.25 (passive regime), obtained for the parameters considered by Costa et al. (2013).

(A)

Radially expanding umbrella cloud with volumetric flow rate, $\dot{V}$

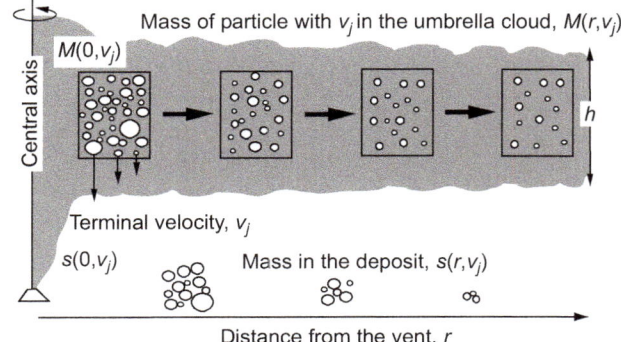

Mass of particle with v_j in the umbrella cloud, $M(r,v_j)$

(B)

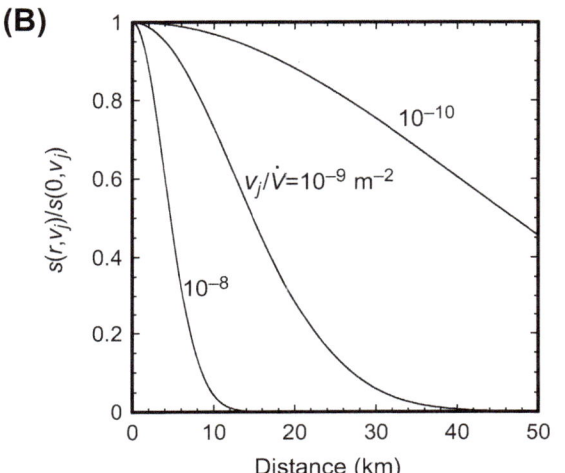

FIGURE 33.7 (A) Schematic illustration of tephra sedimentation from an umbrella cloud spreading as a gravity current (see text and Eqns (33.3) and (33.4) for details). Note that this illustration shows only half of the umbrella cloud, which radially expands from the central axis above the vent. Aggregation processes and sedimentation instabilities are not shown for simplicity. (B) The mass of deposit with terminal velocity v_j is presented as a function of the ratio $v_j/\dot{V}$ on the basis of Eqn (33.3). The volumetric flow rate at the NBL, $\dot{V}$, is approximately proportional to the MER; $\dot{V} = 10^{11}$ m³/s corresponds to a MER of about 10^9 kg/s (value similar to that of the climatic phase of the Pinatubo 1991 eruption). As an example, the line having $v_j/\dot{V} = 10^{-10}$ m^{-2} in this diagram represents a mass of deposit with $v_j = 10$ m/s for an intensive eruption with a MER of about 10^9 kg/s.

3.3. 2D Models

This category of models is based on a simplified solution of the 3D ADS equation:

$$\frac{\partial c_j}{\partial t} + \frac{\partial u_x c_j}{\partial x} + \frac{\partial u_y c_j}{\partial y} + \frac{\partial u_z c_j}{\partial z} - \frac{\partial v_j c_j}{\partial z}$$

$$= \frac{\partial}{\partial x}\left(K_H \frac{\partial c_j}{\partial x}\right) + \frac{\partial}{\partial y}\left(K_H \frac{\partial c_j}{\partial y}\right) + \frac{\partial}{\partial z}\left(K_V \frac{\partial c_j}{\partial z}\right) + S_j + S_{kj}$$

(33.5)

where c_j is the concentration of particle belonging to class j having terminal velocity v_j, u_x and u_y are the horizontal

components of wind velocity, u_z is the vertical wind velocity component, K_H is the horizontal diffusion coefficient, K_V represents the vertical diffusion coefficient, S_j denotes the source term (accounting for the production of particles of class j), and S_{kj} is the sink term (accounting for the removal of particles of class j, e.g. due to aggregation). Introducing some approximations such as (1) negligible vertical wind velocity and vertical diffusion, (2) constant horizontal diffusion, (3) homogeneous wind, and (4) negligible particle–particle interaction (e.g., aggregation), Eqn (33.5) can be simplified into a 2D problem and solved analytically to determine the total mass on the ground at the position (x,y):

$$M_G(x,y) = \sum_i^{N_S}$$

$$\times \sum_j^{N_J} \frac{M_i f_j}{2\pi\sigma_{Gi}^2} \exp\left[\frac{-(x - x_{Gi})^2 + (y - y_{Gi})^2}{2\sigma_{Gi}^2}\right]$$

(33.6)

where N_S denotes the number of discrete source points along the column, N_J indicates the total number of particle classes, f_j is the fraction of that mass belonging to the particle class j, M_i is the total mass emitted from the point source i, and σ_{Gi} is the variance of the mass distribution at the ground of particles belonging to the class j. The diffusion coefficient K_H for this kind of model has to be viewed as an effective horizontal diffusion coefficient that not only describes effects of atmospheric diffusion but also accounts for other physical processes not explicitly considered in these models (e.g., gravitational spreading). The source term S_j (i.e., the eruption column) is typically described as a line source characterized by various mass distribution functions (e.g., Suzuki, 1983). Hence analytical Gaussian models rely on the determination of empirical parameters, such as the diffusion coefficient K_H and column shape parameters introduced for describing the source term S_j. As a result, validation and calibration with field data of specific eruptions is necessary before reliable hazard assessments can be compiled. The advantage of these Gaussian models is the simplicity of the physical parameterization and, therefore, the high computational speed, which allows for a comprehensive probabilistic analysis of the associated inputs and outputs and the solution of inverse problems for the estimation of eruptive parameters, such as total erupted mass, TGSD, and column height (e.g., Connor and Connor, 2006; Volentik et al., 2010). However, they should be applied with caution under certain situations including: (1) near-source regions (as topography and plume dynamics typically are neglected), (2) dispersal from weak plumes (as dispersion from low plumes mainly occurs in the low troposphere where winds are quite variable and vertical components may be not negligible), (3) long-range tephra dispersal (as wind fields over synoptic scales are far from being homogeneous and are likely to change with time over the duration of tephra dispersal), and (4) dispersal from long-lasting eruptions during which conditions in the atmosphere vary with time. This fourth limitation can be partially relaxed by concatenating a series of quasi-steady simulations.

3.4. 3D Models

Some of the limitations of the previously described 2D Gaussian models can be overcome by solving numerically the ADS Eqn (33.5) to obtain time-dependent airborne tephra concentration and ground deposit load/thickness (see Folch (2012) and references therein). Disadvantages include a higher computational cost, which makes 3D models less practical for probabilistic hazard assessment and solving inverse problems. In contrast, these models are very suitable for dispersal forecast and to study past events for which meteorological data are available. The 3D models can be Eulerian, Lagrangian, or hybrid. The Eulerian models solve for variables (e.g., particle mass concentration) at fixed locations or grid points. The Lagrangian models calculate the trajectories of an ensemble of particles (representing moving parcels containing the mass of many real physical particles) and compute mass concentration by averaging over fixed background cells. Neglecting particle inertia and particle—particle interactions, the displacement of a "particle" between two times t_1 and t_2 is given by:

$$r(t_2) = r(t_1) + \int_{t_1}^{t_2} [u(r, t) + u'(r, t) + v_j(r, t)]dt \quad (33.7)$$

where r is the position vector, u denotes the wind velocity vector, u' represents the scale turbulent fluctuations, and v_j is the particle sedimentation velocity. The three terms inside the integral are equivalent to the terms of advection, diffusion, and sedimentation of the ADS Eqn (33.5).

In all cases, 3D tephra dispersal models require (1) a source parameterization defining the particle/mass release with time, (2) a driving atmospheric model (e.g., a numerical weather prediction model), which describes the state and the evolution of the atmosphere, and (3) possibly, parameterization of other processes (e.g., wet and dry deposition, aggregation). Commonly, these models are used by the Volcanic Ash Advisory Centers to forecast the dispersal of volcanic clouds and particle airborne concentration in quasi-real time in order to mitigate aviation risk. As computational capacities are improving, these models have also started to be used probabilistically for solving inverse problems and for long-term hazard assessment.

4. DISCUSSION AND CONCLUDING REMARKS

The transport and deposition of tephra is influenced by many factors including plume dynamics, particle characteristics, sedimentation processes, and atmospheric conditions. Various combinations of these factors can result in different tephra dispersal patterns and tephra deposit features. However, some distinctive characteristics can be identified:

1. Tephra deposits associated with strong plumes vary from nearly circular to weakly elongated geometries, depending on the impact of wind advection, and typically display a certain degree of upwind sedimentation; in contrast, tephra deposits associated with weak plumes are narrow and strongly elongated with negligible upwind sedimentation.
2. Particle shape and density significantly affect particle settling. In particular, the drag coefficient of particles falling within the turbulent regime, i.e., at high Reynolds number, is strongly affected by shape. Nonetheless, travel distance is significantly affected by particle morphology even for particles falling within the laminar regime, i.e., at low Reynolds number. Three main deposit-thinning breaks-in-slope are associated with variations of sedimentation regimes associated with particle Reynolds number, i.e., laminar, intermediate, and turbulent regime.
3. Tephra deposition is affected by various sedimentation processes, such as particle aggregation and gravitational instabilities that make fine ash fall closer to the vent than expected and at a higher sedimentation rate. These processes can act together and are not always evident from the associated tephra deposits. However, field observations, laboratory experiments, and theoretical investigations have shown how particles <100 μm typically fall as aggregates of various types and/or entrapped in sedimentation instabilities that affect the position of thinning breaks-in-slope and can generate double maxima of ground accumulation.

ESPs, i.e., erupted mass, plume height, MER, TGSD, define style and size of volcanic eruptions and are critical inputs for both analytical and numerical models. Given that all models are very sensitive to ESP, associated uncertainty should be quantified accurately in order to interpret model outputs better. In particular, ESP should be determined through the application of many independent strategies in order to identify a range of comprehensive values that could be more representative than any single value. Main strategies include:

1. Integration of various empirical fits (mostly exponential, power-law, Weibull) and inversion of mass/area for the determination of erupted mass from field data; for real-time assessment, erupted mass can be derived by a combination of duration and estimations of MER.
2. Analysis of particle size distribution of the tephra deposits for the calculation of plume height; for real-time assessment plume height can be determined based on various remote sensing strategies and direct observations.

3. MER can be estimated analytically from plume height and wind speed, from geophysical monitoring and/or by combination of erupted mass and eruption duration.

4. Initial TGSD can currently only be derived based on the analysis of granulometry data. The associated fine fraction can also be constrained based on various remote sensing strategies and in situ sampling.

Tephra dispersal and sedimentation need to be described and quantified accurately in order to characterize volcanic explosive eruptions and their associated hazards. Many empirical and analytical models have been developed to quantify ESP, and analytical and numerical models have been developed to investigate tephra transport and assess ground deposition and atmospheric dispersal. The choice of which model to use should be based on the application purpose and the associated assumptions and limitations should be analyzed critically. In particular, the 1D models can be used to investigate tephra sedimentation, tephra deposit thinning, and volcanic cloud spreading; the 2D models are very practical for probabilistic assessment of long-term medial-range ground load accumulation and estimating pivotal ESP by inversion strategies; and the 3D models can be used for real-time tephra dispersal forecasting and for distal-range hazard assessment. Model outputs are associated with a certain level of uncertainty that needs to be quantified in order to interpret the model outcomes accurately. Finally, crucial sedimentation processes, such as the effects of particle morphology on settling velocity, particle aggregation, and gravitational instabilities, still need to be characterized and parameterized accurately in order to be implemented efficiently within tephra dispersal models. Given that both particle aggregation and gravitational instabilities make the sub-100-μm particle fraction fall closer to the vent than expected and increase deposition rate, failing to describe the sink term (i.e., processes that affect the atmospheric residence time of particles) might result in an underestimation of proximal-to-medial hazard related to tephra deposition and an overestimation of the threat to aviation in distal areas. A comprehensive description of particle transport and sedimentation can only be provided by a synergistic combination of geophysical studies together with physical volcanology, experimental investigations, and numerical modeling.

FURTHER READING

Alfano, F., Bonadonna, C., Delmelle, P., Costantini, L., 2011. Insights on tephra settling velocity from morphological observations. J. Volcanol. Geotherm. Res. 208 (3–4), 86–98.

Bonadonna, C., Costa, A., 2013. Modeling of tephra sedimentation from volcanic plumes. In: Fagents, S., Gregg, T., Lopes, R. (Eds.), Modeling Volcanic Processes: The Physics and Mathematics of Volcanism. Cambridge University Press.

Brown, R.J., Bonadonna, C., Durant, A.J., 2012. A review of volcanic ash aggregation. Phys. Chem. Earth. Physics and Chemistry of the Earth, Parts A/B/C 45, 65–78.

Bursik, M., 2001. Effect of wind on the rise height of volcanic plumes. Geophys. Res. Lett. 28 (18), 3621–3624.

Carazzo, G., Jellinek, A.M., 2013. Particle sedimentation and diffusive convection in volcanic ash-clouds. J. Geophys. Res. Solid Earth 118 (4), 1420–1437.

Connor, L.G., Connor, C.B., 2006. Inversion is the key to dispersion: understanding eruption dynamics by inverting tephra fallout. In: Mader, H., Cole, S., Connor, C.B., Connor, L.G. (Eds.), Statistics in Volcanology. Special Publications of IAVCEI. Geological Society London, pp. 231–242.

Costa, A., Folch, A., Macedonio, G., 2010. A model for wet aggregation of ash particles in volcanic plumes and clouds: 1. Theoretical formulation. J. Geophys. Res. Solid Earth 115.

Costa, A., Folch, A., Macedonio, G., 2013. Density-driven transport in the umbrella region of volcanic clouds: Implications for tephra dispersion models. Geophys. Res. Lett. 40 (18), 4823–4827.

Degruyter, W., Bonadonna, C., 2012. Improving on mass flow rate estimates of volcanic eruptions. Geophys. Res. Lett. 39.

Durant, A.J., Rose, W.I., Sarna-Wojcicki, A.M., Carey, S., Volentik, A.C.M., 2009. Hydrometeor-enhanced tephra sedimentation: constraints from the 18 May 1980 eruption of Mount St. Helens. J. Geophys. Res. Solid Earth 114.

Folch, A., 2012. A review of tephra transport and dispersal models: evolution, current status, and future perspectives. J. Volcanol. Geotherm. Res. 235, 96–115.

Ganser, G.H., 1993. A rational approach to drag prediction of spherical and nonspherical particles. Powder Technol. 77 (2), 143–152.

Gilbert, J.S., Lane, S.J., 1994. The origin of accretionary lapilli. Bull. Volcanol. 56 (5), 398–411.

James, M.R., Lane, S.J., Gilbert, J.S., 2003. Density, construction, and drag coefficient of electrostatic volcanic ash aggregates. J. Geophys. Res. Solid Earth 108 (B9).

Koyaguchi, T., Ohno, M., 2001. Reconstruction of eruption column dynamics on the basis of grain size of tephra fall deposits. 2. Application to the Pinatubo 1991 eruption. J. Geophys. Res. 106 (B4), 6513–6533.

Manzella, I., Bonadonna, C., Phillips, J.C., Monnard, H., 2014. The role of gravitational instabilities in deposition of volcanic ash Geology, (in press).

Mastin, L.G., Guffanti, M., Servranckx, R., Webley, P., Barsotti, S., Dean, K., Durant, A., Ewert, J.W., Neri, A., Rose, W.I., Schneider, D., Siebert, L., Stunder, B., Swanson, G., Tupper, A., Volentik, A., Waythomas, C.F., 2009. A multidisciplinary effort to assign realistic source parameters to models of volcanic ash-cloud transport and dispersion during eruptions. J. Volcanol. Geotherm. Res. 186 (1–2), 10–21.

Rose, W.I., Durant, A.J., 2009. Fine ash content of explosive eruptions. J. Volcanol. Geotherm. Res. 186 (1–2), 32–39.

Sparks, R.S.J., Bursik, M.I., Carey, S.N., Gilbert, J.S., Glaze, L.S., Sigurdsson, H., Woods, A.W., 1997. Volcanic Plumes. John Wiley & Sons, Chichester, 574 pp.

Suzuki, T., 1983. A theoretical model for dispersion of tephra. In: Shimozuru, D., Yokoyama, I. (Eds.), Arc Volcanism, Physics and Tectonics. Terra Scientific Publishing Company (TERRAPUB), Tokyo, pp. 95–113.

Volentik, A.C.M., Bonadonna, C., Connor, C.B., Connor, L.J., Rosi, M., 2010. Modeling tephra dispersal in absence of wind: insights from the climactic phase of the 2450 BP Plinian eruption of Pululagua volcano (Ecuador). J. Volcanol. Geotherm. Res. 193 (1–2), 117–136.

Woodhouse, M.J., Hogg, A.J., Phillips, J.C., Sparks, R.S.J., 2013. Interaction between volcanic plumes and wind during the 2010 Eyjafjallajokull eruption, Iceland. J. Geophys. Res. Solid Earth 118 (1), 92–109.

Chapter 34

Pyroclastic Fall Deposits

Bruce Houghton

Department of Geology and Geophysics, National Disaster Preparedness Training Center, University of Hawai'i, Honolulu, HI, USA

Rebecca J. Carey

University of Tasmania, Hobart, TAS, Australia

Chapter Outline

GLOSSARY

componentry Study of the abundances of different individual particle types in a pyroclastic deposit.

dispersal index (D) A measure of the extent of a pyroclastic fall deposit, specifically the area enclosed by an isopach drawn at one-hundredth of the maximum extrapolated thickness of the deposit (introduced by Walker, 1973).

dry explosions Eruptions "powered" by the exsolution and decompression of previously dissolved magmatic gas.

external water Any water phase (e.g., seawater, phreatic water, snow, ice) involved in explosive volcanism that was not originally dissolved in the magma.

fragmentation index (F) A parameter measuring the grain size of a pyroclastic fall deposit, specifically the percentage of ash finer than 1 mm at the point on the dispersal axis corresponding to one-tenth the maximum thickness of the deposit (introduced by Walker, 1973).

grain size half-distance (b_c) The "average" distance over which the maximum clast size in a pyroclastic deposit halves (introduced by Pyle, 1989).

median diameter The midpoint of the grain size distribution, i.e., 50% by weight of the pyroclasts are coarser and 50% are finer than the median grain size, in the assumption of a normal distribution.

isopach The line joining points of equal thickness of a deposit.

isopleth The line joining points where the sizes of the largest clasts (pumice or lithic) are the same.

juvenile clasts Particles derived from fragmentation of fresh erupted magma.

lithic clasts (wall rock) Particles derived from the walls of the conduit or vent.

pumice Microvesicular clasts of high vesicularity. Generally silicic in composition.

pyroclastic fall Sedimentation of pyroclasts through the atmosphere from an eruption jet or plume during an explosive eruption.

scoria Moderately to highly vesicular clasts with abundant millimeter-sized vesicles, typically mafic in composition.

sorting The range or degree of uniformity of grain size in a pyroclastic deposit. Two alternative parameters are described below (note φ_X is the Xth percentile of the grain size distribution.):

Inman graphic standard deviation $\sigma\varphi = \frac{\varphi_{84} - \varphi_{16}}{2}$

Folk and Ward inclusive graphic standard deviation $\sigma_I = \frac{\varphi_{84} - \varphi_{16}}{4} + \frac{\varphi_{95} - \varphi_5}{6.6}$

The Encyclopedia of Volcanoes. http://dx.doi.org/10.1016/B978-0-12-385938-9.00034-1

600   **PART | IV** Explosive Volcanism

thickness half-distance (b_t) The "average" distance over which the thickness of a pyroclastic deposit halves (introduced by Pyle, 1989).

wet explosions Requires an external source for the gas—vapor phase that expands and "powers" the explosions, typically either surface water in lakes or streams or groundwater held in pores and cracks within the shallow rocks that can be flashed into steam.

1. INTRODUCTION

Here we here describe the properties of **pyroclastic fall** deposits on three characteristic length scales (Figure 34.1):

Large scale: those "whole deposit" properties that may be inferred only from integration of data across numerous outcrops across the entire dispersal area.

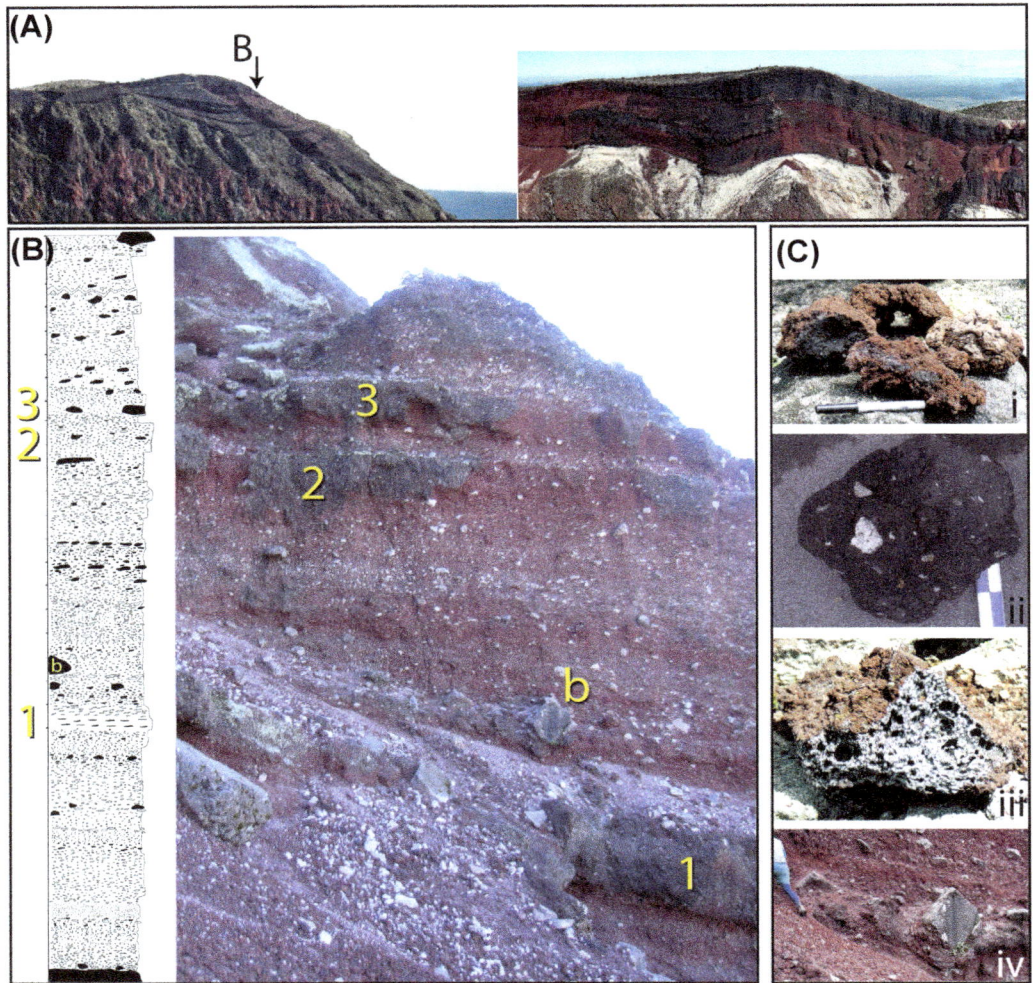

FIGURE 34.1 Montage of images showing (A) large-, (B) intermediate-, and (C) small-scale features for a single fall deposit, the 1886 Plinian fall, Tarawera volcano, New Zealand. (A) Large scale: deposit geometry. Panoramic views of contrasting thinning relationships along the walls of the eruptive fissure. Black and red 1886 fall deposits overlie pyroclastic (white) and effusive (gray/orange) products of earlier rhyolitic eruptions. *Left*: Crater (D) is relatively remote from the major Plinian vents and so beds thin rapidly, suggesting that most clasts were derived from the adjacent craters and explosions of relatively weak intensity. *Right*: Crater (L) contains at least one of the main Plinian vents and proximal beds thin more slowly reflecting that many clasts were falling from the margins of the jet and c. 25-km-high plume. *(After Sable et al. (2006).)* Location of section featured in (B) is labeled with an arrow and the letter "B." (B) Photograph and stratigraphic log of the 1886 fall deposits on the northwestern side of crater D showing decimeter- to meter-scale grading and bedding. Note the concentrations of outsized rhyolitic lithic blocks (e.g., "b"). These largest wall rock clasts are not aerodynamically equivalent to magmatic-derived clasts, and have shapes and dimensions that reflect jointing patterns of underlying dome lavas. The fall deposits show welding corresponding to the times of most rapid tephra accumulation. Examples are labeled as 1, 2, and 3. Note that following convention, juvenile clasts are always shown as white and wall rock pyroclasts as black in stratigraphic logs. *(After Carey and Houghton (2010).)* (C) Diversity seen in representative clasts from the 1886 eruption: (i) ragged scoria typical of the main Plinian phase, (ii) dense, wall rock charged spherical bomb from the upper phreatomagmatic phase, which overlies the main Plinian fall, (iii) partially melted and re-vesiculated rhyolitic lithic clast in welded scoria fall, and (iv) angular rhyolitic wall rock labeled "b" in (B).

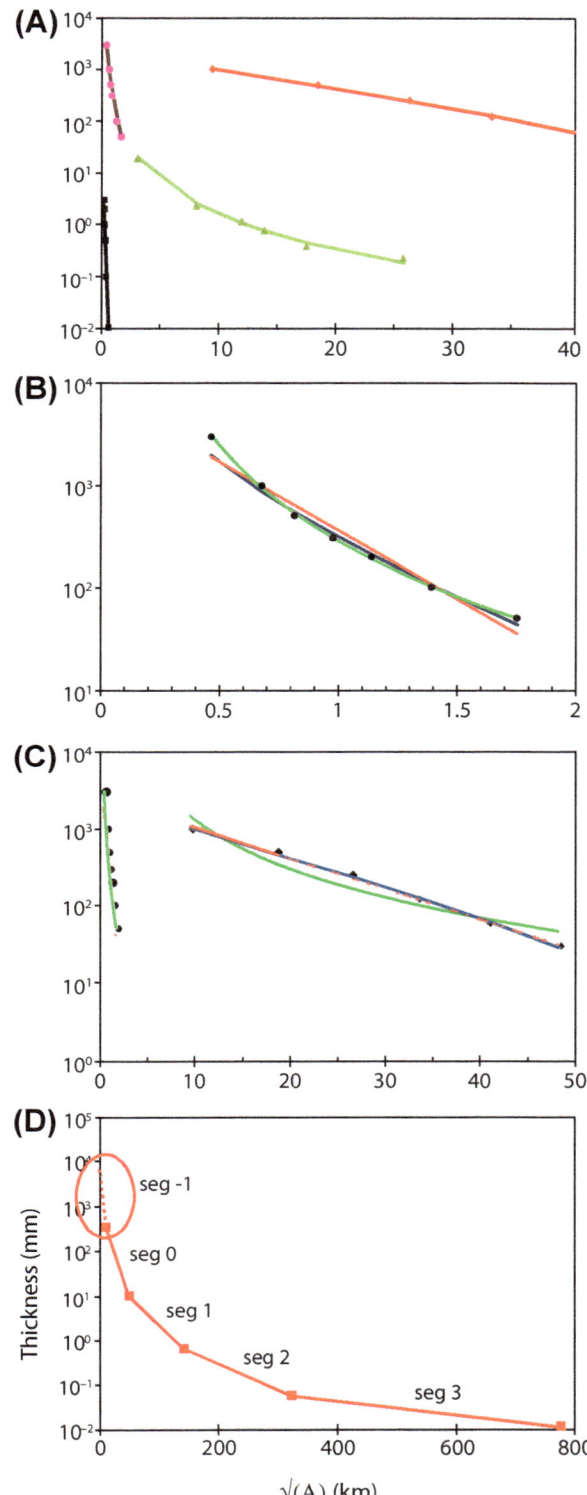

Medium scale: deposit features that may be measured and used on outcrop scale, and vary between outcrops.

Small scale: the properties of the pyroclasts that make up fall deposits.

Large-scale features include the whole geometry of the fallout deposit—the rates at which fall deposits thin and fine away from source—and the whole deposit grain size properties of the entire deposit. Bedding and grading, grain size and **sorting** of single beds, and thermal effects are medium-scale properties. The surface morphology and vesicularity of **juvenile clasts** and the nature of foreign (wall rock) clasts are small-scale features.

2. LARGE-SCALE FEATURES

2.1. Deposit Geometry: Eruption Magnitude and Intensity

Fall deposits are cones or blankets of tephra that thin and fine away from one or more maxima that are usually close to or coincident with the eruptive vent(s). For most eruptions, deposit geometry, eruption magnitude (mass, volume), and intensity (mass/time) are most commonly estimated by constraining the rate of change of tephra thickness (or mass per unit area) over an entire preserved deposit by contouring thickness/mass data to create an **isopach** (or isomass) map (e.g., Figure 6, Chapter 29). The map contours are typically represented in a semilog plot of thickness (t) versus the square root of the isopach area ($\sqrt{(A)}$), and these data points are fit by any of a number of empirical thinning relationships to arrive at dispersal and volume (Pyle, 1989; Fierstein and Nathenson, 1992) (Figure 34.2). Use of $\sqrt{(A)}$ is an effective way of minimizing the difference in shape of the isopachs from different deposits, and hence the influence of wind advection on dispersal.

(Data are fit by Weibull functions following Bonadonna and Costa (2011).) (B,C) Plots showing contrasting exponential (red), power-law (green), and Weibull (blue) fits to thinning data for (B) a typical cone-building deposit, the Hawaiian 1959 Kīlauea Iki fall deposit *(After Klawonn et al. (2014).)*, and (C) a representative sheet-forming deposit, the Plinian 1886 Tarawera lapilli fall. Kīlauea Iki data is inset in (C) to show contrasting thinning trend. The exponential relationship yields the worst fit for cone-forming (Strombolian, Hawaiian) deposits (B). For sheet-forming (Plinian, phreatoplinian) deposits (C) the power-law function typically overestimates thickness in the far field. Inset values are deposit volumes calculated using the fitted relationships. (D) Idealized plot of ln(T) versus $\sqrt{(A)}$ for a Plinian fall deposit showing the possible division into segments corresponding to regimes of contrasting pyroclast transport and sedimentation. Segments 1, 2, and 3 correspond to steady and sustained setting from the umbrella cloud of particles with high, intermediate, and low Reynolds number, respectively. Segment 0 represents sustained sedimentation of particles from the upper margins of the convective plume. Segment −1 reflects larger particles shed from the jet and lower plume. *(Modified after Bonadonna et al. (1998).)*

FIGURE 34.2 (A) Contrasting thinning trends for four basaltic fall deposits related to eruptions of contrasting mass eruption rate. Diamonds: the 1886 Tarawera Plinian deposit featured in Figure 34.1. *(After Walker et al. (1984).)* Triangles: the July 22, 1998, subplinian fall from Mt Etna (Andronico et al., 1999). Circles: the Kīlauea Iki 1959 Hawaiian fall deposit (Houghton, unpublished data) and the deposit from a weakly explosive eruption at Kīlauea on March 19, 2008 (Houghton et al., 2010).

Deposit geometry: The isopach process requires subjective choice by the volcanologist. Results are, however, surprisingly consistent irrespective of the spacing of the field observations, the number and spacing of the contours chosen by the volcanologist, and the degree of smoothing of those contours (Klawonn et al., 2014). At least for cone-building eruptions, data from multiple authors and multiple approaches are consistent, and thus can be grouped for comparative studies.

If data can be fit by straight-line relationships and exponential thinning on a thickness versus distance plot (Figure 34.2), then thinning can be characterized by half-distances, the distance required for the deposit to thin or thicken by a factor of 2×. These half-distances can be measured along single transects or following Pyle (1989) by using a plot of $\ln(t)$ versus $\sqrt{A}$ to eliminate the effect of wind advection to calculate the parameter b_t, "thickness half-distance." b_t is calculated from the formula: $b_t = (\ln 2)/(k\pi^{1/2})$ where k is the slope of the plot of $\ln(t)$ versus $\sqrt{A}$.

The more distal segments of Plinian fall deposits are typically associated with b_t values up to 10–50 km, while the cones formed during Strombolian and Hawaiian eruptions have b_t values generally from 10 to 300 m (Houghton et al., 2013). Most deposits show deviations from straight-line relationships on diagrams of $\ln(t)$ versus $\sqrt{A}$. This is due to (see Chapter 33):

1. Changes in sedimentation region passing from jet to convective plume to umbrella cloud. Extremely proximal products of Plinian events often form a steep, "ultraproximal" segment on a plot of $\ln(T)$ versus $\sqrt{A}$, with b_t values on the order of tens to a few hundreds of meters (segment −1 on Figure 34.2(D)). This is interpreted as intermittent sedimentation from the margin of the jet and perhaps lowermost convective plume (Houghton et al., 2004; Sable et al., 2006) associated with heterogeneities in the mixture of gas and pyroclasts emerging from the vent. For many Plinian deposits, a second inflection point on the $\ln(T)$ versus $\sqrt{A}$ plot, at a few kilometers to tens of kilometers from vent (between segments 0 and 1 on Figure 34.3(C)) reflects a switch in sedimentation from the margins of the buoyant plume to the umbrella cloud (Bonadonna et al., 1998). Typically, b_t is 1–10 km for the proximal portion, and 20–50 km, or more, over the distal portion.

2. Changes in the settling behavior of clasts with decreasing Reynolds number for progressively smaller pyroclasts (Bonadonna et al., 1998). Pyroclast settling velocities vary widely with particle size and density, controlled by the particle Reynolds number, Re. Bonadonna et al. (1998) present three segments on the $\ln(T)$ versus $\sqrt{A}$ plot, which correspond to low, intermediate, and high Re, and laminar, transitional, and turbulent regimes, respectively (Figure 34.2(D)).

3. Ash aggregation in the plume. This influences particle size at medial sites most strongly, but can also produce unusually high thickness/isomass values and "secondary" thickness maxima. Aggregation processes significantly affect the thinning, allowing fine ash to fall in the

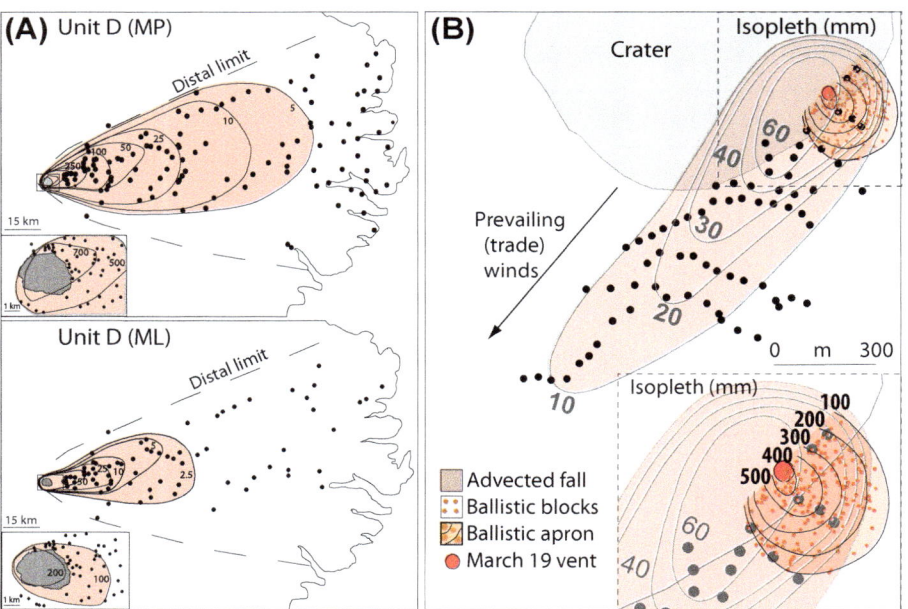

FIGURE 34.3 Isopleth maps for (A) the 1875 Askja D Plinian fall (Carey et al., 2009) and (B) a small explosive eruption at Kīlauea on March 19, 2008 (Houghton et al., 2011). For (B) we show separate contours for the main wind-advected lapilli fall and a ballistic block field (inset). Ballistic sizes are naturally much larger but contours are more narrowly spaced and approximate more closely to circles concentric on the vent (note that data is absent for the inaccessible northwestern sector).

turbulent and intermediate fallout regimes. Thinning breaks-in-slope on ln(T) versus $\sqrt{(A)}$ plots can be shifted either closer to the vent or further downwind depending on the size and density of the aggregates. Aggregates falling in the turbulent regime typically shift breaks-in-slopes closer to the vent, whereas aggregates falling in the intermediate regime typically shift breaks-in-slope further downwind (Bonadonna and Phillips, 2003).

4. Mixed deposits. These result, for example, from multiple source locations such that a single deposit may be derived from two or more vents or from a vent-derived plume and coeval buoyant portions of pyroclastic density currents (pdcs) (Chapter 34). Note that neither reason 3 nor reason 4 will be immediately apparent from a ln(T) versus $\sqrt{(A)}$ plot.

Deposit mass and volume: Data points on ln(T) versus $\sqrt{(A)}$ plots are fitted by a range of continuous functions (Figure 34.3(A) and (B)), such as single-segment exponential, multisegment exponential, power law, and Weibull, to estimate eruptive volume (Bonadonna and Costa, 2012). Piecewise exponential functions (2 or more segments) provide a better fit to the data than a single function, but with additional uncertainty due to the subjective choices of inflection points and number of piecewise functions (Le Pennec et al., 2012). The fitted relationships give the total erupted volume (or mass). A power-law decay function does not require the subjective introduction of inflection points and segments but cannot be integrated between zero and infinity. The user needs therefore to define two arbitrary integration limits to arrive at meaningful volume estimates (see Biass and Bonadonna (2011) and Bonadonna and Costa (2013) for the effect of these choices). Bonadonna and Costa (2012) proposed fitting a three-parameter Weibull function, which does not require manual choice of segments or integration limits. Irrespective of the iteration technique and the geometry of the fall deposit, uncertainty in the volume estimate predictably is least where thickness/isomass data are well constrained widely, and greatest where data is sparse (Klawonn et al., 2014). In practice, this means that a fall deposit can be divided into three fields. Thickness bounds on the field where a deposit can be well constrained are typically between 1 and 5 cm (distally) and 3–10 m (proximally). In the most proximal region, outcrops are sparse and difficult to access due to vent collapse, or difficult to measure because of welding, burial, or extreme thicknesses. In the most distal region, the preservation potential is lowest, the measurement uncertainty becomes a significant proportion of the measurement, and syn-eruptive remobilization by near-surface winds or posteruptive bioturbation can be severe. The implications are different for Plinian (sheet-forming) and Hawaiian/Strombolian (cone-building) eruptions. For cone-forming eruptions a very low proportion of the mass is dispersed

into the most distal region and uncertainty is best reduced by improving data quality and quantity at the most proximal sites. The converse holds for sheet-forming deposits. In well-constrained regions, multisegment exponential, power-law, and Weibull (quadratic residual weighting) functions give very similar results. In poorly constrained regions where a substantial proportion of the mass has been lost to erosion, the exponential function lends conservative estimates; and both power-law and Weibull functions tend to predict larger volumes (Klawonn et al., 2014).

An alternative approach to empirical fitting looks to inversion of thickness/isomass data to determine the optimal set of eruptive parameters that explain best the variation in the field data (Connor and Connor, 2006).

2.2. Deposit Fining: Plume Heights and Mass Eruption Rates

All fall deposits fine (decay in grain size) with increasing distance from vent. Maximum particle size is the chosen parameter for many purposes (especially modeling of plume dynamics, and comparing the "power" of eruptions). However, workers have used many different measurement and averaging strategies (Bonadonna et al., 2013). These include choices of the minimum or maximum ellipsoid, and averages of just the longest, or all three, axes of the 3, 5, 10, or 20 largest clasts at any exposure, and sampling either for an undisclosed time and volume of material, or only over a specified area (0.1, 0.5, 1 m^2) or volume, or for a fixed sampling time over an unspecified area at any exposure (Biass and Bonadonna, 2011). Table 34.1 shows the effect of increasing the number of clasts/axes used for measurement at a single site. The resulting data are plotted as **isopleth** maps (Figure 34.3 (A) and (B)). The major difference in density between vesicular juvenile pyroclasts (**pumice, scoria**) and most wall rock **lithic clasts** means that they are markedly nonequivalent aerodynamically, and so the maximum sizes of juvenile populations (Maximum Pumice (MP)) and lithic populations (Maximum Lithic (ML)) are typically measured separately. In practice having two data sets is redundant. Both clast types can yield apparently conflicting data. The grain size distribution of lithic clasts in a deposit is not always a measure of the efficiency of fragmentation, if it is incorporated from a preexisting unconsolidated tephra or sediment. The size distribution may be effectively inherited, and very large or very small clasts inexplicably under- or overrepresented in the lithic population (Figure 34.1(B)). Likewise, recycling of juvenile or lithic clasts through the vent (see below) can also promote an unusually fine, refragmented clast population that has little relationship to the eruptive conditions at the moment when it finally leaves the vent. Breakage of large

TABLE 34.1 Contrasting ML Values Calculated by Various Formulae, for a Set of Lithic Clasts from One Locality for Subunit 15 of the Taupo Pumice

Clast	Axis	Diameter (mm)
1	a	45
	b	32
	c	29
2	a	33
	b	33
	c	19
3	a	34
	b	32
	c	20
4	a	33
	b	19
	c	19
5	a	23
	b	22
	c	32

Technique	5×3	5×1	3×3	3×1
Average diameter (mm)	28.3	35.4	30.8	37.3

Averages are calculated on the basis of measuring three axes for the five largest clasts (5×3), one axis for the five largest clasts (5×1), three axes for the three largest clasts (3×3), and one axis for the three largest clasts (3×1). The effect of this range of values on estimated maximum column heights using the formula of Carey and Sparks (1986) is relatively small. Using 16-mm isopleths for subunit 15, the 5×1, 3×1, and 3×3 averages gave the same height of 35 km, and 5×3 gave a height of 34 km.

clasts in flight or, more commonly, on impact also leads to an apparent reduction in the grain size of a deposit. Breakage is most commonly in the near field (coarser clasts) and among juvenile clasts. It can be inferred from a jigsaw fit of adjacent clasts in a bed or from a falloff of the MP/ML ratio close to vent (Figure 34.3(C)).

MP and ML data are used to estimate the height of eruption plumes and mass fluxes (Carey and Sparks, 1986), empirically calibrated by data from well-observed eruptions. Clasts disperse from the umbrella cloud, due to diffusion and advection by wind. A higher eruption plume and a higher mass eruption rate will lift a given clast higher, and transport it further from the vent, given identical wind conditions. For identical mass discharge

rates, a higher wind speed will increase the downwind range of a given particle and decrease the upwind range. However, the effects of the wind field can be largely removed by only considering the behavior of particles in the cross-wind regions; e.g., Carey and Sparks (1986) used cross-wind ranges of pyroclasts of fixed radii and density to arrive at estimates of plume height. Most estimates of mass discharge rate are time averages across the lifetime of an eruption episode, and so, if mass eruption rate waxes and wanes, the calculated eruption rate will lie below the peak instantaneous value. Changing the wind field can cause the plume height and mass eruption rate to be overestimated (Figure 34.4).

2.3. Total Deposit Grain Size: Fragmentation Efficiency

The integrated grain size distributions of entire fall deposits (TGSD) reflect only the processes of fragmentation and eruption style, not transport. As such they hold real potential as a parameter for classification of fall deposits and interpretation of eruption dynamics. Unfortunately, few TGSD data sets exist, they have been arrived at by different methodologies, and most are for relatively small historical eruptions. The difficulty lies in the far field due to the low potential for preservation of deposits beyond the 1-cm isopach in subaerial environments and the fact that the distal portions of many fall deposits are deposited offshore. For most prehistoric fall deposits, where distal deposits are not preserved, only crude estimates of the whole deposit grain size can be made. The most common techniques applied to fall deposits are: (1) weighted average of sample grain size distribution for all samples over the whole deposit, (2) various types of arbitrary sectorization of tephra-fall deposits, and (3) the Voronoi Tessellation. Technique (1) has problems with deposits that have irregular distributions of data, whereas (2) is biased due to the arbitrary selection of sectors. The Voronoi Tessellation technique, a well-known method of spatial analysis, provides the better statistical method, dealing with nonuniform data sets without introducing arbitrary sectors. A tephra-fall deposit is divided into cells whose interior consists of all grid points that are closer to a given sample point than to any other point (Bonadonna and Houghton, 2005). The mass per unit area value and the grain size distribution at that sample point are then assigned to its enclosing Voronoi cell. The total grain size distribution is obtained as the area-weighted average of all the Voronoi cells over the whole deposit. Establishing a database of internally consistent TGSDs over the range of mass eruption rates and eruptive styles, and determining separate TGSD for juvenile and wall rock components, are major challenges for physical volcanology.

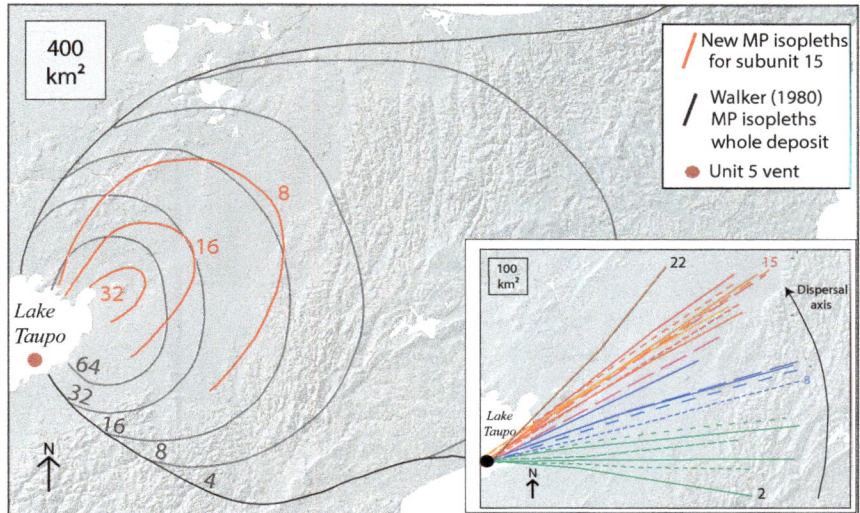

FIGURE 34.4   Selected MP isopleth data for unit 5 of the 1800 a eruption of Taupo volcano. *(After Houghton et al. (2014).)* Isopleths for a single subunit of unit 5 (subunit 15) are superimposed on the full deposit thickness isopleths for unit 5. *(After Walker (1980).)* The apparent large footprint of the entire unit based on full-thickness data is an artifact of previously unrecognized shifts in the wind field during the eruption (inset), from the southeast toward the north (see arrow), rather than extreme eruptive vigor. Isopleths for subunit 15, and all other subunits, enclose much smaller areas than the full-thickness isopleths leading to a marked reduction in the inferred column heights and hence estimates of mass discharge rates during the eruption.

3. INTERMEDIATE SCALE FEATURES

3.1. Bedding: Sustained versus Nonsustained Discharge

The presence or absence of fine-scale stratification in fall deposits (Figure 34.5) helps distinguish between nonsustained versus sustained eruptions, particularly in very proximal sections. The presence of sharp bedding planes, often dividing units of contrasting grain size, implies spasmodic, nonsustained eruption consisting of many short-lived pulses (Figure 34.5). Another possible cause of grain size oscillations, especially in Plinian deposits, is related to the occurrence of partial column collapses. Mass is partitioned between the convective and the collapsing regions so that the convective plume has a reduced heat flux and the column is reduced in height. Steady events

Sustained

Nonsustained

FIGURE 34.5 Photographs and stratigraphic logs showing contrasting bedding for sustained and nonsustained eruptive conditions. (A) The Fontana lapilli is the product of a sustained Plinian eruption in which minor fluctuations in mass eruption rate produce cryptic grain size changes in the deposit but sharp bedding planes are absent. The light-colored, ash-rich intervals represent incorporation of fine ash from an adjacent phreatic vent, and not pauses in the Fontana eruption. (B) A lapilli and ash fall deposit with sharply defined bedding and abrupt shifts in grain size that is inferred to be the rapidly superimposed products of numerous discrete explosive events. The deposit is from another basaltic explosive eruption from the Masaya area, Nicaragua, formerly miscorrelated with the Fontana lapilli. Yellow scale is 1 m in length.

produce massive uniform deposits (Figure 34.5). Ash "partings" in a deposit gives some estimate of the duration of any pauses in the parent eruption, e.g., 10-μm-sized ash can persist in the atmosphere for 2—3 days so beds of this grain size in a sequence imply breaks of at least this duration. Conversely, small amounts of very fine ash erupted in early phases of a prolonged "dry" eruption may not settle until after the close of the last explosive phase of eruption (Fierstein and Hildreth, 1992). With increasing distance from the vent, the definition of partings in a deposit may fade as the period of plume transport becomes significantly longer than the duration of fluctuations in the intensity of the eruption.

Mantle bedding is a characteristic of fall deposits—plane-parallel beds drape and follow preexisting relief on slope angles up to 25° to 30°. On steeper surfaces, material may roll, slide, or avalanche downslope and, in extreme cases, form better-sorted lensoid inverse-graded units on or below any steep slope. Fall deposits tend to lack internal unidirectional bedding features, such as mega-ripples that characterize the deposits of pyroclastic density currents, although near-surface winds and coeval plume-fed dilute density currents occasionally create local cross-bedding and pinch-and-swell structures in fine-grained fall deposits.

3.2. Grading: Steadiness

Many fall deposits consist of alternating coarser and finer grained intervals without discrete bedding planes (Figure 34.6(A)). Such progressive vertical shifts, either in particle size or in density within a bed, are known as grading. *Density grading* is a progressive change in particle density with or without a change in grain size, and is more common in the products of pyroclastic density currents (Chapter 36). *Size grading* is typical of all pyroclastic deposits (Figures 34.1, 34.5, and 34.6) but may be expressed in very subtle ways. Both normal and inverse size grading are common in fall deposits. Size grading is most commonly interpreted in terms of steadiness (fluctuating vigor) of the eruption (Figures 34.6 and 34.7) but can also be produced by wind shifts and/or alteration in the inclination of the eruption column or jet. Figure 34.6 shows examples of a steady Plinian eruption (Novarupta, 1912), and an unsteady, episodic subplinian eruption (c. 700 a Kaharoa eruption).

Submarine and sublacustrine fall deposits have highly distinctive grading patterns. Settling through water promotes size, and particularly, density fractionation.

3.3. Grain Size/Sorting: The Role of Pyroclast Fractionation during Transport

Grain size: The lateral transport of particles in an eruption plume is strongly influenced by their residence time in the

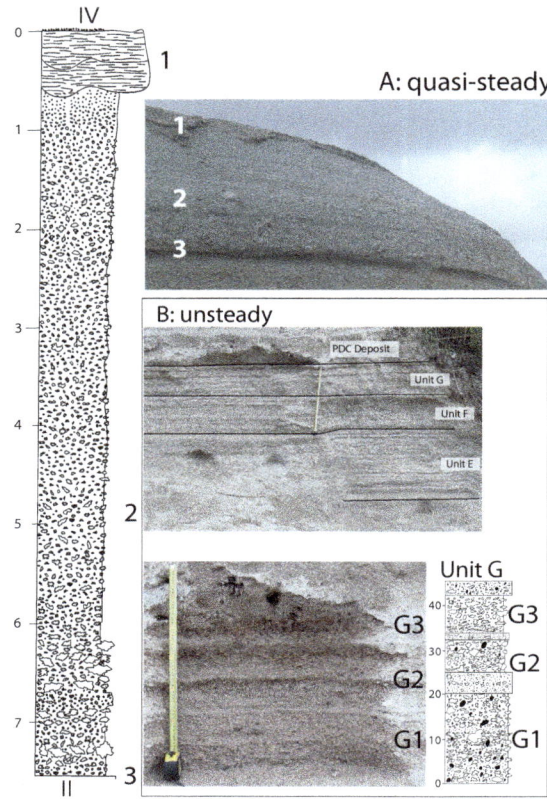

FIGURE 34.6 Photographs and stratigraphic logs for (A) a quasi-steady Plinian eruption phase (Episode III, Novarupta 1912), and (B) three unsteady "subplinian" phases (900 a Kaharoa eruption, Tarawera).

plume and while falling in the atmosphere, and thus by their terminal fall velocities (see Chapter 33) and their ability to form aggregates in flight. The residence time of any clast or aggregate in an eruption plume is inversely proportional to its terminal fall velocity (V_t), which scales with clast diameter (d) and density (ρ) (see Chapter 33). As a consequence, larger particles fall closer to vent and finer particles fall further and, in the absence of aggregation of ash, average grain size fines away from source. If aggregates of ash form through collision and binding of particles (by wet aggregation processes in moist plumes, or by dry, mainly electrostatic aggregation in other plumes), the aggregates will fall with greater velocities than the terminal velocities of their component particles and be deposited "prematurely" (see Chapter 35).

The two commonly used measures are $Md_\phi (= \phi_{50})$ and $M_z (= [\phi_{16} + \phi_{50} + \phi_{84}]/3)$, where the numbers refer to percentiles coarser than the phi size stated and ϕ is the negative log to the base 2 of the grain size in millimeters. In practice, the difference between these two parameters is very slight for most fall deposits. The size of juvenile particles in a deposit at any site is both a measure of the efficiency of the fragmentation processes (i.e., what are the

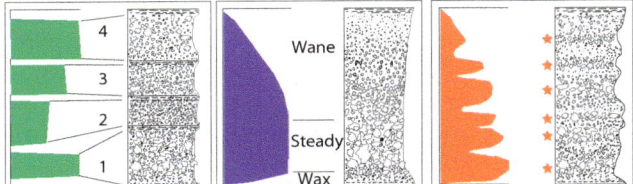

FIGURE 34.7 Schematic cartoon of the relationships between eruptive intensity, the development of bedding, and maximum clast size. On the left (green), four episodes during an eruption produce units delineated by partings of fine to very fine ash, which settle in the pause between episodes. Rapid waxing and waning of each episode is indicated by only limited normal and reverse grading at the top and bottom of the units, respectively. At the center (blue), a single episode in which early steady waxing is followed by an interval of steady peak discharge followed by gradual waning is responsible for the "classical" reverse to normal grading. In reality, few fall deposits show this pattern. On the right (red), a sustained but unsteady pulsating eruption produces a single fall unit characterized by rapid changes in grain size. In a crude way the eruption first waxes and then wanes. In each histogram the vertical axes is normalized time and the horizontal axis is mass discharge rate.

largest particles produced by the explosion) and the vigor of the eruption plume (how high particles of a given size are carried). It is important to realize that fragmentation efficiency and vigor of transport are not always coupled. For example, hydrothermal explosions (Chapter 30), powered by flashing of superheated water to steam, produce highly fragmented assemblages of clasts, because they utilize the available thermal energy very efficiently (down to 100 °C) and often fragment incoherent, intensely altered wall rock. However, they are generally not powerful, as the total energy remaining to eject and transport clasts is limited.

Sorting: Sorting is modulated by the range of grain sizes present in the mixture leaving the vent and, more importantly, by the efficiency of size fractionation during particle transport. It is a measure of the clustering about a mean value. Pyroclastic deposits are poorly sorted in size terms relative to nonvolcaniclastic sediments, reflecting a wide range in pyroclast densities (from 100 to 3000 kg m^{-3}). The two most common measures of sorting, which are highly correlated, are graphic standard deviation ($\sigma\phi = [\phi_{84} - \phi_{16}]/2$) and inclusive graphic standard deviation ($\sigma_I = [\phi_{84} - \phi_{16}]/4 + [\phi_{95} - \phi_5]/6.6$). The latter encompasses a significantly wider portion of the sample but requires definition of the fifth percentile of the deposit, which is often not possible for very coarse-grained falls.

Sorting is often used to separate the products of "dry" (magmatic) and "wet" (phreatomagmatic) explosions. Most dry fall deposits have $\sigma\phi$ of 1–1.5, whereas many deposits of phreatomagmatic eruptions involving abundant water have typical $\sigma\phi$ values of 2 or higher. A general improvement in the degree of sorting beyond the near-vent region is a common feature of many fall deposits.

Md$_\phi$ versus $\sigma\phi$ plots: Grain size studies are among the most widely used tools in studying the origins of pyroclastic deposits. There is a broad correlation of grain size parameters with eruption and transport styles (Walker, 1973). The combination of sorting and absolute grain size was used at an early stage to define fields for "wet" and "dry" fall deposits. This distinction really only held when considering fall deposits of very wet eruptions (with abundant liquid water in the plume). Ash aggregation results in the "flushing" of ash-sized ejecta from the plume and simultaneous deposition of a wide range of grain sizes. Many such very "wet" deposits are poorly sorted and fine grained even close to the vent (Chapter 30), whereas deposits of dry eruptions are coarse grained and better sorted (Chapters 27, 29).

This correlation was originally attributed to a greater fragmentation efficiency of phreatomagmatic explosions. In contrast, there is overlap between the grain size of dry or "magmatic" fall deposits and that of phreatomagmatic deposits from eruptions with low water/magma ratios (Figure 34.8), such that grain size parameters cannot readily separate slightly to moderately "wet" from "dry."

The original defined fields thus now seem rather too restrictive, but use of the Md versus $\sigma\phi$ diagram, in a more subtle way for considering the distinction between different units in one eruptive sequence and particularly trends within a sequence, is still extremely valuable. It is possible to use sorting values to distinguish contrasts in grain size that reflect changing vigor of the explosions or wind shifts, for example, ($\sigma\phi$ remains constant or decreases with decreasing Md), from those where finer-grained beds result from weak magma:water interaction or conduit wall collapse (marked increase in $\sigma\phi$ with decrease in Md).

3.4. Ballistic Pyroclasts

The term ballistic is used for pyroclasts that are too coarse for thermal or mechanical equilibrium with the gas phase, and therefore decouple from the jet and follow independent parabolic trajectories at increasing angles from the vertical. The lower size limit varies with properties of the clasts but is often 20–30 cm for dense lithic clasts. They are proximal phenomena, extending only 1–5 km from the source, depending on the eruption intensity. Their importance is that they are little affected by wind advection (Figure 34.3(B)) and they allow volcanologists to recognize proximity to vent and to calculate approximate minimum ejecta velocities. Ballistic clasts impact into the ground with considerable force to produce impact craters, and crater position and shape indicate the source direction.

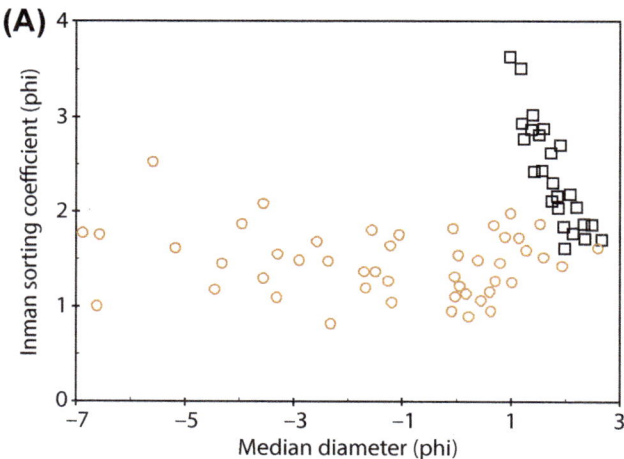

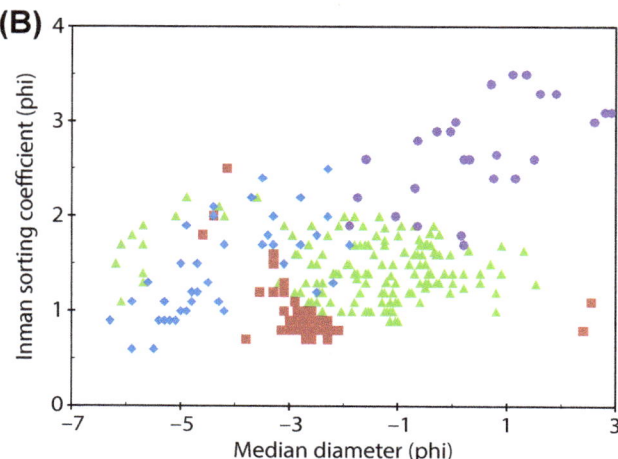

FIGURE 34.8 Contrasting grain size characteristics for a variety of pyroclastic fall deposits from eruptions involving widely different water/magma ratios and mass discharge rates. (A) Contrasting Plinian (orange circles) and phreatoplinian (black squares) samples from Units D and C of the 1875 Askja eruption. *(After Sparks et al. (1978).)* There is a marked contrast in both sorting and median diameter that reflects the role of external water during episode C—an increase in fragmentation efficiency and promotion of "premature" sedimentation of fine—very fine ash as aggregates. (B) Basaltic tephra samples from Strombolian (blue diamonds), Hawaiian (red squares), and weakly phreatomagmatic (green triangles) fall deposits from Kīlauea and Taupo volcanic zone (Houghton, unpublished data) together with samples of PDC (surge) deposits intercalated with the phreatomagmatic falls (purple circles). In contrast to (A), there is little to no distinction between the three families of fall deposits, whereas the surge deposits are both markedly finer and less well sorted.

3.5. Welding and Other Thermal Effects

During fall transport there is prolonged thermal interaction between pyroclasts and air involving cooling of the pyroclasts. Thus, in comparison with many pyroclastic density current deposits (notably ignimbrites, see Chapter 36), fall deposits show less abundant welding/agglutination. These features form a spectrum from thermal coloration through welding and agglutination to clastogenic and rheomorphic lava flows.

The most common and widespread manifestation of high-temperature thermal effects is in coloration of pyroclasts, principally controlled by the growth or alteration of iron oxide mineral species (e.g., hematite), especially as microscopic crystals. Oxidized scoria has a characteristic red color. Such coloration tends to be pervasive on the inner walls of scoria cones and spatter ramparts. Pumice is generally only altered in the interior of large clasts to a wide range of hues from orange through pink, red, purple, and brown. This is interpreted to reflect greater retention of heat by the larger clasts during flight and in the deposit, especially in the case of high sedimentation rate.

At higher retained temperatures pyroclasts will deform viscously and flow on landing. Welding is the slow deformation of individual pyroclasts together, accompanied by development of interclast adhesion under load stresses induced by overlying tephra. Factors that promote welding are the retention of heat in individual clasts (e.g., coarse clast size, short flight time, transport in the denser and higher temperature region of the eruptive jet/fountain, and high initial temperature) and high accumulation rates (to cause rapid burial and loading). Welded fall deposits are most common in deposits of basaltic, peralkaline, or intermediate composition, but are also known from highly silicic compositions. Agglutination is the process by which individual hot, fluid pyroclasts deform, flow, and weld together on impact due to their own momentum. The degree of agglutination varies with the clast viscosity and size, and the flight time, and hence is an important process in deposition from low, fountaining eruptions such as Hawaiian fountains (Chapter 27). These thermal phenomena show a regular increase in thermal intensity toward vents. This contrasts with thermal effects in ignimbrites, which vary more with deposit thickness, which in turn often varies nonsystematically with distance from source. Welding and agglutination in fall deposits are especially useful indicators of proximity to vent. In proximal areas, the distinction between welded deposits of fall versus density current origin is based on the lack of an ash matrix (Figure 34.9(A)) and rapid (decimeter to meter scale) vertical changes in welding intensity in falls (Figure 34.9(B)).

4. SMALL-SCALE FEATURES

4.1. Juvenile Clast Morphology: Cooling Rate and Magma Viscosity

Juvenile pyroclasts have shapes that are related to melt rheology at fragmentation and during transport, and in some cases, deposition. Melt droplets or spray, preserved as fluidal clasts, form from magmas that fragment in a hot fluid state. Achneliths, Pele's tears, and spatter are products of such fragmentation (Figure 34.10(A)–(D)). Angular blocky pyroclasts with planar and curviplanar

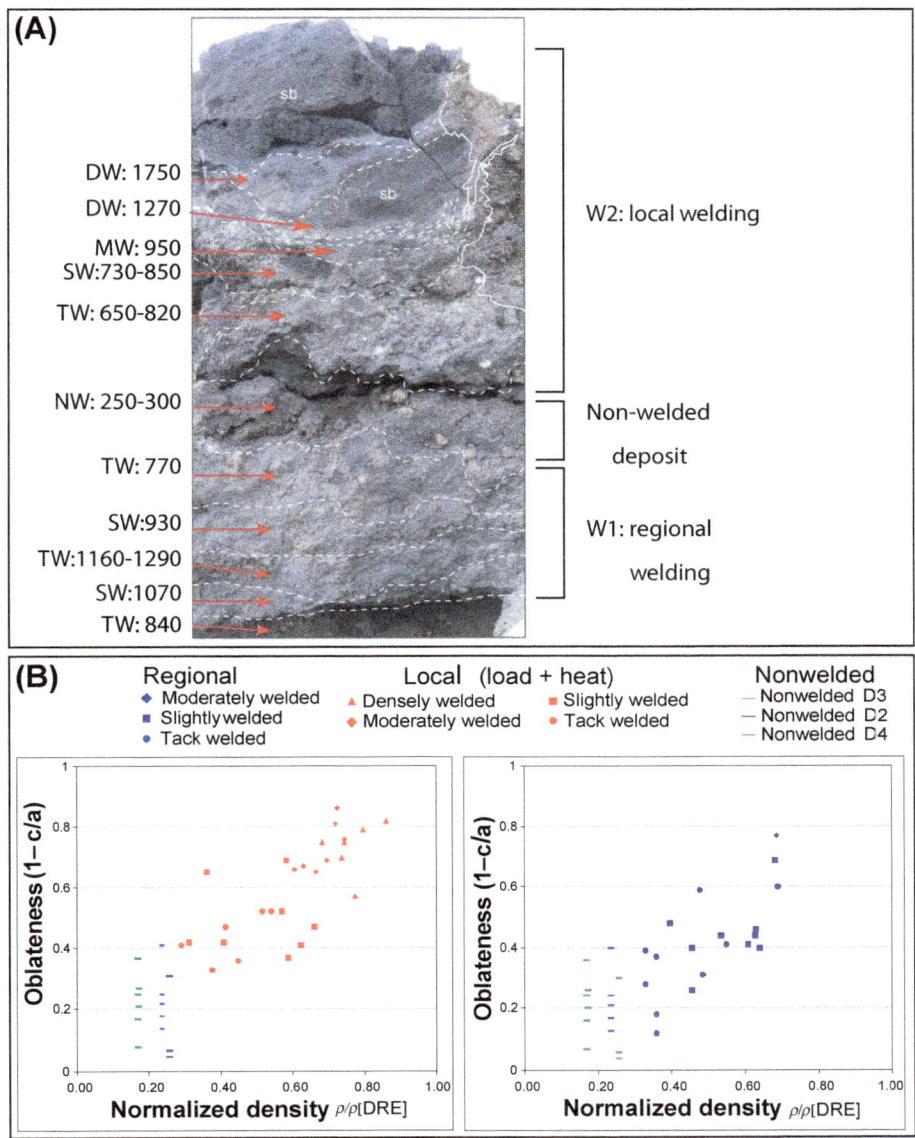

FIGURE 34.9 Welding textures in pyroclastic fall deposits. (A) Profile through welded portions of the 1875 AD fall deposit at Askja volcano, Iceland. At this location there are two welded units. Unit W1 at this location, and laterally over hundreds of meters, has undergone regional welding and shows a limited range in welding grade, deposit bulk density, and clast flattening ratios. Unit W2 is locally welded and strongly influenced by the presence of two large spatter bombs (sb) around which local increases in welding intensity occur as seen in increased flattening ratios of particles and bulk density of the deposits. Loading and transfer of heat from the bombs is responsible for the local increases above the "regional" welding grade. NW, nonwelded; TW, tack welded; SW, slightly welded; MW, moderately welded; DW, densely welded. Density values are in kilograms per cubic meters. (B) Oblateness and normalized density record the strain accumulated during welding and compaction. Oblateness is calculated by combining lengths (a) and heights (c) of flattened lapilli as $[1-(c/a)]$. This figure contrasts the nature of "regional" welding (blue symbols) with local welding due to heating and loading from outsized bombs (red symbols). For regionally welded samples, there is a general trend of increasing oblateness with welding intensity associated with matrix compaction and minor flattening of pumice. For locally welded samples there is a change of slope at normalized density values of 0.6. This change is slope is due to a change of mechanism of strain accommodation, from matrix welding and loss of pore space, to flattening of pumices and only minor matrix pore space.

surfaces (pumice, scoria) reflect brittle failure of magmas that are often cooler and/or more viscous (Figure 34.10(E), (F), and (H)). Brittle fracture can also result from quenching by contact with external water or very high strain rates during fragmentation. The hot interior of large pyroclasts with rapidly quenched rinds may continue to vesiculate and expand, leading to the production of distinctive cooling cracks and "breadcrust" texture (Figure 34.10(G)).

Pyroclast—pyroclast collisions during transport may modify the original morphology of particles. Chipped or rounded edges and corners and hackly fracture surfaces are

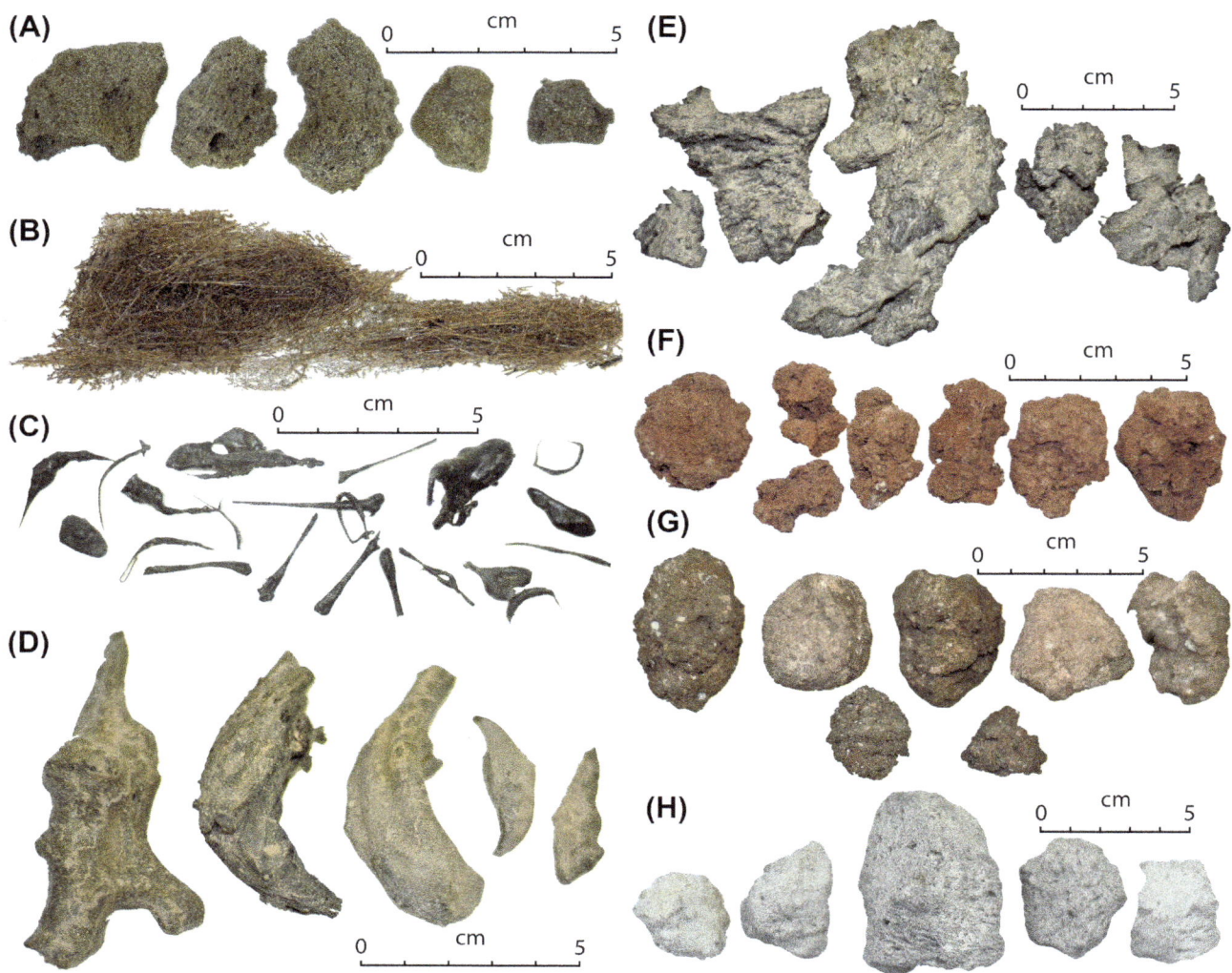

FIGURE 34.10 Examples of juvenile pyroclast morphologies: (A) reticulite, (B) Pele's hair, (C) Pele's tears, (D) fusiform clasts, (E) transitional lapilli (fluidal with ragged exteriors), (F) scoria, (G) dense spherical lapilli, and (H) pumice.

features of transport in a concentrated medium, and may help to distinguish clasts transported by pyroclastic density currents from those of fall deposits.

4.2. Vesicularity: Role of Magmatic Volatiles

Juvenile clast vesicularity for different eruption styles is described in detail in Chapters 27–31. For a given viscosity, increasing number density of bubbles correlates with higher bubble nucleation rates that have been attributed to progressively later onset of bubble nucleation at higher degrees of supersaturation with volatiles. There is a spectrum of bubble maturation as newly formed bubbles first grow unimpeded, and then interact and coalesce. Permeability and outgassing can be achieved in the following two ways. (1) In fluid melts the buoyant ascent rate of large bubbles

may exceed the ascent rate of the host melt and bubbles can escape. (2) In silicic magma permeable pathways of interconnected bubbles may form, often aided by shearing and deformation along conduit margins.

4.3. Ash Aggregates

These features are described in detail in Chapter 30.

4.4. Wall Rock Abundance and Causes

The content of wall rock (foreign) lithic clasts is a sensitive indicator of the competence and stability of the rocks forming the vent/conduit walls, as well as fluctuations of pressure in the conduit and the extent to which the magma has interacted with aquifers. Even among the products of **dry explosions**, the content of foreign lithic clasts correlates with eruptive intensity and with fragmentation depth. Hawaiian fall deposits have probably the lowest lithic content (<1 wt%) of all types of pyroclastic deposit and Plinian deposits often contain 8–20 wt% wall rock lithic clasts. Phreatoplinian deposits, where magma interacts with lake or seawater, typically have foreign lithic contents of 3–20 wt% but many phreatomagmatic deposits associated with aquifer systems have lithic contents of >50 wt%. The contrast is most striking where such deposits are interbedded with Hawaiian, Strombolian, or Plinian units from the same volcano.

In weak explosive eruptions characterized by repeated, nonsustained explosions there is often considerable fallback and recycling of pyroclasts through the vent. These clasts include recycled juveniles that in a thermodynamic sense now behave as lithic clasts, i.e., contribute little, if any, thermal energy to the explosion. Recycled juveniles can be detected by either a mixture of angular and fluidal surfaces, contrasting degrees of surface alteration or coating by fine ash (Houghton and Smith, 1993), or by the presence of thermally induced physical changes in the recycled material (e.g., D'Oriano et al., 2012).

4.4. Wall Rock Componentry: Fragmentation Depth and Vent Position

The nature of wall rock lithic clasts can help both to locate vent position and to constrain the depth over which fragmentation occurs, if the vent and basement geology are well known. A progressive change in the lithic clast assemblage

with time may record either the opening of additional vents in new areas of contrasting basement lithology (Wilson and Hildreth, 1997) or change of the fragmentation region to deeper or shallower levels. Extremely detailed studies of historical eruptions at Vesuvius (Chapter 29) enabled a close match to be made between lithic lithology and the basement stratigraphy. In wet explosive eruptions the nature of the wall rock clasts can help constrain hydrology beneath the vent region. For example, an abundance of well-rounded wall rock clasts may correlate with a gravel aquifer, or faunal assemblages in blocks of unconsolidated marine sediments may help fix water depth for submarine vents. Pulses of shallow-derived lithic fragments in "dry" fall deposits are inferred to accompany intervals of vent widening (Figure 34.1(B)).

5. ATTEMPTS TO CLASSIFY FALL DEPOSITS

Classification of eruptions has been made either on the basis of characteristics of the eruption, observed in real time, or based on the properties of the resulting fall deposits. The classification scheme of Walker (1973) was a major breakthrough permitting the classification of deposits by style based on deposit characteristics alone, thus permitting inclusion of deposits from unobserved eruptions. The classification used two parameters, dispersal and fragmentation indices, which measured the thinning rates and absolute grain size of a deposit (Figure 34.11(A)). Values were inferred to reflect increasing mass discharge

(A)

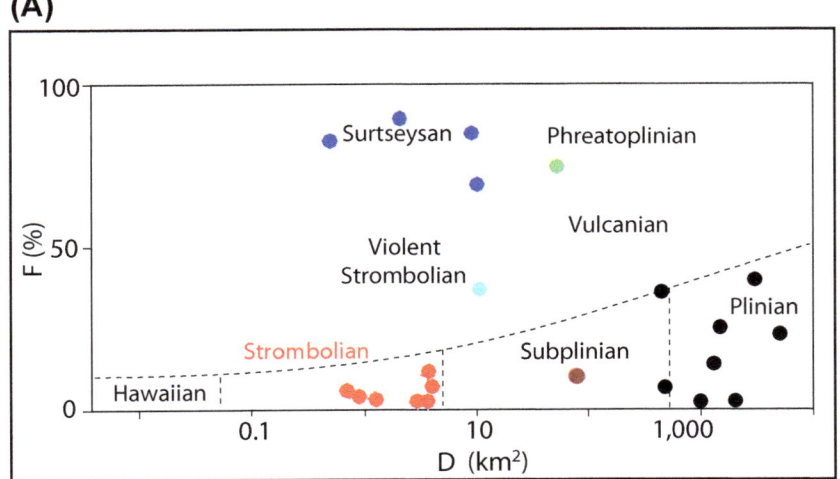

(B)

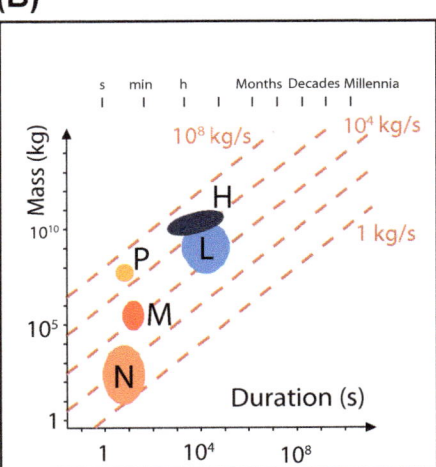

FIGURE 34.11 (A) Plot of pyroclastic fall classification. *(After Walker (1973).)* Points show data used to establish this classification. Walker original term "salic equivalent of Surtseyan" later became "phreatoplinian." Walker used two indices linked to the dispersal of a fall tephra (D) and the efficiency of fragmentation (F). The distinction between Hawaiian, Strombolian, subplinian, and Plinian tephra was largely a function of the D index, assumed to scale with mass discharge rate. (B) Log–log plot of total erupted mass versus duration for well-constrained historical Hawaiian (blue) and Strombolian (red) eruptions. N, M, and P denote the fields of normal, moderate, and paroxysmal Strombolian explosions, respectively, and L and H denote fields of low- and high-fountaining Hawaiian eruptions. Note the major overlap in terms of mass discharge rate (red dashed lines) meaning that dispersal criteria cannot distinguish between these eruptive styles.

rate and the degree of involvement of external water, respectively. The classification relied on measurements from only 26 deposits and no examples of Hawaiian fountaining or any measurements from Stromboli were included. Despite wide acceptance, few deposits have been fully characterized in this fashion. There are clear problems in applying the scheme to eruptions with low mass discharge rates—all documented "normal Strombolian" eruptions fall in the Hawaiian field and the only documented deposit of Hawaiian fountaining lies in the Strombolian field. Subsequent schemes have recognized this; Pyle (1989) includes a Strombolian but not a Hawaiian field and Bonadonna and Costa (2013) group all "small to moderate" eruptions. The Walker (1973) scheme relies principally on dispersal as a proxy for mass discharge rate and calculates this value on a time-averaged basis over the entire eruption span (e.g., Figure 34.4). A priori it cannot distinguish between sustained, long-lived eruptions (Hawaiian, subplinian, Plinian) and series of transient impulsive explosions (Strombolian, Vulcanian) as shown in Figure 34.11(B). Future classifications potentially treating erupted mass and eruption duration as independent variables will solve this issue.

ACKNOWLEDGMENTS

Much of our understanding of fall deposits was gained in the field, on formal excursions and particularly in the close company of a small number of collaborators. In particular, we acknowledge the contributions of Colin Wilson, Thor Thordarson, Costanza Bonadonna, and George Walker. The manuscript was painstakingly reviewed by Costanza Bonadonna, Raffaello Cioni, and Maria Janebo. This work was possible via generous support from the National Science Foundation (USA), GNS Science (New Zealand), and the Australian Research Council.

FURTHER READING

Andronico, D., Del Carlo, P., Coltelli, M., 1999. The 22 July 1998 fire fountain episode at Voragine Crater (Mt Etna, Italy). In: Proceedings of the Volcanic and Magmatic Studies Group Annual Meeting. January 5–6, 1999.

Biass, S., Bonadonna, C., 2011. A quantitative uncertainty assessment of eruptive parameters derived from tephra deposits: the example of two large eruptions of Cotopaxi volcano, Ecuador. Bull. Volcanol. 73, 73–90.

Bonadonna, C., Cioni, R., Pistolesi, M., Connor, C., Scollo, S., Pioli, L., Rosi, M., 2013. Determination of the largest clast sizes of tephra deposits for the characterization of explosive eruptions: a study of the IAVCEI commission on tephra hazard modelling. Bull. Volcanol. 75, 680. http://dx.doi.org/10.1007/s00445-012-0680-3.

Bonadonna, C., Costa, A., 2013. Plume height, volume, and classification of explosive volcanic eruptions based on the Weibull function. Bull. Volcanol. 75, 742.

Bonadonna, C., Ernst, G.G.J., Sparks, R.S.J., 1998. Thickness variations and volume estimates of tephra fall deposits: the importance of particle Reynolds number. J. Volcanol. Geotherm. Res. 81, 173–187.

Bonadonna, C., Houghton, B.F., 2005. Total grain size distribution and volume of tephra-fall deposits. Bull. Volcanol. 67, 441–456.

Burden, R.E., Chen, L., Phillips, J.C., 2013. A statistical method for determining the volume of volcanic fall deposits. Bull. Volcanol. 75, 1–10. http://dx.doi.org/10.1007/s00445-013-0707-4.

Carey, R.J., Houghton, B.F., 2010. "Inheritance": an influence on the particle size of pyroclastic rocks. Geology 38, 347–350.

Connor, L.J., Connor, C.B., 2006. Inversion is the key to dispersion: understanding eruption dynamics by inverting tephra fallout. In: Mader, H.M., Cole, S.G., Connor, C.B., Connor, L.J. (Eds.), Statistics in Volcanology. Special Publications of IAVCEI. Geological Society, London, pp. 231–242.

D'Oriano, C., Pompilio, M., Bertagnini, A., Cioni, R., Pichavant, M., 2012. Effects of experimental reheating of natural basaltic ash at different temperatures and redox conditions. Contrib. Mineral. Petrol. http://dx.doi.org/10.1007/s00410-012-0839-0.

Fierstein, J., Hildreth, W., 1992. The Plinian eruptions of 1912 at Novarupta, Katmai National Park, Alaska. Bull. Volcanol. 54, 646–684.

Fierstein, J.E., Nathenson, M., 1992. Another look at the calculation of fallout tephra volumes. Bull. Volcanol. 54, 156–167.

Houghton, B.F., Wilson, C.J.N., Fierstein, J., Hildreth, W., 2004. Complex and episodic proximal deposition during a sustained Plinian eruption: the 1912 eruption of Novarupta, Alaska. Bull. Volcanol. 66, 95–133.

Houghton, B.F., Swanson, D.A., Rausch, J., Carey, R.J., Fagents, S.A., Orr, T.R., 2013. Pushing the Volcanic Explosivity Index to its limit and beyond: Constraints from exceptionally weak explosive eruptions at Kīlauea in 2008. Geology 41, 627–630.

Houghton, B.F., Carey, R.J., Rosenberg, M.D., 2014. The 1800a Taupo eruption: 'Ill wind' blows the ultraplinian type event down to plinian. Geology 42, 459–461.

Klawonn, M., Houghton, B.F., Swanson, D.A., Fagents, S.A., Wessel, P., Wolfe, C.J., 2014. Constraining explosive volcanism: subjective choices during estimates of eruption magnitude. Bull. Volcanol. 76, 793–798.

Le Pennec, J.L., Ruiz, G.A., Ramón, P., Palacios, E., Mothes, P., Yepes, H., 2012. Impact of tephra falls on Andean communities: the influences of eruption size and weather conditions during the 1999–2001 activity of Tungurahua volcano, Ecuador. J. Volcanol. Geotherm. Res. 217, 91–103.

Newhall, C.G., Self, S., 1982. The volcanic explosivity index (VEI) an estimate of explosive magnitude for historical volcanism. J. Geophys. Res. (Oceans) 87, 1231–1238. http://dx.doi.org/10.1029/JC087iC02p01231.

Pyle, D.M., 1989. The thickness, volume and grain size of tephra fall deposits: Bull. Volcanol. 51, 1–15.

Pyle, D.M., 1995. Assessment of the minimum volume of tephra fall deposits. J. Volcanol. Geotherm. Res. 69, 379–382.

Shea, T., Houghton, B.F., Gurioli, L., Cashman, K.V., Hammer, J.E., Hobden, B.J., 2010. Textural studies of vesicles in volcanic rocks: an integrated methodology. J. Volcanol. Geotherm. Res. 190, 271–289.

Walker, G.P.L., 1973. Explosive volcanic eruptions – a new classification scheme. Geol. Rundsch. 62, 431–446. http://dx.doi.org/10.1007/BF01840108.

Walker, G.P.L., 1980. The Taupo Pumice: product of the most powerful known (ultraplinian) eruption? J. Volcanol. Geotherm. Res. 8, 69–94.

Wilson, C.J.N., Hildreth, E.W., 1997. The Bishop Tuff: new insights from eruptive stratigraphy. J. Geol. 105, 407–440.

APPENDIX: TECHNIQUES FOR CHARACTERIZATION OF FALL DEPOSITS

Thickness/Isomass

Spacing of sampling sites: An ideal isopach/isomass map will have only one contour between any two data points. Our preferred strategy is to space sites at no more than approximately the thinning half-distance apart. Ideally, several passes should be made in the field, initially visiting a sparse number of widely spaced sites to constrain thinning rates crudely, then returning to augment the data by additional points in key regions, where the deposit thins rapidly. The outermost contour is critical in constraining pyroclastic fall deposits, as the margins of a deposit are its most subjective parts. Scattered clasts will land along the margins of coarse-grained fall deposits but are too sparse to form a permanent continuous bed (see Figure 34.5(B) of Houghton et al., 2011). The edges of fine-grained deposits are lost to posteruption erosion and reworking in just a few days under most conditions. Studies made well after an eruption will therefore always underestimate the deposit's "footprint."

Measurement of thinning data: The choice of thickness versus isomass measurements depends on the thickness and degree of induration of a deposit. Ash fall deposits can be constrained accurately to ~1 mm, whereas the realistic error in measuring thickness in a lapilli fall is probably at least 1 cm. Therefore large errors resulting from contouring lapilli deposits below 5 cm thickness can be expected. Strong surface winds have local effects on the deposit both during and after the eruption (Richter et al., 1970). This localized reworking is confined to the top few centimeters of the deposit and has a disproportionate influence on local variability in thickness at the distal sites. Isomass is the preferred technique in distal regions as collecting thin, newly fallen tephras is much easier from man-made surfaces—roads and pavements (Figure 34.12), fence rails, mailboxes, and discarded boxes. In thick deposits (i.e., proximal sites) it is difficult to sample consistently over a fixed area and measurements of thickness are much quicker. If the techniques are combined (proximal thickness data with distal isomass data, for example) it is necessary to have estimates of in situ bulk density of the tephra. This can be done by either inserting a container of fixed volume into the deposit or excavating a known volume in the field (which are preferred approaches), or, in the laboratory by loosely compacting a bulk tephra sample in a measuring cylinder.

Stratigraphic logging: Examples of stratigraphic logs are given in Figures 34.1 and 34.6. By convention large juvenile clasts are shown as white and wall rock lithic clasts as black. A scale should be chosen whereby clasts that are 1 cm in diameter or larger can be drawn to scale. Coarse to fine ash is shown as dots of increasing diameter, and very fine ash as short dashes. Various researchers show finer-grained beds as either receding or protruding; we prefer the latter as finer-grained beds often protrude in outcrop relative to their coarser counterparts, and they contain more delicate bedding structures that are easier to represent on a wider log.

Size of largest clasts (MP, ML): Our preferred solution is to measure the average of the geometric mean of the three orthogonal diameters that define the minimum ellipsoid for the five largest lithic clasts in the unit (Figure 34.13). If a deposit is lithic free or lithic poor, e.g., Hawaiian falls, then we measure the MP instead. The size of a statistically representative ML sample increases with clast size and decreases with increasing wall rock lithic content. Thus, there is no single volume that can be sampled at all sites for any deposit. The diagrams of Bonadonna et al. (2013) have been used both to quantify the uncertainty among different techniques and to define the volume to be sampled depending on the lithic content and medial grain size (cf. Figures 34.4 and 34.8). There are two alternative approaches (1) measure a large number of clasts and calculate the 50th percentile (Bonadonna et al., 2013) or (2) collect large clasts until visual inspection suggests that their mean diameters have converged on a common value and measure the five largest

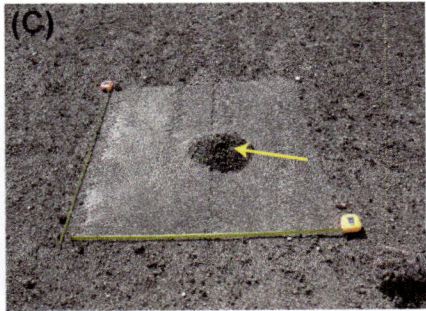

FIGURE 34.12 Images of isomass sampling for the March 19, 2008, explosion of Halemaʻumaʻu crater, Kīlauea. (A) Undisturbed deposit with largest pyroclasts arrowed. (B) Construction of a 1-m-wide square centered on one of the large pyroclasts. (C) Isomass sample prior to being bagged. Largest pyroclast in (B) is conspicuously visible (yellow arrow).

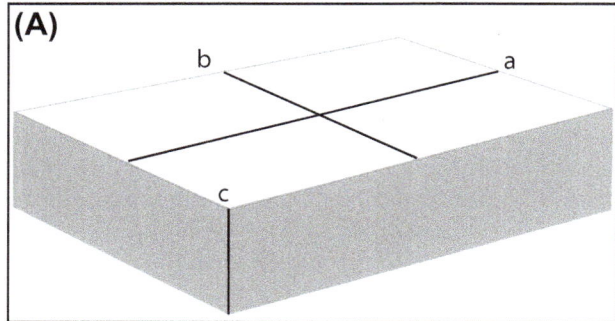

FIGURE 34.13 Measurements of MP and ML data for pyroclasts. (A) Location of minimum ellipsoid for tabular pyroclast. *(After Bonadonna et al. (2013).)* (B) Collection of MP clasts from episode 15 of the 1959 Kīlauea Iki eruption.

clasts. For time efficiency, we prefer technique (2). Where time permits, both values should be determined and reported for future characterizations of tephra deposits. Clasts should be characterized based on the geometric mean of its three orthogonal axes with the approximation of the minimum ellipsoid (Bonadonna et al., 2013).

Grain size sampling: As for MP/ML data, the size of a representative grain size sample scales with clast size. Our preferred strategy is to center a sample on the largest clast within the unit and to collect a sample such that the mass of the largest clast is less than or equal to 5% of the total sample.

This equates to approximately 1–5 g for a sample where the largest clast is 4 mm in diameter, 0.5–2 kg for a sample where the largest clast is 32 mm, and 250–1000 kg if the largest clast is 256 mm. Samples of ash or fine lapilli can be collected en masse in the field and analyzed in the laboratory. Coarse samples require field sieving of the size fractions down to 16 mm, using a technique known as "cone and quartering" to reduce the processed amount of the finer-sized fractions (Figure 34.14). The sample is extracted in buckets and sieved by hand. Clasts larger than −6 phi (>64 mm) are measured by hand and clasts between −4 and −6 phi are removed by sieving at half-phi intervals. All these size fractions are then weighed in the field. The <4 phi (<16 mm) material is weighed and then split to yield a residue of approximately 500 g. This residue is taken to the laboratory, weighed, then dried, and sieved.

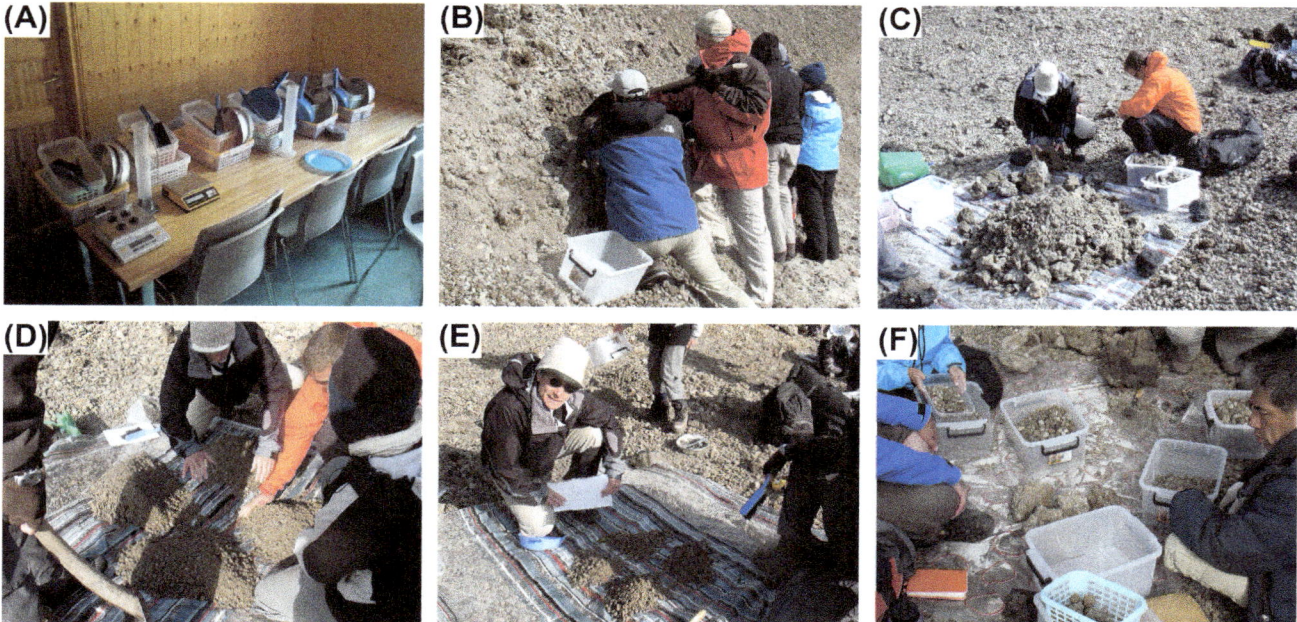

FIGURE 34.14 Field grain size sampling and measurements. (A) Five sets of field sieves and scoops together with portable balances. (B) Collection of bulk sample centered on the largest visible clasts. (C) Bulk sample on tarpaulin during removal and individual measurement and weighing of the largest pyroclasts. (D) and (E) Reduction of sample size for the sub-32-mm fraction via the coning-and-quartering technique. (F) Weighing of the intermediate size fractions under the concerned gaze of a leading modeler.

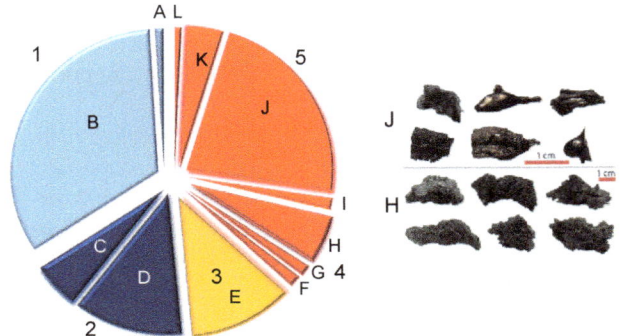

FIGURE 34.15 Componentry of ash erupted on September 28, 2008, from the Overlook vent at Kīlauea. The particles can be divided into 12 categories with five groups. Groups 1 and 2 are altered and fresh wall rock, respectively; groups 3, recycled altered juvenile clasts; and groups 4 and 5, vesicular and outgassed juvenile particles. (A) are intensively hydrothermally altered and (B) weakly altered. Together they represent material collapsed from the superficial exposed part of the vent walls. (C) and (D) are crystal-poor and microcrystalline fresh basalt. They represent the nonexposed portions of the lavas that form the vent walls. (E) are strongly altered but fluidal basaltic lapilli inferred to be formerly juvenile clasts that fell back into the vent and were subsequently reerupted (there are no pre-2008 lapilli fall deposits on the vent walls). (F) are vesicular yet fluidal pyroclasts and (G) are golden pumice. (H) are ragged scoria (see clast population at lower right). (I) are dense but angular glassy fragments. (J) are Pele's tears (see clast population at top right). (K) are small hollow glassy spheres; (L) are Pele's hair. The disparate assemblage of juvenile ejecta ((F) through (L)) reflects a complex pattern of degassing and outgassing in the shallow conduit for these very weak basaltic explosions.

The wet and dry weights are used to calculate moisture correction for the size fractions that are 16 mm and coarser.

In the laboratory samples are sieved at half-phi intervals down to 4 phi (63 μm). The wide size range of pyroclastic rocks makes sieving at quarter-phi intervals impractical. Any field data is then added using a splitting factor. Material finer than 63 μm is analyzed by a range of techniques including sedimentation techniques, laser diffraction methods, photoanalysis, optical or electroresistance counting methods, and acoustic or ultrasound attenuation spectroscopy.

Componentry analysis: Componentry is applied traditionally separately to each size fraction of a grain size sample and the results then integrated. In practice, rapid componentry techniques can only be applied to size fractions larger than 250 μm and in general most workers stop componentry analyses when more than 90 wt% of the sample is characterized. Where a size fraction contains less than 100 grains the entire population is analyzed; for size fractions with larger populations a subpopulation of typically 200 clasts will be used for componentry. The choice of component categories is highly dependent on the diversity of the clast population. We include in Figure 34.15 an example of an unusually complex set of clast types.

Sampling for density and vesicularity: In the field over 100 juvenile clasts of diameter 16—32 mm are collected over a narrow interval of the fall deposit. If the juvenile clasts are diverse the sample size is increased to 200—400 clasts to categorize the end-member vesicularities adequately (Figure 34.16). Sampling lapilli deposits

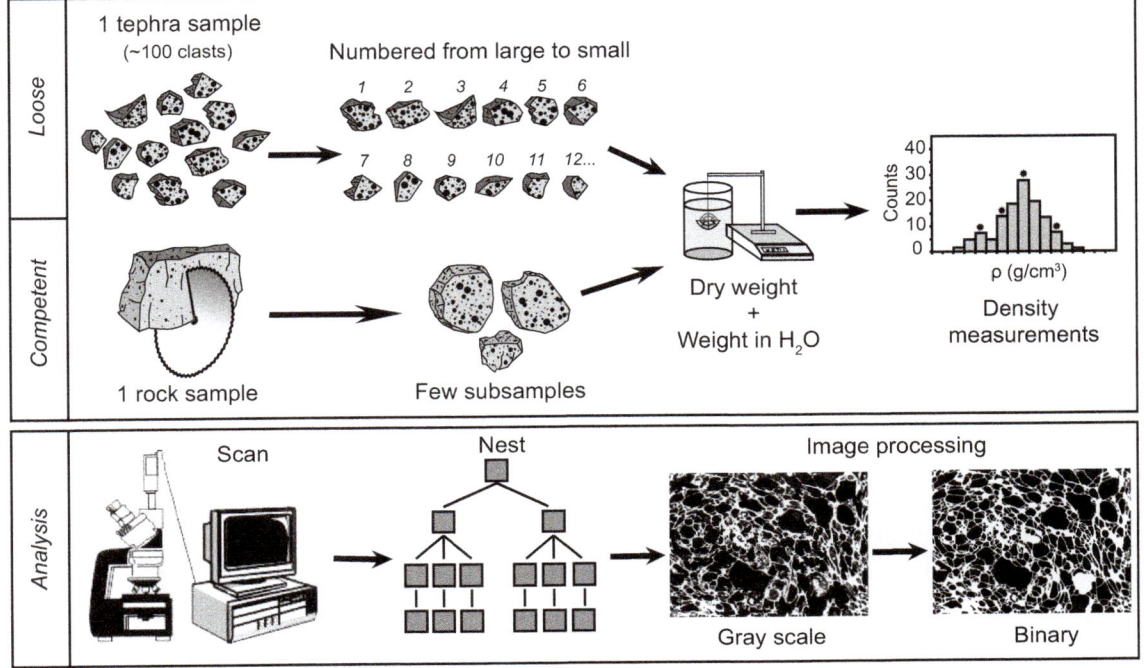

FIGURE 34.16 Quantification of vesicle populations in pyroclasts. Schematic cartoon of sampling procedures, density measurements, scanning electron microscopic photography, and image enhancement modified after Shea et al. (2010).

over a restricted vertical interval of two to three clasts equates to a deposition interval of only minutes. Bombs are often sampled repeatedly in transects across the deposit to establish internal variability (e.g., Gurioli et al., 2014). Clasts or bomb samples are weighed in air (mass ω_{air} in g), and either individually wrapped into polyethylene film (of wet weight ω_{water}^{film}), or made impermeable using water-proofing spray. They are then weighed once more immersed within water (ω_{water}). Density is expressed as: $\rho = \frac{\omega_{air}}{\omega_{air} - \omega_{water} - \omega_{water}^{film}}$.

For buoyant particles, e.g., pumice, the clasts are submerged using a ballast of known wet weight and volume. The density data can then be used as a filter to select representative clasts (Figure 34.16) for measurements of vesicle number density and vesicle size distribution, and for other measurements such as porosity/permeability.

The number and size of vesicles in ejecta can be measured using a scanning electron microscope in two dimensions and converted stereoscopically into three dimensions or measured directly in three dimensions using X-ray microtomography. The former does not deal well with complex bubble shapes or complex connectivity due to coalescence or shearing but does a better job of accurately resolving the size and shapes of the smallest bubbles.

Chapter 35

Pyroclastic Density Currents: Processes and Models

Josef Dufek

School of Earth and Atmospheric Sciences, Georgia Institute of Technology, Atlanta GA, USA

Tomaso Esposti Ongaro

Istituto Nazionale di Geofisica e Vulcanologia, Sezione di Pisa, Pisa, Italy

Olivier Roche

Laboratoire Magmas et Volcans, Université Blaise Pascal-CNRS-IRD, OPGC, Clermont-Ferrand, France

Chapter Outline

GLOSSARY

fluidization Condition when the gravitational force of a bed of particles is balanced by the fluid-particle drag due to upward propagating gas. Fluidization typically occurs in concentrated flows when low permeability results in sustained pore fluid pressure.

Kelvin–Helmholtz instability Shear instability that mixes two adjacent layers of fluid. In the case of a PDC, turbulent mixing is induced between the atmosphere and underlying current when shear is sufficient to overcome the stable density stratification.

lobe-and-cleft instability Entrainment instability in gravity currents where dense material in the leading edge of the current overrides ambient atmosphere due to a no-slip boundary condition, resulting in bulbous fingering at the front of the flow.

pore pressure Fluid pressure in a granular matrix. Pore pressure results from fluid-particle drag and can remain elevated if the granular matrix has low permeability.

1. INTRODUCTION

Pyroclastic density currents (PDCs) are the most hazardous volcanic phenomena for populations living near volcanic edifices due to their rapid propagation along the ground, large dynamic pressures, and high temperatures. PDCs are composed of newly fragmented magma; magmatic and entrained gases; and lithic clasts derived from the vent, conduit, and substrate. PDCs are a type of particle-laden gravity current and as such propagate laterally away from the source region because the current's density exceeds that of the ambient atmosphere. However, while PDCs share much in common with other gravity currents such as turbidity currents and dust storms, they have a wide range of concentrations and forces that can be difficult to quantify from remote observations and deposits.

PDCs encompass a vast array of physical processes operating at many scales of motion, from the granular scale to the scale of the whole current. Some portions of a current may have low particle concentration and are turbulent, while other regions may be dominated by grain–grain contacts and interstitial pore fluid pressure. A specific accounting of forces in PDCs is necessary to predict their dynamics and important quantities such as run-out distance

The Encyclopedia of Volcanoes. http://dx.doi.org/10.1016/B978-0-12-385938-9.00035-3

for flows and inundation patterns. Models and experiments play important roles in quantifying these forces, particularly as direct observations of the internal workings of these flows are hampered by hazardous conditions and their opacity. Deposits contain the most valuable constraints about PDC processes, but this record remains incomplete as many processes are nondepositional and the deposit may only record the conditions in the lower flow boundary immediately prior to deposition. The ultimate goal of developing models and experiments for PDCs is to quantify and describe the forces in these currents to compliment observations made of their deposits.

Both conceptual and numerical models have developed in response to observations and understanding of the physical processes involved in turbulent (dilute) and granular (concentrated) flows. Computational approaches have benefited from improvements in technology and fundamental understanding of the physics of granular, turbulent, and fluidized flows over the past several decades. Improvements in computational approaches have broadly advanced on three fronts: (1) hardware and algorithm development, (2) improvements in understanding of physical processes operating in PDCs, and (3) advances in the description of the natural initial and boundary conditions in the eruptive environment. Hardware advances and improvements in algorithm efficiency have made it possible to improve resolution and consider larger domains, which are more indicative of natural volcanic edifices and can incorporate three-dimensional aspects of PDCs. These computational advances have enabled the exploration of emergent properties of flows such as the feedback between three-dimensional topography and flow concentration.

Understanding the nature of small-scale forces in PDCs is essential for developing fundamental and predictive models of PDCs. Some of these improvements have come from better descriptions of constitutive relationships, such as the rate of energy dissipation during the collision of particles during transport. Others have added new physical terms in the systems of equations, such as incorporating non-inertial acceleration for depth-averaged models interacting with irregular topography and **pore pressure** diffusion. Many of these improvements have come from transfer of ideas from other fields such as turbulent subgrid-scale models in atmospheric sciences and granular flow theory. Finally, improvements of the initial and boundary conditions have come from a variety of techniques from geophysical to petrological. For example, one important example for PDC transport has been the use of vastly improved digital elevation models (DEM) that can greatly improve predictions of flow transport, especially when topographic elements are of comparable size to the dominant flow features.

Experiments have improved our understanding of PDC dynamics by elucidating both the fundamental laws of grain-scale mechanics and the emergent properties of the flow at the large scale. We can roughly distinguish three scales of experiments currently being performed, micro-, meso-, and macroscale, although the boundary between these scales is not sharp. Microscale (or grain-scale) experiments are often focused on local interactions occurring at the same or nearly the same scale as in nature. Because of the similar scale, scaling relationships are most straightforward for this class of experiment. Mesoscale experiments typically scale a specific set of flow features to make PDC dynamics accessible in a laboratory setting. These experiments are aimed typically at explaining emergent behavior that results from the complex interaction between a large number of constituents in PDCs. Even with careful scaling, some parameters are difficult to access in the laboratory, and this has given rise to larger or macroscale experiments that provide an intermediate step between laboratory-scale experiments and natural phenomena.

1.1. Insight from Recent Eruptions

PDCs are produced from volcanic eruptions of many scales, from small-volume events (dome collapse or small column collapse) to caldera-forming eruptions of volumes of 100−1000 cubic kilometers of material (Figure 35.1). Large-scale PDC-forming eruptions typically have evolved magma compositions, ranging from dacitic to rhyolitic. They can produce a range of deposits, related to the depositional temperatures. Some high-silica rhyolitic magmas erupt to form completely nonwelded deposits, while others form ignimbrites that are extremely welded and are difficult to distinguish from lava flows without close inspection. For the largest eruptions, our understanding has been developed principally from studies of the sedimentological and textural characteristic of the deposits and analogy with more frequent, small-volume eruptions.

Direct, but limited observations of smaller volume eruptions (<1 km^3) have yielded key insights into PDC dynamics and have been used to constrain models. These natural examples illustrate the complexity inherent to these types of flows. We describe here three examples of how these eruptions have been used to infer PDC dynamics as an introduction to the general topic of the physics of PDCs. These examples also illustrate the utility of comparing numerical simulation results with well-constrained observations.

1.2. Mount St Helens, 1980

The eruption of Mount St Helens in 1980 profoundly changed the perception and study of PDCs. The cataclysmic eruption on May 18, 1980 started with an edifice collapse that rapidly decompressed a partially crystalline

PDC generation mechanisms

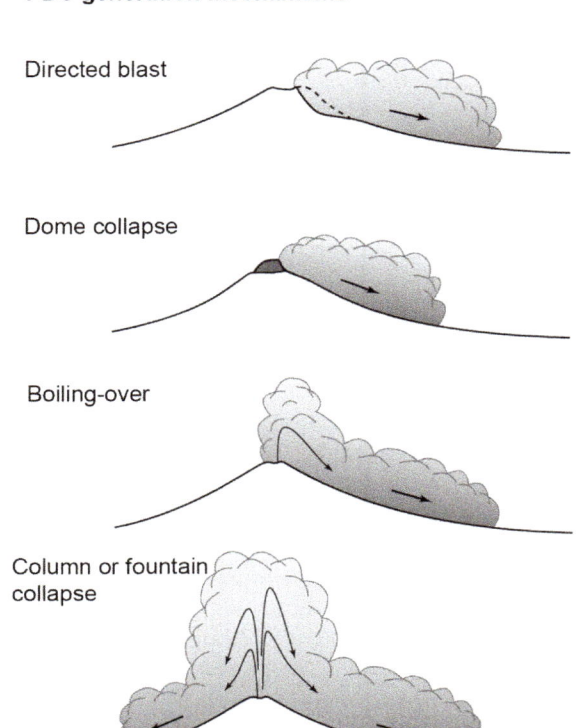

Directed blast

Dome collapse

Boiling-over

Column or fountain collapse

FIGURE 35.1 Summary of PDC generation mechanisms. PDCs are generated when the bulk density of the current exceeds that of the ambient atmosphere and can be generated by several eruption styles, including directed blasts, dome collapse, boiling-over eruptions, and column, or fountain, collapse. From boiling-over to column collapse there is continuum of behavior, but the governing physical mechanism, gravitational collapse of material ejected into the atmosphere, remains consistent.

magmatic body, generating a lateral blast. Instead of being directed vertically, the blast expanded horizontally to the north and overtook the landslide that triggered the eruption. The rapidly expanding current was initially dominated by compressibility effects (Kieffer, 1981). However, recent simulations suggest that beyond an initial burst phase much of the propagation of the current was influenced by gravity as the mixture of pulverized lithics, juvenile material, and gases failed to entrain sufficient ambient atmosphere to become buoyant. In multiphase simulations, this lateral blast-generated PDCs, much like other PDCs, had a high-energy head followed by a depositional body (Esposti Ongaro et al., 2012).

Following the lateral blast, a Plinian eruption continued throughout the day. As the eruption intensity increased, several PDCs were generated due to column collapse. Conceptually, a column collapse occurs when the fragmented mixture exiting the volcanic vent fails to entrain sufficient atmospheric gases to become buoyant. Unlike theoretical end-members where all of the column either collapses or rises buoyantly, the Mount St. Helens eruption

exhibited hybrid or unstable behavior throughout the afternoon of May 18 with initial partial collapse events restricted to the margin of the column. As eruptive flux increased, both the column height increased and more voluminous PDCs were produced. As the eruption waned, smaller PDCs filled in lows in the top of the pumice plain deposits to the north of the volcano.

PDCs at Mount St Helens exhibited several features characteristic of PDC dynamics, e.g., they had long run-out distances, but were confined and directed by topography and were at times erosional. Modulated by topography, PDC deposits thicken in valleys and depressions and become thinner or are absent over topographic highs. Like many PDCs, the ultimate run-out distance was likely impacted by both dissipation of energy in the flow and clast deposition, and also due to reversal of buoyancy. Portions of the flow became less dense than the atmosphere as the currents sedimented clasts and entrained ambient air. These secondary plumes, or co-PDC columns, are a unique feature of PDCs and illustrate the multiphase nature of these flows, as the hot gases that comprise the current only form gravity currents when they are also transporting sufficient amounts of denser volcanic ash, lapilli, and lithic clasts.

1.3. Montserrat 1995–1999

The eruption of Soufriere Hills, Montserrat produced a range of PDCs and extensive study of their deposits and real-time observations have provided a natural laboratory for the study of eruptive dynamics. Transient, short-duration Vulcanian explosions, lava dome collapses, and directed blasts at Montserrat have all produced flows encompassed by the term PDC but display a remarkable diversity of flow types and deposits. Hundreds of lava dome collapses have typically produced block and ash flows that are typified by concentrated granular flows of mostly dense, low vesicularity clasts from relatively degassed lavas. Particularly, the larger collapse events have also produced flows described as pyroclastic surges with finer particles transported in more dilute and turbulent currents. The complexity of even these smaller volume events (10^4-10^8 m^3) and the ability of these flows to transform during transport is particularly emphasized by the recognition that highly concentrated pyroclastic flows can be generated during the sedimentation of these more dilute flows (Druitt et al., 2002).

A particularly large dome collapse in December, 1997 decompressed the dome's interior and produced a lateral blast that flowed over 7 km from the source and continued to propagate for >3 km across the sea. In a manner similar to the Mount St Helens lateral blast, this blast was initiated by the large pressure discontinuity generated by the exposure of a cryptodome. Simulations of the event were consistent with the interpretation that much of the transport occurred after the initial burst phase transformed into a gravity current (Esposti Ongaro et al., 2008). Discrete, Vulcanian explosions at Montserrat also produced abundant PDCs due to fountain collapse. Fountain collapse in this instance refers to the gravitational collapse of material injected into the atmosphere in unsteady or transient explosions.

The forces that contributed to the mobility and transport of the various PDCs at Montserrat have been attributed to a variety of mechanisms. Flows from dome collapse produced abundant impact marks, crushed clasts, and polished surfaces. These features are all consistent with dense granular interaction (Grunewald et al., 2000). Gas-particle drag dominates in collapsing fountains resulting from dome collapse and has been typically inferred for those PDCs described as surges (Esposti Ongaro et al., 2008). Finally, high pore fluid pressures have been inferred in the relatively fines-rich, well-sorted, and mobile pyroclastic flows derived from rapid sedimentation of dilute currents (Druitt et al., 2002), hence reducing the interparticle drag and encouraging long run-out distances.

1.4. Tungurahua, 2006

The recent PDCs generated by eruptions at Tungurahua, Ecuador resulted from low column collapse, or boiling-over, transporting breadcrust bombs, lithic clasts, and ash down the slopes of the volcano (Hall et al., 2013). The boiling-over mechanism has intermediate energy between dome failure and column collapse. Boiling-over occurs when vesiculated magma fountains over the crater rim without forming a convective plume. The boiling-over eruptive style at Tungurahua produces a characteristic type of PDC deposit, which is rich in large breadcrust bombs. High on the slopes of Tungurahua, these breadcrust bombs often form levées and other depositional features usually interpreted to come from concentrated granular flows. The 1—2 m tall levées, oriented parallel to the flow direction and enriched in bombs formed, as the concentrated flow descended the slopes and frictional forces were high at their margins.

These concentrated flows are typically confined to preexisting fluvial channels high on the slopes of the volcano. However, at breaks in slope and channel margins, cross-bedded dunes and deposits of bedded, finer grained material are often observed (Douillet et al., 2013). This provides another example of PDCs transforming in depositional regime, although in this case the PDCs are transformed from concentrated pyroclastic flows to pyroclastic surges. Studies of vesiculated rind formation of the juvenile clasts (Benage et al., 2014), remnant magnetization orientation of both the juvenile and lithic clasts, and variable charring of organic material, all indicate that these flows

are highly heterogenous in temperature with the bombs remaining hot (>540 °C) and the entrained lithics and ash are deposited at much cooler temperatures (<100 °C) suggesting that these flows are efficient at entraining ambient air.

2. PDCS: ENCOMPASSING A RANGE OF PARTICLE CONCENTRATION

The term *pyroclastic density current* has been adopted to refer generically to gravity currents composed of gas and pyroclasts, and encompasses phenomena categorized in the past as *pyroclastic surges* and *pyroclastic flows*. The term *pyroclastic surge* was previously used to typically denote low particle concentration (<1% by volume), where the particles are mostly transported in a turbulent suspension. Inferred deposits from surges have bedforms and stratified deposits, and generally are less confined by topography. However, dilute currents have been interpreted to form massive deposits due to the rapidity of deposition, and a dilute flow need not be strictly linked to stratified deposits. The term, *pyroclastic flow*, in the past has been used to refer to dense particle mixtures that typically produce massive (subtle to no structure) deposits that tend to thicken in topographic lows. The use of the term, pyroclastic density current, is now preferred to refer to all these flows as individual currents can have behavior that was previously attributed to pyroclastic surges and pyroclastic flows. The nature of deposit formation from PDCs has been controversial, in part because of the lack of direct observation. Proposed mechanisms include *en masse* deposition where the dense flow arrests abruptly and the structure of the deposit reflects the structure of the flow during propagation. Conversely, *progressive aggradation* interpretations hold that particles at the base of the flow sediment gradually, and the vertical sections in a deposit reflect the temporal variation at the base of the current as it passes that particular location. Importantly, progressive aggradation implies that the conditions being recorded at the time of deposition reflect conditions in only a small portion of the flow, namely a flow boundary zone immediately above the substrate (Branney and Kokelaar, 2002).

Models and experiments have shown that single PDCs likely have a range of particle concentrations and single flows may exhibit a range of concentrations both in space and time (Burgisser and Bergantz, 2002; Valentine, 1987). The forces on the individual particles that comprise a PDC are in large part determined by the concentration of the particles. In dilute conditions, the dominant force encountered by particles is particle-gas drag. At higher concentrations of particles, interparticle collisions can become important for transferring momentum. These collisions are inelastic, and during each collision some kinetic energy is lost to deformation, breaking, thermal energy, and acoustic energy. At very high concentrations of particles, friction or prolonged contact can also become important in dissipating current energy. However, even in highly concentrated currents, the drag force between the gas and the particle does not always become negligible. When high amounts of fine ash particles create a low permeability matrix, high pore pressure can develop that can partially or fully support the particles weight and reduce frictional forces, creating a fluidized flow.

2.1. Dilute Currents

Low particle concentration is prevalent in the upper portions of PDCs and low particle concentrations can persist toward the base. To examine the transport of these particles, we can consider a simplified equation of motion for a single particle that experiences forces due to interaction with both the gas and other particles. In Eqn (35.1), the change in velocity, or acceleration, is modulated by drag forces (F_{drag}) between the gas and particle per mass of the particle (i.e. acceleration generated by the drag force), forces associated with particle–particle interaction (collisions) per particle mass (F_s), and gravity (F_{grav}),

$$\frac{du_p}{dt} = F_{drag} + F_s + F_{grav} \quad (35.1)$$

Here, and below, the subscript p refers to particles and g refers to the gas phase, and u are the velocity vectors. The momentum exchanged between the gas and particle depends on the drag coefficient that in detail depends on the shape, roughness, and local differential velocity between the particle and gas. This last quantity is encapsulated in a particle Reynolds number, defined as, $Re_p = \frac{\rho_g |u_p - u_g| d}{\eta}$, where ρ_g is the gas density, u_p is the particle velocity, u_g is the gas velocity, d is the particle diameter, and η is the dynamic viscosity of the gas. The Reynolds number describes the relative contribution of fluid inertia (in this case at the scale of the particle) with viscous forcing. The drag per unit mass is

$$F_{drag} = \frac{f}{\tau_p}(u_p - u_g) \quad (35.2)$$

where f is an empirical correction to the drag coefficient sensitive to particle Reynolds number and to particle shape, and

$$\tau_p = \frac{(\rho_p - \rho_g)d^2}{18\eta}, \quad (35.3)$$

is the particle drag response time, based on Stokes drag.

In this way, small particles, with relatively short drag-response times, experience relatively large drag forces and are well coupled to the gas flow. If the particle

response time is short compared to the timescale of fluid flow (τ_f), the particle will accelerate to match the fluid velocity and will closely follow fluid streamlines. A comparison between the fluid timescale, such as the overturn time of an eddy, and the particle drag response time is encapsulated in a nondimensional parameter called a Stokes number,

$$St = \frac{\tau_p}{\tau_f}. \quad (35.4)$$

As the Stokes numbers goes to zero, the particle approximates a perfect tracer of fluid motion (Figure 35.2). When Stokes numbers are nearly unity, particles tend to concentrate at the margins of eddies or are demixed from the average flow. For large Stokes numbers, particles are insensitive to the motion of fluid structures and particle inertia dominates. An example of this includes ballistic particles ejected from volcanic edifices that are influenced only slightly by the surrounding wind field.

Likewise, the force transmitted from a collection of particles colliding with a single particle will depend on the product of the collisional frequency and the force imparted by individual impacts, which depends on the velocity distribution of the particles. While the form of this distribution can be complex, the timescale between collisions (τ_c) is proportional to the inverse of the volume fraction of particles and to the variance of the velocity distribution of the particles. Using this timescale, a collisional Stokes number is defined that ratios the average timescale between particle collisions and the fluid timescale,

$$St_c = \frac{\tau_c}{\tau_f}. \quad (35.5)$$

The collisional Stokes number is large when the timescale between particle collisions is long relative to variation in fluid motion. The ratio between collisional timescale and the particle drag response timescale gives a measure of the transition from dense to dilute flows in PDCs (Burgisser and Bergantz, 2002),

$$D_D = \frac{\tau_c}{\tau_p}. \quad (35.6)$$

If $D_D > 1$, the time between collisions is large relative to the response of the particle to fluid drag and the particles will be influenced mostly by fluid motion, which characterizes the dilute regime. If $D_D < 1$ then the collisional timescale is short relative to particle-fluid drag and the particle motion will be mostly influenced by particle collisions. Depositional regions of the current with $St \sim 1$ and $D_D > 1$ produce bedforms consistent with deposits previously described as surges (Figure 35.3), whereas depositional regions of the current where $D_D < 1$ likely produce massive deposits usually interpreted to come from pyroclastic flows (Burgisser and Bergantz, 2002).

FIGURE 35.2 Particle sorting in eddies produced by the wake behind a blunt object. Particles are seeded in the lee of the object. In both cases, the flow field is identical, but the particle Stokes number varies. (A) St >> 1 particles are relatively impervious to fluid motion, while (B) near unity Stokes number particles are concentrated at the margin of eddies.

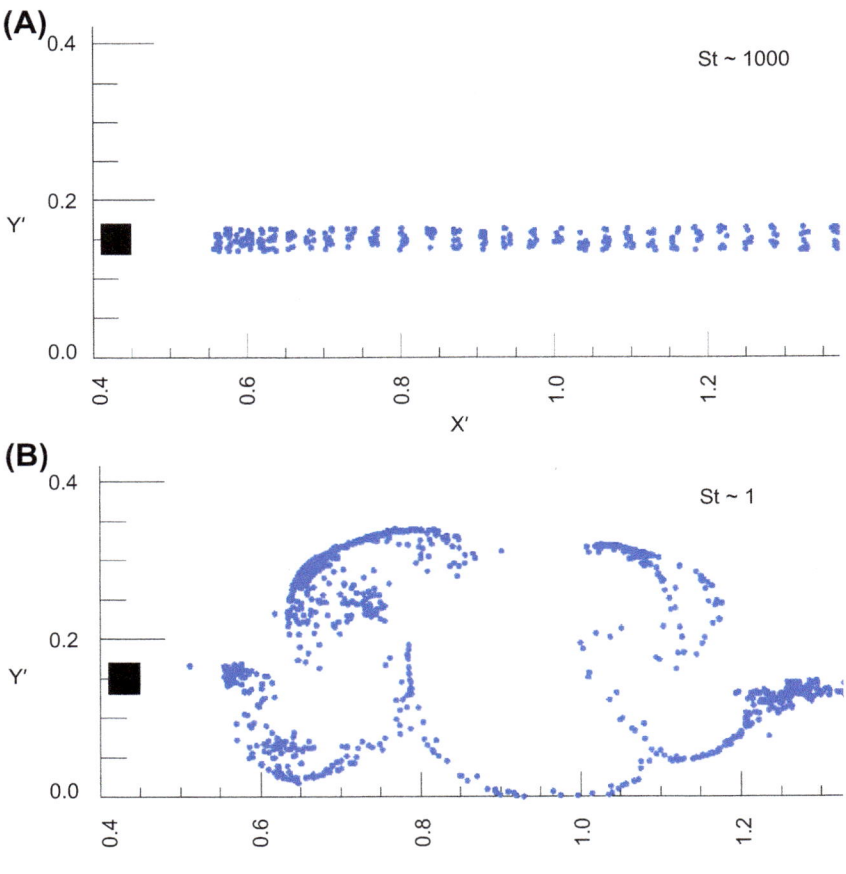

FIGURE 35.3 Idealized channel cross-section for a PDC showing both dilute and concentrated flow. Deposition in dilute $(D_D > 1)$ and low Stokes number flows generates stratified deposits. Deposition in concentrated $(D_D < 1)$ flow generates poorly sorted deposits. *Modified from Burgisser and Bergantz (2002).*

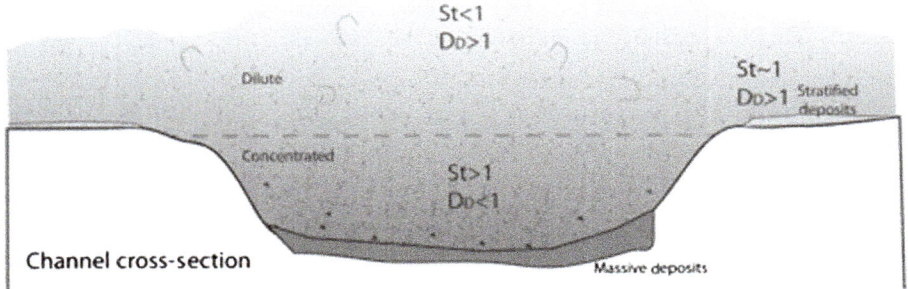

2.2. Concentrated Currents

The basal part of PDCs can often have high particle concentration, reaching values close to that of a deposit. The detailed current dynamics can also be described by dimensionless numbers that account for the ratio of the solid or fluid stresses. In particular, the Bagnold number is defined as the ratio of collisional (inertial) solid stress over fluid viscous shear stress,

$$Ba = \frac{\varepsilon_s \rho_s d^2 \gamma^2}{(1 - \varepsilon_s)\eta\gamma}, \tag{35.7}$$

where ε_s is the solid volume fraction, γ is the shear rate, and η is the dynamic viscosity of the fluid. The Darcy number is defined as the ratio of the approximate fluid drag in a porous media over collisional solid stress, and is given as

$$Da = \frac{\eta\gamma/k}{\varepsilon_s\rho\gamma^2}. \tag{35.8}$$

Here k is the approximate hydraulic permeability. The Darcy number describes the tendency for interstitial pore fluid pressure to buffer particle interactions. The Savage number is defined as the ratio of collisional over frictional solid stresses, which has physical meaning when pore fluid pressure is negligible.

$$Sa = \frac{\rho_s d^2\gamma^2}{(\rho_s - \rho_f)gh}, \tag{35.9}$$

where h is the current depth. Considering mean particle sizes of $\sim 50-500$ μm, representative of the ash matrix of most PDCs, and assuming a volume fraction of particles greater than ~ 0.5 that may be found at the base of the PDC, these dimensionless numbers are in the range $Ba = 10^0-10^2$, $Da = 10^1-10^3$, and $Sa = 10^{-9}-10^{-7}$. That is, in this dense region, the Bagnold number indicates collisional stresses are equivalent to or larger than the fluid viscous stresses. These Darcy number values indicate that pore pressure can dampen particle interactions, and the low Savage numbers in this regime indicate that very close to the close packing limit of particles, frictional interaction will be larger than collisional interactions provided pore fluid pressure is negligible. These values indicate that in this concentrated regime, the interstitial gas may be important in controlling the flow dynamics, notably through the development of pore fluid pressure. When pore fluid pressure is small, solid frictional shear dominates collisions however high pore pressure dampens friction.

Dynamic pore pressure in PDCs arises because of fluid-particle drag as the interstitial gas flows upward and/or particles settle (Figure 35.4). Pore pressure increases with the relative gas-particle velocity and the granular mixture is fluidized if the weight of the particles is fully supported ($P = \rho gh$), which occurs at

$$U_{mf} = \frac{k}{\eta}\frac{P_{mf}}{h}, \tag{35.10}$$

where U_{mf} is the minimum **fluidization** velocity, and P_{mf} is the pore fluid pressure. The small size of the particles constituting the ash matrix of PDCs can reduce permeability down to $k\sim10^{-12}-10^{-11}$ m^2 and U_{mf} as low as ~ 1 mm s^{-1}. Negligible solid friction caused by high pore pressure favors propagation of the concentrated mixture. This is illustrated by analog laboratory experiments showing that granular flows at nearly maximum particle volume fraction with high pore fluid pressure propagate like inertial single-phase fluid flows (Figure 35.4). Pore pressure also favors buoyancy effects in PDCs whose particles can segregate according to their density.

Once pore pressure is generated, it may decrease through pore pressure diffusion so that it does not disappear instantaneously even if there is no longer differential gas-particle motion. Scaling indicates that slow pressure diffusion, favored by low material permeability ($k < 10^{-12}-10^{-11}$ m^2) and large flow thickness ($h > 10$ m), may cause long-lived (typically several minutes) high pore pressure in concentrated PDCs. On the other hand, pore pressure diffuses within only a few seconds in relatively thin ($h < 1$ m) flows with high material permeability ($k > 10^{-10}-10^{-9}$ m^2). When pore pressure is negligible in the concentrated flow, the dynamics are controlled by solid friction as suggested by low values of Sa.

3. MORPHOLOGY AND INTERNAL DYNAMICS

Gradients in the particle volume fraction develop due to the interaction of gravity and the turbulent suspension of different size particles, often resulting in more particle-enriched bases as the particle-laden flows continue to propagate away from the source. The development of gradients in particle volume fraction has important implications for air entrainment and for the development of basal bed-load regions. Detailed measurements of these concentration gradients during transport are not currently possible, but observations and models provide some insight

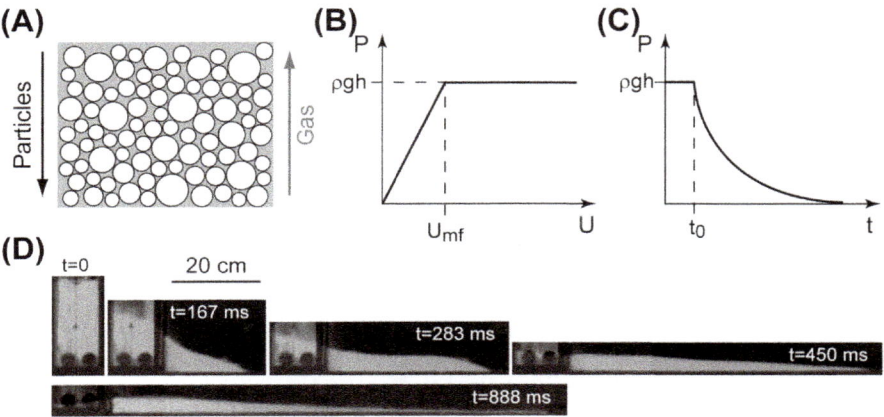

FIGURE 35.4 (A) Relative gas-particle motion as interstitial gas flows upward and/or particles settle. (B) Pore fluid pressure (P) as a function of the relative gas-particle velocity (U), with full bed support (fluidization) at $U > U_{mf}$. (C) Pore pressure diffusion from time t_0. (D) Dense air-particle flow generated from a fluidized granular column released at $t = 0$ and in which pore pressure decreases through diffusion. High pore pressure at $t < 450$ ms damps interparticle friction and causes inertial flow regime. High friction as pore pressure has become negligible at later stages causes flow deceleration and deposition by aggradation ($t = 888$ ms). *Modified from Roche et al. (2008).*

into this structure. In channelized currents, the more concentrated lower portions of the flow will typically follow the channel structure, whereas the more dilute upper portions can form stratified deposits at the flow margin or avulse the channel creating two independent currents (Figure 35.3).

The transport of PDCs over water provides another set of natural boundary conditions to understand what portion of a PDC comes in contact with the bed versus that which is primarily suspended until deposition. A number of interactions may occur at a PDC—water interface as revealed by analog experiments (Freundt, 2003) and numerical simulations (Dufek and Bergantz, 2007). Among these effects, the loss of particles through the water surface, i.e., a leaky boundary, can significantly impact the concentration and run-out distance of a PDC. In the case of the Kos Plateau Tuff, a large rhyolitic eruption in the eastern Aegean, PDCs that travelled over land before deposition transported larger and more abundant lithic clasts than

flows that crossed large expanses of water (Allen and Cas, 2001).

Numerical simulations and analog experiments give some insight into particle concentration stratification in these currents where particles are easily lost from the lower boundary (Freundt, 2003). In Figure 35.5, simulations of two flows with identical initial conditions are shown, where only the bottom boundary condition has been changed. The overland currents have a saltation boundary condition where some energy is lost in every particle-boundary collision, and for overwater conditions all particles that reach the bottom boundary are removed (leaky boundary). Currents with the leaky boundary are more dilute, travel a shorter distance before forming co-PDC plumes, and are finer grained. Currents with the saltating boundary are coarser, develop a concentrated bed-load region, and can, in some cases, even concentrate large, dense lithic clasts into narrow bands as a result of Stokes number sorting in the head of the current. Both

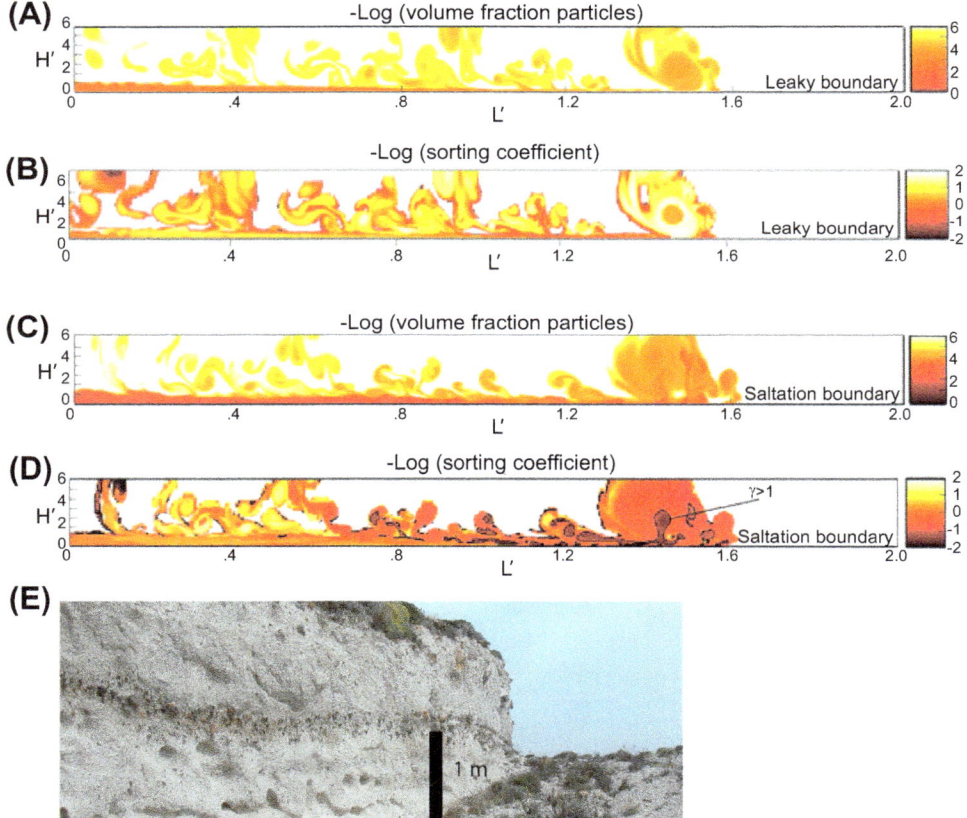

FIGURE 35.5 Example multiphase simulations of the propagation of a PDC with leaky and saltation boundary conditions. All particles that reach the base of the leaky boundary flows are removed from the domain. These simulations included two particles sizes, one with $St \ll 1$ and other with $St > 1$. A sorting coefficient gives the ratio of large St particles to small St in the flow, relative to their initial concentration. (A) Volume fraction of particles in a leaky boundary case (log scale), (B) Sorting coefficient in the leaky flow case (log scale), (C) Volume fraction of particles in the saltation case (log scale), (D) Sorting coefficient in the saltation case, (E) Picture of concentration of lithics in the Kos Plateau Tuff. Saltation boundary flows are more concentrated than the leaky boundary case, develop a bed-load region, and also concentrate larger St particles in the core of the PDC head, and shed these particles in horizontal lithic sheets similar to those seen in deposits of many ignimbrites. *Modified from Dufek and Bergantz (2007).*

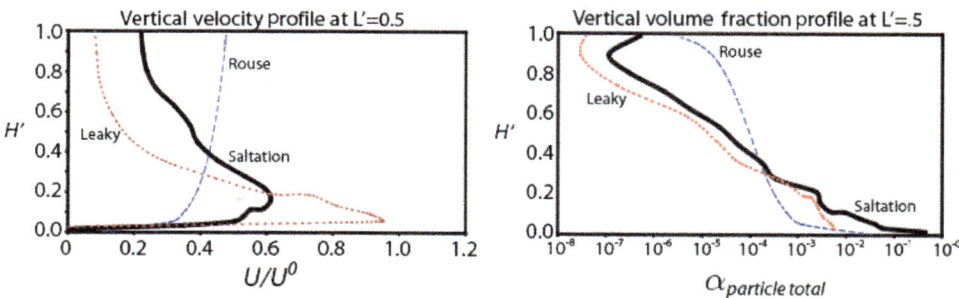

FIGURE 35.6   Vertical velocity and volume fraction profiles for the case of a saltating PDC and a PDC with a leaky boundary. The saltating PDC develops a bed-load region at the base. Note that the peak velocities in the current occur between 10% and 30% of the height of the current. For reference, an advection–diffusion profile of ash (Rouse approach) is shown assuming an open-channel velocity profile (Valentine, 1987).

simulations also illustrate the intense mixing processes occurring at the front and at the top of the PDC. The average vertical concentration and velocity profiles of the saltation currents also indicate a very narrow but concentrated bed-load region (Figure 35.6).

3.1. Buoyancy Reversal

PDCs are driven by their density difference with respect to the atmosphere, resulting from the sum of the bulk density of the interstitial gas and that of pyroclasts transported by the flow. At typical PDC temperatures (300°–700 °C), the interstitial gas density is lower than that of atmosphere. Therefore, when the solid fraction decreases below a critical value, current density becomes lower than atmospheric and horizontal motion stops. Co-PDC plumes eventually form, which convect a dilute gas–pyroclast mixture into the atmosphere.

This process can be described by the *buoyancy* force, B, acting on the unit volume of the current, where

$$B = (\rho_g - \rho_c)g, \qquad (35.11)$$

where the subscripts g refers to gas, and c refers to current. Current motion ceases when buoyancy becomes greater than or equal to zero in Eqn (35.11). Expressed in terms of

the volumetric concentration, ε, and the specific density ρ of its gaseous and solid components, the current density ρ_c is

$$\rho_c = \varepsilon_g \rho_g(T) + \sum_{p=1}^{N} \varepsilon_p \rho_p \qquad (35.12)$$

where the subscripts g and p indicate the gas and dispersed solids, respectively, and T is gas temperature. Given that ρ_s (the density of individual particles) are constant and $\varepsilon_g = 1 - \sum_{p=1}^{N} \varepsilon_p$, current buoyancy depends only on T and particle volume fraction. Since $\frac{\rho_s}{\rho_g} \sim 10^3$, buoyancy reversal may occur below a critical particle concentration $\varepsilon_s \sim 10^{-3}$. Two main mechanisms control the spatial and temporal evolution of particle concentration: sedimentation and turbulent mixing (Figure 35.7).

3.2. Sedimentation

Sedimentation is the main process by which particle concentration progressively reduces until a PDC eventually stops by buoyancy reversal. As a result of particle sedimentation, an idealized current has a stratified transport system with a concentrated basal layer (formed by the continuous sedimentation of particles), a settling zone in which the current is at its initial concentration (in polydisperse currents this region is also layered) and an upper

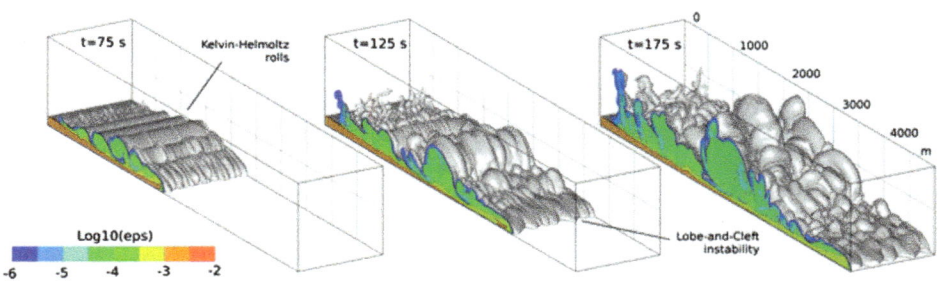

FIGURE 35.7   Three-dimensional numerical simulation of a pyroclastic density currents. The current is fed by a constant inlet (on the left boundary) of 100 m thickness. Initial velocity is 25 m/s, with initial mixture density equal to 3.88 kg/m³, with two particle classes of (30 μm; 2500 kg/m³) and (500 μm; 1000 kg/m³) in equal proportion and a temperature of 300°C.

dilute zone (Figure 35.3). While buoyancy in the intermediate settling zone remains negative, buoyancy reversal occurs initially only in the upper layer, which forms a turbulent convective ash cloud above the PDC (Figure 35.7).

The settling zone progressively thins while particles are gradually conveyed into the basal layer and it finally disappears, leaving a dense undercurrent (or a deposit) overlaid by a hot, dilute, turbulent suspension. At this stage, the current lifts off forming a convective co-PDC plume. In dilute currents ($\varepsilon_s < 10^{-3}$), where particle–particle interactions are negligible, settling velocity in the current body is solely determined by the balance between gravity and gas-particle drag.

3.3. Turbulent mixing

Turbulence is the mechanism by which a gravity current entrains atmospheric air, thus decreasing its bulk density. In PDCs, the entrained air (originally at density $\rho_{atm} > \rho_g$) is rapidly heated by solid particles, and its subsequent adiabatic expansion produces an increase of gas volume and the subsequent decrease of particle concentration and current bulk density, eventually leading to buoyancy reversal (Neri et al., 2002).

Turbulent mixing and entrainment are complex nonlinear phenomena whose dynamical aspects are not understood fully. Although turbulent mixing is a multiscale phenomenon, with relevant spatial scales extending down to the size of a single particle, the rate of entrainment of atmospheric air is imposed by large-scale eddy motions. Laboratory experiments and numerical simulations identify two main large-scale instabilities leading to the onset of turbulence:

1. **Kelvin–Helmoltz (KH)** shear instabilities are generated at the upper interface between the current and the atmosphere and are due to the sharp vertical gradient of the horizontal velocity overcoming the density stratification. This produces vortex rolls whose axis is orthogonal to the propagation direction. Through this instability and the subsequent development of shear turbulence, the interface between the current and the atmosphere is progressively blurred (Figure 35.7). On the upper layer of the current, particle concentration may fall below the critical value and part of the current may lift off forming a convective ash cloud.

2. **Lobe-and-Cleft (LC) instabilities** are generated at the front of a dilute current and are associated with the engulfment of air from the advanced head, producing an unstable region of reversed buoyancy at the base of the current. This is recognized as a key phenomenon in PDC dynamics as it can fluidize the basal, concentrated particle layer and increase the mobility of the current head.

4. BED-LOAD, EROSION, AND DEPOSITION

Complex dynamics at the base of PDCs result from the competition between deposition from the current and erosion of the substrate. Similar to other dynamics in PDCs, the physics of this interaction is mainly dependent on the particle concentration and related flow regime. The observation of abundant erosionally entrained clasts, self-channelization of PDCs into underlying deposits, and unconformities where PDCs have eroded portions of the substrate, all indicate that, while poorly understood, erosion may play an important role in PDC dynamics. Particularly for steep edifices, erosion may play a role in bulking the current and prolonging run-out similar to other granular flows.

4.1. Dilute Regime: The Bed-Load

Bed-load formation is controlled by the rate of sedimentation from the upper regions of the current into a relatively thin basal concentrated region where particle–particle interactions dominate. As shown in Figure 35.6, most of the mass and momentum of the flow can accumulate in basal regions. In many cases, the bed-load is not depositional and its development gives rise to larger average shear stresses at the base of the current. Although conditions for erosion are likely highly variable and are, to date, poorly constrained for PDC substrates, these high shear stresses may provide one mechanism to erode material from the substrate.

4.2. Erosion and Deposition in Concentrated Currents

Dense PDCs commonly erode the substrate over which they propagate. Erosion of rigid coherent substrates is shown by friction marks, suggesting high basal shear stress but limited entrainment. In contrast, a large variety of field observations indicate that a significant amount of granular substrates formed by various types of earlier geological events can be eroded by PDCs, even on horizontal slopes. Direct evidence of material entrainment include substrate-derived blocks found in PDC deposits several decimeters above their base (Figures 35.8 and 35.9) (Roche et al., 2013). Though the origin of blocks of igneous rocks may be arguable since these may have derived from the volcanic vent and may have been transported by the current, blocks from fluvial sediments (for instance) found at distances of tens of kilometers from the vent demonstrate that PDCs can erode and incorporate a granular substrate (Figure 35.9). Indirect evidence of substrate entrainment is provided by underlying pyroclastic fall deposits being thinner than expected according to reconstructed thickness

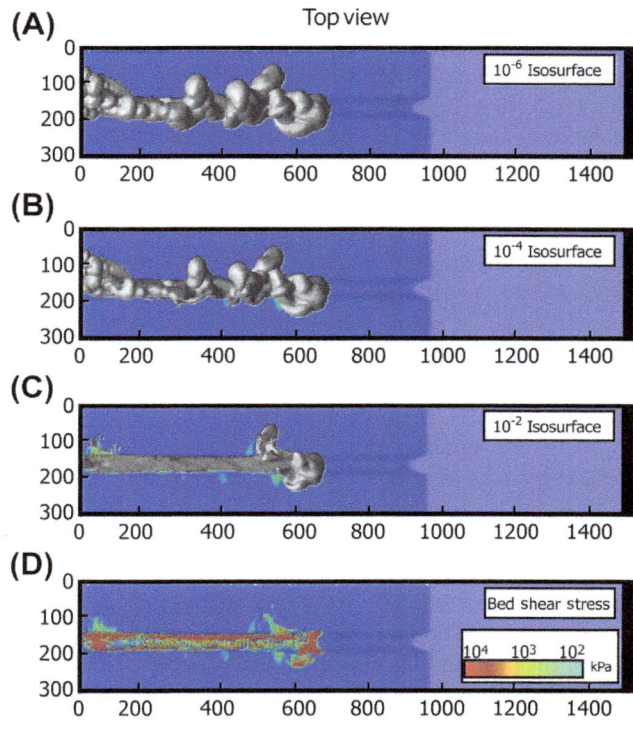

(A) Top view

(B)

(C)

(D)

(E)

FIGURE 35.8 (A–D) Top view of a simulated PDC descending down an idealized channel. (A–C) depict different isosurfaces of particle volume fraction, becoming progressively more concentrated. Most of the concentrated portion of the flow is confined to the channel. (D) shows the bed shear stresses from this continuum calculation. (E) Force chains in an experimental granular flow using a photoelastic technique to visualize stresses. These force chains create transient bed force excursions much greater than predicted by continuum approaches. *Image courtesy J. Estep.*

FIGURE 35.9 (A) PDC deposit (unit 2) with base enriched in blocks entrained from debris avalanche deposit *(Mount St-Helens, WA, photograph B. Brand).* (B) PDC deposit with blocks entrained from fluvial sediments (b) and resting on an undisturbed layer of ash (a); S: fine fluvial sediments *(Peach Spring Tuff, AZ, photograph G. Valentine).*

estimated from isopach data, suggesting that the material missing was eroded by later PDCs.

Many ignimbrites reveal an apparent paradox because underlying substrates of fine particles are often not eroded, whereas substrates of coarser particles (typically coarse-grained fall deposits, lava flow surfaces, debris avalanche deposits, or fluvial conglomerates) are commonly entrained (Figure 35.9). In fact, flat substrates of fine particles have a relatively smooth surface that cause low basal shear

stresses, which does not favor entrainment. In contrast, stresses caused by rough coarse-grained substrates may be sufficiently high to drag particles as large as blocks at the flow base. Considering the mixture of gas and fine ash as a single-phase fluid that creates aerodynamic drag, high dynamic pressures

$$P_d = \frac{1}{2}\rho U^2 \qquad (35.13)$$

exerted on the frontal face of blocks can cause displacement. Analog laboratory experiments suggest that particles set in motion can be uplifted due to underpressure generated at the current base (Roche et al., 2013).

Deposition from dense PDCs shifts the static–flow interface upward (i.e., aggradation). Laboratory experiments show that this occurs progressively up to the top of the current irrespective of the amount of interstitial pore fluid pressure, grain size of the particles, and slope angle (below the repose angle). The aggradation rate of an initially expanded current is equivalent to that of a static bed of the same material, but factors controlling the onset of deposition (irrespective of the material expansion) and rate of aggradation of nonexpanded currents are still not understood. These parameters are critical because they control the timing and duration of the aggradation process and hence the sedimentological characteristics of the deposit regarding possible compositional changes at source. The complex architecture of large-volume PDC deposits such as ignimbrites must be considered in light of both accumulation of successive flow units and variation of eruptive mass flux from waning to waxing phases during long-lived eruptions (Brown and Branney, 2004; Sparks and Wilson, 1976).

Although most approaches for examining bed stresses in PDCs use continuum approaches, which average the interactions between the bed and particles over larger length scales, recent experiments and discrete element simulations have indicated that the discrete nature of the granular portion of the bed-load and concentrated currents may create large heterogeneities in stresses not predicted from continuum methods. These stresses are transmitted along lineaments, or force chains, and can communicate stresses to the bed (Figure 35.8). These transient stresses can vastly exceed the depth-averaged stresses (in some cases by over an order of magnitude), and the stress excursions may enhance erosion, although much work remains to describe these forces for realistic flows and substrates (Estep and Dufek, 2012).

5. MODELS OF PDCS

Numerous conceptual and numerical models of PDC transport have developed over the past several decades. While there is a large diversity of models, they mainly fall into three categories (with some models incorporating

aspects of multiple types): (1) dilute, box model approaches, (2) depth-averaged, concentrated flow models, and (3) multiphase models (Figure 35.10).

The dilute, box model approach was one of the first detailed quantitative approaches to the transport of PDCs. At the core of this approach is the assumption of a turbulent, well-mixed current. These models assume conservation of mass and volume and describe the thickness and the kinematics of the current head by assuming a simple scaling relationship (Dade and Huppert, 1995). Other one-dimensional approaches to dilute currents have included conservation of thermal energy, entrainment, and sedimentation (Bursik and Woods, 1996). These approaches are mostly applicable to dilute currents and provide information on flow run-out and the propensity for such flows to remain denser than the atmosphere before forming co-PDC plumes. This approach has the benefit of being computationally and analytically tractable, and due to the well-mixed assumption is most applicable to dilute currents where particle concentration gradients may have less impact.

Concentrated flow, depth-averaged approaches focus on flows that are near to close-packed. Most of these models compute the local height of the flow and solve for bed shear stresses assuming frictional interaction between the flow and the substrate, but variants have also been modified to account for pore pressure. Internal variations of the flow are

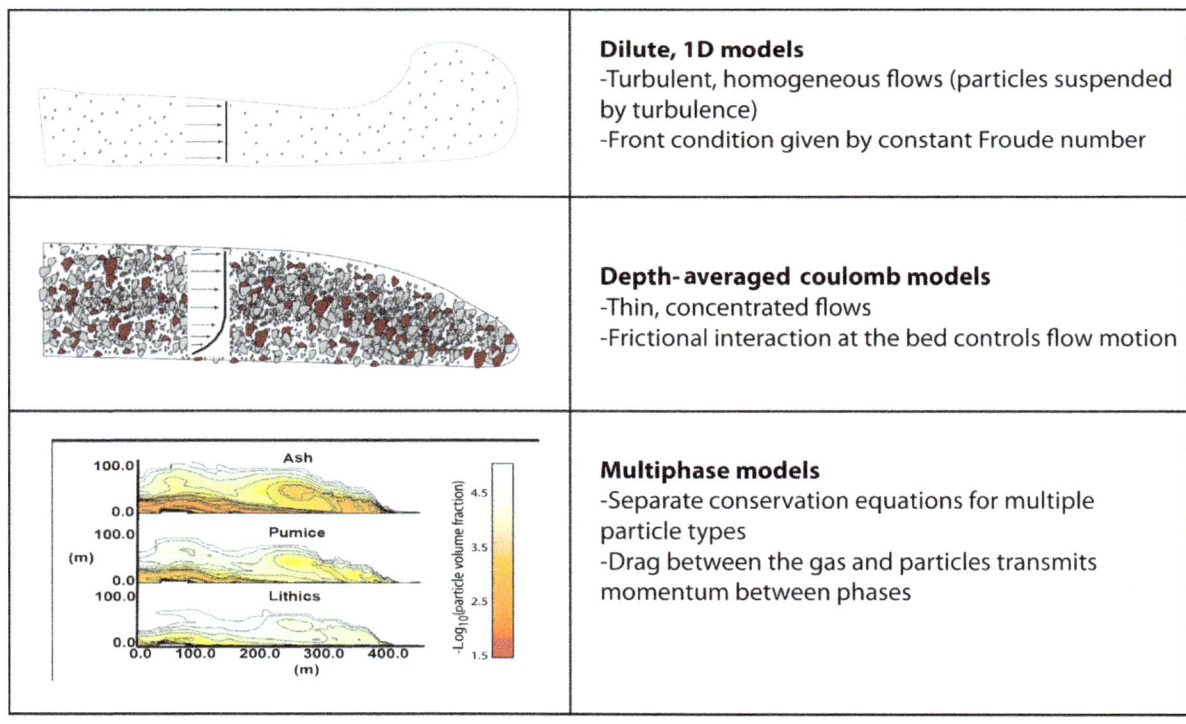

Dilute, 1D models
-Turbulent, homogeneous flows (particles suspended by turbulence)
-Front condition given by constant Froude number

Depth-averaged coulomb models
-Thin, concentrated flows
-Frictional interaction at the bed controls flow motion

Multiphase models
-Separate conservation equations for multiple particle types
-Drag between the gas and particles transmits momentum between phases

FIGURE 35.10 Simplified classification of PDC models.

usually not accounted for, and in this way flows over complex terrain can be solved in pseudo-2D domains efficiently. These simulations have proven adept at calculating the run-out of the dense portions of a flow, and have put renewed focus on the detailed description of topography as such DEM have proven necessary for making robust predictions of flow path.

Multiphase models solve for separate dynamics equations for different particle phases and are thus useful in examining the sorting of different sizes and densities of particles during PDC transport. Variants of this approach have been applied to 2D and 3D problems and have included different force models to account for turbulent interaction, collisions between particles, and frictional interaction. This approach is useful for looking at developing particle stratification in currents, and energy budgets of these currents, although the large computational demands have thus far placed limits on widespread use (Figure 35.10).

FURTHER READING

Allen, S.R., Cas, R.A.F., 2001. Transport of pyroclastic flows across the sea during explosive, rhyolitic eruption of the Kos Plateau Tuff, Greece. Bull. Volcanol. 62, 441–456.

Benage, M., Dufek, J., Degruyter, W., Geist, D., Harpp, K., Rader, E., 2014. Tying textures of breadcrust bombs to their transport regime and cooling history. J. Volcanol. Geotherm. Res. 274, 92–107.

Branney, M.J., Kokelaar, P., 2002. Pyroclastic Density Currents and the Sedimentation of Ignimbrites. The Geological Society, London.

Brown, R.J., Branney, M.J., 2004. Bypassing and diachronous deposition from density currents: evidence from a giant regressive bed form in the Poris ignimbrite, Tenerife, Canary Islands. Geology 32 (5), 445–448.

Burgisser, A., Bergantz, G.W., 2002. Reconciling pyroclastic flow and surge: the multiphase physics of pyroclastic density currents. Earth Planet. Sci. Lett. 202, 405–418.

Bursik, M., Woods, A., 1996. The dynamics and thermodynamics of large ash flows. Bull. Volcanol. 58, 175–193.

Dade, W.B., Huppert, H.E., 1995. Runout and fine-sediment deposits of axisymmetric turbidity currents. J. Geophys. Res. 100, 18597–18609.

Douillet, G., Pacheco, D., Kueppers, U., Letort, J., Tsang-Hin-Sun, E., Bustillos, J., Hall, M.L., Ramon, P., Dingwell, D.B., 2013. Dune bedforms produced by dilute pyroclastic density currents from the August 2006 eruption of Tungurahua volcano, Ecuador. Bull. Volcanol. 75 (11).

Druitt, T.H., Calder, E.S., Cole, P.D., Hoblitt, R., Loughlin, S., Norton, G., Ritchie, L., Sparks, S., Voight, B., 2002. Small-volume, highly mobile pyroclastic flows formed by rapid sedimentation from pyroclastic surges at Soufriere Hills Volcano, Montserrat: an important volcanic hazard. Geological Soc. Lond. Memoirs 21, 263–279.

Dufek, J., Bergantz, G.W., 2007. Suspended load and bed-load transport of particle-laden gravity currents: the role of particle–bed interaction. Theor. Comput. Fluid Dyn. 21 (2), 119–145.

Esposti Ongaro, T., Clarke, A.B., Neri, A., Voight, B., Widiwijayanti, C., 2008. Fluid dynamics of the 1997 Boxing Day volcanic blast on Montserrat, West Indies. J. Geophys. Res. 113 (B03211).

Esposti Ongaro, T., Clarke, A.B., Voight, B., Neri, A., Widiwijayanti, C., 2012. Multiphase flow dynamics of pyroclastic density currents during the May 18, 1980 lateral blast of Mount St. Helens. J. Geophys. Res. 117 (B06208).

Estep, J., Dufek, J., 2012. Substrate effects from force chain dynamics in dense granular flows. J. Geophys. Res. Earth Surf. 117 (F1), F01028.

Freundt, A., 2003. Entrance of hot pyroclastic flows into the sea:experimental observation. Bull. Volcanol. 65, 144–164.

Grunewald, U., Sparks, S., Kearns, S., Komorowski, J., 2000. Friction marks on blocks from pyroclastic flows at the Soufriere Hills volcano, Montserrat: Implications for flow mechanisms. Geology 28 (9), 827–830.

Hall, M.L., Steele, A.L., Mothes, P.A., Ruiz, M.C., 2013. Pyroclastic density currents (PDC) of the 16-17 August 2006 eruptions of Tungurahua volcano, Ecuador: geophysical registry and characteristics. J. Volcanol. Geotherm. Res. 265, 78–93.

Kieffer, S.W., 1981. Blast dynamics at Mt. St. Helens on 18 may 1980. Nature 291, 568–570.

Neri, A., Di Muro, A., Rosi, M., 2002. Mass partition during collapsing and transitional columns by using numerical simulations. J. Volcanol. Geotherm. Res. 115, 1–18.

Roche, O., Montserrat, S., Niño, Y., Tamburrino, A., 2008. Experimental observations of water-like behavior of initially fluidized, unsteady dense granular flows and their relevance for the propagation of pyroclastic flows. J. Geophys. Res. 113, B12203.

Roche, O., Nino, Y., Mangeney, A., Brand, B., Pollock, N., Valentine, G.A., 2013. Dynamic pore-pressure variations induce substrate erosion by pyroclastic flows. Geology 41 (10), 1107–1110.

Sparks, R.S.J., Wilson, L., 1976. A model for the formation of ignimbrite by gravitational column collapse. J. Geological Soc. Lond. 132, 441–452.

Valentine, G.A., 1987. Stratified flow in pyroclastic surges. Bull. Volcanol. 49, 616–630.

Chapter 36

Deposits of Pyroclastic Density Currents

Richard J. Brown

Department of Earth Sciences, Durham University, UK

Graham D. M. Andrews

California State University Bakersfield, Bakersfield, USA

Chapter Outline

GLOSSARY

accretionary lapilli Spherical aggregates of volcanic ash, commonly having a concentric structure and formed by clumping of moist ash. Often used as a general term for all ash aggregates including structureless "ash lumps/pellets."

block-and-ash flow deposit Small volume pyroclastic density current deposit composed of mostly dense to moderately vesicular juvenile blocks in medium to coarse ash matrix. Mostly generated during collapse of lava domes.

co-ignimbrite/co-pyroclastic density current ash cloud Buoyant fine-grained ash plume that rises off the top of moving pyroclastic density currents.

eutaxitic foliation A rock fabric developed in welded ignimbrites when glassy pyroclasts are deformed (flattened and stretched) and aligned.

ignimbrite Pyroclastic density current deposit composed of variable proportions of pumice, ash, and lithic clasts usually used for deposits formed during large explosive eruptions.

lateral blast A rapid decompression of lava domes or cryptodomes on a volcano due to sudden collapse that can result in laterally directed pyroclastic density currents.

pyroclastic density current (PDC) Hot Eruption-derived particulate—gas density current that moves laterally along the ground. This term encompasses *pyroclastic flows and pyroclastic surges*.

rheomorphism Ductile flow of hot ignimbrite during and following deposition. Produces flow folds, stretched vesicles, and rotated crystals and lithic clasts.

welding Process of annealing of hot, ductile, glassy pyroclasts including pumice by compaction and reduction in porosity. Degree of compaction is typically proportional to the thickness of the deposit.

1. INTRODUCTION

Pyroclastic density currents (PDCs) are a diverse range of phenomena generated during explosive volcanic eruptions and by collapse of unstable lava domes. They represent the most visceral and terrifying hazard posed by volcanoes. They are hot, unstoppable, gas—particle mixtures that race across the ground surface, impacting, burning, and burying all that they encounter. They flow across the ground until they become buoyant through deposition of their bed load and through turbulent ingestion and heating of air. The distances traveled can be >100 km and the areas covered

The Encyclopedia of Volcanoes. http://dx.doi.org/10.1016/B978-0-12-385938-9.00036-5

can be enormous ($>20,000$ km^2). In their wake they leave hot, thick pyroclastic deposits that may reach several 1000 km^3 in volume. During explosive volcanic eruptions, PDCs form when the pyroclastic jet exiting the vent fails to become buoyant through the ingestion and heating of the atmosphere. Instead, a hot mixture of gas, ash, pumice, and rock fountains back to the ground and travels away from the volcano as a particulate density current. Collapsing lava domes also generate small volume PDCs with the characteristics of hot, high-particle concentration, granular avalanches (termed "**block-and-ash flows** (BAFs)").

PDCs deposit an extremely diverse range of pyroclastic deposits and this diversity reflects the variability, both between different eruptions and within an eruption, of (1) magmatic properties that influence the nature of the ejected pyroclastic mixture (magma chemistry, volatile content, crystallinity, magma temperature, vesicularity) and (2) topographic and substrate features that determine where PDCs will flow and where and what they will deposit (volcano shape, surrounding topographic relief, presence of water bodies).

Historically, several terms (pyroclastic flows, pyroclastic surges, ash flows) have been widely used to refer to PDCs that were inferred to have distinct characteristics. We use the general term PDC in response to recent studies that have treated the textural and structural variability within PDC deposits as a result of a continuum of density current conditions. Intrinsic current parameters, such as particle concentration gradient, velocity, and turbulence structure are increasingly thought to vary in time and space in response to changes in source conditions, to deposition, and to an irregular and laterally changing substrate. BAFs and PDCs from **lateral blasts** are dealt with separately below.

2. PDC DEPOSITS FROM LARGE VOLUME EXPLOSIVE ERUPTIONS

2.1. Deposit Characteristics

PDCs generated during plinian and subplinian eruptions commonly sediment pumice- and ash-rich deposits called **ignimbrites**. When mapped, ignimbrites show a range of general shapes, from ribbonlike valley-confined deposits through to extensive near-circular sheets that were deposited by currents that traveled radially out from a volcano or caldera (Figure 36.1). They can cover areas up to 20,000 km^2. Particularly extensive ignimbrite sheets include the 28.5 ka BP Taupo ignimbrite, New

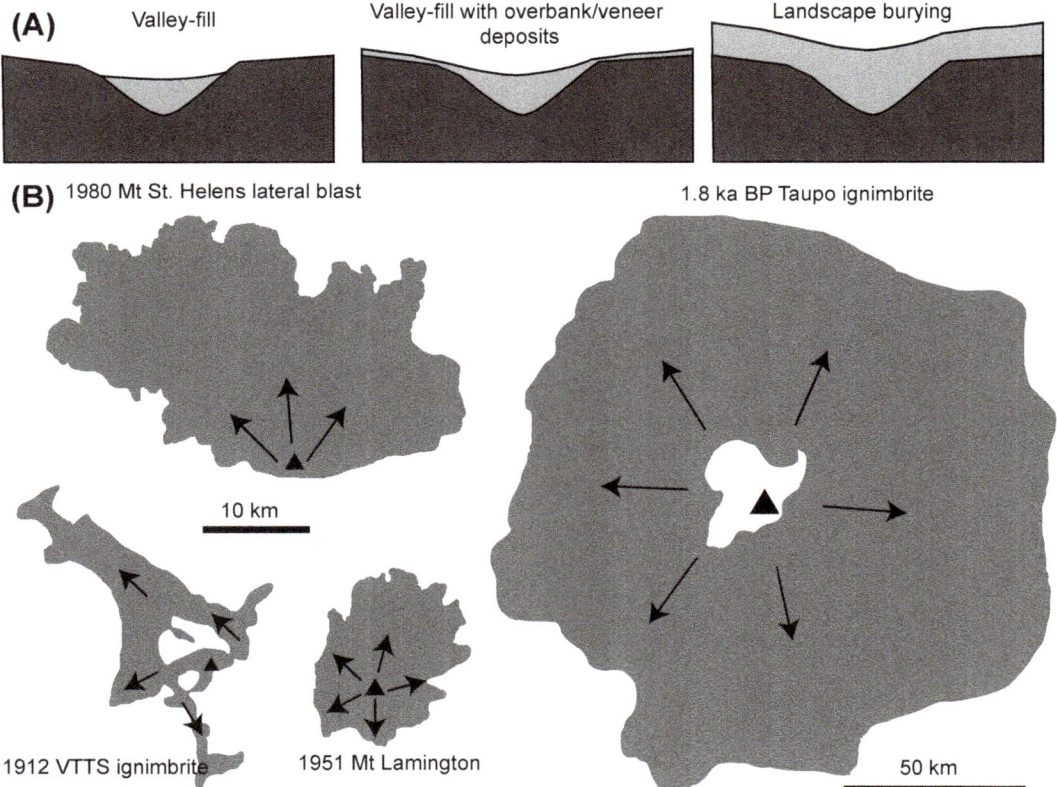

FIGURE 36.1   (A) Simplified cross-sectional patterns of PDC deposits across valleys. (B) Distribution maps of PDC deposits for historic to prehistoric eruptions. Arrows indicate flow directions, and triangles indicate position of vent. VTTS; Valley of Ten Thousand Smokes.

FIGURE 36.2 Aerial view of flat-topped 1912 Valley of Ten Thousand Smokes ignimbrite, Alaska *(Hildreth and Fierstein, 2012)*. Ignimbrite sheet extends 11 km from photographer to distal terminus at foot of mountains in middle ground. *Photo: Wes Hildreth.*

Zealand (Wilson and Walker, 1983) and the 39 ka BP Campanian ignimbrite. The volumes of ignimbrites vary from 0.001 to >1000 km^3: The largest documented ignimbrites, deposited during silicic eruptions at super volcanoes (e.g., Toba caldera and Yellowstone caldera) exceed 3000 km^3.

PDCs are strongly influenced by topography, preferentially chaneling down valleys and swamping topographic depressions (Figure 36.2). All particulate density currents develop density stratification (particle concentration gradients), concentrating more particulate mass, and often coarser, denser clasts, toward their bases as they travel. However, PDCs also travel at high speeds (10s m s^{-1}) and can pass across topographic obstacles like hills, mountains, and ridges and travel great distances over water (10s of km). These characteristics mean that their deposits commonly thin over ridges and thicken into topographic depressions (Figure 36.1)—a result of greater mass being transported at lower levels in the current. Ignimbrite deposited on ridges may be only a few centimeters thick, while ignimbrite deposited in valleys or within subsiding calderas may reach many 100s m thick.

It is useful to think of accumulations of pyroclastic material that are dominated by the deposits of PDCs that build up around volcanoes and calderas during explosive eruptions as *ignimbrite sheets* (e.g., Figure 36.3). Ignimbrite sheets comprise ignimbrite flow units (the products of one discrete PDC) interbedded with co-PDC ashfall deposits, and pumice and ashfall deposits derived from eruption plumes and volcaniclastic deposits (fluvial deposits and lahars) reworked by surface processes during the eruption. Recognition of an *ignimbrite flow unit* requires deposits that record a pause in the passage of PDCs, such as ash or pumice fall deposits or sediments. This is not always simple because PDCs can erode fall deposits resulting in the apparent merging of two flow

units, and because pauses in the passage of a density current may be spatially diachronous (nonuniform). Waxing and waning of a PDC's leading edge during an eruption can mean that ignimbrite that was continually deposited in proximal regions passes downstream into a series of flow units separated, for example, by co-PDC ashfall deposits.

2.2. Compositional Characteristics and Componentry

The chemical composition of juvenile material in ignimbrites covers the full range of magma compositions. Small volume ignimbrites can be of basaltic through to rhyolitic compositions, while large volume ignimbrites are typically of highly evolved compositions. Ignimbrites from ultrabasic kimberlite eruptions are known in the geological record.

PDCs carry a wide variety of material including juvenile clasts (derived from fragmentation of the erupting magma), such as volcanic ash and pumice and scoria lapilli and pyroclasts sourced from the erosion of the volcanic conduit walls, and lithic and debris picked up by the current (rocks, vegetation, soil, sediment, building materials). Ignimbrites display a wide variety of sedimentary structures (massive, planar bedded, stratified, cross-stratified), textures (matrix supported and clast supported), componentry (pumice-rich, lithic-rich, fines-poor), and grading patterns (non-, inverse-, and normal-grading of both pumice and lithic clasts). Commonly recognized lithofacies (parts of a deposit with distinct compositional, structural, and textural characteristics) are summarized in Table 36.1 and illustrated in Figure 36.4 (for a full discussion of common ignimbrite lithofacies see Branney and Kokelaar, 2002).

The particles carried by a PDC have a range of physical properties (size, density, shape). Particles range in size from micron-sized ash particles up to boulder-sized pumice or lithic clasts and tree trunks. Pumice clasts may have densities of 200−1000 kg m^{-3}, while lithic clasts may have densities of >2000 kg m^{-3}; scoria clasts have densities of 1000−2000 kg km^{-3}. Ash particles are commonly angular and/or platy, while pumice and lithic clasts may be more equant in shape. Ash is also generated by attrition and breakage of pumice clasts within moving PDCs. This leads to the rounding of pumice. The ash matrix of ignimbrites can be enriched in phenocrysts relative to pumice. The missing ash is either lofted into co-PDC plumes during transport (as much as 50% of the volume of the ignimbrite can be lost in this way), or is transported to distal sectors of the PDC deposit.

The particulate load in a PDC is supported by a range of different mechanisms (turbulent eddies, particle collisions, saltation, rolling along a substrate, fluid escape). These

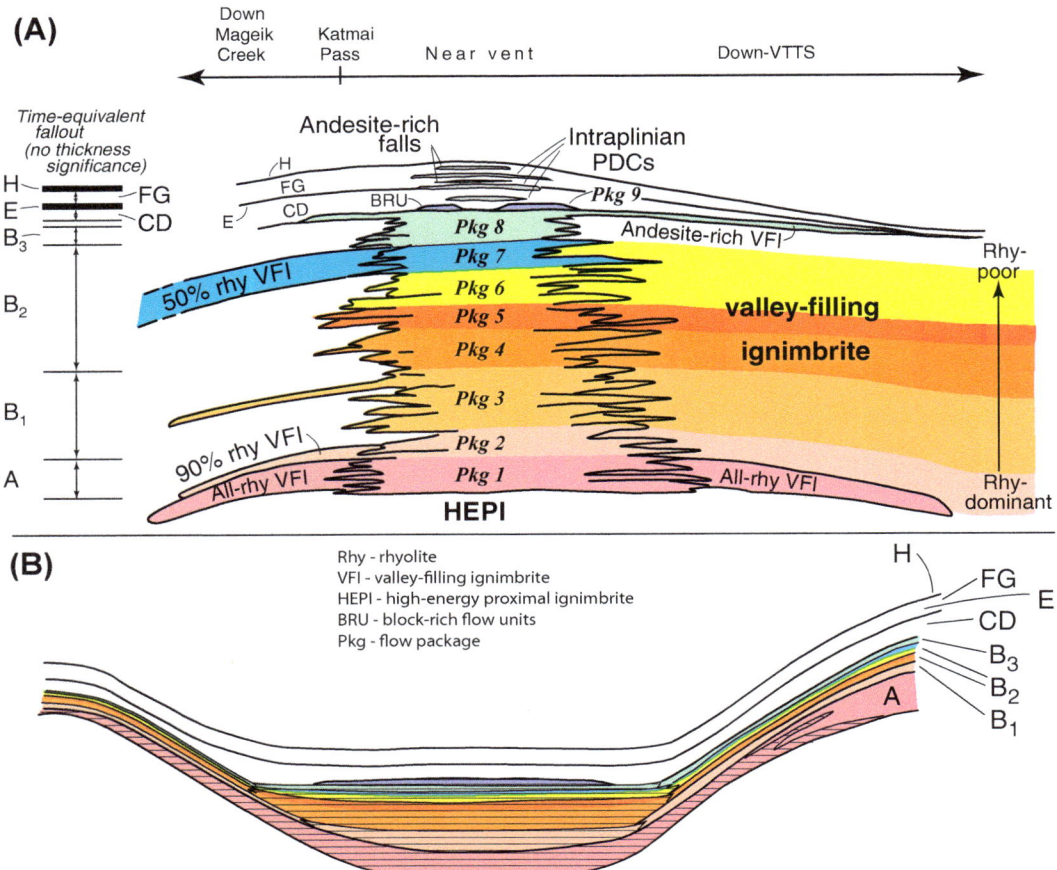

FIGURE 36.3   Schematic time-space sketches of the ignimbrite sheet of the Valley of Ten Thousand Smokes ignimbrite, Alaska. (A) Cartoon summarizing the ignimbrite facies and time-equivalent fall deposits (A—H). (B) Generalized sketch illustrating the influence of topography on fallout and ignimbrite packages. *Modified from Hildreth and Fierstein (2012).*

processes act to segregate particles of different physical properties during transport. In this way, ash, pumice, and lithic clasts erupted from the vent at the same time can have very different transport histories within a PDC and end up in different parts of an ignimbrite. As a current travels out from source its particulate load can become fractionated according to density or size. It should come as no surprise that ignimbrites are commonly compositionally heterogeneous. Lithic clasts tend to sediment early from PDCs resulting in proximal, lithic-rich lithofacies ("lithic breccias"). Fine ash in turbulent suspension can rise buoyantly off the top of PDCs as co-PDC ash plumes. This ash settles widely as thin co-PDC ashfall layers. Low density pumice clasts can travel preferentially toward the distal limits of a current due to the difficulty of depositing through a high-particle concentration shearing dispersion at the base of a PDC. These pumice clasts form clast-supported distal pumice lobes and levees (e.g., the 1993 eruption of Lascar volcano, and the 1980 eruption of Mount St Helens volcano).

Superimposed on the fractionation of the particulate load by the current are variations in clast composition due to changes in the supply of material at source. For example, phases of caldera subsidence or conduit collapse can result in large volumes of lithic clasts suddenly being introduced into a PDC. This will be manifested within the ignimbrite as a lithic clast-rich layer. Progressive evacuation of a chemically zoned magma chamber during an explosive eruption can lead to temporal changes in the chemical composition of the erupted juvenile material (ash, pumice) carried by a PDC (Figure 36.5). There are several well-documented examples of ignimbrites with vertical variations in the chemical composition of juvenile clasts (e.g., Valley of Ten Thousand Smokes (VTTS) ignimbrite, Alaska, Hildreth and Fierstein, 2012; Zaragoza ignimbrite, Mexico, Carrasco-Núñez and Branney, 2005). Chemical zoning and presence of lithic clast-rich layers have proved useful tools for establishing isochrons across extensive ignimbrite sheets (see Branney and Kokelaar, 2002).

TABLE 36.1 Summary Descriptions and Interpretations of Lithofacies Seen in Pyroclastic Density Current Deposits

Lithofacies	Description	Interpretation and Examples
Massive ignimbrite Figure 36.4(A)	Massive, poorly sorted (σ_φ 2–5) ash, and lapilli; median grain size in the ash range; abundance of ash, pumice, or lithic components and crystals varies; pumice clasts are rounded due to abrasion and breakage; clasts may reach block size; may show grading (inverse or normal), clast fabrics, (aligned and imbricated clasts); most common ignimbrite lithofacies; typical in valleys, on plains and depressions.	Deposited from high-concentration flow boundary zones dominated by fluid escape processes; particles experience minimum shear as they are deposited. Examples: Valley of Ten Thousand Smokes, Hildreth and Fierstein (2012).
Massive to stratified lithic breccias Figure 36.4(B)	Massive to stratified, poorly sorted (σ_φ 1.5–4.5) lithic lapilli, minor pumice lapilli; matrix- or clast-supported; matrix may be fines-rich, or fines-poor and dominated by crystal fragments and lithic clasts; lithic blocks are angular to subrounded and up to boulder size (commonly in the centimeter to decimeter size range); lithic clast population is derived from vent erosion, fragmentation of caldera roof rocks during caldera subsidence or picked up by the current during transport; platy clasts may be imbricated; forms massive units up to 10s m thick, or decimeter-thick lenses, stratified layers, and dune bedforms; load and flame structures; common lithofacies in proximal regions.	Rapidly deposited from high-concentration flow boundary zones with variable support mechanisms (fluid escape, traction, granular flow). Examples: Tenerife, Bryan et al. (1998); Santorini volcano, Greece, Druitt et al. (1999).
Massive agglomerate Figure 36.4(C)	Composed of clast- or matrix-supported cognate dense to vesicular scoria or spatter lapilli and blocks; sorting similar to lithic breccias (σ_φ 1–3.3); poorly sorted ash-rich matrix may be present; spatter clasts indicate emplacement while still hot and ductile; clasts can be imbricated; common in proximal settings.	Similar to massive to stratified lithic breccias. Examples: Santorini volcano, Greece, Druitt et al. (1999).
Diffuse-stratified ignimbrite Figure 36.4(D)	Ash and lapilli in subtle cm-scale poorly defined strata; strata may be continuous over decimeters to meters and may show gradual thickening and thinning; low-angle cross-stratification; strata defined by variations in abundance of lapilli-sized clasts; poorly sorted—varies from values typical of massive ignimbrite to values typical of stratified ignimbrite; strata may be non-, inverse-, or normal graded; gradational into massive and stratified ignimbrite; forms packages many meters thick.	Deposited from flow boundary zones characterized by conditions intermediate between fluid escape dominated and traction dominated. Examples: Valley of Ten Thousand Smokes ignimbrite, Alaska, Hildreth and Fierstein (2012); Poris ignimbrite, Tenerife, Brown and Branney (2013).

(Continued)

TABLE 36.1 Summary Descriptions and Interpretations of Lithofacies Seen in Pyroclastic Density Current Deposits—cont'd

Lithofacies	Description	Interpretation and Examples
Stratified and cross-stratified ignimbrite Figure 36.4(E)	Well to poorly sorted (σ_φ 1–3) ash and lapilli with well-defined centimeter scale stratification and cross-stratification; strata may consist of alternating fine-grained and coarse-grained layers; individual strata discontinuous over decimeters to meters; low-angle erosive truncations of strata common; constructional bedforms with strata dipping up to 40°. Typical wavelengths 1–5 m; amplitudes of <1 m; larger scale bedforms composed of poorly sorted ignimbrite; common in ignimbrite deposited over topographic highs (ridges and hills); may occur intercalated with other ignimbrite lithofacies in topographic depressions.	Deposition from traction dominated flow boundary zones. Examples: Taupo ignimbrite, New Zealand, Wilson and Walker, (1985); 1991 Pinatubo volcano, Philippines and the 1980, Scott et al. (1996); Poris ignimbrite, Tenerife, Brown and Branney (2013).
Elutriation pipes Figure 36.4(F)	Subvertical pipes, pods, and sheets enriched in clast-supported crystal and lithic clasts and depleted in fine ash; centimeters to meters long with straight or sinuous margins; may branch upward.	Forms due to aggregative-type segregation of upward-fluxing interstitial gases, during hindered settling, deposition, and compaction; gases derived from air in pore spaces, steam from boiling of wet substrates or combustion of vegetation (Branney and Kokelaar, 2002).
Pumice-rich ignimbrite Figure 36.4(G)	Variably sorted (σ_φ <1->4), clast-supported pumice lapilli and blocks, with or without an ash matrix; may be graded; sharp to gradational boundaries with enclosing ignimbrite lithofacies; can occur at any height in an ignimbrite; common at the base, and at margins of valleys; distal parts of ignimbrites commonly composed of clast-supported pumice-rich lithofacies with steep fronts and levees.	Deposited from currents in which there was efficient segregation of pumice clasts. Examples: 1993 eruption Lascar volcano, Chile, Calder et al. (2000).
Accretionary lapilli-bearing ignimbrite Figure 36.4(H)	Whole and broken concentrically laminated accretionary lapilli matrix supported in massive or stratified ignimbrite.	Accretionary lapilli formed in eruption plumes or co-ignimbrite ash cloud and fell into aggrading ignimbrite at base of current. Examples: Oruanui, New Zealand, Van Eaton and Wilson, (2012).

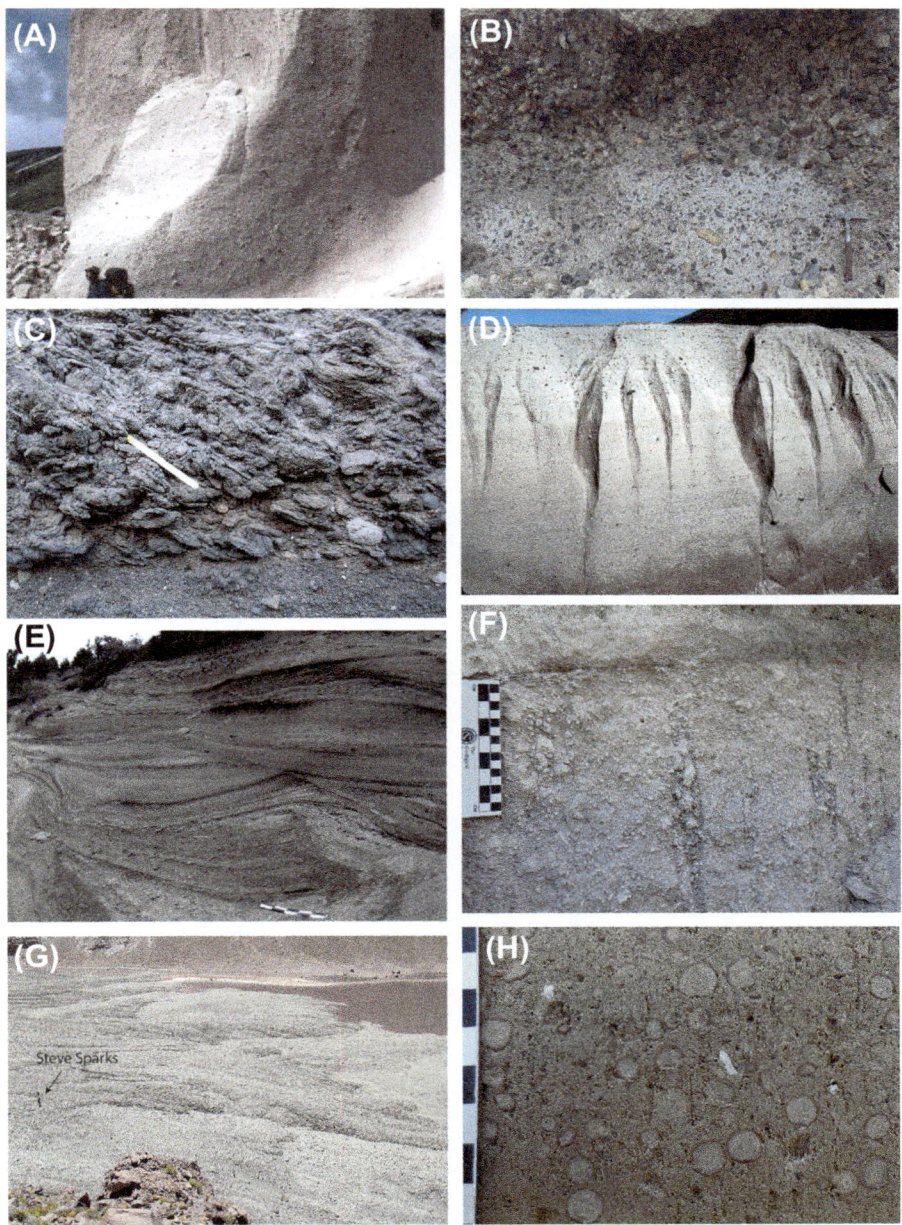

FIGURE 36.4 Representative lithofacies in PDC deposits. (A) Massive ignimbrite, Valley of Ten Thousand Smokes, Alaska. *Photo: Wes Hildreth.* (B) Lithic breccia, Quaternary Fasnia Member, Tenerife. Hammer for scale. (C). Massive agglomerate with imbricated clasts, upper scoria I, Santorini. 30 cm rule for scale. (D) Diffuse-stratified ignimbrite, Valley of Ten Thousand Smokes, Alaska. Exposure is ~8 m thick. *Photo: Wes Hildreth.* (E) Cross-stratified ignimbrite, Gölcük volcano, Turkey. 20 cm divisions on rule. Current direction right-to-left. Foresets in dune bedform dip to left. (F) Elutriation pipes enriched in pumice and lithic clasts and depleted in fine ash relative to host ignimbrite. AD536 Tierra Blanca Joven eruption, Ilopango caldera, El Salvador. 1 cm divisions on rule. (G) Pumice-rich lobes and levees at distal margins of PDC deposits of the 1993 eruption of Lascar volcano, Chile. Steve Sparks for scale. (H) Whole and broken accretionary lapilli matrix supported in ignimbrite from the Phreatoplinian AD536 Tierra Blanca Joven eruption, Ilopango caldera, El Salvador. 1 cm divisions on rule.

2.3. Grain-Size Characteristics

Ignimbrites are composed of clasts of different sizes and in clastic deposits this range is expressed as sorting—the tendency for all grains to be of the same size. Ignimbrites are in general poorly to very poorly sorted (sorting coefficients of $\sigma_\varphi > 2$, where $\sigma_\varphi = 0$ is a deposit with all the clasts of a single size Figure 36.6). Abundant fine ash causes median diameters (Md_φ) to be in the range $Md_\varphi = 0-2$. Different lithofacies can show different

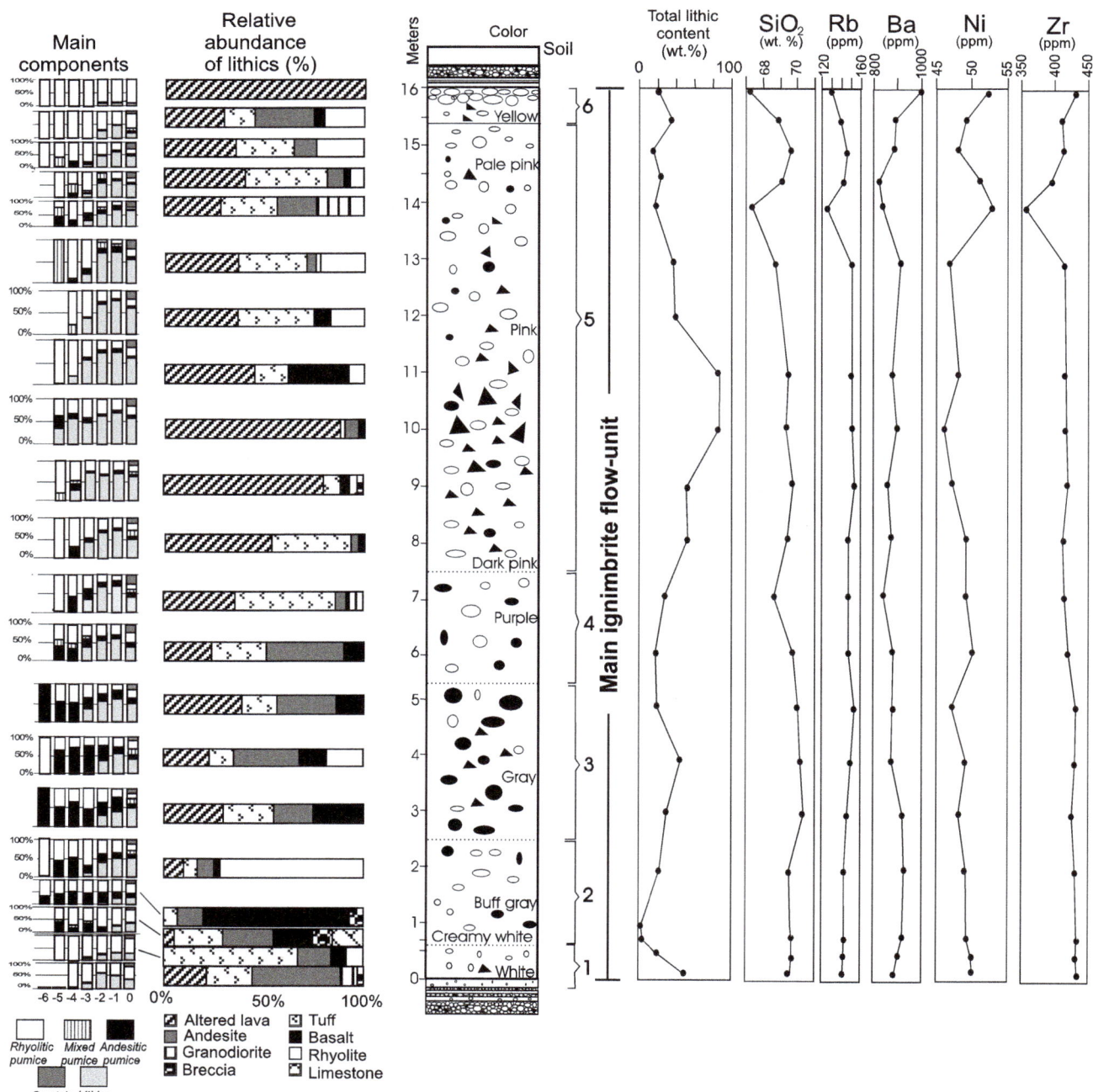

FIGURE 36.5   Measured section through the Zaragoza ignimbrite, Mexico. The log shows six gradational zones (1—6), defined on the basis of overall color (indicated) and compositional zoning. Graphs left of the field log show measured vertical variations in the relative proportions of the main components for size classes larger than 0 φ (1 mm), including the different types of lithic clasts (far left). Plots on the right show vertical variations in lithic clast abundance and in chemistry of the juvenile components. *Modified from Carrasco-Núñez and Branney (2005).*

grain-size characteristics and sorting coefficients (see Table 36.1). Variations in sorting and grain-size characteristics reflect variable segregation of particles of different sizes and physical properties during transport and emplacement, and clast breakage and elutriation processes. In general, with increasing transport distance, the median

grain size decreases, the content of fine ash increases, and sorting is accentuated within ignimbrites. Vent-derived lithic clasts generally decrease in size with distance.

Ignimbrites associated with phreatoplinian eruptions are typically much finer-grained ($Md_\varphi = >1$) than those from plinian eruptions due to the greater degree of

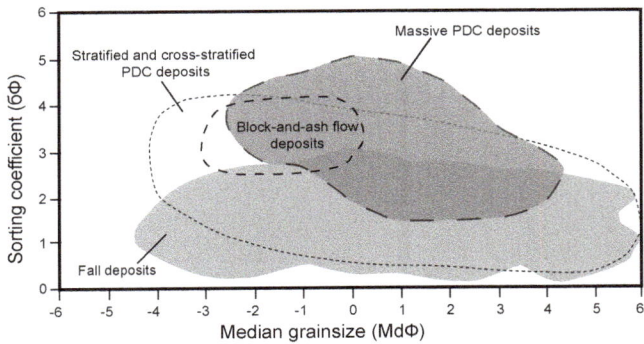

FIGURE 36.6 Indicative Md-σ diagram illustrating the common grain-size characteristics of different types of PDC deposits. Deviations from these fields are known.

fragmentation during magma–water interaction and due to the premature deposition of fine ash through moisture-induced aggregation of fine ash. The ignimbrites are rich in ash and depleted in juvenile lapilli even in proximal

regions. They may contain abundant ash aggregates (**accretionary lapilli** and ash pellets) that formed within eruption or co-ignimbrites plumes and fell into the currents (e.g., the 25.4 ka BP Oruanui eruption, New Zealand, Van Eaton and Wilson, 2012; Figure 36.7).

3. DEPOSITS FROM SMALL VOLUME PDCS

3.1. PDC Deposits of Small Volume Phreatomagmatic Eruptions

Small volume, monogenetic, phreatomagmatic eruptions initiated by explosive mixing between rising magma and ground or surface water at tuff ring or maar volcanoes commonly produce PDCs. Each explosion typically involves only a small volume of fragmented magma as well as abundant fragmented conduit wall rock. The pyroclastic mixture rises hundreds of meters into the air and fountains

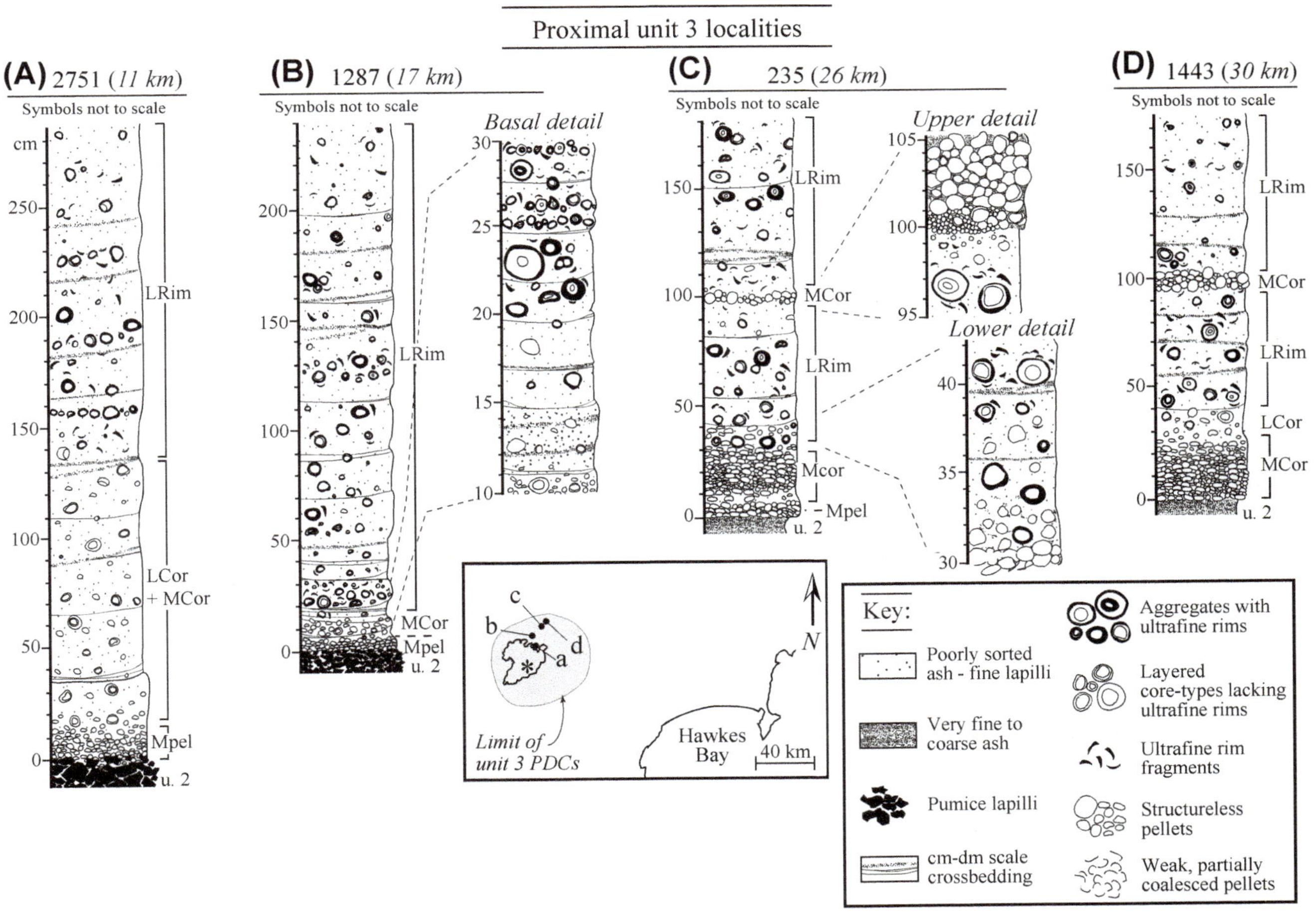

FIGURE 36.7 Stratigraphic logs showing dominant aggregate types in unit 3 deposits of the 25.4 ka BP Oruanui eruption, New Zealand. Proximal localities are dominated by laterally emplaced PDC deposits with fragmented, ultrafine rim-type accretionary lapilli. *Modified from Van Eaton and Wilson (2013).*

back to the ground and travels radially out from the vent as a PDC for distances of <1−10 km. These PDCs are commonly termed "pyroclastic surges" in the literature. Phreatomagmatic eruptions are characterized by numerous closely spaced explosions, each of which may generate a PDC. PDC deposits at maar volcanoes are commonly rich in lithic clasts (up to 90 wt%) because the explosions occur within subsurface aquifers intersected by the rising magma.

The deposits left behind by these PDCs are characteristically thin-bedded, stratified, cross-stratified, or planar bedded. Massive beds up to several meters thick are also common. The deposits have variable but generally poor sorting (σ_φ 1.5−4.5). Lateral lithofacies variations are pronounced and varied: a common sequence is for proximal regions to be characterized by chaotic, massive, and crudely bedded deposits that pass downstream into megaripple bedforms (Figure 36.4(E)), which can exhibit a variety of morphologies (upstream and downstream migrating forms), and then into massive and thinly bedded deposits in distal regions (Figure 36.8, e.g., Chough and Sohn, 1990).

Phreatomagmatic explosions are very efficient at fragmenting magma and their deposits are often rich in fine ash. The generation of abundant fine ash together with the incorporation of water lead to the formation of abundant ash aggregates (e.g., accretionary lapilli or ash pellets). In extreme cases, the water may cool down the erupting pyroclastic mixture sufficiently so that liquid water is present within the PDCs. These cool deposits can plaster onto vertical objects (walls, cliffs) during eruptions (e.g., Cole et al., 2001). Wet PDCs tend to produce more poorly sorted, less well-bedded deposits.

At tuff ring and maar volcanoes, PDC deposits construct low relief cones or tephra rings around the crater. These cones are typically <50 m thick at the crater and thin to centimeters at distances of several 100 meters to several kilometers or more. Beds dip outward at low angles (<20°). Because of the pulsatory nature of the eruptions, co-PDC fall deposits are common within the cones and record the settling of ash and ash aggregates between the passage of currents.

3.2. Deposits of Lateral Blast-Induced PDCs

Rapid decompression of lava domes or cryptodomes due to sudden collapse can result in laterally directed blasts that generate fast moving (80−90 m s^{-1}), highly destructive PDCs (e.g., the 1902 eruption Mount Pelée, Martinique, the 1956 eruption at Bezymianny volcano, Kamchatka, the 1980 eruption of Mount St Helens, and the 1997 eruption of Soufrière Hills volcano, Montserrat). These blasts generate PDCs that are strongly erosive close to source. The deposits they leave behind share many similarities (see Belousov et al., 2007). On Montserrat, streamlined-breccia mounds and hummocks were left in proximal regions, while at greater distances the PDCs deposited two stacked normally graded deposits each composed of a lower normally graded fines-poor deposit rich in lithic clasts, overlain by a fines-rich stratified and cross-stratified layer. Lateral blast deposits commonly contain abundant entrained material (vegetation, soil, and locally derived lithic clasts).

3.3. BAF Deposits

BAFs are a subtype of PDCs that are generated by the gravitational collapse or explosive disruption of typically andesitic to rhyodacitic lava domes, or from the collapse of Vulcanian eruption columns. The volumes of material involved are small (10^3-10^6 m^3), and the flows typically travel only several kilometers away from the volcano, but may reach >10 km. BAFs may be single pulse (lasting for seconds) or composed of multiple pulses reflecting repeated collapse events: such events may be spread over 10s of minutes or several hours. There have been many very well-documented examples of BAFs over the past two decades (e.g., Soufrière Hills volcano, Montserrat; Merapi volcano, Indonesia; Volcán de Colima, Mexico; Unzen volcano, Japan; Mount Sinabung, Indonesia). The velocities at which BAFs travel are largely dependent on the slope angle, but typically range from 5 to 20 m s^{-1}, occasionally up to 100 m s^{-1} (Cole et al., 2002). Emplacement temperatures are high, typically 400−600 °C, and carbonized plant material is common in the deposits where the flows traversed vegetated slopes. BAFs are

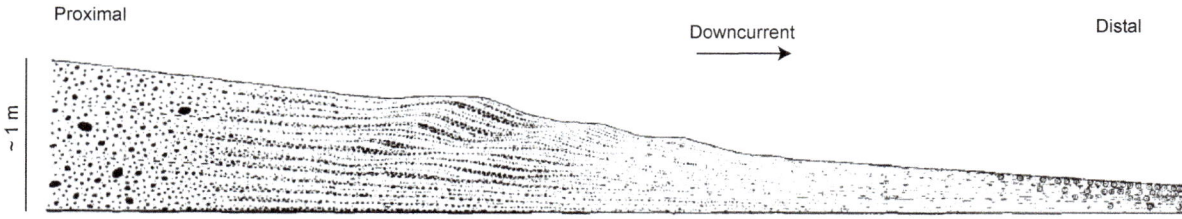

FIGURE 36.8 Typical downstream transitions in pyroclastic density current deposits of the Songaksan tuff ring, Cheju Island, South Korea, over a downcurrent distance of 1−2 km. In proximal regions, deposits are characterized by massive or crudely stratified deposits, these pass laterally into dune-bedded or undulatory deposits and then thin-bedded finer-grained deposits with accretionary lapilli. *Modified from Chough and Sohn (1990).*

strongly controlled by topography, and are commonly confined to valleys and ravines in proximal areas (Figure 36.9(B)), but may spread out on flat-lying plains around volcanoes to form fans. BAF deposits may exceed 100 m thick where confined to valleys, and can weld where thick and hot. They are thinner (centimeters to decimeters) where deposited on ridges (in overbank facies) and on flat-lying aprons, where they are typically 1−3 m thick. Lateral deposits of BAFs (sometimes called *ash cloud surge deposits*) are often finer-grained, sandy, and may be massive or cross-stratified, with wavy or dune bedforms.

BAF deposits are distinct from ignimbrites in being composed predominantly of clast-supported, dense (1700−2700 kg m^{-3}), non- to poorly vesicular juvenile clasts derived primarily from the disintegration of a lava dome. They are poorly sorted and are composed of blocks and boulders up to 15 m in diameter in an ash matrix (Figure 36.9(C); e.g., Cole et al., 2002; Charbonnier and Gertisser, 2011). The juvenile clasts may be prismatically jointed, which indicates hot emplacement, or breadcrusted,

reflecting continued vesiculation after fragmentation. Lithic clasts, derived from the vent area or entrained along the flow path, can account for up to several 10 wt% of the volume of BAF deposits. Juvenile clasts are angular to subangular and may have rounded corners. On Montserrat, large blocks exhibit friction marks on polished slickensided surfaces smeared with thin layers of cataclasite: these features indicate sliding collisions between transported blocks and hard surfaces. BAF deposits may be massive, or have parts that are either inversely graded or normally graded. Large blocks commonly protrude from the top of the deposits. Discontinuous fine-grained lenses may be present along the bases of BAF deposits. Elutriation pipes form where the deposits are emplaced over wet or moisture-rich ground.

Distal parts of the deposits may be enriched in finer-grained particles, may show steep snoutlike accumulations of >1 m diameter boulders, or may be composed of scattered blocks of low-density juvenile material and logs. Decimeter-high ridges and furrows (levees) may be present on the tops of the deposits in flat-lying areas.

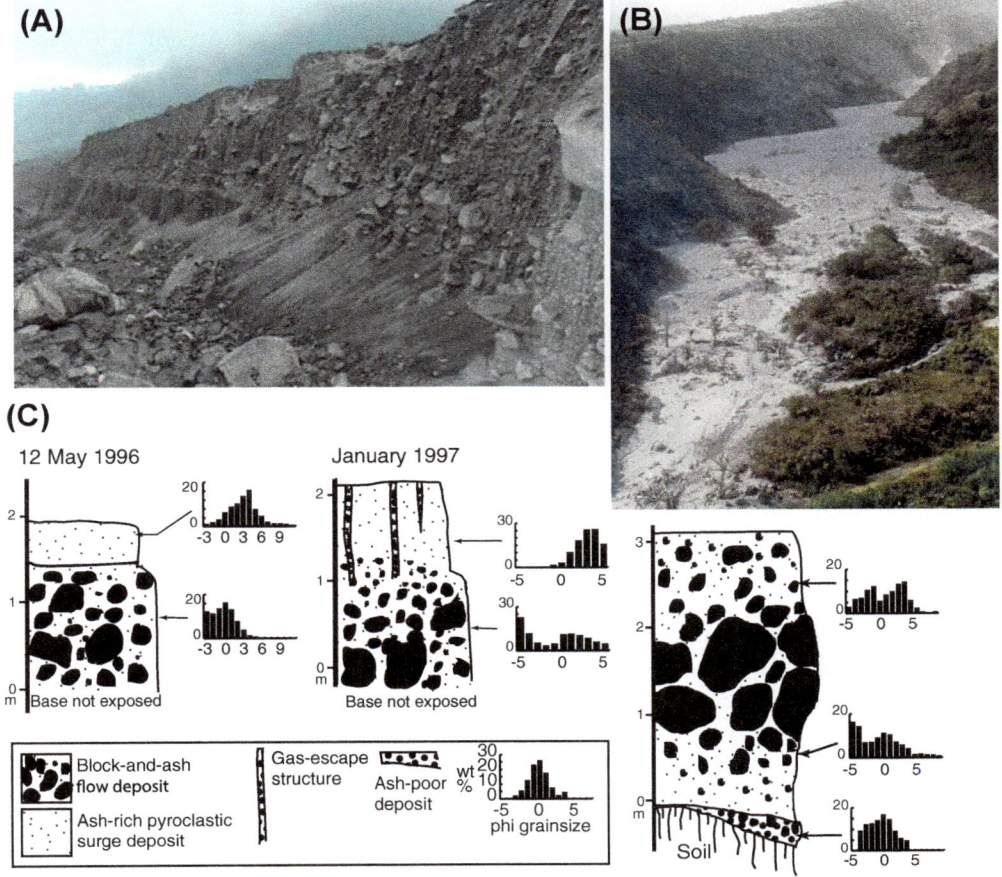

FIGURE 36.9 Block-and-ash flow deposits. (A) Incised block-and-ash flow deposit from 2006 eruption of Merapi volcano, Indonesia. *Photo by Sylvain Charbonnier.* (B) Block-and-ash flow deposits in White River valley, Montserrat immediately following April 11, 1997 dome collapse. Note strong topographic control and standing trees in valley *(Cole et al., 2002).* (C) Measured sections through block-and-ash flow deposits from Soufrière Hills volcano, Montserrat. Histograms show weight% versus phi size. *Modified from Cole et al., (2002).*

BAF deposits are commonly interbedded with lahar and fluvial deposits. Distinguishing between BAF deposits and lahar deposits can be difficult, but paleomagnetic techniques, such as thermoremanent magnetization, can determine whether clasts were emplaced hot and have proved useful.

4. RELATIONSHIP TO TOPOGRAPHY

There are several superb examples of ignimbrites deposited over rough ground that illustrate the influence of topography on PDCs (e.g., 1.8 ka BP Taupo ignimbrite, New Zealand, Wilson and Walker, 1983; 1991 eruption of Pinatubo, Philippines, Scott et al., 1996; Upper Pollara eruption, Salina, Italy, Sulpizio et al., 2008; 273 ka Poris eruption, Tenerife, Brown and Branney, 2013, 1912 eruption of VTTS, Alaska; Hildreth and Fierstein, 2012). In many of these examples, the ignimbrites are thickest in valleys and depressions and thin over high ground, and posteruption dissection of the ignimbrite sheets has allowed their 3-D architecture to be examined. Ignimbrites deposited in valleys tend to be massive and poorly sorted, while coevally deposited ignimbrite on adjacent ridges is often stratified or cross-stratified and may be better sorted and depleted in fines. Ignimbrites that are 10s meters thick in valleys may thin to only a few centimeters on ridges; other ignimbrites may pinch out against valley sides (Figure 36.10). Strata may fade as they are traced from ridges into valleys. Outcrops on Tenerife provide unparalleled cross sections through valley-filling ignimbrites that partially-to-totally buried paleovalleys and paleocanyons (Figure 36.10).

On a regional scale, PDCs accelerate down steep slopes (e.g., the flanks of a volcano) and can bypass (i.e., do not deposit) and be strongly erosive, knocking down and entraining fully grown trees (e.g., Brown and Branney, 2013). Decelerations on lower gradient slopes around volcanoes promote deposition, and thus ignimbrite sheets tend to pinch out upslope against large regional changes in slope. For example, on Tenerife, deposition was induced by currents passing across a regional break-in-slope from slopes of 9°−15° to slopes of 3°−5°.

Paleoflow indicators in PDC deposits are useful in exploring the influence of topography on PDCs (e.g., the 1980 eruption of Mount St Helens). Paleoflow indicators include imbricated platy clasts, and oriented tree trunks and branches. Logs and branches carried by PDCs have become aligned by the flow on deposition. Erosional furrows (flute casts), sheared clasts, and sheared load and flame structures can also indicate paleoflow directions.

5. CO-PDC ASHFALL DEPOSITS

Ash elutriated from PDCs and BAFs during transport can rise buoyantly through the atmosphere as a co-PDC plume.

These plumes differ from eruption plumes centered above vents in that they are enriched in ash (5−25 wt% < 10 μm) and in water due to entrainment of moist tropospheric air and surface water, and they rise from much greater areas (the entire "footprint" of the PDC). Co-PDC ashfall deposits are fine-grained even in proximal regions in contrast to fall deposits from eruption plumes. Co-PDC ash clouds can rise to great heights (>30 km), become dispersed by atmospheric winds, and cover huge areas (10^5−10^7 km^2). During pauses in PDC activity, fine ash settles from co-PDC plumes to form thin layers that mantle the tops of ignimbrite flow units. Some co-PDC ashfall layers contain ash aggregates—commonly these are clast-supported poorly structured ash pellets.

6. EMPLACEMENT TEMPERATURES OF PDC DEPOSITS

During explosive eruptions, magma is erupted at temperatures of 800−1000 °C. Ingestion of atmospheric air by an eruption column and by laterally moving currents can significantly cool the pyroclastic mixture, and the emplacement temperatures of PDC deposits vary widely. Emplacement temperatures can be measured in deposits using a number of methods. For modern eruptions, thermocouples can be inserted into deposits to measure residual temperatures. The most commonly used method for historic and ancient deposits is to analyze paleomagnetic data derived from lithic clasts contained within deposits. This method indicates typical emplacement temperatures of 200−600 °C. Paleomagnetic studies have shown that there can be little decrease in temperature from proximal to distal parts of a deposit and little variation in temperature between different lithofacies. Some phreatomagmatic eruptions can generate cool PDCs, however, the deposits of other phreatomagmatic eruptions show little cooling with respect to magmatic PDC deposits. This may result from the greater degree of fragmentation during phreatomagmatic eruptions which creates a finer-grained particle size distribution that is more efficient at transferring heat to colder lithic clasts. PDC deposits are very good insulators and may remain hot for decades following emplacement, depending on initial emplacement temperature, deposit characteristics (thickness, composition, grain size, porosity), and rate of erosion.

7. WELDING AND RHEOMORPHISM IN IGNIMBRITES

Ignimbrites deposited above at or close to magmatic temperatures are liable to *weld*, whereby viscous compaction of glassy, juvenile pyroclasts reduces interstitial porosity, increases density, and the deposit becomes lithified.

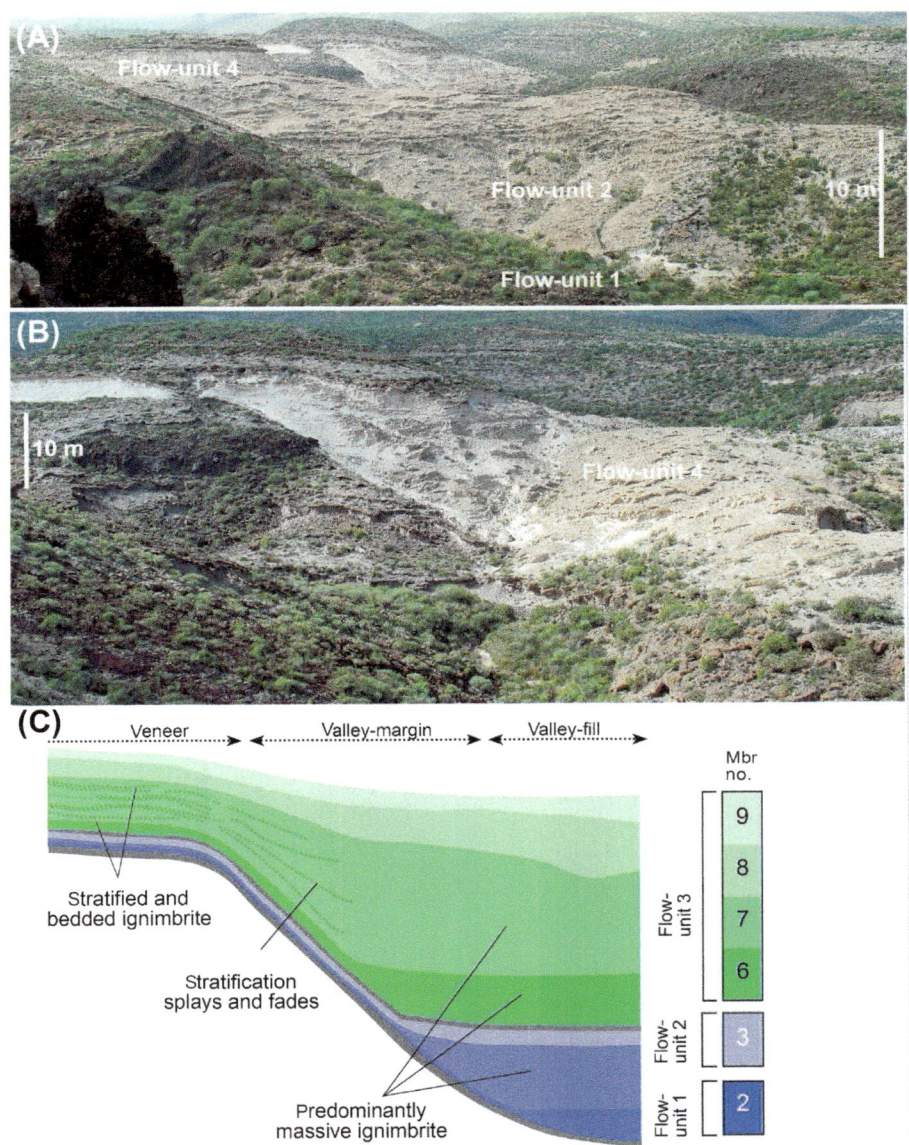

FIGURE 36.10 Effects of topography on the 273 ka BP Poris ignimbrite, Tenerife. (A) Panoramic view of spectacular palaeovalley-filling ignimbrites. Current out of page; view toward the north. Flow units 1 and 2 pinch out against the valley side. (B) Close-up of northern valley-fill in A showing transition from valley-filling facies to veneer facies. Flow unit 4 thins across the paleoridge. (C) Cartoon showing reconstructed transverse-to-current architecture across a palaeovalley. *Modified from Brown and Branney (2013).*

Lithification increases the mechanical strength and preservation potential of tuffs often causing them to be better exposed and overestimated, relative to interstratified non-welded deposits (e.g., Snake River Plain, Idaho). Welded ignimbrites may be low grade or high grade depending on the intensity of welding, and therefore, their emplacement temperature. Vertical profiles through many low-grade welded ignimbrite sheets reveal complex variations in welding intensity that are interpreted as due to discrete *emplacement units*, each representing a single pulse of deposition, that aggraded quickly enough to all cool together in a *composite cooling unit*. Many recent rheological experiments have demonstrated that welding can occur during and immediately after deposition as long as the emplacement temperature is significantly above the glass transition temperature (Tg: e.g., Quane et al., 2009).

During welding, glassy pyroclasts are deformed (flattened and stretched) and aligned to form a characteristic **eutaxitic foliation** (Figure 36.11(A)); the eutaxitic foliation is assumed to be perpendicular to the flattening direction (i.e., gravity) and is typically parallel to the base of the deposit. Pumice lapilli form low-porosity, oblate glassy

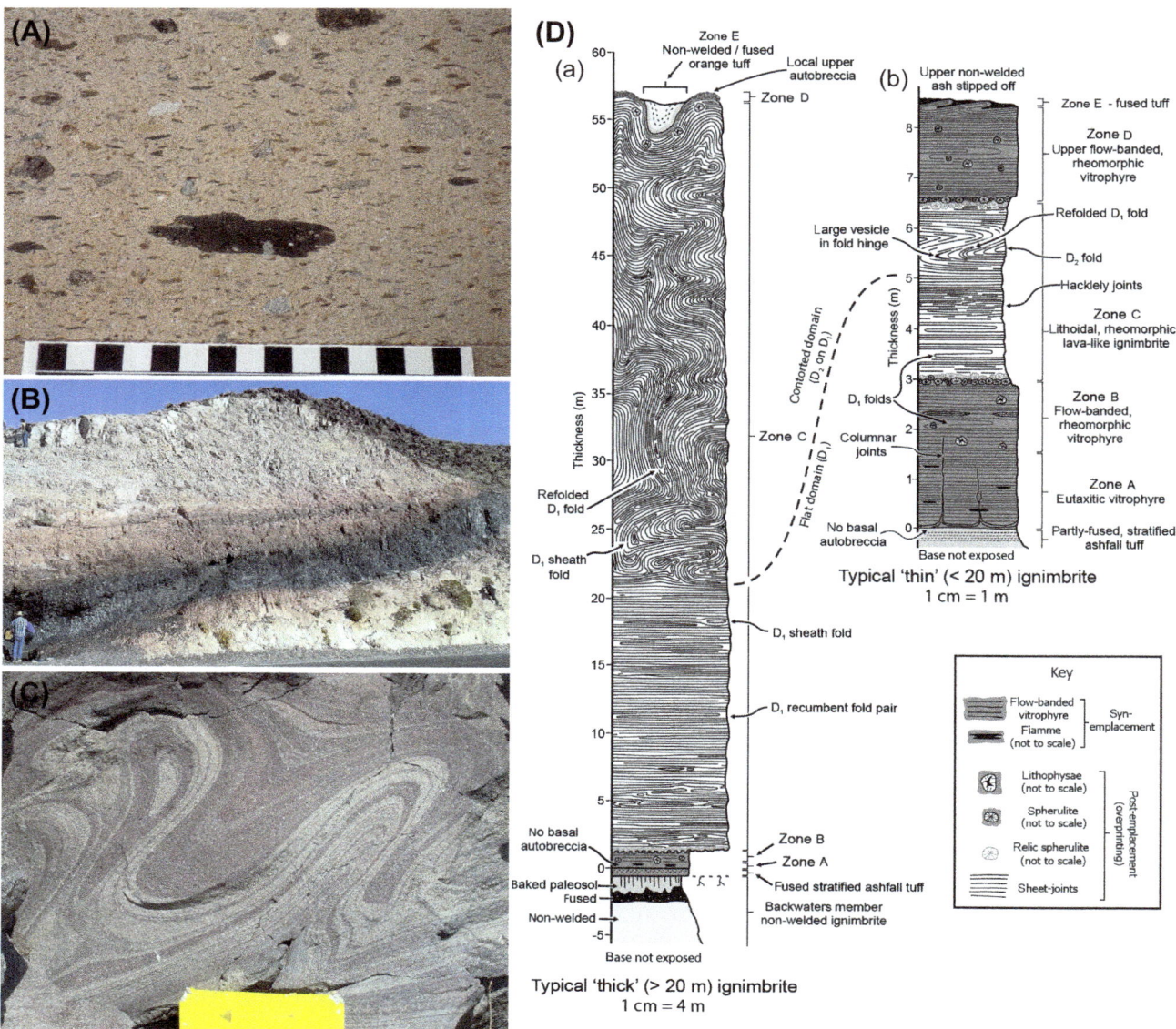

FIGURE 36.11   Welded ignimbrites. (A) Moderately welded Arico ignimbrite, Tenerife with eutaxitic fabric defined by black fiamme. Divisions in centimeters. (B) Welding zonations from nonwelded (base) to intensely welded vitrophyre (black) to strongly welded and lithophysal (purple) in the Resting Springs Pass Tuff, California, USA. (C) Extremely high-grade welded rheomorphic and lava-like Grey's Landing ignimbrite from the Snake River Plain, Idaho, USA. (D) Measured sections through the Grey's Landing ignimbrite. *Modified from Andrews and Branney, 2011.*

disks (lenses in cross-section) called *fiamme*; in contrast, most lithic clasts (e.g., crystalline rocks) are many times stronger than glass under welding conditions, and are not flattened, although they may be rotated or boudinaged. Other characteristic features of welded ignimbrites include columnar jointing and pervasive microcrystallization of the matrix glass (*devitrification*).

The welding process is a direct result of the rheology of the hot pyroclasts and the strain imposed on them. Their rheology is a product of temperature, composition-dependent melt viscosity, dissolved volatile content, and

deformation timescale. The strain is typically the load (mass) exerted by the overlying deposit; therefore strain is usually greatest at the base. However, welding intensity is typically greatest approximately 25–33% up from the base. This is because the base is often deposited at a lower temperature and is quickly cooled by conduction to the substrate; therefore, the ability to weld is progressively retarded closer to the base despite the increasing effective load (strain). Conversely, the upper half of an ignimbrite sheet typically exhibits upward-decreasing welding intensity reflecting the combined effects of decreasing effective load and enhanced cooling by radiation to the atmosphere and advection of heat by through-flowing meteoric water.

7.1. Quantifying Welding

Welding is seldom uniform throughout an ignimbrite sheet, or even a single section (Figure 36.11(B)); instead, welding facies characterize zones that result from combinations of the deposit thickness, composition, and evolving physical properties (e.g., temperature, density, porosity, permeability, and dissolved and exsolved volatile contents). The different welding facies are ascribed to distinct welding intensities related to deposit thickness and thermal regime. Welding intensity can be ranked based on porosity, density, strain, and microtextures (i.e., flattened ash shard) in natural and experimentally derived welded tuffs (e.g., Quane and Russell, 2005). Welding intensity is usually quantified in terms of porosity and density. Primary porosity in an ignimbrite can vary from 50% in the non-welded parts to ~0% in the most intensely welded. Below about 10% primary porosity, the welded ignimbrite is often a low-vesicularity black *vitrophyre* composed of obsidian (Figure 36.11(B)). Finite strain is estimated from porosity and the deformation of pyroclasts (e.g., flattened pumice lapilli).

7.2. High-Grade Ignimbrites

Ignimbrites deposited much hotter than their Tg weld readily and completely, forming a temperature-dependent continuum with agglutinating spatter-fed lavas. Such high-grade and extremely high-grade ignimbrites exhibit features indicative of ductile flow during and immediately after deposition (**rheomorphism**). At high temperatures once primary porosity has been reduced below about 10%, the deposit begins to behave like a coherent, nonparticulate mass (i.e., clast–clast interactions cease as their boundaries cease to be rheologically significant) and exsolution of volatiles induces the growth of postemplacement (secondary) vesicles, thereby increasing the measured porosity. Rheomorphic ignimbrites are characterized by intense welding, flow folding of the eutaxitic fabric, the stretching of vesicles, and prolate strain (Figure 36.11(C) and (D)). Intense welding and rheomorphic deformation modify the eutaxitic fabric into flow banding analogous to that seen in rhyolite lavas, and the lithofacies often become very difficult to distinguish from lavas in the field or under the microscope: *lavalike* ignimbrites often have vitrophyres at both the base and the top, and autobreccias at their upper and distal margins. Once deposited, hot rheomorphic ignimbrites can continue to flow downslope like lavas, producing folds on the order of 10–100s m.

8. TRANSPORT AND DEPOSITIONAL PROCESSES INFERRED FROM PDC DEPOSITS

PDCs are extremely hazardous, unpredictable, and infrequent, and close observation is extremely dangerous. Inferences about the physical characteristics of PDCs (e.g., particle concentration, density stratification) can be made through detailed study of their deposits. Several seminal studies of ignimbrites over the past three decades have shed much light on the nature of PDCs (e.g., Wilson, 1985, 2001; Scott et al., 1996; Branney and Kokelaar, 2002).

On a first-order level, PDCs carry particles of widely differing grain size, shape, and hydrodynamic properties supported within a PDC by a range of mechanisms. Each particle carried in a current must pass through a lower *flow boundary zone* (i.e., the base of the current) before it can be deposited, and adjacent clasts in a deposit may have had different transport and particle support histories. The presence of vertical chemical zoning of juvenile pyroclasts in some ignimbrites (Figure 36.5), which results from changes in the composition of the erupting magma with time, support models of incremental deposition from PDCs (a process termed progressive aggradation, e.g., Branney and Kokelaar, 2002), whereby deposition accompanies transport and the deposit thickness reflects the duration and rate of deposition. Other evidence such as gradual changes with height in grain fabric orientation and lateral transitions from massive to stratified lithofacies also support this model (e.g., Figure 36.8; see Scott et al., 1996).

The massive, poorly sorted nature of many ignimbrites indicates deposition from high-particle concentration, lower flow boundary zones (granular fluid escape flow boundaries), in which particles are supported by grain collisions (in granular flows) and by the upward flux of dusty gas produced as larger particles try to deposit (hindered settling). Hindered settling results in partial fluidization (sedimentation fluidization) and greatly increases a current's mobility. Better sorted, stratified, and cross-stratified deposits result from traction sedimentation from fully dilute, low-particle concentration flow boundaries in which particle interactions have little role in particle support, or particle segregation or in current rheology. Instead, coarser particles are supported either by the turbulence in the surrounding dusty gas or by saltation and rolling along a substrate. These two types of flow boundary can be considered end members of a spectrum of flow boundary types, whose physical properties are influenced by factors such as current velocity and particle concentration, among others. Flow boundary conditions in between the two end members (for example, in terms of particle concentration) will result in deposits that are gradational in character (e.g., diffuse-stratified and bedded ignimbrites; Figure 36.4(D)). Many studies of deposits indicate that within a single PDC the flow boundary conditions can vary spatially at any one instant in time (nonuniformity) and vary temporally at any one location (unsteadiness), e.g., across topography.

The ability of PDCs to travel great distances across water and to cross mountain ranges up to 1500 m high indicates that the greater thickness of PDCs is a dilute, low-particle concentration suspension where particle support is dominated by fluid turbulence. Additional insight into the vertical structure of PDCs can be gained from ignimbrites that were emplaced over rugged topography which indicate that aggradation rates are significantly higher in the valleys and that particle concentrations in the lower flow boundary zone were also higher. What remains unclear for most PDCs, however, is the partitioning of mass between a higher concentration lower flow boundary zone and an overlying dilute and turbulent zone, noting that the vertical structure and the vertical density gradients are likely to vary considerably both between and within PDCs due to current nonuniformity and unsteadiness.

Structural analysis of rheomorphic lavalike ignimbrites has demonstrated synchronous deposition, welding, and rheomorphic deformation at the upward migrating, current-deposit interface as the deposit aggrades (Andrews and Branney, 2011); the shear developed in this narrow zone is sufficient to add up to 200 °C to the tuff due to strain heating sustaining low viscosity and enhancing welding and flow.

Small volume phreatomagmatic eruptions tend to produce transient currents which lose energy rapidly as they travel away from the vent. Close to the vent, material may be dumped rapidly to produce chaotic and massive beds. As the current moves outward, material is deposited and it becomes more dilute: turbulent transport of particles along the substrate can produce dune bedforms. As the current travels away from the vent and decelerates, higher rates of suspended load fallout lead to increasing particle concentrations in the flow boundary zone and the deposition of massive and bedded deposits in distal regions. Variations on these transitions have been seen and probably relate to differences in the initial physical properties of the currents (e.g., particle concentration or velocity).

Transport processes of BAFs are commonly considered to be dominated by granular flow mechanisms. The viscosity of the interstitial fluid and grain-to-grain interactions controls granular flow behavior and ash and gas are important in friction reduction. In these high-particle concentration currents, momentum is largely believed to be managed through grain collisions. Observations at Soufrière Hills volcano, Montserrat, and elsewhere indicate strongly stratified systems composed of granular underflows or "dense basal avalanches" beneath overriding turbulent ash clouds. Flow separation can occur in response to topographic blocking or slowing of the dense lower parts, and the overlying low-particle concentration ash cloud can detach and move independently as a dilute PDC that can swamp topographic highs.

9. DEPOSITS FROM PDCS THAT ENTERED THE SEA

PDCs erupted near the coast or in shallow water may pass across the sea for many 10s km. Debris deposited from PDCs traveling across open water can mix with the water column and transform into widespread submarine volcaniclastic deposits, such as debris flows or turbidites. Welded shallow submarine deposits are known in the geological record indicating significant retention of heat. Thermoremanent studies on submarine volcaniclastic deposits derived from the 1883 Krakatau eruption, Indonesia, indicate emplacement temperatures of >450 °C.

10. POSTDEPOSITIONAL CHANGES TO PDC DEPOSITS

10.1. Erosion

PDCs can rapidly deposit huge volumes of tephra, burying valleys, blocking river drainages and lakes. Erosion can occur instantly through runoff induced by rainfall, or by snow- and ice melt created by hot pyroclastic deposits. Several well-exposed ignimbrite sheets (e.g., VTTS ignimbrite, Alaska, Hildreth and Fierstein, 2012) contain intraeruption fluvial and mudflow deposits, indicating syneruption reworking of newly deposited tephra. Erosion rates increase rapidly after an eruption, carving rills, gullies, and steep-sided gorges through ignimbrite sheets. Erosion rates are high due to the unconsolidated nature of the tephra, the destruction of vegetation by PDCs, and the presence of steep slopes common to volcanic regions. Rapid incision of thick ignimbrite sheets can result in the collapse of steep cliffs and the generation of hot, secondary PDCs, e.g., the 1991 Pinatubo ignimbrite, Philippines. Rapid denudation of PDC deposits produce lahars and can instigate major changes to river systems for many years following an eruption.

10.2. Fumaroles and Phreatic Explosion Craters

Boiling of steam in saturated sediments underneath rapidly deposited, hot ignimbrite can result in phreatic explosions that excavate small craters. Superb examples occur in the VTTS ignimbrite, Alaska (Hildreth and Fierstein, 2012). These craters, 20–60 m wide and up to 20 m deep, are surrounded by rings and fans of poorly sorted cross-bedded ejecta that pockmark the surface of the host ignimbrite.

Rainfall, snowmelt, and rivers and streams can sustain fumaroles within hot ignimbrite sheets for many years after an eruption. Fossil fumaroles vary from narrow joint controlled cracks to funnel shapes up to 5 m wide at the ignimbrites, upper surface (Hildreth and Fierstein, 2012). Many were concentrated above preeruption stream courses. Fumarolic gases result in localized leaching of the ignimbrite, precipitation of vapor-phase minerals, and the early lithification of the deposit.

10.3. Alteration and Lithification

Ignimbrites are prone to alteration due to their hot, porous, permeable nature and the presence of highly reactive volcanic glass. Alteration can result in partial to total lithification as a result of vapor-phase mineralization, clay formation, or zeolitization and this can lead to the development of columnar joints in nonwelded ignimbrites. Deposits with abundant fine ash tend to be more strongly altered because the fine-grain size reduces the reaction time and creates a greater surface area for reactions between permeating fluids and particles.

The first alteration stage is devitrification—the slow crystallization of volcanic glass into microcrystalline arrays of predominantly feldspar (e.g., albite) and anhydrous silica polymorphs (e.g., tridymite, cristobalite). Devitrification occurs by diffusion, and therefore is temperature dependent. It can result in the pervasive crystallization of entire welded ignimbrite sheets, with the exception of the rapidly chilled marginal zones that are preserved as glassy vitrophyres. Localized devitrification, often centered on grain- or crystal-scale heterogeneities, is expressed as spherulites where albite and silica polymorphs typically radiate from a point. Water vapor derived from devitrification of hydrous glass and from heated meteoritic water initiates the redistribution of elements during vapor-phase alteration. This leads to the precipitation of silicate, hydrous silicate, oxide, and other minerals in the interstices between particles, and the removal of soluble elements like potassium. When pore fluid pressure exceeds the strength of the tuff, hydrofracturing creates secondary void spaces and permeability, some of which localize further devitrification (lithophysal cavities); such cavities are often infilled by hydrous silica. Interest in the long-term storage of hazardous waste in welded ignimbrites has chiefly been concerned with the heterogenous nature of porosity and permeability resulting from these diverse, polygenetic processes.

PDC deposits of phreatomagmatic eruptions, or those deposited in lakes and seawater, are particularly susceptible to alteration due to the presence of a condensing vapor phase (H_2O). Typically, reactions between volcanic clasts and moderate pH water will lead to clay formation (smectites) and is followed by zeolite formation. Common replacement and pore-filling zeolite minerals include phillipsite, chabazite, and analcime, which can form gel-like or amorphous microgranular aggregates coating pores and grains. In some cases, zeolite abundances in altered tuffs can reach 75%, with the principal control being the alkalinity of groundwater.

FURTHER READING

Andrews, G.D.M., Branney, M.J., 2011. Emplacement and rheomorphic deformation of a large, lava-like rhyolite ignimbrite: Grey's Landing, southern Idaho. Geol. Soc. Am. Bull. 123, 725−743.

Belousov, A., Voight, B., Belousova, M., 2007. Directed blasts and blast-generated pyroclastic density currents: a comparison of the Bezymianny 1956, Mount St Helens 1980 and Soufrière Hills, Montserrat 1997 eruptions and deposits. Bull. Volcanol. 69, 701−740.

Branney, M.J., Kokelaar, P., 2002. Pyroclastic Density Currents and the Sedimentation of Ignimbrites. Geol. Soc. London Memoir, 27, pp. 152.

Brown, R.J., Branney, M.J., 2013. Internal flow variations and diachronous sedimentation within extensive, sustained, density-stratified pyroclastic density currents flowing down gentle slopes, as revealed by the internal architectures of ignimbrites on Tenerife. Bull. Volcanol. 75, 727.

Bryan, S.E., Cas, R.A.F., Martí, J., 1998. Lithic breccias in intermediate volume phonolitic ignimbrites, Tenerife (Canary Islands): constraints on pyroclastic flow depositional processes. J. Volcanol. Geotherm. Res. 81, 269−296.

Calder, E.S., Sparks, R.S.J., Gardeweg, M.C., 2000. Erosion, transport and segregation of pumice and lithic clasts in pyroclastic flows inferred from ignimbrite at Lascar Volcano, Chile. J. Volcanol. Geotherm. Res. 104, 201−235.

Carrasco-Núñez, G., Branney, M.J., 2005. Progressive assembly of a massive layer of ignimbrite with normal-to-reverse compositional zoning: the Zaragoza ignimbrite of central Mexico. Bull. Volcanol. 68, 3−20.

Charbonnier, S.J., Gertisser, R., 2011. Deposit architecture and dynamics of the 2006 block-and-ash flows of Merapi Volcano, Java, Indonesia. Sedimentology 58, 1573−1612.

Chough, Y.K., Sohn, S.K., 1990. Depositional mechanics and sequences of base surges, Songaksan tuff ring, Cheju Island, Korea. Sedimentology 37, 1115−1135.

Cole, P.D., Guest, J.D., Duncan, A.M., Pacheco, P.-M., 2001. Capelinhos 1957-1958, Faial, Azores: deposits formed by an emergent Surtseyan eruption. Bull. Volcanol. 63, 204−220.

Cole, P.D., Calder, E.S., Sparks, R.S.J., Clarke, A.B., Druitt, T.H., Young, S.R., Herd, R.A., Harford, C.L., Norton, G.E., 2002. Deposits from dome-collapse and fountain-collapse pyroclastic flows at Soufriere Hills Volcano, Montserrat. In: Druitt, T.H., Kokelaar, B.P. (Eds.), the eruption of Soufriere Hills Volcano, Montserrat, from 1995−1999, 21. Geol Soc London Memoir, pp. 231−262.

Druitt, T.H., Edwards, L., Mellors, R.M., Pyle, D.M., Sparks, R.S.J., Lanphere, M., Davies, M., Barriero, B., 1999. Santorini Volcano, 19. Geol Soc. London Memoir, pp. 165.

Hildreth, W., Fierstein, J, 2012. The Novarupta-katmai Eruption of 1912—Largest Eruption of the Twentieth Century: Centennial Perspectives. USGS Prof. Paper 1791, pp. 259.

Quane, S.L., Russell, J.K., 2005. Ranking welding intensity in pyroclastic deposits. Bull. Volcanol. 67, 129–143.

Quane, S.L., Russell, J.K., Friedlander, E.A., 2009. Time scales of compaction in volcanic systems. Geology 37, 471–474.

Scott, W.E., Hoblitt, R.P., Torres, R.C., Self, S., Martinez, M.L., Nillos, T.J., 1996. Pyroclastic flows of the June 15 1991, climactic eruption of Mount Pinatubo. In: Newhall, C.G., Punongbayan, S. (Eds.), Fire and Mud: Eruptions of Pinatubo, Philippines. Philippine Institute of Volcanology and Seismology, Quenzen City. University of Washington Press, Seattle, pp. 545–570.

Sulpizio, R., De Rosa, R., Donato, P., 2008. The influence of variable topography on the depositional behaviour of pyroclastic density currents: the examples of the Upper Pollara eruption (Salina Island, southern Italy). J. Volcanol. Geotherm. Res. 175, 367–385.

Van Eaton, A.R., Wilson, C.J.N., 2013. The nature, origins and distribution of ash aggregates in a large-scale wet eruption deposit: Oruanui, New Zealand. J. Volcanol. Geotherm. Res. 250, 129–154.

Wilson, C.J.N., 1985. The taupo eruption, New Zealand: II. The taupo ignimbrite. Philos. Trans. R. Soc. A. 314, 229–310.

Wilson, C.J.N., 2001. The 26.5 ka Oruanui eruption, New Zealand: an introduction and overview. J. Volcanol. Geotherm. Res. 112, 133–174.

Chapter 37

Lahars and Their Deposits

James W. Vallance and Richard M. Iverson

US Geological Survey, Cascades Volcano Observatory, Vancouver, WA, USA

Chapter Outline

GLOSSARY

bulking The incorporation of solid debris, resulting in increased lahar mass and sediment concentration as the lahar moves downstream.

debris avalanche A flowing mixture of debris, rock, and moisture that moves downslope under the influence of gravity. Debris avalanches differ from debris flows in that they are not water-saturated and in that the load is mostly supported by particle-particle interactions.

debris flow A water-saturated mixture of debris that moves downslope under the influence of gravity, in which the solid and liquid fractions are approximately equal volumetrically and in which the two fractions move downstream approximately in unison.

debulking A process in which a lahar selectively deposits certain particles, owing to their size or density, as it moves downstream. Debulking differs from the general deposition of sediment because it preferentially removes particles, usually large or dense ones, from the flow. It results in decreased sediment concentration.

dilatancy A property of a granular material that enables it to change volume through expansion of pore space when subject to shear strain. Negative dilatancy implies pore space reduction.

hyperconcentrated flow A transitional flow type, between debris flow and streamflow. Unlike streamflow, a hyperconcentrated flow carries very high sediment loads, and unlike debris flow, coarse-grained solids tend to separate vertically from the liquid-and-fine-solids mixture.

granular temperature A scalar measure of random kinetic energy associated with grain agitation. Granular temperature may be interpreted as twice the fluctuation kinetic energy per unit mass of granular solids.

lahar An Indonesian term most commonly defined as a rapidly flowing, gravity-driven mixture of rock, debris, and water from a volcano. A lahar can vary in character with time and distance downstream. It may comprise one or more flow types, which include debris flow, transitional or hyperconcentrated flow, and muddy streamflow or flood flow. Flow-type transitions are commonly defined in terms of solids fraction; however, such transitions are gradational and dependent on other factors such as sediment-size distribution, clay mineralogy, particle agitation, and energy of the flow.

liquefaction A process that enables water-saturated sediment mixtures to flow almost as fluidly as liquids. Liquefaction results from excess pore-fluid pressure that reduces frictional energy dissipation at grain contacts. As pore-fluid pressure exceeds hydrostatic values such water-sediment mixtures become progressively more liquefied.

muddy streamflow or flood flow A type of flow in which fine-grained sediment moves in hydrodynamic suspension (suspended load) and coarse-grained sediment moves along the streambed (bed load). Some floods and muddy streamflows at volcanoes are genetically related to lahar events, but many are not.

stage (of flow) The surface height (above the channel bottom) of a flowing lahar or streamflow at a particular time. Examples

The Encyclopedia of Volcanoes. http://dx.doi.org/10.1016/B978-0-12-385938-9.00037-7

of lahar stages include the initial rising or waxing stage, the peak-inundation stage, and the final long-duration falling or waning stage.

suspension (hydrodynamic) A state in which fine particles remain suspended in a liquid solely as a consequence of buoyancy, liquid viscosity, and Brownian forces. Fluid turbulence can assist hydrodynamic suspension, but it is not a requisite.

turbulence (hydrodynamic) Chaotic fluid movement or deviation of flow from laminar. Turbulent fluid flow characterizes streamflows but not debris flows or sediment-rich hyperconcentrated flows. Turbulence is generated by fluid momentum fluxes and differs from debris agitation that is measured by granular temperature.

1. INTRODUCTION

Lahars occur during volcanic eruptions—or, less predictably, through other processes on steep volcanic terrain—when large masses of water mixed with sediment sweep down and off volcano slopes and commonly incorporate additional sediment and water. Because lahars are water-saturated, both liquid and solid interactions influence their behavior and distinguish them from other related phenomena common to volcanoes, such as **debris avalanches** and floods. The rock fragments carried by lahars make them especially destructive; the abundant liquid contained in them allows them to flow over gentle gradients and inundate areas far away from their sources. People in such distal areas commonly neither expect the danger nor anticipate the destructive power of lahars.

1.1. Historical and Prehistoric Examples of Lahars

Lahars inundate areas surrounding volcanoes and can damage or destroy communities downstream. The 1985 pyroclastic eruption of Nevado del Ruiz in Columbia was a small event (0.01 km^3) but generated lahars 10 times larger ($\sim$0.1 km^3) that flowed as far as 100 km down four of five drainages that head at the volcano (Pierson et al., 1990). These, the deadliest historical lahars worldwide, destroyed more than 5000 homes and killed more than 23,000 people. In the town of Armero, 73 km downstream of the volcano, virtually all structures in the path of a lahar were obliterated and three-quarters of the inhabitants were killed. The interaction of hot pyroclastic flows or surges with glacial ice and snow at the summit of the volcano caused the Nevado del Ruiz lahars. By contrast, the 1980 eruption of Mount St Helens showed that large, destructive lahars form not only owing to the interaction of hot pyroclastic rock with snow and ice but also result from partial or wholesale **liquefaction** of water-laden debris-avalanche deposits (Scott, 1988).

Breakout of crater lakes and volcano-dammed lakes can cause large lahars during or after eruptions. Kelut volcano, on the island of Java in Indonesia, is notorious for producing lahars when explosive eruptions expel its crater lake, and its lahars are notorious for killing people who live on its fertile slopes and for destroying villages within a 40-km radius. Since 1848, the volcano has erupted 10 times. Of these eruptions, seven have expelled the crater lake, and five have produced devastating lahars. The most devastating of these lahars occurred when the largest volumes of water were expelled from the lake. In 1919, an eruption blew 4×10^7 m^3 of water out of Kelut's lake and generated lahars that swept almost 40 km downstream, covered more than 130 km^2, and killed more than 5000 people.

Torrential rains mobilized loose debris and generated hundreds of lahars for years after the 1991 Pinatubo eruption in the Philippines. Although most of the Pinatubo lahars were relatively small, within 6 years they cumulatively remobilized about 2.5 km^3 of the 5.5 km^3 of pyroclastic flows emplaced during the eruption. Filling of downstream channels as well as overbank flow onto surrounding fields and villages inundated more than 400 km^2 and displaced more than 50,000 persons.

In 1998, Hurricane Mitch unleashed torrential rain that caused a slope failure and released a flood of water that generated a lethal lahar at Casita volcano in Nicaragua (Scott et al., 2005). The flood of water and sediment eroded and incorporated three times its volume in sediment as it descended the steep volcano slopes. As it spread across the gentle slopes of the volcano's apron, the lahar destroyed two towns and killed more than 2000 inhabitants.

Studies of prehistoric lahar deposits reveal the potential for even greater disasters. About 5600 years ago, the 3.8-km^3 Osceola mudflow from Mount Rainier, USA began with an edifice collapse that transformed to a lahar. The lahar apparently mobilized because of the enormous volume of water contained in pore spaces and in the volcano's hydrothermal system (Vallance and Scott, 1997). It filled valleys to depths of 80–150 m, flowed more than 120 km down valleys, and continued as far as 20 km underwater in Puget Sound while retaining sufficient coherence to transport large gravel and wood fragments (Crandell, 1971). The 540 km^2 area it inundated is now populated by hundreds of thousands of people.

1.2. Purpose

The chief purpose of this chapter is to summarize what is known about the nature and behavior of lahars on the basis of observations, experiments, theory, and examination of deposits, and further to describe the nature of deposits derived from such events. Because the timing of lahar events is largely unpredictable and working with active flows can be hazardous, much of our present knowledge of flow behavior is inferred from the study of lahar deposits. Nonetheless, key observational, experimental, and theoretical studies have also improved understanding.

2. GENESIS OF LAHARS

Lahars may be primary (syneruptive) or secondary (posteruptive or unrelated to eruptions). Lahar genesis requires (1) an adequate water source; (2) abundant unconsolidated debris, which typically includes pyroclastic-flow and -fall deposits, glacial drift, colluvium, and soil; (3) steep slopes (commonly >25°) and substantial relief at the source; and (4) a triggering mechanism. Water sources include pore or hydrothermal water, rapidly melted snow and ice, subglacially trapped water, crater or other lake water, and rainfall runoff. Water encountered along a lahar's path, either in streams or lakes, or stored within floodplain sediments can also influence downstream dynamics.

2.1. Lahars Caused by Melting of Snow and Ice, Floods, or Heavy Rains

Floods of water moving across loose sediment common on the flanks and aprons of volcanoes readily incorporate that debris and may quickly form lahars (Figure 37.1(A) and (B)). This **bulking** process is critical to all lahars that begin with sudden water releases.

Lahars induced by sudden water release can occur by four principal means. (1) Hot rock avalanches, pyroclastic flows, and surges mix with and melt glacial ice and snow rapidly. Such hot flows may come entirely or nearly to rest, generating meltwater that then runs off, coalesces, and erodes the pyroclastic debris to form water-rich lahars. They may also continue moving across snow or ice, incorporating it continuously to form lahars that often include voluminous slurries of snow, ice, and slush. As they move downstream, the lahars may continue to bulk up with volcanic debris, glacial drift, alluvium, and colluvium so that within a few kilometers to several tens of kilometers they become solids-rich **debris flows**. Lahars of this type are considered primary. (2) Volcanic eruptions can displace large volumes of crater-lake water that form lahars downstream. Crater and caldera lakes and volcanic debris-dammed lakes can also break out months to years after eruptions. Such delayed breakouts occur when water levels gradually rise, then overtop or pipe through, and incise fragile debris dams rapidly. (3) Subglacial eruptions can form subglacial lakes that eventually break out when a section of the ice cap becomes buoyant or pervasively fractured and thereby releases the trapped water. Small-scale outburst floods also occur during periods of glacier ablation and commonly bulk up to form lahars. Huge eruption-driven outbursts cause sediment-rich water floods called jökulhlaups. (4) Lahars that result from intense rainfall often occur after eruptions deposit abundant loose debris from pyroclastic-flow or -fall deposits. Lahars of this type are commonly small but abundant during rainy periods. Size and frequency of rain-induced lahars may increase in the months or years following the primary pyroclastic eruption, then decrease exponentially as drainage networks and vegetation re-establish themselves (e.g., Mount Pinatubo after its 1991 pyroclastic eruption).

Because clay-rich sediment is both uncommon on the flanks or aprons of active volcanoes and resistant to erosion, lahars induced by sudden water release are generally clay-poor and contain less than about 5% clay/(sand + silt + clay) by dry weight.

2.2. Lahars Caused by Collapse of Volcano Flanks

Although most volcano flank collapses behave as debris avalanches, those with sufficient, widely dispersed pore water and hydrothermal water in the precollapse rock may liquefy as the material deforms during collapse. Flank-collapse-induced lahars may have various triggers, including magmatic or phreatic volcanism, volcanic or tectonic earthquakes, hydrothermal groundwater pressurization, and flank bulging caused by magma intrusion.

Hydrothermal alteration of rocks, especially at glaciated volcanoes, increases the probability of edifice-collapse lahars. Acid-sulfate leaching in hydrothermal systems removes mobile elements, adds sulfate, and decomposes framework silicates to form fine-grained silica phases, such as cristobalite and opal, and clay minerals, such as kaolinite and smectite. This process weakens the rock so that it more readily disintegrates during deformation after collapse. Thus, huge blocks of rock typical of debris avalanches are uncommon in lahars of this type. Abundant alteration minerals, especially clay minerals, increase porosity and decrease permeability of the rock and thus, in combination with the hydrothermal system, trap a widely dispersed reservoir of water within the precollapse rock mass. Because of its high water content and its tendency to disintegrate, hydrothermally altered rock, unlike fresh rock, easily liquefies as it deforms. Collapse-induced lahars are commonly clay-rich and most are observed to have greater than 5% clay/(sand + silt + clay) by dry weight (Vallance and Scott, 1997).

Clay-rich, collapse-induced lahars appear to be more common at ice-clad volcanoes than at volcanoes that are free of ice. Glacial erosion tends to expose deeper, potentially more-altered portions of volcanoes, while incised slopes in altered rock are susceptible to failure not only because the rock is weak but also because they are over-steepened. Lastly, melting glacial ice as well as seasonal snowpack provides a slow-release source of water that supplies the hydrothermal system and is important to the efficient operation of the acid-sulfate leaching process.

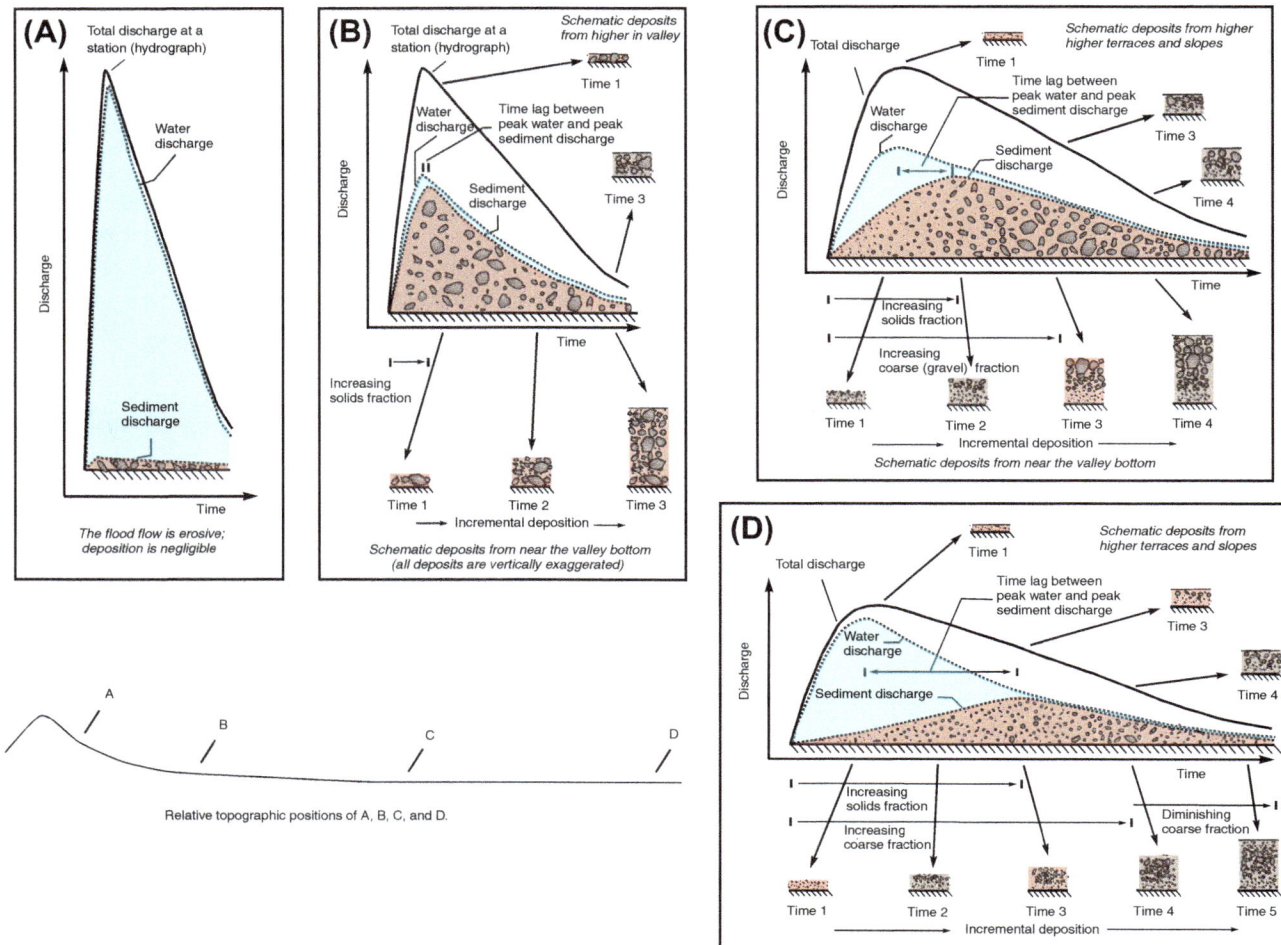

FIGURE 37.1 Schematic hydrographs showing how lahars beginning with water floods are initiated and behave as they undergo downstream dilution. (**A**) Flood flow; (**B**) Debris flow; (**C**) Transitional flow; (**D**) Hyperconcentrated flow. The diagram also illustrates the progressive-aggradation model of inverse grading in panels (**C**) and (**D**).

3. LAHAR BEHAVIOR: A MECHANICAL PERSPECTIVE

Lahars comprise roughly equal volumes of water and granulated rock, and their mechanical behavior can be understood by considering how this two-phase composition influences momentum transfer and energy dissipation (Iverson, 1997). When lahar mixtures flow, they can move almost like liquids—despite having grain concentrations and bulk densities comparable to that of solid ground. Lahar mobility can persist as long as water containing suspended mud-sized particles (silt + clay) reduces frictional energy dissipation by exerting local lubricating forces where large grains contact one another. On a continuum scale that encompasses many grains and adjacent fluid, effects of these local lubrication forces are manifested

by pore-fluid pressures that may cause liquefaction. However, liquefaction diminishes as lahar motion slows and ceases, and lahar deposits ultimately dewater and consolidate until they reach a nearly rigid state (Major and Iverson, 1999). This conspicuous state transition provides clear evidence that lahar rheology is not a fixed property, but is instead dependent on solid-fluid interactions that evolve as a lahar's kinetic energy and composition evolve (Figure 37.2, Video 37.1).

Supplementary video related to this article can be found online at http://booksite.elsevier.com/9780123859389

In a continuum-mechanics context, the chief measure of lahar composition is the mixture bulk density ρ, defined as

$$\rho = \rho_s m + \rho_f (1 - m) \qquad (37.1)$$

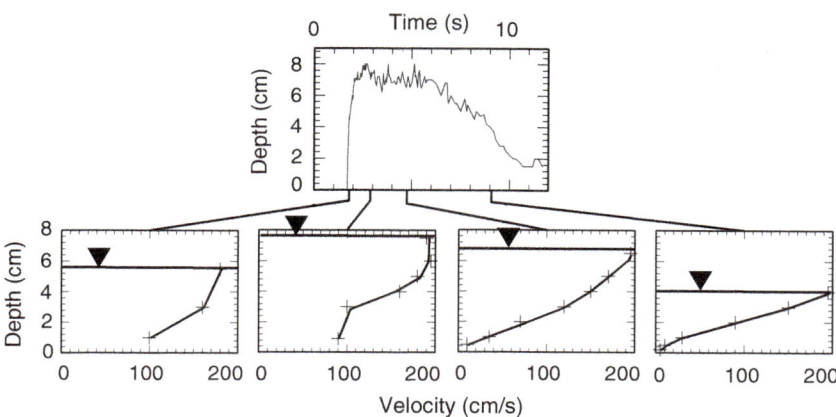

FIGURE 37.2 Variation of flow depths and vertical velocity profiles in experimental debris flow as it passed cross-section 3 m downslope from head of rectangular flume 0.3-m wide and sloped 18°. Variations in velocity profiles imply variation in apparent rheology. Flume bed roughened with grains like those used in flowing debris, 0.8–20 mm. Data extracted from analysis of video of flow past smooth glass walls. Volume fraction 0.5 at flow head, 0.45 flow tail. Black triangles indicate flow surface. *From Iverson and Vallance (2001). See Video 37.1 for an animated version of this figure.*

where ρ_s is the density of solid grains, ρ_f is the density of intergranular fluid (including clay and silt-sized particles held in **hydrodynamic suspension**), and m is the volume fraction occupied by all other solid grains. Typical values of these properties are $\rho_s \sim 2500$ kg/m^3, $\rho_f \sim 1100$ kg/m^3, $m \sim 0.6$, and $\rho \sim 2000$ kg/m^3, but some lahars can have values that differ greatly from these. For example, pumice-rich lahars can have $\rho_s \approx \rho_f$, implying that solid grains can float in the presence of hydrostatic fluid pressure. The values of m and ρ evolve by large amounts in many lahars as a consequence of bulking or **debulking**, while values of ρ_s and ρ_f may evolve in subtler fashion.

Like the bulk density ρ, the velocity $\vec{v}$ of lahar mixtures is defined by a weighted sum:

$$\vec{v} = [\vec{v}_s \rho_s m + \vec{v}_f \rho_f (1-m)]/\rho \qquad (37.2)$$

where $\vec{v}_s$ is the velocity of solid grains and $\vec{v}_f$ is the velocity of adjacent fluid. The weighted sum of the solid-phase momentum $\rho_s \vec{v}_s$ and the fluid-phase momentum $\rho_f \vec{v}_f$ per unit volume of mixture determines momentum per unit volume $\rho \vec{v}$ and dictates the form of (37.2) (Iverson, 1997). Equation (37.2) demonstrates that the mixture momentum $\rho \vec{v}$ can evolve as a result of evolution of m even if $\rho_s \vec{v}_s$ and $\rho_f \vec{v}_f$ remain constant.

Lahar materials generally remain well-mixed, implying that $\vec{v} \approx \vec{v}_s \approx \vec{v}_f$, but small differences between solid- and fluid-phase velocities can have large implications for momentum exchange and energy dissipation. These differences can be summarized by

$$\vec{q} = (\vec{v}_f - \vec{v}_s)(1-m) \qquad (37.3)$$

where $\vec{q}$ is the volumetric fluid flux per unit area in a frame of reference that moves with the adjacent solid grains. Nonzero values of $\vec{q}$ imply the existence of solid-fluid drag that may reduce grain-contact forces. For lahars that contain grains with a great variety of shapes and sizes, grain-scale drag forces cannot be determined precisely.

However, on a bulk continuum scale drag forces can be estimated by using Darcy's law:

$$\vec{q} = -(k/\mu)\nabla p_e \qquad (37.4)$$

where k is the hydraulic permeability of the granular aggregate, μ is the viscosity of intergranular fluid, and p_e is the nonhydrostatic or "excess" fluid pressure. The pressure gradient ∇p_e generated by $\vec{q}$ arises from drag that is inversely proportional to k/μ. Values of k/μ for lahar materials range from about 10^{-4} m^{3}s/kg for dilute mixtures of water, sand, and gravel to 10^{-14} m^{3}s/kg for concentrated, mud-rich mixtures (Iverson, 1997). A mid-range value $k/\mu = 10^{-9}$ m^{3}s/kg in Equation (37.4) implies that a $\vec{q}$ value no larger than 10^{-5} m/s can produce a fluid pressure gradient ∇p_e in excess of a typical hydrostatic pressure gradient, ρ_f g $\approx 10^4$ kg/m^{2}s^2. Recognition that small $\vec{q}$ leads to excess fluid pressure gradients and that excess fluid pressure gradients govern liquefaction is crucial for understanding both the mechanics of lahars and the importance of changes in lahar composition that alter the value of k/μ. Increased proportions of silt and clay greatly diminish permeability, decrease k/μ, and thereby enhance lahar mobility (Video 37.2).

Supplementary video related to this article can be found online at http://booksite.elsevier.com/9780123859389

Excess fluid pressure gradients that counteract gravity-driven settling of grains cause at least partial liquefaction, thereby mimicking the effect of enhanced buoyancy and reducing intergranular friction. Such excess pressure gradients tend to dissipate with time if grain settling proceeds unperturbed, but any process that transiently dilates a lahar mixture can restart the settling process (Iverson and George, 2014). For example, as lahars descend steep channels or encounter obstacles, reduction of solids volume fraction m results from agitation of debris caused by conversion of large-scale translational kinetic energy to disorganized grain-scale kinetic energy (i.e., **granular**

temperature). Provided that sufficient fluid is available to fill intergranular voids that are enlarged by this agitation, the fluid can subsequently pressurize as ρ and m relax toward equilibrium values.

The mobility of lahar mixtures can thus be viewed as the product of an energy cascade that begins with the bulk gravitational potential energy of sediment and water poised at high elevations. The next step of the cascade entails transformation of potential energy into a combination of translational kinetic energy, granular temperature, and elevated fluid pressure. The cascade ends when friction has dissipated all of the original energy as heat and the lahar mixture comes to rest and consolidates (Iverson, 1997).

3.1. Lahar Dynamics and Runout: Statistical Patterns

Lahar mobility and dynamics on a large scale can be gauged from the extent of areas inundated downstream. In this context, the simplest measure of mobility is the ratio H/L, where H is the total vertical elevation lost and L is the total horizontal distance traversed during lahar motion. Although this measure is easy to understand and use, it disregards the important effects of lahar volume and three-dimensional path topography. Alternative measures of bulk mobility account for these effects by using scale-invariant power-law equations, $A = aV^{2/3}$ and $B = bV^{2/3}$, where A is the average vertical cross-sectional area inundated along a lahar path, B is the total planimetric area inundated, V is lahar volume, and a and b are statistically calibrated coefficients (Iverson et al., 1998). Values of a decrease while those of b increase as flows exhibit decreasing frictional resistance and increasing mobility. Indeed, values of a and b that are statistically calibrated using field data provide numerical indices that distinguish the mobility of lahars from less mobile debris avalanches (Figure 37.3). Calibrated values can vary for lahars with differing compositions and origins, but the values $a = 0.05$ and $b = 200$ apply to a broad spectrum of volcanic debris flows. As a result, the empirically calibrated inundation-area equations, $A = 0.05V^{2/3}$ and $B = 200\,V^{2/3}$, have proven useful for assessment of lahar hazards. The same empiricisms also provide a basis for testing some aspects of numerical models of lahar dynamics.

3.2. Lahar Dynamics and Runout: Physics-Based Models

The most powerful tools currently available for understanding lahar dynamics are depth-averaged numerical models. The equations used in these models are derived by applying the laws of mass and momentum conservation, but this application is simplified by assuming that momentum fluxes normal to the bed are negligible in comparison to downstream fluxes. Nevertheless, modern depth-averaged lahar models are sophisticated enough to simulate the primary features of lahar dynamics. For example, they can account for entrainment of bed or bank material that may change total lahar mass and solid volume fraction.

For downstream motion limited to one direction (here denoted by x), depth-averaged conservation of mass within a two-phase lahar that may erode its bed is described by Iverson (2013) as

$$\frac{\partial(\rho h)}{\partial t} + \frac{\partial(\rho h\bar{v})}{\partial x} = \rho_b \frac{\partial z_b}{\partial t} \qquad (37.5)$$

where t is time, h is flow thickness, $\bar{v}$ is the depth-averaged value of the x-component of $\vec{v}$, $\partial z_b/\partial t$ is the local rate of bed lowering due to erosion in the direction normal to x, and ρ_b is the bulk density of bed material subject to erosion. The lahar bulk density ρ may evolve because of differences between ρ and ρ_b, and it may also evolve because of changes in mixture agitation that result in changes in the solid volume fraction m (Iverson and George, 2014).

Depth-averaged conservation of x-momentum in a lahar that obeys (37.5) is expressed by

$$\frac{\partial(\rho h\bar{v})}{\partial t} + \frac{\partial\left(\beta\rho h\bar{v}^2\right)}{\partial x} = \rho g_x h - L_x - \tau_b \qquad (37.6)$$

where g_x is the x-component of the acceleration due to gravity, L_x is a resisting stress proportional to $g_z h[\partial(\rho h)/\partial x]$ that arises when gravity acts on a mass with an uneven weight distribution, τ_b is the resisting basal shear traction, and β is a momentum-distribution coefficient that accounts for deviations of v from $\bar{v}$ but is commonly assumed to equal 1. Importantly, (37.6) contains no term analogous to the bed-erosion term $\rho_b \partial z_b/\partial t$ in (37.5). Such a term is absent because static bed material lacks momentum to contribute to the lahar. However, a crucial and often overlooked implication of (37.6) is that, if $\rho\bar{v}$ is constant, then the value of τ_b in (37.6) must diminish as the rate of erosion increases, for otherwise momentum is not conserved in the two-body system comprising a lahar and its bed (Iverson, 2012). Indeed, relative to τ_b acting in the absence of erosion, τ_b must decrease by exactly the amount $\rho\bar{v}(\partial z_b/\partial t)$ to enable erosion at the rate $\partial z_b/\partial t$ to occur—provided that $\beta = 1$ applies and that the bed-normal momentum flux is zero. This inference helps explain why bed erosion and growth of lahar momentum commonly proceed hand-in-hand, but it does not explain the reasons for reduction of τ_b.

Phenomena that determine the evolving value of τ_b depend on the details of small-scale mixture dynamics and bed composition. Small-scale mixture dynamics generally involves some combination of energy-dissipating collisions and rubbing contacts of grains, as well as viscous fluid flow that simultaneously lubricates grain contacts and dissipates fluid kinetic energy. No complete model of these

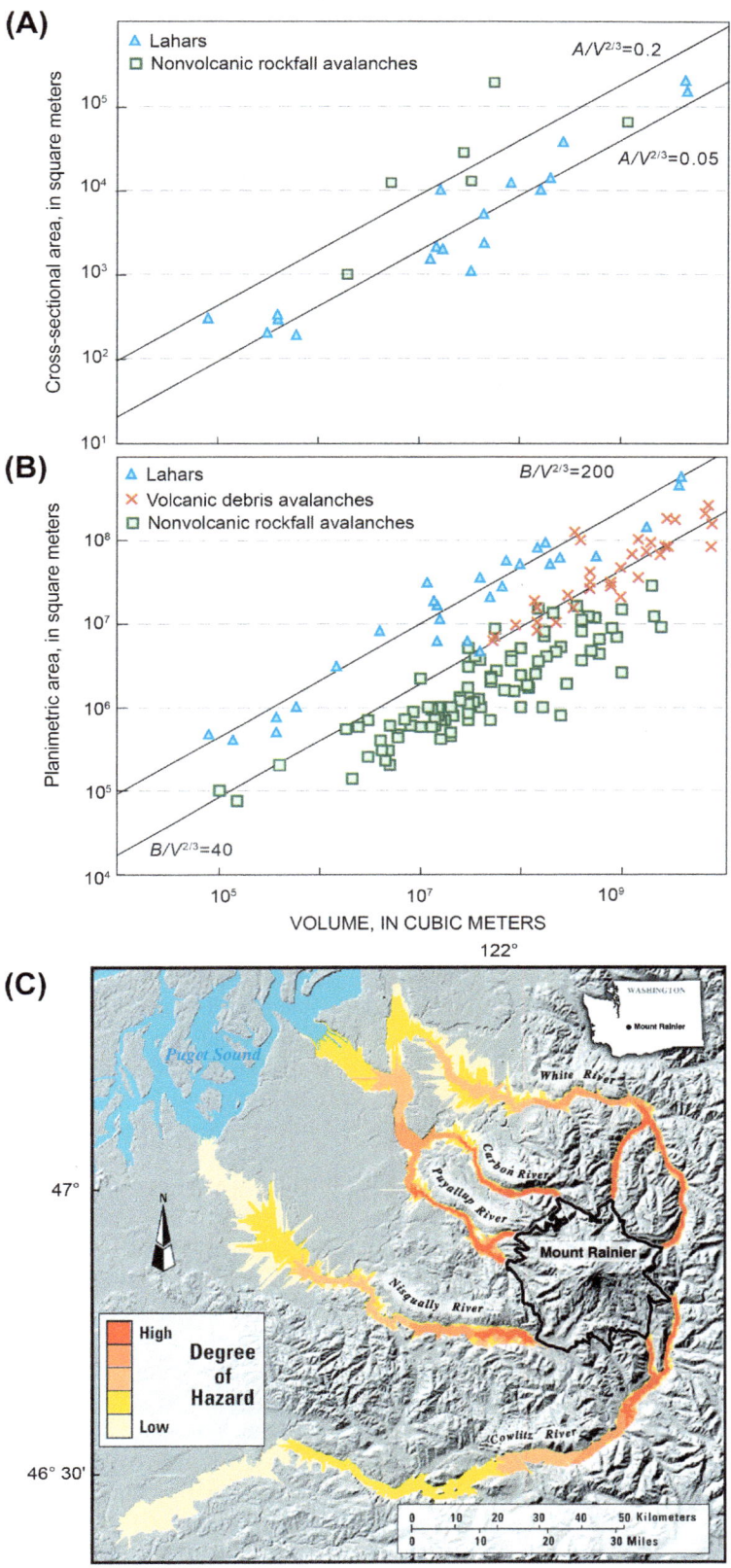

FIGURE 37.3 Plots of (**A**) inundated flow-path cross-sectional area *A* versus flow volume *V*, (**B**) area inundated *B* versus volume, and (**C**) inundation areas calculated for valleys draining Mount Rainier, Washington, USA. *(Details in Iverson et al. (1998).)*

interacting processes in complex lahar mixtures is available. However, where a lahar interacts with its bed, behavior can be approximated by using a modified Coulomb friction equation expressed as

$$\tau_b = \left[(\rho_s - \rho_f)g_z h - p_e\big|_{z_b}\right] \tan \phi(I_v) \qquad (37.7)$$

where g_z is the bed-normal component of gravitational acceleration, $(\rho_s - \rho_f)g_z h$ is the buoyant weight of granular debris acting on the bed, $p_e\big|_{z_b}$ is the excess pore pressure acting on the bed, and $\phi(I_v)$ is a Coulomb friction angle that increases gradually as a function of a dimensionless ratio I_v that compares the viscous resistance to volume change to total normal stress (Boyer et al., 2011). This parameter, defined here as

$$I_v = \frac{\mu(\bar{v}/h)}{(\rho_s - \rho_f)g_z h - p_e\big|_{z_b}}, \qquad (37.8)$$

accounts implicitly for the influences of collisional and viscous energy dissipation on Coulomb friction. These influences grow as the shear rate increases and as the effect of gravity-induced stresses decreases. If $I_v = 0$, then $\phi(I_v)$ equals the static friction angle, which for most sediment ranges from 30 to 40°. As $I_v \to \infty$, values of $\phi(I_v)$ increase but are unlikely to exceed about 60°.

Equations (37.7) and (37.8) imply that a reduction in τ_b accompanying bed erosion might occur for several reasons. For example, undrained loading of wet bed sediment overridden by a lahar might increase $p_e\big|_{z_b}$. In extreme cases such an increase can lead to wholesale bed liquefaction and a drastic reduction in τ_b (Iverson et al., 2011). A subtler but nevertheless important effect can arise when fast-moving, overriding lahar material with a relatively large value of $\phi(I_v)$ interacts with static or slightly moving bed material with a smaller value of $\phi(I_v)$. In this case, τ_b acts on a surface that migrates downward into effectively weaker material as erosion proceeds, thereby diminishing the basal shear resistance.

The mathematical description of lahar motion given by (37.5), (37.6), and (37.7) is incomplete without equations that describe evolution of $z_b(x,t)$, $p_e\big|_{z_b}(x,t)$, and $\rho(x, t)$ (or, alternatively, $m(x, t)$). Various rationales have been offered for formulating such equations, but research has not yet identified the best option. Thus, for the present, all mathematical models of lahar dynamics remain unfinished or, at least, unproven. Many models simply assume that $\rho(x,t)$ is constant and that $z_b(x,t)$ and $p_e\big|_{z_b}(x, t)$ are known independently.

4. LAHAR BEHAVIOR: A GEOLOGICAL PERSPECTIVE

Lahars commonly change character as they travel downstream (Figure 37.1). Floods generated on or near volcanoes can incorporate enough sediment proximally to become **hyperconcentrated flows** or debris flows. In medial or distal reaches, lahars that move down river channels push water ahead of them such that sediment-rich debris flows lag behind water-rich hyperconcentrated flows and flood flows (Figure 37.4). Furthermore, debris avalanches descending the flanks of volcanoes can partly or entirely evolve to debris flows as they move downstream (Scott, 1988; Vallance and Scott, 1997).

4.1. Erosion, Bulking, and Mass Growth

Lahars cause erosion by undercutting steep slopes and terrace scarps and by scouring their beds. Erosion is strongest along steep channel reaches underlain by loose clastic sediment and weakest either along reaches underlain by highly resistant bedrock or reaches with gentle gradients. Along any particular reach, watery sediment floods and water-rich, hyperconcentrated flows are typically more erosive than sediment-rich flows (Video 37.3, Figure 37.4), but local erosion can occur regardless of flow type. The waxing stage of a lahar is likely to coincide with the most widespread and voluminous erosion and bulking. The final, waning stages of a lahar can also be erosive, and commonly result in channel incision of fresh lahar deposits as the flow becomes more watery.

Supplementary video related to this article can be found online at http://booksite.elsevier.com/9780123859389

Erosion at the base of a lahar occurs by piecemeal dislodgment of particles, by liquefaction of loose wet substrate and wholesale incorporation of that sediment, and by rip up of sediment owing to root throw of falling trees. The presence of undisturbed, delicate deposits such as tephra layers at the base of debris-flow deposits suggests that bed erosion is minimal in some topographic settings, especially when the flow type is a sediment-rich debris flow.

An important means by which lahars may incorporate sediment is by compression of loose wet substrate material and subsequent liquefaction of that material. As a lahar overruns wet or water-saturated sediment, it loads, compresses, and shears it slightly, and thereby drives up pore-fluid pressure, causing runaway liquefaction (Video 37.4) (Iverson et al., 2011). Pervasive liquefaction of substrate not only makes that sediment easily eroded and entrained, but also lubricates the base of the flow and allows it to accelerate downstream. Given more than a meter of precipitation during Hurricane Mitch at the end of the 1998 rainy season, the landslide-induced Casita lahar probably owed its downstream growth in volume (times 2−3) and its enhanced mobility to loading and liquefaction of water-saturated sediment in its path.

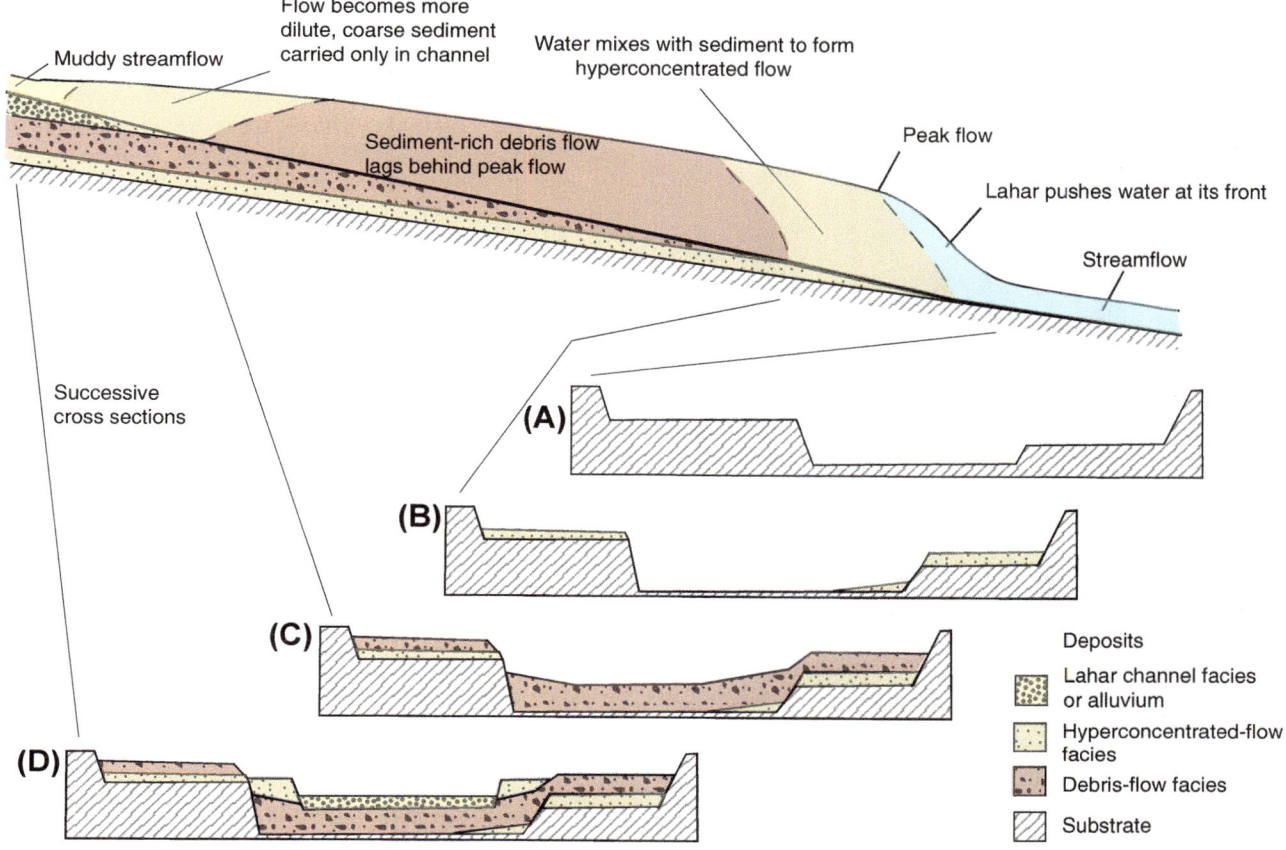

FIGURE 37.4 Schematic illustration of a lahar undergoing downstream dilution from debris-flow phase to hyperconcentrated-flow phase and deposit facies. The model shows the expected sequences of hyperconcentrated- and debris-flow deposits in cross-section ((**A**) through (**D**)).

Supplementary video related to this article can be found online at http://booksite.elsevier.com/9780123859389

Lahars voluminous enough to escape channels knock trees down and incorporate them. Root balls of falling trees drag considerable sediment into the active flow and loosen even more sediment that is then available for erosion. Voluminous lahars that inundate large areas of forested terrain can incorporate considerable quantities of sediment and huge amounts of wood in this way.

Undercutting of steep slopes, fluvial terrace scarps, and active stream banks is one of the most important ways in which lahars erode and incorporate sediment. Undercutting is active during flood flow, hyperconcentrated flow, and debris flow. Large lahars are capable of incorporating megablocks (>10 m across) of unconsolidated sediment, and sometimes even of bedrock, in this way (Figure 37.5). Megablocks may move tens of kilometers downstream before they ultimately fragment into smaller pieces.

Progressive downstream bulking causes downstream changes in lahars. Bulking adds sediment and thereby transforms flood flows and hyperconcentrated flows to more sediment-rich debris flows (Figure 37.1(A) to (B)). If the process continues, both waxing and waning **stages of flow** ultimately become debris flows. With continued downstream motion, debris flows become richer in exotic sediment such as alluvium, colluvium, and glacial drift. Study of deposits indicates that the waxing flow-front and following peak stages of the flow are the most erosive and thus most readily incorporate exotic sediment (e.g., Vallance and Scott, 1997). The sediment-rich falling flow that follows peak flow is less erosive and commonly deposits sediment rather than erodes it (Figure 37.1). The final waning-stage flow is typically more watery, and more erosive, but less voluminous in discharge than the preceding stage. Final waning stages of lahars commonly incise previously emplaced lahar deposits (Figure 37.4).

4.2. Grain-Size and Grain-Density Segregation

Grains in lahars can effectively segregate by density or size, but the most important segregation processes are mediated

FIGURE 37.5 Photograph of cross-sectional view of 20-m thick lahar deposit about 40 km downstream from Cotopaxi, Ecuador, illustrating megaclasts incorporated into lahar by undercutting of channels. Megaclasts are mostly derived from earlier lahar deposits and are rounded from abrasion and rolling during flow. *Photograph by Patty Mothes.*

by the solids fraction m, proportion of coarse grains, and the fluid density ρ_f, the latter being determined by the proportion of mud-sized grains held in hydrodynamic suspension. In dilute lahars, large dense grains may readily settle gravitationally. Through this process, the largest grains collect in the lowest layer of a moving flow and progressively smaller ones collect above that; thus a normally graded flow develops. As larger grains collect toward the base of the flow where velocity is less than average, they progressively lag farther behind the flow front (Figure 37.1(C) to (D)).

In flows with greater solids fraction, grain-to-grain contacts inhibit gravitational settling and may favor preferential rise of large grains, even dense ones. In shearing flows with grains denser than the fluid and with solids fraction $> \sim 0.4$, large particles rise rather than settle. Force imbalances or grain rotations can push the large grains from one layer into another, and smaller grains are likely to fill in beneath larger ones and thus not rise. Once near the flow surface where velocity is greater than at the flow front, large grains migrate forward. Accretion of large grains at flow

perimeters generates bouldery flow fronts (Figure 37.6). The mixture that follows the bouldery front remains fluid so that debris flows moving across fans and channels commonly develop resistant bouldery levee margins and liquefied interiors that govern their behavior (Iverson, 1997) (Videos 37.1 and 37.2, Figure 37.6).

4.3. Downstream Dilution and Flow Transformation

Once off the volcano's slopes, lahars commonly descend river channels that contain significant volumes of water. Lahars, which typically move faster than normal streamflow, push river water ahead of them and gradually, with distance downstream, begin to mix with that water (Pierson and Scott, 1985; Cronin et al., 1997). As the flow front becomes progressively more watery, it loses its capacity to carry larger gravel particles, and these progressively lag behind the flow front (Figures 37.1 and 37.4). With time and distance downstream, a dilution front progresses from the front of the lahar to its middle and eventually the entire lahar becomes more dilute. In lahars that occurred at Mount St Helens in 1980 and 1982, downstream dilution occurred over the course of tens of kilometers and caused a complete transformation from debris flow to hyperconcentrated flow (Figure 37.1(D)). In medial reaches, the hyperconcentrated flow preceded the debris-flow portion of the lahar because the dilution process began at the flow front, then gradually worked backward toward the tail as the flow traveled downstream (Figure 37.1(C)). Although downstream dilution can affect large lahars, the process has little effect on the behavior of lahars so large that their volumes are significantly greater than that of the water in the river being overrun.

5. LAHAR DEPOSITS

5.1. Depositional Processes

An increasing body of evidence indicates that deposits of both hyperconcentrated flows and debris flows aggrade progressively. Stratification in deposits of transitional or hyperconcentrated flows is clear evidence of progressive aggradation. Experimental observation of levees forming during debris flow emplacement shows that levees accrete from bottom to top (Major, 1997). Even in massive lahar deposits, field evidence of progressive aggradation is strong, including (1) strong alignment of elongate clasts parallel to flow directions, (2) imbrication of elongate clasts dipping upstream, (3) strong changes in composition of particles with vertical position in outcrops, especially those that are graded, and (4) marginal deposits with grain compositions similar to those at the base of thick valley-bottom deposits (Vallance and Scott, 1997). Lahars with

(A)

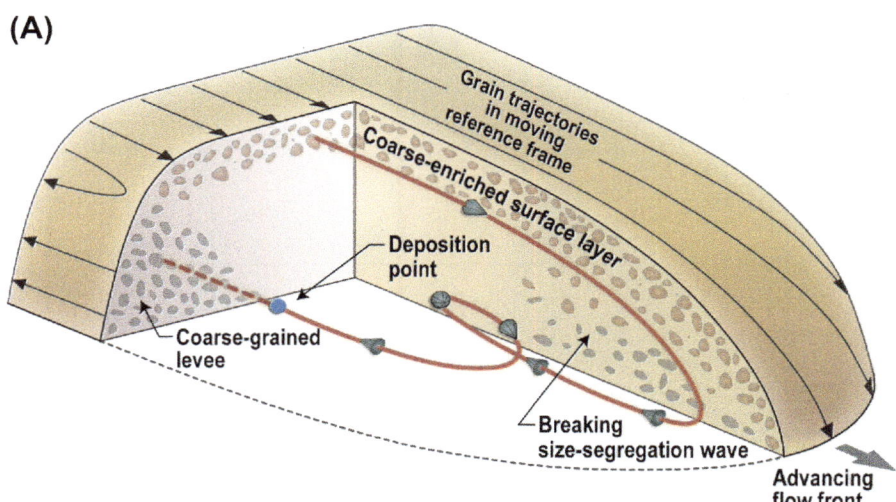

(B)

FIGURE 37.6 (A) Schematic cutaway illustration of the process leading to formation of lateral levees behind a debris flow front. Red path shows how coarse grain near surface migrates toward a levee. Reference frame moves at speed of advancing flow so that grain advancing less rapidly than flow front appears to move backwards (Johnson et al., 2012, gives details). (B) Deposit of experimental debris flow illustrating coarse-grained margins and levees. Width of distal end of deposit is about 2.5 m. White lines are topographic contours.

massive valley-bottom deposits sometimes show bottom-to-top compositional trends that mirror trends in deposits from successively higher to lower positions along valley margins. Such compositional trends imply progressive aggradation of a flow that evolves compositionally with time as it passes. In such cases, marks of peak-flow levels high on valley sides commonly indicate flow depths 5—10 times greater than typical valley-bottom deposit thicknesses.

In lahars, normally and inversely graded deposits generally originate from progressive aggradation of a flow that is compositionally zoned in terms of its grain-size distribution from front to tail of its hydrograph (Figures 37.1 and 37.7) rather than from a vertically graded flow that freezes in place. Figure 37.7(A) and (B) **(Times 1—4)** illustrates schematically how progressive aggradation from

a debris-flow wave with a concentration of large particles at its front can generate a normally graded deposit (c.f., Vallance and Scott, 1997). Accretion occurs for a short time only near inundation limits where grading does not occur (Figure 37.7(B), **Times 1—2**).

Progressive aggradation from a lahar wave that has become more dilute and less capable of carrying larger particles, which lag behind, emplaces inversely graded deposits (Figure 37.1(C) and (D)). Aggradation from a dilute debris flow whose grain-size distribution coarsens from its head toward its tail produces inversely graded deposits, which may be faintly stratified to massive (Figure 37.1(C)). Farther downstream, where the entire lahar has become hyperconcentrated, progressive aggradation produces finer grained beds that may be inversely graded or both inversely graded and normally graded (Scott,

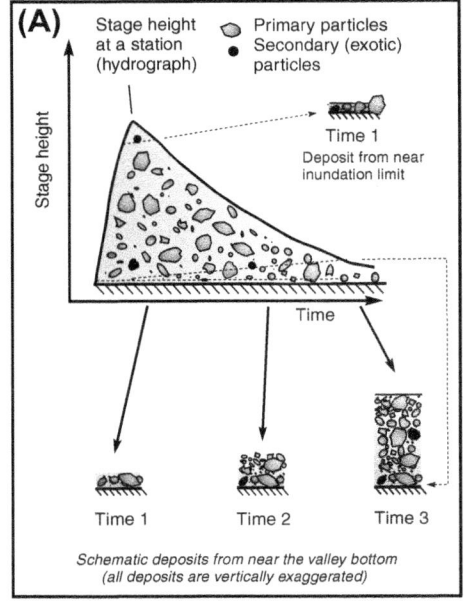

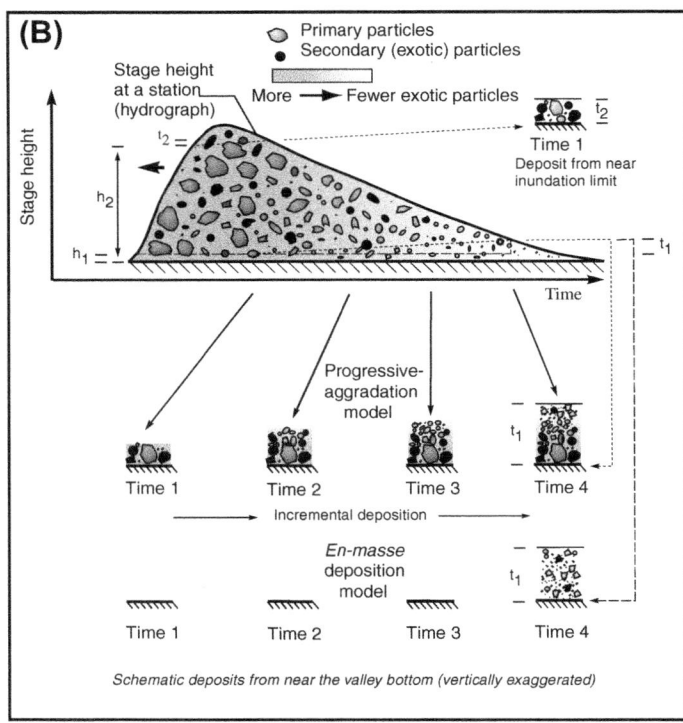

Relative topographic positions of A and B.

FIGURE 37.7 Schematic hydrographs showing the behavior and downstream changes of lahars that begin as avalanches of water-saturated debris. With increasing distance downstream ((**A**) to (**B**)) the lahar incorporates secondary exotic particles, especially near its flow front. Panel (**B**) also illustrates how a flow, coarser grained at its head than at its tail, can accrete incrementally to form a normally graded deposit. *(Adapted from Vallance and Scott, 1997.)* In (**B**), exotic particles are most common at the base of the normally graded deposits and at inundation limits.

1988) (Figure 37.1(D)). Deposits in positions higher on valley walls can also be graded, but often less obviously so.

Small bouldery debris flows that do not undergo downstream dilution commonly form cobble- and boulder-rich margins owing to the size segregation process described above (Figure 37.6). When a debris flow of this type reaches sufficiently gentle slopes, the frictional perimeter slows to a stop and leaves behind a steep-fronted fines-poor margin and a partly liquefied fines-rich interior. Upstream, the fluid interior drains downslope, leaves behind coarse margins, and forms levees (Johnson et al., 2012) (Figure 37.6(A)).

5.2. Characteristics of Deposits

Lahar deposits formed by debris flows, and hyperconcentrated flows that commonly evolve from them, have some similarities, but they have many differences too, and it is useful to characterize each type of deposit separately.

Debris-flow deposits are massive and very poorly sorted to extremely poorly sorted (greater than 2 phi units and typically greater than 4 phi units on the Wentworth scale). Grain-size distributions are commonly bimodal (Figure 37.8). The deposits may be normally (Figure 37.9) or inversely (Figure 37.10(A)) graded throughout or, in some cases, can be inversely graded near their bases and normally graded near their tops (Figure 37.10(B)). Sedimentary fabrics, such as imbrication, are weakly developed. Deposits are extremely compact. Particles found within debris-flow deposits can be monolithologic but are more commonly heterolithologic; they can be rounded to angular, but primary particles are usually subangular to angular. Deposits commonly exhibit vesicles in the matrix, which result from entrapment of air bubbles. Other common constituents include wood fragments, casts of wood fragments, and charcoal. Concentrations of coarse particles, especially low-density particles like pumice, are common at deposit tops.

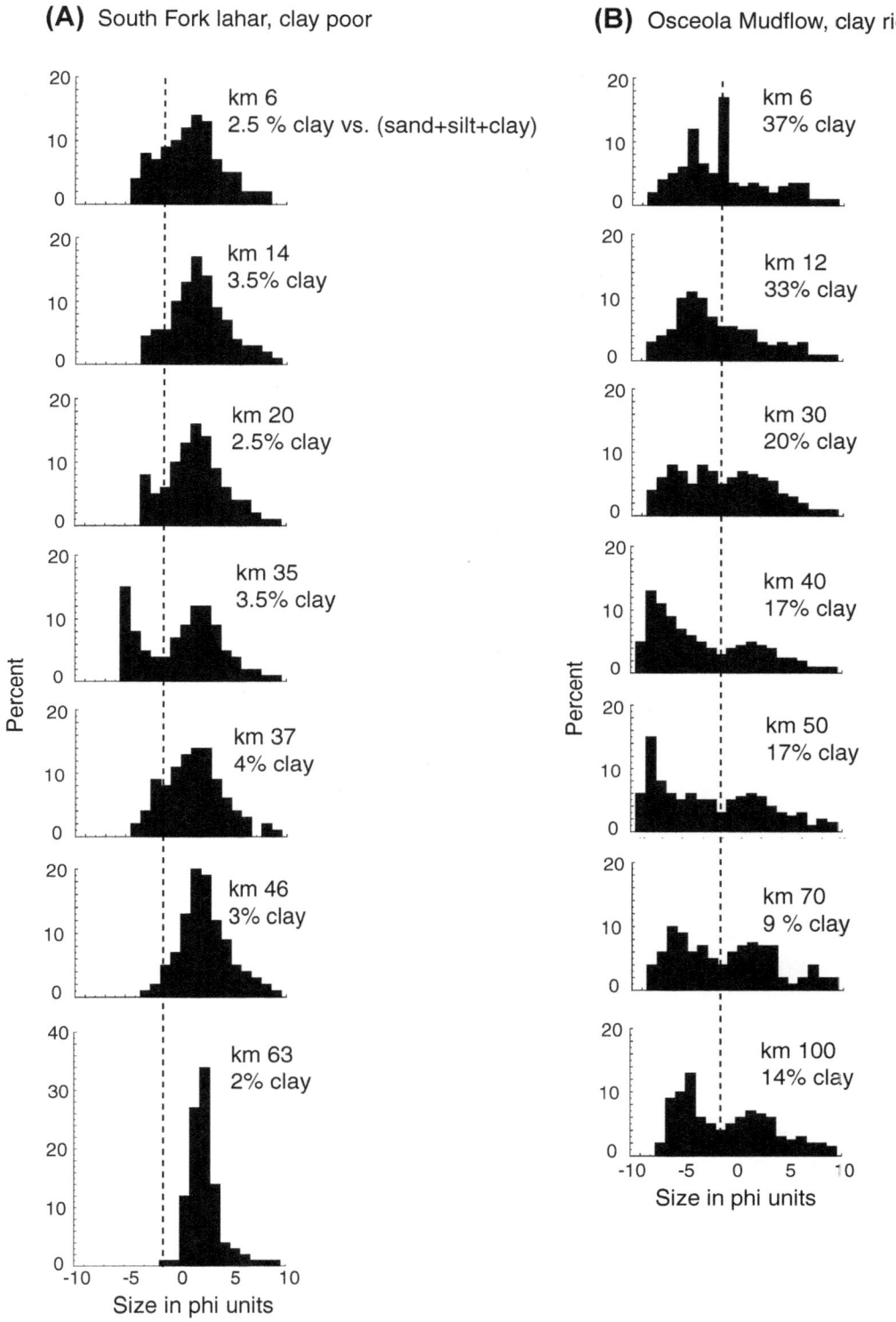

FIGURE 37.8 Grain-size histograms of **(A)** clay-poor South Fork Toutle River lahar. *From Scott (1988)* and **(B)** clay-rich Osceola lahar *From Vallance and Scott, 1997* sediment showing downstream trends. Also shown are proportions of clay in the deposit matrix (sand + silt + clay). In these examples, matrix clay percentage changes little downstream for clay-poor lahar deposit but falls nearly three-fold in clay-rich deposit. Vertical dotted line is at −2 phi units, or 4 mm size, and serves to highlight coarser and finer grained components.

FIGURE 37.9 Photographs of clay-rich lahar deposits caused by volcano flank collapses. (**A**) Trout Lake lahar, Mount Adams, USA and (**B**) Osceola mudflow, Mount Rainier, USA illustrate normally graded debris-flow deposits. (**C**) Hummocks 20−40 m across and 10−15 m high characterize the surface of the Osceola mudflow, derived from Mount Rainier, USA, in the background 70 km upstream. Clay-rich normally graded deposits, locally with hummocks scattered across their surfaces, are common in collapse-induced lahars.

Thicknesses of debris-flow deposits vary from tens of centimeters to tens of meters. Thick fill deposits occur in valley bottoms and lowlands (Figures 37.5 and 37.9(B)). Deposits on higher terraces and slopes within valleys are thinner than those in valley bottoms, and those on steep slopes will drape underlying topography as thin veneers. Both levees and steep terminal flow fronts are common in the deposits of debris flows relatively unaffected by downstream dilution.

Hyperconcentrated-flow deposits have characteristics intermediate between debris-flow and alluvial deposits. They thus have intermediate sorting coefficients (1−2 phi units) and grain sizes. They can be massive (Figure 37.10(C)), but commonly they have weak stratification defined by thin horizontal beds and very low-angle cross bed-sets composed of fine-grained laminae and thicker coarser-grained beds (Figure 37.10(D)). Overbank deposits have grain sizes in the granule, sand, and silt range with the occasional isolated pebble, cobble, or boulder (Figure 37.10(E)). If pumice was an important constituent of the flows, zones of nearly 100% pumice are commonly present at the tops of overbank deposits. These pumice concentrations result from pumice rafts stranded during falling-stage hyperconcentrated flow. Channel-facies deposits commonly exhibit strong bimodality and clast support, with concentrations of cobbles and boulders surrounding granule-sand-silt matrix (Figure 37.10(E)). Vesicles are sometimes present but less obvious than in debris-flow deposits. Deposits are compact except for bouldery channel examples (Figure 37.10(E)). Though rare, dewatering features such as dish structures (Figure 37.10(G)) and pillar structures may be present.

Hyperconcentrated-flow deposits have flat tops and may vary in thickness from a few centimeters to several meters. Flow tops have scattered pebbles and large grains, especially pumice if present; they also commonly have thin layers of fine sand and silt that form during compaction and dewatering.

Lahar deposits fall into clay-rich and clay-poor groups (Figure 37.8), which empirically correspond to origins as avalanches of water-saturated hydrothermally altered rock and origins as flood of water incorporating loose sediment on the volcano. Lahar size, origin, and depositional environment determine the facies that form. Clay-rich lahar deposits at Mount Rainier are solely from avalanches. Such deposits are massive, extremely poorly sorted, and commonly normally graded (Figure 37.9(A) and (B)). The Osceola mudflow for example, has proximal and medial hummocky facies that contain many megaclasts (Figure 37.9(C)). Proximal and medial valley-side facies form thin (0.1−1 m) veneers on steep slopes; and proximal, medial, and distal axial facies form thick (2−20 m) fills with common normal grading in valley bottoms and lowlands. Clay-poor lahar deposits, such as those at Mount St Helens, may have the following facies (Figure 37.10): debris-flow (**A**), flood-plain (**E**), channel (**E**) transition (**B**), hyperconcentrated-flow (**C, D, G**), and stream-flow (**F, H**). The last three of these facies characterize downstream dilution of the lahar. Deposit interiors are more uniform, massive, poorly sorted, and matrix-rich than levees and snouts. Deposits left on steep upland slopes typically form thin lags with concentrations of the coarsest, densest particles, and a deficiency of matrix.

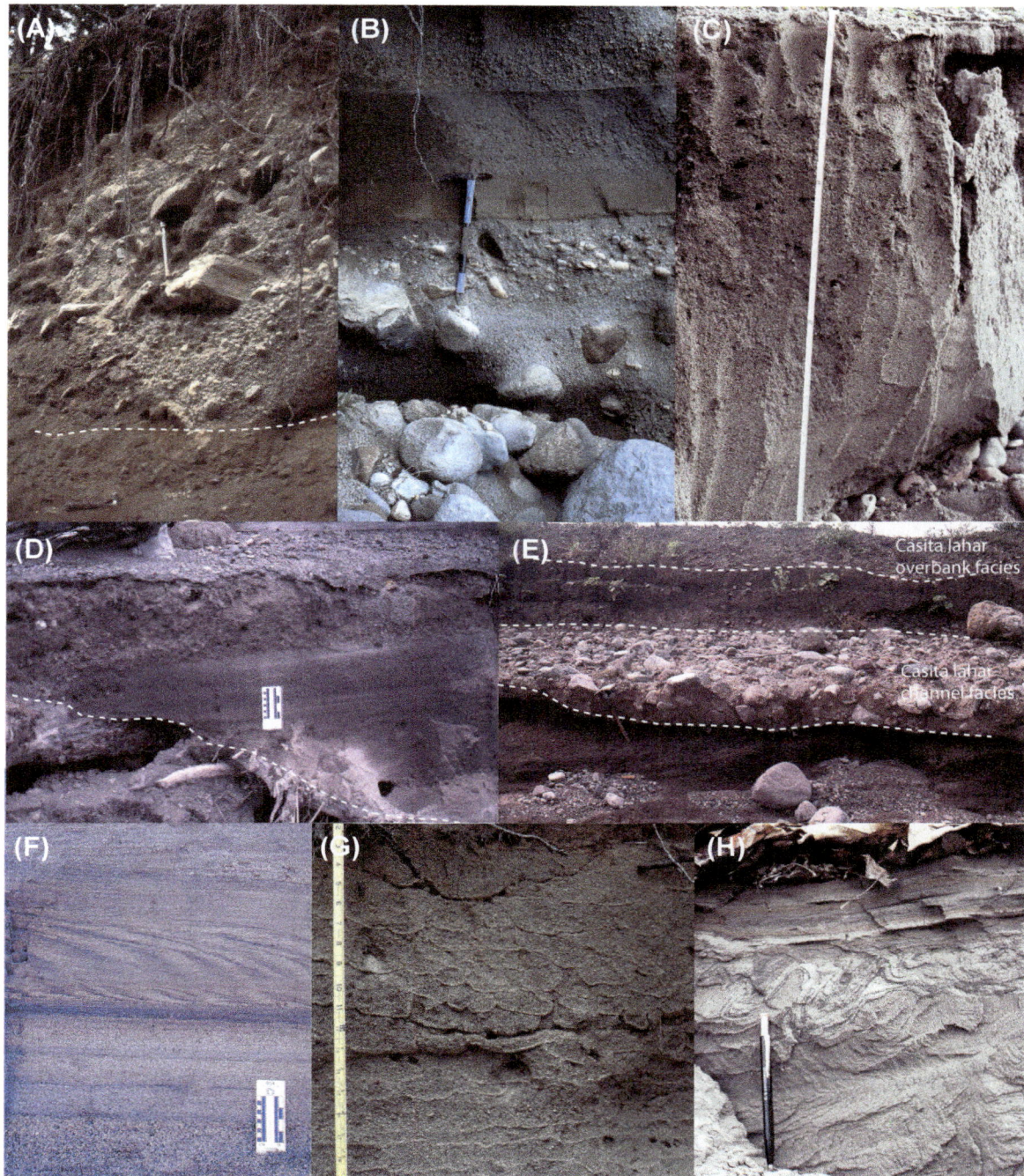

FIGURE 37.10 Photographs showing depositional facies of clay-poor lahars. (**A**) Clay-poor debris-flow deposit at Mount Adams, USA that is inversely graded at its base and both massive and ungraded at its top (dashed line indicates base of deposit; shovel is 50 cm long). (**B**) Transitional debris-flow to hyperconcentrated-flow deposits at Mount Rainier, USA. The basal unit (at the point of the ice axe) contains floating, rounded cobbles and boulders and probably reflects deposition from hyperconcentrated flow in a channel. (**C**) Hyperconcentrated-flow deposit that is inversely graded in the lowest two-thirds of the outcrop and normally graded in the upper third of the outcrop at Mount St Helens in 1982. The deposit comprises silt, sand, granules, and small pebbles with a single mode of coarse sand and granules. Deposit is about 1 m thick. (**D**) Faintly stratified hyperconcentrated-flow deposit and overlying transitional debris-flow deposit at Mount St Helens in 1982. (**E**) 1998 Casita lahar deposit, Nicaragua, illustrating lahar overbank and channel facies. (**F**) Climbing ripple structure transitioning upward to faintly stratified hyperconcentrated-flow facies at Pinatubo, Philippines. (**G**) Dish structure in a hyperconcentrated-flow deposit at Mount Rainier, USA. The structure evolves during or after deposition owing to dewatering during compaction. Structures are more strongly inflected toward the top of the unit. Tape measure for scale. (**H**) Wet-sediment deformation features in stream-flow facies of South Fork Toutle River lahar, 1980, Mount St Helens, USA. (*(C), (G), and (H), photographs by Kevin Scott.*)

5.3. Distinguishing Lahar Deposits from Other Common Diamictons

No single characteristic serves to distinguish lahar deposits from those of nonwelded pyroclastic flows, glaciers, debris avalanches, and landslides. Unlike lahar deposits, nonwelded pyroclastic-flow deposits do not have matrix vesicles, are not as well cemented, and contain mainly juvenile particles. Carbonized wood and magnetically oriented clasts help to distinguish them from deposits of cold lahars (the vast majority) but not necessarily from those of hot lahars. Nonwelded pyroclastic-flow deposits are usually more friable than lahar deposits. Debris-avalanche deposits generally have surfaces that are more irregular than those of lahar deposits. Although lahar deposits, especially clay-rich ones, can have hummocks and lateral levees, these features are more prominent in debris-avalanche deposits. Distal and marginal parts of debris-avalanche deposits can be flat-topped and can contain matrix vesicles, but these features are more typical of lahar deposits. Lateral and terminal moraine landforms or striated boulders and cobbles within deposits distinguish till from lahars. Unlike lahar deposits, till does not contain matrix vesicles, casts of wood fragments, or the wood fragments themselves. Till is also more heterolithologic than most lahar deposits. Landslide deposits generally have more local distribution, and more uniform lithology than lahar deposits do. Mapping distribution and inferring context is perhaps the best way to distinguish among poorly sorted deposits common at volcanoes.

FURTHER READING

Boyer, F., Guazzelli, E., Pouliquen, O., 2011. Unifying suspension and granular rheology. Phys. Rev. Lett. 107, 188301. http://dx.doi.org/10.1103/PhysRevLett.107.188301.

Crandell, D.R., 1971. Postglacial Lahars from Mount Rainier Volcano, Washington. U.S. Geological Survey Professional Paper 677, 75 p.

Cronin, S.J., Neall, V.E., Palmer, A.S., Lecointre, J.A., 1997. Changes in Whangaehu river lahar characteristics during the 1995 eruption sequence, Ruapehu volcano, New Zealand. J. Volcanol. Geoth. Res. 76, 47—61.

Iverson, R.M., 1997. The physics of debris flows. Rev. Geophys. 35 (3), 245—296.

Iverson, R.M., 2012. Elementary theory of bed-sediment entrainment by debris flows and avalanches. J. Geophys. Res. 117, F03006. http://dx.doi.org/10.1029/2011JF002189, 17 p.

Iverson, R.M., 2013. Mechanics of debris flows and rock avalanches. In: Fernando, H.J.S. (Ed.), Handbook of Environmental Fluid Dynamics. CRC Press, Taylor and Francis, pp. 573—587.

Iverson, R.M., George, D.L., 2014. A depth-averaged debris-flow model that includes the effects of evolving dilatency: 1. Physical basis. Proceedings of the Royal Society, London, Series A, 470, 20130819. http://dx.doi.org/10.1098/RSPA20130819.

Iverson, R.M., Reid, M.E., Logan, M., LaHusen, R.G., Godt, J.W., Griswold, J.G., 2011. Positive feedback and momentum growth during debris-flow entrainment of wet bed sediment. Nat. Geosci. 4 (2), 116—121. http://dx.doi.org/10.1038/NGEO1040.

Iverson, R.M., Schilling, S.P., Vallance, J.W., 1998. Objective delineation of lahar-inundation hazard zones. Geol. Soc. Am. Bull. 110, 972—984.

Iverson, R.M., Vallance, J.W., 2001. New views of granular mass flows. Geology 29, 115—118.

Johnson, C.G., Kokelaar, B.P., Iverson, R.M., Logan, M., LaHusen, R.G., Gray, J.M.N.T., 2012. Grain-size segregation and levee formation in geophysical mass flows. J. Geophys. Res. 117, F01032. http://dx.doi.org/10.1029/2011JF002185.

Logan, M., Iverson, R.M., 2007, revised 2013. Video Documentation of Experiments at the USGS Debris-Flow Flume 1992—2006 (Amended to Include 2007—2013). v. 1.3. U.S. Geological Survey, Open-File Report, 2007-1315. http://pubs.usgs.gov/of/2007/1315/.

Major, J.J., 1997. Deposition al processes in large-scale debris-flow experiments. J. Geol. 105, 245—366.

Major, J.J., Iverson, R.M., 1999. Debris flow deposition: effects of pore-fluid pressure and frictions concentrated at flow margins. Geol. Soc. Am. Bull. 111, 1424—1434.

Pierson, T.C., Scott, K.M., 1985. Downstream dilution of a lahar: transition from debris flow to hyperconcentrated streamflow. Water Resour. Res. 21, 1511—1524.

Pierson, T.P., Janda, R.J., Thouret, J.C., Borerro, C.A., 1990. Perturbation and melting of snow and ice by the 13 November 1985 eruption of Nevado del Ruiz, Colombia, and consequent mobilization, flow and deposition of lahars. J. Volcanol. Geoth. Res. 41, 17—66.

Scott, K.M., 1988. Origins, Behavior, and Sedimentology of Lahars and Lahar-runout Flows in the Toutle-Cowlitz River System. U.S. Geological Survey, Professional Paper, 1447-A, 74 p.

Scott, K.M., Vallance, J.W., Kerle, N., Macías, J.L., Strauch, W., Devoli, G., 2005. Catastrophic precipitation-triggered lahar at Casita volcano, Nicaragua—occurrence, bulking, and transformation. Earth Surf. Proc. Land. 30, 59—79.

Vallance, J.W., Scott, K.M., 1997. The Osceola Mudflow from Mount Rainier: sedimentology and hazard implications of a huge clay-rich debris flow. Geol. Soc. Am. Bull. 109, 143—163.

Chapter 38

Landslides, Debris Avalanches, and Volcanic Gravitational Deformation

Benjamin van Wyk de Vries

Laboratoire Magmas et Volcans, Université Blaise Pascal-CNRS-IRD, OPGC, Clermont Ferrand, France

Tim Davies

Department of Geological Sciences, University of Canterbury, Christchurch, New Zealand

Chapter Outline

GLOSSARY

gravitational volcano-tectonics The deformation of a volcanic edifice by the stress of gravitational loading, and accommodated by low-strength rock in the substrata and volcano. The principal types are spreading, where there is outward edifice displacement and sagging where there is inward displacement.

volcano-tectonics The deformation of a volcano through edifice interaction with crustal faults and stress fields.

landslide The outward and downward sliding of a large mass on the flank of a volcano or other slope, where the driving forces exceed the resisting force on the basal layer.

debris avalanche A rapid and catastrophic mass movement originating from a landslide that may travel horizontally several times the fall height. The resistance force on the lower layer is initially very small compared to the driving force.

debris avalanche deposit (DAD) The deposit left by a debris avalanche.

scar The depression left after a landslide has developed into a debris avalanche.

hummock A hill on a debris avalanche deposit surface formed by the brittle extension of the rock mass in a horst and graben style, where the horst forms the hummock.

toreva block A large rotational landslide block at the head of an avalanche deposit.

block facies Parts of a debris avalanche deposit inherited from the original edifice that preserve original stratigraphy and vary from intact to highly brecciated.

matrix facies Highly fragmented parts of a debris avalanche deposit that are mostly derived from disaggregation and fragmentation and mixing of block facies. Mixed facies is a polylithologic variety that may contain significant proportions of substrate, and basal facies is a finer grained and sheared variety.

jigsaw texture A textural result of brecciation in a debris avalanche, where the original fit between the fractured clasts can still be observed.

1. INTRODUCTION

Volcanoes grow incrementally by eruption and intrusion of magma, unlike other mountains that are principally formed by tectonic uplift and erosion. As they grow, volcanoes increase their mass over time, and thus increase the load on underlying rocks. These rocks in turn deform under the imposed weight (Figures 38.1 and 38.2). The gravitational deformation causes the volcanic edifice to sink and/or spread outward. For example (Figure 38.1(A)), the Big Island of Hawaii has caused about 5 km flexure of the oceanic lithosphere (Morgan et al., 2003; see also extensive

The Encyclopedia of Volcanoes. http://dx.doi.org/10.1016/B978-0-12-385938-9.00038-9

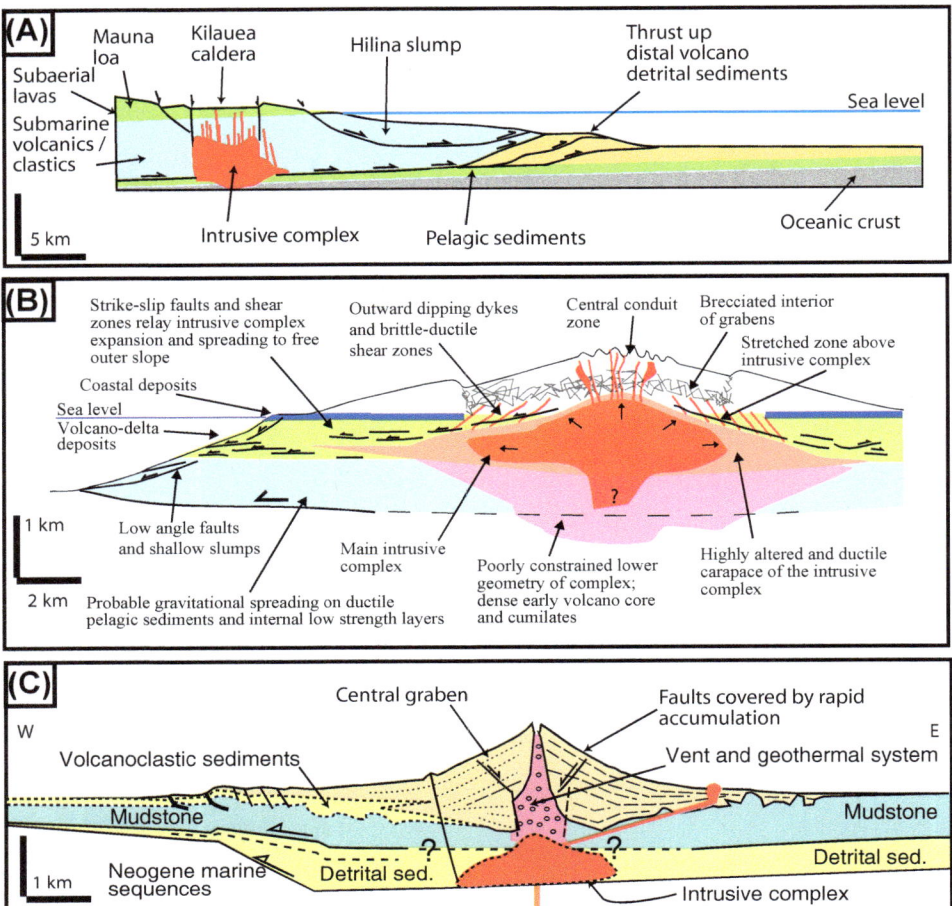

FIGURE 38.1 Examples of gravitational deformation of volcanoes. (A) The Big Island of Hawaii, seen in cross section (Morgan et al., 2003), showing the huge flexure produced by the volcano load on the oceanic crust, and outward sliding on the basal pelagic sediments. (B) Cross sectional of Piton des Neiges volcano, Reunion Island, showing internal structure, deformation and its relationship to structures. (C) Concepcion, Nicaragua, also seen in cross section, showing large-scale gravitational spreading and sagging on weak lake sediments (Borgia and van Wyk de Vries, 2003).

additional bibliography in annex); internal deformation has generated a summit graben complex and flank landslides on Piton des Neiges, Réunion Island (Figure 38.1(B)); and spreading at Concepción volcano, Nicaragua (Figure 38.1(C)) has created a range of folds around its base (Borgia and van Wyk de Vries, 2003).

If outward movement takes place in one main direction, such as for elongated volcanoes or on sloping substrata (Figure 38.2(A)), the volcano may develop landsliding and catastrophic debris avalanches may form. Volcanoes are constructed out of weak rocks, and the weakness is augmented by a combination of hydrothermal alteration, volcano-tectonic deformation, and intrusion-related deformation. The edifice therefore contains many structural weaknesses that can lead to destabilization through increased topographic loading, seismic loading, and progressive strength loss through hydrothermal alteration, deformation, intrusion, and pore pressure increase in hydrothermal systems. The propensity to collapse means

that volcanic landslides are not necessarily related to eruptions and magmatic activity and can occur during dormant periods, or even when a volcano is extinct. Volcanic landslides are commonly termed "sector" or "flank" collapse in the volcanological literature, although here we use the more common term "landslide."

2. VOLCANO SPREADING AND SAGGING

Volcano growth loads substrata and can cause material to flow outward from under the edifice. This outward spreading displaces the volcano flanks outward, creating summit extension and basal constriction. The characteristic structures produced are summit grabens, which are linked at the base to strike-slip faults, thrusts, and folds (Figures 38.1 and 38.2(A)). Alternatively, where there is a thicker ductile layer below the volcano, subsidence may dominate over outward spreading, leading to the central

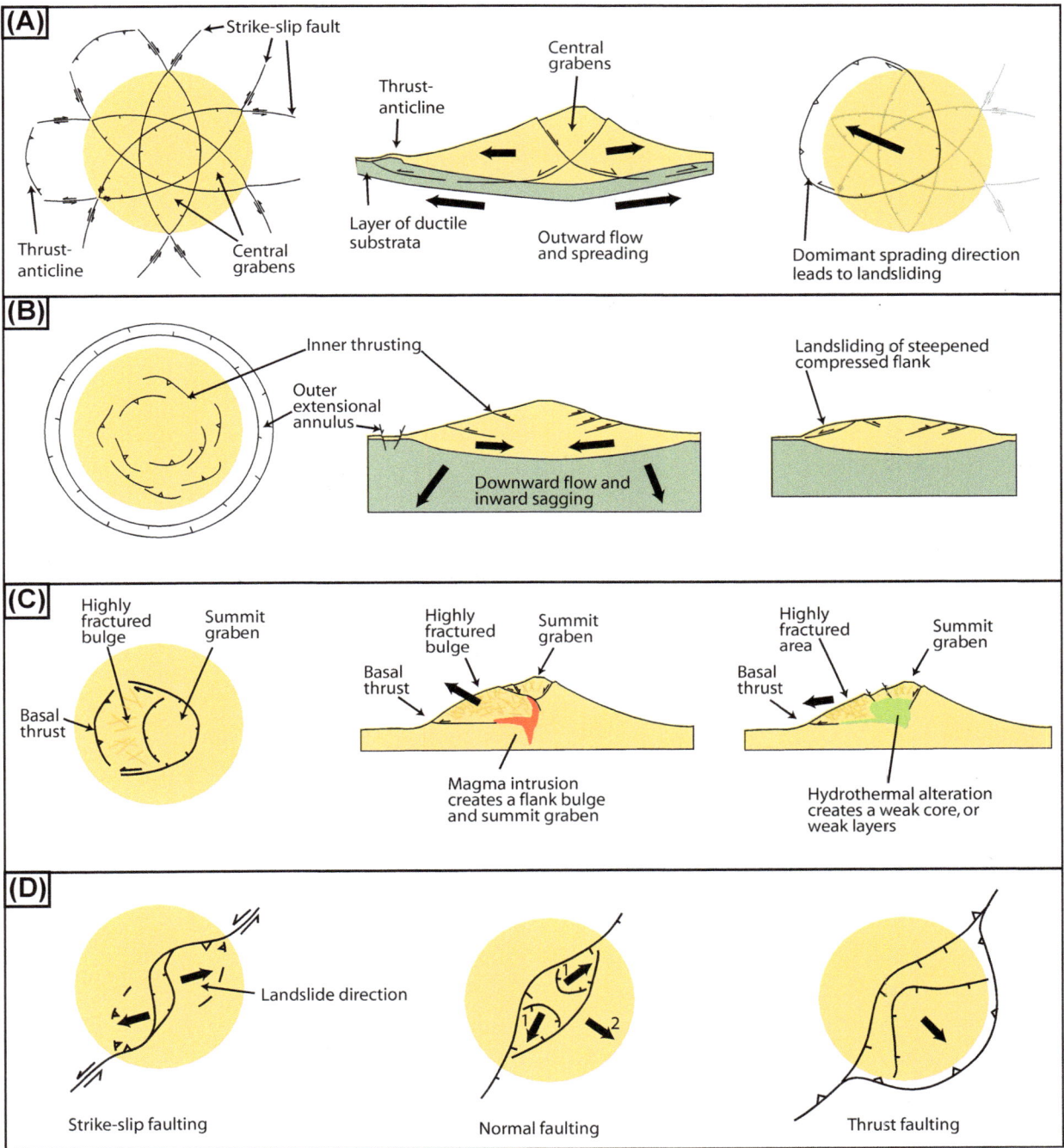

FIGURE 38.2 Styles of volcanic gravitational deformation and landsliding shown in simple sketches. (A) Spreading, with plan view and cross section of ideal fault geometry around a stratocone with central grabens and basal strike-slip or thrust faults. Spreading in a predominant direction can favor landsliding, while radial spreading does not, as the main faults dip toward the center (Borgia and van Wyk de Vries, 2002). (B) Sagging, with plan view and cross section showing the compression created by downward flow in the substrata, and thrusting in the edifice. Constriction on the upper flanks may contribute to landslide formation by increasing slope-parallel stresses. (C) Flank spreading created by magma intrusion or hydrothermal alteration (Cecchi et al., 2005). Plan view and cross sections of structural configurations. (D) Deformation in volcanoes created by regional tectonic faults, showing the consequent landslide directions. For normal faulting, fault-parallel small-scale landslides are created first before large-scale fault normal landslides. For strike-slip faulting, failure directions are slightly oblique to the fault. For vertical faulting and for thrust faults, landslides normal to the fault strike are likely to be generated.

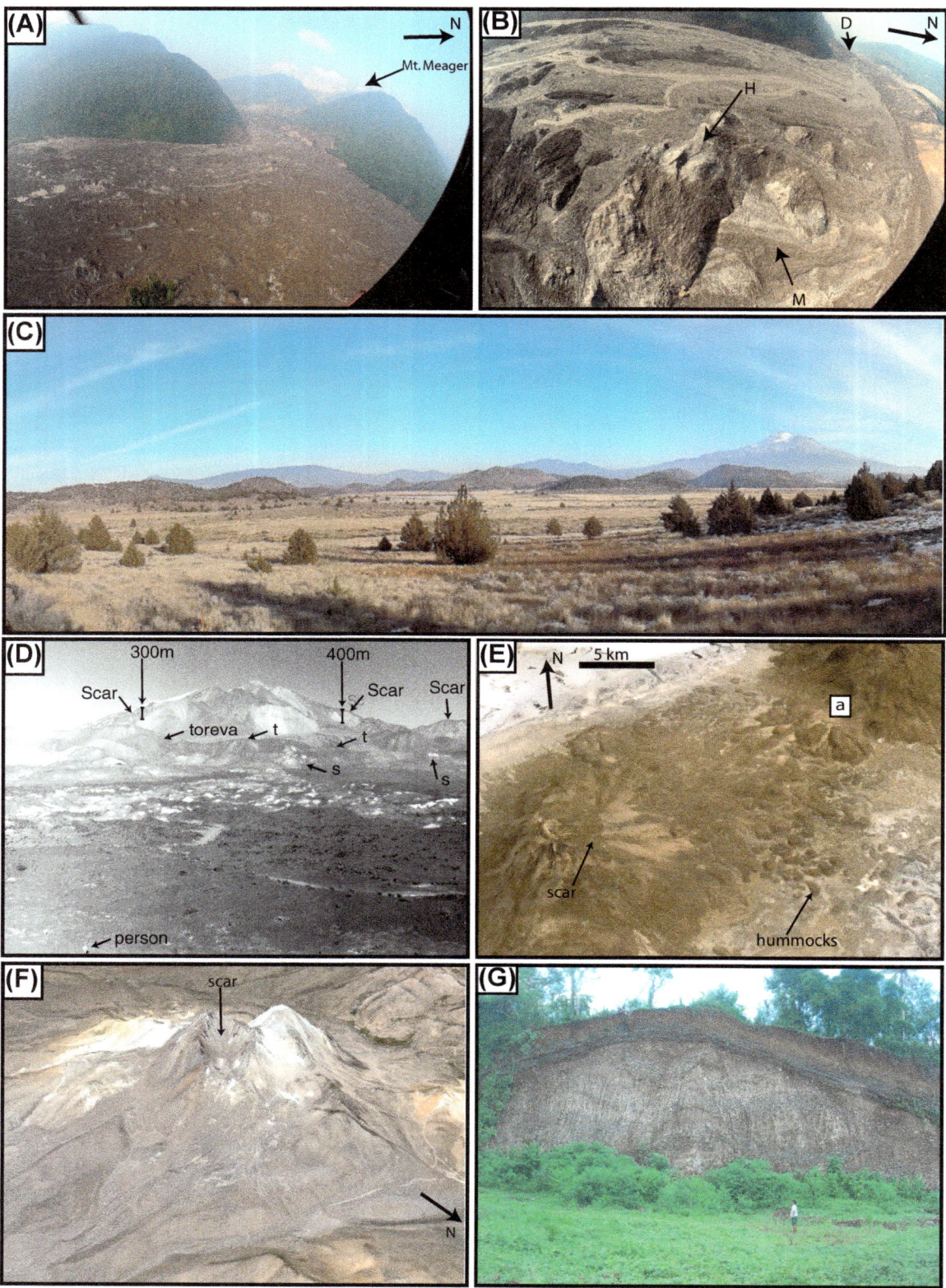

FIGURE 38.3 Examples of volcanic landslides and volcanic debris avalanches. (A) Photograph of the Mount Meager landslide and debris avalanche deposit of 2009 (Photo DB Steers), showing the steep glaciated flanks of the volcano, and the broad deposit on the Lillooet River flood plain. (B) Close-up image of the Mount Meager debris avalanche deposit surface (Photo DB Steers). "H" indicates a hummock with block facies, and "M" a highly stretched interhummock area with matrix facies, including different colored layers of substrata, volcanic and hydrothermally altered units. These are present in the hummocks and highly stretched in the interhummock areas. Note in the middle ground, thin lahars are forming through the dewatering of the deposit. The Meager Creek dam is indicated as "D"; shortly after this image was taken the dam failed and eroded much of the deposit, causing a major flood and changing the Lillooet River course (Guthrie et al., 2012). (C) Photograph of Mount Shasta, California (with snow) and the Shasta debris avalanche deposit.

edifice being constricted while the lower flanks and base are extended (Figure 38.2(B)). This causes thrusting on the upper flanks and normal faulting around the base. The former is termed volcano spreading and the latter volcano sagging.

The deformation and stress states imposed by the gravitational deformation constrain magma intrusion. Magma intrusion, while initially guided by the gravitational stress, may then alter stresses significantly and deform the volcano strongly over short timescales. Sagging will constrict the upper edifice and thus impede magma rise. Spreading, in contrast, will allow rifting and will facilitate magma rise and eruption. A feedback mechanism can develop between gravitational deformation and magma intrusion, which pushes flanks and adds mass. Changes in the composition of magmas and eruptive style have been linked to changes in gravitational stress fields, such as at Concepción (Figure 38.1) and Etna. At Etna, flanks deform continuously in response to gravitational loading, accelerating during intrusive and eruptive periods.

A volcanic edifice is built of primary volcanic deposits that are formed by lava flows (including their autobreccias), fall (tephra), and pyroclastic density currents. Erosion and weathering of these by aeolian, marine, fluvial and glacial activity, and by mass movement, form volcaniclastic sediments that are also incorporated into the edifice. Internal changes, such as hydrothermal alteration, can weaken rocks and make the edifice more susceptible to slope failure. The load of the volcano may thus become greater than the rock's resistance to deformation, failure zones, and slip planes then form and flanks can deform. Such deformation has been termed flank spreading, and may cause a flattening of the summit area, and bulging of the flanks (Figure 38.2(C)). This produces similar structures to volcano gravitational spreading (Figure 38.2(A)), but may be restricted to a single flank while spreading involves the whole edifice.

Regional tectonic stress fields and deformation can combine with the gravitational load of the volcano. In extensional settings the gravitational load may focus rifting and faults onto the edifice, while in strike-slip settings, volcanoes may initiate pull aparts and volcano-tectonic depressions. At volcanic centers in compressional settings, thrusts may be deflected around volcanoes (Figure 38.2(D)). Landslides can be generated with each type of fault movement, and tend to develop distinctive collapse directions (Figure 38.2(D)).

The slow, deep-seated gravitational deformation of volcanoes sets the general conditions for landslides and debris avalanches, which may then be triggered by intrusion, seismic forcing, extra loading from erupted material, or pore pressure changes. Slow deformation may eventually generate landslides and debris avalanches by progressive weakening of the rock mass with no other trigger (van Wyk de Vries and Francis, 1997). In all cases, acceleration of flank deformation would be expected prior to landslide failure, as in nonvolcanic landslides. Hence, monitoring the slow movement of flanks can provide warning of collapse in both active and inactive volcanoes.

3. VOLCANIC LANDSLIDES

The earliest descriptions of volcanic debris avalanches are found relating to the 1888 Bandai-San disaster (Sekiya and Kilkuchi, 1890) and from the collapse and eruption of Bezymianny in 1957 (Gorshkov, 1959). The landslide and debris avalanche that initiated the 1980 eruption at Mount St Helens was the first to be monitored in detail (Lipman and Mullineaux, 1981). Since then, there have been smaller volcanic landslides at Casita, Nicaragua in 1998, Montserrat in 2000, Mount Meager, Canada in 2009 and Askja, Iceland in 2014 that have spawned debris avalanches, debris flows and in the latter case, a small tsunami. The frequency of events suggests that several volcanic debris avalanches may occur worldwide each century.

Volcanic landslide volumes depend partly on the volcano size, as larger edifices have greater volumes, larger flanks, and can deform more. The landslide volume depends also on the local setting. At the smallest scale, landslides of a few thousand cubic meters, such as occur on valley sides, as at Casita Volcano, Nicaragua in 1988, are shallow-slope stability events (Cecchi et al., 2005). However, the example of Casita showed that even such small landslides can be serious hazards, especially if coincident with heavy rainfall. At larger scales, landslides of several million cubic meter involve large valley flanks, and cut deeper into the volcanic edifice to involve hydrothermal systems. An example (Figure 38.3(A) and (B)) is the failure that occurred at Mount Meager, Canada in 2009 (Guthrie et al., 2012). Another example is that at Montserrat in 2000, where an active dome loaded a weak, hydrothermally altered valley side (Sparks et al., 2002). Tutupaca, Peru had two such collapses about 200 years ago (Figure 38.3(F)), leaving deposits, which have been mined for their sulfur

The foreground is a smooth interhummock area, and middle ground shows a 10-km-wide view of hummocks with a low-mountain range in the background. (D) Photograph of the Socompa debris avalanche, N Chile, taken from the Median Scarp (location at star in Figure 38.6(A) and (B)), a large back-thrusted area in the middle of the debris avalanche 20 km from the summit. The Scar (12 km wide) is indicated, and "t" stands for Toreva block, and "s" for white outcrops of the substrate that were involved in the original failure volume. (E) Google Earth Oblique view of the Tetivicha Volcano and its debris avalanche deposit in Bolivia (Shea and van Wyk de Vries, 2008). The 7-km-wide scar has disgorged a hummock-rich debris avalanche onto the flanks of the neighboring volcano (where the hummocks have converged and grown by combining (point "a") and have also spread out and dispersed onto the Salar (salt flat) below. (F) Google Earth oblique image (X2 vertical exaggeration) of the Tutupaca Volcano landslide scar and debris avalanche, Peru, showing the scar, and deposit area. (G) Hummock cut in half by quarrying on the Iriga Debris avalanche, Philippines, showing normal faulting cutting conglomerate bands (Paguican et al., 2014). The original landslide involved the volcano substrata that are preserved in this hummock. Note that this block facies of sedimentary rock would not be easily distinguishable as belonging to a debris avalanche deposit if not for the clear context of this outcrop.

content. The largest scale landslides remove a large sector of an edifice flank and these can involve several cubic kilometers. Examples of these large volume events are Mount Shasta and Socompa (Figure 38.3(C) and (D)).

Most volcanoes have landslides, and collapse to form a debris avalanche at least once during their lifetime, sometimes several times, and such events occur on all types of volcano, be they oceanic, continental, monogenetic, or polygenetic (Figure 38.4). Landslides occur from volcanoes in all geodynamic contexts, including mid-ocean rifts, hot spots, arcs, and intraplate settings. There is also evidence for large volcanic landslides on other planets (Figure 38.4(B)). Collapses may occur at dormant and extinct volcanoes, as at Mayu-yama, Japan in 1792 (Figure 38.4(B)), or after a repose period when a volcano is reactivated, as at Mount St Helens in 1980 (Glicken, 1996) (Figure 38.4(D) and 38.5).

4. VOLCANIC DEBRIS AVALANCHES

A volcanic debris avalanche forms when a volcanic landslide accelerates due to a rapid loss of resistance at the base of the rock mass (Figure 38.5). The result is a fast-moving deforming mass that travels downslope and can extend tens of kilometers from the foot of the edifice (Figures 38.4 and 38.6). Such events have nonvolcanic equivalents and both are able to travel a long distance (L) with respect to the total height loss (H). The H/L ratio is commonly about 0.6 for small valley slope landslides, but can be very low (~ 0.1) for the larger events with volumes over 0.1 km^3. This volumetric influence over distance traveled is shown in Figure 38.6(A), where volcanic, nonvolcanic, and extraterrestrial debris avalanches are compared (Shea and van Wyk de Vries, 2008). The data used in the plot have many sources and involve many geometric and geological factors, which leads to a wide scatter on any such plot of geometric variables.

The area A covered by the debris avalanche is roughly related to its volume V by the relationship $V = C.A^{2/3}$, where C varies according to the geometry and topography of the deposition area and the material rheology. C can be regarded as a mobility coefficient (Figure 38.7) (Dade and Huppert, 1998). Figure 38.7(B) demonstrates the scale-independent relationship between area and volume, where both are cast into dimensionless numbers using H as the denominator (Shea and van Wyk de Vries, 2008).

5. VOLCANIC INSTABILITY PRIMING AND TRIGGERS

5.1. Priming

Landslides initiate when the driving force on a rock mass exceeds the resisting force of the plane or zone of weakness. In and under a volcanic edifice, these low-strength zones can be original weaknesses in the volcanic materials, such as poorly consolidated pyroclastic deposits, lavas that were fractured on cooling, or weathered horizons, soils and breccias. Weak material can be created inside the volcano by hydrothermal alteration, which converts rock to clays, leading to a pervasive loss of strength in the rock mass. Hydrothermal fluids also increase pore pressures, especially when alteration clays form impermeable boundaries. In addition, rock may be fractured, and lose strength, by deformation, intrusion, tectonic movements, or gravitationally induced deformation.

Pore water is an important cause of landslides in granular materials, because it adds mass, and the increase in pore water pressure can reduce rock strength. Thus, the distribution of water in a volcano is an important factor in instability. Volcanoes that hold large volumes of water, especially those in wet climates or with glaciers, may thus have increased landslide susceptibility. For example, the edifice of Mount St Helens had an initial porosity of about 14% and was about 92% water saturated (Glicken, 1996). Weather and climate variability may affect susceptibility, so increased aridity might lead to more stable volcanoes, for example. Sea level changes may also affect the stability of oceanic volcanoes, by changing hydrological conditions in the edifice. Erosion of a volcano (e.g. by glaciers or minor landsliding) can steepen slopes and remove buttressing material, thus diminishing slope stability. Eruptions can progressively load weak zones with new material, and thus can contribute to instability.

5.2. Triggers

Conditions that weaken a volcano may exist or develop over the lifetime of a volcanic edifice (tens to hundreds of thousands of years) with only slow deformation. Such progressive weakening could lead to eventual collapse, but short-term triggers are likely to intervene during these long periods and initiate landslides and debris avalanches.

Shallow intrusion of new magma, such as occurred at Mount St Helens in 1980, can lead to hundreds of meters of deformation (Figure 38.5). This deformation steepens flanks, reduces rock strength by creating shear zones, and by pervasively brecciating the deforming mass.

Changes in the hydrothermal system may trigger collapse by increasing pore fluid pressures, and causing fluid migration as, for example, at Bandai-San in 1888 (Figure 38.4(C)). At Bandai-San, the collapse was preceded by seismic activity and an increase in fumarolic activity, and the collapse was accompanied by a phreatic blast related to the decompression of the hydrothermal system. The hydrothermal system is ultimately driven by magmatic heat, and new intrusions can disturb it by increasing temperature and pressure, and by deforming it, factors, which

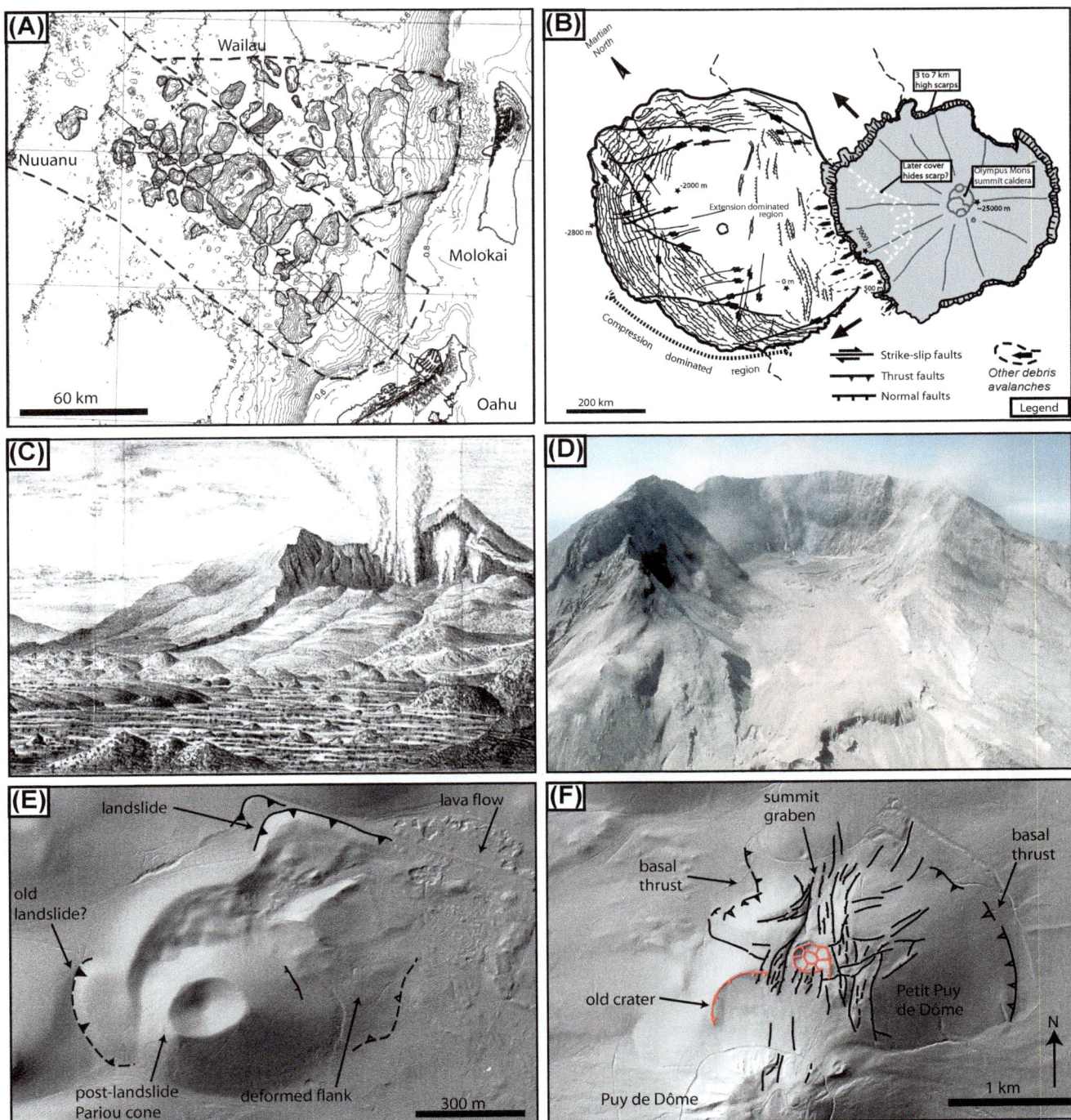

FIGURE 38.4 Examples of different debris avalanche settings. (A) Oceanic volcano debris avalanche deposit: the Nuuanu and Wailau debris avalanches in the Hawaiian Islands (Moore et al., 2000). The Nuuanu deposit is probably the largest one on Earth, and has some hummocks that are tens of kilometers across. (B) The largest extraterrestrial landslide at Olympus Mons (Shea and van Wyk de Vries, 2008); note that the faulted structure of the deposit has close similarities to those at Socompa (Figure 38.6(A)). (C) Original sketch of the 1888 Bandai-San, Japan, debris avalanche deposit, associated with disruption of the hydrothermal system (Sekiya and Kikuchi, 1890). Note the scar, with steam plume rising from the disrupted hydrothermal system, and the hummocky debris avalanche deposit in the middle to foreground. (D) The Mount St Helens scar from the 1980 landslide (Image from USGS). The scar is about 2 km across, and was produced by the landslide shown in Figure 38.5 after 2 months of precursory bulging of the flank by an intrusion, as shown in Figure 38.2(C). (E) Monogenetic volcano collapse at the Pariou volcano, Chaîne des Puys, France. This small volcano has had at least one landslide associated with a voluminous lava flow eruption. (F) Uplifted and deformed flank of the Petit Puy de Dôme, Chaîne des Puys, France, showing an intrusion-generated volcanic landslide that stabilized before developing into a debris avalanche. Shaded relief image from Lidarverne survey.

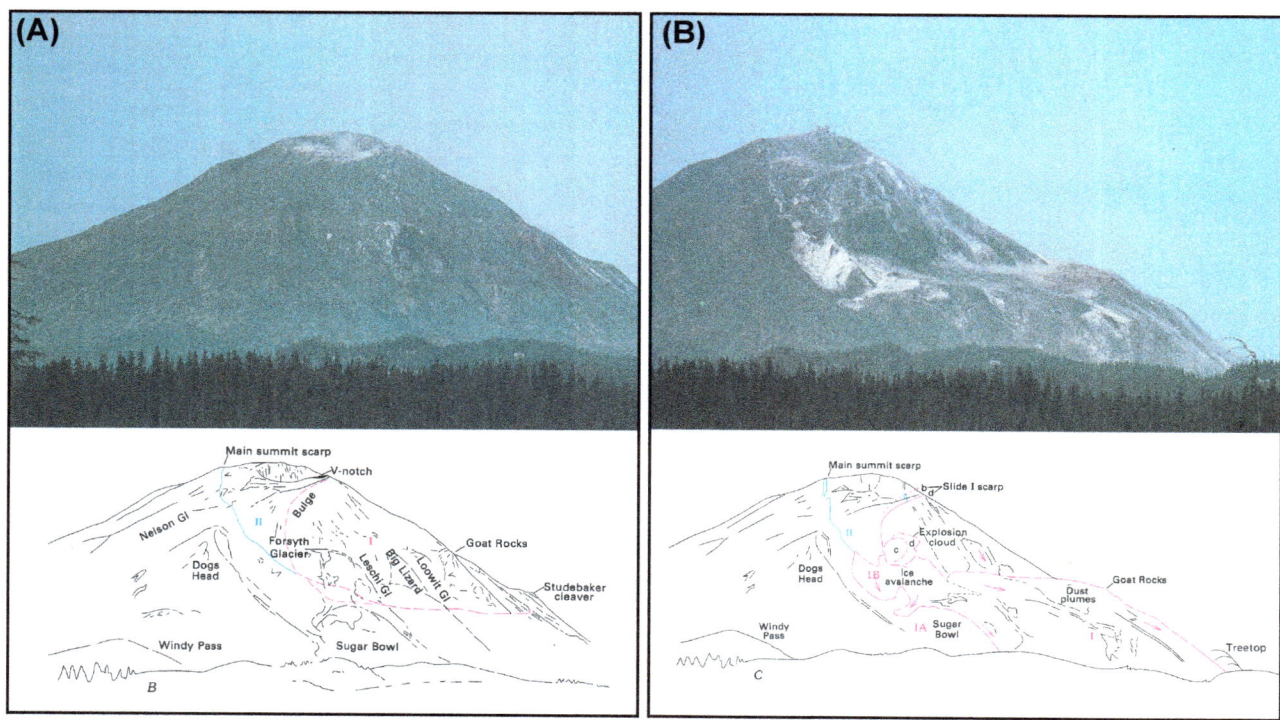

FIGURE 38.5 Two sequential photographs of the first stages of collapse of the 1980 Mount St Helens landslide and debris avalanche (Lipman and Mullineaux, 1981). Photographs from the Gary Rosenquist collection. (A) 5 min before the collapse, with the bulged area on the right hand side. (B) 2.5 s after the landslide commenced, showing the frontal block, containing Goat Rocks and the second slide block with the main part of the bulge. Sketches from Figure 37 in Lipman and Mullineaux (1981).

all lead to strength loss and can trigger a landslide. Hydrothermal systems progressively alter rock and precipitate material; self-sealing, where pores are blocked by precipitation of minerals, can also increase pore fluid pressures and lead to instability.

Large rainfall events, or ice melt, could contribute to triggering smaller collapses, such as at Casita (over 1 m of rain in one day during Hurricane Mitch in 1998) and Mount Meager (ice melt in 2009, Guthrie et al., 2012). Eruptions load the edifice and rapid loading, such as occurs during dome growth, may lead to landsliding, as occurred at Montserrat in 2009 (Sparks et al., 2002) and during the many growth phases of Augustine Volcano, Alaska, and Shiveluch, Kamchatka (Belousov et al., 1999).

Earthquake acceleration (Reid et al., 2010) and movement on regional faults cutting the volcano may trigger failure by causing direct displacement, as at Iriga Volcano, Philippines (Figure 38.6(B)). During caldera formation, stress changes and earthquakes related to the large displacements may generate failures on the volcano flanks, and also within the caldera. Caldera walls then remain landslide-prone, for example the July, 2014 landslide at Askja, Iceland. Eruptions themselves, however, may not have much effect on edifice stability, because their direct energy release is upward into the atmosphere.

6. GENERAL STRUCTURE OF LANDSLIDES

Volcanic landslides differ from nonvolcanic landslides in that they tend to be more deeply seated, involve hydrothermal and magmatic systems, and occur in predominantly granular rocks that can be highly porous. Volcanic landslides have a normally faulted head region, near or surrounding the volcano summit, commonly with a graben. The lateral boundaries of the landslide are defined by strike-slip faults and a thrust-faulted toe zone (Figures 38.8 and 38.9). Volcanic landslides can include the whole flank of a volcano, such as the Hilina slump on Kilauea (Figure 38.1) (Morgan et al., 2003), the eastern flank of Etna, and the south–south east side of Piton de la Fournaise. A landslide may involve a much more restricted area, such as the southeast flank of Casita 1998 (Cecchi et al., 2005), or the landslide that generated the 2009 Mount Meager collapse (Gurthrie et al., 2102), or those at Augustine Volcano, Alaska. Augustine has had multiple collapses produced from rapidly growing domes and cryptodomes that periodically grow on or in the edifice. The combination of rapid dome growth and intrusion has weakened the edifice, in a similar way to Shiveluch (Belousov et al., 1999). Intrusion-related landslides at Mount St Helens in 1980 and

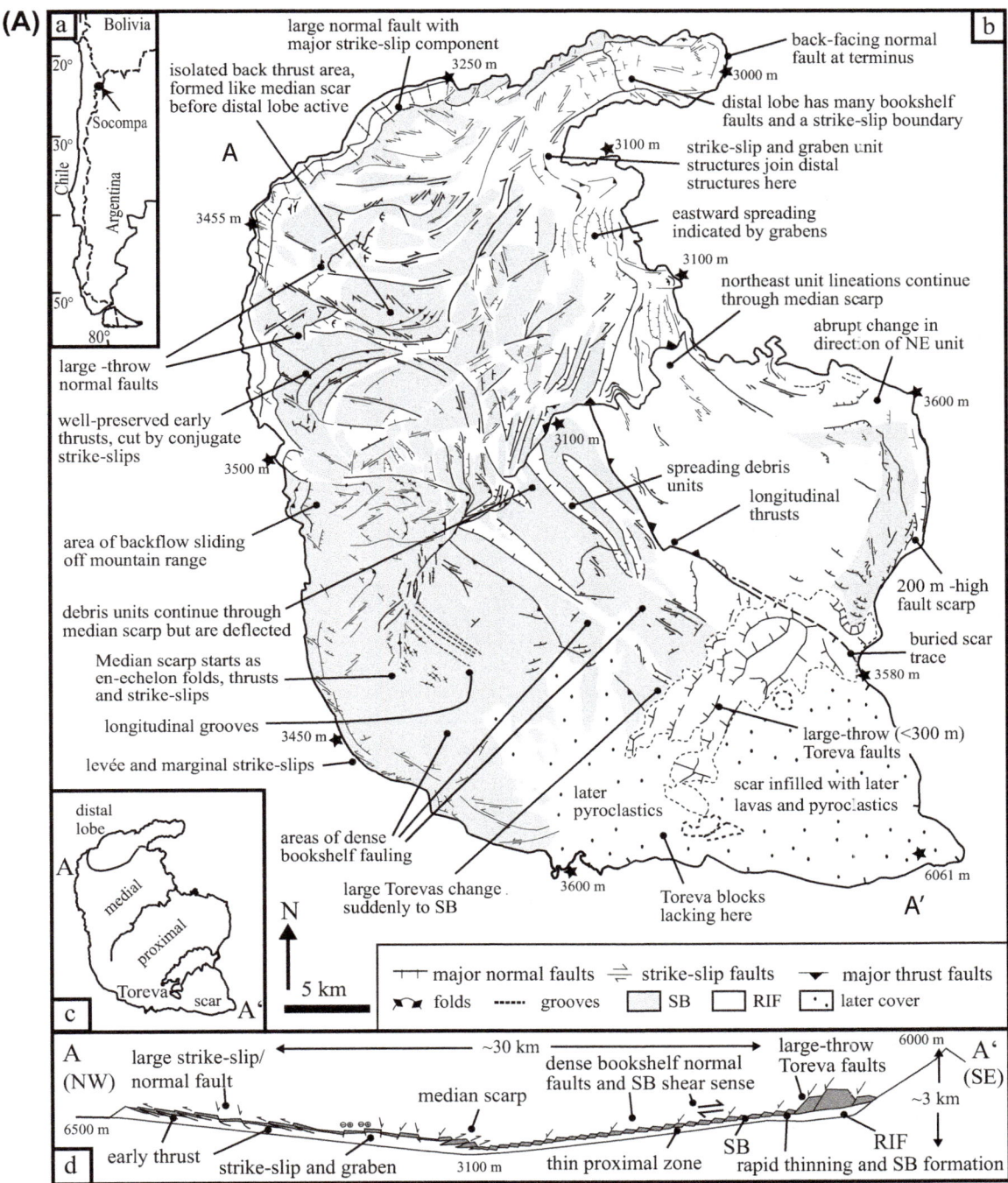

FIGURE 38.6 Maps of two representative volcanic debris avalanches. (A) Socompa debris avalanche deposit, N Chile (Shea and van Wyk de Vries, 2008). a—location of Socompa; b—map of the faulted structure of the deposit, showing the major normal, strike-slip and thrust faults. The main units are SB = Socompa Breccia, derived from the edifice and RIF = Reconstituted Ignimbrite Facies, derived from the volcanic substrate that failed to generate the collapse. The large Toreva blocks are shown filling the scar opening (lower right). The image in Figure 38.3(D) was taken at the star, at 3100 m, in the center of the deposit; c—inset showing the main deposit units from proximal to distal, which can be compared with those in the analogue model in Figure 38.7; d—Cross section of the deposit showing the deposit structure, with the lower substrate level (RIF) and upper volcano layer (SB), which preserves the original stratigraphy of the landslide. (B) Iriga volcanic debris avalanche deposits, Philippines, showing two hummock-rich deposits (Paguican et al., 2014). The volcano is crossed by a major regional strike-slip fault, which has partly guided the landslide directions. The Buhi deposit has dammed a valley to create Lake Buhi. A large number of hummocks have developed out of horsts and grabens as the debris avalanches spread outward and extended.

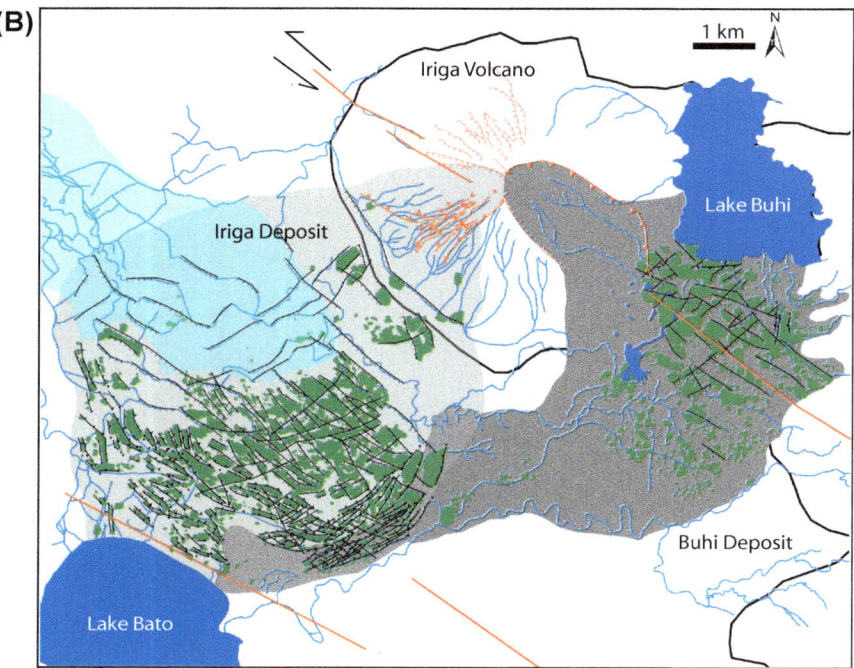

FIGURE 38.6 Cont'd.

FIGURE 38.7 Plots of the geometric relationships of volcanic and nonvolcanic landslides. (A) Height/Length (H/L) versus Volume (V). (B) Area (A) versus volume relationship cast in dimensionless ratios of A/H^2 versus V/H^3 from Shea et al. (2008). The dimensionless ratios allow comparisons over different scales from the largest debris avalanche deposits to the meter-scale laboratory analogue slides. Note that the spread of data indicates that rheological differences and the surface over which the mass transported have a major effect on the area and run out of any particular event.

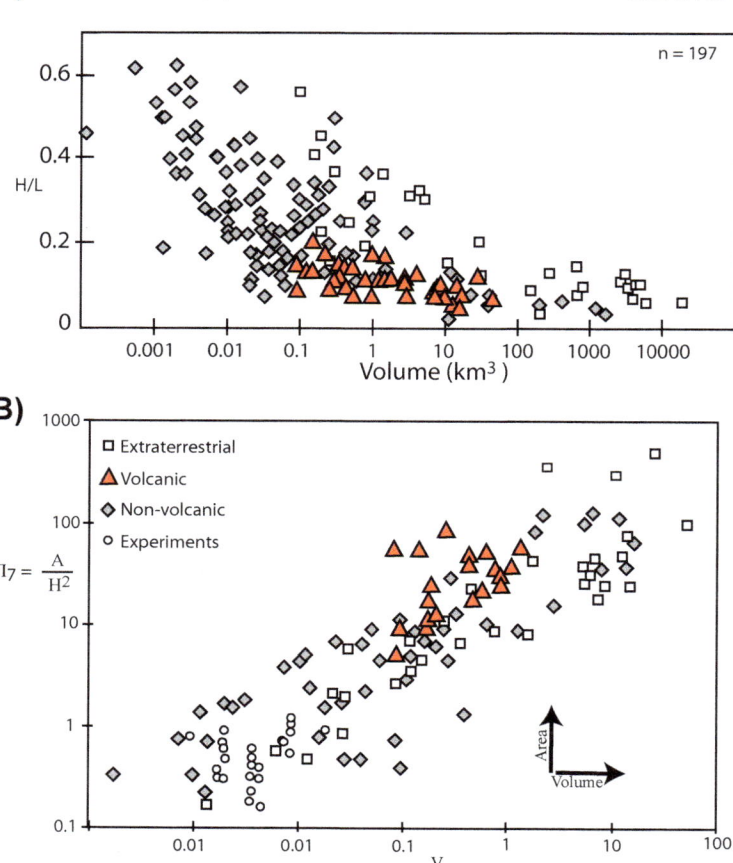

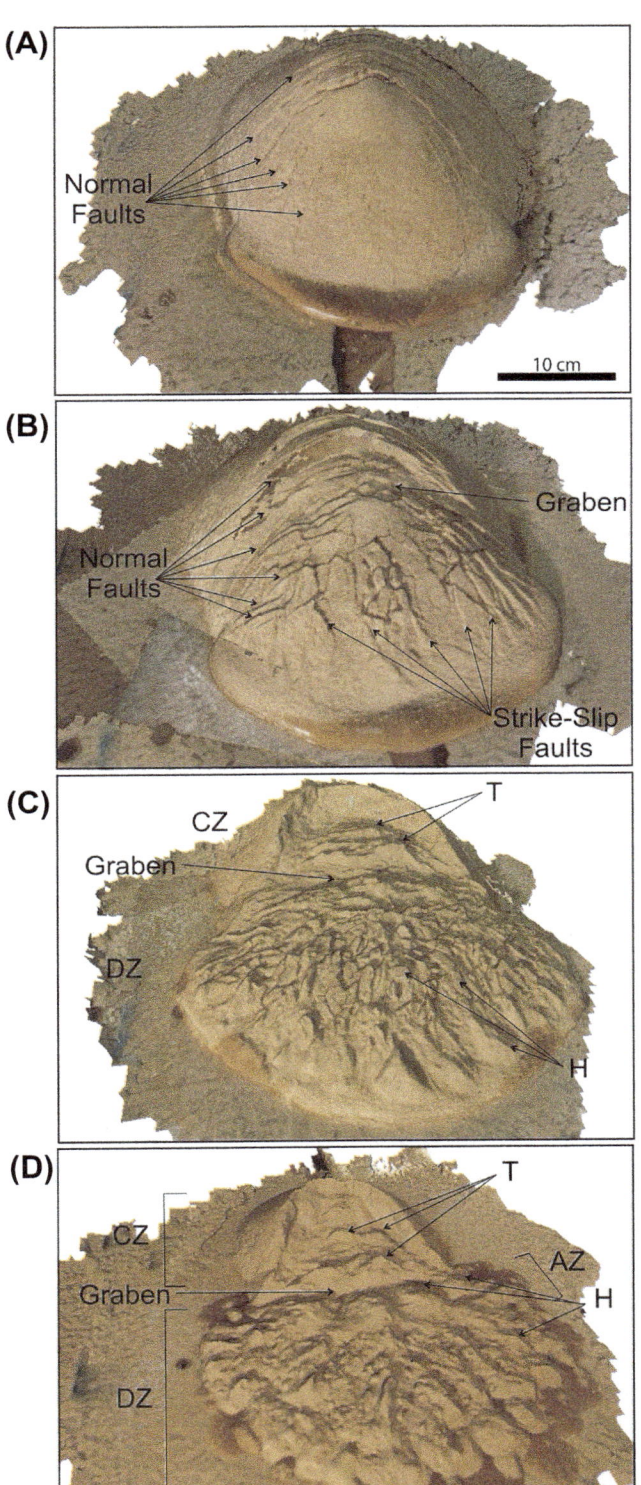

FIGURE 38.8 Example of landslide to debris avalanche evolution provided by an analogue model (Paguican et al., 2014). This model is composed of a sand-plaster mix to simulate brittle rock and a silicone layer to simulate the ductile sliding zone. (A)–(D) Progressive development of structures: (A) The initial landslide phase with principally normal faults at the head and faults forming en-echelon margins with strike-slip movement; (B) The

Bezymianny in 1957 have the same general geometry as purely gravitational landslides, with the difference that initially the central part of the landslide is also uplifted by the intruding cryptodome (Figure 38.5).

7. GENERAL STRUCTURE OF DEPOSITS

The structure of volcanic debris avalanche deposits depends on the initial landslide configuration (Glicken, 1996; van Wyk de Vries and Francis, 1997; Belousov et al., 1999; Morgan et al., 2003). A debris avalanche deposit can be divided into proximal, medial, distal, and marginal regions that roughly correspond to the head, the graben, and the frontal and lateral parts of the original landslide (Figure 38.8). The scar from which the landslide and debris avalanche develop can be horseshoe shaped or triangular, and may cut only the edifice, or extend out into the piedmont and reach into the substrata. Most volcanic avalanche deposits have a hummocky and ridged surface. The hummock size can vary from several kilometers wide and hundreds of meters high down to meter-sized mounds (Figures 38.4 and 38.6).

On the largest landslides, the head is often preserved as Toreva blocks (large rotational blocks), in a proximal region of large hummocks, as at Socompa (Figure 38.3(D), Figure 38.6(A)). The graben area is often seen to be preserved as an area of elongate, transverse-to-motion parallel hummocks in a medial region (Figures 38.8 and 38.9), and the frontal area as a wide zone of variously aligned hummocks, ridges, or fault structures (Figure 38.6(A), 38.8 and 38.9).

Toreva Blocks are backwards-rotated slide blocks originating from the landslide head. They are commonly preserved in the largest volume avalanches that cut deep into the volcano and substrata, but they are not preserved where topography drops steeply away from the flank of the volcano.

Hummocks are characteristic extensional features of debris avalanches and landslides, formed during the spreading of an avalanche mass. They are commonly arranged linearly in the direction of motion. Hummock size generally decreases with increasing distance from source. Two types of hummock have been identified: those formed with a horst and graben structure (Figure 38.3(G)), and those formed by boudinage, where individual blocks of a layer, such as a lava flow, are pulled apart (Figures 38.9 and 38.10). With avalanche spreading and thinning, hummocks may

formation of a rear zone of arcuate normal faults that eventually can become the Toreva blocks; a central graben area, and a distal zone cut by numerous transtensional faults, with a thrusting front. (C) The individualization of hummocks (H), and the spreading of the graben and the formation of a depositional zone (DZ), below the Collapse Zone (CZ). (D) Separate Toreva (T) and hummocks (H) and an accumulation zone (AZ), where large hummocks may merge, as those in front decelerate and stop.

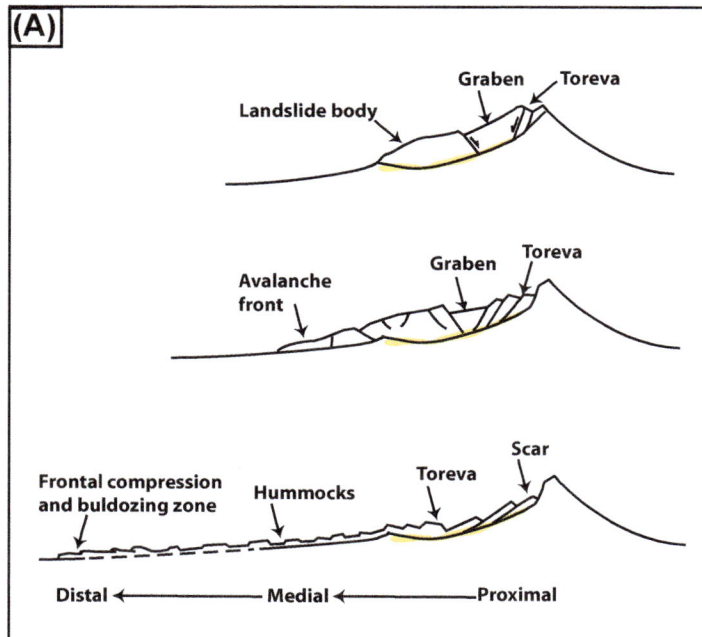
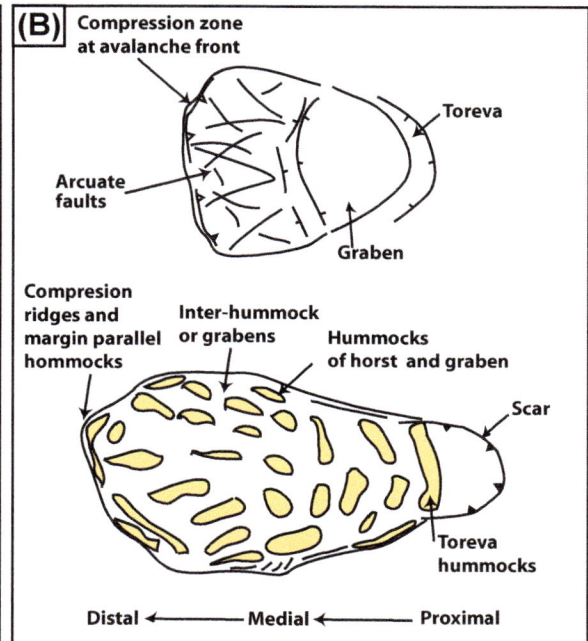

FIGURE 38.9   A general sketch model of a volcanic debris avalanche that has developed from a landslide. Note that this is for a simple deposition surface, while valley deposited landslide structures will be much more complex and will depend on local topography. (A) Cross sections showing the early conditions in the landslide mass, and the development of structures during avalanche spreading. (B) Plan view of the development of hummocks and surface structures as the avalanche spreads.

become grounded and stop moving before the main avalanche mass, or in some cases (like at Parinacota, Chile) blocks may detach from the main mass and travel farther.

Interhummock areas are zones between hummocks in a debris avalanche deposit that generally form a greater proportion of the avalanche surface area than the hummocks. The hummock proportion generally decreases away from source, as hummocks move apart during spreading. These interhummock areas form originally from grabens that separate the hummock horsts, and it is in these grabens that most of the extension is accommodated. Thus, the interhummock areas are zones of intense extensional strain. They are generally composed of significant proportions of matrix and of highly brecciated block facies (see below). Incorporated substrate and original lower layers of the original landslide mass are commonly found. These areas are generally where water is concentrated, and secondary lahars may be generated from them (as at Mount St Helens in 1980). If the water-saturated volume is a large proportion of the debris avalanche, it may transform from a sliding mass into a debris flow, as at Mount Meager in 2009. In such cases the interhummock material may flow around grounded hummocks (Figure 38.3(A) and (B)).

Long ridges may form on avalanches, orientated in the transport direction. These are formed by the production of elongate hummocks bordered by strike-slip faults in lateral areas (Figure 38.6(A)). There may also be a family of

highly elongate ridges some meters high, tens of meters wide, and up to a kilometer long that are may be related to differential granular sorting during high-velocity motion, such as at Shiveluch Volcano (Belousov et al., 1999). Transport-normal ridges are formed in avalanche compression by folds and thrusts (Figure 38.6(A)).

Marginal zones in debris avalanche sides and fronts can be steep-sided and are sometimes lobate (Figure 38.5). The front is often ridged with compressional structures. When the interiors of the fronts of debris avalanches are exposed, the ridges are commonly seen to contain, or entirely comprise, substrate bulldozed at the front of the mass (Belousov et al., 1999).

8. DEBRIS AVALANCHE FACIES

All debris avalanche facies are characteristically brecciated. Jigsaw-crack and jigsaw-fit textures are common (Figure 38.10). Clasts tend to be angular, but shape is dependent on the lithological type, so clay blocks and other ductile materials can be rounded (Figure 38.11). Sorting is poor, but grain size distribution is variable, owing to variations in brecciation (finer material in crush zones) or to original lithological grading (Figure 38.11). As all types of volcano edifice material are incorporated into debris avalanches, both block and matrix facies can have highly variable textures. In the field, volcanic debris avalanche deposits

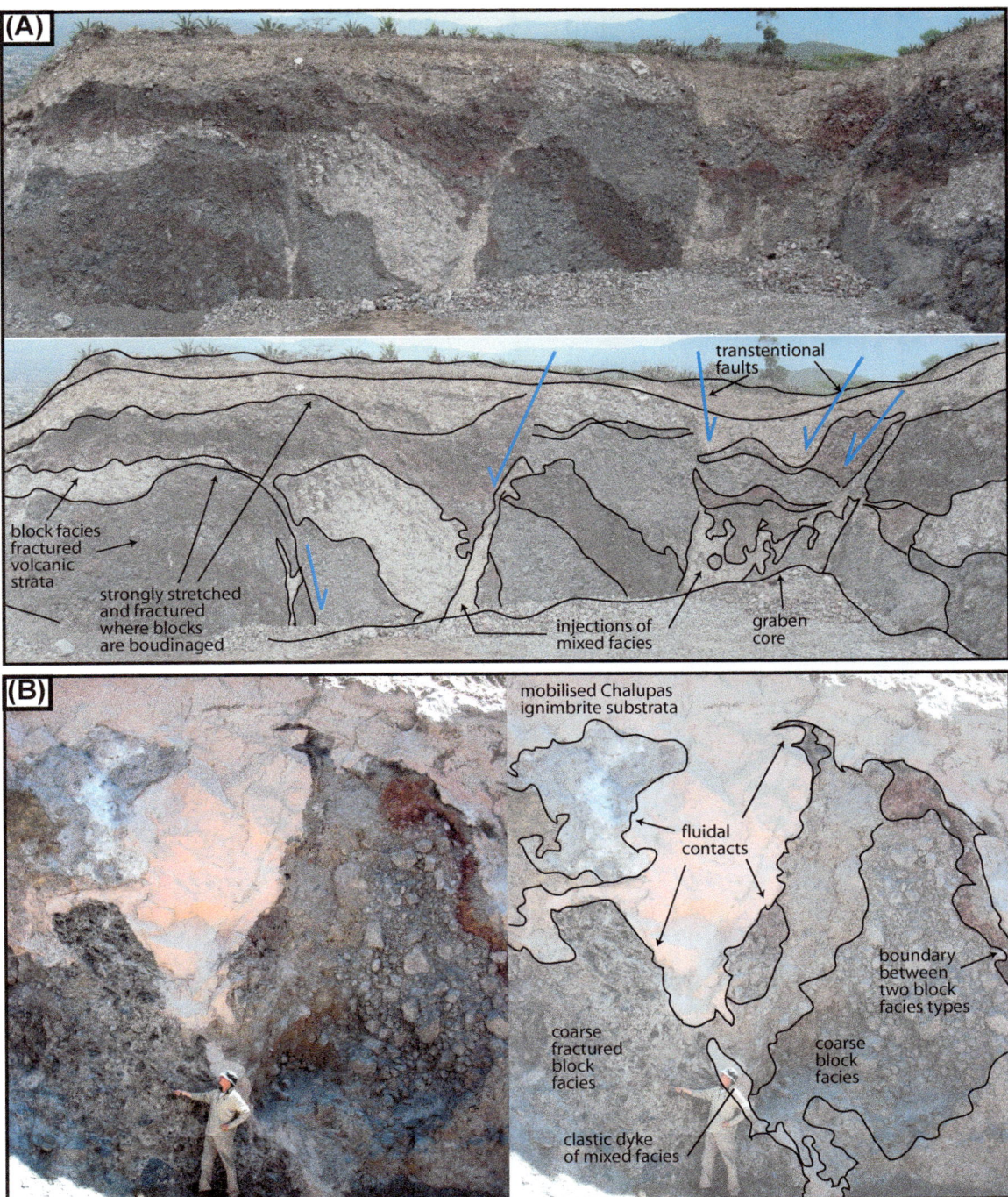

FIGURE 38.10   Examples of debris avalanche deposit textures. (A) Block facies material in a 10-m-high quarry in the Imbabura debris avalanche deposit, Ecuador. This photograph shows the well-preserved bedding of lavas and pyroclastics; note the layer thicknesses change rapidly because of boudinage. Some large fault zones displace bedding by several meters, along which are clastic dykes of light-colored mixed facies. There are also clastic dykes and fluidal textured injections in a small graben. (B) Matrix and block facies and the white incorporated Chalupas ignimbrite in the Chimborazo debris avalanche deposit, Ecuador. In the image are a polylithologic mixed matrix facies, and a highly jigsaw brecciated block facies, and the injected fluidized substrate-derived Chalupas ignimbrite.

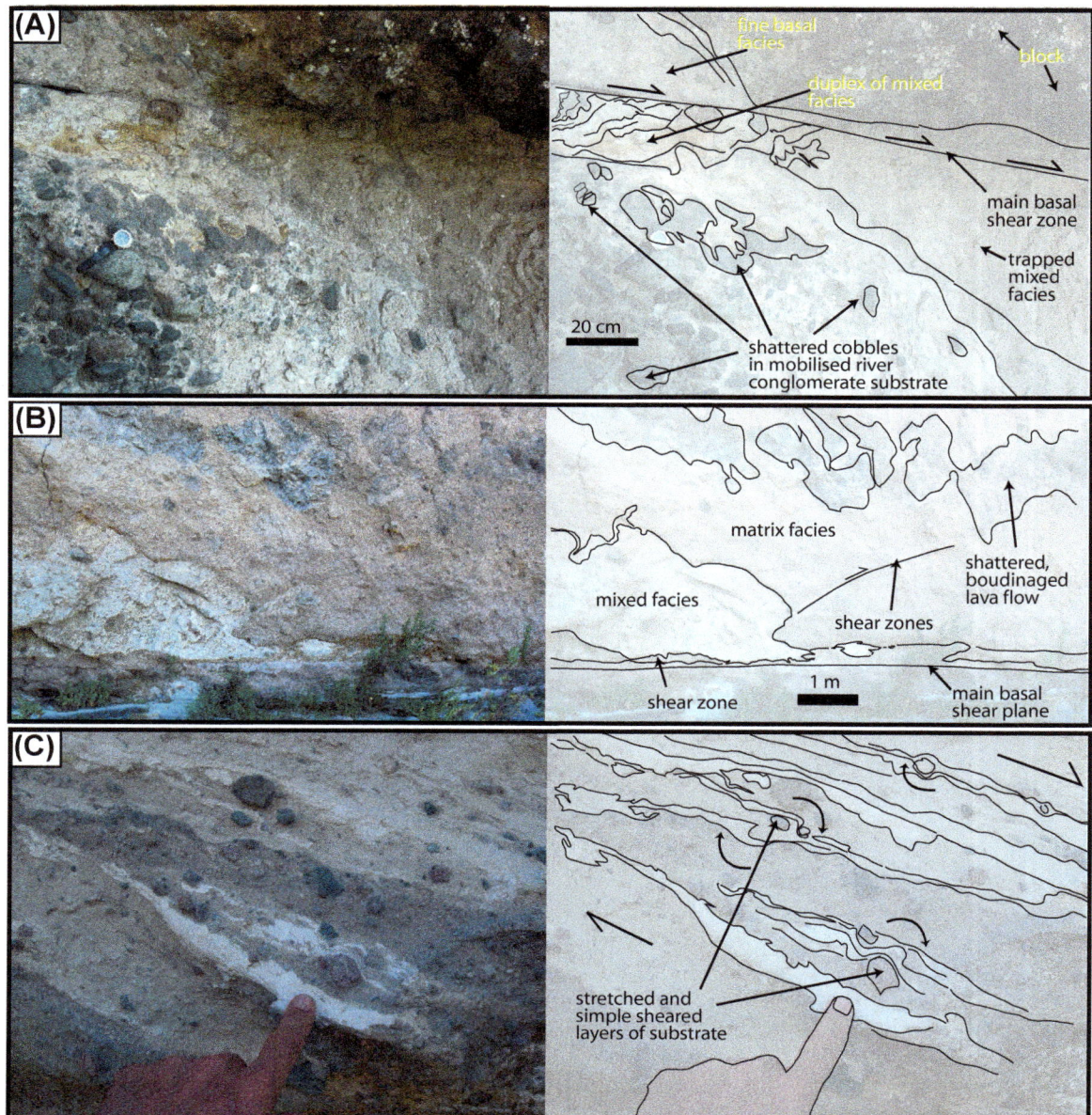

FIGURE 38.11   Basal layers at volcanic debris avalanches. (A) Deformed and mobilized cobbles at the base of the Perrier debris avalanche deposit, France. Note the main shear plane below a large block of lava flow, and the shattered cobbles that are part of a mobilized substrate layer. (B) Highly deformed and planar, 1 m thick, basal shear zone at the Pichu Pichu debris avalanche deposit, Peru. Frictionite is found along the basal contact. Note also the shattered, boudinaged, and dismembered lava flow about 3 m above the base. (C) Close-up of stretched and rotated elements 2 m above the base of the Pichu Pichu debris avalanche deposit.

can be difficult to distinguish from other fragmented in-situ volcanic deposits, and contextual information is generally necessary to determine the origins. For example, in Figure 38.3(F) the sedimentary layers in the hummock are not strongly deformed, and, without the hummock structure, could not be interpreted as being part of a debris avalanche.

Debris avalanche deposits have the following distinct facies (Figures 38.9 and 38.10): toreva and block facies, matrix facies, mixed facies, and basal facies.

Block facies consists of parts of the original slide mass that are intact to highly fractured. The classic texture is that of jigsaw cracks, or jigsaw fit, where the original position of the elements is maintained but the mass is shattered. Block facies can be highly brecciated into a fine sand to silt-sized material and there can be a complete gradation between block facies and matrix facies.

Matrix facies is composed of sand-to-silt-size fragments and isolated blocks (of up to a few meters in diameter). Matrix can be highly polylithologic. The variety of the components may partly be related to a mix of lithologies originating inside the rock mass. It can be also due to strong incorporation of substrate. Matrix facies is commonly highly variable and different matrix compositions can be juxtaposed. Fluidal textures are common features at the boundaries of different units and injections of one matrix into another are common. Such features suggest that the mixed facies can behave as a granular fluidized mass. Faults are also seen, and clastic dykes may cut the matrix (Figure 38.10), indicating brittle behavior at some stages during emplacement. Mixed facies is a term often used to describe a matrix with a highly polylithologic origin, and it is discriminated from matrix produced by milling of block facies originating from one rock type.

Clastic dykes and injected material are commonly seen as matrix and mixed material injections into both matrix and block facies (Figure 38.10), although larger areas of injection can form significant areas within the deposit (Figure 38.10(B)).

In the lowermost meter or so of a debris avalanche deposit the matrix facies is generally finer than the overlying mass (Figure 38.11). In this basal facies, frictionite and abraded surfaces are common, and interleaved lenses or clasts of substrate provide evidence of intrusion into the basal layer. The fine-grained nature and the concentration of structures in this basal layer indicates that much of the shear of the moving mass is concentrated at the base and the fine grain size may be a product of concentrated fragmentation (Davies et al., 2010).

Internal faults and shear zones are common in deposits. In the same way as for clast shape, the nature of these structures depends on the rheology of the materials. For example, contacts between rigid blocks or resistant substrate and matrix can be smooth, or contain gouge zones and striae (Figure 38.11). Alternatively, in the finer, granular deposits, faults are manifest as broad shear zones with no discrete slip surface (Figure 38.10(A)). Folds and rotational strain indicators are common in basal layers of debris avalanches where sediments are incorporated (Figure 38.11). Clastic dykes are commonly found intruded along fault zones (Figure 38.10).

9. DIFFERENT TYPES OF AVALANCHE DEPOSIT

Volcano landslides involving substrata: The initial instability that generates a volcanic landslide may occur under the volcano, and in this case the resulting deposit can include a large proportion of substrata as well as cone material. The Socompa Volcano debris avalanche is an example where up to 60% of the deposit volume is substrate (Figures 38.3(D) and 38.6(A)). Such collapses can form the largest volume volcanic deposits, although domes growing on hydrothermally altered rock can form a smaller volume version, such as at Montserrat (Sparks et al., 2002). Some larger landslides on continental stratovolcanoes, such as those of Parinacota and Chimborazo, attain volumes of over 10 km³ by incorporating large amounts of their basal volcaniclastic aprons and are thus similar in involving sediments lying below the main edifice.

Landslides at oceanic volcanoes can involve the largest volumes (many tens of km³), because these volcanoes are generally the largest, and they are commonly constructed on weak pelagic layers and volcano-derived low-strength strata. Examples are at La Reunion Island (Figure 38.1) and the Hawaiian Nuuanu and Wailau slides from Oahu and Molokai, shown in Figure 38.4(A) (Moore and Clague, 2002).

The Las Canadas, Orotava, and Guimar scars on Tenerife are examples of collapsed sectors of Canary Island oceanic volcanoes. Arc volcanoes in oceanic settings also have large collapses, such as the 1888 collapse of Mount Ritter (Papua New Guinea). Stromboli and the Antilles volcanoes also provide examples of such events.

The Hilina slump (Morgan et al., 2003) on Kilauea, the Enclos Fouqué of Piton de la Fournaise, and the western side of El Hierro, Canaries, are examples of ongoing landslides with slow and intermittent movement that may eventually develop into catastrophic debris avalanches.

Internally generated landslides: Landslides generated from hydrothermally altered cores, and from magma intrusion, will generally involve just the volcano. This limits their volume to the available portions of the edifice affected by the localized deformation. Examples of landslides related to hydrothermal activity are Bandai-San in 1888, and the El Crater landslide at Mombacho, Nicaragua. Examples of magmatic intrusion-driven landslides are Bezimyanny in 1956 and Mount St Helens in 1980. In each case the deposit volume is about 3 km³.

Superficial Landslides: Shallow volcanic landslides may develop in near-surface strata, without involving the volcano core. Such small volume landslides, up to several million cubic meters, include Lastarria (Chile) and Mount Meager (Canada). In the case of Lastarria, slope-parallel scoria beds failed on the flank, leaving a shallow scar and a thin debris avalanche deposit. At Mount Meager deep glacial incision and glacial retreat have eroded and debuttressed an edifice already weakened by hydrothermal alteration, leaving steep slopes that are prone to frequent landslides (Figure 38.3).

Monogenetic: Monogenetic volcanoes may develop landslides and debris avalanches, but being much smaller in volume than their polygenetic counterparts, their avalanches do not travel very far. Landslides may occur in association with cryptodome formation in monogenetic volcanoes and the resulting lava eruption may raft the landslide debris away from the cone to a far greater distance than the landslide would have carried it (Delcamp et al., 2014). Good examples of this are observed at Paracutin, Mexico, Red Mountain, Arizona, and are well preserved in the Lemptégy scoria cone, the Pariou scoria cone, and the Petit Puy de Dôme in the Chaîne des Puys, France (Figure 38.4(E) and (F)). The Petit Puy de Dôme is an example of uplift created by a cryptodome, and show a similar structure to that observed at Mount St Helens before the 1980 collapse. The Petit Puy de Dôme flank did not collapse and the intrusion-related bulge is stabilized in place, providing an excellent example of such deformation (Figure 38.4(F)).

Other planets: Probably the largest, and some of the best-exposed, debris avalanche deposits are preserved in the aureole around Olympus Mons, Mars (Figure 38.4(B)). The scale of such Martian landslides is much greater than on Earth, with several hundreds of kilometers of run out for up to 10 km of height change. As most other volcanoes in the solar system are flat shields, landslides and debris avalanches are restricted to the inner edges of their craters.

10. SECONDARY EFFECTS OF VOLCANIC LANDSLIDES

Landslides and debris avalanches create hazardous secondary effects such as explosive decompression and tsunamis. The most immediate effect may be the destabilization of hydrothermal systems and magmatic systems. Collapses accompanied by eruptions, such as at Mount St Helens in 1980, and Bandai-San in 1988, are examples. In both cases, it is likely that the magmatic and hydrothermal activity played a part in first priming the edifice by weakening the rock mass and then by triggering the landslide. Once failure occurred, decompression caused large blasts with pyroclastic flows and surges. In the cases of Mount St Helens, and Bezymyanny in 1956, explosive decompression led to large plinian eruptions that followed the landslides.

Debris avalanches entering bodies of water may create tsunamis, such as at Mayu-Yama, Japan in 1792; a small tsunami was also created by a minor landslide on Stromboli in 2002. Much larger tsunamis, with hundreds of meters of run-up, are evidenced by deposits on coasts adjacent to oceanic volcanoes, such as on Gran Canaria, Canary Islands, from the collapse of Guimar on Tenerife, or on Mauritius from the collapse of Piton de la Fournaise. Volcanic tsunamis are created by a body that is at most a few kilometers wide entering water and on an ocean scale is effectively a point source, rather than a long fault line as in earthquake-generated tsunamis. This leads to a radial dissipation of energy and thus landslide-generated tsunamis may have extreme local effects, but narrower impact than tectonic ones over the scale of an ocean basin.

Once a landslide and avalanche have occurred, the pressure drop in the hydrothermal system can generate precipitation of epithermal mineral deposits, and deeper decompression may result in the mobilization of magma even down to the mantle, as suggested for ankaramite eruptions in the Canary Islands (Manconi et al., 2009). This suggests that landsliding may result in an increase in eruption rates. Continued volcanic activity leads to a rapid masking of the landslide scar, and many volcanoes with known debris avalanches no longer show obvious source scars.

Volcanic landslides may, like their nonvolcanic counterparts, create dams that can generate floods and debris flows on failure. The 2009 Mount Meager landslide created two dams (Figure 38.3(B)), while the Buhi debris avalanche at Iriga created Lake Buhi (Figure 38.6(B)).

11. TRANSPORT MECHANISMS

The long run-out of volcanic (and nonvolcanic) avalanches remains a fascinating and unresolved problem. Debris avalanches clearly travel much further than would be expected if the motion was constrained by simple granular friction laws alone. However, the processes that cause this increased mobility are still understood poorly. This is partly because the evidence of transport processes is hidden in the hard-to-study deposits, and the appropriate analytical tools (such as sedimentological and textural ones) have yet to be developed or properly applied. Numerical, analytical, and analogue models (Davies et al., 2010; Dade and Huppert, 1998; Kelfoun and Druitt, 2006; Iverson and Denlinger, 2001; Shea and van Wyk de Vries, 2008) thus do not yet have the necessary foundation of observational data to discriminate fully between the possible processes. Numerical models for debris avalanches invoke a retarding stress due to reduced friction on the basal layer to achieve long run-outs. Analogue models use a smooth low-friction base or a ductile layer to simulate the low-friction base, and achieve comparable run-outs to natural events (Shea and van Wyk de Vries, 2008, Paguican et al., 2014). The mechanisms that have been proposed so far for the low friction are: basal air cushioning, lubrication by a watery base, or incorporation of water-rich low-friction clays, fluidization by vibration and acoustic energy. Release of elastic fracture energy during intense basal fragmentation at the base of the mass may also have an important role in providing a dispersive force to keep granular material in a dilated state (Davies et al., 2010).

12. HAZARDS FROM VOLCANIC DEBRIS AVALANCHES

Some volcanoes known to have landslide hazards have large cities at their feet (Popocatapetl—Mexico City, Fuji—Tokyo, Misti—Arequipa, for example), or a park frequented by many tourists (e.g., national parks such as Fuji, Yellowstone, Tongariro) and thus even small landslide events may pose serious risks.

Hazards posed by volcanic landslides and debris avalanches include the direct impact of the rock mass that can cover hundreds of square kilometers in a few minutes. This impact is amplified by the secondary hazards, like tsunamis and triggered eruptions, and because the debris avalanche deposit can be transformed into debris flows that then travel even farther. Hazards can be anticipated by geological mapping that can locate flanks with elevated potential for instability, or which already have evidence of structural deformation. This has been done at Casita (Cecchi et al., 2005), Etna, and Kilauea (Morgan et al., 2012), for example. Geodetic monitoring, especially using GPS and radar interferometry, can also be used to detect and monitor ongoing deformation, such as that detected at Kilauea, Etna or Piton de la Fournaise. At Mount St Helens, photogrammetry was valuable for monitoring the growth of the bulge that eventually failed, and such bulges can be detected with topographic mapping (e.g. Figure 38.4(E)). When a potential landslide is identified, an estimated volume and the local topography can be used as a basis for estimating the area affected using numerical simulations (Iverson and Denlinger, 2001; Kelfoun and Druitt, 2006; Davies et al., 2010).

Monitoring has the potential to detect the onset of instability through the recognition of increased slope movement, the appearance of surface fracturing, changes to the hydrothermal and hydrogeological systems, or increased seismicity. At Mount St Helens, the intrusion-related bulge in 1980 that caused the landslide developed over two months leading up to the failure. At Montserrat in 2000, ground cracking was observed several weeks before the eventual failure. If such symptoms are detected, uncertainty over the possible timescales leading up to a catastrophic failure are a problem for risk mitigation, especially as short-term triggers, such as seismic events (Reid et al., 2010) can cause a sudden loss of stability.

If adequate monitoring systems are in place, then acceleration of ground deformation before any major landslide should be detectable. However, sudden failure could be triggered by abrupt changes in a volcano, such as those occurring in hydrothermal systems, sudden magmatic intrusions, or as a result of seismic shaking. Most volcanoes are not monitored for slope stability, so the likelihood of undetected landslide and debris avalanche events is high. There is a need for a worldwide survey to identify potential landslide zones on volcano flanks, and to plan for both slowly developing failures and rapidly triggered events.

13. SUMMARY

Volcanoes are subject to a wide range of gravitational loading effects due to their growth by eruption and intrusion. Loading causes both substrata and edifice rocks to deform, an effect amplified by hydrothermal alteration, elevated fluid pressures and magma intrusion. Because of loading, volcanoes spread and sag, creating complex fault patterns that may interact with tectonic faulting. Outward-directed deformation can generate landslides, and these can transform into rapid, long run-out debris avalanches. The structure of these debris avalanche deposits is partly inherited from the edifice structure, then modified in the initial landslide and developed by deformation during transport. The structure consists of horst (hummock) and graben (interhummock) areas, and strike-slip zones, with folded and thrusted margins. Debris avalanche deposits comprise a block facies of relatively intact translated edifice rock, grading into fine, matrix facies of highly fractured rock, where different lithologies can combine to produce a mixed facies. The base of the mass usually has a finer-grained, strongly sheared layer that is highly fragmented and can show extreme abrasion, occasional melting, and substrate incorporation.

Volcanoes of all types have landslides and debris avalanches, the largest examples of which originate from the oceanic shield volcanoes, the smallest from monogenetic volcanoes. Debris avalanches can trigger eruptions and generate tsunamis, if they enter bodies of water. Removal of load from the edifice may depressurize the whole hydrothermal and magmatic column down to the mantle, causing renewed magma ascent, eruption, and epithermal mineral precipitation.

The physical processes in volcanic landslides and the mobility of volcanic debris avalanches are still not well constrained and will be a major research challenge for many years. Equally, the prediction of the occurrence and deposit extent of volcanic landslide deposits and debris avalanches will remain an ongoing challenge for hazard and risk assessment.

FURTHER READING

Belousov, A., Belousova, M., Voight, B., 1999a. Multiple edifice failures, debris avalanches and associated eruptions in the holocene history of shiveluch volcano, Kamchatka, Russia. Bull. Volcanol. 61, 324—342.

Borgia, A., van Wyk de Vries, B., 2003. The volcano-tectonic evolution of Concepción, Nicaragua. Bull. Volcanol. 65, 248—266.

Cecchi, E., van Wyk de Vries, B., Lavest, J.M., 2005. Flank spreading and collapse of weak-cored volcanoes. Bull. Volcanol. 67, 72—91.

Dade, W.B., Huppert, H.E., 1998a. Long-runout rockfalls. Geology 9, 803—880.

Davies, T.R.H., McSaveney, M.J., Kelfoun, K., 2010a. Runout of the socompa volcanic debris avalanche, Chile: a mechanical explanation for low basal shear resistance. Bull. Volcanol. 72, 933—944.

Delcamp, A., van Wyk de Vries, B., Petit, S., Kervyn, M., 2014a. The endongenous and exogenous growth of the monogenetic Lemptégy volcano, cahine des puys, France. Geosphere. http://dx.doi.org/10.1130/GES01007.1.

Glicken, H., 1996. Rockslide-Debris Avalanche of May 18, 1980, Mount St. Helens Volcano, Washington. Open-file Report 96—677. US Geological Survey, 90pp.

Gorshkov, G.S., 1959. Gigantic eruption of the volcano Bezymianny. Bull. Volcanol. 20, 77—109.

Guthrie, R.H., Friele, P., Allstadt, K., Roberts, N., Evans, S.G., Delaney, K.B., Roche, D., Clague, J.J., Jakob, M., 2012. The 6 August 2010 Mount Meager rock slide-debris flow, coast mountains, british columbia: characteristics, dynamics, and implications for hazard and risk assessment. Nat. Hazards Earth Syst. Sci. 12, 1277—1294.

Iverson, R.M., Denlinger, R.P., 2001a. Flow of variably fluidized granular masses across three-dimensional terrain 1. Coulomb mixture theory. J. Geophys. Res. 106, 537—552.

Iverson, R.M., Reid, M.E., La Husen, R.G., 1997. Debris-flow mobilization from landslides. Annu. Rev. Earth Planet. Sci. 25, 85—138.

Kelfoun, K., Druitt, T.H., 2006a. Numerical modelling of the emplacement of socompa rock avalanche, Chile. J. Geophys. Res. 110, 12202.

Lipman, P.W., Mullineaux, D.R. (Eds.), 1981. The 1980 Eruptions of Mount St. Helens, Washington. U S Geol Surv, pp. 1—844. Prof Pap, 1250.

Manconi, A., Longpre, M.A., Walter, T.R., Troll, V.R., Hansteen, T.H., 2009. The effects of flank collapses on volcano plumbing systems. Geology 37, 1099—1102.

Moore, J.G., Clague, D.A., 2002. Mapping the Nuuanu and Wailau landslides in Hawaii. In: Takahashi, E., Lipman, P.W., Garcia, M.O., Naka, J., Aramaki, S. (Eds.), Hawaiian Volcanoes: Deep Underwater Perspectives. Geophysical Monograph 128, American Geophysical Union, pp. 223—244.

Moore, J.G., Nomak, W.R., Holcomb, R.T., 1994. Giant Hawaiian landslides. Science 264, 46—47.

Morgan, J.K., Moore, G.F., Clague, D.A., 2003. Slope failure and volcanic spreading along the submarine south flank of Kilauea volcano, Hawaii. J. Geophys. Res. 108, 24—25. http://dx.doi.org/10.1029/2003JB002411.

Paguican, E.M.R., Van Wyk de Vries, B., Lagmay, A.M.F., 2014a. Hummocks: how they from in and how they evolve in rockslide-debris avalanches. Landslides 11, 67—80.

Reid, M.E., Keith, T.E.C., Kayen, R.E., Iverson, N.R., Iverson, R.M., Brien, D.L., 2010. Volcano collapse promoted by progressive strength reduction: new data from Mount St. Helens. Bull. Volcanol. 72, 761—766.

Shea, T., van Wyk de Vries, B., 2008. Structural analysis and analogue modelling of the kinematics and dynamics of large-scale rock avalanches. Geosphere 4, 657—686.

Sekiya, S., Kikuchi, Y., 1890a. The eruption of bandai-san. Trans. Seismol. Soc. Jpn. 13, 139—222.

Sparks, R.S.J., Barclay, J., Calder, E.S., Herd, R.A., Luckett, R., Norton, G.E., Ritchie, L.J., Voight, B., Woods, A.W., 2002. Generation of a debris avalanche and violent pyroclastic density current on 26 December (Boxing Day) 1997 at Soufrière Hills Volcano, Montserrat. The eruption of Soufrière Hills Volcano, Montserrat, from 1995 to 1999. In: Druitt, T.H., Kokelaar, B.P. (Eds.), Geological Society of London, pp. 409—434.

van Wyk de Vries, B., Francis, P.W., 1997. Catastrophic collapse at stratovolcanoes induced by gradual volcano spreading. Nature 387, 387—390.

ANNEXE 1—EXTENDED BIBLIOGRAPHY

Volcano gravitational deformation and volcano-tectonics

Analogue models of volcano spreading

Merle, O., Borgia, A., 1996. Scaled experiments on volcanic spreading. J. Geophys. Res. 101 (B6), 13805—13817.

Extensional tectonics and faulting in volcanoes

van Wyk de Vries, B., Merle, O., 1996. The effect of volcanic constructs on rift fault patterns. Geology 24 (7), 643—646.

Intrusion and stability

Donnadieu, F., Merle, O., 1998. Experiments on the indentation process during cryptodome intrusions: new insights into Mount St. Helens deformation. Geology 26 (1), 79—82.

Intrusion and volcano-tectonics

Galland, O., Hallot, E., Cobbold, P.R., Ruffet, G., de Bremond d'Ars, J., 2007. Volcanism in a compressional andean setting: a structural and geochronological study of tromen volcano (Neuquén province, Argentina). Tectonics 26. http://dx.doi.org/10.1029/2006TC002011.

Regional tectonics and volcano structure

Lagmay, A.M.F., van Wyk de Vries, B., Kerle, N., Pyle, D.M., 2000. Volcano instability induced by strike-slip faulting. Bull. Volcanol. 62 (4—5), 331—346.

Spreading and flexure on pelagic sediments at Hawaii

Nakamura, K., 1980. Why do long rift zones develop in Hawaiian volcanoes—a possible role of thick oceanic sediments. Bull. Volcanol. Soc. Jpn. 25, 255—269.

Spreading and stability at Kilauea

Morgan, J.K., Moore, G.F., Clague, D.A., 2003. Slope failure and volcanic spreading along the submarine south flank of Kilauea volcano, Hawaii. J. Geophys. Res. 108, 24—25.

Spreading linked to landsliding

van Wyk de Vries, B., Francis, P.W., 1997. Catastrophic collapse at stratovolcanoes induced by gradual volcano spreading. Nature 387, 387—390.

Spreading linked to landsliding

van Wyk de Vries, B., Self, S., Francis, P.W., Keszthelyi, L., 2001. A gravitational spreading origin for the Socompa debris avalanche. J. Volcanol. Geotherm. Res. 105, 225—247.

Spreading/sagging

Byrne, P.K., Holohan, E.P., Kervyn, M., van Wyk de Vries, B., Troll, V.R., Murray, J.B., 2013. A sagging-spreading continuum of large volcano structures. Geology 41, 339—342.

Strike-slip tectonics and faulting in volcanoes

van Wyk de Vries, B., Merle, O., 1998. Extension induced by volcanic loading in strike-slip fault zones. Geology 26, 983—986.

Vertical faulting and edifice stability

Merle, O., Vidal, N., van Wyk de Vries, B., 2001. Experiments on vertical basement fault reactivation below volcanoes. J. Geophys. Res. 106, 2153—2162.

Volcanic spreading

van Wyk de Vries, B., Borgia, A., 1996. The role of basement in volcano deformation. In: McGuire, W.J., Jones, A.P., Neuberg, J. (Eds.), Volcano Instability on the Earth and Other Planets, vol.110. Geological Society of London Special Publication, pp. 95—110.

Volcano shape and edifice evolution

Grosse, P., van Wyk de Vries, B., Petrinovic, I.A., Euillades, P.A., Alvarado, G.E., 2009. The morphometry and evolution of arc volcanoes. Geology 37 (7), 651—654.

Volcano spreading

Borgia, A., 1994. The dynamic basis for volcanic spreading. J. Geophys. Res. Solid Earth 99 (B9), 17791—17804.

Volcano spreading

Borgia, A., Delaney, P.T., Denlinger, R.P., 2000. Spreading volcanoes. Annu Rev. Earth Planet Sci. 28, 3409—3412.

Volcano spreading at Concepción

Borgia, A., van Wyk de Vries, B., 2003. The volcano-tectonic evolution of Concepción, Nicaragua. Bull Volcanol 65, 248—266.

Volcano spreading at Etna

Lundgren, P., Casu, F., Manzo, M., Pepe, A., Bernadino, P., Sonsosti, E., Lanari, R., 2004. Gravity and magma induced spreading of Mount Etna volcano revealed by satellite radar interferometry. Geophys. Res. Lett. http://dx.doi.org/10.1029/2003GL018736.

Volcano spreading at Etna

Borgia, A., Ferrari, L., Pasquarè, G., 1992. Importance of gravitational spreading in the tectonic and volcanic evolution of Mount Etna. Nature 357, 231—235.

Edfice instability and volcanic landslides

Bezymianny Volcano

Gorshkov, G.S., 1959. Gigantic eruption of the volcano Bezymianny. Bull Volc 20, 77—109.

Collapse at Mayu-yama, Japan

Miyachi, M., 1992. Geological examination of the two old maps from the Tokugawa Era concerning the Shimabara Catastrophe. In: Yanagi, T., Okada, H., Ohta, K. (Eds.), Unzen Volcano, the 1990-1992 Eruption. Nishinippon & Kyushu University Press, pp. 99—102.

Geotechnical aspects of volcanic landslides

Voight, B., Elseworth, D., 1997. Failure of volcanic slopes. Géothechnique 47, 1—31.

Giant Hawaii landslides

Moore, J.G., Nomak, W.R., Holcomb, R.T., 1994. Giant Hawaiian Landslides. Science 264, 46—47.

Hydrothermal pressurization and collapse

Reid, M.E., 2004. Massive collapse of volcano edifices triggered by hydrothermal pressurization. Geology 32, 373—376.

Hydrothermal system changes after collapse

Silitoe, R.H., 1994. Erosion and collapse of volcanoes: causes of telescoping in intrusion-centered ore deposits. Geology 10, 945—948.

Landlides triggered by caldera collapse

Hürlimann, M., Turon, E., Marti, J., 1999. Large landslides triggered by caldera collapse events in Tenerife, Canary Islands. Phys. Chem. Earth Part A Solid Earth Geodesy 24, 921—924.

Monogenetic collapse, Paracutin Volcano

Luhr, J.F., Simkin, T., 1993. Paracutin: The Volcano Born in a Mexican Cornfield. Geoscience Press p. 427.

Monogenetic volcano collapse

Delcamp, A., Van Wyk de Vries, B., Kervyn, M., 18 August 2014b. The Lemptégy Scoria Cone, Chaîne des Puys France. Geosph. line. http://dx.doi.org/10.1130/GES01007.1.

Montserrat volcano

Sparks, R.S.J., Barclay, J., Calder, E.S., Herd, R.A., Luckett, R., Norton, G.E., Ritchie, L.J., Voight, B., Woods, A.W., 2002. Generation of a debris avalanche and violent pyroclastic density current on 26 December (Boxing Day) 1997 at Soufrière Hills Volcano, Montserrat. The eruption of Soufrière Hills Volcano, Montserrat, from 1995 to 1999 (GSL Memoir 21). In: Druitt, T.H., Kokelaar, B.P. (Eds.), Geological Society, pp. 409—434.

Mount St. Augustine

Beget, J.E., Kienle, J., 1992. Cyclic formation of debris avalanches at Mount St. Augustine volcano. Nature 356, 701—704.

Mount St. Helens

Lipman, P.W., Mullineaux, D.R. (Eds.), 1981. The 1980 eruptions of Mount St. Helens, Washington. U S Geol Surv, pp. 1—844. Prof Pap, 1250.

Progressive strength reduction in edifices

Reid, M.E., Keith, T.E.C., Kayen, R.E., Iverson, N.R., Iverson, R.M., Brien, D.L., 2010. Volcano collapse promoted by progressive strength reduction: new data from Mount St. Helens. Bulletin of Volcanology 72, 761—766.

Red Mountain monogenetic collapse

Riggs, N.R., Duffield, W.A., 2008. Record of complex scoria cone eruptive activity at Red Mountain, Arizona, USA, and implications for monogenetic mafic volcanoes. J. Volcanol. Geotherm. Res. 178, 763—776.

Réunion Island stability

Upton, B.G.J., Wadsworth, W.J., 1965. Geology of réunion island, Indian ocean. Nature 207, 151—154.

Stromboli Volcano

Tibaldi, A., 2001. Multiple sector collapses at stromboli volcano, Italy: how they work. Bull. Volcanol. 63, 121—125. Voight, B., Janda, R.J., Glicken, H., Douglass, P.M., 1983. Nature and mechanics of the Mount St Helens rockslide avalanche of 18 May 1980, Géotechnique, 33, 243—273.

Volcanic landslides and debris avalanche deposits

Bandai-San landslide

Sekiya, S., Kikuchi, Y., 1890b. The eruption of Bandai-san. Trans. Seismol. Soc. Jpn. 13 (2), 139—222.

Blackhawk landslide

Shreve, R.L., 1968. The Blackhawk Landslide. Geological Society of America. Special Paper no.108, p.1—47.

Chimborazo Volcano

Bernard, B., van Wyk de Vries, B., Barba, D., Leyrit, H., Robin, C., Alcaraz, S., Samaniego, P., 2008. The chimborazo sector collapse and debris avalanche: deposit characteristics as evidence of emplacement mechanisms. J. Volcanol. Geotherm. Res. 176, 36—43.

Classic general study of volcanic landslides

Siebert, L., 1984. Large volcanic debris avalanches: characteristics of source areas, deposits, and associated eruptions. J. Volcanol. Geotherm. Res. 22 (3—4), 163—197.

Classic general study of volcanic landslides

Siebert, L., Glicken, H., Ui, T., 1987. Volcanic hazards from Bezymianny- and Bandaï-type eruptions. Bull. Volcanol. 1, 435—459.

Classic general study of volcanic landslides

Ui, T., 1983. Volcanic dry avalanche deposits. Identification and comparison with non-volcanic debris stream deposits. J. Volcanol. Geotherm. Res. 18 (1—4), 135—150.

Colima Volcano

Stoopes, G.R., Sheridan, M.F., 1992. Giant debris avalanches from the colima volcanic complex, Mexico: implication for long-runout landslides (100 km) and hazard assessment. Geology 20, 299—302.

General study of landslide mechanisms

Erismann, T.H., 1979. Mechanisms of large landslides. Rock Mech. Rock Eng. 12 (1), 5—46.

General study on large landslides

Shaller, P.J., 1991. Analysis and implications of large Martian and Terrestrial landslides (Ph.D. thesis). California Institute of Technology, 586pp.

Hummock description in an avalanche

McColl, S.T., Davies, T.R.H., 2011. Evidence for a rock-avalanche origin for "The Hillocks" "moraine", Otago, New Zealand. Geomorphology 127, 216—224.

Landslides from the Antilles volcanoes

Lebas, E., Le Friant, A., Boudon, G., Watt, S.F.L., Talling, P.J., Feuillet, N., Deplus, C., Berndt, C., Vardy, M.E., 2001. Multiple widespread landslides during the long-term evolution of a volcanic island: insights from high-resolution seismic data, Montserrat, Lesser Antilles. Geochem. Geophys. Geosyst. 12 (5), 1—20, 05/2011.

Landslides in mountainous areas

Heim, A., 1932. Zurich, Vierteljahrsschrift. Bergsturz und Menchenleben, 77. no.20, 218 pp.

Mombacho Volcano

Shea, T., van Wyk de Vries, B., Pilato, M., 2008. Emplacement dynamics of contrasting debris avalanches at Volcán Mombacho (Nicaragua), provided by structural and facies analysis. Bull. Volcanol. 70, 899—921.

Mount Meager

Guthrie, R.H., Friele, P., Allstadt, K., Roberts, N., Evans, S.G., Delaney, K.B., Roche, D., Clague, J.J., Jakob, M., 2012. The 6 August 2010 Mount Meager rock slide-debris flow, Coast Mountains, British Columbia: characteristics, dynamics, and implications for hazard and risk assessment. Nat. Hazards Earth Syst. Sci. 12, 1277—1294.

Mount Shasta

Crandell, D.R., 1989. Gigantic Debris Avalanche of Pleistocene Age from Ancestral Mount Shasta Volcano, California, and Debris-avalanche Hazard Zonation. U.S. Geological Survey Bulletin 1861, 32p.

Mount St. Augustine

Swanson, S.E., Kienle, J., 1988. The 1986 eruption of Mount St. Augustine: field test of a hazard evaluation. J. Geophys. Res. 93, 4500—4520.

Mount St. Helens

Glicken, H., 1996. Rockslide-debris avalanche of May 18, 1980, Mount St. Helens volcano, Washington. Open-file Report 96—677. US Geological Survey, 90pp.

Mount St. Helens

Voight, B., Glicken, H., Janda, R.J., Douglass, P.M., 1981. Catastrophic rockslide avalanche of May 18. In: Lipman, P.W., Mullineaux, D.R. (Eds.), The 1980 Eruptions of Mount St. Helens. U.S. Geological Survey, Washington, pp. 347—377. Professional Paper no.1250.

Nevado De Toluca Volcano

Caballero, L., Capra, L., 2011. Textural analysis of Particles from el Zaguán debris avalanche Deposit Nevado de Toluca volcano, Mexico:

evidence of flow behaviour during emplacement. J. Volcanol. Geotherm. Res. 200, 75—82.

Parinacota Volcano

Clavero, J.E., Sparks, R.S.J., Huppert, H.E., 2002. Geological constraints on the emplacement mechanism of the parinacota avalanche, northern Chile. Bull. Volcanol. 64 (1), 40—54.

Ridges of debris avalanche deposits

Dufresne, A., Davies, T.R., 2009. Longitudinal ridges in mass movement deposits. Geomorphology 105 (3—4), 171—181.

Sherman landslide

Shreve, R.L., 1966. Sherman landslide, Alaska. Science 154, 1639—1643.

Shiveluch Volcano

Belousov, A., Belousova, M., Voight, B., 1999b. Multiple edifice failures, debris avalanches and associated eruptions in the Holocene histbory of Shiveluch volcano, Kamchatka, Russia. Bull. Volcanol. 61 (5), 324—342.

Socompa Volcano

Wadge, G., Francis, P.W., Ramirez, C.F., 1995. The socompa collapse and avalanche event. J. Volcanol. Geotherm. Res. 66, 309—336.

Socompa volcano topographic interaction

Kelfoun, K., Druitt, T.H., van Wyk de Vries, B., Guilbaud, M.-N., 2008. The topographic reflection of the socompa debris avalanche. Bull. Volcanol..

Toreva block original definition

Reiche, P., 1937. The Toreva-block, a distinctive landslide type. J. Geology 45 (5), 538—548.

Modeling of landslides and debris avalanches

Analogue model ling of flank spreading

Cecchi, E., van Wyk de Vries, B., Lavest, J.M., 2005. Flank spreading and collapse of weak-cored volcanoes. Bull. Volcanol. 67, 72—91.

Analogue model of landslide formation

Paguican, E.M.R., van Wyk de Vries, B., Lagmay, A.M.F., 2014b. Hummocks: how they from in and how they evolve in rockslide-debris avalanches. Landslides 11 (1), 67—80.

Analogue models of landslide initial structure

Andrade, D., van Wyk de Vries, B., 2010. Structural analysis of the early stages of catastrophic stratovolcano flank-collapse using analogue models. Bull. Volcanol. 72, 771—789.

Analysis of physics of debris avalanches

Kilburn, C., Sorensen, S.-A., 1998. Runout lengths of sturzstroms: the control of initial conditions and of fragment dynamics. J. Geophys. Res. 103 (B8), 17877.

Analytical model of landslides

Dade, W.B., Huppert, H.E., 1998b. Long-runout rockfalls. Geology 26 (9), 803—806.

Distinct element models of landslide structures

Thompson, N., Bennet, M.R., Petford, N., 2010. Development of characteristic volcanic debris avalanche deposit structures: new insights from distinct element simulations. J. Volcanol. Geotherm. Res. 192, 191—200.

General engineering study of landslide mobility

Legros, F., 2002. The mobility of long-runout landslides. Eng. Geol. 63, 301—331.

Model of fragmentation and mobility

Davies, T.R.H., McSaveney, M.J., 2009. The role of dynamic rock fragmentation in reducing frictional resistance to large landslides. Eng. Geol. 109, 67—79. http://dx.doi.org/10.1016/j.enggeo.2008.11.004.

Model of fragmentation and mobility

Davies, T.R.H., McSaveney, M.J., Boulton, C.J., 2012. Elastic strain energy release from fragmenting grains: effects on fault rupture. J. Struct. Geol. 38, 265—277.

Model of fragmentation and mobility

Davies, T.R.H., McSaveney, M.J., Hodgson, K.A., 1999. A fragmentation-spreading model for long runout rock avalanches. Can. Geotech. J. 36 (6), 1096—1110.

Model of fragmentation and reduced basal resistance

Davies, T.R.H., McSaveney, M.J., Kelfoun, K., 2010b. Runout of the socompa volcanic debris avalanche, Chile: a mechanical explanation for low basal shear resistance. Bull. Volcanol. 72, 933—944. http://dx.doi.org/10.1007/s00445-010-0372-9.

Numerical model and experimental tests

Denlinger, R.P., Iverson, R.M., 2001. Flow of variably fluidized granular masses across three-dimensional terrain 2. Numerical predictions and experimental tests. J. Geophys. Res. 106 (B1), 553—566.

Numerical model of landslide-generated tsunami

Kelfoun, K., Giachetti, T., Labazuy, P., 2010. Landslide-generated tsunamis at Réunion Island. J. Geophys. Res. 155 http://dx.doi.org/10.1029/2009JF001381.

Numerical model with "volcflow"

Kelfoun, K., Druitt, T.H., 2006b. Numerical modelling of the emplacement of socompa rock avalanche, Chile. J. Geophys. Res. 110 (B12), 12202.

Numerical model with Coulomb theory

Iverson, R.M., Denlinger, R.P., 2001b. Flow of variably fluidized granular masses across three-dimensional terrain 1. Coulomb mixture theory. J. Geophys. Res. 106 (B1), 537—552.

Numerical modeling

Campbell, C.S., 1989. Self-lubrication for long runout landslides. J. Geol. 97 (6), 653—665.

Numerical modeling

Campbell, C.S., Cleary, P.W., Hopkins, M.J., 1995. Large-scale landslide simulations: global deformation, velocities and basal friction. J. Geophys. Res. 100, 8267—8283.

Particle segregation study

Gray, J.M.N.T., Thornton, A.R., 2005. A theory for particle size segregation in shallow granular free-surface flows. Proc. R. Soc. A 461 (2056), 1447—1473.

Hazards associated with volcanic landslides

Debris flow at Casita

Scott, K., Vallance, K.M., Kerle, J.W., Macias, J.L., Strauch, W., Devoli, G., 2004. Catastrophic precipitation-triggered lahar at casita volcano, Nicaragua - flow bulking and transformation. Earth Surf. Process. Landf. 30, 59—79.

Debris flows

Capra, L., Macías, J.L., 2002. The Cohesive Naranjo Debris-Flow Deposit (10 km3): a dam breakout flow derived from the Pleistocene Debris-Avalanche Deposit of Nevado de Colima Volcano (Mexico). J. Volcanol. Geotherm. Res. 117, 213—235. Elsevier.

Debris flows

Capra, L., Macías, J.L., 2000. Pleistocene cohesive debris flows at Nevado de Toluca Volcano, central Mexico. J. Volcanol. Geotherm. Res. 102, 149—167.

Debris flows

Capra, L., Macías, J.L., Scott, K.M., Abrams, M., Garduño-Monroy, V.H., 2002. Debris avalanches and debris flows transformed from collapses in the Trans-Mexican volcanic Belt, México — behavior, and implication for hazard assessment. J. Volcanol. Geotherm. Res. 113 (1—2), 81—110. Elsevier.

Debris flows

Vallance, J.W., Scott, K., 1997. The Osceola Mudflow from Mount Rainier: sedimentology and hazard implications of a huge clay-rich debris flow. Geol. Soc. Am. Bull. 109, 143—163.

Debris flows from landslides

Iverson, R.M., Reid, M.E., La Husen, R.G., 1997. Debris-flow mobilization from landslides. Annu. Rev. Earth Planet. Sci. 25, 85—138.

Decompressing effects on deep magma plumbing

Manconi, A., Longpre, M.A., Walter, T.R., Troll, V.R., Hansteen, T.H., 2009. The effects of flank collapses on volcano plumbing systems. Geology 37, 1099—1102.

Tsunami

Beget, J.E., Kowalik, Z., 2006. Confirmation and calibration of computer modeling of tsunami produced by Augustine volcano, Alaska. Sci. Tsunami Hazards 24 (4), 257.

Tsunami

Paris, R., Kelfoun, K., Giachetti, T., 2013. Marine conglomerate and reef megaclasts at Mauritius Island (Indian Ocean): evidences of a tsunami generated by a flank collapse of Piton de la Fournaise volcano, Réunion Island? Sci. Tsunami Hazards 32 (4), 281—291.

Tsunami

Tinti, S., Manucci, A., Pagnoni, G., Armigliato, A., Zaniboni, F., 2005. The 30th December 2002 tsunami in Stromboli: sequence of the events reconstructed from the eyewitness accounts. Nat. Hazards Earth Syst. Sci. 5, 763—775.

Extraterrestrial Volcanism

Haraldur Sigurdsson

Graduate School of Oceanography, University of Rhode Island, Narragansett, RI, USA

In the Space Age we have discovered that volcanoes occur on most of the rocky bodies of our Solar System, and future explorations of other worlds will without doubt reveal volcanism as an important feature of planetary evolution in the universe as a whole. The evidence for volcanism on other bodies of the Solar System is reviewed in the five following chapters in Part V of the second edition of the Encyclopedia of Volcanoes, Extraterrestrial Volcanism. Although just beginning, the exploration of extraterrestrial volcanoes has revealed some amazing surprises, as described in this part of the volume.

We begin with the Chapter 39, studying the volcanism of our nearest neighbor. When Apollo 11 landed on the Moon in 1969, it touched down on the surface of a basaltic lava flow. Ever since Galileo aimed his telescope at the Moon in 1610, we have wondered about the origin of its numerous and large craters. The lunar craters, however, turned out to be of impact origin, whereas the dark plains, which are clearly visible with the naked eye from Earth and were termed maria by Galileo, turned out to be made up of great floods of basaltic lavas. Studies of the rocks returned back to Earth laboratories have shown that the Moon has been extinct for over one billion years—a consequence of the small diameter of this rapidly cooling planetary body.

In contrast to our volcanically "dead" Moon, the Jovian satellite Io has a continuous display of spectacular eruptions, as described in the Chapter 43. It is in fact the only body outside the Earth known to have large-scale active volcanism today. Io is indeed a strange and colorful world, with a surface decorated with a wonderful mosaic of yellows, oranges, and reds and sulfurous volcanic plumes that rise over 300 km above its surface. We are accustomed to thinking of volcanic energy as largely derived from primordial heat in the planet, but Io's energy source is different. The intense and continuous volcanism of Io is due to heat from great tidal stresses generated by its giant neighbor Jupiter.

Venus—our nearest planetary neighbor—has volcanic features that are in several respects similar to the activity on Earth, as described in the Chapter 42. While there is no evidence of Earth-style plate tectonics on the Moon, Io, or Mars, or currently on Venus, there is indication that global-scale processes may have been operating there, causing geologically recent global resurfacing, reflected in its extensive volcanic plains. Large central volcanoes dot the surface, but Venus also possesses enigmatic coronae, enormous circular structures that are several hundred kilometers in diameter and found nowhere else in our Solar System.

The Chapter 41 answers the question: how did a planet one-half the size of Earth generate volcanoes that are several times larger than the largest volcanoes on Earth? Mars has widespread volcanic plains, like the Moon, but the characteristic Martian volcanic features are huge central volcanoes that dwarf any terrestrial volcanoes. These immense lava shields are the result of hotspot activity, and the largest is Olympus Mons, 25 km high, 600 km diameter. Is the size of these volcanoes a reflection of a much thicker lithosphere and the absence of plate tectonics?

Volcanism on Mercury is a new chapter in this volume, reflecting the great discoveries made on this planet in recent years, primarily due to the MESSENGER spacecraft mission. These results show that indeed volcanism has been important in shaping the Mercury's surface. There has been extensive flooding of the surface by lava flows to form regional smooth plains.

In the farthest reaches of our Solar System, on the moons of Uranus, Saturn, and Jupiter, a unique form of volcanism occurs, in which geyser-like plumes of nitrogen and water-rich "magmas" are ejected from volcano-like structures. The final chapter of this section, Cryovolcanism in the Outer Solar System, discusses this distinctive form of volcanism. Most of the moons around Jupiter, Uranus, and

Neptune are made up of water ice and other volatiles. From time to time, heat generated within these satellites churns up a slush of ice and volatiles that are erupted at the surface, forming volcano-like manifestations of "dirty ices" mixed with ammonia, silicates, and other impurities. The discovery of nitrogen-rich and geyser-like plumes on Triton, Neptune's distant moon, by Voyager 2 in 1989 and, more recently, the discovery of the Enceladus plume by the Cassini spacecraft, shows that this type of volcanism occurs today.

Chapter 39

Volcanism on the Moon

Paul D. Spudis
Lunar and Planetary Institute, TX, USA

Chapter Outline

GLOSSARY

albedo Reflectivity of an object. Light and dark are high and low albedo, respectively.

ash Very small fragments of volcanic glass, sprayed out at a vent, cooled quickly, and deposited as a blanket of debris.

basalt A dark, fine-grained rock, rich in iron and magnesium, created by solidification of a lava rich in iron and magnesium.

bombardment The repeated collision of a planet with asteroids over a long time.

cinder cone Hill produced by the buildup of ash or other pyroclastic (q.v.) fragments around a volcanic vent.

collapse pit Small depression associated with subsurface collapse, as over a shallow chamber after lava has been drained from it.

dark mantle deposit Large area of the Moon covered by glassy pyroclastic ash deposits.

fire fountain Spray of lava from a vent, producing a dark mantling deposit.

flow front The terminal end of a lava flow.

impact dark halo crater Crater with dark ejecta, caused by the exhumation of buried lava flows in the lunar highlands.

KREEP Acronym for potassium (K), rare-earth elements (REE), and phosphorus (P); a chemical component in lunar rocks, created as the last phase of the magma ocean (q.v.).

lava Liquid magma extruded onto a planetary surface.

lava tube A channel that is partly or completely roofed over to enclose the lava stream; may form a cave after the flow has cooled.

mafic property describing an enrichment in iron and magnesium at the expense of silicon and aluminum.

magma Liquid rock within the interior of a planet.

magma ocean State of the early Moon in which the entire globe was covered by a layer of liquid rock, hundreds of kilometers thick.

mantle The part of a planet below the crust and above the core. Partial melting of the mantle is the source of magma in most planets.

maria (MAR-ee-ah) Dark areas of the Moon; Latin plural for "seas"; singular is mare (MAR-ay).

pyroclastic Literally "fire-broken", meaning fragmental rocks produced in explosive volcanic eruptions, includes ash.

regolith Unconsolidated debris overlying bedrock.

rilles Any relatively narrow depression, whose shape can be linear, arcuate, or sinuous. Linear and arcuate rilles have tectonic origins while sinuous rilles are lava channels or tubes.

shield volcano A broad, low-relief volcanic construct made up of flows of relatively fluid lava, usually basalt.

sinuous rille A lava channel or tube in the lunar maria.

terrae (TER-eye) Latin plural for "land", the cratered highlands of the Moon; singular is terra (TER-ah).

vent Hole in a planet from which volcanic products (lava, ash) may be erupted.

vesicle Hole (frozen bubble) in a sample of lava rock that results from dissolved gas in magma coming out of solution.

viscosity The property of a fluid that causes it to resist flow or movement. Higher viscosity results in slower rates of flow.

xenolith (ZEE-no-lith) Literally "stranger rock", a fragment of rock from great depths carried to the surface as a clast in a lava flow or pyroclastic deposit.

The Encyclopedia of Volcanoes. http://dx.doi.org/10.1016/B978-0-12-385938-9.00039-0

1. INTRODUCTION

In the years before spaceflight, the Moon was thought to be volcanically active. Astronomers are very fond of volcanoes, so an analogy was made between the craters of the Moon and calderas, the large, circular depressions found at the summit of terrestrial volcanoes. Detailed study of lunar craters and terrestrial analogs eventually provided convincing evidence that the abundant craters of the Moon were formed by the collision of solid bodies over geological time.

The few workers studying the Moon largely ignored the dark **maria**, the low-lying smooth plains that make up large fractions of the near side seen through a telescope (Figure 39.1). Most early ideas about the maria associated them with ancient sedimentary deposits, either water-laid or mass-wasted material derived from the surrounding highlands. However, in his landmark 1949 book *The Face of the Moon*, astronomer Ralph Baldwin presented convincing evidence that the maria are made up of floods of **basalt**, a dark **lava**, rich in iron that is abundant on the Earth.

In preparation for the Apollo missions, debate on the origin of the maria raged. The return of samples from the Moon settled the issue. The maria indeed are made up of floods of basaltic lava (Figure 39.2), but what is most striking from the returned samples is the age of these lavas. The basalts returned by Apollo range in age from 4300 to 3100 million years, as old as the very oldest rocks on the Earth.

FIGURE 39.1 Lunar Reconnaissance Orbiter global high-sun image of the Moon showing its two principal terrains: the light, rough highlands (or terrae) and dark, smooth lowlands (or maria). Near side on the left, far side on the right. Study has shown that the maria are made up of basaltic lava flows and constitute about 16% of the lunar surface and <1% of the volume of the crust.

FIGURE 39.2 Hand specimen (top left; markings in centimeters) and thin section (bottom, plane light and xpl; Field of view − 2 mm) of Apollo 17 high-titanium mare basalt 70,017. Lunar basalts are made up predominantly of augite (Ca-rich pyroxene), olivine, plagioclase, and ilmenite; they are depleted in volatile elements and contain no hydrous minerals.

Apollo 17 high-Ti mare basalt 70017

Hand sample scale: 1 cm

Thin section FOV: 2 mm

The lunar lavas have interesting compositions, including a complete absence of hydrous minerals, which often occur in lavas on the Earth. These properties, discovered during the initial examination of the Apollo samples, gave us a first-order understanding of the basic properties of the Moon. We discovered that the Moon has a crust, formed early in its history by global melting (the "**magma ocean**"). This differentiation produced the plagioclase-rich crust and the **mantle** source regions for the later mare basalts. Thus, lunar volcanic rocks contain important information about the composition of the deep interior of the Moon.

2. MARE VOLCANISM ON THE MOON

2.1. Origin of Mare Basalts

The samples of the maria returned by the Apollo missions are a form of lava known as "basalt" (Figure 39.2). Just as on Earth, basalt is created by partial melting in the mantle, composed mostly of the iron-rich and magnesium-rich minerals olivine and pyroxene. From its relatively high density inferred from seismic velocity data, we know that the lunar mantle is largely made up of these same minerals. Radioactive, heat-producing elements, such as uranium, made the early mantle hot enough in some places to partially melt. Blobs of silicate melt coagulate deep in a planet's interior and then slowly migrate upward, where they may force their way to the surface and be extruded onto a planetary surface as a lava flow.

Unlike in the Earth, we see little evidence for the collection of large amounts of **magma** at shallow depths in the Moon's crust and for its retention in subsurface "holding chambers". Most magmas appear to have migrated upward through the mantle and crust and then erupted fairly quickly. The lunar **pyroclastic** glasses are the most extreme examples of this phenomenon, as they appear to be completely unfractionated, suggesting a rapid ascent from the mantle and violent, immediate eruption. Some mare basalts show evidence for the assimilation of highly differentiated crustal material. Such a process does not imply significant near-surface magma storage, as this material (potassium, rare earths, and phosphorus, termed **KREEP**) has a low melting point and could be assimilated easily during brief contact with the basaltic magma.

The chemistry of basaltic magmas tells us approximately where they formed within the Moon (at depths of 150–400 km) along with what processes subsequently affected them. Our study of the mare basalts tells us that many different regions of the mantle underwent melting episodes at several depths over a very long period of time, a period lasting at least 700 million years long and more likely over a time span of one to two billion years. These melted pockets found their way to the surface through cracks that they themselves propagated or via fractures induced by the formation of the giant craters and basins of the highlands. However, only a very tiny volume fraction of the mantle was melted to make basalt. Although the maria appear prominent in areal extent, the lavas are relatively thin compared with the volume of the crust as a whole. It is estimated that the mare basalts cover about 16% of the Moon by area (Figure 39.1) but probably account for less than about 1% of the total volume of the crust.

2.2. Apollo Mare Basalts

Like terrestrial basalts, the lunar mare basalts are made mostly of the minerals pyroxene and plagioclase and are rich in iron and magnesium. The grain size of basalt is very fine (usually <1 mm), a result of rapid cooling. Like some terrestrial basalts, some lunar lavas have small, bubble-like holes in them (**vesicles**), indicating that the magmas contained gas during eruption. As with basalts on Earth, mare basalts are formed by the partial melting of the lunar mantle, made of mostly pyroxene and olivine.

The first basalts returned from the Moon came from the Apollo 11 landing site in Mare Tranquillitatis (Sea of Tranquility). These rocks are remarkable in several respects. Mare basalts are not only devoid of water or any hydrous phase but they are also depleted in all the volatile elements (those that have very low boiling temperatures), including, for example, sodium, zinc, potassium, and phosphorus. Strangely (and surprisingly), the Apollo 11 basalts have large amounts of titanium (Table 39.1), mostly in the form of the mineral ilmenite, an oxide of iron and titanium. The enrichment of the mare basalts in iron and their depletion in aluminum, the exact reverse of the composition of rocks from the lunar highlands, account for the relative darkness (low **albedo**) of the maria as opposed to the **terrae** (lunar highlands).

Lavas from the Moon contain some minor minerals that are not found in Earth rocks. One of these, another iron—titanium mineral, was given the name armalcolite, named in honor of the Apollo 11 crew (the word coming from the first letters in the names of the crew, *Arm*strong, *Al*drin, and *Col*lins). The compositional properties of the lunar basalts reflect the unique chemical environment in which they formed: inside a small planet (resulting in low interior pressures), depleted in volatile elements, containing little or no water (but see below), and erupted onto a low-gravity surface in a vacuum.

Mare basalts from other Apollo missions largely confirm the initial impressions gathered from these studies, but contained some surprises and interesting variations. Lavas from Apollo 12 are lower in titanium (Table 39.1) than the Apollo 11 basalts and are 600–700 million years younger (erupted about 3100 million years ago). Once again, these lavas are low in volatile elements and rich in

TABLE 39.1 Chemical Composition of Typical Mare Basalts and Pyroclastic Glasses (After Taylor, 1982)

	Apollo 11 High-Ti	Apollo 12 Low-Ti	Apollo 15 Low-Ti	Apollo 17 High-Ti	Luna 16 High-Al	Apollo 15 KREEP Basalt	Apollo 15 Green Glass	Apollo 17 Orange and Black Glass
SiO_2	39.8	43.6	44.1	37.8	43.8	50.8	45.2	38.6
TiO_2	10.5	2.60	2.28	13.0	4.90	2.23	0.38	8.81
Al_2O_3	10.4	7.87	8.38	8.85	13.7	14.8	7.5	6.32
FeO	19.8	21.7	22.7	19.7	19.4	10.6	20.0	22.0
MnO	0.30	0.28	0.32	0.27	0.20	0.16	0.26	–
MgO	6.69	14.9	11.3	8.44	7.05	8.17	17.5	14.4
CaO	11.1	8.26	9.27	10.7	10.4	9.71	8.5	7.68
Na_2O	0.40	0.23	0.27	0.36	0.33	0.73	0.13	0.36
K_2O	0.06	0.05	0.04	0.05	0.015	0.67	0.03	0.09
Cr_2O_3	0.25	0.96	0.85	0.41	0.28	0.35	0.53	0.75
P_2O_5	–	–	–	–	–	0.70	–	–
Th (ppm)	1.8	0.75	0.50	0.34	0.9	10.3	0.08	–
Mg/(Mg + Fe)	0.38	0.55	0.47	0.43	0.39	0.38	0.59	0.54

iron. The lower titanium and younger ages of the Apollo 12 basalts confirm that the maria were not erupted as a single, massive flood of lava across the surface all at one time but rather, the formation of the maria was an extended process that involved different batches of magma erupted in different places at different times. In short, the samples told us that the Moon had a complicated volcanic history and a protracted geological evolution.

Mare basalts from the other two mare landing sites extended our picture of mare volcanism. Apollo 15, which landed just inside the rim of the basin containing Mare Imbrium, returned low-titanium basalts (Table 39.1) slightly older than those from Apollo 12; these lavas crystallized about 3300 million years ago. Apollo 17, landing on the edge of Mare Serenitatis, returned very high-titanium basalts (Figure 39.2; Table 39.1), similar to those from Apollo 11, but slightly younger, about 3700 million years old. These results led some to conclude that the Moon had a fairly simple volcanic history, with early eruptions of high-titanium lavas and late eruptions of low-titanium lava. The conclusion was also drawn that the Moon "died" volcanically after the Apollo 12 lavas were erupted at 3100 million years, a totally unwarranted conclusion that even today is widely believed and recounted.

In addition to the basalt samples returned from the mare landing sites, little fragments of basalt have been found in samples from the highlands as well. These basalts occur in two principal ways: as small rocks in the **regolith** from highland sites and as lithic fragments in highland breccias. The former occurrence of mare basalt is likely to result from the deposition of a ray from an impact crater on the distant maria, flung as ejecta to the highland site. Examples include several fragments of mare basalt from the Apollo 16 regolith. Because secondary craters from the large impact crater Theophilus occur near the site, we infer that mare basalts were thrown to the site by the formation of this crater and that they represent a sample of the lava flows of Mare Nectaris, upon which Theophilus was formed. Thus, we can characterize the rocks of distant maria, if plausible candidate craters can be identified.

In contrast, mare basalt fragments found as clasts "within" highland breccias offer clues to the variety and ages of the earliest phases of lunar volcanism. These breccias from the highlands were assembled before 3800 million years ago; therefore, the lava fragments within them must be older than this. Some of these mare basalts large enough to date were erupted well before 3900 million years ago. The oldest mare basalt yet found is about 4200 million years old, only slightly younger than the age of the solidification of the crust. Other fragments display a variety of chemical compositions and date from between 4100 and 3900 million years old. Curiously, most of the ancient mare basalt fragments tend to have relatively high contents of aluminum compared to the basalts from the main phase of mare eruptions, although a few groups of high-alumina basalt date from this later era as well.

3. STYLES OF VOLCANISM AND ASSOCIATED LANDFORMS

3.1. Lunar Lava Flows

From pictures taken by telescopes on the Earth, the maria appear smooth and dark. The impression is that these plains fill in the holes and depressions of the Moon, suggesting a fluid emplacement. At close-up scales, small, lobe-shaped scarps can be seen (Figure 39.3); such scarps or lava fronts are very common in the basalt lava flows found on the Earth. These scarps can even be seen in some of the best telescopic pictures of Mare Imbrium. Thus, we had direct evidence for emplacement of the maria by fluid flow "before" the exploration of the Moon by spacecraft.

We obtained our first detailed look at the maria from the robotic precursor missions sent to scout the Moon for Apollo. The Ranger 7 spacecraft was the first to return detailed, close-up pictures of the Moon. We learned from these images that the scale of impact cratering continues downward to the limits of resolution and that the maria are covered by the regolith, an unconsolidated blanket of debris that overlies the mare lava bedrock. Lunar Orbiter obtained detailed pictures of a wide variety of landforms attributable to volcanism, including a better view of the **flow fronts**, small domes and cones, snakelike **rilles** that served as conduits for molten lava (lava channels and rilles), and irregular craters, whose shapes are difficult to explain by impact origins.

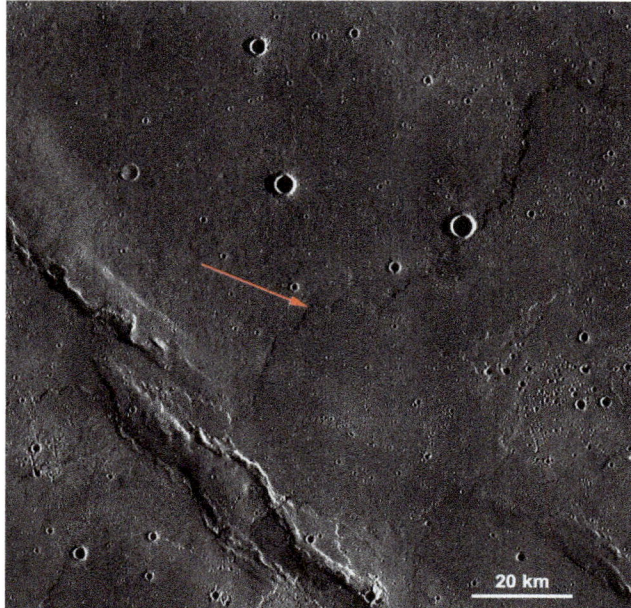

FIGURE 39.3 Flow fronts in Mare Imbrium (arrow). Wrinkle ridges (bottom left) are compressional tectonic features created by loading of the crust by the stack of flood lavas of Mare Imbrium. Flows here are part of a complex that is 800 km long, 20–40 km wide, and 20–60 m thick.

The Surveyor 1, 3, 5, and 6 spacecraft all performed soft landings in the maria. These missions gave us a close-up view of the surface and revealed dark rocks covered with small holes, a morphology typical of lava. The chemical composition of the mare basalts has an interesting side effect. The **viscosity** of the lava is dependent on the composition and temperature of the magma. The low amounts of aluminum and alkali elements and high amount of iron in the lunar magmas, coupled with their relatively high temperature upon extrusion, result in lavas that have an extremely low viscosity. The viscosity of erupted lunar lava was about the same as motor oil at room temperature (about 10 P), a factor of 10 more fluid than terrestrial lava. Such runny, fluid flows spread out great distances, and this property, in addition to the low lunar gravity, accounts for the great lengths (up to hundreds of kilometers) that lava flows can reach on the Moon. Such a fluid character to the lava also explains the tendency of mare lavas to form low, broad structures rather than steep-sided volcanoes and to be erupted in lava channels, as are many basaltic lava flows on Earth, such as on the volcanoes of Hawaii.

Discrete, single flows of mare basalt appear to be rare. Although many units in the maria have a uniform density of impact craters (denoting similar age) and single color, they appear to be made up of many thin, small lava flows. High-resolution photographs sometimes reveal scarps or moats occurring around impact craters that may delineate very thin flow lobes (less than a couple of meters thick). The spectacular lava flows within the Imbrium basin (Figure 39.3) are often used in textbooks to illustrate the volcanic nature of the maria, but these flows are unique on the Moon and their appearance probably indicates a specialized set of eruption circumstances (e.g., the rapid inflation of the crust and discharge of a large magma body over a short period of time). For the most part, the maria appear to be a smooth, nondescript surface and discrete flows are not obvious. It is likely that the maria consist of a complex series of relatively thin lava flows, in which subsequent regolith production and impact erosion have destroyed any original volcanic texture.

Smaller volumes of lava that erupt from a narrow, more localized **vent** (such as a single-pit crater) can produce a variety of other interesting landforms. If the volume of lava cools on timescales similar to the rate of magma supply to the vent, small volcanoes form and may assume a variety of different shapes (Figure 39.4). Typically, lunar volcanoes form domes of low relief, a few hundred meters high and a few kilometers across. Such landforms resemble small basaltic shields found in certain volcanic regions of the Earth, such as Iceland and the Snake River plain of Idaho.

Other domes appear to be slightly larger and steeper. One of the most spectacular areas in the maria are the Marius Hills, a complex area of many small domes in Oceanus Procellarum (Figure 39.5). The domes of the

FIGURE 39.4 Small volcanic features of the lunar maria. Top left: vent area of the late Imbrium flows (Figure 39.3), showing dark mantling and spatter constructs. Bottom left: small shield volcanoes near Hortensius. These features are likely to be small basaltic constructs, erupted from a single vent. Top right: spatter cones aligned along a fissure vent near Hortensius. Bottom right: cones, rilles, and linear vents near the Hortensius volcanic complex.

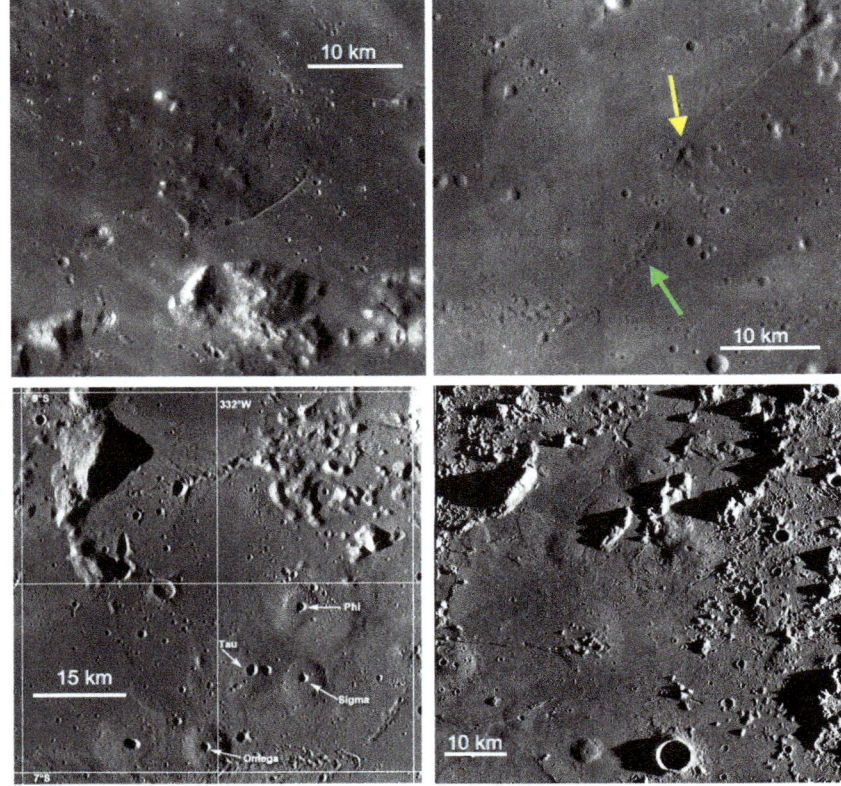

FIGURE 39.5 Marius Hills volcanic complex. Image on the left is the topographic map from stereoimages and laser altimetry showing blister-like shape. Image on the top right is from the Kaguya lunar orbiter, showing numerous small domes and cones occurring on up warped, shield-like surface. Bottom right shows the flanks of the complex in Oceanus Procellarum. This feature has been proposed to be a lunar shield volcano.

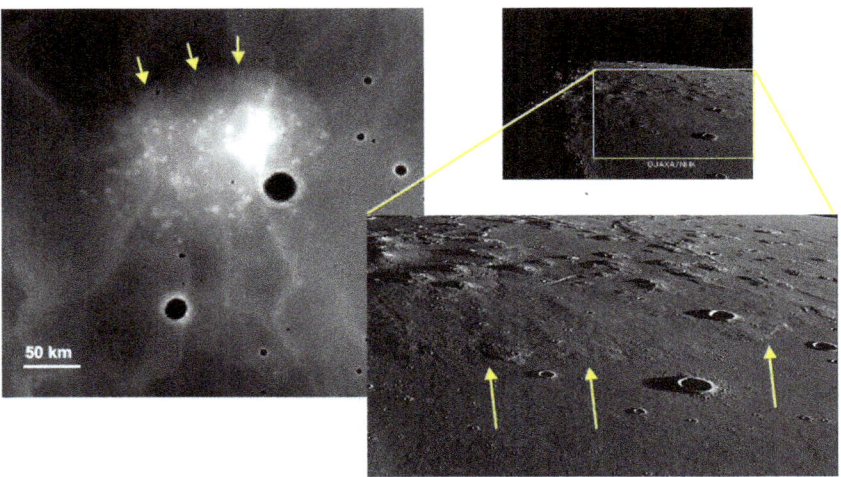

Marius Hills have steeper sides than do the basalt shields mentioned above. On Earth, such differences in shape are caused by differences in the composition of lava, steeper domes containing more silica and less iron and magnesium than the low, broad **shield volcanoes**. The causes for steep domes on the Moon are less well understood but appear to be related to styles and rates of extrusion, rather than to lava composition. Eruptions of shorter duration, possibly mixed with minor interludes of **ash** eruption, would build up a construct with steeper slopes than would the quiet effusion of the very fluid, low-viscosity lavas.

Domes and cones on the Moon seldom are found in isolation, more often occurring as fields of volcanoes within the maria. The Marius Hills display many domes and cones, all of which occur on the summit of a broad topographic swell (Figure 39.5). The complex is several hundred meters high and has a blister-like profile, suggesting that it may be the lunar equivalent of a shield volcano, as are found on the Earth, Venus, and Mars. The Rümker Hills in northern Procellarum is another volcanic complex, similar in appearance to the Marius Hills, but smaller. Small basaltic shields, such as those found near Hortensius (Figure 39.4), occur in several locations near the margins of Maria Imbrium, Nubium, and Serenitatis. The new precision global topographic data reveal that these volcanic fields are all associated with shield-like blisters, possibly reflecting a phase of basaltic shield building. The last type of central-vent volcano on the Moon is the **cinder cone**, typified by irregular dark halo craters found along fractures in some crater floors, such as those of the crater Alphonsus. These craters are surrounded by low albedo ash deposits (as shown by remote-sensing data) and are often associated with the volcanic modification of large craters that originally formed by impact.

3.2. Pyroclastics

Both the Apollo 15 and 17 missions returned some unexpected and surprising material. Small glass beads were found in abundance at both sites: clear emerald green glass at the Apollo 15 site and black and orange glass from the Apollo 17 site (Figure 39.6). These glass beads are homogeneous, with a basaltic composition (Table 39.1), and do not contain mineral debris that characterizes impact-melted glasses from the regolith. The surfaces of these glass beads have small amorphous mounds of a variety of volatile elements, including lead, zinc, and halogens such as chlorine. The Apollo 15 green glasses are very rich in magnesium and extremely low in titanium (an unusual composition for a lunar magma), whereas the orange and black Apollo 17 glass is very rich in titanium (Table 39.1). Once these glasses were recognized from these two sites where they occur in abundance, small spheres of similar material were recognized from every landing site. More than 20 varieties of this type of glass are known.

These glasses are of volcanic, not impact, origin and they represent the end products of low-viscosity lava sprayed into space. During basaltic eruptions on Earth, lava effusion is sometimes accompanied by sprays of magma from the vent. Such eruptions are called **fire fountains** and result in a deposit of ash around the eruptive vent. The ash from Hawaiian eruptions consists of glass that has basaltic composition and is frequently coated by a layer of volatile elements. On the basis of similar characteristics, the lunar glasses represent products of fire fountains that existed on

FIGURE 39.6   Pyroclastic glasses (volcanic ash) from the Moon. At the top is green glass (very low titanium and very high magnesium, age 3.3 Ga) from the Apollo 15 landing site; at the bottom is orange-black glass (very high titanium and very high iron, age 3.7 Ga) from the Apollo 17 landing site. Both types of glass were erupted from deep mantle sources, driven by high volatile content. Glass fragments in each image are the sieved fraction from 90 to 150 μm.

the Moon over three billion years ago. One difference between the lunar and terrestrial ash deposits is that so far no samples of mare basalt lava with compositions corresponding to the lunar ash (presumably derived from the same magma) have been recognized.

Until recently, all lunar volcanic products were thought to be completely free of water. New studies of Apollo 15 green glass revealed a startling relation: the interiors of some of these glasses contain trapped water, dissolved in the glass, indicating that water did exist in the deep interior of the Moon at the time of eruption. Amounts of trapped and dissolved water in the glasses indicate that the lunar mantle may contain up to 700 ppm water, almost as high as some terrestrial mantle values. This water must have been incorporated into the deep lunar interior during accretion and some trapped water may yet remain there.

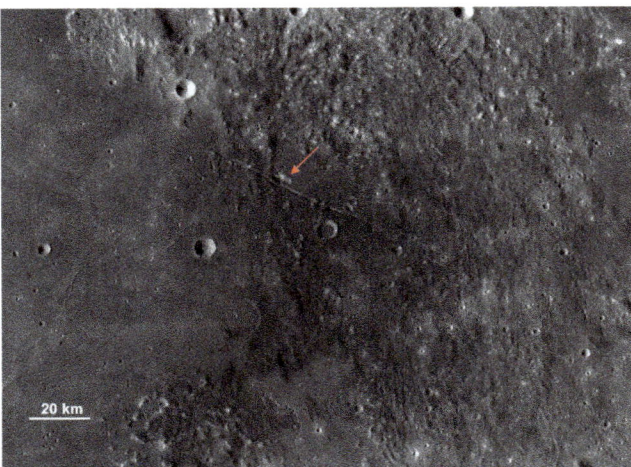

FIGURE 39.7   Regional view of dark mantling deposits near Rima Bode, an irregular rille and vent system (arrow). Such deposits are made up of basaltic glass of pyroclastic origin, such pyroclastic glasses of Figure 39.6.

FIGURE 39.8   The Aristarchus volcanic plateau. This complex displays numerous volcanic landforms, including large sinuous rilles (lava channels), small shields, pyroclastic dark mantling deposits, and flood lavas. Studies have shown the lunar sinuous rilles are lava channels, in some cases roofed over as lava tubes. LRO Wide Angle Camera mosaic.

In addition to pyroclastics in the samples, volcanic glasses are also evident as regional deposits. It was noted during geological mapping that parts of the highlands and maria are blanketed with very dark material (Figure 39.7). Darkness was often equated with geological youth, and these regional dark deposits were thought to represent ash deposits from young volcanic activity. Such a concept was responsible in part for the selection of the Apollo 17 landing site near the margin of one of these regional deposits. It was predicted that this material was volcanic ash because it typically occurs in the vicinity of irregular craters of volcanic origin, indiscriminately covering existing terrain. The principal occurrences of these regional deposits are around the margins of filled mare basins, such as those on the Aristarchus plateau (Figure 39.8), Sulpicius Gallus, Rima Bode (Figure 39.7), and Taurus-Littrow (the Apollo 17 landing site). The Apollo 17 ash is indeed of volcanic origin, but it is old (3700 million years), not young.

The floors of several large craters on the Moon are deformed and fractured. Along some of these fractures, small (typically a few kilometers in size), irregular craters are surrounded by dark, smooth material. These craters are probable volcanic vents, surrounded by ash deposits. They are the lunar equivalent of "cinder cones" found in terrestrial volcanic fields. To create a cinder cone, magma charged with volatiles is squirted out through a very narrow conduit. The release of the low-viscosity lava under high pressure through a small vent causes the lava to spray into a "mist" of liquid rock droplets. The spray of droplets quickly cools in flight and each droplet, thrown on ballistic paths, lands back on the Moon as a small bead of glass. Millions of beads made during an eruption build up a deposit of ash that surrounds the vent.

Phase chemistry allows us to determine which minerals can coexist at certain temperatures and pressures. Studies of the phase chemistry of volcanic glasses have given us a great deal of insight into the very deep lunar interior. It appears that these glasses were generated by the partial melting of an olivine-rich mantle at depths of about 400 km. Moreover, unlike all the mare basalts, the glasses appear to be largely unmodified from their chemical composition at their point of origin. Such a relation indicates that the magmas from which the glasses formed must have ascended very rapidly up through the Moon from deep in the mantle, with little chemical modification from their point of origin. The glasses then erupted into an explosive spray of molten rock at the surface. As such, lunar pyroclastic glass is our best sample of the deep interior and is an important material for determining the bulk composition of the Moon.

Both the fountain nature of the eruption and the small coatings of volatile materials on the glass surfaces indicate that during the main era of mare volcanism over three billion years ago, pockets of gas and other volatile elements existed deep within the Moon. Such an inference is also supported by the presence of the vesicles (holes) that are found in some samples of mare basalt. The principal composition of this gas phase is something of a mystery. Despite the recent discovery of small amounts of water within the glass beads, the reduced chemistry of lunar lavas suggests that the gas phase might have been composed predominantly of carbon monoxide. In any event, we now know that a multicomponent complex of a variety of volatile elements was present in the mare source regions.

3.3. Sinuous Rilles

One common landform of the maria deserves special mention, if only because the Apollo 15 mission (July 1971) was sent specifically to investigate one of them. **Sinuous rilles** are narrow, winding valleys that occur primarily within the maria; in size, they are typically a few hundred meters to a kilometer wide, several hundred meters deep and can be tens of kilometers to over hundred kilometers long. Some originate in highlands terrain (Figure 39.8), but all trend downslope and empty into mare material. Many rilles begin in irregular craters, some of which are surrounded by dark mantling material. Several ideas were advanced for the origin of these features before the Apollo missions, including water-cut stream channels. However, the absence of water on the Moon, the basaltic nature of the maria, and the irregular shapes of their source craters led to the consensus that these features are lava channels, some of which were partly roofed over to form **lava tubes**.

The Apollo 15 mission was sent to Hadley Rille, just inside the rim of the Imbrium basin (Figure 39.9). Hadley is one of the largest sinuous rilles on the Moon, being >140 km in length, 1—3 km wide, and several hundred meters deep. The rille begins in an elongate, irregular crater in the Apennine Mountains and winds its way through the

FIGURE 39.9 Orbital view of the Apollo 15 landing site region. Hadley Rille is long sinuous rille (lava channel), striking mostly NE—SW along the base of the Montes Apenninus, the main rim of the Imbrium impact basin. Light plains on the left are the Apennine Bench Formation, a rare exposure of ancient (3.84 Ga) nonmare KREEP volcanism.

maria, snaking back and forth through the mare, finally becoming shallow and appearing to merge into a complex set of fractures north of the Apollo 15 site. The rille was examined at its rim near the landing site, where the rille is 1.5 km wide and 300 m deep. Samples of basalt collected on the rille edge are probably the only samples in the Apollo collections that were taken from bedrock. A ledge of bedrock seen in the orbital photographs probably consists of this mare basalt unit. Layers of mare lava are exposed in Hadley Rille; at the Apollo 15 landing site, layering is confined to the upper 60 m of the rille wall. Color data from the Clementine mission confirm that the walls of sinuous rilles all over the Moon expose outcrops of mare basalt.

Although all workers agree that sinuous rilles are lava channels and tubes, their exact mode of formation remains contentious. On Earth, lava channels are created when a flow extruded at moderate rates cools from the margins inward; this cooling tends to confine the molten, active part of the flow along a central axis. This axis becomes the channel and, in some cases, is bridged over to form a lava tube. In this mechanism, lava channels are primarily "constructional" features, where the overflow of lava builds up levees and raises the topographic level of the flow axis. Lava can accumulate laterally on the walls of the channel, narrowing the channel width; in fact, such narrowing away from the vent is a common feature both of lava channels on Earth and of sinuous rilles on the Moon. Another concept holds that lava channels are primarily "erosional" features. The claim is that the eruption of a very high-temperature, fluid lava would flow turbulently and would soften, melt, and then remove underlying material, forming a lava channel by erosion. In such a model, the sinuous depression in the maria consists of material removed by the flow of liquid lava and incorporated into the mare deposits.

The occurrence of rille source craters in the highlands that are connected to sinuous rilles by channel segments in terra material indicates that some erosion must have occurred. The eroded segment may have been enlarged by collapse, a process common in lava channels. In general, however, geological evidence indicates that sinuous rilles and terrestrial lava tubes and channels are formed dominantly by construction. The large size of sinuous rilles, argued by some as evidence for erosion, can also be a result of the infilling of preexisting depressions, as clearly shown in the case of Hadley Rille, where the lava channel merges into preexisting valleys north of the Apollo 15 landing site (Figure 39.9).

4. NONMARE VOLCANISM

During the pre-Apollo geological mapping of the Moon, many highland landforms, including smooth light plains, wormy textured materials, and steep domes, were interpreted to be of volcanic origin. The Apollo 16 mission in 1972 was specifically sent to the Descartes highlands to sample these highland (supposedly differentiated) volcanic rocks. However, the rocks from this mission are all impact breccias and the idea of highland volcanism (specifically, the eruption of evolved rocks of higher silica and/or lower iron than the mare basalts) was viewed as discredited by these mission results.

A regional exposure of light plains near the Apollo 15 landing site is the Apennine Bench Formation (Figure 39.9). These plains predate the basalts of Mare Imbrium, but postdate the impact that created the basin. As determined by a variety of remote-sensing techniques, these deposits have composition identical to small fragments of relatively high-alumina basalts of volcanic origin found at the nearby Apollo 15 landing site. These KREEP basalts (Table 39.1) are relatively enriched in potassium (K), the rare-earth elements (REEs), and phosphorus (P). In major-element composition, they are fairly aluminous and iron-poor, compared to most mare basalts. Their origin is still debated, but one idea is that they are partial-melting products of an aluminous source (likely present in the lower crust). If this concept is correct, the Apennine Bench Formation is a major surface manifestation of premare, "highland" (i.e., not mantle-derived) volcanism.

Several isolated domes and mountain-like constructs of very red color occur in the highlands; these features have been suggested to represent a different style of highland volcanism. The blister-shaped Gruithuisen domes are very red and occur along the main rim of the Imbrium basin (Figure 39.10). Spectral data suggest that these features are very low in iron and new spectra in the thermal infrared from the Lunar Reconnaissance Orbiter (LRO) Diviner instrument indicate relatively high silica content. These observations suggest that the domes are piles of extruded silica-rich lavas, similar to terrestrial rhyolite. No such samples have been found in the Apollo collections, but the weight of the new remote-sensing data indicate that the range of volcanic rock types on the Moon may be as great as those found on Earth.

5. THE BEGINNING AND END OF MARE VOLCANISM

The maria were not erupted all at once as a massive flood of lava. The Moon underwent a long and protracted volcanic evolution, characterized by different degrees of interior melting and different types of eruptions of different compositions and places over a long period of time. The production of magma through time is an important basis for reconstructing lunar thermal evolution. Such information allows us to compare the Moon with the other terrestrial

FIGURE 39.10 Gruithuisen Gamma (left) and Delta (right), two highland domes made up of silica-rich lavas. Several highland domes of rhyolite-like material have now been documented, indicating the minor presence of highly differentiated volcanism on the Moon.

planets and to understand the many ways that planets lose their heat.

5.1. Early Phase

The earliest extrusions of lava on the Moon may have been the outpouring of liquid rock onto the still-cooling, crusted-over surface of the early Moon. Indeed, the line between volcanism and crustal formation was probably indistinguishable in early lunar history. The oldest unequivocal volcanism on the Moon is represented by tiny chips of mare basalt found as clasts in the Apollo 14 highland breccias. These fragments represent pieces of a lava flow extruded onto the surface 4200 million years ago, a time so remote that we can only guess at what conditions were like on the Earth. Volcanic eruptions probably were more or less continuous, though sporadic, throughout the period of heavy impact **bombardment** between 4300 and 3800 million years ago. Traces of this volcanism can be found in the tiny fragments of lava in the highland breccias but may also be evident as a chemical signature in cratered terrains. Some regions of the highlands appear to contain relatively large amounts of iron. This iron could represent unknown highland rocks but it is possible that flows of iron-rich mare basalt have been ground up into regolith of the highlands by the intense impact bombardment of early lunar history.

During the final phases of the heavy bombardment (3900–3800 million years ago), several large, well-preserved basins formed. These basins still have recognizable ejecta blankets and the smooth, far edges of their debris layers fill craters and other depressions in many areas. Such deposits have all the compositional properties of highland rocks. The Apollo results tell us that the light plains of the highlands are impact ejecta associated with the large multiring basins. However, some of these plains display small (1- to 3-km diameter) impact craters whose ejecta are relatively dark. Dark halo impact craters are found in some of these light plains and cluster in regional groups, such as those in the Schiller–Schickard impact basin. Spectral observations indicate that their low albedo is caused by mare basalt lava making up the crater ejecta—yet this is an area of impact-generated plains deposits.

These dark halo impact craters are excavating "buried" deposits of mare basalt. Because the plains that bury the lava flows are themselves 3800 million years old, the basalt flows that they cover must be older than this. Thus, light plains on the Moon that display dark halo impact craters are ancient maria—basaltic lavas that were emplaced before 3800 million years ago. This remote-sensing evidence for ancient maria of regional extent complements the sample evidence of tiny fragments of lava in highlands breccias and indicates that the early Moon was a planet of active volcanism as well as subject to intense impact bombardment.

5.2. Main Phase

This stage of mare volcanism began when very high rates of cratering declined to the point where the extruded lava flows were preserved and not destroyed by being ground up into the megaregolith. The impact flux declined very rapidly between 3900 and 3800 million years ago, leveling off after 3800 million years. From that time onward, the extruded lavas were bombarded by impact, but the cratering was not intense enough to destroy the flow surfaces. These lavas make up the visible maria. The earliest lavas from this period, represented by the samples of the Apollo 11 and 17 missions, are the high-titanium basalts of Maria Tranquillitatis and Serenitatis. These flows erupted between 3800 and 3600 million years ago. The complete extent of the early high-titanium lavas cannot be determined because they are partly covered by younger flows; they may be extensive over much of the near side.

A long period of eruption of low-titanium basalt followed, from 3600 million years to an undetermined time, certainly as late as 3100 million years ago, but perhaps much later. Some of the lavas from this time were of the high-aluminum variety, particularly in the eastern maria Crisium and Fecunditatis, as sampled by the Soviet Luna 16 and 24 sample return missions. These basalts date from 3600 to 3400 million years and have moderate to extremely low titanium content. Eruption of low-titanium lavas in Mare Imbrium at 3300 million years ago (Apollo 15) and Oceanus Procellarum at 3100 million years (Apollo 12)

followed. The lavas from the Apollo 12 site are the youngest mare basalts in the sample collection.

Eruptions of mare basalt were rare events, even at the height of lunar volcanic activity, between 3800 and 3000 million years ago. Although the maria appear to dominate the Moon (Figure 39.2), especially on the near side, the visible mare deposits actually make up less than about 1% of the volume of the crust. The total accumulated thickness of lava in most mare deposits varies widely but is typically less than a few kilometers and large areas of basalt may be thinner than a few tens of meters. In part, we know this because of the abundance of highlands debris mixed into the mare soils; this fraction may approach 60–70%. Because most of this debris is derived from rocks beneath the local bedrock, the implication is that the stack of basaltic lava flows is thin.

There is a tendency to think of the maria as a hotbed of geological activity on the Moon, at least in the past. Certainly there has been activity in the maria, but very long time spans separate periods of activity. At the Apollo 11 site, several different lava flows are represented among the samples. The oldest flows are 3860 million years old, but samples of flows of a similar composition have a variety of ages, some as young as 3550 million years. In addition, another group of lavas, of a different composition, are also about 3500 million years old. Thus, at this one site, we have evidence for at least four (and perhaps more) separate lava flows, emplaced over a period of >300 million years. This geologically "active" area of the Moon has been completely quiet for a period of time longer than vertebrate life has existed on the Earth.

5.3. Late Phase

The only other source of information on the age of mare lavas is the relative age data provided by geological mapping. This mapping shows that many different kinds of flows spilled out onto the maria between the dated and sampled eruptions, providing for a more or less continuous infilling of the basins from 3800 million years ago onward. A key piece of information for lunar volcanic history is one we do not have—what is the age of the "youngest" mare basalt eruption on the Moon? This question has an important bearing on the thermal history of the Moon. If we know the age of the last eruption, we know when the Moon "shut down" thermally, at least to the point where magma could no longer get out of its interior.

Many areas of the maria have been identified that have a lower density of impact craters than the sampled flows of the Apollo 12 site (aged 3100 million years). The very young lava flows with well-developed scarps in Mare Imbrium have crater densities a factor of two to three lower than the lavas of the Apollo 12 site. Thus, these Imbrium flows may be 1500–2000 million years old; we cannot estimate their absolute age any more precisely than this. Other relatively young flows are found from Oceanus Procellarum in the west to Mare Smythii in the east. One notable example is the Surveyor 1 landing site (the site of the very first American soft landing on the Moon in 1966) within the lava-filled crater Flamsteed P in Oceanus Procellarum. Crater density suggests that these lavas are some of the youngest flows on the Moon, having an age of about one billion years old. Interestingly, television images returned by Surveyor 1 show that this site has the thinnest regolith of any mare site ever visited, estimated to be between 1 and 1.5 m thick. For comparison, the regolith at the Apollo 12 site (which is underlain by the youngest "sampled" mare basalts) is about 4 m thick. As regolith thickness reflects the age of the bedrock upon which it forms, we have an independent way to estimate the relative age of the Flamsteed basalts from the Surveyor data. These estimated regolith thicknesses are consistent with an age of about 1000 million years for these lavas.

The very last gasps of volcanism on the Moon may have occurred as recently as 800 million years ago. Ejecta from the crater Lichtenberg (20 km in diameter) is covered by a lava flow, which is therefore younger than the crater. Lichtenberg has rays and is a member of the youngest class of craters on the Moon. Estimates of the absolute age of Lichtenberg are uncertain, but craters of this size tend to have their ray systems completely destroyed on timescales of 500–2000 million years. Thus, Lichtenberg and its overlying lava flow are probably "younger" than 2000 million years. This mare unit is the youngest lava flow currently recognized on the Moon.

The Moon has been volcanically active for most of its history. Massive extrusions of lava probably began in dim antiquity, during the era when the Moon's crust was being ground to a pulp by very high rates of bombardment. Extrusion of basalt continued throughout the era of basin formation, including the eruption of the massive amounts of lava that now underlie the highland plains deposited by the impact that made the Orientale basin. As the basins ceased to form, large expanses of mare deposits began to be preserved, forming the complex series of overlapping lava flows and ash deposits that make up the visible maria. Slow, prolonged infilling of the near-side basins continued for several hundred million years, gradually loading the crust and deforming the filled basins by interior compression and exterior extension. Some impact craters

underwent volcanic modification, including interior flooding and the formation of cinder cones on their floors. Large-scale regional deposits of ash were erupted along the margins of some maria. Younger lava deposits tend to be less voluminous than the older deposits, indicating that the intensity of volcanic activity has declined with time. The youngest lava flows are confined to Procellarum and Smythii. Some of these young eruptions may have occurred since one billion years ago, a time that seems remotely old, but, in fact, is "only yesterday" in the ancient and silent world of the Moon.

SEE ALSO THE FOLLOWING ARTICLES

Basaltic Volcanoes and Volcanic Systems ● Lava Flows and Flow Fields ● Volcanism on Io ● Volcanism on Mars ● Volcanism on Venus.

FURTHER READING

Basaltic Volcanism Study Project, 1981. Basaltic Volcanism on the Terrestrial Planets. Pergamon Press, New York, pp. 701–800. http://ads.harvard.edu/books/bvtp/.

Head, J.W., Wilson, L., 1992. Lunar mare volcanism, stratigraphy, eruption conditions, and evolution of secondary crust. Geochimica et Cosmochimica Acta 56, 2155–2175.

Heiken, G.H., Vaniman, D.T., French, B.M. (Eds.), 1991. The Lunar Sourcebook: A User's Guide to the Moon. Cambridge University Press, Cambridge, UK, p. 734. http://www.lpi.usra.edu/publications/books/lunar_sourcebook/.

Jolliff B.L., Wieczorek M.A., Shearer C.K., Neal C.R. (Eds.), 2006. New views of the moon. Reviews in Mineralogy and Geochemistry 60, 365–502.

Spudis, P.D., 1996. The Once and Future Moon. Smithsonian Institution Press, Washington, DC, 308 pp.

Spudis, P.D., 1999. The moon. In: Beatty, K., Petersen, C.C., Chaikin, A. (Eds.), The New Solar System, fourth ed. Sky Publishing Corp. and Cambridge University Press, Cambridge, UK, pp. 125–140.

Spudis, P.D., McGovern, P.J., Kiefer, W.S., 2013. Large shield volcanoes on the Moon. Journal of Geophysical Research 118 (5), 1063–1081. http://dx.doi.org/10.1002/jgre.20059.

Taylor, S.R., 1982. Planetary Science, A Lunar Perspective. Lunar and Planetary Institute Press, Houston, TX. http://www.lpi.usra.edu/publications/books/planetary_science/.

Wilhelms, D.E., 1987. The Geologic History of the Moon, U.S. Geological Survey Professional Paper 1348. U.S. Government Printing Office, Washington, DC, 300 pp. http://ser.sese.asu.edu/GHM/.

Chapter 40

Volcanism on Mercury

James W. Head, III
Department of Earth, Environmental and Planetary Sciences, Brown University, Providence, RI, USA

Lionel Wilson
Lancaster Environment Centre, Lancaster University, Lancaster, UK

Chapter Outline

GLOSSARY

accretional heat The heat generated by the conversion of the kinetic energy of impacting bodies to thermal energy. As the planets formed through the accretion of planetesimals, they heated due to the violence of the collisions.

advective heat transport Advective transport describes the movement of magma and heat from the interior directly to the surface.

albedo The percentage of incident solar radiation reflected back to the observer. Bright and dark surfaces are high and low albedo, respectively.

alkali basalts Volcanic rock characterized by phenocrysts of olivine, titanium-rich augite, plagioclase feldspar, and iron oxides. For similar SiO_2 concentrations, alkali basalts have a higher content of the alkalis, Na_2O and K_2O, than other basalt types such as tholeiites.

anorthositic crust A slowly cooled rock made up almost solely of the calcium-rich mineral plagioclase feldspar and forming much of the lunar highland crust.

basalt A dark-colored fine-grained extrusive igneous rock composed largely of plagioclase feldspar and pyroxene.

cooling-limited lava flows Lava flows that cool sufficiently to cease advancing prior to the end of the eruption. Such flows are often characterized by levee breakouts.

corona Dominantly circular or elliptical structures on Venus consisting of an annulus of concentric ridges or fractures, an interior that may be topographically positive, a peripheral moat or trough, and, frequently, numerous volcanic centers and tectonic landforms in the interior. Coronae may be associated with mantle plumes or hot spots; related volcanism, in the form of large and small volcanoes, is frequently present within the corona or on its margins.

enstatite chondrite A meteorite that is among the most chemically reduced rocks known; most of the iron takes the form of metal or sulfide rather than an oxide. They contain abundant enstatite ($MgSiO_3$).

komatiite An ultramafic rock with a noncumulate texture, presumed to be extrusive.

MERCURY DUAL IMAGING SYSTEM (MDIS) An instrument on the MESSENGER spacecraft consisting of a monochrome narrow-angle camera (NAC) and a multispectral wide-angle camera (WAC).

MESSENGER The MErcury Surface, Space ENvironment, GEochemistry, and Ranging (MESSENGER) mission, a part of NASA's Discovery Program.

oldhamite A calcium magnesium sulfide mineral with the formula (Ca, Mg)S. Ferrous iron may also be present in the mineral.

one-plate planetary body A planet or satellite, such as the Moon, Mars and Mercury, with a globally continuous and unsegmented lithosphere, in contrast to Earth, with its multiple, laterally moving lithospheric plates.

partial melt The igneous process in which a rock begins to melt at the lower end of its melting interval, yielding a magma with a chemical composition different from the bulk composition of the parent rock.

The Encyclopedia of Volcanoes. http://dx.doi.org/10.1016/B978-0-12-385938-9.00040-7

plagioclase feldspar A form of feldspar consisting of aluminosilicates of sodium and/or calcium, common in igneous rocks and typically white.

space weathering A general term used for a number of processes that act on any body exposed to the harsh space environment (collisions of solar cosmic rays, irradiation, implantation, and sputtering from solar wind particles, and bombardment by different sizes of meteorites and micrometeorites).

ultramafic Relating to or denoting igneous rocks composed chiefly of mafic minerals.

volume-limited lava flows Lava flows that cease advancing due to cessation of the eruption rather than cooling.

1. INTRODUCTION: BACKGROUND AND THE QUESTIONS

Volcanism, the eruption of internally derived magma to form surface deposits, provides one of the most important clues to the location of interior heating in space and time and to the general thermal evolution of the planet. Volcanism is a key element in the formation and evolution of secondary crusts (those derived from **partial melting** of the mantle) and tertiary crusts (those derived from remelting of primary and secondary crusts). Volcanism is among the dominant endogenic geologic processes on the Earth, Moon, Mars, and Venus, and has produced significant resurfacing during the evolution of each of these planetary bodies. The geomorphology of surface volcanic features, edifices and deposits, and the stratigraphic relationships that indicate their abundance with time, are thus the key clues to understanding the internal dynamics and thermal evolution of planets. The documentation of the nature, distribution, and abundance of volcanism on Mercury has been a long-standing quest. Unlike the nearby Moon, however, where surface features on the nearside are readily visible, the surface morphology of distant Mercury remained unknown until early in the Space Age.

2. MARINER 10 VISITS MERCURY AND THE POSTMISSION DEBATE

The surface morphology of Mercury was initially revealed by images from the Mariner 10 flybys in 1974−1975. Using a clever set of spacecraft trajectories, Mariner 10 flew by Mercury three times and imaged about 45% of the surface at an average resolution of ∼1 km. These flybys showed that, in contrast to the Moon, where there are distinctive composition-related **albedo** variations between the cratered uplands (relatively high albedo) and the smooth volcanic mare lowlands (relatively low albedo), the albedo of Mercury is relatively uniform across the surface.

Prior to the Apollo 16 mission to the Moon in 1972, a widely distributed smooth, relatively high-albedo plains unit (the Cayley Formation) was mapped in the lunar uplands, lying stratigraphically between the younger, low-albedo maria, and the older, high-albedo impact basins and cratered terrain. One of the purposes of the Apollo 16 mission was to determine the petrology and absolute age of this unit, thought prior to the mission to represent a distinctive premare, highland phase of volcanism. During Apollo 16 surface operations it became rapidly clear that the Cayley Formation consisted of impact breccias, not lava flows, and later assessments suggested that the deposits were a combination of local and regional, basin-related impact ejecta. On the basis of the Apollo 16 results, lunar light plains were subsequently considered by most workers to have been emplaced by impact crater and basin ejecta processes, rather than by extrusive volcanism. Arriving at Mercury shortly after Apollo 16, Mariner 10 revealed the presence of two smooth Cayley-plains-like units alternatively interpreted as representing effusive volcanic deposits or impact basin ejecta. These widespread plains deposits, occurring as relatively smooth surfaces between craters (intercrater plains), and as apparently ponded material (smooth plains), were proposed by some to be volcanic in origin. Others argued that the plains deposits might represent basin ejecta, similar to those found at the lunar Apollo 16 landing site.

Part of the problem concerning interpretation of smooth plains on Mercury as volcanic or impact in origin was the relatively low resolution of the Mariner 10 data. Early on, it was pointed out that the Mariner 10 image data did not have the resolution required to resolve lunar-like volcanic features such as flow fronts, vents, and small domes. Detailed examination of lunar images at resolutions and viewing geometries comparable to those of Mariner 10 readily showed that small shields and cones, elongate craters, sinuous rilles, and flow fronts, all hallmarks of the identification of volcanism on the Moon, would not be resolvable in most of the Mariner 10 images. Furthermore, larger features typical of volcanism on Mars (such as huge volcanic edifices and calderas), and not seen on the Moon, were not observed by Mariner 10 on the ∼45% of Mercury that it observed. Also not observed in the Mariner 10 data were examples of the large (10−30 km diameter) steep-sided domes suggestive of crustal magmatic differentiation processes seen on the Moon and Venus. Lobate fronts exposed at the edge of smooth plains occurrences on Mercury suggested that these might have been volcanic flow margins, but comparisons with marginal basin ejecta deposits on the Moon indicated that such features could also be products of impact ejecta emplacement. Thus, although the surface features observed by Mariner 10 were most similar to lunar plains, there were also fundamental differences between volcanism occurring on the two bodies.

These various observations raised the interesting possibility that there might be **no** identifiable volcanic units on

Mercury. However, crater counts of ejecta facies of the Caloris basin (the largest basin detected on Mercury by Mariner 10) and also on interior/exterior smooth plains deposits indicated that the smooth plains were emplaced after the Caloris basin. On this basis they were interpreted as the product of volcanic eruptions, not contemporaneous ejecta emplacement. Reprocessed Mariner 10 color data, showing some distinctive color boundaries at unit margins, provided additional evidence for the possible volcanic origin of the smooth plains.

Nonetheless, the question of the volcanic origin of the smooth and intercrater plains was hotly debated in the subsequent decades. Could extrusive volcanism not have occurred on Mercury, and if such advective cooling processes did not occur, how did the planet dissipate its **accretional heat** and that from subsequent decay of radioactive isotopes? One possibility is that partial melting of the mantle may have occurred, but that extrusive volcanism did not. Investigating these possibilities, several workers analyzed the ascent and eruption of magma under Mercury conditions for a wide range of scenarios and found that a thick low-density crust could, as with the Moon, inhibit and potentially preclude intruding dikes from reaching the surface and forming effusive eruptions. This, combined with an apparent global compressional net state of stress in the lithosphere observed by Mariner 10, could produce a scenario in which rising magma intruded the crust but did not reach the surface to produce the level of resurfacing or the array of landforms seen on the Moon, Mars, and Venus. Indeed, these workers showed how easy it was, given the range of conditions known to occur in the history of terrestrial planetary bodies, to create a planet with little to no extrusive volcanic activity. Superficially, Mercury looks like the Moon, but Mariner 10 and terrestrial remote sensing data told us that it must be very different in many fundamental respects. Could Mercury be a Moon that did not undergo surface evolution by endogenic processes (e.g., mare volcanism) subsequent to the period of heavy impact bombardment?

The dominant endogenic geologic process on the Earth, Moon, Mars, and Venus is volcanism, characterized by massive extrusions of basaltic lavas, significant resurfacing of their surfaces, and emplacement of large volumes of intrusive magmas. Such fundamental outstanding questions concerning Mercury's volcanism and thermal evolution highlighted the importance of obtaining new spacecraft image, mineralogical, and elemental data. The MErcury Surface, Space ENvironment, GEochemistry, and Ranging (**MESSENGER**) mission was designed to provide a complete picture of Mercury in the context of the early evolution of the terrestrial planets. One of the major goals of the MESSENGER mission was to assess the presence, age, and distribution of volcanic deposits.

3. THE MESSENGER MISSION: THE FLYBYS AND THE ORBITAL PHASE

3.1. The MESSENGER Flybys

In order to place the MESSENGER spacecraft into orbit around Mercury, three flybys were undertaken in 2008—2009. The first MESSENGER flyby of Mercury obtained images of 21% of the surface not seen by Mariner 10, including the center and western half of the Caloris basin and regions near the terminator that show details of the nature of smooth and intercrater plains. These new data immediately helped to address and resolve the series of long-standing questions on the existence and nature of volcanism on Mercury and the distribution of volcanic materials. Data from the **MDIS** on the MESSENGER spacecraft revealed the following observations that were interpreted as supporting the presence of volcanism on the surface, particularly in the smooth plains.

1. **Presence of volcanic vents**: High-resolution images revealed the presence of numerous volcanic vents in the form of irregularly shaped rimless depressions (Figure 40.1). These were observed to be concentrated around the interior edge of the Caloris impact basin.
2. **Pyroclastic eruptions**: Many of these vents were surrounded by bright haloes that were interpreted as representing pyroclastic deposits from explosive eruptions.
3. **Flow margins**: Lobate margins of plains units, seen previously in Mariner 10 data, were documented in MESSENGER images with more clarity and were shown to be often distinctive in morphology and color properties, supporting the interpretation that these features are the edges of lava flow units.
4. **Spectral distinctiveness**: The interior of the Caloris basin was seen to be filled with plains units spectrally distinctive from the rim deposits (Figure 40.2), and comparison with the lunar Imbrium basin and superposed impact crater stratigraphy provided evidence that these units are volcanic in origin; however, the detailed differences in lava flow unit mineralogy, so prominent in the Imbrium basin interior, were not seen in the Caloris basin interior.
5. **Distant plains units**: Some of the smooth plains surrounding the exterior of the Caloris basin at great distances showed distinct differences in color and morphological properties from the surrounding basin textured ejecta. This distinct difference supported a volcanic origin rather than the ponded ejecta origin thought to have formed the lunar Cayley plains at the Apollo 16 site.
6. **Embayment relationships and flooded craters**: Some smooth and intercrater plains units distant from the Caloris basin showed evidence for flooding of small

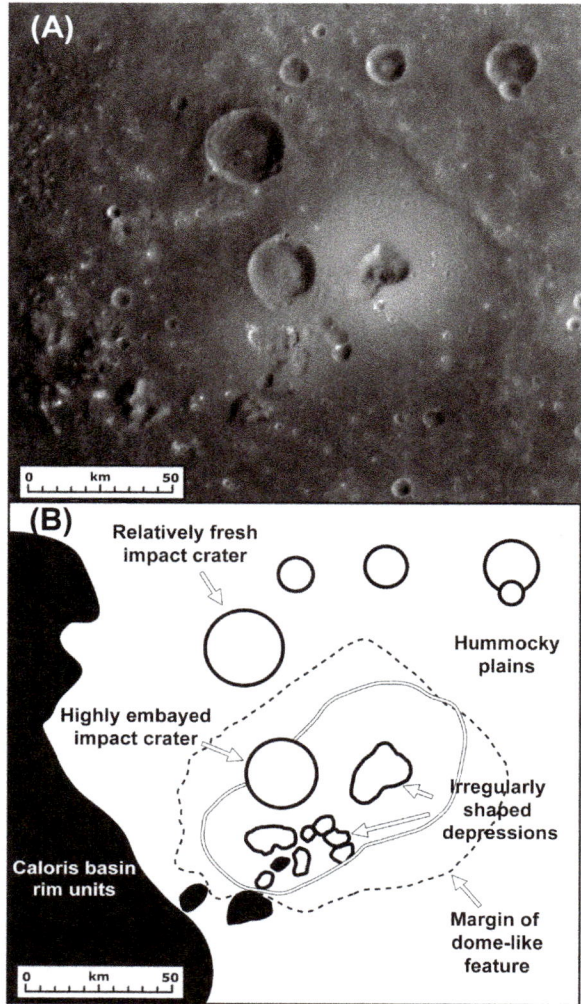

FIGURE 40.1 Central kidney-shaped depression with annulus of bright material interpreted as being of pyroclastic origin located along the southwest margin of the Caloris basin (A) and sketch map illustrating interpretation (B). *From Head et al. (2008). MESSENGER MDIS image.*

9. Counts of impact craters superposed on geologic units: Impact crater size-frequency distributions for the Caloris basin exterior deposits, including relatively high- and low-albedo smooth plains, showed that they are younger than the interior plains of Caloris and thus must be dominantly the product of post-Caloris volcanism. This suggested that there were demonstrable differences in the ages of volcanism subsequent to impact basin formation.

Images from MESSENGER's first flyby of Mercury also showed evidence for several features that are characterized by fractures and graben—rare features on a planet dominated by contractional deformation—that may be linked to intrusive activity. These features included the following:

1. Floor-fractured crater: Interpreted as having been the site of laccolith-like sill intrusions, this feature is similar to some floor-fractured craters on the Moon and shows evidence for individual fractured dome-like uplifts on the floor. Few additional such features have been observed.

2. Radial graben swarm: A large radial graben swarm (Figure 40.4), named Pantheon Fossae, was discovered at the center of the Caloris basin. Characterized by hundreds of individual radiating graben segments ranging from ~5 to ~110 km in length, this feature is unique on Mercury. In the nexus, graben crosscut one another and produce a local polygonal pattern; others curve away from the center as the nexus is approached. Two scales of graben length are observed; the radius of the dense radially symmetric plexus of graben is ~175 km, and a few graben extend to greater radial distances to the north and southwest out to distances that intersect with a ring of generally concentric graben around the outer basin floor. Two width scales of graben are observed; a large graben about 8 km wide emerges from the nexus and extends for ~100 km; most graben are less than half this width. Some graben walls appear cuspate, with convex-outward wall segments that resemble crater-chain segments. One chain of craters with distinctive raised rims parallels the graben. Locally, some graben appear in en echelon patterns, and smaller graben sometimes show crosscutting (superposition) relationships. Abundant impact craters, the most prominent being Apollodorus (Figure 40.4), and secondary crater clusters and chains are superposed on the graben system; there is little evidence that craters greater than 5 km in diameter have been cut by a graben. This relation implies that the graben swarm formed soon after the emplacement of the Caloris floor volcanic plains. These graben are interpreted by many as the surface expression of a radial dike swarm emanating from a subsurface magma reservoir near the

craters and embayment relations unrelated to Caloris ejecta emplacement. Furthermore, local and regional geological and color relationships in these areas supported a volcanic origin for these plains.

7. Stratigraphic relationships: New high-resolution image data of large fresh impact craters, and their more degraded older counterparts, revealed a sequence of embayment of interior floor and exterior ejecta deposits (Figure 40.3) that supported a volcanic origin for the embayment and filling processes.

8. Thicknesses of volcanic deposits: High-resolution images showed crater embayment and flooding relationships that suggested typical thicknesses of volcanic plains of many hundreds of meters, and local thicknesses inside impact craters of up to several kilometers.

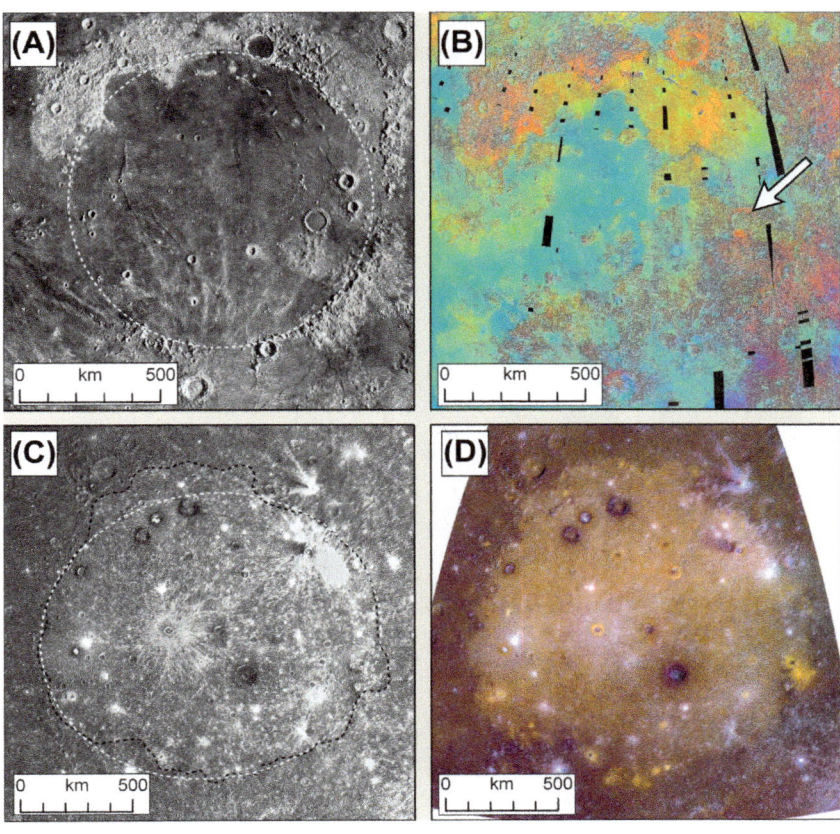

FIGURE 40.2 Comparison of the lunar Imbrium basin and the Caloris basin on Mercury. (A) Consolidated Lunar Atlas Earth-based telescopic view of the Imbrium basin on the Moon. (B) Color-composite Clementine image of the Imbrium basin (red, 750 nm/415 nm; green, 750 nm/950 nm; blue, 415 nm/750 nm). The crater Archimedes, which formed after the basin but was subsequently externally embayed and internally filled by lava, is shown by the arrow. (C) MESSENGER and MDIS mosaic of the Caloris basin. (D) MESSENGER MDIS false-color image of the Caloris basin. Large impact craters in the Caloris interior plains deposits have blue rims similar to the ejecta deposits, and smaller craters excavate orange material similar to the smooth plains and the irregular rimless depressions interpreted as pyroclastic vents around the margins. These relationships argue for the volcanic origin of the plains and superposed crater diameters can be used to estimate volcanic plains thicknesses.

center of the basin. Similar features, in which the dikes contribute to a near-surface stress field that favors radial graben, are known on the Earth, Venus, and Mars. The location of Pantheon Fossae in the center of the Caloris basin suggests that formation of the radial graben structure is linked to basin evolution. Others have proposed an entirely tectonic origin for Pantheon Fossae, related to the dynamic evolution of the basin interior and the formation of the Apollodorus impact crater.

3. **Concentric graben**: A concentric complex of graben was observed inside the peak ring on the floor of the ∼250-km-diameter Raditladi basin and associated with dark plains and possibly embayed by them. This complex may represent an unusual type of floor fracturing associated with deeper intrusions and related ring dikes or cone sheets, or the graben may instead be the product of nonmagmatic deformation of the basin floor.

On the basis of Mercury's high density and proximity to the Sun, it had been thought that the planet would be depleted in volatiles and thus devoid of pyroclastic deposits. One of the biggest surprises of the first MESSENGER flyby was the detection of features that were interpreted as deposits from pyroclastic eruptions. A prominent example is an irregularly shaped depression surrounded by a bright

deposit (Figure 40.1). This candidate pyroclastic deposit has a mean radius of ∼24 km, equivalent in size to the third largest lunar pyroclastic deposit when mapped to lunar gravity conditions. From the extent of the candidate pyroclastic deposit, and assuming primary magmatic volatiles, the eruption parameters of the event that emplaced it could be estimated, including vent exit speed and candidate volatile content. The minimum vent exit speed was estimated to be ∼300 m/s, and the volatile content required to emplace the pyroclasts to an ∼24 km distance is hundreds to several thousands of parts per million (ppm) of volatiles typically associated with pyroclastic eruptions on other bodies (e.g., CO, CO_2, H_2O, SO_2, H_2S). For comparison, measurements of the exsolution of volatiles (H_2O, CO_2, S) from basaltic eruptive episodes at Kilauea volcano, Hawaii, indicate values of ∼1300−6500 ppm for the mantle source. Evidence for the presence of significant amounts of volatiles in partial melts derived from the interior of Mercury was an unexpected result, and it provided a new constraint on models for the formation and early evolution of Mercury.

These new data provided evidence that supported and confirmed earlier hypotheses from Mariner 10 data that volcanism was important in shaping the surface of

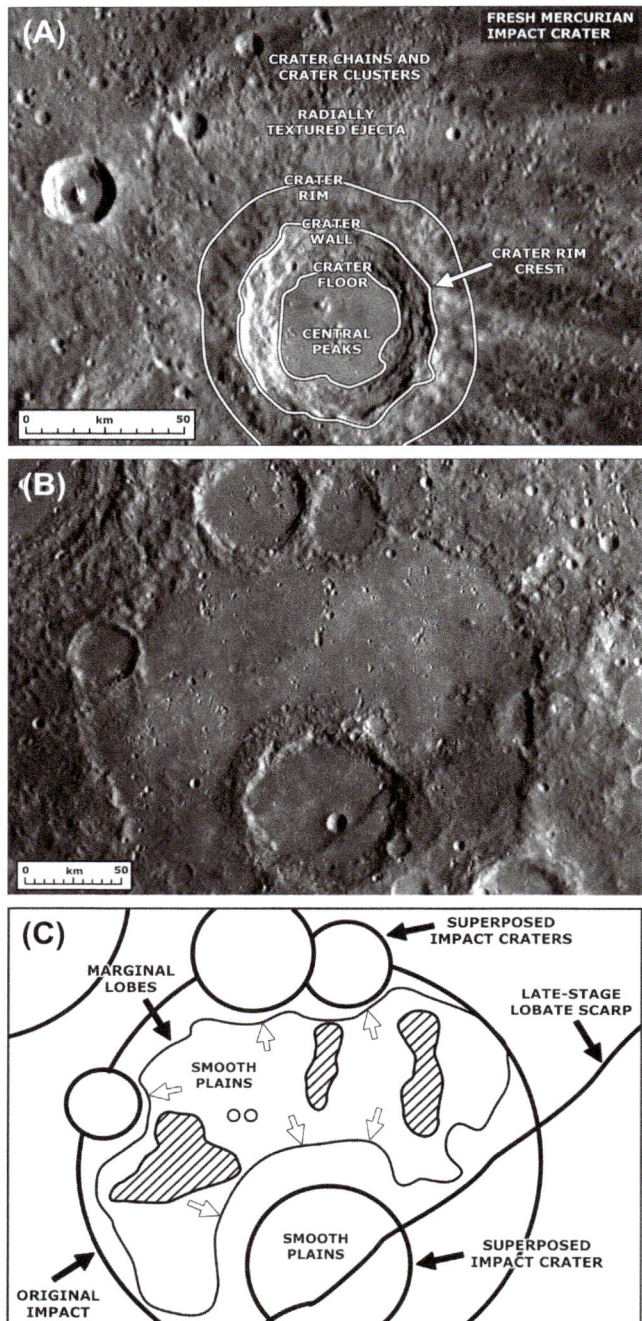

FIGURE 40.3 Fresh craters on Mercury and stratigraphic relationships showing interior and exterior flooding by volcanic plains. (A) Fresh impact crater with major features noted (9.6° N, 125.8° E). (B) Degraded 240-km-diameter crater (2° N, 113° E) that has been modified by subsequent impacts and flooded. (C) Sketch map illustrating major features. The crater floor and exterior of the superposed interior crater have been resurfaced by smooth plains of volcanic origin. MESSENGER and MDIS images. *From Head et al. (2008).*

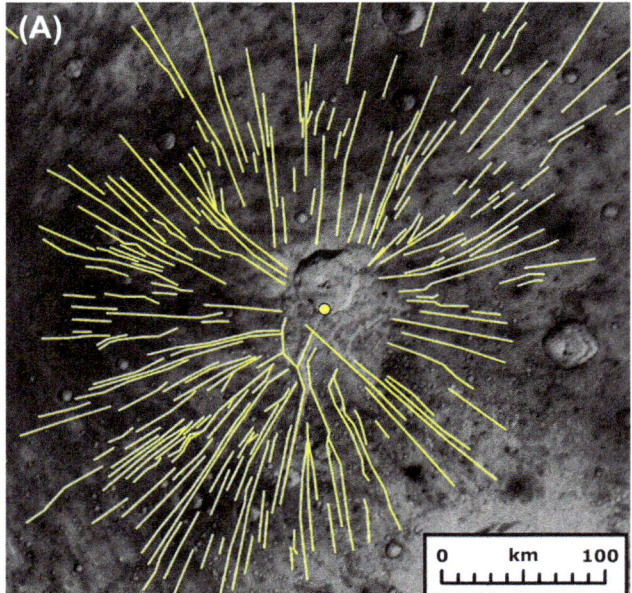

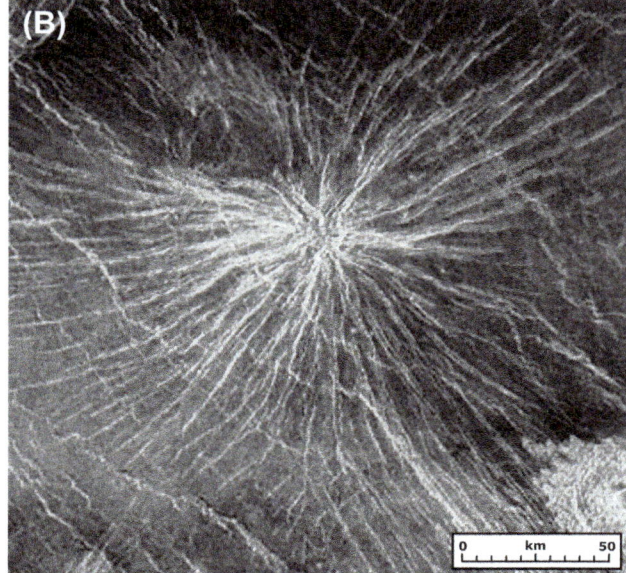

FIGURE 40.4 Pantheon Fossae radial graben in the center of the Caloris basin and comparison with radial graben on Venus. (A) MESSENGER and MDIS image of Pantheon Fossae radial graben (29.9° N, 162.9° E) with position of graben indicated by white lines, on a MESSENGER MDIS image shown as a negative to enhance contrast. (B) Astrum (Becuma Mons) on Venus (34° N, 21.5° E) interpreted as shallow intrusions formed by radial dike emplacement (Magellan image). *From Head et al. (2008).*

Mercury. The emerging picture of the volcanic style of Mercury following these initial MESSENGER flybys was similar to that of the Moon, the other small **one-plate planetary body**; there were no major shield volcanoes and shallow magma reservoirs appeared to be rare (no major calderas). Still undetected was any evidence for surface

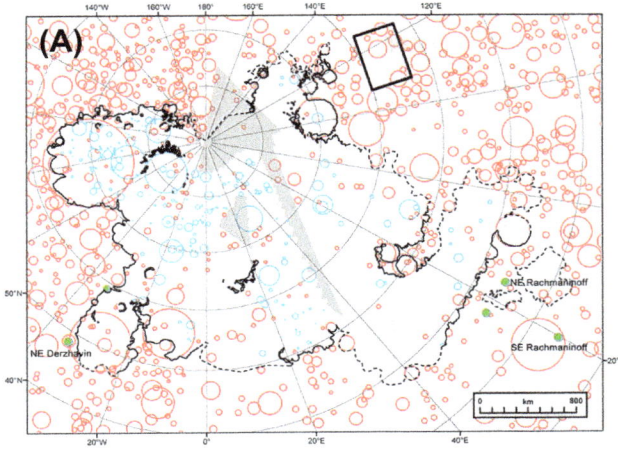

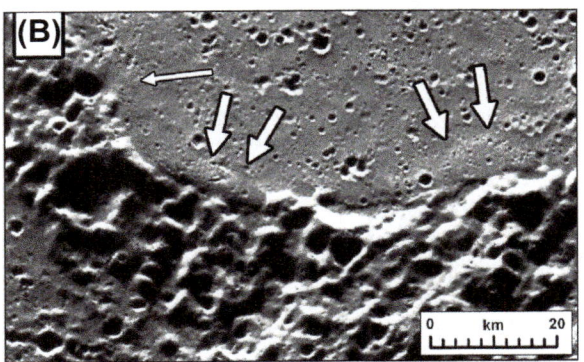

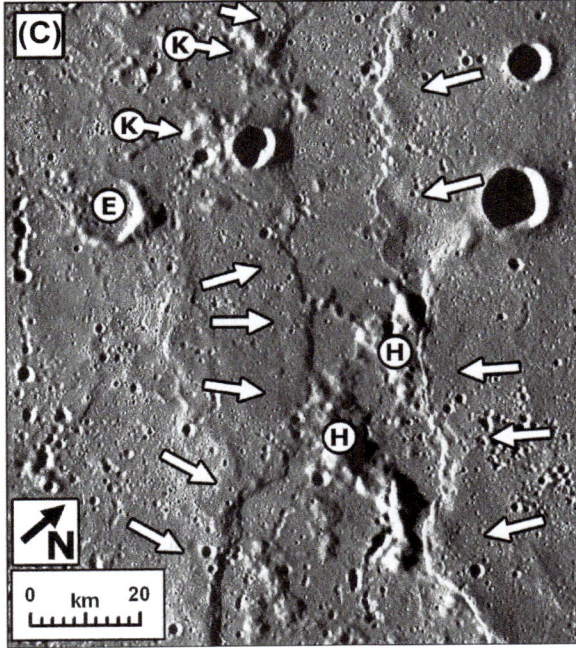

FIGURE 40.5 Flood volcanism in the northern high latitudes of Mercury. (A) Distribution of smooth plains (outlined by heavy black line) and adjacent older units (more heavily cratered). Impact craters >20 km in diameter denoted by red circles, ghost craters within plains by blue circles, and pits and pyroclastic deposits by green dots. The general paucity of impact craters in the smooth plains indicates widespread volcanic

deformation or long-lived volcanic sources related to sites of upwelling mantle, such as the huge rises and volcanoes seen on Mars and attributed to mantle plumes. Furthermore, the close association of volcanic plains and surface contractional deformation features raised the question of the relation between the volcanic flux and the evolving state and magnitude of contractional stress in the lithosphere of Mercury.

3.2. The MESSENGER Orbital Phase

Questions that remained after the three MESSNEGER flybys included (1) the global extent of volcanic plains, (2) their associated features (sinuous rilles, flow fronts, cones, small and large shields, calderas, broad rises, etc.), (3) their style and mode of emplacement, (4) their temporal distribution, (5) their association with other geological features, and (6) their mineralogical and elemental nature. MESSENGER high-resolution images of Mercury and remote sensing data obtained from orbit have provided insight into these issues and questions.

1. **Northern volcanic smooth plains**: MESSENGER observations from Mercury orbit almost immediately revealed the presence of a huge contiguous expanse of smooth plains that covers much of Mercury's high northern latitudes, and occupies more than 6% of the planet (Figure 40.5). These plains are smooth, embay other landforms, are distinct in color from the surrounding more heavily cratered terrain, and show several types of flow features, including distinctive flow-front-like scarps at their margins (Figure 40.5). Observed within the unit is a population of partly-to-wholly buried (ghost) craters that provide evidence for a volcanic origin for these plains and for multiple phases of emplacement. The sizes of these craters indicate lava emplacement to depths in excess of a kilometer. These characteristics, as well as associated features interpreted as having formed by thermal erosion, indicate emplacement in a flood-basalt style. Unlike the situation in mare-filled lunar impact basins, no significant differences in the impact crater size-frequency distributions of superposed craters have been seen within the northern volcanic plains, indicating that they were emplaced over a relatively short time geologically. These observations are also

resurfacing, and the very even distribution of craters superposed on the plains suggests that this resurfacing took place very rapidly. (B) Steep flow margin (broad arrows) within a flooded impact crater (small arrow shows possible flooding of small crater) at the edges of the smooth plains. (C) Candidate lava flow fronts indicated by arrows embaying crater at (E) and flooding hills (H) between flow fronts to form kipukas (K). MESSENGER and MDIS images. *From Head et al. (2011).*

consistent with MESSENGER X-ray spectrometric data indicating surface compositions intermediate between those of **basalts** and **komatiites**, similar to a magnesian alkali-basalt-like composition. The plains formed after the Caloris impact basin, confirming that volcanism was a globally extensive process in the postheavy bombardment era of the history of Mercury.

2. **Massive outflow-like channels**: Although the formation of volcanic lava flow channels and sinuous rilles has not been found to be common on Mercury, a few such examples exist. Near the northern volcanic plains is an unusual assemblage (Figure 40.6) of 5—10-km-diameter pits, teardrop-shaped hills, rough

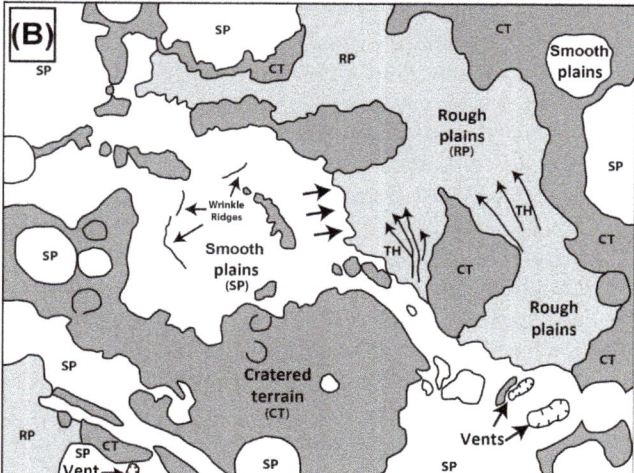

FIGURE 40.6 Image (A) and sketch map (B) showing area adjacent to northern smooth plains (Figure 40.5, black box is location of Figure 40.6) containing an assemblage of vents (lower left-hand corner) and volcanic flow-related features. Blunt arrows show flow-front-like embayment, long arrows show teardrop-shaped hills (TH), suggesting overland flow and thermal erosion by lava. MESSENGER and MDIS images. *From Head et al. (2011). See Byrne et al. (2013) for more details.*

plains, and distal lobate-margined smooth plains, interpreted as representing source vents, lava sculpting of underlying terrain, and distal emplacement of extensive flow lobes. This assemblage is consistent with the rapid emplacement of high-temperature, low-viscosity flood lavas, and the thermal erosion of subjacent terrain. The most striking such landforms are broad channels that host streamlined islands and that cut through surrounding intercrater plains. These are accompanied by narrower, more sinuous channels, coalesced depressions, and evidence for local flooding of intercrater plains by lavas exceeding their channel boundaries. An analysis of lava flow rates suggests that the broad channels define an assemblage of flow features formed by the overland flow of, and erosion by, voluminous, high-temperature, low-viscosity lavas, an interpretation consistent with compositional data suggesting substantial magnesian, iron-poor **ultramafic** lithologies for portions of the crust of Mercury. The proximity of this partially flooded assemblage to the extensive northern volcanic plains (Figure 40.6) suggests that the formation of these flow features may precede total flooding of an area by lavas emplaced in a flood mode.

3. **Sinuous rilles**: A global analysis of images acquired by MESSENGER was undertaken to identify features that may have formed by lava emplacement (lava channels) and lava thermal erosion (sinuous rilles). Despite the presence of the outflow channel-like features near the northern volcanic plains (Figure 40.6), volcanic plains identified on Mercury lack constructional and erosional features that are prevalent on other terrestrial planetary bodies. Several narrow channels observed on Mercury in association with impact craters could be lava channels, but may have formed by impact-melt carving channels into impact ejecta.

4. **Volcanically flooded impact craters**: Images of the northern volcanic plains revealed many dozens of ghost craters, nearly-to-completely buried impact craters with rims whose surface expressions are now visible primarily as rings of deformational features. In the most prominent of these (Figure 40.7), families of troughs, interpreted as graben, are developed on volcanic plains material that largely or completely bury the preexisting craters and basins. These graben, displaying mostly polygonal patterns, are partially-to-fully encircled by rings of contractional wrinkle ridges localized over the rims of the buried impact features to form systems of associated contractional and extensional landforms. This distinctive relationship between wrinkle ridges and graben in buried craters and basins is interpreted as the result of a combination of extensional stresses from cooling and thermal contraction of thick lava flow units

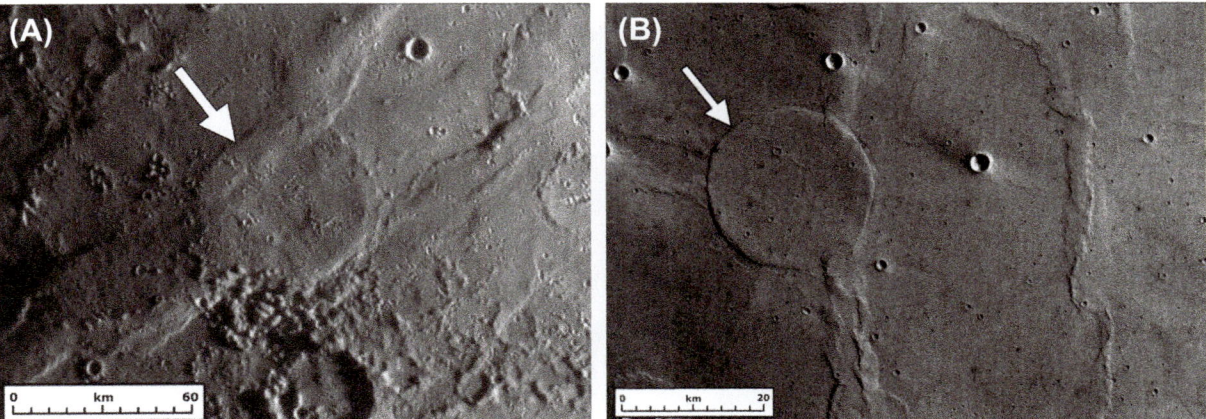

FIGURE 40.7 Volcanically flooded impact craters on Mercury (A) and Mars (B). The circular feature (arrow) on Mercury (centered at 10.0° N, 98.4° E) is interpreted as an impact crater flooded to the brim by lava; linear patterns of wrinkle ridges are disrupted into a circular structure by the presence of the buried impact crater. MESSENGER and MDIS image. A similar circular feature is seen on lava plains in Lunae Planum on Mars (B) (30.0° S, 115° E). High-Resolution Stereo Camera image. *From Head et al. (2008).*

together with compressional stresses from cooling and contraction of the interior of Mercury. Comparison of diameters of ghost craters and fresh crater depth–diameter relationships suggest that the interior of the craters contain volcanic plains material in excess of ~1.5 km in thickness.

5. **Global distribution of smooth plains**: MESSENGER orbital images have permitted a global census of smooth plains deposits and show that ~27% of the surface of Mercury is covered by smooth plains (Figure 40.8). The majority of these (>65%) are interpreted as being volcanic in origin. The spectral characteristics of most smooth plains are similar to those of the northern smooth plains, suggesting that they also share their magnesian alkali-basalt-like composition. A smaller part of smooth plains interpreted as being volcanic in nature has a lower reflectance and shallower spectral slope, suggesting more ultramafic compositions. This implies that high temperatures and high degrees of partial melting in magma source regions could have persisted through most of the duration of smooth plains formation. Although smooth plains are found globally on Mercury, they are more heavily concentrated in the north hemisphere and in the hemisphere surrounding Caloris. A comparison of plains distribution, crustal thickness, and radioactive element distribution found no correlations.

6. **Pit craters**: Global MESSENGER images have revealed more than 100 pit craters (Figure 40.9), rimless depressions ranging from 10 to almost 80 km in widest dimension. The size range of the pits is comparable with that for large calderas on Earth, Venus, and Mars. Evidence in favor of a shallow, noneffusive igneous origin for Mercury's pit craters

includes (1) lack of impact ejecta and raised rims, (2) irregular and often arcuate outlines, (3) frequent association with deposits interpreted as being pyroclastic, (4) spatial association with volcanic smooth plains, and (5) lack of observable lava flows on their rims. Two hypotheses have been proposed to explain pit craters: (1) **Collapse calderas**: This envisions a two-step mechanism beginning with emplacement of a near-surface magma chamber beneath an impact crater floor, followed by withdrawal of magma from or degassing of such a reservoir, and weakening and collapse of the roof to form the pit crater. (2) **Explosion craters**: In this scenario, shallow dike emplacement and stalling results in volatile buildup in the dike tip area, and subsequent explosive venting causes formation of the pit crater. Pits occur in three different geologic settings: (1) on impact crater floors (77%), (2) in intercrater plains (17%), and (3) within impact basins (6%). Lack of floor fracturing that commonly accompanies shallow intrusion below crater floors in lunar floor-fractured craters, and similarities to a large vent and dark ring of pyroclastics in the Orientale basin of the Moon support the dike emplacement and explosion scenario.

7. **Pyroclastic deposits**: MESSENGER flybys 1–3 and orbital data enabled a global survey of candidate pyroclastic deposits on Mercury and a total of 51 deposits have been identified to date (Figure 40.10); these appear to be globally distributed. About 90% of these are found within impact craters and most pyroclastic deposits appear to be unrelated to regional smooth plains deposits. Some deposits, however, cluster around the margins of smooth plains (for example, the Caloris interior plains and the northern volcanic

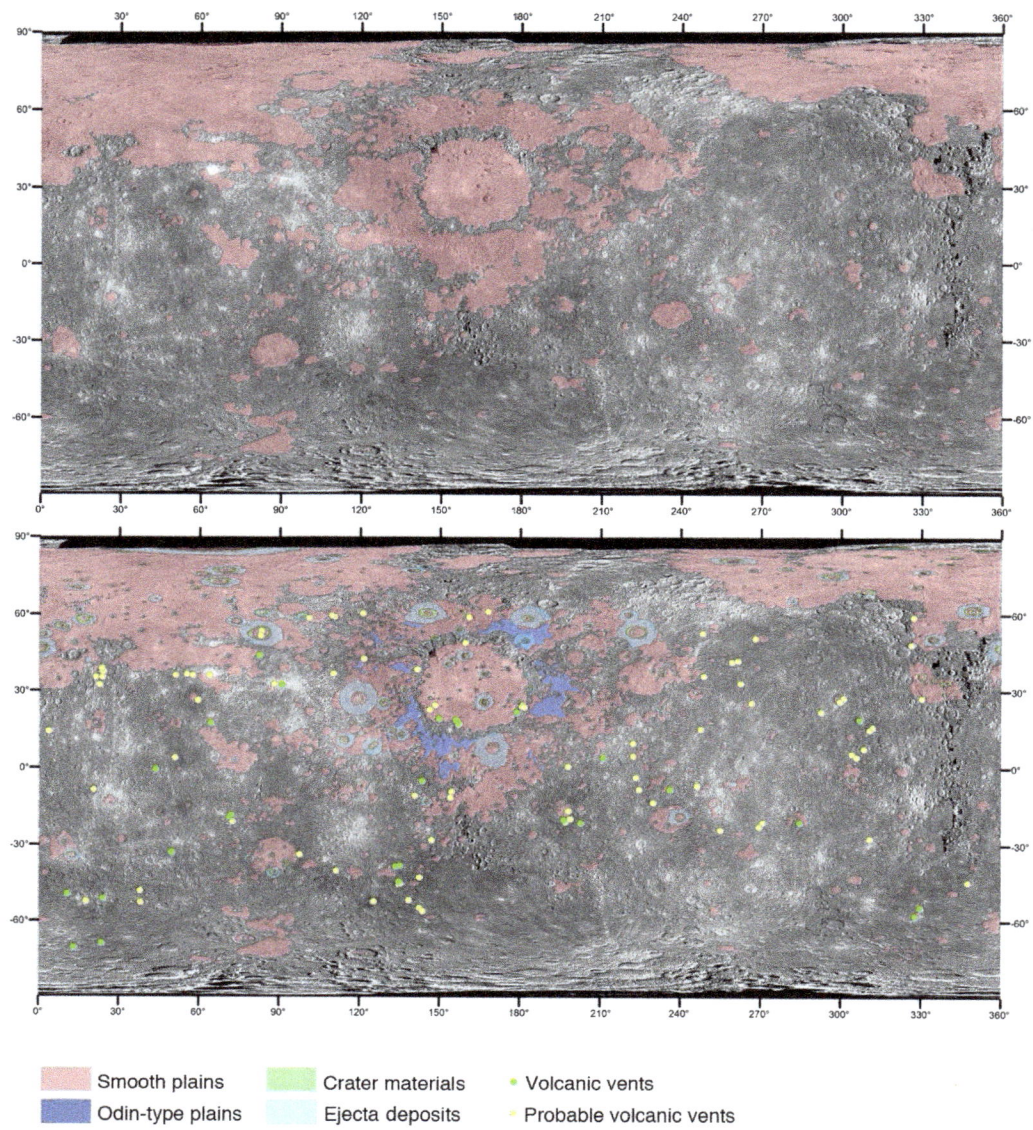

	Smooth plains		Crater materials	•	Volcanic vents
	Odin-type plains		Ejecta deposits		Probable volcanic vents

FIGURE 40.8 Global distribution of smooth plains (tan) on Mercury as mapped in Denevi et al. (2013). The original extent of smooth plains was mapped, regardless of later modification by impact craters. Map is MESSENGER and MDIS mosaic in simple cylindrical projection; central longitude is 180° E. Large circular feature filled with smooth plains ($\sim$30.0° N, 155° E) is the Caloris basin.

plains), similar to the relationship between many lunar pyroclastic deposits and the lunar maria. Deposits are commonly centered on rimless, often irregularly shaped pits, mostly between 5 and 45 km in diameter. Pyroclastic source vent areas range from $\sim$60 to 800 km^2 and depths range from $\sim$1.2 to 2.4 km. When mapped to lunar gravity conditions, the diameters of the deposits are larger than their lunar counterparts, implying that more abundant volatiles were present during typical eruptive processes on Mercury. If these deposits resulted from Hawaiian-style eruptions on Mercury, the volatile contents required would be

between $\sim$1600 and 16,000 ppm CO (or an equivalent value of H_2O, CO_2, SO_2, H_2S, or N_2), a much greater value than predicted by previous compositional models for Mercury. The range of degradation states of the host impact craters suggests that pyroclastic activity occurred on Mercury over a prolonged interval, with some deposits emplaced at least as recently as the Mansurian ($\sim$1$-$3 Ga). These Mercury deposits are generally similar in morphology and absolute reflectance to lunar pyroclastic deposits. Pyroclastic deposits on Mercury are spectrally distinct from their surrounding terrain, as shown by MESSENGER

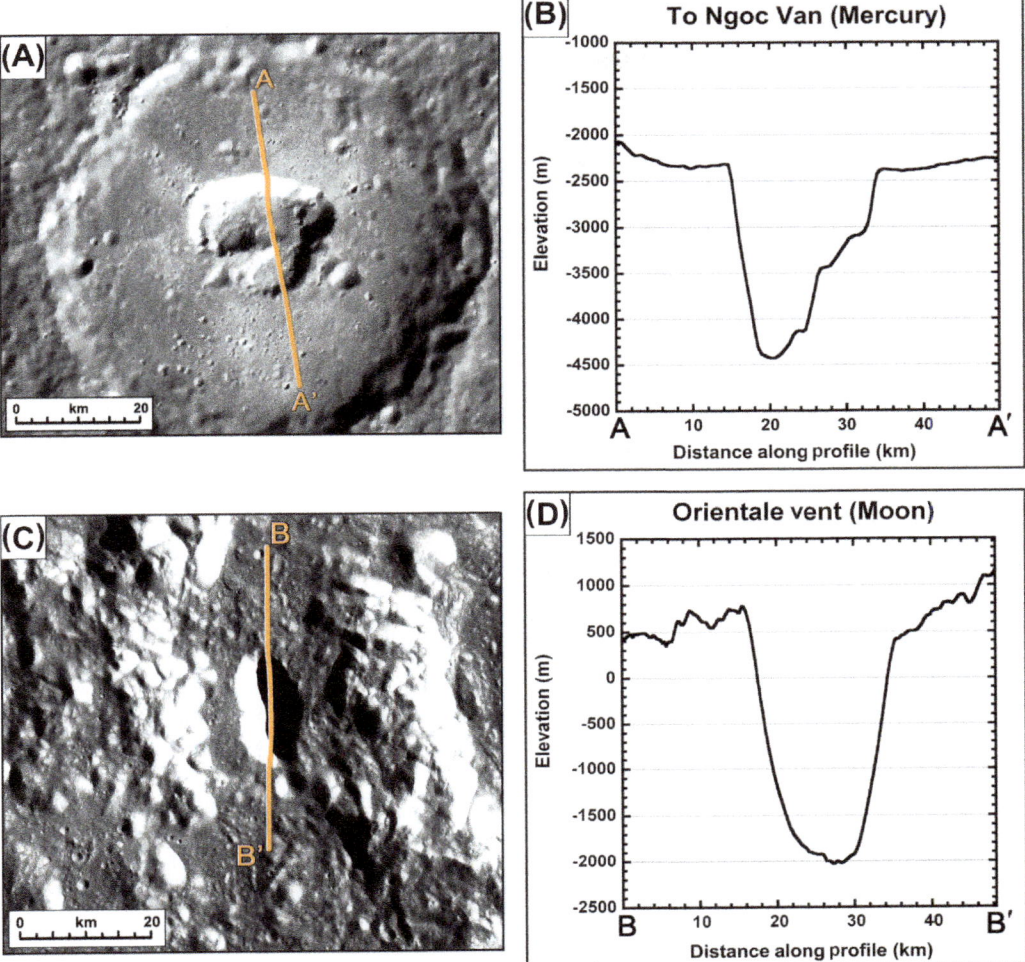

FIGURE 40.9 Pit craters on Mercury and the Moon. Topographic profiles of representative pit craters and pyroclastic source vents on Mercury and the Moon. North is up in all images. (A) The To Ngoc Van pyroclastic source vent on Mercury at 52.8° N, 111.6° E. The locus of a MESSENGER and Mercury Laser Altimeter (MLA) topographic profile is indicated by the orange line, overlaid on the Mercury Dual Imaging System-derived global mosaic. (B) MLA topographic profile of the To Ngoc Van vent. The vent depth is ~2.1 km. Vertical exaggeration is ~12.5:1. (C) The Orientale dark mantling deposit source vent. The location of a Lunar Orbiter Laser Altimeter (LOLA) topographic profile is indicated by the orange line. The image is from a Lunar Reconnaissance Orbiter Camera global mosaic at a resolution of 100 m/pixel. (D) LOLA topographic profile. The vent depth is ~2.6 km. Vertical exaggeration is ~12.5:1. These relationships suggest that volatile fractionation has enhanced volatile content in shallow magmas leading to explosive eruptions of magmatic foams on both bodies, to produce pyroclastic mantles. *From Goudge et al. (2014).*

surface reflectance measurements, with higher albedos, redder (i.e., steeper) spectral slopes, and a downturn at wavelengths shorter than ~400 nm (i.e., in the near-ultraviolet region of the spectrum). Causes for these distinctive characteristics include differences in transition metal content, physical properties (e.g., grain size), or degree of **space weathering** relative to average surface material on Mercury.

8. **Intercrater plains**: Mariner 10 investigators defined three plains types: ancient intercrater plains, intermediate plains (together comprising approximately one-third of the surface area imaged by Mariner 10), and younger smooth plains (Figure 40.11). One hypothesis

attributed plains formation to ponding of fluidized impact ejecta, while another attributed these plains to a volcanic origin. Observations during the MESSENGER orbital mission have been utilized to reevaluate the nature and origin of the intercrater and intermediate plains units. The morphology, spectral properties, impact crater statistics, and topography of intercrater and intermediate plains reveal that the intercrater plains are a highly textured unit with an abundance of secondary craters, whereas the intermediate plains are composed of both intercrater and smooth plains, and can be subdivided into their constituent intercrater and smooth plains units. Various

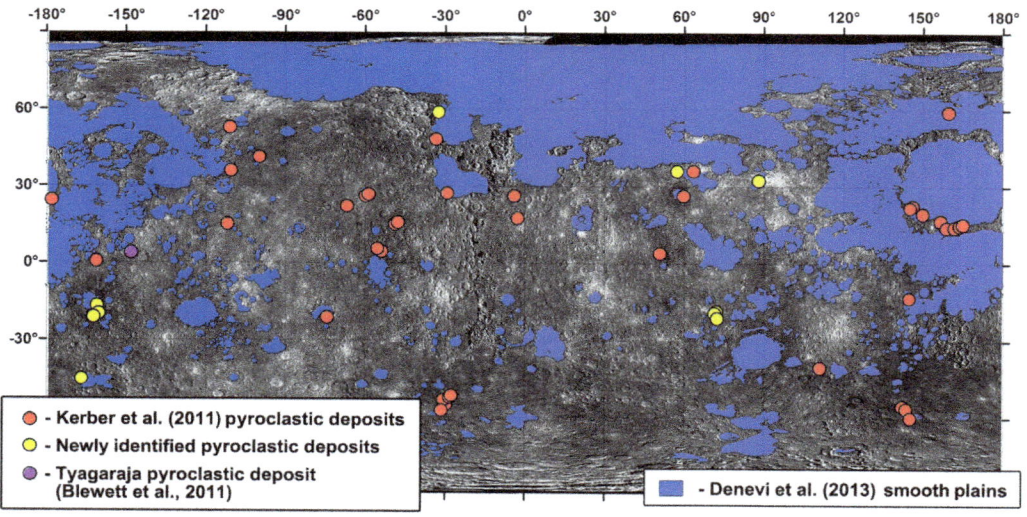

FIGURE 40.10   Global pyroclastic deposit distribution map. Distribution of pyroclastic deposits compared with the distribution of smooth plains deposits mapped by Denevi et al. (2013) (blue regions; see also Figure 40.8). Note that the pyroclastic deposits are either distant from or located on the margins of the smooth plains units. The background is a MESSENGER and MDIS-derived global mosaic. *From Goudge et al. (2014).*

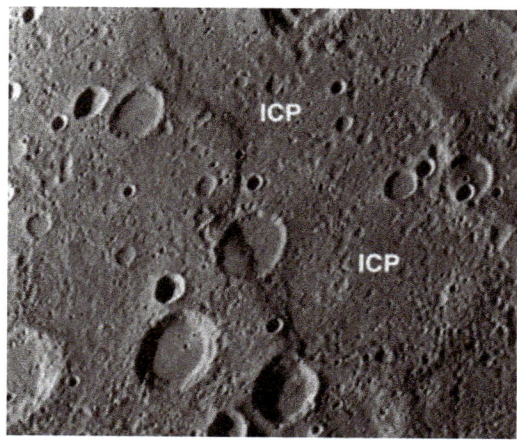

FIGURE 40.11   Example of intercrater plains as initially defined in preliminary geologic mapping using Mariner 10 data, and interpreted by many as being of volcanic origin, predating the smooth plains. In this example, Santa Maria Rupes crosscuts these intercrater plains from the northwest to southeast. The image (Mariner 10 frame 27,448) is ~200-km-across. ICP - Intercrater plains.

lines of evidence suggest that many occurrences of intercrater plains deposits may be of volcanic origin. These lines of evidence include (1) the ability of ejecta from a small number of superposed craters to transform smooth plains deposits of known volcanic origin into a unit indistinguishable from intercrater plains, (2) the different ancient ages for different intercrater plains deposits interpreted from crater size-frequency distributions, and (3) the near-global distribution of intercrater plains compared with the asymmetric distribution of impact basins and their associated ejecta

deposits. Furthermore, several conditions early in the history of Mercury also support a volcanic origin for intercrater plains including (1) the lack of a low-density **anorthositic crust**, which might then favor extrusion of mafic magmas, (2) the extensive nature of young volcanic deposits (smooth plains), suggesting that volcanism might have started early during the period of heavy bombardment, (3) the possibility that Mercury might have been in a net extensional state of lithospheric stress in early Mercury history, (4) the paucity of craters <100 km in diameter relative to the Moon, suggesting volcanic flooding and erasure, and (5) the near-globally continuous distribution of the intercrater plains.

9. **Plains compositions**: Analysis of over 200 spatially resolved MESSENGER X-ray Spectrometer measurements of the surface composition of Mercury shows that the northern smooth plains deposits and the smooth plains interior to the Caloris basin differ compositionally from older terrain on Mercury. Older terrain generally has higher Mg/Si, S/Si, and Ca/Si ratios, and a lower Al/Si ratio than the smooth plains. On the basis of these measurements and related analyses, the surface mineralogy of Mercury is likely to be dominated by high-Mg mafic minerals (e.g., enstatite), **plagioclase feldspar**, and lesser amounts of Ca, Mg, and/or Fe sulfides (e.g., **oldhamite**). The compositional difference between the volcanic smooth plains and older terrain has been interpreted as reflecting different abundances of these minerals. These data have been interpreted as meaning that smooth plains magmas were derived from a more evolved mantle source than those forming the older cratered terrain

crust. A plausible compositional and mineralogical match for much of the surface of Mercury is provided by high-degree partial melts of **enstatite chondrite** material.

10. **Plains ages**: On the basis of Mariner 10 images, the most heavily cratered terrains on Mercury were estimated to be about 4 Gyr old. These terrains exhibit a density of craters <100 km in diameter that is lower than that on the Moon, an observation attributed to preferential volcanic resurfacing of ancient terrains on Mercury, preferentially obliterating smaller craters. MESSENGER data have shown that older regions exist in the portion unobserved by Mariner 10, and these new analyses of the global crater database show that the oldest surfaces formed just after the start of the Late Heavy Bombardment ∼4.0−4.1 Gyr ago. Applying this chronology to the global record of large impact basins yields a similar surface age, and this agreement implies that resurfacing was global, and supports the interpretation that the resurfacing process was volcanism. This resurfacing activity ended during the tail of basin formation from the Late Heavy Bombardment, within ∼300−400 Myr of the emplacement of the oldest terrains on Mercury, suggesting to some that volcanism could have been aided by basin-scale impacts during this bombardment.

4. THE ARRAY OF VOLCANIC FEATURES ON MERCURY: WHAT WE SEE AND WHAT WE DO NOT

Analysis of the generation, ascent, and eruption of magma on the Earth and planets provides substantial information about the geological history and thermal evolution of each body. Here we synthesize the array of extrusive features and landforms seen on the terrestrial planets and those observed to date on Mercury by the MESSENGER spacecraft and explore how they provide insight into eruption styles, lithospheric stress states, and mantle convection on Mercury.

4.1. Volcanic Features and Styles on the Terrestrial Planets

Surface elemental compositions of the terrestrial planets are consistent with a range of mantle compositions, but all are likely to produce mafic to ultramafic melts. The main controls on the types of surface volcanic features and accumulations are expected to be differences in (1) magma compositions and volatile contents, (2) tectonic regimes, (3) crustal densities, (4) crust and lithosphere thicknesses, and (5) mantle convective style. Preferred locations for magma reservoirs appear to be either at a depth within a

planetary interior or relatively shallow within a volcanic edifice. Deeper reservoirs can form near the rheological change at the base of the lithosphere, at upwellings due to pressure-release melting, or at vertical discontinuities in density such as at the base of the crust. Evidence for reservoirs in edifices is seen in calderas. Evidence for deeper magma bodies is seen in giant dike swarms. The position of ascending mantle flow is often marked by broad rises formed from thermal uplift, enhanced crustal construction, and individual edifices built by surface eruptions (e.g., Iceland and Hawaii on Earth, Tharsis on Mars, Beta Regio on Venus). On Venus, volcanic complexes and rises are often accompanied by large annular deformational features (coronae) produced by some combination of uplift and accommodation of intrusive and extrusive loads.

Shallow magma reservoirs are commonly formed within volcanic edifices on Earth, Mars, and Venus. Building a volcanic edifice and reservoir requires multiple pulses of magma to rise frequently within a spatially restricted region over an extended period of time. On the Moon, in contrast, low eruption frequencies and great flow lengths ensure that typical large edifices will not form. Shallow reservoirs form within edifices at levels of neutral buoyancy. Repeated, relatively small-volume eruptions from these shallow reservoirs progressively build shield volcanoes of a range of sizes and aspect ratios on Earth, Mars, and Venus. These shield volcanoes commonly host collapse calderas at their summits, produced when substantial volumes of magma are erupted on the volcano flanks.

Deeper magma reservoirs have certainly existed on Earth, Venus, and Mars. Giant dike swarms can be recognized by eroded outcrops (Earth and Mars), the radial patterns of volcanic vents that they feed (Venus), and/or the graben formation that they cause (Venus and Mars). These giant dikes are close analogs to the large dikes that fed magma to the surface of the Moon from near the base of its crust. When mantle magma rises to a density step at the crust−mantle boundary, a stress regime characterized by net horizontal extension will favor upward propagation of dikes, whereas horizontal compression will favor initial sill formation.

4.2. Evidence for Volcanic Styles from Observed Surface Features on Mercury

On the basis of Mariner 10 and MESSENGER flyby and orbital data, we see no evidence for large shield volcanoes on Mercury such as those on Earth, Mars, and Venus, and we see only a few candidate low shield-like constructs, such as those so common on the Moon and Venus. No evidence has been discerned for extensive centers of volcanism as seen on Mars (e.g., Tharsis, Elysium) or Venus (e.g., Beta and Atla Regiones), or less well-developed ones as seen on the Moon (e.g., Marius and Rumker Hills). Nor has evidence been seen for any Venus-like coronae or related annular deformational features displaying associated volcanism. Only one radial graben structure (Pantheon Fossae; Figure 40.4), centrally located in the Caloris basin, has been documented. Although large pit craters have been described on Mercury, they are often associated with pyroclastic deposits and do not occur in association with edifices or other effusive volcanic deposits.

Observations of Mercury to date also reveal no evidence for several types of volcanic features (cones, leveed flows, or classic examples of lunar-like sinuous rilles). Instead, we see evidence on Mercury for extensive flooding of the surface to form regional smooth plains that appear to be very extensive lava sheet flows, and intercrater plains (found between large, old impact craters) that may also be formed by volcanic eruptions. Volcanic plains filling the interior of the Caloris basin (Figure 40.2) show generally uniform ages and spectral characteristics and are up to several kilometers thick. Exterior plains of volcanic origin have similar to slightly younger ages, but little evidence for flow-related features. Contiguous plains at northern high latitudes cover ∼6% of the surface of Mercury (Figure 40.5), have surface ages and spectral properties that show no resolvable variation, and reveal no specific source regions or associated edifices within them. In one place along their margin are seen large pits and extensive terrains sculpted by outflow-like thermal erosion (Figure 40.6), the one family of features that seem to provide evidence for an effusive volcanic eruption style. Generally the Caloris-related and northern volcanic plains show no signs of broad, rifted rises, constructional landforms (shield volcanoes), or individual linear, leveed flows, and flow fronts. The general characteristics of the plains deposits and features on Mercury strongly suggest that they were emplaced by flood-lava-style eruptions (Figure 40.12) rather than collections of narrow, leveed flows typical of small dike-emplacement events and more limited-volume surface eruptions.

Explosive volcanic activity was not anticipated on Mercury, a high-density, near-Sun planetary body expected to be depleted in volatiles. However, at least 50 deposits morphologically consistent with emplacement by explosive activity have been identified (Figure 40.10). The sources of the deposits are rimless, generally irregular pits, often several tens of kilometers in diameter. Consideration of the energetics of eruptions in a vacuum shows that to reach the observed deposit radii, mainly in the range 20−50 km, requires the erupting magma to contain ∼4000−12,000 ppm CO or the equivalent (inversely proportional to the molecular weight) of other volatiles. Candidate volatiles depend on the oxidation state of Mercury's interior and include CO, N_2, S_2, CS_2, S_2Cl, Cl, Cl_2, or COS (reducing interior, most likely) or CO, CO_2, H_2O, SO_2, or H_2S (oxidizing interior, less likely). Equilibrium release from ascending magmas of up to 12,000 ppm volatiles is not expected given the current understanding of Mercury's composition and oxidation state. Furthermore, the mechanism of formation of the pits associated with the deposits is debated. This suggests that some process may be required to concentrate volatiles into the tops of ascending dikes that fail to breach the surface to form lava flows. The contrast between the small volumes of pyroclastics and the large volumes of flood lavas on Mercury is striking.

5. MAGMA GENERATION, ASCENT, AND ERUPTION

This overview of the range of volcanic and associated tectonic landforms seen on Mercury from Mariner 10 and MESSENGER data indicates little deformational or constructional evidence for localized convective upwelling (e.g., radially/concentrically deformed structures, volcanic rises/edifice concentrations, coronae) or the presence of local shallow crustal magma reservoirs (e.g., large shields, abundant floor-fractured craters, calderas, narrow channelized flows, aggregation of small volcanic constructs). Magma delivery to the surface in the presence of significant

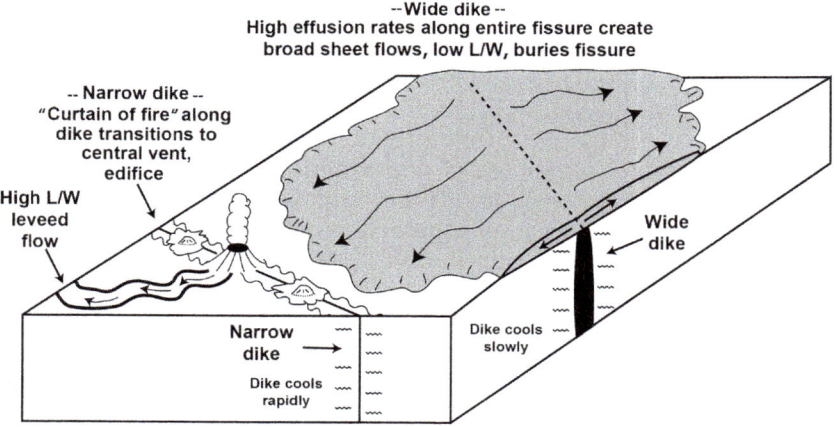

FIGURE 40.12 Illustration of the three-dimensional geometry of narrow and wide dikes. Narrow dikes (<1 m) typically produce a curtain of fire when they erupt to the surface and then centralized along the wide place in the dike as cooling occurs, producing high length/width leveed flows. Wide dikes (>5−10 m) cool slowly and permit very high effusion rates yielding very voluminous eruptions along the entire fissure to produce broad sheet flows, which typically bury the eruptive fissure.

mantle convection occurs as both the host rock and melt rise together in convection cells and encounter more brittle rocks; dikes then transport melt to the surface in the vicinity of a rising mantle diapir. Where convection is suppressed or absent, the process is different; a vertically and laterally extensive melt layer can form beneath a conductively cooled lithosphere and as the amount of partial melting increases, the corresponding volume increase accompanying melting causes an increase in pressure in the growing melt layer. Expansion of the melt layer can exert extensional stresses on the overlying lithosphere, inducing vertical fractures that form dikes through which melt escapes (Figure 40.12). In mantles where the scale length of convection is small, the elastic lithosphere will tend to be thinner and more susceptible to penetrative dike formation than is the case for more vigorously convecting systems, and flood volcanism will be favored. Analysis of magma transport and delivery from depth predicts eruptive fissure widths of ten to several tens of meters and lengths in the 40–90 km range, consistent with flood volcanism.

The typical mode of eruption of magma in the flood lava mode has been explored for a range of mafic mantle melts. The great lengths of fissures and large dike widths cause broad sheet flows rather than long, narrow, leveed and channelized lava flows (Figure 40.12). Lava is released at high-volume fluxes from these long, wide fissures to flow downslope in a turbulent manner. A typical flow will be 1.4–1.9 times thicker than comparable flows on Earth, reaching distances of ~300 km in ~20–60 hours. Such exceptionally high eruption rates would mean that even though large volumes of lava were emplaced, they were emplaced so quickly that there was not enough time to warm up the ground over which they flowed to its melting point; this might help explain the paucity of sinuous rilles.

Flow fronts cease to advance not due to cooling but instead due to cessation (or major reduction) of supply at the vent (flows are volume-limited, not cooling-limited). Under these conditions, the vast majority of lava-flow emplacement events will be in the flood lava mode, and lava distribution will vary as a function of global magma generation history and longer term thermal evolution. Direct eruption from depth will slow and cease as lithospheric horizontal stresses increase. These conclusions hold equally well for both basaltic magma and magma with compositions intermediate between basaltic and komatiitic, because the changing stresses exert more influence on the ability of a dike to remain open through the full vertical extent of the lithosphere than on the speed with which magma can flow through the dike.

In summary, magmatism on Mercury appears to be characterized predominantly by (1) deeper magma sources of large volume, (2) minimal shallow crustal storage of magma, (3) vertically extensive and wide dikes penetrating completely through the lithosphere and crust, and (4) high-volume eruption rates of lava and correspondingly voluminous outpourings producing long/wide lava flows covering extensive areas.

These observations are interpreted as meaning that the ability of magma to reach the surface on Mercury has been strongly influenced by (1) the presence and vigor of mantle convection, and (2) the lithospheric stress state. The comparatively small vertical extent (thickness) of Mercury's mantle (~600 km vs Earth's ~2900 km) may inhibit convection and favor sublithospheric magma buildup and extensional lithospheric stresses on local to regional scales in the early history of the planet. A late-stage tectonic history dominated by horizontally compressive lithospheric stresses could account for the termination of an early period of voluminous volcanic activity. Modeling of the effect of increasing horizontal compressive stress on the ability of dikes to penetrate and remain open through the full thickness of Mercury's lithosphere has shown that lithospheric magma transport would be suppressed when compressive stresses exceeded critical values in the range of several tens of MPa, similar in magnitude to the stresses thought to form the global system of lobate scarps and other contractional landforms currently observed and inferred to have formed at least partly contemporaneously with the late-stage emplacement of volcanic plains.

6. OUTSTANDING PROBLEMS AND FUTURE DIRECTIONS

Among the outstanding problems and unanswered questions in the study of volcanism on Mercury are the following:

1. Is there evidence for multiple phases of smooth plains deposits? Are there recognizable flow fronts in the central parts of the smooth plains deposits?
2. How are the intercrater plains and the smooth plains related in composition, time, and space?
3. What is the nature and origin of the numerous pyroclastic deposits and pit craters? Are they related to smooth and intercrater plains in their composition and times of emplacement?
4. Are pyroclastic deposits formed from primary volatile exsolution or does secondary gas buildup play a role?
5. What is the total time duration of smooth plains emplacement?
6. What is the temporal relationship between the record of extrusive volcanism and the tectonic scarps related to global contractional deformation?

MESSENGER has provided a rich record of the volcanic deposit on Mercury. Upcoming missions, such as the European Space Agency's BepiColombo mission to Mercury, will address many of these outstanding questions.

Supplementary video related to this chapter can be found at http://booksite.elsevier.com/9780123859389. SOM: Global color map in rotation globe.

FURTHER READING

Byrne, P.K., Klimczak, C., Williams, D.A., Hurwitz, D.M., Solomon, S.C., Head III, J.W., Preusker, F., Oberst, J., 2013. An assemblage of lava flow features on Mercury. J. Geophys. Res. 118, 1–20. http://dx.doi.org/10.1002/jgre.20052.

Denevi, B.W., Robinson, M.S., Solomon, S.C., Murchie, S.L., Blewett, D.T., Domingue, D.L., McCoy, T.J., Ernst, C.M., Head, J.W., Watters, T.R., Chabot, N.L., 2009. The evolution of mercury's crust: a global perspective from MESSENGER. Science 324, 613–618.

Denevi, B.W., Ernst, C.M., Meyer, H.M., Robinson, M.S., Murchie, S.L., Whitten, J.L., Head III, J.W., Watters, T.R., Solomon, S.C., Ostrach, L.R., Chapman, C.R., Byrne, P.K., Klimczak, C., Peplowski, P.N., 2013. The distribution and origin of smooth plains on Mercury. J. Geophys. Res. 118, 1–17. http://dx.doi.org/10.1002/jgre.20075.

Fassett, C.I., Head, J.W., Blewett, D.T., Chapman, C.R., Dickson, J.L., Murchie, S.L., Solomon, S.C., Watters, T.R., 2009. Caloris impact basin: exterior geomorphology, stratigraphy, morphometry, radial sculpture, and smooth plains deposits. Earth Planet. Sci. Lett. 285, 297–308. http://dx.doi.org/10.1016/j.epsl.2009.05.022.

Goudge, T.A., Head, J.W., Kerber, L., Blewett, D.T., Denevi, B.W., Domingue, D.L., Gillis-Davis, J.J., Gwinner, K., Helbert, J., Holsclaw, G.M., Izenberg, N.R., Klima, R.L., McClintock, W.E., Murchie, S.L., Neumann, G.A., Smith, D.E., Strom, R.G., Xiao, Z., Zuber, M.T., Solomon, S.C., 2014. Global inventory and characterization of pyroclastic deposits on Mercury: new insights into pyroclastic activity from MESSENGER orbital data. J. Geophys. Res. http://dx.doi.org/10.1002/2013JE004480.

Head, J.W., Chapman, C.R., Domingue, D.L., Hawkins III, S.E., McClintock, W.E., Murchie, S.L., Prockter, L.M., Robinson, M.S., Strom, R.G., Watters, T.R., 2007. The geology of mercury: the view prior to MESSENGER mission. Space Sci. Rev. 131, 41–84. http://dx.doi.org/10.1007/s11214-007-9263-6.

Head, J.W., Murchie, S.L., Prockter, L.M., Robinson, M.S., Solomon, S.C., Strom, R.G., Chapman, C.R., Watters, T.R., McClintock, W.E., Blewett, D.T., Gillis-Davis, J.J., 2008. Volcanism on mercury: evidence from the first MESSENGER flyby. Science 321, 69–72. http://dx.doi.org/10.1126/science.1159256.

Head, J.W., Murchie, S.L., Prockter, L.M., Solomon, S.C., Chapman, C.R., Strom, R.G., Watters, T.R., Blewett, D.T., Gillis-Davis, J.J., Fassett, C.I., Dickson, J.L., Morgan, G.A., Kerber, L., 2009a. Volcanism on mercury: evidence from the first MESSENGER flyby for extrusive and explosive activity and the volcanic origin of plains. Earth Planet. Sci. Lett. 285, 227–242. http://dx.doi.org/10.1016/j.epsl.2009.03.007.

Head, J.W., Murchie, S.L., Prockter, L.M., Solomon, S.C., Strom, R.G., Chapman, C.R., Watters, T.R., Blewett, D.T., Gillis-Davis, J.J., Fassett, C.I., Dickson, J.L., Hurwitz, D.M., Ostrach, L.R., 2009b. Evidence for intrusive activity on mercury from the first MESSENGER flyby. Earth Planet. Sci. Lett. 285, 251–262. http://dx.doi.org/10.1016/j.epsl.2009.03.008.

Head III, J.W., Chapman, C.R., Strom, R.G., Fassett, C.I., Denevi, B.W., Blewett, D.T., Ernst, C.M., Watters, T.R., Solomon, S.C., Murchie, S.L., Prockter, L.M., Chabot, N.L., Gillis-Davis, J.J., Whitten, J.L.,

Goudge, T.A., Baker, D.M.H., Hurwitz, D.M., Ostrach, L.R., Xiao, Z., Merline, W.J., Kerber, L., Dickson, J.L., Oberst, J., Byrne, P.K., Klimczak, C., Nittler, L.R., 2011. Flood volcanism in the northern high latitudes of mercury revealed by MESSENGER. Science 333, 1853–1856. http://dx.doi.org/10.1126/science.1211997.

Hurwitz, D.M., Head III, J.W., Byrne, P.K., Xiao, Z., Solomon, S.C., Zuber, M.T., Smith, D.E., Neumann, G.A., 2013. Investigating the origin of candidate lava channels on mercury with MESSENGER data: theory and observations. J. Geophys. Res. 118, 471–486. http://dx.doi.org/10.1029/2012JE004103.

Kerber, L.A., Head, J.W., Solomon, S.C., Murchie, S.L., Blewett, D.T., Wilson, L., 2009. Explosive volcanic eruptions on Mercury: eruption conditions, magma volatile content, and implications for mantle volatile abundances. Earth Planet. Sci. Lett. 285, 263–271. http://dx.doi.org/10.1016/j.epsl.2009.04.037.

Milkovich, S.M., Head III, J.W., Wilson, L., 2002. Identification of mercurian volcanism: resolution effects and implications for MESSENGER. Meteorit. Planet. Sci. 37, 1209–1222.

Murchie, S.L., Watters, T.R., Robinson, M.S., Head, J.W., Strom, R.G., Chapman, C.R., Solomon, S.C., McClintock, W.E., Prockter, L.M., Domingue, D.L., Blewett, D.T., 2008. Geology of the Caloris basin, Mercury: a view MESSENGER. Science 321, 73–76. http://dx.doi.org/10.1126/science.1159261.

Murray, B.C., Strom, R.G., Trask, N.J., Gault, D.E., 1975. Surface history of mercury: implications for terrestrial planets. J. Geophys. Res. 80, 2508–2514.

Nittler, L.R., Starr, R.D., Weider, S.Z., McCoy, T.J., Boynton, W.V., Ebel, D.S., Ernst, C.M., Evans, L.G., Goldsten, J.O., Hamara, D.K., Lawrence, D.J., McNutt Jr., R.L., Schlemm II, C.E., Solomon, S.C., Sprague, A.L., 2011. The major-element composition of mercury's surface from MESSENGER X-ray spectrometry. Science 333, 1847–1850. http://dx.doi.org/10.1126/science.1211567.

Oberbeck, V.R., 1975. The role of ballistic erosion and sedimentation in lunar stratigraphy. Rev. Geophys. Space Phys. 13, 337–362.

Robinson, M.S., Murchie, S.L., Blewett, D.T., Domingue, D.L., Hawkins III, S.E., Head, J.W., Holsclaw, G.M., McClintock, W.E., McCoy, T.J., McNutt Jr., R.L., Prockter, L.M., Solomon, S.C., Watters, T.R., 2008. Reflectance and color variations on mercury: regolith processes and compositional heterogeneity. Science 321, 66–69. http://dx.doi.org/10.1126/science.1160080.

Spudis, P.D., Guest, J.E., 1997. Stratigraphy and geologic history of Mercury. In: Vilas, F., Chapman, C.R., Shapley Matthews, M. (Eds.), Mercury. Univ. Arizona Press, Tucson, pp. 118–164.

Strom, R.G., Trask, N.J., Guest, J.E., 1975. Tectonism and volcanism on mercury. J. Geophys. Res. 80, 2478–2507.

Strom, R.G., Chapman, C.R., Merline, W.J., Head, J.W., 2008. Mercury cratering record viewed from MESSENGER's first flyby. Science 321, 79–81. http://dx.doi.org/10.1126/science.1159317.

Taylor, S.R., 1989. Growth of planetary crust. Tectonophysics 161, 147–156.

Watters, T.R., Head, J.W., Solomon, S.C., Robinson, M.S., Chapman, C.R., Denevi, B.W., Fassett, C.I., Murchie, S.L., Strom, R.G., 2009. Evolution of the Rembrandt impact basin on mercury. Science 324, 618–621.

Wilhelms, D.E., 1976. Mercurian volcanism questioned. Icarus 28, 551–558.

Wilson, L., Head, J.W., 2008. Volcanism on mercury: a new model for the history of magma ascent and eruption. Geophys. Res. Lett. 35, L23205. http://dx.doi.org/10.1029/2008GL035860.

Chapter 41

Volcanism on Mars

James R. Zimbelman
Center for Earth and Planetary Studies, National Air and Space Museum, Smithsonian Institution, Washington, DC, USA

William Brent Garry and Jacob Elvin Bleacher
Sciences and Exploration Directorate, Code 600, NASA Goddard Space Flight Center, Greenbelt, MD, USA

David A. Crown
Planetary Science Institute, Tucson, AZ, USA

Chapter Outline

GLOSSARY

AMAZONIAN The youngest geologic time period on Mars identified through geologic mapping of superposition relations and the areal density of impact craters.

caldera An irregular collapse feature formed over the evacuated magma chamber within a volcano, which includes the potential for a significant role for explosive volcanism.

central volcano Edifice created by the emplacement of volcanic materials from a centralized source vent rather than from along a distributed line of vents.

composite volcano A volcano that consists of intermixed lava flows and pyroclastic deposits. Flank slopes are typically >10°, or more than twice as steep as the flanks on a typical shield volcano.

fossae Descriptor applied to an aligned series of fractures in a planetary surface.

HESPERIAN The intermediate geologic time period on Mars identified through geologic mapping of superposition relations and the areal density of impact craters.

mons Descriptor applied to a large isolated mountain on a planetary surface.

NOACHIAN The oldest geologic time period on Mars identified through geologic mapping of superposition relations and the areal density of impact craters.

patera Descriptor applied to an irregular or complex crater with scalloped edges.

pseudocrater A "rootless cone" created by the interaction of lava with groundwater or wet sediments beneath a lava flow.

shield volcano A broad volcanic construct consisting of a multitude of individual lava flows. Flank slopes are typically ~5°, or less than half as steep as the flanks on a typical composite volcano.

SNC meteorites A group of igneous meteorites that originated on Mars, as indicated by a relatively young age for most of these meteorites, but most importantly because gases trapped within glassy parts of the meteorite are identical to the atmosphere of Mars. The abbreviation is derived from the names of the three meteorites that define major subdivisions identified within the group: S, Shergotty; N, Nakhla; C, Chassigny.

tholus Descriptor applied to an isolated domical small mountain or hill, usually with slopes that are much steeper than the slopes of a patera.

volcanic plains Planar mappable units interpreted to consist of volcanic materials, often with individual lava flow margins resolvable on the plains surface.

yardang A rounded erosional landform produced by wind-driven sand. Some yardang fields on Mars are interpreted to have formed within wind-eroded pyroclastic deposits.

1. INTRODUCTION

Spacecraft exploration has revealed abundant evidence that Mars possesses some of the most dramatic volcanic landforms found anywhere within the solar system. How did a planet half the size of Earth produce volcanoes like Olympus Mons, which is several times the size of the

The Encyclopedia of Volcanoes. http://dx.doi.org/10.1016/B978-0-12-385938-9.00041-9

largest volcanoes on Earth? This question is an example of the kinds of issues currently being investigated as part of the space-age scientific endeavor called "comparative planetology." This chapter summarizes the basic information currently known about volcanism on Mars.

The volcanoes on Mars appear to be broadly similar in overall morphology (although, often quite different in scale) to volcanic features on Earth, which suggests that Martian eruptive processes are not significantly different from the volcanic styles and processes on Earth. Martian volcanoes are found on terrains of different age, and Martian volcanic rocks are estimated to comprise more than 50% of the Martian surface. This is in contrast to volcanism on smaller bodies such as Earth's Moon, where volcanic activity was mainly confined to the first half of lunar history (see "Volcanism on the Moon"). Comparative planetology supports the concept that volcanism is the primary mechanism for a planetary body to get rid of its internal heat; smaller bodies tend to lose their internal heat more rapidly than larger bodies (although, Jupiter's moon Io appears to contradict this trend; Io's intense volcanic activity is powered by unique gravitational tidal forces within the Jovian system; see "Volcanism on Io"), so that volcanic activity on Mars would be expected to differ considerably from that found on Earth and the Moon.

2. BACKGROUND

The first evidence of the importance of volcanism on Mars came during the Mariner 9 mission, the first spacecraft to be placed in orbit around another planet. Mariner 9 arrived at Mars on November 14, 1971. The first surface features to become visible were four dark "spots" in the region of the planet that telescopic observers called Tharsis. These spots were revealed to each have complex crater assemblages at their summits, which was the first indication that four huge volcanoes are located in the Tharsis region. Mariner 9 eventually mapped the entire Martian surface, showing that the Tharsis region was not the only location with clear evidence of volcanism. This initial Mariner 9 view of Mars has been updated by the steadily improved spatial resolution of cameras and other instruments carried on two Viking orbiters, the Mars Global Surveyor, Mars Odyssey, the Mars Express orbiter, and most recently the Mars Reconnaissance Orbiter.

The distribution of volcanoes on Mars is not uniform; there are several regions where central volcanic constructs are concentrated (Figure 41.1). Many of the largest volcanic constructs have been given names by the International Astronomical Union (Table 41.1), the only organization that can designate official names for features on planetary surfaces. The largest concentration of Martian volcanoes is within the Tharsis region, where the volcanoes are distributed on and around a 4000-km-diameter bulge in the Martian crust, centered on the equator at 250° E longitude. The "Tharsis Rise" represents crust that is elevated ~10 km above the Martian datum level, which is twice the elevation of the highest portions of the cratered highlands that dominate the southern hemisphere of Mars.

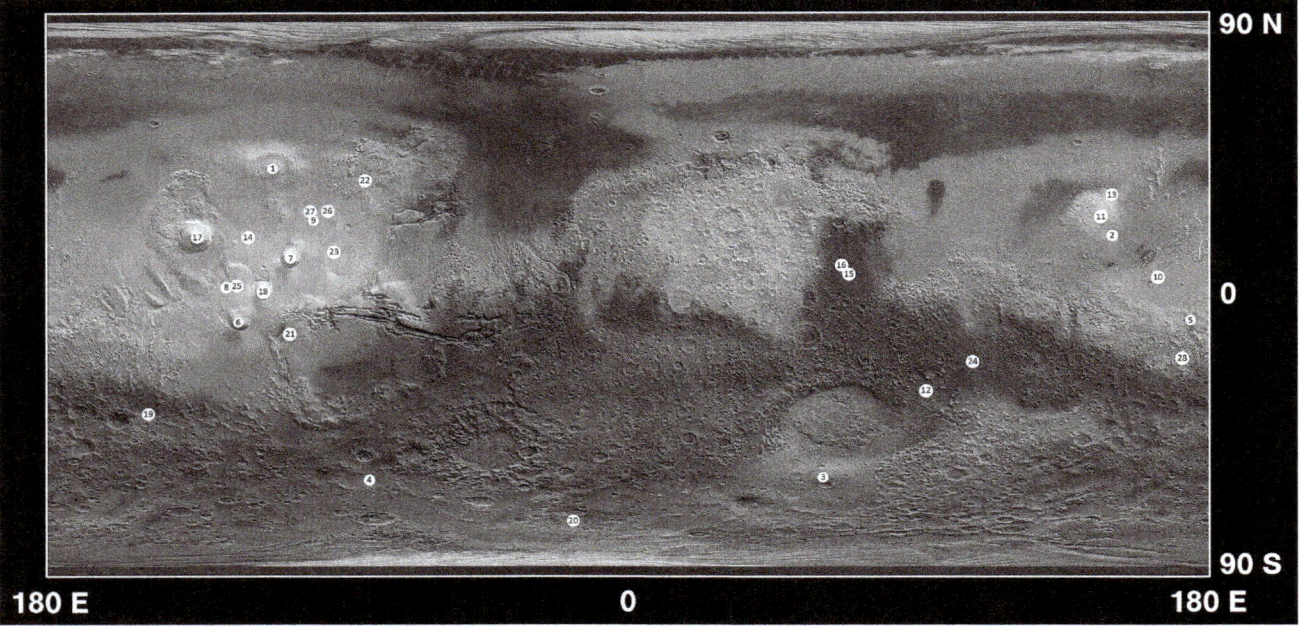

FIGURE 41.1 Location map of major named volcanic features on Mars. Numbers correspond to features listed in Table 41.1. Base map is shaded relief MOLA topography with superposed THEMIS TES albedo. NASA/JPL-Caltech/ASU.

TABLE 41.1 Major Volcanic Centers on Mars

Number[1]	Name[2]	Location[2]	Diameter[2] (km)	Relief[3] (km)
1	Alba Mons	41.1, 249.3	1100	5.8
2	Albor Tholus	18.9, 150.5	160	4.2
3	Amphitrites Patera	−58.7, 60.9	130	0.5−1.5
4	Aonia Tholus	−59.0, 280.0	60	–
5	Apollinaris Mons	−9.1, 174.8	280	5.4
6	Arsia Mons	−8.3, 239.9	470	11.7
7	Ascraeus Mons	11.9, 255.9	460	14.9
8	Biblis Tholus	2.5, 235.6	170	3.6
9	Ceraunius Tholus	24.0, 262.8	130	6.6
10	Cerberus (region)[4]	5, 262	700	–
11	Elysium Mons	25.0, 147.2	400	12.6
12	Hadriacus Mons	−31,3, 91.9	450	1.1
13	Hecates Tholus	32.1, 150.2	180	6.6
14	Jovis Tholus	18.2, 242.6	60	1.0
15	Meroe Patera	7.0, 68.8	50	(<0.5)[5]
16	Nili Patera	9.0, 67.2	70	(<0.5)[5]
17	Olympus Mons	18.7, 226.2	610	21.1
18	Pavonis Mons	1.5, 247.0	370	14.0
19	Sirenum Mons	−38.2, 212.2	120	–
20	Sisyphi Tholus	−75.7, 341.5	30	–
21	Syria Mons	−13.9, 255.7	70	–
22	Tempe (region)[4]	35, 275	300	–
23	Tharsis Tholus	13.3, 269.3	150	7.4
24	Tyrrhenus Mons	−21.6, 105.9	270	1.5
25	Ulysses Tholus	3.0, 238.5	100	1.5
26	Uranius Mons	26.9, 267.9	270	3.0
27	Uranius Tholus	26.3, 262.4	60	2.9
28	Zephyria Tholus	−19.8, 172.9	40	–

[1] See Figure 41.1 for locations on a global map.
[2] Name, center location, and feature diameter (rounded to 10 km) from the Gazateer of Planetary Nomenclature Web site (8/2013).
[3] Relief derived from MOLA data, from Table 1 of Plescia (2004).
[4] Many small volcanic centers occur within a broad region. Cerberus Tholi are several separate small volcanic centers with radiating flows, each 20–50 km in diameter. The Tempe region includes E. Mareotis Tholus (Figure 41.5), Issedon Tholus, N. Mareotis Tholus, and W. Mareotis Tholus as named volcanic features.
[5] Estimated relief of the individual constructs, not of the Syrtis Major region as a whole, as shown in Table 1 of Plescia (2004).

The Tharsis Rise is surmounted by three large volcanoes that are collectively called the Tharsis Montes, aligned along a N40E trend that includes additional volcanic features northeast of the Tharsis Montes. Olympus Mons (Figure 41.2) is larger than any of the three Tharsis Montes, at least in part because it is located on the flank of the Tharsis uplift, but it still reaches heights comparable to that of the tops of the Tharsis Montes. Outside of the Tharsis region, Martian **central volcano**es are present in the Elysium, Syrtis Major, and Hellas regions, all named for broad bright or dark areas visible to astronomers using Earth-based telescopes. In addition to the regions with large

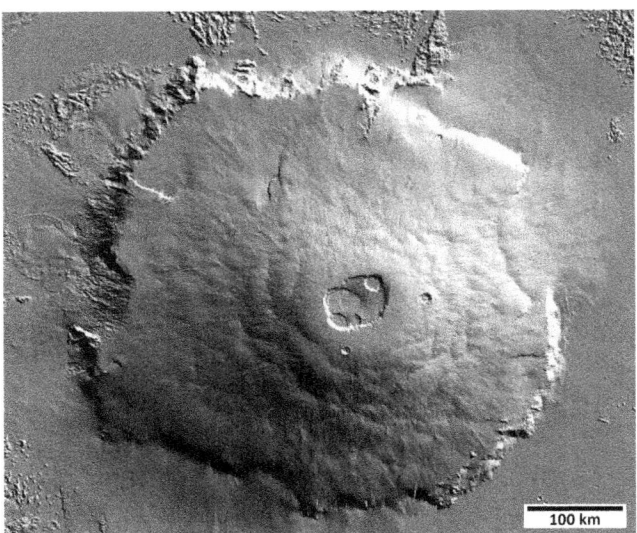

FIGURE 41.2 Olympus Mons, the largest shield volcano on Mars. Shaded relief version of MOLA topographic data. NASA/JPL-Caltech/GSFC.

central volcanoes, broad expanses of regional plains are present across Mars, many of which show compelling evidence that the plains were formed by the emplacement of numerous lava flows.

Through careful observation, a global geologic sequence has been mapped across Mars. The areal density of impact craters on the various material units allows a relative stratigraphy to be established even where units are not in physical contact with each other; recent consensus among the researchers who count impact craters led to absolute ages that can be associated with a statistically significant areal density of impact craters. Three major divisions are now recognized for the geologic history of Mars: the **Noachian** is the oldest era, and it includes materials from the formation of Mars to ∼3.8 Ga; the **Hesperian** is an intermediate era where water was quite abundant across the planet, covering rocks emplaced between ∼3.8 and ∼3.0 Ga; and the **Amazonian** is the youngest era, with rocks emplaced from ∼3.0 Ga to the present. The eventual collection of documented samples for precise age-dating will hopefully refine and calibrate these broad age designations.

Remote sensing instruments provide data about Martian materials that are complementary to the photogeologic interpretation of spacecraft images. Reflected visual and infrared light provide compositional information by the specific wavelengths at which some of the light is selectively absorbed. Earth-based telescopic studies cannot resolve individual volcanoes on Mars, but reflectance spectra indicate that bright regions, where the largest Martian volcanic provinces are found, are covered with dust that includes considerable oxidized iron, whereas the dark regions are less dusty and show evidence of pyroxene-bearing materials

that are common in mafic volcanic rocks such as basalt. Thermal infrared measurements indicate that most of the Martian surface is covered by particulate materials, ranging from the pervasive micron-sized dust to sand-sized particles mixed with larger blocks in the dark regions. Importantly, no thermal measurements from anywhere on Mars have revealed "hot spots" that might be attributable to internally generated heat.

A major advancement in understanding the entire Martian surface was the collection of millions of individual reflected laser points by the Mars Orbiter Laser Altimeter (MOLA). MOLA provided the first global topographic map of Mars where all of the points were tightly constrained by their distance from the center of mass of the planet. Both heights and slopes can be determined accurately from the MOLA measurements, which led to substantial improvements in our understanding of the shapes and topographic characteristics of Martian volcanoes. A shaded relief version of the global MOLA data set is the base map used in Figure 41.1.

Radar signals sent from antennas both on Earth and on spacecraft have shown that some individual Martian volcanoes, like the Tharsis Montes, display very strong scattering behavior. The scattered radar signals indicate the presence of considerable surface roughness at the 10 cm to meter scale on the Tharsis Montes volcanoes, an attribute that may also apply to other Martian volcanoes but the smaller size of other constructs complicates a clear inference of their properties. West of the Tharsis Montes, an area of several million square kilometers displays no reflected radar signal, nicknamed "Stealth" after this extremely low radar reflectance property. Stealth may represent a deposit that is sufficiently thick so that radar signals are absorbed rather than scattered, or Stealth may also have an unusual surface texture that efficiently scatters the radar signals away from the receiving antennas.

3. LARGE CENTRAL VOLCANOES

The four prominent volcanoes in the Tharsis region are some of the largest known volcanoes in our solar system. Olympus Mons (17 in Figure 41.1 and 41.2) and the three Tharsis Montes volcanoes are analogous to **shield** volcanoes here on Earth, except on a much larger scale. The sizes of these volcanoes are so grand (basal diameters are hundreds of kilometers) compared to the overall size of Mars and the slopes along the flanks are so low (≤5°) that an astronaut hiking up the flank would not be able to see the top over the horizon due to the curvature of the planet. Olympus Mons has a basal diameter of 500 km, and 25 km in vertical relief, that would dwarf Mauna Loa volcano here on Earth. The term "**mons**" refers to a large isolated mountain. The four volcanoes contain similar morphologic and structural features that suggest similar processes through their geologic history. A central **caldera** is located at the summit of each volcano. The overlapping nature of the summit pits (which are complex, sometimes nested calderas) indicates that a magma chamber may have been evacuated multiple times within the constructs.

At first glance, the primary flanks of these volcanoes may appear relatively featureless, but high-resolution images from recent missions reveals amazing details of an array of lava flow morphologies (e.g., channeled flows, tube-fed flows, small shields) that have constructed and shaped these shields over time. Irregular channels and canyons are eroded into the shield materials on the northeast, and southwest flanks of Arsia, Pavonis, and Ascraeus Montes (6, 18, and 7 in Figure 41.1) are structural remnants of a series of breaches that created rift aprons comprised of long lava flows and small shields. Lava flows from the rift aprons are hundreds of kilometers long and surround the base of the main flanks as well as extend toward the perimeter of the Tharsis Rise. The volume of material erupted from the rift aprons is generally at least an order of magnitude less than the volume erupted to form the main flanks of the Tharsis Montes, and is more widely spread across the top of the Tharsis Rise. Several researchers have noted morphologic differences between the rift aprons and the main constructs, suggesting that the rift aprons may represent an eruptive episode very distinct from that which formed the bulk of the shield constructs.

The Elysium region is another important volcanic center on Mars, located to the west of the Tharsis Rise. The Elysium volcanic region includes Elysium Mons, Hecates Tholus, and Albor Tholus (11, 13, and 2 in Figure 41.1). Surrounded by the low plains of the northern hemisphere of Mars, Elysium Mons is similar in scale to the Tharsis volcanoes, but with a different topographic profile. The summit of the Elysium Mons construct is relatively steep (slopes up to 12°), giving the profile a more conical appearance, as compared to the low slopes near the summits of the four large shield volcanoes in Tharsis. While the overall profile may be different, high-resolution images show that flow morphologies on Elysium Mons (channeled flows, tube-fed flows) are similar to those observed on the Tharsis volcanoes. Lava flows from Elysium Mons are quite extensive, extending more than 700 km from the summit, which is surmounted by a single caldera about 14 km in diameter, although the floor of the caldera preserves subtle details that suggest repeated collapse occurred prior to the last lava infilling.

Alba Mons (formerly Alba Patera; 1 in Figure 41.1) is a broad volcanic construct north of the Tharsis Montes, but the morphology of the Alba structure is atypical of that of other shield volcanoes on Mars. The diameter of Alba Mons is similar in scale to that of Olympus Mons, but the overall height is lower and slopes are quite gentle (≤1°), giving the volcano a relatively flat, less dramatic profile

than its Martian counterparts. A caldera complex is present at the summit of a central edifice that was constructed on an extensive apron of early-phase lava flows that extend nearly 1000 km from the summit region. High-resolution images and topography show that Alba Mons is comprised of long lava flows and tube-fed flows much like the Tharsis and Elysium volcanoes; however, the Alba eruptions resulted in a very different final morphology. Compelling features present near Alba Mons are the dense swarm of arcuate fractures (graben) that surround this volcano, some of which are partially filled by subsequent eruptions.

4. PATERAE AND THOLI

The descriptor "**patera**" refers to an irregular or complex crater, often with scalloped edges; these complex craters often are surrounded by slopes that are considerably less than the slopes encountered on shield volcanoes. Such features contrast greatly with impact craters or basins of comparable size, so that a volcanic caldera origin is the consensus interpretation. Another distinctive class of volcanic construct, which is consistently smaller than features of the mons type, has the descriptor "**tholus**" that is applied to an isolated domical small mountain or hill, usually with slopes much steeper than the shallow slopes around a patera. Both paterae and tholi (the plural form of these terms) are generally smaller than 200 km in diameter (Table 41.1).

Former paterae in the Tharsis region were recently renamed either Tholus or Mons. Uranius Mons (26 in Figure 41.1) is now recognized by orbital elevation measurements to be considerably larger than the previously named Uranius Patera, which had been applied to features that are now seen to be at the summit of a much broader regional rise that is part of the volcano structure. Tholi are very abundant in the Tharsis region (8, 9, 14, 23, 25, and 27 in Figure 41.1). Slopes on tholi tend to be steeper than those of the large Tharsis shield volcanoes, but the slopes are typically not as steep as those on Elysium Mons. Ceraunius Tholus (9 in Figure 41.1) has sinuous channels carved into the flanks of the volcano, implying that significant effusive (and erosive) flows occurred after the bulk of the construct had formed (Figure 41.3). Some early researchers concluded that such erosive channels on tholi and paterae flanks may be due to pyroclastic flows, while others prefer to interpret the channels to be the result of concentrated fluvial activity that postdates the volcanic activity.

The Elysium region includes two tholi in addition to Elysium Mons (2, 11, and 13 in Figure 41.1); this group represents a substantially lower areal density of volcanic centers than the numerous constructs in the Tharsis region, but Elysium is the second highest volcanic rise on Mars. Hecates Tholus (13 in Figure 41.1) has numerous channels on its flanks and portions of the summit area with few

FIGURE 41.3   Ceraunius (bottom) and Uranius (top) Tholi. Channels have been eroded into the flanks of Ceraunius Tholus, with a possible "lava delta" at the mouth of one channel, emplaced within an elliptical impact crater north of the volcano. Portion of THEMIS Daytime IR mosaic. NASA/JPL-Caltech/ASU.

FIGURE 41.4   Tyrrhenus Mons, a low-profile volcano that likely includes pyroclastic deposits on its flanks. Portion of THEMIS Daytime IR mosaic. NASA/JPL-Caltech/ASU.

preserved impact craters; both characteristics have been interpreted to be the result of pyroclastic activity late in the history of this volcano. Meroe and Nili Paterae (15 and 16 in Figure 41.1) are near the summit of a broad 4-km-high rise that is associated with the classical low-albedo region named Syrtis Major, where remote sensing reveals the presence of both olivine and pyroxene. Apollinaris Mons and Zephyria Tholus (5 and 28 in Figure 41.1) are located on the broad boundary between the southern cratered highlands and the northern lowland plains; both volcanoes are not too far from the landing sites of the Spirit and Curiosity rovers. Six volcanic centers (including paterae, tholi, and mons) have been identified within the intensely cratered (old) southern highlands (3, 4, 12, 19, 20, and 24 in Figure 41.1), so volcanic activity was significant through essentially all of the history of Mars.

5. HELLAS HIGHLAND VOLCANOES

The cratered highlands surrounding the Hellas basin in the southern hemisphere of Mars exhibit a variety of volcanic landforms, most notably ancient, eroded volcanoes (formerly called highland paterae) and vast plains

containing wrinkle ridges that cover low-lying regions in rugged, cratered terrain. Hadriacus Mons and Tyrrhenus Mons (12 and 24 in Figure 41.1) are located northeast of the Hellas basin. Tyrrhenus Mons (Figure 41.4) is found within the ridged plains of Hesperia Planum, interpreted in the early days of Martian exploration to consist of flood basalts analogous to the lunar maria. Hadriacus and Tyrrhenus Montes display central caldera complexes surrounded by relatively flat-lying layered deposits that are heavily dissected by radiating valleys. Recent high-resolution images confirm earlier suggestions that the flanks of these volcanoes are friable in nature and dissected by fluvial processes, although the exact styles and rates of erosion are yet to be determined. Formation of the flanks of these volcanoes has been attributed to large explosive eruptions in Late Noachian (Tyrrhenus) and Early Hesperian (Hadriacus) time that would have emplaced pyroclastic flow deposits in thick sequences around the eruptive vents for hundreds of kilometers. Later effusive volcanic activity may have occurred at the summit regions of both volcanoes and also formed a large field of lava flows that extend from Tyrrhenus Mons to the southwest.

To the south and southwest of the Hellas basin, the ridged plains of Malea Planum bury the basin rim and cover the cratered highlands. Similar in surface morphology and extent to Hesperia Planum, Malea Planum is attributed to voluminous flood volcanism. Within Malea Planum, four major volcanic features have been identified. Amphitrites Patera (3 in Figure 41.1) is the summit depression of a shield-like edifice similar to Tyrrhenus and Hadriacus Montes, with radiating channels characterizing its flank materials. However, in the case of Amphitrites Patera, its surface shows scalloped and pitted textures, and pedestal and ejecta flow craters attributed to periglacial modification of unconsolidated volcanic materials, perhaps similar to the pyroclastic deposits forming Tyrrhenus and Hadriacus Montes. Malea, Peneus, and Pityusa Paterae are large, caldera-like depressions within the ridged plains that also show evidence of surface modification by periglacial activity. Malea and Pityusa Paterae have been compared to large calderas on Earth such as Yellowstone, and have potentially been the source regions for large quantities of lava flows and/or pyroclastic deposits. The volcanic features within Malea Planum are similar in age to those found to the northeast of Hellas, suggesting that this region of Mars was a major volcanic center in ancient times.

6. SMALL CONSTRUCTS

Much knowledge has been gained about small volcanic constructs across Mars from the study of data collected during the post-Viking era. Although Mariner and Viking data revealed tremendous insights into the construction of large volcanoes on Mars, few volcanoes were conclusively identified with diameters of tens of kilometers. However, it was speculated by many researchers that fields of small constructs at this size should exist. Perhaps the most beneficial data to advance our knowledge of these features was the MOLA topography. Even the early, coarser gridded data products showed numerous structures with diameters of tens of kilometers and heights of several hundred meters. Similar structures on Earth have been named low shields due to their low profile and typically form due to dominantly effusive eruptions of lava from point or linear vents. The collection of higher-resolution image data sets over more extensive areas of Mars also reveals steeper-sided cones with diameters of up to several kilometers and heights of several hundred meters (Figure 41.5). Although apparently less common, these features are thought to represent slightly more explosive episodes that produced cones. A third group of small constructs includes fissures with little to no topographic relief, but which appear to link to extensive surface flow units.

Low shields and fissures are now known to be common features across the plains of the Tharsis province, and are also found in very close proximity to the larger central constructs in this region. Both the Tempe Terra and Syria Planum regions of Mars were long known to include several small constructs. Post-Viking era data show that small constructs are more plentiful in these regions than previously thought (Syria includes more than 200 small constructs), and that they are common in other portions of Tharsis, dominantly among and to the east of the Tharsis Montes, southeast of Olympus Mons, and south of Alba Mons. Although each vent represents an eruption point of lava at the surface that built a construct, or "topographic cap" several hundred meters in height and tens of

500 m

FIGURE 41.5 E. Mareotis Tholus, with a 2-km-long summit vent. A circular impact crater is southwest of the elliptical summit vent. Mars Orbiter Camera, MOC2-64 (image 50704fsub2). NASA/JPL-Caltech/MSSS.

kilometers in diameter, the flow fields associated with these features can extend for much greater distances. Where constructs are located in closely spaced groups, the flow fields coalesce to form extensive **volcanic plains**. Impact craters have been counted on the flanks of some of the larger low shields across the Tharsis province, and ages associated with these features place them in the Amazonian, among the youngest volcanic features in the region. However, some low shields are extensively buried, nearly completely, by plains lavas that are considered to be older, and both the Syria and Tempe vent fields are considered to be pre-Amazonian in age. As such, it is not clear if the development of fields of low shields and fissures represents a style of Martian volcanism that was dominant during the latter part of the planet's history or if this style of volcanism has occurred throughout the planet's evolution.

Post-Viking image data have also enabled the study of much smaller domes, cones, and mounds across Mars. These features tend to be hundreds of meters to several kilometers across and hundreds of meters in height. The cones can be generally divided into two groups based upon the interpretation of their formation, including: (1) volcanic cones (cinder and/or spatter), and (2) volcanic rootless constructs (VRCs). Several individual volcanic cones have been identified in high-resolution images across Mars, but only two groups have been identified and studied in more detail. These groups are located in western Tharsis and Utopia Planitia near the north pole of Mars. Ages for these two fields have been estimated at Amazonian and Late Noachian to Early Hesperian, respectively. Some of these features are associated with small lava flows, sometimes from breaches in the structure or from the base. The limited number of these fields suggests that this style of volcanism was never dominant on Mars, or that it was more common in the Noachian and much of the evidence has been buried by younger deposits. VRCs are thought to form when lava flows were emplaced across surfaces that were rich in subsurface ice deposits. As heat is transferred from the active lava through the basal crust, the frozen volatiles begin to melt and can form steam. If pressure is able to build up, local explosive events can occur that excavate the substrate and eject overlying lava to form piles or cones. These cones will be near circular if the lava flow is no longer moving, but can also be fork-shaped if the flow was advancing, as the rootless explosions took place and material was deposited onto a surface that transported it away as new explosions continued. Because these features do not mark the location where a rising magma body reaches the surface they are considered to be "rootless" or not directly connected to a magma body beneath it. Unlike the volcanic cones described previously in this paragraph, VRCs are not typically the source of effusive lava-flow deposits, are located in groups of hundreds to thousands (as opposed to tens of features), are found within a single extensive lava flow, and tend to be smaller in both diameter and height. Because VRCs require the presence of ice-rich materials, their presence and age can be used to infer information about a region's paleoenvironment. As such, it is critical to be able to accurately decipher volcanic cones from VRCs.

The third class of small volcanic constructs on Mars includes fractures or fissures to which channel networks are connected. This type of feature is identified in both Tharsis and Elysium and has been a source of debate in both regions as to their origin. The channel networks are typically sinuous, sometimes branching, and display features such as steep walls, wall terraces, a lack of topographic lava flow levees, and the branching channels can form terraced islands. Due to the similarity of these characteristics to terrestrial fluvial channel systems, some researchers have proposed an origin via the release of subsurface water, possibly by melting of subsurface ice due to interaction with ascending magma, and erosion by overland flow. However, as detailed image and topographic data become available for these features some researchers have suggested an origin associated with volcanic emplacement of sheet lavas. This remains a topic of current study. Yet, in

either style of suggested formation, these features likely represent a site where ascending magma caused the formation of a new feature on the surface of Mars.

7. VOLCANIC PLAINS

Volcanic plains comprise some of the most extensive geologic units on Mars outside the cratered highland terrain. The plains appeared to be quite featureless in moderate to low resolution images from the Mariner and Viking missions, but when viewed in higher resolution images from those missions and post-Viking-era missions a variety of features are observed. Post-Viking observations show that in some cases fields of small constructs, including low shields and low-relief fissures, are the sources for some of the extensive plains units. Although not all plains units have obvious sources, likely due to burial, it seems that the link between fields of small constructs and plains units is likely to have been common, thereby differentiating these two volcanic units from the larger central vent constructs. Perhaps the best example of this relationship is the Cerberus **Fossae** and associated Athabasca Valles lava flows in the Elysium region. Although the origin of Athabasca Valles is thought to involve magmatic melting of ground ice and water release from the Cerberus Fossae, high-resolution image data have shown that this same system was subsequently the site for eruption of flood lavas that flowed down the Athabasca Valles system and resurfaced the Cerberus Palus plains. These volcanic deposits are considered to be Late Amazonian in age, covering $>250,000$ km^2 with >5000 km^3 of lava. Comparable relationships between plains units and small constructs also exist in the Tharsis region and interpretation of the complex morphologies in these flow fields is a current research focus in Martian volcanology, particularly attempting to disentangle the complex relationships between possible fluvial and volcanic features.

The most obvious features in the volcanic plains away from the small constructs are abundant lobate margins surrounding individual packages of material that appear to have been emplaced through flow of fluid materials (Figure 41.6). The lobate margins can be sufficiently thick (various measurements indicate lobe thicknesses of tens of meters to greater than 100 m) that the flowing material is widely accepted to be lava rather than something like water-rich debris flows. Stacked sequences of such flow units likely make up the bulk of the topographic bulges that surround both the Tharsis and Elysium volcanic centers.

Volcanic plains also comprise regionally extensive units in locations that presently do not include major or small constructs. In particular, the plains of the Lunae Planum area east of the Tharsis region and the Hesperia Planum plains northeast of the Hellas basin extend over many thousands of kilometers. These plains units display wrinkle

FIGURE 41.6 Lava flows west of Arsia Mons. Portion of THEMIS VIS frame V46368003. NASA/JPL-Caltech/ASU.

2 km

ridge morphologies that are less abundant or absent from other plains deposits surrounding the Tharsis and Elysium volcanic centers. Both the Lunae Planum and Hesperia Planum volcanic plains have crater densities that suggest they were emplaced during the Hesperian. As such, these plains are significantly older than some of the younger volcanic plains units in and around the Tharsis and Elysium volcanic centers. Thus, substantial outpourings of plains-forming lava flows, likely from vents that are themselves buried by thick stacks of flows, were significant events during the Hesperian and Amazonian periods.

The volcanic plains of Mars display flow features that can be broadly divided into simple and complex flow units. This distinction follows from an extension of George Walker's classification of terrestrial lava flows by the cooling units preserved in the volcanic sequence, with simple flows representing a single cooling unit of great extent (and probable large volume of effusion) and complex flows representing an intermixed sequence of discrete flow lobes that each comprises individual cooling units. Some plains units are nearly featureless in terms of detail visible in even the highest-resolution orbital images, and many of these plains units may eventually turn out to be volcanic in

origin, even though at present they are not readily placed within the simple or complex designation. Many of the volcanic plains that fill the floors of impact basins (like Hellas) or that lack clearly resolvable individual flows over wide areas are considered to be simple in the sense that volcanic emplacement most likely was sufficiently rapid or prolonged to allow the plains to be considered as representing one major cooling unit. The younger volcanic plains around Tharsis and Elysium are primarily complex (Figure 41.6), with numerous finger-like flow lobes traceable in some cases for many hundreds of kilometers. Continued examination of higher resolution image data might show that the older plains display comparable complex morphologies, or these units might be so old that these morphologies are obscured by erosion or burial by sediments.

8. MEDUSAE FOSSAE FORMATION

Some plains units are quite problematic in origin. Such units tend either to be relatively featureless or to have such complicated surface exposures that their origin remains the subject of considerable controversy. The Medusae Fossae Formation (MFF) is the largest such enigmatic deposit that overlies both the northern lowland plains units and the transitional terrains along the margin of the old southern cratered highlands. Spread along the Martian equator south of both the Tharsis and Elysium regions, MFF materials occur within a region that spans more than 100° of longitude (nearly 6000 km). MOLA data show that MFF materials range in thickness from tens of meters in western exposures to more than 3 km in eastern exposures, while orbiting sounding radar investigations reveal that the MFF deposits overlie units conformable with both lowland plains and cratered highlands.

Many alternative hypotheses have been proposed for MFF, but the explanation most consistent with all currently available data is that MFF materials are the result of massive pyroclastic eruptions. MFF exposures consist of layered materials that appear to be very friable, eroded by wind-driven sand into enormous field of **yardangs**, which locally include caprock layers that help to preserve the underlying, more easily eroded materials. MFF materials are clearly easily eroded by windblown sand, but the eroded surface still maintain steep slopes, suggesting that at least some lithification has likely occurred within the materials. These characteristics have been compared to welded and nonwelded zones within many ignimbrite deposits on Earth. A significant issue with the pyroclastic hypothesis is that there is no indication of the possible source vent or vents, which is troubling if explosive eruptions produced more than 1 million km^3 of deposited materials.

The radar "Stealth" region, discussed earlier, has been interpreted to be associated with pyroclastic eruptions from undisclosed vents west of Arsia Mons, which lends support to possible massive ignimbrite eruptions in and around the MFF region. Unfortunately, MFF is within an enormous section of the Martian surface where infrared mapping reveals that a dust mantle, perhaps at least 1 m in thickness, completely masks the underlying rocks, thus effectively hiding MFF materials from remote sensing that could provide some indication of the composition of these friable deposits. It is possible that the Curiosity rover may eventually shed some light on the characteristics of layered materials that might be outliers of nearby MFF materials.

9. COMPOSITIONAL CONSTRAINTS

Compositional information for Mars is available from three different sources: remote sensing data (orbiting spacecraft instruments and Earth-based telescopes), on-site analyses by instruments on landed spacecraft, and the special group of meteorites for which there is compelling evidence that these rocks were blasted off the surface of Mars and eventually fell to Earth. Here we will briefly examine each data type for its implications for volcanic materials on Mars.

Remote sensing studies of Mars began with telescopic measurements of properties of the surface and atmosphere, and since the 1960s, the telescopic data have been augmented with much higher spatial resolution data from numerous instruments on spacecraft that flew by, orbited, or landed on Mars. Earth-based instruments can achieve very high spectral resolution but with only low spatial resolution (typically covering areas hundreds of kilometers across on Mars). Spacecraft have tended to achieve substantially improved spatial resolution, but until recently these data were at relatively limited spectral resolution. The Compact Reconnaissance Imaging Spectrometer for Mars can image the surface of Mars with up to 20-m spatial resolution while also collecting reflectance measurements at hundreds of visual and near infrared wavelengths for each picture element (pixel). The Thermal Emission Spectrometer extended spectral coverage in thermal wavelengths, resulting in global maps of the presence of several minerals across the planet, including olivine and pyroxene in several low-albedo (dark) regions. The primary results of these spectral studies with regard to volcanism are that two types of basalts seem to be common on Mars, a low-silica type similar to tholeittic basalts on Earth and a moderate-silica type that may either be more like terrestrial basaltic andesites or tholeiite-like basalt with a glassy weathering rind with enhanced silica content.

The ubiquitous Martian dust complicates spectral studies of Mars at all spatial and spectral resolutions, masking significant portions of the surface and generally decreasing the contrast of already subtle spectral features being sought. In spite of this dust handicap, spectral studies

have shown that the dark regions of the planet generally have a strong mafic affinity. Unfortunately, only the Syrtis Major volcanoes are located in a low-albedo region, so that the spectral information of nearly every other volcanic center is severely masked by dust. Only tens of microns of dust can strongly affect visual and near infrared reflectance, and only 2 cm of dust can totally obscure even competent bedrock from thermal infrared observations.

The first on-site compositional information for Martian materials came from an X-ray fluorescence instrument on the two Viking landers. In spite of the fact that the two Viking landers were more than 6000 km apart, the composition of Martian fine soils was quite similar at both landing sites. Measurements of the fine soil components at subsequent landing site continued this trend (Table 41.2), which is a strong indication of the homogenizing influence

TABLE 41.2 Chemical Compositions of Selected Martian Materials[1]

Oxide	Spirit Fines[2]	Pathfinder Fines[3]	Spirit Basalt[4]	Pathfinder Basalt[5]	Martian Basalt Meteorites[6]
SiO_2	46.2	49.0	45.7	55.5	49.0–51.4
Al_2O_3	10.2	8.4	10.9	9.1	4.8–12.0
FeO	15.6	16.1	18.8	13.1	17.7–21.4
MgO	8.5	7.9	10.8	5.9	3.7–11.0
CaO	6.4	6.3	7.8	6.6	10.0–11.0
K_2O	0.5	0.2	0.1	0.5	0.06–0.25
TiO_2	1.0	1.2	0.5	0.9	0.8–1.8
SO_3	6.3	5.4	1.2	3.9	0.33–0.80
Na_2O	3.0	3.0	2.4	1.7	1.0–2.2
P_2O_5	1.0	—	0.5	—	0.6–1.5
Cr_2O_3	0.3	—	0.6	—	0.014–0.30
MnO	0.3	—	0.4	—	0.45–0.53
Cl	0.8	0.5	0.2	0.6	0.005–0.013
Total	100.1	98.0	99.9	97.8	

[1]Results are rounded to nearest 0.1% for APXS measurements.
[2]MER-A Gusev mean soil; from Table 4.4, Brückner et al. (2008).
[3]Average of three fines samples (A2, A4, and A5) from Pathfinder; from Table I, Reider et al. (1997). Results are reported as normalized to 98.0%, to allow for unreported P_2O_5, Cr_2O_3, and MnO.
[4]Results for "Adirondack_RAT" from Spirit; from Table 4.1, Brückner et al. (2008).
[5]Results for "Yogi" (A7) from Pathfinder; from Table I, Reider et al. (1997). Results are reported as normalized to 98.0%, to allow for unreported P_2O_5, Cr_2O_3, and MnO.
[6]Basaltic Martian meteorites; from Table 4.4, Brückner et al. (2008).

of the global dust storms. Although repeated attempts were made to collect a rock fragment at the Viking landing sites, no Martian rock was successfully analyzed by the Viking lander instruments (centimeter-sized fragments collected by the sample arm all turned out to be indurated clods of fine soil and dust). This situation changed in July, 1997 when the Mars Pathfinder mission deployed the first mobile vehicle on Mars, the microwave oven-sized Sojourner rover. Sojourner had an alpha proton X-ray spectrometer (APXS) that the mobile rover was able to place directly onto several rocks, as well as the fine-grained materials between the rocks. Subsequently, both Mars Exploration Rovers (MERs) Spirit and Opportunity obtained many dozens of APXS measurements of rocks and soil, and the Mars Science Laboratory rover Curiosity is just beginning to build its own library of APXS measurements (as well as utilize the impressive complement of other instruments available on this latest Mars rover). Table 41.2 lists representative composition results for both fine soils and basalts from the Spirit and Pathfinder landing sites.

Images of the rocks at the Pathfinder site support them being interpreted as being volcanic in origin, with ubiquitous pits thought to be vesicles. Spirit and Opportunity have greatly expanded both imaging and chemistry results for what are definitely volcanic rocks, with several distinct "classes" of rock types identified to date. The high sulfur content of both rocks and fines on Mars (Table 41.2) suggests that a thin soil/dust coating may contaminate measurements of uncleaned rock surfaces. Spirit, Opportunity, and Curiosity all carry instruments that can excavate into rock surfaces in order to obtain fresh material for examination. The Spirit basalt values in Table 41.2 are for an APXS measurement that followed grinding on the surface of the rock Adirondack using the Rock Abrasion Tool (RAT); these results show sulfur and silica contents lower than either the soils or uncleaned Pathfinder targets. Chemistry results from post-RAT APXS measurements by the two MERs, or from drilled rock samples examined by Curiosity, are likely to be least affected by possible contamination from surface dust.

The best chemical information about Martian materials comes from a collection of very special meteorites. Over 100 meteorites are now considered to be from Mars because gases implanted in them during the shock that ejected them from their parent body are identical to the atmosphere of Mars as measured by the Viking landers, and unlike gases obtained from any other terrestrial or extraterrestrial sample. The vast majority of these meteorites are igneous in nature, and they are only about one-third the age (1.3 Ga−180 Ma) of practically all other types of meteorites. Collectively called **SNC meteorites**, the majority are basalts or lherzolites/harzburgites like Shergotty (S), some are clinopyroxenites or wehrlites like Nakhla (N), and Chassigny (C) is the sole dunite. ALH84001 is a special case requiring some additional discussion; this orthopyroxenite is the only Mars meteorite that is as old as the solar system ($\sim$ 4.5 Ga), with veins of carbonate in which are controversial features that may (or may not) be evidence of primitive life from early in Martian history.

The SNC meteorites have told us much about Mars in general, but we unfortunately have no evidence for exactly where on the Martian surface they came from. The abundance of volcanic rocks among the group is consistent with the extensive areas of volcanic plains present on Mars, but their relatively young crystallization ages cannot be used to calibrate the geologic epochs on Mars without knowledge of which specific geologic unit they came from. Some of the nonbasaltic rocks could have come either from slowly cooled cores of thick lava flows where limited fractionation may have taken place, or possibly even from a plutonic body breached by the impact event that ejected the meteorite. ALH84001 likely came from somewhere in the cratered southern highlands of Mars, indicating that at least portions of the highlands must date from near the formation of the planet. There is no good explanation why the highlands, which cover more than half of the surface of Mars, have thus far produced only ALH84001 as an old Mars meteorite. It will require the return of documented samples from Mars before the precise laboratory techniques that are brought to bear on the Mars meteorites can be related to specific geologic materials or units.

10. VOLCANIC HISTORY OF MARS

The oldest terrain on Mars is the Noachian densely cratered southern highlands. If the meteorite ALH84001 is representative of this era, at least some magmatic activity took place during this time of intensive impact events. Isolated volcanic centers developed within the cratered highlands, preserved today only as deeply scoured features scattered throughout the southern highlands. The Noachian materials give way to Hesperian terrains, where massive eruptions of lava produced volcanic plains that covered large portions of the Martian surface. The composition of these plains-forming materials was likely basaltic, although it is possible that more evolved lavas (basaltic andesite) also may have been common in places. Volcanic centers developed on the rim of the enormous Hellas impact basin, as well as within what is now the Syrtis Major region. These volcanoes appear to have involved considerable amounts of ash production, leading to very low overall profiles, as well as intense channelization of the shallow flanks by either fluvial or magmatic liquids. If the massive MFF deposits are the result of pyroclastic eruptions, then ash production was significant from the Late Hesperian well into the Amazonian. Alba Mons (1 in Figure 41.1) may represent a volcano caught in the transition from more ash-rich eruptions typical of the highlands to the more lava-rich eruptions associated with volcanic constructs in the northern hemisphere of Mars.

Late Hesperian to Amazonian epochs involved voluminous eruptions that produced both broad volcanic plains and numerous central volcanic constructs. Volcanic activity concentrated around the Elysium and the much larger Tharsis regions; impact crater densities suggest that effusive activity was most prolonged in the Tharsis region, where some volcanic surfaces are very sparsely cratered and thus may be quite young. Thousands of low-relief craters and domes in the northern lowland plains may be the result of either localized Strombolian activity (cinder cones) or of interaction of lava flows moving over wet sediments (**pseudocraters**). An ignimbrite origin for MFF materials would imply that pyroclastic activity continued well into the Amazonian.

11. FUTURE STUDIES

The recent rover missions have demonstrated the enormous advantage that mobility provides to the exploration of Mars, particularly when supported by consistently improved orbital capabilities. A staggering amount of data have been collected using the diverse instruments on both the rovers and several orbital platforms, so it seems quite likely that the ongoing comparison of results obtained from all of these recent and ongoing missions should provide new insights into volcanism on Mars during the coming years. The question of the abundance and origin of potential pyroclastic deposits on Mars remains unclear at present, but the ongoing orbiter and rover missions may help to shed light on this issue.

The absolute age of all Martian terrains remains a significant uncertainty in studies of the emplacement of volcanic materials on Mars. An improved understanding of the volcanic structures and deposits on Mars will also provide better constraints on the thermal evolution of the planet. The recent identification of possible ancient explosive volcanic centers in the Arabia Terra region suggests that a reinterpretation of some large (>100 km in diameter) depressions throughout the southern highlands may show some presumed impact features may turn out to be calderas, adding additional information about the transition from explosive to effusive eruptions on Mars.

Martian meteorites will continue to provide valuable new insights into the history of Mars, but unfortunately there has yet to be a definitive link between one or more meteorites and a specific bedrock location on Mars. The return of documented samples from diverse localities on Mars would be the most definitive way to answer these and other outstanding questions about Martian volcanism.

FURTHER READING

Brückner, J., Dreibus, G., Gellert, R., Squyres, S.W., Wänke, H., Yen, A., Zipfel, J., 2008. Mars Exploration Rovers: chemical composition by the APXS. In: Bell, J. (Ed.), The Martian Surface: Composition, Mineralogy, and Physical Properties. Cambridge University Press, pp. 58–101.

Carr, M.H., 2006. The Surface of Mars, Volcanism (Chapter 3), pp. 43–76. Cambridge Univ. Press, Cambridge.

Crown, D.A., Bleamaster, L.F., Mest, S.C., 2005. Styles and timing of volatile-driven activity in the eastern Hellas region of Mars. Journal of Geophysical Research 110, E12S22. http://dx.doi.org/10.1029/2005JE002496.

Crown, D.A., Greeley, R., 2007. Geologic Map of MTM -30262 and -30267 Quadrangles, Hadriaca Patera Region of Mars. U.S. Geological Survey Scientific Investigations Series Map 2936, scale 1:1,004,000.

Franics, P., 1994. Volcanoes: A Planetary Perspective. Oxford Univ. Press, New York, pp. 416–425.

Lopes, R.M.C., Gregg, T.K.P., 2004. Volcanic Worlds: Exploring the Solar System's Volcanoes. Springer/Praxis Publishing, Chichester, UK.

McSween Jr, H.Y., 1994. What we have learned about Mars from SNC meteorites. Meteoritics 29, 757–779.

McSween Jr, H.Y., 2008. Martian meteorites as crustal samples. In: Bell, J. (Ed.), The Martian Surface: Composition, Mineralogy, and Physical Properties. Cambridge University Press, pp. 383–395.

Michalski, J.R., Bleacher, J.E. Supervolcanoes within an ancient volcanic province in Arabia Terra, Mars. Nature. doi: 10.1038/nature12482, in press.

Plescia, J., 2004. Morphometric properties of Martian volcanoes. Journal of Geophysical Research 109, E03003. http://dx.doi.org/10.1029/JE002031.

Reider, R., Economou, T., Wanke, H., Turkevich, A., Crisp, J., Bruckner, J., Dreibus, G., McSween Jr., H.Y., 1997. The chemical composition of Martian soil and rocks returned by the mobile Alpha Proton X-ray Spectrometer: preliminary results from the X-ray mode. Science 278, 1771–1776.

Scott, D.H., Tanaka, K.L., Greeley, R., Guest, J.E., 1986. Geologic Maps of the Western Equatorial, Eastern Equatorial and Polar Regions of Mars. Miscellaneous Investigations Series Maps I-1802-A, B, C. U.S. Geological Survey, scale 1:15,000,000.

Smith, D.E., et al., 1999. The global topography of Mars and implications for surface evolution. Science 284, 1495–1503. http://dx.doi.org/10.1126/science.284.5419.1495.

Tanaka, K.L., et al., 2014. Geologic map of Mars. Scientific Investigations Map 3292. U.S. Geological Survey, scale 1:20,000,000.

Taylor, G.J, et al., 2009. Implications of observed primary lithologies. In: Bell, J. (Ed.), The Martian Surface: Composition, Mineralogy, and Physical Properties. Cambridge University Press, pp. 501–518.

Williams, D.A., et al., 2009. The Circum-Hellas volcanic province, Mars: overview. Planetary and Space Science 57, 895–916. http://dx.doi.org/10.106/j.pss.2008.08.010.

Chapter 42

Volcanism on Venus

Mikhail A. Ivanov

Laboratory of Comparative Planetology, Vernadsky Institute, Russian Academy of Sciences, Moscow, Russia; Department of Earth, Environmental and Planetary Sciences, Brown University, Providence, RI, USA

Larry S. Crumpler

Collections and Research Department, New Mexico Museum of Natural History and Science, Albuquerque, NM, USA

Jayne C. Aubele

Education Department, New Mexico Museum of Natural History and Science, Albuquerque, NM, USA

James W. Head, III

Department of Earth, Environmental and Planetary Sciences, Brown University, Providence, RI, USA

Chapter Outline

GLOSSARY

arachnoid A planetary physiographic term for structures characterized by patterns of radial fractures and ridges extending outward for several radii.

canali A planetary physiographic term for long sinuous channels.

chasma A planetary physiographic term for a deep, linear valley with steep sides.

colles A planetary physiographic term for a cluster of small hills or volcanoes.

corona A planetary physiographic term for a circular or elliptical structure outlined by an annulus of concentric ridges or fractures; numerous volcanic centers and tectonic landforms usually occupy the corona interior.

farrum A planetary physiographic term for a flat-topped and steep-sided volcanic dome.

fluctus A planetary physiographic term for an extremely large lava flow field.

fluted dome A type of steep-sided dome (farrum) with ridged or scalloped margins.

magmatic center Refers to any magmatic center in which structural deformation dominates and eruptive morphologies are subsidiary.

patera A planetary physiographic term for a shallow irregular depression.

radar backscatter Characterizes the strength of a radar signal returned from the surface. The lower the backscatter, the darker the surface appears in the images and vice versa.

radial fracture center (nova) A magmatic center in which a pattern of fractures radiates from a common center; annular or circular fractures may be present but are subordinate.

shield clusters Localized concentrations of small (less than 20 km), dominantly shield- to cone-shaped volcanoes.

tessera An elevated and complexly fractured terrain characterized by a mosaiclike pattern resulting from pervasive tectonic structures. **Tessera** massifs are identified as the oldest preserved surfaces on Venus.

The Encyclopedia of Volcanoes. http://dx.doi.org/10.1016/B978-0-12-385938-9.00042-0

1. DEFINITION AND SIGNIFICANCE OF VENUS VOLCANISM

Several specific first-order features of Venus indicate that this planet has great importance in contributing to our understanding of planetary volcanism.

1. Venus is almost as large as Earth and has both thermal and compositional potential for long-lasting volcanic activity, in contrast to the smaller terrestrial bodies (e.g., Moon and Mars), where conductive heat transfer dominates, lithospheres thicken relatively quickly, and volcanic activity is sparse in the last third of solar system history (Head and Solomon, 1981). Comparison of the spatial and temporal patterns of volcanism on Venus and Earth can place important constraints on the geological evolution of the larger terrestrial planets, particularly in relation to early history.

2. Extrusive volcanic materials make up about 80% of the surface of Venus (Ivanov and Head, 2011) and there is a rich variety of volcanic landforms on Venus ranging from small (several kilometers in diameter) volcanic constructs to morphologically homogenous volcanic plains units thousands of kilometers in extent.

3. Atmospheric conditions on Venus preclude the existence of liquid water near the surface due to high temperatures and suppress wind activity. Both these factors enhance the preservation and analysis of volcanic, tectonic, and impact processes. The number of impact craters on the surface of Venus is small (Herrick et al., 1997; Schaber et al., 1998) and they commonly do not obscure landforms of volcanic and tectonic origin.

The style of volcanism on Venus differs from Earth. Heat transfer and volcanism on Earth is dominated by plate tectonics in which upwelling of mantle at spreading ridges and sinking of cold plates at subduction zones liberate much of the heat from the interior. Other planets, specifically Venus, appear not to be dominated by plate tectonics but by a style of volcanism characteristic of the intraplate areas on Earth. This style is referred to as "hot spot" in which heat is released advectively in isolated volcanoes distributed over the surface; the majority of the heat is currently lost by conduction through the lithosphere.

The great size and diversity of volcanic centers on Venus (Crumpler and Aubele, 2000) provide insights into the processes of large magmatic provinces on Earth, many characteristics of which are not recognized because they are not well preserved. Another advantage of studying volcanic phenomena of Venus is that it encourages volcanologists to consider a range of processes and mechanisms not readily apparent on Earth. For example, the very high atmospheric pressure on Venus (see the following "Basic Venus Data" section) inhibits gas exsolution, magma disruption, and pyroclastic volcanism (Head and Wilson, 1992). Another example is the radial fractures that are typical of many magmatic and volcanic centers on Venus.

These fractures may represent radial dike swarms, the only recognizable examples of which on Earth are deeply eroded.

Given the preservation of volcanic landforms in the dry atmosphere, the great abundance of examples, and consistent age relationships established among volcanic landforms at the global scale (Basilevsky and Head, 1998; Ivanov and Head, 2011), Venus offers an opportunity to understand better the details of emplacement of volcanic materials and evolution of styles of volcanism as a function of extended geologic time. This indicates the great importance of Venus volcanism in a comparative planetological context and can serve as one of the keys to understanding terrestrial volcanic landforms.

Basic Venus Data: Despite the large-scale similarity to Earth, surface conditions on Venus differ greatly from Earth. This has potential influence on the eruption styles and morphology of volcanic landforms. The atmosphere consists mostly of carbon dioxide ($\sim$97%) and the atmospheric surface pressure is nearly 100 times that of Earth (or equivalent to pressures at about 1-km depth in the sea). The surface temperatures are hot enough ($\sim$770 K) to potentially increase the time of cooling of magmatic bodies in near subsurface and lava flows on the surface. The dense, opaque atmosphere, capped with clouds requires a special, radar-based technique for the assessment of the geologic details of volcanism on Venus. Venus is the only major planet that does not have a satellite. Its rotation is retrograde, and a single day is 243 Earth-days long. Also unlike Earth, there is no evidence for a magnetic field, a possible consequence of slow rotation on the circulation in the liquid portion of the core necessary to excite a magnetic dynamo in the core, or fundamental differences in the core composition or size.

2. VOLCANIC LANDFORMS

Surveys and analysis of Venera 15/16 and Magellan data identified numerous morphologically varied volcanic landforms ranging from small shields less than 1 km in diameter to extensive lava plains, occurrences of which can cover several millions of square kilometers.

Several specific types of volcanic features have been identified on Venus (Head et al., 1992; Guest et al., 1992; Crumpler et al., 1993; Crumpler and Aubele, 2000; Ivanov and Head, 2013). (1) Features in which effusive characteristics dominate include edifices resembling terrestrial shield volcanoes and low to steep-sided volcanoes that resemble terrestrial volcanic domes, and very long lava channels. (2) Magmatic features that are dominated by tectonic structures with relatively minor effusive characteristics include **coronae**, calderas, **arachnoids**, and novae. Many of these structurally defined annular features are thought to represent the surface manifestation of magma reservoirs or diapiric upwellings.

TABLE 42.1 Areas of the Main Volcanic and Tectonized Units on Venus

Unit	Area, 10^6 km^2	Percent of Venus Surface
Volcanic Units		
psh	84.5	18.5
rp$_1$	150.7	33.0
rp$_2$	44.8	9.8
pl	40.3	8.8
ps	10.3	2.3
Total	330.6	72.4
Tectonized Units Predating psh (Exposed Area)		
t	35.4	7.8
pdl	7.2	1.6
pr	9.6	2.1
gb	39.5	8.7
Total	92.8	20.4
Tectonized Units Synchronous to pl (True Area)		
rz	24.5	5.4

2.1. Volcanic Plains

Volcanic plains are the most abundant and most volumetrically important volcanic features on Venus (Table 42.1) and play a key role in establishing the global stratigraphic scheme of Venus. The abundance of secondary tectonic structures on the surface of the plains and the degree of their deformation define several types of volcanic plains (Figure 42.1).

Strongly deformed plains: Tectonic structures clearly dominate their surfaces and define the final morphology of the units. There are two types of strongly deformed plains.

Densely lineated plains (pdl, Atropos Formation, Figure 42.1(A)): Numerous densely packed, narrow (a few hundred meters wide), short (a few tens of kilometers), and parallel lineaments dissect the surface of this unit. If the lineaments are wide enough, they usually appear as fractures. As a rule, the lineaments are packed so densely that they completely erase the morphology of the precursor materials. In some occurrences of the plains, however, remnants of preexisting material (interpreted as lava plains) are visible between the lineaments. Furthermore, their relatively flat surfaces suggest a volcanic plains origin. Densely lineated plains occupy a small area of the global map, about 7.2×10^6 km^2 (1.6% of Venus, Table 42.1) and are observed as slightly elevated and usually small (tens of kilometers across) patches (Figure 42.2).

Ridged plains (pr, Lavinia Formation, Figure 42.1(B)): Materials of ridged plains have the morphology of lava plains that are deformed by broad (5- to 10-km wide) and long (several tens of kilometers) linear and curvilinear ridges. Often the ridges are collected into prominent ridge belts. Materials of ridged plains are interpreted to be volcanic plains deformed by contractional tectonic structures. Ridged plains and ridge belts occupy about 9.6×10^6 km^2 (2.1% of Venus, Table 42.1) and usually form elongated occurrences hundreds to a few thousands of kilometers long and many tens (to a few hundred) of kilometers wide (Figure 42.2).

Mildly tectonically modified plains show an original morphology that indicates their volcanic nature. There are three types of these plains.

Shield plains (psh, Akkruva Formation, Figure 42.1(C)): The characteristic features of shield plains (Aubele, 1995) are numerous, small (from a few kilometers up to ten kilometers across), shield-like mounds that are interpreted as volcanic edifices (Aubele and Slyuta, 1990; Head et al., 1992; Guest et al., 1992; Aubele, 1995). In many cases, the volcanoes occur close to each other and form clusters listed in the catalog of the volcanic landforms of Venus (Crumpler and Aubele, 2000). Within the plains, the individual shields occur in association with plains material of intermediate **radar backscatter** and may appear to form small clusters that give the plains a hilly appearance. The surface of the shields and the plains between them are morphologically smooth and sometimes deformed by wrinkle ridges and fractures/graben. These structures, however, do not significantly modify the original morphology of the plains. Shield plains cover a significant portion of the surface of Venus, about 84.5×10^6 km^2 or 18.5% (Table 42.1) and typically occur as more or less equidimensional outliers, several tens to hundreds of kilometers across.

The broad and homogeneous areal distribution of shield plains (Figure 42.2) is interrupted by several large areas where the occurrences of the plains are less abundant or absent. These regions correspond to the largest tessera regions (e.g., Fortuna), Lakshmi Planum, the largest chasmata (e.g., Parga Chasma), and some wide lowlands covered by regional plains (e.g., Sedna Planitia). The absence of the plains within the large tesserae and in Lakshmi Planum is consistent with the presence of thickened crust in these regions that may have served as rheological/density barriers inhibiting formation of the plains. Chasmata represent major zones of young extensional structures that may have destroyed occurrences of shield plains. The scarcity of occurrences of shield plains within the lowlands and stratigraphic relationships show that younger volcanic units buried shield plains in these regions (Ivanov and Head, 2004, 2011, 2013). In all seven landing sites, where the chemical analyses of the surface material were taken (Surkov, 1997), the composition of the rocks is consistent with varieties of terrestrial basalts. Analysis of the geology of the Venera-8 landing site has shown that

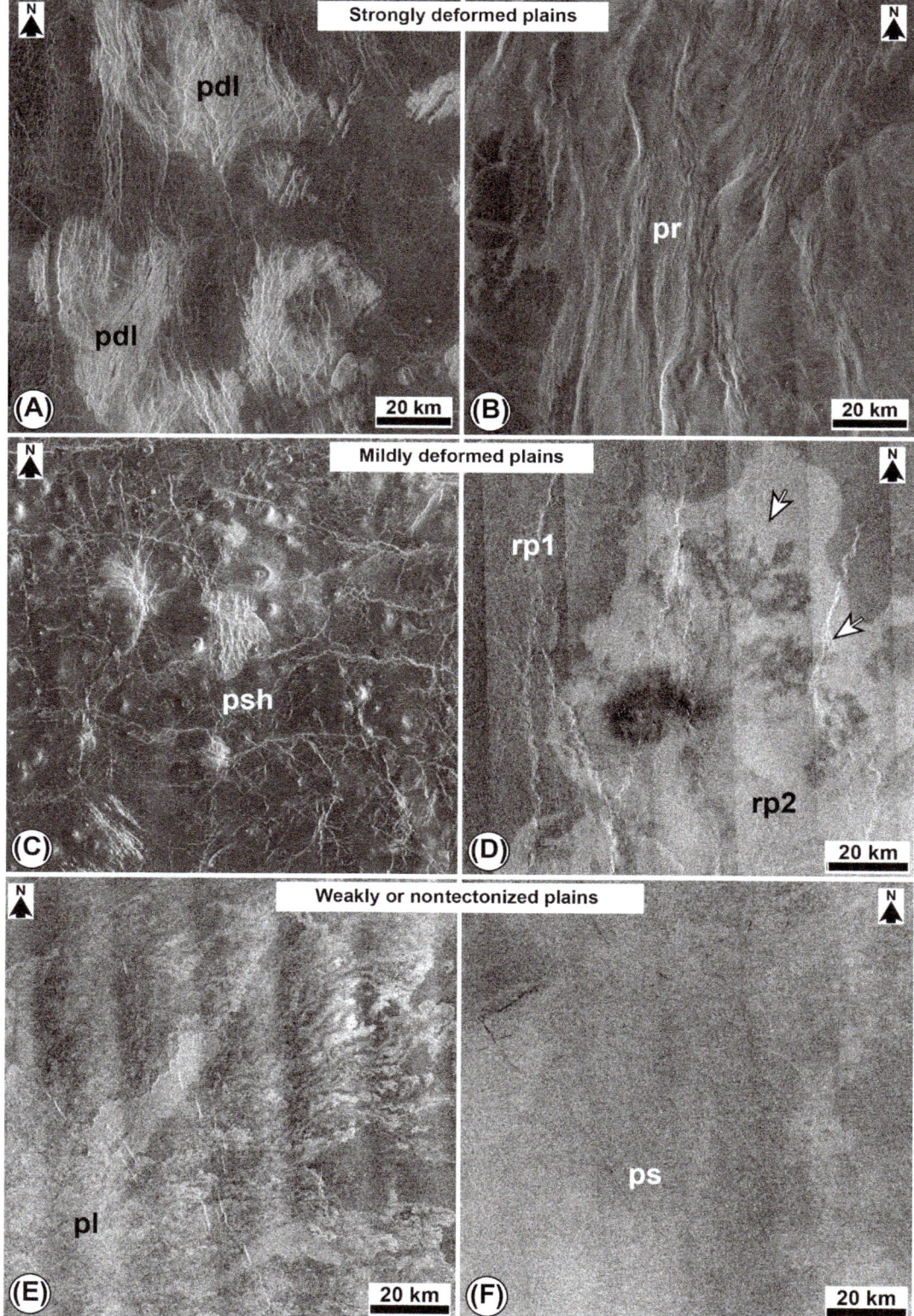

FIGURE 42.1   (A−D): Examples of volcanic plains of Venus. (A) Densely packed and narrow fractures dissect the surface of densely lineated plains (pdl). Center of the image is at 42.5°N, 33.2°E. (B) The surface of ridged plains (pr) is deformed by broad curvilinear ridges that are absent in the surrounding units. Center of the image is at 35.4°N, 157.3°E. (C) Numerous small shield-like mounds interpreted as volcanic constructs populate the surface of shield plains (psh). Center of the image is at 38.7°N, 117.3°E. (D) Two subunits of regional plains (rp$_1$ and rp$_2$) typically have different radar albedo that helps to distinguish these types of plains. The brightness of the lower subunit (rp$_1$) is moderate and uniform, and the surface of the upper subunit (rp$_2$) is brighter and, in places, flow-like features (arrows) are seen on the surface of the unit. Center of the image is at 13.1°N, 171.8°E. E and F: Examples of the weakly and nondeformed volcanic plains of Venus. (E) Numerous brighter and darker lava flows constitute lobate plains (pl). Center of the image is at 10.7°N, 263.9°E. (F) The surface of smooth plains (ps) appears homogeneous and has low to moderate radar brightness. Center of the image is at 3.5°N, 308.1°E.

the lander likely sampled the surface of shield plains (Abdrakhimov, 2005).

Regional Plains (rp): Regional plains represent the most widespread volcanic landforms on Venus (Figure 42.2) and are composed of morphologically smooth, homogeneous plains materials of intermediate-dark to intermediate-bright radar backscatter. Narrow wrinkle ridges form pervasive intersecting networks that cut the surface of the plains. The Venera 9, 10, and 13 and Vega 1 and 2 landers were probably landed on the surface of regional plains (Abdrakhimov, 2005).

Regional plains are subdivided into two subunits on the basis of their radar backscatter characteristics. **The lower subunit of regional plains (rp₁, Rusalka Formation,** Figure 42.1(D)) has a morphologically smooth surface with a homogeneous (locally mottled) and relatively low radar backscatter. These plains are the most abundant and ubiquitous units on Venus (about 150.7×10^6 km² or 33.0% of Venus, Table 42.1). Their extensive fields connect remote regions and can be traced almost continuously around the globe. A very important characteristic of the lower subunit

of regional plains is that lava fronts, volcanic edifices, and sources of the plains are commonly not observed at the resolution of Magellan data. In this respect, regional plains resemble extensive provinces of terrestrial plateau basalts. The lower subunit of regional plains preferentially makes up the floor of the lowlands surrounding the major tessera-bearing uplands (Figure 42.2) and occurs between elevated regions composed of the heavily tectonized units and shield plains.

The upper subunit of regional plains (rp₂, Ituana Formation, Figure 42.1(D)) has a noticeably higher (less frequently, nonuniform, flow-like) radar albedo than the lower subunit. The upper subunit covers about 44.8×10^6 km² or 9.8% of Venus (Table 42.1) and occurs usually as equidimensional or slightly elongated patches of flow-like shape from tens of kilometers to several hundred kilometers across. Occurrences of rp₂ surround some of large volcanic centers and form distal aprons of volcanic materials around them.

Weakly or non-tectonized plains: Their surfaces largely lack tectonic structures. There are two types of such plains.

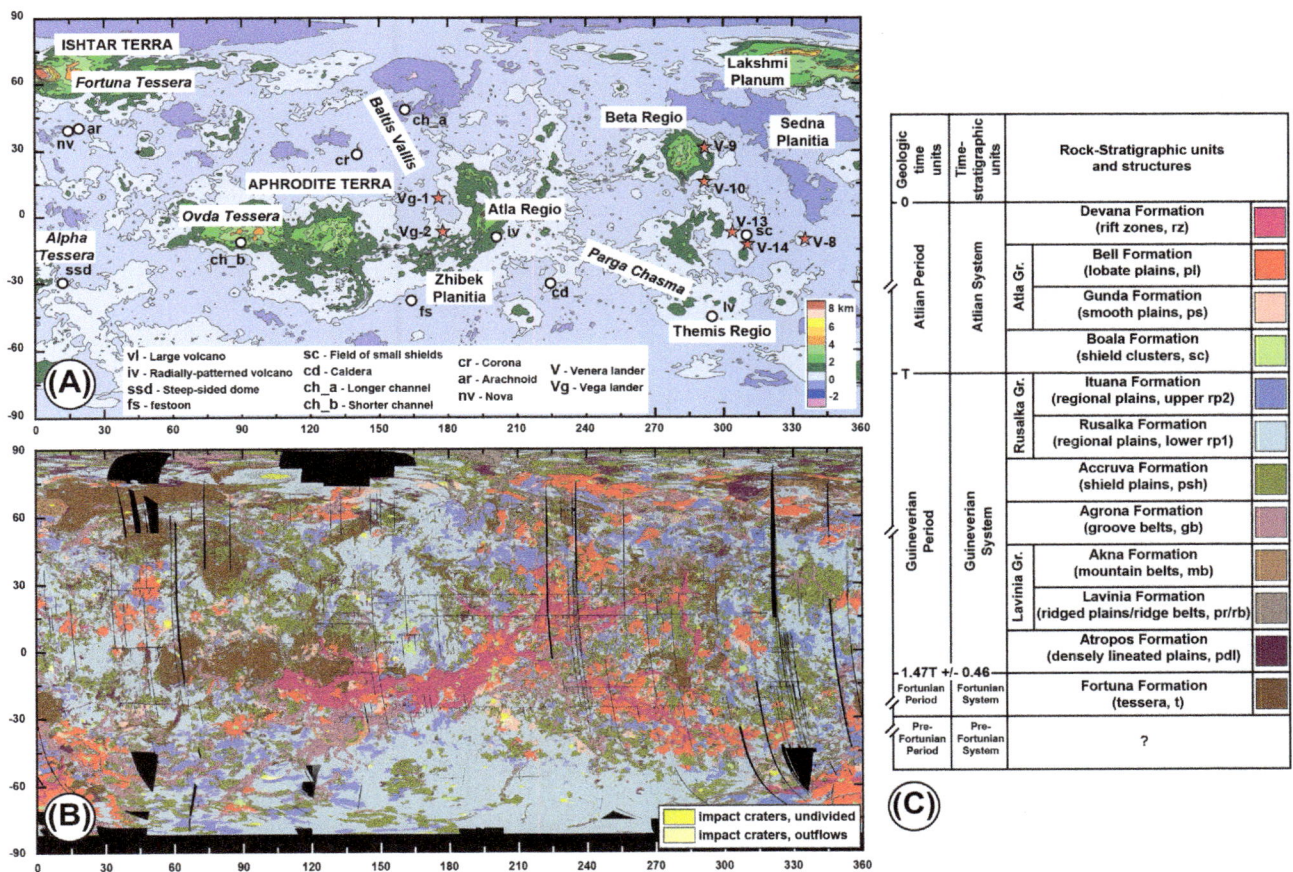

FIGURE 42.2   (A) Topographic map of Venus annotated with the locations of the major physiographic features that are mentioned in the text. Red stars represent landing sites of the Venera and Vega landers; white dots indicate sites with specific volcanic landforms shown in Figures 42.3–42.12. (B) The global geological map. (C) The stratigraphic column of Venus. All maps are in simple cylindrical projection. *(B) and (C) from Ivanov and Head (2011).*

Lobate plains (pl, Bell Formation, Figure 42.1(E)): Occurrences of lobate plains usually have morphologically smooth surfaces, which are occasionally disturbed by a few extensional features (fractures, graben). The most characteristic feature of lobate plains is their nonuniform albedo pattern consisting of numerous bright and dark flow-like features. The flows can be as long as several hundred kilometers and tens of kilometers wide. Lobate plains make up a significant portion of the surface of Venus, about 40.3×10^6 km^2 or 8.8% of Venus (Table 42.1) and their occurrences form distinct equidimensional fields from tens of kilometers up to many hundreds of kilometers across (Figure 42.2). All large complexes of lava flows (**fluctus**) are included in the lobate plains unit.

Lobate plains are usually associated with the large dome-shaped rises (e.g., Atla Regio, Figure 42.2). Large, high-standing, and plateau-shaped tessera regions (e.g., Fortuna, Ovda, Figure 42.2) lack significant occurrences of lobate plains. Within the BAT (Beta–Atla–Themis) region, lobate plains are spatially associated with rift zones. The Venera-14 lander is interpreted to have landed on the surface of one of the occurrences of lobate plains (Abdrakhimov, 2005).

Smooth plains (ps, Gunda Formation, Figure 42.1(F)): The material of smooth plains has a morphologically smooth, tectonically undisturbed, and featureless surface. Areas of smooth plains make up a small portion of the surface, about 10.3×10^6 km^2 or 2.3% of Venus (Table 42.1) and are usually characterized by a low radar backscatter cross-section and appear dark. There are three types of geological settings of smooth plains. (1) Near and within volcanic regions, where the plains are closely associated with fields of lobate plains. (2) Dark plains in spatial association with impact craters may represent remnants of dark parabolas formed due to impact events. (3) Patches of smooth plains within some large tessera regions likely have a volcanic origin.

2.2. Volcanic Centers

The following landforms are classed as volcanoes and are defined as centers of activity dominantly characterized by patterns of radial or circular extruded material.

Large volcanoes (Figure 42.3): These are volcanic centers (diameters $\geq$100 km) characterized by radially dispersed lava flows, frequently associated with concentric or circular central features, radial fracture patterns, and centered on a region of positive topography. The lateral dimensions are measured to the average distal end of the associated digitate lava flows. Complexes of lava flows on flanks of large volcanoes almost exclusively form occurrences of lobate plains. Only a few large volcanoes are associated with the upper subunit of regional plains. Although extensive laterally, most large volcanoes on Venus are relatively low in relief, $\sim$1.5-km high on

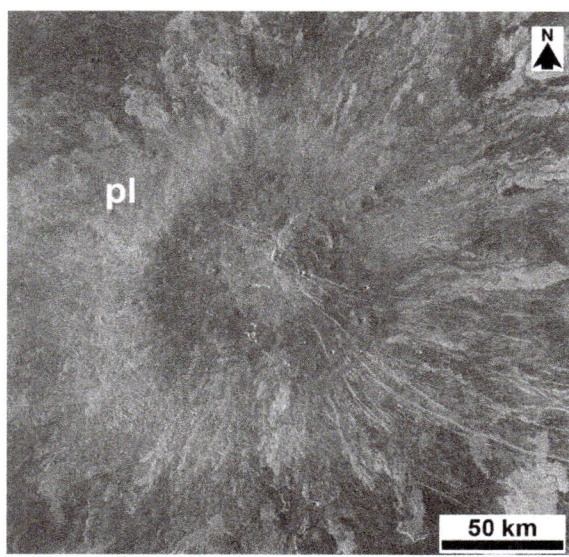

FIGURE 42.3 Chloris Mons, a 300-km diameter large volcano with numerous digitate bright and dark lava flows, radiating fractures, and several small volcanoes with steep-sided dome morphology near the summit. Center of the image is at 38.1°S, 163.9°E. See Figure 42.2(A) for location.

average. Large volcanoes occur preferentially in intermediate to higher elevations, particularly in broad rises and at the junctions of belts of extensive structures.

At the summit of some large volcanoes (e.g., Chloris Mons, Figure 42.3) there are clusters of small volcanoes, many with steep-sided morphology and central pits yielding an overall morphology analogous to many terrestrial shield volcanoes such as Etna or Mauna Kea. The summit region is relatively radar dark due to the presence of either ash or smoother lava flows. The circular shape of the dark area together with faint concentric bands that may be underlying fractures suggests that the dark region may lie within a former caldera-like summit depression.

Other large volcanoes are characterized by complex summit landforms, including multiple and large domical summits. Although the nature of domical volcanoes on Venus is not clear (see discussion of steep-sided intermediate volcanoes below), their presence on the summits of some large volcanoes suggests late-stage silicic volcanism or late-stage volatile-rich pyroclastic-dominated activity.

Intermediate volcanoes: Volcanic centers in the 20- to 100-km diameter range are considered "intermediate" between the diameter of small volcanoes and the more prominent large volcanoes. Several morphologic types of intermediate volcanoes have been identified and include "radially patterned volcanoes," "steep-sided domes," "scalloped domes," "modified domes," or **fluted domes**. Relatively simple volcanoes in this size range are analogous to simple large shield volcanoes.

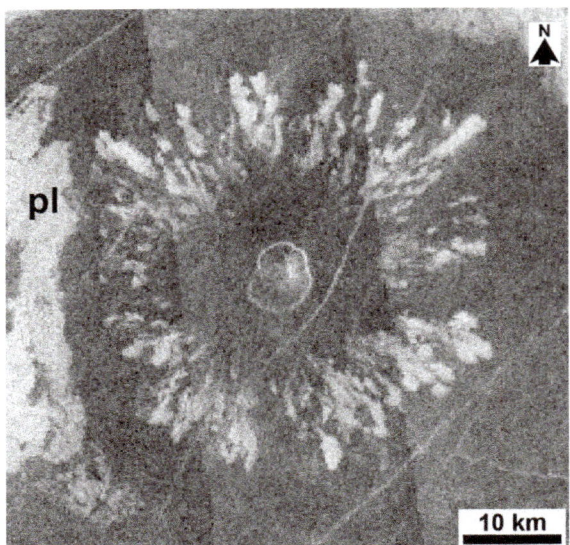

FIGURE 42.4 A simple intermediate volcano about 30 km in diameter with radial bright and dark lava flows and a summit caldera. Center of the image is at 9.6°S, 200.9°E. See Figure 42.2(A) for location.

Radially patterned volcanoes (Figure 42.4) are characterized by multiple radial digitate lava flows of about the same length. Similar to many large volcanoes, radially patterned volcanoes have distinct radial lava flows, typically radar-bright at their termini suggestive of increased roughness at the distal ends.

Steep-sided domes (Figure 42.5(A)) are among the more interesting and enigmatic intermediate volcanoes on Venus. They are also morphologically diverse, including domes with convex, flat, or concave summits. They frequently occur in clusters or appear to be overlapping. Many steep-sided domes on Venus have small summit pits with complexly patterned fractures on their surfaces. The circular form, typical of steep-sided volcanoes, appears to be the result of radial growth of a single extrusive mass, analogous to silicic domes on Earth. However, in size (typically 20–40 km and up to 80 km) and in terms of their remarkable circularity, they are unlike the relatively small and circular-shaped domes associated with silicic volcanism on Earth.

Several origins have been proposed for steep-sided domes. One possibility is that the domes represent lavas with higher silica content due to either fractional crystallization or remelting of crustal substratum or both. Another model suggests that basaltic magmas were oversaturated with volatiles upon eruption. At the high atmospheric pressure on Venus, exsolution of volatiles could result in formation of a froth that can increase the viscosity of lavas. The domes, however, preferentially occur topographically above the mean planetary radius and only a few domes are at lower topographic levels, where the atmospheric

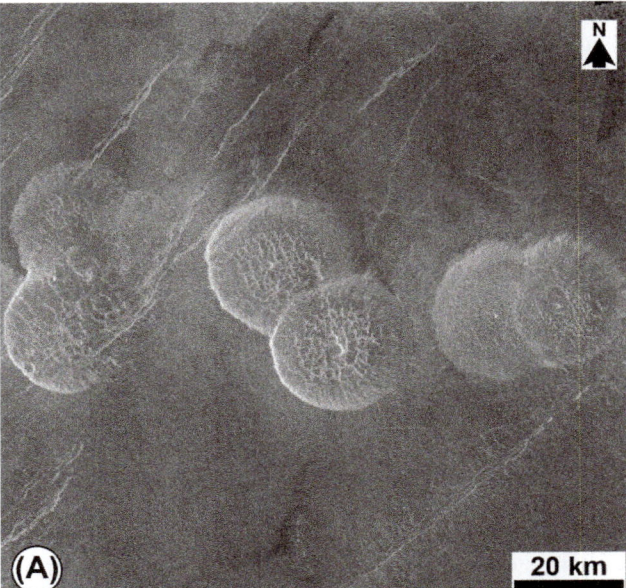

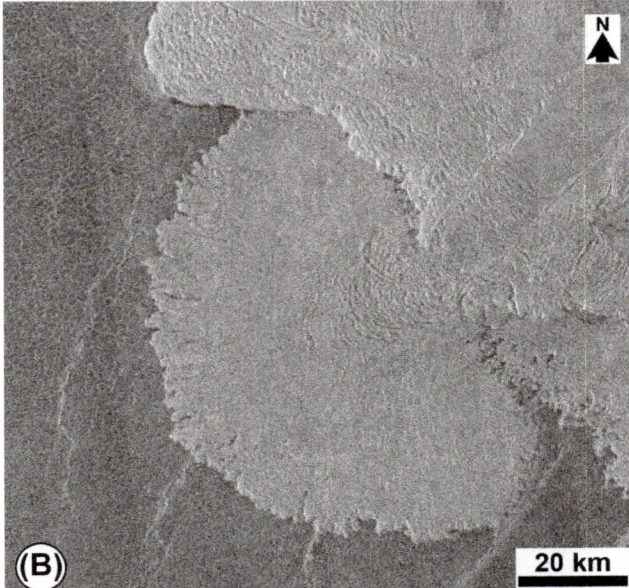

FIGURE 42.5 Examples of volcanic landforms that are related to eruption of lavas with higher apparent viscosity. (A) A group of steep-sided domes to the southeast of Alpha Tessera. The domes occur in pairs of overlapping volcanoes, the surface of which is characterized by radial and concentric fractures. Center of the image is at 29.7°S, 12.0°E. See Figure 42.2(A) for location. (B) The southwestern portion of a large complex of viscous flows (festoon) in Zhibek Planitia. The festoon was built by multiple eruptions of lavas that formed broad fan-shaped lobes with pressure ridges on top and steep-sided frontal scarp. Center of the image is at 38.1°S, 163.9°E. See Figure 42.2(A) for location.

pressure is the highest. Thus, the ambient atmospheric pressure may not be the main controlling factor of formation of steep-sided domes on Venus. Regardless of the mechanism, the widespread occurrence of steep-sided domes over a large range of scales implies that it is a common process on Venus. Both spatially and stratigraphically, steep-sided domes are associated with shield plains (Ivanov and Head, 2011). These associations suggest that the specific volcanic style of shield plains was also favorable for formation of the domes.

The edges of some circular steep-sided domes are scalloped or fluted ("scalloped," "modified," "fluted" domes) due to partial collapse of material along the dome flanks. There are several examples of steep-sided domes that predate nearby impact craters, yet their flanks are relatively undisturbed. This suggests that the gravitational instabilities that caused collapse of the dome edges are part of the emplacement process rather than a degradation of older domes.

Unusual lava flows (festoons) (Figure 42.5(B)) show transverse ridges, troughs, and other fluid-banding characteristics typical of viscous terrestrial lava flow. Three large, lobate, festooned, and thick lava flows occur at isolated points on Venus. The simplest interpretation based on their resemblance to viscous and silicic lavas on Earth is that they are unusually silicic in composition compared with most lava flows on Venus. However, as is the case for steep-sided domes, more mafic composition lavas with unusual viscosity characteristics arising from bubble-rich eruptions are also possible.

Small volcanoes and fields of small shield volcanoes (Figures 42.1(C), 42.6): Small volcanoes are the most abundant volcanoes on Venus and number in the hundreds of thousands. They exhibit many characteristics seen in intermediate and large volcanoes, including steep-sided

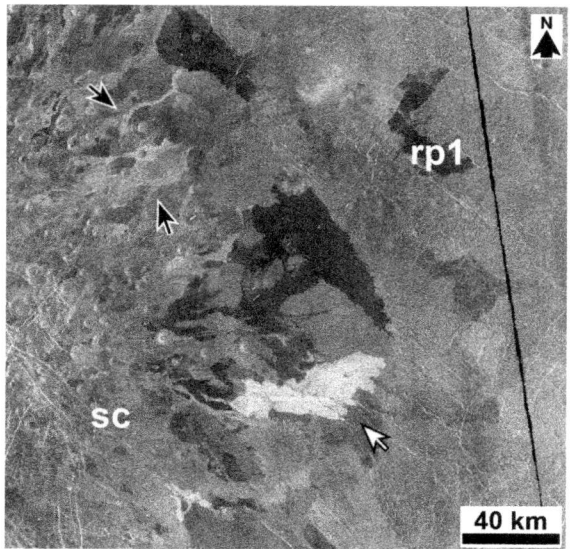

FIGURE 42.6   An example of a field of small shield volcanoes (shield cluster, sc). This volcanic landform is characterized by numerous small shield-like volcanic constructs similar to those in shield plains. In contrast to psh, the constructs of shield clusters often represent the sources of small lava flows (black arrow), some of which superpose the surface of regional plains (white arrows). Center of the image is at 8.9°S, 309.8°E. See Figure 42.2(A) for location.

domes, radial lava flows, and, the most abundant morphology, simple low shield volcanoes. Groups of small volcanic constructs were divided into three types (Crumpler et al., 1997): (1) simple, a cluster of randomly scattered edifices that characterize the surface of shield plains; (2) apron shield fields, clusters of small volcanoes spatially associated with features interpreted to be volcanic flows erupted by the small volcanoes; and (3) companion shield fields, spatially associated with another type of volcanic or tectonic center, usually large volcanic centers. Most of small volcanoes belong to the unit of shield plains (Figure 42.1(C)). Tight groups of small volcanoes (**shield clusters**) unrelated to shield plains occur on the flanks of large volcanoes, in rift zones, and within some coronae (Figure 42.6). These clusters are tectonically undeformed and their materials superpose structures of regional plains. Small shields of the clusters often appear as sources of small but distinct lava flows superimposed on the surrounding plains in contrast to edifices of shield plains. Analysis of the stratigraphic relationships of shield fields cataloged by Crumpler and Aubele (2000) reveals that about 10% of the population of the fields appears to be younger than that of regional plains (Ivanov and Head, 2004). Shield clusters cover a very small fraction of Venus, about 3.3×10^6 km^2 or 0.7% of Venus (Table 42.1).

Shield clusters clearly represent a concentration of numerous small volcanic vents over a localized source supplying magmas at rates insufficient to build a single edifice that can be compared to terrestrial volcanic fields with melt areas of limited extent and low magma rates delivered to the surface. The common occurrence of shield clusters at the summits/flanks of large volcanoes or within the interior of some coronae is consistent with the interpretation that these large hosting structures possessed localized and shallow reservoirs of magma.

Calderas (Figure 42.7): These structures represent topographic depressions that are characterized mainly by concentric patterns of enveloping fractures indicative of a depression formed through collapse of the surface. The most frequent calderas are 60−80 km in diameter with depths up to several kilometers (Figure 42.7). In many cases, the floor is dark and appears to be flooded or partially covered with later flows. Some calderas are the loci of intensely fractured concentric patterns bearing very little evidence of subsequent surface eruptions. This suggests that most of the room for collapse may have been accommodated by lateral withdrawal of magma.

2.3. Channels

Lava channels (Figure 42.8) are linear to sinuous parallel-sided valleys, which range in width from 0.5 to 1.5 km (up to 10 km) that may extend for many hundreds of kilometers. The longest channel on Venus, Baltis Vallis, is about

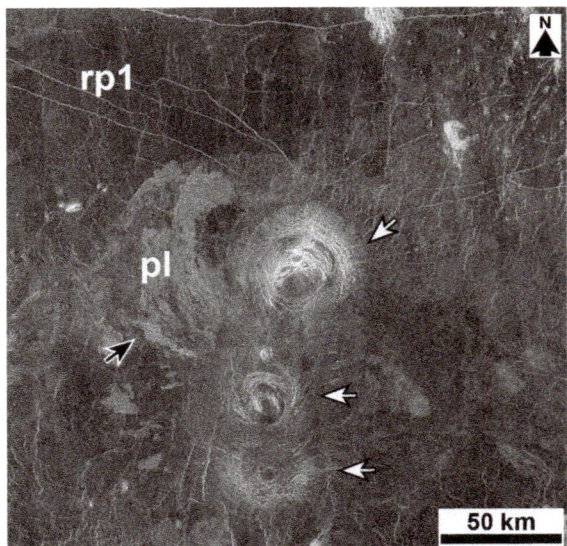

FIGURE 42.7 Examples of calderas (white arrows). Calderas are circular topographic depressions outlined by swarms of concentric fractures. Some of the calderas appear as the sources of lobate plains (black arrow). Center of the image is at 30.5°S, 224.5°E. See Figure 42.2(A) for location.

6800-km long. Several morphologically distinct types of channels have been identified. "Simple" channels are isolated, narrow channels within vast plains. "Complex" channels may have multiple channels, frequently with an anastomosing, braided, and even distributary morphology. "Compound" channels exhibit characteristics of both simple and complex channels in combination. "Integrated" channels are analogous to branching valleys such as those associated with sapping processes on Mars and Earth.

The longer channels ($> \sim 200$ km, Figure 42.8(A)) occur almost exclusively within the lower subunit of regional plains (rp_1). The shorter and more sinuous (rill-like) channels (Figure 42.8(B)) often occur in association with smooth and lobate plains. The great length and occurrence within extensive plains imply extremely fluid materials with viscosities as low as 0.1 Pa s. Potential channel-forming fluids with this viscosity include ultramafic compositions such as komatiites, high-Ti lunar-type basalts, carbonatites, and sulfur flows. The great width of most channels implies that they are not simply collapsed lava tubes. Because our current understanding of Venus excludes the possibility of significant amounts of water now or in the past, channels are interpreted not to be the water-cut features. The likely scenario of channel formation is that the dense atmosphere of Venus resulted in cooling of lavas to form discontinuous, platelike crusts over the fluid filling the channels, thus enabling the fluid to remain molten over great distances with relatively little cooling. Carbonatites and sulfur flows are attractive alternatives, especially due to the abundance of sources for carbon and sulfur on Venus.

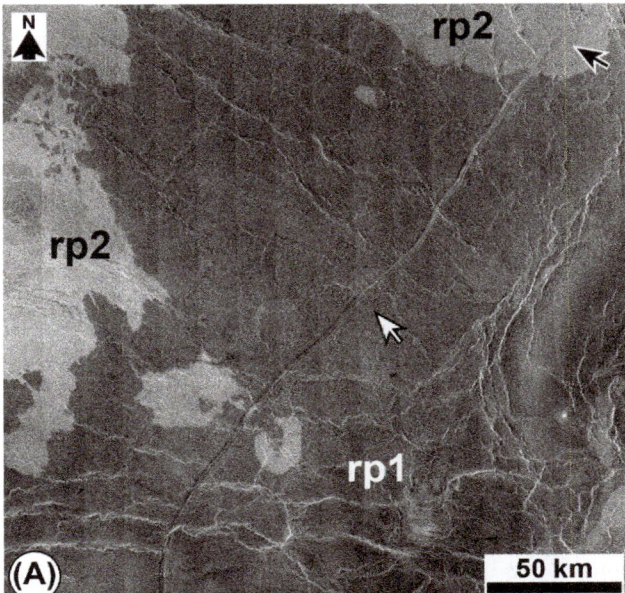

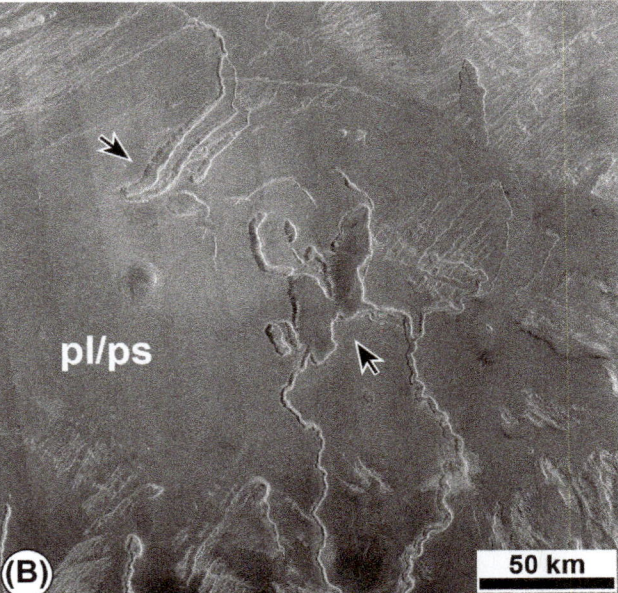

FIGURE 42.8 Examples of narrow sinuous channels. (A) A channel (northern portion of Baltis Vallis) that is associated with the lower subunit of regional plains. The channel cuts the surface of unit rp_1 and appears as a prominent topographic feature (white arrow). Material of unit rp_2 almost completely fills the channels (black arrow). Center of the image is at 48.1°N, 161.3°E. (B) A group of channels that cut the surface of the younger volcanic plains (pl/ps). The channels appear to be much more sinuous compared with those in regional plains and begin in steep-sided and flat-floored depressions (arrows). Center of the image is at 11.7°S, 89.6°E. See Figure 42.2(A) for location.

Another very important characteristic of channels is that they can serve as robust stratigraphic markers and that they record changes in elevation subsequent to their emplacement. Using Magellan altimetry data, topographic undulations with the characteristic wavelength of tens of kilometers were

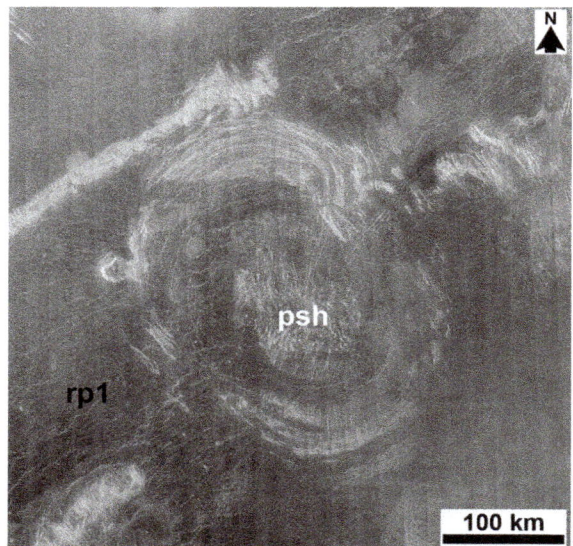

FIGURE 42.9 An example of a corona, Bona Corona, which is outlined by a prominent rim that consists of densely packed fractures. A cluster of small volcanoes (interpreted as an outcrop of shield plains, psh) occupies the corona center. Both the cluster of small volcanoes and the rim are embayed by materials of the lower subunit of regional plains, rp_1. Center of the image is at 27.3°N, 136.5°E. See Figure 42.2(A) for location.

detected along the length of many channels. At the regional scale (hundreds of kilometers), however, the longest channels (e.g., Baltis Vallis) occur at the deepest portion of regional topography. This topographic configuration of channels implies that although the short-wavelength topography changed, the regional-scale topographic pattern remained stable since the emplacement of channels.

2.4. Magmatic Structures

Many features on Venus are characterized more by surface deformation associated with subsurface magmatism on a large scale. In these features, volcanic edifices and lava flows may be relatively minor or subsidiary to tectonic structures. The magmatic structures include the classes of features known as coronae, arachnoids, and **radial fracture centers**.

Coronae are circular to elongate structures consisting of an annulus of fractures (sometimes, ridges) that surrounds topographically either positive or negative interiors with numerous volcanic and tectonic landforms (Figure 42.9). Some coronae possess a peripheral moat and many of them show numerous lava flows radiating away from the annulus; characteristic diameters of coronae are between 200 and 250 km.

Topographically, coronae exhibit a variety of forms, including domical rises, plateaus, plateaus with central depressions, and rimmed depressions. Because of their often circular shape, great size, and associated tectonics and volcanism, they have been widely interpreted as the surface manifestations of mantle plumes, their topography resulting from a combination of uplift of the surface over plume heads and subsequent gravitational relaxation.

Because coronae, arachnoids, and large volcanoes may represent localized areas of mantle upwelling, a map of the distribution of coronae and these features (Figure 42.10) provides clues to the general distribution of regions of mantle upwelling on Venus. Coronae are usually associated with extensive networks of groove belts and rift-like chasmata, suggesting that the lithospheric extension with fracturing and rifting may be closely allied with the processes of upwelling mantle.

Arachnoids (Figure 42.11): These are structures characterized by a combination of concentric and radial pattern of fractures and/or ridges; radial ridges often extend outward for several arachnoid radii. Historically, the term "arachnoid" derives from these ridges that resemble spider

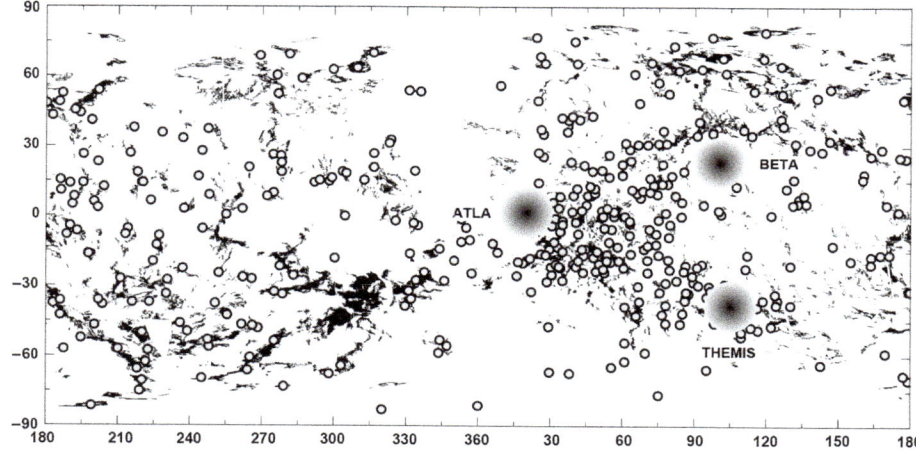

FIGURE 42.10 Spatial distribution of groove belts (black areas) and coronae (white dots). Coronae are spatially associated with the belts, which may suggest mutual development of these features.

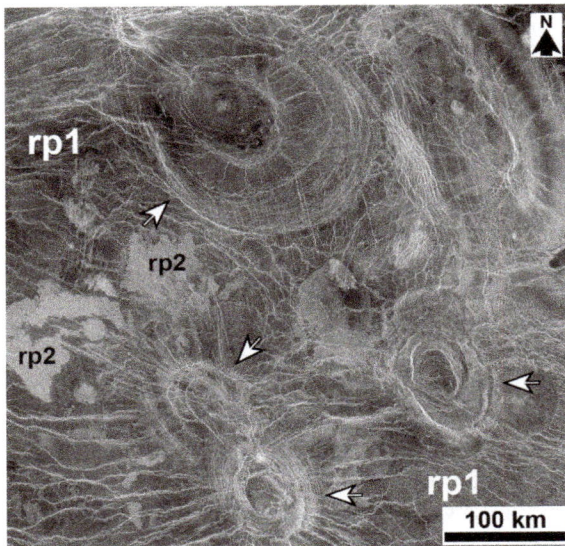

FIGURE 42.11 Examples of arachnoids. These circular and elliptical features (arrows) are surrounded by topographic rims that, at some arachnoids, are cut by swarms of concentric fractures. Wrinkle ridges within the lower subunit of regional plains (lower portion of the image) are converging toward the arachnoids. Graben radiating from these structures appear to be the sources of the flows of unit rp$_2$. Center of the image is at 40.0°N, 19.2°E. See Figure 42.2(A) for location.

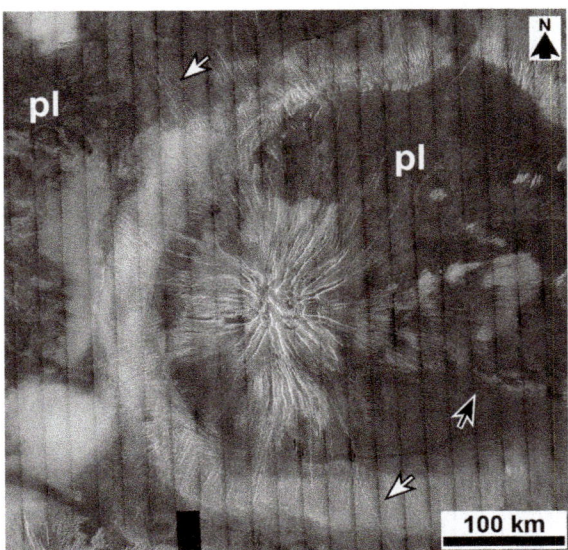

FIGURE 42.12 An example of a nova. The starlike pattern of grooves defines the nova, which is completely inside the rim of Pavlova Corona (radar-bright circular feature). Lobate plains (pl) embay both the nova and the corona rim. Some of the narrow graben radiating from the nova cut both the rim and the embaying plains (white arrows). To the east of the center of the nova, flows of lobate plains seem to emanate from the graben of the nova (black arrow). Center of the image is at 14.5°N, 39.0°E. See Figure 42.2(A) for location.

legs. In many cases, arachnoids include a few interior flows and small shield volcanoes within the interior. The radial fractures frequently merge outward with the linear patterns of the fracture belts on which some arachnoids are arranged. The distinction between arachnoids and coronae is based on the relative abundance of radial versus concentric fractures/ridges, respectively. In general, arachnoid-like circular structures tend to be less than 200 km in diameter.

The relative dearth of extrusive volcanic features associated with arachnoids may imply that magmatism associated with these features is largely intrusive and that the radial ridges are compressional features due to subsidence-related deformation. In most other characteristics, arachnoids are similar to coronae, which suggests that there was a broad spectrum of scales at which coronae and corona-like features formed.

Radial fracture centers (novae, Figure 42.12): These represent swarms of graben radiating from a common point, resulting in a stellate pattern that frequently crosses preexisting structures and surrounding plains; concentric structures are absent. The stellate pattern is frequently centered on a broad domical topographic high that sometimes is surrounded by radial flow patterns or within arachnoid- and corona-like structures. Radial fracture centers occur more frequently in association with coronae and some large volcanoes. In some examples, novae postdate these features and predate them in other examples.

Two different modes of emplacement have been proposed for radial fracture centers: (1) tensile fracturing associated with domical uplift and (2) tensile deformation resulting from shallow dike emplacement. Although both mechanisms are probably influential in the development of radial fracture patterns, domical topography does not characterize all examples of radial fractures centers. Even if the topography has relaxed in such cases, the annular pattern of deformation predicted to occur upon relaxation is not present. In addition, many radial fracture patterns extend for distances much greater than any associated domical rise in the topography. As a result, not all radial fracture centers may be attributed to strain associated with uplift. The recognition that tensile fractures accompany shallow emplacement of dikes on Earth, the similarity to giant radial dike swarms, and the frequent association with clear centers of volcanic activity all suggest that radial fracture centers more likely are the surface manifestation of radial dike swarms.

3. FACTORS INFLUENCING VOLCANIC ERUPTIONS AND VOLCANIC LANDFORMS ON VENUS

Important questions and goals in the study of Venus include determining how the differences in surface and interior conditions are linked to the observed diversity and complexity of volcanic landforms. The influence of the extreme temperature and atmospheric pressure on both extrusive and intrusive magma emplacement is likely to be significant. Past study of volcanism and volcanic rocks on the terrestrial planets and recent studies of magma transport-related processes on Earth suggest that the three most important factors in development of individual volcanic centers on Venus are likely to be the following: (1) magma composition, (2) environmental influences, and (3) magma ascent characteristics, specifically emplacement and replenishment rates.

3.1. Influence of Chemical Composition

Chemical composition is an important influence on the viscosity and volatiles abundances of erupted melt. On Venus, the question arises as to whether all the observed volcanic landforms are the results of silicate melts. As discussed above, the possibility of exotic compositions such as carbonatites and sulfur flows has been considered in order to account for the apparent extreme fluidity of lava channel-forming materials. The high abundance of sulfur (several percent) in the surface soil analyses of Vega 2 lander suggests that sulfur may be important in the surface chemistry. Although observed chemical compositions of materials from Venera landers are consistent with wide range of basaltic compositions, the existence of steep-sided domes may indicate the presence of non-basaltic, silica-rich melts formed either by fractionation of original magma or by remelting of crustal materials.

3.2. Influence of Extreme Surface Temperature and Pressure

The surface temperature (~ 700 K) and pressure (9.2 MPa) at the mean surface elevation of Venus differ substantially from those on Earth and corresponding differences in eruption characteristics and behavior of melts and volatiles on Venus are expected (Head and Wilson, 1986).

Influence on surface eruptions: The influence of surface conditions on eruptions is primarily a result of the extreme pressure rather than temperature. Although the surface temperature is ~ 700 K, this is far below the solidus of common silicate melts. Given the close chemical similarity between measured Venus surface compositions and terrestrial basalts, melt temperatures similar to those on Earth (1700 K) are likely for most volcanic eruptions. Even for rhyolitic eruptions, melt temperatures (1200–1300 K) are far above the surface temperature. Calculations of the rate of cooling of a typical basaltic lava flow erupted onto the surface of Venus show that the temperature at the surface of the flow will be fractionally lower compared with that on Earth soon after eruption. This result merely reflects the enhanced convection in the dense CO_2 atmosphere. The

cooling efficiency is such that it may be compared with a subaqueous eruption. In the longer term, the surface temperature of a lava flow will be greater on Venus as it cools to the higher ambient temperature. A predicted result is that lava crusts will develop more quickly on Venus than on Earth, thereby enhancing the early formation of lava tubes.

The atmospheric density is therefore the primary influence. In fact, the pressure on the surface of Venus is equivalent to the pressures at 1-km depth in Earth's oceans with many of the same predicted consequences. A significant influence of the high pressure is the suppression of gas exsolution, with important consequences for vesiculation of lava flows and pyroclastic disruption of erupting magmas at volcanic vents. For any magma bearing volatiles in concentration comparable to those on Earth ($<2\%$), pyroclastic eruptions and corresponding composite volcanoes and ash flows are unlikely to occur unless volatile contents of eruptions on Venus significantly exceed those common on Earth.

Observations of large increases in sulfur content of the upper atmosphere, originally ascribed to a large Plinian-type eruption process, suggest the possibility of significant volatiles. The occurrence of radar-dark areas around the summits of some large volcanoes, attributed to extensive ash deposits, may be an additional clue to the high volatile abundance. Similarly, vesiculation of lava flows should be suppressed. As discussed in the above section on steep-sided domes, volatiles in the erupting magma may favor extensive vesiculation of lavas, leading to frothy basaltic eruptions. One consequence is that a volcanic event that on Earth would result in an ash-flow eruption might on Venus result in a massive frothy outpouring.

Influence on magma ascent characteristics: The high ambient temperature of the lithosphere is likely to influence the thermal efficiency during magma transport so that magma in dikes and as plutonic bodies are expected to experience less cooling during ascent. The greater abundance of shield fields on Venus than that seen on the other terrestrial planets may be a consequence of the thermal efficiency on Venus of ascending magmas. Fields of volcanoes develop from the ascent of individual bodies of magma directly from a local melting source instead of from the large, shallow reservoirs that are necessary for the development of large **magmatic centers**. This is consistent with calculations of the magma replenishment rates necessary for shallow reservoirs.

Magma ascent may be governed by the presence of variable magma buoyancy with respect to the country rock. The extreme atmospheric pressure may exert a strong influence on magma ascent. Because the crust of Venus is probably largely volcanic, the density in the shallow levels will depend on the density of previously erupted rocks. Given the extreme suppression of vesiculation and the greater abundance of lavas over pyroclastic materials, the density of typical crustal rocks on Venus will be greater than on Earth (Head and Wilson, 1992). In many situations, it is found that magma will ascend buoyantly until the density of the melt approaches the density of the country rocks at which point the magma stalls. Further arrivals of magma will continue to stall at the same level, resulting in the development of reservoirs from which other effects must take over in order for the magma to continue its ascent. The condition of neutral buoyancy may not occur at lower topographic elevations on Venus, where atmospheric pressure is greatest and the volatile exsolution is suppressed, leading to direct ascent and eruption of magmas onto the surface and very little reservoir development. At higher elevations, where atmospheric pressure is lower, shallow reservoirs might be able to form and large magmatic centers characterized by shallow reservoirs might be more common. Many large centers of volcanism on Venus do occur in intermediate and upper elevations.

3.3. Influence of Ascent and Emplacement Rates

Part of the diversity of volcanic morphology observed on Venus may reflect the diversity of rates of magma emplacement. Numerical evaluation of the thermal conditions necessary to establish and maintain an active magma reservoir shows that the most important characteristics necessary for reservoir development under given conditions of magma temperature and ambient crustal rock type are the following: (1) the presence of a neutral buoyancy zone and (2) a rate of magma emplacement great enough so that new magma arrives more quickly before the previously emplaced magma solidifies (in other words, the net thermal flux into the magma emplacement zone is positive).

Volcanological studies on Earth suggest that variations in the morphology of volcanic centers may depend on the rate of magma production, replenishment, and extrusion within individual magmatic systems and volcanic centers. Although the geochemical character of magmas, surface environment of eruption, and buoyancy considerations may be important in controlling the style of eruptions in individual volcanic centers, the total rate of magma eruption, emplacement, and replenishment (the *magma rate regime*) might exert an equal or greater influence on the presence of shallow magma reservoirs, long-lived volcanic centers, and the overall morphology of the volcanic center as a whole. Reservoirs capable of supplying central volcanic edifices, volcanic complexes, or long-lived centers of intrusive or extrusive activity on Earth frequently occur only for a very narrow range of magma transport rates and storage conditions. These conditions are sensitive to lithospheric structure and thermal gradient. As a result, the presence of magma reservoirs on any planet implies specific magma

transfer rates from the mantle and timescales over which magma is stored in the crust.

The largest and shallowest magma reservoirs and most voluminous eruptions on Earth are associated frequently with moderate rates of magma emplacement in midcrustal levels with corresponding development of large magma reservoirs. This rate is approximately equivalent to the rate of magma intrusion and eruption at the Hawaiian hot spot averaged over the past 70 million years. Rates of emplacement lower or higher than this moderate rate tend not to yield well-developed, isolated, long-lived magma reservoirs that evolve chemically and structurally in association with single volcanic edifices. Rates of emplacement lower than this value on Earth frequently result in numerous short-lived volcanic centers and are characteristic of "fields" of small volcanoes. Rates greater than the moderate rate regime frequently result in massive flow fields, incompletely developed volcanic centers and magma reservoirs, and relatively minor associated amounts of evolved magma compositions. Neutral buoyancy is necessary but not a sufficient condition for development of complex dike patterns and long-lived magma reservoirs, and variations in reservoir depth appear insufficient to account for all of the disparate morphologies seen in volcanic centers on Venus.

Results of simple calculations of the influence of replenishment rates on Earth are consistent with observed rates of magma replenishment rates typically associated with reservoir development ($\leq 10^{-3}$ km^3/a). Similar calculations for Venus imply that magma reservoirs are likely to form at volume rates of magma transfer at least a factor of two less than on Earth ($\leq 5 \times 10^{-4}$ km^3/a). This means that magma replenishment rates typical of many fields of small volcanoes on Earth are capable of generating central volcanic centers and magma reservoirs on Venus. Additional details of the process may be significant, including the geometry of the magma emplacement and whether the reservoir is tabular or cylindrical. Rates of magma transport and corresponding style of volcanic centers may also be modulated to a limited extent by magma composition.

4. GLOBAL STRATIGRAPHY OF VOLCANIC LANDFORMS ON VENUS

Age relationships among the main volcanic units: The most areally important volcanic landforms (the main volcanic plains, Table 42.1) on Venus display consistent relationships of relative ages among each other at the global scale. Their stratigraphic relationships (Figure 42.2) provide the possibility to establish a global stratigraphic reconstruction that describes the sequence of major volcanic and tectonic events applicable to the entire planet (Ivanov and Head, 2011).

Morphometric analysis of individual volcanic constructs of shield plains at the contacts with regional plains has shown that the density, characteristic diameter, and height of the shields progressively decrease in the direction away from the edges of contiguous shield clusters toward the interiors of regional plains (Kreslavsky and Head, 1999). These characteristics of the contacts between psh and rp$_1$ strongly suggest embayment and partial burial of the older shield plains by younger regional plains.

Application of a more extensive list of criteria to the global-scale analysis of relative ages of shield plains and surrounding regional plains has shown that in ~70% of analyzed cases regional plains represent an embaying unit and shield plains are older (Figure 42.13(A)). About 10% of analyzed shield fields appear to be synchronous to regional plains and ~10% of shield clusters postdate adjacent regional plains. Shield plains are interpreted as having formed by the eruption from multiple small sources

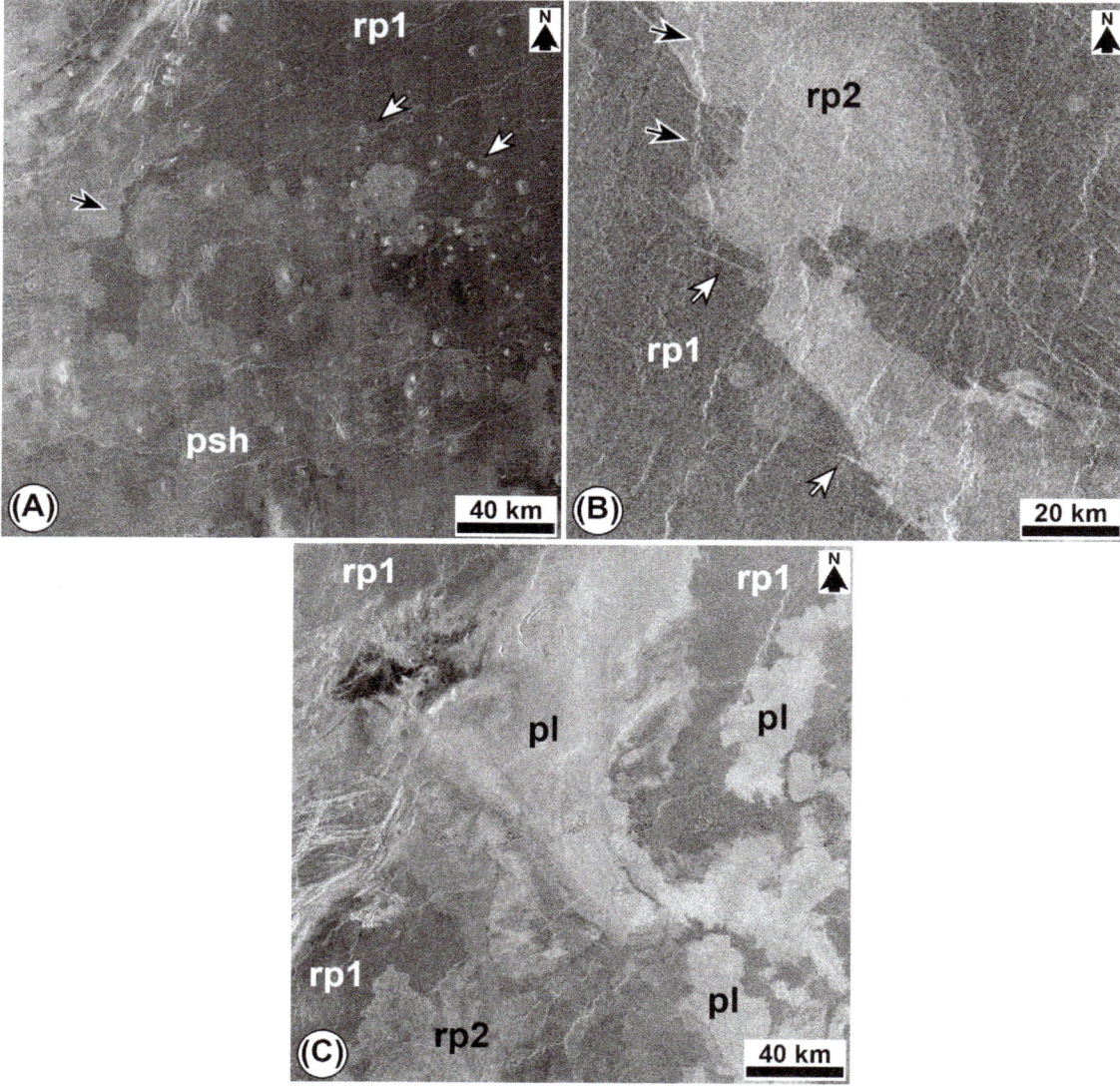

FIGURE 42.13 (A) Relative age relationships between shield plains (psh) and regional plains (rp$_1$). Material of regional plains embays and penetrates into shield plains (black arrow) and separates individual shields from the main exposure of shield plains (white arrows). This is strong evidence for the older relative age of shield plains. Center of the image is at 33.0°N, 352.1°E. (B) Relative age relationships between the lower (rp$_1$) and the upper (rp$_2$) subunits of regional plains. The flow-like occurrence of unit rp$_2$ is brighter and outlined by a sharp boundary. Material of this unit embays and penetrates into narrow fractures that cut the surface of the lower subunit (white arrows). Black arrow indicates a wrinkle ridge that cuts both subunits of regional plains. Center of the image is at 38.4°S, 350.2°E. (C) Relative age relationships between lobate plains (pl) and two subunits of regional plains (rp$_1$ and rp$_2$). Wrinkle ridges cut the surface of regional plains and are embayed by the flows of lobate plains. Center of the image is at 51.5°S, 350.8°E.

over a discrete period of geologic time. Because there are few constraints on emplacement rates of Venus surfaces, it is not known whether shield plains surfaces are produced in a punctuated, catastrophic, or continuous formation. Regardless of the duration of the geologic time period, it is a consistent and constrained time period in the stratigraphy of the mapped area, and therefore it provides evidence for a change in the mechanism of formation of plains units within this region.

These findings imply that the volcanic style of shield plains (formation of small and pervasive volcanic constructs) changed during a specific period of the geologic history of Venus to massive, flood-like, outpouring of lavas that formed regional plains. Most of steep-sided domes are associated with shield plains both spatially and stratigraphically, and none of the domes appear to correlate with regional plains.

The volcanism of small shields, however, was not completely confined to the stratigraphic boundary between shield and regional plains, and some groups of small shields continued to form after emplacement of unit rp_1 (Ivanov and Head, 2004). These occurrences of small shields that postdate regional plains represent a specific unit of shield clusters (Figure 42.6, Boala Formation). The scarcity of the shield clusters ($\sim 0.7\%$ of the surface of Venus, Table 42.1) implies that although the volcanic style responsible for formation of small edifices seems to continue through a large portion of the visible geologic history of Venus, the rate of the small shield volcanic activity dropped significantly (more than an order of magnitude) after emplacement of regional plains.

Shield plains underlie both subunits of regional plains (rp_1 and rp_2). Wrinkle ridges cut the surface of both subunits of regional plains and shield plains as well, thus establishing the upper stratigraphic limit of the emplacement of regional plains. In areas where the contacts between subunits rp_1 and rp_2 are clean and sharp, it is seen that material of the upper subunit (rp_2) penetrates into local lows (e.g., fractures, lava channels) of the lower subunit (Figure 42.13(B)) and embays local highs of unit rp_1. These relationships are observed at the global scale and indicate the relatively young age of unit rp_2. Volcanic flows of the upper subunit of regional plains make up the flanks of some of the intermediate and large volcanoes and sometimes are sourced by arachnoids.

In all places where lobate plains occur in contact with either shield plains or the subunits of regional plains, flows of lobate plains embay wrinkle ridges that deform materials of regional plains and overlap their surfaces (Figure 42.13(C)). These relationships clearly indicate that lobate plains are younger. Lobate plains cover the flanks of the absolute majority of the large volcanoes and represent the volcanic component of some coronae and radial fracture centers.

Age relationships of the main volcanic and tectonic units: In contrast to volcanic units in which the volcanic nature of materials on the surface is evident, the tectonized units/terrains are those in which the surface was strongly modified by tectonic structures that largely erased the characteristic morphologic signatures of underlying material units.

Practically everywhere on Venus, contacts of tessera, densely lineated plains, ridged plains, and the majority of groove belts with surrounding vast volcanic plains (psh, rp_1, rp_2) are very sinuous due to penetration of materials of plains into massifs of the tectonized terrains (Figure 42.14). These characteristics of the contacts provide compelling evidence that both emplacement and tectonic modification of the tectonized terrains were completed before formation of the younger vast volcanic plains. Thus, the absolute majority of the tectonically deformed terrains on Venus (Table 42.1) predate emplacement of both shield and regional plains. Rift zones that represent extensive belts of extension formed broadly contemporaneously with lobate plains. Both these units postdate emplacement of shield and regional plains.

These consistent and globally observed stratigraphic relationships allow division of the visible portion of the geologic history of Venus into three regimes of resurfacing during which specific types of endogenous activity dominated (Figure 42.15).

1. The majority of the tectonized terrains (tessera, densely lineated and ridged plains, and groove belts) that may be related to regional/global mantle convection patterns define the first, *tectonically dominated*, regime. During this time, large regions of thickened crust (tesserae) were formed; a limited contraction and possible underthrusting along specific zones resulted in the formation of ridge and mountain belts. The later phases of the ancient tectonic regime were manifested by the mutual development of groove belts and many coronae. All tectonized terrains of the first regime represent local to regional topographic highs in the background topography.

2. During the second, *volcanically dominated*, regime, the vast plains such as shield plains and both subunits of regional plains were emplaced preferentially in regional lows. The density of craters on regional plains suggests that the first two regimes (tectonic and volcanic) operated during about the first one-third of the observable history.

3. Contemporaneous rift zones and lobate plains define the third, *network rifting-volcanism,* regime. This regime dominated the last two-thirds of the observable geologic history and is linked to the later stages of evolution of the dome-shaped rises.

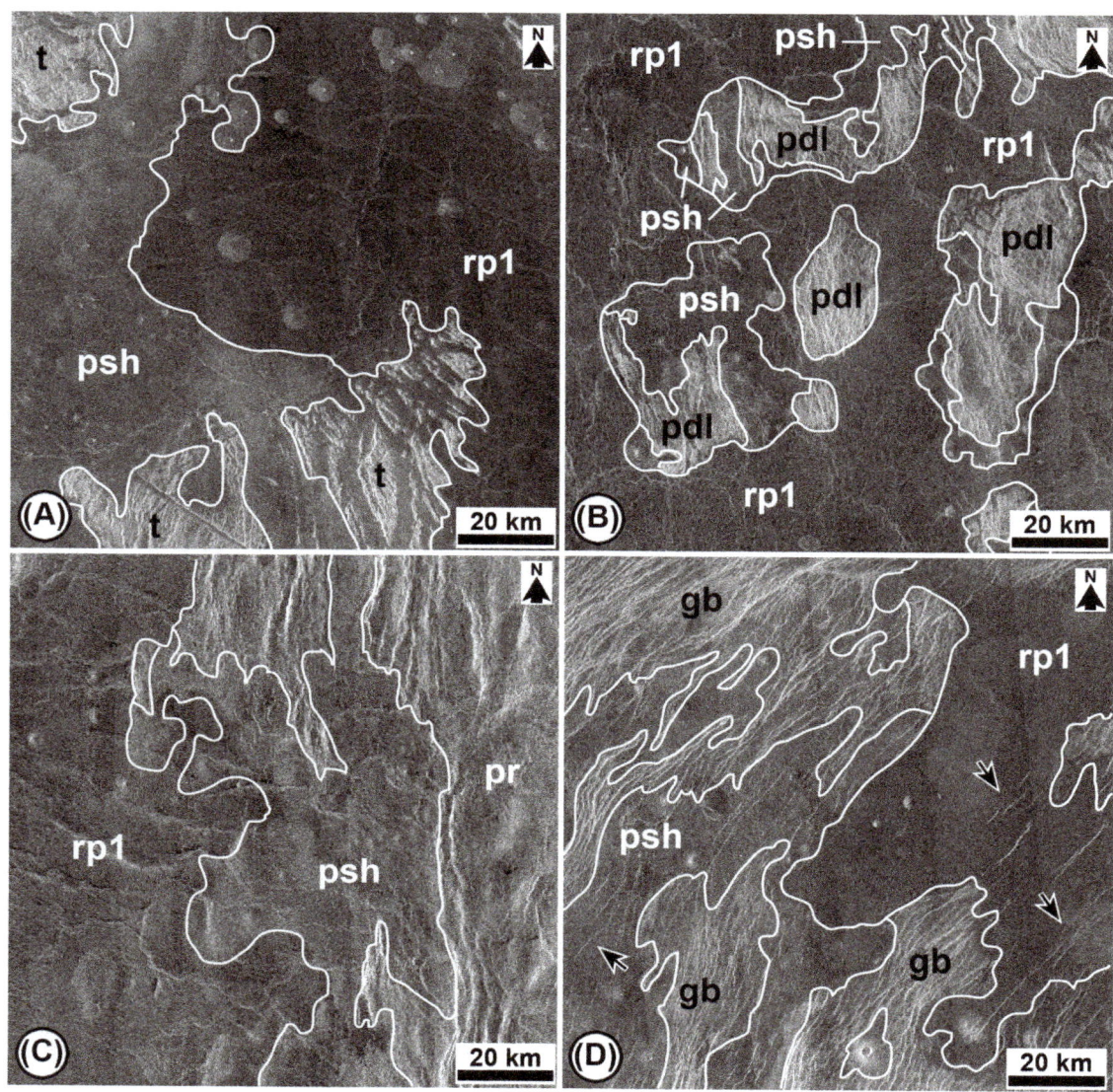

FIGURE 42.14 (A—D): Relative age relationships between the older tectonized units and the vast volcanic plains (psh and rp1). In the cases shown in images A, B, and C, the surface of the vast volcanic plains is mildly deformed and materials of the plains completely superpose all structures of the tectonized units. (D) At the contact of groove belts with either psh or rp1, a few graben of the belts cut the surface of the plains (black arrows), but the absolute majority of the structures of groove belts are embayed by the vast volcanic plains. (A) Center of the image is at 47.3°N, 127.6°E. (B) Center of the image is at 40.9°N, 34.7°E. (C) Center of the image is at 37.5°N, 156.4°E. (D) Center of the image is at 3.0°N, 146.7°E.

5. EVIDENCE FOR ACTIVE VOLCANISM ON VENUS AND FUTURE EXPLORATION

Although volcanic landforms are very abundant on Venus throughout the visible portion of its geologic history, the actual current level of volcanic activity on the planet is unknown. There are several lines of circumstantial evidence suggesting that Venus may have recent or currently active volcanism.

Near the beginning of the ESA Venus Express mission, strong and short bursts of radiation have been detected in the ionosphere of Venus (Russell et al., 2007). These electromagnetic pulses are interpreted as results of lightning in the atmosphere (Russell et al., 2007) that may be related either to purely atmospheric processes or to Plinian-type eruption plumes or to both. Reinterpretation of the radar properties of one of the lava flows on Venus suggests that this flow may have an excess temperature, which is consistent with its young (less than 20 years) age (Bondarenko et al., 2010). The data from the Visible and Infrared Thermal Imaging Spectrometer onboard Venus Express spacecraft reveal significant variations in the

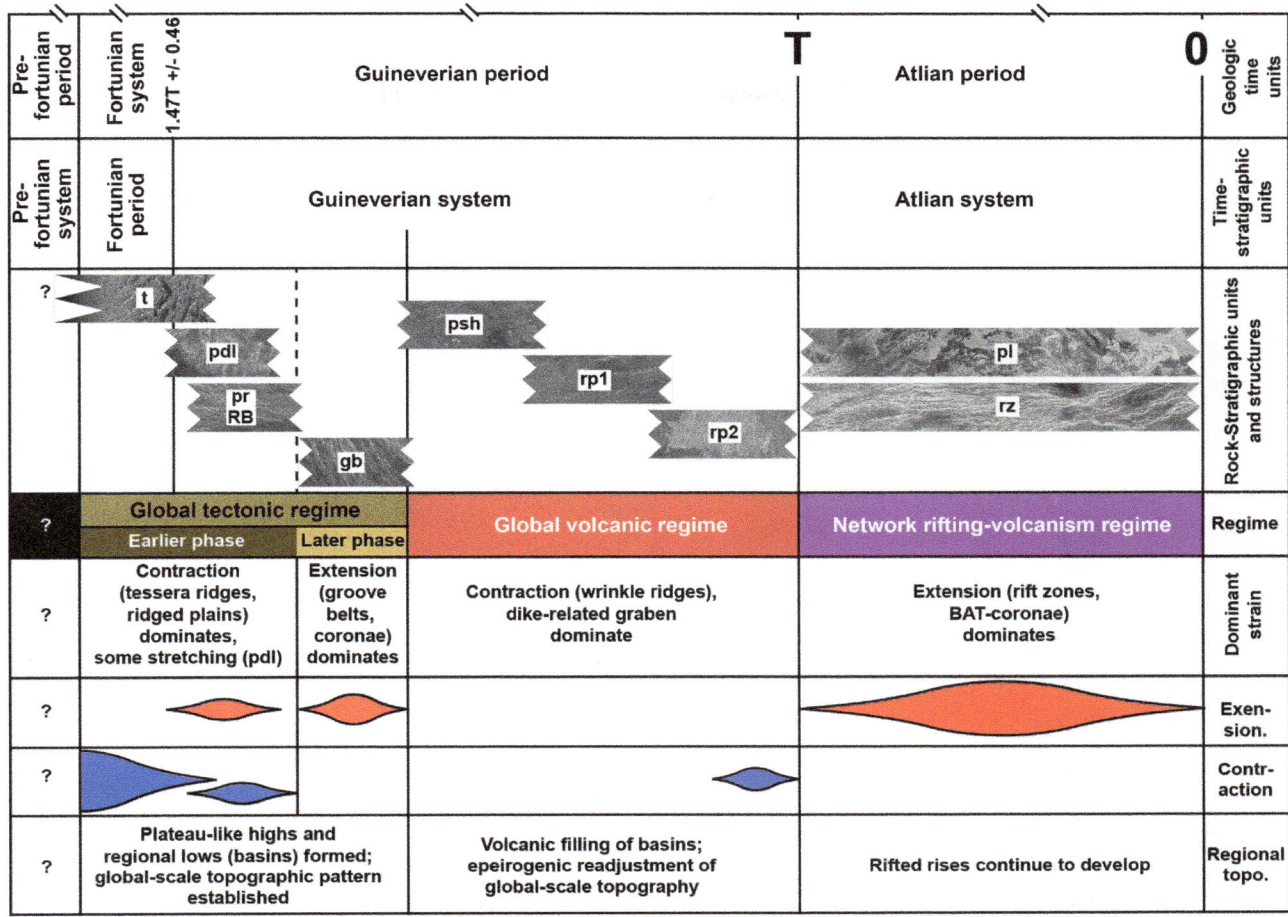

FIGURE 42.15   A global correlation chart that shows the three major regimes of resurfacing on Venus: global tectonic regime, followed by the global volcanic regime, and then by the network rifting-volcanism regime (see text for discussion).

emissivity of the surface of several flow complexes (Helbert et al., 2008) and large volcanic centers in the southern hemisphere of Venus (Smrekar et al., 2010). These emissivity anomalies are consistent with both the variations in composition of lavas (Helbert et al., 2008; Ivanov and Head, 2010) and their relatively young (250,000 years or less) age (Smrekar et al., 2010). Finally, images collected by Venus Monitoring Camera onboard Venus Express spacecraft show several regions within a young rift zone, whose brightness is changed during a series of observations (Shalygin et al., 2014). The variations in the apparent brightness of these spots may correspond to pulses of volcanic activity and related changes of temperature due to eruption of volcanic gases and/or lavas.

All these interpretations are indirect and thus do not provide conclusive evidence for ongoing volcanic activity on Venus. In order to collect evidence for possible current volcanism on the planet, new high-resolution images and topography data are required. For example, with new images, it would be possible to test if patterns of lava flows within the areas of the interpreted young volcanism have

changed compared with the patterns observed by Magellan. Geochemical data are crucially important for the broad understanding of mechanisms of formation of volcanic landforms on Venus and formulation of petrogenetic models. Available data, unfortunately, have very large uncertainties in the determinations of the major petrogenetic and heat production elements, which allow only very general understanding of types of materials that occur on the surface of Venus. A lander-oriented mission with the capability of precise chemical measurements should provide key information on the composition of volcanic materials, mechanisms of their formation, and interaction with the atmosphere.

FURTHER READING

Abdrakhimov, A.M., 2005. Geochemical comparison of volcanic rocks from terrestrial intraplate oceanic hot spots with Venusian surface material. Geochem. Int. 43, 732−747.

Aubele, J.C., Slyuta, E.N., 1990. Small domes on Venus: characteristics and origin. Earth Moon Planets 50/51, 493−532.

Aubele, J.C., 1995. Stratigraphy of small volcanoes and plains terrain in Vellamo Planitia-Shimti tessera region, Venus. Lunar Planet. Sci. 26, 59–60.

Basilevsky, A.T., Head, J.W., 1998. The geologic history of Venus: a stratigraphic view. J. Geophys. Res. 103, 8531–8544.

Bondarenko, N.V., Head, J.W., Ivanov, M.A., 2010. Present-day volcanism on Venus: evidence from microwave radiometry. Geophys. Res. Lett. 37, L23202. http://dx.doi.org/10.1029/2010GL045233.

Crumpler, L.S., Head, J.W., Aubele, J.C., 1993. Relation of major volcanic center concentration on Venus to global tectonic patterns. Science 261, 591–595.

Crumpler, L.S., Aubele, J.C., Senske, D.A., Keddie, S.T., Magee, K.P., Head, J.W., 1997. Volcanoes and centers of volcanism, Venus. In: Bougher, S.W., Hunten, D.M., Phillips, R.J. (Eds.), Venus II Geology, Geophysics, Atmosphere, and Solar Wind Environment. Univ. Arizona Press, Tucson, pp. 697–756.

Crumpler, L.S., Aubele, J., 2000. Volcanism on Venus. In: Houghton, B., Rymer, H., Stix, J., McNutt, S., Sigurdson, H. (Eds.), Encyclopedia of Volcanoes. Acad. Press, San Diego, San Francisco, New York, Boston, London, Sydney, Toronto, pp. 727–770.

Guest, J.E., Bulmer, M.H., Aubele, J., Beratan, K., Greeley, R., Head, J.W., Michaels, G., Weitz, C., Wiles, C., 1992. Small volcanic edifices and volcanism in the plains of Venus. J. Geophys. Res. 97, 15949–15966.

Head, J.W., Solomon, S.C., 1981. Tectonic evolution of the terrestrial planets. Science 213, 62–76.

Head, J.W., Wilson, L., 1986. Volcanic processes and landforms on Venus: theory, predictions, and observations. J. Geophys. Res. 91, 9407–9446.

Head, J.W., Wilson, L., 1992. Magma reservoirs and neutral buoyancy zones on Venus: implications for the formation and evolution of volcanic landforms. J. Geophys. Res. 97, 3877–3903.

Head, J.W., Crumpler, L.S., Aubele, J.C., Guest, J.E., Saunders, R.S., 1992. Venus volcanism: classification of volcanic features and structures, associations, and global distribution from Magellan data. J. Geophys. Res. 97, 13153–13197.

Helbert, J., Muller, N., Kostama, P., Marinangeli, L., Piccioni, G., Drossart, P., 2008. Surface brightness variations seen by VIRTIS on Venus Express and implications for the evolution of the Lada Terra region, Venus. Geophys. Res. Lett. 35, L11201. http://dx.doi.org/10.1029/2008GL033609.

Herrick, R.R., Sharpton, V.L., Malin, M.C., Lyons, S.N., Feely, K., 1997. Morphology and morphometry of impact craters. In: Bougher, S.W., Hunten, D.M., Phillips, R.J. (Eds.), Venus II Geology, Geophysics, Atmosphere, and Solar Wind Environment. Univ. Arizona Press, Tucson, pp. 1015–1046.

Ivanov, M.A., Head, J.W., 2004. Stratigraphy of small shield volcanoes on Venus: criteria for determining stratigraphic relationships and assessment of relative age and temporal abundance. J. Geophys. Res. 109, NE10001. http://dx.doi.org/10.1029/2004JE002252.

Ivanov, M.A., Head, J.W., 2010. The Lada Terra rise and Quetzalpetlatl Corona: a region of long-lived mantle upwelling and recent volcanic activity on Venus. Planet. Space Sci. 58, 1880–1894.

Ivanov, M.A., Head, J.W., 2011. Global geological map of Venus. Planet. Space Sci. 59, 1559–1600.

Ivanov, M.A., Head, J.W., 2013. The history of volcanism on Venus. Planet. Space Sci. 84, 66–92. http://dx.doi.org/10.1016/j.pss.2013.04.018.

Kreslavsky, M.A., Head, J.W., 1999. Morphometry of small shield volcanoes on Venus: implications for the thickness of regional plains. J. Geophys. Res. 104, 18925–18932.

Russell, C.T., Zhang, L., Delva, M., Magnes, W., Strangeway, R.J., Weil, H.Y., 2007. Lightning on Venus inferred from whistler-mode waves in the ionosphere. Nature 450, 661–662.

Schaber, G.G., Kirk, R.L., Strom, R.G., 1998. Data base of impact craters on Venus based on analysis of Magellan radar images and altimetry data. USGS Open File Report 98–104.

Shalygin, E.V., Markiewicz, W.J., Basilevsky, A.T., Titov, D.V., Ignatiev, N.I., Head, J.W., 2014. Bright transient spots in Ganiki chasma, Venus. Lunar Planet. Sci. Conf. 45, 2556 abstract.

Smrekar, S.E., Stofan, E.R., Mueller, N., Treiman, A., Elkins-Tanton, L., Helbert, J., Piccioni, G., Drossart, P., 2010. Recent hot-spot volcanism on Venus from VIRTIS emissivity data. Science 328, 605–608.

Surkov, Y.A., 1997. Exploration of Terrestrial Planets from Spacecraft: Instrumentation, Investigation, Interpretation. Wiley, Praxis Pub, New York, Chichester, p. 446.

Chapter 43

Volcanism on Io

Rosaly M.C. Lopes

Earth and Space Sciences Division, Jet Propulsion Laboratory, California Institute of Technology, Pasadena, CA, USA

David A. Williams

School of Earth and Space Exploration, Arizona State University, Tempe, AZ, USA

Chapter Outline

GLOSSARY

differentiation The process by which planets and satellites develop layers or zones of different chemical and mineralogical composition.

flyby Term used to describe the close approach of a spacecraft to a body, without the spacecraft going into orbit around that body.

Galilean satellites The four largest satellites of Jupiter (Io, Europa, Ganymede, Callisto), named after their discoverer Galileo Galilei.

heat flow Heat that comes out of a body from its interior and is ultimately radiated to space.

hot spots Regions of enhanced thermal emission on Io, a sign of volcanic activity. The term does not imply a particular eruption mechanism, and does not have the same meaning as hot spots on Earth.

IRIS Acronym for Infrared Interferometer Spectrometer, an instrument aboard Voyager.

NIMS Acronym for Near-Infrared Mapping Spectrometer, an instrument aboard Galileo that obtained spectra in the wavelength range $0.7-5.2$ μm. The spectra are used to analyze the composition of Io's surface and to measure the temperatures of its volcanic regions.

Patera A collective term for a variety of saucer-shaped, shallow volcanic constructs that often have a central caldera-like depression.

pyroclastic materials Fragmented materials ejected during a volcanic eruption, including ash, pumice, and rock fragments.

SSI Acronym for solid state imaging system, the CCD camera aboard the Galileo spacecraft.

sulfur allotropes Sulfur cooled rapidly from different temperatures, resulting in different colors.

thermal emission Electromagnetic radiation produced by a body due to its temperature.

volatiles Chemical compounds or elements contained in magmas that are generally released as gases to the atmosphere during a volcanic eruption.

1. INTRODUCTION

Volcanism is a fundamental process that has affected all solid planets and most moons in the solar system and nowhere is this more clearly shown than on Jupiter's moon Io (Figure 43.1). Prior to the Voyager 1 and 2 spacecraft observations of Io in 1979, the Earth was the only planet known to have active volcanic activity. Our views of planetary volcanism were dramatically changed when Voyager 1 revealed active volcanoes on Io, a body about the same size as the Earth's Moon which, according to prior thinking, should have cooled enough to be volcanically dead. Images from Voyager 1 showed plumes up to 300 km in height and a vividly colored surface dominated by large calderas and lava flows.

Io's significance in science dates back centuries. Io is the innermost of the four large satellites of the planet Jupiter, which are known as the **Galilean satellites**. They were discovered by the Italian astronomer Galileo Galilei $(1564-1642)$ in 1610, who soon realized that the objects he initially thought to be stars were, in fact, bodies orbiting Jupiter. These observations became the cornerstone of

The Encyclopedia of Volcanoes. http://dx.doi.org/10.1016/B978-0-12-385938-9.00043-2

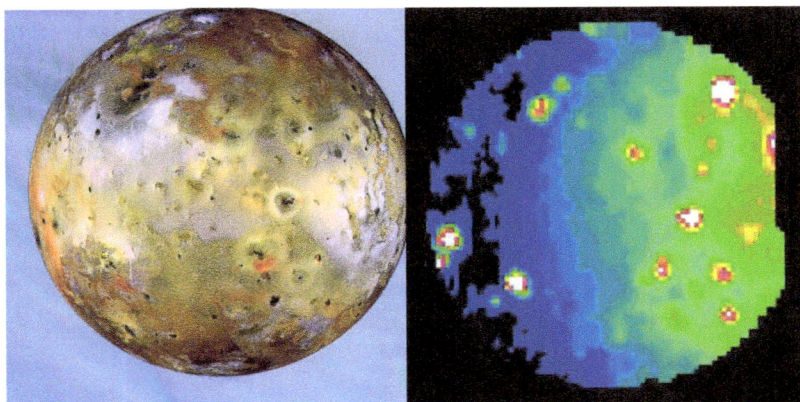

FIGURE 43.1   Io imaged by Galileo's camera Solid State Imaging instrument (SSI) in 1996 (left) and by Galileo Near-Infrared Mapping Spectrometer (NIMS) in 2001 (right). Left: Io's colorful appearance of results from its constant volcanic activity, depositing sulfur compounds and silicates on the surface. The reds and blacks are thought to be the most recent volcanic deposits, probably no more than a few years to decades old. Just to the right of the image center is the Prometheus volcano, one of the most active hot spots on Io. Notice the dark flow and the bright annulus surrounding the volcano, formed by plume deposits. The dark spot on the left edge just below the equator is the volcano Pele, which is surrounded by large red deposits. Jupiter is in the background and appears blue, because it is seen through the camera's infrared filter. On the right, active volcanoes are seen as "hot spots" in this false color image taken using the instrument's 5.0 μm wavelength, showing nearly the same view as the SSI image. Seventy one previously unknown active volcanoes were detected by NIMS during the Galileo mission. (NASA PIA00494 and PIA03535).

evidence that confirmed the Copernican theory, which stated that the Earth and other planets revolved around the Sun, challenging the prevalent view at the time that the Earth was the center of the universe.

Prior to observations from spacecraft, knowledge about Io's surface was inferred from telescopic observations. These showed that Io's brightness varied according to its position in its orbit, suggesting that the moon always keeps one face towards Jupiter. In the 1960s and 1970s, telescopic observations showed that Io differed from the other Galilean satellites because of the absence of water bands in its spectra, and that Io was the reddest object known in the solar system. Ground-based measurements in the mid-1970s also showed the existence of ionized sulfur emission in the inner Jovian magnetosphere; subsequent studies revealed this to be a plasma torus. The **flyby** of the Jupiter system by *Pioneer 10* in 1973 marked the beginning of observations of Io by spacecraft. These space observations detected an ionosphere and thin atmosphere around Io, and a cloud of neutral gasses (sodium, potassium, sulfur, and oxygen) along Io's orbital path. Astronomers now know that Io's volcanic activity is responsible for these phenomena. Nowhere else in the solar system is the effect of volcanism as widespread and far-reaching as on Io.

Io's size and bulk density (Table 43.1) are similar to those of the Earth's Moon and suggest a composition predominantly of silicates. We might have expected Io to be a cratered, dead world much like the Moon. However, even before the first close-up images of Io were returned by Voyager in 1979, there were hints that Io was remarkably different from the Moon. Earth-based telescopic

observations made in 1974 first showed that the overall spectrum of Io closely matched that of sulfur, suggesting that sulfur and sulfur compounds might be abundant on the surface. It was also known that Io's spectrum lacked water ice, unlike those of the other Galilean satellites. Just prior to the Voyager encounters in 1979, two events presaged the discovery of active volcanism. F. Witteborn and colleagues reported a telescopic observation of an intense temporary brightening of Io in the infrared wavelengths from 2 to 5 μm. They explained it, although with some skepticism, as **thermal emission** caused by part of Io's surface being at a temperature of about 600 K, much hotter than the average expected daytime temperature of about 130 K. A few days before the Voyager encounters, a seminal theoretical paper by Stan Peale and colleagues was published in the journal Science. They had studied the tidal stresses generated within Io as a result of the gravitational fields of Jupiter, Europa, and Ganymede. Their calculations showed that the possible heat generated by tidal stresses was in the order of 10^{13} W—much greater than heat that could be released from normal radioactive decay. Their prediction was that Io might have "widespread and recurrent volcanism."

Active volcanoes were not immediately obvious in the first images returned by Voyager 1. The most striking aspect of Io shown in the first images was its colorful surface, with yellows, oranges, reds, and blacks. Scientists on the imaging team nicknamed Io the "pizza moon" and speculated about the presence of large quantities of sulfur on the surface. Another surprising aspect was the absence of impact craters. The obvious conclusion was that Io's surface was very young and the craters must have been

TABLE 43.1 Major Io Characteristics

Orbital period	1.769 days
Rotational period	Synchronous with orbit
Mean radius	1821.6 ± 0.5 km
Bulk density	3528 ± 3 kg m^{-3}
Mass	$8.9320 \pm 0.0013 \times 10^{22}$ kg
Surface gravity	1.80 m s^{-2}
Global average heat flow	>2.5 W m^{-2}
Radius of core	If pure iron, 656 km
	If iron and iron sulfide mixture, 947 km
Geometric albedo	0.62
Local topographic relief	Up to ~17 km
Active volcanic centers	>166
Hot spot temperatures	Up to at least 1500 K
Active plumes	16 observed
Surface composition	SO$_2$ frost mantles surface, other S compounds, silicates
Typical surface temperatures	Away from hot spots: 85 K (night) to 140 K (day)
Crustal thickness (estimated)	30−50 km
Atmospheric pressure	10^{-9} bar or less, higher at locations of plumes
Atmospheric composition	SO$_2$ (main), SO, S$_2$, Na

FIGURE 43.2 Volcanism on Io was first seen on this Voyager 1 image taken on March 8, 1979. The image shows two plumes. The Pele plume is seen on the edge of the disk against the dark sky and it rises about 260 km above the surface. The other plume, Loki, is seen as a bright spot on the nightside of the terminator (the boundary between day and night). The plume is reaching above the darkness of the disk and catching the rays of the rising sun. (NASA PIA00379).

obliterated—but how? The answer came soon after, when a navigation engineer at the Jet Propulsion Laboratory, Linda Morabito, noticed a peculiar umbrella-shaped feature emanating from Io's limb in one of the images that was taken to aid navigation of the spacecraft (Figure 43.2). The pattern turned out to be an eruption plume rising about 260 km above the surface. A second plume was found on the same image, and more plumes were seen upon close examination of various other images. Additional evidence for active volcanism came from another of Voyager's instruments, the **infrared interferometer spectrometer (IRIS)**, which detected enhanced thermal emission from parts of Io's surface—some areas had temperatures of about 400 K, much higher than the rest of the surface, which has noontime equatorial temperatures of about 107−124 K. When one of the hot areas was found to coincide with one of the plumes, there was no doubt that active volcanism was taking place.

Eighteen weeks after Voyager 1's dramatic discovery, the companion spacecraft, Voyager 2, flew close to Io.

Intense activity was still taking place, but significant changes had occurred between the two flybys, including the cessation of the largest plume, Pele, and the altered shape of the deposits associated with this plume. An area of about 10,000 km^2 had been filled in, presumably by fresh material falling down from the plume (Figure 43.3). It became evident that dramatic changes of Io's surface could occur over short timescales. Initial analysis of the Voyager data showed nine plumes and nine "hot spots," though not all plumes coincided with **hot spots** and vice versa. Hot spot is a term used by Io researchers to define a region of enhanced thermal emission, a sign of active volcanism—not the same meaning as a hot spot on Earth. The Voyager IRIS experiment did not observe all of Io's surface, so it was suspected that other active volcanoes existed.

After the two Voyager spacecraft left the Jupiter system on their way to Saturn and beyond, the study of Io's volcanism was continued from Earth, by astronomers using infrared detectors on telescopes. These observations showed that brightenings and fadings of hot spots occur, indicating variations in the level of volcanic activity. These observations showed that Io's most powerful hot spot, Loki, has brightenings that switch on in 1 month or less and last several months before fading. Telescopic observations were also used to analyze the reflected light from Io's surface to determine surface composition, confirming that it was dominated by sulfur dioxide (SO$_2$). Io was also observed by

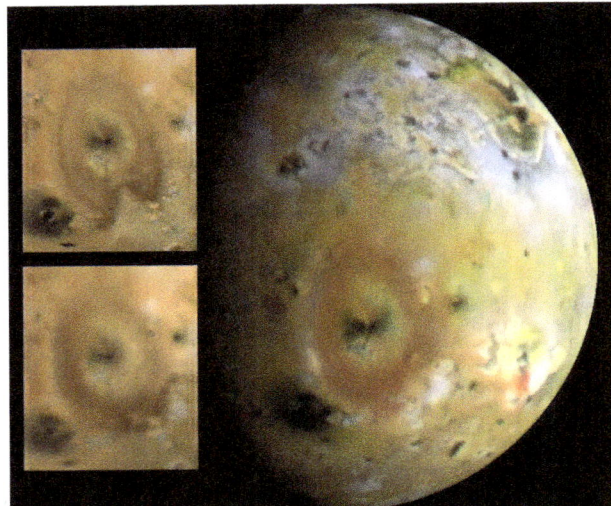

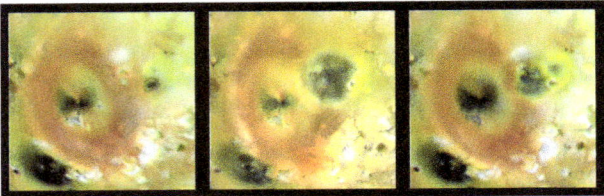

FIGURE 43.3   Io's eruptions cause dramatic changes on the surface. Top: changes in the bright red deposits around the Pele plume. The larger image was taken by the Galileo camera in June 1996 and shows Pele just below the center of Io's disk. The upper inset is an image from Voyager 1 taken in March 1979, while the lower inset was taken by Voyager 2 in July 1979. Note the dramatic change in the shape of the annular deposit in the 4 months between the two Voyager encounters. The differences in the shape of the deposit are thought to reflect the changing shape of the vent from which the plume emerged. (NASA PIA00718). Bottom: changes at Pillan Patera volcano. The image on the left was taken in April 1997. The middle image shows the same area in September 1997 after a huge eruption occurred. The eruption produced the large, dark deposit just above and to the right of the center. The deposit, which is 400 km in diameter, surrounds Pillan Patera and covers part of the bright red ring, which is the deposit from Pele's plume. The image on the right, acquired in July 1999, shows that the red material from Pele has started to cover the dark deposit. This image also shows that a volcano to the right of Pillan has erupted, depositing dark material surrounded by a yellow ring. (NASA PIA02501).

the International Ultraviolet Explorer satellite and by the Hubble Space Telescope.

The first spacecraft to orbit Jupiter was Galileo, from 1995 to 2003. Galileo's instruments were able to image Io and monitor its volcanic activity, revealing variations in volcanic activity, mapping global distribution of volcanism, characterizing eruption styles, measuring lava temperatures, and constraining interior structure, including inferring the existence of a magma ocean under the crust. In 2000, the Cassini spacecraft observed Io on its way to Saturn and, in 2007, the New Horizons spacecraft also made valuable observations of Io on its way to Pluto. Since then, the study of Io's remarkable volcanism has continued

using observations by telescopes both on the ground and orbiting Earth.

2. INTERIOR AND HEAT FLOW

Studies of telescopic and spacecraft data coupled with geophysical modeling of Io suggest that it is differentiated into a metallic core (550–900 km radius if Fe–Fe–S, or 350–650 km radius if Fe), silicate mantle, and crust ($\sim$30 km thick), which likely includes an asthenosphere. Volcanism on Io is powered by tidal heating, induced by a Laplace resonance with Jupiter's moon Europa and Ganymede. For every orbit of Ganymede around Jupiter, Europa orbits twice, and Io orbits four times. This 4:2:1 orbital resonance results in a forced eccentricity in Io's orbit around Jupiter, which causes tidal flexing in Io's crust and mantle ($\sim$100 m every 1.77 days), producing internal friction of such a magnitude that rocks melt to magma. Indeed, this tidal heating results in Io's very high mean global **heat flow**, which is estimated to range from $\sim$1.5 to 4.0 W/m^2 with a mean value of 2.24 $\pm$ 0.45 W/m^2. This is $\sim$20 times larger than the Earth's heat flow. The dominant post-*Voyager* model for heat transport in Io's interior proposes that, rather than undergoing conduction through the lithosphere, heat from Io's interior is advected to the surface via ascending silicate magma in a heat-pipe mechanism that delivers magma to surface hot spots that are contained within a relatively cold lithosphere. This model assumes no global magma ocean within Io, although such oceans have been proposed in the early histories of other solar system bodies like the Moon. However, more recent research by K. Khuruna and colleagues that has reexamined *Galileo* magnetometer data from close flybys of Io in 1999–2001 suggests that Io does have global subsurface magma layer >50 km thick (i.e., a magma ocean) with a rock melt fraction of $\geq$20%. A magma ocean such as this would be akin to a "sponge" with at least 20% silicate melt within a matrix of slowly deformable rock, rather than a completely fluid layer.

Another problem related to understanding Io's interior and heat flow involves comparing the distribution of volcanic features to models of tidal heating. A statistical study of the locations of Ionian **patera** (volcano-tectonic depressions) and hot spots (Figure 43.4) was undertaken by C. Hamilton and colleagues following the publication of the first global geological map of Io. Results showed that the locations of sites of active volcanism are most consistent with asthenospheric heating models, but that volcanic activity is shifted 30–60° E from where models predict. This systematic eastward offset between observed and predicted volcano locations cannot be reconciled with any existing solid body tidal heating models, suggesting that current understanding of Io's tidal heat production and its relationship to surface volcanism is incomplete. Possibilities to

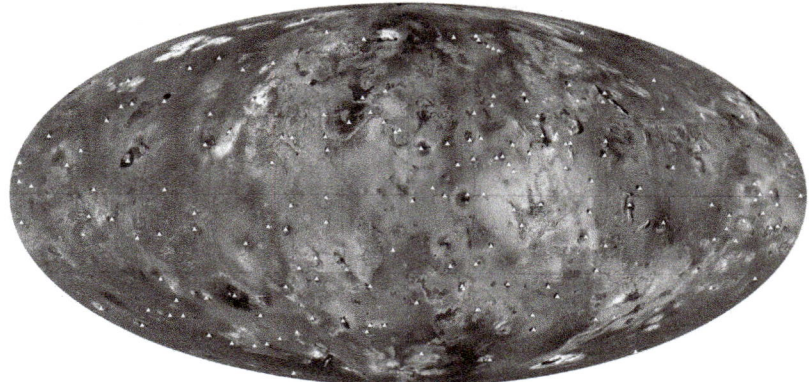

FIGURE 43.4 Map of all hot spots on Io detected by telescopes and the *Voyager, Galileo, Cassini,* and *New Horizons* missions. *Based on data by* Lopes and Spencer *(2007) and* Williams et al. *(2011).*

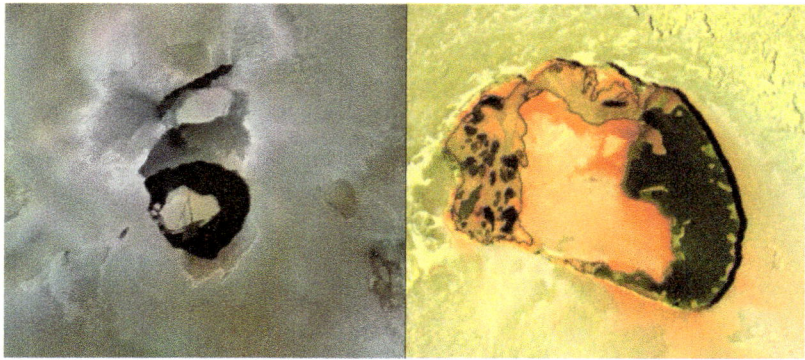

FIGURE 43.5 Two major volcanoes on Io, both of which show a patera (caldera-like structure), which contains dark, active lavas, and "islands," which have cold surfaces where sulfur dioxide is deposited. Left: Loki patera (imaged by Voyager) is about 200 km in diameter and shows dramatic, quasi-periodic, infrared brightenings. Left: Tupan patera (imaged by Galileo) illustrates the result of lava interacting with sulfur-rich materials. Tupan is about 75 km across and is surrounded by cliffs about 0.9 km tall. The central "island" is, like Loki's, presumably higher than the dark floor, as it is not covered by lavas. Much of the area is coated with a diffuse red deposit that is thought to have condensed from sulfur gas escaping from volcanic vents. The floor of Tupan is covered with a surreal pattern of dark black (active lavas), green, red, and yellow materials. The yellow is presumed to be a mix of sulfurous compounds, and the green appears to form where red sulfur has interacted with the dark lavas. (NASA PIA02599).

explain the offset include: (1) a faster than expected rotation for Io; (2) an interior structure that permits magma to travel significant distances to surface eruption sites from where the most heating occurs; or (3) a missing component in existing tidal heating models, like fluid tides from a subsurface magma ocean. The existence of this discrepancy demonstrates the needs for better imaging of Io's surface features and better determination of Io's interior structure.

3. SURFACE EXPRESSIONS OF VOLCANISM

Volcanism on Io manifests itself in several discrete forms of surface features: (1) patera (pl. paterae, caldera-like features), (2) lava flows and flow fields, (3) volcanic "hills," including domes, cones, and possibly shields, and (4)

diffuse deposits (DD) of tephra and frosts derived from condensed volcanic gasses. Paterae are volcano-tectonic depressions that morphologically and structurally resemble the pit craters that occur at the tops of terrestrial shield volcanoes, although on Io they are larger in size (mean diameter 41 km, some >200 km) and more than 90% lack the surrounding shield-like edifice. These caldera-like features are the primary form of volcanic vent on Io, and range in shape from subcircular to very irregular, in which the irregular margins suggest tectonic faults may have influenced their formation. They are called paterae on Io rather than calderas because the connection to subsurface magma chambers typical of terrestrial calderas cannot be confirmed from spacecraft images alone. Nevertheless, high spatial resolution images of Io's paterae (Figure 43.5) suggest they are similar to terrestrial calderas with floors that contain confined lava flows, lava ponds, or lava lakes.

FIGURE 43.6   Lava flow fields on Io imaged by Galileo. Left: the 90-km long Prometheus flow field is surrounded by a bright plume deposit, the result of an apparently ever-present sulfur dioxide plume, centered at the distal end of the flow field. The plume is thought to result from the interaction of hot lava eroding through to a liquid sulfur dioxide substrate, with explosive results. To the right of the image, the main Prometheus patera, shaped like a half-moon, can be seen. Galileo Near-Infrared Mapping Spectrometer (NIMS) thermal data show that the flow is likely tube-fed, and hotter areas are seen near the main vent and at the distal edges. (NASA PIA02565). Right: the Amirani flow field is the longest active flow field in the solar system. The most recent lavas appear darkest because their surfaces are too hot to enable deposition by sulfur compounds from plumes. Thermal maps from the Galileo NIMS instrument show that fresh, hot lavas are coming out of several areas at the northern end of the flow. However, the main vent of the flow field appears to be near the southern end, where a half-moon shaped patera is located. The liquid lava travels under a frozen layer of older lava, breaking out onto the surface only after traveling hundreds of kilometers from the vent. The mosaic shows an area 500 km long and 180 km wide. Amirani (NASA PIA02567).

The variation in color of the floors of Io's paterae (ranging from bright white to yellow-orange to black, often with multiple colored deposits within one patera) suggests that the compositions of patera floor materials include mixes of silicates and various sulfur-bearing compounds, including detection by Galileo's **near-infrared mapping spectrometer (NIMS)** of very pure sulfur dioxide deposits at some localities.

Lava flows and flow fields (Figure 43.6) are typified by their generally elongated morphology (lengths >> widths), crenulate to lobate edges in planetary images, and sharp contacts with the other units. Flows on Io are generally characterized using color and albedo as bright (possibly composed of sulfur), dark (presumably composed of silicates), and undivided (uncertain composition). Albedo variations in the flows are generally thought to be indicative of age on the surface: the freshest dark flows are the darkest black in color; the freshest bright flows are the brightest yellow in color (i.e., brighter than all plains units). Radiation exposure and superposition by DD of various compositions tend to homogenize flow materials to a gray-green color. In fact, it is unclear whether the majority of bright flows are bright because they are composed of sulfur (which cools to a yellow color that changes to white from radiation exposure), or whether they are simply old, cold silicate flows covered by bright, sulfurous plume fallout (there are clear examples of both on Io). Morphologically, the highest resolution Galileo images of flow materials

suggest that they are reminiscent of terrestrial compound pahoehoe flow fields (e.g., Kilauea, Hawaii) or platy ridged lava flows (e.g., Laki, Iceland).

Volcanic hills on Io are called *tholi* (sing.: *tholus*, Latin for "dome"), which is a generalized term for any positive relief volcanic mountain. They are very rare on Io: the *Voyager* spacecraft imaged two low shield-like features (Apis and Inachus Tholi) in 1979, and several much smaller hills were identified during the *Galileo* mission. The relative rarity of volcanic edifices compared to lava flow fields is attributed to the inferred low dynamic viscosities of Ionian mafic to ultramafic lavas, whose emplacement favors widespread, low relief flows rather than the steeper edifices produced by more viscous lavas.

DD refer to the accumulations of frosts (derived from condensed volcanic gasses) and tephra that are found near or around vents and flow margins and that mantle underlying topography. Explosive eruption plumes that have been imaged by spacecraft (Figure 43.7) show that they have shapes, which range from umbrella-like in profile, producing circular haloes around vents, to asymmetric in profile, which produce irregularly shaped deposits that broaden away from the vent or flow margin. DD are the source of the widest range of color variations on Io, in which red, yellow, white, black, and green varieties have been observed. These colors are interpreted to be indicative of key chemical constituents of the deposits: red deposits are interpreted to be composed of short-chain

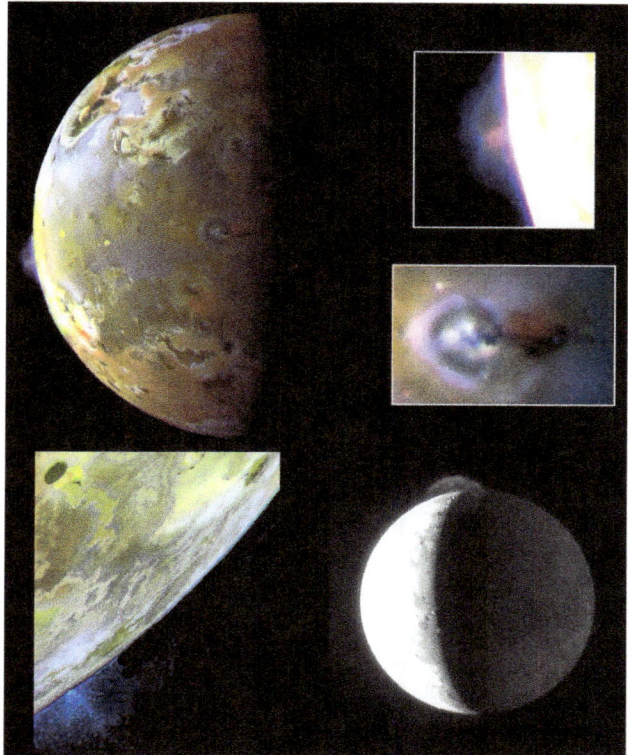

FIGURE 43.7 Volcanic plumes on Io imaged by Galileo (top and left) and New Horizons (bottom right). Top: the Pillan plume is seen against the bright limb in the larger image and also in the upper inset. This plume rose 120 km above the caldera of Pillan Patera volcano in 1997. The second plume in the larger image is erupting from the Prometheus volcano and seen near the terminator, rising about 75 km. The lower inset shows the shadow of this plume extending to the right of the volcano (NASA PIA00703). Bottom left: the Masubi plume is ejected some 100 km above the surface. This plume was seen to erupt from different locations within the Masubi plume, and is thought to be similar to the Prometheus plume, erupting from near the distal edges of the lava flow (NASA PIA02502). Bottom right: the New Horizons spacecraft, which flew by Io in 2007, captured this image of the Tvashtar plume rising some 330 km above the surface near Io's north pole. Io's dayside was deliberately overexposed in this image taken by the Long Range Reconnaissance Imager instrument to bring out details on the nightside and in any volcanic plumes that might be present. Hot, glowing lava at the source of the plume is the bright point of light on the sunlit side of the terminator. (NASA PIA09360).

sulfur $\pm$ sulfur chlorides, yellow deposits are interpreted to be composed of sulfur $\pm$ contaminants, white-gray deposits are interpreted to be composed of sulfur dioxide $+$ contaminants, dark or black deposits are interpreted to be composed of silicate ash, and green deposits are interpreted to be composed of products of silicate—sulfur alteration. Although DD are often ephemeral, fading after a few months when emplaced from a discrete eruption episode, long-term accumulation of red diffuse, yellow diffuse, and white DD from repeated or periodic eruptions could lead to the formation of red-brown plains, yellow bright plains, and white bright plains materials, respectively.

4. SURFACE CHANGES, COLOR, AND COMPOSITION

Comparison between Voyager and Galileo images taken nearly two decades apart showed many areas of surface changes, but not as many as might have been expected given Io's high rate of volcanism. Surprisingly, volcanoes where vigorous activity is known to have occurred, such as Loki, showed little surface change, presumably because the activity is mostly confined to the interior of the caldera. One of the most striking cases of surface change happened on Ra Patera (Figure 43.8), a region that so far has been identified only as an active plume site and not yet as a hot spot, though this could be due to poor spatial resolution of the thermal data over this region. Voyager images showed Ra to have a dark caldera surrounded by narrow flows. The flows showed a sequence of colors along their lengths that was consistent with the interpretation of Carl Sagan that Io's surface was covered by different **sulfur allotropes**, with the supposedly cooler materials further away from the vent. However, this interpretation of composition based on color was criticized on a variety of grounds, including the fact that the exact colors of sulfur can be drastically altered by even small amounts of other materials. Ra Patera became even more of a puzzle when observations from the Hubble Space Telescope showed that a significant brightening occurred in this region between 1994 and 1995. When images of Ra were returned after Galileo's first orbit, they showed that the flows had been covered over by new deposits and that a plume was now erupting from Ra.

The most significant surface changes detected by Galileo were localized changes due to major eruptions. The Pillan eruption of 1997 left a conspicuous "black eye" on Io's surface (Figure 43.3), covering an area of about 200×10^3 km^2 and spreading about 260 km from the source. This dark deposit slowly faded between 1997 and 1999, as it was covered by red deposits from the nearby Pele plume. The Tvashtar eruption that started in 1999 created a large annular red deposit as the result of a plume and the Thor eruption that started in 2001 and resulted in the largest plume so far observed from Io ($\sim$500 km) left a white-gray plume deposit on the surface.

Galileo results brought new insights into what causes the vivid colors of Io's surface, including the colors of the flows that are now gone. Surface colors are the most easily observed expressions of surface changes and can provide insight into composition, however, different illumination angles affect how colors appear in images, so the analysis is not straightforward. Galileo's repeated flybys allowed observations at various illumination angles, and different color deposits were identified. Io's surface has four major color units. About 40% of the surface shows yellow coloration, about 30% appears red to orange, 27% appears white to gray, and black deposits ($\sim$2%) are seen mostly in

FIGURE 43.8 Possible sulfur flows on Io. Top: Voyager 1 image of the Ra Patera volcano on Io taken in 1979. Narrow, lobate lava flows are seen coming out of a caldera-like patera. The colors of the flows were interpreted as colors of different sulfur allotropes. When Galileo imaged Io in 1996, those flows had been covered over by younger deposits. Image is 1000 km across (NASA PIA00361). Right: Emakong patera and bright, channelized flows imaged by Galileo in 1999. The top image, covering an area of about 120 by 40 km, shows a bright flow with a distinct dark channel in the middle. The winding channel and serrated margins suggest a fast-moving flow over shallow slopes. The unusual bright color of the flow may indicate that it is sulfur rather than silicate. The Emakong caldera-like patera is the dark, heart-shaped feature on the bottom part of the picture, about 55 km across. (NASA PIA02518).

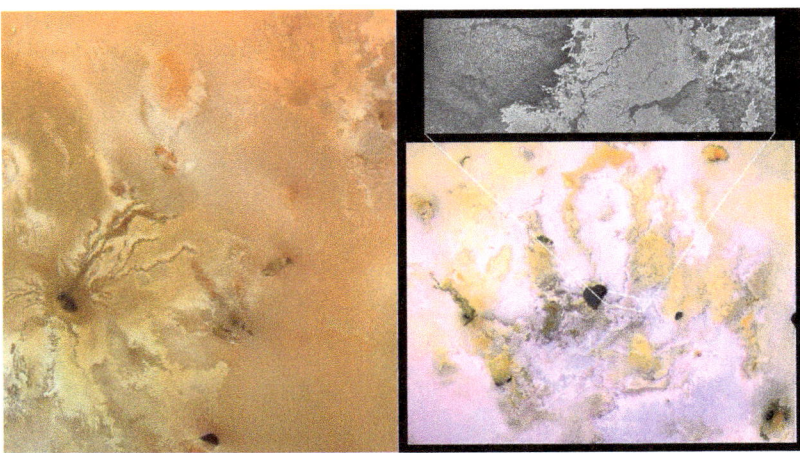

or near active areas. There are also some latitudinal differences. Red and orange materials are thought to be deposits of short-chain sulfur molecules (S_3, S_4) and are found mostly at latitudes higher than $30°$ N and S, where they are thought to result from the breakdown of sulfur (cyclo-S_8) by irradiation from charged particles. Although the time range of observations is limited, it appears that red deposits last longer at higher latitudes, perhaps because the color is the result of irradiation and therefore the origin of the red deposits may be different at higher latitudes. At lower latitudes, red deposits appear to be ephemeral, are associated with hot spots and plumes and probably caused by condensation and fallout of material from sulfur-rich plumes. In some locations, these deposits have been seen to fade to yellow over several years. The vast expanses of equatorial plains are dominated by yellow, white, and gray materials that are generally anticorrelated with active volcanoes, while the black deposits are associated with volcanic activity, occurring in isolated patches, both as DD that are probably pyroclastic in nature and as continuous units confined to caldera interiors.

Sulfur and sulfur compounds are the most viable compositional candidates for the variegated colors of Io, though silicate materials are thought to be exposed in the black areas associated with volcanic activity. Some of the very dark materials that are associated with calderas and active hot spots show an apparent spectral absorption feature at 0.9 μm, seen in the Galileo imaging data and interpreted by P. Geissler and colleagues as being due to orthopyroxene. Although further results are needed to confirm the presence of this spectral feature and to provide a unique mineral identification, the presence of orthopyroxene on Io, a magnesium-rich silicate mineral common in terrestrial mafic and ultramafic rocks, seems likely.

Yellow deposits are interpreted either as cyclo-S_8, possibly with a thin covering of SO_2 frosts deposited by

plumes, or polysulfur oxide and SO_2 without large quantities of elemental sulfur. White and gray materials are interpreted as coarse to moderate grained SO_2 that condensed from plumes and later recrystallized. Perhaps the most intriguing materials on Io's surface are the greenish-yellow deposits seen in a few isolated spots (Figure 43.9). Unlike other terrains on Io, these areas show a negative near-infrared spectral slope, strongly suggesting that a nonsulfur component is present. These green materials could be made up of sulfur contaminated by iron, as suggested by J. Kargel and colleagues, however, the fact that an active hot spot is seen close to some of these areas (such as Balder and Tohil) is consistent with the alternative explanation of silicate deposits such as olivine or clinopyroxene.

The poor spatial resolution of the observations of Io available so far and the thin coating of SO_2 that seems to be prevalent on much of the surface have made it difficult to identify the compositions of the different deposits on Io. The Galileo NIMS instrument detected a broad absorption centered at about 1 μm, but it is still not known what the absorption is due to, a clue is that it is anticorrelated with recently emplaced lavas. Unfortunately, radiation damage to the NIMS instrument during Galileo's close flybys of Io, where high spatial resolution data were obtained, significantly diminished the spectral resolution of the instrument and many questions remain about the composition of Io's surface.

5. HOT SPOT TEMPERATURES AND COMPOSITION OF MAGMA

Shortly after the Voyager mission, the major controversy about Io's volcanism after results from the Voyager mission concerned the nature of the volcanism: sulfur or silicates?

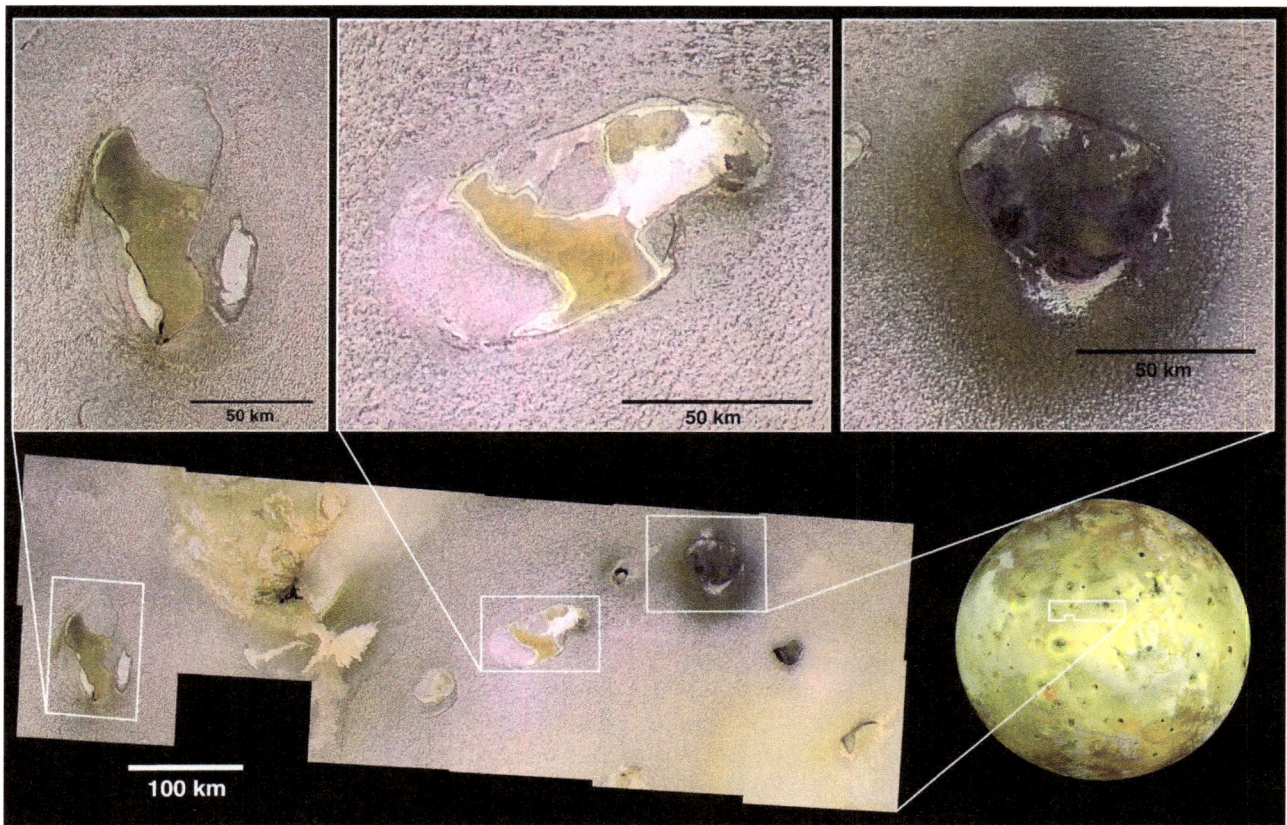

FIGURE 43.9   Colorful volcanic deposits and variety of paterae on Io. The region known as Chaac—Camaxtli was imaged by Galileo in 2000. Chaac patera is the green-floored depression on the upper left. To the right of Chaac is Balder patera with a distinct bright white floor. The Near-Infrared Mapping Spectrometer instrument on Galileo detected revealed that the white is nearly pure SO_2. The middle image shows orange and white materials presumed to be sulfurous deposits, and what may be thin plates of crust that have broken off and rafted over a lava lake. Between these two patera are bright flows that may be sulfur flows. The top right patera, Camaxtli, shows dark deposits indicative of silicates. (Bottom image is PIA02566).

Io's surface colors were initially interpreted as sulfur deposits. The temperatures of the hot spots measured by the Voyager IRIS instrument were relatively low—mostly around 400 K—and could be consistent with either molten sulfur or silicates. Sulfur volcanism could produce temperatures up to ~700 K, while basaltic lavas on Earth range mostly from 1300 to 1450 K. Between the Voyager observations in 1979 and the Galileo observations that started in 1996, several of Io's hot spots were detected by ground-based telescopes. Temperature measurements using infrared detectors mounted on telescopes showed higher temperatures than had been measured by Voyager—up to 1500 K were reported. These measurements are consistent with silicate magmas but not with sulfur volcanism. Sulfur boils vigorously on Io's surface at about 700 K, hence these measurements strongly suggested silicate magmas. Galileo observations, and those that have followed since, showed temperatures consistent with silicate volcanism, although it is possible that sulfur volcanism may be happening in a few areas.

Galileo carried instruments more capable of detecting smaller areas at higher temperatures, including the **solid state imaging (SSI) instrument**, sensitive from ~0.4 to 1.0 μm and the NIMS (sensitive from 0.7 to 5.2 μm). These instruments soon showed temperatures consistent with basalts, however, in 1997, measurements of the violent Pillan eruption showed surprising results. A. McEwen and colleagues showed that temperatures exceeded 1500 K at several hot spots and, in the case of Pillan, for which NIMS and SSI data were combined, lava temperatures were about 1800 K, suggesting ultramafic compositions. This explanation was supported by the discovery of the 0.9 μm spectral absorption feature discussed above, which is interpreted as magnesium-rich orthopyroxene. However, more recent reanalysis of the Pillan data by L. Kesthelyi and colleagues suggest lower temperatures, about 1600 K, which could be due either to ultramafic lavas or to unusually hot (superheated) basalts. Measurements from Cassini and New Horizons at other hot spots showed temperatures in the basaltic range.

The composition of Io's lavas remains unresolved, it is agreed that the Pillan eruption temperature, the highest recorded, could be due to superheated basalts or ultramafic (komatiite-like) lavas that had cooled by a couple of hundred degrees Kelvin at the time the measurements were made. Because there are no direct measurements of the composition of Io's lavas, the question remains open, particularly because small areas of high temperature lavas would not have easily been detected given the spatial resolution of the available observations.

6. STYLES OF VOLCANIC ACTIVITY

Io is a dynamic body—the volcanic plumes turn on and off, the outpourings of magma vary in temperature and extent, and sometimes large-scale changes are seen to occur in a matter of months, completely changing the appearance of a region. Not all volcanic activity on Io produces plumes and plume deposits, or large lava flows. Eruptions can be generally classified as effusive or explosion-dominated, and a classification scheme evolved from observations of individual volcanoes, much like the system used for terrestrial eruptions (e.g., Hawaiian, Strombolian). The major eruption types on Io are flow-dominated (also referred to as "Promethean" after the Prometheus volcano, and accompanying plumes are referred to as "Prometheus-type"), explosion-dominated (or "Pillanian" after the Pillan volcano, and accompanying plumes often referred to as Pele-type plumes), and intra-patera (or "Lokian" after the Loki volcano). Two other types are much rarer, and may involve the eruption of sulfur or SO$_2$.

Flow-dominated eruptions originate either from fissures or paterae (calderas) and produce extensive compound lava flow fields, probably through repeated breakouts of lava similar to compound inflationary flows commonly observed on Hawaiian eruptions on Earth. The two best studied flow fields on Io are Amirani and Prometheus (Figure 43.6). Observations taken about 3 months apart using Galileo's SSI and NIMS instruments showed fresh breakouts of lava in these two flow fields, and temperature profiles obtained by NIMS showed that the flow fields were consistent with largely insulated flows. A curious phenomenon observed first at Prometheus is the movement of the eruption plume. Between Voyager and Galileo observations, the Prometheus plume source and associated deposit had moved ~90 km to the east, advancing with a lava flow. Unlike most terrestrial plumes, these Io plumes do not erupt from the vent but from the distal edges of the flow, a result of hot silicates bury the icy, SO$_2$-rich substrate as proposed by S. Kieffer and colleagues. Smaller jets at the active margins of the flow can also be seen in Galileo images. Although there is no good terrestrial analogue, the plume-generating process on Io is similar to the formation of pseudocraters as lava advances over soggy ground. On

Io, this type of plume shuts off if the lava flow stops advancing, such as happened at the Maui plume and flow. The plumes of these flow-dominated eruptions are relatively small (~100 km high), optically dense, and composed mostly of dust and SO$_2$ gas rather than S gas. An image of the shadow of Prometheus taken by Galileo shows a dense vertical column of dust topped by a mushroom-shaped canopy. The central part may be populated by fine particulates. The resulting plume deposit is white-gray, SO$_2$-rich deposits, rather than the giant red deposits (S-rich) seen at Pele and other locations.

Explosion-dominated eruptions differ from the flow-dominated type in that most of the energy of the eruption is directed into a short-lived, vigorous event that lasts days to weeks. They can originate from either paterae or fissures, but are discreet events compared to the more or less continuous flow-dominated eruptions such as Prometheus. Explosion-dominated eruptions can produce extensive pyroclastic deposits and lava flow fields, and typically a large (>200 km high) plume, thought to originate from the interaction of silicate magma with sulfurous **volatiles**. These large plumes often create large red ring deposits around the source regions (like at Tvashtar and Pele) though the Pillan event of 1997 (Figure 43.3) produced a dark deposit as discussed above. These eruptions often also produce extensive flow fields, but over a shorter period of time than the flow-dominated eruptions. For example, the 1997 Pillan eruption produced a flow field ~3100 km^2 in area, formed in a relatively short period of time (estimated range 52–167 days). The estimated volumetric flow rate is ~1740–7450 m^3 s^{-1}, similar to those calculated for the 1783 Laki eruption. Williams and colleagues analyzed high resolution images (20–30 m/pixel) of the Pillan flow and suggested that the exceptionally rough, disrupted, and platy upper surface was the result of rapidly emplaced flows, possibly turbulent. In contrast, volumetric eruption rates estimated for the Amirani and Prometheus flow fields, based on observed surface changes, range from 50 to 500 m^3 s^{-1}.

The most common type of eruption on Io is the inter-patera, which are confined within paterae (calderas). These eruptions occur with or without associated plumes, and are thought to be lava lakes in a number of cases. Lava lakes on Io were discovered from the thermal signatures of high resolution NIMS images, which revealed hot margins at the edges of paterae (Figure 43.10), similar to terrestrial lava lakes where the crust is broken up as the crusted-over surface of the lava lake hits the crater walls. Most of Io's active hot spots coincide with the interiors of paterae, suggesting that most of the lava resurfacing on Io is confined within these depressions. Io's most powerful volcano, Loki, produces eruptions of this type, which have been monitored from Earth as well as from spacecraft over several decades, revealing reoccurring, almost periodic

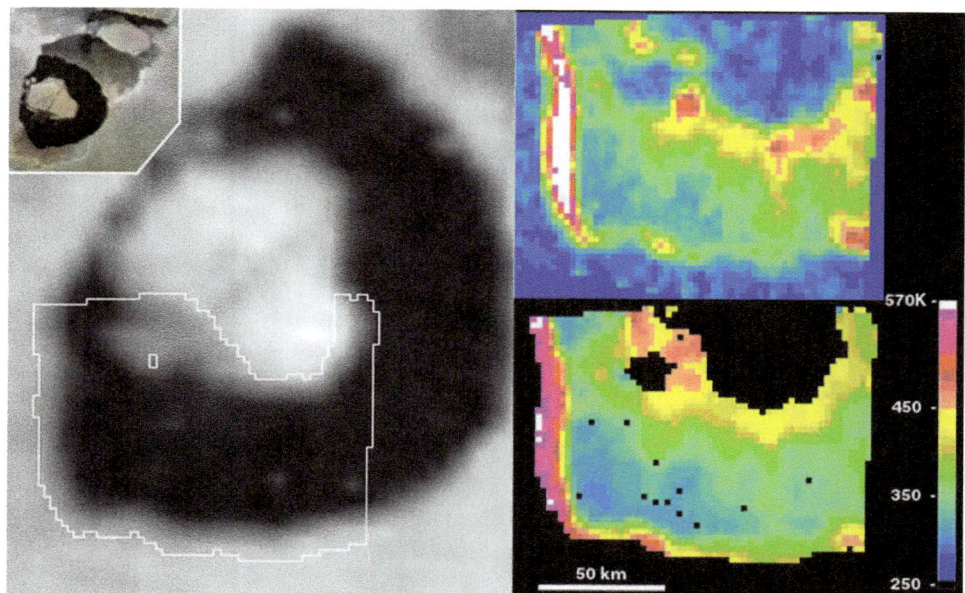

FIGURE 43.10   Thermal emission detected from Loki patera by Galileo's Near-Infrared Mapping Spectrometer (NIMS) instrument. Top left is an image of the ∼200 km across Loki patera from Voyager 1 (see Figure 43.5), bottom left is Loki imaged by Galileo's Solid State Imaging instrument indicating the area at the bottom and most active part of the patera imaged by NIMS in 2001. Top right shows a NIMS image at 2.5 μm. Bottom right is a one temperature fit to the NIMS multiple wavelength data, showing that the hottest lavas are seen near the edges of the patera, indicating a lava lake. (Lopes et al., 2004).

brightenings of Loki in the infrared. Surprisingly, the appearance of Loki did not significantly change in the years between Voyager and Galileo observations, despite many powerful eruptions. The eruption mechanism has been interpreted as either repeated foundering of the crust of a lava lake or, alternatively, repeated flooding of the caldera followed by drainback, in a manner similar to the eruptions from the East Pacific Rise on Earth. If either of these mechanisms is taking place on other paterae, it must be on a much reduced scale, not leading to the large fluctuations in power output as happens at Loki. Other paterae appear to contain lava lakes, however, the formation mechanism for paterae themselves is not well understood, particularly the formation of "islands" in the interior of several of the paterae such as Loki and Tupan. These "islands" are cold regions covered by SO_2 frost that are partially or completely surrounded by lavas and appear to remain remarkably stable over decades.

It is possible that nonsilicate eruptions are taking place on Io, but they are thought to be secondary in nature, due to the remelting and remobilization of crustal sulfurous materials by adjacent silicate heat sources. Examples include, bright flows surrounding smaller volume dark flows at Sobo patera and a yellow and white-gray, 290 km long flow extending from Emakong patera (Figure 43.9). The best evidence for active sulfur volcanism, but not conclusive, is a weak hot spot detected by NIMS at Tsui Goab Fluctus, a bright yellow flow field adjacent to a small, apparently inactive shield volcano in the Culann–Tohil region. The temperature measured using data from Galileo NIMS falls in the range of molten sulfur (∼530 K ± 95 K) and the location coincided with the yellow flow, not dark silicates.

Another possible eruption type of Io is eruption of liquid SO_2 from the substrate. The NIMS instrument detected a strong signature of pure SO_2 confined inside to the floor of Balder patera (Figure 43.9), which visible images show has a homogeneous white floor. It is unclear why the floor should be so SO_2-rich relative to surrounding regions and it is possible that an SO_2 glacial-like flow may have erupted and flooded the floor of the patera. Another possible location for this type of volcanism is near Tohil patera. If these deposits were indeed emplaced by SO_2 coming from the substrate, most would vaporize when exposed to Io's tenuous atmosphere but, given sufficiently large quantities, it is estimated that some could freeze and form a layer at the bottom of paterae.

7. GLOBAL GEOLOGY AND DISTRIBUTION OF VOLCANISM

A global assessment of volcanic features on Io was obtained through production of the first complete global geologic map of Io, which was based on a series of combined *Galileo* and *Voyager* image mosaics that cover Io's surface at 1 km/pixel spatial resolution. A version of the

FIGURE 43.11 Mollweide projection of the 1:15,000,000 global geologic map of Io, centered on the antijovian point (0°, 180° W), from *Williams et al.* (2011). The areal distribution of map units is given in Table 43.2. Map unit abbreviations: p_{bw}, white bright plains; p_{by}, yellow bright plains; p_{rb}, red-brown plains; p_l, layered plains; f_d, dark flows; f_b, bright flows; f_u, undivided flows; pf_d, dark patera floors; pf_b, bright patera floors; pf_u, undivided patera floors; t, tholi; m_l, lineated mountains; m_m, mottled mountains; m_u, undivided mountains; pr, region of poor resolution (likely red-brown plains). Diffuse deposits are not displayed in this version of the map.

map is shown in Figure 43.11. At the global scale, Io's surface can be subdivided into four major classes of units, each with three subclasses: (1) paterae (caldera-like, volcano-tectonic depressions, whose floors are bright, intermediate, or dark); (2) lava flow fields (which are bright, intermediate, or dark); (3) mountains (which are linear, undivided, or mottled); and (4) plains (which are white, yellow, or red-brown). A fifth unit, DD, occur in five colors (red, white, yellow, green, and black) and mantle the other units. Other units that are found on Io include layered plains (which serve as a transition between mountains and plains units) and tholi (which are positive relief features that are volcanic domes or cones). Table 43.2 shows the percent coverage of Io by these units in terms of surface area.

Plains units are thought to be composed of Io's underlying silicate crust overlain by sulfur and sulfur dioxide (SO_2) emplaced via effusive and explosive volcanic processes. Mapping shows that geography constrains the distribution of these plains units, in which red-brown plains occur $> \pm 30°$ latitude, white plains occur mostly in the equatorial antijovian region ($\pm 30°$, $90°-230°$ W), and yellow plains everywhere else. The red-brown plains are thought to result from enhanced alteration of other units induced by radiation coming in from the poles. White plains are thought to be dominated by $SO_2 +$ contaminants, and their restriction to one region is suggestive of a regional cold trap. Yellow plains are thought to be composed of cyclo-octal

sulfur (S_8) that accumulates from sulfur-rich plume deposits. However, there are outliers of white, yellow, and red-brown plains in other areas of Io that appear to result from long-term accumulation of white, yellow, and red DD, respectively.

Bright lava flow fields, possibly composed of sulfur-rich compounds, make up 30% more lava flow fields than dark, presumably silicate lava flow fields (56.5% vs 43.5%). Furthermore, only 18% of bright flow fields occur within close proximity (e.g., within 10 km) of dark flow fields, which would be expected if sulfur flows result from melting of crustal sulfur from adjacent silicate magma chambers (i.e., *secondary* sulfur volcanism). These results suggest that *primary* sulfur volcanism (i.e., eruptions from sulfur magma chambers) could be an important component of Io's recent volcanism. An unusual concentration of bright flows at $\sim 45-75°$ N, $\sim 60-120°$ W could be indicative of more extensive primary sulfur volcanism in the recent past. However, it remains unclear whether most bright flows are bright because they are composed of sulfur flows, or because they are composed of cold silicate flows covered in sulfur-rich particles from plume fallout. Nevertheless, about 29% of all of Io's flow fields are very bright or very dark flows (i.e., young and fresh), suggesting that active lava flow emplacement is occurring in less than one-third of Io's flow fields at the present time.

Global mapping identified 425 paterae (volcano-tectonic depressions), up from 417 previously identified.

TABLE 43.2 Distribution of Geologic Material Units as Percentage of Io's Surface Area

Unit Label	Unit Name	Area (km^2)	Area (%)
Plains Deposits			
p_{rb}	Red-brown plains material	1.41×10^7	33.4
p_{by}	Yellow bright plains material	7.68×10^6	18.4
p_{bw}	White bright plains material	3.75×10^6	8.9
p_l	Layered plains material	1.84×10^6	4.4
	Region of poor resolution (likely red-brown plains material)	7.20×10^5	1.7
	Total plains	2.81×10^7	**66.6**
Mountain Deposits			
m_l	Lineated mountains material	6.40×10^5	1.5
m_m	Mottled mountains material	8.05×10^4	0.2
m_u	Undivided mountain material	5.54×10^5	1.3
	Tholi (domes)	5.25×10^4	0.1
	Total mountain material	1.33×10^6	**3.1**
Patera Floor Deposits			
pf_b	Bright patera floor material	1.84×10^5	0.4
pf_d	Dark patera floor material	1.93×10^5	0.5
pf_u	Undivided patera floor material	6.75×10^5	1.6
	Total patera floor material	1.05×10^6	**2.5**
Flow Deposits			
f_b	Bright flow material	1.80×10^6	4.3
f_d	Dark flow material	1.23×10^6	2.9
f_u	Undivided flow material	8.70×10^6	20.6
	Total flow material	1.17×10^7	**27.8**
Diffuse Deposits			
d_{by}	Yellow bright diffuse material	8.76×10^5	2.1
d_{bw}	White bright diffuse material	2.90×10^6	6.9
d_r	Red diffuse material	3.61×10^6	8.6
d_d	Dark diffuse material	2.68×10^5	0.6
d_g	Green diffuse material	4.09×10^3	0.01

Note: Diffuse deposits are superposed on all other materials, and cover 18.2% of Io's surface. Io's surface Area $= 4.17 \times 10^7$ km^2.
From *Williams et al.* (2011).

Although paterae cover only 2.5% of Io's surface, they correlate with 64% of all detected hot spots. In particular, 45% of Io's hot spots correlate with dark patera floors, which are interpreted to be lava lakes or lava ponds composed of mafic to ultramafic materials, and demonstrate the dominance of active silicate volcanism to Io's heat flow. Of the ~3% of Io's surface that consists of mountains, only 0.1% are tholi, interpreted to be volcanic shields, domes, or cones.

About 18% of Io's surface at any given time is covered by DD, which are thought to be composed of ash and/or frosts derived from condensed volcanic gases from Io's many plume eruptions. About 47% of these DD (by area) are red, presumably deriving their color from condensed sulfur (S_2) gas that recrystallizes to short-chain sulfur (S_3–S_4). In contrast, about 38% of DD are white, presumably dominated by condensed SO_2. The much greater areal extent of gas-derived DD (red + white, 85% of all DD) compared to presumably pyroclast-bearing DD (dark (silicate tephra) + yellow (sulfur-rich tephra), 15% of all DD) indicates that there is effective separation between the transport of tephra and gas in many Ionian explosive eruptions.

Mountains are the dominant structural landforms that are visible on Io and are recognized as steep-sided edifices rising more than ~1 km above the plains, covering ~3% of the surface. Approximately 150 mountains have been identified and mapped. Io's mountains typically rise approximately 6 km in height; the highest (Boösaule Montes) rises >17 km above the surrounding plains. Galileo images show that many mountains are partly or completely surrounded by plateaus, layered plains, and debris aprons. Schenk and Bulmer suggested that a horizontal lithospheric compressive stress is generated because of Io's rapid resurfacing rate, which results in uplift of crustal blocks via thrust faulting. Alternatively, McKinnon and others suggested that sustained reduction in Io's volcanic activity at local, regional, or global scales results in lithospheric heating that causes a large compressive stress at the base of the lithosphere. This stress causes, over time, fluctuating thermally induced stress as resurfacing rates wax and wane, leading to alternating episodes of tensile and compressive faulting. Such repeated episodes of normal and reverse faulting might produce coherent crustal blocks (mountains) that float in a matrix of highly disrupted material, similar to the chaos terrain of Europa. Regardless, the asymmetrical shapes of Io's mountains suggest uplift along thrust faults, implying that compressional uplift is probably the dominant mechanism. Several studies have shown statistical evidence that about 40% of Ionian paterae and mountains are in direct contact with each other, more than random chance would allow (probability ~0.1%). This result suggests a possible genetic link; perhaps the magma exploits the weakness in the lithosphere created by deep orogenic faults. Although a global anticorrelation between the areas with somewhat higher concentrations of observed mountains and the areas with somewhat higher concentrations of observed volcanic centers has been identified, the areas of low mountain concentration also correlate with areas with poor imaging for detecting topographic features. Therefore, it is not yet confirmed that this anticorrelation is not an observational bias.

8. ATMOSPHERE AND TORUS

SO_2 is not only common over the surface of Io but is also the major component of Io's tenuous atmosphere. SO_2 and sulfur are thought to be the volatiles that drive explosive volcanism on Io, as H_2O and CO_2 (the dominant volatiles driving explosive volcanism on Earth) are either absent or highly depleted. SO_2 in the atmosphere is supplied largely by the plumes, but a lesser amount comes from the evaporation of the SO_2 frost deposits on the surface. Some plumes, like Pele, also supply sulfur (S_3 and S_4 have been detected) in addition to SO_2. Observations show that Io's atmosphere is patchy, the average density is low (about 10^{-9} bar) but the density is greater at the location of the active plumes. Io's large volcanic plumes serve as an efficient delivery mechanism for gas and dust particles into Io's atmosphere and beyond. The dynamics of Io's plumes are complex as models have to take into account the very low atmospheric pressure and poorly constrained variables, such as the relative amount of dust and gas in the various plume types. The plumes associated with eruption types, such as at Prometheus and Pele, have been discussed above, but the existence of a third type of plume has been inferred and later supported by Galileo observations. Observations of Io's atmosphere at millimeter wavelengths showed that the amount of SO_2 gas detected would require about 30–50 large plumes to be erupting at one time. Since the two Voyager spacecraft had only detected nine, the question of "missing plumes" arose. Spacecraft observations detect plumes because sunlight is scattered by the small particles of SO_2 within the plumes. The existence of stealth plumes was proposed, that is, plumes composed almost entirely of gas, which would be hard to detect, but that modeling work suggested were possible. These stealth plumes are thought to occur when a hot, molten silicate intrusion encounters a reservoir of liquid SO_2 buried at depths of at least 1.5 km. The high temperature and pressure at the subsurface (>1400 K, >40 bars) results in an eruption that is geyser-like, in which particles do not form through condensation. The existence of stealth plumes has been supported by Galileo observations, including those from the camera. The plume over the hot spot Acala is not seen in sunlight, presumably because of lack of particles in the plume, but it can be seen in observations of Io in eclipse, in which the gas appears to glow (Figure 43.12).

Volcanic outgassing and SO_2 evaporation continuously replenishes Io's atmosphere, but the moon's low gravity allows some material to escape into space, estimated to be at a rate of 1 t s^{-1}. The material forms a corona, neutral clouds, and the Io torus. The corona is a low density shell within Io's gravitational pull populated by atoms and molecules from Io. Neutral clouds of oxygen, sodium, and sulfur extend from the corona to distances of many times the radius of Jupiter. The cloud was discovered from observations by the Pioneer spacecraft that flew by Jupiter in 1973. The most easily observed of the clouds is the sodium cloud, populated by sodium atoms escaping Io at ~2.6 km s^{-1}. Although neutral sodium atoms are more easily detected, Earth-based telescopic studies indicate that

FIGURE 43.12 Io in eclipse (in Jupiter's shadow) showing glowing gases at visible wavelengths (red, green, and violet). The colors, caused by collisions between Io's atmospheric gases and energetic charged particles trapped in Jupiter's magnetic field, had not previously been observed. Bright blue glows mark the sites of dense plumes, and may be places where Io is electrically connected to Jupiter. (NASA PIA01637).

sodium is a minor constituent of the neutral clouds. The primary neutral elements are oxygen and sulfur, which are thought to be dissociated from SO_2. So far, sodium has not been detected either on Io's surface or plumes, but its existence on Io is known from the material in the neutral clouds.

Io's orbit is deep within the Jovian magnetopause, and the interaction of the materials escaping Io with Jupiter's magnetic field is unique in the solar system. The Io plasma torus is a doughnut-shaped trail about 143,000 km wide along Io's orbital path, made up almost exclusively of various charged states of sulfur and oxygen derived from the breakup of volcanic SO_2 and S_2. The ionized particles are held within the torus by Jupiter's magnetic field, in a similar way to the mechanism that holds charged particles in the Van Allen radiation belts around the Earth. It is thought that Io's variable volcanic activity influences the density of the plasma torus and the strength of its interactions with the Jovian magnetic field.

A consequence of the complex interactions between Io and Jupiter's magnetosphere is the existence of an aurora on Io, first detected from observations taken by the Galileo camera while Io was in Jupiter's shadow (Figure 43.12). The vivid colors detected (red, green, and blue) are caused by collisions between Io's atmospheric gases and energetic particles trapped in Jupiter's magnetic field. Observations of Io's aurora were also made by the New Horizons spacecraft in 2007, which observed Io as it went in and out of Jupiter's shadow, showing variations in aurora brightness and morphology.

The effects of Io's volcanic activity are felt throughout the Jovian system, which is a unique interaction in our solar system. Although new in situ observations will only happen when another spacecraft gets to the Jovian system, progress is being made using telescopic observations. Variations in the torus and neutral clouds, particularly of sodium and sulfur, can be measured from Earth and these are thought to respond to Io's variations in volcanic activity, though the nature of the correlations is not simple and still not understood.

9. FUTURE STUDIES

NASA's *Galileo* spacecraft completed its observations of Io in 2002, and the only subsequent spacecraft observations of Io occurred during the February 2007 flyby of NASA's *New Horizons* spacecraft on its way to Pluto. Earth-based telescopic observations of Io continue, but only intermittently. However, new research continues on the existing Io data sets which led, for example, to the publication of the first complete global geologic map of Io in 2012. The global mapping of Io's volcanic materials and the resulting geographic and spatial information is now feeding new studies to investigate Io's interior processes.

The community of Io scientists has produced a blueprint of future Io exploration, which was submitted in the form of a pair of white papers for the 2011 Planetary Science Decadal Survey, *Visions and Voyages*, which included recommendations for future missions with Io as the major target. However, Earth-based telescopic observations and observations from the Hubble Space Telescope continue to yield results that help our understanding of Io. The only mission for which plans are underway to observe Io's surface is the European Jupiter Icy Moons Explorer.

Many important questions remain open about the nature of Io's volcanism. Perhaps the most fundamental is the compositional range of the magma. If ultramafic magmas are indeed present, it raises the question of how these primitive magmas could persist on such a dynamically active body where extreme **differentiation** should be expected. The question of whether sulfur flows exist on Io is still open and, if they do, are they emplaced due to remelting of sulfur deposits by hotter magmas? High spatial resolution composition and thermal measurements are needed to answer this question. Many others questions about how Io's powerful volcanoes work remain. The formation mechanism of the most widespread volcanic landform, the patera (caldera-like depressions) is not understood. Nearly half are found adjacent to mountains, indicating a genetic link between mountains and paterae. Many patera interiors have "islands," which have persisted over decades despite constant and vigorous activity inside the paterae. How the magma is supplied to replenish eruptions, which last for years or decades? Is there indeed a magma ocean under the lithosphere? Most of all, studies of Io's volcanoes can provide insight into how volcanism behaves in an extreme environment, perhaps similar to conditions in the early Earth.

FURTHER READING

Geissler, P.E., McEwen, A.S., Keszthelyi, L., Lopes-Gautier, R., Granahan, J., Simonelli, D.P., 1999. Global Color Variations on Io. Icarus 140 (2), 265–281.

Gregg, T.K.P., Lopes, R.M., 2008. Lava Lakes on Io: New Perspectives from Modeling. Icarus. http://dx.doi.org/10.1016/j.icarus.2007.08.042.

Hamilton, C.W., Beggan, C.D., Still, S., Beuthe, M., Lopes, R.M.C., Williams, D.A., Radebaugh, J., Wright, W., 2013. Spatial distribution of volcanoes on Io: Implications for tidal heating and magma ascent. Earth and Planetary Science Letters 361, 272–286. http://doi.org/10.1016/j.epsl.2012.10.032.

Johnson, T.V., Matson, D.L., Blaney, D.L., Veeder, G.J., Davies, A.G., 1995. Stealth plumes on Io. Geophys. Res. Lett. 22, 3293–3296.

Kezthelyi, L., Jaeger, W., Milazzo, M., Radebaugh, J., Davies, A.G., Mitchell, K.L., 2007. New estimates for Io eruption temperatures: Implications for the interior. Icarus 192, 491–502.

Khurana, K.K., et al., 2011. Evidence of a Global Magma Ocean in Io's Interior. Science 332, 1186. http://doi:1126/science.1201425.

Kieffer, S.W., Lopes-Gautier, R., McEwen, A.S., Keszthelyi, L., Carlson, R., 2000. Prometheus, the Wanderer. Science 288, 1204–1208.

Lopes, R.M.C., Spencer, J.R., 2007. Io after *Galileo*. Springer/Praxis Publishers, Chichester, UK, 342 p.

Lopes, R., Kamp, L.W., Smythe, W.D., Mouginis-Mark, P., Kargel, J., Radebaugh, J., Turtle, E.P., Perry, J., Williams, D.A., Carlson, R.W., Douté, S., 2004. Lava Lakes on Io. Observations of Io's Volcanic Activity from Galileo during the 2001 Fly-bys. Icarus 169 (1), 140–174.

McEwen, A.S., Keszthelyi, L., Spencer, J.R., Schubert, G., Matson, D.L., Lopes-Gautier, R., Klaasen, K.P., Johnson, T.V., Head, J.W., Geissler, P., Fagents, S., Davies, A.G., Carr, M.H., Breneman, H.H., Belton, M.J.S., 1998a. Very High Temperature Volcanism on Jupiter's Moon Io. Science 281, 87–90.

McKinnon, W.B., Schenk, P.M., Dombard, A.J., 2001. Chaos on Io: A model for formation of mountain blocks by crustal heating, melting, and tilting. Geology 29, 103–106.

Peale, S.J., Cassen, P., Reynolds, R.T., 1979. Melting of Io by tidal dissipation. Science 203, 892–894.

Radebaugh, J., Keszthelyi, L.P., McEwen, A.S., Turtle, E.P., Jaeger, W., Milazzo, M., 2001. Paterae on Io: A new type of volcanic caldera? J. Geophys. Res. 106, 33,005–33,020.

Rathbun, J.A., Spencer, J.R., Davies, A.G., Howell, R.R., Wilson, L., 2002. Loki, Io: A truly periodic volcano? Geophys. Res. Lett. 29 (10). http://dx.doi.org/10.1029/2002GL014747.

Retherford, K.D., et al., 2007. Io's Atmospheric Response to Eclipse: UV aurorae observations. Science 318, 237–240.

Sagan, C., 1979. Sulfur flows on Io. Nature 280, 750–753.

Schenk, P.M., Bulmer, M.H., 1998. Origin of Mountains on Io by Thrust Faulting and Large-Scale Mass Movement. Science 279, 1514–1517.

Spencer, J.R., Stern, S.A., Cheng, A.F., et al., 2007. Io volcanism seen by *New Horizons*—A major eruption of the Tvashtar volcano. Science 318, 240–243.

Spencer, J.R., Jessup, K.L., McGrath, M.A., Ballester, G.E., Yelle, R., 2000a. Discovery of Gaseous S_2 in Io's Pele Plume. Science 288, 1208–1210.

Williams, D.A., Keszthelyi, L.P., Crown, D.A., Yff, J.A., Jaeger, W.L., Schenk, P.M., Geissler, P.E., Becker, T.L., 2011. Geologic map of Io. U.S. Geological Survey Scientific Investigations Map 3168 scale 1:15,000,000, 25 p. Available at. http://pubs.usgs.gov/sim/3168/.

Witteborn, F.C., Bregman, J.C., Pollack, J.B., 1979. Io, an intense brightening near 5 micrometers. Science 203, 643–646.

Chapter 44

Cryovolcanism in the Outer Solar System

Paul Geissler

US Geological Survey, Center for Astrogeology, Flagstaff, AZ, USA

Chapter Outline

GLOSSARY

albedo Reflectivity of the surface, ranging from 0 (perfectly absorbing) to 1 (perfectly reflecting), important in determining surface temperature.

apoapsis The farthest point in the orbit of a satellite from its planet.

clathrates Cage structures formed by water ice having 8 cavities for each 46 water molecules that can trap large "guest" molecules of nonpolar volatile gasses.

cryoclastic eruption Eruption of water or other liquids that is driven by the exsolution of more volatile gasses such as CO or CH_4 and may form frosty "ash" deposits as the liquid condenses and expands.

cryovolcanism Eruption of liquid or vapor phases (with or without entrained solids) of water or other volatiles that would be frozen solid at the normal temperature of an icy satellite's surface.

eccentricity A measure of the ellipticity of an orbit, defined as the ratio of the distance between the two foci of the ellipse to the length of its major axis.

inclination Tilt of a satellite's orbit plane with respect to the plane of the planet's equator, important for determining seasonal variations in insolation.

orbital period Time taken for a satellite to complete one orbit of the planet, usually equal to the satellite's spin period (the length of a "day").

orbital resonance A commensurability between the orbits of two or more satellites, such that their orbital periods are related as the ratio of small integer numbers.

photolysis Dissociation of molecules by absorption of energetic photons.

radiolysis Dissociation of molecules by impact of charged particles.

resurfacing Any internal geologic process such as volcanism or tectonism that eradicates older surface features.

solid-state greenhouse Trapping of solar energy by ices that are transparent to visible light but opaque to longer wavelength thermal radiation.

sputtering Ejection of solid surface materials by the impact of energetic charged particles (an efficient erosion mechanism for satellites and rings located within planetary radiation belts).

The Encyclopedia of Volcanoes. http://dx.doi.org/10.1016/B978-0-12-385938-9.00044-4

TABLE 44.1 Physical Properties of the Larger Natural Satellites of the Outer Solar System[1]

Planet	Satellite	Radius (km)	Density (gm cm^{-3})	Gravity (ms^{-2})	Orbit Period (days)	Orbit Eccentricity	Orbit Inclination (deg)
Earth	Moon	1738	3.34	1.62	27.32	0.055	5.15
Jupiter	Io	1815	3.57	1.81	1.77	0.004	0.04
	Europa	1569	2.97	1.30	3.55	0.010	0.47
	Ganymede	2631	1.94	1.43	7.16	0.001	0.19
	Callisto	2400	1.86	1.25	16.69	0.007	0.28
Saturn	Mimas	197	1.17	0.07	0.94	0.020	1.53
	Enceladus	251	1.24	0.08	1.37	0.005	0.02
	Tethys	524	1.26	0.18	1.89	0.000	1.09
	Dione	559	1.44	0.22	2.74	0.002	0.02
	Rhea	764	1.33	0.28	4.52	0.001	0.35
	Titan	2575	1.88	1.35	15.95	0.029	0.33
	Iapetus	718	1.21	0.24	79.33	0.028	7.52
Uranus	Miranda	242	1.26	0.08	1.41	0.003	4.22
	Ariel	580	1.65	0.29	2.52	0.003	0.31
	Umbriel	595	1.44	0.22	4.14	0.005	0.36
	Titania	800	1.59	0.36	8.71	0.002	0.14
	Oberon	775	1.50	0.32	13.46	0.001	0.10
Neptune	Triton	1353	2.06	0.78	5.87	~0	156.8
Pluto	—	1142	2.07	0.63			
	Charon	596	2.07	0.21	6.39	~0	94.3

[1]Not listed are moons smaller than about 200 km in diameter, which number at least 40. Pluto and the Earth's Moon are included for reference.

1. INTRODUCTION AND SCOPE

Cryovolcanism is the eruption of water and other liquid- or vapor-phase volatiles onto the frigid surfaces of the icy satellites of the giant planets. When this chapter was first published in 2000, active cryovolcanism was known to occur on only one of these worlds: Triton, a moon of distant Neptune, where geyserlike plumes of nitrogen were discovered during the explorations of Voyager 2. Since that time, we have learned that cryovolcanic processes are active today in several places in the outer solar system. The Cassini–Huygens mission began exploring the Saturn system in 2004, finding spectacular saltwater plumes erupting from Enceladus and lakes on Titan of methane and other volatiles that probably erupted from the icy interior during past episodes of cryovolcanism. Recent Hubble Space Telescope observations may have detected a water vapor plume from Jupiter's moon Europa. In this chapter, we examine the mechanisms and manifestations of

cryovolcanism, beginning with a review of the materials that make up these unusual "magmas" and the means by which they might erupt and concluding with a volcanologist's tour of the farthest reaches of the solar system.

The Voyager spacecraft encountered a bewildering zoo among the natural satellites of Jupiter, Saturn, Uranus, and Neptune (Table 44.1). Several are larger than our own Moon, and two (Ganymede at Jupiter and Titan at Saturn) are bigger than the planet Mercury and would be considered planets in their own right if not for the fact that they are in orbit about larger worlds. Water ice is the most important constituent of the surfaces of most of these moons, with the notable exception of Io, where water either never was abundant or was driven off by aeons of sustained geologic activity (see "Volcanism on Io"). Many of the icy satellites are heavily cratered, having surfaces that passively recorded the bombardment of impactors, undisturbed by internal activity. Others appear relatively youthful, with areas "resurfaced" so that part of the cratering record was erased by endogenic processes such as tectonism and volcanism. A focus of scientific study of these satellites is the nature of the processes by which they were resurfaced, and in fairness, it should be pointed out that our knowledge of the materials and mechanisms of this geologic activity is still somewhat rudimentary. Much of the uncertainty concerns whether the **resurfacing** materials were in the liquid or solid state at the time they were emplaced or deformed. By confining our attention to cryovolcanism, here defined as the extrusion of liquids and vapors of materials that would be frozen solid at the planetary surface temperatures, we deliberately exclude processes such as icy tectonism, diapirism and solid-state convection, and intrusive magmatism, which probably played important roles in resurfacing the icy satellites. This somewhat arbitrary definition calls into question whether rhyolites should be considered "volcanic," given that they have crystal fractions sometimes exceeding 50%, but it admits an example of cryovolcanic activity here on Earth: the so-called "icefoots" on the shores of Lake Superior, where waves generated over open water in winter pump liquid water through cracks in the ice at the lake margins, forming icy mounds around the "vents." Even Mars hosts current cryovolcanic activity by this definition, erupting plumes of CO_2 vapor from the seasonal polar ice cap as the CO_2 ice sublimates in springtime via a mechanism similar to that which is thought to operate on Triton.

Two conditions must be met for cryovolcanic activity to take place on any icy moon: liquids must be generated in the interior, and the melt must then migrate to the surface of the satellite. These conditions are much more easily met on the larger satellites of the solar system than on smaller objects. A number of heating mechanisms contribute to the generation of fluids in the moons' interiors. The most important energy sources are gravitational (due to accretion and differentiation), radiogenic (through the decay of long-lived radioisotopes), and tidal (frictional heating caused by the daily changes in the shape of the satellite). Solar heating probably drives the nitrogen plumes on Triton, and Io's prodigious heat flow may be fueled in part by electromagnetic induction in addition to the tidal heating known to be important. The planetary factors that determine whether cryovolcanism can occur include size, composition (particularly silicate fraction, which determines radiogenic heating), location in the solar system (related both to composition and surface temperature, which ranges from 38 K at Triton up to more than 165 K in the hottest parts of Callisto), and the history of the moon's orbit from formation through to the present. The orbital history dictates the degree of tidal heating, since the daily tidal distortions are due to noncircular or oblique orbits around the primary planet. Noncircular orbits are commonly induced by gravitational interactions with satellites on neighboring orbits, through **orbital resonances** that gradually increase

the orbital **eccentricity** (Table 44.1). For example, Jupiter's moon Europa is caught in a tidal "tug-of-war" between Io, which orbits nearer to Jupiter with a period exactly half that of Europa, and Ganymede, which orbits farther than Europa with a period exactly twice as long. The orbital eccentricity forced by this "Laplace resonance" raises tides that may be sufficient to maintain a molten layer of liquid water beneath the satellite's surface. Orbital obliquity (**inclination** of the satellite orbit with respect to the planet's equator) is generally the result of collision or capture, extremely energetic events that may have contributed to the melting of Triton, for example. Gravity and composition control the manner of magma ascent, as we shall see later. The morphology of lava flows and other volcanic constructs, once erupted, are influenced by magma viscosity and effusion rate in ways familiar from terrestrial experience but also by planetary factors such as gravity, surface temperature, and atmospheric pressure (for example, in the case of Saturn's moon Titan, which has an atmosphere denser than Earth's).

2. MATERIALS AND MECHANISMS

2.1. Composition of Cryomagmas

Several lines of evidence provide us with incomplete and sometimes contradictory information about the composition of the outer solar system satellites. Direct spectroscopic detection is possible only for optically active materials—those with identifiable absorption features in their spectra—and tells us only about what lies on the uppermost surface, generally after processing by solar and charged-particle irradiation, sublimation, and a host of other "space-weathering" processes. Observations of the presumably "raw" icy materials seen in cometary outflows and in distant extrasolar nebulae provide important added constraints. Satellite densities and moments of inertia determined from gravitational measurements allow us to infer their interior structures and place limits on bulk composition and degree of differentiation. Theoretical modeling, constrained by samples of unprocessed protoplanetary materials in the form of primitive meteorites known as carbonaceous chondrites, provides some of the most valuable clues. Fundamental to the understanding of the icy satellites' composition is the idea of equilibrium condensation: the notion that the constituents of the early protoplanetary system, initially present in cosmic relative abundances, condensed and chemically reacted in accordance with the temperatures and pressures found in their local environments, which varied with distance from the early sun and proximity to the still-forming giant gaseous planets. By these calculations a "frost line" is predicted beyond the orbit of Mars, beyond which water ice is expected to have condensed in abundance. Spectroscopic

TABLE 44.2 Major Ices Detected or Suspected to Be Present on the Outer Solar System Satellites[1]

Name	Formula	Freezing Point (K)	Boiling Point (K)
Water	H_2O	273.2	373.2
Hydrogen peroxide	H_2O_2	272.7	Decomposes
Sulfur dioxide	SO_2	200.5	263.1
Ammonia	NH_3	195.5	239.8
Carbon dioxide	CO_2	194.7	Sublimes
Hydrogen sulfide	H_2S	190.3	211.4
Formaldehyde	H_2CO	181.2	252.2
Methanol	CH_3OH	175.4	337.8
Methane	CH_4	90.7	111.7
Ozone	O_3	80.7	161.3
Carbon monoxide	CO	68.1	81.6
Nitrogen	N_2	63.3	77.4
Oxygen	O_2	54.8	90.2

[1]H_2O, NH_3, CH_4, and H_2S (the reduced forms of the common elements O, N, C, and S) are expected to make up the bulk of ices condensed from a hydrogen-rich nebula in equilibrium. The other species may have been produced by disequilibrium condensation or by photolysis and **radiolysis**. The freezing and boiling points listed are for a pressure of 1 bar (0.1 MPa).

observations confirm that water ice is the dominant constituent of the satellites of Jupiter, Saturn, Uranus, and Pluto. Methane and ammonia should have also been present in the high-pressure nebulae surrounding Jupiter and Saturn, while their oxidized forms (CO, CO_2, and N_2) were favored in the cold outermost solar system, where reactions were too sluggish to proceed to equilibrium. While not predicted by the models, compounds such as methanol (CH_3OH) and formaldehyde (H_2CO) are observed in comets and may be minor components of the most distant satellites. Some properties of these ices are listed in Table 44.2.

As a geologic material, H_2O has a number of unusual properties. Its common hexagonal form (ice Ih) is stable at the low pressures (a few tens of megapascals) attained in the interiors of small satellites such as Enceladus and Miranda, in the thin ice/water crust of Europa, and at depths of up to a few hundred kilometers in the larger satellites. Pressures of hundreds to thousands of megapascals are reached in the interiors of Ganymede, Callisto, and Titan, so that dense high-pressure ice polymorphs should be stable. At the opposite extreme, metastable cubic and amorphous phases of ice can be deposited at the low temperatures (<150 K) and pressures of the icy satellite surfaces. Ice can also form solid cagelike structures known as **clathrates** that can trap adsorbed gasses such as CH_4, CO, and N_2, which are too volatile to condense otherwise, except in the Neptune and Pluto systems. Unlike silicate

minerals and rocks, ice Ih undergoes a volume reduction upon melting so that the liquid is denser than the solid. This presents a profound problem for cryovolcanism, since liquid water magmas should not rise through a pure water ice crust but instead are expected to sink. Water is also an extraordinary solvent, and there is an abundance of highly soluble salts (up to 20% by mass, mostly Mg and Na sulfates and sulfate hydrates) available in the chondritic materials from which the silicate fractions of the icy satellites were formed.

Melting of water-rich magmas can be greatly promoted by adding "antifreeze" in the form of salts or particularly ammonia, which should account for up to 15% of the mixture of ices from a fully equilibrated and condensed primordial nebula. Ammonia has not been detected spectroscopically on any satellite surface but is seen in comets and extrasolar nebulae and may be rapidly photolyzed to nitrogen upon exposure to the space environment. Ammonia has been detected spectroscopically in the plume erupting from Enceladus, however. The ammonia–water system closely parallels the system $MgO–SiO_2$, with ammonia dihydrate ($NH_3 \cdot 2H_2O$) taking the place of enstatite as the minimum-temperature melt product. Addition of ammonia to water ice lowers the melting point of the mixture to 176 K, conceivably attained in the interiors of even modestly sized satellites. Partial melting in the ammonia–water system slightly ameliorates the density problem of pure water magmas, since the melt is more ammonia-rich (28–34%) than its frozen equivalent, making it marginally positively buoyant with respect to the solid ice mixture. On the other hand, addition of salts to water only slightly depresses the melting point and greatly exacerbates the magma density problem, since the eutectic brines tend to sink with respect to solids of comparable composition. Fractional crystallization of brines results in segregation of salts from the solid, similar to the freezing of terrestrial ocean water, and might produce diapirism of the relatively buoyant solid H_2O phase. (Sulfates rather than chlorides are expected to dominate extraterrestrial brines because of the vastly greater primordial S abundance.) Methane does not significantly lower the melting point of ice, but clathrates of methane can break down to form ice Ih and gas or fluid when the vapor pressure of the contained methane exceeds the confining pressure; at low pressures (<0.03 MPa) this can occur when methane clathrates are warmed to only 176 K by an ammonia-rich water magma. Likewise, methane marginally reduces the melting point of nitrogen on Triton to a chilly 62 K.

The plausible candidate cryomagmas cover a range of viscosities and densities that can be compared with those of silicates (Figures 44.1 and 44.2). Briny solutions at their melting points are many orders of magnitude less viscous than even the most fluid silicate lavas and are unlikely to construct observable topography. Ammonia–water liquids attain viscosities similar to those of silicate magmas and may form solid crusts and partially crystalline slurries that aid their ability to build flow fronts and other constructive features found in terrestrial volcanism. The liquid/solid density contrasts that drive silicate volcanism reach up to 15% (in the case of peridotite), whereas a paltry 1% expansion is achieved in the best case by the melting of ammonia dihydrate. Moreover, the surface layers may be rather porous due to fragmentation by repeated impacts. The ascent of cryomagmas must be assisted by increasing the density of the crust, by pressurizing the liquid lavas, or by even more exotic means.

2.2. Eruptive Processes

Cryovolcanism may have been common early in the histories of the icy satellites, while their primordial surfaces were still rich in rocky material prior to differentiation. Even dense brines would have been positively buoyant with respect to sufficiently silicate-rich crusts. Magmas may have migrated upward along fluid-filled fractures and faults, much the same way as crevasses in terrestrial glaciers allow ponded water to sink downward through the ice, the cracks closing again afterward. In such a situation the effects of gravity control the mechanics of the eruption: the low gravity of the small satellites results in slower ascent speeds and lower effusion rates, but greater eruption volumes, than their counterparts on the larger icy moons

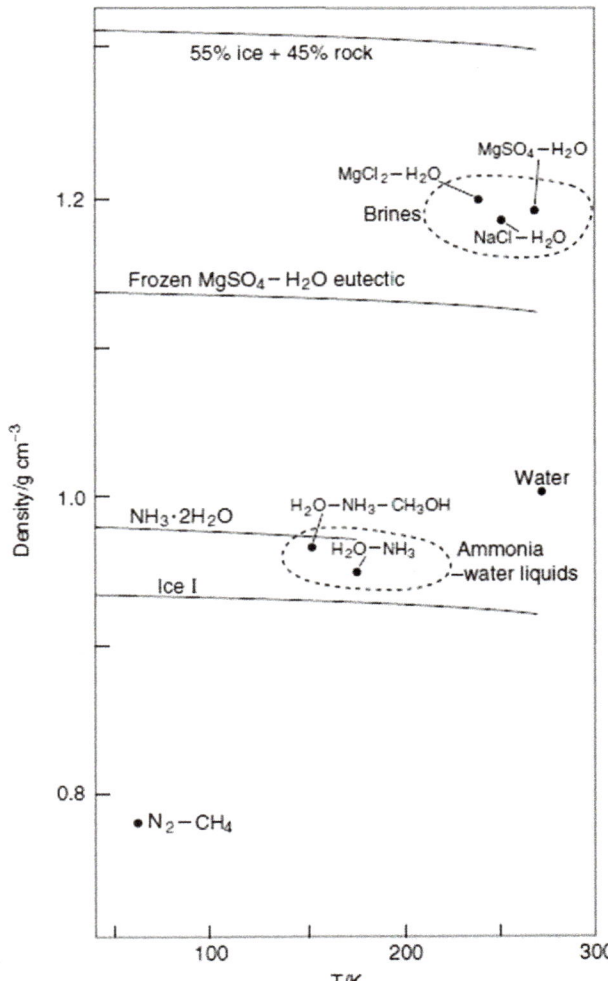

FIGURE 44.1 Densities of candidate cryomagmas and solids as a function of temperature. *From Kargel (1990).*

(because of the larger crack depths possible on smaller objects). This early volcanism would have been a self-limiting process, shutting itself off after crustal composition and density stratification were achieved.

Other ways to overcome the magma density problem are to overpressurize the liquid or add volatiles to provide effervescence for the eruption, as likely takes place on Enceladus. Overpressurization can happen in the late stages of satellite evolution by the cooling and freezing of a subsurface liquid layer. The volume expansion due to freezing of water can then fracture the surface and expel the last melt like an artesian spring. This is an attractive mechanism since much of the tectonism observed on the outer solar system satellites is attributed to global expansion, perhaps in such episodes of freezing. Alternatively, ices can be superheated by volcanism in the underlying silicates or by solar insolation, the latter being the probable explanation for Triton's geysers. The likely incorporation

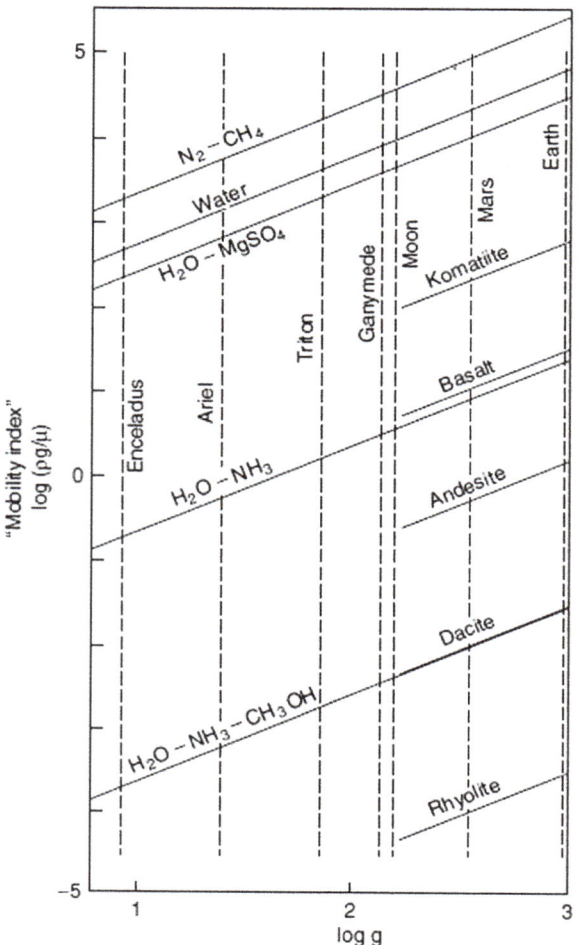

FIGURE 44.2 Mobilities of crystal-free cryomagmas at their liquidus temperatures compared with those of silicate lavas. *From Kargel (1990).*

FIGURE 44.3 Triton. This is a view of the south pole of Neptune's only planet-sized moon. The bright polar cap is made up of relatively mobile N_2 ice, subliming in the summer sunshine. The dark streaks mark the sites of active or recent plumes. *Voyager 2 image mosaic, courtesy of NASA.*

of supervolatiles into solid ice clathrates suggests the possibility of gas-driven cryovolcanism, the icy analog of terrestrial pyroclastic eruptions. The physics of these putative phenomena (dubbed **cryoclastic eruptions**) is similar to explosive volcanism on Earth, with methane and CO taking on the roles of H_2O, CO_2, and SO_2.

Other processes, not yet theoretically explored, have also been suggested to contribute to the widespread resurfacing of the outer solar system satellites described in the next section. Fluids could be derived from pressure-release melting in the course of solid-state convection, provided that the composition of the convecting material differs from water ice Ih (which has the opposite behavior with respect to pressure). A mechanism of mechanical pumping of wet slurries from the interior has been proposed to account for the curious double ridges of Europa. Frictional heating along strike-slip faults is a viable alternative hypothesis. In all cases, the liquids reaching the surface would be subjected to instantaneous freezing and boiling at the

cryogenic temperatures and near-vacuum of the space environment. A solid crust that is meters in thickness would form within hours of such an eruption, but not before the surface surrounding the vent was painted with frost deposits (along with any entrained contaminants) that could be quite extensive under the low gravity of small satellites.

3. TRITON

Active volcanic venting was first observed in the last place one would expect to find it. Triton is the coldest and most distant of the moons yet visited by spacecraft, and the last to be imaged before Voyager 2 was hurled out of the solar system on August 25, 1989. In stereo pictures taken as the spacecraft passed under the south pole of the satellite, at least two plumes can be seen jetting from the surface to heights up to 8 km, with trails reaching 150 km downwind. More than 100 dark windblown streaks decorate the polar cap (Figure 44.3), testifying to widespread previous plume activity. At lower latitudes, an even more interesting episode earlier in the moon's history is suggested by bizarre "cantaloupe" terrain and flowlike features that may have resulted from volcanic flooding of older surface topography. Triton's north polar region was shrouded in darkness from Voyager's cameras and remains unknown.

At 38 K, Triton's surface temperature is well below the freezing point of nitrogen, the material that makes up the south polar cap. It is near the transition temperature for

Neptune. Because of its obliquity, the orbit precesses every 688 Earth-years, causing the latitude of the subsolar point (the tropics) to vary from year to year, with periodic excursions up to 55° from the equator. It happens that Voyager visited on one of these extreme years, while the Sun beat down on the seldom-illuminated polar reaches of Triton.

All of the active plumes identified on Triton are found between 49° S and 67° S, near the latitude of the subsolar point. One model of their eruption suggests a solar **solid-state greenhouse** in which sunlight penetrates a transparent layer of low-thermal-conductivity nitrogen ice but is absorbed and trapped by dark carbon-rich impurities a few meters beneath the surface. This mild heating is enough to induce explosive venting of the gas, since an increase of only 10 K above the surface temperature is sufficient to raise the vapor pressure of N_2 100-fold. The wind streaks are likely caused by circulation in the tenuous atmosphere, as the flow from the south toward the colder northern hemisphere is deflected by coriolis forces eastward, due to the satellite's retrograde rotation. **Photolysis** products of methane might make up the carbon-rich absorbers and the dark particulates of the "cryoclastic" wind streaks.

A young, sparsely cratered surface is visible to the north of the polar cap, particularly on the leading hemisphere. About one-third of the imaged area is imprinted with regular quasicircular depressions bounded by ridges, nicknamed "cantaloupe terrain," which may be the surface manifestation of diapirism or solid-state convection in the interior. In other places, older cratered terrain was resurfaced by seemingly liquid or ductile material that flooded craters and other low-lying topography (Figure 44.5). The source of the material could be four large depressions up to 200 km across that resemble terrestrial calderas. Steep scarps on the flows suggest that they are made up of relatively refractory water ice, since little topography can be supported by solid nitrogen over geologic timescales, even under Triton's low gravity. If these interpretations are correct, an episode of intense heating must have taken place in the past in order to stir up the icy interior and produce water- or ammonia-rich melts. Enormously energetic heating probably occurred when Triton was captured into Neptune's orbit, perhaps by gas drag in Neptune's extended atmosphere or by collision with another satellite. Most of this energy was dissipated within Triton's first orbit around the planet, but a good deal of heating took place soon afterward as its initially elliptical orbit was circularized by tidal friction. Triton's likely origin is from far beyond Neptune in the Kuiper Belt, the outer solar system counterpart to the asteroid belt, home of the Pluto–Charon system and the likely source of short-period comets. Triton could be compositionally similar to Pluto and its large moon Charon and potentially provides us with an indirect look at these so far unseen worlds.

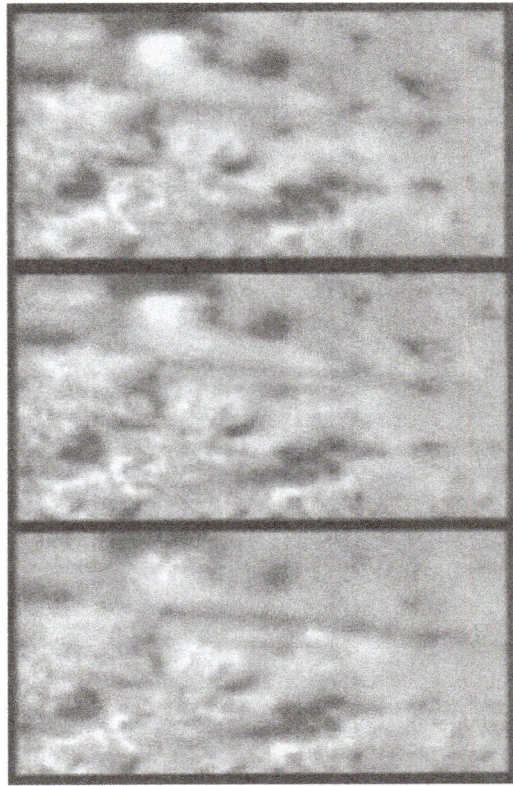

FIGURE 44.4 Nitrogen geyser on Triton. Two active plumes have been definitively identified in Voyager stereo pictures, and several more are possible. This is Mahilani, located at 49° S, 2° E. In these three separate views, Mahilani (left) can be seen to rise 8 km from the surface, with a dark tail that extends 150 km downwind (toward the right). *Voyager 2 images, courtesy of NASA.*

stability of the cubic and hexagonal forms of nitrogen ice, suggesting that seasonal phase changes are possible.

Both N_2 and CH_4 have been spectrally identified on its surface, and CO and CO_2 ices have been detected in minor quantities. H_2O is also expected to be abundant in the crust, but its spectral signature is masked by the more volatile surface constituents. Triton's thin (14 μbar) atmosphere is made up largely of nitrogen vapor with clouds of N_2 ice crystals and traces of CH_4 and CO and is responsible for transporting a couple of meters thickness of nitrogen ice from pole to pole every Triton year, keeping the surface nearly isothermal.

Triton's "geysers" (Figure 44.4) are probably caused by solar sublimation of the polar cap, despite the fact that the satellite is so distant that the Sun would merely appear as a particularly bright star. Triton's year is about 165 Earth-years long, and it was just approaching southern summer solstice at the time of the Voyager flyby. Triton's orbit is oblique with respect to Neptune's equator, and it is retrograde (opposite to the planet's spin direction), indicating that the satellite did not form anywhere in the vicinity of

FIGURE 44.5 Cryovolcanic flows on Triton. An ancient example of cryovolcanism in the solar system, Triton shows evidence of extensive melting, perhaps when the moon was gravitationally captured into orbit about Neptune. This view shows two large caldera-like features near the equator. Rimless pits to the right of the impact crater (in the lower "lake") may be the source of the smooth materials. *Voyager 2 image, courtesy of NASA.*

4. MOONS OF URANUS

Prior to the Voyager encounter in January 1986, the moons of Uranus were thought too small to have any signs of geologic activity on their surfaces. To explain the extensive resurfacing seen on such small satellites as Miranda, only 242 km in diameter, requires exotic compositions and such violent mechanisms as catastrophic disruption. Titania, the largest uranian moon, is rifted by a system of planet-sized fissures that form 20- to 75-km-wide grabens up to 1500 km long with fault scarps 2−5 km high. Umbriel and Oberon are both heavily cratered and show no indications of internal activity. Only about 35% of Ariel was imaged by Voyager 2, but its surface has the lowest crater density of the Uranian satellites. Ariel appears to have been active in the past, with broad grabens similar to those of Titania, suggesting an episode of crustal extension brought about by global expansion, despinning, or relaxation of the satellites' tidal bulges. These grabens appear floored by possible volcanic flows. Smooth deposits within the grabens show flow features such as sinuous medial grooves that appear similar in morphology to familiar lava tubes and channels (Figure 44.6). Material spilling out from the ends of the grabens seems to have flooded older cratered terrain, including a 30-km-diameter crater now half buried. The flows show substantial relief, suggesting that the resurfacing material was fairly viscous at the time it was emplaced.

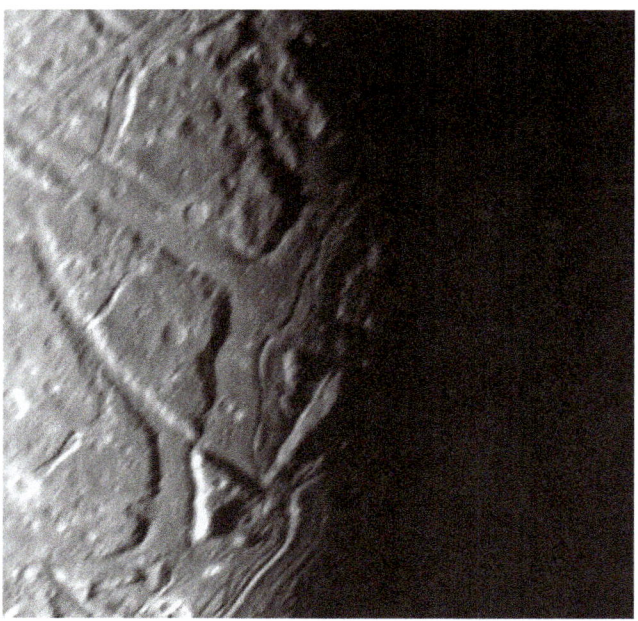

FIGURE 44.6 Ductile material on Ariel. The central channels (right) resemble collapsed lava tubes. *Voyager 2 image, courtesy of NASA.*

Based on the cross sections of the kilometer-thick putative flows, the rheology seems similar to that of terrestrial dacitic to rhyolitic magmas. The flow morphologies would seem to require an ammonia-rich composition, at least partially solidified, perhaps with methanol or some similar material added to thicken the soup.

Miranda's surface is astonishingly complex for such a tiny moon. Three unique tectonic structures are seen on the hemisphere viewed at close range by Voyager (Figure 44.7). These so-called coronae consist of central chaotically deformed regions with patchy **albedo** markings, ringed by broad belts of concentric parallel faults and ridges. A steep fault scarp 10 km high can be seen near Inverness Corona, the chevron-shaped feature near the middle of the picture. All three coronae are large in comparison to the radius of Miranda and seem to reflect some planetary process not active on other solar system objects. One interesting hypothesis is that Miranda might have been disrupted by a catastrophic collision with another satellite and later reaccreted in a haphazard fashion, leaving dense materials on its surface and trapping lighter ices in the interior. Sometime after reassembly, these density differences drove solid-state flow in the ice to produce the observed surface features through diapirism or subsidence as heavy materials sank to the center. Liquids may also have been generated in the process, if two small smooth areas in Elsinore Corona are cryovolcanic flows as has been suggested. Despite the vast difference in size, Miranda resembles giant Ganymede in having ancient heavily cratered terrain that has been partly resurfaced by tectonic deformation on a planetary scale.

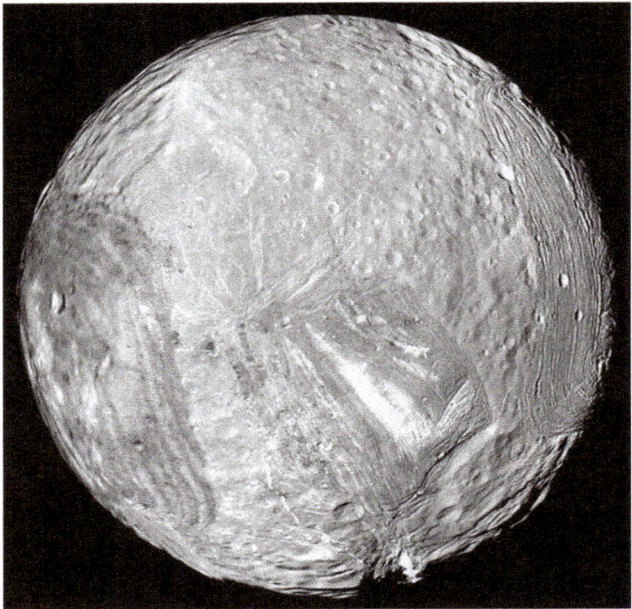

FIGURE 44.7 Miranda. The tectonic features (from left to right: Arden, Inverness, and Elsinore Coronae) appear to be due to density instabilities, despite the fact that the tiny satellite has little gravity to drive such deformation. *Voyager 2 image mosaic.*

5. SATURNIAN SATELLITES

The first definitive evidence of active water cryovolcanism was found on Saturn's moon Enceladus, where the Cassini Orbiter discovered giant plumes of vapor and particles that erupt from the moon's south pole and reach heights of more than 100 km (Figure 44.8). The plumes erupt from fractures nicknamed "tiger stripes" that are bounded by icy ridges and are much warmer (up to 180 K) than the surrounding surface. The plumes vary in brightness with the orbit of Enceladus and are brightest at **apoapsis**, when Enceladus is farthest away from Saturn. This suggests that tidal forces acting on the fractures play a role in the eruption of the plumes.

The precise mechanism of eruption is still widely debated, but some vital clues are provided by the composition of the plumes. Cassini flew through the plumes on three occasions and made direct measurements of the composition of the vapor and the particles. The gas is only 90% H_2O and contains 5% CO_2, 1% methane, and 1% ammonia along with traces of organic molecules. The high CO_2 content has led to speculation that the eruptions are generated by the nucleation of bubbles from dissolved CO_2 when the pressure is relieved by fractures penetrating to a subsurface ocean, much like a shaken carbonated beverage erupts when opened. The dust is made up of frozen saltwater, with an average content of 1% NaCl. The salt in the plumes suggests that the plumes are directly derived from liquid water and not from the icy crust, because salt should be excluded from the ice and concentrated in the water during the freezing of the ice shell. Organic compounds detected in the plumes such as ethane and acetylene have prompted speculation that Enceladus' subsurface sea could be inhabited by living creatures.

Interestingly, Enceladus appears to be the source of the particles that make up the outermost member of Saturn's famous ring system. Most of the gas and up to 10% of the particles erupted from the plumes reach escape velocity and are launched into orbit around Saturn. The faint E ring extends from 3 to 8 Saturn radii from the planet, and the orbit of Enceladus is embedded within the ring's densest part. Ice crystals in the ring are rapidly eroded by charged-particle **sputtering** and are swept up by Enceladus or gravitationally scattered into unstable orbits and escape the ring. The dynamical lifetime of the E ring is thought to be

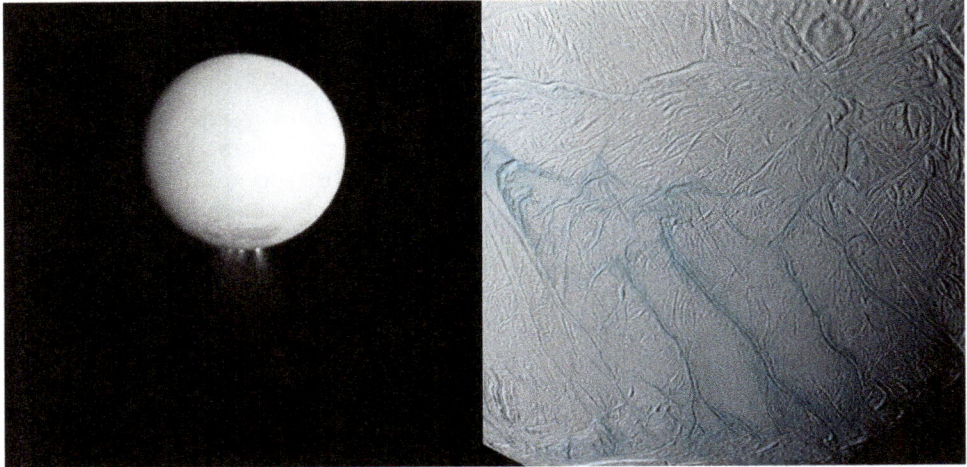

FIGURE 44.8 Enceladus. Saltwater plumes erupt from the south pole (left) from fractures nicknamed "tiger stripes" (right). *Cassini images, credit NASA/JPL-Caltech/Space Science Institute.*

FIGURE 44.9 Titan. View from the Huygens probe of the surface of Titan. *Huygens image, credit ESA/UArizona.*

exceedingly short—on the order of only 10^5 years. The ring would soon disappear without the regular breathing of Enceladus.

The Huygens probe found a fascinatingly familiar landscape when it landed on the surface of Titan in 2005. Huygens landed in a dry stream bed with boulders and cobbles reminiscent of arroyos on Earth, except that the rocks on Titan are made up of water ice and were sculpted by methane rain (Figure 44.9). Titan's nitrogen atmosphere is over four times as dense as Earth's, but because of the low gravity, the surface pressure is only 1.5 bar. At Titan's temperatures, methane behaves much like water on our own planet: liquid lakes of methane and ethane are stable, and methane rain or snow is predicted. Cassini revealed that the tropics of Titan today are deserts dominated by sand dunes, but the poles are home to seasonal lakes of methane and ethane. Extensive lakes of hydrocarbons were found by Cassini at the north pole of Titan (Figure 44.10). The infrared camera on Cassini captured a spectacular specular reflection (sun glint) from one of these lakes, dubbed Jingpo Lacus (Figure 44.11). These volatiles must have been erupted from the icy interior of Titan during volcanic episodes of the past, but identifying the sources of these ancient eruptions has proven problematic.

6. THE JUPITER SYSTEM

Our first close-up looks at icy satellite surfaces were provided by the Galileo mission, which began studying the Jupiter system in 1995. Although hampered by a malfunctioning high-gain antenna, Galileo imaged details on the giant planet's moons as small as a few meters across and revealed a surprising diversity of landforms unsuspected from previous observations. Jupiter's four large "Galilean" satellites exhibit a compositional gradation that mimics that of the solar system, with densities and rock fractions that decrease with distance from the planet. Rocky Io is the innermost of the four moons and due to constant tidal reworking is the most volcanically active body known. Europa is also largely made up of silicates but has a thin outer layer of ice and perhaps liquid water extending to depths of 50–150 km below the surface. Ganymede and Callisto both have large ice fractions, but here their similarity ends. Callisto is an only slightly differentiated mixture of ice and rock with an ancient, heavily cratered surface unbroken by internal geologic activity. Ganymede, on the other hand, has a dense core and an exterior that was extensively rifted in a past episode of activity probably associated with differentiation and core formation.

Prior to Galileo's arrival, much of Ganymede's resurfacing (Figure 44.12) was thought to have been volcanic in nature. As high-resolution images began to arrive from Galileo, the evidence for cryovolcanism eroded. The younger grooved terrain, which appeared smooth at the scales visible to Voyager, could be seen to be intensely shattered. Two scales of tectonic deformation can be recognized in the grooved terrain (Figure 44.13), both likely caused by crustal extension. Broad topographic undulations probably resulted from ductile necking of the crust, while finely spaced fractures were produced by brittle failure. No fluid flow features can be seen issuing from the fractures, so it appears that most of the resurfacing was achieved by tectonism rather than cryovolcanism. (This example should serve to throw considerable "cold water" on our earlier speculations about the satellites of Saturn and Uranus, which are also based on low-resolution images!) Some evidence of early cryovolcanic activity may be preserved in the form of scalloped depressions found in the older cratered regions of Ganymede (Figure 44.14). The origin of these caldera-like features is obscure.

In contrast to its moribund neighbors Ganymede and Callisto, Europa appears to be bristling with signs of recent activity. Europa has long been suspected of harboring a subsurface sea, based, in part, on its lack of topography: its greatest peaks reach heights of only a few hundred meters, and the largest craters excavate to even shallower depths. Relatively few impact craters scar the satellite, indicating that its surface is comparatively young—by some estimates, as little as 100 million years old. Some of the craters look familiar and seem to have formed in a solid target, but the largest impact structures are exceedingly flat and are ringed by evenly spaced concentric fractures unlike any previously seen. These unique features could have been

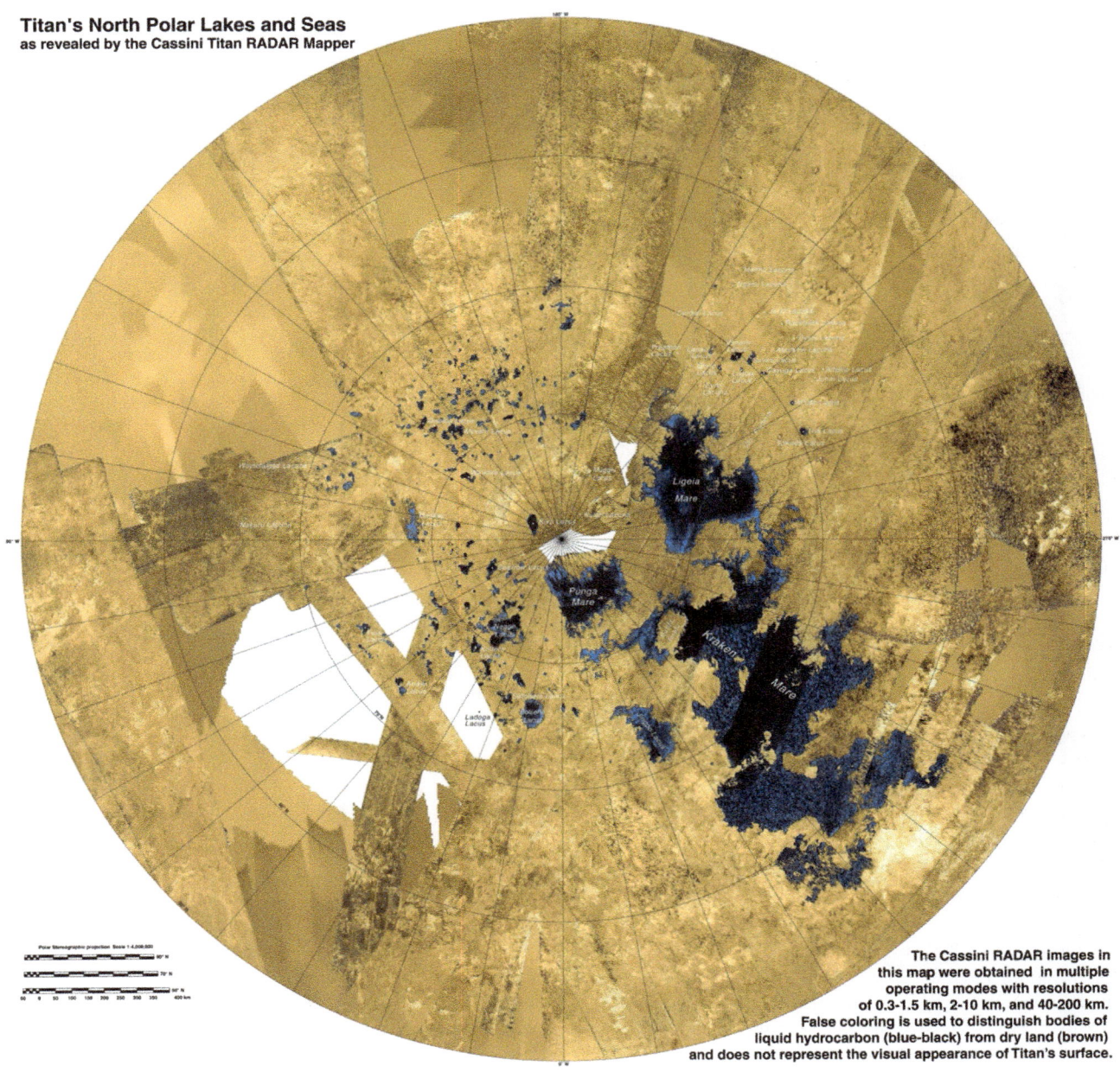

Titan's North Polar Lakes and Seas
as revealed by the Cassini Titan RADAR Mapper

The Cassini RADAR images in
this map were obtained in multiple
operating modes with resolutions
of 0.3-1.5 km, 2-10 km, and 40-200 km.
False coloring is used to distinguish bodies of
liquid hydrocarbon (blue-black) from dry land (brown)
and does not represent the visual appearance of Titan's surface.

FIGURE 44.10   Titan. Map of hydrocarbon lakes (blue) near the north pole of Titan. *Cassini radar mosaic, credit NASA/JPL-Caltech/ASI/USGS.*

created when kilometer-sized impactors penetrated the crust to a fluid layer beneath the surface. Two competing processes act to erase Europa's craters. A dense tangle of ridges resurfaces much of the satellite, representing many generations of repeated ridge-building activity over the course of its short history. In other places, particularly along the equator, the crust has been disrupted by localized heating from below, leaving floe-like fragments that have drifted across the sunken surface (Figure 44.15). In fact, the entire ice shell of the satellite may be drifting gradually to

the east with respect to the silicate interior, due to tidal torques exerted by Jupiter as Europa circles in its eccentric orbit. If so, it implies that the crust is mechanically decoupled from the interior, as if underlain by ductile ice or floating on a liquid layer.

The origins of Europa's perplexing surface features are still being widely debated. Not far below the surface, the ice could be warm enough to flowlike terrestrial glaciers, particularly if it is in thermal contact with liquid water at its melting point. The ridges and thermally disrupted areas

FIGURE 44.11 Titan. Sun glint from hydrocarbon lakes near the north pole of Titan. *Cassini VIMS image, credit NASA/JPL-Caltech/UArizona.*

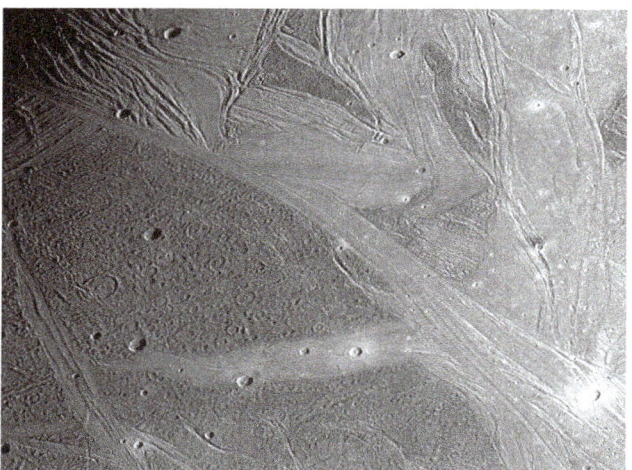

FIGURE 44.12 Ganymede. Terrains of two distinct ages are seen on the solar system's largest moon. Dark, heavily cratered terrain similar in appearance to Callisto is broken by bright young groove lanes. *Galileo image, courtesy of NASA.*

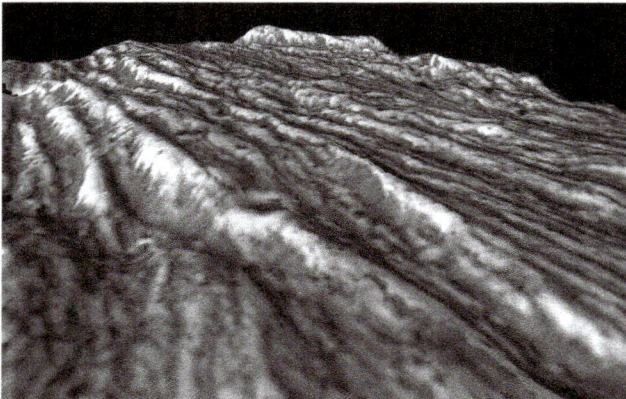

FIGURE 44.13 Topography in Uruk Sulcus, Ganymede. Stereo pictures have been used to render relief on this high-resolution image of a groove lane, formerly thought to be volcanic. *Galileo image, courtesy of NASA.*

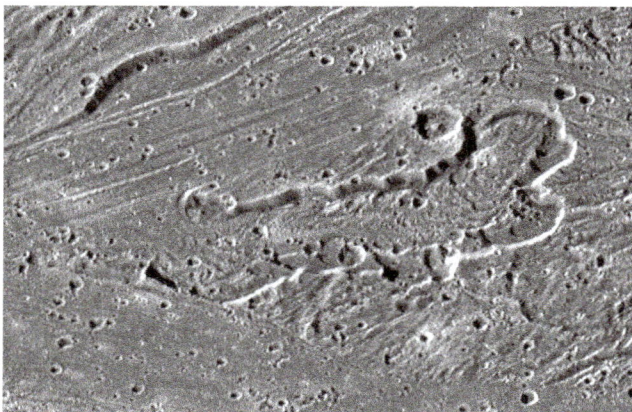

FIGURE 44.14 Scalloped depression on Ganymede. This feature, found on the older heavily cratered terrain, does not appear to be of impact origin. Perhaps it is evidence of early volcanism. *Galileo image, courtesy of NASA.*

might be due to diapirism and convection in the solid ice shell. In these models, the cold ice at the surface is denser than the warmer ice beneath it and tends to sink when disturbed by activity such as solid-state convection taking place in the interior. Buoyant ice from below might also be extruded through fractures in the brittle surface, forming the ubiquitous ridges. Not all of the characteristics of the ridges and disrupted zones are easily explained by solid-state deformation, however. In particular, the last surviving crustal blocks in the areas of thermal disruption are

among the largest and presumably heaviest and appear to have floated on a substrate that was even denser. Thus Europa's smooth spots and "maculae," once thought to have been volcanic flows, are probably places where the icy crust was heated to the melting point from below. An unusual example of a possible zone of melt-through is shown in Figure 44.16. Even small-scale relief has been erased from this 3-km-diameter "puddle," either by thermal topographic relaxation (implying surface temperatures near the melting point) or by fluid extrusion and flow. The source of such energetic and localized heating is a mystery. Europa receives only one-fifteenth of the tidal heating that churns its violent neighbor Io, but it would not be surprising if some high-temperature silicate volcanism was taking place in the subsurface of Europa as well.

Cryovolcanism may help to explain Europa's peculiar ridges (Figure 44.17), which occur in pairs or sets that are

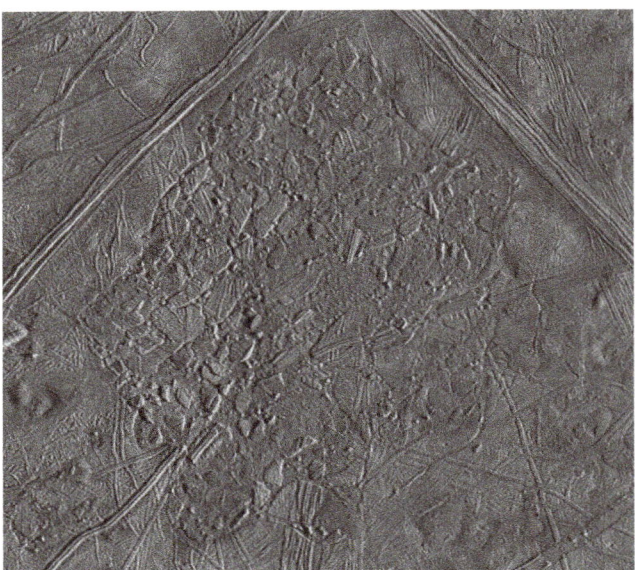

FIGURE 44.15 Conamara Chaos, Europa. Here the icy crust has been melted through by intense heating from below. A few remaining blocks retain the imprint of the original surface. These "floes" have shifted, rotated, and tilted as if floating on a dense water substrate. *Galileo image mosaic, courtesy of NASA.*

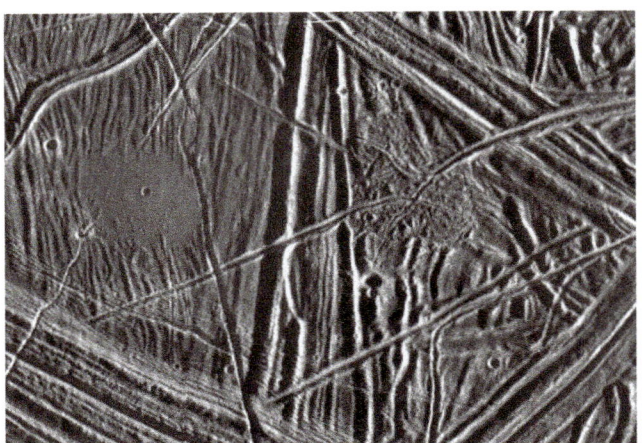

FIGURE 44.16 The "puddle" on Europa at 11° N, 328° W, imaged at 27 m/pixel. Even small-scale relief has been erased from this 3-km-diameter feature left of center, either by thermal topographic relaxation or by fluid extrusion and flow. *Galileo image, courtesy of NASA.*

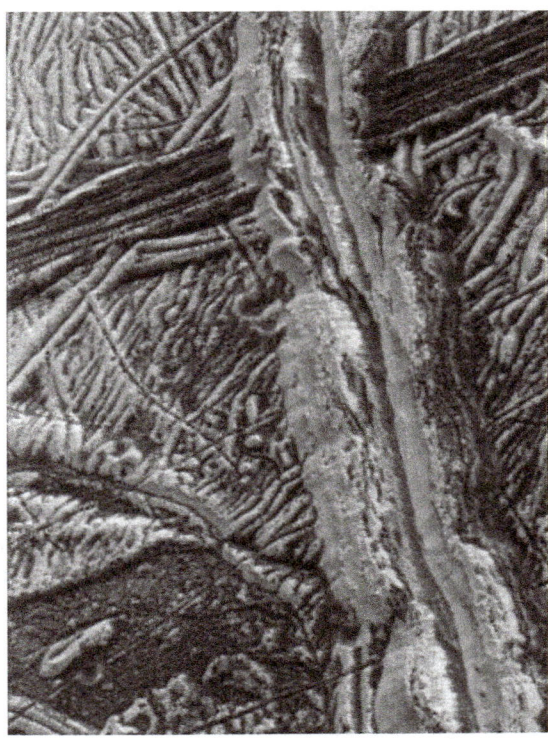

FIGURE 44.17 Freshly formed doublet ridge on Europa. The origin of these ubiquitous ridge pairs is mysterious. Perhaps they formed when icy slurry was pumped up through fractures by the daily tidal flexing. *Galileo image, courtesy of NASA.*

symmetric and uniform over distances often comparable to the radius of the satellite. The margins of many ridge sets are darkened and discolored by a nonice contaminant, perhaps vented from fractures between the ridges. Double ridges are difficult to produce by diapirism and seem to require some cyclic process of extrusion and withdrawal in order to form their central valleys. One possible mechanism is the pumping of fluids along cracks by tidal flexing. If the interior is liquid, tides tens of meters high should be raised each Europan "day" (3.5 Earth-days) as the satellite circles Jupiter. Any fractures that penetrate the icy crust are pulled apart for half of each orbit and slammed shut again for the other half. If liquid can seep into the fractures from below, some slush may be squeezed out of the cracks and onto the surface with each daily cycle. Bright plumes or frost deposits should be produced by this process, but no such telltale sign of activity was detected by Galileo. However, a recent detection of atomic hydrogen and oxygen near Europa's south pole by the Hubble Space Telescope may indicate that plumes of water vapor erupt from Europa as well as Enceladus. Interestingly, the detection was made when Europa was at apoapsis, farthest from Jupiter, where tidal tension was greatest. The plumes appear to reach 200 km in height before falling back onto the surface rather than escaping, consistent with the larger gravity of Europa compared with tiny Enceladus.

7. FUTURE EXPLORATION

Careful observations must be made to follow up on the possibility of water vapor plumes from Europa. It has yet to be confirmed that the hydrogen is produced from dissociation of water vapor, and observations at other wavelengths are needed to detect water vapor directly. If the plumes are confirmed, it will add incentive for the proposed NASA Europa Clipper mission to pass through the plumes and taste the water to determine its composition, much like Cassini did at Enceladus. The European Space Agency-led Jupiter Icy Moons Explorer mission is slated for launch in 2022 and expected to arrive at Jupiter in 2030, yielding new discoveries about each of the icy Jovian moons and Ganymede, in particular, where it will make observations from orbit. Cassini continues to make new discoveries about the Saturn system, and will do so until its fuel is exhausted and it is deliberately crashed into Saturn in 2017.

At the farthest fringe of the solar system, a spacecraft named New Horizons that was launched in 2006 will finally speed past Pluto and Charon in July, 2015. Pluto and Charon are expected to appear quite different from one another. Pluto's surface is dominated by nitrogen ice, with methane, ethane, and carbon monoxide also present whereas Pluto's giant moon Charon has a surface composition of ammonia hydrates and water ice. Pluto and Charon are not expected to be geologically active today, having reached the end of tidal evolution with their rotation locked to their **orbital period**, but significant geological activity may have taken place soon after the Pluto—Charon system formed. Gauging by our past experiences, we may well be surprised by the geologic processes that have gone on in this curious binary system in which the moon is fully one-tenth as massive as the dwarf planet.

SEE ALSO THE FOLLOWING ARTICLES

Migration of Melt ● Volcanism on Io.

FURTHER READING

Burns, J.A., Mathews, M.S. (Eds.), 1986. Satellites. Univ. of Arizona Press, Tucson.

Cruikshank, D.P. (Ed.), 1995. Neptune and Triton. Univ. of Arizona Press, Tucson.

Kargel, J.S., 1990. Cryomagmatism in the outer solar system (Ph.D. dissertation). University of Arizona, Tucson.

Kargel, J.S., 1995. Cryovolcanism on the icy satellites. In: Chahine, M.T., A'Hearn, M.F., Rahe, J. (Eds.), Comparative Planetology with an Earth Perspective. Kluwer Academic, Dordrecht/Norwell, MA, pp. 101–113.

Lewis, J.S., 1995. Physics and Chemistry of the Solar System. Academic Press, San Diego.

Roe, H.G., 2012. Titan's methane weather. Annu. Rev. Earth Planet. Sci. 40, 355–382.

Rothery, D.A., 1992. Satellites of the Outer Planets. Clarendon, Oxford.

Schmitt, B., De Bergh, C., Festou, M. (Eds.), 1998. Solar System Ices. Kluwer, Dordrecht.

Spencer, J.R., Nimmo, F., 2013. Enceladus: an active ice world in the saturn system. Annu. Rev. Earth Planet. Sci. 41, 693–717.

Part VI

Volcanic Interactions

John Stix

Department of Earth & Planetary Sciences, McGill University, Montreal, Quebec, Canada

This section looks at how volcanoes interact with the environment. Volcanoes release unusual types of gases, sometimes in very large amounts and very rapidly, as during a large explosive eruption. This can lead to significant impacts upon global chemical cycles. Volcanoes also produce geothermal systems and ore deposits; these can be exploited for energy and metals.

Chapter 45 addresses volcano degassing and volcanic gases. Volcanoes can release gas in a variety of ways: through fumaroles, as plume emissions, and diffusely through soils. Hence sampling gases is done in different ways depending on the type of sample one needs to obtain. Direct sampling of fumaroles and diffuse emissions can be done fairly easily, but measuring plumes can be difficult, and a variety of remote sensing instrumentation is generally applied to capture the full gas signal of the plume. Volcanic gases are dominated by H_2O and CO_2, with appreciable amounts of S, Cl, and F and other minor constituents as well. Volcanic gases can be characterized isotopically; such information provides clues as to the source or sources of volcanic gases, and can sometimes also shed light on degassing mechanisms. For example, hydrogen and oxygen stable isotopes commonly demonstrate mixing between meteoric and magmatic water. Helium isotopes can reveal relative proportions of mantle and crustal inputs. Carbon isotopes show mixing among mantle, sedimentary, and organic sources of carbon. These complementary approaches can yield valuable information as to the ultimate sources of gases, both for an individual volcano and for groups of volcanoes, e.g., volcanic arcs. Hence regional and even global comparisons can be made. The issue of excess degassing is also discussed, whereby many volcanoes emit an excess amount of gas relative to erupted solid products. The extreme case is a volcano which emits large amounts of gas quiescently but no magma, demonstrating the significant role of unerupted magma in storing and releasing the gas.

Chapter 46 provides a review of geothermal systems associated with igneous systems. Typically such systems are associated with young igneous intrusions, and the geothermal system occurs above the magmatic system and below the groundwater system. Within the geothermal system the rocks are generally porous and permeable, with steam-filled or liquid-filled cracks which transport geothermal fluid. The system is typically convecting but normally isolated from the magma below and groundwater above, except under special circumstances. On stratovolcanoes, meteoric waters can percolate deep into a volcano, gaining heat and various chemical constituents in the process. Magmatic volatiles may be added to the hot fluids, which can rise by convection. Hence in the core of such a volcano, acidic conditions prevail, while at greater distances from the core region, the fluids are diluted by groundwater and are less acidic. Thus the hydrothermal system can be zoned in terms of temperature, pH, and chemistry. Due to the high temperatures and commonly acidic conditions of many geothermal systems, a series of hydrothermal alteration zones develop which are characterized by distinct mineral assemblages. Most geothermal systems are liquid-dominated, meaning that liquid water is present in fractures and other spaces. A few geothermal systems are vapor-dominated, which occurs rarely since a hot source is required as well as efficient isolation from shallow groundwaters.

Chapter 47 looks at hydrothermal venting on the seafloor at arc and back-arc volcanoes. It is only recently that such sites have begun to be explored in a systematic fashion, and many fresh discoveries are being made. For example, the first observations of actively erupting submarine arc volcanoes have been made in the last decade. These observations show emissions of sulfur-rich fluids, sometimes in the form of molten sulfur, with carbon dioxide release also significant. Other volcanoes which are not erupting also exhibit large emissions which can be highly acidic. Once again, elemental sulfur and CO_2 are important constituents. Other seafloor hydrothermal systems show larger amounts of interaction between seawater and the host rock. Compared to hydrothermal venting along mid-ocean ridges, such vents at arcs and back-arcs tend to be shallower and less influenced by normal faulting driven by tectonic processes; instead, volcano-related faults appear to play a greater role. Also, the host rocks are more evolved (e.g., andesite) compared to mid-ocean ridges (basalt).

Chapter 48 examines lakes associated with volcanoes. The vast majority of these lakes are associated with summit craters, and many are acidic owing to the input of magmatic volatiles. Hence the occurrence of an acidic body of water perched at high elevations is an inherently unstable condition, with significant associated hazards. There are four principal types of volcanic lakes: acidic lakes with a large input of magmatic volatiles, geothermal lakes where the fluids have interacted extensively with host rocks, gas-rich lakes which are rare but very hazardous owing to large amounts of dissolved CO_2, and lakes which develop in caldera depressions. Lakes on volcanoes pose a variety of hazards such as sudden releases of toxic waters and gases, phreatomagmatic and phreatic explosions, flank collapses and debris flows, with the potential to generate tsunami waves as well. Hence monitoring is required for many of these lakes, especially when appreciable numbers of people live in close proximity.

Chapter 49 studies the array of ore deposits which form in association with volcanoes and volcanic rocks. Volcanic-hosted massive sulfide (VMS) deposits are found at many submarine volcanoes and are intimately associated with a volcano's volcaniclastic package. Such deposits can form in a wide range of water depths, and the type of metal enrichment depends in part upon the depth. We have a good understanding of these deposits based on comparisons between ancient and actively forming VMS deposits. Nickel sulfide deposits precipitate directly from ultramafic magmas such as komatiites; the nickel is derived from the magma, while much of the sulfur is assimilated from sedimentary rocks. Active and ancient volcanoes commonly host epithermal gold–silver deposits; the shallow permeable interiors of the volcanoes provide ideal conditions for hydrothermal convection and ore deposition. Kimberlite pipes host diamonds which have been incorporated into the kimberlite from the deep mantle. The highly volatile kimberlite is able to rapidly transport the diamonds to crustal levels or the surface.

Chapter 50 looks at gaseous emissions from volcanoes and their contributions to global carbon, sulfur, and halogen biogeochemical cycles. Volcanic gas emissions today mainly come from quiescent volcanoes which continually emit gas, punctuated by explosive eruptions from other volcanoes which emit relatively large quantities of gas into the atmosphere in a short period of time. Super-eruptions, such as those from felsic calderas or mafic large igneous provinces, have the potential of releasing prodigious amounts of gas which can play a significant role in modifying Earth's climate system and global biogeochemical cycles.

Chapter 45

Volcanic, Magmatic and Hydrothermal Gases

Tobias P. Fischer

Department of Earth and Planetary Sciences, University of New Mexico, Albuquerque, NM, USA

Giovanni Chiodini

INGV Osservatorio Vesuviano, Via Diocleziano, Italy

Chapter Outline

GLOSSARY

fugacity A property of real gases similar to vapor pressure in real gases that takes into account the gas density deviations from ideal behavior at high pressures.

gas chromatograph An instrument that is used in gas geochemistry to analyze volcanic gases. In principal it consists of a column that allows the separation of individual gas components from a gas mixture, which are then quantified using a chemical detector.

hydrothermal system A groundwater system that has an area of recharge, an area of discharge, and a heat source. When a magma supplies the heat source and volatiles, the hydrothermal system is termed a magmatic hydrothermal system.

hot spot A volcanic center, 100–200 km across and persistent for at least a few tens of millions of years, that is thought to be the surface expression of a persistent influx of hot mantle material. A hot spot is generally thought to receive volatiles from the lower mantle.

isotope Atoms with the same number of protons, but a different number of neutrons and therefore different masses. Stable isotopes do not decay over time, whereas radionuclides undergo spontaneous decomposition.

magmatic Associated with a magma. Magmatic volatiles are highly volatile elements that have their origin in a magma. Magmatic volatiles exsolve from a magma due to changes in pressure, temperature, or extent of crystallization and form magmatic gases at the surface.

meteoric water Waters of any age that originate from precipitation: rain, snow, ice, river, lake, and most low-temperature groundwaters. They are ultimately derived from the oceans through atmospheric circulation processes.

MORB An abbreviation for midocean ridge basalt, which is erupted at an oceanic spreading center. MORB is commonly used in mantle geochemistry as a compositional end member that represents the upper mantle of the Earth.

subduction A process in geology that describes the sinking of an oceanic plate under a continental or other oceanic plate into the mantle. Subduction transfers materials, in particular volatiles, into the mantle, where they lower the melting point of the mantle rock and cause the generation of arc magmas.

volcanic plume A mixture of water, carbon dioxide, sulfur, and other gases that are emitted from a volcanic crater into the atmosphere.

The Encyclopedia of Volcanoes. http://dx.doi.org/10.1016/B978-0-12-385938-9.00045-6

1. INTRODUCTION

Gas emissions from volcanic and **hydrothermal systems** are the direct result of processes operating in the mantle, the crust, and the hydrosphere. The chemical and isotopic composition as well as the flux of volcanic gases, therefore, provides valuable insights into these processes. It is well-established that globally approximately 70 volcanoes erupt per year. The number of passively degassing volcanoes (without eruption) is still poorly constrained but may be five to seven times higher. Passive degassing likely emits 90–99% of the total volatiles discharged from volcanoes and constraining these emissions is important when assessing mantle–surface element exchange. Fortunately, access to volcanic vents and plumes of passively degassing volcanoes is trivial compared to access to eruption plumes and as a result, the data for gas compositions during passive degassing dominate the literature.

This chapter provides an overview of some of the key characteristics and applications of volcanic, **magmatic**, and hydrothermal gases. It covers sampling and analyses of gases, their chemical and isotopic compositions and implications for volcanic activity and mantle volatile sources, as well as applications of thermodynamics to interpret volcanic gas compositions. We use the term volcanic gases for emissions that are related to degassing volcanoes, as opposed to emissions related to tectonic features, such as faults. Magmatic gases are gases that are directly released from the magma and can be sampled at fumaroles (usually high temperature) in volcanic craters. Hydrothermal gases are gases that are related to volcano-hosted hydrothermal systems and are characterized by compositions that are the result of the interaction of magmatic gases with a liquid phase.

2. SAMPLING AND MEASURING OF VOLCANIC GASES

Volcanic and hydrothermal gases are emitted from a variety of sources. Most notable are gas vents or fumaroles that are located in volcanic craters or on the flanks of volcanoes. These range in temperatures from the boiling point of water to above 1000 °C. Gases also discharge from bubbling springs, mud pots, cold gas vents, and along faults. The location of the gas discharges strongly affects their chemical composition where high-temperature (>400 °C) fumaroles contain acid gases such as SO_2, HCl, and HF that are generally less abundant in medium-temperature (100–400 °C) and absent in low-temperature (<100 °C) bubbling springs. Depending on the temperature, the expected gas composition, and the subsequent analytical protocol, sampling techniques vary slightly. Volcanic gases also discharge from lava lakes or open volcanic craters to form **volcanic plumes**. The composition of volcanic plumes is often measured remotely using spectroscopy. Volcanic plumes can also be sampled for aerosol chemistry and gas compositions using absorbing filters as well as continuously using automated in situ sensors. The goal of any sampling or measurement technique is to provide a means for best characterizing the gases that are emitted from the magmatic or hydrothermal source to the atmosphere.

2.1. Direct Sampling

Direct sampling of volcanic gases discharging from vents has been used since the 1950s. Over the past 60 years various improvements have been made, most notably the use of sodium hydroxide solution in evacuated glass flasks that allows for the dissolution of the abundant acid components (CO_2, SO_2, H_2S, HCl, HF) and concentration of the less abundant and noncondensable gases (N_2, Ar, O_2, H_2, CO hydrocarbons, and noble gases) in the headspace of the bottle. Water, the most abundant component in volcanic gases, condenses in the sample bottle and its concentration is determined by weight difference before and after sampling. This approach enables analyses with relatively simple techniques such as wet chemistry and **gas chromatography**, provides a complete composition of the gases, and was pioneered by Werner Giggenbach. The gas sampling flask that is most commonly used for direct gas sampling and subsequent determination of the complete gas chemistry is often informally termed the "Giggenbach bottle" but other, similar designs have been successfully and widely used by the volcanic gas community. In order to transfer the gases from the vent into the bottle, a titanium or silica glass tube is inserted into the fumarole (Figure 45.1(A)) and connected via silicone or plastic tubing to the glass sampling flask (Figure 45.1(B)). After flushing the tubing, the vacuum valve is opened and gas enters the flask. For the collection and analyses of noble gases, commonly an evacuated lead glass flask is used which has a lower permeability for helium than Pyrex.

2.2. Measurements and Sampling of Volcanic Plumes

Land-based or airborne remote sensing of the composition of volcanic plumes has been performed since the 1970s when volcanologists started using the correlation spectrometer (COSPEC), which was initially developed for measuring SO_2 and NO_x from coal-burning power plants. The instrument measures the absorption of sky ultraviolet (UV) light by SO_2 in the volcanic plume. Through careful calibration and measurement of the sky outside of the plume, the amount of SO_2 in the plume can be determined. The COSPEC is used by either traversing under or, in stationary mode, by scanning through the plume

FIGURE 45.1 (A): Sampling of medium-temperature ($\sim$160 °C) gases at Galeras volcano, Colombia, 1998 using a titanium tube inserted into a volcanic gas vent (fumarole) (Photo S.N. Williams) and connected to an evacuated Pyrex glass flask (often referred to as Giggenbach bottle) partially filled with sodium hydroxide (B).

(Figure 45.2(A)). The result of such a measurement is a concentration profile of SO_2 of the volcanic plume (Figure 45.2(B)) that then can be combined with plume or wind speed measurements to obtain a flux (tons/day) of SO_2 from a volcano.

The original COSPEC that used a chart recorder and heavy car batteries was modified by several research groups to include automated scanning capabilities, lightweight batteries, and computer-based data logging. Since the early 2000s, the COSPEC has been mostly replaced by compact and low-power Universal Serial Bus (USB) UV spectrometers (mini-DOAS, FlySPEC, and others) that are equipped with an appropriate lens and fiber-optics cable, allowing for the collection of UV absorbance spectra of volcanic plumes (Figure 45.3). The collected spectra are analyzed and fitted to calibrated SO_2 spectra to obtain plume SO_2 concentrations. These instruments are operated by laptop and have

resulted in an explosion of the amount of SO_2 flux data collected on volcanoes due to their low cost, low power, and low weight.

For more comprehensive measurements of plume components such as CO_2, H_2O, SO_2, HCl, and HF, an Open-Path Fourier transform infrared spectrometer is used. Similar to UV spectrometers, these instruments collect plume gas spectra, albeit emission spectra in the infrared (IR) portion of the spectrum. Data are reduced by applying atmospheric transfer models and calibration spectra. Fourier Transform infrared (FTIR) instruments are still expensive, bulky, and require significant power and, therefore, have not yet been as widely applied to volcanic degassing studies as the USB spectrometers. FTIR spectrometers are normally pointed at a strong IR light source such as a convecting lava lake which permits continuous collection of detailed data on the degassing behavior of magmas.

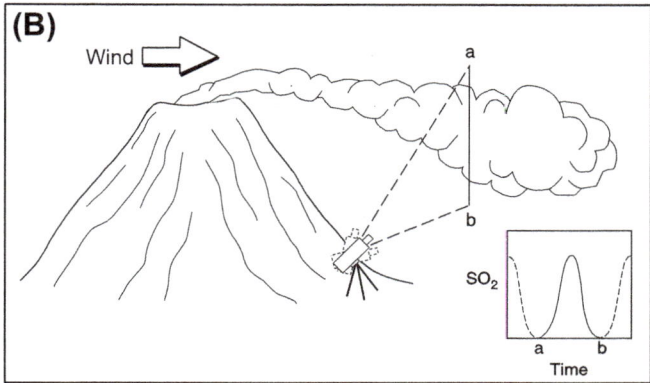

FIGURE 45.2 (A): COSPEC used in stationary scanning mode (USGS/Cascade Volcano Observatory, COSPEC cookbook). (B): Vertical scan results in SO_2 concentration profile of the plume (inset) that can be combined with wind or plume speed measurements to obtain SO_2 flux.

FIGURE 45.3 A mini-DOAS system as used by the University of New Mexico at Oldoinyo Lengai volcano, Tanzania, 2009.

Another approach to continuously collect chemical composition (H₂O, CO₂, SO₂, H₂S, H₂) of volcanic plumes and fumarole fields are low-cost, compact, low-power, and easy-to-use multi-GAS instruments (Figure 45.4) that pump the plume gas and automatically analyze it with a series of chemical sensors. These instruments need to be placed in the plume and can either be operated by walking in the plume to collect concentration and gas ratio data or deployed permanently on the crater rim to semi-continuously measure gas compositions.

2.3. Soil Degassing

Volcanoes may release large amounts of CO_2 and other gases into the atmosphere, during both eruptions and quiescent periods. The CO_2 emission occurs from direct expulsion of fluids at fumaroles and as diffuse soil emanations. In many volcanic areas soil CO_2 degassing correlates well with soil temperature, suggesting that the diffuse degassing is the result of ascent of large amounts of

hot hydrothermal fluids toward the surface. The steam condenses, often near the surface, releasing thermal energy and heating the soil, while the noncondensable CO_2 is released through the soil by diffuse degassing. The amount of volcanic hydrothermal CO_2 discharged by diffuse soil degassing can be significant, compared to the CO_2 released from fumaroles and can in some cases also be comparable to crater plume emissions from active volcanoes. Diffuse soil degassing of deeply derived CO_2 commonly occurs in specific areas (diffuse degassing structures (DDSs)) such as faults, small vents, and steaming ground rather than across the entire volcanic system.

One of the most commonly used methods to measure CO_2 diffuse flux (ϕCO_2) from the soil is based on the accumulation chamber principle (Figure 45.5). Originally, accumulation closed-chamber methods for CO_2 and other gas (e.g., NO_2 and CH_4) soil flux measurements were developed specifically for the determination of soil respiration related to agricultural cultivation. Numerous chamber configurations have been proposed. During the past decade the method was modified according to the method "measure at time 0" (AC0) and adapted to volcanological studies.

The method AC0 consists in the measure of the time variation of the CO_2 concentration (C) inside a chamber with the open side facing the ground. Initially at time 0, ϕCO_2 from soil is a function of the derivative of the curve C versus time (dC/dt):

$$\phi CO_2 = K\left(\frac{dC}{dt}\right)_{t \to 0} \tag{45.1}$$

where the constant K has the dimension of length and depends on the instrumental characteristics. The AC0 method allows rapid direct measurements of ϕCO_2 from soil without significantly altering the natural flux for a wide range of values.

(A)   **(B)**

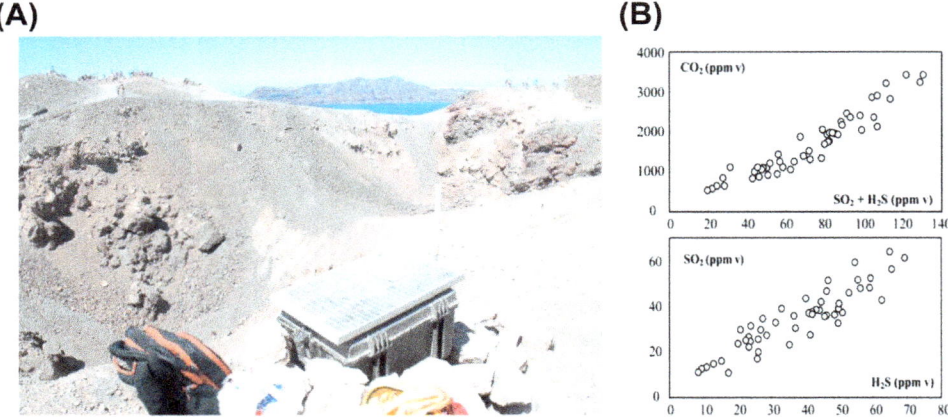

FIGURE 45.4 Multi-GAS instrument deployed permanently at crater rim of Stromboli volcano 2007 (Photo by A. Aiuppa). This instrument collects plume gas compositional data semicontinuously as shown in (B) and (C).

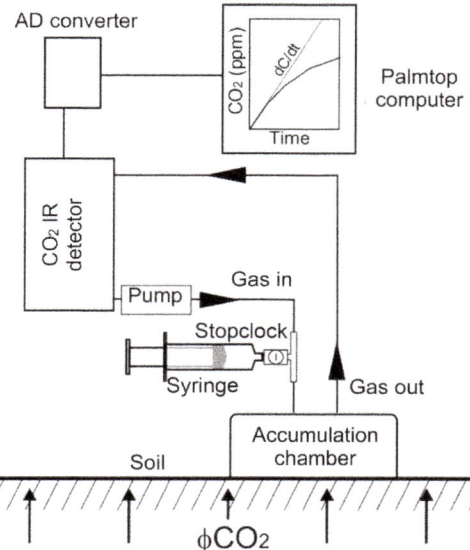

FIGURE 45.5 The accumulation chamber for the measurement of diffuse CO_2 fluxes from the soil. The instrument consists of an inverted chamber, an IR spectrophotometer (CO_2 IR detector), an analog-to-digital (AD) converter, and a handheld computer. The gas is continuously extracted from the chamber through a pneumatic line equipped with a small pump, sent to the IR spectrophotometer, and then reinjected into the chamber. A valve and syringe allows one to sample the gas for carbon isotopic determination.

The mapping of degassing areas (i.e., the definition of the DDSs), the quantification of the amount of released CO_2, and the associated uncertainty are objectives of any study on diffuse degassing. Recently these aims have been achieved using the sequential Gaussian simulation (sGs) approach. The sGs method provides a suitable tool to produce realistic images of the spatial distribution of CO_2 flux that reproduce the histogram and variogram of the experimental data. Furthermore, the sGs procedure permits an estimate of the uncertainty associated with the computed CO_2 output, allowing one to verify if the density of measuring points used in an experiment is adequate to provide a reliable quantification of the total CO_2 output. The sGs procedure considers the CO_2 flux to be a Gaussian stationary multivariate random function. Carbon dioxide fluxes are simulated at locations defined by a grid covering the area of interest. As the sGs procedure needs a multi-Gaussian distribution, measured CO_2 fluxes must be transformed into a normal distribution (normal scores). The transformed data are then used in the simulation procedure. Simple kriging estimate and variance, computed according to the variogram model of normal scores, are used to define a Gaussian conditional cumulative distribution function at each location. A random value is drawn from the conditional cumulative distribution as one "reasonable" simulated value for that location. Once a value is simulated, it is added to the data set and can be used together with the original data to estimate the variable at the next location of the grid. The simulation proceeds to the next grid location and loops until all nodes are simulated. Afterward, the simulated normal scores are back-transformed into values expressed in the original data units, applying the inverse of the normal score transform. Changing the random path of grid nodes visited, N alternative simulations can be performed and N equiprobable realizations can be drawn, each reproducing the sampled data at their locations, the data univariate statistics (e.g., histogram) and the data bivariate properties (e.g., variogram). This information can be used to draw a map of the simulated CO_2 values at any cell obtained through a pointwise linear average of all the realizations or can be used to draw probability maps that represent an alternative method to define the DDSs. In fact, the probability map shows for each cell the probability that the flux is higher than a selected threshold, i.e., the maximum value expected for the background CO_2 fluxes from soil biogenic sources.

2.3.1. The Case of Solfatara Di Pozzuoli

Solfatara di Pozzuoli, a crater located in Campi Flegrei caldera, Italy, is a famous site where techniques for diffuse degassing studies were developed and are now systematically used for monitoring of volcanic activity. We report here the results obtained in a very complete survey performed in March 2007 when 391 ϕCO_2 and 326 $\delta^{13}C_{CO_2}$ soil gas values were determined. Figure 45.6(A) shows the probability map that CO_2 at any location of the surveyed area is higher than 50 g m^{-2} d^{-1}. This cutoff value well discriminates the maximum CO_2 flux fed by shallow biogenic sources (background) from the higher fluxes produced by volcanic hydrothermal sources (Figure 45.6). Sites within the DDS of Solfatara, which are shown in yellow and red in Figure 45.6, are defined as those locations with a greater than 50% probability that the simulated CO_2 fluxes are above the cutoff value. The estimated area within the DDS of endogenous CO_2 is about 0.8 km^2. This large extension of the Solfatara DDS is confirmed by the map of the $\delta^{13}C_{CO_2}$ values shown in Figure 45.6(B), which shows the results of 100 sGss of the spatial distribution of the isotopic composition of CO_2 flux based on the 326 isotopic measurements. Relatively high $\delta^{13}C_{CO_2}$ values ($> -10‰$) separate the sites where flux is mainly sustained by the volcanic hydrothermal source from the zones where background sources of carbon, characterized by more negative carbon isotopic composition, become dominant. Figure 45.6 highlights the complex shape of the Solfatara DDS. Deeply derived CO_2 is released through the soil mainly within the Solfatara crater and along a narrow band about 1 km long and 0.2 km wide located to the east of the crater, which corresponds to an northwest–southeast normal fault system. A second anomaly in the southeastern

(A)

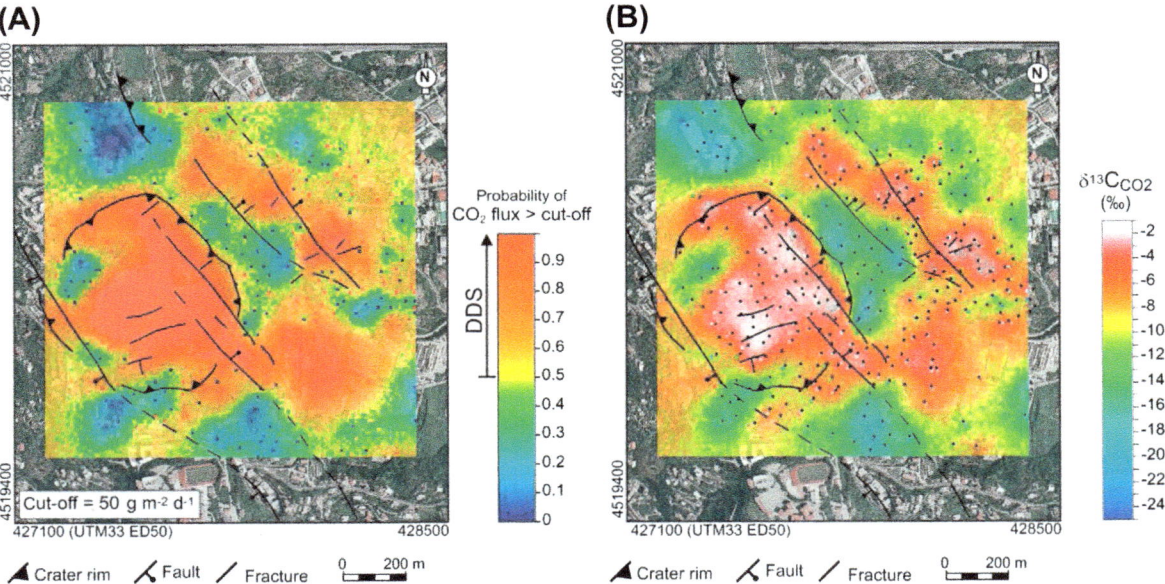

(B)

FIGURE 45.6   (A) Probability map of the CO_2 flux spatial distribution using a cutoff value of 50 g m^{-2} d^{-1}. The color scale shows the probability that at each location the CO_2 flux exceeds the cutoff value. (B) Map of distribution of the isotopic composition of the CO_2 efflux.

sector of the study area is not correlated to any mapped volcanic and tectonic feature, most probably indicating the presence of a hidden volcanotectonic structure. In total, Solfatara crater releases 1000−1500 t d^{-1} of deeply derived CO_2.

3. CHEMICAL COMPOSITION OF VOLCANIC GASES

The chemical and isotopic composition of volcanic and hydrothermal gases is a direct result of the volatile sources and the processes that occur as gases travel from the magma and overlying bedrock to the surface (Figure 45.7). The main sources are the Earth's mantle, magma bodies, crust, hydrosphere, and atmosphere. Mantle and crustal processes result in the generation of magmas in which volatiles are dissolved as governed by solubility relationships. Through lithospheric **subduction**, the mantle itself receives volatiles that at some point originated in the atmosphere or hydrosphere (the ocean). As a result, the mantle contains volatiles that are "primordial" and those that have been added by subduction. Volcanoes are the main pathways that allow mantle volatiles to reach the Earth's surface where they interact with the hydrosphere to form hydrothermal systems, transport precious metals, and modulate the climate. The relative gas abundances, stable and noble gas **isotopes**, and gas fluxes provide information on the sources and processes that operate at depth and at shallower levels. In addition, temporal changes in gas compositions and fluxes provide unique and instant insights into the activity of a volcano that are invaluable for assessing volcanic hazards.

3.1. Chemical Composition

When considering the major components of volcanic and hydrothermal gases, it is useful to distinguish between magmatic and hydrothermal compositions. Figure 45.7 shows a schematic diagram of an active volcano with an overlying hydrothermal system. Here the source of the volatiles is the degassing magma which contains H_2O, CO_2, S, Cl, and F as well as trace gases and noble gases that are not shown. The resulting gases primarily contain H_2O, CO_2, SO_2, H_2S, H_2, and CO. As these high-temperature magmatic gases travel to the surface, they either degas

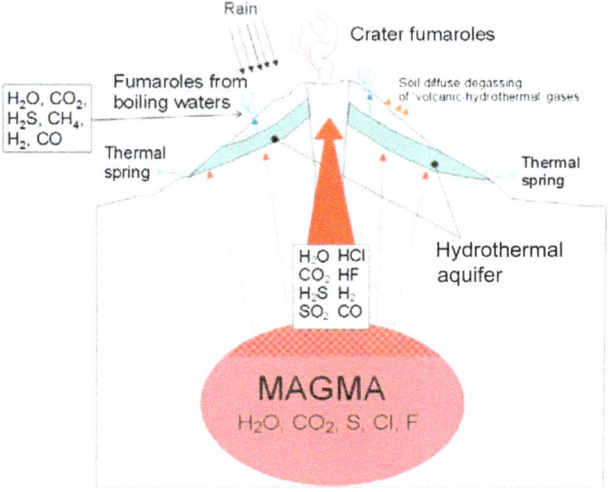

FIGURE 45.7   Schematic diagram of a degassing magma beneath a volcano hydrothermal system.

directly through high-temperature crater fumaroles or interact with an aquifer to form fumaroles from a boiling aquifer or bubbling springs on the volcano's flanks.

Volcanic, magmatic, and hydrothermal gases are dominated by water and CO_2. Characteristically, volcanic and magmatic gases have high SO_2, HCl, HF, and CO contents which are lower or absent in hydrothermal gases due to interaction with aquifers. Hydrothermal gases are generally more water rich and contain CH_4 and H_2S but no SO_2. Oxygen is the result of atmospheric contamination during sampling or the recycling of air at shallow levels, while N_2 and Ar can have both atmospheric and mantle/crustal sources. Helium is dominantly of crustal or mantle origin because of its low concentration in air. Table 45.1 shows the compositions of typical high-temperature magmatic gases and low-temperature hydrothermal gases. In general, samples from volcanic arcs tend to be higher in water contents than samples from continental rifts or hot spots. Arc gases also have higher CO_2/He, N_2/Ar, and N_2/He ratios than rift and **hot spot** gases due to the addition of carbon and nitrogen from subducted slab sources, as confirmed by stable isotopes (discussed below). Figure 45.8 shows a triangular diagram of the samples reported in Table 45.1. The H_2O-CO_2-total S ($SO_2 + H_2O$) diagram is useful to distinguish between magmatic and hydrothermal source of the gas discharges. Magmatic gases generally discharge from crater fumaroles, whereas hydrothermal gases are emitted from the flanks of the volcano after extensive interaction with shallow aquifers of a liquid-dominated hydrothermal system. Magmatic gases have higher total sulfur (S_{total}) contents. H_2O/CO_2 ratios of magmatic and hydrothermal gases are within the same range. CO_2/S_{total} ratios of hydrothermal gases are generally >10, whereas in magmatic gases this ratio is usually <10.

3.2. Noble Gases and Stable Isotopes

Gases that are emitted from volcanoes and hydrothermal vents are a mixture of atmospheric, hydrospheric, mantle, and crustal components. While the relative abundances of major and trace gases can provide general information on gas sources, the use of noble and stable isotopes has the potential to further constrain the ultimate origins of the individual gas species. It is important to note here that each gas component may be a mixture of various sources and that different components may have different sources. For example, water may be predominantly of meteoric (shallow surface) origin, while CO_2 is likely, and to a large extent, derived from the magma and/or the mantle.

Water is the most abundant gas species and its source can be identified using the stable isotopes of oxygen and hydrogen. It was recognized in the early 1990s that the oxygen and hydrogen isotope compositions of volcanic

steam as sampled at fumaroles deviate significantly from that of local **meteoric water**. In a global data set of samples from subduction-related volcanoes, it is clear that the $\delta^{18}O$ and δD values of such samples are shifted toward enriched values of δD and $\delta^{18}O$. Many data points trend toward a common globally almost uniform magmatic end member (Figure 45.9). This end member has been termed "andesitic water" or "arc-type water" to account for its relationship to volcanism occurring in subduction zone settings. The ultimate origin of this arc-type composition is thought to be subducted seawater that has been transported to a depth of arc magma generation by hydrous minerals of the oceanic crust and sediments. The processes related to mineral hydration and subsequent dehydration during subduction generally shift the isotopic compositions of seawater to values in the range of $\delta^{18}O = +10 \pm 2\%$ and $\delta D = -20 \pm 10\%$. Notably, this end member composition is distinctly different from midocean ridge basalt (**MORB**) water which is thought to represent the water isotopic composition of the upper mantle, unaffected by subduction. The oxygen and hydrogen isotope compositions of end member magmatic steam discharging from fumaroles therefore provides direct evidence for the involvement of subducted seawater in the processes of arc magma generation. This involvement also accounts for the high water contents (up to eight wt%) and explosivity of arc magmas. Generally, all subduction zone volcanoes discharge steam that is a mixture of arc-type water and meteoric water derived from the surface or shallow aquifers. As the work on δD and $\delta^{18}O$ compositions of volcanic steam has expanded to include samples from around the globe, it has now become apparent that the original andesitic water box of $\delta D = -20 \pm 10$ does not capture all subduction-related compositions. A more appropriate range, that includes all mixing trends, is $\delta D = -10 \pm 20$, shown as "global arc water" in Figure 45.9. This range is also consistent with δD values measured in olivine-hosted melt inclusions from recently erupted island arc tephras.

Of the noble gases used to interpret the origin of volcanic and hydrothermal gases, helium has received by far the most attention due to its unambiguous crustal and mantle source and low concentration in air (5.4 ppm). Helium has two isotopes: ^{3}He which is primordial and ^{4}He which is an alpha particle and the result of U and Th decay. The ^{3}He/^{4}He ratio in air is 1.4×10^{-6}, attesting to the low abundance of primordial ^{3}He in the atmosphere. The air ratio is commonly used as a reference and expressed as R_A. The upper mantle as sampled by MORB glasses has a globally rather uniform ^{3}He/^{4}He ratio of $8 \pm 1 R_A$ (i.e., 8 ± 1 times the ratio of air). Fluids that discharge from hot spot volcanoes overlap with the MORB ratios but the highest observed ^{3}He/^{4}He ratios measured in such fluids is approximately 28 R_A. Purely crustal volatiles such as those discharging in springs on cratonic shields have ratios that are

TABLE 45.1 Examples of Magmatic and Hydrothermal Gases in mol (%) Total Gas

Magmatic Gases

Volcano	Momotombo	Satsuma Iwojima	Kudryavy	Merapi	White Island	Vulcano	Halemaumau	Erta Ale
Location	Nicaragua	Japan	Russia	Indonesia	New Zealand	Italy	Hawaii, USA	Afar, Ethiopia
Temperature (°C)	747	880	920	803	495	620	307	1130
H_2O (%)	96.25	97.3	95.3	88.7	92.0	86.1	92.1	79.4
CO_2	2.5323	0.51	1.15	5.56	6.0100	11.9000	6.25	10.4
S_{total}	0.2335	0.9820	2.4400	1.1140	1.6000	1.1600	1.5700	6.5000
SO_2	0.2335	0.818	2.05	0.98	1.0000	0.6800	1.57	6.5
H_2S		0.164	0.39	0.134	0.6000	0.4800	n.r.	n.r.
HCl	0.5276	0.5306	0.74	0.608	0.2400	0.4530	0.8	0.42
HF	0.0384	0.0277	0.096	n.r	0.0028	0.1010	0.00269	n.r.
H_2	0.3517	0.6070	0.775	0.501	0.0400	0.1510	0.00439	1.49
CH_4	<0.00002	0.0004	<0.000005	n.r.	0.0014	0.00001	<0.00003	n.r.
CO	0.01122	0.00172	0.00084	0.02350	n.r.	0.0110	<0.00004	0.46
NH_3	0.00007	0.00013	n.r.	n.r.	0.0064	0.0006	n.r.	n.r.
N_2	0.05529	0.008	0.025	3.6047	0.0390	0.1068	0.0026	0.18
Ar	0.00011	0.000058	0.00009	0.0485	0.000056	0.000112	0.00001	0.001
O_2	<0.00003	n.r.	0.0176	0.0180	<0.0001	<0.0002	0.0003	0
He	0.000049	0.000006	0.000014	0.000047	0.000021	0.000115	0.000135	n.r.
CO_2/He	51939	85000	82143	117151	286190	103478	46296	
N_2Ar	503	138	278	74	696	954	260	180
N_2/He	1134	1333	1786	75952	1857	929	19	
$^3He/^4He$ (R_A)	7	7.7	6.8	6.6	6.7	6.0	15.7	
$\delta^{13}C$ CO_2	−2.6	−5.3	−7.3	−4.1	−2.6	−0.5	−3.8	
$\delta^{15}N$	4.9		3.7	1.5				

Hydrothermal Gases

Volcano	Aogashima	Guagua Pichincha	Pacaya	Vulcano	White Island	Papandayan	Kilauea S-Bank	El Chichón
Location	Japan	Ecuador	Guatemala	Italy	New Zealand	Indonesia	Hawaii, USA	Mexico
Temperature	99	86	92.1	97	112	100	97	97
H_2O	99.2	95.9	92.1	90.2	98.8	96.2	96.5	97.2
CO_2	0.772	3.805	7.853	9.437	0.950	2.978	3.323	2.753
S_{total}	0.0050	0.2015	0.0049	0.2665	0.2196	0.6956	0.1201	0.0833
SO_2	n.r.	0.104	n.r.	n.r.	0.007	0.1387	0.1065	0.0009
H_2S	0.0050	0.0975	0.0049	0.2666	0.212	0.5570	0.0136	0.0825
HCl	0.0007	0.0045	n.r.	0.0019	0.0042	0.0283	0.0023	n.r.
HF	0.0002	0.0001	n.r.	0.0003	0.0009	<0.0003	<0.00003	n.r.
H_2	0.0027	0.0015	0.001778	0.07017	0.00024	0.0017	0.0001	0.0016
CH_4	0.0006	0.00005	0.001359	0.01078	0.01332	0.000040	<0.00002	0.000022

TABLE 45.1 Examples of Magmatic and Hydrothermal Gases in mol (%) Total Gas—cont'd

Hydrothermal Gases

Volcano	Aogashima	Guagua Pichincha	Pacaya	Vulcano	White Island	Papandayan	Kilauea S-Bank	El Chichón
Location	Japan	Ecuador	Guatemala	Italy	New Zealand	Indonesia	Hawaii, USA	Mexico
CO	<0.000001	0.000001	n.r.	n.r.	0.000001	<0.0000001	<0.00002	<0.000006
NH_3	0.000072	0.00003	0.000664	0.000314	n.r.	n.r.	n.r.	n.r.
N_2	0.01820	0.88	0.039579	0.083888	0.01224	0.054282	0.006704	0.002802
Ar	0.00011	0.000038	0.000079	0.000666	0.000036	0.000406	0.000061	0.000014
O_2	0.00001	<0.0002	0.000032	0.000049	<0.00004	<0.00012	<0.00015	<0.000006
He	0.000008	0.000011	0.000040	0.000041	0.000004	0.000029	0.000059	0.000009
CO_2/He	91779	345909	198800	229286	255484	101760	56477	294785
N_2Ar	165	23158	501	126	340	134	111	205
N_2/He	2162	80000	1002	2038	3290	1855	114	300
$^3He/^4He$ (R_A)		7.9		5.3		6.2	14.5	7.3
$\delta^{13}C\ CO_2$		−2.8		−3.2		−6.9		−8.1

Note: n.r. is not reported.

<< $1R_A$, and a crustal end member of 0.04 R_A is generally assumed for purely nonmantle helium. The wide range in ratios and low abundance in the atmosphere make helium isotopes a powerful tracer. Of all noble gases, it is the one that always shows the most unequivocal mantle signal. In subduction zone settings, where abundant data are available, the $^3He/^4He$ ratios of discharging fluids are mainly controlled by the crustal thickness through which the gases move to the surface (Figure 45.10).

In addition to providing unambiguous information on the source of helium in volcanic gases, combining helium isotopes with other gas species, notably CO_2 and N_2 and their stable isotopes, is a powerful tool for investigating the sources of those components. Figure 45.11 shows the global data set of $CO_2/^3He$ and $\delta^{13}C$ of CO_2 in volcanic gases. As can be seen from this figure, the MORB mantle is characterized by $CO_2/^3He$ of $\sim 2 \times 10^9$, whereas organic carbon and carbonates have much higher values of around 10^{13}.

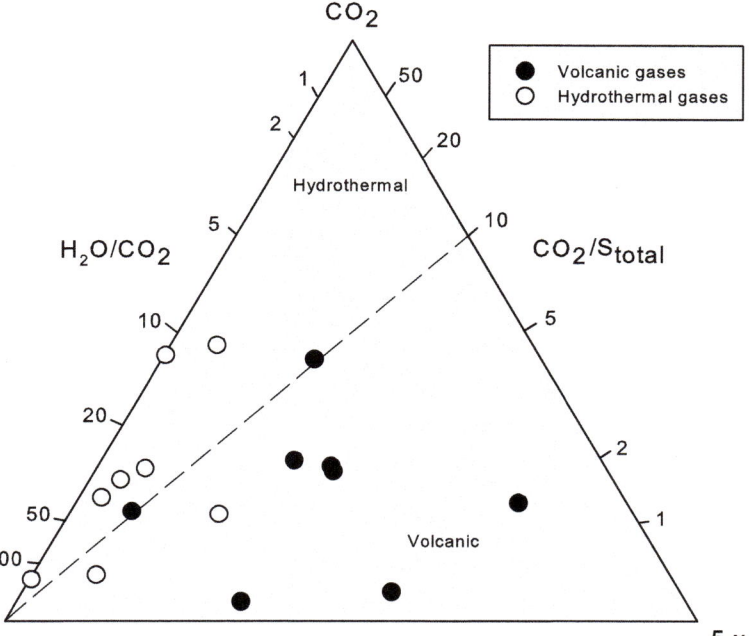

FIGURE 45.8 Triangular diagram showing the relative proportions of H_2O, CO_2, and total S ($SO_2 + H_2S$) of samples in Table 45.1.

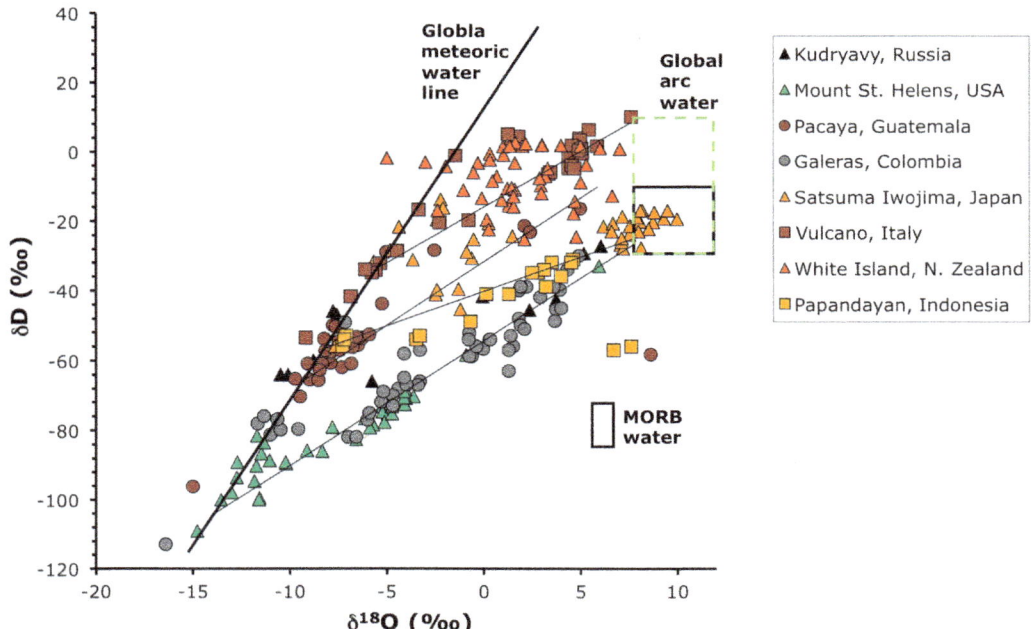

FIGURE 45.9 Oxygen and hydrogen isotope compositions of steam discharging from volcanoes that are related to subduction zones. Also shown are the global meteoric water line, the box defining most arc-type magmatic water and MORB water end members, and mixing lines between local meteoric waters and arc-type water for individual volcanoes. As not all volcanoes degas waters that trend to conventional "arc-type" box, a larger box is shown as dashed lines.

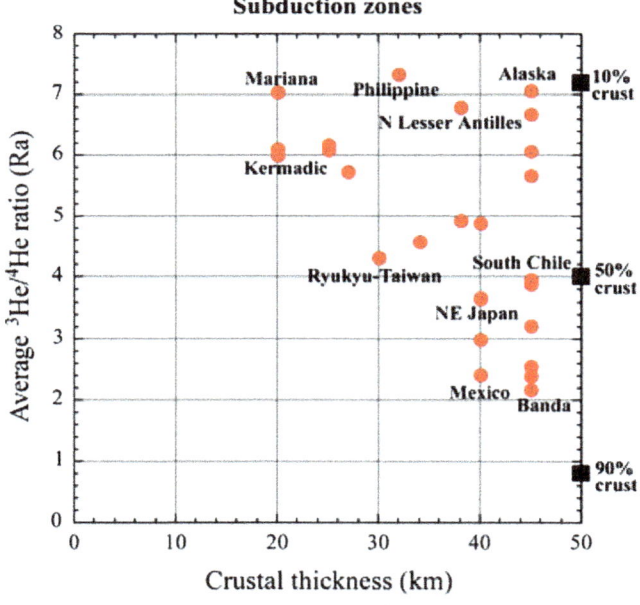

FIGURE 45.10 Relationship of average $^3He/^4He$ sampled from volcanoes and related hydrothermal fluids in an arc segment to crustal thickness. The % crust is calculated using 8 Ra as mantle and 0.02 Ra as crustal end members, respectively. It can be seen that while thin crust results in the highest $^3He/^4He$ ratios, thicker crust does not decrease the ratio in all localities.

The $\delta^{13}C$ value of MORB mantle is $5 \pm 3\%_0$, whereas organic carbon is much lighter at $-30\%_0$ and carbonates are around $0\%_0$. Data points generally fall within the lines that describe mixing of CO_2 from these three sources. Knowledge of the end member compositions allows for the calculation of the relative contributions of mantle, carbonate, and organic sediments to the CO_2 discharges at volcanoes. As seen for water, a significant proportion of CO_2 in subduction zone volcanoes is sourced from the subducting slab that contains carbonates and organic sediments. For investigations of the global carbon cycle, knowledge of these relative contributions, in addition to carbon fluxes, is critical.

As for CO_2, stable isotopes of nitrogen ($\delta^{15}N$) are indicators of source. In volcanic gas samples N_2 is derived from the atmosphere or air-saturated water, the mantle, the crust, or subducted organic sediments. The N_2/He ratio of the upper mantle is approximately 100 with $\delta^{15}N$ of $-5 \pm 3\%_0$, while organic sediments have much higher N_2/He ratios of around 20,000 and $\delta^{15}N$ of $+7\%_0$. Air has N_2/He of 130,000 and $\delta^{15}N$ of $0\%_0$. Application of this approach to gases discharging from the Central American volcanic arc show that the relative contribution of nitrogen from subducted sediment varies along strike of the arc (Figure 45.12). Volcanoes in Nicaragua and

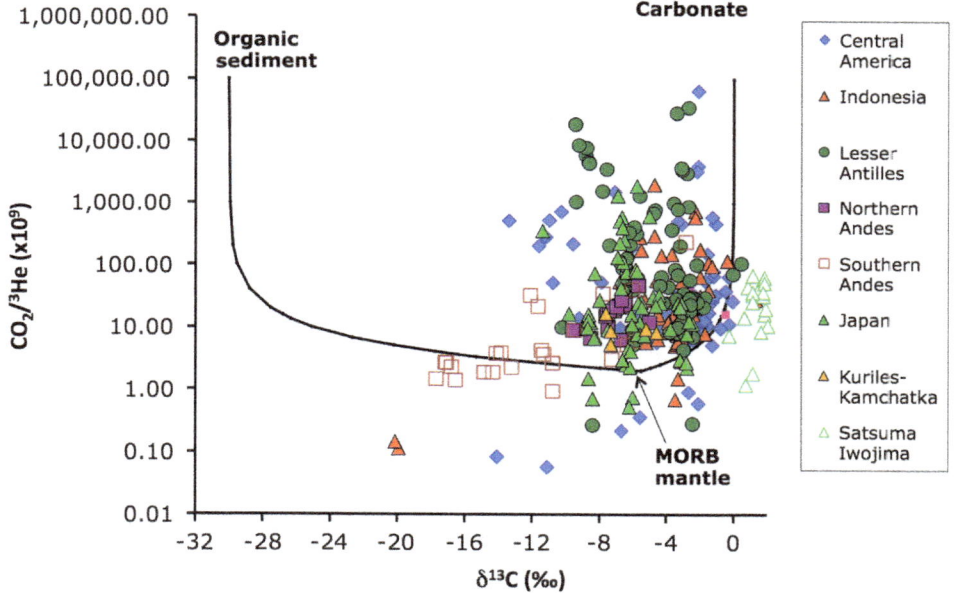

FIGURE 45.11 Global arc data set of CO_2–He systematics of magmatic and hydrothermal gas samples. Knowledge of end members allows for the calculation of the relative contributions of major sources to the discharging CO_2.

Guatemala receive a significantly higher proportion of their N_2 from subducted organic materials than Costa Rican volcanoes. This interpretation is consistent with the idea of sediment removal by off-scraping below the Costa Rica portion of the arc and extensive sediment subduction below Nicaragua and Guatemala due to steeper plate angles.

The multiple oxidation states of sulfur (2^- to 6^+) result in complex behavior in volcanic and hydrothermal environments. Sulfur degassing through volcanoes has been proposed to modify the oxidation state of magmas, while subduction of sulfur may oxidize the subarc mantle. Investigations of sulfur isotopes provide important insights into the global sulfur cycle, the source of sulfur in volcano hydrothermal environments, and magma

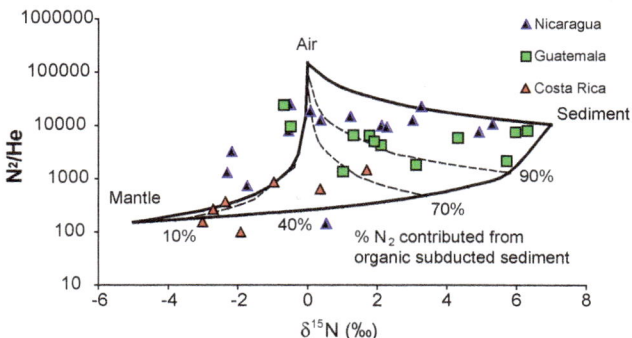

FIGURE 45.12 Application of stable isotopes to determine the source of N_2 in volcanic and hydrothermal gas discharges along the Central American volcanic arc.

degassing processes. Previous measurements on MORB glasses have shown $\delta^{34}S$ values of $0.3 \pm 0.5\%$, but the most recent work using sulfur extraction techniques that allow for the analyses of sulfides and sulfates shows that MORB glasses are lighter at -0.91 ± 0.50. In general high $\delta^{34}S$ values are associated with more oxidized forms of sulfur (sulfates), whereas low $\delta^{34}S$ values are common in more reduced forms (sulfides). Gases from arc volcanoes generally have positive $\delta^{34}S$ values that have been attributed to the subduction of oxidized forms of sulfur, whereas hot spot volcanoes generally exhibit MORB-like $\delta^{34}S$ values (Figure 45.13). The large variations in S isotopes in volcanic gases is due to fractionation processes occurring during degassing, which depend on the temperature of phase separation and the oxidation state of the gas (SO_2 or H_2S) that is emitted from the magma. The more oxidized S species are isotopically enriched in ^{34}S and the more reduced S species are isotopically depleted in ^{34}S. In a magma in which S is primarily dissolved as SO_4^{2-} (S^{6+}), equilibrium degassing of SO_2 would result in a higher $\delta^{34}S$ value in the melt than in the gas. In a melt in which S is primarily dissolved as S^{-2} (sulfide), equilibrium degassing would result in a melt that has lower $\delta^{34}S$ than the exsolved SO_2. Therefore, equilibrium fractionation during degassing can result in evolution of the melt and gas toward higher or lower $\delta^{34}S$ values depending on the dominant S species in the melt. In addition to high-temperature fractionation processes that occur during degassing, low-temperature hydrothermal processes such as precipitation of elemental sulfur,

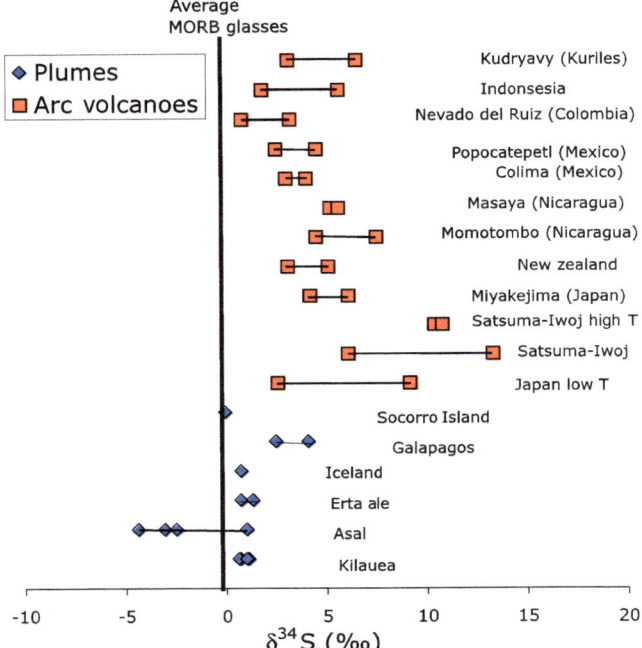

FIGURE 45.13 $\delta^{34}S$ data of fumarolic gas discharges from arc and hot spot volcanoes. Published ranges are given. Arc volcanoes have positive $\delta^{34}S$ values likely due to subduction of seawater sulfate which is not the case at hot spots. The range of $\delta^{34}S$ values measured in MORB glasses is also shown and includes recently measured values that indicate a mean MORB $\delta^{34}S$ value of -0.91 ± 0.50.

sulfides, and sulfates; degassing of SO_2 and H_2S; interaction with liquids; and formation of reduced aqueous species (SO_4^{2-}, S^{-2}) can all significantly affect the sulfur isotope composition of the gas discharges.

4. DEGASSING PROCESSES

Magma degassing begins with the formation of a free volatile phase and the exsolution of bubbles from the melt. The first stage in this process is magma vesiculation resulting in magma with different properties than for bubble-free magma (as described in Chapter 11 on magma vesiculation). The term "degassing" refers to vesiculation followed by separation of the gas phase from the melt and ultimately the release of the gas phase to the atmosphere or hydrosphere. Volatile exsolution is directly related to volatile solubility in melts which depends on pressure, temperature, and magma composition and is discussed in Chapter 7 on magmatic volatiles. The focus here is on volatile saturation and degassing processes. It also needs to be noted that volatile emanations at the surface (in particular CO_2 and He) can occur without the presence of magma at depth and along faults that release mantle volatiles as is the case at the San Andreas Fault, CA, USA, and elsewhere. Commonly, zones of mantle volatile release are characterized by slow seismic P-wave velocities that indicate

areas of higher mantle temperatures. These regions contribute significantly to the global volatile budget.

4.1. Volatile Saturation

The pressure, temperature, and composition of magmas change through time and eventually volatiles dissolved in a magma will reach their points of saturation and enter a vapor phase. Of the major volatile species, CO_2 has the lowest solubility (roughly one order of magnitude lower than water), and CO_2 saturation plays an important role in degassing of magmas at crustal levels. Equally important in early degassing is water because of its high abundance in melts. Upon formation of a H_2O- and CO_2-rich vapor phase, other volatiles (sulfur species, halogens, noble gases, nitrogen) will partition into this vapor phase. The extent of melt–vapor partitioning depends on melt composition, the volatile species and the pressure, temperature, and f_{O_2} (oxygen **fugacity**) conditions. In particular, for degassing of sulfur, magma oxidation state or f_{O_2} is an important variable. Oxidizing conditions promote SO_2 partitioning into the vapor phase, whereas sulfide precipitation is favored at reducing conditions. Overall, pressure exerts the dominant control upon volatile solubility because of the large volume expansion that gases experience when they exsolve from a magma. The control of pressure on solubility is particularly apparent in the CO_2–H_2O system as recorded by melt inclusions. Figure 45.14 shows the CO_2 and H_2O contents measured in melt inclusions of olivine crystals from a variety of tephra samples collected at arc volcanoes. Open-system degassing of magmas will rapidly reduce the amount of CO_2 in the melt while keeping water contents relatively constant. Only at low pressures and

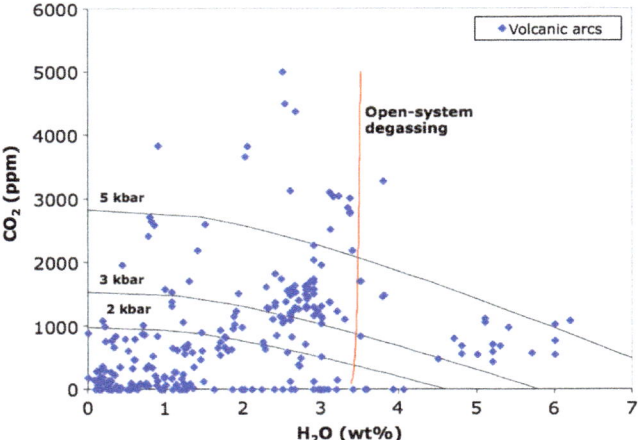

FIGURE 45.14 Water and carbon dioxide contents measured in melt inclusions from arc volcanoes. Isobars represent approximate volatile saturation pressures. The vertical line illustrates an open-system degassing trend, i.e., evolution of melt inclusion compositions as water and carbon dioxide degas from a rising magma.

near-surface conditions will H_2O degas substantially, resulting in a decrease of H_2O content in the melt as recorded by melt inclusions.

Concerning the composition of volcanic gases, the volatile phase first exsolving from an undegassed magma as it rises through the crust should be dominated by CO_2 with only minor contributions from other volatiles such as H_2O and S species. In fact, the CO_2/SO_2 ratio as measured in volcanic gas emissions is an important parameter for assessing the state of activity of a volcano. For example, it has been shown at Stromboli during the 2006–2007 activity that a low CO_2/SO_2 ratio (<10) characterizes a quiescent state and that an increase of this ratio by a factor of about 2 precedes large explosive eruptions (Figure 45.15). The increase in CO_2/SO_2 is attributed to CO_2-rich volatile contributions from a deeper region in the magmatic system. This increase in supply of deeply-sourced CO_2-rich gas likely caused faster convective overturn of magma in the shallow conduit, leading to increased eruptive activity.

4.2. Excess Degassing

It has been recognized since the early 1980s that volcanoes emit more gases during eruptions than can be stored in the amount of erupted magma under realistic pressure and temperature conditions. More recently, this process has been termed "excess degassing" or the "excess sulfur problem." The term "excess degassing" is preferred since this process is not a "problem" but rather an intrinsic characteristic of most volcanoes and has received considerable attention following the eruption of Mount Pinatubo volcano in 1991. At Mount Pinatubo and at other volcanoes the amount of sulfur and other volatiles dissolved in the magma is usually much too low, by a factor of 10–100, than the amount emitted during an eruption (Figure 45.16). The estimation of how much sulfur has degassed from the erupted magma is usually performed by subtracting the amount of volatiles still dissolved in the erupted glass or pumice (posteruptive volatile content) from the amount of volatiles measured in melt inclusions (preeruptive volatile content). Scaling this difference to the amount of magma erupted gives the amount of sulfur (and other volatiles) that could have been released by the erupted magma. All dacitic, andesitic, and rhyolitic volcanoes have emitted excess sulfur but basaltic volcanoes from hot spots generally do not (Figure 45.16). It is likely that preeruptive vapor–melt separation processes (i.e., preeruptive degassing) result in accumulation of a gas phase within the volcanic

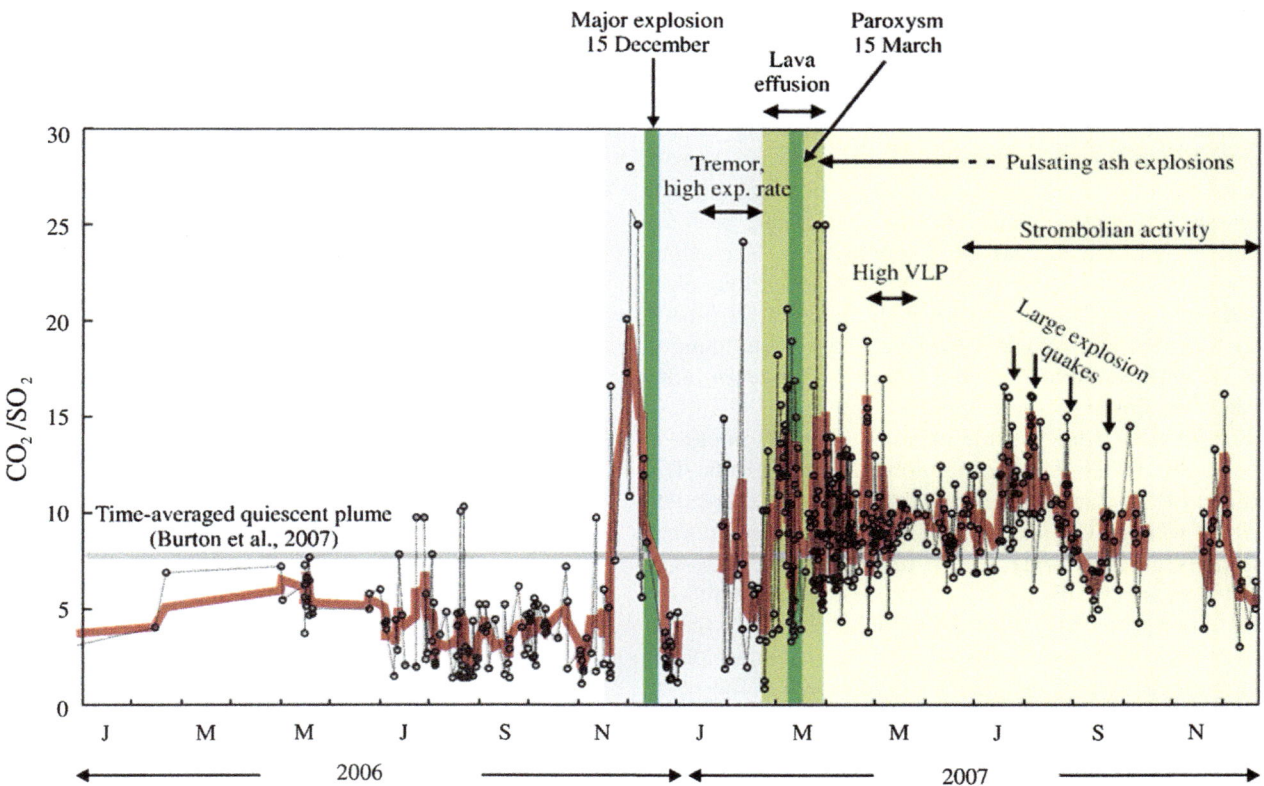

FIGURE 45.15 CO_2/SO_2 ratio variation as measured in the gas plume of Stromboli, Italy, by multi-GAS instrument. Increases in this ratio indicate CO_2 contribution to the magmatic system from deeper levels that result in increased magma convection at shallow levels and eruptions.

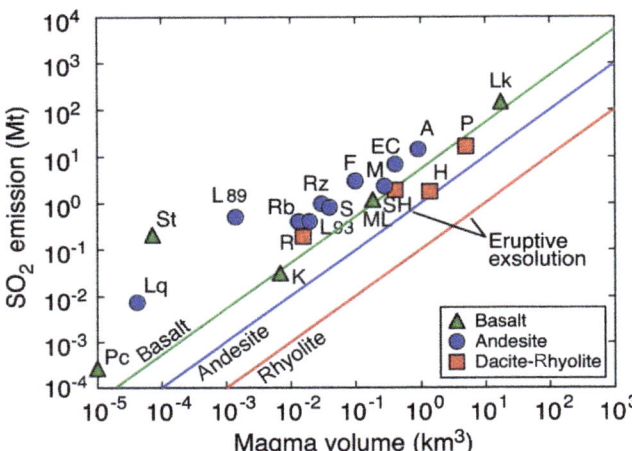

FIGURE 45.16 Volcanic SO_2 emissions versus volume of erupted magma illustrating the excess sulfur that is emitted during volcanic eruptions relative to what can be dissolved in the amount of erupted products, represented by colored lines. SO_2 emissions were measured by remote sensing methods (satellite or ground based). Volcano and eruption labels are: (A) Agung 1963; (EC) El Chichón, 1982; (F) Fuego, 1974; (H) Cerro Hudson, 1991; (K) Kilauea, annual average; (L89) Lascar, 1989; (L93) Lascar, 1993; (Lk) Laki, 1783–1784; (Lq) Lonquimay, 1989; (ML) Mauna Loa, 1984; (M) Mount St Helens, 1980; (Pc) Pacaya, 1972; (P) Mount Pinatubo, 1991; (Rb) Rabaul, 1994; (R) Redoubt, 1989–1990; (Rz) Nevado del Ruiz, 1985; (S) Mount Spurr, 1992; (St) Stromboli, annual average.

edifice on top of the magma chamber. Alternatively, sulfur may also be removed from the silicate magma by an immiscible sulfide melt phase under reducing conditions. When an eruption occurs, accumulated gas is released into the atmosphere, while the magma that degassed at depth to produce this gas phase remains underground. The excess degassing process is particularly obvious at persistently active volcanoes that have open vents, i.e., Sakurajima, Stroboli, Semeru, Karimsky, Popocatepetl, and others, where only minor amounts of ash are emitted explosively but hundreds of tons or more of SO_2 are emitted continuously.

In a broader context, the excess degassing process is also evident when considering global sulfur budgets. The amount of SO_2, and presumably other magmatic volatiles such as water, CO_2, and halogens, that is released during eruptions represents only 1–10% of the global volatile flux from volcanoes. In other words, passive, noneruptive degassing accounts for 90–99% of global volcanic volatile emissions. Therefore, when estimating global emissions it is important to consider passive degassing. On the other hand, new methods are needed to better quantify the amount of an accumulated gas phase from the erupted deposits themselves in order to arrive at volatile budgets that can be extrapolated though time and linked to ancient eruptive products.

5. THERMODYNAMICS OF VOLCANIC AND HYDROTHERMAL GASES

We have described above how inert gas species, for example, noble gases and N_2, can be conveniently used to investigate the origin of the fluids. By contrast other gas species react quickly with the other fluid components and with the minerals in rocks, readjusting their composition to the different pressure–temperature conditions encountered by the gases during their ascent. Reactive gas species include hydrogen (H_2), carbon monoxide (CO), hydrocarbons, and sulfur gases (H_2S, SO_2).

Gas equilibria among reactive gas species in hydrothermal environments have been extensively investigated from 1970 to 2000, mainly for the derivation of gas geoindicators of temperature and pressure resulting from exploration and exploitation of geothermal resources. The $H_2O–H_2–CO_2–CO–CH_4$ system has played a pivotal role in most of these studies, although equilibria involving S-bearing gases, such as H_2S and less frequently COS and N-compounds such as N_2 and NH_3, were also taken into consideration to derive suitable geoindicators. Successively two geothermometers, based on H_2/Ar and CO_2/Ar ratios, were proposed, with the assumption that the relative abundance of Ar in hydrothermal fluids is close to that of air-saturated groundwater.

This section is restricted to hydrothermal gas equilibria in the $H_2O–H_2–CO_2–CO–CH_4$ system.

It is initially assumed that hydrothermal gas species attain chemical equilibrium in a single vapor phase. The following three reactions provide the equilibrium constraints in the $H_2O–H_2–CO_2–CO–CH_4$ system:

$$H_2O = H_2 + 1/2O_2 \qquad (45.2)$$

$$CO_2 = CO + 1/2O_2 \qquad (45.3)$$

$$CO_2 + 2H_2O = CH_4 + 2O_2 \qquad (45.4)$$

Equations (45.2) and (45.3) describe the dissociation of the main components of hydrothermal volcanic gases (H_2O and CO_2), while Eqn (45.4) refers to CH_4 production from CO_2 and H_2O. Production of CH_4 could also take place through reaction of elemental C and H_2. Although elemental C (graphite) may be present in some natural hydrothermal systems such as Cerro Prieto, Mexico, and in Northern Latium, Italy, these are probably peculiar cases rather than the general situation. The equilibrium constants of reactions 2–4 can be written as:

$$\log f_{H_2} = \log K_{H_2} - 1/2\log f_{O_2} + \log f_{H_2O} \qquad (45.5)$$

$$\log f_{CO} = \log K_{CO} - 1/2\log f_{O_2} + \log f_{CO_2} \qquad (45.6)$$

$$\log f_{CH_4} = \log K_{CH_4} - 2\log f_{O_2} + 2\log f_{H_2O} + \log f_{CO_2}$$
(45.7)

The temperature dependence of the thermodynamic constants K_{H_2}, K_{CO}, and K_{CH_4} is given by the following equations, based on thermodynamic data:

$$\log K_{H_2} = -12707/T + 2.548 \qquad (45.8)$$

$$\log K_{CO} = -14955/T + 5.033 \qquad (45.9)$$

$$\log K_{CH_4} = -42007/T + 0.527 \qquad (45.10)$$

Equations (45.5)–(45.7) can be suitably used to compute fugacities of H_2, CO, and CH_4 (f_{H_2}, f_{CO}, f_{CH_4}) at different temperatures once the temperature dependence of f_{H_2O}, f_{CO_2}, and f_{O_2} is known. For a hydrothermal environment, the relationships linking water, carbon dioxide, and oxygen fugacities to temperature are now briefly described.

Water fugacity: Depending on the pressure–temperature distribution, the main component of hydrothermal fluids, H_2O, can be present in different physical states, i.e., liquid, gas ("superheated vapor" in the following) and saturated liquid + vapor. For coexisting vapor and liquid water, changes of $\log f_{H_2O}$ with temperature are closely approximated by the following equation:

$$\log f_{H_2O} = 5.510 - 2048/T \qquad (45.11)$$

where T is in K. Water fugacity is not constrained by temperature alone and becomes an additional external variable both for a single liquid phase at pressures higher than saturation and for a single vapor phase at pressures lower than saturation.

Carbon dioxide fugacity: Carbon dioxide fugacity in "full equilibrium" for hydrothermal systems is fixed, at any given temperature, by univariant reactions involving calcite, a Ca-Al-silicate, K-feldspar, K-mica, and chalcedony. These minerals constitute the thermodynamically stable alteration assemblage resulting from isochemical recrystallization of an average crustal rock. In these conditions the f_{CO_2}–temperature dependence is given by:

$$\log f_{CO_2} = 0.0168 \left(T - 273.15\right) - 3.78 \qquad (45.12)$$

However, it is unlikely that this f_{CO_2} buffer is generally valuable for natural hydrothermal environments, probably because a continuous flux of CO_2 from a magmatic source can occur through the hydrothermal systems and consequently f_{CO_2} acts as an externally fixed parameter.

Oxygen fugacity: Generally, the most suitable parameter describing redox potentials of most natural fluids is as follows:

$$RH = \log\left(f_{H_2}/f_{H_2O}\right) = \log\left(X_{H_2}/X_{H_2O}\right) \qquad (45.13)$$

It has been proposed that the redox potentials of hydrothermal systems are primarily governed by the (FeO)–(FeO$_{1.5}$) couple. The Hydrogen redox (RH) values referring to the (FeO)–(FeO$_{1.5}$) buffer are fixed by the reaction:

$$H_2O + 2(FeO) = H_2 + 2(FeO_{1.5}) \qquad (45.14)$$

the equilibrium constant of reaction (14) has a value of -2.82 ± 0.02, is equal to RH, and is essentially independent of temperature. Considering the formation of water from H_2 and O_2, reaction 14 can be rewritten as:

$$4(FeO_{1.5}) = O_2 + 4(FeO) \qquad (45.15)$$

and $\log f_{O_2}$ can be linked to temperature through the following equation:

$$\log f_{O_2} = 10.736 - 25414/T \qquad (45.16)$$

Another equation relating f_{O_2} and temperature in hydrothermal environments was proposed by D'Amore and Panichi in 1980 and is as follows:

$$\log f_{O_2} = 8.20 - 23643/T \qquad (45.17)$$

Assuming that f_{CO_2} is constrained by Eqn (45.12), f_{O_2} is governed by a suitable buffer, (e.g., Eqn 45.16 or 45.17), and treating f_{H_2O} as a temperature function for coexisting liquid plus vapor (Eqn (45.11)), Eqns (45.5) to (45.7) allow computing the fugacities of H_2, CO, and CH_4 at different temperatures (Figure 45.17). As fugacities of individual gas species are impossible to be determined for fumarolic gases, Figure 45.17 cannot be used to compare theoretical and analytical compositions.

Many of the disadvantages of referring to fugacities of single gas species are overcome by working with isomolar concentration ratios, which can be analytically determined on fumarolic fluids. In order to meet these requirements, the equilibrium constants of reactions 2–4 can be reformulated as:

$$\log\left(f_{H_2}/f_{H_2O}\right) = \log K_{H_2} - 1/2\log f_{O_2} \sim \log\left(X_{H_2}/X_{H_2O}\right)$$
(45.18)

$$\log\left(f_{CO}/f_{CO_2}\right) = \log K_{CO}$$
$$- 1/2\log f_{O_2} \sim \log\left(X_{CO}/X_{CO_2}\right)$$
(45.19)

$$\log\left(f_{CH_4}/f_{CO_2}\right) = \log K_{CH_4} - 2\log f_{O_2} + 2\log f_{H_2O}$$
$$\sim \log\left(X_{CH_4}/X_{CO_2}\right)$$
(45.20)

Example 1: Reaction controlling hydrogen fugacities at Vulcano fumaroles, Italy.

In a $\log(X_{H_2}/X_{H_2O})$ versus temperature (1000/T) diagram we can draw the theoretical values expected for

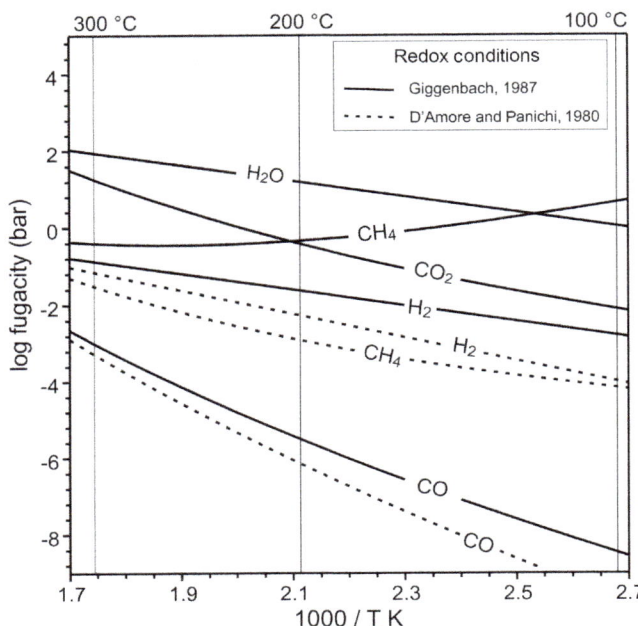

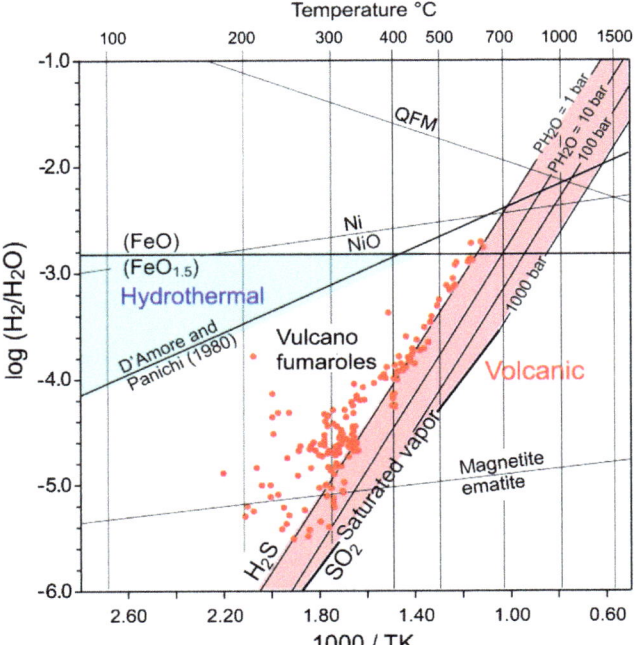

FIGURE 45.17 Fugacities of H_2, CO, and CH_4 at different temperatures, for a hydrothermal vapor phase, calculated assuming that f_{CO2} is constrained by the full equilibrium function (Eqn (45.11)) and f_{O_2} is governed by either the FeO–FeO$_{1.5}$ hydrothermal buffer of Giggenbach (1987) or the f_{O_2} buffer of D'Amore and Panichi (1980). Fugacities of H_2O and CO_2 (Eqns (45.10) and (45.11)) are also shown.

FIGURE 45.18 Plot of log H_2/H_2O versus 1000/T(K). The diagram compares the values measured at high-temperature fumaroles for Vulcano, Italy, with the theoretical compositions computed for various possible buffers of the redox conditions.

different redox buffers and compare them directly with the measured values. This exercise has been done for the high-temperature fumaroles of Vulcano Island (Figure 45.18).

In particular, Figure 45.18 shows the theoretical $\log(X_{H_2}/X_{H_2O})$ values expected for the following redox buffers:

- FeO–FeO$_{1.5}$ (FeO$_{1.5}$) = (FeO) + ½O$_2$
- Hydrothermal empirical (D'Amore and Panichi (DP))
- Hematite-magnetite ($3Fe_2O_3 = 2Fe_3O_4 + ½O_2$)
- Ni–NiO (NiO = Ni + ½O$_2$)
- Quartz-fayalite-magnetite ($2Fe_3O_4 + 3SiO_2 = 3Fe_2SiO_4 + O_2$)
- H_2S-SO_2 (magmatic gas buffer, $SO_2 + H_2O = H_2S + 5/2O_2$)

At temperatures higher than 300–350 °C the values measured at Vulcano fumaroles fit the magmatic gas buffer, i.e., that constituted by the two main sulfur gas species SO_2 and H_2S, at a water pressure of ∼1 bar. This finding suggests that hydrogen fugacity in high-temperature fumaroles of volcanoes, such as those at Vulcano, is controlled by reactions involving the main sulfur gas species at the pressure–temperature conditions in the shallower parts of the fumarolic channels.

Example 2: Reaction controlling hydrogen and carbon monoxide fugacities in hydrothermal systems.

In the Vulcano example above we have seen that hydrogen can quickly react and readjust its concentrations to the discharge conditions of the vents in the high-temperature and relatively oxidizing environment of crater fumaroles. The kinetics of hydrogen, as well as that of other reactive gas species such as CO, becomes slower in the hydrothermal environment, characterized by lower temperatures and by redox conditions generally more reducing than those of volcanic fluids. In this case H_2 and CO may be used to investigate the pressure–temperature conditions governing gas equilibria at depth.

In Figure 45.19 the analytical H_2/H_2O and CO/CO_2 log ratios measured at 21 hydrothermal systems globally are compared with the theoretical compositions expected for redox conditions fixed by the DP empirical function.

The vapor line refers to the composition computed directly by Eqns (45.18) and (45.19) for a pure vapor phase at equilibrium conditions. The other lines refer to the values that $\log(H_2/H_2O)$ and $\log(CO/CO_2)$ assume in the liquid (liquid line) and in the vapor separated during a boiling process at a separation temperature T_s from a liquid at an original temperature T_o. The theoretical compositions are computed assuming adiabatic boiling and considering the vapor–liquid distribution coefficients ($B_i = C_{i,vapor}/C_{i,liquid}$) whose values at different temperatures are known

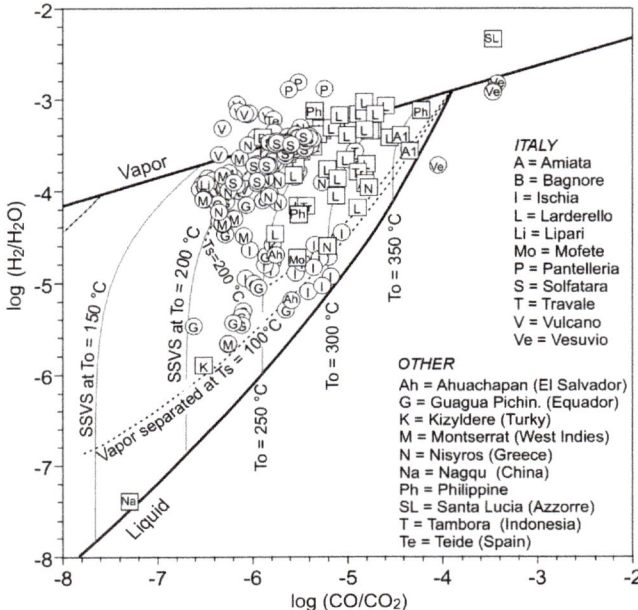

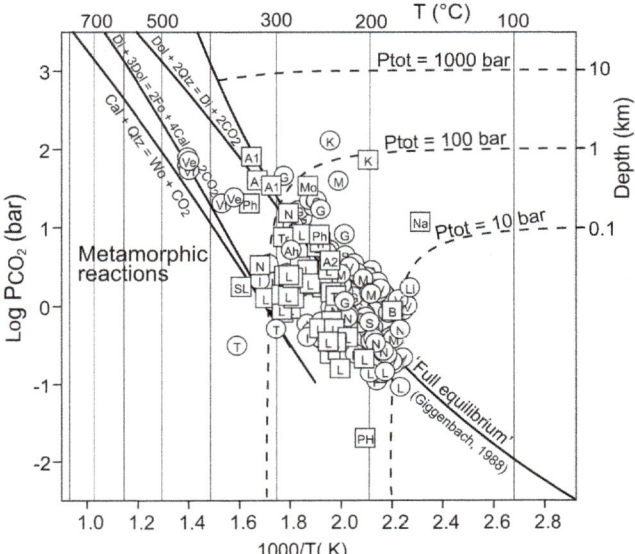

FIGURE 45.19 Plot of $\log(H_2/H_2O)$ versus $\log(CO/CO_2)$. The theoretical grid assumes that redox conditions in the gas equilibration zone are controlled by the f_{O_2} buffer of D'Amore and Panichi (1980). Compositions of both the vapor and liquid phases (thick solid lines) are shown. Compositions of the vapor phase separated in a single step (single-step vapor separation (SSVS)) at different temperatures from a liquid phase initially at T_o of 150, 200, 250, 300, and 350 °C are also shown (thin solid lines) as well as the compositions resulting from SSVS at temperatures of 100 and 200 °C (T_s) starting from any initial temperature (dashed lines). Circles refers to fumarolic vapors, while squares to vapors from geothermal wells.

FIGURE 45.20 Plot of P_{CO_2} values calculated for the hypothetical (equilibrium) single saturated vapor phase vs equilibrium temperatures. Symbols and codes as in Figure 45.19. The full equilibrium functions of Giggenbach (1984, 1988) and $P_{CO_2}-T$ values of relevant metamorphic reactions are also shown.

for each gas species of interest for temperatures from 100 to 340 °C:

$$\log B_{CO_2} = 4.7593 - 0.0109 \times T \qquad (45.21)$$

$$\log B_{H_2} = 6.2283 - 0.0140 \times T \qquad (45.22)$$

$$\log B_{CH_4} = 6.0783 - 0.0138 \times T \qquad (45.23)$$

$$\log B_{CO} = 6.6105 - 0.0151 \times T \qquad (45.24)$$

Most of the experimental points from the various systems plot in the field expected for a vapor phase separated during boiling by liquids originally at 200°C−350 °C, i.e., in the temperature range typical of hydrothermal systems.

Many interpretative diagrams, similar to Fig. 45.19, can be constructed starting from Eqns (45.18)−(45.20) to estimate equilibrium temperature and pressure for fumarolic vents. For example, we could consider equilibria involving CH_4 (Eqn (45.20)), we could check equilibria under different redox conditions (for example, by substituting the f_{O_2}-T function 17 with 16) or we could combine functions 18, 19, and 20 to eliminate the f_{O_2} variable, obtaining a

pressure−temperature geoindicator which is unaffected by different redox conditions.

The final aim is to obtain the most reliable estimates of pressure and temperature conditions governing gas equilibria within the $H_2O-H_2-CO_2-CO-CH_4$ system. In the case of our 21 hydrothermal systems (Figure 45.19), the estimated pressure−temperature conditions are shown in Figure 45.20 which describes typical conditions of most hydrothermal systems: temperatures generally from 150 °C to 350 °C, and total pressures from 10 to 100 bar (i.e., depths of 0.1−1 km assuming hydrostatic pressure conditions). Figure 45.20 also shows the full equilibrium function (see Eqns (45.12)) and CO_2 fugacity−temperature values constrained by relevant thermometamorphic reactions. Most measured points plot along the full equilibrium line, while only a few systems show carbon dioxide fugacities and temperature values consistent with those expected for thermometamorphic reactions. One of these hydrothermal systems is that of Mt Vesuvius where the position of the points suggests CO_2 production through decarbonation of marine carbonates, which are abundant in the basement rocks beneath the volcano. In this case, the past occurrence of decarbonation reactions is documented by skarn and marble ejecta present in Vesuvian volcanic products.

6. CONCLUDING REMARKS

This chapter has reviewed some of the key aspects of volcanic gases and their connection to magma degassing, sources of volatiles in arc and hot spot settings, and the applications of thermodynamics to interpret gas compositions. Although systematic investigations of volcanic and hydrothermal gases has progressed since the mid-1960s, the discipline has plenty of room for new discoveries. New analytical techniques have advanced our understanding of volcanic and hydrothermal gases, as have new in situ and remote sensing tools, both ground and space based. As technology becomes more advanced and more accessible, these techniques will find broader applications opening up gas studies to a larger group of researchers. Volcanic gas data are also finding their way into new databases that allow researchers to correlate gas geochemistry, emissions, and thermodynamic interpretations with other geochemical, geophysical, and volcanological data. New real-time and fast monitoring techniques of volcanic plumes will allow for direct and high-frequency correlations between volcano seismicity and gas emissions. As these techniques become more widespread, we continue to see gas emission monitoring as a standard in the toolbox of volcano monitoring. A likely result of these studies will be scientific discoveries on how volcanoes work and how they move from a state of dormancy through a "critical" state to eruption. Because gases move quickly from depth to the surface and record processes occurring at depth as well as during their ascent, the compositional changes observed are reliable probes into volcanic interiors, and are useful in detecting transitions in volcanic activity.

Sophisticated geothermometers that combine gas chemistry and isotopic compositions will be more extensively tested against actual geothermal well data, as the world is increasingly looking toward using geothermal energy to help meet the planet's growing energy needs. In this case, continuous gas composition measurements as wells are drilled will provide real-time data that can guide energy extraction and sustainability activities.

SEE ALSO THE FOLLOWING ARTICLES

Volatiles in Magmas • Magma Ascent and Degassing at Shallow Levels • Magmatic Fragmentation • Volcanic Plumes • Intrusion-Related Geothermal Systems • Seafloor Hydrothermal Venting at Volcanic Arcs and Backarcs • Volcano-Related Lakes • Volcanic Influences on the Carbon, Sulfur, and Halogen Biogeochemical Cycles • Volcanic Ash Hazards to Aviation • Gas, Plume, and Thermal Monitoring • Volcano Warning Systems.

ACKNOWLEDGMENT

We would like to thank John Stix, who invited us to contribute this chapter to the Encyclopedia. TF would like to thank the US National Science Foundation for supporting his research and that of his students on volcanic, magmatic, and hydrothermal gases through a number of grants since 2001 and the IPA/IRD support during which this chapter was initiated. TF would also like to thank Yuji Sano, Dave Hilton, Yuri Taran, Bernard Marty, Zachary Sharp, Clive Oppenheimer, Sandro Aiuppa, J. Maarten de Moor, and Gary McMurtry for numerous discussions regarding interpretations, sampling, analyses, and detection of volcanic gases. We thank Fraser Goff, John Stix, and an anonymous reviewer for comments that improved the chapter.

FURTHER READING

Aiuppa, A., Burton, M., Allard, P., Caltabiano, T., Giudice, G., Gurrieri, S., Liuzzo, M., Salerno, G., 2011. First observational evidence for the CO_2-driven origin of Stromboli's major explosions. Solid Earth 2, 135−142. http://dx.doi.org/10.5194/se-2-135-2011.

Chiodini, G., Cioni, R., Magro, G., Marini, L., Panichi, C., Raco, B., Russo, M., 1996. Chemical and isotopic variations of Bocca Grande fumarole (Solfatara volcano, Phlegrean Fields). Acta. Vulcanol. 8, 228−232.

Chiodini, G., Marini, L., 1998. Hydrothermal gas equilibria: the H_2O-H_2-CO_2-CO-CH_4 system. Geochim. Cosmochim. Acta 62, 2673−2687.

Chiodini, G., Granieri, D., Avino, R., Caliro, S., Costa, A., Werner, C., 2005. Carbon dioxide diffuse degassing and estimation of heat release from volcanic and hydrothermal systems. J. Geophys. Res. 110, B08204. http://dx.doi.org/10.1029/2004JB003542.

D'Amore, F., Panichi, C., 1980. Evaluation of deep temperatures of hydrothermal systems by a new gas geothermometer. Geochim. Cosmochim. Acta 44, 549−556.

Fischer, T.P., Hilton, D.R., Zimmer, M.M., Shaw, A.M., Sharp, Z.D., Walker, J.A., 2002. Subduction and recycling of nitrogen along the Central American margin. Science 297, 1154−1157.

Giggenbach, W.F., 1980. Geothermal gas equilibria. Geochim. Cosmochim. Acta 44, 2021−2032.

Giggenbach, W.F., 1984. Mass transfer in hydrothermal alteration systems. Geochim. Cosmochim. Acta 84 (2693−2627), 2611.

Giggenbach, W.F., 1987. Redox processes governing the chemistry of fumarolic gas discharges from White Island, New Zealand. Appl. Geochem. 2, 141−161.

Giggenbach, W.F., 1988. Geothermal solute equilibria. Derivation of Na-K-Mg-Ca geoindicators. Geochim. Cosmochim. Acta 52, 2749−2765.

Giggenbach, W.F., 1996. Chemical composition of volcanic gases. In: Scarpa, R., Tilling, R. (Eds.), Monitoring and Mitigation of Volcano Hazards. Springer, Berlin Heidelberg New York, pp. 221−256.

Hilton, D.R., Fischer, T.P., Marty, B., 2002. Noble gases in subduction zones and volatile recycling. In: Porcelli, D., Ballentine, C., Wieler, R. (Eds.), Noble gases in geochemistry and cosmochemistry, Rev. Mineral. Geochem. 47, pp. 319−362.

Labidi, J., Cartigny, P., Birck, J.L., Assayag, N., Bourrand, J.J., 2012. Determination of multiple sulfur isotopes in glasses: a reappraisal of the MORB $\delta^{34}S$. Chem. Geol. 334, 189−198.

Marty, B., 2012. The origins and concentrations of water, carbon, nitrogen and noble gases on Earth. Earth Planet. Sci. Lett. 313−314, 56−66.

Oppenheimer, C., Fischer, T.P., Scaillet, B., 2014. Volcanic degassing: processes and impact. In: Holland, H., Turekian, K. (Eds.), Treatise on geochemistry, second ed, vol. 4. the crust. Elsevier, Amsterdam, pp. 111−179. http://dx.doi.org/10.1016/B978-0-08-095975-7.00304-1.

Taran, Y.A., 2009. Geochemistry of volcanic and hydrothermal fluids and volatile budget of the Kamchatka-Kuril subduction zone. Geochim. Cosmochim. Acta 73, 1067—1094.

Sano, Y., Fischer, T.P., 2013. The analysis and interpretation of noble gases in modern hydrothermal systems. In: Burnard, P. (Ed.), The Noble Gases as Geochemical Tracers. Springer, Berlin Heidelberg New York, pp. 249—317.

Wallace, P.J., 2005. Volatiles in subduction zone magmas: concentrations and fluxes based on melt inclusion and volcanic gas data. J. Volcanol. Geotherm. Res. 140, 217—240.

Chapter 46

Intrusion-Related Geothermal Systems

James Stimac
Stimac Geothermal Consulting, Santa Rosa, CA, USA

Fraser Goff
Earth and Environmental Science, New Mexico Institute of Mining and Technology, Socorro, NM, USA

Cathy J. Goff
Geothermal Consultant, Los Alamos, NM, USA

Chapter Outline

GLOSSARY

boiling point versus depth (BPD) curve A temperature–pressure phase diagram for pure water using hydrostatic depth (weight of a water column to a given depth) as a proxy for pressure. It shows the temperature at which ascending liquid water begins to boil as a function of depth, but is not a strict control on the maximum temperature and pressure observed.

clay cap An impermeable caprock that forms over convecting geothermal systems by alteration of rocks rich in glass and feldspar to abundant smectite and kaolinite.

cupola An "inverted tea cup-shaped" projection from the top of a larger intrusive body that typically acts as a conduit for expulsion of magmatic fluids.

fumarole A hydrothermal manifestation that discharges steam and gas of either magmatic or geothermal origin. Magmatic fumarole discharges are typically >200 °C, whereas geothermal fumarole discharges are typically at the boiling point of water for their elevation, although they may reach 170 °C.

geothermal resource A reservoir of naturally occurring thermal water in the Earth's crust that may be exploited for its energy content.

hydrothermal circulation Movement of water in the Earth's crust resulting from thermal and density gradients.

kaipohan A cold, CO_2-rich gas seep, typically located on high-elevation volcanoes.

magmatic water Water derived from deeply sourced magmas. Magmatic water is a contributing source to many intrusion-related geothermal systems.

The Encyclopedia of Volcanoes. http://dx.doi.org/10.1016/B978-0-12-385938-9.00046-8

meteoric water Water originally derived from precipitation (rain and snow) that is the main source of groundwater. Deeply circulating meteoric water is the dominant fluid in most geothermal systems.

outflow A lateral flow of geothermal fluid that has reached neutral buoyancy or is confined below a low permeability cap, and flows along the hydrologic gradient.

solfatara A volcanic-hydrothermal manifestation that discharges steam rich in CO_2, H_2S, or SO_2, resulting in deposition of large amounts of sulfur and producing acid-sulfate alteration.

steam-heated waters Shallow groundwater of dominantly meteoric origin that has been chemically modified by absorption of geothermal steam and contained gas.

thermal manifestations All hot springs, fumaroles, and related features that discharge fluid from underlying hydrothermal circulation systems.

upflow A buoyant plume of geothermal fluid and contained noncondensable gas that rises from a heat source to a heat sink by advection.

1. INTRODUCTION

Geothermal systems are most prevalent in regions of the Earth that have elevated heat flow and structural settings that support vigorous fluid circulation through fracture networks. These regions are generally located along or near convergent plate margins, near transform plate boundaries, within spreading centers and rifts, and over mantle hot spots (Figure 46.1), where 95% of volcanic activity occurs. Geothermal systems commonly produce hot springs and **fumaroles** as surface **thermal manifestations** of underlying **hydrothermal circulation**. Hot, mineral-rich waters have been universally revered for their therapeutic and mystical qualities. Thermal manifestations also have been exploited for sulfur and other minerals. During the Middle Ages, the commercial value of Italian hot-spring deposits led to wars among local republics. Geothermal electricity was first produced from a steam well at Larderello (Italy) in 1904, and was first marketed commercially in 1912. Since that time, geothermal power has been developed by many countries as an alternative to burning fossil fuels. As of 2010, about 10,900 MW of electric power (MWe) were installed in ~100 geothermal systems worldwide, generating an annual equivalent of about 190 million barrels of petroleum (Table 46.1). In addition, as of 2009, direct utilization of geothermal sources including heat pumps reached 48,493 MW of thermal energy, equivalent to 250 million barrels of oil annually. **Geothermal resources** are not distributed evenly over the globe (Figure 46.1). Not surprisingly, the countries that produce the most geothermal power are located in regions of active or recent volcanism (Table 46.1), while a smaller fraction of geothermal power comes from fault circulation systems and sedimentary basins.

Geothermal resources fall into five categories based on geologic, geophysical, hydrologic, and engineering criteria as follows:

1. Young igneous systems related to fracture networks formed over magma intruded into the shallow crust,
2. Tectonic systems related to fluid circulation along fault and fracture systems,
3. Deep sedimentary aquifers and geopressured systems,
4. Hot dry rock systems or engineered geothermal systems, and
5. Supercritical and/or magma tap systems.

Intrusion-related systems are generally the hottest ($T \geq 220$ °C) and most prolific producers of geothermal electricity worldwide. Extensional and transtensional environments provide open fracture networks that allow extensive upper crustal fluid circulation. Extensional tectonics also promotes rapid accumulation of fragmental rock types with favorable reservoir properties. This chapter focuses on the occurrence and characteristics of intrusion-related systems where the tectonic environment has aided magma ascent to shallow levels, whereas the first edition of the Encyclopedia presented a more generalized chapter entitled *Geothermal Systems*. Chapter 71 of this second edition focuses on utilization of various geothermal resources.

2. INTRUSION-RELATED GEOTHERMAL RESOURCES

2.1. Young Igneous Systems

These systems are associated with Quaternary volcanism and shallow (<6 km) magmatic intrusion. Probably the largest and most abundant high-temperature systems on Earth occur along submarine rifts and transform boundaries, yet we know the least about these systems because they are largely located beyond the current reach of exploitation. Considerably more is known from research and exploitation of geothermal systems on islands and continents. Some commercial developments sit astride active subaerial rift zones (e.g., Reykjanes, Nesjavellir, and Krafla, Iceland; Olkaria, Kenya; Puna in Hawaii, USA), or within calderas or pull-apart basins (e.g., Taupo Zone, NZ; Salton Sea, USA), whereas others are developed on the flanks of stratocones (e.g., Momotombo, Nicaragua) or within dome and lava flow complexes (e.g., Bulalo, Philippines; Salak, Indonesia; San Jacinto, Nicaragua).

Hydrothermal fluid circulation around volcanoes and igneous intrusions connects large mass and energy flows from regions of magma genesis and storage in the Earth to the hydrosphere. Key components of this energy and mass transfer are (1) a heat source, (2) permeable reservoir rocks, and (3) water to carry heat and solutes. Tectonic and

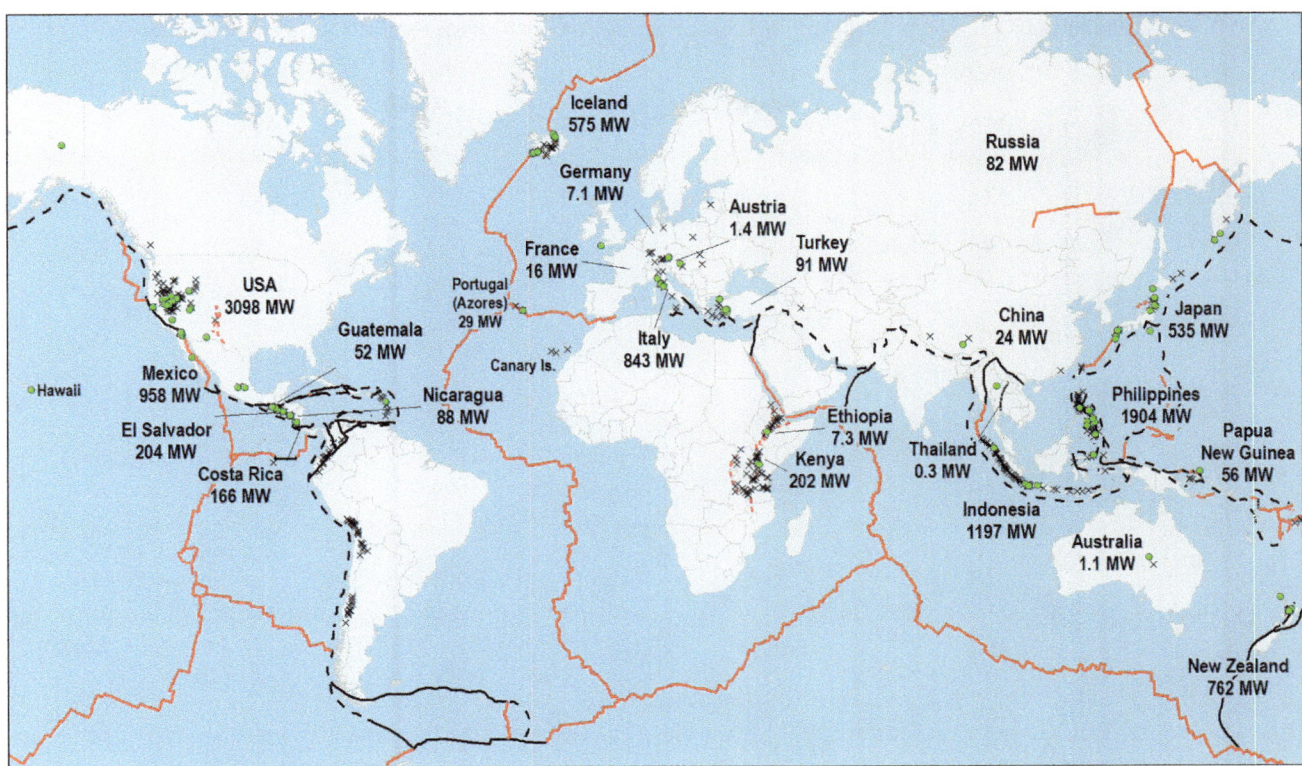

FIGURE 46.1 Map of geothermal systems capable of power production in relation to lithospheric plate boundaries. Geothermal fields producing electricity; green circles, geothermal prospects that are likely to produce power in the future, black x's. Subduction and collision zones, dashed black lines; spreading centers, solid red lines; continental rifts, dashed red lines; and transforms, solid black lines. Installed geothermal capacity is also shown by country in megawatts electric (MWe).

hydrothermal fracturing play the most important role in creating reservoir permeability. **Meteoric water**, mainly consisting of ancient precipitation, may circulate to depths of several kilometers in the crust and is easily set in motion by the large thermal gradients around recently intruded magmas. Seawater may be the main fluid source in some rift and coastal locations. Most geothermal systems also have an overlying **clay cap** that partially isolates them from the near-surface groundwater regime.

Temperatures in intrusion-related geothermal systems generally range from 220 °C to 350 °C, with reservoir depths extending from ≤1 to 3 km below surface. Permeability generally declines with depth, and significant production from >4 km depth is rare. Characteristic assemblages of secondary minerals form in relation to the heat source and to thermal and fluid compositional gradients that develop in the system. Exhumed hydrothermal ore deposits provide fossil analogs for these and other aspects of active systems.

Intrusion-related geothermal systems link the realms of crustal intrusion, *hydrothermal circulation*, and shallow groundwater flow. This is shown schematically in Figure 46.2(A), along with typical temperature and pressure profiles (Figure 46.2(B) and (C)), in which a

region of convective hydrothermal circulation links two regions of lower permeability. While there may be some episodic transfer of **magmatic water** and volatiles directly to the geothermal system, most magma chambers are largely isolated from overlying regions by plastic, low-permeability rocks at temperatures above the *brittle—ductile transition*. Similarly, low-temperature clays formed over the geothermal system limit direct communication with shallow groundwater. Downward circulation of meteoric water peripheral to the top seal provides basal recharge to the system. Within the permeable reservoir, hot fluid flows up as a buoyant plume relative to a hydrostatic gradient. This is referred to as a system **upflow**, where temperatures and pressures may approximately follow the **boiling point versus depth (BPD) curve**. Distal to the main upflow, temperatures are generally below the BPD, but still high compared to areas where heat is transported purely by conduction. In rare cases where the system is strongly isolated from the groundwater realm, steam-dominated conditions may prevail in major fractures. More commonly, fractures are liquid filled, and a hydrostatic gradient, influenced by the temperature of the fluid column with depth, prevails throughout the hydrothermal

TABLE 46.1 Installed Electric Power Generation Capacity and Produced Energy from Geothermal Sources

	2000		2005		2010	
	MWe Installed	Energy (GWh/yr)	MWe Installed	Energy (GWh/yr)	MWe Installed	Energy (GWh/yr)
USA	2228	15470	2534	16840	3098	16603
Philippines	1909	9191	1930	9253	1904	10,311
Indonesia	589.5	4575	797	6085	1197	9600
Mexico	755	5681	953	6282	958	7047
Italy	785	4403	791	5340	843	5520
New Zealand	437	2268	435	2774	762	4055
Iceland	170	1138	202	1483	575	4597
Japan	546.9	3532	535	3467	535	3064
El Salvador	161	800	151	967	204	1422
Kenya	45	366.47	129	1088	202	1430
Costa Rica	142.5	592	163	1145	166	1131
Turkey	20.4	119.73	20	105	91	490
Nicaragua	70	583	77	271	88	310
Russia	23	85	79	85	82	441
Papua New Guinea	0	0	6	17	56	450
Guatemala	33.4	215.9	33	212	52	289
Portugal	16	94	16	90	29	175
China	29.17	100	28	96	24	150
France	4.2	24.6	15	102	16	95
Ethiopia	8.52	30.05	7.3	0	7.3	10
Germany	0	0	0.2	1.5	7.1	50
Austria	0	0	1.1	3.2	1.4	3.8
Australia	0.17	0.9	0.2	0.5	1.1	0.5
Thailand	0.3	1.8	0.3	1.8	0.3	2
Total	7972	49261	8903	55709	10898	67246

regime. However, pressures may be subhydrostatic relative to the surface in high-elevation systems if the deep hydrothermal system is well isolated from groundwaters that reside above the clay cap. A lithostatic gradient generally prevails below the brittle—ductile transition, which serves as a bottom seal of the system.

2.2. Supercritical and Magma Tap Systems

Directly tapping magmatic heat involves drilling into or near shallow bodies of partially molten or recently solidified magma, where pressures and temperatures are in the supercritical regime (above 22.1 MPa and 374.15 °C for pure water, see Figure 46.2(B), and 30.2 MPa and 405 °C for seawater). Supercritical or near-supercritical conditions have been encountered in several geothermal areas including Kakkonda (Japan), Larderello, and Nesjavellir and Krafla at depths from 2 to 5 km. Research wells have also been drilled into lava lakes (e.g., Kilauea Volcano, USA), shallow magma bodies in Iceland and Hawaii, and young volcanic conduits at Unzen Volcano (Japan), and Long Valley Caldera (USA). The Unzen conduit was drilled only 9 years after the last eruptive episode where a dike was intersected at about 2 km depth, but the measured

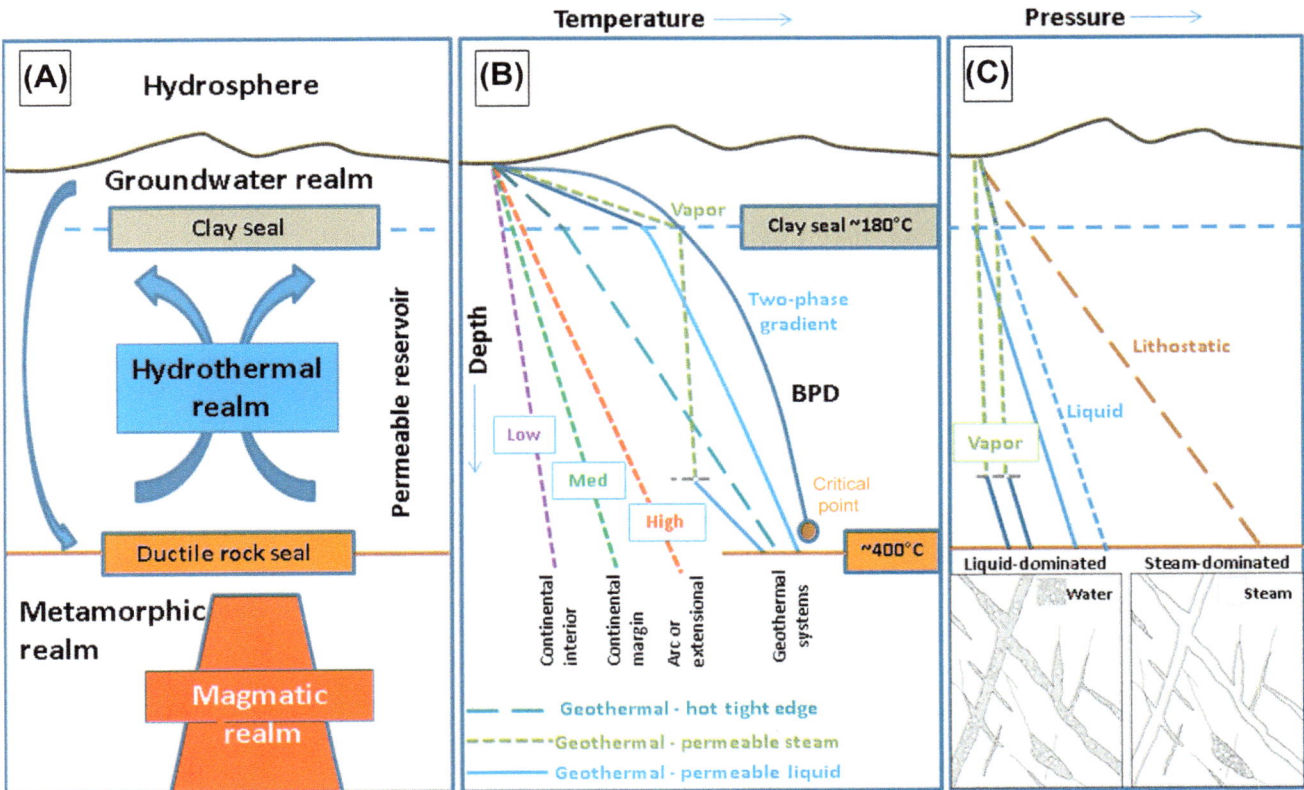

FIGURE 46.2 Links between the realms of crustal intrusion, *hydrothermal circulation*, and shallow groundwater flow. (A) Magmatic intrusion provides heat to drive fluid circulation in the hydrothermal realm. Low-permeability seals at the bottom (ductile–brittle transition, brown line) and top (clay cap, dashed blue line) of the hydrothermal zone limit direct communication between these realms. Downward circulation of meteoric waters occur peripheral to the top seal (down-pointing arrows). Hot fluid flows up as a buoyant plume relative to a hydrostatic gradient (up-pointing arrows). (B) Temperature profiles at different locations in the hydrothermal realm. In the core of the system, temperatures may approach the boiling point verus depth (BPD). Distal to upflow, temperatures are generally below the BPD but high compared to areas where heat is transported purely by conduction. Steam-dominated conditions may form in the shallowest parts of the system if it becomes sufficiently isolated from groundwater. (C) Pressure profiles in different locations in the hydrothermal system. A hydrostatic gradient occurs throughout the hydrothermal regime; however, pressures are commonly subhydrostatic relative to surface in high-elevation systems. A lithostatic gradient prevails below the bottom seal of the system. Liquid or vapor gradients may prevail in the system core depending on its history and extent of isolation from groundwater.

temperature was <200 °C indicating considerable cooling had occurred.

The risk of corrosion related to the acidic nature of volcanic gases has generally discouraged exploration near active conduits of subduction-related volcanoes. Magma tap production may be most feasible in rift settings such as Iceland, East Africa, and possibly New Zealand where the flux of magmatic gas is generally more ephemeral and diffuse than in subduction settings. Extending this concept to greater depths adds exploration and drilling challenges that presently are expensive to solve. Nevertheless, magma is the purest form of natural heat in the shallow crust with expected temperatures from about 650 to 1200 °C.

3. HEAT FLOW IN THE EARTH

Much of the heat in the Earth was generated by attraction, collision, and consolidation of in falling particles due to its

gravitational field. The temperature at the boundary between inner and outer core was recently estimated to be 6000 °C. Additional heat has been generated through time by decay of radioactive elements, particularly U, Th, and K. Today this heat slowly flows through the mantle and crust to the surface. However, magma emplaced in the shallow crust provides the most accessible heat for geothermal production. Here we focus on the heat content of magmas and the role of hydrothermal convection in shaping geothermal systems.

3.1. Heat Content of Magmatic Intrusions and Conductive Heat Flow

Heat content of a magma chamber is dependent mostly on magma volume and temperature. Average silicate magma at 850 °C has latent heat of crystallization of 270 J/g, heat capacity of 1.25 J/g/°C, and mean density of 2.5 g/cm^3. The total heat liberated by magma cooling from 850 to 300 °C is about 960 J/g. If 1 km^3 of magma has a mass of 2.5 × 10^{15} g, the heat liberated by cooling it to 300 °C is 2.4 × 10^{18} J. Large intrusions such as those that produce calderas contain immense quantities of heat. For example, the Bandelier pluton, which created the Valles Caldera (USA), contains roughly 1200 km^3 of magma, releasing roughly 7.2 × 10^{21} J of thermal energy while cooling from 850 to 300 °C.

Conductive heat flow around a magma chamber with time depends especially on emplacement temperature and depth, and on the thermal conductivity of host rocks. A single large intrusion emplaced at 850 °C whose top lies at 5 km depth beneath rocks with average thermal conductivity of 1.5 W/m °C produces a maximum heat flow of about 125 mW/m^2 roughly 400,000 years after emplacement. This value is about two times the average heat flow of the stable crust in the eastern USA (∼55 mW/m^2). Shallower emplacement depth raises maximum heat flow and decreases the time required to achieve the maximum value. Episodic magma replenishment, a common characteristic of plutons and stocks, tends to extend their associated thermal anomalies. Moreover, convection dramatically increases heat flow and shallow thermal gradients, and reduces the time required for cooling of the system.

Typical temperature gradients measured in the shallow crust are summarized in Figure 46.2(B). Conductive gradients produce linear temperature profiles with depth, although the slope may vary when measured over extensive intervals, mainly because of differences in rock thermal conductivities and the localization of magma bodies at certain crustal levels. Shallow thermal gradients in continental areas without young intrusions are typically between 15 and 35 °C/km, whereas background gradients in more tectonically active settings are more variable, ranging from 35 to 120 °C/km (Figure 46.2(B)). Conductive gradients >65 °C/km invariably signal nearby young magmatic intrusions or hydrothermal convection.

3.2. Hydrothermal Convection

Convective heat transport occurs in rocks with sufficient permeability because of the buoyancy effect of heating and consequent thermal expansion of fluids in a gravity field. Cool fluids that access linked magmatic, hydrothermal, and tectonic fractures are heated as they descend through hotter rocks, typically over spans of thousands of years. Here they may mix with magmatic fluids and ascend as a hot *upflow* plume (Figure 46.2(A)). Convection tends to locally elevate shallow temperatures near sources of upflow and depress temperatures in areas of cool downflow relative to the

background thermal gradient. Convection produces steep linear gradients within the top seal and reservoir margins, and nearly isothermal temperature profiles within the reservoir (Figure 46.2(B)). Lateral **outflow** of hot fluid may also occur along avenues of shallow permeability, especially in areas of high topographic relief. Some of the outflowing fluid may reach the surface or be dispersed into regional groundwater regimes.

Convective heat transport is far more efficient than conduction, and is a fundamental characteristic of geothermal systems, sometimes producing spectacular surface manifestations such as geysers, fumaroles, and hot springs (e.g., Yellowstone, USA; El Tatio, Chile). The rate of convective heat removal is highly dependent on circulation depth, reservoir temperature, and mass flow of water through the system. For example, deep conductive heat flow (≥3 km) around larger magmatic systems is about 150−300 mW/m^2, but near-surface heat flow at the largest fumarole areas is commonly ≥5000 mW/m^2. Contrasts such as these illustrate the important role of permeability and convecting fluid to the dynamics and life span of a geothermal system.

The prodigious amount of heat emplaced by shallow intrusion is more than capable of initiating hydrothermal convection, either in existing permeable formations or by hydrothermally mediated permeability enhancement. The minimum permeability required to support hydrothermal convection, on the order of 5 × 10^{-15} m^2 (5 md), can be found in many rock types. The fracture permeability needed in dual porosity simulation models to match observed well behavior is typically 5 × 10^{-14} to 5 × 10^{-13} m^2 (50−500 md). Although large fluid fluxes occur through time, simulation models of producing fields require modest fluid upflows typically on the order of 5−150 kg/s at temperatures of ≥300 °C to match measured temperature patterns.

4. CONCEPTUAL MODELS

The architecture of intrusion-related systems has been determined from studies of exhumed porphyry and epithermal ore deposits, deep geothermal wells, and supporting geophysical data. Figure 46.3 shows temperatures, alteration zones, and sources of deep fluids and their ascent paths in a typical arc-volcanic-intrusive system. Figure 46.4 compares and contrasts intrusion-related geothermal systems that are associated with common volcanic features such as stratocones (Figure 46.4(A)), domes (Figure 46.4(B)), and fissure vents (Figure 46.4(C)). Each volcano-tectonic environment imparts some unique characteristics to the associated geothermal system, although many systems occur in complex volcanic fields with long histories and multiple vent types such as large calderas.

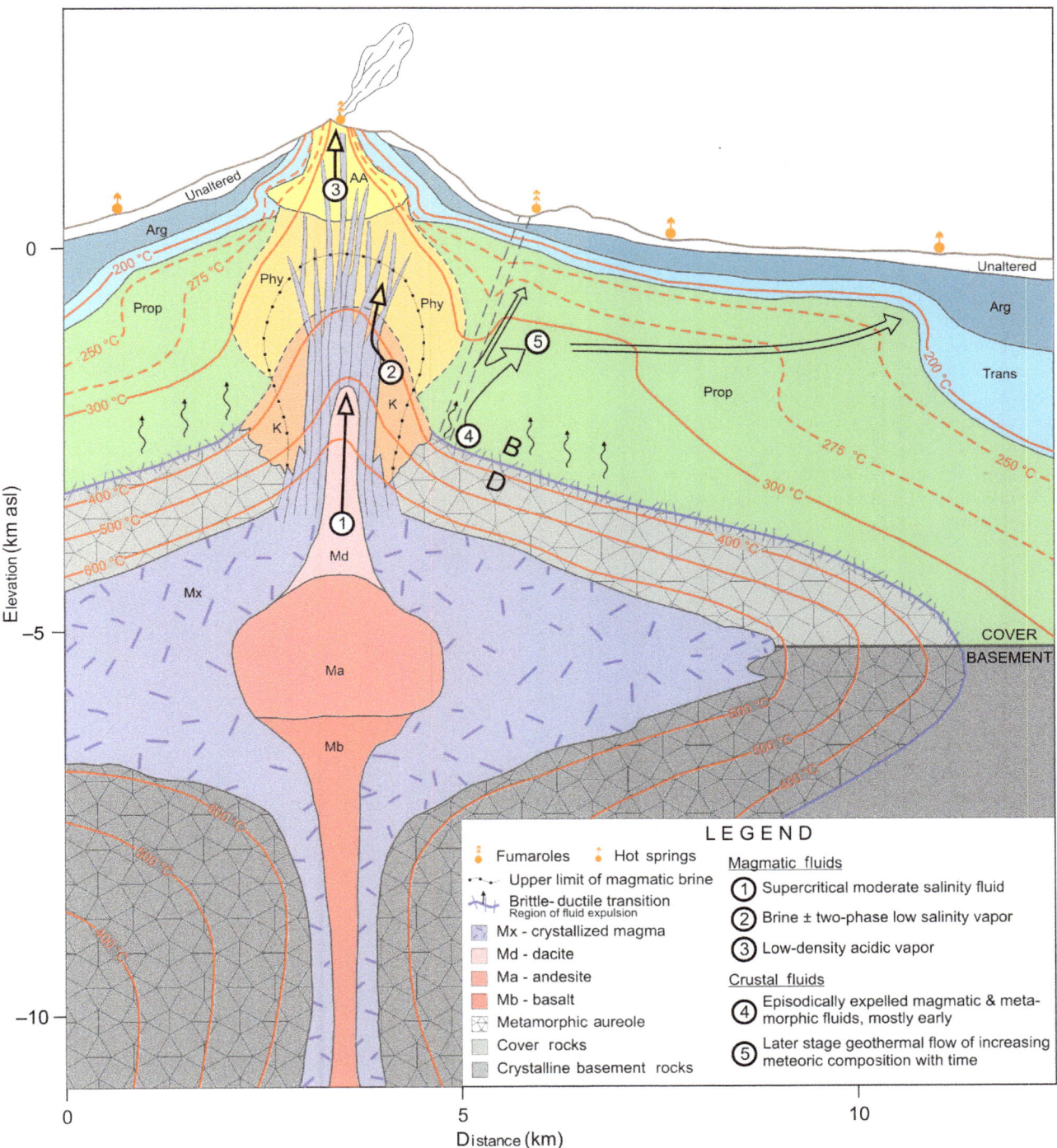

FIGURE 46.3 Schematic cross-section through a typical arc volcanic-intrusive system *(Adapted from Richards (2011))*. Shown are the main alteration zones, isotherms, and sources of deep fluids and their ascent paths (arrows; meteoric fluid descent path not shown). Propylitic (Prop), transition (Trans), and argillic (Arg) alteration formed by long-term circulation of heated meteoric waters affect virtually all the cover rocks above the brittle–ductile transition (B–D, purple line) with greatest intensity close to the intrusion. Metamorphism affects both cover rocks and basement rocks in the ductile regime proximal to the intrusion. Expulsion of metamorphic fluid contributes to brittle fracture during the early stages of the larger pluton but wanes with time as meteoric circulation progressively invades the created fracture network. Potassic (K), phyllic (PHY), and advanced argillic (AA) are more directly related to the flux of magmatic volatiles and circulation of magmatic-hydrothermal fluids that form an envelope around the volcano conduit system. AA may also form more broadly by absorption of H_2S-rich steam by oxygenated meteoric waters.

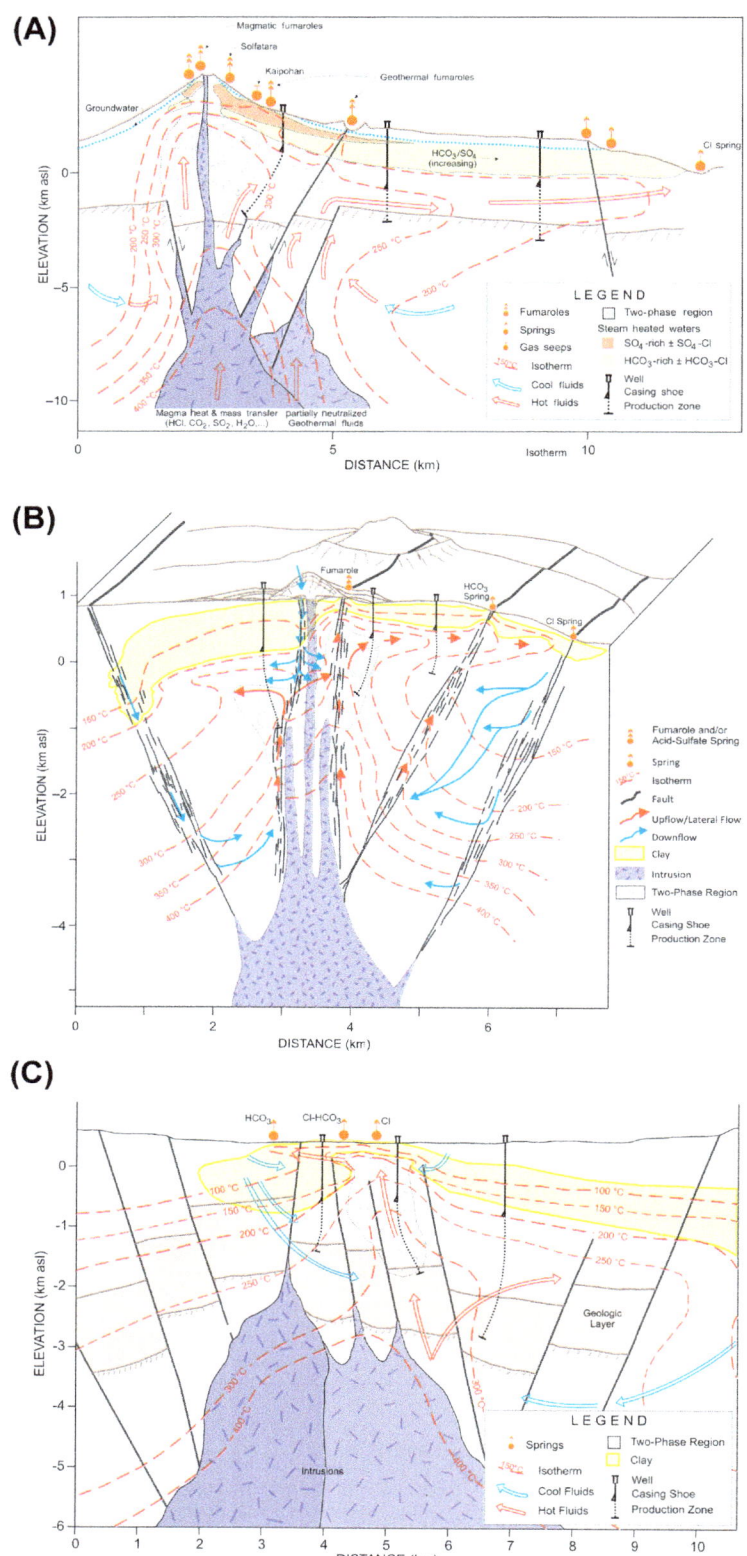

FIGURE 46.4 Conceptual models of geothermal systems. (A) System hosted by an andesitic stratovolcano showing the distribution of major thermal fluid types. The exploitable reservoir is usually found in horizons ≥220 °C. Depth to typical reservoir is ≤1.5 km, whereas depth to intrusions varies from 2 to 10 km. Lateral dimension of reservoir and outflow may exceed 20 km. Hatchered line is faulted basement rocks. (B) System associated with a series of domes of rhyolitic and dacitic composition. This model stresses the interaction of crustal structure and intrusion to localize fluid flow along highly fractured zones. The upper portion of the intrusion should be visualized as a "spine" or dike perpendicular to the section. Complexities in fluid flow paths are indicated by multiple regions of hot fluid upflow and lateral flow, as well as areas of cooler downflow. The interiors of such systems, where hot fluid is upflowing, are commonly at two-phase conditions, whereas more peripheral portions of the reservoir are at liquid-only conditions with some boiling at shallow levels along the outflow path. (C) Low-elevation system, typically hosted within structural basins, with fissure vents, calderas, and dome systems being the most commonly associated volcanic systems.

4.1. Stratocone Systems

Figure 46.4(A) depicts a stratocone system, the types of thermal fluids present, and their distribution relative to the active magmatic conduit and a peripheral geothermal system; these general patterns also apply to high-relief dome and composite volcano systems. Meteoric waters percolating deep into the volcano along permeable zones such as faults, fractures, and volcanic conduits are heated by intrusive bodies. As the hot water circulates, it may become richer in silica, major cations and anions, and soluble metals by mixing with other fluid types and reaction with country rocks. Magmatic volatiles such as H_2O, CO_2, sulfur compounds, HCl, HF, Hg, B, and As may be transferred to the evolving fluid by slow bleeding of gas or episodic magmatic fluid expulsion. The magmatic fluids are believed to be the main source of Cl, F, and S in intrusion-related systems, although these elements may also be derived from leaching of certain sedimentary and volcanic rocks. The deeply circulated fluid is heated and neutralized by reaction with rock. The thermal fluid rises buoyantly due to its decreased density, primarily via major fractures and faults in the volcanic edifice. The core of the system is likely at boiling conditions, and a vapor phase consisting of steam and noncondensable gas is present. Where permeability is sufficient this steam phase may reach the surface, creating fumaroles and areas of steaming ground, sometimes by way of parasitic boiling aquifers that overlie the deep reservoir. When the steam condenses and mixes with shallow meteoric water, H_2S is oxidized to sulfuric acid that chemically alters the rock and produces "acid-sulfate" hot springs.

The composition of thermal springs typically varies systematically with altitude and distance from the volcano summit, with sulfate-chloride and sulfate-rich waters issuing at high elevations, bicarbonate waters issuing at moderate distances and elevations, and chloride waters discharging at lower and more distal locations. Neutral-chloride springs provide direct samples of reservoir fluids that may be diluted en route to the surface, whereas most sulfate and bicarbonate-rich waters are derived by absorption of gas and steam into near-surface aquifers. Sulfate-rich waters are typically acidic, whereas bicarbonate and chloride-rich waters have near-neutral pH. If two water types mix, hybrid sulfate-chloride and bicarbonate-chloride waters may result. Because of stratigraphy, topography, and the hydrologic gradient, the fluids tend to flow laterally away from the volcanic center, forming tabular hydrothermal outflow plumes that reach many kilometers from their deep source. If the intrusive bodies have largely crystallized due to age, small size, or great depth, the contribution of magmatic volatiles to the system may be relatively small or undetectable. This general model describes many features of subduction zone systems such as Rincón de la Vieja and Miravalles (Costa Rica),

Pacaya-Amatitlán (Guatemala), Ahuachapán (El Salvador), and Bacon-Manito and Palinpinon (Philippines).

4.2. Dome Systems

Many geothermal systems are spatially associated with lava domes that occur on the upper flanks of stratocones (Figure 46.4(B)), or in parasitic craters (e.g., Tolhuaca, Chile). Some stratocone systems could also be classified as dome related for this reason (e.g., Bacon Manito and Palinpinon; Lassen Volcano, USA). However, numerous other systems are related to isolated domes, or linear chains or clusters of domes on the distal flanks of stratocones (e.g., Bulalo, Philippines; Wayang Windu, Indonesia), in calderas (Valles; Alcedo, Galapagos, Ecuador), or in extensional (Coso, USA; San Jacinto) or rift settings (Olkaria). Many dome systems are related to magma emplacement in an extensional setting, and thus the combination of shallow intrusion and extensional or transtensional faulting provides more extensive reservoir permeability. Some silicic lava domes may be the only manifestation of more extensive deep to midcrustal magmatism such as at the Salton Sea.

Systems associated with individual domes tend to be relatively small (Acupan and Daklan, Philippines) and if magmatism is still active, can be highly acidic. Larger domes, multiple aligned domes, and dome and lava flow fields tend to host more extensive geothermal reservoirs (Kawerau, NZ: Coso, Bulalo, Wayang Windu, and Olkaria) if emplaced in the past 100,000 years. The fractured conduit system of high-standing domes may provide both a source for cooler fluids and a path for downward migration into geothermal reservoirs. Ingress of cooler fluids along such zones has been documented in a number of cases (e.g., Salak).

4.3. Low-Relief Systems (Caldera Interiors, Extensional and Transtensional Basins, Rifts)

Many geothermal systems occur within areas of local low relief: in collapse calderas, rifts, and various types of tectonic basins (Figure 46.4(C)). Faulted and fractured intracaldera rocks trap meteoric waters that are heated by subjacent magmas (usually ≥ 5 km depth). Because of their large size, calderas (e.g., Yellowstone and the Taupo Zone) host larger, longer lived geothermal systems than those in typical stratovolcanoes. However, some thick intracaldera tuffs have proved to be relatively poor reservoir rocks, possibly due to their high degree of welding, lack of layer-related rheological contrasts, or long alteration histories (e.g., >40 wells drilled in Valles Caldera defined a resource of only 20 MWe).

Low-relief systems include examples from extensional zones in volcanic arcs, such as the Macolod Corridor, Luzon Philippines (Bulalo), most of the Taupo Zone, and many low-standing Icelandic and East African and Afar rift systems. In addition, systems formed in releasing bends of strike-slip faults (many along the Great Sumatra Fault, Indonesia) and in some local rifts or calderas on large shield volcanoes overlying hot spots (Kilauea, Alcedo Volcano) could be considered in this category. In general, these systems have more localized surface expression, and thermal features may be closely grouped. In fault-based systems with ample precipitation (Sumatra, Indonesia), thermal features may be numerous and impressive. Water tables generally reach to the surface, and geothermal fluid pressures slightly exceed a cold hydrostatic gradient. If present, chloride springs may therefore lie directly over the reservoir (e.g., Taupo Zone). However, in rifts with more arid climates thermal features are commonly rather diffuse and chloride springs may be rare to absent (e.g., Kenya Rift).

5. HYDROTHERMAL ALTERATION

Figure 46.3 shows the typical distribution of alteration zones and fluid types of volcanoes, intrusion-related geothermal systems, and exhumed ore deposits. The development of these hydrothermal mineral assemblages are genetically related to (1) the loss of magmatic volatiles from shallow intrusion and magmatic degassing in an active conduit, (2) development of an associated magmatic-hydrothermal fluid envelope, and (3) a related, structurally focused peripheral hydrothermal system dominated by meteoric water with only episodic input of magmatic fluids. The latter is typically the most prospective part of a geothermal system for commercial development.

Some hydrothermal minerals or alteration assemblages are particularly useful in constraining conditions of the hydrothermal systems such as temperature, pH, and redox state. Figure 46.5 summarizes the temperature stability of common hydrothermal minerals comprising the dominant alteration zones. It also shows the typical breakdown products of major igneous minerals as inferred from observed mineral assemblages and temperatures in wells, and the homogenization temperatures of contained fluid inclusions.

The alteration of the most exploitable geothermal systems can be divided into the broadly defined argillic, transition, and propylitic zones (Figures 46.3 and 46.5). Together, the argillic and transition zones form a low-permeability cap by virtue of their abundant clay species, and confine the underlying permeable reservoir. The uppermost argillic zone is the least permeable part of the caprock, and is typically dominated by smectite and kaolinite, with accessory zeolites, chalcedony, amorphous silica, and pyrite. An underlying low permeability transition zone consisting mainly of mixed-layer clays and chlorite,

with accessory zeolites, calcite, chalcedony, quartz, titanite, anhydrite, and pyrite, encloses the permeable reservoir.

The high-permeability reservoir is largely coincident with the propylitic zone in mafic to intermediate host rocks, although this assemblage may be more weakly developed in silicic rocks where an assemblage rich in illite and adularia may dominate. The propylitic alteration assemblage consisting of albite, chlorite, epidote, and pyrite (plus accessory minerals summarized in Figure 46.5) forms over a broad temperature range (220–350 °C) at relatively reducing conditions and near-neutral pH. The roots of igneous geothermal systems are commonly associated with stocks, dike swarms, and magmatic-hydrothermal breccias that are altered to propylitic and potassic mineral assemblages (biotite-adularia) ± amphibole ± tourmaline that formed at temperatures in excess of about 285 °C.

The propylitic assemblage can form with relatively low ratios of water to rock approaching isochemical metamorphism, but a large throughput of hydrothermal fluid is implied in most active geothermal systems over their lifetimes of thousands to hundreds of thousands of years. The highest fluid flux occurs along major faults and fractures where coarse-grained minerals precipitate in open spaces (Figure 46.6). Shearing textures and brecciated vein fragments overgrown and cemented by later mineral growth are clear manifestations of the dynamic nature of fracture growth and filling. Complex sequences of minerals in veins record changes in temperature, permeability, and fluid chemistry along these pathways (Figure 46.7).

In addition to the principal alteration zones described above, advanced argillic alteration forms from acidic, oxidizing fluids. The main zones of acid alteration are kaolinite (up to 120 °C), dickite ± kaolinite (~120–200 °C), dickite ± pyrophyllite (~200–260 °C), and pyrophyllite ± illite (~220–320 °C). Alunite and pyrite may be present in all these zones, whereas diaspore and anhydrite may occur in the latter three. Patches of acid-sulfate alteration may be broadly distributed within the transitional and argillic zones, but it generally crosscuts propylitic alteration at depth, being confined to faults and highly fractured zones. Much of this alteration is produced from shallow secondary sulfate-rich waters that formed by absorption of steam into oxygenated groundwater. These relatively dilute acid fluids may migrate laterally or downward along structural conduits, but tend to remain isolated from neutral pH fluids by precipitation of anhydrite due to its retrograde solubility. Descending acid-sulfate fluids are particularly common in the Philippines where many developed fields (e.g., Tiwi, Bacon-Manito, Mahanagdong, Palinpinon, and Mt Apo) contain sectors with acid fluid production that is spatially associated with this alteration. Other fields contain more pervasive acid fluid zones that were, or still are, associated with direct ascent of magmatic fluids in the near-conduit (e.g., Mt Pinatubo prior

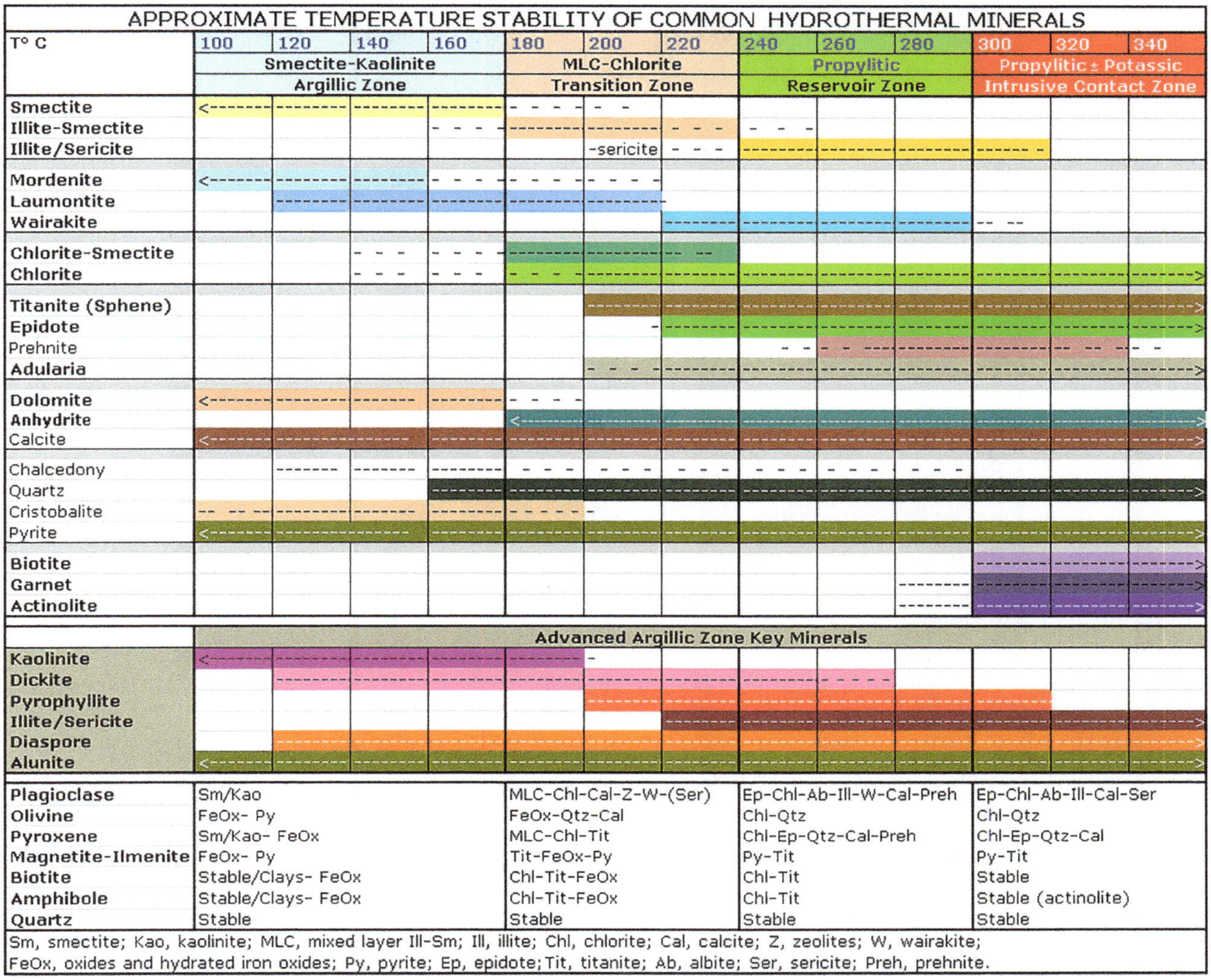

APPROXIMATE TEMPERATURE STABILITY OF COMMON HYDROTHERMAL MINERALS

T° C	100	120	140	160	180	200	220	240	260	280	300	320	340
	Smectite-Kaolinite / Argillic Zone				MLC-Chlorite / Transition Zone			Propylitic / Reservoir Zone			Propylitic ± Potassic / Intrusive Contact Zone		

Mineral stability ranges (approximate):

- **Smectite**, **Illite-Smectite**, **Illite/Sericite** (illite/sericite stable through high temperatures; –sericite noted ~200–220)
- **Mordenite**, **Laumontite**, **Wairakite**
- **Chlorite-Smectite**, **Chlorite**
- **Titanite (Sphene)**, **Epidote**, **Prehnite**, **Adularia**
- **Dolomite**, **Anhydrite**, **Calcite**
- **Chalcedony**, **Quartz**, **Cristobalite**, **Pyrite**
- **Biotite**, **Garnet**, **Actinolite**

Advanced Argillic Zone Key Minerals

- **Kaolinite**, **Dickite**, **Pyrophyllite**, **Illite/Sericite**, **Diaspore**, **Alunite**

	Argillic	Transition	Reservoir	Contact
Plagioclase	Sm/Kao	MLC-Chl-Cal-Z-W-(Ser)	Ep-Chl-Ab-Ill-W-Cal-Preh	Ep-Chl-Ab-Ill-Cal-Ser
Olivine	FeOx- Py	FeOx-Qtz-Cal	Chl-Qtz	Chl-Qtz
Pyroxene	Sm/Kao- FeOx	MLC-Chl-Tit	Chl-Ep-Qtz-Cal-Preh	Chl-Ep-Qtz-Cal
Magnetite-Ilmenite	FeOx- Py	Tit-FeOx-Py	Py-Tit	Py-Tit
Biotite	Stable/Clays- FeOx	Chl-Tit-FeOx	Chl-Tit	Stable
Amphibole	Stable/Clays- FeOx	Chl-Tit-FeOx	Chl-Tit	Stable (actinolite)
Quartz	Stable	Stable	Stable	Stable

Sm, smectite; Kao, kaolinite; MLC, mixed layer Ill-Sm; Ill, illite; Chl, chlorite; Cal, calcite; Z, zeolites; W, wairakite; FeOx, oxides and hydrated iron oxides; Py, pyrite; Ep, epidote; Tit, titanite; Ab, albite; Ser, sericite; Preh, prehnite.

FIGURE 46.5 Summary hydrothermal alteration assemblages, and the temperature stability of common hydrothermal minerals of geothermal systems. Typical alteration products of major igneous minerals are shown for the dominant alteration zones.

to the 1991 eruption). Other well-documented magmatic-hydrothermal envelopes surrounding degassing conduits have proven far too hostile to encourage development (e.g., White Island, NZ; Kawah Ijen, Indonesia).

Most exploited geothermal fluids contain reduced sulfur species and precipitate sulfide minerals, with pyrite being the most common. Other sulfides such as pyrrhotite, galena, sphalerite, and chalcopyrite are present, but are rarely abundant. Native sulfur is typically restricted to near-surface acid-sulfate zones.

6. PHASE DISTRIBUTION

The vast majority of geothermal systems are liquid dominated, meaning they contain mainly liquid water in fractures and pores (Figure 46.2(C)). Two-phase conditions (coexisting water + steam) are most common in areas of vigorous,

hot fluid upflow and in the shallowest portions of well-sealed reservoirs (Figures 46.2 and 46.4). Less commonly, a portion of the reservoir contains steam as the dominant phase in a shallow cap or throughout the entire system.

6.1. Liquid-Dominated Systems

Liquid-dominated systems contain water in all channels and interstitial pores, although bubbles of steam and gas may be present (two-phase regions, Figure 46.2(C)). Groundwater recharge is not as restricted as in vapor-dominated systems (discussed below), and there may be good connections to both shallow and deep meteoric recharge. Hot waters from the reservoir commonly leak at the surface as chloride springs (Figures 46.4 and 46.8(A)). Temperature and pressure increase steadily with increasing depth, either on a liquid or two-phase gradient

FIGURE 46.6   Hydrothermal vein textures from geothermal systems. (A) Brecciated dacite tuff cut by partially filled shear fracture, scale in centimeters. (B) Dacite tuff cut by vuggy open fracture with bridge of cemented rock fragments, scale in centimeters.

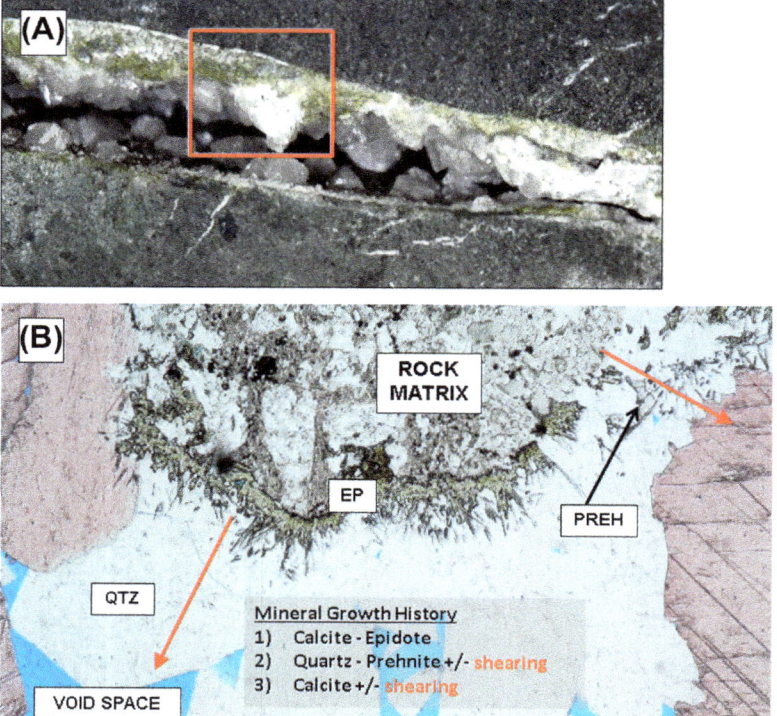

FIGURE 46.7   (A) Andesitic tuff cut by a large open fracture (up to 1-cm aperture) forming part of a fault system. Large calcite and finer epidote crystals are clearly visible. (B) Photomicrograph of vein in Figure 7(A) with red arrows indicating direction of mineral grown from vein walls. Early-formed assemblages are locally sheared indicating cycles of fault movement and sealing over the period of vein mineral formation. Detailed paragenesis consists primarily of early epidote needles and calcite, followed by quartz-prehnite followed by coarse blocky calcite (stained pink). Open space is filled by blue epoxy. Long dimension of photo is about 4 mm.

(A)

(B)

FIGURE 46.8 (A) Chloride spring and surrounding silica sinter near Puchuldiza, northern Chile. (B) Flow test of production well Tolhuaca-4, southern Chile. The well is enclosed in a small building while steam is funneled to a wellhead separator. *Photo courtesy of Anna Colvin, Mighty River Power Chile.*

(Figure 46.2(B) and (C)). If the rate of upflow is rapid, boiling springs or geysers will form and temperature changes with depth will closely follow the boiling curve.

Liquid-dominated reservoirs have maximum temperatures on the order of 370 °C and widely ranging salinities ($\sim$500–125,000 mg/l, averaging $\sim$10,000 mg/l). When fluid is produced from wells, it must be depressurized (flashed) to produce steam for the turbines (Figure 46.8(B)). This is done under controlled conditions with large steam separators. The basic equation that relates fluid enthalpy, h, to steam fraction, X, is

$$h = h_w(1 - X) + h_s \cdot X \qquad (46.1)$$

where h is the enthalpy of the reservoir fluid, h_s is the enthalpy of steam, and h_w is the enthalpy of water at a given separation pressure. Steam tables are used for enthalpy calculations. For a well at 250 °C supplying fluid to a separator at a pressure of 8 bar absolute, X = 0.179, which means that 17.9% by mass of the produced fluid has flashed to steam at a corresponding temperature of 170 °C. A separator receiving 800,000 kg/h of fluid will produce 143,000 kg/h of steam (at 8 bar absolute), which flows into the turbine generator. The remaining 82.1% of the produced fluid is liquid water that must be reinjected. A well supplying 350 °C fluid to the same separator yields X = 0.465 and produces more than double the amount of steam as the previous example. Noncondensable gases such as CO_2, H_2S, and NH_3 partition into the steam phase while the residual brine becomes more concentrated in most other chemical species such as chloride.

6.2. Vapor-Dominated Systems

The occurrence of vapor-dominated regions in the crust is an unlikely state that requires both a potent heat source and isolation from circulating groundwaters. This condition is rarely achieved except at shallow levels because the low pressure exerted by a vapor column relative to a hydrostatic column can only be maintained if the reservoir is sealed from liquid-filled rocks by robust barriers (see Figure 46.2(C)).

Thermal manifestations associated with vapor-dominated systems consist of fumaroles, mud pots, and acid hot springs (Figure 46.9); chloride-rich fluids are absent. Fluid temperature and pressure (T and P) increase with depth in the impermeable cap. However, within the reservoir, T and P are near the maximum enthalpy of "dry" steam ($\sim$240 °C and 3.3 MPa) and remain nearly constant with depth until the bottom of the steam zone is reached (Figure 46.2(B) and (C)). During production, boiling of interstitial water in rock pores and micro-fractures provides additional vapor and latent heat that enter the well. Even though they are not as hot as some liquid systems, vapor-dominated reservoirs are prized because virtually all produced fluid is piped to the power turbines and the expense of phase separation systems and injection wells can be minimized.

Of the $\sim$150 geothermal reservoirs that have been explored by drilling, only a few vapor-dominated systems have been confirmed. These include The Geysers, Larderello, and Kamojang and Darajat (Indonesia). Of these, only Kamojang and Darajat are hosted within young volcanoes. Another system, Matsukawa (Japan), appeared to lie on a liquid pressure gradient at initial conditions, but rapidly evolved to steam as mass was extracted. In this case, low reservoir permeability

(A)

(B)

FIGURE 46.9 Fumarole areas associated with geothermal systems. (A) Boiling pools and warm lake at Kamojang, West Java. (B) Manuk fumarole area, Darajat, West Java, Indonesia. Extensive fumaroles, boiling pools and mud pots are surrounded by clay-altered rock and rock debris.

resulted in rapid pressure drawdown that enhanced boiling in the near wellbore environment to the extent that pure vapor was produced. Thus, a low-permeability system at or near two-phase conditions over a large depth range can evolve to vapor-dominated conditions with fluid extraction.

6.3. Hybrid Vapor-Cap Systems

Some hybrid systems such as Wayang Windu, Patuha, and Karaha-Telaga Bodas (all in West Java, Indonesia) had extensive steam caps underlain by a liquid regime prior to exploitation. In these systems, the liquid regime was in pressure equilibrium with an overlying steam-dominated regime that was sealed on its upper margins. Studies suggest that the Karaha-Telaga Bodas system was driven to shallow steam-dominated conditions by boildown related to eruption and sector collapse of the nearby Galunggung volcano. Fluid inclusion evidence suggests boildown

promoted extreme oversaturation of silica and formation of chalcedony at $T > 250\ °C$.

Shallow initial steam caps or regions of two-phase conditions tend to expand under commercial fluid extraction (e.g., Salak and Tiwi). Small parasitic low-pressure vapor zones extending from near the surface to a few hundred meters may also form (Lassen; Ngatamariki, NZ) and feed fumaroles and acid hot springs similar to those associated with true vapor-dominated systems.

7. GEOTHERMAL FLUID CHARACTERISTICS

Geothermal fluids display wide variations in chemical and isotopic composition that reflect their fluid sources and evolution. Fluid composition is one of the most critical inputs to exploration, and may constrain the type of power cycle that is eventually applied to a particular resource. Waste fluids generally contain constituents that have adverse impacts on the environment if not disposed of properly. Some geothermal fluids contain valuable metals that are sold as by-products of energy utilization. Thus, knowledge of fluid characteristics is important to geothermal development and has applications to understanding volcanic-hosted ore deposits.

7.1. Gas Chemistry

Geothermal gases form by contributions from magmas, reactions of circulating fluids with reservoir rocks, contributions from air-saturated meteoric water, and during thermal breakdown of volatile-producing components in reservoir rocks. Vapor-dominated systems (see Section 6.2) produce more noncondensable gas by weight than typical liquid-dominated system because volatile components are preferentially transported with vapor (steam). However, the gas compositions in the two types of systems are similar (Table 46.2). In contrast to volcanic gases emitted from magma, geothermal gases usually contain more CO_2; less total sulfur, which is invariably in the reduced form (H_2S); and higher contents of CH_4, NH_3, and N_2, and rarely contain acid halides. Gas compositions are sensitive to reservoir rocks. For example, geothermal reservoirs hosted in argillaceous rocks produce gases relatively rich in organic-derived components: H_2S, CH_4, NH_3, and N_2. Reservoirs hosted in carbonates or volcanic rocks tend to be richer in CO_2, and those developed in sedimentary rocks often contain more CH_4 (Figure 46.10; Table 46.2) and NH_3 from thermal breakdown of organic remains.

7.2. Water Chemistry

Geothermal waters range widely in chemical composition (Table 46.3) with variations dictated by temperature,

TABLE 46.2 Gas Compositions from Selected Vapor- and Liquid-Dominated Geothermal Systems Compared with Volcanic Gas Compositions (He, Ar, and O_2, if Present, are Not Reported)

Site	Larderello[1]	The Geysers[2]	Darajat[3]	Ahuachapan[4]	Wairakei[5]	Cerro Prieto[6]	Valles Caldera[7]	Long Valley[8]	Borateras[9]	Galeras[10]	Satsuma I–J[11]
Country	Italy	USA	Indonesia	El Salvador	New Zealand	Mexico	USA	USA	Peru	Colombia	Japan
Reservoir	Vap-dom	Vap-dom	Vap-dom	Liq-dom	Liq-dom	Liq-dom	Liq-dom	Liq-dom	Liq-dom	Magma	Magma
Rocks[12]	Arg/Carb	Arg/Gw/Fel	Volc	Volc	Volc	Arg/Ss	Volc/Ss/Carb	Volc	Volc	Andesite	Rhyolite
Temperature	240	240	240	255	260	≤350	295	95	86.7	360	885
H_2O	98.0	99.4	99.9	99.95	99.94	99.26	–	99.6		89.2	97.5
CO_2	94.1	60.4	84.6	86.8	91.7	86.1	96.9	97.8	85.3	75.1	15.1
H_2S	1.5	5.73	9.85	12.1	4.4	4.43	0.81	0.66	10.4	5.31	1.16
SO_2	–	–	–	–	–	–	–	–	–	10.8	38.6
NH_3	0.7	6.96	0.33	0.7	0.6	0.707	nd	na	na	0.077	0.000
CH_4	1.1	5.77	0.15	0.03	0.9	2.77	0.24	0.023	0.00495	0.000	0.000
N_2	0.6	1.37	1.05	0.05	1.5	1.17	0.29	1.56	2.69	2.18	0.708
H_2	2.1	19.9	4.0	0.126	0.8	4.66	1.85	0.20	0.0054	3.09	19.4
HCl	–	–	–	–	–	–	–	–	–	2.37	21.6
HF	–	–	–	–	–	–	–	–	–	0.346	3.16
Total (dry)	100.1	100.1	100.0	99.8	99.9	99.84	100.1	99.6	98.4	99.27	99.73
T_{gas}	251	254	327	235	239	283	250	208	218	(426)	(427)

Reservoir and volcano fumarole temperatures in °C; values in mol% dry gas except H_2O, which is mol% steam; na, not analyzed; nd, not detected. Parentheses around T_{gas} value mean calculation violates rules of application. All data except columns 1, 3, and 5 are from the files and published reports of F. Goff and C.J. Janik (Goff).
[1] Average of seven production areas.
[2] Average of 10 production wells; some data from high-temperature zone, ≤350 °C.
[3] Average of three wells in upflow zone.
[4] Average from steam separators.
[5] Average well; CH_4 includes all hydrocarbons and N_2 includes Ar.
[6] Average from 35 production wells.
[7] VC-2B scientific well, average of three in situ samples.
[8] Average of nine samples from Casa Diablo fumarole.
[9] Average of two fumarole samples.
[10] Besolima Fumarole, analysis includes 0.4 mol% O_2 + CO.
[11] Ohachi Oku Fumarole.
[12] Arg = argillite, shale, mudstone; Carb = carbonate, marl; Fel = felsite intrusions; Gw = graywacke; Ss = sandstone, conglomerate; Volc = volcanic flows, tuffs, breccias, and sediments.

reactions with reservoir rocks, volume of water relative to rock, residence time, and contributions of components from outside sources (cold groundwater, seawater, magma, etc.). Total dissolved solids in reservoir fluids at temperatures ≥150 °C range from about 1000 to >350,000 ppm (more than 10 times seawater). Liquid-dominated reservoir waters tend to be Na–K–Cl fluids or Na–K–Ca–Cl in very saline cases. Gas analyses indicate that dissolved CO_2 is usually the next major component. Divalent cation contents, particularly Mg, tend to be low due to inverse solubility of carbonates and sulfates with temperature. Silica levels and contents of key trace elements such as As, B, Br,

and Li are relatively high compared to most other groundwaters. The pH generally lies between 6 and 9, although extremely saline fluids may be more acidic.

Neutral-chloride spring and geyser waters usually resemble their parent reservoir fluids. These fluids may be modified during flow to the surface by chemical reactions with wall rocks, by dilution with near-surface groundwater, or by boiling. Such waters are often oversaturated with silica or carbonates at surface conditions and form deposits of sinter or travertine, respectively.

Acid-sulfate springs form where rising steam and volatile compounds condense into near-surface groundwaters,

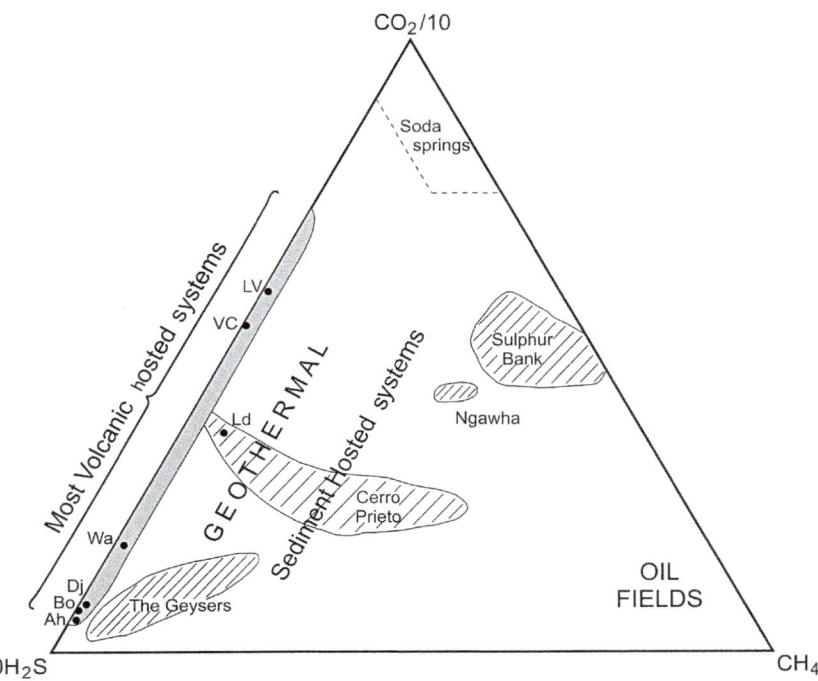

FIGURE 46.10 Ternary plot of $CO_2/10-10H_2S-CH_4$ for geothermal gas data in Table 46.2 (shown as black dots) and four additional sites unusually rich in methane: The Geysers (USA), Cerro Prieto (Mexico), Ngawha (New Zealand), and Sulfur Bank (USA). Such a plot accentuates compositional differences due to varying types of reservoir rocks.

and H_2S is oxidized to form sulfuric acid (Figure 46.4(A)). Additional components are added during acid alteration and dissolution of near-surface rocks. Acid-sulfate waters are comparatively rich in divalent and trivalent cations and low in Na, K, Cl, and most trace elements. Sulfur, acid sublimates, and clays are often deposited in the vicinity of acid-sulfate springs and associated fumaroles.

7.3. Chemical Geothermometers

Temperature-dependent concentrations of some chemical species in geothermal fluids have been applied to exploration and reservoir development. Chemical geothermometers are of two general types: (1) based on absolute concentrations of a constituent in solution, and (2) based on ratios of two or more constituents in solution. Some chemical geothermometers have been calibrated by laboratory experiments under ideal conditions, while others are based on empirical relations involving many fluids produced from different geothermal fields where reservoir temperatures are known. The solubility of silica polymorphs (quartz, chalcedony, cristobalite, etc.) in pure water as a function of temperature represents the prime example of a laboratory calibrated geothermometer. Unique equations describe the solubility of each polymorph. In the second type, one or more chemical ratios (usually cation ratios) from production fluids and associated hot spring fluids are plotted against the

inverse of reservoir temperature, and a linear regression of the data defines the temperature dependence.

Many geothermometers have been developed and refined over the years to meet specific conditions. Calculated subsurface equilibration temperatures for fluids from selected wells and thermal manifestations are listed beneath the analyses in Tables 46.2 and 46.3. The empirical (cation) geothermometers yield excellent results for reservoirs in young igneous systems because these hotter reservoirs have provided most of the data for the regressions, and host rocks consistently have abundant feldspar ± volcanic glass, the main sources for exchange of the alkali elements. Most geothermometers are not applicable for volcanic gas condensates, acid-sulfate waters or their derivatives, or dilute bicarbonate waters.

Application of geothermometer calculations has many potential pitfalls during exploration, mostly because fluid compositions can change during flow from the reservoir to the surface. Dilution, mixing, boiling, precipitation, and other reactions can modify original fluid chemistry as temperature declines along the flow path. Also, many waters (e.g., soda water, fossil seawater, oil field brines) that achieve their initial compositions by reactions at low to moderate temperatures yield calculated temperatures that are too high. Although several schemes exist to evaluate such processes, independent indicators of

TABLE 46.3 Water Compositions from Liquid-Dominated Geothermal Reservoirs Compared with Acid-Sulfate, Acid-Sulfate-Chloride and Neutral-Chloride Hot Spring Compositions

Site	Ahuachapan[1]	Wairakei[2]	Cerro Prieto[3]	Amatitlan[4]	Miravalles[5]	Valles Caldera[6]	Puna (Hawaii)[7]	Salton Sea[8]	Tecuamburro[9]	Miravalles[10]	Borateras[11]
Country	El Salvador	New Zealand	Mexico	Guatemala	Costa Rica	USA	USA	USA	Guatemala	Costa Rica	Peru
Res/Spg	Liq-dom	Liq-dom	Liq-dom	Liq-dom	Liq-dom	Liq-dom	Liq-dom	Liq-dom	Acid-sul	Acid-sul-chl	Neu-chl
Rocks	Volc	Volc	Arg/Ss	Volc	Volc	Qtz Monzonite	Volc	Arg/Ss	Volc	Volc	Volc
Temperature	255	260	≤350	290	240	295	>300	298	77	58.4	≤87
pH	7.1	8.3	7.5	7.22	7.19	4.74	3.77	5.3	2.27	2.76	7.89
SiO_2	537	625	833	845	585	882	1300	774	288	259	278
As	11.3	4.6	1.2	6.62	7.4	2.70	2.68	8.5	0.10	1.5	17.9
B	151	28.9	31.3	73.9	54	29.6	22.6	296	0.31	1.84	88.0
Na	5690	1310	8740	2310	1920	2350	20900	57250	70	64	1325
K	950	212	2150	450	256	700	4890	18920	6.2	4.0	101
Li	17.5	13.5	20.0	11.2	6.91	32.8	15.6	250	0.09	0.16	11.7
Ca	443	19	401	65.6	68	78.5	3600	29900	180	109	57.2
Mg	0.13	0.01	0.38	0.04	0.14	0.76	43.0	38	87.4	49.7	0.59
HCO_3	34.2	28.2	45	32.6	40.3	105	0	117	0	0	69.3
SO_4	34.0	35	4.6	29.5	36.0	7.8	2.5	7.4	3560	2690	74.8
F	1.5	7.8	3.3	1.14	1.24	5.67	1.70	33	0.14	1.4	1.74
Cl	10430	2205	16530	3950	3440	4150	46600	157,600	<0.4	635	2355
Br	43.9	5.7	46.1	15.3	17.5	13.6	na	114	<0.4	0.50	6.10
TDS	18340	4995	28810	7800	6435	6500	77,400	265310	4835	4380	4395
T_{Qtz}	259	274	351	305	267	310	302	296	(206)	(198)	203
$T_{Na/K}$	265	262	309	282	243	333	303	347	(207)	(180)	221
$T_{Na-K-Ca}$	254	254	298	267	235	300	287	316	(142)	(129)	217

Reservoir or spring temperature in °C; other values in parts per million. Reservoir analyses are not reconstructed for steam loss and do not include separated gases. na, not analyzed; TDS, total dissolved solids. Parentheses around geothermometer calculations mean that rules of application are violated. Abbreviations same as Table 46.2. All data except column 2 are from the files and published reports of F. Goff and C.J. Janik (Goff).

[1] Average of nine production wells.
[2] Average of three production wells.
[3] Average of 35 production wells.
[4] AMF-2 production well.
[5] PGM-5 production well.
[6] VC-2B scientific well, in situ at 1765 m.
[7] Kapoho State#8, geothermometers corrected for brine density and steam dilution.
[8] Calif. State#2–14, in situ sample; TDS value does not include significant dissolved Sr, Fe, Mn, NH_4, and trace metals.
[9] Acid spring near Guayabal.
[10] Acid spring on shore of Laguna Ixpaco.
[11] Average of two hot springs on sinter terrace.

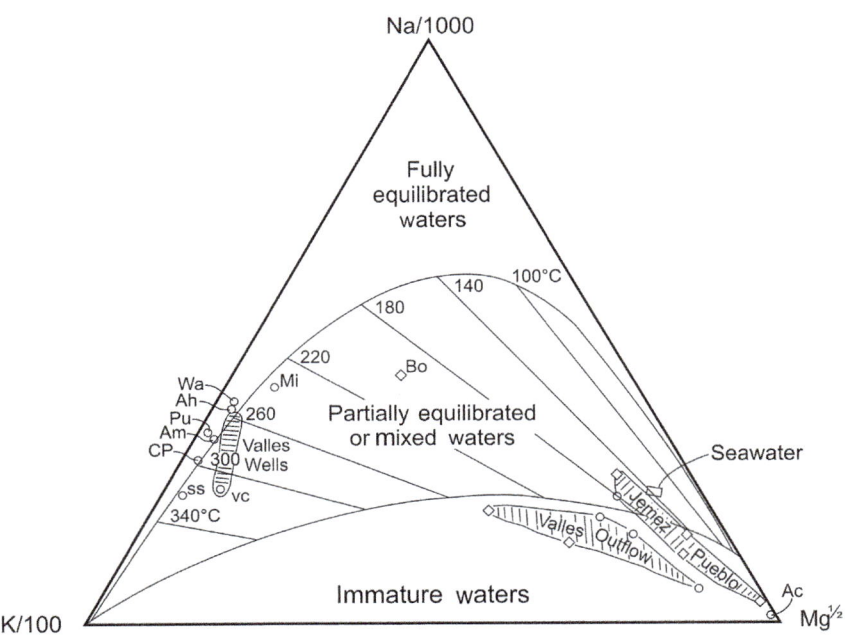

FIGURE 46.11   Ternary plot of Na/1000–K/100–Mg$^{1/2}$ for water data in Table 46.3; open circles are well fluids and open diamonds are spring fluids. High-temperature reservoir fluids hover close to the full (upper) equilibrium line. The sample from Boraterace, Peru (Bo), exemplifies a fluid from a promising reservoir that has not been drilled. Groups of waters from the Valles Caldera region (USA) address the interpretive possibilities of this plot (see text).

high temperature should always be sought. For example, presence of H$_2$S-rich gases, widespread distribution of boiling Cl-rich springs, and spring deposits of sinter (amorphous silica) usually indicate that a reservoir of ≥150 °C lies at depth.

7.4. The Na–K–Mg Ternary

Ternary plots of Na/1000–K/100–Mg$^{1/2}$ (Figure 46.11) compare "mature" geothermal reservoir fluids with other thermal fluids considered to be partially equilibrated, mixed, or "immature." The upper curve on this diagram is defined by a version of the Na/K geothermometer; other interpretations of this ternary plot use different criteria to define temperature. Tie lines pointing toward the Mg axis identify potential mixing trends between mature reservoir fluids and Mg-rich fluids such as cool groundwater or acid sulfate waters. The upper boundary for immature waters is not well defined.

When the data of Table 46.3 are plotted on this diagram, geothermal reservoir fluids (columns 1–8) generally plot at full equilibrium at 240–330 °C and approximately match the geothermometer calculations at the bottom of the table. Neutral-chloride hot spring water from Borateras, Peru, plots in the field for partially equilibrated or mixed waters, yet its position between tie lines suggests the water is derived from a reservoir of about 200 °C. This temperature is in good agreement with

the geothermometers (203–217 °C). In contrast, the two acid fluids of Table 46.3 (labeled Ac) plot toward the extreme Mg apex of the diagram and provide no useful temperature information.

The Na–K–Mg ternary provides valuable information on reservoir outflow processes if a sufficient number of analyzes of various fluids are available for comparison. For example, the Valles Caldera geothermal system discharges a hydrothermal outflow plume whose waters can be sampled from lower elevation springs and shallow wells. Valles outflow waters plot in the field of "immature waters," yet as a group, they trend toward the tie line for 260 °C reservoir fluids. These calculated temperatures are confirmed by deep wells drilled in the Valles parent reservoir that produced fluids from 260 to 300 °C.

Another group of higher chloride thermal waters occurs 30 km southwest of Valles Caldera in the vicinity of Jemez Pueblo and the San Juan Basin. Superficially, the Jemez Pueblo waters have some chemical similarities to Valles outflow waters, yet the former resemble dilute oil field brines and plot along a different trend of cooler temperature (roughly 160 °C) on Figure 46.11. Note that seawater plots at an indicated temperature of 140 °C. As pointed out above, many fluids look encouraging with respect to geothermometers and the Na–K–Mg ternary, but other positive indications of high temperatures at depth should be sought for confirmation.

7.5. Water Origins and System Ages

Although the hottest geothermal systems are associated with magmatism, isotopic studies show that the water itself is derived primarily from ancient to modern precipitation, i.e., meteoric water (Figure 46.12). Some systems contain recycled seawater, fossil seawater and variants, and metamorphic waters (waters released during metamorphism). Positive recognition of magmatic water in geothermal systems was made after studying fluids emitted from active volcanoes. Many commercial geothermal fields in subduction zone settings have stable isotope compositions indicating that magmatic waters comprise up to 15% of the reservoir fluid, but a few systems may contain as much as 40%.

Tritium, ^{14}C, and ^{36}Cl studies show that geothermal reservoir waters are usually older than 1000 years, commonly older than 10,000 years, and rarely older than 100,000 years. Hot springs derived from reservoir fluids sometimes contain younger water due to mixing with near-surface groundwaters.

Life spans of geothermal systems have been estimated by dating of intrusions, alteration minerals in veins, spring deposits, and by geologic relations. Systems such as Valles Caldera have been continuously active for the last 1 million years, even though the present reservoir is not large. Systems found in many arc volcanoes can be no older than the volcanic edifice, often <100,000 year. Volcanic eruptions can create rapid changes in young igneous systems. For example, the present hydrothermal system at Mount St Helens, USA ($\leq$175 °C), formed after the eruptions of 1980 and the existing hydrothermal systems at Mount Pinatubo (Philippines) and El Chichón (Mexico) were extensively modified after recent catastrophic eruptions. These examples underscore that the life spans of hydrothermal systems are dynamically linked to the timescales of local and regional volcanic and tectonic processes.

8. GEOPHYSICS, STRUCTURE, AND ROCK PROPERTIES

8.1. Clay Alteration, Clay Seals, and Means of Detection

A "clay cap" is frequently cited as an impermeable seal on the underlying geothermal reservoir (e.g., Figure 46.2). However, this sealing capacity is a function of many variables including the original rock type; abundance and distribution of clay and other late-stage phases such as carbonates, sulfates, zeolites, and amorphous silica; the state of stress; fault orientation and history of movement; and the recent history of volcanic eruptions. The strong overlap between clay alteration and leakage from geothermal reservoirs shows that the clay seal commonly leaks at limited rates.

Modern magnetotelluric (MT) surveys of many intrusion-related geothermal fields have shown that they possess similar patterns of electrical conductivity with depth that are related to measured temperatures and alteration mineralogy. The most commonly observed pattern is one of the highest electrical conductance (the lowest resistivity) above the reservoir, with a core of higher electrical resistivity where convective fluid circulation prevails. Based on drilling results the lowest resistivity is coincident with low-permeability clay-altered rocks where temperature gradients are steep and conductive. The most electrically conductive zone, typically 10 Ω-m or less, correlates with the abundance of smectite (the argillic zone as defined above). Some authors mark the transition to higher resistivity where chlorite and mixed-layer clays become more abundant, but others have suggested that smectite-rich mixed layer clays contribute to the intensity of the electrical conductor. Thus, the electrical conductor maps the argillic zone, and possibly the upper portion of the transition zone as defined above. Electrical resistivity is considerably higher in the

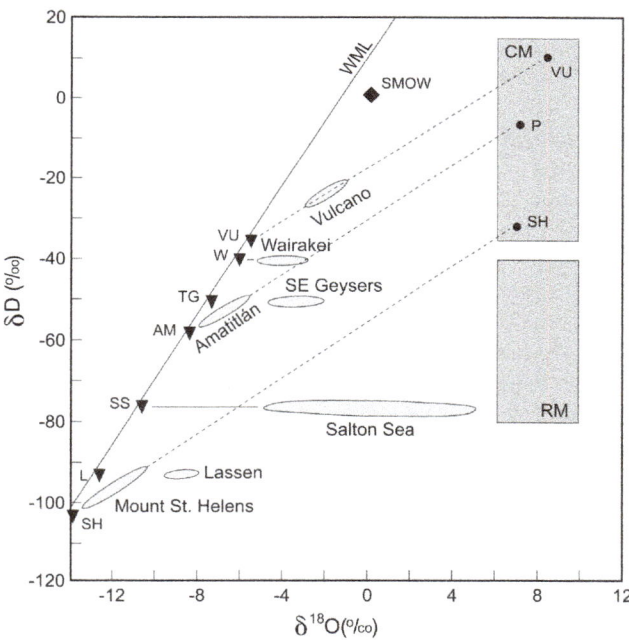

FIGURE 46.12 Plot of deuterium (δD) versus oxygen-18 (δ^{18}O) for meteoric waters (inverted triangles) and reservoir fluids (clouds) at selected geothermal systems (values in per mil relative to Standard Mean Ocean Water, SMOW). Virtually all rocks are isotopically enriched in ^{18}O compared to most waters. Geothermal systems such as Wairakei, The Geysers, Salton Sea, and Lassen contain meteoric waters isotopically enriched in ^{18}O due to high-temperature rock-water isotopic exchange. Systems such as Vulcano (Italy), Pacaya-Amatitlán, and Mount St Helens contain mixtures of meteoric and magmatic waters (dots). WML is the world meteoric line for precipitation. CM is the box for convergent margin (subduction zone) magmatic waters, whereas RM is the box for residual magmatic waters in intrusive bodies.

zone of convective circulation, where epidote and illite alteration occur at temperatures of 240 °C or more. The resistivity of the reservoir zone usually ranges from 30 to 80 Ω-m, and is rarely >180 Ω-m.

The temperature at which smectite alteration gives way to chlorite alteration varies. This transition is reported at 230 °C in Icelandic systems, but as low as at 180–200 °C in many systems hosted by andesitic rocks. Clay types and abundances reflecting the peak temperature of a system are retained even if the system undergoes some cooling, but may be lost at temperatures $\leq$150 °C, possibly as dehydration proceeds. Smectite breakdown and chlorite formation reactions depend on differences in fluid, rock and mineral compositions, and partial pressures of CO_2.

A 3-D inversion of densely spaced MT and TDEM (time-domain electromagnetics) data at Medicine Lake Volcano, USA (Figure 46.13), illustrates the typical shape of the clay cap in an igneous geothermal system. The depth of clay transitions tends to dome upward where hot fluid is ascending, and drape and thicken at the cooler margins of active hydrothermal circulation. Integrated interpretation of the shape of the geophysical conductor and its relation to thermal features provides a useful method of geothermal exploration. However, formations rich in smectite formed by weathering of volcanic ash or epiclastic rocks may have similar resistivity to geothermally altered rocks (e.g., basin fill deposits or crater lake sequences in calderas).

If capping smectite-rich rocks contain cold water, or pockets of steam and gas, this will increase the observed resistivity. In addition, zones of hot water upflow into shallow confined aquifers may generate apparent "holes" or gaps in the smectite cap as they locally convert smectite clays to higher temperature varieties (e.g., Tolhuaca, Chile).

8.2. Microearthquake Monitoring to Detect Faults and Fluid Paths

Microearthquakes (MEQs) have been monitored at the Lassen geothermal systems and at a number of producing geothermal fields. Some fields located along active faults had significant pre-exploitation seismicity (e.g., Salton Sea, The Geysers, Coso), whereas others did not (e.g., Salak). Induced seismicity is primarily associated with reinjection of separated geothermal fluids back into the reservoir; thus, patterns of seismicity can potentially be used to infer movement of injected fluid. Typically, seismicity increases as pressure differentials between production and injection areas increase. Clusters of diffuse events mainly of M < 3 tend to correlate with injection location and rate, and planar arrays of events can sometimes be used to identify probable faults. In some fields, events occur kilometers below the injection points, suggesting a broad, critically stressed, region is accessed by the injected fluid. Both the thermal

contrast (ΔT) between the injected fluid and the fracture faces, and the increased normal stress on them, are factors in triggering events. MEQs associated with fluid production occur predominately where temperature and pressure changes are greatest, for example, in liquid-dominated fields where steam caps develop (e.g., Tiwi, Coso). MEQ locations may thus be helpful in defining the area of productive reservoir.

Slow, vertical ground deformation (bradyseism) occurs naturally in some geothermal areas. Although some cycles of inflation and deflation may be related to magma migration, recent studies have highlighted the role of episodic fluid expulsion from crystallizing magma to overlying hydrothermal regimes at Yellowstone and Campi Flegrei (Italy) in causing this deformation.

8.3. Structural Controls on Reservoir Permeability

Most intrusion-related geothermal systems consist of a dynamic network of interconnected fractures and pores that transmit fluid over distances of many kilometers. The permeability of such fracture networks may be very high compared to typical unfractured volcanic and intrusive rock formations at comparable depths in the crust. Natural rates of fluid flow in such systems are still relatively slow based on initial-state calibration of numerical reservoir models (meters per year), but tracer tests show velocity increases to meters/hour between injection and production wells due to the higher pressure gradients induced by exploitation.

At regions where extensional stress are concentrated, such as stepovers and relay ramps in normal faults, extensional bends in strike-slip faults, fault intersections and fault tips are known to host high permeability in geothermal systems. Fault conceptual models applicable to geothermal reservoirs include division of a mature fault zone into a central "fault core" and a surrounding "damage zone." Most of the fault slip is concentrated in the core, producing gouge and microbreccia. The fault core generally has lower permeability than the surrounding rock due to intensive shear and grain-size reduction, whereas the damage zone is generally more permeable than the surrounding rock by virtue of an array of interconnected shear and extensional fractures. This geometry favors fluid flow parallel to the fault strike rather than across it, and may even result in fault-bounded compartments in a reservoir. The sealing capacity of a fault is strongly influenced by the characteristics of the juxtaposed formations, with fine-grained and clay-rich rocks tending to create ductile seals as they are smeared along offset surfaces. Shear bands and related fractures, which form in more porous formations such as tuffs and volcaniclastics, also tend to enhance fluid flow parallel to their strike and impede it perpendicular to strike. These mechanisms

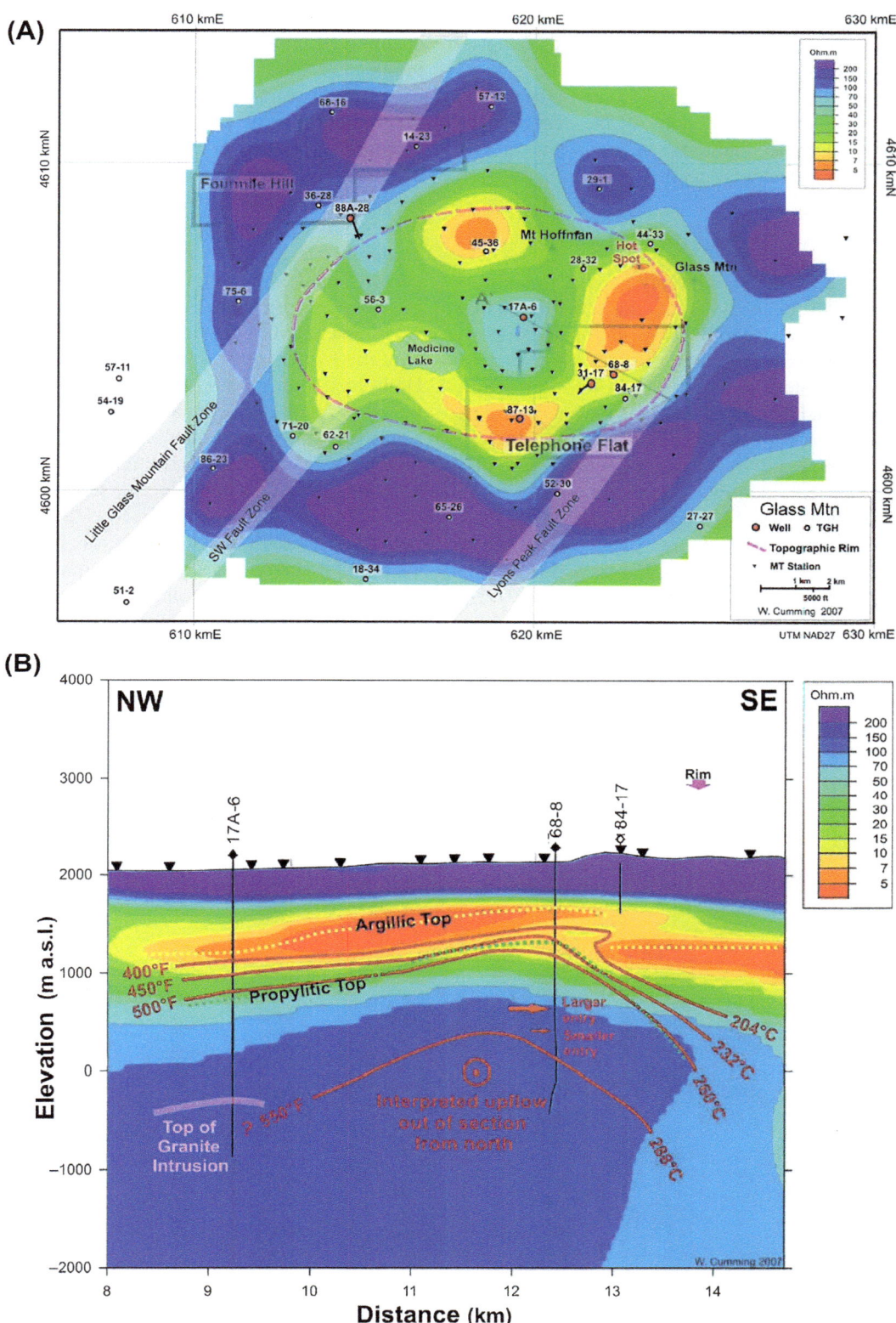

FIGURE 46.13   3D resistivity patterns of the Medicine Lake (Glass Mtn) geothermal system, USA. (A) Map of 3D MT resistivity at 1700 m elevation. (B) NW−SE cross-section line in (A), with contoured temperatures and alteration mineral assemblages. TGH, thermal gradient hole. *From Cumming and Mackie (2007) with permission.*

may explain why a preferred or fast direction of fluid flow is observed in exploited geothermal systems. The direction of enhanced flow is generally aligned with the strike of faults and fractures that are well oriented for failure in the current stress regime, while compartments of the reservoir on opposite sides of these faults are in somewhat poorer communication (e.g., Salak, Bulalo).

8.4. Porosity and Permeability Trends

Although faults and fractures largely control the permeability of intrusion-related systems, the effective matrix porosity and matrix permeability of reservoir rocks plays a secondary role and contributes to the longer term response of the system to commercial fluid extraction. More porous formations mean a higher mass of hot fluid is stored in a given rock volume to support production. Fluid composition, depth of burial, and temperature play the primary roles in mediating changes to reservoir rocks. Fragmental deposits with high porosity and low crystallinity (high matrix glass content) are most prone to alteration, whereas high-crystallinity lavas and intrusions (little or no matrix glass) are relatively resistant. The duration and cyclical behavior of hydrothermal activity is important to determining the intensity of alteration. Protracted fluid circulation leads to lower overall formation porosity, although fracture permeability may be repeatedly regenerated.

Core samples and well logs indicate that the average porosity of hydrothermally altered volcanic rocks declines with depth as a result of mechanical compaction and chemical reaction with hydrothermal fluids (Figure 46.14). Fragmental lithologies such as tuffs and volcaniclastics have higher average initial porosity, but tend to lose pore space more rapidly with burial than lavas and intrusions. The rate of porosity decline tends to decrease with depth of burial, and the range of porosity at 3 km depth tends to be relatively small. Some formations do retain more of their original porosity than others; for example, vapor phase-altered ash-flow tuffs tend to have high initial porosity, moderate strength, and a mineral assemblage that is stable because it was formed in a quasi-hydrothermal environment. Such formations are relatively resistant to both mechanical compaction and hydrothermal alteration. High crystallinity lavas and intrusions with initially low porosity (<4%) may show porosity enhancement as a result of high-temperature alteration (e.g., Coso), and there is some evidence that conditions in upflow zones and adjacent to major structures may tend to locally enhance porosity rather than degrade it. Rock sequences that were rapidly deposited tend to exhibit less hydrothermal alteration and porosity loss than older sequences that have had more time to undergo compaction and alteration events (e.g., The Geysers). Rapidly subsiding basins and rift zones tend to retain higher porosity and be more open to regional recharge than older rock sequences (Salton Sea, Bulalo, Taupo Zone geothermal systems). In fault- and fracture-dominated systems, a significant fraction of total porosity may occur along vuggy, partially sealed fractures and breccias, especially if the matrix porosity is low.

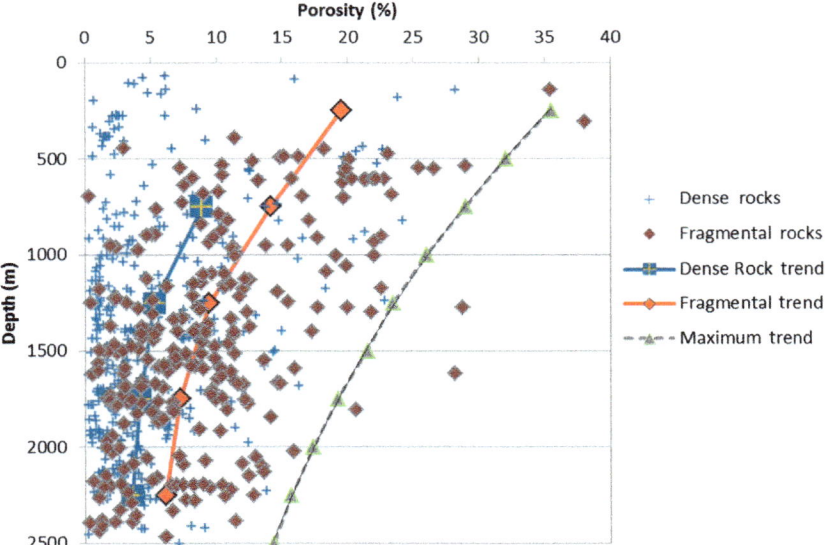

FIGURE 46.14 Core porosity data from volcanic-hosted geothermal systems versus approximate depth of sample. Rocks are divided into dense (e.g., lava, intrusive) and fragmental (breccia, tuff, volcaniclastic) varieties. Trend lines are the mean values for each 500-m interval. Dashed line suggests the maximum porosity that can reasonably be expected as a function of depth.

8.5. Rock Mechanical Properties

Mechanical testing of altered igneous rocks and derived sediments indicates that clay and chlorite content and formation porosity play a key role in determining formation strength, hence its ability to initiate and maintain brittle fracture. Higher porosity and clay-rich lithologies tend to be weak and ductile, leading to formation of shear bands rather than open fractures. Thus, the clay-rich caps of geothermal systems tend to act as ductile, low-permeability lids that inhibit heat and mass transfer (Figure 46.2). Conversely, the more competent and brittle rocks altered at higher temperatures (220–350 °C) support rapid heat and mass transfer through dynamically maintained fracture networks. At temperatures in excess of 350 °C and depths of several kilometers, it becomes increasingly difficult to form and maintain open fractures in most common rocks. Permeability may exist primarily along grain boundaries, microcracks, and connected pore networks at these temperatures. While large fluid volumes may be transported over time, flow rates in this regime are typically too low to support commercial exploitation.

9. EXPLORATION AND APPRAISAL OF HIGH-TEMPERATURE SYSTEMS

The primary objective of exploration is to discover if a given reservoir of steam or hot water can sustain production by virtue of sufficient temperature, permeability, and benign fluid chemistry. Appraisal involves defining those conditions over a sufficiently large area to develop a commercial project. Commercial projects are those that generate enough electricity at a sufficient power price to provide an acceptable rate of return on the capital required for development. Statistically, about 70% of geothermal areas with boiling features will host exploitable systems, but they may not all be developed due to other commercial factors (see Chapter 71).

Regional exploration strategies initially focus on areas of known hot spring and fumarole activity associated with young volcanoes, or areas of known high temperature gradient. Hot aquifers have been discovered by other drilling activities. More detailed exploration of prospective targets includes assessment of the age of the youngest volcanism and the reservoir rocks, structural controls on permeability, potential heat sources, the extent of surface and subsurface clay alteration, and calculation of subsurface reservoir temperatures inferred from fluid geochemical samples (Tables 46.2 and 46.3). Once all this information has been obtained and integrated, one or more conceptual models of the system are developed. These models typically include assessment of (1) reservoir size and thickness, (2) the depth to its top, (3) locations of upflow and outflow zones, (4) the average porosity, and (5) the phase distribution. The model is then tested by drilling an initial well to confirm the presence or absence of an exploitable reservoir (Figure 46.8(B)).

Once the combination of commercial temperature and flow rates are demonstrated with one well, focus shifts to proving the extent of the commercial reservoir. Three to five holes that prove reservoir conditions at depths ≥1.5 km are typically needed to successfully complete the exploration phase. Further appraisal drilling may be required to prove sufficient fluid production to obtain financing for development. Geophysical logs provide temperature-depth profiles, water levels, formation pressures, production zones, and other information. Cores and cuttings provide details on stratigraphy, alteration mineralogy and history, fracture patterns, and porosity of the rocks. Flow tests are conducted to determine preliminary reservoir parameters. Fluid analyses establish background conditions and evaluate corrosion and scaling properties. Well data are then combined with surface studies to refine the conceptual model of the geothermal system in its natural state. Model results are used to create an economic field-development strategy, which is used to obtain financing for development drilling and eventual exploitation of the resource. The entire exploration and development process can be completed in 3–4 years, but has been known to take 10 or more years due to various technical, economic, and political factors.

10. SUMMARY

Geothermal systems are reservoirs of hot water and steam that exist at comparatively shallow depths. The highest temperature systems are associated with shallow intrusion in volcanically and tectonically active regions. Shallowly emplaced magma transfers large mass and energy flows from deep in the Earth to the Hydrosphere via convective circulation of fluids of dominantly meteoric origin. Exploration for geothermal systems relies heavily on interpretation of geothermal fluid chemistry and geophysical surveys that delineate areas of subsurface smectitic clay alteration. Conceptual models, based on well-studied systems and integrating a variety of data, are employed to develop deep drilling targets.

Commercial geothermal fields that mostly produce electricity from intrusion-related reservoirs ≥170 °C now account for about 10,000 MWe of power. Geothermal resources with large fluid withdrawals compared to their size and recharge rates show significant declines from peak production over tens of years, but can be maintained at lower production rates for >100 years with proper management. Worldwide geothermal energy utilization continues to increase yearly because it is an attractive alternative to burning imported and domestic fossil fuel resources.

ACKNOWLEDGMENTS

J. Stimac thanks Geoglobal Energy for financial and logistical support. Fraser and Cathy (Janik) Goff were supported from 1977 to 2004 by the Los Alamos National Laboratory, the U.S. Geological Survey, the U.S. Department of Energy, the U.S. Agency for International Development, and from 2004 to present by the University of New Mexico, and the New Mexico Institute of Technology. Esther Mandeno is gratefully acknowledged for help with illustrations. David Sussman and Greg Raasch provided useful comments on an early draft of this manuscript. We thank Tobias Fischer and John Stix for their reviews.

FURTHER READING

Allis, R., 2000. Insights on the formation of vapor-dominated geothermal reservoirs. In: Proceedings World Geothermal Congress 2000, International Geothermal Association, pp. 2489–2496.

Bertami, R., 2012. Geothermal power generation in the world 2005–2010 update report. Geothermics 41, 1–29.

Caine, J.S., Forster, C.B., 1999. Fault zone architecture and fluid flow: insights from field data and numerical modeling. In: Haneberg, W.C., Mozley, P.S., Moore, J.C., Goodwin, L.B. (Eds.), Faults and Subsurface Fluid Flow in the Shallow Crust, AGU Geophysical Monograph, vol. 113. American Geophysical union, Washington DC, pp. 101–127.

Cumming, W., Mackie, R., 2007. 3D MT resistivity imaging for geothermal resource assessment and environmental mitigation at the glass mountain KGRA, California. Geother. Res. Council Trans. 31, 331–334.

Elders, W.A., Friðleifsson, G.Ó., Albertsson, A., 2014. Drilling into magma and the implications of the iceland deep drilling project (IDDP) for high-temperature geothermal systems worldwide. Geothermics 49, 111–118.

Fournier, R.O., 1999. Hydrothermal processes related to movement of fluid from plastic into brittle rock in the magmatic-epithermal environment. Econ. Geol. 94, 1183–1211.

Giggenbach, W.F., 1988. Geothermal solute equilibria. Derivation of Na–K–Mg–Ca geoindicators. Geochim. Cosmochim. Acta 52, 2749–2765.

Goff, F., McMurty, G.M., 2000. Tritium and stable isotopes of magmatic waters. J. Volcanol. Geother. Res. 97, 347–396.

Grant, M.A., Bixley, P.F., 2011. Geothermal Reservoir Engineering, second ed. Academic Press, Amsterdam.

Lima, A., De Vivo, B., Spera, F.J., Bodnar, R.J., Milia, A., Nunziata, C., Belkin, H.E., Cannatelli, C., 2009. Thermodynamic model for uplift and deflation episodes (bradyseism) associated with magmatic-hydrothermal activity at the Campi Flegrei (Italy). Earth-Sci. Rev. 97, 44–58.

Lund, J.W., Freeston, D.H., Boyd, T.L., 2011. Direct utilization of geothermal energy 2010 worldwide review. Geothermics 40, 159–180.

Menzies, A., Villasenor, L., Sunio, E., Lim, W., 2010. Characteristics of the matalibong steam zone, tiwi geothermal field, Philippines. In: Proceedings World Geothermal Congress. International Geothermal Association, Bali, Indonesia.

Moore, J.N., Allis, R., Renner, J.L., Mildenhall, D., McCulloch, J., 2002. Petrologic evidence for boiling to dryness in the Karaha-Telagaboda geothermal system, Indonesia. In: Proceedings 27th Workshop on Geothermal Reservoir Engineering. Stanford University.

Reyes, A.G., 1990. Petrology of Philippine geothermal systems and the application of alteration mineralogy to their assessment. J. Volcanol.Geotherm. Res. 43, 279–309.

Richards, J.P., 2011. Magmatic to hydrothermal metal fluxes in convergent and collided margins. Ore Geol. Rev. 40, 1–26.

Rowland, J.V., Simmons, S.F., 2012. Hydrologic, magmatic, and tectonic controls on hydrothermal flow, Taupo Volcanic Zone, New Zealand: implications for the formation of epithermal vein deposits. Econ. Geol. 107, 427–457.

Chapter 47

Seafloor Hydrothermal Venting at Volcanic Arcs and Backarcs

Cornel E.J. de Ronde and Valerie K. Stucker

Department of Marine Sciences, GNS Science, Lower Hutt, New Zealand

Chapter Outline

GLOSSARY

acid A molecule or other species that can donate a proton or accept an electron pair in reactions. An acidic fluid has a pH less than 7.

arcs A chain of volcanoes (often referred to as island arcs above sea level) forming an arc shape when seen from above. They result from the subduction of an oceanic tectonic plate under another tectonic plate, and often parallel an oceanic trench.

backarcs Relating to or denoting the area behind an island arc, commonly referred to as backarc basins.

brine Water saturated with, or containing large amounts of a salt, especially sodium chloride.

disproportionation A specific type of redox reaction in which a species is simultaneously reduced and oxidized to form two different products.

end member On a plot of selected elements versus Mg, the Mg-end-member composition of the vent fluid is calculated by fitting a least-squares line through the various measured values for any one element, and seawater, then extrapolating to zero Mg.

exsolve Gases are released from magma through volatile constituents reaching such high concentrations in the magma that they become saturated and are then released by diffusion and phase separation as bubbles.

MORs Mid-ocean ridges are elongate, seismically active submarine ridge systems situated in the middle of an ocean basin that mark the site of the upwelling of magma associated with seafloor spreading.

phase separation The conversion of a single-phase system into a multiphase system; for example, the separation of a solution into two immiscible liquids.

water/rock The interaction between hot (hydrothermal) water and rock resulting in changes to the fluid chemistry.

1. INTRODUCTION

The Earth's surface is covered by $\sim$60,000 km of mid-ocean ridges (**MORs**), or large-scale lineaments otherwise known as spreading centers. This is where newly formed oceanic crust, invariably of basaltic composition, moves inexorably away from the axis of the MORs as the forces of plate tectonics take hold. These spreading centers mark divergent plate margins. Conversely, convergent plate boundaries delineate zones where the oceanic crust, now millions to more than a hundred million years old, is subducted beneath another plate of either oceanic or continental crust. These boundaries are prevalent in regions of the south and western Pacific Ocean and are typically marked by deep trenches on their seaward side, and volcanic **arcs** (or chains of volcanoes) paired with their backarc basins on their landward side (Figure 47.1). These volcanoes sit on top of the overriding plate, typically between $\sim$70 and $\sim$175 km (with a global average of 105 km) above where the subducted plate begins to melt and/or triggers melting of the base of the overriding plate. This fundamental difference in tectonic setting between arcs (and adjacent **backarcs**) and MORs is reflected in the composition of hydrothermal fluids expelled on the seafloor.

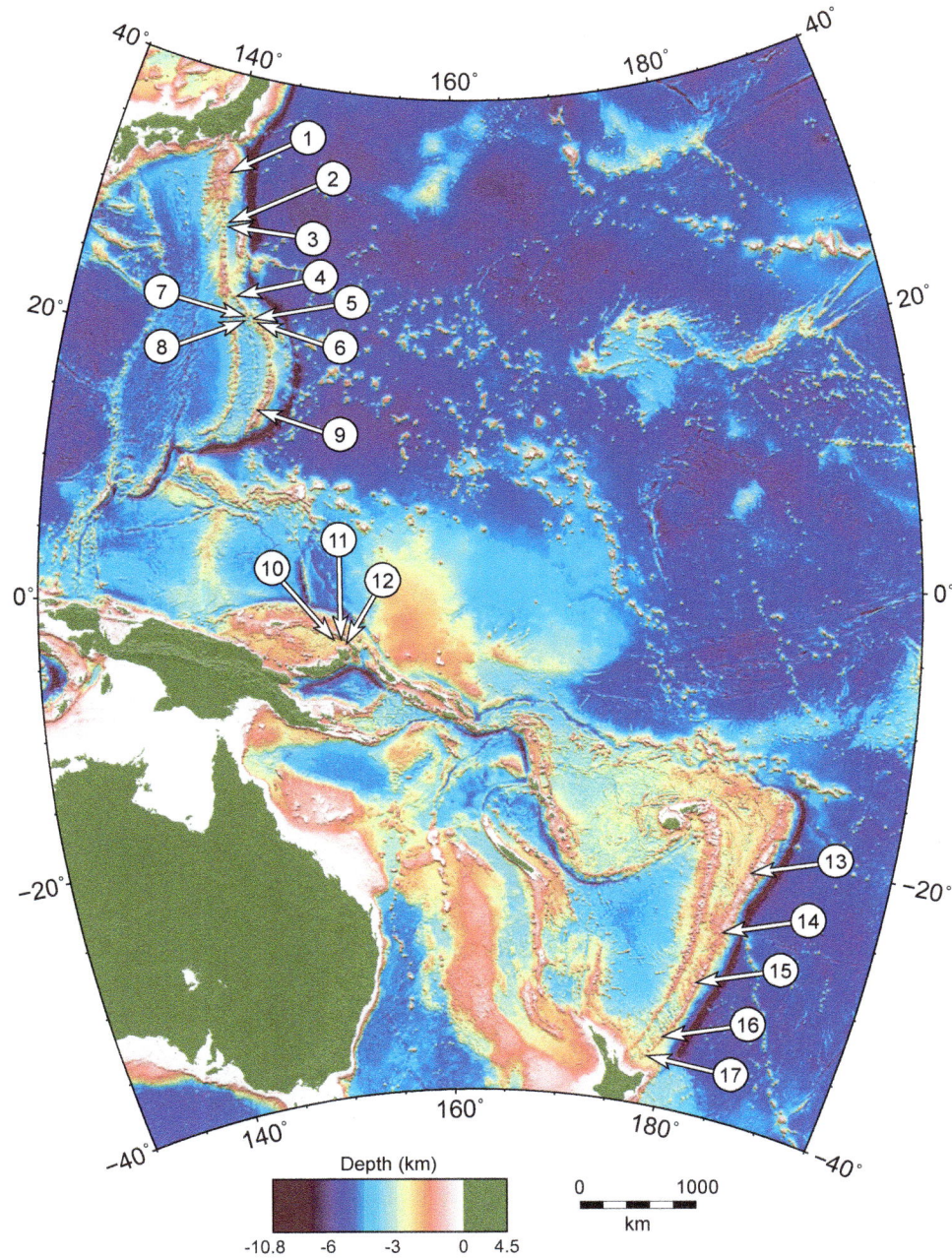

FIGURE 47.1   Map of the central and western Pacific region in a transverse Mercator projection (CM 162°E). This map was created from ETOPO1 data (i.e., a 1 arc-minute global relief model of Earth's surface that integrates land topography and ocean bathymetry; Amante and Eakins, 2009) downloaded from NGDC (http://www.ngdc.noaa.gov/mgg/global/global.html). Hydrothermal vents sites referred to in this paper include: *Izu-Bonin arc:* (1) Myojin Knoll (latitude 139.8500; longitude 32.1000); (2) Suiyo (140.6417; 28.5750); (3) Mokuyo (140.5800; 28.3200); *Mariana arc:* (4) Nikko (142.3333; 23.0833); (5) NW Eifuku (144.0430; 21.4850); (6) Daikoku (144.1940; 21.3240); (7) Kasuga 2 (143.6167; 21.6000); (8) Kasuga 3 (143.6000; 21.3833); (9) NW Rota-1 (144.7750; 14.6010); *Manus Basin:* (10) PACManus (151.6700; −3.7250); (11) DESMOS (151.8717; −3.6958); (12) SuSu Knolls (152.1000; −3.8000); *Tonga arc:* (13) Volcano-1 (−175.7500; −21.1500); *Kermadec arc:* (14) Monowai (−177.1880; −25.8870); (15) Giggenbach (−178.7167; −30.0333); (16) Brothers (179.0667; −34.8667); (17) Clark (177.8400; −36.4490). *Amante, C., Eakins, B.W. (2009). ETOPO1 1 arc-minute global relief model: Procedures, data sources and analysis: NOAA Technical Memorandum NESDIS NGDC-24, 19 pp.*

Volcanic arcs that have a submarine component include both intraoceanic and island arcs and have a combined length of ∼21,700 km—or a third the length of the MOR system—with ∼93% occurring in the Pacific region. Those that are strictly intraoceanic comprise ∼6900 km, or about 32% of the total length (de Ronde et al., 2003). A

compilation of known *hydrothermally* active submarine volcanoes shows that there are at least 170 associated with intraoceanic arcs (Baker et al., 2008); this is a minimum given that large lengths of several arcs have not been mapped in any detail, nor have they been surveyed for their hydrothermal emissions. The density of volcanic centers has been calculated for two of the best-studied intraoceanic arcs. A volcanic center can be an assortment of volcanic edifices comprising a single large volcanic cone or caldera and/or groups of satellite volcanoes with a minimum elevation of 500 m above the seafloor, commonly separated from adjacent centers by several tens of kilometers. The overall volcanic center density for the Mariana arc is 4.4/100 km of arc and that of active centers is 1.9/100 km. An equal length of the Tonga-Kermadec arc holds fewer volcanic centers at 2.9/100 km, but a similar density of active centers at 1.8/100 km (Baker et al., 2008). The lower percentage of active centers on the Mariana arc results from it having both fewer caldera volcanoes—since arc calderas are twice as likely as cones to be active—and a lower percentage overall of active calderas and cones (de Ronde et al., 2001; Baker et al., 2008). Irrespective, intraoceanic arcs are host to multiple sites of seafloor hydrothermal activity that are associated with the full length of subduction zones of the Pacific region, and elsewhere (Figure 47.1).

Submarine arcs are thought to account for ~10% of the global hydrothermal discharge associated with MORs (Baker et al., 2008). Backarcs have received less attention than arcs in recent times, although significant work has been done in the Manus Basin (Figure 47.1), especially where Nautilus Minerals Inc. hopes to be the first to exploit a seafloor massive sulfide deposit at Solwara 1 (http://www.nautilusminerals.com/s/Home.asp). The percentage of global hydrothermal discharge associated with backarcs is unknown.

In this chapter, we concentrate on hydrothermal venting associated with submarine arc volcanoes and to a lesser extent backarcs, so that we might compare and contrast results with those of MORs, which was the focus of Butterfield (2000) in the first edition of *Encyclopedia of Volcanoes*.

2. MOR HYDROTHERMAL VENTING

The discovery of seafloor hydrothermal venting along the Galapagos spreading center in 1978 heralded a new and exciting era of deep-sea research and was the forerunner of many expeditions to hydrothermal systems hosted by MORs. This means a significant database on vent fluid compositions from these sites and a general understanding of the chemical evolution of these systems that involves the convective circulation of seawater in the ocean crust have accrued since that time (e.g., Von Damm et al., 1985; Butterfield, 2000). The general model for MOR hydrothermal systems assumes an increase in temperature as ambient seawater penetrates the seafloor and continues downward via fractures and faults until it gets heated by underlying magma, becomes buoyant, and rises back up through the crust interacting (and exchanging ions) with rock along the way. This fluid is commonly found to phase separate, due to boiling, immediately beneath the seafloor where it can also mix with ambient seawater, ensuring extensive modification of the original seawater composition. The net effect of these processes is the formation of a hydrothermal fluid with

temperatures between ~300 and 400 °C that is depleted in Mg and SO_4, (i.e., Mg and $SO_4 = 0$ in the **end member** fluid), has a pH between 3 and 4, and is enriched in Ca, K, SiO_2, CO_2, H_2S, and dissolved metals such as Cu, Zn, and Pb that form massive sulfide mineralization on, or just below, the seafloor (e.g., German and Von Damm, 2003; Hannington et al., 2005). Furthermore, **phase separation** leads to distinct fluid types, i.e., liquids depleted in Cl relative to seawater, commonly referred to as "condensed vapor" fluids, and those enriched in Cl, commonly termed **"brines"** or more correctly, "low-salinity liquids" and "high-salinity liquids", respectively.

3. STYLES OF ARC AND BACKARC HYDROTHERMAL ACTIVITY

Hydrothermal venting along intraoceanic arcs can be found at depths ranging from just below sea level to around 1800 m, with at least half of all the vent sites occurring in depths $\leq$500 m (de Ronde et al., 2003). Thus, depth plays a crucial role in determining the maximum temperature at which the vent fluid can separate into different phases on, or just below the seafloor. In addition, conductive cooling of the hydrothermal fluid as it rises through the crust and mixes with ambient seawater below the seafloor can affect both the composition and physical characteristics of the fluid as it is expelled. These processes are common at MOR vent sites (Butterfield, 2000) and indeed can also be found at arc sites. However, at arcs there is a much greater influence of magmatic volatiles on the composition of the vent fluids than typically seen at MORs. This is largely due to the compositions of melts beneath arc volcanoes and the relatively shallow depth to the underlying magma chamber, hence their propensity to **exsolve** gases. Taken together, the tectonic setting of arc and backarc hydrothermal systems, the influence of degassing magmas, and the range in depth to the vent sites result in distinct styles of venting.

3.1. Venting Associated with Volcanic Eruptions

During extensive, whole-of-arc surveys for hydrothermal plumes along the Kermadec (e.g., de Ronde et al., 2001) and Mariana (Baker et al., 2008) arcs, followed by detailed seafloor mapping (e.g., Chadwick et al., 2008) and remotely operated vehicle and manned submersible dives (e.g., Embley et al., 2006), it was shown that at least ~5% of arc volcanoes are erupting on the seafloor at any one time. This includes NW Rota-1 (Embley et al., 2006) and Ahyi (W.W. Chadwick, pers. comm., 2014) of the Mariana arc; Monowai (Chadwick et al., 2008), Havre (Carey et al., 2014), and Rumble III (C.E.J. de Ronde, pers. obs., 2011) of the Kermadec arc; Myōjin-shō of the Izu-Bonin arc (Fiske et al., 1998); and Kavachi forearc volcano (i.e., located between the subduction zone and its associated volcanic arc) of the Tabar-Lihir-Tanga-Feni arc (Baker et al., 2002). Similarly, less common surveys of backarcs show that they too have a fairly high incidence for hydrothermal venting (e.g., German et al., 2000; Massoth et al., 2007) and can also include examples of recent seafloor volcanic activity, such as that recorded for the West Mata volcano of the NE Lau Basin (Resing et al., 2011).

Continuous eruption of the basaltic to basaltic andesite NW Rota-1 volcano since 2004 has enabled the style of this eruption and its hydrothermal discharge to be observed and analyzed (Embley et al., 2006; Resing et al., 2008; Butterfield et al., 2011). At a water depth of only 555 m, the ~ 15 m diameter volcanic vent is seen to erupt every hour or so while being observed, ejecting plumes containing droplets of molten sulfur (yellow in Figure 47.2(A)) into the ocean (see http://oceanexplorer.noaa.gov/explorations/06fire/logs/april29/media/movies/nwrota_brimstone12_video.html). Rapid quenching of these sulfur-rich plumes results in the formation of spherical sulfur lapilli that fall out with the ash (Figure 47.2(B)). On other occasions, effusive, blocky lava is spewed onto the seafloor with each block vigorously degassing sulfur gases into the ocean (Figure 47.2(C); see http://oceanexplorer.noaa.gov/explorations/06fire/logs/april29/media/movies/nwrota_brimstone13_video.html). Plumes emanating from the volcanic vent commonly appear to heave then collapse, or "pulse and/or quiver"

rapidly, presumably related to the growth and collapse of bubbles. Inspection of ejected blocks of fresh lava, which have all their vesicles filled by elemental sulfur, further highlights the prevalence of sulfur in these systems (Figure 47.2(D)) and indicates that molten sulfur occurs immediately beneath the seafloor of this volcanic vent. This phenomenon is also seen at other volcanoes. Carbon dioxide is also a major gas species discharged during these eruptions; it can either be expelled simultaneously with the sulfur-rich plumes (Figure 47.2(A)) or discharged separately, forming a "sheet" of bubbles around the margins of the vent and/or is released during periods of volcanic quiescence (see aforementioned video). This suggests that CO_2 and sulfur gases (SO_2 and H_2S) are decoupled, which likely relates to their different solubilities, with C species gases exsolving at higher pressures and depths than S species gases, thus allowing for separate pathways to the seafloor. Alternatively, as these gases are immiscible, they could separate and segregate as they ascend.

3.2. Magmatic-Hydrothermal Venting

Magmatic-hydrothermal systems are thought to make up around 70% of all seafloor hydrothermal systems related to arc volcanoes (e.g., de Ronde et al., 2001, 2003; Baker et al., 2008), and are also represented in backarcs (e.g., Seewald

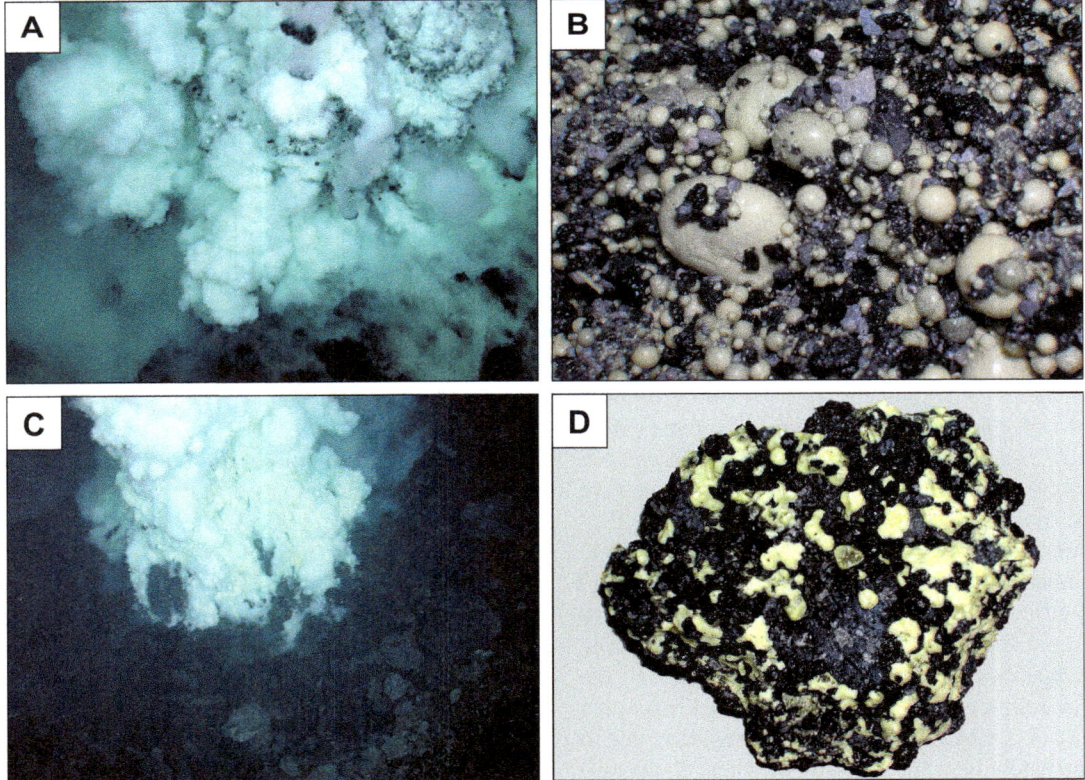

FIGURE 47.2 Examples of venting associated with volcanic eruptions. (A) Plume and ash eruption of the volcanic vent at NW Rota-1 volcano, Mariana arc. A larger block of lava can be seen plummeting down top center, while ash is observed beneath the plumes. The yellow tinge is due to droplets of molten sulfur incorporated within the plume. Bubbles of CO_2 can be seen center left of the photograph. (B) Close-up of sulfur lapilli sitting in ash formed by quenching of molten sulfur droplets within the plumes as they come into contact with seawater. (C) Typical blocky lava form of an effusive basaltic andesite eruption, NW Rota-1. Note the vigorous degassing of the erupted blocks of lava creating steam and molten sulfur-laden plumes (yellow, in center). (D) Fresh, glassy, basaltic andesite lava ejected during the eruption of NW Rota-1, with vesicles filled by sulfur.

et al., 2014). These systems are also dominated by the discharge of magmatic gases but are not associated with volcanic eruptions, although almost certainly they are related to the degassing of magma beneath the seafloor. Oxidation/hydrolysis reactions drive these systems; the **disproportionation** of degassed SO_2 mixed with seawater forms **acid** and sulfate, plus either hydrogen sulfide gas or elemental sulfur (e.g., de Ronde et al., 2005; Resing et al., 2008; Embley et al., 2006; Butterfield et al., 2011):

$$3SO_2 + 2H_2O = S^0 + 4H^+ + 2SO_4 \qquad (47.1)$$

$$4SO_2 + 4H_2O = H_2S + 6H^+ + 3SO_4^- \qquad (47.2)$$

$$2H_2S + SO_2 = 2H_2O + 3S^0 \qquad (47.3)$$

Relatively low H_2S concentrations occur in the vent fluids and plumes in these systems suggesting that reaction (47.2) is not as important a chemical reaction, since H_2S can be oxidized quickly (reaction (47.3)). On the seafloor, these systems are characterized by the expulsion of very acidic fluids (e.g., Resing et al., 2008; Butterfield et al., 2011; de Ronde et al., 2011).

Vent fluid temperatures are typically $<120\,°C$ with the corresponding discharge of white plumes widespread (Figure 47.3(A)). Their milky white color is largely ascribed to the inclusion of sulfur droplets, but also due to particulates of Al-rich sulfates (i.e., alunite or natroalunite), silica, Fe-oxyhydroxides, and some remnant leached host rock (de Ronde et al., 2005; Resing et al., 2008). Similarly, small chimneys, or more commonly low-relief "mounds" formed at these sites, are typically composed of alunite, polymorphs of silica, elemental sulfur, clay (mainly illite), and pyrite, or a mineral assemblage commonly referred to as "advanced argillic," typical of "high-sulfidation" environments (Hedenquist et al., 2000; Figure 47.3(A)).

The discharge of very high concentrations of CO_2 is also common from magmatic-hydrothermal systems. Locally, this can take the form of liquid CO_2 when the depth and pressure are great enough (Figure 47.3(B); see http://oceanexplorer.noaa.gov/explorations/04fire/logs/april10/media/eifuku_champagne_video.html). For example, the Champagne vents of NW Eifuku volcano occur at a depth of 1607 m (equivalent to 2340 lb per square inch, or 16.13 MPa) with temperatures of $2\,°C$, where CO_2 is a viscous, sticky liquid (Lupton et al., 2006; see http://oceanexplorer.noaa.gov/explorations/04fire/logs/april10/media/eifuku_bubbles_video.html).

Elemental sulfur is the most commonly deposited phase seen on the seafloor in these systems. In its most spectacular form, it appears as small lakes of molten sulfur on the seafloor (e.g., Daikoku volcano; Figure 47.3(C)), which effectively act as condensers of the S-rich gases being expelled from the volcano (see http://oceanexplorer.noaa.gov/

explorations/06fire/logs/may4/media/movies/daikoku2_video.html). Temperatures of known sulfur lakes include $187\,°C$ at Daikoku and $197.8\,°C$ at Nikko. These lakes literally slosh backward and forward, driven mainly by discharging CO_2, and contain measurable amounts of metals such as Cu, Au, Te, Se, and Ge (de Ronde et al., 2014a). Vestiges of these sulfur lakes, past and present, can be seen at several volcanoes, including Nikko volcano where remnant pillars of elemental sulfur are testament to a lake $>10\,m$ deeper in the recent past. Macauley Cone in the Kermadec arc was also host to a sulfur lake at least 2 m deep and possibly a lot deeper. The molten sulfur lake immediately beneath the seafloor at Nikko today is 200 m in diameter and was likely substantially larger in the past (de Ronde et al., 2014a). Other forms of elemental sulfur associated with magmatic-hydrothermal systems are remarkable chimneys up to 6 m tall (Figure 47.3(E)) and various examples of spaghetti-like extrusions on the seafloor (Figure 47.3(F)).

3.3. Water/Rock-Dominated Venting

The remaining $\sim 25\%$ of arc seafloor hydrothermal systems are heavily influenced by **water/rock** processes. In other words, they have chemical signatures indicative of hydrothermal fluid interaction with the host rock. Whilst vent-related animals are also found associated with magmatic-hydrothermal systems, they appear more prevalent at water/rock-dominated systems where a number of new and/or rare species have been discovered (Figure 47.4(A)) and in some cases, extensive fields of a dominant animal can be seen (Figure 47.4(B)). Elsewhere, microbes dominate and can cover large tracts of the seafloor near the volcano summits, commonly manifest as soft, gooey organic substances up to $\sim 1\,m$ thick and/or thinner microbial mats (Figure 47.4(C)). They also thrive immediately beneath the seafloor and/or inside small Fe oxyhydroxide chimneys that cover several hundred square meters (Figure 47.4(D)).

Venting at water/rock-dominated systems can be broadly categorized into diffuse and focused venting, as they are at MORs. Diffuse venting is widespread and invariably concentrated at the summit of the volcanoes where permeability is enhanced through fractures and faulting, in contrast to the relative impermeability of the volcano flanks where venting is rarely found more than $\sim 200\,m$ downslope. Temperature of the expelled fluids is usually between a few degrees Celsius above ambient (i.e., $7-10\,°C$) to around $70\,°C$, the result of mixing between a higher temperature end-member fluid and seawater. The occurrence of extensive animal populations (Figure 47.4(B)) and bacterial mats covering the seafloor (Figure 47.4(C)) are tell-tale signs of diffuse venting. In addition, mound and chimney-like structures discharging relatively low-temperature fluids can also be found near the margins of higher-temperature vent fields, including Fe oxyhydroxide-rich $\pm$ amorphous silica

FIGURE 47.3   Examples of magmatic-hydrothermal venting on the seafloor. (A) Chimneys at the Champagne vents site, NW Eifuku volcano, Mariana arc. (B) Liquid droplets of CO_2 seen rising from these same vents. This rarely seen phenomenon is due to the fact that the discharge from this volcano is particularly gas-rich. The high pressure (161.5 bar) at the 1607 m depth of the vents keeps the carbon dioxide in a liquid state. (C) Lake of molten sulfur at a depth of 415 m, Daikoku volcano, Mariana arc. (D) Numerous, horizontally layered pillars of elemental sulfur standing up to a height of 1.5 m above the crater floor at a depth of 469 m, Nikko volcano, Mariana arc. Molten sulfur occurs immediately beneath the seafloor. (E) Sulfur chimneys up to 6 m tall inside a pit crater atop the Cone site (1208 m depth), Brothers volcano, Kermadec arc. (F) Hand sample of extruded molten sulfur, ∼1300 m depth, Seamount X, Mariana arc.

(Figure 47.4(E)), silica-rich (Figure 47.4(F)), and sulfate-rich varieties (Figure 47.4(G)).

Focused venting is defined here as hydrothermal fluid being expelled from a vent field full of chimneys typically, but not exclusively, dominated by sulfides with temperatures in excess of 200 °C. Figure 47.4(H) shows a classic beehive-type structure sitting atop a ∼15-cm-high chimney that itself is perched upon a taller approximately 4-m-high chimney complex at the NW Caldera vent field,

Brothers volcano. Here, fluid temperatures reach 302 °C, which marks the highest temperature recorded at a submarine arc hydrothermal system. This is mainly due to the relatively shallow depths to these seafloor systems when compared to MORs, with the depth-to-boiling-point curve determining the maximum temperature a fluid of largely seawater origin can attain before it boils. In addition, the gas-rich nature of arc vent fluids (see below) means that gases such as CO_2 will contribute to the total vapor pressure

FIGURE 47.4 Examples of water/rock-dominated venting at submarine arc volcanoes. (A) Vent fauna associated with diffuse venting that include rare tube worms and mussels, crabs and rat tail fish, Monowai volcano, Kermadec arc (1152 m depth). (B) Wide swaths of long-necked barnacles associated with diffuse venting at Clark volcano, Kermadec arc (~870 m). Different animals, such as mussels, can dominate hydrothermal systems at other volcanoes along this arc. (C) Bacterial mat covering hummocky seafloor of a silica-Fe oxyhydroxide diffuse vent field, Tangaroa volcano, Kermadec arc. (D) Sampling of bacteria within Fe oxyhydroxide crusts including ~25-cm-high "mound" structures, Healy volcano, Kermadec arc (1373 m). (E) Fe-rich amorphous silica structure, NW Caldera site, Brothers volcano, Kermadec arc (1620 m). (F) Bulbous, ~2-m-tall silica-rich chimney that is host to abundant microbes, Giggenbach volcano, Kermadec arc (186 m). (G) Small, delicate chimneys up to 30 m high comprised mostly of sulfates discharging hydrothermal fluids between 56 and 106 °C, near Marker #14, Clark volcano (885 m). (H) High-temperature black smoker chimney, Marker #4, NW Caldera vent field, Brothers volcano (1665 m). In the main photograph, clear 302 °C hydrothermal fluid can be seen discharging from the top of the main chimney where an ~15-cm-long, narrow chimney has been broken off that previously supported a classic "beehive" structure on top (inset). Within 5 cm of being expelled from the chimney, the hydrothermal fluid mixes with ambient seawater to produce "black smoke."

of the solution so that the fluids can boil at lower temperatures than similar solutions would with lesser amounts of dissolved gases. On rare occasions, boiling has been witnessed at arc vent sites, including Volcano 19 of the Tonga arc (Stoffers et al., 2006) and East Diamante volcano of the Mariana arc (see http://oceanexplorer.noaa.gov/explorations/04fire/logs/april06/media/bubbling_all.html). Significant Cu−Zn−Au mineralization can be found at some of these water/rock-dominated vent sites (e.g., de Ronde et al., 2005, 2011). Backarcs are also host to water/rock-dominated hydrothermal systems, such as those of the PACManus site in the Manus Basin, Papua New Guinea (e.g., Reeves et al., 2011) where Cu−Au mineralization is also prevalent (e.g., the Solwara 1 project of Nautilus Minerals Inc.).

4. ARC HYDROTHERMAL FLUIDS

Published results of vent fluid samples for submarine arc systems are limited to volcanoes from the Izu-Bonin, Mariana, and Kermadec arcs (Table 47.1). A larger data set is available for dissolved (Table 47.2(A)) and pure (Table 47.2(B)) gas samples, including additional volcanoes from the aforementioned arcs and one from the Tonga arc. What is very apparent from the data given in Tables 47.1 and 47.2 is the difference in chemical parameters, and to a lesser degree physical properties, between fluids expelled from magmatic-hydrothermal vent sites and those from water/rock-dominated sites. For example, temperatures for the focused water/rock-dominated sites tend to be >290 °C (e.g., Suiyo and Brothers NW Caldera), whereas those from the magmatic-hydrothermal sites are typically <120 °C (e.g., Kasuga, NW Eifuku, Brothers Cone sites), although temperatures up to 257 °C were recorded at one NW Rota-1 vent, and low temperature, diffuse venting can also occur at the water/rock-dominated sites (e.g., Monowai).

Differences in fluid chemistry clearly distinguish the magmatic fluids from the water/rock fluids. Values of pH are noticeably less for the magmatic-hydrothermal systems, extending to values as low as 1.1, whereas typical values for the water/rock sites are around 3 to 4. Alkalinity appears to correlate with pH and also clearly differentiates these two vent systems. For example, the Brothers NW Caldera and Suiyo water/rock-dominated fluids show a positive correlation between alkalinity and pH ($R^2 = 0.93$), while the Brothers Upper Cone fluids have much higher alkalinity, coincident with magmatic fluids having an order of magnitude greater CO_2 content (Tables 47.1(A) and 47.2(A)).

The most striking chemical difference is the elevated values for Mg and SO_4 relative to seawater seen in the magmatic fluids of NW Rota-1 (Mg up to 55.9 mmol/kg and SO_4 up to 164.7 mmol/kg), Kasuga (Mg up to 65.8 mmol/kg; SO_4 up to 31.8 mmol/kg), and the Cone site of Brothers (Mg up to 53.5 mmol/kg; SO_4 up to 39.4 mmol/kg) when compared to seawater (Mg = 52.3 mmol/kg;

$SO_4 = 28.1$ mmol/kg). Clearly, these ions are being *added* to the hydrothermal fluid, a completely different scenario to water/rock-dominated fluids where these ions are *removed* from the fluid during the interaction between hydrothermal fluid and host rock (e.g., Butterfield, 2000).

Concentrations of Fe are highest in the fluids of NW Rota-1 followed by those from Brothers. More startling are the very high Fe/Mn values for the NW Rota-1 fluids (and to a lesser degree those of Brothers) when compared to the water/rock-dominated fluids of Suiyo and Monowai (Table 47.1(A)). Aluminum is also a good discriminator of these two basic fluid types, with values for water/rock-dominated systems ranging from 2 to 37 μmol/kg for Suiyo Seamount and 3−65 μmol/kg for Brothers NW Caldera, yet reach 2558 μmol/kg for NW Rota-1 and 116 μmol/kg for Brothers Cone.

The total gas measured in the vent fluid samples is significantly higher for magmatic-hydrothermal systems at Nikko, NW Eifuku, Daikoku and NW Rota-1 of the Mariana arc, Volcano 1 of the Tonga arc, and Giggenbach and Brothers Cone of the Kermadec arc than for water/rock-dominated systems (Table 47.2(A)). Indeed, pure gas samples could be collected from these volcanoes and subsequently analyzed (Table 47.2(B)). These samples show that magmatic-hydrothermal systems have very high N_2 and especially CO_2, with up to 2.7 mol/kg CO_2 being expelled from the 103 °C vents at NE Eifuku—the highest ever reported for submarine hydrothermal fluids (Lupton et al., 2006)—and the remarkable occurrence of liquid CO_2 being expelled from the same site (Figure 47.3(B)). The highest concentrations of He and H_2 coincide with the most volcanically active volcano of NW Rota-1 (Table 47.2(A)).

Finally, in the data set for water/rock-dominated vent fluids, Cl concentrations span values both greater and lesser than seawater (i.e., 540 mmol/kg), which has been ascribed to the process of phase separation at MOR systems (e.g., Butterfield, 2000) and which is almost certainly occurring in arc systems as well. For example, Brothers NW Caldera Cl values range from 506 to 724 mmol/kg, or 94−134% seawater values; by comparison, Suiyo fluids range from seawater to 120% seawater values, and Monowai from 95% seawater values to seawater (Table 47.1(A)). Limited Cl data are available for magmatic-hydrothermal systems, although the few analyses from Kasuga and Brothers Upper Cone show essentially seawater values, almost certainly due to mixing with ambient seawater (de Ronde et al., 2011), while Brothers Lower Cone fluids show Cl values 92% that of seawater. In summary, many if not all of the differences mentioned above between fluids expelled from magmatic-hydrothermal vent sites and those from water/rock-dominated sites can be directly attributed to the input of a magmatic fluid, whether adding components such as CO_2 and possibly Mg and Fe, or forming acid and sulfate by condensing magmatic gases and mixing them with seawater.

TABLE 47.1A Vent Fluid Analyses from Submarine Arc Volcanoes

Arc Volcano	Vent	Dive No.^	Sample No.	Depth (m)	T (°C)	pH	Alkalinity (meq/kg)	Mg (mmol/kg)	Cl (mmol/kg)	H₂S (liq) (μmol/kg)	Si (mmol/kg)	Fe (mmol/kg)	Mn (mmol/kg)	Fe/Mn (mol/mol)	Na (mmol/kg)	K (mmol/kg)	Ca (mmol/kg)	SO₄ (mmol/kg)	Al (μmol/kg)	References
Izu-Bonin																				
Suiyo Seamount Vent D W/R		S2-630	1	1370	311	3.8	−0.06	4.2	627											1*
			2	1370	311	3.9	0.08	4.4	626	1461										
			3	1370	311	4.5	0.80	17.2	612											
			4	1370	311	3.8	−0.03	2.0	656											
	Vent F	S2-631	1	1370	300	3.9	−0.12	4.1	632	1558										
			2	1370	300	3.8	−0.14	0.8	627	1655										
			3	1370	300	3.8	−0.12	0.8	633	1363										
			4	1370	300	3.8	−0.16	0.6	642											
	Vent F + G		5	1370		4.3	0.31	10.2	620											
	Vent G		6	1370	296	3.8	−0.12	3.6	633											
			7	1370	296	4.0	0.06	6.3	632											
			8	1370	296	3.9	−0.08	4.1	620											
	ISCS-2	S2-1237	R2	1386	66–198	4.6	0.99	20.2		1300	8.11	0.11	0.43	0.2			53.6		37.0	2
			R4	1386	188–194	4.3	0.62	10.4		1600	10.70	0.13	0.58	0.2			66.5		17.0	
	Mk#230-2		R6	1385	311	3.6	−0.13	1.7		2200	13.60	0.38	0.73	0.5			77.0		33.1	
			R8	1385	311	3.6	−0.10	2.1		2400	13.60	0.40	0.73	0.5			86.2		23.9	
		S2-1238	R2	1374	202–292	4.0	0.20	10.2		2200	12.10	0.30	0.66	0.5			68.6		29.9	
			R4	1371	9–199	5.4	1.83	42.4		500	2.63	0.06	0.13	0.5			25.3		16.3	
	Mk#223		R7	1381	11–27	6.2	2.33	50.2			6.20	0.00	0.03	0.1			13.4		12.9	
	APSK05	HY02	R3	1382	80–100	5.3		29.1	583		3.68	0.16	0.19	0.9			31.0		11.7	
		HY11	R4	1382	260–280	3.9		1.0	646	2300	11.40	0.46	0.61	0.8			81.6		26.5	
	APSK04	HY04	R3	1386	18–21	7.3		49.3	546		0.17	0.00	0.00	1.5			9.8		2.0	
	APSK03	HY08	R2	1374	16–22	6.2		46.6	549	200	0.62	0.02	0.03	0.6			12.6		1.9	
	Shell carpet	S2-1306	P7	1382	5	7.5	2.63	52.5			0.19						10.4			
	APSK01	S2-1307	P3	1375	17	7.2	2.44	49.7			0.16	0.09					11.6			
		S2-1310	P3	1377	40	6.8	2.45	52.2			0.24	0.00	0.01	0.5			10.6			
	Mk#311-2	S2-1311	P6	1374	215	4.5	1.24	20.7		1300	8.37	0.28	0.43	0.7			58.8			
			P7	1374	166	4.7	0.89	27.4		1100	6.20	0.26	0.35	0.7			48.0			
	Sulfide mound	HY32	F2	1384	299–300	3.7		2.0		1461	11.88									3*
	White patch	S2-1388	F2	1376	310	4.0		6.9		1363	10.81									

(Continued)

TABLE 47.1A Vent Fluid Analyses from Submarine Arc Volcanoes—Cont'd

Arc Volcano	Vent	Dive No.^	Sample No.	Depth (m)	T (°C)	pH	Alkalinity (meq/kg)	Mg (mmol/kg)	Cl (mmol/kg)	H2S (liq) (µmol/kg)	Si (mmol/kg)	Fe (mmol/kg)	Mn (mmol/kg)	Fe/Mn (mol/mol)	Na (mmol/kg)	K (mmol/kg)	Ca (mmol/kg)	SO$_4$ (mmol/kg)	Al (µmol/kg)	References
Mokuyo W/R	Mokuyo Seamount				8	7.0	2.50	46.1	509		0.3				461	10.1	10.2	26.3		4*
Mariana																				
Kasuga M-H	Kasuga 2	A-1879		402	39	5.2	31.90	65.8	515	2	1.82	0.01	0.05	0.1	463	9.0	3.6	31.8		5
	Kasuga 3	A-1882		1140	9	7.5	2.58	51.5	536	Nd	0.38	0.02	0.02	1.1	459	9.0	11.3	27.7		
NW Rota-1 M-H	Brimstone Rim	R-783	P20	549	17	3.0		55.9		25	0.39	0.06	0.01	5.6		10.1	11.2	31.0	19.0	6
	Brimstone		B17	566	30	2.1		52.6		64	1.11	0.19	0.04	5.5		10.3	10.5	34.0	67.0	
		R-786	P4	561	26	2.2		50.7		5	1.10	0.22	0.03	8.5		9.8	9.8	34.4	122	
			P5	561	29	2.3		51.6		12	0.38	0.17	0.02	8.5		10.0	10.1	33.3	96.0	
			B8	561	27	2.1		52.9		9	1.29	0.24	0.03	7.2		10.3	10.2	35.8	154	
			BF11	561	30	2.0		51.6		3	1.23	0.28	0.03	9.3		10.0	9.9	36.9	141	
		J2-187	B9	560	28	1.6		54.8		Bdl	Nd	1.62	0.03	50.2		10.2	12.5	67.7	1820	
			PF24	562	95	1.9		48.8		Bdl	0.55	0.12	0.00	25.2		9.0	9.2	47.0	115	
			MajBlu	563	95	1.7		54.9		Bdl	3.19	2.00	0.04	55.8		10.4	9.2	57.3	409	
		J2-188	MajYel	564	110	1.7		50.0		Bdl	1.35	0.48	0.01	50.9		9.7	10.1	63.7	462	
			MajBlu	563	95	1.1		44.3		Bdl	7.06	1.87	0.03	56.2		8.3	11.0	165	2558	
		J2-189	MajWht	559	200	1.4		54.1		Nm		0.81	0.02	51.8		9.7	11.4	44.2	661	
			MajRed	560	257	2.1		50.1		40		0.75	0.02	44.6		9.3	9.2	29.5	376	
		J2-191	PF1	557	42	1.8		52.2		Bdl		0.47	0.01	56.5		9.6	10.3	57.1	814	
		J2-398	PF4	523	81	3.8		50.9			0.10	0.01	0.00	18.5		9.8	9.9		14.6	
			P8	523	31	5.2		52.5			0.09	0.01	0.00	19.6		9.8	9.9		9.9	
			PF2	523	26	5.9		52.0			0.07	0.02	0.00	Nd		9.9	10.2		3.1	
		J2-401	P1	523	47	1.5		51.6			2.55	0.81	0.02	50.6		9.7	11.3		1412	
			PF2	523	58	1.6		51.1				0.60	0.01	52.1		9.8	10.8		1087	
			P3	523	42	1.7		51.4			2.58	0.98	0.07	14.1		9.8	10.8		1192	
			P6	523	207	1.1		46.7			7.03	2.74	0.11	24.0		9.2	11.9		2008	
		J2-403	B24	519	21	4.0		52.8			1.32	0.26	0.02	12.1		10.3	10.5		1890	
			PF4	519	18	4.8		52.5			0.58	0.12	0.01	13.1		10.1	10.4		922	
			P1	519	21	3.7		52.8			1.43	0.30	0.02	13.8		10.2	10.6		2231	

Group	Volcano	ID	Sample															
		J2-405	P3	522	25	2.3	52.0			0.34	0.10	0.00	39.6		9.9	10.0	106	
			P5	522	35	2.3	52.0				0.14	0.00	40.8		10.0	10.1	242	
			PF2	521	45	2.6	52.2			0.27	0.06	0.00	36.9		10.0	10.1	351	
			B22	522	39	2.0	51.9			0.63	0.19	0.01	36.9		9.9	10.1	144	
			B21	521	22	5.7	52.7			0.26	0.03	0.01	4.6		10.0	10.2	64.0	
			P6	521	56	1.5	50.1			4.28	1.11	0.02	53.8		9.5	11.8	2156	
			PF7	521	58	1.5	50.3			4.23	1.15	0.02	54.9		9.6	11.6	1904	
			P8	521	48	1.4	50.2			4.04	1.04	0.02	55.2		9.5	11.3	1936	
			PF9	521	40	1.7	Nd			1.72	0.46	0.01	42.7		Nd	Nd	612	
			BF18	521	35	2.0	52.1			1.30	0.33	0.01	46.2		9.9	10.4	470	
Kermadec																		
Monowai	Monowai Caldera 2	P5-613	MS Blue	1171	56	6.2	51.4	539	684	0.54	0.00	Nm		459	10.0	11.1	7	27.5
W/R	Monowai Ca8	P5-614	HFS 8 bag	1166	44	5.6	49.3	519	3169	0.98	0.00	0.10	0.0	433	9.4	11.0		26.3
			HFS 11 t/bag		43	5.5	49.2	519	3306	1.03	0.00	0.10	0.0	442	9.7	11.0		26.7
			HFS 6 piston		35	5.7	49.9	527	2590	0.83	0.00	0.08	0.0	441	9.6	10.9		27.2
	Monowai Ca9		HFS 14 t/bag	1157	26	5.9	50.4	538	769	0.89	0.07	0.26	0.3	426	9.3	12.2		26.8
			HFS 20 piston		32	5.8	50.4	538	1363	0.97	0.00	0.42	0.0	450	9.9	12.1		26.8
	Monowai Ca10		HFS 1 t/piston	1143	6	7.0	52.1	537	1	0.12	0.00	0.03	0.0	430	9.3	10.4		27.8
			HFS 5 piston		9	6.7	52.1	538	130	0.23	0.00	0.06	0.0	441	9.6	10.7		27.9
	Monowai Ca10b		HFS 18 t/bag	1141	52	5.5	48.3	517	3537	1.43	0.00	0.16	0.0	431	9.4	11.5		26.3
			HFS 16 t/bag		52	5.4	48.1	515	4017	1.43	0.00	0.17	0.0	425	9.3	11.4		26.2
			HFS 19 bag		52	5.4	48.3	518	3869	1.47	0.01	0.17	0.1	425	9.3	11.5		26.2
	Monowai Ca11		HFS 22 piston	1079	25	5.8	50.2	541	1054	0.93	0.00	0.37	0.0	386	8.4	12.4		26.7
	Monowai Ca12	P5-615	MS Blue	1025	13	6.2	52.1	537	685	0.41	0.01	0.03	0.3	456	9.9	11.2		28.4

(Continued)

TABLE 47.1A Vent Fluid Analyses from Submarine Arc Volcanoes—Cont'd

Arc Volcano	Vent	Dive No.^	Sample No.	Depth (m)	T (°C)	pH	Alkalinity (meq/kg)	Mg (mmol/kg)	Cl (mmol/kg)	H2S (liq) (µmol/kg)	Si (mmol/kg)	Fe (mmol/kg)	Mn (mmol/kg)	Fe/Mn (mol/mol)	Na (mmol/kg)	K (mmol/kg)	Ca (mmol/kg)	SO4 (mmol/kg)	Al (µmol/kg)	References
Brothers	Brothers NW Caldera	S6-851	2B	1665	302	3.1	-0.63	3.9	652	1325	12.57	4.80	0.70	6.8	479	62.6	39.4	0.2	64.8	8
W/R			4A	1670	274	3.0	-0.89	5.0	715	1648	12.91	6.61	0.68	9.7	521	67.2	40.5	0.4	54.0	
		S6-852	2A	1656	294	3.2	-0.57	4.5	724	1733	12.13	6.58	0.70	9.4	526	68.0	42.0	0.2	56.8	
			4C	1627	292	2.8	-1.77	5.5	506	2125	11.11	3.77	0.46	8.2	380	44.7	31.2	0.3	54.6	
			Bag	1656	294	3.4	-0.48	16.2	675	387	10.34	4.59	0.53	8.7	504	52.4	34.3	6.0	46.0	
		P5-626	MS Blue	1642	265	6.9	2.26	51.2	545	6	0.30	0.08	0.01	7.9	459	11.2	11.1	27.5		
			MS Yellow	1616	290	2.8		3.4	593	5474	12.97	5.16	0.63	8.2	435	55.2	38.8	6.7	7.1	
		P5-631	MS White	1572	290	3.8	-0.24	22.3	546	2250	7.44	2.70	0.36	7.6	433	35.4	24.5	11.4	3.1	
M-H	Brothers upper cone	S6-853	4B	1222	60	2.3		51.7	547	1703	3.92	0.04	0.14	0.3	451	14.8	9.7	25.4	116	
			Bag	1222	60	3.2	-0.63	50.5	540	701	3.89	0.03	0.08	0.4	439	13.6	9.7		75.9	
		P5-630	MS Blue	1227	67	5.9	1.02	52.4	539	15	0.24	0.04	0.01	4.1	460	10.1	10.4	28.6		
			MS White	1213	122	1.9		53.5	544	600	3.03	1.14	0.17	6.8	452	10.9	10.9	39.4		
	Brothers lower cone	S6-853	2B	1312	68	5.2	3.68	47.8	497	5003	4.32	0.03	0.28	0.1	423	16.6	8.0	24.3	46.7	
		S6-854	2A	1336	67	5.1	4.02	45.5	544	3742	4.25	0.03	0.39	0.1	442	18.5	7.2	25.6	10.0	
			Bag	1336	67	5.1	3.57	46.7	544	2350	3.34	0.04	0.30	0.1	432	16.9	7.9	24.9	36.0	
		P5-630	MS Yellow	1306	46	5.2	2.57	52.3	540	1271	0.94	0.00	0.19	0.0	466	11.0	9.9	28.9		
Background seawater						7.8	2.38	52.3	540	0	0.065	<0.0001			458	10.0	10.1	28.1	0.0	8

Nd—Not detected, Nm—not measured, Bdl—below detection limit, W/R refers to water/rock-dominated hydrothermal systems, and M-H refers to magmatic hydrothermal systems.

[1] Tsunogai, U., Ishibashi, J., Wakita, H., Gamo, T., Watanabe, K., Kajimura, T., Kanayama, S., Sakai, H., (1994). Peculiar Features of Suiyo Seamount Hydrothermal Fluids, Izu-Bonin Arc - Differences from Subaerial Volcanism. Earth Planet. Sc. Lett. **126**, 289–301.

[2] Kishida, K., Sohrin, Y., Okamura, K., Ishibashi, J., (2004). Tungsten enriched in submarine hydrothermal fluids. Earth Planet. Sc. Lett. **222**, 819–827.

[3] Nakagawa, T., Ishibashi, J.I., Maruyama, A., Yamanaka, T., Morimoto, Y., Kimura, H., Urabe, T., Fukui, M., (2004). Analysis of dissimilatory sulfite reductase and 16S rRNA gene fragments from deep-sea hydrothermal sites of the Suiyo Seamount, Izu-Bonin Arc, Western Pacific. Appl. Environ. Microb. **70**, 393–403.

[4] Ishibashi, J.I., Tsunogai, U., Wakita, H., Watanabe, K., Kajimura, T., Shibata, A., Fujiwara, Y., and Hashimoto, J. (1994). Chemical composition of hydrothermal fluids from the Suiyo and the Mokuyo Seamounts, Izu-Bonin Arc. JAMSTEC J. Deep Sea Res. **10**, 89–98.

[5] McMurtry, G.M., Sedwick, P.N., Fryer, P., Vonderhaar, D.L., Yeh, H.W., (1993). Unusual Geochemistry of Hydrothermal Vents on Submarine Arc Volcanoes - Kasuga Seamounts, Northern Mariana Arc. Earth Planet. Sc. Lett. **114**, 517–528.

[6] Butterfield, D.A., Nakamura, K., Takano, B., Lilley, M.D., Lupton, J.E., Resing, J.A., Roe, K.K., (2011). High SO2 flux, sulfur accumulation, and gas fractionation at an erupting submarine volcano. Geology **39**, 803–806.

[7] Leybourne, M.I., Schwarz-Schampera, U., de Ronde, C.E.J., Baker, E.T., Faure, K., Walker, S.L., Butterfield, D.A., Resing, J.A., Lupton, J.E., Hannington, M.D., Gibson, H.L., Massoth, G.J., Embley, R.W., Chadwick, W.W., Clark, M.R., Timm, C., Graham, I.J., Wright, I.C., (2012). Submarine Magmatic-Hydrothermal Systems at the Monowai Volcanic Center, Kermadec Arc. Econ. Geol. **107**, 1669–1694.

[8] de Ronde, C.E.J., Massoth, G.J., Butterfield, D.A., Christenson, B.W., Ishibashi, J., Ditchburn, R.G., Hannington, M.D., Brathwaite, R.L., Lupton, J.E., Kamenetsky, V.S., Graham, I.J., Zellmer, G.F., Dziak, R.P., Embley, R.W., Dekov, V.M., Munnik, F., Lahr, J., Evans, L.J., Takai, K., (2011). Submarine hydrothermal activity and gold-rich mineralization at Brothers Volcano, Kermadec Arc, New Zealand. Miner. Deposita **46**, 541–584.

^A-Alvin, R-ROPOS, H-HyperDolphin, P4-Pisces IV, P5-Pisces V, J2-Jason II, S2-Shinkai 2000, S6-Shinkai 6500, Hy-Hayuko 2000.

*These references reported concentrations per liter. Fluid density of 1.027 kg/L was used for consistent units per personal communication with J-I. Ishibashi, 2014.

TABLE 47.1B Vent Fluid Analyses from Backarc Vents

Arc Volcano	Vent	Dive No.^	Sample No.	Depth (m)	T (°C)	pH	H$_2$S (mmol/kg)	Si (mmol/kg)	Mg (mmol/kg)	Cl (mmol/kg)	Fe (mmol/kg)	Mn (mmol/kg)	Fe/Mn (mol/mol)	Na (mmol/kg)	K (mmol/kg)	Ca (mmol/kg)	SO$_4$ (mmol/kg)	Al (μmol/kg)	References
PACMANUS	Roman Ruins 1	J2-208	IGT8	1677	314	2.3	6.4	13.7	7.3	617	5.56	3.40	1.6	470	71.8	18.4	0.5	8.9	1
W/R			IGT5			2.4	6.5	13.4	7.6	619	5.62	3.28	1.7	485	71.3	18.6	0.6	7.1	
	Roman Ruins 2		IGT2	1675	272	2.3	2.7	16.6	15.9	549		2.26		434	50.6	10.6	2.6	5.6	
			IGT1			2.4	2.9	16.7	16.0	543	0.99	2.27	0.4	435	50.5	10.3	2.5	4.7	
			M2			2.7			27.0	551	0.68	1.70	0.4	443	38.5	10.5		4.1	
	Roman Ruins 3	J2-213	IGT7	1660	278	3.2	2.5	11.6	22.7	948	4.32	2.58	1.7	508	58.8	18.8	10.6	4.6	
			M4			2.5			6.4	708	6.85	4.28	1.6	534	86.4	23.8		15.0	
	Roman Ruins 4	J2-222	IGT1	1680	341	2.7	6.3	17.8	3.6	650	6.47	2.83	2.3	495	77.2	22.3	0.4	6.5	
			M4			2.6		17.3	4.7	647	6.17	2.73	2.3	495	75.5	21.8		5.1	
	Roger's Ruins 1	J2-213	IGT3	1709	320	2.7	3.3	17.0	5.1	635	4.09	2.41	1.7	484	74.5	25.7	1.1	6.1	
			M4			2.7	3.1	16.2	7.6	634	3.89	2.30	1.7	487	70.1	24.4	2.6	5.3	
	Roger's Ruins 2	J2-222	IGT4	1710	274	3.0	1.5	11.0	22.3	606	0.11	1.45	0.1	479	49.9	18.3	9.6	3.4	
			IGT3			2.6	2.5	15.8	9.0	631	0.13	2.15	0.1	481	68.6	22.4	1.5	5.1	
			M2			2.6		15.8	8.6	631	0.14	2.17	0.1	484	68.8	22.4			
	Satanic Mills 1	J2-209	IGT6	1685	295	2.7	7.4	11.8	9.8	523	2.74	2.05	1.3	411	58.2	12.5	2.7	9.5	
			IGT7			2.7	8.1	12.0	9.0	521	2.79	2.14	1.3	409	59.3	12.5	2.2	7.0	

(Continued)

TABLE 47.1B Vent Fluid Analyses from Backarc Vents—Cont'd

Arc Volcano	Vent	Dive No.^	Sample No.	Depth (m)	T (°C)	pH	H$_2$S (mmol/kg)	Si (mmol/kg)	Mg (mmol/kg)	Cl (mmol/kg)	Fe (mmol/kg)	Mn (mmol/kg)	Fe/Mn (mol/mol)	Na (mmol/kg)	K (mmol/kg)	Ca (mmol/kg)	SO$_4$ (mmol/kg)	Al (μmol/kg)	References
Satanic Mills 2			IGT4	1688	241	2.7	3.3	9.4	26.6	478	0.71	1.11	0.6	400	31.0	7.1	12.1	4.5	
			M2			2.4			16.9	455	1.12	1.68	0.7	374	38.6	5.9			
Satanic Mills 3		J2-214	IGT8	1682	288	2.5	8.3	12.3	9.7	510	0.98	1.82	0.5	403	58.1	13.1	2.5	5.2	
			IGT5			2.5	8.3	12.2	9.8	510	0.96	1.82	0.5	402	56.4	13.1	2.4	5.9	
Snowcap 1		J2-210	IGT8	1643	152	4.6	2.9	6.8	30.8	499	0.03	0.99	0.0	419	24.6	6.4	10.8	0.9	
			IGT5			5.0	0.5	1.7	48.5	532		0.21		450	12.9	9.7	24.1	0.2	
Snowcap 2		J2-211	IGT4	1639	180	3.4		9.5	24.2	532	0.14	1.63	0.1	440	34.7	6.8	5.2	1.7	
			IGT3			3.7	1.0	9.3	24.8	536	0.17	1.55	0.1	437	34.0	6.8	5.7	2.4	
			M4			3.4			24.5	530	0.13	1.64	0.1	435	34.4	6.6		2.3	
Tsukushi			IGT7	1660	62	5.9		3.6	44.4	572	0.12	0.57	0.2	477	20.7	12.5	18.2		
		J2-214	IGT2			5.8		3.2	45.2	570		0.56		477	19.8	12.3	22.6		
Fenway 1		J2-210	IGT1	1710	329	2.5	18.5	12.7	6.0	463	7.37	2.31	3.2	347	53.8	14.3	0.9	7.9	
			M4			2.6			5.8	465	7.56	2.56	3.0	340	53.8	14.3		6.4	
		J2-214	IGT1			4.5	4.6	3.7	39.8	520	2.03	0.66	3.1	422	21.2	11.3	20.3	1.9	
Fenway 2		J2-212	IGT8	1707	343	2.7	9.2	13.8	4.9	683	13.30	4.22	3.2	485	86.4	25.5	0.6	8.2	
			IGT5			2.7	9.0	13.8	5.3	685	13.10	4.16	3.1	485	85.8	25.5	0.7	6.9	
Fenway 3			IGT2	1706	358	2.7	15.0	11.9	4.5	589	11.50	3.60	3.2	417	73.9	22.2	1.0	6.1	
			IGT1			2.8	19.3	10.4	4.7	517	9.83	3.09	3.2	377	64.9	20.3	2.5	5.8	
Fenway 4		J2-216	IGT7	1710	284	2.6	11.6	10.9	9.0	527	6.85	2.95	2.3	395	60.2	17.7	2.5	7.5	
			IGT6			2.5	11.6	11.3	8.7	524	6.65	2.94	2.3	392	60.7	17.7	2.2	8.4	
Fenway 5			IGT4	1709	80	5.0	2.5	1.7	45.0	517	0.87	0.33	2.7	437	14.4	13.3	27.6	0.7	
			IGT3			4.9	2.7	1.9	44.0	517	0.57	0.37	1.6	436	15.3	14.1	27.4	1.2	

Group	Region	Dive	Sample																	Ref
DESMOS	Caldera	S6-302	1	1920	88	2.4	5.1	3.4	49.2	485				408	9.9	8.2	29.3	15.6		
	M-H		2			2.2	7.8	4.9	50.3	482				400	10.1	7.7	31.4	20.7		2*
			3			2.1	8.4	5.2	50.1	475				403	10.1	7.4	31.4			
			4			2.1	8.7	5.7	49.1	468				390	9.9	7.1	31.9	24.7		
		S6-306	1	1700	120	2.7	9.4	2.8	46.4	473				393	9.6	8.1	25.8			
			4			2.9		2.3	48.1	506				422	9.8	8.6	26.6			
			5					2.6	46.9	489				406	9.7	8.2	26.1			
			6					1.3	49.2	509				434	9.8	8.5	26.5			
			8					1.4	48.9	516				437	9.7	8.9	26.4			
			TF					5.8	45.2	465				382	10.2	7.0	26.4			
	D1	J2-220	IGT1	1908	113	1.0	0.0	8.3	44.9	492	12.4	0.04	310	391	8.3	9.4	125	480		3
			IGT2		117	1.0	0.0	7.9	45.1	495	11.9	0.04	322	392	8.4	9.4	123	467		
			M4		Nd	1.3	Nd	3.5	50.0	523	5.62	0.03	201	438	9.2	9.8		209		
	D2		IGT4	1908	70	1.4	0.4	5.7	49.3	503	5.56	0.05	121	422	8.7	11.9	55.3	1620		
			IGT3		69	1.4	0.4	5.7	49.2	502	5.46	0.05	119	421	8.8	11.7	54.4	1640		
			M2		Nd	1.4	Nd	5.7	49.4	501	5.53	0.04	129	419	8.8	11.9		1580		
SuSu Knoll	NS1	J2-221	IGT8	1253	47	1.9	0.5	3.0	50.4	527	1.32	0.02	62.9	453	10.0	9.7	35.3	180		
M-H			IGT7		48	1.8	0.6	3.7	49.6	520	1.63	0.03	62.7	447	9.8	9.6	37.4	211		
			M4		Nd	1.9	Nd	3.3	49.8	524	1.42	0.02	64.5	450	9.8	9.6		211		
	NS2		IGT6	1527	206	0.9	Nd	9.9	39.2	443	3.10	0.08	38.3	340	7.8	8.9	132	1080		
			IGT5		215	0.9	Nd	8.6	41.1	456	2.57	0.07	38.4	359	8.0	9.1	120	928		
			M2		Nd	0.9	Nd	8.9	41.1	457	2.47	0.07	36.3	359	8.0	9.0		942		

Nd – Not determined.
W/R refers to water/rock-dominated hydrothermal systems, and M-H refers to magmatic hydrothermal systems.

[1] Seewald, J.S. Reeves, E.P., Bach, W., Saccocia, P.J., Craddock, P.R., Shanks, W.C., Sylva, S.P., Pichler, T., Rosner, M., Walsh, E. (2014) Submarine venting of magmatic volatiles in the Eastern Manus Basin, Papau New Guinea. Geochim. Cosmochim. Ac. (in press).

[2] Gamo, T., Okamura, K., Charlou, J., Urabe, T., Auzende, J., Ishibashi, J., Shitashima, K., Chiba, H., and Shipboard Scientific Party of the ManusFlux cruise. (1997). Acidic and sulfate-rich hydrothermal fluids from the Manus back-arc basin, Papua New Guinea. Geology 25, 139–142.

[3] Reeves, E. P., Seewald, J. S., Saccocia, P., Bach, W., Craddock, P. R., Shanks, W. C., Sylva, S. P., Walsh, E., Pichler, T., Rosner, M. (2011) Geochemistry of hydrothermal fluids from the PACMANUS, Northeast Pual and Vienna Woods hydrothermal fields, Manus Basin, Papua New Guinea. Geochim. Cosmochim. Ac., 75 (4), 1088–1123.

^ J2-Jason II, S6-Shinkai 6500.
*These references reported concentrations per liter. Fluid density of 1.027kg/L was used for consistent units per personal communication with J-I. Ishibashi, 2014.

5. BACKARC HYDROTHERMAL FLUIDS

Hydrothermal fluids expelled from backarc seafloor systems follow a similar pattern to those associated with arc volcanoes. They also show marked differences between magmatic-hydrothermal and water/rock-dominated systems. For example, the magmatic-dominated fluids of backarc vent sites such as DESMOS and SuSu Knolls of the Manus Basin again show high SO_4 values up to 125 and 132 mmol/kg, respectively (Table 47.1(B)). These sites are also distinguished from the water/rock-dominated site of PACManus by having much lower values for pH (down to 0.9), mostly higher Fe concentrations in the fluid and much higher Fe/Mn values, and higher concentrations of Al, as is seen with arc systems. A range in Cl concentrations at PACManus from 463 to 948 mmol/kg, or 86% to 176% seawater values, clearly indicates phase separation as also prominent in backarc systems.

Gases analyzed from the PACManus vent field show relatively high concentrations of CH_4—most likely related to the distillation of organic matter in sediments—as do the water/rock-dominated arc sites, and typically lesser CO_2 than at backarc magmatic-hydrothermal sites, although three of the vent sites, namely Satanic Mills, Snowcap, and Fenway have CO_2 values up to 230, 112, and 58.6 mmol/kg, respectively (not shown in table; see Reeves et al., 2011). By comparison, the North Su magmatic-hydrothermal system has CO_2 values up to 82 mmol/kg (Seewald et al., 2014).

6. COMPARISON WITH MOR VENTING

The universally adopted method of regressing measured values of dissolved ions and gases in MOR vent fluids through $Mg = 0$ to derive "end-member" fluid compositions can also be applied to the water/rock-dominated systems of arcs and backarcs; select data are given in Table 47.3 and the results plotted in Figure 47.5. However, this method is not considered appropriate for calculating the magmatic-dominated (acid-sulfate) fluids commonly found in arc and backarc systems, as here both Mg and SO_4 have been *added* to these fluids, thus making any regression to zero tenuous. Therefore, the fluids least contaminated by seawater from these magmatic-hydrothermal systems are plotted with the other end-member values in Figure 47.5.

Some of the differences between MOR systems and those of arc and backarcs relate to their tectonic environment and include depth, host rock composition, and permeability. Half of arc systems have water depths of ≤ 500 m with ∼95% shallower than 1700 m, similar to most backarc systems, which occur in water depths between 500 and 2000 m. By contrast, the vast majority of MOR systems occur in water depths between 2200 and 3000 m, or deeper (de Ronde et al., 2003; Hannington et al., 2005). Basalts and lesser ultramafic rocks dominate at

MORs, in contrast to the basalt through basaltic andesite, andesite, dacite, and even rhyolite compositions found at arc volcanoes and in backarcs. Permeability at MORs is invariably influenced by seafloor spreading, with variable spreading rates affecting (1) magma supply to the ridges, which is reflected in the nature of the extrusive rocks and hence seafloor topography and (2) incidence of hydrothermal venting (e.g., Baker et al., 1996; German and Von Damm, 2003). At fast-spreading centers, high-temperature fluids circulate to depths of 1–2 km because of the relatively shallow magma chambers. By contrast, at intermediate rate and especially slow-spreading ridge centers, deeply penetrating faults allow circulation of hydrothermal fluids to depths of 5–8 km (Hannington et al., 2005, and references therein). In arcs and backarcs, extension (and accompanying faulting) also plays an important role in the location of individual volcanoes and indeed, in the general location of the hydrothermal systems. However, on a volcano scale, permeability is typically controlled by ring faults at caldera volcanoes, whereby hydrothermal activity is focused along discrete zones (e.g., Embley et al., 2012). By comparison, at cone volcanoes permeability is more widespread (diffuse), with the host rocks commonly dominated by blocky, autobrecciated lavas, pumice and volcaniclastic rocks as a result of gas-driven eruptions in shallow water depths (e.g., de Ronde et al., 2014b).

There are, however, several similarities between MOR (and other tectonic environments) hydrothermal systems and those associated with arcs and backarcs. These include eruptions on the seafloor, such as those at "hot spot" volcanoes, e.g., Surtsey and Macdonald seamounts, while segments of the MOR have erupted in recent times, such as at 9°N of the East Pacific Rise and Axial volcano and the north Cleft portion of the Juan de Fuca Ridge. The effects of these eruptions can be seen in the chemistry of the expelled hydrothermal fluids (see below).

Measured vent fluid pH values show a similar range for MOR, arc and backarc *water/rock-dominated systems*, with MOR and arc end-member fluids bottoming out at around $pH = 3$ and those for backarcs slightly lower (Figure 47.5(A)). By contrast, measured pH values for arc and backarc *magmatic-hydrothermal systems* extend down to $pH = 1.1$ and 0.9, respectively (Table 47.1(A)). Vent fluid temperatures for arc and backarc water/rock-dominated systems peak at 302 °C for a Brothers vent and 358 °C for a PACManus vent, respectively, while those for MORs are up to 400 °C in deeper water systems (Figure 47.5(B)). Again, the magmatic-hydrothermal systems differ, with a maximum vent fluid temperature of 215 °C for the backarc SuSu Knolls system and 256 °C for a single NW Rota-1 vent, which are still considerably lower than the highest values for corresponding water/rock-dominated systems. Even more compelling is the fact that the vast majority of vent fluid temperatures for arc

TABLE 47.2A Gas Compositions of Vent Fluids from Submarine Arc Volcanoes

Arc / Volcano	Vent	Dive No.^	Date Collected	Sample No.	T (°C)	Mg (mmol/kg)	Total Gas (mmol/kg)	R/Ra (Corr)	^{3}He (pmol/kg)	^{4}He (nmol/kg)	Ne (nmol/kg)	CO_2 (mmol/kg)	N_2 (µmol/kg)	H_2 (µmol/kg)	CH_4 (µmol/kg)	Ar (µmol/kg)	References
Izu-Bonin																	
Suiyo Seamount	Vent D	S2-630	07/92	2	311	4.4		8.3	2.9	253		34.0			106		1*
W/R				4	311	2.0		8.3	3.4	292		42.2			193		
	Vent F[a]	S2-631	07/92	2	300	0.8		8.3	2.7	234		37.7			107		
				3	300	0.8		7.9	2.7	243		34.8			94		
				4	300	0.6		8.2	4.2	370		38.4			113		
Mariana																	
Nikko	Station 2	H-496	10/31/05	GT15	112	38.0	68.0	6.8	2.7	290	18.8	58.7	1130	1.1	1.9	20.4	2[b]
M-H	Station 3			GT5	107	50.7	4.1			5.6	9.4	3.53	631	0.6	0.1	12.1	
				GT9	107	51.2				1.7	8.1						
	Deployment site	H-500	11/03/05	GT7	108	40.3	82.5	6.8	3.3	355	17.3	71.7	1320	2.4	1.7	25.3	
				GT6	108	40.7	88.8	6.8	3.7	386	20.5	67.6	15400	4.5	1.7	356	
		H-501	11/04/05	GT15	110	39.9	195	6.8	3.2	336	358	140	13200	4.7	0.4	32.3	
	Southern lone vent	H-500	11/03/05	GT16	90	47.3				154	273						
	Nikko			GT10	61–68	46.8	58.8			44.0	625	26.4	26300	39.6	0.2	407	
	N Nikko	J2-198	05/07/06	GT11	215	46.2	28.6	6.9	0.4	42.0	11.2	25.3	608	0.8	0.1	18.2	
NW Eifuku	Champagne	R-791	04/09/04	GT7	103	52.5	107	7.3	4.8	472	18.2	118	1010	0.1	13.8	19.4	2, 3[b]
M-H				GT9[c]	103	45.1	2296					2308					
				GT11[c]	103	43.9	2625					2711					
		R-793	04/11/04	GT5	68	46.9	255	7.3	3.9	379	32.5	254	1630	0.7	0.2	16.8	
		H-494	10/29/05	GT4	68	47.1	607	7.3	6.2	614	132	564	7350	108	1.3	58.3	
		H-497	11/01/05	GT6	108	46.2	420	7.3	6.2	613	27.0	405	1030	5.9	0.6	12.0	
		H-499	11/02/05	GT7	103	43.5	654	7.2	8.2	819	87.9	591	36000	12.3	1.7	1210	
		H-497	11/01/05	GT5	47	48.6	179	7.3	2.6	254	17.0	174	1010	1.8	0.4	12.6	
	Champagne second site	H-497	11/01/05	GT10	56–63	49.1	144	7.3	1.7	168	12.0	136	690	3.3	0.2	59.9	
				GT16	63	51.5	75.1	7.3	0.7	73.0	14.0	72.3	960	0.9	0.2	14.1	
	Cliff House	R-793	04/11/04	GT7	49	49.2	680	7.3				703					
		H-499	11/02/05	GT15	64	44.9	582	7.3	20.9	2070	29.0	567	1150	3.2	13.6	13.3	
	Sulfur dendrite	R-793	04/11/04	GT11	48	50.4	301	7.2	8.5	843	7.7	306	480	0.2	13.7	7.5	
	Diffuse site	R-791	04/09/04	GT7	49	49.2	107	7.3	4.6	454	30.1	76.0	1560	0.1	11.1	26.5	

(Continued)

Arc Volcano	Vent	Dive No.^	Date Collected	Sample No.	T (°C)	Mg (mmol/kg)	Total Gas (mmol/kg)	R/Ra (Corr)	^3He (pmol/kg)	^4He (nmol/kg)	Ne (nmol/kg)	CO_2 (mmol/kg)	N_2 (μmol/kg)	H_2 (μmol/kg)	CH_4 (μmol/kg)	Ar (μmol/kg)	References
Daikoku	Bottomless pit	H-491	10/26/05	GT7	16	51.1	14.2			205	205	2.5	10600	0.2	0.0	42.7	2[b]
M-H				GT6	16	50.8	7.6			93.0	92.6	2.0	4960	0.3	0.0	53.9	
	Bubble bath	J2-195	05/02/06	GT15	52	46.4	50.9	7.4	0.8	80.0	80.4	38.0	478	0.7	0.3	15.5	
	White smoker second orifice			GT2	210	48.7	15.7	7.4	0.3	33.0	8.9	14.2	543	0.4	0.3	18.8	
	Bubble field	J2-197	05/04/06	GT2	55	48.5	64.2			59.0	710	22.0	33500	11.6	0.2	140	
NW Rota-1	Fault shrimp	R-783	03/29/04	GT9	22	53.4	39.4	8.4	5.9	507	32.0	37.6	1340	0.1	3.4	26.5	2[b]
M-H	Scarp top			GT11	39	49.4	93.5	8.3	13.6	1180	21.0	93.5	1070	58.0	9.9	18.6	
	High flow			GT2	37	52.2	24.4	8.1	3.3	290	106	18.0	4750	3.0	0.6	24.8	
	Iceberg			GT7	45	51.3	81.0	8.3	13.0	1120	14.0	77.7	892	0.2	2.4	16.4	
		J2-187	04/23/06	GT6	50	51.5	22.5	8.3	3.0	263	97.0	21.9	611	0.1	0.8	23.1	
	Brimstone pit	R-786	04/02/04	GT11	28	51.5	25.6	8.2	2.0	172	6.0	25.0	411	154	0.2	10.2	
				GT9	26	51.5	23.5	8.2	3.4	296	15.0	21.5	719	293	0.2	10.4	
		J2-187	04/23/06	GT15	25	51.3	8.6	8.3	0.4	34.0	10.0	7.6	612	65.0	0.0	20.0	
				GT5	95	47.7	20.6	8.3	2.0	175	11.0	18.3	647	1291	0.0	21.9	
		J2-188	04/24/06	GT5	120	48.7	68.7	8.3	7.7	666	12.6	52.2	708	2187	0.0	22.5	
				GT15	95	47.6	110	8.2	4.2	370	82.0	80.3	514	1274	0.1	17.0	
		J2-189	04/25/06	GT6	120	51.2	50.3	8.3	4.9	425	50.0	47.0	314	1192	0.1	12.4	
	Shim sands	R-786	04/02/04	GT5	62	41.6	30.9	8.1	5.0	443	55.0						
Tonga																	
Volcano-1	Vent field	P4-141	06/24/05	GT12	50	50.6	120	6.6	3.6	395	7.9	108	5540		30.6		2[b]
M-H	Mussel field			GT10	68	51.9	137	6.6	4.4	479	8.9	133	649	14.0	21.6	14.9	
	Bubble site	P4-142	06/25/05	GT6	71	52.2	5.7	7.0	0.1	6.4	7.9	3.7	1630		0.5		
	Mussel bowl			GT2	150	51.5	5.3	6.9	0.7	73.0	1.9	4.7	451	0.0	0.0	14.9	
	Marker 43	R-1050	05/11/07	GT10	64	52.1	181	6.6	8.2	892	16.0	177	1167	0.6	26.7	20.5	
	Bubbles vent			GT5	39	54.8	37.5	6.7	0.2	26.0	9.8	37.0	492	0.4	0.3	16.4	
		R-1051	05/12/07	GT15	17	56.0	12.6	6.7	1.0	103	12.1	12.2	620	0.5	0.4	17.0	
	Sulfur vent	R-1053	05/14/07	GT10	36	55.2	44.3	6.7	2.2	242	14.3	42.4	774	0.1	28.6	18.5	
Kermadec																	
Monowai	Monowai	P5-613	04/08/05	GT7	54	47.4	11.5	7.1	1.2	125	13.2						4
W/R				GT10	57	46.5	15.4	7.2	1.7	169	26.8						
		P5-614	04/09/05	GT6	46	45.8	10.4	7.2	1.1	111	34.8						
				GT2	52	43.7	9.3	7.2	0.9	92.0	22.9						
				GT11	52	47.6	10.9	7.2	1.2	118	10.2						

Volcano	Type	Sample description	Sample	Date	ID													Ref
		Monowai caldera mussel	R-1043		GT7	25	50.1	4.3	7.3	0.2	23.0	9.6						
		Caldera mussel mkr 8			GT10	42	50.3	7.8	7.2	0.7	70.0	11.9						
Giggenbach	M-H	Diffuse SW flank	P5-618	04/15/05	GT2	72	47.4	4.6	7.4	0.5	49.0	27300						2[b]
		Summit mussel bed			GT11	70	52.4	4.4	7.5	0.5	48.0	9000	3.82	512	0.1	1.2	17.5	
		Giggenbach mk12	P5-619	04/16/05	GT12	203	34.2	500	7.4	155	15000	54500	361	3430	86.0	16.1	79.9	
					GT6	203	28.2	416	7.4	148	14300	59800						
					GT7	203	30.7	253	7.4	90.6	8760	66500	177	1066	63.0	15.0	45.0	
			P5-620	04/17/05	GT2	203	52.6	2.8	7.5	0.6			2.2	93	0.9	0.5	1.3	
		Giggenbach mk10			GT11	165	51.2	5.0	7.5	0.6	53.0	13100	4.4	656	1.4	0.6	14.9	
Brothers	W/R	NW Caldera	S6-851	10/26/04	1A	302	3.8	42.6	6.9	4.3	448	8.1	39.9	644	14.7	2.8		5
					3A	274	4.4	19.0	6.9	2.5	263	6.1	16.4	466	13.0	2.0		
			S6-852	10/27/04	1C	294	4.1	18.3	6.9	2.4	252	5.5	15.8	421	21.6	1.7		
					3	292	5.1	40.2	7.0	9.3	957	11.4	32.1	993	15.1	6.1		
			P5-626	05/02/05	MS Yellow	290	3.4	33.5	7.1	7.1	717	7.6	25.6	556	67.7	4.7		
					GT7 Black	290	21.8	20.1	7.2	6.5	651	8.0	16.3	587	4.9	3.0		
			P5-631	05/14/05	GT7 Black	290	32.4	16.0	7.2	8.4	818	10.0	13.2	930	3.8	2.6		
	M-H	Upper cone	S6-853	10/28/04	3A	60	45.6	22.0	7.2	0.5	52.0	7.6	19.4	589	4.1			
			P5-630	05/13/05	GT6 Blue	67	51.4	7.38	7.4	0.3	28.0	14.0	6.5	804	0.1			
		Lower cone	S6-853	10/28/04	1A	68	39.8	221	7.2	8.9	893	10.1	206	943	73.5			
			S6-854	11/01/04	1C	67	41.8	140	7.2	9.8	978	9.6	133	844	1.8			
			P5-630	05/13/05	GT10 Red	46	50.8	114	7.4	8.4	818	10.0	108	754	3.0	8.1		
Clark	W/R		P5-623	04/28/05	GT12	185	43.6	8.9	6.8	2.2	224	10.4						6

W/R refers to water/rock-dominated hydrothermal systems, and M-H refers to magmatic hydrothermal systems.

[1] Tsunogai, U., Ishibashi, J., Wakita, H., Gamo, T., Watanabe, K., Kajimura, T., Kanayama, S., Sakai, H., (1994). Peculiar Features of Suiyo Seamount Hydrothermal Fluids, Izu-Bonin Arc - Differences from Subaerial Volcanism. Earth Planet. Sc. Lett. 126, 289–301.

[2] Lupton, J., Lilley, M., Butterfield, D., Evans, L., Embley, R., Massoth, G., Christenson, B., Nakamura, K., Schmidt, M., (2008). Venting of a separate CO_2-rich gas phase from submarine arc volcanoes: Examples from the Mariana and Tonga-Kermadec arcs. J. Geophys. Res. 113, B08S12, http://dx.doi.org/10.1029/2007JB005467.

[3] Lupton, J., Butterfield, D., Lilley, M., Evans, L., Nakamura, K.I., Chadwick, W., Resing, J., Embley, R., Olson, E., Proskurowski, G., Baker, E., de Ronde, C., Roe, K., Greene, R., Lebon, G., Young, C., (2006). Submarine venting of liquid carbon dioxide on a Mariana Arc volcano. Geochem. Geophys. Geosy. 7, Q08007, http://dx.doi.org/10.1029/2005GC001152.

[4] Leybourne, M.I., Schwarz-Schampera, U., de Ronde, C.E.J., Baker, E.T., Faure, K., Walker, S.L., Butterfield, D.A., Resing, J.A., Lupton, J.E., Hannington, M.D., Gibson, H.L., Massoth, G.J., Embley, R.W., Chadwick, W.W., Clark, M.R., Timm, C., Graham, I.J., Wright, I.C., (2012). Submarine Magmatic-Hydrothermal Systems at the Monowai Volcanic Center, Kermadec Arc. Econ. Geol. 107, 1669–1694.

[5] de Ronde, C.E.J., Massoth, G.J., Butterfield, D.A., Christenson, B.W., Ishibashi, J., Ditchburn, R.G., Hannington, M.D., Brathwaite, R.L., Lupton, J.E., Kamenetsky, V.S., Graham, I.J., Zellmer, G.F., Dziak, R.P., Embley, R.W., Dekov, V.M., Munnik, F., Lahr, J., Evans, L.J., Takai, K., (2011). Submarine hydrothermal activity and gold-rich mineralization at Brothers Volcano, Kermadec Arc, New Zealand. Miner. Deposita 46, 541–584.

[6] de Ronde, C.E.J., Walker, S.L., Ditchburn, R.G., Caratori Tontini, F., Hannington, M.D., Merle, S.G., Timm, C., Handler, M.R., Wysoczanski, R.J., Dekov, V.M., Kamenov, G.D., Baker, E.T., Embley, R.W., Lupton, J.E. and Stoffers, P., (2014), The anatomy of a veiled submarine hydrothermal system, Clark volcano, Kermadec arc, New Zealand. Econ. Geol. 109, 2261–2292.

*These references reported concentrations per liter. Fluid density of 1.027kg/L was used for consistent units per personal communication with J-I. Ishibashi, 2014.

^R-ROPOS, H-HyperDolphin, P4-Pisces IV, P5 - Pisces V, J2- Jason II, S2-Shinkai 2000, S6- Shinkai 6500, Hy-Hayuko 2000

[a] applies to Lupton et al. data) Conventional gas chromatograph analyses (CO_2, CH_4, H_2, N_2, Ar) are precise to about ±5%. These ^3He and ^4He concentrations and ratios have had the effect of air addition subtracted out based on the Ne concentration and assuming that the added component had $He/Ne = (He/Ne)_{air} = 0.288$.

[b] Gas phase samples were gathered in one container (for gas composition measurement). The original concentration (C_0) are calculation with the formula $C_0 = C_l + C_g R$. C_l and C_g are liquid and gas phase concentrations respectively, R is liquid/gas volume ratio of sample. This affects CO_2, CH_4 and H_2S values presented.

[c] These samples had gas content more than the extraction system could easily handle, and some fractionation of the gases occurred; only reliable gas compositions are reported.

TABLE 47.2B Compositions of Magmatic-Hydrothermal Submarine Arc Volcano Gas Samples

Arc Volcano	Vent	Dive No.[^]	Date Collected	Sample No.	R/Ra (Corr)	^{4}He (ppm)	Ne (ppm)	CO_2 (%)	N_2 (%)	H_2 (%)	CH_4 (%)	Ar (%)	References
Mariana													
Nikko	*Nikko gas bubble*	H-500	11/03/05	GT17	6.9	18.5	0.11	96.5	1.06	0.0008	0.0033	0.0181	1
NW Eifuku	*Champagne droplets*	H-492	10/27/05	10cc#1	7.3	5.2	0.01	98.7	0.08	0.0003	0.0006	0.0003	1
		H-494	10/29/05	10cc#1	7.3	5.7	0.05	98.6	0.21	0.0002	0.0009	0.0010	
		H-497	11/01/05	10cc#2	7.3	5.4	0.05	99.6	0.23	0.0002	0.0009	0.0011	
		H-499	11/02/05	10cc#1	7.3	7.0	0.02	98.9	0.14	0.0002	0.0009	0.0013	
Daikoku	*Bubble bath*	J2-197	05/04/06	10cc#1	7.4	9.6	0.08	94.0	0.55	0.0018	0.0051	0.0147	1
NW Rota-1	*Brimstone pit*	J2-189	04/25/06	GT11	8.3	38.1	0.03	89.2	0.24	12.6000	0.0002	0.0064	1
	Brimstone	J2-188	04/24/06	GT2	8.3	32.4	0.03	92.5	0.22	10.7000	0.0000	0.0052	
Tonga													
Volcano-1	*Mussel field*	P4-141	06/24/05	Fl22	6.6	12.8	0.28						1
				Fl16	6.6	12.5	0.18	102	1.27	0.0253	0.0520	0.0151	
	Mussel bowl	P4-142	06/25/05	Fl22	6.6	11.3	0.41	100	1.90	0.0253	0.0460	0.0177	
				Fl17	6.6	11.5	0.45						
	Bubbles vent	R-1051	05/12/07	GT7	6.6	19.5	0.12	98.3	1.39	0.0009	0.0051	0.0213	
	Marker 43	R-1053	05/14/07	GT5	6.6	10.3	0.06	97.7	0.73	0.0006	0.0243	0.0088	
Kermadec													
Giggenbach	*Giggenbach mk12*	P5-619	04/16/05	Fl22				84.8	12.90	0.0247	0.0074	0.1290	1
				Fl17	7.5	67.0	3.60						
	Giggenbach	P5-620	04/17/05	GT12	7.4	53.4	1.56						
				GT6	7.4	56.6	1.38	77.5	4.81	0.0330	0.0134	0.0721	
				GT6	7.4	56.6	1.38	88.9	5.45	0.0226	0.0063	0.0725	

Note that some totals are greater than 100% due to accumulated error of analysis (±5%).
[1] *Lupton, J., Lilley, M., Butterfield, D., Evans, L., Embley, R., Massoth, G., Christenson, B., Nakamura, K., Schmidt, M., (2008). Venting of a separate CO₂-rich gas phase from submarine arc volcanoes: Examples from the Mariana and Tonga-Kermadec arcs. J. Geophys. Res. **113**, B08S12, http://dx.doi.org/10.1029/2007JB005467.*
[^] *R-ROPOS, H-HyperDolphin, P4-Pisces IV, P5-Pisces V, J2- Jason II, S2-Shinkai 2000, S6-Shinkai 6500, Hy-Hayuko 2000*

magmatic systems are less than the 120 °C of the Cone site at Brothers (Table 47.1(A)).

Plotting select elements against Cl shows some trends between the different hydrothermal systems. For example, the majority of MOR and water/rock-dominated backarc systems plot with a steeper trend on a diagram of Li versus Cl than do arc water/rock-dominated systems and a few MOR systems; arc and backarc magmatic-hydrothermal system vent fluids contain very little Li (Figure 47.5(C)). Similarly, a plot of K versus Cl shows some distinct trends. Arc and backarc water/rock-dominated systems mostly plot higher and separately from MOR systems, while the arc and backarc magmatic-hydrothermal systems have relatively little K in their vent fluids (Figure 47.5(D)). Calcium shows distinct trends for MOR

and arc water/rock-dominated systems over backarc water/rock-dominated systems that trend similarly, but which have lower concentrations of Ca; the magmatic systems again have the lowest concentrations of all the systems (Figure 47.5(E)). Most of these differences are due to the interaction between hydrothermal fluids and more differentiated rocks found in subduction settings than the typical basalts found at MORs.

The gas content of vent fluids from the various settings also show patterns when plotted against Cl. For example, H_2S is highest in MOR vent fluids that experienced volcanic eruption on the seafloor prior to being sampled, with some water/rock-dominated arc and similar backarc systems having the next highest concentrations. The equivalent MOR systems and arc and backarc magmatic-hydrothermal systems have the lowest concentrations of H_2S (Figure 47.5(F) and Table 47.1(A)). This last observation, when taken in concert with the abundance of sulfur associated with venting at these sites is consistent with reaction (47.1) above being the dominant mechanism of disproportionation at magmatic-hydrothermal systems.

More startling are the results for CO_2 in these different hydrothermal systems, with backarc water/rock-dominated vent fluids containing up to 274 mmol/kg for a PACManus vent (Figure 47.5(G)). At arc magmatic-hydrothermal systems, CO_2 concentrations extend up to 206 mmol/kg at Brothers Cone, 361 mmol/kg at Giggenbach, and five samples between 405 and 591 mmol/kg at NW Eifuku, with another two samples of 2308 and 2711 mmol/kg of CO_2, or more gas than the extraction system could handle at the time (Table 47.2(A)). Carbon dioxide concentrations in MOR vent fluids are at least an order of magnitude less.

Concentrations of Fe for water/rock-dominated arc (7.1 mmol/kg at Brothers NW Caldera; Table 47.3) and backarc (14.6 mmol/kg at PACManus; Table 47.3) systems, together with some magmatic-hydrothermal backarc systems (12.4 mmol/kg at DESMOS; Table 47.1(B)) are generally highest in expelled vent fluids. MOR exceptions are fluids from north and south Cleft of the Juan de Fuca Ridge (Figure 47.5(H)), which had erupted prior to sampling. However, most MOR vent fluids and those from arc magmatic-hydrothermal systems have Fe concentrations <3 mmol/kg (Table 47.1(A)). More striking are the trends for Mn versus Cl with the steepest slope for the Monowai arc water/rock-dominated system, followed by the backarc water/rock-dominated systems, the MOR systems, the remaining arc water/rock-dominated systems and finally the arc and backarc magmatic-hydrothermal systems vent fluids that contain very little Mn (Figure 47.5(I)).

7. DISCUSSION

Both arcs and backarcs are intrinsically tied to the process of subduction. It is here that volatiles are recycled back to the seafloor via volcanic eruption and hydrothermal venting via the emplacement of highly fractionated magmas. Thus, the fluids discharged from many arc and backarc systems will have characteristics indicative of magmatic input. For example, the disproportionation of SO_2 ensures a steady supply of acid, elemental sulfur, and sulfate (see Eqns (47.1)−(47.3) above) to the seafloor, resulting in remarkable manifestations such as lakes of molten sulfur, 6-m-tall chimneys of elemental sulfur, and an abundance of the Al-bearing sulfate natroalunite around the vents, in addition to venting of liquid CO_2 (Figure 47.3).

The release of large amounts of SO_2 and CO_2 also sets apart the magmatic-hydrothermal systems of arcs and backarcs from those at MORs (Figure 47.2; Tables 47.2(A) and (B) and 47.3). For example, mechanisms to produce

a separate CO_2-rich gas phase at the seafloor, as seen at Nikko, NW Eifuku, NW Rota-1, Volcano-1, Giggenbach, and Brothers Cone, require direct injection of magmatic CO_2-rich gas (Lupton et al., 2008). Moreover, calculations show that it is impossible at NW Eifuku to extract sufficient CO_2 from the rock into circulating seawater to form an aqueous fluid saturated with CO_2, indicating extreme CO_2 concentrations requiring direct degassing of CO_2 from a magma chamber, cooling and ascending to the seafloor, as observed by the CO_2-rich and H_2O-rich fluids of Lupton et al (2006). Similarly, δD and $\delta^{18}O$ analyses of vent fluid H_2O from arc and backarc vent sites clearly show a magmatic influence for some of the water, irrespective of the style of the hydrothermal system. For example, all of the backarc basin vent sites and the majority of the arc sites have δD values less than zero, with a clear trend in the data for backarc basin sites at least toward a field of subduction-related volcanic vapor (e.g., Hedenquist and Lowenstern, 1994; de Ronde, 1995). By comparison, the MOR data have values all above zero, with the exception of vent fluid analyses from north Cleft sampled after an eruption (Figure 47.6).

The depth to both arc and backarc hydrothermal vent sites is significantly shallower than at most MOR sites. This affects permeability, particularly at cone volcanoes, where the decrease in depth and pressure promotes degassing of the underlying volatile-rich magma, ensuring there are more explosive eruptions on the seafloor leading to the deposition of highly permeable lithologies, such as pyroclastic rocks. The vast majority of magmatic-hydrothermal systems along arcs and within backarcs are associated with cone volcanoes. The generally deeper caldera volcanoes along arcs more commonly host water/rock-dominated hydrothermal systems. Here, permeability is confined to faults along the caldera walls, with recharge zones to the hydrothermal systems utilizing the same fault system. The focusing of fluids in this setting results in the formation of, in order of increasing vent fluid temperature, Fe-oxyhydroxide, silica-rich, sulfate-rich, and massive sulfide chimneys (Figure 47.4). Depth also affects these systems in that the relatively shallow nature of these sites promotes phase separation, which is further enhanced when the hydrothermal fluids contain significant amounts of gas thereby allowing boiling to occur at lower temperatures (e.g., de Ronde, 1995). Phase separation is clearly reflected in the wide range of Cl concentrations for arc and backarc water/rock-dominated end-member fluids (Table 47.3). Comparison between end-member compositions of MORs and arc and backarc hydrothermal vent fluids show that while these fluids have phase separation and mixing with ambient seawater in common, they can still be differentiated from one another as a direct consequence of their tectonic setting (Figure 47.5).

Phase separation and enhanced volatile flux followed by the discharge of fluids with higher-than-seawater Cl

TABLE 47.3 End Member Compositions of Various Water/Rock-Dominated Vent Fluids

Tectonic Setting Vent/ Location	T (°C)	pH	Alkalinity (meq/kg)	Mg (mmol/kg)	Cl (mmol/kg)	SO₄ (mmol/kg)	H₂S (liq) (mmol/kg)	Si (mmol/kg)	Li (μmol/kg)	Na (mmol/kg)	K (mmol/kg)	Ca (mmol/kg)	Sr (μmol/kg)	Mn (mmol/kg)	Fe (mmol/kg)	Fe/Mn (mol/mol)	CO₂ (mmol/kg)	CH₄ (μmol/kg)	H₂S (gas) (mmol/kg)	³He	References
Submarine Arc Volcanoes																					
Myōjin-shō 1998	280	3.5	0.00	0	545		Nd	10.0		419	33.1	54.1		0.62	0.27	0.4					1*
Myōjin-shō 2012	235	4.0	−0.19	0	541		2.6	12.2		421	27.5	44.8		0.56	0.23	0.4					
Suiyo 1992	300	3.7	−0.20	0	641		Nd	13.2		431	29.0	86.9		0.57	0.39	0.7					
Suiyo 2000–2001	300	3.6	−0.40	0	651		1.4	13.0		425	26.1	87.8		0.64	0.42	0.7					
Suiyo 2007	250	3.7	−0.20	0	644		2.1	12.3		418	Nd	86.2		Nd	Nd						2
Monowai low Cl	52			0	280	4.0	36.7	19.7						2.90	0.00	0.0					
Monowai high Cl	56			0	600	0	23.4	20.2						9.30	0.00	0.0					
Brothers I	265–294	3.0	−1.18	0	737	0	1.6	14.0	828	528	73.0	44.5	186	0.80	7.10	8.9	17.3	2.0	2.2	2.7	3
Brothers II	302	3.1	−0.09	0	661	0	1.4	13.6	750	480	66.8	41.7	168	0.80	5.20	6.5	42.8	3.0	2.2	4.6	
Brothers III	290	2.8		0	597	0	3.9	13.9	682	433	58.3	40.8	163	0.70	5.50	7.9	26.9	5.1	7.2	13.5	
Brothers IV	290	3.8	−2.12	0	561	5.2	5.8	12.8	654	412	56.0	36.6	148	0.60	4.70	7.8	30.5	3.8	4.8	7.6	
Brothers V	290	2.8	−2.25	0	502	0	2.4	12.4	454	370	48.8	33.7	123	0.50	4.20	8.4	35.3	3.8	7.8	10.3	
Back Arc Vents																					
Okinawa JADE	350	4.7	1.83	0	536	0	13.2	12.6	1809	414	70.0	21.7	92.0	0.36	0.03	0.1					4*
Okinawa Hatoma Knoll	240	5.2	4.70	0	381			12.0		285	54.6	17.0	62.0	0.48							5
PACMANUS Roman Ruins	314	2.3		0	632		7.5	15.7	1130	477	81.7	19.8	76.3	4.00	6.50	1.6	17.5				6
	272	2.3					4.0	23.9	895	417	68.5	10.5	48.9	3.30	1.40	0.4	26.5				
	278	2.5					4.4	20.5	1320	541	96.8	25.5	83.7	4.80	7.80	1.6	10.1				
	341	2.6					6.8	19.0	1090	497	82.1	23.1	86.9	3.00	6.90	2.3	10.1				
PACMANUS Roger's Ruins	320	2.7		0	648		3.6	18.8	888	488	81.1	27.1	112	2.80	4.60	1.6	7.31				
	274	2.6		0	650		2.8	19.0	901	485	80.4	24.7	101	2.60	0.17	0.1	7.0				

Location															
PACMANUS Satanic Mills	295	2.6	0	517		9.4	14.5	769	397	69.9	12.8	56.4	2.60	3.40	1.3	212
	241	2.4	0	414		6.8	19.0	627	328	52.4	3.6	6.7	2.40	1.60	0.7	160
	288	2.5	0	503		10.2	15.1	726	387	68.0	13.7	65.7	2.30	1.20	0.5	274
PACMANUS Snowcap	152	4.6	0	441		7.0	16.4	714	343	46.4	0.5	−23.4	2.50	0.08	0.0	268
	180	3.4	0	526		1.9	17.5	893	408	55.9	3.4	24.9	3.00	0.27	0.1	187
PACMANUS Tsukushi	62	5.7	0	749			22.8	1330	505	81.3	23.7	104	3.90	0.81	0.2	24.1
PACMANUS Fenway	329	2.6	0	454		20.8	14.3	724	326	59.3	14.8	95.8	2.70	8.40	3.1	68.7
	343	2.7	0	699		10.1	15.2	1150	486	94.3	26.9	122	4.80	14.60	3.0	25.7
	358	2.7	0	562		18.8	12.2	917	397	76.1	22.3	95.9	3.80	11.80	3.1	56.1
	284	2.5	0	523		14.0	13.3	839	377	70.5	19.1	84.2	3.60	8.30	2.3	63.6
	80	4.9	0	388		17.2	11.3	599	240	42.1	31.8	176	2.30	4.60	2.0	86.4
North Fiji White Lady	285	4.7	0.12	248	0	2.0	13.6	195	204	10.2	6.3	29.2	0.12	0.13	1.1	14.0
Lau Basin Vai Lili	334	2.0	0	769	0	5.0	14.1	607	574	76.9	40.2	19.5	6.90	2.40	0.4	14.6

(North Fiji White Lady marked 7*)

Additional data on figure only:

MOR:

de Ronde, C.E.J. (1995) Fluid chemistry and isotopic characteristics of seafloor hydrothermal systems and associated VMS deposits: potential for magmatic contributions. In: Thompson, J.F.H. (ed) Magmas, fluids and ore deposits. Mineralogical Association of Canada Short Course Series, Victoria, British Columbia, pp 479–509.

Von Damm, K.L., Oosting, S.E., Kozlowski, R., Buttermore, L.G., Colodner, D.C., Edmonds, H.N., Edmond, J.M., Grebmeier, J.M. (1995). Evolution of East Pacific Rise Hydrothermal Vent Fluids Following a Volcanic-Eruption. Nature 375, 47–50.

Charlou, J.L., Fouquet, Y., Donval, J.P., Auzende, J.M., Jean-Baptiste, P., Stievenard, M., Michel, S. (1996). Mineral and gas chemistry of hydrothermal fluids on an ultrafast spreading ridge: East Pacific Rise, 17 degrees to 19 degrees S (Naudur cruise, 1993) phase separation processes controlled by volcanic and tectonic activity. J. Geophys. Res.- Sol. Ea. 101, 15899–15919.

Back arc (magmatic): Seewald, J.S. Reeves, E.P., Bach, W., Saccocia, P.J., Craddock, P.R., Shanks, W.C., Sylva, S.P., Pichler, T., Rosner, M., Walsh, E. (2014) Submarine venting of magmatic volatiles in the Eastern Manus Basin, Papau New Guinea. Geochim. Cosmochim. Ac. (in press).

Arc (magmatic): McMurtry, G.M., Sedwick, P.N., Fryer, P., Vonderhaar, D.L., Yeh, H.W., (1993). Unusual Geochemistry of Hydrothermal Vents on Submarine Arc Volcanos - Kasuga Seamounts, Northern Mariana Arc. Earth Planet. Sc. Lett. 114, 517–528.

[1]Ishibashi, J.I., Nagatomi, K., Takahashi, M., Kodamatani, H., Tomiyasu, T. Takeuchi, A., Yamanaka, T. (2014) Geochemistry of hydrothermal fluids collected from submarine volcanoes in the Izu-Bonin Arc. Presentation SRD45-04 at Japan Geoscience Union Meeting, 28 April – 2 May 2014, Yokohama, Japan.

[2]Leybourne, M.I., Schwarz-Schampera, U., de Ronde, C.E.J., Baker, E.T., Faure, K., Walker, S.L., Butterfield, D.A., Resing, J.A., Lupton, J.E., Hannington, M.D., Gibson, H.L., Massoth, G.J., Embley, R.W., Chadwick, W.W., Clark, M.R., Timm, C., Graham, I.J., Wright, I.C., (2012). Submarine Magmatic-Hydrothermal Systems at the Monowai Volcanic Center, Kermadec Arc. Econ. Geol. 107, 1669–1694.

[3]de Ronde, C.E.J., Massoth, G.J., Butterfield, D.A., Christenson, B.W., Ishibashi, J., Ditchburn, R.G., Hannington, M.D., Brathwaite, R.L., Lupton, J.E., Kamenetsky, V.S., Graham, I.J., Zellmer, G.F., Dziak, R.P., Embley, R.W., Dekov, V.M., Munnik, F., Lahr, J., Evans, L.J., Takai, K., (2011). Submarine hydrothermal activity and gold-rich mineralization at Brothers Volcano, Kermadec Arc, New Zealand. Miner. Deposita 46, 541–584.

[4]Ishibashi, J.I., Tsunogai, U., Wakita, H., Watanabe, K., Kajimura, T., Shibata, A., Fujiwara, Y., Hashimoto, J. (1994) Chemical composition of hydrothermal fluids from the Suiyo and the Mokuyo Seamounts, Izu-Bonin Arc. JAMSTEC J. Deep Sea Res. 10, 89–98.

[5]Kishida, K., Sohrin, Y., Okamura, K., Ishibashi, J., (2004). Tungsten enriched in submarine hydrothermal fluids. Earth Planet. Sc. Lett. 222, 819–827.

[6]Reeves, E.P., Seewald, J.S., Saccocia, P., Bach, W., Craddock, P. R., Shanks, W. C., Sylva, S. P., Walsh, E., Pichler, T., Rosner, M. (2011) Geochemistry of hydrothermal fluids from the PACMANUS, Northeast Pual and Vienna Woods hydrothermal fields, Manus Basin, Papua New Guinea. Geochim. Cosmochim. Ac. 75 (4), 1088–1123.

[7]Ishibashi, J.I. and Urabe, T. (1995) Hydrothermal activity related to arc back arc magmatism in the Western Pacific. In: Taylor, B. (ed) Back arc basins: tectonics and magmatism. Plenum Press, New York, pp 451–495.

[8]Von Damm, K. L., Edmond, J.M., Grant, B., Measures, C.I. (1985). "Chemistry of Submarine Hydrothermal Solutions at 21-Degrees-N, East Pacific Rise." Geochim. Cosmochim. Ac. 49 (11), 2197–2220.

*These references reported concentrations per liter. Fluid density of 1.027 kg/L was used for consistent units per personal communication with J-I. Ishibashi, 2014.

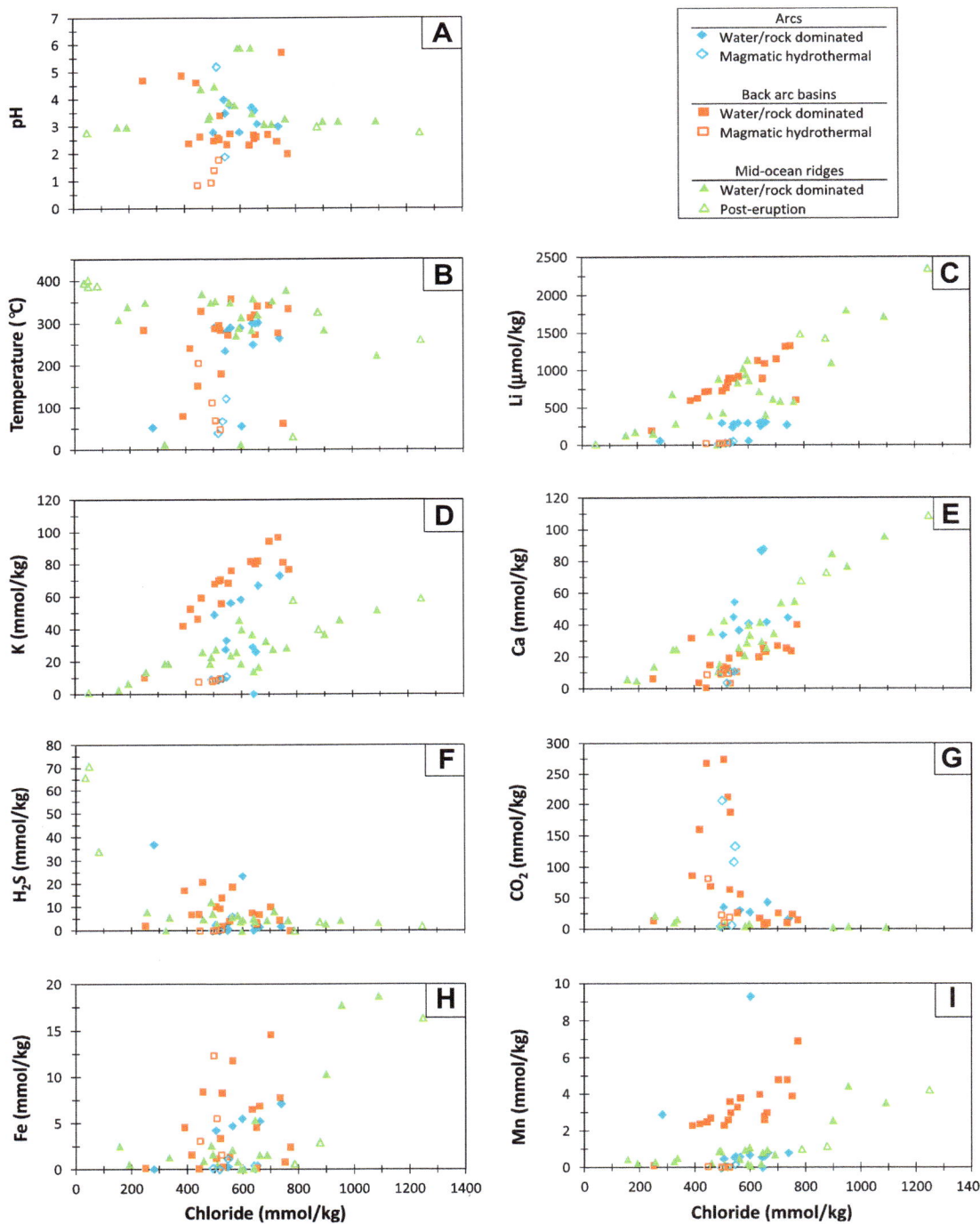

FIGURE 47.5 Representative end-member compositions (solid symbols) of select parameters (A) pH, (B) temperature, (C) lithium, (D) potassium, (E) calcium, (F) hydrogen sulfide, (G) carbon dioxide, (H) iron and (I) manganese versus chloride concentrations. Open arc and backarc symbols are for magmatic fluids that cannot be back calculated to a zero-magnesium end member (see text). For each site, the best sample (i.e., that which contains the most vent fluid) is shown for these vents. Open MOR symbols are for end-member concentrations of post-1991 eruption fluids. NW Rota-1 data are not plotted due to a lack of Cl data reported in the literature (see Table 47.1(A)).

concentrations (>1200 mmol/kg)—thought to temporarily reside beneath the seafloor—has been suggested as a normal evolution of expelled vent fluids following MOR volcanic eruptions on the seafloor (Butterfield et al., 1997).

However, fluids with Cl concentrations of >800 mmol/kg have not been reported for arc or backarc sites, although empirical evidence of Fe-oxyhydroxide crusts with "finger chimneys" expelling fluids denser than seawater have been

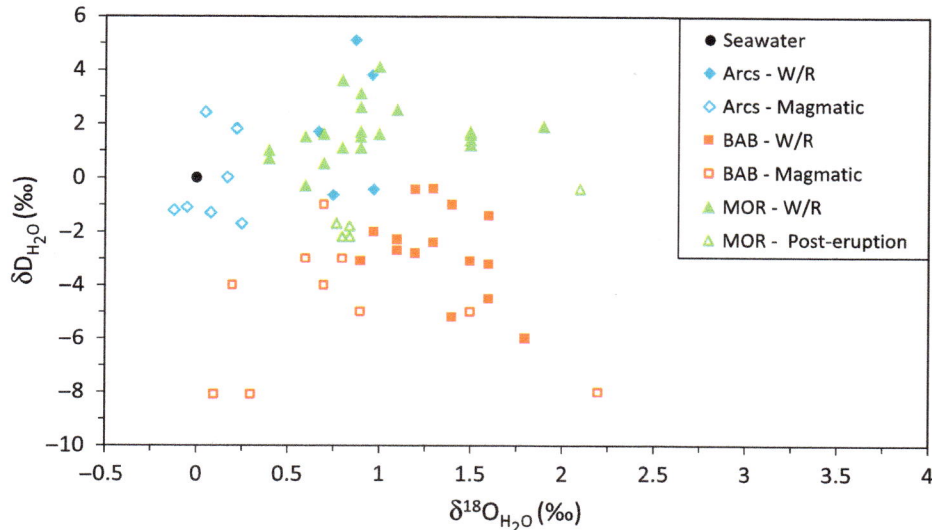

FIGURE 47.6 Plot of δD_{H_2O} versus $\delta^{18}O_{H_2O}$ (calibrated to VSMOW) for seafloor hydrothermal vent fluids. All the backarc and most of the arc vent fluid data have δD values plotting $<0‰$, with the reverse true for MOR data, except for posteruption fluids. The data plotted in the figure can be found in the references listed in the footnotes to Tables 47.1(B) and 47.3. BAB = Backarc basin; W/R = water/rock.

observed at the Brothers Cone site (de Ronde et al., 2005) and from a small chimney atop a sulfur-rich mound at NW Eifuku (C.E.J. de Ronde, pers. obs., 2004). Recently, Gruen et al. (2014) have shown from modeling the Brothers Cone site that metal—sulfide complexes will preferentially ascend during vapor-dominated fluid expulsion in the first few hundred years after emplacement of a hydrous magma. Denser magmatic "brine" with its chloride-complexed base metals is thought to be stored below the seafloor, where together with magmatically derived salt, significant amounts of metals such as Cu, Fe, Pb, and Zn may be also be deposited temporarily.

8. CONCLUSIONS

This chapter is intended to supplement that of Butterfield (2000) in the 1st edition of the *Encyclopedia of Volcanoes* that largely focused on hydrothermal venting at MORs. Here, we focus on venting at submarine volcanic arcs and backarcs (Figure 47.1), two tectonic settings that have received a growing amount of attention from researchers of deep-sea hydrothermal vents.

Arc hydrothermal seafloor vents contribute an equivalent of ~10% of the MOR global flux of hydrothermal emissions into the oceans, while backarcs contribute an unknown but likely similar amount. When combining significantly higher concentrations of gases such as SO_2 and CO_2 with other elements such as K, Mn, and Fe discharged from these sites, subduction-related hydrothermal systems are an important source of chemical emissions into our oceans.

The combination of variable depths and pressures to the arc hydrothermal systems and the input of magmatic volatiles and their ability to transport metals (e.g., Hedenquist and Lowenstern, 1994) has implications for the formation of polymetallic ore deposits. Depth clearly influences the maximum temperature at which the hydrothermal fluid will phase separate—a common mechanism in the precipitation of metals—ensuring a range in mineral deposit types, i.e., Cu- (and Au)-rich mineralization at deeper sites and Zn and Pb at shallower sites. Thus, the potential for significant mineralization to occur below the seafloor in arc volcanoes and in backarc environments is high (e.g., Gruen et al., 2014).

Given the greater range in chemistry found associated with arc and backarc hydrothermal systems compared to MORs, we might expect previously unknown chemoautotrophic microbes and species of associated animals to be found at these sites. Certainly, long-term eruptive activity as seen at NW Rota-1 has produced an unusual and extremely harsh chemical environment with a very unstable benthic habitat exploited by only a few mobile decapods, such as crabs and shrimps (Embley et al., 2006).

There are rare instances where we see both magmatic-hydrothermal and water/rock-dominated systems occurring at the same site. For example, at Brothers volcano, a significant magmatic-hydrothermal vent field (Cone site) occurs within 2 km of two major water/rock-dominated hydrothermal vent fields (NW Caldera and W Caldera sites) with a third, inactive field nearby. This highlights how these systems can coexist, with geophysical and geochemical results showing that they are likely connected at depth and have a common source for some elements.

In summary, the study of arc and backarc systems has clearly identified the role of subduction and the concomitant influence magmatic processes have on seafloor hydrothermal systems in these tectonic environments, ranging from the more common juvenile (or immature) magmatic-hydrothermal systems, through to the more evolved (or mature) water/rock-dominated systems.

ACKNOWLEDGMENTS

We wish to thank J. Ishibashi (Kyushu University), M.D. Lilley (University of Washington), J.S. Seewald (WHOI), and J.E. Lupton (NOAA/PMEL) who provided data from publications in preparation, or unpublished data for tables in this chapter. S.L. Walker and S.M. Merle (NOAA/PMEL) helped make Figure 47.1. Images courtesy of the New Zealand-Japan expedition to Brothers volcano, JAMSTEC and GNS Science; the New Zealand-American Submarine Ring of Fire 2005 Exploration, GNS Science, NOAA/OER/PMEL and NIWA; and the American Submarine Ring of Fire 2006 Exploration, NOAA/OER/PMEL, using the manned submersibles *Shinkai 6500* (JAMSTEC) and *Pisces V* (HURL), and the remotely operated vehicle, *Jason II* (WHOI). This chapter is benefitted from reviews by J. Stix and an anonymous reviewer. The authors were supported by public research funding from the Government of New Zealand.

FURTHER READING

Baker, E.T., Chen, Y.J., Phipps Morgan, J., 1996. The relationship between near-axis hydrothermal cooling and the spreading rate of mid-ocean ridges. Earth Planet. Sci. Lett. 142 (1–2), 137–145.

Baker, E.T., Massoth, G.J., de Ronde, C.E.J., Lupton, J.E., McInnes, B.I.A., 2002. Observations and sampling of an ongoing subsurface eruption of Kavachi volcano, Solomon Islands, May 2000. Geology 30, 975–978.

Baker, E.T., Embley, R.W., Walker, S.L., Resing, J.A., Lupton, J.E., Nakamura, K., de Ronde, C.E.J., Massoth, G.J., 2008. Hydrothermal activity and volcano distribution along the Mariana arc. J. Geophys. Res.: Solid Earth 113, B08S09. http://dx.doi.org/10.1029/2007 JB005423.

Butterfield, D.A., 2000. Deep ocean hydrothermal vents. In: Sigurdsson, H., Houghton, B., McNutt Stephen, R., Rymer, H., Stix, John (Eds.), Encyclopedia of Volcanoes. Academic Press, San Diego, pp. 857–875.

Butterfield, D.A., Jonasson, I.R., Massoth, G.J., Feely, R.A., Roe, K.K., Embley, R.E., Holden, J.F., McDuff, R.E., Lilley, M.D., Delaney, J.R., 1997. Seafloor eruptions and evolution of hydrothermal fluid chemistry. Philos. Trans. R. Soc. Lond. A 355, 369–386.

Butterfield, D.A., Nakamura, K., Takano, B., Lilley, M.D., Lupton, J.E., Resing, J.A., Roe, K.K., 2011. High SO$_2$ flux, sulfur accumulation, and gas fractionation at an erupting submarine volcano. Geology 39, 803–806.

Carey, R.J., Wysoczanski, R., Wunderman, R., Jutzeler, M., 2014. Discovery of the largest historic silicic submarine eruption. Eos, Trans. Am. Geophys. Union 95, 157–159.

Chadwick Jr., W.W., Wright, I.C., Schwarz-Schampera, U., Hyvernaud, O., Reymond, D., de Ronde, C.E.J., 2008. Cyclic eruptions and sector collapses at Monowai submarine volcano, Kermadec arc: 1998–2007. Geochem. Geophys. Geosyst. 9, Q10014. http://dx.doi.org/10.1029/2008GC002113.

de Ronde, C.E.J., 1995. Fluid chemistry and isotopic characteristics of seafloor hydrothermal systems and associated VMS deposits: potential for magmatic contributions. In: Thompson, J.F.H. (Ed.), Magmas, Fluids and Ore Deposits. Mineralogical Association of Canada Short Course Series, Victoria, British Columbia, pp. 479–509.

de Ronde, C.E.J., Baker, E.T., Massoth, G.J., Lupton, J.E., Wright, I.C., Feely, R.A., Greene, R.G., 2001. Intra-oceanic subduction-related hydrothermal venting, Kermadec volcanic arc, New Zealand. Earth Planet. Sci. Lett. 193, 359–369.

de Ronde, C.E.J., Massoth, G.J., Baker, E.T., Lupton, J.E., 2003. Submarine hydrothermal venting related to volcanic arcs. In: Simmons, S.F., Graham, I.J. (Eds.), Volcanic, geothermal, and ore-forming fluids: rulers and witnesses of processes within the earth. Littleton, Colo: Society of Economic Geologists. Special publication / Society of Economic Geologists, 10, pp. 91–109.

de Ronde, C.E.J., Hannington, M.D., Stoffers, P., Wright, I.C., Ditchburn, R.G., Reyes, A.G., Baker, E.T., Massoth, G.J., Lupton, J.E., Walker, S.L., Greene, R.R., Soong, C.W.R., Ishibashi, J., Lebon, G.T., Bray, C.J., Resing, J.A., 2005. Evolution of a submarine magmatic-hydrothermal system: Brothers volcano, southern Kermadec arc, New Zealand. Econ. Geol. 100, 1097–1133.

de Ronde, C.E.J., Massoth, G.J., Butterfield, D.A., Christenson, B.W., Ishibashi, J., Ditchburn, R.G., Hannington, M.D., Brathwaite, R.L., Lupton, J.E., Kamenetsky, V.S., Graham, I.J., Zellmer, G.F., Dziak, R.P., Embley, R.W., Dekov, V.M., Munnik, F., Lahr, J., Evans, L.J., Takai, K., 2011. Submarine hydrothermal activity and gold-rich mineralization at Brothers Volcano, Kermadec Arc, New Zealand. Miner. Deposita 46, 541–584.

de Ronde, C.E.J., Chadwick, W.W., Jr., Ditchburn, R.G., Embley, R.W., Tunnicliffe, V., Baker, E.T., Walker, S.L., Ferrini, V.L., and Merle, S.M. Molten sulfur lakes of intraoceanic arc volcanoes. In: Rouwet, D., Christenson, B., Tassi, F., and Vandelbroulemuck, J. (Eds.), Volcanic Lakes, Springer-Verlag, in press.

de Ronde, C.E.J., Walker, S.L., Ditchburn, R.G., Caratori Tontini, F., Hannington, M.D., Merle, S.G., Timm, C., Handler, M.R., Wysoczanski, R.J., Dekov, V.M., Kamenov, G.D., Baker, E.T., Embley, R.W., Lupton, J.E., Stoffers, P. The anatomy of a buried submarine hydrothermal system, Clark volcano, Kermadec arc, New Zealand. Econ. Geol. 109, 2261–2292.

Embley, R.W., Chadwick Jr., W.W., Baker, E.T., Butterfield, D.A., Resing, J.A., de Ronde, C.E.J., Tunnicliffe, V., Lupton, J.E., Juniper, K.S., Rubin, K.H., Stern, R.J., Lebon, G.T., Nakamura, K., Merle, S.G., Hein, J.R., Wiens, D.P., Tamura, T., 2006. Long-term eruptive activity at a submarine arc volcano. Nature 441, 494–497.

Embley, R.W., de Ronde, C.E.J., Merle, S.G., Davy, B., Caratori Tontini, F., Yoerger, D., 2012. Detailed morphology and structure of an active submarine arc caldera: Brothers volcano, Kermadec arc. Econ. Geol. 107, 1557–1570.

Fiske, R.S., Cashman, K.V., Shibata, A., Watanabe, K., 1998. Tephra dispersal from Myojinsho, Japan, during its shallow submarine eruption of 1952–1953. Bull. Volcanol. 59, 262–275.

German, C.R., Livermore, R.A., Baker, E.T., Bruguier, N.I., Connelly, D.P., Cunningham, A.P., Morris, P., Rouse, I.P., Statham, P.J., Tyler, P.A., 2000. Hydrothermal plumes above the East Scotia Ridge: an isolated high-latitude back-arc spreading centre. Earth Planet. Sci. Lett. 184, 241–250.

German, C., Von Damm, K.L., 2003. Hydrothermal processes. In: Turekian, K.K., Holland, H.D. (Eds.), The Treatise on Geochemistry, vol. 6.07. Elsevier, pp. 181–222.

Gruen, G., Weis, P., Driesner, T., Heinrich, C.A., and de Ronde, C.E.J. Hydrologic controls on different styles of magmatic-hydrothermal activity with implications for ore deposit formation at submarine arc volcanoes. Earth Planet. Sci. Lett. 404, 307–318. http://dx.doi.org/10.1016/j.epsl.2014.07.041.

Hannington, M.D., de Ronde, C.E.J., Petersen, S., 2005. Modern sea-floor tectonics and submarine hydrothermal systems. Econ. Geol. 100th Ann. Vol. 111–141.

Hedenquist, J.W., Lowenstern, J.B., 1994. The role of magmas in the formation of hydrothermal ore deposits. Nature 370, 519–527.

Hedenquist, J.W., Arribas, R., Gonzalez-Urien, E., 2000. Exploration for epithermal gold deposits. Rev. Econ. Geol. 13, 245–277.

Lupton, J., Butterfield, D., Lilley, M., Evans, L., Nakamura, K.I., Chadwick, W., Resing, J., Embley, R., Olson, E., Proskurowski, G., Baker, E., de Ronde, C., Roe, K., Greene, R., Lebon, G., Young, C., 2006. Submarine venting of liquid carbon dioxide on a Mariana arc volcano. Geochem. Geophys. Geosyst. 7, Q08007. http://dx.doi.org/10.1029/2005GC001152.

Lupton, J., Lilley, M., Butterfield, D., Evans, L., Embley, R., Massoth, G., Christenson, B., Nakamura, K., Schmidt, M., 2008. Venting of a separate CO_2-rich gas phase from submarine arc volcanoes: examples from the Mariana and Tonga-Kermadec arcs. J. Geophys. Res. 113, B08S12. http://dx.doi.org/10.1029/2007JB005467.

Massoth, G., Baker, E., Worthington, T., Lupton, J., de Ronde, C., Arculus, A., Walker, S., Nakamura, K., Ishibashi, J., Stoffers, P., Resing, J., Greene, R., Lebon, G., 2007. Multiple hydrothermal sources along the south Tonga arc and Valu Fa Ridge. Geochem. Geophys. Geosyst. 8, Q11008. http://dx.doi.org/10.1029/2007GC001675.

Reeves, E.P., Seewald, J.S., Saccocia, P., Bach, W., Craddock, P.R., Shanks, W.C., Sylva, S.P., Walsh, E., Pichler, T., Rosner, M., 2011. Geochemistry of hydrothermal fluids from the PACMANUS, Northeast Pual and Vienna Woods hydrothermal fields, Manus Basin, Papua New Guinea. Geochim. Cosmochim. Acta 75, 1088–1123.

Resing, J.A., Lebon, G., Baker, E.T., Lupton, J.E., Embley, R.W., Massoth, G.J., Chadwick, W.W., de Ronde, C.E.J., 2008. Venting of acid-sulfate fluids producing a high-sulfidation setting at NW Rota-1 submarine volcano on the Mariana arc. Econ. Geol. 102, 1,047–1,061.

Resing, J.A., Rubin, K.H., Embley, R.W., Lupton, J.E., Baker, E.T., Dziak, R.P., Baumberger, T., Lilley, M.D., Huber, J.A., Shank, T.M., Butterfield, D.A., Clague, D.A., Keller, N.S., Merle, S.G., Buck, N.J., Michael, P.J., Soule, A., Caress, D.W., Walker, S.L., Davis, R., Cowen, J.P., Reysenbach, A., Thomas, H., 2011. Active submarine eruption of boninite in the northeastern Lau Basin. Nat. Geosci. 4, 799–806.

Seewald, J.S., Reeves, E.P., Bach, W., Saccocia, P.J., Craddock, P.R., Shanks, W.C., Sylva, S.P., Pichler, T., Rosner, M., and Walsh, E. Submarine venting of magmatic volatiles in the Eastern Manus Basin, Papau New Guinea. Geochim. Cosmochim. Acta, in press.

Stoffers, P., Worthington, T.J., Schwarz-Schampera, U., Hannington, M.D., Massoth, G.J., Hekinian, R., Schmidt, M., Lundsten, L.J., Evans, L.J., Vaiomo'unga, R., Kerby, T., 2006. Submarine volcanoes and high-temperature hydrothermal venting on the Tonga arc, southwest Pacific. Geology 34, 453–456.

Von Damm, K.L., Edmond, J.M., Grant, B., Measures, C.I., 1985. Chemistry of submarine hydrothermal solutions at 21-Degrees-N, east Pacific rise. Geochim. Cosmochim. Acta 49, 2197–2220.

Chapter 48

Volcano-Related Lakes

Pierre Delmelle

Earth & Life Institute, Environmental Sciences, Université Catholique de Louvain, Louvain-la-Neuve, Belgium

Richard W. Henley

Research School of Earth Sciences, Australian National University, Canberra, ACT, Australia

Alain Bernard

Earth & Environmental Sciences Department, Université Libre de Bruxelles, Brussels, Belgium

Chapter Outline

GLOSSARY

caldera A very large surface depression formed by the collapse of the volcanic edifice above a magma chamber.

composite volcano A relatively large, long-lived constructional volcanic edifice, consisting of intermixed lava flows and pyroclastic material erupted from one or more vents.

hydrothermal system A groundwater system that has a source (or area) of recharge, a source (or area) of discharge, and a heat source. Hydrothermal systems associated with volcanoes are termed volcanic hydrothermal systems.

isotope Atoms with the same number of electrons and protons, but different number of neutrons and therefore, different masses. Most chemical elements occur in nature as two or more isotopes.

maar A maar is a particular type of volcanic crater which has low relief, is roughly circular, and has a flat floor. A maar forms when fissures of magma from beneath Earth's surface meet groundwater, causing volcanic phreatic and phreatomagmatic eruptions.

meteoric water A type of water of any age that originates by precipitation from the atmosphere. They include rain, snow, ice, rivers, lakes, and most low-temperature groundwaters.

phreatic eruption A phreatic eruption is a steam explosion caused by the vaporization of water due to heat generated by proximity to a body of magma or hot rock. A phreatic eruption lacks direct contact with magma and the eruption comprises fragments of preexisting country rock.

phreatomagmatic eruption A volcanic eruption which involves direct interaction between magma and external bodies of water such as the sea, lakes, or groundwater. It differs from a magmatic eruption which is driven by volatiles dissolved within magma.

volcanic hydrothermal fluids A type of fluid that comprises hot water, steam, and volcanic gases in motion in fractured and porous rocks beneath the surface. They receive heat released from a magma at depth, and their composition is governed by magmatic gas inputs, alteration reactions involving the host rock and physical processes such as evaporation and condensation. The surface manifestations of hydrothermal fluids include fumaroles, hot springs, and steaming grounds.

The Encyclopedia of Volcanoes. http://dx.doi.org/10.1016/B978-0-12-385938-9.00048-1

1. INTRODUCTION

According to The Catalog of Active Volcanoes of the World, about 16% of the 714 Holocene ($\sim$10,000 years old) or younger volcanoes contain a lake, but this number is probably underestimated and may also vary with time due to the ephemeral nature of such systems (Rouwet et al., 2014). Volcano-related lakes (Figure 48.1) are found mostly along continental margins (e.g., Central America) or island arcs (e.g., Indonesia), but some, such as Lake Mývatn in Iceland, sit astride the Mid-Atlantic Ridge. Although largely unrepresented in scientific reports, the rift valleys of East Africa also host volcano-related lakes. Volcano-related lakes occur in craters formed by hydrothermal, phreatic, phreatomagmatic, or magmatic explosions and, more rarely, occupy valleys blocked by ash, lava, or debris flows. Large **calderas**, such as the 350 km^2 Yellowstone Lake, USA, act as catchments and drown geothermal features such as hot springs. The warm lake of Grímsvötn, the most active volcano in Iceland, lies under the 250 to 269-m-thick ice sheet of Vatnajökull glacier.

Many volcano-related lakes are extremely hazardous and require constant monitoring. At Kelud, Indonesia, over

FIGURE 48.1 Photos showing the different types of volcano-related lakes: (A) a hyperacidic lake, Ijen lake, Kawah Ijen volcano, Indonesia *(photo courtesy: Dan Kali)*; (B) an acidic lake, Main Crater Lake, Taal volcano, Philippines; (C) a geothermal lake, Champagne Pool, Waiotapu geothermal area, New Zealand *(photo courtesy: Clive Oppenheimer)*; (D) a gas-rich lake, Lake Nyos, Cameroon, (E) a caldera lake, Crater Lake, Mt Mazama, USA *(photo courtesy: Clive Oppenheimer)*.

15,000 people have been killed in separate debris-flow events over the last 450 years due to eruptions within the lake. Over 1700 villagers in Cameroon perished by suffocation in the CO_2-rich cloud expelled from lakes Monoun and Nyos in 1984 and 1986, respectively. Understanding the processes that govern the distinctive character and behavior of each lake therefore underpins the ability of volcanologists to monitor volcano-related lakes and warn of impending hazards.

All volcano-related lakes are connected directly or indirectly to the release of high-temperature gases from magma intrusions within their parent volcanic system. The most spectacular are the hot, very low pH hyperacidic lakes such as Kawah Ijen, Indonesia, with very high concentrations of dissolved material and abundant suspended particles that give rise to distinctive coloration. Other volcano-related lakes can be cooler and contain near-neutral pH waters with thriving ecosystems.

Volcano-related lakes are recognizable through geologic history, and many are closely associated with the formation of some economically important deposits of gold and diamonds. Some also have been mined for native sulfur. As eroded remnants of ancient lake systems, these deposits provide insight to the otherwise inaccessible regions beneath modern lakes and the processes that occur as magmatic gas expands from its source to the surface of a volcano.

There are four broad groupings of volcano-related lakes **(hyperacidic and acidic lakes, geothermal lakes, gas-rich lakes and caldera lakes)** whose characteristics and behavior relate to the relative proportions of volcanic gas and surface-derived waters that enter the lake. In this chapter, we focus on the principal processes that are responsible for their formation and show how these can be linked to monitoring and mitigation of their hazards. Figure 48.2 shows the locations of the volcano-related lakes that are mentioned in this chapter.

2. CHARACTERISTICS OF VOLCANO-RELATED LAKES

The escape of high-temperature gases from intrusions provides the heat and many of the raw materials for the chemistry of lakes hosted by active volcanic systems. The gas mixture released by crystallizing magmas is dominated by water (85—95 mol %) with the remainder made up of carbon dioxide (CO_2) and sulfur dioxide (SO_2), and lower concentrations of hydrogen chloride (HCl), hydrogen sulfide (H_2S), hydrogen fluoride (HF), hydrogen (H_2), and other gases. Most of the heat transferred by this gas mixture is due to the large water component while the acidic gases (SO_2 and HCl) are primarily responsible for the distinctive chemistry of acidic lakes. In lakes associated with geothermal activity in volcano-related **hydrothermal systems**, degassing of magmas occur at depths of several kilometers so that the acidity of this gas mixture is reduced by reaction with rock and dilution with groundwater, resulting in the distinctive characteristics of the lakes which consequently form.

2.1. Hyperacidic and Acidic Lakes

Condensation of gas mixtures near the surface and on entry into a cool lake basin results in the formation of acidic lake water (Christenson and Wood, 1993; Figure 48.3). Sulfur dioxide gas reacts with water to form sulfuric acid (H_2SO_{4aq}) and hydrogen sulfide (H_2S_{aq}) through a simple reaction such as

$$4\,SO_{2(g)} + 4\,H_2O \Leftrightarrow 3\,H_2SO_{4(aq)} + H_2S_{(aq)} \qquad (48.1)$$

Here the subscript "aq" indicates a chemical species that is dissolved in the lake water, and "g" is gas. The acid thus

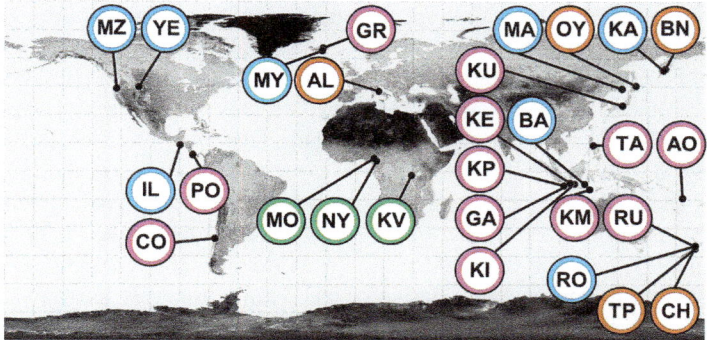

FIGURE 48.2 Geographical locations of the volcano-related lakes mentioned in this chapter. Color code, pink: hyperacidic and acidic crater lakes; orange: geothermal lakes; green: gas-rich lakes; blue: caldera lakes. *AL*, —Lake Albano; *AO*, Lake Voui, Aoba volcano; *BA*, Lake Batur, Mt. Batur; *BN*, Bannoe, Uzon caldera; *CH*, Champagne Pool, Waiotapu geothermal area; *CO*, Copahue, Copahue volcano; *GA*, Galunggung, Mt Galunggung; *GR*, Grímsvötn, Grímsvötn volcano; *IL*, Ilopango, Ilopango caldera; *KA*, Lake Karimsky, Karimsky volcano; *KE*, Kelud, Mt Kelud; *KI*, Kawah Ijen, Ijen volcano; *KM*, Keli Mutu lakes, Keli Mutu volcano; *KP*, Kawah Putih, Mt. Patuha; *KU*, Yugama Lake, Kusatsu-Shirane volcano; *KV*, Lake Kivu, Virunga volcanic chain; *MA*, Lake Mashu; Mt Mashu; *MO*, Lake Monoun, Oku volcanic field; *MY*, Myvátn, Krafla volcano; *MZ*, Crater Lake, Mt Mazama; *NY*, Lake Nyos; Oku volcanic field; *OY*, Oyunuma, Kuttara volcano; *PO*, Laguna Caliente, Poás volcano; *RO*, Lake Rotorua, Okataina volcanic centre; *RU*, Crater Lake, Mt. Ruapehu; *TA*, Main Crater Lake, Taal volcano; *TP*, Rotokawa, Taupo volcanic zone; *YE*, Yellowstone Lake, Yellowstone caldera.

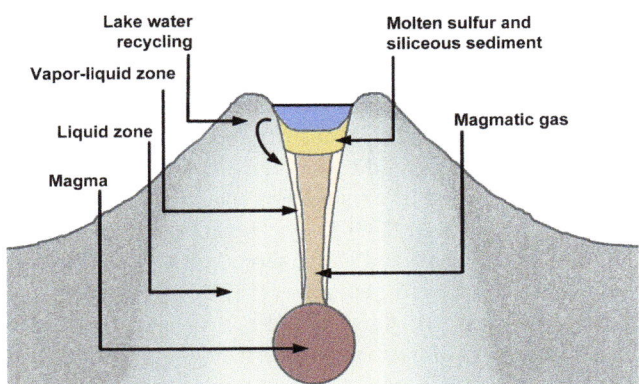

FIGURE 48.3 Cartoon diagram illustrating the association between a hyperacidic crater lake and a hydrothermal system powered by a shallow magmatic intrusion.

formed reacts aggressively with rocks below and around the lake and is responsible for the suspended burden of amorphous silica, sulfur, and anatase (titanium oxide, TiO_2) particles that give rise to spectacular lake coloration (Delmelle et al., 2000; Figure 48.1).

The large sulfur isotopic fractionation associated with reaction Eqn (48.1) typically leads to $H_2SO_{4(aq)}$ that is strongly enriched in the heavy sulfur **isotope** (^{34}S). *Thiobacillus thiooxidans,* a sulfur-oxidizing bacteria, may contribute to sulfur chemistry in active crater lakes, but pH values below one inhibit its growth in the lake waters. Crater lakes that receive a substantial input of hot volcanic gas show a distinctive water isotopic composition, characterized by high deuterium (D) and oxygen18 (^{18}O) contents (Varekamp and Kreulen, 2000). These lakes are usually hot due to intense evaporation and mixing with magmatic water and represent the most acidic and mineralized natural waters on Earth; they can display hyperacidic pH values below one and total dissolved solid (TDS) contents greater than 100,000 mg/kg (Table 48.1). The spectacular TDS values are primarily due to intense water—rock interaction in the subsurface two-phase (liquid plus vapor) region depicted in Figure 48.3, although acid dissolution of fresh rock materials brought to the lake by rock-fall and runoff may also contribute to the dissolved cation load.

The sulfur chemistry of hyperacidic lakes is remarkable (Delmelle and Bernard, 2015), as epitomized by the presence of subaqueous liquid sulfur. Where the total concentration of sulfur in the lake waters is high enough, native sulfur ($S°$) can precipitate as fine particles that settle to form sulfur-rich sediment. This sulfur deposit may be impacted by hot gas flow from below and eventually melt to form a pool on the lake bottom. Molten sulfur bodies in hyperacid crater lakes may also form due to $S°$ deposited directly from the sulfur-bearing magmatic gas phase. Probing of liquid sulfur bodies has been achieved at a few volcanoes, revealing temperatures between ~ 120 and

170 °C. The size of these subaqueous molten $S°$ bodies is poorly constrained. When streaming of hot gas continues through a liquid sulfur pool, small sulfur blisters may develop and float independently to the surface as sulfur globules. Since liquid sulfur can also dissolve $SO_{2(g)}$ and $H_2S_{(g)}$, exsolution of these gases from the sulfur as it cools transforms the blisters into spherical globules that are often seen floating on the lake surface (Figure 48.4). Other evidence for the presence of molten sulfur in volcano-related lakes comes from the recovery of sulfur pyroclasts ejected by **phreatic and phreatomagmatic eruptions**.

In some active volcanoes, groundwater drowns the magmatic heat and gas flow to the surface so that only mild gas inputs to the lake basin are observed. The lake water may be only weakly acidic and the volcano may be interpreted as quiescent or even totally dormant, when in fact intense alteration is occurring inside the edifice. Examples of such lakes are found in the volcanic craters of Taal, Philippines (Figure 48.1), and Kelud, Indonesia.

2.2. Geothermal Lakes

Large-scale geothermal systems, several kilometers in extent, form where magmatic gas is released and dissolved into groundwater-saturated rock at depths of several kilometers. The acidity of the magmatic gas is neutralized by reaction with the rock to form a range of temperature-dependent minerals including clays. The $SO_2(g)$ is converted to $H_2S(g)$ by hydrolysis and reaction with aluminosilicate minerals and in turn reacts with ferrous iron in the rock to form wide zones of disseminated iron sulfide (pyrite, FeS_2). Subsurface geothermal activity is manifested by a variety of thermal features including steaming ground, boiling springs, and low-enthalpy fumaroles, and many such systems have been assessed as possible geothermal energy resources. Catchment lakes are common in geothermal areas but small lakes formed by hydrothermal eruptions also occur. These lakes and pools are formed where CO_2 pressure increases along fault lines leading to eruption when pressure exceeds the overlying rock pressure. They often have streams of CO_2 bubbles, as is particularly spectacular in the apt-named Champagne Pool at Waiotapu, New Zealand (Figure 48.1). Such eruption lakes are commonly surrounded by silica sinters and terraces, and sometimes have distinct layers enriched in arsenic, antimony, mercury as well as gold and silver. Some geothermal lakes host acid sulfate-rich waters when oxidation of aqueous H_2S by atmospheric oxygen or by oxygen dissolved in water generates $H_2SO_{4(aq)}$ (e.g., Oyunuma and Bannoe in Table 48.1). Native sulfur is also produced by aqueous H_2S oxidation in some geothermal lakes such as Rotokawa, New Zealand.

TABLE 48.1 Examples of Water Lake Composition mg/kg for the Different Types of Volcano-Related Lakes

Lake	Ijen	Laguna Caliente	Kelud	Oyunuma	Lake Nyos	Lake Nyos	Crater Lake
Lake Type	Hyperacidic	Hyperacidic	Weakly acidic	Geothermal	Gas-Rich	Gas-Rich	Caldera Lake
Volcano	Kawah Ijen	Poás	Kelud	Kuttara	Nyos	Nyos	Mt Mazama
Sampling depth	0 m	0 m	0 m	0 m	0 m	205 m	200 m
date	August 1996	June 1989	December 1993	October 1995	March 1995	March 1995	1991
Surface area ($\times 10^3$ m^2)	390	~20	240	20	1580		53,000
Maximum depth (m)	170	~5	33	26	209		589
Volume ($\times 10^6$ m^3)	32	~0.03	1.8	–	150		17,300
Temperature (°C)	36	87	41	51	26		9
pH	0.3	–0.4	5.9	2.7	7.1	5.1	7.0
TDS#	106,950	249,500	3430	1030	65	1470	100
Na	1160	2720	720	100	2.8	66	11
K	1470	1730	92	12	1.4	7.0	1.7
Mg	630	2040	55	<1	3.3	90	2.2
Ca	970	650	100	60	3.8	55	6.6
B	53	83	–	22	<0.1	<0.1	<0.5
Al	5410	13,700	<0.1	19	–	–	–
SiO$_2$	160	440	240	156	16	81	20
Fe	2060	5770	<0.1	4.4	<0.1	120	–
Mn	40	130	1.7	0.5	<0.1	1.7	–
SO$_4^{2-}$	71,300	167,000	660	440	<1	<1	8
F$^-$	1050	8960	7	–	–	–	0.1
Cl$^-$	22,630	46,300	1290	221	<2	<1	9.8
HCO$_3^-$	–	–	260	–	38	1040	41
δD	0.8	–3.3	–2.6	–	–10.8*	–11.9*	–7.9
δ^{18}O	8.7	7.3	–1.8	–	–2.6*	–3.0*	–9.8
δ^{34}S$_{SO_4}$	22.3	12.3	15.1	2.2	–	–	–

Isotopic compositions are in delta (δ) units per mil (‰); TDS#: total dissolved solids; *: measured in October 1993. Note that the water composition of a lake may change with time depending on volcanic and meteorological factors. For example, the composition reported here for Poás corresponds to an episode of intense heat and volatile inputs to the Crater lake.

FIGURE 48.4 Sampling of sulfur globules from the shore of Kawah Ijen hyperacidic crater lake. The inset is a scanning electron microscopy photograph (false color) of the inside of a sulfur globule.

2.3. Gas-Rich Lakes

The occurrence of gas-rich volcano-related lakes (Figure 48.1) is not very common. However, Lake Monoun and Lake Nyos, two lake-hosted craters along the Cameroon volcanic line, became tragically famous in the mid-1980s when they suddenly expelled a lethal cloud of CO_2 which engulfed several villages in nearby valleys (Kling et al., 1987). Subsequent geochemical surveys revealed that both lakes are fed by volcanic CO_2-rich springs at the crater floors. The pressure of the overlying water column keeps the gas dissolved and since the lakes are strongly stratified, a situation where the bottom and surface waters do not mix, $CO_{2(aq)}$ can accumulate in tremendous amounts in bottom waters. Lake Nyos has a maximum depth of 209 m and $CO_{2(aq)}$ is overwhelmingly the most abundant chemical species in the lake. Between 1986 and 2001, the annual mean rate of CO_2 accumulation in Lake Nyos was $\sim 5.3 \times 10^6$ kg/year. The presence of $CO_{2(aq)}$ and carbonic acid ($H_2CO_{3(aq)}$) also explains the acidic pH values measured in bottom waters (Table 48.1). Furthermore, anoxic conditions in the lake water column allow iron to exist as ferrous iron and to precipitate as an iron carbonate known as siderite ($FeCO_3$).

Existence of a CO_2-recharge similar in size to that of Lake Nyos has been reported for Lake Mashu, a 1000-year-old caldera lake on Hokkaido, Japan. Fortunately, there is insufficient time for CO_2 to accumulate in the bottom waters since seasonal overturn of the lake expels the gas into the atmosphere twice a year. Lake Kivu in the East African Rift is another example of a gas-rich lake, but on a much larger scale. The predominant gas is magmatic CO_2, but it is accompanied by large quantities of methane (CH_4) originating from bacterial reduction of CO_2 and degradation of settling organic material. Both the CO_2 and CH_4 are currently trapped at the lake bottom; it is estimated that

Lake Kivu contains 1000 times more CO_2 than does Lake Nyos. The lake's dense layers currently prevent vertical mixing, and there is significant concern that a geologic disturbance in the area, or saturation of the water with CO_2 and CH_4 within the next two centuries, could lead to a significant and potentially deadly release of gases.

2.4. Caldera Lakes

There are many large dormant calderas on Earth which are partly or wholly filled by voluminous lakes whose compositions are close to meteoric (e.g., Crater Lake, USA; Batur, Indonesia; Figure 48.1 and Table 48.1). Yet, these waters may still exhibit a dilute volcanic hydrothermal signature. For example, crater lake in the 6800-year-old caldera of Mt Mazama shows near-neutral to slightly alkaline pH and dissolved solids contents typical of fresh water reservoirs (TDS $\sim 80-100$ mg/kg). However, the lake is well mixed to total depth (589 m) and its concentrations of dissolved chloride, sulfate, boron, lithium, and silica are higher than expected. It was shown that warm spring waters discharge on the crater floor; these hydrothermal inputs are probably associated with some remnant of a magma body at depths greater than 7.5 km. The estimated volcanic heat input is in the range $\sim 15-30 \times 10^6$ W.

3. ANCIENT VOLCANO-RELATED LAKES AS ORE-FORMING ENVIRONMENTS

A very large proportion of the world's metal and industrial resources was formed in and around volcanoes throughout Earth's history. For example, 80% of the current world production of copper comes from porphyry copper deposits as well as most of its molybdenum, rhenium, and significant amounts of gold. Extensive geological and geochemical studies have shown that many were formed very close to the paleosurface, from the surface down to about 4 km, but have been deeply eroded with the removal of surface features. However, characteristic features such as lake sediments are sometime preserved. These deposits retain glimpses of the processes that occur below modern volcano-related lakes and that are otherwise inaccessible to the observation.

Some of the world's largest copper and gold resources formed within a few kilometers of the surface of then-active andesitic volcanoes, but have subsequently been exhumed by uplift and erosion. These include Bingham Canyon, Utah, and the 3.2-million-year-old Grasberg deposit in Indonesia. These huge deposits also resulted from the sustained release of volcanic gas from a succession of intrusive bodies beneath the volcanoes and its interaction with groundwater as it expanded toward the surface. Despite extensive erosion, geological and isotope data suggest that acid crater lakes were present at the summit of the volcanoes that hosted ore formation at Bingham and Grasberg.

At Lepanto, Philippines, vestiges of a 1.33-million-year-old hyperacidic crater lake have recently been identified (Berger et al., 2014). This occurred directly adjacent to the rich vein system that was mined extensively for copper and gold from 1948 to 1996. The area of the lake is surrounded by remnants of rock alteration just as is seen in active acidic lakes today. The mined ore is rich in arsenic, which makes copper and gold extraction very difficult and has left a toxic environmental legacy after mine closure. However, arsenic and the suite of other metals that is associated with the ore, including bismuth, antimony, silver, zinc, tin, lead, and tellurium, are similar to the element suite that is commonly encountered in sublimates on active volcanoes. This relationship between ore deposition and lake occurrence suggests that very rapid metal deposition occurred within 1000 m of the surface as volcanic gas expanded through dilations in the fracture system that fed the lake. Similar processes may occur beneath modern acid lakes, with depletion of the metal content of the gas before it reaches lake level. Thus at Kawah Ijen, enargite ($CuAsS_4$) and covellite (CuS) grow as micro-crystalline particles along with barite ($BaSO_4$) inside sulfur globules. Similarly CuS and native gold have been recognized as tiny well-formed crystals in sulfur pools such as on submarine volcanoes. None of these present-day occurrences is likely to be economically exploited, but they provide valuable information as to how very large and valuable ore deposits formed in ancient volcanoes. In the more distal geothermal outflows from volcanic systems, gold and silver deposits are commonly associated with hot spring-fed lakes and hydrothermal eruption craters. The silver-gold deposit at Sulphur (Nevada) preserved silicified reed beds marginal to a former lake, and lakes also formed the base level for geothermal outflows responsible for silver-gold deposits at Bodie (California), Round Mountain (Nevada), and Republic (Washington) mining districts. Major silver-lead-zinc-barium vein deposits, such as at Creede, Colorado, are also associated with evaporative volcano-related lakes into which their outflows fed during their formation. The distinctive metal signature of these deposits may be due to release of metal-rich brines from basement rocks into higher level geothermal systems. In any case, all of these deposits preserve evidence of volcano-related lakes and related paleotopograhy dating back from a few to as many as 50 million years. Famous diamond-bearing kimberlite pipes, such as those in Southern Africa, are now considered to be the deep roots of ancient **maar**-type eruptions. Recent intensive exploration in Northern Canada has identified shallower equivalents of these deposits including the preservation on the Canadian Prairies of the original craters, phreatomagmatic deposits and lake sediments.

4. HAZARDS POSED BY VOLCANO-RELATED LAKES

Volcanoes, while active, are always in a critical state with respect to collapse. Both internal and external events can trigger collapse. The proximity of hyperacidic and acidic lakes to shallow magmatism in active volcanoes provides the ingredients for powerful hydrothermal eruptions (internal trigger) that may themselves result from heavy rainfall and tropical storms (external trigger) . The products of such eruptions provide the greatest hazard, but release of acidic water and toxic gas adds to the risks for surrounding communities.

4.1. Base Surges, Tsunamis, Debris Flows, and Flank Collapses

Eruptions of magma through volcano-related lakes present hazards from phreatomagmatic explosions such as base surges, debris flows, and tsunamis, which may not exist at volcanoes on dry land (Mastin and Witter, 2000). Base surges are ring-shaped clouds that propagate from the base of an eruption column at high velocity along the ground or water surface. While base surges are often associated with phreatomagmatic events, their mode of generation is not fully understood. The eruption of Taal volcano, Philippines, in 1965–1966 that affected its large 267 km^2 caldera lake produced base surges that swept over the landscape, tearing trees from the ground and destroying houses, then flowed across the lake and engulfed fleeing boats before hitting the opposite shores. Base surges also occurred during the 1975 eruption of Ruapehu Crater Lake, New Zealand. A tsunami may also form within a volcano-related lake disturbed by a subaqueous eruption. On New Year's Eve 1996, an eruption through the Karimsky's caldera lake in Russia wiped out the surrounding forest with tsunami waves. This type of hazard is particularly acute for caldera lakes with a populated shoreline such as Lake Taal and Lake Ilopango, El Salvador. During the 1965 eruption of Taal volcano, huge masses of lake water suddenly displaced by explosions generated a tsunami which caused damage up to 6 km from the vent and killed over 200 people.

Eruptions and base surges are not the most common cause of casualties in eruptions through volcano-related lakes; debris flows (slurries of debris, rock, and water) are usually much more deadly because they travel farther and inundate river valleys, which are often inhabited. Volcanic lake-related debris flows usually occur on active **composite volcanoes** containing relatively small (1-km-diameter) crater lakes. Eruptions at these volcanoes can empty the lake in a matter of seconds or minutes, leading to extremely high-water discharges on the order of 10^4–10^5 m^3/s. The steep volcanic flanks favor incorporation of loose debris into the waters and generation of debris flows. Indonesia has suffered dreadfully from crater lake-related debris flows (or lahars), the most devastating of which have occurred at Kelud. The debris flows generated by the eruptions of 1586 and 1919 took the lives of 10,000 and 5000 people, respectively.

Sudden drainage of a volcano-related lake due to crater collapse or crater rim failure/breaching, without precursory activity, can also produce debris flows. In 1953, Tangiwai bridge 35 km downstream from Mt Ruapehu was swept away by a debris flow which formed when the lake water gushed out through the breached crater wall. An express train was derailed and 151 passengers perished in the catastrophe. Debris flows from crater lakes can travel distances of tens of kilometers without diminishing in destructive power. Several catastrophic lake overflows at Lake Albano maar, located a few kilometers from Roma, Italy, during the Holocene produced debris flows that now form a vast plain (De Benedetti et al., 2008). Moreover, when the lake waters are hyperacidic, they carry caustic and toxic waters downstream.

By acting as a trap for magmatic heat and gas flow, acid crater lakes perched high on active volcanoes aggressively attack their surroundings and generate altered rock masses containing minerals such as kaolinite and alunite, with fine-grained amorphous silica. The presence of such hydro-thermally altered rocks within the core and upper flanks of the host volcano, such as at Mt. Rainier, is recognized as a key factor increasing the risk of flank collapse (Reid et al., 2001); it underpins the intrinsic instability of lake-hosted summit volcanic craters.

4.2. Toxic Water and Gas Release

Highly acidic and mineralized waters in summit crater lakes may leak into watersheds on a volcano's flanks, profoundly affecting water quality downstream. A famous example is the Banyuputih River which irrigates over 36 km^2 of cultivated land in the Asembagus coastal plain in East Java, Indonesia. The headwaters of the Banyuputih receive hyperacidic seepages from Kawah Ijen's active crater lake system; as a result low pH values (<4) and elevated concentrations and fluxes of potentially toxic species, notably fluoride, aluminum, and cadmium, are found in the irrigation waters, posing a potential threat for soil fertility and productivity and human health in this agricultural area. Watershed contamination linked to active crater lakes has been reported at other volcanoes including Poás, Costa Rica; Copahue, Argentina; Mt Ruapehu, and Kawah Putih, Indonesia.

The tragedies of lakes Monoun and Nyos in 1984 and 1986, respectively, killed over 1700 people due to asphyxiation by a dense CO_2 cloud. This type of volcano-related lake hazard had never been witnessed by scientists. Even now, the triggering mechanism that led to destabilization of lake stratification and consequent release of the large mass of dissolved CO_2 stored in bottom waters into the atmosphere remains unclear. What is certain is that continuous recharge of CO_2 into the lakes paves the way for future disaster. In the wake of the Monoun and Nyos events, concerns were expressed that several volcano-related lakes worldwide could be "time bombs." However, further studies have revealed that a rare combination of geologic and environmental factors is needed for significant and sustained accumulation of volcanic CO_2 to occur in a lake.

5. MONITORING OF VOLCANO-RELATED LAKES

As well as their scientific interest as windows into active volcanic systems, many volcanic lakes require close monitoring in view of the risk they pose to surrounding communities and infrastructure. Monitoring practices range from simple observations of, for example, changing lake color and temperature, to more refined remote geophysical methods and telemetry.

5.1. Calorimeters and Condensers for Volcanic Heat and Fluids

As well as the cumulative effects of clay alteration upon the walls and foundations of lakes, the hazard potential is readily related to changes in the relative heat and mass inputs and outputs of the lake system. Hyperacidic and acid lakes perched at the summit of active volcanoes act as traps for the volcanic heat and gas flow originating from a magmatic intrusion at depth (Figure 48.3). Volcanologists use a "box model" to relate observed volume and temperature changes in such lakes with estimates of the fluxes of water and volcanic heat into and out of the lake. Mass inputs of water to the lake consist of **volcanic hydrothermal fluids** (i.e., volcanic gas and its derivatives) and **meteoric water** (i.e., precipitation, runoff, streamflow, and groundwater seepage). Thermal and mass outputs are evaporation, lake water seepage and overflow. Heat is derived from the enthalpy of the hot water–gas mixture entering the lake plus solar and atmospheric radiation; it is lost by evaporative, sensible, and radiative fluxes from the lake surface, by seepage and overflow, and in heating precipitation and runoff into the lake. Evaporation is the most important energy loss, and its magnitude depends primarily on the wind velocity and on the difference between the lake surface temperature and that of the ambient air. Figure 48.5 illustrates the influxes and outfluxes used in a typical box model. Seepage of the lake water through the crater floor and recirculation within the gas flow regime complicate this simplified picture and are an important component of the lake's hydrologic budget. This is confirmed by including

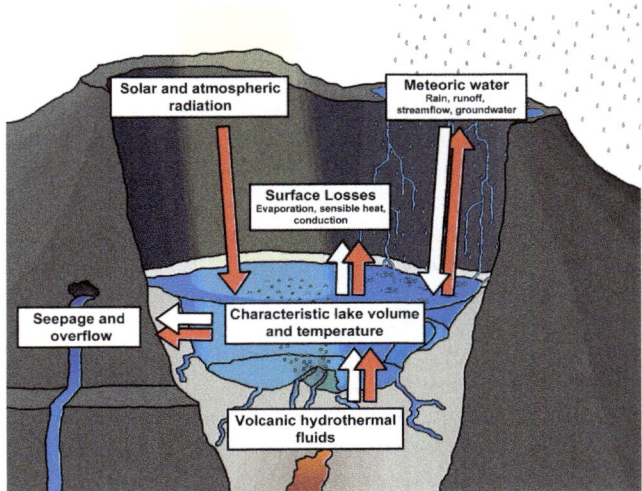

FIGURE 48.5 Cartoon diagram showing fluxes of water (white arrows) and heat (red arrows) into and out of a lake perched at the summit of an active volcano. Volcanologists use this "box model" to equate volume and temperature changes in such lakes and to derive the volcanic water and heat inputs and outputs.

time-series data on the stable isotope composition of water in the heat and mass flow calculations (Ohba et al., 2000).

Pasternack and Varekamp (1997) used the box model approach to conclude that the ratio between volcanic heat input and heat dissipation determines the persistence and the temperature of a lake in an active crater. Since most heat loss occurs at the lake surface by evaporation, small lakes have a limited capacity for heat dissipation, and their temperature is very sensitive to changes in heat inputs; large lakes are better buffered against such variations. On a local scale, each lake size has an upper limit of temperature at which it starts to lose mass. The flux of meteoric water into a lake is also an important parameter in the heat and mass budget, and this depends on the lake's surface area. Calculations suggest that most lakes which maintain both temperature and mass in a steady-state condition have temperatures above 45 °C. This is supported by field observations.

Volcanologists apply the box model to time-series data in order to decipher long-term flux values during the degassing of magmas underlying active crater lakes. Degassing volcanoes which host hot lakes typically have heat outputs ranging from ~20 to 1000×10^6 W and sulfur fluxes ranging from ~50 to 400 metric tons per day, comparable to estimates for other volcanoes which degas freely into the atmosphere. These results, in combination with geologic studies of active volcanoes and their now-eroded ancient equivalents, reveal that volcano-related lakes are in reality only very small expressions of a much larger heat and mass budget within active volcanic systems (Henley and Berger, 2013).

5.2. Lake Color

The color of a volcano-related lake results from absorption and scattering of solar radiation by organic and inorganic materials that are dissolved and/or suspended in the water column. Thus, the color of a volcano-related lake is affected by chemical reactions, particle size, and the vigor of convective circulation in the lake which may resuspend lake sediments. Temporary color changes, from blue to gray, or green to yellow, have been observed in several active crater lakes, most recently at Kelud (Figure 48.6).

At Keli Mutu, Indonesia, the three summit crater lakes reportedly change color between green/blue and red. In some cases, for instance at Mt. Ruapehu, shifts in lake color have been linked to variations in subaqueous hydrothermal discharges. A spectacular and rapid (<2 weeks) color change from light blue to red affected the entire acid crater Lake Voui at Ambae volcano Vanuatu, following phreatomagmatic activity in 2006. This color change event was exceptional considering the large lake volume (40×10^6 m^3). It was suggested that destabilization of the lake stratification by the eruptions led to overturn of the lake waters which brought deep reduced iron-rich waters to the surface, followed by rapid oxidation of iron and consequent iron oxide precipitation.

5.3. Lake Water Temperature

In order for the temperature of a volcano-related lake to increase, the amount of heat entering the system must be greater than the amount that is exiting the system (Figure 48.5). Thus, warming of a volcano-related lake is often regarded as indicative of enhanced subsurface activity. For example, the Main Crater Lake of Taal volcano recorded a 10 °C increase in 2−3 months prior to the September 1965 unrest. At Kelud, the temperature of the crater lake increased from 30 °C in September 1989 to 38 °C several days before the February 10, 1990 eruption. Similarly, temperatures of Yugama Lake at Kusatsu-Shirane volcano, Japan, increased over a period of several months in late 1989 from 9 to 12 °C, although there the heating episode was not followed by an eruption.

Understandably, lake water temperature measurements are normally included in monitoring programs of active crater lakes. However, the demanding logistics of regularly visiting summit crater lakes makes such efforts difficult, and automated methods to record lake temperatures are preferred. The drawback of the latter approach is that the harsh conditions encountered in these environments often cause equipment failure. Thermal satellite imagery offers an attractive alternative for monitoring volcano-related lake temperatures, and a variety of methods have been developed to retrieve reliable data. In particular, encouraging progress has been made to account for atmospheric

FIGURE 48.6 Photos showing the changes in color of the slightly acidic crater lake of Kelud volcano in August—September 2007 which preceded the extrusion of lava dome through the lake. *Photo courtesy: Akhmad Solikhin and Khoirul Huda.*

influences on water reflectance using thermal images taken by the Advanced Spaceborne Thermal Emission and Reflection Radiometer (ASTER; Trunk and Bernard, 2008). However, the number of ASTER images available

for active volcanoes with crater lakes and from which water temperatures can be estimated remains disappointingly small. The new LANDSAT 8 (Land Remote-Sensing Satellite) may alleviate this problem in the near future.

Identifying a forthcoming eruption solely on the basis of lake temperature variation is often not straightforward. The thirty-year record from Ruapehu Crater Lake reveals that a significant decrease of near-surface temperatures over several hours can be associated with rain and snow storms at the summit, whereas the opposite trend may reflect instability of convective heat transfer within the lake rather than changes in volcanic activity. Time-series data on lake temperature can be combined with lake level records to perform a more detailed heat and mass flow analysis. At Ruapehu, this exercise showed that preeruptive periods are characterized by low-heat input to the lake ($<150 \times 10^6$ W compared with a normal average of $\sim 400 \times 10^6$ W). At Poás and Kusatsu-Shirane crater lakes, departures from a steady heat inflow coincided with increase in the lake activity and sometimes with the number of volcanic earthquakes. In both cases, these perturbations were attributed to the formation of fractures in the envelope surrounding a cooling magma body, allowing release of magmatic gases to the overlying hydrothermal system. Upward migration of magma has also been proposed for Poás.

5.4. Lake Water Chemistry

Since acidic magmatic gases expand into lakes on top of active craters, fluctuations of the lake water pH may reflect changes in the composition and/or flux of the volcanic hydrothermal inputs. For example, pH values decreased while the concentration of dissolved Cl increased in the crater lake waters of Kusatsu-Shirane during 1990—1993; this trend was interpreted as reflecting direct degassing of magmatic HCl into the lake. A marked increase in acidity was also reported in the Main Crater Lake of Taal volcano before the 1968—1969 eruptions. The measurement of pH is relatively straightforward although like temperature, regular data collection can be challenging due to field constraints.

New Zealand scientists have routinely monitored the water chemistry of Ruapehu Crater Lake since the mid-1970s. This work led to the analysis of Mg/Cl ratio as a diagnostic tool for monitoring the magmatic activity within the volcano. The rationale is that the Mg/Cl ratio increases in relation to enhanced water—rock interaction beneath the lake due to hydrofracturing or new injection of magma; it decreases with more intense discharge of magmatic HCl-bearing fluids into the lake. In general, eruptions through Ruapehu Crater Lake are preceded by periods of relatively constant Mg/Cl ratios, followed by an abrupt increase and a long-term decline after the eruption.

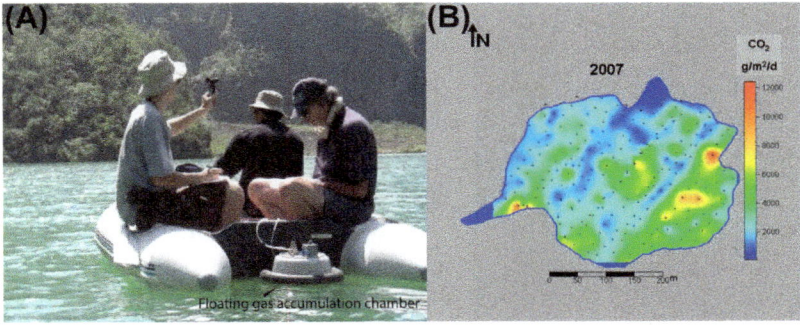

FIGURE 48.7 Measurement of CO_2 flux at the surface of the weakly acidic crater lake of Kelud volcano lake. (A) Photo of the floating gas accumulation chamber in use. (B) Map of the CO_2 flux (g/m²/day) obtained with the floating gas accumulation chamber between July 30 and August 2, 2007.

In addition to requiring post-collection laboratory analysis of the lake water samples, the application of the Mg/Cl method is limited by the assumption that chloride behaves conservatively in the lake waters. However, the vapor pressure of HCl solutions increases with higher acidity and higher temperature. At hyperacidic crater lakes, HCl vapors may escape the lake surface when the water temperature exceeds 50 °C. Such high temperatures are not uncommon in small, active crater lakes, for example at Poás and at Aso volcano in Japan. Another potential difficulty when using the Mg/Cl ratio for monitoring purposes is when seawater contaminates the magma-hydrothermal system that powers the crater lake, like at Taal volcano.

Complex mixtures of polythionates, which are metastable sulfur compounds ($S_xO_6^{2-}$, x = 4, 5, or 6), are commonly found in hyperacidic crater lakes. These chemical species are produced by reaction of $SO_{2(g)}$ with $H_2S_{(g)}$ in acidic solutions. Several studies conducted at Kusatsu-Shirane, Poás, and Ruapehu Crater Lakes in the 1980s and 1990s concluded that polythionate concentrations are coupled to changes in the SO_2/H_2S gas ratio of the magma-hydrothermal inputs (Delmelle and Bernard, in 2014). Notably, sharp declines in polythionate concentrations and corresponding increases in dissolved sulfate are correlated in all cases with renewed phreatomagmatic/seismic activity in these lakes. While polythionate monitoring has provided useful information, important unknowns remain regarding the complex chemistry and temperature-dependent kinetics which govern the cycling of sulfur between its different redox states in a hyperacidic crater lake on top of an active volcano. This gap in knowledge complicates the unambiguous interpretation of polythionate measurements. Further, polythionate analysis is not done routinely in chemical laboratories as it requires special instrumentation.

Sulfur dioxide and CO_2 gas flux monitoring by means of remote sensing and other techniques is commonly used to track changes in volcanic activity. However, such measurements are hampered when the degassing crater is filled with a lake, as most of the volcanic gas emissions of interest are rapidly dissolved in the waters. One exception is CO_2 which can be degassed from the lake surface through bubbling or by diffusion through the water/air interface. In order to estimate the flux of CO_2 from a volcano-related lake, volcanologists use a floating gas accumulation chamber (Figure 48.7).

In 2007, a noticeable increase in CO_2 emissions from Kelud crater lake was measured several months prior to lava dome extrusion. Spectacular intensification of CO_2 degassing from the Main Crater Lake at Taal volcano has been observed repeatedly since 2011. This unusual pattern is believed to relate to changes in magmatic activity rather than to fluctuations in hydrodynamic processes (e.g., fracture opening) within the hydrothermal system.

5.5. Underwater Acoustic Measurements

In volcano-related lake studies, echo sounding methods have traditionally been used to determine bathymetry or to map subaqueous volcanic structures. Since gas bubbles strongly reflect and scatter acoustic energy in water, high-sensitivity hydrophones are also deployed to provide information on the intensity of volcanic hydrothermal fluid discharges at the lake bottom. For example, broad-frequency hydrophones immersed in Kelud crater lake recorded powerful signals of gas bubbles which coincided with the seismic crisis preceding the 1989–1990 eruption. Similar measurements performed at Ruapehu in 1993–1994 showed increased heat transfer to the summit crater lake several days before the lake temperature started to rise.

Recently, echo sounding has been successfully applied to estimate the flux of CO_2 entering the crater lake of Kelud volcano. The method assumes that the gas bubbles, with some defined geometry, contain only CO_2. At Kelud, this type of measurement has revealed that ~70% of the CO_2 enters the crater lake in the dissolved form and ~30% is gaseous (Caudron et al., 2012).

Similar to the floating accumulation chamber method mentioned above, hydroacoustic surveys of this kind offer new perspectives for monitoring changes in CO_2 emissions at crater lakes which may precede other geophysical or geochemical signals indicative of increased volcanic activity.

6. MITIGATION OF VOLCANO-RELATED LAKE HAZARDS

6.1. Drainage

Kelud has erupted 29 times since 1311, with repose intervals between eruptions ranging from one to 75 years. The major volcanic hazard for people living near the volcano is debris flows from the summit crater lake. Aware of the permanent threat posed by this reservoir of volcanic water, the Dutch administration built a dam on the Badak River in 1905 to divert debris flows from reaching the town of Blitar. Unfortunately, the voluminous debris flows from the 1919 eruption swept away the dam. A few months after the deadly 1919 eruption when the lake had not yet reformed, an ambitious engineering project sought to drill a 955-m-long tunnel through the crater wall in order to drain the lake. By 1923, when the crater had filled with 2.2×10^7 m^3 of water, sudden flooding of the tunnel still under construction halted the digging. A new strategy was then adopted to progressively lower the lake level by excavating seven superimposed tunnels while draining the waters with a siphon pipe. In 1926, the lake volume was successfully reduced to less than 2×10^6 m^3. Later eruptions in 1951, 1966, and 1990 only produced minor debris flows whose damage was much more limited, although casualties were reported.

The tunnels of Kelud do not constitute the first example of an engineering endeavor aimed at reducing volcanic hazards. In 394 BC, the Romans excavated a 1.5-km-long tunnel through the rim of Lake Albano maar in order to divert the lake water to the sea. This engineering work was aimed at calming the Poseidon, the God of the sea, earthquakes, and horses, and whose anger was said to be responsible for the catastrophic overflow of the lake in 398 BC. Nowadays, the tunnel remains operational and keeps the lake level about 70 m below the maar rim.

6.2. Artificial Degassing

Since the catastrophic releases of CO_2 in the 1980s, Lake Monoun and Lake Nyos have experienced CO_2 recharge at rapid rates. Gas saturation levels of up to 97% were measured in 2003 and 2001 at Lake Monoun and Lake Nyos, respectively. Alarmed by this situation, scientists sought a permanent solution to mitigate the volcanic hazard, proposing to gradually pipe the gas out of the lake at a controlled rate. In 2001, a 203-m-long pipe was suspended vertically in Lake Nyos, extending down into the gas-rich waters at depth. The principle is relatively simple; initial pumping of the deep waters in the pipe allows CO_2 to come out of solution and form bubbles. As a result, the water–gas mixture can start to rise under its own buoyancy, sucking more water from depth and creating a "soda" fountain at the surface (Figure 48.8). After priming, the system becomes self-sustaining and requires no external energy.

In 2012, measurements of the total gas pressure in the lake showed that the degassing operation is steadily

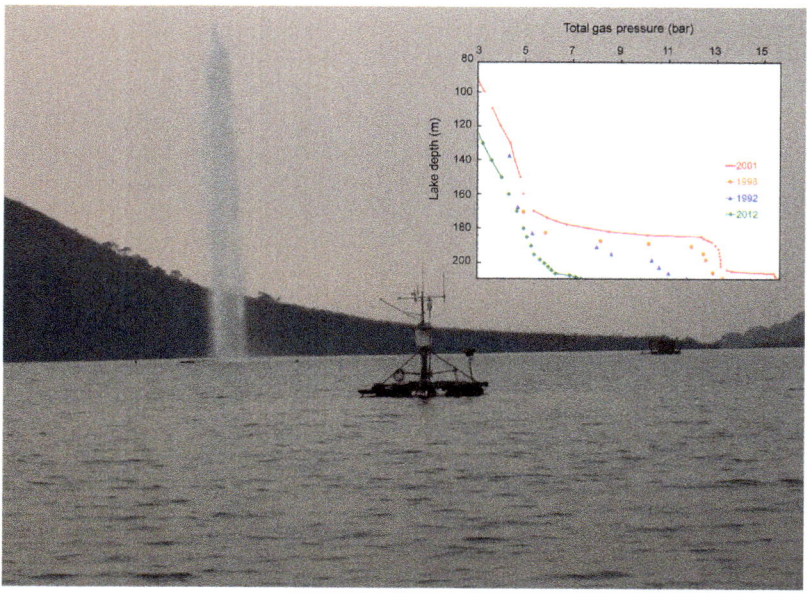

FIGURE 48.8 Photo of the self-sustained CO_2 gas-rich fountain ($\sim$50 m in height) at the surface of Lake Nyos produced by siphoning up the gassy waters from the lake depths. Inset — Profile of total gas pressure ($CO_2 + CH_4$) with depth in the bottom waters of the lake. The gas pressures increased from 1992 to 1998 to a maximum in 2001 when the first degassing pipe was installed. *Data collected as part of the Japanese–Cameroonian SATREPS Project on Safety, Rehabilitation, and Development of the Lakes Nyos and Monoun areas in Northwest Cameroon (http://photos.state.gov/libraries/cameroon/231771/PDFs/nyosreportusembassy2012.pdf), and reproduced by kind permission of T. Ohba and M. Kusakabe.*

reducing the amount of CO_2 gas at the bottom of the lake. Compared to 2001, the gas pressure at the deepest part of the lake has been reduced by ~2.5 times and the total CO_2 in the lake reduced by 40%, from ~710,000 to 425,000 metric tons. However, degassing model results indicate the need for additional pipes at Lake Nyos to reach safe CO_2 levels within a reasonable amount of time (Kling et al., 2005). Artificial degassing of the smaller Lake Monoun started in 2003, and CO_2 concentrations in the lake are now negligible.

7. CONCLUDING REMARKS

Volcanoes are always unstable landscape features due to the threat of catastrophic eruptions and to over steepening of their flanks which may fail when subject to an internal or external trigger. The discharge of magma-derived gas mixtures through the superstructures of active volcanoes inevitably increases their risk to surrounding populations and infrastructure. Entrapment of these gases by lakes within craters or topographic depressions provides a special class of risk that ranges from phreatomagmatic, phreatic, and hydrothermal eruptions to development of unstable stratification in the lake due to CO_2 gas flow. Moreover, crater lakes on top of active volcanoes which act as magmatic gas condensers may promote intense subsurface alteration of the volcanic rocks, thereby weakening rock mass strength and decreasing edifice stability.

Understanding how volcano-related lakes work is crucial to mitigation of hazards and development of effective warning systems. Simple observational monitoring is extremely effective when coupled with effective warning strategies. Experience shows that useful monitoring is achieved by a combination of methods. Considerable advances have been made through monitoring the heat budgets of volcanic lakes and modeling chemical processes such as sulfur reaction with water in relation to the fluxes of reactive gases. However, we have not captured yet the entire complexity of heat and mass transport from a magma intrusion to the crater lake bottom. Detailed hydrological surveys of volcano-related lakes are also needed in order to be able to interpret and predict the dynamics of these systems.

Volcano-related lakes have formed throughout Earth's history but only a few have survived erosion. Their presence, along with extensive rock alteration, records the discharge of gases through ancient volcanic systems and may be related to sites of base and precious metal deposition below the surface to depths of a few kilometers. Volcano-related lakes may also have typified the surface discharge of diamond-bearing kimberlite pipes. As high-energy environments rich in potential nutrients, these lakes may also have played a role in the origin of life and

certainly have been natural laboratories in which new, extremophile organisms have, and will continue to evolve.

In their own right, volcano-related lakes represent some of the most spectacularly diverse physical and chemical settings on Earth, some hosting long-lived molten sulfur bodies. These remarkable phenomena they share with Io, a moon of Jupiter, and possibly the sulfur-rich planets, Venus and Mercury. Rather than feared, volcano-related lakes are unique and extraordinarily dynamic planetary features for us to admire and nurture. There remain plenty of opportunities, for example in Africa, to study them.

ACKNOWLEDGMENTS

The authors are grateful to Paul Ayris for help with the graphics. P.D. acknowledges support from the Fonds National pour la Recherche Scientifique (FNRS) through a MIS-Ulysse grant F.6001.11.

FURTHER READING

Berger, B.R., Henley, R.W., Lowers, H.A., Pribil, M.J., 2014. The Lepanto Cu-Au deposit, Philippines: A fossil hyperacidic volcanic lake complex. J. Volcanol. Geotherm. Res. 271, 70–82.

Caudron, C., Mazot, A., Bernard, A., 2012. Carbon dioxide dynamics in Kelud volcanic lake. J. Geophys. Res. 117 (B05102), 11. http://dx.doi.org/10.1029/2011JB008806.

Christenson, B.W., Wood, C.P., 1993. Evolution of a vent-hosted hydrothermal system beneath Ruapehu Crater Lake, New Zealand. Bull. Volcanol. 55, 547–565.

De Benedetti, A.A., Funiciello, R., Giordano, G., Diano, G., Caprilli, E., Paterne, M., 2008. Volcanology, history and myths of the Lake Albano maar (Colli Albani volcano, Italy). J. Volcanol. Geotherm. Res. 176, 387–406.

Delmelle, P., Bernard, A., 2015. The remarkable chemistry of sulfur in acid crater lakes: a scientific tribute to Bokuichiro Takano and Minoru Kusakabe. In: Rouwet, D., Christenson, B., Tassi, F., Vandemeulebrouck, J. (Eds.), Volcanic Lakes, Advances in Volcanology. Springer, Heidelberg.

Delmelle, P., Bernard, A., Kusakabe, M., Fischer, T.P., Takano, B., 2000. Geochemistry of the magmatic-hydrothermal system of Kawah Ijen volcano, East Java, Indonesia. J. Volcanol. Geotherm. Res. 97, 31–53.

Henley, R.W., Berger, B.R., 2013. Nature's refineries: metals and metalloids in arc volcanoes. Earth Sci. Rev. 125, 146–170.

Kling, G.W., Clark, M.A., Wagner, G.N., Compton, H.R., Humphrey, A.M., Devine, J.D., Evans, W.C., Lockwood, J.P., Tuttle, M.L., Koenigsberg, E.J., 1987. The 1986 Lake Nyos gas disaster in Cameroon, West Africa. Science 236, 169–175.

Kling, G.W., Evans, W.C., Tanyileke, G., Kusakabe, M., Ohba, T., Yoshida, Y., Hell, J.V., 2005. Degassing Lakes Nyos and Monoun: defusing certain disaster. PNAS 102, 14185–14190.

Mastin, L.G., Witter, J.B., 2000. The hazards of eruptions through lakes and seawater. J. Volcanol. Geotherm. Res. 97, 195–214.

Ohba, T., Hirabayashi, J., Nogami, K., 2000. D/H and O-18/O-16 ratios of water in the crater lake at Kusatsu-Shirane volcano, Japan. J. Volcanol. Geotherm. Res. 97, 329–346.

Pasternack, G.B., Varekamp, J.C., 1997. Volcanic lake systematics .1. Physical constraints. Bull. Volcanol. 58, 528—538.

Reid, M.E., Sisson, T.W., Brien, D.L., 2001. Volcano collapse promoted by hydrothermal alteration and edifice shape, Mount Rainier, Washington. Geology 29, 779—782.

Rouwet, D., Tassi, F., Mora-Amador, R., Sandri, L., Chiarini, V., 2014. Past, present and future of volcanic lake monitoring. J. Volcanol. Geotherm. Res. 272, 78—97.

Trunk, L., Bernard, A., 2008. Investigating crater lake warming using ASTER thermal imagery: case studies at Ruapehu, Poás, Kawah Ijen, and Copahué Volcanoes. J. Volcanol. Geotherm. Res. 178, 259—270.

Varekamp, J.C., Kreulen, R., 2000. The stable isotope geochemistry of volcanic lakes, with examples from Indonesia. J. Volcanol. Geotherm. Res. 97, 309—327.

Chapter 49

Volcanic Successions Associated with Ore Deposits: Facies Characteristics and Ore–Host Relationships

Jocelyn McPhie

School of Physical Sciences and CODES, University of Tasmania, Hobart, Tasmania, Australia

Ray Cas

Department of Geosciences, Monash University, Clayton, Victoria, Australia; School of Physical Sciences, University of Tasmania, Hobart, Tasmania, Australia

Chapter Outline

GLOSSARY

autobreccia Clastic aggregate generated as a by-product of lava flowage.

autoclastic facies Clastic facies generated by non-explosive fragmentation accompanying lava effusion and flowage. Autobreccia and hyaloclastite are the two most common kinds of autoclastic facies.

breccia Clastic aggregate composed of relatively coarse (>2 mm) angular clasts.

coherent facies Facies generated by the solidification of molten lava or magma.

diatreme Subvertical breccia pipe that was formerly a feeder or conduit to a volcanic vent.

effusive eruption An eruption style involving passive discharge of magma from a volcanic vent, producing lavas and/or domes.

explosive eruption An eruption style characterized by the explosive expansion of gas, and fragmentation and ejection of magma and/or wall rock from a volcanic vent.

hyaloclastite Clastic aggregate generated by quench fragmentation of hot lava in contact with water and/or ice.

hydrothermal fluid Hot aqueous fluid. The principal component is liquid water and/or steam derived from meteoric, seawater, and/or magmatic sources. High abundances of dissolved ions and fine particles may be present.

phreatic eruption An explosive eruption involving hot wall rock and external (non-magmatic) water; magma is not directly involved.

phreatomagmatic eruption An explosive eruption involving magma that interacts with external (non-magmatic) water; this external water is the principal source of gas.

peperite A facies generated by mingling of molten lava or magma with unconsolidated sediment.

The Encyclopedia of Volcanoes. http://dx.doi.org/10.1016/B978-0-12-385938-9.00049-3

viscosity The ratio between shear stress and strain rate of a substance, or the ease with which a substance deforms and flows in response to applied stress. High-viscosity substances resist flowage (strain) unless a high shear stress is applied.

volcaniclastic facies Facies composed of volcanic clasts or fragments, regardless of the clast-forming process. The main genetic categories of volcaniclastic facies are pyroclastic (clasts formed by explosive eruptions), autoclastic (clasts formed as a by-product of effusive eruptions), and epiclastic (clasts formed by surface weathering).

xenocryst A crystal that did not crystallize from the enclosing lava or magma.

1. INTRODUCTION

Four important ore deposits are commonly hosted in volcanic successions: volcanic-hosted massive sulfide (VHMS) deposits, komatiite-hosted nickel sulfide deposits, epithermal gold—silver deposits, and kimberlite-hosted diamonds. There are different reasons why these ore deposit types occur in volcanic successions. The occurrence of VHMS deposits and komatiite-hosted nickel sulfide deposits in volcanic rocks reflects the strong genetic connection between the volcanism and the ore-forming processes. Such a connection may also link epithermal gold—silver deposits with contemporaneous volcanism, both being products of high heat flow and extension. However, the connection does not always exist for epithermal gold—silver deposits because some of these deposits are significantly younger than the host volcanic succession, or the host succession is not volcanic. In the case of kimberlite-hosted diamonds, kimberlite magmas serve as the medium for transporting diamonds from the mantle to the Earth's surface; the diamonds are not generated by the host kimberlite magmas nor by the processes that produced those magmas. A fifth important ore deposit type, porphyry copper—gold—molybdenum deposits, typically occurs in volcanic arcs and may be hosted by volcanic successions. However, the magmatic-hydrothermal processes that produce these deposits operate exclusively subsurface, and venting of either the fluids or the magmas reduces the metal potential of the system (Pasteris, 1996). Hence, in general, there is no systematic relationship between porphyry deposits and volcanism even in cases where the host succession is volcanic or partly volcanic.

Here, we focus on the characteristics of the volcanic successions that host VHMS deposits, nickel sulfide deposits, epithermal gold—silver deposits, and diamonds, emphasizing the environments and styles of volcanism, volcanic facies associations and architecture, and relationships between the volcanic and mineralizing processes. This knowledge underpins successful exploration for these deposits, and provides the basis for attempts to quantify metal, fluid and magma budgets on local to global scales.

2. VOLCANIC-HOSTED MASSIVE SULFIDE ORE DEPOSITS

VHMS ore deposits are a major source worldwide of copper, lead and zinc, and some examples also produce significant gold and silver (Franklin et al. 2005). The principal metal-bearing minerals are copper sulfide (chalcopyrite, $CuFeS_2$), lead sulfide (galena, PbS), and zinc sulfide (ZnS, sphalerite). Most VHMS contain substantial iron sulfide (pyrite, FeS_2) but are not mined for iron. The deposits are "massive" in the sense of comprising 60% sulfide minerals and very low abundances of gangue.

2.1. Setting of VHMS deposits

VHMS deposits occur in submarine volcanic or volcano-sedimentary settings. Actively forming VHMS deposits have been discovered at many locations in modern oceans, along mid-ocean ridges, in back arc basins, and within the submerged parts of volcanic arcs (Hannington et al., 2005). Deep-sea mining of some modern VHMS deposits is being taken seriously (e.g., Boschen et al., 2013). Sulfides are precipitated at or very close to the seafloor where metal-rich **hydrothermal fluids** vent and mix with cold seawater ("black smokers"). The hydrothermal fluid originates from seawater and seawater is the major source of sulfur so these deposits have a strong link with a submarine setting. There is also a strong link with volcanism, reflecting the requirement of heat energy to drive deep circulation of the hydrothermal fluid. In general, the depth of seawater, either known in the case of modern active examples or inferred for ancient VHMS deposits, is relatively deep, typically deeper than 1000 m. The higher confining pressure of relatively deep-water settings allows hydrothermal fluids to discharge at temperatures in the range of 200—400 °C without boiling.

2.2. Facies and Volcanology

The characteristics of volcanic successions that host VHMS deposits are reasonably well understood from facies architecture research in Japan, Sweden, Australia, Canada, and Portugal (e.g., Horikoshi, 1969; Allen et al., 1996a; Allen et al., 1996b; Allen, 1992; McPhie and Allen, 1992, Paulick and McPhie, 1999; Gibson et al., 1999; Doyle and McPhie, 2000; McPhie and Allen, 2003; Rosa et al., 2008; Rosa et al., 2010). These successions typically comprise complex assemblages of texturally and compositionally diverse volcanic facies that also reflect a wide spectrum in the volcanic environments of massive sulfide ore formation (Figure 49.1). Because the setting is a submarine depocentre, non-volcanic or mixed sedimentary facies are present and may be locally the dominant facies. The essential elements of the facies architecture are lavas and

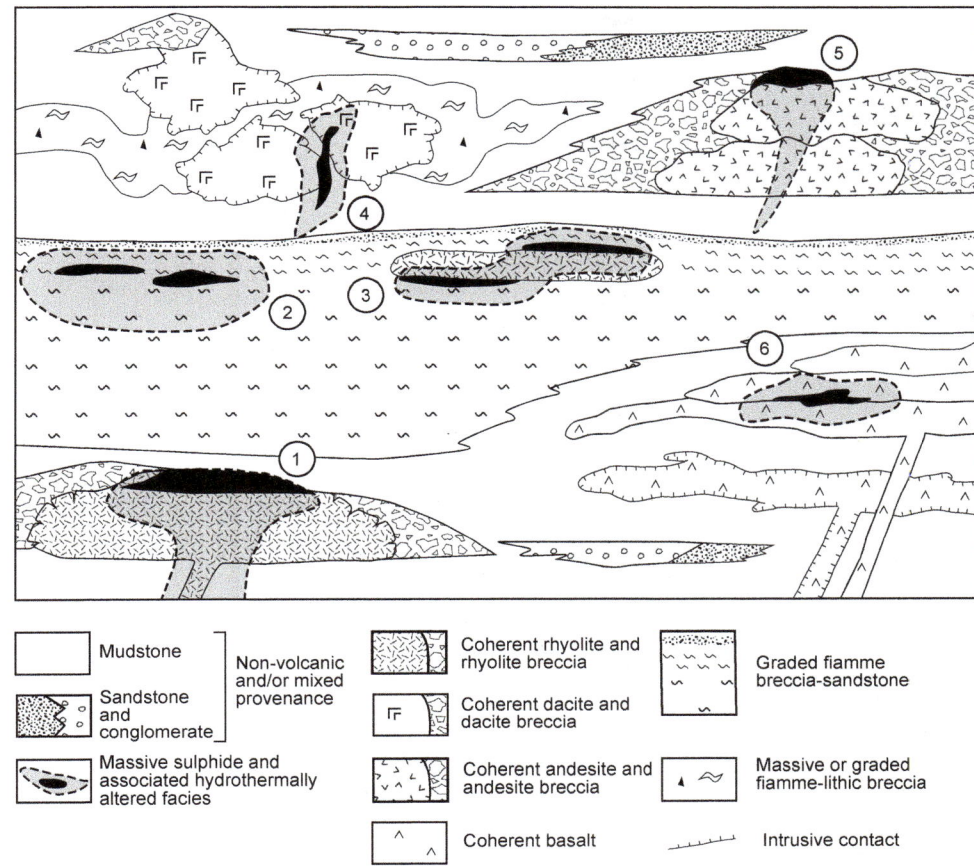

FIGURE 49.1 Schematic facies architecture of a submarine volcanic succession showing the common facies present and possible locations of VHMS deposits. Massive sulfide deposit locations are based on (1) Thalanga, Mount Windsor Volcanics (Paulick and McPhie, 1999) and Neves Corvo, Iberian Pyrite Belt (Rosa et al., 2008); (2) Hercules, Mount Read Volcanics (McPhie and Allen, 2003); (3) Rosebery, Mount Read Volcanics (Gifkins and Allen, 2001); (4) Highway-Reward, Mount Windsor Volcanics (Doyle and McPhie, 2000); (5) Hellyer, Mount Read Volcanics (Waters and Wallace, 1992); (6) FlinFlon greenstone belt, Canada (Syme and Bailes, 1993).

domes, diverse **volcaniclastic facies**, and synvolcanic intrusions. Lavas and domes consist of both **coherent** and **autoclastic** (**autobreccia, hyaloclastite,** pillow fragment **breccia**; Figure 49.2(A) and (B)) facies. Felsic and intermediate lavas and domes can be very thick (tens to hundreds of m), comprising wide (m—tens of m) glassy perlitic zones at the margins and microcrystalline, spherulitic, or micropoikilitic groundmass textures in the interior. Basaltic lavas are relatively thin (<10 m) and commonly pillowed or internally massive, and typically have relatively thin (cm) glassy margins and microcrystalline interiors. **Peperite** may be present along the basal contacts of lavas and domes that overlie volcaniclastic or sedimentary facies.

Although the volcaniclastic facies range widely in textural characteristics, there are four common types. (1) Thick to very thick, massive or weakly graded beds of resedimented hyaloclastite (Figure 49.2(C)) are generated as a by-product of lava or dome effusion, and commonly occur close to lavas or domes of the same composition and texture. (2) Very thick (m to tens of m), massive to graded

beds of rhyolitic or dacitic pumice breccia are the products of **explosive eruption**s driven mainly by magmatic volatiles; in some cases, these eruptions occurred at submarine intrabasinal vents but in others, it appears that the source was a subaerial vent at the basin margin and the products were transported offshore into deep water. In lithified successions, pumice clasts are altered and compacted to fiamme (Figure 49.2(D)). (3) Thick to very thick, graded beds of polymictic volcanic conglomerate- or breccia—sandstone are produced by a variety of mass-wasting and downslope transport processes not necessarily related to eruptions. (4) Massive or laminated shard-rich volcanic mudstone is deposited from relatively dilute suspensions of ash in the water column during periods of otherwise low volcanic aggradation. In successions that include a significant volume of basaltic (or basaltic andesite) lavas, fluidal-clast basaltic breccia (Figure 49.2(E)) may be present though not voluminous. This facies is characterized by fluidal vesicular bombs and lapilli in a matrix of finer basaltic hyaloclastite generated

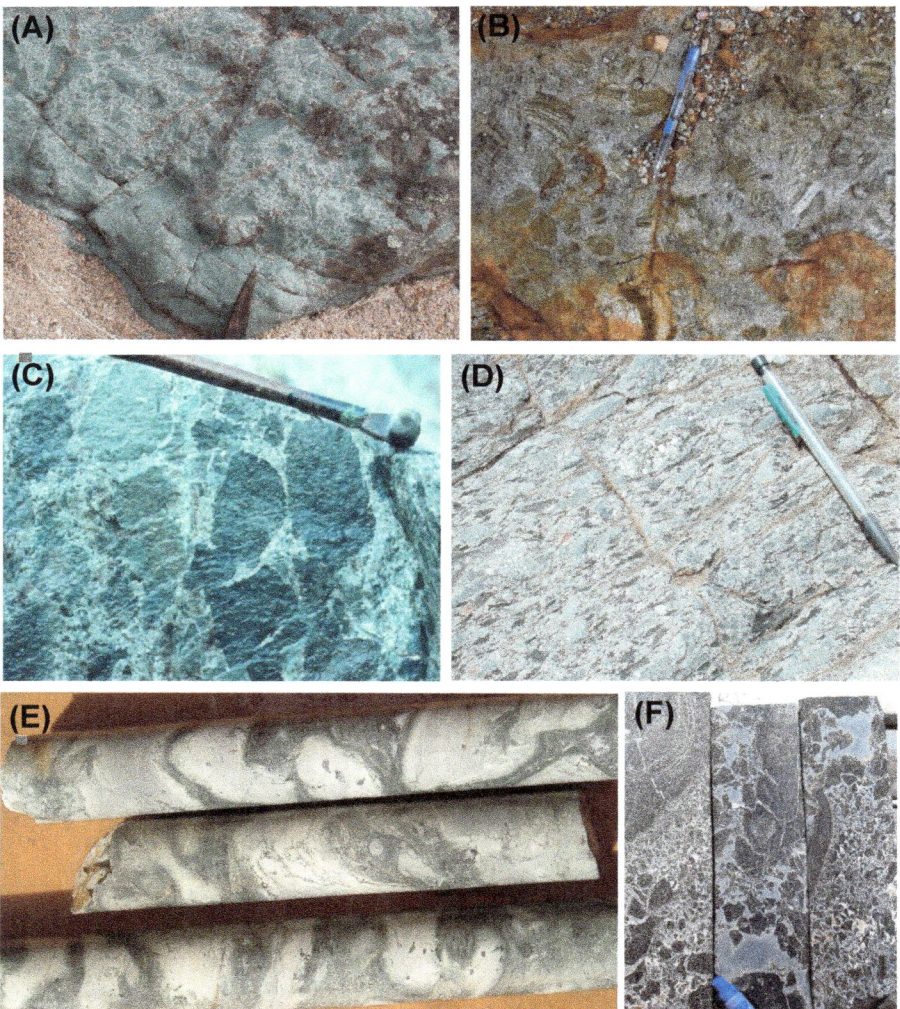

FIGURE 49.2   (A) Monomictic, jigsaw-fit dacite breccia interpreted to be in situ hyaloclastite. Volcano-Sedimentary Complex, Iberian Pyrite Belt, Rio Guadiana, Portugal. (B) Monomictic, clast-rotated rhyolite breccia interpreted to be autobreccia. Central Volcanic Complex, Mount Read Volcanics, Hall Rivulet Canal, Tasmania, Australia. (C) Weakly polymictic, graded volcanic breccia interpreted to be resedimented hyaloclastite. Dark gray dacite clasts have curviplanar margins suggesting they formed via quench fragmentation of coherent dacite. Mount Read Volcanics, Newton Dam Spillway, Tasmania, Australia. (D) Very thick fiamme-lithic breccia. Dark green fiamme (elongate, aligned, chlorite-rich lenses) were originally pumice clasts but have been altered and compacted. Volcano-Sedimentary Complex, Iberian Pyrite Belt, Rio Guadiana, Portugal. (E) Fluidal-clast basalt breccia interpreted to be a basaltic fountain deposit. Jaguar VHMS deposit, Western Australia. (F) Mud-matrix basalt breccia interpreted to be peperite formed by disintegration of molten basalt that intruded wet mud. The basalt clasts have curviplanar surfaces suggesting that quench fragmentation was the main cause of break-up of the intruding basalt. The mudstone matrix is silicified and lacks bedding. Hellyer VHMS deposit, Tasmania, Australia.

by a combination of fountaining and quench fragmentation of low-viscosity lava at an underwater vent (Simpson and McPhie, 2001).

Some of the volcaniclastic facies are clearly eruption-fed, having been generated by a coeval **effusive** or explosive eruption, whereas others are post-eruptive and exhibit evidence for temporary storage and reworking prior to redeposition. Eruption-fed facies are characterized by the dominance of unmodified juvenile components of uniform composition and thick to very thick, mass-flow sedimentation units. Facies (1) and (2) above are both eruption-fed

and related to effusive or explosive eruptions, respectively. In many cases, facies (3) is post-eruptive, especially where clasts are well rounded and sorted. Facies (4) can involve direct supply of pyroclasts to the water column from an eruption followed by a delay before sedimentation takes place, and so falls into a broadly syn-eruptive category rather than being strictly eruption-fed.

Syn-volcanic intrusions are common in VHMS successions because in a submarine setting, rising magma is likely to be denser than wet sediment at the seafloor, and will intrude laterally or updome the sediment instead of erupting (McPhie and Allen, 1992). The internal textures of such intrusions may be closely similar to those of extrusive units; felsic intrusions commonly have glassy brecciated margins, and mafic intrusions may be pillowed. Contact relationships of syn-volcanic intrusions are critical (Allen, 1992), especially the presence of peperite (Figure 49.2(F)) along top contacts. Given the shallow level of intrusion, some emplacement units may show evidence for being both intrusive and extrusive at different locations.

2.3. Relationships between VHMS Deposits and Their Host Facies

Many VHMS deposits are readily understood as sulfide mounds or lenses that formed on and just below the seafloor. Formation on the seafloor results in a marked difference in hydrothermal alteration intensity between the footwall (strongly altered) and the hangingwall (unaltered or weakly altered). For that reason, correct recognition of former seafloor positions is very important in exploration for VHMS deposits. It is also clear that in some cases, the deposition of sulfides occurs somewhat deeper (tens of m) beneath the seafloor, especially where the facies are very porous, permeable and reactive, for example, glassy pumiceous clastic facies or the perlitized coherent facies of domes, lavas and shallow intrusions. In this case, the massive sulfide will not necessarily be a single body, may be disconformable, will have gradational contacts with the host facies, and may have shapes controlled by the shapes of adjacent facies. In addition, both the hangingwall and the footwall will be hydrothermally altered.

Syn-volcanic intrusions can introduce further complications to spatial and temporal relationships between the host succession and the massive sulfide. Syn-volcanic intrusions that predate the mineralizing hydrothermal system can have different physical properties (especially porosity and permeability) from the host succession, and strongly influence the location and shape of the massive sulfide deposit (e.g., Doyle and Huston, 1999). They are likely to be hydrothermally altered and may be mineralized. Syn-volcanic intrusions that postdate the mineralizing hydrothermal system may dismember an initial single massive sulfide deposit into one or more domains, and also modify ore textures and grade. In general, these intrusions will be at most weakly altered or altered only on their margins, and they are not mineralized.

3. KOMATIITES AND THEIR MAGMATIC NICKEL SULFIDE DEPOSITS

3.1. Archaean Komatiites

Komatiites are very low silica (SiO_2 <45 wt%), very high MgO (>18 wt%), low Al_2O_3 (<10 wt%), high CaO (<10 wt%), very low Na_2O (<1 wt%), and extremely low K_2O (<0.1 wt%) ultramafic rocks. Komatiites represent one of the major sources ($\sim$40%) of nickel on Earth. Although some Proterozoic and Phanerozoic examples are known, komatiites formed almost exclusively in the Archaean, probably related to mantle plume events in the Earth's early history. Volcanologically, komatiites are fascinating because they represent one of the lowest **viscosity** magmas known (equivalent in viscosity to carbonatites and sulfur magmas) on Earth. Both komatiite lavas and komatiite intrusions are known.

Komatiite lavas range from tens of cm to tens of m thick and they are variably porphyritic. The most common phenocrysts are forsteritic olivine, chrome-rich pyroxene, anorthite and chromite, giving them a peridotitic mineral assemblage. Komatiite intrusions may also be porphyritic if they are thin, very high-level intrusions. Deeper-seated, thicker (hundreds of m) intrusions of very coarse, holocrystalline, olivine-rich dunites are the intrusive equivalents of komatiite lavas. The rapidly cooled margins of dunitic intrusions are also porphyritic. Being ultramafic, komatiites are easily altered. Common alteration minerals are serpentine, talc, and carbonate.

3.2. Setting

Almost all known Archaean komatiites are preserved as lavas in deep submarine volcanic and sedimentary successions and as intrusions in the subjacent crust that forms part of Archaean cratons. Komatiite lavas were almost certainly erupted in subaerial and shallow marine environments as well, but they have been eroded. Although some vesicular komatiite lavas exist, pyroclastic komatiite successions have been only rarely recognized (Stiegler et al., 2011). These include primary and variably reworked deposits of shards, vesicular pyroclasts, and accretionary lapilli.

3.3. Facies and Volcanology

Komatiite lavas are famous for their unique olivine spinifex and cumulate crystal textures and internal textural zonation. In komatiitic basalts, such textures can also involve pyroxene. In a typical komatiite lava, the textural profile of the lava from top to bottom (Figure 49.3; Arndt et al., 1977) consists of the following:

- A1—This uppermost layer is aphyric or extremely fine grained and only a few mm to a few cm thick.
- A2—Random spinifex texture: Gradationally under A1, there is a layer of apparent needle-like, but actually blade-like, skeletal crystals of olivine with a random orientation (Figure 49.4(A)). These crystals increase in size downward from a few mm to several cm.

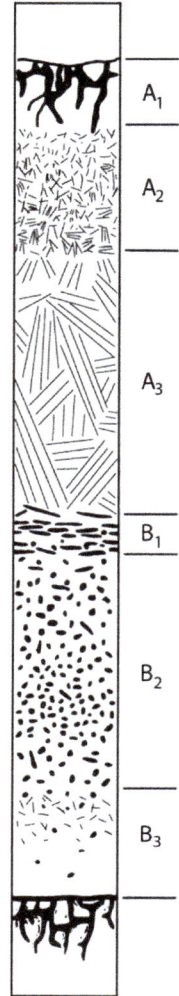

FIGURE 49.3 Profile through a complete komatiite lava, showing the variations in the upper spinifex (A) and lower cumulate (B) textural zones. See text for detailed descriptions of each sub-zone. After Arndt et al. (1977).

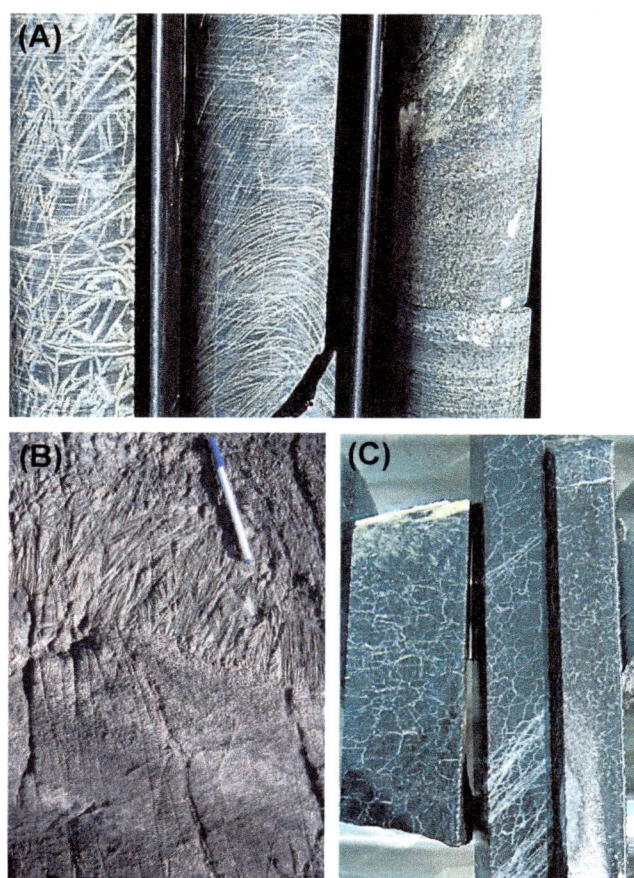

FIGURE 49.4 (A) Drill core through a single komatiite lava showing the textures and textural zonation from random spinifex (left) to blade spinifex (center) to ortho-cumulate (right). Each core rod is 5 cm wide. McCloy deposit, Kambalda, Western Australia. (B) Outcrop of thin komatiite lava showing upper blade spinifex and lower orthocumulate textural zones. Pyke Hill, Ontario, Canada. (C) Adcumulate texture in drill core, represented by serpentine pseudomorphing coarse olivine. Mt Keith mine, Western Australia. The left-hand drill core is 5 cm wide.

- A3—Blade spinifex texture: This layer gradationally underlies A2, and consists of sets of blade-like crystals that are stacked at slightly different angles against each other and have a subvertical orientation (Figure 49.4(A) and (B)). Sets of blades increase in size downward. They can be a cm to tens of cm in length, and in some cases, m in length. The base of this textural zone is sharp, and may be planar to gently undulating.
- B1—Harrisitic texture: This zone is a cm to tens of cm thick and also consists of skeletal crystals, which may be elongate to equant in form. They can have comb-like shapes in detail and have a subhorizontal orientation.

- B2—Cumulate texture: Crystals are equant to slightly elongate, and have a round morphology due to resorption. Olivine phenocrysts are separated by a fine aphyric groundmass (Figure 49.4(C)). Orthocumulate texture comprises phenocrysts wholly suspended in the groundmass. Mesocumulate texture comprises crystals that are just touching each other.
- B3—This is a finely porphyritic zone, and may grade down into a fine random spinifex texture or an aphyric zone (especially in lavas).

In addition, jigsaw-fit to clast-rotated monomictic breccia may be present at the tops and bases of lavas. These breccias are autoclastic and commonly produced by quench fragmentation (hyaloclastite). Thin syn-depositional komatiite sills that were intruded into unconsolidated sediment may have all of the above zones as well as sediment-matrix komatiite breccia (peperite) along the contacts.

Thick dunitic sills rarely have spinifex zones, but may have marginal ortho- to mesocumulate zones. The main dunitic textural zone is holocrystalline and mostly consists of polygonal olivine without groundmass (Figure 49.2(C)) called adcumulate. Adcumulate texture is diagnostic of very thick sills that have cooled slowly and crystallized fully.

Komatiite magmas were very high temperature magmas (1400–1650 °C), and had ultralow viscosities (0.1–10.0 Pa s) and very high density (2800 kg m^{-3}) (Huppert and Sparks, 1985). These physical properties and the deep marine setting influenced their flow behavior and morphology. Active lavas had very thin flow fronts, perhaps only cm thick, and spread as very thin sheets and lobes (Cas et al., 1999). The slightest seafloor slope would have created a hydraulic head effect from vent to flow front. Thicker lava units resulted from ponding in seafloor depressions or inflation under a chilled crust, as happens for modern basalt lavas. Komatiite lavas would have been turbulent mostly near the vent, becoming laminar farther away (Cas et al., 1999), similar to carbonatite lavas at Oldoinyo Lengai volcano in Tanzania. Because of their very high density, komatiite magmas were very prone to forming intrusions, both high-level intrusions into unconsolidated sub-seafloor sediments, and deeper crustal intrusions (Fiorentini et al., 2012).

The spinifex textural zone has been interpreted as a chilled lava crust; the various skeletal spinifex zones reflect extremely rapid chilling of this crust due to the subaqueous environment and the very high temperature gradient between the lava surface and the water. Given that many komatiite lavas are entirely coherent from base to top, an insulating steam layer must have been created at the interface between lava and seawater, preventing quench fragmentation. However, lavas with hyaloclastite at the top imply that steam film collapse occurred and resulted in quench fragmentation.

The underlying cumulate textural zone is interpreted to represent a lava tube insulated from contact with the water by the spinifex-textured lava crust. The lava tube allowed magma to be transferred from the vent to the propagating lava front without losing heat. Slow cooling of stagnant lava in the tube allowed crystallization, forming the ortho- and mesocumulate textural zones. Preserved orthocumulate and mesocumulate textures indicate that crystal settling was not efficient, most likely because of the low density contrast between olivine crystals and the melt.

3.4. Nickel Sulfide Mineralization in Komatiites and Its Origin

Komatiites may be mineralized or barren. Magmatic nickel sulfide deposits occur in both komatiite lavas and intrusions, and were precipitated directly from the ultramafic magmas. However, it is commonly agreed that komatiite magmas had relatively high mantle-derived nickel contents (>400 ppm) but low sulfur contents. Therefore, to precipitate sulfide minerals, komatiite magmas must have assimilated sulfur (Lesher, 1989). Sulfur isotope data indicate the sulfur is not mantle derived. For mineralized komatiite intrusions, the sulfur has to be added by assimilation of sulfidic crustal source rocks through which the magma has intruded. For komatiite lavas, the sulfur may also have been derived from subsurface crustal rock assimilation, and/or through assimilation of sulfidic sea-floor substrate (e.g., sulfidic sediments, VHMS deposits or sulfur-rich, probably felsic volcanic rocks) over which komatiite lavas flowed (Fiorentini et al., 2012). The nickel and sulfur combine to form liquid immiscible, monosulfide drops in the komatiite liquid. These then separate into the component sulfide minerals found in komatiite-hosted nickel sulfide deposits. The principal sulfide minerals associated with komatiites are pyrrhotite (Fe$_{1-x}$S), pentlandite ((FeNi)$_9$S$_8$), and chalcopyrite (CuFeS$_2$). There are two styles of komatiite-hosted mineralization, Type 1 (or "contact ore"), massive nickel sulfide, and Type 2, disseminated nickel sulfide. In komatiite lavas, Type 1 high-grade nickel sulfide is the most important, and most commonly occurs as lenses up to several m thick and several tens of m or more in lateral extent, at the base of the lowest lava in a sequence of komatiite lavas, at its contact with the substrate. Disseminated nickel sulfide in komatiite lavas typically has such a low grade that it is uneconomic, unless it occurs close to the massive nickel sulfide and the global price of nickel is high at the time of mining. Komatiite intrusions can host both Type 1 and Type 2 nickel sulfides although Type 2 disseminated nickel sulfide is more common and can be of high enough grade to be economically viable.

The bases of komatiite lavas are generally planar, although in some successions there are channel-like scours. In the Kambalda region of Western Australia, massive nickel sulfide can occur in such troughs. The troughs indicate physical (Cas et al., 1999) and/or thermal (Lesher, 1989) erosion of substrate sediments by the komatiite lavas, and are considered prospective. Komatiite lavas and intrusions associated with sulfide-bearing felsic (lava) successions are also considered to be prospective for nickel sulfide ore deposits. Volcanic facies analysis and an understanding of emplacement processes in these ancient successions are therefore very important exploration tools for komatiite-hosted nickel sulfide ore deposits.

4. EPITHERMAL GOLD–SILVER ORE DEPOSITS

Epithermal deposits are principally mined as sources of gold and silver. The gold and silver may occur as native elements, electrum, in sulfides (especially pyrite) or in tellurides; a variety of sulfide and sulfosalt minerals are commonly present as well (Simmons et al., 2005). Epithermal deposits form at shallow depths below the surface, generally in the range of hundreds of m to about 1.2 km. Metal-bearing hydrothermal fluids rise and deposit gold and silver in response to boiling of the reduced, near-neutral fluid (low-sulfidation), or through reaction of an oxidized, acidic fluid with the host rocks (high-sulfidation) (White and Hedenquist, 1990). Magmas at depth are thought to provide the heat energy that drives the hydrothermal fluid circulation, and also provide volatile compounds that play a role in transporting metals in solution (primarily H_2S, CO_2, HCl, SO_2). Hydrothermal fluid pathways, and hence also the deposits and the associated altered facies, are typically structurally controlled and make use of highly permeable faults and/or fractures and/or facies. Epithermal deposits may comprise one or a small number of relatively wide (m) veins, sets or clusters of multiple smaller (cm to tens of cm) veins, and/or discordant breccia bodies (m to hundreds of m across). An active epithermal system at depth may be represented at the surface by steaming ground, fumaroles, hot springs, and/or silica or travertine sinter.

The genetic connection between epithermal ore deposits and cooling magmas underpins the common occurrence of these deposits in volcanic successions. Many epithermal deposits of all ages are volcanic-hosted. Of those that are volcanic-hosted, the majority occur in subaerial volcanic successions, reflecting the meteoric origin of the hydrothermal fluid in the case of low-sulfidation deposits, and the lack of dilution of the magmatic-hydrothermal fluid in the case of high-sulfidation deposits. That said, some epithermal deposits are significantly younger than the volcanic rocks in which they are hosted (e.g., Sillitoe et al., 2006), hence clearly not directly linked to active volcanism. Some epithermal deposits occur at locations where deeply penetrating faults, rather than magmas, are responsible for fluid circulation, and in submarine volcanic successions (e.g., Salam et al., 2014, Perkins, 1988).

The hydrothermal alteration that accompanies epithermal mineralization records fluid pathways, especially the typically well-focused upflow zones both beneath and above the areas of mineralization. The effects on volcanic rocks and the new hydrothermal minerals depend on the fluid composition, temperature, and fluid flux (White and Hedenquist, 1990). Common hydrothermal minerals associated with epithermal deposits are quartz, chalcedony, calcite, adularia, illite, kaolinite, pyrophyllite; diaspore

and alunite, and there are important differences in the assemblages and abundances of hydrothermal minerals produced by low versus high-sulfidation hydrothermal systems.

Here, we concentrate on epithermal deposits hosted by subaerial volcanic successions: silicic caldera successions, silicic domes and lavas, andesitic composite cones (aka stratovolcanoes) and silicic maar volcanoes. Regardless of the deposit being low versus high-sulfidation, the range of volcanic host facies is similar and similarly broad, and in general, some combination of host facies characteristics and structure controls the form of the deposits.

4.1. Silicic Caldera Successions

Epithermal deposits can occur in both the intracaldera (e.g., Round Mountain, Henry et al., 1997) and extracaldera (aka outflow; e.g., Cerro Vanguardia, Schalamuk et al., 1997) parts of silicic caldera successions (Figure 49.5). The intracaldera setting lies directly above the magmas that are potential sources of heat and fluids, and is also commonly disrupted by multiple caldera-related and/or tectonic faults that serve as fluid pathways. Diverse facies can be the immediate host, for example, rhyolitic or dacitic lavas or domes, intracaldera ignimbrites, or high-level syn-volcanic intrusions. Ignimbrites generated by large-volume explosive caldera-forming eruptions extend for tens of km radially around the caldera margin. The proximal parts of the outflow environment may be close enough to magma sources for epithermal deposits to form (e.g., Steven and Lipman, 1976). In the medial to distal parts of the outflow environment, caldera-related magmas are too far away and epithermal deposits that occur in these locations probably have more complicated origins. Epithermal deposits that occur in ignimbrite successions may show a strong association with a particular ignimbrite unit (e.g., the Granosa Ignimbrite, Cerro Van Guardia), presumably reflecting the influence of the physical properties and composition of particular units on the hydrothermal processes.

4.2. Silicic Domes and Lavas

Subaerial rhyolitic or dacitic lavas and domes may occur as isolated units or in complexes and are commonly located on faults. They may be part of the intracaldera associations of silicic caldera successions (e.g., Tarawera dome complex, Okataina caldera, New Zealand) or occur independently of calderas (e.g., Taylor Creek Rhyolite, New Mexico). These lavas and domes characteristically have a high aspect ratio that reflects the high viscosity of the silicic lava. The high viscosity also results in the production of a thick carapace of autobreccia, and in ubiquitous flow bands. Single emplacement units are typically very thick (tens of m to ~300 m) and have lateral extents as short as a few

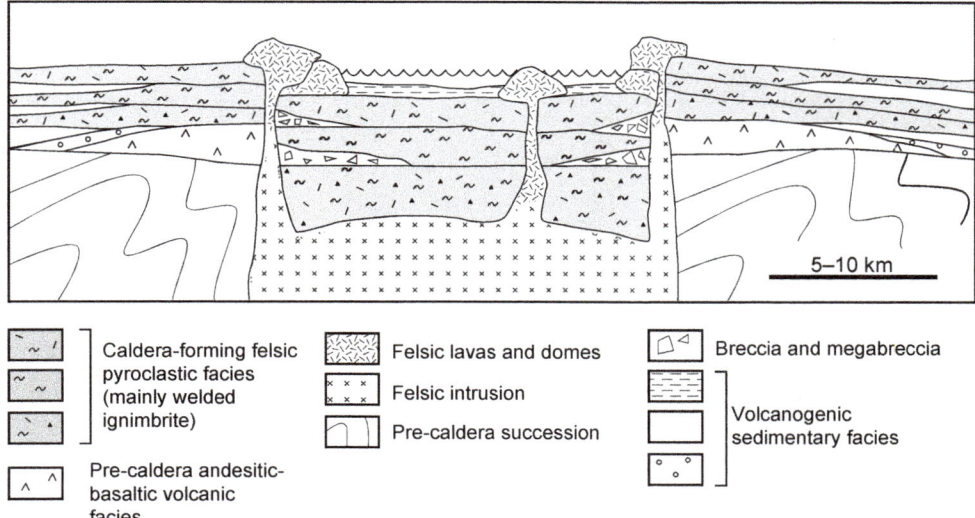

FIGURE 49.5   Simplified cross-section of a subaerial silicic caldera volcano. Epithermal deposits are common in intracaldera settings directly above magmatic sources of heat and fluids. Very thick ignimbrite, felsic lavas and domes, and/or volcanogenic sedimentary facies may be the immediate host rocks to epithermal veins.

hundred m to ~3 km. The groundmass in the outer parts, including the autobreccia carapace, is commonly glassy or partly glassy and highly vesicular (pumiceous). The groundmass in the inner zones includes domains dominated by spherulites, lithophysae, and microcrystalline textures, and is weakly vesicular or non-vesicular. Silicic effusive eruptions may be preceded, accompanied, and/or followed by small-volume explosive eruptions so pyroclastic flow, fall, and surge deposits may be interleaved with the lavas and domes.

4.3. Composite Cones

Composite cones are typically fed by andesitic magmas although significant volumes of basalt, basaltic andesite, and dacite may be present. These volcanoes are characterized by a steep cone around the summit vent, surrounded by a gently dipping ring plain (Hackett and Houghton, 1989). In general, the cone and vent associations are most likely to host epithermal deposits because these associations are directly above sources of magmatic heat and fluids, and because at least in the case of vent associations, there is an established pathway to the surface. Nevertheless, epithermal deposits may be hosted by the ring plain association, especially in cases where structural controls are paramount, and/or the mineralizing hydrothermal system is younger than the volcanism.

Former vents are marked by small-volume (<0.01 km^3) associations of feeder dykes or plugs, domes, shallow intrusions, and coarse breccia (vent-fill talus, proximal pyroclastic deposits), all of which may be hydrothermally altered. Facies geometry and contact relationships are complicated because this location is confined and unstable.

The cone is built by the products of numerous eruptions from one or more central vents and typically dominated by lavas (Figure 49.6). Although explosive eruptions are common, pyroclastic deposits are rarely preserved on the cone because of the steep cone slopes. Andesitic lavas in cone associations may have appreciable primary dip (up to about 20°), are in the range of several to ~20 m thick, and may extend beyond the cone and into the ring plain setting. They comprise a coherent interior encased by coarse autobreccia. In detail, sections through the same cone cannot be easily correlated because lavas are confined within narrow cone segments and do not spread laterally, and small changes in the vent geometry and position greatly affect lava flow directions. The cone stratigraphy can be further complicated by sector collapse events that remove entire portions of the cone instantaneously, creating substantial disconformities within the stratigraphy.

Ring plain associations mainly comprise volcanogenic sedimentary facies generated by mass-wasting, resedimentation, and reworking processes affecting the cone. A wide variety of fluvial, lacustrine, and lahar deposits may be present although all are composed essentially of cone-sourced components. Primary volcanic facies are limited to relatively fine (ash—lapilli), thin (cm) fall deposits and pyroclastic flow deposits. Some ring plain successions include chaotic breccia and megabreccia produced by sector collapse events affecting the cone (debris avalanche deposits).

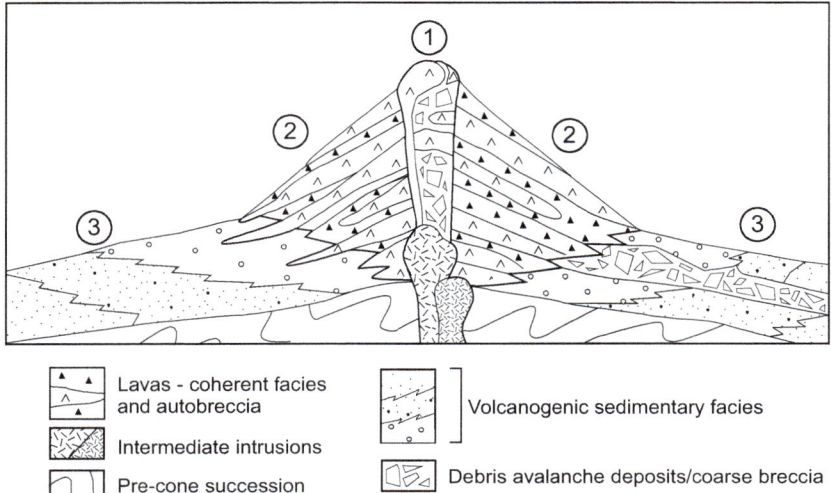

FIGURE 49.6　Facies associations typical of andesitic composite cones: 1. Vent association, dominated by domes, lavas, and coarse breccia. 2. Cone association, mainly comprising lavas. 3. Ring plain association, composed of volcanogenic sedimentary facies derived from erosion and masting events affecting the steep cone.

4.4. Silicic Maar Volcanoes

Maars are small volcanic centers that mainly involve **phreatomagmatic** and **phreatic eruptions** in which external water (or an active hydrothermal system) is the main source of gas (steam) driving explosions. They comprise a deeply excavated crater up to ~2 km across encircled by a rim of pyroclastic surge and fall deposits commonly a few tens of m high (Figure 49.7). Between eruptions and after eruptions cease, the crater may be filled by water within which lacustrine or other volcanogenic sedimentary facies accumulate. Maar craters are widely thought to be underlain by downward-tapering breccia pipes or **diatremes** formed by a variety of pyroclastic, hydrothermal, tectonic, and surface processes. Worldwide, most maars are fed by basaltic magmas. However, silicic maars are known to have been the hosts to a small number of epithermal deposits (e.g., Wau, PNG, Sillitoe et al., 1984; Kelian, Indonesia, Davies et al., 2008). The prime location for epithermal mineralization is within and close to the subsurface diatreme because of the combination of proximity to magmatic heat, permeability and porosity, and connection with circulating groundwater and/or hydrothermal fluids. This location also has the best preservation potential because the loose pyroclastic deposits that form the maar rim beds are easily eroded.

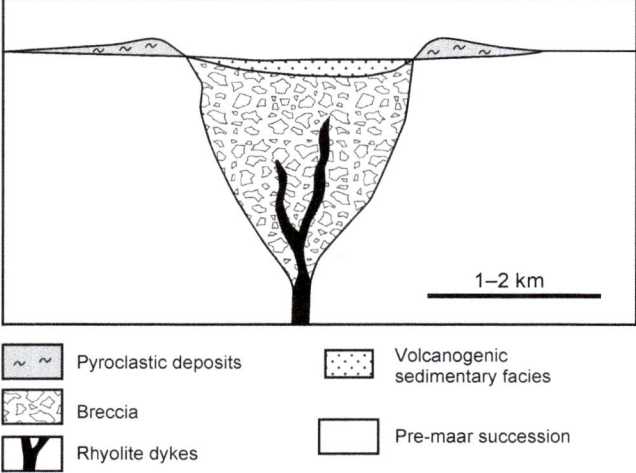

FIGURE 49.7　Simplified cross-section a silicic maar volcano. The wide maar crater surrounds a breccia pipe formed by hydrothermal, phreatomagmatic and mass-wasting fragmentation processes affecting the wall rock and magma. The breccia may be intruded by late felsic dykes that remain unbrecciated.

5. KIMBERLITES, LAMPROITES, AND DIAMONDS

5.1. Kimberlites and Lamproites

Kimberlite and lamproite magmas bring diamonds to the Earth's surface. Diamonds are not magmatic crystals but are **xenocrysts** derived from the deep mantle. Kimberlite magmas are thought to have had very low viscosities (0.1−1 Pa s) and are considered to have risen at very high velocities (>4−20 m s^{-1}) to the Earth's surface (Sparks, 2013). Kimberlite is ultramafic and has extremely low SiO_2 content (<35 wt%), very high MgO (>15 wt%), extremely low Al_2O_3 (<5 wt%), very low Na_2O (<0.5 wt%) and Na_2O/K_2O (<0.5). Lamproite is an undersaturated, ultra potassic basaltic magma. Apart from diamond, the main minerals found in kimberlite are forsteritic olivine, chrome pyrope garnet, chrome diopside, and spinel. Some kimberlites also contain phlogopite. However, kimberlites contain large volumes of xenoliths and xenocrysts, including most of the olivine population. Although there is much debate about the origins and nature of kimberlite magmas, it is currently thought that kimberlite magmas originated as carbonatitic melts that assimilated mantle xenoliths, especially orthopyroxene-bearing xenoliths, thereby transforming them into silicate compositions (Russell et al., 2012). Assimilation would release large amounts of carbon dioxide, which would give the magmas extra buoyancy and help them accelerate to the Earth's surface. Some kimberlite and lamproite magmas were hydrous, as evidenced by the presence of phlogopite, so H_2O also contributed to the ascent and eruption of the magmas. Very few fresh kimberlites have been analyzed, and the actual volatile content at the time of eruption is unknown.

5.2. Setting

Kimberlites and diamond-bearing lamproites occur either within or around the exposed periphery of Archaean cratons. The formation of kimberlite magmas and diamonds therefore relates uniquely to the sub-cratonic mantle. The petrology of mantle xenoliths and the diamond xenocrysts indicate that kimberlite magmas originate from depths of at least 150 km, from the base of the cratonic lithosphere, or deeper.

Kimberlites range in age from Proterozoic to Holocene. The youngest known kimberlite succession is the ~10 ka Igwisi Hills, Tanzania (Brown et al., 2012). The next youngest kimberlites are the early Tertiary Ekati kimberlites, Northwest Territories, Canada. As a result, the original volcanic landforms and surface deposits are rarely preserved. Most kimberlites and lamproites are subvertical pipe-like bodies, which are interpreted to represent the remains of volcanic conduits and vents. The diameter of the pipes ranges from a few tens of m to over 400 m, and they can extend to depths of 2 km. Where they are hosted by older, mechanically competent rock, the pipes are steep-sided (~70°) and taper with depth (e.g., southern African pipes; Lac de Gras kimberlite province, Northwest Territories, Canada; Figure 49.8(A)). However, in the Fort á la Corne kimberlite province of Saskatchewan, Canada, kimberlite bodies are intercalated in contemporaneous Mesozoic shallow marine and coastal plain sedimentary facies that were poorly lithified when kimberlites erupted explosively through them. Because the country rock had little strength, the resulting craters were shallower, more open and flaring in geometry (Cas et al., 2008; Figure 49.8(B)).

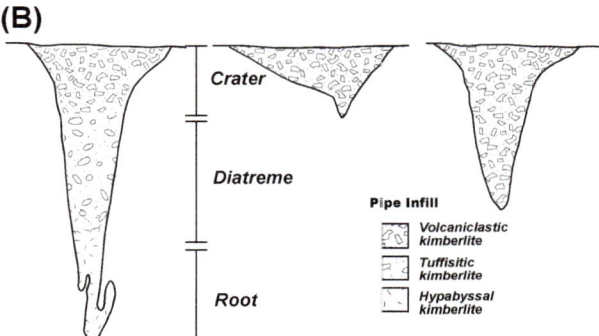

FIGURE 49.8   (A) Aerial view of the open pit into the Koala pipe, Ekati kimberlite field, Canada. The shape of the open pit mimics the pipe shape. (B) Schematic representation of the varying geometry of kimberlite pipes, depending on their geological context. Steep-sided South African and Lac de Gras (Canada) pipes (LHS) are hosted in mechanically strong country rock, whereas Fort á la Corne (Canada) "pipes" (middle and RHS) are hosted by weakly lithified sedimentary rocks are therefore are wider, open and flaring.

5.3. Facies and Volcanology

Most of our understanding of kimberlite volcanic processes comes from study of the deposits that fill kimberlite pipes (Cas et al., 2008, Sparks, 2013) because the volcanic landforms that formed have almost all been eroded. The three Igwisi Hills kimberlite volcanoes are almost complete and consist of relatively steep-sided pyroclastic cones and associated late-stage small-volume lavas, reflecting eruption of residual degassed kimberlite, after an initial phase of explosive volcanism. Craters range from 100 to 200 m in diameter. The cones consist of variably bedded, relatively well-sorted lapilli-stone, lapilli-tuff and tuff, in which the pyroclasts include nonvesicular to poorly vesicular (<20% vesicles) juvenile clasts, pelletal lapilli (lithic or crystal grain in the core coated by a rind of coherent kimberlite), country rock xenoliths, and xenocrystic olivine. Brown et al. (2012) interpreted the pyroclastic deposits to be fall deposits, much like a basaltic scoria cone succession, produced by repeated small explosive eruptions. The lavas are variably vesicular and variably olivine xenocryst-rich, and represent final eruption of degassed kimberlite magma, which is again not uncommon in monogenetic basaltic scoria cones. The small size of the cones and craters indicate low explosive intensity and therefore, low volatile content. All vents in the three Igwisi Hills cones were filled with lava, and were not left as open deep craters, as appears to be the case with many older kimberlite pipes. This fact again indicates that the explosive intensity of the Igwisi Hills volcanoes was relatively weak compared with other kimberlite pipes. That is, kimberlite pipes do not all erupt in the same way. There is no evidence for any phreatomagmatic influence, as has been proposed for the formation of kimberlite pipes by some authors.

The deposits inside older eroded kimberlite pipes are available for study where mined for diamonds or drilled for exploration (Figure 49.8(A)). The principal facies in kimberlite pipes are coherent kimberlite and volcaniclastic kimberlite. Kimberlites are very prone to alteration, and these two facies may be very difficult to distinguish (Hayman et al., 2008). In many kimberlite pipes, most of the pipe consists of syn-eruptive pyroclastic deposits (volcaniclastic kimberlite) and intrusions (coherent kimberlite). However, there are some pipes that are dominated by post-eruptive volcaniclastic, terrigenous, and organic (lignite) sedimentary facies up to hundreds of m thick, indicating that a very deep open vent existed temporarily before being filled post-eruptively. Explosive eruptions excavated the vent by ejecting large volumes of country rock.

Coherent kimberlite commonly intrudes the deep root zone of the pipe, but may also intrude higher into the volcaniclastic pipe fill. It consists of abundant olivine and other xenocrysts of highly variable sizes, mantle xenoliths, and country rock xenoliths, in a groundmass composed of serpentine, carbonate, and clinopyroxene microlites (Figure 49.9(A)). Xenocrysts vary from subangular to round, indicating magmatic resorption. In some coherent kimberlites, there is a population of evenly sized, small, round olivine crystals that may represent magmatic crystals. Coherent kimberlite represents the last phase of degassed kimberlite magma intruding the volcaniclastic pipe-fill succession.

The major volcaniclastic kimberlite facies that occurs in most pipes is massive, poorly sorted, matrix-supported volcaniclastic kimberlite breccia (Figure 49.9(B) and (C)), which ranges from m to hundreds of m thick. The principal components are the same as in coherent kimberlite, except that instead of a groundmass there is a fine (<2 mm) matrix of these components variably altered to serpentine, carbonate, and clinopyroxene microlites. Units of massive volcaniclastic facies can be subhorizontal to

subvertical. This facies is the product of the explosive eruption phase of kimberlite pipe formation.

Cas et al. (2008) noted the similarity of the texture of massive volcaniclastic kimberlite to ignimbrite, and suggested that because of the high density of all pyroclasts in a kimberlite explosive eruption column, these columns would be more prone to gravitational collapse than for any other magma type. If the vents are large, a significant part of the collapsing column would fall back into the vent-producing deposits very similar to ignimbrites. The texture of massive volcaniclastic kimberlite may also be explained by fluidization, but this process is not effective at the scale of entire vents and requires separate stages for vent excavation and vent filling.

In the Fort á la Corne kimberlites, the central massive volcaniclastic facies becomes progressively finer and thinly bedded, laminated, locally normally graded and cross-laminated (Figure 49.9(D)) on the gently sloping flanks and margins of the pipes. Some isolated thin graded beds (<40 cm) of volcaniclastic kimberlite interbedded in the basinal sedimentary facies appear to be distal fall deposits or shelf turbidites (Pittari et al., 2008). Kimberlite pipes and craters can also be filled with volcaniclastic kimberlite resedimented from the crater rim succession, and terrigenous sediment (including carbonaceous mudstone, peat and lignite) eroded and transported into the crater by normal surface processes. In the Ekati field, logs have been found within deep craters. In some pipes, volcaniclastic kimberlite is interbedded with these crater-fill sedimentary facies. These units were not produced by explosive activity from the pipes in which they occur, because the host successions are still intact and not disturbed. They were probably supplied by eruptions from nearby kimberlite volcanoes.

The country rock at the contact with kimberlite pipes varies from unfractured to highly fractured. Shattering was presumably caused by shock waves propagating through the country rock as a result of explosions. Country rock contacts vary from sharp to gradational depending on whether loose, fractured country rock was undisturbed or subsequently removed. Some pipes include megaclasts of country rock (up to tens of m) that collapsed from the vent walls.

Early models for kimberlite pipe formation proposed a "blind" subterranean fluidization-driven process. However, the evidence now indicates that kimberlite pipes vented to the Earth's surface. The volume of country rock xenoliths in the volcaniclastic pipe fill is too small to account for the volume of the country rock that must have been explosively excavated. This deficit of country rock can only be accounted for by explosive ejection from the pipe. In addition, some of the volcaniclastic kimberlite in the pipe fill (especially reworked and resedimented facies) could only have originated under subaerial conditions (Cas et al., 2008). In some cases where pipes are mostly filled by

FIGURE 49.9 (A) Coherent kimberlite from the Koala kimberlite pipe, Canada. Note the very large and irregular nature of most of the large olivine xenocrysts. (B) Poorly sorted massive volcaniclastic kimberlite, Letlehakane kimberlite, Botswana. Lithic clasts are mostly wall-rock basalt. (C) Poorly sorted massive volcaniclastic kimberlite, Jericho kimberlite, Canada. The lithic clasts are granite and limestone. (D) Laminated and cross-laminated fine volcaniclastic kimberlite, Fort á la Corne, Canada.

resedimented volcaniclastic facies, the pipe-forming explosive eruption appears to have been a short-lived major explosive event, whereas in other cases, explosive activity may have been repeated and of intermediate intensity, and in other cases, the explosive activity was relatively weak, as in the Igwisi Hills volcanoes. Such variations in eruption intensity would have been controlled largely by magma ascent rate and volatile contents. Some kimberlite eruptions were probably also phreatomagmatic, but the involvement of external water is not a necessary condition.

5.4. Diamonds in Kimberlites and Lamproites

Not all kimberlite or lamproite pipes are diamond-bearing, either because the mantle through which the magmas passed was not fertile, or diamonds that were incorporated into the magma as xenocryts were resorbed by the magma during ascent. Many diamonds show signs of resorption, which affects their commercial value.

Porritt et al. (2011) assessed the factors that influence diamond grades in the Ekati kimberlites of Canada. They found that higher diamond grades correspond with facies that contain high concentrations of large olivine crystal fragments and low abundances of country rock xenoliths. Olivine is dense and behaves in a similar way to diamonds during transport. They also found that some volcaniclastic kimberlite sourced from a nearby pipe can have a higher diamond grade than the in situ volcaniclastic kimberlite in the pipe, and that large amounts of country rock xenoliths can have the effect of diluting diamond grade. Understanding lithofacies and stratigraphic architecture, eruption, transportation, depositional processes, and the provenance of every deposit in a pipe is therefore critical to understanding grade distribution in a kimberlite pipe.

6. CONCLUSIONS

Volcanic successions are prospective for important resources and in some cases, there is a strong genetic link between volcanism and ore formation. For example, nickel sulfide deposits in komatiite successions are a direct product of the magmatic processes (fractionation, solidification, and assimilation) that operate in komatiitic magmas. The invariable occurrence of VHMS deposits in submarine volcanic successions also reflects a genetic connection between volcanism and ore genesis but a different set of processes and requirements comes into play. For VHMS deposits, volcanism and/or magmatism and hydrothermal systems are both manifestation of high heat flow, but any magma composition and any high heat flow extensional setting may produce these deposits. Apart from heat energy, the only other critical ingredient for the formation of VHMS deposits is seawater because seawater evolves into the hydrothermal fluid and is the main source of sulfur. Furthermore, VHMS deposits are strictly synvolcanic being an "indigenous" component of the volcanic succession in which they occur. The hydrothermal systems that form epithermal deposits commonly operate in contemporaneous subaerial volcanic settings with abundant heat energy and established fluid pathways. However, epithermal deposits may be significantly younger than the host volcanic succession, implying that strongly focused high heat flow, and efficient fluid and heat pathways are more important than regionally extensive high heat flow. Diamonds have a consistent connection with kimberlite magmas but the connection is not genetic. Kimberlite magmas serve to transport xenocrystic diamonds formed in the mantle to the Earth's surface. The rise and eruption of kimberlitic magmas results in diamonds being concentrated in clastic kimberlite facies that fill and surround the vent. Knowledge of the volcanic facies, facies geometry, contact relationships, and internal stratigraphy of the volcanic successions that host ore deposits is essential for understanding ore genesis, alteration patterns, and grade variations.

FURTHER READING

Allen, R.L., 1992. Reconstruction of the tectonic, volcanic, and sedimentary setting of strongly deformed Zn—Cu massive sulfide deposits at Benambra, Victoria. Econ. Geol. 87, 825—854.

Allen, R.L., Lundstrom, I., Ripa, M., Simeonov, A., Christofferson, H., 1996a. Facies analysis of a 1.9 Ga, continental margin, back-arc, felsic caldera province with diverse Zn—Pb—Ag—(Cu—Au) sulfide and Fe oxide deposits, Bergslagen region, Sweden. Econ. Geol. 91, 979—1008.

Allen, R.L., Weihed, P., Svenson, S.A., 1996b. Setting of Zn—Cu—Au—Ag massive sulfide deposits in the evolution and facies architecture of a 1.9 Ga marine volcanic arc, Skellefte district, Sweden. Econ. Geol. 91, 1022—1053.

Arndt, N.T., Naldrett, A.J., Pyke, D.R., 1977. Komatiitic and iron-rich tholeiitic lavas of Munro Township, northeast Ontario. J. Petrol. 18, 319—369.

Boschen, R.E., Rowden, A.A., Clark, M.R., Gardner, J.P.A., 2013. Mining of deep-sea seafloor massive sulfides: a review of the deposits, their benthic communities, impacts from mining, regulatory frameworks and management strategies. Ocean Coast. Manage. 84, 54—67.

Brown, R.J., Manya, S., Buisman, I., Fontana, G., Field, M., MacNiocaill, C., Sparks, R.S.J., Stuart, F., 2012. Eruption of kimberlite magmas: physical volcanology, geomorphology and age of the youngest kimberlitic volcanoes known on Earth (the upper Pleistocene/Holocene Ogwisi Hills volcanoes, Tanzania. Bull. Volcanol. 74, 1621—1643.

Cas, R.A.F., Self, S., Beresford, S., 1999. The behaviour of the fronts of komatiite lavas in medial to distal settings. Earth Planet Sci. Lett. 172, 127—139.

Cas, R.A.F., Hayman, P., Pittari, A., Porritt, L., 2008. Some major problems with existing models and terminology associated with kimberlite pipes from a volcanological perspective, and some suggestions. J. Volcanol. Geotherm. Res. 174, 209—225.

Davies, A.G.S., Cooke, D.R., Gemmell, J.B., Simpson, K.A., 2008. Diatreme breccias at the Kelian gold mine, Kalimantan, Indonesia: Precursors to epithermal gold mineralization. Econ. Geol. 103, 689—716.

Doyle, M.G., Huston, D.L., 1999. The subsea-floor replacement origin of the Ordovician Highway-Reward volcanic-associated massive sulfide deposit, Mount Windsor subprovince, Australia. Econ. Geol. 94, 825—843.

Doyle, M.G., McPhie, J., 2000. Facies architecture of a silicic intrusion-dominated volcanic centre at Highway-Reward, Queensland, Australia. J. Volcanol. Geotherm. Res. 99, 79—96.

Fiorentini, M., Beresford, S., Barley, M., Duuring, P., Bekker, A., Rosengren, N., Cas, R., Hronsky, J., 2012. District to camp controls on the genesis of komatiite-hosted nickel sulfide deposits, Agnew-Wiluna Greenstone Belt, Western Australia: Insights from the multiple sulfur isotopes. Econ. Geol. 107, 781—796.

Franklin, J.M., Gibson, H.L., Galley, A.G., Jonasson, I.R., 2005. Volcanogenic massive sulfide deposits. In: Hedenquist, J.W., Thompson, J.F.H., Goldfarb, R.J., Richards, J.P. (Eds.), Economic Geology 100th Anniversary Volume. Littleton, CO, Society of Economic Geologists, pp. 523—560.

Gibson, H.L., Morton, R.L., Hudak, G., 1999. Submarine volcanic processes, deposits, and environments favorable for the location of massive sulphide deposits. In: Barrie, C.T., Hannington, M.D. (Eds.), Volcanic associated massive sulfide deposits: processes and examples in modern and ancient settings, Reviews in Economic Geology, vol. 8, pp. 13−51.

Gifkins, C., Allen, R.L., 2003. Textural and chemical characteristics of diagenetic and hydrothermal alteration in glassy volcanic rocks: Examples from the Mount Read Volcanics, Tasmania. Econ. Geol. 96, 973−1002.

Hackett, W.R., Houghton, B.F., 1989. A facies model for a Quaternary andesitic composite volcano, Ruapehu, New Zealand. Bull. Volcanol. 51, 433−450.

Hannington, M.D., de Ronde, C.E.J., Petersen, S., 2005. Sea-floor tectonics and submarine hydrothermal systems. In: Hedenquist, J.W., Thompson, J.F.H., Goldfarb, R.J., Richards, J.P. (Eds.), Economic Geology 100th Anniversary Volume. Littleton, CO, Society of Economic Geologists, pp. 111−141.

Hayman, P.C., Cas, R.A.F., Johnson, M., 2008. The difficulties in distinguishing coherent from fragmental kimberlite: a case study of the Muskox pipe (Northern Slave Province, Nunavut, Canada). J. Volcanol. Geotherm. Res. 174, 139−151.

Henry, C., Elson, H.B., McIntosh, W.C., Heizler, M.T., Castor, S.B., 1997. Brief duration of hydrothermal activity at Round Mountain, Nevada, determined from 40Ar/39Ar geochronology. Econ. Geol. 92, 807−826.

Horikoshi, E., 1969. Volcanic activity related to the formation of the Kuroko-type deposits in the Kosaka district, Japan. Miner. Deposita. 4, 321−345.

Huppert, H.E., Sparks, R.S.J., 1985. Komatiites: eruption and flow. J. Petrol. 26, 694−725.

Lesher, C.M., 1989. Komatiite associated nickel sulfide deposits. In: Whitney, J.A., Naldrett, A.J. (Eds.), Ore Deposition Associated with Magmas, Reviews in Economic Geology, vol. 4, pp. 45−101.

McPhie, J., Allen, R.L., 1992. Facies architecture of mineralized submarine volcanic sequences: Cambrian Mount Read Volcanics, western Tasmania. Econ. Geol. 87, 587−596.

McPhie, J., Allen, R.L., 2003. Submarine, silicic, syn-eruptive pyroclastic units in the Mount Read Volcanics, western Tasmania: Influence of vent setting and proximity on lithofacies characteristics. In: White, J.D.L., Smellie, J.L., Clague, D.A. (Eds.), Explosive Subaqueous Volcanism: American Geophysical Union Geophysical Monograph, vol. 140, pp. 245−258.

Pasteris, J.D., 1996. Mount Pinatubo volcano and 'negative' porphyry copper deposits. Geology 24, 1075−1078.

Paulick, H., McPhie, J., 1999. Facies architecture of the felsic lava-dominated host sequence to the Thalanga massive sulfide deposit, Lower Ordovician, northern Queensland. Aust. J. Earth Sci. 46, 391−405.

Perkins, C., 1988. Origin and provenance of submarine volcaniclastic rocks in the Late Permian Drake Volcanics, New South Wales. Aust. J. Earth Sci. 35, 325−337.

Pittari, A., Cas, R.A.F., Lefebvre, N., Robey, J., Kurszlaukis, S., Webb, K.J., 2008. Eruption processes and facies architecture of the Orion Central kimberlite volcanic complex, Fort á la Corne,

Saskatchewan; kimberlite mass flow deposits in a sedimentary basin. J. Volcanol. Geotherm. Res. 174, 152−170.

Porritt, L.A., Cas, R.A.F., Ailleres, L., Oshurst, P., 2011. The influence of volcanological and sedimentological processes on diamond grade distribution in kimberlites: examples from the EKATI Diamond Mine, NWT, Canada. Bull. Volcanol. 73, 1085−1105.

Rosa, C.J.P., McPhie, J., Relvas, J.M.R.S., Pereira, Z., Oliveira, T., Pacheco, N., 2008. Facies analyses and volcanic setting of the giant Neves Corvo massive sulfide deposit, Iberian Pyrite Belt, Portugal. Miner. Deposita. 43, 449−466.

Rosa, C.J.P., McPhie, J., Relvas, J.M.R.S., 2010. Type of volcanoes hosting the massive sulfide deposits of the Iberian Pyrite Belt. J. Volcanol. Geotherm. Res. 194, 107−126.

Russell, J.K., Porritt, L.A., Lavallee, Y., Dingwell, D.B., 2012. Kimberlite ascent by assimilation fuelled buoyancy. Nature 481, 352−356.

Salam, A., Zaw, K., Meffre, S., McPhie, J., Lai, C.-K., 2014. Geochemistry and geochronology of the Chatree epithermal gold-silver deposit: implications for the tectonic setting of the Loei Fold Belt, central Thailand. Gondwana. Res. 26, 198−217.

Schalamuk, I.B., Zubia, M., Genini, A., Fernandez, R.R., 1997. Jurassic epithermal Au-Ag deposits of Patagonia, Argentina. Ore Geol. Rev. 12, 173−186.

Sillitoe, R.H., Baker, E.M., Brook, W.A., 1984. Gold deposits and hydrothermal eruption breccias associated with a maar volcano at Wau, Papua New Guinea. Econ. Geol. 79, 638−655.

Sillitoe, R.H., Hall, D.J., Redwood, S.D., Waddell, A.H., 2006. Pueblo Viejo high-sulfidation epithermal gold-silver deposit, Dominican Republic: a new model of formation beneath barren limestone cover. Econ. Geol. 101, 1427−1435.

Simmons, S.F., White, N.C., John, D.A., 2005. Geological characteristics of epithermal precious and base metal deposits. In: Hedenquist, J.W., Thompson, J.F.H., Goldfarb, R.J., Richards, J.P. (Eds.), Economic Geology 100th Anniversary Volume. Littleton, CO, Society of Economic Geologists, pp. 485−522.

Simpson, K.A., McPhie, J., 2001. Fluidal-clast breccia generated by submarine fire fountaining, Trooper Creek Formation, Queensland, Australia. J. Volcanol. Geotherm. Res. 109, 339−355.

Sparks, R.S.J., 2013. Kimberlite volcanism. Ann. Rev. Earth Planet. Sci. 2013 (41), 497−528.

Steven, T.A., Lipman, P., 1976. Calderas of the San Juan Volcanic Field, southwestern Colorado. USGS Professional Paper 958, 35.

Syme, E.C., Bailes, A.H., 1993. Stratigraphic and tectonic setting of volcanogenic massive sulfide deposits, Flin Flon, Manitoba. Econ. Geol. 88, 566−589.

Waters, J.C., Wallace, D.B., 1992. Volcanology and sedimentology of the host succession to the Hellyer and Que River VHMS deposits, northwestern Tasmania. Econ. Geol. 87, 650−666.

White, N.C., Hedenquist, J., 1990. Epithermal environments and styles of mineralization: variations and their causes, and guidelines for exploration. J. Geochem. Explor. 36, 445−474.

Chapter 50

Volcanic Influences on the Carbon, Sulfur, and Halogen Biogeochemical Cycles

Pierre Delmelle and Elena Maters

Earth & Life Institute, Environmental Sciences, Université Catholique de Louvain, Louvain-la-Neuve, Belgium

Clive Oppenheimer

Department of Geography, University of Cambridge, Cambridge, United Kingdom; Institut des Sciences de la Terre d'Orléans, Universite d'Orléans, Orléans Cedex, France

Chapter Outline

GLOSSARY

anaerobic A state or process without oxygen.

ash Solid particles less than 2 mm in diameter ejected from a volcano during an eruption. The primary ash production mechanism is magma fragmentation during volcanic explosive activity.

basalt Dark-colored extrusive mafic rock characterized by a composition relatively low in silica (~ 50 weight%).

burial diagenesis The set of physical and chemical processes involving the transformation of unconsolidated sediment into rock at various timescales, and depending on factors such as sediment deposition rate, temperature, and pressure.

contact metamorphism The process by which preexisting rock undergoes changes in composition, mineralogy, and/or texture as a result of exposure to a hot intrusive magma body, lava flow, or pyroclastic flow.

dike A body of igneous rock which intrudes preexisting rock, cutting across layers typically in a vertical orientation.

fumarole A volcanic vent from which hot gases and vapors are emitted.

lithology A particular set of characteristics, such as composition and texture, shared by a rock unit.

mafic A magma or igneous rock relatively poor in silica and rich in magnesium and iron. Basalt, dolerite, and gabbro are the most common mafic rocks.

petrological analysis An approach commonly used to estimate the degassing budget of past volcanic eruptions. It involves comparing the volatile content in posteruption material such as

The Encyclopedia of Volcanoes. http://dx.doi.org/10.1016/B978-0-12-385938-9.00050-X

glassy lava or rock fragments to that in glass inclusions trapped within crystals in such material which are representative of the preeruption melt.

silicic A magma or igneous rock relatively rich in silica and poor in magnesium and iron. Granite and rhyolite are the most common silicic rocks.

sill A body of igneous rock which intrudes preexisting rock, running parallel to layers typically in a horizontal orientation.

stratosphere The part of the atmosphere which extends from the upper boundary of the troposphere (i.e., the tropopause) at a height of approximately 7−16 km to a height of approximately 50 km above the Earth's surface.

troposphere The lowest part of the atmosphere which extends from the Earth's surface to a height of approximately 7−16 km above the Earth's surface, at the lower boundary of the stratosphere.

volatiles The chemical species in magmas that occur in a gas or vapor phase at near atmospheric pressure and high temperature observed in magmas. The most important magmatic volatile species are water and carbon dioxide.

1. INTRODUCTION

Volcanic activity and its associated outputs of lava, **ash**, and gas are the ultimate result of complex geological processes occurring in the Earth's interior. The significance of such outputs for the biogeochemical cycles of carbon, sulfur, and halogens has long been acknowledged, but it is only recently that quantitative studies have become available to determine the extent to which volcanoes affect, directly and indirectly, the Earth's carbon, sulfur, and halogen cycles on various spatial and temporal scales. New evidence has also emerged that volcanic disturbances on individual biogeochemical cycles may reverberate throughout the system and affect other cycles.

Various types of volcanic activity ranging from explosive eruptions to nonexplosive gas emissions occur on Earth. Present-day activity is characterized by eruptions of ~50−60 subaerial volcanoes worldwide annually. Thanks to the capabilities of satellite remote sensing, most of these eruptions are detected and for some, their sulfur emissions can be measured spectroscopically. In addition to these short-lived eruptions, a number of volcanoes release gases quiescently into the atmosphere, i.e., nonexplosively and without the discharge of lava flows. Such activity can last for years, decades or even centuries, and these volcanoes serve as natural laboratories for assessing the contribution of contemporaneous volcanic carbon, sulfur, and halogen emissions to global cycles.

It is also known that the Earth's volcanoes sporadically produce enormous explosive eruptions greatly exceeding in size any recorded in human history. These so-called supereruptions involve **silicic** magmas with high contents of volatile constituents. The most recent supereruption occurred 74,000 years ago at Toba in Indonesia. Supereruptions not only inject vast quantities of sulfur into the

atmosphere, they also have the potential to produce widespread (up to several million km^2) and thick ash fallout deposits on land and in the ocean. There is much uncertainty concerning the intensity and duration of the atmospheric and environmental impacts of such eruptions, but the effects may induce significant perturbations in the global carbon cycle.

Large igneous provinces (LIPs) represent another class of exceptionally large volcanic events. They are associated with repeated intrusion and eruption, probably spaced many hundreds to thousands of years apart, of gigantic volumes (likely exceeding 2000 km^3) of magmas over short periods of geologic time (<1 Ma, Ma = million years). The best preserved LIPs are from the Mesozoic (~251−66 Ma ago) and Cenozoic (~66 Ma ago to present). In terms of composition, LIPs are predominantly **mafic** and form series of voluminous basaltic lava flows (LIP eruptions are also referred to as to flood **basalt** eruptions), but some contain significant quantities of silicic rocks. LIP activity has been linked to severe disruptions of the Earth's carbon cycle, and severe repercussions for the environment and life have been speculated.

In this chapter we provide an overview of the carbon, sulfur, and halogen emissions by present-day volcanic activity as well as supereruptions and LIPs. We then discuss how the release of volcanic carbon, sulfur, and halogens may impact the biogeochemical cycles in the atmosphere, ocean, and/or terrestrial ecosystems. Finally, we highlight some feedbacks between the individual cycles.

2. VOLCANIC ACTIVITY AND INFLUENCE ON THE CARBON CYCLE

Volcanic activity is connected with both the short- and long-term geological carbon cycle. The short-term cycle operates on a timescale of days to thousands of years and is governed by exchanges between the near-surface reservoirs of the Earth, i.e., the atmosphere, oceans, biosphere, and sedimentary rocks. In contrast, the long-term cycle operates on a timescale of millions of years and involves the deep Earth's interior, where carbon is buried in the mantle during subduction and returned to the near-surface by volcanic and metamorphic activity and loss of volatile carbon species during **burial diagenesis** of sediments.

2.1. Volcanic Carbon Emission Output

2.1.1. Present-Day Volcanic Activity

Carbon dioxide (CO_2) is usually the dominant carbon compound quiescently or explosively released by active volcanoes. Quantifying the present-day global volcanic CO_2 emission is important to understand fully the carbon cycle and isolate the impact of additional CO_2 emission by anthropogenic activity. The majority of volcanic CO_2 input to the atmosphere is from quiescently degassing activity, with episodic emissions from erupting volcanoes considered comparatively negligible (Burton et al., 2013). For instance, the estimated CO_2 output from the four largest eruptions of the last two centuries (Tambora, 1815 and Krakatoa, 1883 in Indonesia, Katmai, 1912 in Alaska, Mt Pinatubo, 1991 in the Philippines) equals less than 1% of CO_2 output from quiescent degassing over the same period. The most recent global emissions estimate for subaerial volcanism is $\sim$540 Tg CO_2 a^{-1} (Tg = 10^{12} g; a^{-1} = per year; Figure 50.1) which includes emissions from quiescently degassing volcanoes, volcanic lakes, tectonic, hydrothermal and geothermal volcanic areas, and flanks of historically active volcanoes (i.e., soil degassing) (Burton et al., 2013). Volcanoes also emit other carbon-bearing gases including methane (CH_4), carbon monoxide (CO), and carbonyl sulfide, but their impact on the total volcanic carbon output is generally considered negligible compared to that of CO_2. For example, the estimated global CH_4 flux from volcanic activity is 0.34 Tg a^{-1}, although inclusion of output from soil degassing, hot springs and **fumaroles** in volcanic geothermal areas puts the estimate in the range of 2.5−6.3 Tg CH_4 a^{-1}.

The contribution of submarine volcanism at mid-ocean ridges and subduction zones to the Earth's carbon degassing remains mostly unconstrained due to the difficulty of performing emission measurements at the ocean floor. A handful of studies have placed the global CO_2 flux at mid-ocean ridges in the range 50−100 Tg a^{-1}, but reported values of up to $\sim$800 Tg a^{-1} reflect the large uncertainty in these estimates (Burton et al., 2013 and references therein).

Similarly, poor knowledge of the extent of submarine volcanism at subduction zones hinders extrapolation of individual flux estimates to the global scale. A study at submarine volcano NW Rota-1 in the Mariana arc indicates a CO_2 flux of 0.4 Tg a^{-1}, comparable to the lower range values for CO_2 emissions from continuously erupting subaerial volcanoes. Given that most volcanic activity on Earth occurs at the ocean floor, more measurements of CO_2 release by submarine volcanoes are needed in order to quantify the submarine contribution to the total volcanic CO_2 output. In addition, the possibility of unrecognized submarine volcanic phenomena such as cold liquid CO_2 emissions has also been raised, but their significance is not known.

At present, CO_2 fluxes from biological and more notably anthropogenic sources vastly overshadow those from volcanic sources; anthropogenic emissions of $\sim$35 Pg (1 Pg = 10^{15} g) CO_2 a^{-1} exceed volcanic emissions by a factor of more than 60. To put this figure into context, 700 eruptions exhibiting comparable CO_2 release to that of one of the largest eruptions of the last century (Mt Pinatubo 1991; $\sim$0.05 Pg CO_2) would be required every year to match the present annual anthropogenic CO_2 flux (Gerlach, 2011). In order to match current annual anthropogenic CO_2 emissions, the volume of magma erupted would have to be roughly 210 times the current output of 4 km^3 a^{-1}. Clearly, the influence of present-day volcanism is trivial compared to the impact of anthropogenic activity on atmospheric CO_2 concentrations.

2.1.2. Supervolcanoes and LIP Eruptions

Fluxes of CO_2 from supereruptions are poorly constrained but the explosive activity of Toba, the largest eruption of

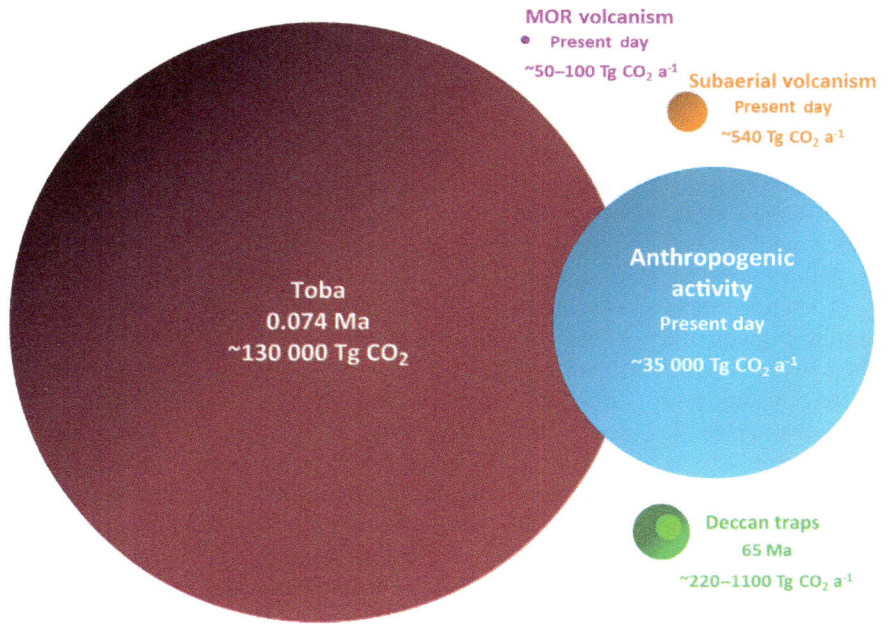

FIGURE 50.1 Diagram depicting volcanic (large igneous provinces, supereruptions, and mid-ocean ridge and subaerial activity) and anthropogenic CO_2 sources with the area of each circle scaled according to the source emission strength. *Values are taken from Burton et al. (2013), Rose and Chesner (1990) and Self et al. (2006).*

MOR volcanism
Present day
$\sim$50−100 Tg CO_2 a^{-1}

Subaerial volcanism
Present day
$\sim$540 Tg CO_2 a^{-1}

Toba
0.074 Ma
$\sim$130 000 Tg CO_2

Anthropogenic activity
Present day
$\sim$35 000 Tg CO_2 a^{-1}

Deccan traps
65 Ma
$\sim$220−1100 Tg CO_2 a^{-1}

the last 2 Ma, is estimated to have released 130 Pg of CO_2 to the atmosphere. This instantaneous output of CO_2 exceeds by more than two orders of magnitude the amount of CO_2 emitted annually by present-day subaerial volcanic activity (Figure 50.1).

As noted above, a number of researchers suggest that LIP eruptions had a major impact on the global carbon cycle, globally marked by a negative carbon isotope excursion recorded in marine and terrestrial carbonates and organic matter. Such perturbation may have served as a trigger for global warming episodes during Phanerozoic history ($\sim$543 Ma ago to present; Wignall, 2001). Interest in investigating the impact of LIPs on the carbon cycle also lies in the apparent association between LIP eruptions and mass extinctions in the Phanerozoic. For instance, the most significant mass extinction in Earth's history (in which $\sim$96% of marine species were eliminated) coincides, within error, with the Siberian Traps eruptions in Russia at the end of the Permian period ($\sim$252–251 Ma ago).

The CO_2 content of LIP mafic magmas is not well known but petrological studies of flood basalts from the Deccan Traps eruptions ($\sim$65 Ma ago) in India reveal that up to 14 Tg CO_2 per km^3 of magma may have been degassed into the atmosphere during each individual eruption (Self et al., 2006). This roughly corresponds to as much as $\sim$1 Pg CO_2 degassed per year assuming a 10-year eruption duration, i.e., about twice the present-day annual global CO_2 output from subaerial volcanism. Despite these staggering rates, CO_2 emissions from LIP magma degassing were probably not sufficient to exert a long-term effect on the mass of CO_2 already present in the contemporaneous atmosphere.

Aside from the volatile content of LIP magmas, the nature of the country rocks (e.g., carbonates, shales, evaporites, coal-bearing sediments) through which the magma intrudes also dictates carbon release to the atmosphere during LIP formation. It has been shown that the mass of CO_2 produced by decarbonation of sedimentary rocks during **contact metamorphism** with **sills** and **dikes** at temperatures below 650 °C can greatly exceed the mass of magmatic CO_2. Importantly, subsurface contact metamorphism of organic-rich shale and coal around magma bodies during LIP eruptions may dramatically enhance the release of CH_4, a greenhouse gas $\sim$20 times more effective than CO_2. Thus, the global warming potency of a given LIP is greater if it intrudes a carbon-rich host crust compared to a carbon-poor crust. Modeling results also predict that assimilation of carbonaceous rocks into mafic magmas at high temperature (1200 °C) generates a magmatic gas phase strongly enriched in CO. Evidence of interaction between LIP magma intrusions and coal-bearing sedimentary rocks has been found for the Emeishan ($\sim$258 Ma ago, China) and Siberian Traps eruptions (Figure 50.2).

FIGURE 50.2 The emissions of carbon, halogen, and halocarbon compounds into the atmosphere during emplacement of the Siberian Traps' flood basalts (shown in the background) were greatly enhanced due to interaction between the hot magmas and carbon-rich (coal) sedimentary rocks (shown in the foreground). Other types of sedimentary rocks were also involved in the eruption. *Photo courtesy of Linda Elkins-Tanton, Carnegie Institution for Science, USA; reproduced by kind permission of the author.*

In a recent study, Rothman et al. (2014) proposed a new mechanism by which the Siberian Traps eruption may have impacted the Earth's carbon cycle. Carbon isotope, archaeal genome, and sediment records in South China show that this event was coeval with the emergence of a nickel-limited microbial metabolic pathway, which allowed efficient conversion of marine organic carbon to CH_4. The nickel limitation may have been alleviated by copious amounts of nickel released into the atmosphere during degassing of the LIP magmas. The methanogenic expansion may then have acted to perturb CO_2 and oxygen levels in the oceans, possibly leading to **anaerobic** oxidation of CH_4 followed by toxic release of hydrogen sulfide (H_2S) to the atmosphere. This chain of reactions may have caused mass extinctions in the oceans and on land.

Finally, it has been argued that changes in the subaerial volcanic CO_2 output on millennial timescales may be partly responsible for glacial/interglacial variations in atmospheric CO_2 concentrations (Huybers and Langmuir, 2009). There is good evidence that the last deglaciation (between 12,000 and 7000 years ago) increased terrestrial volcanism two to six times above background levels. The causal mechanism may have been intensification of magma production owing to the mantle decompression initiated by ablation of glaciers and ice caps. This probably led to a significant rise in volcanic CO_2 emissions and imbalances in the atmospheric carbon budget, which persisted for thousands of years. Thus, volcanism may forge a positive feedback between glacial variability and atmospheric CO_2 concentrations: deglaciation increases volcanic eruptions, which raise atmospheric CO_2, thereby promoting global warming and more deglaciation. Along other oceanic processes, such a positive feedback may contribute to the rapid passage from glacial to interglacial periods.

2.2. Volcanic Activity-Related Carbon Sinks

2.2.1. Removal of Atmospheric Carbon Dioxide by Volcanic Rock Weathering

Volcanism releases CO_2 into the atmosphere but also contributes to CO_2 withdrawal from the atmosphere via chemical weathering of continental volcanic silicate rocks. A general weathering reaction of volcanic rock containing the feldspar mineral albite ($NaAlSi_3O_8$) is,

$$2NaAlSi_3O_{8(s)} + 2CO_{2(g)}$$
$$+ 11H_2O_{(aq)} \rightarrow Al_2Si_2O_5(OH)_{4(s)} + 2Na^+_{(aq)}$$
$$+ 4H_4SiO_{4(aq)} + 2HCO^-_{3(aq)}$$

where the subscripts g, aq, s refer to the gas, aqueous, and solid phases, respectively. The dissolved cations (here Na^+), silica (as silicic acid, H_4SiO_4) and bicarbonate (HCO_3^-) produced by this reaction are exported by rivers to the ocean basins, with HCO_3^- eventually being captured in carbonate sediments on the ocean floor. In the long-term, ocean floor subduction returns the carbon buried as carbonates to the Earth's interior before it is outgassed near the surface once again, mainly by volcanic activity. Overall, silicate weathering results in the net removal of atmospheric CO_2 and, on geological timescales, it roughly compensates for the addition of volcanic CO_2 to the atmosphere.

The rate of silicate weathering, and thereby of atmospheric CO_2 consumption, at the Earth's surface depends principally on **lithology**. The much higher HCO_3^- concentrations measured in rivers draining basaltic rock compared to rivers draining granitic rock simply reflect the higher weathering rate (up to five times) of basalt compared to other igneous rocks such as granite and gneiss. In an influential study, Dessert et al. (2003) utilized river geochemistry data from basalt regions across contrasting climates to conclude that basalt weathering consumes ~ 180 Tg CO_2 a^{-1} and accounts for $\sim 30\%$ of the total continental silicate weathering flux.

Based on the above considerations, the idea has emerged that enhanced weathering following the voluminous eruption of silicate rocks during LIP formation led to long-term negative feedback between volcanic activity and the carbon cycle. For instance, it is suggested that the Deccan Traps eruption initially resulted in an increase in atmospheric CO_2, but emplacement of fresh basaltic rocks across an area of 500,000 km^2 amplified weathering over the following 1 Ma which in turn, resulted in net drawdown of atmospheric CO_2. Continental silicate weathering was similarly enhanced by a factor of 1.5 due to the eruptions of Central Atlantic Magmatic Province basalts in eastern North America ~ 201.5 Ma ago. The eruption lasted <1 Ma but was accompanied by a decline in atmospheric CO_2 over the subsequent 1.5 Ma to a CO_2 concentration below preeruptive levels.

In active volcanic regions, silicate rock weathering can be driven largely by hydrothermal fluids containing sulfuric and hydrochloric acids. Under these conditions, the role of atmospheric CO_2 in the weathering reactions is reduced. For example, $\sim 25\%$ of the total riverine cation flux exiting the Kamchatka Peninsula can be accounted for by hydrothermal fluid–rock interaction. These findings prompt caution when equating the dissolved rock-forming element flux in areas hosting volcanic hydrothermal systems with the consumption of atmospheric CO_2.

Finally, the contribution of extensive continental ash deposits, blanketing areas of up to several millions of square kilometers, which may be emplaced during supereruptions to atmospheric CO_2 consumption via weathering is not known but could be significant. However, water and wind erosion may quickly remove the deposits such that the impact of ash weathering on atmospheric CO_2 may be relatively short-lived.

2.2.2. Removal of Atmospheric Carbon Dioxide due to Ocean Fertilization by Volcanic Ash

Ocean chemistry essentially sets the concentration of CO_2 in the atmosphere on millennial timescales. Oceanic primary productivity, i.e., biomass production by photosynthetic phytoplankton, affects atmospheric CO_2 by exporting both organic carbon and calcium carbonate ($CaCO_3$) from the surface ocean to depth. The former lowers atmospheric CO_2 while the latter modestly raises it. Oceanic primary production accounts for about half of the carbon fixation occurring on Earth, yet phytoplankton growth, and therefore primary productivity, is strongly reduced in $\sim 30\%$ of the ocean due to extremely low concentrations of dissolved iron, a key micronutrient involved in photosynthesis. These iron-limited regions are known as high nutrient low chlorophyll (HNLC) zones.

There is now ample evidence that primary productivity in HNLC zones is contingent upon iron delivery via atmospheric deposition of continental aerosols such as wind-blown mineral dust from arid and semiarid regions. The hypothesis has emerged that the strength of atmospheric CO_2 withdrawal and carbon fixation by oceanic primary productivity is modulated by the flux of mineral dust to HNLC zones. The discovery of an inverse relationship between levels of atmospheric CO_2 and iron recorded throughout the geological past in ice sheet cores lends credence to this idea.

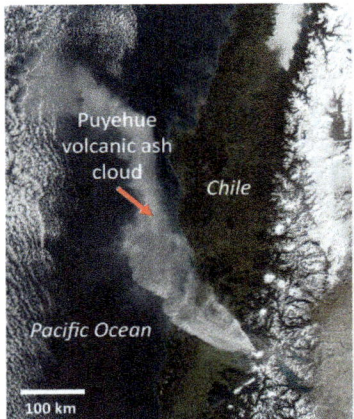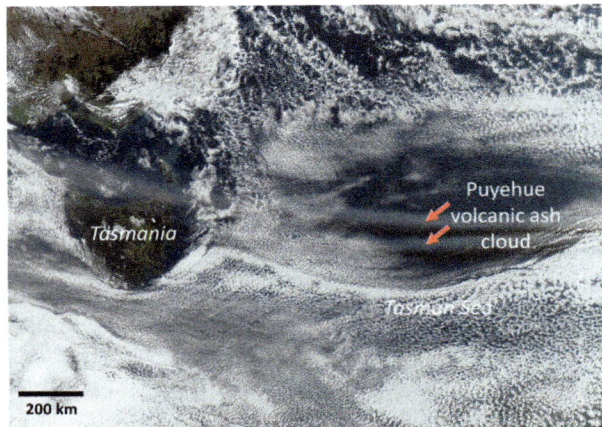

FIGURE 50.3 The long-range atmospheric transport of a volcanic plume can deliver iron-bearing ash to the open ocean and enhance phytoplankton growth and atmospheric CO_2 withdrawal. The photo shows the volcanic ash plume from the 2011 eruptions of Puyehue volcano, Chile, drifting toward the northwest over Chile and the Pacific Ocean on July 2 (left) and toward the east in the jet stream over the Tasman Sea and Tasmania on June 13 (right). Images were captured by Moderate Resolution Imaging Spectroradiometer (MODIS) onboard NASA's Aqua satellite. *Courtesy of the NASA, available from http://earthobservatory.nasa.gov/IOTD/view.php?id=51316&eocn=related_to&eoci=related_image and http://earthobservatory.nasa.gov/ NaturalHazards/view.php?id=50985.*

A connection between volcanic ash deposition and phytoplankton growth in the ocean has similarly begun to be recognized (Figure 50.3). Volcanic ash contains iron in variable concentrations and mineralogical forms and has been found in considerable quantities in ocean sediments, demonstrating ash input to the ocean from eruptions throughout Earth's history. A recent assessment of the millennial deposition flux of ash and dust to the Pacific Ocean and their respective iron release rate in seawater concluded that these two sources provide comparable iron inputs to the marine environment. Furthermore, a single volcanic eruption has the capacity to emit large quantities of iron-bearing particles over very short timescales. For example, ash fallout from the 1991 eruption of Mt Hudson, Chile, deposited an equivalent amount of iron to the sub-antarctic South Atlantic Ocean as ∼500 years of mineral dust fallout from Patagonia. Indirect evidence of past ocean fertilization by ash is further provided by the association of ash deposit and marine diatom (a dominant species of phytoplankton) sequences in deep ocean sediments from the Southern Ocean and in marine deposits off the coast of Denmark.

The first direct proof of ocean fertilization by volcanic ash emerged only recently following the eruption of Kasatochi volcano, Alaska, in August 2008. Researchers detected an anomalous phytoplankton bloom in the Gulf of Alaska, a well-studied HNLC zone, and realized that the area of increased chlorophyll concentrations inferred from satellite-derived ocean color images matched the area that received the Kasatochi ash fallout. Further in situ measurements corroborated the notion that the ash input boosted primary production in the iron-limited surface waters. These results agreed with those of earlier laboratory bioassay experiments. Other recent volcanic eruptions believed to have stimulated phytoplankton growth via oceanic ash deposition include Anatahan, Mariana Islands, in 2003 and Eyjafjallajökull, Iceland, in 2010.

While the Kasatochi fertilization event provided evidence for the ability of volcanic ash to impact ocean biogeochemistry on a basin-wide scale, the magnitude of the resultant oceanic carbon uptake remains uncertain. Nevertheless, the strong temporal correlation between major explosive silicic eruptions and cold climate periods throughout the Phanerozoic could reveal significant strengthening of oceanic CO_2 uptake due to repeated injection of large volumes of iron-rich ash into the world's oceans. Cather et al. (2009) posited that large volcanic ash fluxes into the oceans may be linked to the onset of the Cenozoic icehouse (∼65 Ma ago), the termination of global warming at the Paleocene—Eocene boundary (∼54.8 Ma ago), and the start of Antarctic glaciation at the Eocene—Oligocene boundary (∼33.7 Ma ago).

2.2.3. Sequestration of Organic Carbon in Volcanic Soils

Volcanic soils are formed through weathering of fresh volcanic material, mostly ash and to a lesser extent lava (see Chapter 72 on Volcanic Soils). They exhibit a distinct capacity to store large amounts of organic carbon derived from biological activity, a property relating to repeated burial of the soil surface by ash deposition, which inhibits further microbial decomposition of soil organic matter. Additionally, interaction of organic matter with the poorly crystalline and noncrystalline secondary mineral phases (i.e., iron and aluminum oxyhydroxides and allophanes) often found in volcanic soils protects organic carbon against its conversion to CO_2 during decomposition. Overall, volcanic soils act as a sink for organic carbon on timescales of hundreds to thousands of years and globally sequester organic carbon at a rate of ~ 20 Tg C a^{-1}, equivalent to $\sim 14\%$ of the total CO_2 flux released into the atmosphere by subaerial volcanoes (~ 540 Tg CO_2 a^{-1}) and $\sim 41\%$ of the CO_2 flux consumed by chemical weathering of basalts (~ 180 Tg CO_2 a^{-1}). These figures highlight the significant role that volcanic soils play in the global carbon cycle despite occupying only $\sim 1\%$ of the continental land surface.

3. VOLCANIC ACTIVITY AND INFLUENCE ON THE SULFUR CYCLE

With varying redox states and four stable isotopes, sulfur compounds play diverse, complex, and vital roles in biogeochemical processes. In the Earth's surface reservoir, most sulfur resides in sulfides (e.g., pyrite, FeS_2) and sulfates (e.g., gypsum, $CaSO_4.2H_2O$) in continental sedimentary rocks and as dissolved sulfate in the oceans. Sulfate is the most stable form of sulfur on today's oxic (i.e., where oxygen is present) Earth. Similar to carbon, sulfur is released from the Earth's interior by volcanism, metamorphism, and gas loss during burial diagenesis of sediments. In the oceans, the circulation of seawater at spreading ridges extracts sulfur from hot igneous sulfides and sustains a significant sulfur flux at hydrothermal vents. Sulfur is returned to the mantle at subduction zones via sediments and sulfur mineralization of crustal rocks by seawater. Since most sulfur on Earth is locked up in rocks and in the oceans, sulfur transport occurs mainly through the atmosphere. Major perturbations of the global and atmospheric sulfur cycling have occurred as a result of fossil fuel combustion and metal ore smelting, both of which emit sulfur dioxide (SO_2). Similarly, magma degassing and concomitant atmospheric release of SO_2 and to a lesser degree H_2S at terrestrial volcanoes can affect the sulfur cycle.

3.1. Volcanic Sulfur Emission Output

3.1.1. Present-Day Volcanic Activity

Inventories of volcanic sulfur emissions to the atmosphere reveal that a significant proportion of the total volcanic sulfur load at any one time may be attributed to a mere handful of quiescently degassing volcanoes. In recent decades, these have included Ambrym, Vanuatu; Anatahan, Mariana Islands; Bagana, Papua New Guinea; Mt Etna, Italy; Kilauea, Hawaii, United States; Miyakejima, Japan; Nyiragongo, Democratic Republic of Congo; and Popocatepetl, Mexico.

According to Oppenheimer et al. (2011), the total global sulfur flux from all volcanic sources amounts to ~ 10.4 Tg a^{-1} (~ 20.8 Tg SO_2 a^{-1}), based on observations made over a 25 year period up to 1997. Averaged on centennial timescales, an estimated 1 Tg of sulfur (2 Tg SO_2) reaches the stratosphere annually. However, such estimates have significant margins of error because of the challenges of arriving at meaningful time-averages for the sporadic but large magnitude releases of SO_2 to the stratosphere from explosive eruptions, and of extrapolating field data for a comparatively small number of observed tropospheric volcanic plumes to the global volcano population. This is reflected in an uncertainty of around a factor of two in the global figure. Nevertheless, the majority of volcanic sulfur output worldwide comes from continuous volcanic degassing.

Another significant source of uncertainty in the global volcanic sulfur source is that estimates are based almost exclusively on observations of SO_2 emissions, which may neglect a potentially substantial contribution from H_2S degassing. The relative proportions of SO_2 and H_2S in volcanic emissions reflect temperature, pressure, and redox conditions in the source magma, although interaction between magmatic gases and a hydrothermal system, or sedimentary country rocks, can modify the composition of emissions reaching the surface. A more recent estimate of the total volcanic sulfur flux to the atmosphere attests to these uncertainties, reaching a figure of $\sim 9-46$ Tg a^{-1} of sulfur (Oppenheimer et al., 2011). This estimate is comparable to the present-day sulfur output of ~ 76 Tg a^{-1} from anthropogenic activities (Figure 50.4).

3.1.2. Supereruptions and LIP Eruptions

There are considerable uncertainties in the sulfur emission estimates from ancient supereruptions but values on the order of ~ 200 and 1400 Tg of sulfur have been suggested for Yellowstone in the United States (~ 2 Ma ago) and Toba, respectively, exceeding by far annual global sulfur emissions ($\sim 9-46$ Tg) from present-day volcanoes. The magnitude of sulfur release also appears to depend on the redox state of the magma; a small volume of oxidized magma can give off more sulfur than a larger volume of reduced magma.

Sulfur gas releases from ancient LIP eruptions, which emplaced huge volumes of mafic lavas are also poorly constrained, but detailed petrological studies of the Columbia River, United States (~ 15 Ma ago) and Deccan Traps flood basalts reveal that ~ 1.8 Tg (~ 3.6 Tg SO_2) and ~ 3.5 Tg (~ 7 Tg SO_2), respectively, of sulfur per cubic kilometer of lava may have been injected into the atmosphere. This translates into enormous total sulfur outputs compared to the background atmospheric sulfur concentration of <1 Tg (Self et al., 2006). An estimated 500 Tg of

FIGURE 50.4 Diagram depicting volcanic (large igneous provinces, supereruptions, and subaerial activity) and anthropogenic sulfur (S) sources with the area of each circle scaled according to the source emission strength. *Values are taken from Beerling et al. (2007), Oppenheimer et al. (2011), Self and Blake (2008), Self et al. (2006) and Thordarson and Self (1996).*

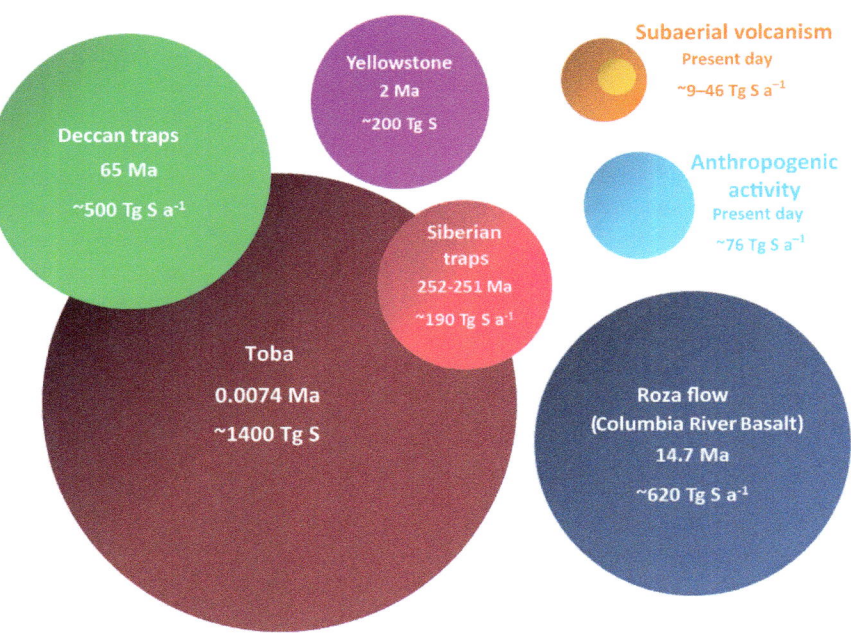

FIGURE 50.4 Diagram depicting volcanic (large igneous provinces, supereruptions, and subaerial activity) and anthropogenic sulfur (S) sources with the area of each circle scaled according to the source emission strength. *Values are taken from Beerling et al. (2007), Oppenheimer et al. (2011), Self and Blake (2008), Self et al. (2006) and Thordarson and Self (1996).*

sulfur (1 Pg SO_2) released per year during LIP eruptions corresponds to an emission rate 10—50 times greater than the present-day annual volcanic sulfur flux. As noted previously for CO_2, the type of crust through which magma intrudes also influences the sulfur output during LIP eruptions, such that contact metamorphism of sulfur-rich evaporites and dolomites may produce sediment-derived SO_2 release several times that of magmatic SO_2.

3.2. Atmospheric, Climatic and Environmental Impacts of Volcanic Sulfur Emissions

3.2.1. Volcanic Forcing of Climate

The impacts of volcanic activity on sulfur in the atmosphere depend on numerous factors but can be broadly divided according to the intensity of activity; minor eruptions and quiescent degassing typically influence the **troposphere** while more powerful eruptions can release sulfur directly into the stratosphere. Release of volcanic sulfur gases causes significant changes to the normal atmospheric concentrations of SO_2, H_2S, and sulfate aerosol, particularly in the stratosphere where these species normally occur in extremely low abundances. Since the atmosphere is oxidizing, all volcanic sulfur that is emitted will ultimately form sulfate aerosols unless it is previously removed by rapid processes of dry and/or wet deposition to the Earth's surface. The main oxidant in the atmosphere is the hydroxyl radical (OH), which reacts with almost all gases including sulfur emissions.

The oxidation of SO_2 to sulfate is slow in the gas phase, occurring on a timescale of ~ 2 weeks in the troposphere. However, the atmospheric lifetime of SO_2 is considerably reduced to days or even hours when its oxidation takes place in the liquid phase. This is because SO_2 is readily taken up into aqueous aerosol particles or cloud droplets where it reacts with other oxidants (e.g., ozone and hydrogen peroxide) and with oxygen in the presence of iron and manganese acting as catalysts. The primary fate of H_2S in the atmosphere is oxidation to SO_2 by $OH\cdot$, followed by formation of sulfate aerosols as described above. In-plume H_2S oxidation could also be mediated at a comparatively faster rate by halogen (chlorine and bromine) radicals.

The present-day volcanic contribution to the tropospheric sulfate budget is not fully understood, yet it is clear that it is disproportionally important in terms of sulfate aerosol formation compared to other sulfur sources including anthropogenic emissions (~ 76 Tg a^{-1}), biomass burning (~ 2.2 Tg a^{-1}), and dimethyl sulfide production (~ 25 Tg a^{-1}) because of the generally higher altitude of emission and thus, longer atmospheric residence time, of volcanic sulfur. However, volcanogenic sulfate aerosols in the troposphere have a much shorter residence time ($\sim$ one week) than in the stratosphere (several years) as they are washed out by cloud formation and rain. Interest in volcanic sulfur emissions in the stratosphere stems largely from the importance of the radiative effects of sulfate aerosols formed by oxidation of gaseous SO_2 and H_2S. As an airborne particulate, sulfate reflects and absorbs sunlight, thereby leading to cooling of the atmosphere below (the so-called "direct effect") and local heating, respectively. For example, the 1991 eruption of Mt Pinatubo resulted in the production of ∼30 Tg of stratospheric sulfate aerosol over a period of months which, in turn, led to a decrease in the global surface temperature of ∼0.4 °C in the year immediately after and ∼0.25 °C in the subsequent year. Huge stratospheric sulfur emissions associated with supereruptions in the past also likely led to severe global cooling episodes. However, it should be emphasized that the magnitude of such cooling does not scale linearly with the volcanic sulfur release into the stratosphere due to the strong influence exerted by particle size distribution on the radiative properties of volcanogenic sulfate aerosols and atmospheric residence times (larger particles sediment more rapidly to the surface). Sulfate aerosols can also act as cloud condensation nuclei, which enable the condensation of water vapor to form clouds. This results in an "indirect effect" on radiation by generating clouds or modifying the properties such as albedo, precipitation rate, and lifetime of existing clouds by introducing additional cloud condensation nuclei. Chapter 53 provides a detailed account of the climatic impacts linked to volcanic activity.

The voluminous and sustained releases of sulfur, dominantly as SO_2, associated with LIP mafic eruptions likely increased sulfate aerosol concentration in the ancient atmosphere. However, the altitude reached by sulfur gases emitted during lava emplacement likely determined the magnitude and duration of the atmospheric effect. In contrast to large explosive volcanic eruptions, which inject sulfur directly into the stratosphere, much of the sulfate aerosol loading associated with LIP eruptions may have been in the mid to upper troposphere where deposition processes, and hence removal, are faster. Additionally, the production of sulfate aerosols from sustained sulfur release was probably self-limited by oxidant (e.g., OH•) availability in the atmosphere. In order to fully understand the magnitude and duration of atmospheric impact from LIP eruptions, further work is required to better constrain the temporal spacing of individual eruptions and the quantity and vertical distribution of sulfur gases emitted into the atmosphere.

3.2.2. Feedback of Volcanic Forcing of Climate on the Carbon Cycle

The direct impact of volcanogenic sulfate aerosols on solar radiation and climate can lead to indirect perturbations to the carbon cycle by influencing CO_2 exchange between the atmosphere and terrestrial and/or oceanic systems. Since measurements began in 1950, several declines in the rates of increase of atmospheric CO_2 abundance have been recorded within an overall trend of increasing atmospheric CO_2, which cannot be accounted for by natural variability or changes in anthropogenic emissions, yet coincide with large volcanic eruptions (e.g., Mt Agung, Indonesia, in

1963, El Chichón, Mexico, in 1982, Mt Pinatubo in 1991) (Jones and Cox, 2001).

A number of hypotheses have been evoked to explain the apparently rapid impact of volcanic activity on the carbon cycle, including an increase in seawater CO_2 solubility, a decrease in soil and plant respiration and an increase in primary productivity on land. A coupled climate-carbon model was used to investigate the impact of the Mt Pinatubo eruption on the carbon cycle. This study attributed the posteruption decline in atmospheric CO_2 to a reduction in the rate of terrestrial CO_2 release. Lower temperatures reduced soil and plant respiration globally and, in combination with variable precipitation changes, reduced primary productivity in temperate regions and increased primary productivity in tropical regions. These effects resulted in a net terrestrial CO_2 uptake dominated by changing CO_2 fluxes in tropical regions in the years immediately following the eruption (Jones and Cox, 2001).

A similar model applied to simulate carbon cycle impacts of volcanic eruptions on a thousand-year scale highlighted a slower return to preeruption carbon fluxes over several decades from land relative to from the ocean. This finding demonstrates the longer term biologically mediated capacity of soil to accumulate and retain carbon than the physically mediated capacity of seawater to do the same, with carbon cycle effects relating to both systems lasting longer than climate effects induced by volcanic sulfate aerosol loading of the stratosphere (Frölicher et al., 2011). The atmospheric CO_2 anomaly typically persists for twice as long as the global temperature anomaly following large explosive eruptions.

It has also been suggested that an increase in diffuse solar radiation resulting from volcanic sulfate aerosol in the stratosphere can enhance photosynthetic CO_2 drawdown due to the greater efficiency with which plants utilize diffuse radiation relative to direct radiation. However, the positive effect of increased diffuse radiation on primary productivity may be offset by the negative effect of decreased direct radiation such that radiative property disturbances by volcanic aerosols have effectively no influence on primary production-mediated CO_2 fluxes.

3.2.3. Suppression of Methane Emission in Wetlands Subject to Volcanogenic Sulfate Deposition

Sulfur loading of the troposphere during volcanic activity can enhance the flux of sulfur returned to the ground via dry and wet deposition processes, as demonstrated locally downwind of several quiescently degassing volcanoes. The effect can also be felt at the regional scale; in Japan, Indonesia, and the Philippines, ∼30, 50% and 20%, respectively, of the total SO_2 deposition have been attributed to volcanic degassing. A modeling study of the impact of the

1783–84 Lakagigar fissure eruption in Iceland on atmospheric composition revealed that spreading of the SO_2 gas plume led to increased sulfate deposition over much of the Northern Hemisphere. Enhanced deposition of volcanic sulfur onto terrestrial ecosystems may contribute to soil acidification but the response, if any, will be highly variable depending on soil mineralogy and drainage conditions.

Perhaps a more important consequence of volcanogenic sulfate addition to terrestrial ecosystems relates to perturbation of the carbon cycle in wetland areas. Many wetlands worldwide are situated within transport distance of gas emissions from active volcanic regions, e.g., in Patagonia, Alaska, Kamchatka, and Iceland. In Indonesia, 50–80% of the wet-deposited sulfur in peat swamp areas is thought to be of volcanic origin. Wetlands constitute the largest single source of CH_4 in the world, releasing $\sim$100-200 Tg a^{-1} to the atmosphere. Methane production in these environments is due to decomposition of organic matter by archaebacteria known as methanogens. Temperature and water table level are two key variables controlling methanogen activity in wetlands. In addition, sulfate availability impedes CH_4 production due to competition for essential organic substrates between methanogens and sulfate-reducing bacteria. Based on this idea, Gauci et al. (2008) predicted that volcanic sulfur inputs to northern peatlands following the 1783–84 Lakagigar eruption stalled CH_4 emissions by up to $\sim$9 Tg a^{-1}, corresponding to a 6% reduction in wetland CH_4 output globally. However, the authors attributed an even greater negative impact on wetland CH_4 emissions to climate cooling due to volcanogenic sulfate aerosol loading of the upper troposphere/lower stratosphere.

4. VOLCANIC ACTIVITY AND INFLUENCE ON THE HALOGEN CYCLES

The halogens (fluorine, chlorine, bromine and iodine) are reactive elements that are distributed between the different Earth system reservoirs. Despite their low concentrations in the Earth's mantle, halogens play an important role in the evolution of magmatic systems and the transport of metals from magmas to hydrothermal systems and the atmosphere, with important implications for understanding the genesis of ore bodies and the interaction between volcanic gases and the Earth's atmosphere.

4.1. Volcanic Halogen Emission Output

4.1.1. Present-Day Volcanic Activity

Halogens released to the atmosphere by volcanoes are degassed as halogen halides, i.e., hydrogen chloride (HCl), hydrogen fluoride (HF), hydrogen bromide (HBr), and hydrogen iodide (HI). While other trace halogen species such as halocarbons and halogen oxides have been detected

in volcanic gases, they generally represent a negligible fraction of the halogen halides. The most recent global volcanic halogen flux estimates are $\sim$4.3 Tg a^{-1} of HCl, $\sim$0.5 Tg a^{-1} of HF, 5–15 Gg (Gg $= 10^9$ g) a^{-1} of HBr, and 0.5–2 Gg a^{-1} of HI (Pyle and Mather, 2009), but these figures are poorly constrained due to the limited number of observations. Halogen releases from submarine volcanism are thought to be comparatively small.

Volcanic halogen loading to the atmosphere is dominated by background emissions from continuously degassing arc volcanoes, but explosive eruptions can inject substantial quantities of halogens into the atmosphere instantaneously. For some magma compositions, emissions of chlorine may be significantly greater than those of sulfur, as appears to have been the case for the Minoan eruption of Santorini in Greece and the Millennium Eruption of Changbaishan (also known as Paektu) on the Chinese/North Korean border.

4.1.2. LIP Eruptions

LIP activity is associated with substantial outputs of halogen gases in the form of HCl and HF from magma degassing. Based on petrological analyses of Columbia River (Roza flow, $\sim$14.7 Ma ago) and Deccan Traps LIP eruptions, halogen emissions from flood basalts are estimated to be $\sim$0.5–1 Tg HCl and $\sim$1.3 Tg HF per km^3 of lava. The lava volume of such eruptions exceeded 1000 km^3 and resulted in the release of hundreds to thousands of Tg of halogen gases to the atmosphere over periods of years to decades (Thordarson and Self, 1996; Self and Blake, 2008). For instance, an estimated $\sim$3400–8700 Pg Cl and $\sim$7100–13 600 Pg F were released during the massive Siberian Traps eruptions, which lasted for several hundred thousand years (Black et al., 2012). These eruptions additionally emitted an estimated 980–10 000 Pg of halocarbon compound (as methyl chloride) via contact metamorphism of Siberian sedimentary country rocks during magma intrusion (Beerling et al., 2007). In all cases, at least half of the halogen gases released by LIP eruptions were injected into the upper troposphere/lower stratosphere with the remainder being released from lava flow into the lower to mid troposphere.

4.2. Atmospheric and Environmental Effects of Volcanic Halogen Emissions

4.2.1. Atmospheric Ozone Destruction

Chlorine chemistry in the stratosphere, particularly over the polar regions, is known to deplete ozone and create so-called "ozone holes." By providing additional reactive surfaces, sulfate aerosols emitted into the stratosphere during powerful explosive volcanic eruptions may enhance

FIGURE 50.5 Volcanic gas plumes in the troposphere, such as the one from Erebus volcano, Antarctica, are highly reactive chemically and have been shown to cause ozone depletion downwind. See text for further details.

the chlorine-mediated ozone depletion reactions. Moreover, such eruptions may supply additional chlorine in the form of HCl for ozone destruction, although anthropogenic chlorofluorocarbon compounds remain the primary reservoir of reactive halogen species. Some models suggest that significant stratospheric injection of volcanic HCl is hampered due to very efficient removal in precipitation. However, more recent simulations predict that interaction of volcanic gases with ice particles in the rising eruption plume enables transfer of up to 25% of the emitted HCl to the stratosphere. Thus, large explosive volcanic eruptions may still have a profound but episodic impact on ozone concentration in the stratosphere.

Another type of volcanism that may lead to stratospheric ozone level disturbances relates to ancient LIP eruptions, where copious releases of halogens and halocarbon compounds into the atmosphere originated from the interaction between hot magmas and organic carbon-rich sedimentary rocks. For example, emplacement of the Siberian Traps flood basalts may have resulted in ozone depletion ranging from moderate to total in both the Northern and Southern Hemispheres. This effect may have hastened the catastrophic end-Permian mass extinction (Black et al., 2012).

The emission of HBr from volcanoes has garnered attention because it can rapidly oxidize to bromine monoxide (BrO) in the atmosphere with consequences for ozone abundance (von Glasow, 2010). Bromine monoxide has been measured in tropospheric emissions just downwind of several volcanoes (e.g., Soufrière Hills, Montserrat; Mt Etna, Italy; Masaya, Nicaragua; Ambrym, Vanuatu) and in eruption plumes in the upper troposphere/lower stratosphere (e.g., the eruptions of Kasatochi volcano in 2008 and Eyjafjallajökull volcano in 2010). The rapid generation of BrO in the atmosphere through heterogeneous chemical processes is known as the "bromine explosion," which produces reactive bromine. The bromine explosion involves in-plume uptake of emitted volcanic

HBr by aerosols and light-dependent reactions. It also requires an acidic environment, which is readily available in volcanic emissions. In polar regions, the bromine explosion triggers chemical destruction of atmospheric ozone during springtime. Recent field measurements indicate that a similar process takes place in tropospheric volcanic plumes (Figure 50.5). Interestingly, reactive bromine in volcanic emissions may also play a role in the rapid oxidation of volcanic and background atmospheric mercury and, consequently, may enhance mercury deposition downwind of degassing volcanoes.

4.2.2. Environmental Effects

Compared to HBr, volcanic HCl and HF emissions in the troposphere are both unreactive. However, HCl can be taken up by and acidify aerosols, thereby enhancing the acid deposition flux in the vicinity of a degassing volcano (e.g., at Masaya and Mt. Etna volcanoes). As a weak acid, volcanic HF does not contribute to aerosol acidity, but rapid HF deposition can contaminate plants, soil surfaces, and water bodies downwind. Overexposure of animals and humans to fluorine-contaminated food and water may be associated with negative health consequences; several volcanic eruptions and some quiescently degassing volcanoes have been held responsible for fluorosis in animal and human populations with varying degrees of severity.

5. CLOSING REMARKS

Notwithstanding the numerous uncertainties and doubts regarding the intensity, timing, and duration of present-day and ancient volcanic outputs of lava, ash, and gas to the atmosphere, land and/or ocean, it is important to recognize that these emissions interact with the biogeochemical cycles of carbon, sulfur, and halogens in numerous complex ways. The sulfate aerosol veil formed by oxidation of volcanic sulfur gases released to the atmosphere during

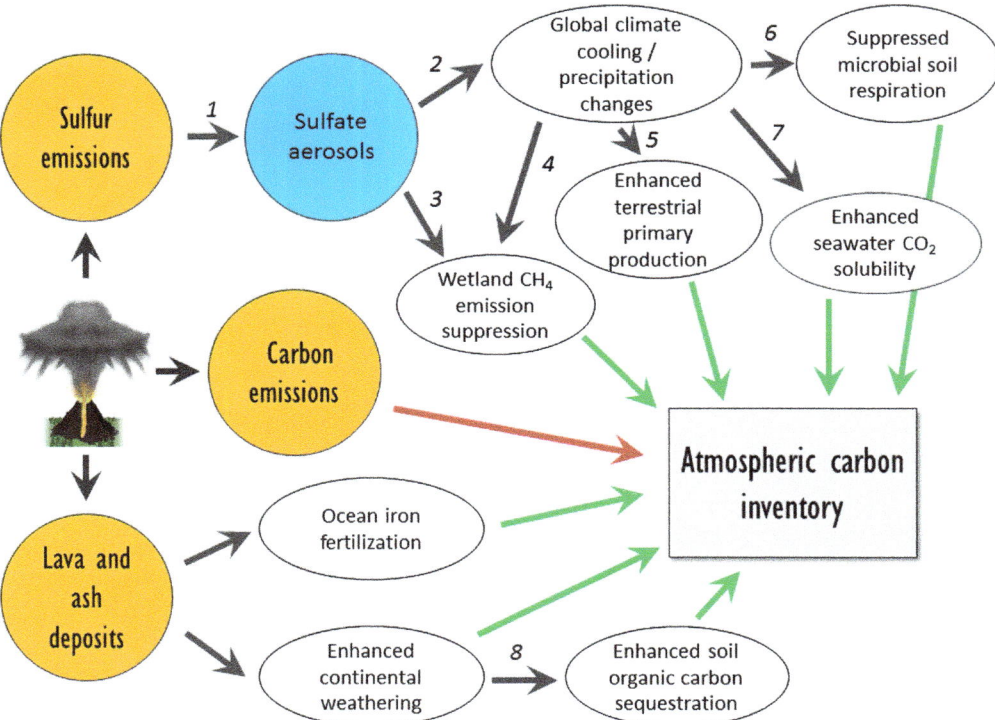

FIGURE 50.6 Conceptual diagram depicting interactions among various volcanic processes and their impacts on the atmospheric carbon inventory. The red arrow signifies a resulting increase in atmospheric carbon while the green arrows signify a resulting decrease in atmospheric carbon. The numbers refer to the following processes: 1. Volcanic sulfur emissions including SO_2 and H_2S are rapidly oxidized in the atmosphere to produce sulfate aerosols; 2. Sulfate aerosols injected into the stratosphere can bring about a decrease in global temperatures as well as regionally varying changes in precipitation; 3. Sulfate aerosols can act to suppress wetland CH_4 emissions directly due to an inhibitory effect of sulfate deposition on methanogenic bacteria activity; 4. A decrease in global temperatures can act to suppress wetland CH_4 emissions indirectly by exerting an inhibitory effect on methanogenic bacteria activity; 5. Sulfate aerosols injected into the stratosphere can bring about regionally varying changes in temperature and precipitation resulting in a net global increase in terrestrial primary production; 6. A decrease in global temperatures can inhibit the rate of microbe-mediated soil organic matter decomposition and plant cellular respiration; 7. A decrease in global temperatures and thereby, sea surface temperatures results in an increase in the solubility of CO_2 in seawater; 8. Weathering of volcanic material produces soils, which exhibit a high-organic carbon sequestration capacity due to interaction with poorly crystalline and noncrystalline secondary mineral phases, which shield organic carbon from microbial decomposition.

powerful explosive eruptions is a primary cause of volcanic forcing of climate. Such sudden climate perturbations may also instigate longer term disturbances of the global carbon cycle, with complex feedback on the CO_2 content of the atmosphere and hence on climate (Figure 50.6).

Counter-intuitively, large-scale explosive eruptions that deliver huge quantities of silicate particles onto the land and into oceans have the potential to strip CO_2 from the atmosphere and perhaps cause global cold snaps on Earth. On land, this phenomenon involves the rapid chemical weathering of ash and lava, and relatedly, the production of volcanic soils with high-carbon sequestration capacity. The mechanism evoked in the ocean involves a biological route where iron (and possibly other elements) in ash plays a determining role. LIP eruptions in the Phanerozoic, which produced voluminous gas emissions and gigantic outpourings of lava also likely created severe environmental stresses through direct or indirect disturbances of the carbon cycle. In addition, new studies suggest that some LIP

events led to collapse of the stratospheric ozone layer, with likely consequences for biota at the surface. Returning to the present day, there is renewed appreciation of the role that volcanic halogen emissions play in atmospheric chemistry.

In order to understand better the impacts of past and present volcanism on the biogeochemical cycles of carbon, sulfur, halogens, and possibly other elements, improved quantitative data on volcanic emissions as well as information on linkages between cycles are needed in a variety of volcanic settings. Addressing this interdisciplinary challenge will contribute to our understanding of the intimate links between volcanism, the environment and life on Earth.

ACKNOWLEDGMENTS

P.D. and E.M. acknowledge support from the Fonds National pour la Recherche Scientifique (FNRS) through a MIS-Ulysse grant F.6001.11.

FURTHER READING

Beerling, D.J., Harfoot, M., Lomax, B., Pyle, J.A., 2007. The stability of the stratospheric ozone layer during the end-Permian eruption of the siberian Traps. Phil. Trans. R. Soc. A 365, 1843–1866.

Black, B.A., Elkins-Tanton, L.T., Rowe, M.C., Peate, I.U., 2012. Magnitude and consequences of volatile release from the siberian Traps. Earth Planet S. C. Lett. 317–318, 363–373.

Burton, M.R., Sawyer, G.M., Granieri, D., 2013. Deep carbon emissions from volcanoes. Rev. Mineral. Geochem. 75, 323–354.

Cather, S.M., Dunbar, N.W., McDowell, F.W., McIntosh, W.C., Scholle, P.A., 2009. Climate forcing by iron fertilization from repeated ignimbrite eruptions: the icehouse silicic large igneous province (SLIP) hypothesis. Geosphere 5, 315–324.

Dessert, C., Dupre, B., Gaillardet, J., Francois, L.M., Allere, C.J., 2003. Basalt weathering laws and the impact of basalt weathering on the global carbon cycle. Chem. Geol. 202, 257–273.

Frölicher, T.L., Joos, F., Raible, C.C., 2011. Sensitivity of atmospheric CO_2 and climate to explosive volcanic eruptions. Biogeosciences 8, 2317–2339.

Gauci, V., Blake, S., Stevenson, D.S., Highwood, E.J., 2008. Halving of the northern wetland CH_4 source by a large icelandic volcanic eruption. J. Geophys. Res. 113, G00A11, doi:10.1029/2007JG000499.

Gerlach, T., 2011. Volcanic versus anthropogenic carbon dioxide. Eos 92, 201–202.

Huybers, P., Langmuir, C., 2009. Feedback between deglaciation, volcanism, and atmospheric CO_2. Earth Planet S. C. Lett. 286, 479–491.

Jones, C.D., Cox, P.M., 2001. Modeling the volcanic signal in the atmospheric CO_2 record. Glob. Biogeochem. CY 15, 453–465.

Oppenheimer, C., Scaillet, B., Martin, R.S., 2011. Sulfur degassing from volcanoes: source conditions, surveillance, plume chemistry and impacts. Rev. Min. Geochem. 73, 363–421.

Pyle, D.M., Mather, T.A., 2009. Halogens in igneous processes and their fluxes to the atmosphere and oceans from volcanic activity: a review. Chem. Geol. 263, 110–121.

Rose, W.I., Chesner, C.A., 1990. Worldwide dispersal of ash and gases from earth's largest known eruption: toba, Sumatra, 75 ka. Palaeogeogr. Palaeocl. 89, 269–275.

Rothman, D.H., Fournier, G.P., French, K.L., Alm, E.J., Boyle, E.A., Cao, C., Summons, R.E., 2014. Methanogenic burst in the end-Permian carbon cycle. Proc. Natl. Acad. Sci. 111, 5462–5467.

Self, S., Widdowson, M., Thordarson, T., Jay, A.E., 2006. Volatile fluxes during flood basalt eruptions and potential effects on the global environment: a deccan perspective. Earth Planet S. C. Lett. 248, 518–532.

Self, S., Blake, S., 2008. Consequences of explosive supereruptions. Elements 4, 41–46.

Self, S., Blake, S., Sharma, K., Widdowson, M., Sephton, S., 2008. Sulfur and chlorine in late cretaceous deccan magmas and eruptive gas release. Science 319, 1654–1657.

Thordarson, Th, Self, S., 1996. Sulfur, chlorine and fluorine degassing and atmospheric loading by the roza eruption, Columbia River Basalt Group, Washington, USA. J. Volcanol. Geoth. Res. 74, 49–73.

von Glasow, R., 2010. Atmospheric chemistry in volcanic plumes. Proc. Natl. Acad. Sci. 107, 6594–6599.

Wignall, P.B., 2001. Large igneous provinces and mass extinctions. Earth-Sci. Rev. 53, 1–33.

Part VII

Volcanic Hazards

Hazel Rymer

Faculty of Science, The Open University, Walton Hall, Milton Keynes, UK

Volcanic eruptions are major natural hazards on Earth. The hazard from a volcanic eruption depends on the type of volcano, time since the last eruption of that volcano, geographical location, local climate, and time of year. The longer a volcano is dormant between eruptions, the larger the eruption tends to be. Thus, while Kilauea (Hawaii) and Etna (Sicily) are active volcanoes that erupt almost continuously, the direct risk to humans from their eruptions is negligible. The risk to property and the environment is not diminished, however. In this section, the various types of hazards posed by volcanoes are discussed mostly from the point of view of their effects on people and the environment. The processes that give rise to these hazards are discussed in Part III (Effusive Volcanism) and Part IV (Explosive Volcanism). The location and timing of a volcanic event affect the risk, as wind direction and strength in the upper atmosphere vary with latitude and season, so that ash from an explosive eruption might be transported around the globe at one time but remain confined to a small region at another time. The 2010 eruption of Eyjafjallajokull is an example of this, where a relatively small ash eruption in southern Iceland caused havoc to Europeans air traffic because of the prevailing wind direction at the time.

The area affected by volcanic activity may range up to several tens of square kilometers, depending on the eruption type. For example, lavas may flow between 1 m and 30 km per hour for distances of several tens of kilometers. Explosive eruptions may eject large quantities (0.1–100's of km^3) of volcanic ash into the atmosphere, much of which falls out locally. The collapse of an explosive eruption column may form a glowing cloud of volcanic debris (nuees ardentes) or pyroclastic flows. Most eruptions are accompanied by sulfur dioxide gas release. This may escape into the upper atmosphere to form a volcanic aerosol that can have global consequences, or it may combine with water vapor at lower elevations and form aerosols that cause local acid rain. Eruptions occurring on volcanic islands or close to the sea may produce tsunamis (tidal waves) with devastating effects reaching tens or even hundreds of kilometers.

The section on Hazards begins with an introduction to Probabilistic Volcanic Hazard Assessment by Chuck Conner and colleagues. This methodology seeks to quantify the probability of a particular volcanic phenomenon. Short-term and long-term probabilistic forecasts have very different uses, but both are increasingly essential as world population continues to grow and more people and the infrastructure on which they rely are at risk.

Passenger airliners that encounter volcanic ash clouds are subject to complete engine failure, which can have catastrophic effects. In Chapter 52, Fred Prata and Bill Rose document several cases of eruption-induced aviation incidents and explain why this hazard is becoming increasingly important. The release of gases to the atmosphere is one of the hazards associated with volcanic eruptions. The several million tons of gases released during large eruptions can be of global significance. Etna volcano (Sicily) emits about 13 Mt (million tons) of CO_2 per year, equivalent to a 1000-MW coal-fired power station in terms of greenhouse gas production. Etna also produces some 1.4 Mt of SO_2 per year. The 1991 eruption of Pinatubo (Philippines) ejected 15 Mt of SO_2, causing surface temperatures to fall globally by about 0.1 °C. The aerosol formed from this gas contributes to global cooling by absorbing incoming solar radiation. In his chapter "Climatic Impacts of Volcanic Eruptions", Alan Robock documents unusual climatic features caused by volcanic eruptions as far back as the seventeenth century BC. It also considers the effects, unintended and otherwise of the various geoengineering solutions that have been proposed to alleviate the impact of climate change. The following chapters cover the other hazards from volcanic eruptions, from pyroclastic density currents to lightning.

The speed and great extent of pyroclastic flows make them by far the greatest and most lethal volcanic hazard.

Lava flows represent a much more sedate volcanic eruption process, but the effects, are still significant.

Volcanic eruptions can trigger secondary events if significant quantities of groundwater, snow, or ice are involved. Risk may then extend beyond the immediate region of the eruption. A relatively minor eruption of Nevado del Ruiz (Colombia, 1985) resulted in the loss of 23,000 lives as the glacier on the top of the volcano melted. Water flowed down the river valleys, removing the loose material from the banks, and the muddy torrent rushed downslope at 30 km/h to cover the village of Armero to a depth of 3 m. A similar muddy torrent (lahar) resulted in the death of about 900 people after the eruption of Mount Pinatubo in 1992. The Hazard from Lahars and Jokulhlaups illustrates how these devastating events can continue to occur even without further volcanic activity, sometimes for decades after the eruption has ceased.

Hazards of Volcanic Gases (Chapter 57) illustrates the effects that various volcanic gases have on people, but perhaps more importantly when these gases fall as acid aerosols on crops, vegetation, and communities downwind. Persistent gas release may cause asphyxiation by inhalation locally and asthma more regionally. Acid rain may also produce skin disorders. All these phenomena are devastating to wildlife and farmland for the duration of the eruption and sometimes far longer.

Most of the world's volcanoes are below the sea, and many are located in coastal areas. Explosion or collapse of a volcano can therefore often cause a large body of water to move suddenly, creating a tsunami. The chapter on Volcanic Tsunamis presents several mechanisms that are thought to produce tsunamis and the substantial number of volcano-related fatalities attributed to tsunamis.

Volcanic earthquakes have different characteristics from the deep and tectonic seismic events that occur along great faults. They are considerably less devastating and may continue at a detectable level for weeks or months prior to eruption. As the Chapter 59 demonstrates, they are the most important window to the processes going on inside a volcano before eruption, and their study has considerable scientific value and application to risk mitigation. As human populations continue to grow and expand, many people live close to active or potentially active volcanoes. The impact on human health of volcanic activity may at first seem obvious, but as Peter Baxter and Claire Horwell show in Chapter 60, the health consequences of natural disasters in developing countries can be reduced by integrating disaster recovery and resilience with poverty reduction and improved disaster management. They present a graphic account of the nature of injuries and the cause of death from various types of eruptions.

While human life is only one of the casualties in the face of an eruption, it is striking just how resilient life is in the face of volcanic activity. Eruptions do not completely eradicate life and many species survive; even newly formed volcanic islands soon become home to a wide range of opportunists.

A look back through the geological record reveals considerable evidence for huge eruptions. If we scale up the observed effects of recent eruptions, the global consequences of large eruptions are awesome. Mike Rampino and Steve Self provide compelling evidence in Chapter 61 for a correlation between mass extinctions and the largest basalt lava flows on Earth. Whether these flood basalts, mass extinctions, and asteroid impacts are related in some way is a matter of ongoing debate.

Finally in this section, we consider volcanic lightning. This is not a new phenomenon, but it is only in recent years that it has been studied systematically. Some of the most stunning pictures of eruptions include volcanic lightning and it is a significant risk for people on the ground. Thus, volcanic hazards have always impinged on life and the environment, and an understanding of their past and present effects is vital for effective monitoring, prediction, and mitigation.

Chapter 51

Probabilistic Volcanic Hazard Assessment

Chuck Connor

School of Geosciences, University of South Florida, Tampa, FL, USA

Mark Bebbington

Statistics Group and Volcanic Risk Solutions, Massey University, Palmerston North, New Zealand

Warner Marzocchi

INGV—Istituto Nazionale di Geofisica e Vulcanologia, Italy

Chapter Outline

GLOSSARY

akaike information criterion (AIC) Comparatively good models have greater likelihood and fewer parameters. Therefore, the relative quality of a statistical model can be assessed based on the AIC, which depends on the number of parameters and maximum estimated likelihood of the model: $AIC = 2m - 2\ln L$, where m is the number of model parameters and L is the maximum estimated likelihood.

Bayesian methods Although model parameters may be unknown, usually there is prior information about the potential distribution of these parameters. Bayes' theorem allows additional prior information to be included explicitly in probabilistic hazard assessments, even if this prior information is somewhat subjective.

hazard curve The potential distribution of event magnitudes, such as the thickness of tephra accumulation or the dynamic pressure of a lahar, may be represented as a histogram, a cumulative distribution function or as a complementary distribution function. When shown as a complementary distribution function, the curve represents the probability of an event magnitude exceeding a specific value, and hence is referred to as a hazard curve, or survival curve.

Markov model A statistical model that includes some dependence on previous states of the system, either in time or space.

parameter An unknown quantity in any mathematical model. Model parameters are estimated using data and sometimes additional information.

point process Any model for events occurring at random in time and/ or space. The chance of occurrence may depend on the history of the process, and the points may have "marks" corresponding to specific characteristics (e.g., volume, rock-type). Poisson and renewal models are special cases.

Poisson model One standard steady-state model for discrete random variables, such as volcanic events, in which the occurrence of any event does not inhibit or encourage the occurrence of additional events. Therefore, the hazard curve for a Poisson model is exponentially decreasing, i.e., memoryless.

probabilistic eruption forecast A statement of the likelihood of an eruption and some indication of the uncertainty in this estimate, including windows for the onset time, range of possible eruption location, size, style, and the probability of the next eruptive event falling within the specified windows.

probability density function A function of a continuous random variable from which it is possible to retrieve a measure of the likelihood of any specific interval of values.

random variable A function whose value changes depending on random chance, that is, the value of the variable will change in successive experiments due to complex changes in the experimental system. Features of volcanic systems, such as the timing of eruptions or the thickness of tephra accumulation are quantitatively studied as random variables.

renewal model A statistical model, usually applied to event recurrence rates, in which the repose interval between events is

The Encyclopedia of Volcanoes. http://dx.doi.org/10.1016/B978-0-12-385938-9.00051-1

described by a renewal function, which may attempt to model variations in rate with elapsed time. Renewal models are used to describe event clustering, periodic behavior, and similar processes.

stationary A statistical process whose properties (mean, variance, autocorrelation) remain constant over its domain. Processes that are characterized by trends are termed nonstationary.

Weibull model A two-parameter standard statistical model that is a generalization of the exponential distribution and is sometimes used to describe accelerated failure with time. For example, if stress is applied to a volcanic system, rock mechanics suggests that the system is more likely to fail with passing time. Such a system can be described by a Weibull model.

1. INTRODUCTION

The goal of volcanic hazard assessment is to provide quantitative, probabilistic estimates of the occurrence and magnitude of potential volcanic phenomena. In this context, volcanic hazard is defined as the probability of a specific event (e.g., volcanic eruption of a specific size) or phenomenon (e.g., the occurrence of lahar inundation of a specific location during a specific time interval). Worldwide, tremendous progress has been made in recent decades in forecasting volcanic events, such as episodes of volcanic unrest, timing of eruptions, and the potential impacts of volcanic eruptions. Practitioners of volcanic hazard assessment strive to quantify the probabilities of such events using volcanological data, statistical models, and numerical simulations of specific volcanic phenomena.

People, of course, have been concerned with the frequency and magnitude of volcanic activity for a very long time. The Greek philosopher Strabo (approximately 64 BC−AD 24) wrote that volcanoes are "safety valves" for the Earth and that we derive our understanding of hazards from observation of things like deluges, earthquakes, and volcanic eruptions. Pliny counted new islands formed in the Mediterranean as a result of volcanic activity, presumably to get a sense of the number of volcanoes formed through time. Although their observations lack any probabilistic context, Strabo and Pliny certainly had empirical views of these phenomena and their frequency. During 1900s−1920s, both Thomas Jaggar and Giuseppe Mercalli studied the frequency of volcanic eruptions, at Hawaiian volcanoes and Vesuvius, respectively, with an eye toward forecasting the potential for future eruptions. Reginald Daly considered physical mechanisms leading to eruption periodicity during this time. The advent of widespread radiometric age dating of volcanic rocks in the 1960s−1970s allowed volcanologists to develop a long-term perspective on rates of volcanic activity. For example, radiometric age determinations allowed Dwight Crandell and Donal Mullineaux to constrain and compare rates of activity at Cascades volcanoes for the first time.

Increased resolution of the geological record was mirrored by development of statistical models of rates of volcanic activity. For example, Frans Wickman applied alternative statistical models of volcano repose intervals during this period, using **Poisson** and **Markov models**. These developments marked the beginning of the understanding that volcanological data and statistical models might be used together to quantify rates of volcanic activity, and hence to assess volcanic hazards.

Experience with volcanic disasters such as the 1902 eruption of Mount Pelée, Martinique clarified the need to understand the potential magnitudes of volcanic events and the potential areas affected by them, in addition to their frequency. An excellent example of an early, qualitative hazard assessment is the work of Thomas Jaggar completed in the 1930s and 1940s on lava flow hazards to Hilo Harbor, Hawaii. A central theme of that study was the frequent activity and "periodicity" of effusive eruptions of the Kilauea−Mauna Loa volcano system. Like other studies of the time, Jaggar used the record of past lava flow activity and current topography to assess hazards to Hilo Harbor. The study is highly qualitative by modern standards because of its lack of statistical modeling or numerical simulations of lava flows, which were unknown at the time of his hazard assessment. Nevertheless, Jaggar's work represents an important step toward assessment of hazards. During the last 20−30 years, numerical simulation of volcanic processes has enabled studies that consider the magnitude of specific phenomena, like the potential volume of lava flows, and estimation of the probability of inundation of specific areas in far greater detail than was previously possible.

Thus, the rise of probabilistic volcanic hazard forecasts has relied on developments in several aspects of volcanology. First, vastly improved quality of available data has enabled forecasts. These are geologic data, such as high-resolution stratigraphy calibrated by radiometric age determinations, and geophysical data associated with volcano monitoring. Second, numerical simulation of volcanic phenomena was virtually unknown more than 30 years ago, but has rapidly developed through our understanding of the physics of volcanic processes, and our ability to run simulations using ever more sophisticated and faster codes and computers. Third, statistical models are evolving at an equally rapid rate, allowing application of improved methods for understanding, for example, uncertainty in volcanic hazard, extreme events, and incorporation of disparate data sets into forecasts using event trees, **Bayesian methods**, and expert elicitation.

Today, probabilistic forecasts of volcanic activity are conveniently divided into two categories. Short-term forecasts are prepared in response to unrest at volcanoes, rely on geophysical monitoring and related observations, and have the goal of forecasting events on timescales of hours to weeks to provide time for evacuation of people, shutdown of facilities, and implementation of related safety measures. Long-term forecasts are prepared to better understand the potential impacts of volcanism in the future and to plan for potential volcanic activity. Long-term forecasts are particularly useful to better understand and communicate the potential consequences of volcanic events for populated areas around volcanoes. In the following, we present general methods in probabilistic volcanic hazard assessment for forecasting long-term recurrence rate, location, and magnitude of potential volcanic events—the basic components of any comprehensive long-term volcanic hazard assessment. Other chapters in this volume present details of specific monitoring techniques and numerical simulation of specific volcanic phenomena.

2. RECURRENCE RATES

Long-term volcanic hazard assessment requires estimation of the recurrence rate of volcanic activity, defined as the expected number of volcanic events within a given time interval and often within a specific range of eruption size. Recurrence rate estimates of volcanic events are most often based upon the frequency with which these events (eruptions or other evidence of volcanic unrest) occurred in the past for a specific volcanic system. These estimates are then used to forecast future volcanic activity. A key component of long-term hazard assessment is the assumption that future rates of activity can be extrapolated from the past rate, at least throughout some timescale of interest. In other cases, the primary concern is the duration of quiescence, or lack of activity, over a prolonged period of time. Some volcanic systems can be argued to be "extinct." That is, there is no credible potential for future eruptions from the volcanic system. Although specific volcanoes certainly go extinct, it is often quite ambiguous whether the system is truly extinct, or merely experiencing a prolonged period of inactivity. Data on the timing of past volcanic events are required, and statistical models must be developed, in order to assess recurrence rate of volcanic activity or the significance of the duration of volcano quiescence on any timescale.

Consider a sequence of volcanic events (eruptive events) associated with a volcanic system. An estimation of the current recurrence rate at time t, $\lambda_{H_{S,T}}(t)$, is derived from the history of the volcano, $H_{S,T}$, between times S and T. The early limit, S, might be defined as the age of the oldest event in the entire volcanic system, the known date at which observation started, or the time at which the volcano "woke from slumber," that is, the onset of some recent episode of activity. The later limit T is typically the time at which the calculation of the recurrence rate was done, which could be the time of the youngest event in the sequence, or perhaps

the present time. The number of events known to have occurred during the interval between S and T is denoted $N(S,T)$. The probability of renewed eruptions during some time period, u (e.g., the next 100 years) is:

$$Pr[N(T, T + u) \geq 1] = 1 - \exp\left[-\int_{T}^{T+u} \lambda_T(t)dt \right].$$

$$(51.1)$$

Then the probabilistic analysis becomes a matter of estimating at time T the recurrence rate, $\lambda_T(t)$ for times $t > T$. The recurrence rate $\lambda_T(t)$ is also referred to as the **point process** intensity, or the instantaneous recurrence rate. Most generally, the recurrence rate depends on the history of the volcanic system, $H_T = H_{S,T}$:

$$\lambda_T(t|H_T) = \lim_{\Delta t \to 0} \frac{Pr(N(t, t + \Delta t) = 1|H_T)}{\Delta t} \quad (51.2)$$

where the history H_T is reduced to the onset times of a sequence of eruptive events $S \leq t_0, t_1, t_2 \ldots, t_n < T$. Equivalently, the history of the volcano can be considered in terms of repose intervals, defined as $r_i = t_i - t_{i-1}$ for $i = 1, 2, \ldots, n$. These events may additionally be considered in terms of eruption volume, v_i, eruption magnitude or explosivity, m_i, or composition, x_i. Consideration of volume eruption rate is important because many volcanic systems follow a volume–time-predictable model. Long-term trends or cycles in eruption magnitude or composition are also inferred for many volcanic systems.

Although this general model structure is straightforward, there is nearly always ambiguity in how events, t_i, and the onset, or observation, of activity, S, are defined. For historically observed eruptions, the timing of the onset of eruptions is usually reported, but the duration of eruptions is often not reported. This becomes problematic, for instance, when the repose interval between eruptions is comparable to the duration of eruptions. To address this, and to attempt to clarify the nature of events, volcanologists use terms such as eruptive phase, to characterize and differentiate one part of an eruptive sequence from another. Eruptive episodes are sometimes defined as sequences of eruptions that are relatively closely spaced in time. In the geological record, it is often difficult to distinguish eruptive phases, eruptions, or eruptive episodes without very high precision geochronology. Often, eruptions are defined in the geological record in terms of mappable units (e.g., lava flows) or stratigraphic units (e.g., tephra layers). In practice, it is most important to use a consistent definition of terms such as eruption, eruptive episode, and events when estimating recurrence rate for a volcanic system. The implicit units or dimensions of the hazard are then those of the data used to construct the model.

2.1. Data Requirements for Estimating Recurrence Rates

Determining the long-term recurrence rate of volcanic activity for a specific volcanic system requires detailed study of the volcano. The most fundamental data required are geologic maps and stratigraphic studies that may be used to delineate the sequence of eruptive events in the volcanic system. Stratigraphic relationships within volcanic systems are often quite complex and ambiguous. Mapping efforts, therefore, must focus on the identification of individual eruptive units and their stratigraphic relationships. Of primary importance is the recognition that the goal of geologic mapping and stratigraphic analyses done for hazard assessments may not be the same as the goals of such studies for more general purposes.

There are numerous factors to consider at the outset of data collection for recurrence rate estimates. For many volcanic systems only the most recent volcanic events are known. In some cases the timing of these events is known from historical accounts of eruptive activity. For many volcanoes, the most recent eruptive events are the best known, simply because the products of these eruptions mantle the topography and comprise the most accessible rock outcrops. As a result, in the geological literature and volcano databases, often only the ages of the most recent activity have been reported. Conversely, the oldest volcanic units are often not exposed or are unrecognized, so the onset of activity, S, and the total duration of volcanic activity are often uncertain. Similarly, eruptive data tends to be biased toward the largest eruptive events, simply because these large eruptions leave copious deposits with a greater chance of being preserved in the rock-stratigraphic record. It is often unclear if the youngest ages for volcanic systems are associated with the largest recent eruptions or are significantly younger, as the record of smaller eruptions may go unmapped or may be completely removed by erosion. Overall, the number of eruptions identified in the geological record decreases with time into the past. For example, the number of known eruptions of a given magnitude decreases exponentially with time for the geologic record of the last 2 Ma in Japan (Figure 51.1). In fact, for this data set, 50% of the total known eruptions occurred within the last 65 ka. The trend for volcanic explosivity index (VEI) 4 eruptions indicates that 97% of these eruptions older than 200 ka are missing from the geologic record (eroded or unrecognized). This same trend exists for even the largest eruptions (VEI 6 and 7). As the tephra stratigraphy for Japan is among the best known in the world, these figures are a clear indication that recurrence rate estimates based on known events must be biased toward lower rates if underreporting is not properly addressed.

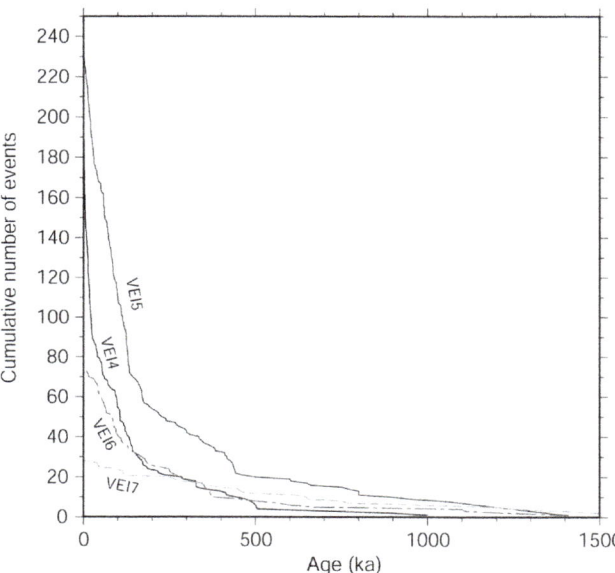

FIGURE 51.1   The number of explosive volcanic eruptions known to have occurred in Japan during the last 2 Ma, shown by volcanic explosivity index (VEI). The decrease in frequency of explosive eruptions within increasing time into the past indicates that many eruptions are missing from the geologic record. Either these eruptive deposits are eroded away or are unrecognized. This creates bias in recurrence rate estimates. *Figure courtesy of Koji Kiyosugi.*

Sequences of volcanic eruptions often are, or appear to be, episodic on widely varying timescales. Periods of activity in the volcanic system may persist for tens to thousands of years, with periods of inactivity between episodes of thousands, or even tens of thousands of years duration. Especially for volcanoes that appear to be in prolonged periods of inactivity, analyses may need to consider the likelihood of renewed episodes of volcanic activity, rather than recurrence rate of individual eruptions. Other volcanic systems have nonstationary behavior. There are documented cases of volcanic systems increasing recurrence rate by one order of magnitude or more. Similarly, volcanic systems may decrease in rate of activity over time. Finally, many volcanoes appear to exhibit a frequency—magnitude relationship, as large-volume eruptions are much less frequent than small-volume eruptions in many volcanic systems.

Given these realities, even with the most detailed mapping, the stratigraphic sequence of activity at volcanoes is generally incompletely known. This leads to uncertainty in recurrence rate estimates of volcanic activity, which may span several orders of magnitude. This incomplete record of activity may also result in bias in recurrence rate estimates, as undetected events lead to lower recurrence rate estimates. It is a typical experience that long-term, detailed study of volcanic systems reveals additional evidence of the frequency of volcanic eruptions, which invariably increases

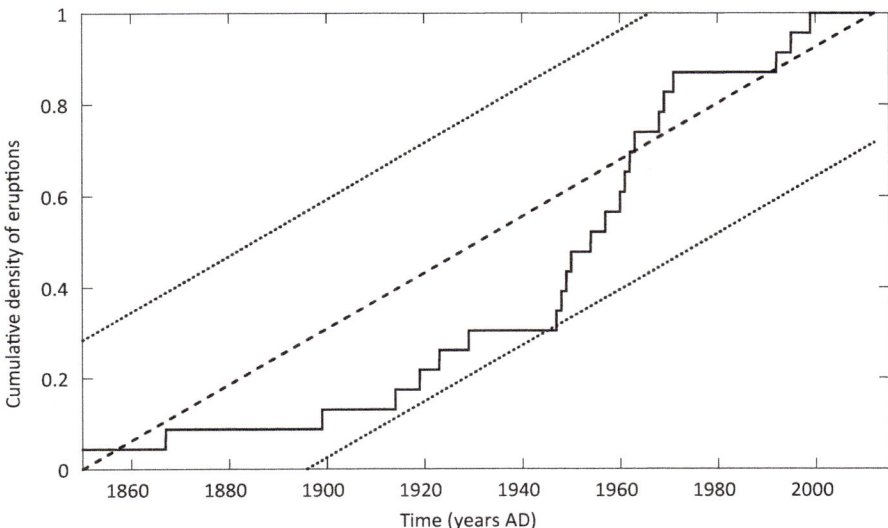

FIGURE 51.2 Cumulative distribution function for eruptions of Cerro Negro volcano, Nicaragua, since its formation in 1850 (solid line). The dashed line indicates a steady-state model with the confidence interval for this steady-state model shown by finer dashed lines. The Kolmogorov—Smirnov test indicates that this series not in steady state, with eruptions occurring at a significantly increased frequency after approximately 1947.

the estimated recurrence rate of events. For example, prior to the 2008 eruption of Chaíten volcano, Chile, this volcanic system was considered to be long dormant, perhaps not having erupted during the last 9 ka. Following the 2008 eruption, there was renewed interest in stratigraphic studies of the volcano, which revealed more frequent Holocene activity than previously known. Some of these uncertainties are irreducible. But in all cases, clear documentation of the data set and discussion of its completeness is required in any volcanic hazard assessment.

3. STEADY-STATE MODELS

Whether or not some statistical models (e.g., Poisson process or **renewal models**) can be applied to estimate recurrence rate and probability of eruption within a given time period for a given set of eruption sequences depends on determining if this sequence is in steady state. Often, eruption sequence data over some period of time, say the Holocene record, may appear to be in steady state, while the same data set may appear over a longer (or shorter) time period to not be in steady state. Lack of steady state may reflect sampling bias, as described previously, or may reflect the episodic nature of volcanic eruptions in some systems.

If a sequence is in steady state, then the level of activity in the future is well-represented by activity in the past. Mathematically, this means that estimated probabilities of future events will be independent of t, the time at which they could occur. If t is a factor in the probability forecast, then the system is not in steady state.

The Kolmogorov—Smirnov test may be used to determine if the eruption sequence is steady state. This test is ideal because usually there are few events comprising the

eruption sequence and because the test is simple. For the cumulative number of volcanic events in the sequence:

$$F_n(t) = \frac{\#(t_i \le t)}{n}, \quad i = 1 \ldots n, \quad S < t < T \quad (51.3)$$

then the Kolmogorov—Smirnov test statistic is:

$$D_n = \max_{t \in [S,T]} \left[\left| F_n(t) - \frac{t - S}{T - S} \right| \right]. \quad (51.4)$$

In other words, in the Kolmogorov—Smirnov test statistic, the maximum deviation from the expected frequency of events determines whether the process is steady state or not. The 95% confidence bounds are given as $1.36/\sqrt{N(S,T)}$.

Consider the history of eruptions at Cerro Negro volcano, Nicaragua, since its formation in 1850. For this series, $N(S,T) = 23$, and $D_n = 0.29$, at 1947 (Figure 51.2). The cumulative distribution function for Cerro Negro eruptions shows significant deviation from steady state, based on the Kolmogorov—Smirnov test at the 95% confidence level. Eruptions prior to approximately 1947 occurred at a significantly lower rate than subsequent eruptions. This suggests that a change in eruptive regime occurred at Cerro Negro at approximately this time, and a different recurrence rate applies to these two different regimes.

4. STATISTICAL MODELS

Numerous statistical models have been applied to describe patterns of volcanic activity through time. Renewal models include the Poisson process, where volcanism can be described as steady state. Non-Poisson renewal models are invoked to describe varying degrees of clustering or regularity in eruption onsets.

More elaborate models are required to describe waxing or waning rates of volcanism, which are observed in many well-studied volcanic systems. The simplest of these is the monotonic **Weibull** (or Power-law) process, where activity decreases asymptotically towards zero, or increases at a decelerating rate. Increase at an accelerating rate is also possible, but not necessarily realistic in the limit. More complex cases of wax-and-wane or cyclic behavior have been modeled using hierarchical models, where steady-state episodes themselves occur in a Poisson process, or the trend renewal process, where nonstationary behavior is embedded through a parametric timescaling. Although recurrence rate is basically estimated using the number of volcanic events in a specific time range, there is a clear dichotomy between steady-state and nonsteady-state models. We will consider steady state first.

The simplest approach to steady-state recurrence rate estimation is to average the number of events that have occurred during some arbitrary time period. An alternative approach, especially appropriate when the total number of known events is small, is to calculate recurrence rate using the repose-time method, in which the time range is restricted by the estimated ages of youngest and oldest events:

$$\widehat{\lambda}_T = \frac{N-1}{T-S} \tag{51.5}$$

where $\widehat{\lambda}_T$ is the recurrence rate (identical for all t), N is the number of events, occurring between S, the age of the first (oldest) event, and T, the age of the most recent (youngest) event. This is the Poisson process estimate, and is actually a maximum likelihood estimate, provided that an event occurs at time S. If the distribution of repose intervals in a volcanic system show Poisson behavior, then the probability of a given number of events, k in some time interval u is:

$$Pr\left[N(t+u) - N(t) = k\right] = \frac{e^{-\lambda_T u}(\lambda_T u)^k}{k!}, \quad k = 1\dots n \tag{51.6}$$

for all t (the definition of stationarity), and the probability of zero events in the time interval, u, is:

$$Pr\left[N(t) = 0\right] = e^{-\lambda_T u} \tag{51.7}$$

indicating that repose intervals between eruptive events have an exponential distribution, the memoryless property of which implies that there is no possibility of using patterns of activity in the past history to improve on the forecast.

The Poisson process specifies that events occur at random, i.e., that we know nothing about the likely time to the next event based on the time elapsed since previous events. Renewal processes are a generalization that assumes

that the length of the current repose (only) is informative. In a non-Poissonian renewal process, the time between events is usually assumed to have at least two **parameters**, which can be fitted by method of moments. To illustrate, suppose that the renewal distribution is a gamma(α,ν) distribution, with mean repose time α/ν and variance α/ν^2. Then the parameters α and ν can be estimated as the solution of the two equations

$$\frac{\alpha}{\nu} = \frac{N-1}{T-S} \tag{51.8}$$

and

$$\frac{\alpha}{\nu^2} = \frac{\sum_{i=1}^{N}(r_i - \bar{r})^2}{N} \tag{51.9}$$

where the $\{r_i\}$ are the reposes, and $\bar{r}$ their mean. Note that in order to model clustering or regularity in this way, we must have data on past clustering or regularity.

Given this framework, alternative statistical models can be constructed using alternative models of recurrence rate for an eruption sequence, t, or repose intervals, r. In the latter case, where u is defined as the time elapsed since the last eruption, the recurrence rate $\lambda_T(u)$ is equivalent to the hazard function.

It is useful to compare these models fit to the same eruption sequence data set using the **Akaike Information Criterion (AIC)**. AIC is extremely useful for identifying the better statistical model when the eruption sequence is precisely known, or can be rigorously sampled. In practice, there may be insufficient information available to distinguish among these different statistical models. Usually a simpler (e.g., one parameter) model is justified over more complex (e.g., two parameter) models if the major uncertainties are geological, and lie with the determination of the eruption sequence. The AIC does not tell one whether a model is actually a reasonable description of the data, that requires goodness of fit tests.

Models for nonsteady-state behavior naturally require more subjectivity, due to the infinite ways in which the data sequence may be fit with nonsteady-state statistical models. The simplest approach is to parameterize the time-varying behavior. In a nonhomogeneous **Poisson model**, the recurrence rate $\lambda_T(t)$, is allowed to vary with time t, but events still occur at random. In other words, the absolute time t carries information about the likely time to the next event, but the past events and their timings do not. At time t,

$$Pr\left[N(t+u) - N(t) = k\right] = \frac{e^{-\lambda_T(t)u}(\lambda_T(t)u)^k}{k!},$$

$$k = 1\dots n. \tag{51.10}$$

Note that this depends on t, but not at all on what may have happened prior to t, whether it be 9 kyr of quiescence, or centuries of Strombolian behavior. These factors only enter into the process of estimating $\lambda_T(t)$.

For a nonhomogeneous (power-law) Poisson model:

$$\lambda_T(t) = \alpha t^\beta. \qquad (51.11)$$

In this case, the parameters α and β must be estimated in order to use the distribution, usually using maximum likelihood. Note that estimates of $\widehat{\beta} > 1$ are possible, although not necessarily physically reasonable, and that the trend is relative to some arbitrary time denoted $t = 0$, not easily defined.

Any steady-state renewal model can be generalized to nonsteady state using the trend renewal process formulation. This replaces the reposes $r_i = t_i - t_{i-1}$ by the difference $r_i^* = \Psi(t_i) - \Psi(t_{i-1})$, where Ψ is a monotonically increasing function. In this way, models incorporating clustering, trend, wax-and-wane and cycles can be fitted, and the results compared using AIC. Note that the scaled reposes remain statistically independent.

Frans Wickman was the first to suggest that the activity of a volcano might not change continuously; rather that it may be at any given time in one of a discrete number of steady-state "regimes." This has been substantiated by many subsequent investigations. This makes for a very attractive treatment in terms of recurrence rates, provided that we can, first, identify the breaks in regime and, secondly, provide a stochastic model for changes in regime. Many methods have been proposed to answer the former problem. Essentially these boil down to determining a graphical "envelope" within which the record of the volcano, be it eruptive volume, number of onsets or size of event, should track: excursions beyond the envelope are considered to indicate a change in regime. In a regime model, the future activity is a function of (and differs with) the regime, and hence mixture models are not regime models, as the possible future activity is probabilistically identical at all times. Hence, forecasting in a regime model needs to encompass both future activity in the current regime, and possible changes to other regimes. This can be done through a hierarchical scheme or, more generally, in a hidden Markov model framework.

Modeling future onsets requires at a minimum the use of previous onset data. It is possible that other data, formally termed covariates, from the eruption record may be available that may carry predictive value. The classic example in long-term forecasting is the erupted volumes of previous eruptions. In a load-and-discharge type model, the likelihood of an eruption is assumed to increase with the amount of unerupted magma volume, which is itself decreased by eruptions. Hence large(er) eruptions should be followed by long(er) reposes for magma recharge,

which is the basis of the time-predictable model. Many models and tests for time-predictability exist, and it can be incorporated in a renewal function framework. Eruption duration, usually correlated with volume, can also be used as a covariate. It is also possible to weaken the renewal (steady state) formulation by tracking the level of the inferred magma chamber. In the volume-history (dependent) model:

$$\lambda_T(t) = \exp\{\alpha + \nu[\rho t - V(t)]\}, \qquad (51.12)$$

where $V(t) = \sum_{k:t_k<t} v_k$ is the cumulative volume erupted prior to time t, and $\alpha, \nu > 0$, $\rho > 0$ are parameters to be estimated through maximum likelihood. The term α incorporates the unknown state of the volcano at time S.

Explanatory variables need not be limited to measures of past eruption size. For example, accumulated extensional strain may be a covariate. Geochemical composition has also been correlated with repose length, and can be incorporated using a proportional hazards approach into a renewal model, or, indeed, any other model that can be parameterized via a recurrence rate. A similar approach has been used to test for triggering of eruptions by large earthquakes.

While temporal forecasting through recurrence rates appears to be on a firm footing, the forecasting of eruption "size" is not. Analogous to the time-predictable model is the "size-predictable model," where long(er) repose intervals are supposed to be followed by large(er) eruptions. Although data on eruption size is limited, size-predictability appears to be much rarer than time-predictability, a pattern that continues when size is measured by duration. Hence, the usual approach is to forecast eruption size independently of onset time, using a power-law or extreme value distribution. Improving on this is one of the most important challenges facing volcanic hazard analysis.

5. LOCATION OF ERUPTIVE VENTS

In addition to forecasting when a volcano will erupt, the question of where the volcano will erupt is fundamental. Nearly all volcanoes have distributed vents. Sometimes volcanoes erupt from a central vent; other times they erupt from lateral or distributed vents. Some volcanic systems, such as monogenetic volcanic fields, lack central vents altogether. The question of where vents form is critical because many flow phenomena, lava flows, pyroclastic flows, and the like, can impact completely different areas even with small changes in vent location. This is particularly the case on steep-sided composite volcanoes, where a small change in vent location can lead to the flow descending a completely different flank of the volcano. Probabilistic methods developed in spatial statistics are amenable to forecasting the probability of vent formation in a given area. Spatial density, showing the map distribution of probability of new vent formation, is a conditional probability, depending on the assumption that a new vent forms, and integrates to one across the region of interest (the new vent must form somewhere).

How can vent spatial density be estimated? A simple approach relies on the distribution of mapped vent locations that have erupted in the past. Like in probability models of the timing of future eruptions, this assumption assumes a steady-state model for the distribution of new volcanic vents is appropriate. Kernel density estimation is a nonparametric method for estimating the spatial density of future volcanic events based on the locations of eruptive vents. Two important parts of the spatial density estimate are the kernel function and its bandwidth, or smoothing parameter. The kernel function is a **probability density function** that defines the probability of future vent formation at locations within a region of interest.

A two-dimensional elliptical kernel with a bandwidth that varies in magnitude and direction is given by:

$$\widehat{\lambda}(\mathbf{s}) = \frac{1}{2\pi N \sqrt{|\mathbf{H}|}} \sum_{i=1}^{N} \exp\left[-\frac{1}{2}\mathbf{b}^T\mathbf{b}\right] \tag{51.13}$$

where,

$$\mathbf{b} = \mathbf{H}^{-1/2}\mathbf{x}.$$

The local spatial density estimate, $\widehat{\lambda}(\mathbf{s})$ is based on N total events (vent locations). The bandwidth, $\mathbf{H}$, is a 2×2 element matrix that specifies the bandwidth as an ellipse, with major and minor axis lengths, and a rotation. This bandwidth matrix is both positive and definite, important because the matrix must have a square root; $\mathbf{x}$ is a 1×2 distance matrix, $\mathbf{b}$ is the cross product of $\mathbf{x}$ and $\mathbf{H}^{-1/2}$. The resulting spatial density at each point location, $\mathbf{s}$, is usually distributed on a grid that is large enough to cover the entire region of interest.

A special case of the kernel function is a two-dimensional radially symmetric Gaussian kernel:

$$\widehat{\lambda}(\mathbf{s}) = \frac{1}{2\pi h^2 N} \sum_{i=1}^{N} \exp\left[-\frac{1}{2}\left(\frac{d_i}{h}\right)^2\right] \tag{51.14}$$

that depends on the distance, d_i, to each event location, N, from the point of the spatial density estimate, $\mathbf{s}$, and the smoothing bandwidth, h that is constant in all directions.

The bandwidth is selected using some criterion, often visual smoothness of the resulting spatial density plots. Bandwidths that are narrow focus density near the locations

of mapped vents. Conversely, a large bandwidth may oversmooth the density estimate, resulting in unreasonably low density estimates near clusters of past events, and overestimate density far from past events. This dependence on bandwidth can create ambiguity in the interpretation of spatial density if bandwidths are arbitrarily selected. Several methods have been developed for estimating an optimal bandwidth matrix based on the locations of the event data, such as the modified asymptotic mean integrated squared error method. Such optimization methods are extremely useful because, although they are mathematically complex, they find optimal bandwidths using the actual data locations, removing subjectivity from the process. That said, alternative optimization methods result in alternative spatial density maps (Figure 51.3), so an important part of the hazard assessment is evaluation of the sensitivity of vent spatial density to different model assumptions.

A potential disadvantage of these kernel functions and similar methods is that they are not inherently sensitive to geologic boundaries and to ancillary data. For example, the distribution of earthquake hypocenters or zones of high strain observed during magma intrusion should inform and update the spatial density estimate as this information becomes available. Various schemes have been developed to weight spatial density estimates in light of geological or geophysical information in a **Bayesian** framework. A difficulty with such weighting is the subjectivity involved in recasting geologic observations as density functions. Furthermore, geologic insight is not always consistent with event distributions. Developing such weighting schemes remains an important area of research. In addition, note that this statistical structure uncouples the temporal forecasting problem from the spatial forecasting problem. That is, estimates of the recurrence rate are not mathematically linked to estimates of the spatial density. This assumption may be inappropriate on some volcanoes where there is migration of activity. An area of significant future research is the formulation of spatiotemporal models, which make this link explicit.

6. PROBABILISTIC VOLCANIC HAZARD CURVES AND MAPS

Many people living near volcanoes are less concerned with the probability of a volcano erupting, or the probable locations of future vents, than with the probability that the products of these eruptions might reach their homes, or destroy their city and its infrastructure. The role of **hazard curves** and maps is to quantify potential hazards and to communicate these potential hazards to stakeholders. Probabilistic hazard maps accomplish this

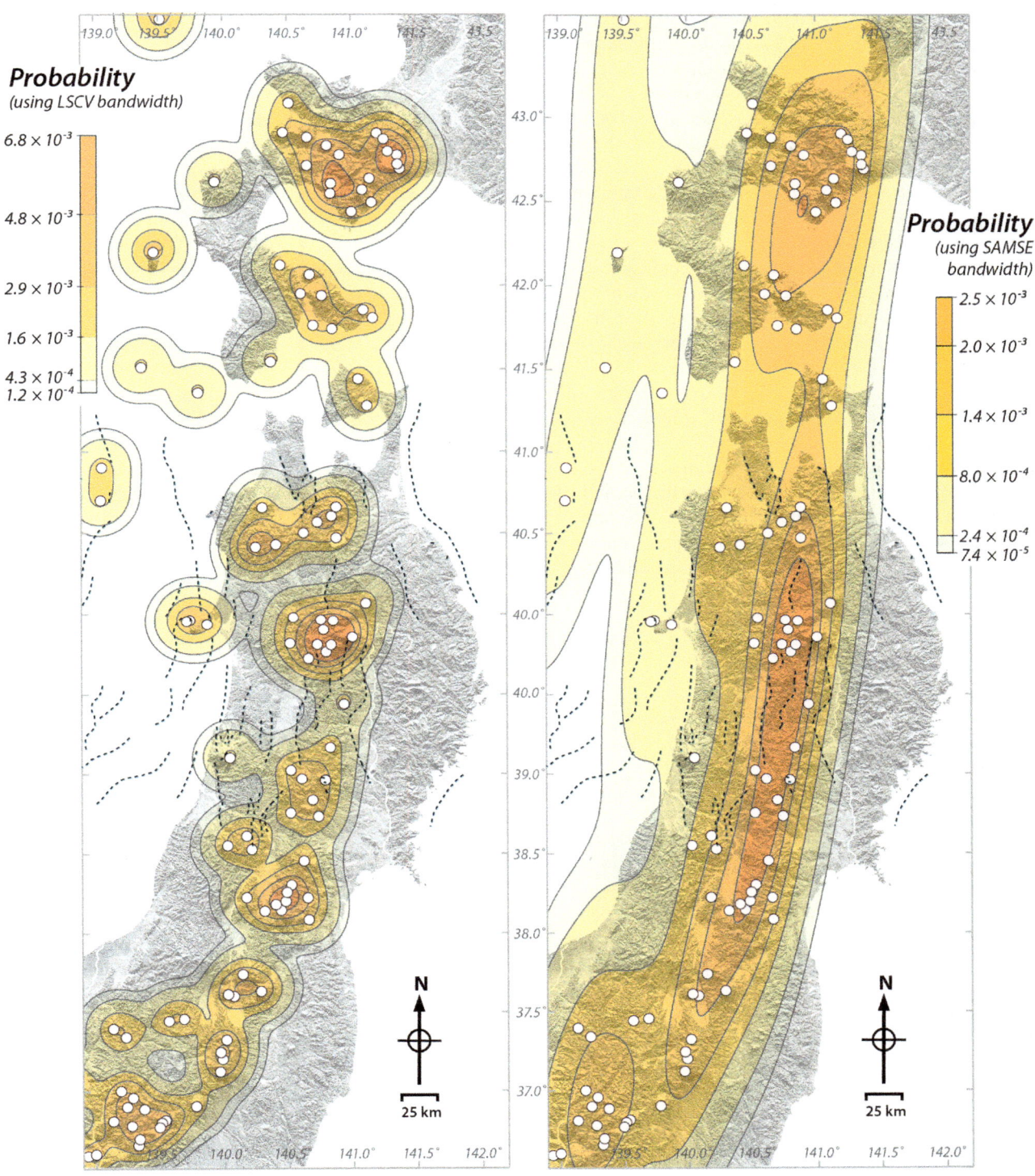

FIGURE 51.3 Alternative spatial density models of Quaternary volcano distribution in the Tohoku arc, northern Honshu, Japan. Both maps are prepared using kernel density estimation, with alternative kernel bandwidths estimated using different optimization algorithms. An isotropic, short bandwidth calculated with least squares cross-validation (LSCV, left panel) emphasizes clustering along the volcanic arc, possibly in response to along-arc variations in magma productivity and heat transfer in the mantle wedge. An anisotropic, long bandwidth calculated using smoothed asymptotic mean squared error (SAMSE) methods smooths volcano distribution, and emphasizes relatively uniform melt production along the arc (right panel). Hazard assessments need to consider such alternative models, which are linked to our understanding of volcanological processes on various scales. *Figure courtesy of Laura Connor.*

in a quantitative way. That is, probabilities are assigned to the potential for a given location to be inundated by flow phenomena, such as lava flows or lahars, or by tephra fallout, or other products of volcanic eruptions. These probabilities might be represented as conditional probabilities—the probability that a given location will experience a specific volcanic phenomena given that an eruption occurs—or as total probabilities accounting for the likelihood of a volcanic eruption of different sizes. Such probabilistic hazard maps are different from empirical hazard maps, which tend or represent a qualitative assessment of probable impacts based on past patterns of activity, or maps that depict thresholds or zones based on interpretation of probabilistic hazard maps and related data.

Once temporal and spatial estimates have been made for event timing and location, respectively, it becomes possible to forecast potential event impacts and to prepare hazard curves for specific sites, or hazard maps for regions. Generally, hazard curves and maps rely on numerical models to simulate volcanic processes. For example, Figure 51.4 shows a hazard curve for tephra fallout from Cerro Negro volcano for the center of the city of Leon,

Nicaragua, located approximately 30 km from the volcano. The curve is a survivor function of the mass loading or thickness of tephra fallout expected in Leon, given a VEI 3 eruption of Cerro Negro, and given the distribution of wind speed and direction observed historically in the region. The curve indicates the conditional probability, also named exceedance probability, that given a VEI 3 eruption of Cerro Negro tephra accumulation will equal or exceed a given value.

The procedure for running such a model, which is identical to the procedures used to model a wide variety of volcanic hazards, requires several steps. The first is to define a model—a numerical abstraction of the eruptive phenomenon. In this case, an analytical solution to the advection—diffusion equation is used to estimate tephra fallout, given eruption and meteorological conditions. This model must be tested and calibrated, a task usually accomplished by applying the model to well-known or observed eruptions. The second task is to define eruption conditions to be modeled. This involves a forecast of eruption magnitude, often characterized by the VEI. Eruption magnitude must then be cast in terms of distributions of specific model input parameters. In this case, these parameters include eruption column height, volume, total grain-size distribution, and additional parameters. The variability in tephra accumulation shown on the graph indicates the natural variability in tephra accumulation, due to factors such as variation in eruption column height, or mass flux, variation in eruption duration, tephra grain-size, which affects particle settling velocity, and meteorological conditions.

Once these input parameter distributions are defined, the model can be run repeatedly, in Monte Carlo fashion, sampling the ranges of input parameters, generally on the order of 10,000 times or more. Taking these factors into account, and assuming the numerical model is correct, yields a range of model outcomes represented by the conditional hazard curve. Note that potential correlation among input parameters, such as between eruption column height and grain-size distribution, is important to consider in constructing and evaluating such models. The exceedance probability can be made unconditional by multiplying by the probability that an eruption of given magnitude will occur.

Similarly, probabilistic hazard maps can be defined by contouring probability across a region. Often the probability is a conditional probability related to the hazard curve, such as the probability that tephra accumulation will exceed some specific mass loading value of interest. For other phenomena, such as lava flows (Figure 51.5) or pyroclastic flows, the main issue is whether a location is potentially inundated by the phenomena, and this probability is usually shown as a conditional probability.

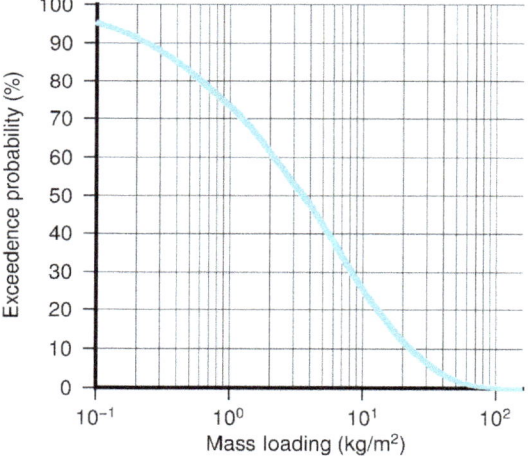

FIGURE 51.4 A hazard curve showing the conditional probability of tephra accumulation in the city of Leon, Nicaragua, given a volcanic explosivity index (VEI) 3 eruption of Cerro Negro volcano, located approximately 30 km from the city. The curve shows the exceedance probability, that is, the probability that tephra accumulation will be equal to, or exceed, a given value, given that the VEI 3 eruption occurs. Calculations are based on 10,000 simulations using a numerical solution to the advection—diffusion equation, using the computer code Tephra2. Variation in tephra accumulation is caused by variation in eruption parameters, such as eruption column height and volume, and by variation in meteorological conditions, including wind speed and direction. Such hazard curves may be prepared for specific regions and a wide variety of volcanic phenomena. *Figure courtesy of Laura Connor.*

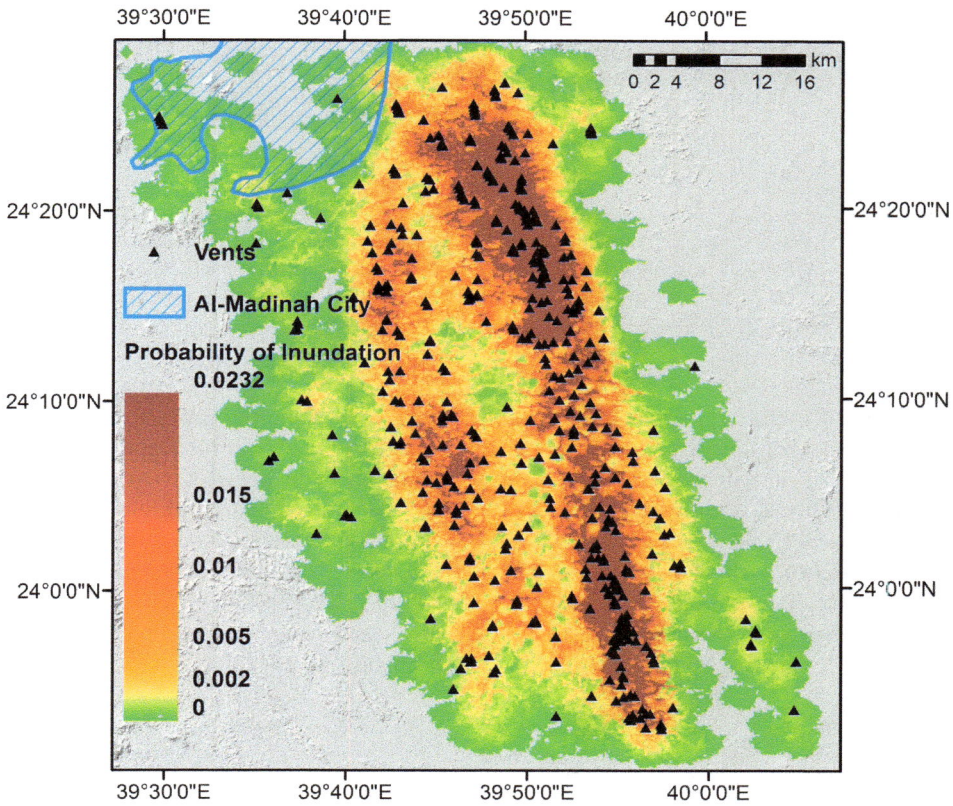

FIGURE 51.5 An example probabilistic volcanic hazard map, for lava flow inundation associated with potential new eruptions from the Harrat Al-Madinah, an active basaltic volcanic field in Saudi Arabia. Probability, conditional on an eruption, that a point is inundated by lava, based on 10,000 iterations of a lava flow simulation code, with vent location drawn from an optimized kernel function based on the previous vent locations. Mapped vents are shown as black triangles. *Figure courtesy of Mel Runge.*

7. THE EVENT TREE SCHEME FOR LONG-TERM HAZARD ASSESSMENT

As described in previous sections, a full and quantitative volcanic hazard analysis requires the probability estimation of a wide range of possible phenomena like eruption occurrence, the arrival of a pyroclastic flow at a specific site, the amount of tephra expected at a specific area, and so on. Each of these probabilities needs the estimation of other quantities like, for example, the eruption recurrence rate, and the space-time propagation of tephra and pyroclastic flows in areas surrounding the volcano. A suitable procedure to represent the full complexity of the volcanic hazard analysis is based on the event tree concept. The event tree is a branching graph representation of events in which individual branches are alternative steps from a general prior event, state, or condition, and which evolve through time into increasingly specific subsequent events. Eventually the branches terminate in final outcomes representing specific hazards or risks that may transpire in the future. In this way, an event tree attempts to graphically display all relevant possible outcomes of volcanic unrest in progressively higher levels of detail. Points on the graph where new branches are created are referred to as **nodes**.

The event tree of one specific case is shown in Figure 51.6, in which the probability is calculated of at least one pyroclastic flow reaching more than 10 km in the NE sector of a given volcano within the next 50 years. Such a probability is

$$P = P(eruption) \times P(explosive\ activity | eruption)$$

$$\times P(PF | explosive\ activity) \times P(sector\ NE | PF)$$

$$\times P(distance > 10\ km | PF\ in\ NE\ Sector)$$

$$= 0.1 \times 0.7 \times 0.3 \times 0.25 \times 0.1 = 5.3 \times 10^{-4}$$

$$(51.15)$$

where $P(\cdot | \cdot)$ is a conditional probability, and each one of the probabilities on the right-hand side of Eqn (51.15) represents the probability of one node in the event tree (Figure 51.6).

The Bayesian event tree (BET) represents the evolution of the event tree concept. BET includes two main features

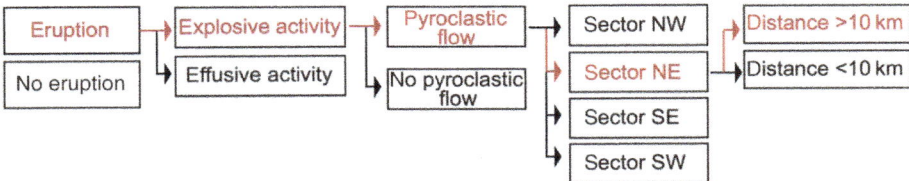

FIGURE 51.6 General scheme of an event tree for pyroclastic flows. The selected path is represented in red, illustrating the events corresponding to probabilities given in Eqn (51.15). For example, the probability of eruption in the next 50 years is 0.1; the conditional probability of explosive activity, given an eruption has occurred, is 0.7 and so on. Each of these numbers can be obtained by expert elicitations, output of one or more pyroclastic flow models, and/or data of the past activity of the volcano (or analog volcanoes). Although this example is a simplified version of the volcanic hazard assessment for a real volcano, it captures the essence of the use of the event tree.

that extend the concept of a classical event tree. First, it uses Bayesian logic to merge information coming from expert opinion, models, and data of the past activity. For example, the prior probability at each node can be set up using the outcome of a model, and then updated to a posterior probability including data about past activity using Bayes' rule. Evaluation of long-term volcanic hazard is based on these posterior probabilities. Second, in BET, each probability is not one single number but is a distribution.

Current volcanic hazard assessment is entangled by an intrinsic complexity of the system, scarce data, and relatively poor knowledge of the physical processes. Cumulatively, these difficulties involve two different types of uncertainty—aleatory variability and epistemic uncertainty—that are of primary importance in hazard and risk studies. Aleatory variability arises from the intrinsic complexity of the system and produces an irreducible stochasticity (randomness) in outcomes, regardless of our physical knowledge of the system. Epistemic uncertainty is associated with limitations in our knowledge and is, in principle, reducible by increasing the number or quality of data and/or improving our knowledge of the physical system. For example, while quantitative, the outcomes of the probabilistic hazard assessments described so far depend on model and data assumptions. These assumptions may have a profound influence on the hazard assessment. For instance, the alternative spatial density models of volcanic vent distribution for Tohoku (Figure 51.3) are both statistically valid models, yet yield significantly different probabilities for parts of the arc. Thus, it is clear that alternative models and data must be considered in probabilistic volcanic hazard assessment.

The use of a distribution instead of a single value for the probability at each node allows us to formally account for both types of uncertainty. Interpreting the probability as an expected long-term frequency of events under the same physical conditions, the average of the distribution represents the best guess for the aleatory variability, and the dispersion of the distribution around the central value

represents the epistemic uncertainty. For example, the hazard curve shown for tephra fallout (Figure 51.4), illustrates aleatoric variability associated eruption magnitude and meteorological conditions. One might also generate alternative hazard curves using alternative data or numerical models to simulate eruptions. Those alternative models would represent the epistemic uncertainty associated with the hazard estimate. One convenient distribution for the probability θ at each node is the beta distribution:

$$[\theta] = Beta(\alpha, \beta) = \frac{\Gamma(\alpha + \beta)}{\Gamma(\alpha)\Gamma(\beta)} \theta^{\alpha-1}(1-\theta)^{\beta-1} \quad (51.16)$$

where $[\cdot]$ indicates the probability density function, $\Gamma(\cdot)$ is the gamma function. The average and variance are

$$E(\theta) = \frac{\alpha}{\alpha + \beta} \quad (51.17)$$

and

$$V(\theta) = \frac{\alpha\beta}{(\alpha + \beta)^2(\alpha + \beta + 1)} = \frac{E(\theta)(1 - E(\theta))}{\alpha + \beta + 1} \quad (51.18)$$

The beta distribution is particularly suitable for our purpose because it is defined for **random variables** in the range [0,1], it is unimodal, and it is the conjugate prior distribution in the binomial model. The choice of the beta distribution is subjective and other distributions can be used, such as the Gaussian distribution for the logistic transformation of the probability. In practice, the differences associated with the use of reasonable distributions are not usually significant, and the beta distribution is the most used in practical applications. On the other hand, we argue that using a single number instead of a distribution is certainly a much stronger subjective choice.

Using the beta distribution, the passage from a prior to a posterior distribution is straightforward. Consider the case in which we set a prior beta distribution for θ (e.g., the probability to have one type of event hitting one specific location) with the parameters α_{prior} and β_{prior}. In case the state of the art does not allow us to estimate a reasonable prior model, we can express the state of maximum

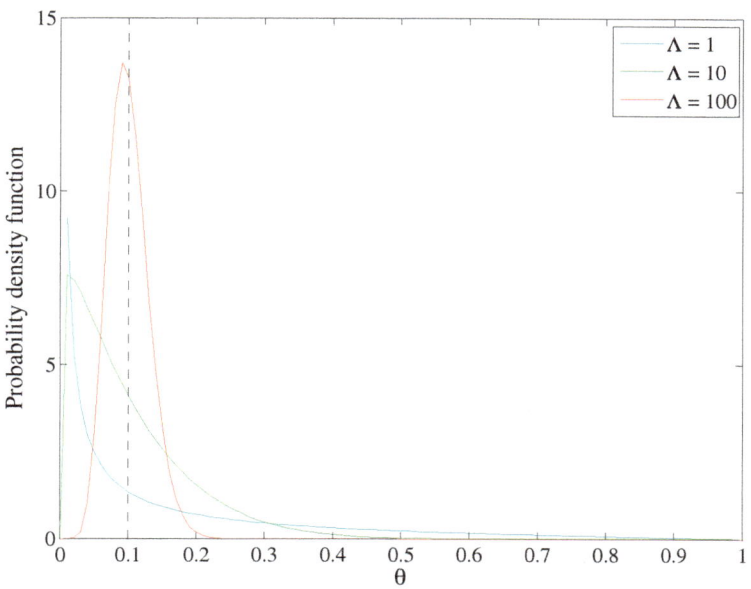

FIGURE 51.7 Probability density function for a beta distribution having average equal to 0.1 and different **equivalent number of data**.

ignorance using the uniform distribution that corresponds to a beta distribution with $\alpha_{prior} = \beta_{prior} = 1$ (any value of θ is equally likely). Then, we collect a set of data Y of the past activity, n data with y occurrences. The posterior distribution for θ is given by

$$[\theta|Y] = Beta(\alpha_{post}, \beta_{post})$$
$$= Beta(\alpha_{prior} + y, \beta_{prior} + n - y) \qquad (51.19)$$

For example, consider one specific node in Figure 51.6, i.e., $\theta \equiv P(\text{distance} > 10 \text{ km}|PF \text{ in NE Sector})$. In our example of the event tree, this number was set to 0.1. In the BET approach, this average of the beta distribution can be derived from expert elicitation or from numerical modeling, e.g., $E(\theta) = 0.1$. In order to have a complete description of the prior distribution, we need to quantify the dispersion around this central value. Conveniently, this equation may be rewritten as:

$$V(\theta) = \frac{E(\theta)(1 - E(\theta))}{\Lambda + 2} \qquad (51.20)$$

where $\Lambda = \alpha + \beta - 1$ can be interpreted as the **equivalent number of data** that mimics the degree/amount of information contained into the prior. In fact, in Eqn (51.19), we can see that the sum of α and β is related to the number of data available. In practice, we can set Λ according to our confidence in the accuracy and precision of the information used to set the prior distribution. For example, if we set $\Lambda = 1$, it means that very few data can significantly alter the prior, or, more pragmatically, it means that the model we are using is very rough. If we set $\Lambda = 10$, it means that only tens of data can alter the prior significantly. Finally, if we

set $\Lambda = 100$ we strongly rely on our model and only hundreds of data can influence the prior. In Figure 51.7, we show the distribution using $E(\theta) = 0.10$ and $\Lambda = 1, 10$ and 100. The choice of Λ is subjective, but is much less subjective than using a single value for the probability, even when using models with completely different reliability. Once the prior distribution has been set, we can include the number of data observed in the field. For example, we count that in the past only y' pyroclastic flows in the NE sector out of a total number of n' reached 10 km from the cone. These numbers may be used to update the prior distribution using Eqn (51.19).

This example represents one possible (simple) case among the wide variety of possible hazard assessments. The use of event trees and BET can be generalized to much more complicated situations incorporating, for example, the output of different models and/or different types of volcanic phenomena. This may require the use of a more complicated probabilistic calculations, but the essence of the problem is the same.

8. FUTURE DEVELOPMENTS

So far, the volcanological community has relatively limited experience with probabilistic volcanic hazard assessments, given that these methods have only been employed widely for the last couple of decades. We suggest several future developments, likely among many. First, an important goal is to use a consistent framework for volcanic hazard assessments. Implementation of standardized outputs of the probabilistic volcanic hazard assessment is needed in order to compare the hazards of different volcanoes and different kind of risks in specific areas. BET is one example of a tool that may provide a comprehensive estimation of the volcanic hazard and associated uncertainties in a proper way, but other strategies are conceivable. Second, the current generation of statistical models is relatively static. They tend to depend mostly on the distribution of past events and prior experience. Future models will be more adept at incorporating disparate geological and geophysical observations of volcanic systems to create ensemble forecasts. Such improvements are likely required, for example, to successfully use statistical models to forecast eruption magnitude. Third, the current generation of numerical models used to simulate volcanic phenomena will likely see dramatic improvements. This is crucial in order to improve uncertainty analysis of hazard curves and maps. For example, the current generation of pyroclastic flow models is quite computationally expensive relative to current computing power; this makes it exceptionally difficult to build probabilistic maps for pyroclastic flows even for the most vulnerable communities living near active volcanoes. Fourth, our views of the consequences of volcanic activity are rather anecdotal, based on relatively few albeit sometimes very well-documented, observed catastrophes. Additional detailed study of specific impacts, on people and infrastructure, will certainly help the community to improve the focus and use of probabilistic hazard assessments.

As far back as the 1930s, Thomas Jaggar proposed bombing active lava flows to remove crust, accelerate cooling, and arrest the downhill progress of the flow. Probabilistic hazard assessment methods will reach fruition when we have the confidence in our models to test alternative mitigation strategies and use these models to weigh the cost and benefit of specific actions in response to volcanic activity.

FURTHER READING

Baxter, P.J., Aspinall, W.P., Neri, A., Zuccaro, G., Spence, R.J.S., Cioni, R., Woo, G., 2008. Emergency planning and mitigation at Vesuvius: a new evidence-based approach. J. Volcanol. Geotherm. Res. 178 (3), 454—473.

Bayarri, M.J., Berger, J.O., Calder, E.S., Dalbey, K., Lunagomez, S., Patra, A.K., Pitman, E.B., Spiller, E.T., Wolpert, R.L., 2009. Using statistical and computer models to quantify volcanic hazards. Technometrics 51, 402—413.

Bebbington, M.S., 2007. Identifying volcanic regimes using hidden Markov models. Geophys. J. Int. 171, 921—942.

Bebbington, M.S., Lai, C.D., 1996. On nonhomogeneous models for volcanic eruptions. Math. Geol. 28 (5), 585—600.

Bebbington, M., Cronin, S.J., 2011. Spatio-temporal hazard estimation in the Auckland Volcanic Field, New Zealand, with a new event-order model. Bull. Volcanol. 73 (1), 55—72.

Burt, M.L., Wadge, G., Curnow, R.N., 2001. An objective method for mapping hazardous flow deposits from the stratigraphic record of stratovolcanoes: a case example from Montagne Pelee. Bull. Volcanol. 63, 98—111.

Connor, C.B., Hill, B.E., 1995. Three nonhomogeneous Poisson models for the probability of basaltic volcanism: application to the Yucca Mountain region, Nevada. J. Geophys. Res. 100 (12), 107—110, 125.

Connor, L.J., Connor, C.B., Meliksetian, K., Savov, I., 2012. Probabilistic approach to modeling lava flow inundation: a lava flow hazard assessment for a nuclear facility in Armenia. J. Appl. Volcanol. 1 (3) http://dx.doi.org/10.1186/2191-5040-1-3.

Coppersmith, K.J., Jenni, K.E., Perman, R.C., Youngs, R.R., 2009. Formal expert assessment in probabilistic seismic and volcanic hazard analysis. In: Connor, C.B., Chapman, N.A., Connor, L.J. (Eds.), Volcanic and Tectonic Hazard Assessment for Nuclear Facilities. Cambridge University Press, pp. 593—611.

De la Cruz-Reyna, S., 1991. Poisson-distributed patterns of explosive eruptive activity. Bull. Volcanol. 54, 57—67.

Garcia-Aristizabal, A., Selva, J., Fujita, E., 2013. Integration of stochastic models for long-term eruption forecasting into a Bayesian event tree scheme: a basis method to estimate the probability of volcanic unrest. Bull. Volcanol. 75 (2), 1—13.

Hill, B.E., Connor, C.B., Jarzemba, M.S., La Femina, P.C., Navarro, M., Strauch, W., 1998. 1995 eruptions of Cerro Negro volcano, Nicaragua, and risk assessment for future eruptions. Geol. Soc. Am. Bull. 110 (10), 1231—1241.

Jaggar, T.A., 1945. Protection of harbors from lava flow. Am. J. Sci. 243, 333—351.

Klein, F.W., 1982. Patterns of historical eruptions at Hawaiian volcanoes. J. Volcanol. Geotherm. Res. 12, 1—35.

Martin, A.J., Umeda, K., Connor, C.B., Weller, J.N., Zhao, D.P., Takahashi, M., 2004. Modeling long-term volcanic hazards through Bayesian inference: an example from the Tohoku volcanic arc, Japan. J. Geophys. Res. 109 (B10208).

Marzocchi, W., Bebbington, M.S., 2012. Probabilistic eruption forecasting at short and long time scales. Bull. Volcanol. 74, 1—29.

Marzocchi, W., Sandri, L., Selva, J., 2010. BET_VH: a probabilistic tool for long-term volcanic hazard assessment. Bull. Volcanol. 72, 705—716.

Mulargia, F., Gasperini, P., Tinti, S., 1987. Identifying regimes in eruptive activity: an application to Etna volcano. J. Volcanol. Geotherm. Res. 34, 89—106.

Newhall, C., Hoblitt, R., 2002. Constructing event trees for volcanic crises. Bull. Volcanol. 64 (1), 3—20.

Pyle, D.M., 1998. Forecasting sizes and repose times of future extreme volcanic events. Geology 26 (4), 367—370.

Wickman, F.E., 1966. Repose-period patterns of volcanoes. I. Volcanic eruptions regarded as random phenomena. Arch. Mineral. Geol. 4, 291—367.

Chapter 52

Volcanic Ash Hazards to Aviation

Fred Prata

Nicarnica Aviation AS, Kjeller, Norway

Bill Rose

Michigan Technological University, Houghton, MI, USA

Chapter Outline

GLOSSARY

ash mass loading Total amount of ash sensed in a column per unit area. The units are usually given in g m^{-2}.

BTD Brightness temperature difference.

glass transition temperature Temperature at which the amorphous content of volcanic ash changes from brittle to plastic behavior (always less than the melting temperature).

hydrometeor Atmospheric water or ice particle formed from sublimation or condensation of water.

IVATF International Volcanic Ash Task Force.

NFZ No-fly zone.

reverse absorption Absorption phenomenon exhibited by SiO_2 where absorption of infrared radiation decreases with wavelength between 7 and 13 μm.

VAAC Volcanic Ash Advisory Centre.

VAFTAD Volcanic Ash Forecast and Dispersion Model.

VASAG Volcanic Ash Scientific Advisory Group.

very fine volcanic ash Volcanic ash with particle radii < 30 μm.

In April 2010, a little-known volcano located in the southeastern part of Iceland had a small but prolonged ash-rich eruption. Between 15 April and 23 May 2010, Eyjafjallajökull volcano erupted between 6 to 10 Tg of fine ash (radii < 63 μm) into the mid-troposphere (3−8 km) and, because of the prevailing westerly and north-westerly winds during the period, the ash reached continental Europe, where it caused the largest shutdown of commercial aviation since World War II (see *News report on the effects of the Eyjafjallajökull volcanic ash on European travel*—Movie 1).

The global economic impact of this disruption to commercial aviation has been estimated at US$ 5 billion. In the last edition of the Encyclopedia, Miller and Casadevall noted that between 1980 and 1998 the cost of damage to aviation engines, airframes, and avionics was US$ 250 million. While the problem of jet aircraft encounters with airborne volcanic ash remains a serious safety issue, the economic impact, particularly in the crowded European, US, and fast-growing Asian air corridors, has refocussed efforts to develop robust and timely means for detecting, forecasting, and mitigating the hazard.

This chapter provides a timely update on the excellent discussion provided by Miller and Casadevall and we look in more detail into the methods and technologies being employed to mitigate the ash/aviation hazard. This has largely been made possible since the recent eruptions of Eyjafjallajökull (March−May 2010) and Grímsvötn (May 2011) in Europe, the hemispheric-scale ash/SO_2 eruption of Puyehue Córdon-Caulle in southern Chile, and the SO_2-rich eruption of Nabro in Eritrea. There have also been several large eruptions in the North Pacific (NOPAC) region, e.g., Okmok, Kasatochi, Sarychev Peak, and Mt Redoubt. Because of the significance of the Eyjafjallajökull eruption, we take a detailed look at this event and use it to assess current practice for detecting, modeling, and responding to the volcanic ash hazard to aviation.

The Encyclopedia of Volcanoes. http://dx.doi.org/10.1016/B978-0-12-385938-9.00052-3

The problem for aviation is that ash and other volcanic emissions, notably SO_2 gas, can reach high into the atmosphere and intersect airspace at all flight levels. Figure 52.1 shows a schematic of the main processes of importance for the aviation problem. Commercial aircraft at cruise altitude often fly just below the tropopause. Violent eruptions are able to reach the tropopause, shown at 40,000 ft here and even penetrate into the stratosphere. If SO_2 is present in the violent eruption, then at least some of it will reach the tropopause or higher, and form stable layers where it will be chemically transformed into sulfate. At lower levels, there may be mixtures of ash and SO_2, or separated ash and SO_2 layers, transported in thin layers by the wind field. At the lowest levels, ash will fall out or rain out, most prominently close to the volcano. A particularly insidious problem, first recognized by Rose et al. (1995), is the presence of ice in volcanic clouds, and ice-coated ash particles. These almost always occur in vigorous tropical eruptions, which reach high into a cold, moist atmosphere. Ash coated with ice is very difficult to discriminate from ice-only clouds. While ice clouds themselves pose a threat to aviation, ash coated in ice is likely to be just as hazardous. Volcanic emissions remain in the atmosphere for different amounts of time, depending mostly on the height they initially reach. In the stratosphere, volcanic aerosols may have lifetimes of a few years, whereas in the lower atmosphere, ash and SO_2 are removed quickly, within a few hours, by gravitational settling, wet, and dry aggregation that enhance fallout and by atmospheric processes such as rainout. Recent studies have shown that ash and SO_2 often form in thin vertical layers ($\sim 1-2$ km) as they are transported by the winds. Very little is known about the size distribution of ash in these layers, other than that the physics demands that they are small (radii $< \sim 10\ \mu m$) if they have traveled far (distances greater than approximately hundreds to thousands of kilometers).

Volcanic ash is defined as pyroclasts (volcanic fragments) that are less than 2 mm in diameter. The product of eruptions, volcanic ash is fragmented magma, characterized by its mineralogy, texture and shape, and size distribution. There are many types of volcanic ash, reflecting variable magma compositions, eruption styles, atmospheric/weather conditions, and distance from the volcano. A typical ash consists of volcanic silicate glass (usually

FIGURE 52.1 Schematic showing some of the processes important for aviation after a violent volcanic eruption of ash and SO_2 gas.

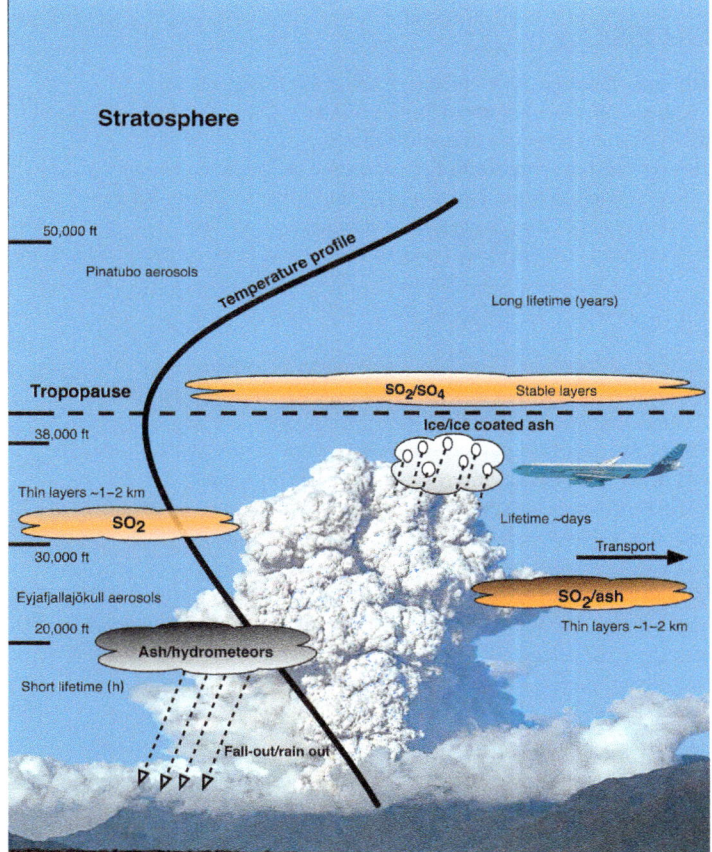

30—100% of the total mass; 45—75 wt% SiO_2) and from one to several minerals (mainly silicates), reflecting the explosion of a partly solidified magma. The shape of the ash reflects the fragmentation mechanism of the eruption, usually either explosive vesiculation (bursting gas bubbles), water—magma (phreatomagmatic or fuel—coolant) interactions, or comminution within a pyroclastic flow or volcano conduit. The volcanic ash size distribution may include a wide range of diameters, and is typically sorted during its atmospheric residence by gravitational settling. Large (<1-mm diameter) particles fall quickly through the atmosphere (atmospheric residence < 1—30 min) in the turbulent regime of high ($r_e > 500$ μm) particle Reynolds numbers. Fine ash, less than 1 mm in diameter, falls in the intermediate flow regime (0.4 μm < r_e < 500 μm), while very fine ash of diameter less than 30 μm falls in the laminar flow regime ($r_e > 0.4$ μm). An eruption produces a broad size distribution of pyroclasts and if these are spread into the atmosphere, sorting occurs during fallout as large particles separate quickly, while finer ones remain and disperse, carried by winds and buoyed by air resistance. Thus, fine (atmospheric residence > 30 min) and especially very fine ash (atmospheric residence from three hours to several days) are agents for volcanic cloud aircraft hazards. Because very fine ash that is often dispersed large distances from the source volcano can fall out in low, sometimes barely detectable amounts, estimates of the actual amount of ash produced are unreliable. Fine and very fine ash fallout is apparently controlled largely by meteorological processes, where very fine ash acts as cloud condensation and ice nuclei and becomes part of the **hydrometeors**. Volcanic clouds have been directly sampled on some rare occasions, giving us insight on their nature. Silicate particles dominate into the very fine ash size ranges and glass often is more dominant as size decreases. Particles with diameters less than about 5 μm include nonsilicates, probably incrustation and sublimate minerals (metallic oxides, halides, sulfides, native metals, sulfates, and acids). Volcanic gases (H_2O, CO_2, SO_2, HCl, HF) are also part of the volcanic cloud, and as these components aggregate they may react with each other.

Aircraft are affected when they encounter volcanic clouds. Ash enters jet engines and the glass pyroclasts may melt when they encounter temperatures in excess of about 800 °C, which causes engine failures. Ash may also cause damage to many other parts of the aircraft because of its abrasive effects. Sulfuric acid aerosol may be dispersed widely in the lower stratosphere where jets typically spend most of their flight time and even in very small concentrations produces crazing in aircraft windshields. Volcanic clouds, particularly when they are carried away from the volcano, look like meteorological clouds and to prevent aircraft encounters, their detection is of major interest. In this regard, remote sensing has been a major tool in the detection and monitoring of volcanic clouds. Because SO_2 is released in most volcanic clouds, yet is present in only low concentrations in ambient atmosphere, its detection using remote sensors was an initial focus. The use of SO_2 sensing methods employing both ultraviolet and infrared detectors still provides valuable input to ash cloud/aircraft mitigation efforts. SO_2 seems to frequently stay strongly correlated with ash in many drifting clouds, although there are spectacular examples of volcanic clouds where separation of SO_2 and ash occurs. IR methods of direct volcanic ash detection grew out of applications using weather satellites. Prata (1989) first noticed that ash present in clouds seemed to correlate in weather satellite data measured using dual and multiple thermal IR channels within the spectral wavelength range 8—12 μm. This discovery led to many applications to volcanic clouds demonstrating successful detections under varying environmental conditions and some problematic cases were identified, and eventually increased understanding of detection parameters and retrieval methodologies. Direct detection of ash is related to the reststrahlen effect, where semiconductor materials (in this case silicates) of small scale (in the Mie region where IR wavelength is similar to ash particle size) show changes in refractive index concurrent with absorption bands. Weather satellites typically have two or more spectral "channels" in the 8- to 12-μm wavelength region, where silicate reststrahlen occurs. Measurements of refractive index of the various components of volcanic ash, both real and imaginary, at IR wavelengths are sparse but sufficient enough to allow development of retrievals. Researchers used IR weather satellite measurements of volcanic clouds to estimate ash particle size, optical depth, and particle mass, and this approach has been improved and applied by many since. Assumptions in these retrievals are pragmatic and problematic:

1. Particles are assumed to be spherical—unlikely.
2. The volcanic cloud has planar geometry—rarely well known.
3. The background of the cloud as seen from space is homogeneous—often not.
4. The atmosphere nearby is a "clear window"— meteorological clouds cause interference.
5. Size distribution is assumed to be simple—unlikely.

Volcanic ash is not the only silicate-bearing cloud type—dust from desert regions also produces detections that resemble volcanic clouds. In recent years, retrievals have been improved to include cloud height estimates to address an important gap.

1. HAZARDS

Volcanic ash that affects jet aircraft occupies only a very small fraction of the mass of material erupted—typically

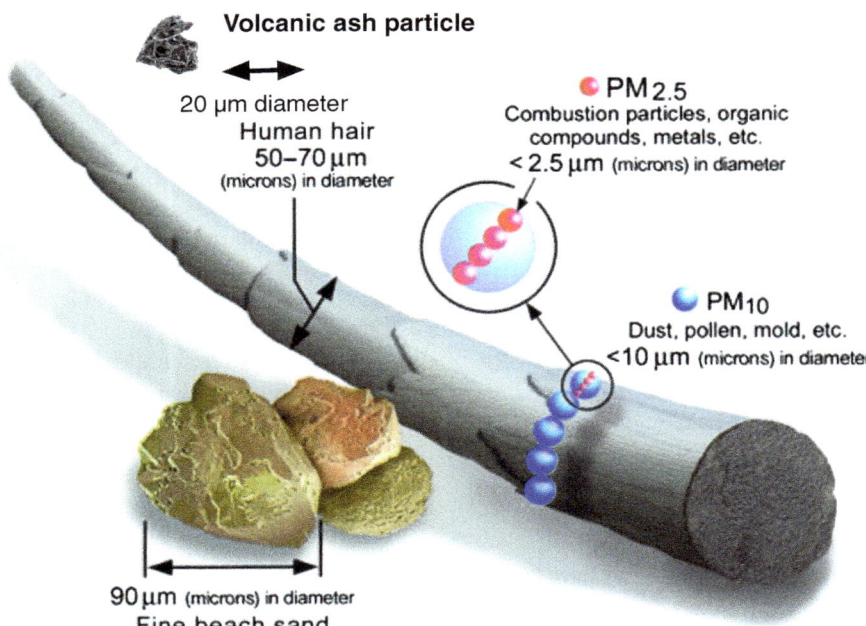

Volcanic ash particle

20 μm diameter
Human hair
50–70 μm
(microns) in diameter

PM2.5
Combustion particles, organic
compounds, metals, etc.
< 2.5 μm (microns) in diameter

PM10
Dust, pollen, mold, etc.
<10 μm (microns) in diameter

90 μm (microns) in diameter
Fine beach sand

FIGURE 52.2   Electron micrograph of a single ash particle shown together with some other common materials. PM, particulate matter. *Image courtesy of the US EPA.*

less than a few percent, with the remainder falling out near the crater so quickly that aircraft do not encounter it. The particles that travel long distances and therefore have more chance of entering airspace are small, typically 1–30 μm in diameter. These particles may take several months to fall out of the atmosphere, the residence time depending initially on the height of injection into the atmosphere and on atmospheric processes such as rainout, aggregation, and vertical updrafts and downdrafts. Particles may also remain longer in the atmosphere if their shapes are less aerodynamic, by increasing the drag, but it is highly unusual for particles with diameters larger than ∼30 μm to travel more than a few hundred kilometers and reside in the atmosphere for more than a few hours.

Figure 52.2 shows an electron micrograph of an ash particle alongside some other common materials, including a human hair. The particle is jagged, glassy, and has smaller material attached to it. As it represents material that is frozen by sudden contact with the Earth's atmosphere at temperatures <25 °C, it usually contains glass and high-temperature minerals, including feldspar, olivine, pyroxene, quartz, and others.

The glassy components of volcanic ash become "sticky" as they pass through their **glass transition temperatures** (always lower than the melting temperature) where their properties change from brittle to plastic at temperatures ranging from about 700 to 1100 °C, depending on various factors, including impurities and the proportion of minerals that require higher temperatures

to melt the silicate. The combustion region of modern jet turbine engines are maintained at temperatures of 1200–1500 °C and so if volcanic ash finds its way into the hot parts of the engines, glass and a small fraction of the minerals may also melt (see How ash damages engines. Movie 2).

Glass is analogous to a frozen liquid, already polymerized and structurally similar to liquid. The hot, sticky, and glassy material can clog the air bleed holes and may adhere to metallic surfaces. When engine temperatures are lower, as when the engine is powered off, the glass will cross the transition temperature back to brittle behavior and may fracture, thus clearing the metal surfaces. Very little is known about the real behavior of crystal-rich ash in this state, but volcanic ash particles without glass can migrate through the engine without liquefying, block inlets/outlets, and abrade parts of the engine. Several researchers have investigated ash ingestion in experiments using jet engines and also modeled the trajectories of particles as they pass through. These conclude that the exact mechanisms of deposition depend on composition and the size distribution. Other deleterious effects that have been identified include clogging of the bleed filter system with consequent loss of pressurization, short circuit and intermittent failure of electronic components, and obstruction of the pitot-static system leading to unreliable speed indications. Windows, leading edges of the wings, and other exposed edges can be abraded and damaged; see Casadevall (1994) and references therein.

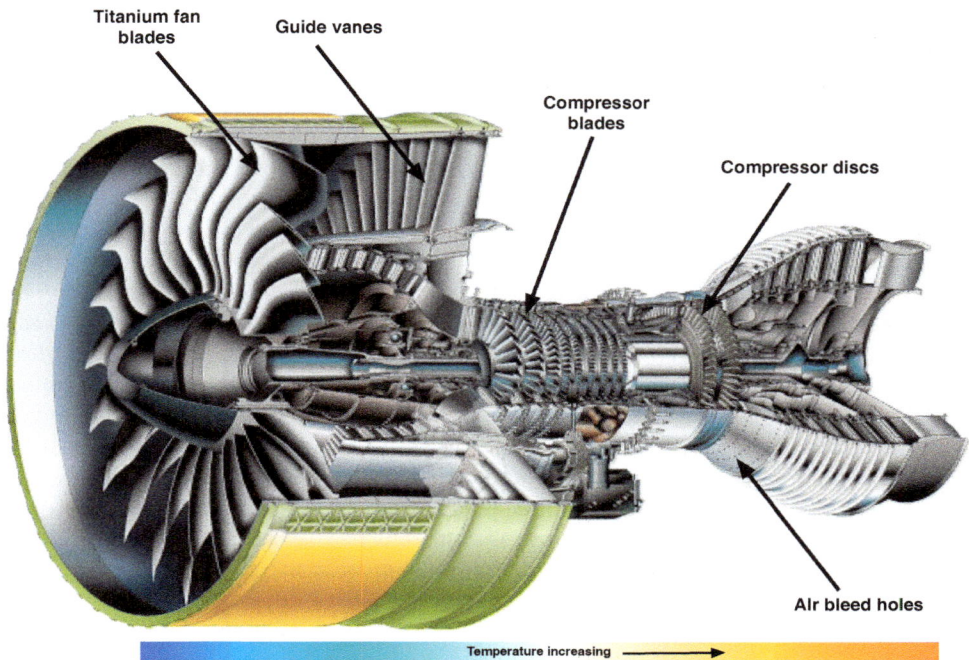

FIGURE 52.3 Schematic of a gas turbofan jet engine. The main components are shown by the arrows and the color scale gives an indication of the temperature gradient through the engine from cooler (blue) to hotter (red).

Needless to say, many of these effects can lead to catastrophic consequences for jet aircraft, with loss of engine power and potentially loss of aircraft. Figure 52.3 shows a diagram of a gas turbofan jet engine with the main parts labeled. An important factor in acerbating the problem is the presence of a temperature gradient through the engine core from cooler regions near the entrance to much hotter regions toward the compressor blades and exit. Damage to engine parts has been recorded in several incidents, the most well known being the British Airways Flight 9 (BA9, also known as Speedbird 9) into an ash cloud from Galunggung volcano, Indonesia on 24 June 1982. A photograph of damage to one of the Rolls—Royce engines is shown in Figure 52.4. Ash particles can also be highly irregular in shape and some researchers believe that the asperities on ash particles can result in greater abrasive damage to the leading edges of the aircraft and more severe sandblasting of the windows.

2. IMPACTS

When Eyjafjallajökull produced an ash-rich eruption on 14 April 2010, experience of ash-related aviation problems in Europe was very limited. Safety concerns based around the BA9 incident in Indonesia and the 1989 KLM Redoubt encounter caused the Civil Aviation Authority (CAA), the London Volcanic Ash Advisory Centre (**VAAC**), and other regulatory bodies to take a very cautious approach. Consequently, there followed the largest shutdown of

aviation since World War II. The global economic impact of the Eyjafjallajökull eruption has been estimated by Oxford Economics to be ~US$ 5 billion, with about ~US$ 2 billion of direct impact on the aviation industry. It is of some interest and importance, with the spyglass of hindsight, to document the events that led to this unprecedented effect on commercial aviation. Prior to the April—May 2010 Eyjafjallajökull eruptions, Europe had not experienced any significant aviation disruptions from volcanic ash. The Grimsvótn 2004 eruption had caused some minor flight disruptions in northern Europe, but this

FIGURE 52.4 Engine damage to one of the BA9 engines involved in an encounter with ash from Galunggung volcano on 24 June 1982.

cloud was predominantly composed of SO_2 and no aircraft incidents or aircraft damage was reported. The global community involved in this problem meet regularly (every 3 years) to discuss all aspects of monitoring and forecasting volcanic ash movement—the sixth meeting of this group was held in Citeko, Indonesia in March 2013 and the fifth meeting was held in Santiago, Chile in March 2010, just prior to the Eyjafjallajökull eruptions. It is generally agreed in hindsight that the closure of European airspace during April and May 2010 was an overreaction. It can be argued that for safety concerns and because the London VAAC relied on a model that did not utilize observations, a highly conservative approach was necessary and the airspace closures justified. The industry and regulatory response based on discussion and scientific input provided a "three-zone approach" aimed at permitting aircraft to fly in areas contaminated by different concentrations of ash, provided various safety and operational procedures were implemented. These zones are depicted in Figure 52.5.

Outside of the zones, where ash concentrations were measured or forecast to be less than 0.2 mg m^{-3}, normal operations could be adopted. Inside the red-colored zone, where concentrations were measured or forecast to be higher than 0.2 mg m^{-3} but less than 2 mg m^{-3}, an enhanced procedures zone (EPZ) was designated; a gray-colored region, also denoted as a "buffer or time-limited zone" marks a transition ($\sim$60 nm) between the EPZ and a no-fly zone. Ash concentrations of 4 mg m^{-3} are deemed likely to cause economic damage to aircraft engines and

aircraft are not permitted to fly within this zone of measured or forecast ash concentrations $\geq$4 mg m^{-3}. The rationale is explained in documentation by the CAA of the United Kingdom, where it is emphasized that there has been a degree of arbitrariness used to establish the zones but that there have been no reported incidents of economic engine damage in the EPZ. Prata and Prata (2012) showed that satellite observations suggest that the ash clouds over parts of Europe were highly heterogeneous, and although mean concentrations were relatively low (<2 mg m^{-3}), maximum concentrations exceeded 4 mg m^{-3} in patches. Models are unlikely to be able to forecast and pinpoint in space and time these small, highly concentrated patches.

The general guidance for aircraft operating in ash-affected airspace is that if the ash is visible, then it should be avoided. This guidance is flawed for two reasons: it is unclear what concentration of ash is actually visible (to the human eye) because of variations in illumination and viewing geometry and it is unclear what level of ash concentration is dangerous (causes an unacceptable safety risk). In the congested European airspace, the approach taken was to designate large areas of the atmosphere as ash-contaminated, as both the amount of ash erupted and the amount dispersed across Europe were uncertain.

In the 3 years since the Eyjafjallajökull event, much has been learned and there have been further eruptions to study, and practice exercises have been undertaken.

Discussions and meetings concerning improvements in the coordination of the VAACs in order to provide consistent warnings to the aviation industry are ongoing. Following the Eyjafjallajökull 2010 event, there have also been aviation issues with the May 2011 Grímsvötn eruption, Iceland; the June 2011 Puyehue Córdon-Caulle eruption, Chile; the SO_2-rich eruption of Nabro, Eritrea also in June 2011; and the February 2014 eruption of Kelut, Java, Indonesia. A summary of all known ash—aviation encounters can be found in Guffanti et al. (2010). In terms of impacts, the best documented and most severe ash encounters with commercial aircraft number only five. The main characteristics of these encounters are listed in Table 52.1. An interesting observation to note is that all of these encounters occurred in low visibility or at night, suggesting that the ash was not visible to the human eye.

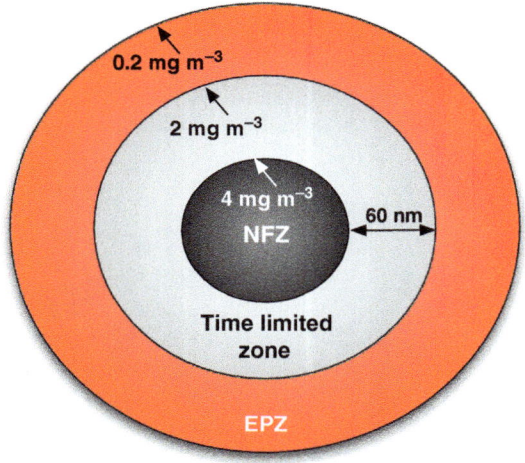

FIGURE 52.5 The three-zone system introduced by the UK Civil Aviation Authority in May 2010 in response to the Eyjafjallajökull ash crisis. The enhanced procedures zone (EPZ) permits aircraft to fly in ash concentrations either measured or forecast up to 2 mg m^{-3}, while a time-limited zone (gray-colored) was introduced as a "buffer" between the EPZ and a no-fly zone (**NFZ**), where aircraft are not permitted to fly in measured or forecast ash concentrations of 4 mg m^{-3} or higher. Outside the three zones, normal operation procedures apply.

3. QUANTIFICATION

Since the recognition of the problem that dispersing ash clouds present to commercial jet aircraft, means for detecting and quantifying volcanic ash in the atmosphere have been sought. Identification at source is problematic because of the danger that erupting volcanoes present to humans (and equipment), and in any case it is the fine-ash content of the erupted material that is of most concern to

TABLE 52.1 Five Ash Encounters with Commercial Aircraft That Are Known to Have Caused Engine Damage. In These Cases There is Documented Engine Damage. The Duration of the Incident is Approximate and the Ash Mass Loadings Have Been Derived from Satellite Information, When Available. The Time of the Encounter is Given as Approximately at the Midpoint of the Encounter. "?" Indicates Not Known or Unreliable

Eruption	Volcanic Explosivity Index	Location (Lat Lon Height)	Date	Time (UTC)	Duration (Minutes)	Time of Day	Altitude Range (Feet)	Maximum Mass Loading (g m^{-2})
Galunggung, Java, Indonesia	4	7° 15' 24" S 108° 04' 37" E 2168 m	24 June 1982	15:00	12–16	Night	12,000–37,000	>20
Galunggung, Java, Indonesia	4	7° 15' 24" S 108° 04' 37" E 2168 m	13 July 1982	13:00	?	Night	21,000–29,000	1–5
Soputan, Sulawesi, Indonesia	2	1° 06' 29" N 123° 43' 48" E 1784 m	19 May 1985	17:00	7–8	Night	37,000	>10
Redoubt, Alaska, USA	3	60° 29' 07" N 152° 44' 35" W 3108 m	15 December 1989	20:46	9–15	Morning	13,300–35,000	?
Kelut, Java, Indonesia.	4	7° 55' 48" S 1 12° 18' 29" E 1731 m	13 February 2014	22:15	~10–15	Night	20,000–35,000	1–10

Volcanic ash also impacts operations at airports close to volcanoes. A good description of the problems including a list of vulnerable airport can be found in Guffanti et al. (2009).

aviation. In situ measurements, that is, instruments carried on board the aircraft to monitor particle (and/or gas) intake have also been considered, but such data give no advance warning of or information on what to do if there is an ash encounter. Since the advent of Earth-observing satellites in the late 1960s, it has been recognized that they have great potential for monitoring the movement of clouds, and satellites dedicated just for meteorological purposes have been developed. Finally, much work and progress has been made in the use of so-called Volcanic Ash Forecast Transport and Dispersion (**VAFTAD**) models. These are modified versions of atmospheric dispersion models (Eulerian and Lagrangian) that require input parameters (a source term) to provide the necessary a priori information, which is used together with forecast three-dimensional wind fields. In the following, we briefly review the various options described above for quantification of dispersing volcanic ash clouds.

3.1. Measurements at Volcanoes

Information at the source is needed in order to establish the amount and height of ash (and gas) emissions into the atmosphere as a function of time. The total size distribution and mass eruption rate are of primary interest; both parameters may be complex functions of height and time. The exact composition of the erupted material is also of importance, but for the ash—aviation problem it is secondary. At most volcanoes none of these data are available in real time and often they are indirectly inferred only later, for example, by examining the ashfall field soon after the end of activity. Examination of fall deposits is extremely useful, as the data can often reveal hitherto unknown processes concerning the erupted material and its interaction with the atmosphere. For example, following the Mount St Helens eruption of 18 May 1980, it was found that the fall deposit downwind of the eruption exhibited a double maximum in the size distribution, suggesting that some aggregation of particles had occurred in the dispersing cloud. Such effects are important for the aviation problem, as aggregation can accelerate the removal of material from the cloud.

Standard geophysical variables, such as seismic measurements, gas composition measurements, ground deformation measurements, infrasonic data, and Webcam imagery, are of great value to detect the onset of an eruption and provide information on its size and duration. During the Eyjafjallajökull eruption in April—May 2010, Webcams were established to continuously monitor the ash plume —see *Webcam footage taken from Hvolsvöllur of ash erupting from Eyjafjallajökull in April 2010*—Movie 3. These standard tools are being supplemented by sophisticated remote sensing technologies, including ground-based UV and IR cameras to measure gases and particles and also ground-based radars, normally used for meteorological purposes. In the early stages of the eruption, radar is very

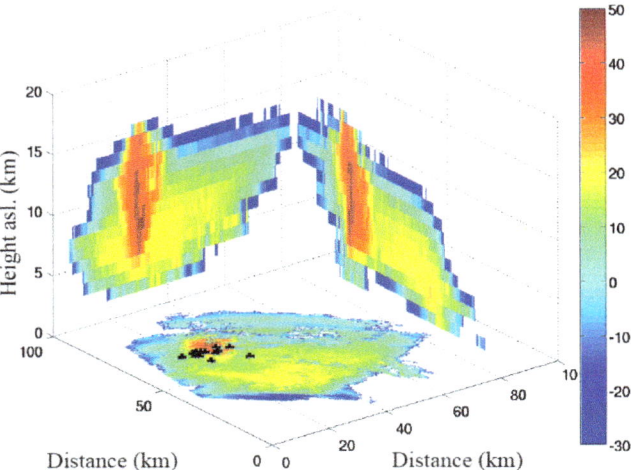

FIGURE 52.6 X-band radar observation of the Grímsvötn volcanic eruption on May 22 2011 (Iceland). Horizontal Vertical Maximum Indicator Zhh reflectivity (dBZ) image at 06:12 UTC, where the maximum of radar reflectivity volume along the vertical and horizontal directions is considered around the volcano vent. The X-band radar, in the position (0,0), is located at about 70 km from the volcano. The 15 detected lightning positions, projected to the ground, are also plotted as black symbols, showing that lightning electrical activity may be intense for Plinian plumes. *After Marzano et al. (2013).*

useful because the microwave energy can penetrate the optically thick plume and provide estimates of the column height and the mass eruption rate. Figure 52.6 shows an example of the three-dimensional structure of radar reflectivity of an eruption of Grímsvötn in Iceland. These instruments operate at millimeter wavelengths, so their sensitivity to micron-size particles is low, but radar observations have become increasingly important at volcanoes because the waves penetrate clouds and provide height estimates. Other novel approaches are also being tested at volcanoes, including the use of unmanned aerial vehicles carrying sensitive gas and particle probes, measurements of the electrification in plumes, and direct sampling of plumes using light aircraft armed with optical particle counters.

3.2. Satellite Measurements

The discovery that dual-band infrared measurements of volcanic ash clouds from space could be used to provide real-time discrimination from normal meteorological clouds led many VAACs to incorporate fast satellite data processing systems and automated schemes in order to assist in the production of Volcanic Ash Advisories (VAAs). Table 52.2 provides a summary of the detection schemes in use up to the present (2014). The evolution of the algorithms (see Table 52.2) has followed improvements in the understanding of the interaction of infrared radiation with ash particles but also with improvements in technology. It should be noted that there has not yet been a

TABLE 52.2 Summary of Ash Detection Algorithms and Techniques Used with Satellite Infrared (IR) and Visible Channel Data

Name	Principle	Comment
RA	2-band IR (11 and 12 μm)	Widely used at VAACs
Ratio	2-band IR (11 and 12 μm)	Similar to RA
4-Band	IR + visible	Daytime only
TVAP	3-band IR (3.9, 11, and 12 μm)	Need correction for reflected sunlight in 3.9-μm channel
PCI	Multiband principal components	Statistical basis
WVC	2-band IR + water vapor correction	Improved RA
RAT	3-band IR (3.5, 11, and 12 μm)	Climatological database needed
3-Band	3-band (IR and visible)	Daytime only
Concavity	High-spectral (IR)	Spectral data needed
β-	Multiband (IR)	Improved RA; physical basis
OE	Multiband (IR)	Improved RA; physical basis
CID	Multiband (IR)	Improved detection
UA	High-spectral (IR)	Spectral data needed; physical basis

RA, **reverse absorption**; TVAP, three-band volcanic ash product; PCI, principle components; RAT, ratio method; WVC, water vapor correction method; OE, optimal estimation; CID, cloud identification; UA, unified algorithm.

satellite instrument developed for the purpose of ash detection and retrieval, and researchers have relied on suboptimal measurements, at least until the arrival of the high-spectral resolution spectrometers (e.g., AIRS) and interferometers (e.g., IASI). The most advanced algorithms use optimal estimation schemes that incorporate spectral databases on the scattering properties of volcanic ash particles and retrieve effective particle radii, column mass loading, and cloud-top height simultaneously. Remote sensing technology exploits gaseous and particle absorptions and scattering throughout the electromagnetic spectrum. Figure 52.7 provides an illustration of the regions of the electromagnetic spectrum suitable for measuring properties of gases and particles using remote sensing techniques.

The "reverse absorption" infrared method for detecting and quantifying volcanic ash in the atmosphere is a basic part of all of the algorithms that utilize broadband infrared sensors (e.g., the MODerate resolution Imaging Spectrometer (MODIS), or the National Polar-orbiting partnership (NPP) sensors). Geostationary infrared instruments are well suited to monitor ash clouds using this method because they provide a near-continuous view of the atmosphere and the Earth below. Although coverage is best in the equatorial and tropical regions, they can also monitor eruptions as far as 60° N and 60° S and this is illustrated in the movie loop (*Movie loop of ash detections from SEVIR*—Movie 4) that

shows a stream of ash from Eyjafjallajökull during May 2010. The detailed methods for retrieving effective radius and mass loading from 2-band IR data have been further explained in several papers. Absorption of infrared radiation is greater at shorter wavelengths in the IR window (7−14 μm) for silicate-containing particles (e.g., ash), while the opposite is true for water droplets and ice particles. An illustration of the method is given in Figure 52.8, and an example of the retrieval of ash mass loadings from the eruption of Eyjafjallajökull on 6 May 2010 is shown in Figure 52.9.

The unpredictable nature of volcanic activity, the global distribution of active volcanoes, and the potentially far-reaching spread of ash clouds in the atmosphere make satellites ideal for monitoring them. These factors also mean that continuous observation is needed (day and night) and at relatively high temporal frequency. Spatial resolution is not too much of a demand because modern satellite instruments have 1- to 10-km^2 footprints, which are adequate for monitoring dispersing ash in the atmosphere. Vertical resolution is a concern, as passive remote sensing is unable to provide accurate height information. The Caliop lidar on board the polar-orbiting CALIPSO platform is capable of providing very good vertical resolution (∼30 m) but with very limited temporal resolution (repeat cycle of 14 days). Obtaining height and thickness of dispersing volcanic clouds is of great value to aviation.

FIGURE 52.7 The electromagnetic spectrum showing regions where remote sensing of gases and particles is possible. (a) Ultraviolet, visible, and near-infrared parts of the spectrum indicating regions where gases absorb and the locations of the bands used by various satellite instruments. The region between 280 and 340 nm is used by several satellite instruments to measure volcanic SO₂. (b) The middle infrared part of the spectrum. Water vapor and CO₂ have major absorption bands, but there are gaps where remote sensing of the surface to infer volcano "hot spots" is possible. Thermal alerts generated by satellites are useful early indicators of volcanic unrest. (c) The thermal infrared part of the spectrum from ~5- to 20-μm wavelength. This is the principal region used to measure ash particles in the atmosphere. There are also important absorption bands of SO₂ in this region that are routinely used to measure this volcanic gas.

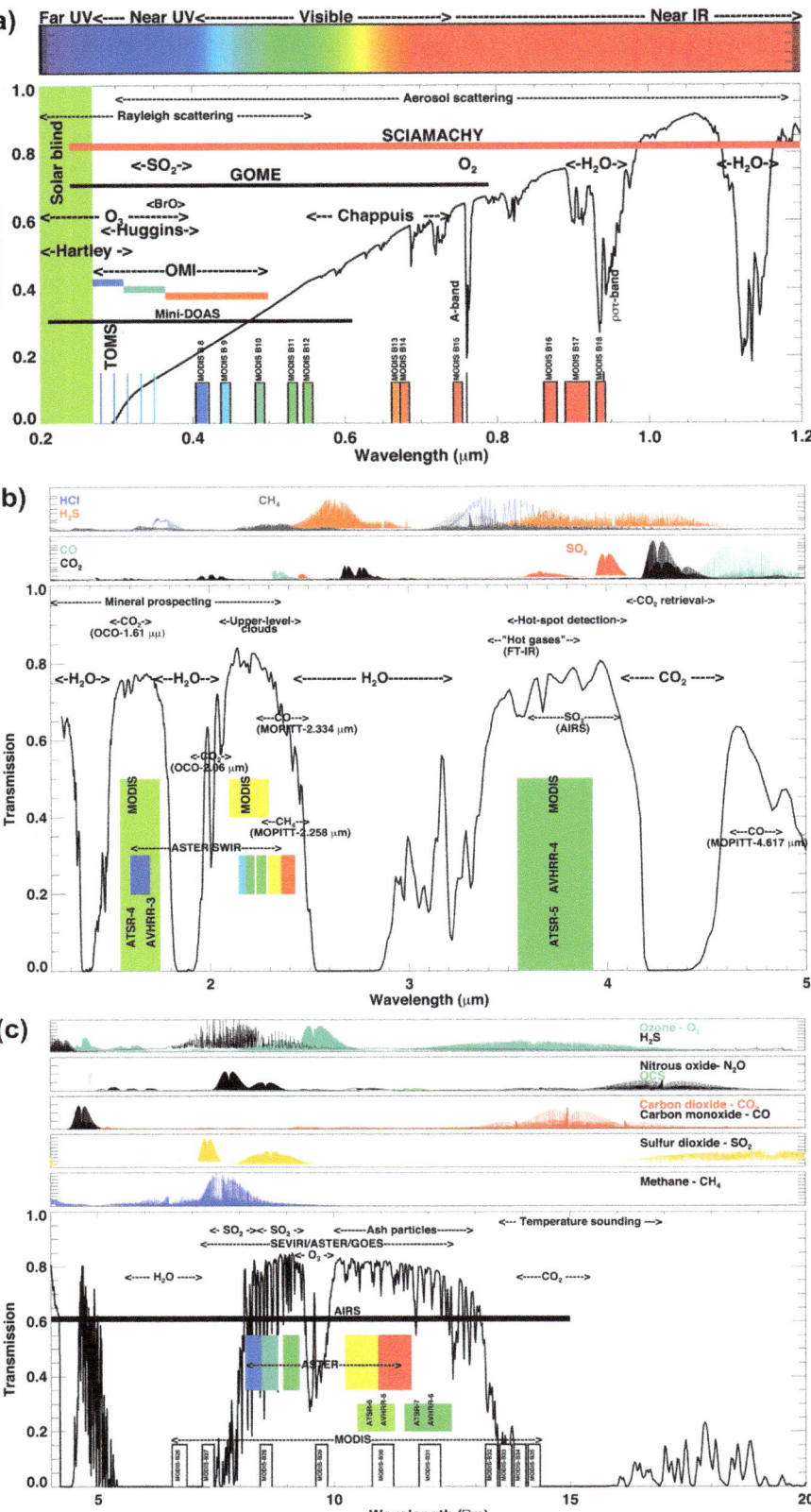

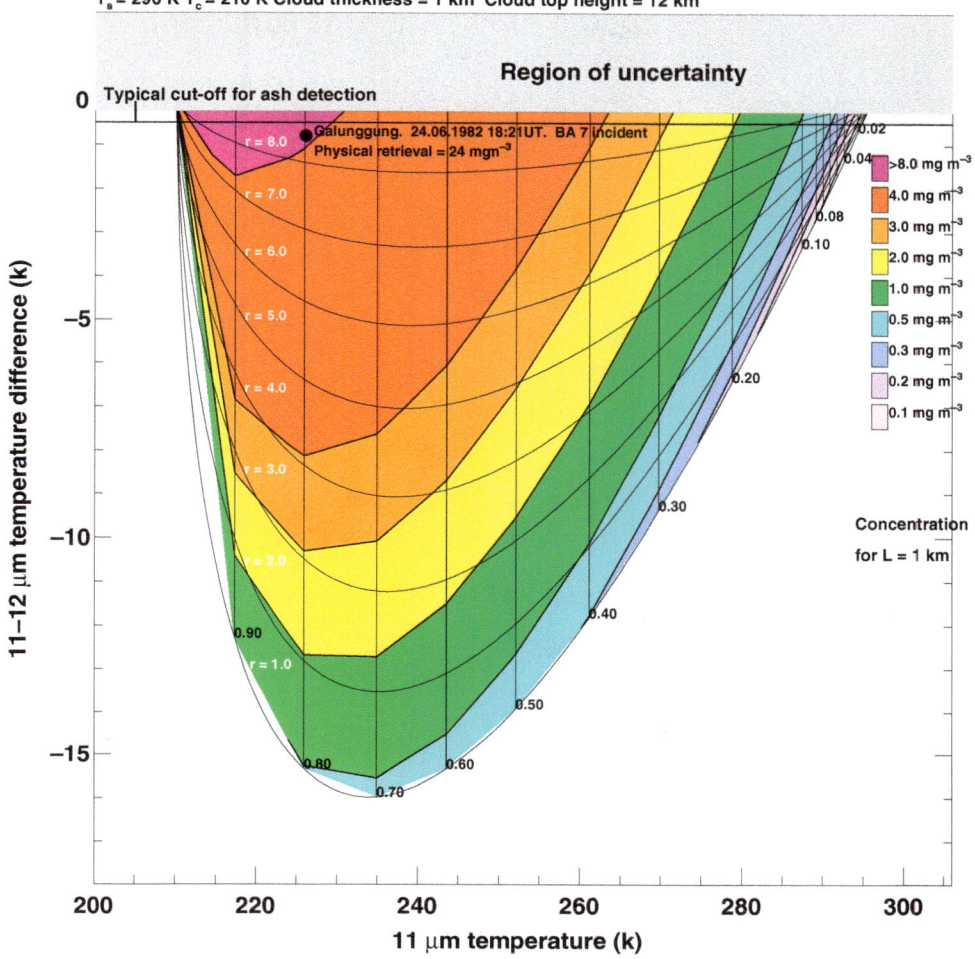

FIGURE 52.8 Schematic illustrating the principle behind the "reverse absorption" method for determining ash mass loadings from two-channel broadband infrared data. The ordinate shows the temperature difference between the two channels, while the abscissa shows the temperature of one of the channels. The distribution of measurements falls into a "U" shape, where at one extreme the difference is close to zero and the cloud is transparent, while at the other extreme the difference is also zero while the cloud is now opaque. The vertical lines show lines of constant optical depth, while the curved lines are the lines of constant effective particle size. Wen and Rose (1994) describe the diagram as a "radiant net." The colors show the regions of constant concentrations, assuming a cloud of 1-km uniform depth.

SO_2 is also a common component of dispersing volcanic clouds and is readily detectable from space using both ultraviolet and infrared sensing. In fact, it turns out to be easier to detect volcanic SO_2 than ash and the accuracy of retrieval is also better. Thus, it makes sense to utilize SO_2 as a tracer for volcanic ash whenever they travel together, and there are many examples of cases detecting SO_2, especially using UV sensors such as TOMS (before 2000), OMI (after 2001), SCIAMACHY, and GOME/GOME-2. The capability of satellite sensors to monitor SO_2 dispersion in the atmosphere is illustrated in Figure 52.10, which shows the path of SO_2 emitted by an eruption from Hekla volcano in 1980.

There have been some studies of the coincidence of ash and SO_2 in the dispersing eruption clouds from Eyjafjallajökull in April/May 2010 and these find that most of the time the constituents traveled together, but sometimes they

do not. Figure 52.11 shows a composite of ash and SO_2 retrievals from different sensors, for the eruption clouds from Grímsvötn between 22 and 25 May 2011. In this case, the separation of SO_2 and ash was spectacular: the SO_2 went high into the atmosphere (>10 km) and traveled northward before spreading east and west, while the ash traveled southward in a low-altitude (<4 km), ground-hugging cloud that eventually reached the southern parts of Scandinavia, where elevated values of PM10 were recorded at air quality stations at the surface. Ash concentrations were low (<1000 $\mu g\ m^{-3}$) and unlikely to cause a hazard to aviation.

Apart from the volcanic ash hazard, there remains an issue of exactly what hazard is posed by dispersing volcanic SO_2. Concentrations in dispersing volcanic clouds are not usually high enough to cause a health hazard, while any

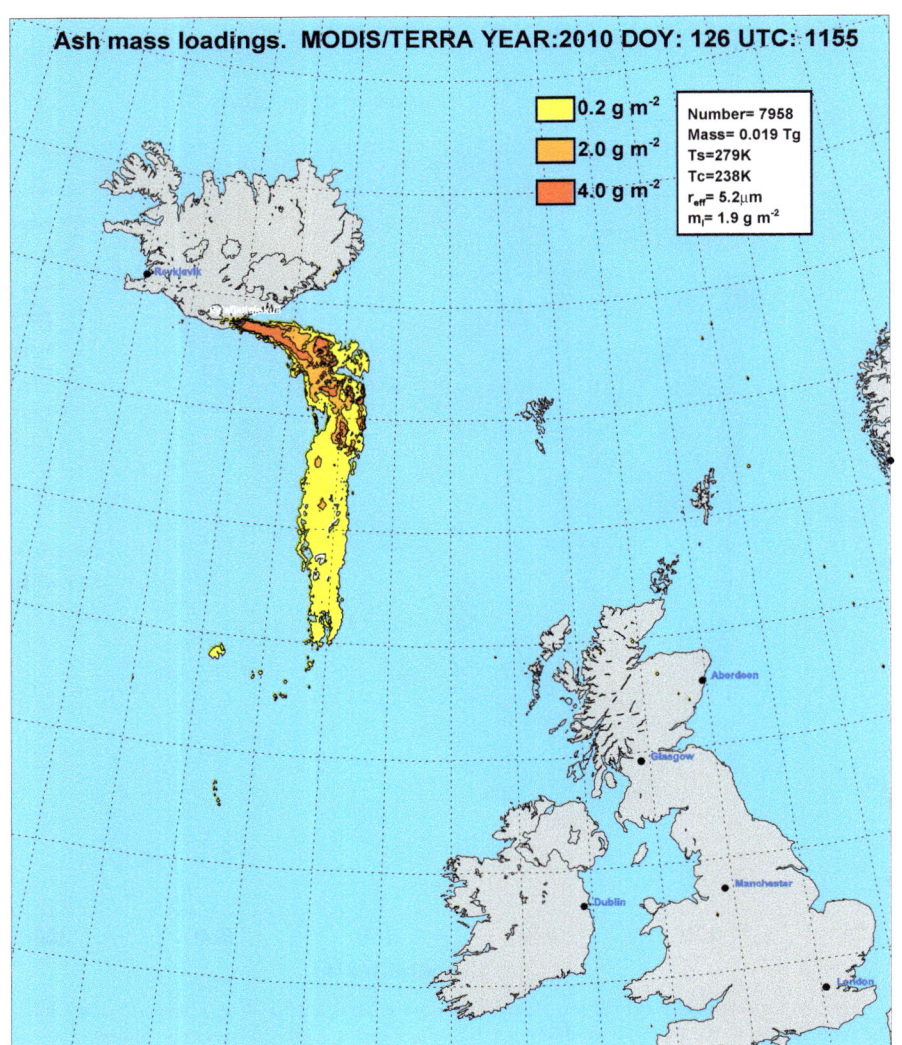

FIGURE 52.9   An example of a retrieval of fine-ash mass loading based on measurements from the MODIS for an eruption of Eyjafjallajökull, Iceland on 6 May 2010. The colors have been assigned to show levels at 0.2, 2, and 4 g m^{-2}, which correspond to ash concentrations of 200, 2000, and 4000 μg m^{-3} for an ash cloud 1-km deep.

damage to aircraft is likely to be secondary and may not be critical to flight operations. SO$_2$ undergoes chemical conversion into H$_2$SO$_4$, which can corrode parts on aircraft and hence there are potential long-term effects on aircraft from SO$_2$.

Ice in volcanic clouds, especially in the tropics, has been discussed previously and may be particularly problematic for aviation because the usual satellite detection techniques are unable to isolate the ash signal. As vigorous eruptions can easily reach 40,000 ft, aviation encounters with ash concealed by ice or concealed within ice are possible. Aircraft radar should be able to detect these volcanic ice clouds but avoidance, especially in the tropics, may not always be feasible nor warranted if the cloud presents itself as a cumulonimbus cloud with ice particles. This topic is the subject of ongoing research.

The difficulties associated with discriminating ash from ice and water in infrared satellite data are compounded by the lack of high spatial resolution and the lack of any vertical resolution. On occasion, satellite data can be rendered less useful because the ash can be obscured from view by meteorological cloud above the ash or because the opacity of the cloud is so great that the IR methods can no longer discriminate them from meteorological clouds. In addition, there is a time lag between acquiring the satellite data, processing it, sending out a volcanic ash product to the Meteorological Watch Office, and the issuance of a VAA. As an example of the problems associated with early detection and cloud cover, the 19–20 May 1985 Soputan eruption provides an important case study, as this eruption caused damage to the engines of a Qantas B747-combi flight en route from Hong Kong to Sydney, Australia. The

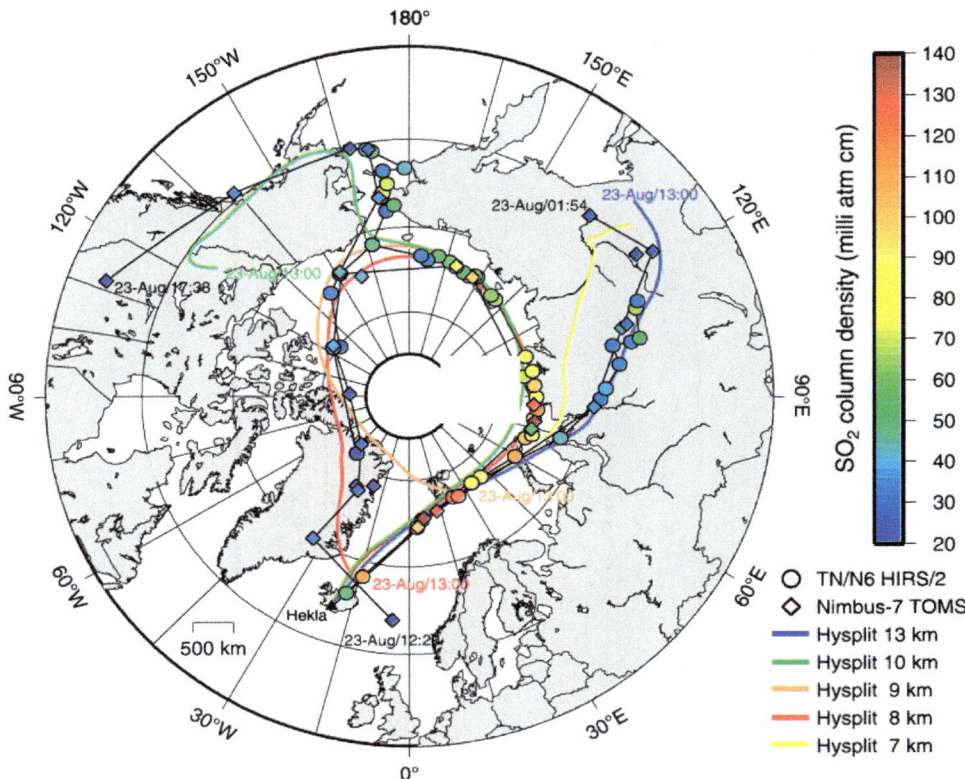

FIGURE 52.10   Peak SO$_2$ column amounts measured by N7/TOMS and TN/N6 HIRS/2 during atmospheric residence of the 1980 Hekla volcanic cloud, with best-fit HYSPLIT forward trajectories superimposed. Dates and times (UTC) of the final satellite observations and model trajectory end points are indicated.

aircraft encountered volcanic ash in an area very close to Soputan volcano; St Elmo's fire was observed, ash entered the cabin (samples were collected), and all four engines were removed and inspected. In 1985, the main satellite data available for ash detection and quantification were from the AVHRR's on board the NOAA operational polar-orbiting satellites. A processed thermal image around the time of the incident is shown in Figure 52.12 together with the aircraft track and an HYSPLIT (see the next section) showing the trajectory of ash at 16.6 km, close to the cloud top.

Several other parameters are plotted on this image. These are as follows:

- The possible track of the Qantas flight. The reported route is airway A61D, which lies to the north of the track shown; however, the reported location of the incident is not on A61D. The track shown is a part of the great circle between Hong Kong and Sydney airports.
- The location of the reported incident, marked by a filled circle and indicated by the time stamp 16:58 UTC.
- A section of the track between two asterisks that has been estimated to be the extent of the main cloud over the region and consequently a value for the distance

spent by the aircraft inside the cloud. According to this, the distance in the cloud is ∼110 km, suggesting a ∼7.3-min transit at a cruise speed of 900 km h^{-1}.

- An HYSPLIT trajectory starting at Soputan at 10:00 UTC on 19 May from a height of 16.6 km. The ambient temperature along this trajectory is indicated at regular intervals.
- Ellipse A: the region of the cloud mass near Soputan and extending southward.
- Ellipse B: the region of a second cloud mass, well away from Soputan, and presumed to be a normal meteorological cloud system.

The data suggest a very large, cold (high) cloud in the vicinity of Soputan with a morphology similar to eruption clouds observed in satellite imager. The lowest cloud-top temperatures are in the region of 193—194 K, which suggests that these clouds have reached the tropopause, or perhaps penetrated into the stratosphere (nearby radiosonde data confirm this). The two ellipses have been chosen to include data for the suspect ash cloud and a meteorological cloud system nearby in order to compare brightness temperatures and brightness temperature differences (**BTDs**). The BTD image is shown in Figure 52.13.

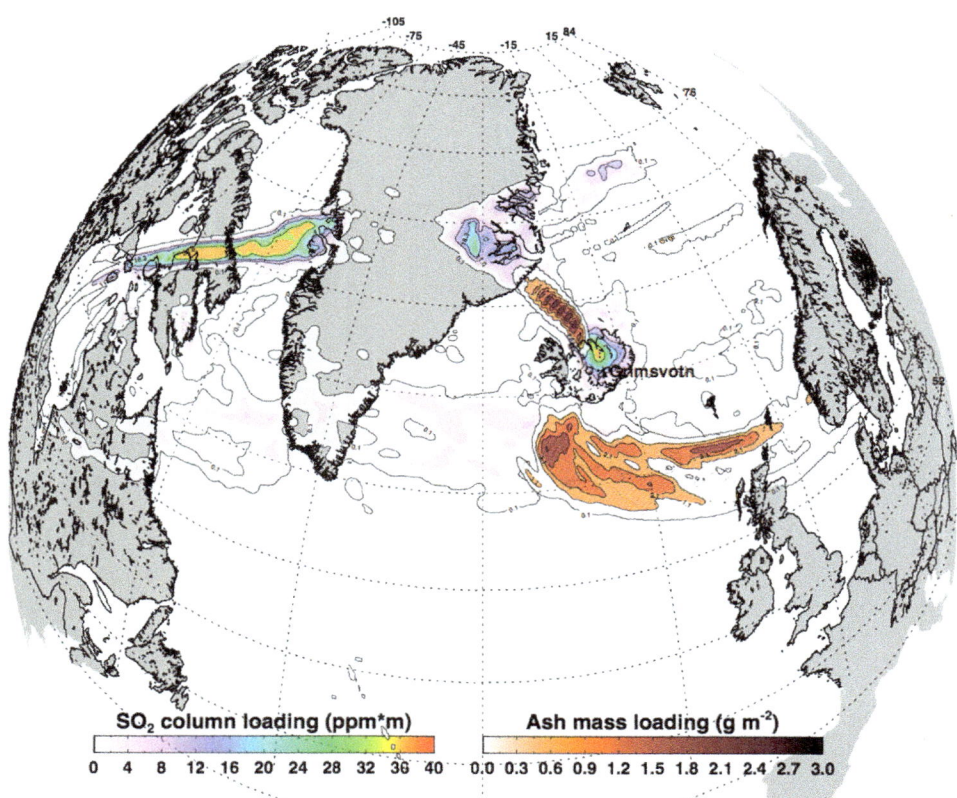

FIGURE 52.11   Ash and SO$_2$ column amounts determined from the SEVIRI and AIRS instruments, respectively. The retrievals have been composited for the days 22–25 May 2011 to show the extent and movement of the different branches of the dispersing volcanic emissions from Grímsvötn.

Satellite imagery obtained on 20 May 1985 from the TOMS instrument showed a region of increased absorption of UV light, suggesting the presence of absorbing particles, likely to be ash, but AVHRR data (Figure 52.14) show no evidence of ash. A possible reason for the discrepancy could be that the AVHRR is no longer sensitive to the ash, as concentrations are now too low to be detected or that water vapor is masking the detection.

The conclusion from this case study is that the aircraft probably flew directly into the ash cloud without any prior warning in conditions that did not permit visual identification of the cloud. Satellites had difficulty identifying the ash because it was below the main umbrella of the developing ash column and possibly the ash was concealed by a coating of ice. The ash particles encountered were too small to be detected by the on-board radar.

One potential way to circumvent the problems of time delays in communication and the inability of satellites to see through opaque clouds is to offer a forward-looking remote sensing instrument to monitor the skies ahead of the cloud. The instrument must be able to work during the night as well as in daylight and provide sufficient warning time ($\sim$5 min) to enable the aircraft to take evasive action. Recent tests (see *Airbus tests ash detector*—Movie 5) on detection of an

artificially generated ash cloud over the Bay of Biscay have demonstrated the proof-of-concept of an infrared imaging device (AVOID, Airborne Volcanic Object Imaging Detector) that was able to correctly identify an ash layer of $\sim$200-m thickness with mean concentrations of $\sim$1 mg m^{-3} from a distance of $\sim$70 km. The technology is being tested by Airbus with the support of the budget European airline easyJet, with plans to install a device on its aircraft operating between the UK and Iceland in the near future.

3.3. Dispersion Modeling

Satellite data are indispensable for locating and monitoring volcanic ash clouds as they are transported around the globe by atmospheric winds. VAACs (see Section 4) utilize this information (among many other sources) to verify forecasts of the movement of ash clouds to advise aviation of the potential hazard and to help with flight scheduling. All VAACs are associated with or housed within meteorological centers and generally have access to current weather information and three-dimensional wind fields as well as the models that produce weather forecasts. The shorter time-scales and distances involved in volcanic ash forecasting

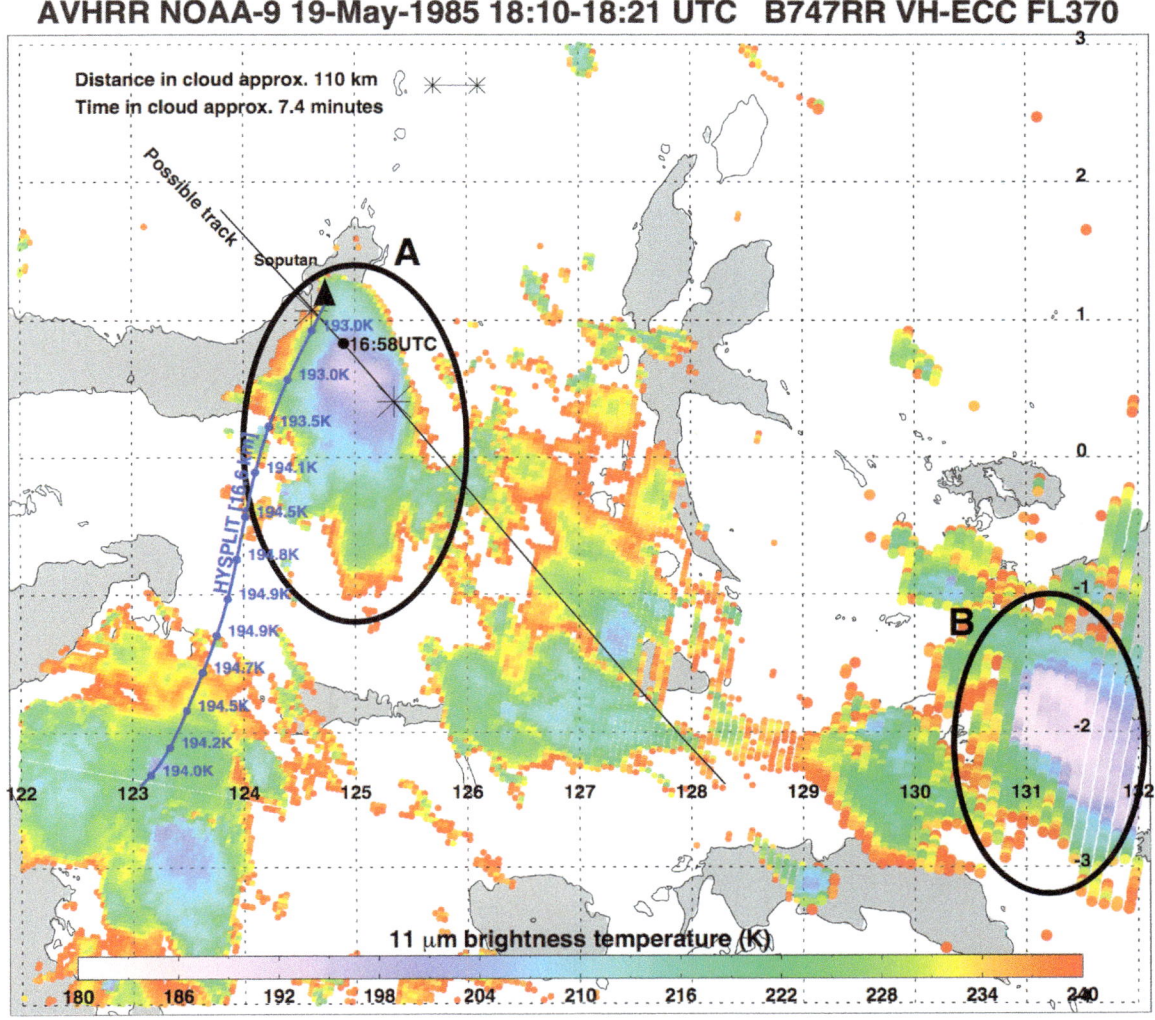

FIGURE 52.12   Top-of-atmosphere (toa) brightness temperatures in Kelvin derived from the AVHRR-2 11-µm channel. No corrections have been applied apart from that required for calibration. The data are at 4 × 4-km resolution and are reprojected without resampling (hence the gaps between pixels). Only pixels with toa temperatures between 180 and 240 K are shown.

make using atmospheric dispersion models more suitable than global weather forecasting models.

These models have been used for many years and were originally developed for pollution forecasting, especially for nuclear accidents. The models are classed as advective-diffusion models, where the atmospheric pollutant (e.g., volcanic ash) is considered as a passive tracer and spreads through advection by the atmospheric wind field and disperses through diffusive processes based on vertical and horizontal diffusion coefficients. Additional processes included are dry and wet deposition, and sedimentation, but most models do not include aggregation or particle breakup or chemistry. The University of Alaska has provided a series of animations using the PUFF model (see Table 52.3) for volcanic eruptions in the NOPAC region. One such simulation for Cleveland volcano during 2001 is shown in

PUFF dispersion model animation—Movie 6. Table 52.4 provides a list of some of the main VAFTAD models and the processes that are included.

There are many efforts currently underway to improve VAFTAD models, including multidisciplinary work on characterizing the source term, using remote sensing data to constrain forecasts by inverse modeling, making use of ensemble forecasting methods, and improvements in operational modeling for providing information to airlines.

4. AVIATION REGULATORY ENVIRONMENT

After the eruptions of Mt Pinatubo, Philippines in June—July 1991, International Civil Aviation Organization

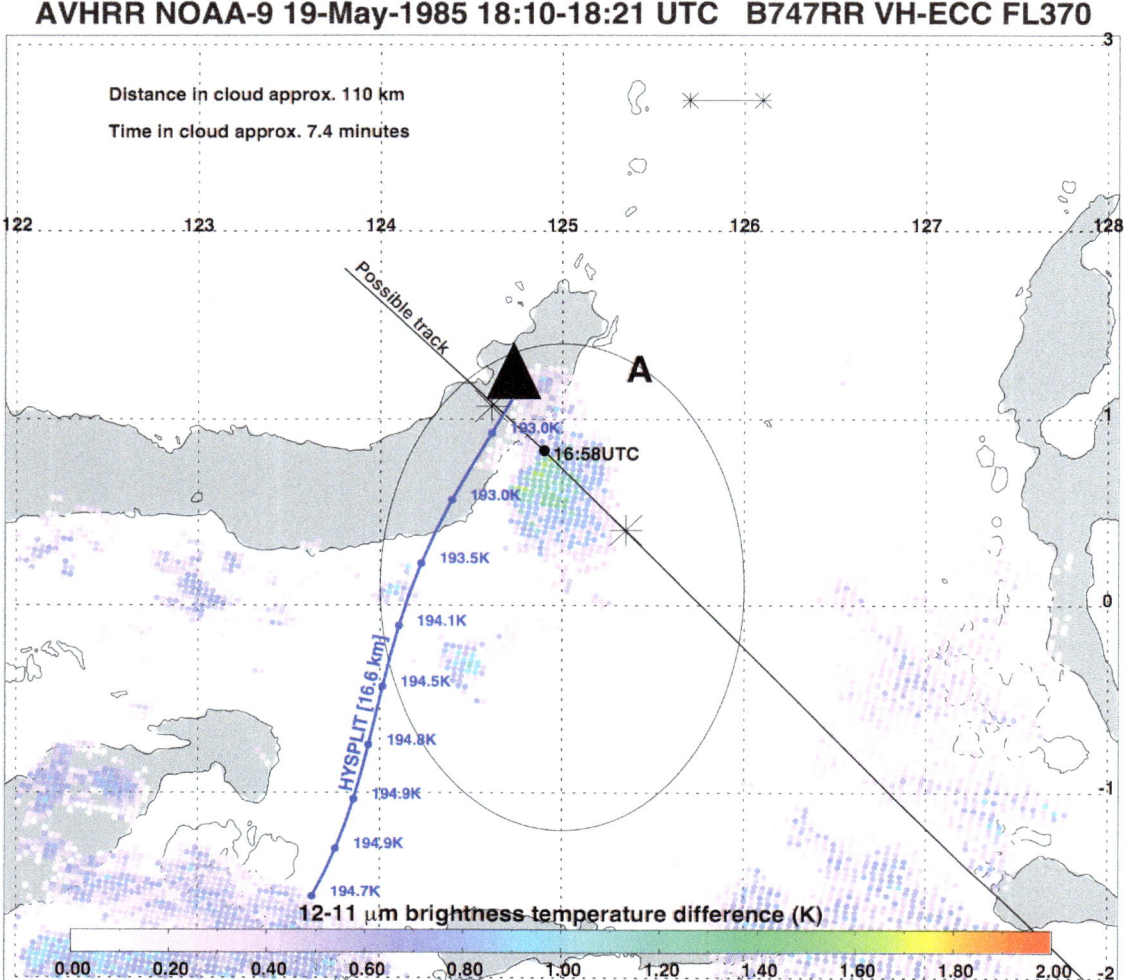

FIGURE 52.13   Brightness temperature difference (BTD) image for 19 May 1985 at 18:10−18:21 UTC (1.5 h after the incident). The BTD is computed as 12−11 μm differences; usually this is done the other way around. Notice that the cloud near Soputan contains many pixels with BTD > 0 K, which is indicative of ash, but the BTD curves suggest that these positive differences are very much suppressed by the presence of ice crystals. Note that only positive differences are shown, making the cloud mass appear much lower. The negative differences are associated with ice particles in this case because the cloud has penetrated into the stratosphere, where temperatures now increase with height.

(ICAO) established a network of nine VAACs to provide advice and timely warnings for aviation concerning volcanic clouds. Figure 52.15 shows the locations of the VAACs and their regional boundaries. Notice that some parts of the world are not covered by these boundaries and there is an informal agreement among the VAACs with regard to warnings in these regions. Ash clouds tend to drift across the boundaries, and in recognition of the need to provide consistent and accurate information on the ash hazard, VAACs tend to cooperate extremely well and have regular meetings. In addition, VAACs are informed by the latest scientific developments through a mechanism established under the auspices of ICAO and WMO. The Volcanic Ash Scientific Advisory Group (**VASAG**) is an ad hoc group of scientists working within academia that meet

every 18 months with representatives of the VAACs, volcanologists, and aviation meteorologists to discuss activities related to the ash−aviation problem. The main topics include satellite remote sensing, volcano observation, eruption warnings, and dispersion modeling. During the Eyjafjallajökull crisis, the International Volcanic Ash Task Force (**IVATF**) was established to quickly advise governments and industry on ways to mitigate the hazard and provide information on better ways to designate safe airspace during ash cloud events. The VASAG provided advice to one of the working groups of the IVATF. Unlike the VASAG, the IVATF was a temporary arrangement and it has now been formally disbanded, while the VASAG continues to report and meet. Perhaps, the most important development arising from the deliberations of the IVATF is

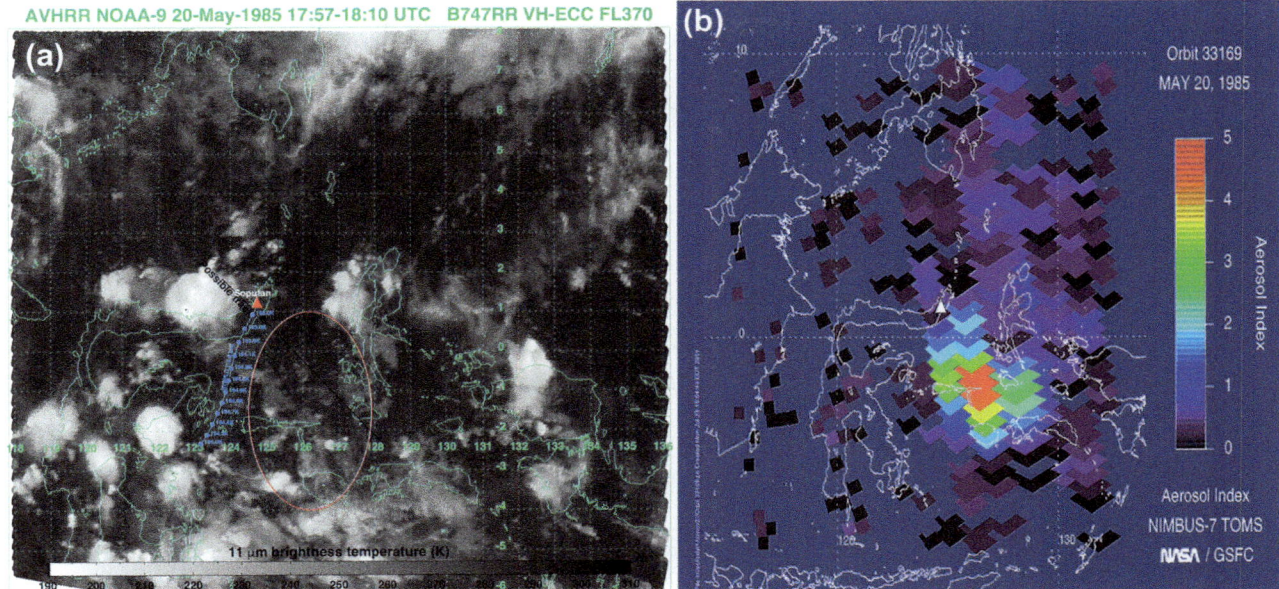

FIGURE 52.14 (a) 11-µm brightness temperatures for the AVHRR image acquired at 17:57 UTC on 20 May 1985. Notice that there are many high (cold) cloud systems in the vicinity of Soputan. The AVHRR BTD image (Figure 52.13) shows no evidence of ash, which may be due to the masking effect of water vapor—likely in this moist atmosphere to be as much as 2—4 K. Note that the region is relatively cloud-free. The red-colored ellipse corresponds roughly to a region identified in the TOMS image. (b) TOMS image showing regions of high absorbing index.

TABLE 52.3 Locations of the Nine Volcanic Ash Advisory Centers (VAACs), Geostationary Satellites Used to Generate Ash Products, and the Principal Dispersion Model(s) Utilized. The Current (2014) URL is Also Given, Where Available

VAAC	Region Lon (°E)	Lat (°N)	Geostationary Satellites	Dispersion model
Anchorage[1]	[150, −135]	[50, 90]	GOES	PUFF, HYSPLIT, CANERM
Buenos Aires[2]	[−90, −10]	[−90, −10]	MSG-3/GOES-E	
Darwin[3]	[75, 160]	[−90, 10]	MTSAT	HYSPLIT
London[4]	[−30, 60]	[45, 90]	MSG-3	NAME
Montreal[5]	[−135, 0]	[45, 90]	GOES-E/GOES-W	CANERM
Tokyo[6]	[90, 165]	[15, 60]	MTSAT	
Toulouse[7]	[−30, 90]	[−90, 70]	MSG-3	MEDIA/MOCAGE
Washington[8]	[−150, −40]	[−10, 45]	GOES/POES	HYSPLIT
Wellington[9]	[160, −140]	[−90, 0]	MTSAT	HYSPLIT

[1] http://vaac.arh.noaa.gov/
[2] http://www.meteofa.mil.ar/vaac/vaac.htm
[3] http://www.bom.gov.au/info/vaac/
[4] http://www.metoffice.gov.uk/aviation/vaac/index.html
[5] http://meteo.gc.ca/eer/vaac/index_e.html
[6] http://ds.data.jma.go.jp/svd/vaac/data/index.html
[7] http://ds.data.jma.go.jp/svd/vaac/data/index.html
[8] http://www.ssd.noaa.gov/VAAC/washington.html
[9] http://vaac.metservice.com/vaac/

TABLE 52.4 Capabilities of 12 Volcanic Ash Forecast and Transport Models Currently in Use by Volcanic Ash Advisory Centres and Research Centers

	ASH3D	ATHAM	FALL3D	FLEXPART	HYSPLIT	JMA	MLDP0	MOCAGE	NAME	PUFF	TEPHRA2	VOL-CALPUFF
Operational Approach	▮	▮	▮	▮	▮	▮	▮	▮	▮	▮	▮	▮
Method	E/H	E	E	L	H	L	L	E	L	L	E	H
	N	N	N	N	N	N	N	N	N	N	A	S
Coverage	LRG	L	LR	LRG	LRG	G	LRG	G	LRG	LRG	L	LR
Physics												
Topography												
H-wind advection												
V-wind advection												
H-Atm. diffusion								See note‡				
V-Atm. diffusion												
Particle sedimentation												
Other dry deposition												
Wet deposition												
Dry part. aggr.												
Wet part. aggr.												
Particle shape												
Gas species												
Chemistry												
Granulometry												
Variable size class												
Variable grain size dist.												
Variable size limits												
Mass distribution	Ln	O	ALL	PS/Li U/P/O	PS/Li U/P/Ln	PS/Li U/P/Ln	PS/Li U/Lni	PS/Li	PS/Li O	PS/Li U/P	Li/U Ln	PS/BP PS/BP

L=Lagrangian. E=Eulerian. H=Hybrid. A=Analytical. S=semi-analytical. N=Numerical. L=Local. R=Regional. G=Global.

PS=Point Source. Li=Linear. U=Umbrella type. P=Poisson. Ln=Log-normal. BP=Buoyant Plume. O=Other.

‡Neglected. Diffusion of numerical origin appears to be sufficient wit particularly good results at 0.5° resolution.

Based on an analysis by A. Folch.

the recognition of the need for clarity on the term *visible ash*. This term is used by ICAO in advice to aviation for avoidance of areas of airspace contaminated with ash. To circumvent the need to specify ash concentrations and provide ash tolerances for engines, aviation is able to fly as long as ash is not visible. There are obvious problems with this definition, not least about what to do at night. The IVATF has adopted a compromise approach by providing a new term and clarifying the definitions for both of these terms. The fourth (and final) report of the IVATF states,

"…the task force agreed to take the following initial definitions forward to the IAVWOPSG:

a) ***Visible ash:*** *volcanic ash that can be observed by the human eye; and*

b) ***Discernible ash:*** *volcanic ash that can be detected by defined impacts on the aircraft or defined in-situ and/or using remote-sensing techniques."*

The report does provide a recommendation based on scientific studies that assert,

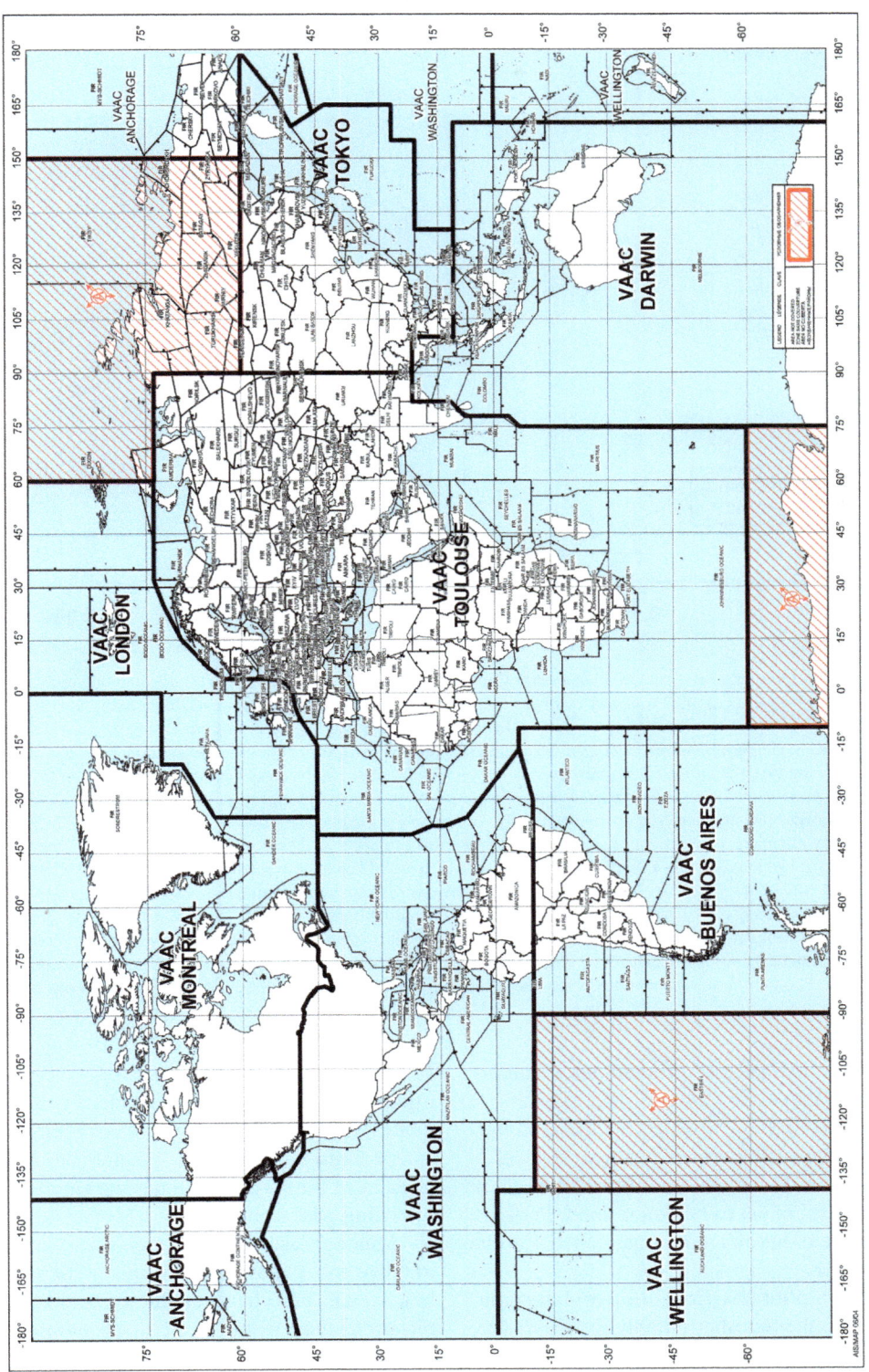

FIGURE 52.15   The nine Volcanic Ash Advisory Centre (VAAC) regions covering most of the global airspace. Parts that are not currently covered are shaded in red.

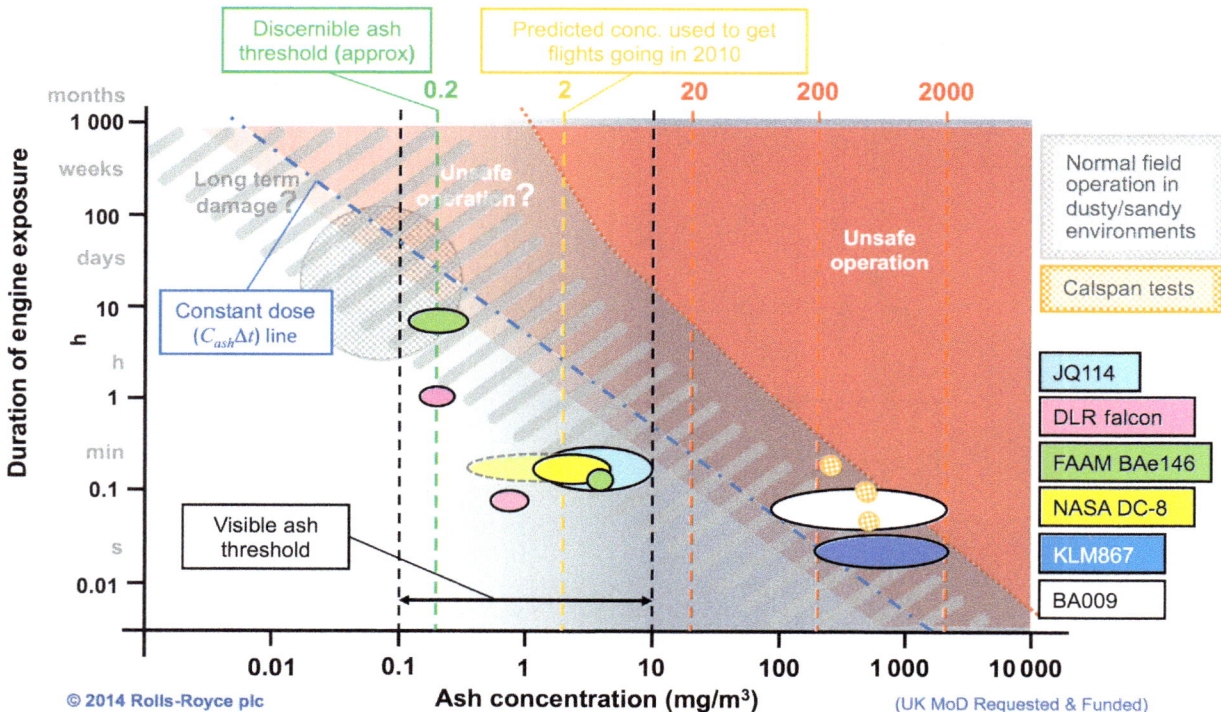

FIGURE 52.16   Chart demonstrating the relationship between exposure duration and ash concentration with relation to safe operating levels. The chart is essentially a sketch of the available data, plotted in a way that tries to illustrate the important considerations when flying in volcanic ash-contaminated airspace. Plotting ash concentration (c_{ash}) against exposure time (Δt) has proved the most instructive way of presenting the information. The event bubbles (BA009, KLM867, NASA DC-8, etc.) are positioned using the best data available on the exposure duration and ash concentration of the event. The BA009 and KLM867 concentrations are taken from Rolls–Royces latest study of these events. The NASA DC-8 concentration was taken from Witham et al. (2012). The 2010 DLR and Dornier/BA flights concentrations were taken from various Internet sources and from Weinzierl et al. (2012). The red "Unsafe operation" region is the optimists view of where you do not want to fly. The curved lower boundary line of this region reflects the possibility that some of the damage mechanisms may attenuate at lower concentration—something that is plausible. The pink "Unsafe operation?" region would be the pragmatists view, assuming that damage is essentially a function of the mass of ash ingested and it is accepted that the BA009 and KLM867 events occurred at around ash concentrations of 100–200 mg m^{-3}; its lower bound is a constant-dose extrapolation from these events. The additional pink region below the constant-dose line covers the increased susceptibility of modern engines compared with those from the early mid-1980s. The gray-hatched "Long-term damage?" region illustrates that completing a flight safely is not the only consideration for an operator; damage may have been done to the engines, which only becomes apparent tens or hundreds of flights later. This is essentially economic damage, but there are airworthiness implications if the damage is not detected early enough. The "?" on all these regions reflects the uncertainty in their extent and position. *Courtesy: R. Clarkson, Rolls–Royce.*

"*…the current best estimate of the minimum satellite detection threshold for ash mass loading is 0.2 g m^{-2}, with a standard error of ±0.15 g m^{-2} under favourable conditions using the most advanced retrieval methodologies ….*" Whether these new definitions and recommendations will be adopted and be useful is yet to be determined. Perhaps, the next large volcanic cloud reaching continental Europe will provide the answer.

As well as the contributions from the regulatory authorities and academia, the industry, notably Rolls–Royce, Airbus, and easyJet, have invested time and funds to improve knowledge concerning the impact of ash on engines, airframes, and safe flying. Rolls–Royce has developed a "Safe to Fly" chart, which succinctly shows under

what circumstances airspace may be hazardous (see Figure 52.16). The idea of the chart is that it be used as a guide and there is an implicit recognition that definitive safety limits or dosages cannot be established easily because of the variations in turbine engine performance and operating characteristics.

Although charts of this type could in principle be used to judge safe exposure times, this particular chart cannot; it is a sketch. For charts of this type to be usable for flight planning, they would need to be bespoke charts for ash concentration, ash type, and engine design, and take other factors into consideration (operating conditions). The current chart implies that flying faster, thus shortening your exposure time, is safer; it is not. The duration of

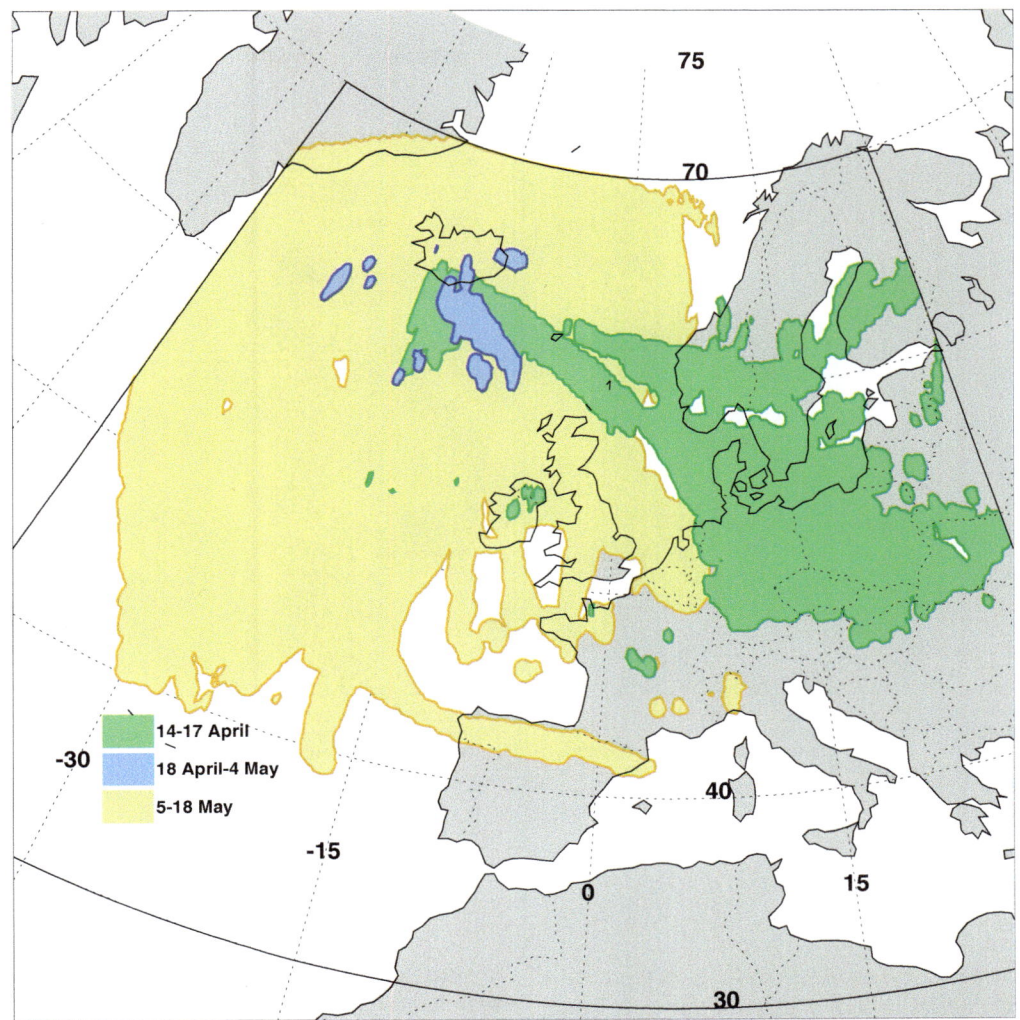

FIGURE 52.17   A synthesis of satellite observations of ash from the three main periods of the eruption of Eyjafjallajökull in 2010.

exposure should be converted to "flight path length" (L) by multiplying time spent in the ash (Δt) by velocity (v), which conveniently converts the dose to $g\,m^{-3}$, or $c_{ash}\ \Delta t V = x$ (where x is the dose to be avoided). The gray "normal field operation in dusty/sandy environments" bubble is included to give an indication for repeated long-term exposure to very low concentrations. Sand/dust is less damaging than volcanic ash, but effective economic damage is seen on engines that regularly operate from such airports after many tens of hours of exposure (usually takeoffs). The discussion and progress on the potential hazard of SO_2 gas to aircraft is at an early stage and no definitive statements can be made concerning the potential hazard. SO_2 in high concentrations is harmful to humans and therefore encountering a concentrated volcanic SO_2 plume where concentrations $\sim$5 ppm in flight would present a dangerous situation. At concentrations of $\sim$100 ppm, SO_2 presents an immediate danger to life or health. However, it is extremely unlikely

that a commercial aircraft would encounter such a highly concentrated cloud and the more likely hazard is to passengers with a preexisting health condition that makes them more susceptible to SO_2 at lower concentrations. Pilots are trained to recognize odors associated with SO_2 as indicators of an encounter with a volcanic cloud.

The regulatory environment for the ash-aviation problem is complex, involving many national and international agencies and authorities. Over Europe during the Eyjafjallajökull eruption in April and May 2010, the ash was widely dispersed (Figure 52.17). Volcanic ash advisories issued by the London VAAC of the kind shown in Figure 52.18 for conditions on 16–17 April 2010, prompted authorities to close large areas of airspace, without specifying concentration levels. A new graphic (Figure 52.19) shows areas of predicted ash concentrations based on two levels of severity. Often decisions are taken by the operators rather than by the regulators, and it

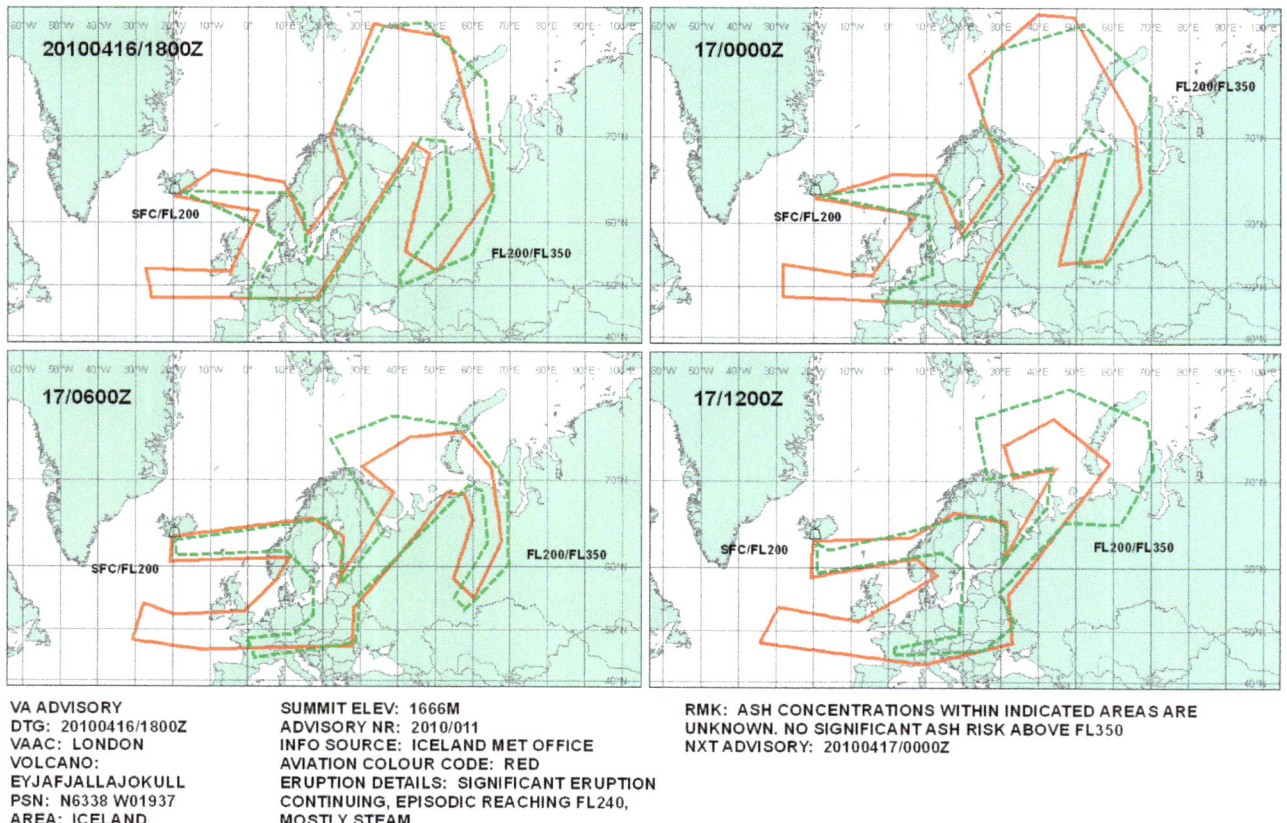

VA ADVISORY
DTG: 20100416/1800Z
VAAC: LONDON
VOLCANO:
EYJAFJALLAJOKULL
PSN: N6338 W01937
AREA: ICELAND

SUMMIT ELEV: 1666M
ADVISORY NR: 2010/011
INFO SOURCE: ICELAND MET OFFICE
AVIATION COLOUR CODE: RED
ERUPTION DETAILS: SIGNIFICANT ERUPTION
CONTINUING, EPISODIC REACHING FL240,
MOSTLY STEAM

RMK: ASH CONCENTRATIONS WITHIN INDICATED AREAS ARE
UNKNOWN. NO SIGNIFICANT ASH RISK ABOVE FL350
NXT ADVISORY: 20100417/0000Z

FIGURE 52.18 Volcanic ash graphic issued by the London Volcanic Ash Advisory Centre (VAAC) on 16 April 2010 at 17:50 UTC at the start of the shutdown of aviation over European airspace.

seems that the system appears to be working, as there have been no known fatalities or loss of aircraft due to volcanic ash. To facilitate airline operators within European airspace, it was decided that operators could submit *safety cases* to their relevant aviation authority, which would allow them to fly in conditions where forecast ash concentrations were below 2 mg m^{-3}. Safety cases must include information sources and pathways that help the operator decide whether or not they can fly safely through a forecast area of ash-contaminated airspace. Third-party data (i.e., not sourced from a VAAC) may be used. The case must also outline inspection and maintenance procedures to be followed and a risk analysis. To date it is known that at least five European carriers have approved safety cases. This is a significant change to the situation that existed prior to April 2010 and the industry appears to be strongly supportive of the procedure. It is not yet clear whether the rest of the world will adopt this procedure.

LIST OF ACRONYMS

AIRS Atmospheric Infrared Sounder
ATZ Air Traffic Zone (air traffic management)

AVHRR Advanced Very High Resolution Radiometer
BTD Brightness Temperature Difference
CAA Civil Aviation Authority
CAeM Commission for Aeronautical Meteorology (WMO)
Caliop Cloud−Aerosol Lidar with Orthogonal Polarization
CALIPSO Cloud−Aerosol Lidar and Infrared Pathfinder Satellite Observations
ESA European Space Agency
ESP Eruption Source Parameter
EUR/NAT Europe/North Africa Region (ICAO)
FAA US Federal Aviation Authority
GEO Geostationary Earth Orbit
GOES Geostationary Operational Environmental Satellite
GOME Global Ozone Monitoring Experiment
IACVEI International Association of Volcanology and Chemistry of the Earth's Interior
IATA International Airline Transport Association
IASI Infrared Atmospheric Sounding Interferometer
IAVW International Airways Volcano Watch system (ICAO)
IAVWOPSG International Airways Volcano Watch Operations Group (ICAO)
IUGG International Union of Geophysics and Geodesy
ICAO International Civil Aviation Organization
LEO Low Earth Orbit
MODIS MODerate resolution Imaging Spectroradiometer

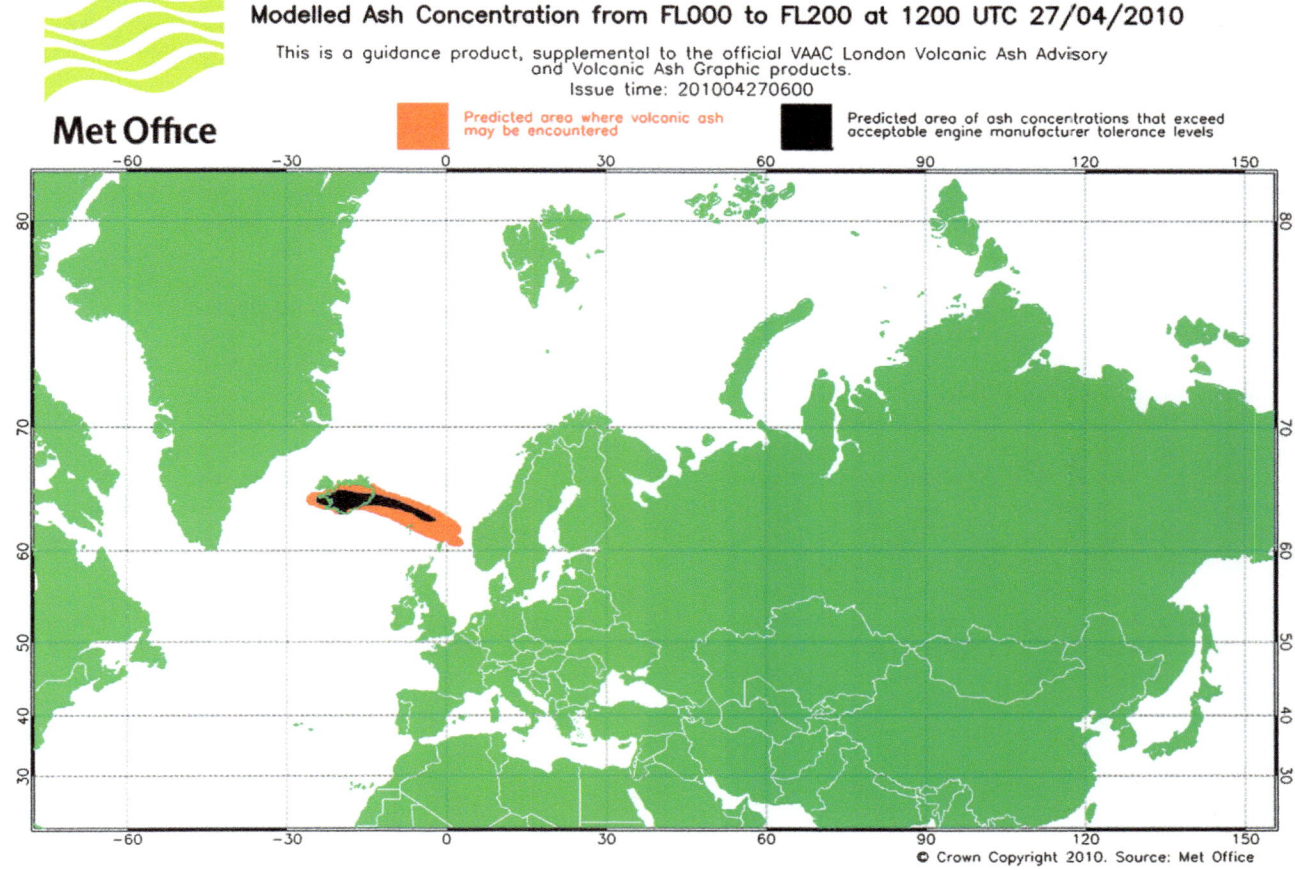

Modelled Ash Concentration from FL000 to FL200 at 1200 UTC 27/04/2010

This is a guidance product, supplemental to the official VAAC London Volcanic Ash Advisory and Volcanic Ash Graphic products.
Issue time: 201004270600

Predicted area where volcanic ash may be encountered

Predicted area of ash concentrations that exceed acceptable engine manufacturer tolerance levels

© Crown Copyright 2010. Source: Met Office

FIGURE 52.19 New ash graphic issued by the London Volcanic Ash Advisory Centre (VAAC) on 27 April 2010 at 06:00 UTC using concentration levels.

MSG Meteosat Second Generation
MWO Meteorological Watch Office
NAME Numerical Atmospheric-Dispersion Modeling Environment
NOAA National Oceanic and Atmospheric Administration
NPP National Polar-orbiting Platform
OMI Ozone Monitoring Instrument
SEVIRI Spin-stabilized Enhanced Visible and Infrared Imager
SIGMET Notice of SIGnificant METeorological Phenomena (ICAO)
USGS United States Geological Service
VAAC Volcanic Ash Advisory Centre
VAFTAD Volcanic Ash Forecast Transport and Dispersion
VASAG Volcanic Ash Science Advisory Group
VEI Volcanic Explosivity Index
VONA Volcano Observatory Notice for Aviation
WMO World Meteorological Organization
WOVO World Organization of Volcano Observatories
WRF Weather Research and Forecasting

Supplementary videos related to this chapter can be found at http://booksite.elsevier.com/9780123859389

FURTHER READING

Casadevall, Thomas J., 1994. Volcanic Ash and Aviation Safety: Proceedings of the First International Symposium on Volcanic Ash and Aviation Safety, vol. 2047. DIANE Publishing.

Clarisse, L., Coheur, P.-F., Prata, F., Hadji-Lazaro, J., Hurtmans, D., Clerbaux, C., 2013. A unified approach to infrared aerosol remote sensing and type specification. Atmos. Chem. Phys. 13 (4), 2195–2221.

Guffanti, M., Mayberry, G.C., Casadevall, T.J., Wunderman, R., 2009. Volcanic hazards to airports. Nat. Hazards 51 (2), 287–302.

Guffanti, M., Casadevall, T.J., Budding, K.E., 2010. Encounters of Aircraft with Volcanic Ash Clouds: A Compilation of Known Incidents, 1953–2009. US Department of Interior, US Geological Survey.

Lin, John, 2013. Lagrangian Modeling of the Atmosphere, vol. 200. John Wiley & Sons.

Marzano, Frank S., Picciotti, Errico, Montopoli, Mario, Vulpiani, Gianfranco, 2013. Inside Volcanic Clouds: Remote Sensing of Ash Plumes Using Microwave Weather Radars. Bull. Am. Meteorol. Soc. 94 (10).

Miller, Thomas P., Casadevall, Thomas J., et al., 2000. Volcanic ash hazards to aviation. In: Encyclopedia of Volcanoes, 915–930.

Pavolonis, M.J., Heidinger, A.K., Sieglaff, J., 2013. Automated retrievals of volcanic ash and dust cloud. J. Geophys. Res. Atmos. 118, 1436–1458.

Prata, A.J., 1989. Infrared radiative transfer calculations for volcanic ash clouds. Geophys. Res. Lett. 16 (11), 1293−1296.

Prata, A.J., Prata, A.T., 2012. Eyjafjallajökull volcanic ash concentrations determined using Spin Enhanced Visible and Infrared Imager measurements. J. Geophys. Res. 117, D00U23.

Prata, A.J., Tupper, A., 2009. Aviation hazards from volcanoes: The state of the science. Nat. Hazards 51 (2), 239−244.

Rose, W.I., Delene, D.J., Schneider, D.J., Bluth, G.J.S., Krueger, A.J., Sprod, I., McKee, C., Davies, H.L., Ernst, G.G.J., 1995. Ice in the 1994 Rabaul eruption cloud: implications for volcano hazard and atmospheric effects. Nature 375 (6531), 477−479.

Rose, W.I., Durant, A.J., 2009. Fine ash content of explosive eruptions. J. Volcanol. Geotherm. Res. 186 (1−2), 32−39.

Rose, W.I., Bluth, G.J.S., Ernst, G.G.J., 2000. Integrating retrievals of volcanic cloud characteristics from satellite remote sensors: a summary. Phil. Trans. R. Soc. Lond. A 358 (1770), 1585−1606.

Rose, W.I., Gu, Y., Watson, I.M., Yu, T., Blut, G.J.S., Prata, A.J., Krueger, A.J., Krotkov, N., Carn, S., Fromm, M.D., Hunton, D.E., Ernst, G.G.J., Viggiano, A.A., Miller, T.M., Ballenthin, J.O., Reeves, J.M., Wilson, J.C., Anderson, B.E., Flittner, D.E., 2003. The February−March 2000 Eruption of Hekla, Iceland from a Satellite Perspective. Volcanism and the Earth's Atmosphere, 107−132.

Shinozaki, Maya, Roberts, Kevin A., van de Goor, Bennie, Clyne, T. William, 2013. Deposition of ingested volcanic ash on surfaces in the turbine of a small jet engine. Adv. Eng. Mater. 15 (10), 986−994.

Stohl, A., Prata, A.J., Eckhardt, S., Clarisse, L., Durant, A., Henne, S., Kristiansen, N.I., Minikin, A., Schumann, U., Seibert, P., Stebel, K., Thomas, H.E., Thorsteinsson, T., Tørseth, K., Weinzierl, B., 2011. Determination of time- and height-resolved volcanic ash emissions and their use for quantitative ash dispersion modeling: the 2010 Eyjafjallajökull eruption. Atmos. Chem. Phys. 11 (9), 4333−4351.

Thomas, H.E., Prata, A.J., 2011. Sulphur dioxide as a volcanic ash proxy during the April and May 2010 eruption of Eyjafjallajökull Volcano, Iceland. Atmos. Chem. Phys. 11 (14), 6871−6880.

Webley, P.W., Dean, K., Peterson, R., Steffke, A., Harrild, M., Groves, J., 2012. Dispersion modeling of volcanic ash clouds: North Pacific eruptions, the past 40 years: 1970−2010. Nat. Hazards 61 (2), 661−671.

Weinzierl, Bernadett, Sauer, Daniel, Minikin, Andreas, Reitebuch, Oliver, Dahlkötter, Florian, Mayer, Bernhard, Emde, Claudia, Tegen, Ina, Gasteiger, Josef, Petzold, Andreas, et al., 2012. On the visibility of airborne volcanic ash and mineral dust from the pilot's perspective in flight. Phys. Chem. Earth 45, 87−102. Parts A/B/C.

Wen, S., Rose, W.I., 1994. Retrieval of sizes and total masses of particles in volcanic clouds using AVHRR bands 4 and 5. J. Geophys. Res. 99 (D3), 5421−5431.

Witham, Claire, Webster, Helen, Hort, Matthew, Jones, Andrew, Thomson, David, 2012. Modelling concentrations of volcanic ash encountered by aircraft in past eruptions. Atmos. Environ. 48, 219−229.

Chapter 53

Climatic Impacts of Volcanic Eruptions

Alan Robock

Department of Environmental Sciences, Rutgers University, New Brunswick, NJ, USA

Chapter Outline

GLOSSARY

aerosol Any suspended particle in the atmosphere. An aerosol can be a liquid droplet, such as sulfuric acid, or a solid, such as ash or tephra.

Arctic Oscillation A strengthening or weakening of the polar vortex in the Northern Hemisphere, the main pattern of oscillation of winter variation of extratropical atmospheric circulation. For a positive mode, with a strong vortex, winter temperatures are above normal over Northern Hemisphere continents.

El Niño A periodic warming of the tropical Pacific Ocean that increases global average temperature and produces patterns of positive and negative regional temperature and precipitation variations.

geoengineering Ideas for deliberate, large-scale manipulation of the environment to counteract global warming. The technology does not currently exist, but suggestions include producing a permanent stratospheric aerosol cloud in the stratosphere.

lidar Light detection and ranging. Sending a pulse of laser light into the atmosphere and measuring the reflection from aerosols. The timing of the reflected pulse tells the distance, and the strength of the reflection gives the amount. Using multiple wavelengths can give information about the size distribution of the aerosols.

nuclear winter The climate response to nuclear war. Smoke from targeted cities and industrial areas would rise into the stratosphere, block sunlight, and cool the surface. A nuclear war between the United States and Russia could produce temperatures below freezing in the summer over land, killing all crops and sentencing most people on the Earth to starvation.

ozone depletion Destruction of stratospheric ozone from reactions with anthropogenic chlorine. Aerosols in the stratosphere serve as surfaces for heterogeneous reactions, enhancing ozone depletion.

troposphere The lowest layer of the atmosphere from the surface up to 8 km at the poles and 18 km in the tropics. Weather takes place in the troposphere, including clouds and precipitation. The troposphere is well mixed, and the average lifetime of aerosols is about 1 week.

stratosphere The second layer in the atmosphere from the boundary with the troposphere (the tropopause) up to about 50 km. The stratosphere has little vertical motion and no precipitation, so the average lifetime of volcanic sulfate aerosols is 1–2 years.

1. INTRODUCTION

Explosive volcanic eruptions affect climate by injecting gases and **aerosol** particles into the **stratosphere**. In most cases, the eruption directly injects SO_2 into the stratosphere, but for the 2011 Nabro eruption, the largest since Mt Pinatubo in 1991, the eruption put most of the sulfur into the upper **troposphere**, and the summer Asian monsoon pumped the sulfur into the stratosphere over the next couple months. Only if the eruption cloud is rich in SO_2, will the eruption produce a long-lived aerosol cloud of sulfate aerosols that form over the next few weeks. Otherwise, explosive eruptions that only produce large ash particles, such as the 1980 Mount St Helens eruption, can produce a large local weather perturbation but do not have long-lasting climatic effects. Some volcanoes, such as Kilauea and Etna, produce large tropospheric emissions of sulfate aerosols, and only if there is a dramatic change in these emissions, will climate be changed. Stratospheric aerosol clouds last for several years, reflecting sunlight and cooling the surface. These clouds also absorb both solar (near infrared) and terrestrial radiation, heating the lower stratosphere. Volcanic aerosols also serve as surfaces for heterogeneous chemical reactions that destroy stratospheric ozone, which lowers ultraviolet (UV)

The Encyclopedia of Volcanoes. http://dx.doi.org/10.1016/B978-0-12-385938-9.00053-5

TABLE 53.1 Major Volcanic Eruptions of the Past 250 Years. Volcanic Explosivity Index (VEI; Newhall and Self, 1982) Data are from the Smithsonian Global Volcanism Program, http://www.volcano.si.edu/world/largeeruptions.cfm. Stratospheric Loading is from Gao et al. (2008) and Sato et al. (1993), Updated at (http://data.giss.nasa.gov/modelforce/strataer/). The VEI is a Logarithmic Measure of the Explosivity and not of the Potential to Cause Climate Change

Volcano	Year of Eruption	VEI	Global Stratospheric Sulfate Loading (Tg)
Laki (or Lakagigar), Iceland	1783–1784	4+	93[1]
Unidentified	1809	6?	54
Tambora, Sumbawa, Indonesia	1815	7	110
Cosiguina, Nicaragua	1835	5	40
Krakatau, Indonesia	1883	6	22
Okataina (Tarawera), North Island, New Zealand	1886	5	2
Santa María, Guatemala	1902	6?	4
Novarupta (Katmai), Alaska, the United States	1912	6	11[1]
Gunung Agung, Bali, Indonesia	1963	5	17
Mount St Helens, Washington, the United States	1980	5	0
El Chichón, Chiapas, Mexico	1982	5	14
Mt Pinatubo, Luzon, Philippines	1991	6	30
Cerro Hudson, Chile	1991	5+	?

[1]Northern Hemisphere only.

absorption and reduces the radiative heating in the lower stratosphere, but the net effect is still heating. This also allows more UV radiation to reach the surface. As this chemical effect depends on the presence of anthropogenic chlorine, it has only become important in recent decades. Tropical eruptions produce asymmetric stratospheric heating, producing a stronger polar vortex and associated positive mode of the **Arctic Oscillation** in tropospheric circulation. This pattern is one of enhanced warm advection over Northern Hemisphere (NH) continents in winter, producing winter warming after large tropical eruptions. Although observations and some climate models show this response, many modern climate models still do a poor job of simulating this dynamic response to tropical volcanic eruptions. There is no evidence that volcanic eruptions can produce **El Niños**, but El Niño/Southern Oscillation variations must be considered when searching the climatic record for volcanic signals, as they have similar amplitudes and timescales.

There have been several large volcanic eruptions in the past 250 years (Table 53.1), and each has drawn attention to the atmospheric and potential climatic effects. The 1783 Laki eruption in Iceland was followed by a very warm summer and then a very cold winter in Europe, causing Benjamin Franklin, the U.S. Ambassador to France, to publish the first paper on the subject in more than

1800 years. The 1815 Tambora eruption, combined with the effects of the unidentified 1809 eruption, produced the "Year Without a Summer" in 1816 and inspired *Frankenstein*, written by Mary Shelley on the shores of Lake Geneva, Switzerland, that summer. The 1883 Krakatau eruption was the largest explosion ever heard, and the sound wave was tracked on microbarographs for four complete circuits of the Earth, taking 33 h for one circuit. The Royal Society (UK) report on this eruption published 5 years later (which misspelled the volcano name as Krakatoa) remains the most extensive report on the atmospheric effects of a volcanic eruption (Symons, 1888). The observations of the westward propagation of the resultant volcanic cloud were a hint at strong easterlies in the stratosphere, an atmospheric layer yet to be discovered. The 1963 Agung eruption produced the largest stratospheric aerosol cloud in more than 50 years and inspired many modern scientific studies. The subsequent 1982 El Chichón and 1991 Mt Pinatubo eruptions produced very large stratospheric aerosol clouds and large climatic effects.

2. VOLCANIC EMISSIONS

Quantification of the size of volcanic eruptions is difficult, as different measures reveal different information. For example, one could examine the total mass ejected, the

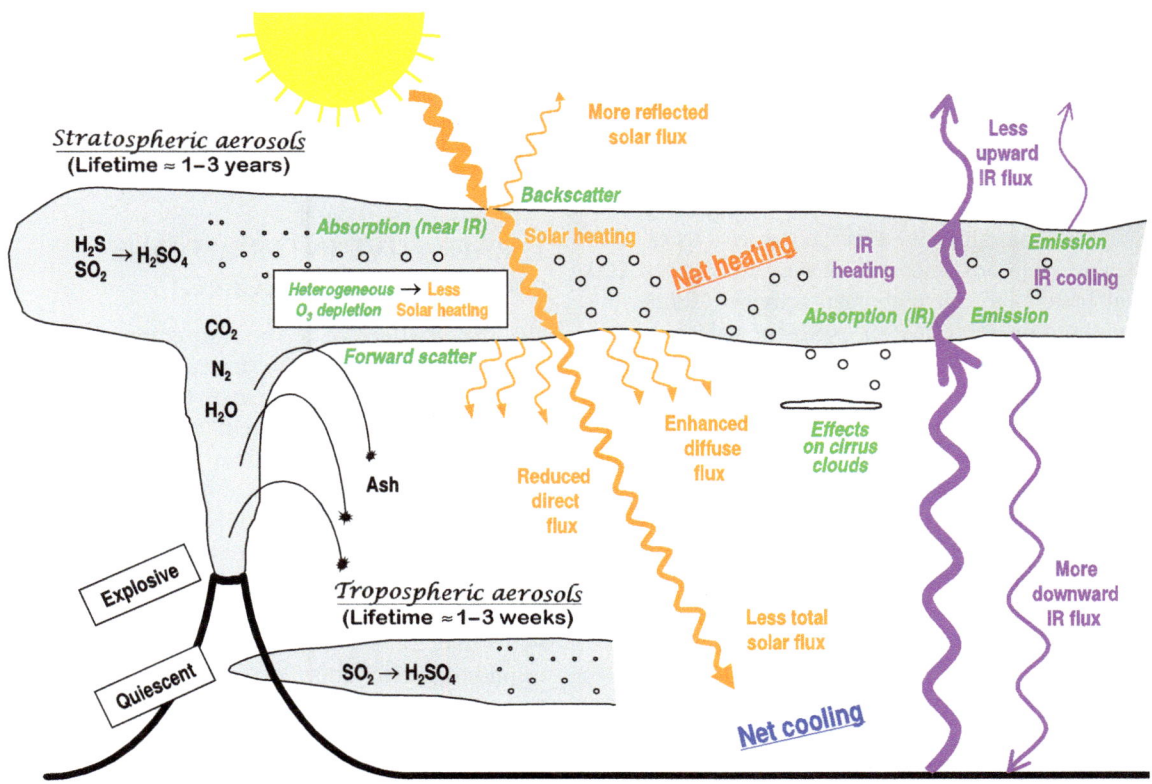

FIGURE 53.1 Schematic diagram of volcanic inputs to the atmosphere and their effects. *Plate 1 of Robock (2000),* [C]*Copyright, American Geophysical Union.*

explosiveness, or the sulfur input to the stratosphere. While the first two of these are of great interest to volcanologists, as they relate to the geology and the local hazards, it is the third that determines the climatic impact. Thus, you will find climatologists and volcanologists disagreeing on the largest volcanic eruptions in the past. For example, although the 1980 Mount St Helens eruption was a tremendous blast and was assigned a Volcanic Explosivity Index of 5, the blast was lateral, full of solid material, and injected negligible amounts of sulfur to the stratosphere. It had a large impact on local weather for a few days (Robock and Mass, 1982), but no global climatic impact (Robock, 1981).

Volcanic eruptions inject several different types of particles and gases into the atmosphere (Figure 53.1). The gases can be assessed based on measurements from active, but not explosive, eruptions, but it is not clear that these would be the same from strong, explosive eruptions. In situ measurements from balloons and airplanes have been used to measure the bottoms of the resulting stratospheric aerosol clouds, but remote sensing from **lidar**, radiometers, and satellites is needed for global coverage. Still, there are gaps in our ability to monitor volcanic clouds in detail because the thickest parts of the clouds require lidar and satellite orbits prevent continuous monitoring. Several

satellites routinely monitor SO_2, allowing us to directly measure stratospheric injection of gases from eruptions.

The major component of volcanic eruptions is the matter that emerges as solid, lithic material or solidifies into large particles, which are referred to as ash or tephra. These particles fall out of the atmosphere very rapidly, on time-scales of minutes to a few days, and thus have no climatic impacts but are of great interest to volcanologists, as seen in the rest of this encyclopedia. When an eruption column still laden with these hot particles descends down the slopes of a volcano, this pyroclastic flow can be deadly to those unlucky enough to be at the base of the volcano. The destruction of Pompeii and Herculaneum after the AD 79 Vesuvius eruption is the most famous example.

Volcanic eruptions typically also emit gases, with H_2O, N_2, and CO_2 being the most abundant. Over the lifetime of the Earth, these gases have been the main source of the Earth's atmosphere and ocean after the primitive atmosphere of hydrogen and helium was lost to space. The water has condensed into the oceans, the CO_2 has been changed by plants into O_2 or formed carbonates, which sink to the ocean bottom, and some of the C has turned into fossil fuels. Of course, we eat plants and animals, which eat the plants, we drink the water, and we breathe the oxygen, so each of us is made of volcanic emissions. The atmosphere is now mainly composed of N_2 (78%) and O_2 (21%), both of which had sources in volcanic emissions.

Of these abundant gases, both H_2O and CO_2 are important greenhouse gases, but their atmospheric concentrations are so large (even for CO_2 at only 400 ppm in 2013) that individual eruptions have a negligible effect on their concentrations and do not directly impact the greenhouse effect. Global annually averaged emissions of CO_2 from volcanic eruptions since 1750 have been at least 100 times smaller than those from human activities. Rather the most important climatic effect of explosive volcanic eruptions is through their emission of sulfur species to the stratosphere, mainly in the form of SO_2, but possibly sometimes as H_2S. These sulfur species react with H_2O to form H_2SO_4 on a timescale of weeks, and the resulting sulfate aerosols produce the dominant radiative effect from volcanic eruptions.

The 1982 El Chichón eruption injected 7 MT of SO_2 into the atmosphere. There has not been a large stratospheric injection since 1991, when Mt Pinatubo in the Philippines put about 20 MT of SO_2 into the lower stratosphere. In 2008, Kasatochi (in the Aleutian Islands of Alaska); in 2009, Mt Sarychev (in the Russian Kamchatka Peninsula); and in 2011, Nabro (in Eritrea), each put about 1.5 MT of SO_2 into the lower stratosphere, and these eruptions contributed to the reduced global warming of the past decade. The Eyjafjallajökull eruption in Iceland in 2010, while very disruptive of air traffic for weeks, had so little SO_2 and with a short lifetime of a week or so in the troposphere that it had no impact on climate.

Once injected into the stratosphere, the large aerosol particles and small ones being formed by the sulfur gases are rapidly transported around the globe by stratospheric winds. Observations after the 1883 Krakatau eruption showed that the aerosol cloud circled the globe in 2 weeks. Both the 1982 El Chichón cloud and the 1991 Pinatubo cloud circled the globe in 3 weeks. Although El Chichón (17°N) and Pinatubo (15°N) are separated by only 2° of latitude, their clouds, after only one circuit of the globe, ended up separated by 15° of latitude, with the Pinatubo cloud straddling the equator and the El Chichón cloud extending approximately from the equator to 30°N. Subsequent dispersion of a stratospheric volcanic cloud depends heavily on the particular distribution of winds at the time of eruption. For trying to reconstruct the effects of older eruptions, this factor adds a further complication, as the latitude of the volcano is not sufficient information.

Quiescent continuous volcanic emissions also add sulfates to the troposphere, but their lifetimes there are much shorter, although longer than anthropogenic sulfates, as they are emitted from the sides of mountains rather than at the surface. The local pollution produced by the emission of the Kilauea crater on the Big Island of Hawaii is called "vog" (volcanic fog). Global sulfur emission of volcanoes

to the troposphere is about 15% of the total natural and anthropogenic emission, producing cooling at the surface. Only if there is a long-term trend in these emissions, will they be important for climate change; nevertheless, they must be considered when evaluating the effects of anthropogenic sulfate emissions.

3. RADIATIVE INTERACTIONS AND CLIMATE FORCING

The major effect of a volcanic eruption on the climate system is the effect of the stratospheric cloud on solar radiation (Figure 53.1). Some of the radiation is scattered back to space, increasing the planetary albedo and cooling the Earth's atmosphere system. The sulfate aerosol particles (typical effective radius of 0.5 μm, about the same size as the wavelength of visible light) also forward scatter much of the solar radiation, reducing the direct solar beam but increasing the brightness of the sky. After the 1991 Pinatubo eruption, the sky around the sun appeared more white than blue because of this. After the El Chichón eruption of 1982 and the Pinatubo eruption of 1991, the direct radiation was significantly reduced, but the diffuse radiation was enhanced by almost as much. Nevertheless, the volcanic aerosol clouds reduced the total radiation received at the surface.

As the sun sets, the yellow and red light (because Rayleigh scattering removes the shorter wavelengths in the process that produces the blue sky) is reflected from the bottom of stratospheric volcanic clouds, producing a characteristic yellow and red sky 1/2−1 h after the time of sunset (Figure 53.2). This effect has been used in the past to detect distant eruptions and to estimate the height of the aerosol cloud and its extent. But such sunsets also inspire artists. After seeing the brilliant volcanic sunsets from the 1883 Krakatau eruption over Oslo, the Norwegian artist

FIGURE 53.2 Volcanic sunset over Lake Mendota in Madison, Wisconsin, in July 1982, 3 months after the El Chichón eruption. Photograph by Alan Robock. *Plate 4 from Robock (2000),* ©*Copyright, American Geophysical Union.*

Edvard Munch painted *The Scream* 10 years later. As he wrote in his diary, "I was walking along a path with two friends—the sun was setting—suddenly the sky turned blood red—I paused, feeling exhausted, and leaned on the fence—there were blood and tongues of fire above the blue-black fjord and the city—my friends walked on, and I stood there trembling with anxiety—and I sensed an infinite scream passing through nature."

4. CLIMATIC IMPACT OF VOLCANIC AEROSOLS

Stratospheric aerosol clouds from volcanic eruptions cool the Earth's surface for several years but produce winter warming over the continents in the NH. These and other effects are summarized in Table 53.2. Volcanic aerosols can be important causes of temperature changes for several years following large eruptions, and even on a millennial timescale they can be important when their cumulative effects are taken into account. This is very significant in analyzing the global warming problem, as the impacts of anthropogenic greenhouse gases and aerosols on climate must be evaluated against a background of continued natural forcing of the climate system from volcanic eruptions, solar variations, and internal random variations from land—atmosphere and ocean—atmosphere interactions.

Individual large eruptions produce global or hemispheric cooling for 2 or 3 years, but the winter following a large tropical eruption is warmer over the NH continents, and this counterintuitive effect is due to a nonlinear response through atmospheric dynamics. The winter warming pattern is illustrated in Figure 53.3, which shows the global lower-tropospheric temperature anomaly pattern for the NH winter of 1991−1992 following the 1991 Mt Pinatubo eruption. This pattern is closely correlated with the surface air temperature pattern where the data overlap, but the satellite data allow global coverage. The temperature over North America, Europe, and Siberia was much higher than normal and that over Alaska, Greenland, the Middle East, and China was lower than normal. In fact, it was so cold that winter that it snowed in Jerusalem, a very unusual occurrence. Coral at the bottom of the Red Sea died that winter because the water at the surface cooled and convectively mixed the entire depth of the water. The enhanced supply of nutrients produced anomalously large algal and phytoplankton blooms, which smothered the coral. This coral death had only happened before in winters following large volcanic eruptions. At the tropopause, the boundary between the troposphere and stratosphere, the strongest winds are found in the midlatitudes in the winter and are called the jet stream or polar vortex. The strength of the jet stream depends on the temperature difference (gradient) between the tropics and the polar region, which is largest in the winter when the polar regions cool. For a tropical eruption, the stratospheric heating from volcanic aerosols is larger in the tropics than in the high latitudes, producing an enhanced pole-to-equator temperature

TABLE 53.2 Effects of Large Explosive Volcanic Eruptions on Climate

Effect/Mechanism	Begins	Duration
Stratospheric warming	1−3 months	1−2 years
Stratospheric absorption of shortwave and longwave radiation		
Global cooling	Immediately	1−3 years
Blockage of shortwave radiation		
Global cooling from multiple eruptions	Immediately	Up to centuries
Blockage of shortwave radiation		
Winter warming of Northern Hemisphere continents	½−1½ years	1 or 2 winters
Differential stratospheric heating, dynamical interaction with troposphere		
Reduced tropical precipitation	Immediately	~1 year
Blockage of shortwave radiation, reduced evaporation		
Reduction of Asian and African summer monsoon	½−1 year	1 or 2 summers
Continental cooling, reduction of land—sea temperature contrast		
Ozone depletion, enhanced UV	1 day	1−2 years
Dilution, heterogeneous chemistry on aerosols		

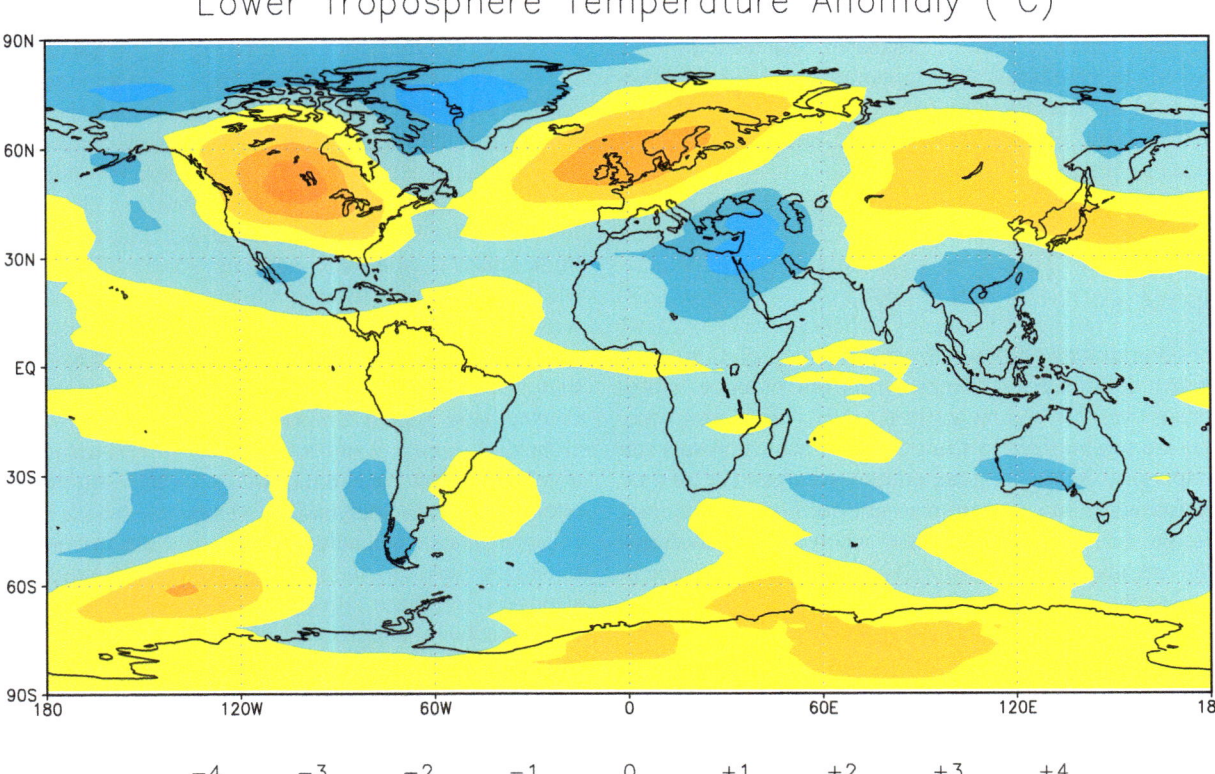

FIGURE 53.3   Winter (DJF) lower-tropospheric temperature anomalies (with the nonvolcanic period of 1984–1990 used to calculate the mean) for the 1991–1992 Northern Hemisphere winter (DJF) following the 1991 Mt Pinatubo eruption. This pattern is typical of that following all large tropical eruptions, with warming over North America, Europe, and Siberia, and cooling over Alaska, Greenland, the Middle East, and China. Data from Microwave Sounding Unit Channel 2R, updated courtesy of J. Christy and now called Channel 2LT. *Plate 8 from Robock (2000),* ©*Copyright, American Geophysical Union.*

gradient, and in the NH winter, a stronger polar vortex and winter warming of NH continents. The stronger jet stream produces a characteristic wind pattern in the troposphere, which warms some regions and cools other ones. This pattern is called the "Arctic Oscillation" and is the dominant mode of tropospheric variability. Tropical eruption clouds push the atmosphere into the positive phase of this natural variation. This indirect advective effect on temperature is stronger than the radiative cooling effect that dominates at lower latitudes and in the summer.

Because volcanic aerosols normally remain in the stratosphere no more than 2 or 3 years, the radiative effect of volcanoes is interannual rather than interdecadal in scale. A series of volcanic eruptions could, however, raise the mean optical depth significantly over a longer period and thereby give rise to a decadal-scale cooling. If a period of active volcanism ends for a significant period, such as for the 51-year period from 1912 to 1963 when global climate warmed, the adjustment of the climate system to no volcanic forcing helped produce the warming. Furthermore, it

is possible that feedbacks involving ice and ocean, which act on longer timescales, could transform the short-term volcanic forcing into a longer-term effect. The current century is the warmest of the past 10, with the previous several centuries called the Little Ice Age due to their coldness. Studies show that the planet cooled at the end of the thirteenth century, following the most volcanic period in the past 1500 years, including the largest eruption of the entire period, the 1257 Samalas eruption, which produced 258 Tg of sulfate aerosols in the stratosphere (Gao et al., 2008). Miller et al. (2012) showed that in a climate model, these eruptions cooled the Earth so much that an Arctic sea ice feedback caused the cooling to persist until the recent global warming, which started in the nineteenth century, and thus that these volcanic eruptions were the cause of the Little Ice Age. This agreed with evidence from vegetation that had been killed and then preserved by ice sheet advances on Baffin Island following the largest eruptions of the period, the 1257 Samalas eruption and Kuwae in 1452. The large warming of the past century, however, can only partially be explained by these natural causes, and in fact the second half of the twentieth century would have cooled due to volcanic eruptions if there had been no human emissions. The large warming of this period can only be explained by including the effects of warming from anthropogenic greenhouse gases.

While Haslam and Petraglia (2010) showed that a 1000-year glacial period started just before the 74 ka BP Toba eruption, disproving the theory that the eruption produced the ice advance, the potential for a human genetic bottleneck produced by the death of most humans in a volcanic winter just after the eruption is still not resolved. This eruption on the island of Sumatra left a caldera about 86-km long and 30-km wide, with a large island inside, the resurgent block of the caldera. It erupted 1000 times more rock than the 1980 Mount St Helens eruption and injected approximately 100–300 times the amount of SO_2 into the stratosphere than the 1991 Pinatubo eruption. Depending on the assumptions about the properties of the resulting sulfate particles in the stratosphere, climate model simulations produce global average coolings of 3–15 °C, lasting for a decade or more. The larger range would have certainly been devastating for many species. However, there are no observations or paleo-reconstructions with a high enough time resolution to tell whether there was a several-year volcanic winter.

5. OZONE IMPACTS

Volcanic aerosols have the potential to change not only the radiative flux in the stratosphere but also its chemistry. The most important chemical changes in the stratosphere are related to O_3, which has significant effects on UV and longwave radiative fluxes. The reactions that produce and destroy O_3 depend on the UV flux, the temperature, and the presence of surfaces for heterogeneous reactions, all of which are changed by volcanic aerosols. The heterogeneous chemistry responsible for the ozone hole over Antarctica in October every year occurs on polar stratospheric clouds of water or nitric acid, which only occur in the extremely cold isolated spring vortex in the Southern Hemisphere. Conditions in the NH are now changing and small O_3 depletions are being observed in spring there now, too. Reactions on polar stratospheric clouds make anthropogenic chlorine available for chemical destruction of O_3. However, sulfate aerosols produced by volcanic eruptions can also provide these surfaces at lower latitudes and at all times of the year. In fact, after the 1991 Pinatubo eruption, column O_3 reduction of about 5% was observed in midlatitudes, ranging from 2% in the tropics to 7% in the midlatitudes. Therefore, **ozone depletion** in the aerosol cloud is much larger and reaches 20%. The chemical ozone destruction is less effective in the tropics, but lifting of low-ozone-concentration layers with the aerosol cloud causes a fast decrease in ozone mixing ratio in the low latitudes.

Decrease of the ozone concentration following volcanic eruptions causes less UV absorption in stratosphere, which modifies the aerosol heating effect. The net effect of volcanic aerosols on the surface UV flux is to increase it, as the aerosols back scatter less UV than the subsequent O_3 depletion allows through. The reduced O_3 absorption of shortwave and longwave radiation reduces the stratospheric heating effect and can affect the winter warming phenomenon described above.

The volcanic effect on O_3 chemistry is a new phenomenon, depending on anthropogenic chlorine in the stratosphere. While we have no observations, the 1963 Agung eruption probably did not deplete O_3, as there was little anthropogenic chlorine in the stratosphere. Due to the Montreal Protocol and subsequent international agreements, chlorine concentration has peaked in the stratosphere and is now decreasing. Therefore, for the next few decades, large volcanic eruptions will have effects similar to Pinatubo, but after that, these O_3 effects will go away and volcanic eruptions will have a stronger effect on atmospheric circulation without the negative feedback produced by O_3 depletion.

6. DISCUSSION

There is no evidence that volcanic eruptions produce El Niño events, but the climatic effects of El Niño and volcanic eruptions must be separated to understand the climatic response to each. It had been suggested that the simultaneous appearance of the large 1982–1983 El Niño and the 1982 El Chichón eruption and the 1991 smaller El Niño and the Pinatubo eruption suggested a cause-and-effect relationship. However, no plausible mechanism has been suggested and further research into the oceanography of those El Niños shows that they started before the volcanic eruptions. Examination of the entire record of past El Niños and volcanic eruptions for the past two centuries also shows no significant correlation.

As volcanic eruptions and their subsequent climatic response represent a large perturbation to the climate system over a relatively short period, observations and the simulated model responses can serve as important analogs for understanding the climatic response to other perturbations. While the climatic response to explosive volcanic eruptions is a useful analog for some other climatic forcings, there are also limitations. For example, successful climate model simulations of the impact of one eruption can help validate models used for seasonal and interannual predictions. But they cannot test all the mechanisms involved in global warming over the next century, as long-term oceanic feedbacks are involved, which have a longer timescale than the response to individual volcanic eruptions. Theory tells us that volcanic eruptions also will produce multidecadal impacts on oceanic heat content and sea level, but these impacts are small and cannot be separated from other factors in observations.

The theory of "**nuclear winter**," the climatic effects of a massive injection of soot aerosols into the atmosphere from fires following a global nuclear holocaust, includes upward injection of the aerosols to the stratosphere, rapid global dispersal of stratospheric aerosols, removal of the aerosols, heating of the stratosphere, ozone depletion, and cooling at the surface under this cloud, still possible with the current global nuclear arsenal (Robock et al., 2007a; Toon et al., 2008). The use of only 100 nuclear weapons could produce climate change unprecedented in recorded human history (Robock et al., 2007b), which could produce significant decreases of agriculture in the main grain-growing regions of the world, the United States and China. As this theory cannot be tested in the real world, volcanic eruptions provide analogs that support these aspects of the theory.

Recent suggestions that we consider using **geoengineering** to control global climate through the creation of a permanent stratospheric aerosol cloud have used volcanic eruptions as an analog (e.g., Robock et al., 2013). Volcanic eruptions teach us that a stratospheric aerosol cloud would indeed cool the surface, reducing ice melt and sea level rise, and increase the terrestrial carbon sink. But volcanic eruptions also teach us that a stratospheric aerosol cloud would produce ozone depletion, allowing more harmful UV radiation at the surface, reduce summer monsoon precipitation and even produce drought, produce rapid warming if geoengineering were suddenly stopped, reduce solar power, damage airplanes flying in the stratosphere, and degrade surface astronomical observations and remote sensing (Robock et al., 2013). This raises many issues about the wisdom of geoengineering, and there are many other reasons why geoengineering may be a bad idea (Robock, 2008).

Given our current understanding of the climatic impact of volcanic eruptions, we can safely predict that following the next large tropical eruption, there will be global cooling for about 2 years and winter warming of the NH continents for 1 or 2 years. There will also be reduced summer monsoon precipitation over Asia and Africa. A large NH high-latitude eruption, if it occurs in spring or summer, will also produce a weak summer monsoon.

FURTHER READING

Franklin, B., 1784. Meteorological imaginations and conjectures. Manchester Literary and Philosophical Society Memoirs and Proceedings 2, 122 [Reprinted in Weatherwise 35, p. 262, 1982.].

Gao, C., Robock, A., Ammann, C., 2008. Volcanic forcing of climate over the past 1500 years: an improved ice-core-based index for climate models. Journal of Geophysical Research 113, D23111. http://dx.doi.org/10.1029/2008JD010239.

GRL Special Issue, 1983. Geophysical Research Letters 10, 989–1060 [Studies of the 1982 El Chichón eruption].

GRL Special Issue, 1992. Geophysical Research Letters 19, 149–218 [Studies of the 1991 Mt. Pinatubo eruption].

Haslam, M., Petraglia, M., 2009. Comment on "Environmental impact of the 73 ka Toba super-eruption in South Asia" by Williams, M.A.J., et al. [284 (2009) 295–314]. *Palaeogeography, Palaeoclimatology, Palaeoecology* 296, 199–203. http://dx.doi.org/10.1016/j.palaeo.2010.03.057.

Lamb, H.H., 1970. Volcanic dust in the atmosphere; with a chronology and assessment of its meteorological significance. Philosophical Transactions Royal Society of London A266, 425–533.

Miller, G.H., et al., 2012. Abrupt onset of the Little Ice Age triggered by volcanism and sustained by sea-ice/ocean feedbacks. Geophysical Research Letters 39, L02708. http://dx.doi.org/10.1029/2011GL050168.

Newhall, C.G., Self, S., 1982. The Volcanic Explosivity Index (VEI): An estimate of explosive magnitude for historical volcanism. Journal of Geophysical Research 87, 1231–1238.

Robock, A., 1981. The Mount St. Helens volcanic eruption of 18 May 1980: Minimal climatic effect. Science 212, 1383–1384.

Robock, A., 2000. Volcanic eruptions and climate. Reviews of Geophysical 38, 191–219 [From which this chapter was condensed and updated.].

Robock, A., 2008. 20 reasons why geoengineering may be a bad idea. Bulletin Atomic Scientists 64 (2), 14–18. http://dx.doi.org/10.2968/064002006, 59.

Robock, A., Mass, C., 1982. The Mount St. Helens volcanic eruption of 18 May 1980: Large short-term surface temperature effects. Science 216, 628–630.

Robock, A., Oman, L., Stenchikov, G.L., Toon, O.B., Bardeen, C., Turco, R.P., 2007a. Climatic consequences of regional nuclear conflicts. Atmospheric Chemistry and Physics 7, 2003–2012.

Robock, A., Oman, L., Stenchikov, G.L., 2007b. Nuclear winter revisited with a modern climate model and current nuclear arsenals: Still catastrophic consequences. Journal of Geophysical Research 112, D13107. http://dx.doi.org/10.1029/2006JD008235.

Robock, A., MacMartin, D.G., Duren, R., Christensen, M.W., 2013. Studying geoengineering with natural and anthropogenic analogs. Climatic Change 121, 445–458. http://dx.doi.org/10.1007/s10584-013-0777-5.

Sato, M., Hansen, J.E., McCormick, M.P., Pollack, J.B., 1993. Stratospheric aerosol optical depths, 1850-1990. Journal of Geophysical Research 98, 22,987–22,994.

Siebert, L., Simkin, T., Kimberly, P., 2011. Volcanoes of the World, third ed. University of California Press, Berkeley, Ca. 568 pp.

Symons, G.J. (Ed.), 1888. The Eruption of Krakatoa, and Subsequent Phenomena. Trübner, London, England, 494 pp.

Timmreck, C., 2012. Modeling the climatic effects of large explosive volcanic eruptions. WIREs Climate Change 3, 545–564. http://dx.doi.org/10.1002/wcc.192.

Toon, O.B., Robock, A., Turco, R.P., 2008. Environmental consequences of nuclear war. Physics Today 61 (12), 37–42.

Chapter 54

Hazards from Pyroclastic Density Currents

Paul D. Cole

School of Geography, Earth and Environmental Science, Plymouth University, Plymouth, UK

Augusto Neri

Istituto Nazionale di Geofisica e Vulcanologia, Sesione Di Pisa, Pisa, Italy

Peter J. Baxter

Institute of Public Health, University of Cambridge, Cambridge, UK

Chapter Outline

GLOSSARY

block and ash flow Small-volume deposits of pyroclastic density currents formed by gravitational collapse of lava dome or lava flow.

dome collapse Retrogressive gravitational collapse of intermediate to silicic lava domes, either partial or total, to form a single or more typically a series of pyroclastic density currents. Generally dome collapses occur over a few hours but may last a few tens of minutes. Sometimes referred to as "Merapi type."

ignimbrite Deposit of pumiceous pyroclastic density currents, term generally used for those with volumes >1 km³, although smaller volume examples might appear identical. Fragments may be fused together or welded and some examples from the geological record are >1000 km³.

Monte Carlo simulation Repeated running of computer models, often thousands of times, using different system parameters to obtain representative variations of uncertainty of inputs.

Plinian and Subplinian eruptions Sustained explosive eruption typically lasting several hours. Eruptions columns up to several tens of kilometers (Plinian) and 10 to 20 km (Subplinian) are typical.

pumice flow Small-volume pumiceous pyroclastic density currents. Deposits are typically lobate and sinuous with levee and bulbous snouts.

pyroclastic flow High particle concentration pyroclastic density current. Term also frequently used for all types of pyroclastic density current.

pyroclastic surge Low particle concentration, dilute pyroclastic density current. Density varies widely with some pyroclastic surges typically having densities less than 1 kg m⁻³ as they have been observed traveling over water. Typically generated from, and may travel on top of or detach from, dense PDCs (pyroclastic flows), in the case of detached surges or may travel on top of a more concentrated basal layer.

scoria flow Small-volume pyroclastic density current composed of scoria (mafic to intermediate vesicular juvenile material). Deposits typically contain bulbous, cauliform blocks of juvenile scoria that show evidence of having plastic deformation and being Semi-molten during transport.

vulcanian explosion Short-lived explosion caused where gas build-up is released by sudden failure of an impermeable cap rock.

1. INTRODUCTION

Pyroclastic density currents (PDCs) are probably the least predictable and the most dangerous of all volcanic hazards. As a consequence they have been responsible for most

The Encyclopedia of Volcanoes. http://dx.doi.org/10.1016/B978-0-12-385938-9.00054-7

deaths in volcanic eruptions in recent times and they present the most important challenge of all volcanic hazards for disaster planners at volcanoes in densely inhabited regions. Their extreme destructiveness first came to scientific attention in two major eruptions, in the West Indies in 1902, on the islands of St. Vincent (May 7) and Martinique (May 8). The most notorious catastrophe was on Martinique, in the city of St. Pierre, where the PDC produced at Mont Pelée caused the death of almost all of the 28,000 inhabitants. An even more powerful blast on May 20 destroyed the buildings still standing, leaving an eerie and lifeless cityscape, which the scientists who had just arrived from overseas compared to the ruins of Pompeii and Herculaneum, but who were unable to explain the eruptive phenomenon.

In this chapter, we summarize some of the recent progress in our understanding of PDCs and their hazards, including advances in PDC modeling for hazard mapping and mitigation purposes.

2. DEFINITIONS

PDCs are laterally moving, buoyantly expanding mixtures of hot gases and fragmental particles (ash, lapilli, blocks, and boulders). They were first called **nuées ardentes**, or glowing clouds, after their striking appearance to early observers. The terms **pyroclastic flows** and surges have been used frequently by authors in the past, but the term PDC has now become preferred within the academic community as it includes all types of PDC, irrespective of their particle concentration or origin. The older terms (surge and flow) are effectively two end members, with pyroclastic flows at the dense end, with a high concentration of particles and **pyroclastic surges** at the dilute end. The same PDC may exhibit properties of both end members during the course of its runout. They are hazardous because of their high temperatures and velocities, making them capable of devastating the areas they inundate.

3. PDC GENERATION MECHANISMS

There are three main ways in which PDCs are produced in different types of eruption, namely the collapse of the vertical eruption column of gas and ash, the collapse of a lava dome and in a lateral or directed blast. In all, regardless of their mode of generation, the PDC is denser than the surrounding air and moves under the influence of gravity as it travels down the volcano's flanks. From the geological record we know that PDCs are able to travel distances of more than 100 km from their source, but typical historical examples of PDCs have generally not gone beyond 20 km. One exception to this was the 1883 eruption of Krakatau where PDCs traveled more than 40 km over the sea and were still able to fatally injure people. The volumes of material involved range over many orders of magnitude,

from 10^{-5} to $>10^3$ km^3 for the largest known examples in the geological record, termed **ignimbrite**.

3.1. Eruption Column Collapse (Fountain Collapse)

Explosive eruptions of different sizes, varying from small **Vulcanian explosions** through to large sustained **Plinian eruptions**, typically generate PDCs by the mechanism known as column collapse. With this mechanism PDCs are typically radially distributed around the volcanic edifice. Where the height of collapsing column is low, e.g., a few hundred meters, then this process is frequently referred to as fountain collapse.

Observers of small PDC-forming events often make reference to similarities with "a pot of rice boiling over." The actual mechanism is unclear, but it is likely to be a low energy form of fountain collapse in which the PDCs are largely contained within the crater or can only just surmount the crater rim.

3.2. Lava Dome and Lava Flow Front Collapse

Gravitational collapse of viscous (intermediate to silicic) lava domes is a common method of PDC generation. In most cases this is considered to be solely related to the gravitational instability of the lava dome since there are no evidence for any explosive activity. Nevertheless, some lava **dome collapses** (in particular large examples $>10^6$ m^3) are associated with explosions that are probably related to excavation of deeper and more gas-rich parts of the lava dome.

Collapse of viscous lava flow fronts on steep slopes (close to or exceeding the angle of repose) is also a typical method for generation of small-volume PDCs, e.g., Colima, Mexico, and Santiaguito, Guatemala. Associated with this origin is another less common mechanism, which is crater wall collapse where the wall of a summit crater can fail, forming a series of small PDCs or rock avalanches as has occurred several times at Arenal, Costa Rica, or recently at Mt Etna, Italy.

PDCs formed by lava dome collapse and lava flow front collapse are almost always unidirectional, flowing down one sector of the volcano at a time, but different sectors can be impacted subsequently as the dome continues to grow and collapse in different directions.

3.3. Lateral Blasts

The catastrophic failure of the flank of Mt St Helens on May 18, 1980, which unroofed a cryptodome growing inside the volcano, generated a laterally directed blast and a PDC of wide extent, high velocity, and immense destructive power that flowed as far as 28 km over a 180° sector north of the crater. Smaller, directed blasts have also occurred at Pelée, Martinique 1902, Lamington, Papua New Guinea 1951, Bezymianny, Kamchatka 1956, Soufrière Hills, Montserrat 1997, and Merapi, Indonesia in 2010, for example, and elsewhere in the geological record as inferred from the deposits at many other volcanoes.

From a hazard perspective, lateral or directed blasts are important because of the high velocity of the PDCs generated and the extensive areas that may be impacted, which can vary by more than two orders of magnitude from a few to over 600 km².

3.4. Other Mechanisms

There are several other less common mechanisms that can generate PDCs and many of these can present significant hazards:

Surge derived PDCs—where a dilute PDC interacts with strong relief, rapid sedimentation can occur and drain to form highly mobile dense, high-concentration PDCs, e.g., at Soufrière Hills Volcano, Montserrat in the June 25, 1997 dome collapse event.

Remobilization of unconsolidated ignimbrite—observed at Mt Pinatubo after the 1991 eruption from the walls of deep canyons once the ignimbrite had been dissected by fluvial action.

Interaction of a lava flow with snow and ice—e.g., at Mt Etna on 11 February 2014 in Sicily, but the PDCs are small and have limited runouts <1 km.

Hydrovolcanic explosions of PDC entering the ocean—at Soufrière Hills Volcano, Montserrat large-volume dome collapses in 2003 and 2006 sent high flux PDCs into the ocean. Hyrdovolcanic explosions were formed as a result, sending dilute PDCs (see Edmonds and Herd, 2005) inland for several kilometers.

4. VELOCITIES OF PDCs

Accurate velocity measurements of PDCs (using video) are rare, although a few examples of generally small-volume PDCs have been documented where runout distances are less than 10 km. Alternatively, estimates of velocities have been made from the interpretation of PDC deposits at locations of superelevation where PDCs had flowed around bends, or by determining the heights they manage to climb over topography.

Small-volume PDCs associated with fountain collapse or dome collapse that have been observed in recent decades, typically have velocities <30 ms^{-1} and rarely in excess of 60 ms^{-1}, although close to the source velocities may have exceeded this. Within the last few kilometers of travel of some PDCs, velocities are low, of the order of 1−2 ms^{-1}, as evidenced by the many buildings and trees

that remain standing in the peripheral areas (See Figure 54.1(A) and (B)).

Lateral blasts are associated with considerably higher velocities. Early studies of the May 18, 1980 lateral blast of Mt St Helens suggested that it began moving at around 100 ms^{-1}, but numerical modeling subsequently indicated that it may have been initially internally supersonic (with respect to speed of sound of the particulate mixture), with velocities up to 150 ms^{-1} (Esposti Ongaro et al., 2011). Other smaller directed blasts have mean velocities inferred from the deposits of around 90 ms^{-1} (e.g., Soufrière Hills Volcano) and maximum velocities of approximately 100 ms^{-1} as reckoned for the violent PDCs produced on October 26 and November 5, 2010 at Merapi, though these fell rapidly to around 50 ms^{-1} within 7 km of the vent.

Studies of examples from the geological record indicate that some PDCs attained very high velocities over 300 ms^{-1}, for example, the eruptions forming the Table Rock tuff ring, USA, and the Taupo ignimbrite AD 186. If there is no evidence of a lateral blast, then eruption column collapse from a considerable height and involving a large fraction of the erupted mass is likely to be responsible for such high velocities. Radial distribution of the deposits strongly supports a column collapse origin. Hence, at the eruption of Mt Lamington in 1951 a high velocity, lateral blast PDC was initially considered responsible for the scenes of devastation, but the radial distribution of the deposits is perhaps more suggestive of such a column collapse mechanism. This is an important consideration for hazard assessors at other volcanoes where collapse of large portions of the plume from high up in eruption columns could generate similar high velocity PDCs.

5. DYNAMIC PRESSURES AND BUILDING DAMAGE

The dynamic pressure of a PDC is related to the velocity and density of the moving current and is given by the following simple equation:

$$P_{dyn} = 1/2\rho v^2$$

where $\rho =$ is the density of the moving current and $v =$ is the velocity. The units are typically expressed in kilopascals (kPa).

Dynamic pressure is the kinetic energy per unit volume of the flow and can be represented as the lateral pressure of the PDC, which, when applied to structures, may be highly damaging and is analogous to the dynamic pressure wave in the blast from a nuclear or conventional bomb explosion, except with PDCs it lasts for minutes and is typically not accompanied by a shock front. It may be conceptualized as being struck by a dense, powerful wind traveling at hurricane force. Assumptions have to be made about the

(A)

(B)

FIGURE 54.1 (A) Pyroclastic density currents (PDC) during the final minutes of the small dome collapse of June 25, 1997 at Soufrière Hills Volcano, Montserrat. At this point the PDC was moving at velocities of a $\sim$1–2 ms^{-1}. (B) Deposits formed by the PDC shown in (A) viewed the following day. The house on the left is visible in both photographs. Note trees still standing in the peripheral/distal areas and houses partially buried. Destruction is in this case caused by the heat of the PDC igniting buildings and large boulders transported by it rather than the dynamic pressure. Photos: P. Cole

density of the moving PDC as it cannot be ascertained directly. Maximum dynamic pressures vary widely between PDCs, depending on their mechanism of formation and the velocity they attain, exceeding 35 kPa at the PDC front not far from source and decreasing along its path as the PDC loses energy and slows down. The dynamic pressure of the PDC erupted at Merapi on November 5, 2010, attenuated rapidly from an estimated 15 kPa at 5.75 km from the crater where the weakly constructed buildings were totally destroyed to less than 1 kPa where there was minimal building damage at 1.25 km further down the flank near the end of the runout (Jenkins et al., 2013) (Figure 54.2).

Examples where the dynamic pressures associated with PDCs vary considerably owing to variations in velocity and particle density have been well illustrated showing the impact of PDCs with buildings at Soufrière Hills Volcano, Montserrat (e.g., Baxter et al., 2005). On December 26, 1997, a high velocity lateral blast occurred impacting a large area to the southwest of the volcano. Buildings situated along its central axis were completely destroyed. However, the level of damage decreased rapidly over a

distance of 3 km towards the lateral edge of the PDC. This variation corresponded to the decline in velocity of the PDC towards the periphery.

On February 11, 2010, a large partial collapse of the lava dome led to a series of PDCs that impacted the northern flanks of Soufrière Hills Volcano, Montserrat, and the village of Harris 3 km to the north–northeast of the volcano for the first time during the eruption. The dilute PDC (equivalent to a pyroclastic surge) swept through the whole village. Marked variations in the damage to buildings could be observed in the village, with those houses in lower topographic positions being completely destroyed, whereas those in higher locations, e.g., on the ridge, suffered less structural damage (Figure 54.3).

These variations in damage may be interpreted as being caused by density stratification of the PDCs. High density parts of the current remained in topographically lower areas (valleys) generating higher dynamic pressures and resulting in extensive building damage, whereas the less dense parts of the current swept over the ridge generating lower dynamic pressures and resulted in less damage. This highlights the point that dilute PDCs, as detailed in the

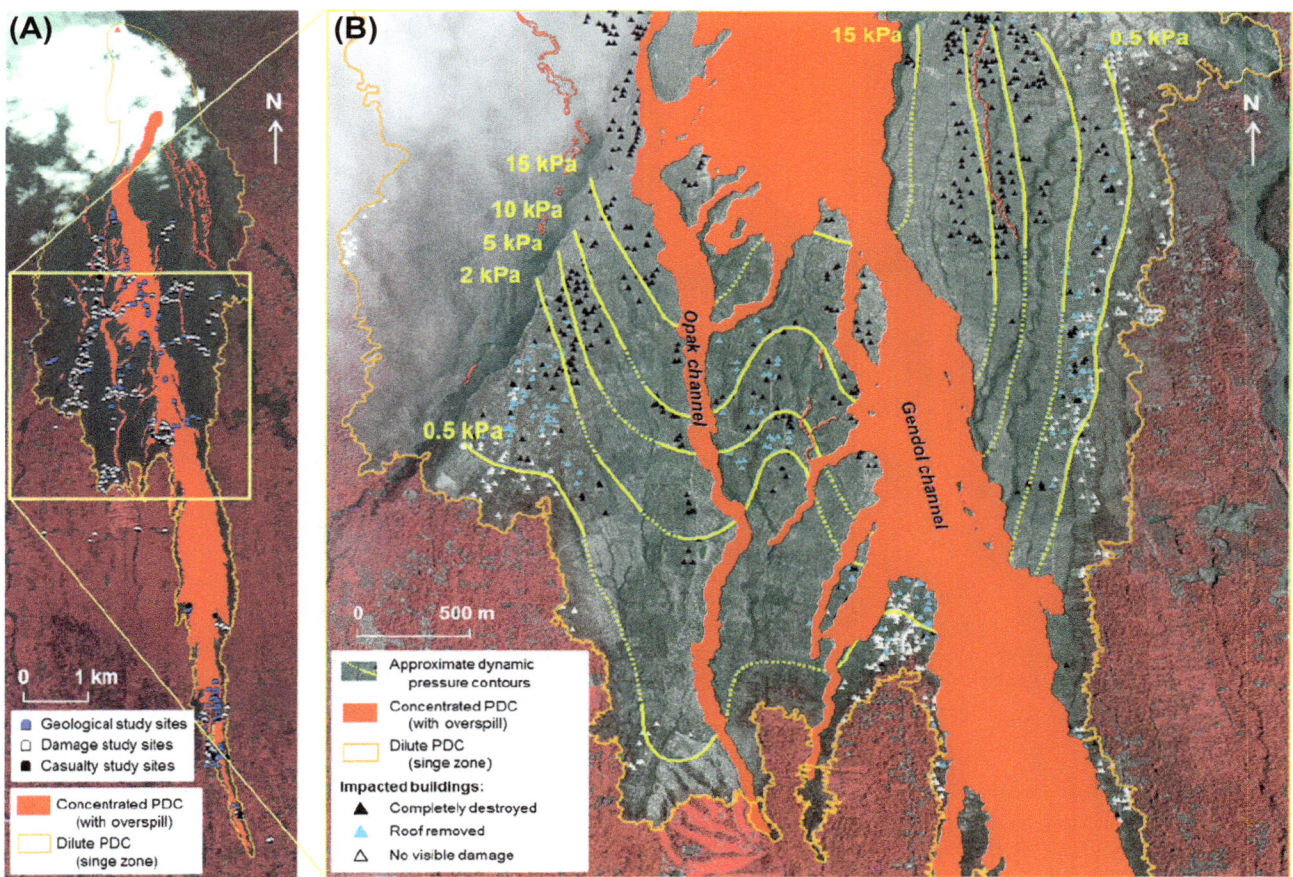

FIGURE 54.2 Map showing inferred dynamic pressures formed as the result of the November 5, blast-like pyroclastic density currents (PDC) at Merapi, Indonesia. *After Jenkins et al. (2013).*

FIGURE 54.3 Building damage in Harris village Soufriere Hills Volcano, Montserrat resulting from pyroclastic density currents caused by dome collapse on February 11, 2010. Current direction is from left to right (A) buildings in the lower part of the village. These have suffered nearly total destruction (foundations are still visible) whereas those in the upper part (B) have suffered less structural damage with walls standing and some roofs still intact. Photos: P. Cole

next section, are particularly dangerous as they are less constrained by topography and are often able to travel further or in different directions than their dense PDC counterparts.

An additional hazard linked to the dynamic force of the PDC is from the turbulent behavior of the moving current. Chaotic internal flow patterns in the body and tail of the PDC can add substantially to the impact of the dynamic pressure of the flow front at the advancing head. Thus buildings can have their walls blown down by the PDC front and then the remaining contents fragmented by the turbulent current. These effects were characteristic of the PDC on the November 5, 2010 at Merapi mentioned above, which was violent but not hot enough to ignite fires.

6. DETACHMENT/DECOUPLING OF PDCs

As already mentioned, most PDCs develop density stratification, with a dense, high particle concentration lower part (pyroclastic flow) and low particle concentration upper part (pyroclastic surge). It has long been recognized that this differential can lead to decoupling with the two parts moving independently and in different directions. The late Richard Fisher was one of the first volcanologists to recognize this process (see Fisher (1995)) and to show how the decoupling was accentuated by topographic features causing the denser parts of the PDCs to be obstructed or deflected, while the upper dilute part continues along its path without being impeded by topography.

Important examples of PDCs detaching in this way and giving rise to fatalities include the eruptions of Unzen, Japan, 1991, Soufrière Hills Volcano in 1997, and, probably, at Mt Pelée in 1902. Also, at Merapi on November 5,

2010, a PDC ran 15.5 km along the Glendol valley and a dilute PDC detached from it at a bend before a Sabo dam and flowed over the bank into Bronggang village, leading to 54 deaths among villagers who had been unaware of the imminent danger while they were attempting to evacuate (Figure 54.4).

An uncommon behavior of dilute PDCs that only became fully recognized on June 25, 1997 at Soufrière Hills Volcano was when a detached dilute PDC decelerated and rapidly sedimented material that drained from the steep topography to form a dense, but highly mobile PDC, which took a new course and traveled a further 3 km into a populated area (Loughlin et al., 2002).

These kinds of unexpected behavior of PDCs occur frequently in major eruptions and they highlight the dangers of building special structures in an attempt to deflect, or shelter people from, PDCs. Such attempts at mitigation measures may also have unforeseen consequences and show that, at this time, there is no substitute for a wide and timely evacuation from areas at risk of inundation by PDCs. The construction of PDC proof buildings (bunkers) is therefore generally not recommended, though they have been occasionally used as an absolute last resort.

7. TEMPERATURES OF PDCs

The temperatures of PDCs can vary widely. Large-volume ignimbrites commonly display welding, where particles of the juvenile magma have been fused together in the high temperature of the PDC that formed them, which must have been not far below magmatic temperatures ($>800\,°C$) for this to occur. In contrast, those formed during hydro-volcanism may typically be close to the boiling point of water. Small-volume PDCs are also often cooler than large

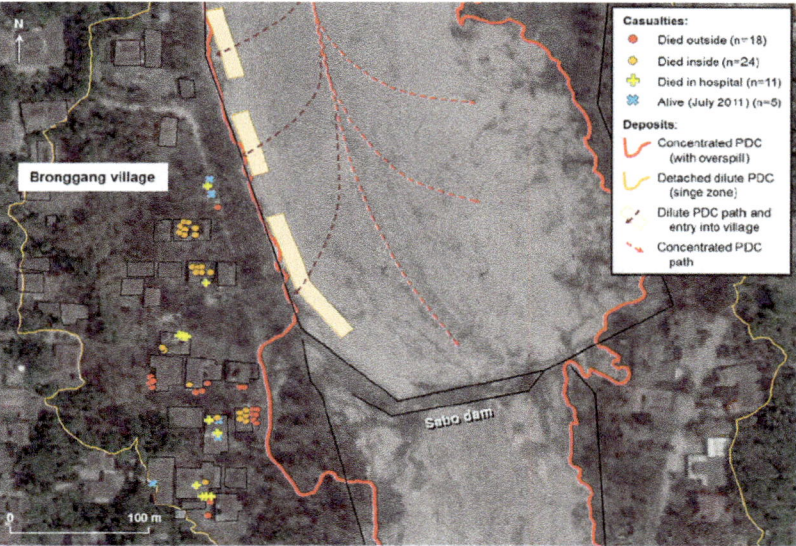

FIGURE 54.4 Map showing area of Bronggang village, Merapi, where dilute portions of pyroclastic density currents (PDC) detached before the Sabo dam and flowed over three places on the bank as the parent flow moved along the valley (after Jenkins et al. (2013)). 54 villagers were killed some inside and others outside their houses.

ones, although studies have shown that this is a variable finding depending on their mode of formation, as shown in the following case studies.

7.1. Vesuvius, Italy

Detailed magnetic studies of the AD 79 deposits on the flanks of Vesuvius indicate that the PDCs formed by column collapse had temperatures in the range 250–400 °C, similar to findings on the PDCs formed during the smaller volume AD 472 eruption. Moreover rock magnetism studies of the AD 79 PDC deposits have shown that buildings in the city of Pompeii influenced the transport and movement of PDCs significantly, and local topographic effects were associated with the temperature variations, possibly due to mixing with surrounding air.

7.2. Soufrière Hills Volcano, Montserrat 1995–2010

Temperatures of PDCs at Soufrière Hills Volcano have been measured directly using industrial temperature patches and also by thermocouples inserted in deposits within hours after emplacement. They have also been inferred from their effects on materials accidentally incorporated within them during their emplacement.

At Soufrière Hills Volcano, the largest PDCs formed by dome collapse were found to have temperatures exceeding 400 °C as measured from the deposits. Some localized deposits were measured at over 600 °C. At the blast-generated PDC on December 26, 1997, studies of charcoal and carbonized vegetation suggested temperatures ranging between 300 and 425 °C. Temperatures of deposits of pumice-rich PDCs were cooler and did not char paper, or melt foam mattresses, indicating temperatures less than 200 °C. These PDCs were apparently cooled significantly by the entrainment of air during the fountain collapse process.

7.3. Merapi, Indonesia

The paroxysmal PDCs generated at Merapi volcano in November 2010, did not generate widespread fires or combustion of furnishings inside houses indicating that temperatures were lower than 300 °C (Jenkins et al., 2013). Although the reasons are unclear, the lower temperatures at Merapi might possibly be related to greater air entrainment associated with these small-volume, turbulent PDCs.

In summary, recent studies show that the temperatures of PDCs formed by different generation mechanisms can vary widely. In addition, it is evident that temperatures may fluctuate significantly within a single PDC and be related to local flow conditions and the entrainment of air, the volume of the PDC and the topography over which the current travels.

7.3.1. Heat Hazards

Hot ash associated with PDCs convects as turbulent clouds as it moves because of its high temperature compared to the ambient air. The ash particles in the PDC radiate heat internally in the cloud and also conduct heat when they come into contact with the human body or form a deposit inside a house. The overall measure of hazard when engulfed by a PDC is therefore the sum of the convective heat and radiant heat fluxes and the additional element of conducted heat from the hot ash settling on the surfaces of the body or furnishings.

In the open the dynamic pressure adds to this lethal total heat flux, making survival almost impossible, even in slow-moving dilute PDCs. Indoors, in buildings, which remain intact, or outside yet protected in some fortuitous way from the force and heat of the PDC, a few individuals have lived only to die later in hospital from complications of severe burns to the skin and frequently to their airways and lungs as well. Survival in PDCs is discussed elsewhere (see Chapter 60 in this book).

The hot ash deposits may or may not ignite fires after the PDC has dispersed. The atmosphere inside a PDC is low in oxygen and it is only after the plume has lofted away and drawn in fresh oxygen-rich air that fires will break out if the ash deposit is hot enough to ignite flammables. This was the case in the dome collapses that generated PDC events on Montserrat in 1997. Initially it was thought that the burnt out roofs were ignited by the contact with the hot PDC as it flowed over the wooden structure covered in asphalt tiles. But field surveys showed that the furniture left behind after rapid evacuations of the areas around the volcano had ignited first and the fires had burnt into the wooden roofs from below. In contrast, the houses in minor to severe states of destruction by blast in the main PDCs in the Merapi 2010 eruption did not show signs of fire, but had more minimal heat effects, such as slight charring of some exposed timbers or external wooden structures caused by the heat flux from the moving PDC.

Studies on buildings impacted by PDCs on Montserrat (Baxter et al., 2005) provided clear evidence that ash from a PDC at or over 300–400 °C had first to gain entry inside robust buildings to set them on fire. Since Hurricane Hugo struck Montserrat in 1989, when many buildings on the island were badly damaged, houses have been built to more hurricane resistant standards. Many concrete block buildings stood up to the PDCs remarkably well in the more peripheral areas where the impact was least and in so doing showed how the PDCs could easily enter the tropical housing through the windows and doors. The resistance to dynamic pressure of the aluminum slats installed in most windows was low, just as glazed windows without protective shutters in temperate climates would offer little resistance, too. Small missiles such as dense clasts entrained in the PDC would also add to the hazard. But there were several examples of houses where the simple expedient of covering the windows with plywood boards—a standard form of protection against hurricanes in the West Indies—was sufficient to protect the house from the penetration of hot ash in the PDC and thus from catching fire.

Houses in many countries may have dry vegetation surrounding the house, or keep materials such as hay and straw for pets and even livestock. Kindling material for wood fires may also be kept in outside lean-to kitchens. At Merapi in 2010, isolated fires were ignited in these materials by firebrands formed from the huge number of felled trees burning inside the main PDC and carried in the detached dilute part of the current into the village. Some of these fires then spread inside the houses, whereas the ash deposits were not hot enough to trigger fires on their own.

There is as yet no way of determining the temperature of PDCs in advance of an eruption, but it is likely that urban areas would be at special risk of major fires with the potential for these spreading even beyond the PDC impact zone if the ash were hot enough to cause multiple and widespread ignitions.

8. PDC HAZARD MODELING AND MAPPING

Assessment of the hazard posed by PDCs is particularly challenging due to the variety of generation mechanisms described above, and the properties and dynamics of the currents, as well as the significant uncertainty in several of the parameters controlling the characteristics and evolution of the phenomenon. Therefore, hazard maps and scenarios that delineate areas that are likely to be impacted by PDCs have been developed in a number of ways and using quite different approaches.

These methods can be subdivided into either deterministic (i.e., scenario-based) or probabilistic types. Deterministic hazard maps typically define the areas that are inundated once specific eruptive conditions (i.e., PDCs generation mechanisms) have been assumed and according to a limited number of specific past or foreseen events, but probabilistic approaches are often the preferred technique as they are able to incorporate uncertainty in the physical phenomena by using numerous multiple computer runs as in a "Monte Carlo" simulation approach. This methodology, by varying input parameters across likely ranges, explores the effect of modifying key, but uncertain, variables. In both approaches the maps generated can then be used, for instance, in a geographical information system for volcano crisis management and in urban planning.

Hazard maps have traditionally been based on the maximum extent of PDCs as derived from previous eruptions. In most cases these were based on the extent of deposits from prehistoric eruptions within the geological record. However, studies of PDC-forming eruptions in the last few decades (e.g., Unzen 1991; Soufrière Hills Volcano 1995−2010) have highlighted that PDC deposits, particularly those of dilute PDCs, leave deposits that are only a

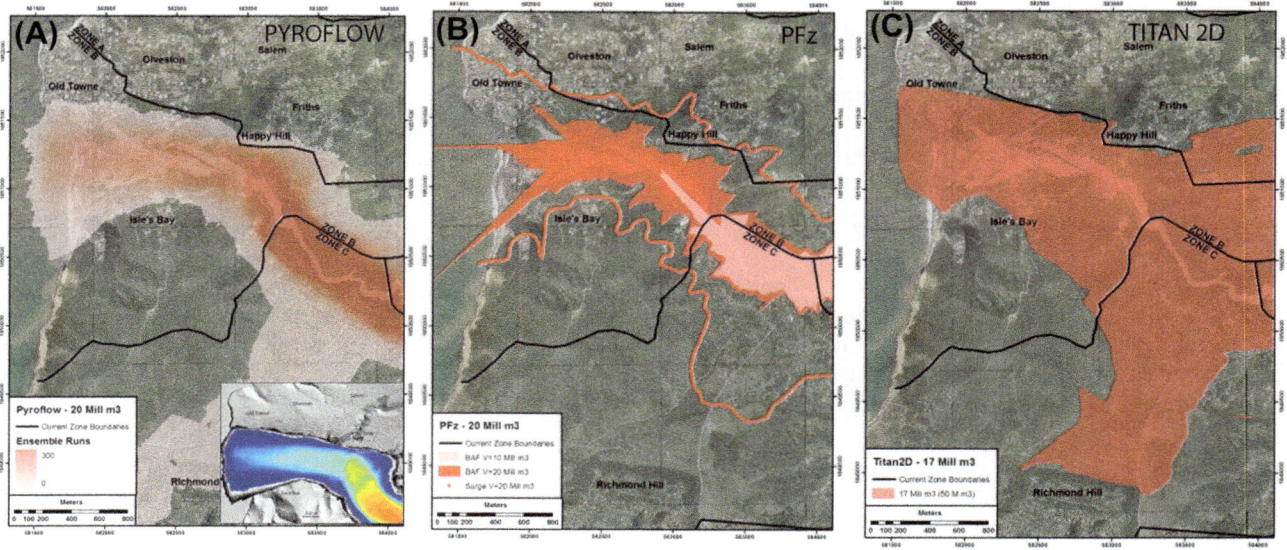

FIGURE 54.5 Examples of three different computer modeling methods for hazard mapping in use at Soufrière Hills Volcano to assess the pyroclastic density currents hazard around the Belham Valley in Montserrat (after SAC 15 2010). (A) Pyroflow (Wadge et al., 2009) (B) PFz (Widiwijayanti, 2008) (C) Titan2D (Patra et al., 2005). The results of these models were collectively used to determine the location of hazard zone boundaries.

few centimeters thick and may are likely to be quickly removed by erosion. As a consequence the geological preservation potential of such deposits is typically extremely poor and will also add to the uncertainties of estimating the extent of past runout distances. Moreover, volcanic phenomena are determined by many parameters of the volcanic system (nature and location of the magmatic reservoir, volcano topography, etc.) that vary with time and therefore volcanic events never occur in exactly the same way as previous events. Thus hazard maps solely based on the geological record may provide only a limited perspective and should be treated with extreme caution as they are likely to seriously underestimate the true extent of future PDCs.

As physical modeling of volcanic processes improves in accuracy the use of numerical models and computer simulations for hazard mapping is also becoming more common. Although volcanologists are still someway away from realistically simulating PDCs, the use of physical models allows them to describe quantitatively several aspects of their dynamics and also to investigate the sensitivity of them by varying the different input parameters. In some cases, probabilistic hazard maps are able to show the likelihood of PDC invasion or the distribution of a specific hazard variable associated with the current (e.g., dynamic pressure, temperature, ash-in-air concentration, etc.). A few examples of these models and maps are briefly discussed below.

For instance, computer programs such **Pyroflow** (Wadge et al., 1998) enable volcanologists to estimate the PDC path and runout based only on the digital elevation model of the volcano and a simple 3-fold parameterization of current acceleration. In fact, during the Soufrière Hills Volcano crisis, this model, once calibrated with field data, was able to provide first-order estimates of the maximum distance reached by the granular portion of the PDC generated by gravitational collapse of the dome (Wadge et al. 2009). An example of this model as applied in the crisis at Soufriere Hills Volcano in 2010, using a Monte Carlo methodology is shown in Figure 54.5(A).

Similarly, the **PFz** (Pyroclastic Flow Z) model (Widiwijayanti et al., 2009), which is based on empirical relationships between the volume of the material forming the PDC and the associated area invaded, can provide a statistically based correlation of propagation over a realistic terrain (Figure 54.5(B)). But this model, like Pyroflow, although useful to assess the likely propagation of the concentrated component of the current, does not explicitly model the flow dynamics, and particularly in particular the dilute (Pyroclastic Surge) component.

Another family of computer models able to describe the propagation of PDCs over a realistic topography are those based on a transient 2D shallow-water representation of the current. **Titan2D** (Patra et al., 2005 and Figure 54.5(C)) and **Volcflow** (Kelfoun and Druitt, 2005) are examples, which are used to define scenarios for hazard mapping of dense PDCs and rock avalanches. These models have significant drawbacks in simulating some of the complexities of PDCs, i.e., they make quite restrictive assumptions on the nature of the flow such as its homogeneity and rheology, but they do incorporate some aspects of the physics of the phenomena, which, once calibrated against field data, allow the

Ash concentration

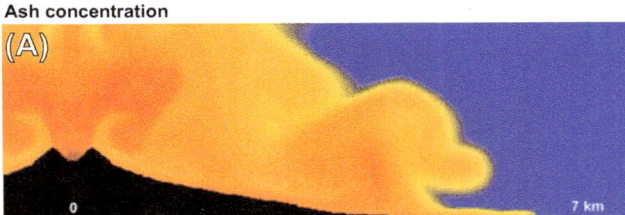

Temperature

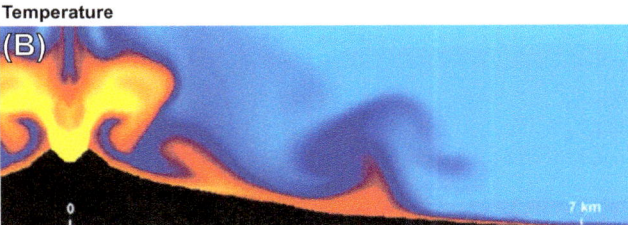

Dynamic pressure

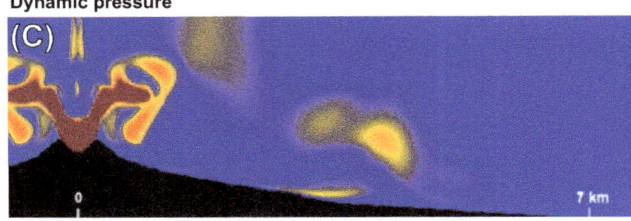

FIGURE 54.6 This numerical simulation of an eruption of a pyroclastic density currents (PDC) at Vesuvius illustrates the spatial distribution of three fundamental variables of pyroclastic flows 10 min after the occurrence of the collapse of the volcanic column that generated them. The topographic profile assumed in the 2D axisymmetric simulation corresponds to the mean profile of the south sector of Vesuvius (between Torre del Greco and Torre Annunziata). The mass flow rate corresponds to a sub-Plinian eruption (5×10^7 kg/s) dispersed over a sector of 90°. (A) The ash-in-air concentration profile corresponds to volume fractions of about 10^{-2} (red) and 10^{-8} (yellow) on a blue background of undisturbed air. The formation of a new batch of pyroclastic mixture detaching from the fountain and feeding the flows is evident with a thermal instability in the flow and the generation of a coignimbritic or **phoenix** cloud at 4.5-km-runout. The PDC head has reached the coastline (7 km from the vent). (B) The flow temperature profile corresponds to 950 °C above the vent (yellow), and 400 (red) and 100 °C (blue). Thermal instabilities at about 2 and 4.5 km from the vent generate convective plumes (thermals and **phoenix** clouds). At about 7 km, the conditions are unsurvivable outdoors (over 200 °C). (C) The dynamic pressure profile shows 50 kPa in the fountain region (red) and about 3 kPa at 4.5 km from the vent, corresponding to the flow wave feeding the **phoenix** cloud (yellow). At the more distal portion of the PDC, the dynamic pressure is around 1−2 kPa, capable of breaking unprotected windows but leaving most structures intact. *Modified from Todesco et al. (2002).*

description of their transient and multidimensional dynamics. They also allow user input of a range of parameters, such as the volume of the starting pyroclastic mixture or friction and internal angles, for producing a range of scenarios and probabilistic invasion maps.

More complex models of PDCs are those based on the solution of the fundamental transport equations of the multiphase mixture (e.g., Dobran et al., 1993). 2D/3D computer models such as the **PDAC** code (e.g., Esposti Ongaro et al., 2007) simulate some complexities of the physics of the dilute portion of PDCs, such as the generation mechanism (e.g., column collapse or lateral blast), flow stratification, air ingestion during PDC propagation and formation of

instabilities leading to the generation of buoyant convection and coignimbrite (or **phoenix**) plumes from the current. This type of model can also provide oblique and immersive three-dimensional visualizations of the process, which are often better than traditional two-dimensional hazard maps, that are sometimes misunderstood. These models typically run on supercomputers and are used to investigate specific eruptive scenarios in order to improve our understanding of the physics of the phenomenon under realistic system conditions, but they are not generally suitable for the rapid development of probabilistic hazard maps in crisis situations. Their applications will be discussed more in the next section.

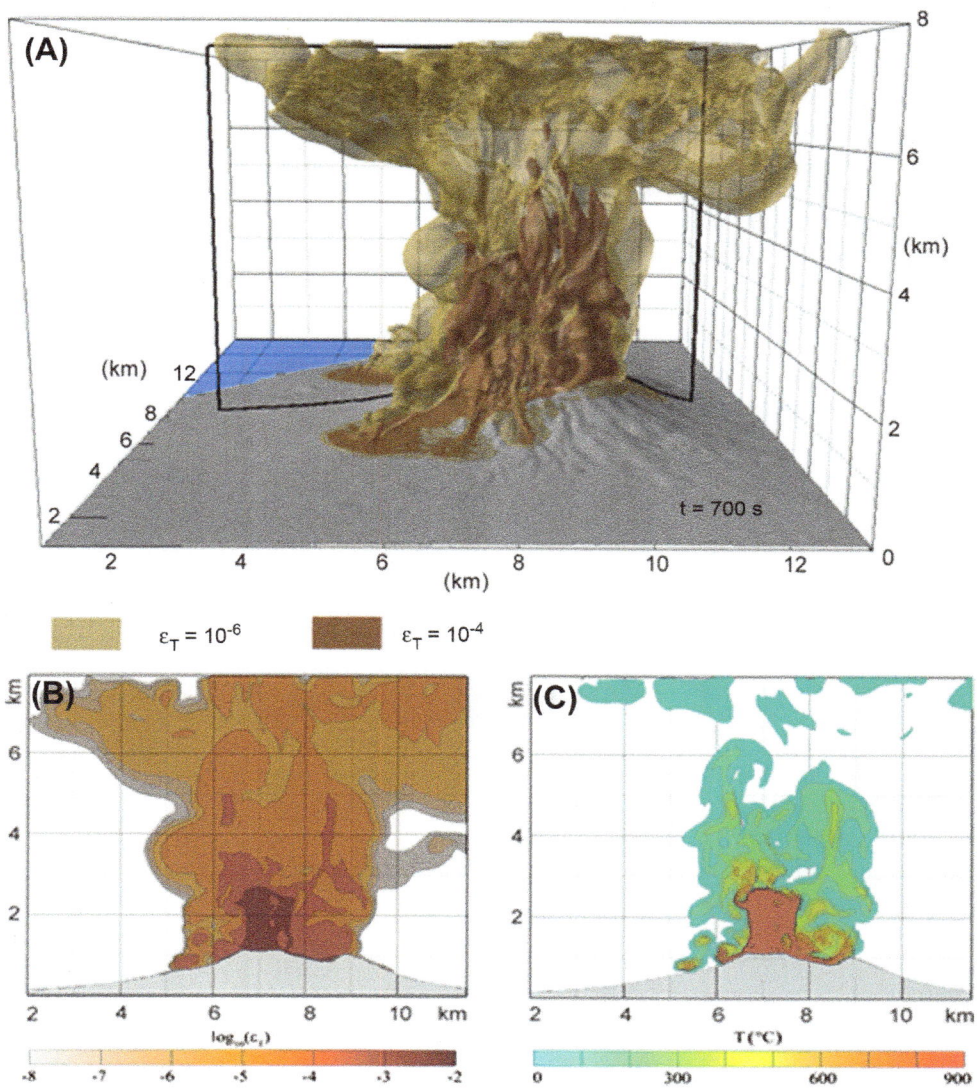

FIGURE 54.7 Simulation of a sub-Plinian event at Vesuvius characterized by the collapse of the volcanic column and generation of pyroclastic density currents. (A) 3D view of 10^{-4} and 10^{-6} isosurfaces of total volume particle fractions at 700 s, (B) color contours of the base 10 logarithm of the total volume particle fraction, and (C) gas temperature at 700 s along a N–S section of the column (shown in (A)). See color legends for concentration and temperature values. *From Esposti Ongaro et al. (2008).*

9. VESUVIUS HAZARD MAPPING CASE STUDY

Mt Vesuvius is considered one of the most dangerous volcanoes in the world due to its explosive nature and the large population living on and around its flanks. The generation of PDCs has typically been associated with its Plinian and sub-Plinian eruptions, thus representing the primary source of risk for more than 0.5 million inhabitants living within a radius of about 7 km from the summit. PDCs were responsible for the majority of casualties during the AD 79 eruption as well as during the last sub-Plinian event in 1631. The spatial distribution of these PDCs has been reconstructed to some extent based on stratigraphic studies and frequency of occurrence maps have been produced as a first hazard assessment (Gurioli et al., 2010). However, such studies may seriously underestimate the runout of PDCs, as has been mentioned above.

Despite large epistemic uncertainties over the future input parameters, 2D/3D numerical models have been developed at Vesuvius in the last decade, as a means of quantifying the hazard from PDCs. Properties of the magmatic system and conditions at the vent were assumed representative of past or future events. Dynamic pressure, temperature, velocity, and density were then computed by the PDAC numerical simulation model for a sub-Plinian eruption to study flow dispersion along selected 2D axisymmetric lines (Figure 54.6) (e.g., Todesco et al., 2002; Esposti Ongaro et al., 2002) representative of the southern and northern slopes of Vesuvius as well as the fully 3D topography of the volcano (Esposti Ongaro et al., 2008). Due to the resolution of the numerical grid (typically of the order of several meters), these simulations were representative of the dynamics of the low particle concentration portion of the PDC (pyroclastic surge) and do not consider the high particle concentration part (Pyroclastic flow). Time-dependent and peak curves of the main hazard variables were thus produced at specific locations as a function of the defined scenario. Similarly, hazard maps of PDC invasion as well as the distribution of the limits of the hazards, were produced by the transient 3D simulations. In this latter case, simulations such as those represented in Figure 54.7 specifically quantified the effect of Mt Somma—the former volcano edifice still present on the northern side of Mt Vesuvius—as well as the local topography on the propagation of the current and therefore on the hazard.

Simulated output variables were also used to make deterministic estimates of building damage in test scenarios, including the infiltration of the hot gas and particles inside buildings for estimating survival times of occupants and where the openings may have resisted the PDC due to sheltering or other reasons. In fact, based on the Montserrat findings, dynamic pressures of few kilopascals would break most glazed windows in the area and allow entry of the hot ash, with the potential of destroying at least parts of the buildings by fire. At moderate and high dynamic pressures (>2 Kpa) large building structural damage would be expected with additional damage added by missiles transported by the PDC and strong sheltering effects due to the proximity of buildings (Zuccaro et al., 2008). The model also illustrated the additional hazard presented by the isotropic pressure of the PDC, which acts over the building envelope due to its weight and the large-scale atmospheric circulation

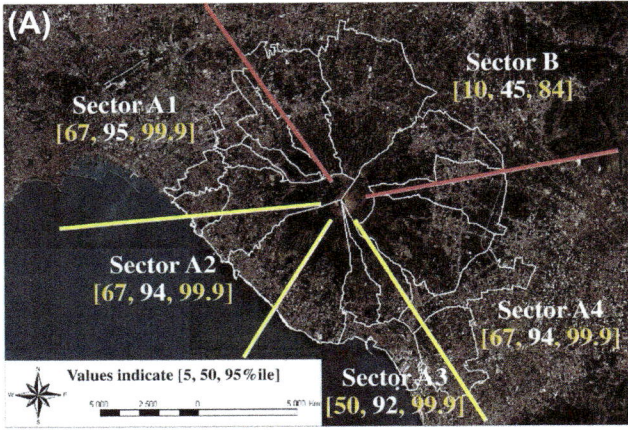

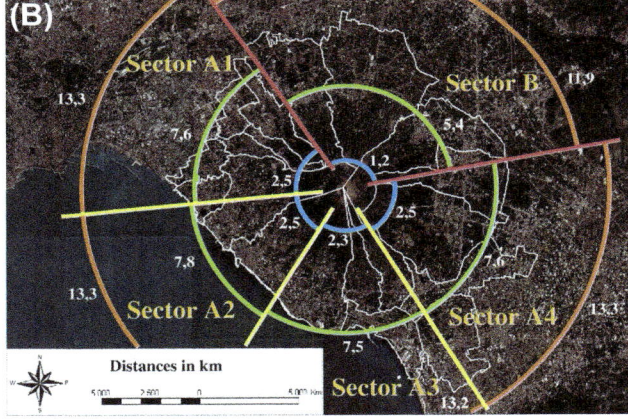

FIGURE 54.8 Probabilistic map of pyroclastic density currents (PDC) flow invasion at Vesuvius as produced by expert elicitation techniques. (A) Broad segmentation of area around Vesuvius recognizing the first-order effect of Mt Somma topography in determining areas that might be invaded by PDCs as the result of a sub-Plinian eruption. The bracketed values in each sector show elicited modal probabilities that a PDC will affect that sector (expressed in percentage terms) together with the corresponding credible intervals, in quantile form [5 percentile, 50 percentile, 95 percentile]. (B) As (A), showing elicited estimates of maximum runout distances (in km) for PDCs occurring during a sub-Plinian eruption, by sector. Inner arcs (blue) are 95% confidence levels for exceeding distance shown (e.g., 2.5 km for Sector A1), central arcs (green) are expected (50%) values, and outer arcs (orange) are the runout distances assessed has having only a 5% chance of being exceeded. *From Neri et al. (2008).*

produced by the dispersal process as well as the fire hazard due to the high temperature of the PDCs.

At Vesuvius, field reconstructions of PDC dispersal and outcomes of the 3D simulations of PDCs produced by sub-Plinian events were also combined to define first-order probabilistic hazard maps for PDCs produced by a sub-Plinian eruption with a crater location coinciding with the present Gran Cono (Neri et al., 2008). This was a first attempt to incorporate outcomes from the different methodologies adopted as well as to account for the different sources of uncertainty involved. By using structured expert elicitation techniques, the various strands of evidence were summarized by subdividing the volcano in different sectors and by assigning probabilities that each sector would be affected by PDCs and to the extent of the runout. Each probability was assigned by a group of experts as a modal value and a credible interval in quantile form (5% and 95% percentile). Resulting maps such as those illustrated in Figure 54.8, although related to a single eruptive scenario and in need of continuous updating as more information becomes available, allowed appreciation of the remarkable uncertainty affecting the hazard assessment process, even for a well-known volcano such as Vesuvius.

10. SUMMARY

Understanding of PDC dynamics and assessment of their hazards still represent real challenges for modern volcanology. PDCs can be formed in a number of different ways and show many complex and often unpredictable processes. Some of the key variables characterizing the dynamics of PDCs such as the velocities, temperature, dynamic pressure, and potential damage have been described. Techniques for mapping potential inundation areas of PDCs have developed significantly in the last decade and typically involve the use of simulation models and probabilistic techniques in a variety of ways. Several of the techniques used have been outlined and examples of some of the most sophisticated computer models and how these have been used for hazard mapping are described. However, despite such recent advances, understanding the real limitations of our present ability to quantitatively describe the dynamics of PDCs and accurately forecast zones at risk from their inundation remains a key concern.

FURTHER READING

Baxter, P.J., Boyle, R., Cole, P., Neri, A., Spence, R., Zuccaro, G., 2005. The impacts of pyroclastic surges on buildings at the eruption of the Soufrière Hills volcano, Montserrat. Bull. Volcanol. 67 (4), 292–313.

Cioni, R., Gurioli, L., Lanza, R., Zanella, E., 2004. Temperatures of the A.D. 79 pyroclastic density current deposits (Vesuvius, Italy). J. Geophys. Res. 109 (B02207). http://dx.doi.org/10.1029/2002JB002251.

Dobran, F., Neri, A., Macedonio, G., 1993. Numerical simulation of collapsing volcanic columns. J. Geophys. Res. 98, 4231–4259.

Edmonds, M., Herd, R.A., 2005. Inland-directed base surge generated by the explosive interaction of pyroclastic flows and seawater at Soufrière Hills volcano, Montserrat. Geology 33 (4), 245–248.

Esposti Ongaro, T., Neri, A., Todesco, M., Macedonio, G., 2002. Pyroclastic flow hazard assessment at Vesuvius by using numerical modelling. 2. Analysis of flow variables. Bull. Volcanol. 64, 178–191.

Esposti Ongaro, T., Neri, A., Menconi, G., de'Michieli Vitturi, M., Marianelli, P., Cavazzoni, C., Erbacci, G., Baxter, P.J., 2008. Transient 3D numerical simulations of column collapse and pyroclastic density current scenarios at Vesuvius. J. Volcanol. Geotherm. Res. 178 (3), 378–396. http://dx.doi.org/10.1016/j.jvolgeores.2008.06.036.

Esposti Ongaro, T., Widiwijayanti, C., Clarke, A., Voight, B., Neri, A., 2011. Multiphase-flow numerical modeling of the 18 May 1980 lateral blast at Mount Saint Helens, USA. Geology (6), 535–538. http://dx.doi.org/10.1130/G31865.1.

Fisher, R.V., 1995. Decoupling of pyroclastic currents: hazards assessments. J. Volcanol. Geotherm. Res. 66, 257–263.

Gurioli, L., Sulpizio, R., Cioni, R., Sbrana, A., Santacroce, R., Luperini, W., Andronico, D., 2010. Pyroclastic flow hazard assessment at Somma–Vesuvius based on the geological record. Bull. Volcanol. 72, 1021–1038. http://dx.doi.org/10.1007/s00445-010-0379-2.

Jenkins, S., Komorowski, J.-C., Baxter, P.J., Spence, R., Picquout, A., Lavigne, F., Surono, 2013. The Merapi 2010 eruption: an interdisciplinary impact assessment methodology for studying pyroclastic density current dynamics. J. Volcanol. Geotherm. Res. 261, 316–329.

Kelfoun, K., Druitt, T.H., 2005. Numerical modeling of the emplacement of Socompa rock avalanche, Chile. J. Geophys. Res. 110 (B12202.1–12202.13). http://dx.doi.org/10.1029/2005JB003758.

Loughlin, S.C., Calder, E.S., Clarke, A., Cole, P.D., Luckett, R., Mangan, M., Pyle, D.M., Sparks, R.S.J., Voight, B., Watts, R.B., 2002. Pyroclastic flows and surges generated by the 25 June 1997 dome collapse Soufrière Hills volcano, Montserrat. In: Druitt, Kokelaar (Eds.), The 1995–1998 Eruption of Soufrière Hills Volcano, Montserrat. Memoir, vol. 21, pp. 191–210.

Neri, A., Esposti Ongaro, T., Macedonio, G., Gidaspow, D., 2003. Multiparticle simulation of collapsing volcanic columns and pyroclastic flow. J. Geophys. Res. 108. http://dx.doi.org/10.1023/2001JB000508.

Neri, A., Aspinall, W.P., Cioni, R., Bertagnini, A., Baxter, P.J., Zuccaro, G., Andronico, D., Barsotti, S., Cole, P.D., Esposti Ongaro, T., Hincks, T.K., Macedonio, G., Papale, P., Rosi, M., Santacroce, R., Woo, G., 2008. Developing an Event Tree for probabilistic hazard and risk assessment at Vesuvius. J. Volcanol. Geotherm. Res. 178 (3), 397–415. http://dx.doi.org/10.1016/j.jvolgeores.2008.05.014.

Patra, A.K., Bauer, A.C., Nichita, C.C., Pitman, E.B., Bursik, M., Sheridan, M.F., Rupp, B., Webber, A., Stinton, A.J., Namikawa, L.M., Renschler, C.S., 2005. Parallel adaptive numerical simulation of dry avalanches over natural terrain. J. Volcanol. Geotherm. Res. 139, 1–21.

Scientific Advisory Committee (SAC) Technical report 15. Available from: http://www.mvo.ms/pub/SAC_Reports/SAC15-Technical.pdf.

Todesco, M., Neri, A., Esposti Ongaro, T., Papale, P., Macedonio, G., Santacroce, R., Longo, A., 2002. Pyroclastic flow hazard assessment at Vesuvius by using numerical modelling. 1. Large scale dynamics. Bull. Volcanol. 64, 155–177.

Wadge, G., 2009. Assessing the pyroclastic flow hazards from dome collapse at Soufrière Hills Volcano, Montserrat. In: Thordarson, T., Self, S., Larsen, G., Rowlands, S.K., Hoskuldsson, A. (Eds.), Studies in Volcanology: The Legacy of George Walker, vol. 2. Spec. Pub. IAVCEI, pp. 211–224. Geol. Soc. London.

Widiwijayanti, C., Voight, B., Hidayat, D., Schilling, S.P., 2009. Objective rapid delineation of areas at risk from block-and-ash pyroclastic flows and surges. Bull. Volcanol. 71, 687–703. http://dx.doi.org/10.1007/s00445-008-0254-6.

Zuccaro, G., Cacace, F., Baxter, P.J., Spence, R., 2008. Impact of explosive eruption scenarios at Vesuvius. J. Volcanol. Geotherm. Res. 178, 416–453.

Chapter 55

Lava Flow Hazards and Modeling

Christopher R.J. Kilburn

Department of Earth Sciences, University College London, London, UK

Chapter Outline

GLOSSARY

aa lava Lava flows with extremely irregular surfaces, usually covered by fragments of broken crust that are typically decimeters thick. The thickness of the surface crust is controlled by cooling (compare blocky lavas).

blocky lava Lava flows with fractured surfaces, usually covered by debris up to meters across. The size of surface fragments is controlled by the rheology of the lava interior, rather than by the thinner surface crust (compare aa lavas).

crust The outer part of a lava flow that has solidified after losing heat to the exterior.

flow A discrete body of lava emplaced as a dynamically continuous unit.

flow field A collection of lava flows produced by the same effusion.

lava Molten or partially molten rock erupted at a planet's surface.

pahoehoe lava Lava flows with smooth, continuous surfaces.

1. INTRODUCTION

On Saturday, 18 March, 1944, San Sebastiano was an unassuming Italian village in the foothills of Mount Vesuvius. On Tuesday, 21 March, it had ceased to exist. In less than 48 h, **lava** streams from a new eruption had descended the volcano's flanks and crept through the village, slowly and inexorably eating their way from one building to the next. Today, San Sebastiano is flourishing again as a popular residential area 10 km east of Naples. Yet the 1944 **flows** were the third set to have overwhelmed the settlement in less than 100 years. After both previous eruptions, in 1855

and 1872, the village and its neighbors were rebuilt around the very lavas that had caused their destruction. Not even the menace from Vesuvius could break the lure of the familiar.

The resilience, or perhaps stubbornness, that keeps populations returning to hazardous districts is the chief reason why lava flows are a common threat to human settlements, even though they are produced by one of the least powerful styles of eruption. Between 1973 and 2014, lava effusions worldwide caused property losses of hundreds of millions of US dollars. One of the most dramatic examples occurred in the Democratic Republic of Congo in January 2002, when lava flows destroyed the center of Goma, a city at the foot of Nyiragongo volcano, and triggered the evacuation of as many as 450,000 people. Even though the losses from effusions are small compared to those due to major explosive eruptions, they have long-lasting repercussions on local economies and are a key force driving investigations into how lavas behave.

A second motive for studying lavas is that they are the single most common feature on the surfaces of the terrestrial planets. They cover 90% of Venus, 50% of Mars, at least 20% of the Moon, and some 70% of the Earth, where they are mostly hidden from view on the ocean floor. Understanding how lavas are emplaced is thus crucial to reconstructing the surface evolution of the inner planets and to investigating the conditions below the surface that favor effusive eruptions.

The Encyclopedia of Volcanoes. http://dx.doi.org/10.1016/B978-0-12-385938-9.00055-9

TABLE 55.1 Major Historic Lava Flows

Composition	Typical Length (km)	Mean Thickness (m)	Mean Discharge Rate (m^3 s^{-1})	Flow (& Flow Field) Volume (km^3)
SiO$_2$ < 55 wt% (basalts, basaltic andesites)	<10 (basalts to 50)	3−20	10−100 (basalts to 1000)	0.01−0.1 (<1−2)
For example, Kilauea & Mauna Loa, Hawaii; Etna & Vesuvius, Italy; Iceland; Piton de la Fournaise, Réunion Is; Lanzarote & Tenerife, Canary Is; Arenal, Costa Rica; Parícutin, México.				
SiO$_2$ > 55 wt% (andesites, dacites, trachytes)	<5 (some to 15)	20−300	1−10 (andesites to 100)	0.01−1.0 (<10−20)
For example Lonquimay, Chile; Nea Kameni, Santorini, Greece; Hibok−Hibok, Philippines; Trident, Alaska.				

The main factors controlling how lava flows develop are the rate at which lava is effused from the ground, the lava's physical properties, and the local environment (such as ground slope, topography, and whether the eruption occurs on land, or below water or ice). Each of these factors can vary greatly among eruptions, as well as during a single effusion, and it might be expected that lavas should show a wide range of behavior. In fact, the contrary holds and common lava types evolve along only a few, clearly defined trends, which link the morphology of lava surfaces to styles and rates of flow advance. Only by understanding why these natural evolutionary sequences occur will it be possible to improve strategies for mitigating lava hazard.

2. WHAT ARE LAVA FLOWS?

Lava flows are outpourings of molten rock, or magma. On Earth, the overwhelming majority have silicate compositions, for which common melting temperatures are in the range 800°−1200 °C; lavas of sulfur (e.g., Siretoko-Iosan volcano in Japan, and Lastarria volcano in Chile) and of carbonate compositions (e.g., Ol Doinyo Lengai volcano in Tanzania) also occur at lower temperatures (about 150 °C for sulfur and 600 °C for carbonatite), but these are extremely rare and are not important as far as general hazard studies are concerned. First applied at Vesuvius, the word "lava" is derived from the Italian *lavare* (to wash); ironic since the washing normally meant cleaning away the fruits of human labor.

Flows are distinguished from lava domes by their extreme elongation downslope. Historically, the volumes produced by single effusions range from minor dribbles to outpourings of a few cubic kilometers (the latter include Etna, Sicily, 1614−1624; Lanzarote, Canary Islands, 1730−1736; Lakagigar, Iceland, 1783−1785; and Pu'u 'Ō'ō, Hawaii, continuing since 1983). The resulting **flow fields** can extend for tens of kilometers, spread kilometers

across, and reach thicknesses of hundreds of meters, although most are more modest in size (Table 55.1). The durations of single eruptions also cover a large range and, while some may reach decades, the majority lie between days and months. Rates of flow lengthening rarely exceed a brisk walking pace, so that it is usually possible for people to escape immediate danger. Exceptions occur during the start of effusions, when lavas can sometimes advance as fast as a galloping horse. In the saddest example, on Nyiragongo at 10.15 in the morning of 10 January, 1977, a fluid lava swept some 5−6 km downslope from the lower end of a flank fissure in 20 min (an average advance rate of about 15−20 km per hour), catching a small village unawares and roasting 70 people alive.

The Nyiragongo tragedy highlights the need to prepare against lava invasion even before an eruption begins. To identify the most vulnerable districts, it is necessary to recognize probable locations of future eruptions and to forecast the likely travel distance of at least the initial lava flow. The first task is achieved by applying statistical analyses to the known distributions of vents at a volcano. The second requires a model that can link probable flow length to factors that can be measured at the start of eruption, and it is here that the evolutionary sequences of lava flows assume a fundamental importance.

3. EVOLUTIONARY SEQUENCES AMONG LAVA FLOWS

Most lavas are crystallizing as they erupt, owing to chemical imbalances induced in magma as it approaches the surface from below. They continue to solidify during effusion, aided by loss of heat to the ground and to the atmosphere. As a result, flows begin to form channels or tubes (Figures 55.1 and 55.2) that concentrate motion along only a small number of paths, so that subsequent lava is transported more efficiently from the vent to the front.

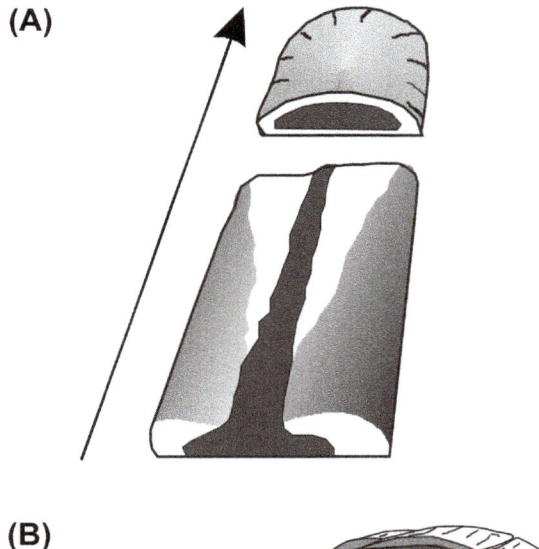

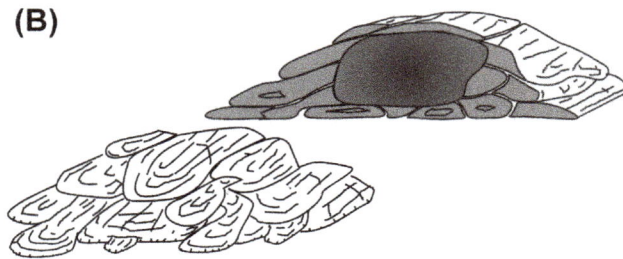

FIGURE 55.1 Major flow structures. (A) In aa and blocky flows, open channels (below) typically feed lava to simple fronts (above). Motion, in direction of arrow, is concentrated in the black zones. For major flows, the fronts are commonly ∼100 m wide and tens of meters thick. In long-lived eruptions, the channels may evolve into tubes *(After Lipman and Banks (1987))*. (B) The fronts of pahoehoe flows are normally a complex of intermingling tongues and toes (up to meters across and tens of meters long) fed by lava from a tube system.

A flow initially forms a tube or a channel according to whether the lava surface can develop a continuous **crust**. When exposed to the atmosphere, a fresh lava surface chills to a strong, solid crust within minutes. At the same time, the new crust is pulled forward by more mobile lava. If the forward pull is large enough, the crust continually breaks into fragments and so, being unable to form a stable roof, the flow develops an open channel for containing the lava. If the forward pull is too small, a continuous crust can

develop across the flow and this, anchored to the flow margins, gives birth to a tube.

A similar battle between crustal growth and disruption occurs at flow fronts. When disruption dominates, the front moves forward as a single unit, controlled by the properties of the frontal interior. When crust formation dominates, the front advances by oozing small tongues of lava through localized punctures in the crust.

From observation alone, it is clear that crustal structure can be linked directly to the early formation of lava channels and tubes, and to the style with which a flow front moves forward. The link with crustal structure is convenient since only the outer parts of an active flow are normally accessible to investigate, either by direct observation on the ground or by remote monitoring using aircraft or satellites. Difficulties in studying the rest of a flow arise because lavas are hot and viscous. To measure the temperatures, velocities, and fluidities of flow interiors, it is necessary to stand close to poorly crusted lava, where it is easiest to insert monitoring equipment. The most active parts of flows have mean surface temperatures commonly between 475 °C (when their color is a very dull red) and 1100 °C (when they are golden yellow), a range sufficient for radiant heat to cause serious burns even meters away. Protective clothing, although cumbersome, can normally overcome this difficulty. However, even at the start of the eruption, the most fluid lavas are frequently a million times more viscous than water (Table 55.2) and it is extremely difficult both to force a measuring device into a flow and, if successful, to retrieve it again.

As a result, virtually no data are available from direct measurements of active lava interiors. Investigations must instead rely upon theoretical studies or upon indirect evidence from solidified lava interiors (exposed by natural collapse or by artificial excavations, such as road cuttings or quarries). In the first case, no theoretical model has yet been completely verified, precisely because the necessary field data are unavailable. In the second, the features preserved by solidified lava interiors (such as crystal size distributions, which depend on a lava's cooling history) may not reflect conditions that prevailed while the flow was still active, because these features (1) may have been dominated by changes (e.g., further crystallization) that

FIGURE 55.2 (A) A well-developed channel between static banks of aa lava. This example, from the 1983 eruption on Mt Etna, is a few meters wide, but channel widths may reach tens of meters in aa and more in blocky flows. Note the haze due to hot gases (mostly steam) escaping from the lava, moving from left to right *(Photo: C. R. J. Kilburn.)*. (B) Extensive lava tubes are essential features of pahoehoe flow fields and may also form in aa flow fields during long-lived eruptions. *(Photo: H. Pinkerton.)*

TABLE 55.2 Physical Properties of Lava Flows

Composition	Eruption Temperatures °C	Density at Eruption Temperatures kg m^{-3} (without Vesicles)	Viscosities at Eruption Temperatures Pa s[1]
Basalt	1050–1200	c. 2600–2800	10^2–10^3
Andesite	950–1170	c. 2450	10^4–10^7
Rhyolite	700–900	c. 2200	10^9–10^{13}
Komatiite	>1600?	c. 2800	<1?
Water at Earth's surface	20	c. 1000	10^{-3}
	Temperature °C (K)	Color of lava surface	
	1150 (>1423)	White	
	1090 (1363)	Golden yellow	
	900 (1173)	Orange	
	700 (973)	Bright cherry red	
	600 (873)	Dull red	
	475 (748)	Lowest visible red	

[1] Newtonian approximation at low shear rate.

occurred after the flow had come to rest or (2) may have formed at different times during flow advance, and so cannot simply be related to the state of the active flow at any particular moment. Thus, until more reliable data are available for active flow interiors, the links between crustal structure and flow dynamics offer the best prospect of quantifying flow behavior.

4. FIELD CLASSIFICATION OF LAVA FLOWS

Crustal appearance provides the basis for classifying lava flows on land into three major categories, **pahoehoe, aa,** and **blocky**. Pahoehoe and aa are Hawaiian terms introduced in the late nineteenth century to describe the common lava types found on Mauna Loa and Kilauea, but they apply equally to other lavas with silica contents less than about 50–55 wt% (basalts and some basaltic andesites), as well as to the rare flows of sulfur and carbonatite. Blocky flows are common among lavas with silica contents greater than 55 wt% (basaltic andesites to rhyolites).

Diagnostic features are most evident when viewing a crust over distances of decimeters and meters (Figure 55.3; Table 55.3). At these distances, pahoehoe surfaces are smooth and, though occasionally broken, are normally continuous, while aa surfaces are extremely irregular, frequently fractured, and usually covered by rough, contorted fragments with typical dimensions of centimeters

and decimeters. **Blocky lavas**, like aa, also have fractured surfaces and a covering of debris; they differ from aa flows in that their fragments are smooth and angular with common dimensions from decimeters to meters. From a distance, indeed, both aa and blocky surfaces look as exciting as piles of rubble on a building site.

4.1. Aa and Blocky Lavas

Aa and blocky flows show simple evolutionary trends. As might be anticipated from their broken surfaces, their fronts tend to advance as single units and it is rare for one part of a front to move far ahead of neighboring sections. Blocky fronts crumble to produce a snout of debris from the early stages of emplacement. Aa fronts show a greater range of behavior, often starting as fluid sheets, but finishing as near-solid masses that fragment throughout their thickness to maintain advance; between these limits, they move by various combinations of fracture and flow.

Fronts thicken while advancing, often growing to more than 10 times their initial thickness. Final thicknesses are typically about 20 m or less for aa fronts, but several tens of meters for blocky flows; their maximum lengths are measured in tens of kilometers and in kilometers, respectively. Major flows can achieve volumes of 1–100 million cubic meters, and tend to be emplaced within days when they are aa but within months when they are blocky.

FIGURE 55.3 The surfaces of aa and blocky flows are covered by broken surface fragments. (A) Aa fragments (1991—1993 flow, Mt Etna, Sicily) are contorted and initially appear black and spinose but, during later stages of advance, the surface breaks to yield rounded and abraded rubble. (B) Blocky fragments (Nea Kameni, Santorini, Greece) are angular and have planar faces. (C) Pahoehoe surfaces (1736 flow, Lanzarote, Canary Islands) are smooth and often billowy over distances of meters. *Photos: C.R.J. Kilburn.*

TABLE 55.3 Common Features of Flow Surfaces

Feature	Description
1. Aa Lava	Surface is covered by a jumble of irregular crustal fragments.
Cauliflower	Crust twists upward as cauliflower-like protrusions. These break to give fragments up to decimeters across. Surfaces are gray-black, often glassy, and rough and spinose at the millimeter scale.
Rubbly	Crust fractures downward to yield rounded rubble up to meters across, often with an ochre-black granular surface, millimeters deep.
2. Blocky Lava	Surface is covered by broken lava, containing fragments up to meters across with smooth, planar, and angular surfaces.
3. Pahoehoe Lava	Surface is smooth and continuous, often with a millimeter-scale texture of interweaved lava threads or filaments.
Entrail	Dribbles of lava yield convoluted surfaces reminiscent of entrails.
Ropy (or corded)	Flexible crusts ruck into tight folds before chilling. Surface resembles segment of coiled rope. Each "rope" can be centimeters thick.
Shelly	Highly vesicular, fragile crusts. Often associated with skins, centimeters thick, over hollow lava blisters. The skins break underfoot, giving the impression of walking on egg shells.
Slabby (sometimes slab aa)	Slabs of broken crust, up to meters across and centimeters thick.
4. Toothpaste Lava	Protrusions of viscous lava squeezed through gaps in flow crust. They may be tens of meters long and their cross-sections often mimic the shape of the source gap, like toothpaste emerging from its container.

Both flow types initially develop channels to feed lava to their fronts. Channels form when a flow stops widening and concentrates motion downhill. The fronts themselves grow during advance because they decelerate as they solidify and allow faster lava to accumulate from upstream. For example, during the opening stage of emplacement (when they are fastest), major aa fronts may advance a few kilometers a day (occasionally 10—30 km in the first 24 h),

although the velocity of lava near the vent may be at least 10 times greater.

At one extreme, advance and thickening continue until a flow stops being fed by new lava, whereupon the front slows to a halt as remaining lava drains from the feeding channel. At the other extreme, effusion continues into the flow even though the front has come to rest. Lava begins to pile up within the channel, starting at the front and working its way backward. Such thickening of a flow is also described as inflation. As it thickens, the channel lava exerts an increasing pressure on its margins, and these may eventually breach to form an outlet through which the channel lava can escape. If the breach is too small, it may be able to heal itself through cooling or by being plugged with crustal debris. Otherwise, the breach may become a permanent outlet from which a major new flow can develop. The new flow, in turn, may halt and thicken until the cycle is repeated. In such a way, a flow field, the final product of one effusive eruption, may evolve with time from a single flow to a collection of interconnected flows (Figure 55.4).

Breaching is more common among aa than blocky lavas. Apart from high channel pressures, the propagation of new flows by breaching requires channel lava that is much more fluid than its lateral margins. Only in this case can the channel lava easily escape through a breach; if, instead, the channel lava is almost as solid as the margins, breaching simply allows channel lava to spread into the breach and heal it. By virtue of its chemical composition, channel lava in aa flows tends to be more fluid than its blocky equivalent, so that conditions are more favorable in **aa lavas** for the propagation of new flows.

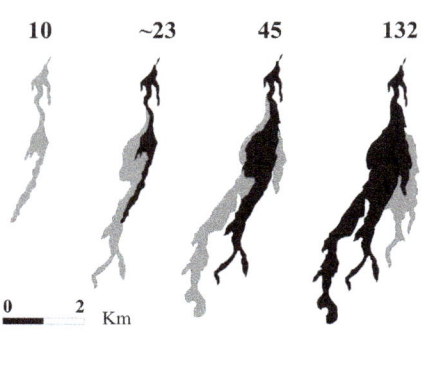

10   ~23   45   132

0   2
            Km

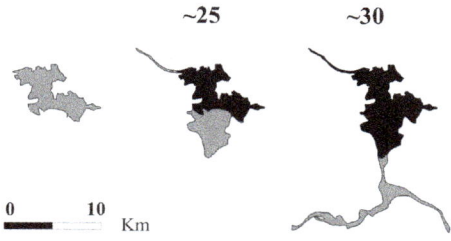

~25   ~30

0   10
            Km

FIGURE 55.4 Top. Aa flow fields grow as they propagate a sequence of flows, new flows (gray) extending from the sides of earlier streams (black). This example shows the evolution of Etna's 1983 flow field at 10, 23, 45, and 132 days after the start of effusion. Note how the final length of the flow field on day 132 is only about 30% greater than the length of the initial flow on day 10. *(Data from Frazzetta and Romano (1984)).* Bottom. Pahoehoe flow fields tend to continue growing throughout the course of an eruption. This example is based on a reconstruction of the 1736 flow field from Montaña de las Nueces on Lanzarote, in the Canary Islands. *(Data from Solana et al. (2004).)*

Although breaching may occur anywhere along a flow, it most commonly occurs somewhere along the upstream half of a flow's length. Newly propagated flows may extend downslope beyond an earlier stream, but rarely do they increase the length of a whole flow field by more than half the length of the initial flow. A result of such behavior is that the propagation of flows tends to widen, rather than to lengthen, the area covered by new lava (Figure 55.4). It also means that, when addressing the hazard from aa and blocky flows, the first goal is to estimate the probable maximum length of the initial flow.

As well as promoting flow breaching, flow thickening can trigger the overflow of fluid channel lava across its lateral margins. The margins can thus build themselves up as a series of superposed overflows, which, under favorable conditions, develop an inward overhang across the channel surface. As the exposed channel surface narrows, it becomes easier for adjacent segments of crust to congeal together and to form a continuous roof over the flow. In this way, a lava channel can evolve with time into a tube, better insulating the lava beneath and allowing it to travel further before solidification sets in.

Tubes form in aa and blocky flows only after a channel has become well established, and weeks may be required for a major tube system to be developed. When they occur, therefore, tubes are found along those parts of flows that remained active for long periods, typically behaving as feeders to sites of breaching downstream. As with the propagation of flows by breaching, the growth of lava tubes is favored by a large contrast between the fluidity of internal lava and its external margins. Accordingly, tubes are more common in aa than in blocky flows. Should lava drain from a tube toward the end of effusion, it may leave behind a tunnel perhaps kilometers long that is large enough at least to crawl through.

4.2. Pahoehoe Lavas

Pahoehoe flows have final dimensions similar to aa flows, but normally advance up to 10 times more slowly. Their cooling surfaces resist extensive tearing and, though feeding channels can develop, it is not unusual for continuous crusts to form across a whole flow from the early stages of emplacement. Modern descriptions of active **pahoehoe lavas** have been dominated by observations from Kilauea, on Hawaii. The crust remains continuous around flow margins and fronts, whose initial thicknesses are usually of only a few decimeters. Since early flows are thin, they are easily retarded by the crust and spreading occurs by a combination of lava leaking out as small tongues through breaks in the crust, and by inflating as new lava from upstream burrows beneath the tongues and slowly lifts the front upward. The front thus appears as a collection of intermingling tongues, each much narrower (by 10−1000 times) than the width of the whole flow (Figure 55.1).

Pahoehoe tongues may extend by some 100 m, and even start to develop feeding channels, before crusting over and extruding small lava toes, typically a few meters or less in length, analogous to the lava pillows among submarine flows (see Chapter 19). As a result, pahoehoe fronts and margins soon develop as a complex of budding tongues and toes, all of which are gently uplifted by newly arriving lava. The surfaces of pahoehoe flows thus evolve a curious hummocky form, involving swells that extend distances from decimeters to hundreds of meters. The largest of these seem almost flat to the casual gaze; the smallest appear curiously grotesque and intestinal (Figure 55.3; Table 55.3).

Lava from the vent continually raises the surface by inflation and, within weeks, a flow field may have thickened to several meters, while thin lava tongues continue to emerge from around its edges and from occasional surface ruptures. Although many tongues and toes stagnate after halting, others remain connected beneath the crust to form a network of distributary tubes, casually resembling an underground river system. Compared with the open channels in aa and blocky flows, lava tubes reduce the rate of lava cooling, especially when they are partly drained so that hot gases can collect beneath the tube roof and keep the flowing lava surface at temperatures close to its initial value. Thus, although pahoehoe flows commonly advance more slowly than their aa counterparts, their interiors remain fluid for much longer periods and it is normal for pahoehoe flow fields to continue extending for as long as they are supplied with fresh material (Figure 55.4). The greater lengthening time often dominates the slower velocity, so that pahoehoe flow fields can achieve lengths greater than aa flow fields of similar volume.

Historical pahoehoe flow fields have extended several tens of kilometers, the best examples being found on the Big Island of Hawaii. Some prehistoric flow fields, however, have been traced for more than 100 km, notably those in Queensland, Australia. Ancient flood basalts (such as the Columbia River Basalts in the USA), which have lengths of several hundreds of kilometers, may also have developed as enormous pahoehoe flow fields. The drained tube systems in such flows are truly impressive, reaching tens of meters across, up to 10−20 m high, and extending for several kilometers at least; the prehistoric Australian flow fields, for example, contain a tube system about 100 km long.

5. THE LAVA SPECTRUM

In the simplest of cases, a flow maintains the same type of surface morphology throughout emplacement. Frequently, though, a flow surface evolves through more than one type with distance downstream. Among basaltic lavas, downstream transitions occur from pahoehoe to aa

morphologies, while the change from aa to blocky is found among some basaltic andesites; both transitions are unidirectional, so that blocky surfaces do not become aa, and aa surfaces do not become pahoehoe (although, as discussed later, special field conditions may give the false impression that reverse transitions can occur). The pahoehoe, aa, and blocky morphologies are thus not independent entities, but parts of a continuous spectrum of lava types. The association of the two transitions with different lava chemistry shows that composition is one underlying control. However, since each transition occurs between lavas of similar composition, other nonchemical factors must also be involved.

Both pahoehoe and aa surfaces are created during the formation of surface crust. At the start of eruption, the lavas are often too fluid to break before cooling, so that the degree of rupture must be controlled by the solidifying surface layers. To break a surface before it has chilled to its maximum strength, the rate of energy supplied predominantly by gravity to a unit volume of crust must be greater than a critical value. The critical value depends on how quickly a crack can be healed by chilling the newly exposed lava beneath. If the critical value is exceeded, a flow breaks its crust more quickly than existing cracks can be healed and so evolves a fragmented aa surface; otherwise, a flow cannot tear its crust quickly enough and develops a continuous pahoehoe surface.

The rate of energy supply is measured by the product of the stress applied to the crust and the rate at which the crust deforms. The larger the pull, or applied stress, the smaller will be the deformation rate at which critical conditions are reached. More viscous lavas require larger applied stresses to induce a given rate of deformation. As a result, more crystallized, and, hence, more viscous, basalts cross the threshold from pahoehoe to aa under larger applied stresses and slower rates of deformation (Figure 55.5).

Faster rates of energy supply are also associated with faster rates of advance down a particular slope. The threshold advance rate U_{cr}, above which lava becomes aa, is given by $[(\epsilon S/\rho g)(\rho c_p/\Sigma \sigma \Theta^3)]/\sin \beta$, where β is the angle of slope and the remaining physical quantities are defined in Table 55.4. For basalts, the expression simplifies to $U_{cr} \approx 0.0006/\sin \beta$, for advance rates measured in meters per second (Figure 55.6). Pahoehoe flows can therefore advance more quickly on shallower slopes before they are transformed into aa.

Blocky surfaces are associated with stronger and more viscous lavas that break before they have cooled significantly after leaving the vent. They can be considered almost as magmatic glaciers, slowly moving sheets that flow near their bases but break at their surfaces. The aa−blocky transition thus occurs when the lava interior becomes too crystalline to move forward only by flowing,

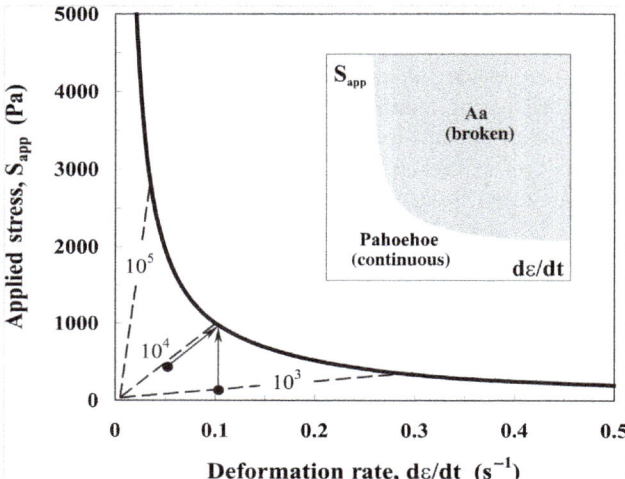

FIGURE 55.5 The evolution of pahoehoe and aa surfaces depends on the rate at which energy is supplied to deform cooling crust. The rate of energy supply per unit volume is given by the product of applied stress and deformation rate ($S_{app}\ d\varepsilon/dt$). Below a critical energy flux, which is controlled by the cooling and strengthening of lava exposed between cracks and can be expressed as $(\Sigma S/\kappa)(\varepsilon\sigma\theta^3/\rho c_p)^2$ (about 100 J m^{-3}), the surface chills as a continuous crust to form pahoehoe. When this flux is exceeded, the surface breaks persistently to form aa. The dashed lines show lavas with different Newtonian viscosities (in Pascal Second). The viscosity increases with crystallization. The arrows illustrate two simple conditions for the transition from pahoehoe to aa: one for increasing deformation rate at a constant viscosity (in this case, of 10^4 Pa s) and another for increasing crystallization and viscosity under a constant deformation rate (0.1 s^{-1}). Once crust has been broken to form aa, it cannot return to pahoehoe. Blocky lavas are not shown on this diagram, because they are strong before eruption and, as if they were magmatic glaciers, they must break their surfaces to advance, independent of cooling effects.

at which stage its upper levels, including the surface, also fracture as they deform.

The transitions between surface morphologies are irreversible; pieces of broken crust cannot recombine to form a continuous surface (hence aa surfaces do not evolve into pahoehoe), while subcrustal lava does not become less solid with time (hence blocky surfaces do not evolve into aa). On occasion, however, it might appear that a reverse transition has taken place, especially between pahoehoe and aa types. The deception occurs under two main sets of conditions. The first corresponds to changing conditions of effusion from the vent. As an eruption decays to its close, so the rate of effusion also decreases. While early, fast lava may produce an aa surface, later lava may emerge slowly enough to form a continuous pahoehoe crust. This change, though, does not represent a transition from aa to pahoehoe, since the later pahoehoe crust formed across an independent flow and did not evolve from lava already erupted with an aa surface. The second condition involves the escape from aa channels of hot internal lava, the result either of a channel overflow or of a margin being breached. In this case, the escaping lava may be less crystallized and, if advancing more slowly than the parent flow, may form a pahoehoe crust. Once again, such a situation does not correspond to an evolution of the crust from aa to pahoehoe, because the escaping lava comes from deep within the parent flow and has followed a crystallization history independent from that of the near-surface lava layers that formed the earlier crust.

TABLE 55.4 Symbols for Physical Properties

Symbol	Meaning	Units	Nominal Value for Basalt
β	Slope angle	Degrees	2–10
c_p	Specific heat capacity	J kg^{-1} K^{-1}	1150 (During crystallization)
D	Thickness	m	1–10 m at flow fronts
ϵ	Extension before failure	–	Min. 10^{-3} (for chilled crust)
ϵS	Energy per unit volume for failure	J m^{-3} (or Pa)	2×10^4 (during crystallization)
g	Gravitational acceleration	m s^{-2}	9.81 (for planet Earth)
κ	Thermal diffusivity	m^2 s^{-1}	4.2×10^{-7}
L_m	Maximum potential length of flow	m	Commonly 10^3–10^4
Q	Mean discharge rate along flow	m^3 s^{-1}	Commonly 1–10^3
Θ	Eruption temperature (absolute)	K	1350–1400
ρ	Density of lava crust	kg m^{-3}	2200 ($\sim$20% vol. Vesicles)
S	Tensile strength	Pa	Max. 10^7 (For chilled crust)
Σ	Surface emissivity	–	1
σ	Stefan–Boltzmann constant	J m^{-2} s^{-1} K^{-4}	5.67×10^{-8}
t	Time	s	

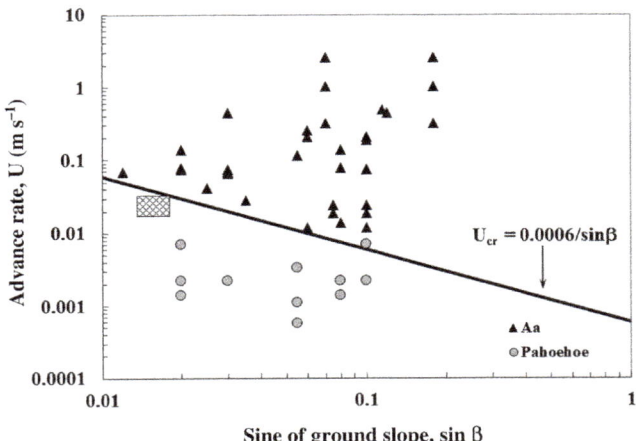

FIGURE 55.6 On a particular slope, flows must advance more quickly than a critical value if they are to break their crusts persistently and develop as aa. For basaltic lavas, the critical velocity increases as slope decreases, according to the relation $U_{cr} = 0.0006/\sin \beta$, for U_{cr} in meters per second (solid line). Note that the scales are logarithmic. *Field data are from Hawaii (triangles and circles for aa and pahoehoe, respectively; Rowland and Walker (1990)) and Lanzarote (cross-hatched rectangle for pahoehoe; Solana et al. (2004)).*

6. LAVA DYNAMICS AND INTERNAL SOLIDIFICATION

Because of its high viscosity, lava can rapidly reduce accelerations in a flow. As a result, flow fronts tend to settle into a steady dynamic state. A simple example is shown by the preference of flows to advance at nearly constant velocities for extended periods of time (Figure 55.7). Since inertia is not important, flow growth is controlled by how a lava's rheological resistance is overcome by gravity (pulling lava downslope) and by pressure differences due to local variations in flow thickness (notably at the flow periphery).

Another consequence of low inertia is that lava strives to maintain laminar flow, whereby adjacent packets of lava tend to flow past each other, rather than intermingling, as would occur if motion were turbulent. Without intermingling, a lava interior can diffuse heat only by conduction (i.e., interactions between neighboring molecules), so that cooling effects tend to migrate inward from the flow exterior, itself maintained at a low temperature by radiation to the atmosphere, by cooling wind, and by precipitation. Lava, however, is a very poor conductor. The time needed for cooling to penetrate a depth D into a flow is given approximately by $D^2/4\kappa$, where the thermal diffusivity (κ) of lava is between 10^{-7} and 10^{-6} m^2 s^{-1}. Thus, it takes minutes to chill a layer centimeters deep, but weeks for a layer meters thick.

To give an example of a lava flow 10 km long, the time for lava to travel from the vent to the front may be only hours for pahoehoe and aa flows, and days for blocky flows. (Recall that this travel time is less than the interval since effusion began: the former is given by [flow length]/[mean velocity along a flow]; the latter is [flow length]/[mean velocity of the flow front].) The corresponding mean flow thicknesses will be measured in meters or tens of meters, implying conductive cooling times of weeks or longer. It might thus be expected that the interiors of pahoehoe and aa flow fronts will contain lava almost as fluid as it was near the vent. This condition is typical among pahoehoe flows, but not among aa flows, which can develop solidified fronts within days.

Precisely why the interiors of aa flows can solidify so quickly is still not fully understood. One possibility is that, compared with other lava types, aa lavas are initially richer in volatiles. All magmas contain some volatiles, mostly water, that remain dissolved at depth, where the high pressure from surrounding rock prevents them from forming bubbles. As the magma ascends, the imposed pressure decreases and bubbles can eventually develop, just as they

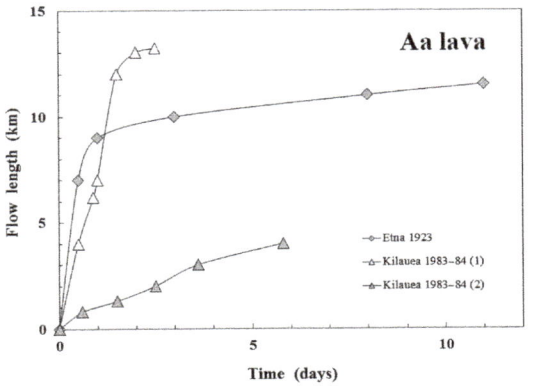

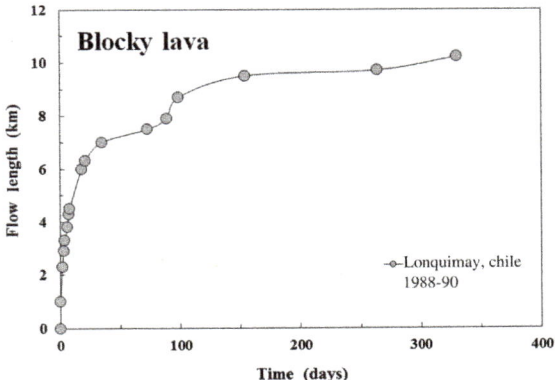

FIGURE 55.7 Most flow lengthening for aa and blocky lavas occurs during the early stages of effusion and at nearly constant rates of advance. This may be followed by a rapid drop in advance rate, which, though continuing for a long time, does not extend a flow significantly. *The examples of aa flows are from Etna (Ponte, 1923) and Kilauea (Wolfe et al., 1988); the blocky lava is from Lonquimay (Naranjo et al., 1992).*

appear in a bottle of soda water when pressure is released as the bottle is opened. The loss of volatiles upsets the chemical balance in the liquid and can trigger crystallization at a rate that increases with the amount of gases originally dissolved in the magma (and now exsolved as bubbles), but decreases with increases in liquid viscosity. Thus, compared with aa lavas, pahoehoe interiors may crystallize more slowly due to smaller amounts of initial gas, while crystallization in blocky interiors is retarded by high lava viscosity.

Another explanation appeals to crustal entrainment. Pieces of cold crust are dragged back into the lava interior, accelerating solidification by forcing internal lava to intermingle. Importantly, the intermingling in aa is a result of crustal entrainment and not of natural turbulence; when possible, the flow again seeks a simple laminar motion. This mechanism is less common in pahoehoe flows, since they form continuous crusts, and also in blocky flows, because their very large viscosity impedes incorporation of surface debris.

7. FORECASTING THE BEHAVIOR OF AA FLOWS

The interaction of factors controlling flow behavior may be complex in detail and so is ideally modeled by computer simulations, using numerical models that are based on well-established relations for describing the fluid dynamics, rheology and thermal behavior of fluids. In practice, however, a major limitation to applying such models is uncertainty in how to quantify a lava's change in rheology as it solidifies. Although excellent numerical models have been developed, more comprehensive field data on lava rheology are required before they can produce fully realistic results. Until such data are available, an alternative approach is to use the patterns of flow-field growth to identify limiting conditions of emplacement.

The three main types of lava surface are associated with distinct styles of flow growth. Since lava surfaces are easily monitored in the field, it would be convenient if the conditions for producing a specific surface type could also provide limits upon the likely distance and velocity a given flow might travel. Such a state of affairs does exist for aa flows, a happy circumstance since aa flows are the lava type that has most frequently threatened human activity, and they are also intermediate between pahoehoe and blocky flows. Understanding the limits to aa behavior thus provides clues to the limiting behavior of the other flow types.

The aa surface criterion can be used to estimate the advance velocity needed to keep breaking lava crust. The time for which a lava flow can continue to advance is controlled either by the duration of eruption (for short-lived aa flows) or, during long eruptions, by the time needed for the flow to acquire a solid front. Combining the

requirement for a lava front to solidify (which gives the longest time for advance) with the fracture criterion to form an aa surface, it is possible to link the maximum potential length (L_m) of an aa flow to lava properties and mean rate of discharge (Q):

$$L_m \approx [3\epsilon S/\rho g\kappa]^{1/2} Q^{1/2} \qquad (55.1)$$

where the remaining symbols are defined in Table 55.4. The terms in the square brackets describe the physical properties of the lava crust and are approximately constant for a given lava composition. For basalts, in particular, Eqn (55.1) can be simplified to $L_m \approx 2.5\,Q^{1/2}$, for length in kilometers and discharge rate in cubic meters per second. The maximum potential length is thus expected to increase with discharge rate, in agreement with observation (Figure 55.8).

The length-discharge rate trend may be used for short-term forecasts of maximum aa flow length while an eruption is in progress. During the 1991—1993 effusion on Etna, for example, secondary aa flows began to overtop an artificial barrier and descend toward Zafferana, a town almost 3 km further downslope. Initial observations suggested that it was unlikely for the mean discharge rate to exceed $1\,m^3\,s^{-1}$ along any flow. From Eqn (55.1), a maximum flow length of less than 2.5 km was anticipated, indicating that the immediate threat to Zafferana was small. A forecast was issued to that effect and, in the event, the longest secondary flow traveled only 1.8 km beyond the barrier.

The success of Eqn (55.1) is remarkable not only for the simplicity of its underlying assumptions, but also because it does not include slope directly and uses mean values of discharge rate to determine length. In reality, surface slopes

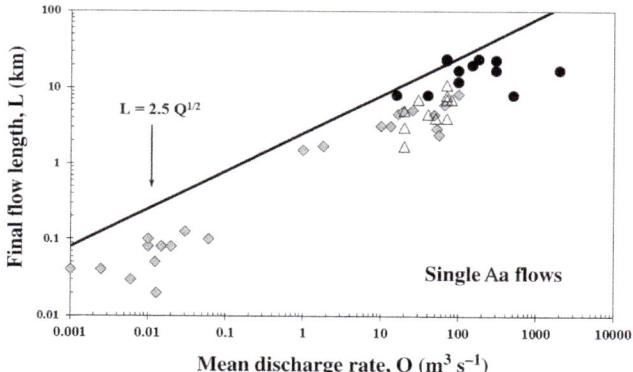

FIGURE 55.8 The conditions that (1) aa surfaces must break before they cool to their maximum strength and (2) flow fronts advance until solidified yield criteria for relating the maximum potential length of a single aa flow to the mean rate of discharge along the flow (Q). The solid line shows the theoretical trend from Eqn (55.1). Note that the scales are logarithmic. The limiting trend agrees well with field data from Etna (diamonds), Mauna Loa (circles) and Kilauea (triangles).

are uneven, while discharge rate changes with time and position along a flow. The agreement between theory and observation thus suggests that local variations have normally a secondary effect on maximum potential flow length, so that mean values are sufficient for most purposes. Extreme variations, such as flow over a cliff or through a very narrow ravine, may induce significant changes from the model results and, until more sophisticated analyses are available, these rare cases must be evaluated individually.

The expression for maximum length assumes that an aa flow travels until its front has solidified. The maximum time required is approximately given by $[(\epsilon S/\rho g)(\rho c_p/\Sigma\sigma\Theta^3)^{1/2}]/\sin\beta$. As before, the combination of physical properties (Table 55.4) in the square brackets is roughly constant for a particular lava composition and, for typical angles of $2°-10°$ for the slopes of basaltic volcanoes, yields lengthening times of about $1-10$ days (the longer times corresponding to smaller slope angles). During a long-lived eruption, therefore, major breaching and the propagation of new flows is most likely to commence within days of the start of effusion. Since new flows rarely extend a flow-field by more than half the length of the first major flow, another forecasting problem is to assess the final width of a flow field, which may be as large as several kilometers.

An aa flow field can widen until halted by topography or by the end of effusion. The topographic control depends on local conditions and must be assessed on a case-by-case basis. The probable duration of an effusion can be estimated by comparison with previous eruptions, but even on well-studied volcanoes, such as Etna, Kilauea, and Mauna Loa, such estimates are rarely better than inspired guesswork. Improvements will follow when it becomes possible either to estimate the quantity of magma below ground that is available for eruption, or to forecast the long-term decay in rate of effusion as an eruption proceeds.

8. FORECASTING THE BEHAVIOR OF PAHOEHOE AND BLOCKY FLOWS

It is possible before an eruption to estimate the maximum length of an aa flow because the surface criterion provides a velocity, and the solidification control yields a time. Such a combination is not available either to pahoehoe or to blocky flows and so forecasts of their growth is inherently more difficult.

For pahoehoe flows, the requirement to form a continuous crust can be used to constrain maximum rates of advance to about $1-2$ km per day on slopes typical of basaltic volcanoes (Figure 55.6). Unfortunately, the burrowing mechanism, by which new lava intrudes into the front after traveling through insulated tubes, has so far defied realistic estimates of maximum cooling times. As it

happens, observed pahoehoe flows have frequently lengthened for the whole of an eruption (Figure 55.4), even when this has continued for several months. In practice, therefore, pahoehoe emplacement is often limited by the available supply of magma, and so forecasting flow behavior is subject to the same problems encountered when estimating the final width of an aa flow field.

Among blocky lavas, surface fracturing appears to be controlled by the strength of lava beneath the chilled crust. As a result, fracturing is not required to be faster than the rate of surface chilling, and so no surface criterion is available for forecasting mean rates of advance. Velocity forecasts must therefore rely on knowing the rheological state of a lava and the rate at which it is effused, factors that can only be estimated by analogy with previous effusions from the same volcano.

Since the lava is viscous and slow moving, a blocky front is expected to solidify at a rate controlled by conductive heat transfer to the breaking surface. From standard conduction theory, a crude estimate of the solidification time for common final thicknesses of $10-30$ m is from 6 months to over 4 years, lengths of time that are greater than the usual durations for eruptions of blocky lava. Once again, the volume of magma available appears crucial for determining how far a lava can travel, and estimating this volume remains a fundamental obstacle to improving preeruptive forecasts of blocky flow growth.

9. THE CONTROL OF LAVA FLOWS

Techniques for controlling the growth of a flow-field seek to change the environment around a flow or the structure of a flow itself. The first approach involves the construction of retaining dams or diversion barriers; the second aims to redirect a flow by piercing its margins, or to slow its advance by artificial cooling. The preferred choice depends on the prevailing topography and on the value of the land surrounding a flow. Dams are most effective when sealing the exit from a natural depression toward which a flow is already heading. The aim is to contain the areal coverage of the lava behind the dam until the eruption ceases. Diversion barriers, in contrast, are more effective when a flow can be allowed to spread over low-value terrain.

Changing a flow's structure takes advantage of the natural restraining effect of a lava's solidified margins. The earliest recorded attempt occurred during the 1669 effusion on Mt Etna, when an aa lava stream threatened the city of Catania on the southeast coast of the volcano. Diego Pappalardo and some 50 others climbed to the vent, at about 950 m above sea level, and attempted to divert the flow by breaching its chilled margins with makeshift picks and axes, using water-soaked animal hides as protection from the heat. The ambitious group were successful, but at the expense of a new lava being directed toward the

TABLE 55.5 Selected Topics for Further Reading

Topic	Reference
Overviews of lava behavior.	Macdonald (1972); Chester et al. (1985); Kilburn and Luongo (1993); Kilburn (1996).
Flow-field emplacement.	Rowland and Walker (1990); Kilburn and Lopes (1991).
Field descriptions of lavas. (a) aa lavas; (b) pahoehoe lavas; (c) blocky lavas; (d) lava tubes.	(a) Ponte (1923); Frazzetta and Romano (1984); Lipman and Banks (1987); Wolfe et al. (1988); Solana (2012). (b) Hon et al. (1994); Solana et al. (2004). (c) Naranjo et al. (1992). (d) Calvari and Pinkerton (1998).
Solidification of lava during emplacement.	Crisp and Baloga (1994); Applegarth et al. (2013).
Forecasting flow behavior.	Kilburn (1996, 2004).
Numerical models of lava flows.	Harris (2013).
The control of lava flows.	Barberi and Carapezza (2004).

neighboring town of Paternò. Unsurprisingly, some 500 armed Paternesi encouraged the first group to desist. The breach healed and the lava resumed its course toward Catania, eventually destroying a large part of its suburbs. Modern attempts have adopted the same approach, replacing picks and axes with aerial bombing on Mauna Loa in 1935 and 1942, and with explosives inserted directly into flow margins on Etna in 1983 and 1992.

Cooling by water spraying has also been used in attempts to halt flows by accelerating their rates of solidification. The most celebrated effort occurred in 1973, when lavas from the Eldfjell volcano on Heimaey island threatened the adjacent settlement of Westmannaeyjar, the principal fishing center in Iceland. Water spraying was continued for several weeks and successfully inhibited the lava's progress, although it is uncertain whether their shortened final lengths were much different from the values they would have reached if they had been left undisturbed.

10. SUMMARY

Lava flows on land evolve along a small number of trends, each of which links surface structure to styles of advance and to modes of flow-field growth. These trends correspond to particular combinations of the forces driving and resisting flow motion. One combination, corresponding to aa lavas, provides the possibility for forecasting the maximum potential lengths of individual flows. Emplacement of the remaining flow types, pahoehoe and blocky, appears to need prior knowledge of the amount of lava available, a quantity that is not yet possible to assess before an eruption has finished. Improved forecasts are expected from the new generation of computer-assisted numerical models of flow development, combined with enhanced data

on changes in lava rheology during emplacement. Such improvements will aid more effective strategies for mitigating lava hazard, from direct interventions on flow advance to the more timely evacuation of threatened communities.

FURTHER READING

Much of the vast literature on lava flows would not have been possible without the pioneering observations of Gordon Macdonald, George Walker, and John Guest. Selected references are listed in Table 55.5. In addition to addressing key topics, they provide comprehensive bibliographies for further information.

Applegarth, L., Tuffen, H., James, M., Pinkerton, H., 2013. Degassing-induced crystallization in basalts. Earth-Sci. Rev. 116, 1—16.

Barberi, F., Carapezza, M.L., 2004. The control of lava flows at Mt Etna. In: Bonnacorso, A., Calvari, S., Coltelli, M., Negro, Del, Falsaperla, S. (Eds.), Mt Etna: Volcano Laboratory, vol. 143. American Geophysical Union Monograph, pp. 357—369.

Calvari, S., Pinkerton, H., 1998. Formation of lava tubes and extensive flow field during the 1991—1993 eruption of Mount Etna. J. Geophys. Res. 103, 27,291—27,301.

Chester, D.K., Duncan, A.M., Guest, J.E., Kilburn, C.R.J., 1985. Mount Etna. The Anatomy of a Volcano. Chapman and Hall, London.

Crisp, J.A., Baloga, S.M., 1994. Influence of crystallization and entrainment of cooler material on the emplacement of basaltic aa flows. J. Geophys. Res. 99, 7,177—7,198.

Frazzetta, G., Romano, R., 1984. The 1983 Etna eruption: event chronology and morphological evolution of the lava flow. Bull. Volcanol. 47, 1079—1096.

Harris, A.J.L., 2013. Lava flows. In: Fagents, S.A., Gregg, T.K.P., Lopes, R.M.C. (Eds.), Modeling Volcanic Processes. Cambridge University Press, Cambridge, pp. 85—106.

Hon, K., Kauahikaua, J., Denlinger, R., McKay, K., 1994. Emplacement and inflation of pahoehoe sheet flows — observations and measurements of active lava flows on Kilauea volcano, Hawaii. Geol. Soc. Am. Bull. 106, 351—370.

Kilburn, C.R.J., 2004. Fracturing as a quantitative indicator of lava flow dynamics. J. Volcanol. Geotherm. Res. 132, 209−224.

Kilburn, C.R.J., Lopes, R.M.C., 1991. General patterns of flow field growth: aa and blocky lavas. J. Geophys. Res. 96, 19,721−19,732.

Kilburn, C.R.J., Luongo, G. (Eds.), 1993. Active Lavas: Monitoring and Modelling. UCL Press, London.

Kilburn, C.R.J., 1996. Patterns and predictability in the emplacement of subaerial lava flows and flow fields. In: In Scarpa, R., Tilling, R.I. (Eds.), Monitoring and Mitigation of Volcano Hazards. Springer, Berlin, pp. 491−537.

Lipman, P.W., Banks, N.G., 1987. Aa flow dynamics, Mauna Loa 1984. In: Decker, R.W., Wright, T.L., Stauffer, P.H. (Eds.), Volcanism in Hawaii. U. S. Geological Survey, pp. 1527−1567. Prof. Paper 1350.

Macdonald, G.A., 1972. Volcanoes. Prentice-Hall, Englewood Cliffs.

Naranjo, J.A., Sparks, R.S.J., Stasiuk, M.V., Moreno, H., Ablay, G.J., 1992. Morphological, structural and textural variations in the 1988-1990 andesite lava of Lonquimay Volcano, Chile. Geol. Mag. 129, 657−678.

Ponte, G., 1923. The recent eruption of Etna. Nature 112, 546−548.

Rowland, S.K., Walker, G.P.L., 1990. Pahoehoe and aa in Hawaii: volumetric flow rate controls the lava structure. Bull. Volcanol. 52, 615−628.

Solana, M.C., 2012. Development of unconfined historic lava flow fields in Tenerife: implications for the mitigation of risk from a future eruption. Bull. Volcanol. 74, 2,397−2,413.

Solana, M.C., Kilburn, C.R.J., Rodriguez Badiola, E., Aparicio, A., 2004. Fast emplacement of extensive pahoehoe flow-fields: the case of the 1736 flows from Montaña de las Nueces, Lanzarote. J. Volcanol. Geotherm. Res. 132, 189−207.

Wolfe, E.W., Neal, C.A., Banks, N.G., Duggan, T.J., 1988. Geologic observations and chronology of eruptive events. In: Wolfe, E.W. (Ed.), The Puu Oo Eruption of Kilauea Volcano, Hawaii: Episodes 1 through 20, January 3, 1983, through June 8, 1984. U.S. Geological Survey, pp. 1−97. Prof. Paper 1463.

Chapter 56

Hazards from Lahars and Jökulhlaups

Magnús T. Gudmundsson
Institute of Earth Sciences, University of Iceland, Reykjavík, Iceland

Chapter Outline

GLOSSARY

debris flow A wet, flowing slurry of solid particles or rocks, commonly defined as flows with less than 20% mass percentage of water (40% by volume), with remaining mass being solid material. Unlike for normal water flow, the transfer of momentum by impact of particles/rocks is a major factor in the flow with the water providing lubrication. Debris flows commonly display laminar behavior, advancing as a plug, but sluggish turbulent behavior is also observed. Debris flows occur in a variety of settings, not only on volcanoes.

hydraulic routing A method of estimating the timing, depth, velocity, and distribution of bodies of water in open channels. Routing is usually based on the St Venant equations or equations describing the translation and behavior of dynamic or translatory waves. Hydraulic routing is increasingly used to predict the impact and reach of lahars, although the processes of bulking (accumulation of sediment) and debulking (deposition of sediment) make the routing of lahars more challenging than for water flows. Routing of flowing water has been used extensively to estimate sizes of areas that can be expected to be inundated by major jökulhlaups in Iceland.

hyperconcentrated flow The flow of a mixture of water and sediment intermediate in behavior between a debris flow and normal fluvial streamflow (water flow). A hyperconcentrated flow is often defined as having a solid part between 40% and 80% of the flow mass (20–60% by volume). Hyperconcentrated flows look similar to normal sediment-rich streamflow (water flows) but turbulence is reduced and standing waves are common.

ice cauldron Circular or elongated depressions in glaciers, often bound by a set of concentric crevasses or vertical ice walls. Ice cauldrons form as result of basal heating caused by subglacial hydrothermal activity or when volcanic eruptions bring magma to the base of glaciers. Ice cauldrons are common in Iceland and are found on ice-covered volcanoes worldwide. Sometimes ice cauldrons store water that is released in jökulhlaups. Such jökulhlaups, some of which may be sediment-rich lahars, are common in eruptions in ice-covered volcanoes.

jökulhlaup Floods released from glaciers regardless of how they originate. Jökulhlaups arising from volcanic eruptions are sometimes termed volcanogenic. The term comes from Icelandic and no distinction is made between water flows and lahars of high sediment concentration.

lahar The term is Indonesian and refers to debris flows or hyperconcentrated flows in volcanic regions, usually caused by mobilization of pyroclastic material on the slopes of volcanoes. Lahars make heavy noise as they roar down the slopes. The largest lahars occur during eruptions. They are often called primary lahars. Secondary lahars result from heavy rain in areas of abundant tephra, and can be very serious. Primary lahars can have high temperatures resulting from mobilization of still hot pyroclastic material. The term lahar is usually reserved for the process of sediment-rich flow in a volcanic region; it is not used as a term for the deposits.

The Encyclopedia of Volcanoes. http://dx.doi.org/10.1016/B978-0-12-385938-9.00056-0

mudflow A category of sediment-rich flows where the solids are chiefly small sized particles or muds. Some, but not all lahars are mudflows.

muds Mixtures of silts (particle diameters 2−62 μm) and clays (particle diameters <2 μm). The term is commonly used for the fine-grained flood deposits resulting from lahars initiating from the mobilization of pyroclasts deposited by ashfall or pyroclastic density currents. Post-eruption alteration of the pyroclasts mobilized in mudflows is often the main source of clay in muds.

sabo dam/check dam A dam used to trap sediments in lahars, constructed over a drainage channel. The dam generates a temporary reservoir where parts of the solids settle. Several sabo dams are usually placed in a single drainage channel. The space above each dam has to be cleared of deposits in between lahar occurrences in order for the dam to work as an effective trap. Sabo dams have to be very sturdy to withstand the force of full scale debris flows. They are used extensively in Japan and are increasingly being constructed in lahar prone areas elsewhere.

water flow/normal streamflow Newtonian flow of water as observed in a normal river. Sediment concentrations are usually proportionally low but can reach up to 40% by mass (20% by volume) but the sediments do not influence much the flow behavior.

1. INTRODUCTION

Lahars are fast flowing mixtures of solid particles and water are one of the principal hazards resulting from explosive volcanic activity in tropical and warm temperate regions. Sometimes formation of lahars is caused by the sudden bursting of a water body such as a lake in a caldera or a crater. But lahars also occur when unconsolidated tephra and pyroclastic flow deposits are mobilized by heavy rain, resulting in slurries of water and debris roaring down the slopes of volcanoes. Unlike most other volcanic hazards, lahars can occur without a direct link to a particular eruption, all it takes is heavy precipitation and abundance of unconsolidated volcanic debris. However, the most severe lahars usually occur during or soon after explosive eruptions. Although the hazard potential of lahars is greatest in densely populated warm temperate to tropical regions, they also occur in cooler climates provided tephra or other unconsolidated debris on slopes is being mobilized by rain, glacial meltwater, or other sources of surface water. **Jökulhlaups** caused by volcanic activity happen in regions with ice-covered active volcanoes. The frequency and severity of volcanogenic jökulhlaups are greatest at higher latitudes, in places such as Iceland and Alaska. These are usually sparsely populated areas in comparison with, e.g., the fertile volcanic regions in Indonesia, the Philippines, or Japan. Jökulhlaups have claimed relatively few lives, but they pose a very real threat to several communities and can lead to heavy damage (Major and Newhall, 1989, Gudmundsson et al., 2008).

Table 56.1 lists some important examples of lahars and jökulhlaups and Figure 56.1 shows the location of volcanoes with notable reported lahar or jökulhlaup events in the twentieth century. Records of fatalities resulting from volcanic eruptions in the twentieth century show that lahars are only surpassed by pyroclastic density currents as the most deadly volcanic hazard. A recent compilation puts the death toll by lahars at 30,700 people in the period 1900−1999, or about one-third of the total of volcanically induced fatalities (Witham, 2005). Pyroclastic density currents are responsible for about 45,000 deaths. The disaster of the town of Armero, Colombia, in 1985, when 23,000 people were killed weighs heavily in these statistics.

This chapter focuses on the hazards resulting from lahars and jökulhlaups caused directly or indirectly from volcanic activity. Details of the physical principles of lahars are discussed in a separate chapter and will only be described here as needed. We will first look into the characteristics of lahars and jökulhlaups. This discussion is followed by a section on hazard, where case histories of both lahars and jökulhlaups are presented to illustrate the various settings that occur in volcanic regions around the world. In the final section, hazard mitigation is briefly examined and summarized.

2. LAHAR AND JÖKULHLAUP CHARACTERISTICS

Flows of water and volcanic debris down the slopes of volcanoes and over the surrounding areas occur where sufficient amounts of water are available to form or initiate flows. Figure 56.2 shows examples of lahars and jökulhlaups and illustrates the range observed in style and magnitude. The amount of solid material in the flows is highly variable and dependent on the surface conditions at the volcano. The sources of water can be glacier ice or snow cover on the slopes of volcanoes. In some cases where large glaciers dominate the volcanic terrain, the flows can be principally made of meltwater; this is the case for many jökulhaups, especially those that drain subglacial reservoirs or lakes formed by sustained geothermal activity. Such subglacial lakes can be very large. During most of the twentieth century, the subglacial caldera lake in Grímsvötn (Iceland) drained every 5−10 years producing floods ranging in size from 1 to 5 km^3 (Björnsson, 2003). The Grímsvötn subglacial lake is sustained by geothermal activity that melts the glacier ice. Most subglacial reservoirs are smaller than Grímsvötn. Another potential source of what begins as principally water floods are naturally dammed crater or caldera lakes. Breaching of the natural dams occurs as the lake level rises to unstable levels resulting in swift floods. This has happened with the crater lake at Mt Ruapehu in New Zealand and was the cause of the disaster at Christmas of 1953, when a railway bridge

TABLE 56.1 Notable Lahars and Jökuhlaups and Their Triggering Mechanism

Mechanism	Volcano, Occurrences	Fatalities	Total Volume, km³	Maximum Discharge, m³/s
Contemporaneous with Eruption				
Crater lake expulsion	Galunggung, Indonesia, 1822	3600		
	Kelut, Indonesia, 1919	5110		
Crater lake breakout (wall failure)	Pelée, Martinique, 1902	25		
Subglacial eruption (jökulhlaups)	Öræfajökull, 1727	3	<1	<100,000
	Katla, Iceland, 1918	0	2–4	300,000
	Gjálp, Iceland, 1996	0	3.6	40–50,000
	Eyjafjallajökull, 2010	0	<0.1	3000
Surface melting of snow and ice	St Helens, USA, 1980	6		
	Nevado del Ruiz, Colombia, 1985	23,000	0.04	27,000–48,000
	Redoubt, 2009, April 4	0	0.2	60,000–150,000
Pyroclastic flows entering streams	Asama, Japan, 1783	550		
	Santa Maria, Guatemala, 1929			
	St Helens, USA, 1980	6		
Landslide	White Island, New Zealand, 1914	11		
Avalanches into lakes and streams	St Helens, USA, 1980	6		
Rain during eruption	Agung, Indonesia, 1963	200		
During Volcanic Quiescence				
Avalanche	White Island, New Zealand, 1914			
Crater lake failure	Ruapehu, New Zealand, 1953	151	0.002	850
Crater lake failure	Parker, Philippines, 1995	>60		
Heavy posteruptive rain	Irazu, Costa Rica, post-1964			
	Mayon, Philippines, post-1984			
	Pinatubo, Philippines, post-1991			
	Semeru, Indonesia, 1981			
	Merapi, Indonesia, post 2010			
Lake breakouts	Santa Maria, Guatemala, post-1902			
	Pinatubo, Philippines, post-1991			
Earthquake-induced avalanche on extinct volcano	Ontake, Japan, 1984			
Drainage of subglacial lake (jökulhlaup)	Grímsvötn, Iceland 1954		3	10,000
	Skaftá, Iceland, 1984		0.4	1500

References: Witham (2005), Rodolfo (1999), Major and Newhall (1989), Gudmundsson et al. (2008), Pierson et al. (1990), Waythomas et al. (2013), Björnsson (2003).

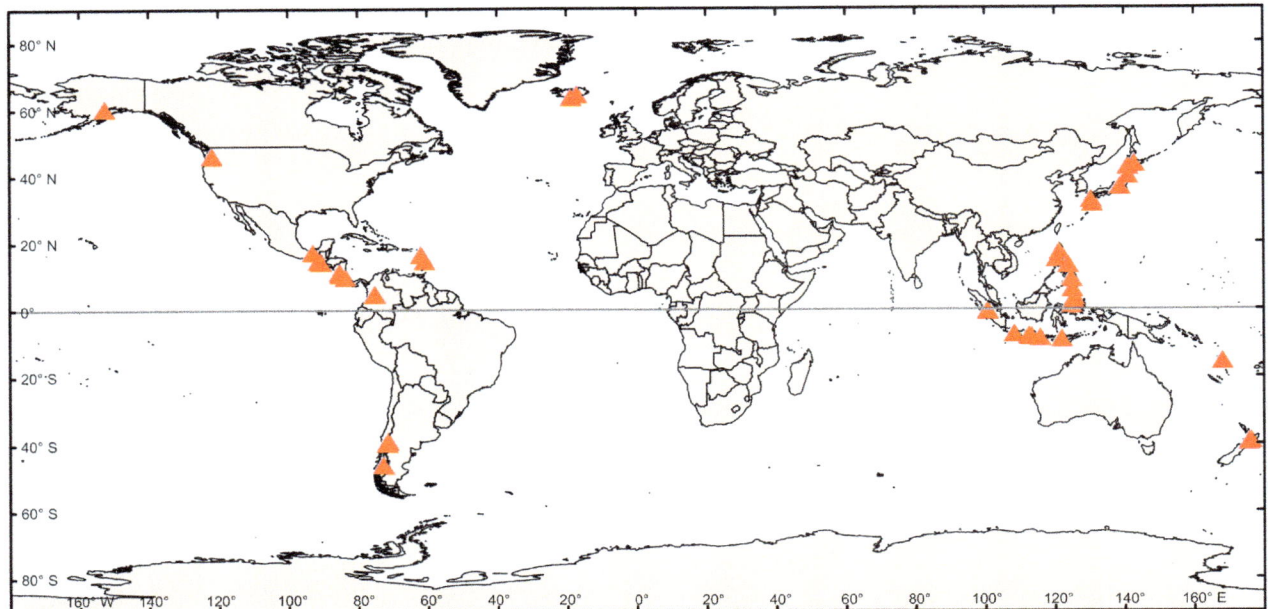

FIGURE 56.1   Distribution of volcanoes with reported lahars or jökulhlaups causing accidents or damage since AD 1900. *Based on Witham (2005) and other sources.*

collapsed and six carriages of the night train from Wellington to Auckland were engulfed in the lahar, leading to the death of 151 people. The crater lake of Gunung Kelut on Java Indonesia was expelled from the crater in the eruption of 1919, with the resulting lahar killing about 5000 people. However, most lahars do not require the breaching of a preexisting source of standing water. Precipitation during eruptions is sufficient to mobilize the tephra on the slopes, forming in some cases **debris flows** or **hyper-concentrated flows**. Explosive eruptions generating large amount of fine ash can form thick tephra layers of limited permeability, leading to water saturation and eventually liquefaction of the tephra.

2.1 Lahars

The term lahar comes from Indonesia and refers to **mud-flows** or debris flows down the slopes of volcanoes and down river valleys. All lahars have a component of water as part of the flow. A lahar may begin as a debris flow (solid mass >80% of total flow mass) down steep slopes but through sedimentation (debulking) become hyper-concentrated (mass of solid material between 40% and 80%, Beverage and Culbertson, 1964) and as sedimentation progresses on flat ground turn into a **water flows** (solids <40%). They can travel several tens of kilometers down canyons and river valleys where the topographic confine-ment and sufficient bed slope prevents significant debulk-ing. Lahars are often of short duration, but their discharge can be very high, orders of magnitude above normal river

flow. However, due to the typically short duration (minutes, tens of minutes) the total volume of solid material trans-ported by a single lahar is usually moderate. Bank erosion of lahars can be extremely high, e.g., at the bends in lahar channels carved into unconsolidated sediments. Peak discharge can be rapidly attenuated as the lahar reaches lowlands and can expand without constrictions. Devas-tating lahars that roar out of canyons onto flat ground may have all but disappeared some kilometers from the mouth of the canyon.

Due to their high density and momentum, the impact of lahars on structures is often devastating. For the same reason a lahar need not be deep to float cars or people attempting to ford what looks like a small, relatively innocent looking muddy stream.

Primary lahars often result from interaction of water with pyroclastic flow deposits. As a result the lahars may be very hot (Figure 56.2(B)); temperatures close to boiling have been recorded. This can increase the hazard as the high temperatures lead to scalding of people and animals caught in the lahar. The principal cause of death in lahars is, however, suffocation as people get caught up in the flow and are rapidly buried during rapid sedimentation.

2.2 Jökulhlaups

Probably the most common type of a jökulhlaup worldwide is where a glacier dams a tributary valley leading to the formation of a marginal lake. Usually such events have nothing to do with volcanoes. Subglacial reservoirs have

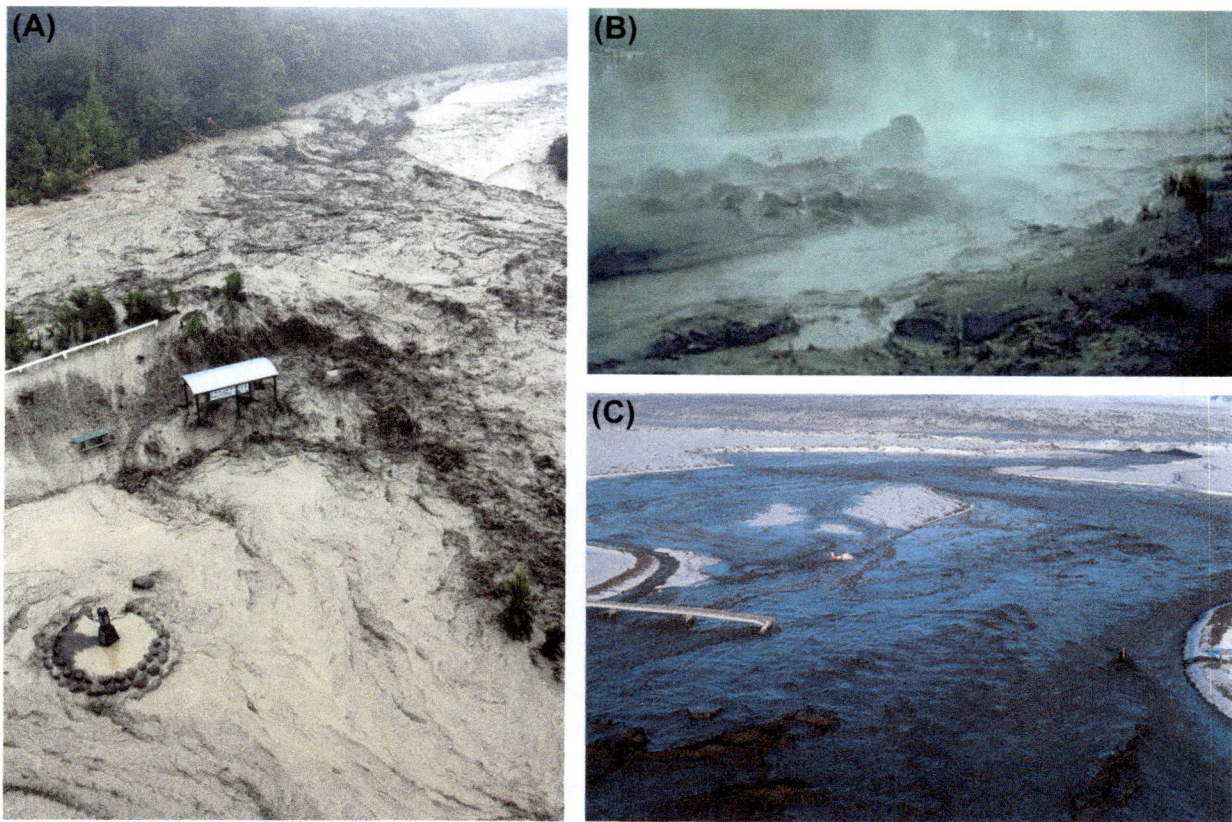

FIGURE 56.2 (A) The 2007 Ruapehu lahar, New Zealand, running over the Tangiwai rail disaster memorial site. In this disaster 151 people lost their lives as a consequence of a rail bridge being swept away by a lahar in 1953. (Photo by Geoff Mackley.) (B) A lahar carrying large blocks of rock down the flank of Mayon volcano in the Philippines on September 14, 1984. The high temperature of about 80 °C caused the fumes rising from the lahar. (Photo by Ernesto Corpuz, Philippines Institute of Volcanology and Seismology.) (C) The bridge over the river Gígjukvísl collapsing during a rapidly growing jökulhlaup on November 5, 1996. The flood carried ice blocks of 1000−2000 tons, broken from the edge of the glacier. The cause of the jökulhlaup was a 13 day long subglacial eruption that occurred about a month earlier, leading to accumulation of 3.6 km³ of meltwater within the Grímsvötn caldera before its release in the jökulhlaup.

also been identified under Antarctica and Greenland. These are nonvolcanic in origin and are not considered further here. Volcanogenic jökulhlaups are principally of two types: Firstly, the jökulhlaups coming from subglacial lakes formed over geothermal areas by sustained melting at the base of the ice, and secondly, the jökulhlaups that result from direct melting of ice in volcanic eruptions.

Geothermally sustained subglacial lakes are semi-permanent features of glaciers in volcanic regions. Such subglacial lakes are usually drained periodically. They are common in Iceland and often associated with calderas. The smallest jökulhlaups are minor events with duration of hours to a day, and a peak discharge of order 100 m³/s. Larger reservoirs exist, storing 0.1−3 km³ of water and draining periodically every few years resulting in jökulhlaups with peak discharge of order 1000−10,000 m³/s. These jökulhlaups are water flows but the amount of H_2S released can be dangerous to people close to where the waters emerge from the edge of the glacier. These events usually do not cause any major hazard.

Belonging to the second type of volcanogenic jökulhlaups are those directly associated with volcanic eruptions (Figure 56.2(C)). These floods usually have higher discharge and sediment load than the geothermal ones, as volcaniclastic debris is carried with the meltwater. Some of these jökulhlaups are hyperconcentrated and may therefore be called lahars, usually also carrying large blocks of ice broken from the glacier. Flow of jökulhlaups on the surface of glaciers can lead to large-scale entrainment of snow and firn, especially if the eruptions occur in winter. The resulting floods may therefore be slurries of water, slush, tephra, and ice blocks. As such a flood progresses after reaching gently sloping and unconfined terrain, debulking will occur leading to the dilution and transition to water flow. The melting region in a volcanic eruption within a glacier is often at the base of the ice, where the eruption melts the glacier from below. A different scenario is common in eruptions on tall stratovolcanoes with partial ice cover at the summit. Pyroclastic density currents produced by collapsing eruption column and by partial collapse of a lava dome can flow over glaciers, melting their surface rapidly from above through entrainment and mechanical mixing.

The total amount of water involved in jökulhlaups associated with eruptions is often much greater than for lahars that do not initiate through the expulsion of water from a region of rapid melting or a preexisting lake. This is because of the very large energy transfer that occurs when a large fraction of the thermal energy of a subglacial eruption is dissipated extremely fast by ice melting.

3. LAHAR AND JÖKULHLAUP HAZARDS

The source regions of jökulhlaups are usually more remote from densely inhabited areas than that of many lahars. However, large jökulhlaups that occur infrequently can have major consequences for rural areas and towns in some regions of the world. With increased tourism, however there is real prospect of a considerable number of people being located within danger zones during outbreak of eruptions and initiation of jökulhlaups in once remote regions. This is, for example, the case for now popular hiking routes in the vicinity of the Katla volcano in Iceland.

3.1 Nevado del Ruiz 1985—A Worst Case Scenario

The most serious volcanic disaster since the eruption of Mt Pélee in 1902 occurred when 23,000 people were killed in a few minutes on November 13, 1985, as the town of Armero in Colombia was destroyed by a lahar (Figure 56.3). The cause of the lahar was a relatively modest (volcanic explosivity index, VEI 3) short-lived eruption of the volcano Nevado del Ruiz (Pierson et al., 1990).

The eruption generated pyroclastic density currents that traveled across the 10-km²-summit ice cap (Figure 56.4(A)). Entrainment and mechanical mixing of snow and firn with the hot pyroclasts lead to partial melting and removal by diluted water flow and avalanches down the steep slopes of the volcano. Further entrainment of tephra and sediment in the narrow river valleys radiating out from the volcano lead to bulking of the floods, turning them into debris flows. The largest lahar and the one that engulfed Armero traveled down the Rio Azufrado valley at a velocity of 6–10 m/s. The maximum discharge varied along the flowpath. The highest discharge of 48,000 m³/s is considered to have occurred about 10 km from the crater. As the lahar reached the mouth of the narrow river valley where it hit the town of Armero, 74-km-downstream from the summit, the peak discharge had dropped to 27,000 m³/s. The eruption occurred at 21:08 and the lahar reached Armero at 23:35. It arrived as several pulses. The first one was the largest, engulfed most of the town and caused most of the fatalities. It lasted for 10–20 min. The next pulse arrived at about 23:50 and lasted 30–35 min. The total volume of the Armero lahar was 40 × 10⁶ m³, whereof only ~15% was meltwater from the summit ice cap; the bulk of the volume was solid material (Pierson et al., 1990). Lahars also came

FIGURE 56.3 The town of Armero on December 9, 1985, about a month after the lahar resulting in the loss of life of 23000 people. The flood marks in the river valley above are apparent and in the foreground the thick sediments covering what was once the center of the town of Armero. *Photo by J. Janda, USGS.*

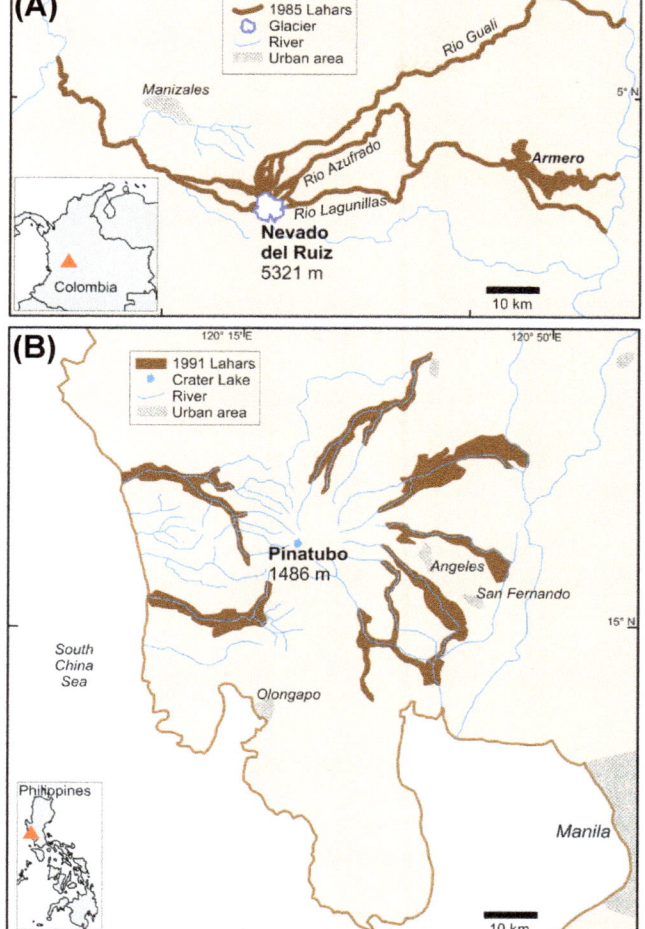

FIGURE 56.4 (A) Nevado del Ruiz and the paths of lahars on the evening of November 13, 1985 (based on Pierson et al. (1990)). (B) Pinatubo volcano and the areas affected by lahars during the eruption in 1991 and in the years that followed (based on Newhall et al. (1997)).

and ice at the summit ice cap showed that in eruptions generating pyroclastic density currents the surface area can be more important in determining melting potential than ice thickness. Thirdly, the knowledge that Armero was highly vulnerable to lahars had been established by Colombian geologists and it was known that it was built on the deposit of an even larger lahar that occurred in 1595. However, this information had not been registered by officials or the local population. As a result, no effective response or evacuation plan existed to get the inhabitants out of harm's way in the event of an eruption. Fourthly, no lahar detection system existed for the areas around Nevado del Ruiz. This may have been of secondary importance, since swift evacuation orders simply based on the occurrence of a summit eruption would have averted most if not all loss of life at Armero. A time lag of 2.5 h between the onset of eruption and the arrival of a lahar should be quite sufficient for decision-making and evacuation out of possible lahar paths if a predefined response plan exists.

3.2 Eyjafjallajökull 2010—Jökulhlaups and Lahars

The eruption of Eyjafjallajökull, Iceland, in 2010, is known for the unprecedented disruption it caused to aviation in Europe and the Northern Atlantic with cascading aviation effects worldwide. Locally, however, the principal concern in the first few days was the jökulhlaup hazard. Eyjafjallajökull is located on the south coast of Iceland. It rises about 1600 m above the surrounding lowlands. The lowlands are alluvial plains formed by sedimentation of glacial rivers, and to a considerable extent by jökulhlaups during the end of the last glaciation and the Holocene. Large parts of the lowlands are agricultural areas with the nearest farm on the south side only 7 km from the summit. The upper part of Eyjafjallajökull is covered by an ice cap, about 200-m-thick in the 2.5-km-wide summit caldera but thinner on the slopes.

The eruption of 2010 (14 April—22 May) occurred in the summit caldera, melting **ice cauldrons** in the glacier with the meltwater draining mostly toward north, down the steep 4-km-long Gígjökull outlet glacier. Ice melting was not particularly intense but amounted to an average of 300—500 m³/s in the first 2 days (Magnússon et al., 2012). This meltwater was released in repeated jökulhlaups that inundated the 2- to 3-km-wide and uninhabited valley of the Markarfljót River (Figure 56.5). By the time the jökulhlaups reached the bridge on the main road over Markarfljót 20-km-downstream from the outlet glacier their discharge was 2000—3000 m³/s. In the hours before the arrival of the first jökulhlaup at the bridge on the main road along the south coast, to prevent its destruction bulldozers were used to cut the road on both sides of the bridge, allowing parts of the floods to bypass it. A system of levees,

down other rivers but they were smaller and towns or villages were not as exposed as Armero. About three-fourth of the inhabitants of Armero drowned or suffocated in the lahar, which broke down concrete buildings and destroyed everything in its path. The first wave of the lahar carried some large boulders, but most of the sediment was fine grained (a mean particle diameter <1 mm, Pierson et al., 1990). Most of the solid material had sedimented from the flow 15-km-downstream from Armero. The volumetrically minor water component turned into a relatively harmless water flood.

Several lessons have been drawn from this devastating event. Firstly, it showed that modest eruptions can lead to major lahars in ice-covered volcanoes and that lahars can bulk up and grow for tens of kilometers from the source, provided that the flow is confined to a narrow valley, that prevents debulking. Secondly, the mode of melting of snow

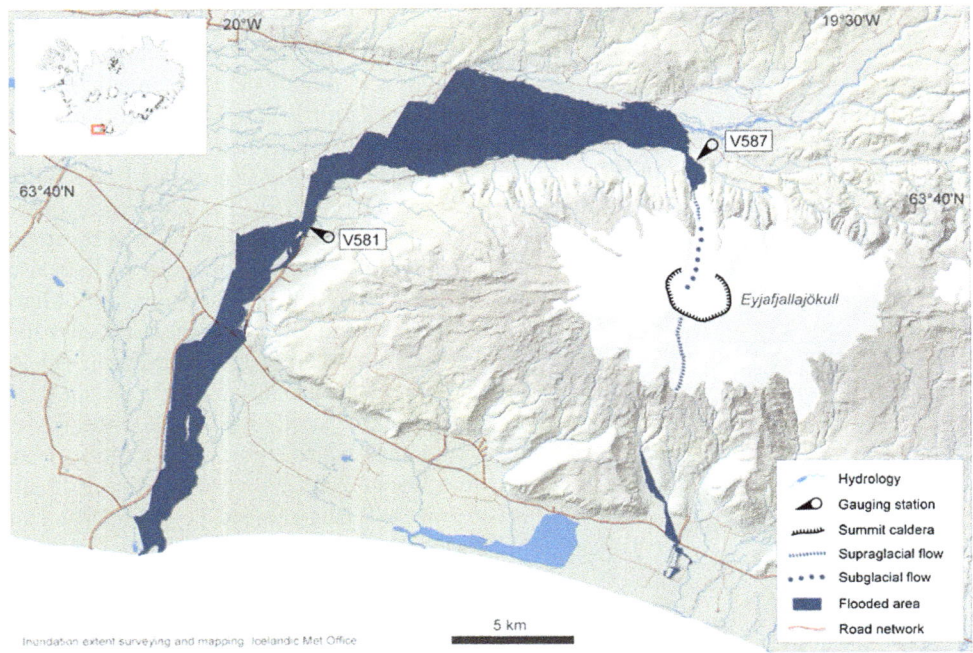

FIGURE 56.5 The jökulhlaup paths in the eruption of Eyjafjallajökull, Iceland, on April 14–15, 2010. *After Karlsdóttir et al. (2012).*

built to contain normal meteorological floods within the wide river bed, mostly withstood the jökulhlaups. This prevented the floodwaters from destroying agricultural lands on both sides of the river. Large morphological changes occurred where the hyperconentrated jökulhlaups reached the lowlands. A glacial lagoon, about 0.8-km-long, 0.4-km-wide, and 30-m-deep was filled by volcanic debris in the first 24 h. After the lagoon had been filled, the last of the main jökulhlaups occurred, in the late afternoon of April 15. The mixture of water, tephra, and ice blocks had the characteristics of a hyperconcentrated flow for the first few kilometers after it issued from the glacier (Figure 56.6). Sedimentation from the flow within the wide gently sloping river valley led to a transition to normal water flow behavior within 5 km from the former lagoon. A small jökulhlaup, causing little damage, also occurred on the south side of the volcano on the first day of the eruption (Figure 56.5).

A few years prior to the eruption in 2010, a comprehensive assessment of the jökulhlaup hazard had taken place for Eyjafjallajökull and the western part of the Katla caldera. This assessment involved mapping of volcanic fissures on the slopes of both volcanoes, mapping of flood marks from earlier, mostly prehistoric (1200–7000 years BP) jökulhlaups, estimates of the melting potential of eruptions in both volcanoes, mostly based on ice thickness and experience of earlier eruptions, and the **hydraulic routing** of jökulhlaup paths using reasonable worst case scenarios. Importantly, the hazard assessment was followed by extensive response planning, principally aimed at

developing realistic schemes for evacuation of the large areas in danger of inundation by volcanogenic floods in the event of eruptions in either of the two volcanoes. A crucial element in the planning was consultation with local people, involving a large number of meetings between the local population, officials, and scientists where the hazard and the plans were introduced. A part of the consultation was input from the locals to weed out errors and unworkable plans and replace them with other, more realistic procedures. The plan was tested in 2006, 4 years before the eruption of Eyjafjallajökull, with an exercise where most of the

FIGURE 56.6 The hyperconcentrated jökulhlaup of April 15, 2010 from Eyjafjallajökull overtopping a levee, about 2 km from where the flood cascaded down to the lowlands. *Photo by Þórdís Högnadóttir.*

inhabitants in the vulnerable areas took part by evacuating their homes for a part of a day. These preparatory measures were put to the test in the Eyjafjallajökull eruption. Due to jökulhlaup danger a large area with several hundred inhabitants was temporarily evacuated three times during the eruption. Luckily, very minor damage occurred to farmland or people's dwellings in these jökulhlaups. The response to the eruption is considered to have been a success at a local level. An important part in this success was the trust built up between local people, officials, and scientists during the years before the eruption, through the hazard assessment and the development of the response plans.

The hazards were not over by the end of the eruption (Figure 56.7). Large bodies of tephra had accumulated on the glacier on the southern slopes of the volcano during April and May. Until late May, mostly dry northwesterly winds had prevailed. However, during the last days of the eruption the winds turned southerly and on May 19, in heavy rain these masses were mobilized for the first time, flowing down the river gullies as lahars. This situation occurred repeatedly over the next 2 years, requiring work by bulldozers for extended periods to clear the main river channels and prevent the rivers from overflowing, destroying surrounding farmland, and closing the road along the south coast, the main transport route between western and eastern Iceland. By 2012, 2 years after the eruption, the situation had stabilized, as most tephra that could be mobilized had been removed by the lahars.

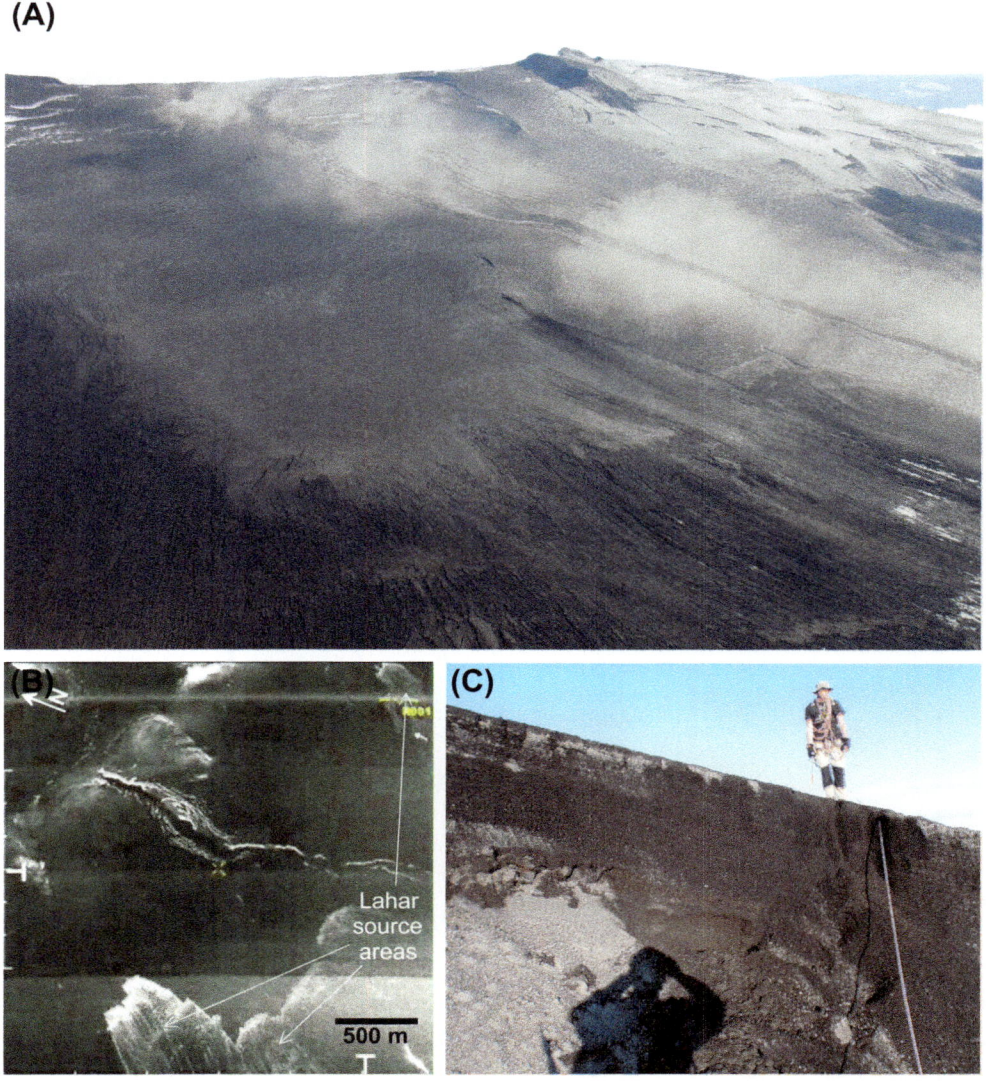

FIGURE 56.7 The source area of lahars on the southern slopes of Eyjafjallajökull after the eruption in 2010. (A) The southern slopes on June 11, 2010, with the lahar source areas in the lower part of the picture. (B) A synthetic aperture radar (SAR) image obtained from aircraft of the Icelandic Coast Guard on May 19, 2010. (C) Erosional escarpment in the unconsolidated tephra after removal by lahars in May 2010. *Photo by Björn Oddsson.*

3.3 Katla 1918—The Largest Jökulhlaup

Preparation of authorities and local communities for eruptions in Katla on the south coast of Iceland has been the single most important effort in volcanic hazard mitigation in Iceland in recent years. This is principally because of the potential of Katla to generate major jökulhlaups as testified by its recent eruption history. The eruptions occur within the Katla caldera which is 100 km² in area and filled with 400–700 m thick ice. Eruptions in Katla have occurred on average about once every 50 years over the last millennium. Large eruptions in Katla probably generate the largest volcanogenic jökulhlaups that occur presently on Earth (Figure 56.8). The last eruption of Katla in 1918 produced large volumes of basaltic tephra. The jökulhlaup that occurred in the first day of the eruption is considered to have had a peak discharge of about 300,000 m³/s, carrying with it hundreds of millions of cubic meters of tephra (Tómasson, 1996). Systematic earthquake monitoring in Iceland had not started in 1918 and no instrumental records exist of the eruption. Photographs, eye witness accounts, and the contemporaneous written records nevertheless give a clear picture of the event. The first warning of an imminent eruption on October 12 was an earthquake at around 1 PM, followed by repeated tremors. Roughly 2 h later an eruption column was seen rising over the glacier filling the caldera. At a similar time, a jökulhlaup was observed, flowing from the glacier onto Mýrdalssandur, a jökulhlaup-generated outwash plain to the east and southeast of the Katla volcano. This flood grew in magnitude and inundated an area of approximately 500 km² in the course of 2–3 h. This flood gradually receded and had fully stopped at daybreak the following morning. After the jökulhlaup a new peninsula, 2- to 3-km-long had formed and 5–10 m of volcaniclastic sediment had been deposited over large areas the sandur outwash plain. It was a coincidence that shepherds were not passing the sandur plain with flocks of sheep on the day of the jökulhlaup. Several shepherds had a narrow escape on the northern part of the sandur plain as they had to flee on their horses from the advancing jökulhlaup, leaving the sheep they were hoarding behind. A large number of livestock perished, mainly sheep and horses.

The fact that no fatalities resulted from the Katla 1918 jökulhlaup was mainly because the floodplain is an uninhabitable sand desert. The road across this desert is now driven daily by hundreds of people in winter and thousands in the summer. In response to a similar event in the near future detailed evacuation plans have been made, prepared in collaboration with local people. The sizes of the evacuation areas were based on expected inundation of the largest jökulhlaups from hydraulic routing and history of previous floods. An increasing concern is the growing number of tourists in the area, since they sometimes camp in the sandur outwash plain, a practice hardly occurring with Icelanders due to the common knowledge of the swiftness and short warning times of Katla eruptions and jökulhlaups. Reaching out to tourists with education on hazards is regarded as an important element in the effort of preventing fatalities in a future Katla eruption.

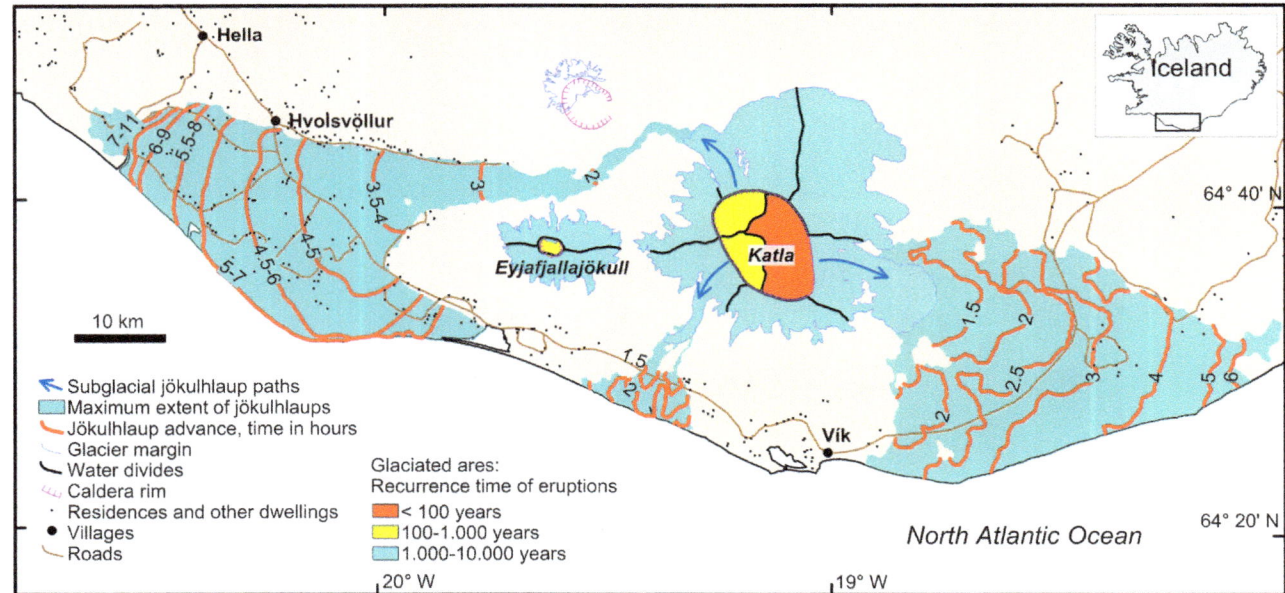

FIGURE 56.8 Katla and Eyjafjallajökull volcanoes on the south coast of Iceland. The color of the glacier covered areas indicates the recurrence times of eruptions. The expected maximum extent of jökulhlaups is based on flood modeling using the magnitude of the Katla 1918 as a worst case scenario. The extent of floods along all three expected pathways out of the Katla caldera are shown. The jökulhlaup advance time refers to the time in hours since the onset of a subglacial eruption. *Modified from Gudmundsson et al. (2008).*

3.4 Posteruption Lahars, Pinatubo 1991 and Merapi 2010

The eruption of Pinatubo in the Philippines in 1991, reached a climax on June 15. This VEI 6 eruption is considered the second largest in the twentieth century, producing a bulk volume of tephra of about 10 km³. In the weeks preceding the climactic eruption in June 15, large areas had been evacuated, since volcanologists feared that the escalating precursors and initial eruptive activity would lead to a major Plinian eruption. These evacuations turned out to be well founded and saved thousands of lives. Major lahars occurred, as heavy precipitation caused by the typhoon Yunya struck the Philippines simultaneously with the climactic phase of the eruption. The pyroclastic density currents and the tephra fallout lead to the deposition of thick layers of pyroclasts.

After the eruption it was clear that a major threat was posed by the vast tephra deposits on the slopes of the volcano, and that during rainy seasons these deposits would be mobilized as lahars. Warning systems were put in place where watchmen on the slopes sent messages to places downslope of the advancing lahars. This measure saved hundreds of lives. The government of the Philippines decided not to abandon towns susceptible to lahar hazard since it would be safer and more economical to protect the towns with dams and levees containing and diverting lahars. These measures were largely a failure (Rodolfo, 1999) as the difficulties and costs of effective measures were greatly underestimated. The scale of the problem can be seen from the fact that by 1997, the amount of material carried to the lowlands since the eruption by lahars was 3 km³ (Newhall et al., 1997). Levees and dams were often constructed from the only readily available material, tephra from the eruption itself. These structures turned out to be totally inadequate and were repeatedly washed away by the lahars. Several villages and towns were destroyed in the years that followed, leading to loss of life and destruction to property. The town of Bacolor was hit by repeated lahars (Figure 56.9), with the lahar of October 1, 1995 being the worst, when large parts of the town were buried in several meters of sediment. The death toll of this event is uncertain but over 100 people lost their lives (Crittenden and Rodolfo, 2002) (Figure 56.9).

The VEI 4 eruption of Merapi in Java, Indonesia, in late 2010, was the largest in the volcano since the nineteenth century. The lower slopes of Merapi have a high population density and the Indonesian authorities responded to the eruption with the evacuation of about 400,000 people. Despite this major effort several people were killed by pyroclastic density currents. The amount of pyroclastic material deposited on the slopes was far greater than in previous recent eruptions and rain-triggered lahars in the period since the eruption have been numerous. A large effort has been put into clearing lahar pathways, with thousands of people taking part in the clearing work (Bélizal et al., 2013). **Check dams (sabo dams)** help in containing lahars by forming temporary breaks in the flow, allowing sedimentation from the lahars above the dams. However, for these to be effective, the sediments have to be

FIGURE 56.9   The town of Bacolor suffered frequent lahars after the Pinatubo eruption and was badly hit by the October 1, 1995 lahar. *Published with permission from Tips Images.*

FIGURE 56.10   A lahar from Mt Redoubt's eruption on April 4, 2009, advancing down the Drift River and towards the Drift River Oil Terminal captured by the Advanced Land Imager on board the EOS-1 satellite (NASA's Earth Observatory).

cleared before the next lahar arrives, in order for the trapping nature of the dams to be preserved. An important part of the success of this effort has been a community-based early warning system. Volunteers have been trained to recognize the meteorological conditions that trigger lahars and warning signals are sent quickly to vulnerable areas in the community, allowing evacuation. This early warning system has contributed to limiting the losses; three fatalities were reported for the rainy season of 2010−2011 (Bélizal et al., 2013).

It is not simple to make direct comparisons between Pinatubo and Merapi and the responses to these two events. The scale is not the same; the Pinatubo eruption was orders of magnitude larger and the problems correspondingly greater. However, it seems that progress in posteruption hazard management was considerable in the two decades that passed between Pinatubo and Merapi.

3.5 Redoubt 2009

The Redoubt stratovolcano by the Cook Inlet in Alaska has had four moderate eruptions since 1900, the latest occurring in 2009. Lahars are generated in these eruptions due to partial melting of the glacier covering the upper parts of the volcano. The eruptions have been vulcanian in character, melting the glacier through minor pyroclastic density currents and hot debris avalanches originating from the partial collapse of the dome. Such events lead to debris flows down the Drift glacier with melting of ice occurring through mixing of hot pyroclasts with snow and ice. When this happens, large lahars or hyperconcentrated jökulhlaups flow down the glacier and out toward the ocean down the Drift River valley. An oil storage and transfer station is located by the Cook Inlet, 50 km from downstream from

Redoubt's crater and in the pathway of lahars. The largest lahars during the eruption of 2009 reached the oil terminal (Figure 56.10) with the largest event having peak discharge estimated in the range of 100,000 m^3/s in the upper part of the Drift valley (Waythomas et al., 2013).

Lives were not in danger at Redoubt but an important facility was threatened. Through seismic monitoring and the use of Webcams in the upper reaches of the flood path, the progress of lahars could be monitored and their duration estimated. The largest lahars could be detected on the seismic network for almost 2 h, as they propagated down the volcano and the Drift River valley.

4. HAZARD MITIGATION

The examples given in the previous section highlight some key issues that need to be addressed if measures to significantly reduce fatalities and the loss of property due to lahars and jökulhlaups are to be successful.

4.1 Knowledge of the Geology and Eruptive History

A proper recognition of the existence of a lahar or jökulhlaup hazard is needed, both within the communities at risk and for officials at both local and national level. In regions where eruptions do not happen often this recognition may be lacking, since neither written or oral records exist of previous eruptions in a long dormant volcano. This illustrates the importance of basic geological mapping, since it should reveal a history of previous events. The response plans in place for Katla in Iceland are based on the past history of jökulhlaups. This history is partly based on written sources covering the jökulhlaups toward east and

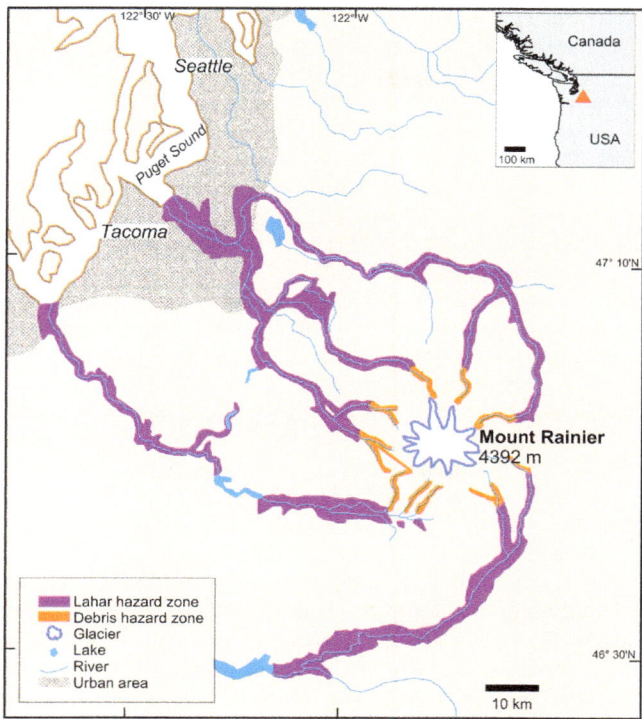

FIGURE 56.11 Mt Rainier in Washington, USA is located close to the Seattle—Tacoma urban area, the home of almost 5 million people. The upper parts of the volcano are covered by large glaciers and eruptions can melt large volumes of ice generating lahars. Landslide generated lahars are also a threat, e.g., due to intrusion of magma leading to destabilization of the flanks. *Based on Iverson et al. (1998).*

includes the last several hundred years. However, it is only through geological mapping of deposits and old river channels that the possibility of floods down paths toward west and south become apparent. For Mt Rainier (Figure 56.11), geological mapping of the lowlands to the west of the volcano revealed that major prehistoric mudflows had inundated large areas forming deposits that are several tens of meters thick; hydrological routing of the path and inundation zones of possible lahars is shown in Figure 56.11 (Iverson et al., 1998).

4.2 Monitoring

In regions susceptible to flooding, a monitoring system should be in place. Such a system can provide early warning of imminent volcanic activity. In some regions the onset of an explosive eruption with associated pyroclastic density currents will lead to lahars. In ice-covered volcanoes such an eruption can be taken as a sure sign of jökulhlaups/lahars occurring within hours in one or several lowland areas around the volcano. As for other volcanic hazards seismic monitoring is of major importance. Automated river gauges, Webcameras and seismic networks can then track the

advance of a lahar or jökulhlaup. Progress is being made toward recognizing advancing lahars in automated systemic networks and this holds considerable potential. However, there is a long way to go before such systems adequately cover dangerous volcanoes, especially in developing countries, many of which are found in tropical regions of high population density. Simple and robust systems may rely on lahar watchmen on the slopes of volcanoes, relaying through radio information of advancing lahars.

4.3 Conditions for Jökulhlaup or Lahar Initiation

A necessary condition for any warning system to be effective is to know the precursors and events likely to spark of lahars or jökulhlaups. This includes knowledge of the level of precipitation required to initiate secondary lahars. For jökulhlaups, knowledge of the size of known subglacial water reservoirs helps in predicting the likely maximum discharge. Another aspect of this is knowledge of the approximate size of the expected inundation area, likely lahar/flood depths, velocities, and travel times. For regions where only prehistoric geological information exists, modeling of lahar or jökulhlaup propagation down expected flood channels is sometimes used. Avalanche models have been used with some success in Iceland for such modeling of floods flowing down steep slopes. On flat ground, where large volumes of especially floodwater may spread out, hydraulic routing based on solving the 2-D Saint Venant's equations is required. The inundation zones shown for possible Katla jökulhlaups (Figure 56.8) are obtained through such modeling.

4.4 Response Plans

A proper and realistic response plan is required and it needs to be used by officials for evacuation and other measures meant to prevent fatalities, reduce the risk of injury and loss of property. A key element in successful execution of any such plan is that local people are involved in the planning, that knowledge of the plan is widespread and it is accepted by the communities (Bird and Gísladóttir, 2012).

4.5 Vulnerability Reduction Measures

Engineering projects such as construction of deflection levees for diversion of mostly water floods and checking dams (sabo dams) for smaller lahars are proving to be a part of reducing community vulnerability (Figure 56.12). Another important and relatively simple measure is the construction and maintenance of shelters on high ground, within easy reach of farms and villages. These can be of major importance in saving lives.

FIGURE 56.12 A network of dams and levees built to contain posteruption lahars on Unzen volcano, Japan, after the 1991–1995 eruption. The flowpath is periodically dammed with so-called sabo dams, which provide a sediment trap behind them. These dams are effective at stopping smaller lahars and slowing down and reducing larger ones. *Photo: SABO.*

4.6 Land Development Regulations

The final and probably the most difficult step, and one that few countries have as yet achieved fully, is proper incorporation of lahar and other volcanic hazards into planning and development regulations, especially for residential areas near volcanoes. With time, implementation of such measures should reduce greatly the vulnerability of the affected societies. However, the sheer scale of VEI 6 and even larger eruptions is such that a community hit by an event of that magnitude will always be seriously affected. Nevertheless, proper planning regulations coupled with strong monitoring systems and sound, workable response plans are key stages in increasing the resilience to the lahar hazard.

FURTHER READING

Bélizal, E., Lavigne, F., Hadmoko, D.S., Degeai, J.P., Dipayana, G.A., Mutaqin, B.W., Marfai, M.A., Coquet, M., Mauff, B.L., Robin, A.K.,

Vidal, C., Cholik, N., Aisyah, N., 2013. Rain-triggered lahars following the 2010 eruption of Merapi volcano, Indonesia: a major risk. J. Volcanol. Geotherm. Res. 261, 330–347.

Beverage, J.P., Culbertson, J.K., 1964. Hyperconcentrations of suspended sediment. J. Hydraul. Div., Am. Soc. Civ. Eng. 90, 117–128.

Bird, D.K., Gísladóttir, G., 2012. Residents' attitudes and behaviour before and after the 2010 Eyjafjallajokull eruptions-a case study from southern Iceland. Bull. Volc 74, 1263–1279.

Björnsson, H., 2003. Subglacial lakes and jokulhlaups in Iceland. Global Planet. Change 35, 255–271.

Crittenden, K.S., Rodolfo, K.S., 2002. Bacolor town and Pinatubo Volcano, Philippines: coping with recurrent lahar disaster. In: Torrence, R., Grattan, J. (Eds.), Natural Disasters and Cultural Change. Routledge, pp. 43–65.

Gudmundsson, M.T., Larsen, G., Höskuldsson, Á., Gylfason, Á.G., 2008. Volcanic hazards in Iceland. Jökull 58, 251–268.

Iverson, R.M., Schilling, S.P., Vallance, J.W., 1998. Objective delineation of lahar inundation hazard zones. GSA Bull. 110, 972–984.

Karlsdóttir, S., Gylfason, Á.G., Höskuldsson, Á.H., Brandsdóttir, B., Ilyinskaya, E., Gudmundsson, M.T., Högnadóttir, þ.H., þorkelsson, B. (Eds.), 2012. The 2010 Eyjafjallajökull Eruption, Iceland, IMO, IES and NCIP-DCEPM Report to ICAO, p. 209.

Magnússon, E., Gudmundsson, M.T., Sigurdsson, G., Roberts, M.J., Höskuldsson, F., Oddsson, B., 2012. Ice-volcano interactions during the 2010 Eyjafjallajökull eruption, as revealed by airborne radar. J. Geophys. Res. 117, B07405.

Major, J.J., Newhall, C.G., 1989. Snow and ice perturbation during historical volcanic eruptions and the formation of lahars and floods. Bull. Volcanol. 52, 1–27.

Newhall, C., Stauffer, P.H., Hendley, J.W., 1997. Lahars of Mount Pinatubo. U.S. Geological Survey, Fact Sheet, Philippines, 114–97.

Pierson, T.C., Janda, R.J., Thouret, J.C., Borrero, C.A., 1990. Perturbation and melting of snow and ice by the 13 November 1985 eruption of Nevado-del-Ruiz, Colombia, and consequent mobilization, flow and deposition of lahars. J. Volcanol. Geotherm. Res. 41, 17–66.

Rodolfo, K.S., 1999. The hazard from lahars and jökulhlaups. In: Sigurdsson, H., Houghton, B., McNutt, S., Rymer, H., Stix (Eds.), Encyclopedia of Volcanoes. Academic Press, pp. 973–995.

Tómasson, H., 1996. The Jökulhlaup from Katla in 1918. Ann. Glaciol. 22, 249–254.

Waythomas, C.F., Pierson, T.C., Major, J.J., Scott, W.E., 2013. Voluminous ice-rich and water rich lahars generated during the 2009 eruption of Redoubt Volcano, Alaska. J. Volcanol. Geotherm. Res. 259, 389–413.

Witham, C.S., 2005. Volcanic disasters and incidents: a new database. J. Volcanol. Geotherm. Res. 148, 191–233.

Chapter 57

Hazards of Volcanic Gases

Glyn Williams-Jones
Department of Earth Sciences, Simon Fraser University, Burnaby, BC, Canada

Hazel Rymer
Faculty of Science, The Open University, Walton Hall, Milton Keynes, UK

Chapter Outline

GLOSSARY

aerosol A colloidal dispersion of liquid particles in a gas, e.g., SO_2, which will react with the OH radical in the atmosphere to form tiny droplets of sulfuric acid (H_2SO_4).

hazard In a volcanic context, hazard refers to the phenomena produced by a volcanic event and is directly related to *Risk*, where *Risk = Hazard × Vulnerability*, with *vulnerability* referring to the consequences for population and infrastructure.

PEL The recommended permissible exposure limit to a given chemical compound above which health risks may occur. The exposure limit is generally averaged over an 8-h day, 40-h week. It is measured in parts of compound (e.g., CO_2) per million parts of air (ppm).

Gas solubility The maximum amount of a gas that can be dissolved in a given amount of water at 20 °C; measured in g/L.

δ_{gas} The vapor density of a gas relative to air (density = 1); measured in g/L.

1. INTRODUCTION

Gases are the invisible yet often continuous products of volcanic activity. Even volcanoes in a state of quiescence, not actually erupting or showing signs of unrest through seismic activity, are able to degas continuously. Eruptions can produce lethal quantities of toxic gases, but long-term exposure to a lower dose also can pose a significant **hazard**. Although volcanic gases are only directly responsible for 1−4% of volcano-related deaths, they are nevertheless hazardous and responsible for deaths every year. They have an important effect on the regional and global environment and may contribute greenhouse gases to the atmosphere. Indirectly, through the destruction of crops, volcanic gas emissions have resulted in starvation and disease (40% of volcano-related deaths between 1600 and 1982).

The composition of volcanic gases depends on the type of volcano and its eruptive state. However, the most common volcanic gases in order of abundance are water (H_2O, 30−90 mol%), carbon dioxide (CO_2, 5−40 mol%), sulfur dioxide (SO_2, 5−50 mol %), hydrogen (H_2, <2 mol%), hydrogen sulfide (H_2S, <2 mol%), and carbon monoxide (CO, <0.5 mol%). Some of these, when emitted from active vents (Figure 57.1), react in the atmosphere or volcanic plume to form **aerosol**s, the most important being hydrochloric acid (HCl), hydrofluoric acid (HF), and sulfuric acid (H_2SO_4).

2. TOXICITY OF VOLCANIC GAS SPECIES

It is difficult to determine accurately the contribution of gases to volcano-related deaths, since much of the data reflect deaths during eruptive periods, whereas the majority of gas-related deaths occurred during noneruptive periods. The long-term health effects of volcanic gases are poorly understood; they may be responsible for or accelerate epidemic diseases because of their irritant and depressing

The Encyclopedia of Volcanoes. http://dx.doi.org/10.1016/B978-0-12-385938-9.00057-2

FIGURE 57.1 Degassing vent of Santiago crater at Masaya volcano, Nicaragua, 2006.

effects, which reduce the resistance of ocular, respiratory, and digestive systems to microbial attack. From a health perspective, the most important volcanic gases and aerosols are CO_2, SO_2, Rn, H_2S, HCl, HF, and H_2SO_4 (Table 57.1). Exposure to these has been the cause of the majority of volcanic gas-related fatalities.

TABLE 57.1 Toxicology of Volcanic Gases and Aerosols

Carbon dioxide (CO₂)

Characteristics	Colorless odorless gas. Irritation of eyes, nose, and throat only at high concentrations. Vapor density relative to air (δ_{gas}) = 1.52 g/L. **Gas solubility** in water at 20 °C = 0.14 g/L. Permissible exposure limit (**PEL**) averaged over 8 h = 5000 ppm in air; 9000 mg/m³ of air.
Effects of overexposure	
Short-term	A simple asphyxiant, symptoms appear only when such high concentrations are reached that there is insufficient oxygen to support life. Inhalation may cause rapid breathing and increase heart rate (at >7.5%), headache, sweating, dizziness, shortness of breath, muscular weakness, mental depression, drowsiness, and ringing in the ears. Concentrations at >11% result in unconsciousness in 1 minute or less. Convulsions may occur at concentrations of >25%. Rapid recovery occurs on removal from exposure.
Long-term	Prolonged exposure to concentrations of >10% may result in unconsciousness.

TABLE 57.1 Toxicology of Volcanic Gases and Aerosols—cont'd

Sulfur dioxide (SO₂)

Characteristics	Colorless gas or liquid (<−10 °C) with characteristic pungent odor. Perceptible odor at 0.3−1.0 ppm and easily noticeable at 3 ppm. δ_{gas} = 2.26 g/L. Gas solubility = 10 g/L. PEL = 5 ppm in air; 13 mg/m³ of air.
Effects of overexposure	
Short-term	Inflammation and irritation of the eyes and respiratory tract resulting in burning of the eyes, coughing, and difficulty in breathing. Approximately 90% of inhaled SO_2 is absorbed in the upper respiratory tract, where it forms sulfurous acid which then oxidizes to form sulfuric acid. Concentrations of 6−12 ppm cause immediate irritation of nose and throat. Exposure to >20 ppm causes irritation of the eyes, while concentrations of 10,000 ppm irritate moist skin within minutes.
Long-term	Prolonged exposure to low concentrations may be dangerous for persons with preexisting cardiopulmonary diseases.

Hydrogen sulfide (H₂S)

Characteristics	Colorless, flammable gas with offensive odor (rotten eggs). Characteristic odor perceptible at 0.77 ppm and easily noticeable at 4.6 ppm. δ_{gas} = 1.19 g/L. Gas solubility = 2.9 g/L. PEL (averaged over 10 min) = 20 ppm in air; 28 mg/m³ of air.
Effects of overexposure	
Short-term	Inhalation of 20−150 ppm may cause eye irritation, while slightly higher concentrations cause irritation of upper respiratory tract. In low concentrations, exposure may result in headache, fatigue, dizziness, excitement, staggering gait, diarrhea, followed sometimes by bronchitis and bronchopneumonia. In small amounts the gas acts as depressant and as stimulant in larger amounts. Very large amounts result in paralysis of the respiratory center and death; exposure to 1000−2000 ppm may cause coma after a single breath.
Long-term	Prolonged exposure to concentrations as low as 50 ppm may cause pharyngitis and bronchitis, while concentrations >250 ppm may result in pulmonary edema.

TABLE 57.1 Toxicology of Volcanic Gases and Aerosols—cont'd

Radon (Rn)

Characteristics	Colorless, odorless, tasteless, radioactive gas, formed from the radioactive decay of uranium. $\delta_{gas} = 9.73$ g/L. Gas solubility = 51 g/L. PEL = 200 Bq/m3.

Effects of overexposure

Short-term	There is no information on the acute noncancerous effects of radionuclides in humans; however, animal studies have reported inflammation in the nasal passages and kidney damage from acute inhalation exposure to uranium.
Long-term	Chronic exposure by inhalation has been linked to respiratory disorders, such as lung disease and lung cancer, in humans. Smokers exposed to radon are at ~10–20 times greater risk for lung cancer than nonsmokers.

Hydrochloric acid (HCl)

Characteristics	Colorless gas or colorless fuming liquid with an irritating pungent odor. Detectable odor by most people between 1 and 5 ppm. $\delta_{gas} = 1.27$ g/L. Gas solubility = 62 g/L. PEL = 5 ppm in air; 7 mg/m^3 of air.

Effects of overexposure

Short-term	Irritation of the mucous membranes of the eyes and respiratory tract with burning, choking, and coughing. At concentrations >35 ppm, there is irritation of the throat after only short exposure. Severe breathing difficulties may occur along with skin inflammation or burns. Concentrations >100 ppm will result in pulmonary edema and often laryngeal spasm.
Long-term	Repeated or prolonged exposure may lead to erosion of the teeth and skin rash.

Hydrofluoric acid (HF)

Characteristics	Clear, colorless, fuming corrosive liquid or gas with strong irritating odor. Irritation of nose and eyes at <5 ppm. $\delta_{gas} = 0.7$ g/L. Gas solubility = miscible in all proportions. PEL = 3 ppm in air, 2 mg/m^3 of air.

(Continued)

TABLE 57.1 Toxicology of Volcanic Gases and Aerosols—cont'd

Effects of overexposure

Short-term	Extreme irritation and corrosion of the skin and mucous membranes. Contact with the eyes will cause deep-seated burns, and if the chemical is not removed immediately, permanent visual impairment or blindness may result. Skin exposure produces severe burns, which are slow to heal. Subcutaneous tissues may be affected becoming blanched and bloodless, which may result in gangrene. A severe irritant to the nose, throat, and lungs, inhalation of the vapor may cause ulcers of the upper respiratory tract; concentrations at 50–250 ppm are dangerous even for brief exposures.
Long-term	Repeated or prolonged exposure to lower concentrations may cause changes in the bones as well as chronic irritation of the nose, throat, and lungs.

Sulfuric acid (H$_2$SO$_4$)

Characteristics	Colorless to dark brown, oily, odorless liquid. Irritation of nose and eyes at low concentrations. $\delta_{gas} = 3.4$ g/L. Gas solubility = miscible in all proportions. PEL = 20 ppm in air, 1 mg/m^3 of air.

Effects of overexposure

Short-term	Irritation of eyes, nose, and throat. Severe burns with rapid destruction of tissue and erosion of teeth may occur. Inhalation may also lead to difficulty in breathing and inflammation of upper respiratory tract.
Long-term	Repeated or prolonged exposure to the vapor may cause erosion of the teeth, chronic irritation of the eyes, nose, throat, and lungs.

3. HAZARDS TO POPULATION AND THE ENVIRONMENT: CASE STUDIES

The relative degree of hazard from volcanic gases is dependent upon the type of gas emitted. Some gases are poisonous, while others are dangerous only if present in such high concentrations that they block oxygen respiration. The dispersion of a given gas species is also directly related to the hazard. Emission of gases at high elevations (e.g., Mt Etna, Italy) will have less direct impact on population than low-level gas plumes (e.g., Masaya, Nicaragua; Kilauea, USA). Relatively short-lived eruptions may eject significant amounts of gases into the stratosphere with short-term global consequences (e.g., Mt Pinatubo, Philippines). Persistently active volcanoes, however, degas continuously and may present a long-term hazard (e.g., Ambrym, Vanuatu). In some instances, even dormant volcanoes can pose a threat to human health and the local environment (e.g., Long Valley, USA).

3.1. CO_2 Hazards

The most lethal CO_2-related events have been those in which CO_2, which is colorless, odorless, and denser than air, flowed downhill as a density current, collecting in low-lying areas and asphyxiating all lives in its path. The first recorded incident of this occurred in the Dieng Volcanic Complex (or Dieng Plateau), Indonesia: a complex of volcanic centers forming a large depression, approximately 14-km long and 6-km wide. On the morning of February 20, 1979, the inhabitants of the village of Batur felt three seismic shocks at 2:00 AM, 3:30 AM, and 4:00 AM and then observed a phreatic eruption at 5:15 AM, which ejected a dark gray cloud from the Sinila crater (a small water-filled vent, 3 km northeast of the village). This eruption, which formed a 90-m wide and 100-m deep crater, was accompanied by the ejection of blocks and mud, steam and gas, and a hot lahar (or mudflow) that flowed 3.5 km downslope. At 6:45 AM, a second minor eruption occurred 300 m west of Sinila and resulted in the formation of a new crater, Sigludung. Many of the villagers from Koputjukan fled west toward Batur and were killed by a gravity current of gas (probably CO_2 and H_2S), which was emitted from multiple small vents and fissures to the south and west of the two craters. Others, having witnessed the deaths along the Koputjukan—Batur road, retreated to a nearby elementary school, where they were also killed. In total, the gas killed at least 149 people and injured over 1000 people.

Some volcanic lakes can also pose significant risk due to the accumulation of CO_2 and CH_4 in thermally stratified waters, which when perturbed can lead to limnic eruptions or catastrophic gas discharge from the lakes. Lake Kivu (on the border between Rwanda and the Democratic Republic of Congo), Lake Monoun, and Lake Nyos (Cameroon) are the best examples, with Monoun and Nyos being the only two with recorded events. The CO_2 is believed to have originated from cold springs degassing below the lakes, which being thermally stratified, allowed for the gas accumulation. Although still somewhat controversial, it is generally believed that these two events may have been triggered by landslides, which caused the deep gas-rich layer to rise to a point in the lakes where the hydrostatic pressure was insufficient to keep the CO_2 in solution. Being denser than air, the CO_2 gas that was released formed a large density current, which flowed over the crater rim and downslope.

On August 15, 1984, at approximately 11:30 PM local time, a cloud of concentrated CO_2 burst from the crater in Lake Monoun and flowed down valley several hundred meters, where it settled along the Panke River depression. The emission was heralded by an explosion heard by the people in Njindoun village (1 km north of the lake) and by seismic shocks felt 6-km north in the village of Mbankouop. At approximately 3 AM, the majority of the victims (39 in total) had left the village of Njindoun and were heading south toward the Foumbot market. The location of the bodies indicated that the victims encountered and succumbed to the dense gas cloud near the bridges over the Panke River. When police arrived at the scene ($\sim$6:30 AM), a whitish smoky cloud still covered the area, drifting with the wind off Lake Monoun. They were only able to enter the area when the cloud had finally dissipated at approximately 10:30 AM. It was then that they noted that the bodies of the victims were covered with reddish first-degree burns and blisters and that mucous and blood had frothed from their noses and mouths. The surrounding vegetation was bleached yellowish and withered, yet the victims' clothes were unaffected. The carcasses of domestic and wild animals were also found nearby.

Between 9:00 PM and 10:00 PM (local time) on August 21, 1986, just 2 years after the Lake Monoun event, a cloud of concentrated CO_2 was emitted from the lowest point of the Nyos Crater Lake and spread into the surrounding valleys, affecting an area approximately 20-km long and 15-km wide. The gas emission appears to have been preceded by the rise of a gas column at 4:00 PM and a weak explosion at 8:00 PM and then by two or three violent explosions between 9:00 and 10:00 PM. No seismic events were recorded. There was little or no damage to vegetation or housing, although taller vegetation was flattened in some areas between the lake and Lower Nyos. As with the Lake Monoun event, the carcasses of many domestic and wild animals were found in the affected area (Figure 57.2). The exact number of casualties is unknown, but according to government sources, at least 1700 people died, while

FIGURE 57.2   Cattle asphyxiated by the dense CO_2 cloud emitted from Lake Nyos, Cameroon, 1986. *Photograph by M.L. Tuttle, U.S. Geological Survey.*

approximately 5000 people in the affected area escaped exposure or survived its effects.

Another type of CO_2-related hazard is that seen at Mammoth Mountain, California, a large dacitic volcano located on the southwestern rim of the 760,000-year-old Long Valley caldera. Beginning in 1990, extremely high levels of CO_2 soil degassing (20–90% CO_2) on the flanks of Mammoth Mountain have resulted in tree-kill areas, which cover >500,000 m^2. The CO_2, which kills the trees during the winter by inhibition of root function and oxygen deprivation, originates from the degassing of intruded magma and gas release from magmatically heated carbonate metasedimentary rocks beneath the caldera. The CO_2 flux from Mammoth Mountain was estimated at 1200 metric tons per day (t/d) in 1995 dropping down to ~10 t/d in 2010, occurring principally in the tree-kill areas. This diffuse degassing may also be producing catastrophic acidification of the soil and mobilization of toxic Al^{3+} into local aquatic ecosystems. Furthermore, although CO_2 generally dissipates when it leaves the ground, its relatively high density causes it to collect in hollows, wells, and confined places, where it creates a serious asphyxia hazard. During the winter months, the relatively impermeable snow around Mammoth Mountain prevents CO_2 from escaping, dramatically increasing the hazard; a number of cases of asphyxia or near-asphyxia have been reported after skiers fell into snow-covered fumaroles or took shelter in small snow caves or snow-covered cabins.

Global CO_2 emission from subaerial volcanoes has been calculated, using CO_2/SO_2 ratios and SO_2 flux estimates, to be approximately 0.15–0.26 billion metric tons (Gt) of CO_2 per year, with over half of that coming from passively degassing volcanoes. Although only a fraction of that of anthropogenic emissions (36 Gt of CO_2 in 2013), volcanoes nevertheless contribute large amounts of CO_2 to the atmosphere. In earlier geologic times, catastrophic volcanic event such as the eruption of the Deccan Traps (India) may have added significant amounts of CO_2 to the atmosphere and therefore may be partially responsible for increased global warming.

3.2. SO_2 Hazards

When sulfur dioxide is released to the atmosphere, it oxidizes with the OH radical in air to form sulfurous acid (SO_3), which then reacts with water to produce sulfuric acid particles. At Masaya volcano, a basaltic complex of nested calderas and craters located in northwestern Nicaragua, the long-term degassing activity from the currently active crater, Santiago (Figure 57.1), has had a significant impact on the surrounding vegetation: high concentrations of SO_2 disturb stomatal respiration and cause necrosis. Sulfur dioxide fluxes from the crater have been measured at as much as 2500 t/d. The presence of this gas, as well as HCl and HF,

FIGURE 57.3 A continuous gas plume rises from the Pu'u 'O'o vent on Kilauea volcano, Hawaii, USA. The prevailing winds from the northeast often lead to accumulation of vog or volcanic fog against the southwestern flank of Mauna Loa and the Kona coast, 1983. *Photograph by R.W. Decker, U.S. Geological Survey.*

has led to extensive fumigation and contamination of >1200 km^2 downwind of the volcano. Concrete and metal fences, telephone wires, and other metal equipment are also severely damaged in the affected areas (Figure 57.3). Since the nineteenth century, the significant economic impact from the degassing has led local coffee farmers to even demand the capping or destruction (by aerial bombardment) of the degassing vent; none of these attempts proved successful. Mixing of rain from Hurricane Mitch (late 1998) with the volcanic gas plume produced concentrated acid rain, which local farmers blamed for the damage to young palm trees and even the complete destruction of a soya field during a single afternoon.

On June 6, 1912, Novarupta, a dacitic volcano located on the Alaska Peninsula, explosively erupted ~13 km^3 of magma and triggered the collapse of Mt Katmai volcano. The Katmai–Novarupta eruption released substantial amounts of gas, which created acid rains that appear to have greatly impacted the local environment. As with Masaya, acid rain was reported to have dissolved the metal work of buildings 400-km northeast at Seaward, while polished brass was tarnished 1100 km away at Cape Spencer. A month after the eruption, housewives in Vancouver, Canada (2400 km south), reported that clothes left out to dry had "turned to shreds" upon being ironed.

The conversion of volcanic sulfur dioxide into aerosol particles is often responsible for formation of an acid smog or "vog" (e.g., Kilauea, Hawaii; Figure 57.4). This acid vog is slowly neutralized by ammonia in the atmosphere to form a haze that may block solar ultraviolet rays from reaching the lower atmosphere, which in turn may lead to a cooling effect (e.g., Mt Pinatubo, Philippines). The volcanic haze may also act as cloud condensation nuclei and perhaps even as sites for the catalytic destruction of ozone with obvious global implications.

FIGURE 57.4 A metal gate, 15 km downwind of Masaya volcano, Nicaragua, severely damaged by prolonged exposure to acid gases, 1999.

3.3. H₂S Hazards

Hydrogen sulfide is an extremely toxic gas, which has been responsible for at least 46 fatalities (since the early twentieth century) at Rotorua (New Zealand) and a number of volcanoes in Japan, where it is believed to be the most common cause of volcanic gas accidents. This has led many volcano observatories in Japan to install H_2S detectors and automated warning systems in areas frequented by the public.

One such incident occurred in 1971 on the flanks of Kusatsu-Shirana volcano, Honshu, when six downhill skiers died almost instantly after passing through a depression filled with H_2S. Being denser than air, the gas may accumulate in snow-covered depressions or caves which are breached from time to time, releasing lethal concentrations to the surface. Another more recent accident occurred on Adatara volcano, Honshu, a basaltic-to-andesitic volcano, which forms part of the volcanic front of northeastern Japan. On September 15, 1997, 4 hikers, from a party of 14, were killed after inhaling volcanic gases on the floor of the Numano-taira crater. The group had become disoriented because of fog and left the trail, which had signs warning of volcanic gas hazards in the area. Three of the hikers fell into the crater, where they succumbed to noxious gases that had accumulated on the crater floor due to calm wind conditions. The fourth hiker died, after succumbing to the gases while attempting a rescue. Scientists from the Kusatsu-Shirane Volcano Observatory reported that the fumarolic gas from the southwest rim of the crater was composed of 0.5% SO_2, 33–37% CO_2, and 60–65% H_2S.

3.4. HCl Hazards

Although there are few cases of hydrochloric acid being solely and directly responsible for volcano-related fatalities, it is nevertheless an important component because of its effects on the environment. HCl is highly soluble in water and is therefore easily removed by rain from a volcanic plume, resulting in low-pH acid rains (e.g., Kilauea, USA; Masaya, Nicaragua). Significant amounts of HCl may also have been injected into the stratosphere during the cataclysmic eruption of El Chichon (Mexico) in 1982 with potentially serious environmental implications. In Hawaii, clouds of lava haze or "laze" are formed from the boiling and vaporization of seawater when lava flows enter the sea. These large white laze plumes contain a mixture of HCl (up to 10–15 ppm) and seawater and produce acid rains (pH 1.5–2), which pose a hazard to the local population.

3.5. HF Hazards

Hydrogen fluoride is highly soluble and with excessive intake leads to dental and skeletal degradation and is indirectly responsible for the most lethal gas-related volcanic event. The Laki fissure (Iceland), a NE-trending 27-km-long row of cones and craters, was the source of the Grímsvötn caldera eruption, which lasted from June 8, 1783 to May 26, 1785. The second largest historic basaltic fissure eruption since the ∼AD 935 Eldgjá eruption, it was preceded by tremors and earthquakes starting on May 15. The eruption involved at least 19 km³ of basaltic magma of which 15 km³ was erupted as lava and tephra covering 565 km². Lava fountains are thought to have risen as high as 1400 m and were accompanied by convecting eruption columns, which rose to a maximum altitude of ∼15 km, resulting in an atmospheric loading of 219 Mt of SO_2, ∼7.0 Mt of HCl, and 15.0 Mt of HF over an 8-month period. A low-altitude gas haze caused grasses contaminated by fluorine to be stunted, leading to the loss of over 50% of Iceland's grazing livestock. This consequently lead to the "Haze Famine," which, in combination with various diseases and two severe winters, caused the death of 10,521 people, 22% of Iceland's population. The injection of ash, gases, and aerosols into the lower stratosphere also affected parts of Western Europe, North Africa, and Western Asia and resulted in cooling of the Northern Hemisphere by 1–2 °C.

Ambrym is a persistently and vigorously degassing island volcano (>17,000 t/d SO_2; ∼950 t/d HCl; 3400 t/d HF) in Vanuatu, which is having a significant impact on the health of its 9000 residents. Since rainwater is used by the majority of residents for cooking and drinking, fluoride contamination is important with concentrations in rainwater tanks reaching 9.5 ppm F; the WHO-recommended F concentration in drinking water is 1 ppm. This has led up to 96% of the population suffering from moderate to extreme dental fluorosis, and it is known that prolonged exposure to F concentrations of 4–6 ppm can lead to skeletal fluorosis.

4. GAS HAZARD MITIGATION

While it is impossible to stop volcanoes from degassing, volcanic gases may be monitored and studied using a variety of techniques ranging from direct sampling of fumaroles (e.g., Giggenbach bottles, MultiGAS) to remote sensing techniques (e.g., FLYSPEC, mini-DOAS, OMI). Ideally, gas monitoring is carried out sufficiently frequently by volcano observatories in order to characterize the baseline or background activity for every volcano. Following the example of those working on industrial-related gas emissions, volcanologists have begun to develop and apply models of gas dispersion and deposition downwind from the volcanic source (e.g., Vulcano, Italy; Poás, Costa Rica). Based on this information, alert levels and hazard maps can then be developed and used for risk management. However, it is often the case that areas at risk lack the necessary funding and infrastructure to support such studies.

In only a few instances, such as the artificial degassing being carried out at Lake Nyos, Cameroon, is it possible to reduce or eliminate the gas hazard physically. In most cases, monitoring of the volcano and limitation of access by the public to affected areas are the only means of reducing the risk. On a short timescale, the impact of volcanic gases can be reduced by limiting exposure time, not overexerting oneself, and resting frequently. Where possible, one should remain to the windward side of the gas source and wear full- or half-face gas masks (respirators) with appropriate absorbers/filters. Where gas masks are unavailable, a wet cloth held over the face can partially reduce the amount of water-soluble gases entering the lungs (Figure 57.5).

In some cases, evacuation is necessary, but for extended long-term activity such as at Masaya or Ambrym, it is clearly not feasible. Education of the population at risk is therefore critical, and some excellent resources are now available. The International Volcanic Health Hazard Network (http://www.ivhhn.org) has prepared a comprehensive set of guidelines and pamphlets for the public and emergency managers. The International Association of Volcanology and Chemistry of the Earth's Interior (http://www.iavcei.org) also has safety recommendations for the public online and has produced video on "Understanding Volcanic Hazards and Reducing Volcanic Hazards."

5. SUMMARY

Volcanic gases, although a relatively minor hazard in comparison with other volcanic phenomena, can have important short- and long-term impacts on people and the environment. Gases and resulting acid rains may sometimes be detected over 1000 km from the volcanic source but are generally directly responsible for relatively few deaths. Their effects on buildings are rarely totally destructive except in the case of long-term exposure (e.g., Masaya, Ambrym). Sulfur dioxide emissions from the 1783 Laki fires were responsible in part for global temperature decreases, while CO_2 from subaerial volcanoes may contribute to global warming. Although some attempts have been made to physically reduce gas hazards (e.g., artificial degassing), gas monitoring and education of the public are the most effective means of reducing the hazard from volcanic gases.

FIGURE 57.5 A sulfur miner at Kawah Ijen volcano, Indonesia, carrying 70–80 kg of native sulfur. Without access to gas masks and filters, miners use a wet scarf in their mouth to limit gas entering their lungs, 2006.

FURTHER READING

Allibone, R., Cronin, S.J., Charley, D.T., Neall, V.E., Stewart, R.B., Oppenheimer, C., 2012. Dental fluorosis linked to degassing of Ambrym Volcano, Vanuatu: a novel exposure pathway. Environ. Geochem. Health 34, 155–170.

Baxter, P.J., 2005. Human impacts of volcanoes. In: Marti, J., Ernst, G.G.J. (Eds.), Volcanoes and the Environment. Cambridge University Press, New York, pp. 273–303.

Blong, R.J., 1984. Volcanic Hazards. Academic Press, Orlando, Florida.

Cantrell, L., Young, M., 2009. Fatal Fall into a volcanic fumarole. Wilderness Environ. Med. 20, 77–79.

Farrar, C.D., Sorey, M.L., Evans, W.C., Howle, J.F., Kerr, B.D., Kennedy, B.M., King, C.-Y., Southon, J.R., 1995. Forest-killing diffuse CO_2 emission at Mammoth Mountain as a sign of magmatic unrest. Nature 376, 675–678.

Gerlach, T., 2011. Volcanic versus anthropogenic carbon dioxide. EOS Trans. Amer. Geophys. Union 92, 201–202.

Hansell, A., Oppenheimer, C., 2004. Health hazards from volcanic gases: a systematic literature review. Arch. Environ. Health 59, 628–639.

International Volcanic Health Hazard Network. Volcanic Gases and Aerosols Guidelines. http://www.ivhhn.org.

Le Guern, G., Tazieff, H., Faivre-Pierret, R., 1982. An example of health hazard: people killed by gas during a phreatic eruption, Dieng Plateau (Java, Indonesia), February 20th 1979. Bull. Volcanol. 45, 153–156.

Le Guern, G., Tazieff, H., 1989. Lake Nyos. J. Volcanol. Geotherm. Res. 39 special issue.

Schmid, M., Halbwachs, M., Wehrli, B., Wüest, A., 2005. Weak mixing in Lake Kivu: new insights indicate increasing risk of uncontrolled gas eruption. Geochem. Geophys. Geosyst. 6, Q07009.

Sigurdsson, H., Devine, J.D., Tchoua, F.M., Presser, T.S., Pringle, M.K.W., Evans, W.C., 1987. Origin of the lethal gas burst from Lake Monoun. Cameroon. J. Volcanol. Geotherm. Res. 31, 1–16.

Scarpa, R., Tilling, R.I. (Eds.), 1996. Monitoring and Mitigation of Volcano Hazards. Springer-Verlag, Berlin.

Thordarson, Th, Self, S., 1993. The Laki (Skaftár Fires) and Grímsvötn eruptions in 1783–1785. Bull. Volcanol. 55, 233–263.

Thordarson, Th, Self, S., Óskarsson, N., Hulsebosch, T., 1996. Sulphur, chlorine, and fluorine degassing and atmospheric loading by the 1783–1784 AD Laki (Skaftár Fires) eruption in Iceland. Bull. Volcanol. 58, 205–225.

Chapter 58

Volcanic Tsunamis

Simon J. Day

Department of Earth Sciences, Institute of Risk and Disaster Reduction, University College London, London, UK

Chapter Outline

GLOSSARY

amplitude The vertical separation of adjacent crest and trough in a tsunami or other wave. Not necessarily twice the height of the crest above the initial water level before arrival of the tsunami. In tsunamis, increases with decreasing water depth. See also run-up, below.

bore A long-period wave with a steep, turbulent front, and a sustained flow of water behind. Bores are regularly generated by high tides propagating into narrow channels, but tsunami waves propagating into shallow water also often transform into bores.

dispersion The evolution of a tsunami wave train during propagation in which an initially small number of large, nonsinusoidal waves contain components that travel with slightly different velocity and so separate over time into many smaller waves. Frequency dispersion, whose strongest effect is on shorter-period tsunami waves including most volcanic tsunamis, is the most common type of dispersion.

impulse wave A wave produced primarily by the transfer of kinetic energy to a water body, rather than the transfer of potential energy by seabed uplift. Some types of volcanic tsunamis are impulse waves, particularly those produced by the entry of landslides, pyroclastic flows, and lahars into the ocean.

inundation distance The distance between the initial shoreline and the inland limit of the area flooded by a tsunami. Depends on the amplitude of the tsunami waves at the shoreline, the wave period, the topography of the inundated area (see run-up, below), and the roughness of the inundated land, which affects the drag exerted by the land on the flood-like bore of water.

period The time interval between passage past a fixed point of successive crests in a wave train. The reciprocal of wave frequency (number of wave crests per second). Largely conserved during tsunami propagation, although in complex wave trains components with different periods may separate out producing apparent changes in wave period of the dominant waves in the train.

phreatomagmatic (synonym for hydromagmatic) Eruptions involving interaction between hot magma and water, with transfer of heat to the water leading to steam explosions. Explosions of this type are a common cause of small tsunamis.

run-up The highest point within the inundation zone at a site inundated by a tsunami. Owing to upslope surges of rapidly flowing water, especially where focused into gullies on hillsides, the run-up may be two or more times the height of the tsunami wave at the shoreline.

seiche (meteotsunami) A tsunami-like long-period wave produced by the action of atmospheric forces such as winds and pressure waves upon a water body, causing it to resonate. Seiches may also form in small water bodies due to the passage of earthquake waves.

tsunami A Japanese word that translates as "long wave seen in a harbor," originally describing waves associated with earthquakes but now applied to all long-period waves: such waves have the characteristic of increased amplitude as they propagate into shallow water such as harbors.

wavelength The distance between successive wave crests in a wave train. Unlike wave period, wavelength is not conserved during wave propagation but in the case of tsunamis decreases with decreasing water depth.

The Encyclopedia of Volcanoes. http://dx.doi.org/10.1016/B978-0-12-385938-9.00058-4

1. INTRODUCTION: CAUSES AND CONSEQUENCES OF VOLCANIC TSUNAMIS

Tsunamis or seismic sea waves (commonly, but incorrectly and dangerously, also called tidal waves) are most often produced by seabed displacements due to earthquake fault ruptures. However, they may also be produced by many other mechanisms, which range from anthropogenic nuclear and chemical explosions to submarine landslides and asteroid impacts into the ocean. Among these other mechanisms, perhaps the most widely known and second most common (after nonvolcanic landslides) are the diverse group of tsunamigenic processes associated with volcanic activity. About 100 notable tsunamis of volcanic origin have been produced in the world's oceans during the past three centuries in which historical records of these events are reasonably complete, in addition to local tsunamis confined in caldera lakes and other lakes on or near volcanoes.

The fundamental process involved is the transfer of energy from the volcano to the ocean or other water body (the following discussion refers to the ocean for brevity, but the same physical processes are involved in generation of tsunami waves in other water bodies). This can be, for example, release of gravitational potential energy from a high edifice into the kinetic energy of a landslide, whose interaction with the ocean produces kinetic and potential wave energy in a tsunami wave train; or conversion of the thermal energy of hot magma into the kinetic energy of an expanding steam bubble, which again interacts with the ocean to produce the kinetic and potential energy of a wave train. Most volcanic tsunamis are therefore **impulse waves**. The source events involved are numerous and varied but can be divided into the following broad categories:

- Submarine volcanic explosions, typically **phreatomagmatic** in nature, in which the ultimate energy source is the thermal energy of hot magma that is transferred to the ocean via a working medium in the form of steam produced by heating of ocean water or subsurface pore water, in one of various types of fuel—coolant interaction. Rarely, subaerial volcanic explosions produce such intense atmospheric pressure waves and high-velocity winds that coupling between ocean and atmosphere generates tsunami-like **seiche** waves, especially in confined bodies of water that resonate in response to the atmospheric phenomena.
- Entry of volcanic flows into the ocean and their subsequent movement. These maybe hot pyroclastic flows, in which case the displacements of the ocean due to movement of the flow are commonly accompanied or followed by tsunamigenic steam explosions, cold syneruptive flows, or posteruptive lahars.

- Landslides involving unstable but initially coherent volcanic material, which maybe newly emplaced or built up during an extended **period** of volcano growth: these range from small events triggered by the collapse of lava deltas or pyroclastic fan deltas at the shoreline, through the collapses of summit dome complexes and thin-skinned flank failures, to large-scale sector collapses in which volumes ranging from cubic kilometers to hundreds of cubic kilometers of material maybe involved.
- Sudden ground deformations of volcanic origin, which range from inflations and deflations at the start of eruptions, to earthquakes of volcanic origin (for example, at those oceanic island volcanoes where episodic flank deformation involves significant earthquakes), to caldera subsidences.

An important distinction between most volcanic tsunamis (and indeed, most tsunamis produced by processes other than earthquakes)—except for the very largest—and those of seismic origin, arises from the different dimensions of the sources. Volcanic tsunamis are typically generated in regions a few hundred meters to, at most, a few tens of kilometers across; in contrast, earthquake-generated tsunamis have sources tens of kilometers across at minimum, to hundreds of kilometers wide by more than a thousand kilometers long in the case of those tsunamis produced by the largest subduction zone earthquakes such as Chile 1960 and Sumatra—Andaman 2004.

The significance of this distinction is that the dominant wave period of the tsunami is related to the dimensions of the source, and in turn produces a difference in wave behavior. Long-period tsunami waves, with periods of 500—5000 s, show little dependence of wave velocity upon wave period, velocity being dependent almost entirely upon the square root of ocean depth. Therefore, a tsunami wave train dominated by components with long periods will show only very slow separation of the initial sequence of a few large-amplitude waves into a longer sequence of many smaller-amplitude waves, with the effects of **dispersion** only being important after the waves have traveled over transoceanic distances of order 10,000 km. In contrast, most volcanic tsunamis, generated at sources a few kilometers across or less, are typically dominated by waves with periods in the range of 100—300 s. These disperse into wave trains consisting of many smaller waves over distances of order 100—1000 km, so most volcanic tsunamis present mainly local to regional-scale hazards, albeit at greater distances than most other volcanic hazards other than airfall tephra, aviation hazards, and climate-related gas and aerosol hazards. Important exceptions to this rule are those tsunamis generated by large explosive eruptions, and perhaps also by the largest volcano lateral collapse landslides. In the case of Krakatau 1883, in addition to the source being as much as a few tens of kilometers across, long-period tsunami waves may have been produced by resonance of the partially confined waters of Sunda Strait as discussed further below. The question of whether sector collapses at oceanic island volcanoes generate tsunami waves with sufficiently long periods to cross oceans with only limited dispersion is more controversial: this is addressed further in the final section of this chapter.

Notwithstanding their comparatively limited range, the very large **amplitudes** of many volcanic tsunamis near source (Table 58.1) mean that they present major local to regional hazards. This is compounded by the fact that they often occur with little warning. Some of the largest death tolls in individual volcanic eruptions (most notably Krakatau 1883, Tambora 1815, and Unzen 1792; Table 58.1) have been caused by volcanic tsunamis. Note, however, that it is important to distinguish deaths caused by tsunamis from deaths caused by other processes in the same eruptions. Overall, as much as a quarter of all deaths due to volcanic activity in the last three centuries may have been caused by volcanic tsunamis.

Evidence from oral traditions and archaeology indicates that coastal communities and civilizations may have been affected by volcanic tsunamis throughout human history. Archaeological sites in Alaska, in the South West Pacific and most famously on Crete show indications of the impact of tsunamis that have been linked to volcanic eruptions. Oral traditions of tsunamis associated with disappearing islands and volcanic eruptions are widespread in the South West Pacific. However, the interpretations of these traditions are often ambiguous. Different versions of the same tradition—for example, those related to the disappearance of Yomba Island in the Bismarck Sea region of Papua New Guinea—may contradict one another on such basic points as whether or not the lost island was a volcano or not. Such contradictions should not be regarded as invalidating the traditions as it may be that, over the generations since the events that they describe, the traditions have telescoped or combined events happening at different times and different places into a composite narrative. However, they do point to the need to evaluate the traditions critically and to test them against archaeological and geological evidence.

The problems involved in such work are well illustrated by the controversy over the relationships between the Bronze Age caldera-forming explosive eruption of Santorini c.1628 BC, associated tsunami waves around the southern Aegean Sea, and the evolution of the Minoan and Mycenaean civilizations in that area. This eruption undoubtedly had a major direct impact upon Santorini, exemplified by the Minoan settlement of Akrotiri that was buried beneath its proximal deposits. In addition there is an increasing amount of geological evidence of syn-eruptive deposits from tsunami waves with **run-ups** of as much as tens of meters both on Santorini itself, adjacent islands, and on the north coast of Crete. However, the hypothesis that the Minoan

civilization was replaced or conquered by Mycenaeans who took advantage of the destruction caused by tsunami waves is both difficult to reconcile with some archaeological evidence, that has been interpreted in terms of gradual change at a later date, and uncertain because the effects of the tsunami waves upon Minoan coastal communities are poorly known. As with the South Pacific examples, integrated geological, geochronological, and archaeological studies are needed to address these questions.

2. EXAMPLES OF DIFFERENT VOLCANIC PROCESSES WHICH CAN PRODUCE TSUNAMIS

2.1. Submarine Eruptions

Although there are many more submarine than subaerial volcanoes, there are few instances in which tsunamis can be linked with confidence to particular volcanic eruptions. This is partly because a large proportion of submarine volcanoes are at depths too great—usually, more than 150–300 m below the ocean surface—to form efficient tsunami sources except in very large eruptions or landslides, but also because of the limited evidence for the source events except in very recent years. Among the first written accounts of tsunami waves from a submarine volcanic eruption are those from the AD 1650 Kolumbo eruption northeast of Santorini; perhaps the first photographs are those taken by Stehn of waves generated by submarine explosions in the first stages of the Anak Krakatau eruption in 1928–1929.

The 1952–1954 submarine explosive eruption of Myojin-Sho, at a submarine vent on the rim of a shallow submerged caldera in the Izu-Bonin arc, is notable both for the instrumental records of the eruption and its tsunami waves, and for the loss of the *Kaiyo Maru 5*, a Japanese research vessel sent to investigate the eruption. In the worst loss of life ever to occur in the course of volcanological research, the ship and all 29 people on board were lost as it crossed over the vent. Subsequent finds of floating debris from the ship, charred and with embedded volcanic scoria matching that from other phases of the eruption, suggest that it was engulfed in the high-velocity jets of tephra and steam generated by a violent submarine explosion of Surtseyan type. This explosion and many like it produced tsunami waves, recorded at a tide gauge on an adjacent island some 100 km away, and underwater sound waves recorded on hydrophone arrays across the Pacific. The vicinity of the vent has remained a maritime exclusion zone ever since, but recent research has confirmed the presence of a phreatomagmatic tephra deposit covering large parts of the adjacent caldera.

Another well-known tsunamigenic submarine volcano is the Kick 'em Jenny submarine volcano northeast of Grenada in the Lesser Antilles volcanic arc of the eastern

TABLE 58.1 Characteristics of Selected Volcano Tsunamis

Location	Date	Volcano Type	Cause of Tsunami	Maximum Wave Heights/run-ups	Travel Distance of Significant Waves	Impact of Tsunami
Santorini, Aegean sea	c. 1628 BC	Island caldera (partly submerged)	Large explosive eruption (mechanism uncertain)	10–50 m? (uncertain)	Up to 500 km? (uncertain)	Uncertain
Komagatake, Hokkaido, Japan	AD 1640	Coastal stratovolcano	Lateral collapse landslide	8 m run-up	50 km (Limited by bay shores)	About 700 deaths
Kolumbo, Aegean sea	AD 1650	Submarine volcano	Submarine explosive eruption	15 m run-up	About 100 km	About 20 deaths
Oshima–Oshima, sea of Japan	AD 1741	Island stratovolcano	Lateral collapse landslide	14 m maximum observed run-up	1200 km	About 2000 deaths
Unzen, Ariake sea, Japan	AD 1792	Coastal lava dome complex	Dome collapse and debris avalanche into sea	10–55 m run-ups?	<50 km (limited by bay shores)	About 10 000 deaths
Tambora, Sumbawa, Indonesia	AD 1815	Island caldera volcano	Entry of large pyroclastic flows into sea	>10 m run-up	>100 km	About 10 000 deaths
Krakatau, Sunda Strait, Indonesia	AD 1883	Partly submerged island caldera	Large explosive eruption (mechanism uncertain)	c. 40 m run-up	3000 km; waves detected at >8000 km	About 33 000 deaths
Mount St Augustine, Cook Inlet, Alaska	AD 1883	Island stratovolcano	Debris avalanche into sea	7–9 m run-up	>100 km	Limited damage (occurred at low tide); no fatalities
Ritter Island, Bismarck sea, Papua New Guinea	AD 1888	Island stratovolcano	Lateral collapse landslide removing most of island	14 m measured run-up (perhaps >50 m close to island?)	>500 km	More than 1500 deaths likely to have occurred
Paluweh Island, Indonesia	AD 1928	Island stratovolcano	Debris avalanche into sea	10 m	>100 km	More than 150 deaths
Kick 'em Jenny, Grenada, Caribbean	AD 1939 onward (many events)	Submarine volcano	Submarine explosive eruptions; possible landslides in some events	<5 m	<10 km	No known fatalities to date
Iliwerung, Indonesia	AD 1979	Coastal stratovolcano	Debris avalanche into sea	9 m	>100 km	More than 500 deaths
Mount St Helens, Washington, USA	AD 1980	Inland stratovolcano	Debris avalanche into lake; lateral blast	260 m on lake shore	4 km	No known fatalities (tsunami within lateral blast area)
Karymskoye, Kamchatka	AD 1996	Caldera lake	Phreatomagmatic eruptions	30 m on lake shore	5 km on lake shore; flooding downriver from overflow	No known fatalities
Stromboli, Italy	AD 2002	Island stratovolcano	Subaerial and submarine landslides	8 m on Stromboli	170 km	No fatalities; minor damage on Stromboli

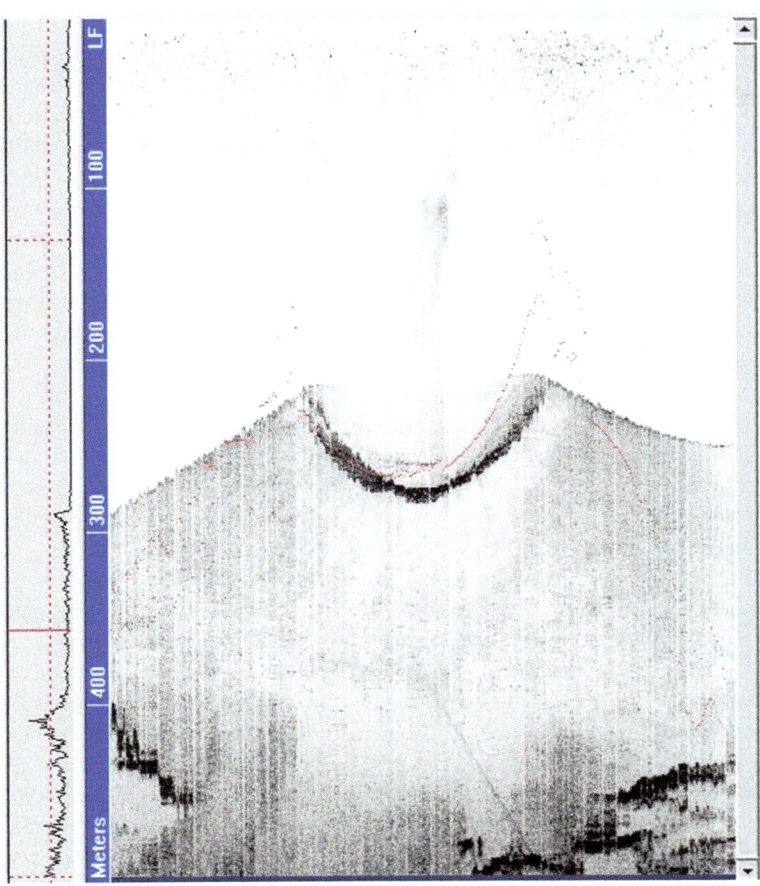

FIGURE 58.1 Backscatter Intensity sonar image of the Ritter Island submarine vent in December, 2006 showing plume of water or gas bubbles rising from the crater. Figure rotated for ease of interpretation: ship track at top, vertical scale is distance from sonar.

Caribbean. From 1939 onward, repeated eruptions at a vent around 150-m below the ocean surface have produced many small tsunami waves that form a persistent hazard for passing yachts and local shipping, and occasionally impact the beaches of adjacent islands. Most of these waves appear to have been generated by small (volcanic explosivity index = 0−1) phreatomagmatic eruptions, and others by rising, cavitating steam bubbles and plumes of hot water. Repeated bathymetric surveys and (sometimes hazardous) investigations by research submarines indicate that the volcano is growing at as much as 4 m/year. This raises the possibility that its eruptions maybecome more violent in future as its summit approaches the sea surface and the hydrostatic pressures that confine and to some extent suppress submarine explosions diminish. Increasing eruption violence will most likely increase the tsunami hazard presented by Kick 'em Jenny. Conversely, the vent is on a steep submarine slope and another possibility is that a future eruption may trigger a lateral collapse landslide that could also produce a significant tsunami.

Other shallow submarine volcanoes that are known to produce frequent tsunamigenic eruptions include Kavachi in the Solomon Islands and Ritter Island in Papua New Guinea, where the submarine vent that has grown in the scar left by the lateral collapse landslide of AD 1888 (discussed below) has produced a number of small tsunami waves that have struck the beaches of adjacent islands despite the vent summit currently being about 300-m below the ocean surface. Figure 58.1 shows an active side-scan sonar image of the vent and a plume of hot water or bubbles rising from it. This dates from December, 2006 when there was no activity evident at the surface, but eruptions in 2007 and 2009 produced small local tsunamis. It is however likely that minor tsunamis produced by activity at many other such submarine volcanoes go unrecognized, and publicly available monitoring data are very limited.

2.2. Tsunamis Produced by Eruptions Through Lakes

Many volcanoes have crater lakes, ranging in size up to tens of kilometers across in the case of caldera lakes, and others have blocked drainage systems leading to the formation of adjacent lakes; while rift valley and other tectonically formed lakes often have volcanoes located within them or on their shores. These lakes are especially prone to the occurrence of local tsunamis that may often attain extreme run-ups on the lake shores due to the proximity of these shores to the tsunami sources. If the waves exceed the height of the lowest points on crater or caldera rims, sudden outflow floods may also result, analogous to floods from reservoir dam overtopping events such as that which occurred at Vajont, Italy, in 1963. As a result of these distinctive hazards it is useful to distinguish tsunamis in volcanic lakes from those in the open ocean, although the various source event processes are similar.

These source events can be landslides from crater walls or sudden gas releases from within or beneath the lake (as at Lake Nyos, Cameroon in 1986), but are perhaps most commonly eruptions beneath lake surfaces or on islands in the lakes concerned. Notable examples where such tsunamis have been generated include eruptions at Taal caldera volcano in the Philippines, in 1910 and 1965, and at Karymskoye Lake in the Akademia Nauk caldera, Kamchatka, in 1996. In the latter case many tsunami waves were produced by phreatomagmatic explosions over a whole day, with the largest waves up to 30-m high causing extensive stripping of soil and vegetation from the lake shores and overtopping the lowest point on the rim to cause extensive flooding along the Kamchatka river.

Risks from lake tsunamis maybe anomalously high, relative to the size of events, because of the lack of tides and large storm waves in most lakes which means that intensive development of lake shores may extend right down to lake level. Tsunamis comparable to those developed at Karymskoye Lake in 1996, if they occurred at caldera or other volcanic lakes with densely populated shorelines such as Taal, Toba in Sumatra, Managua in Nicaragua, Towada and Toya in Japan, Taupo and Rotorua in New Zealand, and Yellowstone in the USA, could cause major damage and casualties. Against this it has to be noted that the short distances from source vent to impacted shorelines mean that the latter would also be vulnerable to other volcanic hazards from the same eruptions, such as pyroclastic surges traveling over the surface of the lake. Surges were by far the most deadly components of the 1910 and 1965 Taal eruptions, for example.

2.3. Tsunamis Produced by Sudden Ground Deformations Associated with Volcanic Eruptions, and by Volcanic Earthquakes

Eyewitness accounts of the onset of a few volcanic eruptions at coastal volcanoes indicate that tsunamis were generated at the very start of the eruptions, before pyroclastic flows or landslides impacted the surface of the ocean or lakes. These include the 1631 eruption of Vesuvius, where a gradual withdrawal of the ocean by some hundreds of meters in the minutes before the start of the eruption at the summit of the volcano, was followed by a sudden return wave as magma breached the surface and the eruption

started. On a smaller scale, a small wave some 30-cm high was generated in Lake Tarawera in 1886, at the start of the nearby eruption of the same name. More ambiguous cases may have occurred at the start of eruptions at Rabaul caldera in 1937 and 1994, where there was certainly rapid uplift of the caldera floor immediately prior to the eruption.

The mechanism involved maybe that of rapid ground deformation as vesiculating, expanding magma within volcanic conduits is in the final stage of its ascent to the surface. At this point the overpressure of the magma may cause rapid deformation of the wall rocks and surrounding volcanic edifice, and hence uplift the surface; once this overpressure is released as the magma breaches the surface and gas exsolving from the magma can escape more freely, the edifice moves back toward its original position, causing a sudden subsidence of the surface. Such a reversal of deformation would account in particular for the gradual withdrawal and sudden return wave seen at Vesuvius in 1631.

A related but distinct tsunamigenic phenomenon, seen at those few volcanoes sufficiently large to experience progressive flank deformation in seismogenic slip events, is that of tsunamigenic volcanic earthquakes. These occur on discrete fault structures within or at the base of volcanic edifices, as part of their ongoing flank deformation, and are distinct from the complete (or catastrophic, in the materials sense) failures of volcanic flanks that produce the lateral collapse landslides discussed below. The most notable historic examples of volcano flank earthquakes occurred at Kilauea and Mauna Loa volcanoes in Hawaii, in the 1868 Great Kau and 1975 Kalapana earthquakes. The latter had an instrumentally determined moment magnitude of 7.2, while the Great Kau earthquake may have had a magnitude of as much as 8. These earthquakes were associated with one or more destructive tsunami waves up to 10-m high on adjacent coastlines. Detailed modeling of the 1975 Kalapana event indicates, however, that while the first tsunami wave to impact the coast was produced by seabed deformation directly due to the felt earthquake, the second and larger wave was generated by a seismically inefficient event further offshore on or in the lower submarine flank of Kilauea. This may have been either a slow-rupturing, seismically inefficient fault slip event on the base of the volcanic edifice or a slump-type submarine landslide.

2.4. Tsunamis Produced by Volcano Landslides and Debris Avalanches

Volcanoes produce landslides (in the broadest sense) on a wide range of scales, from rockfalls with volumes of cubic meters or tens of cubic meters to lateral or sector collapses with volumes of up to tens of cubic kilometers on volcanic arc stratovolcanoes to thousands of cubic kilometers in the case of the largest oceanic island volcano flank collapses.

Where these landslides enter water bodies, or are partly to wholly submerged at their initiation, they displace water and generate tsunami waves by rapidly transferring varying combinations of potential (due to deformation of the floor of the water body) and kinetic (due to landslide motion) energy to the water body. Especially where these waves are generated in open water, the waves can then transmit to the energy of the landslide to impact far distant shorelines. The efficiency of this energy transfer, which determines the size of the tsunami waves produced by a given landslide volume, is influenced by a multitude of factors that are in many cases poorly understood due to a combination of:

- The lack of direct observations of the source landslides (especially in the case of the largest, lowest-frequency events, and of submarine landslides), that forces reliance upon indirect evidence from landslide deposits;
- The potential for feedbacks between landslide motion and tsunami wave generation (most obviously, through the effect of the waves upon the motion of the landslide, but also upon its shape in the case of deforming landslides);
- The complexities of modeling the subsequent propagation and impact of the typically shorter-period, frequency-dispersive tsunami waves that are the typical products of landslide sources: such modeling is a key step in inverse modeling of shoreline observations of tsunamis to infer characteristics of the source, and the uncertainties that result need to be carefully estimated and accounted for.
- The scale dependence of landslide and tsunami processes, which make extrapolation from small-scale laboratory experiments (and the numerical models that use outputs from them) to much larger-scale natural events a difficult exercise that results in large uncertainties in experiment-based predictions.

Nevertheless the most important factors affecting tsunami generation by volcano landslides are generally accepted to be as follows:

Volume of the landslide appears to be an obvious factor that can be well defined from both the landslide scar and from the volume of the deposits, but needs to be carefully considered as a deposit initially considered to be from a single event may on closer examination be found to be the product of multiple landslides. Furthermore, if a landslide deposit has been formed by a gradual slope failure (for example, as a retrogressive failure beginning at the base of the slope and propagating back up the slope) or in a series of discrete events, then the duration of the failure or the time separation of the discrete events needs to be considered. Critically, if these times are less than the dominant wave period of the resulting tsunami, then their effect is limited, and certainly much less than if the gradual failure or a series of discrete events are spread over a longer time and so

produce series of more or less independent tsunami wave trains, rather than a single wave train of larger waves. In such a situation, accurate post-event modeling based on a landslide deposit requires not only identification of subunits within the deposit, but also determination of the time intervals between emplacement of successive subunits and estimation of the volumes of individual subunits recognizing that these may differ by orders of magnitude (for example, in the case where a large initial landslide is followed by a series of smaller landslides from the newly formed and unstable landslide headwall). A further problem in the interpretation of submarine landslide deposits is that the landslide may entrain sediments from the deep seafloor, particularly from submarine basin plains floored by weak, easily eroded sediment, to produce deposits that are much greater in volume than the volume of the initial collapse scar (which may have been infilled by later volcanic activity, making an independent check difficult) and, critically, greater than the volume of the landslide at the point when it was generating its largest tsunami waves.

Thickness of the landslide, both in absolute terms and relative to the depth of water in which it moves. A notable difference between volcano collapse landslides and almost all nonvolcanic landslides is that the former are much thicker than the latter, with maximum thicknesses of 0.5–3 km. Experimental data clearly indicate that landslides are most efficient as tsunami sources—or, put another way, lose energy most rapidly to the wave that they are generating—when their thicknesses are comparable to water depth: that is to say, when they are fully submerged but are not yet in water that is deep compared to their thickness. Thus, it would be expected that volcano collapse landslides would continue to be efficient tsunami sources as they descended into much deeper water than, for example, continental slope sediment landslides: these are much wider but even in the largest examples mostly have thicknesses of a few hundred meters at most. The problem in applying this rule to natural cases is that the near-ubiquitous development of different morphological facies in different parts of the resulting deposits indicate that large volcano landslides deform continuously during their motion, beginning as large discrete sliding blocks that fragment during motion to produce a matrix that continues motion long after the larger blocks begin to sediment out. Consequently, in the absence of eyewitness observations such as exist for the subaerial collapse of Mount St Helens in 1980 but have no counterparts for comparable ocean-entering let alone submarine volcano landslides, estimates of the thicknesses and shapes of these landslides at the time of maximum tsunami generation are highly uncertain. The problem is compounded by the potential for feedback between wave generation and landslide deformation, with wave resistance at the front of the landslide compressing and thickening it.

Velocity of the landslide, both in absolute terms and relative to tsunami wave velocity in the water depth that the landslide is moving. Faster landslides obviously have more energy to transfer to the ocean, but if landslide velocity exceeds wave velocity—a condition that in reality mainly applies to fast subaerial landslides as they enter the ocean—then the landslide tends to initially generate an impact cavity and a splash, rather than coherent tsunami waves, although once it has slowed it will generate waves. Such cases are exemplified by a nonvolcanic event, the 1958 Lituya Bay rockslide and tsunami waves, but could also occur as a result of water entry by subaerial volcanic landslides. Velocities of submarine landslides of volcanic origin have almost never been determined by direct observation. Although the timing of submarine telecommunications cable breaks by nonvolcanic submarine landslides and turbidity currents has been used to estimate their velocities, no such data exist for volcanic landslides. A recent determination of landslide velocities on West Mata submarine volcano in the Lau Basin using acoustic interference effects in the noise produced by the landslides, as recorded on a temporary scientific hydrophone array, yielded an estimate of between 10 and 25 m/s landslide velocity. These were small landslides in deep water and would not have produced significant tsunami waves, but the method would also apply to larger tsunamigenic landslides if such arrays were more widely deployed. In the absence of such data, indirect methods based upon modeling the relationship between landslide velocity and landslide run out, which in subaerial landslides are linked through their common dependence upon basal friction forces, have been used. However, the results from these methods are ambiguous when they are applied to submarine and ocean-entering landslides partly because entrainment of sediment substrate may greatly reduce basal friction during landslide motion—producing greater landslide deposit lengths for given peak landslide velocity—but primarily because landslides moving in water are subject to velocity-dependent drag and wave resistance forces as well as basal friction forces. The involvement of more than one significant force resisting motion means that it is only possible to model velocity histories of submarine landslides from their final deposit distributions when an independent constraint on landslide velocity at some point in their motion is known: for example, from the interaction of the landslide with submarine topography such as where it was deflected by or ran over an obstacle in its path, providing respectively maximum and minimum velocity estimates for at least a significant part of the landslide. A few key examples of this are known and are discussed below.

Understanding the effects of these and other factors upon the efficiency of tsunami generation by volcano landslides is further complicated by interdependencies between the factors. Thus, for example, landslides from bigger volcanoes can obviously be larger and thicker, but because they may also descend greater heights before reaching more or less flat seabed, have a longer path over which to accelerate. Similarly, thicker landslides are not only efficient sources in deeper water (and therefore after traveling greater distances), they also have an increased ratio of accelerating gravitational forces (depending on their weight and therefore volume) to drag forces (dependent on their area). Furthermore, larger landslides tend to produce longer-period waves as noted above, and therefore their initially larger waves will also not disperse as fast as waves from smaller landslides.

Notwithstanding all these uncertainties and complications in modeling and understanding volcano landslide-generated tsunamis, sufficient observational data sets exist, especially for the relatively small volcano collapse landslides that have occurred in historic time and for which we have at least written and other eyewitness accounts of the tsunami waves if not the source landslide, that it is evident that volcano collapse landslides are an important source of tsunamis from volcanoes. A number of such events appear in Table 58.1, and it is worth emphasizing that every volcano collapse landslide with a volume of more than about 0.1 km^3, for which we have historic records, produced tsunami waves that led to at least local, and often regional-scale, destruction on coasts exposed to the waves. Some authors have linked a large lateral collapse landslide scar on Tinakula volcano, Solomon Islands, to accounts of an event that produced a small local tsunami in 1966, and have argued that this is an exception to the rule; however, descriptions of the topography of the island dating back to the late 1950s show that the landslide scar was present then and is likely a much older prehistoric feature, and that the 1966 tsunami was therefore produced by a relatively small landslide or rockfall within the collapse scar.

Similar events producing locally destructive tsunamis occur with annual to decadal frequencies, with various causes: historic examples include syn-eruption dome collapse events such as that which occurred at Montserrat in 1997; thin subaerial and submarine slope failures in stratified sequences on steep volcano flanks, such as that at Stromboli in 2002; landslides from the steep seaward faces of growing lava deltas such as have occurred several times since 1983 from the active Pu'u O'ou lava deltas on the south coast of Kilauea; and landslides from growing volcanic domes and cryptodomes such as occurred from the Yenkahe block on Tanna Island, Vanuatu, during an episode of intrusion-driven uplift in 1878. Submarine landslides from the seaward faces of coastal alluvial and volcaniclastic fans, analogous to the delta-front landslides that are common sources of local tsunamis in nonvolcanic settings, may also be an important category of small but locally significant volcanic tsunami sources, because such fans are often important areas of flat land on the coasts of volcanic islands and so tend to be densely populated.

The two largest volcano lateral collapses to have occurred at island volcanoes in historic time certainly produced regionally destructive tsunami waves. These are the 1741 Oshima—Oshima collapse in the Sea of Japan, with a poorly constrained volume of around 2.5 km^3 (the collapse scar has been largely filled by later activity, and the submarine deposits have not been completely mapped), and the 1888 Ritter Island collapse in the Bismarck Arc, Papua New Guinea, with an initial collapse volume of between 4 and 7 km^3 (but likely at the lower end of this range, with a best estimate of 4.5—4.7 km^3); the Ritter landslide appears to have entrained large volumes of substrate sediment toward the end of its travel over the deep seafloor and the total deposit volume, including debris flow as well as debris avalanche facies deposits, maybe as much as several times this volume. Accounts of the Ritter tsunami waves by German colonial officials and other colonists are particularly important because they include timing information as well as information about tsunami wave heights (up to several meters even at distant sites, but in excess of 20 m according to oral traditions from sites close to Ritter), **inundation distances** (up to 1.5 km inland), and run-up heights (measured up to 14 m at inland sites, but likely greater on steeper island coasts). They show that there was a single tsunami wave train, with dominant wave periods around 3 min, whose duration increased from around 30 min at the nearest observation points to the source, to more than 2 h at the most distant observation points several hundreds of kilometers away. Such data are critical to models of the dispersive, short-period tsunami waves produced by landslide sources, and make 1888 Ritter Island a key test case for models of tsunami generation by volcano landslides. They also show that the Ritter Island collapse can, at least as far as the tsunami is concerned, be considered to be a single-event landslide in which initial motion of the entire slide mass began within a time span short compared to the dominant tsunami wave period: the significance of this is discussed further below. The Ritter landslide deposits have also been mapped in detail, and have the important feature of having been deflected by the submarine slopes of adjacent islands at more than one point in their trajectory: as noted above, these deflections provide additional constraints upon models of landslide motion and therefore of tsunami generation by the Ritter landslide.

Other historic volcano lateral collapse tsunamis at island arc volcanoes include those produced at Komagatake, Hokkaido, in 1640, and at Augustine volcano in Cook Inlet, Alaska, in 1883. The latter is notable as the most recent in a series of lateral collapses that have occurred at short intervals at Augustine: as many as 12 in the last 2000 years, based upon geological evidence. This reflects the relatively small volume of the Augustine collapses, which have typically affected only the dome complex that has repeatedly grown and collapsed at the summit of the volcano: the consequence of this has been that the landslides have spread out to form relatively thin although likely fast-moving debris avalanches before they entered the water. In contrast, the Ritter Island 1888 lateral collapse cut down to the base of the edifice some 900-m below sea level, and almost completely removed precollapse Ritter Island, which had been a cone rising some 800-m above sea level and is now reduced to a crescent-shaped island no more than 150-m high, representing the highest part of the rim of the largely submerged collapse scar. Although as noted above, an active submarine volcanic cone is growing within the collapse scar, it is unlikely that this will generate a collapse and collapse-generated tsunami comparable to that of 1888 in the near future. Comparable steep-sided islands formed by the summits of stratovolcanoes elsewhere in the Bismarck arc may present more immediate volcano tsunami hazards.

The coastlines impacted by the Oshima—Oshima and Ritter tsunamis were relatively sparsely populated at the time: the Oshima—Oshima tsunami claimed around 2000 lives, while the Ritter tsunami wiped out at least 10 villages on adjacent islands (suggesting a death toll of at least 1500, based on the typical population of 150 that is found in traditional villages throughout Papua New Guinea coastal areas). Similar events occurring today in the same or similar locations would likely produce much greater casualties, owing to the vast increases in coastal populations. This is also seen in smaller historic events that have impacted nearby densely populated coastlines, most notably the 1792 Unzen collapse landslide into the largely enclosed bay of the Ariake Sea, Kyushu, Japan. This collapse of an old dome complex had an estimated volume of 0.34 km^3, not all of which entered the bay, but it generated run-ups on the facing coastline of up to 20 m or more. Around 5000 people were killed by burial under the landslide, and another 10,000 or so by tsunami waves that would have taken only 10 min or so to cross the bay.

Tsunamis produced by the largest volcano collapse landslides at oceanic island volcanoes maybe far larger still than those produced by collapse landslides at island arc volcano such as Ritter Island. Mapped landslide units around many ocean island groups, such as the Canary Islands, Cape Verde Islands, and around Reunion Island in the Indian Ocean, have estimated volumes of several tens to hundreds of cubic kilometers; while the largest landslide units of all, around the Hawaiian Islands, are considered to have volumes up to 5000 km^3. Furthermore, the latter have run out distances of hundreds of kilometers and also extend upslope by hundreds of meters onto the Hawaiian Arch topographic high around the islands, indicating high velocities; and the great thicknesses of at least some of the deposits, with individual blocks up to 2 km thick, also imply a capacity to generate exceptionally large tsunami waves. The large horizontal dimensions of the collapses, tens of kilometers wide and long (in the direction of travel) also imply that the dominant wave periods would be as much as 10 min and that the waves would also be less dispersive than the waves from the much smaller collapse landslides that have occurred in historic time; therefore that they would propagate over transoceanic distances with only slow reduction in height by dispersion. These inferences have however proved controversial, because no such giant tsunamigenic landslide events have been observed in historic time: the controversy and possible routes to its resolution are discussed further in the final section of this chapter.

The relationships of volcano landslides to eruptions vary widely. Some are not associated with eruptive activity, and are thus particularly difficult to predict, but many including most of the larger events, are associated with at least phreatic or phreatomagmatic explosive activity. A common pattern seen on island volcanoes as at inland volcanoes is that collapses often occur at the beginning of, and may trigger, magmatic eruptions; however, the collapses may follow precursory activity such as ground deformation, rising seismicity, anomalous gas and groundwater discharges, and initial phreatic and phreatomagmatic explosions. Conversely, the 1792 Unzen collapse landslide followed an extended period of eruption at the adjacent Fugendake volcano, associated earthquakes, and according to some reports, episodes of anomalous discharge of hot groundwater and block sliding around the Unzen old volcanic dome itself. Such associations indicate that volcano collapse landslides are linked to destabilization of the volcanic edifices by phenomena associated with eruptions or shallow intrusions, for example, pressurization of ground water: this may explain the greater thickness of volcano collapse landslides as compared to most nonvolcanic landslides of comparable thickness, and so provide a basis for the expectation that being thicker future volcanic landslides will be more efficient tsunami sources. They also suggest a broad strategy for mitigation of volcano landslide tsunamis, discussed in the following section.

2.5. Tsunamis Associated with Large Explosive Eruptions

This category includes arguably the largest and certainly the most destructive volcanic tsunami waves of the past three centuries, those associated with the 1883 eruption of Krakatau and produced in the climactic phases of the eruption on August 25 and 26, 1883. Four series of major tsunami waves, the last by far the largest and most destructive, were generated in the space of less than 24 h.

Tsunami run-ups of as much as 40 m were recorded on the coast of Java to the east, some 60 km distant, with inundation distances of 3 km or more inland from the coast in flat-lying areas. Islands closer to the Krakatau volcano itself may have experienced even greater inundations, from the evidence of tsunami deposits interbedded with the pyroclastic deposits of the eruption. Uniquely among historic volcano tsunamis, long-period tsunami waves were generated in the fourth series of waves associated with the largest explosions of all, and propagated over transoceanic distances. These long-period waves, with wave periods of 30 min to 1 h, caused damage and one fatality in Sri Lanka, over 3000 km from Krakatau, and were recorded on tide gauges all around the world. Around the Sunda Strait itself, the tsunami waves caused over 90% of all the deaths associated with the Krakatau eruption, with around 33,000 deaths in some 250 towns and villages on the coasts of Java and Sumatra.

The Krakatau tsunami waves are also scientifically notable as the subject of the first coordinated and systematic scientific studies of a tsunami. These were carried out in particular by R.D.M. Verbeek, a civil engineer and surveyor in the Dutch East Indies colonial service, who collected eyewitness accounts from close to Krakatau and conducted the first detailed and quantitative post-event tsunami damage survey as part of his monumental report into the eruption and its effects in Indonesia; and by W.J.L. Wharton, a Royal Navy captain and Fellow of the Royal Society, who contributed a compilation of reports of the tsunami from around the world to the Royal Society report into the eruption. Unfortunately, these two studies came to different conclusions about the tsunami. Verbeek's primary data (Figure 58.2) showed that the size and effects of the fourth and largest tsunami wave train varied around Sunda Strait, with the largest run-ups and inundations to the northwest, north, and east, and significantly smaller values from section of coast in the arc from the west round to the southeast of Krakatau. This pattern is still more striking when it is recognized that large tsunamis, capable of producing the observed run-ups, traveling in these directions would have had to propagate for tens of kilometers in water depths not more than twice the amplitude of the waves, with the resulting potential for energy loss by wave breaking and seabed friction; whereas waves propagating to the west in particular would have traveled for the most part in water depths of hundreds of meters before striking southernmost Sumatra. Wharton, despite using essentially the same data, chose to dismiss the variation and asserted his belief that an average wave height of "50 feet" (16 m) was developed all around Sunda Strait. Unfortunately, in a summary table of both near and distant tsunami wave heights he then attached this value to each and every one of the sites investigated by Verbeek. This error has misled many tsunami modelers who have used Wharton's

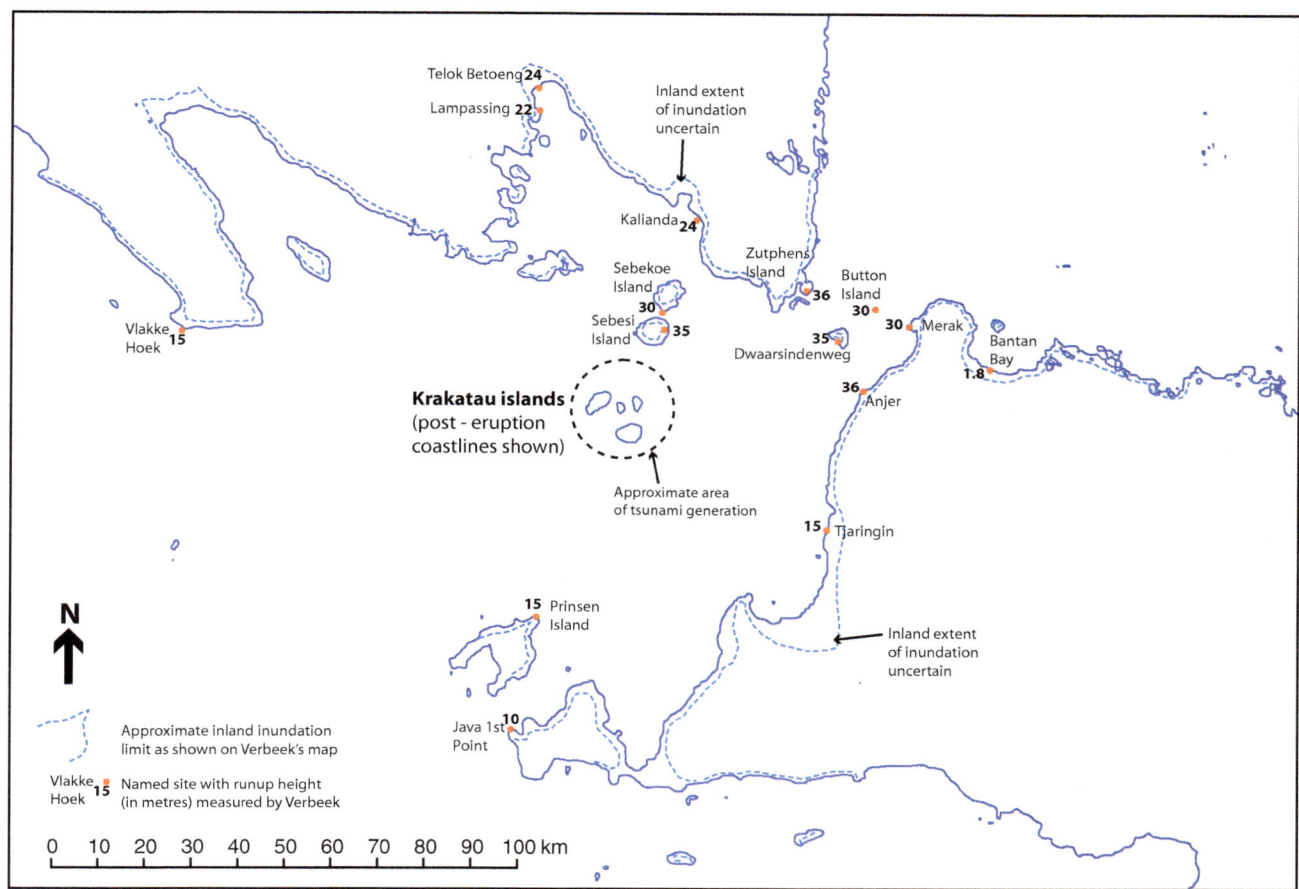

FIGURE 58.2 Simplified map of Sunda Strait map showing the locations of run-up height measurements made by R.D.M. Verbeek (Verbeek, 1885) and the approximate limit of the inundation zone mapped by Verbeek. Significant uncertainties exist in the map of the inundation limit both because of incomplete data and because of inaccuracies in the base map used by Verbeek.

summary table as the basis for their models, which have with few exceptions therefore favored a radially symmetric tsunami source such as a submarine explosion or crater excavation. As a result these models and their developers have been at odds with volcanologists studying the Krakatau eruption, who have for many decades used Verbeek's data, along with later geological studies, as the basis for their investigations of the origin of the tsunami waves.

These volcanological investigations, beginning with Verbeek's own interpretations of the eruption, have identified at least five different mechanisms for generation of the tsunami waves despite the large number and range of constraints provided by the detailed timing information from tide and pressure gauges, the numerous eyewitness accounts, and an abundance of geological, topographic, and bathymetric data. In part this reflects the complexity of the eruption itself, which likely involved many potentially tsunamigenic phenomena. These mechanisms, some of which correspond to the different categories of process discussed earlier in this chapter, are:

- Magmatic and phreatomagmatic explosions, some of which may have been submarine producing cavities in the ocean that then collapsed to produce tsunami waves. Such explosions are likely to have been located within the partly flooded Krakatau caldera.
- Coupling between the atmospheric pressure waves produced in these explosions and the ocean, producing seiche-like tsunami waves.
- Flooding of the Krakatau caldera as it subsided further during the climactic phases of the eruption. If sufficiently rapid this might have generated a rebounding **bore** that flooded out of the caldera again.
- Collapses of the walls of the deepening caldera, particularly from the unstable face of the Rakata cone at the south of the island that was bisected by the caldera wall, would have produced landslides and therefore landslide-generated tsunamis.
- Rapid emplacement of dense pyroclastic flows into the ocean, that are known from recent submarine geological investigations to have formed an extensive sheet of

pyroclastic deposits around the islands and were recognized by Verbeek as having formed an arc of transient islands or banks of pumice to the northwest, north, and east of the preexisting islands of the Krakatau group. Such flows would have been capable of pushing the ocean ahead of them as they advanced into the ocean, so generating tsunami waves.

Recent tsunami modeling work, that uses Verbeek's original run-up and inundation data, argues against some of these mechanisms but in favor of the last in particular. Tsunami sources within the caldera (the first, third, and fourth above) are unlikely to have produced much wave energy outside the caldera; furthermore most of the energy that would have escaped would have done so through deep radial grabens to the southwest and southeast, inconsistent with the pattern of strongest waves to the northwest, north, and east. The atmospheric coupling mechanism, which may have a significant role to play in generation of the long-period, far-field tsunamis waves, would produce a symmetric wave field in Sunda Strait. Only the fifth mechanism seems able to produce the observed pattern of tsunami wave directionality, since the arc of transient islands that indicates the location of thickest pyroclastic flow deposits corresponds to the arc of directions of the largest tsunami run-ups from the source area; furthermore this mechanism generates tsunami waves outside the caldera, avoiding the problem of trapping of wave energy within the caldera, and is also consistent with data on the timing of wave generation relative to the explosions once it is recognized that the pyroclastic flows were likely to be initially traveling faster than tsunami wave velocities in the shallow water around the Krakatau islands, resulting in the tsunamis arriving earlier at sites with well-constrained arrival times than would be the case if the waves were generated within the Krakatau caldera itself. Although it should be recognized that most of the data, and therefore most studies based on that data, relate to the tsunami waves generated by the fourth and largest explosion in the 1883 eruption (with the implication that the earlier tsunami wave trains may have been generated by different mechanisms), it therefore seems that the balance of evidence favors generation of the fourth tsunami wave train in particular by entry into the ocean of large and fast-moving dense pyroclastic flows.

The extended controversy and remaining uncertainties about the mechanism or mechanisms of origin of the Krakatau tsunamis, despite the comparative wealth of relevant data, underlines the difficulty of interpreting the origins of tsunamis generated by large-volume explosive eruptions. This difficulty is likely to be an inherent one, because of the complexity and variety of these eruptions. The tasks of understanding tsunami generation in other geologically recent but less well documented caldera-forming eruptions

for which there are indications of associated tsunamis from oral traditions or tsunami deposits, such as the c. AD 1452 eruption of Kuwae volcano in Vanuatu, the Minoan eruption of Santorini, and the c. 7300 years BP eruption of Kikai in southern Japan, are likely to be still more difficult than in the case of Krakatau.

3. MONITORING OF TSUNAMIGENIC VOLCANOES AND THE MITIGATION OF HAZARDS FROM VOLCANIC TSUNAMIS

Major international efforts have been directed, since the late 1940s, into instrumental detection and real-time warning of tsunamis generated by large earthquakes: initially by detection, location, and interpretation of the source earthquakes but more recently by direct detection of tsunami waves while still in deep water. While effective in mitigation of transoceanic tsunamis these systems have had only partial success, despite large financial and technical investments, in protecting populations against the effects of near-field tsunamis: those that are generated close to the coastal populations concerned and take only a few tens of minutes to arrive. This problem, seen most notably in the 2011 Tohoku earthquake and tsunami disaster, indicates that a different approach is needed to address the hazards associated with volcanic tsunamis. These, even more than earthquake-generated tsunamis, are primarily a threat to populations on coasts within several hundred kilometers of the source with the possible exception, discussed in the following section, of tsunamis generated by the largest volcano collapse landslides.

In contrast to the major earthquakes, for which there has been little success in identifying reliable precursors, most volcano tsunamis (with the exception of those generated by spontaneous or externally triggered slope failures producing large landslides) occur within sequences of events including potential precursor phenomena such as ground deformation, seismic swarms, gas and groundwater discharges, that can be indicative of impending tsunamigenic events such as eruptions and volcano flank instabilities. Therefore, whereas mitigation of earthquake-generated tsunami hazards is largely centered around responsive mitigation (that is to say, based upon actions taken in response to the source event), anticipatory mitigation based upon actions resulting from interpretation of precursory phenomena may be more relevant to mitigation of volcanic tsunamis. Development of monitoring techniques specifically as a basis for anticipatory mitigation of volcanic tsunamis needs to take account of three main situations that present different problems:

- Where a coastal volcano or lake volcano that is a potential tsunami source is located close to densely populated coastlines: examples include Unzen, Sakurajima, and Taal volcanoes.

- Where a relatively unpopulated, relatively remote island volcano has the potential to produce tsunamis that threaten more distant populated coastlines, particularly by volcano lateral collapse: examples include Oshima—Oshima volcano, and numerous island volcanoes in the South West Pacific volcanic arcs.
- Where a submarine volcano has the potential to generate tsunamis that threaten adjacent coastlines: examples include Kick 'em Jenny and the post-1888 vent in the Ritter Island collapse scar.

In the first case, it is likely that significant resources and monitoring capacity are already available for the volcanoes concerned. The principal additional requirements for effective monitoring and mitigation of volcano tsunami hazards are a capacity to detect precursors to even quite small tsunamigenic events, that may produce locally damaging tsunamis; and efforts to educate shoreline populations in particular of the need to respond to anticipatory warnings of potential tsunamis, because tsunami travel times from source to populated shorelines are likely to be short.

In the second case, while only larger and rarer events such as volcano collapse landslides are likely to produce tsunamis large enough to cause major damage to more distant coastlines, and the population on the source island volcano maybe small or nonexistent, the key problem is that resources to detect precursory phenomena maybe very limited. This applies in particular to remote island volcanoes in developing countries such as Bam and Manam volcanoes in Papua New Guinea. In these cases even strong indicators of potential flank instability (Figure 58.3) may not produce a corresponding effort to monitor potential precursors, because of the logistic difficulties of maintaining monitoring infrastructure on remote islands and because of a lack of human, technical, and financial resources.

In the third case, even for those submarine volcanoes close to populated coastlines in developed countries, technologies for monitoring precursors are limited in scope: in particular, it is very difficult to monitor surface deformation underwater (apart from point measurements of vertical deformation using seabed pressure gauges). New and alternative technologies are therefore needed. The detection of the Myojin-Sho eruptions of 1952—1954 on hydrophones at great distances across the Pacific Ocean suggests that more modern hydrophone arrays should have the capacity to detect and locate the sound sources formed by submarine eruptions, but the capacities of such arrays are highly classified because of their military applications. Temporary deployments of hydrophones for scientific purposes have detected submarine landslides on volcanoes as discussed above, further indicating their potential for monitoring and hazard mitigation at submarine volcanic vents. It should however be noted that most submarine volcanoes have summits well below the surface and are likely to be inefficient tsunami sources, due to limited volcanic explosivity and because landslides from them have to be large in order to attain a thickness/water depth ratio close to unity.

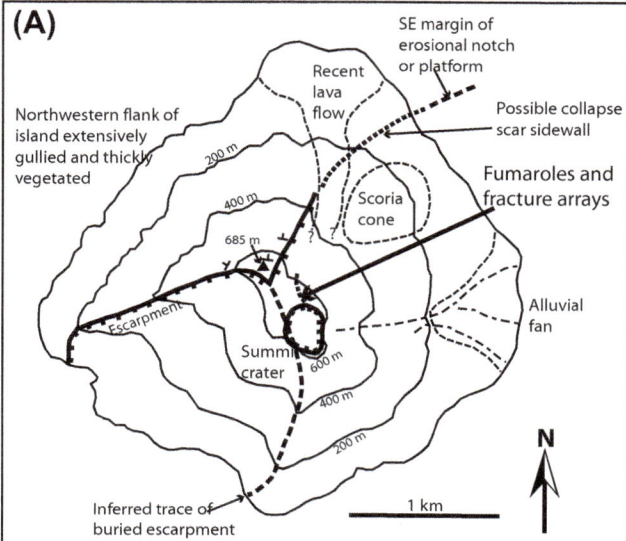

FIGURE 58.3 (A): Sketch map of Bam island showing the location of fissures with fumarolic activity in 2004—2005 along a possible incipient flank failure headwall scarp. (B): Photograph of the fissures taken by the author in August, 2005, some 12 months after the first report of activity. Audible sounds of boiling and steam escape indicated that the fissures were deep, extending to the water table at hundreds of meters depth. Lack of monitoring of such incipient landslides at remote volcanoes presents a major problem for mitigation of volcano tsunami hazards.

Submarine volcanoes with summit vents just below the surface are a greater cause for concern and have produced a number of small tsunamis as noted above; however, these are more easily recognized due to their propensity to produce phreatomagmatic explosions.

In all three cases, monitoring efforts are of limited value unless linked to practical plans for mitigative actions, for which planning, education, and infrastructure preparations are likely to be needed. There may also need to be an acceptance of the costs of mitigation and a preparedness to act despite uncertainty. For example, in the first category described above, evacuation of densely populated coastlines adjacent to an incipiently unstable erupting volcano in a bay containing important port facilities might lead to an extended period of economic disruption not just to the area directly threatened but to the entire economic hinterland that depends upon imports and exports passing through those port facilities.

4. CURRENT AND FUTURE RESEARCH DIRECTIONS

The low frequency of occurrence but high destructive potential of volcanic tsunamis, coupled with the variety and complexity of the processes that generate them, presents major problems for research into the hazards that they present and thus for mitigation of those hazards. Key areas for future research are:

- Models of the tsunami generation and propagation processes, respectively specific to the different generation processes discussed above and to the shorter wave period dispersive volcano tsunamis, that can be used in conjunction with data and volcanological models of potential source volcanoes to produce at least predictive scenarios if not probabilistic estimates of the tsunami hazards that these volcanoes may present in future.
- Understanding precursory phenomena to tsunamigenic events at coastal and island volcanoes, and the mechanisms that underlie them. There is in particular a need for an understanding of how volcano flanks may become destabilized and how they deform prior to collapse, as a basis for developing and implementing methods of anticipatory warning and mitigation such as evacuation of vulnerable coastal areas in response to detection and analysis of these precursory phenomena.
- Methods of recognizing and quantitatively interpreting geological evidence of past volcano tsunamis, to provide a basis for estimating the frequency of occurrence, spatial distribution and intensity of impact of past volcano tsunamis beyond the short historical record, that is at most a few centuries long in many areas vulnerable to volcanic tsunamis. This includes both

means of recognizing tsunami deposits within the complex volcanic sequences of large explosive eruptions, and of distinguishing deposits from volcanic tsunamis that typically have many shorter-period waves, from deposits from earthquake-generated tsunamis that typically have a few very long-period waves. A related problem is that of quantitatively characterizing the tsunami source events and correlating them with the resulting tsunami deposits: the latter problem is especially difficult in the case of far-field tsunami deposits, where the link requires exceptionally accurate radiometric dates or other geochronological work.

These problems are especially critical in the cases of the largest volcanic tsunamis, mainly those produced in large-volume explosive eruptions and by large volcano collapse landslides, as the large size and longer wave periods of these tsunamis mean that they are likely to cause greater damage at greater distances from the source volcano, and may therefore represent a disproportionately large component of the overall volcano tsunami hazard relative to their frequency of occurrence. Furthermore, these groups of events have been the subject of substantial scientific debates and disagreement in recent decades. The controversies over the origin of the 1883 Krakatau tsunamis have been discussed above and may be approaching some degree of resolution, although much further work is needed to understand the physics of tsunami generation by entry into the ocean of large-volume, dense pyroclastic flows. Tsunami generation by volcano collapse landslides, especially the hypothesis of generation of giant transoceanic tsunamis by the largest lateral collapse landslides at oceanic island volcanoes, has been and remains another controversial subject. The main areas of debate can be summarized under the headings noted above.

Existing models of tsunami generation by giant landslides disagree on how the landslides should be treated, either as sliding blocks, progressively disintegrating blocks that transform into frictional flows, or as viscous or viscoplastic fluids. As a result there is little agreement on the initial acceleration history, peak velocity, and history of deformation of shape change in the model landslides: as discussed above, all of these are critical to the efficiency of the landslides as tsunami sources. With few exceptions, the models have not been tested against experimental data (which give the additional problem of scaling the models up from the laboratory to the giant landslide scale), and even fewer have been tested against the incomplete but nevertheless important data sets from historic volcano collapses and resulting tsunamis such as 1741 Oshima—Oshima and 1888 Ritter Island. Important disagreements also exist about how to model the extremely large waves likely to be generated, since these may experience nonlinear behaviors such as wave breaking close to source; although equally it is not known how much effect the spilling or surging type of breaking that is likely to result will have on the eventual sizes and periods of waves that propagate into deep water, since our understanding of wave breaking is mainly based upon waves propagating into shallow water. A final area of disagreement between existing models, that greatly affects their outputs, is that of whether oceanic island volcano collapses should be treated as single giant landslides or as a succession of smaller landslides, but this derives directly from disagreements about how to interpret the landslide deposits and is discussed further below.

With respect to the propagation of very large tsunami waves that maybe generated by oceanic island volcano collapses, repeated claims have been made that even the largest volcano landslides will produce relatively short-period waves that will disperse rapidly and have little impact at transoceanic distances. However, the data from Oshima–Oshima and Ritter Island tsunamis, coupled with studies of other short-period tsunami waves such as those generated by nuclear weapons tests, indicate that while such dispersion will be significant, the result is likely to be that oceanic island collapse landslides will produce extended wave trains with waves of amplitudes of a few tens of meters at transoceanic distances—comparable in size to those generated near source by the largest submarine earthquakes and by the 1883 Krakatau eruption, but potentially affecting coastlines up to tens of thousands of kilometers long, rather than only a few hundreds of kilometers. The Krakatau tsunami waves are also relevant to another model-based prediction, that wide continental shelves will protect coastlines from the transoceanic waves from oceanic island volcano collapses by absorbing the energy of these waves through friction on the shallow seabed. As discussed above, large tsunami waves from Krakatau propagated to the north and east for several tens of kilometers through the shallow waters of Sunda Strait before impacting coastlines with run-ups of up to 40 m, while having less impact to the south and especially the west, despite the intervening sea being deeper in those directions. This asymmetry is difficult to explain if seabed friction is really a significant mechanism for absorbing tsunami wave energy during propagation through waters of order 100 m depth.

Mechanisms of lateral collapse at oceanic island volcanoes are important both for the recognition and interpretation of potential precursors, and for the resulting controls on the geometry and initiation of the collapses themselves. A key feature of these collapses appears to be that, like the smaller lateral collapses that occur at stratovolcanoes, they are very much thicker than most nonvolcanic landslides and typically have a wedge shape in longitudinal section defined by the surface slope of the volcano and a much lower angle basal failure surface. The mechanism for producing this shape and particularly the deep, low-angle basal failure surface is not fully understood but may be related to volcano destabilization mechanisms involving pressurization of pore fluids within the core of the edifice, for example, by heating of low-compressibility pore water, and forces exerted by pressurized magma in dikes and sills that are also concentrated in the cores of the edifices. Equivalent thermal and magma-generated forces do not operate in nonvolcanic landslides such as continental slope sediment failures, which are typically thinner in relation to their area and have slope-parallel basal failure surfaces, and also have very different permeability distributions and internal structures. Arguments that oceanic island volcano lateral collapses are likely to involve sequential or retrogressive failures, beginning at the base of the slope and propagating up to the summits, based on analogy with the inferred behavior of continental slope sediment failures, are therefore problematic. While analogies with the smaller collapses at island arc stratovolcanoes such as Ritter and Oshima–Oshima are useful, it may also be that different oceanic island volcanoes show different mechanisms of destabilization and failure. A few such volcanoes, most notably Kilauea among active examples, show continuous flank deformation on well-developed faults and creeping structures, presenting the problem of how they can exhibit both such quasi-continuous deformation and rare sudden flank failures. Most oceanic island volcanoes show much less evidence for quasi-continuous deformation and usually lack active flank faulting, but do exhibit evidence of past lateral collapses. It is for this reason that even slight indications of flank destabilization on such volcanoes, such as long-term structural reconfigurations and even small flank deformations as are seen on oceanic island volcanoes such as the Cumbre Vieja (La Palma, Canary Islands) and Fogo (Cape Verde Islands), should be of concern and the subject of intensive monitoring.

Investigations of the deposits from past lateral collapses, particularly around the Canary Islands but also around Hawaii, have also been used to address the question of whether these collapses involved single giant landslides or a number of much smaller events of more limited tsunami-genic potential. The critical evidence comes not from the proximal debris avalanche deposits, but from turbidite sequences at the far distal end of the deposits hundreds of kilometers from the source islands. These sequences contain multiple turbidite beds, inferred to correspond to discrete landslide events. The upper beds are certainly distinct, with interbed contacts that can be correlated between cores and appear to represent significant time intervals. However, the lower beds in each sequence are less clearly separated and represent the bulk of the volume of the sequence in those cases where volumes have been determined. Such a sequence can therefore be interpreted in a different way, as recording a single giant tsunamigenic landslide followed by a series of smaller landslides from the collapse scar headwall. Unfortunately, corresponding studies have not been carried out at the time of writing on the distal end of deposits from historic island arc volcano lateral collapses known to have produced single tsunami wave trains, such as 1888 Ritter Island and 1741 Oshima–Oshima, to see if they also contain sequences of multiple turbidite beds.

An important alternative approach to the problem of whether or not past giant lateral collapses produced correspondingly large tsunamis is of course to look for such deposits either on the source islands or on adjacent islands. Unfortunately, the steep coasts of most oceanic island groups are not good site for tsunami deposit preservation. Nevertheless, a number of unusual marine fossil bearing conglomerate deposits found up to a few hundred meters above contemporary sea level on islands in Hawaii and the Canary Islands in particular, have been interpreted as deposits from such tsunamis, but have proved controversial. This is in part because of extensive reworking by alluvial processes since initial deposition, that has destroyed any sedimentary structures indicative of deposition by tsunamis and leaves only the marine provenance of fossils and other clasts as secure evidence of the tsunamigenic origin of these deposits. Distant flat-lying coastlines maybe better sites for tsunami deposit preservation without reworking and also offer the potential for demonstrating transoceanic propagation of tsunamis from oceanic island volcano collapses, but present the problem of determining the ages of both the deposits and of the source collapse events with sufficient accuracy to enable convincing correlation between them.

Overall, therefore, volcano tsunamis and the resulting hazards present many problems requiring further research. While as discussed above these problems are particularly difficult in the case of the rarest, largest classes of event, there is also a need for further investigations of the phenomena that can lead to smaller, more locally damaging volcano tsunamis both because these phenomena occur more frequently but can also, as in the case of island arc volcano lateral collapses and the resulting tsunamis, provide important insights into the rarer, larger events, and their tsunamis.

FURTHER READING

Antonopolos, J., 1992. The great Minoan eruption of Thera volcano and ensuing tsunami in the Greek Archipelago. Nat. Hazards 5, 153–168.

Béget, J.E., Kienle, J., 1992. Cyclic formation of debris avalanches at Mount St. Augustine volcano. Nature 356, 701–704.

Béget, J.E., 2000. Volcanic tsunamis. In: Encyclopedia of Volcanoes, first ed., pp. 1005–1014.

Belousov, A., Voight, B., Belousova, M., Murveyev, Y., 2000. Tsunamis generated by subaquatic volcanic explosions: unique data from 1996 eruption in Karymskoye Lake, Kamchatka, Russia. Pure Appl. Geophys. 157 (6–8), 1135–1143.

Caplan-Auerbach, J., Dziak, R.P., Bohnenstiehl, D.R., Chadwick, W.W., Lau, T.-K., 2014. Hydroacoustic investigation of submarine landslides at West Mata volcano, Lau Basin. Geophys. Res. Lett. 41 http://dx.doi.org/10.1002/2014GL060964.

Carey, S., Sigurdsson, H., Mandeville, C., Bronto, S., 2000. Volcanic hazards from pyroclastic flow discharge into the sea: examples from the 1883 eruption of Krakatau, Indonesia. In: McCoy, F.W., Heiken, G. (Eds.), Volcanic Hazards and Disasters in Human Antiquity, Geological Society of America Special Paper, vol. 345, pp. 1–14.

Day, S.J., Watts, P., Grilli, S.T., Kirby, J.T., 2005. Mechanical models of the 1975 Kalapana, Hawaii earthquake and tsunami. Mar. Geol. 215, 59–92.

Fiske, R.S., Cashman, K.V., Shibata, A., Watanabe, K., 1998. Tephra dispersal from Myojinsho, Japan, during its shallow submarine eruption of 1952–1953. Bull. Volcanol. 59, 262–275.

Hunt, J.E., Wynn, R.B., Masson, D.G., Talling, P.J., Teagle, D.A.H., 2011. Sedimentological and geochemical evidence for multistage failure of volcanic island landslides: a case study from Icod landslide on north Tenerife, Canary Islands. Geochem. Geophys. Geosy. 12 (12), Q12007. http://dx.doi.org/10.1029/2011GC003740.

Kienle, J., Kowalik, Z., Murty, T.S., 1987. Tsunamis generated by eruption from Mount St. Augustine volcano, Alaska. Science 236, 1442–1447.

Latter, J.N., 1982. Tsunamis of volcanic origin: summary of causes with particular reference to Krakatoa, 1883. Bull. Volcanol. 44, 467–490.

Maeno, F., Imamura, F., 2011. Tsunami generation by a rapid entrance of pyroclastic flow into the sea during the 1883 Krakatau eruption, Indonesia. J. Geophys. Res. 116, B09205. http://dx.doi.org/10.1029/2011JB008253.

Novikova, T., Papadopoulos, G.A., McCoy, F.W., 2011. Modelling of tsunami generated by the giant Late Bronze Age eruption of Thera, South Aegean Sea, Greece. Geophys. J. Int. 186, 665–680.

Paris, R., Wassmer, P., Lavigne, F., Belousov, A., Belusova, M., Iskandarsyah, Y., Benbakkar, M., Ontowirjo, B., Mazzoni, N., 2014. Coupling eruption and tsunami records: the Krakatau 1883 case study, Indonesia. Bull. Volcanol. 76 (814) http://dx.doi.org/10.1007/s00445-014-0814-x.

Satake, K., 2007. Volcanic origin of the 1741 Oshima-Oshima tsunami in the Japan Sea. Earth Planets Space 59, 381–390.

Sigurdsson, H., 1989. Submarine investigations in the crater of Kick 'em Jenny volcano. Bull. Sci. Event Alert Network 4, 45–49.

Silver, E., Day, S.J., Ward, S.N., Hoffmann, G., Llanes-Estrada, P., Driscoll, N., Appelgate, B., Saunders, S., 2009. Volcano collapse and tsunami generation in the Bismarck Volcanic Arc, Papua New Guinea. J. Volcanol. Geoth. Res. 186 (3), 210–222. http://dx.doi.org/10.1016/j.jvolgeores.2009.06.013.

Simkin, T., Fiske, R.S., 1983. Krakatau 1883: The Volcanic Eruption and Its Effects. Smithsonian Institution, Washington, DC, p. 464.

Smith, M.S., Shepherd, J.B., 1993. Preliminary investigations of the tsunami hazard of Kick 'em Jenny volcano. Nat. Hazards 7 (3), 257–278.

The eruption of Krakatau, and subsequent phenomena. In: Symons, G.J. (Ed.), 1888. Report of the Krakatau Committee of the Royal Society. Publ. Trubner, London, p. 494.

Verbeek, R.D.M., 1885. Krakatau. Batavia Government Press, p. 495.

Ward, S.N., 2001. Landslide tsunami. J. Geophys. Res. 106 (6), 11,201−11,215.

Ward, S.N., Day, S.J., 2001. Cumbre Vieja volcano: potential collapse and tsunami at La Palma, Canary Islands. Geophys. Res. Lett. 28, 3397−3400.

Ward, S.N., Day, S.J., 2003. Ritter Island Volcano − Lateral collapse and tsunami of 1888. Geophys. J. Int. 154, 891−902.

Chapter 59

Volcanic Seismicity

Stephen R. McNutt

School of Geosciences, University of South Florida, Tampa, FL, USA

Diana C. Roman

Department of Terrestrial Magnetism, Carnegie Institution of Washington, DC, USA

Chapter Outline

GLOSSARY

explosion earthquake Events that are recorded as seismic waves traveling through the ground (including P waves and S waves) often followed by air waves. The air waves travel through the air and are coupled back into the ground near the recording site.

high-frequency (HF) earthquake Discrete events with clear P waves and S waves. The dominant energy is above 5 Hz (synonyms are A-type earthquake or volcano-tectonic (VT) earthquake).

low-frequency (LF) earthquake Discrete events with weak P and no discernable or distinct S waves. Recordings show monotonic, low frequency (1—5 Hz) waveforms that resonate for many cycles (synonyms are B-type event and long-period (LP) event).

magnitude A measure of earthquake size, usually determined by measuring the highest-amplitude waves and correcting for distance and instrument type. The scale is logarithmic, so each increase of one unit corresponds to an amplitude increase of a factor of 10.

P wave Primary wave, a compressional elastic wave with primarily longitudinal particle motion parallel to the propagation direction. P waves travel faster than other waves and are thus the first to appear on seismograms.

S wave Secondary wave, a shear elastic wave with particle motion perpendicular to the direction of propagation. S waves are usually the most prominent phases on a seismogram; they travel more slowly than P waves and thus arrive later on seismograms; they cannot pass through liquids.

swarm A group of many earthquakes of similar magnitude occurring closely clustered in space and time with no dominant mainshock.

volcanic tremor Continuous seismic signal with regular or irregular sine wave appearance and characterized by low frequencies (0.5—5 Hz). Harmonic tremor has very uniform appearance, whereas spasmodic tremor is pulsating and consists of higher frequencies with a more irregular appearance.

The Encyclopedia of Volcanoes. http://dx.doi.org/10.1016/B978-0-12-385938-9.00059-6

Volcano seismology is the study of earthquakes of volcanic origin as well as of velocity structure, attenuation, and other physical properties of the Earth materials that affect the passage of seismic waves at volcanoes. Volcanic earthquakes may be defined as earthquakes that occur at or near volcanoes, generally within 15 km, or that are related to volcanic processes. Volcanoes are places where heat and mobile fluids are concentrated, so the number of earthquakes per unit time is high when compared with normal crust. Most volcanic earthquakes take place at shallower depths (1−9 km) than tectonic earthquakes on typical faults (generally to depths of about 15 km in the crust; as deep as 700 km in subduction zones). Volcanic events also differ in their patterns of occurrence: they often occur in **swarms**, which are groups of many small events with similar **magnitudes** and locations. This is in contrast to the typical mainshock−aftershock sequences characteristic of tectonic earthquakes. Volcanoes produce different types of earthquakes that are thought to represent different physical processes.

Among these earthquakes are **high-frequency (HF)** (also known as volcano-tectonic (VT) earthquakes), **low-frequency (LF)** (also known as long-period (LP) events), explosions, and **volcanic tremor**. These typically occur in increasing numbers prior to or during eruptions. Seismicity also increases during intrusions—episodes of magma ascent that do not culminate in an eruption—and even though the eruptive outcomes differ in these cases, the processes that cause various types of volcanic earthquakes are similar. The source mechanisms of some volcanic earthquakes are not yet well understood.

1. INTRODUCTION

Volcanoes are the sources of a great variety of seismic signals that differ from those produced by tectonic earthquakes. Nearly every recorded volcanic eruption has been preceded by an increase in earthquake activity beneath or near the volcano and accompanied and followed by varying levels of seismicity. For this reason, seismology has become one of the most useful tools for eruption forecasting and monitoring. At present, approximately 200 of the world's volcanoes are seismically monitored, although the number and quality of stations at each volcano varies considerably. This represents about one-third of the 541 volcanoes that have erupted in historic times. Over the last several decades, 55−70 individual volcanoes have erupted each year. Because erupting volcanoes draw attention both publicly and scientifically, over half of these are seismically monitored (see Chapter 63).

This chapter reviews some of the developments in volcano seismology over the last 10 years that have led to an improved understanding of volcanoes and the volcanic processes that cause earthquakes and other types of seismicity. The chapter also describes several patterns and relationships in volcano seismology that form the physical basis of contemporary monitoring and forecasting, although these topics are covered in greater detail elsewhere.

1.1. Some Famous Early Eruptions

Throughout history, the relationship between earthquakes and volcanoes has been close, although there has been confusion about causal relations between the two. This is because both volcanoes and earthquakes form parallel belts at subduction zones, where 95% of the Earth's subaerial volcanoes are located. It is common in the historic and recent record to find an increase in reports of regional earthquakes around the times of eruptions. For example, the reports of the eruption of Vesuvius in AD 79 tell of numerous local earthquakes preceding the eruptive activity. People living near Mount Nuovo, Italy, in 1538 and Mount Usu, Japan, in 1663 fled from the areas before the eruptions because of many felt earthquakes. Early observatories at Vesuvius starting in 1856, Usu in 1910, and Hawaii in 1912 recorded different types of earthquakes and began the systematic study of eruption precursors. Three weeks of felt earthquakes heralded the formation of the new volcano Paricutin in Mexico in 1943. These and many other examples illustrate that volcanoes are seismically active features.

1.2. Brief Review of Earthquake Seismology

In order to appreciate the unique features of earthquakes at volcanoes, it is first necessary to review some features of typical earthquakes. Most earthquakes, which volcanologists often refer to as "tectonic earthquakes" to distinguish them from volcanic earthquakes, are caused by shear failure of rock. That is, rock masses on either side of a planar fracture slip past each other in a shearing motion, generating heat, deformation, and elastic waves. It may surprise most readers to learn that only about 1% of the energy of an earthquake is released in the form of elastic waves, which is the part that is felt or recorded by instruments as ground shaking. Most of the energy is liberated as heat and deformation. While rocks are fractured and faulted at all scales, it is only where they are organized into large planar features that we can identify them as faults. Faults can be the sites of large earthquakes, those that involve a large surface area of the fault, as well as many smaller ones that rupture only a small surface area. In general, there are many more small earthquakes than large ones; in fact, the distribution of earthquake numbers versus magnitude follows a power law, known to seismologists as the frequency−magnitude distribution. The relation is

$$\log_{10} N = a - bM$$

where M is magnitude, N is the cumulative number of earthquakes greater than or equal to M, and a and b are constants. The slope, b, is known as the b-value, and usually takes on values of 0.9—1.0 for tectonic earthquake sequences. This means that for every M3 earthquake, there are ten M2 earthquakes, one hundred M1 earthquakes, and so on.

There are several different measures of earthquake size. The most widely known is magnitude, which is the amplitude of the elastic waves recorded on a particular instrument and corrected for distance. Magnitude was originally developed by C. Richter using the Wood—Anderson torsion seismometer. A more physical measure of earthquake size is seismic moment, which is the product of fault area, slip, and the rigidity or strength of the rock. The ranges of these parameters and how they relate to each other are called scaling relations, and they permit the estimation of one earthquake parameter from another. Table 59.1 shows typical dimensions and parameters for earthquakes of different magnitudes.

Once an earthquake occurs, the elastic waves travel away from the source (the fault) along different paths and eventually reach the seismometers. Characteristics of the seismogram (the way the earthquake appears on a paper record or other media) may be generated at the source, along the path, and at the recording site; thus, a major task for seismologists is to determine, for each seismogram, the relative contributions of source, path, and site effects. Once these are known, a basis exists to demonstrate that a particular earthquake is LF, harmonic, etc., and physical models are then constructed to evaluate the likely causes of peculiar (and normal) waveforms.

Earthquakes are located using a velocity model of the Earth's structure. Most velocity models are 1-D models consisting of horizontal layers of varying thicknesses and velocities that generally increase with depth. Velocities in the shallow crust (0—10 km) typically have values of ~5 km/s, while velocities near the Earth's surface (upper

1 km) have lower values of ~3 km/s near volcanoes. To locate an earthquake, a trial location and depth are assumed and rays are traced from the earthquake to the various stations using Snell's law according to the velocity model. The theoretical travel times are then compared with observed travel times. If the waves arrive at the station too early (or too late), then the earthquake is moved away from (or toward) that station by a small amount and the rays are traced again. Each step is known as an iteration, and computer programs are designed to iterate with predefined small increments until the travel time errors (difference between observed and theoretical time) reach a minimum. At this point, further iterations do not result in reduced errors, and the event is considered to be located. Note that an earthquake location is a model-dependent interpretation and not a fact. If the velocity model is changed or arrival time picks are modified, then the location changes as well.

Seismology is also concerned with the distribution of velocities within the Earth. Seismic waves have much higher velocities than what we typically deal with on a day-to-day basis and are measured in kilometers per second. For example, a car traveling on a highway at 60 mph is only going about 0.027 km/s, whereas air waves (sound waves) travel at 0.331 km/s (740 mph) and typical **P waves** at volcanoes travel at about 3—5 km/s (6710—11,180 mph) or faster. Seismic wave velocities, like all wave velocities, depend on the characteristics of the material they are passing through; for example, seismic waves travel more slowly through molten rock than they do through solid rock. Therefore, seismic tomography, similar to medical tomography, uses variations in the speed of elastic waves to recover information about where magma is located. With this brief overview as a reference frame, we now turn our attention to some of the features of volcanic earthquakes.

2. TYPES OF VOLCANIC EARTHQUAKES AND TERMINOLOGY

Active volcanoes are the sources of a great variety of seismic signals. Traditionally, volcanic earthquakes have been classified based on seismogram appearance into four different types: (1) HF, VT, or A-type; (2) LF, LP, or B-type; (3) explosion quakes; and (4) volcanic tremor. This general classification scheme works well at a large number of volcanoes. Other local signals, such as those caused by glaciers or landslides, are also recorded at volcanoes. Although the original classification included restricted depth ranges for various events, these have often been relaxed because of improvements in location accuracy and better understanding of source and propagation effects. This section also provides additional discussion on terminology. Examples of volcanic events from several volcanoes are shown in Figure 59.1.

TABLE 59.1 Typical Earthquake Parameters (Circular Fault, 30-bar Stress Drop)

Magnitude	Radius	Slip	Moment[1] (Nm)	Duration[2] (s)
−2	0.85 m	0.06 mm	4×10^6	0.00053
0	8.5 m	0.6 mm	4×10^9	0.0053
2	85 m	6 mm	4×10^{12}	0.053
4	0.85 km	6 cm	4×10^{15}	0.53
6	8.5 km	60 cm	4×10^{18}	5.3
8	85 km	6 m	4×10^{21}	53

[1]Rigidity $= 3 \times 10^{10}$ N/m^2.
[2]Assumed $\beta = 3.18$ km/s.

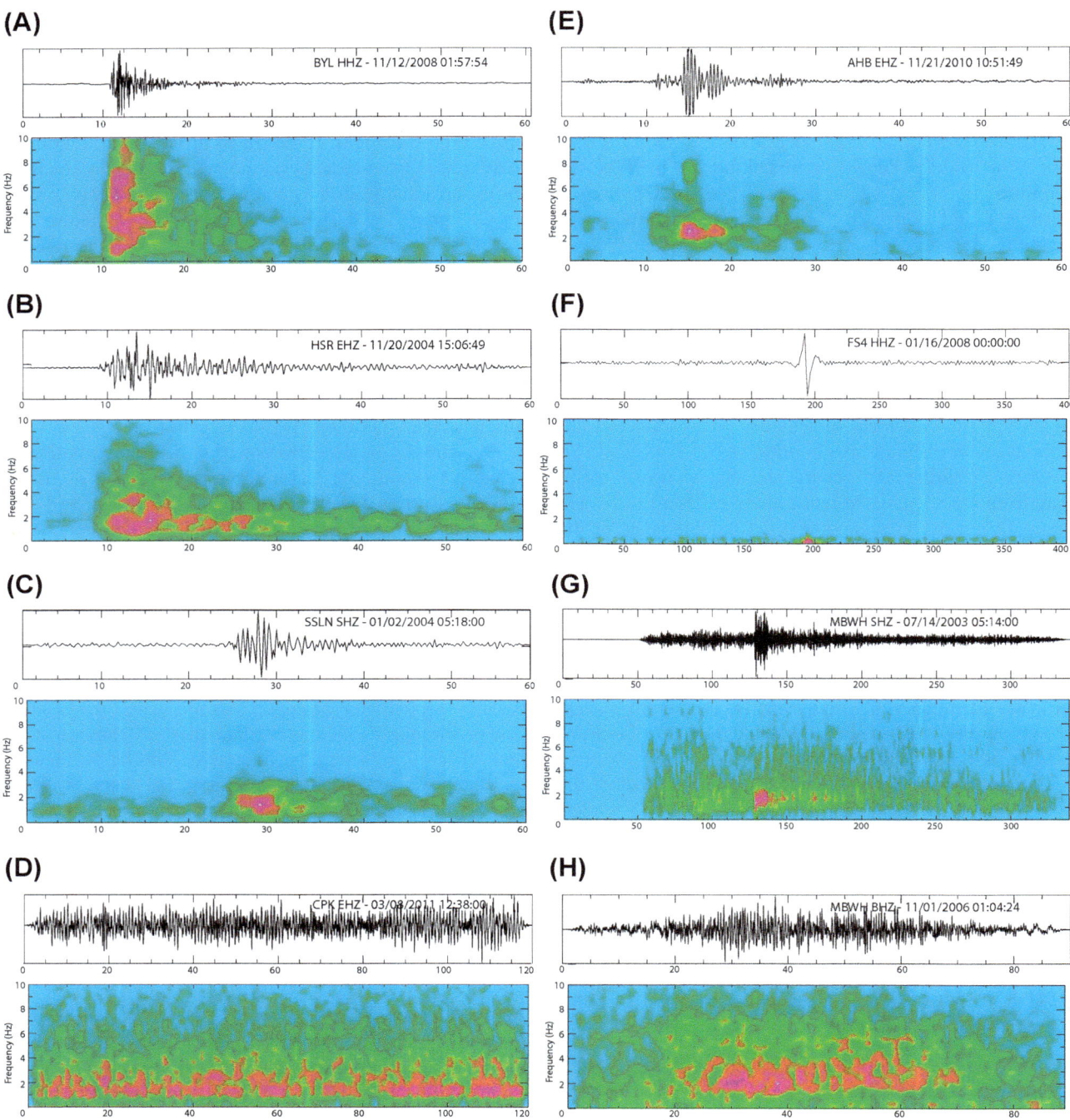

FIGURE 59.1 Example waveforms (top) and spectrograms (bottom) of volcanic seismicity: (A) Volcano-tectonic earthquake recorded at Kilauea volcano, Hawaii. *(Data courtesy HVO.)* (B) Hybrid event recorded at Mount St Helens, Washington. *(Data courtesy CVO.)* (C) Long-period event recorded at Shishaldin volcano, Alaska. *(Data courtesy AVO.)* (D) Volcanic tremor recorded at Kilauea volcano, Hawai'i. *(Data courtesy HVO.)* (E) Deep long-period earthquake recorded at Akutan volcano, Alaska. *(Data courtesy AVO.)* (F) Very long-period earthquake recorded at Fuego volcano, Guatemala. *(Data courtesy G. Waite.)* (G) Explosion earthquake recorded at Soufrière Hills volcano, Montserrat. *(Data courtesy MVO.)* Note that this event begins with a dome collapse — the explosion begins at approximately 130 s. (H) Rockfall event recorded at Soufrière Hills volcano, Montserrat. *(Data courtesy MVO.)* All signals have been bandpass filtered between 0.1 and 10 Hz, with the exception of (F), which was bandpass filtered between 60 and 12 s periods. Note different timescales (given in seconds at the bottom of each plot). Station name, channel code, and event date and time are given on each waveform. HVO, Hawaiian Volcano Observatory; CVO, Cascades Volcano Observatory; AVO, Alaska Volcano Observatory; MVO, Montserrat Volcano Observatory.

2.1. HF Earthquakes

Most HF earthquakes are thought to be caused by shear failure or slip on faults and differ from their tectonic counterparts only in their maximum magnitudes and patterns of occurrence. At volcanoes, earthquakes typically occur in swarms rather than mainshock—aftershock sequences. A swarm may be defined as a group of many earthquakes clustered in space with no dominant shock. In practical terms, the difference in magnitude between the largest and second largest event of a swarm is 0.5 magnitude unit or less, as opposed to 1.0 magnitude unit or more for most mainshock—aftershock sequences. HF earthquakes have clear P- and S-wave onsets, and dominant frequencies are 5—15 Hz. Higher frequencies are generated at the source but are not recorded because of instrumental limitations and high local attenuation. Their magnitudes are typically lower than M = 3.0. A typical HF (VT) earthquake is shown in Figure 59.1(A).

2.2. LF Events

Most LF events are thought to be caused by fluid pressurization processes, such as bubble formation and collapse, and also by shear failure or tensile failure of the rock, or nonlinear flow processes that occur at very shallow depths for which attenuation and path effects also play an important role. These events often have emergent P waves, lack distinct S waves, and have dominant frequencies between 1 and 5 Hz, with 2—3 Hz being the most common. Examples of LF events, including LP and deep LP events are shown in Figures 59.1(C) and (E).

Prior to eruption, magma must move from depth to the Earth's surface through conduits, dikes, sills, reservoirs, or various combinations of these. Thus, models of LF volcanic earthquakes often select a suitable geometry, such as a rectangular crack or a cylindrical pipe, and then attempt to reproduce seismograms by appropriate choice of length, width, and velocity of the material within it, which can be either magma or water (seismicity similar to that at volcanoes is found at geysers and geothermal areas). A source of mechanical energy is needed; this can be a small earthquake adjacent to a conduit, a flow transient or pressure fluctuation within a conduit, gas bubbles expanding or contracting, a shock wave from choked flow, or other causes. Some researchers consider the source to be the mechanical energy alone, while others treat the source as the ensemble of the mechanical energy and the resonant response of the magma or water in conduits or dikes.

2.3. Hybrid Events

Some earthquakes share attributes of both HF and LF events. These are called hybrid events. For example, the event shown in Figure 59.1(B) displays the HF onset of an HF event, but the later part of the signal (the coda) is similar to LF events. It is thought that such events may represent a mixture of processes such as an earthquake occurring adjacent to a fluid-filled cavity and setting it into oscillation. Others suggest that hybrids are shallower than LF events and thus preserve most of the HF energy that is attenuated for deeper events.

2.4. Volcanic Tremor

Volcanic tremor is a continuous signal with duration of minutes to days or longer. The dominant frequencies of tremor are 1—5 Hz (2—3 Hz is the most common), similar to LF events, and many investigators have concluded that tremor is a series of LF events occurring at intervals of a few seconds. Harmonic tremor and spasmodic tremor are two special cases of more general volcanic tremor. Harmonic tremor is an LF, often single-frequency sine wave with smoothly varying amplitude, or sometimes it consists of a fundamental frequency with a set of harmonics. Spasmodic tremor is an HF, pulsating, irregular signal. An example of volcanic tremor is shown in Figure 59.1(D).

It has previously been stated that earthquakes obey a power law, known in seismology as the frequency—magnitude relation. There are more small earthquakes than large ones, and the relative numbers of events form a power law distribution. Volcanic tremor, however, obeys a different type of relation, which is an exponential law distribution (Figure 59.2). For tremor, which is a continuous signal, we cannot count the number of events, so we compute the duration at different elevated amplitudes. Synthetic experiments have shown that this is equivalent to counting events because large and small earthquakes have different durations (large ones shake longer). The exponential law has implications for the source of volcanic tremor. For earthquakes, both the fault area and the slip increase as magnitude increases, but for tremor, either the source size (such as conduit length) or the magnitude of pressure fluctuations (analogous to slip) must remain constant to produce the observed exponential scaling. The available evidence suggests that the source (conduit) size remains constant.

2.5. Very-Long-Period Events

Over the past two decades, new types of seismometers known as broadband seismometers have been deployed at volcanoes. These have the ability to detect ground motions over a wider frequency band, particularly at the LF end (down to f = 0.016 Hz, or periods of 60 s and lower). Not surprisingly, a new class of events called very-long-period (VLP) events has been observed at some volcanoes using broadband seismometers. Observations to date show events with periods of

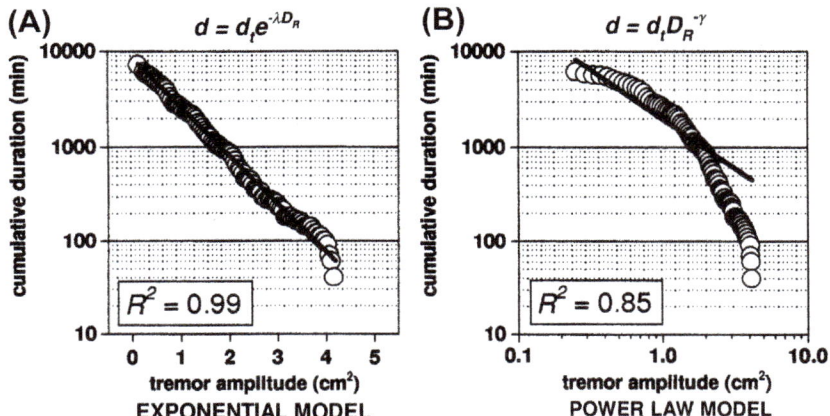

FIGURE 59.2 Volcanic tremor data from Mount Spurr, Alaska, comparing an (A) exponential and a (B) power law model. Note that the vertical axis is logarithmic in both cases, but the horizontal axis is linear for the exponential model and logarithmic for the power law model. The circles show the duration of tremor observed at various amplitudes; strong tremor occurs less often than weaker tremor. The lines are weighted least squares mathematical fits of the data to the two models shown. The data fit better using the exponential model as shown by the higher value of the correlation coefficient R^2. *After Benoit et al. (2003).*

3–50 s originating from shallow depths of 1.5 km or less under several active volcanoes, including Kilauea, Stromboli, Aso, Sakurajima, Fuego, Augustine, and Satsuma-Iwojima. The VLP events at these volcanoes have been associated with either eruptions or vigorous fumarolic activity, and the observed events have fairly small amplitudes. An example of a VLP event is shown in Figure 59.1(F).

Normally, it takes a large structure, such as a long fault, to produce significant waves with these long periods. However, many of the VLP events at volcanoes appear to come from small source zones despite the long wavelengths. Some current models suggest that these events may be produced as magma moves through a flapper valve (such as a restriction in a dike or sill) in distinct pulses. Additionally, theoretical calculations show that it is possible for a crack or tube filled with fluid to produce long-wavelength (and LP) waves if the acoustic velocity in the fluid is slow compared with the rigid walls. The interaction between the fluid and walls produces low phase velocities along the interface. Such waves are called crack waves or tube waves.

During its climactic eruption on June 15, 1991, Mt Pinatubo, Philippines, produced strong seismic waves with periods of 228–270 s. These are thought to be caused by a type of oscillation in the atmosphere in which thermal energy from the vent travels up and then reflects off the top of the ash column (in this case, in the stratosphere). These waves also coupled into the ground and appeared on seismic stations thousands of kilometers away, lasting for over 2 h. During the strong initial phase of the 2008 eruption of Okmok volcano, Alaska, a nearby broadband seismometer recorded several hours of similar waves with periods of 540 s.

Broadband sensors have been deployed at a number of volcanoes that produced no observed VLP signals, even though several of these were erupting. They include Mt Spurr, Montserrat, Arenal, and Akutan volcanoes. VLP events have been observed as precursors to only a few large eruptions, one example of which was the August 2008 eruption of Kasatochi, Alaska. There VLP signals were associated with some of the large precursory earthquakes (M = 4–5.8) even though the nearest broadband seismometer was 70 km away. The VLP signals made visible by the new broadband instruments are an exciting part of volcano seismology, and it is anticipated that many new advances will be made in this area over the next few years.

2.6. Explosion Earthquakes

Explosion quakes accompany explosive eruptions, and many are characterized by the presence of an air-shock phase on the seismogram. There is a partitioning of energy at the source: part of the energy travels through the ground as seismic waves and part travels through the air as acoustic or air waves. The air wave then couples back into the ground and is detected by the seismometer. An example of an explosion earthquake is shown in Figure 59.1(G).

The air waves also show up clearly on microphones or infrasound sensors. Infrasound data have been used increasingly to study volcanoes and are a very useful complement to seismic data. The instruments are typically deployed in small arrays (a few hundred meters across) and instruments may be colocated with seismic stations. The data recording and telemetry are similar, using sample rates of 20–100 samples per second and recording pressure instead of ground velocity. Infrasound signals have the advantage that they travel through a simple medium (the air) and, at least for local stations, do not suffer from many of the complications present in seismic data such as reflections, refractions, and other path and site effects (see also Chapter 63).

2.7. Surficial Events

Seismometers located on volcanoes may record a variety of local signals caused by shallow processes. These include nonvolcanic processes, such as glacial events, shore ice movement, and landslides, as well as volcanic processes, such as outburst floods, lahars, pyroclastic flows, and rockfalls, from crumbling lava domes. Some of the seismograms are similar, so they must be recognized and properly treated by analysts.

- *Glacial events*. Volcanoes at high latitudes and tall volcanoes are frequently covered by glaciers, which produce seismic events along the ice—rock boundary and within the ice when cracks and crevasses form. Because the seismic wave velocity in ice is lower than that in rock, the seismic waves may become partially trapped in the ice and set up oscillations of the ice itself. Seismograms from such events resemble those for volcanic LF events. Glacial events are more common in the summer months, when the glaciers are moving faster.
- *Shore ice*. Ice forms from seawater when the air temperature falls below 10 °F (−12 °C). Large blocks of such ice gather on the shores of Augustine Island, Cook Inlet, Alaska, and move around as the tides ebb and flood. When the blocks collide or break, they produce seismic events because the blocks are coupled to the ground. The events are shallow (surface), so the seismic waves are mostly LF surface waves traveling in thin sediment layers. The individual ice events resemble volcanic LF events, and the events occur in swarms of a few hours duration because of the tides. Both of these features are similar to seismicity preceding eruptions, so thermometers are needed to measure air temperature in such situations!
- *Landslides*. Avalanches and landslides of various sizes occur on volcanoes. Those associated with partial melting of ice and snow mainly occur in the spring and summer. The avalanche or landslide generates a seismic signal of several minutes' duration, depending on the size and run-out. The amplitude varies with the amount of material, with larger avalanches generating stronger seismic signals. Some landslides on volcanoes (and elsewhere) are preceded by small discrete seismic events, whose rate of occurrence increases up to the time of the main event. This is similar to a common pattern of earthquakes before some eruptions, so again,

additional information is needed to distinguish between volcanic and nonvolcanic causes.

- *Rockfalls*. Rockfalls are a special type of landslide in which one or a few pieces of bedrock become detached and free-fall. They are especially common at the edges of volcanic domes and at volcanoes with very steep slopes or cliffs. In the summer of 1995, workers at Mt Spurr, Alaska, observed rocks more than 1 m in diameter falling several hundred meters and impacting on the crater floor. The corresponding seismograms showed strong local signals lasting more than 1 min. An example of a seismic signal from a rockfall is shown in Figure 59.1(H).
- *Pyroclastic flows*. Pyroclastic flows are products of explosive eruptions and consist of particles of hot rock and gases that move rapidly over the ground surface. Large-scale collapses of volcanic domes can also produce pyroclastic flows. These latter events are characterized by complex seismograms that have been modeled as a sequence of forces. First, rock is removed from the dome (the dome moves up slightly in response); second, the falling rock collides with the slope below (an impact); and third, the rock breaks apart and the fragments continue to move down the slope as an irregular flow. These seismograms last several minutes, they may contain high frequencies, and the larger flows are associated with larger-amplitude signals. Pyroclastic flows have also been associated with another seismic feature: At Pinatubo volcano, Philippines, pyroclastic flows traveled down a prominent valley between two seismic stations, one of which transmitted its signal to the other. As the ash clouds rose into the air, the telemetry path was interrupted for several tens of minutes after which the signal returned. Thus, the temporary absence of the telemetry could be used to infer that pyroclastic flows were occurring.
- *Outburst floods and lahars*. The surface melting of ice and snow, or heavy rains, may cause floods and lahars (volcanic mudflows). Similarly, a volcanic eruption under a glacier will melt the ice, and the flooding that subsequently occurs is called a jokulhlaup. All of these floods or lahars cause long-duration seismic signals that resemble volcanic tremor, except that the signal is stronger near the flow channel as opposed to tremor, which is stronger nearest the volcanic vent. The floodwater travels downstream, so the source is moving. Also, the recorded frequencies may be high if the seismometer is near the flow channel, even though this may be far from the vent.

Different groups of investigators and various observatories have used several different local terminologies for volcanoseismic event types, and no consensus has yet emerged about an appropriate global terminology. The lack of a global terminology makes it difficult to evaluate and

TABLE 59.2 Selected Volcano Seismology Terminology

This Article	Minakami (1960)	Latter (1979)	AVO[1]	Other names	Example (Figure 59.1)
High-frequency (HF)	A-type	Tectonic, volcano-tectonic	Volcano-tectonic (VT)	Short-period earthquake	(A)
Low-frequency (LF)	B-type	Volcanic	Long-period (LP)	Long-coda event, tornillo[2]	(C), (E)
Hybrid	—	Medium-frequency	Hybrid	Mixed-frequency	(B)
Explosion quake	Explosion quake	Volcanic explosion	Explosion	—	(G)
Volcanic tremor	Volcanic tremor	Volcanic tremor	Volcanic tremor	Harmonic tremor, Spasmodic tremor	(D)

[1]Alaska Volcano Observatory.
[2]Tornillo is the Spanish word for "screw." The codas of these events resemble a wood screw in profile.

compare observations from different volcanoes and creates confusion in the literature. Table 59.2 compares several of the more common local terminologies. Different mechanisms acting at the seismic source are responsible for many of the features of the different events, and path and site effects, which are extreme at many volcanoes, may greatly modify the signals. Four principal seismic sources are shear failure, also called "double-couple" mechanisms; tensile failure, or other "non-double-couple" mechanisms; single forces resulting from explosive injection of material into the atmosphere; and passive and active fluid involvement in producing LF events and volcanic tremor. Nonlinear flow has also been modeled as a source for volcanic tremor. Some authors have investigated differences between volcanic and tectonic earthquakes by comparing factors such as magnitude using different waves, and seismic moment versus fault length. Such comparisons reveal that many volcanic earthquakes are enriched in low frequencies and occur on smaller structures (faults) than tectonic earthquakes of the same magnitude. Intermittent or "banded" tremor occurs in regular, periodic bursts separated by periods of quiescence of uniform duration. The resulting pattern looks like stripes or bands on seismograms recorded on rotating drum recorders. Many other patterns are observed and various terms are used as adjectives to describe them, but this does not necessarily mean that different source processes are required.

3. GENERAL FEATURES OF EARTHQUAKES AT VOLCANOES

The goals of volcano seismology include monitoring the present state of a volcano, forecasting eruptions and other changes in a volcano's activity (the end of an eruption, for example), estimating the size of eruptions in progress, locating magma chambers, and understanding the physical processes that are occurring within the magmatic system. The following topics clarify many features that are common to earthquakes at volcanoes.

3.1. Seismicity Rates and Background

Volcanoes nearly always have a nonzero background level of seismicity caused by heat; movement of groundwater, volatiles, or magma; glaciers (if present), landslides, and rockfalls; and reactions to stresses from regional tectonics, tides, or other forcing functions. Thus, a baseline of measurements of at least a few years duration is necessary to characterize and understand the background seismicity. The background may consist of several different types of events that are related to different processes. Typical background rates of seismicity are a few to a few tens of events per day, depending on the locations and sensitivity of the seismic stations. In contrast, the rates of seismicity before and during eruptions are typically several tens to several hundreds or more events per day and include larger-magnitude events. Examples of rates and magnitudes of seismicity prior to selected eruptions are shown in Table 59.3.

Many successful eruption forecasts have been made because of an increase in seismicity above previously recorded background levels. This is based on the idea that the level of seismicity reflects the level of volcanic activity and suggests that there is a constant long-term probability of eruption, usually assumed to be a Poisson process. This means that eruptions behave as a random variable and that volcanic systems have no memory (mathematically) of previous eruptions. The average rate of eruptions is also assumed to be constant, and the probability for a new eruption increases when earthquake swarms occur. Commonly, the outcome of an increase in seismic activity is magma intrusion without an eruption. Therefore, there will always be the possibility of a false alarm because some intrusions remain at depth, whereas others reach the surface and erupt.

TABLE 59.3 Seismicity Rates for Selected Earthquake Swarms

Location	Detected[1]	M > 1	M > 2	M > 3	M > 4	M_{max}	LP[2]/tremor	Eruption
Augustine								
Feb 1986	>5000/day					M = 2.5	No	Yes
Jan 2006	80/hour	17/hour				M = 1.7	Yes	Yes
Campi Flegrei								
Oct 1983	>300/day					M = 4	No	No
Mar 1984	≈500/day			5/day		M = 4	No	No
El Hierro								
Aug 2011	640/day					M = 4.6	Yes	Yes
Eyjafjallajökull								
Apr 2010	360/day					M = 3.5	Yes	Yes
Fuego								
Jan 1977	2000/day	70/day	≈5/day			M = 2.8	No	No
Galapagos								
Jun 1968				90/day	33/day	M = 5.2	?	No
Kasatochi								
Aug 2008	530/day				12/day	M = 5.8	Yes	Yes
Kilauea								
Jan 1983	>1100/day	92/day	18/day	1/day		M = 3.3	Yes	Yes
Long Valley								
May 1980					≈15/day	M = 6(4)	No	No
May 1982				7/3 days		M = 4.1	No	No
Nov 1982	≈100/day		≈9/day			M = 2.8	No	No
Jan 1983	≈800/day			25/hour	≈10/day	M = 5.3(2)	No	No
Jun 1989	≈40/day	≈25/day	2/day			M = 3	No	No
Mar 1990	>300/day					M = 3	No	No
Matsushiro								
Nov 1965	≈2000/day			>200/day		M = 5.0	No	No

(Continued)

TABLE 59.3 Seismicity Rates for Selected Earthquake Swarms—cont'd

Location	Detected[1]	M > 1	M > 2	M > 3	M > 4	M_{max}	LP[2]/tremor	Eruption
Medicine Lake								
Sep 1988	80/hour			2/hour		M = 4.2	No	No
Mt Hood								
Jul 1980	20/hour		14/hour			M = 2.8	No	No
Mt St Helens								
Mar 1980	≈600/day		>70/day	≈50/day	1–4/day	M = 4	Yes	Yes
Ofu-Ito								
Jul 1989	≈400/day			40/day		M = 5.5	Yes	Yes
Okmok								
Jul 2008	32/hour	24/hour				M = 2.4	Yes	Yes
Pavlof								
Apr 1986	800/day	400/day				M = 2.1	Yes	Yes
Rabaul								
Apr 1984	≈1700/day					M = 4.8	No	No
Redoubt								
Dec 1989	≈150/hour					M < 2	Yes	Yes
Mar 2009	120/hour					M = 3.4	Yes	Yes
Shishaldin								
May 2003	1505/day	247/day				M = 1.2	Yes	No
Usu								
Aug 1977	200/hour			5/hour		M = 3.8	Yes	Yes

[1]Minimum magnitude not specified but generally M < 1.
[2]LP: Long-period earthquake.

3.2. Seismicity Associated with Large versus Small Eruptions

In general, large eruptions are relatively easier to forecast than smaller ones. The amount of magma involved is large, and precursory earthquakes tend to be numerous as well as distributed over a large volume. Other precursors such as deformation and steaming are usually observed in conjunction with earthquake swarms. Recent examples include Pinatubo, Philippines, 1991; Spurr, Alaska, 1992; Augustine, Alaska, 2006; Kasatochi, Alaska, 2008; Redoubt, Alaska, 2009; Eyjafjallajökull, Iceland, 2010; and Merapi, Indonesia, 2010. Small eruptions and many phreatic (water-driven) eruptions, in contrast, involve much smaller amounts of magma and have subtle precursors or sometimes no observable precursors. They are, therefore, generally harder to forecast. Recent examples include Galeras, Colombia, January–June 1993; Arenal, Costa Rica, August 1993; White Island, New Zealand, February 1992; and Ontake, Japan, September 2014.

Occasionally, "unexpected" eruptions occur; often these take place during later phases in a sequence of eruptions. Even though these may be large, they may have subtle or undetectable seismic precursors. This is a consequence of the vent remaining "open" or mechanically weak following an initial stage of activity, preventing buildup of stresses that would cause earthquakes. A recent example is the second eruption of Crater Peak (Mt Spurr), Alaska, in August 1992, which occurred without immediate detectable precursors.

3.3. Locations of Volcanic Earthquakes

Generally, volcanic earthquakes occur beneath the point of eruption (i.e., in proximal clusters). This helps reduce the monitoring problem to that of estimating the time of eruption, since the place is considered known. For example, seismicity preceding the 2004 eruption of Mount St Helens, Washington, occurred in a small shallow volume approximately 1 km below the eruptive vent. However, several cases have occurred in which the locations of initial precursory earthquake swarms did not coincide with the eruptive vent (i.e., distal clusters). For instance, the initial earthquake swarms prior to the dacitic eruption of Mt Pinatubo, Philippines, in 1991, occurred about 5 km northwest of the volcano. Twelve days before the climactic eruption, activity of these swarms declined and vigorous shallower swarms began beneath the eventual eruption site. Similarly, prior to the onset of eruption at Soufrière Hills, Montserrat, the first recorded earthquakes were located ~4 km west-northwest of the volcano. After 2 days, the distal earthquake swarm ceased and a swarm began directly under the Soufrière Hills vent. These and other cases demonstrate that earthquakes occur where stresses are concentrated, which is not necessarily where the magma is located. It is expected that earthquake locations would migrate upward toward the Earth's surface as magma rises from depths prior to eruption. There are, however, surprisingly few well-documented cases of such a trend associated with eruptions. This overall pattern was first shown for Kilauea volcano, Hawaii, in 1960; a more recent example is a seismic swarm that migrated upward over a period of 35 h, preceding the 1998 eruption of Piton de la Fournaise volcano, la Reunion. One spectacular case of lateral migration of hypocenters was observed at Krafla volcano, Iceland, in September 1977; when the earthquakes passed a geothermal borehole, fresh pumice actually erupted from the hole! Most well-documented examples of hypocenter migration involve noneruptive swarms, e.g., at Mammoth Lakes, California; Upptyppingar, Iceland; and Paricutin, Mexico, suggesting that migrating swarms are caused by a process other than preeruptive magma ascent such as hydrothermal fluid circulation.

3.4. Large Regional Earthquakes Near Volcanoes

Large regional earthquakes are believed to occasionally trigger eruptions. Proposed examples include the eruption of Puyehue, Chile, 48 h after the M9.5 earthquake of 1960 and the 1999 eruption of Cerro Negro, Nicaragua, 3 days after a series of nearby M6.5 earthquakes. However, such examples of "static" triggering are difficult to prove conclusively, and the exact mechanism of eruption triggering is also not well understood. Proposed mechanisms include local stress changes near the volcano induced by the nearby large earthquake (analogous to squeezing a tube of toothpaste) and several mechanisms involving the displacement or growth of bubbles by passing seismic waves. Most well-documented examples of "dynamic" triggering involve triggered episodes of noneruptive volcanic unrest. For example, the only documented example of volcanic unrest triggered by the 2004 M9.1 Sumatra–Andaman earthquakes was at Wrangell volcano, Alaska (11,000 km away), which experienced a series of 14 volcanic earthquakes during the passage of surface waves through the volcano. Immediately after the 2011 M9.0 Tohoku earthquake in Japan, a swarm of over 1600 earthquakes was recorded at the quiescent Hakone volcano, 450 km away, that included multiple felt earthquakes. Proposed examples of static triggering of noneruptive unrest also abound: for example, it has been demonstrated convincingly that the June 1992 M7.3 earthquake in Landers, California, remotely triggered earthquake swarms at 17 volcanic areas in the western United States at distances of 165–1250 km. Although no eruptions occurred, these swarms demonstrated that small transient stress changes can trigger significant changes in seismicity in volcanic areas.

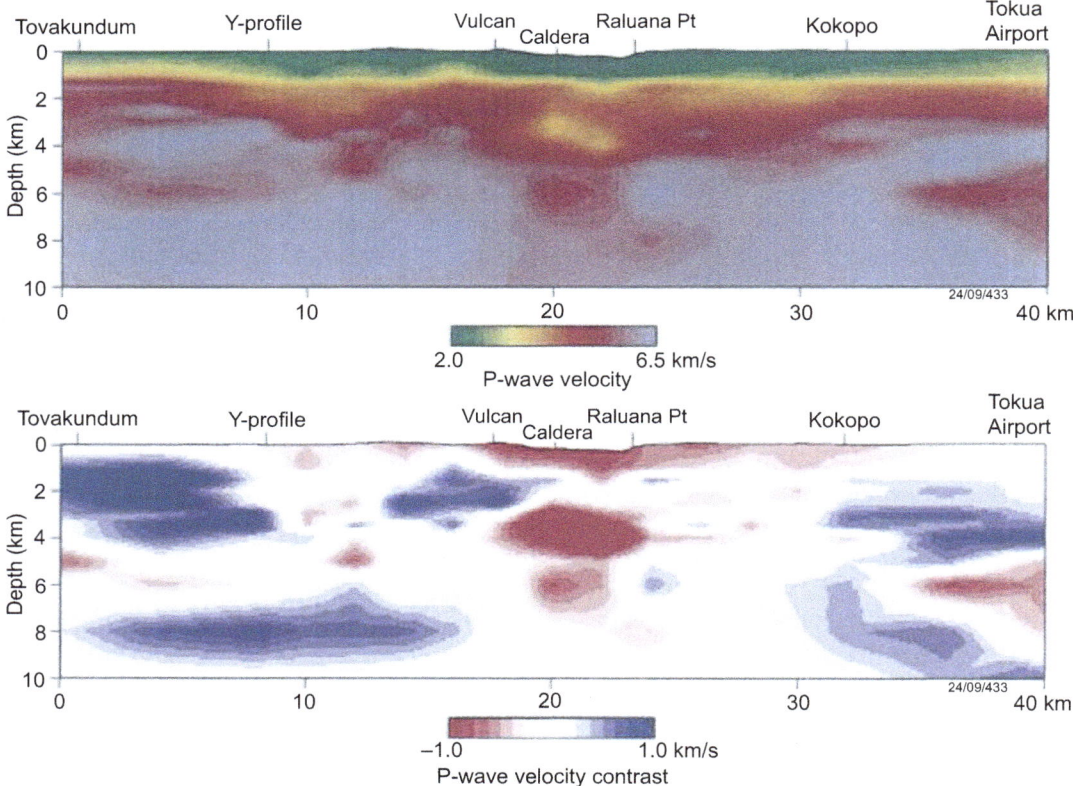

FIGURE 59.3 Tomographic images of P-wave velocity along an NE—SW profile at Rabaul caldera, Papua New Guinea. (Top) P-wave velocity and (bottom) residual velocity difference after subtracting the regional one-dimensional velocity from the top. Note the significant low-velocity region under the center of the caldera, interpreted to be a region of high-temperature magma accumulation. *After Finlayson et al. (2003).*

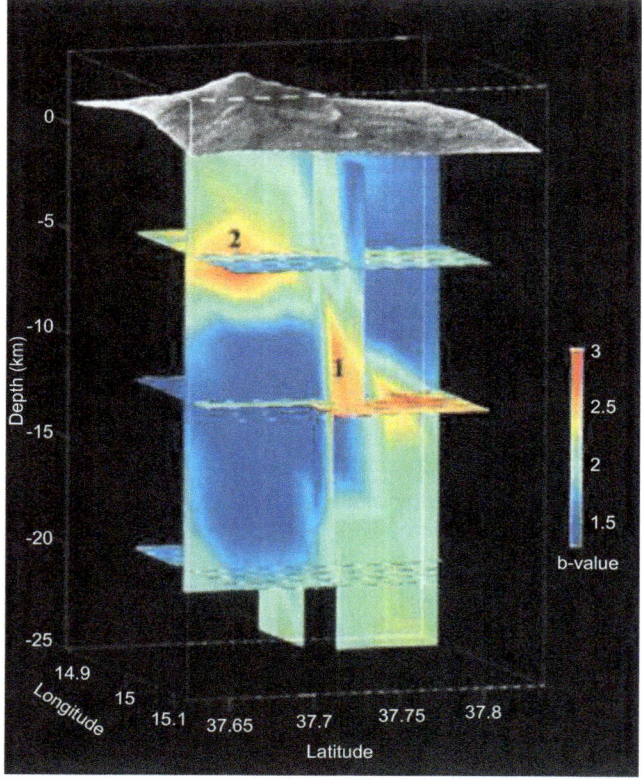

3.5. Caldera Earthquakes

Large earthquakes often occur at large structures such as calderas. For example, M6 earthquakes have occurred at or near calderas in Long Valley, California; Yellowstone, Wyoming, Montana and Idaho; and Aso, Japan. M5.1 earthquakes have occurred at Rabaul caldera, Papua New Guinea, in 1982 and 1984. While small eruptions have occurred frequently at the central vent of Aso, no eruptions have occurred at Long Valley or Yellowstone. Rabaul finally erupted on September 19, 1994, following 13 years of caldera-wide seismicity and 27 h of intense local seismicity. It is normal for large calderas to show frequent signs of unrest, the vast majority of which are not precursors to eruptions. This adds to the uncertainty in dealing with earthquake swarms at calderas: false alarms are more likely, but large eruptions are also possible.

FIGURE 59.4 Three-dimensional map of b-values (here expressed as mean earthquake magnitude) under Mt Etna. Red volumes mark locations characterized by a lack of larger earthquakes. These volumes are interpreted as places of magma storage. b-values are estimated every 1 km by samples of N = 50 events, which occur in radii between 2.5 and 6 km. The period covered by the data is 1990 through 1997—1999, and using M > 2.5 earthquakes. The numbers 1 and 2 mark a stronger and weaker anomaly, respectively. *After Murru et al. (1999).*

3.6. Seismicity at Volcanoes with Long Repose Times

Large eruptions often follow long repose periods, and the precursory earthquake swarms may include relatively large events. Unfortunately, because large eruptions are relatively rare and volcanoes with long repose times are usually unmonitored, very few observations exist to date. Recent examples include Chaitén volcano, Chile, 2008 (volcanic explosivity index (VEI) 4, repose 9400 years; M = 5.2) and Kasatochi volcano, Alaska, 2008 (VEI 4, repose 248 years; M = 5.8). At Mount St Helens, 1980 (VEI 5, repose 123 years), an M = 5.0 earthquake that occurred seconds before observation of the cataclysmic eruption may have been a short-term precursor or a manifestation of summit failure. Large eruptions commonly result in caldera formation, which may also result in large-magnitude syn- or posteruptive earthquakes. For example, at Mt Pinatubo, Philippines, in 1991 (VEI 6, repose 600 years), two M = 5+ earthquakes occurred shortly after the start of the cataclysmic eruption. In 1912, approximately 11 h into a large-volume eruption at Katmai volcano, Alaska, a series of fourteen M = 6–7 earthquakes accompanied caldera formation. The presence of large structures such as caldera-bounding faults may result in large earthquakes not directly related to an eruption. For example, a sequence of four M = 6 earthquakes at Long Valley caldera, California, in May 1980, and some of the later seismicity at Long Valley may have been related to intrusions of magma at depth.

3.7. Synchronous Volcanic and Tectonic Activity

A 40-km-wide belt of volcanoes and frequent shallow crustal earthquakes is located above the subduction zone in Central America. While the general structural features are similar to most other subduction zones, the Central American tectonic seismicity is more abundant and appears to be more closely linked to volcanism. Both eruptions and damaging earthquakes have occurred in isolation, but occasionally the two have been coincident. For example, at Boqueron volcano, El Salvador, an M6.2 earthquake occurred alone in May 1965, while in June 1917 and September 1659, eruptions and destructive (M6) earthquakes occurred simultaneously. These may be cases of one phenomenon modifying or triggering the other, or an external agent may be the cause of both.

3.8. Seismic Features of Magma Chambers

Magma is thought to be stored in some type of chamber or reservoir underground, and seismology is useful for determining the size, shape, and location of such chambers. Magma chambers at stratovolcanoes appear to be generally equant in shape and on the order of 1–20 km^3 in volume. Those at calderas are larger. The majority of depths to the tops are mapped at about 5–20 km or deeper based on tomography, S-wave screening, and posteruption seismicity. Seismic tomography, similar to medical tomography, requires seismic waves to travel through the target region. Thus, a good distribution of earthquakes and seismic stations is needed; both local and distant earthquakes have been used as sources for such studies. The technique exploits the fact that P waves speed up in competent rock and slow down in unconsolidated materials or magma. An example of a tomographic image of a magma chamber under Rabaul caldera, Papua New Guinea, is shown in Figure 59.3. S-wave screening exploits the fact that S waves cannot pass through liquids. Thus, the target region will be illuminated by the presence of P waves but lack of S waves for those earthquake rays that pass through magma. Posteruption seismicity is often concentrated at depths of 5–20 km. Here, the upward removal of magma during eruption causes stress changes at depth that induce earthquakes. The idea is that the magma chamber walls collapse inward, and earthquakes are concentrated where these processes are most pronounced. A technique that requires well-distributed seismicity (as opposed to earthquakes concentrated at one spot) looks for systematic spatial variation in the b-value, discussed earlier. It has been shown in the laboratory that heat, high pore pressure, low applied stresses, or heterogeneous materials all produce high b-values. These are also the conditions that would be expected in the vicinity of magma bodies. Although this technique cannot uniquely identify which process is dominant, strong anomalies have been found at depths of 3–4 km and 7–9 km beneath more than 10 volcanoes studied. Examples of b-value anomalies thought to represent magma chambers beneath Mt Etna volcano, Italy, are shown in Figure 59.4. The parameter plotted is mean magnitude: If the b-value is high, there will be many small events and the mean magnitude will be low, whereas if the b-value is low, the sample will have more number of larger events and hence a high mean magnitude.

3.9. Source versus Path and Site Effects

A long-standing issue in volcano seismology is the relative contribution of source and propagation (path and site) effects on the resulting seismogram. Two examples illustrate why this issue is important. The first shows three-component seismograms from three small earthquakes near Mammoth Mountain, California (Figure 59.5). Both the P and S waves are distinct for the events that occur at a depth of approximately 4 km and epicentral distance of 3 km (Figure 59.5(A)). The station MMB is located on an

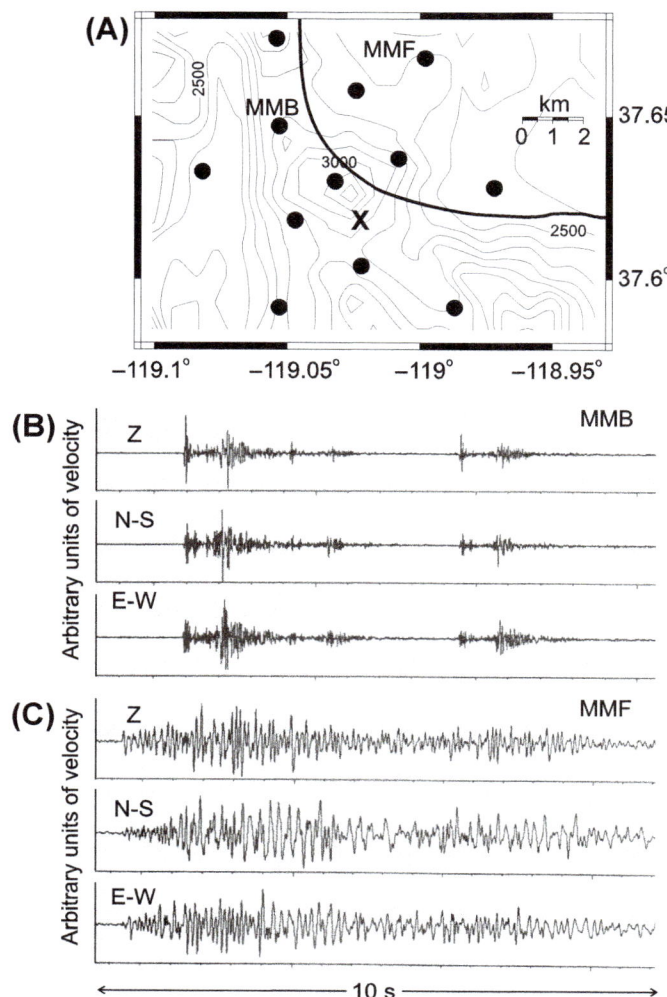

FIGURE 59.5 (A) Map of temporary seismic stations (dots) deployed near Mammoth Mountain, California, in 1989. *(After Julian et al., 1998.)* The edge of Long Valley caldera is the heavy curved line. The approximate location of the earthquakes is shown as a cross. (B) Vertical, north—south, and east—west seismograms from station MMB. Data for 10 seconds are shown. Clear P and S waves can be seen for three small earthquakes beginning roughly 1.5, 3, and 7 s into the record. (C) The same events for vertical, north—south, and east—west seismograms from station MMF. Note that the seismograms look like a single low-frequency event. The third event can be identified with close scrutiny, but the small second one cannot be seen. This is a clear example of site effects.

old dome and is essentially a bedrock site. Figure 59.5(C) shows the same events at an adjacent site that is approximately at epicentral distance of 4 km. The seismograms look completely different. Whereas the earthquakes appear to be ordinary HF events at station MMB (Figure 59.5(B)), on station MMF (Figure 59.5(C)) the events look like a single LF event with extended coda. The third event can be discerned barely, but the small second one cannot even be identified. Station MMF is a soft sediment site. The network geometry was designed to record optimally for focal mechanism studies. This example clearly demonstrates the high impact of path and site effects on the resulting waveforms.

The second example shows that when seismometers are right on top of the source (0—2 km), LP events can appear as simple pulses, whereas at distances of 3—10 km, the seismograms show typical long codas (Figure 59.6(A)). Because of steep topography, most stations at volcanoes are in the range of 3—10 km. The simple pulse can be caused by slow rupture failure in unconsolidated sediments (such as tephra), and the long codas are caused by propagation effects such as resonance in layers of the volcanic pile. Thus, fluids need not be involved. Simple pulses have been observed at close distances at several volcanoes and at Mt Etna during several eruptions (Figure 59.6(B)). These observations demonstrate that alternative explanations need to be considered. No sweeping generalizations can be made regarding source versus site and path effects, and it is recommended that each case be evaluated independently.

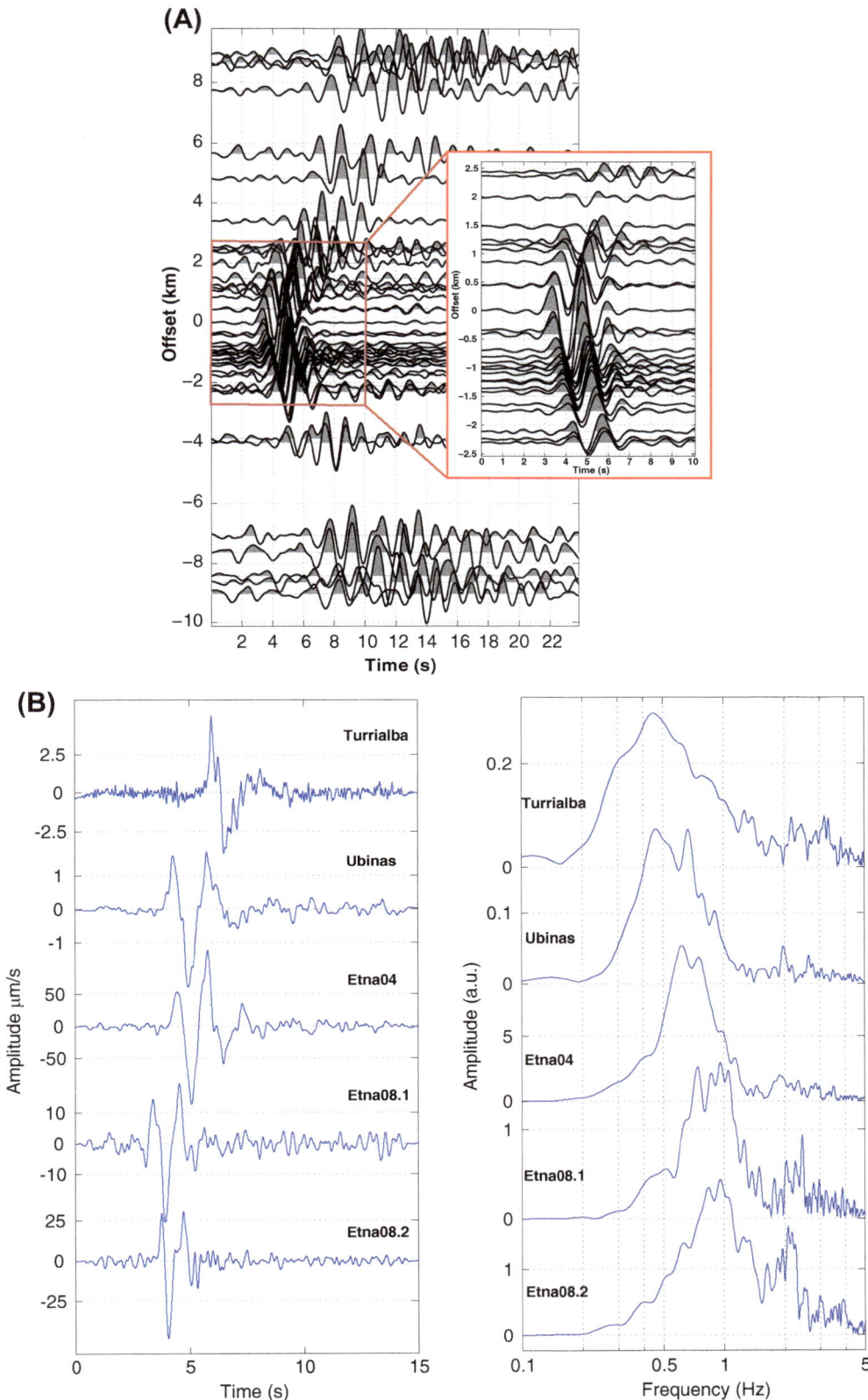

FIGURE 59.6 (A) Spatial distribution of waveforms for a long-period (LP) event family at Mt Etna in 2008. Each seismic trace is a stack of about 60 events in the family. Events are located beneath the summit at depths of <800 m. Normalized vertical component traces are plotted as a function of the station's distance from the volcano summit. (B) Left panel: Shallow pulse-like LP events (vertical component) detected on near-summit stations at Turrialba, Costa Rica (2009); Ubinas, Peru (2009); and two different time periods on Mt Etna (2004 and 2008). Right panel: amplitude spectra for the data shown in the left panel. a.u. stand for arbitrary units. *After Bean et al. (2013).*

TABLE 59.4 Parameters of Volcanic Activity: Seismic Case Histories

Volcano Name	Date	VEI[1]	Swarm Duration (days)	Times since previous eruption (years)	Event type[2]	Maximum Magnitude
Mount St Helens	Sep 2004 to Jan 2005	2	130	13	H,L,T	2.9
Soufrière Hills	Mid-1995 to present	3	>6700	365	H,L,T,R	3.3
Okmok	Jul 12, 2008 to Aug 19, 2008	4	0.21	>800 (11)[3]	H,L,T	2.4
Augustine	Apr 30, 2005 to May 2006	3	256	20	H,L,T,R	1.7

[1]Volcanic Explosivity Index.
[2]H, high-frequency earthquakes; L, low-frequency earthquakes; T, volcanic tremor; R, rockfalls.
[3]Okmok volcano erupted in 1997 from a different vent (Cone A), located 4 km W of the 2008 event.

4. CASE HISTORIES

We consider the earthquake activity at four volcanoes to illustrate the variety of seismic activity. The four examples include small and large eruptions and systems dominated by basalt, andesite, and dacite magmas. The volcanoes chosen all had high-quality seismic data. Parameters of the four case studies are summarized in Table 59.4. The Further Reading section at the end of this chapter gives at least one recent reference for each of the case histories.

4.1. Mount St Helens, 2004—2008

Mount St Helens, Washington, USA, experienced a 4-year-long, near-continuous eruption involving gradual extrusion of dacitic magma beginning in October 2004.

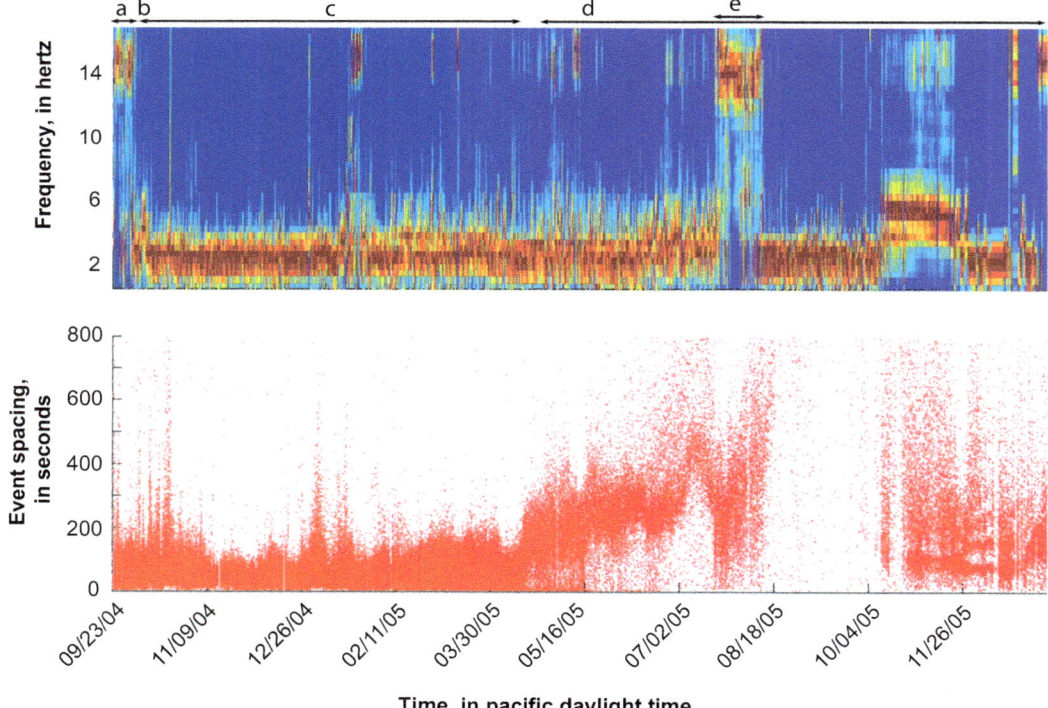

FIGURE 59.7   Plot showing Earthquake Spectral Amplitude Measurement (top) and interevent spacing (bottom) for detected events at Mount St Helens on station HSR between September 23, 2004 and December 31, 2005. Roughly 370,000 events were detected during this time period, including occasional noise glitches and other false triggers. Note the brief initial swarm of volcano-tectonic earthquakes, followed by several months of extremely regular and repetitive "drumbeat" events (Figure 59.1(B) shows an example of a "drumbeat" event from Mount St Helens), and the gradual slowing of the rate of seismicity beginning in April 2005. See text for additional details. *After Moran et al. (2008), courtesy of S. Moran/CVO.*

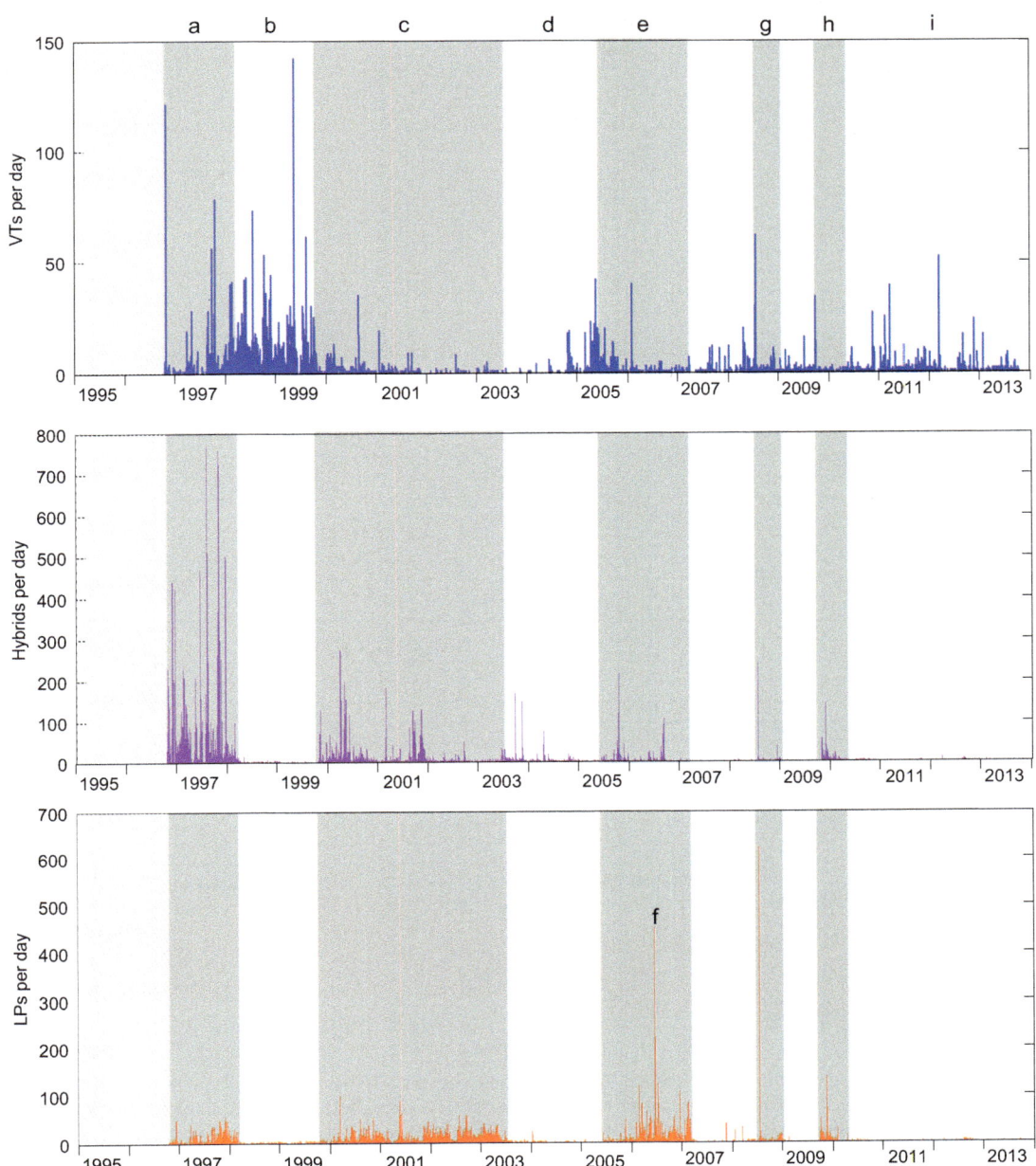

FIGURE 59.8 Histograms of the number of volcano-tectonic (VT) (top), hybrid (middle), and long-period (LP) (bottom) events per day at the Soufrière Hills volcano, Montserrat, from 1996 to 2013 (period of operation of the MVO digital seismic network). The precursory phase and the first year of the eruption, which preceded the installation of the digital seismic network, are indicated by the gray stippled areas. Solid gray areas indicate periods during which Soufrière Hills was in eruption, and white areas indicate intereruptive periods. Note that hybrid and LP activity occurs primarily during eruptive (gray) periods, while VT activity declines during eruptive periods and is strongest during inter-eruptive periods and immediately before the onset of each eruptive period. See text for additional details. MVO, Montserrat Volcano Observatory. *Data courtesy P. Smith/MVO.*

Seismicity preceding the eruption began on September 23, 2004, with a shallow swarm of VT earthquakes beneath the 1980 dome (Figure 59.7: phase a). VT activity increased on September 24, the first LP event was recorded on September 25, and the first hybrid event was recorded on September 26. Seismicity continued to increase in intensity until the first phreatic explosion on October 1. From October 1–5, recorded seismicity included one M2+ earthquake every minute, with tremor bursts beginning on October 2. Magma reached the surface and began to erupt from the dome on October 11 in a series of spines, accompanied by

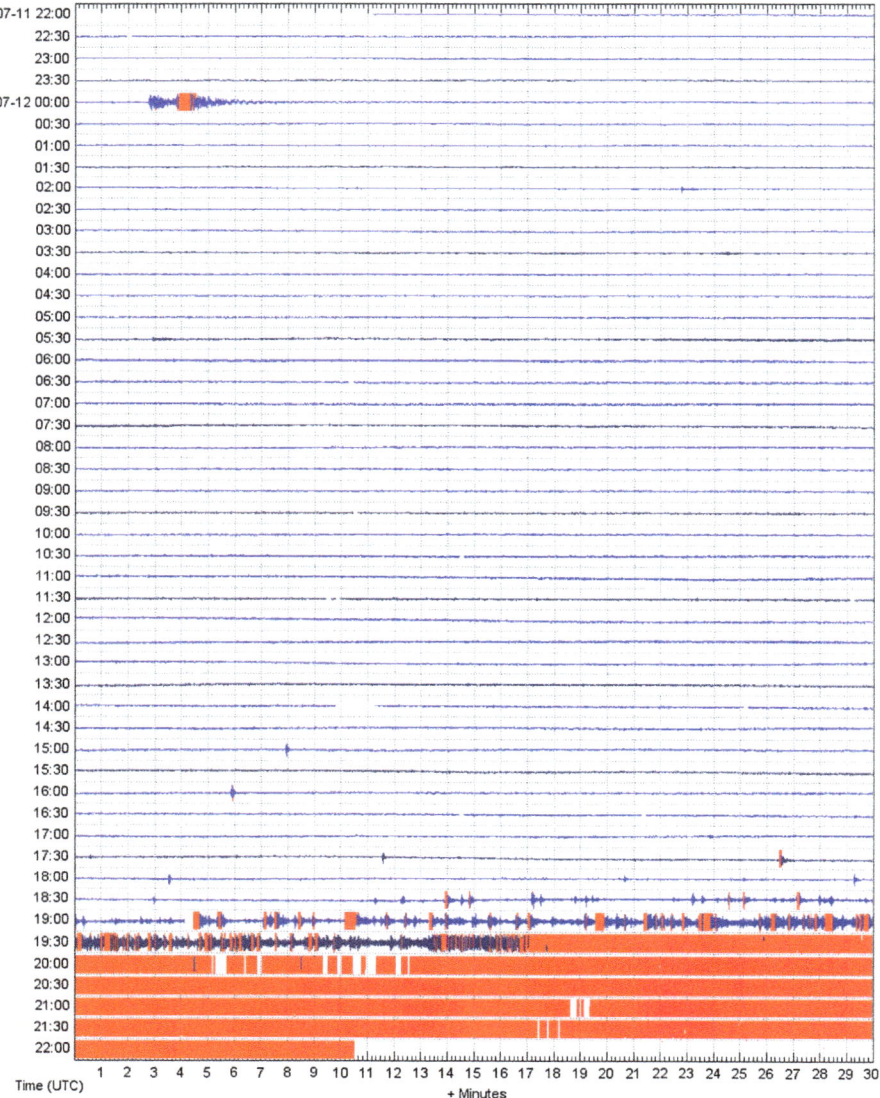

FIGURE 59.9   Seismogram of 24 h of data from station OKWE, 10 km NW of the vent of the 2008 eruption of Okmok volcano. Each line shows 30 min of data, and time increases from top to bottom. A regional earthquake occurs at 00:02 UTC (upper left). The first visible precursors at this station are small earthquakes at 15:08 and 16:05 UTC. Precursors intensify at 18:41 UTC, and the eruption begins with continuous tremor at 19:43 UTC, which increases in strength at 19:47 UTC. The strong initial phase of the eruption lasted for 10 h. Red color indicates strong signal saturating the plot. Gaps represent telemetry dropouts.

strikingly regular "drumbeat" earthquakes beginning on October 16 (Figure 59.7: phase b). Drumbeats had HF onsets and predominantly LF codas, with many having nearly identical waveforms. An example waveform from a drumbeat earthquake is shown in Figure 59.1(B). Drumbeat seismicity continued throughout the dome-building phase of the eruption, at a remarkably steady rate until April 2005 (Figure 59.7: phase c), then at a declining rate through the remainder of the eruption (Figure 59.7: phase d). A swarm of VT earthquakes

occurred in July—August 2005 (Figure 59.5: phase e), coincident with the disintegration of the largest dacitic spine extruded during the 2004—2008 eruption. On January 16, 2008, steam was observed seeping from the lava dome, coincident with an M2.9 earthquake, followed by a small episode of tremor that lasted nearly 90 minutes, and an M2.7 earthquake. By the end of January 2008, the lava dome growth stopped along with elevated seismic activity. Overall more than one million earthquakes were recorded during this eruption.

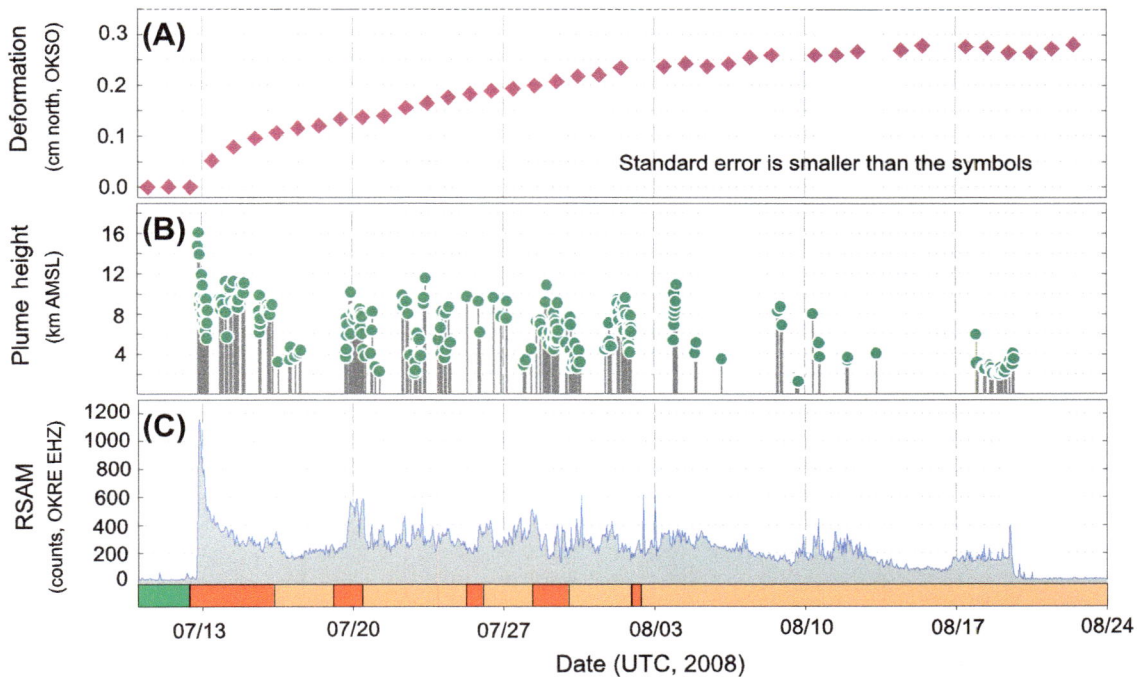

FIGURE 59.10   Summary monitoring data from Okmok volcano, July—August 2008 eruption. (A) Deformation (up (N) corresponds to deflation) from GPS station OKSO, 12 km SW of the vent. (B) Plume height in km. (C) Real-time Seismic Amplitude Measurement (RSAM) from station OKRE. *After Larsen et al. (2009).*

4.2. Soufrière Hills, 1995—Continuing

In the century preceding its current eruption, the Soufrière Hills volcano, Montserrat, experienced three major seismic crises that occurred approximately every 30 years (1897—1898, 1933—1937, and 1966—1967). The current eruption began in mid- to late 1995 following approximately 3 years of escalating seismic activity, with a series of VT earthquake swarms beginning in January 1992, which intensified in 1994. As of mid-2013, the eruption has been ongoing for over 17 years and has consisted of five discrete phases of extrusive activity to date (Figure 59.8). A phreatic explosion on July 18, 1995 marked the onset of the phreatic phase of the eruption, which continued until the first emergence of lava at the volcano's summit in September 1995. Soufrière Hills then remained in continuous eruption until March 1998, accompanied by high rates of VT, LP, and hybrid earthquakes (Figure 59.8: phase a). The exact appearance of the VT, LP, and hybrid events varied by station. An inter-eruptive pause then lasted approximately 20 months and was accompanied by the highest levels of VT earthquakes recorded to date and almost no LP or hybrid seismicity (Figure 59.8: phase b). The second phase of the eruption began in November 1999 and lasted until July 2003 and was characterized by dome growth punctuated by episodes of dome collapse and explosions, reflected by moderate rates of LP seismicity and declining rates of VT and hybrid seismicity (Figure 59.8: phase c). After a second inter-eruption pause characterized by minimal LP and hybrid activity and increasing VT seismicity (Figure 59.8: phase d), eruptive activity resumed approximately 2 years later with the onset of the third phase of lava extrusion in August 2005 (Figure 59.8: phase e), which lasted until early 2007. Dome growth during this phase was initially sluggish, with extrusion rates increasing significantly in February—March 2006. Significant explosions occurred in May 2006 and January 2007, and a major dome collapse occurred on May 20, 2006, accompanied by a strong swarm of LP events (Figure 59.8: phase f). Fourth and fifth phases were relatively short (August 2008—December 2009 and October 2009—February 2010, respectively, Figure 59.8: phases g and h). As of mid-2014, the volcano is showing low levels of surface activity but is still seismically active, with a moderately high rate of VT seismicity (Figure 59.8: phase i) and undergoing inflation. In general, during each phase of eruptive activity, seismicity is dominated by LP/hybrid events, and pauses between eruption phases are dominated by VT seismicity. Overall, there has been a gradual decline in VT seismicity throughout the eruption, with subsequent extrusive phases preceded by weaker and shorter VT swarms.

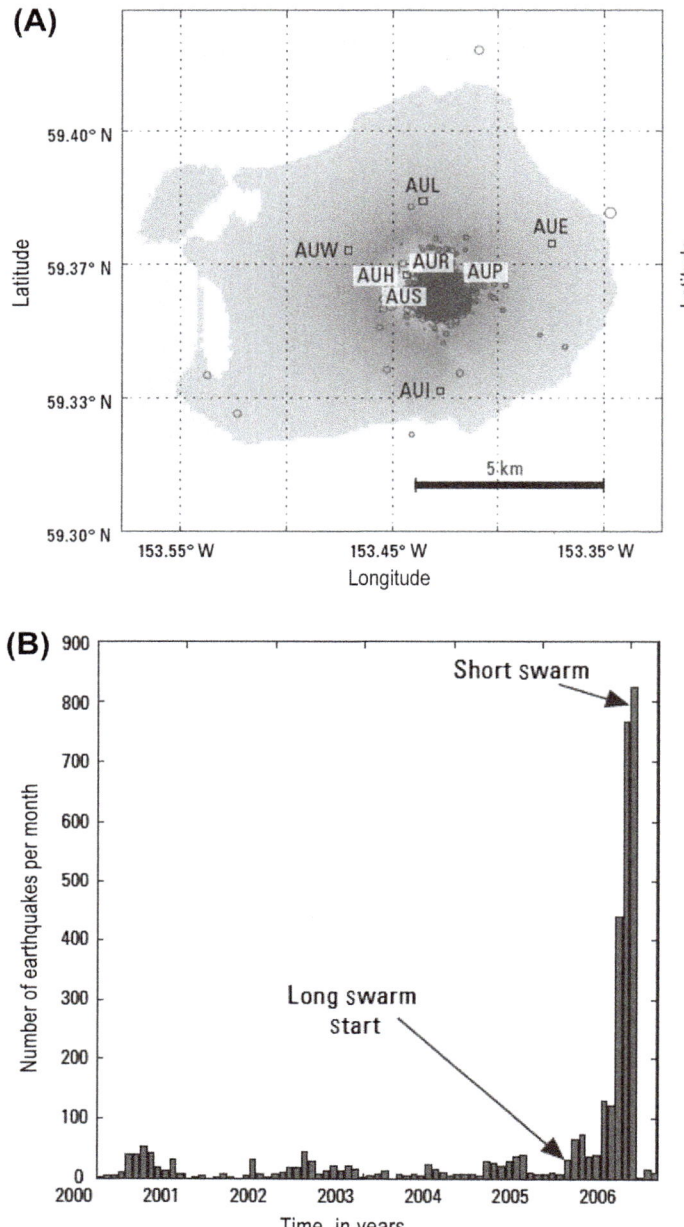

FIGURE 59.11   (A) Map of Augustine volcano showing the permanent AVO monitoring network stations, including broadband seismic station AUL and the pressure sensor colocated with a short-period seismic station at AUE (small squares). The active vent is located at the summit. The four summit stations and AUL were destroyed over the course of the eruption. Nearly all the earthquakes were located at shallow depths under the summit in a zone 2-km across. (B) Number of earthquakes per month from 2000 to 2006. From a variable background, the long swarm started 8.5 months before the eruption, and the short swarm started 10 h before the eruption onset. *After Jacobs and McNutt, (2010).*

4.3. Okmok, 2008

Okmok volcano, Alaska, erupted on July 13, 2008 following a 5-h-long earthquake swarm that intensified approximately 1 h before the eruption onset (Figure 59.9). A new vent was opened at a location where no previous eruptions had occurred for over 800 years. The onset was accompanied by volcanic tremor that became stronger about 5 min later (Figure 59.9). The eruption quickly formed a 16-km-high ash column (Figure 59.10(B)) and was sensed by several satellites. The initial main phase lasted about 10 h as determined from seismic and infrasound data (Figure 59.10(C)). Continuous tremor was observed on stations up to 160 km away. During the

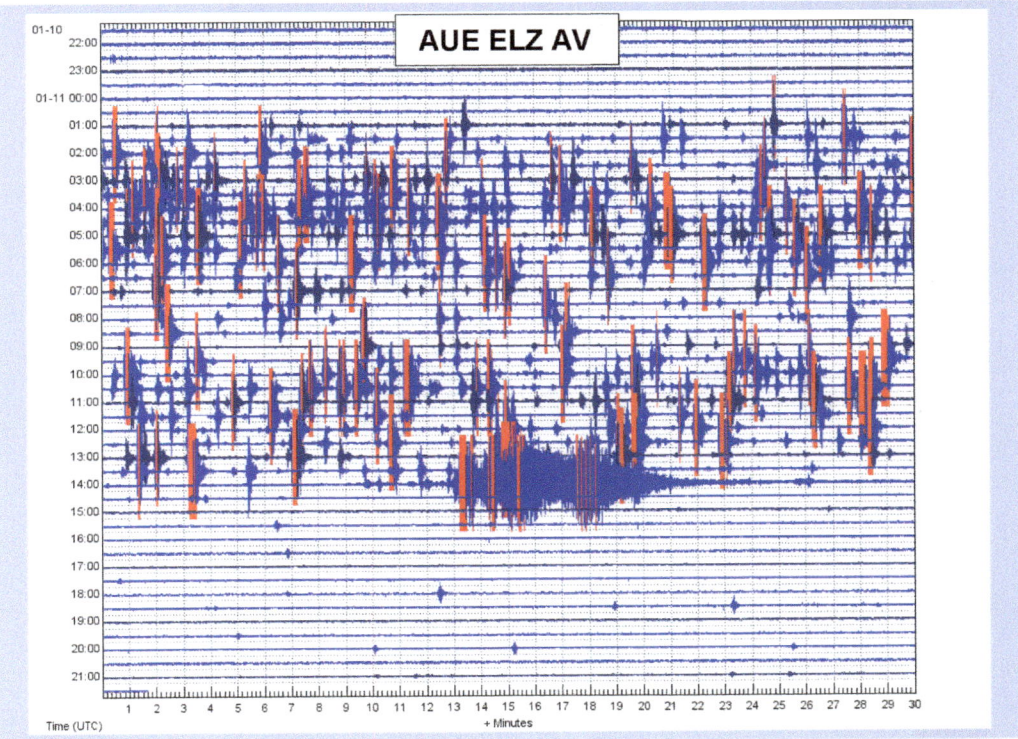

FIGURE 59.12 Seismogram of 24 h of data from station AUE, 3 km E of the vent of the 2006 eruption of Augustine volcano. Each line shows 30 min of data, and time increases from top to bottom. Over 700 earthquakes occurred in the 10 h preceding the eruption. Two eruptions are shown in the central part of the figure at 13:44 and 14:12. Because they are 30 min apart, the traces overlap. Red color indicates signal saturation.

strongest portion, gravity waves with a period of 9 min (540 s) were also observed on a broadband seismometer 14 km east of the vent. Later analyses of InSAR and GPS geodetic data showed deflation at the time of eruption (Figure 59.10(A)) followed by inflation as the magma chamber began to refill. The eruption lasted 5 weeks with variable intensity and seismicity. The plume height varied and there was a rough correlation with the strength of the seismic signal as shown by RSAM (real-time seismic amplitude measurement) data (Figure 59.10(B) and (C)). Given the eruption size and the fact that a new vent was formed, the seismic precursors were remarkably small and short lived.

4.4. Augustine, 2006

Augustine volcano, Alaska (Figure 59.11), began to erupt on January 14, 2006, following an 8.5-month-long earthquake swarm, inflation of the edifice, and increased steaming. A distal cluster of earthquakes was also observed 25 km NE of Augustine at the same time as the later stages of the long-lasting precursory swarm. The local earthquake swarm intensified in the 10 h immediately preceding the eruption with 722 events recorded (Figure 59.12). Of the 722 events, 221 or 31% had VLP

energy as revealed by filtering. This suggests magma injection in a series of pulses during the final ascent of magma before the eruption. A series of strong explosive eruptions from January 14 to 28 produced ash columns from 4 to 14 km (Figure 59.13(A)). Each was accompanied by strong seismic and infrasound signals (Figure 59.13(B)) as well as lightning in the ascending ash plumes. Ash clouds were tracked for many hours in satellite images. Following the explosive phase, dome growth occurred for several months and persistent small ash clouds were produced. Vigorous dome growth was accompanied by repetitive small earthquakes, which were similar to the "drumbeats" observed at Mount St Helens. Numerous rockfalls produced secondary deposits and accompanying clouds; these also produced seismic signals and were recorded on low-light cameras. Two previous eruptions of Augustine, in 1976 and 1986, had similar precursory seismic sequences (8.5 and 9 months, respectively) as well as similar eruptions including both explosive (4–18 days) and dome growth (several months) stages. The close similarities suggest that Augustine has a characteristic eruptive activity. Thus, a useful background investigation for other volcanoes would be to determine the parameters of typical eruptions at candidate volcanoes. If there are no

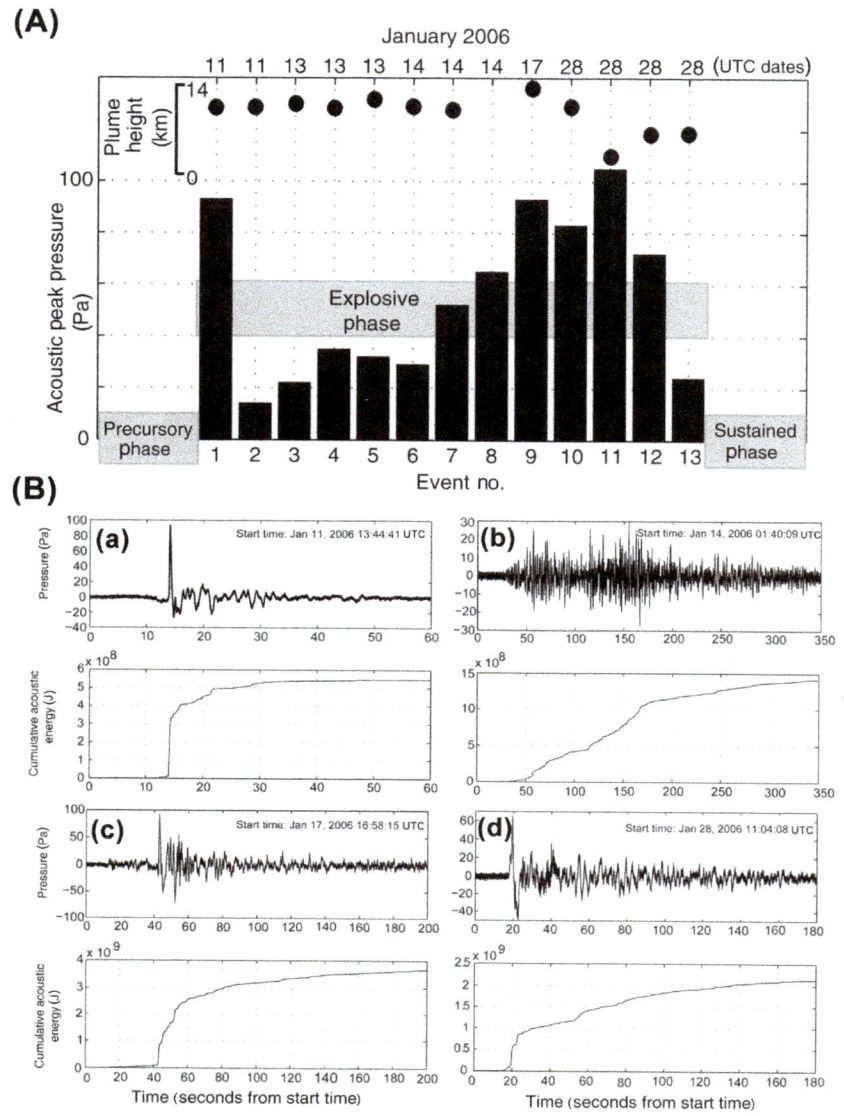

FIGURE 59.13   (A) Histogram of infrasound data from Augustine volcano, showing zero-to-peak pressures of the 13 acoustic signals associated with the explosive eruption phase (January 11–28, 2006). Preliminary plume heights determined by National Weather Service are indicated by dots. Horizontal axis is event number, with dates shown at the top. (B) Examples of traces and cumulative energies of acoustic signals associated with the Augustine January 2006 explosive eruptions recorded at AUE. The signals are high-pass filtered with a corner frequency of 0.1 Hz (a, c, and d) are "impulsive events"; (b) example for an "emergent event." Note that impulsiveness of individual degassing events is clearly visible as the vertical part in the cumulative energy plots. *After Petersen et al. (2006).*

previous eruptions at a candidate volcano, then a systematic search for activity at similar volcanoes may be helpful.

5. CONCLUSIONS

This chapter has discussed many general features of earthquake activity at volcanoes, and it has been shown that activity varies widely. This is true because the volcanoes and their eruptions vary, and also because the stresses differ and volcanoes have different geometries as well as combinations of rock, magma, water, and gases within them. Although seismic data are plentiful, the complexities of the volcanic environment often make interpretations ambiguous. Different processes can generate similar seismograms, and similar processes can generate different seismograms, although detailed diagnostics can often distinguish between them. This keeps volcano seismology challenging, but also suggests that seismology alone may be a somewhat limited tool.

There is a close link between earthquakes and volcanoes. This is true at a large scale in the parallel belts of volcanoes and earthquakes at subduction zones. It is also true at the smaller scale of individual volcanoes, where rates of local seismicity often mimic eruptive activity. Although seismology is often viewed as a separate subject, seismology is an integral component of volcanology. Similarly, volcanology may be viewed simply as a component of the larger study of tectonics. Volcano seismology has helped to answer some fundamental questions: (1) Where is the magma? Tomography and b-value anomalies have identified large (kilometer scale) magmatic or near-magmatic structures, while LF events and volcanic tremor are associated with small-scale movement of magma or water. (2) What is "normal" earthquake activity at volcanoes? The case studies and general features sections showed that activity varies widely, but with a number of systematic trends. Successful separation of source, site, and path effects of seismograms helps to elucidate the causes of specific activity. (3) What will happen next? Understanding of physical processes and study of case histories are the two main tools used to evaluate this question, which is treated more fully elsewhere. It is also possible, and indeed likely, that new high-quality data will lead to fresh insights about the new as well as older examples of earthquake activity at volcanoes.

SEE ALSO THE FOLLOWING ARTICLES

Calderas • Hawaiian and Strombolian Eruptions • Lahars • Magma Chambers • Migration of Melt • Plate Tectonics and Volcanism • Seismic and Infrasonic Monitoring • Synthesis of Volcano Monitoring.

FURTHER READING

Battaglia, J., Ferrazzini, V., Staudacher, T., Aki, K., Cheminée, J.-L., 2005. Pre-eruptive migration of earthquakes at the Piton de la Fournaise volcano (Réunion island). Geophys. J. Int. 161, 549–558.

Bean, C.J., De Barros, L., Lokmer, I., Métaxian, J.-P., O' Brien, G., Murphy, S., 2013. Long-period seismicity in the shallow volcanic edifice formed from slow-rupture earthquakes. Nat. Geosci. http://dx.doi.org/10.1038/NGEO2027.

Benoit, J.P., McNutt, S.R., Barboza, V., 2003. The duration-amplitude distribution of volcanic tremor. J. Geophys. Res. 108 (B3), 2146. http://dx.doi.org/10.1029/2001JB001520.

Chouet, B.A., Matoza, R.S., 2013. A multi-decadal view of seismic methods for detecting precursors of magma movement and eruption. J. Volcanol. Geotherm. Res. 252, 108–175.

De Angelis, S., McNutt, S.R., Webley, P.W., 2011. Evidence of atmospheric gravity waves during the 2008 eruption of Okmok volcano from seismic and remote sensing observations. Geophys. Res. Lett. 38. http://dx.doi.org/10.1029/2011GL047144.

Diez, M., La Femina, P.C., Connor, C.B., Strauch, W., Tenorio, V., 2005. Evidence for static stress changes triggering the 1999 eruption of Cerro Negro Volcano, Nicaragua and regional aftershock sequences. Geophys. Res. Lett. 32, L04309. http://dx.doi.org/10.1029/2004GL021788.

Finlayson, D.M., Gudmundsson, O., Itikarai, I., Nishimura, Y., Shimamura, H., 2003. Rabaul volcano, Papua New Guinea: seismic tomographic imaging of an active caldera. J. Volcanol. Geotherm. Res. 124, 153–171.

Gardine, M., West, M., Cox, T., 2011. Dike emplacement near Paricutin volcano, Mexico in 2006. In: Roman, D.C., Moran, S.C., Newhall, C.G., (Eds), Failed eruptions: Late-stage cessation of magma ascent. Bull. Volcanol. 73(2), 123–132.

Jacobs, K.M., McNutt, S.R., 2010. Using Seismic b-values to Interpret Seismicity Rates and Physical Processes During the Augustine 2005–2006 Pre-eruptive Earthquake Swarm. U.S.G.S. Prof. Paper 1769.

Julian, B.R., Pitt, A.M., Foulger, G.R., 1998. Seismic image of a CO_2 reservoir beneath a seismically active volcano. Geophys. J. Intl. 133, F7–F10.

Larsen, J., Neal, C., Webley, P., Freymueller, J., Haney, M., McNutt, S., Schneider, D., Prejean, S., Schaefer, J., Wessels, R., 2009. Eruption of Alaska Volcano breaks historic pattern. EOS Trans. Amer. Geophys. Union 90, 173–174.

Lees, J., 2007. Seismic tomography of magmatic systems. J. Volcanol. Geotherm. Res. 167, 37–56.

Manga, M., Brodsky, E., 2006. Seismic triggering of eruptions in the far field: volcanoes and geysers. Annu. Rev. Earth Planet. Sci 34, 263–291.

Moran, S.C., Malone, S.D., Qamar, A.I., Thelen, W., Wright, A.K., Caplan-Auerbach, J., 2008. Seismicity associated with renewed dome-building at Mount St. Helens, 2004–2005. In: Sherrod, D.R., Scott, W.E., Stauffer, P.H. (Eds.), A Volcano Rekindled: The Renewed Eruption of Mount St. Helens, 2004–2006. U.S.G.S. Prof. Paper 1750, pp. 27–60.

Murru, M., Montuori, C., Wyss, M., Privitera, E., 1999. The locations of magma chambers at Mt. Etna, Italy, mapped by b-values. Geophys. Res. Lett. 26, 2553–2556.

Neuberg, J.W., Tuffen, H., Collier, L., Green, D., Powell, T., Dingwell, D., 2006. The trigger mechanism of low-frequency earthquakes on Montserrat. J. Volcanol. Geotherm. Res. 153, 37–50.

Power, J.A., Lalla, D.J., 2010. Seismic observations of Augustine Volcano, 1970–2007. In: Power, J.A., Coombs, M.L., Freymueller, J.T. (Eds.), The 2006 Eruption of Augustine Volcano, Alaska. U.S. Geological Survey Professional Paper 1769, pp. 3–35.

Power, J.A., Stihler, S.D., White, R.A., Moran, S.C., 2004. Observations of deep long-period (DLP) seismic events beneath Aleutian arc volcanoes; 1989–2002. J. Volcanol. Geotherm. Res. 138, 243–266.

Petersen, T., De Angelis, S., Tytgat, G., McNutt, S.R., 2006. Local infrasound observations of large ash explosions at Augustine Volcano, Alaska, During January 11–28, 2006. Geophys. Res. Lett. 33, L12303. http://dx.doi.org/10.1029/ 2006GL026491.

Roman, D.C., Cashman, K.V., 2006. The origin of volcanotectonic earthquake swarms. Geology 34, 457–460.

Roman, D.C., De Angelis, S., Latchman, J.L., White, R., 2008. Patterns of volcanotectonic seismicity and stress during the ongoing eruption of the Soufrière Hills Volcano, Montserrat (1995–2007). J. Volcanol. Geotherm. Res. 173 (3–4), 230–244.

Ruppert, N.A., Prejean, S., Hansen, R.A., 2011. Seismic swarm associated with the 2008 eruption of Kasatochi Volcano, Alaska: earthquake locations and source parameters. J. Geophys. Res. 116, B00B07. http://dx.doi.org/10.1029/2010JB007435.

West, M., Sanchez, J.J., McNutt, S.R., 2005. Periodically-triggered seismicity at Mt. Wrangell Volcano following the Sumatra-Andaman Islands earthquake. Science 308, 1144–1146.

White, R.A., Power, J.A., 2001. Distal volcano-tectonic earthquakes (DVT's): diagnosis and use in eruption forecasting. EOS Trans. AGU 82 (47).

Yukutake, Y., Ito, H., Honda, R., Harada, M., Tanada, T., Yoshida, A., 2011. Fluid-induced swarm earthquake sequence revealed by precisely determined hypocenters and focal mechanisms in the 2009 activity at Hakone Volcano, Japan. J. Geophys. Res. 116, B04308. http://dx.doi.org/10.1029/2010JB008036.

Chapter 60

Impacts of Eruptions on Human Health

Peter J. Baxter

Institute of Public Health, University of Cambridge, Cambridge, UK

Claire J. Horwell

Institute of Hazard, Risk and Resilience, Department of Earth Sciences, Durham University, Durham, UK

GLOSSARY

acid rain Pure water in equilibrium with atmospheric carbon dioxide has a pH of 5.6, but with pollution by strong acids, the rain pH ranges from 2 to 4.

asphyxia Death from lack of oxygen.

crystalline silica The three main crystalline forms of silicon dioxide, namely, quartz, cristobalite, and tridymite.

fluorosis Dental fluorosis is a permanent hypomineralization of enamel that manifests itself in the permanent teeth as small white areas in the enamel or, in more severe forms, as obvious staining and pitting of the teeth. In humans it is most commonly encountered in regions where drinking water from groundwater sources has a raised fluoride level. In sheep, spiked teeth may grow in time in opposition to soft teeth and make grazing impossible. Acute fluorosis in livestock is the condition of acute poisoning by an excess intake of fluoride from eating high-fluoride-bearing ash on grass. It is often fatal without urgent treatment. Chronic fluorosis or skeletal fluorosis in livestock is when the bones are deformed by repeated ingestion of fluoride and may arise in a just a few months. As in humans, who can develop a similar condition after many years of exposure to raised fluoride in drinking water and living in tropical or subtropical regions, the main feature is increased bone growth causing the bones to thicken and deform.

gas gangrene A soft-tissue infection caused by contamination of wounds by gas-producing organisms (*Clostridium welchii*) that thrive in the absence of oxygen.

necrotizing fasciitis A soft-tissue infection characterized by rapidly spreading inflammation and destruction of connective tissue (muscle fascia) and subcutaneous fat.

silicosis Chronic silicosis is a fibrous (scarring) reaction of the lungs caused by inhalation of crystalline silica. A common occupational disease of quarry workers, miners, and stone workers if dust controls are inadequate. It is progressive over years and there is no specific treatment. Very high exposures can cause fatal disease in a few months—acute silicosis.

tetanus A soft-tissue infection caused by contamination of wounds by the bacterium *Clostridium tetani*. The bacteria produce a toxin that affects the nervous system, resulting in involuntary muscle spasms. There is no specific treatment and the condition is almost invariably fatal.

traumatic asphyxia Fixation of the chest by external pressure so the victim is unable to breathe.

tympanic membrane The eardrum, or the thin membrane between the outer and middle ear, which is in contact with the air and transmits sound energy to the auditory apparatus.

Risk reduction in natural disasters is a part of the discipline of public health and for volcanoes it also overlaps with global and planetary health in its scope. For years, governments have tended to ignore the importance of risk reduction compared to disaster relief, but we are now living in an era of mega disasters in which it is hard to ignore the scale of the human and economic impacts, and the specific threats they may pose to global security. By contrast, recent studies show that the health consequences of natural disasters in developing countries can be reduced by integrating disaster recovery and resilience with poverty reduction and improved disaster management.

Experience in dealing with major volcanic eruptions in densely populated regions has been limited owing to their rarity compared to earthquakes, floods, and windstorms, and the evidence base on their health and traumatic impacts in humans is correspondingly still rudimentary. The main reason for general concern is that the science of forecasting the timing, type, size, and duration of the most dangerous explosive events, even with the best scientists and technology available, is also not a mature science, yet it has to be relied upon by decision makers for the timely evacuation of whole populations when a volcano in repose moves into a state of unrest and begins to threaten to erupt. Unlike in other natural disasters, the life-threatening activity and eruptive events may continue over weeks, months, and even years. Over this time, the uncertainty surrounding the eruptive behavior, the credibility and reliability of warnings, the art of communication to the public, and planning the actions that follow to protect the population at risk present one of the greatest challenges in disaster management.

Knowledge of the causes of death, injury, and disease during and after volcanic eruptions has grown alongside the recent advances made by volcanologists in the studies of eruptive phenomena. A whole range of injury agents are involved in eruptions and these are described for pyroclastic flows and surges, lahars, explosions, tephra falls, and gas emissions. Burgeoning human populations have formed high-density settlements in volcanic areas, some of which were barely populated 100 years ago. Recent eruptions have revealed how explosive events inflict death and injury, and provide insights into the mitigation measures needed for the management of volcanic crises. This chapter summarizes these observations, many of which have been made during and since the eruption of Mount St Helens, USA, in 1980.

1. EARLY STUDIES OF THE HUMAN IMPACTS OF ERUPTIONS

Curiosity over the fate of victims in volcanic eruptions has always been at the heart of popular interest in the excavations at Pompeii and Herculaneum, ever since they began in the eighteenth century. The younger Pliny's account of the AD 79 eruption of Vesuvius is the first recorded description of an eruption and it includes the first written report describing the death of a victim, his uncle, Pliny the Elder, who was probably caught on the edge of a pyroclastic surge. In 1863, Italian archaeologists devised a method of preserving the positions of Pompeiians in the moment of death; it involved forming a cast by pouring plaster of Paris into the cavity that had once contained the victim's body. These eerie replicas led archaeologists to speculate on causes of death and it was widely believed that victims had died by being buried by flowing mud or a heavy ashfall. In 1902, two eruptions occurred within a day of one another in the Caribbean and would alert scientists to a newly reported and highly destructive volcanic phenomenon. The May 8 eruptive blast from Mont Pelée on Martinique devastated the city of Saint Pierre, killing 28,000 people and leaving only two survivors in the city itself. In Saint Vincent, over 1500 people died in another eruption of a "glowing cloud" from La Soufrière containing intensely hot ash and gases. Scientists were so puzzled by the phenomena at Pelée that they undertook the first investigations of how people died and buildings were razed, but it was not until the eruption of Mount St Helens in 1980 that volcanologists were able to extensively study, at first hand, the eruption of what, by then, was known as a pyroclastic surge (see below). The eruption of Soufrière Hills volcano, Montserrat, West Indies, which began in 1995, was the next event after Mount St Helens that provided the largest stimulus to study, in detail, the health effects of volcanic ash.

Mount St Helens also renewed scientific interest in lahars, but it was the tragedy at the eruption of Nevado del Ruiz in 1985 (see below) that demonstrated the scale of the threat posed by these phenomena and the importance of devising adequate emergency planning and warning measures in localities at risk.

2. INJURY AGENTS

We can categorize injury agents, in general, as various forms of energy, all of which may inflict harm in volcanic eruptions. These agents (and examples) are mechanical (lahars and rock/debris avalanches, ballistic ejecta, and tephra falls), thermal (pyroclastic flows and surges), chemical (gases and aerosols), electrical (lightening), and ionizing radiation (radon gas) or just **asphyxia** by volcanic gases (carbon dioxide, hydrogen sulfide). Volcanoes can also present hazards in repose periods, however. People living on the flanks of volcanoes can be subjected to soil gas emissions that can flow into houses, or they may be in areas subject to fumigation by gases discharged in the plume from the crater or active fumaroles. Much more deadly are the hazards of lahars, as they can form without warning even in the absence of eruptive activity, as in the rapid mobilization by heavy rainfall of loose material on the volcano flanks.

3. DEATHS IN VOLCANIC ERUPTIONS

In 2005, Witham reported that 91,724 people died as a result of volcanic eruptions in the twentieth century, with most loss of life occurring in two major events, namely, the eruption at Mont Pelée, Martinique, and Nevado del Ruiz, Colombia. Dilute pyroclastic density currents (PDCs) were responsible for the most deaths (e.g., Mont Pelée; see Table 60.1) and lahars caused the most injuries (e.g., Nevado del Ruiz). The Philippines, Indonesia, and Southeast Asia, as a region, were the most affected. These data should be regarded with caution, as accurate mortality figures are sometimes not even obtainable today in developing countries. The eruptions of the twentieth century are not representative of the whole range of eruption types and minimize the real potential threat from a massive eruption, especially today when concentrations of populations around some volcanoes are much greater than even several decades ago.

Estimates suggest that over 500 million people, or 10% of the world's population, are living in areas of active volcanism (Small and Naumann, 2001). In the Vesuvius area alone, over 500,000 people could be killed in a major eruption in the absence of adequate warning and evacuation measures. Thus, a range of vulnerability factors and mitigation measures need to be considered when estimating mortality risks, and it is quite deceptive to use total deaths in the past as a reflection of the importance of volcanoes in a future league of natural hazards. The dominance of windstorms, floods, and earthquakes in terms of numbers of deaths in the twentieth century, with only a relatively small number attributable to volcanic eruptions, ignores the fact that volcanoes are capable of

unleashing the most destructive forces on earth. It is indeed fortunate for our world that the recurrence rate of major eruptions in populated areas is substantially lower than that of other natural hazards.

Deaths from lava flows are rare. The flows of most effusive eruptions move sufficiently slowly for people to avoid them, but vents may open unexpectedly at low elevations on a volcano's flanks, and rapidly threaten settlements. In the 1928 eruption of Mount Etna, the town of Mascali was destroyed in just over one day, the lava moving silently at about 2 m/min over a 400 m front. Although eruptions of high-effusion, low-viscosity flows of lava are rare, two have occurred on Hawaii and two in Zaire (now the Democratic Republic of Congo) during historical times, with velocities of 40–100 km/h. Possibly 600 people were killed in the Zaire event in 1977, when the lava lake at Nyiragongo rapidly emptied through fissures that opened up on its flanks, with villagers engulfed in the fast-moving flows; a similar eruption occurred in 2001 when nearly 200 people died. Onlookers may also face the dangers of small explosions when flows come into contact with seawater and by the ignition of hydrocarbon gases formed in the pyrolysis of organic matter beneath the moving lava, and escaping through porous old rock at, and beyond the flow margins.

4. CAUSES OF DEATH AND INJURY

By studying eruptive phenomena and their impacts on people, we can learn to understand the types of injuries or health effects that can be caused by eruptions, as well as the special problems that may arise in the pre-and posteruptive

TABLE 60.1 Casualties due to Dilute Pyroclastic Density Currents in Classical Twentieth-Century Explosive Eruptions

Eruption	Number of Deaths	Dead: Injured Ratio	Survivors after Treatment
Pelée, 1902	28,000	230:1	163 treated, 123 survived
La Soufrière, 1902	1565	11:1	194 treated, 120 survived
Taal, 1911	1335	10:1	Not known
Lamington, 1958	2942	44:1	70 treated, 67 survived
St Helens, 1908	57	16:1	130 airlifted; 9 required treatment, 7 of whom survived
Unzen, 1991	43	5:1	17 treated, 4 with minor burns
Merapi, 1994	94	3:1	81 treated, 11 dead on arrival at hospital
Soufrière Hills, Montserrat, 1997	19	4:1	7 treated, all survived

phases that themselves can last for prolonged periods. Of course, such information is invaluable for health professionals, especially emergency workers, facing a volcanic crisis, who may be responsible for planning the management of casualties. Of much greater importance, in terms of preventing loss of life, is the application of this information to all aspects of volcanic emergency management, especially community preparedness and evacuation planning. Health sector involvement in volcanic risk management is, therefore, critical to any risk assessment process which, with hazard assessment, is a key step in setting priorities in emergency planning and long-term mitigation measures. The most common hazards to arise in explosive eruptions are now described, together with their consequences for human safety and health.

4.1. Pyroclastic Density Currents

Until recently, volcanology was taught around a few catastrophic eruptions, like Krakatau in 1883 and the destruction of Pompeii and Herculaneum as well as the obliteration of the city of St Pierre, Martinique, in 1902, when the Mont Pelée volcano erupted a "column of fire," a surge that swept over the city, toppling over and igniting the buildings in a fireball that killed virtually the whole population of 28,000 men, women, and children within 3 min. For decades after, the violent blast was considered unique in the history of volcanic eruptions, until the cataclysmic eruption of Mount St Helens, USA, in 1980, produced an even greater surge that flowed into the wilderness of the Cascades range. This event sparked a new awareness of the destructive hazards of PDCs and the need for an improved understanding of their behavior and impacts for disaster planning.

The distinction between dilute (surges) and concentrated PDCs (flows) is important because anyone meeting a pyroclastic flow traveling down a deep valley, for example, will not survive to tell the tale and any buildings are likely to be totally destroyed and even buried. Violent, dilute surges of small to moderate volume, though, have been responsible for most human volcanic disasters in recent times and these should be at the top of the list of hazards in populated areas threatened by an explosive eruption. Their destructiveness emanates from their high temperatures, which can be extreme enough to cause built-up areas to catch fire, combined with their dynamic pressure acting as a lateral force to smash or topple structures, but they can be less damaging than concentrated PDCs and even survivable in some instances, as will be described below.

Human beings enveloped by a surge in the open can withstand moderately raised dynamic pressures but are very vulnerable to the heat, which may be intense enough to cause sudden death, and asphyxia due to the high concentrations of fine ash and the effects of heat on lung tissue and the airways, as victims attempt to breathe in an intolerable atmosphere short of air and its oxygen. Being indoors may provide some protection, but raised dynamic pressure can blow in windows and cause structures to collapse on the occupants.

4.1.1. Pompeii and Herculaneum

Surges were responsible for the destruction of Pompeii and Herculaneum in AD 79 by Vesuvius, though how this happened was not fully understood until after the eruption of Mount St Helens in 1980. The excavated cities are the most graphic examples ever preserved of the impacts of surges in a populated area and, 2000 years later, the eruption continues to enthral and deeply move the huge number of visitors to the sites. As previously mentioned, a remarkable feature of Pompeii is the preservation of skeletons in cavities in the excavated eruption deposit, which, when filled with plaster of Paris and allowed to set hard, yield a mould of the body at the time of death. In some, the detail is sufficient to show that the volcanic material closely sealed itself around the body to reveal even the togas they were wearing, as well as the posture of the body. Many victims died with their limbs fixed by the intense heat of the surge in a flexed position known as the "pugilistic attitude," as if warding off an attacker. This posture is assumed by victims in fires, but there were few, if any, large fires ignited at Pompeii, as the temperatures of the surges were not high enough.

The raised dynamic pressures toppled the Roman buildings and the heat of the surges instantly killed the remaining inhabitants who were inside and outside their houses at the time and, for unknown reasons, had not already left with most of the population to safety. At Herculaneum, most people had also left the city leaving hundreds to take refuge in caves by the beach and these men, women, and children all died as the first surge to enter the city ran out as far as the sea and swept into the caves. The surge temperatures were hotter than at Pompeii, as shown by the examples of carbonized furniture that survive to this day while, in Pompeii, most wood has rotted away over time. Only skeletons without their surrounding body cavities, with some showing signs of slight carbonization, have been found preserved at Herculaneum. In the past, archaeologists and volcanologists had mistakenly attributed the building damage to the heavy ash and pumice falls, and most of the deaths at Pompeii to suffocation in the ash during these events in the earlier stages of the eruption.

4.1.2. Mount St Helens, USA

At Mount St Helens, the area devastated by the pyroclastic surge extended northward in a 180° sector as far as 28 km from the summit, far beyond what anyone had expected in the hazard assessment. A blast zone within 8 km of the summit was covered by flow and collapsed flank material, but the rest was devastated by the main surge. The images communicated through the news media made a huge and immediate impact on scientists and the general population the world over. Anyone lucky enough to fly over the volcano after the eruption looked out over an awe-inspiring, lifeless scene of flattened forest and ash stretching as far as the eye could see. What if, instead of an unpopulated wilderness, the once perfectly shaped summit cone had cast its shadow over a city and the millions of downed trees had been people?

Fifty-seven people, who happened to be in the area, died as the surge swept over them and the local coroner took the unusual step in a natural disaster to arrange for autopsies to be conducted on the 19 bodies the search and rescue teams were able to find. The area was traversed by volcanologists keen to study the still-hot deposits and their characteristics, recording and reconstructing what had transpired. The findings were the first in-depth scientific analysis of a volcanic disaster since the eruption of Mont Pelée and marked the birth of modern volcanology.

The autopsies showed how, in some instances, the hot ash in the fast-moving surge had been pushed into the mouth and airways leading to rapid suffocation, but death had probably been virtually instantaneous when the victims were struck by the force and heat of the surge as well. By contrast, four loggers survived in the tree blow-down area, who had been cutting young trees at the time and saw the surge coming toward them. They felt the air becoming hard to breathe and the intense heat, but both passed quite quickly, as they were not far from the end of the PDC run-out and the surge temperature had dropped to less than about 150 °C. Three were lucky enough to be rescued and treated in the hospital. They should all have survived with treatment for their skin burns, which were only moderately severe, but two of the three men who had clinical evidence of thermal lung injury eventually died.

4.1.3. Soufrière Hills Volcano, Montserrat

The anatomy of highly destructive surges became more evident years later in two well-studied and separate eruptive surge events at the Soufrière Hills Volcano on the Caribbean island of Montserrat in 1997. The growing activity of the volcano, which had commenced in July 1995, was beginning to cause increasing concern by mid-1997. The housing on Montserrat had been built to be hurricane-resistant after the damage wrought by Hurricane Hugo in 1989, and no one could have foreseen then that one day the buildings would prove capable of standing up to the much higher dynamic pressures of a surge and give scientists the opportunity to evaluate the forces involved and produce a building damage scale for volcanic eruptions. By contrast, the forested area around Mount St Helens had provided little resistance to dynamic pressure, with trees being

uprooted or broken off where the pressure was highest and blown down over most of the remaining impact zone.

On June 25, 1997, the deadliest PDC of the Montserrat eruption emanated from the summit and headed toward the sea but, at a turn in the valley, a surge decoupled and flowed into the Streatham area in the evacuated Exclusion Zone, where local farmers had gone back, against official advice, to work on their fields. Nineteen people were killed before they could escape from its path and protect themselves inside their houses, but many of the latter caught fire where the surge gained access through openings and ignited furniture. The clothes of the rescued bodies had burnt off in the surge, an invariably fatal occurrence. Decoupled surges may move at much lower velocities than the parent PDC and, by being more dilute, have lower and less damaging dynamic pressures while still retaining enough heat to be highly lethal to human beings. Predicting the formation and behavior of decoupled surges is not straightforward, but always needs to be considered in emergency planning during volcanic crises.

The damage to buildings in the Boxing Day (December 26, 1997) surge ranged from minimal at the margins to total loss from the highest dynamic pressure in the central axis of the impacted zone (Baxter et al. 2005). The houses had already been evacuated when the danger from the volcano had escalated and since the people had an expectation that they would eventually return to their homes, they left their furniture behind. The furnishings were immediately ignited by the hot ash deposits as the surge forced its way in through windows and other openings facing the summit. The fires then rapidly burnt into the wooden roofs leaving many buildings gutted.

4.1.4. Merapi, Java, 1994 and 2010 Eruptions

Dilute, decoupled surges may also have low enough temperatures and dynamic pressures for people to survive and be treated in the hospital, as occurred in two recent eruptions at Merapi, the most active volcano in Indonesia, where the temperatures of the surges were much lower than in the two Soufrière Hills examples. The eruption in 2010, in particular, has provided new evidence for the thermal effects of surges in humans, which we summarize next (Jenkins et al. 2014).

In the eruption in 1994, a decoupled surge, arising from an unexpected eruption of a small block and ash flow along a nearby valley, ran into a group attending a wedding ceremony on Turgo Hill. Twenty-four of the group were killed outright, as were 30 people in the valley, but 81 were taken to the main hospital in Yogyakarta for emergency treatment of their burns. In October/November 2010, another much larger eruption climaxed on November 5, with a concentrated PDC that traveled for 17 km down the Gendol valley, causing most of the 200 deaths due to the eruption in an area that was still in the process of being evacuated. In two separate surge events striking the villages of Kinarejo (October 26) and Bronggang (November 5), 25 were rescued alive with burns and were sent to the same major hospital in Yogyakarta. Altogether, when we combine the patient data for the 1994 and 2010 eruptions, 104 victims, including some children, were rescued with three-quarters suffering from extensive burns (over 40% total body surface area). The prognosis was surprisingly good for mass burns casualties with 38% patients eventually leaving hospital.

This is the largest series of hospitalized surge victims to date which, together with our field work at previous eruptions, has shown how people enveloped in a dilute, slow-moving surge may be injured and killed through direct thermal injury to the body from the radiant and convective heat transferred by the moving cloud of intensely hot particles. A surprising degree of protection to the skin itself from conducted heat—transfer of heat by contact of the skin with the hot ash particles—can be provided by even a thin layer of light clothing, such as a T-shirt and cotton trousers, for example, but being inside a sealed, resistant building that can keep the surge out is the only way to avoid severe burns to the exposed skin. Unfortunately, most eruptions occur in hot climates, as at Merapi, where the houses are made as open as possible to allow a good flow of air, so people inside can be very badly burned when the surge finds a way in. A slow-moving surge will have lower dynamic pressure and produce little in the way of blast damage, but may still be hot enough to be lethal. This was clearly shown in the detached surge event at Bronggang village, where the ventilation openings in the traditionally designed buildings allowed easy access to the surge, even though it was slow moving, so people located inside as well as outside their houses were badly burnt or killed; 54 died and there were only 5 survivors.

We have outlined how the human impacts of dilute PDCs may vary according to their temperatures and velocities (dynamic pressures) in several very different eruption scenarios. These can range from surge temperatures high enough to cause instant, widespread ignitions and a fireball in a city (St Pierre), houses and victims' clothing catching fire (Soufrière Hills), to survival at lower temperatures albeit with serious burns near the end of a cataclysmic surge run-out (Mount St Helens) and in two decoupled surges at low dynamic pressures (Merapi). The lessons of these landmark eruptions should be incorporated in managing future volcanic crises and in devising long-term mitigation measures in volcanic regions.

4.2. Lahars

At some erupting volcanoes, like Merapi in Indonesia, lahars are a common and much feared occurrence, for they can occur with little warning for those living alongside rivers on the lower flanks. Bursts of heavy rain at the summit area, which may pass unheeded in villages lower down, readily remobilize ash deposits from recent ashfalls or PDCs that then flow down valleys as fast flowing streams of slurry, picking up boulders and other debris on the way. Villages for many kilometers downstream are at risk of sudden inundation or even demolition as the lahar may erode and even overwhelm the river or stream channels where the valleys narrow or turn. People who cannot escape in time will be killed or severely injured by multiple trauma from the impacts.

Just as the cataclysmic eruption of Mount St Helens revealed much about the impact of pyroclastic flows, the eruption of Nevado del Ruiz, Colombia, in 1985 demonstrated to the world's media the most destructive features of lahars and produced on a much larger scale than had been seen hitherto. The eruption melted glacial ice on the summit and the mobilized water-entrained debris scoured out valley bottoms to produce a huge mudflow that ran under gravity toward the towns of Chinchina and Armero. As much as 85% of the town of Armero was left covered in 2—4 m of mud. Official figures put the death toll in the town at 21,015 out of a total population of 29,170. Altogether, there were 22,942 deaths and 4470 injured survivors in this event. In the rescue operation over the following days, 1244 survivors, mainly from Armero, were admitted to hospitals (138 subsequently died). The main lahar struck Armero at around 11:30 PM and was preceded by a river of water that flowed along the streets and was sufficiently fast and deep in places to overturn cars and sweep away people. The height of the lahar as it left the Langunillas River and headed toward Armero was 30 m above the riverbed. It traveled at an estimated velocity of 12 m/s and lasted for 10—29 min, during which time most of the town was devastated as buildings collapsed and broke up. Survivors clung to moving pieces of debris or were miraculously swept along on top of the mud. Overall, the inundation of mud lasted about 2 h, with two slower moving major pulses and several smaller pulses over this period.

The head of the lahar would have been in turbulent motion and contained cobbles and boulders. Most people would have been killed immediately by the severe trauma caused by the collapse of buildings, flying debris, and burial by the slurry mass. In addition to the risk of being engulfed, bodies would have been driven against stationary objects, or contorted and crushed by entrained debris such as trees and collapsed parts of buildings, resulting in mutilation and fractures of limbs and skull bones. Stones and other sharp objects would cut into the skin, causing deep lacerations. Mud was inhaled as it forced its way into the eyes, mouth, ears, and open wounds. The pressure of the mud against the chest would have inhibited breathing in those buried to the neck and caused some deaths by **traumatic asphyxia.**

This type of flow arising after a volcanic eruption is described as being noncohesive and, in Armero, became mostly granular or sandy in consistency, which would explain the presence of so many survivors trapped in the mud. Video film shows how the soft consistency made access to the injured so difficult, with many having to be airlifted from the mud by helicopter. The most common lesions in the hospitalized patients were lacerations, penetrating wounds, and infections including **gas gangrene**, **tetanus**, and a form of **necrotizing fasciitis**, which occurred almost exclusively in victims rescued after being in the mud for 3 days. This dreaded complication was due to normally nonpathogenic soil organisms replicating in wounds in the absence of oxygen, and is resistant to medical treatment.

The Ruiz disaster has many implications for disaster planners. Emergency workers need to be aware of the special problems of trying to rescue people from lahars, including the danger to rescuers being themselves engulfed in further pulses of material that might come down 1 or 2 h after the initial wave. Evacuation is the only safe precaution in the face of a continuing threat from an active volcano, but community preparedness should include last-resort advice to move quickly to higher ground if a lahar is on its way. A warning system located on a volcano might give a warning of an hour or more of a moving debris flow. Such advice, however, is going to be of little value for the elderly, the sick, patients in hospitals, etc. who are unlikely to be able to flee, or be moved in time.

Finally, the lahar hazard may last for years after the eruption and cause much disruption to livelihoods, as well as deaths. The eruption of Mount Pinatubo in 1991 left huge quantities of tephra on the flanks of the volcano, and repeated noncohesive lahars are formed in the wet season every year when heavy rains readily mobilize the ash deposits. Six years later, the death toll had exceeded 500 and over 40,000 families remained evacuated from their at-risk lowland areas.

4.3. Explosions

Ballistic ejecta and blast/sound waves come under this heading. Minor explosions on volcanoes can be steam driven, or phreatic, with rocks being hurled without warning from fumaroles or craters. Small explosions may also occur without warning at lava domes or plugged vents, which can be lethal to volcanologists taking gas samples, as happened at the crater of Galeras, Colombia, in 1994, when six scientists met their deaths. Very powerful blasts usually only occur in explosive eruptions.

Most explosive eruptions do not produce shock or blast waves since the release of energy is too slow. The sound of an eruption may not be heard at all close by and instead may undergo atmospheric reflection and concentration to be heard loud enough to rattle windows 100—300 km away, as occurred at Mount St Helens in 1980. In some eruptions, initial shock velocities exceeding the speed of sound arise when ejecta are released at subsonic velocities and a shock wave can be produced. In some Vulcanian eruptions, rocks may be hurled from vents with ejection velocities as high as the speed of sound. The shock waves from these eruptions are unlikely to directly cause injuries unless an individual is close to the erupting center, when they are most likely to be killed in the eruption itself by being struck or obliterated by erupted material.

Loud sound waves can have enough energy to rattle and even break windows, as when a plane flying overhead breaks the sound barrier, but there should be no serious harm to hearing. Eruptions on Stromboli can rattle windows as far as 5 km away from the crater. The ear is the most sensitive organ to shock wave damage, but the blast required to rupture eardrums or cause more serious internal injuries to other organs will also cause major structural damage to buildings as well. Blast waves of this magnitude, at some distance from the eruptive center, are only rarely reported in eruptions. A possible example of this was at Mount Asama, Japan, in 1958, which produced terrific detonations and air shocks that were heard in an area 8—18 km away from the crater and caused damage to windows of houses as far out as 15 km. Seven persons were injured by flying glass. At Galeras, historic eruptions were reported to have generated shock waves that knocked down people as far as 13 km from the crater. In summary, injuries to the **tympanic membrane** of the ear (eardrum), or the hearing apparatus, from loud noise or blast waves in eruptions are rare, as are deaths and injuries caused by the effect of blast waves on major organs. Blast waves with overpressures of 7 kPa may break windows, and the threshold for rupture of the normal eardrum is 35 kPa.

In contrast, ballistic ejecta thrown out in explosions present a more common hazard. To be truly ballistic, the diameter of clasts of typical density has to be at least 15—20 cm for them to have the momentum to be driven against air resistance in an explosion. Rocks (clasts) of this size, unless they are made of light pumice, are easily capable of causing lethal skull injuries, lacerations, and internal chest and abdominal injuries on impact at their terminal fall velocities. Nevertheless, reports of people being killed from such fallout are rare. Mount Sakurajima, Japan, has been erupting frequently for years, and sometimes school children on Sakurajima Island have to wear hard hats to walk to school; concrete shelters have also been built in fields and public places. A potentially more serious risk to life is the danger of fire from ballistics since large clasts can readily penetrate galvanized steel or wooden board roofs and shatter into small, hot pieces inside. The maximum range for ballistic hazards is about 5 km from the crater.

4.4. Tephra Fall

Although it has been recognized that the weight of ash from the fallout in volcanic eruptions can bring down roofs and cause serious damage, as well as injury, it was not until the eruption of Mount Pinatubo in 1991 that the potential scale of this hazard became fully appreciated. Most of the 300 deaths in this eruption occurred outside the 30- to 40-km radius of evacuation as a result of roofs collapsing under the weight of ash (Spence et al. 1996). The ash on the roofs had been made even denser by the addition of rain from Typhoon Yunya on the afternoon of June 15, when the climactic eruption occurred. At the city of Olangapo, about 35 km from the summit, widespread damage to roofs occurred as a result of a deposited depth of ash of 15 cm which, when wet, gave a loading of some 2.0 kN/m^2 (approximately 200 kg/m^2). Typical designed loads for pitched roofs would rarely approach this level. Longer span roofing fared worse than domestic roofs. Although the causes of death were not studied at Pinatubo, the collapse of roof components may cause direct injury to the skull and the body, and the inflowing ash may partially bury and suffocate those inside unless they are able to rapidly extricate themselves. Future eruptions with ashfalls into cities will present a similar hazard, especially as heavy rainfall is known to be triggered by Plinian eruptions. Unfortunately, the best way to mitigate the risk is not clear at present, as shoveling ash off tall house roofs, for example, is not an advisable option for householders during a heavy ashfall when visibility, even during daytime, is drastically reduced and the roofs are slippery from the ash and dangerous to climb on.

Much concern is usually voiced over the respiratory effects of ash after eruptions when the levels of fresh, fine ash particles resuspended in the air reduce visibility and are unpleasant to inhale. Ashfalls from explosive eruptions usually contain an abundance of particles less than 10 μm in aerodynamic diameter, including fine particles (less than 2.5 μm) that are also present, and these are the size ranges of most concern in health studies of air pollution (PM_{10} and $PM_{2.5}$, respectively). Inhaling such ash in sufficient amounts will irritate the lungs and provoke symptoms in asthma sufferers. In the weeks after the Mount St Helens eruption in 1980, increased numbers of asthma sufferers in the impacted areas attended hospital for treatment, and many patients with chronic respiratory disease experienced frequent aggravation and worsening of their condition while the ash remained visibly present in the environment for months, but no deaths that could be obviously linked to the ash were identified (Baxter et al. 1981). Since then, the regional mortality data have not been retrospectively submitted to time series analysis, but the impacts on people's health were nothing like what were initially feared. Rainfall, in general, has a dramatic effect of clearing the air and stabilizing ground deposits and, hence, reducing the concentrations of suspended fine ash in the air.

PM_{10} and $PM_{2.5}$ (which are thought to trigger acute and chronic lung disease and cause premature cardiorespiratory mortality) are routinely monitored in cities in developed countries because of pollution from vehicle emissions. For months after an ash eruption, the air quality standards and guidelines for PM_{10} and $PM_{2.5}$ over a 24-h averaging period can be regularly exceeded unless the ash is removed from the environment by rain and wind and human clearance operations, and, in the least favorable conditions, the PM_{10} concentrations can be in the milligram per cubic meter range until it rains.

An important new respiratory hazard was encountered at the eruption of Mount St Helens in 1980 and then again at the Soufrière Hills Volcano, in 1995. The fine, respirable ash from these volcanoes that fell in populated areas contained substantial amounts of cristobalite, a form of **crystalline silica** that can cause **silicosis** (Baxter et al. 1999), a disabling and potentially fatal lung disease typically found in miners and quarry workers exposed to high concentrations of siliceous dust (Figure 60.1). At Soufrière Hills, exposure was long lived, as the eruption has lasted over 15 years creating, over this period, continuing concerns about the respiratory health risk in the general population, especially in children (Baxter et al. 2014). The cristobalite was formed in the hot lava dome as it grew on and off in the crater and was almost invariably present in the ash clouds generated whenever portions of the dome collapsed. The amount of cristobalite in the ash appears to have been related to the rate of magma extrusion into the dome and the amount of time that the dome was in place prior to collapse (Horwell et al. 2014).

FIGURE 60.1 Cristobalite forming in the lava dome extruded into the Soufrière Hills crater. Microscopic crystals of potentially harmful cristobalite are seen growing among a bed of plagioclase feldspar crystals on the inside surface of a vesicle in Soufrière Hills dome rock *(Image: David Damby)*. When part of the lava dome collapses, the rock is ground down in the moving pyroclastic density current and forms a plume of fine ash particles containing cristobalite.

A full risk assessment for silicosis at Soufrière Hills, based on air quality data, ash characteristics, and toxicological analyses, found that the risk of the local population developing silicosis after 15–20 years of cumulative exposure to cristobalite-laden ash was significant if measures were not undertaken to substantially reduce the population's exposure (Hincks et al. 2006). This meant regularly removing ash deposits from homes and public places, and minimizing heavy exposure to ash by such measures as wearing recommended masks when required and only to shovel ash after rainfall, or wetting down, to prevent the resuspension of fine ash.

Several epidemiological and clinical studies were carried out after the eruptions of Soufrière Hills and Mount St Helens but no long-term disease associated with ash inhalation was identified. Whether this is due to the success of the preventative measures to reduce exposure, or the ash is simply not as toxic as we thought, or both explanations are true, remains to be seen. The toxicological assays that have been performed have, on the whole, suggested that cristobalite-rich ash is not as toxic as might be supposed (Horwell and Baxter, 2006). Physicochemical evidence indicates that the cristobalite toxicity may actually be masked by other associated mineral phases as well as the cristobalite itself being an impure form, a consideration that would also apply to crystalline silica in industrial exposures to some common mineral dusts (Horwell et al. 2012).

High levels of respirable ash in the ambient air are not, in themselves, a reason for medically advising the evacuation of a population from an area of heavy ashfall, unless the conditions are persistently arid and dry. Current advice is that people should stay indoors or wear lightweight industrial masks during very dusty periods, but children and even adults with bad asthma and other lung disorders might need to be advised to leave temporarily as a precaution. In tropical areas, the housing and school buildings are not usually built to keep the air out, so they may offer little protection against the ash. Repeated or persistent exposure to airborne ash containing potentially harmful levels of crystalline silica would be an indication for considering evacuation from, or not returning to, an area, if high and unlimited exposure was likely to continue over years (as in a protracted dome-forming eruption like the one that occurred at Soufrière Hills). In all instances, a full risk assessment, based on monitoring airborne ash levels (and their crystalline silica content) in the populated areas, needs to be undertaken before the most appropriate advice can be given.

Planning for major ash falls in states of unrest at explosive volcanoes is an essential part of volcano crisis management. Public health officials and experts in lung diseases must be involved in eruption planning and in the response, and work closely with scientists monitoring the volcanic activity.

It is essential to exclude raised levels of crystalline silica as an urgent priority after a heavy ashfall by sending ash samples to a laboratory with experience of undertaking the analysis. If raised levels are suspected, confirmation will be needed, which is best done by sending split samples for analysis in different laboratories and under strict scientific protocols. Specialist advice on risk assessment may be needed for reassuring the population and providing guidance on measures to reduce exposure to ash to safe limits, particularly outdoor workers who may be most exposed, as well as children, who may be most susceptible to developing silicosis. Regular measurement of exposure to the ash will need to be undertaken for risk assessment purposes by an experienced team of occupational or environmental scientists (Baxter et al. 2014).

Ash clouds may affect areas over hundreds of kilometers away from the volcano in large eruptions, and even cross national borders, raising further public anxiety over air pollution. Fears may also arise over the presence in erupted ash of fluoride and other toxic elements, and the impacts on the environment and animal health, even with fine or sparse deposits from dispersing plumes. Most laboratories are not used to the analyses required to assess the toxic hazard and the main danger is from alarmist, but erroneous, results being disseminated to a public, their politicians, and media demanding rapid answers (see also below).

We have learnt from some recent eruptions that the most disruptive types of eruption are not necessarily the largest, but relatively small ones like the continuous, open-vent eruptions (e.g., Eyjafjallajökull 2010 that lasted 6 weeks) or the long-lasting dome growth and collapse type (e.g., Soufrière Hills, 1995 to present). These contrast with the major one-off eruptions like Mount St Helens, 1980, and Pinatubo, 1991, where the visible ash deposits may persist for many months or even a few years but are not replenished by frequently repeated episodes of emissions or continuous venting of ash and gases. Being unable to escape the ever-present ash landing on surfaces and equipment interferes with daily life and working, and the disruption can become intolerable. Another scenario arises where one-off heavy ash falls occur in semiarid areas, for example, in Patagonia after large eruptions of Hudson in 1991 and Puyehue Cordón-Caulle in 2011. Abandonment of farm areas may eventually occur, where huge ash deposits lie downwind of the Andean volcanoes and are regularly blown over farmland by very strong winds for many years, causing economic losses compounded by anxieties over the human health effects of such high, repeated exposures.

An important recent example of the effects of continuous, small-scale eruption in a densely populated area was at the port of Rabaul, which for 2 years (2007–2008) was subjected to continuous exposure to gases, mainly sulfur dioxide, and freshly erupted fine ash from the Tavurvur cone every day for 6 months, until the seasonal prevailing winds reversed the direction of the plume away from the populated area of 70,000 people. In the next year, a drought lasted for 3 months in the middle of the 6-month period and the disruption nearly made the main hospital close because the ash infiltrated wards and even the operating theatre, closed schools because the ash made it difficult to conduct lessons and triggered asthma attacks in a proportion of the school children, and some adults also developed asthmatic symptoms. The authorities had to consider evacuating Rabaul Town, mainly for the risk to the health of the population. Fortunately, the eruption stopped and ordinary life became possible again. Rabaul is an active caldera and other populated calderas could experience similar small but persistent eruptions that are capable of causing severe air pollution and disruption to infrastructure and transport, especially in vulnerable modern cities (a potential example is Naples and the currently nonerupting Campi Flegrei caldera).

4.5. Gases

The principal volcanic gases are water vapor, carbon dioxide, sulfur dioxide, hydrogen chloride, hydrogen fluoride, and hydrogen sulfide, with minor amounts of radon, helium, hydrogen, and carbon monoxide (Hansell and Oppenheimer, 2004). Exposure to these gases in active craters or discharging fumaroles can present a hazard to volcanologists monitoring volcanoes unless adequate respiratory equipment is worn.

Sudden loss of consciousness and death are hazards of hydrogen sulfide exposure if its concentration in the air goes above 1000 ppm, especially as its presence is not readily detectable by smell at dangerous concentrations. Fortunately, the usually more abundant and highly irritant acid gases ward off scientists from receiving too much exposure to this lethal gas, and no cases of asphyxiation are known to have been reported among them. However, there are sundry cases of tourists, especially, being overwhelmed by hydrogen sulfide while visiting hydrothermal areas such as the hot springs in Japan, or the town of Rotorua, New Zealand. Hydrogen sulfide is heavier than air (as is carbon dioxide) so tends to pond in low depressions and can fill confined spaces.

People living close to degassing volcanoes can suffer from their pollution if local topography and weather result in the grounding of plume emissions. Sulfur dioxide is the main gas of interest as it can provoke asthma attacks in asthma patients at the low concentrations that can be found at long distances from the source of emissions. The classic example is Masaya in Nicaragua, where an extensive area as far as 30 km from the crater is denuded of normal vegetation as a result of the prevailing winds blowing the plume into a populated area during years of repeated

degassing. **Acid rain** also has an important impact on soils and vegetation at Masaya (Delmelle et al. 2002). Another example of this type of eruption is the persistent degassing that has occurred since 2007 at Turrialba Volcano in Costa Rica. Around volcanoes the principal agent in the formation of acid rain is hydrogen chloride, which is very soluble and immediately taken up by rainwater falling through the plume (this is in contrast to acid rain formation in industrialized countries where it is mainly due to sulfur dioxide, which is slowly oxidized to eventually form the more soluble sulfuric acid; nitrogen oxides and nitric acid may also contribute).

In Hawaii, the abundant white clouds that are formed when the outpouring lava from Kilauea runs into the sea comprises water vapor, hydrogen chloride gas, and hydrochloric acid, which can be blown downwind toward residential areas where acid rain can fall. The visible lava-induced haze in the air is known locally as "laze." Sulfur dioxide emissions from the eruptive site of Pu'u O'o have also been of concern at times in the current phase of activity, which began in 1983; the haze of gas, sulfuric acid, and ammonium sulfate aerosols is called "vog" (volcanic smog).

Poas volcano in Costa Rica went through a period of increased emissions starting around 1986 and lasting until 1995. The crater lake began to dry out in the dry season and refill with rainwater at other times of the year. The already acidic lake became even more so when its level fell and gas emissions through the lake carried with them highly concentrated acid aerosols that damaged vegetation and apparently provoked respiratory complaints in the local villages exposed to the prevailing wind, though elevated concentrations of sulfur dioxide were also present. The crater lakes of active volcanoes can therefore make the air pollution hazard worse under some circumstances, and so it is essential to study the crater activity, in addition to monitoring the plume and its impacts downwind. In addition to its impact on humans, volcanic air pollution can impair the health of farm animals since they may spend most of their lives outdoors, and have other potential consequences such as marked deterioration to the outer fabric of buildings and impairing local infrastructure.

In 2000, a protracted eruption on Miyakejima Island, 180 km south of Tokyo, led to the evacuation of the island's population of nearly 4000 people due to the high levels of sulfur dioxide emissions that occurred. The evacuation order was not lifted until February 2005, when about 70% of the population returned after the emissions had subsided, but were not over. An island-wide monitoring system was established with a control center and a real-time warning system that triggered lights and auditory warnings if levels of sulfur dioxide rose to above guideline heath levels, when inhabitants and visitors are advised to put on gas respirators or even stay indoors if the short-term sulfur dioxide levels exceed 3 ppm. Some large buildings were fitted with gas filters for their ventilation systems and individual free-standing filtration systems were placed inside some houses. The management of this episode has been successful, but at great financial cost, and it has set a precedent for living with future gas pollution episodes from volcanoes.

Occasionally, in the past, deaths from acute exposure to sulfur dioxide have been reported among tourists who were suddenly enveloped by the plume when they incautiously approached the rim of a crater to capture the view, e.g., Hawaii and at Mt Usu, Japan. They were probably already severe asthma sufferers, for such patients are known to be very sensitive to the action of sulfur dioxide.

Gas bursts of carbon dioxide from the ground or from lakes have been agents of disaster in volcanic areas. In 1979, at the Dieng Plateau in Java, 139 people died fleeing from a phreatic eruption when a gas cloud flowed out from a crater above their route of escape. The release had probably been triggered from the ground hydrothermal system by earthquake activity associated with the eruption. In 1986, an estimated quarter of a million tons of carbon dioxide was suddenly released from Lake Nyos, Cameroon, in West Africa, on a still night in August as local villagers were going to bed. About 1700 people and countless numbers of birds and animals died as the denser than air gas flowed down the adjacent valleys. A similar overturning of the lake waters had occurred at Lake Monoun, only 100 km south of Lake Nyos, in 1984, when 35 people died. The source of the gas at Lake Nyos is believed to be a soda water spring at the bottom of the 200-m-deep lake. This recharges the lower, well-stratified water layers where the gas is held in solution under high pressure until some unknown trigger provokes an overturning of the lake, releasing the gas rather like opening a champagne bottle. Apart from Lake Kivu in East Africa, close to the active Nyiragongo volcano, no other deep lakes in the world have yet been positively identified to be in this very hazardous category, but evidence of past overturning has been found at Lake Albano, near Rome.

Soil gas emissions are another way in which carbon dioxide from the underlying magma can present an ever-present hazard in volcanic areas. Studies at Vulcano (Italy), Furnas (Azores, Portugal), Mammoth Mountain (California, USA), and Colli Albano (near Rome) have shown that enough gas may be released into structures to present an asphyxia hazard to occupants by accumulating in confined spaces or parts of buildings below ground level. Carbon dioxide will induce distressed breathing and eventual loss of consciousness at levels of 5—10% in the air, but it is odorless, and at concentrations above 10—15% can cause immediate loss of consciousness after a few breaths and before the individual becomes aware of any danger. At higher concentrations, the gas pouring from ground vents is pungent and hard to breathe in at all.

Carbon dioxide is a carrier of radon gas from the magma and, at Furnas, levels of radon daughters can also accumulate in the indoor air in sufficient concentrations to be hazardous to health. Radon is a cause of lung cancer and, as there is no known threshold to the risk, it may be advisable to take remedial measures in the building, especially if indoor national or international guideline levels for radon are being exceeded.

The ground emissions of carbon dioxide may increase without warning during periods of magma unrest or unfelt seismic activity, and so the hazard should be regularly monitored. In pre- and posteruptive phases of Vesuvius, there are historic reports of people having been overcome by carbon dioxide in cellars or caves on the volcano's flanks.

Undoubtedly, more attention needs to be given to the hazard of carbon dioxide in areas of volcanic and hydrothermal activity. Where volcanic gas emissions are considered to be a health and safety hazard the concentration should be carefully monitored as part of a risk assessment. The hazard of gases in eruptions has tended to be ignored by volcanologists since gas bursts, such as that at Dieng, have so far fortunately proved to be rare events. However, gas releases may have occurred in past eruptions and in some unexplained incidents but gone unrecognized.

4.6. Gas and Ash Interactions in Human and Animal Health

Acid gases and toxic metals may adsorb onto the surfaces of fine ash particles. Volcanic gases adhere to ash, forming salts and sublimates at high temperatures as the ash is erupted; at lower temperatures, in the plume, acids adhere directly to volcanic glass surfaces and form aerosol or fluid droplets such as sulfuric acid and halogen acids. Owing to the vast surface areas presented by these particle and aerosol surfaces in the erupting plume, the amounts may be enough to cause toxic hazards to animals grazing on ash-coated grass and drinking from surface waters contaminated by deposits of ash.

In practice, the element of major concern in eruptions has always been fluoride and, although it does not arise as a major problem at most eruptions, surface waters used as sources of drinking water for humans or available for animals should be routinely tested for fluoride and a range of other toxic elements at the same time (Stewart et al. 2006). It is common in low-economy countries, in large eruptions with widespread ash falls, for livestock and other animals to be seen dying in fields, or looking sick, and casual observers may rush to the conclusion that the ash is poisonous. The usual explanation is that the animals are short of food when the grass becomes covered with ash as most animals will not, or are unable to, graze and the ash may also have covered over their usual sources of water. Pregnant sheep (ewes) are very susceptible to the added stress caused by such conditions and will soon die, again raising alarms that the ash contains poisonous levels of fluoride. Ruminant animals will also succumb if hunger makes them eat large amounts of ash in the absence of any digestible intake; a recognized finding in such animals is a bolus of ash lying heavily in the stomach at autopsy, suggestive of it being the cause of death, but there could be other explanations as well in these disasters. In advanced economy countries, many thousands of livestock may need to be evacuated to ash-free areas to prevent them from dying.

All ash samples should be routinely subjected to water and acid leachate analyses, the latter being the laboratory equivalent of the acid conditions in the animal stomach. Once again, a specialized water testing laboratory should be used for the task and preferably one that follows the protocols of the International Volcanic Health Hazard Network (www.ivhhn.org).

Acute fluoride toxicity and chronic **fluorosis** following eruptions of Hekla volcano have been reported in sheep, horse, and cattle in Iceland since 1693. In 1783, the livestock and vegetation were decimated by repeated emissions of fluoride-bearing ash from the Laki fissure eruption, which lasted for 8 months and, in the ensuing famine, about 20% of the Icelandic population died. Abnormal sheep bones, affected by the Hekla eruption in 1845 and preserved into the twentieth century, were finally analyzed and confirmed to have a high fluoride content by a Danish physician who was able to identify, for the first time, similar bone changes in workers in an aluminum factory engaged in a process using cryolite, which released hydrogen fluoride into the factory atmosphere. This disease, chronic skeletal fluorosis, was later identified as a crippling condition in parts of the world where there are raised concentrations of fluoride in ground water used for drinking, where the contamination comes from local geological strata. Severe dental fluorosis will also be found in such areas, usually associated with poor living conditions. The bone disease has not been confirmed in human beings after volcanic eruptions, however, including during the Laki eruption in 1783 when widespread fluoride contamination occurred. Dental fluorosis has been linked to gaseous volcanic emissions on Ambrym Island, Vanuatu, where the plume from the chronically degassing volcano regularly contaminates the rainwater collected by the islanders and used as their main source of drinking water (Allibone et al. 2012). Some degree of dental fluorosis is a commonplace finding in every part of the world, as fluoride is present in many water supplies and in various sources such as toothpaste, and arises during the formation of the permanent teeth from excess exposure to fluoride in childhood. The bone disease, however, arises in later life after many years of consumption of heavily contaminated water.

In the "dry" stage of the eruption of Eyjafjallajökull in 2010, when the glacier around the vent had melted away, the concentration of fluoride as measured by water leachate studies became very high (>1000 ppm). It was springtime and the livestock were due to be released on to the pastures after being kept inside their barns over the winter, but the farmers in the ash fall area nearest to the volcano were warned not to let them out or they would perish in the fields from acute fluorosis as soon as they ate the ash-covered grass. When the eruption stopped a few weeks later, the fluoride on the ash deposits was rapidly leached off by the rain and it was possible for the animals to graze normally. The regular water to the farms from springs was tested and found to be uncontaminated.

5. CONCLUSION

Understanding the human impacts of eruptions requires interdisciplinary studies of the behavior of volcanoes in periods of repose as well as during and after eruptive events. Computer numerical simulation models are becoming important tools for predicting injury impact in PDCs (see Chapter 54) and for forecasting ash falls. With the continuing rise of global urbanization it will be the management of volcanic crises in cities and other densely populated regions that will present the major challenges for scientists and disaster workers in the years ahead.

FURTHER READING

Allibone, R., Cronin, S.J., Charley, D.T., Neall, V.E., Stewart, R.B., Oppenheimer, C., 2012. Dental fluorosis linked to degassing of Ambrym volcano, Vanuatu: a novel exposure pathway. Environ. Geochem. Health. 34, 155–170.

Baxter, P.J., 1990. Medical effects of volcanoes. 1. Main causes of death and injury. Bull. Volcanol. 52, 532–544.

Baxter, P.J., Ing, R., Falk, H., French, J., Stein, G., Bernstein, R.S., Merchant, J.A., Allard, J., 1981. Mount St Helens eruptions, May 18 to June 12, 1980: an overview of the health impact. J. Am. Med. Assoc. 248, 2585–2589.

Baxter, P.J., Bonadonna, C., Dupree, R., Hards, V.L., Kohn, S.C., Murphy, M.D., Nichols, A., Nicholson, R.A., Norton, G., Searl, A., Sparks, R.S.J., Vickers, B.P., 1999. Cristobalite in volcanic ash of the Soufrière Hills volcano, Montserrat, British West Indies. Science 283, 1142–1145.

Baxter, P.J., Boyle, R., Cole, P., Neri, A., Spence, R., Zuccaro, G., 2005. The impacts of pyroclastic surges on buildings at the eruption of the Soufrière Hills volcano, Montserrat. Bull. Volcanol. 67, 292–313.

Baxter, P.J., Aspinall, W.P., Neri, A., Zuccaro, G., Spence, R.J.S., Cioni, R., Woo, G., 2008. Emergency planning and mitigation at Vesuvius: a new evidence-based approach. J. Volcanol. Geotherm. Res. 454–473.

Baxter, P.J., Searl, A., Cowie, H.A., Jarvis, D., Horwell, C.J., 2014. Evaluating the respiratory health risks of volcanic ash at the eruption of the Soufrière Hills Volcano, Montserrat, 1995–2010. In: Wadge, G., Robertson, R., Voight, B. (Eds.), The Eruption of

Soufriere Hills Volcano, Montserrat from 2000 to 2010, vol. 39. Geological Society, London, pp. 405–423.

Cole, P.D., Neri, A., Baxter, P.J. Hazards from pyroclastic density currents. Encycl. Volcanoes, Chapter 54, this volume in press.

Delmelle, P., Stix, J., Baxter, P.J., Garcia-Alvarez, J., Barquero, J., 2002. Atmospheric dispersion, environmental effects and potential health hazard associated with the low-altitude gas plume of Masaya volcano, Nicaragua. Bull. Volcanol. 64, 423–434.

Hansell, A., Oppenheimer, C., 2004. Health hazards from volcanic gases: a systematic literature review. Arch. Environ. Health 59, 628–639.

Hincks, T.K., Aspinall, W.P., Baxter, P.J., Searl, A., Sparks, R.S.J., Woo, G., 2006. Long term exposure to respirable volcanic ash on Montserrat: a time series simulation. Bull. Volcanol. 68, 266–284.

Horwell, C.J., Baxter, P.J., 2006. The respiratory health hazards of volcanic ash: a review for volcanic risk mitigation. Bull. Volcanol. 69, 1–24.

Horwell, C.J., Hillman, S.E., Cole, P.D., Loughlin, S.C., Llewellin, E.W., Damby, D.E., Christopher, T., 2014. Controls on variations in cristobalite abundance in ash generated by the Soufrière Hills volcano, Montserrat in the period 1997–2010. In: Wadge, G., Robertson, R., Voight, B. (Eds.), Memoir of the Geological Society of London, pp. 397–404.

Horwell, C.J., Williamson, B.J., Le Blond, J.S., Donaldson, K., Damby, D.E., Bowen, L., 2012. The structure of volcanic cristobalite in relation to its toxicity: relevance for the variable crystalline silica hazard. Part. Fibre Toxicol. 9, 44.

The International Volcanic Health Hazard Network (www.ivhhn.org), the umbrella organization for volcanic health-related research and dissemination, has produced pamphlets and guidelines on volcanic health issues for the public, scientists, governmental bodies, and agencies.

Jenkins, S., Komorowski, J.-C., Baxter, P.J., Spence, R., Picquout, A., Lavigne, F., Surono, 2013. The Merapi 2010 eruption: an interdisciplinary impact assessment methodology for studying pyroclastic density current dynamics. J. Volcanol. Geotherm. Res. 261, 316–329.

Small, C., Naumann, T., 2001. Holocene volcanism and the global distribution of human population. Environ. Hazards 3, 93–109.

Spence, R.J.S., Pomonis, A., Baxter, P.J., Coburn, A.W., White, M., Dayrit, M., Field Epidemiology Training Program Team, 1996. Building damage caused by the Mount Pinatubo eruption of June 15, 1991. In: Newall, C.G., Punongbayan, R.S. (Eds.), Fire and Mud: eruptions and lahars of Mount Pinatubo, Philippines. Philippine Institute of Volcanology and Seismology, University of Washington Press, Seattle, pp. 1055–1061.

Stewart, C., Johnston, D.M., Leonard, G., Horwell, C.J., Thordarsson, T., Cronin, S., 2006. Contamination of water supplies by volcanic ashfall: a literature review and simple impact modelling. J. Volcanol. Geotherm. Res. 158, 296–306.

Witham, C.S., 2005. Volcanic disasters and incidents: a new database. J. Volcanol. Geotherm. Res. 148, 191–233.

Chapter 61

Large Igneous Provinces and Biotic Extinctions

Michael R. Rampino

Department of Biology, New York University, New York, NY, USA; Department of Environmental Studies, New York University, New York, NY, USA

Stephen Self

Department of Environment, Earth, and Ecosystems, The Open University, Milton Keynes, MK, UK; Department of Earth and Planetary Science, University of California, Berkeley, CA, USA

Chapter Outline

GLOSSARY

aerosol A suspension of liquid droplets or fine particles in a gas.

biocalcification The ability of some marine organisms to extract calcium carbonate from seawater to produce internal and external skeletons.

fire fountain The fountains of lava created by the thrust of escaping gases at volcanic vents.

greenhouse effect Warming of the atmosphere as a result of greenhouse gases (such as CO_2, CH_4, H_2O, and SO_2) that absorb outgoing radiation.

large igneous province (LIP) Great outpourings of primarily basaltic magma on land and in the oceans, which eventually cover millions of square kilometers and constitute millions of cubic kilometers in volume.

mantle plume Upwelling material in the Earth's mantle driven by differences in temperature and density. Plume heads are thought to cause LIP volcanism as they approach the Earth's surface.

mass extinction A widespread and rapid decrease in the amount of life on Earth. Such an event is identified by a sharp change in the diversity and abundance of (mainly) macroscopic life forms.

ocean anoxic event (OAE) Periods in Earth history when large portions of the oceans became depleted in oxygen, marked by the deposition of organic-rich sediments.

paleocene−eocene thermal maximum (PETM) Brief period in time about 56 million years ago when Earth's temperature was several degrees warmer than present.

1. INTRODUCTION

It has been estimated that as much as 99.9% of the species that have ever lived are extinct. Extinctions of individual species commonly occur during mass extinction events, where a significant number of species die out in a relatively short period of time. Mass extinctions are important factors in the history of life on Earth; they are often followed closely by major adaptive radiations of surviving species. Studies beginning in the 1980s discovered that mass extinctions were typically widespread and rapid events, and catastrophic causes were suggested. Two catastrophic processes have been invoked (Rampino, 2010): (1) impacts of asteroids and comets (Alvarez et al., 1980), and (2) massive volcanism, in the form of eruptions of large igneous provinces (LIPs)—continental flood basalts on land and oceanic plateaus in the sea (Rampino and Stothers, 1988; Courtillot and Renne, 2003). For example, on the one hand, the end-Cretaceous (66 Ma) mass extinction (the Cretaceous−Paleogene or K−Pg event) has been convincingly linked to the impact of a large asteroid or comet (Schulte et al., 2010). On the other hand, the near-coincidence of the K−Pg extinctions and the voluminous Deccan flood basalt province in India led to suggestions that the

The Encyclopedia of Volcanoes. http://dx.doi.org/10.1016/B978-0-12-385938-9.00061-4

volcanic eruptions might have contributed to the extinction event (see Courtillot, 1999).

In recent years, as radiometric dating has improved and as the LIPs are more closely studied, a general correlation has been proposed between LIPs and extinctions, and also with ocean anoxic events (OAEs), periods when the deep oceans became depleted in oxygen (Courtillot and Renne, 2003) (Table 61.1; Figures 61.1 and 61.2). It has also been suggested that the coincidence of both a large impact and an ongoing flood basalt province-forming event might be necessary to cause severe mass extinctions—the press-pulse hypothesis. However, only one such coincidence is well established, at the end of the Cretaceous. Some researchers have even proposed that large impacts might in some way trigger or enhance LIP volcanism.

A major question regarding any possible relationship between LIP volcanism and extinction events involves the nature and severity of the environmental effects caused by the eruptions and the potential impact on life (Wignall, 2001). Furthermore, LIP episodes seem to be related to the inception of mantle plume activity, and thus LIPs may represent only one facet of a host of geological factors (e.g., changes in seafloor-spreading rates, rifting events, tectonism, and sea-level variations) that tend to be correlated, and may have contributed to climatic and faunal and floral changes.

2. THE RECORD OF LIP EVENTS, EXTINCTIONS AND OAES

Twenty five continental and oceanic LIPs over the past 580 Ma (an average of about 1 every 23 Ma) have been recognized, with correlative biological and geological events (Rampino and Stothers, 1988; Courtillot and Renne, 2003; Kravchinsky, 2012) (Table 61.1; Figures 61.1 and 61.2). Using the results of the $^{40}Ar/^{39}Ar$ geochronological method (and, to a lesser extent, U—Pb age determinations), and paleomagnetic studies, it has been proposed that the bulk of LIP episodes was erupted over quite brief periods of geologic time, less than a few hundred thousand years in some cases (Kelley, 2007), and perhaps much less (Chenet et al., 2009). The total extent of older LIPs is difficult to establish because of erosion and burial of continental flood basalt provinces, and the subduction and destruction of oceanic LIPs.

In particular, the 4 most severe mass extinction events in the last 260 million years—the end-Guadalupian event at 260 Ma, the end-Permian event at 252 Ma, the end-Triassic event at 201 Ma, and the end-Cretaceous event at 66 Ma—are closely correlated in time with the peaks of 4 LIP events—the Emeishan basalts, the Siberian Traps, the Central Atlantic Magmatic Province (CAMP), and the Deccan Traps, respectively (Figure 61.1). For LIP eruptions prior to 260 Ma (Table 61.1; Figure 61.2) good correlation can be found with stratigraphic boundaries marked by faunal extinctions.

Furthermore, close correlations exist between the ages of the other known LIPs and those of OAEs such as the mid-Cretaceous Cenomanian—Turonian OAE at 94 Ma, (also near the Caribbean Plateau at 89 Ma and the Ontong Java Plateau at 90 Ma), the Toarcian OAE in the Jurassic at 183 Ma (the Karoo basalts at 183 Ma), and the end-Permian widespread anoxic event at 252 Ma (which accompanied a major extinction event), with the Siberian Trap basalts at the same time. There is also a correlation between the Paleocene—Eocene Thermal Maximum (PETM) at 56 Ma (not a mass extinction but a time of extreme warmth and incipient ocean anoxia) and the North Atlantic Igneous Province (Table 61.1).

In some cases, the LIPs, mass extinctions, and OAEs can be correlated rather precisely. For example, in the case of the CAMP basalts at 201 Ma, recent work in Morocco and Eastern North America on rocks near the Triassic—Jurassic boundary shows that the lava flows can be dated (using precise U—Pb age determinations and astrochronology) exactly to the time of the end-Triassic extinction event (Blackburn et al. 2013). For the Siberian basalts at 252 Ma, the paleontological evidence has been interpreted as placing the time of the end-Permian mass extinction within the lowest lava suite (Wignall, 2001), and the most reliable age determinations on the basalts agree closely with U—Pb radiometric ages associated with the latest Permian extinctions in the type section in China (also dated at 252 Ma). In the case of the Deccan Traps province, there seem to have been several major eruptive phases. Paleomagnetic results and radiometric dating suggest that the flows were extruded rapidly during three discrete intervals that occurred both before and after the K—Pg extinction (Chenet et al., 2009) alternating with times of quiescence.

3. THE NATURE OF LIPS

Before assessing the effects of LIPs as possible causes of mass extinctions and OAEs, it is worth examining the style of flood basalt volcanism and the amounts of climatically active gases these volcanic events can release into the atmosphere. LIPs are exceptional volcanic events in Earth history because of the large volume of basaltic magma emitted during individual eruptions (thousands to tens of thousands of cubic kilometers), and the total volume of magma released during the formation of an entire LIP (up to several million cubic kilometers), erupted over a relatively brief period of time. The combination of large erupted volumes and relatively frequent eruptions led to the rapid construction of extensive lava plateaus ranging from 1 to 3 km in thickness. The eruptive style of flood basalt lavas

TABLE 61.1 LIPs of the Last 580 Ma Compared to Stratigraphic Boundaries, Faunal Extinction Events, and OAEs

Flood Basalt Episode	Age (Ma)	Volume (10⁶ km³)	Paleolatitude	Duration of Peak Pulse (Ma)	Stratigraphic Boundary	Age (Ma)
Columbia river	16	0.25	45°N	≤1 (for 90%)	Early/Mid-Miocene	16
Ethiopian	30 ± 1	≤1.0	10°N	≤1	Early/Late Oligocene	28.1
North Atlantic (NAIP)	56	>1.0	65°N	≤1	Paleocene/Eocene (OAE?)	56
Deccan	65.5	>1.3	20°S	≤1	Cretaceous/Tertiary (**E**)	66
Madagascar	87 ± 3	?	45°S	?	Turonian/Coniacian	89.8
Caribbean	89 ± 1	>2	10°N	?	Cenomanian—Turonian (OAE)	93.9
Rajmahal	116 ± 1	?	50°S	≤2	Aptian/Albian (OAE)	113
Ontong Java	90 ± 2	20—57	20°S	?	Cenomanian/Turonian	93.9
Ontong Java	122 ± 1.5				Barremian/Aptian	126
Paraná/Etendeka	133 ± 1	>1.0	40°S	≤1 or ~5?	Valanginian/Hauterivian	132
Antarctica (Ferrar)[1]	183 ± 1	>0.5	50—60°S	≤1?	Pliensbachian/Toarcian (OAE)	183
Karoo	183 ± 1	>2.0	45°S	0.5—1	Pliensbachian/Toarcian (OAE)	183
Central Atlantic (CAMP)	201 ± 1	>2.0	~30°N	?	Triassic/Jurassic (**E**)	201
Siberian	252	>4.0	55—75°N	≤1	Permian/Triassic (**E**) (OAE?)	252
Emeishan	259 ± 3	0.4	4°N	?2	End-Guadalupian (**E**)	260
Skagerrak-centered, Northern Europe	297 ± 4	0.5	11°N	?	End-Carboniferous	299
Barguzin-Vitim, Mongolia	275—310	0.3	60°N	?	End-Carboniferous	299

(Continued)

TABLE 61.1 LIPs of the Last 580 Ma Compared to Stratigraphic Boundaries, Faunal Extinction Events, and OAEs—Cont'd

Flood Basalt Episode	Age (Ma)	Volume (10⁶ km³)	Paleolatitude	Duration of Peak Pulse (Ma)	Stratigraphic Boundary	Age (Ma)
Villuy	350–380	>1	30°N	?	Tournasian—Visean	347
Pripyat-Dnieper-Donets	364–367	>1.5	5°N	?	End-Frasnian	372
Kola/Kontogero	363–370	>0.1?	45°N	?	End-Frasnian	372
Altai-Sayan	393–408	0.1	15°S	?	End-Silurian?	419
Ogcheon area, Korea	430–480	?	?	?	End-Ordovician (**E**)?	444
Central Asian intraplate volcanism	470–510	0.025	15°S	?	?	
Kalkarindji	503–510	>1	10–30°N	?	End age 4?	509
Volyn	545–580	>0.2	40–60°N	?	End-Ediacaran?	541

Bold (**E**) indicates a major extinction event, otherwise stratigraphic boundary coincides with environmental changes and/or less severe extinction.
Ma, million years.
OAE indicates Ocean Anoxic Event.
NAIP, North Atlantic Igneous Province.
CAMP, Central Atlantic Magmatic Province.
[1]Antarctica (Ferrar basalts) considered to be part of the Karroo by sc.
After Courtillot and Renne (2003), Kelley (2007), Kravchinsky, (2012), and sources therein. Updated timescale for extinctions/geological boundaries nt (2012) GSA Geologic Time Scale.

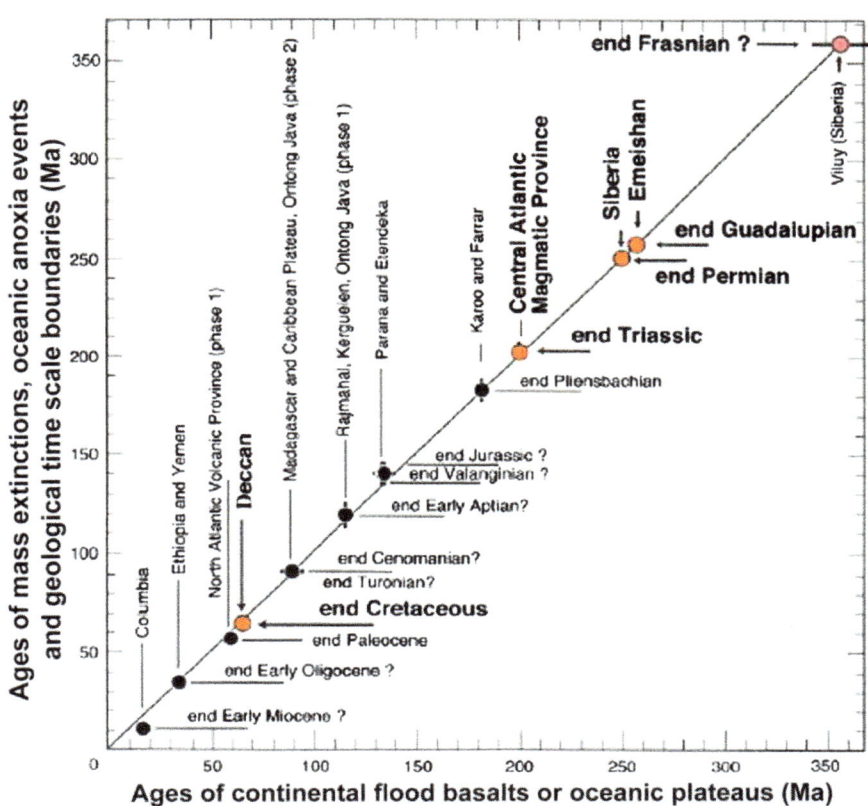

FIGURE 61.1 Ages of LIPs versus times of mass extinctions, geologic timescale boundaries, and ocean anoxic events over the last 350 Ma. *(After Courtillot and Renne (2003).)* The red dots show coincidences of LIPs with strong extinction events.

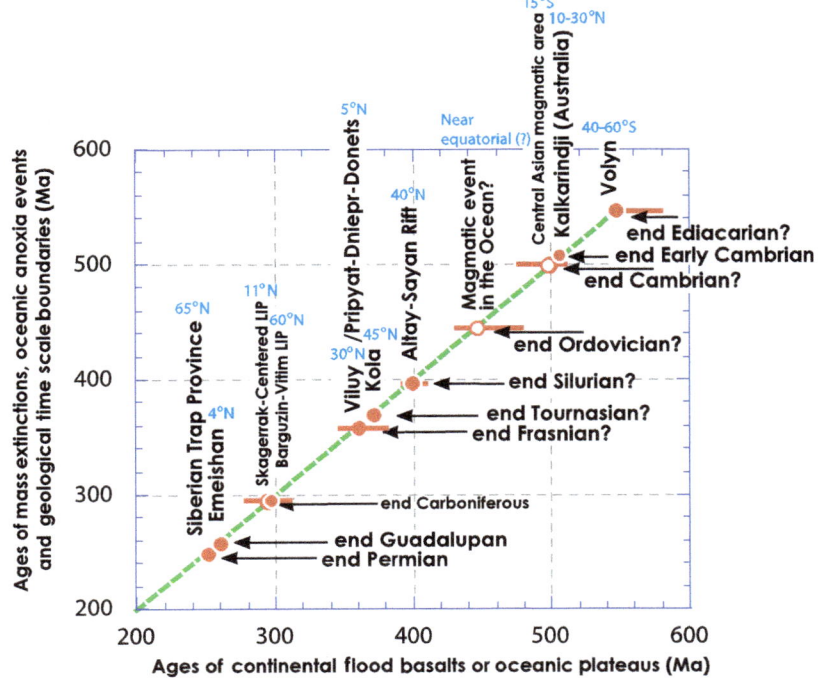

FIGURE 61.2 Ages of LIPs versus times of mass extinction, geologic timescale boundaries, and ocean anoxic events over the period from 250 Ma to 550 Ma. *After Kravchinsky (2013).*

is important in controlling the atmospheric and climatic effects. Flood basalt lavas were originally envisioned to flow as turbulent sheets 10–100 m thick that covered large areas in a matter of days. It is now believed that flood basalts are erupted mainly as fissure-fed pāhoehoe flows that inflate to their greatest thicknesses during the time in which they reach their great extent (Self et al., 1997). This suggests emplacement of the individual flows over periods of time ranging from years to decades, or perhaps longer. For example, based on the peak output rates of a large historical lava flow eruption, the Laki (Iceland) event in AD 1783, (about 5×10^3 kg per second (or 1.5–1.8 m^3) per meter length of fissure, for a 2 to 3-km-long fissure), the time required for eruption of one large LIP lava flow ($\sim$1000–10,000 km^3) would be in the range of tens to hundreds of years (Self et al., 2014 and papers therein).

A major unanswered question is whether the individual flood basalt eruptions were frequent enough for the gases released during the sequence of eruptions to have had a cumulative effect on the environment. An entire 1–2 million cubic kilometer LIP could be erupted in $\sim$30–50 thousand years, without a significant pause in eruption, if eruption rates along kilometer-long fissure systems were similar in magnitude to those observed at the peak of the Laki eruption ($\sim$12 km^3 erupted in a few months). However, the emplacement of lava flows in LIPs is known to be episodic, but accurately determining the length of the quiescent periods (hiatuses) remains challenging due to the precision limitations of current dating methods compared with the duration of volcanism. The timescale of eruptive hiatuses thus plays a role in the global carbon dioxide (CO_2) cycle, as this greenhouse gas has an average residence time of over 100 years in the atmosphere.

The tempo of eruptions is also especially critical for sulfur gases (sulfur dioxide, SO_2, and to a lesser extent hydrogen sulfide, H_2S), which can create sulfate (sulfuric acid, H_2SO_4) aerosols in the troposphere and stratosphere, but with a relatively short atmospheric residence time of a few years. Thus, a prolonged eruption producing a large LIP lava flow could give rise to increased sulfate aerosols lasting many years. In fissure-fed basaltic lava flow eruptions, the volumetric eruption rate is a critical parameter in determining the height of fire fountains over active vents, which controls the convective rise of the volcanic gas plumes in the atmosphere, and hence the geographical distribution of the SO_2 gas released by the eruption and the resulting aerosols. Based on high eruption rates and fire-fountain height estimates, it is likely that plumes from these large basaltic eruptions would have reached the stratosphere (a maximum of 17 km in Earth's present climatic regime, decreasing with higher latitude) in many cases. Furthermore, in some cases (e.g., the Siberian Traps and the Etendeka basalts), the basaltic eruptions were apparently accompanied by more explosive silicic eruptions that could lead to stratospheric height plumes.

4. HOW MIGHT LIP VOLCANISM AFFECT THE ATMOSPHERE AND ENVIRONMENT?

Despite the apparent coincidence of LIP events with extinctions and OAEs, there is no general agreement on how LIP volcanism could have caused these die-offs and oceanographic changes (Bryan and Ferrari, 2013). LIP eruptions have been suggested to produce several kinds of direct environmental effects, including (1) greenhouse warming from CO_2 and methane (CH_4) emissions (and to a lesser extent possibly SO_2), (2) ocean surface acidification from releases of CO_2 and SO_2, (3) climatic cooling from sulfuric acid aerosols in the troposphere and especially the stratosphere, and (4) widespread acid rain from emissions of SO_2, hydrochloric acid (HCl), and hydrofluoric acid (HF). Indirect environmental effects include changes in ocean chemistry, circulation, and oxygenation, especially from LIP volcanism associated with large submarine oceanic plateaus. In this case, the underwater LIP volcanism such as that which created the Ontong Java Plateau (phases about 90 and 122 Ma ago) or the Caribbean volcanic province (about 89 Ma ago) might have altered ocean chemistry through hydrothermal reactions. Moreover, global warming from increased greenhouse gas concentration in the atmosphere is expected to disrupt and slow down ocean circulation, and favor anoxic conditions leading to OAEs.

In most cases, the environmental effects of these eruptions are ascribed to the gases, mainly CO_2 and SO_2 (along with lesser amounts of HCl and HF) that are released directly from the magma during the volcanic events. Another source of gases is from interactions of the intrusive LIP magma with carbon-rich deposits (e.g., coal) releasing large amounts of thermogenic CH_4, which is a strong greenhouse gas that converts rapidly in the atmosphere to CO_2 (Svensen et al., 2009 and references therein). In support of this idea, several LIPs are associated with marked decreases in the $\delta^{13}C$ of contemporaneous marine sediments, suggesting a source of light carbon, such as thermogenic CH_4, to the atmosphere–ocean system. Another possibility is interaction with anhydrite ($CaSO_4$) deposits that can yield excess SO_2 gas. Such was the case at the end-Permian extinction, which is tentatively correlated with an extreme hot greenhouse climate (Sun et al., 2012), although the model results of this study have been challenged. A distinct negative carbon-isotope anomaly in end-Permian marine sediments coincides with the eruption of the Siberian Traps, which are known to have interacted with coal, halite, and anhydrite, and may thus have triggered a release of light thermogenic carbon to the atmosphere.

5. RELEASE OF GREENHOUSE GASES

Greenhouse warming caused by large emissions of CO_2 from LIP volcanism has been suggested as the primary cause of climatic change leading to mass extinctions and OAEs. But the climatic effects depend critically on the timing of the erupted lavas. For example, a large amount of CO_2 could have been produced by the eruptions of an entire LIP. Self et al. (2014) estimate the total amount of CO_2 produced during the entire Deccan eruption sequence to have been more than 10^6 to 10^7 Mt (a megaton equals 10^6 tons or 10^{12} g), but these gases would have been released over relatively long periods of time; slowly enough for the global carbon cycle to remove excess CO_2 from the atmosphere. Thus, it is important to know the duration of the eruptions and the spacing between the large volcanic events that produced flow fields.

Based on historic eruptions, it has been estimated that approximately 10 Mt of CO_2 could be released for every cubic kilometer of basaltic magma erupted. Thus, the release from an erupted flow with a volume of 1000 to 10,000 km^3 of lava (large LIP flows) would be about $10-100$ Gt (A gigaton equals 1000 Mt) of CO_2. Although this is a large volume of CO_2, it should be noted that it represents only a fraction of the CO_2 present in the modern atmosphere (about 600 Gt), and that atmospheric CO_2 is estimated to have been as much as 10 times greater in the geologic past when the eruptions occurred. Degassing of CO_2 during a 1000 to 10,000 km^3 eruption over a period of tens of years would yield at most a few Gt per year, and produce a relatively small increase in atmospheric CO_2 concentration (present anthropogenic CO_2 production is about 35 Gt per year). Despite the size of LIP eruptions, direct release of CO_2 from the basaltic magma seems to be less than would have been required to change climate significantly (Caldeira and Rampino, 1990; Self et al., 2014).

A significantly larger source of greenhouse gases (both CO_2 and CH_4) can come from interactions between the magma and organic-rich sedimentary deposits, which seems to be a common occurrence during sill intrusions by LIP volcanism (Svensen et al., 2009). The suggested mechanism of gas release is through extensive hydrothermal vent complexes associated with intrusions that are now recognized as a characteristic component of LIPs. The release of several thousand Gt of isotopically light carbon gases (mostly methane) from interaction of magma and organic-rich sedimentary deposits has been proposed to be the cause of negative shifts in $\delta^{13}C$ in the ocean and atmosphere, and spikes of warming that

occurred at the Permian—Triassic boundary (252 Ma), Triassic—Jurassic boundary (201 Ma), the Toarcian OAE in the Jurassic (183 Ma) and the PETM (56 Ma). The resulting climate changes are extensively documented by proxy data, which may include a global warming of $5-9\ °C$ lasting a few hundred thousand years, accompanied by anoxic conditions in the oceans and in some cases severe extinctions.

An initial volcanically induced greenhouse warming could lead to the release of additional CH_4 from unstable methane hydrates in high-latitude shelf areas. The CAMP volcanism, for example, seems to have coincided with a large release of isotopically light carbon as CH_4 into the atmosphere, probably from thermogenic and methane hydrate sources. Using carbon-isotope data from paleosols, the CAMP basalts are estimated to have released enough CO_2 to raise the atmospheric concentration from ~ 2000 ppm to ~ 4400 ppm, in less than 300 thousand years (Schaller et al., 2011), which would seem to require more CO_2 (~ 9000 Gt) than could be supplied by magmatic sources alone.

6. OCEAN ACIDIFICATION

Addition of large amounts of CO_2 and SO_2 into the ocean/atmosphere system over a relatively short period of time can cause changes in seawater pH and carbonate saturation. This is happening at the present time as a result of anthropogenic release of CO_2 (Silverman et al., 2009). Berner and Beerling (2007) used carbon-cycle model calculations to estimate the degree of under-saturation for carbonates in the surface ocean that might be produced by the CAMP basalts degassing of CO_2 and SO_2. They conclude that CO_2 and SO_2 from CAMP eruptions could have produced an undersaturated surface ocean that persisted for $20-40$ thousand years, but only if the total degassing took place in less than $50-100$ kyr. Thus, at the Triassic—Jurassic boundary (201 Ma ago), doubling of atmospheric CO_2 concentration over less than 100 thousand years may have created a biocalcification problem with reduced carbonate accumulation (especially aragonite and high-magnesian calcite) in shallow- and deep-water shelf sediments, while coral reefs experienced a near total collapse.

Other extinction episodes such as the end-Permian extinction are similarly marked by preferential survival of organisms with resilience to reduced carbonate saturation, and a major extinction of more vulnerable reef builders. The Toarcian OAE (early Jurassic) at the time of the Karoo eruptions was also associated with a reef crisis and a decrease in the production of calcareous nanoplankton. Further work is needed to establish whether oceanic undersaturation of carbonates was a general accompaniment to LIP volcanism.

7. POSSIBLE EFFECTS OF SULFUR GAS RELEASE

Climatic cooling at Earth's surface attributed to volcanic eruptions is primarily a result of the formation and spread of sulfuric acid (sulfate) aerosols. These small droplets are formed from sulfur volatiles (largely SO_2) injected into the upper atmosphere by convective eruption columns and plumes rising above volcanic vents and fissures. The SO_2 is quickly oxidized to form sulfuric acid aerosols. The aerosols have a short residence in the troposphere, but those that reach the stratosphere may persist for several years, where they primarily backscatter incoming sunlight, cooling the planet.

Studies of sulfur degassing of basaltic magmas suggest that up to $\sim 75\%$ of volatile sulfur in the magma is released (largely as SO_2) into the atmosphere at the eruptive vents. In historic time, the large volume ($\sim 15\ km^3$), 8-month-long Laki eruption of 1783—1784 released about 120 Mt of SO_2 delivered in quasi-continuous emissions with maximum fluxes of about 6 Mt per day, and averages of 3 Mt per day over the first few months (Thordarson et al., 1996).

Observations and model calculations suggest that the high eruption rates during the Laki eruption produced high fire fountains (up to 1500 m), so that the convective plume rise above the fountains could have attained altitudes of up to 13 km above sea level (Thordarson et al., 1996; Figure 61.3) into the upper troposphere and the lower stratosphere. The atmospheric effects of the Laki eruption were severe and quite widespread, supporting the creation of at least some stratospheric aerosols, which agrees with eruption models (Oman et al., 2006). Haziness and dimming of sunlight were noticeable in 1783 in Europe, and a so-called "dry fog" was reported as Far East as China. Climatic cooling followed, and the winter of 1783—1784 was the coldest recorded in the Eastern United States from that date forward.

The observed sulfur content of basaltic magmas and the evidence for degassing of the lava flows suggest that the sulfur gas release from a large flood basalt eruption (such as the Roza flow of the Columbia River basalts) could be about 10 Gt of SO_2, along with significant amounts of fluorine and chlorine gases (Thordarson and Self, 1996; Figure 61.3). Using the maximum eruption rates for Laki,

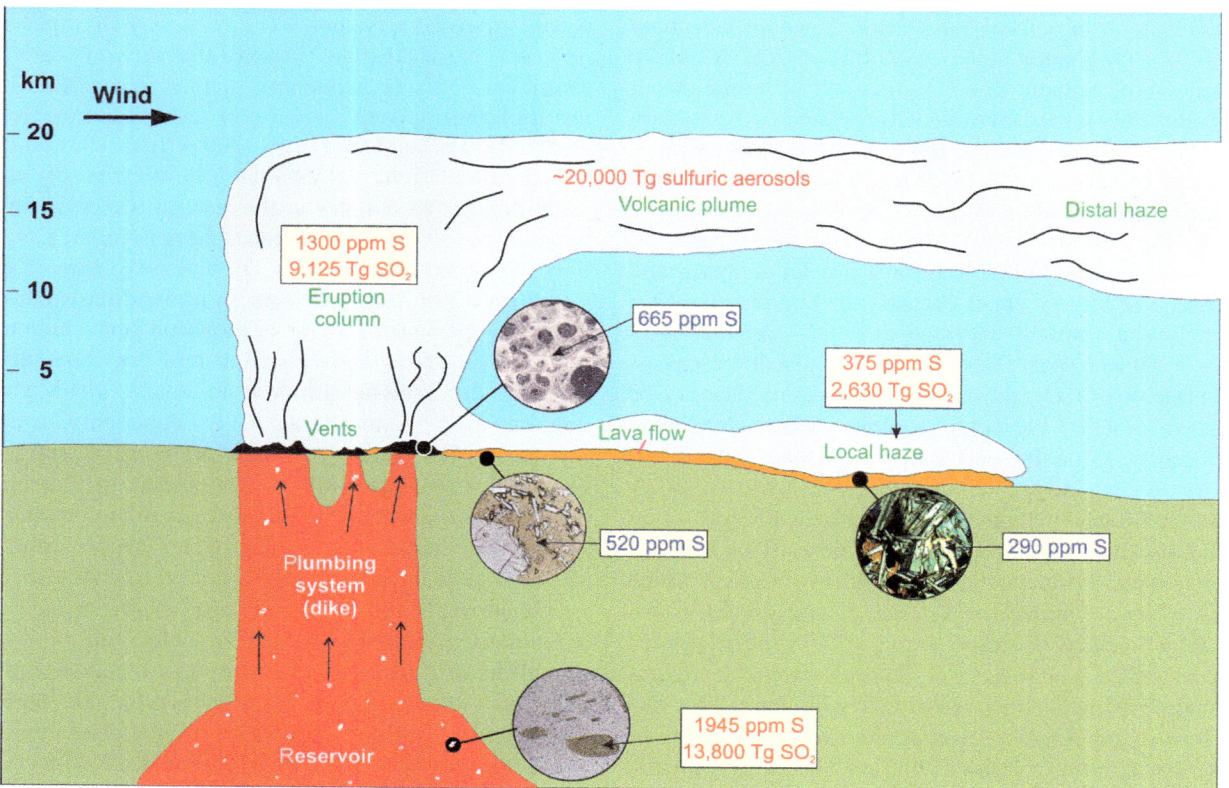

FIGURE 61.3   Cartoon showing generalized eruptive style and atmospheric dispersal of gas and aerosol particles (with minor ash) associated with basaltic fissure eruptive activity such as a flood basalt eruption, based on studies of the Roza eruption of the Columbia River basalt province, Washington, USA. *(After Thordarson and Self (1996a); see also Self et al. (2014).)* Tg is terragrams (10^{12} g). Total amount of SO_2 released into the atmosphere is estimated to have been almost 12,000 Tg, ~ 12 Gt (gigatons), with ~ 9000 Tg (9 Gt) emitted from the vents with the eruption columns.

by analogy the large Roza eruption would have produced high lava fountains and a convective column rising more than 15 km above the volcanic vents. If the Roza eruption continued over 10 years, a great deal of aerosols could have been generated in the atmosphere, with a portion going into the lower stratosphere.

Some atmospheric scientists have suggested, however, that physical and chemical effects in dense aerosol clouds might limit the mass of volcanic aerosol that persists in the stratosphere during and after an eruption. Further modeling of dense stratospheric aerosol clouds, and their effects on atmospheric dynamics and chemistry, is needed before we can predict the climatic changes that would occur from the release of SO_2 during a long-duration flood basalt eruption.

By absorbing and reflecting incoming solar radiation, volcanic aerosols at all altitudes should cool Earth's surface, with maximum effects stemming from low-level tropospheric aerosols. However, tropospheric aerosols have a very short lifetime of about a week before they are washed out of the lower atmosphere. Yet, massive gas release from a long-lasting flood basalt eruption could generate such large amounts of volcanic aerosols that a regional steady state aerosol cloud might be maintained in the troposphere despite this washout, while a portion of this aerosol mass might enter the stratosphere.

Studies using a global climate model coupled with the expected oceanic response suggest that if flood basalt eruptions were continuous at an average activity level for decades, then significant cooling of climate might be possible. However, with hiatuses of perhaps thousands to tens of thousands of years between individual flood basalt eruptions during the formation of an LIP province, it has not yet been shown that such large releases of sulfur gases would have a cumulative and lasting effect on climate and the environment.

Modeling studies by Schmidt et al. (2010) showed that Laki also had the potential to dramatically impact global cloud condensation nuclei (CCN) concentrations with an increase in the total particle concentrations in the upper troposphere by a factor of 16 over large parts of the Northern Hemisphere during the first 3 months of the eruption. In this simulation, volcanic particles could affect cloud formation even as far away as Asia and perhaps into the Southern Hemisphere. An increase in CCNs should cause more small cloud droplets, which increases cloud albedo and contributes to surface cooling. This effect might be much amplified during large LIP events, and the effects on climate could be significant.

Releases of SO_2, a strong greenhouse gas, into the lower atmosphere by flood basalt eruptions are a possible source of climate warming. During the summer of 1783 in Europe, while the sulfurous dry fog generated by the Laki eruption hung in the air, the weather was stifling. Historical reports of acrid odor, difficult breathing, dry deposition of sulfate,

and vegetation damage indicate especially high SO_2 concentrations in the lower atmosphere from mid-June to early August (Schmidt et al., 2012). The July 1783 temperatures in Western Europe were up to 3 °C warmer than the long-term average. While it is not certain that these effects were caused by the Laki gases, the influence of low-elevation volcanic plumes downwind from basaltic eruptions is a topic that should be further investigated.

8. ACID RAIN

Yet another possible environmental factor is acid rain derived not only from the SO_2 but also from chlorine and fluorine gases released from flood basalt eruptions. For the historic Laki eruption, the local haze over Iceland contained high concentrations of H_2SO_4, HF, and HCl, causing skin lesions in animals and humans. Grass growth was stunted, trees were killed, and 50% of livestock perished, most likely from fluorine poisoning. The resulting "haze famine" led to the death of 20% of the Icelandic population, and acid rain effects were also reported from Northern Europe, where growth in peat bogs was affected. Scaling up from Laki, it seems clear that acid rain could be a serious problem during LIP eruptions. The end-Permian extinction, for example, is marked by the global die-off of terrestrial vegetation that might have been related to acid rain created by the massive Siberian Traps eruptions.

9. SUMMARY

The radiometrically determined ages of LIPs show a correlation with the ages of mass extinction events, OAEs, and stratigraphic boundaries (Figures 61.1 and 61.2). Several lines of evidence suggest that gases such as CO_2 and SO_2 released by LIP volcanism and associated intrusions may be capable of causing significant environmental effects that may lead to climatic effects, faunal/floral crises and extinctions. Factors that contribute to the environmental impact of LIP volcanism include (1) rapid global warming resulting from greenhouse gas emission, especially from related magma intrusions into carbon-bearing sedimentary rocks; this led at times to OAEs, (2) ocean acidification from CO_2 and SO_2 releases, (3) cooling resulting from conversion of SO_2 to sulfuric acid aerosols in the troposphere and lower stratosphere, and (4) acid rain from SO_2, HCl, and HF emissions.

With regard to CO_2, direct emissions from the magma during flood basalt events seem to be insufficient to cause the global warming effects associated with many LIPs, and isotopically too heavy to cause the observed carbon-isotope anomalies. A key factor is the type of rocks with which the magma interacts. There is, however, potential for significant volatile release when the intrusive component of LIPs (sills and dikes) comes in contact with sedimentary rocks such as coal or other organic-rich deposits, and evaporite deposits, releasing thermogenic CH_4 and SO_2. An important method of gas release is through hydrothermal vent complexes that are being recognized as a characteristic component of LIPs. Further release of CH_4 can come from destabilization of methane hydrates (clathrates) on continental margins as a result of the initial global warming, providing a strong positive feedback mechanism.

More precise correlations may be discovered with increasing precision of radiometric dates of extinctions and the massive basaltic volcanism. There is thus a good case for a link between LIP volcanism and environmental and biotic changes, although the precise mechanism(s) remains unclear. Direct evidence of causality is difficult to demonstrate, and there is a continuing debate on the cause or causes of the mass extinction events and OAEs.

FURTHER READING

Alvarez, L.W., Alvarez, W., Asaro, F., Michel, H.V., 1980. Extraterrestrial cause for the Cretaceous—Tertiary extinction: experimental results and theoretical interpretation. Science 208, 1095—1108.

Berner, R.A., Beerling, D.J., 2007. Volcanic degassing necessary to produce a $CaCO_3$ undersaturated ocean at the Triassic-Jurassic boundary. Palaeogeogr. Palaeoclimatol. Palaeoecol. 244, 368—373.

Blackburn, T.J., Olsen, P.E., Bowring, S.A., McLean, N.M., Kent, D.V., Puffer, J., McHone, G., Rasbury, E.T., Et-Touhami, M., 2013. Zircon U-Pb geochronology links the end-Triassic extinction with Central Atlantic Magmatic Province. Science 340, 941—944.

Bryan, S.E., Ferrari, L., 2013. Large igneous provinces and silicic large igneous provinces: progress in our understanding over the last 25 years. Geol. Soc. Amer. Bull. 125, 1053—1078. http://dx.doi.org/10.1130/B30820.1.

Caldeira, K., Rampino, M.R., 1990. Carbon dioxide emissions from Deccan volcanism and a K/T boundary greenhouse effect. Geophys. Res. Lett. 17, 1299—1302.

Chenet, A.-L., Fluteau, F., Courtillot, V.E., Gerard, M., Quidelleur, X., Khadri, S.F.R., Subbarao, K.V., Thordarson, T., 2009. Determination of rapid Deccan eruptions across the Cretaceous—Tertiary boundary using paleomagnetic secular variation: 2. Constraints from analysis of eight new sections and synthesis for a 3500-m-thick composite section. J. Geophys. Res. Solid Earth 114, B06103. http://dx.doi.org/10.1029/2008JB005644.

Courtillot, V.E., 1999. Evolutionary Catastrophes: The Science of Mass Extinctions. Cambridge University Press, New York.

Courtillot, V.E., Renne, P.R., 2003. On the ages of flood basalt events. Comp. Rendus Geosci. 335, 113—140.

Kelley, S.P., 2007. The geochronology of large igneous provinces, terrestrial impact craters, and their relationship to mass extinctions on Earth. J. Geol. Soc. 164, 923—936.

Kravchinsky, V.A., 2012. Paleozoic large igneous provinces of Northern Eurasia: correlation with mass extinction events. Glob. Planet. Change 86—87, 31—36.

Oman, L., Robock, A., Stenchikov, G.L., Thordarson, T., Koch, D., Shindell, D.T., Gao, C., 2006. Modeling the distribution of the volcanic aerosol cloud from the 1783—1784 Laki eruption. J. Geophys. Res. 111, D12209. http://dx.doi.org/10.1029/2005JD006899.

Rampino, M.R., 2010. Mass extinctions of life and catastrophic flood basalt volcanism. Proc. Natl. Acad. Sci. U.S.A. 107, 6555—6556. http://dx.doi.org/10.1.073/pnas.1002478107.

Rampino, M.R., Stothers, R.B., 1988. Flood basalt volcanism during the past 250 million years. Science 241, 663—668.

Schaller, M.F., Wright, J.D., Kent, D.V., 2011. Atmospheric pCO_2$ perturbations associated with the Central Atlantic Magmatic Province. Science 331, 1404—1407.

Schmidt, A., Carslaw, K.S., Mann, G.W., Wilson, M., Breider, T.J., Pickering, S.J., Thordarson, T., 2010. The impact of the 1783—1784 AD Laki eruption on global aerosol formation processes and cloud condensation nuclei. Atmos. Chem. Phys. 10, 6025—6041.

Schmidt, A., Thordarson, Th, Oman, L.D., Robock, A., Self, S., 2012. Climatic impact of the long-lasting 1783 Laki eruption: inapplicability of mass-independent sulfur isotopic composition measurements. J. Geophys. Res.: Atmos. 117 (D23), D23116. http://dx.doi.org/10.1029/2012jd018414.

Schulte, P., Alegret, L., Arenillas, I., et al., 2010. The Chicxulub asteroid impact and mass extinction at the Cretaceous—Paleogene boundary. Science 327, 1214—1218.

Self, S., Schmidt, A., Mather, T.A., 2014. Emplacement characteristics, timescales, and volatile release rates of continental flood basalt eruptions on Earth. In: Keller, G. (Ed.), Geol. Soc. Amer. Special Paper 505, Flood basalts, impacts and extinctions. http://dx.doi.org/10.1130/2014.2505(16).

Self, S., Thordarson, T., Keszthelyi, L.P., 1997. Emplacement of continental flood basalt lava flows. In: Mahoney, J.J., Coffin, M.F. (Eds.), Large Igneous Provinces: Continental, Oceanic, and Planetary Flood Volcanism, Geophysical Monograph Series, vol. 100. American Geophysical Union, Washington, DC, pp. 381—410.

Silverman, J., Lazar, B., Cao, L., Caldeira, K., Erez, J., 2009. Coral reefs may start dissolving when atmospheric CO_2 doubles. Geophys. Res. Lett. 36, L05606. http://dx.doi.org/10.1029/2008GL036282.

Sun, Y., Joachimski, M.M., Wignall, P.B., Yan, C., Chen, Y., Jiang, H., Wang, L., Lai, X., 2012. Lethally hot temperatures during the Early Triassic greenhouse. Science 338, 366—370.

Svensen, H., Planke, S., Polozov, A.G., Schmidbauer, N., Corfu, F., Podladchikov, Y.Y., Jamtveit, B., 2009. Siberian gas venting and the end-Permian environmental crisis. Earth Planet. Sci. Lett. 277, 490—500.

Thordarson, T., Self, S., 1996. Sulfur, Chlorine, and Fluorine degassing and atmospheric loading by the Roza eruption, Columbia River Basalt Group, Washington, USA. J. Volcanol. Geotherm. Res. 74, 49—73.

Thordarson, T., Self, S., Óskarsson, N., Hulsebosch, T., 1996. Sulfur, chlorine, and fluorine degassing and atmospheric loading by the 1783—1784 Laki (Skaftár Fires) eruption in Iceland. Bull. Volcanol. 58, 205—225.

Wignall, P.B., 2001. Large igneous provinces and mass extinctions. Earth-Sci. Rev. 53, 1—33.

Chapter 62

Volcanic Lightning

Stephen R. McNutt

School of Geosciences, University of South Florida, Tampa, FL, USA

Ronald J. Thomas

Department of Electrical Engineering, New Mexico Institute of Mining and Technology, Socorro, NM, USA

Chapter Outline

GLOSSARY

BLM Alaska lightning detection system The U.S. Bureau of Land Management (BLM) in Alaska operates a low-frequency lightning location network for forest fire information and supplies the Alaska Volcano Observatory with lightning location data. The location capability does not extend into the Aleutian Islands.

discharge A sudden elimination of electrical charge and stored energy from a charged region in the form of a spark to a large lightning flash. The energy stored in the static electrical field is released as heat and light during the discharge.

leader A lightning flash has many kilometers of conducting channel made of very hot gas in which the electrons are free to move and carry current. The hot channels are created by leaders formed by small discharges in the very high electric fields at their tips. The leader carries a high electrical field that ionizes air, allowing for extension of a channel.

lightning flash A large electrical discharge in the atmosphere. The lightning flash forms a large interconnected double-ended treelike structure with many kilometers of very hot conducting channels. A flash may or may not contact the ground and is between and propagates throughout two charge regions or a charge region and the ground. The lightning flash reduces the static electric field and effectively decreases the charge. A flash typically is about half a second long but can last 1 s or more. Lightning to the ground is called cloud-to-ground (CG) lightning.

LMA The Lightning Mapping Array (LMA) is a very-high-frequency lightning location and mapping network that locates hundreds or thousands of sources produced by lightning leaders in each flash.

For each lightning flash, these source locations give a 3-D map of the lightning flash.

near-vent lightning Volcanic lightning starting at or near the erupting vent. This lightning goes upward and is often smaller and shorter lasting than thunderstorm lightning.

plume lightning Volcanic lightning in the plume of the erupting volcano. This volcanic lightning is very much like thunderstorm lightning. It can range in size from subkilometer to tens of kilometers.

polarity The sign of charge transferred vertically downward by a lightning flash. A lightning flash can transfer either positive or negative charge to ground or downward. A negative cloud-to-ground (CG) lightning flash transfers negative charge to ground.

stroke A large pulse of current along a lightning channel to ground. A large lightning flash can have a dozen or more strokes with the current going to zero and the channel to dark between strokes.

vent discharges Very small volcanic lightning that occurs at the vent during the eruption. It is estimated to be of the order of 10–100 m in length.

WWLLN World Wide Lightning Location Network, operated by the University of Washington, is a very-low-frequency lightning location network, which can locate lightning globally.

1. INTRODUCTION

Volcanic lightning is relatively common but has only recently been studied systematically with modern instrumentation. Lightning is important because it reveals electrical properties of processes such as fragmentation and

The Encyclopedia of Volcanoes. http://dx.doi.org/10.1016/B978-0-12-385938-9.00062-6

collision during jetting and convection, affects ash particle clumping, and shows the effects of water and ice in rising ash columns. Lightning is also a hazard in its own right. People have been killed by volcanic lightning flashes at Paricutin (Mexico) and Vulcan (Papua New Guinea) volcanoes.

As of December 2012, lightning occurrence has been documented for 394 eruptions at 152 volcanoes. This is about 28% of the 540 known historically active volcanoes. The data include photographs, videos, reports from pilots, and ground observers as well as instrumental lightning data from **WWLLN** (World Wide Lightning Location Network), **BLM Alaska Lightning Detection System** (Bureau of Land Management), **LMA** (Lightning Mapping Array), and others.

Of the 131 eruptions for which the time of day was tabulated, it was found that 44% occurred during daylight hours and 56% at night. This suggests that lightning is more easily seen against a dark background and may be missed during bright daytime conditions. It may also help explain why lightning is more common with larger eruptions— these often produce so much ash that they "turn day to night" in the vicinity of the eruption.

Lightning has been observed associated with the full spectrum of magma compositions from basalt to rhyolite. The composition of erupting magma is of interest in studies of volcanic lightning because this characteristic is known to influence the dissolved water content in the magma (with important effects on electrification in its own right), with important consequences for the style of the eruption (i.e., Strombolian, Plinian, etc.). The broad range for chemical composition in eruptions goes from basalt to andesite to dacite to rhyolite (48−77% SiO_2), with a systematic decline in equilibrium magma temperature from about 1200−800 °C over this range. Water contents range from about 0.1−6.5 wt% and are systematically higher with increasing silica content. High silica content also generally means more explosive eruptions producing more fine ash as well as higher plumes.

The volcanic explosivity index (VEI) values for eruptions with lightning span six orders of magnitude. VEI values were estimated for 177 eruptions, and a histogram of these events is shown in Figure 62.1 (bottom). Note that heights are estimated above the vent for eruptions with VEI = 2 and smaller, but above sea level for eruptions with VEI = 3 and larger following the standard conventions (see Chapter 13). For comparison, a histogram of all known VEI is shown in Figure 62.1 (top). There are many more small eruptions than large ones, so the numbers increase from right to left. The drop-off for VEI = 1 is likely due to underreporting.

The number of occurrences of volcanic lightning at various VEI values is also shown in Table 62.1 along with all known VEI and the percent of cases showing lightning versus all eruptions. The percent of eruptions with lightning is nearly the same for VEI = 3, 4, 5, and 6 at about 13−16%. This suggests a standard reporting efficiency. Large eruptions attract attention and are generally well reported. The heights of the ash plumes are >10 km for VEI ≥4, so these are taller than most thunderclouds (also referred to as deep convection by atmospheric scientists), and we infer that similar ice-contact charge generation or separation mechanisms may be acting.

FIGURE 62.1 Histogram of volcanic explosivity index (VEI) for all eruptions in *Volcanoes of the World*, by Simkin and Siebert, 1994 (A) and for eruptions accompanied by lightning (B) using the most complete data available as described in the text. Note that the eruptions accompanied by lightning are skewed toward higher VEI values. The plume heights for VEI 3 and larger are similar to the heights of thunderstorms.

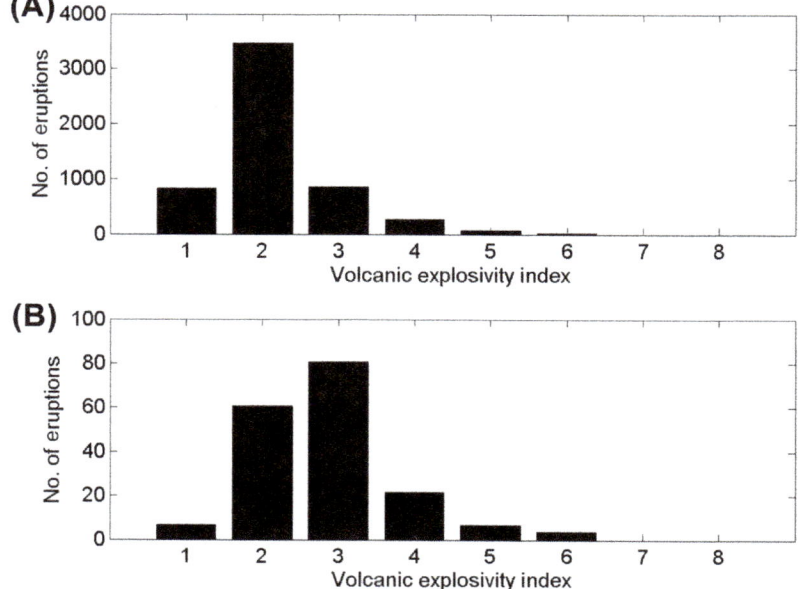

TABLE 62.1 Volcanic Lightning and Global Volcanic Explosivity Index (VEI) Data

VEI	No. of Cases with Lightning	Total No. of Eruptions	Percent
0	1	–	–
1	9	845	1.07
2	124	3477	3.57
3	126	869	14.50
4	45	278	16.19
5	12	84	14.29
6	5	39	12.82
7	0	4	0.00
8	0	0	–

The percentage of eruptions with recorded lightning with VEI = 2 and smaller drops off sharply. This behavior suggests a systematic reporting problem, or that lightning is simply less common in these smaller eruptions, and charging processes that occur do not make enough charge to cause lightning.

Volcanic lightning has been observed associated with eruptions having a wide range of ash column heights from as little as 200 m to as high as 33 km. Height is one component of VEI that offers finer resolution in terms of deducing underlying charging processes. A histogram of the number of cases versus height shows two peaks (Figure 62.2(a)), one from 1 to 4 km and another around 10 km.

Eruptions with VEI = 3 have ash plume heights of 3–15 km (see Chapter 13); these straddle typical thunderstorm heights, and we find many cases with lightning and also many without it. For VEI = 1 and 2 eruptions, there are relatively fewer cases with lightning, but these are important because the ash plume heights are <5 km, less than summertime thunderclouds (tops at 7–20 km above sea level), and they suggest that some other lightning-producing mechanisms may be acting. A histogram of ash plume heights is shown in Figure 62.2(a) for the 142 cases for which data are available. The histogram shows a bimodal distribution with a broad peak in the range of 7–12 km (maximum at 10 km) and another broad peak in the range of 1–4 km. The higher-altitude peak represents the typical heights of ordinary thunderstorms. The low-altitude peak, however, is significantly lower than thunderstorm values.

Another systematic observation is that there is generally more lightning when there is more fine ash. This comes from observation of volcanoes such as Yasour, where Strombolian eruptions producing mainly clots of lava produce no lightning, whereas jetting of fine ash from the same vent produces lightning. The ash production and jetting process are seen to be efficient charging mechanisms.

In addition to being visually spectacular and helping to reveal electrical processes in plumes, the occurrence of lightning also helps to confirm that ash-producing eruptions are occurring. Thus, the occurrence of lightning may be used as part of a warning system to verify the onset and development of ash-producing eruptions.

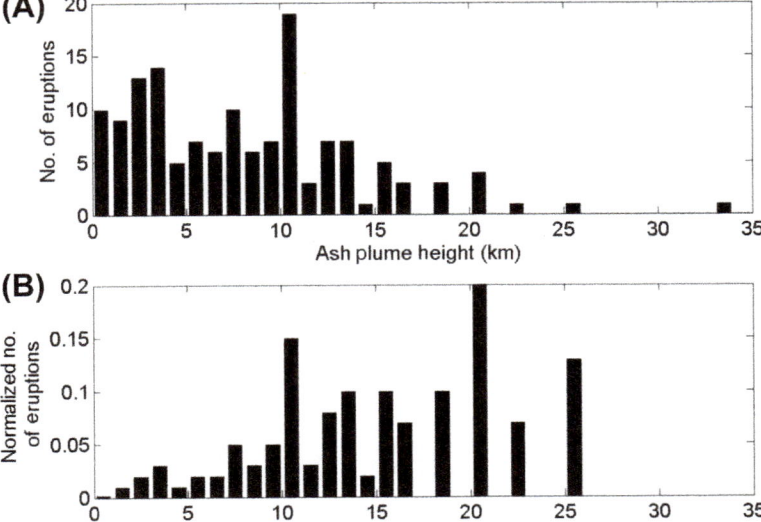

FIGURE 62.2 (A) Histogram of ash column heights for eruptions accompanied by volcanic lightning. The broad peak >7 km (maximum at 10 km) includes heights of similar dimension and larger than typical thunderclouds. This suggests that similar mechanisms may be acting. Also note the second peak from 1 to 4 km. These heights are significantly smaller than thunderclouds and suggest a possible second mechanism or mechanisms. (B) The same data plotted as percent of expected eruptions for each height interval.

2. INSTRUMENTS AND OBSERVATIONS

Lightning during volcanic eruptions is often observed and reported by people in the area and recorded with photographs and videos. It is now easy to find spectacular photos and videos of volcanic lightning on the Internet. Much of the lightning and electrical activity is often hidden from view by bad weather or the ash plume and is best studied with techniques developed and used for thunderstorm research. The buildup and **discharge** of the electrical field can be measured by electric field mills and electric-field change sensors. By very accurately measuring the times of field changes due to lightning at several locations, the lightning can be located in a manner similar to a seismic network.

Measurement of the electric field on the ground near the volcano would seem to be a good measure of the charges in the plume; however, thunderstorm research has shown that this measurement is difficult to interpret, as the measured field could be due to many different charge configurations. The electric field is due to all charges in and around the plume, with the nearby ones weighted the most (the field decreases as $1/r^2$). In thunderstorms, the charges inside the cloud are often shielded by charges of the opposite sign on the edges. Since the changes in the field due to lightning are only due to charges involved with the lightning, field change measurement is more useful.

While visual observations including photos and videos are very interesting, they miss much of the lightning. Clouds, remote locations, and the opaqueness of the plume obscure most of the optical flashes of lightning. In thunderstorms, lightning is routinely detected and located by the longer-wavelength electromagnetic radiation (radio waves) that the lightning generates. This radiation is not blocked by the clouds or ash. There are three different lightning location systems that use different parts of the radio frequency spectrum. All these systems have been used to detect and locate volcanic lightning. Each system has different characteristics in the type and property of lightning detected, the detection efficiency, area coverage, and the location accuracy.

The lowest frequency systems can locate lightning globally because the signals can travel around the Earth in a waveguide between the surface and the ionosphere. These very-low-frequency signals are generated by large currents that flow in larger lightning. Examples of this type of system are WWLLN (operated by the University of Washington), ATDnet (Arrival Time Difference Network operated by the UK Met Office), and GLD360 (Global Lightning Dataset operated by Vaisala). These systems detect about 10–70% of the lightning to ground with an accuracy generally better than 10 km. Each system is unique with its own specifications. WWLLN has been programmed to issue an alert when lightning is detected in the real-time data at the location of an active volcano where thunderstorm lightning is not normally seen. They had their first successful early alert for "volcano ash cloud lightning" with the successful identification of the explosive ash cloud for Sheveluch volcano (2011), Kamchatka, over 1 h prior to any other detection confirmation of the eruption.

Systems that use low-frequency radio waves can detect the signals from lightning hundreds of kilometers away and locate up to 90% of the lightning to ground and 10–20% of the in-cloud lightning. The accuracy is generally better than 1 km. Examples of these systems are the NLDN (National Lightning Detection Network operated by Vaisala in the United States and Canada), the BLM Alaska Lightning Detection System, and EUCLID (EUropean Cooperation for LIghtning Detection). Each **lightning flash** may generate several cloud detections and several ground strikes (ordinary thunderstorm lightning flashes may have 10–20 **strokes** to ground).

At higher frequencies (very high frequency (VHF), generally ~60 MHz), the lightning signals are produced by much smaller events, which are caused by the formation of the lightning channels, not current flowing along them. One system, the LMA (developed by New Mexico Tech) detects hundreds to thousands of locatable sources in each lightning flash, giving a 3-D picture of each lightning flash. The LMA sees all the flashes including very small lightning events that occur near the vent as well as the smaller **vent discharges**. The very small discharges are not detected by the other systems. Table 62.2 shows a comparison of the

TABLE 62.2 Lightning Statistics for LMA, WWLLN, and BLM Networks for Three Representative Eruptions at Redoubt Volcano, Alaska in 2009

Date	Time (UTC)	Plume Height (km)	LMA Regular Discharges	WWLLN Discharges	BLM Discharges
March 23	09:38	13.6	1975	172	79
March 28	01:34	15	534	3	39
March 28	09:19	15	445	39	28

WWLLN, World Wide Lightning Location Network; LMA, Lightning Mapping Array; BLM, Bureau of Land Management.
Data sources: LMA—S. Behnke. WWLLN—R. Holzworth. BLM—T. Weatherby. Plume Height—Alaska Volcano Observatory.

lightning associated with eruption of Redoubt volcano on three systems: LMA, WWLLN, and the BLM Alaska Lightning Detection System. This underscores that various lightning measurement systems measure different properties of lightning and are complementary to each other.

Lightning is detected from space by the optical flashes. Currently, the Lightning Imaging Sensor, developed by NASA Marshall Space Flight Center, is flown on the TRMM satellite. Since this instrument is in a low Earth orbit and is only over any location for a few minutes each day, it is not very useful for observing volcanic eruptions. The Global Lightning Mapper on GOES-R is scheduled for launch in 2015 and will be in a geostationary orbit. It will be useful for observing volcanic lightning at low and mid-latitudes.

3. TYPES OF LIGHTNING

The volcanic lightning has been divided into three different types depending on the flash size, location, and duration. While conceptually each type is distinct, there is likely a continuum between them. The smallest and shortest are **vent discharges**, then **near-vent lightning**, and the largest is thunderstorm-like **plume lightning** (Figure 62.3).

3.1. Vent Discharges

Vent discharges last less than a millisecond, and thousands are detected each second. They have been detected with the LMA during explosive eruption events at Augustine, Redoubt, and Eyjafjallajökull volcanoes. Each source located by the LMA appears to be independent of other sources. This differs from normal lightning, where the located sources group in both time and position. The

sources can occur so often (1–10 per millisecond) that the signal appears to be continuous. Since the measured radio frequency power is similar to that of **leader** steps in a thunderstorm, which are 10–100 m in length, we assume that these are a similar size. At Augustine, the vent discharges began within 1 s of the beginning of the eruption as determined by the infrasound detector. This indicates that the discharges occur very close to the vent and that the tephra is charged as it emerges. All the eruptions where the vent discharges have been detected were erupting ash and gases. Photos of lightning at the vent during an eruption show many small lightning flashes of the size close to 10–100 m and may produce the signals we receive (Figure 62.4). The vent discharges appear to be a good indicator that ash and gas are erupting. The rate of these discharges may be related to the rate of tephra discharge from the vent, but more research is needed to verify this.

3.2. Near-Vent Lightning

Some of the vent discharges develop into longer lightning channels. These start on or near the vent and propagate upward into the developing plume. These are small compared with most thunderstorm lightning and are a few kilometers in length lasting up to 30 ms. Near-vent lightning is detected along with vent discharges while the explosive event is active, generally starting within 1–2 min of eruption onset. They are also seen at a low rate when the ash eruption is too weak to produce any vent discharges. Measurements from both Augustine and Eyjafjallajökull show that this lightning is negative breakdown propagating upward into a region of positive charge. A photograph of a typical near-vent lightning flash is shown in Figure 62.5.

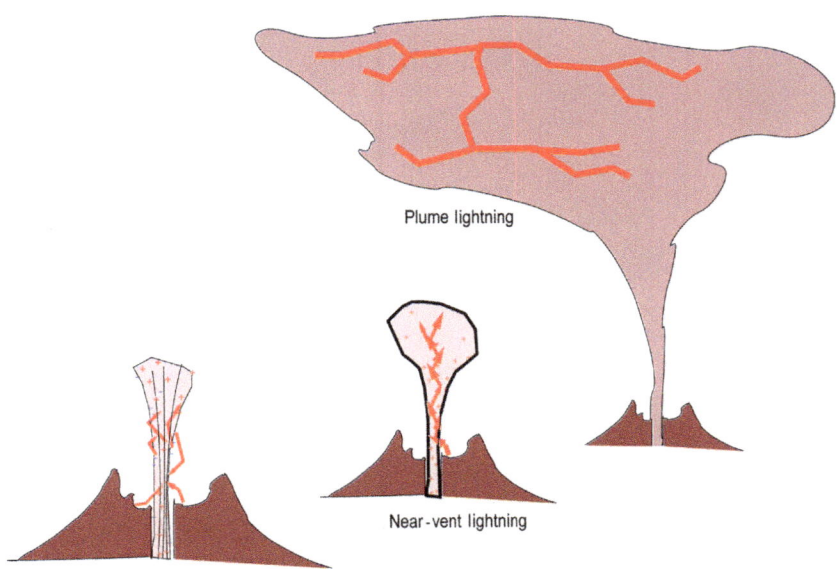

FIGURE 62.3 Schematic of the three types of volcanic lightning or electrical discharges. As the explosive eruption is under way, the small vent discharges occur at a rapid rate near the vent. When the erupting column is high enough, larger discharges begin. These near-vent lightning flashes are still small compared with thunderstorm lightning. In the final part, the thunderstorm-like lightning is seen in the well-developed plume.

Plume lightning

Near-vent lightning

Vent discharges

FIGURE 62.4   Time exposure photograph of an eruption at Tavurvur volcano, Papua New Guinea, showing likely vent discharges. No instrumental data were available at the time of the photograph; however, 17 small discharges with scale lengths of tens of meters are visible (purple and light green streaks). The vent diameter is approximately 100 m. *National Geographic, September 2007, p. 14–15.*

FIGURE 62.5   Photograph of a small eruption at Karymsky volcano, Kamchatka, on October 24, 2007, showing likely near-vent lightning. The ash column is about 2-km high, so the lightning flash is hundreds of meters to 1 km in length. *Photograph by V. Yudin.*

3.3. Plume Lightning

Once the plume has developed, several minutes after the eruption onset, lightning very similar to that seen in thunderstorms begins (Figure 62.6). As winds move the plume away from the volcano, the plume lightning can go with it.

As in thunderstorms, the lightning is either between two regions in the plume or between the plume and ground. The CG lightning can lower either positive or negative charge to the ground. This lightning is generally larger (subkilometer to >20 km) than the near-vent lightning and lasts longer

FIGURE 62.6   In this time exposure of one explosive event in the 2009 eruption of Redoubt volcano, lightning is seen throughout the plume. Composite photograph of in-cloud and cloud-to-ground lightning during the explosive eruption at 07:19 UTC on March 28, 2009. Volcano (lower left) is 3 km high; top of image is about 14 km a.s.l. *Photographs taken from Seldovia, Alaska, 125 km to the south-southeast, by B. Higman; http://www. groundtruthtrekking.org/ blog/?p=849.*

(10 ms−1 s). As the plume matures, the lightning becomes larger and less frequent, and the largest flashes can occur farthest from the vent. This reflects the structure of the plume, which initially is very turbulent and slowly becomes more stable and stratified.

4. ELECTRIFICATION PROCESSES

4.1. Background

Electrical phenomena are widespread and important, even though people may not be aware of them. For example, the electric field in normal undisturbed atmosphere is approximately 100 V/m (vertically). It can be substantially higher in and near a thunderstorm. Electrical breakdown occurs when the electric field exceeds the dielectric breakdown. A familiar small-scale example is walking on a carpet. An electric charge develops on the person, and bringing a finger near a metal object such as a doorknob causes a spark to jump across the air gap. The relevant parameters are the charge magnitude (moderate in this case), the distance (short), and the permittivity of air (nearly constant). Basic physics theory shows that the maximum electric field at the surface of a uniformly charged spherical volume of radius R is $R\rho/3e_0$, where ρ is the charge density and e_0 is the permittivity of free space. For ordinary thunderstorm lightning, the charges are on the order of tens of Coulombs and the distances a few kilometers. The permittivity again is constant, while the breakdown decreases slowly with increasing altitude. Hence, the lengths of lightning flashes

in ordinary thunderstorms are many kilometers. Lightning flashes propagate at speeds of about 10^5 m/s (much slower than the speed of light, which is 10^8 m/s), so there is a systematic relation between the duration of a flash and its length. Vent discharges occur with very short durations and hence short-scale lengths. Because permittivity is constant, this means that the charge density in jets is much higher —an order of magnitude or more—than that of ordinary thunderstorms. Presumably, this occurs because the ash particles are charge carriers and they are found at high concentrations in the rising ash columns. In the main part of the plume, the charge density is much like a thunderstorm.

4.2. Processes at Volcanoes

Several mechanisms contribute to the creation and generation of charges at volcanoes. The first is fracto-emission. When rock materials are broken, the irregular surfaces have different abilities to hold charge, so there is a net charge transfer and some particles end up positively charged and others negatively charged. In the case of magma, the dissolved gases (mainly water) expanding violently drive the fracturing of the magma into tephra particles of various sizes. The particles may be rock or crystal fragments, or glass. It is not known how the charges are distributed among the various components. A second charging process is called tribo-electrification. When particles collide, there is a transfer of charge because some electrons are only loosely held on the surface of the particles and they move if the thermodynamic conditions are favorable. This process has been studied in laboratory settings using uniform glass spheres and shows that smaller particles generally acquire negative charges and larger particles positive charges. Small-scale lightning has also been observed in the laboratory and is more readily produced when there are both large and small particles. A third mechanism is called streaming potential. This occurs when a solid body moves through an ionized fluid and charges build up on the surface of the body. This occurs, for example, when airplanes fly through air and is one of the reasons why planes are electrically grounded before they are unloaded or refueled. At volcanoes, the rock fragments are the solid bodies and the air is the fluid. The relative contributions of these three processes are not well known at volcanoes, but it is thought that fracto-emission and tribo-electrification are more important than streaming potential.

Once charged particles have been created, they must be redistributed into regions of like charges with sufficient charge density to create large electric fields and promote electrical breakdown. Here, the likely mechanism is gravitational separation, with small particles rising with respect to large ones. The implication then is that there is systematic **polarity** based on particle size, as indicated above for tribo-electrification. In rising ash clouds, turbulent flow likely supplies the mechanical energy. This is mainly caused by turbulence at the edges of the column in the jetted part and throughout the column in the convective portion. This is analogous to "deep convection" in ordinary thunderstorms but likely occurs more energetically.

Volcanic ash columns are very water rich. Calculations show that in some cases the mass of water per unit volume of air in ash columns is much greater than for ordinary thunderstorms and is greater than for saturation conditions. This suggests an important role for water in generating lightning. As ash columns rise to heights of 4—7 km, ice begins to form when the temperatures fall to −10 to −20 °C. Ice particles are small and tend to rise in the column. Ice is also the most electropositive substance known, so these rising ice particles carry a positive charge. The larger liquid water and graupel particles preferentially carry a negative charge and tend to concentrate lower in the cloud. Hence, the processes in ash columns are quite analogous to those in ordinary thunderstorms. Indeed, some investigators have referred to volcanic ash clouds as "dirty thunderstorms." This mechanism involving water and ice is thought to be the dominant mechanism for producing plume lightning.

The idea that magmatic water is dominant over entrained water from the atmosphere has some implications for latitudinal and seasonal dependence of volcanic lightning. Ordinary thunderstorm lightning is concentrated in the tropics and drops gradually with increasing latitude so that there is very little lightning above 60° latitude. This occurs because the ability of air to hold water is a function of temperature, with higher temperatures coinciding with higher moisture contents. The distribution of volcanic eruptions with lightning, however, shows high values at high latitudes (Figure 62.7), suggesting that sources and processes within the ash clouds are more significant than the small amount of water entrained from the surrounding air.

One other effect of electrification of ash clouds is the clumping of ash particles. It has been widely reported that ash particles clump together to form lapilli (see also Chapter 33). The effects of water are also important

FIGURE 62.7   (A) Percent of volcanoes with lightning versus latitude. (B) Percent of eruptions with lightning versus latitude. For each plot, both north and south latitudes are combined. High values in (B) between 30—35° and 60—65° represent observations from Japan and Alaska, respectively. These distributions are essentially flat, suggesting that entrainment of meteorological water does not play a major role in volcanic lightning occurrence.

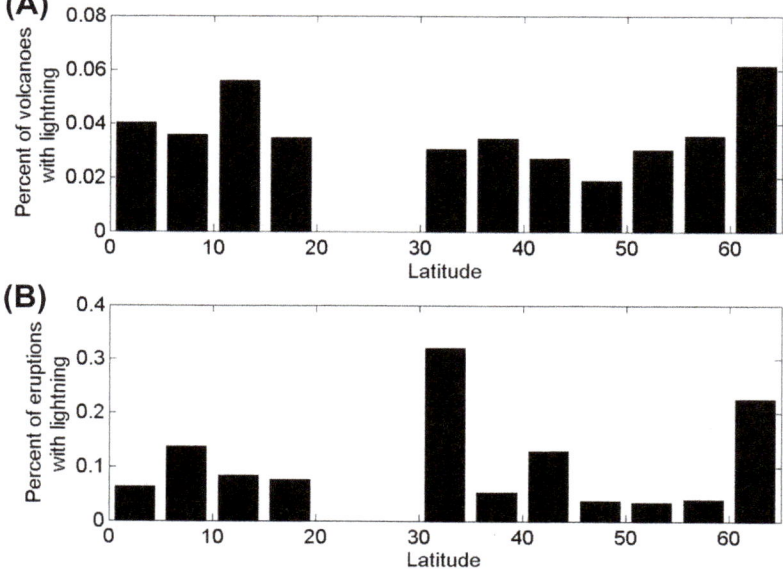

because water makes the particles sticky. The clumped particles are larger and heavier than individual particles, so they tend to fall out of the ash clouds more quickly. Dry aggregation can form low-density particles that stay aloft longer. This has several implications: (1) it is a mechanism to help redistribute charges within the ash clouds, (2) patterns of ashfall on the ground are affected, and (3) the number and size distribution of ash particles at altitude are affected, which has implications for both hazards and ash cloud microphysics.

As shown here, volcanic lightning is fairly common with nearly 400 documented eruptions with lightning. First-order studies have been completed and show the existence of three types of lightning: vent discharges, near-vent lightning, and plume lightning. The latter is strong enough to show up on several different types of detection systems, whereas the former two have only been studied in a few cases with LMA networks. Three processes likely contribute to charge creation: fracto-emission, tribo-electrification, and streaming potential. Detailed studies are needed to determine the relative efficiencies of these in producing volcanic lightning. Lightning at volcanoes is a hazard in its own right, and the occurrence of lightning can be used as one component of a multisensor system to help confirm that ash-producing eruptions are in progress.

FURTHER READING

Anderson, R., Björnsson, S., Blanchard, D.C., Gathman, S., Hughes, J., Jónasson, S., Moore, C.B., Survilas, H.J., Vonnegut, B., 1965. Electricity in volcanic clouds. Science 148, 1179–1189.

Arason, þ., Bennet, A.J., Burgin, L.E., 2011. Charge mechanism of volcanic lightning revealed during the 2010 eruption of Eyjafjallajökull. J. Geophys. Res. 116, B9. http://dx.doi.org/10.1029/2011JB008651.

Behnke, S.A., Thomas, R.J., McNutt, S.R., Schneider, D.J., Krehbiel, P.R., Rison, W., Edens, H.E., 2012. Observations of volcanic lightning during the 2009 eruption of Redoubt Volcano. J. Volcanol. Geotherm. Res. http://dx.doi.org/10.1016/j.jvolgeores.2011.12.010.

Behnke, S.A., Thomas, R.J., Edens, H.E., Krehbiel, P.R., Rison, W., 2014. The 2010 eruption of Eyjafjallajökull: Lightning and plume charge structure. J. Geophys. Res. Atmos. 119, 833–859. http://dx.doi.org/10.1002/2013JD020781.

Bennett, A., Odams, P., Edwards, D., Arason, þ., 2010. Monitoring of lightning from the April–May 2010 Eyjafjallajökull volcanic eruption using a very low frequency lightning location network. Environ. Res. Lett. 5, 44013–44022.

Blanchard, D.C., Björnsson, S., 1967. Water and the generation of volcanic electricity. Mon. Wea. Rev. 95, 895–898.

Bruning, E., MacGorman, D., 2013. Theory and observations of controls on lightning flash size spectra. J. Atmos. Sci. http://dx.doi.org/10.1175/JAS-D-12-0289.1.

Hoblitt, R., 1994. An experiment to detect and locate lightning associated with eruptions of Redoubt Volcano. J. Volcanol. Geotherm. Res. 62, 499–517.

Houghton, I.M.P., Aplin, K.L., Nicoll, K.A., 2013. Triboelectric Charging of Volcanic Ash from the 2011 Grımsvotn Eruption. Phys. Rev. Lett. 111 http://dx.doi.org/10.1103/PhysRevLett.111.118501.

James, M.R., Lane, S.J., Gilbert, J.S., 2000. Volcanic plume electrification—Experimental investigation of fracture-charging mechanism. J. Geophys. Res. 105, 16641–16649.

James, M.R., Wilson, L., Lane, S.J., Gilbert, J.S., Mather, T.A., Harrison, R.G., Martin, R.S., 2008. Electrical charging of volcanic plumes. Space Sci. Rev. 137, 399–418.

Lane, S.J., Gilbert, J.S., 1992. Electric potential gradient changes during explosive activity at Sakurajima volcano, Japan. Bull. Volcanol. 54, 590–594.

Mather, T.A., Harrison, R.G., 2006. Electrification of Volcanic Plumes. Surv. Geophys. 27, 387–432. http://dx.doi.org/10.1007/s10712-006-9007-2.

McNutt, S.R., Davis, C., 2000. Lightning associated with the 1992 eruptions of Crater Peak, Mount Spurr Volcano, Alaska. J. Volcanol. Geotherm. Res. 102, 45–65.

McNutt, S.R., Williams, E., 2010. Volcanic lightning: global observations and constraints on source mechanisms. Bull. Volcanol. 72, 1153–1167.

Miura, T., Koyaguchi, T., Tanaka, Y., 2002. Measurements of electric charge distribution in volcanic plumes at Sakurajima Volcano. Japan. Bull. Volcanol. 64, 75–93. http://dx.doi.org/10.1007/s00445-001-0182-1.

Rodger, C.J., Brundell, J.B., Dowden, R.L., 2005. Location accuracy of VLF World Wide Lightning Location (WWLL) network: post-algorithm upgrades. Ann. Geophys. 23, 277–290.

Thomas, R.J., Krehbiel, P.R., Rison, W., Hunyady, S., Winn, W.P., Hamlin, T., Harlin, J., 2004. Accuracy of the Lightning Mapping Array. J. Geophys. Res. 109, D14207. http://dx.doi.org/10.1029/2004JD004549.

Thomas, R.J., Krehbiel, P.R., Rison, W., Edens, H.E., Aulich, G.D., Winn, W.P., McNutt, S.R., Tytgat, G., Clark, E., 2007. Electrical activity during the 2006 Mount St. Augustine volcanic eruptions. Science 315, 1097.

Thomas, R.J., McNutt, S.R., Krehbiel, P.R., Rison, W., Aulich, G., Edens, H.E., Tytgat, G., Clark, E., 2010. Lightning and electrical activity during the 2006 eruption of Augustine Volcano. In: Powers, J., Coombs, M., Freymueller, J. (Eds.), The 2006 Eruption of Augustine Volcano: Alaska. U.S. Geological Survey, pp. 579–608. Professional Paper 1769.

Williams, E.R., McNutt, S.R., 2005. Total water contents in volcanic eruption clouds and implications for electrification and lightning. In: Pontikis, C. (Ed.), Recent Progress in Lightning Physics: Research Signpost, pp. 81–94.

Part VIII

Eruption Response and Mitigation

Stephen R. McNutt

School of Geosciences, University of South Florida, Tampa, FL, USA

Scores of volcanoes will continue to erupt (about 55—70 being active each year) and will thus continue to pose a significant risk as the world's population grows and we rely increasingly on technology for the basic needs of civilization. However, some of the same technologies that improve life, such as the Global Positioning System, have also led to improvements in forecasting of eruptions. Better data, plus improvements in the use of data, have led to increasingly effective responses to eruptions and mitigation of their hazards.

One of the main tools for monitoring is seismology. Nearly every eruption is preceded by an increase in earthquake activity, and eruptions are accompanied by a continuous vibration of the ground known as volcanic tremor. Once gases or tephra breach the surface, infrasound signals are also produced. These provide a direct means of assessing the eruptive state of a volcano. The opening chapter of this section, *Seismic and Infrasonic Monitoring*, discusses the use of seismology and infrasound in eruption forecasting and monitoring in general. The chapter includes case studies of eruptions at Redoubt, Eyjafjallajokull, Tungurahua, Sarychev Peak, Cleveland, Soufriere Hills, and El Hierro volcanoes.

Deformation of the ground, such as bulging, doming, or the formation of cracks, is also frequently found prior to eruptions. Further, if new magma is injected beneath a volcano, gravity changes may occur because new mass is being added. Finally, when rock is heated near its Curie temperature, its magnetization decreases. The following chapter, *Ground Deformation, Gravity, and Magnetics*, discusses modern techniques of measuring deformation, gravity, and magnetic signals and their use in monitoring volcanoes and eruption forecasting. Case histories include Sakurajima, Krafla, Mount St Helens, Soufriere Hills, Augustine, and Okmok volcanoes for deformation; Campi Flegrei, Poas, Etna, and Kilauea for gravity; and Mount St Helens and White Island for magnetic studies.

In addition to earthquakes and deformation, volcanoes typically show signs of increased heat or changes in degassing prior to eruptions. These may be measured from the Earth's surface or from satellites if the signal is strong enough. During eruptions, plumes of ash and gases rise and may then travel great distances downwind. The next chapter, *Gas, Plume, and Thermal Monitoring*, examines the monitoring of gases, heat, and plumes and illustrates these with examples from Etna, Fuego, Nyiragongo, Nabro, Reventador, Kelud, Nyamuragira, Shishaldin, Erta'Ale, and others.

Each type of data may be handled and interpreted individually, but experience has shown that much more can be learned when various data sets are combined. Indeed, closely monitored eruptions with many data sets provide much of the basis for scientific progress in volcanology. The chapter, *Synthesis of Volcano Monitoring* describes how various types of data contribute to an evolving understanding of the activity of volcanoes. The sequence of geological, geochemical, and geophysical events varies for different types of eruptions. These ideas are illustrated using a detailed case study from Merapi volcano, Indonesia.

Once scientists agree about the meaning of the data (and even if they disagree), a series of communications need to occur between scientists and those responsible for emergency response. The actions that need to be taken appropriate for the current or expected hazards, need to be identified and assigned to the right groups of people. The content and distribution of warning messages, maps, videos, alert levels, etc., are the focuses of the chapter, *Volcano Warning Systems*.

Although most eruptions are small, any eruption can pose a risk if people or structures are close to the volcano. Large eruptions, in contrast, affect nearby populations with certainty and have many effects at great distances. In each case, the perception of the events and the response of society will dictate whether an eruption is benign or whether a disaster will occur. The next chapter, *Volcanic Crises Management*, discusses elements of volcanic crises management, drawing on experiences at Eyjafjallajokull, Soufriere Hills, La Soufriere, Popocatepetl, and others.

When an eruption occurs or is about to occur, the need for information suddenly becomes acute. Much of the needed information can be prepared ahead of time; indeed, the in-between periods last much longer than the eruptive periods. Effective strategies for education and intervention are the subjects of the chapter, *Social Processes and Volcanic Risk Reduction*. Case studies include lessons learned at Nevado del Ruiz, Eyjafjallajokull, Nyiragongo, and Mayon volcanoes.

Past eruptions have left deposits that can be studied to learn about the type and distribution of specific volcanic hazards. Further, contemporary eruptions at similar volcanoes may be studied to learn about the general effects of hazards. Once these are identified, coherent strategies may be developed to help mitigate risk. Some of these are direct, such as building dams to protect areas from mudflows. Some are longer term and less direct, such as land use zoning. Others are indirect and strategic, such as setting appropriate rates for insurance. All are influenced by society's understanding of and tolerance for risk. These and related topics are discussed in the final chapter of this section, *Volcanic Risk Assessment*.

Although volcanoes will continue to erupt, knowledge of their behavior and effects and adequate monitoring programs, can greatly reduce the risks from future eruptions. People can coexist with volcanoes provided they do so intelligently.

Chapter 63

Seismic and Infrasonic Monitoring

Stephen R. McNutt and Glenn Thompson
School of Geosciences, University of South Florida, Tampa, FL, USA

Jeffrey Johnson
Boise State University, Boise, ID, USA

Silvio De Angelis
University of Liverpool, Liverpool, Merseyside, UK

David Fee
University of Alaska Fairbanks, Fairbanks, AK, USA

Chapter Outline

GLOSSARY

anomaly An observation that differs from normal or expected values.

blast wave Explosion-generated bipolar-shaped pulse, which relates to acceleration of eruption materials from the vent.

deterministic A statement of the likelihood of an event based on a physical model and measurements. See also Probabilistic.

effective sound speed Combined intrinsic sound speed (function of temperature) and advection (function of wind), which determines how infrasound refracts in the atmosphere.

explosion earthquake An eruption that produces both seismic radiation and infrasound radiation that couples to the ground as a slower phase.

forecast A general description of future events, including rough estimates of time, location, and likely activity. See also Prediction.

infrasound Sound waves below 20 Hz, the region where the majority of volcano acoustic energy is concentrated.

infrasound array Distribution of three or more sensors with spacing that is similar to the sound wavelength and small compared with the source—receiver propagation distance; used to distinguish between signal and noise and determine wave parameters such as bearing to the source.

infrasonic tremor Sustained infrasound indicating a long-duration open-vent source that may be stationary or fluctuate in intensity.

local, regional, and global infrasound Atmospheric propagation distances corresponding to direct arrivals, first stratospheric refractions, and ducting in the stratosphere and thermosphere, respectively.

precursor A geological event that occurs prior to an eruption and is related to the preparation processes of the forthcoming eruption.

The Encyclopedia of Volcanoes. http://dx.doi.org/10.1016/B978-0-12-385938-9.00063-8

prediction A specific description of future events, including time, size, type, location, and formal errors for each. See also Forecast.

probabilistic A statement of the relative likelihood of an event based on study of a population of similar events that have occurred in the past. See also Deterministic.

reduced displacement The amplitude of a displacement seismogram multiplied by a geometrical spreading correction factor (ignoring other path and site effects). Typically expressed in units of cm^2. Reduced displacement data are often sampled once per minute, as for RSAM data.

reduced pressure The infrasound pressure scaled to a common reference distance (typically 1 km).

RSAM data A downsampled version of continuous seismicity which can be rapidly computed, analyzed, and fed into an alarm system. Typically the average raw seismic amplitude is computed and stored for each minute of continuous data.

seismic network A group of seismic stations which are deployed in a region for the purpose of detecting and locating earthquakes.

seismic station A set of instruments including a seismometer (buried in the ground), cables, batteries, solar panels, amplifiers, radios, and antennae located in an enclosure nearby.

spectrogram A plot of frequency in hertz (Hz) of seismic or infrasound signals as a function of time. Typically the strength of the signal is shown as a color or shading.

swarm A group of many earthquakes of similar size occurring closely clustered in space and time with no dominant main shock.

telemetry The set of instruments and communications that send a data signal from their source to a distant data acquisition system. May include radio, microwave, telephone, and satellite components.

Seismology and **Infrasound** are important and effective tools for monitoring volcanoes and **forecasting** eruptions. In the past two decades there have been over 25 successful forecasts. Well-monitored volcanoes have six or more local **seismic stations** within 15 km and several regional stations (>15 km) which are able to detect volcanic earthquakes of M ~ 0 under the volcano. Ongoing analyses of the data provide the basis for determining the eruptive state of the volcano. **Local infrasound** stations are becoming common on volcanoes but are far from ubiquitous.

The recording of several years of background seismicity is important to establish a baseline for evaluation of possible **precursors**. Five main types of events are recorded: high-frequency earthquakes, low-frequency earthquakes, **explosion earthquakes**, very-long-period (VLP) events, and volcanic tremor. Surface events such as rockfalls, avalanches, and mudflows are also recorded. Both explosion earthquakes and other surficial phenomena produce atmospheric propagating sounds that may be recorded with infrasound sensors. Volcanic earthquakes prior to eruptions generally occur in **swarms**; have maximum sizes of approximately M 5; include many events with similar waveforms; have high b values (up to 3.7); increase in numbers before eruptions; and occur beneath or near the site of eruptions. Many swarms follow a regular sequence consisting of the following components: background,

swarms of high-frequency events, relative quiescence after the peak rate, low-frequency events, volcanic tremor, eruption, and deep earthquakes following eruption. This sequence reflects systematic changes in locations of earthquakes and in the dominant physical processes.

Infrasonic monitoring complements seismic monitoring in that it provides direct and unambiguous records of surface activity that are largely "uncontaminated" by internal volcano sources. Erupting volcanoes accelerate the atmosphere and produce a variety of pressure waves at infrasonic frequencies (<20 Hz). Discrete infrasound events, including explosions, and continuous infrasound, in the form of a variety of tremors, are used to continuously track and assess the style of an eruption or the openness of a volcanic vent. Eruptions are well detected by arrays of infrasound sensors that may be deployed local to a volcano or at regional and global distances. Most volcano acoustic studies focus on infrasound in the band 0.1—20 Hz, as this is the band of most intense volcanic sounds and these long wavelengths propagate with very little attenuation.

1. INTRODUCTION—HISTORY AND ORGANIZATION

Seismic monitoring is the systematic observation of the status of seismic activity at a volcano. Monitoring attempts to associate key diagnostics, such as rates, types of earthquakes, and changing locations, with physical processes involved in the movement or state of magma. This can be done **probabilistically** by comparing current activity at the volcano of interest with comparable past activity at similar volcanoes. It can also be done **deterministically** when a particular process can be associated with seismic signals with a high degree of confidence. Seismic monitoring can help determine if a volcano (1) is in a background or restful state; (2) is restless; (3) shows precursory activity; (4) has begun to erupt; (5) is actively erupting; (6) is ending an eruption; or (7) is returning to a background state. Seismic monitoring is fairly effective alone, but is better utilized as one component of an integrated monitoring effort.

Nearly every recorded volcanic eruption has been preceded by changes in seismic activity beneath or near the volcano. For this reason, seismology is one of the most useful tools for eruption forecasting. Early observatories (Table 63.1) included seismometers that were at first crude mechanical devices, but were later upgraded to more modern electromechanical types (see Figure 63.1) that have higher magnification and higher dynamic range. At the present time, approximately 200 of the world's volcanoes are seismically monitored (about 37 percent of the 541 volcanoes that have erupted in historic times). Volcanoes that are actively erupting draw attention both publicly and scientifically, and over half of these are seismically

TABLE 63.1 Selected Early Volcano Observatories and Observations with Seismometers

Volcano	Country	Date	Type of Instrument
Vesuvius	Italy	1856	Mechanical
Usu	Japan	1910	Electromechanical
Kilauea	USA	1912	Electromechanical
Sakurajima	Japan	1914	Electromechanical
Aso	Japan	1930	Electromechanical
Merapi	Indonesia	1924	Electromechanical

TABLE 63.2 Some Successful Forecasts of Explosive Volcanic Eruptions, 1980–1999

Volcano and Location	Year of Forecast
Mount St Helens, Washington	1980–1986
Galunggung, Indonesia	1982
Merapi, Indonesia	1984
Izu Oshima, Japan	1986
Banda Api, Indonesia	1988
Kie Besi, Indonesia	1988
Tokachi-dake, Japan	1988
Galeras, Colombia	1989
Izu-Tobu, Japan	1989
Redoubt volcano, Alaska	1989–1990
Kelut, Indonesia	1990
Unzen, Japan	1990–1991
Mount Pinatubo, Philippines	1991
Colima, Mexico	1991
Lonkon, Indonesia	1991
Mount Spurr, Alaska	1992
Mayon, Philippines	1993
Mount Klyuchevskoi, Kamchatka	1994
Rabaul Caldera, Papua New Guinea	1994
Popocatepetl, Mexico	1994
Montserrat, West Indies	1995–1997
Pavlof, Alaska	1996
Shishaldin, Alaska	1999
Mount St Helens, Washington	2004
Augustine, Alaska	2006
Pavlof, Alaska	2007
Kasatochi, Alaska	2008
Redoubt, Alaska	2009

Note: Compiled by J. Power, C. Newhall, T. Casadevall, and S. McNutt.

FIGURE 63.1 Seismic station PN7 at Pavlof volcano, Alaska. The seismologist is checking to make sure the seismometer is level in the bottom of the hole in which it will be buried. A cable carries the signal to a custom-designed hut which houses the batteries, antenna, and electronics. Solar panels are mounted on the top left of the hut, which faces south. *Photo by S. McNutt.*

monitored. A good example of this is Soufriere Hills Volcano, Montserrat, West Indies. There have been a number of generally successful forecasts during the past two decades (Table 63.2), although the exact criteria for claiming success and the levels of precision have been variable.

The violent Krakatau eruption of 1883 demonstrated that erupting volcanoes generate high-intensity sounds, including low-frequency oscillations. Krakatau produced audible sounds out to 3500 km and observable barometric changes that propagated around the earth up to seven times. Subsequent large eruptions, including Bezymianny in

1956, Mount St Helens in 1980, and El Chichon in 1982, also excited the atmosphere with ultra long period oscillations known as acoustic gravity waves (T ~ 300 s) that propagate and can be recorded globally and in some cases with multiple circumglobal transits. Higher frequency atmospheric acoustic (infrasound) airwaves from eruptions were not initially recorded with dedicated microphones, but with seismometers, which registered ground shaking due to atmospheric pressure waves impinging upon the free surface and shaking the solid earth (termed ground-coupled airwaves).

In the early 1980s, dedicated infrasound monitoring gained adherents who used pressure transducers sensitive to the infrasound band. Early studies at Mt Erebus, Antarctica, and several volcanoes in Japan demonstrated the utility of recording infrasound and seismic data on separate channels. These dedicated instruments permitted identification of small explosion transients, which would have been overlooked through analysis of seismicity alone. These instruments also identified many examples of continuous tremor-like signals, which are often produced by open-vent but nonexploding volcanoes. Today numerous volcano observatories around the world use infrasound to monitor volcanoes.

The figurative "explosion" of volcano infrasound study since the first edition of Encyclopedia of Volcanoes is testament to both the scientific and monitoring benefit of recording the atmospheric wavefield near active volcanoes. The reader is referred to Johnson and Ripepe (2011) and Fee and Matoza (2013) for reviews of, and literature reference to, a spectrum of studies. These reviews summarize how volcano infrasound has been used to continuously monitor explosive activity, precisely locate active vents and depth of source, quantify source motions, track gravity-driven rockfalls and pyroclastic flows, quantify resonant phenomena of the vent/conduit region, understand atmospheric sound propagation, and probe atmospheric structure. In this chapter we introduce the basic characteristics of infrasound from volcanoes and describe how they can be applied to volcano monitoring.

2. SEISMIC INSTRUMENTS AND NETWORKS

Volcano observatories vary widely in location and sophistication. The simplest observatories may be small observation huts with a single seismograph recorded on paper on a revolving drum. Many larger observatories are located near volcanoes at distances of several tens to several hundreds of kilometers, and seismic data from local and regional stations are telemetered to them by cable, radio, phone, satellite, and/or Internet. Some observatories are colocated with universities or other research organizations. These may be hundreds of kilometers from the volcanoes of interest, yet the vast array of modern communications

available today makes such facilities common and economical. They are, further, safe from large eruptions that may affect closer facilities. There are approximately 77 members of the World Organization of Volcano Observatories (WOVO) as of this writing, most with some kind of seismic monitoring capability.

Many seismometers deployed at volcanoes are short-period (1 s) vertical instruments that record useful data between about 0.7 and 30 Hz. They consist of a magnetic mass suspended on a spring, surrounded by a coil of wire. When the mass moves in response to ground motion, a current is generated in the coil. Modern three-component, digitally telemetered, high-dynamic range, broadband seismometers record over a much wider band of frequencies (0.02–50 Hz). These instruments also have a mass and spring, but the coil that senses the motion is connected to a feedback circuit that sends a current through another coil exactly to oppose the motion. Thus the mass is held fixed, and the feedback current is proportional to the ground motion. Such broadband seismometers also have a much higher dynamic range, up to 145 dB, compared with about 40 dB for analog **telemetry**. Temporary deployments of these broadband instruments have become common since the early 1990s, and permanent deployments of 1–10 instruments at well-monitored volcanoes are now widespread.

Most seismometers used to monitor volcanoes are located on the flanks at distances within 15 km (Figure 63.2(B)). About a third of the monitored volcanoes have at least one station within 1 km of the active vent. Modern networks generally consist of six or more stations. Station spacing of a few kilometers enables better detection of small earthquakes and generally small location errors (~0.5 km). Data are usually telemetered to a common site and recorded digitally on computers. A number of volcanoes are monitored indirectly because a nearby (>15 km) seismic station is part of a larger regional **seismic network**. Maps showing distributions of seismic stations near several volcanoes in Alaska are shown in Figure 63.2. Topography, telemetry, and ease of access often determine the exact locations of stations.

The distance from an earthquake to a seismic station and the magnification of the station are the main factors that determine the magnitude detection threshold, that is, the smallest signals that can be recorded. Typical magnifications at volcanoes are 5000–50,000, and the corresponding magnitudes of detection are about M 0 to M 1 for the 1–15 km distances involved. It is desirable to record the smallest earthquakes possible, but because volcanoes often have high noise levels, this is usually achieved by putting stations closer to the vent rather than by operating more distant stations at higher magnifications. At such close distances, short-period seismometers will usually saturate ("clip") at less than M = 3 for signals generated at or near

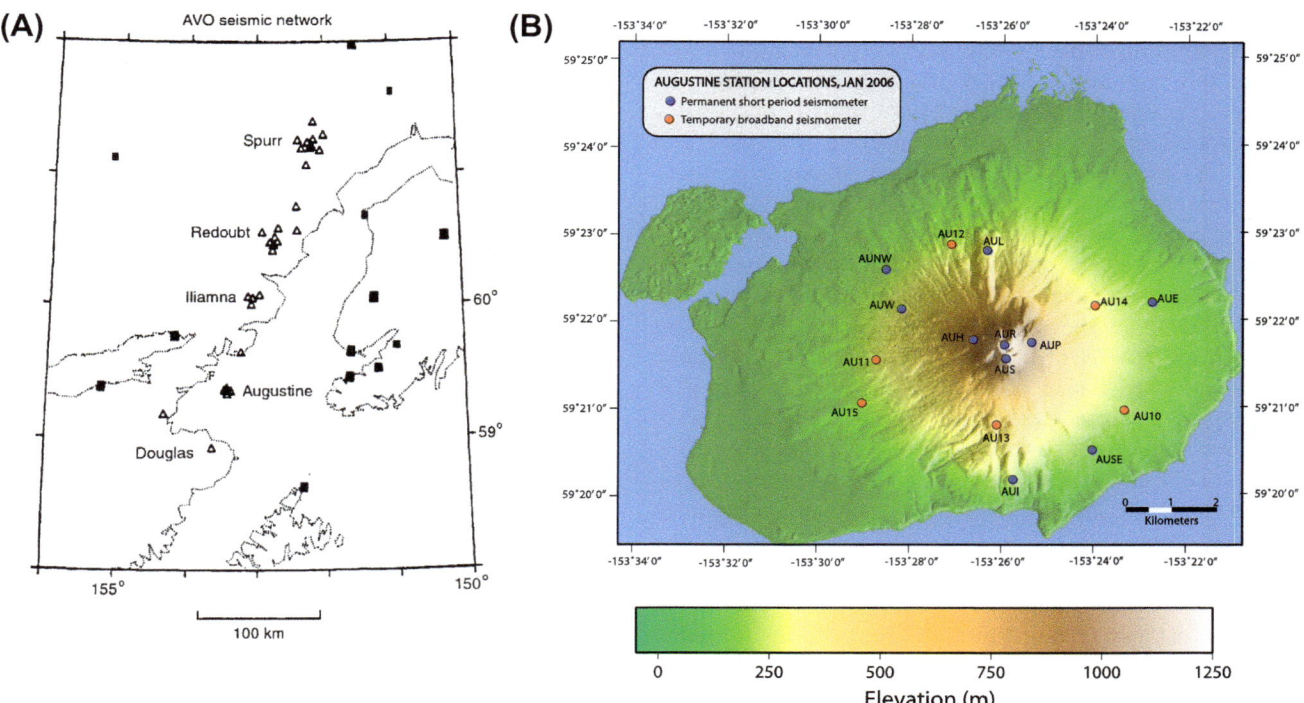

FIGURE 63.2 (A) Map of local (open triangles) and regional (solid squares) seismic stations in the Cook Inlet region of Alaska. Clusters of stations are located near each of the five active volcanoes. (B) Map of local seismic stations near Augustine Volcano. Most well-monitored volcanoes have similar networks of 6—10 telemetered stations. Six campaign broadband seismometers were installed in fall 2005 prior to the January 2006 eruption.

the vent. Given the much greater dynamic range of broadband seismometers, signals remain on-scale up to around M = 6 at these same distances. It is important to keep track of the amplitude of the ground shaking, so calibration information for each seismic station is stored and used to help determine earthquake magnitudes and the **reduced displacement** of continuous signals such as volcanic tremor. These "normalized" measures allow comparison between different volcanoes. Similarly, several observatories compute **spectrograms**—plots of the frequency of ground shaking versus time—to help identify low-frequency events and to help distinguish between earthquakes, noise, and other types of signals.

A seismometer in the vicinity of a volcano will typically record seismic signals from four sources: the volcano, local earthquakes (<15 km), regional earthquakes (>15 km), and teleseisms (generally >600 km). Part of the job of the seismologist is to recognize and discriminate among these. It is especially helpful to record data from several stations at regional distances of 50—200 km to aid in the identification process. Two examples illustrate why these are helpful. At Nevado del Ruiz volcano, Colombia, in November 1985 and later, many earthquakes from the Bucaramanga nest, a persistent deep source of earthquakes beneath Colombia, appeared on seismograms from stations near the volcano. The Bucaramanga event seismograms resembled many of the volcanic events, and their occurrence created confusion.

At Spurr volcano, Alaska, in 1992, the seismic signal for two eruptions began gradually but the third began sharply and its onset looked like that of a regional earthquake. However, the pattern of arrival times and relative amplitudes at local versus regional stations quickly revealed that the event was an eruption at the volcano.

Some volcanic signals have gradual onsets or very long durations (hours to days or more). For these events, traditional location methods, which depend on measuring arrival times for P and S waves, cannot be used. New techniques use data from dense networks or "arrays." Here, the stations are very close, a few tens to hundreds of meters apart, so it is possible to see the same wave coherently on all the stations (the wave need not be a first or prominent arrival). Because the physics of waves are well known, it is possible to determine what type of wave is present, which direction it is coming from, and how fast it is moving. All these provide information about the sources. Thus, dense arrays can be used like antennae to discern certain features of specific types of seismic waves even when the signals are weak or noise is present. Array methods look at different features of seismograms and thus extend the information available for analyses.

Many tasks are automated and run in near-real time, which means the data are available for viewing a few minutes after the fact. For example, when an earthquake occurs, there is an abrupt increase in the signal level at several stations at nearly the same time. This feature is exploited by algorithms that compute the ratio of short-term average (STA) and long-term average (LTA) and declare an event when this rises above a threshold. Hypocenters and magnitudes are typically determined automatically in near-real time too. These events are later reviewed, classified, and their locations refined by scientists. A decade ago, it was common to archive only the data corresponding to these events because data storage was expensive. Now it is practical for volcano observatories to store continuous data too. This is crucial for volcano-seismic monitoring, because emergent signals like tremor are usually entirely missed by STA/LTA algorithms. Researchers can explore the continuous data to find other signals such as VLP events, to analyze earthquake swarms in detail, or to perform studies such as ambient noise tomography.

Modern earthquake monitoring and data analysis systems are built around relational database management systems. Schema have been developed which fully describe everything related to producing earthquake catalogs: earthquake origins, magnitudes and the arrivals used; a full history of instrument deployments and their response characteristics; station coordinates; and the location of archived waveform data files. Scientists are insulated from the overhead of organizing the data themselves, and can easily combine, subset, and sort data from different tables to create unique ways of analyzing data. Databases have greatly facilitated the storage and manipulation of large data sets, which in turn have facilitated the use of data-intensive forms of analysis.

Institutions such as Incorporated Research Institutions in Seismology make data readily available to the entire scientific community, and benefit monitoring agencies by providing offsite backup. Automation of many tasks has improved the ability to analyze many types of data in near-real time, which leads to better methods of both forecasting and characterizing eruptions in progress. The software package SWARM produces digital helicorder plots and has rendered paper-based helical drum recorders obsolete. In 1998 the Alaska Volcano Observatory (AVO) began routinely posting on a Web site near-real-time spectrograms and reduced displacement plots for all volcano-seismic stations in Alaska, allowing scientists to monitor volcanic seismicity from anywhere with an Internet connection. Both of these produce time series (see below) that are routinely interpreted in terms of changing volcanic processes. Continuous spectrograms help scientists recognize low-frequency events, tremor, and gliding spectral lines.

2.1. Infrasound Instruments and Deployment Strategies

Data acquisition requirements for volcano infrasound are similar to those for volcano seismology in that most volcano sound energy is beneath 20 Hz, permitting relatively low-sample-rate data acquisition. Continuous data recording with Global Positioning System (GPS) time synchronization is standard and necessary for typical analyses. Signals may either be recorded with stand-alone sensors and digitizers, which is common for many short-duration campaign science studies, or can be relayed via telemetry to observatories for monitoring applications. For both scientific and monitoring applications 24-bit resolution is desirable given the wide range (6 orders of magnitude) of volcano infrasound amplitudes that are recorded.

A variety of infrasound sensors may be used to record excess pressure time series, which are records of atmospheric pressure oscillations above and below the ambient atmospheric pressure. These pressure changes are typically reported in units of Pascals (Pa). Infrasonic waves passing over these sensors will produce measurable changes in pressure propagating at acoustic velocities (~ 340 m/s at the ground surface). Consumer-grade audio microphones often have sensitivity extending down to the near-infrasound band, but these instruments' responses decay at low frequencies. While they may serve as low-cost detectors for explosion counting, a proper infrasonic microphone is necessary to perform detailed waveform analysis and source characterization. Both high-fidelity microbarometers and low-cost MicroElectroMechanical Systems transducers are capable of measuring typical volcano infrasound down to periods of several tens of seconds and with amplitudes as low as a few mPa at 1 Hz. At the other end of the amplitude spectrum, large and/or impulsive eruptions can produce excess pressure signals exceeding 10^3 Pa at 1 km; thus a dynamic range of more than 20 bits (120 dB) is ideal.

Ambient noise is an issue with infrasound data collection and is most often attributed to pressure fluctuations from wind turbulence. Because wind can obscure important signal, it is critical to deploy infrasound instrumentation away from high-wind environments such as the upper slopes of a volcano. In many instances this means deploying the sensors several kilometers from the vent, preferably in areas of thick vegetation to reduce wind noise. Shallow burial in porous media, like snow or tephra, or connection to wind-noise reduction manifolds, are additional techniques that are used to combat wind noise and are common in long-term deployments.

Single sensor deployments can be useful for monitoring, particularly in near-vent applications with high signal-to-noise ratios. Large explosions are clearly recorded as high-amplitude transients. However, wind noise complicates the interpretation of time series data from a single pressure transducer, and nonvolcanic sources can be difficult to differentiate from volcanic sources. It is often convenient and cost effective to colocate seismic and infrasonic sensors or telemetry resources. Colocating seismic and acoustic sensors may permit the distinction of separate seismic and acoustic sources, and recent work has shown how cross-correlation between colocated seismic and acoustic sensors permits identification of coherent acoustic signals and ground-coupled energy onto seismometers (Ichihara et al., 2012).

Network and array deployments of infrasound sensors are used to effectively identify small amplitude signals and/or signals in noisy environments. They also allow infrasonic source location. As in seismology, an **infrasound array** configuration involves three or more sensors, where the spacing of sensors is small compared with propagation distances (from the vent) and is generally tuned to the wavelength of the expected signal. Typical 1 Hz infrasound has a wavelength of ~330–350 m so that many volcano infrasound arrays have apertures of 75–150 m (approximately a quarter to half wavelength). Permanent global nuclear monitoring arrays of the International Monitoring System (IMS) have apertures about 10 times as large. Infrasound array deployments permit the identification of volcano infrasound sources through effective discrimination between volcanic signals and other infrasound and/or ambient noise. Wind noise, for instance, will be incoherent across the array. Numerous other infrasonic sources, such as pervasive ocean microbaroms (with amplitudes of tenths of Pa and peak frequency around 0.2 Hz; Bowman et al., 2005) and other cultural, geophysical, or weather-related sources (e.g., chemical blasts, vehicles, thunder, avalanches, meteors, storms) can be properly differentiated from volcano signals, by using an array distribution of sensors.

The infrasonic array is a powerful tool for distinguishing between signal, noise, and nonvolcanic signal through a variety of signal processing techniques. Three techniques are most commonly used to identify the propagation velocity and azimuth of coherent plane waves traversing the array. The first technique finds the sound propagation vector that is associated with maximum semblance, a measure of the power associated with stacked pressure waves. For each candidate slowness vector, array pressure waveforms are time shifted and summed (Neidell and Taner, 1971). When pressure waveforms sum in phase, signal associated with that slowness vector is considered coherent. Another common technique measures the lag times between pairs of array elements associated with maximum cross-correlation (Cansi, 1995) and finds the best fitting sound propagation vector associated with all inter-array time lags. Least-squares fitting of plane waves traversing the array is also common, and relies on waveform cross-correlation between sensors to determine time delays (Szuberla and Olson, 2004). This technique can also be used to derive uncertainties in the wave parameters.

Arrays of infrasound sensors are capable of recording volcano signals at local distances (on the flanks of a volcano out to about 15 km), at regional distances (out to several 100 km), and at global distances for very large

eruptions. Local distances in this context correspond to mostly straight-line sound propagation between the volcano and the recording site. Because the volcanic vent is generally above the background topography, sound propagates down from the vent. Regional distances are associated with refraction of sound in the stratosphere (~50 km a.s.l.) and are common when prevailing stratospheric winds parallel the sound propagation direction. Global propagation includes both ducted stratospheric waves, which refract multiple times, and/or thermospheric returns where sound propagates to altitudes more than 100 km a.s.l.

Networks of infrasound sensors are also common at local volcano installations. Although coherence of infrasound across a network, which may span many kilometers, is not usually as pronounced as for array infrasound, networks of pressure sensors can be used to precisely locate and distinguish vents or track gravity-driven flows. Networks are particularly valuable for locating activity at multivent volcanoes. Unlike seismic signals, the relatively simple structure of the atmosphere permits relatively robust correlation of infrasound signal across a network. Furthermore, the comparatively slow propagation velocity of sound may permit high-resolution (to within a few tens of meters) source localization.

3. TERMINOLOGY AND TYPES OF VOLCANIC EARTHQUAKES

Active volcanoes are the sources of a great variety of seismic signals. Traditionally, volcanic earthquakes have been classified based on seismogram appearance into four different types: high-frequency or A-type, low-frequency or B-type, explosion quakes, and volcanic tremor (Minakami, 1960). This classification scheme works very well at a large number of volcanoes. For monitoring, the exact terminology is less important, as it is more important to quickly recognize different types of events. Other local signals, such as those caused by glaciers or landslides, are also recorded at volcanoes, and need to be distinguished from the volcanic events. Although the original classification included restricted depth ranges for various earthquakes, these have often been relaxed because of improvements in location accuracy and better understanding of source and propagation effects. Examples of volcanic events are shown in Figure 63.3.

1. High-frequency events

Most high-frequency events are thought to be caused by shear failure or slip on faults, and differ from their tectonic counterparts only in that they occur predominantly in swarms at volcanoes. High-frequency events have clear P and S waves, and dominant frequencies are 5–15 Hz. A typical small high-frequency event is shown in Figure 63.3(A).

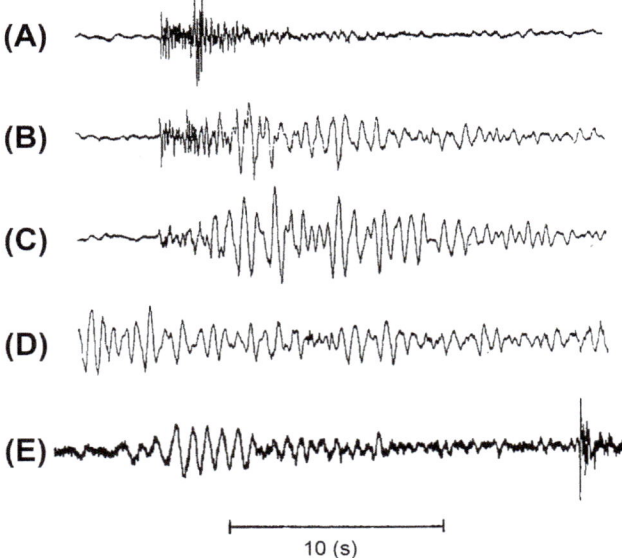

FIGURE 63.3 Typical waveforms of volcanic earthquakes. (A) High-frequency or volcano-tectonic earthquake, 6.8 km depth, Redoubt volcano, station RED, 8 km from the vent. (B) Hybrid or mixed-frequency event, 0.6 km depth (0.6 km above sea level), Redoubt volcano, station RED. (C) Low-frequency or long-period event, 0.4 km depth, Redoubt volcano, station RED. (D) Volcanic tremor, Redoubt volcano, station RED. (E) Explosion quake, Pavlof volcano, station PVV, 8.5 km from the vent (note prominent airwave arrival). *After McNutt (2000). p. 1101, Encyclopedia of Volcanoes, First Edition.*

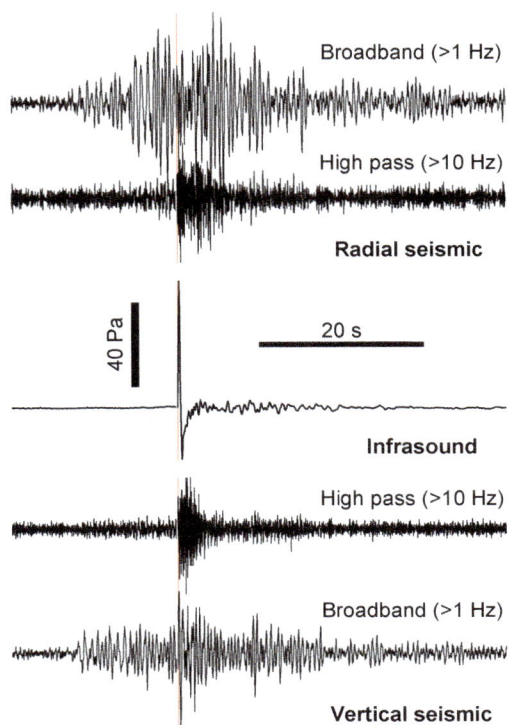

FIGURE 63.4 An example of a volcanic explosion earthquake recorded with colocated seismometer and infrasonic microphone 6.5 km from Tungurahua in 2009. Emergent broadband seismicity precedes the air phase due to higher seismic wave speeds. Air-to-ground coupling is especially evident at higher frequencies for both transverse and radial seismograms and is less pronounced in the transverse seismogram (data not shown). Peak-to-peak amplitude of infrasound pulse is ~80 Pa.

2. Hybrid earthquakes

   Some earthquakes share attributes of both high- and low-frequency events. These are called hybrid events, and may represent a mixture of processes. For example, the event shown in Figure 63.3(B) displays the high-frequency onset of a high-frequency event, but the later part of the signal (the coda) is similar to low-frequency events.

3. Low-frequency events

   Many low-frequency events are thought to be caused by fluid pressurization processes such as bubble formation and collapse, and also by shear failure or tensile failure in rock, or nonlinear flow processes which occur at very shallow depths for which attenuation and path effects play an important role. These events often have emergent P waves, lack S waves, and have dominant frequencies between 1 and 5 Hz, with 2−3 Hz most common. An example of a low-frequency event is shown in Figure 63.3(C).

4. Volcanic tremor

   Volcanic tremor is a continuous signal with duration of minutes to days or longer. The dominant frequencies of tremor are 1−5 Hz (2−3 Hz is most common), similar to low-frequency events, and many investigators

have concluded that tremor is a series of low-frequency events occurring at intervals of a few seconds. Harmonic tremor and spasmodic tremor are two special cases of more general volcanic tremor. Harmonic tremor is a low-frequency, often monotonic sine wave with smoothly varying amplitude, and spasmodic tremor is a higher frequency, pulsating, irregular signal. An example of volcanic tremor is shown in Figure 63.3(D).

5. Explosion earthquakes

   Explosion quakes accompany explosive eruptions, and many are characterized by the presence of an air-shock phase on the seismograms. There is a partitioning of energy at the source; part of the energy travels through the ground as seismic waves, and part travels through the air as acoustic or airwaves. The airwaves then couple back into the ground near the seismometer. The airwave also shows up clearly on microphones, barographs, and infrasound sensors. An example of an explosion quake is shown in Figure 63.3(E). An explosion quake recorded with both seismic and infrasound sensors is depicted in Figure 63.4.

6. VLP events

Since the 1990s, broadband seismometers have become widespread at volcanoes. These have the ability to detect a wider frequency band, particularly at the low end (down to a period of 120 s). A new class of events called VLP or very-low-frequency events have been observed at some volcanoes. Observations to date show events with periods of 3–200 s originating from shallow depths of 1.5 km or less under several active volcanoes including Kilauea, Stromboli, Aso, Sakurajima, and Satsuma-Iwo-jima. The VLP events at these volcanoes have been associated with either eruptions or vigorous fumarolic activity, and the observed events have fairly small amplitudes. (See also Figure 59.1 F of Chapter 59.)

7. Surficial events

Seismometers located on volcanoes may record a variety of local signals caused by shallow processes. These include nonvolcanic processes such as glacial events, shore ice movement, and landslides, as well as volcano-related processes such as outburst floods and lahars, pyroclastic flows, and rockfalls from crumbling lava domes. These are described in Chapter 59 of this Encyclopedia.

3.1. Quantification of Volcanic Seismicity

Seismic and infrasonic monitoring require regular checks of data at times of interest, and preparation of plots and maps to display data so that anomalies may be identified and interpreted. For example, plots of the number of earthquakes per unit time, and of cumulative energy per unit time, are often routinely generated for each type of earthquake over many different timescales (e.g., last week, last year) to help scientists recognize escalations in event seismicity: these are especially helpful to identify small events that increase in numbers before eruptions. Maps and cross-sections are made of all locatable earthquakes at periodic intervals such as a day or week. Locatable events must be large enough to show up on three or more stations, so these are fewer in number than countable events.

Real-time Seismic Amplitude Measurement (RSAM) is a rapid analysis tool developed by the US. Geological Survey (Endo and Murray, 1991). The RSAM system provides consecutive 1- or 10-min averages of absolute seismic amplitude for each seismic station, regardless of event type. **RSAM data** are generally plotted as time series for each individual station, and because of their low sampling rates they are useful for quantitatively determining the gross features of seismicity automatically and quickly during crises when detailed manual analyses are not possible. RSAM used to be a stand-alone system, but is mainly used now as an optional module for the Earthworm automatic seismic data processing system (Olivieri and Clinton, 2012). RSAM data have units of raw counts—the direct electronic output of the seismometer. Reduced displacement is another parameter often computed on a 1-min or 10-min timescale, but differs from RSAM in that the instrument response is removed, the signal integrated to displacement, and a correction is made for geometrical spreading. RSAM and reduced displacement plots can help distinguish storm noise from tremor, and to track tremor once it starts.

The largest events in swarms preceding most eruptions typically have magnitudes of M = 2 to M = 3. However, many larger events have occurred at volcanoes, especially calderas (Table 63.3), and some of these are large enough to be hazardous in their own right. Magnitude 4–5 earthquakes are common at volcanoes with large deformations in the form of doming, bulging, and sector collapses; events associated with the latter may reach M = 5.5. To date, only weak correlations have been found between the magnitude of preceding earthquakes and the Volcanic Explosivity Index (VEI) of the resulting eruption (Benoit and McNutt, 1995).

In general, there are many more small earthquakes than large ones. This is a consequence of the fact that the size distribution of earthquakes obeys a power law, which in seismology is of the form $\log_{10}(N) = a - bM$ (where N is the cumulative number of events, M is magnitude, and a and b are constants) and is known as the Gutenberg–Richter relation. The slope, or b value, determines the relative number of large versus small events. While for tectonic earthquakes the b value has typical values of 0.9–1.0 worldwide, more extreme values have been observed at volcanoes. High-frequency events have values ranging generally between 0.6 and 1.3, but low-frequency events have b values of up to 3.7. This means, for low-frequency events, that small earthquakes dominate in both numbers and energy, and also implies that the source size is restricted. Determination of the b value for volcanic earthquakes is an important diagnostic tool for identifying changes in seismicity.

At volcanoes, most earthquakes occur at depths of 10 km or less. Deeper events are common, but the number of events is considerably smaller. A trend recently observed at several volcanoes is that deeper earthquakes (10–40 km) occur after major phases of eruption. It is thought that these events are occurring in response to stress changes caused by the removal of magma at depth. The cumulative energy of deep quakes may be proportional to the volume of magma erupted.

All the seismic data are combined with other data on a regular basis, and attempts are made to develop a working hypothesis of what is physically occurring to cause

TABLE 63.3 Selected Large Earthquakes Near Volcanoes

Volcano	Year	Mmax	Comment
Arenal	1968	4.5	Preeruption
Asama	1916	6.3 MS	No eruption
Aso	1975	6.0	No eruption
Bandai	1888	5	Triggered eruption
Bezimianny	1956	5.0	Preeruption
Fernandina	1968	5.7 MS	Caldera collapse
Gorely	1985	6	Posteruption (8 months)
Ito-oki	1989	5.5	Preeruption
Jan Mayen	1954	5.8	No eruption
Kalapana	1975	7.1 MS	Shallow thrust
Katmai	1912	7.0	Posteruption
Kasatochi	2008	5.8 MW	Preeruption
Long valley	1980	6.2 ML	4 earthquakes with M 6, no eruption
Miyake-jima	1962	5.9	Strong swarm
Miyake-jima	1982	6.4 MJMA	Swarm 20 km deep
Miyake-jima	1983	6.2 MS	During eruption
Mt St Helens	1980	5.2 MS	Triggered eruption
Mt Usu	1910	5.5	Preeruption
Mt Usu	1943–1944	5.0	Preeruption
Mt Usu	1977	3.7	Preeruption
Mt Usu	1977–1980	4.3	Emplacement of cryptodome
Pinatubo	1991	5.7	Coeruption
Sabancaya	1991	5.6 MS	10 km NE of summit
Sakurajima	1914	7.0 MS	10.5 h posteruption, depth 13 km
Sheveluch	1964	5.5	Preeruption
Showa-Shinzan	1944–1945	3.8	Dome emplacement
Unzen	1792	5–5.5	Accompanied collapse

Note: Mmax, maximum magnitude; MS, surface wave magnitude; MW, moment magnitude; ML, local (Richter) magnitude; and MJMA, Japan Meteorological Agency magnitude.

the observed signals. Such hypotheses or conceptual models are continually revised as conditions change. These are the links between the raw data and the science. In the remainder of this chapter it is assumed that the seismic events have been properly identified and classified.

3.2. Terminology and Types of Infrasound Events

Volcano infrasound signals may be broadly characterized as short-duration transients or long-duration tremors, and the nomenclature generally follows that of volcano seismology. The first category often corresponds to short-duration explosive emissions while the latter corresponds to open-vent or longer term eruptive activity. Often an eruptive event begins with a short-duration, high-amplitude pulse associated with an explosion(s), and continues with extended duration tremor indicative of continuing emissions and/or open-vent degassing.

Short-duration infrasound transients from a range of different volcano types exhibit some common attributes (Figure 63.5). Typically, the pressure waveform begins impulsively with a compression that lasts from a few tenths of a second to several seconds. Next, a lower amplitude rarefaction with somewhat longer period occurs. In some circumstances these *bipolar pulses* are nearly symmetric, while in other cases the rarefaction amplitude is almost negligible. The rise rate and shape of the initial compressional pulse is also variable, ranging from a near sinusoidal shape to an *N-shaped wave*, which has very rapid rise followed by a linear pressure time decay. The latter waveform is reminiscent of a chemical explosion **blast wave** and may be associated with volcanic eruptions that produce supersonic shocks. Examples of various short-duration eruption transients are shown in Figure 63.5. Many high-amplitude infrasound transients are followed by damped sinusoidal oscillations suggestive of crater or vent resonance. Occasionally, high-amplitude infrasound pulses are immediately preceded by a subtle pressure increase (lasting a few tenths to a few seconds) that is attributed to distension of the vent region or lava lake prior to the primary explosion.

In many cases infrasound signals are absent in the minutes prior to an explosion, indicating the vent is acoustically quiet and/or physically sealed. After the transient explosion, the signal often continues as lower amplitude **infrasonic tremor** that may taper off over tens of seconds to minutes. This suggests that the vent has been reamed open and that degassing continues. Sustained amplitude infrasonic tremor is also common in the absence of explosions and is common in many open-vent volcanoes.

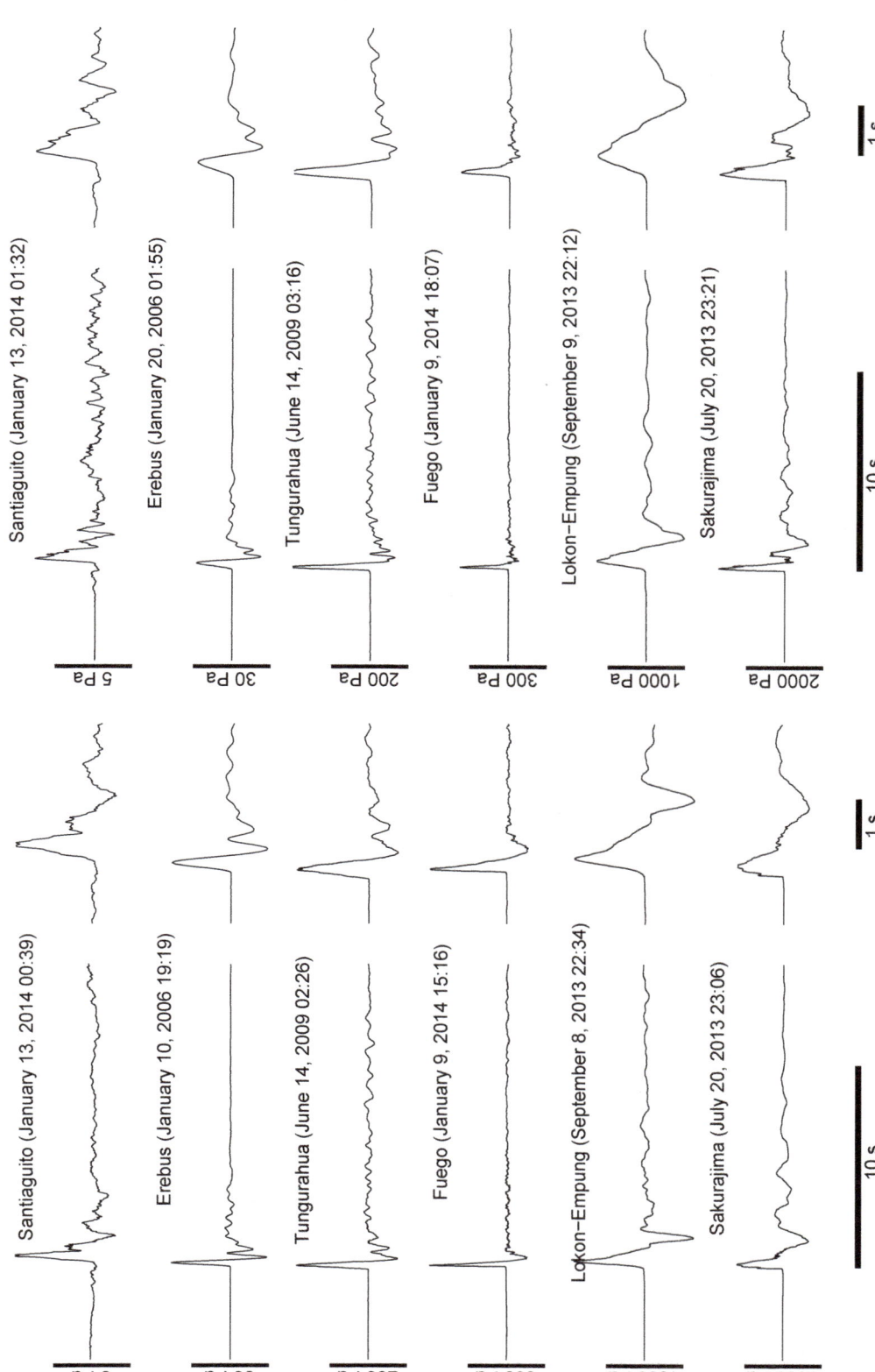

FIGURE 63.5 Volcano explosion infrasound pulses (and detail) are shown for a spectrum of volcanoes that exhibit episodic explosive activity. Excess pressures are reduced to a distance of 1 km for comparison and signals are displayed in order of increasing peak-to-peak pressures. Signals presented correspond to "slow emission" pyroclastic-laden explosions from Santiaguito (Guatemala), gas slug bursts from the Erebus (Antarctica) lava lake, and ash-rich impulsive explosions from Tungurahua (Ecuador) and Fuego (Guatemala). More vigorous vulcanian eruptions are represented by the signals at Lokon-Empung (Indonesia) and Sakurajima (Japan). Two explosions are shown for each volcano to demonstrate similarity in typical signal from given volcano for the given time period. Timing of individual events is shown in parentheses.

Tremor may be classified according to its observed spectral and envelope characteristics (Figure 63.6). In many cases, geneses of the various infrasound tremors are inferred because other direct observations of the vent (e.g., degassing rate or vent geometry) are known. There are several types of infrasonic tremor:

- *Broadband infrasonic tremor* is often broadly peaked throughout the infrasound band. It is typically associated with vigorous and continuous ash and gas venting. Some broadband tremor has been hypothesized to represent a low-frequency form of the sound produced by jet engines (jet noise), which results from turbulence within the jet itself. Thus the volcanic source of jet noise would be in the lowermost portion of the volcanic jet. Volcan Tungurahua's sub-Plinian eruptions generated broadband tremor signals, but lesser intensity episodic eruptions, such as Strombolian events at Karymsky Volcano, also produce sustained, broadband infrasound signals.
- *Harmonic infrasonic tremor* possesses a peak fundamental frequency (often near 1 Hz) and integer overtones at multiples of this frequency. Time series features may appear as a sequence of periodic pulses with timing that is coincident with audible "chugs" and coincident seismic response. Chugging harmonic tremor has been attributed to pulsed emissions for activity at Volcan Sangay, Karymsky, Arenal, and Reventador volcanoes.
- *Monotonic (or monochromatic) infrasonic tremor* is characterized by a pronounced spectral peak (within a few octaves of 1 Hz) and absence of frequency overtones. The time series manifestation is a sinusoidal sustained infrasound. Stability of the spectral peak suggests one of several potential resonant phenomena in the crater/vent or in the uppermost conduit. Intense monotonic tremor at Kilauea and Villarrica volcanoes has been attributed to Helmholtz-type resonant phenomena.

The envelope of infrasound tremors provides further constraints on the nature of near-vent activity. Commonly observed infrasound tremors include

- Stationary tremor possesses relatively constant amplitude and constant spectral content signal indicative of a stable open vent and degassing regime. Open-vent volcanoes such as Villarrica, Turrialba, Masaya, and Telica maintain a stationary tremor that appears related to vent/conduit dimensions.
- Spasmodic tremor is pulsatory, in that its amplitude rises and decays slowly with time scales of tens of seconds, to minutes, or hours. Spasms often occur regularly with time. Kilauea's Halemaumau Vent has experienced extended intervals of spasmodic tremor related to increased/decreased lava lake degassing.

- Episodic tremor begins and ends rapidly and seemingly at random, suggestive of a vent that is intermittently closed and open. This behavior has been observed at Reventador Volcano.
- Gliding tremor is a type of harmonic tremor in which a fundamental tremor frequency, and associated integer overtones, vary smoothly with time. Arenal and Reventador volcanoes have exhibited seismo-acoustic harmonic tremor with frequencies that glide over time.
- Cigar-shaped tremor—infrasonic tremor that begins and ends emergently and is often relatively short (a few to few tens of seconds) in duration. Broadband nature, relatively small amplitude, and envelope are often suggestive of debris flows. Cigar-shaped infrasonic tremor corresponding to pyroclastic flows at Soufriere Hills Volcano has been observed.

3.3. Quantification of Volcano Infrasound

Compared with the dramatic structural variability inherent within a volcanic edifice, the overlying atmosphere is relatively homogenous. At local distances infrasound multipathing and refraction are minimal. Combined with the fact that only compressional waves are supported in the atmosphere, the recorded infrasound is (to first order) representative of source processes. For local propagation distances out to ~ 15 km it is often assumed that the atmosphere is isotropic, acoustic wavefronts are composed of spherical shells, and that excess pressure falls off linearly with distance. Considerations for propagation beyond a few kilometers are outlined below in the section entitled Propagation Considerations.

The following metrics assume an isotropic atmosphere and pressure amplitude decay that falls off with the inverse of distance:

- *Reduced pressure* is best calculated for a common reference distance, typically fixed at 1 km because it is generally outside of the lumped vent/crater source region and also beyond the elastic radius. Typically peak reduced pressures are given although peak-to-peak reduced pressures are also common. This metric is analogous to seismic reduced displacement.
- *Sound pressure level (SPL)*—Although infrasound by definition is subaudible, it is nonetheless illustrative to convert reduced pressures to a logarithmic scale to which humans can intuitively relate. SPL is referenced to the threshold of audibility (20×10^{-6} Pa at 1000 Hz). Volcanoes such as Sakurjima, Redoubt, or Tungurahua that are especially "loud" may have bimodal pulses exceeding 200 Pa at 1000 m, implying an SPL of 140 dB, equivalent to a jet engine at 50 m.

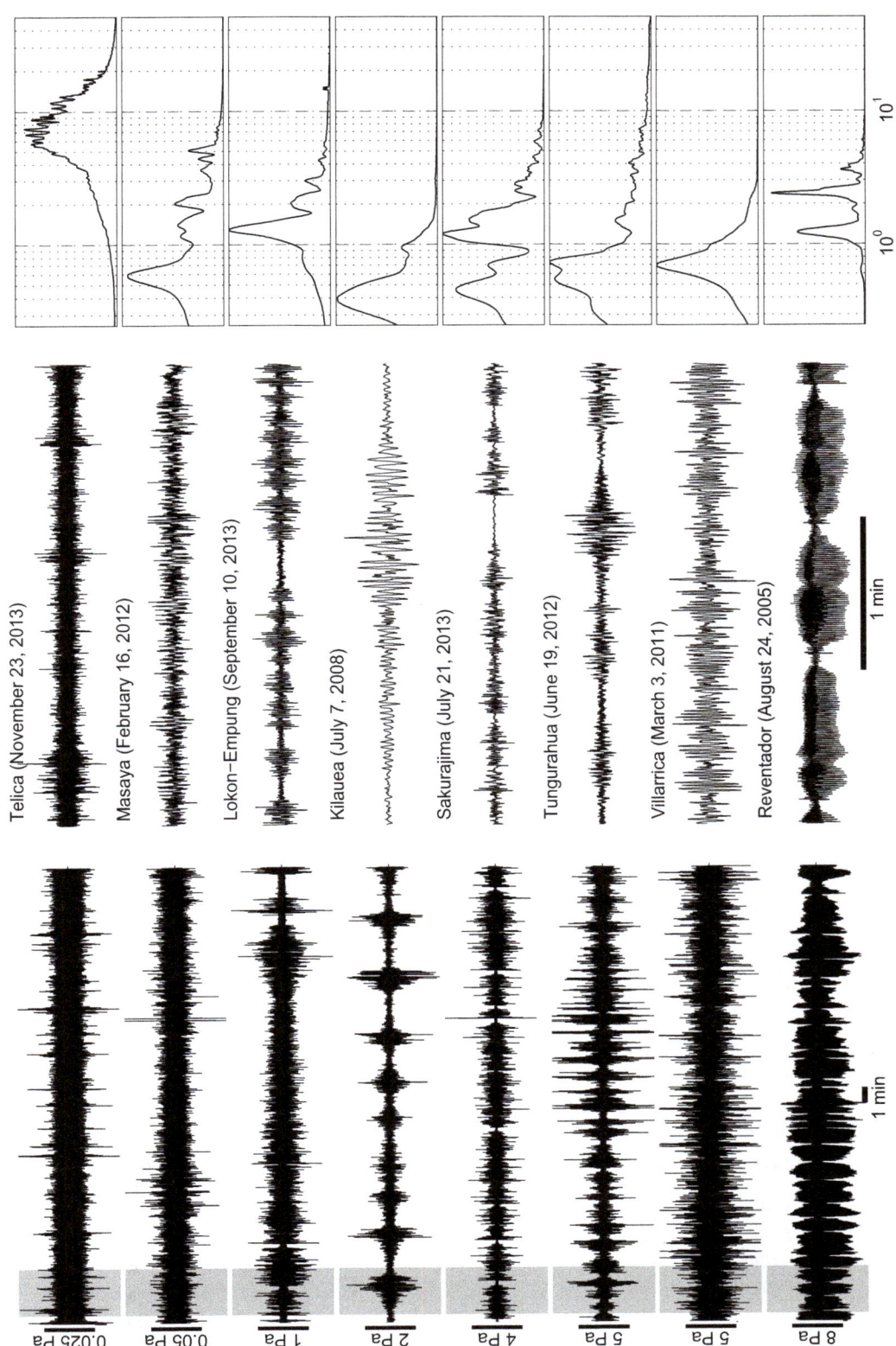

FIGURE 63.6 A spectrum of infrasonic tremor types is shown in order of increasing tremor amplitude calculated for a reduced distance of 1 km. Corresponding activities include continuous gas venting at Telica (Nicaragua) and Masaya (Nicaragua). These "fumarole" sounds contrast with the vigorous ash-venting at Lokon-Empung (Indonesia), Sakurajima (Japan), and Tungurahua (Ecuador). Lava lake degassing and induced crater resonance is shown for Kilauea (USA) and Villarrica (Chile). Open-vent vigorous degassing of an ash-poor column is associated with high-amplitude infrasound activity at Reventador (Ecuador). It is notable that stationary tremors are evident at Telica, Masaya, and Villarrica whereas intermittent or spasmodic tremor dominates at the other volcanoes where the vent is intermittently open. Power spectral density corresponds to the coherent portion of signals recorded across arrays or networks of sensors. Examples of monotonic and harmonic tremors are evident at Villarrica and Reventador, respectively.

- *Infrasound intensity* in the far field is the product of pressure and particle velocity, which is equal to squared pressure divided by impedance. Units of intensity are W/m^2 and are expected to vary spatially in cases where radiation is asymmetric (e.g., for volcanic jets) or where atmospheric structure is nonhomogeneous.
- *Infrasound power* is spatially integrated intensity associated with a common source origin time. For a subset of isotropic infrasound sources, including compact volumetric sources, infrasound power is estimated by calculating intensity integrated over a hemispherical shell. In some cases, where sound is radiated into a region larger than a half space, the corresponding solid angle may be adjusted accordingly.
- *Infrasound energy* is time-integrated infrasound power and gives a measure of total radiated energy. It is often useful to quantify infrasound energy over a specified duration interval, such as over the course of an explosion signal.
- *Volcano acoustic seismic ratio (VASR)* is the ratio of infrasound energy to coincident seismic energy during a volcanic eruption and provides a measure of relative energy partitioning into the atmosphere relative to energy into the ground. High VASR implies efficient ensonification of the atmosphere, which may occur for especially shallow explosions and/or explosions with minimal overburden.
- *Peak frequency* is a useful measurement for comparison of different phases of an eruption sequence or comparison between different volcanoes. It also provides a measure of the source size radiation process. Smoothed spectra or power spectra computed using one of several periodogram methods are common.
- *Spectral band*—Different volcanoes produce infrasound with different characteristic tones and with different peakedness. It is therefore beneficial to specify the spectral band containing the primary contribution to power. One method to quantify the band is the energy percentile where 90 or 99% energy content is indicated by low and high corner frequencies. Special care must be taken to analyze recordings if they include contamination from nonvolcanic infrasound.

3.4. Propagation Considerations

When quantifying infrasound signals it is imperative to consider the influence of propagation effects, including atmospheric focusing, attenuation, scattering, and potential multipathing. In general, propagation effects become more pronounced with increasing distance; the assumption of pressure amplitude inversely decaying with distance loses validity beyond a few kilometers. Beyond local distances, atmospheric shadow zones exist where there are no infrasound ground returns. Care must be taken to avoid deployment in these shadow zones (between a few tens to few hundreds of kilometers from the volcano) where infrasound sensors become "deaf" to potential volcanic activity. Significant local topography or crater morphology may also perturb the signal (Figure 63.7).

Atmospheric structure and corresponding infrasound radiation is inherently dynamic as it depends upon the changing atmospheric temperature and wind structure. To first order these parameters vary primarily with altitude and they can be modeled or sampled with meteorological tools. As with seismic energy in the earth, infrasound energy can be treated as a ray that is refracted in the atmosphere according to a ray parameter, which is conserved along an acoustic raypath. The ray parameter, or horizontal slowness, is related to the sine of the incidence angle divided by the **effective sound speed**.

$$p = \frac{\sin i}{c_{eff}} = \frac{\sin i}{c + \vec{n} \cdot \vec{w}} = \frac{\sin i}{\sqrt{\gamma R T} + \vec{n} \cdot \vec{w}} \quad (63.1)$$

The effective sound speed is the sum of the adiabatic sound speed (a function of specific heat ratios (gamma), gas constant R, and ambient temperature T) and the projected wind vector on to the direction of propagation

$$n \cdot w \rightarrow \vec{u} \cdot \hat{r}$$

where $\vec{u}$ is wind direction and $\hat{r}$ is propagation direction normal.

Both winds and temperatures tend to vary most significantly with altitude. Equation (63.1) predicts that infrasound ray paths return to the earth when they reach regions of the atmosphere with high temperature and/or strong winds. Ray theory is useful for modeling propagation paths and travel times, but is limited in that it is a high-frequency approximation that does not take into account diffraction and scattering. More sophisticated modeling techniques, such as the parabolic equation and finite difference, are able to simulate diffraction and scattering and provide detailed amplitude information. Figure 63.8 gives an example ray tracing simulation.

For typical atmospheric stratification the temperatures and adiabatic sound speeds decrease in the tropopause (up to ~ 17 km), increase up to the stratopause (~ 50 km), decrease again toward the mesopause (~ 80 km), and increase dramatically through the thermosphere. This general structure facilitates ducting in the stratosphere for favorable winds and always in the thermosphere. Zonal (east–west) stratospheric winds have been shown to be responsible for controlling the majority of long-range infrasound detections. These winds are seasonally and latitudinally dependent, meaning that stratospheric returns may be common to the east during a given season and to the west during the rest of the year (e.g., see Figure 63.8). Tropospheric wind jets, such

(A)

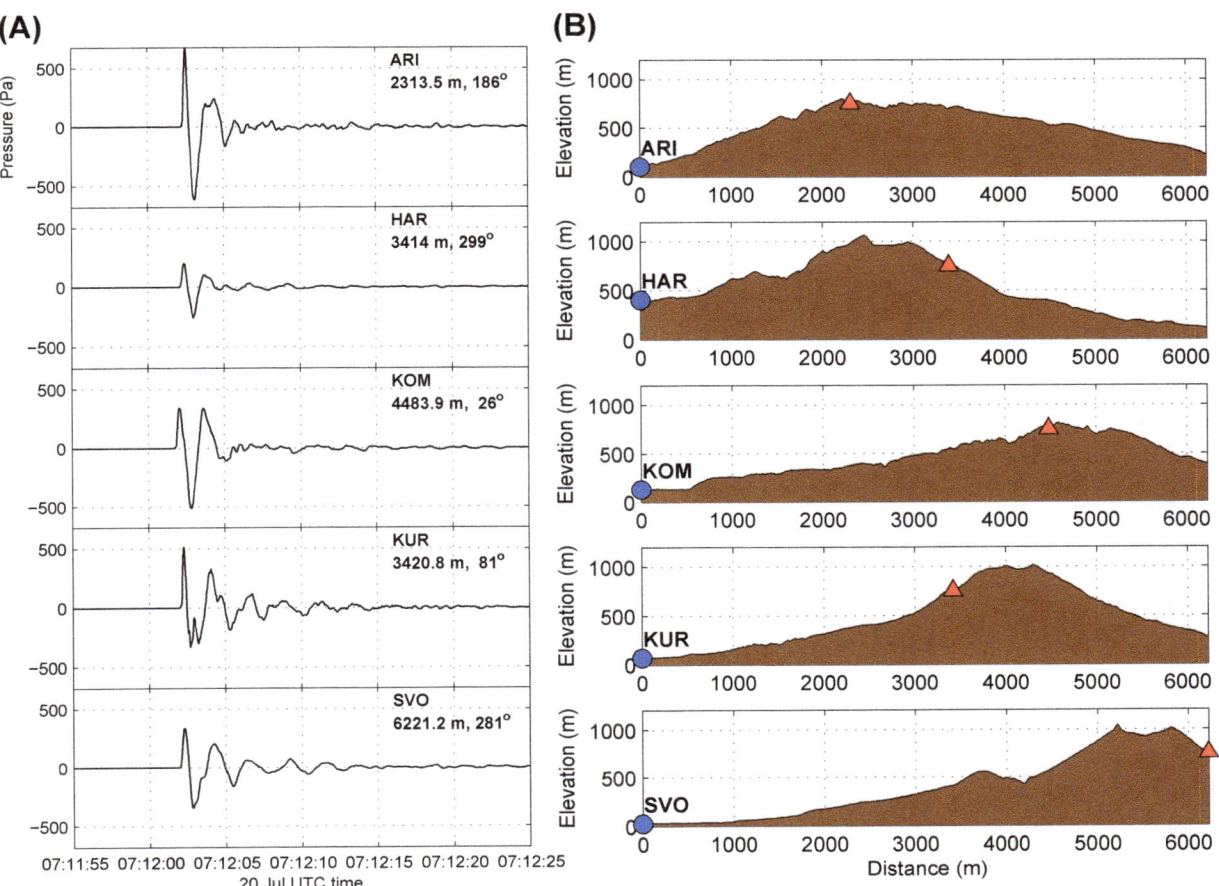

(B)

FIGURE 63.7 Infrasound waveforms and topographic profiles for an infrasound network surrounding Sakurajima Volcano, Japan. (A) Waveforms for a typical vulcanian explosion, with amplitudes reduced to a common distance of 1 km (assuming 1/r spreading). Station name, distance to vent, and azimuth are listed above each waveform. There is significant variability in the recorded infrasound, both in waveform shape and amplitude. (B) Topographic profiles between each station and the active vent at Show a Crater. The variations in recorded infrasound are likely due to topography, as multiple stations do not have direct line of sight with the active vent, in which case sound must diffract around the topography. Reflections off topography may also contribute. *Figure after Fee et al. (2014).*

as the jet stream at ~10 km height, can also significantly guide **regional and global infrasound**.

Although near-surface temperature inversions are occasionally present and facilitate ducting in the lower troposphere, it is more common that temperature structure facilitates upward refraction of infrasound relatively near to the volcano. Beyond local distances it becomes complicated to quantify infrasound reduced pressures or acoustic energies as 1/r pressure decay does not apply. Even at 8 km a study of 0.65 Hz infrasound at Volcan Villarrica revealed that sound amplitudes routinely varied by a factor of two to three depending upon wind conditions (Johnson et al., 2012). For this reason local infrasound is loosely associated with radiation out to about 10–15 km; beyond this distance recorded signals are referred to as regional. Global propagation implies multiple refractions in the stratosphere and/or thermosphere. At further propagation distances intrinsic attenuation needs to be also considered, especially for

refractions in the thermosphere where ambient density is very low. In general, infrasound wavelengths are very long (hundreds of meters) implying nearly negligible intrinsic attenuation throughout local distances.

Waveform comparison, modeling, and interpretation of infrasonic signal shape require consideration of diffraction around topographic features and/or site response, even at local distances. Particularly strong influences can be observed for infrasound generated by a vent that is recessed within a deep crater, or where significant topography obstructs the direct line of sight (Figure 63.7). In these cases ray theory is insufficient, and full-waveform propagation modeling is required. Very large amplitude pressure sources, or supersonic eruption velocities, must also be considered as these sources may produce airwaves that propagate nonlinearly within an elastic radius. Many studies that employ waveform modeling from bipolar pulses to estimate explosive gas flux, or preeruption surface

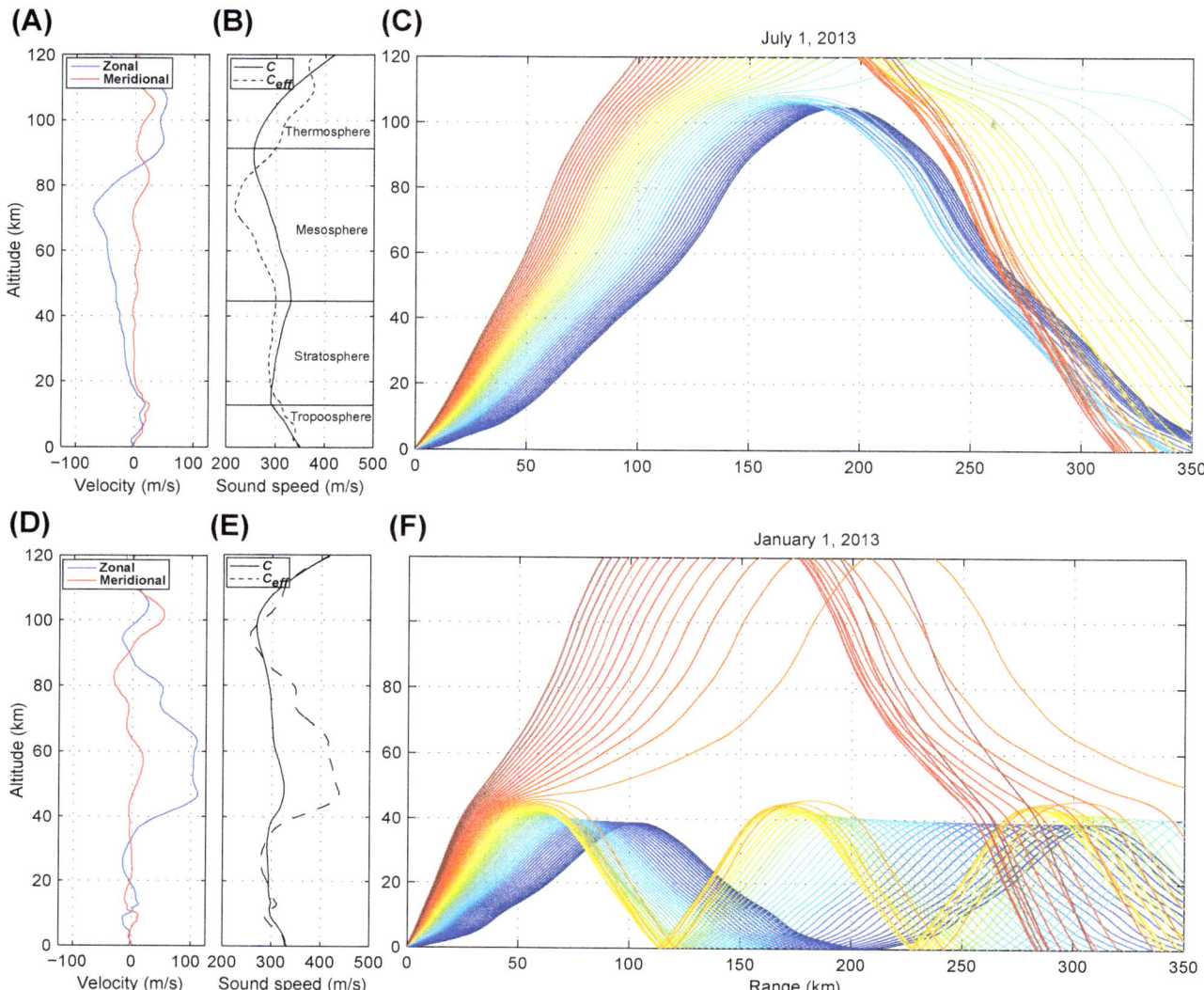

FIGURE 63.8   Long-range infrasound propagation examples for summer (top) and winter (bottom) in the northern hemisphere. (A) Zonal (east—west, positive for easterly blowing) and meridional (north—south, positive northerly) winds and (B) sound speed and effective sound speed (C_{eff}) for propagation to the east on July 1, 2013. The dominant wind feature is the westerly zonal wind peaking at ~70 km altitude. (C) Ray tracing for sound propagating to the east using the profiles in (A) and (B). Sound is refracted down in the thermosphere around 110 km, illustrating a large shadow zone during this time of year out to ~300 km. (D) and (E) Wind and sound speed profiles for propagation to the east on January 1, 2013. Note the strong easterly stratospher-ic—mesospheric wind peaking at ~110 m/s between 45 and 65 km height. (F) Ray tracing for sound propagating to the east using the profiles in (D) and (E). Sound is refracted by the strong stratospheric winds at ~40 km. Rays are colored by the take-off angle.

distension, require care to make sure that the analyzed signal is attributable to the source, and not to propagation phenomena. With this said, high infrasound waveform integrity is possible even at distances of hundreds of kilometers under the right circumstances.

4. FORECASTING ERUPTIONS

The main goals of seismic and infrasonic monitoring of volcanoes are to know the present status of a volcano, to forecast eruptions, to estimate the size of eruptions in progress, and to understand the physical processes that are occurring. Seismic and infrasonic monitoring can help determine and establish time frames for whether a volcano (1) is in a background or restful state; (2) is restless; (3) shows precursory activity; (4) has begun to erupt; (5) is actively erupting; (6) is ending an eruption; or (7) is returning to a background state. A primary focus of this article is forecasting eruptions. For this, goals are estimating the date and time, the place, the size and type, and the likely duration of eruptions. A variety of obser-vations have a bearing on these problems.

Infrasound monitoring is a vital, low-bandwidth technology for tracking the upper portion of the conduit: it tracks when a volcano is exploding; when it is venting; and when it is plugged. It complements visual tools, such as video observations of the volcano vent, because it is functional at night, during poor weather, and/or when the vent is obscured by gas or ash. It complements seismic observations, which often provide ambiguous constraints on whether a particular seismic event occurred at the vent (and was indeed an explosion), or was internal. Most importantly for monitoring applications, continuous tracking of volcano infrasound permits improved understanding of the trends of an eruption sequence. Larger and/or more continuous emissions are reflected qualitatively in the infrasound records, which can be used to assess whether a volcano is building toward an eruption. Steady-state open or episodically open activity is well tracked through infrasound surveillance.

4.1. Seismicity Rates and Background

Volcanoes nearly always have a high background level of seismicity. Thus, a baseline of measurements of at least a few years duration is desirable to characterize and understand background seismicity. The background may consist of several different types of events (see Section 3), which are related to different processes. Typical background rates of detectable seismicity are a few to a few tens of events per day, depending on the locations and magnifications of the seismic stations, and noise levels. In contrast, the rates of detectable seismicity before and during eruptions are typically several tens to several hundreds or more events per day, and include

larger magnitude events. Many successful eruption forecasts have been made based on an increase in seismicity above previously recorded background levels. However, the main physical processes that cause seismicity are those associated with magma intrusion. Thus, there will always be a false alarm rate because some intrusions remain at depth whereas others reach the surface and erupt.

In terms of trying to assess the type of eruption prior to its onset, previous eruptive activity is a better guide than any seismic feature yet identified. This is true for all of the case studies discussed in Section 4. For the formation of new volcanoes, such as Paricutin, Mexico (1943), Surtsey, Iceland (1963), or Novarupta, Alaska (1912), geology of the surrounding area is a better guide than earthquake activity to determine the *type* of impending eruption.

4.2. Statistics of Volcanic Seismicity

Statistical information on many volcanoes, swarms, and eruptions provides a broad perspective to evaluate potential precursors (e.g., Figure 63.9 and Figure 63.10 and Table 63.3). It is helpful to have such information available during a seismic crisis to have an idea of the possible range of activity. A comparison of seismic activity at Kilauea and Krafla demonstrates how the information may evolve over time. Both of these volcanoes produce basalt, and fissure eruptions are common at each. For Kilauea, of 73 swarms studied, 31% accompanied eruptions, 42% accompanied intrusions, and 27% were associated with inflation episodes. Except for harmonic tremor generated near an erupting vent, intrusions and

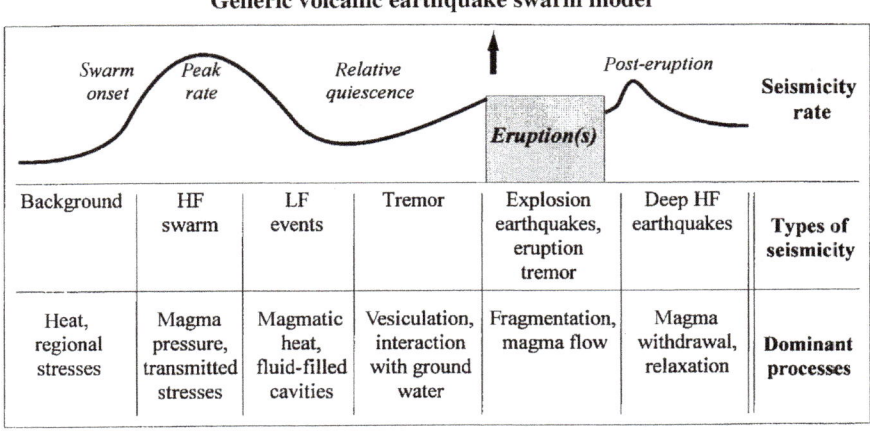

Generic volcanic earthquake swarm model

Background	HF swarm	LF events	Tremor	Explosion earthquakes, eruption tremor	Deep HF earthquakes	**Types of seismicity**
Heat, regional stresses	Magma pressure, transmitted stresses	Magmatic heat, fluid-filled cavities	Vesiculation, interaction with ground water	Fragmentation, magma flow	Magma withdrawal, relaxation	**Dominant processes**

Time ⟶

FIGURE 63.9 Schematic diagram of the time history of a generic volcanic earthquake swarm model. Seismicity rates are shown along with the main types of events observed at each stage. Some ideas about dominant processes are presented at the bottom of the figure. *After McNutt, S. (1996). Page 99–146 in "Monitoring and Mitigation of Volcanic Hazards" (Scarpa/Tilling, eds.). Springer-Verlag, Berlin/Heidelberg.*

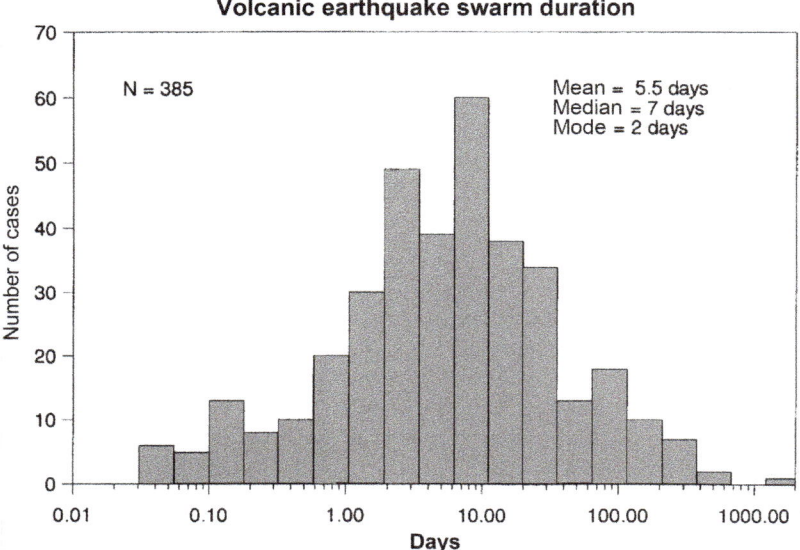

FIGURE 63.10   Histogram of volcanic earthquake swarm durations. The horizontal axis is logarithmic, and the numbers shown represent the middle value of each bin. The data form a log-normal distribution with a mean and median of 5.5 and 7 days, respectively. *After Benoit, J., and McNutt, S. (1996). Annali Geophys. 33, 221—229.*

eruptions were seismically indistinguishable. This suggests that the only difference between an intrusion and an eruption is the magma level compared with ground level. At Krafla, similar numbers were obtained for 20 swarms that included volcanic tremor and occurred in association with deflation episodes; 9 swarms (45%) accompanied eruptions and 11 (55%) occurred during intrusions. However, the first five were intrusions and the last six were eruptions; thus the statistics changed substantially as a function of time. Here the magma level was presumably lower for the intrusions and higher for the eruptions.

4.3. Volcanic Tremor

Volcanic tremor (Figure 63.3(D)), which has several possible mechanisms, is an especially common short-term precursor to eruptions. Tremor was the most common short-term precursor in a study of 132 historical cases of phreatic eruptions (Barberi et al., 1992). This suggests a close link between tremor and hydrothermal processes (tremor known to be related to hydrothermal processes is weaker than that associated with magmatic activity, although frequencies are similar). Tremor was the most consistent short-term indicator of impending eruptions in a separate study of 200 calderas worldwide (Newhall and Dzurisin, 1988). While many tremor episodes accompany eruptive activity, about 20% precede eruptions on a time frame of 10 days (McNutt, 1992). While neither of these studies distinguished between different tremor mechanisms, volcanic tremor is a common and important precursor to many eruptions.

4.4. Generic Volcanic Earthquake Swarm Model

Common observations suggest a general model of earthquake activity at volcanoes. The model, which is termed a generic volcanic earthquake swarm model, is shown as a time series in Figure 63.9. Seismicity rates, types of events, and dominant processes are indicated in the commonly observed temporal sequence. Variations in the involvement of groundwater, as well as variations of parameters such as magma composition, temperature, rate of flow, and volatile content, can explain many of the observed differences from this model of seismic activity at different volcanoes.

In the generic model, high-frequency earthquakes reflect shear fracture of the country rock in the vicinity of volcanoes in response to increasing magmatic pressure. Since magma movement at depth may occur over large distances and long periods of time, the durations of high-frequency swarms are long. At shallow depths of 1—3 km, several significant changes occur. First, gases begin to exsolve from the magma, changing its mechanical behavior and, hence, its ability to transmit or reflect seismic waves. For example, exsolution of volatiles will increase the viscosity and lower the acoustic velocity. A seismic event generated by resonance of a magma-filled cavity would thus be richer in lower frequencies. Second, the magma may encounter groundwater, modifying the shallow hydrothermal system and removing some of the available magmatic heat by conduction and advection. Most low-frequency events and tremor originate at shallow depths, and several models for their origin involve fluids, either water, magma, exsolved volatiles, or all three. Tremor caused by boiling of water is generally weaker than tremor involving magma movement. Third, open cracks are found above 3 km depth in the earth, permitting venting of excess pressure. This may explain the relative quiescence after the peak rate of a swarm as well as the inverse relation between high-frequency events and tremor.

The upper 1−3 km are small distances compared with depths to magma chambers (5−20 km) or crustal thicknesses (about 35 km). Thus, if vertical magma ascent rates are constant, then durations of low-frequency event swarms and tremor sequences will be shorter than those of high-frequency events simply because the distances are smaller. Deep earthquake swarms following eruptions probably reflect stress changes associated with partial evacuation of magma chambers deeper than 5 km.

Relative seismic quiescence following an initial seismicity peak occurred in 25% of a worldwide sample of 192 swarms. While quiescence 25% of the time may seem too low a percentage to be included in a generic model, it is included because of its importance in terms of eruption forecasting and identifying physical processes. Factors that may cause quiescence include strain hardening; high temperatures near the brittle−ductile transition; increased water content (e.g., groundwater) lowering effective stress; or a reduction in strain rate. It is common for low-frequency events and tremor to occur after seismicity rate peaks, during times of relative quiescence. This may be explained by gradual shoaling of the source, particularly magma interacting with groundwater or exsolution of volatiles.

Each volcano is unique in its seismic behavior, but the observations suggest that variations in a few parameters such as those discussed above may account for the main differences. Minor variations in each of several parameters may greatly affect one observed parameter, such as swarm durations. The durations of earthquake swarms are characterized by a log-normal distribution with a mean of 5 days (Figure 63.10). There is a weak correlation between the duration of a swarm and the VEI (Benoit and McNutt, 1996). However there is considerable scatter, and large ranges of sizes, rates, compositions, and types of structures observed.

Parameters of 700 volcanic earthquake swarms have been compiled into the Global Volcanic Earthquake Swarm Database (Benoit and McNutt, 1996). The database has one table of basic information on each volcano, such as type of lava, height, and location; one table on eruptions, including volume, time of occurrence, and type; and a table of swarm parameters such as duration, number of earthquakes, depths, and magnitudes. Using database queries it is possible to find all known cases of earthquake swarms that fit a particular profile. For example, we can search for the cases that last 3 weeks, have 60 events per day, and a largest event of M 3.3. The "answer" is a list of similar swarms that

can then be analyzed to determine whether eruptions occurred, how long they lasted, etc. These answers are probabilistic in that they look for a likelihood of something happening based on past occurrences. Such information is complementary to simultaneous efforts to determine and to constrain what is happening physically at a volcano. A large-scale effort called WOVOdat is under development (Ratdomopurbo et al., 2012).

4.5. Curve Fitting Techniques

The Failure Forecast method is an analysis technique based on the materials science approach of estimating failure criteria. An observable parameter, such as strain or seismicity, is fit to the equation

$$\dot{\Omega}^{-\alpha}\ddot{\Omega} - A = 0, \tag{2}$$

where A and α are constants, Ω is the observed quantity, and the dot refers to differentiation with respect to time (Voight, 1988). This method has been shown to be successful in several volcano applications in retrospect, but not beforehand. These included the March 1982 eruption of Mount St. Helens, the April 1960 eruption of Bezymianny, and the 1987 summit eruption of Izu-Oshima volcano. The method predicts the time of the eruption onset.

RSAM data from four volcanoes have been analyzed using an automated curve-fitting scheme (Endo et al., 1996). The same exponential equation was objectively selected to best fit preeruption normalized RSAM data from dome-building eruptions at Mount St Helens and Mount Pinatubo, suggesting that a common process is acting regardless of the details of the seismicity. Selected portions of premonitory seismicity at Redoubt, Mount Spurr, and Pinatubo volcanoes also showed exponential increases with different durations. The best-fit equation is of the form

$$R(t) = a + b\exp(-t/c) \tag{3}$$

where R is the normalized RSAM value at time t, and a, b, and c are parameters determined by the curve-fitting program. This approach aims at identification of an aggregate process that does not depend on specific types of earthquakes or their relative times of occurrence.

4.6. Alarm Systems

According to the generic volcanic earthquake swarm model, volcanic eruptions are expected to be preceded by earthquake swarms or tremor. Recognizing swarms and tremor in near-real time (seconds to minutes) is therefore crucial. Volcanic activity can escalate rapidly, for example, the July 12, 2008, eruption of Okmok volcano occurred after less than 5 h of precursory seismicity. Periodic data checks can be made to catch such escalations in activity, but can fatigue observatory staff quickly. Another strategy is to man the observatory 24 h a day, which requires a large team and is therefore expensive. The development of effective volcano-seismic alarm systems is therefore a viable alternative and a priority for many volcano observatories.

A tremor alarm module is included in the RSAM system (now part of Earthworm; Olivieri and Clinton, 2012). For a station to trigger, the amplitude must exceed a predefined RSAM threshold for a predefined duration (usually several minutes). If multiple stations trigger in the same time window, a tremor alarm is declared. Another alarm system works in a similar way, but uses the reduced displacement between 0.5 and 15 Hz as the amplitude metric, instead of RSAM. This has the advantage of being comparable between different stations and volcanoes. Tremor alarm systems have proved useful during the eruptions of Redoubt volcano from 1989 to 1990 and 2009, and Soufriere Hills Volcano, Montserrat, from 1999 to 2010 (the Operations Room was permanently manned 24/7 from 1995 to 1998). False alarms do occur, however, due to storms moving through the region, other noise or station malfunctions.

A swarm alarm system can be based on a near-real time catalog of located events, a product already available at many observatories. If the rate of earthquakes in a volcanic region exceeds a threshold, a swarm start alarm is declared. Alarms can also be declared when a swarm escalates, deescalates, and ends. The University of Alaska operated a swarm alarm system that successfully detected the six main swarms of the 2009 eruption of Redoubt volcano. A similar system is included in the Advanced National Seismic System Quake Monitoring System, which is an extension of the Earthworm system. During swarms, events will sometimes coalesce into a continuous (tremor) signal, especially at distal stations, at which point the STA/LTA detector used will break down and an algorithm based on a real-time event catalog will conclude that the swarm has ended, when in fact activity is continuing to escalate. Running swarm and tremor alarm systems in parallel might therefore be the best strategy.

Volcano-seismic alarm systems are exciting areas of research that have the potential to forecast escalations in activity. However, detecting all significant changes in seismicity remains a great challenge. While catalogs of volcanic earthquakes are available, observatories do not typically catalog swarms, tremor, or the signals corresponding to volcanic hazards such as explosions, pyroclastic flows, or lahars in any systematic way. This is a barrier to designing detection algorithms for these different signal types. Furthermore, an alarm system is only useful if it also produces a low-false alarm rate. This is a difficult goal to reach if data are afflicted by frequent spikes and drop-outs, or if data are clipped: these introduce step functions into the data stream which are hard to distinguish from seismicity changes and wreak havoc with automated algorithms. Seismic networks need to be engineered to robustly capture high-quality continuous data if seismic alarm systems are going to be a successful part of an observatory's monitoring strategy.

5. CHARACTERIZING ERUPTIONS IN PROGRESS

Forecasting eruptions has received dominant focus as a goal of volcano monitoring, especially when the volcano is located near population centers. However, characterization of eruptions in progress, including estimates of tephra volume, height of ash column, and end time of eruption, is also of importance for hazard mitigation, particularly regarding aviation. Remote or recently inactive volcanoes may also have limited to no local monitoring networks, thus forecasting eruptions at these volcanoes are unlikely and rapid characterization of their activity is essential. VEI can be estimated based on the reduced displacement (D_R) measured during eruptions (Figure 63.11). The computed relation is

$$\log_{10}(D_R) = 0.46(\text{VEI}) + 0.08 \qquad (4)$$

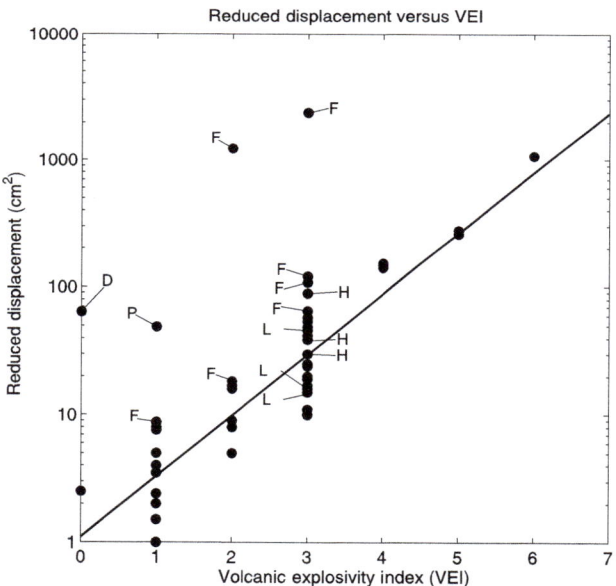

FIGURE 63.11 Reduced displacement, a normalized measure of amplitude, versus the Volcanic Explosivity Index (VEI) for 50 eruptions at 31 volcanoes. The regression line is from McNutt (1994) based on a smaller data set and is shown for comparison. Fissure eruptions are labeled F; a phreatic eruption is labeled P; deep (40 km) tremor from Kilauea is labeled D; and three pairs of values from eruptions with high and low gas content are labeled H and L, respectively. *After McNutt, S.R. (2005). Annu. Rev. Earth Planet. Sci. 33, 461–491.*

Reduced displacement is a method of normalizing volcanic tremor to a common scale. For body waves (e.g., P waves and S waves) the formula is

$$D_R = rms(D) \cdot r \qquad (5)$$

For surface waves (e.g., Rayleigh waves, Love waves, and PL waves) the formula is

$$D_R = rms\left(D\right) \cdot \sqrt{(\lambda r)} \qquad (6)$$

where $rms(D)$ is root-mean-square displacement amplitude in centimeters; r is distance from source to seismic station in centimeters; and λ is tremor wavelength in centimeters. Tremor is generally stronger from fissures as opposed to cylindrical conduits, and stronger when the gas content of the magma is higher (Figure 63.11). A handful of cases suggest that phreatic eruptions also produce stronger tremor when compared with other eruption parameters.

Information on tremor D_R is helpful for estimating plume heights and tephra volumes. The ends of eruptions can be estimated as coinciding with the end of strong tremor associated with eruptions. In many cases tremor declines more or less exponentially on a timescale of minutes to tens of minutes and returns to background levels. In some cases, eruptions of ash end but weak tremor continues for days in association with hydrothermal activity or degassing. The ends of eruption sequences, which may last months to years, are more difficult to determine than the ends of individual eruption pulses, which last a few hours to days. Generally volcanic tremor ceases altogether and the number of earthquakes per unit time decreases regularly. Volcanic earthquake swarms decline with a similar power law decay to aftershocks, except that the timescale is often longer.

Infrasound is key to providing direct information on what is occurring at the vent. Large explosions are often easily detected on infrasound sensors, and sensor arrays and networks permit identification and location of even low-energy eruptions. Changes in the eruptive style can be detected easily with infrasound. For example, the cessation of infrasonic tremor prior to an eruption may represent a decrease in degassing due to conduit sealing. The low amount of attenuation at infrasonic frequencies also permits long-range detection and characterization of explosive eruptions, which may provide the most detailed information available for remote eruptions. Recent work has also found positive correlations between infrasound pressure and energy with plume heights, as well as an inverse relationship between peak frequency and plume heights. Future work shows promise in using infrasound to quantitatively determine critical eruption source parameters, such as mass eruption rate, in near-real time.

6. CASE HISTORIES

Case histories provide specific examples of the concepts and methods described above. Each was chosen to illustrate a sequence of events in which seismic and/or infrasound data were used to constrain processes and hence influence decision making. Monitoring is not a "one size fits all" endeavor. Rather, broad processes and behaviors are inferred from the nuances and details of seismic and infrasonic observations, considering all other available data as well.

6.1. Redoubt, February—April 2009

Redoubt Volcano, Alaska, had a prior eruption from December 1989 to April 1990 that was seismically well monitored. The 1989 precursory sequence was short and intense, lasting only 23 h and including over 4000 long-period (LP) events. There was an expectation that this was typical activity for this volcano. However, the precursors to the 2009 eruption lasted months and included several different types of events.

After 6 months of unrest, Redoubt volcano explosively erupted at least 19 times between March 23 and April 4, 2009, producing ash clouds as high as 19 km (McNutt et al., 2013). Several low-amplitude tremor bursts were recorded in late September 2008, and there were reports of anomalous gas emissions and noises coming from the volcano. LP earthquakes were recorded in December 2008 at about 30 km depth. Higher amplitude tremor was recorded on January 25, 2009, prompting the AVO to strengthen the seismic network and establish an arsenal of near-real-time monitoring tools. On January 30, high-amplitude tremor occurred, coincident with the onset of shallow tectonic earthquakes. Sustained tremor occurred throughout February, escalating and then ending abruptly on February 26, and followed within hours by the first swarm that lasted 31 h and contained almost 900 events. Tremor was observed again on March 15, coincident with a short-lived phreatic eruption. On March 20 a more intense 64-h swarm began, containing 2000 earthquakes, during which the first evidence of dome growth was observed. The latter part of this swarm coincided with rapidly escalating tremor that culminated in the first magmatic explosion on March 23, destroying the first dome. Further swarms occurred on March 27, 29 and from April 2—4, culminating in the final explosive eruption. A final swarm occurred from May 2—8, containing over 7000 events. These swarms stand out clearly in an earthquake rate plot (Figure 63.12). Note the differences between manual event counts (Figure 63.12, top) and automatic event counts (Figure 63.12, bottom). The automatic counts are always higher (and crucially they are available in near

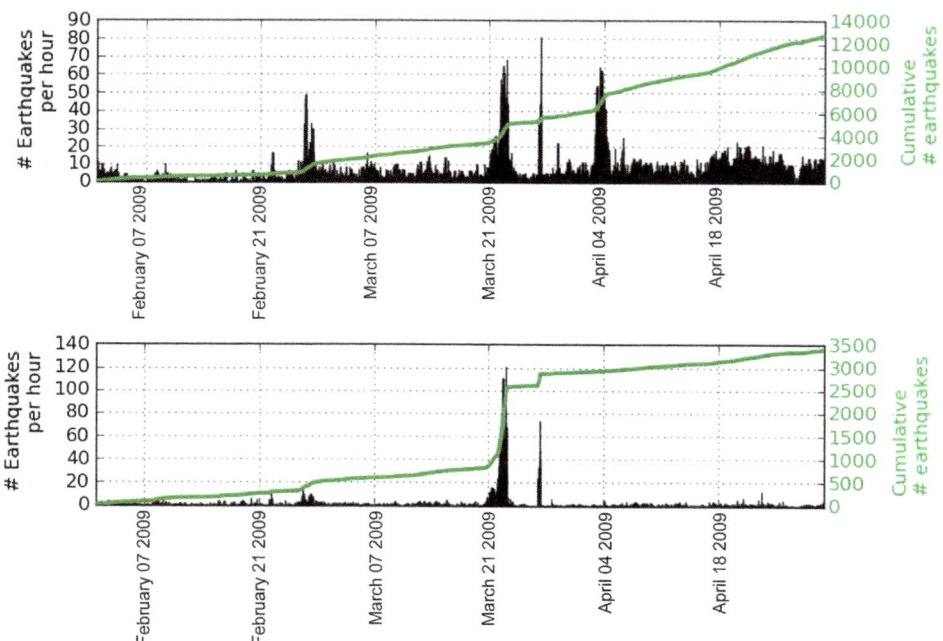

FIGURE 63.12   Number of earthquakes versus time for Redoubt Volcano using automatically picked data (top) and manually detected data (bottom). (left axis) Number of earthquakes per hour within 20 km of Redoubt Volcano summit from February 1, 2009, to May 1, 2009. (right axis) Cumulative number of earthquakes for same data set. All earthquakes detected in real-time using Antelope (bottom), see Thompson and West (2010) for details. Earthquake swarms on February 26, March 21–23, March 27 and April 2–4 are clearly visible. The absolute numbers of events are different, but the main episodes of seismicity are visible in both data sets.

real time), and help to identify subtle changes in seismicity that are not obvious in manual plots. The relative sizes and importance of the various rate changes are subject to interpretation.

Between January and May 2009 a total of 37,000 seismic events were detected (Thompson and West, 2010). The start and end of each swarm were detected, as were the sharp increases in overall seismicity corresponding to the onset of most of the explosions. Gliding spectral lines occurred in the minutes prior to several of the explosions and could be seen in near-real time web-based spectrogram plots, one of AVO's core monitoring tools. A unique feature of these gliding lines compared with those seen at other volcanoes is that they glided to frequencies as high as 30 Hz (Hotovec et al., 2013).

6.2. Eyjafjallajokull March to May 2010

Eyjafjallajokull volcano, Iceland, erupted effusively from a flank vent from March 20 to April 12, then explosively from its summit vent from April 14 to May 22, 2010 (Tarasewicz et al., 2012). The latter activity disrupted thousands of airline flights to and from Europe over the course of a week. The volcano had only three previously known eruptions with the most recent activity in 1821–23, thus nothing was known about its likely precursors. Once the flank eruption was underway, the

volcano was known to be erupting, and the focus of monitoring shifted to tracking the ongoing activity. The onset of the summit eruption was quite rapid and somewhat unexpected.

Three major earthquake swarms occurred in 1994, 1996, and 1999–2000. Swarm activity renewed in March 2009, strongest from June to August 2009 with the largest event an M = 2.4 earthquake. A partial decline in activity occurred for 4 months, then the swarm increased again in late December 2009. Six temporary three-component broadband seismometers were deployed on March 5 to augment the eight closest regional network stations. Similar to previous swarms, the main activity was clustered east of the summit at depths of 9–11 and 2–4 km (Figure 63.13(A)). In late February to early March seismicity migrated toward the SSE from 4 to 9 km depth with the largest event M = 3.0. Volcanic tremor was first recorded on March 4 and the flank eruption began late on March 20 following 2 days of declining earthquakes at an ice-free site called Fimmvorduhals Pass, 4 km E of the summit (Figure 63.13(B)). Eruption tremor accompanying fire fountaining was strongest from 07:00–08:00 UTC March 21. Earthquake activity continued with the strongest event of M = 3.6 in early April (Figure 63.13(C)). The precursors to the summit eruption were very short, lasting just a few hours before the onset of the second eruption beneath the ice

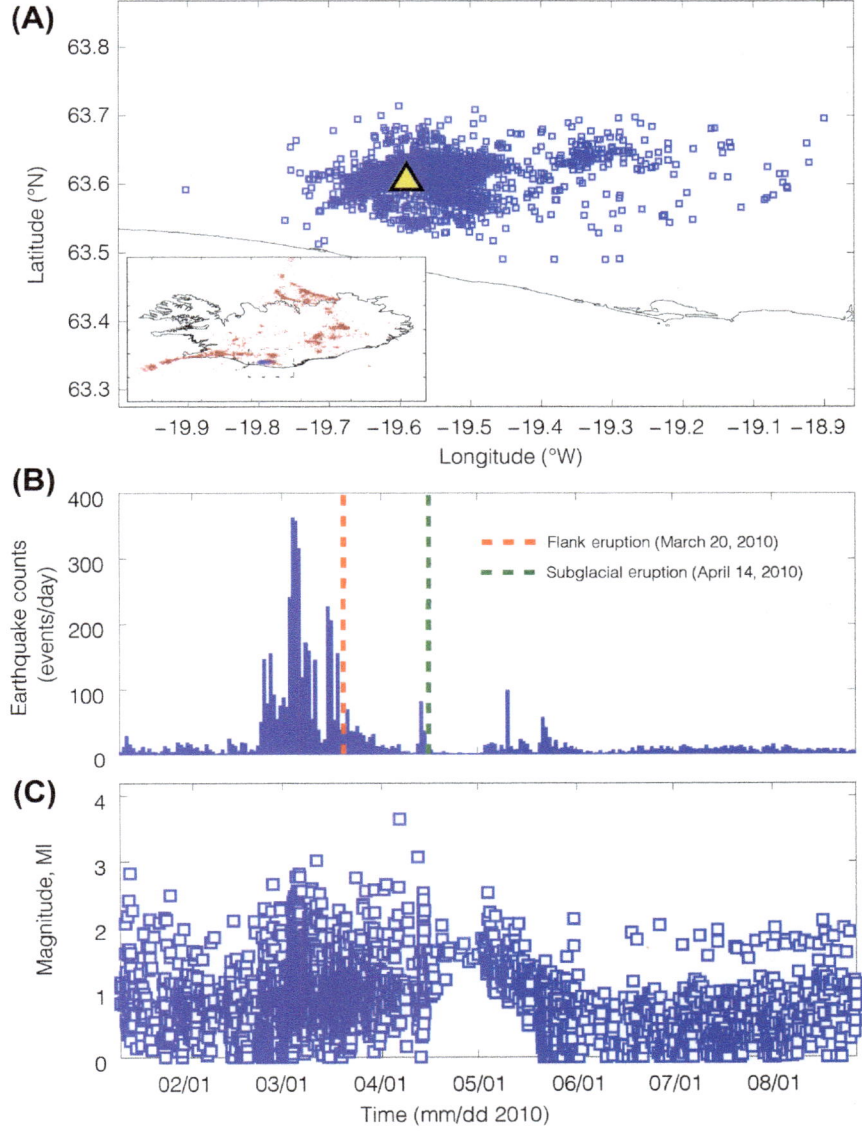

FIGURE 63.13 Seismicity associated with the eruption of Eyjafjallajokull volcano. (A) Large panel: Seismicity between October 1, 2009, and June 1, 2010, in the Eyafjallajokull region. Inset Iceland seismicity between 2006 and 2010 (red) and Eyafjallajokull earthquakes (blue). Yellow triangle is Eyafjallajokull. Earthquake catalog from Icelandic Meteorological Office (IMO). (B) Earthquake rate in the Eyafjallajokull region around the time of eruptions (Jan–Aug 2010). (C) Magnitude/time plot of earthquakes in the Eyafjallajokull region around the time of eruptions (January to August 2010).

cap on April 14 (Figure 63.13(B)). Earthquakes were clustered at depths of 5.5–7 and 0–3 km just south of the summit crater. In late April an increase in volcanic tremor occurred that was not correlated with the eruptive activity; it was later determined that this signal was caused by melting of ice and snow producing lahars near the seismic station. Seismicity dropped after the onset of the summit eruption, then increased again on May 3 when deep activity was recorded, followed by shoaling and an increase in ash production on May 5. The explosive phase of the eruption lasted about 5 weeks with variable intensity, and vigorous eruptive activity ended

on May 22. Subsequent investigations suggest that basaltic magma intruded from depth to a shallow magma chamber at about 3–5 km depth. Basaltic magma was erupted in the flank eruption, and the basalt likely triggered the trachy-andesite explosive summit eruption.

6.3. Tungurahua Volcano, Ecuador, 2006–2008

Tungurahua Volcano, Ecuador, regularly produces high-amplitude infrasound signals and provides an

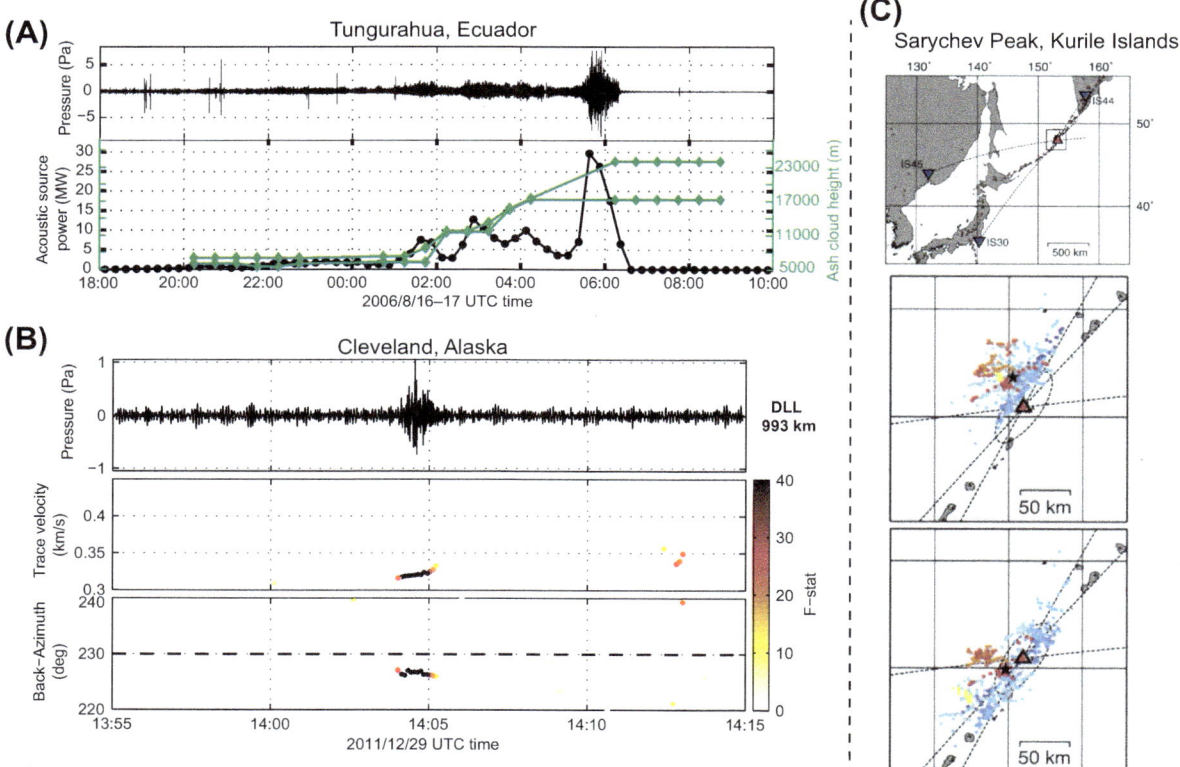

FIGURE 63.14 Examples of regional and global infrasound monitoring of explosive volcanic eruptions. (A) Infrasound from the August 2006 plinian eruption of Tungurahua Volcano, Ecuador. The top panel shows the sustained, high-amplitude infrasound from the eruption recorded at 37 km. Bottom panel shows the correlation between acoustic power (black line) and ash cloud height (green lines). *(Figure from Fee et al. (2010)).* (B) Infrasound array processing results from an explosion at Cleveland Volcano, Alaska, on December 29, 2009, recorded in Dillingham, Alaska at 993 km distance. The top panel shows the 0.1−5 Hz beamed waveform, middle panel the trace velocity, and bottom panel back-azimuth. Trace velocity and back-azimuth estimates are colored by the Fisher statistic. Infrasound from Cleveland is clearly detected beginning ∼14:04 UTC. *(Figure after De Angelis et al. (2012)).* (C) Infrasound source localization from the 2009 eruption of Sarychev Peak, Kuril Islands. The explosive eruption of this remote volcano was detected by multiple infrasound arrays (top panel). Source localization was performed using back-azimuth cross-bearings from three arrays (middle panel) as well as after applying a correction due to deflection from cross-winds (bottom panel). *Figure modified from Matoza et al. (2011).*

opportunity to examine the relationship between acoustic signals and different eruption styles. In 2006 an infrasound array was deployed 37 km from Tungurahua as part of the Acoustic Surveillance of Hazardous Eruptions (ASHE) project. The goal of the ASHE project was to assess volcano monitoring using infrasound arrays at regional distances; here the propagation time was about 111 s. This array recorded extensive, diverse infrasound from both short-duration explosions and longer duration tremor and jetting. Fee et al. (2010) used this data to design an automated system that detected over 20,000 short-duration explosions and large, sustained eruptions in July and August 2006 and February 2008. They found that acoustic energy from Tungurahua broadly scaled with eruption intensity and ash cloud height during the large subplinian−plinian eruptions (Figure 63.14(A)). Using this information the automated system was successful in near-real time detection of the onset, change in intensity, and cessation of a large, hazardous eruption

in February 2008. Infrasound array signal detection permitted unambiguous identification and characterization of large eruptions. Additionally, the frequency content of the sustained, energetic infrasound correlated with the eruption size, and exhibited characteristics similar to that of jet engine noise (Matoza et al., 2009; Fee et al., 2010).

6.4. Sarychev Peak, Kuril Islands, June 2009

An explosive eruption sequence from the remote Sarychev Peak in the Kuril Islands was well documented using remote infrasound arrays located 640−6400 km from the volcanic vent (Matoza et al., 2011). Due to the remoteness of the volcano, no local seismic stations were available to detect precursory activity or track this eruptive sequence, thus real-time observations were limited to remote sensing. In this case, infrasound arrays located at global distances were shown to be uniquely capable of characterizing the eruptive sequence with a higher level of detail than remote sensing. Although remote, this eruption had important hazard impacts for the well trafficked North Pacific air corridor.

Sarychev eruptions during June 2009 were detected with multiple arrays in the IMS and with infrasound arrays operated by the Korea Institute of Geoscience and Mineral Resources (Figure 63.14(C)). Explosions were detected on a subset of the infrasound stations as early as June 11. Approximately 17 h later a continuous signal, punctuated by discrete pulses of energy, was suggestive of vulcanian–plinian activity that endured for 10 h. Satellite observations showed plumes that routinely extended more than 500 km from the vent during this sequence. For the next few days at least 10 eruptive events were detected on multiple arrays, with infrasound durations lasting from tens of minutes to over an hour. The end of the primary eruptive activity tracked by infrasound arrays was around June 15.

This eruption demonstrated the capabilities of global monitoring of large remote eruptions. Although recorded signals at these distances were relatively weak (~ 0.1 Pa peak pressure at 643 km), the ability of multiple arrays to provide accurate detection and localization proves the utility of remote infrasound monitoring. Not all IMS arrays detected the entire sequence of eruptions, and arrays located East of the Kurils were generally in the acoustic shadow zone, but more than eight infrasound arrays contributed to identification of an eruptive chronology that would otherwise have been poorly known. The propagation time to the nearest arrays was about 32 min.

6.5. Cleveland Volcano, Alaska

The Aleutian Island Arc is another remote and volcanically active region where infrasound monitoring has been shown effective. Monitoring volcanoes in this region is particularly challenging, as the remoteness makes local networks difficult to establish and maintain, and satellite remote sensing is limited by frequent cloud cover and relatively low-temporal resolution of the imagery. The AVO has thus employed regional infrasound arrays to help monitor this region with promising results. Cleveland Volcano is particularly remote and difficult to monitor, with the nearest seismic station ~ 75 km away. However, it is also the most consistently active volcano in the arc. De Angelis et al. (2012) documented explosive activity from Cleveland between December 2011 and August 2012 and showed how regional–global infrasound arrays and ground-coupled airwaves on regional seismic networks permitted detections of 20 explosions during this period (Figure 63.14(B)). Although the latency of these detections was occasionally up to 1.5 h,

only 7 of these 20 events were detected with satellite imagery, suggesting 13 explosions would have gone undetected without the use of infrasound. These promising results have led to the expansion of infrasound monitoring by AVO in the region.

6.6. Soufriere Hills Volcano, Montserrat, July 1995 to February 2010

The Soufriere Hills Volcano on Montserrat had five phases of andesitic dome growth between November 1995 and February 2010. Each lasted 6 months to 3.8 years. These phases began with hybrid earthquake swarms, and were followed by increasing levels of rockfall and pyroclastic flow seismicity, and irregularly spaced long period earthquakes as the growing dome became increasingly unstable. These phases of growth were punctuated by several catastrophic dome collapses with some the largest being on March 20, 2000; July 29, 2001; July 12, 2003; May 20, 2006; July 28, 2008; and February 11, 2010. Pauses in dome extrusion were marked by sharp drops in low-frequency seismicity and rockfalls, and an increase in volcano-tectonic earthquakes (see Figure 59.6 of Chapter 59). GPS data showed inflation of the volcanic edifice during these pauses, suggesting continued magma supply at depth. The most useful prognostics of renewed dome extrusion identified so far have been rotation of fault plane solutions of these volcano-tectonic events, providing as much as 6 months warning. Examples and details of seismic data are given by Green and Neuberg (2006) and Luckett et al., (2002).

The dominant volcanic hazards on Montserrat are pyroclastic flows, surges, and lahars. Once the eruption was underway, techniques were developed that responded to these hazards. In 2000 a rockfall location system was developed (Jolly et al., 2002). The dome was shrouded in cloud most of the time, but often changed its direction of growth. Once the dome began growing, dome failure would begin to generate rockfalls and pyroclastic flows of increasing size and runout in that direction. Switches toward the northwest were of particular concern because dome failure events could then follow the Belham Valley toward the largest settlement on the island. So scientists set out to detect switches in dome growth direction by routinely locating and mapping rockfall activity based on signal amplitude across the network. Using the method, seismic monitoring could be used to identify switches in dome growth as much as a few days before it could be visually confirmed (Figure 63.15). Scientists also routinely estimated the direction of larger events in real-time based on RSAM data. Additional seismometers were also deployed at two sites within tens of meters of the Belham Valley to enhance identification of lahars.

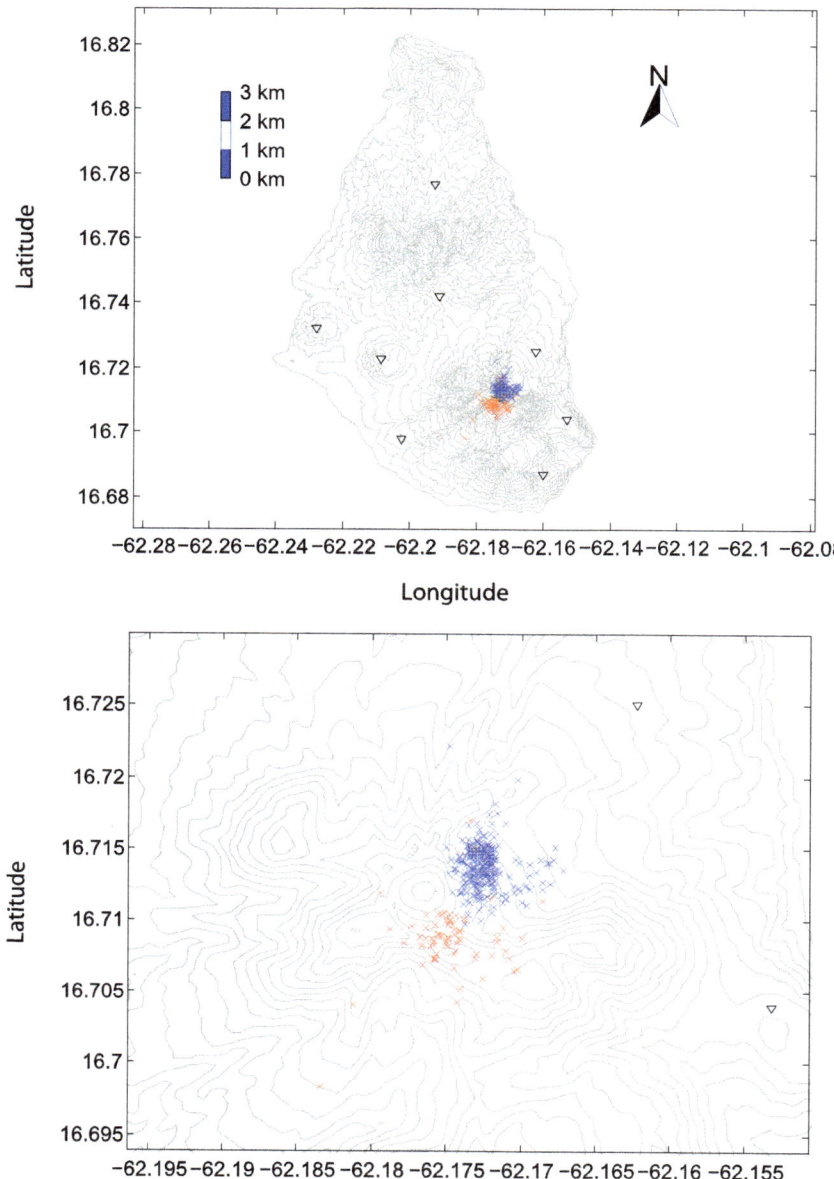

FIGURE 63.15   Seismicity from rockfalls at Soufriere Hills Volcano. (top) Island of Montserrat. Seismic stations are triangles. (bottom) Enlargement of the volcano. Each cross denotes the average location of a rockfall signal, based on the amplitude distribution across the digital seismic network, as determined by the Montserrat Volcano Observatory rockfall location system. The blue crosses are the average rockfall locations February 6 to March 3, 2001. The red crosses are the same for March 4 to April 1, 2001. February 26 to March 3 corresponded to a period of banded tremor/hybrid swarms, following dome collapses to the northeast and southeast February 24–25. These results suggest the direction of lava extrusion shifted from the northern and eastern sides of the dome, to the southern and western sides of the dome in late February or early March.

6.7. El Hierro, July to December 2011

When seismic activity began to occur in the vicinity of El Hierro, there were fundamental questions concerning whether the activity was tectonic or magmatic, whether it would erupt, and if so, where? There were no historic eruptions to serve as a guide, and three rift zones had been identified by geological studies. The seismicity as initially determined spanned the entire island (Figure 63.16).

Five phases of activity were identified in later studies (Lopez et al., 2012): Phase I (7–18 July 2011)—slight deformation; Phase II (July 19 to September 3, 2011)—beginning of the unrest; Phase III (4–26 September 2011)—migration to the south; Phase IV (September 27 to October 7, 2011)—acceleration of the process; and Phase V (8–10 October 2011)—forthcoming eruption. Increased seismic activity was detected beginning on July 17, 2011. The seismic monitoring network was increased in density

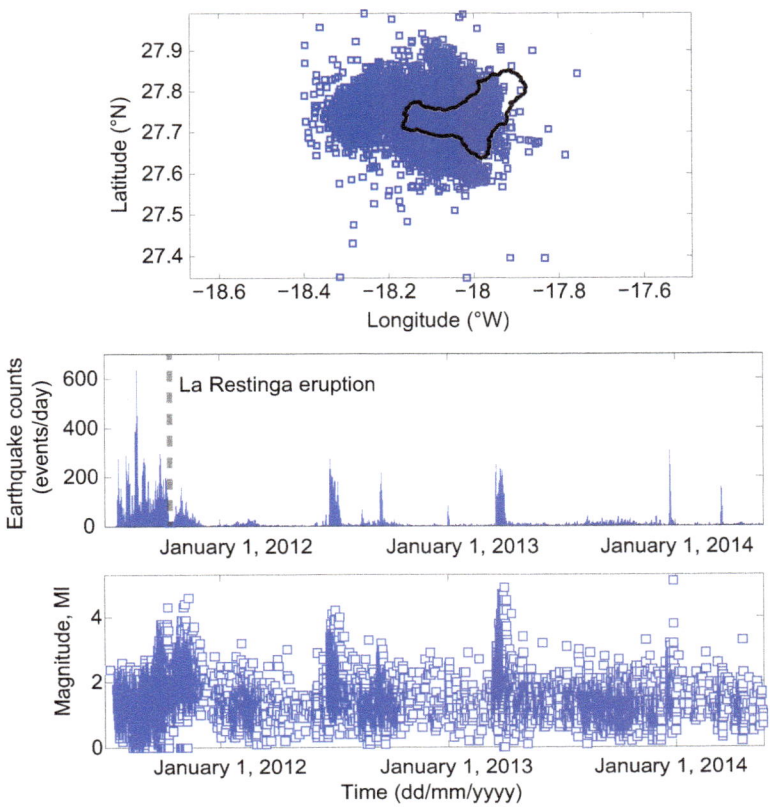

FIGURE 63.16 Seismicity associated with the eruption of El Hierro volcano. Top panel: Seismicity between July 1, 2011, and May 1, 2014, in the El Hierro region. The outline of the island is shown in black. The eruption site was on the sea floor just south of the central part of the island. Middle panel: Earthquake rate in the El Hierro region from July 1, 2011, to May 1, 2014. Bottom panel: Magnitude/time plot of earthquakes in the El Hierro region from July 1, 2011, to May 1, 2014.

on July 21 to allow better detection and location of the seismic events. There was an earthquake swarm with more than 400 minor earthquakes between 20 July and 24 July; by 27 July a further 320 earthquakes had been recorded (Figure 63.16). Geodetic instruments showed that some horizontal deformation had been detected. At that time, the total number of earthquakes had exceeded 4000. By the end of September, the earthquakes had increased in frequency of occurrence and magnitude, with experts fearing landslides affecting the town of La Frontera, and also a small possibility of a volcanic eruption through a new vent. Emergency services evacuated several families in the areas at most risk, and made plans to evacuate the island if necessary. On 8 October the largest earthquake of M = 4.3 occurred near the south part of the island and about 6 km NW of the eventual eruption site. Between 4:15 and 4:20am on October 10, 2011, the earthquake swarm changed behavior and produced harmonic tremor. A small submarine eruption began, 7 km south of La Restinga. As of November 7, 2011, a confirmed Surtseyan type of eruption phase was underway at the fissure on the sea floor. The eruption occurred from October 10, 2011, to March 5,

2012, with vigorous phreatic bubbles emerging, occasional lava fragments, and discoloration visible at the sea surface. Volcanic tremor was recorded with variable intensity during the eruption, and seismicity varied from as few as 6 to as many as 135 earthquakes located per week. Seismicity declined once the eruption was over (Fig. 63.16).

These case histories show a great variety of seismic and infrasonic behavior. The Redoubt case showed the differences between newer (2009) and previously observed (1989) precursors. The Redoubt and Tungurahua cases demonstrated the advantages of using automated data streams especially when hundreds to thousands of events occurred. The Eyjafjallajokull and El Hierro cases show the impact of using seismic data to constrain activity when there was no previous basis for comparison. The Soufriere Hills case demonstrated how observatories frequently need to innovate new techniques as seismicity and hazards change; determining the trajectories and amplitudes of dome failure events provided improved estimates of dome evolution. The Tungarahua, Sarychev, and Cleveland cases showed that eruptions could be characterized and detailed chronologies developed using infrasound data. Indeed without such data, we would not even know that some of the eruptions had occurred.

7. CONCLUSIONS

Although seismology is a powerful tool in terms of eruption monitoring, it is nowhere near as powerful alone as it is when combined with other techniques. Infrasound in particular is an important complement. Additional methods include visual observations, geodesy, gas sampling, satellite images, and others. Nevertheless, seismology is the backbone of most volcano observatories, and a strong seismic monitoring program generally results in effective hazard mitigation. Future efforts will undertake to provide a more physical basis for eruption forecasting, and to continually search new and existing high-quality data sets for better diagnostics. Such data, and well-trained and experienced scientists, will help reduce the potential for disasters at dangerous volcanoes.

Infrasound complements seismic monitoring in its capability to robustly detect and characterize surface activity. Joint infrasound observations permit discrimination of surface from subsurface seismicity. Infrasound records permit quantification of explosive output, qualification of eruptive style, and can be used to track locations of active volcanic vents and determine whether those vents are open or closed. Because atmospheric structure is relatively simple, and the source of infrasound is primarily at a known location (a volcanic vent), infrasound signal evolution and comparison of infrasound from various volcanoes can be very robust. Ongoing study of infrasound includes a focus on relating and modeling characteristic volcano infrasound radiation to other metrics, such as gas and thermal flux, ejection velocity, and seismic radiation. Scientists have now demonstrated the merits of infrasound data for elucidating eruption physics. As more volcano observatories implement multidisciplinary approaches that incorporate infrasound, more long-term monitoring infrasound case studies will become available. As with seismic records, infrasound requires low-bandwidth data acquisition hardware and is suitable for telemetry. It is a low-cost, high-potential technology that allows volcano observatories to monitor both the atmospheric and ground propagating elastic wavefield. The global IMS network is an effective tool for identifying large eruptions, but will be ideally complemented by local and regional installations capable of tracking early and less violent stages of eruptive sequences.

Volcano seismology and infrasound have matured in the past two decades, helped by dramatic improvements in the use of computers and detailed study of many eruptions of different types and sizes. And yet, despite the extraordinary increase in amount and precision of knowledge, there is no guaranteed method for forecasting eruptions. Rather, volcanoes remain complex structures for which a great variety of behaviors are possible. Several general patterns of seismic and infrasound activity and physical models have been discussed here, yet these may be overlooked during crises. This happens for two main reasons. First, during crises there simply is not time to perform many detailed analytical or research tasks. For this reason, most monitoring tasks are designed to be simple and reliable, and hence less prone to error. Many are now routinely automated. Second, scientists tend to be conservative in their decision making, and crises are difficult times to test new hypotheses. Rather, the best observations, which are built on the most secure, previously published foundations, will guide response. This trend sometimes delays response, but prevents unacceptably high-false alarm rates.

Several of the observations and conclusions presented in this chapter deserve additional emphasis. First, it is very important to have a sample of background seismicity and infrasound of several years duration to serve as a benchmark for comparison of possible precursors. This requires establishing seismic and infrasound networks or at least a single station before a volcano shows signs of reawakening. Second, a variety of recording methods are necessary to emphasize different features of the data. These may include visual recorders, event-triggered computers, continuous digital recording, spectrograms, and reduced displacement plots or RSAM. Third, three-component and broadband seismometers are necessary to properly analyze many signals. These provide improved monitoring at the time, and much needed data for later analyses to understand the physical processes. Broadband, high-dynamic-range seismometers are particularly important because short-period instruments always saturate for large eruptions, and because energy at long periods (>1 s) is simply not seen on standard short-period seismometers. Thus our current understanding does not incorporate either the full spectrum or high-amplitude near-source effects. Fourth, monitoring strategies need to be flexible and should not rely exclusively on a single process or type of signal. In spite of better understanding and greatly improved data quality and precision, volcanoes are still capable of new and unexpected activity. Finally, false alarms are inevitable. These may be reduced as patterns and processes become better understood, but the fundamental processes that are monitored using seismology and infrasound are common to both intrusions and eruptions. Progress toward probabilistic methods may help overcome this obstacle.

SEE ALSO THE FOLLOWING ARTICLES

Gas, Plume, and Thermal Monitoring ● Synthesis of Volcano Monitoring ● Volcanic Ash Hazards to Aviation ● Volcanic Seismicity ● Volcano Warnings.

FURTHER READING

Barberi, F., Bertagnini, A., Landi, P., Principe, C., 1992. A review on phreatic eruptions and their precursors. J. Volcanol. Geotherm. Res. 52, 231–246.

Benoit, J.P., McNutt, S.R., 1996. Global volcanic earthquake swarm database 1979–1989. U.S. Geol. Surv. 96–69, 334. Open-File Report.

Bowman, J.R., Baker, G.E., Bahavar, M., 2005. Ambient infrasound noise. Geophys. Res. Lett. 32, L09803.

Cansi, 1995. An automatic seismic event processing for detection and location; the P.M.C.C. method. Geophys. Res. Lett. 22, 1021–1024.

De Angelis, S., Fee, D., Haney, M., Schneider, D., 2012. Detecting hidden volcanic explosions from Mt. Cleveland Volcano, Alaska with infrasound and ground-coupled airwaves. Geophys. Res. Lett. 39, L21312. http://dx.doi.org/10.1029/2012GL053635.

Endo, E.T., Murray, T., 1991. Real-time seismic amplitude measurement (RSAM): a volcano monitoring and prediction tool. Bull. Volcanol. 53, 533–545.

Endo, E.T., Murray, T.L., Power, J.A., 1997. A comparison of pre-eruption real-time seismic amplitude measurements for eruptions at Mount St. Helens, Redoubt Volcano, Mt. Spurr, and Mount Pinatubo. In: Newhall, C., Punongbuyan, R. (Eds.), Fire and Mud. U. of Washington Press.

Fee, D., Garces, M., Steffke, A., 2010. Infrasound from Tungurahua volcano 2006–2008: strombolian to plinian eruptive activity. J. Volcanol. Geotherm. Res. 193, 67–81.

Fee, D., Matoza, R.S., 2013. An overview of volcano infrasound: from hawaiian to plinian, local to global. J. Volcanol. Geotherm. Res. 249, 123–139.

Fee, D., Yokoo, A., Johnson, J.B., 2014. Introduction to an open community infrasound dataset from the actively erupting Sakurajima Volcano, Japan. Seismological Research Letters 85 (6). http://dx.doi.org/10.1785/0220140051.

Green, D.N., Neuberg, J., 2006. Waveform classification of volcanic low-frequency earthquake swarms and its implication at Soufrière Hills Volcano, Montserrat. J. Volcanol. Geotherm. Res. 153 (1), 51–63.

Hill, D.P., et al., 1993. Seismicity remotely triggered by the magnitude 7.3 Landers, California, earthquake. Science 260, 1617–1623.

Hotovec, A.J., Prejean, S.G., Vidale, J.E., Gomberg, J., 2013. Strongly gliding harmonic tremor during the 2009 eruption of Redoubt Volcano. J. Volcanol. Geotherm. Res. 259, 89–99.

Ichihara, M., Takeo, M., Yokoo, A., Oikawa, J., Ohminato, T., 2012. Monitoring volcanic activity using correlation patterns between infrasound and ground motion. Geophys. Res. Lett. 39, L04304.

Johnson, J.B., Ripepe, M., 2011. Volcano infrasound: a review. J. Volcanol. Geotherm. Res. 206, 61–69.

Johnson, J.B., Anderson, J., Marcillo, O., Arrowsmith, S., 2012. Probing local wind and temperature structure using infrasound from Volcan Villarrica (Chile). J. Geophys. Res.: Atmos. 117, D17 (1984–2012).

Jolly, A.D., Thompson, G., Norton, G.E., 2002. Locating pyroclastic flows on Soufriere Hills Volcano, Montserrat, West Indies, using amplitude signals from high dynamic range instruments. J. Volcanol. Geotherm. Res. 118 (3), 299–317.

López, C., Blanco, M.J., Abella, R., Brenes, B., Cabrera Rodríguez, V.M., Casas, B., Villasante-Marcos, V., 2012. Monitoring the volcanic unrest of El Hierro (Canary Islands) before the onset of the 2011–2012 submarine eruption. Geophys. Res. Lett. 39 (13).

Luckett, R., Baptie, B., Neuberg, J., 2002. The relationship between degassing and rockfall signals at Soufriere Hills Volcano, Montserrat. Geol. Soc. Lond. Memoirs 21 (1), 595–602.

Matoza, R., Fee, D., Garces, M., Seiner, J.M., Ramon, P.A., Hedlin, M.A.H., 2009. Infrasonic jet noise from volcanic eruptions. Geophys. Res. Lett. 36, L08303.

Matoza, R., Le Pichon, A., Vergoz, J., Herry, P., Lalande, J.-M., Lee, H.-i., Che, I.-Y., Rybin, A., 2011. Infrasonic observations of the June 2009 Sarychev Peak eruption, Kuril Islands: implications for infrasonic monitoring of explosive volcanism. J. Volcanol. Geotherm. Res. 200, 35–48.

McNutt, S.R., 1992. Volcanic tremor. In: Encyclopedia of Earth System Science, vol. 4, 417–425.

McNutt, S.R., 1994. Volcanic tremor amplitude correlated with eruption explosivity and its potential use in determining ash hazards to aviation. In: Proc 1st int symp on volcanic ash and aviation safety. US Geol Surv Bull, vol. 2047, pp. 377–385.

McNutt, S.R., Thompson, G., West, M.E., Fee, D., Stihler, S., Clark, E., 2013. Local seismic and infrasound observations of the 2009 explosive eruptions of Redoubt Volcano, Alaska. J. Volcanol. Geotherm. Res. 259, 63–76.

Minakami, T., 1960. Fundamental research for predicting volcanic eruptions, part 1. Bull. Earthq. Res. Inst. 38, 497–544.

Neidell, N., Taner, M.T., 1971. Semblance and other coherency measures for multichannel data. Geophysics 36, 482–497.

Newhall, C.G., Dzurisin, D., 1988. Historical Unrest at Large Calderas of the World. USGS Bull, 1855.

Olivieri, M., Clinton, J., 2012. An almost fair comparison between Earthworm and SeisComp3. Seismol. Res. Lett. 83 (4), 720–727.

Ratdomopurbo, A., Widiwijayanti, C., Win, N.T.Z., Chen, L.D., Newhall, C., 2012. WOVOdat progress 2012: installable DB template for volcano monitoring database. In: EGU General Assembly Conference Abstracts, vol. 14, p. 6742.

Szuberla, C.A.L., Olson, J.V., 2004. Uncertainties associated with parameter estimation in atmospheric infrasound arrays. J. Acoust. Soc. Am. 115, 253–258.

Tarasewicz, J., Brandsdóttir, B., White, R.S., Hensch, M., Thorbjarnardóttir, B., 2012. Using microearthquakes to track repeated magma intrusions beneath the Eyjafjallajökull stratovolcano, Iceland. J. Geophys. Res. 117 (B9) (1978–2012).

Thompson, G., West, M.E., 2010. Real-time detection of earthquake swarms at Redoubt Volcano, 2009. Seismol. Res. Lett. 81, 505–513.

Voight, B., 1988. A method for prediction of volcanic eruptions. Nature 332, 125–130.

Chapter 64

Ground Deformation, Gravity, and Magnetics

Jeffrey T. Freymueller

University of Alaska Fairbanks, Fairbanks, AK, USA

John B. Murray and Hazel Rymer

Faculty of Science, The Open University, Walton Hall, Milton Keynes, UK

Corinne A. Locke

University of Auckland, New Zealand

Chapter Outline

GLOSSARY

absolute gravity A direct measurement of the acceleration of gravity, usually by tracking and rise and fall of a mass in a vacuum, with an accuracy of $1-10 \times 10^{-8}$ m s^2.

CGPS An abbreviation for a continuously recording Global Positioning System (GPS) site, that is, one permanently mounted to record for months to years, as compared to a survey marker that is measured on an episodic basis.

Earth tides The displacement of the Earth's surface associated with the change in gravity caused by the relative movement of the Sun and Moon. Earth tide displacements are removed from deformation measurements, and the tidal gravity changes are removed from gravity measurements.

GNSS Global navigation satellite system, a general term for satellite positioning systems such as GPS. Currently active GNSS include Beidou/Compass (China) DORIS (France), Galileo (EU), GLONASS (Russia), GPS (USA), IRNSS (India), and QZSS (Japan). As of this writing, GPS remains the most mature and commonly used of these systems. New generation receiver hardware can track satellites from multiple GNSS systems.

GPS Global Positioning System, a satellite positioning system that uses signals from a constellation of artificial satellites to determine position on the Earth's surface.

The Encyclopedia of Volcanoes. http://dx.doi.org/10.1016/B978-0-12-385938-9.00064-X

Gal, milligal, μGal Units of gravity and gravity change, named after Galileo. 1 Gal = 1 cm s^{-2}, so 1 milligal = 10^{-5} m s^{-2} and 1 μGal = 10^{-8} m s^{-2}. Milligals are commonly used units for gravity anomalies (variations from the reference gravity field), and μGal are commonly used units for gravity changes.

gravimeter Sensitive geophysical instrument used for measuring the relative acceleration due to gravity.

InSAR Interferometric Synthetic Aperture Radar (InSAR) is a remote sensing technique for measuring topography or displacements of the Earth's surface. It requires imaging the surface by radar at least twice from almost the same position in space. The differences in phase of the backscattered radar signal between two images produce interferometric fringes that will represent both the topography of the surface and any changes in position of the surface between the acquisition of the two images.

microgravity Technique using gravimeters at the limit of precision to measure gravity differences of the order 10^{-6}–10^{-8} m s^2.

Tesla, nanoTesla SI units of magnetic field strength or magnetic flux per square meter. The Earth's magnetic field is more conveniently expressed in units of nanoTesla or 10^{-9} T.

piezomagnetic Magnetic properties of rock produced by stress.

Plate Boundary Observatory (PBO) A network, mainly comprised of continuous GPS sites, established by the U.S.A. National Science Foundation to study deformation across the Pacific–North America plate boundary zone. PBO includes networks on several volcanoes, including Mt Augustine and Mt St Helens.

proton precession magnetometer A geophysical instrument which exploits the property that the magnetic moment vector of a proton aligns itself parallel to an applied field. When the applied field is switched off, the magnetic moments precess (like a spinning top) about the remaining (Earth's) magnetic field. The frequency of the precession is proportional to the strength (magnitude) of the field, which provides a measure of the prevailing Earth's magnetic field strength.

strain The spatial gradient of horizontal displacement, strains measure the internal deformation of a body. Linear strains are the fractional change in length, equal to the change in length divided by the original length, and shear strains are related to changes of angles within the body.

tilt The change in slope of the ground surface due to relative vertical motions. Significant tilting is common at volcanoes due to the vertical movements caused by pressurization or depressurization sources. Mathematically, tilt is correlative to strain, being the spatial gradient of the vertical displacement. Tilt is often measured in terms of an angle, with units of microradians (μrad).

Volcanic eruptions are preceded by the upward migration of magma, and sometimes by the long-term accumulation of magma in the shallow subsurface. Upward migration of magma causes pressure changes within magma or hydrothermal fluid reservoirs, which cause ground deformation that can be measured as displacement at the surface or as tilt or strain. Deformation measurements are made using Global Positioning System (GPS) and InSAR (interferometric synthetic aperture radar), and sometimes also using terrestrial surveying techniques. Displacements from volcanic sources can range from millimeters to a few meters. Tiltmeters and strainmeters are far more sensitive than GPS and InSAR to small deformations, but suffer from instrumental drift and require the removal of Earth tides and many smaller nonvolcanic signals; however, they can measure rapid changes in deformation very well. Local changes in subsurface density may be produced by magma movements (intrusion/withdrawal) and by variations in the degree of magma vesiculation. These changes can be detected by measuring variations in the acceleration due to gravity. Variations on the order of 10–100s of μGal are typical. Magmatic processes may also induce changes in the total magnetic field strength. Magnetometers can measure field strength to 0.1 nT and fluctuations of a few nanoTesla to a few thousand nanoTesla are typical at volcanoes. Measurements of gravity, magnetic fields, and ground deformation can be made during and between eruptions, providing information on the processes going on inside a volcano when it is not erupting.

1. PRINCIPLES OF GROUND DEFORMATION MONITORING

For many volcanoes, ground deformation is among the most useful indicators of the general state of the volcano and of any impending eruption. Magmatic pressure increases cause the ground surface to move upward and away from the pressure source, with the reverse pattern occurring with a pressure decrease. Thus in simple terms volcanoes swell up or inflate before erupting and deflate during eruption. The details of this process and the timescales involved vary from volcano to volcano, and the time history of pressure changes and deformation can be complex. Some volcanoes accumulate large volumes of magma in the shallow subsurface between eruptions, causing large deformation and gravity change signals observable over years, while in other cases relatively small changes are seen before eruption. Deformation measurements are valuable both for understanding where magma resides at depth and for monitoring during volcanic crises.

The use of ground deformation for studying and monitoring volcanoes dates back more than a century, for example, at the Hawaiian Volcano Observatory in the USA and the Osservatorio Vesuviano in Italy. Before the 1990s, deformation was measured by surveying methods such as leveling (which measures relative vertical heights), distance measurement and theodolite angular measurement (which give horizontal measurements of position), and various types of tiltmeters (which measure ground tilt only). Since then, GPS has overtaken terrestrial surveying as the fastest and most precise way of measuring positions in situ. Leveling remains more precise than GPS vertical over the distances typically used for volcanic studies, but is much more time consuming. However, all of these techniques require that an instrument be taken to a site in the field, which may be costly and sometimes impossible due to unsafe conditions. InSAR requires no presence on the ground, only satellite imagery, and has become an essential tool for volcano deformation studies, especially for remote volcanoes. The chief limitation of InSAR has been that its measurements are sparse in time (typically every few weeks), although new satellite missions promise to allow measurements to be made more frequently. Cross-correlation of optical or radar imagery is also used to provide measurements of surface displacements, and recently developed software tools now make this straightforward.

Surface deformation measurements are based on repeat measurements of positions on the surface, and detection of displacements due to ground deformation depends upon repeat measurement of exactly the same point on the ground. In situ measurement points are marked by a permanent GPS antenna mount, permanent survey markers in the ground, or other unambiguous indicators like a specific pattern of drill holes in rock. InSAR, on the other hand, measures the motion of one or more dominant scattering objects within each pixel, rather than the motion of a survey marker. The deformation can be expressed as displacements or velocities (displacement rates).

All deformation measurements are made relative to some reference. For example, terrestrial measurements require a local reference station. GPS measures the relative positions of two GPS sites, although one site can be located far from the volcano. Using today's high-precision orbit and clock products, GPS positions can be measured relative to a global reference network, which makes them nearly like absolute positions. InSAR measures motions relative to a reference pixel, or such that the average deformation is zero (or a similar constraint). As a general rule, the reference point should be as close to the area studied as possible without being involved in the movement under investigation. If the reference point is located too far away, nonvolcanic deformation due to tectonic motions may be included in the deformation. If the reference point is too close, it may also be moving and this needs to be accounted for in modeling.

In situ ground deformation measurements can be made episodically or continuously. The former mode was the earliest approach and remains common, mostly because many sites with a high spatial density can be measured with a small number of instruments and personnel. However, continuous measurements are becoming more common and a combination of the two is often employed now.

The earliest continuous measurements of deformation came from tiltmeters and strainmeters. Continuous GPS (CGPS) measurements are now common, as the cost of the instrumentation has became low enough. Tiltmeters or strainmeters measure strain or ground tilt continuously.

Such records provide immediate information on events within the interior of the volcano. Episodes of magma injection, for example, can be followed in detail as they unfold, and increasing ground tilt can be used to help predict the time of an impending eruption. CGPS stations have been installed on many volcanoes and provide continuous information on how the edifice is deforming with time. If enough of these are installed, then they can provide measurements that are dense in both space and time, thus providing an ideal monitoring system.

Field surveying methods take varying amounts of time. A kinematic GPS traverse can be completed within several hours if rapid transportation is available, but more precise GPS surveys take a few days, and a complete survey of many points using GPS or a terrestrial surveying method can take several days to weeks. It is therefore not usually practicable to measure the network more than once a month; financial and logistic considerations often mean that once or twice a year is nearer the norm for many volcanoes, and less frequently for remote or logistically challenging volcanoes.

2. TECHNIQUES USED IN THE FIELD

2.1. GPS and Other GNSS Positioning

The GPS is now the most accurate way of measuring horizontal and vertical positions. As of 2014, it is the most mature and reliable global navigation satellite system (GNSS). A great advantage of GNSS (Figure 64.1) is that it can be carried out in any weather, including rain, snow, fog, and high winds, which would preclude conventional surveying altogether. Atmospheric path delays (especially the effects of water vapor) are estimated along with the site positions and (largely) removed in the processing of the GNSS data. Residual atmospheric errors depend on the amount and spatial/temporal variability of water vapor in the atmosphere and are greater in the tropics.

GNSS provides a constellation of satellites (currently about 30 for GPS) that broadcast radio navigation signals used by a receiver to determine its position and precise time. The fundamental GNSS observations are a set of distances to each visible satellite, derived from correlation of timing codes on the satellite signal and from phase changes on the underlying carrier signals. To derive positions, these measurements are combined with external products: ephemerides for the satellite orbits and estimates of the satellite clock errors. The precision and accuracy of positioning is dependent mainly on the quality of these products. The satellites broadcast ephemerides and clock corrections sufficient for single receiver point positioning with an accuracy of a few meters. Far more accurate ephemerides and clock corrections are available from the

FIGURE 64.1   Example of a campaign GPS site, a Trimble dual-frequency GPS receiver in use inside the caldera of Okmok Volcano, Aleutian Islands, Alaska. The antenna is set up directly above a ground marker and the data stored in the receiver (not shown). *Photo by J. Freymueller.*

International GNSS Service (IGS), among other sources. Data from two simultaneously measured sites can be combined to measure the vector between the sites; this provides high-precision measurements and does not require knowledge of satellite clock corrections. Multi-GNSS receivers, which observe both GPS and other GNSS, are becoming common, and before 2020, we can expect that use of multi-GNSS signals will become routine, with a substantial improvement in precision and accuracy over the use of GPS alone.

Positions can be averaged over a day or several hours (static mode), or as a series of independent positions at each epoch of data (kinematic mode). As of 2014, postprocessed static daily point positions have an accuracy of ~ 1 mm for horizontal and ~ 3 mm for vertical components over most of the world, when the best products and analysis techniques are used, and somewhat more accurate measurements of the vector between two sites can be made over distances up to a few hundred kilometers. Vertical positions are never as accurate as horizontal positions because the satellites are only on one side (above the instrument) so the observing geometry is weaker, and because the vertical position is strongly correlated with atmospheric path delays that also must be estimated in the processing. Postprocessed kinematic positions are at least an order of magnitude less accurate than daily static positions, although displacements over short time periods (seconds to minutes) can be measured at the ~ 10 mm level or better,

and a similar accuracy can be reached for relative kinematic positions over distances of tens to a few hundred kilometers. Larson et al. (2010) discuss data processing and filtering approaches for the application of kinematic GNSS positions to volcano monitoring. Kinematic surveys can operate in a "stop and go" mode, in which the position is tracked continuously but averaged while the receiver is stopped at a site. Real-time kinematic surveys use two nearby receivers connected by a radio (or the Internet), so that the moving (rover) station can use the data from the base station to compute the vector between the sites in real time. In this case, the data processing is done inside the receiver.

Real-time GNSS processing is becoming increasingly robust and accurate. Accurate ephemerides are now available for GPS in real time from the IGS and commercial vendors, and the accuracy of real-time clock products is increasing. Real-time point positioning techniques currently provide positions with an accuracy of $\sim 20-50$ mm at best, limited primarily by the precision and accuracy of real-time satellite clock corrections. These have improved substantially in recent years, by increasing the number of real-time continuous stations and improving the reliability of real-time processing, and before 2020, it is likely that real-time point positioning accuracy will approach the 10 mm level globally.

The IGS, an international cooperative service, offers access to a global set of stations, and extremely precise orbit ephemerides and satellite clock corrections. Standard GPS processing using IGS products give coordinates in the International Terrestrial Reference Frame (ITRF), an accurate and stable global coordinate system. Changes in position over time can be expressed as displacements or velocities in ITRF, relative to the motion of a tectonic plate whose motion in ITRF is known, relative to a particular measurement site, or relative to the average of a set of sites. Motions relative to a site or set of sites are often used in volcanic deformation studies so that tectonic or other nonvolcanic motions are not included in the data.

Over the last two decades, an increasing number of permanent or semipermanent CGPS stations (Figure 64.2) that record measurements continuously have been established for surveying, scientific, and other purposes. The total number of installed CGPS stations globally numbers well over 10,000, with hundreds installed on or near-active volcanoes. Many volcanoes are remote and pose logistical difficulties, so repeated temporary survey measurements of survey markers in the ground are frequently used to augment CGPS installations. The temporal coverage and resolution of CGPS data make continuous measurements more powerful than occasional or episodic measurements, so CGPS data are preferred when possible.

2.2. Interferometric Synthetic Aperture Radar

InSAR can provide information on ground deformation at any point in the world without the need to deploy any instrumentation on the ground. Imaging radars send a radar pulse from satellite to ground and record the backscattered radar energy. Two satellite radar images taken at different times but from almost the same point in space are then compared. The differences in phase between the two images produce interferograms (Figure 64.3), interferometric fringes that represent both the topography of the surface and any changes in position of the surface during the period between the acquisition of the two images, that is, any deformation that has taken place. The interferogram also includes phase variations due to additional factors that processing tries to minimize, including orbital error, ionospheric and tropospheric path delays, and noise. For most of the world, the topography is known well from sources like the Shuttle Radar Topography Mission or other digital elevation models (DEM). The effect of topography can be removed from the interferograms as long as the topography and satellite orbits are well known. The end product of InSAR is essentially a map of the volcano showing contours of displacement in the satellite look direction between two radar images.

FIGURE 64.2 Examples of continuous GPS sites. (left) Continuous GPS site OKNC on Okmok volcano, Aleutian Islands, Alaska. The GPS antenna is mounted on a braced structure anchored in rock, with a separate hut for the receiver and batteries (behind). Solar panels run the instrument and its colocated seismometer. The ash deposits here are from a 2008 eruption from the cone in the background (see case study). Photo by M. Kaufman. (right) A "spider" continuous GPS instrument is deployed in the crater of Mt St Helens by a US Geological Survey (USGS) scientist. Spiders are temporary continuous instrument packages, featuring a GPS receiver and other instruments with a radio for communications. GPS, *USGS Photo.*

FIGURE 64.3 Interferogram showing the 1997 coeruptive deflation from Okmok Volcano, Aleutian Islands, Alaska. Each fringe corresponds to 28 mm, or half the C-band wavelength, of deformation in the LOS direction. The inset shows the region inside the caldera. A. Original data. B. Inset of area within the caldera. C. Model fit to the data using a combination of a Mogi model and a sill (see section 3). D. Inset of model predictions. LOS, line of sight. *After Mann et al. (2002).*

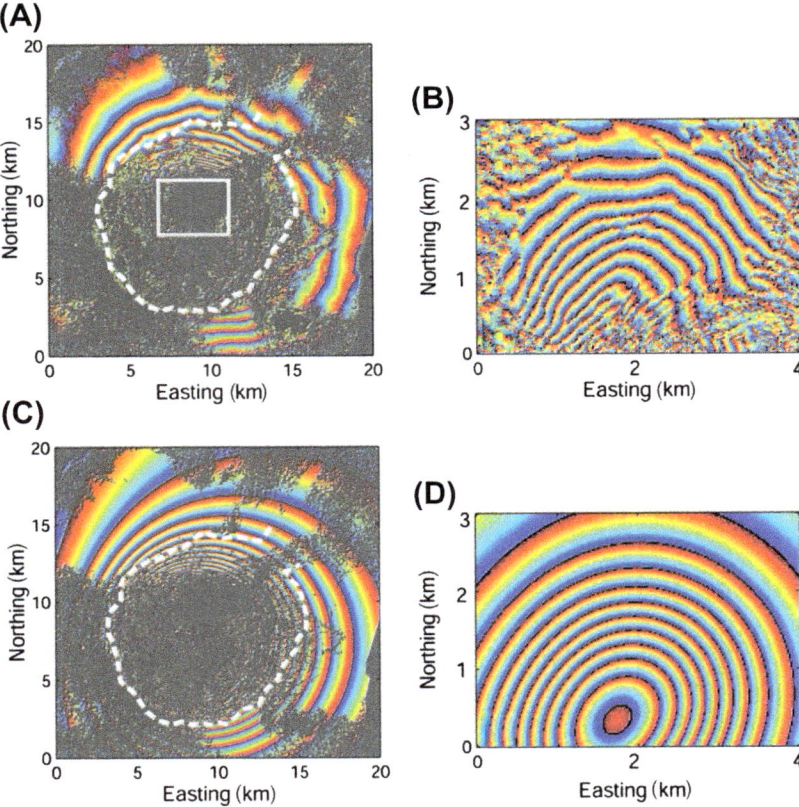

This technique is particularly important for remote or rarely monitored volcanoes where no other means of deformation monitoring are possible. Lu and Dzurisin (2014) demonstrated the power of the InSAR technique for the study and monitoring of remote volcanoes in the Aleutian arc, Alaska. Most past and present radar satellites have used an orbit repeat interval of on order 1 month and a variety of wavelengths for the radar signal, most commonly C-band (~6 cm) and L-band (~20 cm) and X-band (~3 cm). Short wavelength radars are more prone to signal loss due to surface decorrelation than are long wavelength radars. Longer wavelength radars can more easily see through vegetation (see the discussion of decorrelation, below), which is of great importance on many volcanoes. However, longer wavelength radars are more sensitive to ionospheric path delay errors.

InSAR displacements are measured in one component only, the satellite look direction, generally called the line of sight (LOS). The look direction is ~30−40° off-vertical, depending on the satellite involved. Thus InSAR tends to measure vertical deformation very effectively (although combined with one horizontal direction), and is completely blind to displacements in the orthogonal horizontal direction. Sometimes interferograms from different passes, having significantly different LOS directions, can be combined to separate the horizontal and vertical deformation. InSAR deformation measurements are biased by tropospheric path delays (mainly due to water vapor but also pressure, etc.), and no usable data are recovered in places where changes in the ground surface, vegetation, or snow cover cause the surface backscattering characteristics to change from one satellite pass to the next (decorrelation).

Decorrelation results when the phase changes of adjacent pixels are inconsistent with each other; it is mainly caused by scattering objects within each pixel moving relative to each other. Short wavelength radars decorrelate more easily because the waves scatter off of smaller objects, which are more prone to random motion. For example, short wavelength radars may scatter off of objects like leaves, which will generally change position from pass to pass. Future planned radar satellites will use wider ground swaths, steerable antennas, and/or multiple satellites to image the ground more frequently to improve temporal resolution and mitigate the problem of decorrelation.

Although in principle the measurement precision of the phase changes is at the millimeter level, the accuracy of the measurements is substantially worse due to the impact of signal path delays through the atmosphere and residual topographic and orbit errors. Accuracy in many cases is at the centimeter level, but biases at the decimeter level can occur when dense concentrations of water vapor are present. Unlike with GPS, atmospheric path delays can only be removed from interferograms by filtering, although attempts have been made to remove them through external calibration using tropospheric path delays estimated by GPS or weather models. In extreme cases, ionospheric path delays can distort the image enough to make it essentially unusable for measurements of deformation.

InSAR time series techniques use a large set of radar images and construct many interferograms that overlap in time. The set of interferograms is then used to estimate a time series of deformation at each coherent pixel. These techniques attempt to use filtering to remove biases from atmospheric path delays and separate them from deformation signals (e.g., Hooper et al., 2012).

Terrestrial laser or radar scanners are starting to be used to map out deformation at volcanoes. These devices use a similar principle to satellite-based InSAR, but are mounted at an observation point on the ground. Currently they are more commonly used to map out the shape of the ground surface, but repeat measurements can measure deformation in the direction of the laser or radar scan. This tool requires an accessible location for the instrument with LOS to the volcano, but provides a way to measure rapid deformation or motion because many images can be acquired quickly. These measurements are also affected by the atmospheric path delay on the LOS to the volcano, which depends on the distance to the target (it is negligible for nearby targets).

2.3. Tiltmeters and Strainmeters

Various kinds of tiltmeters and strainmeters have been deployed on volcanoes, the first being a horizontal pendulum seismometer that proved to be an effective tiltmeter installed at Kilauea volcano, Hawaii, in 1912. Other tilt sensors used over the years have included water-tube (hydrostatic) tilt instruments and borehole tiltmeters. Strainmeters are designed to measure one or more components of the strain tensor and are commonly borehole instruments. Tilt and strain sensors have been installed at many volcanoes to provide a continuous record of deformation. They are very sensitive to both deformation and environmental changes such as temperature and rainfall, usually suffer a "running-in" period of rapid instrumental drift after initial installation and often suffer from long-period instrumental drift.

Nevertheless, these instruments can be extremely valuable due to their sensitivity and high-time resolution. Processes such as the final ascent of magma at the beginning of a large explosion can be measured and modeled accurately, giving important constraints on the timing and rate of magma ascent. Because the number of such instruments at any volcano is usually small, their placement is critical and may require knowledge of the expected pattern of deformation so that data can be readily interpreted.

2.4. Precision Leveling and Conventional Ground Surveying

Leveling is the most accurate method of deriving changes in relative heights over short distances. It was first used for volcanic studies in Japan, to investigate the 1910 eruption of Usu volcano and the 1914 eruption of Sakurajima volcano. In its simplest form, leveling involves the use of a vertical graduated measuring staff and an instrument called a level, consisting of a telescope with a central crosshair setup to read exactly horizontally. Using this technique, relative height accuracies of 0.8 mm over 1 km are obtainable. Leveling is time-consuming and labor-intensive, which makes its application costly relative to GPS. Today it is used mostly for specialty applications, where vertical displacements occur over short distances and submillimeter accuracy is required. For measuring vertical motions over distances of more than a few kilometers, it has largely been displaced by GPS.

Trigonometric leveling, triangulation, and trilateration (determining angles or distances between sets of markers) were in use to measure horizontal and three-dimensional changes in position of networks of benchmarks at volcanoes prior to the mid-1990s. Distances were measured by electronic distance measurement (EDM) and horizontal or vertical angles read with a theodolite. A total station combines an EDM and a theodolite and these instruments are still used to survey or map out topography in 3D. These measurements require LOS to all targets, which makes their use time-consuming and labor-intensive and thus expensive. Advances in GPS techniques and the present low cost of GPS receivers have made these methods of deformation measurement obsolete except where dangerous activity prevents occupation of a remote station. In such cases continued measurements of a previously installed target can still be valuable.

2.5. Dry Tilt

Dry tilt is the colloquial name given to a method of measuring changes in tilt of the ground, which employs the same optical instruments as leveling. The technique came to be known as "dry tilt" by way of contrast with the alternative of water-tube tiltmeter measurements. In its simplest form, a dry tilt station consists of at least three benchmarks arranged in a roughly equilateral triangle with sides of about 50 m. The level is set up in the center and the staff placed on each of the three benchmarks in turn with their relative altitudes measured to a high precision. Next time the station is measured, any tilt of the ground in the intervening period will show up as a slight change in relative altitudes of the three benchmarks.

Dry tilt stations are cheap and quick to measure and can be occupied several times a day during periods of crisis. For poorly funded or infrequently visited volcanoes, they are a simple means of providing rudimentary information on deformation.

3. MODELING DEFORMATION

The work of the Japanese scientist K. Mogi, published in 1958, marks the beginning of modern deformation studies on volcanoes because he developed a simple mathematical model to compute the deformation due to a volcanic source. His model described the effect at the surface of a change in pressure within a small, spherical source inside an elastic half-space (Figure 64.4), simulating the effect of increasing magma pressure within a buried magma chamber. Despite the fact that volcanoes are not strictly elastic in behavior, nor are their surfaces flat and horizontal, many volcanoes show behavior surprisingly close to Mogi's model. Mogi applied his model successfully to Sakurajima volcano, Japan and Kilauea volcano, Hawaii. Common volcanic source models used today include (1) Mogi or similar pressure sources to represent magma bodies/chambers at depth or pressure changes in hydrothermal systems; (2) injection of tabular bodies like a dike or sill; (3) subsidence due to lava loading, gravitational settling, or spreading of the whole volcano; and (4) slope movement caused by slope creep prior to failure or by magma pressure variations on steep slopes. Combinations of these sources can represent a complex pattern of deformation. More complex and realistic rheological models are also used in modeling including layered elastic spaces or fully 3D elastic variations.

The Mogi model has a particularly simple analytical solution and is described below. Pressure changes in a spherical or spheroidal source volume produces deformation with a roughly radial symmetry, and most examples of volcanic deformation involve sites moving toward or away from the volcano. The deformation due to a dike or sill can be modeled using an opening mode dislocation within an elastic half-space, using the formulation of Okada (1985). Very long dikes can be approximated using simpler 2D dislocation formulae. For a vertical dike, points on either side of the dike will be displaced away from the dike, normal to its long axis, as it inflates. A more complete summary of volcanic source models, along with a useful library of computer code, can be found in Battaglia et al. (2013). The mathematical basis for many of these models is developed in Segall (2010).

As illustrated in the case study of Krafla below, it is important to recognize when the assumptions behind the models are be valid. For example, strains in volcanic regions can exceed the elastic limit of the rocks resulting in permanent and pervasive ground cracking.

The Mogi model predicts that horizontal displacements will be exactly radially symmetric (Figures 64.4 and 64.5), with a simple form for the vertical and horizontal displacements:

$$\Delta d = \frac{3a^3 P d}{4\mu (f^2 + d^2)^{3/2}}$$

$$\Delta h = \frac{3a^3 P f}{4\mu (f^2 + d^2)^{3/2}},$$

where

- a = radius of the source sphere
- P = change in pressure in the sphere
- f = depth to the center of the sphere
- μ = Lamé's constant, shear modulus
- d = radial distance on the surface from point above source
- Δd = radial horizontal displacement of a point at the surface
- Δh = vertical displacement of a point at the surface

The multiplicative constants common to both of these expressions are often combined into a single constant commonly called the source strength, which is proportional

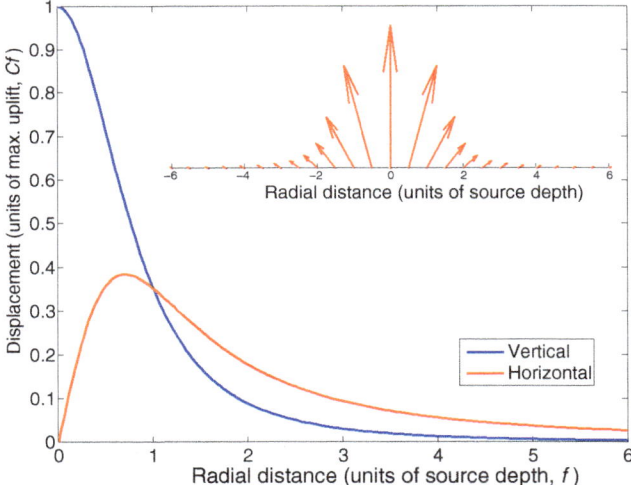

FIGURE 64.4   Mogi model displacements. The blue and red curves show the vertical and horizontal displacements, respectively. The x-axis is scaled in units of source depth f, while the y-axis is scaled in units of the maximum vertical displacements Cf. The inset shows the displacement vectors along a radial cross-section.

(A)

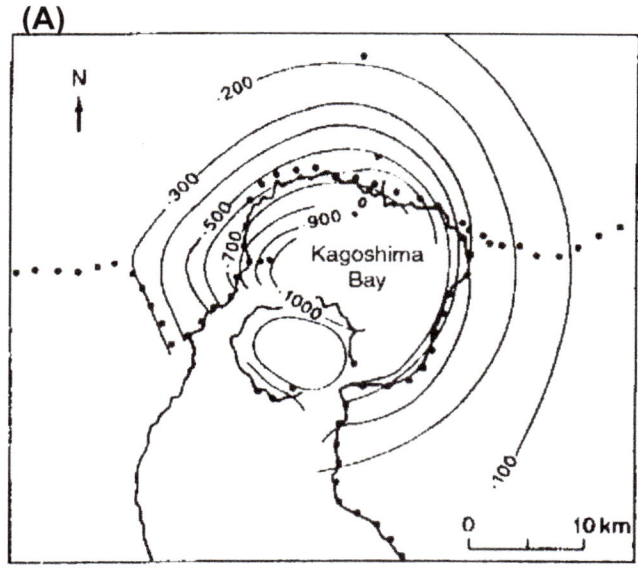

(B)

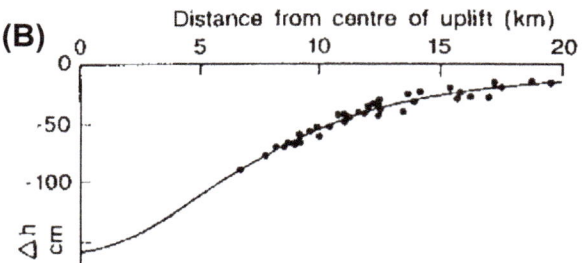

FIGURE 64.5 (A) Map and diagram of the 1914 deflation episode at Sakurajima volcano, Japan, from Mogi (1958). The map above is overlaid with contours of deformation at 100-mm intervals, interpolated between values of vertical displacement measured at benchmarks around the volcano. A benchmark distant from the volcano has been used as a zero reference. Below, vertical displacement is plotted against distance from the volcano in kilometers. The black dots show displacement at individual benchmarks, while the curves show the theoretical displacement expected from a buried pressure source according to Mogi's (1958) model. The center of deflation lies not at the volcano itself, but in the center of Kagoshima Bay, the adjacent caldera. The curve in (B) is for a pressure source at a 10-km depth.

to the volume of the body and the pressure change, and inversely proportional to the shear modulus:

$$C = \frac{3a^3 P}{4\mu}.$$

The volume change of the subsidence bowl or uplift bulge at the surface is equal to $2\pi C$. The volume change of the magma body at depth is proportional to the source strength.

$$\Delta V = \frac{4\pi}{3} C = \frac{\pi a^3 P}{\mu}.$$

Note that this volume change represents the change in volume of the cavity rather than the volume change of magma at ambient pressure. If magma contains exsolved

volatiles it may have a significant compressibility, so the relationship between volume change ΔV and magma mass change ΔM is not always simple.

Models such as the Mogi model are typically used to describe kinematic snapshots of the deformation at a volcano, but physics-based models for magma flow are becoming more prevalent. Segall (2013) discusses two main categories of such models, those that model pressure-driven fluid flow between a magma body and conduit, and those that attempt to integrate volcano deformation and seismicity through changes in stress. In the absence of magma recharge, a simple fluid flow model predicts that the magma volume flux (or intrusion) should decay exponentially with time such that the total erupted volume will be proportional to the initial overpressure. Such models have been borne out by recent observations (Hreinsdóttir et al., 2014).

4. CASE HISTORIES OF DEFORMATION

The examples given here were selected to illustrate the range of deformation measurement tools introduced in this chapter. Recent studies of volcanic deformation are heavily dominated by GPS and InSAR.

4.1. Sakurajima Volcano (Japan, 1914)

Six months after a large eruption in 1914, leveling benchmarks around Sakurajima volcano, which had been installed and leveled in 1895 as part of a countrywide survey, were resurveyed. The displacements showed a broad, roughly circular subsidence bowl of more than a meter amplitude and at least 60-km wide, not centered on the volcano itself, but roughly concentric with the center of Kagoshima Bay, an old caldera. In the 1950s, Mogi plotted the displacement against the distance from the center of subsidence, and found that the data fitted very closely to the deformation expected from elastic deformation caused by a buried source, from which he was able to develop his model. Mogi found that the Sakurajima data were consistent with a depth to source (f) of 10 ± 1 km (Figure 64.5).

4.2. Krafla Volcano (Iceland, 1978)

The Krafla fissure swarm in northern Iceland began erupting in December 1975 after 230 years of dormancy. There were nine eruptions in the following 9 years. The accompanying deformation patterns showed repeated inflation and deflation sequences, essentially of the Mogi type, centered on the 9-km-wide Krafla caldera. It appears that a magma chamber was being supplied with magma over this 9-year period, and that the consequent steady inflation of the chamber was punctuated by about 20 rapid deflation events. During each deflation event, vertical fissures crossing the area widened and filled with magma from the reservoir, forming dikes, which extended rapidly to distances up to 50 km from the caldera center. Eleven of these dikes did not reach the surface, but remained as intrusions, and these were detected through the deformation they caused.

Characteristic displacements perpendicular to the strike of the dike are expected on the ground surface above a vertical or near-vertical dike during emplacement. The vertical displacement increases rapidly with distance from the dike, reaches a peak value, and then decreases slowly toward an asymptote of zero displacement (Figure 64.6(A)). In the case of vertical dikes, this displacement is symmetrical about the center of the dike. Analysis of vertical displacements measured by leveling associated with one of the intrusive events at Krafla (that of January 7, 1978) indicated that the intrusion was consistent with a vertical dike 5.5-km in height, whose center lay at a depth of 3 km. The measured values in Figure 64.6(A) depart from the theoretical curve only near the dike itself, where the theoretical horizontal stresses are tensile (Figure 64.6(B)), and many faults and cracks were visible at the surface. Pervasive ground cracking indicates that the ground in this zone was not behaving elastically and thus would not be expected to conform to the model.

4.3. Mt St Helens (1980 and 2004–2008)

When a volcano builds up by the steady accumulation of lavas and pyroclastics, its slopes become progressively steeper and prone to slope failure, where a large section of the volcanic edifice may peel off and slide downslope in a sector collapse. Flank collapses can also be triggered by the expansion of pressure sources within a volcano, leading to a flank collapse and lateral blast. The most famous example of this kind of event was the Mt St Helens eruption of May 18, 1980, where nearly 3 km^3 of the northern slopes of the volcano collapsed and slid off the volcano, taking the pressure off the subsurface magma body and provoking a large pyroclastic eruption. A similar event occurred at Bezymianny volcano, Kamchatka, in 1956. Before such catastrophic events occur, slope creep and other precursory movements may take place on the unstable slope and may be capable of indicating both where and when failure will take place.

Slope movements prior to the 1980 Mt St Helens lateral blast were so great that they were first detected from air photographs, where the changes were plainly visible. Air photographs had been taken the previous year, on August 15, 1979, and were taken again on April 7 and 12, 1980 and May 12, 1980. Between each of these dates, there was subsidence close to the summit and broad inflation about 1-km downslope on the northern flank. The amount of subsidence had totaled more than 70 m by May 12, the

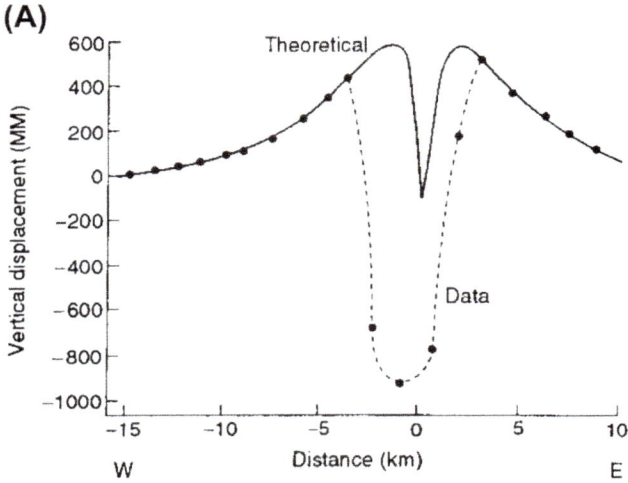

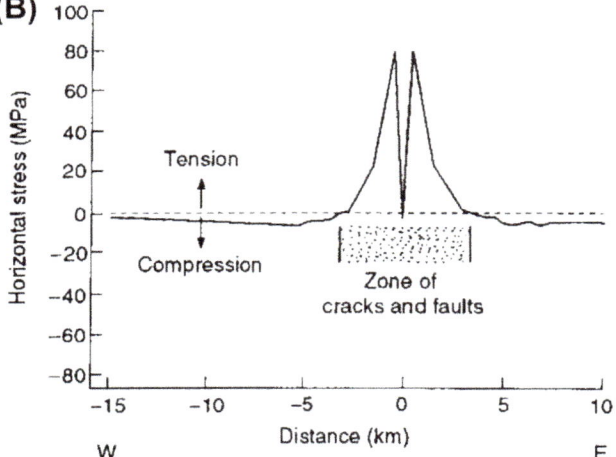

FIGURE 64.6 (A) Plot of vertical displacement against distance from a dike emplaced during a rifting event north of Krafla volcano, Iceland, on January 7, 1978. Data points closely follow the theoretical curve (solid line) for a vertical dike whose center lies at a 3-km depth, and whose height is 5.5 km. The fit is good apart from close to the dike itself, where tensional stress is calculated to be very high (see (B)). (B) Calculated horizontal stress in MPa plotted against distance from the dike modeled in (A).

subsiding area forming an expanding trough or graben on the north side of the summit aligned approximately east–west. This graben was 1.5-km long and grew from 360 m wide on April 7 to about 600 m wide on May 12. The inflation further down eventually raised some areas more than 150 m above the preeruption topography (Figure 64.7), forming a prominent bulge nearly 2 km in diameter on the northern flank, which became increasingly obvious with time.

Dry tilt stations were measured frequently during the crisis, and in some cases showed large short-term fluctuations in tilt, measured at up to 50 μrad per hour on April 10. At the same time, distance measurements to reflectors installed at several points on the upper slopes of the

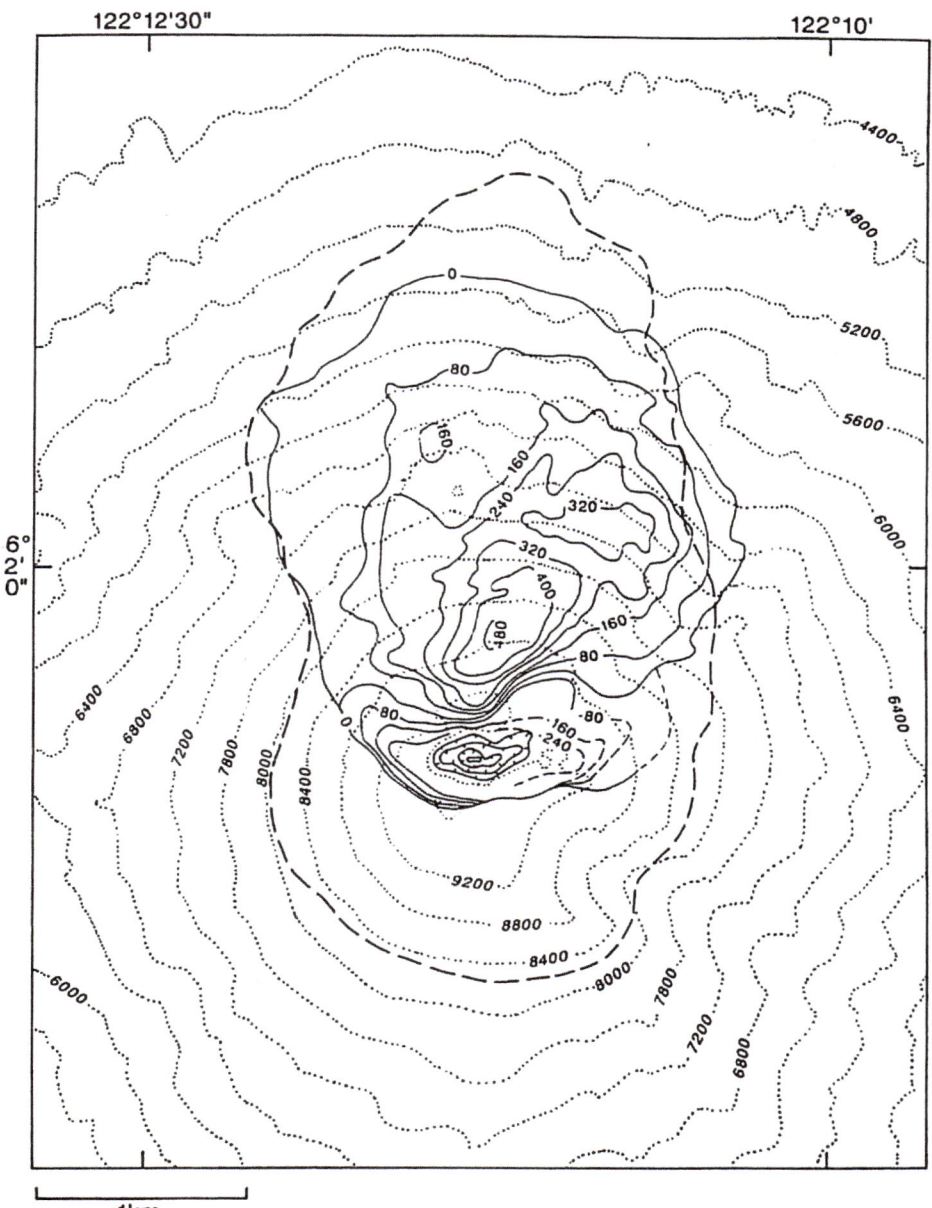

122°12'30"

122°10'

6°
2'
0"

1km

FIGURE 64.7 Map showing topography (dotted lines) and elevation changes in feet (solid lines) between August 15, 1979 and May 12, 1980 at Mt St Helens volcano, USA, before the catastrophic slope failure and pyroclastic eruption of May 18. The bold dashed line shows the position of the crater rim that formed in the May 18 eruption.

mountain showed that stations on the bulge were moving north at a fairly constant rate of up to 2.5 m per day. Many reflectors were in fact moving in an almost horizontal direction, which carried them north into regions well above the preeruption slope that made them appear as inflated areas on the topographic difference maps. The enormous rates of ground movement began only 8 weeks before the eruption and continued at a fairly constant rate until the moment of eruption. This indicates a different behavior from slope creep observed before some types of slope failure, where a progressively increasing rate of deformation is observed that can be used to predict the time of failure.

Sector collapses and their associated pyroclastic flows are among the most dangerous and destructive types of volcanic activity as they occur suddenly and progress very rapidly, and are capable of wiping out towns as far as 100-km distant from the volcano in some cases. The Mt St Helens eruption of May 18, 1980 is the only example of sector collapse on a volcano where monitoring has been carried out both before and after the event, and as such contains important information to bear in mind when facing possible future sector collapses on other volcanoes.

After the cataclysmic 1980 eruption, St Helens erupted until 1986, with a series of episodes that together extruded a significant lava dome ~300 m high within the collapse crater. A new eruption commenced in September 2004 and renewed lava dome extrusion continued until January 2008. According to the US Geological Survey, about 7% of the crater volume has been refilled since the 1980 eruption.

The 2004–2008 eruption began with essentially no measured deformation precursors. At the time of the eruption, the closest CGPS site was located ~ 20 km away. Eleven new CGPS sites were installed within a few weeks of the start of the eruption by the US National Science Foundation's Plate Boundary Observatory (PBO) project, and by the end of the eruption 25 CGPS sites were in place. Ground deformation associated with the 2004–2008 eruption was subtle outside of the crater, only a few centimeters maximum. Displacements were inward (radial) and downward, consistent with the deflation of a magma source at ~ 7-km depth. A vertically elongated source fit the data better than a spherical source. Inside the crater, displacements were extremely large. A series of disposable sensor packages, called "spiders" due to their multilegged appearance, were deployed inside the crater. The spiders included a low-cost CGPS instrument to measure deformation and other monitoring sensors (Figure 64.2, right).

4.4. Montserrat (2001–2008)

Soufriére Hills Volcano on the island of Montserrat in the Lesser Antilles began to erupt in 1995. The eruption featured a series of dome-building episodes and subsequent dome collapses and pyroclastic flows, which resulted in the destruction of the capital city and the evacuation of most residents of the island. The eruption was studied from its early stages by campaign and semicontinuous GPS measurements and later a CGPS network. In general, the volcano deflated during episodes of dome-building or other lava effusion, and inflated during periods of apparent repose; several such cycles occurred over the duration of eruption. The inflation and deflation source was found to be located at ~ 10-km depth for most of these episodes.

Three borehole instrument sites were installed in 2002–2003, which included a Sacks-Evertson dilatometer, a seismometer, and a tiltmeter. CGPS stations were installed at the surface of each site. These instruments observed a particularly large dome collapse in July 2003, along with an explosion on March 3, 2004 and several other events. The 2003 dome collapse resulted in pressurization of a magma source at ~ 5-km depth below sea level, detected by its strain signal. This was interpreted as being due to rapid exsolution of bubbles in a subsurface magma body, triggered by the abrupt removal of the load of the dome. Strainmeter data from the March 3, 2004 explosion event required a compound source, involving a reduction of pressure in a magma body at ~ 5-km depth and

opening/propagation of a dike upward from this magma body to within ~ 1.4 km of the surface (Figure 64.8). Both of these changes occurred over durations of a few hundred seconds, and could be measured because of the high precision and time resolution of the strainmeters.

4.5. Augustine (2006)

The 2006 eruption of Mt Augustine, Alaska provided an outstanding example of how near-real-time GPS data can provide important monitoring information during an eruption. Augustine erupted in January 2006 after 8.5 months of precursory unrest. Eruptive activity began with a series of short explosions starting on January 11. With each explosion, the critical question for the Alaska Volcano Observatory (AVO) was what is likely to happen next? Did an explosion mark a release of pressure that would be followed by a lull in activity, or was it the beginning of a larger sustained eruption?

During the precursory period, AVO implemented an hourly GPS solution, using data from the PBO CGPS sites. Solutions were run ~ 30 min after each hour for the previous hour's data. Despite the scatter of the individual hourly solutions, over the next 17 days the episodic explosions caused no change in the overall inflation trend of the volcano, which was interpreted to indicate that the volcano continued to build to a possible larger eruption. After the sustained activity began on January 28, the GPS positions indicated deflation; the change in deformation was clear within a few hours. The sustained explosive activity and deflation continued for 2 weeks, followed by a 3-week hiatus during which slow inflation was observed. Deflation of a shallow source began again when the volcano went into an effusive phase in early March. Although the deformation signals were subtle and exceeded 2–3 cm only at one site near the summit, the strong correlation between the inflation and deflation observed in the GPS data and the progress of the eruption meant that the GPS data played a critical role in the rapid assessment of the eruption progress.

4.6. Okmok (1993–2010)

Okmok Volcano is a ~ 10-km diameter caldera located in the eastern Aleutian Islands, Alaska. It erupted from an intracaldera cone in 1997, producing basalt lava flows up to ~ 50-m thick, based on differencing DEMs before and after the eruption. Total eruptive volume estimates ranged from 0.05 to 0.015 km^3, based on InSAR and DEM differencing. Modeling of the coeruptive deformation provided a smaller volume change estimate, ~ 0.05 km^3. InSAR showed evidence for ~ 1.7 m of coeruptive subsidence at the center of the caldera (Figure 64.3). Okmok erupted explosively in 2008, producing a volume of ash ~ 2 times larger than the

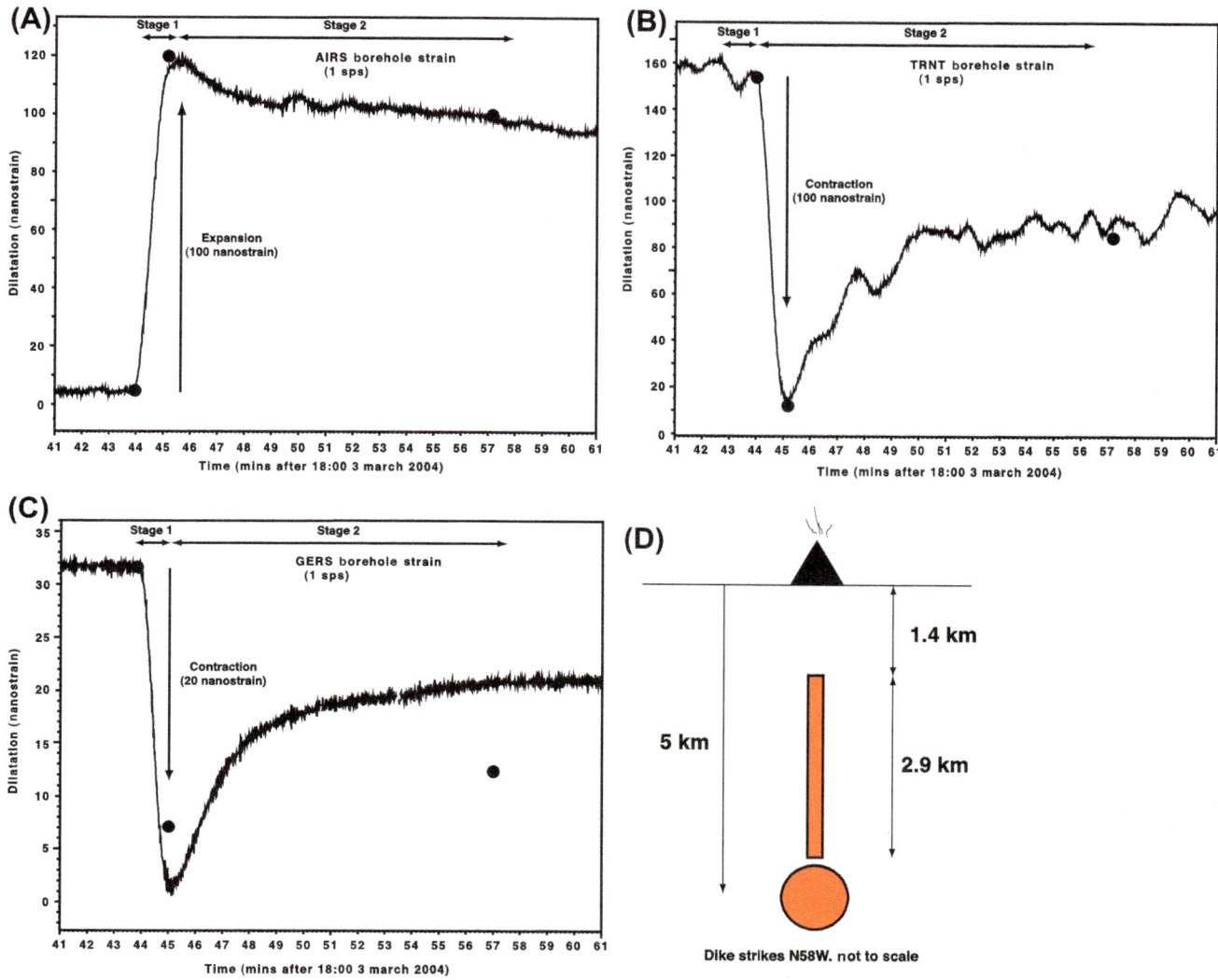

FIGURE 64.8 Data and model from the March 3, 2004 explosion of Montserrat. (A)—(C) Strainmeter records from 3 strainmeters arranged around the volcano showing the time-dependent strain signal associated with the explosion. Large dots show model calculated values. (A) Site AIRS (Air Studios). (B) Site TRNT (Trants). (C) Site GERS (Gerald's Yard). (D) Interpretive model for the magma body and dike inferred by Linde et al. (2010).

lava volume from the 1997 eruption. Between the two eruptions, Okmok underwent substantial inflation that was measured extensively with GPS and InSAR, producing >0.5 m uplift in the center of the caldera. The post-1997 inflation occurred consistently until 2006, with major pulses in 1997—2000, 2002, and 2004 (Figure 64.9). The 1997—2006 posteruptive inflation was associated with a total volume change of 0.035 km^3. The inflation source was remarkably consistent in location throughout this entire period, located in the geometric center of the caldera 2.5-km below sea level.

Okmok began to inflate again in early 2008, culminating in an explosive eruption that began on July 12, 2008, which lasted ~6 weeks. The eruption blasted a new vent within the caldera, and its high explosivity and entrainment

of a large amount of water in the ash cloud indicated interaction between magma and ground and surface water. The 2008 magma was significantly more evolved than the 1997 basalt, and both GPS and InSAR data indicate that it erupted from a shallower depth (by 500—700 m) than the preeruptive inflation source, but also beneath the center of the caldera.

The end of the 2008 eruption demonstrated how quickly a volcano could begin reinflating after an eruption. Renewed posteruptive inflation at Okmok due to new magma accumulating at <2-km depth began within 3 weeks or less from the cessation of eruptive activity. A CGPS site inside the caldera was repaired at that time and immediately showed evidence of rapid inflation. It is possible that renewed inflation at Okmok began essentially

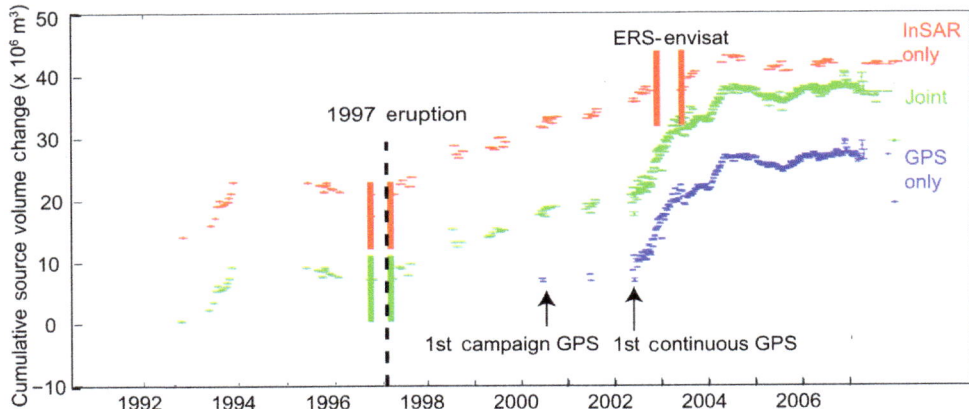

FIGURE 64.9　Volume change estimates for the shallow magma storage beneath Okmok volcano for 1997–2006, based on a joint inversion of InSAR and GPS data by Biggs et al. (2010). The green points show the joint inversion results, red are InSAR only and blue are GPS only. The InSAR only and GPS only curves have been offset for comparison. Vertical bars indicate data gaps in the time series due to the 1997 eruption and a change from ERS-1/2 to Envisat satellites. Several prominent pulses of rapid inflation are evident in the data.

immediately after the cessation of eruption, either because magma had been supplied from depth continuously during the eruption, or because the eruption itself depressurized a shallow magma body enough to allow new magma to rise in response.

5. PRINCIPLES OF GRAVITY MONITORING

The acceleration due to gravity at the Earth's surface is approximately 9.8 m s^{-2}. Modern gravimeters can measure gravity differences (either between different places or at one place through time) to about 1 part in 10^8; variations as small as 1–10 μGal can be detected, depending on the instrument. Spatial variations of that magnitude or larger occur due to lateral heterogeneities in the subsurface density. Thus buried dikes, feeder pipes, and even magma chambers can be identified by making a map of the spatial variations in gravity, provided there is a density contrast between the structures of interest and the surrounding rock. Gravity also varies with elevation (gravity decreases vertically at roughly 308 μGal m^{-1} = 3 × 10^{-6} m s^{-2}; this is called the free-air gradient (FAG)). Changes in gravity with time result from changes in the subsurface mass and/or changes in elevation. At volcanoes, these changes may result from intrusions of magma, or from the eruption of magma that had previously been stored in the subsurface. As shown in section 3, pressure changes in volcanic systems lead to changes in elevation, which must be accounted for in analyzing the gravity changes. However, gravity changes can distinguish changes in pressurization resulting from changes in mass from those that reflect exsolution of volatiles or other chemical processes.

The most common type of gravimeter used on volcanoes basically comprises a mass on a spring (Figure 64.10). A complex array of gears provides an accurate measure of the force that needs to be applied to the mass to return the spring to its original length within the prevailing gravity field. Calibration of the instrument allows differences in the measured force to be expressed in terms of gravity differences. Absolute gravimeters measure the acceleration of gravity directly, rather than differences in gravity from one

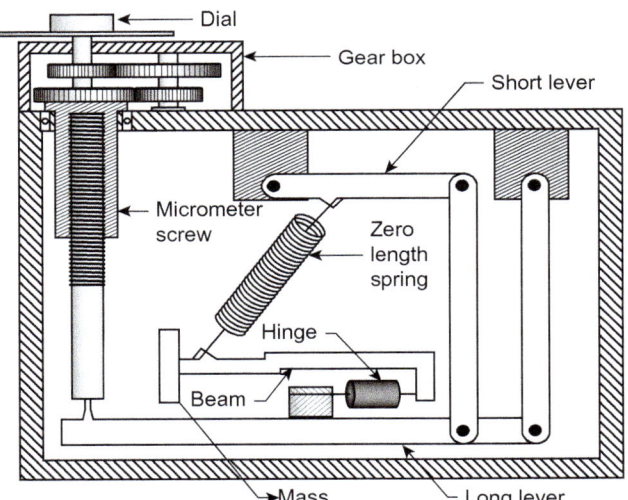

FIGURE 64.10　Basic operating mechanism of a LaCoste and Romberg relative gravimeter. A key element is the "zero length spring," a specially designed coil spring that is pretensioned so that the force it exerts is exactly proportional to its length, over a sufficient range of lengths (the plot of force exerted vs. extension of the spring passes through the origin). The name arises because if no force is applied to the spring it would collapse to zero length (that is a physical impossibility, but the spring behaves as such an idealized spring within its operating range).

site to another, but are much larger and more expensive than relative gravimeters. Absolute gravimeters can also measure gravity with greater accuracy than relative gravimeters.

Measured gravity values need to be corrected to remove the effects of changes in latitude and elevation and the Earth tides. For any point on the Earth's surface, Earth tides vary by up to 400 μGal peak to peak with a maximum rate of change of about 60 μGal hr^{-1} (1.8×10^{-10} m s^{-2}). The Earth tides are well known and theoretical tides can be computed for any time and location. Any deviations from the tide, called residual gravity changes, are due to instrumental effects or external processes such as ground deformation and volcanic activity. Changes in material properties can result in measurable changes in the tidal response. Many modern gravimeters can measure continuous variations in gravity.

6. TECHNIQUES USED IN THE FIELD

6.1. Microgravity

Microgravity monitoring involves the measurement of small changes in the value of gravity at a network of stations, using small, portable instruments such as the LaCoste and Romberg gravimeter (Figure 64.10) or similar instruments made by other manufacturers. Microgravity is a valuable tool for mapping out the subsurface mass redistributions that are associated with volcanic activity. As with all types of gravity work, any changes in elevation must be measured at the same time as the gravity measurements are made. Microgravity measurements are usually made by making one or more loops around the network, always returning to the starting station or making multiple repeat occupations of stations. This is needed to remove instrumental drift. Gravity values are usually expressed relative to the value of gravity at a base or reference station, which is preferably outside the area of volcanic activity. The network is then remeasured sometime later, or after some volcanic event has occurred. The corrected gravity difference at each station relative to the base station can be related to subsurface density or mass changes, such as the addition or removal of magma.

6.2. Absolute Gravity

The most common method for determining the absolute value of gravity involves measuring the rate of fall of a mass. A laser interferometer is used to track the vertical rise of a corner cube reflector propelled vertically upward and its subsequent fall. Most absolute gravity instrumentation is bulky and heavy and only transportable by vehicle. It is also complex to use, taking a day or so for one value to be measured. However, the advantage of absolute gravity values is that, unlike relative gravity values, there is no dependence on reference to a base station that may or may not be stable, and no need to control for instrumental drift. Absolute gravimeters can measure gravity with an accuracy of ∼1 μGal, although the effect of local groundwater variations is often larger than this. In recent years, newer model absolute gravimeters have become much more portable, at the cost of some accuracy (∼10 μGal for smaller, more portable instruments). Absolute gravimeters remain extremely expensive.

7. CASE HISTORIES OF GRAVITY CHANGES

Microgravity surveys have been made at active volcanoes since the 1950s, but tidal gravity, with its requirements of continuous power, accurate time keeping, and large volumes of data output, has only been a realistic option for volcano monitoring since the mid-1990s. Absolute gravity is still too cumbersome and expensive to use regularly at most volcanoes, but has been measured at the oldest volcano observatory in the world, the Osservatorio Vesuviano at Vesuvius, Italy. The first absolute gravity measurement there was made in 1986, and was repeated several times since then, with measurements accuracy of 3–6 μGal each time. The absolute gravity site is also included in a microgravity network using relative gravimeters; those measurements are less precise but more frequent in time. The measured gravity value decreased by ∼100 μGal from the early 1980s until 1994. Gravity decreased a further ∼20 μGal after a seismic swarm in 2000, then recovered about half of the total decrease between 2003 and 2004. These variations are interpreted in terms of changes in fluid mass (not magma), with the gravity decrease in 2000 thought to be caused by opening of cracks and pore spaces resulting from the seismicity.

The typical compositions of eruptive products at a volcano provide a basis for classification of the nature of precursory microgravity changes, as each type tends to behave differently.

7.1. Microgravity and Deformation at Calderas: Campi Flegrei (1969–1984)

Large silicic calderas, where long-term unrest is common, are characterized by periods of uplift and subsidence with residual gravity changes indicating subsurface mass increases (Figure 64.11). Gravity changes differing from those predicted from the elevation changes have been interpreted in terms of magma movements and hydrothermal fluid migration. Observed elevation changes have been

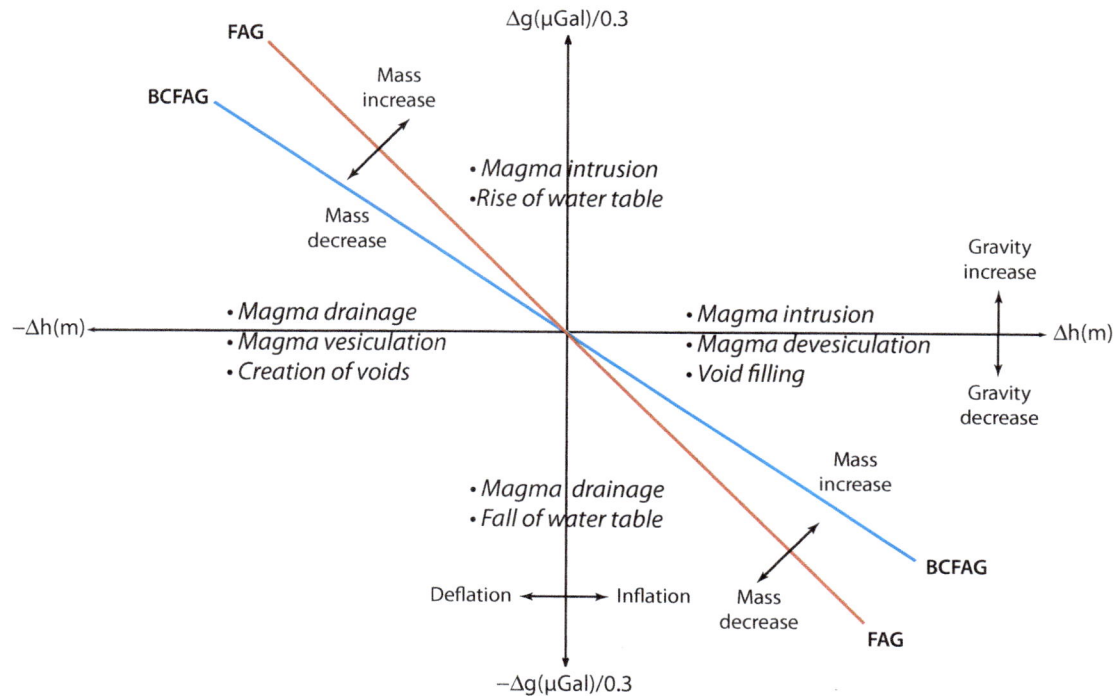

FIGURE 64.11 Graph of height changes (horizontal axis) against gravity changes (vertical axis, scaled by free-air gradient (FAG), of 0.308 milligal/m). Measured gravity and elevation changes may be plotted on this graph and the interpretations read off. The diagonal line with a 1:1 slope given this scaling is the FAG. Data plotted above this line on the right-hand side (positive Dh, i.e., inflation) represent net gravity increases and therefore subsurface mass increases, which, on volcanoes, are most likely caused by magma intrusion, magma devesiculation, and void filling. Data falling in other regions of the graph have correspondingly different interpretations as marked. The line labeled BCFAG marks the Bouguer-corrected free-air gradient.

large at these calderas (ca. 1 m) and the gravity data fall close to, but not always on, the FAG.

In the center of Campi Flegrei caldera, the ruins of an ancient Roman market (Serapeo; see Figure 64.12) bear witness to long-term caldera unrest. Mollusk borings in the stone columns indicate earlier periods below sea level and recent measurements indicate that during two crisis periods (1969–1972 and 1982–1984), uplift of some 3.3 m and a gravity decrease of 330 µGal occurred there. This gravity change was less than predicted by the FAG, suggesting a subsurface mass increase. Inflation and gravity increases were seen over a large area and were thought to be caused by a magmatic intrusion of around 2×10^{11} kg at some ~3-km depth. From 1985 to 2001, Serapeo slowly subsided and gravity increased, but these changes were in accord with those expected from the FAG, showing no further change in mass.

7.2. Microgravity at Andesitic Volcanoes: Poas Volcano (1985–1989)

The explosive nature of many andesitic eruptions and the possibility of risk mitigation make this category of volcanoes the most important to monitor effectively in the short to medium term. The intrusion of magma into poorly consolidated material and even the buildup of magmatic gas pressure are sometimes detectable because these processes can occur close to the surface and the density contrasts are large.

A characteristic of andesitic volcanoes is that the gravity—height change data tend to plot close to the Δg axis of Figure 64.11 (the height changes are usually rather small, as at Augustine, section 4.5). One of the longest time series of microgravity data for a volcano is from Poas in Costa Rica. A network of stations in and around the active crater revealed increases and decreases in gravity associated with phreatic eruptions and the gradual disappearance and final reestablishment of a crater lake. Increases at stations in the southern part of the crater bottom began in 1985, preceding the phreatic activity and loss of the lake by 4 years (Figure 64.13). By integrating the magnitude of the residual gravity changes over the area of their extent (the principle of excess mass), the subsurface mass increase can be deduced: 10^8 kg of new mass appeared beneath the crater between 1985 and 1989. This new mass may have been a magmatic intrusion, deposition of sulfur, the redistribution of the hydrothermal system, or some combination of these.

FIGURE 64.12   The ruins of the market of Serapeo at the center of Campi Flegrei caldera (Naples, Italy) provide a classic example of long-term caldera unrest. When built in ancient Roman times, the market must have been above sea level, yet by 1834 (left), it had sunk below sea level, and the mollusk borings more than a meter up the columns indicate lengthy periods even further below sea level prior to this. However, recent surveying measurements show that the region rose by 3.3 m between 1969 and 1984, raising the whole market well above sea level again by 1991 (right).

7.3. Microgravity at Basaltic Volcanoes: Mt Etna (1990−1993) and Kilauea (1983−Present)

Comprehensive surveillance programs have been carried out at highly active basaltic volcanoes such as Kilauea on the island of Hawaii and Etna on Sicily. An important characteristic of these volcanoes that influences the observed microgravity changes is their tendency to develop flanking rift systems. Magma may be intruded passively into dikes and can feed flank eruptions exploiting these rifts. Thus very large mass changes can occur without elevation changes (data plot near the Δh axis on Figure 64.11), but large elevation changes can also occur (during rifting events, for example) and during these events gravity data fall close to the FAG. Separate deflation events on Kilauea had gradients between -171 and -607 μGal m^{-1}, thought to indicate magma migration away from the rift zone leaving voids, and then later passive intrusion into these void spaces.

Microgravity increases as large as 400 μGal were found between June 1990 and June 1991 at a network of stations near the summit of Etna volcano, Sicily (Figure 64.14). There was no eruptive activity at this time, and the ground deformation was less than a few centimeters, the microgravity increases were therefore caused by subsurface mass changes. A model for the gravity increases (Figures 64.15 and 64.16) shows the change in the shallow plumbing system at this time. The level of magma within the summit feeder pipe (modeled with a radius of 25 m) rose at

sometime between the two sets of observations and a dike was apparently intruded in a south−southeasterly direction trending away from the summit. The point at which the projection of this modeled dike intersects the topography became the site of the vent for the December 1991−April 1993 eruption. In this case it would appear that the gravity increases indicated rising magma, which later fed a large eruption.

Kilauea volcano, Hawaii, has been erupting since 1983, with the eruption having gone through many distinct phases. The eruption has been centered along Kilauea's East Rift Zone, building a cone called Pu'u 'O'o and adjacent shields. During the eruption, the south flank of the volcano, south of the East Rift Zone, has moved rapidly seaward. This seaward motion was accompanied by extension of the rift zone, presumably due to dike intrusions. As of the end of 2012, about 4 km^3 of magma had erupted as part of this eruption. In 2008, a new vent opened at the Kilauea summit, and an active lava lake remains within this summit vent. Vigorous degassing from the summit vent has continued since then, and may contribute to the reduced rates of lava extrusion from the East Rift Zone since then.

Most eruptive episodes since 1983 have followed a similar pattern, with changes in tilt at the summit, centered about a point in the southern part of the summit caldera, being a leading indicator of changes in the eruption. Increasing summit tilt results from accumulation of new magma there and decreases of summit tilt occur when magma begins to move down the rift zone toward Pu'u 'O'o. Throughout this period, there was a net mass

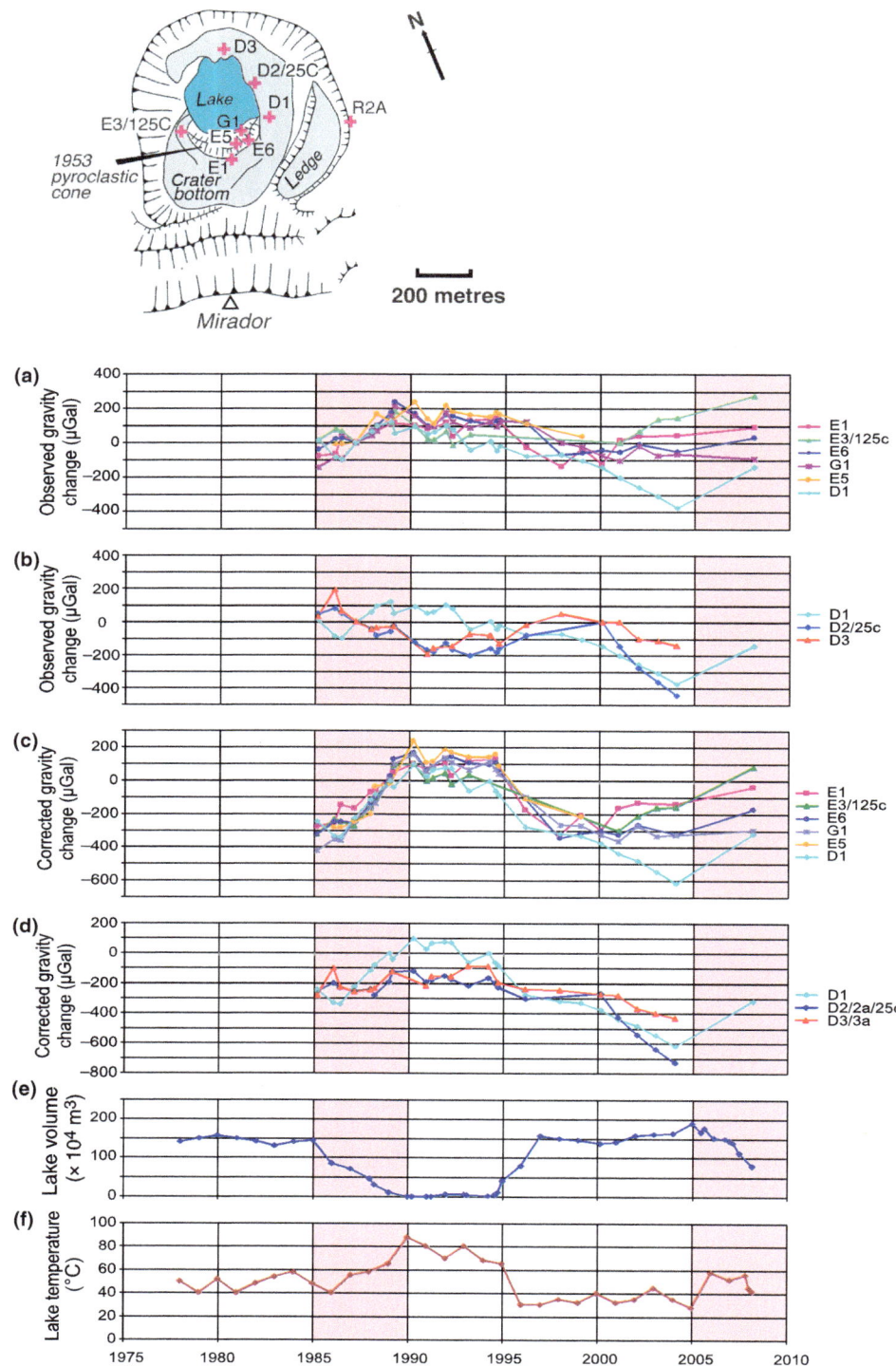

FIGURE 64.13 Microgravity, and lake volume and temperature changes at stations in and around the active crater at Poás. A map of the summit area is shown at top, including station locations. Panels a and b show observed gravity variations, panels c and d show gravity corrected for variations in lake level and water table, and panels e and f show lake volume and temperature. Red shading indicates time periods of intrusions inferred from these data. Geyser activity began at the lake in 1987–1978 and the lake disappeared in 1989. It was reestablished in 1997. Throughout this time period, deformation near the summit was small. *After Rymer et al. (2009).*

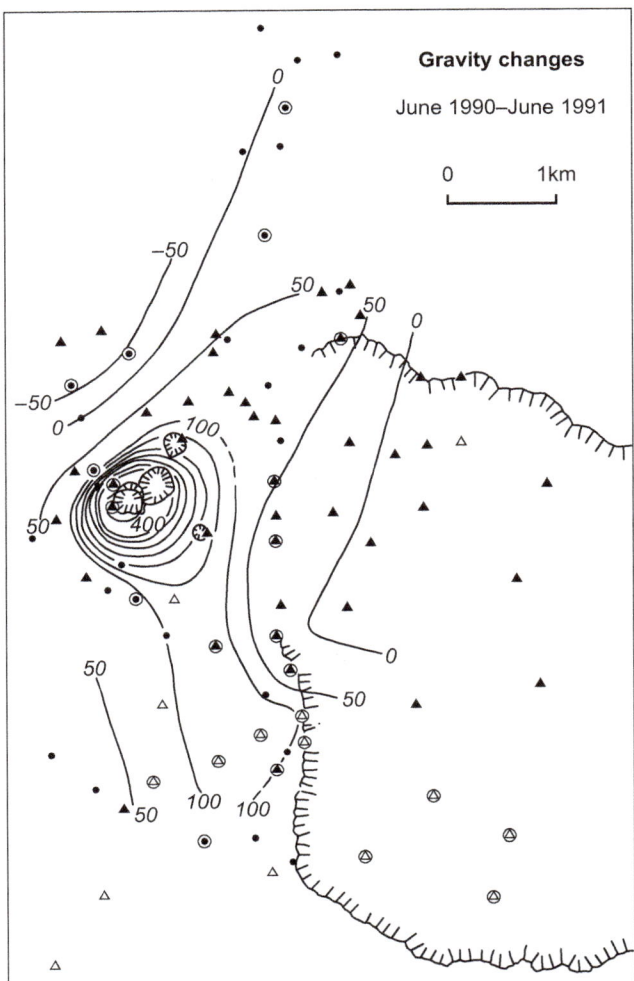

Gravity changes

June 1990–June 1991

0   1km

FIGURE 64.14 Map of microgravity increases at Etna volcano (in μGal) measured between June 1990 and June 1991. The largest increases are close to the summit craters and reflect an increase in magma level within the feeder system. Gravity increases were also seen in an elongated south–southeasterly trending region close to fractures that opened up at the end of an eruption in 1989. This was probably caused by the intrusion of a dike.

increase of $6-33 \times 10^{10}$ kg from 1975 to 2008 beneath the eastern part of the Halema'uma'u crater at the summit (the location of the 2008 summit vent). The mass increase was not accompanied by significant uplift. This was interpreted to mean new magma progressively filled cracks and pore spaces created by extension from the 1975 Kalapana earthquake on the south flank. Continuous gravity measurements made at a site only 150 m from the vent, combined with measurements of the height of the lava lake, have demonstrated the surprising result that the magma density in the lava lake is only 950 g cm^{-3}, less than that of water! This indicates that the lava lake consists mainly of a gassy froth rather than purely liquid magma. Shelly

pahoehoe overbank deposits at the margins of other lava lakes on Kilauea provide evidence that gassy, low-density magma is not restricted to this one example.

8. PRINCIPLES OF MAGNETIC MONITORING

Magnetic monitoring of volcanoes is relatively underdeveloped compared with ground deformation and microgravity monitoring. Magnetic changes associated with volcanic activity have been successfully observed, some of which were precursory to eruption and hence identify the potential of magnetic monitoring methods. However, magnetic effects are generally more complex and difficult to interpret than gravity or deformation data. A total magnetic field map can be drawn using data from a network of stations measured either near the ground surface or from an airplane or helicopter. Variations through time may also be monitored using a stationary continuously recording magnetometer or by repeating measurements at a network of stations, if stations can be precisely relocated.

The magnetic field at the surface of the Earth ranges in intensity from approximately 25,000–70,000 nT, depending mainly on latitude. About 99% of this field is generated by circulation in the Earth's partially molten outer core, with minor contributions from the magnetic effect of near-surface rocks themselves. The magnetic field measured on the surface also includes the effects of the Earth's external field, resulting from charged particles in the ionosphere. This component is generally removed from magnetic measurements by differencing between magnetometers. However, shock waves in the ionosphere induced by volcanic eruptions can cause observable changes in the external magnetic field, and in ionospheric measurements made using GPS.

There are two main types of rock magnetism: remanent magnetism and induced magnetism, both of which depend on the occurrence of magnetic or magnetizable minerals in the rock. Remanent magnetization, the essentially permanent magnetization in rocks, is acquired at the time of rock cooling or chemical recrystallization. Induced magnetization is a temporary magnetization generated in all rocks by the present day ambient magnetic field of the Earth. Rock magnetism only occurs at temperatures below the Curie temperature (ca. 550 °C).

The intensity of magnetization in a rock depends largely on the abundance of magnetic minerals such as magnetite or hematite in the rock. Volcanic rocks are often strongly magnetized, strong enough to divert a compass needle in some cases. The intensity of magnetization in more basic rocks (e.g., basalt) is in general stronger than that in less basic rocks (e.g., andesite). Surveys of the spatial

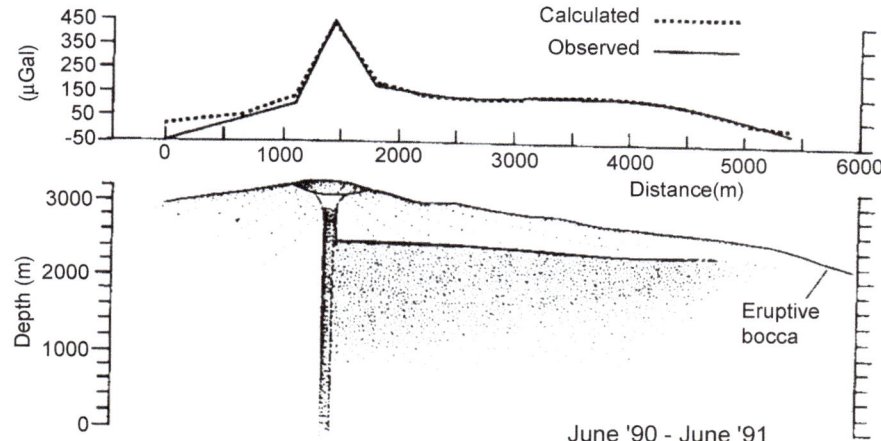

FIGURE 64.15 Model of the subsurface magma movements thought to have caused the microgravity changes seen in Figure 64.14. The south—southeasterly trending dike is only 4-m wide, but the summit feeder pipe is 25 m in radius, which accounts for its larger gravitational effect.

FIGURE 64.16 Portable proton precession magnetometer in use on the south flank of Mt Etna, Sicily. Precise centering over a permanent marker is essential where repeat measurements of a network are to be made.

distribution of magnetic effects of volcanic rocks can be used to generate models of structures within volcanoes or to determine the depth and volume of lava flows. Magnetic surveys are also used to detect zones of hydrothermal alteration (hematite is a particularly important mineral in this case), which can also be zones of structural weakness. In magnetic monitoring of active volcanoes, measurements of the magnetic field are repeated, either continuously or intermittently, at the same location over a period of time in order to detect changes in the magnetization of the subsurface rocks.

"Volcanomagnetic" effects are caused by a number of mechanisms, the most important being thermal demagnetization or remagnetization, and piezomagnetic effects. Thermal demagnetization occurs as the rock heats through its Curie temperature. The rock can be remagnetized when it cools again. Piezomagnetic effects are magnetic changes due to changes in stress regime such as magma intrusion or depressurization, which are generally irreversible. Other possible mechanisms for volcanomagnetic effects include (1) chemical changes to magnetic minerals, for example, alteration by hydrothermal activity, (2) mass redistribution such as intrusion or major landslides, or (3) electrokinetic effects due to fluid flow in the rocks.

9. TECHNIQUES USED IN THE FIELD

Only the intensity of the total magnetic field is generally measured, though it is possible to measure its constituent parts (e.g., horizontal and vertical components). However, making vector magnetometer measurements requires precisely leveled instruments, which can be difficult to maintain in areas undergoing significant ground deformation and impossible to maintain in an airborne survey. Instruments used are either proton precession magnetometers (accurate to 0.1 nT) or cesium/rubidium vapor magnetometers (accurate to 0.01 nT). Measurements must be corrected for secular variations (i.e., changes with time) in the Earth's magnetic field itself and induced external fields, which is achieved by monitoring magnetic field changes at a base station outside the region influenced by the volcano. Each site on the volcano may respond differently to these ambient field changes but this can be calculated mathematically and subtracted from the observed data.

10. CASE HISTORIES OF MAGNETIC OBSERVATIONS

Most commonly, magnetometers are set up at a site on a volcano and left to record "continuously" (i.e., measurements are recorded every 1−10 min). Alternatively, measurements may be repeated intermittently (e.g., every few days or weeks) at monumented sites using a portable magnetometer (Figure 64.16). In this case it is essential that the magnetometer be relocated in precisely the same position (horizontally and vertically) for each measurement. In some studies, combinations of both continuous and intermittent sites have been used.

10.1. Mt St Helens Volcano (USA, 1980)

Magnetic changes were observed on the west side of the edifice of Mt St Helens volcano at the times of the three major eruptions on May 18, May 25, and June 12, 1980. No magnetic changes were observed immediately before the catastrophic May 18 eruption (Figure 64.17); this is consistent with the triggering of the eruption by a sudden depressurization of the magma by a landslide. The increase of 9 ± 2 nT occurring during the 12 h following the May 18 eruption has been interpreted as a piezomagnetic effect resulting from the release of stress during the eruption. Since the net magnetic change is permanent, electrokinetic and thermal demagnetization mechanisms cannot have played a major role, and also the magnetic effects are in the opposite sense than could be attributed to the accompanying mass redistribution. This interpretation has been endorsed by mathematical models, which show that a magnetic change with the same amplitude and sense would be generated by a pressure release of ~ 100 MPa in a cylindrical or spherical magma chamber ~ 1 km in diameter at a depth of ~ 5 km. There was some indication of magnetic disturbances for several hours before the May 25 (Figure 64.18) and June 12 events, and even greater effects were recorded after these eruptions. These posteruption changes are considered to reflect the effect of eruptive shock waves on the ionosphere and were also recorded at least 1000 km away from the mountain. Shock waves from several other large volcanic explosions also have been detected from ionospheric disturbances measured from GPS data.

10.2. White Island (New Zealand, 1974)

Magnetic measurements have been made at White Island in surveys repeated at approximately 4-month intervals since 1968. Localized changes of up to 400 nT were measured within the crater between 1976 and 1982; both increases

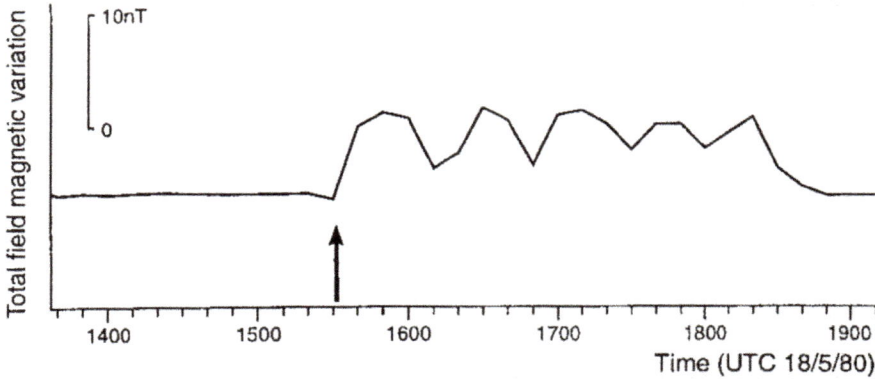

FIGURE 64.17 Variations in total magnetic field recorded at a site on the west of Mount St. Helens volcano before and after the catastrophic eruption of 1980 May 18 at 0832 PDT (1532 UTC). Time of eruption denoted by arrow.

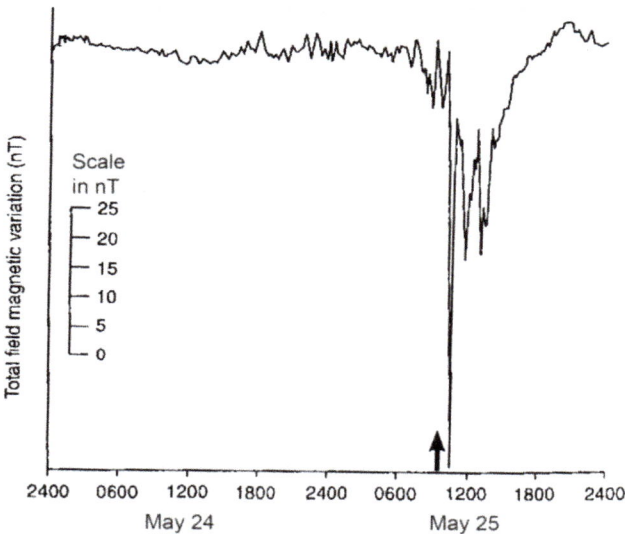

FIGURE 64.18 Variations in total magnetic field recorded at a site on the west side of Mt St. Helens before and after the eruption on May 25. Time of eruption denoted by arrow.

11. CONCLUDING REMARKS

Ground deformation has long been used as a tool for studying, monitoring, and predicting volcanic activity. By combining the method with seismic monitoring, monitoring of gas and water chemistry and gravity and magnetic potential field techniques, it becomes a more powerful and rigorous indicator of subsurface processes. Not all volcanoes lend themselves equally well to each monitoring technique. Nevertheless, these methods have been used on volcanoes ranging from young basaltic cones to andesitic stratocones to silicic calderas to detect changes before, during, and after magma movement or eruption.

The cost of regular visits to a volcano to make routine measurements and the hazard to human life are such that there is an understandable drive to develop reliable, automated instrumentation with a telemetering capacity to send data in real time or at discrete intervals. InSAR is a valuable ground deformation early warning tool for these reasons, although more rapid access to data are needed. In situ measurements are still needed because InSAR measurements still lack the temporal resolution to study all of the important processes, and because atmospheric path delays and loss of coherence can cause problematic systematic errors or spatial gaps in data. GPS techniques are now the essential ground deformation measurement tool, and continuously recording GPS can provide exceptional time resolution and accuracy for deformation measurements, if an adequate number of sites can be installed. Dense arrays of continuously recording instruments, with data telemetered to a remote operations center, would constitute an ideal surveillance system. The development of more and more Earth observing satellites means that such data will soon be combined with thermal infrared, gas, and even real-time InSAR to provide real-time volcano monitoring.

and decreases were observed during this time. Increases occur after eruption sequences and indicate cooling, while decreases are associated with eruptions and result from heating of the rocks at depths of 500−1000 m. For example, the magnetic field changes observed between February and August 1974 (Figure 64.19) were interpreted as resulting from localized heating below Donald Mound. This was confirmed by the opening of a vigorous fumarole in that vicinity in September 1974. Modeling has shown that variations of 200−300 °C are required to account for the observed field changes. The rapid occurrence of these changes indicates convective (rather than conductive) transport mechanisms to be active in this case, such as the movement of hot gases or hydrothermal fluids, or the ingress of cold rainwater.

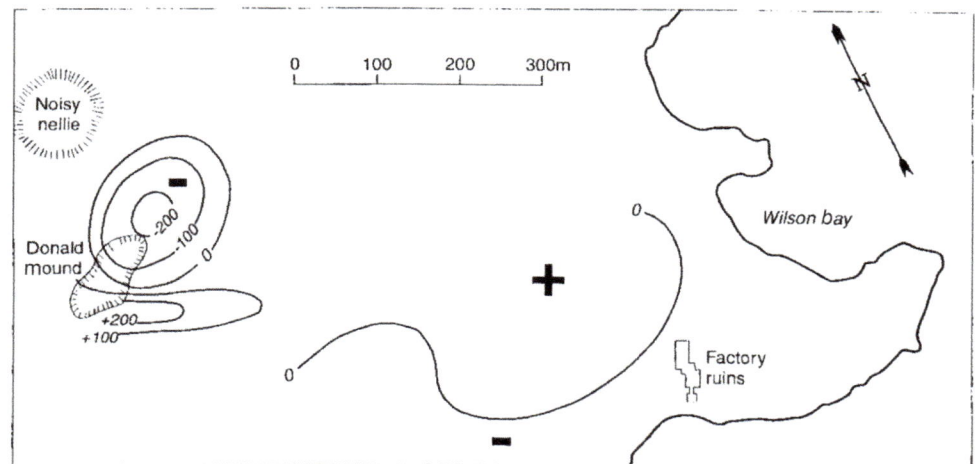

FIGURE 64.19 Map of the total field magnetic changes within the crater of White Island volcano between February and August 1974. Contour interval is 100 nT.

SEE ALSO THE FOLLOWING ARTICLES

Gas, Plume, and Thermal Monitoring ● Seismic and Infrasonic Monitoring ● Volcanic Hazards and Risk Management ● Volcano Warnings.

Supplementary video related to this chapter can be found at http://booksite.elsevier.com/9780123859389.

FURTHER READING

Battaglia, Maurizio, Cervelli, P.F., Murray, J.R., 2013. Modeling Crustal Deformation Near Active Faults and Volcanic Centers—A Catalog of Deformation Models. U.S. Geological Survey Techniques and Methods book 13, chap. B1, 96 p. http://pubs.usgs.gov/tm/13/b1.

Biggs, J., Lu, Z., Fournier, T., Freymueller, J.T., 2010. Magma flux at Okmok Volcano, Alaska, from a joint inversion of continuous GPS, campaign GPS, and interferometric synthetic aperture radar. J. Geophys. Res. 115, B12401. http://dx.doi.org/10.1029/2010JB007577.

Dzurisin, D., 2006. Volcano Geodesy: New Geodetic Monitoring Techniques. Springer-Praxis, ISBN 978-3-540-49302-0, 441pp.

Fournier, T., Freymueller, J., Cervelli, P., 2009. Tracking magma volume recovery at Okmok volcano using GPS and an unscented Kalman filter. J. Geophys. Res. 114, B02405. http://dx.doi.org/10.1029/2008JB005837.

Gottsmann, J., Rymer, H., Berrino, G., 2006. Unrest at the Campi Flegrei caldera (Italy): a critical evaluation of source parameters from geodetic data inversion. J. Volcanol. Geotherm. Res. 150, 132–145.

Hooper, A., Bekaert, D., Spaans, K., Arıkan, M., 2012. Recent advances in SAR interferometry time series analysis for measuring crustal deformation. Tectonophysics 514–517, 1–13. http://dx.doi.org/10.1016/j.tecto.2011.10.013.

Hreinsdóttir, S., Sigmundsson, F., Roberts, M.J., Björnsson, H., Grapenthin, R., Arason, P., Árnadóttir, Th, Hólmjárn, J., Geirsson, H., Bennett, R.A., Gudmundsson, M.T., Oddsson, B., Ófeigsson, B.G., Villemin, T., Jónsson, Th, Sturkell, E., Höskuldsson, Á., Larsen, G.,

Thordarson, T., Óladóttir, B.A., 2014. Volcanic plume height correlated with magma-pressure change at Grímsvötn Volcano, Iceland. Nat. Geosci. 7, 214–218. http://dx.doi.org/10.1038/ngeo2044.

Larson, K.M., Poland, M., Miklius, A., 2010. Volcano monitoring using GPS: developing data analysis strategies based on the June 2007 Kilauea Volcano intrusion and eruption. J. Geophys. Res. 115, B07406. http://dx.doi.org/10.1029/2009JB007022.

Lu, Z., Dzurisin, D., 2014. InSAR Imaging of Aleutian Volcanoes. Springer-Praxis, ISBN 978-3-642-00348-6, 390 pp.

Mann, D., Freymueller, J.T., Lu, Z., 2002. Deformation associated with the 1997 eruption of Okmok Volcano, Alaska. J. Geophys. Res. 107 (B4), ETG 1-7—ETG 7-12, April 2002 2001JB000163.

Mogi, K., 1958. Relations between the eruptions of various volcanoes and the deformations of the ground surfaces between them. Bull. Earthq. Res. Inst. Univ. Tokyo 36, 99–134.

Murray, J.B., Pullen, A.D., Saunders, S., 1995. In: McGuire, W.J., Kilburn, C.R.J., Murray, J. (Eds.), Monitoring Active Volcanoes. UCL Press, London, pp. 113–150.

Rymer, H., 1996. In: Scarpa, R., Tilling, R.I. (Eds.), Monitoring and Mitigation of Volcanic Hazards. Springer, Berlin/Heidelberg, pp. 169–198.

Rymer, H., Locke, C.A., Borgia, A., Martinez, M., Brenes, J., van der Laat, R., Williams-Jones, G., 2009. Long-term fluctuations in volcanic activity: implications for future environmental impact. Terra Nova 21, 304–309. http://dx.doi.org/10.1111/j.1365-3121.2009.00885.x.

Segall, P., 2010. Earthquake and Volcano Deformation. Princeton University Press, ISBN 9780691133027, 456 pp.

Segall, P., 2013. Volcano Deformation and Eruption Forecasting, vol. 380. Geological Society, Special Publications, London. http://dx.doi.org/10.1144/SP380.4.

Williams-Jones, G., Rymer, H., Mauri, G., Gottsmann, J., Poland, M., Carbone, D., 2008. Towards continuous 4D microgravity monitoring of volcanoes. Geophysics 73 (6), 19–28.

Zlotnicki, J., 1995. In: McGuire, W.J., Kilburn, C.R.J., Murray, J. (Eds.), Monitoring Active Volcanoes. UCL Press, London, pp. 275–301.

Chapter 65

Gas, Plume, and Thermal Monitoring

Simon A. Carn

Department of Geological and Mining Engineering and Sciences, Michigan Technological University, Houghton, MI, USA

Chapter Outline

GLOSSARY

active remote sensing A remote sensing technique using an artificial source of electromagnetic radiation, typically mounted on the same platform as the detector (e.g., light detection and ranging [LiDAR] and radio detection and ranging [radar]).

atmospheric window A region of the electromagnetic spectrum with high atmospheric transmission, usually exploited for remote sensing of the Earth's surface.

electromagnetic radiation Radiation comprising orthogonal electric and magnetic fields; its interaction with atoms and molecules in matter is the basis of all remote sensing techniques. Travels at the speed of light, and can be transmitted through a vacuum.

fumarole A crack or fissure in a volcanic edifice through which volcanic gases are emitted.

hyperspectral Measurements providing contiguous coverage of a region of the electromagnetic spectrum at high spectral resolution.

multispectral Measurements at discrete, non-contiguous channels at specific wavelengths of the electromagnetic spectrum.

passive remote sensing A remote sensing technique that exploits natural sources of electromagnetic radiation (e.g., the Sun, thermal emission from the Earth or the atmosphere).

remote sensing The measurement of some property of an object, by an instrument that is not in physical contact with that object.

retrieval The derivation of the abundance (e.g., column amount) of an atmospheric constituent (gas or particles) from spectra of electromagnetic radiation intensity.

spatial resolution The ability of a remote sensing system to resolve objects or phenomena of a given spatial dimension.

spectral resolution The ability of a remote sensing system to discriminate wavelengths of electromagnetic radiation.

spectroscopy The measurement of electromagnetic radiation intensity as a function of wavelength.

stratosphere The layer of Earth's atmosphere directly above the troposphere, extending up to ∼50 km above sea level (ASL), characterized by stability (due to a temperature inversion above the tropopause) and the ozone layer.

thermal infrared The region of the EM spectrum corresponding to peak thermal emission from objects at terrestrial temperatures, extending from ∼4 to 15 μm wavelength, used for remote sensing of several volcanic gases and volcanic ash/aerosols.

troposphere The lowest, most dynamic layer of Earth's atmosphere, containing 99% of its water vapor and extending to altitudes of up to 20 km ASL in the tropics, or 7 km ASL in polar regions. The top of the troposphere is the tropopause.

ultraviolet The region of the EM spectrum whose major source is solar radiation, extending from ∼10 to 400 nm wavelength, the

The Encyclopedia of Volcanoes. http://dx.doi.org/10.1016/B978-0-12-385938-9.00065-1

300—400 nm band of which is used for remote sensing of sulfur dioxide (SO_2), ozone, and reactive halogens.

volcanic cloud A discrete body of volcanic gases and particles drifting in the atmosphere.

volcanic degassing The separation or release of volcanic gases from their magma source, preceded by exsolution of volatile species to form bubbles. Often classified as passive (quiescent) or eruptive/explosive.

volcanic plume A cloud of gases and particles entrained by the prevailing wind but still attached to the volcanic vent at its source.

1. INTRODUCTION

Eruptions are relatively rare in the life cycle of most volcanoes. At any given time, most subaerial volcanic activity occurring on Earth consists of persistent, passive emissions of gases and aerosols, either in the form of focused gas plumes from summit vents (e.g., Mt Etna, Italy) or as diffuse emissions from **fumaroles** and volcano flanks (e.g., Vulcano, Italy), and thermal features associated with the heat transported to the surface from depth by the emitted gases. The ultimate origin of many volatile species in volcanic gas emissions, and surface geothermal manifestations, is magma stored at depth beneath volcanoes, and volcanic emissions provide the only direct chemical link to subsurface magma that could fuel eruptions. Volcanic gas, plume, and thermal monitoring is therefore one of the linchpins of the tripartite approach to volcano surveillance that typically also includes seismic and geodetic monitoring, with the overarching goal of forecasting volcanic eruptions. Furthermore, significant amounts of current volcanological research are devoted to measuring, interpreting, and understanding the origins of temporal trends in volcanic gas and heat emissions, and to developing new tools to quantify volcanic gas and heat fluxes (e.g., Kern et al., 2010a; Harris, 2013).

In addition to the relatively benign or "passive" releases of volcanic gases described above, monitoring of **volcanic clouds** and plumes is also crucial during eruptions (e.g., Surono et al., 2012). The large quantities of volcanic gases and aerosols (e.g., volcanic ash) discharged during major explosive eruptions can have significant effects on climate (e.g., the 1815 Tambora eruption), Earth's atmosphere, the natural and built environment, and the health of humans and other living organisms. Airborne volcanic ash is also a major hazard to jet aircraft, and accurate tracking of drifting volcanic ash clouds is essential for aviation safety (see Chapter 52). **Remote sensing** of volcanic gases and ash from the synoptic perspective of Earth-orbiting satellites plays a key role in these efforts.

This chapter summarizes the state-of-the-art in volcanic gas, plume, and thermal monitoring, describing the significant advances in this field that have occurred in the past decade. We focus on remote sensing from various platforms (ground based, airborne, and space-borne), since the vast majority of volcanic gas, plume, and thermal monitoring employs this technique. Although the theoretical principles of remote sensing remain essentially unchanged, the available technology and instrumentation for volcano remote sensing, and available computational resources for data processing and dissemination, have advanced greatly in recent years, permitting significant advances in measurement automation and sensitivity, instrument portability and data availability.

2. GAS MEASUREMENTS AND MONITORING

2.1. Principles

The key role of volatiles in driving volcanic activity has been appreciated since the early days of volcanology. At the great depths (i.e., pressures) of basaltic magma generation in the mantle, volatile species such as water (H_2O), carbon dioxide (CO_2), and sulfur species (sulfur dioxide, SO_2 and hydrogen sulfide, H_2S) are completely dissolved in the melt. As magmas ascend to lower pressures, volatile solubility tends to decrease and volatiles begin to exsolve from the melt and form bubbles, thus initiating the process of magma degassing and lowering bulk magma density. In low-viscosity basaltic magmas these bubbles, being less dense than the surrounding melt, will typically rise relative to the melt and may escape continuously to the surface through fracture networks in the crust and/or the magma column itself, or they may become trapped at some intermediate depth. If the gases escape to the surface, their detection and measurement, and variations over time, may represent an eruption precursor and provide indications of the quantity, composition, and storage conditions (pressure, temperature) of magma stored at depth. Gas release in a more sporadic, explosive manner in the form of rising gas pockets or slugs drives the classic Strombolian activity observed at many basaltic volcanoes (e.g., Stromboli, Italy).

In higher viscosity, silicic magmas, exsolved bubbles may be unable to separate from the melt, or may do so much more slowly. In such cases there may be little or no precursory release of gas prior to an eruption. As the bubble population increases, magma density decreases and this drives magma ascent toward the surface. At sufficiently high bubble fractions, the magma may fragment into a low-density mixture of gas and melt fragments (e.g., ash and pumice) and accelerate rapidly upward as an explosive volcanic eruption plume (or plinian eruption), instantaneously releasing large quantities of volcanic gas. Alternatively, if gas loss is permitted (e.g., through permeable networks of bubbles or via collapse of bubble-rich foams), the resulting degassed melt may erupt nonexplosively in an effusive eruption and form a lava dome or flow. Since volatile loss (particularly dewatering) impacts magma rheology by increasing melt viscosity (by increasing melt polymerization and undercooling), which in turn impedes subsequent degassing, there can be complex feedbacks between degassing and magma ascent on various time-scales. Monitoring volcanic gas emissions with high temporal resolution (e.g., hourly and daily) can assist efforts to recognize these processes and thus contributes to volcanic hazard assessment.

While their relative proportions depend on tectonic setting and magma composition, the major volatile species comprising volcanic gases are invariably H_2O, CO_2, sulfur species (SO_2, H_2S), and halogen species such as hydrogen chloride (HCl) and hydrogen fluoride (HF). In addition, volcanic gases contain many other minor and trace components, such as hydrogen (H_2), nitrogen (N_2), the noble gases helium (He) and argon (Ar), and trace metals (lead, zinc, gold). The abundances or isotopic ratios of some of these components (e.g., He) can be characteristic of the volatile source region (e.g., crust vs. mantle), and therefore of considerable diagnostic value. Although the details vary and are beyond the scope of this chapter, CO_2 is typically the least soluble volatile species in magma, SO_2 has intermediate solubility, and water and the halogen species (HCl, HF) are the most soluble. Among the sulfur species, H_2S is associated with higher pressures (i.e., greater depths) and lower temperatures than SO_2, and thus the former (recognized by its characteristic "rotten egg" odor) is the dominant sulfurous gas in fumarolic emissions at dormant volcanoes. SO_2, on the other hand, is thermodynamically favored in high-temperature magmatic gases at low pressures, and thus predominates in eruptive gases and emissions sourced from shallow magma reservoirs. For this reason, and because it is the easiest gas to measure using remote sensing techniques (see below), most volcanic gas monitoring is focused on SO_2.

Another important attribute of volatile species is their reactivity; the species that form strong acids (e.g., SO_2, HCl, and HF) are more water-soluble and susceptible to dissolution in hydrothermal fluids or groundwater (e.g., hydrolysis), whereas other volatile species (e.g., CO_2, H_2S, and noble gases) are more inert. Measurements of the latter are particularly important at "wet" volcanoes with extensive hydrothermal systems, ground- or surface water, for example tropical or glaciated volcanoes, volcanoes hosting crater lakes, or volcanic systems awakening after an extended period of dormancy. The depletion of magmatic volatiles prior to emission via dissolution, gas—water or gas—water—rock reactions (forming precipitates) within the volcanic edifice is generally referred to as "scrubbing." SO_2 is particularly prone to hydrolysis and disproportionation to sulfide (H_2S), sulfur (S), or sulfate (H_2SO_4), and hence SO_2 degassed from magma at depth will only reach the surface if a dry pathway is established. Prior to this, volatile species such as CO_2 and H_2S (along with H_2O of both magmatic and hydrothermal origin) may dominate in volcanic emissions. Thus, monitoring of multiple volatile species (e.g., CO_2, SO_2, H_2S, HCl, and HF) and their ratios is deemed more useful than surveillance of a single species for prognostic purposes, particularly at reawakening volcanoes.

Apart from scrubbing, the other major factor that determines whether volatiles degassed from magma will reach the surface is permeability. Volatiles may migrate upward through the magma column as bubbles, or through fractures in the surrounding crust and volcanic edifice. Broadly speaking, **volcanic degassing** regimes may be characterized as open-system, where permeability is high and most exsolved volatiles can freely escape to the surface, or closed-system, where gas loss is impeded (e.g., due to high melt viscosity or fracture sealing by precipitates). In the latter case, high gas pressures may develop, possibly triggering an explosive release of trapped gas or an eruption. Transitions between open- and closed-system degassing may be recognized by monitoring volcanic gas emission rates in conjunction with other geophysical data (e.g., seismic data; Carn et al., 2008).

Once released into the atmosphere as a **volcanic plume**, the fate of volcanic gases is species-dependent. The complex atmospheric chemistry of volcanic plumes is a major focus of current research. Understanding the depletion rates of volcanic gases is critical for accurate interpretation of remote sensing data collected at varying distances downwind of the emission source. Depletion of volcanic gases can occur by dissolution in condensed water (SO_2 and HCl are particularly soluble) and wet deposition, via gas—or aqueous-phase oxidation (e.g., SO_2 is oxidized to sulfate) and conversion to aerosol, or via dry deposition in grounded plumes. Reactive halogen species (BrO, ClO) can be generated in volcanic plumes from primary halogen species (e.g., HBr) via photochemical halogen-catalyzed ozone depletion reactions. Conversion of volcanic SO_2 to sulfate aerosol, which has significant effects on shortwave (UV—visible) and long-wave (IR) radiation and hence can impact climate, is the most important and well-understood atmospheric effect of volcanic emissions (see Chapter 53). However, in the current era of climate change and rising anthropogenic CO_2 emissions there is a renewed interest in the relative magnitude of volcanic CO_2 emissions, which are poorly constrained on a global scale. As discussed below, volcanic CO_2 is challenging to measure using remote sensing, but CO_2 does have the advantage of being relatively inert and hence immune to the processes that can deplete other volatile species in volcanic gases, either prior to or after emission.

2.2. Direct/In situ Sampling of Volcanic Gases

2.2.1. High-Temperature Fumarolic Discharges from Active Craters

Direct, in situ sampling of volcanic gases at the point of emission remains the only technique capable of providing the complete chemical and isotopic composition of volcanic gases (after laboratory analysis of the collected samples). Gases are most commonly collected by inserting a titanium or glass tube into a fumarole and bubbling the gases through an evacuated bottle of alkali solution (NaOH), which absorbs the acidic gases and allows other species to collect in the headspace (the "Giggenbach bottle" technique). The value of volcanic gas samples generally correlates with gas emission temperature, since hotter emanations will be more representative of magmatic gases and less contaminated by air or hydrothermal system interaction. The desire to sample the hottest and most highly pressurized fumaroles within active craters renders direct gas sampling inherently hazardous. Another drawback is the typically low temporal resolution of such measurements, particularly at more active volcanoes or those with less accessible craters and fumaroles. Furthermore, the most abundant volcanic gas, H_2O, can be difficult to quantify due to condensation in the collection tube. For further information on the collection and measurement of volcanic gases via direct sampling, the reader is referred to Chapter 45 on "Volcanic Gases."

2.2.2. In situ Measurements of Volcanic Plumes

Recent improvements in sensor technology have stimulated the development of portable sensor packages capable of measuring concentrations of multiple gas species in volcanic plumes (H_2O, CO_2, SO_2, H_2S, HCl, HF, and H_2), commonly termed Multi-gas sensors (e.g., Aiuppa et al., 2007). These can be deployed on crater rims to sample young volcanic plumes as they begin to drift downwind, permitting monitoring of gas ratios with high temporal resolution. In situ measurements are advantageous since they yield actual mixing ratios (e.g., in parts per million or ppm) of volatile species in volcanic plumes, which are often difficult to obtain from remote sensing data unless the physical dimensions of the plume are known (see below).

Multi-gas packages typically incorporate a combination of nondispersive infrared (NDIR) spectrometers (for CO_2, CO, and H_2O measurement), electrochemical sensors (for SO_2, H_2S, HCl, and HF), and semiconductor sensors (for H_2). Quantification of H_2O is most challenging since volcanogenic water must be distinguished from the large background atmospheric water vapor abundance, and it can only be measured in the vapor phase, whereas volcanic plumes invariably begin to condense immediately upon emission. Cross-sensitivity to various gas species can also be an issue with some detectors. The inevitable degradation of sensors in the acidic environment of a volcanic plume may also preclude long-term autonomous deployments.

Figure 65.1 shows an example of real-time multi-gas observations of H_2O, CO_2, and SO_2 applied to the forecasting of eruptions at Etna volcano (Aiuppa et al., 2007). The transition from normal, quiescent degassing behavior to explosive activity is presaged by an increase in the CO_2/SO_2 ratio, which is interpreted as the result of replenishment of the volcanic plumbing system with CO_2-rich magma rising from depth.

In situ measurements of volcanic plumes can also be made by mounting sensors on piloted aircraft, unmanned aerial vehicles (UAVs), balloons, or kites and penetrating the airborne gas plume aloft. In addition to gas mixing ratios, emission rates of SO_2, CO_2, and H_2S can also be derived from airborne in situ measurements by flying gas sensors in a ladder traverse across a volcanic plume. The resulting data are interpolated to produce a gas concentration map, which is then integrated over the plume cross-sectional area and multiplied by wind speed to yield gas flux. Data collection typically takes at least ~ 1 h and therefore stable wind conditions are required in order to avoid errors in the flux calculation, but in ideal conditions such techniques are complementary to, and perhaps more accurate than, flux calculations derived from remote sensing data.

Recently, several attempts have been made at in situ volcanic plume measurements conducted from UAVs (e.g., McGonigle et al., 2008), and this is likely to become more widespread in the future. The use of UAVs for volcano monitoring is becoming more attractive due to the significantly reduced cost and risk, as well as increased accessibility with lightweight, portable, low-cost UAVs now available, including relatively stable heli-, quad, or octo-copters. However, UAVs often require experienced pilots, can be difficult to control in windy conditions, and although becoming more affordable are not yet truly expendable. Furthermore, widespread adoption of UAVs (or unmanned aircraft systems, UAS) for volcano surveillance is likely to depend on regulations for their integration into national airspace shared with manned aircraft, which will vary between nations. At the extreme end of the UAV spectrum are platforms such as NASA's Global Hawk, which has a range of 11,000 nautical miles, a 32 h flight time, and can carry a 1500-pound payload, and could potentially be used to intercept and sample drifting volcanic clouds in remote regions. Such in situ airborne measurements are valuable for validation of satellite remote sensing **retrievals** of volcanic gases.

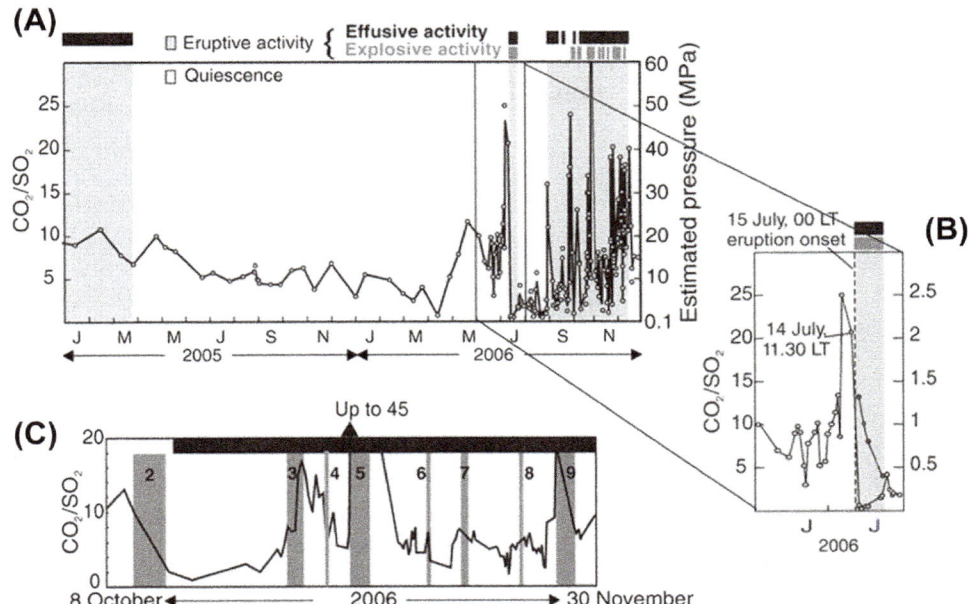

FIGURE 65.1 (A) Time evolution of CO_2/SO_2 molar ratios in Etna's central crater plume, measured by in situ multi-gas sensors. Right axis shows estimated pressures (in MPa) evaluated by combining volcanic gas data with model results. Timing of 2004—2005 and 2006 eruptions are also shown. (B) Detail of June—July 2006 period. Gray dots refer to composition of plume released at eruptive vent (right scale). LT—local time. (C) Detail of the October—November 2006 period. Dark gray bars indicate timing of Strombolian events 2—9 that occurred at southeast summit crater (SEC). *Reprinted from Aiuppa, A., Moretti, R., Federico, C., Giudice, G., Gurrieri, S., Liuzzo, M., Papale, P., Shinohara, H. and Valenza, M., 2007. Forecasting Etna eruptions by real-time observation of volcanic gas composition. Geology. 35 1115—1118, doi:10.1130/G24149A.1. Copyright 2007, with permission from Geological Society of America.*

2.2.3. Soil Gas (CO_2 Emissions)

In addition to the conspicuous, focused volcanic emissions from summit crater plumes and fumaroles, volcanoes also release significant amounts of gas (particularly CO_2) through their flanks in the form of diffuse or soil gas emissions. Since soil gases must traverse considerable thicknesses of crust after release from magma reservoirs at depth, they are dominated by inert gases that are not scrubbed by groundwater, such as CO_2 and noble gases (He, Rn). The global volcanic flux of diffuse CO_2 is poorly constrained but believed to be quite significant: perhaps ∼25% of the total global subaerial volcanic CO_2 flux from passive crater degassing, diffuse degassing, and volcanic lakes (Burton et al., 2013).

Several techniques have been used to quantify diffuse CO_2 emissions. Direct methods include the accumulation chamber (or closed-chamber) method and eddy covariance or alternately eddy correlation (EC). The static accumulation chamber method, originally used in agriculture to determine soil respiration, was adapted to measure volcanogenic soil CO_2 emissions. An open-bottomed chamber of known volume is inverted on the surface and the increase in CO_2 concentration is measured; the initial rate of change of the concentration is proportional to the CO_2 flux. Advantages of the accumulation chamber method include

immunity to assumptions about soil characteristics or the flux regime (e.g., advective/diffusive). The main challenge with such localized measurements is extrapolation to an entire volcanic edifice, with the usual approach involving multiple measurements to achieve a representative mapping of CO_2 flux, followed by estimation of the total CO_2 release using interpolation algorithms. Floating accumulation chambers have also been deployed on volcanic lakes to determine CO_2 fluxes.

More recently, the micrometeorological eddy covariance technique has been tested as a method to monitor diffuse volcanic CO_2 emissions. EC derives the CO_2 flux at the surface from the covariance between fluctuations of the vertical component of the wind and the fluctuations of the atmospheric CO_2 concentration. The EC provides advantage of being an automated, time-averaged, and area-integrated technique with a spatial scale significantly larger (square meter to square kilometer) than that of the accumulation chamber method. However, volcanic environments may be too heterogeneous for EC application in terms of the spatial and temporal variability of surface fluxes and surface morphology.

Diffuse CO_2 degassing measurements require careful interpretation since CO_2 fluxes are influenced by surface soil permeability and atmospheric pressure changes that can mask signals from deeper magma reservoir or conduit degassing. Soil gas permeability is dependent on soil moisture content; hence, tropical volcanoes are particularly susceptible.

2.3. Remote Sensing Measurements of Volcanic Gases

The use of remote sensing to measure volcanic emissions offers several advantages over direct sampling and in situ measurements. Remote sensing is less hazardous, as data can be collected from relatively safe distances, and data can also be collected more rapidly, more frequently, and over larger areas (e.g., an entire volcanic plume) than is achievable with direct sampling. The principal drawbacks of remote sensing are the limited range of volcanic volatile species that can be reliably quantified, and that the technique yields integrated (or "column") amounts of gases along the optical path between the sensor and the source of EM radiation (i.e., gas mixing ratios cannot be determined unless the plume thickness is known). To be measurable by **passive remote sensing**, utilizing a natural source of electromagnetic (EM) radiation (e.g., the Sun, skylight, or heat from an active volcanic vent), a volatile species must satisfy several criteria: it must absorb EM radiation in a waveband that provides adequate signal at the Earth's surface (usually in an **atmospheric window** with high transmittance), it must be present in significant concentrations in the volcanic plume (i.e., above detection limits), and its abundance in the ambient atmosphere surrounding the plume must be low (or ideally, zero). The two most commonly exploited wavebands for passive remote sensing of volcanic gases are the **ultraviolet** (UV) band (wavelengths of $\sim 0.3-0.4$ μm; Figure 65.2) and the infrared (IR) band (wavelengths of $\sim 2-10$ μm; Figure 65.2).

The only major component of volcanic plumes that usually satisfies the above criteria is SO_2, which has strong absorption bands in the UV (at $\sim 310-340$ nm; Figure 65.2) and IR (at ~ 4, ~ 7.3 and ~ 8.6 μm; Figure 65.2), and low ambient concentrations away from pollution sources (e.g., power plants and metal smelters). Minor volatiles quantifiable by passive remote sensing include CO, HCl, HF, halogen oxides (bromine monoxide (BrO), chlorine monoxide [ClO], and chlorine dioxide (OClO)), carbonyl sulfide (OCS), and nitrogen dioxide (NO_2). Notably, the most abundant volcanic gases (H_2O and CO_2) are notoriously difficult to measure by remote sensing due to their high atmospheric abundance ($\sim 1-4$ vol% and ~ 400 ppmv (parts per million by volume), respectively), although this problem can be surmounted by maximizing concentrations of volcanic gases in the optical path, or via **active remote sensing** using an artificial source of EM radiation. The latter technique has also been used to measure volcanic H_2S emissions, using a

deuterium UV lamp to supply a source of the short-wavelength UV radiation ($0.2-0.25$ μm) where H_2S absorbs. Active sensing techniques can be used to artificially extend the optical path and detect volatile species present in very low concentrations (e.g., OClO).

Passive UV remote sensing typically uses scattered sunlight as the source of EM radiation, restricting the technique to daylight hours, ideally under relatively cloud-free conditions and low solar zenith angles (around local noon). It is often assumed that the solar UV radiation, after scattering by air molecules, follows a direct (i.e., straight) path through the volcanic plume. However, determination of the actual UV optical path can be highly complex when using a scattered light source due to multiple scattering of UV radiation by air molecules and aerosols (e.g., clouds) either outside or within the volcanic plume. This can be a major source of error in UV remotely sensed measurements of volcanic gases. Direct sun measurements, made by pointing the sensor at the Sun, are also possible in the UV and circumvent the optical path problem since the path can be deduced straightforwardly from geometric principles. Path-length issues are generally negligible in IR remote sensing of volcanic gases due to less significant scattering at the longer IR wavelengths.

2.3.1. Ground-Based and Airborne Gas Measurements

There have been significant advances in remote sensing technology since the adoption of the UV correlation spectrometer (COSPEC) for ground-based SO_2 measurements in the late 1960s. The COSPEC is a bulky, analog "black box" instrument that provides accurate SO_2 measurements under certain conditions but provides scant opportunity for data reprocessing or evaluation of radiative transfer effects on the measurements. In the past decade, the COSPEC has been largely superseded by compact fiber-optic UV spectrometers that record continuous spectra at high **spectral resolution** in the UV wavelength range (Galle et al., 2003). These spectrometers can be deployed in exactly the same configurations as the COSPEC: by traversing beneath a volcanic plume by road, sea, or air to obtain a plume cross-section, or by scanning the sensor field of view (FOV) through a plume horizontally or vertically from a stationary position. However, the significant reduction in physical size, cost, and power consumption of miniature UV spectrometers relative to a COSPEC has also permitted novel applications such as walking traverses (using a helmet-mounted spectrometer), UAV deployment, and automated scanning spectrometer networks (e.g., to ensure that volcanic plumes can be measured autonomously under multiple wind direction regimes). Dual-beam and dual-FOV spectrometer configurations have also been developed in a bid to reduce the largest source of error in SO_2 flux

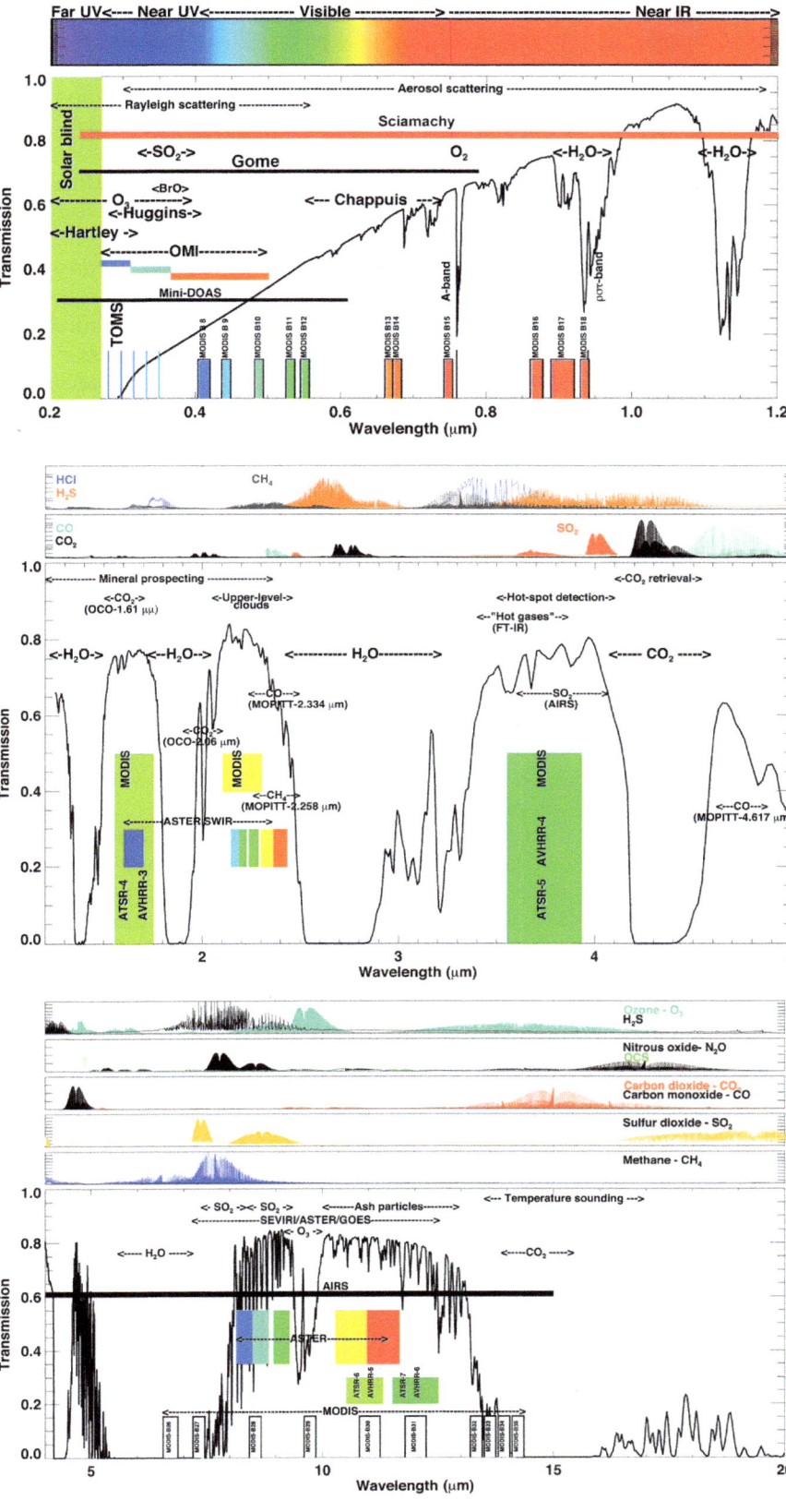

FIGURE 65.2 The electromagnetic spectrum showing regions where remote sensing of gases and particles is possible, and typical levels of atmospheric transmission (*black line*). Top panel: Ultraviolet (UV), visible, and near infrared (NIR) part of the spectrum indicating regions where gases absorb and the locations of the bands used by various satellite instruments (Table 65.1). The region between 280 and 340 nm (0.28 and 0.34 µm) is used by several satellite instruments to measure volcanic SO_2. Middle panel: The middle infrared part of the spectrum. Water vapor and CO_2 have major absorption bands, but there are gaps where remote sensing of the surface to detect volcanic hot-spots is possible. Thermal alerts generated by satellites are useful early indicators of volcanic unrest. Bottom panel: The TIR part of the spectrum from ∼5–20 µm wavelength. There are important absorption bands of SO_2 in this region that are routinely used to measure this gas. This is also the principal region used to measure ash particles in the atmosphere. Figure courtesy of Fred Prata, NILU.

measurements, which is invariably the determination of plume speed (usually assumed to equal the wind speed at plume altitude). In turn, these developments have stimulated the deployment of autonomous spectrometer networks at a greater number of degassing volcanoes (e.g., the Network for Observation of Volcanic and Atmospheric Change or NOVAC project (Galle et al., 2010)) and are improving our knowledge of global volcanic SO_2 emission rates.

Reduction of UV spectrometer data typically utilizes the differential optical absorption **spectroscopy** (DOAS) technique, widely used in the atmospheric science community, and capable of retrieving path amounts of SO_2, BrO, ClO, OClO, and NO_2 from UV–visible spectra of volcanic plumes. DOAS exploits the characteristic narrow-band UV–visible absorption structures of trace gas molecules, and hence requires measurements at high spectral resolution. Retrievals can be performed in real-time using current laptop computers, and collected spectra can be archived and reprocessed using improved retrieval algorithms. Corrections for radiative transfer effects (e.g., multiple scattering) can also be implemented using DOAS measurements, improving the accuracy of volcanic SO_2 retrievals (e.g., Kern et al., 2010b).

UV spectrometers are also deployed on aircraft to measure SO_2, although the cost of aircraft time usually prohibits routine airborne plume surveillance, except in remote regions where it may be the only option. Airborne plume traverse measurements of SO_2 with UV spectrometers can also be combined with in situ measurements (e.g., CO_2 and H_2S) to derive fluxes of other volcanic gases. Currently, this technique is probably the most robust for measuring volcanic CO_2 fluxes (which cannot be readily quantified using remote sensing techniques), since airborne SO_2 measurements have the advantage of proximity to the volcanic plume, minimizing the effect of light dilution between the plume and the instrument. Airborne measurements of volcanic CO_2 emissions from Kilauea have also been achieved by measuring IR transmission through the volcanic plume with the AVIRIS **hyperspectral** radiometer.

2.3.2. UV Imaging Cameras

A more recent development in UV volcanic plume remote sensing has been the adoption of UV imaging cameras for SO_2 measurements. Although the concept of volcanic plume imaging was first posited some time ago, only recently has appropriate UV imaging hardware become readily available. Originally designed for astronomical observations, commercially available UV cameras utilize charge-coupled devices (CCDs) that provide sensitivity at UV wavelengths where SO_2 absorbs radiation. The advantage over conventional UV DOAS techniques is that

UV cameras provide two-dimensional images of volcanic plumes, and when collected at high temporal resolution (≤ 1 Hz is typical) such imagery can be used to visualize plume dynamics and obtain plume speed information, which is required for SO_2 emission rate calculations. Hence SO_2 emission rates can be measured at high temporal resolution, permitting correlation with other geophysical data such as seismic or infrasound measurements (e.g., Figure 65.3).

Disadvantages of UV cameras are that they have low spectral resolution, since wavelengths must be selected using relatively broadband filters that cannot resolve the detailed spectral variation of the SO_2 absorption cross-section. This also necessitates calibration, either using SO_2 gas cells of known concentration or using independent spectral information from a co-aligned UV spectrometer. Hybrid techniques such as Imaging-DOAS (I-DOAS) circumvent the calibration issue by physically scanning one row of a 2D CCD array across an FOV to form an image (with one CCD dimension acquiring spectral information while the other acquires vertical spatial information), performing a DOAS trace gas retrieval for each scan increment, at the expense of temporal resolution. Despite the disadvantages of UV cameras, the potential insights into plume dynamics and volcanic conduit processes (such as gas accumulation and release) provided by UV cameras, and their increasing affordability, have stimulated major interest in these instruments among the volcano remote sensing community.

UV camera technology for volcanological applications continues to improve, with more sensitive back-thinned CCDs now available that increase the signal-to-noise in UV camera imagery, along with systems lacking mechanical shutters with higher frame rates and longer lifetimes. Calibration using a co-aligned spectrometer is likely to remain the best approach for SO_2 retrieval from UV camera imagery, due to the limitations on filter bandwidths and the trade-off between spectral resolution and signal-to-noise (narrower filter bandwidth reduces the number of photons transmitted). These advances render UV cameras more suitable for long-term autonomous operation, and many volcano observatories will likely incorporate UV cameras into their monitoring networks over the coming decade to measure SO_2 emissions.

2.3.3. IR Spectroscopy

Although UV spectroscopy is more commonly used for remote sensing of volcanic plumes, IR spectroscopy offers several key advantages. It can be used by day or night, and many volcanoes inherently provide a strong, natural source of IR radiation (e.g., a hot vent, lava lake, or lava flow) that can be exploited for measurements. Nonvolcanic natural sources of IR radiation, such as direct solar radiation from

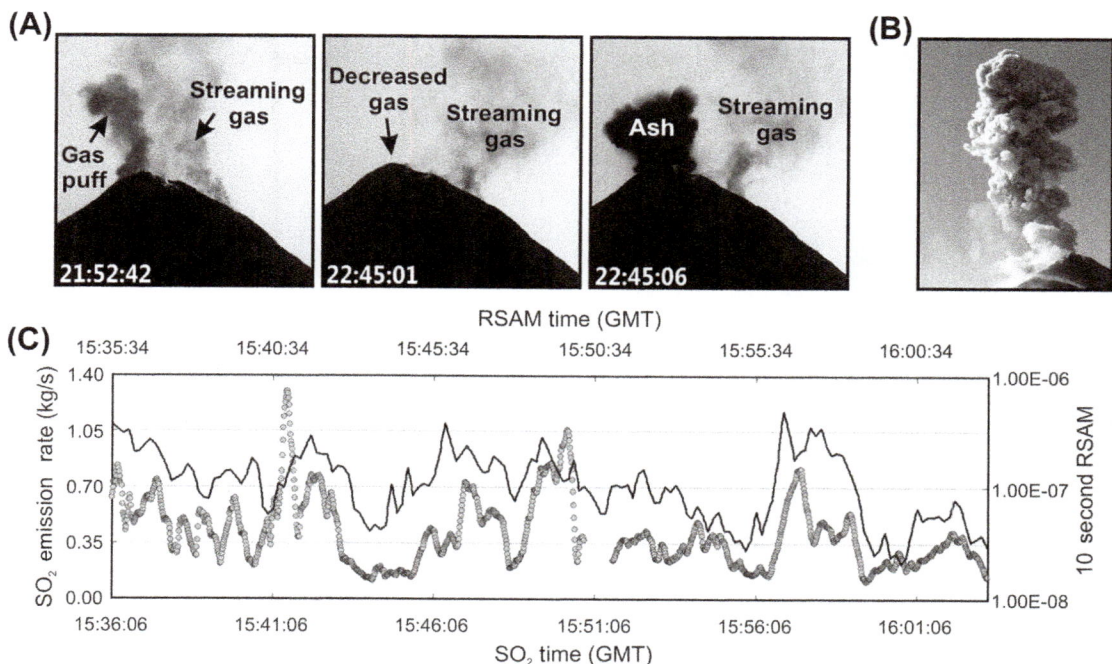

FIGURE 65.3 (A) Ultraviolet (UV) camera imagery of variable gas emissions from two vents at Fuego volcano, Guatemala, prior to an explosion. Times are UTC on 21 January 2009; (B) Example of an ash-rich explosion plume (~ 450 m in height above summit); (C) UV camera SO_2 emission rates from Fuego on 21 January 2009 (*circles*) and low-frequency 10 s average seismic amplitude (RSAM; *black line*). Note the strong correlation between these high-frequency datasets, with a slight temporal offset between them (x-axes are offset by 32 s to compensate for overall lag). *From Nadeau et al. (2011).*

the Sun or reflected solar radiation from clouds or the Moon, or artificial sources (e.g., an IR lamp), can also be used. The Fourier-Transform IR (FTIR) spectroscopy technique permits simultaneous measurement of several important volcanic gas species, including H_2O, CO_2, SO_2, HCl, and HF, providing valuable constraints on gas ratios, which can be diagnostic of the pressure (i.e., depth) of origin of volcanic gases (e.g., Allard et al., 2005; Figure 65.4). The usual requirement for FTIR measurements is that the volcanic gas be at a lower temperature than the source of background IR radiation, so that gas abundance can be determined using principles of absorption spectroscopy, which is always the case for solar occultation FTIR measurements. Since volcanic gases cool rapidly to sub-magmatic temperatures upon emission, these conditions are also usually satisfied when observing gas emissions against a background of lava lakes (e.g., Figure 65.4), lava fountains, or hot vents. Alternatively, an artificial source of IR radiation such as an IR lamp can be employed in active-mode FTIR spectroscopy. Volcanic gases can also be observed in emission, e.g., against a cloudy or clear-sky background at a lower temperature, although in these cases absolute radiometric calibration of the FTIR using a blackbody radiation source is essential.

Although FTIR spectrometers may never be as compact as fiber-optic UV spectrometers, reasonably rugged and portable FTIR instruments are available for field deployment at volcanoes, with thermoelectric or Stirling-cycle detector cooling available to avoid the use of liquid nitrogen. Most existing FTIR volcanic gas data have been collected during short field campaigns at degassing volcanoes. However, in 2008 a remote-controlled FTIR instrument was installed at the summit of Stromboli volcano (Italy), to permit more frequent measurements of volcanic gas ratios (e.g., SO_2/CO_2 and SO_2/HCl) in emissions from individual vents on the volcano's crater terrace. Measurements have shown systematic increases in CO_2/SO_2 ratios prior to major explosions at Stromboli, and are thus valuable for hazard assessment. Imaging FTIR instruments, which collect both spectral and spatial information, are also available but remain prohibitively expensive for most volcano monitoring applications, although their use may increase in the future.

2.4. Satellite Remote Sensing of Volcanic Gases and Aerosols

The first satellite measurements of SO_2 emitted by explosive volcanic eruptions, made by the Total Ozone Mapping Spectrometer (TOMS) instrument in the 1970s, provided a new perspective on volcanic degassing. The large quantities of SO_2 released in some cases, such as the ~ 20 Tg measured after the 1991 eruption of Pinatubo (Philippines),

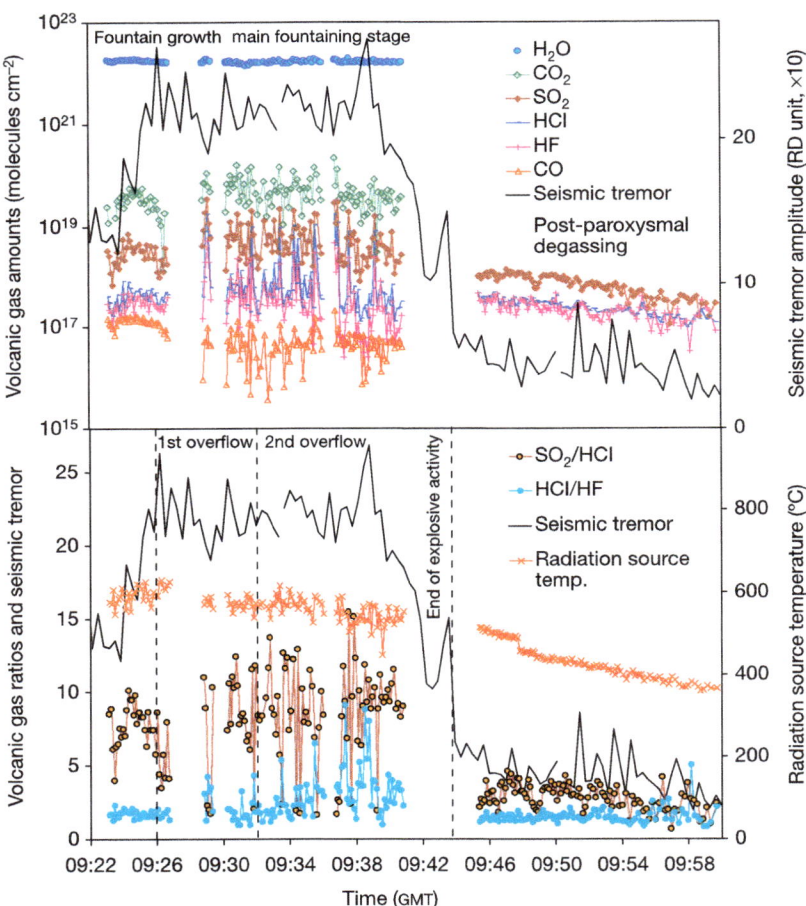

FIGURE 65.4   Top: Fourier-Transform infrared (FTIR) spectrometer measurements at Nyiragongo volcano (DR Congo) in June 2007. The active lava lake in the central pit crater, at magmatic temperatures, acts as the source of infrared radiation for the FTIR spectrometer (*white box in foreground*). Volcanic gases cool rapidly to near-ambient temperatures upon emission, permitting measurement via infrared absorption spectroscopy. Bottom: Temporal evolution of volcanic gas amounts and ratios measured by FTIR during and after the 14 June 2000 lava fountain at Etna (Sicily). Upper plot shows a time series of measured path amounts of volcanic gases (left-hand log scale), together with volcanic tremor amplitude (right-hand scale, RD indicates reduced displacement, in cm^2) as an indicator of eruption intensity. Lower plot shows the evolution of SO_2/HCl and HCl/HF ratios (left-hand scale) during and after the fountaining, and evolution of the radiating source temperature retrieved from FTIR spectra (right-hand scale). Note the higher SO_2/HCl ratio during the fountaining phase, indicative of a deeper gas source. Left and center dashed lines indicate the onset of bursts of lava overflowing the upper south crater rim; right dashed line indicates end of explosive activity. *From Allard et al. (2005).*

often vastly exceeded petrological estimates of degassing and spawned new research into the origin of this "excess sulfur." Space-borne SO_2 measurements also facilitated more quantitative studies of the effects of volcanic emissions on the Earth's atmosphere and climate, the latter of which are largely due to sulfate aerosol derived from SO_2. Satellite sensors remain the only viable means to detect and monitor emissions from remote or unmonitored volcanoes, or from volcanoes in eruption that are too hazardous for ground-based or airborne surveillance. Over the past two decades, satellite remote sensing has evolved from a tool used exclusively to study large volcanic eruptions to one that can also be used to monitor passive volcanic degassing of SO_2 and small eruptions, and thus plays a greater role in routine volcano monitoring (e.g., Carn et al., 2013). The increased sensitivity to volcanic SO_2 is the result of improvements in the spectral and **spatial resolution** of satellite instruments.

2.4.1. Ultraviolet Measurements of Volcanic Gases

Like suborbital UV remote sensing, satellite UV measurements use scattered sunlight as a radiation source and are predominantly used to measure volcanic SO_2, but the reactive halogen species BrO and OClO have also been detected in volcanic clouds after some large eruptions (e.g., Kasatochi in 2008 and Puyehue-Cordón Caulle in 2011; Theys et al., 2014). Since solar UV radiation at longer wavelengths ($>0.31-0.32$ μm) is not absorbed by O_2 and O_3 in the upper atmosphere, it is transmitted to Earth's surface, providing sensitivity to volcanic SO_2 at low altitudes (i.e., passive degassing) in the absence of thick overlying meteorological clouds. The increase in atmospheric transmission of UV radiation with wavelength has also been exploited to retrieve the altitude of volcanic SO_2 directly from UV satellite measurements, rather than inferring cloud height indirectly (e.g., by comparing cloud drift with winds aloft). Constraining SO_2 altitude is crucial for climate modeling and aviation hazard mitigation.

Four UV TOMS missions provided a near-continuous, long-term record of volcanic SO_2 emissions from major eruptions between 1978 and 2005. There are currently three operational, polar-orbiting UV satellite sensors continuing this long-term record: the Ozone Monitoring Instrument (OMI) on NASA's Aura satellite, the Global Ozone Monitoring Experiment 2 (GOME-2) on MetOp-A/B, and the Ozone Mapping and Profiler Suite (OMPS) aboard Suomi-NPP (Table 65.1). OMI, GOME-2, and OMPS are hyperspectral instruments, providing higher spectral (and in the case of OMI, spatial) resolution than the **multispectral** TOMS. This allows OMI, GOME-2, and OMPS to detect lower tropospheric volcanic SO_2 plumes released by passive degassing (e.g., Figure 65.5), in addition to

volcanic SO_2 clouds in the upper **troposphere** and lower **stratosphere** (UTLS; e.g., Figure 65.6; also see related multimedia content), and hence quantify volcanic SO_2 emissions on a broader scale than was previously possible. Although reconciling ground-based and satellite measurements of volcanic SO_2 emissions has proved challenging at some volcanoes (likely due to the significant spatial averaging of volcanic plumes in satellite observations), these measurements are providing improved constraints on the spatial and temporal variability of volcanic degassing (e.g., Carn et al., 2013). They can also be used in conjunction with other geophysical datasets to gain insight into volcanic processes, even when ground-based SO_2 measurements are unavailable (e.g., Figure 65.7).

UV satellite measurements of volcanic SO_2 clouds in the UTLS are very robust, and sensitive instruments such as OMI, OMPS, and GOME-2 can detect SO_2 emissions from a wide range of eruption magnitudes. One constraint on eruption detection is timing: since UV measurements are daytime only and at specific overpass times (Table 65.1), nighttime eruptions may not be captured if SO_2 discharge is low and disperses before the following day (but IR sensors measure at night—see below). Some of the largest SO_2 releases measured over the past decade have been from Kasatochi in August 2008 (~ 2 Tg), Sarychev Peak (Kurile Islands) in June 2009 (~ 1 Tg), and Nabro (Eritrea) in June 2011 ($\sim 3-5$ Tg; Figure 65.6), although the proportion of SO_2 reaching the stratosphere (where it has most impact on climate) varies between eruptions. The Nabro eruption was notable as the pattern of atmospheric SO_2 dispersion after the initial injection led to suggestions that the Asian Monsoon anticyclonic circulation played a role in lofting the gas into the stratosphere, rather than direct injection above the volcanic vent, although the relative significance of these processes remains unclear. However, it does highlight other potential pathways for tropospheric volcanic SO_2 to enter the stratosphere and impact climate. In addition to these relatively large SO_2 emissions (although none approach the 1991 Pinatubo SO_2 loading), there have been regular injections of $\sim 0.1-0.3$ Tg SO_2 into the UTLS from smaller eruptions. Recently it has been posited that these frequent, small injections may play an important role in maintaining the stratospheric sulfate aerosol layer at above-background levels, partly offsetting the warming effects of tropospheric greenhouse gases.

2.4.2. Thermal IR and Microwave Measurements of Volcanic Gases

Numerous thermal IR (TIR) satellite sensors also have the capability to measure volcanic SO_2 (Table 65.1; Figure 65.2), including multispectral instruments (e.g., MODIS, ASTER, SEVIRI, HIRS, and VIIRS) and hyperspectral sounders (e.g., AIRS, IASI, and CrIS). The

TABLE 65.1 Currently Operational (June 2014) and Near-Future Satellite Missions with Volcanic Plume/Thermal Monitoring Capabilities

Satellite[1]	Sensor(s)[1]	Launch Date	Volcanic Features Monitored[2]	Overpass Time (Local)[3]	Resolution Spatial (nadir)[4]	Resolution Temporal[5]	Websites[6]
Polar-orbiting (LEO)							
NOAA-15	AVHRR/3 HIRS/3	13 May 1998	Hot spots, ash UTLS SO₂	7:30 am	1 km 18 km	Daily Daily	USGS: http://volcview.wr.usgs.gov/
Landsat 7	ETM+	15 Apr 1999	Hot spots	9:45 am	15–60 m	16 days	Landsat science: http://landsat.gsfc.nasa.gov/
NASA Terra	MODIS ASTER MISR MOPITT	18 Dec 1999	Hot spots, SO₂, ash, aerosols Hot spots, SO₂, ash, aerosols Ash/gas plumes (altitude) CO	10:30 am	250 m–1 km 15–90 m 250–275 m 22 km	2 × daily 16 days 9 days	MODVOLC: http://modis.higp.hawaii.edu/ AVA: http://ava.jpl.nasa.gov/ MISR at JPL: http://www-misr.jpl.nasa.gov/ NCAR: https://www2.acd.ucar.edu/mopitt
NASA EO-1	ALI Hyperion	21 Nov 2000	Hot spots Hot spots	10:00 am	10–30 m 30 m	16 days 16 days	USGS: http://eo1.usgs.gov/
NASA Aqua	MODIS AIRS	4 May 2002	Hot spots, SO₂, ash, aerosols UTLS SO₂, ash, aerosols	1:30 pm	250 m–1 km 13.5 km	2 × daily 2 days	MODVOLC: http://modis.higp.hawaii.edu/ SACS: http://sacs.aeronomie.be/nrt/
NASA Aura	OMI MLS	15 Jul 2004	SO₂, BrO, OClO, ash UTLS SO₂, HCl, CH₃Cl	1:45 pm	13 × 24 km 1.5 × 3 × 300 km	~Daily 16 days	NASA: http://so2.gsfc.nasa.gov/ https://earthdata.nasa.gov/labs/worldview/
NOAA-18	AVHRR/3 HIRS/4	20 May 2005	Hot spots, ash UTLS SO₂	1:30–2:30 pm	1 km 10 km	Daily Daily	USGS: http://volcview.wr.usgs.gov/
CALIPSO	CALIOP	28 Apr 2006	Ash/aerosol cloud altitude	1:31 pm	333 m (horizontal)	16 days	http://www-calipso.larc.nasa.gov/
CloudSat	CPR	28 Apr 2006	Hydrometeors, ash aggregates	1:31 pm	1.5 km	16 days	http://cloudsat.atmos.colostate.edu/
EUMETSAT MetOp-A	GOME-2 IASI AVHRR/3 HIRS/4	19 Oct 2006	SO₂, BrO, OClO, ash SO₂, H₂S, CO, ash, aerosols Hot spots, ash UTLS SO₂	9:30 am	80 × 40 km 12 km 1 km 10 km	~Daily 2 × daily Daily Daily	SACS: http://sacs.aeronomie.be/nrt/ http://cpm-ws4.ulb.ac.be/Alerts/index.php http://www.nsof.class.noaa.gov/
NOAA-19	AVHRR/3 HIRS/4	6 Feb 2009	Hot spots, ash UTLS SO₂	1:30–2:30 pm	1 km 10 km	Daily Daily	USGS: http://volcview.wr.usgs.gov/

Platform	Instrument	Launch date	Overpass time	Species	Spatial resolution	Temporal resolution	URL
NASA/NOAA Suomi NPP	OMPS / VIIRS / CrIS	29 Oct 2011	1:30 pm	SO_2, ash / Hot spots, SO_2, ash / UTLS SO_2, aerosols	50 km / 375–750 m / 14 km	Daily / 2 × daily / Daily	NASA: http://so2.gsfc.nasa.gov/ http://viirsland.gsfc.nasa.gov/index.html http://npp.gsfc.nasa.gov/cris.html
EUMETSAT MetOp-B	GOME-2 / IASI / AVHRR/3 / HIRS/4	17 Sep 2012	9:30 am	SO_2, BrO, OClO, ash / SO_2, H_2S, CO, ash, aerosols / Hot spots, ash / UTLS SO_2	80 × 40 km / 12 km / 1 km / 10 km	~Daily / 2 × daily / Daily / Daily	SACS: http://sacs.aeronomie.be/nrt/ http://cpm-ws4.ulb.ac.be/Alerts/index.php http://www.nsof.class.noaa.gov/
Landsat 8	OLI, TIRS	11 Feb 2013	10:00 am	Hot spots, ash	15–100 m	16 days	USGS: http://landsat.usgs.gov/landsat8.php
NASA OCO-2	OCO-2	2 Jul 2014	1:15 pm	CO_2	1.3 × 2.3 km	16 days	http://oco.jpl.nasa.gov/
ESA Sentinel-2	MSI	2014	10:30 am	Hot spots	10–60 m	5 days	
ESA Sentinel-5 precursor	TROPOMI	2015	1:35 pm	SO_2, BrO, OClO, ash	7 × 7 km	Daily	http://www.tropomi.eu/TROPOMI/Home.html
Geostationary (GEO)							
EUMETSAT Meteosat-7	MVIRI	9 Feb 1997	n/a	Hot spots	2.5–5 km; 57°E	30 min	http://volcano.ssec.wisc.edu/
CMA FY-2E	S-VISSR	19 Oct 2004	n/a	Hot spots, ash	1.25–5 km; 105°E	30 min	http://www.ssec.wisc.edu/data/geo/
EUMETSAT Meteosat-9	SEVIRI	22 Dec 2005	n/a	Hot spots, ash, SO_2	1–3 km; 9.5°E	5 min	http://volcano.ssec.wisc.edu/ http://fred.nilu.no/sat/ HotVolc: http://www.obs.univ-bpclermont.fr/SO/televolc/hotvolc/HVOS/
JMA MTSAT-2	Himawari-7	18 Feb 2006	n/a	Hot spots, ash	1.25–5 km; 145°E	30 min	http://volcano.ssec.wisc.edu/
CMA FY-2D	S-VISSR	15 Nov 2006	n/a	Hot spots, ash	1.25–5 km; 87°E	30 min	http://www.ssec.wisc.edu/data/geo/
GOES-14 (E)	Imager	27 Jun 2009	n/a	Hot spots, ash	1–4 km; 75°W	1 min	http://volcano.ssec.wisc.edu/
GOES-15 (W)	Imager	4 Mar 2010	n/a	Hot spots, ash	1–4 km; 135°W	1 min	http://volcano.ssec.wisc.edu/
KMA COMS-1	MI	26 Jun 2010	n/a	Hot spots, ash	1–4 km; 128°E	10 min	http://www.ssec.wisc.edu/data/geo/

(Continued)

TABLE 65.1 Currently Operational (June 2014) and Near-Future Satellite Missions with Volcanic Plume/Thermal Monitoring Capabilities—cont'd

Satellite[1]	Sensor(s)[1]	Launch Date	Volcanic Features Monitored[2]	Overpass Time (Local)[3]	Resolution Spatial (nadir)[4]	Resolution Temporal[5]	Websites[6]
CMA FY-2F	S-VISSR	13 Jan 2012	Hot spots, ash	n/a	1.25–5 km; 113°E	30 min	
EUMETSAT Meteosat-10	SEVIRI	5 Jul 2012	Hot spots, ash, SO₂	n/a	1–3 km; 0°	15 min	http://volcano.ssec.wisc.edu/ http://fred.nilu.no/sat/
JMA MTSAT	Himawari-8	2014	Hot spots, ash, SO₂	n/a	0.5–2 km; 145°E	2.5 min	http://mscweb.kishou.go.jp/himawari89/
GOES-R	ABI	Early 2016	Hot spots, ash, SO₂	n/a	0.5–2 km; 75°/137°W	30 s	http://www.goes-r.gov/
L1 Lagrange libration point							
NOAA DSCOVR	EPIC	2015	SO₂, ash	n/a	8 km; sunlit Earth disk	90 min	http://www.osd.noaa.gov/DSCOVR/dscovr.html

[1]Satellite and sensor acronyms: NOAA: National Oceanic and Atmospheric Administration; NASA: National Aeronautics and Space Administration; EO: Earth Observing; CALIPSO: Cloud-Aerosol Lidar and Infrared Pathfinder Satellite Observations; EUMETSAT: European Organization for the Exploitation of Meteorological Satellites; NPP: National Polar-orbiting Partnership; ESA: European Space Agency; CMA: China Meteorological Administration; FY: Feng-Yun; JMA: Japan Meteorological Agency; MTSAT: Multi-functional Transport SATellite; GOES: Geostationary Operational Environmental Satellite; KMA: Korean Meteorological Administration; COMS: Communication, Ocean, and Meteorological Satellite; DSCOVR: Deep Space Climate ObserVatoRy; AVHRR: Advanced Very High-Resolution Radiometer; HIRS: High-resolution Infrared Radiation Sounder; ETM: Enhanced Thematic Mapper; MODIS: MODerate resolution Imaging SpectroRadiometer; ASTER: Advanced Space-borne Thermal Emission and Reflection radiometer; MISR: Multi-angle Imaging SpectroRadiometer; MOPITT: Measurements Of Pollution In The Troposphere; ALI: Advanced Land Imager; AIRS: Atmospheric InfraRed Sounder; OMI: Ozone Monitoring Instrument; MLS: Microwave Limb Sounder; CALIOP: Cloud-Aerosol Lidar with Orthogonal Polarization; CPR: Cloud Profiling Radar; GOME: Global Ozone Monitoring Experiment; IASI: Infrared Atmospheric Sounding Interferometer; OMPS: Ozone Mapping and Profiler Suite; VIIRS: Visible Infrared Imaging Radiometer Suite; CrIS: Cross-track Infrared Sounder; OLI: Operational Land Imager; TIRS: **Thermal Infrared** Sensor; OCO: Orbiting Carbon Observatory; MSI: Multispectral Imager; TROPOMI: TROPOspheric Monitoring Instrument; MVIRI: Meteosat Visible and Infrared Imager; S-VISSR: Stretched Visible and Infrared Spin Scan Radiometer; SEVIRI: Spinning Enhanced Visible and InfraRed Imager; MI: Meteorological Imager; ABI: Advanced Baseline Imager; EPIC: Earth Polychromatic Imaging Camera.

[2]Only those phenomena that can be monitored quantitatively are listed (i.e., visible imaging of volcanic plumes is excluded).

[3]Only the daytime overpass of each satellite is given. Infrared sensors also collect data at night.

[4]Spatial resolution typically varies between spectral bands.

[5]The maximum temporal resolution is listed. Some LEO spacecraft/sensors are pointable (e.g., NASA's EO-1, Terra/ASTER), which can increase temporal resolution. Note that geostationary imagers typically have a nominal temporal resolution of 15–30 min, but higher resolution is possible in special observation modes (e.g., GOES Super Rapid Scan).

[6]URLs provided are single examples of sites where instrumental datasets (if available) can be viewed or accessed. Note that website URLs are subject to change but were active at the time of writing.

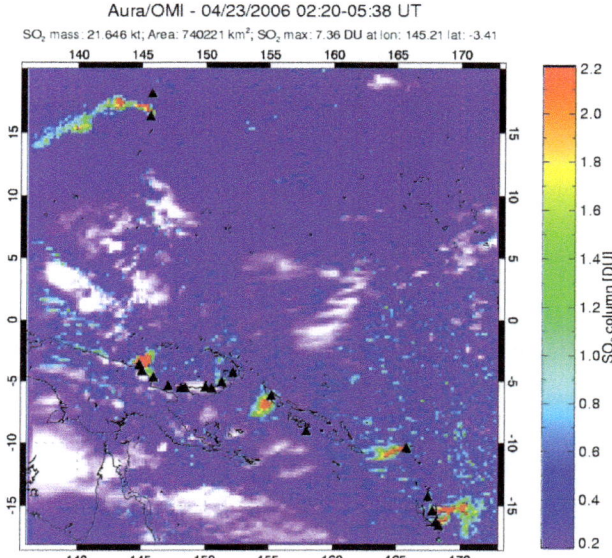

Aura/OMI - 04/23/2006 02:20-05:38 UT

SO₂ mass: 21.646 kt; Area: 740221 km²; SO₂ max: 7.36 DU at lon: 145.21 lat: -3.41

FIGURE 65.5 Composite OMI SO₂ and reflectivity image of the southwest Pacific region on April 23, 2006. In this single image, tropospheric SO₂ plumes can be discerned issuing from seven volcanoes (from north to south): Anatahan (CNMI), Manam (PNG), Ulawun (PNG), Bagana (PNG), Tinakula (Solomon Islands), Aoba (Vanuatu), and Ambrym (Vanuatu). *From Carn et al. (2013).*

latter provide higher sensitivity to SO₂, though spatial resolution is also a factor. The TIR sensors complement UV instruments by measuring at night and by offering different sensitivity to SO₂. Unlike in the UV, the main factor governing detection of SO₂ in the IR is the temperature difference, or thermal contrast, between the SO₂ layer and the background (which could be the ground surface or a layer of meteorological clouds). All other factors being equal, IR sensitivity to SO₂ increases with thermal contrast, thus it tends to peak for volcanic clouds located at the cold-point (or thermal) tropopause—usually the coldest level of the atmosphere. Detection of SO₂ at high altitudes can be compromised when it is viewed against a cold background (e.g., polar regions; strong temperature inversions; and so on) due to the lack of thermal contrast. The other major constraint on IR SO₂ measurements is water vapor abundance, particularly when the 7.3 μm absorption band of SO₂ is exploited, due to reduced atmospheric transmission in this region of the EM spectrum (Figure 65.2). However, SO₂ can be quantified using this waveband if the gas is located in the UTLS, above the lower tropospheric peak in water vapor concentration. Combined use of the three IR SO₂ absorption bands (ν_1 at ~8.6 μm, ν_3 at ~7.3 μm, and $\nu_1 + \nu_3$ at ~4 μm) can provide constraints on SO₂ altitude owing to their distinctive vertical sensitivity. Detection of passive volcanic degassing in the IR is

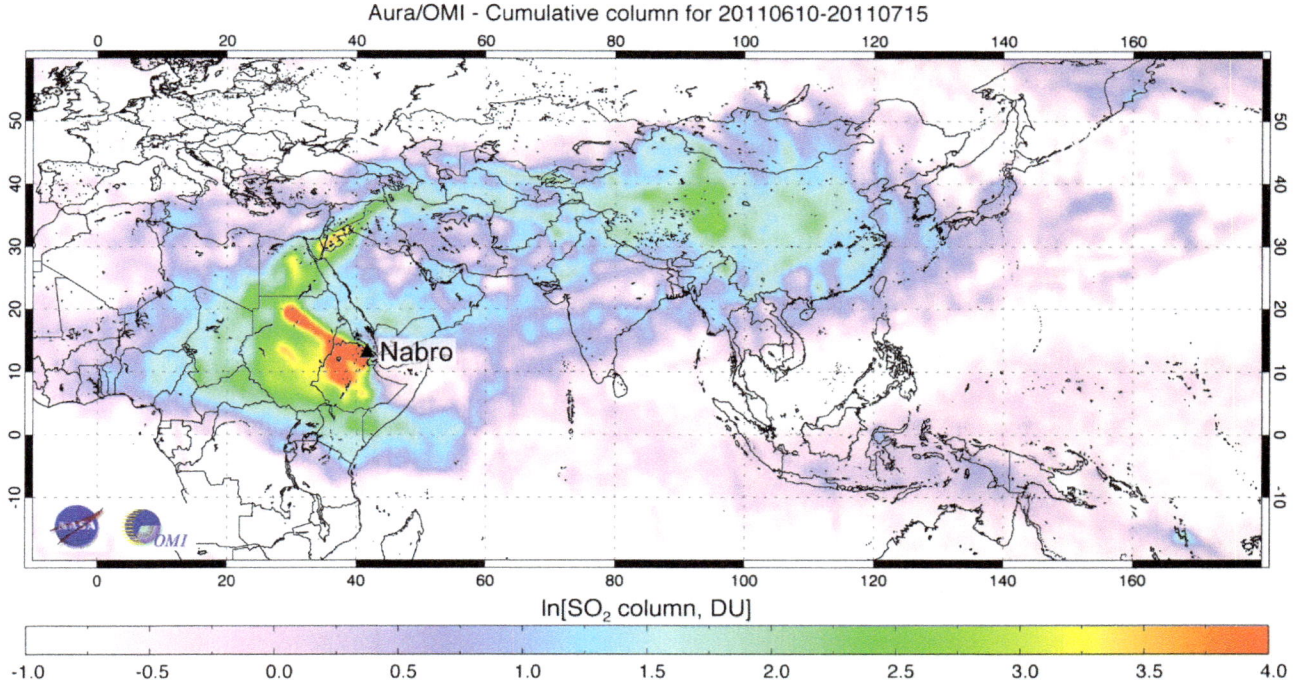

FIGURE 65.6 Cumulative SO₂ column amount (note logarithmic color scale) measured by OMI in the Nabro volcanic cloud from June 10 to July 15, 2011.

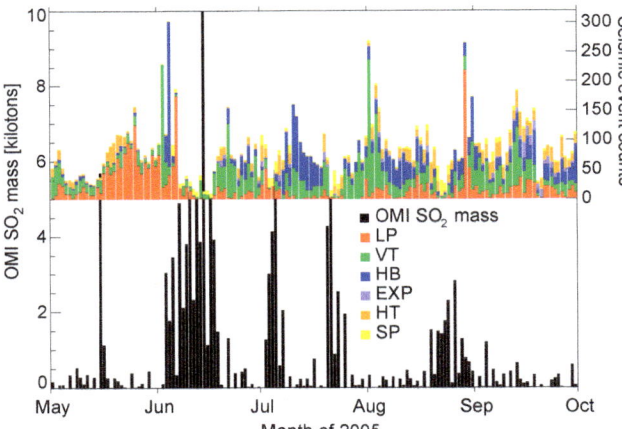

FIGURE 65.7 OMI SO$_2$ burdens (lower panel) and seismic event counts (upper panel; courtesy of the Instituto Geofísico–Escuela Politécnica Nacional (IG–EPN), Ecuador) for Reventador volcano, Ecuador in May–September 2005. Seismic events shown are long-period (LP), volcano-tectonic (VT), hybrid (HB), explosion signals (EXP), harmonic tremor (HT), and spasmodic tremor (SP). Note the anticorrelation between OMI SO$_2$ and seismicity, indicating alternation between open- (elevated SO$_2$; low seismicity) and closed-system (low SO$_2$; elevated seismicity) degassing at Reventador. *From Carn et al. (2008).*

possible using 8.6 μm data, but requires high spatial resolution (e.g., MODIS or ASTER; Table 65.1).

An advantage of the TIR is that several other volcanic gas species exhibit absorption features in this spectral band, in addition to SO$_2$ (Figure 65.2). Although more challenging, and hence less common than SO$_2$ retrievals, such measurements provide unique constraints on the budgets of other volcanic volatile species. Recently, H$_2$S has been detected and quantified in volcanic plumes from Kasatochi and Grimsvötn (Iceland) using the IASI instrument (Clarisse et al., 2011), providing the first direct observations of the abundance of this important sulfur species in explosive eruption clouds. IASI and MOPITT (Table 65.1) also detected volcanic CO in the emissions from Grimsvötn (in 2011) and Eyjafjallajökull (in 2010); the first time CO of volcanic origin has been detected from space (Martinez et al., 2012).

Arguably the most useful volatile species to measure, from the standpoint of eruption forecasting, would be CO$_2$, but volcanic CO$_2$ is among the most challenging gases to detect using satellite remote sensing due to the high atmospheric background ($\sim$400 ppmv), which dominates the signal. Even at the strongest volcanic CO$_2$ sources, such as Etna, the volcanic CO$_2$ anomaly may be <10 ppm above the background close to the crater, imparting stringent demands on the precision and accuracy of satellite CO$_2$ measurements. The Japanese Greenhouse Gases Observing Satellite (GOSAT) mission, launched in 2009, carries one of the first satellite sensors with the capability to detect CO$_2$ from point sources such

as volcanoes and power plants. This can be achieved by pointing the sensor at the emission source, with the goal of filling the $\sim$10 km diameter FOV with emitted CO$_2$ to maximize the signal. NASA's recently launched Orbiting Carbon Observatory 2 (OCO-2; Table 65.1) offers even greater potential for volcanic CO$_2$ detection due to its smaller footprint (1.3×2.3 km). Although neither GOSAT nor OCO-2 provide adequate temporal resolution or spatial coverage to be effective volcano monitoring tools, such measurements will certainly improve constraints on the sources and sinks of CO$_2$.

Beyond the TIR region, the microwave region of the EM spectrum is also used to measure volcanic gases. The Microwave Limb Sounder (MLS) on NASA's Aura satellite detects thermal microwave emission from Earth's limb (looking horizontally through the atmosphere instead of downward in the nadir direction) by scanning vertically from the surface to $\sim$90 km altitude. MLS can measure vertical profiles of volcanic SO$_2$ and HCl (and potentially other species such as CH$_3$Cl) after volcanic eruptions, provided that the gases reach the UTLS or above (e.g., Figure 65.8).

2.4.3. Ash and Aerosol Measurements

The eruption of the relatively obscure (at the time) Eyjafjallajökull volcano in southeast Iceland in April 2010 was a landmark event in volcanology. The eruption, although modest in magnitude, produced ash-rich volcanic plumes that had a major impact on aviation in Europe and subsidiary effects worldwide. Although the volcanic ash hazard to aviation was recognized by the scientific community prior to 2010, the economic impact of the Eyjafjallajökull eruption arguably brought the issue to worldwide public attention for the first time, and also gave renewed impetus to studies involving volcanic ash measurements and monitoring. Ash is usually a minor component of tropospheric volcanic plumes and effusive eruption clouds, but ash concentrations can be very high in fresh explosive eruption clouds, especially in the first few hours after emission. For a detailed discussion of volcanic ash hazards to aviation, and current remote sensing techniques used to detect volcanic ash in the atmosphere, the reader is referred to Chapter 52 on "Volcanic Ash Hazards to Aviation."

Aerosols such as volcanic ash, desert dust, and smoke absorb UV radiation (the principal UV-absorbing components in ash and dust are iron oxides), while sulfate aerosol, sea salt, and ice crystals are only weakly absorbing; thus UV satellite measurements can be used to detect, quantify, and characterize atmospheric aerosols. However, satellite-based ash detection more commonly utilizes TIR wavelengths at $\sim$11 and 12 μm in the

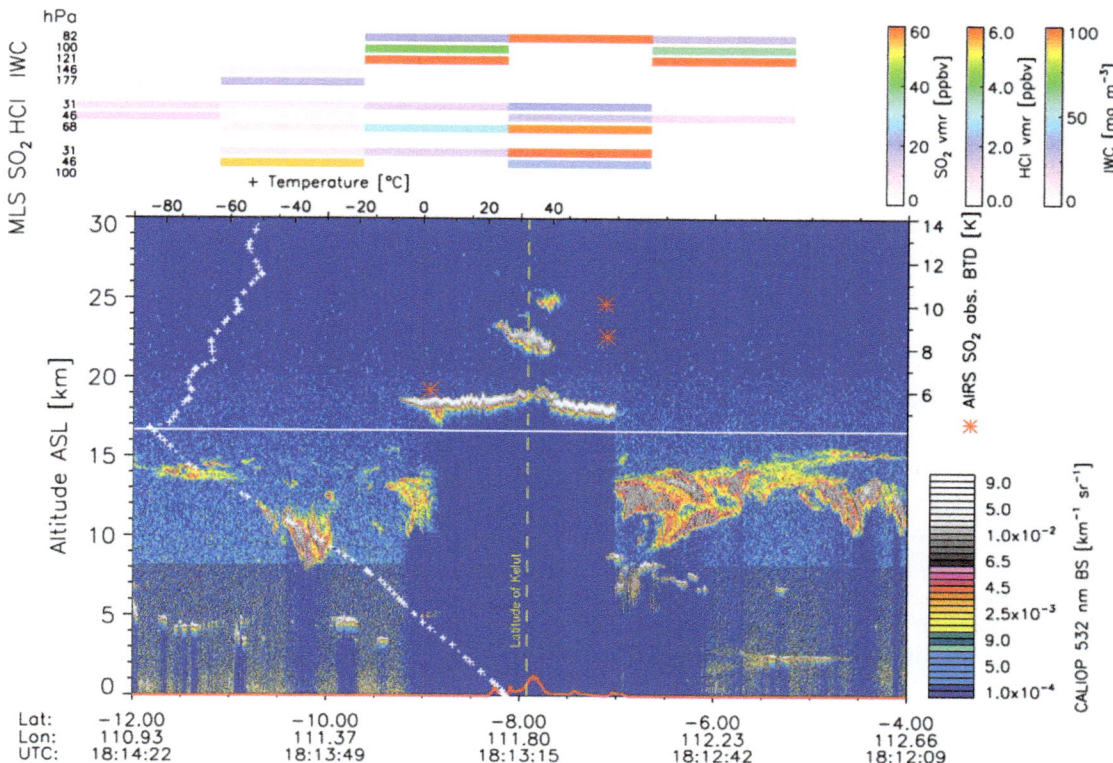

FIGURE 65.8 Vertical profile through the stratospheric volcanic cloud produced by the eruption of Kelud (Java) in February 2014, observed by sensors in NASA's A-Train satellite constellation at 18:13 UTC on February 13, just ∼2 h after the eruption. The central image shows a vertical profile of aerosols observed by the CALIOP lidar on CALIPSO (532 nm total attenuated lidar backscatter); the volcanic umbrella cloud (at ∼18−19 km) and two levels of overshooting top (at 22−26 km) are clearly apparent. The *horizontal white line* is the tropopause. *Red stars* indicate AIRS SO_2 brightness temperature differences (BTDs; restricted here due to cloud opacity preventing SO_2 detection in the TIR over much of the cloud). *White crosses* denote an atmospheric temperature profile from the Surabaya-Juanda rawinsonde sounding at 00Z on February 14. Above the CALIOP image are collocated stratospheric SO_2, HCl, and IWC (ice water content) retrievals from Aura/MLS at several pressure levels (in hPa, indicated top left). The MLS data show clear signs of stratospheric injection of SO_2 and HCl, with SO_2 detected as high as 31 mb (equivalent to an altitude of ∼23−24 km, which is consistent with the CALIOP data).

"split-window" technique, based on the inverse wavelength dependence of ash and water/ice cloud absorption in the TIR. Numerous operational, polar-orbiting, and geostationary multispectral IR satellite sensors possess the TIR window channels required to detect volcanic ash (e.g., AVHRR, GOES, SEVIRI, MTSAT, and MODIS; Table 65.1), and these data can also be used to quantify ash column abundance and determine ash concentrations if cloud thickness is known. Hyperspectral IR sensors such as IASI and AIRS can be used to identify the characteristic spectral signatures of ash and other aerosol particles (e.g., sulfate) and determine the composition of airborne particles. This was demonstrated after the rhyolitic eruption of Chaitén volcano (Chile) in 2008 (Gangale et al., 2010).

The most critical factor determining the impact of a volcanic eruption on climate or aviation is plume altitude. Direct measurements of ash or gas altitude using passive UV/IR remote sensing techniques are possible but inherently imprecise, whereas active sensing

techniques (e.g., LiDAR and radar) provide very accurate altitude information. The 2006 launch of the Cloud-Aerosol Lidar with Orthogonal Polarization (CALIOP) aboard the CALIPSO satellite, and the Cloud Profiling Radar (CPR) aboard CloudSat (Table 65.1), into NASA's polar-orbiting A-Train spacecraft constellation was a major boon to volcanological remote sensing. CALIOP can provide vertical profiles of particles (including ash, sulfate, and hydrometeors) in volcanic clouds (e.g., Figure 65.8), although the chances of sampling fresh volcanic clouds are slim owing to LiDAR's poor spatial coverage. CPR is a millimeter-wavelength radar that is sensitive to water/ice particles in clouds and precipitation, and also coarser volcanic particles such as ash aggregates. Working in concert, CALIOP and CPR can yield information on particle abundance, phase (e.g., liquid or solid) and size in volcanic clouds, in addition to cloud altitude and thickness. The latter can be used to calculate concentrations of SO_2 and ash in volcanic clouds.

2.4.4. Satellite Sensor Synergy

Since a single satellite has a limited payload of sensors, obtaining coincident measurements at multiple wavelengths and/or using multiple techniques requires coordinated observations from a group (or constellation) of spacecraft flying in close formation along the same orbital track. NASA's A-Train (http://atrain.nasa.gov/), which currently contains the OCO-2, GCOM-W1, Aqua, CALIPSO, CloudSat, and Aura satellites (Table 65.1), is the largest such constellation in operation. Several A-Train sensors make near-coincident (within ~ 15 min) observations of volcanic cloud constituents, including OMI, MLS, AIRS, MODIS, CALIOP, and CPR (Table 65.1). Combining data from these sensors has yielded new insight into the composition and vertical distribution of gases and particles in volcanic clouds. A spectacular example of A-Train observations following the February 2014 eruption of Kelud (Java, Indonesia) is shown in Figure 65.8. Satellite constellations such as the A-Train represent a new paradigm for remote sensing of dynamic phenomena such as volcanic eruptions, and will likely be replicated in the future.

Other potential satellite sensor synergies exist between geostationary and polar-orbiting (or Low Earth Orbit; LEO) satellites. Most current satellite measurements of SO_2 are made from LEO spacecraft (Table 65.1). The only geostationary sensor currently capable of measuring volcanic SO_2 is the IR SEVIRI on the Meteosat satellites observing Europe and Africa, whereas most geostationary instruments can detect volcanic ash (Table 65.1). Future IR SO_2 measurements from geostationary platforms (e.g., Himawari-8, GOES-R ABI; Table 65.1) will offer greater temporal resolution (~ 15 min) than current LEO data, and cover East Asia and the Americas, both regions with significant volcanic activity. The availability of multiple sources of satellite SO_2 measurements is useful for sensor intercomparisons and validation. For volcanic SO_2, the total mass measured using different LEO sensors has been observed to differ by more than a factor of two, owing to different instrument characteristics, sampling, and algorithms.

3. THERMAL MEASUREMENTS AND MONITORING

3.1. Principles

Volcanoes can be prodigious sources of heat, and remote thermal surveillance from the ground, air, or space using the shortwave IR (SWIR), mid-IR, and TIR regions of the EM spectrum is one of the most effective and widely used volcano monitoring techniques, largely superseding in situ temperature measurements using thermocouples. Remote

sensing measurements in the TIR rely on the detection of **electromagnetic radiation** emitted by the target, and the theoretical bases of all such measurements are the radiation laws (Planck, Stefan–Boltzmann, and Wien). Planck's law defines the relationship between the absolute temperature (T) of a body and the emitted spectral radiance (with units of power per unit area per unit solid angle per unit wavelength, e.g., W m^{-2} sr^{-1} μm^{-1}; Figure 65.9), which is the fundamental quantity measured by IR sensors, though it is usually converted to an equivalent brightness temperature (by inverting Planck's law). Integrating Planck's law with respect to wavelength yields the Stefan–Boltzmann (or Stefan's) law ($F = \sigma T^4$; where σ is the Stefan–Boltzmann constant), which defines the total power per unit area (F; W m^{-2}) emitted by a blackbody at temperature, T, and stresses the very strong temperature dependence of the emitted energy. Differentiating Planck's law once with respect to wavelength yields Wien's law ($\lambda_{max}T = 2898$), which defines the important inverse relationship between temperature and the peak wavelength (λ_{max}) of thermal emission. Planck's law curves (Figure 65.9) have a characteristic shape, with a sharp decline in spectral radiance at shorter wavelengths, a peak (defined by Wien's law) and a slower decline toward longer wavelengths. Planck's and Wien's law tell us that active lava at magmatic temperatures (~ 700–1300 °C or 1300–2400 °F; e.g., uncrusted lava lakes, or flows) can emit significant amounts of radiation at

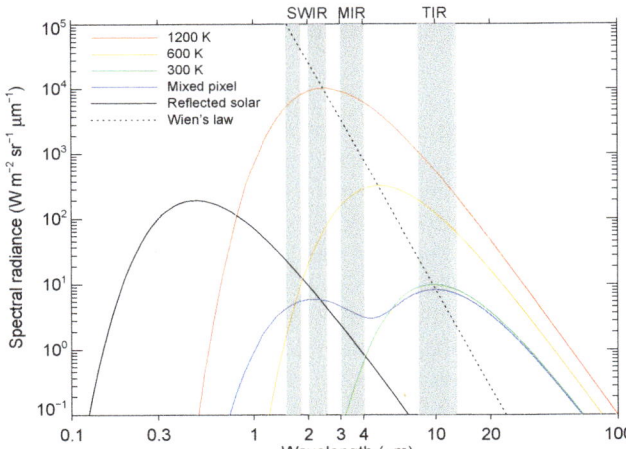

FIGURE 65.9 Spectral radiance (W *m-2* sr-1 μm-1) from blackbodies at a range of temperatures (300, 600, and 1200 K) as a function of wavelength (μm), calculated using Planck's law. *Mixed pixel* represents the radiance from a sub-pixel sized area of active lava (0.03% of the pixel at 1100 °C) within a background at ambient temperatures (15 °C). The *Reflected solar* curve shows the reflected solar radiance for a surface reflectivity of 10%. Spectral radiance maxima from blackbody radiators lie along the *Wien's law* curve. *Gray bands* indicate commonly exploited atmospheric windows in the SWIR (~ 1.6 μm, 2–2.5 μm), mid-IR (MIR, 3–4 μm), and TIR (8–13 μm) spectral bands.

visible wavelengths, thus explaining the red-orange glow of freshly erupted lava.

Wien's law also indicates that, in order to achieve maximum sensitivity to the full range of ambient to magmatic surface temperatures, sensor channels in a range of TIR wavebands are preferable. Since water vapor and clouds strongly attenuate TIR radiation, sensor channel placement is largely dictated by the presence of atmospheric windows where transmission of radiation is highest. Some of the most commonly exploited SWIR, mid-IR, and TIR windows are at wavelengths of ~1.6 μm, 2–2.5 μm, 3–4 μm, and 8–13 μm (Figure 65.2, Figure 65.9) and most IR satellite instruments have channels in at least one of these spectral regions. Due to its proximity to the peak emission from high-temperature targets, and sensitivity to rising temperatures relative to longer wavelength channels, the mid-IR channel at ~3.5–4 μm (e.g., on the AVHRR and GOES Imager) is among the most useful wavebands for volcano monitoring (e.g., Figure 65.10), particularly at nighttime when reflected solar radiation (which can be significant at wavelengths of ≤4 μm; Figure 65.9) is absent (e.g., Harris, 2013).

An important consideration in IR remote sensing, particularly from satellite platforms, is the "mixed pixel" effect. In volcanic observations, the instrument's Instantaneous Field of View (IFOV) will typically contain surfaces at a range of temperatures, and the emitted spectral radiance will be an area-weighted sum of the radiance from each thermal component. This effect depends on the spatial resolution of the sensor (~10 m–1 km; Table 65.1), satellite viewing geometry (which varies from day-to-day) and the nature of the target. Typically, with the exception of major effusive eruptions, features at magmatic temperatures occupy very small fractions of the instrument IFOV, although the profuse spectral radiance from such bodies permits their detection even for very small fractional areas (e.g., Figure 65.9). A further advantage of 3.7 μm mid-IR data is its sensitivity to high-temperature features of sub-pixel extent, even in a 1 km AVHRR pixel, due to the strong thermal emission in this waveband (Figure 65.9, Figure 65.10; e.g., Steffke and Harris, 2011). For example, Figure 65.9 shows the significant increase in spectral radiance at 3.7 μm due to a sub-pixel sized area of active lava covering only 0.03% of the pixel. Combining mid-IR data with TIR measurements less sensitive to high temperatures (e.g., at 11 μm) in "dual-band" techniques permits quantitative analysis of the spatial extent of hot volcanic features, though the solution becomes better constrained when data at multiple wavebands are available. However, since satellite sensors are not optimized for the exceptional temperatures encountered in active volcanic contexts, sensor saturation is a common problem and this prohibits quantitative analysis. The AVHRR 3.8 μm channel saturates at pixel-integrated temperatures of ~325 K,

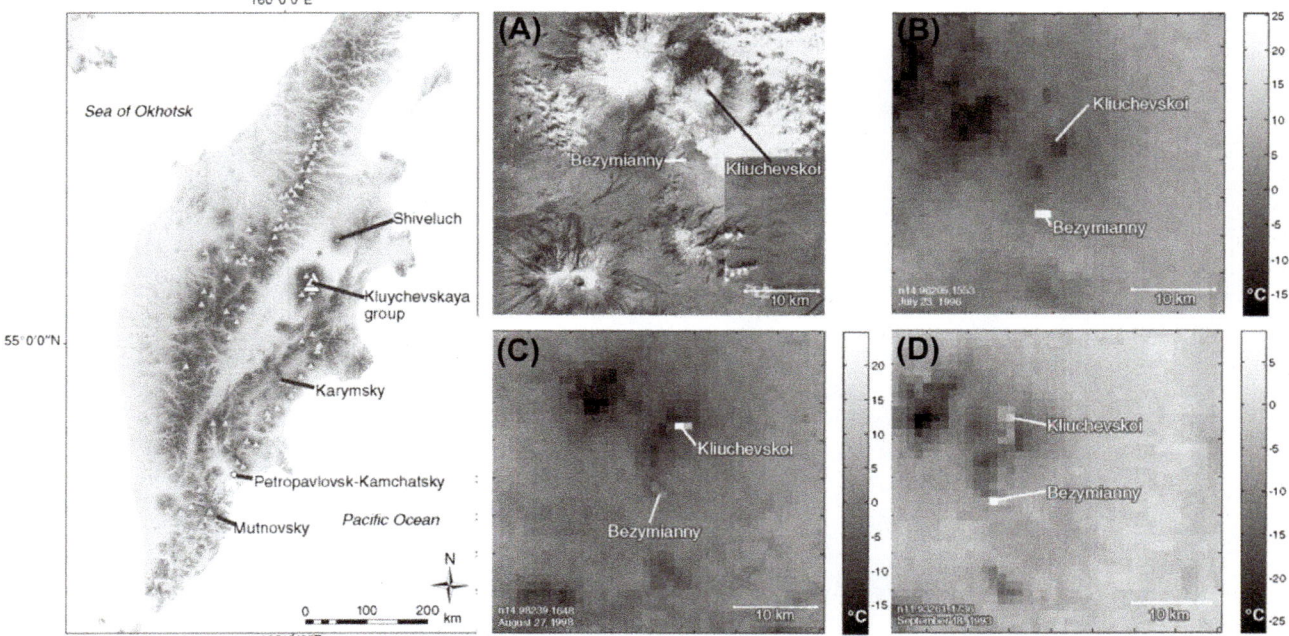

FIGURE 65.10 Shuttle Radar Topography Mission (SRTM) digital elevation model (DEM) mosaic of the Kamchatka Peninsula and selected TIR (3.55–3.93 μm) Advanced Very High Resolution Radiometer (AVHRR) data. All images are north oriented to top, scale bar is in °C, and satellite image numbers are noted at bottom. (A) Google Earth image showing Klyuchevskaya group. (B) Thermally elevated pixels at summit of Bezymianny. (C) Thermally elevated pixels at Kliuchevskoi. (D) Contemporaneous thermal activity at both volcanoes. *From van Manen and Dehn (2009).*

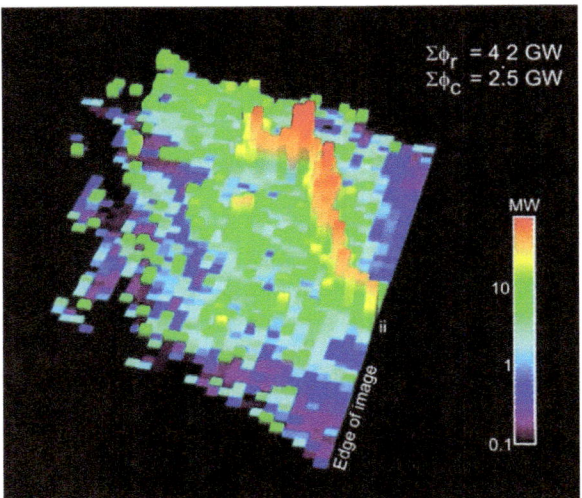

FIGURE 65.11   Heat flux (due to radiational and convective heat loss) from an 'a'a lava flow erupted from Nyamuragira volcano, Democratic Republic of Congo, observed by NASA's EO-1 Hyperion sensor on 21 May 2004. North is down. The height of each colored column is proportional to the heat flux and a logarithmic color scale has been applied. *From Wright et al. (2010).*

unsaturated data over a wider range of high-temperature targets than 3–4 μm channels. However, the strong reflected solar radiation contribution can easily mask any volcanic signal in daytime 1.6 μm data (Figure 65.9), so observations at this wavelength are best made at night. A further advantage of nighttime data is that any spectral radiance recorded at 1.6 μm can be attributed entirely to volcanic phenomena since, in contrast to the 3–4 μm waveband, thermal emission from surfaces at ambient temperatures is negligible at 1.6 μm (Figure 65.9). Hyperspectral imagers providing 100s of contiguous channels (e.g., NASA's EO-1 Hyperion; Table 65.1) provide the best opportunity for unsaturated data and detailed thermal characterization of volcanic surfaces (e.g., lava flow cooling; Figure 65.11), though saturation can still be an issue.

3.2. The Origin and Detection of Volcanic Thermal Anomalies

The range of thermal features present on a volcanic edifice can include fumarole fields with temperatures slightly above background, warm crater lakes, hot vents heated by emitted gases, and active lava lakes, flows, or domes (e.g., Figure 65.10, Figure 65.11, Figure 65.12, Figure 65.13). TIR monitoring techniques are employed to detect surface manifestations of volcanic unrest, such as increased fumarole or vent temperatures due to magma ascent from depth. They are also used in a more

although some sensors (e.g., MODIS) mitigate the impact of saturation by including channels with larger dynamic ranges to record higher temperatures. At shorter wavelengths, a 1.6 μm channel (e.g., available on AVHRR, ASTER, MODIS, Landsat ETM+, and OLI) can provide

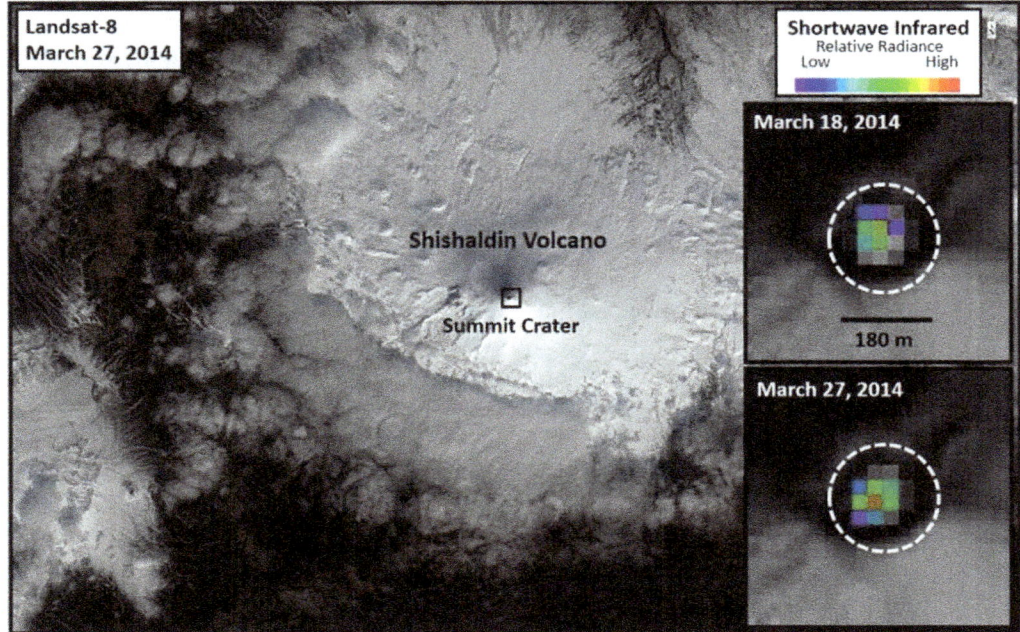

FIGURE 65.12   Landsat 8 satellite imagery showing thermal anomalies due to lava in the summit crater of Shishaldin volcano, Alaska, on March 18 and 27, 2014. Image courtesy of Alaska Volcano Observatory/US. Geological Survey (created by David Schneider). URL: http://www.avo.alaska.edu/images/image.php?id=57861.

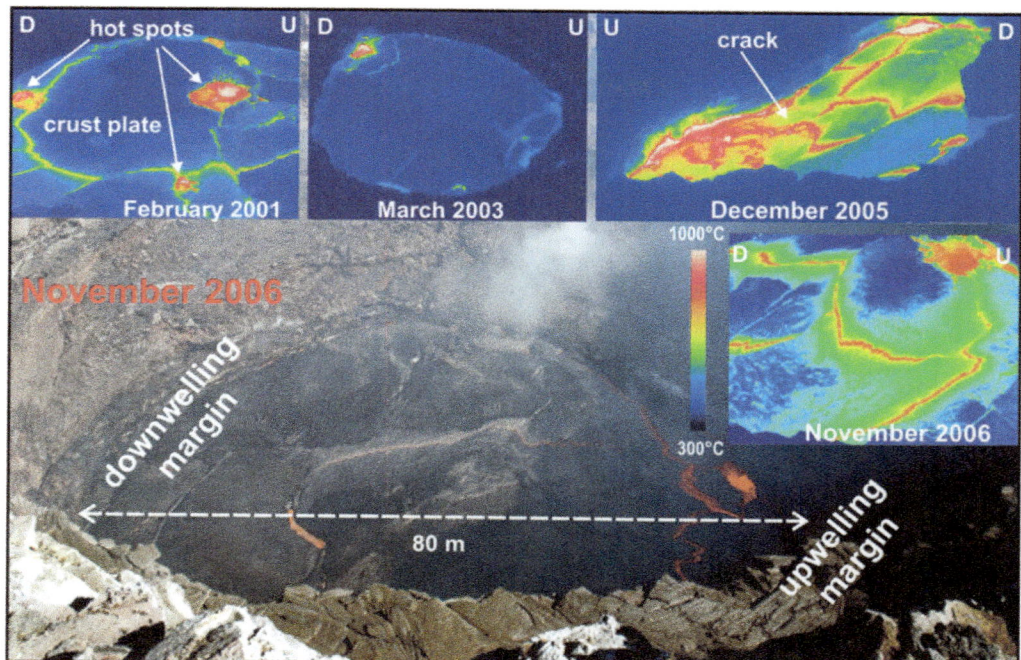

FIGURE 65.13 The photograph shows the surface of the active lava lake at Erta'Ale (Ethiopia) in November 2006. The inserts are thermal camera images of the lava lake collected in February 2001, March 2003, December 2005, and November 2006. The thermal images reveal details of the lake surface such as incandescent cracks, crustal plates, and bursts of hot gas bubbles at hot spots. U and D mark the locations of the upwelling and downwelling margins, respectively. *From Spampinato et al. (2011).*

quantitative way to determine the spatial extent of active lava flows or lakes, or to estimate lava effusion rates during effusive eruptions (e.g., Harris et al., 2007).

The principal goal of thermal surveillance is the detection of volcanic thermal anomalies (e.g., Steffke and Harris, 2011), though quantitative analysis is also desirable. Effective space-based monitoring systems should permit detection of active volcanism on kilometer scales or smaller, and provide repetitive, frequent observations (requiring use of low spatial, high temporal resolution data) with low noise and at low cost. As described above, detection of small volcanic heat sources can be very effective and timely using night-time mid-IR (e.g., 3–4 μm) band data from high temporal, and low-to-moderate spatial resolution sensors such as AVHRR, GOES, or MODIS (Figure 65.10; e.g., Harris, 2013). The use of operational meteorological satellites (e.g., NOAA's Geostationary Operational Environmental Satellites (GOES) and Polar-Orbiting Environmental Satellites (POES)) offers the additional benefits associated with assets designed to monitor weather conditions, that is, 24/7 operations and near real-time (NRT) data availability. However, temporal variations in spectral radiance (or derived pixel brightness temperatures) can rarely be unambiguously attributed to volcanic processes, due to the modulating effects of variable IFOV size, viewing geometry, meteorological cloud cover, solar radiation, surface

emissivity, atmospheric attenuation, and other factors; thus interpretation of trends can be challenging. High spatial resolution sensors (e.g., Landsat ETM+, OLI, and TIRS; ASTER; Table 65.1) provide more detailed characterization of volcanic thermal features (e.g., Figure 65.12), but less frequently (and sometimes at significant cost), hence integration of these observations with high temporal resolution data is the best approach. Identification of long-term trends in heat flux is generally more valuable than measurements of absolute temperature and thermal distributions at any given time (e.g., van Manen and Dehn, 2009).

3.3. Acquisition of IR Measurements

Remotely sensed IR data can be collected from the ground, air, or space. Airborne IR surveys of active volcanoes using aircraft-mounted sensors are uncommon and not discussed further here. Ground-based and airborne IR measurements are now predominantly made using thermal cameras, which capture a 2D image rather than the single spot measurement collected by an IR radiometer. Detector technology for thermal cameras has greatly improved in recent years, with most devices using either a focal plane array (FPA) of IR-sensitive pixels or microbolometers (whose resistance depends on the intensity of incident radiation). FPAs can operate at cryogenic or ambient temperatures, and microbolometers are uncooled, but cooled detectors are more sensitive due to lower noise levels (e.g., IR emission from the detector itself). However, uncooled sensors are more compact, less expensive and require less power, and are usually adequate for monitoring high-temperature volcanic targets. Currently available portable thermal cameras operate in either the 3−5 μm or 7.5−13 μm wavebands (Figure 65.9), and offer variable image resolution (640 × 480 or 1024 × 1024 pixels) and sampling rates (60−132 Hz). Temperatures of up to 2000 °C can be measured, thus avoiding saturation when imaging active lava and permitting quantitative analysis, although surface emissivity is often unknown (but can usually be assumed to be close to unity). As for satellite instruments, the use of the 3−5 μm and 7.5−13 μm wavebands is primarily due to the high atmospheric transmission in these spectral regions (although volcanic gases such as SO_2 and water vapor do absorb some radiation at these wavelengths) and their proximity to radiance maxima from bodies at magmatic and ambient temperatures, respectively (Figure 65.9). While most thermal cameras used for volcano monitoring are broadband instruments, it is also possible to equip TIR imaging cameras with multiple narrow band (0.5−1.0 μm) spectral filters, rotated sequentially into the camera FOV with a filter wheel, to select specific wavelengths for measurements of SO_2 and volcanic ash.

Owing to their portability, low power requirements and relatively low-cost, thermal cameras and analogous instruments have been employed in a broad range of volcano monitoring applications (e.g., Ramsey and Harris, 2013). These include thermal mapping of hydrothermal areas and fumarole fields, imaging active lava flows and domes, visualizing convective motion on the surface of lava lakes (e.g., Figure 65.13), analyzing the dynamics of Strombolian eruptions, measuring lava effusion rates, lava flow velocity and cooling rates, and deriving plume rise speeds. Regular thermal surveys can be used to detect changes in temperature related to subsurface magma migration. Lava effusion rates are a critical parameter for lava flow hazard mitigation. Remote measurement of effusion rates requires several assumptions, resulting in significant errors, but results comparable to field data have been obtained.

Numerous space-borne sensors collect TIR data at varying spatial, spectral, and temporal resolution and at different overpass times (e.g., AVHRR, Landsat ETM+, MODIS, ASTER, SEVIRI, GOES, EO-1 ALI, Hyperion, and VIIRS; Table 65.1; Figure 65.10, Figure 65.11, Figure 65.12). TIR sensor channel locations are dictated by the presence of atmospheric windows (Figure 65.2), hence channels at 3−4, 11 and 12 μm are ubiquitous, permitting detection of volcanic hot-spots. The large number of space-borne TIR datasets provides useful redundancy for operational volcano monitoring, and high temporal resolution even from LEO satellites, especially in high-latitude volcanic regions such as Alaska, the Aleutian Islands, and

Kamchatka. Useful synergies can be exploited between high spatial, low temporal resolution data (e.g., from AS-TER or Landsat), which permit detailed thermal characterization of volcanic features, and more frequent overpasses by low-to-moderate spatial resolution assets (e.g., AVHRR and MODIS) for timely hot-spot detection. This synergy is also exploited in autonomous spacecraft observing systems such as NASA's Volcano Sensor Web (VSW), which monitors volcanic activity alerts from diverse sources (e.g., satellite-based hot-spot detection, volcanic ash advisories, and in situ monitoring networks) and uses them to trigger data acquisition by the ALI and Hyperion sensors aboard the EO-1 spacecraft (Table 65.1). The VSW has been successfully implemented in several volcanic crises (e.g., Nyamulagira in 2006 and Eyjafjallajökull in 2010), often pre-empting manual triggering, and collecting valuable observations that would not otherwise be made (e.g., Davies et al., 2013).

4. WEBSITE DISSEMINATION AND ALERT SYSTEMS

A major advance over the past decade has been the increasing dissemination of remote sensing data via the Internet, particularly satellite measurements. There are now numerous websites providing publicly available satellite measurements of SO_2, volcanic ash, and TIR hot-spots (e.g., MODVOLC, SACS, and HotVolc; Table 65.1). In addition to the worldwide expansion of the Internet, mobile devices, and increased computational power, a major driver for the development of these services has been the need to issue timely warnings of drifting volcanic clouds to aviation. Rapid growth, and interest, in web-based volcanic SO_2 and ash services has occurred since the volcanic ash crisis caused by the April 2010 eruption of Eyjafjallajökull (Iceland), although many existed beforehand and the volcanic ash hazard to aviation (see Chapter 52) has been recognized for several decades. TIR hot-spot monitoring websites have also been operational for some time, the most widely used of which is the MODVOLC system driven by MODIS data (http://modis.higp.hawaii.edu; Wright et al., 2004).

The time latency of the data presented on these websites (i.e., the delay between data acquisition on the spacecraft and display on the website) varies, and although many services aim to provide data in "'near real-time'" (NRT), the delay can be a few hours for a polar-orbiter. Many satellites offer direct broadcast (DB) capabilities, where satellite data can be received directly at a ground receiving station as the spacecraft flies overhead, rapidly processed and displayed within ∼5−30 min, and a few systems exploit this capability (e.g., the OMI/OMPS Satellite Measurements from Polar Orbit (SAMPO) service: http://sampo.fmi.fi/volcanic.

html; NASA's Direct Readout Laboratory: https://directreadout.sci.gsfc.nasa.gov/). Receiving stations are mostly located at high latitudes to maximize the "visibility" of polar-orbiting satellites. Even these low latencies are arguably not rapid enough to enable a jet aircraft traveling at cruising speed to divert around a fresh volcanic eruption cloud, but capabilities are improving.

Several websites also offer eruption "alerts," based on TIR, SO_2, or volcanic ash measurements. Typically, an alert is issued if the detected radiance, or SO_2 or ash abundance exceeds a predefined threshold, although the latter are often set at low levels in order to capture both small and large eruptions. Alerts based on SO_2 data may also be triggered by periods of vigorous volcanic degassing, in the absence of an eruption. The provision of timely, accurate eruption alerts, with no false alarms, is a desirable but challenging goal (e.g., Steffke and Harris, 2011), and more accurate alert thresholds could perhaps be derived from data mining and analysis of prior satellite measurements of eruptions.

In addition to satellite measurements, real-time or NRT web-camera (webcam) imagery from many volcanoes can now be accessed online, usually via volcano observatories. Webcams typically operate at visible wavelengths, providing qualitative monitoring of volcanic gas and ash plumes and weather conditions. Visible wavelength cameras are ineffective at night unless very high-temperature phenomena are present, but near-IR or TIR webcams (e.g., the thermal cameras described above) are also in use by some observatories (e.g., at Etna, Italy; http://www.ct.ingv.it/en/webcam-etna-en.html), providing nighttime surveillance.

5. FUTURE DIRECTIONS

Although funding for costly satellite missions is ultimately subject to the vagaries of federal budgets, capabilities for space-borne monitoring of volcanic plumes and hotspots are likely to remain uninterrupted over the next decade (e.g., Table 65.1). Several instruments capable of measuring volcanic SO_2 emissions are currently in orbit (e.g., OMI, OMPS, and IASI) or planned for launch in the near future (e.g., TROPOMI). MODIS-like TIR measurements will be continued by VIIRS on the Suomi NPP and JPSS satellites. Improved iterations of high spatial resolution assets are also being launched (e.g., Landsat-8), but high temporal and low spatial resolution data are likely to remain the mainstay of operational volcanic gas, plume, and thermal monitoring from space (e.g., Ramsey and Harris, 2013). However, although there is usually a trade-off between spatial and temporal resolution, advanced sensors such as TROPOMI (daily or better temporal resolution and 7 km spatial resolution) attempt to optimize both. NASA's Hyperspectral Infrared Imager (HyspIRI) sensor, one of the several missions recommended by the last National Research Council's Earth Science Decadal Survey for launch in the 2020 timeframe, will provide 60 m spatial resolution and 5 days revisit time in the TIR. In the nearer term, the NASA/NOAA DSCOVR mission (Table 65.1) will offer the prospect of volcanic SO_2 measurements from the unique perspective of the L_1 Earth–Sun Lagrange (or libration) point at 1.5 million kilometers from Earth, where the gravitational forces of the Sun and Earth are balanced. Positioning a satellite at L_1 permits a continuous view of the sunlit hemisphere of the Earth.

The next decade is also likely to see an increased focus on quantifying CO_2 emissions from space, including the volcanic contribution. Motivation for this includes the need to better quantify the global volcanic budget of deep carbon, and understand the role of volcanic degassing of greenhouse gases in climate change. Several imminent or planned satellite missions, such as NASA's OCO-2, JAXA's GOSAT-2, and ESA's CarbonSat, will offer the potential to measure volcanic CO_2 emissions from space, although this is never likely to become as routine as volcanic SO_2 measurements due to the challenges of detecting the volcanic CO_2 signal (a few ppmv) against the atmospheric background ($\sim$400 ppmv). It will likely be restricted to the strongest volcanic CO_2 degassing sources, such as Mt. Etna, major eruptions notwithstanding. High spatial resolution ($\sim$1–2 km) is essential in order to maximize the volcanic signal in the sensor's IFOV, hence the temporal resolution of satellite CO_2 measurements is low (e.g., 16 days for OCO-2; Table 65.1).

To assess the feasibility of space-borne volcanic CO_2 detection, we can consider the situation at a strong CO_2 source such as Etna, which emits CO_2 at an average rate of $\sim$190 kg s^{-1}. If Etna's plume was aligned along the full 2.3 km length of an OCO-2 IFOV, thus maximizing the amount of volcanic CO_2 in the scene, then for a wind speed of 5 m s^{-1} the total volcanic CO_2 loading in the IFOV would be $\sim$87,000 kg. Converting to molecules of CO_2 and averaging over the 1.3 × 2.3 km OCO-2 nadir IFOV yields a CO_2 column amount of $\sim$3.5 × 10^{19} molecules cm^{-2}, which is $\sim$0.6% of the atmospheric background CO_2 column at Etna ($\sim$6 × 10^{21} molecules cm^{-2}). The volcanic CO_2 contribution would be higher if the wind speed were lower, and vice versa. The required accuracy for OCO-2 is 0.2% or better, in which case detection of strong volcanic CO_2 sources would be feasible under optimal conditions (although note that the significant impacts of meteorological clouds are ignored here).

Major developments in ground-based observations are likely to arise in the field of infrared laser absorption spectroscopy, using techniques such as Differential Absorption LiDAR (DIAL) and cavity ring-down spectroscopy (CDRS) to analyze volcanic gases such as H_2O and CO_2. The advantage of laser spectroscopy is that it allows very precise wavelength selection, which can be tuned to the absorption lines of individual isotopes in volcanic gases (e.g., $^{12}CO_2$ and $^{13}CO_2$). Isotopic ratios can yield unique information on the origin of volatile species (e.g., Chiodini et al., 2011); e.g., they could be used to distinguish between CO_2 derived from magmatic and non-magmatic sources (e.g., biogenic, thermal decomposition of carbonates) and identify potential eruption precursors. The main barriers to volcanological application of these techniques are the very high precision required on isotopic ratios and the size, robustness, and power consumption of the instrumentation.

Ramsey and Harris (2013) review the potential developments in thermal remote sensing between 2010 and 2020. They suggest a somewhat pessimistic outlook for space-borne thermal monitoring, due to the lack of dedicated volcano remote sensing missions relative to the previous decade, and launch delays for key assets such as HyspIRI. Data gaps, particularly in high spatial resolution TIR observations (e.g., as currently provided by ASTER), are likely. However, Ramsey and Harris (2013) propose that the impending satellite data gaps may invigorate research into ground-based thermal monitoring techniques, such as applications of TIR cameras, and result in increased integration of such data into monitoring networks. This will be assisted by the ever-decreasing cost and increasing portability of thermal imaging devices. Data processing algorithms will also have to adapt to ingest expanding volumes of input data as the spatial and temporal resolution of sensors increase.

Other probable areas of future development include exploitation of the considerable potential of UAVs for low-risk volcano surveillance. Given the strengths and limitations of each observing platform (satellite, ground-based, suborbital, manned, and unmanned) and remote sensing technique, the most effective approach will be to combine gas, plume, and thermal monitoring synergistically in an integrated, remote volcano monitoring system.

FURTHER READING

Aiuppa, A., Moretti, R., Federico, C., Giudice, G., Gurrieri, S., Liuzzo, M., Papale, P., Shinohara, H., Valenza, M., 2007. Forecasting Etna eruptions by real-time observation of volcanic gas composition. Geology 35, 1115–1118.

Allard, P., Burton, M.R., Mure, F., 2005. Spectroscopic evidence for a lava fountain driven by previously accumulated magmatic gas. Nature 433 (7024), 407–410. http://dx.doi.org/10.1038/nature03246.

Burton, M.R., Sawyer, G.M., Granieri, D., 2013. Deep carbon emissions from volcanoes. Rev. Mineral. Geochem. 75, 323–354.

Carn, S.A., Krueger, A.J., Krotkov, N.A., Arellano, S., Yang, K., 2008. Daily monitoring of Ecuadorian volcanic degassing from space. J. Volcanol. Geotherm. Res. 176 (1), 141–150. http://dx.doi.org/10.1016/j.jvolgeores.2008.01.029.

Carn, S.A., Krotkov, N.A., Yang, K., Krueger, A.J., 2013. Measuring global volcanic degassing with the Ozone Monitoring Instrument (OMI). In: Pyle, D.M., Mather, T.A., Biggs, J. (Eds.), Remote Sensing of Volcanoes and Volcanic Processes: Integrating Observation and Modelling, 380. Geological Society, London, Special Publications. http://dx.doi.org/10.1144/SP380.12.

Chiodini, G., Caliro, S., Aiuppa, A., Avino, R., Granieri, D., Moretti, R., Parello, F., 2011. First $^{13}C/^{12}C$ isotopic characterisation of volcanic plume CO_2. Bull. Volcanol. 73 (5), 531–542.

Clarisse, L., Coheur, P.F., Chefdeville, S., Lacour, J.L., Hurtmans, D., Clerbaux, C., 2011. Infrared satellite observations of hydrogen sulfide in the volcanic plume of the August 2008 Kasatochi eruption. Geophys. Res. Lett. 38, L10804. http://dx.doi.org/10.1029/2011GL047402.

Davies, A.G., Chien, S., Doubleday, J., Tran, D., Thordarson, T., Gudmundsson, M.T., Höskuldsson, Á., Jakobsdóttir, S.S., Wright, R., Mandl, D., 2013. Observing Iceland's Eyjafjallajökull 2010 Eruptions with the Autonomous NASA Volcano Sensor Web. J. Geophys. Res. 118, 1936–1956. http://dx.doi.org/10.1002/jgrb.50141.

Galle, B., Oppenheimer, C., Geyer, A., McGonigle, A.J.S., Edmonds, M., Horrocks, L., 2002. A miniaturized ultraviolet spectrometer for remote sensing of SO_2 fluxes: a new tool for volcano surveillance. J. Volcanol. Geotherm. Res. 119, 241–254.

Galle, B., Johansson, M., Rivera, C., Zhang, Y., Kihlman, M., Kern, C., Lehmann, T., Platt, U., Arellano, S., Hidalgo, S., 2010. Network for Observation of Volcanic and Atmospheric Change (NOVAC)— A global network for volcanic gas monitoring: Network layout and instrument description. J. Geophys. Res. 115, D05304. http://dx.doi.org/10.1029/2009JD011823.

Gangale, G., Prata, A.J., Clarisse, L., 2010. The infrared spectral signature of volcanic ash determined from high-spectral resolution satellite measurements. Remote Sens. Environ. 114, 414–425.

Harris, A.J.L., 2013. Thermal Remote Sensing of Active Volcanoes: A User's Manual. Cambridge University Press, Cambridge, UK.

Harris, A.J.L., Dehn, J., Calvari, S., 2007. Lava effusion rate definition and measurement: a review. Bull. Volcanol. 70, 1–22.

Kern, C., Kick, F., Lübcke, P., Vogel, L., Wöhrbach, M., Platt, U., 2010a. Theoretical description of functionality, applications, and limitations of SO_2 cameras for the remote sensing of volcanic plumes. Atmos. Meas. Tech. 3, 733–749. http://dx.doi.org/10.5194/amt-3-733-2010.

Kern, C., Deutschmann, T., Vogel, L., Wöhrbach, M., Wagner, T., Platt, U., 2010b. Radiative transfer corrections for accurate spectroscopic measurements of volcanic gas emissions. Bull. Volcanol. 72, 233–247. http://dx.doi.org/10.1007/s00445-009-0313-7.

Martínez-Alonso, S., Deeter, M.N., Worden, H.M., Clerbaux, C., Mao, D., Gille, J.C., 2012. First satellite identification of volcanic carbon monoxide. Geophys. Res. Lett. 39, L21809. http://dx.doi.org/10.1029/2012GL053275.

McGonigle, A.J.S., Aiuppa, A., Giudice, G., Tamburello, G., Hodson, A.J., Gurrieri, S., 2008. Unmanned aerial vehicle measurements of volcanic carbon dioxide fluxes. Geophys. Res. Lett. 35, L06303. http://dx.doi.org/10.1029/2007GL032508.

Nadeau, P.A., Palma, J.L., Waite, G.P., 2011. Linking volcanic tremor, degassing, and eruption dynamics via SO_2 imaging. Geophys. Res. Lett. 38, L01304. http://dx.doi.org/10.1029/2010GL045820.

Prata, A.J., Bernardo, C., 2007. Retrieval of volcanic SO_2 column abundance from Atmospheric Infrared Sounder data. J. Geophys. Res. 112, D20204. http://dx.doi.org/10.1029/2006JD007955.

Ramsey, M., Harris, A.J.L., 2013. Volcanology 2020: How will thermal remote sensing of volcanic surface activity evolve over the next decade? J. Volcanol. Geotherm. Res. 249, 217–233.

Spampinato, L., et al., 2011. Volcano surveillance using infrared cameras. Earth Science Reviews 106 (1−2), 63−91.

Steffke, A.M., Harris, A.J.L., 2011. A review of algorithms for detecting volcanic hot spots in satellite infrared data. Bull. Volcanol. 73 (9), 1109−1137. http://dx.doi.org/10.1007/s00445-011-0487-7.

Surono, Jousset, P., Pallister, J., Boichu, M., Fabrizia Buongiorno, M., Budisantoso, A., Costa, F., Andreastuti, S., Prata, F., Schneider, D., Clarisse, L., Humaida, H., Sumarti, S., Bignami, C., Griswold, J., Carn, S., Oppenheimer, C., Lavigne, F., 2012. The 2010 explosive eruption of Java's Merapi volcano—A '100-year' event. J. Volcanol. Geotherm. Res. 241-242, 121−135.

Symonds, R.B., Gerlach, T.M., Reed, M.H., 2001. Magmatic gas scrubbing: implications for volcano monitoring. J. Volcanol. Geotherm. Res. 108, 303−341.

Theys, N., De Smedt, I., Van Roozendael, M., Froidevaux, L., Clarisse, L., Hendrick, F., 2014. First satellite detection of volcanic OClO after the eruption of Puyehue-Cordón Caulle. Geophys. Res. Lett. 41, 667−672. http://dx.doi.org/10.1002/2013GL058416.

Van Manen, S., Dehn, J., 2009. Satellite remote sensing of thermal activity at Bezymianny and Kliuchevskoi from 1993 to 1998. Geology 37, 983−986.

Wright, R., Flynn, L.P., Garbeil, H., Harris, A.J.L., Pilger, E., 2004. MODVOLC: near-real-time thermal monitoring of global volcanism. J. Volcanol. Geotherm. Res. 135, 29−49.

Wright, R., Garbeil, H., Davies, A.G., 2010. Cooling rate of some active lavas determined using an orbital imaging spectrometer. J. Geophys. Res. 115, B06205. http://dx.doi.org/10.1029/2009JB006536.

Chapter 66

Synthesis of Volcano Monitoring

John Pallister
Volcano Disaster Assistance Program, U.S. Geological Survey and U.S. Agency for International Development, USA

Stephen R. McNutt
School of Geosciences, University of South Florida, Tampa, FL, USA

Chapter Outline

This chapter synthesizes the monitoring of active volcanoes using a wide range of techniques and instrumentation. We cover definitions, concepts of instrument design and use, patterns and processes that occur before, during, and after eruptions. Ideas, trends, and the interplay between monitoring and research are discussed for both under-monitored and well-monitored "laboratory volcanoes," including a recent case history of the 2010 eruption of Merapi Volcano, Indonesia. We conclude with a discussion of: the importance of using multiple techniques and databases in forecasting, the reliability of forecasts, and future directions in volcano monitoring.

1. INTRODUCTION

Volcano monitoring serves two key functions: it provides basic scientific data to develop our understanding of the structure and dynamics of volcanoes, and it is crucial for hazard assessment, eruption forecasting and warnings, and risk mitigation at times of volcanic unrest. Monitoring provides the means to address questions of vital interest to communities affected by impending eruptions, such as when and where will the volcano erupt? Which areas are safe or dangerous? When will eruptions cease? Optimal interpretation of data from monitoring, especially for the purpose of warning and forecasting, depends critically on an adequate scientific understanding of volcano structure and processes, both in general and for each specific volcano. Thus, volcano monitoring at its most effective is a synergy between basic science and mitigation of risk.

A volcanic disaster on one side of the world can now have a significant economic impact on countries on the other side. For example, the economic cost of the 2010 eruption of Iceland's Eyjafjallajökull volcano is estimated

The Encyclopedia of Volcanoes. http://dx.doi.org/10.1016/B978-0-12-385938-9.00066-3

at approximately \$US 5.0 billion, with the majority of losses in GDP shared by nations in Europe, the Americas, Africa and the Mid-East, and Asia (Oxford Economics, 2010). It is therefore of interest to all nations that volcanoes are monitored so that eruptions can be forecast and mitigating action taken. Successful mitigation of the local effects of an eruption (such as evacuations and removal of movable assets, such as aircraft) requires a detailed understanding of the eruption process, for which monitoring data are a vital component. When successful, such efforts can save tens of thousands of lives, as was the case at Pinatubo volcano, Philippines in 1991 and at Merapi volcano, Indonesia in 2010. More widespread effects can be mitigated to some degree with early warning, for example, rerouting of aircraft, preparing airports, electrical grids, and critical facilities for ashfall. Similarly, an appreciation of the climatic effects both locally and globally requires data from surveillance before, during, and after eruptive activity.

Volcano monitoring comprises the systematic checking of various types of geological, geochemical, and geophysical data at a volcano. Each type of data (such as measurements of earthquakes, deformation, gas emissions, and temperature) provides information about physical processes that may be related to movement of molten rock or other eruption precursory phenomena. All the available data should be synthesized in a coherent way so that interpretations may be arrived at that are consistent with all the observations and with information about past eruptions. This interpretation then serves as a guide to anticipate what may happen next and provides an opportunity to take appropriate mitigation measures. Monitoring is complicated by the fact that data come in at different times and in different formats. Some data can be viewed 'raw,' but other data must be processed first, causing delays. For example, a seismic record may show that an earthquake has occurred, but additional processing is necessary to determine magnitude and location. Fortunately, automated data processing is decreasing the delays, and more and more monitoring data of all types is moving into the real-time domain.

Monitoring results may be ambiguous. For example, thermal anomalies may occur if lava is erupted, but can also result from fires and from solar heating of the ground. Similarly, ground deformation may be caused by the injection of magma prior to eruption, but also by movement of ground water or hydrothermal fluids (Dzurisin, 2007). Modern techniques of volcano surveillance involve the use of a wide range of physical and chemical measurements, some of which require lengthy laboratory work to analyze and some of which provide instant results. Each technique has advantages and drawbacks of some kind, but the more methods that can be used at one time, the better the chances of detecting precursory variations and therefore of being able to provide a warning, provided these data are effectively synthesized and interpreted. The most appropriate

techniques to use depend on the type and scale of the hazard envisaged. In the last two decades, a revolution in monitoring has taken place. Completely new technologies have become available including the use of satellites to monitor explosive SO_2 emissions and deformation with global positioning system (GPS) and SAR interferometry (InSAR). In addition, the expansion of the Internet and digital data has enabled easy transmission and sharing of data from remote locations to computers in observatories and laboratories for automatic analyses and interpretation, dramatically improving the resolution and forecasting capabilities at many volcanoes. Thus, the potential exists to use better and more complete data to monitor volcanoes, but with higher initial costs and more equipment to maintain.

Monitoring a volcano has some similarities with monitoring the health of a patient. Some tests, such as checking a pulse, are simple and direct. Others, such as blood tests, take time to perform and to analyze, but provide fuller results. Like volcanoes, people can show similar symptoms that may be related to a wide variety of illnesses or conditions. For example, a simple skin rash may be caused by a minor and temporary allergic reaction, or it may herald a terminal disease. New fumaroles may open up at a volcano simply because the previous path for passive degassing has become blocked by precipitation of minerals, or because a new batch of gas-rich magma has been intruded to shallow depth. The first situation may not be too serious, but the second certainly is. Monitoring techniques were formerly based on the most easily measured phenomena; other effects were either not recorded or were treated as noise. Future progress will be enhanced by taking account of these more subtle or complex effects and by the more comprehensive acquisition and real-time analysis of continuous data sets over extended periods. In addition, the development of global databases of eruptive phenomena and use of Bayesian statistics (Marzocchi and Bebbington, 2012; Newhall and Pallister, 2015) provide a means to evaluate the value of various monitoring data streams for forecasting and to estimate the probability of eruptions and their impacts.

2. INSTRUMENTS AND NETWORKS

The instruments used to monitor volcanoes vary widely in type, sophistication, cost, and effectiveness. Some instruments are permanent and telemetered; once installed they collect and relay data in real time to observatories for long periods of time (e.g., permanent seismic, infrasound, continuous GPS, webcams, and in situ or scanning gas sensor networks). Other instruments are usually deployed temporarily (e.g., seismic arrays, gravimeters, ground-based radar) and some can be driven or flown in the vicinity of volcanoes to make measurements (e.g., vehicle-based or airborne gas monitoring). Despite considerable progress over the past decade in building real-time and multiparametric monitoring networks, it is important to realize that only a fraction of the world's active and potentially active volcanoes are being monitored in real time today, and of these most still have only one or two seismometers and lack multiparametric networks. However, several dozen volcanoes now have extensive real-time measurements of multiple parameters, and these 'laboratory volcanoes' are the ones about which the most is known (e.g., Etna, Vesuvius, Kilauea, Mount St. Helens, Yellowstone, Long Valley, Popocatépetl, Soufriere Hills, Rabaul, Merapi, Mayon, Bezymianny, Sakurajima, Unzen, Fuji, and Augustine).

2.1. Seismic Monitoring

A century of seismological data testifies that seismic unrest in the form of earthquakes and tremor almost always precedes and/or accompanies volcanic activity at all types of volcanoes. Seismic activity is considered to be the best indicator, and often a reliable short-to midterm (days to weeks) forecaster, of the level and evolution of volcanic activity. Real-time seismic monitoring is one of the most common surveillance tools used by geophysicists and the equipment is fairly cheap and simple to install, although data collection and transmission may be difficult in remote areas. Traditionally, real-time telemetered networks have utilized robust low-cost single-component vertical seismometers (geophones), which are capable of measuring the vertical components of seismic waves with frequencies ranging from about 0.7 to 30 Hz and they have utilized relatively low-frequency VHF radio telemetry. Such networks still have advantages in some situations; however increasingly, digital broadband seismic networks are being installed because of their broader range of frequencies (ranging from 0.2 to 50 Hz), higher dynamic range, ability to resolve all three components of the seismic waveform, 'Internet-ready' circuitry, and availability of 'plug-and-play' digital UHF radios for networking and interface with data processing systems. With the proliferation and reduction in cost of very small aperture terminals for satellite Internet access, and with the increasing coverage of cellular telephone networks, even remote sites are now becoming more accessible.

Once the background seismicity has been characterized for a particular volcano, changes in the level or intensity, or appearance of different types of events, may herald the onset of activity. However, the resolution of the technique is not always very good, and source locations (such as those caused by movement of volcanic gases or cracking of rocks as fresh magma is intruded) are difficult to distinguish unless there are several seismic receivers spread on and around the volcano. Successful forecasts of eruption have been made using seismology, and tens of thousands of lives have been saved by evacuations around Pinatubo (Philippines) and Merapi (Indonesia) following seismic and other precursors to renewed activity. However, small explosions at 'open-system' volcanoes that are frequently active have fewer and smaller precursory earthquakes. In some of these cases, precursory earthquakes either did not take place or were not observed due to lack of nearby seismometers. In other such cases, precursory events were recognized only as changes in high background rates of seismicity, e.g., at Telica volcano, Nicaragua (Rodgers et al., 2013). Consequently the seismic warnings are of shorter duration and more subtle as was observed before a series of explosive eruptions at Soputan volcano, Indonesia by Kushendratno and others (2012). Although our ability to detect precursors and issue short-term eruption warnings at open-system volcanoes is improving, small phreatic or hydrothermal explosions occasionally take place with little warning and result in fatalities, principally to tourists and volcanologists who venture into hazardous crater areas (e.g., Galeras volcano near Pasto, Colombia in 1993; Semeru volcano, near Malang, East Java, Indonesia in 2000; Mayon volcano, near Legazpi City, Philippines in 2013; or Mount Ontake, Honshu, Japan in 2014).

Seismic sources at volcanoes are highly complex and involve the interaction of gases, melts, and solids. The role of the melt and gas may be either (1) active, giving rise to pressurized intrusions of magma into preexisting or newly formed zones of weakness, or to sustained vibration of the magma and host rocks or (2) passive, where brittle failures and the consequent stress readjustments modify the distribution of melts in the crust. Because volcanic media comprise systems of pores, fractures, and faults at all scales, sudden modification of the local stress field may induce seismic failure independent of melt propagation. Further, the complexity of the typical volcanic geology affects the transmission of seismic energy between source and receiver, e.g., 'path effects' can absorb energy and filter high frequencies, thereby making a volcano-tectonic (VT) earthquake representative of rock breakage appear at the seismic receiver to be a lower frequency or 'hybrid' event usually associated with gas or fluid movement. These factors contribute to the substantial degree of ambiguity in the interpretation of volcanic seismology. Interpretation of the source mechanisms of volcanic earthquakes is complex and still the subject of intensive study and debate [e.g., whether the ~1 million hybrid events at Mount St. Helens in 2004–2006 were produced by stick-slip faulting or by resonance of a fluid-filled crack (Iverson et al., 2006; Waite et al., 2008; Pallister et al., 2013)]. However, despite the uncertainty as to exact source mechanism, common precursory patterns of seismicity are well known and are used effectively in eruption forecasting world wide (e.g. White and McCausland, in review; Zoback et al., 2013).

In addition to the real-time telemetered seismic networks described above, recording networks or arrays of seismometers (typically consisting of broadband instruments) are frequently deployed to image subsurface magmatic systems using seismic tomography and to study the source processes for volcanic earthquakes (e.g., Waite et al., 2008). Although lacking real-time data transmission and hence, not appropriate for eruption warnings, these experiments do provide important insights into magmatic processes and plumbing — insights that subsequently inform hazard analyses.

2.2. Infrasound

A relatively new field of volcano monitoring utilizes sound waves below about 20 Hz, which are generated by the coupling of volcanic degassing events (explosions, degassing bursts, jetting, eruption tremor) with the atmosphere (Garces et al., 2013). These sound waves are similar to the P-waves as used in seismology. Specialized sensor arrays are used to detect coherent acoustic waves from these sources and to determine their locations, acoustic pressures, and frequency characteristics. From these parameters, it is possible not only to detect eruptions, but also to characterize intensity and type of eruptive behavior and intensity and to interpret source characteristics (see Chapter 63 for more details).

2.3. Ground Deformation

Ground deformation is another common technique for volcano monitoring. Repeated measurements of relative vertical height and/or horizontal distance are made, and changes often precede a change in volcanic activity. Traditionally measurements have been made with levels and electronic distance measurement (EDM) theodolites, but GPS measurements using a constellation of 24—32 satellites combined with synthetic aperture radar interferometry (see below) have largely replaced EDM methods for monitoring volcanic deformation in most developed countries (Dzurisin, 2007). GPS provides a measure of changes in the distance from a satellite to ground stations and is conducted in both campaign and continuous modes. In campaign mode one or more ground-based receivers are moved from place to place during a survey; thus, a large ground deformation network can be covered. In continuous mode, permanent GPS stations are installed at key locations surrounding a volcano and networked to provide near-real-time position data that is processed to reveal the subtle changes in the ground surface that accompanies the intrusion or withdrawal of magma or hydrothermal fluids. The advantage of GPS methods over the conventional ground deformation techniques is that they are quicker and less labor intensive, and not weather dependent.

A satellite-based technique that has revolutionized ground deformation monitoring is synthetic aperture radar interferometry (InSAR), in which radar images of large areas (such as a whole volcano) are made by scanning a radar beam over the area, and then comparing the resulting image to those from previous passes of the satellite. The interval between times the satellites pass over a site limits their applicability for short-term monitoring. However, this issue is improving. With the current constellation of satellites, repeat pass inteferograms can be constructed at intervals of ∼12 days for equatorial volcanoes and ∼4 days for high-latitude volcanoes (such as those in the Aleutians), and it is expected that with newly planned satellites, daily repeat coverage for high priority targets will be possible within a decade. Differences between the images reflect ground deformation, and the method is sensitive to movements of less than a centimeter under favorable conditions. There are limitations: atmospheric anomalies can delay the radar beam and produce false signals, areas with snow cover and dense vegetation do not produce coherent signals, although longer-wavelength radars have proven more effective in penetrating vegetation. In addition to InSAR, the radar images themselves may be useful for detecting and quantifying changes in a volcano's morphology, such as opening of fractures and growth of lava domes — changes that can be important for assessing hazards and issuing warnings. As an active-source method, radar can produce images night and day and it 'sees' though clouds — thereby enabling observations of cloud-shrouded volcanoes, as is commonly the case in the tropics and the Aleutian Islands.

It is usually assumed that a volcanic edifice behaves elastically or viscoelastically as pressure builds up within or beneath it; the volcano expands in a predictable way to accommodate the increased stress. However, brittle fracture rather than elastic/viscoelastic deformation also occurs at volcanoes and the effect is that ground deformation may be abrupt and contemporaneous with eruption, rather than gradual and precursory. In addition, deformation may take place without eruption, or may precede an eruption by many years. Thus, like seismic monitoring, this method is unreliable on its own for the forecasting of volcanic activity. Further, even though instruments are in place, some eruptions have little if any measurable deformation associated with them. This is commonly the case at 'open-system' volcanoes, which erupt frequently and have hot pathways (conduits) for magma ascent.

2.4. Gas and Thermal Monitoring

Volcanic gas analysis and temperature measurements are also commonly made. Precursory variations have been observed prior to and accompanying eruptions and these are used both in forecasting and to improve understanding of magmatic processes. Gas monitoring, that was initially done manually and at great risk using evacuated bottles to extract gas from fumaroles, followed by laboratory analysis, is now also accomplished with safer remote aerial surveys and *in situ* telemetered instruments that utilize UV spectrometers and chemical sensors to generate real-time or near-real-time data (Aiuppa et al., 2007; Gerlach and McGee, 2008; Galle et al., 2010). Satellite measurements of SO_2 emissions are also useful when gas plumes reach high into the atmosphere. The Total Ozone Monitoring Satellite was used to determine SO_2 emissions from a number of eruptions in the 1980s and 1990s. Currently, the most important satellite systems for eruption detection and volcanic emission monitoring are the Ozone Monitoring Satellite (OMI) and the Atmospheric Infrared Sounder, which return quantitative estimates of sulfur dioxide emissions into the upper atmosphere on a daily global basis (Carn et al., 2009). The gases SO_2, H_2S, and CO_2 are also monitored from aircraft or from ground-based gas sensors using a combination of spectrographic and direct chemical sensors (e.g., Gerlach and McGee, 2008). These airborne techniques are complementary to the satellite monitoring because they detect multiple gas species at low levels of the atmosphere and close to the volcano. As opposed to the satellite measurements, which for volcano hazards are used mainly for eruption detection and ash-cloud tracking, the airborne methods are also used in eruption forecasting.

Thermal monitoring is used for eruption detection by detecting hot spots at volcanoes, using sensors such as the Advanced Very High Resolution Radiometer (Dehn et al., 2000). However, the source (a vent in a volcanic crater or a new lava flow or dome) is often too small for quantitative analysis using the available satellite thermal sensors. Instead, quantitative thermal analysis is currently done mainly using hand-held or airborne infrared sensors. Such thermal measurements have proven valuable in identification of zones of extrusion and in documenting the evolution of lava dome eruptions.

2.5. Gravity and Magnetic, Electromagnetic Monitoring

The strength of a potential field varies with distance from the source, but never falls away to zero. Thus observation of gravity and magnetic fields can be made at relatively safe distances from the source, and provided they are made over a wide enough area, can reveal precursory variations. Magnetic precursors to eruption have been observed at the basaltic volcano Mount Etna, where shallow intrusions have produced significant changes in the magnetic field (Del Negro and Napoli, 2004). However, the technique has not been widely adopted at other volcanoes, likely due to the complexity of the magnetic field in these areas. Gravity measurements are ambiguous without simultaneous elevation (ground deformation) observations, but the combination has been used effectively for volcano monitoring. The combined technique, which has become known as microgravity monitoring, is capable of detecting precursory changes within a volcano that other techniques have not resolved. As with other techniques described here, microgravity monitoring can be carried out in both campaign and continuous modes. Examples of effective use of microgravity for volcano monitoring include studies that documented magmatic intrusion beneath Long Valley caldera, California (Battaglia et al., 2003) and helped image the progressive intrusion of a magmatic dike at Etna volcano, Sicily (Carbone et al., 2006). Additional electromagnetic methods, such as transient electromagnetic, self-potential, and magnetotelluric are not utilized for routine monitoring, but they may reveal information about a volcano's hydrothermal system.

Explosive eruptions with gas and ash plumes generate significant electrical charges, resulting in lightning, which at times can be intense and spectacular, as during the large explosive eruption of Chaitén volcano, Chile, in 2008. These electrical discharges create very low-frequency radio waves, which can be detected at distances of up to several thousand kilometers, using a low-cost network of radio frequency sensors. The World Wide Lightning Location Network (WWLLN; wwlln.net) was formed to detect lightning of all types, and since 2010 WWLLN has been operating an experimental ash cloud monitoring project, which has detected ash clouds from explosive eruptions, especially for volcanoes at high latitudes where thunderstorms are rare (see also Chapter 62).

2.6. Water-Level Monitoring

As magma rises to shallow levels below a volcano it disturbs the ground-water system. Rising magma and gas may pressurize aquifers, resulting in expulsion of water from springs and rising or falling water levels in wells. Additionally, deformation preceding eruptions will affect surface water levels. Changes in groundwater resulting from pressurization may be of large magnitude, with meter-scale changes in water levels in wells, such as those that took place preceding the 1993 eruption of Mayon volcano, Philippines and the 2000 eruption of Usu volcano, Japan. In addition, this effect may result in the expulsion of vast quantities of water (sufficient to create lahars) such as the one that took place at Mount Pelee in 1902 and preceded the eruption of Huila volcano, Colombia in 2007. Although not commonly included in volcano observatory operations, where appropriate, networks of water-level sensors can provide a relatively low-cost and sensitive monitoring method. Similarly, tide gauges, such as those used at

Rabaul Caldera, New Guinea, or the natural tide gauge provided by the Roman ruins at Macellum, Pozzouli (Campe Flegrei caldera, Italy) can serve as effective indicators of intrusion-related deformation.

2.7. Observations and Other Types of Monitoring

The most basic and most traditional of monitoring techniques is visual observations and sampling. As with the geophysical techniques, major advances in observational monitoring have come about in recent decades. These include advances in near real-time photogrammetry to quantify changes in morphology and extrusion rate. Similarly, sampling and rapid analysis during nonexplosive eruptions enable 'petrologic monitoring' of changes in magma chemistry, viscosity, and vesicularity — parameters that strongly affect explosivity. Although direct sampling of ballistic or flowage deposits from ongoing explosive eruptions is too dangerous to be practical, sampling of tephra can provide a safe means to evaluate these magmatic characteristics. In addition, ex post facto petrologic studies of eruptive products provide improved understanding of the subvolcanic and eruptive plumbing of volcanoes and of the processes that the real-time geophysical techniques monitor. In this manner, petrologic analysis of eruptive deposits after an eruption, as well as studies of previous eruptive products provide important conceptual frameworks in which to interpret real-time monitoring data and to make forecasts.

3. MONITORING BEFORE ERUPTIONS

There is a period of time between the end of dormancy and the onset of eruptive activity that may last days, weeks, years, or even decades during which processes go on inside and beneath a volcano in 'preparation' for eruption. These processes can be detected by making measurements at the surface and are termed 'precursors.'

Before magma or gases are erupted at the surface, they must either be transported from a deep storage region to a shallow region for temporary storage or simply intruded into pipes and dikes feeding to the surface without storage (hence the difference in 'preparation' time described above). Often the first observed signal is deformation, sometimes (but not always) accompanied by microearthquakes. If the magma movement is deep, the signals are small and spread out over a wide area. If it is shallow, then the deformation signal is stronger and is concentrated in a smaller area. The same processes that cause deformation also frequently cause small fractures to occur below ground level that are detectable as earthquakes. The earthquakes may occur in swarms at areas of stress concentrations — such as along nearby faults — and in the immediate vicinity of magma where pressure changes are occurring. The types of earthquakes typically change

during magma ascent, from high-frequency distal VT events that occur nearby faults to shallow low-frequency or hybrid (mixed high- and low-frequency) events as the magma shallows and gases are exsolved and the resulting magmatic fluid moves through conduits to the surface. Finally, the earthquakes merge into continuous tremor as the eruption ensues. This pattern of changing earthquake types has been observed preceding many eruptions and is widely used in combination with the other monitoring techniques, but typically as the primary tool, in forecasting eruptions (White and McCausland, in review).

When magma is ascending the gases dissolved in it change their relative concentrations because different gases have different solubilities. Carbon dioxide (CO_2) has very low solubility so it is the first to exsolve. Explosive eruptions that have been preceded by days to several weeks by increases in CO_2 emissions have been documented at several volcanoes including Etna (Aiuppa et al., 2007), Merapi (Surono et al., 2012), and Redoubt (Werner et al., 2013). Other gases such as chlorine (Cl) and sulphur (S) exsolve at shallower depths, and measurements of the C/S and Cl/S ratio can help determine whether fresh magma is present at shallow depths. The total amount of SO_2 also typically increases and may reach several hundred or even many thousands of tons per day when magma reaches very shallow depths. Further, decrease in SO_2 accompanied by continued or increasing seismicity, as at Pinatubo volcano in 1991, can signal sealing of conduits and building pressure prior to an explosive eruption. Another common precursor is increased heat. Typically the temperature at fumaroles increases, and the average ground temperature may increase as well; this is especially easy to spot on snow-covered volcanoes because the snow melts.

The precursors to an eruption will depend on the type of volcano, although virtually all eruptions are preceded by seismicity and emission of gases. Basaltic volcanoes tend to erupt low-viscosity freely flowing lavas, and the precursors might be increases in gas emissions and changes in the ratio of gas species, swelling of the volcano associated with the intrusion of magma into the edifice and a corresponding mass increase (observed as a gravity increase). Andesitic, dacitic, and rhyolitic volcanoes tend to erupt high-viscosity magma explosively and so in addition to the changes in earthquake types noted above, other precursors may include buildup in volcanic gas pressure beneath the surface accompanying intrusion of magma within the edifice. This would be seen as a decrease in gas emissions coincident with continued or increased seismicity, ground deformation, and possibly in changes in microgravity. Calderas erupt large volumes explosively, and although a large caldera eruption has not been observed, such an eruption may be triggered by the intrusion of hot basaltic magma into a reservoir of more silicic magma. The basaltic magma would be cooled and liberate volatiles and the silicic magma would become superheated and the dissolved gases would rapidly expand and escape to the surface. The bubbles would entrain hot magma that itself would degas rapidly, resulting in a runaway process that would erupt a large volume of the magma reservoir and lead to caldera collapse. A process such as this was documented from petrologic work on the caldera-forming eruption of Pinatubo volcano in 1991.

At a volcano that has not been historically active, or one where ground surveillance is not available, the first information that an eruption is imminent may come from 'remotely sensed' or satellite thermal infrared data, or from reports of earthquakes, gas emissions, or other changes by local people. The area near the vent may appear in satellite thermal imagery as a 'hot spot,' an area that is significantly warmer than its surroundings.

The appearance of a new lava lake or dome in a crater can be seen using these techniques, although strictly speaking, an eruption has already begun by this stage. When magma is very close to the Earth's surface, incandescence may be seen in cracks, areas of steaming ground appear, lakes (of water) dry up, cracks and bulges may be evident, and volcanic tremor occurs on seismometers. Many or all of the monitoring techniques available can detect processes that occur before an eruption.

It can be expensive to install and maintain monitoring networks. Consequently, during the 1990s, the U.S. Geological Survey conducted analyses of vulnerability, gaps in monitoring, and benefit—cost of early warnings and published a method for prioritization of national monitoring networks (Ewert et al., 2005). As a result, monitoring network resources in the USA are increasingly being prioritized for the highest-risk volcanoes including not only those that erupt frequently, but also those that pose high risk from infrequent eruptions. Similar prioritizations are underway in a number of other nations.

4. MONITORING DURING ERUPTIONS

Interestingly, and perhaps surprisingly, the most important yet ambiguous monitoring data are those collected during an eruption. The processes that occur during an ongoing eruption involve the movement of magma and gases from a shallow storage or feeder region to the surface. During an eruption, the feeder system remains open and variations in geophysical phenomena that can be measured (seismicity, potential fields, temperatures, etc.) are limited as the process is continuous and some of the changes in physical properties that are exploited by monitoring techniques are not detectable. However, there are changes during an eruption, such as in the ratio and flux of gases species as detected with spectrographic or aerial plume sampling, magma flux determined from satellite radar or other observations, pulses of energy from explosions, and intensity

of tremor. These observations provide valuable information about the next stage in the eruption (e.g., whether it is becoming more or less explosive).

We next approach what happens during eruptions by considering several different types of eruptions.

4.1. Explosions

An explosion is a very short-lived (a few seconds) but energetic event. When a strong explosion occurs at a volcano, seismic waves are generated in the ground, acoustic waves in the atmosphere, and a plume starts to rise. Seismic waves travel at several km/s, so the seismic ground waves are often the first signals recorded on monitoring instruments. Air waves travel more slowly, about 330 m/s, so these take several tens of seconds to reach instruments on the flanks of a volcano, either microphones, barographs, or seismometers (the air wave couples back into the ground). Gas is suddenly expelled during the explosion. Gas and ash plumes rise at rates of 50—100 m/s, so these may be observed locally and detected by satellites as they rise and spread laterally with the wind.

4.2. Sustained Explosive Eruptions

A sustained explosive eruption may occur on a timescale of a few minutes to a few hours. Such events are highly energetic and produce primarily ash, tephra, and rapidly moving pyroclastic density currents. The onset of such an event produces seismic waves and acoustic waves as above, and a plume rises as well. Lightning may form within an ash plume, as early as tens of seconds for Surtsey volcano but more typically 4—20 min after charge separation has occurred, as seen in spectacular fashion in the 2008 eruption of Chaitén volcano in Chile. Deformation may include static strain steps as well as permanent changes in the shape of the vent, such as the excavation of a large crater. When sustained, the seismic disturbance is continuous and long lasting (volcanic tremor) and the strength of the signal is proportional to the intensity of the eruption. This typically changes as a function of time within the sustained eruption. Acoustic waves are continuous or episodic, and typically occur in bursts during strong pulses of explosions. In some cases acoustic waves can be heard 200 km or more away from the volcano. Lightning continues to occur for as long as the eruption lasts. Some lightning extends from the ash cloud to the ground, and some extends between charged regions within the ash cloud (intracloud lightning). If the plume is injected into the stratosphere, the amount of SO_2 can be estimated by ultraviolet and infrared satellite measurements. Plumes may travel hundreds of kilometers downwind, connected to the vent, and when the eruption stops, the plume continues to travel as a discrete mass carried by the winds. Some plumes have remained hazardous several days after the eruptions that produced them have ended. Large plumes that reach the stratosphere have been known to circle the Earth and the largest and most sulfur-rich may result in temporary global cooling. Pyroclastic density currents from sustained explosive eruptions (along with lahars created by remobilization of the pyroclastic and ash-fall deposits) can scour and then bury vast regions surrounding volcanoes, as at Pinatubo in 1991.

4.3. Effusive Eruptions

Effusive eruptions produce lava flows or domes instead of tephra. They are thus not explosive, although such eruptions sometimes follow initial explosive phases, and collapse of accumulating summit lava domes may also produce explosions, as well as convective gas and ash columns. The seismic signal produced by effusive eruptions is volcanic tremor, which may have a gradual onset. The strength of the tremor varies with the rate of effusion of lava. Ash plumes from effusive eruptions are typically small, typically rising to only a few km, so lightning is rare. (Ash plumes from effusive eruptions may account for a small fraction of the total erupted volume.) Deformation is of several types. New lava flows cover and change the shape of the land surface. Effusive eruptions often start at fissures, and the flow later localizes to one or more cylindrical vents. Each of these has a characteristic pattern of deformation that occurs before and during the eruption. The filling and emptying of subsurface dikes and feeders affects the gravity (causing increases and decreases, respectively). Magnetic intensity will increase over a lava flow or dike as it cools, but decrease over a dike or feeder as it drains.

Although lava flow eruptions destroy property, because of relatively slow flowage, they rarely take human lives. However, lava dome-forming eruptions are among the most frequently deadly eruptions on Earth, as the gravitational collapse of lava domes unleashes internal gas pressure in the lava, resulting in rapidly moving pyroclastic density currents, surges, and blasts. Seismic signals accompanying dome-forming eruptions vary widely and may include phases rich in VT earthquakes and tremor, however; a characteristic seismic signature of dome extrusion is the occurrence of repetitive hybrid earthquakes, such as that which accompanied the 2004–2008 eruption of Mount St. Helens (Iverson et al., 2006).

5. MONITORING AFTER ERUPTIONS

5.1. General Cases

It is not easy to predict the beginning of an eruption, but it is even more difficult to predict the end. Partly this is a problem of definition. A volcano never looks quite the same after an eruption as it did before and so it cannot return to exactly the state it was previously in. Therefore the definition of the end of an eruption may be rather subjective. It is clear when magma is no longer coming out of a vent, but there may be gases still being emitted, and lava flows and domes take weeks, years, or more to cool to ambient temperatures. In some cases, the end of an eruption means simply the end of a particular phase of or type of activity. Here we define an eruption as the flux (mass flow per unit time) of magma through a vent including steam and gases. This is an operational definition to help put boundaries on the subject at hand; other definitions appear throughout this encyclopedia.

Most explosive eruptions, those that produce tephra, generate volcanic tremor as the eruption is under way. In particular, the strength of the tremor is proportional to the intensity of the eruption as measured by ash-column height or flux. Thus when an eruption wanes and approaches its end, the tremor decreases and reaches background values. Tremor generally accompanies each phase of eruption, and may occur between phases and after the last eruption as well. Earthquake swarms often occur throughout eruptive sequences, with some clusters of events associated with eruptive phases and some as indications of changing stresses in the vicinity. As eruptions near their ends, the swarms typically decay in rate with a power law, similar to aftershock sequences following large earthquakes. Thus the two chief seismic indicators, earthquake swarms and tremor, both decay from high levels to background levels as eruptions are ending.

At the end of an eruptive period when intruded magma is cooling and solidifying in situ, the ground may begin to subside slightly due to thermal contraction and or magma withdrawal. At the same time, as magma cools it will acquire a magnetic signature measurable at the surface and the density may increase, raising the values of the magnetic and gravity fields at the surface. Degassing decreases also, although frequently it decays slowly and sporadically. Gas may continue to be released months or even years after an eruption, although typically at lower levels. Tephra eruptions deposit large amounts of unconsolidated material on steep slopes. If water is abundant, either rain or glacial, then lahars may be easily formed and may continue to occur for months or years after eruptions end. Seismometers or microphones and web cameras placed near flow channels may serve as effective lahar sensors. Several seismometers along channels at intervals of a few kilometers, or an array of microphones, are needed to be able to tell if a lahar is moving and to estimate its speed. Although the eruption of lava or pyroclastic material onto the ground may end, some of the finest ash, gases, and aerosols that were injected to high altitudes may remain suspended in the atmosphere. If the suspended particles reach the stratosphere, they may be carried by the jet stream, carried around the globe. In addition to climatic effects such as global cooling, which can have widespread impacts on agriculture, fine particles and droplets of sulfuric acid cause abrasion to airliner leading edges and produce 'crazing' of windows, resulting in more frequent replacement. Thus, the economic effects of eruptions can last for years.

5.2. Restless but No Eruption

A large-magnitude caldera-forming eruption has not taken place since volcano monitoring techniques have been developed. However, although a large caldera eruption is less likely to occur than a small caldera-forming eruption, such as the one that occurred at Pinatubo, its effect would be so devastating that large calderas in regions of dense population are monitored closely. There were crises during the 1980s at two such calderas, Rabaul (Papua New Guinea) and Campi Flegrei (Naples, Italy). In both cases, the ground rose up by over a meter and gravity decreased (as would be expected for an elevation increase), but not by as much as it should if the gravity change could be attributed solely to the height change. Thus, in both cases, an overall increase in subsurface mass occurred (approximately 10^{11} kg), but in neither case did an eruption follow in the short term. Thousands of microearthquakes occurred beneath and adjacent to each caldera, but virtually no volcanic tremor was detected. Thus while it seems that magma was added to the reservoir of silicic magma beneath the calderas, there was not enough to cause the catastrophic superheating expected to precede an explosive eruption. Although the amount of ground uplift was large enough to be alarming, the calculated increased volume of magma was small (a few percent), and so the risk in fact was small, although in both cases people were evacuated temporarily. Rabaul eventually erupted in 1994, but this was nearly 10 years after the seismic and deformation crisis. This implies that appreciable amounts of time may pass between the time when the magma reservoir is replenished and the time when the volcano finally erupts. Additional episodes of inflation, as well as deflation, have taken place at Long Valley and Yellowstone calderas, leading to the conclusion that the larger the magmatic reservoir, the more capable it is to receive new inputs of magma without triggering eruptions. In addition, large distant tectonic earthquakes can cause seismic unrest, but to date no eruptions at these caldera systems. For example, triggered microearthquake swarms at Long Valley and Yellowstone calderas following the M7.9 Denali Fault earthquake in 2002. The current consensus among most volcanologists and seismologists is that although large magnitude earthquakes are known to trigger seismic activity at nearby and even distant volcanoes, only in cases where a volcano is already

pressurized and ready to erupt will a distant tectonic earthquake trigger an eruption.

6. CASE HISTORY: MERAPI 2010

In the first edition of this chapter for the Encyclopedia, case histories for Mount Etna and Soufrière Hills Volcano were summarized. These are laboratory volcanoes for which extensive multiparameter monitoring data are available. The reader is referred to the first edition, to the 2002 Special Volume of the Geological Society of London on Soufrière Hills, and to the subsequent review paper by Alparone et al. (2004), as well as to numerous other papers in the international literature on monitoring and research at these two volcanoes.

In this edition, we focus on a new case study that involved explosive and dome growth and destruction phases: the 2010 eruption of Merapi volcano, Indonesia, an anomalously large '100-year' eruption at a volcano known mainly for small dome eruptions and lithic pyroclastic density currents. The case study reported below is drawn from the overview paper by Surono et al. (2012) and from data, interpretations, and references published in the 23 papers that constitute a Special Issue of the *Journal of Volcanology and Geothermal Resources* on the 2010 Merapi eruption (Jousset et al., 2013).

6.1. Introduction

The 2010 eruption of Merapi volcano, near Yogyakarta, Indonesia, was the largest eruption in more than 100 years at a volcano that is better known for small eruptions, which occur on average every 4—6 years. Typical eruptions of the twentieth century produced summit lava domes, which collapsed to produce block-and-ash pyroclastic flows and surges, known as 'Merapi-type' pyroclastic density currents. In contrast, the 2010 eruption did not only extrude lava domes, but it also produced powerful explosions and numerous pyroclastic density currents that extended into populated areas at distances of up to 16 km from the summit.

Because of its frequent activity and hazard to the population of Yogyakarta region (pop. 3.4 million), Merapi has long been an international 'laboratory volcano.' Observatory functions began in the late 1800s and the first seismograph was installed at Merapi in 1924. Currently, the Indonesian Center for Volcanology and Geologic Hazard Mitigation (CVGHM, formerly known as the Volcanological Survey of Indonesia) operates a technology development center (Balai Penyelitikan dan Pemgambangan Teknologi Kegunungapian) and Merapi Volcano Observatory (MVO) in Yogyakarta to monitor and study Merapi volcanic activity. MVO operates five observatory outposts ('Pos' stations) at strategic locations on the flanks of

Merapi volcano. The mission of these entities is to forecast eruptions, to improve knowledge of volcanic processes, and to develop new volcano monitoring technology.

Continuous research efforts have been underway at Merapi for decades, with numerous Indonesian and international projects directed at understanding the structure and the mechanisms of dome-collapse eruptions and at improving eruption forecasting. An important lesson from past work is that the volcano is capable of much larger and more dangerous eruptions than those of the twentieth century — a lesson that was relearned in 2010.

6.2. 2010 Eruption Summary

Eruptions at Merapi are so common and their precursors and effects have been so predictable in recent decades that when Merapi began to experience seismic swarms and inflation in late 2009 and early 2010, the MVO anticipated another similar eruption. For example, during the last eruption (in 2006) a lava dome grew at rates $2-4$ $m^3 s^{-1}$ and over a period of several months, it added $\sim 5 \times 10^6$ m^3 of basaltic-andesite to the complex of pre-existing lava domes at the summit. The 2006 dome collapsed repeatedly; the largest pyroclastic density current descended 7 km down the south flank and killed two persons.

In early October 2010, when the cumulative seismic energy release approached levels comparable with those of past eruptions, MVO forecasted an eruption for mid-October. However, during the third week of October, the seismic energy and extent of deformation exceeded forecast values and CO_2 emissions spiked. As seismicity and summit deformation continued to increase rapidly, CVGHM concluded that a much larger eruption was coming. The 2010 eruption started with a powerful phreatomagmatic blast and pyroclastic density current on 26 October, which killed 34 people. The eruption continued and reached a climactic phase during the night of 4–5 November, with a vertical ash column to 17 km altitude and with pyroclastic density currents that swept a broad region on the southern flank of the volcano toward Yogyakarta. Satellite remote sensing aided forecasting during the eruptive crisis. These data revealed a new lava dome at the summit, which grew at exceptionally rapid rates (>25 $m^3 s^{-1}$ on average) and reached 5×10^6 m^3 volume just before the climactic eruption (Surono et al., 2012; Pallister et al., 2013). In addition to VT and multiphase (MP, i.e. hybrid) swarms, very long-period and long-period (LP) earthquakes preceded the eruption and were linked to magma ascent. The 4–5 November eruption was approximately 10 times larger and more explosive than eruptions of the past several decades, and it validated the concern that had long been apparent at Merapi — that much larger and more hazardous eruptions are a continuing threat at the volcano.

6.3. Crisis Response

The Geological Agency of Indonesia monitored the eruptive activity, interpreted the monitoring data, and issued warnings and recommendations for areas to be evacuated. The Indonesian national emergency response agency (BNPB) and their provincial and local counterparts managed the evacuations. Teams from Europe, the USA, and Japan assisted the Indonesian scientists during the crisis and provided new monitoring equipment, remote sensing information, and consultation regarding potential hazards. Space agencies from several nations provided remote sensing data (e.g., Canada, Europe, Italy, Japan, and USA), including support through the International Charter for Space and Natural Disasters, which was activated and managed by the U.S. Geological Survey on behalf of the Government of Indonesia.

Evacuations took place in stages as the monitoring parameters escalated in intensity and the scientists' concerns of a larger-than-normal eruption increased. A restricted zone with a radius of 10 km from the summit was established on 25 October, just 35 h before the initial explosive eruption. The radius of the restricted zone on the volcano's south flank was then increased in rapid succession from 10 to 15 km on 3 November and to 20 km late on 14 November only a few hours before the climactic eruption took place at midnight. The official count of evacuees increased commensurate with the radii of the zones from 22,599 on 26 October to a total of almost 400,000 people by 13 November. Although rushed, recognition that a large eruption was imminent and the resulting rapid evacuations saved approximately 15,000 lives.

The challenge following the eruption has been to determine what triggered this 'larger-than-normal' 2010 eruption, what monitoring parameters were most important, and how to improve forecasting of such anomalous large events in the future. Indonesian and international teams have worked to answer these questions.

6.4. Magmatic Plumbing

Tomographic seismic modeling of data from the MERapi AMPhibious EXPperiment, a broad seismological experiment conducted in 2004, revealed a large ($>50,000$ km^3) low-velocity body that extends to upper mantle depths beneath Central Java. This anomaly is considered to be the pathway for ascent of magmas erupted at Merapi and at the other active volcanoes in the region (Figure 66.1). The anomaly is unique in size and amplitude compared to other arcs and it suggests that this segment of the arc has a large proportion of fluids or melts (13–25%) and a high magma flux. Detailed analysis of seismic data associated with Merapi's eruptions reveal a shallow low-velocity zone below the summit and a deeper low-velocity zone that

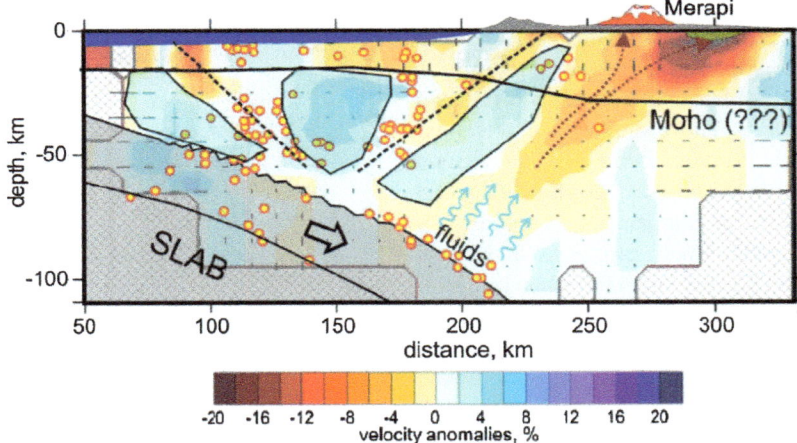

FIGURE 66.1 P-wave seismic velocity anomalies (colored zones) and earthquake locations below the island of Java in a section that transects the island perpendicular to the arc near the longitude of Merapi volcano. In this interpretation, fluids are liberated from the downgoing slab flux melting in a seaward dipping low-velocity zone beneath the arc. *Figure modified from Luehr et al. (2013).*

extends to depths of more than 10 km. Modeling of GPS and tilt data indicate depths of 6−9 km for the source reservoir for magmas erupted in the 1990s. However, petrologic data indicate that magmas erupted at Merapi contain crystals that were derived from a wide range of depths, ranging from the near-surface conduit environment down to about 30 km, and high-resolution gravity modeling suggests a complex plumbing structure with multiple solidified magma intrusions beneath the summit of the volcano. Together, these data suggest a magmatic plumbing system consisting of a plexus of relatively small but interconnected reservoirs (Figure 66.2).

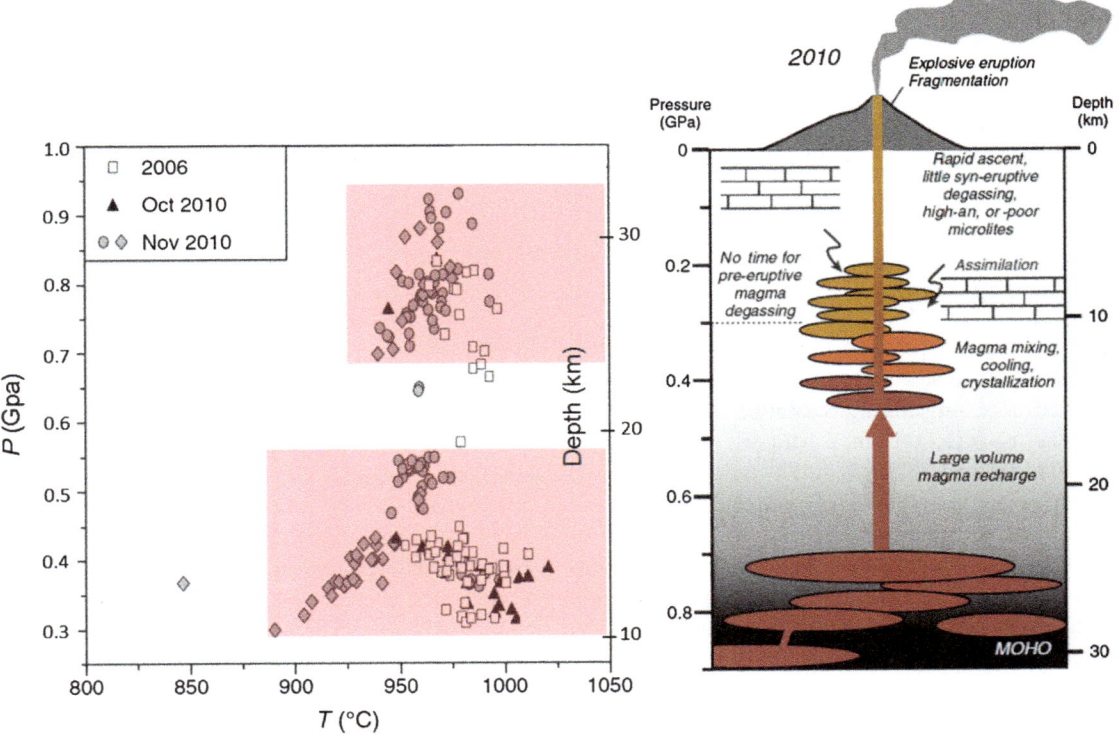

FIGURE 66.2 Diagram showing the equilibration pressures and temperatures for amphiboles from the 2006 and 2010 eruptions of Merapi on the left and the resulting conceptual model of the magmatic plumbing system on the right. *Figure modified from Costa et al. (2013).*

The working petrologic model for eruptions at Merapi is that batches of reduced mafic magma from the lower crust are typically slowed and trapped by a crystal-rich 'eruption filter' — a plexus of shallow crustal magma reservoirs and partially solidified intrusions. These shallow reservoirs are connected to the surface through conduits that remain partially molten between eruptions over much of its extent (as suggested by Merapi's high-temperature magmatic gas fumaroles). Frequent 'normal-sized' eruptions of Merapi take place when magmatic pressure in the shallow reservoir system exceeds load pressure in the conduit. Increases in magmatic pressure result from addition of new batches of magma and from exsolution of a supercritical fluid, which unmixes to a brine and vapor during ascent, accompanied by addition of CO_2 from reacting limestone wallrock inclusions. Crystal-rich magma then rises, and as seen during the 2006 eruption, it may push a plug of nearly solidified conduit magma upward, accompanied by localized summit deformation and earthquake swarms. Once the solidified plug is pushed out, deformation and VT seismicity typically decline as less-viscous magma is then extruded to form a summit lava dome. Extrusion of dome lava may continue for weeks to months, feeding block and ash pyroclastic density currents as the growing domes collapse down the steep slopes from the summit. Although most aspects of this model also apply for 2010, petrologic studies of 2010 eruptive products explain the explosive character of the eruption as resulting from a rapid rise of an anomalously large batch of volatile-rich magma from a deep (~ 30 km) reservoir below the volcano. Because of the large volume and volatile load, this batch of magma was able to rise more quickly and power the large, but short duration explosive 2010 eruption.

6.5. Seismic and Deformation Monitoring Data Used to Forecast Eruption Occurrence and Magnitude

Merapi has long been monitored using seismology, deformation, gas emission data, and petrology. Under noneruptive conditions, the rate of inflation/deflation (measured as change in lengths of EDM lines between the volcano's summit and flanks) is ~ 0.003 md^{-1}; the cumulative seismic energy release is less than 35 MJd^{-1} with daily averages of five MP earthquakes and one VT earthquake; the baseline SO_2 flux is $\sim 50-100$ Mgd^{-1}, the long-term eruption rate is 0.038 m^{3}s^{-1} (1.2×10^6 m^3 yr^{-1}), and the typical magma erupted in all but a few of the two dozen eruptions of the twentieth century has been basaltic andesite.

Peak cumulative seismic energy release preceding eruptions of Merapi during the twentieth-century eruptions was generally in the range of 20,000 to 30,000 MJ (Figure 66.3). Precursory seismic unrest typically begins

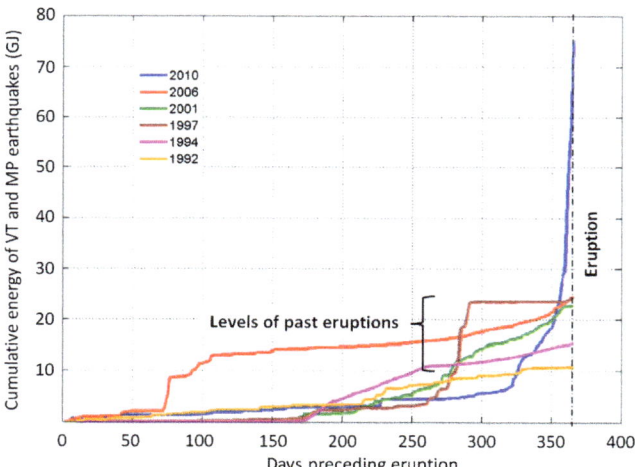

FIGURE 66.3 Cumulative seismic energy release of volcano-tectonic (VT) and multiphase (MP) earthquakes for eruptions of Merapi in 1997, 2001, 2006, and 26 October 2010. Cumulative energy is plotted relative to days preceding the peak eruptive activity during each eruptive episode (for the pre-2010 activity, the peak activity is defined as the largest dome collapse event among a series of smaller collapses and is designated as 'Eruption' in the figure). *Figure modified from Budi-Santoso et al. (2013).*

with small VT earthquake swarms that start months to as much as a year before normal eruptions, increase in number and magnitude, and are joined by MP (hybrid) events as the time of eruption approaches. Similarly, summit deformation typically begins months in advance and escalates in rate before eruptions begin. As noted above, once a 'normal' eruption begins, pressure within the conduit is reduced and deformation and seismicity decrease in intensity (e.g., as represented by the flattening of the seismic energy curves in Figures 66.3 and 66.4).

Merapi's seismic energy trends emphasize how knowledge of an individual volcano's monitoring history plays a fundamental role in forecasting future eruptions of a similar type and magnitude. However, as previously noted, it has also long been recognized that Merapi is also capable of much larger eruptions. Evidence of larger eruptions includes stratigraphic relationships and historic accounts of events that took place in the late 1800s and early 1900s. In particular, eruptions in 1930 and 1872 were much larger and more explosive than those of more recent decades. Consequently, volcanologists working at Merapi have worried that the next eruption could be 'the big one' and have struggled to distinguish precursors for such an event. In 2010 Merapi provided precursory warnings in rapid succession, and fortunately, the Indonesian volcanologists recognized that these were warning of a large eruption and they responded quickly. Now and in retrospect, the 2010 eruption provides an important case study of precursors for anomalously large eruptions at frequently active volcanoes.

A remarkable feature of the 2010 eruption was the rapid increase in rates and magnitudes of seismicity and

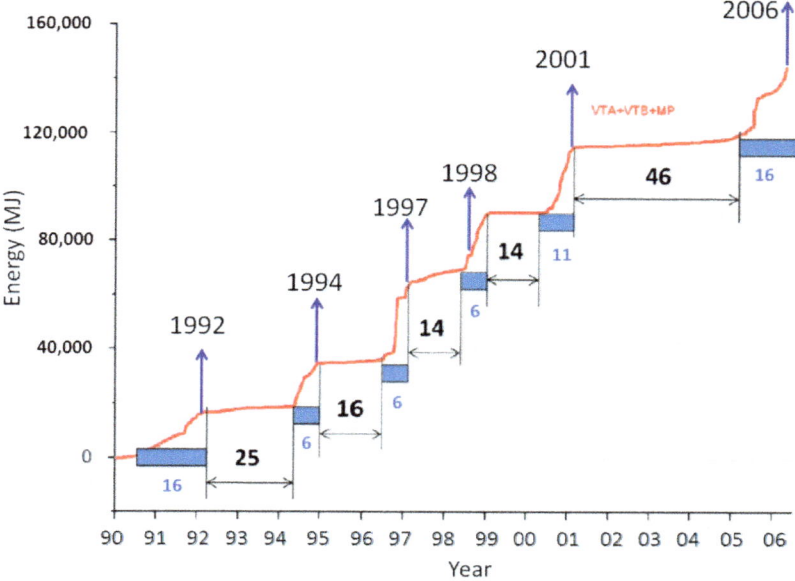

FIGURE 66.4 Cumulative seismic energy curve for volcanic earthquakes at Merapi volcano for the period 1990 through 2006. Arrows indicate times when eruptions began. Blue boxes indicate intereruptive periods with times in months indicated in blue. Eruptive periods (characterized by prolonged dome growth and collapse) indicated by horizontal arrows with times in black. *Figure modified from Ratdomopurbo et al. (2013).*

deformation that took place over the week to 10 days preceding the eruption. As seen in Figure 66.3, the cumulative seismic energy release reached a level several times greater than previously recorded by the time of the initial eruption on 26 October. This was a result of the much larger and more frequent VT and MP events during the 10 days preceding the 2010 eruption (Figure 66.5).

Seismicity decreased after the initial phreatomagmatic eruption on 26 October as the number of VT and MP events dropped sharply immediately following this eruption, reflecting a decrease in pressure in the conduit associated with initial vent opening, as had been seen in previous eruptions. However, during the period between 26 October and the climactic magmatic eruption on 4–5 November, a number of other unprecedented events took place. This was a time of rapid lava dome extrusion, accompanied by modest size explosive eruptions on 29 and 31 October, and 1 and 3 November. More than 150 large LP earthquakes (with dominant frequencies ~2 Hz) also took place between 29 October and 3 November. Posteruption modeling indicates that these LP events were the result of movement and resonance of gas and magma within the shallow part of the edifice. High-frequency tremor then began on 3 November and was associated with frequent pyroclastic density currents. The tremor became continuous by midday on the 3rd and during the 4th short-period stations became saturated. The climactic eruption began on 5 November at 00:01 local time (4 November, 17:01 UTC), lasted 27 min, and generated an ash column to 17 km altitude and pyroclastic density currents that extended 16 km down the Gendol drainage on the south flank of the volcano.

Simultaneously with the initial phase of rapidly increasing seismic energy, the rate and magnitude of deformation increased dramatically. An EDM line on the south flank decreased in length by as much as 3 m (Figure 66.6). However, EDM lines on other flanks of the volcano showed no such changes, and posteruption InSAR analysis revealed that only the summit area was deforming and shifting downslope to the south. In effect, the localized downslope movement of the summit led to collapse, which opened the upper conduit and triggered the climactic eruption.

Posteruption analysis of earthquake locations after the eruption reveal a vertical and roughly cylindrical distribution extending to ~5 km depth below the summit, with most of the located earthquakes within the uppermost kilometer (Figure 66.7). A shift from mainly deep to shallow VT seismicity was recorded as magma ascended and fractured its way to the surface leading to the climactic eruption. Similarly, analysis of LP seismicity indicates progressive excitation of three shallow fluid-filled conduits in the summit region as the magma rose.

6.6. Satellite Monitoring and Probabilistic Methods Used to Forecast the 2010 Climactic Eruption

Through the International Charter for Space and Major Disasters, a variety of satellite data were used to monitor the 2010 eruption, including Synthetic Aperture Radar (SAR) from the COSMO SkyMed, RADARSAT-2, and

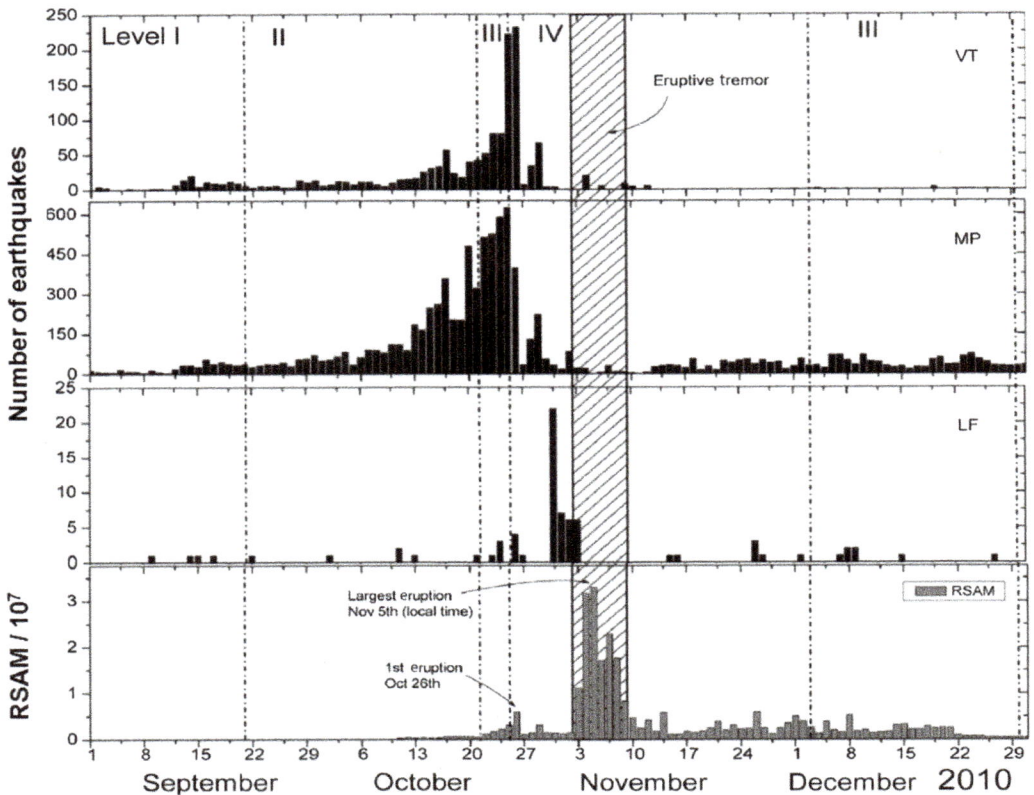

FIGURE 66.5 Histograms showing the daily number of recorded volcano-tectonic (VT), multiphase (MP), and low-frequency (LF) earthquakes and the real-time seismic amplitude measurement (RSAM) unit counts. Dash-dot lines divide periods at alert levels I, II, III, and IV. Diagonally lined box represents the period of intense tremor that accompanied the climactic eruption. *Figure modified from Surono et al. (2012).*

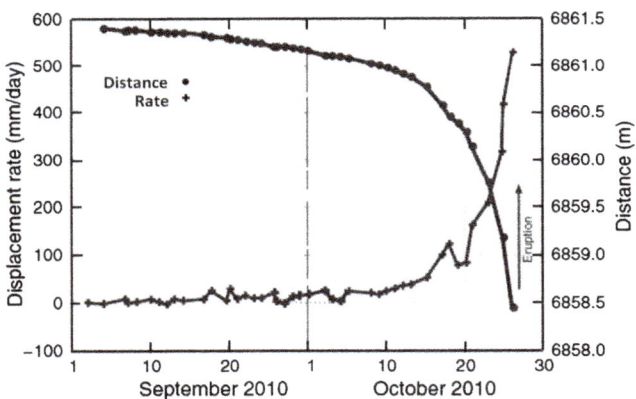

FIGURE 66.6 Diagram showing the change in distance and rate of displacement along an EDM line measured from the south flank (Kaliur-ang station) to the summit of Merapi during September and October of 2010. Notice the rapid change in distance and rate during the week pre-ceding the initial eruption on 26 October (when the EDM summit reflector was destroyed). *Figure modified from Surono et al. (2012).*

TerraSAR-X sensors, and when weather and orbits permitted, thermal infrared from the ASTER sensor and high-resolution visible and near-infrared data from the GeoEye 1 and WorldView-2 sensors. Cloud cover limited

use of data from optical sensors. However, the radar sat-ellites supplied frequent and detailed images of the summit crater, rapidly growing lava domes, vent features, and pyroclastic density current deposits. Images were analyzed and critical data and analyses were delivered to the response team in Yogyakarta each day or in some cases twice a day during the crisis.

During the early eruptive phase of the 2006 eruption of Merapi, the petrologic hypothesis described previously was used to develop a probability tree to forecast the magnitude and potential outcomes of the eruption (Newhall and Pallister, 2015). The basis of this forecast is the assumption that the rate of lava extrusion is proportional to pressure and ascent rate within the magmatic conduit. To determine a precursor for a more explosive eruption, the observed rate of extrusion was compared to Merapi's long-term average extrusion rate and to rates at other similar volcanoes where transitions from effusion to explosive fountaining have been observed (e.g., typically seen when rates exceed $10\ m^3 s^{-1}$). These comparisons, along with other moni-toring precursors were used to assign probabilities for various outcomes. Factors that increased the probability of an explosive eruption included a marked increase of hybrid or LP earthquakes (especially monochromatic LP's), a

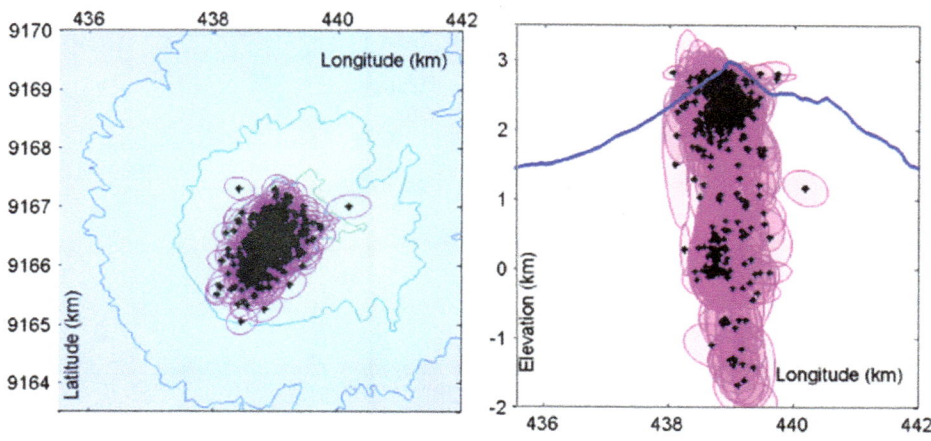

FIGURE 66.7 Map (left, in UTM units, UTM Zone 49S, WGS84 projection) showing epicenters and east–west cross section showing hypocenters of precursory VT earthquakes recorded at Merapi volcano in 2010. Confidence intervals (67%) are indicated by ellipses. *Figure modified from Budi-Santoso et al. (2013).*

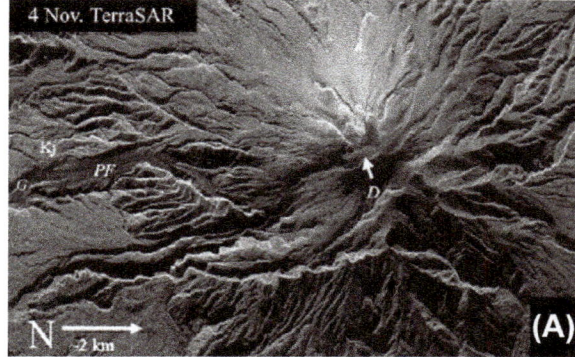

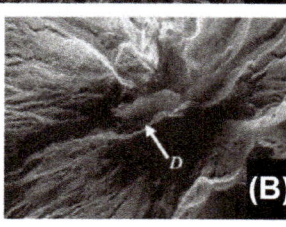

FIGURE 66.8 TerraSAR radar image of the Merapi volcano (A) and enlargement of the summit area (B). The new $5 \times 10^6 \, m^3$ lava dome (D) extruded during the preceding 4 days is seen at the summit and extending into the head of the Gendol drainage (G). Recent pyroclastic flows (PF) extend to the vicinity of the village of Kinorejo (Kj), where 34 people were killed by overbank surges from a pyroclastic flow during the initial eruption on 26 October.

sudden increase in deformation, rate of extrusion, or SO_2 emissions (or a sudden decrease in SO_2 accompanied by increased seismicity), sharply accelerated rates of any of these monitoring parameters, and any evidence suggestive of potential summit collapse (such as the one that last took place in 1930), such as water expulsion along structural discontinuities.

During the period 1–4 November 2010, a new lava dome was extruded at Merapi's summit. Satellite radar imagery was used to estimate the volume and rate of lava extrusion. The rate ($>25 \, m^3 s^{-1}$) was remarkable; it produced a $5 \times 10^6 \, m^3$ lava dome in just a few days

(Figure 66.8). Taken in the context of the 2006 probability tree, these high rates and the large volume of new lava perched at the summit raised concerns of an imminent and much larger eruption. Consequently, when the seismic tremor increased greatly on the evening of 4 November, CVGHM extended the restricted zone to 20 km distance several hours before the climactic eruption ensued.

6.7. Gas Monitoring Using Ground and Satellite Sensors

SO_2 burdens in the plume were available daily from satellite sensors. Plume altitudes were reported by the Darwin VAAC and were used to assign the appropriate altitude for OMI retrievals (~ 17 km for 4–5 November, and altitudes in the $\sim 5-8$ km range after 5 November). Subtracting the SO_2 burdens from two consecutive images allowed a mean SO_2 flux to be estimated. The SO_2 flux rates corresponded directly with the explosive phases of the eruption and even the smaller eruptions greatly exceeded the SO_2 fluxes measured during the 1992–2007 eruptions (Figure 66.9). The SO_2 cloud was carried by high-level winds west across Java and the southwestern Pacific Ocean (Figure 66.10). The plume from the climactic eruption was tracked and advisories were issued to aviation interests by the Darwin Volcanic Ash Advisory Center.

The total SO_2 emission during the 2010 Merapi eruption was estimated at 0.44 Tg, indicating that at least 0.1 km^3 of basaltic andesite was degassed during the eruption (assuming ~ 1000 ppm S in the melt phase) and suggesting a magnitude and Volcanic Explosivity Index for the eruption of about 4 (Surono et al., 2012). However, field-based estimates of juvenile magma within the 2010 deposits are considerably smaller (0.02 to 0.05 km^3) and suggest the presence of a significant excess gas phase to power the explosive phases of the eruption.

FIGURE 66.9   SO$_2$ fluxes during the 2010 Merapi eruption as determined from various ground and satellite sensors and compared to real-time seismic amplitude measurement (RSAM) values (in arbitrary units). (*Modified from Surono et al. (2012)*). Symbols: E = explosive eruption, L = lahar. The flux during previous eruptions is based on measurements by correlation spectrometer (COSPEC).

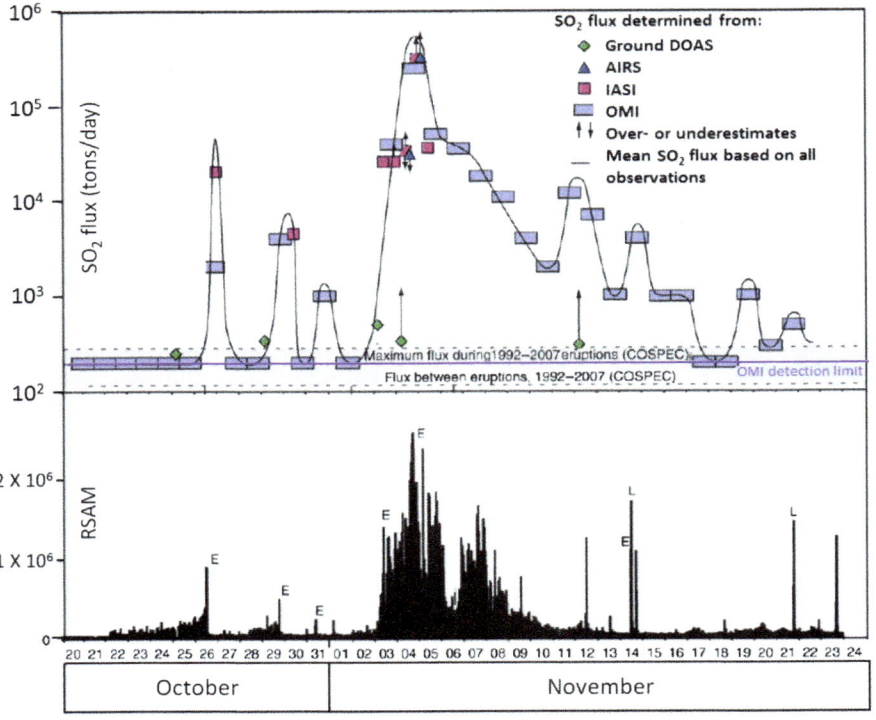

FIGURE 66.10   Composite image from the Aura/Ozone Monitoring Instrument (OMI) aboard the Aura satellite showing the distribution of the SO$_2$ cloud from the 2010 eruptions of Merapi volcano from 4 through 8 November 2010. *Image courtesy of NASA (see http://aura.gsfc.nasa.gov/science/feature-20120305a.html).*

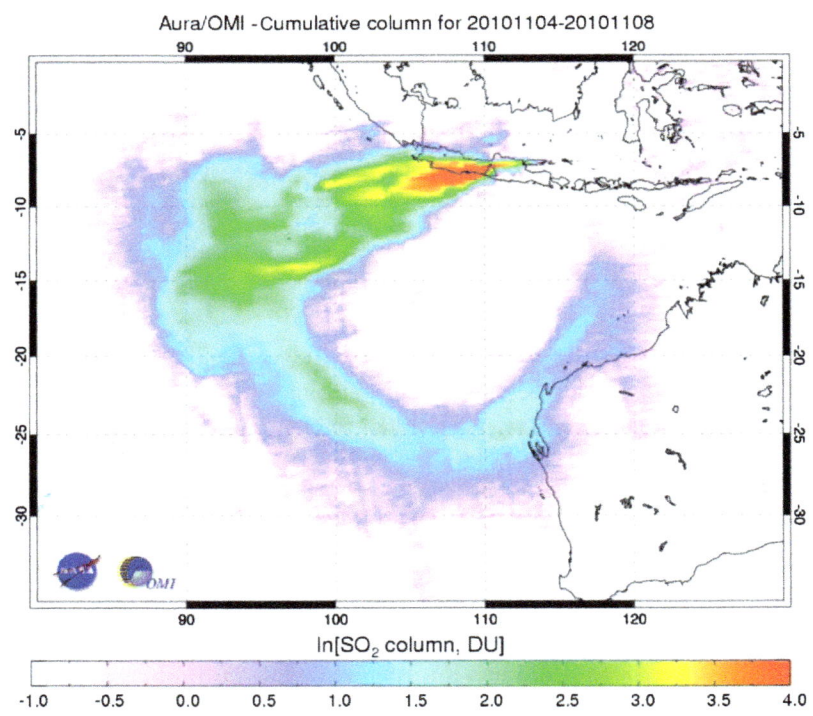

6.8. Carbon Dioxide: A 'Smoking Gun' Precursor for Large Eruptions at Frequently Active Volcanoes?

In situ monitoring of volcanic gas emissions (H_2O, SO_2, CO_2, H_2S, CO, HCl, H_2, O_2, and CH_4) has been carried out for decades at Merapi by regularly collecting samples from the Woro solfatara at summit. Traditional methods are utilized (e.g., collection by bubbling the gas through NaOH solutions in evacuated flasks and analysis using spectrometric and volumetric methods).

Because of the pressure dependency of CO_2 solubility in silicate melt, delivery of a large volume of mafic magma from depths of 30 km to a much shallower crustal reservoir (as indicated by the petrology) should release significant amounts of CO_2. As it happened, a gas sampling team from the Merapi observatory climbed the volcano on 20 October, only 5 days before the initial explosion. Given the circumstances, this was a highly risky and unwise expedition. Regardless, the gas samples that were collected from the high-temperature summit Woro fumarole field revealed unusually high CO_2 abundances and ratios of CO_2/SO_2. These results are consistent with the recent arrival of new gas-rich magma from great depth. This result and similar precursory CO_2 emissions at Etna and Redoubt volcanoes suggest that such emissions may be effective 'early warnings' for ascent of gas-rich and potentially explosive magma. Unfortunately, at Merapi none of the fumarole gas samples were retained for isotopic analyses. Consequently, the distinction between CO_2 derived from degassing of mafic magma versus assimilation of crustal limestone remains a question. Yet, regardless of the source of the CO_2, the prominent spike in abundance preceding the initial eruption is a strong argument for enhanced remote multigas monitoring of volcanoes.

7. DISCUSSION

7.1. Techniques Alone versus in Combination

Most of the monitoring techniques discussed in this chapter can give ambiguous data or, in worst cases, can give misleading data or even fail to give needed data. For example, earthquake swarms may happen during eruptions or when no eruption is occurring. If a large regional earthquake happens to occur, it will swamp the seismometer for several minutes so that any volcanic signals cannot be seen. Further, instruments are subject to noise and physical damage; radio and electronic interference can degrade many types of telemetered data, and eruptions themselves, or related phenomena like lightning, can damage or destroy instruments. On practical considerations alone, it is generally best not to rely on only one instrument or one type of data, but rather to consider data from several different types of instruments simultaneously. In addition, the combination of different types of monitoring data provides a better understanding of magmatic processes, and therefore aid in forecasting. For example, decreased gas emissions coupled with increased seismicity signal increasing pressure and a greater probability of an explosive eruption, as at Pinatubo in 1991. Similarly, unusually rapid rates of deformation combined with unusually high levels of seismic energy release or with unusually rapid rates of dome extrusion increase the probability of a larger and more explosive phase of an ongoing eruption, as at Merapi in 2010. Conversely, deformation without seismicity, as during the first 8 years of dramatic InSAR-detected uplift at Three Sisters volcano in Oregon, USA, indicated deep magmatic intrusion (below the brittle–ductile transition) and consequently, extremely low probability of an imminent eruption.

7.2. Reliability, False Alarm Rate, and Safety Margin

Many types of data occur over various size ranges, and the small signals are 'safe'; whereas, the larger ones indicate significant hazard. Thus a threshold must be determined which will serve as a guide to judge the appropriate level of concern. Examples would be earthquakes of a certain number or cumulative energy release or deformation of a certain amount. Once a threshold is agreed upon, however, several issues arise. It is possible to make the threshold too sensitive: this means the dangerous signals will definitely be registered, but with a high false alarm rate. Alternatively, the threshold may be set too high: this will keep the false alarm rate low, but may miss the onset of dangerous activity. A few minutes extra warning time may not seem like a lot, but could be significant for a commercial jetliner traveling at eight miles per minute. The ideal threshold minimizes the false alarm rate while providing the earliest warning. Thus it gives the highest reliability and safety margin. Appropriate threshold estimates may be made in advance of eruptions, but then need to be fine-tuned as the activity evolves, because physical changes may occur at the volcano. Also, because the history of human observation is short compared to the lifespan of volcanoes, and because geologic data show that many volcanoes have had much larger than 'normal' eruptions in the past, volcanologists must be ever vigilant for potential precursors for larger-than-normal eruptions.

7.3. Intrusions versus Eruptions

One of the main large-scale problems of monitoring volcanoes is that many of the signals regarded as precursors, such as earthquakes and deformation, occur with both eruptions and intrusions. The reason the signals look the same is that the processes are the same at depth, differing only in whether or not the magma reaches the Earth's surface. Thus, not every geophysical event is regarded as a precursor; they often simply drive an additional inquiry or checking of other data. Calderas are notorious for having high levels of unrest, so there are lots of potential precursors. For example, Long Valley caldera in eastern California has experienced more than 25 earthquakes of $M = 5$ or larger, uplift of 60 cm, degassing of CO_2, LP earthquakes, local deformation, and increased heating, all in the 19-year period of 1981–2000. No eruption has occurred, but it is thought that several intrusions of magma to depths of about 5 km have occurred. Thus, it is clear that intrusions do not always result in eruptions; the challenge is to decide if an intrusion will progress to an eruption or not.

The experience of the USGS–USAID Volcano Disaster Assistance Program in responding to dozens of volcanic crises world-wide is that less than one in 10 episodes of volcanic unrest results in an eruption, and most such eruptions that ensue are small. Further, as volcano monitoring networks expand, the incidence of recorded volcanic unrest is increasing. This is at once good and bad news, as although the probability of a highly explosive and deadly eruption during a period of unrest is typically small, repeated examples of unrest without eruption tends to foster public complacency, or worse, a lack of confidence in scientists' warnings. One of the greatest future challenges for volcanologists and emergency managers is not only to continue to improve our ability to forecast eruptive activity (or lack thereof), but to find ways to clearly communicate and foster effective public responses to warnings of low-probability, high-impact events.

8. VOLCANO MONITORING IN THE FUTURE

While monitoring solely for the purpose of risk mitigation may reduce loss of life or property, its effectiveness can be greatly improved through the concurrent development of scientific methods and enhanced understanding of volcanic processes. Given the average interval between eruptions at volcanoes in most nations, volcano monitoring tends not to be a high spending priority of governments, yet major volcanic eruptions can be national (and international) social and economic disasters.

Future developments in volcanic monitoring will depend on many factors including technological advances, improvements in analytical and interpretation methods, the ability to successfully integrate multiple data sets, and having detailed knowledge of volcanic structure and processes. The continued development of robust, inexpensive, and effective real-time technologies for monitoring volcanoes is a vital and sometimes urgent requirement for

volcano-affected countries. Advances in electronics in the period between 1970 and 1990 resulted in cheap expendable sensors and analogue radio telemetry, thus reducing the inherent financial risk of losing expensive monitoring equipment and greatly benefiting the monitoring of volcanoes in developing countries. The current transition to broadband seismic systems and digital networks has been a mixed blessing for these countries; as such 'off-the-shelf' systems are Internet-ready, but at increased cost and investment in information technology. The Internet has added a new standard for communication and it is now routine at many volcano observatories to run analysis tasks automatically and to post the results on web pages. These can then be checked on demand by anyone with a computer, Internet access, and the appropriate permissions. The web protocols have made it no longer necessary for each observatory to develop its own unique way of handling data and new software is beginning to enable real-time visualization and interpretation of multiparametric data streams. This is an area that is ripe for advances over the next decade: development of common software for analysis of volcanic monitoring data, wide access to these data and software via the Internet, as well as to archived data that are important as analogues for use in forecasting. Several database projects are underway aimed at these goals, e.g., WOVOdat (http://www.wovodat.org), VHUB (https://vhub.org/), as well as to aid in analysis of volcanic risk VOGRIPA (http://www.bgs.ac.uk/vogripa/index.cfm).

The importance of a good knowledge of the geological and volcanic history and the nature of background activity at each particular volcano cannot be overemphasized; if what is 'normal' is not known then the 'abnormal' cannot readily be recognized. By analogy to medicine, we know that a person's normal temperature is 98.6 °F, so this provides the basis for identifying a fever. Many of the world's potentially active volcanoes, however, are not currently monitored. When these volcanoes become active, it is necessary to quickly survey the geology of the region to determine the character and explosivity of past eruptions and to draw analogies from other similar volcanoes world wide. This is one of the operating principals of the Volcano Disaster Assistance Program (VDAP) program, which played an important role in the successful response of the Philippine Institute of Volcanology and Seismology and VDAP to the 1991 eruption of Pinatubo. In such cases, as well as for evaluating the probability of abnormal events at well-known volcanoes, use of global databases of past eruptions, such as the Global Volcanism Program of the Smithsonian Institution (http://www.volcano.si.edu/index.cfm) coupled with precursor databases, such as being constructed by WOVOdat, will be keys to advancing probabilistic eruption forecasting. Although such forecasts are often described as 'pattern-based' and 'empirical,' most are in fact based on physical and chemical constraints on magmatic evolution and ascent, and on geologic models of magma reservoir and conduit systems. Over the next decade, we anticipate more advanced development of physical—chemical models of volcanic systems. Such models have already been inverted to estimate volumes, overpressures, and volatile abundances of crustal magma reservoirs. The next frontier will be in quantitative forward modeling utilizing the determined physical parameters and a new generation of higher-precision real-time monitoring data to make deterministic forecasts. Such deterministic forecasts become probabilistic when they take into account past activity (global and local) as well as thresholds and uncertainties in the modeled parameters (c.f., Marzocchi and Bebbington, 2012; Newhall and Pallister, 2015).

Perhaps the area of greatest advance during the past decade has been in monitoring of Earth surface deformation. The expansion of the GPS satellite constellation and the development and deployment of thousands of high-precision GPS ground stations world-wide have given us unprecedented insights into plate and microplate tectonics, faulting, magmatic intrusion, and eruption. The complementary and synoptic views of InSAR have further advanced our understanding of these earth processes. Future expansion of the constellation of SAR satellites, along with airborne and ground-based SAR systems, will surely continue these advances over the coming decade, and provide rapid repeat high-precision temporal digital elevation models, which will continue the volcano geodesy revolution and enable modeling of extrusion and flow modeling at a level never before possible, as well as contributing to deterministic forward models and eruption forecasting.

Gas monitoring, that was initially done manually and at great risk using evacuated bottles to extract gas from fumaroles, followed by laboratory analysis is now accomplished with safer remote aerial surveys and telemetered instruments that utilize UV spectrometers and chemical sensors to generate near-real-time data. These techniques have benefitted from research and development investment driven by air pollution and climate change problems. New miniature ultraviolet spectrometers, designed to monitor industrial emissions, have been adapted to monitoring SO_2 in addition to gases involved in air pollution, such as nitrogen dioxide (NO_2), formaldehyde (HCHO), bromine monoxide (BrO), and ozone (O_3). Such sensors have largely replaced the larger and more expensive correlation spectrometer (COSPEC) that was formerly the mainstay of remote gas monitoring. Similarly, newly developed small and low-cost commercial chemical sensors for CO_2, H_2O, H_2S, and SO_2 are now available and are being combined with miniature air pumps, digital data recorders, and microprocessors in suitcase-size packages. These 'multigas sensor' packages can be deployed downwind of a volcano in telemetered networks or flown through a dilute plume on an aircraft to provide abundance and ratios of these key volcanic gases. Using these instruments, as well as Fourier transform infrared spectrometers, to remotely measure the halogen gases has already revolutionized the utility of volcanic gas monitoring. The future is bright for continued development and deployment of real-time multigas networks, which will improve understanding of the chemistry and dynamics of magmatic degassing — the ultimate driver of explosive eruptions. Of particular interest is the measurement of CO_2, SO_2, and the halogen gases and the ratios between them, as these species vary in solubility in silicate melts and therefore exsolve at progressively lower pressures. CO_2 is difficult to measure directly due to its high abundance in the atmosphere; however, it can be determined by ratio to the more readily measured sulfur species. This is of importance to hazard analysis and forecasting, as a shift from low to high CO_2/SO_2 ratios can indicate ascent of magma from depth to the near surface. To address the consequent early warning potential of CO_2 emissions and to fingerprint sources of the carbon, new instruments are now being tested to enable direct measurement of CO_2 as well as isotopes of carbon in the field.

Potential developments in analysis and interpretation of seismic data include improvements to automatic detection and processing (ADP) such that the large volume of data generated by an intense seismic swarm can be processed, event types classified, and that real events can be more reliably distinguished from noise. The effectiveness of ADP depends on the density and geometry of the array compared with the extent and depth of the active zones; data from recent eruptions suggest an array density of one seismometer per km^2 is required to resolve events associated with shallow magmatic intrusion. The Integrated Mobile Volcano Monitoring System developed by the USGS for use in volcanic crises was used to very good effect in the 1991 Pinatubo eruption; however, portable seismometers do not constitute an adequate alternative to permanent recording stations. While they provide satisfactory monitoring during volcanic events, they have no role in establishing background seismicity or in recording the onset of seismic activity. In addition, these 'during the crisis' installations pose significant risks to those involved in responding to the crisis. As a consequence, the USGS has shifted to a strategy of installing seismic networks in advance of any crisis activity at all potentially active and high-risk volcanoes (Ewert et al., 2005). The same approach is being taken in many other nations, and even the USGS—USAID VDAP is focused on helping developing countries build multidisciplinary monitoring networks in advance of crises, as opposed to the former 'mobile volcano observatory' approach. Advances in interpretation will require more detailed knowledge of the structural complexities and heterogeneities typical of volcanoes in order to develop adequate models. Recent advances in nonlinear, high-resolution three-dimensional seismic tomography, for example, may allow the detailed interpretation of broadband seismic data. Such broadband data with a period greater than a few seconds are considered to be associated with mass transport and are currently the subject of much research into their causative mechanisms.

Although seismology will remain the most important monitoring tool for the foreseeable future, the integration of seismic data with results from other methods, for example, those based on geodesy, infrasound, gas and thermal flux, as well as gravity and electrical signals, will be far more powerful. 'Smart' systems using artificial neural networks and sequential data assimilation techniques are being developed to integrate the vast quantities of data resulting from technological developments to produce the simplest best-fitting models consistent with as much of the data as possible. Such systems, however, are only as good as the input data, and hence the need for detailed structural models and background information.

We anticipate that the revolution in remote volcano monitoring will continue. New satellite radar missions will increase the repeat time for radar image acquisitions of volcanoes, enabling radar imaging and InSAR to become more effective near-real-time monitoring tools. Innovative new techniques, such as cosmic-ray muon radiography, offer promise in imaging the internal structure and areas of low density (e.g., magma or hydrothermal fluids) within volcanoes, and a new generation of remote gas monitoring methods are moving volcanic emission detection and quantification to the realm of real-time techniques and offer promise for early warning of magmatic recharge. Petrology, active- and natural-source seismology, and other geophysical imaging techniques will remain of great value for understanding volcanic plumbing and magmatic processes. New petrologic methods that measure ascent rates of magmas based on reaction kinetics are improving our understanding of magma storage, eruption triggers, and the dynamics of magma ascent (e.g. Costa et al., 2013) and near-real-time analysis of tephra are beginning to also move petrologic characterization into the realm of near-real-time monitoring. In addition, new investigations to link seismic signals to observed field structures offer promise to improve our understanding of seismic sources for volcanic earthquakes (Pallister et al., 2013).

Important questions in contemporary volcanology include the following: What are the processes that trigger eruptions? How can forecasts of the timing and impacts of an eruption be improved? Is it possible to determine in advance the explosivity and magnitude of an eruption and its likely duration? These issues present pressing problem (both to basic science and to affected societies), and are being studied at a number of 'laboratory volcanoes' using a combination of geophysical and geochemical techniques. The priority now in volcanic monitoring is the safe acquisition of real-time multiparameter data and the development of rapid and reliable methods to collate, analyze, and interpret them. Our understanding of the mechanisms and processes operating within the plumbing systems below volcanoes is improving, but we now need to expand this knowledge by integrating the experience gained from decades of independent research. In addition to technological and scientific advances, it remains a simple fact that cannot be emphasized enough that effective communication and interaction between volcanologists, civil authorities, and the affected populace are required to prevent future volcanic disasters.

SEE ALSO THE FOLLOWING ARTICLES

Gas, Plume, and Thermal Monitoring ● Ground Deformation, Gravity, and Magnetics ● Seismic and Infrasonic Monitoring ● Volcanic Gases ● Volcanoes and Tourism ● Volcano Warnings - Volcanic Lightning.

FURTHER READING

Aiuppa, A., Moretti, R., Federico, C., Giudice, G., Gurrieri, S., Liuzzo, M., Papale, P., Shinohara, H., Valenza, M., 2007. Forecasting Etna eruptions by real-time observation of volcanic gas composition. Geology 35, 1115−1118. http://dx.doi.org/10.1130/G24149A.1.

Alparone, S., Andronico, D., Giammanco, S., Lodato, L., 2004. A multidisciplinary approach to detect active pathways for magma migration and eruption at Mt. Etna (Sicily, Italy) before the 2001 and 2002−2003 eruptions. J. Volcanol. Geotherm. Res. 136, 121−140.

Battaglia, M., Segall, P., Roberts, C., 2003. The mechanics of unrest at long valley caldera, California. 2. constraining the nature of the source using geodetic and micro-gravity data. J. Volcanol. Geotherm. Res. 127, 219−245.

Budi-Santoso, A., Lesage, P., Dwiyono, S., Sumarti, S., Subandriyo, J., Surono, Jousset, P., Metaxian, J.-P., 2013. Analysis of the seismic activity associated with the 2010 eruption of Merapi volcano, Java. J. Volcano. Geotherm. Res. 261, 153−170.

Carbone, D., Zuccarello, Saccorotti, G., 2006. Analysis of simultaneous gravity and tremor anomalies observed during the 2003−2003 Etna eruption. Earth Planet. Sci. Lett. 245, 616−629.

Carn, S.A., Krueger, A.J., Krotkov, N.A., Yang, K., Evans, K., 2009. Tracking volcanic sulfur dioxide clouds for aviation hazard mitigation. Nat. Hazards 51, 325−343. http://dx.doi.org/10.1007/s11069-008-9228-4.

Costa, F., Andreastuti, S., Bouvet de Maisonneuve, C., Pallister, J., 2013. Petrological insights into the storage conditions, and magmatic processes that yielded the centennial 2010 Merapi explosive eruption. J. Volcanol. Geotherm. Res. 261, 209−235.

Dehn, J., Dean, K., Engle, K., 2000. Thermal monitoring of North Pacific volcanoes from space. Geology 28, 755−758.

Dzurisin, D., 2007. Volcano Deformation − Geodetic Monitoring Techniques. Springer/Praxis Publishing, Chichester, UK, 441 p.

Ewert, J.W., Guffanti, M., Murray, T.L., 2005. An Assessment of Volcanic Threat and Monitoring Capabilities in the United States: Framework for a National Volcano Early Warning System NVEWS. USGS. Open-File Report 2005-1164, 62 p.

Galle, B., Johansson, M., Rivera, C., Zhang, Y., Kihlman, M., Kern, C., Lehmann, T., Platt, U., Arellano, S., Hidalgo, S., 2010. Network for Observation of Volcanic and Atmospheric Change (NOVAC)—A global network for volcanic gas monitoring: Network layout and instrument description. J. Geophys. Res. Atmos., 115, D05304. http://dx.doi.org/10.1029/2009JD011823.

Garces, M.A., Fee, D., Matoza, R., 2013. Volcano Acoustics, (Chapter 16). In: Fagents, S.A., Gregg, T.K.P., Lopez, R.M.C. (Eds.), Modeling Volcanic Processes: The Physics and Mathematics of Volcanism. Cambridge University Press, New York.

Gerlach, T.M., McGee, K.A., 2008. Emission rates of CO_2, SO_2 and H_2S, scrubbing, and preeruption excess volatiles at Mount St. Helens, 2004—2005. In: Sherrod, D.R., Scott, W.E., Stauffer, P.H. (Eds.), A Volcano Rekindled: The Renewed Eruption of Mount St. Helens, 2004—2006. U.S. Geol. Survey Professional Paper 1750, pp. 543—571.

Iverson, R.M., Dzurisin, D., Gardner, C.A., Gerlach, T.M., LaHusen, R.G., Lisowski, M., Major, J.J., Malone, S.D., Messerich, J.A., Moran, S.C., Pallister, J.S., Qamar, A.I., Schilling, S.P., Vallance, J.W., 2006. Dynamics of seismogenic volcanic extrusion at Mount St. Helens in 2004—05. Nature 444, 439—443. http://dx.doi.org/10.1038/nature05322.

Jousset, P., Budi-Santoso, A., Jolly, A.D., Boichu, M., Surono, Dwiyono, S., Sumarti, S., Hidayati, S., Thierry, P., 2013. Signs of magma ascent in LP and VLP seismic events and link to degassing: an example from the 2010 explosive eruption at Merapi volcano, Indonesia. J. Volcanol. Geotherm. Res. 261, 171—192.

Kushendratno, Pallister, J.S., Kristianto, Bina, F.R., McCausland, W., Carn, S., Haerani, N., Griswold, J., Keeler, R., 2012. Recent explosive eruptions and volcano hazards at Soputan volcano — a basalt stratovolcano in north Sulawesi, Indonesia. Bull. Volcanol. 74, 1581—1609. http://dx.doi.org/10.1007/s00445-012-0620-2.

Luehr, B.–G., Koulakov, I., Rabbel, W., Zschau, J., Ratdomopurbo, A., Brotopuspito, K.S., Fauzi, P., Sahara, D.P., 2013. Fluid ascent and magma storage beneath Gunung Merapi revealed by multi-scale seismic imaging. J. Volcanol. Geotherm. Res. 261.

Marzocchi, W., Bebbington, M.S., 2012. Probabilistic eruption forecasting at short and long time scales. Bull. Volcanol. 74, 1777—1805. http://dx.doi.org/10.1007/s00445-012-0633-x.

Newhall, C.G., Pallister, J.S, 2015. Using multiple data sets to populate probabilistic volcanic event trees, p. 203—232 (Chapter 8). In: Papale, P., Eichelberger, J., Loughlin, S., Nakada, S., Yepes, H. (Eds.), Volcanic Hazards, Risks and Disasters. in: Hazards, Risks, and Disasters. Elsevier, Amsterdam, 505 pp.

Oxford Economics, 2010. The Economic Impacts of Air Travel Restrictions Due to Volcanic Ash — A Report Prepared for Airbus. Oxford Economics, Oxford, UK, 12 p. http://www.airbus.com/company/environment/documentation/?docID=10262&eID=dam_frontend_push.

Pallister, J.S., Cashman, K.V., Hagstrum, J.T., Beeler, N.M., Moran, S.C., Denlinger, R.P., 2013. Faulting within the Mount St. Helens conduit and implications for volcanic earthquakes. Geol. Soc. Am. Bull. 125, 359—376. http://dx.doi.org/10.1130/B30716.1.

Ratdomopurbo, A., Beauducel, F., Subandriyo, J., Nandaka, A., Newhall, C.G., Suharna, Sayudi, D.S., Suparwaka, H., Sunarta, 2013. Overview of the 2006 eruption of Mt. Merapi. J. Volcanol. Geotherm. Res. 261, 87—97.

Rodgers, M., Roman, D.C., Geirsson, H., LaFemina, P., Muñoz, A., Guzman, C., Tenorio, V., 2013. Seismicity accompanying the 1999 eruptive episode at Telica Volcano, Nicaragua. J. Volcanol. Geotherm. Res. 265, 39—51.

Surono, Jousset, P., Pallister, J., Boichu, M., Buongiorno, M.F., Budisantoso, A., Costa, F., Andreastuti, S., Prata, F., Schneider, D., Clarisse, L., Humaida, H., Sumarti, S., Bignami, C., Griswold, J., Carn, S., Oppenheimer, C., 2012. The 2010 explosive eruption of Java's Merapi volcano — a '100-year' event. J. Volcanol. Geotherm. Res. 241-242, 121—135.

Waite, G.P., Chouet, B.A., Dawson, P.B., 2008. Eruption dynamics at Mount St. Helens imaged from broadband seismic waveforms: Interaction of the shallow magmatic and hydrothermal systems. J. Geophys. Res. 113, B02305. http://dx.doi.org/10.1029/2007JB005259.

Werner, C., Kelly, P.J., Doukas, M., Lopez, T., Pfeffer, M., McGimsey, R., Neal, C., 2013. Degassing of CO_2, SO_2, and H_2S associated with the 2009 eruption of Redoubt Volcano, Alaska. J. Volcanol. Geotherm. Res. 259, 270—284.

White, R.A., McCausland, W.A., in review 2015. Estimating intrusive volumes and forecasting eruptions using distal volcano-tectonic earthquakes. J.Volcano. Geotherm. Res.

Zoback, M.L., Geist, E., Pallister, J., Hill, D.P., Young, S., McCausland, W., 2013. Advances in natural hazard science and assessment, 1963—2013. Geol. Soc. Am. Special Paper 501, 81—154.

Chapter 67

Volcano Warning Systems

Chris E. Gregg

Department of Geosciences, East Tennessee State University, Johnson City, TN, USA

Bruce Houghton

Department of Geology and Geophysics, National Disaster Preparedness Training Center, University of Hawai'i, Honolulu, HI, USA

John W. Ewert

Cascades Volcano Observatory, US Geological Survey, Vancouver, WA, USA

Chapter Outline

GLOSSARY

aviation-based alerts Alerts targeting people in aircraft or linked to the aviation sector (commercial and military).

digital age or information age The present time, when most information is stored, disseminated, and received in digital form, especially concerning web and cellular smart phone technologies.

ground-based alerts Alerts targeting people on the ground.

instrumental monitoring The use of hardware and usually corresponding computer software to detect and monitor physical and chemical changes of a volcano.

milling behavior The act or acts of people seeking (searching for) information in response to a perceived threat or risk communication message in order to find answers to unresolved questions or exchange information with others.

notifications An all-inclusive term encompassing the range of risk communication messages (i.e., alert, information statement, advisory, watch, and warning) disseminated by volcano observatories or emergency management agencies.

table-top exercises A type of training where agency representatives role play their individual and organizational responses to a real or potential threat without mobilizing resources (cf., drills, which involve mobilization of personnel and resources).

volcano alert level systems A subsystem within volcano warning systems that involves development and communication of alert messages. It is synonymous with status levels, condition levels, or color codes.

warning confirmation process A model that describes people's response to a risk communication messages as a sequence of outcomes that depend on the content and style of a message and on characteristics of the receiver.

volcano warning system A system designed to detect impending danger and convey pertinent information to people at risk, enabling them to make protective action decisions.

1. INTRODUCTION

The existence of clear precursors to many volcanic eruptions facilitates volcano warnings because precursors are detected by monitoring instrumentation and associated software systems used at volcano observatories as part of an official **volcano warning system**. Such systems are designed to provide people with important information about the hazards associated with volcanic unrest, eruption, and post-eruption processes. Information often includes a description of the level of unrest/activity at a volcano, the expected hazards, and a time frame for specific activity (forecasts), but may also include information about human and structural vulnerability and generic recommendations for protective actions, among other things. Volcano observatories disseminate information not only to various end-users, including agencies linked to emergency management/civil defence and the aviation industry, but also to the public and media. These end-users are not just recipients of information, but also sources. Each source, including observatories, both disseminates and receives information through multiple channels (e.g., radio, television, internet, and cell phone), but observatories disseminate messages to two distinctly different human populations— those on the ground around volcanoes and those in aircraft who are at risk to ash in plumes that may reach thousands of kilometers from the eruptive source. These two distinctly different audiences require separate warning processes. The warning process for the aviation-based population is highly structured and streamlined because **notifications** are issued to an audience that has specific standard operating procedures (SOPs—action plans) for responding. In contrast, the warning process for ground-based populations is much less structured because, unlike the observatories and emergency management agencies, generally the public and media do not have SOPs. These differences lead to drastically different responses to notifications. A lack of routine training for volcanic crises in most of the global population at risk from volcanic hazards is exacerbated by the great diversity in characteristics of volcanic unrest and eruption and of post-eruption hazards.

Durations of volcanic unrest, eruption, and post-eruption impacts differ greatly and range from hours to decades, which distinguishes the warning process for volcanoes from other common hazardous events such as floods and severe weather. Also, most large eruptions have precursors which can be used by scientists to alert people of potential hazards through the warning process. This means that volcano warning and alerting systems (such as the Volcano Activity Alert-Notification System of the US Geological Survey) have a real chance to reduce risk effectively by notifying people of the status of volcanic activity and the urgency with which protective actions need to be taken.

The occurrence of precursors is an important and distinguishing characteristic of volcanic eruptions compared to many other hazards, and forms the basic premise for volcano early warning systems built around detection and monitoring systems. However, sustaining effective response to volcano warning notifications is a real challenge in some instances because some volcanoes experience elevated levels of unrest for years before an eruption occurs, if ever; some eruptions last for years to decades; and some eruptions produce long lasting hazards such as remobilized ash and lahars for many years after an eruption ends.

Figure 67.1 compares the durations of precursors and events of three volcanic crises with several other common hazardous events— floods, hurricanes, earthquakes, and tsunamis. Note how all three volcanic events had precursors, but how Kilauea volcano's eruption has no distinct ending compared to Mount St. Helens' 2004—8 eruptions. Also, two events, Phlegraean Fields and Long Valley, have not (yet) culminated in an eruption. While there has not yet been a recent eruption at Long Valley, early dissemination of information between scientists, emergency management, and the media and general public during the period of unrest contributed to considerable social, political, and economic impacts in the affected communities.

The durations of precursory time before eruption affect the window of opportunity during which information about an eruption is disseminated by officials (scientists and emergency managers) to the media and the public. Figure 67.2 is a model that illustrates this relationship between precursory activity of a volcano and the expected surface and social manifestations accompanying unrest and eruption, including the expected alert levels. In Figure 67.2, we adopted US alert levels— Normal, Advisory, Watch, and Warning. In this model, magma is shown in the upper left corner of Figure 67.2 to be in shallow storage beneath a

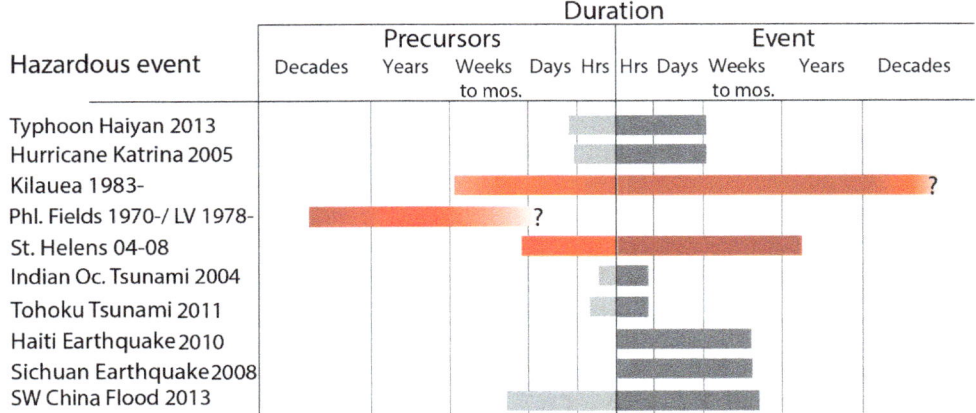

FIGURE 67.1 Comparison of durations of precursors and some common hazardous events, including volcanic eruptions (red), hurricanes, earthquakes, tsunamis, and floods (gray). Phl = Phlegaean. LV = Long Valley.

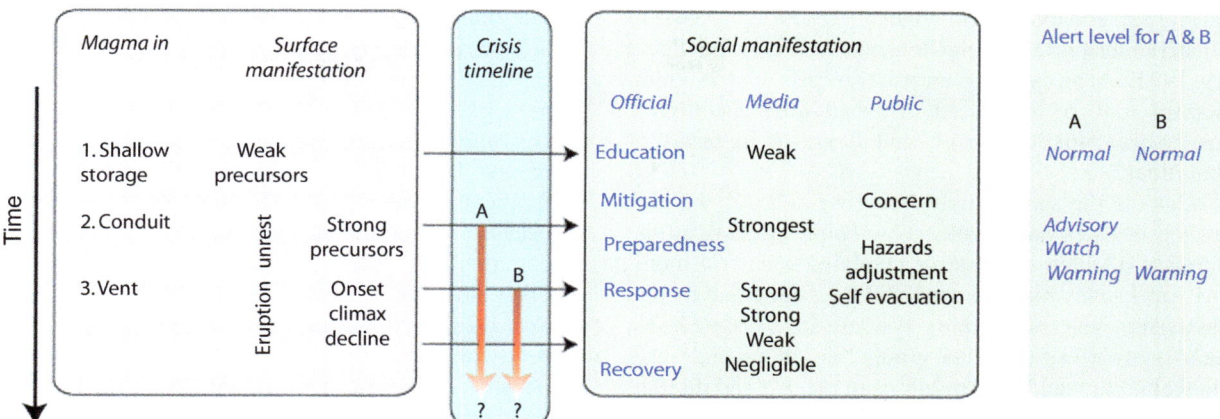

FIGURE 67.2 Physical and social manifestations and alert levels for two crisis timelines (Crisis A and Crisis B) associated with three levels of magma and their associated precursors at the surface. Figure 67.2 indicates that as the volcano becomes restless and magma migrates toward the surface, precursors strengthen. Conditions change for officials and for the media and the public. At this stage, response of the media peaks, officials begin or intensify mitigation and preparedness actions, and the public and media become increasingly interested. At this time broad forecasts may be made that indicate when and where an eruption might occur.

volcano. Reading the figure from left to right, one discerns that the magma is close to equilibrium and typically creates only weak precursors that prompt little to no interest or reactions by media or the public. During this time officials engage in intensified educational programs to raise awareness and preparedness to reduce longer term vulnerability and risk to a future eruption (e.g., such as happened at Nevado del Huila, Columbia, 2007−2008 and Rabaul, Papua New Guinea, 1983−1994). However, in the absence of strong signals of an eruption, the educational information often competes with other demands for the public's attention (e.g., Phlegraean Fields, Italy, 1972 to present and Long Valley, USA, 1978 to present). While outreach increases with shallow emplacement of magma, education is a continuous process and it continues throughout phases of mitigation, preparedness, response, and recovery, as shown in Figure 67.2. Factual statements are issued by scientists describing the state of the volcano, but forecasts are not made for the long or intermediate time period (several months to years). Such conditions seldom attract much attention from the media or public, despite the intensified flow of information. A low alert level (Normal) exists, as depicted on the right side of the figure.

Two crisis timelines (Crisis A and Crisis B) in the middle of Figure 67.2 define the timing of onset of eruption associated with the progressively shallower movement of magma. In one case, stronger precursors associated with magma in the conduit (Crisis A) allow scientists time to delay issuing higher levels of alert (Warning) and instead establish intermediate levels of alert such as Advisory and Watch. This is done because of the uncertainty surrounding the exact timing of the eruption, but broad forecasts of an eruption (to occur in weeks to months or years) may still be made. However, if no strong precursors occur, or there is

little or no **instrumental monitoring** in place well in advance of an eruption that seems imminent (Crisis B), then scientists move from a Normal background alert level to Warning directly, without issuing intermediate alert levels of Advisory or Watch. Such conditions may shorten the window of time associated with any forecasts since the certainty of eruption is greater than in the Watch/Advisory stage. The events of 2014 at Kelut Volcano on the island of Java, Indonesia, did not totally bypass intermediate alert levels, but it is an example of where rapid unrest at a monitored volcano led to scientists quickly, but incrementally, raising alert levels. The alert level was raised from 1 to 2 on 2 February; from 2 to 3 on 10 February; and from 3 to 4, with eruption imminent/underway, on 13 February. Situations like this, and situations where intermediate alert levels may be bypassed, are most likely to occur at volcanoes with high eruption frequencies (like Kelut) and where the local population is knowledgeable about the volcano and the hazards it presents (again, like Kelut, but also including Merapi; Manam, Papua New Guinea; Sakura-jima, Japan; San Cristobal, Nicaragua; Tungurahua, Ecuador; and Nevado del Ruiz, Columbia).

When alert levels of Watch and Advisory are established, the public may engage in actions to protect property, but when the situation reaches the alert level of Warning, self-evacuations generally increase. However, some people may choose to self-evacuate before being advised to do so by officials or they may choose to delay their evacuation beyond the recommended time. Some even fail to evacuate at all, despite being advised to do so, but the reason(s) for each of these behaviors do not only reflect levels of awareness of hazard, vulnerability, and risk, but also involve more complex socio-demographic (e.g., household characteristics, social ties, and concerns about protection of property) and economic (e.g., financial resources, access to alternative transportation, and housing) factors (Lindell and Perry, 1992). At any stage of unrest, forecasts of an eruption or hazard may be expressed in broad semi-quantitative terms (years, months, weeks, and days) or in terms of probabilities.

Following the subdivision of Sorensen (2000), we now turn to a description of volcano warning systems as the integration of three subsystems involving scientific monitoring, emergency management, and the public and media. In this sense, volcano warning systems are discussed as a dynamic network of agencies, groups, and individuals each using technology and human senses to receive and disseminate information and make protective action decisions.

2. VOLCANO WARNING SYSTEMS

Researchers generalize the warning process as three extended and distributed subsystems of individuals and organizations connected by communications systems. The subsystems are designated: *scientific hazard detection and monitoring*, *emergency management*, and *public response*. However, warning systems have traditionally focused on the hazard detection and emergency management subsystems, and much less on the public response subsystem and the role of media in influencing public and organizational response. Not surprisingly, people often conceptualize volcano warning systems as a volcano observatory's ability to monitor, detect, and forecast volcanic eruptions and associated hazards and emergency management's ability to coordinate and manage a crisis. However, this view ignores the important roles that the public and the media play in the warning process and its outcomes. This chapter provides an overview of all the subsystems, with a focus on describing how official agencies (observatories, emergency management), and the media and public act as both information senders and receivers during volcanic unrest.

Information about volcanic hazards is communicated by multiple sources through multiple channels. There are four primary sources of information from which risk information is derived and only one is derived from official agencies. These four sources and their roles in the warning process are as follows (points 3 and 4 are collectively referred to as social warnings):

1. *Officials*: These are authoritative agencies, such as volcano observatories and emergency management agencies. They issue official notifications about hazards, vulnerability, risk, and protective action.
2. *Environmental cues*: These are naturally occurring phenomena associated with volcanism, or other hazards, that can be detected by one or more of the human senses— seeing, hearing, smelling, tasting, and touching.

They alert people of a real threat (e.g., by seeing a steam or tephra plume) or potential threat (e.g., seeing ground cracking, hearing a phreatic explosion, smelling H_2S or smoke, or feeling seismicity).

3. *Informal*: These are communications between members of the lay public and media. They are extremely frequent but highly variable in their quality, primarily because most people are not trained in how to communicate pertinent information in a real or potential crisis situation.
4. *Social cues*: These are observations of other people's behavior, such as seeing people congregating (e.g., in unusual places or numbers or at unusual times) or evacuating.

Official scientific notification of volcanic unrest or eruption is initiated at volcano observatories once a threshold of precursor activity detected by monitoring instrumentation and associated software is exceeded, or occasionally from direct observation of the early stages of volcanic activity (see Figure 67.3). Figure 67.3 illustrates the general flow of information for ground- and aviation-based messages once the official volcano warning system is triggered by precursors, or by direct observation. Two separate paths are shown for ground- and aviation-based messages. Aviation messages follow a "linear" path where information flows from observatories to VAACs to MWOs and then to the aviation sector. Ground-based messages are also usually linear (where there is a progression of information dissemination from observatory to emergency management to media and public end-users), but they may also be non-linear (where all end-users receive messages simultaneously). The delay in communication of notifications by

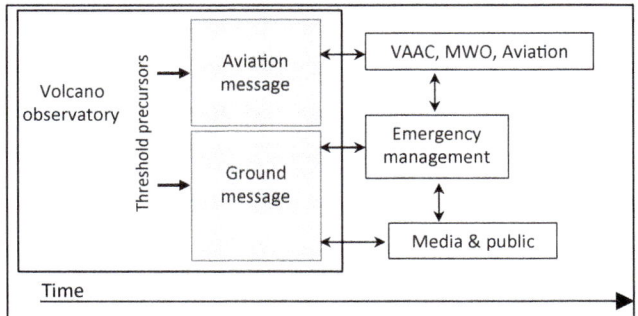

FIGURE 67.3 Basic flow of primary messages in a volcano warning system for aviation-based and ground-based notifications. There is a delay in a volcano observatory's dissemination of a message to the media and public due to a duration of time when observatory scientists first discuss a change in volcanic activity with "official" agencies in emergency management. The duration could last for minutes to days depending on the severity of the precursors and risk. Double arrows reflect two-way flow of information since each entity (volcano observatory, media and public) may be both senders and receivers. VAAC = Volcano Ash Advisory Center; MWO = Meteorological Watch Office.

the observatory to the media and public is due to the standard practice of observatory scientists first discussing changes in a volcano's status with emergency management as an intermediate step before disseminating the information broadly to the media and public. The duration of this delay is to some degree proportional to the threat. The delay would be very short or non-existent if there was an immediate need to notify everyone simultaneously, but longer if there was no urgency. Whether this intermediate step is taken or not, digital technologies allow modern observatories to disseminate messages simultaneously to many entities (e.g., emergency management, media, and public).

Instrumental monitoring connected to an observatory almost always provides the first notice of unrest to the public. Historically, environmental and social cues and informal warnings provided first warnings until technological advances in the mid-twentieth century. Recent examples of first warnings based on instrumentation include: Sinabung, Indonesia, 2010 and 2013; Chiles-Cerro Negro de Mayesquer, Ecuador–Colombia border region, 2014; Merapi, Indonesia, 2010; Nevado del Ruiz, Colombia, 2012; and Kelut, Indonesia, 2014.

While instrumental monitoring may lead to release of an official notification, precursors to eruptions or environmental cues (e.g., steam and other gas venting, phreatic explosions, felt seismicity) are often very conspicuous to local populations, prompting them to act and disseminate their observations to other people through informal communications. Observations of environmental cues may occur before individuals receive notification from an official source (or any other source) for several reasons. Possible reasons include when a volcano is insufficiently instrumented; instrumentation does not perform as intended; instrumentation performed as intended but precursors have not yet reached a threshold that prompts a scientific response to disseminate information to the public and media; or the threshold has been reached but scientists have not yet disseminated information to either emergency management or the public and media, or both. Alternatively, an official notification may have been issued, but specific people may not have received it before they receive environmental cues, or even social cues or informal warnings.

Regardless of the reason for variations in the order and timing of receipt of information from different sources, all sources of information are important in protective action decision making because of their role in the **warning confirmation process**. Any one source of information received through any one or more channels may supplement information already received and in circulation among message receivers. Access to information may relate to factors such as the degree to which technology is or is not available to both monitor a volcano and disseminate information, or the speed with which volcano observatories and emergency management agencies disclose information that may have high degrees of uncertainty but minor to extreme human and financial consequence. Furthermore, there is great variability in the characteristics of people receiving warning information, including but not limited to overall experience with volcanic hazards; access to pre-event information describing volcanic hazards, vulnerability, and risk; access to technology; and socio-demographic, cultural, and cognitive factors that influence the availability of information and people's interpretation and response to it. For these reasons, a discussion of volcano warnings and volcano warning systems is very multi-disciplinary.

Research findings demonstrate how the public are not passive recipients of information, but pursue it actively by milling information and exchanging information with other people. They do this to search for answers to unanswered questions and to confirm characteristics of information received in relation to themselves and the people and things they value. Occasionally, people progress directly to protective action behavior such as evacuating or seeking shelter without passing through the intermediate warning confirmation process, but this behavior is infrequent and the factors influencing such behavior are not well known. If pertinent information is either not provided by authorities or is inconsistent then the public increase their level of milling. Official notifications are therefore needed early during volcanic unrest and messages need to be disseminated through all available channels (e.g., television, radio, SMS text, email, and Twitter). Accurate and consistent updates need to be issued frequently to help increase belief in the message and reduce milling by satisfying public demand for information. To date, important factors in determining rapid response to warning messages involve the frequency with which messages are disseminated (i.e., a proxy for number of messages received, which facilitates confirmation of the warning message and belief—the more messages received the better), the message content (e.g., hazard characteristics, affected/unaffected areas, and timing) and message style (e.g., the need for specific, consistent, and internally accurate messages from multiple sources disseminated through multiple channels).

2.1. Scientific Monitoring Subsystem

Methodologies for monitoring volcanoes are numerous and vary considerably in cost and effectiveness. Monitoring involves the use of direct and indirect measurements by ground- and space-based instruments and direct observation. Instrumental methods with robust track records involve seismic, ground deformation, and gas monitoring (see Chapters 63–66 on monitoring).

Observation of volcanic successions (deposits) and current volcanic processes through field studies are also critical to identifying the range in physical characteristics of past volcanic eruptions and their associated hazards. Pre-eruption field mapping of deposits and hazard assessment are therefore critical components of a volcano observatory's ability to issue accurate notifications about the potential characteristics of a future eruption. Such characteristics include the styles of past eruptions and their magnitude, intensity, and duration in addition to the "footprint" of associated hazards, including run-out distances and velocities. So, while instrumentation and corresponding software systems are critical to detect unrest and to initiate the warning process, interpretation of prior eruptive events from observations of deposits provides context to interpret and understand the unrest and expected hazards. This information must be synthesized with monitoring data to provide comprehensive alert notifications that have sufficient detail, to be useful for managing risk.

Notifications issued by many volcano observatories designate a specific alert level for a volcano and outline the expected hazards for ground-based and aviation-based recipients using separate, dedicated messages, but both populations are not necessarily placed on the same alert levels for a given eruptive scenario. Generally speaking, if an eruption is expected to be effusive then only ground-based populations are alerted, but if an eruption is expected to be explosive then both ground- and aviation-based populations would be alerted. However, in practice it is not possible to know a priori whether the opening phase of an eruption will be explosive or effusive in style, and, even if the eruption is effusive, there is typically some explosive behavior at the onset of the eruption. As a result of this uncertainty, both ground and aviation notifications are likely to be issued until the eruption is in progress and it is known which style the eruption is taking. At this point decisions can be made to issue supplemental ground- or aviation-based notifications before issuing a cancellation message.

2.2. Emergency Management Subsystem

Numerous agencies are involved directly and indirectly with warning systems as a component of emergency management. Each agency and its personnel must function as a team that links to and interacts with numerous agency representatives, the public and the media (see Chapter 68). There are three primary functions of emergency management in the warning process. These are disseminating information about protective actions, coordinating and managing official and community response, and disseminating scientific and response information as an authoritative local source for all pertinent information (including responding to requests for information). Emergency management may, for example, elaborate on scientific alert notifications by providing more detailed recommendations

for protective actions associated with evacuation and sheltering. Such recommendations are usually omitted in scientific alert notifications because they are made principally by elected and appointed officials with formalized roles in emergency management rather than by scientists in observatories.

Effective interagency communication that conveys warning information or helps manage a population's response is highly dependent on the development of inter-personal and inter-organizational trust, in addition to creation of positive relationships with community members to develop trust and confidence (see Chapter 69 on the perception and communication of volcanic hazards). Developing and maintaining organizational trust requires that teams form and work together before a crisis (see Chapter 70). Such trust is needed to manage SOPs that specify organizational response to alert notifications and to develop shared mental models (Paton and Gregg, 2013). The term "mental model" refers to how an individual or group of persons (e.g., agency personnel) thinks about and mentally represents or conceptualizes objects, events, relationships, and responsibilities in their lives. Mental models facilitate people's ability to adapt to their environment. While people generally force new information to fit into their existing mental model, models evolve as people accumulate experience, especially when new information is discordant with or contradicts current beliefs.

Table-top exercises and full scale drills provide training that strengthens operational responses and shared mental models. Given the diversity of volcanic crises, eruption scenarios involving a range of expected hazards and corresponding responses help establish the need for inter-organizational communication. For example, scenarios involving large magnitude and high intensity explosive volcanism produce short-term but widespread vulnerability to tephra fall and pyroclastic density currents. They require rapid decision making to effect evacuation and sheltering behaviors in the affected population. In contrast, long-term lahar hazards that may follow such eruptions require periodic short-term, but rapid, decision making over years to decades after the eruption. The long-term threat of post-eruption lahars to people and property following the 1991 eruption of Mount Pinatubo was captured by Newhall and Punongbayan (1996), who reported that: "... direct and indirect damage from lahars has probably exceeded that from the eruption several times over."

Sorensen (2000) described how principles of emergency management suggest that coordination of organizational response in a crisis is improved when organizations within the network: (1) remain flexible, (2) establish and understand interagency relationships, (3) know the responsibilities of each agency, and (4) know who is supposed to do what in different scenarios. Interagency coordination plans are developed to serve as the basis for conducting table-top exercises that support all four. In contrast, the most common causes of poor warning dissemination were reported to be human error and equipment failure. Miscommunication could be considered a common human error.

2.3. Public Response Subsystem

Information during volcanic unrest is communicated by numerous sources through multiple channels, and a number of models have been proposed to explain human decision making and response to short- and long-term risk communications. Such communications encourage the adoption of long-term preparedness actions before volcanic unrest, and short-term warnings during volcanic unrest and eruption. Three leading models developed in the USA by Lindell and Perry (2012) and Mileti and Sorensen (1990) and in Australia by Paton (2003) delineate the most salient factors influencing people's response to warnings identified to date. The US models are largely based on studies of earthquake, hurricane, flood, volcano, and technological hazards on US populations, while the Australian model is more international in its application and inclusion of non-western societies.

The Paton social-cognitive model involves a three step process whereby people must first learn of a hazard and perceive a risk, then develop intentions to either take action or seek additional information. This is followed by the decision to perform the actions. The decision-making process is mediated by social, cognitive and cultural factors, and situational impediments or facilitators that increase or decrease an individual's likelihood of either developing intentions to take protective actions or translating intentions into action. Some of these include *trust* in the message source, *self-efficacy* (self-appraisal of one's ability to take action) and *outcome expectancy* (self-appraisal of society's ability to take any action), and *perceived responsibility* (if they accept responsibility for their personal preparedness or transfer responsibility from themselves to others, usually official agencies such as observatories or emergency management). For example, an individual may have an accurate understanding of both the prevailing hazards, their vulnerability and their risk, but low-self-efficacy may translate to a lack of positive actions to reduce risk because the individual lacks belief in their ability to take the action. This could include, for example, actions to mitigate effectively the respiration of fine ash or infiltration of ash into a building.

The warning response model of Mileti and Sorensen (1990) characterizes response to warnings as a sequence of events that involve: perceiving the warning (hearing, seeing), understanding the contents of a warning message, believing that the message is credible and accurate,

personalizing the warning to oneself, confirming that the warning is true, and that others are complying with its recommendations, and responding by taking protective action(s). The protective action decision model of Lindell and Perry (2012) describes response as a sequence of steps involving people's interface with environmental or social cues or social warnings, which are influenced by characteristics of the receiver (age), the source of the message, and access to and preference for specific channels. These interactions elicit centralized perceptions of the environmental threat, variety of protective actions available, and perceptions of agencies. Perceptions provide the basis for decisions concerning protective actions and they interact with situational facilitators and impediments to produce a behavioral response. The decision to take protective action involves a five-stage process whereby a person asks: is there a threat? (risk identification), do I need to take protective action? (risk assessment), what can be done to achieve protection? (protective action search), what is the best method of protection? (protective action assessment), and does protective action need to be taken now? (protective action implementation).

All three of these models make similar predictions about people's behavior in response to warnings. Each acknowledges that all people do not progress through all stages of a model— some may skip one or more stages and advance immediately to take protective action. Furthermore, each of these models acknowledge the importance of **milling behaviors** in responding to risk information. The models differ in their lack of attention to cultural factors such as indigenous knowledge that may significantly affect response to hazards and warnings and the relative attention given to the different warning sources.

Each model also accounts for the observation that people do not wait passively for information once a threat is initially received and that receivers actively seek new information through the warning confirmation process. In general, people will seek information about a threat until their immediate questions are answered. A question might be "do I need to take protective action now or can I delay until the time is more appropriate?" such as when family would be together or day versus night. The warning confirmation process often leads to delays in initiating protective actions, because, for example, the recipient does not personalize the message or the message content is ambiguous or inconsistent with other messages. The warning confirmation process itself is modeled as a sequence of events and the human outcomes depend on the content and style of the message and characteristics of the receiver. The content pertains to the hazard, the authoritative and trusted source, timing (whether it is night or day), and guidance (recommended actions). Style pertains to the specificity, consistency, certainty, clarity, accuracy, and sufficiency of the message and the channel(s) through which the message is delivered. Finally, the characteristics of the receiver pertain to a person's social setting (whether alone or with others), social ties (responsibility for others), social structure (are they a head of household or a dependent), and cognitive factors. Not surprisingly, decisions to respond to warnings or other alerts involves a complex interplay between their environment and broad receiver characteristics— a person's membership in a particular socio-demographic group or their cognitive ability to receive and process information, in addition to characteristics of the message.

For well monitored volcanoes linked to an observatory, official warnings are likely to be issued and people will process the information along with other information received informally and from environmental and social cues. In contrast, at unmonitored or poorly monitored volcanoes in rural areas where there is no official warning system tied to volcanoes, environmental and social cues and informal warnings provide the only source of information for people to interpret changes in their environment and make decisions about protective actions. The number of sources from which an individual may receive a warning message and the number of channels through which that information may reach them varies considerably, but there is no single source or channel that is more or less likely to trigger a response.

In contrast to the limited number of sources of information, there are substantially more channels through which information is disseminated than there are sources. Official alerts are issued through dedicated and third party systems and involve a mix of one- and two-way channels. Informal messaging also involves both one- and two-way channels. For environmental cues and natural warnings they include sensory experiences such as hearing (an explosion), seeing (a plume), or feeling (an earthquake). The total number of channels people have available to receive and disseminate information depends on access to traditional and emerging digital technology and its redundancy and functionality in a crisis. People's access will also depend on their proximity to the hazard source. For example, weather may obscure visual cues, earthquakes may not be felt at distance, and sulfur smells are rapidly diluted. Consequently, cues are more likely to attract the attention of people near their source than further away. A person in a catchment subject to lahar hazards and further removed from the cues, for example, may not receive them.

Digital technologies have created a shift in how risk information is communicated in terms of available channels and the style and content of information (e.g., text only versus text plus graphics). Channels change as new technologies emerge and others become obsolete, but an emerging technology in one area may be obsolete in more technologically advanced areas, so no two warning systems are identical. Traditional channels (e.g., face-to-face,

television, radio, and telephone) will continue to be good dissemination channels of information for the foreseeable future, but they are being supplemented by new digital technologies including handheld devices (e.g., smart cell phones and tablets), internet, email, and social media (Twitter, Facebook) and online blog sites.

Social media is effective for disseminating and receiving risk information in some events, but little is known about the cyber-psychology of protective action behavior. Since some cities with volcanic hazards have sophisticated technology and other, often rural, areas do not, globally there is a dichotomy between countries or regions with advanced versus basic technology to deliver notifications of alert and for end-users to receive and subsequently disseminate information. Furthermore, while the digital technologies have resulted in our ability to rapidly develop and disseminate useful information, they have also produced much chaff, such as inaccurate and misleading information produced by self-proclaimed experts and people who promote conspiracies. Simply put, there is a lot more information available for the technologically-wired public to mill through in digitally connected areas, but the process of disseminating, seeking, and receiving risk information is not necessarily better as a result in every case.

One common thread in prevailing socio-behavioral and cognitive models describing response to warnings is the finding that most people seek information rapidly after receipt of an initial risk communication. Figure 67.4 illustrates how people milled for web-based information from USGS during volcanic unrest at Mount St. Helens in 2004. The figure plots Real-time Seismic Amplitude Measurement (RSAM) against average daily requests for web pages. While there is a sharp increase in information seeking paralleling the initial increase in seismicity beginning on about 09/29/2004, peaks in requests for information do not increase on days of highest seismicity or on weekends when individuals have more discretionary time. Instead, they were found to correlate generally with times immediately following publicized changes in alert level and release of related information statements, updates, and press conference-derived media coverage.

3. SPATIAL AND TEMPORAL CLASSIFICATION OF NOTIFICATIONS

The great diversity of volcanic hazards associated with the range of eruption styles and magnitudes possible at a volcano must be kept in mind when considering the temporal and spatial classification of notifications for eruptions. Volcano warnings can be expected to change considerably in terms of their content, style, and audience and in terms of frequency with which supplemental messages are provided as a volcanic crisis waxes and wanes and hazards emerge and subside.

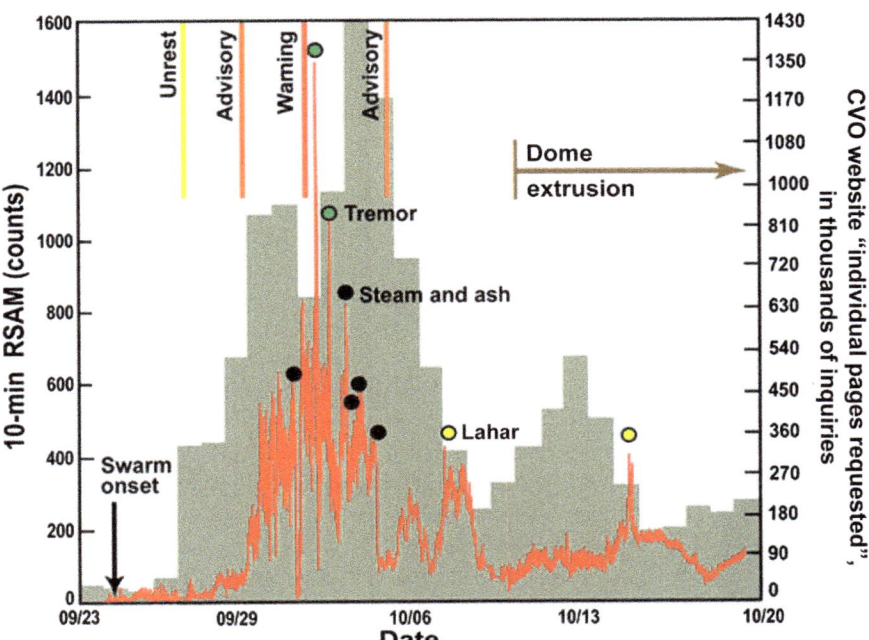

FIGURE 67.4 Information seeking and milling behavior illustrated by daily average requests for web pages (gray logs) at the Cascades Volcano Observatory, US (in thousands of pages) during volcanic unrest at Mount St. Helens, 2004. Red trace is RSAM. *Modified after Driedger et al. (2008).*

3.1. Spatial Classification of Notifications

Volcano warnings target people on the ground or those connected to aviation. Aviation-based populations include people operating, traveling in, or scheduled to travel in aircraft, because aircraft are vulnerable to potentially catastrophic damage from fine ash carried in plumes produced from explosive volcanism (see Chapter 52).

Different notification schemes are needed for each of the ground-based populations and the aviation-based populations. Elevated aviation warnings are only associated with explosive volcanism or the threat of explosive volcanism. For example, lava flows may pose a high hazard to ground populations, warranting an elevated alert level, but if no eruption plume poses a hazard to aviation then the aviation warning can be relatively low. Ground-based populations are at risk from the great diversity of volcanic hazards while the aviation-based population is only at risk from airborne ash. Notifications for the ground-based population are directed to subgroups of society ranging from emergency management and business and industry to lay people, while aviation notifications address the commercial and military aviation sector specifically.

The linear path in warning dissemination describes aviation warnings very well. Here, the monitoring subsystem, which includes volcano observatories and Volcano Ash Advisory Centers (VAACs), issue alerts and these are disseminated to aviation end-users. The aviation sector is highly controlled, is accustomed to receiving and reacting to short-notice hazard situations, such as severe weather,

and has a robust warning delivery system involving VAACS, Meteorological Watch Offices (MWOs), Significant Meteorological Information (SIGMETs), and Notices to Airmen (NOTAMs). VAACs and MWOs use highly structured notifications to convey the location and duration of the hazardous condition through SIGMETs and NOTAMs. Consequently, there is limited milling for informal information and confirmation of the emergency alert once these are disseminated; authoritative information is delivered and it is acted on by the recipient. However, despite this structured approach in response to aviation alerts, in practice there is some milling of information when volcano color codes are at yellow and orange. Typically the milling is done by airline and air force dispatchers who need to be concerned with en route weather issues or other circumstances which have potential to affect a flight plan. Such a system, even with limited milling, contrasts starkly with most ground-based populations, where warning confirmation and milling is common behavior.

Relatively speaking, for a given explosive eruption, the areas included in **ground-based alerts** are much smaller than the areas included in aviation alerts. This is because the threshold for vulnerability of aircraft engines to suspended ash is very low. However, the number of people directly affected by ground and aviation alerts varies considerably. The number of people directly affected by ground-based alerts reflects characteristics of the eruption and particularly the number of people in exposed communities, whereas the number of people affected by aviation alerts reflects characteristics of the eruption and how the plume intersects aviation flight paths. The number of people directly affected by volcanic ash during aviation alerts will be relatively small, as it affects only people in aircraft. However, the number of people indirectly affected can be orders of magnitude greater, as aircraft become grounded in airports, passengers' flights delayed or cancelled, and air cargo is disrupted. The 2010 eruption of Eyjafjallajökull, Iceland, for example, had major effects on commercial and military aviation.

3.2. Temporal Classification of Notifications

The content and style of notifications disseminated during a volcanic crisis and the urgency with which decisions must be made depend on the state of restlessness of the volcano and the likely style, magnitude, and intensity of the expected hazards. Notifications will also differ depending on whether the eruption has ended and there is no further threat of ongoing hazards, or if the eruption has ended but hazardous conditions still exist. Consequently, in the first edition of the Encyclopedia, Newhall (2000) divided volcano warnings into three time periods: (1) pre-eruption warnings, (2) syn-eruption warnings, and (3) post-eruption warnings. These time periods are summarized below from his original contribution.

Pre-eruption warnings are further subdivided into long-, intermediate-, and short-range forecasts and each reflects decreasing uncertainty about the expected phenomena. Long-range forecasts consist of broad statements about the long-term likelihood of an eruption over many years. They may be exemplified by maps designating areas of a volcano that are likely to experience specific hazards such as tephra fall, pyroclastic density currents, lava flows, gas emissions, and lahars. In contrast, intermediate-range forecasts are issued when a volcano exhibits a change in behavior where it could become hazardous. For example, forecasts could be issued stating that "The volcano is showing signs of increased activity and might erupt within weeks or months and should be monitored closely." This level of notification usually prompts scientists to increase monitoring activity and emergency management to review their SOPs for managing a crisis. Public information about volcanism, hazards, vulnerability, and risk-reduction actions is disseminated widely and attention from the news media is heightened. This is also when just-in-time or crisis education is critical because it is a teachable time window when at risk people learn how they may be at risk in the short term and what they can expect. Short-term forecasts of an eruption, in contrast, are provided once certain indicators of unrest have been reached, such as a threshold of seismicity or tilt, explosions releasing new magma, all indicating that magma has made its way close to the surface. Estimates of the type and size of the eruption,

locations of vents and areas expected to be affected by specific hazards are sometimes addressed in these notifications. They commonly trigger evacuation notifications by emergency management and heightened mitigation and preparedness actions.

For *syn-eruption warnings*, Newhall (2000) described the need for two subdivisions of warnings based on the immediate hazards associated with the eruption and the intermediate-range outlook about the future course and duration of the eruption. For the immediate hazards potentially affecting communities, the focus is on detection and forecasting their characteristics. These include the travel direction, arrival times, and intensity and duration of specific hazards. For the aviation community, the focus is on detecting and forecasting the location of the dispersed plume and modeling or forecasting its future location, and ash concentrations. Volcanic ash Temporary Danger Zones are established around the areas of high concentrations of ash and provide for an area of modeling uncertainty through inclusion of a vertical buffer zone beneath the high contamination area depicted on the graphic VAAC charts. The buffer zone allows for uncertainty in modeling of eruption strengths and meteorological conditions. The science of ash impacts on jet aircraft is still relatively young, and new "no-fly" or "fly with caution" policies can be expected in the coming years. In contrast to immediate hazards, the intermediate-range outlook focuses on forecasting what can be expected from an eruption beyond the immediate time frame (i.e., over the next few days to weeks or months). This has applications to both ground- and aviation-based populations.

Unlike almost all other natural hazards, except drought, volcanic eruptions last for weeks to months, or even years or decades, and this makes forecasting durations of eruption difficult. From Simkin and Siebert's (2000) analysis, less than 10% of some well-documented 3301 eruptions had durations of less than 1 day (some 28% lasted 1 month to 6 months). However, of the 252 moderate to powerful eruptions, over 40% reached their peak intensity in less than 1 day. This means that the total duration of warnings for unrest and eruption in general can be very long, measured in days to years, but it also suggests that the duration of notifications for the climactic phase of moderate to powerful explosive eruptions and associated hazards should be comparatively short. Furthermore, once an eruption has begun and the composition of erupted magmas compared to historical analogues, intermediate-range forecasts can be refined to inform people about how the eruption is expected to proceed over the following weeks to months.

Post-eruption warnings apply primarily to explosive eruptions because the watersheds surrounding volcanoes are disturbed and unconsolidated pyroclastic material is re-transported by water from upland areas to areas of lower elevation where populations of people are usually concentrated. The threat of lahars and floods produces two categories of warning that Newhall outlined: short-range warnings of lahars and intermediate- and long-range estimates of sediment yield, shifts in stream channels, and related flooding. Short-range lahar warnings are urgent warnings of immediate flooding that are normally triggered by heavy rainfall or natural dam failure. These may provide only minutes to hours of advanced warning to downstream communities. The intermediate- and long-range warnings address the hazards created when abundant pyroclastic debris disrupts drainage systems. A disturbed landscape with abundant new erodible material at the surface causes channels to avulse to new paths and transport sediment swiftly downstream. This raises riverbeds, which leads to flooding. Although ground-based hazards usually have much longer durations than aviation-based hazards, aviation warnings can remain in effect for several days longer than warnings for ground populations, since hazardous conditions may have ended in the vicinity of the volcano, but dangerous concentrations of ash may still be found in downwind plumes in distal areas. As recently as 2010, aviation exclusion zones resulting from zero tolerance policies for aircraft in areas containing airborne ash in airspace above Europe caused enormous aviation losses due to protracted durations of grounded planes and stranded passengers. The exclusion zone was later relaxed when policy changes permitted flying in airspace with some concentration of ash.

4. AN EFFECTIVE WARNING SYSTEM

Society's ability to respond effectively to volcanic eruptions requires consideration of more than linear technological solutions to detecting hazards and subsequent dissemination of scientific alert notifications to emergency management, media and the public. Effective response requires consideration of social, cognitive, political, and cultural factors at individual, organizational, and group levels which influence people's receipt, dissemination, and interpretation of social warnings and environmental and social cues and their responses to them. This is true for most of society because the public generally do not practice response to the official warning process as do scientists and agencies with ties to emergency management.

Effective warning systems motivate the desired response in at risk-populations. As such, they generally require considerable work before the official warning process is initiated during a volcanic crisis. This includes planning for the various hazards associated with effusive and explosive eruptions of varying magnitudes and intensities, consultation and discussion between community stakeholders, participation in education and drills and exercises for some hazards. All this should be supported by on-going research and evaluation to ensure that the potential performance of a warning system is not over-estimated. See Leonard et al. (2008) for a more detailed discussion.

5. THE FORM OF WARNINGS

Warning information can be presented in many different formats, including text only, or text and graphics. Newhall (2000) distinguished between concepts of factual statements, eruption forecasts, and probability trees and these are summarized here. Factual statements are commonly used by scientists to explain what is simply known about the past or present physical state of a volcano from what they expect to occur in the future. Factual statements are derived from studies of a volcano's eruptive history and from current monitoring and they form the foundation of eruption forecasts. Eruption forecasts, in contrast, provide information about whether an eruption or specific hazards associated with eruption are expected to happen at a volcano and when and where. The forecasts may be developed from consideration of a single explosive or effusive event or an eruption scenario that includes multiple events (e.g., several moderate explosions separated by days). Frequently, scientists can estimate how likely or unlikely a volcano is to erupt, but they lack sufficient data to state the style, magnitude, intensity, or duration the eruption will take. In these instances, probability trees may be established to deduce the probability that a specific event is to occur. Probabilities are established by individual scientists and the concept of expert elicitation has been advanced from the crisis at Montserrat to constrain better the probabilities expressed by scientists. See Neri et al. (2008) for a discussion of probability event trees and expert elicitation at Vesuvius and Donovan et al. (2012) for a recent evaluation of these methods for the Montserrat crisis. However, complex probability event trees are not usually made public during a crisis because scientists believe that the public might think they are guessing about the volcano's expected behavior. Instead, they are used to help scientists develop forecasts by allowing them to focus on understanding how a particular unrest episode is developing. Event trees may also help scientists discuss possible outcomes with professional emergency managers or other authorities.

A variety of **volcano alert level systems** can be found internationally for ground-based populations (Fearnley et al., 2012). Worldwide, there is considerable variation in the behavior of individual volcanoes and local and regional monitoring capabilities. These differences, along with variations in the needs of specific end-user populations (e.g., experience with volcanic hazards, symbolism of numeric and color codes for alert levels) have led to diversity in alert schemes. Most alert schemes begin with a background or normal activity level or color code (0 or 1 and a cold or soft color) and progress to higher levels of unrest or eruptive activity (e.g., 1, 2, 3, 4, 5, or yellow, orange, red colors, which correspond to conditions of increasing unrest or likelihood of eruption). Recently, New Zealand, on July 1, 2014, abandoned its two schemes of alert levels for frequently active and reawakening volcanoes in favor of a single scheme for all volcanoes. The new scheme recognizes a background level (0) with two levels of unrest (1, 2—previously there was only one level of unrest for all volcanoes), followed by 3 levels of activity (3, 4, and 5).

While there is variation in alert schemes for ground-based populations, the same is not so for the aviation sector, where there is a single color-coded scheme with four colors that has been adopted internationally. This consists of a base color alert level, two levels of unrest or unrest and mild eruption and one level of imminent eruption or eruption with significant ash into the atmosphere.

6. SUMMARY

Volcanoes around the world display a great diversity and complexity of unrest, eruption, and post-eruption behavior. Human development on and around volcanoes is equally complex and there are vast differences in the extent to which individual volcanoes are monitored and connected to early warning systems for volcanic eruptions. These complexities contribute to challenges in developing volcano warning systems and providing effective warnings, especially for ground-based populations. The paths of information dissemination differ for ground-based population and aviation-based population. Information dissemination for the aviation sector follows a linear model where official notifications originate from observatories and VAACs and are disseminated to specific end-users in the aviation sector, with relatively little milling for information. That is, a specific response or set of responses is recommended and subsequent action is taken. In contrast, information dissemination for ground-based populations involves notifications being issued by volcano observatories to emergency management, then to the public and media, or alternatively, all simultaneously.

While there is certainly a trend in warning systems toward developing non-linear dissemination paths that notify all end-users nearly simultaneously, it is very likely that, in most instances, observatory scientists will continue to initially discuss eruption precursors and the likely volcano outcomes (e.g., hazards) with emergency management and selected government agencies prior to going public with official alert notifications. In this sense, the dissemination path continues to be linear as scientists disseminate alerts to emergency management, then to the public and media. The real change is the rapidity with which scientists initially discuss conditions with selected agencies and subsequently disseminate alerts to all end-users, usually through numerous digital technologies. This trend is a result of both increasing technological innovation in monitoring, forecasting, and information dissemination and receipt, but it is also the result of the public and media's increased demand for near-real time information. The trend at observatories to notify emergency management and the media and the public of alerts nearly simultaneously improves trust between the media and public and science and emergency management agencies. Rapid notification also helps to accommodate the human tendency to confirm risk information through information milling and warning confirmation behaviors. Thus, rapid official alert notifications serve to provide the public with authoritative confirmation of information they may have received initially from informal sources or from environmental or social cues.

In the age of digital information, where information can be disseminated widely in text and/or graphic form, volcano observatories, emergency management and any related agencies must be prepared to support the warning confirmation process and public milling for information by providing pertinent and consistent information from numerous sources through multiple channels. Notifications need to be updated frequently to satisfy information demands from all parties. In this case, information dissemination and information seeking are iterative, cyclical processes that contribute to dynamic community and organizational protective action decisions, not necessarily one-off actions.

FURTHER READING

Donovan, A., Oppenheimer, C., Bravo, M., 2012. The use of belief-based probabilistic methods in volcanology: Scientists' views and implications for risk assessments. J. Volcanol. Geother. Res. 247–248, 168–180.

Driedger, C., Neal, C., Knappenberger, T., Needham, D., Harper, R., Steele, W., 2008. Hazard Information Management During the Autumn 2004 Reawakening of Mount St. Helens Volcano, Washington. In: Sherrod, D., Scott, W., Stauffer, P. (Eds.), A Volcano Rekindled: The Renewed Eruption of Mount St. Helens, 2004–2006. US Geological Survey Professional Paper 1750, 505–519.

Fearnley, C.J., McGuire, W.J., Davies, G., Twigg, J., 2012. Standardisation of the USGS volcano alert level system (VALS): analysis and ramifications. Bull. Volcanol. 74, 2023–2036.

Leonard, G.S., Johnston, D.M., Paton, D., Christianson, A., Becker, J., Keys, H., 2008. Developing effective warning systems: Ongoing research at Ruapehu volcano, New Zealand. J. Volcanol. Geother. Res. 172, 199–215.

Lindell, M., Perry, R.W., 1992. Behavioral Foundations of Community Emergency Planning. Hemisphere Publishing Company, New York, 309 pp.

Lindell, M., Perry, W., 2012. The Protective Action Decision Model: Theoretical Modifications and Additional Evidence. Risk, 32 (4), 616–632.

Mileti, D.S., Sorensen, J.H., 1990. Communication of Emergency Public Warnings: A Social Science Perspective and State-of-the-art Assessment. ORNL-6609, Oak Ridge National Laboratory, Oak Ridge.

Neri, A., Aspinall, W.P., Cioni, R., Bertagnini, A., Baxter, P.J., Zuccaro, G., Andronico, D., Barsotti, S., Cole, P.D., Ongaro, T.E., Hincks, T.K., Macedonio, G., Papale, P., Rosi, M., Santacroce, R., Woo, G., 2008. Developing an event tree for probabilistic hazard and risk assessment at Vesuvius. J. Volcanol. Geother. Res. 178 (3), 397–415.

Newhall, C., 2000. Volcano warnings. In: Sigurdsson, H., Houghton, B.F., McNutt, S.R., Rymer, H., Stix, J. (Eds.), Encyclopedia of Volcanoes. Academic press, California, pp. 1185–1198.

Newhall, C.G., Punongbayan, R.S., 1996. Fire and Mud: Eruptions and Lahars of Mount Pinatubo, Philippines. University of Washington Press, Seattle, 1126.

Paton, D., 2003. Disaster Preparedness: a social-cognitive perspective. Disaster Prevention and Management 12, 210–216.

Paton, D., Gregg, C.E., 2013. Mental models. In: Penuel, K.B., Statler, M., Hagen, R. (Eds.), Encyclopedia of Crisis Management. SAGE, pp. 312–314.

Simkin, T., Siebert, L., 2000. Earth's volcanoes and eruptions: an overview. In: H Sigurdsson, B.H., McNutt, S., Rymer, H., Stix, J. (Eds.), Encyclopedia of Volcanoes. Academic Press, San Diego, pp. 249–261.

Sorensen, J., 2000. Hazard Warning Systems: Review of 20 Years of Progress. Natural Hazards Review, May: 119–125.

Chapter 68

Volcanic Crisis Management

Gill Jolly

GNS Science, Wairakei Research Centre, New Zealand

Servando de la Cruz

Centro Nacional de Prevencion de Desastres, SEGOB, Mexico, Instituto de Geofisica, UNAM, Mexico, Instituto de Ingeniera, UNAM, Mexico

Chapter Outline

GLOSSARY

alert A condition of heightened watchfulness or preparation for action. A warning method or system to make people aware of impending danger.

alert level or code A set of rules or a system used for transmitting alert messages requiring brevity and clarity. The symbolic arrangement of instructions in a warning system.

consensus General agreement in the diagnostics and the prognosis of the majority of scientists involved in the study of an episode of volcanic activity. Consensus of the involved scientists is important for the decision-making of authorities, however, if consensus cannot be reached, then the range of views should be provided to decision-makers.

contingency or emergency management plan A plan developed to respond to an outcome outside of normal business as usual. These are often devised to deal with an exceptional risk that might have catastrophic consequences.

crisis An unstable situation of increased danger. A crucial stage in the course of volcanic unrest, when the threat of an eruption, or the possibility of an eruption, exceeds some reference or threshold level. Defining the reference level prior to the onset of volcanic unrest is important for clear communication of the threat.

disaster risk reduction Reducing the damage of natural hazards through prevention or mitigation of impact, rather than reaction to a disaster.

disaster An event produced by a phenomenon capable of breaking the fabric of society.

pre-event Period of time when there is significant evidence that a disaster could occur.

preparedness A state of readiness or preparation for use or action in the case of volcanic activity or any other threat. It involves a clear understanding by the population and authorities of the natural phenomena and their destructive effects.

recovery The period of time following a volcanic eruption, when actions are required for returning a community to a former (or improved) state of activity.

resilience The ability of a sector of society impacted by disaster to recover fully from the event.

response The period of time during which decision-makers and a community at threat react to the situation in order to mitigate loss of life and/or property.

risk assessment Determination of the value of risk related to a specific hazard. This may be either quantitative or qualitative and requires an assessment of the likelihood of the hazard occurring and the magnitude of any potential losses incurred.

risk management cycle A sequence of events before, during, and after a disaster whereby decision-makers and communities prepare for potential future disasters to reduce their risk, respond to disastrous events, and then recover from the impacts of a disaster.

The Encyclopedia of Volcanoes. http://dx.doi.org/10.1016/B978-0-12-385938-9.00068-7

volcano monitoring Undertaking observations and measurements to be able to assess the state of the volcano.

vulnerability The inability of physical and human systems to withstand the impacts of a natural hazard. It can be expressed as a probability of damage, or as the proportion of the total exposed assets expected to be affected by a given manifestation.

Potentially destructive phenomena such as volcanic eruptions occur independently of any human action. Volcanic disasters may occur when a social group fails to respond to a threatening situation resulting from volcanic activity. Society should thus react to a potential threat and reduce the future impacts of a disaster. Volcanic crisis management is a framework whereby scientists, emergency managers (civil protection), and communities work together to develop and implement a set of preparedness and response measures aimed toward the mitigation of the effects of an eruption. In this chapter, we outline some general principles for volcanic crisis management.

1. INTRODUCTION

A natural hazard is a naturally occurring phenomenon that might have a negative impact on people or the environment. A **disaster** can occur when a natural hazard impacts on a community that is not well prepared to respond to the effects of the phenomena. The risk from natural hazards is related to the likelihood of the phenomena occurring, the magnitude of losses that might be incurred, and the **vulnerability** of potentially impacted people or environment. To reduce the risk from disasters, the United Nations International Strategy for Disaster Reduction (UNISDR) developed the Hyogo Framework for Action (HFA) in 2005. The desired outcome of the HFA (2005–2015) was stated as "the substantial reduction of disaster losses, in lives and in the social, economic and environmental assets of communities and countries." Five priorities for action were outlined as follows:

1. Ensure that **disaster risk reduction** is a national and a local priority with a strong institutional basis for implementation.
2. Identify, assess, and monitor disaster risks and enhance early warning.
3. Use knowledge, innovation, and education to build a culture of safety and **resilience** at all levels.
4. Reduce the underlying risk factors.
5. Strengthen disaster **preparedness** for effective **response** at all levels.

A volcanic eruption is an example of a natural hazard that can cause significant impact on communities at local, regional, national, or even transnational level. During the twentieth century, over 90,000 people were killed by volcanic activity and over 5 million people were evacuated from their homes (Witham, 2005). The small eruption of Eyjafjallajökull in 2010 resulted in the loss of over US$1 billion for the airline industry. As populations increase in cities and regions close to active volcanoes, the exposure of communities increases and hence there is a critical need for clear and coordinated volcanic risk reduction.

Here we discuss the identification, assessment, and monitoring of volcanic hazards to reduce risk. Together with appropriate mitigation measures, these elements increase preparedness of a community. We then discuss the factors that contribute to a successful **crisis** response and subsequent **recovery** from a volcanic eruption. These actions constitute volcanic crisis management.

Volcanic crisis management is a partnership between the vulnerable sectors of society, the scientists in charge of monitoring volcanic activity and assessing the volcanic hazard, and the government through its civil protection bodies. Without the engagement of all parties, crisis management cannot be effective. Such a partnership should include a clear definition of the responsibilities of each partner. Ideally, an appropriate legal framework, which may be regarded as a crucial component of the preparedness, should define those responsibilities.

Volcanic disaster risk reduction and crisis management involve very complex and uncertain situations, on both the scientific and the management sides. A volcanic eruption may generate a number of different hazards—over different time scales—that can affect different sectors of society in a variety of ways, and with different degrees of intensity. The number of variables involved sometimes makes it difficult even to define the problem for which the crisis management plan is to be designed. Although a disaster may derive from an unexpected intensity of a given phenomenon (which may be an outcome of a lack of understanding or preparedness), disasters, or at least serious crisis management problems, often also result from the unsatisfactory responses of different social sectors. Dealing with a crisis may prove to be rather difficult, particularly when a large population is at risk.

One way to mitigate volcanic hazards is to construct realistic scenarios based on the study of previous episodes of activity and on computer simulations, and then to implement relocation programs to reduce the threatened population to zero—or at least to a minimum—before any eruptive activity develops. However, the attachment of people to their homes, the difficulty (oftentimes impossibility) of finding places to relocate refugees, and many other cultural, economic, and social factors force scientists and officials to look for solutions that permit large populations to live together with a potentially active volcano. Some level of risk acceptance should thus be defined and specific emergency management plans should be developed to be put into action when that level is exceeded. The level of risk deemed to be acceptable will be different across various cultures. Ideally, community engagement prior to an eruption will determine the acceptable level of risk; this can then lead to good decision-making under pressure and lack of the "blame game" during or after a disaster. Discussion with the public during quiescent periods is crucial for developing awareness of volcanic hazards and risk and for promoting appropriate crisis response.

A specific methodology for management of eruptive crises is not possible due to the great variety of situations deriving from differences in physical settings and in social, cultural, economic, and political conditions. However, some basic components of a crisis management plan can be identified, together with some steps necessary for its successful implementation. These steps result from the careful analysis of past volcanic crises and the reasons for the success or failure of their management.

Volcanic crisis management may be considered as an evolutionary process involving three phases:

- **Pre-event** preparedness
- Crisis response
- Recovery

The quality of a crisis management plan, measured by its success in mitigating risk, depends on the extent to which different sectors of society react to the countermeasures. Throughout the **risk management cycle**, good communication between the different organizations and sectors of the population is critical. This is particularly important for volcanic crises, where there can be a high degree of uncertainty in the outcomes of scenarios.

At all stages of crisis management, lessons learned from analyzing the effectiveness of actions should be documented and used to inform improvements to future responses.

2. PRE-EVENT PREPAREDNESS

Preparedness may be defined as a state of readiness prior to volcanic activity. It involves a clear understanding, by the population and authorities, of the volcanic phenomena and its potentially destructive effects. It also involves the design of preventive measures based on that understanding and the development of response plans, in case of major eruptive activity.

To increase the level of the preparedness of a community, some basic components are needed as follows:

- Long-term volcanic hazards and **risk assessment**; associated mitigation measures;
- A **volcano monitoring** system;
- A structure for the provision of scientific advice;
- A volcanic **alert level** system;
- A **contingency or emergency management plan** to respond to changes in activity and exercising of that plan.

2.1. Long-term Hazards and Risk Assessments; Associated Mitigation Measures

The first step of a preparedness plan for disaster risk reduction must be done well before any expected eruptive crisis. This is the assessment of risk, understood as a combined measure of hazard and vulnerability. Once the risk is identified some long-term mitigation methods can be implemented.

Initially, the potentially active volcanoes should be identified. Although this may seem like an obvious first step, in many communities, where there has been little volcanic activity for decades or even centuries, the local population may not be even aware of a volcanic threat. For example, in Montserrat, prior to the onset of the eruption in 1995, many of the local people did not understand that the summit of the Soufrière Hills was a volcano, rather they incorrectly identified an area of hot springs and fumaroles on the southern flank of the volcano as the vent.

After identifying the volcanoes that have a potential to produce eruptions, a thorough knowledge of the characteristics of each volcano and of its past activity allows an assessment of eruptive scenarios with their associated probability of occurrence. For a frequently active volcano, an assessment of past historic activity is informative. For quiescent volcanoes, geological mapping of older eruptive deposits can provide information on likely future activity. Where there is limited geological or historical knowledge, the use of analog volcanoes can inform potential scenarios. Computer simulations can help to assess future impacts such as likely ashfall depths and pathways for lahars or other mass flows.

One product of these studies is a hazard map depicting the areas that could be reached by each type of eruption. The large amount of information contained in hazard assessments makes it difficult to handle as a conventional printed map. However, a balance needs to be struck between a comprehensive representation of hazard for decision-making and a comprehensible visualization of volcanic hazards for the communities. In the former, a geographic information system could be designed that incorporates all aspects required by emergency managers such as location of critical infrastructure, distribution of population, and likely impact zones. In the latter, simple perspective views of the volcano may communicate hazard better to populations unfamiliar with maps. It is also very useful to divide each hazard area into sectors and consider the possibility of setting different levels of **alert** in different sectors for particular scenarios (Figure 68.1).

Increasingly, probabilistic hazard and risk assessments of volcanic activity are being used to determine the likelihood of different scenarios. These can be used both in the long term for land-use planning and in the short term for

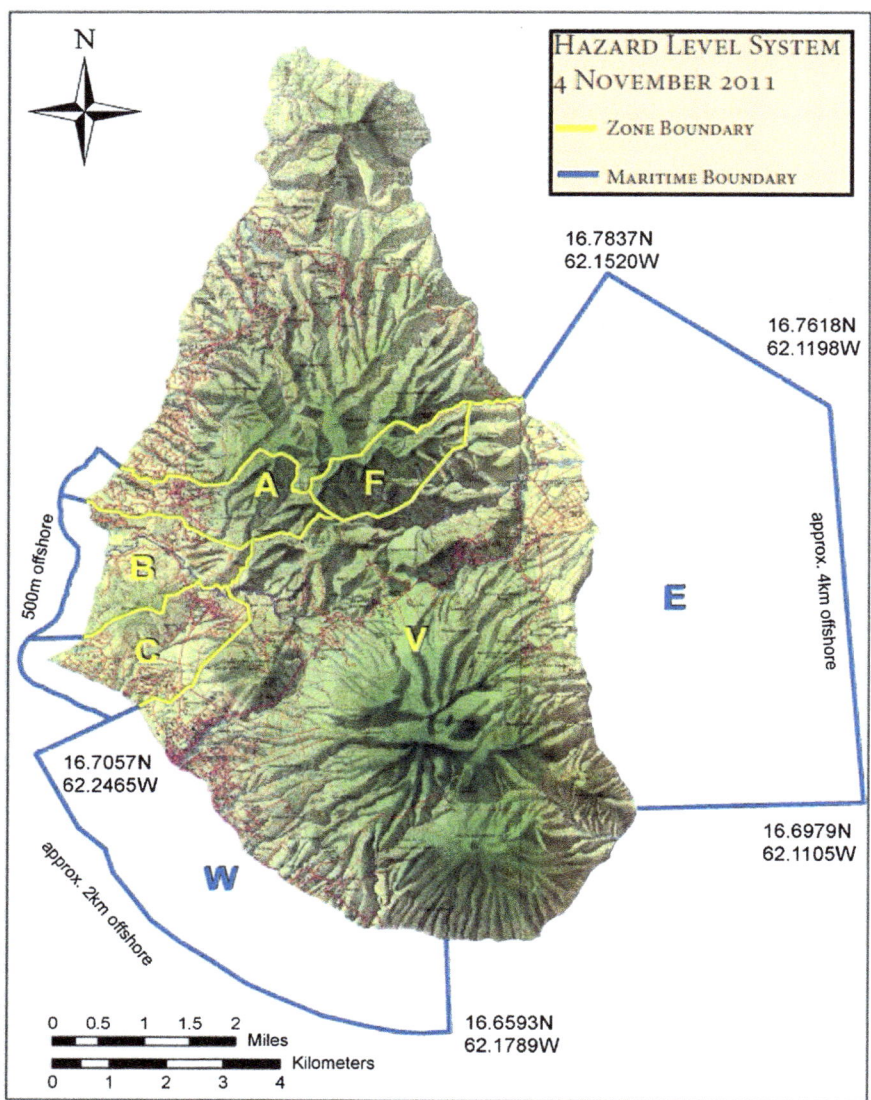

FIGURE 68.1   The hazard level system on Montserrat from 2011. The areas around the volcano are split into sectors and depending on the activity of the volcano, access to different sectors is allowed.

decision-making on evacuations. A probabilistic event tree assigns a probability to each eruptive scenario, and also describes the possible evolution and concatenation of scenarios. Assigned probabilities must be updated as a crisis evolves.

Risk is defined as the likelihood of a hazard occurring multiplied by the vulnerability of people or infrastructure to be impacted by the hazard. Because natural phenomena and their associated hazards cannot generally be modified, risk mitigation is mainly related to the reduction of the hazard exposure and of the vulnerability of the exposed systems. Vulnerability is defined as the inability of physical and human systems to withstand the impacts of a natural hazard. Formally, it can be stated in terms of a probability

of damage, or as percentage of the total exposed assets expected to be affected by a given peril.

It is essential to assess vulnerability and expected damage for each possible eruptive scenario so that mitigation measures and emergency management plans can be designed and implemented. A thorough collection of the physical and demographic information on the region at risk is required to assess vulnerability and to design the most effective measures of risk reduction. Specifically, among the most effective measures for volcanic vulnerability reduction are those directly affecting the public. These measures should aim to achieve a proper response from the population in the threatened area, mainly through building a collective consciousness about the existing hazard and the implementation of a policy of land development (discouraging new settlements and encouraging emigration into safer areas).

At this point, continuing education of the population is required to ensure that awareness of the potential threat is kept current, especially if a volcano has not erupted for many years. Dissemination of information about the phenomenon, its reach and effects, and the mitigation measures should thus be one of the top priorities in any preparedness actions. Specifically, the main measures of self-protection, such as self-evacuation, preparation of emergency kits, or development of community or household plans, and response to a volcanic emergency must be widely explained. There are many methods of dissemination of messages, and an assessment of the most effective methods for communication of hazard information should be undertaken. In developing countries with basic infrastructure, this may take the form of community meetings and face-to-face discussions; in areas with highly developed communication, other measures such as social media, Internet, and television may be available.

Some difficulties in this dissemination process may be anticipated. For instance, in highly developed, industrialized countries, the possibility of land devaluation may sometimes pose a problem in the handling of this information.

Within a developing country, there are many requirements for limited resources. For example, if a volcano has not erupted recently, and communities are subjected to social deprivation issues such as high levels of crime or lack of suitable housing, it is likely that the volcanic mitigation measures are seen as much lower priority than the provision of basic sanitation or crime prevention. In this case, communities and the authorities together need to decide on the best use of available resources. Sometimes political problems, particularly when the volcano is located at or near a border region, may arise.

In these cases, important aspects of preparedness include the construction and dissemination of a document outlining organizational responsibilities, resulting from a discussion and agreement among the involved parties.

Another important step for risk reduction is the implementation of infrastructural measures, related mainly to engineering works aimed at protecting structures and its inhabitants. Although not much can be done in this respect for the safekeeping of structures from the most severe eruptive phenomena, such as pyroclastic and lava flows, protection at reasonable costs from the fall of tephra can be achieved with proper roof structure.

Some of the most destructive volcanic phenomena are mudflows or lahars. The energy and range of lahars may be reduced through barriers to retain or divert solid materials in the mudflows. These constructions are rather complex and expensive. However, they have been extensively implemented in several countries; for example, in Japan, a general technology (SABO Engineering) has been developed for this purpose.

The economic impact of volcanic eruptions on aviation has been increasingly apparent with the eruptions of Eyjafjallajökull, Iceland in 2010 and Puyehue-Cordon Caulle, Chile in 2011. These two eruptions illustrated how relatively minor eruptions can cause major disruption of air traffic across the globe. Since 2010, considerable work has been undertaken to monitor and quantify ash clouds. A better understanding of ash dispersion characteristics and the vulnerability of aircraft to small concentrations of ash in the atmosphere is required to reduce the risk to aviation.

2.2. Volcano Monitoring

A volcano monitoring system aims to gather the maximum relevant information about the volcanic activity in order to detect precursory signals of an impending hazardous event. For that purpose, it is important that the monitoring system is in place well in advance of the eruptive crisis in order to establish a baseline of activity against which anomalies can be compared. These systems consist of the instrumentation of the volcano with specialized measuring equipment adequate for timely detection of any geophysical or geochemical changes of the volcano for correct decision-making and management of a possible emergency.

Here we define "minimum monitoring" as basic instrumentation that should allow the detection of the first symptoms of reactivation. This is subjective and will depend on the activity of the volcano. In some countries, an overview of different levels of monitoring for different levels of volcanic risk has been undertaken to guide how to best focus resources. Monitoring technology is developing rapidly, and the ability to install basic, cheap, and robust monitoring instrumentation is improving. In addition, remote monitoring through satellite observations can be a cost-effective way of observing changes at multiple volcanoes if there are insufficient resources to implement ground-based monitoring.

Once eruption precursors increase, the monitoring system should be augmented. More complete monitoring will encompass a wider set of geophysical or geochemical parameters, with data analyzed in real time (s) or near real time (min). Campaign monitoring can be added where staff and expertise is available. Ideally, a monitoring system should be designed with the ability to add more instrumentation quickly and easily during a crisis response.

All high-risk volcanoes should have a comprehensive and long-term monitoring system, especially those volcanoes that are frequently active; however, resource limitations might mean that a country will need to rely on an approach of minimum monitoring in the long term, with the ability to augment the networks during a crisis response.

Globally, it has been recognized that many volcanoes have little or no monitoring in place and very few volcanoes have robust long-term monitoring systems. To reduce global risk from volcanic activity, improved baseline monitoring of all high-risk volcanoes is a fundamental requirement.

Volcano monitoring is usually the responsibility of a volcano observatory (VO), although the constitution of a VO may differ markedly from country to country. In some cases, the VO is a single government institution and may work only on one volcano; in other countries, a VO might have responsibility for multiple volcanoes; elsewhere, a group of organizations, including academic institutes, may collaborate to deliver coordinated monitoring advice. The basic function of a VO, however, is to operate and maintain a volcano monitoring system.

A significant initial expense is required to set up, maintain, and update a volcano monitoring system. Maintaining a continuous operation in the long term, which requires a stable institution and trained personnel, is also expensive. When setting up a monitoring system, it is important to ensure that it is sustainable in the long term, i.e., there is sufficient funding and staffing to maintain continuous operation. Hence to ensure its proper operation, volcano monitoring must be included in a nation's development planning as part of comprehensive disaster prevention programs. In view of other high-priority needs, the conditions for permanent monitoring are often difficult to achieve in developing countries without technical and economic international cooperation.

The monitoring system for a given volcano depends on its hazard and risk levels, its eruptive history, the availability of economic and technical resources, the location and access difficulties of the volcano, and other aspects. It is generally accepted that the reactivation of a volcano usually involves a progressive appearance of certain unrest signals or precursors. Precursor signals are sometimes strong enough to be perceived by people living close to the volcano, for example, noises, vibrations, gas emissions or fumaroles, water level changes in wells and springs, deformations, and opening of fractures. In some countries, community science is used to expand the available technological measurements. For example, in Ecuador, local volunteers communicate observations to the scientists. At many volcanoes, the signals are very small and can only be detected by sensitive instruments. Among the most important symptoms of unrest that can be measured with a basic monitoring system are changes of the seismicity, ground deformation, and chemical changes in springs or fumaroles.

A minimum monitoring system should consist of one, or preferably several, seismic and deformation sensors located close to the crater of the volcano. A good choice is starting with something simple and inexpensive and gradually upgrading the monitoring system as experience is gained, and more and better data are required for specific scientific purposes. Detailed discussion of several measurement techniques is beyond the scope of this article and is found in other chapters, but will be summarized below. A complete description of instruments and techniques used by the USGS to monitor volcanoes may be found in the work by Moran et al. (2008). In general terms, the type of instrumentation that is convenient for monitoring purposes is not necessarily the same as that which would be needed for research on basic volcanic behavior. However, if there is funding for detailed research on volcanic activity achieved with advanced instrumentation, this can, in turn, lead to better, smarter monitoring in the longer term.

The most common and widely used types of high-profile monitoring systems will be described briefly here.

Visual surveillance. This consists of continuous observation and recording of the volcano by means of video and photographic cameras, as well as frequent visual reconnaissance missions from the ground and air (including satellite pictures) from which any conspicuous anomaly can be detected. Infrared or low-light imagery is increasingly being used to allow observation of thermal features and to provide continual imagery during the night. Lava, debris and mudflows, ash emissions, fumaroles, dome growth, landslides, deformation, ice and snow changes, and weather conditions, among others, are phenomena which can be observed and quantitatively assessed. Webcams are now commonly used in many different VOs around the world, some of which include thermal imagery (Figure 68.2). Such imagery can be used both for immediate observations of volcanic activity, but also analyzed retrospectively for purposes such as dome volume changes.

Seismic monitoring. Seismic stations provide fundamental information about the internal structure of the volcano, its state of activity, and the way this changes. Recording of seismicity can be done locally, using autonomous portable instruments at lower-risk locations, or remotely, through radiotelemetry links into a central recording station. Ideally, real-time information will be available at the VO, so that rapid assessments of activity can be made (see Chapter 63 for more details).

Geodetic monitoring. Ground deformation is a second physical parameter that provides fundamental information about the internal state of the volcano and supplies additional insight on the expected dimension of the developing process. Usually, its measurement is done using conventional geodetic methods, although increasingly interferometric synthetic aperture radar is used to assess whole edifice deformation (see Chapter 64 for more details).

Geochemical monitoring. This includes the chemical analyses of gases, water from springs and crater lakes, volcanic ash and rocks, fumaroles, and, in general, all volcanic materials that may provide information about the evolving state of the volcano. Many of the methods require

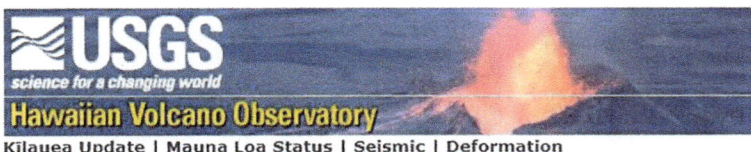

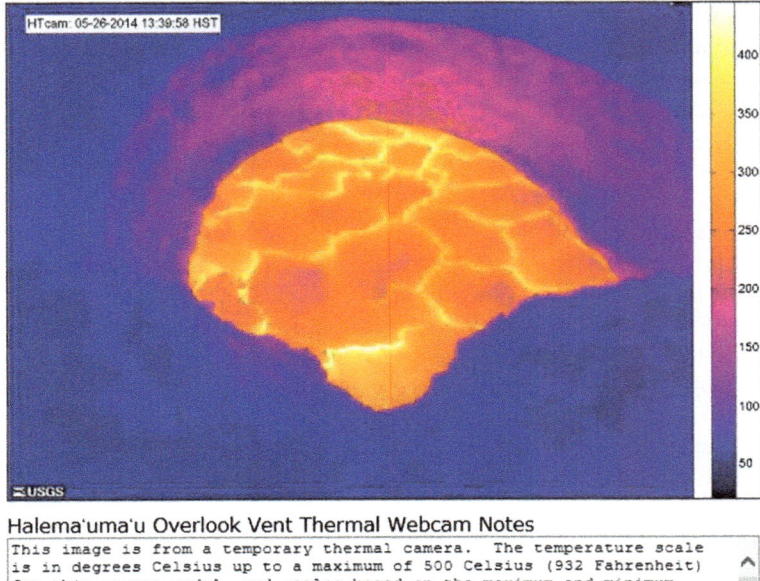

Halema'uma'u Overlook Vent Thermal Webcam Notes

This image is from a temporary thermal camera. The temperature scale
is in degrees Celsius up to a maximum of 500 Celsius (932 Fahrenheit)
for this camera model, and scales based on the maximum and minimum
temperatures within the frame. Thick fume, image pixel size and other
factors often result in image temperatures being lower than actual
surface temperatures.

FIGURE 68.2   A real-time infrared image from the Hawaiian Volcanoes observatory web page showing a lava lake at Halema'uma'u. Similar web sites may be found for other volcanoes of the world.

campaign measurements or analysis of samples in a laboratory, so data may not be available for several weeks or even months. However, remote sensing by means of satellite imagery and ground or airborne spectroscopy provide concentrations or fluxes of gases such as SO_2 and CO_2 in near real time (see Chapter 65 for more details).

Hydrometeorologic monitoring. Monitoring weather conditions (temperature, barometric pressure, wind speed and direction, rainfall, etc.) on a volcano is very important for the forecasting and detection of ash plumes and mudflows produced by rain or melting of glaciers and snowcaps. Ground-based radar is being used in different sites to detect ash emissions as well as heavy rainfall. C-, X-, and S-band radar have all been used on different volcanoes such as the 2011 Grimsvotn and 2011 Mt Etna eruptions. Flow and lahar detection systems, installed along different drainage paths of the volcano, are methods which should definitely be considered if there is

any indication that potentially destructive flows might be triggered by the volcanic activity, as occurred at the Nevado del Ruiz volcano in 1985. Fast-response microbarometers for detecting and measuring infrasound pressure waves from volcanic explosions are also useful in eruption detection systems as at Mt Ruapehu, New Zealand (see Chapter 63 for more details on infrasound).

Other measuring and monitoring systems. There are many other physical and chemical properties of a volcano that can be monitored. Detection and study of gravity and geomagnetic anomalies, concentrations of soil gases such as radon and other tracers, and groundwater levels provide additional and important information on the state of a volcano. Some of these may involve a higher number of support staff and sophisticated and costly measuring equipment, and therefore cannot always be widely implemented.

Data processing and alarm systems. A well-equipped and centralized data collecting and processing facility is essential for a VO, ideally for display and analysis of multiple data sets in real time. For some volcanoes, alarm systems are in place whereby if thresholds of a data stream are surpassed, an alert is communicated. Such alarm systems are key elements for personnel on duty in an observatory and for other scientists, authorities, emergency agencies, or the threatened population during a potentially hazardous event. A good example of this is the Ruapehu, New Zealand eruption detection system, whereby seismic and acoustic signals are compared to a data from a previous eruption. If the thresholds are exceeded, an alert is sent to cell phones, pagers, and a siren system on a ski field on the volcano to warn of possible lahars. In this case, warnings need to be promulgated within a couple of minutes (the time taken for the lahars to reach the pistes), so skiers can take immediate action. This can lead to false alarms, but there is an understanding by the ski field operators that it is better to have a few false alarms than an increased risk of fatality. In many other situations, there is more time for the scientists to view the data and make a judgment on whether an eruption has occurred. In these instances, response actions are not linked automatically to instrumental alerts.

During the process of selection and installation of a monitoring system, several factors should carefully be considered: reliability of the equipment, protection, site selection (geology and noise levels), access, maintenance, and so on. Also, one has to take into account that these systems might need to be operative during long periods of time (most likely years), sometimes in harsh environments, before the volcano of interest shows any sign of reactivation. This permanent monitoring and instrument maintenance can become tedious. The absence of volcanic activity could induce lack of motivation, lead to budget cuts for operation and maintenance, or even complete abandonment of the system with the possible consequence of failing to detect any critical precursor activity that might lead to an short-term eruption. It must be emphasized that once an eruption is in progress, it is usually too late or impossible to implement the desired monitoring system, and therefore, it is better to have some basic instrumentation than none at all.

There are many examples in the literature of volcanic disasters that could have been prevented if minimum monitoring systems had been operating. One such example was El Chichón volcano, located in the southeast part of Mexico. It erupted in 1982 causing a death toll estimated to be around 2000 (Figure 68.3). Had proper monitoring systems been installed on this volcano in the early stages before the eruption, precursor activity could have been detected and adequate countermeasures implemented to save the lives of many people. However, it is important to

FIGURE 68.3 Pyroclastic flow deposits at the foot of El Chichón volcano, near the town of Francisco León. The flows produced during the eruptions of April 3 and 4, 1982, descended along the valley and devastated several towns. The flow deposits also dammed the river shown in the photograph, and the accumulated water and debris produced lahars 6 weeks later.

point out that, by the definitions accepted at that time, El Chichón was considered a low-risk volcano.

Priority criteria for setting monitoring instruments on different volcanoes should thus be carefully reviewed at regular intervals.

2.3. A Structure for the Provision of Scientific Advice

Prior to the onset of a crisis, there needs to be clear understanding of how scientific advice is provided to the authorities. Hence the constitution of a scientific group in charge of hazard assessment and responsible for providing specific recommendations to the authorities must be included in any preparedness planning.

It is common that volcanologists, especially when they are specialized in different aspects of the volcanic activity, such as geology, seismology, and geochemistry, have rather distinct views about the actual hazard and give different opinions to authorities and media. Great confusion can arise from such a situation, causing a general distrust about the capability of scientists to make a sound assessment of the hazard or a reliable forecast of an eruption, eventually leading to a poor response to warning and evacuation orders. A way to discuss activity and convey scientific **consensus** is thus essential. If consensus cannot be reached, then the range of different opinions needs to be communicated so that the authorities and the public can understand the uncertainty associated with forecasting volcanic eruptions.

A framework for scientific discussion must be established in advance of any volcanic crisis. This may help to preclude the possibility of scientific groups divided by competition for the attention of the media and authorities, which can be particularly harmful. It is important that the lines of communication are established well in advance of a crisis, so that trusted relationships can be developed between the science community, the public, and the authorities. In any communication, a balance must be struck between communicating uncertainty and providing confusing messages.

An example of a situation when two scientific groups were providing conflicting advice is the case of La Soufrière volcano on the eastern Caribbean island of Guadeloupe in 1976, where the differences of opinions among groups of scientists, greatly emphasized by the media, caused confusion, misunderstandings, and incorrect decisions. About 72,000 people were evacuated for several months at a very high cost without a clear definition of the actual level of risk. No major eruption occurred, although some minor phreatic activity impacted areas close to the summit of the volcano (Fiske, 1984).

In many cases, there is an official agency responsible for monitoring and risk evaluation. This may be the VO or it may be a separate group that includes representation of VO scientists. It is necessary that all external specialists channel their findings and opinions through such an official body. The conduit for official scientific advice differs from country to country. In some countries, a scientific advisory committee is the main forum for discussion on hazards and risk. Herein recognized specialists, local and visiting, can present the results of their studies and discuss their opinions with the purpose of reaching a consensus about the evaluation of hazard and risk for determining possible forecasts. Subsequent recommendations are then provided to the authorities. An example of this mode of operation is on Montserrat where a Scientific Advisory Committee meets annually to discuss the latest activity and produces a quantitative risk assessment that is presented to the authorities for medium to long-term planning purposes. Over shorter time frames, during eruptions, the Montserrat VO provides advice directly to a local civil protection committee.

In other jurisdictions, a single government agency, most commonly the VO, is charged with monitoring the volcano, undertaking hazard assessments and communicating that information to the authorities. This is the situation in New Zealand, where the organization tasked with monitoring the volcano provides information on the current state of volcanic activity and future scenarios direct to civil protection and other responding agencies as well as the public; it is then the responsibility of the responding agencies to make decisions on closure of areas to the public. Depending on the time available, external scientists are encouraged to attend scientific meetings and their input is used to assess hazard and risk. However, if an eruption has suddenly occurred, the core monitoring organization will provide public information rapidly without wider consultation.

In both cases, it is important to establish the roles and responsibilities of the scientists and the authorities, so that it is clear where the line lies between science advice and making decisions on the welfare of the population. The responsibilities differ between countries. In some cases, VOs may go as far as recommending evacuations, based on their understanding of the volcanic hazard, whereas in other jurisdictions, a committee including both science and civil protection personnel might have a joint discussion before making a decision. In other cases, the VO may only be responsible for reporting current data to civil protection, and then decisions are made by civil protection alone.

2.4. Volcanic Alert Level Systems

Another essential part of volcanic risk preparedness is the definition and implementation of an alert level. The basic function of an alert level is to define the status of a volcano for the authorities and for the public, so they better understand the level of potential threat. The authorities need an efficient way to combine scientific opinion and the requirements of decision-makers so that the vulnerable population can react appropriately. Usually, the scientific consensus and the authorities' decisions are communicated to the population through an alert level based on colors, numbers, or words, generally with between three and six levels. A small number of levels could make it difficult to define the proper hazard level of a given volcano, especially as the actual quantitative hazard or risk is a continuum. Nonetheless, a small number of levels are more understandable to the population, reducing the chances of inaction derived from an ambiguous perception of the risk.

Recent experiences in Mexico show the convenience of using a combined **code** in which a six-level scale is used to communicate the hazard between the scientific advising body and the civil protection or civil defense authorities. The vulnerable population is advised through a simplified three-level color scale. In this way, the first multilevel scale of the code reflects the state of the volcano, whereas the second scale reflects the level of alertness of the population.

Elsewhere in the world, reviews of alert level systems in the United States and New Zealand have led to two different approaches. First, in the United States, a four-level system using words, similar to those used in weather alerts, is used and has some element of prediction (Table 68.1). Second, in New Zealand, a six-level system using numbers is based on the current activity of the volcano (Table 68.2). It is unlikely that a single common system for all volcanoes will be possible as the hazards and needs of civil authorities differ markedly from country to country.

In the United States, a color code system was also developed for use by the aviation sector to provide warnings of the ash in the atmosphere (Table 68.3). This system has been adopted by the International Civil Aviation

TABLE 68.1 USGS Volcano Alert Level (Developed in 2006)

Normal	Volcano is in typical background, noneruptive state or, *after a change from a higher level*, volcanic activity considered to have ceased, and volcano reverted to its normal, noneruptive state.
Advisory	Volcano is exhibiting signs of elevated unrest above known background activity *or, after a change from a higher level*: Volcanic activity has decreased significantly but continues to be closely monitored for possible renewed increase.
Watch	Volcano is exhibiting heightened or escalating unrest with increased potential of eruption activity, time frame uncertain, or Eruption is underway that poses limited hazards
Warning	Hazardous eruption is imminent, underway, or suspected.

Organization and is now in place in several countries around the world.

The actual process of communication of the alert level itself represents a difficult problem, similar to the dissemination of a hazards map, which must be addressed in the preparedness phase. Public campaigns of information, not only on the volcanic phenomena but also on the alert level and the preventive actions, should be started in regions neighboring potentially active volcanoes in quiet periods. Improved confidence in the management may be obtained through direct contact of scientists and government officials with local authorities and communities in the hazard area and persons who have influence on public opinion. Regular meetings to discuss preparedness and develop joint contingency plans help to strengthen relationships, foster trust between all parties, and create an environment that results in more efficient and effective crisis responses.

Ongoing outreach and education is critical to maintain awareness of the volcanic hazard and risk. This can take the form of public meetings, development of outreach materials, surveys of the level of knowledge of the population,

TABLE 68.2 New Zealand Volcanic Alert Level System (Updated in 2014)

New zealand volcanic alert level system

	Volcanic alert level	Volcanic activity	Most likely hazards
Eruption	5	Major volcanic eruption	Eruption hazards on and beyond volcano*
	4	Moderate volcanic eruption	Eruption hazards on and near volcano*
	3	Minor volcanic eruption	Eruption hazards near vent*
Unrest	2	Moderate to heightened volcanic unrest	Volcanic unrest hazards, potential for eruption hazards
	1	Minor volcanic unrest	Volcanic unrest hazards
	0	No volcanic unrest	Volcanic environment hazards

An eruption may occur at any level, and levels may not move in sequence as activity can change rapidly.

Eruption hazards depend on the volcano and eruption style, and may include explosions, ballistics (flying rocks), pyroclastic density currents (fast moving hot ash clouds), lava flows, lava domes, landslides, ash, volcanic gases, lightning, lahars (mudflows), tsunami, and/or earthquakes.

Volcanic unrest hazards occur on and near the volcano, and may include steam eruptions, volcanic gases, earthquakes, landslides, uplift, subsidence, changes to hot springs, and/or lahars (mudflows).

Volcanic environment hazards may include hydrothermal activity, earthquakes, landslides, volcanic gases, and/or lahars (mudflows).

***Ash, lava flow, and lahar (mudflow) hazards may impact areas distant from the volcano.**

This system applies to all of New Zealand's volcanoes. The Volcanic Alert Level is set by GNS Science, based on the level of volcanic activity. For more information, see geonet.org.nz/volcano for alert levels and current volcanic activity, gns.cri.nz/volcano for volcanic hazards, and getthru.govt.nz for what to do before, during and after volcanic activity. Version 3.0, 2014.

TABLE 68.3 Aviation Color Code Focussed on Ash Hazards in Accordance with International Civil Aviation Organization Regulations (Developed 2006)

Green	Volcano is in normal, non-eruptive state. *or, after a change from a higher level:* Volcanic activity considered to have ceased, and volcano reverted to its normal, non-eruptive state.
Yellow	Volcano is exhibiting signs of elevated unrest above known background levels. *or, after a change from higher level:* Volcanic activity has decreased significantly but continues to be closely monitored for possible renewed increase.
Orange	Volcano is exhibiting heightened unrest with increased likelihood of eruption. *or,* Volcanic eruption is underway with no or minor ash emission *[specify ash-plume height if possible].*
Red	Eruption is forecast to be imminent with significant emission of ash into the atmosphere likely. *or,* Eruption is underway with significant emission of ash into the atmosphere *[specify ash-plume height if possible].*

and publicity campaigns to encourage personal preparedness, e.g., having a family emergency response kit or personal plan for relocation.

The most important factor for the successful transmission of volcanic risk information is clear and consistent communication between all players, leading to the reduction of misunderstandings. This can be achieved by establishing a communication routine consisting of periodic bulletins briefly explaining, with simple words, the state of the volcano and the level of alert for the population. This develops among the population an awareness of the relation between the level of hazard of the volcano and their level of alertness. Bulletins containing the same information may be issued in different formats and should be handed to authorities and media, and can also be posted at sites easily available to the general public, such as on Internet sites, on local radio, or in public places such as community halls and libraries.

2.5. Contingency or Emergency Management Plans and Exercising

To ensure appropriate crisis response actions, an emergency management or contingency plan should be developed well in advance of a crisis. A contingency plan is generally the responsibility of civil protection and other responding agencies, although scientific groups should also develop their own response plans. Science response plans should include standard operating procedures (SOPs) including a list of key responding agencies to contact at different stages of a crisis and a set of actions for scientists to follow when making decisions and communicating them to the authorities. The latter is becoming

more important in the wake of the legal judgments on the 2009 L'Aquila earthquake. In this case, scientific advisors were convicted of involuntary manslaughter for downplaying the likelihood of a large earthquake and for providing inaccurate and incomplete statements on risk. If a scientific group has clear SOPs and they are followed correctly, then the likelihood of missing or shortcutting important steps in hazard and risk communication becomes lower.

Ideally, contingency plans should be shared among all responding agencies so that all parties are aware of different actions being undertaken at various times in a response and the requirements of information from the scientific groups are understood. To ensure that plans are current and effective, regular exercises should be undertaken involving all parties potentially involved in a future event. Exercises can be very small scale using only a desktop approach to test responding agency plans or they may include significant involvement of potentially affected communities. The latter has a twofold benefit in terms of (1) testing crisis response plans and (2) increasing awareness of the hazards within the communities.

Summarizing, alert and warning procedures can be done in different ways, according to specific social, economic, and cultural conditions. However, efficient and effective communication of information and development of robust plans will underpin a successful crisis response.

3. CRISIS RESPONSE

Crisis response is the phase of the risk management cycle when actions are required to mitigate risk. For volcanic eruptions, there are several critical areas that need clarification, depending on how the eruption progresses. They are:

- Identification of the onset of a crisis
- Role of scientists during a crisis response
- Role of the authorities during a crisis response
- Ongoing eruption responses

3.1. Identification of the Onset of a Crisis

The decision-making process that defines a crisis response starts with monitoring and evaluation of the volcanic activity and an assessment of how the hazard and risk has changed. For a long-dormant volcano, this is usually manifested by anomalous data signifying increased unrest. However, for a frequently active volcano, the crisis might be triggered by a sudden escalation in eruptive activity.

Communicating the status of the volcano to the authorities is required and the need for emergency measures, such as an evacuation or closure of critical infrastructure, can then be assessed. The responsibilities for different stages in the evaluation of hazard, risk, and emergency actions will be different according to emergency management legislation in each country. Matters become more complicated when authorities of different ranks and jurisdictions become involved, at municipal, state, or federal levels.

A timely response is essential to the proper functioning of this stage as activity might escalate rapidly from background. For example, it is considered that the rhyolitic Chaiten, Chile eruption in 2008 was initiated ca. 36 h after the first precursory seismicity; the remarkably similar Cordón Caulle, Chile eruption in 2011 also started after two days of intense seismicity. In other cases, precursory activity might last months or even years, such as the 1994 eruption of Rabaul, Papua New Guinea that followed almost 20 years of earthquakes and ground deformation. This range of time frames should be communicated to the authorities.

The decision-making criteria for establishing a crisis response depend on different factors. A fundamental one is the degree of risk acceptance defined by the contingency plans. This level should be the result of a consensus between the scientific group, the community, and the civil protection authorities. In general terms, the level of risk acceptance would be directly related to the quality of the monitoring, the efficiency of the warnings, and the logistic capabilities of authorities and the population.

Therefore, the roles and responsibilities in the decision-making process for the onset of a crisis response must be established well in advance of an event, and must be clearly specified in the contingency plans. Should a situation of slow buildup of the volcanic activity persist, the monitoring network should be enhanced to acquire the necessary information needed, first, to better understand the physical processes and, second, to manage any eventual emergency.

There are several recent examples of successful identification of the onset of a crisis, resulting in appropriate emergency management responses being put in place. For the 1991 Pinatubo eruption, although precursory activity fluctuated, the evacuation of communities living on the flanks of the volcano as a result of correct assessment of increased hazard saved the lives of many thousands of people. At the 2010 Merapi eruption, the observation of a rapidly growing dome on the summit of the volcano prompted the increase in alert level and subsequent evacuation of several villages (see Chapter 66 for more details). The comparatively small eruption of Tongariro in 2012 had some minor precursors, which nevertheless prompted the development of an evacuation plan that was used on the night of the eruption.

3.2. Role of Scientists during a Crisis Response

When activity increases, the problem arises of how to respond to this alert and decide what action to take within the VO. Following the SOPs that should have been developed prior to the crisis is crucial to an effective response. Thus, a consistent set of actions takes place and information is disseminated rapidly to the appropriate organizations and personnel at the right times, allowing them to trigger their own response plans. A key end-user sector that needs timely warning is aviation, so that flights can be diverted to avoid ash clouds. Hence, SOPs should include contact details of the Volcanic Ash Advisory Centre responsible for the airspace around the volcano.

Judging whether a signal is of no serious consequence or a critical precursor indicating the onset of a more important event is a crucial problem, and a few basic guidelines can be proposed. First, it is assumed that an adequate monitoring system exists, with sufficient instrumental diversity, redundancy (in case of instrument failure) and reliability to be able to detect and correlate different signals and patterns detected on a volcano. Second, it is assumed that a moderate to large eruption would produce significant changes in a range of different parameters. Therefore, a major change in the state of equilibrium of an active volcano will, in most cases, be detected by different monitoring devices. Third, it is assumed that there is a scientific team charged with monitoring the volcano and providing hazard assessments. If a volcano is showing major signs of unrest, it is worthwhile to consider scheduling a shift system of the monitoring team to enable 24-h-surveillance, so that changes in the volcanic activity are rapidly noticed and assessed. How to organize such a group and implement a specific action plan depends on many local circumstances and workplace policies.

It is expected that a comprehensive monitoring system should detect early volcanic manifestations that could be interpreted as precursors to hazardous eruptive activity. In general terms, it also may be expected that the earlier this interpretation is made, the higher the probability of having a false alarm. On the other hand, an inefficient warning procedure or insufficient logistic capabilities may result in a long delay in the implementation of response actions. Looking for a compromise among these factors is a task that should be addressed from the earliest stages of the detection of increased activity, and must be updated according to the evolution of the activity.

Once a potentially hazardous situation is identified, the consensus of the scientists must be transmitted to the authorities in a clear and unambiguous way. Even when a consensus has been attained in that respect, the opinions of individual scientists may have important differences. A delicate balance must be maintained between the right of individual specialists to make public their own views by direct access to the media and the need of giving to the population a unified position. Ideally, consistent messages should be provided by a single authority or spokesperson, but this should be accompanied by an explanation of the uncertainty inherent in volcanic activity and the range of possible scenarios. It is particularly important that the position be expressed in a language that can be understood by the great majority of the population with a minimum content of technicalities. The risk to life and wealth is such a critical issue to a threatened population that messages can be easily misinterpreted, rumors can spread rapidly especially through social media channels such as Facebook or Twitter, and contending opinions dispersed, leading to confusion, disbelief, and inaction.

The high credibility of the scientific group in charge of the risk evaluation is essential for a quick response to warning and for the success of the emergency plan. The failure of the warning about the great mudflow produced by a relatively moderate eruption of Nevado del Ruiz in 1985 has been mainly attributed to the disbelief or lack of understanding of people and authorities about a recommendation by the scientific group to evacuate areas that could be affected by a large lahar (Figure 68.4). The communication of the hazard was not made clear and strong enough, and the perception of the risk by the population differed from that of the scientific community.

Whether the spokesperson for the scientific group should advise directly the population at risk or whether the alert to the population be issued by the civil protection authorities after a recommendation of the scientific group depends on the design of the responsibility matrix and the local legal framework. Since major decisions involving

FIGURE 68.4 Over 21,000 people were killed when mudflows devastated the city of Armero (Colombia) in November 1985. The disaster occurred despite the existence of basic volcanic emergency management elements and resulted from misunderstanding and indecision. *Photo courtesy of Omar D. Cardona.*

displacements of people involve aspects other than science, such as availability of shelter, provision of food, sanitary conditions, and so on, a careful balance must be achieved to define this responsibility, and is better understood by the authorities.

A particularly sensitive issue in this regard is the position of international specialists determining risks of eruption in a developing country. Their opinions are sometimes given more credit by the media, the authorities, and the public than those of the local scientists in charge of the problem, and can again give rise to confusion and lack of credibility. The International Association of Volcanology and Chemistry of the Earth's Interior commissions have prepared a document on the professional conduct of scientists during volcanic crises in which some of these subjects are addressed (IAVCEI Subcommittee for Crisis Protocols, 1999).

3.3. Role of the Authorities during a Crisis Response

The civil protection authorities and other responding agencies are responsible for making decisions affecting the communities, based both on scientific advice and on their knowledge of the vulnerability of the local population. There is a broad body of emergency planning literature covering the role of civil protection authorities for a range of natural hazards. A review of this is beyond the scope of this chapter, but a few key points relating to volcanic crises are outlined below.

A key factor in the response to a volcanic crisis is the geographic, economic, and urbanization situation around the volcano, which imposes different conditions on the decision-making and emergency management. For instance, about 2.7 million people live within a radius of 15 km around Vesuvius (a similar number would be at risk in the case of volcanic reactivation of the Campi Flegrei), 1 million within 40 km from Popocatépetl, over 1 million within 30 km of the Auckland Volcanic Field and about 530,000 within 20 km of Sakurajima. Mobilization or a general evacuation of such a large number of people requires a very complex operation. In addition, these cases face particular problems. For instance, while the population at Pozzuoli or near Vesuvius is mainly urban and high income, and it can rely to a great extent on alert communication channels and its own resources for an evacuation, the high population density adds special difficulties of automobile traffic flow in an emergency. In contrast, the population around Popocatépetl is mainly rural and low income, scattered in hundreds of small villages, with great differences in culture and even language, thus making the dissemination of information and alert communications particularly difficult, and requiring great support for the evacuation. The emergency plans cannot be easily transferred from one volcano to the other, and the decision-making criteria should take into account the social and economic conditions, as well as the extent of the population. In Auckland, an eruption of a new vent in the volcanic field might occur with very short precursory activity; the time to evacuate might differ considerably from day to night, depending on the likely location of the vent.

Once an increased level of alert has been declared by the scientific group and communicated to the authorities, the decision of declaring an emergency condition to the population must be addressed by the authorities in terms of the contingency plan, defining all tasks involved in each operational phase, such as priorities of evacuation (what places should be evacuated first), rescue, and relief.

It must be clearly defined who is in charge of each task, how to mobilize the necessary resources for the plan's implementation, and how to coordinate with other groups. The main parts of the plan are those related to warnings, transportation, shelter, and health and food services. In this case, a map with the hazard areas divided in sectors may be helpful in allocating levels of risk to each sector and assigning specific tasks and responsibilities.

3.4. Ongoing Eruption Responses

The way the volcanic activity develops and reaches a critical stage has a great influence on the management of the situation. If the activity shows a clear trend toward a major event, with convincing evidence of impending hazard, it is easier for scientists to obtain the support of the authorities, and thus the support of the population and task forces for the implementation of the emergency plan. The 1991 Pinatubo, 2010 Merapi, and the 1994 Rabaul crises are good examples of situations in which the clear consensus of scientists and authorities led to successful management of the emergency.

In some cases, the increased activity fluctuates for a time that may extend for several years, with contradictory signs of a possible evolution toward a major eruption. In a situation like this, there is enough time to organize, test, and improve the emergency management plan, but on the other hand, it is difficult to maintain the consciousness and preparedness of the population and the authorities. There is an increased possibility of false alarms and of growing differences of opinion among scientists that may result in loss of credibility by the population. Several cases of slow development of activity and of misleading signals of an imminent eruption exist, as happened at La Soufrière in Guadeloupe in 1976.

Popocatépetl volcano had a sharp increase in activity in late 1994 and since then has shown large fluctuations, several times reaching stages that were close to an evacuation order, causing great unrest among the population. The persistency of a state of intermediate alert has also caused some loss of interest in the warning signals among the population.

Media play an increasingly important role in the communication of hazard and risk information throughout a volcanic crisis. Scientists should be trained for interaction with the media and there should be an understanding of the different drivers for journalists. Their job is to develop interesting or controversial stories, so there may be a tendency to overemphasize any discrepancy among scientists, or between scientists and authorities. However, if a trusting relationship between the scientific group and the media is developed before a crisis, then the media can become a conduit for correct information to be disseminated. The roles of the spokesperson for the scientific group and for the authorities become crucial in transmitting the best estimate of the hazard condition and the existing countermeasures. When transmitting this information, the speakers should take into account that the attitude of local media is generally different from those having a national or international reach. News intended for people external to the problem tends to highlight the most negative aspects of the situation and to exaggerate the extent of risk, thus causing serious consequences for people involved in the crisis. For instance, in the crisis of Popocatépetl (1994–present) there has been a clear trend by national and international media to overemphasize the risk to Mexico City, which is located farther than 65 km from the volcano, thus giving more emphasis to the most catastrophic scenarios, with a low probability of occurrence, than the more probable ones affecting the vicinity of the volcano.

An effective way to keep a stream of reliable information to the media is establishing a more regular routine of daily (or more frequent) bulletins with different formats but containing the same information. Understanding media deadlines and providing information to meet the media requirements will help the media, for example, morning radio news bulletins may require information by a certain time and if the scientific group can provide this on a regular basis through written updates, it will reduce the need for reactive interviews that can be time consuming. In this way, unexpected calls or interviews from media, which sometimes have to be attended by staff of the permanent watch that may not be experienced in dealing with media, can always be addressed to the official report. The bulletins should contain information about the scientific group's assessment of the state of the volcano and the level of alertness of the population. If there is critical information, a joint media release from the scientists and the civil protection can demonstrate that the two groups are well aligned and strengthen the message. Additionally, keeping a frequently updated Web page with the same information as the bulletins has proven to be very effective in dealing with the large demand for information.

4. RECOVERY

A decision as difficult to make as declaring the start of a volcanic emergency is declaring the end of the emergency. Once a major eruption has started, it may be followed by periods of quiescence that may not represent the end of the eruption. There is an increased difficulty in assessing the state of the volcano, for the remaining activity may obscure precursors to new eruptive episodes. It is important during this phase to continue to monitor at a heightened level.

Additionally, some social pressure may build up from the population and authorities to return to a normal state, particularly if the eruptive event was not very intense, few or no external manifestations can be viewed by the public, or people have remained in shelters for a long time. This pressure may be high enough to exceed the authorities' actions and recommendations about remaining in safe areas. On the other hand, the decision of allowing the return of evacuated people can only be made in the context of the type of eruption and its evolution. This may result in conflicting situations in the way in which the information on the existing hazard is transmitted to the affected population.

To deal with this, in some cases, it is possible to define a "return" alert code in which the actions recommended to the authorities and population are equally simple as those in the "direct" alert code, but different, and strongly dependent on whether the mobilized population is urban or rural. Although such a "return" code may be designed in the pre-event phase, or during the onset of the critical phase, it is difficult to establish a program and disseminate it to the population in those stages. A more practical solution would be to work out a version of the code for the authorities in which some actions of partial or gradual return may be defined, under additional security conditions.

This "return" code has to take into account the current degree of acceptable risk. If some level of hazard has been defined, the only way to stay within the level of acceptable risk is by reducing vulnerability through the reduction of exposure. For instance, if a rural area was evacuated, and the eruptive activity decreases, but there is no evidence that it has ended, a "return" alert code may define a condition under which it is possible to allow part of the population to return and work during the daytime in specific relatively low-hazard areas. This would help to reduce the degree of economic effects of the region, a factor that must also be taken into account in the definition of acceptable risk.

In these kinds of actions, all scenarios must be carefully considered, and if there is any evidence that hazardous events may occur without any precursory activity (as for instance, in dome collapse), the return should be delayed.

Once a crisis has ended, the authorities should implement a recovery phase. This may include cleanup of volcanic deposits to enable the population to return to their homes in safety or for normal agricultural practice to resume. It should also include debriefs of the response to the activity, both from a science and from a civil protection perspective. This will provide lessons on what went well and how a response could be improved in the future. It is important that these lessons are then embedded in operational response plans so that the same mistakes are not repeated.

5. SUMMARY

Managing a volcanic crisis requires the implementation of an organized set of actions as that volcanic activity develops. The preeruption phase includes scientific work aimed at the identification of hazardous volcanoes, the elaboration of risk scenarios, and the installation and operation of monitoring devices. Clear protocols for the recognition and communication of an increased risk should be developed and responsibilities between scientists and civil protection should be agreed upon. It also includes education and outreach work involving communication procedures that develop a clear understanding, by the population and authorities, of the natural phenomena and its destructive effects, and establishing an operational plan in which the responsibilities of scientists, authorities, and the population are defined and the actions to be taken for different scenarios are clearly specified.

Should an eruption develop, or clear evidence of precursory activity be detected, a warning of the potential for hazard must be declared and communicated through an appropriate alert level. Alarm methods have to be developed to make sure that the information through the communication chain is clear and unambiguous. In reply to the alarms, response plans should be implemented and the programmed measures taken according to the responsibility matrix defined in the operational plan. When the volcanic activity stops or declines, a transition to a recovery phase has to be planned. It is important to learn lessons from any volcanic crisis and build them into future response plans.

FURTHER READING

Araña, V., Ortiz, R., 1993. In: Martí, J., Araña, V. (Eds.), La Volcanología Actual. CSIC, Madrid, pp. 277–385.

Auker, M.R., Sparks, R.S.J., Siebert, L., Crosweller, H.S., Ewert, J., 2013. A statistical analysis of the global historical volcanic fatalities record. J. Appl. Volcanol. 2 (2).

Barberi, F., Carapezza, M.L., 1996. In: Scarpa, R., Tilling, R.I. (Eds.), Monitoring and Mitigation of Volcano Hazards. Springer-Verlag, Berlin/Heidelberg, pp. 771–786.

Blong, R.J., 1984. Volcanic Hazards: A Sourcebook on the Effects of Eruptions. Academic Press, Orlando.

Cardona, O., 1997. Management of the volcanic crisis of Galeras volcano: social, economic and institutional aspects. J. Volcanol. Geotherm. Res. 77, 313–324.

CENAPRED/SINAPROC-SEGOB/UNAM, 1995. Volcán Popocatépetl. Estudios Realizados durante la Crisis de 1994–1995. CENAPRED, Mexico.

Cepeda, H., Michael, J., Murcia, A., Parra, E., Salinas, R., Vergara, H., 1985. Mapa de Riesgos Volcánicos Potenciales del Nevado del Ruiz. INGEOMINAS, Medellin, Colombia.

Crandell, D.R., Booth, B., Kusumadinata, K., Shimozuru, D., Walker, G.P.L., Westercamp, D., 1984. Sourcebook for Volcanic-Hazard Zonation. UNESCO, Paris.

Ewert, J.W., Swanson, D.A. (Eds.), 1992. Monitoring Volcanoes: Techniques and Strategies Used by the Staff of the Cascades Volcano Observatory, 1980–90. U.S. Geol. Surv. Bull., 1966.

Fiske, R.S., 1984. Volcanologists, journalists, and the concerned local public: a tale of two crises in the eastern Caribbean. In: Boyd, F. (Ed.), Explosive Volcanism: Inception, Evolution and Hazards. Studies in Geophysics. National Academy Press, Washington DC, pp. 170–176.

Fournier d'Albe, E.M., 1979. Objective of volcanic monitoring and prediction. J. Geol. Soc. Lond. 136, 321–326.

IAVCEI Subcommittee for Crisis Protocols, 1999. Professional conduct of scientists during volcanic crises. Bull. Volcanol. 60, 323–334.

Miller, C.D., Mullineaux, D.R., Crandell, D.R., 1981. In: Lipman, P.W., Mullineaux, D.R. (Eds.), The 1980 Eruptions of Mount St. Helens, Washington. U.S. Geol. Surv. Prof. Paper, pp. 789–802, 1250.

Moran, S.C., Freymueller, J.T., LaHusen, R.G., McGee, K.A., Poland, M.P., Power, J.A., Schmidt, D.A., Schneider, D.J., Stephens, G., Werner, C.A., White, R.A., 2008. Instrumentation Recommendations for Volcano Monitoring at U.S. Volcanoes under the National Volcano Early Warning System. U.S. Geological Survey Scientific Investigations Report 2008–5114, 47 p.

Peterson, D.W., 1996. In: Scarpa, R., Tilling, R.I. (Eds.), Monitoring and Mitigation of Volcano Hazards. Springer-Verlag, Berlin/Heidelberg, pp. 701–718.

Punogbayan, R.S., Newhall, C.G., Bautista, M.L., García, D., Harlow, D.H., Hoblitt, R.P., Sabit, J.P., Solidum, R.U., 1996. In: Newhall, C.G., Punongbayan, R.S. (Eds.), Fire and Mud: Eruptions and Lahars of Mount Pinatubo, Philippines. Univ. Washington Press, Seattle, pp. 67–85.

Shimozuru, D., 1996. In: Scarpa, R., Tilling, R.I. (Eds.), Monitoring and Mitigation of Volcano Hazards. Springer-Verlag, Berlin/Heidelberg, pp. 787–806.

Sparks, R.S.J., 2003. Forecasting volcanic eruptions. Earth Planet. Sci. Lett. 210, 1–15.

Tilling, R.I., 1989. Volcanic hazards and their mitigation: progress and problems. Rev. Geophys. 27 (2), 237–269.

UNDRO/UNESCO, 1985. Volcanic Emergency Management. UNDRO/UNESCO.

Voight, B., 1996. In: Scarpa, R., Tilling, R.I. (Eds.), Monitoring and Mitigation of Volcano Hazards". Springer-Verlag, Berlin/Heidelberg, pp. 719–769.

Witham, C.S., 2005. Volcanic disasters and incidents: A new database. J. Volcanol. Geotherm. Res. 148, 191–233.

Chapter 69

Social Processes and Volcanic Risk Reduction

Jenni Barclay

School of Environmental Sciences, University of East Anglia, Norwich, UK

Katharine Haynes

Risk Frontiers, Department of Environment and Geography, Macquarie University, Sydney, NSW, Australia

Bruce Houghton

Department of Geology and Geophysics, National Disaster Preparedness Training Center, University of Hawai'i, Honolulu, HI, USA

David Johnston

Joint Centre for Disaster Research, GNS Science/Massey University, Lower Hutt, New Zealand

Chapter Outline

GLOSSARY

extensive risk (*definition drawn from the United Nations Office for Disaster Risk Reduction, UNISDR*) The widespread risk associated with the exposure of dispersed populations to repeated or persistent hazard conditions of low or moderate intensity, often of a highly localized nature, which can lead to debilitating cumulative impacts.

hazard (UNISDR) A dangerous phenomenon that may cause loss of life, injury, or other health impacts, property damage, loss of livelihoods and services, social and economic disruption, or environmental damage

information milling The process of confirming a warning or other message. It takes the form of searching and interacting with formal or informal information sources.

intensive risk (UNISDR) The risk associated with the exposure of large concentrations of people and economic activities to intense hazard events, which can lead to potentially catastrophic impacts involving high mortality and asset loss

Eruption Response and Mitigation. http://dx.doi.org/10.1016/B978-0-12-385938-9.00069-9

livelihoods The means by which individuals can provide for themselves with the essentials for living (food, shelter, clothing, water) and the capacity to secure those and the other assets needed (e.g., medicine, fodder) for secure living. A *sustainable* livelihood in a volcanic context is one that is capable of withstanding the physical, economic, and social shocks associated with volcanic eruptions **without** the loss of access to those essentials or the means to restore and improve them following activity. More generally, a sustainable livelihood is also one that secures essentials without irreversibly depleting resource for future generations.

preparedness Actions taken in advance of (volcanic) activity that improve response and recovery once it occurs. This includes preventative and mitigative actions.

recovery Actions taken by the population at risk following hazardous (volcanic) activity as a direct or indirect consequence of the impacts of the activity including improved preparedness for future events.

resilience The capacity of a system, community, or society to absorb, accommodate, and recover from the impacts of a volcanic hazard in a timely and effective manner. Some definitions focus on the ability to restore structures and functions to their initial state but others acknowledge a desire in many settings to have individuals and communities "build back better."

response Actions taken by the population at risk during the course of hazardous (volcanic) activity. Response can act to both decrease (seen as desirable) and increase individual or societal risk. A response of "no action" is not necessarily neutral in terms of risk consequence in this situation.

risk perception a measure of the likelihood people give that a hazard will impact upon them, their lifestyle or **livelihood**. Analysis involves the examination of individuals', societal, or cultural groupings' attitudes, beliefs, judgments, and feelings toward risks.

risk tolerance Level of risk accepted in order to achieve a specific objective.

social capital A measure of social relationships that have productive or positive benefits before, during, and after a volcanic eruption. Chiefly, this is the value that comes from strong supportive networks that provide emotional, physical, and direct or indirect economic resources during a volcanic crisis.

uncertainty Situation where the current state of knowledge is such that (1) the order or nature of things is unknown; (2) the consequences, extent, or magnitude of circumstances, conditions, or events is unpredictable; and (3) credible probabilities to possible outcomes cannot be assigned. Uncertainty is a feature of every facet of volcanic crises. This uncertainty may be either epistemic (lack of knowledge) or aleatoric (intrinsic randomness or irreducible uncertainty).

volcanic risk the combination of the likelihood of a volcanic event, or combination of events, and its negative consequences.

vulnerability (UNISDR) The characteristics and circumstances of a system, community, or society and their surroundings (for example, critical infrastructure and buildings) that make it susceptible to the impact of a volcanic hazard.

1. INTRODUCTION

By definition, **volcanic risk** is a consequence of both physical **hazard**s and the social, economic, and political landscape into which the volcano erupts. The wider social factors, such as poverty, weak or good governance, and effective or inappropriate land-use planning modulate the risks due to eruptions. Volcanic impacts are often felt disproportionally by the most vulnerable who have the least capacity to reduce their risks, cope, and recover. Actions and reactions to eruptions are also influenced by individuals' perceptions of and prior exposure to risk, acceptability, and tolerance, which are again related to underlying issues of vulnerability and associated life factors.

Risk communication is the dialogue between the monitoring agencies (e.g., observatories), risk managers, and populations at risk. The effectiveness of this information exchange is modulated by complex social and political issues. These can enhance or impede the effective translation of warnings or technical information into actions to reduce risk.

Studies of, and a dialogue with, populations at risk provide two immediate benefits: (1) an improved ability to characterize and understand the social, economic, and cultural processes that act to promote and constrain risk reduction in volcanic regions and (2) insights into issues that inhibit the transformation of hazard knowledge into individual, group, or institutional actions that act to reduce risk. This includes both the analysis of communication processes and the identification of the most salient components of hazard information for specific physical and social contexts.

This chapter starts by describing desired and actual social outcomes in **response** to differing volcanic activity. These are a function of societal awareness, readiness, and pre-existing or newly introduced vulnerabilities. The typical methods associated with the communication of risk are then also catalogued. In the second part, the most important social factors that act to amplify or attenuate volcanic risk, including the observed consequences of the perception of the threat, are analyzed.

The chapter concludes by providing case studies of social pathways during and after volcanic crises that exemplify the ways in which these factors compete and interact. The case studies provide an illustration of eruptions where the impact of the physical hazard was intensified or reduced by some of the social factors identified. Analysis of the social understanding and modulation of risk promises dividends not only for better communication processes but also in helping populations at risk adapt to and cope with volcanic activity. Currently, there are many knowledge gaps in this field.

2. PART I—VOLCANIC ACTIVITY: SOCIAL OUTCOMES, INDIVIDUAL BEHAVIOR, AND COMMUNICATION PROCESSES

2.1. Volcanic Activity in a Social Context

Volcanic eruptions and their associated hazards are diverse, complex, and highly variable in space and time.

The modal duration of an eruption is between 1 and 6 months but variance spans individual pulses of activity that last less than a few minutes to ongoing activity of decades or even centuries. An eruption may be preceded by less than a day or many decades of heightened unrest and **uncertainty**. The remobilization of deposited material as lahars can persist for many years after eruption and volcanic edifices are prone to externally triggered instabilities ranging from small landslides to sector collapse.

From a social perspective, this range in volcanic activity can be framed by different states of ideal population response in parallel with different types of activity (Table 69.1):

Affected populations will have different reactions according to the stage, intensity, and overall duration of activity (Table 69.1). When compared to other natural hazards, volcanic eruptions have a tendency to be prolonged and to switch rapidly between different styles and intensities of behavior during eruptions. Volcanic activity can invoke both **intensive** and **extensive risk**. In considering the social impacts of eruptions, **uncertainty** in forecasting is as important as physical parameters, such as eruptive intensity, duration, and variability. The "acute" phase predominantly involves emergency response; at this time, scientists, risk managers, and populations will ideally place a stronger emphasis on early warning and the saving of lives (Table 69.1). Most of the other phases of activity and population state introduce an additional emphasis on protecting **livelihoods** and assets via longer term planning. However, a weaker short-term response to emergency warnings can be driven by longer term social issues relating to livelihoods and assets. This is covered in Part II.

2.2. Outcomes for Affected Populations in a Volcanic Crisis

Potential **outcomes** (both negative and positive) for affected populations as a result of the broad eruptive stages detailed in Table 69.1 are summarized by the typology shown in Figure 69.1. This characterizes the variance in style of activity and some of the chief characteristics that populations affected by volcanic activity might possess and how these can control the impact or outcome of the activity.

It provides a summary of the social outcomes associated with the transformation of volcanic crises into disasters for each stage of an eruptive cycle. Social analysis can help identify the means by which populations have the capacity to act to reduce their volcanic risk (moving from amplification to attenuation of the impact of volcanic hazards). A particular challenge for risk managers arises from identifying actions that can improve outcomes for populations or sectors of a population that have several characteristics associated with high vulnerability. During the course of an eruption, this may require the improvement of warning and forecasting, communication processes, and actions that improve or offset underlying vulnerabilities. "Improvement" may not be possible during the short timescale of an acute volcanic crisis and positive outcomes are thus reliant on good, well-communicated warnings and a resilient population (Figure 69.1).

It is unrealistic to expect populations to completely and permanently isolate themselves from volcanic terrains to minimize risk. The negative outcomes associated with volcanoes and their activity are partially balanced by the resources and positive benefits they bring during quiescence. However, a particular challenge is to encourage the treatment of the volcano as a threat as well as a resource,

TABLE 69.1 Eruptive Stages, Population Reaction, and Typical Timescales of Volcanic Activity

Volcano "Stage"	Ideal Population Reaction	Typical Time "Stage" Duration
Dormancy or inactivity	Preparedness (longer term planning and adaptation)	Months to centuries
Unrest	Preparedness (short-term planning and readiness)	Minutes to decades
High-intensity activity	Acute emergency response	Hours to weeks
Low-intensity activity	Emergency response	Hours to decades
Secondary hazards (posteruptive activity)	Recovery and response	Months to decades
Dormancy or inactivity	Recovery, rehabilitation, and adaptation	Months to millennia

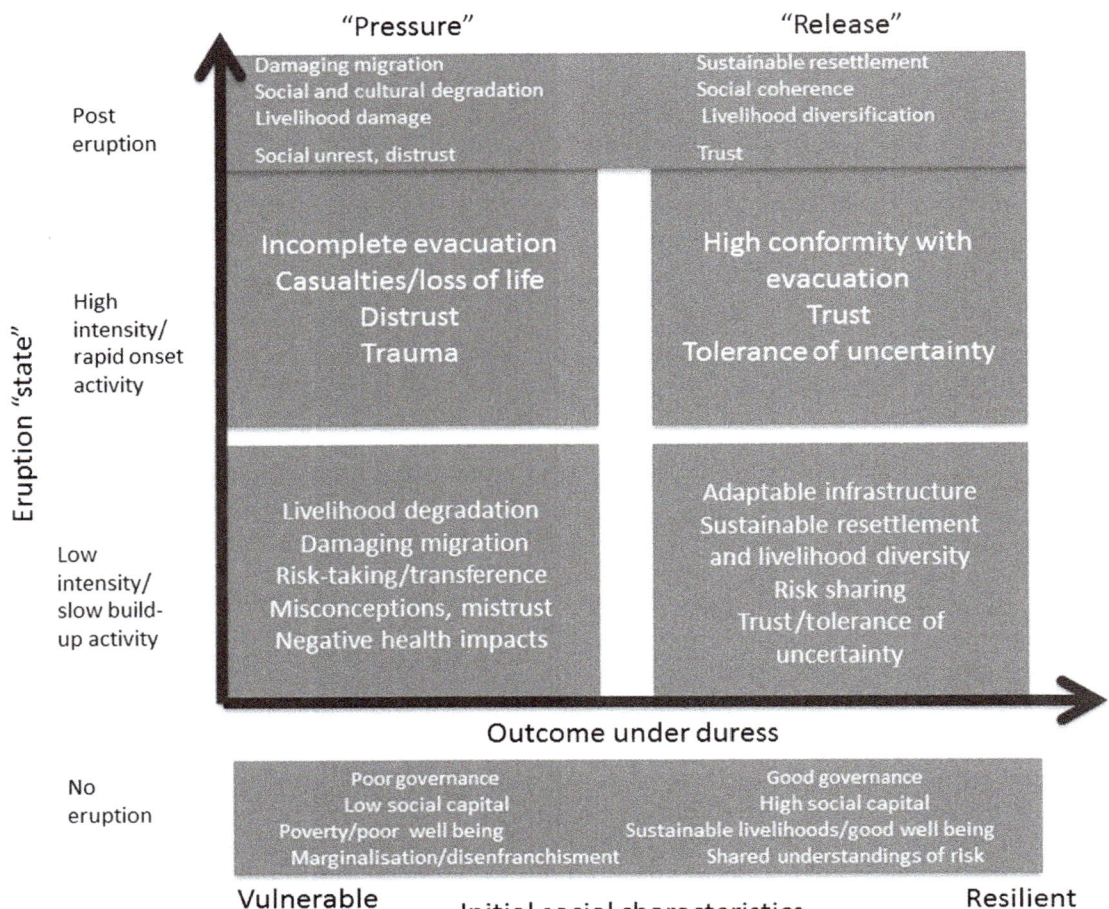

FIGURE 69.1 Summary of outcomes associated with the interaction of two end-member population types with volcanic hazards at differing stages of an eruptive cycle (see Table 69.1). The "response" phase of volcanic crises has been subdivided according to intensity and duration. Low intensity, slow-building, or long-duration activity (e.g., persistent small explosions or lahars) contains most characteristics of extensive risk for affected populations, whereas the higher intensity activity (e.g., Plinian explosions, sector collapse) has impacts more consistent with intensive risk. "Pressure" and "release" draws an analogy with the Pressure and Release Model for vulnerability and disaster risk from Wisner et al. (2004). This model considers disaster risk as the result of two opposing forces: vulnerability and hazard. Positive characteristics of these forces thus alleviate (or "release") the disastrous consequences of an event while negative characteristics reinforce (or "push") them. This allows for a more sophisticated reading of the contributing factors toward risk and disaster, drawn on in this diagram.

and in some instances to adjust rapidly to this as it moves through the eruptive cycle. This type of adaptable view of volcanic settings is embedded into some of the resilient outcomes described in Figure 69.1 and recognizes the desire to minimize risk while living near to episodically active volcanic settings.

2.3. "Unhelpful" Volcanic Behavior

Even the most tolerant, well-governed, and prepared societies can be tested by the extremes of volcanic behavior. High-intensity activity can locally and temporarily overwhelm populations and lead to the disastrous loss of life. The test for resilient populations under these extremes is a minimization of this loss and reduction of long-term trauma and instability.

The extremes of uncertainty associated with poorly felt or observed but relatively high-risk activity (e.g., intense subsurface deformation or movement that is detected by monitoring instruments alone) or low-grade persistent ash emission have all been identified as drivers for longer term changes in the social characteristics of populations with an attendant change from resilient to vulnerable outcomes (Figure 69.1).

2.4. Communication Pathways, Scientific Uncertainty, and Encouraging Risk-Reducing Behavior

The differing states of volcanic activity identified in Table 69.1 clearly drive different purposes for the communication of volcanic risk. Volcanic risk communication can be categorized as follows:

1. *Emergency warnings*—to achieve a short-term **preparedness** and an immediate **response** during a crisis, e.g., evacuation messages

2. *Information provision and education*—to promote long-term **preparedness** and mitigation, e.g., preparedness materials, hazard and risk maps, location of community shelters. This information is usually independent of any event but requires extended time, either that associated with the preparedness or **recovery** phases of a volcanic eruption or as part of a strategy during long-lived eruptions to cope with change or raise awareness.

3. *Partnership and dialogue*—with communities to build consensus and shared meanings over controversial issues, e.g., deciding on acceptable levels of risk, appropriate mitigation and preparedness strategies, appropriate changes of policy to reduce risks (land-use planning and building codes), and understanding who is already self-reliant and who will need greater support to become resilient. This process is independent of any specific threat, but can be performed most usefully during the **preparedness** and **recovery** phases or in the development of strategies to cope with high-intensity or rapid-onset activity.

The prevailing approach to hazards education has been to provide the public with "objective" risk information, based on an assumption that if people are informed they will then take action to protect themselves. Research has demonstrated, however, that awareness of hazards does not necessarily translate into protective actions (Haynes et al., 2008a; Sims and Baumann, 1983). Sims and Baumann (1983) carried out a detailed review of natural hazard communication and education programs and found a mass universal failure of these programs to bring about any behavior change. The work identified that even those who are fully informed and appreciate the risk, may not take actions to protect themselves:

*it doesn't necessarily follow that because information is given it is received or because education is provided there is learning; nor does it follow that even if a public is informed of a risk and **does** know what to do, it therefore **will** do what it knows it could or should do.*

Sims and Baumann, 1983 p167, emphasis in original

At worst, a proportion of those receiving information may even feel less concerned about the hazard as a result as they perceive others are taking control ("risk transference," see Figure 69.1). It should be pointed out that this does **not** imply there is no place for *hazard education* in risk communication. Hazard education forms the foundation for risk communication: the more the people know about the hazard the easier it becomes to explain the important facts that guide decisions to reduce personal, institutional, or community risk. Effective *communication* builds on education by also understanding the decisions to be made and the competing issues to be resolved in order that information is appropriate and timely.

Further, research shows that receipt of a warning is followed by a period in which recipients **mill** this information, often seeking confirmation from alternative sources. Time delays during this process may be critical in preventing an efficient response, and if confirmation is sought from ill-informed unofficial sources, the original message may become diluted or distorted. Figure 69.2 illustrates the competing messages and differences in perception of communication pathways between those at risk and those in charge of transmitting messages. People also often wait for visible evidence of a threat before taking action.

Thus, there is a shift in emphasis toward partnership and dialogue at active volcanoes (Barclay et al., 2008). This is enriched by knowledge of the social and cultural context as well as analysis of **risk perception** (see Part II). Here, effective risk communication is not simply information provision and education; this dialogue takes into account

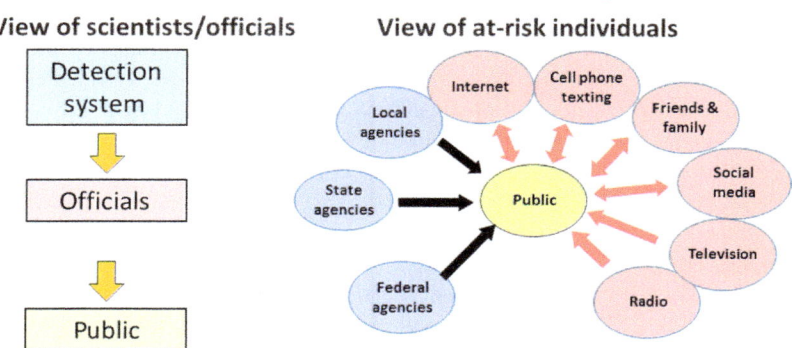

Warning confirmation: milling

View of scientists/officials   View of at-risk individuals

FIGURE 69.2 Comparison of perceptions of information pathways between those delivering information and those "at risk." Official agencies often have a simple source-oriented model of the warning process. By contrast, a receiver-oriented model of the warning process shows people being bombarded by information from a wide range of sources. In the diagram, official sources are local agencies, state agencies, federal agencies, and initiatives like "Reverse 911." Informal sources include the Internet, cell phones, direct contacts with friends and family, texting, social media, television, and radio. The public will attribute real significance to both types of communication, and for this reason it is critical that the informal "conversations" reflect an accurate picture of the situation. Official sources are in blue, informal sources are in pink.

and addresses the wider factors that underpin people's capacities and intentions to take action and their existing knowledge and perspectives. Evidence is now emerging that this provides improved understanding of and toleration for the scientific uncertainty that inevitably accompanies volcanic crises. These are further explored in Part II. Nonetheless, these methods are resource- and time intensive and not necessarily appropriate in all circumstances, for example, clear simple statements of knowledge may be the most appropriate during high-intensity activity and acute emergencies. This change in emphasis also throws out some provocative and important questions in defining the role of the volcanologist as a full participant in the decision-making processes associated with disaster risk reduction and risk management (see Chapter 70).

New methods of visualization, rolling 24-h news and pervasive social media coverage present both opportunities and challenges in acute phases of volcanic activity and in engendering social cohesion during recovery phases. It is necessary to use multiple consistent sources across differing media to convey warning messages and dynamic tracking and earlier identification of inappropriate information sources is critical (see "competing messages" below). GoogleEarth, Geographical Information Systems, and high-resolution imagery from remote sensing provide rapid means to survey, update, convey, and share changing risk information as situations develop, even with rapidly changing information or evolving hazards. Analysis of the impact of these types of media is only just emerging.

3. PART II — SOCIAL FACTORS THAT MODULATE VOLCANIC RISK

Research within the social sciences has demonstrated that the decisions, actions, and ultimately outcomes for those at risk are driven by a myriad of social, psychological, economic, and political factors (Wisner et al., 2004). These not only shape people's capacities to respond, cope, recover, and adapt to volcanic impacts but also their accepted level of exposure in the first place (i.e., who lives and works in high-risk zones). For example, in some countries, land tenure rights on and around volcanoes are often passed down through families over generations, which can generate a strong sense of tradition and place. The impact on vulnerability can be both negative and positive dependent on social circumstances. In the worst situations, people may feel compelled to live and work or return to high-risk locations to sustain their **livelihoods**, often without adequate protection from government (e.g., through land-use planning and emergency management systems).

Further, exposure and vulnerability to hazard impacts are unevenly distributed among any population; no geographic group is uniformly empowered, wealthy, or

healthy. Volcanic eruptions, like all hazard events, often exacerbate the sociopolitical problems within communities (Figure 69.1). The processes that drive both these negative and positive outcomes in the face of different types of activity are summarized below. The case histories described in Part III illustrate the manner in which these different drivers reinforce and interact with one another in differing types of volcanic crises.

3.1. Risk Perception, Knowledge, and Risk-Reducing Behavior

An inability or lack of willingness to act in the face of damaging volcanic risk, poor toleration of scientific uncertainty, and a lack of trust in any, some, or all of those involved in the risk management process (Figure 69.1) are some of the principal means by which social circumstances can amplify volcanic risk, inducing negative outcomes. These are rooted in how individuals and institutions interpret and act upon risk information. Understanding this is a complex function of individual, social, and cultural predispositions and experience as well as the economic and political landscape.

Risk perception provides quantitative measures of the likelihood people give that a hazard will impact them, their lifestyle, or livelihood. Risk perception research usually involves the examination of individuals' attitudes, judgments, and feelings toward risks and the role this plays in formulating preferences and making decisions under conditions of uncertainty, with the aim of improving risk communication and motivating protective behavior.

A number of heuristics or measures have been identified that influence the judgments of all involved in milling risk information, for example (see Slovic, 2000):

- **Law of randomness**: The hazard occurs only once and therefore it would not occur again.
- **Normalization bias**: Where a past experience of coping with a hazard may condition people to be complacent, expecting only the experienced.
- **Availability bias**: The probability of an event is judged by the ability to recall it from memory.
- **Anchoring and adjustment**: People think they have a better picture of the truth than they really do. People anchor on a salient value and then begin adjusting their confidence bounds. However, often their original anchor value is inaccurate and based on availability bias.
- **Affect**: This can be described as the "knee-jerk" reaction to external stimuli where the "attractiveness" of the choices under consideration may influence behavior. In other words, the greater the instinctive benefit of that choice, the lower the perception of the associated risk will be. For example, relocating to a crowded,

impersonal shelter may lead to subconscious underestimation of the volcanic risk associated with staying within an exclusion zone.

- **Outcome expectancy**: Whether people perceive the impacts of a hazard to be surmountable or insurmountable.
- **Self-efficacy**: People's belief in their own coping ability.

These heuristics, according to Slovic et al., (2000) *"may have profound implications for the judgments about the risks associated with natural hazards or the benefits of plans for coping with those hazards"* (p17). In particular, the "affect" heuristic and the inverse relationship between perceived risk and perceived benefit it can cause is evident in a number of volcanic risk situations.

In line with other domains of natural hazard research a strong link between volcanic hazard knowledge alone (i.e., good perception of the hazards), preparedness, and most importantly action has rarely been seen (e.g., Haynes et al., 2008a). Recent research acknowledges the important role that underlying vulnerabilities, social and cultural norms, and communication pathways play in moderating behavior (e.g., Eiser et al., 2012).

Nonetheless, the analysis of risk perceptions adds considerably to evaluations such as vulnerability assessments as it provides insights into the knowledge, attitudes, and feelings of individuals and how they interpret risks and choose actions in the face of that risk. The perception of risk, like many social attributes, can be dynamic on timescales similar to volcanic activity itself and understanding how and what drives positive change is important.

3.2. Governance

The majority of volcanic disasters have contributing social and/or political causes that may include one or more of the following: unstable governance; social and economic inequality; a lack of investment in, or capacity for, risk reduction; prioritization of economic interests over safety; and politicization of the crisis to advantage powerful individuals or groups ("Poor governance" in Figure 69.1). Developing countries are disproportionately impacted by weak governance and limited capacity for risk reduction. However, volcanic crises in all countries are often politicized with economic interests competing with those of safety and long-term investment (See Part III for examples). This can be a particular problem under conditions of high uncertainty and during eruptions of long duration.

In this context, "good governance" (Figure 69.1) recognizes the need for long-term planning to invest in adaptable infrastructure, and to develop shared understandings of risk that reduce social and economic inequalities and guards against the marginalization or disenfranchisement of particular community groups (using "Partnership and dialogue," identified in Part I). In turn, these long-term strategies improve the ability to react and mobilize populations during short-term acute crises. Currently, there is very little detailed analysis that considers the best means by which this can be achieved.

3.3. Livelihoods and Poverty

There is a strong link between livelihood insecurity, poverty, and vulnerability to volcanic hazards. Many of those who suffer the greatest impacts are people who rely directly on the volcano and its surroundings for their livelihoods. For example, many of those returning to the highest danger zone around Merapi during the 2010 crisis were engaged in animal husbandry during both the 2010 eruptions of Merapi (Mei et al., 2013).

For smaller volcanic island nations or more isolated regions, volcanic activity can devastate the country's economy as demonstrated on Montserrat through the destruction of the capital city Plymouth, the airport, and port.

Livelihoods can be an important indirect driver of impact in well-developed countries due to the economic fallout of threatened or actual activity on particular industries. For example, the impact of unrest in the Long Valley Caldera on the Mammoth Mountain ski resort and the disruption to air travel and associated global economic impact following the 2010 Eyjafjallajökull eruptions (Case Study 2; Part III).

Kelman and Mather (2008) outline a sustainable livelihood approach to living with volcanic risk by maximizing the social and economic benefits while reducing risks through livelihood diversification, one of the key positive outcomes recognized in a resilient recovery in Figure 69.1. Encouraging sustainable livelihoods also contributes to the strengthening of well-being. Another important factor is **social capital** (Figure 69.1) and, although some communities may be economically impoverished, they can possess strong social networks that help to reduce the impacts of an eruption and improve recovery prospects. Wilson et al., (2012) examined the differential recovery of Argentinean farming communities affected by heavy ash fall during the eruption of Mt Hudson, Chile, in 1991. The recovery process was influenced not only by the initial ash impacts and proximity of the farm to the volcano, but also by the financial and psychosocial capacity of the farmers.

3.4. Culture and Religion

Knowledge and action to reduce risk is mediated through cultural contexts such as languages, traditions, meanings, and beliefs. Cultural or religious beliefs can influence risk perception and in turn actions to reduce risk, e.g., impacting self-efficacy. Strong cultural groupings can provide high social capital in the face of a volcanic crisis and a number of volcanic events have demonstrated the importance of the cultural and religious belief context in determining the outcome of a crisis.

For example, in 2006, residents around Merapi in Indonesia refused to evacuate when the volcano entered a new phase of dome growth as they had not received traditional warnings from the spiritual creatures termed Makhluk Halus. Dome collapse produced pyroclastic flows that traveled within 300 m of inhabited areas, although only two fatalities resulted from this eruption. Merapi erupted again in late 2010, with flows extending further, leading to 367 fatalities, including the spiritual caretaker of the volcano, Mbah Maridjan, and 34 other people who remained in his village, some trying to persuade others to leave (Mei et al., 2013). As well as cultural beliefs, the need to protect and look after livestock and crops was a compounding issue in their reluctance to evacuate. This again highlights the importance of livelihood sustainability to volcanic risk reduction and the multifaceted nature of vulnerability (Mei et al., 2013).

A number of positive accounts also exist where the social networks associated with strong cultural and spiritual beliefs have reduced risks. For example, during the 1991 eruption of Pinatubo, a local group of nuns, the "Franciscan Sisters of Mercy," were used as intermediaries to help communicate with rural groups living in the mountains. In many accounts of volcanic eruptions, a positive and strong message from respected community leaders to these social networks assists very strongly in risk communication. This is discussed further in Section 3.6.

3.5. Age, Gender, Disabilities

Women, the very young and old, and those with disabilities are often among the most vulnerable during disasters and are overrepresented in death tolls (Wisner et al., 2004). They are often more vulnerable due to their limited mobility, as well as lower ability to take direct action from warnings and other disaster information (Wisner et al., 2004). All of these groups, but, in particular, children, may have little or no influence over family and community decisions about disaster preparedness and response. Directly involving these groups in risk reduction activities has been shown to be an effective way to reduce household and community risks. Gender is now a mainstream issue in disaster risk reduction and increasingly the rights and unique capacities of children, the elderly, and disabled are being recognized. However, the examination of these groups within a specific volcanic context remains largely unstudied.

3.6. Trust and Competing Messages

Trust is a fundamental factor in all aspects and phases of risk reduction and has been shown to influence how people interpret the information and advice they receive (Haynes et al., 2008b). The development and loss of trust influences, and is influenced by, individual and institutional decision making (governance), risk perception, risk tolerability, and ultimately actions to reduce risk. Trust can be difficult to build but easy to lose. Volcanic crises are often uncertain and their effective management requires a delicate balance between precaution and reassurance. Dealing with uncertainty, by means of a precautionary attitude and long-term evacuations, can lead to the perception of false alarms, undermining the credibility of the risk management team and reducing the impact of risk communication messages (Figure 69.2). On the other hand, managers pressured to maintain a more relaxed policy in the face of uncertain information, may have to deal with the consequences of an "unpredicted" disaster and a significant loss of trust. In terms of managing uncertainty, the International Association of Volcanology and Chemistry of the Earth's Interior (IAVCEI) guidelines for volcanologists recommend that the scientists and authorities understand each other's expectations, methods, and limits ("Partnership and dialogue" in Part I). Discussion is encouraged among scientists, officials, and the public about hazards, uncertainty, and tolerable risk with the aim of increasing scientific credibility and by inference trust. Where possible, building trust, via dialogue and knowledge exchange, is even more effective well in advance of a crisis situation.

Where trust has been explored during volcanic crises some communities place high value on their own informal social networks and if used well these can be a helpful means for the population to personalize and make sense of information ((Figure 69.2). Unfortunately, informal communications can become competing messages, particularly during periods of high risk but low activity, when scientific and public observations differ (Haynes et al., 2008b). During challenging times of social and financial loss, such competing messages are inevitable. However, informal and trusted networks can be utilized in order to reinforce official messages and receive feedback from the community and some successful case studies demonstrate the value of the instinctive use of this type of communication. For example, on Montserrat during the peak of activity church leaders and radio presenters became very effective "translators" and intermediaries.

Increasingly, social media, such as Facebook and Twitter, have become very effective official communication tools during disasters. They provide opportunities for official agencies to engage with communities producing a blend of official and informal messages, including feedback and discussion. The data can be updated continually as events unfold, supplying real-time and locally specific messages. Community-based communications can be monitored via Joint Information Centers (not controlled by any one managing organization), acted upon if assistance is required, and if necessary corrected when information is inaccurate. The opportunities to utilize social media during volcanic disasters need further research in order to ensure they can be used to their full potential.

4. PART III—CASE STUDIES

The following case studies provide examples of some of the concepts and drivers described in Parts 1 and 2.

4.1. Case Study 1: Communication, National and Local Governance Processes: The Nevado del Ruiz Eruption of 1985

The eruption of Nevado del Ruiz in November 1985 was the world's fourth largest volcanic disaster; the ensuing lahars killed over 23,000 people, in what was, a "forecast" event. Given the statement above, there are many essential messages and lessons that can be learned from this disaster. Questions are why did this eruption lead to a tremendous loss of life and why was there inadequate awareness and preparedness of officials and the public?

Unrest at the volcano began in December 1984 and intensified in February 1985, weak explosions occurred in September 1985. The main explosion on 13 November, 1985 followed 3 days of recorded continuous seismic tremor. At 15:06 local time the first small gas and ash explosions occurred. The main explosive eruption began at 21:08, and hot ejecta melted the snow cap generating lahars. The lahars destroyed the village of Chinchina at 22:40 and Armero at 23:35. Any of the events prior to 23:35 could have triggered a warning to evacuate Armero but seismic data were not immediately telemetered at that time.

Previous eruptions in 1595 and 1845 caused 636 and 1000 fatalities in Armero, from lahars that took a similar path as those in 1985. Although triggering eruptions were relatively modest, the lahars generated were large. In 1985, Armero had a population of ~30,000 people. However, residents were completely unaware of these lahars, and most had no experience with natural or volcanic hazards. Despite almost a year of precursors, there was no unified official voice to at-risk residents. Seismic recording at the volcano had began in July 1985 and a detailed hazard map was presented to government in October, when a United Nations Disaster Relief Organization report stated "Ashfall and lahars are very likely and within near future. Protect populations."

The risks were thus known early, but provincial and national government made conscious decisions not to evacuate villages unless, and until the precise moment, that danger was *guaranteed*. For many officials, the political and economic costs of a false alarm were too high and would have caused enormous logistical problems and political risks for a financially strapped government. The Armero mayor stated shortly prior to the event that many residents did not know whether to stay or leave, the local emergency committee lacked information and resources, and people had lost confidence in the veracity of the information made available. On the day of the lahar, a series of provisional alerts that were later retracted (when conditions appeared to return to normal) also confused residents.

Many citizens did not believe that the threat was real or that they were in any real danger, an opinion that was shared by some officials at all levels. This was compounded by widespread illiteracy and poverty in the areas most at risk. Furthermore, the society was deeply religious, and early on the day of the eruption "stay calm" messages were issued by commercial radio and the village priest. In any case, the local population was not prepared to act, and no official evacuation order was issued.

The more complete elaboration of the lessons is in Voight (1996). Some clear messages, which contribute to a very different system of volcanic hazard management in Colombia today, are as follows:

1. Lack of a unified chain of command and a single voice contributed mixed messages at critical times in the crisis.
2. In the moments of crisis, complex decision-making processes that involve a bureaucratic chain-of-command, hours of committee discussion, or that assume rapid, unstressed communications linkages, were not effective or even feasible.
3. Short timescales for preparedness of agencies or the populace led to critical errors.

4.2. Case Study 2: Global Dispersal, Global Governance, Science in Decision Making, and Livelihood Preservation: Eyjyafjallajökull 2010

"Celebrity chefs prepared lavish barbeques at the edge of the new lavafield" and "Britain's TopGear [television] program sent a team with a specially customized jeep to drive on the hot rocks, much to the chagrin of the local police."

Benediktsson et al. (2011)

An initial fissure eruption started on the ice-free portion of Eyjafjallajökull on March 20. On April 14, a new prolonged and more explosive eruption was initiated, this time focused under 300—500 m of ice. A series of <10-km-high phreatomagmatic plumes was generated with fine ash dispersed across much of mainland Europe and the United Kingdom during an unusual period of unfavorable wind patterns. The initial fissure eruption caused only minor localized disruption and in fact served as a tourism fillip. It excited considerable global media interest, optimistically deemed the "tourism friendly eruption" (Benediktsson et al., 2011). The onset of the second phase heralded unprecedented disruption to global airspace and in modern times to the local Icelandic population. Emergency management planning in Iceland was well prepared for the impact of jokullhlaups with ~800 people evacuated. Although roads, infrastructure, and farmland were destroyed, no lives were lost. However, the long-term impacts of volcanic ash on local people and livestock were not well anticipated and the effects were locally sometimes severe (Bird and Gísladóttir, 2012).

At the time of the eruption, the existing protocol of the aviation industry amounted to a "zero tolerance" of ash in airspace, based on historical precedents of short-lived eruptions in relatively uncluttered airspace. It was untested by the prolonged but weak activity experienced in Iceland, and the congested airspace in Europe. It resulted in the closure of airports and the cancellation of thousands of flights. The consequences as reported by the various aviation authorities are as follows: Air Council International Europe: 313 airport closures; International Air Transport Association: 100,000 flights disrupted with 19,000 canceled on 18 and 19 April alone losing $1.7 billion (USD) of airline revenue peaking at $400 million (USD) a day from 17–19 April. EUROCONTROL (European Organisation for the Safety of Air Travel) reported some 10 million passengers were unable to travel.

By April 21, a new policy was formulated and implemented, with the formation of concentration-dependent flying "zones." This policy reduced the impact of later ash dispersal.

As the mass flux of material peaked during Eyjafjallajökull's second phase, the relatively unusual prevailing wind conditions and the extended eruption interval were undoubtedly a key factor in the disruption. This was considerably exacerbated by the lack of an international protocol that was "fit for purpose" or in step with modern air traffic. There was no knowledge of tolerable flying limits or ash dosages and no preparedness for the potential disruption associated with the closure of crowded European airspace. Nor was there a clear chain of command for decision-making responsibilities within the aviation industry. As early as 1997 the scientific (volcanological) community had called for studies to determine those safe limits but without impact on the aviation authorities. The current flying zones still remain only indirectly linked to scientific data and the concentrations cited in current protocols are not detectable with current instrumentation.

Tourism in Iceland also initially suffered (compounded by an eruption from Grimsvötn in May 2011) but an immediate £3.7 m government-sponsored global campaign

entitled "Inspired by Iceland" featuring prominent use of Web cameras and visual imagery to emphasis the ash-free natural landscape in an effort to counteract this.

The references contained in Bird and Gísladóttir (2012) and Benediktsson et al. (2011) provide a fuller discourse but this case study provides a demonstration of

1. The economic expediency of incorporating scientific evidence into the governance of volcanic hazards, in advance of crises;
2. The manner in which volcanic hazards cross international and political boundaries and the rate at which decision-making processes accelerate when economic as well as scientific pressure is brought to bear.
3. The increasing role that the modern media can play in disseminating information and influencing individual behavior, both positively and negatively.

4.3. Case Study 3: An Unmanageable Crisis? The Eruption of Nyiragongo (Democratic Republic of Congo) in 2002: High-Intensity Volcanic Activity, Acute Crisis Modulated by Inherent Governance and Poverty Characteristics

A disaster where political instability was a significant cause was the eruption of Nyiragongo, within the Democratic Republic of Congo (DRC) on 17 January, 2002. Lava flows have inundated the nearby city of Goma in 1977 and 2002. The region around the volcano has suffered from a complex series of conflicts since the early 1990s. This includes civil war, the 1994 genocide in neighboring Rwanda, and struggles for power and control of DRC's mineral wealth with Uganda and Rwanda (Wisner et al., 2004). The city's population more than doubled over this time from an estimated 173,000 in 1993 to 550,000–600,000 in 2003 (Verhoeve, 2004). The level of suffering and poverty within the city and surrounding camps was demonstrated by the occurrence of one of the world's worst cholera epidemics during the height of the Rwanda crisis in 1994[1] (Baxter et al., 2002).

The risk posed by the volcano in 2002 was not unknown. In 1991, Nyiragongo was designated a Decade Volcano, one of 16 such volcanoes chosen by IAVCEI. Night glows from the lava lake are visible from inhabited areas and the 2002 eruption was preceded by many months of precursory activity monitored by Goma Volcano Observatory scientists. However, the pre-existing humanitarian crisis and lack of formal institutions made any kind of emergency management and warning extremely difficult. On 18 January, 2002, with armed rebels in control of the

---

1 At this time approximately 800,000 Rwandan refuges sought sanctuary in and around Goma (Wisner et al., 2004).

city, lava flowed through the heart of Goma. Up to 160 were killed,[2] the homes of 120,000 people were destroyed, and 80% of the city's economic facilities were destroyed (Wisner et al., 2004; Baxter et al., 2002; Verhoeve, 2004). Approximately 75% of the population evacuated, mostly to nearby Rwanda, but returned within days when it became possible to cross the cooled lava safely (Baxter et al., 2002). The residents' rapid return to the still dangerous, largely destroyed, and overcrowded city was to protect their property from looting and bigger fears for their safety in the Rwandan relief camps (Wisner et al., 2004).

The local people in Goma continue to face a multitude of risks in everyday life, many of which may be more significant than the imprecisely defined threat of volcanic activity. Years of conflict in the DRC had reduced Goma to a lawless city without functioning institutions, making disaster management almost impossible.

This provides an end-member demonstration of the important role that pre-existing social, economic, and political factors (Figure 69.1) play in a disaster, many of which cannot be addressed on the timescale of a volcanic crisis. Nonetheless, scientific information that reduces uncertainty or forecasts zones of inundation has a role to play in reducing the immediate impacts of volcanic activity such as this. Allowing volcanological knowledge to feed into the roots of longer term actions primarily aimed at improving governance, reducing conflict and vulnerability (in this instance) could help to improve effectiveness of response during a future crisis.

4.4. Case Study 4: Mayon Volcano 2006—Risk Reduction through Resettlement: an Example of Shared Decision Making and livelihood Diversification

Resettlement, adaptive building design, and **livelihood** diversification can all significantly reduce volcanic risks. Permanent resettlement can be challenging, and numerous examples from the literature document how it can increase vulnerability (Usamah and Haynes, 2012). In some cases, volcanic hazards may devastate a community so completely that resettlement becomes the only option. On 30 November, 2006, severe lahars were generated around Mayon volcano in the Philippines by the passage of a typhoon. About 1400 people were killed and thousands were left homeless and destitute (Usamah and Haynes, 2012). The predominant group impacted was farmers, who relied on the fertile soils close to the volcano. The area is

densely populated, with almost half of the population living below the poverty line. Impacts were further heightened due to the passage of an earlier typhoon, which had partially filled drainages leading from the volcano. Warnings were also limited because the earlier typhoon had caused power outages in many communities and inhibited the communication capabilities of the local emergency management agency.

Following the disaster, the local government and numerous development agencies collaborated to permanently relocate the high-risk communities and approximately 1160 houses were built (Usamah and Haynes, 2012). The highest risk and most vulnerable households were relocated first. The government conducted public consultation through community meetings and seminars. A risk assessment of the new site was undertaken by the government to ensure that the risks posed by geological and meteorological hazards were low. The new buildings are designed to reduce the risks from strong winds and ash loading.

Overall, research has demonstrated that people feel much happier and safer at the resettlement site, although a number of challenges exist (Usamah and Haynes, 2012). Residents noted that their involvement in decision-making processes was minimal and that homes were not designed with cultural needs in mind. Most importantly, resettlers did not think enough had been done to ensure livelihood sustainability. With limited skills, funds, and support to diversify, the majority of the resettlers returned to their old homes to farm and look after livestock, often staying overnight. Although this was not an intended outcome, and lives and livelihoods remain at risk, it does allow fertile farmland to be cultivated while providing safe homes during times of heightened volcanic activity. One positive livelihood outcome at the resettlement site has been a microfinance scheme popular with women who had begun grocery or handcraft businesses (Figure 69.1 "livelihood diversification"). Previously it was uncommon for women to engage in economic activities separate to their husbands and family.

Although the resettlement program has not been an outright success, it has certainly reduced risks and promoted greater gender equality. Future schemes must do more to involve residents in decision-making processes and ensure that livelihoods are secure.

This case study demonstrates the following:

1. The value of a long-term collaborative approach (between scientists, engineers, local government, and development agencies) to problems induced by volcanic activity.

2. That balanced or satisfactory outcomes to the problems posed by hazardous activity often contain elements of compromise between all parties (i.e., at Mayon farming

---

2 The exact numbers killed, especially among those who lived on the slopes close to the volcano, are unknown.

continues on the volcano's slopes but farmers have safe, alternative accommodation during an acute crisis).

3. Monitoring or evaluation of these collaborative approaches can document unintended consequences of mitigative measures (negative and positive) and in so doing improve future analysis and mitigative actions.

4.4.1. Concluding Remarks

This chapter illustrates the means by which social, cultural, and political processes act to modify outcomes for populations impacted by volcanic activity. Part II has highlighted some of the key drivers of social vulnerability in disaster causation while Part III has explored these issues through four volcanic case studies. The outcomes described can be clearly traced to underlying issues of disadvantage in some cases but also serve to illustrate that pressures from social, cultural, and political forces can modulate (in both a negative and positive manner) volcanic risk in any society.

Where volcanic outcomes are more positive they serve to illustrate the benefits that vulnerability reduction and the proactive sharing of risk information can bring to difficult and often uncertain volcanic crises acting to improve the **resilience** of affected populations. Disaster risk reduction, which aims to address underlying vulnerabilities and enhance preparedness, is growing to complement the more traditional activities of emergency management and disaster response. In terms of volcanic risk reduction, this involves augmenting the reactive approach that relies on emergency services, evacuations, and rehabilitation with the more complex but worthwhile task of proactively integrating monitoring with community development. While these case studies provide important lessons and insights, detailed studies are still relatively few and far between and the study of the social modulation of volcanic risk is a field of research that is ripe for significant new advances in the coming years.

FURTHER READING

Barclay, J., Haynes, K., Mitchell, T., Darnell, A., et al., 2008. Framing volcanic risk communication within disaster risk reduction: finding ways for the social and physical sciences to work together. In: Liverman, D.G., Pereira, C.P.G., Marker, B. (Eds.), Communicating Environmental Geoscience. Geological Society Special Publication, vol. 305. Geological Society, London.

Baxter, P., Allard, P., Halbwachs, M., Komorowski, J.-C., Woods, A., Ancia, A., 2002. Human health and vulnerability in the Nyiragongo volcano eruption and humanitarian crisis at Goma, Democratic Republic of Congo. Acta Vulcanol. 14 (1−2), 109−114, 2002 − 15 (1-2), 2003.

Benediktsson, K., Lund, K.S., Huijbens, E.H., 2011. Inspired by eruptions? Eyjafjallajokull and icelandic tourism. Mobilities 6, 77−84.

Bird, D.K., Gísladóttir, G., 2012. Residents' attitudes and behaviour before and after the 2010 Eyjafjallajökull eruptions − a case study from southern Iceland. Bull. Volcanol. 74, 1263−1279.

Eiser, J.R., Bostrom, A., Burton, I., Johnston, D.M., McClure, J., Paton, D., van der Pligt, J., White, M.P., 2012. Risk interpretation and action: a conceptual framework for responses to natural hazards. Int. J. Disaster Risk Reduct. 1, 5−16.

Haynes, K., Barclay, J., Pidgeon, N.F., 2008a. Whose reality counts? Factors affecting the perception of volcanic risk. J. Volcanol. Geotherm. Res. 172 (3−4), 259−272.

Haynes, K., Barclay, J., Pidgeon, N.F., 2008b. The issue of trust and its influence on risk communication during a volcanic crisis. Bull. Volcanol. 70 (5), 605−621.

Kelman, I., Mather, T.A., 2008. Living with volcanoes: the sustainable livelihoods approach for volcano-related opportunities. J. Volcanol. Geotherm. Res. 172 (3−4), 189−198.

Mei, E.T.W., Lavigne, F., Picquot, A., deBelizal, E., Brunstein, D., Grancher, E., Sartohadi, J., Cholik, N., Vidal, C., 2013. Lessons learned from 2010 eruptions at Merapi volcano. J. Volcanol. Geotherm. Res. 261, 348−365.

Newhall, C., Aramaki, S., Barberi, F., Blong, R., Calvache, M., Cheminee, J.-L., Punongbayan, R., Siebe, C., Simkin, T., Sparks, S., Tjetjep, W., 1999. Professional conduct of scientists during volcanic crises. Bull. Volcanol. 60, 323−334.

Sims, J.H., Baumann, D.D., 1983. Educational programs and Human response to natural hazards. Environ. Behav. 15, 165−189.

Slovic, P., 2000. Perceived risk, trust and democracy. In: Slovic, P. (Ed.), The Perception of Risk. Earthscan.

Usamah, M., Haynes, K., 2012. An examination of the resettlement program at Mayon Volcano: what can we learn for sustainable volcanic risk reduction? Bull. Volcanol. 74 (4), 839−859.

Verhoeve, A., 2004. Conflict and the urban space: the socioeconomic impact of conflict on the city of Goma. In: Vlassenroot, K., Raeymaekers, T. (Eds.), Conflict and Social Transformation in Eastern Congo. Academia Press, Ghent.

Voight, B., 1996. The Management of Volcano Emergencies: Nevado del Ruiz. In: Scarpa, R., Tilling, R.I. (Eds.), Monitoring and Mitigation of Volcano Hazards. Springer, Berlin Heidelberg New York.

Wilson, T., Cole, J., Johnston, D., Cronin, S., Stewart, C., Dantas, A., 2012. Short- and long-term evacuation of people and livestock during a volcanic crisis: lessons from the 1991 eruption of Volcán Hudson, Chile. J. Appl. Volcanol. 1 (2).

Wisner, B., Blaikie, P., Cannon, T., Davis, I., 2004. At Risk: Natural Hazards, People's Vulnerability and Disasters. Routledge, London.

Chapter 70

Volcanic Risk Assessment

Willy Aspinall

Aspinall & Associates, Tisbury, and Bristol University, Bristol, England, UK

Russell Blong

Aon Benfield Asia-Pacific, Sydney, and Risk Frontiers, Macquarie University, Sydney, Australia

Chapter Outline

GLOSSARY

ALARP The principle behind a legal requirement (in the UK) for organizations to reduce risks, to workers and to persons externally, "as low as reasonably practicable." For a risk to be ALARP it is necessary to demonstrate that any cost involved in further reducing that risk would be grossly disproportionate to the benefit gained.

cost–benefit analysis (CBA) In volcanic risk management, CBA provides one comparative dimension for objective decision support by enumerating the costs of losses (life or other assets) and comparing these with the costs of mitigation.

***F–N* curve** This is a probability plot that shows the chance a given number of fatalities (or injured persons) in a population may be exceeded in a given time period due to the impact of a volcanic hazard or hazards (see also societal risk).

IRPA Individual risk per annum is a measure of the chance of a person (usually a hypothetical individual) being killed by a volcanic hazard, expressed as an annualized probability; for some decision purposes, volcanic risk IRPA can be compared with other risk exposure levels, such as due to earthquakes or hurricanes (see also quantitative risk assessment).

mass casualty incident A mass casualty incident is one in which emergency and medical services resources can be overwhelmed by the number or severity of casualties. On a small island, as few as five volcano burns casualties could represent a mass casualty event, for example.

quantitative risk assessment (QRA) This involves making a numerical estimate of the probability that a defined harm will result from the occurrence of a particular event (see also volcanic risk management).

societal risk This is a single measure of the chance of an event that could harm a significant number of people in one incident; given limited emergency response and medical capacities, the likelihood of a "mass casualty" emergency can be inferred from a societal risk analysis. For volcanic events, the range of casualty numbers and probabilities of exceedance can be expressed in the form of an *F–N* curve (see above; see also Quantitative Risk Assessment); while this is an analogue measure of that used for industrial ALARP (see above), criteria for volcanic risk ALARP are not established.

volcanic risk management (VRM) Concerns the recognition of risks due to volcanic hazards, the assessment of their likelihoods of occurrence, and the selection of measures to reduce or remove harmful outcomes; increasingly, VRM makes use of objective QRA measures of risk, such as individual and societal risk exposures, and of decision-informing concepts, such as CBA and ALARP.

1. INTRODUCTION

Although the two themes are inextricably linked, this chapter is about quantitative volcanic risk assessment, not volcanic hazard assessment; the latter is addressed in Part

The Encyclopedia of Volcanoes. http://dx.doi.org/10.1016/B978-0-12-385938-9.00070-5

VII. The present contribution seeks to outline how, in current practice, risk assessment procedures can inform **volcanic risk management (VRM)**. The latter concerns the recognition of risks, the assessment of likelihoods of occurrence, and the selection of measures to reduce or remove harmful outcomes. Ultimately, responsibility for risk mitigation and management lies with government, but also with communities and individual members of society.

The responsibilities of volcanologists in this context are—or should be—circumscribed: their function is to provide scientific advice to decision-makers. Thus it should not be the role of a volcanologist to decide whether to evacuate people (Newhall et al., 1999). Such decisions require consideration of multiple social, economic, and political factors—responsibility lies with the political directorate. This said, it is often the case in a crisis that volcanologists are able to make more informed judgments about risks than persons in positions of civic responsibility, and so are well placed to assist them firsthand.

Within proper limits, how can volcanologists best contribute to risk management? The present chapter focuses on recent developments in several aspects of risk assessment, many of which come from the 1995 to present dome-building eruption of Soufrière Hills Volcano, on Montserrat (Druitt and Kokelaar, 2002; Wadge et al., 2014). That protracted crisis has provided an opportunity, not usually afforded by short-lived eruptions or other natural disasters, to appraise the roles of scientists and of decision-makers in such situations, and to analyze the **quantitative risk assessment (QRA)** process that has run now for most of the eruptive episode (Wadge and Aspinall, 2014). However, this particular case history is exceptional in certain features, and thus may differ from the realities of eruption crises elsewhere.

2. VOLCANIC RISK: LANDMARK ERUPTIONS

Certain eruptions from the second half of the twentieth century onwards represent landmarks in relation to the development of volcanic risk assessment. Table 70.1 lists a

TABLE 70.1 Landmark Eruptions for Volcanic Risk Assessment from Second Half of Twentieth Century Onwards

Volcano	Date	Remarks
Mt Lamington	1951	Violent explosive eruption with major death toll (about 2950 killed)
Tristan da Cunha	1951	Evacuation of whole population from small, remote South Atlantic island
Sakurajima	1955	Ongoing many minor eruptions promote development of well-practiced population response; in effect promoting risk management that involves "living with the volcano"
Mt Agung	1963	Significant death toll from pyroclastic flows (about 1150 killed) and later lahars (additional 200 deaths)
La Grande Soufrière, Guadeloupe	1976	Large-scale population evacuation in a failed eruption
Mt St Helens	1980	Large, economically costly explosive eruption in the western United States; 57 direct deaths
El Chichón	1982	Large death toll (about 1900) and massive release of SO_2 into stratosphere
Galunggung	1982	Eruption demonstrated susceptibility of modern jet aircraft engines to volcanic ash
Nevado del Ruiz	1985	At least 23,000 killed by massive lahar, despite credible lahar maps
Pinatubo	1990	Largest explosive eruption of period, with successful evacuation response saving thousands of lives; huge injection of SO_2 into stratosphere produced detectable lowering in global temperatures
Mt Unzen	1991	12,000 people evacuated; 43 scientists and journalists killed
Galeras	1993	Nine killed (including six volcanologists) in crater by a small dome explosion
Rabaul	1994	Multiple vent explosive eruption, close to Rabaul city
Montserrat	1995	Ongoing small island population living with pyroclastic flow hazards from a protracted and repetitive dome-building eruption
Eyjafjallajökull	2010	Ash cloud from a small eruption massively disrupts air travel and commerce over a wide area for several days
Merapi	2010	About 20,000 people evacuated from within a radius of 10 km, but an exceptional pyroclastic flow traveled farther; more than 350 people died

few selected events: these are not necessarily the biggest eruptions of the modern era and the list is not exhaustive, but these episodes were noteworthy for some of the challenges they presented for risk management, or for advances in volcanic risk assessment.

Against the background of these recent historical losses and demonstrated vulnerability of modern life to even small eruptions—and in the light of plausibly larger future eruptions and greater disasters at major population concentrations—improved techniques for volcanic hazard and risk assessment have been pursued by volcanologists, in parallel with endeavors to achieve better eruption forecasting, risk management, and threat mitigation.

3. VOLCANIC RISK: ASSETS AND VULNERABILITIES

In natural hazard studies, risk is usually expressed as a function of three elements—the probability of the hazard occurring (with given intensity and spatial footprint), the assets exposed to the hazard, and the vulnerability of those assets to the hazard. One of the most challenging aspects of VRM is that there can be a lot of different hazards, some of which can occur simultaneously or sequentially, sometimes all in one place, other times at differing locations around an erupting volcano. Broadly, we can recognize five classes of assets and social capital exposed to the various volcanic hazards: **people** (in terms of lives or personal livelihoods), **economic activity**, **agriculture**, **buildings**, **and other infrastructure**. Some might recognize another broad class of physical asset—the environment—but we have chosen not to consider that special topic here.

These asset classes represent a varied and diverse mix of effects and interactions. The complexities are such that, in practice, it is difficult to construct a single uniform framework that can accommodate all the various combinations, their likelihoods, and all possible consequences in a rational and probabilistically coherent way.

To go with the diversity of volcanic hazards, there is also a wide range in the vulnerabilities (or fragilities) of assets to those hazards. Vulnerabilities can vary, too, and might depend, for instance, upon time of day or season of the year.

We can recognize five broad VRM options for responding to the exposures of asset classes and to the range of vulnerabilities:

1. Accept risk: do nothing—by far the most commonly adopted option. This might mean the risk is not understood or recognized, is accepted or that it is regarded officially as sufficiently low in some sense, such as "as low as reasonably practicable" (**ALARP**—see Glossary).

2. Reduce hazard—almost impossible in the case of volcanic eruptions, except for minor interventions such as draining crater lakes to reduce the lahar hazard

3. Reduce vulnerability (increase resilience)—for example, by evacuating people, strengthening buildings, or constructing levees to route lahars away from the most valuable assets

4. Adapt—for example, by early harvesting of crops or by simply accepting interruptions to electric power or water supplies

5. Transfer risk—to an insurance company, a government, or an international aid organization.

This said, the fivefold division of asset classes and the five risk management options summarize the main possibilities appropriate for a community or a government to consider. Most of these options can be implemented only at official level, but some may also be relevant for personal choices, and some might be acted upon in both contexts, although not necessarily in the same direction! With volcanic hazards, societal tolerability of risk, public policy—and political willingness to invest in mitigation—can conflict substantially with individual responses and preferences (see, e.g., Loughlin et al., 2002).

Most risk management studies focus on negative aspects of risk, but there can be positive sides to volcanic activity, which may bestow benefits or opportunities (for example, a fall of volcanic ash may reduce insect pests more than it damages crops; tourism may flourish after the eruption subsides).

4. RISK METRICS, CRITERIA, AND TOLERABILITY

The aim of VRM is to deal with the risks associated with eruptions and, where possible, to reduce those risks. One important question is how much time, effort, and money should be spent to reduce risk? Is it reasonable to spend $10 million to save one human life? Is it sensible to spend the same amount to reduce the risk of the local school (and community center) being overwhelmed by a lahar? Obviously the answers depend on a range of attributes including consequences, the resources available to the community, political priorities, perceptions of risk, and how the particular risk is regarded, relative to other risks.

The way in which risk is assessed and classified varies from country to country, and also can depend on the relevant scientific domain and its worldview. There are four main approaches to risk assessment. The first is the "technical" or "statistical-probability" approach, founded in science and engineering, and used for risk assessment in some recent volcanic crises. The second is the "economic" or "cost–benefit" approach, which, in our context, has to place a value on a human life (we come back to this issue, later). The third, the "psychological" approach, emphasizes the individual's response to risk, while the fourth, the "sociological" approach, seeks to access group behavior theory, treating risk against and within the context of other environmental and social factors. Critically, understandings and characterizations of uncertainty can differ greatly across these domains—sometimes irreconcilably—and remains an important problem that is too often ignored (as discussed further in Chapter 69).

Human risk metrics for large-scale hazards, such as those associated with major natural disasters, can be contemplated in two basic ways. Although various alternative characterizations have been used for industrial risks (e.g., harmful doses of radiation or chemicals, as well as fatal effects), in the context of volcanic eruptions the usual definition of *individual risk* relates to the probability of a person being killed; up to now, injuries and long-term health effects have been secondary concerns, if considered at all. While criteria for tolerability levels of individual risk exposure may exist in industrialized societies, similar criteria are rarely established for natural hazards, and certainly not uniformly or universally across the world.

In the same context, the situation regarding **societal risk**, as a metric for decision purposes and tolerability considerations, is even more tricky. One significant difficulty is that there is no clear basis (in law) for taking into account society's aversion to multiple fatalities in large disasters (known as "scale aversion," Box 70.1) or how this aspect might be incorporated into a risk management strategy. In some cases, policy is predicated—when making judgments about mitigating societal risk—on whether proposed or suggested measures are deemed grossly disproportionate, economically, in relation to what is reasonably practicable. Ultimately, this is a political, not a volcanological, matter.

Although scale aversion is not measurable in a literal sense, some guidelines are needed as to what additional weighting should be given to extreme disasters in order to allow practical societal risk mitigation (see Box 70.2). That, too, is a problem for politicians, not for volcanologists.

Considerations of tolerability of risk generally entail occurrence probability levels that sit between those associated with risks that are clearly unacceptable and those that are broadly acceptable. However, that range typically includes risk levels that many people are prepared to tolerate individually, in order to secure or retain personal benefit. Some people expect that the nature and level of risks will be objectively assessed, that such findings are made available as public advice and that, in addition, they will be used to determine official mitigation measures (in this regard, the 2009 L'Aquila earthquake disaster comes to mind).

Ideally, both individual risk and societal risk ought to be taken into account when deciding whether a situation is acceptable or not. However, from experience when the danger is from a volcano, it seems that either the one or the other comes to dominate in official thinking, not a joint combination of both.

In industrial risk assessment, residual risk is often ascribed to one of the following tolerability bands:

"Intolerable" risk: if the risk is in this category (whether for individual or societal risk), then ALARP cannot be demonstrated and action must be taken to reduce the risk, irrespective of cost.

"Tolerable if ALARP" risk: if the risk falls in this category, then a case-specific ALARP demonstration is required, proportionate to the level of risk.

"Broadly acceptable" risk: if the risk can be shown to be in this category, then the ALARP demonstration may be based on adherence to codes, standards, and established good practice.

Figure 70.1 shows these three bands of risk tolerability and their risk thresholds of death per annum for an individual (see Box 70.2 for an illustration of this concept applied to societal risk tolerability). Note that the expectation is that the worker, in pursuit of his or her livelihood, will tolerate a higher threshold level than a member of the public.

A similar concept for presenting risk exposures to decision-makers and the public was adopted, for a time, in Montserrat during the eruption ongoing since 1995 (Figure 70.2). However, this graphical format was criticized for appearing too complicated for lay consumption and, subsequently, a simpler layout was adopted, which omitted the risk category descriptors and the triangular depiction of elevated risk.

These volcanic risk level estimates were determined by formalized QRA using a probabilistic methodology, adjunct to normal observatory activities such as activity monitoring and interpretation.

BOX 70.1 Scale Aversion

Governments and their agencies usually seek to achieve risk-neutral attitudes in decision-making. However, for low-probability high-consequence events, many decision-makers tend to be risk averse because of the potentially catastrophic nature of the hazard. "Scale aversion" is a term defined as "the tendency to want greater protection where consequences are high …," focusing on the notion that people are more averse to large-scale disasters than more frequent but smaller losses (Health & Safety Executive, 2009). In cost—benefit or integrated risk analysis approaches, scale aversion can be implemented in calculations as a power-law exponent, which modulates the numerical evaluation of impact.

BOX 70.2 Risk Integral, Tolerability, and Volcanoes

For assessing industrial disaster risks, the UK Health & Safety Executive uses a concept called the "risk integral," which, in decision theory, represents one form of disutility function (i.e., cost equation). A measure of societal risk exposure can be derived by calculating the annualized expectation value (EV) of the number of fatalities in an accident—in essence, EV is the sum of all possible toll numbers from 1 to some maximum N_{max}, multiplied by the probability of occurrence. This produces an overall number that can be compared with criteria determined on the same basis. There are various versions of this measure, including the approximate risk integral (ARI) (Hirst and Carter, 2002), which incorporates a scale aversion factor (see Box 70.1). Here, the basic EV concept, without aversion

In the cases of the past volcanic disasters in the Table, with known death tolls, the event probabilities (per day) were not estimated a priori, so illustrative values are used here (marked with a ?). The corresponding volcanic EV number is given, and related to the industrial *intolerable* threshold by a ratio; the volcanic event probability that would match this threshold is also given, recognizing that a quite different threshold value might be invoked for volcanic risks. The table also shows volcanic EVs for a hypothetical future major volcanic disaster, with an arbitrary but not implausible total of 100,000 deaths; the first example is based on a long-term event probability for the volcano when it is in a quiescent state, and the second when event probability is elevated because of significant unrest.

Event	Eruption Probability	N_{max}	Volcanic EV	Volcanic EV/Industrial ARI Threshold	Event Probability for ARI Intolerable Threshold
Montserrat (June 25, 1997)	?10^{-2} per day	19	2.7×10^8	540	10^{-6} per day
Mt St Helens (May 18, 1980)	?10^{-3} per day	57	1.1×10^8	211	5×10^{-6} per day
Mt Pelée (May 8, 1902)	?10^{-3} per day	29,000	1.3×10^{11}	250,000	4×10^{-9} per day
Hypothetical major eruption state: quiescent	?5×10^{-5} pa	100,000	6×10^7	120	1.5×10^{-7} pa or 10^{-10} per day
Hypothetical major eruption state: unrest	?10^{-6} per day	100,000	4×10^8	895	10^{-9} per day

scaling (i.e., the ARI aversion power-law factor $a = 1$), is illustrated for three volcano case histories, and for an extreme hypothetical future major eruption, such as a caldera-forming event, occurring in a large populated area (see Table). The tabulated *Volcanic EVs* are compared numerically with ARI = 500,000, the value at which an industrial risk is considered to become *Intolerable*.

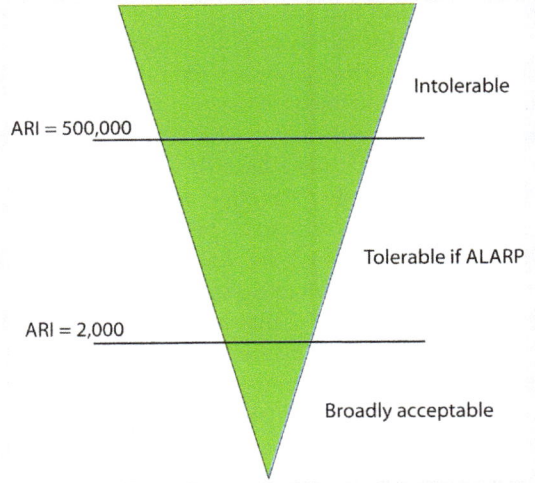

UK industrial *ARI* societal risk tolerability thresholds.

These approximate numbers are not intended to be in any way definitive or authoritative, or suggestive of possible criteria—their purpose is to illustrate that, with a volcano, a risk manager could be decision-making in the face of event occurrence probabilities that are fundamentally miniscule and, even when restlessness is manifest, any short-term increase in event probability may be very difficult to determine with any confidence. Worse yet, if scale aversion were compounded into the equation (e.g., $a > 2.5$, for massive casualty tolls), the event probability picture easily becomes orders of magnitude more extreme, relative to "ordinary" societal risks.

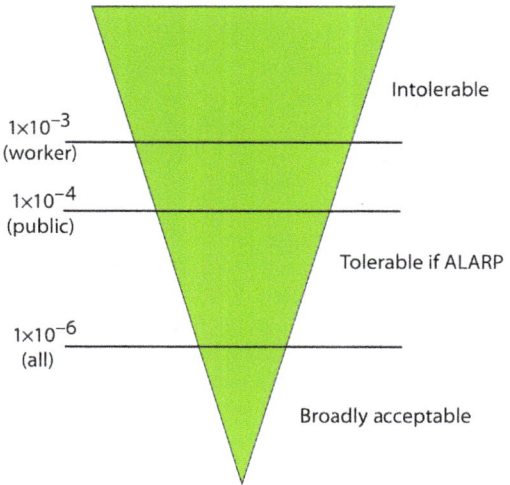

FIGURE 70.1 Industrial risk tolerability criteria for individuals. ALARP, as low as reasonably practicable.

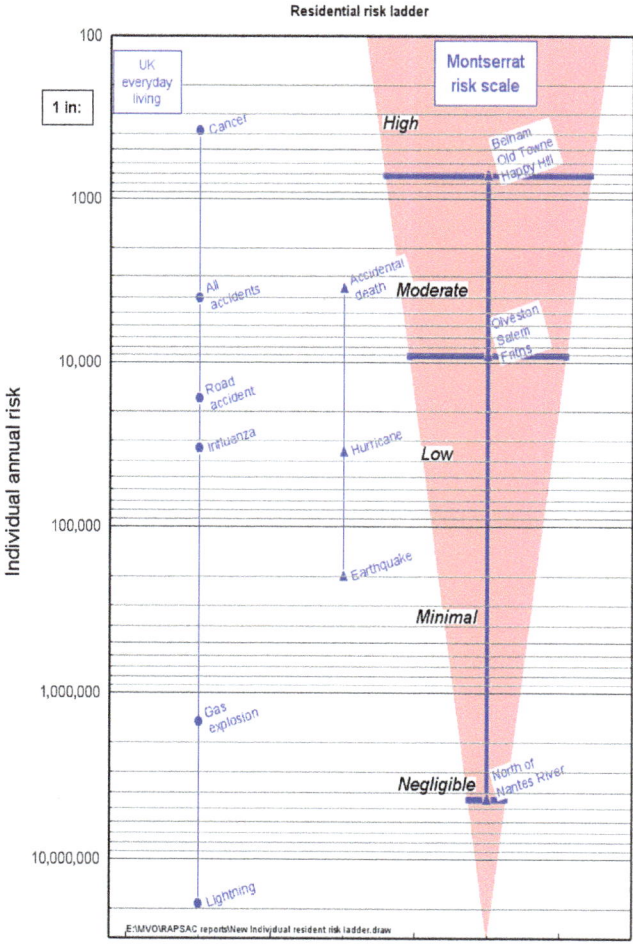

FIGURE 70.2 Risk ladder representation of volcanic risk exposure (shaded area) of individual residents in different localities on Montserrat in 2008, with comparisons to everyday risks (available for UK) and accidental death and long-term hurricane and earthquake risks in Montserrat.

5. CHARACTERIZING VOLCANIC RISK

There are a variety of ways to represent risk exposure and risk levels to decision-makers, the media, and public. Here we outline the two main methods used in current practice.

5.1. Risk Matrices

One commonly used approach for articulating different risk levels, an intermediate stratagem between qualitative appraisal and full QRA analysis, is the "risk matrix." In this paradigm, risk is interpreted as the combination of consequence (severity) and likelihood (frequency), with the risk matrix enabling different combinations to be represented graphically as a method for visualizing the spread of risk. The main advantages claimed for it are its easy depiction of different risk levels, and minimal amount of compilation time compared with a quantitative analysis. A graphical layout is promoted in early versions of the AS/NZS ISO 31,000 Risk Management Standard, in the form of a simple "traffic light" color-coded likelihood/consequence risk matrix, as shown in Figure 70.3.

However, the basis of risk estimation for the matrix is usually qualitative, although it can be semiquantitative (for either consequences or frequencies, or both). This version recognizes five levels of likelihood and five levels of severity of the expected consequences, but the actual number used can vary with the needs of the study. Cells are then categorized, with E implying "Extreme risk, immediate action required"; H "High risk, attention needed"; and so on, each category to be managed by preordained or routine procedures.

In Figure 70.3 the traffic light colors do not correspond uniformly to the E, H, M, and L categories but follow Cox (2008), who argues that any configuration of ratings other than that shown by the three colors is likely to result in illogical rankings. That study has identified serious mathematical defects and inconsistencies in most risk matrix formulations: different risk assessors may assign vastly different ratings to the same hazard; scatter and variability can be high, even after lengthy deliberations, and the underlying drivers of disparate ratings relate to fundamentally different worldviews, beliefs, and a panoply of psychosocial factors, seldom explicitly acknowledged. Not least among the issues, risk matrices have a limited ability to reproduce risk ratings estimated in formal quantitative models because their definitions and scales are usually geared to risk management planning and risk governance at different levels of government. As a consequence, risk matrices often are so generic they have questionable operational value in a crisis and, moreover, do not easily convey risk-shifting time frames, which, short-term, can change dramatically for certain hazards.

Likelihood	Consequences				
	Insignificant 1	Minor 2	Moderate 3	Major 4	Catastrophic 5
A (almost certain)	H	H	E	E	E
B (likely)	G	H	H	E	E
C (moderate)	L	M	H	E	E
D (unlikely)	L	L	M	H	E
E (rare)	L	L	M	H	H

FIGURE 70.3 A simple Likelihood/Consequence Risk Matrix based on AS/NZS ISO 31,000 Risk Management Standard, with categories denoted by: E, Extreme; H, High; M, Moderate, and L, Low (see text). The colors, which do not correspond uniformly with risk letters, and the significance of such schemes are discussed in text.

Taking a longer-term perspective on volcanic hazards, the MIAVITA (2012) threat matrix, shown in Figure 70.4, endeavors to overcome some of these difficulties. The "frequency scale" is semilogarithmic with numeric equivalents assigned to each qualitative description. Similarly, the intensity scale uses qualitative descriptions such as "high," which refers to a maximum human impact of severe injuries potentially leading to deaths. In Figure 70.4, frequency and intensity values are multiplied together to give a "threat matrix" score, where "negligible hazards" (or risks) are shaded gray, indicating no associated recommendations for normal permanent human settlements; "low" (yellow) implies no specific associated recommendations but vigilance is required; "moderate" (orange) suggests permanent human settlement is possible but with specific recommendations; "high" (red) indicates permanent human settlement is inadvisable unless major precautions are taken. "Very high" (dark red) proposes that permanent human settlement should be avoided (MIAVITA, 2012, pp. 101−102).

While MIAVITA's Threat Matrix is a substantial advance on the qualitative matrices of many civilian risk schemes, it still suffers from using verbal descriptors such as "numerous deaths" and "high financial losses." While such "fuzzy" terms are intended to acknowledge the reality of uncertainty in many risk assessments, their use often makes difficult realistic comparisons or risk rankings.

Another problem with Figure 70.4 is that it also expresses resulting risks as numbers, and thus may appear unduly quantitative for many people. As a consequence, it can have limitations as a communication tool. Lastly, use of the risk matrix concept for decision purposes, whether in a crisis or for planning purposes, may entail some, as yet untested, legal ramifications.

Table 6: : Threat Matrix Resulting from Crossing Intensity and Frequency Baseline Scales

Frequency (Q_F)	Intensity (E_1) 0.5	2.5	15	60	100
0.01	0.005	0.025	0.15	0.6	1
0.02	0.01	0.05	0.3	1.2	2
0.1	0.05	0.25	1.5	6	10
0.2	0.1	0.5	3	12	20
1	0.5	2.5	15	60	100
2	1	5	30	120	200
10	5	25	150	600	1000

FIGURE 70.4 Volcanic Threat Matrix proposed by MIAVITA (2012, Table 6 at p. 102).

5.2. *F*−*N* Curves

In order to determine the statistical "expected" number of fatalities in a given time interval, societal risk is often expressed quantitatively as an ***F*−*N* curve**. The cumulative exceedance number of fatalities N that can result from a major incident—ranging from zero or one to some maximum value N_{max}—is related to the frequency or probability F with which that number of fatalities is estimated to occur. This relationship is usually presented graphically, which is useful for comparing situations from different periods with changing activity levels, and for assessing potential "mass casualty" scenarios for preparedness purposes.

Idealized *F*−*N* curves are occasionally proposed as criteria for judging the tolerability of societal risk from an installation or for national safety criteria. Plots such as Figure 70.5 are adopted, delineating three regions within which the risks are described as "unacceptable," "tolerable if ALARP," and "broadly acceptable." However, there are different ways of using such criterion lines, and the intended method of use often is not stipulated.

One way of using a risk criterion line would be to require that for a situation to be acceptable its *F*−*N* curve should be everywhere below the criterion line. Alternatively, it might be sufficient that the *F*−*N* curve is below the criterion line in some integral sense, allowing it to be above the line in places, provided it is sufficiently far below the line elsewhere.

The first way of using the criterion line is the simplest—as an absolute barrier—but it has been shown

(Evans and Verlander, 1997) that this requirement can lead to decisions that appear unreasonable and, more seriously, to some decisions that are inconsistent across a range of similar risks. For a volcanic risk example, Figure 70.6 is a plot of an *F*−*N* curve from Montserrat, together with two companion curves showing how societal risk can be reduced by selectively evacuating different areas.

Uncertainty analysis of the Monte Carlo societal risk modeling in this case furnishes confidence bounds on these alternative volcanic risk *F*−*N* curves (Figure 70.6), indicating that real, and meaningful, risk reductions could be achieved by a strategy of selective location evacuation.

It may also be noted that, in contrast with typical *F*−*N* plots for industrial risks, which are generally linear, the Montserrat societal risk curve exhibits several humps. These arise because an erupting volcano can have a variety of ways of posing a threat to a population, with differing levels of hazard intensity and spatial extent potentially causing disparate total magnitudes of casualty numbers, all with different probabilities of occurrence. Such complexity in the *F*−*N* curve serves to make decision-taking about public safety very much more challenging than in the industrial risks case, especially when the risk findings are

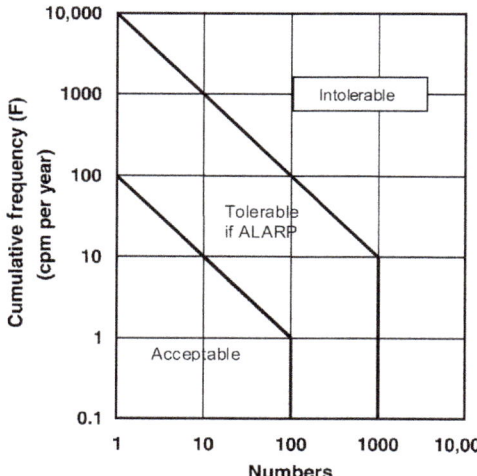

FIGURE 70.5 An *F*−*N* plot depicting societal risk tolerability criteria (e.g., for UK hazardous installations, Hirst and Carter, 2002). The vertical axis has units of "cpm per year," that is, "chances per million per year," sometimes expressed in terms of annualized probability of occurrence. "Numbers" on the horizontal axis usually refers to deaths or injuries. ALARP, as low as reasonably practicable.

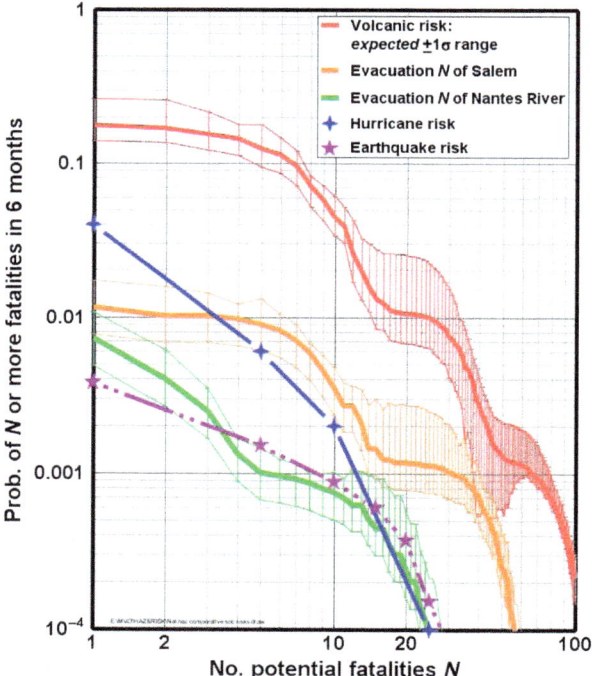

FIGURE 70.6 Montserrat societal risk (*F*−*N*) plot showing mean risk curve in May 2003 (continuous red line) and effects on risk levels if people were evacuated to north of Salem (orange curve) or to north of Nantes River (green). Also shown are the ±1 standard deviation uncertainty spreads on these curves, estimated by the Monte Carlo method, and indicative estimates of hurricane and earthquake risk levels in Montserrat, based on historical experience. *Reproduced from Wadge and Aspinall (2014), with permission: © Geological Society of London.*

convolved with the substantial scientific uncertainties associated with volcano forecasting.

For comparison, rudimentary $F-N$ curves for hurricane and earthquake risk on Montserrat are also shown on Figure 70.6. The latter are based on historical experience and not event forecasting, as is the case for the volcanic risk exposure estimates. It is very likely, for instance, that societal risk exposures to future hurricanes and earthquakes are both reduced nowadays due to the higher building standards that have been adopted following the impact of Hurricane Hugo in 1988.

While volcanic risk $F-N$ curves have been applied almost exclusively with fatalities as variable, the concept could be extended readily to apply to injuries and to effects of chronic exposure, and also to threats to physical assets, such as buildings and critical infrastructure.

6. COST—BENEFIT ANALYSIS IN VRM

On their own, hazard event probabilities are somewhat abstract and do not easily translate into immediate, readily comprehended factors for decision-making. For generic risk management, **cost—benefit Analysis (CBA)** is seen as providing a more explicit comparative dimension for decision support by enumerating costs of risk (in terms of lives or other assets) against costs of mitigation.

In recent work, Marzocchi and Woo (2009) have developed the argument for utilizing CBA for VRM, demonstrating the way in which the trade-offs involved in taking mitigating action in the interests of public safety can be studied within an economic decision framework (Box 70.3).

Viscusi (1992) has reviewed situations where the avoidance of fatalities is a sensitive public policy issue.

BOX 70.3 Principles of Cost—Benefit Analysis

Cost—benefit analysis may be used in a situation where a decision-maker has to choose between two actions: either (1) protect or (2) do not protect. The cost of protection is C. In the absence of protection, a loss L is incurred, which exceeds C, if an adverse hazard state arises. Let the probability of the adverse hazard state arising, within a specified time window, be denoted by P. If the expected expense is to be minimized, then the optimal policy is

To protect, if $P > C/L$;

Not to protect, if $P < C/L$.

The minimal expense is then $min\{C, PL\}$.

In a volcanological context, protection may involve evacuation, which carries a cost of C. The adverse hazard state here is one of volcanic eruption, for which a decision not to protect carries a large loss penalty of L, measured in human fatalities.

A quantitative CBA requires the evaluation of costs and benefits in the same common unit, the most convenient being monetary value. A standard practical approach (Viscusi, 1992) is to determine the amount of money authorities are willing to pay to save a human life. While there is natural public antipathy to the notion of placing monetary value on human life, this value is indirectly measurable by the "willingness to pay," to avoid danger.

As an example, Marzocchi and Woo (2009) consider the important mitigation action that has been taken by authorities to reduce long-term risks at Mt Vesuvius (VesuVia, literally meaning "away from Vesuvio"). Based on the historical experience of the past several millennia, the likelihood of a long-term resident of the Red Zone dying in a Vesuvius eruption is of the order of 0.01 (see Appendix 1 of Marzocchi and Woo, 2009). They show that the willingness to pay to save a human life in this Italian context is the ratio between the money offered to people to move away from the Red Zone (8600 Euros per person) and the probability of the deadly event (0.01), i.e., about 860,000 Euros. Allowing for an uncertainty margin in the underlying assumptions of threat level and future life expectancy, a lower bound value for a life might reasonably be taken to be 1 million Euros.

While this monetary worth is consistent with values used in other risk studies in affluent industrialized nations, the financial offer by the Italian government was, seemingly, *not* based on formal CBA. For a purely political judgment, the correspondence between it and the quantitative CBA is remarkable. This said, in reality the risk decision calculus is more complicated: for instance, the long-term cost of dealing with severely injured, long-term disabled persons is vastly more than the nominal value of a life lost.

From a decision-maker's perspective, CBA may be viewed as a logical framework to define key probability thresholds. Evacuation is by no means the only difficult decision which CBA can help rationalize. At lower levels of threat, CBA can justify preparatory measures in advance of a crisis: these might include actions such as putting in place transportation logistics and emergency services preparedness; and, for long-term purposes, providing a reasoned basis to justify land-use planning and to refine the geography of emergency planning.

The adoption of a systematic procedure using the CBA concept could provide a transparent audit trail for decisions, which should reduce the degree of subjective judgment—and the temptation to furnish such opinions—in volcanological communications with civil authorities. Skeptics, however, might criticize the approach by asserting that CBA is simply rebranding what is (in the beginning) a qualitative expert judgment process: determining the critical "value" parameter is based on some subjectivity as to what is considered, and what is not. In other words,

CBA dresses up a subjective process in quantitative clothing.

Despite volcanic risk having been defined quantitatively more than 30 years ago, more often than not managing the repercussions in any particular set of circumstances has proceeded without an effective way of measuring risk. Marzocchi and Woo (2009) argue that recent progress in quantifying eruption probabilities paves the way for a new era of rational risk management, using "volcanic risk metrics." They put forward basic principles, in effect coupling probabilistic volcanic hazard assessment and eruption forecasting with CBA. This approach offers a strategy for rationalizing decision-making across a broad spectrum of volcanological questions. When should the call for evacuation be made? What areas should be covered by an emergency plan? During unrest, which zones around a volcano should be evacuated, and when? A set of quantitative and transparent decision rules can be established in advance of a crisis.

While such analyses do not provide definitive risk tolerability criteria, the decision principles should make the comparative merits of different choices more comprehensible to civil authorities. In principle, this would be applicable to all asset classes at risk, not just human lives. Ideally, the merits and detriments of such decision choices would benefit from public deliberation at an early stage—all societies find it necessary to strike a balance between what is desirable and what is affordable and feasible.

6.1. Societal Risk CBA

The essence of societal risk estimates (e.g., $F-N$ curves, above) can also be expressed as a statistical value: that is, the "expected" number of fatalities in a given time interval. Typically, for a period of 3 months (which is the sort of time interval population displacement measures might have to be in force), the statistically expected number of fatalities in the danger zone among a population of 1000 people might be close to 1, if the probability of the hazard is 0.001 over the same period.

For most aspects of life, public safety policy requires an assessment of the costs of reducing the chance of fatal accidents, implying some nominal value can be ascribed to the prevention of a "statistical fatality" or, its more acceptable equivalent, achieving a "statistical life saved." An alternative perspective is to look at a minimal decision-specific equivalent value of a life, as indicated by other, accepted, economic loss measures. Although criticized, the usual indicator for socioeconomic loss is the per capita average annual gross domestic product (published in World Bank reports and expressed in terms of US dollars). As with road safety measures, willingness to pay for wider public safety measures varies quite widely from one risk context to

another, and from one country to another. An example CBA calculation of this type is given as Example 1 in Box 70.4. Of course, more costly measures can be invoked, according as the value attributed by the authorities to preventing a statistical fatality.

As mentioned in Box 70.1 and in the previous section, societal aversion to multiple casualties generally increases with the total number of lives lost, following some power-law factor. This implies that consideration has to be given to the value of preventing multiple fatalities—and hence investment in protective measures—that significantly exceeds any lower bound figure for the monetary worth an individual life.

Ultimately, which mitigation measures to apply are decisions for the civil authorities; volcanologists help inform that decision process by providing quantification of probabilities and expected casualty numbers. Both factors can be input to cost–risk analysis in the ways illustrated here, and can track how volcanic hazard factors vary through time with the state of the volcano, and how

---

BOX 70.4 Societal and Individual Risk CBA

Example 1: case for evacuating 1000 residents from a given locality due to volcanic risks for 3 months.

If the hazard probability level in 3 months is assessed such that the statistical expected (mean) number of fatalities in a population of 1000 is 1, and the cost of evacuation were (as little as) $250,000 then it appears that evacuating 1000 people for 3 months would be unequivocally justified against a yardstick of $1 M for a statistical life saved (even in circumstances where the "expected" number of fatalities averted is small). If a higher value of life is appropriate, the case is even more compelling.

Example 2: balance of economic worth to an individual of not evacuating from danger for 3 months.

Say, for a hypothetical person in a danger zone, the probability of being killed is 1 in 4000 in a 3-month period (commensurate with the previous example, where societal risk was the issue). If the person decides not to leave, this is equivalent to saying that he is not prepared to pay a notional "life insurance premium" of 1/4000th of (some) value of his or her life. If, to provide for his or her family in the event of his or her death in normal circumstances, the person is willing to buy commercial insurance for dependents' protection in the typical sum of $500 k on his or her life, then declining to be evacuated for 3 months is tantamount to saying he or she is not prepared to pay a notional premium of $500 k/ 4000 = $125 to avoid death by being temporarily displaced. If he or she chooses to stay in danger it appears to be "worth more" in economic terms to this person than having their life and their dependents' income protected by way of a one-off payment of $125 (which cost might take the form of increased living expenses due to evacuating, say).

population numbers vary. Given their knowledge base, volcanologists are perhaps uniquely placed to carry out the computation of the hazard-related part of a CBA calculation. In principle, CBA can provide an objective basis that could contribute to societal risk decision-making, but care is needed to ensure that the separate roles and responsibilities of calculator and decision-maker are not mixed up.

6.2. Individual Risk CBA

While societal risk is undoubtedly important for a political directorate, individual risk can be a more influential measure for expressing exposure: it is the means by which many people—but not all—base their own responses to assessed threats. In Montserrat, for instance, individual risk per annum (**IRPA**) is a quantitative risk measure that has been used extensively, expressed as personal odds of death due to the volcano (e.g., "1-in-1000" per annum). This metric was supplemented originally by an associated descriptive medical risk scale (e.g., "low" risk), then by an alphabetic, nonadjectival version (e.g., "risk category C"). Another innovation has been to estimate a numerical risk increase factor, which indicates the increase in an individual's background annual risk of accidental death, such increase being due to volcanic hazard exposure (e.g., "4.6× background"). Publishing comparative IRPA equivalents for familiar everyday hazards and occupations, relevant to the country and society concerned, help people understand such personal metrics of volcanic risk.

As with societal risk CBA, personal risk choices can also be framed in terms of a monetized analysis of cost or detriment versus benefit. Example 2 in Box 70.4 illustrates a cost–benefit calculation of this type However, life insurance is just one possible cost–benefit yardstick—most individuals would, inevitably, value their own life and its preservation at much more than some level of insurance cover. Being obliged to move away from home in an evacuation induces different degrees of stress and worry in different people; for some it can be considerable, particularly if the accommodation they are forced to live in, even temporarily, is much less comfortable than their own home. Such people would put a high additional value on not having to undergo that stress, even if evacuation had a guaranteed end after 3 months.

The pair of simple numerical examples in Box 70.4, which are indicative, not normative in any sense, illustrate how CBA principles might be applied with the kinds of volcanic hazard probabilities and quantitative risk estimates that can be provided by volcanologists before or during an eruption crisis.

These complementary examples show how a gap can open up between an individual's perception of the risk–cost–benefit inequality, and typical governmental or

political attitudes to the prospect of significant loss of life. The nettle which the politicians need to grasp is that of deciding what levels of risk are acceptable, tolerable, and intolerable to a government and to people living near a volcano. Volcanologists can devise credible scenario definitions and perform the quantitative risk analyses, but the ultimate tolerability issue is a sociopolitical one.

Traditionally, volcanologists have focused their risk interests and active advice roles on minimizing immediate loss of life from eruptions. Thus nearly all the foregoing discussion has been concerned with quantifying potentially lethal risks associated with volcanic hazards. Clearly, acute or minor injuries to persons can be regarded as losses, too, in much the same way as damage to other assets. Some aspects of mitigation against hazards that are not life-threatening are starting to receive attention from policy makers.

7. RELATIVE RISKS AND RISK MANAGEMENT FOR NONLETHAL HAZARDS

Nowadays, governmental and political anxiety in some countries is expanding to include the potential for wider social and other impacts due to volcanic activity. For instance, following the 2010 disruption to European air travel caused by the Eyjafjallajökull eruption, the UK government has added volcanic activity to its National Risk Register and its civil contingencies planning; potential scenarios include the prospect and societal implications of a repeat of the terrible Lakagígar 1783 eruption.

As a generalization, populations and communities—and insurers—are less and less inclined to accept the consequences of a volcanic eruption as an "act of God" or an unpredictable "force of nature." Responsibility, blame, and even culpability for "losses" are more readily ascribed to key actors *post hoc*. Here, briefly, are some elements in this emerging domain of concern for nonlethal VRM.

Livelihoods and the economy: Widespread disruption to ordinary life can have significant economic impacts and potentially could compromise incomes, employment, cause business interruption, and halt productive activities such as agriculture or manufacture. A minor example, but disastrous for the employees in Montserrat, was an electronics assembly plant forced to close by low concentrations of ash in the air.

Buildings and structures: A comprehensive model to assess the impact of a range of different volcanic hazards on building structures, the first of its kind, has been developed for Vesuvius and Naples; the implications, and how to handle them, are still being worked out.

Public health: Certain types of eruption, e.g., a gas-rich episode like Lakagígar 1783, have the potential to affect people's health deleteriously in various ways across a very wide area, potentially shortening lives and increasing demands on health services.

Infrastructure: As experienced in NW Europe in 2010, even quite small eruptions can impact utilities regarded as essential for modern life (e.g., power generation and distribution; water supplies; communications; transport; travel). Military systems and logistics are another, perhaps less obvious, branch of life that may be susceptible to volcanic eruption effects.

Social unrest: It is not too far-fetched or alarmist to be concerned about a breakdown in social fabric if a volcanic eruption and its effects intrude massively into day-to-day lives for a prolonged period. In many developed countries the chain of food and other staple supplies is often very short, only a few days, and a sudden absence of stock on supermarket shelves or fuel at pumps could easily engender panic buying, civil commotion or worse.

Contingency planning for such prospective scenarios, and their management, would benefit from the best possible and most informative assessment of hazards and risks; better data, information, and robust QRAs are needed to inform and guide appropriate strategies.

In work to extend volcanic risk assessment to include infrastructure and other effects, in addition to life-threatening risks, Magill and Blong (2005) described a methodology for determining relative risk for a range of volcanic hazards expected in Auckland, New Zealand. The ranking relies on a thorough understanding of the record of past eruptions affecting Auckland (including volcanoes more than 200 km distant) to define hazard frequency and hazard extent. The expected consequences for two sets of asset—human life and buildings—when exposed to each hazard required a set of vulnerability estimates. Based on a meta-analysis of earlier studies, a human life was valued at US$2.2 million (see also CBA, above), giving an aggregate asset class value for the human "inventory" approximately

50 times greater than that of the building inventory in the Auckland region.

Figure 70.7 shows the relative risk rankings for the 11 hazards considered, emphasizing that tephra fall and base surge are easily the most important risks. Notably, the probability of a distal volcanic event impacting the Auckland region was judged to be more than four times the probability of an eruption from the local Auckland volcanic field itself (Figure 70.7).

Figure 70.8 provides an example of tephra fall thickness exceedance probability curves for nine capital cities in the Asia—Pacific region. The probability of an eruption volcanic explosivity index ≥ 4 producing an ashfall in a city is based on Smithsonian Institution records for all volcanoes in the Asia—Pacific region and 1000 simulations of eruption characteristics, eruption column height, tephra volume, and wind directions at multiple levels in the atmosphere for each of the 190 volcanoes in the region (Jenkins et al., 2012). The curves can be read in the following way, for example: for Jakarta a tephra fall of 10 mm or more can be expected at least once in 1000 years on average, over time. Over a similar time frame, Tokyo can expect a fall of 100 mm, once on average.

Figure 70.8 compares only ashfall hazard in the nine cities. Jenkins et al. derived an estimate of exposure for 18 cities, based on the mean urban population density within 10 km of each city. In the absence of city level data, vulnerability to natural hazards was based on country values of the United Nation's Human Vulnerability Index, calculated as (1—HDI, where HDI is the UN Human Development Index), normalized to the highest value (Papua New Guinea). With this approach, the risk to each of the 18 cities from a 1 mm or greater fall of tephra can be compared (Figure 70.9), and categorized by their risk scores: very high (>90)—Manila; high (9—90)—Jakarta, Surabaja, Kainantu, and Tokyo; moderate (0.3—9)—Busan,

FIGURE 70.7 Normalized relative risk rankings by hazard for combined risk (humans and buildings) PDC, pyroclastic density currents (Magill and Blong, 2005).

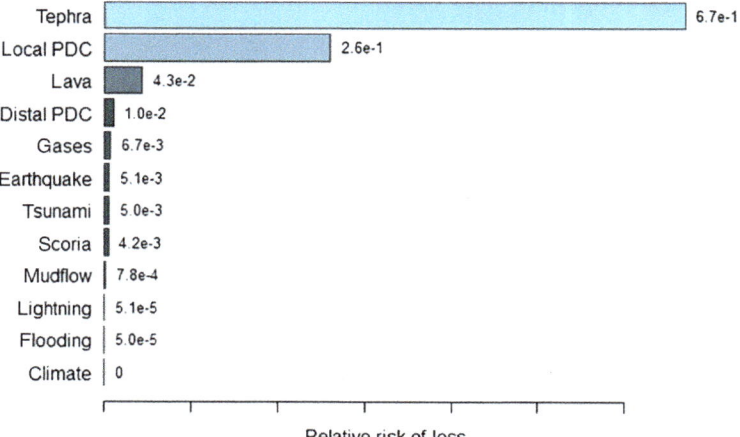

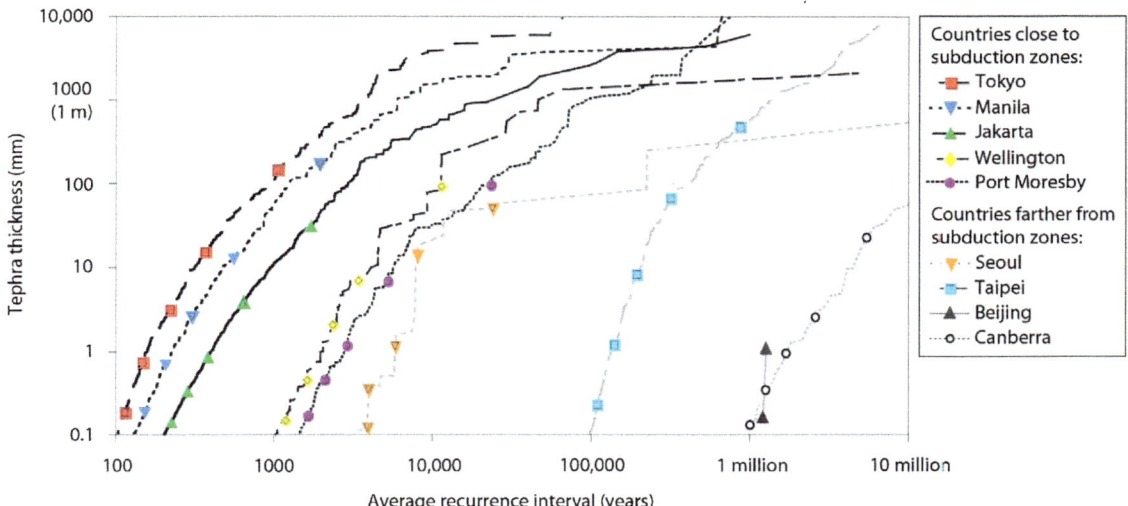

FIGURE 70.8 Exceedance probabilities for ashfall of specified thicknesses (or greater) for nine Asia—Pacific capital cities (Jenkins et al., 2012).

Port Moresby, Nagasaki, Cagayan de Oro, Auckland, and Seoul; and low risk (<0.3)—the remaining seven cities.

Figure 70.9 shows the influence of each of the risk components to total risk for each city from a 1 mm or greater fall of tephra, highlighting the different contributions made by hazard, exposure, and vulnerability. Tokyo's risk is dominated by the hazard component, as 54 surrounding volcanoes contribute a total 67% of the risk score. At Busan, population exposure (61% of the score) is the most significant determinant of risk, while Port Moresby's risk is controlled, almost entirely, by vulnerability (87%; Jenkins et al., 2012).

This level of understanding of risk forms a more dependable basis for risk management than a simple qualitative risk matrix, but it is also requires greater technical expertise to develop and more time to complete.

This said, more sophisticated risk assessment methodologies and metrics, used in safety-critical industries and in the insurance industry to create criteria for risk, have considerable promise; they allow ready comparisons across a range of hazards, accumulation of potential losses, and pricing of risk.

Before leaving this brief visit to volcanic risk implications that are not directly concerned with sudden life-threatening dangers, it is important to mention also "climate change." Volcanoes are the big imponderable factor when it comes to future climate change trends. A decent-sized eruption, bigger than historical experience but on the scale of some relatively unexceptional eruptions in the very recent geological record, could easily have the power to change weather patterns and climate trends for very long periods of time.

Thus, there is a need for volcanology to extend its hazard and risk assessment capabilities to help address the scales and minute likelihoods of such extreme events. A start has been made, for instance, with the creation of the LaMeve database for large magnitude Quaternary eruptions (Crosweller et al., 2012), building upon that unique, ineffable, and indispensible data resource, the Smithsonian Institution's Volcanoes of the World (Siebert et al., 2010).

8. EXPERT ELICITATION FOR RISK ASSESSMENT

Whatever the form of a volcanic risk assessment—qualitative, quantitative; risk matrices, relative risk rankings, or exceedance probability curves—elements of expert judgment are involved in the analysis of information, whether by single experts or a group. Structured and formalized procedures for expert elicitation are another area where volcanologists have made substantial advances in support of natural hazards risk management.

For example, Figure 70.10 shows a range graph for an expert group elicitation on the probability that "nothing significant" will happen at the Soufrière Hills volcano, Montserrat, in the following year (at the time, 2001, the bulk of a large new dome remained in the crater, but there had been no surface magmatic activity for nearly 3 years). The group comprised 12 scientists and observers, all with long experience of the ongoing eruption, and their judgments were pooled in two ways: the bar labeled "EqWts" shows the result of pooling the experts' distributions with equal weights, while the lowest bar, "PerfWts" shows the solution obtained when individual distributions are combined with weights that are derived from expert calibration using the classical model (Cooke, 1991). The latter are based on individual statistical performance (hence

FIGURE 70.9   Risk = f (Hazard × Exposure × Vulnerability) for 18 cities in Southeast Asia, in terms of 1 mm ashfall hazard exceedance (Jenkins et al., 2012).

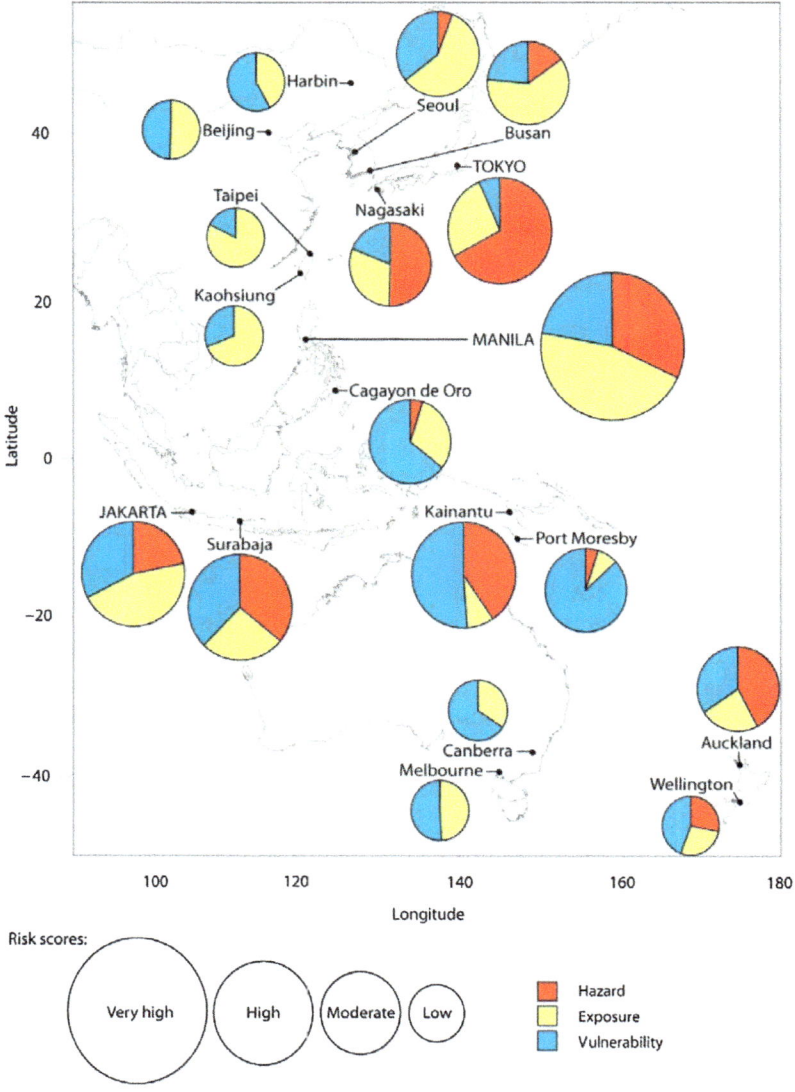

PerfWts) on a set of "seed" questions; each item has a determinable value that an expert is not expected to know precisely but can be expected to capture by providing an informative credible range. A basic hypothesis test of the individual expert's statistical accuracy over the set of seeds, compounded with a measure of their overall informativeness, represents the evidence for differentially weighting their judgments on target questions (Cooke, 1991). The subject matter of the seed questions must be closely related to that of the target questions so that skill in quantifying uncertainty, measured on seed items, can be presumed to map to uncertainty judgment on target questions.

Thus, for the two combination solutions in Figure 70.10, while the medians are quite similar at about 40% probability that "nothing significant" will happen in the next year, the uncertainty range for the performance-based solution (PerfWts) is narrower than the "democratic" EqWts spread—although not as narrow as some individual experts' credible ranges. These are typical traits and outcomes when this method is used to pool expert judgments in situations of scientific uncertainty, not just in volcanology.

Large individual or collective uncertainties, such as those shown, are common in forecasting volcanic activity, and reflect both the stochastic nature of eruption processes and our limited capability to conceptualize the behavior of a complex dynamic system. The first-order interpretation of this range graph and the PerfWts distribution in particular is that the group considers it slightly more likely than not that there will be a significant event, such as restart of magmatic

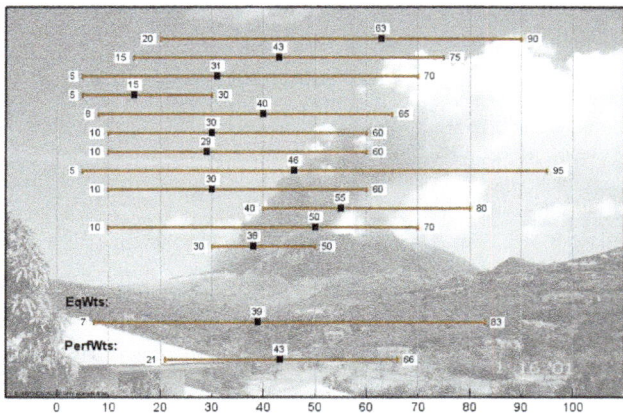

% prob: Nothing significant happens within next year

Labels show 5th, 50th and 95th percentiles for experts and pooled solutions

FIGURE 70.10 Example of an expert elicitation range graph for short-term future behavior of Montserrat volcano, early 2001 (see text). "Nothing significant" was taken to mean no new magmatic extrusion or magmatic explosion, nor a major collapse of the dome in the next 12 months (except for minor passive mass wasting by rockfalls). The upper bars show the credible ranges provided by 12 experts, expressed in terms of their individual judgments on suitable 5th percentile, median and 95th percentile values (e.g., Expert one gives: 20%; 63%; 90% probability as their distribution markers). The bar labeled "EqWts" shows the result of pooling all these experts' distributions with equal weights; the lowest bar "PerfWts" shows the solution obtained when individual distributions are combined with weights derived from expert calibration (see text).

activity, in the following 12 months but, given all the uncertainties, the odds of another year of quiescence could be as high as 2-to-1 on (66%). A major challenge for scientists is how to convey uncertainty of this extent to decision-makers. The goal of the elicitation is not to remove uncertainty but to express it as informatively as our understanding and data allow.

This underscores the need to canvass the views of several experts rather than rely on the judgment of a single person, such as a Chief Scientist, and to utilize structured, auditable elicitation procedures at relevant stages of volcanic risk assessment.

9. FUTURE DEVELOPMENTS

In recent decades, there has been rapid accumulation in knowledge relating to natural disasters but, despite this, losses due to natural hazards, in terms of economic capital and human lives, are increasing. Effective management of natural hazards is a key priority in both the underdeveloped and, increasingly, the developed world. The challenge therefore is for the scientific community to continue to develop robust, scientifically informed risk advice that is both dependable and timely so that decision-makers can implement rational strategies and policies.

Risk criteria for public safety and the levels at which risks might be deemed tolerable are not anywhere

established in relation to volcanic hazards—a significant vacuum exists. Given this, how can risk mitigation capabilities be strengthened from the perspective of the science? One avenue is in combining fundamental research into volcanic phenomena with state-of-the-art probabilistic risk assessment and risk mitigation models. Industrial criteria for risk tolerability represent one possible reference basis for articulating and contextualizing risks from volcanic hazards. However, the unsatisfactory inference from the Montserrat eruption is that, currently, even basic quantitative expressions of risk do not interface easily with political decision-making, nor bridge gaps to public risk perception. These challenges are not restricted to volcanic risks: MacKenzie (2014) gives a useful discussion of multiple issues associated with risk matrices, risk indices, and risk communication.

In contrast, public involvement in the risk management process is increasingly recognized as an important ingredient for maintaining trust in the science—and the scientists—behind mitigation decisions. Experience suggests that, on both sides of the dialogue, communication of risk improved on Montserrat over the course of the crisis, but scope remains for further investigation of this crucial element in VRM and resilience.

Public risk communication challenges are exemplified by eruption false alarms. These can have direct costs and intangible costs, some of the latter relating to loss of credibility in scientists because society may overestimate their prediction capabilities (e.g., Haynes et al., 2008). Other difficulties can arise because some volcanologists exhibit strong "epistemic risk aversion" (a fear of being proved wrong, influencing their advice), and others worry about "crying wolf." The latter concern may be something of a myth, perpetuated by social scientists who, by subscribing to the linear communication chain status quo, ascribe disasters to scientists' "failure to communicate." There is good evidence from research into deliberative approaches that public concern about false or unfulfilled alarms diminishes when a population feels included in risk decisions—such issues tend to recede if risk communication works. So, are social scientists—or volcanologists trained in social sciences—essential for effective risk assessment? Several aspects of the present chapter seem to make this case.

This said, recent natural disasters worldwide, both great and small, and fatalities in eruptions at Merapi (2010), Sinabung (2014) and Mt Kelud (2014) have demonstrated repeatedly that risk assessment and communication continue to involve stubborn difficulties for Earth scientists. Achieving optimal risk mitigation effectiveness within wider society remains elusive, especially for rare, extreme events. Those recent examples involved risks in small eruptions—what about the situation if we had a really large eruption? What is the risk calculation for a caldera-forming eruption on Kyushu, Japan? About 36 million people live in

Kyushu, and the chance of a caldera-forming eruption, while very small annually, is not zero. Is risk assessment infeasible for geologically rare, continental scale volcanic activity, or simply unwanted by society?

In the context of assessing extreme events, one promising topic for research may be the notion of disaster "near misses," and how these can inform probabilistic low-probability hazard assessments through formal techniques of counterfactual analysis (Woo, 2011). For volcanoes, a weakness of an elementary historico-statistical approach is that it does not account for failed but incipient eruptions. The simplistic dyadic view of history—an event either happened or it did not—can lead to underestimation of the chances of a future disaster if such near-miss evidence is ignored. Reasoning like this carries the danger of being falsified *post hoc* by events and, as increasingly happens with major disasters, any limitations in risk assessment can be exposed if subjected to intense legal scrutiny.

Thus, analysis of "close calls" offers a means to reduce disastrous surprises—the lesson for volcanology to be taken from the 2011 Tōhoku, Japan, earthquake and tsunami. With statistical limitations in volcanic data and intrinsic uncertainties in risk models inevitable, the corollary is that expert judgment has to be exercised for viable, and defensible, risk assessment.

Recalling that risk management is about "the identification, assessment, and prioritization of risks and involves the coordinated and economical application of resources to minimize, monitor, and control the probability and/or impact of unfortunate events," this chapter has endeavored to highlight where significant advances have been made toward more robust assessments of volcanic risk. Knowledge of these steps forward, recognition of shortcomings, and a more comprehensive understanding of societal needs by volcanologists should help improve eruption risk management. However, some state-of-the-art aspects may not be feasible in every society, applicable in every culture or appropriate for all political environments.

SEE ALSO THE FOLLOWING ARTICLES

Chapter 66: Synthesis of Volcano Monitoring ● Chapter 67: Volcano Warning Systems ● Chapter 68: Volcanic Crisis Management ● Chapter 69: Social Processes and Volcanic Risk Reduction.

FURTHER READING

Aspinall, W.P., 2006. Structured elicitation of expert judgment for probabilistic hazard and risk assessment in volcanic eruptions. In: Mader, H.M., Coles, S.G., Connor, C.B., Connor, L.J. (Eds.), Statistics in Volcanology. Special Publications of IAVCEI, 1. Geological Society, London, pp. 15–30.

Aspinall, W.P., Cooke, R.M., 2013. Expert elicitation and judgement. In: Rougier, J.C., Sparks, R.S.J., Hill, L. (Eds.), Risk and Uncertainty Assessment in Natural Hazards. Cambridge University Press, pp. 64–99 (Chapter 4).

Aspinall, W.P., Sparks, R.S.J., 2004. Volcanology and the law. IAVCEI Newslett. 1, 4–12.

Baxter, P.J., Aspinall, W.P., Neri, A., Zuccaro, G., Spence, R.J.S., Cioni, R., Woo, G., 2008. Emergency planning and mitigation at Vesuvius: a new evidence-based approach. J. Volcanol. Geotherm. Res. 178, 454–473.

Blong, R.J., 1984. Volcanic Hazards: A Sourcebook on the Effects of Eruptions. Academic Press, Sydney, 424 pp.

Cooke, R.M., 1991. Experts in Uncertainty: Opinion and Subjective Probability. Oxford University Press, New York, 321 pp.

Cox Jr., L.A.T., 2008. What's wrong with risk matrices? Risk Anal. 28, 497–512.

Crosweller, H.S., et al., 2012. Global database on large magnitude explosive volcanic eruptions (LaMEVE). J. Appl. Volcanol. 1 http://dx.doi.org/10.1186/2191-5040-1-4.

Druitt, T.H., Kokelaar, B.P., 2002. The Eruption of Soufrière Hills Volcano, Montserrat, from 1995 to 1999, vol. 21. Geological Society, London, Memoirs. http://dx.doi.org/10.1144/GSL.MEM.2002.021.01.32, 645 pp.

Evans, A.W., Verlander, N.Q., 1997. What is wrong with criterion $F-N$ lines for judging the tolerability of risk? Risk Anal. 17, 157–168.

Haynes, K., Barclay, J., Pidgeon, N.F., 2008. The issue of trust and its influence on risk communication during a volcanic crisis. Bull. Volcanol. 70, 605–621.

Health & Safety Executive, 2009. Evidence or Otherwise of Scale Aversion: Public Reactions to Major Disasters. Reference 0091699-TN03 Rev 4. http://www.hse.gov.uk/societalrisk/evidence-or-otherwise-of-scale-aversion.pdf.

Hirst, I.L., Carter, D.A., 2002. A "worst case" methodology for obtaining a rough but rapid indication of the societal risk from a major accident hazard installation. J. Hazard. Mater. 92, 223–237.

Jenkins, S., Magill, C., McAneney, J., Blong, R., 2012. Regional ash fall hazard II: modelling results and implications. Bull. Volcanol. 74, 1713–1727.

Magill, C.R., Blong, R.J., 2005. Volcanic risk ranking for Auckland, New Zealand. II. Hazard consequences and risk calculation. Bull. Volcanol. 67, 340–349.

Loughlin, S.C., Baxter, P.J., Aspinall, W.P., Darroux, B., Harford, C.L., Miller, A.D., 2002. Eyewitness accounts of the 25 June 1997 pyroclastic flows at Soufrière Hills Volcano, Montserrat, and implications for disaster mitigation. In: Druitt, T.H., Kokelaar, B.P. (Eds.), The Eruption of Soufrière Hills Volcano, Montserrat, from 1995 to 1999, vol. 21. Geological Society, London, Memoir, pp. 211–230.

MacKenzie, C.A., 2014. Summarizing risk using risk measures and risk indices. Risk Anal. http://dx.doi.org/10.1111/risa.12220. Article first published online: June 10, 2014.

Marzocchi, W., Woo, G., 2009. Principles of volcanic risk metrics: theory and the case study of Mount Vesuvius and Campi Flegrei, Italy. J. Geophys. Res. 114, B03213. http://dx.doi.org/10.1029/2008JB005908.

MIAVITA, 2012. Handbook for Volcanic Risk Management: Prevention, Crisis Management, Resilience, Mitigate and Assess Risk from Volcanic Impact on Terrain and Human Activities (MIAVITA). European Commission, 204 pp.

Newhall, C., Aramaki, S., Barberi, F., Blong, R., Calvache, M., Cheminée, J.-L., Punongbayan, R., Siebe, C., Simkin, T., Sparks, S., Tjetjep, W., 1999. Professional conduct of scientists during volcanic crises. Bull. Volcanol. 60, 323–334.

Siebert, L., Simkin, T., Kimberly, P., 2010. Volcanoes of the World, third ed. Smithsonian Institution, University of California Press, Washington, DC. 551 pp.

Sparks, R.S.J., Aspinall, W.P., Crosweller, H.S., Hincks, T.K., 2013. Risk and uncertainty assessment of volcanic hazards. In: Rougier, J.C., Sparks, R.S.J., Hill, L. (Eds.), Risk and Uncertainty Assessment in Natural Hazards. Cambridge University Press, pp. 364–397 (Chapter 11).

Viscusi, W.K., 1992. Fatal Tradeoffs: Public and Private Responsibilities for Risk. Oxford University Press, New York, 306 pp.

Wadge, G., Aspinall, W.P., 2014. A review of volcanic hazard and risk assessments at the Soufrière Hills Volcano, Montserrat from 1997 to 2011. In: Wadge, G., Robertson, R.E.A., Voight, B. (Eds.), The Eruption of Soufrière Hills Volcano, Montserrat from 2000 to 2010,

vol. 39. Geological Society, London, Memoirs, pp. 439–456. http://dx.doi.org/10.1144/M39.24 (Chapter 24).

Wadge, G., Robertson, R.E.A., Voight, B., 2014. In: The Eruption of Soufrière Hills Volcano, Montserrat from 2000 to 2010, vol. 39. Geological Society, London, Memoirs. http://dx.doi.org/10.1144/M39.0, 501 pp.

Woo, G., 2008. Probabilistic criteria for volcano evacuation decision. Nat. Hazards 45, 87–97.

Woo, G., 2011. Calculating Catastrophe. Imperial College Press, London, 355 pp.

Part IX

Economic Benefits and Cultural Aspects of Volcanism

Stephen R. McNutt

School of Geosciences, University of South Florida, Tampa, FL, USA

Most people think of volcanoes either as objects of awe and fascination or as agents of destruction. Indeed, at times they can be both. However, volcanoes also contribute to commerce and culture in myriad ways. The eight chapters in this section explore a few of the ways in which volcanoes enrich our lives and contribute to economic activity.

Modern life is highly dependent on energy to heat buildings, cook food, and run automobiles and machinery. In many volcanic regions geothermal resources are exploited either directly for heat or indirectly for the generation of electricity. Geothermal energy may be renewable, provided that it is withdrawn at a lower rate than the natural heating. It is also clean compared with the burning of fossil fuels. These issues and other aspects of geothermal energy are discussed in the opening chapter of this section, Utilization of Geothermal Resources.

Volcanic ash and lava from eruptions often cover large expanses of ground and are detrimental to life, but over time these same materials weather to form soils. The resulting volcanic soils have unique physical and chemical features, which affect properties such as moisture retention. Some mountainous volcanic regions are noted for their production of coffee or wine, owing to volcanically derived soils. So, if you start your day with a cup of coffee and end it with a glass of wine, volcanoes may have been indirectly responsible. The following chapter, Volcanic Soils, details the properties, distribution, and formation of such soils.

When volcanoes erupt, the products of eruption may dramatically alter the local conditions for plants and animals. Over time, life forms are reestablished as soils form and other changes occur in deposits. The chapter on Volcano Ecology: Disturbance Characteristics and Assembly of Biological Communities, discusses the changes for plants and animals in the context of the geographic distribution of volcanoes.

Volcanic rocks are the most plentiful rocks in many areas and are often used as building materials for structures and road beds and in some types of cinder blocks, landscaping, and walls. The altered volcanic clay, bentonite, is commonly used in diverse products from drilling mud to cosmetics. Some minerals concentrated in volcanic rocks are important for components of cell phones and other modern electronic devices. The next chapter, Volcanic Materials in Commerce and Industry, describes many products derived from volcanoes and their usage and availability.

Hot springs and fumaroles are often found in close association with volcanoes. This combination has led to the development of many spas and resorts at volcanoes. The attractive mountainous terrain of volcanic areas is also suitable for hiking, camping, and climbing as well as for viewing the various surface manifestations of volcanic activities. These and related topics are discussed in the following chapter, Volcanoes and Tourism.

Ironically, volcanic deposits are wonderful preservers of ancient buildings and artifacts, even though at the time of eruption they were undoubtedly viewed as destructive. Much of what we know of some ancient civilizations was instantly preserved by the devastating force of volcanic eruptions. Examples include the well-known Roman cities of Pompeii and Herculaneum and the Bronze Age Akrotiri in the Aegean. Other less well-known sites near Ilopango in El Salvador, Arenal in Costa Rica, Sunset Crater in Arizona, Mount Rainier in Washington, Mount St. Helens in Washington, and sites in the Yukon Territory have yielded clues showing the widespread effects of eruptions on the life of early peoples. The Chapter Volcanoes, Ancient People and Their Societies explores how archaeologists have harvested information about early cultures.

Volcanoes are often used as metaphors for power or unpredictability. Volcanoes are also frequently viewed as windows to the interior of the earth. These ideas show up repeatedly in popular culture and often make their way into books, movies, cartoons, fine art, and other forms of expression. Two famous examples that began as books and were later made into films are the novel The Last Days of Pompeii by George Bulwer-Lytton (1834) and the science fiction work Journey to the Center of the Earth by Jules Verne (1864). This section of the encyclopedia, and the entire text, concludes with the Chapters 77 and 78. These chapters provide brief descriptions and synopses of many volcano books, movies, and art works.

Readers may be surprised by the far-reaching benefits of volcanoes outlined in this section and their broad impact on our lives. Although we tend to think about volcanoes primarily while they are erupting, their products and images are found in many places in both ancient and modern life. A world without volcanoes would be a duller, less economically viable, and certainly less interesting place.

Chapter 71

Utilization of Geothermal Resources

Stefán Arnórsson

Institute of Earth Sciences, University of Iceland, Reykjavík, Iceland

Sverrir Thórhallsson

Iceland GeoSurvey, Reykjavík, Iceland

Andri Stefánsson

Institute of Earth Sciences, University of Iceland, Reykjavík, Iceland

GLOSSARY

appraisal well A well drilled in a prospective wellfield within a geothermal field that has been roughly delineated by step-out wells with the purpose of quantifying its production characteristics and proving hot fluid.

direct use (of geothermal heat) Extraction of heat from hot fluid to heat buildings, dry vegetables, etc.

earth energy sources There are only five sources of energy on the Earth: energy generated by (1) nuclear fusion, (2) nuclear fission, (3) pull of gravity, (4) energy stored in chemical bonds (fossil fuel, biomass), and (5) primordial heat. About half of this heat still remains in the Earth.

energy resource Something that can be used to do work, such as moving vehicles, and produce heat or electricity with present-day technology.

enhanced geothermal system Geothermal system in hot-dry (impermeable) rock, where permeability is artificially created by fracturing the rock at depth.

geothermal energy Energy stored in the form of heat below the surface of the solid Earth.

geothermal exploration well A well drilled for the purpose of discovering a prospective wellfield within a hydrothermal system.

geothermal field (area) An area with finite boundaries, usually defined by the distribution of hot springs and/or fumaroles.

geothermal fluid Hot water and/or steam hosted in a hydrothermal system. Geothermal fluids contain dissolved gases and solids.

geothermal system Any body of hot rock with or without hot fluid in the uppermost part of the Earth's crust. Some geothermal systems are hydrothermal systems but others are not.

heat pump A device to transfer energy in the form of heat from a cooler body to a warmer one.

hot-dry rock Geothermal system characterized by impermeable rock.

hydrothermal field See geothermal field.

hydrothermal reservoir A body of permeable hot rock in the uppermost part of the Earth's crust that contains hot fluid, which can be extracted from the reservoir and brought to the surface through drillholes.

hydrothermal system A body of permeable hot rock with hot fluid that underlies a hydrothermal field. See also geothermal field.

The Encyclopedia of Volcanoes. http://dx.doi.org/10.1016/B978-0-12-385938-9.00071-7

injection well A well used for disposing of spent hydrothermal fluid from production wells/power plants.

production well Any well that produces steam/hot water.

renewable energy resource An energy resource that is replenished at a rate equal to or higher than it is consumed. It cannot be exhausted. Examples: direct sun energy, wind, and hydropower.

resistance heating Use of electricity for house heating that involves passing electric current through radiators. When there is resistance to the flow of this current through the radiator, the electric energy is converted into heat energy, hence resistance heating.

scaling Formation of mineral deposits in wells and surface equipment by precipitation of solids from the flowing fluid.

step-out well Well drilled in the vicinity (~ 1 km) of a successful exploration well with the purpose of delineating the size of the anomaly discovered by the exploration well.

1. INTRODUCTION

Geothermal energy is defined as energy in the form of heat below the Earth's solid surface. **Geothermal systems** represent bodies of hot rock with or without hot fluid in the upper part of the Earth's crust. Many classifications have been proposed for such systems. The one used here is that of Goff and Stimac (see Chapter 46), which is based on the geological environment in which these systems occur. The most important types are systems in young igneous settings, tectonic systems, sedimentary basins, and impermeable **hot-dry rock** (also called **enhanced geothermal systems** (EGS)). The first three types have also been termed **hydrothermal systems**, and this terminology is used here. Their development has been proved to be technically and economically feasible. Thus, they represent an **energy resource**. The heat stored in hot-dry rock within drillable depths in the Earth's crust could also be classified as an energy resource if the current technology can be refined to extract heat economically from such rock.

Examples of exploited fields in young igneous settings include Wairakei in New Zealand and Palinpinon in Philippines. Laugarnes and Laugarland in Reykjavík and Akureyri, respectively, in Iceland represent tectonic systems. The former is an old high-temperature system of $> 250\,°C$ as indicated by the hydrothermal alteration mineralogy, but at present reservoir temperatures are $130-140\,°C$. Szentes in the south Hungarian plains provides an example of a sedimentary basin hydrothermal system. The hot-dry rock of the Cooper basin in Australia represents an EGS, where a 25 MW_e demonstration plant is being developed.

Globally, geothermal energy is not an important energy resource. Today, it accounts for only about 0.22%, ~ 11 GW (gigawatts), of the total installed electric power capacity worldwide (~ 5000 GW). Despite this, it is very important for many countries with abundant hydrothermal

activity. For example, within a few years, almost half of Kenya's electric power will be generated by geothermal steam.

The heat transported from the Earth's interior through its surface is $\sim 50,000$ GW. Accordingly, the heat flowing through every meter square of the Earth's surface is around 100 mW (milliwatts). In comparison, the radiation from the Sun that reaches the Earth is 342 W/m^2 or 3420 times larger than the energy flow from inside the Earth. The heat flow passing through ice-free dry land areas is only about twice the installed electric power. By looking at all the above numbers, it should be evident that the future source of renewable energy for mankind is the Sun. Further, it also seems impossible to make any significant use of the renewable conductive heat current from the Earth's interior.

The amount of energy stored within the Earth in the form of heat, 10^{31} J, is enormous. So far, however, use of this energy has been technically possible and economically feasible only where geological and hydrological conditions are favorable for the formation of hydrothermal systems. Volcanic and tectonic systems are most important. They are characterized by high permeability allowing groundwater circulation to depths of a few kilometers, where it heats up by contact with hot rock. Subsequently, it ascends due to buoyancy. Use of geothermal water in sedimentary basin has also proved to be economic. Their water may be connate, at least in part, i.e., of the same age as the host sediments, and fluid convection is not significant. Heat is recovered from hydrothermal systems by drilling, thus bringing the hot fluid to the surface.

Several attempts have been made to extract heat from hot-dry rock. They involve drilling of "injection" and **production wells**, and hydrofracturing of the rock to create permeability. Subsequent production consists of pumping cold water into the **injection well** and extracting this water from the production well. As the water flows between the two wells through the artificially created fractures, it gains heat from the rock. So far, this technology has not resulted in projects of economic importance. If current technology can be improved, it may become feasible to extract heat from such impermeable rocks in which case an enormous mine of heat has become a resource.

Geothermal energy is sometimes classified among the **renewable energy resources**. However, renewability of the different types of geothermal systems is highly variable. It is a good approximation to regard tectonic systems, sedimentary basins, and hot-dry rock as nonrenewable resources (see O'Sullivan et al., 2010). The renewability of systems in young volcanic settings is variable, uncertain, and affected by the extent of heat withdrawal by the exploitation relative to the natural heat output (see Sanyal, 2005).

In the present chapter, the main focus is on the development and use of hydrothermal systems and the rapid growth in **heat pump** usage. Hot-dry rock systems are not discussed further.

2. USE OF HYDROTHERMAL FLUIDS

2.1. Energy Usage

Modern usage of geothermal energy started at Larderello, Italy. On 4 July 1904, geothermal steam was used for the first time to generate electric power. In 1912, this was followed by the installation of the first turbine generator unit powered by underground steam. By 1944, the installed power had increased to 127 MW. In the 1920s, hydrothermal power explorations were carried out in California and Japan but at the time did not result in power developments. Since the 1960s, many countries have been active in developing hydrothermal resources for various types of direct uses of the heat stored in the **geothermal fluid** such as house heating and greenhouse farming in addition to power production. During the 2000–2010 period, the direct use of geothermal energy increased much, by 23,300 TJ annually (12.2%). The most striking feature is in the rapid increase in the use of geothermal heat pumps, which rose on average by 76% annually in the 2000–2010 period. The increase in installed capacity for electric power generation has been slower over the same period, 9.2% annually corresponding to a total of 2743 MW_e. Direct use of geothermal energy in the form of heat utilization is summarized in Table 71.1. Figure 71.1 shows installed power capacity in 2010 and planned capacity by 2015. In the past, however, similar plans for capacity increases have been made but not realized.

Power generation by geothermal steam involves passing it through a turbine in much the same manner as steam produced by burning coal. Geothermal steam is produced from high-temperature hydrothermal systems ($T > 200\,°C$). Direct use of heat in geothermal fluids is generally based on exploiting geothermal systems with lower temperatures (50–150 °C) and even lower in the case of swimming pools and heat pumps. The growth of the heat pump technology has occurred mostly in countries, or in areas within countries, with very limited or no hydrothermal activity, such as Sweden, Germany, France, the Netherlands, and Norway, and is a reflection of energy policies. They make more efficient the use of electricity for space heating than direct use of the electricity, i.e., **resistance heating**.

Hydrothermal resources account for only a very small fraction of the world's total installed electric capacity, or ~11 GW, out of the total estimate of ~5000 GW for 2009, i.e., 0.22%. The annual production of electric energy from geothermal plants is, however, most likely higher than these numbers indicate because they have higher load factor than fossil fuel plants and most hydropower plants. A load factor of 100% means that the power plant is running at full capacity all year round.

2.2. Extraction of Chemicals

Hydrothermal fluids have been utilized on a small scale as a source of various chemicals. At Larderello, Italy, boron was extracted from geothermal steam before electric power generation started. In Japan, cesium has been extracted from hot spring waters and mercury from hot spring deposits at Sulfur Bank in California. Production of various metals from the hot brine in the Salton Sea area in California (e.g., lithium) has been tested and a new plant is under construction. Many hydrothermal fluids are rich in lithium, which could be extracted by the use of ion exchange resins if the silica in the fluid could be removed. In New Zealand, a technique was developed to extract silica from the waste brine of the Wairakei geothermal power plant by coprecipitating it with lime to form a useful calcium silicate for the building industry. Extensive research has been carried out that may lead to production of high-purity silica polymers from geothermal brines. Common salt was for a time produced in a factory on Reykjanes, Iceland, from geothermal seawater and the steam used in the evaporation process. Carbon dioxide originating from geothermal steam is liquefied or turned into dry ice in factories in Turkey and Iceland.

2.3. Factors Affecting Energy Usage

The principal factors that determine potential use and economics of hydrothermal resources are reservoir temperature, reservoir size, formation permeability, and the chemical composition of the fluid. The location of the resource in relation to the market is also of importance, at least when using the heat in hot water or steam directly, i.e., for greenhouses. For financial and technical reasons, transport of hot fluid over long distances is not possible. In this respect, the value of geothermal energy is inferior to fossil fuel, but for power production it is on pair with hydropower.

Baldur Líndal, an Icelandic chemical engineer, assessed the potential use of geothermal water and steam on the basis of their temperature (Figure 71.2). From this figure, it is clear that thermal water with temperatures as low as 20–30 °C constitutes a useful energy resource. Sites with temperatures even lower than this are still a viable resource in the case of heat pumps.

The chemical composition of hydrothermal fluids is highly variable with respect to gases and dissolved solids ranging from very dilute to hypersaline brines. The chemical composition can significantly affect the economy of resource exploitation by causing operational problems, such as deposition of solids from the flowing fluid in wellbores and surface equipment (**scaling**), and corrosion. High gas content in the steam may affect its quality for power generation. Disposal of spent brine and condensed

TABLE 71.1 Direct Use of Geothermal Heat in 2010 by Country

Country	Installed Capacity (MW$_t$)	Annual Energy Usage (TJ)	Load Factor
Albania	11	40	0.11
Algeria	67	2099	1.00
Argentina	307	3907	0.40
Australia	33	235	0.22
Austria	663	3728	0.18
Belarus	4.5	44	0.31
Belgium	118	547	0.15
Bosnia and Herzegovina	22	255	0.37
Brazil	360	6622	0.58
Bulgaria	98	1370	0.44
Canada	1126	8873	0.25
China	8898	75,348	0.27
Columbia	14	287	0.63
Croatia	67	469	0.22
Czech Republic	216	1290	0.19
Denmark	200	2500	0.40
Estonia	63	356	0.18
Finland	994	7966	0.25
France	1345	12,929	0.30
Georgia	27	689	0.82
Germany	2485	12,764	0.16
Greece	135	938	0.22
Hungary	655	9767	0.47
Iceland	1826	24,361	0.42
India	265	2545	0.30
Iran	42	1064	0.81
Ireland	138	692	0.16
Israel	82	2193	0.84
Italy	867	9941	0.36
Japan	2100	25,698	0.39
Jordan	153	1540	0.32
Kenya	16	127	0.25
Korea (South)	229	1955	0.27
Lithuania	48	412	0.27
Macedonia	47	601	0.40
Mexico	156	4023	0.82
The Netherlands	1410	10,699	0.24

TABLE 71.1 Direct Use of Geothermal Heat in 2010 by Country—cont'd

Country	Installed Capacity (MW$_t$)	Annual Energy Usage (TJ)	Load Factor
New Zealand	393	9552	0.77
Norway	1000	10,800	0.34
Poland	281	1501	0.17
Portugal	28	386	0.44
Romania	153	1265	0.26
Russia	308	6144	0.63
Serbia	101	1410	0.44
Slovak Republic	132	3067	0.74
Slovenia	116	1015	0.28
Spain	141	684	0.15
Sweden	4460	45,301	0.32
Switzerland	1061	7715	0.23
Tunisia	44	364	0.26
Turkey	2084	36,886	0.56
Ukraine	11	119	0.35
The United Kingdom	187	850	0.14
The United States	12,611	56,552	0.14
Vietnam	31	92	0.09
Other countries (22)[1]	93	1344	0.38[2]
Total	48,523	423,374	

[1]Armenia, Caribbean Islands, Chile, Costa Rica, Ecuador, Egypt, El Salvador, Ethiopia, Guatemala, Honduras, Indonesia, Latvia, Mongolia, Morocco, Nepal, Papua New Guinea, Peru, Philippines, South Africa, Tajikistan, Thailand, Venezuela, Yemen, all with <10 MW$_t$.
[2]Weighed average.

steam into drillholes is practiced in many geothermal fields worldwide in order to minimize the environmental impact of utilization. The geochemical environment and fluid temperatures mostly control fluid composition in each type of hydrothermal systems. The most common features that have delayed and, in some cases, even inhibited the development of drilled hydrothermal systems are inadequate permeability and foreseen operational/environmental problems as determined by the chemical characteristics of the hydrothermal fluid.

2.4. Power Generation, Direct and Multiple Uses

Although geothermal steam is principally used in the same way to generate electric power as is steam formed in boilers using fossil fuel or nuclear energy, there is one major difference. In the case of fossil fuel and nuclear energy, the steam inlet pressure for the turbines can be chosen to maximize efficiency. For geothermal steam, the characteristics of the reservoir determine, at least to some extent, the optimum turbine inlet pressure, which is usually in the range of 4—11 bar gauge. The efficiency of geothermal steam for electric power generation is poor. Only about 20% of the thermal power of the flowing steam can be converted into electricity, depending on turbine inlet pressure and efficiency. The efficiency is even less relative to the total thermal power (steam and water) of wells. As an example, consider a reservoir with liquid water at 260 °C and a steam separation pressure of 6 bar abs. (this is equal to a temperature of 159 °C for saturated steam). Boiling by a pressure drop from 260 to 159 °C will cause 22% (steam fraction by weight) of the reservoir water to be converted

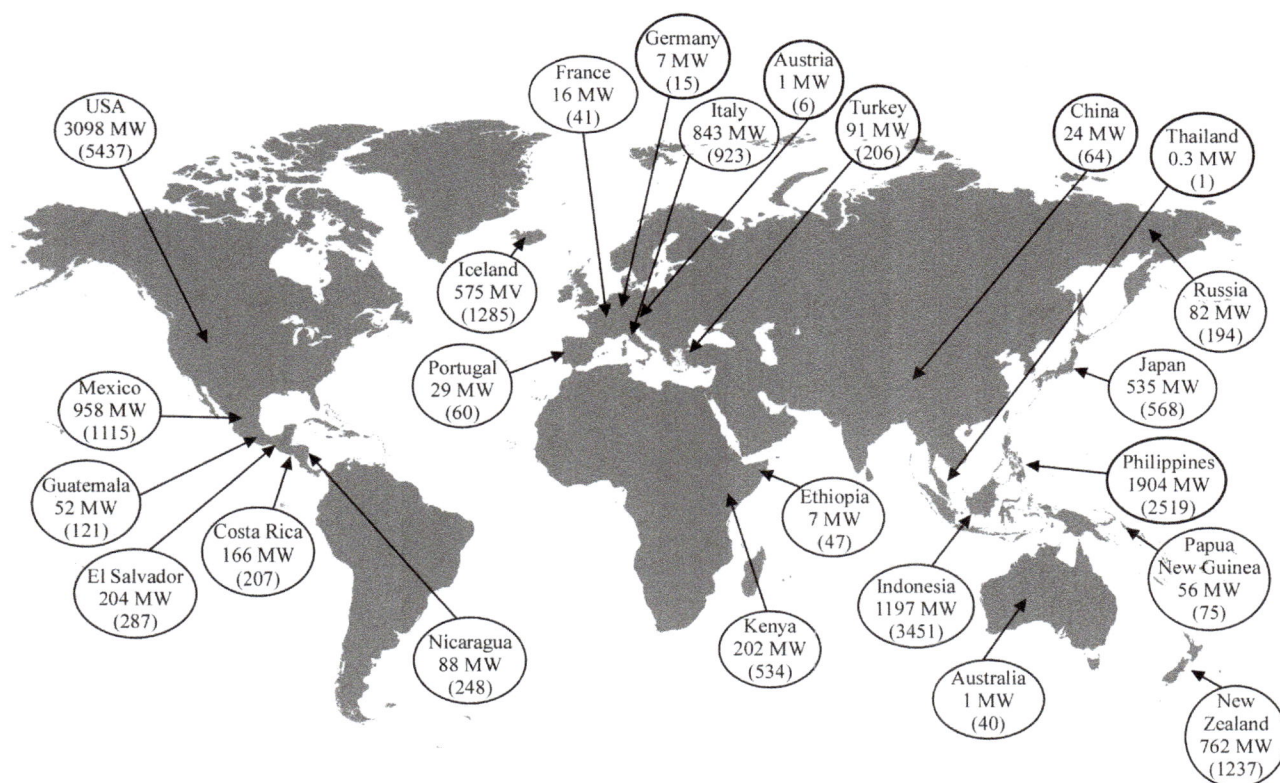

FIGURE 71.1   Installed geothermal power by country in 2010 and planned increase by 2015 (in parenthesis). Twenty-two countries that did not produce electricity from geothermal fluids in 2010 plan to have an installed capacity of 1120 MW_e by 2015.

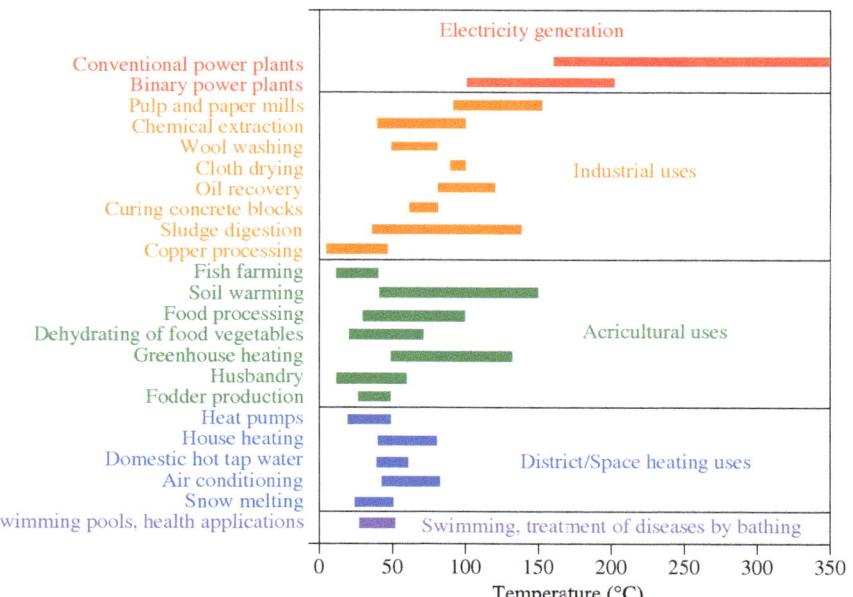

FIGURE 71.2   The Líndal diagram. This diagram is named after an Icelandic chemical engineer, Baldur Líndal. It depicts potential uses of steam and hot water depending on their temperature. The present figure is a shortened version of the initial diagram.

into steam. Only the steam fraction can be utilized to generate power, while the remaining 78% of the produced mass is injected back into the reservoir, so the heat is not totally lost. The wastewater contains 46% of the total heat coming from such a well. Assuming that the wastewater could be utilized by cooling it further to 35 °C, the remaining heat of 46% is reduced to 10%. A steam flow of 1.7−2.0 kg/s is required to generate 1 MW of electric power in a geothermal turbine.

Direct use of hot water/steam for heating is much more efficient than the use of steam only for power generation. For space heating, heat can easily be extracted in radiators by cooling it down to 35 °C and even lower in floor-heating systems. As an example, for an initial hot-water temperature of 80 °C and assuming useful heat as that in excess of 35 °C (20 °C) gives an efficiency factor of 56% (75%).

The poor utilization of heat in hydrothermal fluids by conventional electric power generation has led to technological improvements. Two methods have mainly been pursued. One involves a so-called binary cycle (bottoming cycle) and the other utilization for multiple purposes from the same field such as power generation, house heating, bathing, etc. (Figures 71.2 and 71.3). Multiple utilization of this kind may also involve extraction of useful chemicals from geothermal fluids. The binary cycle involves heating a low-boiling point secondary fluid in a closed loop of a heat exchanger. That fluid is then used to drive the turbines or expanders. Heat recovered from turbine condensers is then used as the first stage in the heating of freshwater, e.g., for space heating. The water separated from the steam in steam separators provides additional heating in heat exchangers to the final temperature. In this way, the geothermal resource can be more effectively utilized (Figure 71.3(B)). Problems with efficient use of high-temperature geothermal fluids may arise as they become supersaturated with minerals at low temperatures. Cooling of the fluid sometimes leads to amorphous silica scaling in production wells and surface equipment, and scaling of calcium carbonate and anhydrite in injection wells. Such scaling can impede smooth operation of a geothermal installation by clogging pipelines and reducing heat transfer in exchangers, thus increasing the operational costs. The possibilities of multiple uses of hydrothermal fluids depend on the availability of a local market for the heated freshwater.

2.5. Heat Pumps

As already mentioned, the use of heat pumps has grown rapidly during the last two decades or so. In principle, a heat pump is used to extract heat from a cooler body and transfer it to a warmer body. Heat pumps can be switched between operating in a heating mode or as air conditioners.

There are basically two configurations for heat pumps, one that extracts heat from air (air-source heat pumps) and other from the ground (ground-source heat pumps). In the latter case, the heat may be extracted either from the soil directly or from deeper strata in the ground using drillholes. The latter are geothermal heat pumps. The ultimate source of heat utilized by a heat pump that uses air and soil is solar. Geothermal heat pumps may also use some solar heat if

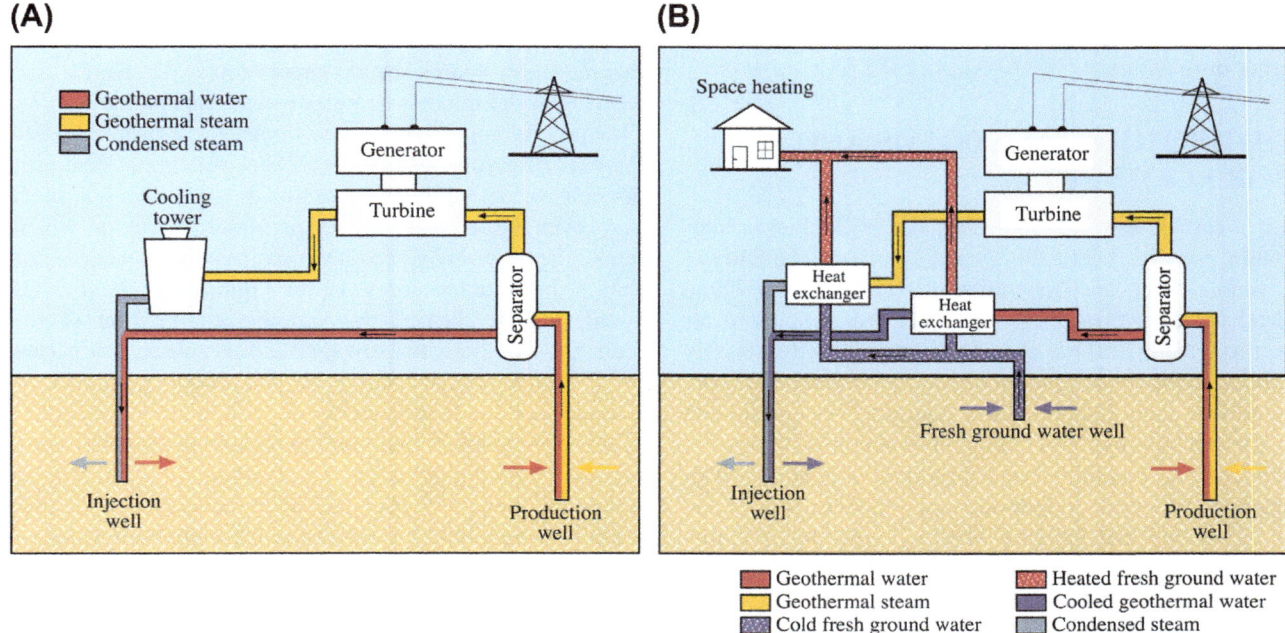

(A)

(B)

FIGURE 71.3 (A) Conventional power generation. (B) Power generation combined with space heating.

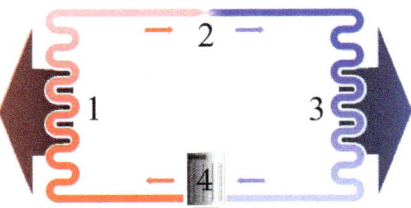

FIGURE 71.4 Simplified illustration of how heat pumps work. (1) Flow of heat from warm liquid in red loop into house. (2) Expansion valve. Liquid vaporizes here and cools extensively. (3) Cooled gaseous fluid picks up heat from surroundings. (4) Compressor. Gaseous fluid becomes liquid and warm.

they extract heat from downward-percolating groundwater and also when they are used for cooling buildings in summer time and the heated water warms up rock around drillholes.

Heat pumps are based on the so-called Joule—Thompson or Joule—Kelvin effect, discovered in 1852 (William Thompson later became 1st Baron Kelvin). This effect describes that all gases except hydrogen and helium cool upon rapid expansion. The development of the first heat pump is attributed to an Austrian, Peter von Rittinger (1811—1872). Essentially, a heat pump consists of a fluid that circulates in a closed loop, which includes a compressor and an expansion valve (a pressure-lowering device) (Figure 71.4). When fluid passes through the expansion valve, it is extensively depressurized, leading to severe cooling (from −20 to −30 °C). The cooled fluid gains heat from the environment (air, ground) by conduction. When the gaseous fluid reaches the compressor, it is compressed into liquid, which is warmer than the liquid that entered the expansion valve. The compressor uses only one-fourth to one-third of the electricity needed for with resistance heating in radiators.

3. HYDROTHERMAL SYSTEMS AND THEIR ROOTS

Their volume but also many other factors, including temperature, porosity, permeability, and the chemical quality of the fluid, affect the generating capacity of individual hydrothermal systems, both for power production and for direct use. Most of these characteristics can be reasonably well quantified by drillings into the geothermal reservoir. This is, however, not so when it comes to system volume because the base depth of fluid convection is not known.

In order to account for the high natural heat output of many high-temperature hydrothermal systems, it is necessary to assume that the heat source is very hot and that temperature gradient from this heat source to the base of the convecting fluid must be very high, 10 °C/m, even more in the case of magma or very hot rock that represents recently consolidated magma. Modeling studies indicate that fluid convection over very hot heat sources may be one-dimensional, at least if permeability above it is sufficiently high. One-dimensional convection involves descent of liquid water through very small pores in the rock and along mineral grain surface and rise of vapor along more permeable fractures, both occurring right above the heat source. The ascending vapor may form by complete evaporation of the descending liquid water. The rising steam may condense partly or completely in shallower groundwater, heating it in the process.

After fluid convection has started in a newly born hydrothermal system, the ascending hot fluid will gradually heat the rock above the heat source and probably also to the sides by lateral hot-water flow. When the system has matured, considerably greater proportion of the heat in the system will be stored in the rock rather than in the hot fluid depending on rock porosity of course. Exploitation of such fields will involve enhanced natural recharge of groundwater that ultimately was mostly if not solely nonthermal (meteoric water or seawater). This recharge makes it possible to extract heat from the hot rock of the system, its effectiveness depending on its porosity the spacing of permeable fractures in the reservoir and the extent of exploitation.

In recent years, holes drilled into volcanic systems have penetrated magma, in Hawaii in 2005, two times at Krafla in Iceland in 2008—2009, and very likely at Menengai in Kenya in 2011. At Krafla and Menengai, a zone of super-heated steam overlies the magma body but underlies a two-phase (water + steam) reservoir. At Alto Peak in the Philippines, it is known from drillings that a vapor zone underlies a two-phase reservoir although drillholes have not penetrated magma. It remains to be verified by deeper drillings in new areas whether or not two-phase **hydrothermal reservoirs** are typically separated from their magma heat source by a vapor zone. At Krafla and Menengai, the magma bodies were as shallow as ∼2 km. This sets an upper limit to the thickness of usable hydrothermal reservoir to around 1500 m assuming that production will be confined to depths of more than 500 m. It may even be thinner. Magmatic gases, some of which render the fluid acid when they dissolve in it, may cause this fluid to become too corrosive for exploitation. Acid fluids could also be formed by complete evaporation of the recharging liquid water to vapor and its subsequent partial or complete condensation in shallow groundwater. Gradually, water—rock interaction will destroy the acidity of this deep fluid, as many of the rock-forming minerals in common rock types act as bases when they dissolve in water, i.e., their dissolution involves consumption of protons. If superheated steam forms in the roots of volcanic hydrothermal systems, its depressurization may lead to intense silica scaling. The solubility of silica in steam can be considerable at high pressure, but upon depressurization the solubility decreases strongly and intense deposition sets in. Such deposits may render the steam unexploitable, at least with respect to present-day technology.

For environmental purposes, it is common practice to inject spent hydrothermal fluids into the ground. Injection of the spent fluid back into the reservoir has been considered to be beneficial because it will counteract reservoir pressure drawdown. Such injection may, on the other hand, reduce the effective withdrawal of heat from the reservoir rock, at least in the case of high-temperature two-phase systems. Liquid water in micropores may be immobile due to its adhesion onto mineral grain surfaces, but if pressure drawdown in larger permeable fractures causes sufficient cooling of the fluid by depressurization boiling, it may create sufficient temperature difference between fluid and rock to cause the immobile water to boil. The vapor so formed could flow into the larger pores and ultimately into production wells due to its better flowing properties as compared to liquid water.

In order to explain the relatively large heat output from some tectonic systems in nonvolcanic regions, it must be assumed that their heat source is hot rock in their roots. Mining of heat occurs under natural conditions, and exploitation that involves enhanced fluid withdrawal will enhance this mining of heat from the rock.

Some hydrothermal systems are vapor-dominated, others are liquid-dominated. In the latter, steam forms by depressurization boiling as the liquid water flows through the aquifer and ascends in the well. The quantity of steam that forms depends on the initial temperature of producing aquifers and the magnitude of pressure drop. As an example, 41% of 300 °C liquid water is transferred into vapor by depressurization boiling to 100 °C. It is for this reason that wells drilled into liquid-dominated systems discharge a two-phase mixture of water and steam or steam only if the boiled liquid water is retained completely in the aquifer due to its adhesion onto mineral grain surfaces. Segregation of the two phases in this way is affected by their relative volumes in the flow but also by the surface area between rock and fluid. Due to its properties, liquid water forms a thin layer on mineral surfaces but vapor does not "wet" these surfaces.

4. RENEWABILITY OF GEOTHERMAL SYSTEMS

Possibly the best definition of a renewable energy resource is that given by Girardet and Mendonça (2009): "An energy resource is renewable if it cannot be exhausted." It is frequently stated that geothermal energy is a renewable energy resource. Indeed, the European Union and the Department of Energy in the United States classify geothermal energy among renewable energy resources despite the fact that almost all geothermal energy is not renewable, certainly not on a timescale that matters to

mankind. The reason is likely political. In their estimates of geothermal potential for power production, the U.S. Geological Survey and the National Energy Authority in Iceland evaluate this potential by referring to 30- and 50-year production periods, respectively. From this, it may be deduced that both organizations view geothermal resources as mines of heat.

The renewability of geothermal systems depends on the type of their heat source and the extent of exploitation that generally outpaces the natural heat output considerably. Opinion among scientists is divided on to what extent individual geothermal systems are renewable. If the heat that renews the system is by conduction alone from deeper and hotter levels in the Earth's crust, it is, however, clear that it is a good approximation to regard these systems as nonrenewable, as recovery time after 100 years of production may be on the order of 1000−100,000 years or even more depending on heat extraction rates by the exploitation. In some tectonic systems, deep drilling indicates that mining of heat from the rock in the roots of these systems occurs under natural conditions (Figure 71.5) and it is this heat mining that sustains the system. Exploitation will enhance the heat mining and conductive heat transfer from below and possibly also from the sides presents no significant short-term contribution, making these systems a transient phenomenon, hence nonrenewable.

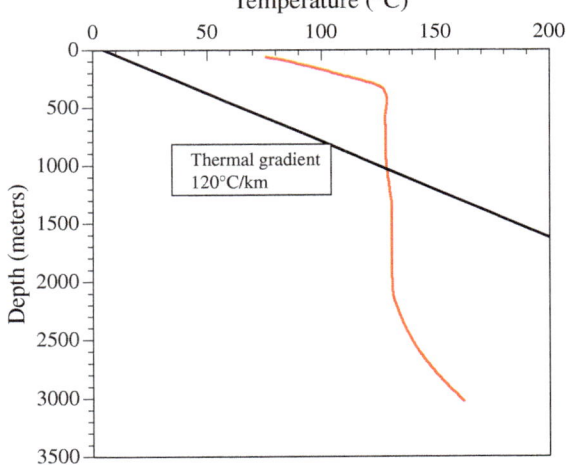

FIGURE 71.5 Temperature profile (red curve) in a deep well at Laugarnes field in Reykjavík, Iceland. The black curve shows temperature gradient in wells outside the geothermal field. Above the point of intersection of the two curves, the temperature is higher within the hydrothermal system than in the rock enveloping it. Below this point, on the other hand, the temperature within the geothermal system is lower at any depth than in the surrounding rock. The temperature distribution in the well is considered to reflect that the convecting water has mined heat from the rock (under natural conditions) in the roots of the system and transported it to shallower depth levels due to its density-driven convection, thus leveling out temperature over a great depth range within the geothermal system. The temperature distribution in the well also indicates that recharge is from above, not from the sides.

Geothermal systems with a magmatic heat source may have higher renewability than other types of geothermal systems. It is, however, also affected by the extent of exploitation, how frequently magma is intruded into their roots, and to what extent the heat flow from the heat source is lost to the surface after production is halted and how much goes into heating up the rocks that cooled during the production period as a consequence of enhanced cold water recharge. It is not possible to generalize about the renewability of these systems. Data on the flow of new magma into their roots are also lacking, except possibly in a few cases. To be able to estimate the renewability of any magmatic hydrothermal system, data are needed that are specific for each system and the outcome may turn out to be no more than a skilled guess.

Recent publications on the subject of renewability of magmatic hydrothermal systems do not address sufficiently the concept of time. This is, however, important. Such systems are born, they develop, cool down, and become extinct. All existing geothermal systems in young igneous settings need not have a magma heat source today although they might have had one earlier in their life. At diverging plate boundaries, like in Iceland, a volcanic hydrothermal system will eventually be displaced from its magma heat source as it drifts out of the volcanic zone at which time it will become practically nonrenewable and it will gradually cool down and become extinct. Some 40 fossil volcanic hydrothermal systems are known in Iceland. A mature hydrothermal system without magma or very hot igneous intrusive body in its roots can have high natural heat output over a long time on the human timescale, at least if permeability is sufficient just as is the case with some tectonic systems.

The heat stored in a given volume of rock within a hydrothermal system depends largely on its porosity and temperature (Figure 71.6). Measurement of rock porosity in

cores from drillholes allows an estimation to be made of the heat stored in a given volume of rock with reasonable accuracy. For example, in a reservoir with a temperature of 250 °C and a porosity of 10%, the heat stored in basaltic rock is 88% of the total heat in excess of 150 °C. To extract this heat from the rock through recharge of 5 °C groundwater, it is necessary to replace the original liquid water 4.5 times. However, extraction of heat from the rock may not be very effective. It has been estimated to be 5—20% in system under exploitation depending on the spacing of permeable fractures.

5. DRILLING FOR HYDROTHERMAL FLUIDS

Drilling of deep wells is required to reach geothermal reservoirs. They have to be of large diameter to bring economic quantities of geothermal fluids to the surface. The targets for geothermal wells are (1) desired temperature and (2) permeable strata, and both have to be present for a successful well. For low-temperature wells, the temperature target is in the range 50—130 °C; for medium-temperature wells, 130—210 °C; and for high-temperature wells, >210 °C. Low-temperature wells are suitable for direct use of the heat in the geothermal fluid, medium-temperature wells for binary generation, and high-temperature wells for steam power plants. If the temperature is below 210 °C, it is not certain that a high-temperature well can sustain self-flow of the boiling fluid, and thus the last casing string needs to be deep enough to block inflow of colder fluids. The permeability, aquifer temperature and pressure, and well diameter dictate the maximum flow rate. For low-temperature wells, it should be 30—90 L/s ($\sim$5—15 MW$_t$; megawatts thermal for space heating) and for high-temperature wells, around 10 kg/s of steam (40—60 kg/s total flow corresponding to $\sim$5 MW$_e$; megawatts electric). This requires drilling of wells to a depth of 1000—3000 m in the case of high-temperature resources. For low- and medium-temperature wells, the range is in depths of 400—6000 m.

The drilling equipment and technology is mainly derived from the petroleum industry with some modifications in the equipment selection such as mud-cooler and greater mud-pumping capacity. Currently, the number of drilling rigs in the world engaged in geothermal drilling is about 1% of that for petroleum.

Some drilling procedures as well as casing designs have been modified to meet different reservoir conditions. The drilling tools and methods are, in general, similar to petroleum drilling. The drill bits are selected for hard rock, mainly tricone bits with tungsten carbide inserts but also some polycrystalline diamond compact bits (PDC) and hybrids of the two have found application. The steel grade

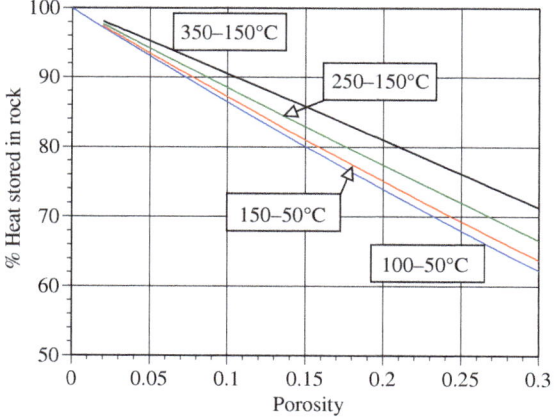

FIGURE 71.6 Percentage of heat stored in basalt as a function of porosity. Four scenarios are shown corresponding with different reservoir temperatures. The numbers on the left in each box represent the reservoir temperature and the ones on the right the lower limit of usable temperature.

in the casings, e.g. API K55 or L80 is selected with respect to the presence of H_2S and H_2 and the technical standards are the same as for oilwell casings.

Nowadays, most geothermal wells are drilled directionally, rather than vertically. The trajectory of the directionally drilled geothermal wells is similar in most fields worldwide. The special directional drilling tools employed are a steerable mud motor just above the drill bit with logging equipment (MWD, measurement while drilling) next to it to relay information to the surface on the angle of inclination and azimuth (direction). The inclination relative to vertical is changed gradually from 1.5 to 3 per 30 m at a depth of 300–400 m until it reaches 30–45° according to the directional plan. The final inclination is reached at 700 to 800 m depth, and continued drilling is usually straight in the desired direction and at fixed inclination until the total depth is reached. Thus, the bottom of a typical directional well is about 1000 m displaced from vertical (Figure 71.7).

The drilling fluid is rather simple for geothermal drilling. While drilling for the casings, the drilling fluid is mainly a low-solid drilling mud (bentonite about 50–60 kg/m^3) with some caustic soda added to stay above pH 9. In the open hole section, the drilling fluid is water only, but often compressed air and drilling soap are added for "pressure balance drilling" after permeable horizons have been intersected. Permeability shows up as loss zones, where the drilling fluid enters the reservoir due the lower pressure in the reservoir than in the fluid-filled well. Large fluid losses often encountered in geothermal wells give rise to drilling problems such as cleaning cuttings from the well and achieving good cementing of the casings. The drill string may also get stuck in the hole because of fluid losses, especially in the case of multiple loss zones. Although fluid losses cause many drilling problems, they are precisely what is required for a productive well. An indication of future well productivity can be obtained with the rig still in place with a step-rate injection test and monitoring of the pressure change downhole at (or near) the main loss zone. For the test, water is pumped into the well in 3 hour long steps of 20, 40, and 60 L/s and the downhole pressure change is recorded. This usually produces a linear relationship of flow vs pressure and the result is reported as the Injectivity Index in units of (kg/s)/bar. Generally a Injectivity Index less than 2 (kg/s)/bar indicates a poor well, around 5 (kg/s)/bar is fine, but values above 10 (kg/s)/bar indicate a very permeable and a good well.

The drilling equipment for production-size wells is large and heavy, requiring 50–100 truckloads to transport. The drill rigs have a hook load rating of 200–450 tons that have a capacity to drill to a depth of 3000–7000 m. A few smaller truck-mounted rigs exist in the 60 to 100 ton range that drill slimholes to 1500–2000 m for exploration and also small wireline coring rigs for 1000–1500 m used by the mineral exploration industry.

For high-temperature wells, the casing design usually calls for three cemented casing strings of different diameters that extend to successively greater depths (Figure 71.7). The selection of drill bit and casing diameters usually follows a recommendation by the American Petroleum Institute and three different casing programs presently dominate. The "regular" high-temperature well casing program has a production casing outside diameter of 244.5 mm (9 $\frac{5}{8}$″) and ends in a 177.8 mm (7″) holed liner. This is the most common diameter size selection. The "large" well design has a 339.7 mm (13 $\frac{3}{8}$″) production casing that ends in a 244.5 mm (9 $\frac{5}{8}$″) liner. In addition, a few slimhole designs have a 193.7 mm (7 $\frac{5}{8}$″) production casing and a 114.3 mm (4 ½″) holed liner. These casing programs are the same for vertical or/and directional drilling. The "regular" well is designed for production, whether drilled for the purpose of exploration or other purposes. In very permeable and productive reservoirs, the well output may be restricted by the casing diameter. Then the "large" casing program, which can produce up to twice as much, is preferred.

The surface casing is typically set at a depth of 60–80 m to support the well through the soil and loose overburden, and the first blowout preventer is attached to it. The blowout preventers are of the same type as for petroleum drilling. They have different sizes depending on the section of the well being drilled. The blowout preventer stack typically consists of a double-gate ram preventer, an annular preventer, and a rotating head preventer. The casing is cemented in place with a high-temperature cement mix that is blended with water to produce a slurry, which is pumped down inside the casing, out of the bottom of the pipe, and back up to the surface via the annulus between the casing and open hole. It is extremely important that all casing strings are fully cemented from top to bottom. The cementing operation is one of the most critical procedures for longevity of the well. The second casing string is the anchor casing, which is landed at a depth of 300–400 m. The innermost casing is the production casing, set to 800–1100 m depth in a typical 2000–2500 m deep well. All of these three casing strings reach back to the surface and are cemented in place along their full length. The "open hole" section of the well is then drilled to the target depth. Finally, a liner with drilled holes is landed in the well, which allows the reservoir fluid to flow into the well (up to 100 holes per meter of liner, each with a diameter of 20 mm). For low-temperature wells in volcanic regions like Iceland that typically reach 1000–2000 m depth, well designs are similar in diameter but the production casing and liner are omitted. In continental settings where the geothermal gradient is near normal, wells may reach 3000–6000 m depth. Such wells require more casing strings and have well screens in the open hole section. All low-temperature wells have downhole pumps installed inside the production casing, either of the shaft-driven type

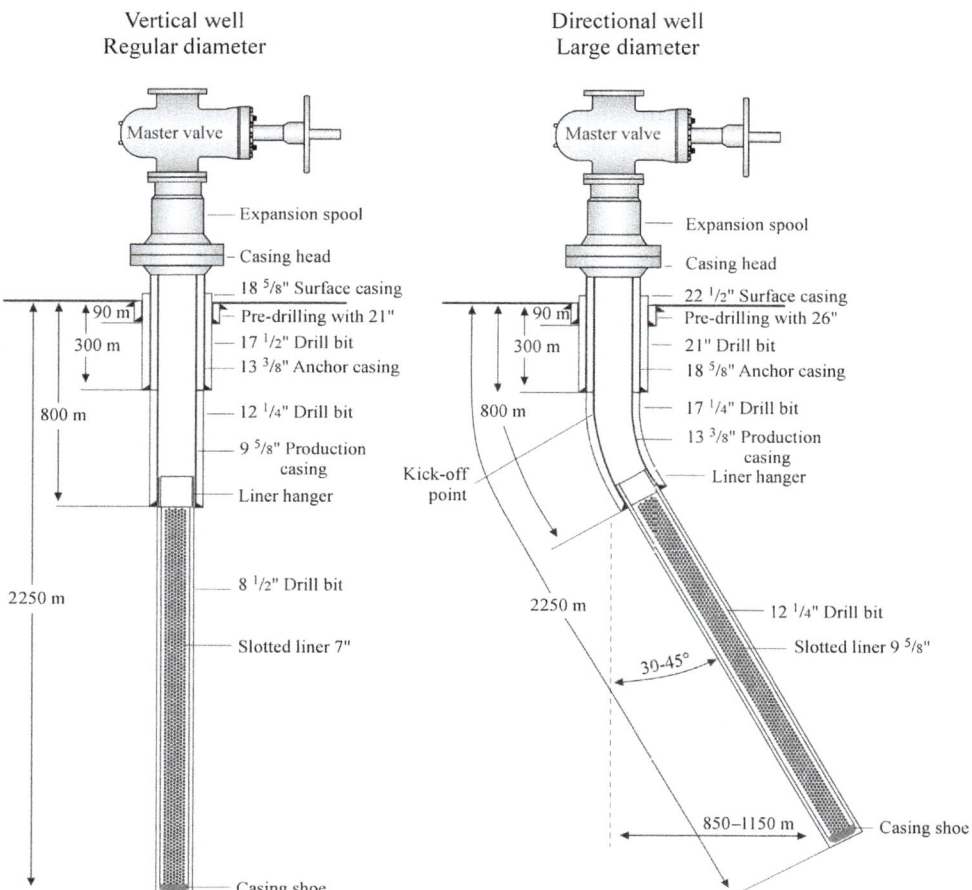

FIGURE 71.7   Schematic layout of high-temperature geothermal wells drilled vertically and directionally. The casing program shown on the left is for a "regular" diameter well but on the right of a "large" diameter well. Casing design is the same, irrespective of the trajectory but the directional wells require special tools to deviate the well while drilling. Logging operations for each well section, as practiced in Iceland, are shown.

or electrical submersible pump. Some wells can be so productive that the maximum flow is limited by the diameter of the pump body that can be placed inside the casing.

6. STRATEGY IN GEOTHERMAL DEVELOPMENT AND USE

6.1. General

The presence of hydrothermal reservoirs is generally manifested by hot springs and/or fumaroles and, in some cases, by hydrothermal alteration and sinter deposits. Exploration has, however, revealed hidden reservoirs. Sometimes the distribution and intensity of surface hydrothermal activity and the lateral extent and productivity of the underlying reservoir are not strongly related.

As with exploitation of the Earth's mineral resources, uncertainty is always involved with the success and economy of developing hydrothermal reservoirs at the time when the decision is made to go ahead with exploration. The uncertainty is due to limited information that is

available about the characteristics and size of the anticipated resource. Development, therefore, always requires risk capital. Sometimes, exploration and subsequent development work reveals that utilization of the resource would be uneconomic. The most common causes for this are inadequate permeability and unfavorable reservoir fluid chemical composition.

Due to the uncertainty in predicting the success of developing a hydrothermal resource, it is common to divide exploration and subsequent development into phases. For each phase, the aim is to minimize cost and maximize information. At the end of each phase, a decision is made whether to continue or terminate the project. The phases are as follows:

1. Surface exploration
2. Exploration drilling
3. Appraisal drilling
4. Feasibility study and preliminary plant design
5. Additional production drilling to recover the amount of fluid required

The first four phases listed above correspond to comparable phases used in the exploration and development of subsurface mineral resources. The mining industry calls the different phases as anomaly, indicated deposit, proven deposit, and economic deposit. An environmental impact assessment is part of the early development phases. A more detailed description of the steps involved in geothermal exploration and development is shown in Figure 71.8.

6.2. Surface Exploration

Surface exploration is used to locate favorable geothermal fields within a particular region or country and to aid in skillful siting of exploration wells within them. It includes geological studies, geochemical and geophysical surveys, and sometimes hydrological balance assessment also.

Geological studies focus on mapping geological formations and the distribution and type of thermal manifestations. An attempt should always be made to relate the distribution of thermal manifestations with local structures such as faults and volcanic edifices. Frequently, the overall distribution of geothermal activity can be linked with larger geological structures, such as calderas, in which case, favorable drilling targets might be confined to the caldera, or by fractures/faults with the purpose of intersecting these permeable structures. Calderas, which form in response to rapid emptying of a magma chamber during a major explosive volcanic eruption, define approximately the lateral extent of the underlying magma chamber that was emptied when the caldera formed. New magma is likely to flow along the path of earlier magma that formed the magma chamber. Thus, the lateral extent of the magma heat

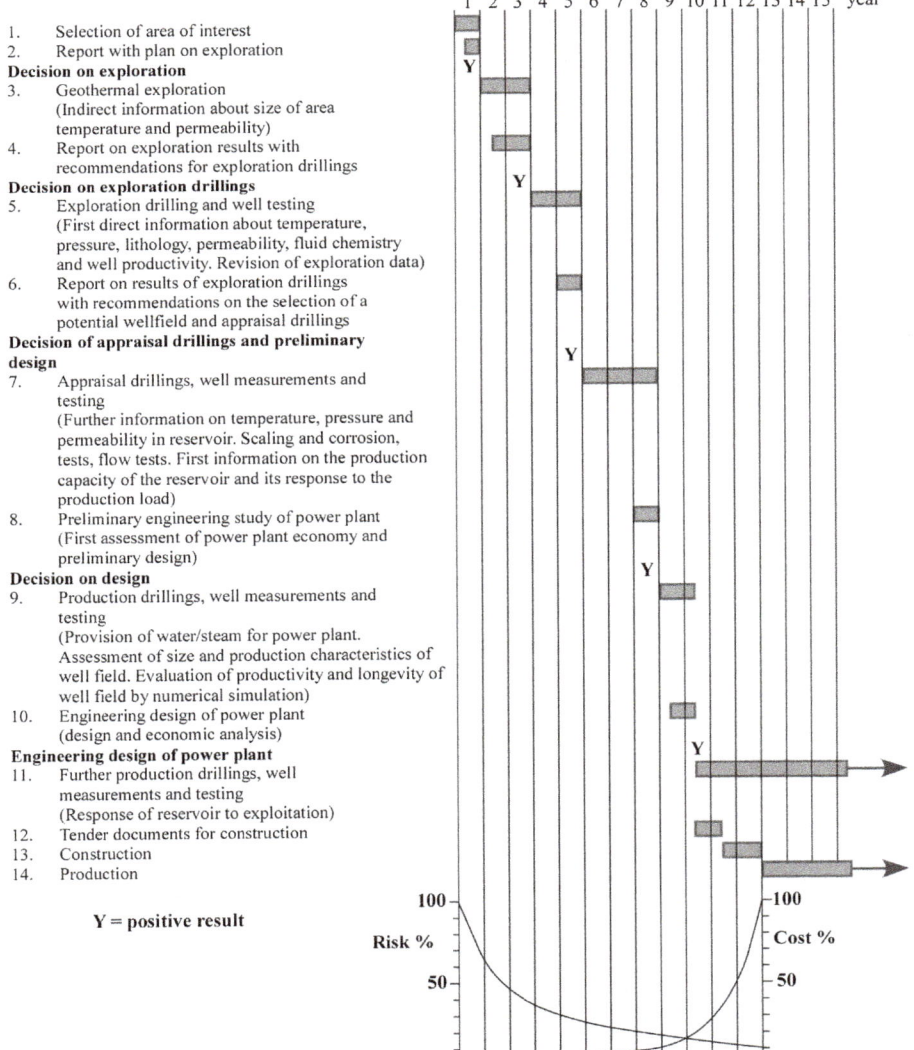

1. Selection of area of interest
2. Report with plan on exploration
Decision on exploration
3. Geothermal exploration
   (Indirect information about size of area temperature and permeability)
4. Report on exploration results with recommendations for exploration drillings
Decision on exploration drillings
5. Exploration drilling and well testing
   (First direct information about temperature, pressure, lithology, permeability, fluid chemistry and well productivity. Revision of exploration data)
6. Report on results of exploration drillings with recommendations on the selection of a potential wellfield and appraisal drillings
Decision of appraisal drillings and preliminary design
7. Appraisal drillings, well measurements and testing
   (Further information on temperature, pressure and permeability in reservoir. Scaling and corrosion, tests, flow tests. First information on the production capacity of the reservoir and its response to the production load)
8. Preliminary engineering study of power plant
   (First assessment of power plant economy and preliminary design)
Decision on design
9. Production drillings, well measurements and testing
   (Provision of water/steam for power plant. Assessment of size and production characteristics of well field. Evaluation of productivity and longevity of well field by numerical simulation)
10. Engineering design of power plant
    (design and economic analysis)
Engineering design of power plant
11. Further production drillings, well measurements and testing
    (Response of reservoir to exploitation)
12. Tender documents for construction
13. Construction
14. Production

Y = positive result

FIGURE 71.8 Strategy in geothermal exploration and development. Work involving environmental impact assessment is not shown. It varies from country to country when this work is carried out during the development work. Likely, this is affected by legislation in each country.

source to the hydrothermal system would not be expected to extend beyond the caldera.

The principal purpose of geochemical surveys is to estimate temperatures in the underlying hydrothermal system with the aid of geothermometers. Sometimes carbon dioxide and radon are measured in soil gases to locate thermal anomalies. Data on hydrogen and oxygen isotopes in the water provide information on the origin of the hydrothermal fluid and subsurface flow directions. Studies in many geothermal fields have shown that mineral—solution equilibria exist at depth in hydrothermal reservoirs. This is an important observation, as equilibrium greatly constrains fluid compositions, making it possible to predict fluid properties with respect to their scaling tendencies and corrosion. Application of geothermometers assumes that specific mineral—solution equilibria prevail in the reservoir and that chemical reactions in upflow zones insignificantly modify the reservoir fluid composition, making it possible to use the chemical composition of hot-spring waters and fumarole steam to estimate subsurface temperatures.

Geophysical surveys have been classified into two groups: one providing information on geological structures and another supplying temperature information. The most widely used geophysical exploration tool maps electrical resistivity of bedrock. Hydrothermal systems are typically identified as resistivity lows. Resistivity surveys have proved useful to map the lateral extent of hydrothermal systems. Gravity surveys are sometimes useful in locating faults and intrusive bodies. Magnetic surveys are sometimes employed because hydrothermal alteration can cause decay of magnetite, leading to a negative magnetic anomaly.

Shallow ground temperature surveys (1 m) and drilling of shallow gradient holes (from few tens to few hundreds of meters deep) have also been used to locate upflow zones in hydrothermal systems. Ascending hot water/steam produces anomalously steep thermal gradient near the surface.

All data from exploration surveys are synthesized into a first conceptual reservoir model, which shows predicted subsurface temperatures, structural control of fluid flow, lateral extent of the **hydrothermal field**, and possibly other geological features such as caldera structure. If the outcome of the exploration survey is favorable, the report describing the results should include proposed sites for exploration drillholes.

6.3. Exploration Drilling

Exploration drilling provides the first direct data about the hydrothermal reservoir. It is much more expensive than the earlier surface exploration. The most important data obtained from exploration drillings include temperature, permeability, fluid chemical composition, and well's thermal output (mass flow and enthalpy). Successful results

from exploration drillings lead to the selection of a prospective well field within a particular hydrothermal area. Following the drilling of a successful exploration well, drilling of **step-out wells** is usually done as a next step to get preliminary information about the extent of the anomaly (temperature, permeability) discovered by the exploration well.

In a hydrothermal area that is tens of square kilometers in size, it is common to drill several exploration holes in one go. These holes are drilled at possible production well fields on the basis of the surface exploration results, and the best prospect is selected for drilling step-out wells.

The main purpose of exploration drilling is to locate a promising production well field. However, in the case of small fields, it may be justified to drill wells near the anticipated field margins to verify its boundaries, especially when size and capacity are of concern for the intended utilization. As a rule, the principal aim of drilling any deep borehole is to prove hot fluid. New information from exploration drillings and later drilling and well testing requires updating the conceptual reservoir model.

6.4. Appraisal Drilling

Having located a promising production well field through exploration and step-out wells, the next development phase involves appraisal drillings to quantify the production characteristics of the identified hydrothermal reservoir. Appraisal drilling provides the necessary information for design and economic analysis of the project and, last but not least, the recovery of fluid for the intended utilization. For that reason, appraisal drilling is often termed as production drilling. However, any productive well is appropriately termed a production well. Having delineated a production well field, additional production wells are drilled between the **appraisal wells** to recover the needed amount of fluid for the production.

6.5. Feasibility Study and Plant Design

There are no fixed rules as to how much fluid should be proved (recovered) before a firm decision is made to construct a plant, whether for direct use of heat or for power generation. In a new area, the amount of steam to be proved for a power plant may be as much as 80% of the total steam required. However, if a new production field is being developed within an already exploited hydrothermal area, the number may be considerably lower (~50%) because understanding of the field production characteristics is already likely high based on information and experience from the previous development.

It used to be common practice to discharge all productive wells drilled in a particular well field simultaneously over a period of time (several months) to test interference and measure decline in output with time. For environmental reasons, such tests may be limited to shorter periods and each new well is flow-tested shortly after it has been completed. Through numerical simulation, the flow-test results are used to predict the generating capacity of the reservoir as well as decline in well output and the longevity of the well field. When production history is short, the prediction carries considerable uncertainty but becomes more reliable as production history becomes longer.

Details of the strategy of geothermal exploration and development depend on the size of the plant to be constructed. The scheme described above was drawn up for a 50- to 100-MW electric power plant, which is a common size for such plants.

The electric power generating capacity per unit area of a wellfield depends on many factors such as temperature, permeability, and porosity. A useful working rule, when information is limited, is to assume a capacity of 15 MW_e/km^2 of well field. Thus, for a 75 MW plant, a wellfield of 5 km^2 size would be required plus some reserve area for drilling additional wells to replace the anticipated decline in output from initial production wells. Assuming the thickness for a hydrothermal reservoir of 1.5 km^2, rock porosity of 10%, and average reservoir temperature of 250 °C, the amount of fluid stored in 1.5 km^2 area of rock suffices to generate 15 MW of electric power over a period of 35 years, i.e., a period more than sufficient to depreciate the power plant.

Geothermal wells are never bonanzas like oil and natural gas wells. Oil wells produce a commodity measured in dollars, whereas wells drilled into hydrothermal systems produce a commodity measured only in cents. For this reason, exploration and development of such system demands a high ratio of successful wells. To generate electric power from geothermal steam in a particular area would be economically very attractive only if the average well yield was ~ 10 kg/s of steam (equivalent to $\sim 5\ MW_e$) over a period of at least 10 years or proportionally longer if the yield is lower. The average steam yield per drilled well into volcanic hydrothermal systems worldwide is 4—5 MW_e. The poorest average yield of wells drilled into exploited hydrothermal reservoirs is a little less than 2 MW_e.

7. MONITORING STUDIES

During the early days of exploitation of hydrothermal systems, it was considered sufficient to prove the required steam flow rate for the intended use by short-term testing (a few months or even less) of producing wells and extrapolation of the results into the future. Experience, however, has shown that fluid pressures may decline considerably in exploited hydrothermal reservoirs because recharge is less than the rate of fluid extraction. This leads to decreased flow from wells and enhancement of cooler groundwater inflow from the surrounding rock. This

necessitates monitoring of the response of hydrothermal reservoirs to the production load. Therefore, geoscientific work is not complete at the time of commission of a power plant.

Overexploitation of hydrothermal reservoirs is not uncommon. It leads to rapid drawdown of reservoir pressure and a corresponding decline in well output and power generation. This happened, for example, at the Geysers vapor-dominated field in California. The reason for rapid pressure drawdown was low porosity of reservoir rock and the correspondingly low quantity of liquid water in the reservoir that would generate steam by its boiling. To counteract pressure decline, water has been injected into the reservoir at the Geysers.

Today, monitoring of the response of hydrothermal reservoirs to the production load has become the general practice worldwide. Data on the long-term response of a hydrothermal reservoir to production are necessary for specifying how the reservoir should be best exploited both economically and environmentally. Data to be collected are both physical and chemical. They involve regular measurement of flow rate and discharge enthalpy of wells and sampling and analysis of well fluid. Reservoir pressure also needs to be monitored at depth in the nonproducing wells. The chemical data are useful for assessing any changes in scaling tendencies but more importantly to map cold water recharge into the reservoir that is enhanced by production. Monitoring data are valuable for timing and siting of new wells that need to be drilled to counteract the decline in flow from earlier production wells. The data are further used to update the reservoir production model to improve predictions about the long-term generating capacity of the reservoir.

8. ENVIRONMENTAL ASPECTS

8.1. General

Most often geothermal energy can be regarded as a relatively clean resource and it is so in comparison with fossil fuels. The environmental impact of exploitation, however, can be considerable. It varies by the fluid's chemical composition, the strength of rock overlying the reservoir and its stress field, and the magnitude of exploitation. Pollution caused by harnessing hydrothermal reservoirs can be (1) visual, (2) thermal, and (3) chemical. Additional adverse environmental impacts include the following: (1) extinction of surface thermal activity, (2) spoiled scenery, (3) noise pollution, (4) soil erosion and damage to vegetation, (5) land subsidence, and (6) induction of or enhancement of seismic activity. Volcanic hydrothermal fields often occur in areas of highly valued scenic beauty and they may, therefore, be of economic value not only as energy resources but also for tourism.

8.2. Physical Impact and Pollution

The principal change resulting from exploitation of hydrothermal systems is reservoir pressure drawdown. This may cause hot-water springs to dry out but enhance fumarole activity. Also, pressure drawdown will enhance flow of recharging cold groundwater into the reservoir. The magnitude of these changes depends on the intensity of exploitation, i.e., the rate of fluid withdrawal from the hydrothermal system.

Power plants, steam supply systems, and roads to individual wells and along the steam pipelines all degrade the scenery and disturb the land surface, particularly in rugged terrain. A combination of heavy rains and steep slopes of incompetent rock can cause landslides and high soil erosion rates. Steam flow from the powerhouse (steam ejectors, cooling towers), steam separators, and wells leads to visual and some thermal pollution. If wastewater is disposed of on the surface, shallow groundwater and streams and other surface waters can become thermally polluted. Noise pollution occurs by wells, steam separators, steam ejectors, and cooling towers.

If the geological formation on top of geothermal reservoirs is incompetent, as is, e.g., the case at the Wairakei and Ohaaki-Broadlands fields in New Zealand, lowering of the water level in the reservoir causes emptying of fluid from pore spaces. As a consequence, they collapse, leading to land subsidence. This can cause problems in relatively densely populated areas and in flat-lying areas where rivers are flowing. Power production at Wairakei started in 1958, and by 2005, subsidence at Wairakei amounted to as much as 14 m.

Injection of waste fluid from geothermal power plants into the ground causes the groundwater table around an injection well to rise, thus increasing the hydrostatic head. This may affect the stress field in the rock sufficiently to cause it to fracture, thus creating earthquakes. The likelihood for this depends on the initial stress field.

Utilization of low-temperature fields ($<150\,^{\circ}C$), which are developed for direct use of the geothermal heat, generally has much less environmental impact than exploitation of volcanic hydrothermal systems for power generation. This is so partly because low-temperature fields tend to occur in areas of lower relief. Hot-water pipelines require little maintenance and can, therefore, be subsurface. Also, the hot water is generally pumped from the wells and the quantity pumped equals the usage, so no water is wasted. For example, in the low-temperature fields exploited by Reykjavík Energy in Iceland, no visual impact exists except for small shelters that cover each wellhead.

8.3. Chemical Pollution

The environmental impact of geothermal energy utilization generally of the most concern is chemical pollution due to air-borne and water-borne pollutants. Geothermal waters have, as a rule, much higher salt content than nonthermal groundwater. Many elements can occur in concentrations much higher than acceptable for domestic use, irrigation, stock watering, various industries, and aquatic life. This is particularly the case for arsenic, boron, hydrogen sulfide, and overall salt concentration. Furthermore, the concentrations of some trace elements, especially in saline fluids, may be unacceptably high. These include elements such as arsenic, mercury, lead, manganese, and zinc. Aluminum concentrations may be quite high in dilute high-temperature water and may far exceed permitted concentrations in domestic water and is too high for aquatic life.

Geothermal steam always contains some gases and other volatiles, including carbon dioxide (CO_2), hydrogen (H_2), hydrogen sulfide (H_2S), methane (CH_4), atmospheric gases, as well as mercury (Hg) and sometimes boron (B). Generally, the most abundant gases are CO_2, H_2S, and H_2. The source of the geothermal gases may be the magma heat source, the rock with which the geothermal fluid has interacted, or they may form by chemical reaction such as CH_4 by reduction of CO_2. The greenhouse gases, CO_2 and CH_4, are present in highly variable concentrations. The CO_2 concentrations are sometimes equilibrium-controlled but sometimes source-controlled. Methane and boron may be present in very high concentrations in steam in high-temperature systems associated with marine sediments such as at Ngawha in New Zealand. At least some volcanic geothermal systems in the Eastern Rift Valley of Kenya contain fluids rich in CO_2 (up to 10% in the steam).

The gas in geothermal steam that is usually of the greatest concern from an environmental perspective is H_2S. This gas is highly poisonous, is corrosive, and has a noxious odor. It is common to remove H_2S from the noncondensable gases extracted from the turbine condensers to reduce atmospheric pollution. Many removal methods are known, but the ones most commonly used involve oxidation of the H_2S into either native sulfur or sulfate.

8.4. Mitigating Measures

In addition to various mitigation measures such as plant design and effluent treatment, basically two types of measures have been implemented to reduce the environmental impact: directional drilling and injection of spent fluid.

Today, directional drilling is very common in hydrothermal fields, mainly for environmental reasons but also to reach difficult targets under mountains or rugged terrain. When drilling directionally, several wells can be drilled from the same platform requiring fewer roads and rig platforms. The success of directional drilling depends on the depth of producing horizons. If aquifers are abundant in the reservoir just below the production casing, the directionally drilled wells will be closely spaced at the depth level of these aquifers and interference between them is likely. Directional drilling works best when productive aquifers occur at deep levels in the reservoir. It seems logical to first drill vertical wells in a new field to obtain information on the depth level of producing horizons and subsequently deviated wells when, at least, several appraisal wells have been drilled in a prospective wellfield.

During the early years of geothermal energy utilization, the waste fluid was generally disposed of by the least expensive method available. This most often involved surface disposal into the nearest stream or into ponds constructed for this purpose. To reduce harmful chemical pollution, injection of all waste fluid into wells is the norm today. Injection may involve the drilling of special wells, either within the hydrothermal reservoir or outside of it. Alternatively, wells drilled for the purpose of production that turn out to be nonproductive or very poor may be used for injection. In any case, if injection is intended, this has to be taken into account during phase 3 (see Figure 71.8) in the development strategy.

Injection of spent hydrothermal fluids was seriously discussed for the first time at the UN Conference on Geothermal Energy held in Pisa, Italy, in 1970. There were two basic reasons for the interest to inject spent geothermal fluid. One is that injection helps maintain reservoir pressures, thus reducing temporal decline in well flow. The second reason is to eliminate the environmental impact caused by surface disposal of water with high concentrations of undesirable chemicals.

The waste fluid may be injected into shallow or deep wells. Injection back into the reservoir is generally considered to be the best solution assuming that scaling is not troublesome. Caution is, however, advised on such injection because of the risk of thermal breakthrough, i.e., the relatively cool injected water may flow rapidly into the aquifer of producing wells and cause their performance to deteriorate. Because permeability within hydrothermal reservoirs is variable and anisotropic, it is not possible to predict with confidence how an injection well will perform with respect to its capacity or to which way the injected fluid will flow once in the reservoir. Although injection will have a positive effect on reservoir pressure, it may reduce mining of the heat stored in the reservoir rock. Extensive depressurization boiling in a high-temperature reservoir that occurs as a consequence of the reservoir pressure drop will cause cooling of the depressurized flowing fluid. This creates a positive temperature gradient from rock to fluid that could enhance boiling of immobile water in micropores (vesicles in volcanic rocks), but the steam formed would flow due to its better flowing properties and end up in production wells.

A problem sometimes associated with injection is deposition of minerals from the fluid in the well, in the receiving formation, or in both. The mineral that is most often of concern is amorphous silica. Other common scale-forming minerals are calcite and anhydrite. The solubility of amorphous silica decreases with decreasing temperature. For this reason, deposition of this phase is only expected to occur in surface equipment and possibly in wells, if cooling is sufficient to make the initially quartz-saturated fluid oversaturated with more soluble amorphous silica. Calcite and anhydride have retrograde solubility with respect to temperature, so they may deposit in the receiving aquifer, where the wastewater gains temperature. Deposition will reduce the well capacity, which can be restored by acid cleaning. Sulfide minerals have both prograde and pH-dependent solubility and deposit readily from solution. Scales of sulfides invariably form from high-temperature fluids when they cool. These are minor from dilute waters but may be severe from very saline fluids. It is possible to predict with reasonable confidence the temperature range in which deposition of different minerals does not occur or is minimal in order to reduce mineral deposition. Injection plans should take this into consideration.

FURTHER READING

Arnórsson, S., 2004. Environmental impact of geothermal energy utilization. In: Gieré, R., Stille, P. (Eds.), Energy, Waste and the Environment: A Geochemical Perspective. Geological Society of London, pp. 297–336. Special Publication 236.

Arnórsson, S., Axelsson, G., Saemundsson, K., 2008. Geothermal systems in Iceland. Jökull 58, 269–302.

Arnórsson, S., Stefánsson, A., Bjarnason, J.Ö., 2007. Fluid-fluid interaction in geothermal systems. In: Liebscher, A., Heinrich, C.A. (Eds.), Reviews in Mineralogy and Geochemistry, vol. 65, pp. 259–312.

Bertrani, R., 2012. Geothermal power generation in the world 2005–2010 update report. Geothermics 41, 1–29.

Coumou, D., Driesner, T., Heinrich, C.A., 2008. Heat transport at boiling, near-critical conditions. Geofluids 8, 208–215.

Davis, J.H., Davis, R.D., 2010. Earth's surface heat flux. Solid Earth 1, 5–24.

Duffield, W.A., Sass, J.H., 2003. Geothermal Energy — Clean Power from the Earth's Heat. U.S. Geological Survey Circular 1249.

ENGINE Coordination Action, 2008. Best Practice Handbook for the Development of Unconventional Geothermal Resources with a Focus on Enhanced Geothermal Systems. BRGM Editions, Orleans, ISBN 978-2-7159-2482-6. Collection Actes/Proceedings.

Finger, J., Blankenship, D., 2010. Handbook of Best Practices for Geothermal Drilling. SANDIA Report, SAND 2010-6048, 84 pp.

Girardet, H., Mendonça, M., 2009. "A Renewable World — Energy, Ecology, Equality." A Report for the World Future Council. Green Books Ltd, Darlington, UK.

Glover, R.B., Mroczek, E.K., 2009. Chemical changes in natural features and well discharges in response to production at Wairakei, New Zealand. Geothermics 38, 117–133.

Grant, M.A., Donaldson, I.A., Bixley, B.F., 1982. Geothermal Reservoir Engineering. Academic Press, New York.

International Geothermal Association, 2010. In: Proceedings of the World Geothermal Congress 2010, Bali, Indonesia, 24–29 April, 2010.

Lund, J.W., Freestone, D.H., Boyd, T.L., 2010. Direct use of geothermal energy 2010 worldwide review. Geothermics 40, 159–240.

Mannington, W.I., O'Sullivan, M.J.O., Bullivant, D.P., Clotworthy, A.W., 2004. Reinjection at Wairakei-Tauhara: a modelling case study. In: Proceedings Twenty-ninth Workshop on Geothermal Reservoir Engineering, Stanford University, Stanford, California, 26–28 Jan., 2004. STP-TR-175.

O'Sullivan, M., Yeh, A., Mannington, W.I., 2010. Renewability of geothermal resources. Geothermics 39, 314–320.

Rybach, L., Muffler, L.P.J., 1981. Geothermal Systems: Principles and Case Histories. Wiley, Chichester.

Sanyal, S.K., 2005. Sustainability and renewability of geothermal power capacity. In: Proceedings World Geothermal Congress 2005, Antalya, Turkey abstract 0520.

Schwarzschild, B.M., 2011. Neutrinos from Earth's interior measure the planet's radiogenic heating. Phys. Today 64, 14–17.

Stefánsson, V., 2000. The renewability of geothermal energy. In: Proceedings of the World Geothermal Congress 2000, Kyushu – Tohoku, Japan, 28 May to 10 June 2000, pp. 883–888.

Williams, C.F., Reed, M.J., Mariner, R.H., DeAngelo, J., Galanis Jr, S.P., 2008. Assessment of Moderate- and High-temperature Geothermal Resources of the United States. U.S.G.S. – science for a changing world.

World Energy Council, 2010. 2010 Survey of Energy Resources.

Chapter 72

Volcanic Soils

Pierre Delmelle, Sophie Opfergelt and Jean-Thomas Cornelis
Earth & Life Institute, Environmental Sciences, Université Catholique de Louvain, Louvain-la-Neuve, Belgium

Chien-Lu Ping
School of Natural Resources and Agricultural Sciences, University of Alaska Fairbanks, Fairbanks, AK, USA

Chapter Outline

GLOSSARY

bulk density The mass of dry soil per unit bulk volume, including the air space.

clay mineral Naturally occurring crystalline silicate mineral found in soils, the particles being of clay size, that is, <0.002 mm in diameter.

chronosequence A sequence of related soils that differ, one from the other, in certain properties primarily as a result of time as a soil-forming factor.

horizon A layer of soil, approximately parallel to the soil surface, differing in properties and characteristics from adjacent layers below or above it.

layer silicate clay A combination in silicate clays of tetrahedral silicon and octahedral aluminum sheets. One distinguishes 1:1 clay minerals (one tetrahedral sheet for each octahedral sheet) from 2:1 clay minerals (two tetrahedral sheets for each octahedral sheet).

permeability The ease with which gases, liquids, or plant roots penetrate or pass through a bulk mass of soil or a layer of soil.

porosity The volume percentage of the total bulk soil not occupied by solid particles.

primary mineral A mineral that has not been altered chemically since deposition from volcanic eruption.

profile A vertical section of the soil through all its horizons and extending into the parent material.

secondary mineral A mineral resulting from the decomposition of a primary mineral or from the reprecipitation of the products of decomposition of a primary mineral.

soil order The category at the highest level of generalization in US Department of Agriculture Soil Taxonomy. The properties selected to distinguish the orders are reflections of the degree of horizon development and the kinds of horizons present.

soil taxonomy A basic system developed by the United States Department of Agriculture to provide a hierarchical classification of soils.

tilth The physical condition of soil as related to its ease of tillage, fitness as a seedbed, and impedance to seedling emergence and root penetration.

1. INTRODUCTION

Volcanoes have wrought destruction on the Earth in many ways and will continue to do so in the future. But disaster sometimes brings long-term fortune. Most soils that form in volcanogenic materials, i.e., volcanic soils, have a high agronomical potential and as such support up to ~10% of the world's population, despite accounting for only ~0.7% of the Earth's surface. These soils are widely distributed but are invariably located near active or dormant volcanoes. Their natural fertility is due to a set of recognizable properties, which are seldom observed in soils derived from other parent materials. They are light and fluffy, have a low **bulk density** and remarkable water-holding capacity, contain a clay fraction (solid material <2 μm) dominated

The Encyclopedia of Volcanoes. http://dx.doi.org/10.1016/B978-0-12-385938-9.00072-9

by poorly crystalline and noncrystalline minerals, and tend to accumulate organic matter.

Volcanic soils with these unique characteristics are termed Andisols, one of the 12 **soil orders** of **Soil Taxonomy** (Soil Survey Staff, 2010). In the FAO World Reference Base classification, they are known as Andosols (IUSS Working Group WRB, 2006). The name Andisol or Andosol relates to the Japanese words *anshokudo*, meaning "dark colored soil" (*an*, dark; *shoku*, color or tint; *do*, soil), and *ando*, meaning "dark soil," which both denote the high content of dark organic matter usually occurring in these soils. Most volcanic soils worldwide are Andisols. They occur under various environmental conditions, but the presence of appreciable amounts of silicate glass in the parent material is highly influential in their formation. Furthermore, the intermittent deposition of fresh tephra from nearby volcanic eruptions on these soils counteracts the effects of physical erosion and chemical weathering, thereby enabling maintenance of their distinctive attributes over relatively long periods of time. However, in more stable and older landscapes, Andisols can evolve toward other soil types and acquire new characteristics.

Volcanic soils, particularly the Andisols, fulfill important functions. They are generally good media for plant growth and display a range of agricultural productivity. In the Andes, they act as vast sponges, storing and releasing water. They have the ability to slow or filter the transfer of pollutants to groundwater reservoirs. However, like other soils, volcanic soils are fragile and degradable if not properly utilized and managed. Since volcanic soils constitute a source of livelihood for millions of people who live in the shadow of volcanoes, understanding their physical, chemical, and mineralogical properties and how they function is central to the development of the best practice for their sustainable use during volcanic repose and for their rehabilitation after an eruption.

The purpose of this chapter is to provide an introduction to the formation, evolution, distribution, properties, utilization, and environmental implications of volcanic soils. There exist several excellent compilations on volcanic soils, which interested readers may wish to consult for further details (Arnalds et al., 2007; Dahlgren et al., 2004; McDaniel et al., 2011; Mizota and van Reeuwijk, 1989; Shoji et al., 1993; Ugolini and Dahlgren, 2002; Wada, 1986).

2. FORMATION AND EVOLUTION

2.1. Parent Material

The large majority of volcanic soils form from volcanic pyroclasts and/or epiclasts, but some also derive from lava flows. Pyroclasts include the material deposited by pyroclastic density currents, close-to-source ballistic bombs and scoria, and widely spread tephra fallout. Epiclasts correspond to volcaniclastic materials that have been remobilized and redeposited on the landscape after their emplacement; they include deposits from volcanic debris avalanches, volcanic debris flows, alluvium on the flanks of a volcano, and wind-blown tephra. The majority of volcanic soils originate from tephra (Shoji et al., 1993), which is the most widespread volcanic product.

The volcanic parent material on which volcanic soils develop varies in composition, from low-silica basalt to high-silica rhyolite. Basaltic tephra does not commonly cover extensive areas, and therefore, most volcanic soils correspond to andesitic, dacitic, and rhyolitic compositions. However, volcanic soils derived from lavas are usually associated with the low-silica end-member because basaltic lavas tend to cover larger areas than the more silicic ones. Regardless of composition, all volcanic parent materials contain various proportions of silica-based glass, i.e., an amorphous solid consisting of a disordered arrangement of silicate oxides, with various proportions of aluminum and other elements.

2.2. Formation

Soil formation is effected through weathering of a parent material with time. The interaction of environmental variables such as climate, vegetation, and topography establishes the progression of soil-forming processes. The unique properties (see Section 4) of most volcanic soils emphasize the importance of the parent material lithology on which these soils have formed. Specifically, the ubiquitous presence of silica-based glass in the parent material is central to volcanic soil genesis. Volcanic glass is a fast-weathering source of silicon and aluminum, as well as other chemical elements, for **secondary mineral** formation. The results of laboratory experiments demonstrate that silicate glass of a given composition dissolves more rapidly than its crystalline equivalent. It is also known that glass dissolution rates, and thus weathering, increase with decreasing silica content. Besides crystallinity and chemical composition, two other variables that influence the weathering rate of volcanic parent material are grain size and vesicularity, both of which impact on the surface area available for reactions. Thus, highly vesicular pumice weathers more rapidly than dense rock of identical composition.

Volcanic material subject to weathering generally releases silicon, aluminum, and iron in solution at rates that lead to rapid oversaturation with respect to secondary mineral products. The mineralogy of the solids that can precipitate from such solution is strongly influenced by leaching conditions, as dictated by the amount of rainfall (Churchman and David, 2011). When weathering takes place under high rainfall, intense leaching gives rise to

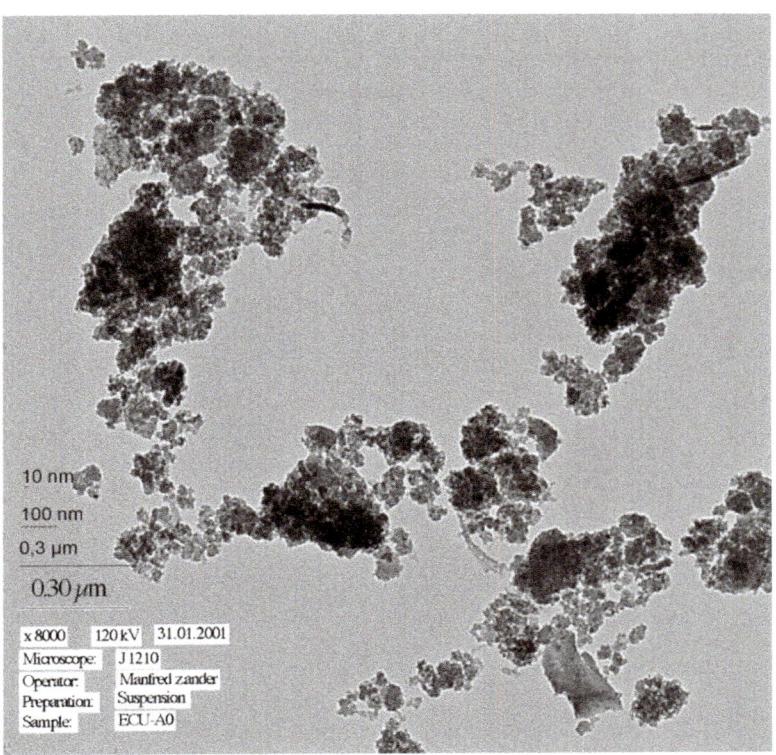

FIGURE 72.1 Transmission electron microscopy micrograph of allophane. The globular particles are a few nanometers in size. *Image by Stephan Kaufhold, Federal Institute for Geosciences and Natural Resources, Hannover, Germany; reproduced by kind permission of the author.*

relatively low dissolved silicon concentrations, which, in turn, favor formation of allophane and to a lesser degree imogolite. Allophane and imogolite are both poorly crystalline aluminosilicates (Parfitt, 2009). Allophane has a globular structure (Figure 72.1) and is represented by the chemical formula $Al_2O_3.SiO_2.nH_2O$, but its aluminum-to-silicon atomic ratio varies (from 2:1 to 1:1) depending on the composition of the soil solution from which it precipitates.

Imogolite has the chemical formula $SiAl_2O_5.2H_2O$ and generally occurs in association with allophane, albeit in minor quantities. Where very strong leaching occurs, i.e., in humid climates, little silicon may be left in solution during weathering, and a secondary aluminous mineral known as gibbsite ($Al(OH)_3$)) can form instead of allophane and imogolite. In contrast, when weathering occurs in dry climate areas, limited leaching assists the buildup of a high silicon concentration in solution. Under such conditions, halloysite, a crystalline aluminosilicate **clay mineral** with the empirical formula $Al_2Si_2O_5(OH)_4$, is the stable weathering product that precipitates. Previously, the presence of halloysite in volcanic soils was assumed to be the result of allophane/imogolite aging. However, this mechanism of halloysite generation in volcanic soils is now largely discounted (Parfitt, 2009). A conceptual model illustrating the effect of soil drainage

and leaching conditions on allophane/imogolite versus halloysite formation is shown in Figure 72.2.

Studies of young volcanic soils found along an altitudinal gradient and derived from a common parent material confirm the critical role of climate, and more specifically leaching regime, in controlling secondary aluminosilicate mineral formation in the early process of weathering (e.g., Rasmussen et al., 2007). At high elevations, where climate is more humid, allophane and imogolite dominate the soil weathering mineral assemblage. On the contrary, halloysite is the major secondary mineral phase occurring in soils at low altitudes, which receive comparatively less rainfall.

In the above model of volcanic material weathering, aluminum is assumed to be unlimited. However, this is not always the case in environments where organic matter is abundant and pH values ≤ 5 can arise. Under such circumstances, organic matter competes effectively for aluminum to form stable aluminum—organic matter complexes (Parfitt, 2009; Shoji et al., 1993). As a consequence, less aluminum in solution is available for coprecipitation with silicon to form allophane and imogolite (Figure 72.2). Typically, allophane in volcanic soils is lower in the soil surface layer, which is continuously replenished with organic matter by decomposition of plant and animal residues, than in the subsoil. The term "anti-allophanic effect"

FIGURE 72.2 Simplified model depicting the effect of silicon and aluminum leaching conditions on the formation of secondary minerals during weathering of volcanic tephra, and the likely occurrence or not of Andisols as a result. ± indicates that the mineral may also be present. *Modified after Churchman and David (2011) and McDaniel et al. (2011).*

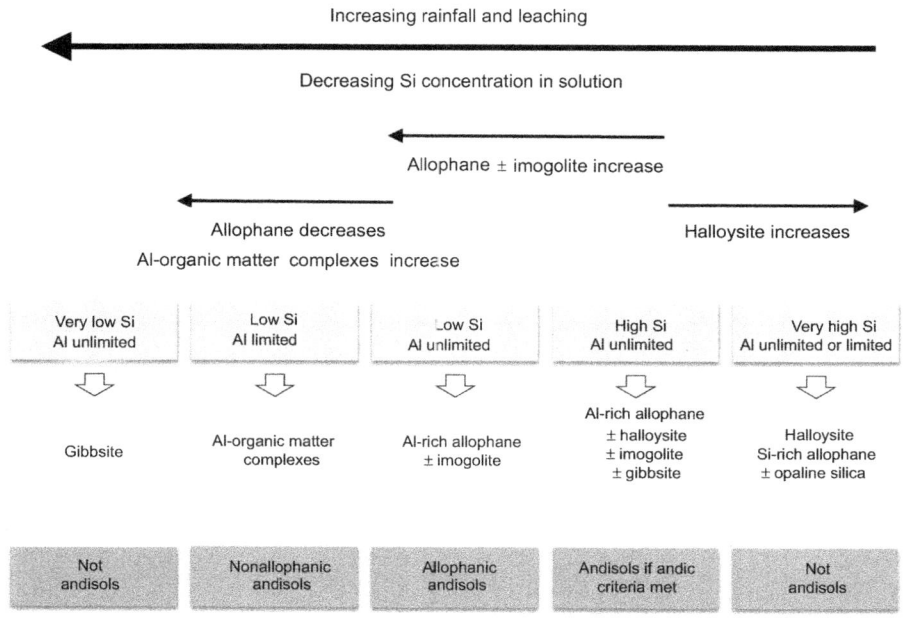

has been coined to describe preferential incorporation of aluminum into organic matter complexes rather than into allophane/imogolite materials.

The presence of crystalline 2:1 **layer silicate clays**, such as smectite and vermiculite, in the soil-forming environment may similarly give rise to an anti-allophanic effect because these mineral phases also bind aluminum in solution. The origin of 2:1 layer silicate clays in volcanic soils is not entirely clear. These clays probably originate mainly from eolian dust addition or in relation to the presence of hydrothermally altered rock fragments in the parent material (Dahlgren et al., 2004). However, smectite can also be a product of basalt weathering in humid tropical and subtropical climates. Combined with unfavorable leaching conditions, the presence of organic matter and/or 2:1 layer clay minerals during soil genesis favors the buildup of high concentration of silicon in solution, eventually leading to precipitation of opaline silica (SiO_2, Figure 72.2) (Dahlgren et al., 2004).

Similar to silicon and aluminum, iron is readily released during weathering of volcanic materials. The main sources of iron are the mafic **primary minerals** such as pyroxenes, amphiboles, biotite, and olivine, but glass also contains iron. Once iron is brought into solution, it will quickly precipitate as ferrihydrite, a noncrystalline iron oxy-hydroxide with the general chemical formula $Fe_5(OH)_8 \cdot 4H_2O$. The formation of more thermodynamically stable crystalline iron oxides such as goethite and hematite in the early weathering stage of the volcanic parent material is believed to be inhibited due to the presence of organic matter as well as silicate and phosphate

ions. The type of iron oxides that are formed in volcanic soils probably depends on the same environmental conditions that influence which particular secondary alumino-silicates are stable (Churchman and David, 2011). For example, Rasmussen et al. (2007) reported that with increasing elevation—and hence, higher rainfall and lower temperature—on the western slope of the Sierra Nevada, California, crystalline iron oxides in the soil are found in lesser abundance than poorly crystalline and noncrystalline iron-bearing minerals. Finally, it has been shown that noncrystalline ferrihydrite formed in young volcanic soil converts to crystalline materials such as goethite as weathering advances with time.

2.3. Evolution

As noted and emphasized above, volcanic soils include all soils derived from material of volcanic origin, but the large majority of volcanic soils worldwide are Andisols. Volcanic Andisols generally originate from fragmented and unconsolidated ejecta produced during explosive eruptions (rather than effusive ones). The key attribute of these soils is the presence of poorly crystalline and noncrystalline secondary minerals. Volcanic Andisols can form under a broad range of environmental conditions, and therefore, their properties vary quite widely. Andisols have andic properties, which generally correspond to low bulk densities, high levels of organic carbon to depth, and high phosphorus retention. Rigorously, to qualify for an Andisol, a soil has to have andic properties in 60% or more of the thickness of soil material within 60 cm of the mineral soil

FIGURE 72.3 Andisol profiles: (left) soil under grassland from southeast Iceland. Several thin, dark tephra layers are visible and (right) soil under banana culture from the Mungo Plain in Cameroon. A paleosol derived from older volcanic tephra is buried at ~75 cm depth.

surface. One distinguishes between the allophanic Andisols, where allophane and minor imogolite outweigh any other aluminosilicates, and the nonallophanic Andisols in which aluminum—organic matter complexes and minor 2:1 layer silicate clay minerals dominate. The *Keys to Soil Taxonomy* (Soil Survey Staff, 2010) describes the diagnostic criteria and identification procedures for Andisols.

The generation of the andic character in volcanic tephra deposits can be rapid; 200—300 years may be sufficient to form an Andisol from fresh tephra under a humid weathering regime. In temperate regions, Andisols can reach maturity in 2000—3000 years. Typical Andisols in Japan are 4000—7000 years old (Shoji et al., 1993). In Iceland, most of the Andisols range in age from a few hundred years to over 10,000 years. In general, the drier and cooler the environment, the longer it takes for an Andisol to form (McDaniel et al., 2011).

The Andisol character in volcanic soils can be maintained as a relatively stable condition over time due to intermittent deposition of volcanic pyroclasts (mainly tephra) and consequent rejuvenation of soil formation processes. During periods of volcanic repose between major eruptions, soil development takes place through normal top-down weathering of the volcanic parent material, a process that progressively gives rise to differentiation of the soil material into subsurface **horizons** with distinct physical and chemical properties. However, when the land surface is completely buried under a thick deposit of fresh tephra, soil formation processes are suddenly terminated.

As a result, the buried soil becomes a paleosol (i.e., a soil that is isolated from the range of most soil-forming processes) and a new soil will begin to form in the freshly deposited tephra. The accretion at a particular site of multiple tephra deposits from eruptions of one or more volcanoes is responsible for the distinctive layered **profiles** of most Andisols (Figure 72.3).

While Andisols are the major type of volcanic soils, they actually depict an intermediary stage of soil formation between the volcanic parent material and one of the more "mature" soil orders. Poorly developed volcanic soils come under the Entisols order, which in *Keys to Soil Taxonomy* (Soil Survey Staff, 2010) represents relatively young soils occurring under low-temperature and low-moisture conditions. In older and more stable landscapes not subject to repeated burial under new volcanic ejecta, weathering can progress with time. Thus, some volcanic soils have developed into more "mature" soil orders such as Mollisols, Alfisols, Spodosols, Ultisols, Vertisols, or Oxisols. Often, this transformation involves gradual conversion of the poorly crystalline secondary aluminosilicate minerals to more stable crystalline layer clays and aluminum hydroxides. The composition of these new phases is controlled largely by climate conditions. In warm and humid regions having a short dry season, kaolinite and vermiculite prevail and the resulting volcanic soil belongs to the Inceptisols, Alfisols, or Ultisols order in *Key to Soil Taxonomy*. In contrast, soils containing expanding 2:1 layer silicate clays (e.g., smectite) known as Vertisols form when weathering of Andisols occurs in regions with a distinct dry season. Under wet tropical environments, an Oxisol dominated by kaolinite and aluminum/iron oxides can evolve from an Andisol with the passage of time.

A beautiful example of volcanic soil evolution can be observed in the Hawaiian Islands, where soil weathering degree correlates with the age of the basaltic parent material. There, soils younger than 20,000 years still contain some primary volcanic minerals such as plagioclase, pyroxene, olivine, and silicate glass, but these components of the parent material are progressively converted to allophane, imogolite, and ferrihydrite as weathering progresses. In soils older than 400,000 years, the poorly crystalline and noncrystalline materials as well as the more resistant primary minerals are no longer stable, and they are gradually replaced by crystalline mineral phases—these soils are Ultisols. The weathering products of soils with age of 1,000,000 years consist nearly exclusively of halloysite, kaolinite, and crystalline iron/aluminum oxyhydroxides, and the soils belong to the Oxisols order.

Under a cool to cold-humid weathering regime such as in Southcentral Alaska and the mountainous areas of the Pacific Northwest, most volcanic soils have developed into Spodosols due to moderate to strong leaching under conifer forests (Ugolini and Dahlgren, 2002). The Spodosols are characterized by a "bleached" (grayish white) layer strongly depleted in aluminum and iron, which may grow with time. This layer (called the E horizon) overlies a "spodic" horizon in which organic matter mobilized from the surface organic-rich horizon formed complexes with iron and aluminum leached from the E horizon. These andic Spodosols share similar physical and chemical properties with Andisols and undergo similar management practices. It may take a few thousand years for an Andisol to transform into a Spodosol, although in some instances,

Spodosols form directly through weathering of the volcanic parent material in less than 1000 years. Finally, some volcanic soils are classified as Mollisols and Gelisols (Soil Survey Staff, 2010). Volcanic Mollisols are typically found in old basaltic terrains under grassland cover in temperate and humid climate areas, whereas Gelisols represent weathered volcanic deposits underlain by, or partially within, permafrost.

3. DISTRIBUTION

The proportion of the Earth's ice-free land surface covered by volcanic soils is not known. A lower bound is set by considering the global distribution of Andisols, which dominate volcanic soils. According to the Soil Survey Staff (1999), 910,000 km^2 or ~0.7% of the world's land area is covered by Andisols, with the majority of Andisols in high-rainfall regions. Approximately 50% of Andisols occur in the tropics, with the remaining half being split between boreal and temperate regions. Importantly, there are large land areas, for example, in the United States, with soils that have been influenced by volcanogenic materials, but which do not qualify as Andisols. These soils with >18 cm but <35 cm of andic material considerably increase and perhaps double the ice-free land area endowed with volcanic soils.

Andisols are found across a broad range of climatic conditions, from cold to hot and wet to dry, revealing that climate is less important to the genesis than is proximity to a volcanic parent material source. The distribution of Andisols (Figure 72.4) closely parallels that of active and dormant volcanoes. The major areas of Andisols by far occur along the Circum-Pacific Ring of Fire, where oceanic plate subduction produces extensive rhyolitic and andesitic volcanism. Volcanic Andisols are common in Chile, Peru,

FIGURE 72.4 Global distribution of Andisols. *(Courtesy of USDA, Natural Resources Conservation Soils, available from http://www.nrcs.usda.gov/wps/portal/nrcs/main/soils/use).* The soil map shows the distribution of the 12 soil orders, including Andisols, according to *Keys to Soil Taxonomy. Soil Survey Staff (2010).*

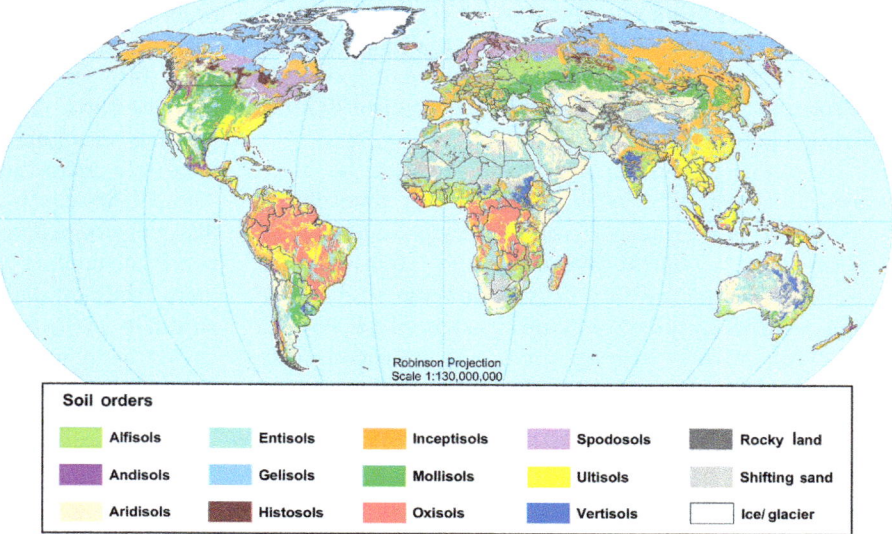

Ecuador, Colombia, Central America, the United States, Kamchatka, Japan, the Philippines, Indonesia, Papa New Guinea, the North Island of New Zealand, and the islands of the Southwest Pacific. Andisols and volcanic ash-derived Spodosols occupy 2.5% of the United States land area, including some very productive forests in the Pacific Northwest and southern Alaska regions. The second major area of Andisols extends along active continental rift zones, notably the East African Rift Valley. They also occur throughout the Mediterranean region (e.g., Sicily and Greece), where tectonic plates converge, and in association with spreading centers along the Mid-Atlantic Ridge, as in Iceland and the Azores. Andisols dominate in Hawaii and La Réunion, where hot mantle plumes pierce through the lithosphere. Cameroon and the Canary Islands are other areas covered with Andisols. In China, volcanic Andisols occur in the northeastern and southern provinces, inner Mongolia, Xinjiang, Hainan, and Taiwan. France, Germany, Portugal, Slovakia, Hungary, and Romania all have

several thousand square kilometers of land covered with Andisols, which are linked to relatively old periods of volcanism. Iceland represents the largest area of Andisols in Europe and accounts for ~5–7% of the world's Andisols.

The frequency of volcanic eruptions and the dispersal pattern, texture, and thickness of the deposits determine the distribution of volcanic Andisols at the landscape scale. These soils are usually the dominant soil in young volcanic terrains; they are encountered on all types of topography and at various elevations, from sea level to altitudes above 3000 m. In general, less-evolved and coarse-textured volcanic soils occur on the flanks of volcanoes, whereas at more distal sites and at long distances downwind, there may be a wide variety of landscapes upon which finer tephra has accumulated over thousands of years to form thick deposits. Thus, deep (>20 m) fertile Andisols can occur in lowlands surrounding active and dormant volcanic centers (Figure 72.5).

In contrast, in hilly areas tephra accumulation may be of variable thickness due to past erosional histories, notably during the Last Glacial Maximum (about 19,000–26,500 years ago), resulting in shallow parent material and thinner soil profiles. Also, water can rework tephra such that extensive flat areas of Andisols are found in intermountain valleys; a good example of this is the inter-Andean valley of Ecuador.

4. PROPERTIES

4.1. Charge Characteristics, Nutrient Availability, and Aluminum Toxicity

Andisols, the most abundant volcanic soil type, display remarkable chemical properties that are linked to the presence of allophane/imogolite, ferrihydrite, and aluminum/iron–organic matter complexes. This fine material gives rise to a very large surface area (Shoji et al., 1993). To illustrate, a single teaspoon (~5 g) of allophane has the surface area of a rugby field (Parfitt, 2009). Moreover, the surface of this fine mineral and organic material carries an electrical charge that varies with pH. Thus, there is a pH value of the soil solution above which the surface charge of allophane/imogolite and ferrihydrite is negative and below which the surface charge is positive. Organic matter has a charge, which remains negative in a wide pH range, but which becomes gradually more negative with increasing pH (i.e., decreasing acidity). The existence of a variable surface charge in volcanic Andisols greatly affects the behavior of ions, including nutrients. In general, these soils have a greater propensity to retain cations (positively charged ionic compounds) at higher pH values, i.e., when the surfaces of poorly crystalline and noncrystalline minerals and organic matter develop a larger negative charge.

FIGURE 72.5 A thick (~16 m) allophane-rich Andisol overlying a halloysite-rich volcanic soil. This formation was recently discovered in northwest Ecuador and covers >4000 km² (Kaufhold et al., 2009). The allophane layer is thought to derive from Quaternary volcaniclastic materials. *Photo by Stephan Kaufhold, Federal Institute for Geosciences and Natural Resources, Hannover, Germany; reproduced by kind permission of the author.*

Conversely, their capacity to hold anions (negatively charged ionic compounds) increases as the pH becomes more acidic.

Such interaction is demonstrated by considering phosphorus, one of the key macronutrients required by plants. The plant-available form of phosphorus is the phosphate anion. Phosphate may be released from tephra upon its addition. In more evolved soil–plant systems, the large accumulation of organic matter acts as the main source of phosphate for plants. However, there is strong competition for phosphate in Andisols due to strong retention of the anion on positively charged mineral surfaces. Much of this retention is irreversible, making phosphate largely unavailable to plant roots. As a consequence, farming on Andisols in many areas necessitates heavy application of phosphorous fertilizers. Sulfur deficiency in some volcanic Andisols is attributed to similar interactions with the variable charge components of the soil.

The availability of plant nutrients in volcanic soils closely parallels the chemical composition of the parent material, most often tephra. A peculiar characteristic of tephra is its ability to provide a variable array of macro- and micronutrients depending on its chemical composition. While tephra deposited on soil may release agronomically significant amounts of nutrient cations along with sulfur and phosphorus, during the early stage of weathering (Ayris and Delmelle, 2012), a range of elemental deficiencies is reported in Andisols, especially in those formed from silica-rich parent materials, as they have low-nutrient contents. Such situations may pose health problems for animals grazing on pasture land. A classic example is the low cobalt in volcanic soils developed on Taupo Tephra in the central North Island of New Zealand. In the early part of the twentieth century, cobalt shortage in herbage led to a vitamin B12-related stock wasting disease known as "bush sickness" in ruminants. Selenium deficiency in Andisols is also common and is cited as a cause of selenium-related disorders in livestock feeding on such soils.

Nitrogen is virtually absent from the volcanic parent material, but large quantities of nitrogen, primarily in the form of the nitrate anion, are required by most agricultural plants throughout their growth period. Nitrogen is also the major nutrient limiting the establishment of plants in newly emplaced volcanic deposits. Vegetation regeneration in buried landscapes can occur when plant roots are able to tap nitrogen pools from the underlying soil. In areas heavily impacted by tephra, plants may not be able to access this source of nitrogen, and plant recolonization of the barren landscape is limited by spreading of nitrogen-fixing species such as lupine (Figure 72.6). In areas with maritime climate, the quick return of sea birds to the sterilized island covered with thick fresh tephra often provides a source of nitrogen and phosphate for early establishment of plants.

FIGURE 72.6 The first plant to colonize the land buried by pyroclastic deposits from the 1980 eruption of Mount St Helens, United States, was the prairie lupine, a hardy nitrogen-fixing subalpine plant. *Photo by John Bishop, Washington State University Vancouver, United States; reproduced by kind permission of the author.*

In more stable landscapes seldom impacted by volcanic eruptions, Andisols have accumulated large quantities of nitrogen-containing organic matter; it is the ease with which this material is converted to inorganic nitrogen (a process known as mineralization), which will determine the amount of nitrogen available to plants. Usually, nitrogen mineralization is slower in allophanic Andisols than in nonallophanic Andisols because organic matter interacts with and is stabilized by allophane in the former soil type (Parfitt, 2009).

Potassium is another key plant nutrient. Although low contents are measured in basaltic rocks, potassium is normally present in the volcanic parent material in amounts adequate for plant growth. Nonetheless, the availability of potassium in volcanic soils is reduced in nonallophanic Andisols and generally decreases with the advance of soil weathering. This simply reflects incorporation of potassium in 2:1 layer silicate clay minerals, a process that does not occur in allophanic Andisols (Dahlgren et al., 2004).

4.2. Interaction with Organic Matter

Both allophanic and nonallophanic volcanic Andisols share the capacity to accumulate relatively large quantities of organic matter. The typical organic carbon content of an allophanic Andisol is ∼8–12%, whereas nonallophanic soils may contain up to 25% of this material (McDaniel et al., 2011; Shoji et al., 1993). Despite constituting a relatively minor proportion of the Earth's land surface, volcanic Andisols represent up to 5% of the global soil organic carbon, i.e., ∼2344 billion metric tons (Dahlgren et al., 2004).

The excellent ability of Andisols to store organic carbon underlines their unique mode of formation. Organic carbon accumulation in these soils is partly due to repeated "top dressing" of the land with tephra; a new tephra layer on soil isolates organic matter from soil macro- and mesofauna and reduces soil aeration. As a result, the capability of microbes to degrade organic material is severely diminished, thereby leading to its preservation. High organic carbon content can persist in deep soil layers for several thousand years after burial by tephra. Organic matter in Andisols can also be physically protected from decomposing organisms by encapsulation within abundant soil aggregates and small pores (see Section 4.3).

Studies of volcanic soils of different age that have evolved under similar conditions of vegetation, topography, and climate (such soil sequence is called a **chronosequence**) have highlighted a positive relationship between organic carbon accumulation and the abundance of poorly crystalline and noncrystalline minerals. The observed dependency points to interaction between organic matter and the surface of allophane/imogolite and/or ferrihydrite. The residence time of soil organic carbon associated with these mineral phases (i.e., organomineral associations) is considerable, on the order of 160 000−175 000 years for volcanic soils from Hawaii and La Réunion, revealing an efficient mechanism of protection against microbial degradation. However, the capacity of Andisols for storing organic carbon evolves with time, as poorly crystalline and noncrystalline minerals are converted to more crystalline, less reactive minerals with the advance of weathering. On Hawaii, such mineralogical changes have been documented along a chronosequence that tracks the transition from Andisol to Oxisol. As illustrated in Figure 72.7, organomineral associations, and hence soil organic carbon, rose to

a maximum after 150,000 years of soil development, and then decreased by 50% over the next 4 million years when stable crystalline minerals accumulated at the expense of the poorly crystalline and noncrystalline minerals.

Organic carbon in volcanic soils is also stabilized via binding of organic matter with aluminum and iron, as the so-formed complexes are less susceptible to microbial attack. This mode of interaction dictates organic carbon accumulation in nonallophanic Andisols and in the organic-rich surface layer in allophanic Andisols. The respective contributions of aluminum/iron−organic matter complexes and organic matter−mineral associations to organic carbon stabilization in Andisols probably depend on the rates at which poorly crystalline secondary minerals are formed and organic matter compounds are added. It is also worth noting that organic carbon accumulation in young volcanic soils proceeds faster than it declines during the later stages of weathering.

4.3. Physical Properties and Interaction with Water

Volcanic Andisols have a high **porosity** and a wide range of pore size distributions. This relates primarily to structural assemblages of the poorly crystalline and noncrystalline secondary minerals into stable (sand- and silt-sized) aggregates. Accumulation of organic matter in nonallophanic Andisols also forms highly porous aggregates. The well-developed particle aggregation explains the typical low bulk-density values (usually <0.9 g/cm^3) of both allophanic and nonallophanic Andisols. These soils typically display a strong resistance to water erosion and dispersion of soil aggregates due to a high **permeability** that reduces runoff. Among the mineral soils, the volcanic allophanic Andisols are also unrivaled for their capacity to retain large amounts of water thanks to the presence of high-surface-area components, including allophane/imogolite and ferrihydrite (Shoji et al., 1993). Recent work has highlighted that some nonallophanic Andisols exhibit a similar property. In these soils, the exceptionally large accumulation of organic matter is responsible for their high porosity and water retention capacity.

Another notable feature of Andisols that are rich in allophane/imogolite is their ability to release water when disturbed with gentle pressure, even if they appear barely moist. This behavior, known as "thixotropy," distinguishes allophanic Andisols from their nonallophanic equivalents. Thixotropy in allophanic Andisols is explained by the fact that much water is held in the small pores created by aggregates and whose geometry is easily altered by pressure. Thus, these Andisols can reach the liquid limit (i.e., the water content at which the soil begins to behave as a liquid and begins to flow) and become smeary upon mechanical

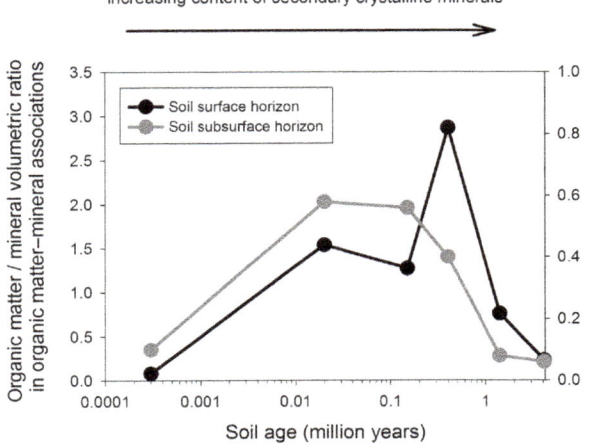

FIGURE 72.7 Graph illustrating the decreasing capacity of the soil to store organic matter (i.e., decreasing organic matter-to-mineral ratio in organomineral associations) as poorly crystalline secondary minerals transform to crystalline minerals with time. This evolution is shown for the surface (left axis) and subsurface (right axis) horizons of soils along a chronosequence in Hawaii. The scale of the x-axis is logarithmic. *Data from Mikutta et al. (2009).*

FIGURE 72.8 Oblique aerial view of the landslide that buried Colonia Las Colinas, El Salvador, in 2001. This landslide occurred on a slope covered by allophanic Andisols. *Photo by Ed Harp, US Geological Survey, reproduced by kind permission of the author.*

FIGURE 72.9 Volcanic Andisols in active volcanic regions are often intensively cultivated. The fields shown in this photo are regularly affected by the ongoing eruption of Tungurahua volcano, Ecuador.

disturbance. In steep terrains, allophanic Andisols may be quite susceptible to failure with subsequent formation of hazardous debris flows (Basile et al., 2003) (Figure 72.8).

Nevertheless, from an engineering point of view, there is indication that allophanic Andisols can also perform remarkably well. For example, in Indonesia and other tropical countries, terraced rice fields exist on very steep (up to 40°) slopes in areas of allophanic Andisols. These soils remain stable despite permanent saturation with irrigation water, which flows from terrace to terrace. Moreover, many water-retaining structures have been successfully constructed from allophanic Andisols.

5. UTILIZATION AND ENVIRONMENTAL IMPLICATIONS

5.1. Utilization

Volcanic soils, and more particularly the dominant Andisols order, are usually excellent media for crop production (Figure 72.9). This is due to their superior physical properties, including high water-holding capacity, high porosity, low bulk density, and stable structure, which make them resistant to drought stress and water erosion and render them excellent plant-rooting media with favorable **tilth**. Moreover, in some localities, periodic additions of fresh tephra can provide an adequate supply of nutrients and maintain favorable fertility levels. Thus, volcanic Andisols have occupied a central role in the history of humanity, notably by supporting the earliest human race in the Rift Valley in East Africa and by providing the Mayan and Inca civilizations in Central and South America with fiber, food, and grazing ranges. In Japan, Indonesia, and the Philippines, both native crop and rice fields were cultivated in Andisols for more than ten centuries. Vineyards have

flourished in Andisols in southern Italy and nourished the Mediterranean culture for hundreds of years.

The use of volcanic Andisols depends on a suitable climate but is tempered to some extent by the unique properties of these soils. Nowadays, Andisols support more than 10% of the total world population, a disproportionate utilization, since they represent slightly less than 1% of the total land surface. Andisols are cultivated for a wide variety of crops, including wheat, barley, rice, sweet potatoes, peanuts, vegetables, fruits, etc., in warmer regions, and potatoes, wheat, sugar beets, various beans, etc., in cooler regions. They support large quantities of coniferous forests in southern Alaska and along the coastal regions of the Pacific Northwest of the United States and provide fertile grazing ground for animals in New Zealand. Volcanic Andisols are also increasingly used for growing high-value crops such as flowers (e.g., roses, tulips) and fruits (e.g., bananas, pineapples, kiwis), for example, in Japan, Ecuador, and Costa Rica.

5.2. Environmental Issues

Contrary to popular opinion, Andisols are not systematically "rich and fertile" despite their generally favorable physical and chemical properties. Thus, volcanic soils must be properly managed in order to maintain high productivity while preventing deterioration and the occurrence of undesirable environmental effects. As described in Section 4.1, phosphorous deficiency occurs in most allophanic and nonallophanic Andisols. Heavy phosphorus addition is therefore carried out in many agricultural areas to sustain production. Phosphate fertilizers are produced from phosphate rocks, which also contain fluorine. In New Zealand, the long-term application of phosphate to agricultural Andisols is held responsible for increased fluorine concentration in these soils. There is concern that the continued accumulation of soil fluorine poses a health risk to grazing ruminants and could also affect groundwater quality in the future (Loganathan et al., 2001).

Like most other soils, nitrogen is added to volcanic soils under culture. In the groundwater-dependent Central Valley of Costa Rica, intensive use of nitrogen fertilizer in coffee plantations is a suspected cause of rising groundwater nitrate concentrations. However, subtle mineralogical differences in volcanic soils have a significant impact on nitrate mobility. Soils containing aluminum-rich allophane showed significant nitrate retention and thus, delayed nitrate leaching, whereas more weathered soils with silicon-rich allophane and halloysite were essentially ineffective in retaining nitrate. This finding emphasizes the importance of evaluating volcanic soil properties at local and landscape scales in order to develop appropriate nitrogen fertilization programs.

In Martinique and Guadeloupe, volcanic Andisols and Oxisols have been used for intensive banana cropping. Until its ban in the early 1990s, the toxic pesticide chlordecone was applied to protect banana plantations from root borers. The affinity of this chemical for poorly crystalline and noncrystalline minerals and organic matter has considerably retarded its leaching, particularly in Andisols. As a result, the agricultural land that supported banana growth more than two decades ago is still heavily loaded with chlordecone. This poses serious and long-term environmental issues, as the pesticide is slowly released to surface- and groundwater. Further, human exposure to this pollution is a cause for concern because crops grown nowadays on the contaminated soils contain potentially harmful levels of chlordecone. Removal of the top soil to remediate this situation is not recommended, as previous tillage practices led to redistribution of the pesticide to deeper levels. Soil removal would also drastically reduce soil fertility and enhance soil erosion.

While most allophanic Andisols are acidic, with soil pH values ranging from 4.8 to 6.0, they can effectively neutralize acid deposition by rapid chemical weathering reactions and pH buffering by variable-charge materials. Nevertheless, long-term nitrogen fertilizer inputs may induce acidification of allophanic Andisols, with serious consequences for crop production. In the case of non-allophanic Andisols and volcanic Spodosols, critical pH values below 5.0 are not uncommon. Such conditions may give rise to a high concentration of aluminum in solution, which, in turn, may severely impair root proliferation and inhibit crop growth. Compared to allophanic Andisols, the higher susceptibility of nonallophanic Andisols to aluminum toxicity reflects the presence in these soils of crystalline 2:1 layer silicate clays. The surface of such clays is known to retain a large amount of aluminum that can be released into solution upon acidification (Dahlgren et al., 2004). The abundance of aluminum−organic matter

FIGURE 72.10 Páramo landscape in the Ecuadorian Andes. The volcano in the background is Antisana. *Photo by Mark Horrell, http://www.markhorrell.com; reproduced by kind permission of the author.*

complexes in nonallophanic Andisols also contributes to aluminum toxicity problems.

Both allophanic and nonallophanic Andisols are known to undergo irreversible physical changes upon drying due to structural degradation of allophane/imogolite minerals and organic matter. This behavior makes them particularly fragile when they are used inappropriately. The Andean páramos are cold and moist grasslands and shrublands that cover the mountain sides of the northern Andes from Venezuela to Peru, at elevations of between 3000 and 5000 m (Figure 72.10). Most páramos are supported by volcanic soils, including nonallophanic Andisols, Entisols, Inceptisols, and Histosols. Like glaciers, páramos act like vast sponges, storing and releasing water. As such, they play a key role in the hydrology of the continent, notably by feeding many of the largest tributaries of the Amazon basin. Páramos are also the main water source for the Andean highlands and for a vast area of arid and semiarid lowlands. However, there is growing concern that the intensification of human activities such as cattle grazing, cultivation, and pine planting in the páramo irreversibly damages the ability of the soil to retain water, with potential long-term negative consequences for the regional hydrology (Buytaert et al., 2006).

The high permeability and large water-holding capacity of Andisols reduce runoff and render these soils more resistant to water erosion than other soils. However, because of their low bulk density, they may be very susceptible to erosion when vegetative cover is removed (McDaniel et al., 2011). It has been shown that both traditional and intensive agricultural practices may lead to structural degradation in both allophanic and nonallophanic Andisols, giving rise to a higher susceptibility to erosion. Increased population pressure on land intensifies this problem. In Mexico, and particularly along the Trans-Mexican Volcanic Belt, tillage erosion has exposed an indurated soil layer (or pan), known locally as "tepetate," to the surface. The presence of a hard pan layer in volcanic soils is usually attributed to compaction and cementation of tephra by weathering products such as calcite, precipitated clays, allophane, and silica. Tepetate has a low fertility due to reduced porosity, high compaction, and low carbon and nitrogen contents. Consequently, its formation requires intensive labor to restore the attractive agricultural potential of the original volcanic soil. A similar problem exists in northern Ecuador, where the hardened tephra layer is called "cangahua."

6. CONCLUDING REMARKS

Volcanic soils, in particular those belonging to the Andisols order, are amongst the highest value soils on the Earth. They have unique properties and generally good fertility, but akin to other soils they are a limited resource. While the majority of research on Andisols to date has focused on measuring their characteristics, less is known about their susceptibility to natural and anthropogenic disturbances. More work should be conducted to determine critical thresholds in andic soil properties above which plant growth and hydrologic performance are adversely affected. Addressing this gap in knowledge will entail long-term detailed study, and successful management will ultimately require an understanding of the mechanisms involved.

The bountiful gift of volcanic soils given by nature encourages population growth in the shadow of volcanoes and in previously unsettled volcanic regions; yet, ironically, with this comes a higher volcanic hazard to both people and property. Such an interplay will always exist, and in order to enable farmers to produce crops sustainably, efforts should be made to fully recognize the economic and social significance of proper soil management in both active and dormant volcanic regions.

ACKNOWLEDGMENTS

P.D. acknowledges support from the Fonds National pour la Recherche Scientifique (FNRS) through a MIS-Ulysse grant F.6001.11. S.O. and J-T. Cornelis contributed to this work as FNRS-funded postdoctoral researchers.

FURTHER READING

Arnalds, Ó., Óskarsson, H., Bartoli, F., Buurman, P., Stoops, G., García-Rodeja, E., 2007. Soils of Volcanic Regions in Europe. Springer, Berlin.

Ayris, P., Delmelle, P., 2012. The immediate environmental effects of tephra emission. Bull. Volcanol. 74, 1905–1936.

Basile, A., Mele, G., Terribile, F., 2003. Soil hydraulic behaviour of a selected benchmark soil involved in the landslide of Sarno 1998. Geoderma 117, 331–346.

Buytaert, W., Célleri, R., De Bièvre, B., Cisneros, F., Wyseure, G., Deckers, J., Hofstede, R., 2006. Human impact on the hydrology of the Andean páramos. Earth-Sci. Rev. 79, 53–72.

Churchman, G.J., David, J.L., 2011. Alteration, formation, and occurrence of minerals in soils. In: Huang, P.M., Li, Y., Summer, M.E. (Eds.), Handbook of Soil Sciences. CRC Press, pp. 1–72.

Dahlgren, R.A., Saigusa, M., Ugolini, F.C., 2004. The nature, properties and management of volcanic soils. Adv. Agron. 82, 113–182.

IUSS Working Group WRB, 2006. World Reference Base for Soil Resources. World Soil Resources Reports No. 103, second ed. FAO, Rome, Italy.

Kaufhold, S., Kaufhold, A., Jahn, R., Brito, S., Dohrmann, R., Hoffmann, R., Gliemann, H., Weidler, P., Frechen, M., 2009. A new massive deposit of allophane raw material in Ecuador. Clays. Clay. Miner. 57, 72–81.

Loganathan, P., Hedley, M.J., Wallace, G.C., Roberts, A.H.C., 2001. Fluoride accumulation in pasture forages and soils following long-term applications of phosphorus fertilisers. Environ. Pollut. 115, 275–282.

McDaniel, P.A., Lowe, D.J., Arnalds, O., Ping, C.-L., 2011. Andisols. In: Huang, P.M., Li, Y., Summer, M.E. (Eds.), Handbook of Soil Sciences. CRC Press, pp. 29–45.

Mikutta, R., Schaumann, G.E., Gildemeister, D., Bonneville, S., Kramer, M.G., Chorover, J., Chadwick, O.A., Guggenberger, G., 2009. Biogeochemistry of mineral–organic associations across a long-term mineralogical soil gradient (0.3–4100 kyr), Hawaiian Islands. Geochim. Cosmochim. Acta 73, 2034–2060.

Mizota, C., van Reeuwijk, L.P., 1989. Clay Mineralogy and Chemistry of Soils Formed in Volcanic Material in Diverse Climatic Regions. International Soil Reference and Information Centre (ISRIC), Wageningen.

Parfitt, R.L., 2009. Allophane and imogolite: role in soil biogeochemical processes. Clay Miner. 44, 135–155.

Perret, S., Dorel, M., 1999. Relationships between land use, fertility and Andisol behaviour: examples from volcanic islands. Soil Use Manage. 15, 144–149.

Rasmussen, C., Matsuyama, N., Dahlgren, R.A., Southard, R.J., Brauer, N., 2007. Soil genesis and mineral transformation across an environmental gradient on Andesitic lahar. Soil Sci. Soc. Am. J. 71, 225–237.

Shoji, S., Nanzyo, M., Dahlgren, R.A., 1993. Volcanic Ash Soils. Genesis, Properties and Utilization. In: Developments in Soil Science, vol. 21. Elsevier, Amsterdam.

Soil Survey Staff, 1999. Soil Taxonomy - a Basic System of Soil Classification for Mapping and Interpreting Soil Surveys. US Department of Agriculture—Natural Resources Conservation Service, Washington, D.C.

Soil Survey Staff, 2010. Keys to Soil Taxonomy. US Department of Agriculture—Natural Resources Conservation Service, Washington, D.C.

Ugolini, F.C., Dahlgren, R.A., 2002. Soil development in volcanic ash. Global Environ. Res. 6, 69–81.

Wada, K., 1986. Ando Soils in Japan. Kiyushu University Press, Fukuoka, Japan.

Chapter 73

Volcano Ecology: Disturbance Characteristics and Assembly of Biological Communities

Charles M. Crisafulli

USDA Forest Service, Pacific Northwest Research Station, Olympia Forestry Sciences Laboratory, Mount St Helens National Volcanic Monument, Amboy, WA, USA

Frederick J. Swanson

USDA Forest Service, Pacific Northwest Research Station, Corvallis Forestry Sciences Laboratory, Corvallis, OR, USA

Jonathan J. Halvorson

USDA Agricultural Research Station, Northern Great Plains Research Laboratory Mandan, ND, USA

Bruce D. Clarkson

Environmental Research Institute, University of Waikato, Hamilton, New Zealand

Chapter Outline

GLOSSARY

biological assemblage A group of species occurring together in the same geographical area.

biological assembly The patterns and processes that occur in species accumulation and community organization at a site following an ecological disturbance.

biological community Organisms of a variety of species that co-occur in the same habitat or area and interact through trophic and spatial relationships; typically characterized by reference to one or more dominant species.

biological legacy Organisms, propagules, and organic structures and spatial patterns that remain following a disturbance; includes surviving plants, animals, seeds, spores of microorganisms, and plant or animal organic matter, including above- and below-ground wood that affect ecological and soil processes and properties.

colonization The successful invasion of a habitat by a species.

defaunated Depleted of animals.

dispersal Spread of organisms or propagules from their point of origin or release; one-way movement of organisms from one home site to another; the outward extension of a species range, typically by a chance event.

The Encyclopedia of Volcanoes. http://dx.doi.org/10.1016/B978-0-12-385938-9.00073-0

disturbance Any relatively discrete event in time that disrupts ecosystem, community, or population structure and changes resources, substrate availability, or the physical environment.

ecosystem A community of organisms and their physical environment interacting as an ecological unit.

establishment A recently arriving plant or animal growing and reproducing successfully in a given area.

primary succession Ecological succession commencing on a substrate that does not support any organism or contain any biological remnants from previous life.

propagule(s) Any part of an organism, produced sexually or asexually, that is capable of giving rise to a new individual (e.g., seeds, spores).

refugium (Refugia, pl) An area that has escaped ecological changes imposed by disturbance in surrounding areas, and so provides suitable habitat for relict species.

succession The process of gradual replacement of one species population by another over time and the concurrent change in **ecosystem** properties after a site has been disturbed. The concept can be extended to the replacement of one kind of community by another, the progressive changes in vegetation and animal life that **may** culminate in dominance by a community that attains dynamic equilibrium until the next disturbance. Succession commonly refers to changes that are scaled to the longevity of dominant organisms and not to seasonal changes in populations and communities.

trophic levels The sequence of stages in a food chain or food pyramid, from producers to primary, secondary, or tertiary consumers, and decomposers.

1. INTRODUCTION.

Volcanic activity has profoundly disturbed ecological systems since the earliest days of life. Volcanoes are broadly distributed around the world, particularly around the Pacific Rim, and several dozen volcanoes erupt during any given year. Eruptions, in all their various types and intensities, affect contemporary natural ecological systems as well as important natural resources and even climate in many ways, and often are the most extreme form of natural disturbance to ecosystems. As a result, volcanism is of great importance to human societies and especially to scientists who use post-eruption landscapes as exemplary living laboratories for basic and applied research. Volcanoes exhibit a tremendous range of interactions with the Earth's ecosystems: volcanic processes, products, and structures support distinctive ecosystems, such as the incredible "black smoker" and other bizarre communities at deep-sea spreading centers; and volcanoes create new landforms, deliver production-limiting nutrients to marine and freshwater systems, disturb terrestrial and freshwater ecosystems, and trigger change on a gradient from minimal to profound.

The study of the interactions of volcanoes and ecosystems, here termed volcano ecology, focuses on the ecological responses of organisms and biological processes to eruption events. It is a subdiscipline of the broader field of disturbance ecology, which addresses ecosystem responses to a wide variety of perturbations. Ecologists define "disturbance" as "any relatively discrete event in time that disrupts ecosystem, community, or population structure, and changes resources, substrate availability, or physical environment."

Research conducted at volcanoes, particularly at iconic locations such as Krakatau, Indonesia (1883); Surtsey, Iceland (1963); and Mount St Helens, USA (1980), has profoundly influenced our understanding of ecosystem dynamics.

Eruptions are certainly remarkable for their impetuous, fiery, and explosive displays, but perhaps much less appreciated for their role in stage-setting of ecological processes that play out over decades, centuries, and even millennia. Initial questions addressed through direct observation and experimentation in the weeks to a few years after an eruption include assessment of severity and spatial distribution of impacts on ecological components and measurements of the earliest stages of **succession**. Trajectories of ecosystem assembly over decades to centuries, termed "succession" by ecologists, are tracked by continuous or periodic sampling, as in cases such as Islas Revillagigedo (Brattstorm, 1990), Krakatau (Whittaker and Jones, 1994), Long Island, Papua New Guinea (Ball and Glucksman, 1981), Mount St Helens (Dale et al., 2005), and Surtsey (Jakobsson et al., 2009). On the scale of centuries to millions of years, ecologists employ chronosequence techniques to examine ecosystem dynamics over the course of soil and landform development (Vitousek, 2004).

In our treatment of volcano ecology, we emphasize response of terrestrial ecosystems to volcanic disturbance and consider areas undergoing **primary succession**, where the substrates have never been previously inhabited, and also areas undergoing secondary succession, where at least some survivors of the pre-eruption biological communities are present.

In this chapter, we begin with a discussion of the specific types of volcanic processes and describe characteristics of their intensities and mechanisms of impact on biota. Further, we examine the physical and chemical characteristics of newly emplaced volcanic products (e.g., lava, tephra) and discuss these materials relative to soil development and to the assembly of plant and animal communities. We then present information on the current state of knowledge regarding both initial and midterm (i.e., successional) responses of plant and animal species and assemblages to volcanism and present examples of a few salient ecosystem processes and biotic interactions that have been observed following volcanic disturbance. We highlight the importance of the context under which volcanism is taking place, which so profoundly influences the pace and pattern of ecological responses and includes key factors such as geographic setting; climate; volcanic disturbance type, intensity, and spatial extent; and stochastic properties influencing assembly processes (see glossary).

2. CONCEPTUAL MODELS OF SUCCESSION: BIOTIC ASSEMBLY AND ECOSYSTEM CHANGE OVER TIME

Succession—A volcanic disturbance event can initiate succession, which can be described as biotic change over time, including species **establishment** and loss. Primary succession is the establishment and change of vegetation on newly formed or exposed land surfaces comprising new parent materials (e.g., a recently erupted lava flow), rather than on a developed or modified soil. Sites of primary succession contain no biological or physical legacies (residuals) persisting from the predisturbance ecological system, including surviving or dead organisms, vegetative **propagules**, seeds, spores, organic matter derived from previous vegetation, litter, or soil. True primary succession after volcanic disturbance is rare, but newly erupted islands such as Anak Krakatau in Indonesia (1930) and Surtsey in Iceland (1963) were undoubtedly sterile surfaces when they first emerged from the ocean (Whittaker et al., 1989; Jakobsson et al., 2009), as are areas subjected to extensive lava flows (e.g., flood basalts) and pyroclastic flows. **Secondary succession** is a much more frequent occurrence, involving the presence of at least some surviving organisms following a disturbance event (e.g., a tephra fall or lahar) and begins with a more or less mature soil, which contains an established seed bank and vegetative propagules (Glenn-Lewin et al., 1992). However, this mature soil is often buried beneath newly emplaced volcanic deposits. Secondary succession is a common form of succession following volcanic disturbance because the disturbance intensity of most eruptions is not great enough to exterminate all life. Although there are some parallels to plant succession, animal ecologists generally view animals in terms of community assembly as a framework to describe changes following a disturbance event. Here, we strive to provide roughly equivalent structure in our treatment of plant and animal responses to volcanism.

We provide two conceptual models as a framework for structuring information presented in our chapter. The first is a simple generalized model that characterizes the overarching geological and ecological processes of an ecosystem that has been subjected to volcanic disturbance. The second model focuses more narrowly on salient processes and factors that occur during **biological assembly** in the absence of legacies and forms the basis for presenting information in the plant and animal sections below.

General Response to Disturbance Conceptual Model—This model (Figure 73.1) begins with a pre-eruption ecosystem that is subject to a generic primary volcanic disturbance, which causes not only loss of some biophysical components of the predisturbance ecological system but also retention of some organisms

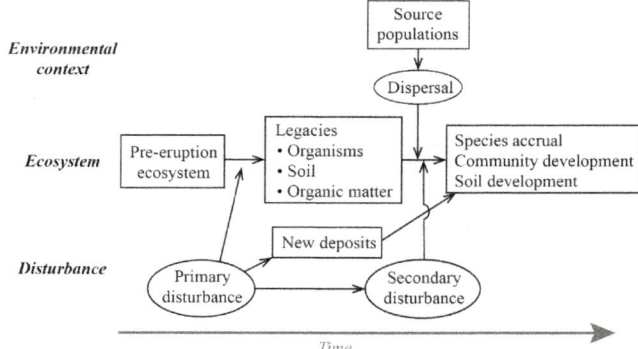

Ecological response to volcanic disturbance

FIGURE 73.1 Generalized model depicting key biological and physical factors influencing succession following volcanic disturbance. In this conceptual model a pre-eruption ecosystem is subjected to not only some form of volcanic disturbance that results in loss of some biota but also the retention of some biophysical components of the predisturbance ecosystem (e.g., living and dead organisms, soil). These legacies are joined by newly arriving organisms from adjacent source populations that must disperse to the recently disturbed area and collectively the survivors and colonists become important components in assembly processes and community development. Volcanic disturbance universally results in new deposits that are lacking crucial nutrients for plant and microbial growth, and these conditions must be contended with by plants colonizing the surface of these deposits. Primary volcanic disturbances predispose landscapes to a number of potential secondary disturbances, such as increased flooding (lahars), landslides (deep tephra fall), and wildfire (blast scorched forests).

(living and dead), soil, and new volcanic deposits. Biophysical carryover of predisturbance ecosystem components and the arrival of new colonists and propagules from source populations initiate the accrual of species, community assembly, and soil development. Primary disturbance predisposes the ecosystem to subsequent secondary disturbance (e.g., surface erosion, landslides) that influences biological assembly and soil development. These ecosystem processes have a temporal and spatial aspect that is highly context-dependent.

Community Assembly Conceptual Model—In probably all but the most benign cases, volcanism reduces plant and animal populations through direct mortality and also through changes in important resources that influence persistence and growth of individuals. In the aftermath of eruptions, biological assembly proceeds and is a central theme of volcano ecology. Assembly begins with a regional pool of species that are potentially available to colonize a recently volcanically disturbed site (Figure 73.2). In order for these organisms and propagules to be part of the new local community, they must contend with three filters related to processes of **dispersal**, establishment, and community development, as well as associated deterministic and stochastic factors. Successful dispersal is influenced by the distance from the source population to the disturbed site; the condition of the intervening landscape/waterscape,

Community assembly model

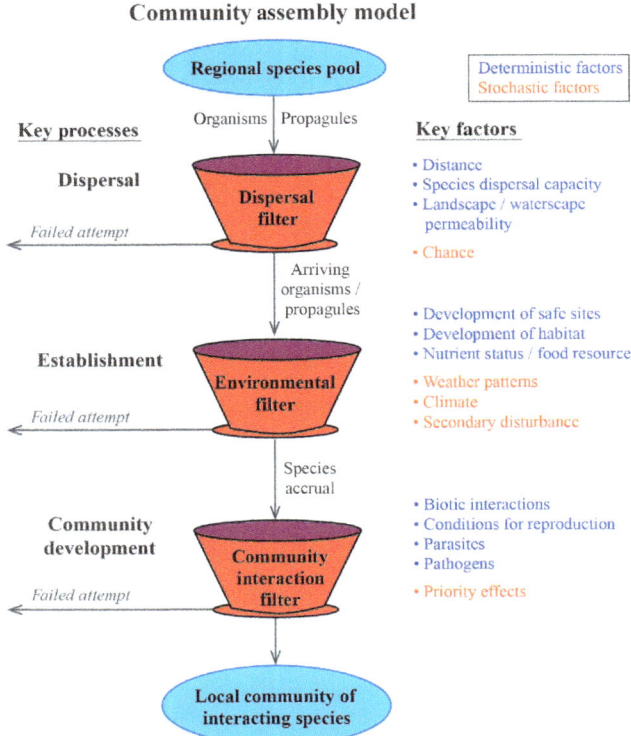

FIGURE 73.2 Conceptual model showing principle factors of **biological community** assembly following volcanic disturbance that initiates primary succession. In this conceptual model intense volcanic disturbance (e.g., pyroclastic density current, lava flow) removes (or deeply buries) all or most of the pre-eruption biota and soil, and the assembly process begins from a clean slate. There are three key ecological processes: dispersal, establishment, and community development that are integral to biological assembly, and each of these have several factors, some of which are deterministic and others that are stochastic that govern the pace at which each process and assembly occurs. Note that there are a number of factors that can preclude the success of any key process and this may result in a failed attempt in community assembly. For example, an individual small mammal may successfully disperse from a source population to a disturbed site but appropriate habitat (e.g., food, cover) is lacking and the individual either dies or is forced to move on and does not successfully establish at the disturbed site. See text for details on key processes and their associated factors.

referred to as permeability; and factors related to chance. Establishment of individuals in the disturbed site is influenced for plants by the presence of sites where seeds and spores can successfully establish, referred to as safe sites, and for animals by development of habitat; nutrient status (plant) and food resources (animal); and stochastic factors of weather patterns, climate, and secondary disturbances. Site amelioration occurs as early establishing plants, and physical processes (e.g., weathering of deposits) create conditions favorable for establishment. Species accrual describes the phase of assembly where species are arriving at a site and filling in both physical and ecological space, eventually reaching some threshold at which species begin

interacting and community development ensues. Simple ecological interactions occur very early in the accrual process, but the number and the strength of interactions increase as assembly proceeds. Community development is influenced by a broad suite of biological interactions among species, including predation/herbivory, pollination, facilitation, competition, symbioses, and others; conditions for reproduction; parasites; pathogens; and the order in which species arrive in a disturbed site, referred to as priority effects.

3. VOLCANIC PROCESSES AND INITIAL IMPLICATIONS FOR SOIL AND ECOSYSTEM DEVELOPMENT

Volcanic and associated hydrologic processes involved in eruptions are highly varied in their abilities to disturb terrestrial, freshwater, and marine ecosystems; to deliver new deposits; and to create new landforms, making it possible for new types of ecosystems to develop where they had not existed before. To aid interpretation of disturbance impacts, each type of volcanic process (e.g., tephra fall, lava flow) can be viewed as composed of one or more disturbance mechanisms, such as heating (to the point of exceeding the physiological tolerance of organisms, scorching, or combustion), erosion/deposition, abrasion, chemical toxicity, suspended particles in air or water, canopy loading, and impact force (Dale et al., 2005, Chapter 3; Swanson et al., 2013). Many processes may involve multiple mechanisms, such as the impact force, abrasion, heating, and deposition of materials (burial) characteristic of pyroclastic flows. Ultimately, organisms respond to specific mechanisms of disturbance, and there are thresholds at which they perish or their health or fitness is compromised. These mechanisms of disturbance and their intensities (temperature, burial depth, etc.), acting singularly or in combination, determine both initial (direct) effects on populations and longer-term patterns of ecological response.

Explosive eruptions commonly involve a diverse suite of destructive processes, such as pyroclastic density currents (PDCs), tephra fall, lahars, and extreme sediment runoff.

Consequently, the post-eruption landscape is a mosaic of partially overlapping disturbance zones of various origins and degrees of alteration. The 2008 eruption of Chaiten volcano (Chile) provides an example of such a diversity of types and extents of zones influenced by various combinations of disturbance mechanisms (Table 73.1). Notably, deposition is the most pervasive disturbance mechanism, and airfall tephra is the most extensive; these are common features of explosive eruptions globally.

TABLE 73.1 Characteristics of forest disturbance mechanisms for disturbance types observed in the 2008–2009 Chaitén (Chile) eruption, modified from Swanson et al. (2013) (see that reference for further details). Depth of burial varies with position within the geographic distribution of a deposit; these values are for sample plots in this study. Estimates of moderate heat (50–200°C), are based on presence of charred wood in deposit, scorching of foliage in the scorch zone, and associated laboratory studies. Lateral abrasion ranged in severity from very high (removal of epiphytes, bark, impact scars on the wood surface), to moderate (removal of epiphytes and some bark), to very low. Abrasion by vertical fall of gravel tephra was of high severity (removal of foliage, canopy epiphytes, and twigs). Impact forces ranged from very high (trees uprooted and moved downslope); through high (force sufficient to topple trees); to moderate to very low (gradient of force laterally from the axis of flow to thin, tranquil flow near margins). Canopy loading was precluded or obscured by canopy removal in several cases, but was high, with sufficient mass entrained in canopy to break branches up to 15 cm diameter at the point of attachment to the trunk, in areas with fine tephra >10 cm thick. "—" = not significant

Disturbance type	Extent (km^2)	Burial depth (m)	Heat	Abrasion	Impact force	Canopy loading
Blast Zone						
Tree removal	0.8	~1–2	Moderate	Lateral, very high	Very high	Precluded
Toppled tree	2.5	<0.1–1	Moderate	Lateral, high	high	Precluded
Scorch	0.7	<0.08	Moderate	–	–	–
Tephra Fall						
Gravel (>5 cm thick)	50	<0.27	–	Vertical, high	–	Precluded
Fine tephra (>10 cm thick)	480	<0.5	–	–	–	High
Pyroclastic flow	1.5	<6–8	Moderate	Lateral, very high to moderate	Very high	–
Fluvial deposition	7	~1–2	–	Lateral, very low	Moderate to very low	–

Development of soil ecosystems on volcanic substrates is influenced by the initial texture and composition of the deposits and subsequent physical and chemical changes brought about by weathering and biological activity. Low-silica magma, for example, commonly produces basalt and andesite lava flows with temperature, depth of burial, and impact force great enough to severely disrupt ecosystems. The surface textures of resulting deposits are likely to impede plant establishment, but the chemical and mineral compositions may be more susceptible to weathering and vegetation establishment over the long term than silica-rich material with its lower nutrient content. Silica-rich magmas are more likely to produce explosive eruptions with attendant tephra fall and pyroclastic flows of dacite and rhyolite, resulting in fragmental deposits with a texture more amenable to plant and animal life, but lower availability of mineral nutrients may limit the rates of ecosystem and soil development. Lava flows and deep, expansive tephra or pyroclastic deposits may lack the legacies that link their patterns of reassembly to those of the pre-eruption ecosystem (Figure 73.1).

Some volcanic processes and resulting deposits obliterate ecosystems that previously occupied the area, but rich and diverse biological legacies commonly persist in areas of moderate disturbance (Figure 73.3). Deposits that are thick (several meters) or impenetrable to roots descending from above or stems attempting to emerge from below may effectively erase the influence of plants and animals that previously occupied the area. At the other extreme of disturbance intensity, a few centimeters of tephra may permit most species to persist, and actually serve as water-retaining mulch, enhancing plant growth. Intermediate intensities of disturbance and the different combination of mechanisms involved in different disturbance types can result in a variety of forms of "editing" of plant and animal communities. An overview of common volcanic processes in terms of the disturbance mechanisms involved, areal extent, and characteristics of deposits and landforms

FIGURE 73.3 Examples of plant and animal survivors following volcanic eruptions in the Americas. A lone fern (*Blechnum magellanicum*) survived an explosive blast surge at the Chaiten volcano, Chile (A). An overstory tree (*Caldcluvia paniculata*) that was leveled by the blast surge at Chaiten resprouts from its bole (B). The fully aquatic form (i.e., neotene) of the Northwestern Salamander (*Ambystoma gracile*) survived under a protective cover of ice and snow in many high-elevation lakes at Mount St Helens, USA (C). Western Toads (*Bufo boreas*) that survived beneath the ground in winter hibernacula successfully dug through the overlying blast and tephra deposits and bred in lakes during the first post-eruption summer at Mount St Helens, USA (D). The Northern Pocket Gopher (*Thomomys talpoides*), a denizen of subterranean habitats, survived throughout the blast area of Mount St Helens, USA (E). Western Carpenter Ants (*Camponotus modoc*) survived within large downed logs on the forest floor at many locations in the blast area and tephra fall zones at Mount St Helens. Shown here is the frass, or wood filings produced by their tunneling, that has accumulated on top of blast deposits (F).

created provides background for discussion of volcano ecology.

- Lava flows: Effusive eruptions of lava flows commonly obliterate pre-existing ecosystems by burial and extreme heat leading to incineration, and cover areas 100 m^2 to a few 10 s km^2. Lava flow deposit surfaces may be inhospitable due to extreme porosity (e.g., aa) or hard, continuous surface (pahoehoe). Flood basalts create exceptional circumstances in volume, thickness,

extent, and magnitude of impacts, possibly being a major component of four of the Earth's five global extinctions of 50—90% of species found in the fossil record. Plant growth and microbial activity important to soil formation are often constrained on lava flows because of environmental extremes in temperature and moisture availability and because they lack critical nutrients, especially nitrogen. Weathering of lava flow surfaces leading to the creation of soils is a slow process that may require thousands to millions of years, but more rapid pedogenesis and biotic assembly can occur if imported materials, including guano, loess, tephra, or sediment, are deposited on top of lava surfaces (Figure 73.4, See examples below Section 7, Deligne et al., 2013). This may occur extensively or very locally, such as in surface depressions or cracks, forming "resource islands" that allow establishment of isolated plants or small communities of vegetation. The resultant organic matter supplied by the plants supports microbial activity and further accumulation and transformation of nutrients, all of which promote biological assembly.

- Tephra fall: Explosive eruptions of tephra are particularly common and extensive, with plumes that can circle the globe and, in extreme cases, alter climate for several years. Tephra fall disturbance mechanisms include burial, canopy abrasion and loading, chemical toxicity, and asphyxiation of animals. Disturbance to ecosystems ranges from very high plant and animal mortality at tephra thickness >50 cm to minimal damage by tephra fall of only a few millimeters. However, thin tephra

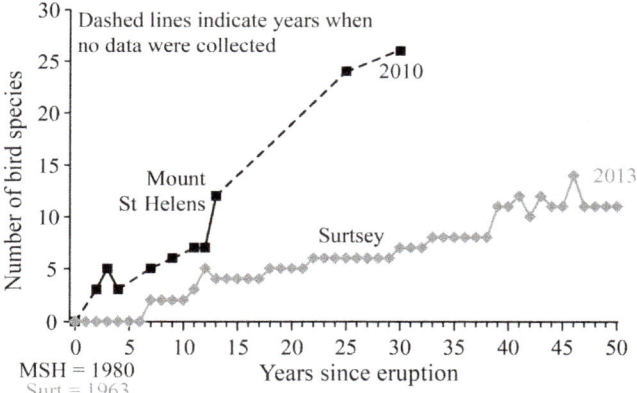

FIGURE 73.4 Bird species accrual at Surtsey and Mount St Helens (MSH). At both MSH blast area and on Surtsey, bird community assembly began from zero. However, the habitat structure differed dramatically between these sites. At Surtsey, colonization began on barren rock, whereas at MSH there was tremendous legacy material left from the former forest in the form of down wood, snags, and sparse plant cover. Colonization at MSH during the first 10 post-eruption years was dominated by ground nesters and foragers, air salliers, and bole foragers. Later, there was a sharp increase in species accrual in response to the establishment of dense riparian and patchy upland shrub cover. In contrast, bird community assembly on Surtsey has been slow, with only minor increases in richness for the first 37 post-eruption years. At that point, the establishment of gull colonies promoted the development of dense herbaceous vegetation that ushered in the recruitment of several song-bird species. The data for Surtsey include the entire island, which ranged from 1.4–2.6 km² during the period of record, whereas the MSH data were based on an effective survey area of 0.2 km². The disparity in sample area only reinforces the importance of biological habitat legacies and also the role of isolation in the rate of biological reassembly. Dashed lines between data points (MSH) indicate intervals when no data were collected.

burial can cause selective damage to certain ecosystem components (Dale et al., 2005, Chapter 4). Tephra texture and bulk density affect ecosystems; for example, the fall of gravel-sized, lithic tephra may abrade foliage and epiphytes, but low-density pumice of the same size distribution may cause minor damage to tree and shrub canopies. Further, tephra falling through vegetation may strip foliage and fine twigs from limbs and incorporate this organic material in the tephra strata, thus increasing the levels of nutrients and organic matter. Shallow tephra deposits (<20 cm) may have little impact on soil functions or may even provide some short-term benefits to existing vegetation by suppressing competitors. However, rain and snowmelt on fresh tephra may release leachates of chlorine, fluorine, or sulfur compounds that contaminate animal forage and lead to conditions toxic to soil organisms and plants. Moreover, bioturbation caused by animals can lead to rapid amelioration through mixing of nutrient-rich buried soil with overlying nutrient-impoverished tephra. Deeper or fine-grained deposits of tephra, however, may smother or damage standing vegetation and inhibit emergence of shoots from buried seeds. High-porosity pumice may be broken down readily through mechanical weathering, leading to accelerated soil formation.

- PDCs: Explosive pyroclastic density currents or flows containing juvenile pumice are extremely destructive, characterized by high velocity and impact force and extreme heat (to >700 °C). Resulting deposits from individual flows may be a few to tens of meters thick and commonly multiple flows are stacked on top of one another. Pyroclastic surges (low concentration of rock particles) may also be highly abrasive and impose high impact force and moderate temperature (e.g., 50–200 °C). Laterally directed surges into forest ecosystems create a gradient of impact signatures from forest removal close to the surge source, to trees toppled in place, to a "scorch zone" of standing, dead trees around the perimeter of the severely impacted area. Events of this type in recorded history have generally ranged from a few to hundreds of square kilometers in extent. Weathering of pyroclastic density current deposits can occur relatively rapidly, and commonly results in formation of fertile soils (Andisols) that are often layered as a result of multiple deposition events and characterized by metastable, noncrystalline minerals and accumulation of soil organic carbon in complexes with metals such as Al and Fe (organometallic) (Ugolini and Dahlgren, 2002).

- Lahars: Relatively cool, water-rich flows that may be triggered by interactions of eruption processes with groundwater, surface water, snow, and ice. The resulting lahar may race down hillslopes and river channels, damaging aquatic and riparian habitats for tens of kilometers. Lahar deposit thickness is commonly less than several meters, but abrasion and burial may be extreme close to the channel. Flows may be tranquil along their margins and within inundated floodplains, leading to survival of both plants and animals. Lahar deposits are typically unsorted and unstratified inorganic material with minor components of organic material incorporated into the mixture, derived from soils and organisms that were plucked from the valley walls or bottom as the lahar traveled downslope. Consequently, lahar deposits commonly have some carbon, nitrogen, and other organic constituents available for surviving and colonizing organisms, thus initiating soil formation. Lahar deposits in narrow valley bottoms are well situated to accumulate nutrients, propagules, and organisms from adjacent valley walls, which can accelerate soil development and community assembly.

- Debris avalanche: Massive debris avalanches resulting from collapse of major sections of volcanic cones, such as at Mount St Helens (1980, USA) and Bandai-san (1888, Japan), obliterated the previous ecosystems with deposits tens of meters thick composed of generally cool, fragmented, former mountain-top rock. The resulting hummocky deposit surface may be pock-marked with new ponds, and lakes may be formed by damming of tributary streams. These disturbance processes and the resulting deposits affect areas on the scale of a few to $50 + km^2$. As with lahars, some organic material derived from soil and organisms incorporated into the flow, but unless these components are near the surface they contribute little to soil formation and biological assembly.

Volcanic events can profoundly alter the form and surface properties of landscapes, accentuating conditions that allow wind and water to redistribute new deposits. These processes can help ameliorate site conditions, and also act as secondary disturbance processes that may impede plant establishment by displacing seeds and scouring and removing vegetation, and also kill animals and destroy their habitat. In some cases, such as in the blast area at Mount St Helens, small gullies eroded into the fresh, nutrient-deficient blast deposits liberated propagules in the pre-eruption forest soil and enhanced plant recovery and establishment (Dale et al., 2005, Chapter 20). Gradual accumulation of plant cover on new deposits eventually suppresses overland flow and surface erosion. Consequently, biological and physical properties of the land surface interact intensively over a period of years to decades following the primary disturbance. Particularly protracted secondary disturbance occurs where primary processes, such as debris avalanches and lahars, inundate river valleys with thick (tens of meters), unconsolidated

volcanic deposits, which can yield extreme sediment loads for decades, creating complex issues for both downstream ecosystems and human societies.

4. GEOGRAPHY OF VOLCANISM: IMPLICATIONS FOR VOLCANO ECOLOGY

Geographic context and landscape conditions are paramount in determining the pace and pattern of succession, community assembly, and soil development following volcanic disturbance, and these constraints are accentuated in **defaunated** or newly created volcanic landforms, where assembly is dependent on arrival of organisms from distant source populations into areas devoid of life (Figure 73.2). Anak Krakatau and Surtsey are prime examples of small, remote volcanic islands that emerged from the sea and are dependent on distant source populations for community assembly (Thornton, 1996; Jakobsson et al., 2009). At these sites, assembly can be protracted because of isolation and permeability issues related to species limitations to disperse across expanses of sea water. On the other hand, sea currents are vectors for delivering flotsam containing organic material (e.g., logs, plants, animal carcasses) and numerous arthropods and other living animals to new volcanic islands, and often so-called "strand communities," found in a narrow band along the shoreline, are the first to develop. In sharp contrast, mainland eruption sites typically have surviving organisms present to initiate assembly and are also within easy dispersal distance of members of the regional species pool; thus isolation per se is often not an obstacle to community assembly.

Climate plays a critical role in the pace and pattern of soil development and biological assembly following volcanic disturbance. Extremes in temperature (hot or cold) and aridity may result in extremely protracted timelines, as seen in the arid, continent-interior climate settings of Craters of the Moon volcanic site (Idaho, USA), and Wudalianchi Volcanic Field (China). Temperature extremes, particularly in the context of recently emplaced volcanic substrates, pose serious challenges for species' tolerance limits, productivity, and soil development. In sharp contrast, mild and wet conditions lead to a surprisingly rapid response time in community assembly and initial soil development, as has been observed at equatorial sites such as Pinatubo (Philippines), and also at temperate, midlatitude locations like Mount St Helens (Figure 73.5) and Chaiten (southern Chile). In such locations, abundant moisture facilitates plant establishment, and thus animal resources such as food, cover, and sites for reproduction, and increases the permeability for dispersal across otherwise-inhospitable terrestrial environments. Moreover, high precipitation can contribute to atmospherically

FIGURE 73.5 Photo chronosequence showing changes in stream channel conditions and riparian plant community development at Herrington Creek from 1981 through 2012 following the 1980 eruption of Mount St Helens, USA. Herrington Creek did not exist before the 1980 eruption and was created during post-eruption watershed evolution. Photo credits: Brian Franzen and Pete Bisson.

delivered nutrient inputs and also to increased onsite production, leading to more rapid soil development.

Establishment and development of biotic communities can be strongly influenced by stochastic patterns of environmental conditions and limited by availability of nutrients and water sufficient to support plant growth

(Figure 73.2). More deterministic patterns of soil development in volcanic deposits accompany chemical and physical weathering of the parent materials, favored under warm, moist climate regimes and in unconsolidated deposits of ash or tephra that contain glassy material through which water can percolate freely.

5. DISTRIBUTION OF VOLCANO ECOLOGY STUDIES ACROSS BIOMES OF THE WORLD

The status of the field of volcano ecology is illuminated by assessing studies of plant and animal responses to eruptions in relation to the distribution of active volcanoes among terrestrial biomes of the globe. In many respects the famous 1883 eruption of Krakatau initialized volcano ecology as a formal field of inquiry and serves as a good reference point to evaluate progress in this field. According to the Smithsonian Institute/USGS catalog of volcanoes (http://www.volcano.si.edu/search_volcano.cfm, access date March 20, 2014), 404 volcanoes have erupted subaerially since 1883. A plot of these volcanoes on the World Wildlife Fund's global map of terrestrial biomes (http://www.worldwildlife.org/publications/terrestrial-ecoregions-of-the-world, accessed 2008) reveals information on the types of terrestrial ecosystems affected by processes proximal (within a few 10 s km) to source volcanoes (Figure 73.6). Overall, only a small percentage (11%) of the volcanoes that have erupted since 1883 have been the subject of published ecological study, but these are broadly distributed around the globe. At least one volcano in all of the biome types received study, and the range of study attention varied from 4—30% of the volcanoes in each biome. Volcanoes along the Pacific margins of North and South America and the Aleutians have received more ecological study than understudied regions such as the northeastern and southeastern Pacific, Africa, and Indonesia.

The number of volcanoes with recent eruptions in each biome and the number per million km^2 provide an indication of opportunity to study ecological effects in the various biomes (Table 73.2). For the seven biome types considered, the number of volcanoes ranges from 10 (Mediterranean forest/woodland/scrub) to 175 (tropical and semitropical

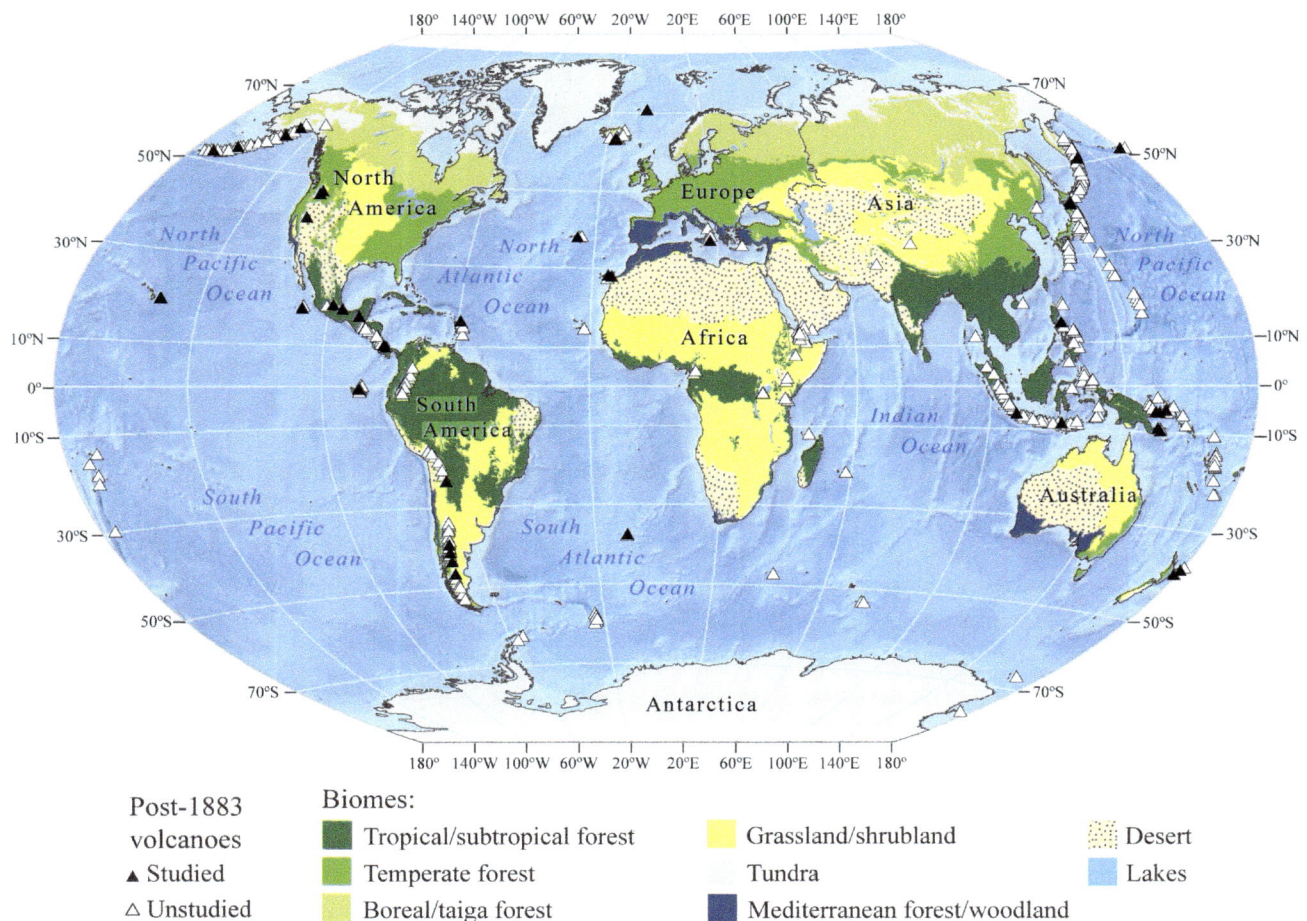

Post-1883 volcanoes
- ▲ Studied
- △ Unstudied

Biomes:
- ■ Tropical/subtropical forest
- ■ Temperate forest
- ■ Boreal/taiga forest
- ■ Grassland/shrubland
- □ Tundra
- ■ Mediterranean forest/woodland
- ▨ Desert
- ■ Lakes

FIGURE 73.6 Map of the world with volcanoes that have erupted subaerially since 1883 according to the Smithsonian Institute/USGS catalog of volcanoes plotted on a subset of the World Wildlife Fund's biomes of the globe. In this figure, several of the WWF biome types, given the limited number of volcanoes subject to ecological studies, have been combined. Volcanoes that have received ecological study are indicated with white triangles and those that have gone unstudied are indicated with black triangles. See text for additional information.

TABLE 73.2 Distribution and number of volcanoes by biome type, and biome area, and associated number of volcano-ecology studies of three ranks of study intensity (see text for explanation). The volcanoes tallied are in the Smithsonian Institution/US Geological Survey Global Volcanism Program catalog reported to have erupted since 1883, excluding those noted as "submarine." Locations of these volcanoes were plotted onto the World Wildlife Fund's (WWF) map of terrestrial biomes of the globe. Some of the 14 biome types in the WWF system were combined (e.g., three types of tropical and subtropical forests are shown as one and four types of grassland/shrubland are combined), and several involving very small numbers (<10) of volcanoes and/or small area are not reported (i.e., flooded grasslands; mangrove; lake, and rock/ice). The Mediterranean type refers to Mediterranean forest, woodland, and shrubland. For volcanoes designated as "ocean" biome, but not "submarine," we assigned a terrestrial biome type based on published ecological literature from the volcano or surrounding area, web-based searches on vegetation of the target area, and visual inspection of Google Earth imagery

Biome	No. active volcanoes	Area (10^6 m^2)	Volcanoes/biome area (no./10^6 km^2)	Number of volcanoes with ecology studies, by study category			Volcanoes studied (%)
				A	B	C	
Tropical forest	175	24.08	7.27	1	5	9	8.6
Temperate forest	63	16.90	3.73	1	3	8	19.0
Boreal forest	49	15.13	3.24	1	0	2	6.1
Mediterranean	10	3.22	3.11	0	0	3	30.0
Grassland/shrubland	34	35.99	0.94	0	0	4	11.8
Desert	23	27.89	0.82	0	0	1	4.3
Tundra	39	11.65	3.35	1	0	5	15.4

forest), and the density of active volcanoes in tropical and subtropical forest biomes (7.27 per 10^6 km^2) is more than twice than in other forest biomes; grassland/shrubland (0.94) and desert (0.82) biomes have the lowest density of active volcanoes, despite having relatively great extent. Volcanoes experiencing eruptions since 1883 affecting forest biomes account for 73.7% of all volcanoes erupting in terrestrial settings.

A tally of the number of studies, types of plants and animals (i.e., taxonomic scope) investigated, and duration of study for each volcano provides an index on the extent to which individual volcanoes have contributed to our knowledge of volcano ecology (Table 73.2). We rank volcano ecology study intensity at three levels: level A: multiple or continuous sampling over time, intensive, multiple topics and taxa; level B: several samplings over time, multiple topics and taxa; level C: single to a few sampling efforts, limited subject matter and depth of investigation. The majority (72%) of volcanoes were the subject of very limited study (level C), and only four (0.9%) of the 404 — Krakatau (1883), Surtsey (1963), Mount St Helens (1980), and Kasatochi (2008) — have been intensively studied on a sustained basis with commitment to long-term investigation. Interestingly, these four sites are representative of four different biomes, which should provide opportunities for interesting cross-biome comparisons in ecological responses to eruptions.

Ecological studies have taken place across all types of volcanic disturbance, but various forms of pyroclastic density currents, tephra falls, and lava flows have received most attention, likely because of their relative prevalence. Animal studies have been conducted at 28 (64%) of the 44 volcanoes that have erupted since 1883. Our extensive review of 302 scientific publications on animal studies following these eruptions have shown a near even attention given to invertebrates (47%) versus vertebrates (43%), and the remaining studies (10%) included both groups. Vegetation investigations have taken place at no fewer than 26 of the 44 volcanoes that have erupted since 1883. Although the numbers of volcanic sites studied for animals and plants are similar, there is only 11 (25%) of the 44 volcanoes where both groups were studied.

The limited proportion of active volcanoes studied is probably a result of several factors: not all eruptions did enough damage to ecosystems to draw the attention of ecologists and ecologists appear to be preferentially drawn to large eruptions or newly emerged volcanic islands; many of the volcanoes are in remote areas, making access difficult and expensive; sustained funding for ecological research is uncommon in most countries; unlike the volcanology community, the ecological research community is not organized and funded to promptly address volcanoes in states of unrest or to undertake post-eruption study.

6. PLANT SUCCESSIONAL PROCESSES FOLLOWING VOLCANISM

6.1. Key Processes of Plant Community Development

Early stages of succession are strongly influenced by persistence and growth of survivors (Figure 73.3(A) & (B)); immigration, establishment, and growth of colonists (Figure 73.6); and interactions among these colonists. Dispersal of species to a disturbed site is commonly thought to be determined by the distance to source populations, but complexities of disturbance process and patterns and state of the affected ecological system make such simplistic, distance-related interpretations unrealistic (Dale et al., 2005, Chapter 1). Moreover, dispersal capabilities of species vary tremendously, with some species being very mobile and readily dispersed long distances and others poorly equipped to disperse beyond the immediate area of the mother plant.

Plants exploit many means of dispersing their seeds or spores, including passive dispersal by wind and water, or by animals either in their gut tract or attached to fur or feathers. Wind dispersal is the most probable way for a new species to reach an isolated disturbance site, but large-seeded, poorly dispersed species can be more likely to establish if they do reach a site. Because many annuals, perennial forbs, and grasses are dispersed by wind, they are often the first plant species to arrive on volcanic deposits following a severe disturbance, but some pioneer trees, such as *Metrosideros*, also have light wind-dispersed seeds. Many species with poor wind-dispersal mechanisms can be transported great distances by hitchhiking in or on animals or capitalizing on the influence of gravity. Dispersal by birds, for example, is a common mode of plant species arrival onto volcanically disturbed sites in New Zealand (Clarkson, 1990). Once a plant propagule disperses into a new location, its ability to establish, grow, and reproduce is determined by prevailing climate, site conditions, previously established organisms, and the organism's own requirements and tolerances (Dale et al., 2005, Chapter 1). Once early colonists start to produce seed, local dispersal becomes possible and the rate of vegetation change increases, sometimes exponentially. Establishment from seed, comprising the phases of germination, seedling survival, and subsequent growth, is a particularly vulnerable phase of the plant lifecycle. Following a severe site-altering disturbance event, substrates such as lava and tephra are typically deficient in nutrients, particularly phosphorus and nitrogen, and can have very low water-holding capacity, but light is readily available. This commonly observed negative correlation led to the resource-ratio hypothesis as a general model for explaining plant succession, focused on the concept that succession results from a gradient through time in the relative availability of limiting nutrients and light resources, and also the associated competitive ability of coexisting species to secure requisite resources. Site amelioration is an important process in succession that can involve changes in soil conditions, microclimate, and microtopography. Primary forms of amelioration include creation of shade, organic matter, and nutrient inputs; increase in water holding capacity; and weathering of substrates. As a site ameliorates, the rate of succession gradually accelerates as plants establish and spread, species interact, and a community develops. For example, on lava flows in New Zealand and the Hawaiian Islands, soil and microclimate conditions are greatly improved below the canopy of the pioneer tree *Metrosideros*, which increases humidity, provides shade, and contributes to the development of soil—vital factors for the arrival of later-successional species (Clarkson, 1990; 1997). Pocket gophers (*Thomomys talpoides*) played a surprisingly important role in site amelioration on Mount St Helens, by digging through tephra and mixing it with buried soil complete with organic matter, soil microbes and macrofauna, and seeds, and depositing this mixture on the surface in the form of a mound (Figure 73.3 and, Dale et al., 2005, Chapter 14).

6.2. Factors Influencing the Pattern and Pace of Succession

Volcanic disturbance events can greatly influence vegetation cover, structure, and species richness and abundance. Accrual is the community assembly process whereby survivors grow and new colonists arrive, resulting in an accumulation of species, biomass, and structural complexity (Dale et al., 2005, Chapter 20). In early primary succession, species may accumulate slowly without interacting with one another, but as vegetation cover increases, interactions become inevitable and the pace of succession generally, but not always, increases. In the course of succession, some species fail because they cannot reproduce in the emerging environment, whereas others will be eliminated by competition. The net result is species turnover, one way to recognize succession (Dale et al., 2005, Chapter 7).

Various types of interactions among species drive community development, and these processes may change in their relative importance over the course of succession. The three most widely discussed mechanisms of plant succession relate to how species interact with the environment and with later-arriving species to either facilitate (relay floristics), inhibit (initial floristic), or have minimal effects on (tolerance) subsequent colonizers. Facilitation involves positive interactions between species and the improvement of a site's physical condition and is thus more common under stressful environmental conditions such as during primary succession. Early colonists create shade, alter soil moisture, and ameliorate soil texture and nutrient conditions on new volcanic substrates, facilitating the establishment of later-successional species, which otherwise could not have coped. Plants with the ability to fix atmospheric nitrogen (frequently legumes) are a prime example of facilitation during succession, such as *Lupinus*, which colonized on Mount St Helens (Figure 73.6 and, Dale et al., 2005, Chapter 17), or *Coriaria* on Mt Tarawera (Walker et al., 2003). In contrast to facilitation, inhibition occurs where resources such as space, water, or nutrients are so intensively used by one or more species that they are not available for other species and therefore end up inhibiting growth and slowing successional development. In this case, the species that arrive first at a site (initial floristic composition) can have a strong competitive advantage and strongly influence the trajectory of succession. Tolerance is intermediate between facilitation and inhibition, whereby the species best able to tolerate the prevailing site conditions are favored, and species replacement is not affected by the present residents. However, succession is highly variable, and in most cases these three mechanisms, plus others, occur simultaneously and can all be applicable to an individual at some stage in its life cycle. Although the relatively persistent stage of successional species composition that may eventually be achieved is often referred to as the climax community, this is better replaced with the notion of equilibrium or stabilization. Some successions can also be cyclic, never reaching equilibrium, or in some instances, the reverse of succession can occur—the degradation of ecosystems known as retrogression (Glenn-Lewin et al., 1992).

6.3. Examples of Plant Succession Following Volcanic Disturbance

Surtsey—Surtsey emerged from the sea in 1963, and the first plant had colonized by two years later (1965). Since that time, plant succession has been studied annually through the present (2013), making it one of the most thorough and long-term studies of post-eruption succession in the world (Jakobsson et al., 2009, pgs. 57–76, Magnusson et al., 2014). The pattern of plant establishment on the island occurred in five stages (Figure 73.8). The first decade was characterized by the establishment of several shoreline plants species, presumably dispersed by sea water. During the second decade species richness changed little, with the arrival of relatively few species new to the island. The third decade began a period of accelerated species accrual that was mediated by the formation of a large gull colony (*Larus* spp.), which resulted in the transport of marine-derived nutrients to land and also the dispersal of seeds from adjacent islands to Surtsey. During this period, particularly during 1990−1998, two to five new plant species established on the island each year. From 1999 through 2005 the pace of establishment declined to zero to three new species each year, followed by a sharp increase from 2006 to 2007, when five new species were added each year. From 2008−2013, species accrual stabilized, with no new species added, presumably because most of the species in the regional pool had already colonized or conditions were not suitable for establishment of arriving propagules.

Mount St Helens—Vegetation responses to the 1980 eruption of Mount St Helens have been studied extensively (Dale et al., 2005). Here we provide one example of this large body of work that was conducted on the debris avalanche deposit (Dale et al., 2005, Chapter 5). The pattern of plant establishment on the debris avalanche occurred in five stages (Figure 73.7). The first stage included the presence of only a few surviving plants, all of which were sprouting from plant fragments that happened to come to rest on or near the deposit surface; 20 species were known to have survived this way. The next stage was characterized by a rapid accrual during the first two post-eruption growing seasons that resulted in about 68 species present. This was followed by a much slower increase in establishment that lasted from post-eruption year three through nine, and ended with 90 species present. The fourth stage (year 10−14) was characterized by rapid recruitment of an additional 60 species (total ∼150 species). The final stage in this period of study was a sharp decline, as about 40 species were lost between post-eruption year 14 and 20. This decline in species is thought to be tied to the development of forest canopy that altered the light regime, resulting in the loss of light-dependant species, which had dominated the precanopy plant communities.

Comparing the trajectories of plant species accrual at Surtsey and Mount St Helens illustrates the importance of the context in which volcanism occurs and ecosystems respond. Surtsey, a recently emerged, small, isolated island with new lava surfaces, poses serious constraints on the pace of ecological responses related to a small and relatively distant regional species pool, dispersal limitations due to being surrounded by sea, lava flows that are slow to weather and have hard surfaces limiting rooting, and a relatively cold climate. This resulted in a slow but steady accrual for the first 30 years. With the development of a large gull colony, the inhospitable conditions for plant establishment were ameliorated and the pace of plant accrual accelerated. On the debris avalanche deposit at Mount St Helens, the accrual process was rapid due to the presence of some surviving plants (albeit at low densities), close proximity and connectivity to large regional source populations, valley-bottom topography that facilitated arrival of propagules, numerous mesic environments that promoted trapping of seeds and ample moisture for

establishment, unconsolidated deposits that allowed root penetration, and a wet and mild climate.

7. ANIMAL RESPONSES TO INITIAL ERUPTION AND SUBSEQUENT ASSEMBLY PROCESSES

It should come as no surprise that volcanic eruptions can negatively affect individuals, populations, and assemblages of animals and that these effects may be driven by both direct (killed by primary disturbance) and indirect causes (mortality caused by changes in vital resources). The ways in which animals are affected by eruptions vary by a multitude of factors, including, type, intensity, and spatial extent of volcanic disturbance; stochastic factors, such as timing of the event; characteristics of the fauna; and extent of habitat alteration. Animals ultimately are either resistant to disturbance, and individuals persist through an eruption, or at least many species demonstrate remarkable abilities to colonize defaunated or newly created landscapes and aquatic systems. One only has to visit a volcanic area after a few years, or in some cases after only a few days, have elapsed since the last eruption to witness surviving animals from the pre-eruption ecosystem and the arrival of new colonists (Figure 73.3(C−F), Figure 73.9(A−F)).

7.1. Factors Influencing Animal Survival and the Importance of Survivors

Of the several factors influencing animal survival, disturbance intensity is probably most important. Animals perish when subjected to extreme temperature, impact force, or burial associated with lava flows, PDCs, lahars, or deep (>50 cm) tephra, under almost any circumstance. However, rarely are volcanic disturbances uniform, and complex disturbance mosaics or gradients, occurring at a variety of spatial scales, are the norm, which lead to heterogeneity in survivorship across impacted landscapes. Hence, under most eruption scenarios residual species are observed. Indeed, following the 1980 eruption of Mount St Helens, survival of animals was widespread throughout much of the 600-km² blast area (Figure 73.3, and Dale et al., 2005, Chapters 9−14) and after what appeared to be an annihilating eruption in 2008 on Kasatochi Island (Alaska, USA), at least a few arthropod species were observed and thought to have survived within cracks and crevices of rocks. **Refugia**, or places within a disturbed landscape where organisms survive, can range from small outcrops to entire mountainsides or even whole lakes. Refugia can be created by factors that reduce the intensity of disturbance mechanism(s) experienced by an animal and thus allow survival within a broader area of defaunation (Figure 73.3(C−F)). In all cases, they involve some form of protection from

FIGURE 73.7   Examples of plant colonizers following eruptions at several volcanoes. Plants gain a foothold in cracks and crevices on pahoehoe lava, Hawaii, USA (A). Lava flows colonized by herbs, shrubs, and coniferous trees on Mt. Etna, Italy, with more recent flows burying and incinerating vegetation on the slope in the background (B). Nalca (*Gunnera magellanicum*), a plant that has a symbiotic relationship with cyanobacteria located on its roots and gains nitrogen to support growth on nutrient-impoverished lahar deposits at the Orsorno volcano, Chile (C). An arid-land cushion plant grows on scoria deposits at the La Payunia Volcanic Complex, Argentina (D). The early succession, wind-dispersed herb, *Gamochaeta spicata*, rapidly spread throughout the blast area during the third post-eruption growing season at the Chaiten volcano, Chile (E). The Prairie Lupine (*Lupinus lepidus*) is a primary colonizer on pyroclastic flow, debris avalanche, and lahar deposits at Mount St Helens, USA. It has a symbiotic relationship with bacteria on its roots that enables it to flourish on nutrient-impoverished substrates (F).

volcanic forces and include examples such as topography that offers topographic shielding or presence of snow or ice, soil, sediment, rock crevices, interstices within rotten logs, or water.

Animal size is inversely related to survivorship because small animals can more easily find protection in refuges, whereas larger animals may perish. This was well documented at Mount St Helens, where many small-bodied mammals survived, yet all five species of large mammals present in the pre-eruption landscape perished in the 600-km² blast area. Similarly, habitat associations favor the survival of some animals. For example, species living in subterranean environments, like certain small mammals, survived, whereas animals living in exposed habitats, such as resident birds, perished. Animal life-history strategies have been documented as important in leading to

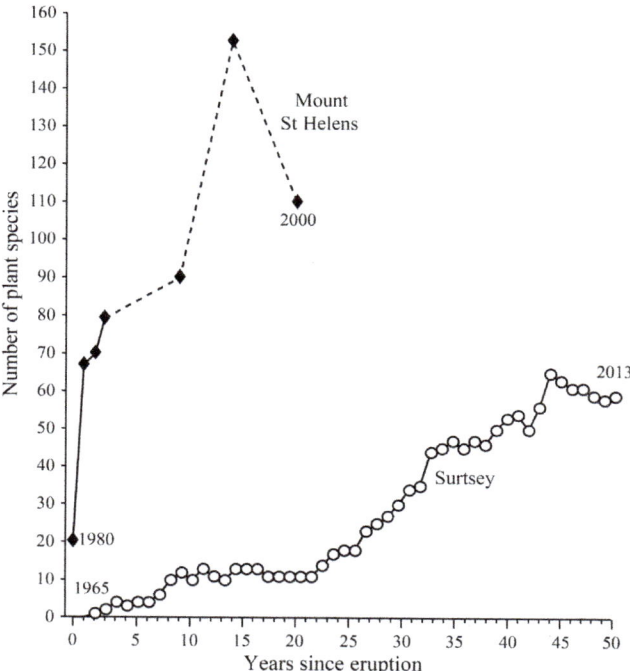

FIGURE 73.8 Plant species accrual at Surtsey and Mount St. Helens (MSH) volcanoes. At Surtsey, colonization began on barren lava surfaces. Species accrual has been a relatively slow process because of inhospitable substrates for plant establishment and a small regional pool of potential colonists that are somewhat dispersal limited because the small island is surrounded by sea water. The eventual establishment of breeding gull colonies led to a rapid increase in accrual and spread of vegetation. Species accrual on the debris avalanche at MSH began with the survival of a few isolated plant fragments that took root during the first post-eruption growing season. This was followed by a rapid invasion of plant species due to favorable surface conditions for plant establishment and a rich local seed source. However, the development of closed forest on parts of the deposit reduced light conditions that presumably led to the sharp decline of plants found among the long-term study plots. Data from Surtsey provided by Borgthor Magnusson, Icelandic Institute of Natural History. Data from MSH provided Virginia Dale, Oak Ridge National Laboratory.

survivorship and include biphasic organisms, such as certain insects and amphibians (Figure 73.3(C)), that have aquatic larval forms and terrestrial adults, which afford protection by spreading the risk of mortality (e.g., habitats possessing different life-stages are disturbed to differing extents), and anadromous fish, which migrate between fresh and salt water, have several age cohorts at sea and thus avoid mortality of those population segments when freshwater habitat is severely damaged. Finally, timing of an eruption, a stochastic factor, can influence survivorship at a number of time scales, including diel, seasonal, and decadal. For example, an eruption occurring during the daylight hours may allow nocturnal animals to survive in the protection of their day time retreats. Seasonally, migratory species may not be at an eruptive site at the time of the eruption, or the conditions, such as snowpack, may protect organisms that live in soil or downed wood. Finally,

at decadal scales, the successional stage of a system influences the types and abundance of survivors, and this is probably most apparent for plants that show different responses to disturbance by their ability to send shoots up through volcanic deposits.

7.2. Factors Influencing Animal Community Assembly

The pace and pattern of animal community assembly following volcanism is strongly influenced by presence of surviving individuals, several aspects of landscape characteristics including geographic context (see Section 4 above), species characteristics, habitat development, biotic interactions, and stochastic factors, such as weather (Figures 73.1 and 73.2).

Roles of surviving animals—Surviving animals influence the pace and pattern of community assembly and, more broadly, succession in several important ways. By having in situ survival, the need for **colonization** from potentially distant source populations and the challenges associated with dispersal are negated, and if sexually mature individuals are present, reproduction may ensue immediately, leading to population growth. These survivors or their progeny may serve as important source populations for adjacent defaunated sites. However, in some cases animals surviving an eruption succumb shortly afterward because of indirect effects related to changed resource conditions, such as limited food, water, and protective cover, and stressful microclimate. If this is the case, their carcasses decompose, release vital nutrients, and ameliorate impoverished substrates. On the other hand, surviving animals may exploit the newly created conditions under reduced competition or predation, and their populations may burgeon. Such responses have been documented for pond-breeding amphibians following the 1980 eruption of Mount St Helens (Dale et al., 2005, Chapter 13) and for other taxa at other eruption sites. Surviving animals interact with other ecosystem components that may either hasten or retard ecological processes. For example, seed dispersal and fecal inputs can accelerate succession, whereas herbivory and bioturbation, such as burrowing, slow succession. Overall, increased presence and abundance of survivors should result in a more deterministically based assembly, whereas under circumstances where survivors are absent stochastic processes will play a more prominent role.

Following the primary disturbance, animal assembly processes, such as establishment, are often tightly coupled with habitat genesis—a form of site amelioration. The three-dimensional architecture of vegetation forms habitat conditions crucial to establishment of plant-eating insects, small mammals, and birds, and often the development of animal assemblages follows in lock-step with vegetation

FIGURE 73.9   Examples of animal colonizers following the 1980 eruption of Mount St Helens, USA. North American elk (*Cervus elaphus*) perished in the 600 km² blast area, but within weeks of the eruption returned to the area and have influenced plant community dynamics in numerous ways (A). The deer mouse (*Peromyscus maniculatus*), an extreme generalist in both habitat and diet, was among the most successful small mammal colonists following the eruption (B). White-crowned sparrows (*Zonotrichia leucophrys*), ground foragers and shrub nesters, became ubiquitous in the post-eruption landscape, exploiting the open terrain (C). The common nighthawk (*Chordeiles minor*) is a ground nester and aerial screener and is common on the barren pyroclastic flow surfaces (D). Several grasshopper species and shield-backed katydids (*Steiroxys* sp.) colonized pyroclastic flow deposits following the development of an herbaceous plant layer (E). Darwin beetles (*Chiasognathus grantii*) colonized the blowdown forest zone at the Chaiten volcano, Chile, presumably in response to the enormous recruitment of dead wood, which the larvae forage on.

development, reflecting controls on both habitat structure and food resources (Figure 73.9(A–F)). Hence, volcanic landscapes present many examples where animals arrive to a site, but fail to establish because of slow rates of plant succession (Dale et al., 2005, Chapter 14). On the other hand, there are examples where transitions from low-statured herb communities to tall-shrub habitats result in a major accrual of small mammal and songbird species (Dale et al., 2005, Chapters 14 and 20).

Recently disturbed volcanic sites may contain unusually food-rich habitats that have developed unexploited by consumers, with few (or no) predators, parasites, or pathogens (the three 'Ps'). Eventually these habitat patches may be colonized by an animal species whose population grows extraordinarily large and either exploits all resources and wanes or is found by one of the three 'Ps', and a boom-and-bust cycle ensues. Boom-and-bust cycles are likely a common feature of assembling communities on severely disturbed volcanic landscapes. Such patterns and processes have been observed regularly at Mount St Helens and at the Chaiten volcano, Chile (Dale et al., 2005). As succession proceeds and species interactions become more tightly coupled and involve more complexity and participants, these oscillations are expected to diminish.

Species dispersal is a key factor in assembly processes, with two important components that warrant consideration. First is the mobility of different animal species; second is the propensity for a particular animal species to disperse through different landscape configurations composed of vegetation patches and boundaries. These factors can strongly influence establishment sequences and hence ultimately community structure and potential biotic interactions. "Aerial plankton" of arthropods readily rains down on recent volcanic sites, and birds and bats can quickly colonize a site, if appropriate resources exist (Section 8, below). However, less-mobile taxa (e.g., pedestrian species) may take years, decades, or even centuries to establish and spread. For example, at Mount St Helens, the highly mobile Deer Mouse (*Peromyscus maniculatus*) established in an area undergoing primary succession one year post-eruption, presumably arriving from sites several kilometers away, whereas the fossorial Northern Pocket Gopher that survived on adjacent hill slopes, but digs its way to new locations, was not observed until 12 years post-eruption (Figures 73.3 and 73.9). There are examples of volcanic landscapes in faunal disharmony resulting from the interplay between eruption frequency (return interval) and species-specific dispersal capacity, such that the volcano is erupting at a pace that outstrips the movement capacity of the species. This may create an extensive zone where certain species in a region are absent because of repeated volcanic disturbance.

Development of animal community structure is rarely linear, but instead is characterized by a series of steps or thresholds often related to changing resources and biotic interactions. These two factors can lead to major shifts in both species richness and in the specific members of an animal community during the assembly process.

8. ECOSYSTEM PROCESSES AND BIOLOGICAL INTERACTIONS

Soils and plants processes and interactions—Soil is the product of many biological and geophysical processes. Lack of soil organic matter and critical nutrients, notably nitrogen and phosphorus, often shape and limit ecosystem development in volcanic landscapes. Soil organic matter, composed mainly of carbon, is generally considered to include plant residues and living microbial biomass, detritus, and stable or humified material. The amount and type of soil organic matter affects a large number of

fundamental biological, chemical, and physical soil functions, including its ability to support microbial activity, supply nutrients to plants, resist chemical change, transport and hold water, and resist erosion. Typically, heterotrophic soil microorganisms are limited by the availability of carbon substrates, while nitrogen (N) needed by plants may be in limited supply, especially in early primary successional sites, which begin in the absence of biological and physical legacies. In addition to nitrogen, plant-available phosphorus (P) may become an important limiting nutrient on older volcanic soils. Thus, early colonizers of new land surfaces include vascular and nonvascular plants and microorganisms that are able to avoid or tolerate environmental extremes, overcome nutrient limitations, and modify the physical and chemical characteristics of the substrates through various strategies.

Vascular plants fix the majority of the carbon (C) needed by heterotrophic organisms, while soil microorganisms drive nutrient acquisition during early succession and later become the drivers of nutrient recycling. In general, increasing inputs of exudates and accumulation of soil organic matter derived from plants are associated with larger, more active, and more complex microbial communities but specific interactions among plants and soil biota are complex and can vary over space and time.

Nutrient acquisition strategies for plant colonists of low-fertility volcanic substrates include specialized anatomical features such as cluster roots, whose exudates can mobilize nutrients such as P from parent material, or development of specialized, coevolved relationships whereby plants divert a portion of their photosynthetic output to support symbiotic relationships with microorganisms. Examples of the latter include fixation of atmospheric N by *Rhizobium* in legumes or *Frankia* in alder (*Alnus* spp.) and scavenging of P from substrates where it is in short supply by arbuscular mycorrhizal fungi in roots. Plants with the ability to fix atmospheric nitrogen (frequently legumes) are a prime example of facilitation during succession, such as *Lupinus*, which colonized on Mount St Helens (Figure 73.7 and Dale et al., 2005, Chapter 17), or *Coriaria* on Mt Tarawera (Walker et al., 2003). Some inputs of organic matter to aggrading soils are the result of free-living photosynthetic microorganisms such as cyanobacteria, or of nonphotosynthetic microorganisms called chemoautotrophs that use light or the energy produced by chemical reactions to fix CO_2 and form organic compounds even below the soil surface. Besides nutrient inputs, communities of such microorganisms can form crusts on unconsolidated substrates that stabilize surfaces and reduce erosion by wind and water. Soil microorganisms affect other important soil physical attributes like aggregate formation and water movement by producing compounds such as polysaccharides, glycoproteins, and hydrophobins and shape the plant community structure by producing compounds that can inhibit other microorganisms or plants.

Early patterns of biological reassembly and soil genesis are both spatially and temporally context-dependent and may appear as a gradual process, occur in a stepwise fashion characterized by periods of relative stability separated by critical thresholds, or even appear stochastic. Short-term or small-scale patterns of plant colonization may not be tightly coupled to the distribution of soil properties, but instead may depend on unpredictable patterns of environmental conditions, survivors, and colonists. Some soil characteristics such as hydraulic or thermal conductivity are strongly influenced by the physical characteristics of deposited material, such as color or texture, and thus substrates can demonstrate some soil-like functions before organic matter and nutrients accumulate or diverse biotic communities evolve. The physical characteristics that influence water availability or temperature extremes can act as strong selective filters of early colonists, but young substrates may not exhibit the resilience and consistency associated with a more completely developed soil ecosystem. With time, plant communities and soil properties become more tightly linked (Laliberté et al., 2013). Synoptic ecosystem patterns emerge and become more predictable but, because they remain dynamic, can be difficult to study (Walker et al., 2010).

Animal processes and interactions—Flying animals, such as birds and bats, or those passively dispersed by wind, referred to as "aerial plankton," including many arthropod species, are often the first animals to arrive in isolated volcanic landscapes. These animals play many pivotal ecological roles and often precede the establishment of plants. Hence, the first biological communities may very well be animal-based and dominated by predators and scavengers eager to consume other aerial plankton, rather than the commonly assumed sequence of plants, primary consumers, secondary consumers, and so forth. Moreover, the majority of passively dispersed arthropods that arrive on primary successional surfaces are unable to establish and thus perish. These unsuccessful "derelicts of dispersal" provide important sources of imported nutrients to impoverished volcanic substrates (Thornton, 2000; Dale et al., 2005, Chapter 9).

On Surtsey, sea birds, particularly four species of gulls, have profoundly influenced the development of invertebrate and plant communities (Figure 73.8), as well as the establishment of terrestrial birds, through transport of marine-derived nutrients in their feces, which are deposited in localized areas on the island, greatly enriching the volcanic substrates with high nutrient concentrations and increasing biological productivity (Figure 73.8). Frugivorous bats have played similar roles on Anak Krakatau, where they have dispersed seeds of fruit trees from the mainland to the remote island (Thornton, 2000). At mainland eruption sites, landscape context has also been identified as an important factor of animal assembly (Dale et al., 2005, Chapter 20). For example, small mammal and bird assemblages are tightly coupled to "oasis" habitats of complex vegetation architecture embedded within a vast "sea" of relatively barren vegetation (Dale et al., 2005, Chapter 14). Thus, there are parallels in community assembly between oceanic islands and mainland habitat islands.

9. SUMMARY

Volcano ecology focuses on the initial types, intensities, and mechanisms of volcanic disturbance to biota and their habitats and resources, on the longer-term processes of succession and biological assembly, and on soil development. Volcanic disturbance is highly variable in its effects on ecosystems and impacts range from minor and short-term to profound and enduring. Nonetheless, life is either tenacious or resilient, and eventually soil develops and plant and animal communities assemble. Focusing on specific disturbance mechanisms, such as burial, abrasion, heating, impact force, and chemical toxicity, provides an important framework for evaluating biotic responses to volcanic disturbances and also provides opportunities to compare volcanic disturbance to other forms of natural and human disturbance.

Rarely does volcanic disturbance result in complete annihilation of the biota, and survivorship is the general rule ranging from most of the biota persisting to only a few individuals of some species surviving in isolated refugia (Figure 73.3). High levels of survival lead to a more deterministic assembly process, whereas when life is starting anew, chance is expected to play a larger role, and predictability of the pace and pattern of assembly is expected to decline. In cases of newly emerged volcanic islands and in the context of extreme disturbance on sites that supported life beforehand, there are examples of landscapes devoid of life. Within minutes to hours of cessation of volcanic activity, organisms and propagules initiate the processes of succession and biological assembly. The pace and pattern of these processes will be strongly dictated by the presence of survivors; texture and nutrient characteristics of the new deposits; climate; factors influencing dispersal, establishment, and growth of individuals and populations; and a diverse array of biotic interactions.

Although there has been substantial research on volcanic landscapes, the vast majority of this work has focused on only a few iconic locations, limiting the inferences that can be made to volcanism globally. Plants and animals have been studied at roughly the same number of volcanoes, and within the animal kingdom invertebrates and vertebrates have received roughly equivalent attention. In addition, a dearth of work has addressed the patterns and rates of soil development at volcanic sites, and truly well-integrated interdisciplinary work is largely absent. We encourage ecologists to establish a research agenda that integrates soil development, plant succession, and animal assembly, and that they do so immediately following an eruption so that the initial disturbance characteristics can be mapped and described and questions of survivors versus colonists can be accurately assessed. Ecologists are also encouraged to integrate their work with that of geologists, who can aid in interpreting the volcanic events and resulting deposits. Apart from being a natural driver of damage or rejuvenation of ecosystems, volcanism provides a rich source of study to better understand Earth's ecological systems.

ACKNOWLEDGMENTS

Our understanding of volcano ecology has benefitted greatly from discussions with members of the Mount St Helens Ecological Research Community and also our collaborators at the US Geological Survey-Cascades Volcano Observatory and from working with our colleagues on volcanoes in Chile, Alaska, and China. We thank Kathryn Ronnenberg for creating figures and tables and for providing copy editing. Kelly Christiansen graciously provided GIS support and figure production. Jackson T. Efford is thanked for his contribution compiling literature for the plant ecology section. Support for CMC and FJS was provided by the US Forest Service-Pacific Northwest Research Station and the US National Science Foundation (DEB-0614538, NSF 0917697 and NSF 0823380).

FURTHER READING

Ball, E.E., Glucksman, J., 1981. Biological colonization of a newly created volcanic island and limnological studies on New Guinea Lakes, 1972–1978. Nat. Geo. Soc. Res. Reports 13, 89–97.

Brattstorm, B.H., 1990. Biogeography of the Islas Revillagigedo, Mexico. J. Biogeogr. 17, 177–183.

Clarkson, B.D., 1990. A review of vegetation development following recent (<450 years) volcanic disturbance in North Island, New Zealand. New Zeal. J. Ecol. 14, 59–71.

Clarkson, B.D., 1997. Vegetation succession (1967–89) on five recent montane lava flows, Mauna Loa, Hawaii. New Zeal. J. Ecol. 22, 1–9.

Dale, V.H., Swanson, F.J., Crisafulli, C.M. (Eds.), 2005. Ecological Responses to the 1980 Eruption of Mount St Helens. Springer, New York.

Deligne, N.I., Cashman, K.V., Roering, J.J., 2013. After the lava flow: The importance of external soil sources for plant colonization of recent lava flows in the central Oregon Cascades, USA. Geomorphology 202, 15–32.

Edwards, J.S., 2005. Animals and volcanoes: Survival and revival. In: Marti, J., Ernst, G.G.J. (Eds.), Volcanoes and the Environment. Cambridge University Press, New York, pp. 250–272.

Glenn-Lewin, D.C., Peet, R.K., Veblen, T.T. (Eds.), 1992. Plant Succession: Theory and Prediction. Chapman & Hall, London.

Jakobsson, S.P., Magnusson, B., Gunnarsson, K., 2009. Surtsey Research Society. Surtsey Research 12. Reykjavik, Iceland.

Laliberté, E., Grace, J.B., Huston, M.A., Lambers, H., Teste, F.P., Turner, B.L., Wardle, D.A., 2013. How does pedogenesis drive plant diversity? Trends Ecol. Evol. 28, 331–340.

Swanson, F.J., Jones, J.A., Crisafulli, C.M., Lara, A., 2013. Effects of volcanic and hydrologic processes on forest vegetation: Chaiten volcano, Chile. Andean Geol. 40, 359–391.

Thornton, I., 1996. Krakatau: The Destruction and Reassembly of an Island Ecosystem. Harvard University Press, Cambridge, MA.

Ugolini, F.C., Dahlgren, R.A., 2002. Soil development in volcanic ash. Glob. Environ. Res. 6, 69–81.

Vitousek, P.M., 2004. Nutrient Cycling and Limitation: Hawai'i as a Model System. Princeton University Press, Princeton, NJ.

Walker, L.R., Clarkson, B.D., Silvester, W.B., Clarkson, B.R., 2003. Colonization dynamics and facilitative impacts of a nitrogen-fixing shrub in primary succession. J. Veg. Sci. 14, 277–290.

Walker, L.R., Wardle, D.A., Bardgett, R.D., Clarkson, B.D., 2010. The use of chronosequences in studies of ecological succession and soil development. J. Ecol. 98, 725–736.

Whittaker, R.J., Jones, S.H., 1994. The role of frugivorous bats and birds in the rebuilding of a tropical forest. J. Biogeog. 21, 245–258.

Whittaker, R.J., Bush, M.B., Richards, K., 1989. Plant recolonization and vegetation succession on the Krakatau Islands, Indonesia. Ecol. Monogr. 59, 59–123.

Chapter 74

Volcanic Materials in Commerce and Industry

Jonathan Dehn

Geophysical Institute, University of Alaska Fairbanks, Fairbanks, AK, USA

Stephen R. McNutt

School of Geosciences, University of South Florida, Tampa, FL, USA

Chapter Outline

GLOSSARY

absorbent A material that takes up, assimilates, or incorporates liquids through pores or interstices.

aggregate Mineral material composed of a mixture of different sizes that is separable by mechanical means. Aggregate is passed through sieves and other sorters to grade its size. Graded aggregates are mixed with cement or bituminous material to form mortar or concrete.

basalt A fine-grained, sometimes glassy igneous rock. Basalts have low SiO_2 contents of 45–50%. They are usually found in lava flows, which may be of variable size. They are frequently erupted from fissures and less often from central vents.

bentonite A special assemblage of clay minerals often formed from weathering of volcanic ash. Typically, it is composed of non-refractory clays rich in montmorillonite and is off-white in color. When the principal ion in the clay is sodium, bentonite is highly absorbent and subject to large volume changes.

cinder block A type of brick made from compressed aggregate material, often volcanic.

drilling mud A mud mixture of clays and water with a high density, viscosity, and strength. Acts as a lubricant, coolant, and sealant and floats drill cuttings up and out of the hole.

montmorillonite A clay mineral common as a weathering product of volcanic material.

obsidian A black, wholly glassy rock that displays conchoidal fracture. Most obsidians are of rhyolitic composition. Because glass is thermodynamically unstable, most obsidians undergo a process of devitrification and form crystal structures.

perlite A glassy devitrified rock of rhyolitic composition, which contains many curved or spherical cracks caused during cooling and contraction. These may be so well developed that the rock easily breaks into irregular pieces. Contained spherical pieces of unaltered obsidian are known as "Apache's tears." Perlite is used as an absorbent and insulator.

pozzolan A siliceous volcanic ash used to create hydraulic cement.

pumice A highly vesicular pyroclastic rock produced during explosive eruptions. It tends to accumulate as rounded lumps and also as abraded fragments called shards. Densities are typically less than 1 g/cm^3, so pumice often floats on water.

refractory A material that does not significantly deform or change chemically at high temperatures. Bricks of such material in various shapes are used to line furnaces.

tephra All fragmental volcanic ejecta, which are produced during explosive volcanism, including ash, cinders, lapilli, scoriae, pumice, bombs, etc.

The Encyclopedia of Volcanoes. http://dx.doi.org/10.1016/B978-0-12-385938-9.00074-2

trap rock An older term for compact, fine-grained igneous rocks such as lava flows, which have been broken into roughly uniform fragments. Often used for roadbeds, running tracks, and the like.

tuff A general name for consolidated ash or tephra, usually with largest clasts less than 2 cm. The material is erupted and is still hot when it hits the ground, resulting in consolidation and partial welding. Tuffs may be crystal-rich (crystal tuffs), lithic-rich (lithic tuffs), glass-rich (vitric tuffs), or lapilli-rich (lapilli tuffs).

The products of volcanic eruptions have provided useful raw materials for man throughout history. In many regions of the world volcanic rock is ubiquitous, and the inhabitants of these regions have developed ingenious uses for this natural resource. Ranging from early weapons and implements to building materials, volcanic rocks have been sought out because of their physical properties for use in manufacturing processes, as insulators, **absorbents**, or abrasives. Volcanic materials play an important role in our lives, regardless of whether or not we live near volcanoes, although most of us do not realize their wide variety and applications.

1. INTRODUCTION

The use of volcanic materials most certainly sprang from necessity, but it was quickly learned that many volcanic materials are ideal for a variety of uses. They have a unique set of physical properties, and their natural occurrence makes them relatively inexpensive and thus ideal for construction and manufacturing. For example, **pumice** has a relatively high strength, yet in some cases a low enough density to float on water, and provides excellent insulation and absorption characteristics. Most explosive eruptions produce pumice and spread it over a large area. Volcanic rocks are present almost everywhere in the world and annually $6-8\,km^3$ of new volcanic material is erupted, nearly $2\,km^3$ subaerially, and about 80% of that is particulate material from pyroclastic flows, debris avalanches, and ashfalls. This is enough to cover 1600 football or soccer fields a kilometer deep in **tephra**.

Before the industrial age, volcanic rocks were used for a wide range of purposes and provided a tradable commodity and a basis for local economies. Volcanic soils are fertile, and to this day many countries use volcanic ash as a soil supplement in agriculture (see "Volcanic Soils"). The use of volcanic materials has increased as additional applications have been found, strengthening the practical and economic relationship between mankind and volcanoes. Few data exist on the amount of volcanic material used or the "size" of the affected industries. Hence we make a comparison with several other industries and commodities to help get an idea of the number of volcano-related products available. We determined the number of pages in the 1994 *Thomas Register of American Manufacturers: Products and Services Section* devoted to several volcano-related topics and the number of products listed in the online version of the register in 2013 (Figure 74.1). We then added several familiar everyday examples to provide a basis for comparison. These include a small niche market (surfboards), medium-sized markets (air conditioners and refrigerators), and a large market (batteries). We found that one page or less of the *Thomas Register* was devoted to sulfur, monuments and memorials, pumice, **bentonite**, and additives; however, each of these volcano-related products had more entries than surfboards. Thus, these are all relatively small markets. Abrasives, on the other hand, are a substantially larger industry than air conditioners, and represent a medium market. Absorbents, polishers, and refractories, of which volcanic materials are components, offer a moderate number of products, slightly smaller than refrigerators. All the examples given have fewer manufacturers than batteries, suggesting that volcanic materials in total are small to medium components in the marketplace. These ratios remained consistent over the following 19 years, although a new entry, "volcanic" was added as global volcanic awareness rose following the 2010 eruption of Eyjafjallajökull. Although crude, these estimates give a sense of the growing and significant positive impact volcanic materials have on our everyday lives.

2. TYPES OF MATERIALS

Volcanic products can be divided into several categories based on their use and chemical composition. These materials are summarized in Table 74.1. The density and strength of each type of material depends on its chemistry and eruptive style, and in turn determines the ideal use for each product. As a result of the eruption process, explosive volcanic products tend to be in the form of **aggregates** and have lower densities. In addition, they are often vesicular, giving them excellent insulating qualities. Many industrial processes, for example, the manufacture of **perlite**, are designed to enhance these characteristics. Volcanic materials often have very high strengths relative to their densities, particularly for massive deposits, such as lava flows or welded ignimbrites. This makes them cost-effective building materials. In the case of aggregate material, the irregular angular shapes allow particles to stick together, yielding very high angles of internal friction. This makes these aggregates ideal for use in constructing roadbeds or in laying railroad tracks. Because volcanic glass fragments are made of silica, a very hard substance, and the shards are often in the form of bubble-wall junctions, or small sharp tetrahedra, ashfall deposits are a good source of abrasives. Volcanic rocks are not readily soluble in water nor do they take on water and increase in volume (with the exception of volcanic clays, see following discussion). This makes them more resistant to weathering than limestone or sandstone. Volcanic glass is resistant to corrosion and forms the

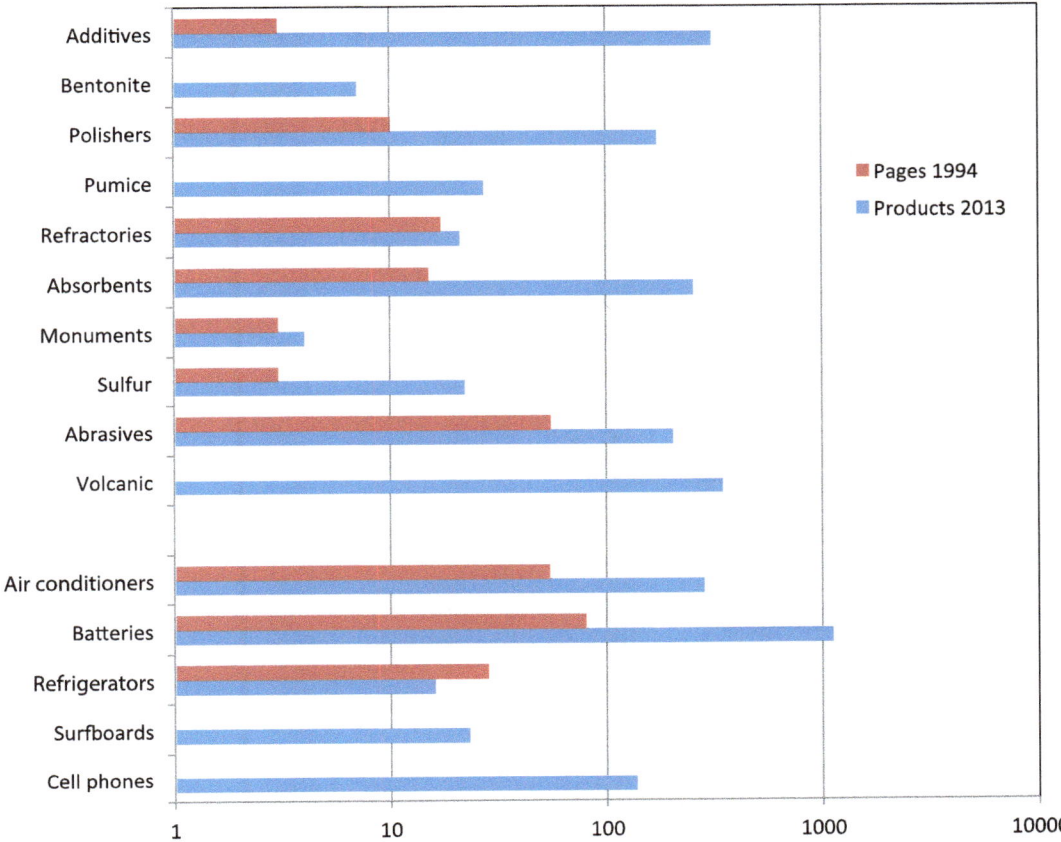

FIGURE 74.1 A histogram showing the number of pages of the 1994 *Thomas Register of American Manufacturers: Products and Services Section* devoted to various volcano-related topics and the number of products listed in the Thomas register online in 2013. Several familiar everyday examples of other products are shown in the lower part of the chart to provide a basis for comparison.

groundmass of most volcanic rocks. Most volcanic rocks have very small coefficients of thermal expansion (on the order of 1×10^{-7} cm deformation/cm distance/°C or less), and thus will not change in size or shape even under very large changes in temperature. This makes them very good refractories. Concrete fabricated from volcanic cinders is a better insulator by a factor of 3−4 than normal concrete (thermal conductivities of 0.35−1.5 kcal/m h deg). This is because of the high vesicularity of the cinders used in its manufacture. The small vesicles resist the transfer of heat, making it more difficult for thermal energy to move through the block. Igneous rocks (e.g., granite) are also used in construction and have similar strengths compared to volcanic materials, but have higher thermal conductivity and thus do not insulate as well. Because these rocks were typically created at some depth, and thus pressure, the crystals are tightly pressed to one another, giving very good heat conduction across these boundaries. These materials often have larger crystals, making them useful for aesthetically pleasing facades. This chapter will not go into depth on the use of igneous materials, although in many ways they parallel similar volcanic products.

3. PREHISTORIC USES

Volcanic glass has been used for early cutting implements, as well as microcrystalline volcanic rocks, such as fine-grained **basalts** or andesites. Arrowheads, scrapers, and knives made from volcanic products are common finds at prehistoric sites near volcanic regions (Figure 74.2). How our ancestors used volcanic rocks is nearly as varied as how we use them today. Pumice can be used as a float in fishing, and as an excellent abrasive, like sandpaper, to fashion other tools. Nonvesicular volcanic rocks are also ideal for crushing and have been used to grind wheat, by hand and in more carefully fashioned millstones.

Blocks of **tuff** are good refractories, reflecting and holding in heat, thus ideal for the construction of oven walls, chimneys, and kilns. The clays produced by the weathering of volcanic ash are ideally suited for making pottery (Figure 74.3). A culture that had a good source of volcanic rocks nearby was not only threatened by possible eruption but also had a strong economic base. In Mexico, the early empire of Teotihuacan (preninth century) was founded on the manufacture and export of **obsidian** blades.

TABLE 74.1 Physical Properties of Useful Volcanic Materials: Typical Grain Size in a Volcanic Occurrence, Density, Strength (Compressive), Thermal Conductivity, Heat Capacity, and Common Uses

Volcanic Product	Color	Typical Grain Size	Density (g/cm³)	Strength (kbar)	Thermal Conductivity (kcal/m h deg)	Common Uses
Raw Products						
Basaltic scoria	Black to red	Coarse aggregate <3 cm	1.2–2.5	0.6–1.6	<1.0	Road construction, cinder blocks, moderate insulator
Basaltic lava	Black/gray	Massive	2.4–3.1	<2	2.0–3.0	Construction, decorative purposes, moderate insulator
Rhyolite ash	Light gray/brown	Fine aggregate <2 mm	1.5–2	<0.1	<1.0	Abrasives, creation of perlite, a good **refractory**, insulator
Pumice	Light gray/brown	Aggregate 0.2–10 cm	0.5–1.5	<0.5	<0.75	Absorbent, abrasives, good insulator
Silicic ignimbrite	Light to dark brown	Massive	2.1–2.8	<1	2.0–3.0	Decorative uses, construction, poor to moderate insulator
Rhyolite lava	Brown to gray/black	Massive	2.1–2.8	<25	2.0–3.0	Decorative uses, construction, poor to moderate insulator
Obsidian	Clear to black	Small lenses or tears (cm)	2.0–2.5	<11	2.7–3.5	Decorative uses, cutting implements, poor insulator
Native sulfur	Yellow	Microcrystalline	1.95–2.1	<0.1	0.13	Chemical additive, needed to "vulcanize rubber"
Bentonite clays	Light brown	<0.005 mm	1.8–2.6	<0.1	Varies widely	Additive to drilling muds, good insulator and sealant
Man-Made Products						
Perlite	White to light gray	Coarse aggregate <3 cm	0.3–1.2	<0.1	<1.0	Absorbent, insulator, light-weight concrete
Cinder concrete	Gray	Blocks (man-made)	2.0	<0.5	3.0–4.5	Construction, insulation

Where appropriate, a range of values has been given. The values given are typical for the materials shown but there can be considerable variance. For example, basaltic scoria has been observed up to meters in size, though these are typically not used for raw materials. The thermal character of the materials is greatly dependent on physical factors, such as vesicularity, or for aggregates their grain size and spatial relationship to one another.

FIGURE 74.2 Implements have often been created using volcanic materials. The arrowheads are made of obsidian and the hammer of basalt. Such weapons maintained a keen edge after several uses and are easily as sharp as modern implements. *These artifacts are from St. Lawrence Island, courtesy of the University of Alaska Fairbanks Museum.*

FIGURE 74.4 Bottom: Entrance to a tunnel cut into volcanic tuff from Rabaul caldera, Papua New Guinea. Hundreds of miles of tunnels were dug by the Japanese military during World War II. Top: Inside view of one of the larger tunnels. This one was used to store boats and barges to protect them from Allied bombing raids. The boats were moved using winches, sometimes with trolleys and rail tracks. This tunnel is about 100 m from the water's edge.

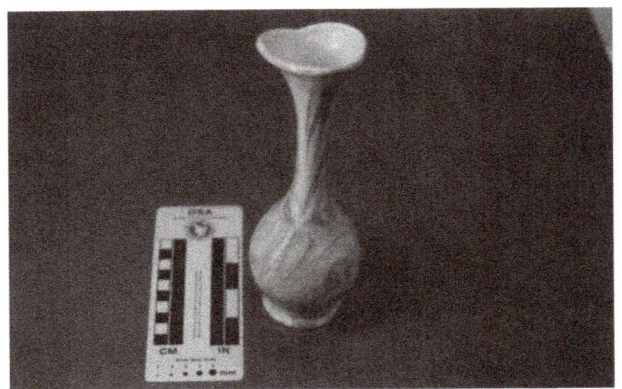

FIGURE 74.3 A vase decorated with volcanic ash from the 1980 eruption of Mt St Helens. The ash, which occurs in different colors ranging from tan to gray to black to red, is used to add color and texture to pottery. Here the potter's clay is white, and brown ash is added to the exterior to make the swirl pattern.

Volcanic rocks are also good building materials, not only for stand-alone structures but for caves and tunnels as well. Large ignimbrites have been excavated by the local inhabitants in Japan, Italy, and Turkey for living spaces, storage, and emergency shelters. A modern example was during the World War II, when the Japanese military dug over 300 miles of tunnels in ignimbrites in Rabaul, Papua New Guinea (Figure 74.4). These were used for shelter, to hide weapons and to provide bomb-proof storage for ammunition.

Because of the sharp color contrasts of volcanic rocks (e.g., basalt to sandstone or limestone), volcanic rocks have also been used for decorative purposes. Monuments, temples, and churches throughout Asia and Europe have used volcanic rock for sculpture and decoration.

4. MODERN USES

Since prehistoric times more sophisticated uses for volcanic materials have been developed through the growth in world population and modern industry. These uses range from the mining of volcanogenic minerals for manufacturing aids, lubricants, and abrasives to construction materials. Volcanic products have also been used for artistic purposes (e.g., in gardens or as monuments). Often the applications are local, although certain types of volcanic minerals and rocks have become important exports for many countries.

5. ADDITIVES

Additives cover an extremely wide range of products, and we review several of the main volcanic contributors here. Volcanic ash and clays are common additives to many industrial materials. Whereas pumice and ash are excellent abrasives, volcanic clays can be used as lubricants and additives.

5.1. Bentonite

Bentonite is an off-white **montmorillonite** clay formed from altered volcanic ash. It has a sheet-silicate structure and is especially notable for the way in which it absorbs and loses water and for its base-exchange properties. Sodic bentonite can absorb up to 10 times its own weight in water and can swell to 18 times its dry volume. Thus, it can be applied where its colloidal properties can be exploited. Bentonite has a wide variety of uses including foundry, animal feed, **drilling mud**, absorbent, industrial, and specialty uses.

About 70% of the world's known supply of Western or sodium-type bentonite occurs in the state of Wyoming (notably in the Mowry Formation), with additional significant deposits elsewhere in the Western United States. Volcanoes of the Yellowstone area of Wyoming were active for long periods in the early Cretaceous (c.100 million - years) and deposited ash layers of a few centimeters up to 15 m thick into nearby shallow seas. The ash weathered with time, forming the bentonite.

Early uses of bentonite in the 1800s, included lubricant for wagon wheels, sealant for log cabin roofs, and soap. Modern uses are highly variable. In foundry or metal casting, bentonite is mixed with sand to make the molds. The bentonite forms a pliable bond with the sand grains, so the mixture holds its shape well. Molten metal is then poured into the impression to make the casting. When complete, the bentonite and sand mold can then be broken away from the casting and reused. Bentonite is used as a binder for food pellets for animals and poultry and as a carrier for food supplements. Bentonite also absorbs excess moisture and oils, helping to prevent malodor, lumping, and caking of the pellets. Some recent research has shown that the base-exchange properties of bentonite allow it to help remove certain toxins and ammonia. It also reduces feed flow through the intestines so that nutrient uptake is improved, indirectly aiding such factors as milk production, wool growth, and the quality and production of eggs.

Bentonite is widely used as drilling mud. It is mixed with water to form a slurry that is pumped through the drill string and out through the drill bit. The drilling mud helps to cool and lubricate the drill bit, and due to its high viscosity and yield strength, it extracts drill cuttings, carrying them to the surface, and helps to seal the drill hole and to prevent wall cave-ins. Bentonite is used to fill and seal abandoned drill holes. Similarly, bentonite has been used for sealing other waterways such as ponds, irrigation ditches, reservoirs, and industrial lagoons and has sometimes been used as a grout.

Because of its ability to absorb water and odors, bentonite has become a major component of the "clumping"

types of cat litters. (Nonclumping or regular cat litter is made from Georgia clay and other similar non-montmorillonite clays.) The bentonite clumps are easy to handle, and the rest of the unused material remains intact, so the litter lasts longer between changes. With 74 million domestic cats in the United States alone, this could be a significant impact! Materials very similar to cat litter, both clumping and nonclumping, are used in industry to clean up liquid spills.

Bentonite has been used as a binder for taconite, a low-grade iron ore. The taconite is ground finely, mixed with bentonite to make pellets or balls, and then briefly sintered in kilns to give the pellets a hard surface. The pellets can then be easily handled and shipped to steel mills.

Bentonite additives have been used to emulsify oils in asphalt and detergents; to improve strengths of ceramics and bricks; to retain microparticles in papermaking; to clarify water, wine, and juice; to treat wastewater, deink paper, and control airborne stack emissions; and to gel cosmetics and pharmaceuticals. It has also been used as an ingredient in cleansers, polishes, and sprays of various types. For obvious reasons, bentonite has been called the product of 1000 uses. The uses are dependent on its absorption characteristics, which are in turn dependent on the chemical composition of the clay.

5.2. Pumice

Pumice is a porous, bubble-rich volcanic glass with low density. It is easily handled because of its low density and, when broken into finer particles, is widely used as an abrasive. The presence of the many bubbles means that the small pieces produced upon breaking are sharp and irregular. The particles are typically graded by size and backed with paper or other materials to produce sandpaper, grinding wheels, and other tools. Some pumice also had a highly specialized abrasive use that has declined recently. Pumice that floats on water may be used in large washing machines to gently abrade cloth such as "stone-washed jeans." Eventually the pumice became waterlogged and sank, but it could be dried and used again. The need for pumice in such processes increases and decreases as fashions evolve.

Pumice is included in small amounts in certain soaps (e.g., "Lava" hand soap among others, Figure 74.5) to provide a mildly abrasive action to aid in cleaning. Thus, it adds a mechanical component to the cleaner that enhances the soap's chemical action.

6. AGGREGATES

Aggregates are broken pieces of rocks which range in size from fine sand to several centimeters in diameter. Volcanic aggregates, cinders, or scoria have been used for road

FIGURE 74.5 Soap made using volcanic ash. Most of the soap is made from standard materials, but small amounts of fine-grained volcanic ash were added to provide a mild abrasive action in addition to the chemical action. This particular soap was sold in tourist shops soon after the famous Mt St Helens eruption of May 18, 1980.

construction, and the production of building materials (e.g., concrete and **cinder blocks**). In addition, they have widespread use as ballast, and as a component in plaster and related materials. In general, the use of aggregates falls into three categories; loose aggregates, refractories, and absorbents.

Both perlite and pumice are also used as aggregates, especially in applications in which low weight is desired, such as lightweight precast masonry or concrete floors. The main cost of aggregate is its transportation, so pumice is widely used wherever it is the closest available source of aggregate. Many countries, however, export pumice, taking advantage of its low density to minimize shipping costs. Both perlite and pumice are also used in certain types of insulation and acoustic tiles.

6.1. Perlite

Perlite is a highly hydrated volcanic glass. The hydration occurs along conchoidal fractures that pervasively criss-cross the material, so that perlite breaks up easily into small pieces. Many of these have round surfaces and a waxy luster of typically light gray color, although greenish, brown, blue, and red also occur. The pieces resemble pearls, hence the name perlite.

Perlite is used generally as a lightweight aggregate but also has some specialty uses. It is a substitute for sand in lightweight wall plaster and is also used in ceramic products, fillers, and filters. In agriculture, perlite is mixed with peat from 25% to 75% and used to grow many types of plants. Because of its high fracture density and high surface area, perlite is ideal for simultaneously holding water and air pockets. This dual ability of aeration and water retention is necessary for plant roots to function properly. Even when wet, a perlite—peat mixture weighs about four times less

than wet sand, so the perlite is easier to handle and much less expensive to ship. Economic deposits of perlite are found in California, Nevada, and New Mexico in the United States as well as in Russia and Greece.

6.2. Cinder

Mining of scoria or cinder cones for loose aggregate is common in Central Europe, the Southeastern United States, and Japan. In the United States alone, c.1.5 tons of aggregate per person per year are used. Cinders are primarily used in the fabrication of concrete and cinder blocks. This was also one of earliest industrial uses of volcanic aggregates, developed by the Romans, who had convenient sources of many types of volcanic products from the nearby volcanoes (e.g., Colli Albani near Rome and Vesuvius near Naples).

Today, aggregates are still an important resource. It is inconceivable to think of modern construction without the use of concrete or pressed cinder blocks.

In road construction, a flexible bed of high strength and with good insulating properties is important to build a long-lasting roadway. Volcanic cinders are ideal for this purpose. The irregular surface allows particles to stick together, providing high angles of internal friction (up to 45°). The insulating properties ensure that the ground beneath the roadway is protected from extreme temperatures, and the associated deformation. In Alaska roadbeds like these are sometimes used to minimize the effect of annual frost heave and permafrost on roads. In addition to the thermal insulation, cinders provide a sound insulation as well, and individual vesicles can act as tiny cushions, absorbing vibrations.

Cinder can also hold water in open vesicles, or in the pore space between aggregates. It is often used to hold water around plants. In addition, cinder sorted to specific sizes becomes an effective barrier for tunneling pests, such as termites. The material is too well anchored to be moved, and large enough that a termite cannot grasp or carry each particle. The cinders are too strong to be broken, are resistant to weathering, yet are small enough and so closely packed that the termite cannot fit between individual particles.

6.3. Refractories

Refractories are materials that are not deformed or damaged by high temperatures, thus they are used to make firebricks, crucibles, insulation, furnace linings, and furnaces themselves for metallurgy and glass making. Pumice and clays are sometimes ingredients of refractories, and bentonite is used as a binder during production. Ignimbrites or welded tuffs can be cut into bricks and are used directly as refractories.

7. CONSTRUCTION

Volcanic stone has been used worldwide due to its stability, strength, coloring, availability, and resistance to weathering. Its uses vary from simple garden walls to baroque cathedrals to modern skyscrapers.

7.1. Massive Stones

Massive volcanic stone provides a good building material because lavas and ignimbrites have high strengths and good insulating qualities. Such stone has been used throughout Europe and Britain for monuments and churches. One of the best examples of this is the Cologne Cathedral, which is composed of 12 different types of stone, over 300,000 tons, more than half of which are volcanic, derived from the nearby Eifel volcanic field in central Germany. The Clermont-Ferrand Cathedral (Figure 74.6), is of similar construction, built with largely from a nearby trachyandesite. Its construction began in 1248 but was not finished until 1902. As with aggregate, a significant cost of this building material is its transport, and thus local sources are often preferred over imported materials. Lavas tend to dominate the use of massive volcanic rocks in construction,

though less resistant rocks have also been used. In the past, ignimbrites were much easier to cut and shape but are more susceptible to weathering. As the science of masonry has progressed, harder and more resistant stone could be just as easily shaped or fabricated. In the case of the Cologne Cathedral, many of the older stones were made of more malleable softer rock and are now being replaced with harder, more massive stone (e.g., basalt).

Volcanic material is also ideal for use in monuments, markers, or gravestones. This is the most common use of volcanic rock throughout Asia. In Japan, one of the most volcanically active countries in the world, lanterns, signposts, carved idols and icons are typically made from one of the many volcanic stones of the region (Figure 74.7). Often the dark colors of volcanic rock provide a somber presentation, appropriate for many religious ceremonies.

Larger scale construction projects in Japan are seldom made of stone because these buildings tend to fracture and collapse in this earthquake-prone region. Strict regulations are imposed to ensure that the supporting frame of a building is made of a more flexible material (e.g., wood or steel). Today, even with improved construction techniques, the use of massive volcanic stone is limited to tiles, steps, sidewalks, and facades.

7.2. Concrete and Cinder Blocks

Concrete is one of the oldest uses of volcanic products and was originally developed by the Romans as a nonwater-soluble alternative to the gypsum-based mortar used by the Egyptians. Primarily developed to tile and seal the aqueducts, this strong new material was used in many of the classic Roman structures, a number of which are still standing today. This material is known as **pozzolan**, and is formed using fine-grained siliceous ash as the basis for a hydraulic cement. The precise technique used by the

FIGURE 74.6   The spires of the Clermont-Ferrand Cathedral. Volcanic stone was often used during the Middle Ages, not only because it was local and easily acquired but also because it was resilient, afforded great strength for elaborate construction, and the dark colors were very imposing.

FIGURE 74.7   Lanterns at a temple in Kyoto, Japan. These were created using local andesite. Volcanic rocks are often used in monuments and idols.

Romans is still a mystery, and modern concrete has not yet achieved the durability shown by these early structures. Although concrete can be made from any aggregate material, volcanic aggregate provides enhanced strength and insulation, making it a preferred material.

Today pumice and cinders are used in creating concrete prefabricated parts for buildings. In addition to its strength and insulation characteristics, pumice concrete can withstand flexure better than its counterparts. Although it has limited flexibility, it has a small coefficient of thermal expansion, and when reinforced, it can withstand significant thermal shock, as in the case of a fire. For example, when heated to 1000 °C, a temperature where a metal door or window frames would melt, and then suddenly quenched with cold water, pumice concrete shows little or no damage whereas concrete made of nonvolcanic material rapidly and severely fractures. This difference is the direct result of the very small coefficient of thermal expansion of volcanic material. Buildings made of pumice concrete or cinder blocks are less likely to collapse when on fire than buildings made with limestone concrete or even steel.

This material also has the advantage of weighing less than normal aggregates or massive blocks. When shear strength is not the dominant factor in the design of a building, a lightweight concrete is a good low-cost alternative.

7.3. Tunneling

Soft volcanic rocks, such as ignimbrites have often been used for tunneling, storage, and housing. Examples of this can be found in Japan, Ischia, Herculaneum, and New Guinea (Figure 74.4). In Japan, the soft volcaniclastics, which were formed submarine and are now subaerial, are tunneled for roads, although there is often a danger of collapse because these rocks are porous and have low strengths, only a little stronger than loose aggregate material. In the winter, water trapped in the rock freezes and expands and can cause fatal tunnel collapses like the one in the Fall of 1995 on Japan's northern island Hokkaido.

7.4. Mining

Sulfur is the ninth most abundant element in the universe and one of the four most basic chemical commodities. The chief sources of free sulfur are volcanoes, fumaroles, and certain sedimentary deposits. Sulfur occurs naturally in several forms, of which rhombic sulfur is the most common at volcanoes (often as flowers of sulfur). Important compounds of sulfur include hydrogen sulfide (H_2S), which has the familiar rotten egg smell, and is used extensively in chemical plants and laboratories as an analytical agent. Sulfur dioxide (SO_2) is a precursor of sulfuric acid; it is also used as a bleach and as an industrial reducing agent. In small amounts, sulfur dioxide is used as a food preservative and to aid the ripening of fruits. Sulfur dichloride (S_2Cl_2) is used in the manufacture of chemical products such as lubricants and mustard gas. Sulfur hexafluoride (SF_6) is used as an insulator in many electrical devices. Sulfurous acid (H_2SO_3) forms the salt sodium sulfite (Na_2SO_3), which is used in the manufacture of paper pulp, in photography, and in removing oxygen from boiler feed water. Sulfuric acid (H_2SO_4) is used in the manufacture of explosives, detergents, dyes, drugs, pigments, fertilizers, and many inorganic compounds. Synthetic organic sulfur compounds are components of pharmaceuticals, insecticides, and solvents and are also used to help vulcanize rubber and to produce rayon. The first recorded therapeutic use of sulfur was by Philippus Aureolus Theophrastus, Bombast of Hohenheim, in the late fifteenth century, not only as a bath additive but also in various compounds taken internally to relieve minor ailments.

8. DECORATIVE USES

Volcanic rocks are often aesthetically pleasing and can take many forms. Minerals in the rocks can often reach sizes useful for decorative purposes and can even be used as precious or semiprecious stones. Volcanic products can supply the materials used to create works of art, such as decorative pottery (Figure 74.3). Polished banded rhyolites, or basalts, are commonly used as veneer for countertops, walkways, or stair steps and facades for buildings. In some cases, volcanic rocks are exported for this purpose from famous localities, such as the Clermont-Ferrand region of France. Volcanic rocks often yield semiprecious stones. Olivines are common in basalts, as are pyroxenes. Even more exotic minerals can be found in lavas (e.g., hauene, a rare pyroxene mineral, in the trachytes of Central Germany and topaz in rhyolites of the Southwestern United States

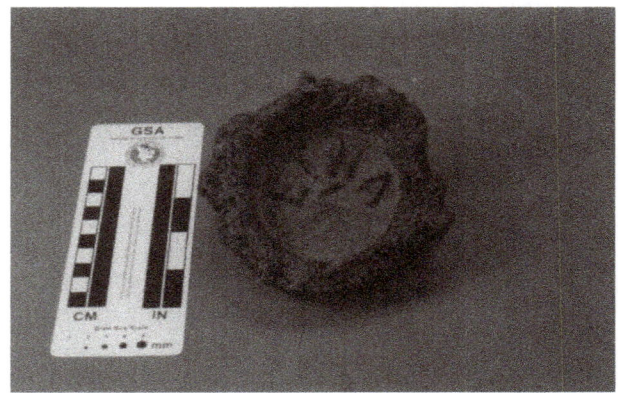

FIGURE 74.8 An ash tray made from lava at Mt Etna. In perhaps the most direct use of red hot lava, the still molten rock is scooped up and placed into a mold and squeezed immediately. The ash trays are then set aside to cool and are ready for sale a few hours later.

and Mexico). Over time, volcanic ash weathers into fine clays, which are preferred for delicate pottery in Japan. The fresh glass can be used as glaze due to its low melting temperature. Volcanic glass can also be melted and molded to make synthetic "gem" stones. Decorative uses of volcanic rocks can be more coarse; different colored cinders are often used in gardens or in more elaborate arrangements such as in Japanese rock gardens. Fresh lava is pliable, and local vendors in Italy gather the fresh lava and create souvenirs for tourists (Figure 74.8). This is the most direct use of lava known to us.

9. CONCLUSIONS

Volcanic raw materials have been used by man throughout history. The oldest uses relied on the fact that volcanic rocks were plentiful locally and were easy to fashion into tools, such as the shaping of obsidian arrowheads. Modern uses include those for which volcanic rocks area local source of aggregate, such as roadbeds, but also a variety of new specialty uses. For example, most uses of bentonite were developed in the last 100 years, and perlite has been exploited only since the1950s. Volcanic rocks have some unusual properties, particularly density and thermal conductivity, so it is likely that new uses will continue to be found. These materials may have uses in combination with reinforced platinum, carbon fibers, or other new products to make unique composite materials for construction. Volcanic eruptions occur at more or less steady rates on a global scale, thus there will always be replenishment of volcanic raw materials from new eruptions. Therefore, in addition to

the excitement, awe, and scientific value of eruptions, in a very substantial way volcanoes will continue to contribute raw materials to sustain the industrial needs of modern life.

SEE ALSO THE FOLLOWING ARTICLES

Earth's Volcanoes and Eruptions: An Overview ● Ignimbrites and Block-and-Ash Flow Deposits ● Mineral Deposits Associated with Volcanism ● Volcanic Soils.

FURTHER READING

Bordaz, J., 1971. Tools of the Old and New Stone Age. David & Charles Newton Abbot, Devon.

Carmichael, R.S. (Ed.), 1989. Practical Handbook of Physical Properties of Rocks and Minerals. CRC Press, Boca Raton, FL.

Chestermann, C.W., 1956. Pumice, Pumicite and Volcanic Cinders in California. State of California Press, San Francisco.

Fischer, R.V., Heiken, G., Hulen, J.B., 1997. Volcanoes: Crucibles of Change. Princeton University Press, Princeton.

Heiken, G. (Ed.), 2013. Tuffs—Their Properties, Uses, Hydrology, and Resources. Geol. Soc. Amer., Boulder, p. 131. Special Paper 408.

Heiken, G., Funiciello, R., DeRita, D., 2005. The Seven Hills of Rome—A Geological Tour of the Eternal City. Princeton University Press, Princeton, 288 pp.

Heiken, G., 2005. Industrial uses of volcanic materials. In: Marti, J., Ernst, G.G.J. (Eds.), Volcanoes and the Environment. Cambridge University Press, pp. 387—403. http://dx.doi.org/10.1017/CBO9780511614767.

Thomas Publishing Company, 1994. The Thomas Register of American Manufacturers: Products and Services Section, eighty fourth ed. Thomas Publishing Co., New York http://www.thomasnet.com/.

Chapter 75

Volcanoes and Tourism

Patricia Erfurt-Cooper

School of Business, James Cook University, Cairns, QLD, Australia

Haraldur Sigurdsson

Graduate School of Oceanography, University of Rhode Island, Narragansett, RI, USA

Rosaly M.C. Lopes

Earth and Space Sciences Division, Jet Propulsion Laboratory, California Institute of Technology, Pasadena, CA, USA

Chapter Outline

GLOSSARY

volcanic hazards The risk to life and property from volcanic activity, which includes eruptions, ash fall, gas emissions and toxic fumes, lahars and mudflows, lava flows and pyroclastic flows, extreme hot springs, and other geothermal features such as geysers and fumarolic vents.

volcanic heritage Active and dormant areas included in national parks and other protected sites to provide controlled access and to minimize environmental impact. Remnant volcanic landforms and their unique features are increasingly preserved as natural heritage for future generations.

volcanic risk management Identification, assessment, and implementation of suitable strategies to minimize the potential for accidents and injuries. It should include education of tourists about the potential hazards from volcanic activity and how to stay safe.

volcanic springs In a health and wellness tourism context, natural volcanic and nonvolcanic hot springs are used for therapeutic benefits based on their mineral content. Volcanic hot springs such as geysers are also tourist attractions for their visual appeal.

volcano tourism The exploration and study of active volcanic and geothermal landforms. Volcano tourism also includes visits to dormant and extinct volcanic regions where remnants of activity attract visitors interested in geological heritage (Erfurt-Cooper, 2010).

Active and dormant volcanoes provide a valuable natural resource for various tourism sectors, including geotourism, adventure tourism, and ecotourism. Destination development in many countries and regions has profited from the close proximity of volcanoes and geothermal springs with historical links to hot spring spa tourism or thermal tourism. Volcanic regions are major tourist destinations that are visited by people of all age-groups for a variety of reasons. These can include plain curiosity, a scientific interest in volcanoes, the ambition to climb a volcano, taking photos, or just visiting as part of a guided tour. The interesting combination of volcanic attractions and geothermal phenomena presents visitors with an experience of scenic beauty in close contact with nature while learning about the surrounding geodiversity.

1. WHAT IS VOLCANO TOURISM?

With over 1500 volcanoes classed as active worldwide, (Siebert et al., 2011) for many regions **volcano tourism** represents an opportunity to gain independence from changing seasons and climate, as well as an economic advantage. Volcanoes in Japan, for example, are marketed with a special focus on each of the four seasons to encourage repeat visits, e.g., skiing in winter and hiking the same mountain in summer. The development of tourism, including volcano tourism, has evolved from a minority activity of the elite in the eighteenth century into the mass tourism we see today. A global tourism boom, a major consequence of the economy of the 1990s, has continued into our current millennium, despite a brief interruption during the global financial crisis in 2009, when many other economies failed. The last two decades of the twentieth century have seen a strong trend towards nature-based tourism with a focus on ecotourism, geotourism, and adventure travel, generating infrastructure development and economic profits at many destinations. The focus of global tourism destinations has shifted from Europe toward East Asia and the Pacific Region, as people travel farther than in the past. The total economic contribution of travel and tourism during 2012 was USD$6.6 trillion, supporting 260 million jobs worldwide, either directly in the tourism industry or in related sectors (e.g., air travel, accommodation, service industry). The future outlook for tourism is optimistic based on an average annual growth of 3.4% during the 10-year period from 2000 to 2010 (World Travel & Tourism Council, 2013).

Geotourism is a fast-growing sector within the nature-based tourism industry and takes place in some of the world's most spectacular volcanic regions. In the seventeenth and eighteenth centuries, the northern European aristocracy visited Vesuvius and Etna as part of the Grand Tour. Today, people from all social classes visit volcanic regions for their scenery and for their educational, ecological, and adventurous aspects. And while volcanic geotourism has continued to draw visitors to Vesuvius, Mt St Helens, or Fuji, lesser known volcanic destinations such as Ethiopia, Comoros, Reunion, Vanuatu, and Kamchatka have become increasingly popular and easier to access. To save their geological heritage, many nations protect their unique volcanic tourist attractions for future generations in national parks, geoparks, and World Heritage sites.

Presently, between 450 and 500 million people live within close proximity to volcanoes classed as active. Millions of visitors every year spend time near active volcanoes; sometimes not aware of any potential danger while they learn about these fascinating natural wonders. Large cities in the neighborhood of active volcanoes include Naples in Italy (Vesuvius), Tokyo in Japan (Fuji), Mexico City (Popocatepetl), Jogjakarta in Indonesia (Merapi), Manila in the Philippines (Pinatubo), Seattle in Washington, USA (Mt Rainier), Nagasaki and Kagoshima in Japan (Unzen and Sakurajima), Goma in the Democratic Republic of Congo (Nyiragongo), Auckland in New Zealand (Auckland Volcanic Fields), as well as numerous others (Erfurt-Cooper, 2014). While some of these volcanoes may have entered a quiet phase, most of them have the potential to add to their record of destructive activity in the future. However, significant progress was made in monitoring volcanoes for signs of unrest over the last three decades. Many locations are now able to communicate to residents and visitors whether eruptive activity is imminent.

2. VOLCANOES IN HISTORY, CULTURE, AND RELIGION

Visiting places of interest dates back to the first pilgrims and the documentation of their journeys. Since then tourism has developed with the advancement of technology and the increasing amount of information available. Historically, the Romans, besides engaging in extensive and well-documented "military travel", used volcanic locations such as the Campi Flegrei and the island of Ischia for recreation because of their preference for natural hot springs. Roman scholars documented all destinations they "visited" most accurately, thereby providing information for posterity.

It was in Europe, particularly in Great Britain, that the tourist industry was invented, refined, and developed as the Grand Tour for the young, well-educated, and wealthy members of the elite. The Grand Tour had its onset around 1660, but increased considerably in the 1760s. At the heart of the Grand Tour was Italy with Naples and Vesuvius as a main attraction. Excursions to the archaeological excavations of Pompeii and Herculaneum began in 1755 and 1738, respectively, while adventurous visitors climbed Vesuvius to see the "fireworks" up close. From the Grand Tour, travel developed with further industrialization and enhanced transportation, particularly the advance of the railways, as well as the growth of an increasingly leisured middle class. These changes in travel and transportation lead to a great leap forward in tourism. In England, Thomas Cook started tours for the masses in 1841 and to Europe in 1855, marking the beginning of mass tourism (Whitey, 1997). Cook was aware of the attraction of volcanoes for tourists and by 1864 he was arranging tours to Pompeii and Vesuvius (Figure 75.1).

In an attempt to further develop volcano tourism, Cook acquired the Vesuvius Railway (Funicolare) that took tourists directly up to the volcano (Figure 75.2). This presented serious competition to the local guides and porters who had earned a living by carrying people to the top in sedan chairs. In protest they damaged the railway, threw the carriages down the volcano, and set fire to the station. Cook

FIGURE 75.1   **Nineteenth-century tourists visiting Vesuvius had three choices.** Either they could walk, be carried to the crater rim in a chair (left), or be dragged with a rope by Neapolitan porters (right). *Woodcuts from The Illustrated London News, 1872.*

overcame these initial problems and convinced the local porters to allow the railway to operate peacefully while profiting in other ways from the tourists who used it. Over the centuries, the funicular railway was destroyed by lava flows three times, and after the damage from the eruption in 1944 it was not rebuilt again. In its place a chairlift was constructed to carry tourists up to the crater rim, which is also no longer operating. Visitors today have to walk up to the crater rim and visualize the events of the past. The increasing knowledge available online provides access to the original footage from the 1944 eruption, an amazing video that lends a perspective to this major eruption of Vesuvius (Vesuvius Eruption, 1944).

Documentaries, books, and movies about the unleashed natural forces of volcanoes have also contributed to the growing interest in famous volcanic destinations such as

FIGURE 75.2   **In 1880, the Thomas Cook Company opened the Funicolare to the top of Vesuvius, an event that transformed tourism to Vesuvius and put many of the volcano's guides out of work.**

Pompeii and Herculaneum (Italy), Yellowstone and Mt St Helens (United States), and Krakatau (Indonesia). Prior to the eruption of Krakatau in 1883, visitors were already fascinated by spectacular displays of smaller eruptive events of the mountain with many observations documented in writing. Today tour groups and individual travelers can visit Anak Krakatau, the new mountain that is replacing the original volcano. Around the same time as the Grand Tour era in Europe, Iceland's volcanic landscapes and the original Geysir in Haukadalur were already visited by tourists.

Times have certainly changed since the turn of the last century, when Englishman Tempest Anderson, a photographer and amateur volcanologist, spent two decades exploring volcanic regions, including Vesuvius, Etna, Lipari Isles, Auvergne, Eifel, Canaries, and Iceland, as well as numerous volcanoes of the Cascades and Yellowstone. In May 1902, when La Soufrière (St. Vincent) and Mont Pelée (Martinique) erupted in the Lesser Antilles, Anderson recorded these events in pictures and gave lectures about his journeys. His published work includes a volume with photographs of volcanic features (Anderson, 1903) and some of his publications about volcanic history have been reprinted recently (Volcanic Studies in Many Lands, 2009).

Volcanic eruptions (Table 75.1) are generally accompanied by change, both good and bad. In Japan, eruptions of Mt Unzen (1991–1995) have caused significant new developments for tourism, including infrastructure such as volcano museums and memorial sites for public education (Figure 75.3). Equally, the eruption of Mt St Helens in 1980 launched a new era of tourism in volcanic environments in the years following the event. In Iceland, a growing demand by visitors for information about the eruption of Eyjafjallajökull in 2010 brought about the establishment of a volcano museum (þorvaldseyri) close to the eruption site.

TABLE 75.1 Historic Volcanic Eruptions

Chaitén, **Chile** (2008)
Puyehue-Cordón Caulle, **Chile** (2011)
Nevado del Ruiz, **Colombia** (1985)
Arenal, **Costa Rica** (1968, 1988, 1993, 2000, 2010)
Nyiragongo, **Democratic Republic of the Congo** (1877, 2002)
Cotopaxi, **Ecuador** (1744, 1877, 1907)
Santorini, **Greece** (1610, 1650, 1866, 1950)
Santa Maria, **Guatemala** (1902, 2008, 2011, 2013)
Askja, **Iceland** (1875, 1921, 1926, 1961)
Eyjafjallajökull, **Iceland** (1612, 1821, 2010)
Grimsvötn, **Iceland** (1783, 1903, 1983, 1996, 1998, 2004, 2011)
Hekla, **Iceland** (1554, 1766, 1845, 1947, 2000)
Krafla, **Iceland** (1975, 1984)
Laki, **Iceland** (1783–1785)
Agung, Bali, **Indonesia** (1808, 1943, 1963)
Kelud, **Indonesia** (1586, 1641, 1826, 1919, 1956, 1961, 1990, 2008)
Krakatau, **Indonesia** (1883, 1999, 2009, 2010, 2011, 2012)
Merapi, **Indonesia** (1822, 1872, 1961, 2001, 2006, 2010)
Tambora, **Indonesia** (1812–1815, 1880)
Etna, **Italy** (2012, ongoing activity)
Vesuvius, **Italy** (AD 79, 1875, 1944)
Fuji, **Japan** (1560, 1700, 1707)
Sakurajima, **Japan** (1471, 1779, 1914, ongoing activity)

Unzen, **Japan** (1663, 1690, 1792, 1990–1995, 1996)
Usu, **Japan** (1663, 1769, 1822, 1853, 1943, 1977, 2000)
Mont Pelée, **Martinique** (1851, 1902, 1929)
Soufriere Hills, **Montserrat** (1995, 2009, 2010)
Ruapehu, **New Zealand** (1881, 1895, 1945, 1953, 1995, 2007)
Taranaki (Egmont), **New Zealand** (1655)
Tarawera (Okataina), **New Zealand** (1310, 1886)
Taupo, **New Zealand** (c.1800 years ago or AD 200)
Mayon, **Philippines** (1814, 1897, 1968, 2006, 2009, 2010, 2013)
Pinatubo, **Philippines** (1450, 1991)
Taal, **Philippines** (1716, 1749, 1754, 1965, 1977)
Colima, **Mexico** (1622, 1818, 1890, 1913, 1975)
El Chichón, **Mexico** (1982)
Popocatépetl, **Mexico** (1994, ongoing activity)
Bezymianny, Kamchatka, **Russia** (1956, 1997, 2012, ongoing activity)
Shiveluch, Kamchatka, **Russia** (1944, 1990, 1999, 2005, 2010)
La Soufriere, **Saint Vincent** (1902, 1979)
Kīlauea, Hawai'i, **USA** (1983 to present)
Augustine Volcano, Alaska, **USA** (1883, 1976, 1986, 2006)
Novarupta, Alaska Peninsula, **USA** (1912)
Redoubt Volcano, Alaska, **USA** (1902, 1967, 1989, 2009)
Mt St Helens, **USA** (1480, 1482, 1800, 1980)

The table shows examples of eruptions from the last 2000 years. Some volcanoes erupt on a daily basis for years (Sakurajima, Japan), others erupt frequently (Etna, Italy), and some reawaken after several thousand years of quiescence (Chaitén, Chile) and surprise the world.

2.1. Mythology and Mountain Spirits

Volcanoes have a fixed place in many cultural and religious practices. Vulcan was worshiped by the Romans as their god of fire and Pele to this day is considered by many people as the goddess of fire, and features in traditional

FIGURE 75.3 The Unzen Disaster Memorial Hall complex (Gamadas Dome) includes a volcanic rock garden; a geopark information center; a library, museum shop, and restaurant; conference facilities; and, most importantly, the volcano museum showcasing the Unzen eruption and its aftermath in great detail. The monument in the foreground reflects the "power of the universe." *Photo: P. Erfurt-Cooper.*

Hawai'ian legends. Ancient mythologies claim volcanic craters as entrances to the underworld; a world where deities and demons dwell. Hence, volcanoes were worshiped and sacrifices were offered to appease the fiery mountain spirits. The practice of sacrificial offerings is still maintained in Indonesia, Japan, Hawai'i, and several other countries, although the rather ineffective custom of human sacrifice is fortunately no longer acceptable. Appeasement ceremonies in Indonesia include processions to active craters lead by a priest in charge with offerings carried up the mountain. The gifts include flowers, money, live animals, and food items and in some regions have become an event for tourists. Mt Merapi, for example, is an additional attraction for tourists who visit the World Heritage listed Borobudur Temple, close to Jogjakarta on the Indonesian Island of Java.

The list of volcanoes, taking into account ancient remnants of volcanism, that feature in legends and mythology is long. Some examples include Mayon (Philippines), Popocatépetl and Ixtaccíhuatl (Mexico), Etna in Sicily (Italy), Fuji (Japan), Crater Lake in Oregon (USA), Tongariro, Tarawera, and Ruapehu (New Zealand), Glass House Mountains, Seven Sisters, and Mt Warning (Australia), Hekla and Katla (Iceland), Giant's Causeway (Ireland), Arthur's Seat (Scotland), and Poás and Arenal (Costa Rica). There are many more volcanoes, both active and dormant, that are intricately connected to cultural and religious traditions worldwide with enough legends to fill several books.

In Japan, volcano tourism involves the acknowledgment of different mountain spirits, and shrines can be found all over the countryside. Torii gates, a common sight, lead to places of worship on mountain sides and volcanic hot springs throughout Japan (Figure 75.4). The city of Beppu on Kyushu Island (Japan) is famous for its over 4000 hot

FIGURE 75.4 An old stone Torii gate leads to a shrine near the summit of Mt Unzen. *Photo: P. Erfurt-Cooper.*

FIGURE 75.6 Torii Gates at the Umi Jigoku are leading to a small shrine. The turquoise pond in the foreground is a crater lake, which emerged after a volcanic explosion 1200 years ago. *Photo: P. Erfurt-Cooper.*

springs and onsen resorts. A unique attraction for Beppu's visitors are the **jigoku**, which means hell in Japanese. The "jigoku meguri" is a tour or pilgrimage to individual small geoparks showcasing different types of geothermal phenomena. Here visitors can observe boiling lakes of different colors, gushing geysers, bubbling mud ponds, small mud volcanoes, and hissing steam vents (Figure 75.5). While for foreign visitors this "pilgrimage" is not necessarily religious, for Japanese tourists each "hell" has several shrines for prayer (Figure 75.6). Jigoku are also a common tourist attraction in other active volcanic areas in Japan.

New Zealand's ancient mythology has its own interpretation of volcanoes and their creation. For the Māori, volcanic landscapes are of special cultural and spiritual significance, and the peaks of the mountains are considered sacred. According to their belief, some of New Zealand's volcanoes including Tongariro, Taranaki, Ngauruhoe, and Ruapehu were once warriors and gods who fought with each other, using volcanic eruptions and earthquakes.

Other legends state that the volcanoes were created by fire sent from heaven. In 1993, the Tongariro National Park was listed as a World Heritage site because of the cultural and religious significance of the mountains for the Māori people and their spiritual connection to the natural environment.

Another significant destination for volcano connoisseurs is the Snæfellsjökull National Park in the west of Iceland. Snæfells, a stratovolcano, has historically been linked to the book *Journey to the Center of the Earth* by the French author Jules Verne, who chose this volcano as the entrance to the underworld. Visitor and heritage centers throughout the region provide information about the geology and cultural history of the Snæfellsjökull National Park. According to legend, the nearby located volcanic mountain Stapafell (Figure 75.7) is home to the "hidden

FIGURE 75.5 At the geothermal park Umi Jigoku in Beppu a boiling hot spring is feeding a rock pool with terracotta colored water. *Photo: P. Erfurt-Cooper.*

FIGURE 75.7 Stapafell (526 m), an eroded mount of palagonite in the Snæfells National Park. *Photo: P. Erfurt-Cooper.*

people" or elves that feature in folk tales throughout Iceland. Other volcanoes associated with legends include Katla, Askja, Hekla, and Laki, all commonly linked to the gates of hell and to meeting places for witches, trolls, and demons.

2.2. Volcanic Hot Springs and Their Cultural Significance

Volcanic hot springs have a tradition as centers of cultural activity and socializing in many countries. The significance of such springs in a spiritual (cultural) sense stems from the belief that natural springs are divine gifts to human beings for the purpose of purification and renewal, which must be honored and worshiped. In countries such as Iceland, hot spring water has on occasion been used for baptism since the introduction of Christianity. Today, hotels and guesthouses in Iceland offer their visitors hot tubs and outdoor baths fed by local hot springs (e.g., Viking Pool in Leirubakki with Mt Hekla in the background), and spiritual significance is purely personal to the individual tourist.

Many other volcanic areas offer tourists the experience of volcanic hot springs; especially health resorts and spas are taking full advantage of the cultural significance of this natural phenomenon. Japanese hot spring resorts (Onsen) have maintained a time-honored atmosphere for centuries, based on a distinct preference for authentic hot spring water. The close cultural and historical connection of active volcanoes with social cohesion and recreation in Japan has provided a sustainable foundation for the tourism industry. Over 90% of tourists in Japan are of domestic origin, and visitors from other prefectures usually combine onsen tourism with visits to the nearby volcanic attractions. For the Japanese people the famous Mt Fuji (Figure 75.8) has immense cultural and spiritual significance, which was officially recognized by the

UNESCO in 2013 by awarding World Heritage status. More than 100 million people visit the area around Mt Fuji every year throughout the four seasons. The threat of a possible future eruption has not affected tourism in the large Fuji—Hakone—Izu National Park or in the many resort towns located around Mt Fuji.

Japan with over 20,000 hot springs of volcanic origin attracts over 150 million predominantly domestic visitors every year. Resorts and health clinics are built in close proximity to the natural springs, which have been used for centuries to improve health and provide relaxation after work. Traditional onsen resorts in quiet rural settings with a reputation as pleasant, peaceful, and healthy getaways are highly sought after and can be found in historical onsen towns such as Toyako, Yufuin and Kusatsu. Each of these locations is located close to a volcano that offers attractive activities like hiking, climbing, crater viewing, and skiing. Apart from supplying health spas with natural hot spring water, many areas in Japan have established geothermal parks or jigoku that feature various types of geothermal phenomena, where visitors are offered warm foot and hand spas, as well as various types of food cooked over steam vents.

Turkey is another country with an immense **volcanic heritage** that includes an abundance of hot springs, which have a long history of being used for health purposes. Between 1680 and 1193 BC, the Hittite Empire is said to have used natural hot springs for recreation and therapeutic treatments (Özgüler & Kasap, 1999). Today, approximately 1300—1500 volcanic hot springs are used for health and wellness tourism in over 700 areas throughout Turkey. One rather unusual hot spring though has only a mythological attraction; the **Plutonium** at the ancient city of Hierapolis (above the travertine terraces of Pamukkale) in Turkey's southwest, is a sacred cave, which is located among the ruins of the city. Here the Roman god Pluto is said to have entered the underworld domain. Undoubtedly, toxic fumes and the volcanic origin of this hot spring have contributed to the mythology and "sacredness" of the Plutonium. To prevent people from entering the cave the entrance was closed a long time ago, leaving only a small opening (Figure 75.9).

Extensive research into the cultural use of geothermal resources has shown that most countries with volcanic hot springs are using them successfully as unique tourist attractions and as a foundation for sustainable destination development (Erfurt-Cooper & Cooper, 2010).

2.3. The Health and Wellness Benefits of Volcanic Hot Springs

The key geological processes causing the natural discharge of hot springs involve the rise of water to the surface from unconfined aquifers, assisted by pressure, friction, and the

FIGURE 75.8 During a short climbing season of only 2 months during summer around 300,000 people hike up Mt Fuji (3776 m). *Photo: P. Erfurt-Cooper.*

FIGURE 75.9 The Plutonium, also known as "gate to hell," is connected to a cave below, where toxic fumes are said to have poisoned people and animals in the past. While the sound of gushing water can be heard, no toxic fumes were experienced during research on site. *Photo: P. Erfurt-Cooper.*

geothermal temperature gradient while passing through faults and fissures. In areas with volcanic activity the water temperature increases through circulation close to magma reservoirs below the surface and can reach boiling point or even flash into steam, causing geyser eruptions (Figure 75.10).

Volcanic hot springs are highly mineralized due to increased water temperatures dissolving the surrounding host rock, thereby releasing high concentrations of minerals and trace elements into the rising spring water. The amount of dissolved minerals is influenced by the flow rate, flow path, and length of time the water is in contact with the surrounding rock. Other factors affecting the mineral concentration include the type of rocks the water passes

FIGURE 75.10 Litli Geysir is located close to the original great Geysir (Iceland), which only infrequently sends up giant spouts of boiling water. Today's crowd pleaser is Strokkur, which erupts every few minutes. *Photo: P. Erfurt-Cooper.*

through as well as the possible addition of seawater, if near an ocean, or the quality of the recharge water (see Chapters 46 and 71 for more details).

The benefits of volcanic hot springs for health and wellness from their mineral content have been known in many countries for several thousand years. Often referred to as "medicinal waters," these natural resources are traditionally linked to health improvements. Based on their unique properties and their occurrence worldwide, hot springs are a major resource in health and spa tourism in many countries. Ongoing research by scientists in Asia, Oceania, and Europe (e.g., Taiwan, Japan, France, Germany, and New Zealand) confirms the benefits of minerals and trace elements in assisting the healing process and rehabilitation of a large number of health conditions. The curative effect is said to be directly related to the ability of the human body to absorb minerals and metallic trace elements contained in the hot spring water (Erfurt-Cooper, 2012; Ghersetich and Lotti, 1996).

Even dormant volcanic regions such as the Auvergne (France) and the Vulkaneifel (Germany) are rich in volcanic hot springs, which have been utilized as health and wellness destinations since the Romans built the first thermal baths. In the Auvergne, approximately 100 dormant volcanoes provide the heat source for 109 hot springs, including the spa town Vichy. In the Vulkaneifel, also referred to as a "therapeutic landscape," the Romans already used the volcanic hot springs 2000 years ago to relax after battle. Other examples of European hot spring destinations include Budapest (Hungary) and Ischia (Italy), both destinations also popular with the Romans. During their occupation of various Mediterranean countries the Romans established hot spring baths wherever they discovered geothermal resources. The volcanic heritage of Greece, Turkey, the Middle East, and several north African countries has left a legacy of countless historical hot springs that are still used as tourist destinations today.

Long after the Romans, in the eighteenth century Sir William Hamilton used to ride from Naples to La Solfatara to enjoy the hot spring baths. Destinations such as the mud baths in nearby Ischia and Vulcano in the Aeolian Islands have remained popular to this day and have kept their reputation of fulfilling the needs of health-conscious visitors who are seeking cures and restoration.

The use of volcanic hot springs for healing purposes was practiced by the native Indian tribes of America who were aware of the beneficial mineral content and considered sites with volcanic hot springs as sacred and neutral ground. In Mexico the Aztecs enjoyed hot springs (e.g., Agua Hedionda) for medicinal reasons as did the Maya in Central America and the Incas of Peru. In Oregon the Modoc Indians used Klamath Hot Springs for bathing and to build sweat huts (Theodoratus et al., 1990), while Shoshone Indians believed in the healing powers of the hot springs of Yellowstone. In Costa Rica, the Tabacon thermal spa resort claims to have the perfect combination of water volume, temperature, and mineral content for health purposes with attractions such as steaming waterfalls and thermal pools heated by Arenal's magma chamber.

The therapeutic use of volcanic hot springs in New Zealand also has a very long social and cultural history. For centuries the Māori traveled to Waiwera just north of Auckland, to heal themselves in the therapeutic warm waters. The Māori as well as the European settlers would dig holes along the beach and soak in the hot springs that seeped through the sand. Volcanic valleys, geothermal wonderlands, and hot spring spas are the main attractions of Rotorua (New Zealand), a city that is known for its rather unpleasant sulfur smell but still attracts over three million visitors every year. The city's Queen Elizabeth Hospital has a long tradition of using hot springs and volcanic mud for health treatments and rehabilitation. The Polynesian Spa offers therapies based on Rotorua's hot springs and thermal mud with 26 pools at different temperatures and mineral compositions. At "Hells Gate," one of the geothermal nature parks near Rotorua, again reference is made to the underworld. In addition to its visual "hellish" attractions, the park also operates a health spa with hot spring pools and mud baths.

Very similar to New Zealand, Iceland's abundance of volcanic hot springs has been used for tourism, as well as domestic and industrial purposes since the early 20th century. In Hveragerði, the NLFÍ Spa and Medical Clinic specializes in rehabilitation after accident or illness using mud from the local hot springs. The Blue Lagoon (Figure 75.11) is one of Iceland's most visited destinations and has a reputation for effectively treating psoriasis and other health conditions. This popular geothermal spa is located between Keflavik and Reykjavik, directly next to the Svartsengi geothermal power station. This makes the large lagoon, which is fed by surplus geothermal water from the power plant, a truly unique attraction, although the

FIGURE 75.11 The Blue Lagoon is a major stopover for tourists between the international airport in Keflavik and the capital Reykjavik. *Photo: P. Erfurt-Cooper.*

Mývatn Nature Baths in Iceland's northeast are a similar tourist destination.

Taiwan also has an established health tourism industry, based on more than 130 hot springs originating from the Datun volcano group, an active volcanic system, with the highest concentration of springs found in the north of the island. While this is not volcano tourism as such, the natural resources originate from volcanic activity and have provided not just health benefits but also economic advantages through a booming hot spring tourism industry. The Datun volcano group is part of the cultural and natural landscape of the Yangmingshan National Park, where visitors can hike the multipeak trail or the Datun Nature Park trail. Nearby, the Beitou District north of Taipei City has developed into a famous hot spring destination with over 30 resorts, where local and international visitors enjoy the many volcanic hot springs. A former public bath house, which was established in 1913 under Japanese rule, is now a "Hot Spring Museum."

3. VOLCANIC NATIONAL PARKS, GEOPARKS, AND WORLD HERITAGE SITES

Volcanic regions throughout the world have been designated as protected sites such as National Parks, Geoparks, and World Heritage listed areas. Other conservation settings include National Heritage Areas, National Landscapes (Australia), Quasi National Parks and Prefectural Parks (Japan), National Geoparks or State Parks, Global Geoparks, Volcanic Island Reserves, and private properties. Unlike national parks, geoparks are a rather new entity (Table 75.2). They frequently include a number of national parks and other protected sites within their boundaries, as well as townships and other urban areas. Currently China has the highest number of global geoparks (27 areas officially recognized by UNESCO) with several of them containing volcanic heritage.

More common however are national parks with the first park in the world created in 1872, when the spectacular geologic and natural wonders of the Yellowstone caldera were protected for future generations. In Japan, the first 12 national parks were established between 1934 and 1936 and included the volcanic areas of Unzen, Kirishima, and Aso in Kyushu, as well as Akan and Daisetsusan in Hokkaido. In effect, 11 of the initial 12 Japanese national parks were created around volcanoes such as Fuji-Hakone, Towada, Yoshino-Kumano, Daisen, Nikko, and the Japanese Alps. National parks are also frequently included in World Heritage listed areas as preexisting protected sites of outstanding value.

In comparison, the Vesuvius National Park in Italy was only established in 1995 and the French Réunion National

TABLE 75.2 Volcanic National and Global Geoparks

Active

Katla Geopark, **Iceland**
Batur Global Geopark, **Indonesia**
Toya Caldera and Usu Volcano Global Geopark, **Japan**
Unzen Volcanic Geopark, **Japan**
Aso Geopark, **Japan**
Azores Geopark, **Portugal**

Dormant or Extinct

Shimonita Geopark, **Japan**
Itoigawa Global Geopark, **Japan**
Yehliu Geopark, **Taiwan**
Shishan Volcanic Geopark, Hainan, **China**
Tengchong Volcanic Global Geopark, **China**
Wudalianchi Geopark, **China**
Hexigten Geopark, **China**
Sai Kung Volcanic Rock Region, Hongkong, **China**
Vulkaneifel Geopark, **Germany**
Jeju Island Geopark, **Korea**
Magma Geopark, **Norway**
Shetland Geopark, **UK**
Lochaber Geopark, **UK**
Bohemian Paradise Geopark, **Czech Republic**
Kanawinka Geopark, **Australia**
Geopark Shetland, **UK**

Active and dormant volcanic areas are included in geoparks, either on a national basis, incorporated in the European Geoparks Network or as a Global Geopark.

FIGURE 75.12 Mt Damavand in northern Iran is a popular tourist destination, although climbing requires several days and special equipment to cope with extreme temperatures at higher elevations. *Photo: P. Erfurt-Cooper.*

Park in the southern Indian Ocean was established in 2007. France has also set up a number of "regional natural parks," one of which includes the volcanoes of the Auvergne (Parc Naturel Régional des Volcans d'Auvergne) and dates back to 1977. Turkey's extensive volcanic heritage is protected in national parks including Mt Nemrut and Göreme/Cappadocia, which are also listed as World Heritage sites based on their unique landscape and rock formations.

In fact, most volcanoes worldwide, whether dormant or active, are located in protected areas. The legendary Krakatau, which erupted in 1883 and destroyed itself in the process, created a large caldera in the Sunda Strait between Sumatra and Java. Today the Krakatau Nature Reserve is part of the Ujung Kulon National Park, which is also inscribed as a World Heritage site. The growing vent of Anak Krakatau (Child of Krakatau) maintains its reputation as a potentially dangerous mountain, which nonetheless attracts volcano tourists who can join special tours. However, access to the volcano itself depends on the state and level of its activity.

In the Iranian Alborz Mountains north of Tehran the volcano Damavand (5671 m) is the highest peak in the Middle East. Although there are no recent recorded eruptions of Mt Damavand, the volcano is classed as potentially active with numerous hot springs and fumaroles on the

mountain sides. In 2008, Damavand was placed on the tentative UNESCO World Heritage list and one year later was registered as a national heritage site to protect the unique landscape and ecology, as well as the natural resources of the region (Figure 75.12). According to ancient stone reliefs the volcano was used by Persian kings for recreational purposes including climbing and hunting (Vafadari, 2010). Today trekking and climbing tours are promoted online for groups (beginner to advanced) and led by experienced mountain guides.

In Asian countries like Indonesia and the Philippines, volcanoes have become major tourist attractions. Indonesia is a volcanic island arc with close to 130 active volcanoes. Mt Merapi on the island of Java is especially worth mentioning because this active volcano is promoted as a tourist destination despite an eruption history that has claimed thousands of lives. However, the close location of Merapi to the World Heritage site of Borobudur makes it a popular destination, activity status permitting. On the Indonesian island of Bali, volcano tourism includes climbing Mt Batur in the early hours of the morning to watch the sunrise as a worthwhile adventure (Figure 75.13). Although accidents happen occasionally as mentioned by local tour guides, the geodiversity of Bali's volcanoes presents an enormous potential for the local tourism industry. In fact, so enormous that in 2012, the geological heritage of the Mt Batur caldera was officially recognized as a UNESCO Global Geopark.

On the African continent, particularly in the Great Rift Valley, volcanoes are included in national parks and other protected areas. The Virunga National Park (Mt Nyiragongo) is World Heritage listed and the Ngorongoro Crater (Tanzania) is a conservation area. Other protected volcanic sites include Mount Kenya National

FIGURE 75.13 On the Indonesian island of Bali "sunrise trekking" to the summit of Mt Batur is part of the popular day tours. Tourists can see the sunrise above Lake Batur with Mt Abang and Mt Agung to the right. In the middle distance, Mt Rinjani (Lombok Island) was erupting together with the sunrise, making it a very special experience. *Photo: P. Erfurt-Cooper.*

TABLE 75.3 Volcanic World Heritage Listed Properties

Active	Dormant or Extinct
Heard and McDonald Islands, **Australia** (1997)	Gondwana Rainforests of **Australia** (1994)
Area de Conservación Guanacaste (**Rincon de la Vieja**), **Costa Rica** (1997)	Lord Howe Island Group, **Australia** (1982)
Virunga National Park, **DR Congo** (1979)	Macquarie Island, **Australia** (1997)
Morne Trois Pitons National Park, **Dominica**, Lesser Antilles (1997)	Wet Tropics of Queensland, **Australia** (1988)
Galápagos Islands, **Ecuador** (1978; Ext. 2001)	Fernando de Noronha, **Brazil** (2001)
Sangay National Park, **Ecuador** (1983)	Rapa Nui National Park, **Chile** (1995)
Pitons, Cirques and Remparts of Reunion Island, **France** (2010)	Mt Huangshan, **China** (2012)
Surtsey, **Iceland** (2008)	Cocos Island National Park, **Costa Rica** (1997)
Bali Cultured Landscape, **Indonesia** (2012)	Þingvellir National Park, **Iceland** (2004)
Borobudur Temple (**Merapi**), Java, **Indonesia** (1991)	Komodo National Park, **Indonesia** (1991)
Ujung Kulon National Park (**Krakatau**), **Indonesia** (1991)	Giant's Causeway, **Ireland** (1986)
Aeolian Islands, **Italy** (2000)	Taï National Park, **Ivory Coast** (1982)
Mt Etna, Sicily, **Italy** (2013)	Shiretoko National Park, Hokkaido, **Japan** (2005)
Pompeii, Herculaneum and Torre Annunziata (**Vesuvius**), **Italy** (1997)	Mt Kenya Natural Park, **Kenya** (1997; Ext. 2013)
Mt Fuji, **Japan** (2013)	Jeju Volcanic Island and Lava Tubes, **Korea** (2007)
Tongariro National Park, **New Zealand** (1993)	Pitons Management Area, St. Lucia, **Lesser Antilles** (2004)
Pico Island Vineyard Culture, Azores, **Portugal** (2004)	El Pinacate and Gran Desierto de Altar Biosphere Reserve, **Mexico** (2013)
Garajonay National Park, La Gomera, **Spain** (1986)	Rock Islands Southern Lagoon, **Palau** (2012)
Teide National Park, Tenerife, **Spain** (2007)	Old and New Towns of Edinburgh (**Arthurs Seat**), **Scotland** (1995)
Volcanoes of Kamchatka, **Russia** (1996; Ext. 2001)	Kilimanjaro National Park, **Tanzania** (1987)
Hawai'i Volcanoes National Park, **USA** (1987)	Ngorongoro Conservation Area, **Tanzania** (1979; Ext. 2010)
Yellowstone National Park, **USA** (1978)	Göreme National Park Cappadocia, **Turkey** (1985)
	Hierapolis-Pamukkale, **Turkey** (1980)
	Yosemite National Park, California, **USA** (1984)

Active, dormant, and extinct volcanic areas contribute at least 5% to the current total of 981 World Heritage sites. The year of the original listing is included in brackets.

Park and Meru National Park (Kenya), Volcans National Park (Rwanda), Kilimanjaro National Park (Tanzania), Rwenzori National Park (Uganda), Mount Cameroon National Park (Cameroon), and the Fogo National Park on the Cape Verde Islands.

Apart from the famous Yellowstone, the United States offers a wide range of volcanic national parks. The entire west coast with the Cascades mountain range has been recognized as a unique volcanic area, and includes four national parks. In 1902, Crater Lake National Park was established to preserve the spectacular Mazama caldera, its unique lake, and the surrounding environment. Other national parks in the Cascades include Lassen National Park (established 1916), which includes a vast hydrothermal area and a dozen smaller volcanoes. The creation of Mt St Helens National Monument was prompted by the cataclysmic events of 1980 that changed the local landscape. The aim to preserve a unique volcanic setting was also behind the creation of the Sunset Crater National Monument in Arizona.

Many countries have followed various models of preserving spectacular volcanic regions, with a significant increase in the number of national parks alone over the last three decades. However, developing countries frequently have limited resources to either establish or maintain protected sites including World Heritage listed areas (Table 75.3).

In Latin America, several of Costa Rica's volcanoes are protected in national parks. Ecuador also preserves many of its volcanic regions as national parks, including the Galápagos Islands and Cotopaxi, whereby the Galápagos Islands are also World Heritage listed. Chile and Argentina have established national parks to protect the Andean volcanic environments including Villarica, while Colombia's notorious Nevado del Ruiz is located within the Los Nevados National Park.

3.1. Protected Sites and Their Educational Value

Protected areas such as national parks, geoparks, and World Heritage sites play an important role in volcano tourism by offering education about the geological heritage of active and dormant areas, as well as showcasing the overall geo-diversity. The quest for knowledge is a driving force in tourism and volcanic environments are an ideal place to learn about geology. Volcanoes attract people of all ages with visitors making use of educational displays in museums and information centers (Table 75.4) or hiking the trails leading to fumaroles, crater lakes, and other volcanic features.

The Hawai'i Volcano National Park (established 1916) preserves a large fraction of Mauna Loa and most of Kīlauea, two of the world's most active volcanoes. The spectacular scenery and frequent eruptions attract on average 4000 visitors per day, who hike the park's well-laid out trails or use stopover viewing points when driving. A large number of plaques with explanations are placed at points of interest along the trails and roads. Located inside the park is the Thomas A. Jagger Museum, an educational facility with informative displays that opened in 1987.

TABLE 75.4 Volcano Museums—Education, Information, Interpretation

Penshurst Volcano Discovery Centre, **Australia**	Sakurajima Volcano Museum, Kyushu, **Japan**
Volcano Museum Huilo Huilo, **Chile**	Toyako Science Visitor Centre, Hokkaido, **Japan**
Arenal Observatory Lodge Museum, **Costa Rica**	Unzen Volcano Museum (several), Kyushu, **Japan**
Volcano House, Reunion Island, **France**	Volcano Museum Martinique, **Caribbean**
Volcano Park of Lemptégy, **France**	Volcanic Activity Centre Taupo, **New Zealand**
Vulcania Theme Park and Museum, **France**	Volcanic Forces—Auckland Museum, **New Zealand**
Maarmuseum Manderscheidt, **Germany**	Te Papa Museum, Wellington, **New Zealand**
Vulkanmuseum Daun, Vulka-neifel, **Germany**	Masaya Volcano Visitor Centre, **Nicaragua**
Eldfjallasafn Volcano Museum, Stykkishólmur, **Iceland**	Vulcão dos Capelinhos Exhibition, Azores, **Portugal**
þorvaldseyri Visitor Centre, **Iceland**	Visitor Centre Teide National Park, Tenerife, **Spain**
Volcano House Reykjavik, **Iceland**	International Museum of Volcanoes, Lanzarote, **Spain**
Batur Volcano Museum Bali, **Indonesia**	Jaggar Museum, Hawai'i Volcanoes National Park, **USA**
Merapi Volcano Museum, **Indonesia**	
Aso Volcano Museum, Kyushu, **Japan**	Mt St Helens Visitor Centers, Washington, **USA**

Volcano museums have been established worldwide and are important educational facilities, which are part of the visitor experience in national parks, geoparks, and world heritage areas.

4. VOLCANIC ERUPTIONS—UNIQUE TOURIST EXPERIENCES

Volcanic eruptions can have a large impact on the tourism industry of a region, and often the aftermath of a violent eruption will bring increased numbers of visitors. Once a catastrophe is over and the tourism industry is on the path of recovery, tourists arrive in growing numbers to witness the destruction and also to provide an economic boost for the affected region. A famous example in more recent times (2010) was the eruption of Eyjafjallajökull in Iceland—the event that closed European airspace with consequences that affected passenger movement all over the globe. While the eruption initially attracted tourists to watch the fiery fissures from as close as possible, the next eruption stage generated vast ash clouds, a major hazard to aviation, which effectively prevented aircraft at many European airports from takeoff or landing. This was a forceful reminder of how exposure to a natural disaster can cause widespread damage and economic loss, while at the same time putting an area on the map and increasing visitor numbers after the event.

The unexpected explosive eruption of Chaitén volcano in Chile in 2008 after thousands of years of quiescence caused significant damage from pyroclastic flows and lahars. Although there is now a realistic expectation of future eruptions of the Chaitén volcano, tourism in this region has seen an increase of visitors interested in the volcano and the damage it caused. Other well-known examples of areas that benefit from past eruptive events include the Greek island of Santorini, which is a thriving tourist destination surrounded by physical evidence from one of the largest eruptions in history (Thera). In Sicily, Mt Etna's ongoing activity has influenced the lives of people who live in its vicinity (Figure 75.14). Although lava flows have caused disruptions and threatened not just local towns such as Linguaglossa, Nicolosi, Zafferana, and Randazzo but also Catania's large population (during the 1669 eruption), Mt

FIGURE 75.14 Mt Etna was listed as a UNESCO World Heritage Site in 2013 because of its great scientific interest and its value as a key destination for education and research. *Photo: P. Erfurt-Cooper.*

Etna has remained an attractive destination for volcano tourists who enjoy hiking and skiing.

Other European volcanic regions that attract tourists with their spectacular scenery include the Canary Islands (Spain) and the Azores (Portugal). Although the volcanoes of these two island groups are not currently erupting, they are nevertheless classed as active based on their recent history. Volcanic attractions feature throughout the tourism industry with popular tours to the Teide caldera Las Cañadas on Tenerife, volcano walks in Timanfaja National Park on Lanzarote, or rock climbing on the volcanic walls in the Azores.

4.1. Eruptive Activity—Awe-Inspiring Attractions

Heightened volcanic activity encourages volcano tourists to explore volcanic features such as Strombolian eruptions, lava flows, geysers and hot springs, lava lakes, crater lakes, fumaroles, boiling mud ponds, hot rivers, and travertine terraces created by volcanic hot springs and their mineral deposits. These volcanic features are generally included as attractive highlights for volcano tourists.

Volcanic eruptions are among nature's most awesome spectacles; however, while an active volcano can contribute significantly to the economy of a region, it can also destroy infrastructure and negatively affect the local tourism industry. From a tourism point of view, the best types of volcanic activity are long-lived Hawai'ian or Strombolian eruptions. These eruptions allow tourists to observe the activity from a reasonably close range and in relative safety. If the eruption lasts for months or years, the local economy can truly prosper from the sustained number of visitors. However, other factors—notably, ease of access—are important for turning an active volcano into a major tourist destination. Stromboli, for example, which has been continuously active for over 2000 years, is one of the world's top destinations for those who want to see a live volcano. However, its remote location and its dependence on the weather (access restricted to summer) make it far less popular than many other volcanic regions around the world.

Mt Yasur (Tanna Island, Vanuatu) is another example of a volcano that sustains long-lived Strombolian activity and has attracted growing numbers of visitors in recent years, despite its rather remote location in the South Pacific. According to tourism statistics for the last quarter of 2012, Vanuatu registered a 14% increase in international visitor arrivals (Vanuatu National Statistics Office, 2013). Although not all of these visitors may be volcano tourists, the figures clearly indicate that destinations, formerly avoided as too remote or inaccessible, receive growing numbers of international visitors. Online video footage of Yasur erupting is used as a major attraction in tourism marketing. To see a boiling lava lake on Vanuatu's Ambrym Island, visitors can take a helicopter flight to view the active craters of Benbow and Mbewelesu.

In fact, using helicopters is a common way of accessing unique volcanoes in remote locations; i.e., visitors of the Kamchatka volcanoes (Russia) rely on a fly-in-fly-out system as do tourists in Ethiopia who want to have quick access to the lava lake of Erta Ale. Scenic flights by helicopter are also used at Mt Aso in Kyushu (Japan), especially during times when the crater emits toxic fumes and access to the summit by car or ropeway is suspended. Helicopter tours are also popular in Hawai'i, Costa Rica, Chile, and New Zealand. The volcano White Island, located 50 km off the coast of New Zealand's north island, is accessed by boat and by helicopter with both services continuing during periods of heightened activity. Moreover, the tour operators and visitors agree that the best time to visit White Island is when volcanic tremors and degassing are accompanied by minor eruptive activity.

However, volcanoes do not have to be active to attract adventure-seeking tourists. The unique rock formations of central Turkey are visited all year round, although the area is not promoted directly as a volcanic destination but as an outstanding geological and cultural experience that is also World Heritage listed. Located between the dormant volcanoes Erciyes, Melendiz, and Hasan, the plains of Cappadocia are attracting growing numbers of tourists with a 15% growth in January 2013 compared to the previous year. The unusual landforms were created by erosional forces, which carved the bizarre tuff towers capped with basalt layers out of alternating layers of welded tuff and basaltic lava flows. The resulting "Fairy Chimneys" are remnants of the up to 150-m-thick tuff beds that were laid down during historic volcanic eruptions (Figure 75.15).

FIGURE 75.15 Cappadocia's Fairy Chimneys are part of the unique volcanic landscape around Gőreme (Central Turkey), a World Heritage listed area. *Photo: P. Erfurt-Cooper.*

Some volcanoes are so visually spectacular that this alone makes them major tourism destinations. Crater Lake in Oregon (USA) is surrounded by forests and volcanic peaks and lies inside the Mt Mazama caldera. After a cataclysmic eruption, the caldera developed and was eventually filled with water from rain and snow, forming a lake 589 m (1932 ft) deep with the volcanic cinder cone of Wizard Island at the western end of the lake. The purity of the water and the great depth of the caldera are responsible for its startling indigo blue. Another "Blue Lake" is part of the Mt Gambier maar complex in South Australia. The crystal clear water of this Blue Lake changes its color every November from steel blue to brilliant turquoise and back again in March to steel blue. While the Blue Lake provides the water supply for the city of Mt Gambier, the crater lakes are a State Heritage Area and are located within the boundaries of the Kanawinka Geopark, which includes large parts of the Newer Volcanic Province in western Victoria.

One of the most popular active volcanoes in the world is Kīlauea on Hawai'i. Kīlauea has attracted tourists since the early twentieth century, when a lava lake filled Halemaumau crater. A hotel, Volcano House, was built in 1846 on the edge of Kīlauea caldera, with full view of Halemaumau. Another leading example of spectacular volcanic scenery is the vast summit crater of Haleakalā (on the island of Maui), a breathtaking sight and an impressive geologic wonder. The most magnificent sight offered by this volcano is early in the morning, when tourists hike to the summit, starting their ascent in darkness to watch the sunrise. The Pu'u 'Ō'ō-Kūpaianaha eruption, ongoing since 1983, is the longest eruptive event on record in Hawai'i. The sustained activity has benefitted local tourism and has spawned businesses ranging from new hotels to helicopter tours.

Before the 1980 eruption of Mt St Helens, visitors were attracted to Spirit Lake and its magnificent setting. When the lake and its surrounding forests were destroyed, it seemed that tourism in the area would not recover. However, the devastated landscape became a major attraction with many people wanting to see the eruption's effect, which has made the area considerably more popular than it was before 1980. A landmark event was the creation of the Mt St Helens National Monument in 1982. In 1993, the Coldwater Ridge Center was opened, and during that year alone received about one million visitors. The Center was permanently closed in 2007 due to federal budget constraints, but reopened in 2012 as the "Science and Learning Center at Coldwater." The climb to Mt St Helens' crater rim, from where one can see the impressive dome growing inside the gaping crater, is still one of the most popular activities in the Cascades.

Costa Rica's Arenal was not recognized as an active volcano until it started to erupt in 1968. The ongoing eruption (until 2010) contributed to the development of Arenal's surrounding areas into one of the most popular destinations in Costa Rica. Hotels and lodges with views of the volcano were built to watch Arenal's activity at times of eruption. Many visitors stayed up all night to observe glowing lava flowing down the volcano's steep flanks. However, Arenal has not erupted since 2010 and although in September 2013 some activity including steam plumes and rock falls was observed, by January 2014 the volcano failed to show further signs of reawakening. And while there is a reasonable expectation that Arenal will erupt again in the future, tourists will include the volcano in their travel plans.

Finally, one of the most unique, although remote volcanic destinations is Deception Island in Antarctica, the only place in the world where cruise ships sail directly into a volcanic caldera. Entering the center of the horseshoe-shaped island is only possible under calm conditions through Neptune's Bellows, a narrow channel at the caldera entrance. This major stopover is hailed as a highlight by visitors and provides the opportunity to learn about volcanic history, local wildlife, and aspects of the unique Antarctic ecology. Apart from penguin colonies, a barren volcanic landscape, and black volcanic sand beaches, steaming hot springs emerge through the black sand at the beach and mix with cold seawater; a unique opportunity for cruise guests to soak in a hot spring in Antarctica. Owing to Deception Island's eruption history, it is classed as a restless caldera with the potential for further eruptions at any time. Deception Island is one of the most frequently visited sites in Antarctica by tourists (Deception Island, 2012), which shows a growing demand for remote volcanic locations.

5. ACTIVITIES IN VOLCANIC ENVIRONMENTS

Tourism in volcanic environments incorporates a number of elements ranging from geotourism to ecotourism and adventure tourism. The study of geological phenomena in extreme environments requires education and interpretation, which is provided via on-site information centers and volcano museums. Tour guides and sign boards present additional learning experiences while sightseeing and exploring. Depending on the volcanic activity level, a sense of adventure plays an important role for many volcano tourists. It is common in most regions where volcanic geotourism is practiced, that local history and culture are part of the visitor experience, adding to the overall geodiversity. The aspect of sustainability also plays a significant role in nature-based (volcano) tourism with the awareness about environmental protection at an ever-increasing level.

Volcanic geoheritage, which includes active and dormant landscapes, is often co-located with natural hot springs, which are also included as volcanic attractions in sightseeing tours as shown above. Due to easier access to remote locations and more affordable air travel, the number of visitors is constantly increasing. In Sicily, eruption viewing during the night is available in villages and townships surrounding Mt Etna. In New Zealand, the Chateau Tongariro Hotel is located in the Tongariro National Park in the foothills of Mt Ruapehu; in Costa Rica, the Arenal Observatory Lodge offers volcano views; and in Kagoshima, hotel rooms with views of Sakurajima in eruption are highly sought after.

5.1. Recreation and Volcanoes

Apart from hiking, trekking, backpacking, climbing, and mountain biking, some rather unusual sporting activities have emerged in recent years. On the slopes of the Cerro Negro volcano in Nicaragua "volcano surfing" is a favorite with adventure tourists, who rank it as a major attraction. After climbing to the summit of Cerro Negro, adventure-minded visitors use toboggans to race down the coarse cinder slopes at speeds of up to 80 km/h. The potential for injury is readily accepted, as well as the possibility of an unexpected eruption of the active volcano. According to the Huffington Post (2011), volcano surfing "is the 'latest hot extreme sport' among the travel adventure community, with over 10,000 people having tried it" despite the fact that Cerro Negro was last active in 1999 with a potential for renewed activity at any time. Another extreme activity has groups of tourists running down the slopes of the Cerro Negro and El Misti volcanoes, in a similar fashion to scree running in nonvolcanic environments. For the extreme sports enthusiast, volcano abseiling into an active volcanic crater (e.g., Ambrym, Vanuatu) offers a certain adrenaline rush, although the need for special equipment restricts this activity to a minority of visitors.

Possibly one of the most exciting, although dangerous tourist attractions is bungee jumping from a helicopter above the active crater of Chile's Villarica volcano. So far no casualties have been reported but the excitement for dedicated adventure seekers comes with a price tag of around USD$10,000 for a 5-day trip to Pucon, which includes this unusual highlight. Alternatively, hot air balloon tours can prove fatal. An accident in Turkey in 2013 claimed 3 lives and injured 21 passengers in a midair collision of two balloons. Because Turkey's volcanic heritage is completely integrated in the tourism industry, many activities include hiking, trekking, climbing, and skiing the volcanic landscape. However, most famous are the above mentioned hot air balloon trips, which have taken place since 1997. Cappadocia is known as one of the most important hot air balloon centers worldwide, and over the

FIGURE 75.16 Hot air balloon flights in Göreme start early in the morning before the wind can become a hazard to the many individual balloons and their passengers. The eroded tuff landscape and the cave buildings of Cappadocia are major attractions and are best viewed from above. *Photo: P. Erfurt-Cooper.*

last two decades this activity has grown in popularity. Early in the morning, colorful balloons rise to the skies and take tourists over the bizarre volcanic landscape. Passengers can choose varying levels of height to accommodate people who are not comfortable at higher levels (Figure 75.16). As with any adventure activity there is a risk involved that people should be aware of when they decide to participate, although the view of the spectacular landscape of Cappadocia is definitely worth it.

In the French Auvergne, hot air ballooning is promoted as "a truly magical meeting with nature," where passengers can view the dormant volcanic landscape from above. In Australia, hot air balloons take visitors on sunrise flights to experience the breathtaking view of the volcanic caldera of Mt Warning (Wollumbin), also known as the Green Cauldron, one of Australia's "National Landscapes" and part of the World Heritage listed Gondwana Rainforest on the border of Queensland and New South Wales. Here, the remnants of an ancient shield volcano are presented together with lush rainforests and the unusual Australian wildlife, as well as the cultural aspects of the local indigenous heritage. Further north in the "Wet Tropics" World Heritage area, balloon rides take off from Mareeba to fly above the volcanic landscape of the Atherton Tablelands. For volcano tourists visiting this area, the Undara Volcanic National Park, 320 km west of Cairns, lets visitors explore some of the largest and best preserved lava tubes worldwide (Figure 75.17); for geotourists this is an amazing destination surrounded by the Great Australian Outback.

In Spain, hot air balloons take flight above the landscape of the Garrotxa Volcanic Zone Natural Park; in Costa Rica, ballooning includes views of Arenal volcano as a special attraction; and in Southern California, tourists have the

FIGURE 75.17 Some of the Undara lava tubes have caved in over time and access to the remaining caves is only possible with a guided tour. The Undara Volcanic National Park is part of the Wet Tropics of Queensland World Heritage area. *Photo: P. Erfurt-Cooper.*

FIGURE 75.18 A number of houses were buried by lahars from Unzen volcano, Japan, and have been covered to protect them from erosion and to make them available to visitors all year round. *Photo: P. Erfurt-Cooper.*

opportunity to enjoy the panoramic views of ancient volcanoes below. Scenic flights by airplane or helicopter are a popular way in Oceania to access volcanoes such as White Island, Tongariro, Ruapehu, and Tarawera in New Zealand and Lopevi, Ambrym, and Yasur in Vanuatu. In the northern hemisphere, "flightseeing" includes the national parks of the Alaskan Ranges and the Aleutians with a number of active volcanoes and spectacular glaciers as visitor experience.

5.2. Volcanic Disaster Tourism

The revenue raised through volcano tourism is an opportunity to counteract economic shortfalls caused by natural disasters such as eruptions that can cause widespread losses. In response to visitor expectations, information centers in volcanic areas frequently show video footage as forceful reminders of the destruction caused by different types of eruptions. Consequently, volcano tourism can sometimes overlap into what is known as "disaster tourism" or "dark tourism." After a volcanic eruption, affected regions on the path to recovery need a boost to the economy, and the tourism industry often incorporates the devastating changes to the landscape to attract visitors. Well-known disaster tourism sites are Pompeii and Herculaneum in Italy, with some modern day examples located in Iceland (Pompeii of the North) and in Japan (Mt Unzen). In the city of Shimabara, posteruption lahars from Mt Unzen were caused by a disastrous combination of thick layers of volcanic ash and heavy rainfall, and consequently buried numerous buildings that were located in the flow path (Figure 75.18). Some of these buildings are under cover to protect them from the elements and are known as the Mizunashi River Memorial Park, an important stopover for tourists. At the seafront of

Shimabara, built on volcanic debris, and surrounded by open spaces, the Unzen Disaster Memorial Hall (Unzendake Saigai Kinenkan) provides vital information about the effects of exposure to pyroclastic flows and lahars. During excavations, several cameras buried by pyroclastic flows were found; one of them with a few seconds of footage showing a superheated cloud of volcanic gas and ash heading toward a group of 43 people, who perished in the immense heat.

5.3. Potential Dangers and Visitor Safety

Volcano tourists who are venturing into the mountains are facing potential dangers that are common in many mountain areas. These dangers include hiking accidents; getting lost; sudden weather changes causing rain, fog, or snow; rock falls, mud slides, and snow avalanches; as well as altitude sickness and hypothermia. While people can prepare for these risks to a certain degree, there are other hazards related directly to volcanic environments. Unexpected eruptions, gas emissions and toxic fumes, earthquakes, lahars, steam vents, pyroclastic flows, and thermal burns from lava flows or hydrothermal springs add a host of potential risks that should be given some serious thought before embarking into an area where help may not be available. Potential problems can stem from a lack of preparedness that is frequently related to a lack of safety instructions for visitors, as these are not available at every volcanic destination. Existing safety guidelines are usually available for national parks and local communities, but this information is not always made available to temporary visitors in active volcanic environments. Information about the local area including escape routes and directions to shelters and emergency phones in several major languages is essential. Visitors who are not familiar with the local terrain also need a basic hazard map to keep them in safer zones. Especially important is that interpretive signage and warning signs should be in more than one language with images or pictograms for immediate visual recognition.

5.4. Personal Risk Acceptance and Emergency Situations

Volcano tourism is growing in popularity, and in Iceland, Japan, Costa Rica, and New Zealand geotourism in volcanic regions already has the dimensions of mass tourism. Back in 2007, the Philippines advertised Mt Pinatubo as the "hottest" tourist destination in Luzon and encouraged ecotourism groups to visit the devastation area around the mountain (Reyes, 2007). That this can in some instances have tragic outcomes was evident when Mt Mayon (also in the Philippines) erupted in May 2013. An explosion caused the death of five tourists, with several more injured. A permanent exclusion zone of several kilometers around Mt Mayon either was not adequate or was ignored due to the assumed quiescence of the volcano. It is therefore important that volcano tourism is made as safe as possible by raising awareness about potential hazards in these extreme environments, as people often underestimate possible dangers and overestimate their own abilities to cope with challenging situations. In many active volcanic and geothermal areas, existing safety standards or guidelines do not include tourists, whose risk perception frequently can be impaired without appropriate safety advice from local authorities.

6. SUMMARY

Millions of visitors every year travel to the most active volcanoes of the world to experience the added element of adventure with volcanic features used as attractions throughout the tourism industry. While some volcanic areas are popular mainly for their magnificent scenery, the combination of volcanoes, adventure, risk, vegetation, and unusual wildlife can bring visitors to the remotest island destinations. Volcanoes in countries such as Indonesia, the Philippines, and Vanuatu have become major tourist attractions, despite their often dangerous activity levels. Even more remote regions such as Antarctica draw cruise ships to volcanic attractions such as the active caldera of Deception Island, while on the Russian Kamchatka helicopter tours provide access to the volcanic wilderness.

Throughout the world, active and dormant volcanic regions have been protected as national parks, geoparks, and World Heritage areas. Protected sites such as these play an important role in volcanic geotourism, as they offer education about the geological heritage, as well as a variety of recreational activities including adventure and extreme sports. For the tourism sectors of geotourism, adventure tourism, and ecotourism, volcanoes are priceless natural resources. The educational aspect of volcanoes in terms of their role in nature and society is one of the major attractions for volcano tourists, especially for those who are exploring and studying volcanic and geothermal landforms. Volcanoes have also played a significant role in history and culture, as well as in religious legends and in the mythology of many regions.

Volcanic geoheritage is often co-located with natural hot springs. While some hydrothermal features like geysers and boiling lakes are only for visual enjoyment, the pleasant addition of volcanic hot springs for bathing purposes has traditionally supported health and wellness tourism. Many destinations take full advantage of the therapeutic use of this natural phenomenon and offer tourists the experience of health resorts and spas based on volcanic hot springs and their beneficial mineral content. This is particularly evident in countries such as Japan, New Zealand, Iceland, Germany, France, and Taiwan, where the close proximity of volcanoes and hot springs has historical links to health and wellness tourism.

Volcanic eruptions can have a significant impact on the tourism potential of a region, as heightened volcanic activity can encourage visitors to stay longer to explore exciting volcanic features. Depending on the level of activity a sense of adventure plays an important role for many volcano tourists. As a result, volcano tourism is frequently linked to disaster tourism or dark tourism, with Pompeii and Herculaneum as classic examples. To have a safe experience and to avoid accidents and injuries, volcano tourists need to have sufficient up-to-date information before they embark on their adventure. The main objective in volcano tourism, especially in active environments, is the safety of all visitors.

FURTHER READING

Anderson, T., 1903. Volcanic Studies in Many Lands. C. Scribner's, New York. Reprinted 2009 by HP from the collections of Californian Libraries.

Aramaki, S., Barberi, F., Casadevall, T., McNutt, S., 1994. Report of the IAVCEI sub-committee for reviewing the safety of volcanologists. Bull. Volcanol. 56 (2), 151—154.

Deception Island, 2012. Management Plan for Antarctic Specially Managed Area No 4 Deception Island, South Shetland Islands, Antarctica. Online: www.ats.aq/documents/recatt/Att512_e.pdf (accessed 18.06.13.).

Erfurt-Cooper, P., Cooper, M., 2009. Health and Wellness Tourism: Spas and Hot Springs. Channel View Publications, Bristol, UK.

Erfurt-Cooper, P., 2010. Introduction to volcano and geothermal tourism. The context of volcano and geothermal tourism. In: Erfurt-Cooper, P., Cooper, M. (Eds.), Volcano and Geothermal Tourism: Sustainable Geo-Resources for Leisure and Recreation. Earthscan, London, pp. 3—31.

Erfurt-Cooper, P., 2012. An Assessment of the Role of Natural Hot and Mineral Springs in Health, Wellness and Recreational Tourism (Unpublished Ph.D. thesis). James Cook University.

Erfurt-Cooper, P., 2014. Volcanic Tourist Destinations. Geoheritage. Geoparks and Geotourism Series. Springer Verlag, Berlin, Heidelberg.

Ghersetich, I., Lotti, T.M., 1996. Immunologic aspects: immunology of mineral water spas. Clin. Dermatol. 14 (6), 563–566.

Huffington Post, 2011. Volcano Surfing: The "Latest Hot Extreme Sport" (VIDEO). Online: http://www.huffingtonpost.com/2011/11/20/volcano-surfing_n_1093122.html (accessed 22.05.13.).

Krafft, M., 1991. Volcanoes — Fire from the Earth. Thames and Hudson, London.

Life and Death in Herculaneum HD, 2013. Documentary. Online: www.youtube.com/watch?v=edgpma-xeK4 (accessed 16.05.13.).

Lopes, R., 2005. The Volcano Adventure Guide. Cambridge University Press, Cambridge, UK.

Lopes, R., 2010. Volcanoes: A Beginner's Guide. Oneworld Publications, Oxford, UK.

Özguler, M.E., Kasap, A., 1999. The geothermal history of Anatolia, Turkey. In: Cataldi, R., Hodgson, S.F., Lund, J.W. (Eds.), Stories from a Heated Earth. Geothermal Resources Council, International Geothermal Association, Sacramento, CA.

Pompeii: The Last Day. Movie. Online: www.youtube.com/watch?v=xS-N_M_QI4E (accessed 16.05.13.).

Reyes, N., 2007. Volcano Tourism. Online: www.bloggernews.net/110955 (accessed 14.06.13.).

Rosi, M., Papale, P., Lupi, L., Stoppato, M., 2003. Volcanoes: A Firefly Guide. Firefly Books, New York.

Siebert, L., Simkin, T., Kimberly, P., 2011. Volcanoes of the World, third ed. Smithsonian Institute/University of California Press.

Theodoratus, D.J., Ashman, M.M., McCarthy, H., Genetti, D.L., 1990. Klamath River Canyon Ethnology Study. Prepared under Contract for the United States Department of the Interior by Theodoratus Cultural Research, Inc, Fair Oaks, California.

TripAdvisor, 2013. Mount Etna Catania, Sicily, Italy. Online: www.tripadvisor.com.au/Attraction_Review-g187888-d195063-Reviews-Mount_Etna-Catania_Province_of_Catania_Sicily.html (accessed 22.05.13.).

Vafadari, K., 2010. Volcano tourism in Iran: Mt Damavand, the highest peak in the Middle East. In: Erfurt-Cooper, P., Cooper, M. (Eds.), Volcano and Geothermal Tourism: Sustainable Geo-Resources for Leisure and Recreation. Earthscan, London, pp. 180–186.

Vanuatu National Statistics Office, 2013. Quarterly Statistical Indicator (Quarter 4, 2012). Online: www.vnso.gov.vu/index.php (accessed 30.05.13.).

Vesuvius Eruption, 1944. Eruption of Mt Vesuvius 1944. Online: www.youtube.com/watch?v=1bsmv6PyKs0 (accessed 16.05.13.).

Whitey, L., 1997. Grand Tours and Cook's Tours. William Morrow.

World Travel & Tourism Council, 2013. Global Travel & Tourism Industry Defies Economic Uncertainty by Outperforming the Global Economy in 2012 — and Predicted to Do it Again in 2013. Online: www.wttc.org/news-media/news-archive/2013/global-travel-tourism-industry-defies-economic-uncertainty-outpe (accessed 12.05.13.).

Chapter 76

Volcanoes, Ancient People, and Their Societies

Payson Sheets

Department of Anthropology, University of Colorado, Boulder, CO, USA

Chapter Outline

GLOSSARY

australopithecus An early human ancestor that lived in sub-Saharan Africa at least from 4 to 2.5 million years ago, and had the beginnings of culture. They walked upright like humans today, but had small brains like those of the apes.

egalitarian society A human society without inherited inequality of power, wealth, or social status. Made up of families, and decisions in the group were made by consensus. Decisions on how to react to suddenly changed circumstances are made rapidly.

homo habilis Our human ancestor that immediately followed the Australopithecines in Africa, and lived until about 1.6 million years ago. They made the earliest stone tools known, by percussion flaking, and used them in various cutting and scraping tasks.

lahar A mudflow or debris flow down the slopes of a volcano that can come right after a volcanic eruption, or much later.

mesoamerica The ancient culture area from highland central Mexico to Guatemala, Belize, Honduras, and El Salvador, where complex society (full civilization) developed. Much of the area was and still is volcanically active.

pyroclastic flow A very hot dense cloud of volcanic gasses, and rock and ash particles that travel downhill at high speeds and are devastating to flora, fauna, and people.

tephra Any size of rock fragment blasted into the air during a volcanic eruption, ranging from tiny particles of volcanic ash to large blocks and lava bombs that explode upon impact. The larger particles fall near the vent, while the smallest ash particles in big eruptions can encircle the globe.

1. INTRODUCTION

Volcanoes have provided hazards and opportunities for people as long as they have been on Earth. When the hazards manifested themselves as eruptions, our ancestors often died or had to relocate to refuge areas until the environment recovered. With soil formation and recovery of flora and fauna, the environment was often superior in life-support capacity compared to preeruption conditions. For well over 99% of our existence on Earth, our ancestors were hunter-gatherers with a high degree of residential mobility from one campsite to another. That mobility, coupled with very low population densities and knowledge of resources at a distance, often facilitated relocation when impacted by unanticipated stresses. **Egalitarian societies** frequently were resilient; even when the circumstances suddenly changed, they recovered and continued their sustainable lifestyles. However, as populations increased in the last 10000 years and complex societies emerged, people and societies often were more vulnerable to volcanic eruptions. Much of their vulnerability derived from their dependence on intensive agriculture, fixed facilities such as irrigation and storage structures, redistributive economies, large-scale architecture, strong sense of place, and, often, defensive and offensive facilities for warfare. And within societies, the elite often inhabited higher locations within their territories, with the lower classes in low-lying areas that were more vulnerable to lava flows, **pyroclastic flows**, and **lahars**, in addition to flooding and, often, earthquake vulnerability.

The focus of this chapter is the relationships between volcanoes and people. Volcanoes present hazards and opportunities, and our ancestors have been dealing with both for as long as we have been on Earth. Researching these relationships requires careful dating of volcanic events and people, and researchers must be careful to avoid seeing a

The Encyclopedia of Volcanoes. http://dx.doi.org/10.1016/B978-0-12-385938-9.00076-6

correlation of natural event and culture change as a causal relationship. Fortunately, more recent volcanic and cultural events, those that occurred in the past two millennia, are more precisely dated and causality can be more compellingly explored (Table 76.1). Here, we begin in the very distant past, and move up in time. An excellent book examining volcanism and its effects on our ancestors is *Eruptions that Shook the World*.

Prior to the scientific understanding of volcanic activity, native cultures consistently explained volcanic eruptions in supernatural terms. Deities, spirits, and gods were responsible for initiating eruptions, and people could propitiate the causes by their religious activities. Cultures often embedded their perceived risks into oral traditions and religious beliefs, which often helped them in dealing with the volcanic hazards in their homelands. Thus, people often reacted to the volcanic eruptions with resilience and creative responses, and passed the knowledge of how to handle their suddenly changed circumstances down to younger generations.

TABLE 76.1 Major Volcanic Eruptions that Affected People and Their Societies in the Past

Volcano/Location	Date	Society/Human Ancestors	Effects	References
Rift valley, Africa	3.6 million years ago	Australopithecines	Habitat changes	Raichlen et al., 2010
O'a Caldera, Ethiopia	240,000 years ago	Archaic *Homo sapiens*	Deaths, stress, survival	Oppenheimer, 2011: 175
Toba, Indonesia	73,000 years ago	Anatomically modern *H. sapiens*	Climate change, may have stressed Neanderthals	Rose and Chesner, 1987
Papua, New Guinea	40,000 years ago to recent	Egalitarian hunter-gatherers and horticulturalists	Creative reactions to numerous eruptions	Grattan and Torrence, 2007
Laacher See, Germany	13,000 years ago	Hunter-gatherers, egalitarian. Bromme culture.	Isolation, simplification	Schmincke et al., 2009; Oppenheimer, 2011
Campi Flegrei, Italy	40,000 years ago	Hunter-gatherers	Culture change, stress on Neanderthals	Oppenheimer, 2011: 210
Vesuvius, Italy	AD 79	Roman civilization: Pompeii and Herculaneum	Burial and preservation of cities	Oppenheimer, 2011
Thera (Santorini), Aegean Sea	1600 BC	Minoan civilization	Debilitation of society, favoring Mycenaeans	Oppenheimer, 2011; Warburton, 2009
Popocatepetl, Mexico	AD 50	Cholula—Puebla civilization	Devastation, migration to Teotihuacan	Evans and Webster, 2001
Xitle, Mexico	AD 10—100	Buried Cuicuilco	Favored Teotihuacan	Evans and Webster, 2001: 723
Ilopango, El Salvador	AD 535	Devastated nearby Miraflores Maya	Northern hemisphere drought, cold, societal collapses	Dull et al., 2010; Dull et al., 2001
Arenal, Costa Rica	2000 BC—AD 1450	Hunting/gathering-horticultural villages	Resilience to 10 major and many minor eruptions	Sheets and McKee, 1994; Cooper and Sheets, 2012.
Baru, Panama	AD 800	Barriles chiefdoms	Had to immigrate	Cooper and Sheets, 2012
Laki, Iceland	AD 1783	Iceland, Europe	Agricultural failures	Oppenheimer, 2011
Tambora, Indonesia	AD 1815	Local agricultural societies, Europe, and North America	Proximal societies eliminated, agricultural failures at distances	Oppenheimer, 2011

2. VOLCANIC ACTIVITY AND ANCIENT HUMAN ANCESTORS

The earliest human ancestors with sufficient samples to be relatively well understood, and that are fairly well dated are the **australopithecines** of Africa, from about 4 to 2.5 million years ago. Many australopithecines inhabited the volcanically active landscape of the rift valley of eastern Africa and must have suffered often from eruptive effects, but benefited in the long run. More importantly, the topographically and environmentally complex rift valley would have provided varied habitats for our small ancestors, and they exploited them successfully. At Laetoli in Tanzania, actual footprints of australopithecines were discovered, preserved in hardened volcanic ash, dated to 3.6 million years ago, which proved they walked upright. Following ecological recovery of soils, plants, and animals, the surviving australopithecines must have benefited from the environment enriched in food sources. However, dating of volcanic events and skeletal remains is insufficient to document any individual cases that are this ancient. And volcanism was one of many factors resulting in marked environment and climate changes, encouraging flexible adaptability and resilience by our ancestors. Only during the past few thousand years are researchers able to investigate causal relationships between volcanic eruptions and people, and even then the research is challenging.

The earliest member of our genus is *Homo habilis*, from about 2.5 to 1.8 million years ago (Feder, 2007), who made stone tools and presumably had similar beneficial and detrimental experiences with volcanoes in the same area. Our earliest ancestors to move out of Africa were *Homo erectus*, who colonized Asia with more developed culture shortly after their emergence at about 1.8 million years ago, and they had expanded into Europe by about 800,000 years ago. The earliest *Homo sapiens* developed by about 400,000 years ago, with more sophisticated culture and more widespread adaptations and also in volcanically active areas taking advantage of the beneficial aspects and suffering the consequences of explosive eruptions. Vast numbers of eruptions surely affected people by altering their environments and food sources during the past 4 million years, but the details of when and how are elusive, and few eruptions have been studied and dated.

An example of an eruption that must have caused many deaths near it, and difficulties at greater distances, is that of the O'a Caldera in Ethiopia, about 240,000 years ago. The archaic *H. sapiens* individuals would have been almost modern biologically, and had a relatively sophisticated culture and adaptation.

Anatomically modern *H. sapiens* with a Middle Stone Age culture begin migrating out of sub-Saharan Africa sometime around 90,000 years ago through the Near East, India, and into Asia. Arrival in Europe was later, about 50,000 years ago. Between these times a gigantic eruption of Toba volcano in Sumatra (Indonesia) occurred, dated to 73,000 years ago. The eruption was so great, along with two much earlier eruptions, that it is called a "super-eruption." It inevitably caused great stress among people in South and Southeast Asia, and may have contributed to the isolation of human subgroups, and the biological and cultural diversification that occurred at about this time.

Grattan and Torrence documented many volcanic eruptions in Papua New Guinea over the past 40,000 years. They note that the focus of virtually all literature on volcanic human interaction is on the immediacy of the disaster. In refreshing contrast, they discovered significant human resilience to the stresses, and creativity in dealing with the eruptions and their aftermaths. They suggest that volcanic events can be conceived as natural selection phenomena that can stimulate innovative coping strategies by people.

Although, we normally do not associate Germany with volcanic activity, it was the location of a sizable eruption about 13,000 years ago. The blast and pyroclastic flows devastated people, flora, and fauna near the vent, but most of the regional effects were from airborne volcanic ash and gasses of sulfur and chlorine. The widespread hunting and gathering peoples largely headed to refuge areas in southern Europe, leaving an isolated small group called Bromme in Denmark. That isolated group suffered a deculturation in their material culture, a simplification of their stone tools, perhaps resulting from the volcanically caused fragmented habitat and lack of the former widespread cultural contacts.

3. VOLCANIC ERUPTIONS AND COMPLEX SOCIETIES

Certainly the most famous eruption that affected people is that of Vesuvius that buried the Roman city of Pompeii in AD 79. It is so well known that it need not be mentioned in detail here. However, an earlier eruption of the adjacent Campi Flegrei almost 40,000 years ago was vastly larger than Vesuvius, dropping **tephra** all over southern Europe, Turkey, and well into Russia. Many scholars believe that the stress caused by the eruption resulted in accelerated change in culture (from Middle to Upper Paleolithic) and in human biology, even accelerating the demise of Neanderthals.

The eruption of Thera (Santorini) and Minoan civilization in the eastern Mediterranean is the earliest example of a complex society affected by a large volcanic eruption, and is a cautionary tale of the importance of precision in dating natural and cultural phenomena in trying to understand causality. Minoan civilization was based on the island of Crete and thrived because of sophisticated ships plying the sea in an elaborate trading network. Minoans built elaborate palaces for their rulers and multistory homes for other people, they developed a graceful art style and intensive agriculture to feed the masses. Thera erupted about 1600 BC and devastated a city on the island. The tephra that landed on Crete would have caused some difficulties, but the major destructive element was a tsunami that struck the island and destroyed the shipping fleet. Minoan civilization never fully recovered from the eruption, and their land-based competitor, Mycenaean civilization, took advantage of their weakness and expanded geographically, politically, and economically. Many scholars believe in this causal chain, while others point out that dating the eruption and the cultural changes are not even accurate to a century. More research is necessary for this possible causal sequence to be demonstrated.

Mesoamerican cultures (ancient Mexico into Central America) experienced numerous volcanic eruptions. For example, Popocatepetl volcano erupted about AD 50 and buried an extensive area to its east, magnificently preserving farms and households for archaeologists to excavate but also causing the migration of many tens of thousands of refugees eastward into the nascent city of Teotihuacan, carrying with them a sacred architectural style that became standard for religious buildings there.

The interaction of volcanoes with the fortunes of Teotihuacan (Figure 76.1), in highland central Mexico, illustrates the beneficial and detrimental aspects of the relationship. Teotihuacan was the largest and most influential city of Mesoamerica during the Classic period (AD 300–900). It began as a small town about 2000 years ago until Xitle volcano erupted in the southern basin of Mexico and devastated its competitor, Cuicuilco. Its population grew steadily and then experienced a huge influx of refugees from the above-mentioned eruption of Popocatepetl. The city's rulers, perhaps a city council, successfully put the masses of people to work, building the most immense pyramids known to that date in Mesoamerica. The city thrived economically, politically, and religiously for some five centuries and

impacted societies all the way from Chaco in the north in New Mexico to the Maya cities of Tikal in Guatemala and Copan in Honduras in the far south. The demise of Teotihuacan was dramatic, with deliberate burning of temples and many palaces, and destruction of religious statues, while lower-class residences were abandoned without deliberate destruction. This appears to be a failure of belief in the power structure of the city that could only have been caused by a powerful event. The revolution occurred about AD 550, which suggests it may have been caused by the effects of a cataclysmic volcanic eruption in El Salvador. The decade or more of cold and drought from that event would have caused crop failures and starvation, and therefore the loss of confidence in the religious and political authority of Teotihuacan. So, two volcanic eruptions played major roles in the emergence of the great city, and the climatic effects of a huge eruption played a key role in its demise.

The sixth-century AD eruption of Ilopango volcano (Figure 76.2) devastated the southeastern Maya of El Salvador by its direct effects. It was the greatest eruption in Central America in the last 84,000 years, and the combination of vast amounts of fine tephra and abundant sulfur aerosols apparently was the cause of the worldwide cold and drought period that began in AD 536. It caused some 75% of the people of the Chinese Wei Dynasty to die from crop failures, starvation, and disease, and the emperor lost the mandate of heaven. That resulted in major political, economic, and religious changes, much like what happened at the same time in Teotihuacan. Historians in Europe, North Africa, and the Middle East recorded the same phenomena at exactly the same time. The tephra and sulfur could have caused the drastic climatic change only for a few years; it is likely that the polar snow and ice fields grew in those years to sufficient extents to reflect solar radiation and extend the problem for almost two decades. The phenomenon is also recorded in tree rings and a great sulfur

FIGURE 76.1 Teotihuacan, the ancient city in the basin of Mexico. All the pyramids around the plaza and the long street had wood-and-thatch temples that were deliberately burned in the mid-sixth century.

FIGURE 76.2 Lake Ilopango, in central El Salvador. Under this west end of the lake is the vent, the source of the immense amount of tephra and sulfur gas from the mid-sixth century eruption.

spike in Greenland ice cores, and in a smaller spike in Antarctic cores, which indicates that the cause was a volcano slightly north of the equator. The most likely candidate is Ilopango volcano in El Salvador.

The greatest immediate impacts of the eruption were within a 100-km radius of the Ilopango caldera, where villages and cities of the southeast Maya were buried by a few to many meters of tephra. Most people were killed outright, along with all flora and fauna, as the lush tropical landscape was turned into a white desert. Ecological and then human recovery required many decades.

The most vulnerable societies to the cold and drought caused by the eruption were the urban agricultural groups with dense populations that were living at high elevations or high latitudes close to the frost line in various parts of the world. The Wei Dynasty and Teotihuacan in Mexico are prime examples of civilizations with insufficient resilience to withstand many sustained years of stress. Although, it may seem counterintuitive, civilizations closer to the eruption, but beyond the zone of deep tephra burial, fared better because they were well below the frost line. Most Maya city states in Northern Guatemala, Belize, and the Yucatan Peninsula of Mexico were able to withstand the climate change.

Farther south in tropical Costa Rica, native societies also survived the stresses of the AD 536 phenomenon, but what is notable is that the direction of outside cultural influences reversed at that time. External contacts prior to the sixth century AD by native Costa Ricans were to the north with Mesoamerica, as seen in ceramics and in the abundant jade that was traded in from the source in the middle of the Maya area. The Ilopango eruption effectively quashed the jade access and trade. The shift southward was dramatic, as external influences changed completely toward Panama and Colombia and elites changed from using jades as status indicators to goldwork.

Costa Rican societies have also been impacted by local eruptions. Arenal volcano (Figure 76.3) erupted some 10

FIGURE 76.3   Arenal volcano in northwestern Costa Rica, with Lake Arenal in the foreground.

times in the past 4000 years with sufficient magnitude to leave tephra layers interspersed with soils and evidence of human habitation. What is remarkable about native peoples in the Arenal area is their extraordinary resilience to these eruptions, in marked contrast to eruption's effects on more complex societies. The Arenal natives lived in small egalitarian villages (no permanent leaders, no inherited differences among families in power or wealth), with some agriculture of maize and beans but most of the food gathered from the rich resources of the tropical rain forest. Families were almost entirely self-sufficient, and when danger was perceived they could migrate to a refuge area without waiting for an authority to tell them what to do. The regional populations were low, so refugees could move into uncontested terrain and live off of forest resources until they could reoccupy their original villages. They did go back to their same villages after environmental recovery and reestablished the exact same pilgrimage routes to their distant graveyards to get back into contact with the spirits of their deceased ancestors.

The Arenal families living in the egalitarian villages contrast sharply with more complex societies in Mesoamerica. In Mesoamerican civilizations, as with their brethren in other culture areas, the populations were dense and dependent on intensive agriculture and dietary staples of domesticated species. That agriculture often involved irrigation canals, dams, or other fixed facilities. All civilizations have monumental constructions for elite residences and often massive religious and economic buildings. Political and religious authorities largely reside in the elite, so major decisions in stressful situations are top–down. The areas surrounding civilizations routinely are also densely populated, and often hostile, as groups are competitive for territory. Refuge areas were not readily available in hostile territories. The contrast with Arenal is striking, as nonhostile areas were available as refuges. When disaster struck the Panamanian chiefdom and the Mesoamerican societies, commoners awaited decisions as to plans of action, and if elites were also incapacitated, inaction ensued. Decision making under sudden duress in hierarchical societies is slower than in egalitarian groups. The policy implications are clear for present-day planners and authorities: prior to emergencies, do planning at the local level and delegate authority so decisions can be made efficiently.

Arenal villages also contrast with the modestly complex chiefdoms of nearby Panama. The Barriles and adjoining chiefdoms expressed monumentality in terms of life-size stone sculptures, and they were at least minimally competitive in that they captured an occasional opponent and decapitated him with a specially manufactured axe. Their population densities were fairly significant, concentrated on the terraces of rivers. Baru volcano erupted about 1000 years ago. The eruption was smaller than those of

Arenal volcano, but it caused greater societal impacts. The reasons include the greater population density in Panama along with societal hostility denying easy refuge and the heavy reliance on agriculture. The Baru eruption created stress greater than the Barriles chiefdom could handle, and the people had to migrate over the continental divide and down onto the Caribbean slope. The new environment was so different that they had to change their adaptation and lifeways fundamentally, and they never returned to their original homeland.

4. RECENT LARGE-SCALE VOLCANIC IMPACTS

Explosive eruptions, such as those considered above, have affected human societies in often dramatic ways. But so can eruptions that are primarily lava flows. The 1783—1784 eruption of Laki in Iceland is a good example. Huge amounts of sulfur-rich lava, perhaps 6—14 cubic kilometers, flowed from complex fissures for some 8 months, devastating settlements in its path and creating a volcanic fog. Approximately one-fourth of the people of Iceland perished, and when the fog reached Europe, it continued to wreak havoc. The fog dramatically lowered temperatures in the winter and summer, causing famine, exceptionally high death rates, and lowered birth rates. Although Benjamin Franklin observed the phenomenon while he was living in Paris and understood the relationship of the fog with the cooling, most authorities did not understand that and did not react in ways that could have buffered the effects. The eruption unfortunately has attracted some extreme environmental determinists who claim that the eruption's effects caused public unrest in France, leading to the French Revolution of 1789. Even a cursory examination of the data show the eruption, its effects, and the French Revolution were quite unconnected.

The most recent big explosive eruption to cause worldwide deleterious effects occurred in Indonesia. Tambora was a towering volcano that erupted with cataclysmic force in April, 1815. The ashfalls and pyroclastic flows instantly killed all people along with fauna and flora for huge distances around the volcano, and the sounds were heard by people even a thousand miles away. Some 100,000 people were killed in Indonesia alone. The volcanic ash and especially the sulfur in the upper atmosphere became widely distributed in 1816 and caused a "year without a summer" with devastating distal effects on agriculturally based societies. The summer months in Europe, the US, and Canada never experienced summer temperatures, resulting in crop failures, malnutrition, and famine. The weakened human and animal populations were susceptible to disease, and outbreaks of plague, dysentery, and typhus, often followed by death, were devastating to people in Europe and into the Near East.

5. VOLCANOES AND PEOPLE IN THE FUTURE

Of course volcanoes will continue to erupt in the future and affect people. Applying lessons learned from past eruptions can assist in disaster preparations and land-use zoning, and there are policy suggestions in a book *Surviving Sudden Environmental Change*. The greatest dangers are to the world's biggest cities that are close to volcanoes, such as Seattle, Mexico City, Tokyo, Manila, and Jakarta, and those cities could suffer greatly even with a relatively minor eruption, at least compared to those considered in this chapter. A very large eruption, on the scale of Ilopango, Tambora, Thera, O'a, or Campi Flegri, would cause almost unimaginable disruption to worldwide societies, particularly because of the lack of warning, dense populations in the billions that would suffer, and the cost of maintaining mitigating measures such as immense food and water storage is beyond what any nation is willing to invest.

FURTHER READING

Birkmann, J. (Ed.), 2006. Measuring Vulnerability to Natural Hazards: Towards Disaster Resilient Societies. United Nations University Press, New York.

Cooper, J., Sheets, P. (Eds.), 2012. Surviving Sudden Environmental Change: Answers from Archaeology. University Press of Colorado, Boulder.

Dull, R., Southon, J., Sheets, P., 2001. Volcanism, ecology and culture: a reassessment of the volcan Ilopango TBJ eruption in the southern Maya realm. Lat. Am. Antiq. 12, 25—44.

Dull, R., Southon, J., Kutterolf, S., Freundt, A., Wahl, D., Sheets, P., 2010. Did the Ilopango TBJ eruption cause the AD 536 event?. In: American Geophysical Union Fall Meeting, San Francisco. Abstract V13C-2370.

Evans, S., Webster, D. (Eds.), 2001. Archaeology of Ancient Mexico and Central America: An Encyclopedia. Garland Publishing, NY.

Feder, K., 2007. The Past in Perspective: An Introduction to Human Prehistory. McGraw Hill, Boston.

Friedrich, W., 2009. Santorini: Volcano, Natural History, and Mythology. Aarhus: Aarhus University Press, Copenhagen.

Grattan, J., Torrence, R., 2007. Beyond gloom and doom: the long-term consequences of volcanic disasters. In: Grattan, J., Torrence, R. (Eds.), Living under the Shadow: Cultural Impacts of Volcanic Eruptions. Left Coast Press, Walnut Creek, California, pp. 1—18.

Gunn, J. (Ed.), 2000. The Years Without Summer: Tracing AD 536 and Its Aftermath. BAR International Series, vol. 872. Archaeopress, Oxford, England.

Harrington, C. (Ed.), 1992. The Year Without a Summer? World Climate in 1816. Canadian Museum of Nature, Ottawa.

King, G., Bailey, G., 2006. Tectonics and human evolution. Antiquity 80, 265—286.

Linares, O., Ranere, A. (Eds.), 1980. Adaptive Radiations in Prehistoric Panama. Peabody Museum of Archaeology and Ethnology, Harvard University, Cambridge, MA.

Linares, O., Sheets, P., Rosenthal, J., 1975. Prehistoric agriculture in tropical highlands. Science 187 (4172), 137—144.

Oppenheimer, C., 2011. Eruptions that Shook the World. Cambridge University Press, Cambridge.

Raichlen, D., Gordon, A., Harcourt-Smith, W., Foster, A., Haas, W., 2010. Laetoli footprints preserve earliest direct evidence of human—like bipedal biomechanics. PLoS ONE 5 (3), e9769. http://dx.doi.org/10.1371/journal.pone.0009769.

Rose, W., Chesner, C., 1987. Dispersal of ash in the great Toba eruption, 75 kyr. Geology 15, 913—917.

Schmincke, H., Park, C., Harms, E., 2009. Evolution and environmental impacts of the eruption of Laacher see volcano (Germany) 12,900 a BP. Quat. Int. 61, 61—72.

Snarskis, M., 1981. The archaeology of Costa Rica. In: Abel-Vidor, S. (Ed.), Between Continents/Between Seas: Precolumbian Art of Costa Rica. Abrams, NY, pp. 15—84.

Sheets, P., McKee, B. (Eds.), 1994. Archaeology, Volcanism, and Remote Sensing in the Arenal Region, Costa Rica. University of Texas Press, Austin.

Warburton, D., 2009. Time's Up! Dating the Minoan Eruption of Santorini. Aahus: Aarhus University Press, Copenhagen.

Chapter 77

Volcanoes in Art

Haraldur Sigurdsson

Graduate School of Oceanography, University of Rhode Island, Narragansett, RI, USA

Chapter Outline

GLOSSARY

classicism The type of art that was characteristic of Greek and early Roman culture. It was revived in Florence in the fifteenth century, when it was often regarded as the highest point that art could reach.

edo painting The style of painting typical in Japan up to the mid-nineteenth century, before the arrival of a strong western influence in art.

genre A type of painting showing realistically scenes from everyday life; a category or style of art.

iconography The study of the symbolic meaning of objects, persons, or events depicted in art.

impressionism A type of realism in art, with the aim of showing the immediate impression of the scene on the artist, mainly in terms of the representation of light.

neoclassicism An art form that was the result of the revival of interest in classical forms and values in the eighteenth century, largely due to the discoveries of ancient Greek and Roman classical art works during the eighteenth century excavations in Pompeii and Herculaneum.

picturesque In the late eighteenth century, European people began to travel specifically to enjoy the scenery, seeking out spectacular natural scenery, ruined castles, and unusual sights. Picturesque tourism coincided with the rise of landscape painting and a romantic celebration of the landscape in art. It is an art form that is first and foremost an expression of our worship of Nature.

romanticism An attitude in art and literature that stems from the nineteenth century philosopher Rousseau, where moral passion counted for more than intellectual analysis, and where principal values were sincerity, authenticity, and moral exaltation.

sublime Nature overwhelming the onlooker and filling him with awe for its unpredictable force. The epitome of the sublime in art is the terrifying spectacle of a volcanic eruption.

Ukiyoe Traditional Japanese woodblock color prints, generally portraying the ever-changing world of the city or village. This form of print popularized the landscape in Japanese art in the eighteenth century.

1. INTRODUCTION

Volcanoes create the most beautiful landscapes in the world. No one who has viewed the symmetry of Fuji volcano in Japan will forget its aesthetic profile, defined by a perfect catenary curve. Volcanic mountains are associated with a great geologic process which has resulted in the construction of a fascinating variety of landforms, but the spectacle of volcanoes goes far beyond their pleasing landscape form. Their eruptions are exciting and thrilling because they are dangerous and exceed all natural catastrophes known to man in their explosive release of energy; they are without doubt the most spectacular natural phenomenon witnessed by humans. In terms of the cultural context of the nineteenth century, volcanic eruptions are **sublime**, invoking an expression of terror, awe, fascination, and savage beauty. They are also among the most stunning of visual phenomena, as they display an overabundance of fire, light, energy, and terror on the grandest scale. It is therefore not surprising that volcanic eruptions have been a subject for landscape

The Encyclopedia of Volcanoes. http://dx.doi.org/10.1016/B978-0-12-385938-9.00077-8

artists in the Western World from the sixteenth century to our time. Similarly, because of their purity of form and mythological significance, volcanoes have also featured prominently in Oriental art, particularly in the treatment of Japanese landscape pioneered by Hokusai in the early nineteenth century. This chapter discusses the artists' vision of volcanic eruptions in painting from the earliest examples to our time. The topic is vast in scope, and only a few examples of the principal and better known works can be given here. Table 77.1 lists over 90 artists who have painted volcanic eruptions or related phenomena. The portrayal of volcanic eruptions in art through the ages is best displayed in the unique collection of Eldfjallasafn, the volcano museum in Stykkisholmur, Iceland. Many of the images shown and discussed in this chapter can be viewed there.

2. THE BEGINNINGS OF LANDSCAPE PAINTING

Artists began to paint volcanoes and other landscape forms in the Renaissance, the period of artistic creativity which began in the fifteenth century and brought us geniuses such as Copernicus and Leonardo da Vinci. In Renaissance culture, art and science were closely associated and were in fact almost interchangeable types of mental activity, as both were stimulated by a new vision of the Earth, humans, and the world as a whole. Through its representation of objects in their true proportions, for example, Renaissance art opened the door for a new methodology in the empirical sciences, whereby scholars could accurately demonstrate their findings in drawing or illustration. The impact of art on anatomy and on the biological sciences is well-known.

TABLE 77.1 Artists and the Volcanoes They Painted

Artist	Volcano
Cesare Ripa (c.1550), Italian	Etna
Pieter Bruegel (1528—1569), Dutch	Vesuvius 1558, 1561, Monte Nuovo
Annibale Carracci (c.1594), Italian	Etna
Filippo D'Angeli (1587—1640), Italian	Vesuvius (1631)
Didier Barra (1590—c.1650), French	Vesuvius
Guercino (Giovanni Francesco Barbieri (1591—1666))	Vesuvius
Claude Lorrain (1600—1682), French	Vesuvius (1631)
Micco Spadaro (Domenico Gargiulo (1609—1675)), Italian	Vesuvius (1631)
Scipione Compagno (c.1670), Italian	Vesuvius (1631)
Salvator Rosa (1615—1673), Italian	Etna (Empedocles)
Thomas Wijck (1616—1677), Dutch	Vesuvius
Johannes Lingelbach (1622—1674), German	Vesuvius (1669)
Jan Wyck (seventeenth century), Dutch	Vesuvius (1689)
Antonio Joli (1700—1770), Italian	Vesuvius
Claude-Joseph Vernet (1714—1789), French	Vesuvius (1748)
Carlo Bonavia (eighteenth century), Italian	Vesuvius (1757)
Katsukawa Shunsho (1726—1793), Japanese	Fuji
Pierre-Jacques Volaire (1729—1802), French	Solfatara, Vesuvius (1777)
Joseph Wright of Derby (1734—1779), English	Vesuvius (1774)
Tomasso Ruiz (eighteenth century), Italian	Vesuvius (1737)
Jacob Philipp Hackert (1737—1807), German	Vesuvius (1774, 1779)
Michael Wutky (1739—1822), Austrian	Vesuvius
Pietro Antoniani (1740—1781), Italian	Vesuvius (1767)
Pietro Fabris (1784), Italian	Vesuvius (1776), Solfatara, volcanic rocks
Jacob More (1740—1793), English	Vesuvius AD 79
Angelica Kauffmann (1741—1807), Swiss	Vesuvius AD 79

TABLE 77.1 Artists and the Volcanoes They Painted—cont'd

Artist	Volcano
Jean-Louis Desprez (1743—1804), French	Vesuvius (1779)
Alessandro D'Anna (1749—1810), Italian	Vesuvius (1794), Etna (1766)
Xavier Della Gatta (1827), Italian	Vesuvius (1794)
Jean-Baptiste Genillion (1750—1829), French	Vesuvius
Giovan Battisa Lusieri (1755—1821), Italian	Vesuvius (1787, 1794)
Saverio Della Gatta (1755—1827), Italian	Vesuvius
Odoardo Fischetti (late eighteenth century), Italian	Vesuvius (1805)
Camillo de Vito (c.1760—c.1830), Italian	Vesuvius (1794, 1820, 1822, 1828)
John Webber (1751—1793), English	Krakatau
Pierre-Henri de Valenciennes (1750—1819), French	Vesuvius (AD 79)
Katsushika Hokusai (1760—1849), Japanese	Fuji
Joseph Franque (1774—1833), French	Vesuvius (AD 79)
Joseph Mallord William Turner (1775—1851), English	Vesuvius, Soufriere of St Vincent
John Vanderlyn (1775—1852), American	Vesuvius (AD 79)
Martin Johnson Heade (nineteenth century), American	Omotepe, Nicaragua, (1867)
William Hodges, American	Mauna Loa
Jacob Munch (1815), Norwegian	Vesuvius
Francois-Joseph Heim (1787—1865), French	Vesuvius allegorical
Johann Christian Dahl (1788—1857), Norwegian	Vesuvius (1820—1823)
John Martin (1789—1854), English	Vesuvius (AD 79)
Keisai Eisen (1790—1848), Japanese	Fuji
Sebastian Pether (1790—1844), English	Destruction of Pompeii AD 79
Jean-Baptiste Louis Gros (1793—1870), French	Popocatepetl
Utagawa Hiroshige (1797—1858), Japanese	Fuji, Sakurajima
Carl Rottmann (1797—1850), German	Etna, Santorini
Karl P. Briullov (1799—1852), Russian	Vesuvius (Pompeii AD 79)
Titian Ramsay Peale (1799—1885), American	Kilauea
James Baker Pyne (1800—1870), English	Vesuvius (1856)
Thomas Cole (1801—1848), American	Etna, Vesuvius
Johann Moritz Rugendas (1802—1858), German	Colima
Robert S. Duncanson (1802—1872), Afro-American	Vesuvius and Pompeii
Paul Kane (1810—1871), Canadian	Mount St Helens
Raden Saleh (1816—1880), Indonesian	Merapi (1865), Vesuvius (1839)
James Hamilton (1819—1878) American	Vesuvius (Pompeii AD 79)
Frederic Edwin Church (1826—1900), American	Cotopaxi, Sangay, Chimborazo
Louis Remy Mignot (1831—1870), American	Cotopaxi
Archibald D. Willis, New Zealand	Tarawera (1886)
Albert Bierstadt (1830—1902), American	Vesuvius (1868)
Degas, Edgar (1834—1917), French	Vesuvius (1890)

(Continued)

TABLE 77.1 Artists and the Volcanoes They Painted—cont'd

Artist	Volcano
Charles Furneaux (1835–1913), American	Kilauea, Mauna Loa (1881)
Jose Maria Velasco (1840–1912), Mexican	Popocatepetl, Iztaccihuatl, Orizaba
Auguste Renoir (1841–1919), French	Vesuvius
Masayoshi Tokonami (1842–1897), Japanese	Sakurajima (1895)
D. Antonio Munoz Degrain (1843–1924), Spanish	Vesuvius
Jules Tavernier (1844–1889), French	Kilauea (1887)
D. Howard Hitchcock (1863–1943), American	Kilauea, Mauna Loa (1896)
Charles H. Woodbury (1864–1940), American	Montagne Pelee
Seiki Kuroda (1866–1924), Japanese	Sakurajima (1914)
Dr Atl (Geraldo Murillo; 1875–1964), Mexican	Paricutin, Popocateptl
Asgrimur Jonsson (1876–1958), Icelandic	Laki, Grimsvotn
Hiroshi Yoshida (1876–1950), Japanese	Fuji
Kawase Hasui (1883–1957), Japanese	Fuji
Johannes Kjarval (1885–1972), Icelandic	Volcanic landscapes
Diego Rivera (1886–1957), Mexican	Paricutin (1943)
Saturnino Herran (1887–1918), Mexican	Legend of volcanoes
Ruyzaburo Umehara (1888–1986), Japanese	Asama, Sakurajima (1935)
David Alfaro Siqueros (1896–1974), Mexican	Volcanic eruption
Gudmundur Einarsson (1895–1963), Icelandic	Grimsvotn (1934)
Ichien Somiya (1893–1994), Japanese	Sakurajima (1957)
Alfredo Zalce (1908–2003), Mexican	Paricutin (1956)
Misao Yokoyama (1920–1973), Japanese	Sakurajima (1956)
Walter Womacka (1925–2010), German	Vesuvius
Matazo Kayama (1927–2004), Japanese	Sakurajima (1961)
Andy Warhol (1928–1987), American	Vesuvius
Emilia Cersosimo (1944–present), Costa Rican	Arenal, Poas, Irazu
David Haste (contemp.), English	Etna
Robert Hawkins (contemp.), American	Surreal volcanoes
Gudni Hermansen (1928–1989), Icelandic	Surtsey (1963), Heimaey (1973)
Bill Martin (contemp.), American	Imaginary volcanic landscapes
Sandra Del Valle (contemp.), Mexican	Myths of Popocatepetl and Iztaccihuatl
Roger Brown (1941–1997), American	Mount St Helens (1980)
Bill Sullivan (contemp.), American	Cotopaxi
Wahru (contemp.), Indonesian	Galunggung (1982)
Vaino Kola (contemp.), Finnish	Icelandic volcano landscapes
Beth Neville (contemp.), American	Volcanoes as metaphors of change
Dennis Bayuzick (contemp.), American	Volcano symbolism and surrealism
Mario Schifano (contemp.), Italian	Vesuvius
Tony Foster (1946–present), English	Mount St Helens, Soufriere Hills

Similarly, in geology, the storing of information and observations through the new dimension in art led to advances in the study of fossils and minerals by simply enabling communication through graphic representation of natural objects.

The study of landscape advanced much more slowly. There are glimpses of natural scenery in some Medieval manuscripts, and, in the Renaissance, admirable landscapes were portrayed by some artists, such as Leonardo da Vinci's graphic panoramic drawing of the Arno Valley in Italy, but on the whole the study of landscape did not captivate artists and scholars until late in the seventeenth century. By that time, the surveying of the land and the development of geography had opened the realm of landscape to the human mind and made the study of the Earth's wrinkled surface a worthy and manageable topic for art and science. During the age of **Romanticism**, which began in the late eighteenth century, admiration for the wilder natural phenomena, such as mountains, storms, and waterfalls, is abundantly expressed in landscape art. Another phenomenon common in Romantic landscapes are volcanic eruptions, which also inspired exhilarating terror and most dramatically expressed the sublimity of nature.

After their common origins in the Renaissance, science and art have steadily and inexorably diverged from one another on their separate cultural paths. This regrettable trend has developed to the point of schism in our modern society, where these two cultural pursuits are in practice often mutually exclusive, in contrast to their unity in the Renaissance age. The gradual parting of the ways was of course an inevitable consequence of the cultural specializations which our structured society imposes on its members, and in our age, the division has reached a level where most artists and scientists are largely illiterate in each other's fields. It is, for example, ironic and symptomatic that Roger Brown's painting of the Mount St Helens eruption in 1980 (Figure 77.13) is styled after an industrial oil refinery fire, displaying surprising ignorance of a volcanic process, and harking back to the Medieval concept that volcanic eruptions were the consequence of the combustion of fossil fuels in the Earth. In contrast, many of the seventeenth century paintings of Vesuvius display details of eruption, pyroclastic surges, and lava flows which indicate a high degree of understanding of the volcanological process, based on direct observation by an eyewitness.

The earliest visual representation of volcanic eruption dates to one of the oldest village cultures on Earth, which thrived on the Anatolian plateau in southeastern Turkey more than 8000 years ago. Excavation of this site, known as Çatal Hüyük, revealed a Stone Age town, with a long and rich cultural and artistic history, that flourished between 6500 and 5700 BC. Numerous shrines found at the site attest to the development of religion and display a wealth of decoration, reliefs, and wall paintings. In their work the early artisans of Çatal Hüyük used rich colors and had a full range of pigments at their disposal, derived from a variety of minerals. They also manufactured fine artifacts of highly polished obsidian. A wall painting of an erupting volcano was found in one of the older shrines at a level carbon-14 dated to 6200 BC. In the foreground is a group of domino-like rectangular houses, which represents the town, and in the background rises a twin-peaked volcano, ochre in color, and spotted with black dots, with an eruption spewing from the higher cone. The eruption is portrayed as curved lines or spray above the crater, and the dots on the cone and outside its outline may represent falling volcanic "bombs" or large pieces of semiliquid lava. The lines emerging from the base of the volcano may represent lava flows. Today the plain surrounding Çatal Hüyük is indeed dominated by the twin volcanic peaks of Hasan Dag and Karaca Dag, and it is very likely that this ancient painting represents an eruption of Hasan Dag. It has been proposed that the painting had religious meaning, but several factors may have motivated the artist: perhaps it was a response to the threat from the erupting volcano, a work to appease the powers of the underworld? Was it inspired by gratitude to the volcano, the source of valued obsidian? A large number of obsidian spearheads and other implements have been found in the excavations, and it has been suggested that the town derived much of its wealth from the obsidian trade. Or was the painting simply inspired by fear in witnessing Earth's greatest display of power?

Most of the known examples of early Roman landscapes have been unearthed at Pompeii and Herculaneum, the cities buried by eruptions of Mount Vesuvius. These landscapes show a variety of views, natural forms, cultivated lands, wooded groves, streams, and harbors populated by shepherds, farmers, nymphs, and other mythological figures. Several Roman wall paintings of Vesuvius have survived in houses excavated from the AD 79 volcanic deposits in Herculaneum and Pompeii. A painting from the lararium of the House of the Centenary in Pompeii shows the steep volcanic cone as it appeared before the AD 79 eruption (Figure 77.1). The mountain is covered in vineyards and other vegetation to the top. In the foreground stands Dionysus (Bacchus) clad in a mantle of grapes. Thus the few images which can be identified as Vesuvius show it as a benign mountain, its slopes covered with an abundance of the vines which thrive in the rich volcanic soil. Of course, these were painted before AD 79, but even after this time there is no evidence of a painter attempting to portray the awesome spectacle of the eruption. A violent eruption or any other catastrophe would be quite unsuitable as a wall decoration! Pliny the Younger's vivid and stirring description of Vesuvius erupting would have to wait some 1700 years before its expressive, visual qualities could be appreciated.

FIGURE 77.1   This is one of the oldest surviving paintings of a volcano, and it was preserved for posterity due to the volcanic burial of a city. Before the eruption of AD 79, Vesuvius was covered with forests, vineyards, and other vegetation up to the summit crater, as shown in this Roman wall painting, found in the House of the Centenary in Pompeii. The figure draped in bunches of grapes is the god Dionysus (Bacchus), and the snake represents the household gods. Archaeological Museum of Naples.

During the Middle Ages, the only images associated with volcanoes are vivid, but symbolic pictures of flaming hell, devils, and burning rocks found in illuminated manuscripts and frescoes. After 1400, with the Renaissance revival of classical culture, a new world view emerged as people began to examine the natural world around them and to consider mankind's place within it, but the first portrayals of a real volcano were not seen until around 1550. In Cesare Ripa's *Iconologia,* a handbook of symbolic personifications used by painters, "Sicily" is described as represented by a seated woman holding attributes that identify her as this particular island. Behind her is a representation of Etna volcano shown simply as a truncated, irregular pyramid topped by flames. This symbolic image does not attempt to show the actual appearance of Etna in particular or any volcanic eruption as observed in nature. A 1541 woodcut of Etna erupting in the *Cosmographia Universa* of Sebastian Münster indeed shows the volcano as an icon (Figure 77.2). The volcano, with the appearance of

FIGURE 77.2   An eruption of Etna volcano, shown in a woodcut from the *Cosmographia Universa* of Sebastian Münster, from 1541 (author's collection).

a block of wood, emits flames, with rounded boulders flying through the air, threatening the nearby town of Catania.

One of the earliest attempts at showing a more realistic view of Etna erupting was made in 1594 by the two Bolognese painters Annibale and Agostino Carracci, who decorated the Farnese Palace in Rome with fresco scenes representing moral actions taken from classical literature. In one scene, we see an erupting Etna in the background. In the foreground two young men flee from the flames and smoke, carrying their mother and father on their backs to safety. This story of the legendary Catanian brothers was intended as a moral allegory, for unlike the Catanians who attempted to save material possessions and perished, the brothers showed filial devotion by rescuing their aged parents. The primary subject of the painting is the moral action of the brothers, and the volcano is relegated to the background providing a reason for this heroic deed but stimulating little interest in itself. This portrayal of an eruption is unusual at this time, for neither Etna nor Vesuvius appear in a landscape for another 40 or 50 years.

Volcanic eruptions have been witnessed and described by mankind for thousands of years in many parts of the world, but it is mainly within the past 300 years, since the Renaissance, that the subject has appeared in the visual arts. In fact, most paintings of volcanic activity have been produced only since the latter half of the eighteenth century when the birth of scientific investigation and systematic observation of the workings of nature stimulated an interest in unusual and marvelous phenomena, which most people would never actually see in their own lives. Before artists could represent volcanic eruptions and the human reaction to them, they had to create a visual vocabulary for such depictions. It was therefore necessary to invent an **iconography** of volcanoes, which until the eighteenth century had rarely been seen in painting. Volcanoes were not part of the standard schema for landscape painting, which had developed during the Renaissance. Most European artists would have been totally unfamiliar with the appearance of an eruption. Furthermore, volcanoes had no place in the range of conventional subjects of the arts: Biblical scenes, mythological themes and **genre**, that is, views of common folk at work or play.

The European exploration of distant lands, such as the East Indies in the sixteenth century, revealed many new volcanoes, whose topography and eruptions were at times documented in art. Such is the case with the early seventeenth century eruption of Gunung Api in the Banda Islands in Indonesia (Figure 77.3).

3. VESUVIUS 1631 ERUPTION

In 1631 Vesuvius erupted with a violence it had not demonstrated since ancient times. This catastrophe killed over 4000 people and devastated the population of Naples, as has been recorded in several vivid eyewitness accounts. The event is described not only as frightening but also spectacular to see. By this time, Italian artists had developed a full range of techniques for portraying landscape images with convincing naturalism, and there are several pictorial records of this eruption. As a result of this

eruption, a new interest and direction in the depiction of volcanic eruptions arise in the seventeenth century. Rather than purely documentary or religious, the volcano itself becomes the focus as a powerful, awe-inspiring phenomenon. A sketch by Filippo D'Angeli shows a view of the 1631 eruption from the harbor, with people fleeing, and a lightly rendered view of Vesuvius with line suggesting flames and lava (Figure 77.4). The main focus of the drawing, however, is Saint Gennaro (St Januarius), the patron saint of Naples, in the heavens. Januarius gestures toward Vesuvius while three angels pour down water in an attempt to quench the flames. In essence, this is still a symbolic representation expressing the hope of averting the eruption, rather than recording its spectacle. The image of the saint is unremittingly tied to the volcano, and the saint's duty, in the eyes of the populace, is to avert volcanic catastrophe. If the saint fails in his duty, the people may take strong measures, which include attacking the images of the saint himself. During the unusually violent Vesuvius eruption of 1872, the people of Campania felt totally abandoned by the patron saint, and attacked the statue of Saint Gennaro in anger (Figure 77.5). A detailed and finished contemporary view of the 1631 disaster can be seen in a painting by Micco Spadaro, who also witnessed this eruption. A city square is shown from above to display the crowds in a procession, to invoke the protection of St Januarius whose relics are carried by the clergy. The saint and angels can again be seen hovering in the clouds, while in the distance Vesuvius belches forth flows of ash. Another contemporary artist, Scipione Compagno also painted two views of the eruption, which focus more on Vesuvius itself. What is particularly interesting in these paintings is the phenomenon of ground-hugging surge clouds, that sweep

FIGURE 77.3 Engraving from 1631 of an eruption of Gunung Api volcano in the Banda Islands in the South Moluccas. An early seventeenth century example of the representation of a volcanic eruption (author's collection).

FIGURE 77.4 As seen by the contemporary Neapolitan artist Filippo D' Angelo, the 1631 Vesuvius eruption brought out Saint Gennaro, the patron saint of Naples, and his helpers, serving as a fire brigade, but the eruption also caused panic among the citizens. Janos Scholz collection, Pierpont Morgan Library, New York.

FIGURE 77.5 The Neapolitan population attacking the statue of Saint Gennaro, the patron saint of Naples, when their appeals to the saint failed to stop the April 1872 Vesuvius eruption (*Harper's Weekly,* June 1872).

down the slopes of Vesuvius—a process that these artists clearly witnessed. If the early twentieth century volcanologists had only bothered to look at these paintings, they would have appreciated that the deadly phase of the eruption was associated with surges and pyroclastic flows and not with mud flows.

4. EMPEDOCLES LEAPS INTO ETNA

A remarkable volcanic painting by the Neapolitan painter Salvator Rosa also dates from this time. It portrays the Greek philosopher Empedocles hurling himself into the flaming crater of Etna in order to commit suicide (Figure 77.6). According to ancient legends he had done this so that the followers of his philosophy would believe he had disappeared from Earth and thus gained godlike immortality. The truth became known when the philosopher's followers found one of his sandals at the crater's edge. Rosa's Empedocles hurtles headlong into the turbulent, steaming crater, his fluttering robes and outflung arms echoing the shapes of the rocks and clouds. He seems to merge with the volcano, a small man lost in the immensity of nature's turbulence. Painting toward the end of his life, in 1666, Rosa, who was something of a poet and philosopher as well as painter, intended the volcano as a metaphor for uncontrolled creative inspiration which can overwhelm the artist. Although he took the idea from a poem by Horace, this is the first time the image appears in painting. The volcano has taken on new meaning, and its violent spectacle, unpredictability, and deadly might are exploited for their aesthetic qualities. The idea that an eruption, even while destroying life, could contain beauty worthy of artful

FIGURE 77.6 Empedocles leaping into the crater of Etna. The Greek philosopher Empedocles lived in Sicily and studied the activity of Etna. He had novel ideas about volcanic activity, but legend has it that he took his life by secretly jumping into the crater of Etna, in order to convince his pupils that he had ascended to heaven. Brown ink and chalk drawing by Salvator Rosa, c.1666 (Galleria Palatina, Palazzo Pitti, Florence).

stenches, and sudden danger that evoke emotions of terror, fear, and pain. This notion suggests an earthquake, terrible storm, or a volcanic eruption as an ideal sublime image and experience filled with the sights and sounds of uncontrolled forces, immensity, darkened skies, stinking fumes, harsh lights, burning, and perhaps death. The sublime is not seen as the opposite of the beautiful, but as an alternative aesthetic taste. This taste soon became widespread among the cultured, art-loving classes, an audience now attuned to seeing the terrifying as pleasing—and for them a volcanic eruption was the ultimate sublime experience.

5. THE NEAPOLITAN SCHOOL

From the middle of the eighteenth century, the subject of Vesuvius in eruption became increasingly popular in painting, gouache, and prints. These were frequently produced as souvenirs for the English, German, and French gentlemen and ladies who came to Italy on the Grand Tour, a highlight of which was a visit to Naples and Vesuvius. Luckily for the travelers, Vesuvius became very active once again and was in frequent eruption from the 1730s through the end of the century. It was a great tourist attraction for those who could safely view the fiery lava and clouds of smoke from a distance, a scene they would later recall back home in London or Paris—where they could share their sublime experience with friends through these colorful works of art.

Claude-Joseph Vernet arrived in Naples in 1734, where he had a profound and pioneering influence on the painting of Vesuvius, establishing a formula for the representation of its eruptions. In his numerous works, the fiery glow of molten lava is shown silhouetted against the night or ash-darkened sky, with the steely blue cool light of the moon in strong contrast. In the foreground, relatively tiny human figures gesticulate toward the eruption in a mixture of wonder, terror, and admiration. Many contemporary artists adapted this scheme of volcano painting, such as the English Joseph Wright of Derby, the German Jacob Philip Hackert, the Austrian Michael Wutky, and later the Norwegian Johann Christian Dahl. Vernet's most prominent pupil, Pierre-Jacques Volaire, stayed in Naples for 20 years and established himself as Vesuvius' visual biographer. His painting of the *Eruption of Vesuvius in 1771* epitomizes this genre. He plays the burning red against the ice cold blue, a fiery lava fountain illuminates red clouds of steam and a river of molten lava contrasts with the silvery moonlight and the tranquil Bay of Naples on the other side. In earlier times Vesuvius, the powerful, earth-shaking giant, might have been represented as a Cyclops at the forge, a fiery, muscular figure, but now the mountain itself takes on the characteristics of sublime beauty. The dark side of nature had become worthy of portrayal, and these artists arranged their compositions and choice of

speech or painting was developed by Rosa and others in the seventeenth century, and it helped to create a taste for the marvelous, unusual, and awesome in nature. The style that captured these qualities was termed the sublime, which was great and inspired feelings of awe, fear, and amazement in the spectator.

In a treatise on the art of painting published in 1715 by Jonathan Richardson, the superiority of visual expression over verbal description was exemplified by an eruption of Etna. The poets since classical times had described eruptions, but how much more effective a picture could be when the whole scene opened up at once! The taste for the sublime became most popular in England after the 1757 publication *A Philosophical Enquiry into the Origin of Our Ideas of the Sublime and the Beautiful* by Edmund Burke. He defined the "beautiful" as something that is well-proportioned, graceful, smooth, clear, quiet, and harmonious, a prospect which elicits an emotional response of calm, security, and joy. In striking contrast Burke's "sublime" contains darkness, noise, overwhelming magnitude, colors to convey both the actual appearance as they observed it and the dramatic artistic effects. No longer does St Januarius hover protectively over Naples; now the figures are excited spectators who study and enjoy the phenomenon.

Natural catastrophes and landscape were among the subjects which were considered potentially sublime in late eighteenth century painting in Europe. In no subject matter, however, did the catastrophic possibilities become as manifest as in the painting of volcanic eruptions, as was demonstrated by the paintings by Joseph Wright of Derby of the eruptions of Vesuvius. Wright had climbed the volcano in 1774 and witnessed its eruption, and his pictures of this subject are startling for their vividness of colors and tumultous movement of the volcanic emissions. His views of the volcano were both at the very edge of the crater, from within the crater, or from a great distance, such as in *Vesuvius from Posillipo,* creating scenes that were almost too intense to be natural. One of the paintings of Vesuvius from this time, and most pleasing to modern taste, is a gouache study of 1774 by Wright of Derby. The reddened mountain and billowing gray smoke are made in a sketchy, soft manner with emphasis on the color and shapes. Here there are no small fleeing figures or scientific observers but simply the volcano in splendid isolation.

The German artist Jacob Philipp Hackert was also attracted to Italy by the strength of the seventeenth century Roman landscape tradition. He became the court painter for King Ferdinand IV in Naples, where he had ample opportunity to paint Vesuvius in action. Many of his views of Vesuvius erupting are remarkable because of the close-up of the volcano in action and are closer to a geological study than most other paintings of the mountain, in particular his painting of *The Eruption of Vesuvius in 1774* (Figure 77.7). This emphasis on scientific process may have been influenced by Hackert's close association with the pioneer volcanologist William Hamilton, also a resident of Naples.

6. NEOCLASSICISM

When Herculaneum was discovered out of the ashes of Vesuvius in 1709, and Pompeii in 1748, European artists became fascinated by the great eruption of AD 79 and by the cities which had been buried in layers of ash, as well as the newly revealed classical culture. The recovered buildings and artifacts quickly promoted a taste for classical styles in furniture, clothing, and art, and the evidence of the inhabitants' swift deaths, and the immensity of destruction further stimulated fascination with the sublime. Paintings which depicted scenes of this ancient catastrophe were now made, along with the landscapes of Vesuvius as it looked in modern times.

In 1785 the Swiss painter Angelica Kauffman made a painting based on the events of August 24, AD 79 as described by Pliny the Younger. The young Pliny and his mother are shown in the courtyard of their villa at Misenum where a friend urges them to flee before it is too late. In the distance, framed within the archway of a gate is Vesuvius, an outline on the horizon shooting forth lava, pumice, and ash. The sky is dark and the waves on the usually calm sea are agitated. The only suggestions of sublime emotion here

FIGURE 77.7 *The Eruption of Vesuvius in 1774* by the German artist Jacob Philipp Hackert (1737–1807), the court painter for King Ferdinand IV in Naples. The painting is an unusual close-up view of an erupting volcano, with pioneer volcanologists clambering right up to the vent on the volcano's flank to get a closer look. The Hessen State Collection, Germany.

are the expression and gestures of the two frightened women on the steps down to the sea. The painting has the cool, detached crispness, and clarity typical of **Neoclassicism**. The artist has recreated a historical moment in which the human figures dominate, rather than the volcano.

A more emotional version on the subject of the AD 79 eruption of Vesuvius can be seen in a spectacular painting by Pierre Henri de Valenciennes in 1813. Now in the Romantic era, the sublime becomes even more powerful through emotional and imaginative imagery to emphasize the horror of the moment in the appearance of the sky, brilliant color, and small-scale figures. Valenciennes also used Pliny's letters as a source, in a scene where his uncle Pliny the Elder is carried by two slaves as he succumbs to choking clouds of fumes issuing from the volcano. This subject had been painted as early as 1780 by Jacob More in *Mount Vesuvius in Eruption: The Last Days of Pompeii,* but the subject became increasingly popular through the nineteenth century in many versions that stressed the physical

agony, fear, and other sublime emotions in romanticized paintings. This was especially true after the 1834 publication of Bulwer-Lytton's tremendously popular novel *The Last Days of Pompeii.* Vesuvius in eruption was interpreted as divine retribution for wicked mankind; corrupt civilization was contrasted with the purity of nature.

In 1833 the Russian artist Karl Briullov painted a huge composition titled *Last Day of Pompeii,* which was inspired by the opera by Pacini: *Ultimo giorno di Pompeia,* performed in Naples 1827. Briullov had visited the excavations of Pompeii, and the sight of the city buried by layers from the great AD 79 eruption astounded the artist. He became captivated by the idea of sudden interruption of life by this great natural catastrophe. Briullov made many sketches of the composition and then began a study of historical documents of the excavations of Pompeii as well as Pliny the Younger's description of the AD 79 eruption, in order to create authenticity and to follow the Romantic principle of historical truth (Figure 77.8).

FIGURE 77.8 In 1832 the Russian artist Karl Briullov painted *The Last Day of Pompeii,* a representation of the catastrophic AD 79 eruption and its effects on Pompeii. It is the quintessential example of Neoclassicism, with Greek and early Roman architectural and physical forms juxtaposed to the terror of the volcanic eruption. The State Russian Museum, St Petersburg.

7. TURNER'S FIRE

By far the most notable artist to paint volcanic eruptions was Joseph Mallord William Turner (1775—1851). Turner was the greatest British painter of his age and probably the greatest painter England has ever produced. He was both by training and inclination a true landscape painter in the tradition of the **picturesque** sublime, and his choice of subject matter included apocalyptic and catastrophic topics. He had a brilliant gift to translate onto canvas the most fleeting effects of light, such as a passing storm, sunrise, or dissolving mist. In 1817, Turner painted *Vesuvius in Eruption,* creating a work that has a sense of volcanic energy and the spectacle of light that is unique to this artist. However, Turner's basic treatment of the eruption follows the eighteenth century formula: a white to red-hot fire fountain out of the central cone, reflected on the Bay of Naples, a cool blue-gray sky with a crescent moon in the background, contrasting with the conflagration around the volcano. Lightning and the arched trails of glowing volcanic bombs are shown by generous use of a scraping-out technique.

Turner's earliest painting (1815) of a volcanic eruption is *The Eruption of the Soufriere Mountains, St Vincent.* In April 1812, the Soufriere volcano on St Vincent in the Caribbean erupted explosively. The eruption resulted in ashfall on the island of Barbados to the east and produced deadly hot ash flows down the flanks of the volcano. The eruption blasted out a new crater in Soufriere's summit and killed many of the Carib Indians living on the volcano's slopes. News of the eruption spread with the British colonialists to England, and these reports stimulated Turner to put his impression of this volcanic eruption on canvas. Turner wrote these lines in the catalogue published at the first Royal Academy, showing in 1815:

Then in stupendous horror grew

The red volcano to the view.

And shook in thunders of its own,

While the blaz'd hill in lightning shone,

Scattering their arrows round.

As down its sides of liquid flame

The devastating cataract came,

With melting rocks, and crackling woods,

And mingled roar of boiling floods,

And roll'd along the ground.

In 1815 another volcanic eruption occurred in far-off Indonesia which may have influenced Turner's art in a fundamental way. The volcano Tambora erupted in 1815, lifting into the atmosphere more ash and volcanic gases than any other eruption in history. The atmospheric effects of the eruption included spectacular twilights of yellow and red skies in the years following the event. Turner undoubtedly experienced this marvelous coloration of the skies caused by the aerosols or fine particles injected into the atmosphere after the eruption. Is it possible that Turner's fiery skies, for which he is so justly famous, were in this way inspired by the effects of a distant volcanic eruption of which he was unaware and would never hear of in his lifetime?

8. HOKUSAI AND THE SPELL OF FUJI

At the time when Turner was executing his rich paintings of Vesuvius, there was at work in Japan an artist gifted with extraordinary powers of draftsmanship and design, who devoted much of his career to the imagery of Fuji volcano near Tokyo. Like Turner, Hokusai was the most eccentric of men and dedicated to his art with the passion and single-mindedness of a saint to his religion. Hokusai, like Minsetsu, Hiroshige, and other Japanese artists, was drawn to Fuji volcano, not because of its spectacular eruptions, but because of the veneration and idolatry for the volcano by the Japanese people. From 1823 to 1829 Hokusai created *Thirty-Six Views of Fuji,* often regarded as among his finest landscapes. The series of landscape color prints includes the incomparable *Fuji in Clear Weather.* Here the calm selection of lines and graduation in colors bears the imprint of the woodblock printer's art (Figure 77.9). In 1834—1835, he published another set of prints entitled *One Hundred Views of Fuji.* The earlier 36 landscapes are dwarfed by the eternal and monumental cone of Fuji, which hints at the transitoriness of human life. Is it the gentle curve of the cone, with its near-perfect geometric form, sweeping the eye to the sky, which has led to Japanese idolatry of this volcano? Or was it based on fear because of the history of Fuji's deadly eruptions? The phenomenon is outside our western experience, and thus we miss some of the special significance that Houksai's views of Fuji have for his countrymen. Already in the eighth century, Fuji was revered with superstitious awe, as shown for example in Manyoshu, one of the earliest poems of the Japanese language:

No words may tell of it

No name know I that is fit for it

But a wondrous deity it surely is!

It is the peace giver, it is the god, it is the treasure.

On the peak of Fuji, in the land of the Suruga

I never weary of gazing.

FIGURE 77.9 *Fuji in Clear Weather.* Color woodcut by the Japanese artist Katsushika Hokusai (1760—1849). *From the series* Thirty-Six Views of Fuji, *c. 1830.*

In the later 100 views, the curious and unexpected glimpses of Fuji, reflected in the sake cup held by a tippler or seen through a bamboo thicket, transmute the ordinary into the memorable.

To my knowledge, Hokusai only executed one picture featuring a volcanic eruption, which is *The Upthrow of Mount Hoei* in his great book of woodblock prints *One Hundred Views of Fuji* published in 1834. In the winter of 1707, the great volcano exploded in one of the largest eruptions of the historical period. In this flank eruption, Fuji created a new cone or "hump" on the slope of the volcano that was named Hoei. In the woodcut *The Upthrow of Mount Hoei* Hokusai lets his imagination run free in what is an image of the effects of the eruption on the people and an utter demolition of their world (Figure 77.10). The plate shows the terror, the eruption and its associated earthquake created among the villagers living near the volcano. Rocks and stones are hurled from the new crater and showered down in destructive abundance over the village. Men, women, and children are tossed about the air, together with the debris of their ruined homes. Some villagers try to escape, dragging after them their fainted wives. The terrible eruption has created panic, destruction, and despair among the helpless folks. Fuji itself is not to be seen in this picture of chaos, but its effects are devastating. In the woodcut *Sono ni* in the same volume, Hokusai shows us the quiet volcano after the eruption (Figure 77.11). The great eruption of 1707 did not occur from the central crater but rather from a lateral vent on the southeast slope of the volcano, forming Hoeizan, an unsightly lump on the otherwise perfect profile of Fuji.

Hokusai's color prints achieved a compromise between representation of actuality on one hand and creation of an abstract design in color and line. His style of brushwork was consistent with the highest ideals of Japanese calligraphy. Hokusai's prints lie between two opposing styles—the Japanese ancient Edo school, suggesting a landscape of the imagination, and the school of Western landscape artists, approaching as near as the oil or water color mediums will allow to a realistic representation of a certain place. His work had significant influence upon Western art in the direction of a new formalism and simplification (e.g., in the French Impressionists). Hokusai's work was, however, most influential in Western book illustration and design.

A large number of other Japanese artists, who had mastered the color woodblock, were captivated by Fuji, including Keisai Eisen (1790—1848), Utagawa Hiroshige (1797—1858), Hiroshi Yoshida (1876—1950), and many others listed in Table 77.1. In addition to Fuji, another volcano has received much attention among Japanese artists, particularly in the twentieth century. This is Sakurajima in Kagoshima Prefecture in south Japan. Early studies of this volcano are shown in the oil paintings of Masayoshi Tokonami (1842—1897), who was a pioneer in making the transition from Edo style painting to a more modern form of Western painting. Seiki Kuroda (1866—1924) also painted a number of works showing the 1914 eruption, which he witnessed. The dramatic Sakurajima landscapes of Ryuzaburo Umehara (1888—1986) are impressionistic works, but he studied under Renoir in France and was acquainted with Picasso. He successfully fused the traditions of Japanese and European painting, and his Sakurajima works brought Japanese art into another phase. Ichien Somiya (1893—1994) painted a close-up of the Sakurajima crater during the 1956 eruption. Perhaps the most striking works of Sakurajima is the 1961 painting *Island of Fire* by Matazo Kayama (1927—2004), where a

FIGURE 77.10 *The Upthrow of Hoei.* Woodcut by the Japanese artist Katsushika Hokusai (1760—1849), showing the effects of the 1707 eruption of Fuji on the local people. *From the series* One Hundred Views of Fuji, *c. 1830.*

modern approach to Japanese painting is combined with a traditional sense of design that turns the volcanic landscape into a fantastically designed fabric screen.

9. AMERICAN EXPLORATION OF UNKNOWN VOLCANOES

With the opening of the American continent in the nineteenth century came the discovery of new worlds, which included volcanic mountains greater and more terrifying than any of the European experience. In the 1840s, a young Canadian artist, Paul Kane, was employed by the Hudson's Bay Company to aid in the documentation of the Pacific Northwest. During his travels, he stumbled upon Mount St Helens volcano, which he represented in a series of paintings, mostly in vigorous eruptions. In the painting *Mount St Helens Erupting,* Kane shows a fiery Strombolian-like fountain rising out of a crater on the north flank and a

dark eruption cloud billows away from the eruption column, watched by awestruck Indians in a canoe. The eruption site appears to be the Goat Rocks dome, a parasitic vent on the north flank, which was about 2000 feet lower than the summit of Mount St Helens. Today there is no trace remaining of Goat Rocks: the entire north flank of the volcano was completely obliterated during the explosive eruption on May 18, 1980. Although Kane's view appears to be accurately observed and was intended as a scientific record, it is composed using the conventional formulae for the depiction of volcanoes that had developed during the eighteenth century in conjunction with Vesuvius. He shows the darkened sky, the silhouetted volcano, the fiery reflection on the water, and the small foreground figures who react with expressive gestures.

The most prominent American artist in the early nineteenth century belonged to the Hudson River School, which still contained residual elements of European influence. Thomas Cole (1801—1848) was the founder of the Hudson

FIGURE 77.11   In the woodcut *Sono ni* Hokusai shows the changed profile of Fuji after the 1707 eruption, which took place from a lateral vent on the southeast flank of the volcano and created the cone of Hoei.

River School of painting and one of the leading figures in the development of a national tradition of American art. His work was a mixture of devotion to natural scenery on the one hand and illustration of moral truths on the other hand. Gradually, however his natural landscape themes gave way to the moral and allegorical themes and strong religious influences.

Several American landscape painters of the nineteenth century chose volcanic eruptions as subjects for painting, and Thomas Cole painted *View of Mt Etna* in 1844. The development of tourism as an aspect of leisure and education may have had a profound and stimulating effect on appreciation of landscape painting in America in the early nineteenth century. The appreciation of the wilderness grew with the developing middle class and urban elite, and people sought pleasure through tourism and developed a demand for views of the familiar and picturesque landscape. Cole, as one of the greatest landscapists in North America, was perfectly suited to meet this new demand, when fine art was becoming popular in the emerging

American culture at large. Cole was the first American artist to tackle the wilder and untamed features of the scenery of American mountains. By 1825, his Romantic pictures of the Hudson Highlands and the Catskill Mountains had received attention, particularly because of their sublime representation of American mountain tops and inaccessible peaks.

Cole was a close friend of the geologist Amos Eaton, and Cole was also acquainted with the pioneer American geologist Benjamin Silliman. It is likely that Cole's interest in volcanoes may have been influenced by these friendships. Cole spent several years in Italy, including an extended stay at the foot of Vesuvius volcano in 1832. He ascended Mount Etna volcano in Sicily in 1842, but the ascent of the volcano took a strong hold of Cole's imagination and formed the subject of four pictures. The famous painting of Mount Etna from Taormina took 5 days to paint. It was then sold for $500 to the Hartford Lyceum, which was "pretty good for five days work," as Cole stated in a letter to his wife. The painting produces the effect of a huge and lofty mountain, whose summit is white in perpetual snow, while its slopes below are basking in perpetual summer. In this distant view from the northeast, a gentle steam plume rises from the main crater, and the northeast crater is also visible on the upper slopes. The height and steepness of the top cone are exaggerated in this painting to attain the sublime effect.

In the early nineteenth century, biblical ideas of a volcanic judgment day appear frequently in American poetry and art. In 1828, Cole painted the *Expulsion from the Garden of Eden*, where he introduced the more terrible objects of nature including a volcano in violent eruption. Here Adam and Eve leave the sunlit paradise, to enter desolation and darkness, with all the elements raging against them—including a volcanic eruption towering overhead. Cole had not yet seen a volcano at this stage but borrowed the image from the conventional view of an eruption of the time.

The work of many early nineteenth century American artists reflects a viewpoint and a social evolution that has been referred to as Manifest Destiny—the Euro-American dream of discovering, exploring, and conquering an unspoiled and undeveloped terrain, toward creating an authentic American Civilization. The works of Cole and his brilliant disciple Frederic Church reflected this spirit. Frederic Edwin Church (1826–1900) was a central figure in nineteenth century American landscape painting and prominent in the group of artists associated with the Hudson River School. But he was first and foremost an adventurer–explorer artist. As a former pupil of Thomas Cole, Church received the mantle of American landscape painting from the master who had carried it for an entire generation.

Church replaced the subjectivity of his romantic teacher with an objectivity perceived by a mind steeped in geology. He was particularly influenced by the writings of the great German geographer Alexander von Humboldt whose descriptions of South American volcanoes drew Church on his first expedition to Cotopaxi in 1853. Church was, however, primarily a product of the golden era of prosperity and cultural nationalism just before the Civil War, the years of Manifest Destiny, which fostered a New World authentic movement of painting and heroic art. Taken to its extreme, the faith of Manifest Destiny was the faith that natural history had dictated Anglo-Saxon domination of the great American continent and regeneration of the whole world. Thus Church's art was premised on the Grand Design that the Creation existed for man.

Church's famous volcano painting *Cotopaxi* (1862) was painted early in the Civil War and dramatizes the struggle between the forces of darkness, represented by the belching volcano, and the forces of light. This work resulted from Church's travels to South America in 1857, when he was at the base of Cotopaxi volcano in Ecuador. In his painting

Cotopaxi, he invokes the moment of creation itself. It is painted with the fidelity of a naturalist, yet the style suggests Earth's terrible power and mankind's insignificance. His monumental South American scenes started with the *Heart of the Andes* in 1859. They attempt to combine into a single colossal scene the geological, meteorological, and botanical history of the continent, and his volcanic paintings metaphorically suggest a new world in the making. On his second journey to South America in 1857, Church made a trip to Sangay volcano in Ecuador. Traveling in the pathless wilderness of the high Andes mountains, Church ascended a ridge to obtain a view of the volcano, which rewarded him with a series of explosions:

Above a serrated, black, rugged group of peaks which form the crater, the columns arose, one creamy white against an opening of exquisitely blue sky, delicate white, cirrus formed, flakes of vapor hung about the great cumulus column and melted away into the azure. The other, black and sombre, piled up in huge, rounded forms cut sharply against the dazzling white of the column of vapor and piling up higher and higher, gradually was diffused into a yellowish tented smoke through which would burst enormous heads of black smoke which kept expanding, the whole gigantic mass gradually settling down over the observer in a way that was appalling.

This experience resulted in his painting *Distant View of Sangay Ecuador.*

The great public following and admiration for the landscape works of Church and other American artists went into rapid decline in the late nineteenth century for reasons that can be attributed to an advance in science. By the 1880s, the Scientific Revolution sparked by Darwin's world-shaking theory of evolution had triumphed in the intellectual world. The concept of the Grand Design was shattered, and man's place in nature shifted from the heroic to the unheroic. Church was one of the victims of this revolution, and by the 1990s his work was all but forgotten.

Albert Bierstadt (1830–1902) was one of the artists who discovered the grandeur of the American West and was the principal rival of Frederic Church. He became well-known for his paintings of the spectacular mountainous vistas of the Rocky Mountains. Bierstadt's interest in volcanoes was kindled when he visited Naples in 1857 and climbed Vesuvius, resulting in the painting *Mount Vesuvius and the Bay of Naples*. His second treatment of the subject was painted in 1868, showing the volcano in full eruption, as seen from the upper slopes of Vesuvius, northwest of the crater, in a gaudy, almost fluorescent view.

10. HAWAIIAN ERUPTIONS

The painting of Hawaiian volcanoes begins with the visit of Robert Dampier to Hawaii in 1825. Another important interpreter of Hawaii's volcanoes was Titian Ramsay Peale, son of the noted American painter Charles Willson Peale, who painted the famous portrait of George Washington. The younger Peale traveled widely and was selected to participate as an illustrator in the US Exploring Expedition of the South Seas (1838–1842) under the command of Lieutenant Charles Wilkes. On the fleet of six ships were a large number of naturalists and scientists, who collected a wealth of specimens which were returned to Washington, D.C., and formed the nucleus of the National Museum of the United States in the Smithsonian Institution. One of the scientists on board was the American geologist James Dwight Dana. In September 1840, the expedition reached Hawaiian waters and remained near the volcanic islands until April 1841. A major eruption of Kilauea volcano had partly filled in the caldera with lava in 1839, which solidified to form the Black Ledge. During their climb on the flanks of Mauna Loa volcano, Peale sketched a view of the distant Kilauea, which he later used for the painting *Volcano Kaluea Pele.* The painting shows Peale's companions hiking across lava formations in the foreground, and in the background is the shield of Kilauea, fuming from the summit region and from the site of a recent eruption on the lower flank. When Peale reached the summit of Kilauea volcano, the Black Ledge was still about 200 m below the caldera rim and there was an inner pit, about 100 m deep below the Black Ledge, with several erupting lava pools on the floor. From this vantage point, he painted West crater "Kaluea Pele" from "Black Ledge," showing the outer caldera wall, inner Black Ledge, and the main fire pit (Halemaumau).

Charles Furneaux (1835–1913) became the first of Hawaii's resident volcano painters. He arrived in Hawaii in 1880 with the volcano enthusiast William T. Brigham. At the time, an eruption of Mauna Loa produced a major lava flow, which advanced toward the town of Hilo. Furneaux took advantage of this opportunity to the fullest, and painted over 40 views of the advancing lava and other landscapes in connection with the eruption.

Jules Tavernier arrived in Hawaii in 1884 and can be credited with starting the Hawaiian "Volcano School" of painting. He was born in Paris, and traveled widely in Europe and North America as an illustrator and writer. Upon arriving in Hawaii, Tavernier immediately went to view Kilauea, and from that day forward, the volcano became his chief subject, painting about 100 views of it during his stay in Hawaii. At this time, the caldera of Kilauea was flooded by a lava lake, and Tavernier devoted much effort to capturing this spectacle on canvas. Many of these paintings are based, however, on the Italian iconography of Vesuvius painting developed by Vernet in the eighteenth century, with the caldera shown at night, the glowing lava lake sluggishly rolling and flashing steam, surrounded by massive walls of lava cliffs, and the cold silvery moon emerging from the clouds in the background.

One of Jules Tavernier's art students was the Hawaiian-born David Howard Hitchcock, whose output of Hawaiian volcano landscapes was enormous.

11. DR ATL—VOLCANOLOGIST AND ARTIST

In the early twentieth century, the visual arts turned away from attempts at rendering the visible world and turned instead toward the representation of abstract forms. This may have been brought about in part by the invention of photography. Thus interest in landscape painting declined, and along with it, volcanoes vanished as part of the painted natural scene. One notable exception is the Mexican artist Dr Atl, who showed the same devotion to Paricutin volcano that Volaire had shown to Vesuvius in the eighteenth century. In February 1943, a fissure opened in a farmer's cornfield in the state of Michoacan in western Mexico. A volcanic eruption out of the fissure threw up scoria and lava, which during the following 9 years built up the 410 m high volcanic cone known as Paricutin. In his vibrant, expressionistic paintings, Atl has recorded the actual birth of Paricutin producing a unique pictorial biography of the volcano. Atl was a representative of **Impressionism** in Mexican art, but his simplified forms and details and brilliant coloring made his art highly personal and new.

Dr Atl was one of the outstanding figures in Mexican cultural and political life in the first half of the twentieth century. He was one of the original creators of Mexican landscape painting and a leading figure in the revolution in Mexican art. Thus he fought for the identity of national and racial consciousness in his nation's art, against the Spanish colonial past. The modern vision of Mexican landscape created by Dr Atl is a reflection of the enthusiastic temperament, travels, and explorations of this man of culture and wide artistic experience. He spent whole seasons on the snow-covered peaks of the Mexican volcanoes, which he painted many times.

He made major contributions in art, literature, philosophy, poetry, history, revolutionary politics, and the science of volcanology. Jose Gerardo Francisco Murillo was born in Guadalajara, Mexico, in 1875. When he came of age, he renounced the Spanish name and took an Indian pseudonym, changing his name to *Atl,* meaning *water* in the Aztec language, to declare his separation from the dreary influences of the secondhand European culture into which he was born. He knew instinctively that the arts in Mexico were counterfeit, and he set out to rediscover or create a contemporary native art—striving for a renaissance of Mexican and Pagan culture. At the age of 21, he was already noted in Mexico as an outstanding artist and writer. He traveled widely during a 7-year stay in Europe, Egypt, and China, but upon his return to Mexico in 1907, he took up mountaineering, which led to his lifelong affair with the volcanoes of his country. In Mexico City, he set about to revolutionize the Academy of San Carlos, where he lectured about the art of the Impressionists, but he also encouraged his students—including the young Diego Rivera—to paint out of doors and appreciate the drama of the Mexican landscape.

In 1912, he traveled to Italy, where he studied volcanology for a time under the expert guidance of Immanuel Friedlander and Frank A. Perret at the University of Naples—the leading volcanologists of the time. From this period, he began to combine his two primary interests in a unique manner—painting and volcanoes. Prior to the work of Dr Atl, the only Mexican artist to paint the volcanoes was Jose Maria Velasco, who painted more than 200 Mexican landscapes, noted for their poetic realism and sensibility. He explored Popocatepetl before Dr Atl's time and published texts and lithographs and engravings of the revered mountain. Several Mexican volcanoes had been painted earlier by European visitors, such as the French artist Jean-Baptiste Louis Gros, whose 1833 work shows in a vivid manner the sulfur encrustation of the interior of the crater of Popocatepetl, and numerous dramatic paintings of Colima volcano were executed by the German artist Johann Moritz Rugendas (1834).

Dr Atl required a new medium to capture the striking colors of the Mexican volcanoes. He therefore invented the Atl-colors, which were, like pastel, for use on surfaces like wood, paper, plaster, fabric, and board. The colors were composed of wax, dry resins, oil color, and gasoline, manufactured in small bars like the sticks of sealing wax. The color itself is apparently unalterable, although the sticks can be melted like wax or dissolved in gasoline for spreading with a brush or palette knife.

The Mexican revolution at the beginning of the twentieth century not only transformed the political and social fabric but also created a new tremendous flowering of artistic and intellectual creative energy that changed the course of Mexican culture. Intense nationalism reacted to the largely European-oriented culture of the regime of the dictator Porfirio Diaz. In 1910, Dr Atl organized an exhibition of nationalistic paintings in honor of the centenary of Mexican independence, which has often been regarded as the beginning of the revolution in Mexican art. A few months later, the political revolution broke out, and Dr Atl became a leader of the revolutionary intelligensia, persuading other artists to devote their energies to journalism and political propaganda in the cause of the revolution. Among his assistants were the young David Alfaro Siqueiros and José Clemente Orozco, who later became Mexico's best known painters. Prior to the Mexican revolution, many artists and writers had felt intellectually oppressed, forced to admire the foreign and to scorn the native and the ordinary. With the revolution, they were able

to share these feelings and become fully absorbed in the revolutionary movement—and for many, the volcano was a potent symbol. This sentiment is illustrated in the novel *Los de Abajo* (Those from Below) by Mariano Azuela, when one of the characters states: "You ask me why I am still a rebel? Villa, Obregon, Carranza? What's the difference? I love the revolution like a volcano in eruption; I love the volcano because it is a volcano, the revolution because it's the revolution! What do I care about the stones left above or below after the cataclysm? What are they to me?"

Following the revolution, Dr Atl spent much time exploring and painting the Mexican volcanoes. His interest in volcanic activity was further aroused when Popocatepetl erupted in 1919. He had earlier written of his beloved volcano:

For millions of years it slept in the silence of death, for millions of years the wind lashed it, for millions of years the forces of nature have tried to destroy it. They closed its mouth, gnawed at its vertebra, they shook its formidable mass and tore its lips which in other times vibrated from thundering eloquence. But one day in its venerable old age, its entrails stirred and from the decayed lips of the Colossus, fire shot forth again. Oh marvelous teaching of nature! Nothing is old and nothing has died: in the end of the Destruction of Life.

In 1939 he published a monograph on this volcano, *Volcanes de Mexico: La actividad del Popocatepetl*. In it he put forward the curious view that the eruption of 1919 had "acquired an importance of the first order in the history of geology, because it was the result of purely artificial action." Sulfur deposits around fumaroles in the bottom of the volcano's crater had been mined since the sixteenth century. When the volcanoes or sulfur miners blasted the fumarole deposits with dynamite, it was apparently Dr Atl's belief that the dynamite blasting by the volcanoes had triggered the 1919 eruption. Popular tradition in Mexico claimed that the by-now legendary Dr Atl lived in a cave in the side of Popocatepetl volcano, where he sustained himself by eating locusts and honey. Legend also has it that he was erupted out of the volcano into this world:

Subio al Popocatepetl,

Bajo del Ixtaccihuatl,

Y es hombre de gran valer el inquieto Doctor Atl.

Dr Atl also used to good effect the curvilinear perspective in his landscapes and flew in an airplane over his beloved volcanoes to paint them from a new vantage point; a method he called aeropainting. The finest examples of Dr Atl's Aeropainting give a panoramic view of Mexico's most famous volcano, Popocatepetl. He uses this extraterrestrial view repeatedly when painting his favorite volcanoes. In a 1960 painting, Popocatepetl pokes its majestic snow-clad crater above the clouds, framed by the curving horizon and the starlit Milky Way. For all its twentieth century realism, Atl retains the romanticized vision with which earlier ages viewed volcanoes. He enhances this mood by juxtaposing the volcanoes to our contemporary sublime, the vastness and mystery of the stars and space. The volcano as part of the Earth has become one of the last natural phenomena that we can still awe. In Atl's paintings, there are no small, frightened figures, no evidence of human devastation, no moralizing overtones, just the volcano itself.

During the 1920s and 1930s, Atl dropped out of the postrevolutionary government of Mexico and began to wander the world, chiefly in the study of volcanoes. When the volcano Paricutin was born in a cornfield in 1943, the aging Dr Atl disappeared from Mexico City and moved to a shack at the foot of the growing volcano and began to paint. He produced 130 drawings, 11 paintings, and a large book on the activity, documenting all stages of its evolution. Births of new volcanoes is a common experience in the history of the Mexican people. In 470 BC, the birth of Xitle volcano destroyed the Mexican city of Cuicuilco. Again in 1759 the volcano El Jorullo was born in the state of Michoacan.

In 1950, Dr Atl published his observations of the eruption in a unique volume, including a large number of paintings and sketches of the activity, which document the growth of the volcano, in *Como nace y crece un volcano: El Paricutin*. The illustrations in brilliant color capture and preserve the spectacle of the eruption and the artist's excitement in the presence of this awesome display of nature's force. The text includes a detailed chronological description of the eruption and his conclusions regarding the fundamental mechanisms of volcanic activity. He rejects the ideas that volcanic heat is derived from reactions between iron and sulfur or from the combustion of petroleum or indeed the modern concept of heat generation by the decay of radioactive elements. Instead, he proposes that the volcanic energy is residual heat left over from the primitive globe, and only a small amount of additional heat from chemical reactions or from radioactive decay are needed to trigger an eruption. As a fundamental driving force for volcanism, Dr Atl embraced Alfred Wegener's theory of continental drift: "It is possible to conclude, as a consequence, that this abundance of volcanic structures is due to innumerable fractures caused by continental drift." Atl was not alone in painting Paricutin; his most illustrious student, the famous Mexican painter Diego Rivera, was also fascinated by the activity and created *Volcano Erupting* in 1943.

In *Crater en Llamas,* we see a full view of the volcano in the height of activity, with fiery fountains of red scoria mixed with black ash clouds billowing out of the crater, while a lava flow creeps along the ground toward the observer (Figure 77.12). In *Paricutin Violeta al Am* (1946),

FIGURE 77.12   The painting *Crater en Llamas* (Crater on Fire) by the Mexican artist and volcanologist Dr Atl shows the Paricutin volcano in 1943 in the height of activity.

the activity produces mostly ashfall to one side of the volcano with the entire landscape draped by a uniform layer of gray ash. In *Falda Incendiada del Paricutin* (1948), the lava flow creeps forward, oozing red-hot molten lava at its base. The waning phase of the eruption is shown in *Paricutin en Eruption* with a wispy black plume of ash rising from the crater. These paintings dominate the materials, colors, and movements of the volcano, but the artist uses the expressive qualities of the paint to render the scene as experienced by human eyes rather than with the objective clarity of the camera.

In *Paricutin en Noche Estrellada,* we see the eruption from a vantage point characteristic of Atl's works. As if gliding by Earth in a spaceship, hundreds of kilometers above the planet's surface, we have a full view of the erupting volcano against the strongly curved horizon of the planet in the background.

12. THE MODERN VOLCANO LANDSCAPE

Until Mount St Helens reminded us of nature's fury, strength, and indifference in 1980, our modern fears had been those we created ourselves—wars and the horror of nuclear devastation. What most Americans had forgotten was both the power and beauty of natural cataclysm, the coming together of earth and sky, brilliant color contrasted with uniform gray, the sight of all those trees laid flat in parallel lines, and the emotional impact of lives disrupted or gone. Artists like Dr Atl had followed earlier painters in giving form to these observations so that we can appreciate and understand volcanoes as both real phenomena and as symbols of our own states of mind and emotions. Toward the end of the twentieth century, a new interest is emerging among artists in capturing on canvas the spectacle of volcanic eruption.

In the 1980 painting *First Continental Eruption* by Chicago artist Roger Brown, the residents of two high-rise apartment buildings gaze out their windows at the erupting volcano and its billowing clouds of ash (Figure 77.13). The painting cleverly expresses the surprise of the American nation over this volcanic eruption in their own backyard, and the effrontery of Mount St Helens volcano to intrude into their lives in the spring of 1980. After all, volcanic eruptions belonged to Hawaii and Italy and were definitely not part of the safe continental scene! In this painting, the stylized volcano is erupting a black plume which rises sinuously like smoke toward the sky. The nasty-looking eruption plume spreads horizontally on top, with black to gray mammary-like clouds of ash and steam marching across the sky and fading in the distance. At the tip of the volcano is a red-hot flame which feeds the black smoke plume rather like the flame which continuously burns off waste products at an oil refinery. The artist has reached back to the Middle Ages in his interpretation of volcanic activity, when eruptions were thought to be the burning of combustible rocks in the earth. Roger Brown's paintings have been grouped with the art movement of the 1980s

FIGURE 77.13 The 1980 painting of the *First Continental Eruption* by American artist Roger Brown was inspired by the Mount St Helens eruption.

referred to by some as Survival Art. The politics of survival in his paintings address the idea of a definitive upheaval from which none will escape, such as in this volcanic eruption. The gloom is, however, tempered by a jazzy, parodic idiom with a vaudeville touch. In *Mania* (1984), the American artist Robert Hawkins conveys a dark, whimsical view of reality. Hawkins often selects the most menacing gifts of nature and blows them up to a frightening scale. In his aerial view of a volcano, the active crater has an inner pool of glowing, steaming lava, while several rows of volcanoes spew out ashes in the background.

The contemporary American artist Bill Sullivan has the temerity to take on the panoramic tradition of nineteenth century American art. Like his nineteenth century pre-decessors Church and Heade, some of Sullivan's pictures have been inspired by the volcanoes of the Ecuadorian Andes, particularly Cotopaxi. In painting volcanoes, he plays with the theatrical qualities of the subject, height-ening it with extravagant color combinations and a sense of elegance that keeps his work from turning into kitsch. In the *Last Refuge,* the painting is of Cotopaxi, the highest active volcano on Earth, and the Lord of Terror of the Incas. The volcano is framed between snow-covered rocks, across a vast gray plain. Cotopaxi is almost completely covered with snow, and a full moon breaks through the clouds, punctuating the majestic stillness.

In 1985, the Museo di Capodimonte in Naples in Italy mounted an exhibition of the work of American artist Andy Warhol, commissioned especially on the theme of *Stermi-nator Vesevo* or Vesuvius the exterminator. The hallmark of Andy Warhol's pop art was to take a mass media image out of context, be it a soup can label or a photo of Marilyn Monroe, and to formulate a new meaning for it. An erupting Vesuvius is such a well-known artistic icon that it too ap-proaches a mass media image. In 1980, Warhol spent some time in Naples and returned in 1985 to work on the image of Vesuvius and transferred it to his realm of fantasy. For Warhol, a Vesuvius "eruption is an overwhelming image, an extraordinary happening and even a great piece of sculpture." His product from the stay in Naples was a great one: 16 hand-painted canvases and 26 silk-screen images of Vesuvius erupting, each one developing a highly distinctive color scheme. The basic image of an eruption is nearly identical, however, with the symmetrical cone of Vesuvius rising out of the ruins of Monte Somma, whose ragged outline is apparent on the left (Figure 77.14). Warhol could not witness a Vesuvius eruption—the last one was in 1944—but he has created a very powerful and fiery Plinian eruption, probably based on his impressions from the widely distributed photographs of the great 1980 eruption of Mount St Helens in the state of Washington, or from the common images of erupting Vesuvius by Pierre Jaques Volaire and his contemporaries in the late eighteenth century.

FIGURE 77.14   *Vesuvius 1985* by the American artist Andy Warhol. Silk screen on cardboard (author's collection).

13. ARTISTIC ILLUSTRATION

Pictorial representation of volcanoes, eruptions, and other geologic phenomena is found in the domain of the artists and illustrators of books and media as well as among the works of geologists. A spectacular example of an early volcanic illustration is the Japanese color woodcut of the July 1783 eruption of Asama in Japan, which resulted in a large loss of life and was associated with the Tenmei famine (Figure 77.15). Earth science has a visual language, which relates to the communication of ideas and information between geologists by visual materials, especially maps, diagrams, and landscapes. Geologists are not only concerned, however, with the visual representation of topography but also with the penetration of the topography to form a three-dimensional picture of the structure of Earth. The main medium for geologic illustration in the eighteenth century was the copper engraving, which provided an excellent means of reproducing several 100 copies of the kinds of illustrations presented in early texts on the subject of Earth. The process was, however, very expensive, and only rarely was it used to its full effect. However, the copper engraving was a complex process, consisting of several steps, that could result in a finished product significantly different than that conceived by the naturalist who commissioned the illustration. Between the naturalist and the reader stood not only the artist but also the engraver and printer. Few scholars who explored volcanoes were in the position of William Hamilton (1776–1779), who had at his disposal a fine artist Pietro Fabris. At the request of Hamilton, Fabris lavishly illustrated his volcano monograph *Campi Phlegrei* and showed the volcanic features of Vesuvius and other craters in southern Italy in superb hand-colored engravings (Figure 77.16). He is also the only artist who has turned humble rock specimens of obsidian, lava, and sulfur into pure works of art.

In the early part of the nineteenth century, the invention of lithography had a far-reaching impact on geologic illustration, as it cut out the engraver as a middle man between artist and printer and enabled greater precision, a finer shading, and more subtle expression of detail. Lithography began to be exploited widely for scientific purposes in the 1820s, and it had the great advantage of being much cheaper than copper engraving, and thus allowed authors to illustrate their publications relatively lavishly. In the eighteenth century, an artistic tradition of realism had developed in the presentation of geologic landscape. In a way, this may be related to the development of neoclassical movement in art, which encouraged a realistic and clear landscape style, although the

FIGURE 77.15   A contemporary Japanese color woodcut showing the great July 1783 Asama eruption, which took many lives, both as a result of pyroclastic flows as well as a famine.

FIGURE 77.16 In 1768 the Neapolitan artist Pietro Fabris illustrated the great Italian volcano monograph *Campi Flegraei* for William Hamilton. Among the illustrations is this colored engraving of Stromboli erupting.

neoclassical artist would be generally scorned for painting an identifiable scene and, therefore, preferred to invent one that suited that overall composition. Similarly, the depiction of wild mountain landscapes was a common theme adopted by the artists of the Romantic movement, but they generally exaggerated the vertical scale greatly in order to heighten the Romantic impact of the work, with little regard for realism or accurate representation of the actual landscape. Therefore, the accurate representation of topography or landscape was looked upon with undisguised scorn by the principal spokesmen of eighteenth century art; consequently, most of the illustrators of geologic and other natural history subjects were from a socially humbler level of documentary topographical drawing and painting. The topographic art or accurate prepresentation of landscape was referred to by Henry Fuseli as "that last branch of uninteresting subjects, that kind of landscape which is entirely occupied with the tame delineation of a given spot." Similarly, Thomas Gainsborough declined to paint "*real Views* from Nature in this country." Thus topography was held in very low esteem by academic artists, and Twyman, for example, argued that "the whole concept of the picturesque is at variance with the needs of topography." Fortunately, this was not the view of all artists, and we are thankful to them for bringing us several wonderful and informative views of the volcanic eruptions of the past.

SEE ALSO THE FOLLOWING ARTICLES

Archaeology and Volcanism ● Plinian and Subplinian Eruptions ● Volcanoes and Tourism ● Volcanoes in Literature and Film

FURTHER READING

Atl, Dr, 1950. Como Nace Y Crece Un Volcan?: El Paricutin, Mexico. Editorial Stylo, Mexico City.

Boime, A., 1991. The Magisterial Gaze. Manifest Destiny and American Landscape Painting c. 1830–1865. Smithsonian Press, Washington, DC.

Crausaz, W., 1985. Dr Atl: Pioneer mexican volcanologist. In: Geological Society of America Centennial Special, vol. 1, pp. 251–256. Boulder, CO.

Forbes, D.W., Kunichika, T.K., 1983. Hilo: A Century of Paintings and Drawings (Lyman House, Hawaii).

Helm, M., 1968. Modern Mexican Painters. Books for Libraries Press, Freeport, NY.

Huntington, D.C., 1966. The Landscapes of Frederic Church. G. Braziller, New York.

Luhr, J.F., Simkin, T., 1993. Paricutin, the Volcano Born in a Mexican Cornfield. Geoscience Press, AZ.

Luna Arroyo, A., 1992. Dr Atl. Hachette, Mexico.

Manthorne, K., 1985. Creation and Renewal: Views of Cotopaxi by Frederic Church. National Museum of American Art and Smithsonian Institution Press, Washington, DC.

Mellaart, J., 1967. Catal Huyuk: A Neolithic Town in Anatolia. McGraw-Hill, New York.

Myers, B.S., 1956. Mexican Painting in Our Time. Oxford University Press, New York.

Nicolson, M.H., 1959. Mountain Gloom and Mountain Glory. Cornell University Press, Ithaca, NY.

Parry, E.C., 1981. Acts of god, acts of man: geological ideas and the imaginary landscapes of Thomas Cole. In: Schneer, C.J. (Ed.), Two Hundred Years of Geology in America. University Presses of New England, pp. 53–71.

Rudwick, M.J.S., 1976. The emergence of a visual language for geological science, 1760–1840. Hist. Sci. 14, 149–195.

Wallach, A., 1981. Thomas Cole and the aristocracy. Arts Mag. 56, 94–106.

Vesuvius by Warhol, Exhibition Catalogue, 1985. Museo di Capodimonte, Electa, Napoli.

Chapter 78

Volcanoes in Literature and Film

Haraldur Sigurdsson

Graduate School of Oceanography, University of Rhode Island, Narragansett, RI, USA

Rosaly M.C. Lopes

Earth and Space Sciences Division, Jet Propulsion Laboratory, California Institute of Technology, Pasadena, CA, USA

Chapter Outline

GLOSSARY

neptunists A group of geologists who maintained that basalt layers and related formations were formed by sedimentation in the ocean, and not by solidification from magma.

Plutonian theory The opinion maintained by a group of nineteenth century geologists that some controversial rock formations were of volcanic and magmatic origin, but not derived from sedimentation in the ocean, as maintained by the Neptunists.

pyroclastic flows Currents of volcanic ash, gases, and heated air that cascade down the slopes of a volcano, at high temperature and high velocity.

sublime An event depicted in art or literature that is both terrifying as well as beautiful.

1. INTRODUCTION

The beauty and power of erupting volcanoes, as well as the often tragic outcomes of violent eruptions, have long inspired writers and filmmakers. The fictitious Mt Doom is probably one of the most famous volcanoes in the world. In 1954, the British philologist and writer JRR Tolkien (1892–1973) completed his three-volume work entitled *The Lord of the Rings*. The novel set the stage for modern fantasy literature and featured as one of the central and key elements the volcano Mt Doom or Orodruin ("Mountain of red flame"). This trilogy is one of the most popular works of fiction in the twentieth century and has been named the book of the millennium by many critics and readers. The Orodruin volcano in Mordor was the destination of the Fellowship of the Ring, in the quest to destroy the One Ring, leading to the climax of a long and complex epic struggle. This Tolkien tale became the source of an enormously successful film trilogy directed by Peter Jackson: *The Fellowship of the Ring* (2001), *The Two Towers* (2003), and *The Return of the King* (2003). The filming of the trilogy was a massive undertaking that took 8 years, but resulted in a product that is so far the highest grossing film series of all time.

The roots of the tale created by Tolkien lie in Germanic and Old Norse mythology and the Icelandic sagas. When Tolkien graduated from university in 1915, he had the Old Norse language as a special subject and later he became professor of Anglo-Saxon at Oxford. In Old Norse mythology, as in the *Fellowship of the Ring*, the idea of

The Encyclopedia of Volcanoes. http://dx.doi.org/10.1016/B978-0-12-385938-9.00078-X

Ragnarök or Armageddon is a central theme, which gives much scope for heroes battling the monsters at the center of the Earth. Mt Doom volcano thus literally becomes the scene of "the crack of doom" when the announcement of the last day is made. The fateful Ring could only be destroyed by throwing it into the red-hot churning magma pool in the Crack of Doom, the fissure at the mouth of Mt Doom volcano. As a result of Tolkien's writings and the successful film trilogy, Mt Doom has become the universal icon or at least the stereotype of an erupting volcano with the general public worldwide. The fame of Mt Doom has now spread even beyond Earth and to other worlds. Planetary scientists have given this name also to a volcano on Titan: Doom Mons. Volcanoes are terrifying, **sublime**, and awesome, and their eruptions represent the ultimate exterminator. It is therefore natural that they have been frequently used in literature and film, both as a metaphor and as a plot device. In one of the better if lesser known novels, *Gentlemen in England* (1985), the potency implied by erupting volcanoes allows the Victorian volcanologist to win over the younger woman he is courting. In the most famous volcano film, Rossellini's *Stromboli* (1950), the erupting volcano is used as an image of the despair felt by the young bride, unwelcome and trapped in a foreign society on a tiny island.

This chapter reviews several of the principal works of fiction and film that feature volcanoes and their eruptions. Many of the novels and films are based on or strongly influenced by historical eruptions, whereas others are in the genre of science fiction or fantasy, such as the pioneering work of Jules Verne in *The Journey to the Center of the Earth*. The various works differ dramatically in how accurately they portray volcanoes and volcanic eruptions. Some novels and films attempt to portray volcanoes in a credible, realistic way. For example, *Juan of Paricutin* was written by a geologist and can be considered a mostly historical account of the eruption. Authors of some other novels have taken considerable care to describe volcanoes in a realistic light. Some authors, such as Jules Verne, present scientific views that are no longer accepted, but their novels reflect the prevailing thinking of their times. Other authors use the power and fury of volcanoes purely for dramatic effect, and their works can only be considered fiction.

Most films that prominently feature volcanoes unfortunately fall into the dramatic effect category or, as they are known in the industry, "disaster movies." Yet even these differ in their attempts to present reality; for example, the 1997 film *Dante's Peak* portrayed well the difficult decision of whether or not to evacuate a threatened town—a dilemma that is often faced by volcanologists and public officials, but little known to the general public.

The volcanoes that have most often inspired works of fiction, either in literature or film, are those that either erupt frequently, such as Kilauea in Hawaii, or those whose eruptions brought tragedy to the people living near them, such as Vesuvius in AD 79, Krakatau in 1883, and Mt Pelee in Martinique in 1902. All these events are major catastrophes that still capture the imagination of the public and are rich, exciting themes for stories about the interaction between volcanoes and the people who live around them.

The authors acknowledge that the literary works discussed here are predominantly from Western sources and in English, and that the volcano-related literature from the volcanic regions in Russia, Japan, the Philippines, Indonesia, and other parts of Asia, as well as Africa, are as yet largely unexplored by Western volcanologists.

2. PRETWENTIETH-CENTURY NOVELS

Probably the earliest Western novel that features volcanic activity is *Montalbert*, a Gothic novel by Charlotte Smith (1795). The heroine Rosalie undertakes a journey from England to Italy, where she witnesses the activity of Etna volcano and is captivated by its sublime beauty and force: *We are never surprised at the sight of a small Fire that burns clear, and blazes out on our own private Hearth, but view with Amaze the celestial Fires, tho' they are often obscured by Vapours and Eclipses. Nor do we reckon any thing in nature more wonderful than the boiling Furnaces of Etna, which cast up Stones and sometimes whole Rocks from their labouring Abyss, and pour out whole Rivers of liquid and unmingled Flame.*

2.1. The Last Days of Pompeii

The first major novel that features a volcano as a central theme is *The Last Days of Pompeii* by the English writer George Bulwer-Lytton (1803—1873). The novel, dealing with the catastrophic eruption of Vesuvius in AD 79, was first published in 1834 and became an immediate and spectacular success, with numerous editions published up to our day. It is a work that is based upon substantial research by Bulwer-Lytton in Pompeii and Herculaneum and elsewhere around the volcano. The novel is particularly important because the author attempts to describe the volcanic processes on the basis of the eyewitness report of Pliny the Younger, it is thus the earliest recreation of a volcanic eruption.

Bulwer-Lytton's novel coincides with the rise of awareness for volcanoes among the European intellectuals. It is virtually certain that he was aware of the volcanological research of his British contemporaries, such as the experiments of Sir Humphry Davy (who gave frequent popular lectures on volcanoes in London) and the pioneering fieldwork of George Poulett-Scrope on the volcanoes of Central France, as well as the writing of Charles Lyell.

The author with his wife arrived in Naples in 1832, to spend a winter abroad and to write. He explored the ruins of Pompeii and Herculaneum with the guidance of the archaeologist Sir William Gell, whose major work *Pompeiana* (1852) described in great detail the topography, edifices, and ornaments of Pompeii.

Bulwer-Lytton got the idea for the subject matter of the Pompeii novel—and the title—from the painting *The Last Days of Pompeii* that he had seen in Milan. This was the huge 1831 work of the Russian Karl Bryullow. In Italy, Bulwer-Lytton pored over literature sources about Vesuvius volcano, the eruption of AD 79 and the Roman cities, reading Vitruvius, Strabo, Livy, Tacitus, Seneca, and Dio Cassius, as well as the letters of Pliny the Younger. He returned to England in 1833, after having written most of the novel in Naples.

While exploring Pompeii, Bulwer-Lytton heard a suggestion that during the blackness in the city at the time of the eruption, a blind person would have a better chance of escape than one who had full sight. This remark led to the creation of Bulwer-Lytton's blind flower girl Nydia, the main heroine of the story, who leads her friends to safety in the total darkness beneath the eruption cloud (Figure 78.1).

The novel has been justly criticized for turgid oratory and lush style, pseudopoetical excesses, pretentious striving for wit and profundity, melodrama, ham philosophy, and moralizing. These are said to be features of style that match well Bulwer-Lytton's personality: ostentation, pomposity, and hauteur. Still, *The Last Days of Pompeii* triumphed over all its deficiencies, and made an immediate and spectacular conquest of the readership, becoming by far the most popular of all Bulwer-Lytton's works. His accuracy to history was justly praised, as was his description of Pompeiian life and architecture. In this sense, the novel is one of the great recreations of the past. Perhaps the novel's most notable achievement is in the mounting suspense of its plot, beginning as a dark cloud hangs over the summit of Vesuvius. It reaches a climax in the last scenes of the earthquake and the eruption, with terrified crowds, crashing buildings, darkness, the falling of hot rocks and ash over the city, destruction, and chaos.

The threat of Vesuvius is gradually brought into the story and is used effectively to create suspense.

Bulwer-Lytton even makes an attempt at a volcanic hazard assessment when he describes the population sprawl up the slopes of the volcano. The fact that the residents of Pompeii may have noticed some atypical behavior of the volcano is brought into the lovers' conversation. *Could that mountain have any connection with the last night's earthquake? They say that, ages ago, almost in the earliest era of tradition, it gave forth fires as Etna still. Perhaps the flames yet lurk and dart beneath.* Bad omens continue: *From the summit of Vesuvius, darkly visible at the distance, there shot a pale, meteoric, livid light—it trembled an instant*

FIGURE 78.1 The blind Nydia leading Glaucus and Ione to safety out of Pompeii during the eruption of Vesuvius in AD 79. *From the historic novel* The Last Days of Pompeii *by Bulwer-Lytton, 1850 edition.*

and was gone. An earthquake from Vesuvius later saves the hero Glaucus, as he is about to be killed by the villain Arbaces the Egyptian, who also lusts for Ione: *…at that awful instant, the floor shook under them with a rapid and convulsive throe—a mightier spirit than that of the Egyptian was abroad!*

Glaucus is saved once again by the volcano on the city's last day. Arbaces has framed Glaucus for murder, and the Greek hero is sentenced to go before the lions in the Amphitheater that fateful day, armed only with a dagger. In a suspenseful scene, the grating is lifted and the lion jumps into the arena. But miraculously it shows no interest in attacking Glaucus, which the crowd takes as a proof of his innocence. Glaucus is saved by a witness, who pins the crime on the Egyptian and wants to throw him to the lions. Arbaces has noted, however, that the volcano has become active. In describing the next stage of the eruption, Bulwer-Lytton takes the common view that mud flows overwhelmed the cities around Vesuvius: *Amidst the other horrors, the mighty mountain now cast up columns of boiling water. Bent and kneaded with the half-burning ashes, the streams fell like seething mud over the streets in frequent intervals.*

Bulwer-Lytton describes the climax of the eruption by citing Pliny's Letters in footnotes, in what is perhaps the earliest (1834) volcanological interpretation of the eyewitness account. Glaucus and Ione escape the horror of the eruption with help from their friend and guide, the blind Nydia, who alone can find the way in the total darkness below the eruption cloud. They reach the shore and find safety on a boat at sea: *Meanwhile the showers of dust and ashes, still borne aloft, fell into the wave, borne by winds, those showers descended upon the remotest climes, startling even the swarthy African; and whirled along the antique soil of Syria and of Egypt.* Here Bulwer-Lytton cites, in a footnote, the writings of Dio Cassius as evidence for the ash fallout in Africa and the Levant. At the very end of the novel, Nydia, unable to attain the love of Glaucus, slips out of the boat and into the sea while the lovers sleep, and drowns below the waves.

In a footnote, Bulwer-Lytton discusses the various theories that have been advanced for the mode of destruction of Pompeii. He adopts the one *…which, upon inspecting the strata, appears the only one admissible by common sense; namely, a destruction by showers of ashes, and boiling water, mingled with frequent eruptions of large stones, and aided by partial convulsions of the earth. Herculaneum, on the contrary, appears to have received not only the showers of ashes, but also inundations from molten lava.* Where did he get his relatively advanced volcanic ideas? Was he a friend of his geological contemporaries, the pioneer volcanologist George Poulett-Scrope and the geologist Charles Lyell? Although he is close to the mark on Pompeii, his assertion that lava flows inundated Herculaneum, on the contrary, is wrong and is probably based on the misidentification of the outcrops of consolidated pyroclastic flow deposits in that city as lava flows.

2.2. The Crater

The American author James Fenimore Cooper, famous for his Leatherstocking Tales, published a novel *The Crater*, or *Vulcan's Peak*, in 1847. The story begins in the style of Robinson Crusoe, with a shipwreck on a desert island. The hero is Mark Woolston, who becomes a sailor for the East India trade in 1793. The vessel is shipwrecked, and the only two survivors land on a reef-fringed volcanic island, which Cooper describes in considerable geologic detail. Mark and his companion Bob take advantage of the volcanic formations in many ways, using hollows in the lava as water catchment cisterns, the good soil inside the crater as a natural garden in fertile volcanic ash, and the sheltered crater as their home. Bob is driven off to sea in a small boat during a storm and feared lost, but Mark continues to eke

out a living in solitude on the volcanic island for many months.

One day Mark wakes up to terrifying tremors and a violent volcanic eruption, with suffocating gases. Smoke and ashes fill the air. A large part of the surrounding reef and his island has been thrust up above the surrounding sea, by a sudden elevation of the Earth's crust. The volcanic outburst came from a new crater offshore. A new mountain—Vulcan's Peak—has risen overnight to at least 1000 ft above the ocean. Mark climbs the new mountain and sees other islands in the distance and, much to his joy, a boat approaching, with Bob, a native from a distant island, and a group of new settlers. Bob had returned to America and had told the news of their shipwreck. Then he returned to the Pacific in the company of Mark's wife and several other settlers—the Craterinos. Soon a virtual Utopia is established on the new volcanic island and they begin to multiply, ruled by the author's concept of the ideal government (Scudder, 1947). Years later, Vulcan's Peak sinks back into the ocean, and their Utopia is destroyed in a catastrophe, leading to disintegration and physical destruction of the Craterinos.

The Crater is a story of survival by courage and ingenuity, with precise description and explanation of many geological phenomena. The central scientific idea of *The Crater* is the craters of elevation theory of the German geologist Leopold von Buch (1774–1853). He was probably the most illustrious student of the arch-Neptunist Leopold Werner and one of the most traveled of all field geologists. Leopold von Buch proposed a volcanic origin of oceanic islands from his studies in the Canary Islands in 1815. There he devoted much energy to the study of the form of volcanic mountains and developed his craters of elevation theory, which stated that volcanic mountains were thrust up by upheaval of the Earth's crust. Volcanoes, he noted, were generally conical in form, and their strata dipped in all directions away from the central crater at the summit.

Curiously, he did not recognize that this was due to the pouring out of lava and ash from the summit but attributed their form to uplift, accounting for the craters through the bursting and collapse of a bubble-like mass at the summit. In order to support his theory, he proposed cavernous openings beneath the volcanoes, separated by walls or arches, and hypothesized that these caverns were periodically subject to influx of great quantities of gases and lava. Although von Buch abandoned the Neptunist theory, he never abandoned the Craters of Elevation theory. The theory was shattered, however, by a volcanic eruption in the Mediterranean in 1831, creating a new island variously called Graham Island, Isola Julia, Ferdinanda, Sciacca, Corrao among others, depending on the nationality of the name-giver. Charles Lyell was quick to point out that the dip of strata on Graham Island was due to deposition, not upheaval. The island has since disappeared beneath the waves—not by subsidence but rather by erosion.

Cooper's contemporary Herman Melville also featured geology in several of his novels, notably in *Mardi* (1849), *Moby Dick* (1851), and *The Encantadas* (1854), a tale of the Galapagos Islands. Melville was most likely influenced by the popular and widely read writings of the British geologist Charles Lyell, particularly his *Elements of Geology* (1841). In *Mardi*, the philosopher Babbalanja offers his theory of volcanism and the evolution of his island. Foster (1945) has proposed that Melville, with the elevation of strata by "volcanic throes," was here parodying the **Plutonian theory**, particularly the ideas of Buffon and Hutton.

2.3. Journey to the Center of the Earth

The French writer Jules Verne (1828—1905) is the undisputed founder of the genre of science fiction. *Journey to the Center of the Earth* (1864) was the third in his series *Les Voyages Extraordinaires*, where he spins a tale of high adventure around a credible scientific hypothesis—in this case a geological theory about the physical conditions in the interior of the Earth. Verne was clearly very familiar with the conflicting ideas about the origin of heat in the Earth, which basically reflected the debate between the **Neptunists** and the Volcanists of the later eighteenth century. One hypothesis centered on primordial heat in the Earth and predicted a gradual increase in temperature with depth, all the way down to the Earth's core. The competing hypothesis was that heat in the Earth was superficial and caused by combustion at or near the surface, involving heat-giving reactions of the alkali metals on contact with air and water—a hypothesis championed by Sir Humphry Davy. Verne's German hero, Otto Liedenbrock, is firmly in the Davy camp and insists that there is no proof that heat increases greatly with depth. He is basically a Neptunist at heart—the typical position of German scholars of his time—but he is always on the lookout for evidence that can support his theory.

The professor is therefore ecstatic when he and his nephew Axel discover a soiled slip of ancient parchment with runic lettering, in a volume of the twelfth-century chronicle of Heimskringla by the Icelandic scholar Snorri Sturluson. After much effort, they manage to decipher the message: "Descend, bold traveler, into the crater of Snaefellsjokull, which the shadow of Scartaris touches before the kalends of July, and you will attain the center of the Earth; which I have done, Arne Saknussemm." The professor and his nephew take the first ship to Iceland, in order to reach the volcano's summit before the kalends of July. Upon arriving in Reykjavik, they hire Hans, an Icelandic guide, who is intimately familiar with the terrain and will assist them in carrying their instruments into the bowels of the Earth, including thermometers, aneroid barometer, chronometer, compasses, and Ruhmkorff's apparatus (based on a Bunsen pile and the bichromate of potash) to give an abundant supply of crucial electric light.

As they approach the glacier-capped volcano, Axel becomes increasingly worried about the prospects of descending into a crater that could erupt at any moment. The professor reassures him and insists that there is no cause for concern, pointing to the widespread fumaroles and jets of hot springs around the volcano: *You see all these volumes of steam, Axel: well, they demonstrate that we have nothing to fear from the fury of a volcanic eruption.*

As they start their journey down the volcano, the Professor makes geological observations, and after a near-vertical descent of thousands of feet he declares, *The further I go the more confidence I feel. The order of these volcanic formations affords the strongest confirmation to the theory of Davy. We are now among the primitive rocks, upon which the chemical operations took place which are produced by the contact of elementary bases with water. I repudiate the notion of central heat altogether. We shall see further proof of that very soon.*

The travelers are pleased to find that, even at a depth of 1100 ft, the temperature has scarcely increased. They make good progress downward in an inclined lava tube during the next few days, through a succession of fossil-bearing geologic periods. After 20 days, they reach a depth of 16 leagues, at the base of the crust, but the temperature has increased only slightly. Among the wonders conceived by Verne is a subterranean sea in a vast cavern, which the professor promptly names the Liedenbrock Sea. The travelers fashion a raft and set sail across the sea until they are caught in a violent torrent or waterfall that carries their raft instantly to great depths. The torrent reverses direction, and now they are carried upward at even greater speed, and the heat becomes unbearable. The walls of the shaft are burning hot, and the professor concludes that they are being carried up the throat of an erupting volcano! The water column below their raft is replaced by red-hot lava lifting them rapidly up the conduit. Suddenly they are ejected violently from the crater, and, badly shaken but not injured, find themselves lying on the crater rim of an erupting volcano—Stromboli, thousands of miles from Iceland.

After Jules Verne's death, a number of manuscripts were found by his family among his papers. One of these was a manuscript of a novel titled *Le Volcan d'or* or the Golden Volcano. An altered and highly edited version of the novel was published by his son Michael in 1906, but the first complete translation in English of the original version of the novel did not appear until 2008. It is an adventure tale dealing with gold fever in the Klondike region of the Yukon Territory in Canada, hardships, action, and adventure, in a story that anticipates in style the Arctic sagas of Jack

London and the westerns of Ben Harte. Two Canadian cousins inherit a mining claim in the Klondike. During their expedition they learn of a fabulous gold-filled volcano near the Arctic Ocean. Rather than wait for the volcano to erupt their riches, the miners excavate a tunnel and divert a river into the magma chamber, causing a violent steam explosion. Unfortunately, the volcano discharges its load of gold nuggets into the deep Arctic Ocean, out of reach of the Canadian cousins.

3. TWENTIETH-CENTURY NOVELS

A large number of twentieth-century novels feature a volcano, either as a central theme or as a symbol. Perhaps the most famous modern novel that features a volcano in the title is Malcolm Lowry's *Under the Volcano* (1947), which has been referred to as having the best account of a "drunk" in fiction. This modern classic is regarded as one of the major novels of our time, but its relevance to volcanic activity is little or none, except for the memorable title—and perhaps the use of the volcano as a metaphor for the mental and physical hell that the British consul, the novel's main character, descends into. Similarly, do not look for wisdom about eruptions in the massive novel *The Volcano God* by Philip Freund (1959). It is an epic profile of our age and, again, has no relation to volcanism except for the enigmatic title.

Modern novels about volcanoes can be roughly divided into literature and adventure genres, with works in the latter being by far the most numerous. Perhaps the most popular adventure writer of our time, Patrick O'Brian, puts a volcanic eruption to good use in *The Wine-Dark Sea*. When naval Captain Aubrey is chasing a privateer in the South Pacific Ocean, amazing natural phenomena occur, as the sky turns wine-red, a tidal wave swamps his ship, and violent explosions emerge from the deep. When daylight breaks, they witness the eruption of a new volcanic island that has devastated their prey with ashfall and bombardment of volcanic ejecta.

The novels in the literary genre often take real volcanoes and their eruptions as a backdrop for the plot but not as the main theme. The 1902 eruption of Montagne Pelee is featured in several of these novels, which is not surprising, considering the strong human appeal of the story. In addition to novels, a few literary accounts of volcanic eruptions have been written. Two noteworthy historical accounts of the catastrophic Pelee eruption are *The Tragedy of Pelee* by George Kennan (1902) and *The Day the World Ended* by Gordon Thomas and Max Witts (1969). Similarly, *Krakatoa* by Rupert Furneaux (1964) is an accurate historical account of the great 1883 eruption. However, in the category of historical accounts of volcanic eruptions, *The Night of Purnama* by Anna Mathews (1965) is unique. Shortly after the author and her husband took up residence in the

village of Iseh on the south flank of Mt Agung in Bali, the volcano burst into a furious series of eruptions, producing numerous and deadly **pyroclastic flows**. Her book is a tense and thrilling account of the impact of the lethal eruption on the Balinese village community, providing much insight on the relationship between the Balinese and their gods—the most powerful of which reside in Agung.

In the theater, a recent work based on the 1902 eruption of Montagne Pelee is the play *The Prisoner of St. Pierre* by Pat Gabridge, which describes the real-life ordeal of the prisoner Ciparis, a black man sentenced to death. When the eruption took place, Ciparis was being held in a dungeon below the city, which protected him from the worst effects of the scalding hot pyroclastic surge. He became the most famous survivor of the catastrophe, and the play describes the hallucinations he experiences while waiting for rescue. Ciparis later made a living by touring in the United States as a member of the Barnum and Bailey Circus. It is perhaps surprising that his fantastic story has not yet been featured as the central theme of a novel or a film.

3.1. The Violins of Saint-Jacques

Patrick Leigh Fermor traveled in the Caribbean after World War II and as a result published *The Violins of Saint-Jacques* in 1953. This beautiful and haunting novel was immediately hailed as a rare and exotic seep of color across the drab monochrome of the postwar years. It is set on a Caribbean island of romantic passion, tropical luxury, and European decadence, and the story captures the delicacy of high society entanglements and the unforeseen drama of the volcanic forces beyond human control.

The narrator first encounters the Caribbean volcanic island in a painting during his visit to the artist Berthe. It is a picture of a colorful town, above which "a tropical forest rises in a cone, hiding to its crater the steep and concave flanks of a volcano from whose blunt apex curled a languid blue-gray banner of smoke." It is inscribed by the artist and the year 1902. Berthe tells him she had spent 6 years as a governess to the Sheridans, a patrician family on the island of Saint-Jacques, in the shadow of the ever-smoldering Salpetriere volcano.

Every year on the last day of carnival, a great ball was traditionally held in the Count Sheridan household in Saint-Jacques. On this occasion, in the days when preparations were under way for the ball, the Salpetriere had been burning for a week or so with unaccustomed vigor. The ball, attended by the governor of the island, turns out to be a tremendous success, but interrupted by an unexpected event.

The description of the eruption of Salpetriere is largely based the May 8, 1902, eruption of Montagne Pelee, when ships in the harbor were set aflame by fiery hot surges passing over the ocean, while ships further out in the bay were singed, but limped away to safety. Berthe, who is safe on a schooner outside the harbor, watches in horror as the citizens are halted in their flight and swept down to the ground by the descending gas cloud that invaded the city when the mountainside opened up. The cloud rolls into the harbor and over the bay, as ships caught on fire. She is terrified that the deadly cloud will reach the schooner, but *The cloud's soft progress over the water halted and its edges began to creep backwards again as though the ghostly mass were drawing in its skirts.*

The captain orders all sails set and steers the schooner due west, out of harms way. They look back and see that the island is literally coming apart, with the erupting fissure opening in a fiery yawn. Hills near the coast collapse, sending up a huge tidal wave that wildly tosses the schooner on the sea. Saint-Jacques is sinking into the ocean as the schooner speeds away. Soon only the summit of la Salpetriere remains above the waves, with huge billows of steam around. Then only the crater rim remains visible, with an immense plume of fire, but at last that too subsides into the sea, *and the water flowed in and snuffed it almost in silence.*

In the following years, ships sailing over the location of Saint-Jacques find no physical trace of the island. However, fishermen from Dominica and Guadeloupe claim that when they sail over the submerged city at carnival time, they can hear the faint sound of violins coming up through the water, as though a ball were in full swing at the bottom of the sea. They call it "the violins of Saint-Jacques."

3.2. The Sleeper of the Moonlit Ranges

In the novel *The Sleeper of the Moonlit Ranges* by Edison Marshall (1925), the native Alaskan Breed Bert from Pavlof village becomes a guide for rich and spoiled Americans in the Alaskan volcanic and rugged wilderness. It is a story that is strongly influenced by the Katmai eruption of 1912, and makes frequent references to the rapidly changing face of the Earth as a result of volcanic activity. Bert and beautiful Grace fall in love, and their journey eventually turns into a deadly love triangle, involving Grace's jealous fiance Paul. But Pavlof Mountain begins to erupt, at first with a ruddy fountain of lava from the summit crater. The native Alaskans begin a dance to pacify the god of evil in the volcano, and Paul joins them in a frenzied dance. In a trancelike condition, Paul demands that the natives sacrifice both Grace and Bert to the fire god, but they escape—at least for a while—by running straight up the flank of the volcano. But now they are caught in a vast volcanic outburst, as Pavlof explodes and they loose consciousness in the cataclysm. When they come to, Bert discovers that one side of the volcano has been blown out, the village has been destroyed, and lakes of molten lava are forming in the crater and flowing toward them. The lovers

escape miraculously and begin a new life as pioneers of the Alaskan frontier.

Human sacrifices to the volcano gods have been a highly popular topic in a number of other novels, including *The Phantom City* by William Westall (1870), where a young maiden is sacrificed to the fire god in an erupting volcano in the Mayan region of Mexico.

3.3. Paricutin

Paricutin is a truly historical volcano novel, written by Bernice I. Goodspeed (1945) and published immediately after the birth and growth of Paricutin volcano in Mexico. The central theme of the novel is the mental struggle among the native Tarascan Indians in response to the devastating eruption. It is a struggle over whether to follow the Christian ways and pray or to pay homage to their ancient fire gods and make human sacrifices to appease their wrath. It is thus a story about a people caught at the crossroads of two cultures as they react to a natural disaster.

Some of the elders of Paricutin village still secretly worship the fire god Ome Teotl. Thus when Mazatl witnesses an erupting fissure opening up in his field, he is convinced that someone must have offended Ome Teotl and that more tributes to the god are required (Figure 78.2). The priest of their village, however, tries to persuade his people that the eruption is an act of nature, and not due to the actions of the Christian god or of Ome Teotl. But the elders recall the creation of the Jorullo volcano nearby, that had severely affected their ancestors. When all prayers had failed at Jorullo, the elders gathered in the temple, and decided that only a human sacrifice would pacify the god of fire. A youth was chosen, and he climbed up the cone to the very brink of the erupting crater, where great tongues of the

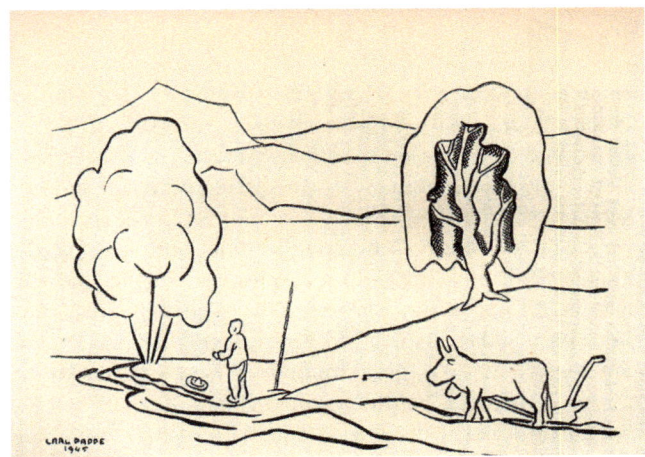

FIGURE 78.2 When Mazatl plows his field, a fissure opens up in the ground, marking the beginning of a violent eruption and the birth of Paricutin volcano in Mexico. *The illustration is from the historic novel* Paricutin *by Bernice I. Goodspeed.*

flames reached out and swept him down into the fiery abyss of Ome Teotl. After a final explosion, the Jorullo crater subsided, and it appeared that their sacrifice had brought them peace.

As the eruption of Paricutin intensifies, the elders become increasingly worried, convinced that the prayers to the Christian god are of little or no value in calming their fire god. Mazatl and his two friends become convinced that only a human sacrifice will save their community from destruction, and select their teenage children, the virgin Zolintzin and young Tochtil, to sacrifice to the flames. They decide, however, that the sacrifice must be made to Popocatepetl—father of all volcanoes—the principal residence of Ome Teotl. The children are convinced that the sacrifice is a necessity and set out on the long journey to Popocatepetl with one of the elders. As they near the crater, the elder is killed in a fall into a canyon, and the frightened youngsters are taken under the protection of an old forester and a Padre in a nearby village. They are torn in a great conflict between the promise they have made to the elders to sacrifice themselves in the volcano crater, and what they hear from the forester and the Padre. Their new guardians gradually convince them that such a sacrifice is not essential for the well-being of their village. The teenagers settle down in the village at the foot of Popocatepetl, marry, and have a child.

Meanwhile, the eruption of Paricutin continues and lavas engulf more fields and villages. Mazatl is now convinced that the sacrifice of his children has been in vain, the ancient god has let him down, and he is heartbroken. The youngsters return to Paricutin 1 year later, where they are welcomed by all, except Mazatl, the father of Zolintzin, who is furious over the breach of promise of the sacrifice. He runs up the side of Paricutin's erupting cone to sacrifice himself to Ome Teotl.

3.4. Volcano

As a land where volcanoes are a dominant part of the landscape and often threaten lives and property, Japan is the source of a number of volcano novels, including Shusaku Endo's *Volcano* (1959). In this carefully elaborated Japanese thriller, the aging volcanologist Jinpei Suds and the Catholic priest Dehorned become obsessed by the Akadake volcano. For both, the question whether the volcano is still active or not becomes the overriding concern in their lives. Suds has spent 15 years studying Akadake and is convinced that it is dying. This opinion is very good news to a tycoon who is building a lavish hotel resort on the volcano's slopes. Dehorned is equally certain that the volcano will erupt again, pouring streams of lava down over the religious Christian sanctuary being built at its foot. To Dehorned, the volcano is a symbol of evil and a judgment on the people who cannot understand the significance of sin

and redemption: *Akadake is absolutely going to explode. Because evil is a volcano that will never be extinct.* To Suds, the volcano must remain dormant, or else it will wreck the scientific thesis on which he has spent much of his life and on which the book he is about to publish, *The History of the Eruptions of Akadake*, is based. The book is the result of 15 years of research, with the financial support of the business tycoon.

The novel's volcano has had numerous historical eruptions, the latest at the end of the nineteenth century, but since Suda's arrival at the observatory, only a plume of thin smoke rises from the crater. Professor Koryama of Kyoto University has declared that this is sign of dormancy; the volcano was stricken with old age: *What a mount of heartache it is. A volcano resembles human life. In youth it gives reign to the passions, and burns with fire. It spurts out lava. But when it grows old, it assumes the burden of past evil deeds, and it turns as quiet as a grave.*

Shortly after the ailing Suds is admitted to the hospital, Akadake suddenly explodes.

The observatory monitoring the volcano stresses that there was no cause for alarm and no imminent danger, even though the eruption was totally unexpected. Investigators report that no tremors of any kind, that might be portents of volcanic activity, have been recorded before the event. They also report that the eruption has not produced any new magma, merely a discharge of gases trapped in the crater mouth. After a few days, life is back to normal, and it seems as if people have completely forgotten about the event—Akadake appears to have resumed its old forlorn appearance.

Suds had earlier observed that, historically, the volcano has had a peculiar behavior. Successive eruptions always originate at a higher elevation on the flanks than the previous one: It is *an upward moving volcano*. He is relieved when he finds out that the new explosion has taken place from the very top—it is just the last gasp of a dying volcano. This is challenged by a young Tokyo University professor who discovers that the gas emissions include enormous amounts of halogen elements. He predicts the coming of a big eruption at the 900-m level on the volcano's eastern flank, which would devastate the planned hotel complex. Suds feels humiliated, and his pride is hurt by this novelty hypothesis concocted without evidence by a greenhorn scholar. He reassures the tycoon that all is well; *Everybody wants to publish novel ideas. Look it's safe, I tell you. Akadake is impotent with old age.* When Suds regains some of his strength, he accompanies the tycoon on an excursion to inspect the volcano. They are both pleased to find that the hotel site appears normal and safe—until Suds notices a dead field mouse and withered vegetation. He remains quiet about his find and terminates the investigation, fearing what he might find. A few days later, when he visits the observatory, he discovers to his dismay that ominous tremors have occurred during the night, suggesting an imminent explosion. He returns to the flanks of the volcano, where he finds unmistakable warning signs, such as increased temperature of the ground. Still keeping his findings secret, he returns in silence and in deteriorating health, dying a few days later. His son carries a crock with his ashes to the slopes of the volcano for burial. Shortly afterward, Father Dehorned also dies, but a single plume of smoke continues to rise from Akadake. Will it erupt tomorrow, or is it truly dormant? The reader is left to wonder at the novel's end.

This is a geologist's novel in many ways. It is full of field observations, historical volcanological data, and lucid descriptions of landscape and Earth processes. The author Shusaku Endo had great affection for Kyushu Island, the setting for his novel. Thus the city in the novel is based on Kagoshima, and Akadake volcano is inspired by Sakurajima, one of Japan's most active volcanoes. It is reported that the author's curiosity about the volcano was so great that he had himself lowered from a helicopter into the Sakurajima crater.

3.5. The Volcano Ogre

The Volcano Ogre by Lin Carter (1976) is an example of the disaster/science fiction/fantasy novel. The action takes place on the slopes of the fictitious Mt Rangatoa volcano on a Polynesian island. In its innards, the volcano harbors a grotesque, humanlike monster, covered with glowing hot lava, who occasionally lumbers out of the crater to threaten the local inhabitants. After the lava devil kills again, the islanders send a plea of assistance to the all-powerful Prince Zarkon and his crime-fighting Omega Crew. The hideous volcano monster, wading through red-hot molten lava as if it were tepid sea water, turns out to be an American geologist, wrapped in protective plastic boron-fiber insulation against the heat of the lava. There is no knowledge of volcanoes to be gained from this novel, although the idea of a geologist as an evil volcano monster is no doubt appealing to some people.

3.6. Genesis Rock

In this fantasy novel, Edwin Corley (1980) spins a lively tale about a volcanic eruption in the heart of New York's Central Park. It is indeed a far-fetched concept but an entertaining story that drops the names of several contemporary volcanologists and makes abundant references to famous eruptions and geological processes. Janet McCoy, a beautiful and brilliant volcanologist who works for the U.S. government, is recalled to New York from Iceland, where she has been studying methods to harness geothermal energy. The reason for Janet's recall is that she has been picked by NASA for a mission to explore the

geothermal energy resources of the planet Mars. Passing through Central Park, she is amazed to come across a chunk of anorthosite lying by the path—a Genesis Rock. In her opinion, it is a very rare rock type that is only found at the surface when hurled out of the Earth by cataclysmic upheavals, such as volcanic explosions. She also finds out that Central Park is riddled with dikes.

A homeless man, who has made shelter in a cave in Central Park, notices hot steam venting out of a fissure in the rock and makes use of it to cook squirrels for lunch. Events unfold as mysterious bubbles of sulfur-rich gases emerge from the lake, and a construction team blasting a subway tunnel through the park encounters steam veins in the solid rock. When a violent steam explosion scalds a worker to death and injures several others, the news spreads rapidly that something very unusual is afoot below Central Park.

Janet assembles a team of Earth scientists to study this strange phenomenon, and they are startled to find that infrared aerial photos show a "hot" line that strikes from the Baltimore Trench, goes across the continental shelf and terminates near Staten Island. Is this thermal anomaly associated with a magmatic intrusion at depth? Or is it just a stringer from the Gulf Stream? Meanwhile, the fissure erupts again, forming a geyser that shoots up steam to a great height. Janet McCoy informs the mayor that the birth of a new volcano may very well be taking place in Central Park, and that the entire New York region may soon be covered with 4 ft of ash from a Krakatau-size explosion; alternatively, it may result in an Icelandic-type eruption, with widespread lava flows covering the park and large parts of the city. Meanwhile, her geophysicist colleagues are exploring ways to head off a volcanic eruption, possibly with the detonation of a nuclear bomb. They discuss an alternative scenario of diverting lava flows by pumping water on them, as was done successfully on Heimaey in Iceland in 1973.

The subway tunnel construction crew decides to pump large volumes of water down the fissure to cool things off at depth before they resume drilling—but with disastrous results. *With an incredible belch, the solid rock heaved and began to reject the liquid diet.* Several steam explosions occur, ejecting boiling water and superheated steam all over the park and into the city. A thick mudflow rushes down the East Drive and into the East River, resulting in many casualties and destruction of buildings, including the mayor and his residence. Huge fountains of steam and water rise to thousands of feet above the boiling reservoir. When most of the water is dissipated, the top of a glowing lava mass begins to emerge (Figure 78.3).

Janet predicts that worse is to come and demands that the city be evacuated before more deaths occur. The president agrees, and, at Janet's suggestion, the Joint Chiefs of Staff plan to bomb the hot vein of magma that runs across the continental shelf to New York with a nuclear device, in order to relieve the pressure below the city. Meanwhile, the

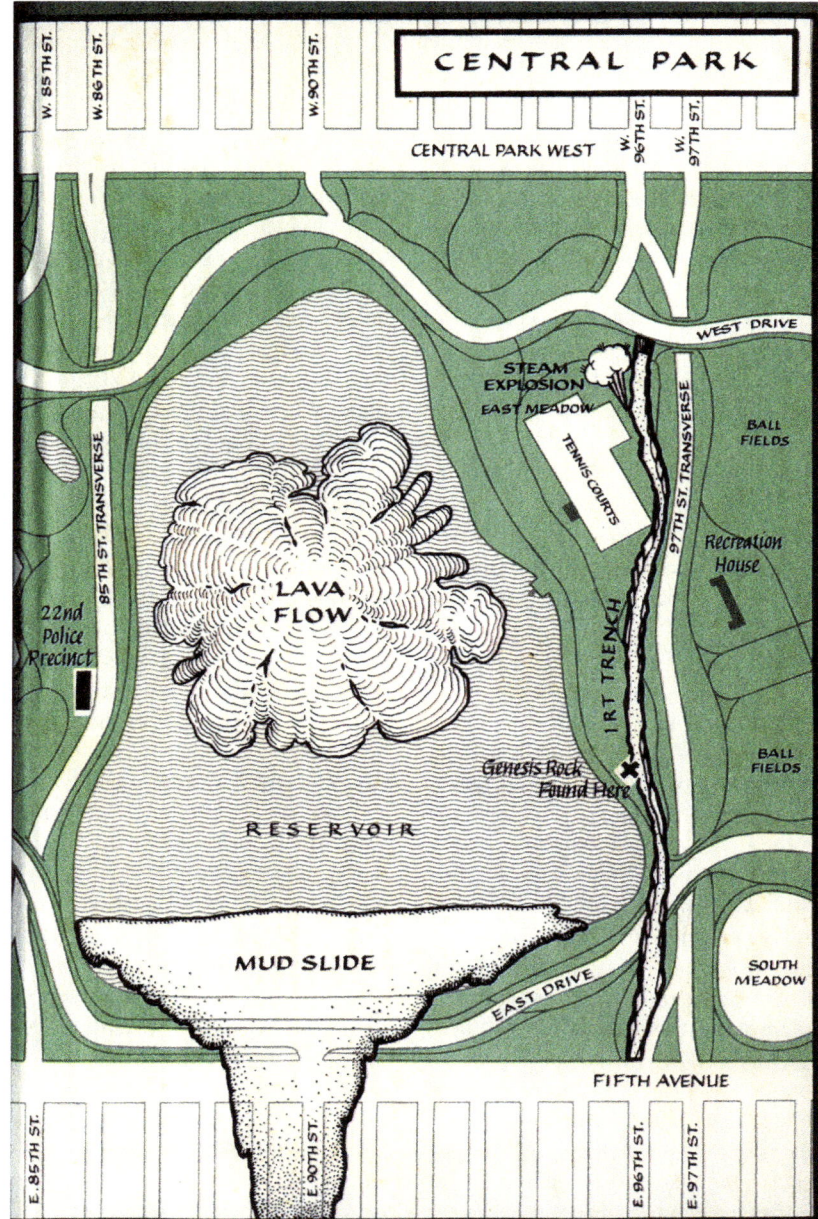

FIGURE 78.3   In the novel *Genesis Rock*, a new volcano bursts forth in Central Park, New York.

reservoir has nearly filled with lava and it is about to burst out of the park. A submarine delivers the nuclear bomb to a spot above the magmatic sill in the ocean floor, and shortly after it is detonated, the magma extrusion rate in the new Central Park volcano slows down to a trickle. Janet McCoy has saved New York City.

The author of this novel came up with an extremely unlikely scenario, but he did some homework, making use of ideas behind the lava diversion in Heimaey in 1973, as well as early bombing of lava flows on Kilauea—though nuclear devices were not used.

3.7. Gentlemen in England

A.N. Wilson (1985) has written a novel of repressed Victorian life and manners, set in London in 1880. *Gentlemen in England* is not a novel about volcanoes, but it features a volcanologist as one of the main characters. The plot centers around young Charlotte, who is courted by the much older Professor Horace Nettleship, an enormously respected volcanologist. She is impressed by the potency that is implied by his fieldwork on volcanoes, and they marry. Soon however, the marriage is in trouble, largely because her husband confines all his passion to the three volumes he is writing about his research. As the plot develops, husband and wife have not spoken privately for 15 years, but keep up the appearance of a happy marriage in public—a caricature of Victorian society. The author makes the perfect statement—indeed the epitaph—of the typical volcanologist: *Volcanoes his expertise, he found out about the world by peering at little pieces of it through a microscope, by hammering little chunks of it, and keeping specimens of it in boxes.* The volcanologist's alienation from society is also perfectly captured. The professor is too absorbed in his work: *He has been working on a great book about volcanoes ever since my daughter married him…All their little holidays as far as I can recollect have been spent in places of particularly volcanic interest.*

3.8. Texaco

Martinican writer Patrick Chamoiseau won the prestigious Prix Goncourt, France's most celebrated literary award, for *Texaco* (1992), a novel about his French Caribbean island home, written in the French Creole oral storyteller tradition. Like *The Violins of Saint-Jacques*, the deadly 1902 eruption of Montagne Pelee is a subplot within the novel, but here it is seen from the perspective of a former slave. *Texaco* is a mythic history of the rise of the Creole nation in Martinique from its African and French roots. The title refers to a gas station at the outskirts of Fort-de-France, around which a shantytown developed.

Esternome Laborieux is born into slavery on a plantation near the city of Saint-Pierre, at the foot of Montagne Pelee. The slaves knew that *…the lumpish mountain overlooking Saint-Pierre was in reality a raging bull.* When slavery is abolished, Esternome abandons the plantation and takes up carpentry in Saint-Pierre. Here he meets beautiful Ninon, and, together, they leave the city and squat on abandoned land in the hills, living the life of subsistence farmers. The idyll is shattered when Ninon runs off to Saint-Pierre with a musician. Esternome lives on in sadness, and in this daze he did not *see how from time to time the horizon became roaring and how ash from the mountain suddenly floured the land, more and more often, for a longer and longer time.* One morning a huge bang blogodoom shook him into consciousness. He came out of his thousand-year stupor with a ball of hope. He calls for Ninon while "manic fleering shook the quarter." *He wants to rescue Ninon but he discovered a lump of coal instead of Saint-Pierre. Despite his barbecued heart, it stunned him…Soon he couldn't take another step forward. His eyes were burning. Even his gray hair burned him. The grass, the trees had withered…And he moved on and on and on. So that he was the first to enter Saint-Pierre.*

He went from cadaver to cadaver and examined every blackish thing, but of course he never found Ninon.

The tragedy of Montagne Pelee, so vividly descried here, is a recurring theme in novels about the Caribbean. Chamoiseau's later novel, *Solibo Magnificent* (1998), also refers to the eruption, with discussion of Mt Pelee, the ash of Le Precheur, and other geological scenes on Martinique.

3.9. The Volcano Lover

In 1992, there appeared a romanticized biography or a historical novel about the life of Sir William Hamilton, by the celebrated American writer Susan Sontag. In this novel, the life of the central character is intertwined intimately with the past and present volcanic activity of Vesuvius. The plot is a well-known tale: Hamilton, an elderly diplomat and nobleman, is the British ambassador to the Kingdom of Naples, where he becomes fascinated by the activity of nearby Vesuvius. His interest in the volcano gradually evolves from amateurish enthusiasm to a fully fledged research project, which culminates in a series of important and lavishly illustrated scientific publications by this first volcanologist. He is tireless in studying the object of his fascination, traversing its lava flows, repeatedly climbing to the fire-spitting summit crater, and facing danger for the sake of more information about the workings of the inner Earth. He collects rock samples and ships them back to England for chemical analysis. When his wife dies, the lonely and aging nobleman begins a relationship with Emma, a young and strikingly beautiful former mistress of his nephew, and soon the unlikely pair weds. The celebrated couple participate in the rediscovery of the cities of Pompeii and Herculaneum, unearthed from the ashes of the AD 79 eruption of Vesuvius; the ambassador acquires a fine collection of antiquities, and the wife develops a new performance art form, with her "attitudes" or posing inspired by the recovered Roman paintings. When the great British naval hero Lord Nelson appears in Naples, he is nursed back to health by Emma, and here begins a somewhat scandalous long-term relationship that the ambassador manages to ignore.

The Volcano Lover is a formidable romantic biography that is based on meticulous research. All volcanologists are familiar with Hamilton's pioneering work on Vesuvius and on the entire volcanic region of Campi Phlegraei in Italy, and it is fascinating to read an account where this historic figure of volcanology is brought back to life in vivid detail.

3.10. Swimming in the Volcano

In *Swimming in the Volcano*, a love story and sophisticated political novel by Bob Shacochis (1993), the American economist Mitchell Wilson becomes unwittingly involved in a struggle between two factions of the government of a small Caribbean island, against the backdrop of volcanic activity. The novel's smoldering volcano is Mt Soufriere on the Caribbean island of St Catherine, and it is of course based on La Soufriere volcano on St Vincent. The volcano is correctly described as dormant since 1902, with a hot brown lake surrounding a steaming magmatic island in the crater.

In the novel, forest ranger Godfred Ballantyne of the Ministry of Agriculture is dispatched weekly up to the crater to monitor any change in activity. The topographic descriptions of the volcano are excellent, and many details about the monitoring of the Soufriere in the 1970s are brought to life. Because of Mitchell's liberal, leftist leanings and a nonconformist lifestyle, he is viewed with suspicion by the authorities and politicians on both sides. Some suspect that he may have connections with a group of guerrillas that are operating from camps on the remote flanks of La Soufriere. When Mitchell leads a group of friends on an ill-fated climb of the volcano, they are intercepted by a group of policemen. During the questioning, Mitchell's girlfriend is shot by an unknown assailant—a guerrilla or a government agent? She dies shortly afterward and Mitchell is jailed without trial, though eventually released.

Although this is not primarily a novel about a volcano or an eruption, the author did his homework on Soufriere, and his descriptions of the volcano and its monitoring are excellent.

4. JUVENILE FICTION AND COMICS

Volcanoes are a favorite topic with children and many authors of children and teenage fiction bring in volcanoes and their eruptions as a subtopic, but rarely are volcanoes featured as a central theme. Probably the best known children's novel to feature volcanoes is *The Little Prince*, written and illustrated by Antoine de Saint-Exupery. First published in 1943, it quickly became one of the world's favorite books. Although written as a story for children, it contains an urgent humanitarian message for people of all ages. Volcanoes are brought into the story as the prince describes his home, the asteroid B-612, to his newly found terrestrial friend. There are two active volcanoes on the asteroid, and the prince finds them convenient for heating up his breakfast in the morning. He is careful to clean the volcanoes out regularly.

The illustration of the Prince, straddling his tiny planet and cleaning out a volcano with a long poker, is an unforgettable image for children and adults alike.

4.1. The Twenty-One Balloons

The Twenty-One Balloons, a humorous story by William Pene du Bois (1947), tells of a fabulous voyage around the world in a balloon, undertaken by Professor Sherman from San Francisco in August 15, 1883. The date is important, because his only stop on the journey is on the island of Krakatau. The cataclysmic eruption of Krakatau actually occurred on August 26, 1883. In the story, the professor makes an emergency landing on a small volcanic island, after his balloon is punctured by seagulls. He soon finds out that this is Krakatau, and that it is inhabited by a handful of highly sophisticated American settlers, wealthy beyond belief because they have discovered a fabulous diamond mine under the volcano. The island is ruled by the Gourmet Government, whose motto is *Not new things, but new ways*. The only major drawbacks of the island are the frequent earthquakes and rumblings of the volcano. To avoid earthquake damage, the Krakatoans build their houses on foundations of solid diamond boulders.

The professor settles into a life on the island and joins the Krakatoans in short trips in their own balloon. One trip takes them directly over the crater, where the professor notes that a lake of molten lava seethes in the bottom, emitting clouds of sulfur. The islanders are acutely aware of the danger of living on Krakatau: *This whole island is like a turkey stuffed with nitroglycerin.* As an escape mode for a fast getaway in case of a violent eruption, they have constructed a flying platform to carry the 20 families to safety, lifted by 21 balloons.

On August 26, during the professor's lecture, a violent earthquake strikes the island, cracking walls, and breaking windows. The order is given to all to dash to the balloon platform for evacuation. They take off, but unfortunately the wind carries them directly toward the volcano and over the crater. Here they *were greeted by a roaring swift upward blast of hot air which catapulted the balloon far up into the sky*. Trapped above the fiery red crater, they bob up and down on top of the eruption column for the entire night, in hot sulfurous fumes. Fortunately, they eventually drift off the eruption column, toward Java, just before the entire island of Krakatau blows up in a tremendous explosion. A thick blanket of ash settles on the balloon, and the Krakatoans are deadly afraid that diamonds from the mine will fall and puncture their balloons. They drift away, over the Indian Ocean, the Mediterranean, and Europe. Finally the professor makes a controlled crash of the balloon into the North Atlantic Ocean. This is an unusual novel because it uses a major volcanic eruption as the central theme for a lighthearted story, with delightful illustrations of the volcanic activity.

4.2. Juan of Paricutin

The children's novel *Juan of Paricutin* (1953) is unusual in that it was written by a geologist—Marion Isabelle Whitney, who witnessed the eruption and spent considerable time with the local people. Her book can thus be properly regarded as a historical novel. It is a tender and sympathetic story of the hardships that a family in the village of Paricutin experiences as a result of the eruption.

When Juan and his grandparents visit a nearby village, a strong earthquake strikes, swaying the church tower and toppling another tower into the plaza. Rumblings from the Earth and quakes are common during the next few weeks. One day Juan and his family accompany his uncle Dionisio to help him prepare a field. While they get ready, Paricutin village is struck by the strongest earthquake yet. When they arrive in the field, they hear loud rumbling and tremors underground and see a pit opening in the cornfield right before their eyes! A part of the field appears to sink into the Earth, as trails of white smoke curl out of the fissure. A column of black dust rises from the center of the pit, and stones are ejected, as the deep rumble changes to a loud roar. They rush back to the village, where others have seen the activity from afar. An incessant roar comes from uncle Dionisio's cornfield during the night, and a fountain of brilliant red stones plays above the fissure, higher and higher. When morning comes, Dionisio and Juan return to the cornfield and find a terrible sight. A huge cone of rock and black ash had formed around the pit, and the cornfield was ruined. Back in the village, black ash spreads over everything, killing crops and making life nearly unbearable. The volcano grows in height and is up to 450 ft high after 12 days, and 1000 ft high after 2 months, still showering the countryside with rocks and dusty ash. Leaves wither on the trees and corn and all other plants die. Their livestock also die in the fields.

Tourists of all nations are now flocking into Paricutin to see the new volcano. Juan's father sets up a stand to sell soft drinks to the thirsty visitors, and the family manages to eke out an existence on this trade. The roof of their house falls in due to the thick ashfall, and eventually they are forced outdoors to set up tents. Most other villagers flee the area. One day a violent explosion blows the top off the cone, and a new vent opens up on its side, sending out a river of red-hot lava. After the initial fright, Juan begins to marvel at the fascinating spectacle and even at the beauty of sparkling crystals in the cooling lava. Gradually the river of hot lava begins to cover some of the houses of Paricutin village. But the American tourists keep coming, and the demand for bottled drinks is high. Juan's father gets the bright idea of selling the tourists donkey rides to the volcano, which enables his family to survive in the face of the disaster.

One day when young Juan is guiding American visitors, he witnesses a great outbreak of lava, which threatens to engulf the rest of the village. Soon the village is totally covered by thick lava, and not even the church is spared—only the church tower stands, sticking out of the black flow. One of Juan's customers, Bill, is a knowledgeable American volcanologist who teaches Juan many things about the workings of the volcano and the nature of the rocks they produce. In turn, Juan tells Bill about the

sequence of events during the first day of activity—the birth of a new volcano.

There are some excellent passages in the book that are highly educational for young readers interested in volcanoes. Juan cries over the loss of his village, but he can also see another side of his situation: *I don't like that old volcano. He makes me very unhappy. But he is beautiful.*

4.3. The Comics

The beauty and excitement of volcanic eruptions make them excellent material for comics. Two notable examples are *The Volcano with Hiccups*, one of the Flintstones stories, and *Donald Duck in Volcano Valley. The Volcano with Hiccups* (Horace J. Elias, 1982) is Mt Firepot, about 10 miles outside of Bedrock. It has not erupted for a long time but huffs and puffs a little once in a while. A couple of visitors, masquerading as the Flintstones' "lost cousins" show up in Bedrock for a few days, and, coincidentally, the very day they leave town, the banks are robbed. Fred Flintstone suspects that the robbers have sought refuge and a hiding place for the loot in the crater of Mt Firepot. When he and the police approach the volcano, it huffs and puffs, and spews out rocks, brushes, and 50 suitcases of money!

In *Donald Duck in Volcano Valley* (1973), Donald and his three nephews acquire a government-surplus airplane. Major Mañana, a stranger from the Latin American republic of Volcanovia, offers Donald 300,000 pesos for the plane. The major takes them for a test flight to Volcanovia, where they narrowly avoid crashing into several erupting and fire-spitting volcanoes. When attempting to land, however, they end up crashing on the summit of one of the active volcanoes. The situation turns serious when that volcano, Old Ferocio, gets ready to explode. Donald and his nephews get the bright idea to fill the crater with corn in order to stop an eruption. This results, however, in an explosion that spreads a thick layer of popcorn over the entire country of Volcanovia.

5. VOLCANO FILMS

5.1. Early Films and Movie "Classics"

It is appropriate that the first volcano film was *The Last Days of Pompeii* (1897), a British silent film inspired by Bulwer-Lytton's novel. Another black and white film of the same title but with a different plot was made in 1935, starring Basil Rathbone. This version is notable for its special effects, which are considered remarkable for its time. Italian filmmakers have also worked on this theme, first in the silent film *Gli Ultimi Giorni di Pompeii* (1913) and again in 1960, when writer Sergio Leone and director Mario Bonnard made an Italian—Spanish film with this

FIGURE 78.4   A poster advertising *Stromboli*, the movie about a beautiful foreign woman who settles with her fisherman husband on his remote, isolated, and active volcanic island.

title, featuring Steve Reeves and Fernando Rey. The theme's appeal is lasting; in 1984, Bulwer's novel was the basis of a television miniseries, starring Ernest Borgnine and Laurence Olivier. In 2013, the movie "Pompeii" also used the AD 79 eruption of Vesuvius as the theme, though its plot was different from Bulwer-Lytton's novel.

The most important film that relates to a volcano is the American–Italian production *Stromboli* (1950). The film is remembered in part for its artistic merit and in part because of the American establishment's fervently puritanical

reaction to the love affair between the actress, Ingrid Bergman, and the film's Italian director and producer, Roberto Rossellini (Figure 78.4).

Rossellini wrote the script for Stromboli. In 1948, he wrote a personal letter to Ingrid Bergman, then a major Hollywood movie star, about the plot he had conceived: A young and beautiful Latvian woman, victim of the horrors of war, seeks freedom by marrying an Italian soldier. Her simple husband takes her to his home on the volcanic island of Stromboli. Rossellini explains in his letter to Bergman that the Lipari Islands consist of seven volcanoes in the Tyrrhenian Sea, with Stromboli continuously active. The islands "earned sad fame during fascism, because it is there that the enemies of the Fascist Government were confined." On this island, Rossellini wishes Bergman to act out "the life of the Latvian girl, so tall, so fair, in this island of fire and ashes, amidst fishermen, small and swarthy, amongst the women with the glowing eyes, pale and deformed by childbirth, with no means to communicate with these people of Phoenician habits, who speak a rough dialect." The girl "is stranded in this savage island, all shaken up by the vomiting volcano, and where the earth is so dark and the sea looks like mud saturated with sulfur."

In April 1949, Bergman, Rossellini, and their film crew boarded a fishing schooner from Sicily to Stromboli. They sailed past Volcano Island, where Anna Magnani, Rossellini's former mistress, was starring with Rossano Brazzi in the film *Volcano* (1953).

The crew settled in the small village clinging to the side of the volcano and began shooting a movie. It was not without incident, however. One crew member was overcome by the volcanic fumes and died of a heart attack (Miss Bergman paid the 60,000 lira funeral expenses and set up a 600,000 lira fund for his family).

In her autobiography, Bergman said of Stromboli: *And the volcano…the volcano! The first time I had to try to climb it, we slogged for four hours upward. After I climbed for two hours, I just sat down, gasped and said, 'I'm sorry, I can't make it.' But after a rest I did make it, and at the top I could have just lain down and died.* During the filming on the island, a full-scale and highly publicized love affair blossomed between the married star and Rossellini. Puritan America was not yet ready for this display of free love, however, and the establishment attacked the famous actress publicly. *Stromboli* may be the only volcano film that can be considered a classic in movie history. However, a more recent film that has attained high praise is *Joe Versus the Volcano* (1990), described as a "story of love, lava, and burning desire." In this lighthearted story by John Patrick Shanley, Joe (played by Tom Hanks) is diagnosed with a destructive brain disease—a brain cloud—and given 6 months to live. A business tycoon makes Joe a peculiar offer: he can have an unlimited expense account, if he agrees to jump into the smoldering volcano of Big Woo on the South Pacific island of Waponi Woo. The local Waponis believe that there is an angry fire god in the volcano that will destroy their island home in a violent eruption unless they toss a human into the smoldering crater, a sacrifice that must be carried out every 100 years. In return for a volunteer, the Waponis offer the American tycoon exclusive rights to mine bubureau, a mineral that is only found on Waponi Woo and is essential for the manufacture of superconductors.

Joe agrees to the "bargain," not realizing that he has been fooled into believing he will die. He sails off on a private luxury yacht across the Pacific Ocean, with the tycoon's beautiful daughter, Patricia (played by Meg Ryan), as an escort. On the way, the two develop a friendship, are shipwrecked, drift for days atop Joe's expensive luggage, and eventually are washed up on the beaches of Waponi Woo. Patricia declares her love to Joe and pleads with him not to jump into the crater. However, the volcano is showing increasing unrest, and Joe insists on keeping his part of the bargain. Patricia decides to die with him but, as they jump into the crater, a violent eruption ejects them off the island and into the ocean—not only miraculously alive, but again able to use Joe's expensive luggage as a raft. As the lovers float away, they see the island explode and subside into the ocean. *Joe* has become something of a "cult" movie, probably because of its zany plot and dialogue. The theme of a volcano in the South Pacific and of natives who believe in human sacrifice to appease their volcano god has appeared in several other films.

An unusual type of volcano film is Akira Kurosawa's *Dreams*. In 1990, Japan's most celebrated film director, Akira Kurosawa, made a film consisting of a number of short stories, based on the director's dreams. The film was produced by Steven Spielberg, with special effects by George Lucas. Two of the short stories involve volcanoes. *Mount Fuji in Red* is a nightmare about the future, where the meltdown of a nuclear plant causes Fuji to erupt. Six nuclear reactors explode in succession as fireballs rise from the volcano, and a panicked crowd swamps the roads. The "hero" watches Mt Fuji turn incandescent, while multi-colored smoke drifts down its slopes. The colored smoke symbolizes different types of radioactive gas. People all around commit suicide by throwing themselves off a cliff. This is followed by a dialogue between the hero, a woman, and an older man about the lies about nuclear plants and their safety. The older man turns out to be one of the plant's engineers, excuses himself politely, and also jumps off the cliff. The hero stands alone, trying to protect the woman and her baby from the radioactive fumes, in vain, as they become surrounded by the colored gas.

5.2. Science Fiction and Disaster Movies

Jules Verne's classic science fiction novel, *Journey to the Center of the Earth*, was made into a major Hollywood production in 1959, starring Pat Boone, James Mason, and Arlene Dahl. The novel was the subject of two other films, in 1976 and in 1989, and of a television movie in 1993. None of the versions has been rated particularly highly by critics, although the special effects of the 1959 film were considered impressive for its time. Another volcano film that can be considered a science fiction thriller is *Crack in the World* (1965). The plot centers around a terminally ill villain, played by Dana Andrews, who is tapping geothermal energy from the Earth's interior by detonating a thermonuclear device at great depth. The explosion causes a crack to form that spreads through the Earth's crust and threatens to split the planet in two if not stopped in time. The film's slogan was, appropriately, "Thank God it's only a movie!"

Most other volcano films fall squarely into the disaster movie category. *Volcano* (1942) is a Superman animation movie, where a lethal eruption is predicted on the island of Monokoa. *The Daily Planet* reporters Clark Kent and Lois Lane rush to the scene, and immediately Lois is threatened by hot lava. In *The Devil at 4 O'Clock* (1961), a priest, played by Spencer Tracy, and three convicts land on a South Sea Island where a volcanic eruption threatens the inhabitants. The priest and the convicts, one of whom is played by Frank Sinatra, rescue the situation. The epitome of a volcano disaster movie is *Krakatoa East of Java* (1969). The all-star cast, including Maximillian Schell, Rossano Brazzi, and Diane Baker, battle the elements at sea and on land during the Krakatau eruption of 1883. For some reason, the producers could not figure out on which side of Java the volcano sits. The 1980 production *When Time Run Out*, also known as "Earth's Final Fury" is another disaster movie featuring famous stars, including Paul Newman, Jacqueline Bissett, and William Holden. The theme is a tired one, also seen in *Jaws* and ultimately based on Ibson's *Enemy of the People*: A volcano in a tropical island becomes active, and evil resort developers insist there is no danger. The resort's architect, played by Newman, leads a small party across the island to safety and the resort is eventually destroyed by the volcanic eruption.

More recent major volcano movies are the 1997 productions *Dante's Peak* and *Volcano*. The two films differ substantially in terms of plot and in how they portray a volcanic crisis. In *Dante's Peak*, a volcano in the Cascades threatens to wake up. The movie producers used three volcanologists from the U.S. Geological Survey as consultants, and the result is a disaster movie that is more true to life than most, although it still uses abundant artistic license. The plot centers around a dashing but brooding volcanologist (Harry Dalton, played by Pierce Brosnan), who believes the volcano is about to blow up; his boss from the Geological Survey, who sees no concrete evidence for an impending disaster; and the town's mayor, an embattled single mother whose kids eventually need to be rescued by the hero. One of the movies' major accomplishments is the portrayal of the difficult decision of whether or not to evacuate a town when the signs of an impending eruption are dubious. Harry's boss points out that evacuation of the town will lead to economic disaster for the residents— even if there is no eruption.

Dante's Peak eventually erupts and the usual disaster movie ingredients are brought in. The volcano gives out the widest possible range of products, from Hawaiian-type lavas to an acid lake to a pyroclastic flow. The pyroclastic flow simulation near the end of the movie is very impressive, and several real-life volcanologists have pointed out that it is highly educational. Relatively few people are aware that this type of flow exists and how totally deadly it is. The more recent movie "Pompeii" was also noteworthy because of its special effects depicting a pyroclastic flow.

Dante's Peak provoked something of a flurry in the volcanological community, probably because it can be considered the least inaccurate of all volcano disaster movies. It was generally well received by scientists and even reviewed in the scientific periodical *Nature* (Fink, 1997). The Harry Dalton character is credible as a volcanologist and will probably be remembered by professionals for his line, *I move around a lot, wherever there is a volcano with an attitude.*

The second volcano disaster movie of 1997 was *Volcano* ("the Coast is toast"), a film written by Jerome Armstrong, directed by Mick Jackson (1997). The story seems to be loosely based on the plot from the *Genesis Rock* novel, but transferred from New York to Los Angeles. Like *Genesis Rock*, the principal character is a female volcanologist, and an eruption in the city leads to the growth of a volcano on the streets and brings about a terrifying catastrophe. The main characters, Mike Roark (Tommy Lee Jones) of the Office of Emergency Management and geologist Amy Barnes (Anne Heche), team up to save the city. The plot unfolds as seven men perish in a hot cloud of yellowish sulfur-bearing gas in the manholes below the street. The La Brea tar pits begin to steam and bubble-like a pot on the stove. Mike Roark, investigating the deaths in the manhole, just manages to escape before a hot yellow gas cloud explodes through the tunnel system. *We'd better call a geologist*, he says. Here enters Dr Amy Barnes, driving a Humvee that is a science lab on wheels. She concludes that magma is moving below the city, but the Office of Emergency Management is highly skeptical—until manhole covers start rocketing into the sky, borne on top of columns of white steam, and a huge ball of a fiery explosion from the La Brea tar pits sends white ashfall over the city. A violent volcanic eruption has begun in the heart of Los Angeles. Magma breaks up to the surface, and a thick river of lava flows along the streets, setting fires, choking off traffic, and hindering evacuation. The volcano (later named Mt Wiltshire) quickly grows to 200 ft in height, threatening large parts of the city. Emergency workers build up barricades of cars and earth to try to contain the lava, which begins to form a huge pool. Suddenly Amy has an idea: bring in all available fire trucks, tankers, and helicopters to dump water on the lava pool. It works! A brittle crust begins to form on the lava. The lava diversion idea is used again—this time by the explosive demolition of a large new building. The city is saved as the lava flows harmlessly into the Pacific Ocean.

The lava diversion concept is probably the only credible idea in this movie. Explosives have been used to breach and divert lava on Mt Etna, and the cooling of lava using water hoses was successful in Heimaey, Iceland. Real-life volcanologists were quick to point out that the only real-eruption footage used in the movie was in the final scene, when lava pours into the sea.

The two 1997 movies were welcomed, by the professional community because, despite their artistic license, they succeeded in bringing the potential threats of volcanoes to the attention of the public and, perhaps more importantly, because scientists were portrayed as heroes in both movies. *Dante's Peak* can even be considered generally educational. These and many of the earlier movies contain special effects that were difficult to create at the time—an article in the *New York Times* (1997) discusses how "flowing lava is one of the most difficult illusions to pull off in special effects." Even though the work in creating these effects has to be commended, it is disappointing that the large resources and tremendous creativity available in the movie industry have been limited to making disaster-type films about volcanoes. The fantastic real-life stories of human endurance, courage, and tragedy during volcanic eruptions have not yet been seen in the movies.

SEE ALSO THE FOLLOWING ARTICLES

Archaeology and Volcanism ● Hazards from Pyroclastic Flows and Surges ● The History of Volcanology ● Mantle of the Earth ● Risk Education and Intervention ● Volcanoes and Tourism ● Volcanoes in Art.

FURTHER READING

Bergman, I., Burgess, A., 1980. Ingrid Bergman, My Story. Delacorte Press, New York.

du Bois, W.P., 1947. The Twenty-One Balloons. Viking Press, New York.

Bulwer-Lytton, E., 1850. The Last Days of Pompeii. Routledge and Sons, New York. Carter, L. (1976).

Chamoiseau, P., 1997. Texaco. Random House, New York.

Chard, C., 1990. Rising and sinking on the Alps and Mount Etna. J. Philos. Vis. Arts, 61—69.

Cooper, J.F., 1962. The Crater, or Vulcan's Peak. Belknap Press, Cambridge.

Corley, E., 1980. The Genesis Rock. Doubleday, New York.

Dean, D.R., 1981. The influence of geology on American literature and thought. In: Schneer (Ed.), Two Hundred Years of Geology in America. University Press of New England, pp. 289—303.

Endo, S., 1980. Volcano. Taplinger, New York.

Fermor, P.L., 1985. The Violins of Saint-Jacques. Oxford University Press, New York.

Fink, J., 1997. Cinemagmatic mayhem. Nature 386, 33.

Foster, E.M., 1945. Melville and geology. Am. Literature 27, 50—65.

Goodspeed, B.I., 1945. Paricutin. The American Book and Printing Co., Mexico.

Kennan, G., 1902. The Tragedy of Pelee. Outlook Co., New York.

Marshall, E., 1925. The Sleeper of the Moonlit Ranges. Cosmopolitan, New York.

Mathews, A., 1965. The Night of Purnama. Jonathan Cape, London.

O'Brian, P., 1993. The Wine-Dark Sea. Norton, New York.

Scudder, H.H., 1947. Cooper's the crater. Am. Literature 19, 109—126.

Sontag, S., 1992. The Volcano Lover. Farrar Strauss, New York.

Spiller, R.E., 1955. Literary History of the United States. Macmillan, New York.

Tolkien, J.R.R., 1954—55. Lord of the Rings, 3 vols. George Allen and Unwin.

Verne, J., 1966. A Journey to the Center of the Earth. Heritage Press, New York.

Verne, J., 2008. The Golden Volcano. U of Nebraska P.

Westall, W., 1886. The Phantom City. Cassell, New York.

Whitney, M.I., 1953. Juan of Paricutin. Steck Co., Austin, Texas.

Wilson, A.N., 1985. Gentlemen in England. Hamish Hamilton, London.

Common Units and Conversion Factors

1. MATHEMATICAL PREFIXES OF UNITS

Prefix	Symbol	Multiple
Tera-	T	10^{12}
Giga-	G	10^{9}
Mega-	M	10^{6}
Kilo-	k	10^{3}
Hecto-	h	10^{2}
Deka-	da	10
Deci-	d	10^{-1}
Centi-	c	10^{-2}
Milli-	m	10^{-3}
Micro-	μ	10^{-6}
Nano-	n	10^{-9}
Pico-	p	10^{-12}
Femto-	f	10^{-15}
Atto-	a	10^{-18}

2. MEASUREMENT UNITS AND CONVERSION FACTORS

Unit	Conversion Factor
Length Units	
Ångstrom	$1\ \text{Å} = 0.1\ \text{nm} = 10^{-8}\ \text{cm} = 10^{-10}\ \text{m}$
Nanometer	$1\ \text{nm} = 10\ \text{Å} = 10^{-7}\ \text{cm} = 10^{-9}\ \text{m}$
Micron (micrometer)	$1\ \mu\text{m} = 10^{-6}\ \text{m} = 0.00003937\ \text{in}$
Millimeter	$1\ \text{mm} = 1000\ \mu\text{m} = 0.1\ \text{cm} = 0.03937\ \text{in}$
Centimeter	$1\ \text{cm} = 10{,}000\ \mu\text{m} = 10\ \text{mm} = 0.3937\ \text{in}$

(Continued)

—cont'd

Unit	Conversion Factor
Meter	1 m = 100 cm = 39.37 in = 3.2808 ft = 1.0936 yd
Kilometer	1 km = 3280.8399 ft = 1000 m = 10^5 cm = 0.6214 mi
Inch	1 in = 2.54 cm = 0.0833 ft
Foot	1 ft = 30.48 cm = 12 in = 0.3048 m
Yard	1 yd = 91.44 cm = 36 in = 3 ft = 0.9144 m
Statute mile	1 mi = 5280 ft = 1760 yd = 1609.344 m = 1.6093 km
Nautical mile	1 n.m. = 1852 m = 1.852 km = 1.1508 mi
Area Units	
Square centimeter	1 cm^2 = 100 mm^2 = 0.155 in^2
Square meter	1 m^2 = 10,000 cm^2 = 10.7639 ft^2 = 1.1960 yd^2
Hectare	1 ha = 10^8 cm^2 = 10,000 m^2 = 2.4711 a = 0.01 km^2 = 0.003861 mi^2
Square kilometer	1 km^2 = 10^6 m^2 = 100 ha = 0.3861 mi^2
Square inch	1 in^2 = 6.4516 cm^2
Square foot	1 ft^2 = 929.0304 cm^2 = 144 in^2 = 0.0929 m^2
Square yard	1 yd^2 = 1296 in^2 = 9 ft^2 = 0.8361 m^2
Acre	1 ac = 43,560 ft^2 = 4840 yd^2 = 4046.8564 m^2 = 0.4047 ha
Square mile	1 mi^2 = 640 ac = 2.5900 km^2
Volume Units	
Cubic centimeter	1 cm^3 = 0.0610 in^3
Cubic meter	1 m^3 = 35.3147 ft^3 = 1.3080 yd^3 = 1000 l = 264.1721 gal
Cubic kilometer	1 km^3 = 10^{15} cm^3 = 10^9 m^3 = 0.2399 mi^3
Cubic inch	1 in^3 = 16.3871 cm^3
Cubic foot	1 ft^3 = 1728 in^3 = 0.02832 m^3
Cubic yard	1 yd^3 = 27 ft^3 = 0.7646 m^3
Liter	1 l = 10 dl = 1000 cm^3 = 61.0237 in^3 = 1.0567 qt = 0.2642 gal
Deciliter	1 dl = 0.1 l = 100 cm^3 = 6.1023 in^3
Pint	1 pt = 28.875 in^3 = 0.4732 l
Quart	1 qt = 57.75 in^3 = 2 pt = 0.473 l
The U.S. gallon	1 gal = 231 in^3 = 8 pt = 4 qt = 3.7854 l
Weight Units	
Kilogram	1 kg = 1000 g = 2.20461 lb
Gram	1 g = 0.03527 ounces avoirdupois
Metric ton	1 t = 106 g = 2204.6226 lb = 1000 kg
Teragram	1 Tg = 10^{12} g = 10^6 t
Gigaton	1 Gt = 10^{15} g = 10^9 t
Energy Units	
Electron-volt	1 eV = 1.6022×10^{-19} J

—cont'd

Unit	Conversion Factor
Erg	$1 \text{ erg} = 10^{-7} \text{ J} = 0.23885 \times 10^{-7} \text{ cal} = 1 \text{ g cm}^2/\text{sec}^2$
Joule	$1 \text{ J} = 10^7 \text{ erg}$
Calorie	$1 \text{ cal} = 4.1868 \text{ J}$
Kilogram-meter	$1 \text{ kg-m} = 2.344 \text{ cal}$
Kilocalorie	$1 \text{ kcal} = 1000 \text{ cal} = 3.9683 \text{ BTU}$
Force Units	
Dyne	$1 \text{ dyn} = 10^{-5} \text{ N} = 1 \text{ g cm/sec}^2$
Newton	$1 \text{ N} = 1 \text{ kgm/s}^2 = 10^5 \text{ dyn}$
Gravity Units	
milligal	$\text{mGal} = 1 \text{ cm/s}^2$
Frequency Units	
Hertz	$1 \text{ Hz} = 1 \text{ cps}$
Power Units	
Watt	$1 \text{ W} = 1 \text{ J/s} = 0.05687 \text{ BTU/min} = 0.01433 \text{ kcal/min} = 0.001341 \text{ hp} = 107 \text{ erg/sec}$
Horsepower	$1 \text{ hp} = 745.6999 \text{ W}$
Kilowatt	$1 \text{ kW} = 1000 \text{ W} = 1.341 \text{ hp}$
Megawatt	$1 \text{ MW} = 10^3 \text{ kW} = 10^6 \text{ W} = 1341.0221 \text{ hp}$
Pressure Units	
Nanobar	$1 \text{ nbar} = 10^{-9} \text{ bar}$
Centimeter of mercury	$1 \text{ cm Hg} = 0.01316 \text{ atm}$
Pound per square inch	$1 \text{ psi} = 0.06805 \text{ atm} = 0.07031 \text{ kg/cm}^2$
Kilogram per square centimeter	$1 \text{ kg/cm}^2 = 0.9678 \text{ atm}$
Bar	$1 \text{ bar} = 10^5 \text{ N/m}^2 = 0.9869 \text{ atm} = 10^6 \text{ dyn/cm}^2$ $= 10^6 \text{ g/cm/sec}^2 = 10^5 \text{ Pa}$
Kilobar	$1000 \text{ bars} = 100 \text{ MPa}$
Atmosphere	$1 \text{ atm} = 1.01325 = 10^5 \text{ N/m}^2 = 76 \text{ cm Hg} = 29.9213 \text{ in Hg} = 14.6959$ $\text{psi} = 1.0332 \text{ kg/cm}^2 = 1.01325 \text{ bar} = 1.013 \times 10^5 \text{ Pa}$
Megapascal	$1 \text{ MPa} = 10^6 \text{ Pa} = 10 \text{ bar}$
Gigapascal	$1 \text{ GPa} = 10^9 \text{ Pa} = 10 \text{ kbar} = 10^3 \text{ bar} = \text{ca. 30 km depth in Earth's mantle}$
Torr	$1 \text{ Torr} = 1 \text{ mm Hg} = 1.3332 \times 10^{-3} \text{ bar}$
Time Units	
Year	$1 \text{ yr} = 1 \text{ a} = 3.15567 \times 10^7 \text{ s} = 5.2560 \times 10^5 \text{ min} = 8760 \text{ h}$
Ma	One million years before present, in terms of absolute geologic time
Ga	Billion years $= 10^9$ years, in terms of absolute geologic time
m.y.	Million years, in terms of duration; a relative term
Temperature Units	
Degree celsius	$^\circ\text{C} = ((^\circ\text{F} - 32) \times 5/9) = \text{K} - 273.15$

(Continued)

—cont'd

Unit	Conversion Factor
Kelvin	$K = {}^{\circ}C + 273.15 = (({}^{\circ}F - 32) \times 5/9) + 273.15$
Degree fahrenheit	${}^{\circ}F = {}^{\circ}C \times (9/5) + 32$

Gravimetric Factors

$Fe_2O_3 \times 0.89985 = FeO$

$FeO \times 1.1113 = Fe_2O_3$

$FeO \times 0.77731 = Fe$

$Fe_2O_3 \times 0.69944 = Fe$

$K_2O \times 0.83015 = K$

Viscosity Units

Unit	Conversion Factor
Poise	1 P (After poiseuille) = 1 g/cm/s = 10^{-1} Pa s = 100 centipoise (cp) = 0.1 Pa s
Pascal second	1 Pa s
Perfect gas constant	R = 8.31441 J K mol
Density	1 g/cm^3 = 1000 kg/m^3
Velocity	1 m/s = 3.6 km/h = 2.2369 mi/h = 0.003017 mach number

Catalog of Earth's Documented Holocene Eruptions

Lee Siebert, Elizabeth Cottrell, Edward Venzke and Benjamin Edwards

U.S. National Museum of Natural History, Smithsonian Institution, Washington, DC, USA

This two-part appendix begins with an alphabetical listing of all volcanoes with documented Holocene eruptions. Each name is followed by the volcano's Catalog number, to facilitate its rapid location in the regionally sequenced Catalog listing Holocene eruptions that forms the second and main part.

The Catalog is a condensed and updated version of the 165-page "Directory" table in *Volcanoes of the World*, 2010 (see Chapter 12 for this and other references), to which the reader is referred for more information. It includes known Holocene eruptions through the end of September 2014, but excludes uncertain eruptions, and does not include month/day dates, exact vent locations, eruptive characteristics, or precise durations. Sequence is regional, proceeding along generally linear volcanic belts and following the numbering system established decades ago by the International Association of Volcanology and Chemistry of the Earth's Interior (IAVCEI) and used in *Volcanoes of the World*, 2010, but later modified in the Smithsonian's online *Volcanoes of the World* electronic Catalog (www.volcano.si. edu). Regional sequencing starts in Europe, and then moves south and east to New Zealand, generally clockwise around the Pacific, then down the Atlantic from Iceland to Antarctica. A geographic sequence allows the perspective of viewing a wide region's volcano data at once, but readers interested in finding a single, unfamiliar volcano may do so through the alphabetical name index. Users will note that Japanese volcano names have been modified from those in *Volcanoes of the World,* 2010 to follow the Japan Meteorological Agency and Geological Survey of Japan in attaching generic terms such as *san, zan,* or *yama* for mountain, *dake* or *take* for peak, and *jima* or *shima* for island to the volcano name. Noteworthy volcanoes meeting criteria including eruption size, frequency, or fatalities are highlighted in **boldfaced** type (and these are the volcanoes named on the endpaper world map).

Coordinates (in *italics*) are given in decimal degrees, and elevations in meters above/below sea level. Volcano edifice type follows *Volcanoes of the World* (and Chapter 12) and subsidiary morphologic features are not included. Unusually large explosive eruptions, as measured by the volcanic explosivity index or VEI (see Chapter 12), are highlighted by typography (see header) allowing swift location of these larger events, keeping in mind that the eruptive record becomes more incomplete with passing time and many documented Holocene eruptions have not been studied in sufficient detail to calculate tephra volumes used to assign eruption magnitudes.

Uncertainty is unavoidable in the volcanic record, and the following conventions are used to display this aspect of eruptive histories. A question mark *after* a year indicates that the date is uncertain, and commercial "at" sign (@) *before* a year indicates that the location of the eruption is uncertain (it may have been a neighboring volcano). When an eruption is known to have begun sometime prior to a particular year, the "less than" symbol (<) appears after that date. If an eruption was continuing when last observed, but the actual stop date is unknown, then the "greater than" symbol (>) follows the observation date. BC (BCE) eruptions are distinguished from AD (CE) eruptions by showing a "-" before the eruption year.

Nonhistorical eruptions are distinguished by a dating method code preceding the eruption date as follows: **A** = anthropology, **C** = radiocarbon (uncorrected), **D** = dendrochronology (tree ring), **E** = surface exposure, **F** = fission track, **G** = radiocarbon (corrected), **H** = hydration rind, **I** = ice core, **K** = potassium–argon (K–Ar), **L** = lichenometry, **M** = magnetism, **N** = thermoluminescence, **R** = argon–argon (Ar–Ar), **S** = SOFAR (hydrophonic), **T** = tephrochronology, **U** = uranium series, and **V** = varve count. The dating uncertainty in years is shown after many eruptions and can be quite large for some dating techniques, but many other eruptions for which the precise uncertainty is not known are followed only by a question mark. Users should keep in mind that these eruptions also may have large uncertainties and should pay particular attention to distinguishing corrected from uncorrected radiocarbon dates (see *Volcanoes of the World*, 2010 and other resources for discussion of these or other dating method techniques). Volcanoes for which the only historical eruptions are questionable are not included in the Catalog but appear on the endpaper maps as the smallest red symbol.

In addition to the date uncertainties attached to many eruptions in this Catalog, users must be aware of the significant underrepresentation of smaller-volume eruptions throughout much of Holocene and even historical time as well as the widely variable completeness (regionally as well as within regions) of the geologic record that reflects relative effort and resources available for field studies of individual volcanoes. It is these detailed stratigraphic studies of Holocene and earlier Pleistocene eruptions that are essential to define the eruptive history of individual volcanoes and allow better informed assessment of volcanic risk.

Volcano Index: Name with Regionally Sequenced Volcano Number

Abu	283001	Ambang	266020	Atacazo	352021	Barkhatnaya Sopka	300084
Acatenango	342080	Ambitle	254020	Atitlán	342060	Barren Island	260010
Acigöl-Nevsehir	213004	Ambrym	257040	Atka	311160	Barrier, The	222030
Adams	321040	Amukta	311190	Auckland Field	241020	Barú	346010
Adams Seamount	333050	Anatahan	284200	Augustine	313010	Barva	345050
Adatarayama	283170	Andahua-Orcopampa	354004	Avachinsky	300100	Batur	264010
Agrigan	284160	Aniakchak	312090	Awu	267040	Bayuda Volcanic Field	225060
Agua de Pau	382090	Antillanca Group	357153	Axial Seamount	331021	Belknap	322060
Aguilera	358062	Antisana	352030	Azufral	351090	Berlin	390022
Agung	264020	Antuco	357080	Azul, Cerro	353060	Besar	261250
Ahyi	284141	Aoba	257030	Azul, Cerro (Quizapu)	357060	Bezymianny	300250
Aira (Sakurajima)	282080	Aogashima	284060	Azumayama	283180	Biliran	272080
Akademia Nauk	300125	Apagado	358024	Babuyan Claro	274030	Billy Mitchell	255011
Akan	285070	Api Siau (Karangetang)	267020	Bachelor	322090	Black Butte Crater Lava Field	324010
Akita-Komagatake	283230	Apoyeque	344091	Bagana	255020	Black Peak	312080
Akita-Yakeyama	283260	Ararat	213040	Baitoushan (Changbaishan)	305060	Black Rock Desert	327050
Akutan	311320	Ardoukôba	221126	Bakening	300123	Blanco, Cerro	355210
Alaid	290390	Arenal	345033	Baker	321010	Bliznetsy	300552
Alamagan	284180	Arenales	358059	Bam	251010	Blue Lake Crater	322030
Alayta	221112	Arhab, Harra of	231090	Bamus	252110	Bogoslof	311300
Alcedo	353040	Arjuno-Welirang	263290	Banda Api	265090	Bolshoi Semiachik	300150
Aliso	352031	Arshan	305011	Bandaisan	283160	Bolshoi-Kekuknaysky	300360
Almolonga	342040	Asamayama	283110	Banua Wuhu	267030	Boomerang Seamount	234000
Alney-Chashakondzha	300450	Askja	373060	Baransky (Sashiusudake)	290080	Bouvet	386020
Alutu	221270	Asosan	282110	Bárcena	341020	Bravo, Cerro	351012
Amak	311390	Asuncion	284150	Bárdarbunga	373030	Brennisteinsfjöll	371040

Volcano Index: Name with Regionally Sequenced Volcano Number—cont'd

Bristol Island	390080	Cherpuk Group	300273	Dabbahu	221113	Egon	264160
Buckle Island	390010	Chichinautzin	341080	Daisetsuzan (Taisetsuzan)	285060	Ekarma	290270
Bulusan	273010	Chichón, El	341120	Dakataua	252040	Elbrus	214010
Burney, Monte	358070	Chiginagak	312110	Dalaffilla	221070	Elovsky	300590
Buzzard Creek	315001	Chikurachki	290360	Dallol	221041	Emuruangogolak	222051
Cabalían	272050	Chillán, Nevados de	357070	Dama Ali	221141	Endeavour Segment	331010
Caburgua-Huelemolle	357112	Chimborazo	352071	Damavand	232010	Epi	257060
Cagua	273090	Chirinkotan	290260	Dana	312050	Erciyes Dagi	213010
Calatrava Volcanic Field	210040	Chirippusan (Chirip)	290090	Dar-Alages	214080	Erebus	390020
Calbuco	358020	Chirpoi	290150	Darwin	353030	Erta Ale	221080
Callaqui	357091	Chokaisan	283220	Davis Lake	322100	Es Safa	230050
Cameroon	224010	Churchill	315030	Dawson Strait Group	253060	Esan	285011
Camiguin	271080	Chyulu Hills	222130	Deception Island	390030	Escanaba Segment	331040
Camiguin de Babuyanes	274010	Cleft Segment	331030	Dempo	261230	Etna	211060
Campi Flegrei	211010	Cleveland	311240	Descabezado Grande	357050	Etorofu-Atosanupuri	290050
Campi Flegrei Mar Sicilia	211070	CoAxial Segment	331020	Dhamar, Harras of	231120	Etorofu-Yakeyama (Grozny)	290070
Candlemas Island	390100	Cobb Segment	331011	Diamond Craters	322170	Eyjafjallajökull	372020
Canlaon (Kanlaon)	272020	Cofre de Perote	341096	Didicas	274020	Falcon Island	243050
Carlisle	311230	Colima	341040	Dieng Volcanic Complex	263200	Farallon de Pajaros	284140
Carrizozo	327110	Colo (Una Una)	266010	Diky Greben	300022	Fayal	382010
Carrán-Los Venados	357140	Concepción	344120	Don Joao de Castro Bank	382070	Fentale	221190
Cayambe	352006	Conchagüita	343120	Doña Juana	351070	Fernandina	353010
Cayutué-La Viguería	358012	Copahue	357090	Dotsero	328010	Fisher	311350
Ceboruco	341030	Corcovado	358050	Dubbi	221100	Flores	382001
Cendres, Ile des	275060	Cosigüina	344010	Dukono	268010	Fogo	384010
Cereme	263170	Cotopaxi	352050	Eastern Gemini Seamount	258001	Fonualei	243100
Chacana	352022	Crater Lake	322160	Ebeko	290380	Fort Portal	223001
Chachadake (Tiatia)	290030	Craters of the Moon	324020	Ebulobo	264100	Fournaise, Piton de la	233020
Chachimbiro	352002	Cuicocha	352003	Ecuador	353011	Fourpeaked	312260
Chaine des Puys	210020	Cumbal	351100	Edgecumbe	315040	Fremrinamur	373070
Chaitén	358041	Cumbres, Las	341098	Edziza	320060	Fuego	342090
Changbaishan (Baitoushan)	305060	Curacoa	243102	Egmont (Taranaki)	241030	Fueguino	358090

(Continued)

Volcano Index: Name with Regionally Sequenced Volcano Number—cont'd

Katmai	312170	Krakatau	262000	Leonard Range	271031	Malinche, La	341091
Kavachi	255060	Krasheninnikov	300190	Leroboleng	264200	Maly Semyachik	300140
Kazbek	214020	Krísuvik	371030	Lewotobi	264180	Mammoth Mountain	323150
Kelimutu	264140	Kronotsky	300200	Lewotolo	264230	Manam	251020
Kelut	263280	Ksudach	300050	Liamuiga	360030	Manda Hararo	221115
Kerinci	261170	Kuchinoerabujima	282050	Lipari	211042	Manda-Inakir	221122
Ketoi	290200	Kuchinoshima	282043	Little Sitkin	311050	Marapi	261140
Khangar	300272	Kueishantao	281031	Ljósufjöll	370030	Marchena	353080
Kharimkotan	290300	Kujusan	282120	Llaima	357110	Marion Island	234070
Khaybar, Harrat	231060	Kunlun Volcanic Group	304030	Llullaillaco	355110	Mariveles	273081
Khodutka	300053	Kupreanof	312060	Loihi	332000	Maroa	241061
Kick 'em Jenny	360160	Kurikomayama	283210	Lokon-Empung	266100	Marra, Jebel	225030
Kikai	282060	Kurile Lake	300023	Lolobau	252130	Martin	312140
Kikhpinych	300180	Kusatsu-Shiranesan	283120	Loloru	255030	Maruyama	285061
Kilauea	332010	Kussharo	285080	Long Island	251050	Masaya	344100
Kinenin	300551	Kuttara	285034	Longaví, Nevado de	357063	Mashu	285081
Kirishimayama	282090	Kuwae	257070	Longgang Group	305050	Matthew Island	258010
Kiska	311020	Kverkfjöll	373050	Longonot	222100	Matutum	271020
Kita-Fukutokutai	284121	Kyejo	222170	Lonquimay	357100	Mauna Kea	332030
Kita-Ioto (Kita-Iwojima)	284110	La Palma	383010	Lopevi	257050	Mauna Loa	332020
Kizimen	300230	Lamington	253010	Lunayyir, Harrat	231040	Mayon	273030
Klyuchevskoy	300260	Lamongan	263320	Lvinaya Past (Moekeshiwan)	290041	Mayor Island	241021
Kolbeinsey Ridge	375010	Langila	252010	Maca	358056	Mayotte Island	233005
Kolokol Group	290120	Langjökull (Hveravellir)	371080	Macauley Island	242021	McDonald Islands	234011
Komarov	300220	Lanín	357122	Macdonald	333060	Meager	320180
Kone	221200	Lanzarote	383060	Machín	351040	Medicine Lake	323020
Koniuji	311140	Larderello	211001	Madeira	382120	Medvezhia (Moyorodake)	290100
Korovin	311161	Láscar	355100	Mageik	312150	Megata	283262
Koryaksky	300090	Lassen Volcanic Center	323080	Mahawu	266110	Meidob Volcanic Field	225050
Koshelev	300020	Late	243090	Maipo	357021	Melbourne	390015
Kostakan	300122	Lautaro	358060	Makaturing	271040	Melimoyu	358052
Kozushima	284030	Lawu	263260	Makian	268070	Mendeleev (Raususan)	290020
Krafla	373080	Lengai, Ol Doinyo	222120	Makushin	311310	Menengai	222060

(Continued)

Volcano Index: Name with Regionally Sequenced Volcano Number—cont'd

Name	Number	Name	Number	Name	Number	Name	Number
Mentolat	358054	Nabukelevu	245030	Numazawa	283151	Penguin Island	390031
Merapi	263250	Nakanoshima	282040	NW Rota-1	284211	Perbakti-Gagak	263040
Merbabu	263240	Namarunu	222040	Nyamuragira	223020	Peuet Sague	261030
Meru	222160	Nantai	283141	Nyiragongo	223030	Pico	382020
Methana	212020	Naolinco Volcanic Field	341095	Nylgimelkin	300650	Pico Fracture Zone	381040
Metis Shoal	243070	Naruko	283200	Ofu-Olosega	244010	Picos Volcanic System	382081
Michael	390090	Nasudake	283150	Ojos del Salado, Nevados	355130	Piip	300271
Michoacán-Guanajuato	341060	Nazko	320140	Okataina	241050	Pilas, Las	344080
Midagahara	283080	Negra, Sierra	353050	Okmok	311290	Pinatubo	273083
Mikurajima	284041	Negro de Mayasquer, Cerro	351110	Olca-Paruma	355050	Pinta	353070
Milos	212030	Negro, Cerro	344070	Olkaria	222090	Planchón-Peteroa	357040
Minami-Hiyoshi	284131	Nejapa-Miraflores	344092	Omanago Group	283142	Plat Pays, Morne	360110
Minchinmávida	358040	Nemo Peak	290320	Ontakesan	283040	Pleiades, The	390013
Miravalles	345030	Nemrut Dagi	213020	Opala	300080	Poás	345040
Misti, El	354010	Newberry	322110	Öraefajökull	374010	Pogromni (Westdahl)	311340
Miyakejima	284040	Newer Volcanics Province	259010	Orizaba, Pico de	341100	Popa	275080
Mocho-Choshuenco	357130	Ngauruhoe (Tongariro)	241080	Oshima-Oshima	285010	Popocatépetl	341090
Moekeshiwan (Lvinaya Past)	290041	Ngozi	222164	Osorezan	283290	Porak	214090
Moffett	311111	Nightingale Island	386011	Osorno	358010	Prestahnukur	371070
Momotombo	344090	Niigata-Yakeyama	283090	Ostry	300680	Prevo Peak	290190
Monaco Bank	382110	Niijima	284020	Pacaya	342110	Protector Shoal	390140
Mono Craters	323120	Nikko-Shiranesan	283140	Pagan	284170	Pululagua	352011
Mono Lake Volcanic Field	323110	Nila	265060	Pago (Witori)	252080	Puntiagudo-Cordón Cenizos	357160
Monowai Seamount	242050	Niseko	285031	Paka	222053	Puracé	351060
Montagu Island	390081	Nishinoshima	284096	Pali-Aike Volcanic Field	358080	Putana	355090
Moua Pihaa	333030	Nisyros	212050	Palinuro	211031	Puyehue-Cordón Caulle	357150
Moyorodake (Medvezhia)	290100	Niuafo'ou	243110	Paluweh	264150	Qualibou	360140
Muria	263251	NNE of Iriomotejima	282010	Pantelleria	211071	Quetrupillan	357121
Musuan	271070	Norikuradake	283060	Papandayan	263100	Quill, The	360020
Mutnovsky	300060	North Gorda Ridge Segment	331031	Parinacota	355012	Quilotoa	352060
Myojinsho	284070	Northern EPR-Segment RO2	334020	Parker	271011	Quimsachata	354000
Myokosan	283100	Northern EPR-Segment RO3	334021	Pavlof	312030	Quizapu (Azul, Cerro)	357060
Nabro	221101	Novarupta	312180	Pelée	360120	Rabaul	252140

Volcano Index: Name with Regionally Sequenced Volcano Number—cont'd

Ragang	271060	Salton Buttes	323200	Shasta	323010	St. Paul	234002
Rahat, Harrat	231070	San Cristóbal	344020	Sheveluch	300270	St. Paul Island	314010
Raikoke	290250	San Francisco VF (Sunset Crater)	329020	Shikotsu (Tarumaesan)	285040	Stromboli	211040
Rainier	321030	San Jorge	382030	Shiretoko-Iozan	285090	Sumaco	352040
Ranakah	264071	San José	357020	Shishaldin	311360	Sumbing	261180
Raoul Island	242030	San Martín	341110	Shisheika	300511	Sumbing	263220
Rasshua	290220	San Miguel	343100	Sibayak	261070	Sumisujima	284080
Raung	263340	San Pablo Volcanic Field	273060	Silali	222052	Sundoro	263210
Rausudake	285082	San Pedro	355070	Simbo	255050	Suoh	261270
Raususan (Mendeleev)	290020	San Salvador	343050	Sinabung	261080	Suphan Dagi	213021
Reclus	358063	Sanbesan	283002	Sinarka	290290	Supply Reef	284142
Redoubt	313030	Sand Mountain Field	322040	Sirung	264270	Suretamatai	257010
Reporoa	241060	Sangay	352090	Slamet	263180	Suwanosejima	282030
Reventador	352010	Sangeang Api	264050	Snaefellsjökull	370010	Taal	273070
Reykjanes	371020	Santa Ana	343020	Snowy Mountain	312200	Taapaca	355011
Rincón de la Vieja	345020	Santa Isabel	224004	Soche	352001	Tacaná	341130
Rinjani	264030	Santa Isabel	351021	Socompa	355109	Tafu-Maka	243120
Rishirizan	285041	Santa María	342030	Socorro	341021	Tair, Jebel at	221010
Ritter Island	251070	Santiago	353090	Sollipulli	357111	Taisetsuzan (Daisetsuzan)	285060
Robinson Crusoe	356020	Santorini	212040	Soputan	266030	Takaharayama	283143
Rocard	333020	Sarychev Peak	290240	Sorikmarapi	261120	Takahe	390027
Romeral	351011	Sashiusudake (Baransky)	290080	Soufrière Guadeloupe	360060	Takawangha	311090
Roundtop	311380	Savai'i	244040	Soufrière Hills	360050	Talang	261160
Ruang	267010	Savo	255070	Soufrière St. Vincent	360150	Tambora	264040
Ruapehu	241100	Sawâd, Harra Es-	231160	South Island	222020	Tanaga	311080
Ruby	284202	Sedankinsky	300520	South Sarigan Seamount	284193	Tandikat	261150
Ruiz, Nevado del	351020	Seguam	311180	Southern EPR-Segment I	334140	Tangkubanparahu	263090
Rumble III	241130	Semeru	263300	Southern EPR-Segment J	334130	Tao-Rusyr Caldera	290310
Rungwe	222166	Semisopochnoi	311060	Southern EPR-Segment K	334120	Tara, Batu	264260
Saba	360010	Serua	265070	Spokoiny	300671	Taranaki (Egmont)	241030
Sabancaya	354006	Sete Cidades	382080	Spurr	313040	Tarumaesan (Shikotsu)	285040
Sakurajima (Aira)	282080	Seulawah Agam	261020	St. Andrew Strait	250010	Taryatu-Chulutu	303010
Salak	263050	Severny	300700	St. Helens	321050	Tatun Group	281032

(Continued)

Volcano Index: Name with Regionally Sequenced Volcano Number—cont'd

Taunshits	300160	Tongkoko	266130	Veer	300121	Yavinsky	300021
Taupo	241070	Tor Zawar	232080	Veniaminof	312070	Yellowstone	325010
Taveuni	245010	Torfajökull	372050	Vestmannaeyjar	372010	Yokoate-jima	282021
Tavui	252150	Toshima	284011	Vesuvius	211020	Yokodake	283031
Teahitia	333010	Towada	283271	Victory	253030	Yonemaru-Sumiyoshiike	282081
Tecuamburro	342120	Toya (Usu)	285030	Viedma, Volcan	358061	Yoteizan	285032
Telica	34440	Traitor's Head	257090	Villarrica	357120	Yucamane	354050
Telong, Bur ni	261050	Trident	312160	Vilyuchik	300083	Yufu-Tsurumi	282130
Tenduruk Dagi	213030	Tristan da Cunha	386010	Visoke	223050	Yunaska	311210
Tenerife	383030	Trois Pitons, Morne	360100	Vsevidof	311270	Zaozan	283190
Tengchong	275110	Tromen	357072	Vulcano	211050	Zavaritsky	300124
Tengger Caldera	263310	Tseax River Cone	320100	Vulsini	211003	Zavaritzki Caldera	290180
Teon	265050	Tskhouk-Karckar	214100	Vysoky	300221	Zavodovski	390130
Terceira	382050	Tullu Moje	221250	Waiowa	253040	Zheltovsky	300040
Terpuk	300512	Tungurahua	352080	Wapi Lava Field	324030	Zhupanovsky	300120
Theistareykjarbunga	373090	Tupungatito	357010	Watt, Morne	360101	Zitácuaro-Valle de Bravo	341061
Three Sisters	322070	Turrialba	345070	Wells Gray-Clearwater	320150	Zubair Group	221020
Thule Islands	390070	Ubehebe Craters	323160	West Crater	321060	Zuni-Bandera	327120
Tianshan Volcanic Group	304020	Ubinas	354020	West Eifel Volcanic Field	210010		
Tiatia (Chachadake)	290030	Udokan Plateau	302030	West Mata	243130		
Ticsani	354031	Ugashik-Peulik	312130	Westdahl (Pogromni)	311340		
Tinakula	256010	Uinkaret Field	329010	White Island	241040		
Tinguiririca	357030	Ukinrek Maars	312131	Witori (Pago)	252080		
Titila	300560	Ulawun	252120	Wolf	353020		
Tjörnes Fracture Zone	373100	Ulreung	306030	Wrangell	315020		
Tofua	243060	Una Una (Colo)	266010	Wudalianchi	305030		
Tokachidake	285050	Unzendake	282100	Wurlali	265040		
Tolbachik	300240	Ushishur	290210	Yakedake	283070		
Tolima, Nevado del	351030	Ushkovsky	300261	Yantarni	312100		
Tolmachev Dol	300082	Usu (Toya)	285030	Yanteles	358049		
Toluca, Nevado de	341070	Uwayrid, Harrat	231020	Yar, Jabal	231080		
Tomariyama (Golovnin)	290010	Uzon	300170	Yasur	257100		
Tongariro (Ngauruhoe)	241080	Vailulu'u	244000	Yate	358022		

VNUM	VOLCANO NAME	LAT	LONG	ELEV	TYPE	ERUPTION YEARS (VEI≤3 <u>VEI=4</u> **VEI=5 <u>VEI=6-7</u>**)

Western Europe

210010 West Eifel Volcanic Field.......... *50.17N* *6.85E* *600* Maars .... C-8300±300, G-8740±150

210020 Chaine des Puys......................... *45.78N* *2.97E* *1464* Lava domes.......... C-4040±150, N-5760?, C-6020±150, N-6250?, C-6550?, C-7020±100, N-7740?, C-7840±200

210040 Calatrava Volcanic Field .... *38.87N* *4.02W* *1117* Pyroclastic cones.. G-3600±500

Italy

211001 Larderello .... *43.25N* *10.87E* *500* Explosion craters.. <1282

211003 Vulsini... .... *42.60N* *11.93E* *800* Caldera................. -0104

211010 Campi Flegrei .... *40.83N* *14.14E* *458* Caldera................ 1538, 1198, <u>T-1650?</u>, <u>C-1870±50</u>, <u>T-2000±150</u>, T-2040?, T-2080±75, **C-2150±500**, C-2220±50, T-2330±150, C-2440?, <u>T-2500?</u>, C-2580±50, C-2890±50, C-6300±50, T-6490?, <u>C-6650±100</u>, C-7590±50, T-7980±500, <u>G-8480±100</u>

211020 **Vesuvius** .... *40.82N* *14.43E* *1281* Somma................ 1913-44, <u>1875-1906</u>, 1870-72, 1864-68, 1854-61, 1841-50, 1824-39, 1796-1822, 1783-94, 1770-79, 1764-67, 1742-61, 1732-37, 1701-30, 1696-98, 1685-94, 1682, 1654-80, 1637-52, **1631-32**, <1570-72±1, 1500, 1150, 1139, 1037, 1006, 0999, 0991, <u>0968</u>, M 0900±40, M 0860±50, 0787-88, <u>0685</u>, 0536, <u>0512</u>, **0472**, 0379-95, 0222-35, <u>0203</u>, 0172, **0079**, -0217-16, A <-0600, <u>G-0880±50</u>, <u>G-1430±300</u>, <u>G-1550±75</u>, **G-2420±40**, **G-6940±100**

211030 Ischia..... .... *40.73N* *13.90E* *789* Complex............... 1302, G 1290?, G 0820±300, G 0540±150, 0295±10, 0145?, 0080±1, G 0040±75, -0006±20, -0091, M-0200±200, M-0370±150, -0470?, M-0490±150, T-0700=75, G-0930±150, M-1480±300, K-2350±1400, T-2700±1750, K-3050±1000, G-3580±300, M-3880±200, K-7550±2400

211031 Palinuro . ... *39.48N* *14.83E* *-70* Submarine .... G-8040±100

211040 **Stromboli**................................. *38.79N* *15.21E* *924* Stratovolcano .... 1934-2014, 1910-32, <1558-1907, 0950±50, 0550±50, 0250±50, 0150±50, 0050±50, -0050±50, -0210±10, -0350±50, C-4050?, M-4250?, M-4550?, M-4800±300, M-5050?, M-5550?, M-5800±300, M-6050?

211042 Lipari..... .... *38.48N* *14.95E* *602* Stratovolcanoes.... M 1230±40, G 0780±100, G-5820±75

211050 Vulcano . ... *38.40N* *14.96E* *500* Stratovolcanoes.... 1888-90, 1886, 1873-79, 1780, 1771, 1731-39, 1727, 1688, 1651, 1631, 1626, 1618, 1550, 1444, M 1230±20, M 1200±75, M 1040±75, 0925±25, 0729, 0526?, 0144, 0050±50, -0024±5, -0091, -0126, U-0150±300, <u>-0183</u>, -0215, -0300, -0360±10, -0475?, U-0950±500, T-1300?, K-2650±1700, K-3550±1300, K-6350±1600, K-6550?

211060 **Etna**....... .... *37.73N* *15.00E* *3330* Stratovolcanoes.... 1966-2014, 1955-64, 1949-52, 1928-47, 1878-1926, 1874, 1868-69, 1863-65, 1852-53, 1842-43, 1838-39, 1791-1833, <u>1787</u>, 1780-81, 1776, 1732-67, 1727-28, 1702, 1693-94, 1688-89, 1682, 1669, 1651-56?, 1646-47, 1643, 1634-38, 1614-24, 1603-10, 1579-80?, 1566, 1540-41, 1536-37, 1493?-1500?, 1446-47, 1444, 1408, 1381?, 1350, 1333, 1329, 1284-85, 1250, 1224, 1194, 1164?, 1160, 1157, 1063±1, 0417?, 0252, T 0100±100, 0039±1, 0010?, -0020?, -0032, -0036-35, -0044, -0049, **-0122**, -0126, -0135, -0141, -0396?, -0425-24?, -0479-75?, -0695±2, C-1050±75, G-1420±75, **-1500±50**, C-1980±50, C-2330±100, C-3050±150, C-3390±50, C-3510±150, C-4150±150, C-5150±150, C-6190±200

211070 Campi Flegrei Mar Sicilia.......... *37.10N* *12.70E* *-8* Submarine .... 1867, 1863, 1846, 1831, 1632. -0253±12

211071 Pantelleria................................. *36.77N* *12.02E* *836* Shield................... 1891, C-1080±300, G-4430±200, M-4550±300, C-5610±100, C-6130±75, K-7050?

Greece

212020 Methana. .... *37.62N* *23.34E* *760* Lava domes.......... -0258±18

212030 Milos .... *36.70N* *24.44E* *751* Stratovolcanoes.... G 0140±300

212040 **Santorini**................................. *36.40N* *25.40E* *367* Shields .... 1950, 1939-41, 1928, 1925-26, 1866-70, 1707-11, <u>1650</u>, 1570-73, <u>0726</u>, 0046-47, -0197, **G-1610±14**

212050 Nisyros .. .. *36.59N* *27.16E* *698* Stratovolcano 1888, 1873, 1871, 1422

Turkey

213004 Acigöl-Nevsehir .... *38.57N* *34.52E* *1689* Caldera................. A-2080±200, C-2370±200, C-3500?, C-6230?, C-7810?

213010 Erciyes Dagi .... *38.52N* *35.48E* *3916* Stratovolcano .... G-6880±40

213020 Nemrut Dagi............................ *38.65N* *42.23E* *2948* Stratovolcano .... 1650, <1597, 1441, V 1402?, 1111, V-0531?, V-0657±25, V-0787±25, V-1396?, V-1662?, V-4055±50, V-4321?, V-4615?, V-4849?, V-4938±75, V-5085?, V-5152?, V-5242±75, V-5320?, V-5745?, V-6213?, V-6471?, V-7087?, V-7579?, V-7769?, V-9950±150

213021 Suphan Dagi .... *38.92N* *42.82E* *4158* Stratovolcano .... T-8050?

213030 Tenduruk Dagi......................... *39.37N* *43.87E* *3584* Shield................... 1855, C-0550?

213040 **Ararat**... .... *39.70N* *44.30E* *5165* Stratovolcano .... 1840, A-0550, A-2450±50

Western Asia

214010 Elbrus .... *43.35N* *42.44E* *5642* Stratovolcano .... T 0050±50

214020 Kazbek... .... *42.70N* *44.50E* *5050* Stratovolcano .... T-0750±50, C-4000±50

214070 Ghegam Ridge......................... *40.28N* *44.75E* *3597* Volcanic Field...... A-1900±750

214080 Dar-Alages.............................. *39.70N* *45.54E* *3329* Pyroclastic cones.. A-2000±1000?

214090 Porak .... *40.02N* *45.78E* *2800* Stratovolcano .... A-0778±5, T-4510±300

214100 Tskhouk-Karckar...................... *39.73N* *46.02E* *3000* Pyroclastic cones.. T-3000±300

VNUM	VOLCANO NAME	LAT	LONG	ELEV	TYPE	ERUPTION YEARS (VEI≤3 <u>VEI=4</u> **VEI=5** **<u>VEI=6-7</u>**)
Africa (northeastern) and Red Sea						
221010	Tair, Jebel at	15.55N	41.83E	244	Stratovolcano	2007-08, 1883, 1863, 1833, 1750±50
221020	Zubair Group	15.05N	42.18E	191	Shield	2011-12, 1824
221041	Dallol	14.24N	40.30E	-48	Explosion craters	2011, 1926
221070	Dalaffilla	13.79N	40.55E	613	Stratovolcano	2008
221080	**Erta Ale**	13.60N	40.67E	613	Shield	<1967-2014, 1960, 1940, 1906
221100	**Dubbi**	13.58N	41.81E	1625	Stratovolcano	1861, 1400
221101	Nabro	13.37N	41.70E	2218	Stratovolcano	<u>2011-12</u>
221112	Alayta	12.88N	40.57E	1501	Shield	1915, 1907
221113	Dabbahu	12.60N	40.48E	1442	Stratovolcano	2005, R-3450±1800, R-4450±2700, R-5850±4300
221115	Manda Hararo	12.17N	40.82E	600	Shields	2009, 2007
221122	Manda-Inakir	12.38N	42.20E	600	Fissure vents	1928
221126	Ardoukôba	11.58N	42.47E	298	Fissure vents	1978
221141	Dama Ali	11.28N	41.63E	1068	Shield	@1631
221190	Fentale	8.98N	39.93E	2007	Stratovolcano	1820?, 1250±50
221200	Kone	8.80N	39.69E	1619	Calderas	1820±10
221250	Tullu Moje	8.16N	39.13E	2349	Pyroclastic cone	1900?, A 1775±25
221270	Alutu	7.77N	38.78E	2335	Stratovolcano	C-0050?
Africa (eastern)						
222020	South Island	2.63N	36.60E	800	Stratovolcano	1888
222030	Barrier, The	2.32N	36.57E	1032	Shield	1921, 1917, 1897, 1895, 1888, 1871±3, M 1090±50, M 1050±150, M 1030±150, T-7710±200
222040	Namarunu	1.98N	36.43E	817	Shield	T-6550=1500
222051	Emuruangogolak	1.50N	36.33E	1328	Shield	M 1910±50, C 1700±100, M 1300±150, M 1230±150, M 1160±150, M 1120±150, T-6550±1500, C-8050±1000
222052	Silali	1.15N	36.23E	1528	Shield	R-5050±2000, R-6050±3000, R-7050±2000
222053	Paka	0.92N	36.18E	1697	Shield	<u>R-7550±2000</u>
222060	**Menengai**	0.20S	36.07E	2278	Shield	**C-6050?**, T-7350±500
222090	Olkaria	0.90S	36.29E	2434	Pyroclastic cones	C 1770±50, T-4050?, T-6050?
222100	Longonot	0.91S	36.45E	2776	Stratovolcano	A 1863±5, C-1330±100, C-7200±100
222120	**Lengai, Ol Doinyo**	2.76S	35.91E	2962	Stratovolcano	<2011-13, 2007-10, 1994-2006, 1983-93, 1967, 1960-66, 1958, 1955, 1954, 1940-41, 1926, 1921, 1916-17, 1914-15, 1913, 1907±3-10, 1904, 1882-83, 1880, T 1350?, T 0700?, T-0050?, T-1550±1500
222130	Chyulu Hills	2.68S	37.88E	2188	Volcanic Field	A 1855±5, C 1470±200
222160	Meru	3.25S	36.75E	4565	Stratovolcano	1910, 1886?, 1878±1, <u>C-5850?</u>
222164	Ngozi	8.97S	33.57E	2622	Caldera	G 1450±40, **G-8250?**
222166	Rungwe	9.13S	33.67E	2956	Stratovolcano	G 1250±40, <u>G-0050±100</u>, **T-2050?**
222170	Kyejo	9.23S	33.78E	2175	Stratovolcano	1800?
Africa (central)						
223001	Fort Portal	0.70N	30.25E	1615	Tuff cones	C-2120±100, C-2750±75
223020	**Nyamuragira**	1.41S	29.20E	3058	Shield	2014, 2010-12, 2006, 2004, 2000-02, 1998, 1996, 1991-94, 1986-89, 1984, 1980-82, 1976-77, 1971, 1967, 1956-58, 1954, 1951-52, 1948, 1920-40, 1912-13, 1907-09, 1904-05, 1901-02, 1899, 1896, 1894, 1882, 1865, G 1550±100
223030	**Nyiragongo**	1.52S	29.25E	3470	Stratovolcano	2002-14, 1994-96, 1982, 1927-77, 1920-21, 1918, 1911, 1905-06, 1898-1902, 1894, @<1891, 1884
223040	Karisimbi	1.50S	29.45E	4507	Stratovolcano	K-8050?
223050	Visoke	1.47S	29.49E	3711	Stratovolcano	1957, @1891
Africa (western)						
224004	Santa Isabel	3.58N	8.75E	3007	Shield	1923, 1903, 1898?
224010	Cameroon	4.20N	9.17E	4095	Stratovolcano	1999-2000, 1989, 1982, 1959, 1954, 1925, 1922, 1909, 1871, 1868, 1865-66, 1852, 1838, 1825±10, 1807±8, 1650±50, -0450±50
Africa (northern)						
225030	Marra, Jebel	12.95N	24.27E	3042	Volcanic Field	<u>G-2000?</u>
225050	Meidob Volcanic Field	15.32N	26.47E	2000	Pyroclastic cones	N-2950±500, <u>T-3000?</u>, <u>T-3050?</u>, <u>T-4150±1450</u>, N-5250±500, N-6050±1600
225060	Bayuda Volcanic Field	18.33N	32.75E	670	Pyroclastic cones	C 0850±50

VNUM	VOLCANO NAME	LAT	LONG	ELEV	TYPE	ERUPTION YEARS (VEI≤3 <u>VEI=4</u> **VEI=5** <u>**VEI=6-7**</u>)

Middle East (eastern)

230020	Unnamed	36.67N	37.00E		Volcanic Field	1222
230040	Unnamed	33.00N	36.63E	1050	Volcanic Field	G-2670±200
230050	Es Safa	33.25N	37.07E	979	Volcanic Field	1850±10
232010	Damavand	35.95N	52.11E	5670	Stratovolcano	U-5350±200

Middle East (southern and western)

231020	Uwayrid, Harrat	27.08N	37.25E	1920	Volcanic Field	A 0640?
231040	Lunayyir, Harrat	25.17N	37.75E	1370	Volcanic Field	<1000
231060	Khaybar, Harrat	25.00N	39.92E	2093	Volcanic Field	0650±50
231070	Rahat, Harrat	23.08N	39.78E	1744	Volcanic Field	1256, 0641
231080	Yar, Jabal	17.05N	42.83E	305	Volcanic Field	1810±10
231090	Arhab, Harra of	15.63N	44.08E	3100	Volcanic Field	0500±100, A >0200
231110	Haylan, Jabal	15.43N	44.78E	1550	Volcanic Field	A >-1200
231120	Dhamar, Harras of	14.57N	44.67E	3500	Volcanic Field	1937
231160	Sawâd, Harra Es-	13.58N	46.12E	1737	Volcanic Field	1253
232080	Tor Zawar	30.48N	67.49E	2237	Fissure vents	2010

Indian Ocean (western)

233005	Mayotte Island	12.83S	45.17E	660	Shield	T-2050±500, T-3550±500, T-4050±500
233010	Karthala	11.75S	43.38E	2361	Shield	2005-07, 1991, 1977, 1972, 1965, 1956, 1952, 1948, 1928±2, 1918, 1910, 1904, 1883-84, 1880, 1876, 1872, 1865, 1857-60, 1855, 1850?, 1848, 1833, 1830, 1828, 1821, 1814, 1808, 1050±150
233014	Itasy Volcanic Field	19.00S	46.77E	1800	Pyroclastic cones	C <-6050, C-7130±100
233020	**Fournaise, Piton de la**	21.24S	55.71E	2632	Shield	2014, 1998-2010, 1990-92, 1983-88, 1981, 1979, 1975-77, 1972-73, 1963-66, 1950-61, 1941?-48, 1929-39, 1924-27, 1920-21, 1917, 1915, 1913, 1909-10, 1907, 1897-1905, 1894, 1889-91, 1884, 1878, 1874-76, 1871, 1869?, 1868, 1863-65, 1858-61, 1842-52, 1832, 1830, 1824, 1820-21, 1812-17, 1809-10, 1807, 1800-02, 1797?, 1794-95, 1791-92, 1789, 1784-87, 1774-76, 1771-72, 1768, 1766, 1759-60, 1753, 1751, 1734, 1733?, 1721, 1708-09, <1703-05, 1671-72, 1669, 1649, 1640, G 1600?, G 1440?, G 1340?, G 0960?, G 0600?, C 0460?, G 0120?, C-1790±100, G-2700?, C-2800±150

Indian Ocean (southern)

234000	Boomerang Seamount	37.72S	77.83E	-650	Submarine	U 1995
234002	St. Paul	38.72S	77.53E	268	Stratovolcano	1793
234010	Heard	53.11S	73.51E	2745	Stratovolcano	2012-14, 2006-08, 2003-04, 2000-01, 1992-93, 1985-87, 1950-54, 1910
234011	McDonald Islands	53.03S	72.60E	230	Complex	2005, 2001, 1996-97, 1992
234070	Marion Island	46.90S	37.75E	1230	Shields	2004, 1980

New Zealand

241010	Kaikohe-Bay of Islands	35.30S	173.90E	388	Volcanic Field	C 0400±300
241020	Auckland Field	36.90S	174.87E	260	Volcanic Field	C 1350?
241021	Mayor Island	37.28S	176.25E	355	Shield	**G-5060±200**, C-6050±75
241030	Taranaki [Egmont]	39.30S	174.07E	2518	Stratovolcano	1854, T 1800?, D 1755?, C 1700±50, <u>D 1655?</u>, C 1590±40, C 1570±40, C 1560±40, C 1550±40, C 1500±30, C 1480±50, C 1400±50, C 1340±40, C 1300±50, C 1070±40, C 0970±30, C 0820±30, T 0550?, C 0520±150, C 0390±40, T 0150?, C 0100±40, C-0040±75, C-0150±30, C-0420±30, T-0590±500, T-1130±200, C-1190±40, T-1250?, C-1560±40, **C-1700±100**, T-2150?, C-2400±40, T-2450±300, T-2700?, T-2850±300, T-3250?, C-5120±50, T-6050?, C-7000±100, C-7270±50, T-7330?, C-7650?
241040	**White Island**	37.52S	177.18E	321	Stratovolcanoes	2012-13, 1998-2001, 1986-95, 1976-84, 1974, 1966-71, 1962, 1957-59, 1955, 1947, 1933, 1930, 1928, 1926, 1924, 1922, 1909, 1885-86, 1836±2, 1826
241050	**Okataina**	38.12S	176.50E	1111	Lava domes	1981, 1978, 1973, 1951, 1926, 1924, 1917-20, 1912-15, 1910, 1908, 1900-06, 1896, **1886**, G 1310±12-15?, T 0180?, T-0300?, G-1330±75, <u>G-1750?</u>, **G-3580±50**, T-5550?, **G-6060±50**, **G-7560±18**, T-8050?
241060	Reporoa	38.42S	176.33E	592	Caldera	T 1180?
241061	Maroa	38.42S	176.08E	897	Calderas	T 0180?, T-7050?
241070	**Taupo**	38.82S	176.00E	760	Caldera	T 0260?, **G 0233±13**, G-0200?, T-0800?, G-1010±200, G-1050?, G-1250?, **G-1460±40**, G-2500?, <u>G-2600?</u>, T-2800?, G-2850?, <u>G-2900?</u>, G-3070?, T-3120?, <u>G-3170±200</u>, G-3420?, G-4000?, <u>G-4100?</u>, <u>G-4700?</u>, G-5100?, **G-8130±200**, <u>T-9210?</u>, **G-9240±75**, **G-9460±200**
241080	**Tongariro [Ngauruhoe]**	39.16S	175.63E	1978	Stratovolcanoes	2012, 1972-77, 1968-69, 1962, 1958-59, 1948-56, 1939-40, 1937, 1934-35, 1931, 1928, 1924-26, 1917, 1913-14, 1909-11, 1904-07, 1896-98, 1892, 1885±1-87, 1883, 1881, 1878, 1869-70, 1862-65, 1859, 1857, 1855, 1844-45, 1841, 1839, T 1500±50, **C-0550±200**, **T-9350?**, **G-9450?**, **T-9650?**, G-9850?
241100	**Ruapehu**	39.28S	175.57E	2797	Stratovolcano	2006-07, 1995-97, 1984-92, 1966-82, 1959, 1956, 1950-52, 1944-48, 1942, 1940, 1934-36, 1925, 1921, 1918, 1906, 1903, 1895, 1889, 1861, C 1210±150, T-5550±2500, C-7590±100, C <-7840, T-9650?, G-9850?

VNUM	**VOLCANO NAME**	**LAT**	**LONG**	**ELEV**	**TYPE**	**ERUPTION YEARS** (VEI≤3 <u>VEI=4</u> **VEI=5** <u>**VEI=6-7**</u>)
241130	Rumble III	35.75S	178.48E	-220	Submarine	2008±1, S 1986, S 1973, S 1970, S 1963-66, S 1958-62
241140	Healy	35.00S	178.97E	-980	Submarine	G 1360±75

Kermadec Islands

242005	Havre Seamount	31.11S	179.04E	-720	Submarine	2012
242021	**Macauley Island**	30.20S	178.47W	238	Caldera	**C-4360±200**
242030	**Raoul Island**	29.27S	177.92W	516	Stratovolcano	2006, 1964-65, 1886, 1870, 1814, <u>C 1720±50</u>, <u>C 1630±50</u>, T 1450?, <u>T 0850?</u>, T 0700?, <u>C 0550?</u>, <u>T 0400?</u>, <u>T 0100?</u>, T-0050?, **C-0250±75**, <u>C-1200±150</u>, <u>C-2000±100</u>

Tonga Islands

242050	Monowai Seamount	25.89S	177.19W	-132	Submarine	S 2012, S 2008, S 2002-06, S 1995-99, S 1990-91, S 1988, S 1986, S 1982, S 1977-80
243010	Unnamed	21.38S	175.65W	-500	Submarine	1932, 1907
243030	Unnamed	20.85S	175.53W	-13	Submarine	1999, 1923, 1911
243040	Hunga Tonga-Hunga Ha'apai	20.57S	175.38W	149	Submarine	2009, 1988, 1937, 1912
243050	Falcon Island	20.32S	175.42W	-17	Submarine	1936, 1933, 1927-28, 1885-86, 1877
243060	Tofua	19.75S	175.07W	515	Caldera	2004-14, 1958-60?, 1906, 1885, 1854, 1792, 1774
243070	Metis Shoal	19.18S	174.87W	43	Submarine	1995, 1991, 1979, 1967-68, 1886, 1878, 1858, 1851, <1781
243080	Home Reef	18.99S	174.78W	-10	Submarine	2006, 1984, 1852
243090	Late	18.81S	174.65W	540	Stratovolcano	1854, 1790
243091	Unnamed	18.33S	174.37W	-40	Submarine	2001
243100	Fonualei	18.02S	174.33W	180	Stratovolcano	1957, 1951, 1939, 1906, <u>1846</u>, 1791
243102	Curacoa	15.62S	173.67W	-33	Submarine	1979, 1973
243110	Niuafo'ou	15.60S	175.63W	260	Shield	1985, 1946, 1943, 1935-36, 1929, 1912, 1887, <u>1886</u>, 1867, 1853, 1814
243120	Tafu-Maka	15.37S	174.23W	-1400	Submarine	2008
243130	West Mata	15.10S	173.75W	-1174	Submarine	2008-09

Samoan and Wallis Islands

244000	Vailulu'u	14.22S	169.06W	-592	Submarine	2003±2. S 1995, S 1973
244010	Ofu-Olosega	14.18S	169.62W	639	Shields	1866
244040	**Savai'i**	13.61S	172.53W	1858	Shield	1905-11, 1902, 1760, G 1610±200, G 1350±50, G 1310±50, G 1240±30, G 1040±150, G 0170±100, G-0480±300, G-1150±150, G-1990±150

Fiji Islands

245010	Taveuni	16.82S	179.97W	1241	Shield	G 1550±100, G 1420±20, T 1350±75, G 1160±150, T 1020?, T 0880?, T 0770?, G 0640±40, G 0520±100, G 0480±75, G 0400±50, T 0350?, G 0320±100, T 0270?, G 0220±100, G-0090±200, T-0200?, G-0330±75, G-0400±20, T-0600?, T-0680?, T-1020?, T-1100?, T-1200?, T-1300?, T-1450?, T-1700?, G-3200±150, G-3580±200, G-4800±100, G-5230±100, G-5920±75, G-6560±75, G-8040±50
245030	Nabukelevu	19.12S	177.98E	805	Lava domes	G 1660±30, G 0340?, G-0580±300

Admiralty Islands

250010	St. Andrew Strait	2.38S	147.35E	270	Complex	1953-57, 1883, C 0350?, C-0240±100
250030	Unnamed	3.03S	147.78E	-1300	Submarine	S 1972

Northeast of New Guinea

251010	Bam	3.61S	144.82E	685	Stratovolcano	1954-60, 1947, 1944, 1924, 1913, 1907-09, 1877, 1874, 1872±4
251020	**Manam**	4.08S	145.04E	1807	Stratovolcano	2010-14, <u>2004-09</u>, 1974-2004, 1956-66, 1953-54, 1946-47, 1936-39, 1932-34, 1925-28, 1920-22, <u>1919</u>, 1917, 1909-14?, 1904, 1899±1, 1887-95, 1885, 1877, 1830, 1700, 1643, 1616
251030	Karkar	4.65S	145.96E	1839	Stratovolcano	2013-14, 1979, 1974-75, 1895, 1885, 1643, C 1070±200, C 0730?, C 0520±100, C-0870±75, <u>C-7140±150</u>
251050	**Long Island**	5.36S	147.12E	1280	Complex	1993, 1976, 1973-74, 1968, 1953-55, 1943, 1938, 1933, **C 1660±20**, **C-2040±100**
251070	**Ritter Island**	5.52S	148.12E	140	Stratovolcano	2006-07, 1974, 1972, 1888, @1887, 1793, 1700

New Britain

252010	**Langila**	5.53S	148.42E	1330	Complex	2012, 2002-10, 1960-2000, 1958, 1954-56, 1907, 1900, 1890, 1884, 1878
252040	**Dakataua**	5.06S	150.11E	400	Caldera	A 1895±5, **C 0800±50**
252070	Garbuna Group	5.45S	150.03E	564	Stratovolcanoes	2008, 2005, C 0150?

VNUM	VOLCANO NAME	LAT	LONG	ELEV	TYPE	ERUPTION YEARS (VEI≤3 VEI=4 **VEI=5** <u>**VEI=6-7**</u>)
252080	**Witori [Pago]**	5.58S	150.52E	742	Caldera	2012, 2007, 2002-03, 1933, 1920±2, 1911-18, <u>T 1800?</u>, T 1730±25, <u>T 1550?</u>, T 1450?, T 1050±100, T 0950?, **C 0710±75**, **G 0690±100**, **G 0310±100**, C-0640±300, <u>**G-1370±100**</u>, <u>**G-4000±200**</u>, C-7510±150
252100	Hargy	5.33S	151.10E	1148	Stratovolcano	C 0950?, C-5050?
252110	Bamus	5.20S	151.23E	2248	Stratovolcano	1886±8, C 1650±50, C-0270±50, C-0350±75
252120	**Ulawun**	5.05S	151.33E	2334	Stratovolcano	2010-13, 2001-07, <u>2000</u>, 1999, 1993-94, 1989, 1983-85, 1980, 1978, 1973, 1970, 1967, 1960-63, 1958, 1941, 1927, 1918-19, 1915, 1898, 1878, 1700
252130	Lolobau	4.92S	151.16E	858	Caldera	<u>1911-12</u>, <u>1904-05</u>, C 1100±30
252140	**Rabaul**	4.27S	152.20E	688	Pyroclastic shield	2013-14, 2010-11, <u>2006-10</u>, 1995-2006, <u>1994-95</u>, 1940-43, <u>1937</u>, 1878, 1850?, 1791, 1767, T 1450±150, **C 0540±100**
252150	Tavui	4.12S	152.20E	200	Caldera	**N-5150±2000**

New Guinea and D'Entrecasteaux Islands

VNUM	VOLCANO NAME	LAT	LONG	ELEV	TYPE	ERUPTION YEARS
253010	**Lamington**	8.95S	148.15E	1680	Stratovolcano	<u>1951-56</u>, C-4850±300, C-5980±300
253030	**Victory**	9.20S	149.07E	1925	Stratovolcano	1890?-1935±5
253040	Waiowa	9.57S	149.08E	640	Pyroclastic cone	1943-44
253060	Dawson Strait Group	9.62S	150.88E	500	Volcanic Field	H 1350?

New Ireland

VNUM	VOLCANO NAME	LAT	LONG	ELEV	TYPE	ERUPTION YEARS
254020	Ambitle	4.08S	153.65E	450	Stratovolcano	C-0350±100

Bougainville and Solomon Islands

VNUM	VOLCANO NAME	LAT	LONG	ELEV	TYPE	ERUPTION YEARS
255011	**Billy Mitchell**	6.09S	155.23E	1544	Pyroclastic shield	**C 1580±20**, C 1030±25
255020	**Bagana**	6.14S	155.20E	1855	Lava cone	<2000-14, 1970-95, 1961-68, 1959?-60, 1956, 1953, <u>1952</u>, 1945-51, 1943, 1937-39, 1908, 1899, 1897, 1894-95, 1865±3-83, 1842
255030	Loloru	6.52S	155.62E	1887	Pyroclastic shield	C-1050?, C-1260±300, C-2150?, C-3150?, C-4150?, C-6950?
255050	Simbo	8.29S	156.52E	335	Stratovolcanoes	A 1910±10
255060	**Kavachi**	9.02S	157.95E	-20	Submarine	2014, 2007, 1999-2004, 1997, 1991, 1985-86, 1980-82, 1974-78, 1972, 1969-70, 1961-66, 1957-58, 1950-53, 1942?, 1939
255070	**Savo**	9.13S	159.82E	485	Stratovolcano	1835±5-47?, G 1650±20, <1568

Santa Cruz Islands

VNUM	VOLCANO NAME	LAT	LONG	ELEV	TYPE	ERUPTION YEARS
256010	Tinakula	10.38S	165.80E	851	Stratovolcano	2006-12, 1999-2002, 1995, 1989-90, 1984-85, 1971, 1965-66, 1951, 1909, 1897, 1886, 1871, 1869, 1857, 1855, 1840?, 1797, 1768, 1595, C-1050?

Vanuatu

VNUM	VOLCANO NAME	LAT	LONG	ELEV	TYPE	ERUPTION YEARS
257010	Suretamatai	13.80S	167.47E	921	Complex	1965-66, 1861?, 1856?
257020	Gaua	14.27S	167.50E	797	Stratovolcano	2009-11, 1980-82, 1976-77, 1973-74, 1971, 1965-69, 1962-63
257030	**Aoba**	15.40S	167.83E	1496	Shield	2011, 2005-06, 1995, 1870?, A 1670?, C 1530?
257040	**Ambrym**	16.25S	168.12E	1334	Pyroclastic shield	1996-2014, 1994, 1988-91, 1983-86, 1979-81, 1957-77, 1952-55, <u>1950-51</u>, 1942, 1935-38, 1929, 1912-15, 1908-10, 1898, 1894-95, 1888, 1886, 1883-84, 1871, 1863-64, 1820?, 1774, **C 0050±100**
257050	**Lopevi**	16.51S	168.35E	1413	Stratovolcano	2003-07, 1998-2001, 1982, 1978-80, 1974-76, 1967-72, 1962-65, 1960, 1939, 1933, 1922, 1908, 1898, 1892-93, 1883-84, 1874, 1863-64
257060	Epi	16.68S	168.37E	833	Stratovolcanoes	2004, 2002, 1999?, 1979±5, 1960, 1958, 1953, 1920
257070	Kuwae	16.83S	168.54E	-2	Caldera	1974, 1971, 1958-59, 1952-53, 1948-49, 1923-25, 1897-1901, G 1430?
257090	Traitor's Head	18.75S	169.23E	837	Stratovolcano	1881
257100	**Yasur**	19.53S	169.44E	361	Stratovolcano	<1774-2014, C 1150?, C 0850±300, C 0550?

Pacific Ocean (southwestern)

VNUM	VOLCANO NAME	LAT	LONG	ELEV	TYPE	ERUPTION YEARS
258001	Eastern Gemini Seamount	20.98S	170.28E	-80	Submarine	1996
258010	Matthew Island	22.33S	171.32E	177	Stratovolcano	1956±2, 1954, <1949
258020	Hunter Island	22.40S	172.05E	297	Stratovolcano	1903, 1895, 1841, 1835
258030	Unnamed	25.78S	168.63E	-2400	Submarine	S 1963-64

Australia

VNUM	VOLCANO NAME	LAT	LONG	ELEV	TYPE	ERUPTION YEARS
259010	Newer Volcanics Province	37.77S	142.50E	1011	Shields	C-2900±150, F-3000±500, C <-5290, C-5850?

VNUM	VOLCANO NAME	LAT	LONG	ELEV	TYPE	ERUPTION YEARS (VEI≤3 <u>VEI=4</u> **VEI=5** <u>**VEI=6-7**</u>)

Andaman Islands

260010 Barren Island .... *12.28N* *93.86E* *354* Stratovolcano .... 2013-14, 2005-11, 1994-95, 1991, 1832, 1803-04, 1795, 1789, 1787, T-8060±100

Sumatra

261020 Seulawah Agam .... *5.45N* *95.66E* *1810* Stratovolcano .... 1839, 1510±10

261030 Peuet Sague .... *4.91N* *96.33E* *2801* Complex.............. 1998-2000, 1991, 1986, 1979, 1918-21

261050 Telong, Bur ni............................ *4.77N* *96.82E* *2617* Stratovolcano 1937, 1919, 1856, 1839, 1837

261070 Sibayak.. .... *3.23N* *98.52E* *2212* Stratovolcanoes.... 1881

261080 Sinabung .... *3.17N* *98.39E* *2460* Stratovolcano 2013-14, 2010

261120 **Sorikmarapi** .... *0.69N* *99.54E* *2145* Stratovolcano .... 1986, 1970, 1917, 1893, 1892, 1879, 1829?

261140 **Marapi**.. .... *0.38S* *100.47E* *2891* Complex.............. 2014, 2011-12, 2004, 1999-2001, 1987-94, 1982-84, 1975-80, 1973, 1970-71, 1966-67, 1954-58, 1949-52, 1943±5, 1932, 1929-30, 1927, 1925, 1913-19, 1910-11, 1907-08, 1904-05, 1888-89, 1885-86, 1883, 1876-77, 1871, 1863, 1861, 1854-56, 1845, 1833-34, 1822, 1807, 1770

261150 Tandikat. .... *0.43S* *100.32E* *2438* Stratovolcanoes.... 1924, 1914, 1889

261160 Talang.... .... *0.98S* *100.68E* *2597* Stratovolcano .... 2005-07, 2001, 1967-68, 1963, 1876?, 1845, 1843, 1833

261170 **Kerinci**.. .... *1.70S* *101.26E* *3800* Stratovolcano .... 2007-09, 2004, 2001-02, 1998-99, 1996, 1990, 1966-70, 1963-64, 1960, 1952, 1936-38, 1923, 1921, 1908-09, 1887, 1878, 1874?, 1842, 1838

261180 Sumbing .... *2.41S* *101.73E* *2507* Stratovolcano 1921, 1909

261220 **Kaba** .... *3.52S* *102.62E* *1952* Stratovolcano 2000, 1956, 1950-52, 1939-41, 1907, 1873-92, 1868-69, 1853, 1833-4

261230 Dempo... .... *4.03S* *103.13E* *3173* Stratovolcanoes.... 2009, 2006, 1994, 1973-74, 1964, 1939-40, 1936, 1934, 1926, 1923, 1921, 1908, 1905, 1900, 1895, 1884, 1879-81, 1853, 1839?, 1817

261250 Besar...... .... *4.43S* *103.67E* *1899* Stratovolcano? 1940

261270 Suoh...... .... *5.25S* *104.27E* *1000* Calderas .... <u>1933</u>

262000 **Krakatau** .... *6.10S* *105.42E* *813* Caldera................ 2014, 2007-12, 1999-2001, 1992-97, 1988, 1978-81, 1975, 1972-73, 1965?, 1958-63, 1955, 1952-53, 1949-50, 1927-47, <u>**1883**</u>, 1684, 1680-81, 1530, 1320?, 1150±50, 1050±50, 0950±50, 0850±50, <u>0416</u>, 0250±50

Java

263040 Perbakti-Gagak .... *6.75S* *106.70E* *1699* Stratovolcanoes.... 1938-39, 1935-36, 1929, 1923, C-6450?

263050 Salak.... .... *6.72S* *106.73E* *2211* Stratovolcano 1938, 1935, 1919, 1902-03, 1780

263060 Gede .... *6.78S* *106.98E* *2958* Stratovolcano 1956-57, 1947-49, 1909, 1899, 1891, 1886-88, 1870, 1866, 1852-53, 1847-48, 1845, 1843, 1840, 1832, 1761, 1747-48

263090 Tangkubanparahu .... *6.77S* *107.60E* *2084* Stratovolcano .... 2013, 1983, 1969, 1967, 1965, 1961, 1957, 1952, 1929, 1926, 1910, 1896, 1846, 1842, 1829, 1826, C-7500±50, C-8020±50

263100 **Papandayan**.............................. *7.32S* *107.73E* *2665* Stratovolcanoes.... 2002, 1942, 1923-25, 1772

263130 Guntur ... . *7.14S* *107.84E* *2249* Complex.............. 1847, 1843, 1840-41, 1832-36, 1827-29, 1825, 1818, 1815-16, 1809, 1807, 1803, 1800, 1780, 1777, 1690

263140 **Galunggung**.............................. *7.25S* *108.06E* *2168* Stratovolcano .... 1984, <u>1982-83</u>, 1918, 1894, **1822, C-2250±150**

263170 Cereme .. .. *6.89S* *108.40E* *3078* Stratovolcano 1951, 1937-38, 1805, 1775, 1772, 1698

263180 **Slamet**... .... *7.24S* *109.21E* *3428* Stratovolcano 2014, 2009, 1999, 1988, 1973, 1969, 1966-67, 1960-61, 1957-58, 1955, 1951-53, 1948, 1943-44, 1939-40, 1937, 1932-33, 1926-30, 1923, 1904, 1890, 1885, 1875, 1860, 1849, 1847, 1835, 1825, 1772

263200 **Dieng Volcanic Complex** .... *7.20S* *109.92E* *2565* Complex.............. 2009, 2003, 1996, 1993, 1986, 1981, 1979, 1964, 1956, 1953-54, 1943-44, 1939, 1928, 1883-84, 1847, 1825-26, 1786, 1375±75, C 1180±100, C-0050±100, C-0500±75, C-6590±150

263210 Sundoro . ... *7.30S* *109.99E* *3136* Stratovolcano .... 1971, 1906, 1902-03, 1887, 1882-83, 1818, 1806?, C 0470?, C 0230?

263220 Sumbing .... *7.38S* *110.07E* *3371* Stratovolcano .... 1730?

263240 Merbabu .... *7.45S* *110.43E* *3145* Stratovolcano 1797, 1560

263250 **Merapi**.. .... *7.54S* *110.44E* *2968* Stratovolcano .... 2013-14, <u>2010-12</u>, 2006-07, 1992-2002, 1972-90, 1967-70?, 1961, 1953-58, 1948, 1942-45, 1939-40, 1930-35, 1924, 1920-22, 1918, 1915, 1902-13, 1897, 1891-94, 1883-89, 1878-79, 1872-73, <u>1872</u>, 1861-71, 1849, 1846-47, 1840, 1832-38, 1828, 1812-23, 1810, 1807, 1797, <1791, 1786, 1768, 1755, 1752, 1745, 1677-78, 1672, 1663, 1658, 1587, 1584, 1560, 1554, 1548, G 1480±300, G 1440±100, G 1380±300, G 1300±75, G 1230±200, G 1190±30, G 1140±150, G 1090±100, G 1010±25, G 0940±100, G 0870±100, G 0680±200, G 0630±30, G 0540±50, G 0480±75, G 0410±150, G 0280±150, G 0190±300, G 0120±75, <u>G 0020±300</u>, G-0340±500, G-0700±150, G-1010±200, <u>C-1180±75</u>, G-1410±50, G-1770±75, G-1890±55, G-2910±150, G-4690±75, G-7310±300, G-8780±150

263251 Muria..... .... *6.62S* *110.88E* *1625* Stratovolcano .... C-0160±300

263260 Lawu...... .... *7.63S* *111.19E* *3265* Stratovolcano 1885

263280 **Kelut** .... *7.93S* *112.31E* *1731* Stratovolcano .... <u>2014</u>, 2007-08, <u>1990</u>, 1967, <u>1966</u>, <u>1951</u>, 1920, <u>1919</u>, 1901, 1864, 1851, 1848, 1835, <u>1826</u>, 1825, 1811, 1785, 1776, 1771, 1752, 1716, <u>1641</u>, **1586**, 1548, 1481, 1462, 1450-51, 1411, 1395, 1385, 1376, 1334, 1311, 1000, C-0230±300

VNUM	VOLCANO NAME	LAT	LONG	ELEV	TYPE	ERUPTION YEARS (VEI≤3 <u>VEI=4</u> **VEI=5** <u>**VEI=6-7**</u>)
263290	Arjuno-Welirang	7.73S	112.58E	3339	Stratovolcano	1952, 1950
263300	**Semeru..**	8.11S	112.92E	3676	Stratovolcano	1967-2014, 1950-64, 1945-47, 1941-42, 1915±1, 1907-13, 1903-05, 1884-1901, 1877-79, 1872, 1867, 1865, 1860, 1856-57, 1851, 1848, 1844-45, 1842, 1838, 1836, 1832, 1829-30, 1818
263310	**Tengger Caldera**	7.94S	112.95E	2329	Stratovolcanoes	2010-12, 2004, 2000-01, 1995, 1983-84, 1980, 1972, 1955-56, 1950, 1948, 1939-40, 1935, 1928-30, 1921-22, 1915-16, 1906-10, 1896, 1893, 1890, 1885-87, 1877, 1865-68, 1857-60, @1856, 1842-44, 1835, 1829-30, 1825, 1822-23, 1820, 1815, 1804, C 1590±50, C 0330±50, C 0190±50, <u>C-0830±50</u>, C-5260±780
263320	Lamongan	7.98S	113.34E	1651	Stratovolcano	1898, 1896, 1893, 1883-91, 1877, 1874, 1869-72, 1864, 1861, 1859, 1856, 1849, 1847, 1841-44, 1838, 1829-30, 1826, 1824, 1821-22, 1817-18, 1808, 1806, 1799
263340	**Raung**	8.13S	114.04E	3332	Stratovolcano	2007-08, 2002, 1999-2000, 1997, 1993-95, 1985-91, 1982, 1973-79, 1971, 1955-56, 1953, 1943-45, 1936-41, 1933, 1927-29, 1924, 1921, 1915-17, 1913, 1902-04, 1896-97, 1890, 1885, 1881, 1864, 1859-60, 1849, 1838, <u>1817</u>, 1815, 1812-14?, 1804±4, 1793±6, 1730, <u>1638</u>, 1597, **1593**
263350	Ijen	8.06S	114.24E	2799	Stratovolcanoes	1999, 1993-94, 1952, 1936, 1917, 1817, 1796, C-0640±50

Lesser Sunda Islands

VNUM	VOLCANO NAME	LAT	LONG	ELEV	TYPE	ERUPTION YEARS
264010	Batur	8.24S	115.38E	1717	Caldera	1998-2000, 1994, 1970-74, 1968, 1963-66, 1921-26, 1904-05, 1897, 1888, 1854, 1849, 1821, 1804
264020	**Agung**	8.34S	115.51E	3142	Stratovolcano	**1963-64**, 1843, 1808
264030	**Rinjani..**	8.42S	116.47E	3726	Stratovolcano	2009-10, 2004, 1994, 1965-66, 1953, 1949-50, 1944-45, 1915, 1909, 1906, 1900-01, 1884, 1847, <u>I 1257</u>, C-0600?
264040	**Tambora**	8.25S	118.00E	2850	Stratovolcano	1967±20, 1880±30, 1819, <u>**1812-15**</u>, C 0740±150, C-3050?, C-3910±200
264050	Sangeang Api	8.20S	119.07E	1949	Complex	2014, 1997-99, 1985-88, 1964-66, 1953-58, 1927, 1911-12, 1860, 1821, 1715, 1512
264071	Ranakah	8.62S	120.52E	2350	Lava domes	1991, 1987-89
264080	Inierie	8.88S	120.95E	2245	Stratovolcano	C-8050?
264090	Inielika	8.73S	120.98E	1559	Complex	2001, 1905
264100	Ebulobo	8.82S	121.18E	2124	Stratovolcano	1969, 1941, 1938, 1924, 1910, 1888, 1830
264110	Iya	8.90S	121.65E	637	Stratovolcano	1969, 1953, 1882, 1871, 1867-68, 1844, 1671?
264140	Kelimutu	8.77S	121.82E	1639	Complex	1968, 1938, 1865±5
264150	**Paluweh**	8.32S	121.71E	875	Stratovolcano	2012-14, 1984-85, 1980-81, 1972-73, 1963-66, 1928, 1650±50
264160	Egon	8.67S	122.45E	1703	Stratovolcano	2008, 2004-05
264180	Lewotobi	8.54S	122.78E	1703	Stratovolcanoes	2002-03, 1999, 1990-92, 1968-71, 1939-40, 1935, 1932-33, 1921, 1914, 1909-10, 1907, 1889, 1868-69, 1865, 1861, 1675±25
264200	Leroboleng	8.36S	122.84E	1117	Complex	2003, 1881, 1876, 1873
264220	Iliboleng	8.34S	123.26E	1659	Stratovolcano	1993, 1991, 1986-87, 1982-84, 1973-74, 1948-51, 1944, 1927, 1925, 1909, 1904, 1888, 1885
264230	Lewotolo	8.27S	123.51E	1423	Stratovolcano	2012, 1951, 1920, 1899, 1864, 1852, 1849, 1819, 1660
264250	**Iliwerung**	8.53S	123.57E	1018	Complex	2013, 1999, 1993, 1983, 1973-74, 1952, 1951, 1950, 1949, 1948, 1928, 1910, 1870
264260	Tara, Batu	7.79S	123.58E	748	Stratovolcano	2007-14, 1847-52
264270	Sirung	8.51S	124.13E	862	Complex	2012, 1970, 1964-65, 1960, 1953, 1947, 1934

Banda Sea

VNUM	VOLCANO NAME	LAT	LONG	ELEV	TYPE	ERUPTION YEARS
265030	Gunungapi Wetar	6.64S	126.65E	282	Stratovolcano	1699, 1512
265040	Wurlali	7.13S	128.68E	868	Stratovolcano	1892
265050	Teon	6.92S	129.13E	655	Stratovolcano	1904, 1693, 1663, <u>1660</u>, 1659
265060	Nila	6.73S	129.50E	781	Stratovolcano	1968, 1964, 1932, 1903
265070	Serua	6.30S	130.00E	641	Stratovolcano	1921, 1919, 1858-59, 1846, 1844, 1694, <u>1693</u>, 1687, 1683
265090	Banda Api	4.53S	129.87E	640	Caldera	1988, 1901, 1890, 1825?-31, 1824, 1820, 1816, 1778, 1775, 1773, 1765-66, 1762, 1749, 1722, 1712, 1690-96, 1683, 1635, 1632, 1615, 1609, 1598-1602, 1586

Sulawesi

VNUM	VOLCANO NAME	LAT	LONG	ELEV	TYPE	ERUPTION YEARS
266010	Colo [Una Una]	0.17S	121.61E	507	Stratovolcano	<u>1983</u>, 1938±10, 1898-1900?
266020	Ambang	0.75N	124.42E	1795	Complex	2005, 1845±5
266030	**Soputan.**	1.11N	124.73E	1784	Stratovolcano	2011-12, 2000-08, 1991-96, 1989, 1984-85, 1982, 1973, 1970-71, 1966-68, 1953, 1947, 1923-24, 1917, 1915, 1906-13, 1901, 1890, 1845, 1833?, 1819, 1785, 1450±10
266100	**Lokon-Empung**	1.36N	124.79E	1580	Stratovolcano	2011-13, 2000-03, 1991-92, 1986-88, 1973-80, 1969-71, 1961-66, 1958-59, 1951-53, 1949, 1942, 1930, 1893-94, 1829, 1775±25, 1375±25
266110	Mahawu	1.36N	124.86E	1324	Stratovolcano	1977, 1958, 1952, 1904, 1846, 1789, <1788
266130	Tongkoko	1.52N	125.20E	1149	Stratovolcano	1880, 1843-46, 1821, 1801, 1694, 1683, **1680**

VNUM	VOLCANO NAME	LAT	LONG	ELEV	TYPE	ERUPTION YEARS (VEI≤3 <u>VEI=4</u> **VEI=5** <u>**VEI=6-7**</u>)

Sangihe Islands

267010 **Ruang** ... . *2.30N 125.37E* *725* Stratovolcano .... <u>2002</u>, 1949, 1914-15, 1904-05, 1889, 1874, 1870-71, 1856, 1840, 1836?, 1808

267020 **Karangetang [Api Siau]** .... *2.78N 125.40E* *1784* Stratovolcano 1995-2014, 1991-93, 1982-89, 1970-80, 1965-67, 1961-63, 1952-53, 1947-49, 1940-41, 1935, 1930, 1926, 1924, 1921-22, 1905, 1899-1900, 1892, 1886-87, 1883, 1864, 1825, 1712, 1675

267030 Banua Wuhu .... *3.14N 125.49E* *-5* Submarine 1918-19, 1904, 1895, 1889, 1835

267040 **Awu**...... .... *3.67N 125.50E* *1320* Stratovolcano .... 2004, 1992, <u>1966</u>, 1930-31, 1921-22, 1913, 1892-93, 1885, 1883, 1875, 1856, <u>1812</u>, 1711, 1646±5

Halmahera

268010 **Dukono** . ... *1.68N 127.88E* *1335* Complex .... 1933-2014, 1901, 1868?, 1719±150, 1550

268030 Ibu .... *1.49N 127.63E* *1325* Stratovolcano 2008-14, 2004-05, 2001, 1998-99, 1911

268040 Gamkonora .... *1.38N 127.53E* *1635* Stratovolcano .... 2007, 1987, 1981, 1949-52, 1926, 1917, 1911, 1885±5, **1673**, 1564

268060 **Gamalama** .... *0.80N 127.33E* *1715* Stratovolcanoes.... 2011-12, 2003, 1996, 1993-94, 1990, 1988, 1983, 1980, 1962-63, 1938, 1932-33, 1923, 1918, 1911, 1907, 1897-98, 1895, 1884, 1871, 1868-69, 1864-65, 1862, 1849-50, 1846-47, 1842-43, 1838-40, 1835, 1833, 1831, 1814, 1811-12, 1770-75, 1763, 1739, 1737, 1686-87, 1676, 1659, 1653, 1648, 1635, 1608, 1605, 1561, 1538, 1510±10

268070 **Makian**.. .... *0.32N 127.40E* *1357* Stratovolcano .... 1988, 1890, 1863-64, <u>1861-62</u>, <u>1760-61</u>, <u>1646</u>, <1550

Sulu Islands and Mindanao

270010 Jolo....... .... *6.01N 121.06E* *811* Pyroclastic cones.. 1897

271011 **Parker**... .... *6.11N 124.89E* *1824* Stratovolcano .... **1640-41**, G 1380±75, C-1920±40

271020 Matutum .... *6.37N 125.07E* *2286* Stratovolcano .... C 1290±40, C-0170±75, C-0400±50

271031 Leonard Range............................ *7.38N 126.05E* *1080* Stratovolcano .... C 0120±100, C-0080±50, C-4090±100

271040 Makaturing .... *7.65N 124.32E* *1940* Stratovolcano 1882, 1865

271060 Ragang... .... *7.70N 124.50E* *2815* Stratovolcano 1873, 1871, 1858, 1856, 1840, 1834, 1765

271070 Musuan.. .... *7.88N 125.07E* *646* Lava dome 1886

271080 **Camiguin** .... *9.20N 124.67E* *1552* Stratovolcanoes.... 1948-53, 1871-75, 1862, 1827

Central Philippines

272020 Kanlaon [Canlaon]..................... *10.41N 123.13E* *2435* Stratovolcano 2005-06, 2002-03, 1996, 1991-93, 1985-89, 1980, 1978, 1969-70, 1932-33, 1927, 1904-06, 1902, 1893-94, 1866

272050 Cabalían. .... *10.29N 125.22E* *945* Stratovolcano .... C 1820±30

272080 Biliran.... .... *11.52N 124.54E* *1301* Compound............ 1939

Luzon

273010 Bulusan.. .... *12.77N 124.05E* *1565* Stratovolcanoes.... 2010-11, 2006-07, 1994-95, 1988, 1983, 1978-81, 1933, 1928, 1918-22, 1916, 1894, 1892, 1889, 1886, C 0950?, C-3050?

273030 **Mayon**... .... *13.26N 123.69E* *2462* Stratovolcano .... 2013-14, 2008-10, 1999-2006, 1993, 1984, 1978, 1968, 1947, 1943, 1941, 1938-39, 1928, 1902, 1900, <u>1897</u>, 1895-96, 1890-93, 1885-88, 1881-82, 1876, 1871-73, 1868, 1857-62, 1855, 1853, 1851, 1845-46, 1839, 1834-35, 1827-28, <u>1814</u>, 1800, 1766, 1616, C 0470±75, C-3100±300

273042 Isarog..... .... *13.66N 123.38E* *1966* Stratovolcano .... G-3500±125

273060 San Pablo Volcanic Field........... *14.12N 121.30E* *1090* Stratovolcano .... A 1350=100

273070 **Taal**....... .... *14.00N 120.99E* *311* Caldera................ 1976-77, 1966-70, <u>1965</u>, 1911, 1903-04, 1878, 1873-74, 1842, 1825, 1808, 1790, <u>1754</u>, <u>1749</u>, 1731, 1729, <u>1716</u>, 1715, 1709, 1707, 1645, 1641, 1634-35, 1608±3, 1591, 1572, **C-3580±200**

273081 Mariveles................................. *14.52N 120.47E* *1388* Stratovolcano .... C-2050?

273083 **Pinatubo** .... *15.13N 120.35E* *1486* Stratovolcano .... 1992-93, <u>**1991**</u>, G 1450±50, <u>**G-1050±500**</u>, **G-3550?**, G-7030±300, <u>**G-7460±150**</u>

273090 Cagua..... .... *18.22N 122.12E* *1133* Stratovolcano 1860

North of Luzon

274010 Camiguin de Babuyanes .... *18.83N 121.86E* *712* Stratovolcano .... <1857

274020 Didicas... .... *19.08N 122.20E* *228* Compound............ 1978, 1969, 1952-53?, 1900, 1856-60, 1773

274030 Babuyan Claro .... *19.52N 121.94E* *1080* Stratovolcanoes.... 1924, 1917-19, 1907, 1860, <u>1831</u>, 1652

274050 Unnamed .... *20.33N 121.75E* *-24* Submarine............ 1854, 1850, 1773

274060 Iraya....... .... *20.47N 122.01E* *1009* Stratovolcano .... 1454?, C 0470±50, C 0250±200

Southeast Asia

275001 Hainan Dao............................... *19.70N 110.10E* Pyroclastic cones.. 1933, 1883

275060 Cendres, Ile des .... *10.16N 109.01E* *-20* Submarine............ 1923

VNUM	VOLCANO NAME	LAT	LONG	ELEV	TYPE	ERUPTION YEARS (VEI≤3 VEI=4 **VEI=5** **VEI=6-7**)
275080	Popa......	20.92N	95.25E	1518	Stratovolcano	A-0442
275110	Tengchong...............................	25.23N	98.50E	2865	Pyroclastic cones..	U-5750±1000

Taiwan

281030	Unnamed	24.00N	121.83E		Submarine............	1853
281031	Kueishantao..............................	24.85N	121.92E	401	Stratovolcano	1785±10
281032	Tatun Group	25.17N	121.52E	1120	Lava domes	G-4100±40

Ryukyu Islands and Kyushu

282010	Unnamed NNE of Iriomotejima.	24.57N	123.93E	-200	Submarine............	<u>1924</u>
282020	Io-Torishima.............................	27.88N	128.22E	212	Complex..............	1967-68, 1959, 1903, 1868, 1855, 1829, 1796, 1664
282021	Yokoate-jima.............................	28.80N	129.00E	495	Stratovolcanoes....	1835±30
282030	**Suwanosejima**	29.64N	129.71E	796	Stratovolcanoes....	1999-2014, 1956-97, 1949-54, 1940, 1938, 1925, 1921-22, <u>1889</u>, 1884-85, <u>1877</u>, <u>1813-14</u>, T 1600?
282040	Nakanoshima.............................	29.86N	129.86E	979	Stratovolcanoes....	1914
282043	Kuchinoshima.............................	29.97N	129.93E	628	Stratovolcanoes....	G 1190±40, G 0750±50, C-0900?, G-6750±50
282050	Kuchinoerabujima	30.44N	130.22E	657	Stratovolcanoes....	2014, 1980, 1976, 1972-74, 1968-69, 1966, 1945, <u>1933-34</u>, 1931, 1914, 1841, <1840, G 1560±100, G 1470±50, G 1440±50, G 1110±75, G 1100±100, G 0970±75, G 0600±75, G-1140±150, G-1450±75, G-3480±150, G-9520±300
282060	**Kikai**	30.79N	130.31E	704	Caldera................	2013, 1997-2004, 1988, 1934-35, C 1430±75, C 1340±30, C 1030±40, C 1010±40, C 0830±40, T 0750?, C 0390±100, C-0280±75, C-1090±100, C-1830±75, T-2450±840, C-3250±75, **C-4350?**
282070	Ibusuki Volcanic Field [Kaimon]	31.22N	130.57E	924	Calderas	<u>0885</u>, <u>0874</u>, <u>T 0770?</u>, <u>T 0720?</u>, <u>T 0660?</u>, <u>T 0600?</u>, T 0550?, T 0270?, <u>T 0150?</u>, <u>T 0130?</u>, T 0030?, <u>T-0080?</u>, <u>T-0270?</u>, <u>T-0650?</u>, T-0700?, T-1450?, <u>T-1500?</u>, T-1550?, T-1610?, <u>T-1780?</u>, <u>T-2010?</u>, **C-2690±75**, T-5050?
282080	**Aira [Sakurajima]**....................	31.59N	130.66E	1117	Caldera................	1955-2014, 1950, 1948, 1946, 1938-42, 1935, <u>1914-15</u>, 1860, 1799, 1797, 1794, 1790-91, 1785, 1782-83, <u>1779-81</u>, 1756, 1749, 1742, 1706, 1678, 1642, 1478, **1471-76**, 1468, 0778, 0766, <u>0764</u>, 0716-18, 0708, C-0650?, M-1050?, M-2050?, <u>C-2900?</u>, <u>C-3050?</u>, C-4800?, C-5400?, C-5950?, **C-6050?**, **C-6200±1000**, C-6350?, C-7750?
282081	Yonemaru-Sumiyoshiike	31.77N	130.59E	40	Maars	T-6200?, T-6250?
282090	Kirishimayama	31.93N	130.86E	1700	Shield..................	2008-11, 1991-92, 1979, 1971, 1959, 1923, 1913-14, 1903, 1894-1900, 1891, 1887-89, 1880, 1832, 1822, 1771-72, 1768-69, 1719, 1716-17, 1706, 1690, 1677-78, 1659-64, 1637-38, 1628, 1620, 1613-18, 1598-1600, 1595, 1587-88, 1585, 1576-78, 1574, 1566, 1554, 1524, <u>1235</u>, 1184, 1167, 1112-13, T 1000?, 0945, 0857-58, 0843-48, 0837-39?, <u>0788</u>, 0742, T 0700?, C-2050?, <u>C-2650?</u>, T-3050?, T-3550?, <u>C-4350?</u>, T-5700±1350, T-7050±2350
282091	Fukue.....	32.66N	128.85E	315	Shields	G-0400?
282100	**Unzendake**..............................	32.76N	130.30E	1483	Complex..............	1996, 1990-95, 1792, 1663, N-1150±500, N-1450±500, N-2150±800, G-2640±500, G-2720±300, F-4050±3000
282110	**Asosan**...	32.88N	131.10E	1592	Caldera................	2014, 2011, 2008, 2003-05, 1988-95, 1983-85, 1953-81, 1943-51, 1932-41, 1925-30, 1923, 1918-20, 1916, 1914, 1906-12, 1897-99, 1894, 1884, 1872-74, 1856, 1854, 1837-38, 1835, 1826-32, 1814-16, 1806, 1804, 1772-88, 1765, 1709, 1691, 1683, 1675, 1668-69, 1649, 1637, 1631, 1620, 1611-13, 1598-99, 1592, 1587, 1582-84, 1576, 1573-74, 1564, 1562, 1558-59, 1542, 1533, 1522, 1505-06, 1485, 1473-74, 1438, 1434, 1387, 1375-77, 1343, 1340, 1335, 1331-33, 1324, 1305, 1286, 1281, 1271-74, 1269, 1265, 1239-40, 0867, 0864, G 0440±75, C-0630±50, C-1270±75, G-1350?, G-1650?, C-1830±75, G-2050?, G-2150?, C-2350±75, T-2550?, T-2850?, C-3610±50, R-8050?
282120	**Kujusan**	33.09N	131.25E	1791	Stratovolcanoes....	1995-96, 1675, 1662, G 0370±40, <u>G-0100±300</u>, <u>G-0990±940</u>, T-1720±300, <u>G-2440±300</u>, T-3110±500, <u>G-3780±500</u>, T-4490±500, T-7180±2640, <u>G-9160±1190</u>
282130	Yufu-Tsurumi...........................	33.28N	131.39E	1584	Lava domes	0867, 0771, <u>C-0200±50</u>

Honshu

283001	Abu.........	34.50N	131.60E	641	Shields	N-6850?
283002	Sanbesan	35.14N	132.62E	1126	Stratovolcano	G 0650±50, <u>T-1920?</u>, G-3550=50
283010	Izu-Tobu	34.90N	139.10E	1406	Pyroclastic cones..	1989, T-0750?, <u>G-1150±50</u>, T-2050?, <u>G-2100±100</u>, T-8050?
283020	Hakoneyama.............................	35.23N	139.02E	1438	Complex..............	G 1170±100, T-0050?, T-1050?, G-1200?, G-1400±100, G-3700±100, G-6000±100
283030	**Fujisan**..	35.36N	138.73E	3776	Stratovolcano	**1707-08**, 1700, 1511, 1435, 1083, 1033, 0999, 0937, 0932, 0870, 0864-66, 0830, 0826, <u>0800-02</u>, 0781, C 0720±100, T 0530?, C 0520±100, C 0470±100, T 0400?, C 0370±200, C 0350±300, T 0300?, T 0250?, C 0240±150, T 0220?, T 0200?, T 0100?, T 0050?, C-0100±150, T-0190±100, G-0520±300, G-0780±500, **C >-0930**, G-1010±100, <u>T-1030?</u>, G-1300±150, **T-1350?**, T-1450±100, G-1510±100, C-1850±150, C-2050?, C-2450±500, T-2550?, T-2800±300, C-3050?, C-3690±100, T-4120±300, T-4730±500, C-5070±200, C-5540±200, T-6050?, C-6240±300, C <-6580, C-7310±500, C-7530±300, C-7820±200, C >-8540
283031	Yokodake	36.09N	138.32E	2480	Stratovolcanoes....	G 1200±100, G >-0400
283040	Ontakesan.................................	35.89N	137.48E	3067	Complex..............	2014, 2007, 1991, 1979-80
283050	Hakusan.	36.16N	136.77E	2702	Stratovolcano	1658-59, 1582, 1579, 1554-56, 1547-48, 1239?, 1042, 0706, G 0500±100, T 0200?, <u>G-0200?</u>, G-2550±150, T-3550±500, G-3900±200, C-5000?, G-6550±50, T-7050±500, G-7550±50

VNUM	VOLCANO NAME	LAT	LONG	ELEV	TYPE	ERUPTION YEARS (VEI≤3 VEI=4 **VEI=5** VEI=6-7)
283060	Norikuradake	36.11N	137.55E	3026	Stratovolcanoes	T-0050?, G-7250±150, G-7700±150
283070	**Yakedake**	36.23N	137.59E	2455	Stratovolcanoes	1995, 1962-63, 1939, 1935, 1929-32, 1907-27, 1746, 1585, T 1460?, T 1440?, T 1270?, 0686, T 0630?, G-0350?, T-0400?, T-0850?, T-2550?, T-7450?
283080	Midagahara	36.57N	137.59E	2621	Stratovolcano	1839, 1836, T-0900?, T-3200±2100, T-7300±1000
283090	Niigata-Yakeyama	36.92N	138.04E	2400	Lava dome	1997-98, 1989, 1987, 1983, 1974, 1962-63, 1949, 1852-54, 1773, 1361, 0989, 0887, 0813?, G-0700±100, G-1750?, T-1900±1050
283100	Myokosan	36.89N	138.11E	2454	Stratovolcano	T-0750, T-1200?, T-2100±500, G-2750±100, T-2900?, T-3450?, T-3700?, T-4000?, T-4300?, **G-4750±300**
283110	**Asamayama**	36.41N	138.52E	2568	Complex	2008-09, 2003-04, 1990, 1982-83, 1973, 1965, 1961, 1958-59, 1949-55, 1944-47, 1927-42, 1924, 1919-22, 1906-17, 1904, 1899-1902, 1894, 1891, 1889, 1879, 1875, 1869, 1815, 1803, 1783, 1776-77, @1769, 1762, 1754-55, 1752, 1731-33, 1728-29, 1720-23, @1719, 1717-18, 1713, 1708-11, 1706, 1703-04, 1695, 1669, 1655-61, 1651-52, 1647-49, 1644-45, 1609, 1605-06, 1600, 1595-98, 1590-91, 1582, 1527-28, **1108**, @0685, A 0350±10, T-1130±1430, A-2550±500, C-3450?, G-4200±100, T-4700±500, T-6400±1050, G-7200±200
283120	Kusatsu-Shiranesan	36.62N	138.53E	2165	Stratovolcanoes	1982-83, 1976, 1958, 1937-42, 1934, 1932, 1927, 1925, 1905, 1902, 1900, 1897, 1882, 1805, C 1470±150, T 0050?, T-0550?, C-1120±150, G-3750?, C-6270±200, G-6550?
283122	Harunasan	36.48N	138.85E	1449	Stratovolcano	**A 0550±10**, A 0520±10, T 0450±50
283131	Hiuchigatake	36.96N	139.29E	2356	Stratovolcano	1544, T-6050?
283140	Nikko-Shiranesan	36.80N	139.38E	2578	Shield	1952, 1889-90, 1872-73, 1649, 1625, T 0800?, T-0400?, T-2000?, G-4150±200
283141	Nantai	36.76N	139.49E	2486	Stratovolcano	G-9540±500
283142	Omanago Group	36.79N	139.51E	2367	Lava domes	C-3050?
283143	Takaharayama	36.90N	139.78E	1795	Stratovolcano	G-4570?
283150	**Nasudake**	37.13N	139.96E	1915	Stratovolcanoes	1963, 1960, 1953, 1881, 1846, 1410, 1408, 1404, 1397, G 0330±200, G 0250±200, T-0250?, G-0700±200, T-1440±500, T-2000±1450, T-4350±950, G-5550±500, T-6050?, G-7850?, T-8550±1500
283151	Numazawa	37.44N	139.57E	835	Shield	**G-3400?**
283160	**Bandaisan**	37.60N	140.07E	1816	Stratovolcano	1888, 1808?, <1787, 0806, T-0550?, T-1800±1250, T-3850?, T-4650?, T-5050?, T-6350?, T-7450?
283170	Adatarayama	37.65N	140.28E	1728	Stratovolcanoes	1996, 1899-1900, T 0950±50, T-0050±900, G-0590±200, T-1550±1100, G-2600±50, T-4300±850, G-6150±100, G-6650±100, T-8050?
283180	Azumayama	37.74N	140.24E	1949	Stratovolcanoes	1977, 1950, 1893-95, 1800?, 1711?, 1331, G 0600±200, G-0150±200, G-0950±100, G-1800±50, G-2750±200, G-3000±50, G-4150±500, T-4550±1000, G-5400±200, G-5700±50
283190	Zaozan	38.14N	140.44E	1841	Complex	1940, 1905, 1894-96, 1873, 1867, 1833, 1830-31, 1821-22, 1809, 1806, 1804, 1796, 1794, 1694, 1668-70, 1641, 1630, 1622-24, 1620, G 1400±100, 1230, 1227, 1183, 0884?, G 0300±200, G-1600?, T-2000±500, G-2300±75, G-2600±200, G-3350±50, G-3850±200, G-4150±200, G-5500?, G-5600?, G-7600?
283200	Naruko	38.73N	140.73E	470	Caldera	0837, G <-0800, G-1350±50, T <-1400, G >-4400
283210	Kurikomayama	38.96N	140.79E	1627	Stratovolcano	1950, 1946, 1944, 1744, 1726±10, T 1450±50, T-3540±2480
283220	Chokaisan	39.10N	140.05E	2236	Stratovolcanoes	1974, 1834, 1821, 1800-04, 1740-47?, 1659-63±1, 0939, 0871, 0830, 0817±7, 0711±3, T-0450?, C-0650?, T-1050?
283230	Akita-Komagatake	39.76N	140.80E	1637	Stratovolcanoes	1970-71, 1932, 1902, 1890-91, T 1100?, 0807?, T 0400±500, G-0050±200, G-0200±200, G-0350±200, G-1450±50, G-5950±200, T-6150±300, G-6350±200, T-7100±900, G-7850±200, T-8300±300, T-8800±300
283240	Iwatesan	39.85N	141.00E	2038	Complex	1919, 1732, 1686, C 1450±100, T 1300±50, C 0150?, C-0350?, G-0450?, C-1150?, G-1250?, T-1500±300, G-1650?, G-2000?, G-2050±200, T-2700±200, G-2950±50, C-3050?, T-3250±500, T-3750±100, T-4350±500, T-4450±500, G-4850±50, G-4900±100, G-5650±50, G-6300±100, T-6450±1600
283250	Hachimantai	39.96N	140.85E	1613	Stratovolcano	T-5350?, G-7900±150
283260	Akita-Yakeyama	39.96N	140.76E	1366	Stratovolcano	1997, 1957, 1948-51, 1929, 1890, 1887, 1867, 1678, G 1390±75, T 0570?, G-1250±200, N-3050?
283262	Megata	39.95N	139.73E	291	Maars	T-2050?, T-7050?
283270	Iwakisan	40.66N	140.30E	1625	Stratovolcano	1863, 1844-45, 1790, 1782-83, 1618, 1604, 1600, T-0550±500, T-1050?, T-4050?, K-8050?
283271	**Towada**	40.51N	140.88E	1011	Caldera	0915, G-0750?, **G-4150?**, G-5550?, G-6250?, **G-7250?**, **G-8250?**, G-9490?
283280	Hakkodasan	40.66N	140.88E	1585	Stratovolcanoes	G 1550±100, G 1340±75, G 0450?, G-0050?, G-1150?, G-2250?, G-2850?
283290	Osorezan	41.28N	141.12E	878	Stratovolcano	<1787

Izu, Volcano, and Mariana Islands

VNUM	VOLCANO NAME	LAT	LONG	ELEV	TYPE	ERUPTION YEARS
284010	**Izu-Oshima**	34.72N	139.39E	758	Stratovolcano	1990, 1986-88, 1974, 1962-71, 1956-60, 1953-54, 1950-51, 1937-40, 1933-35, 1928, 1922-23, 1919, 1912-15, 1910, 1876-77, 1870, 1846, 1822-24, V 1821?, 1803, 1792, 1789, 1783-86, 1777-79, 1695, 1684-90, 1637-38, 1634, 1623, 1612, 1600, 1552, V 1471?, 1442-43, 1421, 1415-17?, 1338, V 1307?, V 1245?, V 1183?, 0886?, 0854?-56?, 0838?, V 0822?, V 0713?, V 0700?, 0684?, 0680?-81, 0654?, V 0625?, V 0600?, V 0580?, G 0340±200, T 0250±100, G 0150±50, T-0150?, T-0600±500, G-0900±100, T-1050±200, T-1200?, T-1450?, T-2550?, T-3650?, T-3750?, T-4000±300, T-4250?, T-4450?, T-4920±500, T-5450±75, T-5550?, T-6050±500, T-6550?, T-6650?, T-7150±500, T-7650?, T-8050±500, T-8450?

VNUM	VOLCANO NAME	LAT	LONG	ELEV	TYPE	ERUPTION YEARS (VEI≤3 VEI=4 **VEI=5** VEI=6-7)
284011	Toshima.	34.52N	139.28E	508	Stratovolcano	T-4550±2550
284020	Niijima...	34.40N	139.27E	432	Lava domes	0886, 0856?-57?, T-1250?, T-4350?, T-5950?
284030	Kozushima	34.22N	139.15E	572	Lava domes	0838, H-0100±950, H-0750±700, T-8050?
284040	Miyakejima	34.09N	139.53E	775	Stratovolcano	2008-10, 2004-06, 2000-02, 1983, 1962, 1940, 1874, 1835, 1811, 1763-69, 1712-14, 1643, 1595, 1535, 1469, 1154, 1085, 0850?, 0832, G 0750±50, G 0500=50, T 0320±500, T 0260±500, G-0050±50, T-0600±500, T-0950±200, T-1250?, T-1450±500, T-1800±50, G-2000±100, T-2900±500, T-6450±500
284041	Mikurajima	33.87N	139.60E	851	Stratovolcano	G-4100±100, T-5050?
284050	**Hachijojima**	33.14N	139.77E	854	Stratovolcanoes	1605-06, 1518-23, 1487, G 0850±200, G-0150±50, T-0350±500, T-1150±700, T-1250±800, **T-2050?**, T-2450±500, T-2550±500, G-2700±100, T-3350±1300, **T-4650?**, T-5020±370, T-7650±3000, T-8020±2640
284060	**Aogashima**	32.46N	139.76E	423	Stratovolcano	1780-85, 1670-80, 1652, C-0600±200, T-1100±300, G-1200±50, G-1800±100
284070	Myojinsho	31.89N	139.92E	11	Submarine	1970, 1957-60, 1952-55, 1946, 1934, 1915, 1906, 1896, @1871, @1869
284080	Sumisujima	31.44N	140.05E	136	Submarine	1916, 1870
284090	**Izu-Torishima**	30.48N	140.30E	394	Stratovolcano	2002, 1975, S 1965, 1939, 1902, 1871
284096	Nishinoshima	27.25N	140.87E	25	Caldera	2013-14, 1973-74
284100	Kaitoku Seamount	29.23N	140.75E	-95	Submarine	1984, 1543
284110	Kita-Ioto [Kita-Iwojima]	25.42N	141.28E	792	Stratovolcano	1930-45, 1880-89, 1780
284120	Ioto [Iwojima]	24.75N	141.29E	169	Caldera	2012, 2001, 1999, 1994, 1982, 1980, 1978, 1976, 1969, 1967, 1957, 1944, 1935, 1930?, 1922, 1889, G-0850±50
284121	Kita-Fukutokutai	24.42N	141.42E	-73	Submarine	1953-54
284130	Fukutoku-Oka-no-Ba	24.29N	141.48E	-29	Submarine	2010, 2005, 1992-93, 1986-87, 1973-75?, 1914, 1904-05
284131	Minami-Hiyoshi	23.50N	141.94E	-107	Submarine	1975
284133	Fukujin ..	21.93N	143.47E	-217	Submarine	1973-74, 1968, 1951
284134	Kasuga...	21.77N	143.71E	-598	Submarine	1959
284140	Farallon de Pajaros	20.54N	144.90E	360	Stratovolcano	S 1967, 1951-53, 1947, 1943, 1941, 1939, 1936, 1934, 1932, 1928, 1925, 1912, 1900?-01, 1874?-76, 1864
284141	Ahyi	20.42N	145.03E	-75	Submarine	S 2014, S 2001
284142	Supply Reef	20.13N	145.10E	-8	Submarine	@1989, S 1969
284150	Asuncion	19.67N	145.41E	857	Stratovolcano	1906
284160	Agrigan..	18.77N	145.67E	965	Stratovolcano	1917
284170	Pagan	18.13N	145.80E	570	Stratovolcanoes	2010-12, 2006, 1996, 1992-93, 1987-88, 1981-85, 1925, 1923, 1917, 1909, 1873?, 1864, 1825±5, C 1800±50, 1669, G 1340±100
284180	Alamagan	17.60N	145.83E	744	Stratovolcano	C 0870±100, C 0540±75
284190	Guguan ..	17.31N	145.85E	287	Stratovolcano	1883±1
284193	South Sarigan Seamount	16.58N	145.78E	-184	Submarine	2010
284200	Anatahan	16.35N	145.67E	790	Stratovolcano	2003-08
284202	Ruby	15.62N	145.57E	-230	Submarine	1995, S 1966
284211	NW Rota-1	14.60N	144.78E	-517	Submarine	2003?-10

Hokkaido

VNUM	VOLCANO NAME	LAT	LONG	ELEV	TYPE	ERUPTION YEARS
285010	**Oshima-Oshima**	41.51N	139.37E	732	Stratovolcano	1790, 1759, 1741-42, C 0250±150, C-0800±100
285011	Esan	41.81N	141.17E	618	Lava domes	1874, 1846, T 1350?, G-0440±46, T-1050?, G-3900±100, G-5770±924, G-6670±27
285020	**Hokkaido-Komagatake**	42.06N	140.68E	1131	Stratovolcano	2000, 1998, 1996, 1942, 1937, 1929, 1928, 1922-24, 1919, 1905, 1888, 1856, 1694, **1640**, G-4350?, T-4500±150, **G-4600±50**
285030	**Toya [Usu]**	42.54N	140.84E	733	Stratovolcano	2000-01, 1977-82, 1944-45, 1910, 1853, 1822, 1769, T 1690±10, **1663**, 1638, 1626, G-4600±200, G-6550±1500
285031	Niseko ...	42.88N	140.63E	1308	Stratovolcanoes	T-4900?
285032	Yoteizan	42.83N	140.81E	1898	Stratovolcano	T-1050?, T-3550?
285034	Kuttara...	42.49N	141.16E	549	Stratovolcanoes	T 1820±100, G 0200±75, T-8050?
285040	**Shikotsu [Tarumaesan]**	42.69N	141.38E	1320	Caldera	1981, 1978-79, 1953-55, 1951, 1944, 1936, 1933, 1931, 1928-29, 1926, 1923, 1917-21, 1909, 1894, 1885-87, 1883, 1874, 1867, 1804-17, **1739**, T 1707±30, **1667**, G 1550±75, G 1500±150, T-0050?, G-0100±100, **C-0550?**, **C-6950?**
285041	Rishirizan	45.18N	141.24E	1721	Stratovolcano	C-5830±300
285050	**Tokachidake**	43.42N	142.69E	2077	Stratovolcanoes	2004, 1988-89, 1985, 1961-62, 1958-59, 1956, 1954, 1952, 1931, 1925-28, 1889, 1887, 1857, G 1570±100, G 1250?, T 1050±200, G 0950±50, T 0600±300, G 0350±100, G 0001±50, G-1350?, G-1750±50, G-2650±200

VNUM	VOLCANO NAME	LAT	LONG	ELEV	TYPE	ERUPTION YEARS (VEI≤3 <u>VEI=4</u> **VEI=5** <u>**VEI=6-7**</u>)
285060	Taisetsuzan [Daisetsuzan]..........	43.66N	142.85E	2291	Stratovolcanoes....	T >1739, T-0550±500, G-1450±50, G-2800±100, G-3200±75
285061	Maruyama...................................	43.45N	143.04E	2013	Stratovolcanoes....	1898, G-1700?
285070	Akan......	43.38N	144.01E	1499	Caldera.................	2008, 2006, 1998, 1996, 1988, 1984, 1964-66, 1962, 1954-60, 1808?, 1780±75, T 1600±50, T 1550?, T 1250±300, T 0950?, T 0600±300, G 0250?, T 0100±100, <u>G-0050?</u>, T-0300±100, T-0550?, C-0800±250, C-2050?, C-3050?, C-3550±500, C-4550±500, <u>C-7050?</u>
285080	Kussharo	43.61N	144.44E	574	Caldera.................	T 1320±300, T 0700?, T 0450?, T-1550±2000, T-3550?, T-5800±2250
285081	**Mashu**...	43.57N	144.56E	857	Caldera.................	**G 1080±100**, G 0350±100, <u>G 0150±100</u>, <u>G-2050±40</u>, T-2800±750, <u>G-3550±40</u>, <u>**G-5550±100**</u>
285082	Rausudake	44.08N	145.12E	1660	Stratovolcano	T 1800±50, G 1350±100, <u>G 0550±100</u>, G 0080±50, G-0270±100
285090	Shiretoko-Iozan	44.13N	145.16E	1562	Stratovolcano	1935-36. 1889-90, 1876, 1857-58, T 0850±500

Kuril Islands

VNUM	VOLCANO NAME	LAT	LONG	ELEV	TYPE	ERUPTION YEARS
290010	Tomariyama [Golovnin]	43.84N	145.50E	535	Caldera.................	1848, T 1290±20, T-4550?
290020	Raususan [Mendeleev]...............	43.98N	145.73E	882	Stratovolcano	1880, C-2270±50
290030	Chachadake [Tiatia]...................	44.35N	146.25E	1822	Stratovolcano	1981, 1978, <u>1973</u>, 1812
290041	**Moekeshiwan [Lvinaya Past]** ..	44.61N	146.99E	528	Stratovolcano	<u>**C-7480±50**</u>
290050	Etorofu-Atosanupuri	44.81N	147.13E	1206	Stratovolcano	1932, 1812
290070	Etorofu-Yakeyama [Grozny]	45.01N	147.87E	1158	Complexes	2012-13, 1989, 1973, 1970, 1968
290080	Sashiusudake [Baransky]...........	45.10N	148.02E	1125	Stratovolcano	1951, C 1570±30, C 1460±30
290090	Chirippusan [Chirip]..................	45.34N	147.92E	1587	Stratovolcanoes....	1860?, 1843
290100	Moyorodake [Medvezhia]..........	45.39N	148.84E	1124	Somma.................	1999, 1958, 1883, 1778, T-0050?
290120	Kolokol Group.........................	46.04N	150.05E	1328	Sommas	1973, 1970, 1952, 1946, 1940±6, 1924, 1894, 1845-46, 1780±10
290150	Chirpoi...	46.53N	150.88E	742	Caldera.................	2012-14, 1982, 1960, 1879, 1857, 1854, 1811, 1790±20, <u>1712</u>
290160	Unnamed	46.47N	151.28E	-502	Submarine............	S 1972
290170	Goriaschaia Sopka	46.83N	151.75E	891	Stratovolcano	1914, 1883, 1881, 1849, 1842
290180	Zavaritzki Caldera	46.93N	151.95E	624	Caldera.................	1957, 1923±8
290190	Prevo Peak...............................	47.02N	152.12E	1360	Stratovolcano	1825±25, 1765±5
290200	Ketoi.......................................	47.35N	152.48E	1172	Stratovolcano	1960, 1924, 1843-46
290210	Ushishur	47.52N	152.80E	401	Caldera.................	1884, >1769, 1710±10, <u>C-7450±50</u>
290220	Rasshua..	47.77N	153.02E	956	Stratovolcano	1957, 1846
290230	Unnamed	48.08N	153.33E	-150	Submarine............	1924
290240	Sarychev Peak	48.09N	153.20E	1496	Stratovolcano	<u>2009</u>, 1989, 1986, 1976, 1965, 1960, 1954, <u>1946</u>, 1930, 1928, 1927, 1923, 1879, 1805, 1765±5
290250	Raikoke..	48.29N	153.25E	551	Stratovolcano	<u>1924</u>, <u>1778</u>, 1765±5
290260	Chirinkotan..............................	48.98N	153.48E	724	Stratovolcano	2013-14, 2004, 1986, 1979-80, 1955?, 1900±10, 1884±6
290270	Ekarma	48.96N	153.93E	1170	Stratovolcano	2010, 1980, 1767-69
290290	Sinarka...	48.88N	154.18E	934	Stratovolcano	<u>1872-78</u>, 1855, 1846, 1725±25
290300	Kharimkotan............................	49.12N	154.51E	1145	Stratovolcano	**1933**, 1931, 1883, 1848, 1846, 1713
290310	**Tao-Rusyr Caldera**	49.35N	154.70E	1325	Stratovolcanoes	1952, **C-5550±75**
290320	Nemo Peak	49.57N	154.81E	1018	Caldera.................	1938, 1906, 1710±10, T 1350?, T 0750?, T-0550±100, T-1850?, T-3050?, T-5550?, T-7050?, T-7550?
290340	Fuss Peak.................................	50.27N	155.25E	1742	Stratovolcano	1854, T 1250?, G-1590±70, T-1850?, T-3150?, T-3850?, T-4850?, T-5250?, G-5340±30
290350	Karpinsky Group	50.15N	155.37E	1326	Cones	1952
290360	Chikurachki	50.32N	155.46E	1781	Stratovolcanoes....	2007-08, 2005, 2002-03, <u>1986</u>, 1973, 1967, 1964, 1961, 1957-58, 1853-59, <u>T 1690±10</u>, <u>T-1500±250</u>, T-1750?, <u>T-1950?</u>, <u>G-6310±70</u>, T-7550?, T-8500±950, G-9360±80
290380	Ebeko.....	50.69N	156.01E	1103	Somma.................	2009-10, 2005, 1987-91, 1969, 1967, 1963-65, 1934-35, 1859, 1833, 1793, C 1670?, C 1650?, C 1600?, C-0390±75
290390	Alaid......	50.86N	155.57E	2285	Stratovolcano	2012, 1996, 1986, 1982, <u>1981</u>, 1972, 1933-34, 1894, 1860, 1854, <u>1790-93</u>

Kamchatka Peninsula

VNUM	VOLCANO NAME	LAT	LONG	ELEV	TYPE	ERUPTION YEARS
300010	Kambalny	51.31N	156.88E	2116	Stratovolcano	C 1350?
300020	Koshelev	51.36N	156.75E	1822	Stratovolcano	1690±10, C-1350?, T-4050?, C-4550?
300021	Yavinsky	51.57N	156.60E	705	Stratovolcano	T-4050?
300022	**Diky Greben**	51.45N	156.98E	1040	Lava domes	T 0350±300, C-2250?, T-3050?, T-5700±100

VNUM	VOLCANO NAME	LAT	LONG	ELEV	TYPE	ERUPTION YEARS (VEI≤3 VEI=4 **VEI=5** VEI=6-7)
300023	**Kurile Lake**	51.45N	157.12E	81	Caldera	**G-6440±25**, T-7550±500
300030	Iliinsky	51.50N	157.20E	1555	Stratovolcano	1901, C 0050?, C-2050?, **C-2850?**, C-4550?, T-5700±50
300040	**Zheltovsky**	51.58N	157.33E	1926	Stratovolcano	1923, **C-3050?**, T-6050?, **T-7050±1000**
300050	**Ksudach**	51.84N	157.57E	1005	Stratovolcano	**1907**, G 1750?, G 1000±50, T 0700?, G 0350?, **G 0240±100**, T-0200?, T-3000?, T-4100?, T-4550?, T-4750?, **G-4900?**, **G-5200?**, G-5600?, **G-7900?**
300053	Khodutka	52.06N	157.71E	2039	Stratovolcanoes	T-0300±300, **G-0930±100**, T <-1050
300060	Mutnovsky	52.45N	158.20E	2288	Complex	2000, 1960-61, 1945, 1938-39, 1927-29, 1916-17, 1904, 1898, 1852-54, 1848, 1750±50, 1650±50, T 1300?, T 0950?, T 0750?, T 0250?, T 0050?, T-0100?, T-0200?, T-0450?, T-2050?, T-2150?, T-2900?, T-3650?, T-4050?, T-4550?, T-4650?, T-4700?, T-5000?, T-5050?, T-5250?, T-5350?, T-5450?, T-5800?, T-5900?, T-6000?, T-7550?
300070	Gorely	52.56N	158.03E	1799	Caldera	2010, 1984-86, 1980-81, 1961, 1947, 1929-31, 1869, 1832, 1828, T 1750±50, T 1330±25, T 1030±25, T 0550?, T 0250?, T 0200?, T 0050?, T-0350?, T-0700±850, T-2000?, T-2050?, T-2200?, T-2250?, T-2450?, T-2750?, T-3450?, T-3550?, C-3580±100, T-3900?, T-3950?, T-4150?, T-4300±150, T-4350?, T-4450?, T-4500?, T-4600?, T-4650?, T-4700?, T-4750?, T-4950?, T-5150?, T-5300?, T-5450?, T-5500?, T-5650?, T-5950?, T-6050?, T-7250?, T-7400±150
300080	Opala	52.54N	157.34E	2439	Caldera	1776, **G 0610±50**, C-1550?, T-3500?
300082	Tolmachev Dol	52.63N	157.58E	1021	Pyroclastic cones	T 0300±150, C-2650?
300083	Vilyuchik	52.70N	158.28E	2173	Stratovolcano	T-8050?
300084	Barkhatnaya Sopka	52.82N	158.27E	870	Lava domes	T >-3550
300090	Koryaksky	53.32N	158.71E	3430	Stratovolcano	2008-09, 1956-57, 1926, 1890±3, C-1550?, C-1950±300, T-5050?
300100	**Avachinsky**	53.26N	158.84E	2717	Stratovolcano	2001, 1991, 1945, 1938, 1926-27, 1909, 1901, 1894-95, 1881, 1878, 1851-55, 1828, 1827, 1779, 1772, 1737, C 1550?, C 1400?, T 1200?, T 1100?, T 0900?, T 0700?, T 0400?, T 0100?, **C-1350?**, **G-1500?**, T-1700?, T-2100?, T-2300?, T-2500?, G-2530±300, T-2650?, T-2900?, T-2950?, **G-3200±150**, T-3400?, T-3500?, T-3700?, G-3790±100, T-4050?, T-4200?, T-4250?, **G-4340±75**, T-4400?, G-4460±100, G-4550±200, T-5450?, T-5500?, T-5600?, T-5700?, **G-5980±100**, T-6100?
300120	Zhupanovsky	53.59N	159.15E	2899	Compound	2013-14, 1959, 1956-57, 1940, 1929, 1925, 1882, 1776, T 1000±500, T-0050?, C-0220±50, T-3050?, T-5050?
300121	Veer	53.75N	158.45E	520	Pyroclastic cones	T 0390±75
300122	Kostakan	53.83N	158.05E	1150	Pyroclastic cones	T 1350?, T 1200±50, T 1000±50, T 0800±50, T-6550±500, T-8050±1000
300123	Bakening	53.91N	158.07E	2278	Stratovolcano	T-0550, T-1550?, T-6300±300, T-6550±500, T-7550±500
300124	Zavaritsky	53.91N	158.39E	1567	Pyroclastic cones	T-0800±500, C-0850?
300125	Akademia Nauk	53.98N	159.45E	1180	Stratovolcanoes	1996, G-0950?, G-3850?, G-5500±500
300130	**Karymsky**	54.05N	159.44E	1513	Stratovolcano	1996-2014, 1970-82, 1960-67, 1955-57, 1952-53, 1945-47, 1943, 1940, 1938, 1932-35, 1929, 1925, 1923, 1921, 1915, 1911-12, 1908, 1854, 1852, 1830, 1771, G 1730±25, G 1550?, G 1450?, G 1150?, G 1050?, G 0950?, G-0850?, G-1100?, G-1400?, G-2050?, G-2250?, G-2350?, G-3200?, G-3450?, G-4150?, **G-6600?**
300140	Maly Semyachik	54.14N	159.67E	1527	Caldera	1952, 1945-46, 1851-52, 1804, G 1550?, G 1400±50, G-0550?, G-0650?, G-0850?, G-1800±50, G-2250?, G-2450?, G-3500±50, G-4500±50, G-4650?, G-5050?, G-5450?, G-5750?, G-5850?, G-6150?, C-6950?, C-7550?
300150	Bolshoi Semiachik	54.32N	160.02E	1720	Stratovolcanoes	G-4450?, C-6800±300
300160	Taunshits	54.53N	159.80E	2301	Stratovolcano	C-0550?, C-5800±50
300170	Uzon	54.50N	159.97E	1617	Calderas	T 0200±300, T-1550?, C-5700±50, T-5750?
300180	Kikhpinych	54.49N	160.25E	1506	Stratovolcanoes	T 1550?, C 1350?, C 0900±50, T 0830±25, C 0650?, G 0550?, G-2780±25, G-2850?
300190	Krasheninnikov	54.60N	160.27E	1816	Caldera	T 1550?, T 1350?, G 0850?, G 0750?, T 0650?, G-0150?, G-0250?, G-0350?, G-0650?, G-0850?, G-1000±50, G-1050?, G-1150?, G-1350?, G-1650?, G-2250?, G-2950?, G-3250?, G-3550?, G-4450?, C-4850?, C-5050?, G-5250?, C-5450?, C-5800±50, C-6000±50, C-6250?, C-6350?, C-6550?, C-7250?, C-8050?
300200	Kronotsky	54.75N	160.53E	3482	Stratovolcano	1922-23, T-0050?
300210	Gamchen	54.97N	160.70E	2539	Complex	T-0550?, T-1650?
300220	Komarov	5.03N	160.73E	2065	Stratovolcano	C >0950, C 0450?
300221	Vysoky	55.06N	160.77E	2129	Stratovolcano	C-0550?
300230	**Kizimen**	55.13N	160.32E	2334	Stratovolcano	2010-13, 1927-28, T 0850?, T 0700±50, C 0350±75, T-1010?, T-4050?, T-4450?, T-5800±50, **C-6400±50**, **T-8050?**
300240	**Tolbachik**	55.83N	160.33E	3611	Shield	2012-13, 1975-76, 1954-70, 1947, 1939-41, 1936-37, 1931-32, 1904, 1793, 1788-90, 1769, 1739-40, T 1550?, T 1050?, T 1000?, T 0950?, T 0900?, T 0550?, T 0450?, T 0400?, T 0350?, T 0250?, T 0150?, T 0050?, T-0100?, T-0200?, T-0700?, T-0750?, T-0800?, T-1650?, T-1750?, T-2050?, T-4550?, T-5450?, T-5600?, T-5650?, C-6050?, G-7600?
300250	**Bezymianny**	55.97N	160.60E	2882	Stratovolcano	1976-2014, 1958-74, **1955-57**, T 0950?, T 0850?, T 0700±50, T 0600?, T 0250?, T 0150?, T 0050?, G-0450?, T-1350?, T-1550±500, T-2750±500, T-5050?, T-7050±2000

VNUM	VOLCANO NAME	LAT	LONG	ELEV	TYPE	ERUPTION YEARS (VEI≤3 VEI=4 **VEI=5** VEI=6-7)
300260	**Klyuchevskoy**	56.06N	160.64E	4754	Stratovolcano	2007-13, 2002-05, 1991-2000, 1986-90, 1977-86, 1956-74, 1953-54, 1951, 1948-49, 1944-46, 1935-39, 1932, 1931, 1929, 1925-26, 1922-23, 1915, 1913, 1909-11, 1907, 1904, 1896-98, 1890, 1882-83, 1877-79, 1865, 1852-54, 1848, 1840, 1829, 1819-22?, 1812-13, 1807, 1787-91, 1785, 1772, 1770, 1767, 1762, 1740, 1737, 1727-31, 1720-21, 1697-98, C 0550?, T-1050?, C-3950?
300261	Ushkovsky	56.11N	160.51E	3943	Compound	1890, C-6670±150, C-7550?
300270	**Sheveluch**	56.65N	161.36E	3283	Stratovolcano	1999-2014, 1997-98, 1984-95, 1980-81, 1964, 1944-50, 1928-30, 1905, 1897-98, 1879-83, **1854**, 1739, T 1700?, **G 1652±11**, T 1550?, G 1430?, T 1150?, **G 1034±23**, T 1020±40?, G 0970±80?, T 0750?, T 0700?, **G 0650±40**, T 0630?, **G 0600?**, T 0580?, T 0530?, G 0500?, G 0380?, T 0250?, T 0230?, G 0170±20, G 0120?, T 0100?, T-0010?, T-0150?, G-0300?, T-0400?, G-0500?, T-0650?, G-0780±300, T-0900?, **G-0950?**, T-1010?, G-1330±200, T-1500?, T-1650?, T-1700?, **G-2000?**, **G-2100?**, T-2150?, T-2200?, G-2490?, T-2530?, **G-2620±300**, T-2750?, T-2900?, T-3050?, T-3200?, **G-3500?**, **G-3650?**, T-4250?, G-4350?, **G-4400?**, G-4530?, T-4900?, T-5400?, **T-5500?**, T-6000?, T-6100?, T-6200?, T-6350?, T-6380?, G-6400±150, T-6500?, T-6600?, **T-6800?**, T-7000?, **T-7150?**, **T-7300?**, **G-7400±150**, T-7500?, T-7550?, T-7600?, T-7630?, T-7700?, T-7750?, T-7850?, T-7950?, T-8100?, T-8200?, T-8350?, T-8450?, T-8500?
300271	Piip	55.42N	167.33E	-300	Submarine	T-5050?
300272	**Khangar**	54.76N	157.41E	1967	Stratovolcano	G 1500±40, G 1000±16, G-0350±30, G-2700±25, G-5500±25, **G-5700±16**, G-6400±75, G-7100±100, G-8250±100, G-9500±300
300273	Cherpuk Group	55.55N	157.47E	1868	Pyroclastic cones	C-4550?
300280	Ichinsky	55.68N	157.72E	3596	Stratovolcano	1740, G 1300±200, G 0800±300, T 0550?, G 0050±300, T-0600?, T-1200?, G-1950±300, G-2850±300, **T-5400?**, T-5650?, T-5850?, G-6150±50, T-6950?
300360	Bolshoi-Kekuknaysky	56.47N	157.80E	1401	Shields	C-5310±100
300450	Alney-Chashakondzha	56.70N	159.65E	2598	Stratovolcano	C 1600?, C-0650±75, C-0660±75
300511	Shisheika	57.15N	161.08E	379	Lava dome	C-2240?
300512	Terpuk	57.20N	159.83E	765	Shield	C-0800±300
300520	Sedankinsky	57.27N	160.08E	1241	Shield	C-7050±1000
300550	Gorny Institute	57.33N	160.20E	2125	Stratovolcano	C 1250?, C 1000?, C-4250?
300551	Kinenin	57.35N	160.97E	583	Maar	C 0850±50
300552	Bliznetsy	57.35N	161.37E	265	Lava cone	C-1060±40
300560	Titila	57.40N	160.10E	1559	Shields	C-0550?
300590	Elovsky	57.55N	160.53E	1381	Shields	T-7550±500
300650	Nylgimelkin	57.97N	160.65E	1764	Shields	C-3550?
300671	Spokoiny	58.13N	160.82E	2171	Stratovolcano	C-3450?
300680	Ostry	58.18N	160.82E	2552	Stratovolcano	C-2050?
300700	Severny	58.28N	160.87E	1936	Shield	C-1550?

Russia (southeastern)

302030	Udokan Plateau	56.28N	117.77E	2180	Pyroclastic cones	C-0220±50, C-2670±100, C-5990±100, C-6210±100, C-7290±100
302060	Jom-Bolok	52.70N	98.98E	2077	Volcanic Field	G-5180±140

Mongolia

303010	Taryatu-Chulutu	48.17N	99.70E	2400	Volcanic Field	C-2980±150

China (western)

304020	Tianshan Volcanic Group	42.50N	82.50E		Volcanic Field	0650±50, 0050±50
304030	Kunlun Volcanic Group	35.52N	80.20E	5808	Pyroclastic cones	1951

China (eastern)

305011	Arshan	47.50N	120.70E		Pyroclastic cones	C 0000±150
305030	**Wudalianchi**	48.72N	126.12E	597	Volcanic Field	1776, 1720-21
305040	Jingbo	44.08N	128.83E	1000	Volcanic Field	C-0520±100, C-1540±150, C-3550?
305050	Longgang Group	42.33N	126.50E	1000	Pyroclastic cones	C 0350?
305060	**Changbaishan [Baitoushan]**	41.98N	128.08E	2744	Stratovolcano	1903, 1898, 1702, 1668, **G 0942±4**, C-0180±75, C-2160±100

Korea

306030	**Ulreung**	37.50N	130.87E	984	Stratovolcano	C-2990±40, G-6450?, **G-8750?**
306040	Halla	33.37N	126.53E	1950	Shield	1007, 1002, C-2050±200, C-2830±50

Aleutian Islands

311020	Kiska	52.10N	177.60E	1220	Stratovolcano	1990, 1969, 1964, 1962
311050	Little Sitkin	51.95N	178.54E	1174	Stratovolcano	1828-30, 1776

VNUM	VOLCANO NAME	LAT	LONG	ELEV	TYPE	ERUPTION YEARS (VEI≤3 VEI=4 **VEI=5** **VEI=6-7**)
311060	Semisopochnoi	51.93N	179.58E	1221	Stratovolcano	1987, 1873
311070	Gareloi	51.79N	178.79W	1573	Stratovolcano	1989, 1987, 1982, 1980, 1950-52, 1929-30, 1922, 1873, 1790-92
311080	Tanaga	51.89N	178.15W	1806	Stratovolcanoes	1914, 1829, C 1550?, C 1050?, T-0550±2500, R-1050?
311090	Takawangha	51.87N	178.01W	1449	Stratovolcano	C 1550?
311110	Kanaga	51.92N	177.17W	1307	Stratovolcano	2012, 1994-95, 1942, 1906, 1904, 1783?-87?, G 1400±50, G 1150±100, G 0850±200, G 0200±150, G-1550±100, G-1900±300, G-2150±200, G-2300±150, G-4700±150, G-7300±500
311111	Moffett	51.94N	176.75W	1196	Stratovolcano	G-1600?, G-3750?, G-7850?
311120	Great Sitkin	52.08N	176.13W	1740	Stratovolcano	1974, 1949-50, 1945, 1933, 1792
311130	Kasatochi	52.18N	175.51W	314	Stratovolcano	**2008**, 1760
311140	Koniuji	52.22N	175.13W	273	Stratovolcano	R-1150±1900, R-2650±2000, R-3850±3100
311160	Atka	52.33N	174.14W	1451	Stratovolcanoes	1987, 1812
311161	Korovin	52.38N	174.15W	1533	Stratovolcanoes	2004-07, 2002, 1998, 1987, 1973, 1907
311180	Seguam	52.32N	172.51W	1054	Stratovolcanoes	1992-93, 1977, 1902, 1891-92, 1786-90, U 0250±500, U-4050±4000, U-5100±2000, **R-7300±2250**
311190	Amukta	52.50N	171.25W	1066	Stratovolcano	1996-97, 1987, 1963, 1878, 1786-91
311210	Yunaska	52.64N	170.63W	550	Shield	1937, 1830, 1824
311230	Carlisle	52.89N	170.05W	1620	Stratovolcano	1828, 1774
311240	Cleveland	52.83N	169.94W	1730	Stratovolcano	2005-14, 2001, 1997, 1994, 1986-87, 1984, 1944, 1938, 1932, 1897, 1893
311260	Kagamil	52.97N	169.72W	893	Stratovolcano	1929
311270	Vsevidof	53.13N	168.69W	2149	Stratovolcano	1878, 1830, 1817
311290	**Okmok**	53.43N	168.13W	1073	Shield	2008, 1997, 1986-88, 1983, 1981, 1960-61, 1958, 1945, 1943, 1938, 1931, 1899, 1878, 1824-30, 1817-20, 1805, **C-0100±50**, C-6310±500
311300	Bogoslof	53.93N	168.03W	150	Submarine	1992, 1931, 1926-28, 1909-10, 1906-07, 1883-95±2, 1806-23, 1796-1804
311310	Makushin	53.89N	166.92W	1800	Stratovolcano	1993-95, 1987, 1980, 1951, 1938, 1926, 1907, 1883, 1865, 1826, 1802, 1768-69, T 1150±500, T-0550±900, C-1750?, T-3650±1850, **C-6100±50**, C-6650±200
311320	**Akutan**	54.13N	165.99W	1303	Stratovolcano	1986-92, 1982-83, 1980, 1976-78, 1973-74, 1962, 1953, 1951, 1946-48, 1931, 1927-29, 1911, 1907-08, 1896, 1892, 1887, 1883, 1867, 1865, 1852, 1848, G 1420±100, C 0550?, **C 0340?**, C-4150?, G-7620±300
311340	Westdahl [Pogromni]	54.52N	164.65W	1654	Stratovolcano?	1991-92, 1978, 1964, 1827-30, 1820, 1796, 1795
311350	**Fisher**	54.65N	164.43W	1112	Stratovolcano	1830, 1826-27, <1795, C 0400±50, **C-3170±75**, **C-7420±200**
311360	**Shishaldin**	54.76N	163.97W	2857	Stratovolcano	2014, 2004, 1995-99, 1993, 1986-87, 1978-79, 1975-76, 1967, 1963, 1955, 1953, 1951, 1946-48, 1932, 1927-29, 1925, 1922, 1901, 1898, 1883, 1842, 1838, 1824-30, C 0950?, C-7050?, C-7550?
311380	Roundtop	54.80N	163.59W	1871	Stratovolcano	**T-7600±500**
311390	Amak	55.42N	163.15W	488	Stratovolcano	1796, 1700-10, T-2550±500

Alaska Peninsula

VNUM	VOLCANO NAME	LAT	LONG	ELEV	TYPE	ERUPTION YEARS
312030	**Pavlof**	55.42N	161.89W	2519	Stratovolcano	2013-14, 2007, 1996-97, 1990, 1986-88, 1983, 1980-81, 1973-77, 1966, 1960?-63, 1958, 1950-54, 1936-48, 1929-31, 1922-24, 1917, 1914, 1906-11, 1901, 1894, 1892, 1886, 1880, 1866, 1845-46, 1838, 1825, 1817, 1790, @1762-86
312050	Dana	55.64N	161.21W	1354	Stratovolcano	**C-1890?**
312060	Kupreanof	56.01N	159.80W	1895	Stratovolcano	1987
312070	**Veniaminof**	56.17N	159.38W	2507	Stratovolcano	2013, 2008, 2002-06, 1993-95, 1987, 1983-84, 1956, 1944, 1939, 1930, 1892, 1874, 1838-39, **C-1750?**
312080	**Black Peak**	56.55N	158.79W	1032	Stratovolcano	**T-1900±150**
312090	**Aniakchak**	56.88N	158.17W	1341	Caldera	1931, C 1560±50, C 1550?, C 1390?, T 1220±150, T 1050?, T 0700±300, C 0460?, T 0200, C-0350?, **I-1645±10**, T-2550±500, **T-5250±2700**
312100	Yantarni	57.02N	157.19W	1345	Stratovolcano	**T-0800±500**
312110	Chiginagak	57.14N	156.99W	2221	Stratovolcano	1998, 1971
312130	Ugashik-Peulik	57.75N	156.37W	1474	Stratovolcano	1814, C 1050±100, C-5850?, C-6550?
312131	Ukinrek Maars	57.83N	156.51W	91	Maars	1977, C-0350?
312140	Martin	58.17N	155.36W	1863	Stratovolcano	@1953, @1951, C-0800±50, C-1750?
312150	Mageik	58.20N	155.25W	2165	Stratovolcano	C-0500±50, C-0550?, C-0650?, C-1650?, C-1950±100, C-4400±300, G-7380±150, G-8670±300
312160	Trident	58.24N	155.10W	1864	Stratovolcano	1974, 1967-68, 1966?, 1957-64, @1956, 1953-54, @1950, 1949, 1913

VNUM	VOLCANO NAME	LAT	LONG	ELEV	TYPE	ERUPTION YEARS (VEI≤3 <u>VEI=4</u> **VEI=5** **VEI=6-7**)
312170	Katmai...	58.28N	154.96W	2047	Stratovolcano	1912
312180	**Novarupta**	58.27N	155.16W	841	Caldera	**1912**
312190	Griggs	58.35N	155.09W	2317	Stratovolcano	G-1790±40
312200	Snowy Mountain	58.34N	154.68W	2162	Stratovolcanoes	G 1710±200
312250	Kaguyak	58.61N	154.03W	901	Lava domes	**C-3850?**, C-4060±150
312260	Fourpeaked	58.77N	153.67W	2105	Stratovolcano	2006

Alaska (southwestern)

313010	Augustine	59.36N	153.43W	1252	Lava domes	2005-06, <u>1986</u>, <u>1976-77</u>, 1971, 1963-64, 1935, 1908, <u>1883-84?</u>, 1812, T 1650±100, <u>G 1540±100</u>, G 1230±50, G 0930±150, G 0570±150, G 0340±40, G-0120±40, G-0310±100, T-0350±200, G-1820±300, G-2040±300, G-4150±100, G-5420±50
313020	Iliamna...	60.03N	153.09W	3053	Stratovolcano	1876, 1867, 1778-79, C 1650?, C-0450?, <u>C-2050?</u>, <u>C-5050?</u>
313030	Redoubt .	60.49N	152.74W	3108	Stratovolcano	2009, 1989-90, 1966-68, 1902, V 1650±75, V 1530±75, V 1510±75, V 1400±75, V 1360±75, V 1200±75, V 1160±75, V 1120±75, V 1030±75, V 0720±75, V 0640±75, V 0580±75, V 0520±75, V 0440±75, V 0290±75, V 0190±75, C 0110±50, V 0070±75, V-0090±75, V-0100±75, V-0160±75, T-0210±200, V-0220±75, V-0270±75, V-0350±75, V-0380±75, V-0390±75, V-0400±75, V-0420±75, V-0450±75, V-0470±75, V-0600±75, V-0610±100, V-0620±100, V-0670±100, V-0770±100, V-0790±100, V-0830±100, V-0870±100, V-1040±100, V-1050?, T-1080±150, V-1100±100, V-1150±100, V-1210±100, V-1310±100, V-1350±100, V-1510±100, V-1540±100, C-1550±150, V-1560±100, V-1690±100, V-1700±100, V-1810±150, V-1870±150, V-2040±150, V-2230±150, V-2350±150, V-2830±200, V-2860±200, V-2920±200, V-3030±200, V-3150±200, V-3180±200, V-3600±150, V-3650±150, V-3690±150, V-3730±150, V-3890±150, V-4010±150, V-4360±150, V-4510±150, V-4810±150, V-4900±150, V-4960±150, V-5240±150, V-5580±150, V-5820±150, V-5910±150, V-5940±150, V-5950±150, V-6050±150, V-6150±150, V-6190±150, V-6280±150, V-7210±200, V-7270±200, V-7380±200, V-7430±200, V-7800±200, V-9310±300
313040	Spurr	61.30N	152.25W	3374	Stratovolcano	<u>1992</u>, <u>1953</u>, T 1650±50, C-3250?, T-4050?, C-5110±100, C-6050?
313050	Hayes	61.64N	152.41W	3034	Stratovolcano	T 1200±300, **T-1550?**, T-1850?

Alaska (western)

314010	St. Paul Island	57.18N	170.30W	203	Shield	C-1280±40
314060	Imuruk Lake	65.60N	163.92W	610	Shields	C 0300?

Alaska (eastern)

315001	Buzzard Creek	64.07N	148.42W	830	Tuff rings	C-1050?
315020	Wrangell	62.00N	144.02W	4317	Shield	2002, 1999, 1969, 1911-12, 1902, 1899-1900, <u>C 0190±200</u>
315030	**Churchill**	61.38N	141.75W	5005	Stratovolcano	**G 0800±100**, **C 0060±200**
315040	Edgecumbe	57.05N	135.75W	970	Stratovolcanoes	C-2220±100, C-3810±75, C-7220±150

Canada

320060	Edziza....	57.72N	130.63W	2786	Stratovolcano	F 0950±6000, C 0610±150, T-0750±100, C-6520±200
320080	Hoodoo Mountain	56.78N	131.28W	1850	Subglacial	T-7050?
320090	Iskut-Unuk River Cones	56.58N	130.55W	1880	Pyroclastic cones	C 1800?, C 1590±50, C-0620±150, C-1830±300, C-3450±150, C-4700±300, C-6830±150
320100	Tseax River Cone	55.12N	128.90W	609	Pyroclastic cone	G 1690±150, C 1330±75
320140	Nazko	52.90N	123.73W	1230	Pyroclastic cones	C-5220=100
320150	Wells Gray-Clearwater	52.33N	120.57W	2015	Pyroclastic cones	D 1550?, C-5650?
320180	Meager...	50.63N	123.50W	2680	Complex	**G-0410±200**
320200	Garibaldi	49.85N	123.00W	2678	Stratovolcano	C-8060±500

USA (Washington)

321010	Baker	48.78N	121.81W	3285	Stratovolcanoes	1880, 1870, 1863, 1858-60, 1852-54, 1843, 1820?, G-4550?, G-7850?
321020	Glacier Peak	48.11N	121.11W	3213	Stratovolcano	T 1700±100, C 0900±50, <u>C 0200±50</u>, C-0850?, C-3150?, C-3550?
321030	Rainier...	46.85N	121.76W	4392	Stratovolcano	1894, G 1450±100, G 0910±500, G 0440±100, T-0150?, <u>G-0250±200</u>, T-0400±50, T-0500±50, G-0610±100, T-0650±50, T-0700±50, C-2550?, C-2750?, G-3650?, G-3850±200, G-4850?, G-5050?, G-5350?, G-5550?, G-7800±300, G-8050?
321040	Adams ...	46.21N	121.49W	3742	Stratovolcano	T 0950?, T 0200?, T-0300?, T-0400?, T-0550±1000, T >-1850, T-2650±300, T-2950±100, T-3250±300, T-3550?, T-3800±1950, T-4050±500, T-4550?, T-5150±500, T-7050±1000
321050	**St. Helens**	46.20N	122.18W	2549	Stratovolcano	2004-08, 1989-91, **1980-86**, 1857, 1853-54, 1850, 1847-48, 1842-45, 1835, 1831, **D 1800**, G 1610±40, D 1525±25, **D 1482**, **D 1480**, C 0780±300, G 0420?, G 0270?, T 0230?, G 0190?, G 0100?, G-0100?, G-0220?, T-0250?, G-0280?, **G-0530?**, G-0800?, C-0830±75, G-1010?, G-1100?, G-1180?, G-1610?, G-1680?, **T-1770±100**, <u>G-1860?</u>, T-2100±300, **G-2340?**

VNUM	VOLCANO NAME	LAT	LONG	ELEV	TYPE	ERUPTION YEARS (VEI≤3 VEI=4 **VEI=5 VEI=6-7**)
321060	West Crater	45.88N	122.08W	1329	Volcanic Field	C-5750?, C-6110?
321070	Indian Heaven	45.93N	121.82W	1806	Shields	C-6250±100

USA (Oregon)

VNUM	VOLCANO NAME	LAT	LONG	ELEV	TYPE	ERUPTION YEARS
322010	Hood	45.37N	121.70W	3426	Stratovolcano	1865-66, 1859, D 1781-1801?, G 0480±37, C-4940±150
322020	Jefferson	44.67N	121.80W	3199	Stratovolcano	V 0950?, V-4500±50
322030	Blue Lake Crater	44.41N	121.77W	1230	Maar	G 0680±200
322040	Sand Mountain Field	44.38N	121.93W	1664	Pyroclastic cones	G 0070±150, G-0800±300, G-0900±100, G-1740±300, G-2290±300
322060	Belknap	44.29N	121.84W	2095	Shields	G 0480?, G-0800±300, G-1030±300
322070	Three Sisters	44.17N	121.77W	3074	Complex	G 0440±150, G 0040±200, C-0050?, C-0350?, T-0800?, T-7350±2700
322090	Bachelor	43.98N	121.69W	2763	Stratovolcano	M-5800±750
322100	Davis Lake	43.57N	121.82W	2163	Volcanic Field	C-2790?
322110	Newberry	43.72N	121.23W	2434	Shield	G 0690±100, G 0490±100, H-1450?, H-4450?, G-4690±150, G-4770±75, G-4860±150, G-4960±100, G-5070±150, G-5260±150, G-9210±1200
322160	**Crater Lake**	42.93N	122.12W	2487	Caldera	G-2850?, T-5250?, T-5550?, **I-5680±150**, **G-5900±50**
322170	Diamond Craters	43.10N	118.75W	1435	Volcanic Field	G-5610±470
322190	Jordan Craters	43.15N	117.46W	1473	Volcanic Field	C >-1250

USA (California)

VNUM	VOLCANO NAME	LAT	LONG	ELEV	TYPE	ERUPTION YEARS
323010	Shasta	41.41N	122.19W	4317	Stratovolcano	1786, C 1250?, C 1200?, C 0850?, C 0150?, T 0050?, C-0150?, T-0550±500, C-0650±800, C-0850?, C-1150?, C-2050?, C-2550?, T-3050±1000, C-4050?, C-6050?, T-6650?, T-7250?, T-7350?, C-7420±300, C-7650±100, C-7750?, T-8050?
323020	**Medicine Lake**	41.61N	121.55W	2412	Shield	M 1080±25, C 0890±100, T 0830±25, M 0800?, M 0720?, M-0050?, C-0780±100, C-1080±50, C-2410±100
323080	Lassen Volcanic Center	40.49N	121.51W	3187	Stratovolcano	1914-17, D 1666?, G 0980±300, G 0880±300, G 0800±300
323110	Mono Lake Volcanic Field	38.00N	119.03W	2121	Pyroclastic cones	T 1790±75, T 1550±300, H 1150±200, T 0350±100
323120	Mono Craters	37.88N	119.00W	2796	Lava domes	D 1350±20, H 1000±200, G 0620±27, C 0490±100, C 0440±100, G 0320±200, G 0010±200, H-0700±800, H-3850±1160, H-6750±1740
323130	Inyo Craters	37.69N	119.02W	2629	Lava domes	G 1380±50, G 0290±50, H-4050?
323150	Mammoth Mountain	37.63N	119.03W	3369	Lava domes	G 1260±40, C-6960±500
323160	Ubehebe Craters	37.02N	117.45W	752	Maars	A-4050?
323170	Golden Trout Creek	36.36N	118.32W	2886	Volcanic Field	T-5550±2500
323200	Salton Buttes	33.20N	115.62W	-40	Lava domes	H-6450?

USA (Idaho)

VNUM	VOLCANO NAME	LAT	LONG	ELEV	TYPE	ERUPTION YEARS
324010	Black Butte Crater Lava Field	43.18N	114.35W	1478	Shield	C-8400±300
324020	Craters of the Moon	43.42N	113.50W	2005	Pyroclastic cones	C-0130±50, C-0260±25, M-0350?, C-1680±150, C-2560±100, C-4070±50, T-4250?, C-4600±100, C-5470±150, C-5890±150
324030	Wapi Lava Field	42.88N	113.22W	1604	Shield	T-0300?
324040	Hell's Half Acre	43.50N	112.45W	1631	Shield	C-3250±150

USA (Wyoming)

VNUM	VOLCANO NAME	LAT	LONG	ELEV	TYPE	ERUPTION YEARS
325010	Yellowstone	44.43N	110.67W	2805	Calderas	G-1350±200, T-3050?, T-6050?, G-7400±860

USA (Utah)

VNUM	VOLCANO NAME	LAT	LONG	ELEV	TYPE	ERUPTION YEARS
327050	Black Rock Desert	38.97N	112.50W	1800	Volcanic Field	C 1290±150

USA (Colorado)

VNUM	VOLCANO NAME	LAT	LONG	ELEV	TYPE	ERUPTION YEARS
328010	Dotsero	39.66N	107.04W	2230	Maar	C-2200±300

USA (Arizona)

VNUM	VOLCANO NAME	LAT	LONG	ELEV	TYPE	ERUPTION YEARS
329010	Uinkaret Field	36.38N	113.13W	1555	Volcanic Field	A 1100±75
329020	San Francisco VF [Sunset Crater	35.35N	111.68W	3850	Pyroclastic cones	M 1075±25

USA (New Mexico)

VNUM	VOLCANO NAME	LAT	LONG	ELEV	TYPE	ERUPTION YEARS
327110	Carrizozo	33.78N	105.93W	1731	Pyroclastic cones	E-3250±500
327120	Zuni-Bandera	34.80N	108.00W	2550	Volcanic Field	G-1170±300, G-8710±300

Pacific Ocean (northern)

VNUM	VOLCANO NAME	LAT	LONG	ELEV	TYPE	ERUPTION YEARS
331010	Endeavour Segment	47.95N	129.10W	-2050	Submarine	U-3490?, U-6930?

VNUM	VOLCANO NAME	LAT	LONG	ELEV	TYPE	ERUPTION YEARS (VEI≤3 VEI=4 **VEI=5** **VEI=6-7**)
331011	Cobb Segment	46.88N	129.33W	-2100	Submarine	U-1180?
331020	CoAxial Segment	46.52N	129.58W	-2400	Submarine	1993, 1986±5
331021	Axial Seamount	45.95N	130.00W	-1410	Submarine	2011, 1998, 1976±6, G 1650±117, G 1400±71, G 1300±91, G 1260±72, G 1230±76, G 1000±98, G 0800±107, G 0410±123
331030	Cleft Segment	44.83N	130.30W	-2140	Submarine	1986, <1982, U-0270?
331031	North Gorda Ridge Segment	42.67N	126.78W	-3000	Submarine	1996, U-3020?, U-4840?
331040	Escanaba Segment	40.98N	127.50W	-1700	Submarine	U-2260?

Hawaiian Islands

VNUM	VOLCANO NAME	LAT	LONG	ELEV	TYPE	ERUPTION YEARS
332000	Loihi	18.92N	155.27W	-975	Submarine	1996, K-0050±3000, K-5050±5000, K-7050±7000
332010	**Kilauea..**	19.42N	155.29W	1222	Shield	1983-2014, 1982, 1979-80, 1977, 1967-75, 1965, 1959-63, 1954-55, 1952, 1934, 1929-32, 1927, 1902-24, 1896-97, 1823-94, 1820?, 1790. 1750?, M 1700±25, T 1650±50, C 1610±50, C 1510±50, G 1500?, G 1490±16, C 1460±50, G 1410?-70?, A 1340±40, G 1140±75, C 1050±75, T 0900±50, G 0850±150, C 0680±75, G 0540±200, M <0450, T 0420±20, M 0150±300, G-0050±150, G-0200±150, G-0270±75, C-0410±100, G-0800?, C-1550?, C-1650?, G-2080?, T-2200±500, C-2850?, G-3300?, C-4650?
332020	**Mauna Loa**	19.48N	155.61W	4170	Shield	1984, 1975, 1949-50, 1942, 1940, 1935-36, 1933, 1926, 1919, 1914-16, 1907, 1903, 1899, 1896, 1892, 1887, 1879-81, 1871-77, 1868, 1865-66, 1859, 1855-56, 1851-52, 1849, 1843, 1832, 1750?, C 1730?, C 1685?, C 1680?, C 1650?, C 1640?, C 1540?, C 1510?, C 1500?, C 1470?, C 1440?, C 1390?, C 1370?, C 1360?, C 1310?, C 1190?, C 1170?, C 1130?, C 1070?, C 1040?, C 0940?, C 0830?, C 0810?, C 0680?, C 0630?, C 0600?, C 0550?, C 0480?, C 0450?, C 0350?, C 0300?, C 0200?, C 0150?, C 0100?, C 0050?, C-0030?, C-0060?, C-0080?, C-0200?, C-0300?, C-0400?, C-0500?, C-0600?, C-0950?, C-1300?, C-1650?, C-1700?, C-1750?, C-1800?, C-1900?, C-2000?, C-2050?, C-2150?, C-2250?, C-2350?, C-2750?, C-3250?, C-3350?, C-3750?, C-4250?, C-5350?, C-5650?, C-5850?, C-6250?, C-6550?, C-6650?, C-7150?, C-7350?, C-7550?, C-7850?, C-8050?
332030	Mauna Kea	19.82N	155.47W	4205	Shield	C-2460±100, C-2540±200, C-2750±200, C-3370±150, C-3680±200, C-5150±150
332040	Hualalai .	19.69N	155.87W	2523	Shield	1800-01, 1784±7, C 1650±50, C 1240±150, T 1150?, C 1050±100, C 0920±50, C 0770±200, C-0080±75, C-0350±75, C-0400±75, C-0440±50, C-0720±75, C-1080±200, C-1150±75, C-1650±200, C-2040±75, C-2440±75, C-2770±75, C-4410±100, C-6820±200, C-7540±200
332060	Haleakala	20.71N	156.25W	3055	Shield	A 1750?, C 1460±75, G 1420±100, G 1360±100, G 1350±150, G 1200±75, C 1080±40, C 1020±50, C 1010±40, C 0990±50, C 0980±50, C 0910±40, C 0790±50, C 0080±40, G-0290?, C-0390±40, C-0580±50, C-1140±30, G-1240?, G-1310?, C-1800±50, G-1850?, C-1900±50, C-1940±40, C-2120±50, C-2210±40, C-2260±40, G-2470±50, C-3070±40, C-4760±40, C-5860±40, C-6030±40, C-6220±50, C-6700±100, C-6760±40, G-7210?, C-7450±300, C-7570±75, G-7950±200
332090	Unnamed	23.58N	163.83W	-4000	Submarine	1955

Pacific Ocean (central)

VNUM	VOLCANO NAME	LAT	LONG	ELEV	TYPE	ERUPTION YEARS
333010	Teahitia..	17.57S	148.85W	-1400	Submarine	S 1982-85
333020	Rocard ...	17.64S	148.60W	-2100	Submarine	S 1971-72, S 1966
333030	Moua Pihaa	18.32S	148.67W	-160	Submarine	S 1969-70
333050	Adams Seamount	25.37S	129.27W	-39	Submarine	K-0050±1000, K-1050±1000, K-4050±2000, K-5050±1000
333060	Macdonald	28.98S	140.25W	-39	Submarine	1987-89, S 1986, S 1979-84, S 1977, S 1967, @1936, @1928

Pacific Ocean (eastern)

VNUM	VOLCANO NAME	LAT	LONG	ELEV	TYPE	ERUPTION YEARS
334020	Northern EPR-Segment RO2	16.55N	105.32W	-2700	Submarine	M-0050±1000
334021	Northern EPR-Segment RO3	15.83N	105.43W	-2300	Submarine	M-0050±1000
334040	Unnamed	10.73N	103.58W		Submarine	2003
334050	Unnamed	9.83N	104.30W	-2500	Submarine	2005-06, 1991-92, 1988±1, M 1950?, M <1875, U 1650±100, U 1600±150, U 1200±300, U 0950±2000, U 0850±200, U-0050±2000, U-1050±2000, U-2050±2000, U-3050±2000, U-4050±2000, U-5050±8000
334070	Galapagos Rift	0.79N	86.15W	-2430	Submarine	1996±6, 1972
334100	**Unnamed**	8.27S	107.95W	-2800	Submarine	1964?-69?
334120	Southern EPR-Segment K	17.44S	113.21W	-2566	Submarine	1990±2, M 1965?, M 1930?, M 1840?, M 1625?
334130	Southern EPR-Segment J	18.18S	113.35W	-2650	Submarine	M 1890?, M 1820?
334140	Southern EPR-Segment I	18.53S	113.42W	-2600	Submarine	M 1915±40, M 1860?, M 1620?

Mexico

VNUM	VOLCANO NAME	LAT	LONG	ELEV	TYPE	ERUPTION YEARS
341020	Bárcena..	19.30N	110.82W	332	Pyroclastic cones..	1952-53
341021	Socorro ..	18.78N	110.95W	1050	Shield	1993-94, 1951, C-3090±500
341030	**Ceboruco**	21.13N	104.51W	2280	Stratovolcano	1870-75, 1567, 1542, **C 0930±200**

VNUM	VOLCANO NAME	LAT	LONG	ELEV	TYPE	ERUPTION YEARS (VEI≤3 VEI=4 **VEI=5** VEI=6-7)

341040 **Colima** .....   *19.51N 103.62W 3850*   Stratovolcanoes.... 2013-14, 1997-2011, 1994, 1991, 1985-87, 1975-82, 1957-70, 1926±4-31?, **1913**, 1908-09, 1891-1906, <u>1889-90</u>, 1869-87, 1819, <u>1818</u>, 1806-09, 1804, 1794-95, 1780, 1769-71, 1743-44, 1711, 1690, <u>1622</u>, 1611-13, <u>1606</u>, 1590, <u>1585</u>, 1576, 1560, 1519-23, G 1110±200, C 0730±100, G 0540±150, G-0650±200, C-1140?, G-1170±200, C-1320?, C-1450±100, C-1890±75, G-1940±300, <u>C-2370±150</u>, G-2800±100, C-3030±50, G-3180±100, T-3270?, G-3350±300, G-3510±200, G-3600±200, G-4110±100, G-4430±300, C-4500±200, C-4960±200, G-5880±200, C-6320±200, C-7420±500, C-7690±500

341060 **Michoacán-Guanajuato**........... *19.85N 101.75W 3860*   Pyroclastic cones.. <u>1943-52</u>, <u>1759-74</u>, G-1140±865, C-1880±150, A-2050?, C-2750±200, G-4140±300, G-5940±335, C-6480±300, C-7350±300

341061 Zitácuaro-Valle de Bravo........... *19.40N 100.25W 3500*   Caldera................ K-3050±2000

341062 Jocotitlán .... *19.73N 99.76W 3900*   Stratovolcano .... C 1270±75, C-7740±75

341070 Toluca, Nevado de.................... *19.11N 99.76W 4680*   Stratovolcano .... C-1350?

341080 **Chichinautzin**........................ *19.08N 99.13W 3930*   Volcanic Field...... G 0400±100, G 0200±100, G-2238±1413, C-4250±75, C-5840±500, G-7340±1050, <u>G-7370±300</u>, C-7930±500

341090 **Popocatépetl**.......................... *19.02N 98.62W 5426*   Stratovolcanoes.... 1994-2014, 1947, 1942-43, 1933, 1919-27?, 1802-04, 1720, 1697, 1663-67, 1642, 1592-94, 1590, 1580, 1571, 1548, 1542, 1539-40, 1530, 1528, 1518-23?, 1512, 1504, 1488, 1354, 1345-47, <u>I 0823</u>, C 0250?, <u>C-0200±300</u>, C-1890±75, C-2370±75, **G-3700±300**, <u>C-5150?</u>, T-6250±500, <u>C-7150?</u>

341091 Malinche, La........................... *19.23N 98.03W 4461*   Stratovolcano .... C-1170±50, C-5580±300, C-5870±100, C-6120±100, C-6310±75, C-6710±200, C-6890±500

341093 Humeros, Los .... *19.68N 97.45W 3150*   Calderas .... G-4470?

341095 Naolinco Volcanic Field........... *19.67N 96.75W 2000*   Pyroclastic cones.. G-1200±100

341096 **Cofre de Perote** .... *19.49N 97.15W 4282*   Shields .... G 1150±100

341098 Cumbres, Las........................... *19.15N 97.27W 3940*   Stratovolcano .... C-3920±50

341100 Orizaba, Pico de...................... *19.03N 97.27W 5675*   Stratovolcano .... 1846, 1687, 1613, 1569-89, 1566, 1545-55?±10, C 1260±50, A 1175, C 0220±75, C 0140±50, C 0090±40, C 0040±40, C-0780±50, C-1500±75, C-2110±100, <u>C-2300±75</u>, C-2500±75, C-2780±75, C-4690±300, C-6220±75, **C-6710±150**, <u>C-7030±50</u>, C-7530±40

341110 San Martín .... *18.57N 95.20W 1650*   Shield.................. 1794-96, <u>1793</u>, 1664, C 0890±40, C 0480±50, T 0380±75, C 0120±200, T-0150±300, C-0750±40, C-1320±300, C-2130±50, C-3440±50

341120 **Chichón, El**............................ *17.36N 93.23W 1150*   Lava domes.......... **1982**, A 1850?, **G 1360±100**, <u>G 1190±150</u>, **G 0780±100**, G 0590±100, G 0480±200, G 0190±150, G-0020±50, G-0700±200, G-1340±150, **G-2030±100**, G-6510±75

341130 Tacaná ... . *15.13N 92.11W 4060*   Stratovolcano .... 1986, 1949-50, 1878, C 1030±40, <u>G 0070±100</u>, G-1080±150, G-4740±200, G-5720±200, C-5940±500, G-9450±150

Guatemala

342030 **Santa María**........................... *14.76N 91.55W 3772*   Stratovolcano 1922-2014, 1903-13, **1902**

342040 Almolonga............................... *14.82N 91.48W 3197*   Stratovolcano .... 1818, 1765, G 0800±50

342060 Atitlán.... .... *14.58N 91.19W 3535*   Stratovolcano .... 1853, 1843, 1837, 1833, 1826-28, @1717-21, 1663, 1579?-81, 1505?, 1469, C-1020±150

342080 Acatenango............................... *14.50N 90.88W 3976*   Stratovolcano .... 1972, 1924-27, A 1450±50, C 0090±100, C-0260±75, C-0370±200, C-2710±75

342090 **Fuego** .... *14.47N 90.88W 3763*   Stratovolcano .... 2002-14, 1999-2000, 1987, 1977-79, 1975, <u>1974</u>, 1973, 1971, 1966-67, 1962-63, 1957, 1955, 1953, 1949, 1947, 1944, <u>1932</u>, 1896, <u>1880</u>, 1860, <u>1857</u>, 1855-56, 1829, 1826, 1799, <u>1737</u>, 1732, 1730, <u>1717</u>, 1710, 1705-06, 1702, 1699, 1686, @1685, 1629-32, 1623, 1620, 1617, 1614, 1585-87, <u>1581-82</u>, 1551-52, 1542, @1541, 1531, 1524, C 0970±50, C 0900±75, C 0590±75, C-1580±75

342110 **Pacaya**... .... *14.38N 90.60W 2552*   Complex............... 2013-14, 2004-10, 1965-2002, 1961, 1885, 1846, 1805, 1775, 1699, 1693, 1690, 1687, 1678, 1674, 1671, 1668-69, 1664, 1655, 1651, 1623?, 1565, C 1360±75, C 1160±75, T 0880±500, C 0400±50

342120 Tecuamburro .... *14.16N 90.41W 1845*   Stratovolcano .... C-0960±75

El Salvador and Honduras

343020 Santa Ana .... *13.85N 89.63W 2381*   Stratovolcano .... 2005, 1904, 1884, 1879-80, 1874, 1734, 1722, 1576, 1570?, 1524, 1521

343030 Izalco..... .... *13.81N 89.63W 1950*   Stratovolcano .... 1966, 1939-57, 1937?-38?, 1933-34, 1930-31, 1924-28?, 1920-21, 1912-16, 1902-05, 1887-1900, 1878-85, 1872-73, 1863-69, 1856-60, 1854, 1850, 1844, 1842, 1838-40, 1836, 1825, 1817, 1805-07, 1802-03, 1798, 1793, 1783, 1772?, 1770

343050 **San Salvador** .... *13.73N 89.29W 1893*   Stratovolcano .... 1917, 1658-71, 1575, <u>A 1200?</u>, G 0640±30

343060 Ilopango .... *13.67N 89.05W 450*   Caldera................ 1879-80, **G 0450±30**

343100 **San Miguel**............................. *13.43N 88.27W 2130*   Stratovolcano .... 2013-14, 2002, 1997, 1995, 1985-86, 1976-77, 1970, 1967, 1966, 1964, 1954, 1939, 1929-31, 1919-25, 1890-91, 1884, 1882, 1867-68, 1862, 1857, 1855, 1844-48, 1819, 1787, 1769, 1762, 1699, 1510±5

343120 Conchagüita.............................. *13.23N 87.77W 505*   Stratovolcano 1892

VNUM	VOLCANO NAME	LAT	LONG	ELEV	TYPE	ERUPTION YEARS (VEI≤3 <u>VEI=4</u> **VEI=5** <u>**VEI=6-7**</u>)

Nicaragua

344010 Cosigüina.................................... *12.98N   87.57W   872*   Stratovolcano .... 1859, 1852, **1835**, 1709?, C 1500?

344020 **San Cristóbal**............................ *12.70N   87.00W   1745*   Stratovolcano .... 1999-2014, 1997, 1976-77, 1971, 1684-85, 1680, 1528±1

344040 **Telica** .... *12.60N   86.85W   1061*   Stratovolcanoes.... 2011, 2004-08, 1999-2002, 1994, 1987, 1981-82, 1975-78,
1969-71, 1965-66, 1962, 1951, 1948-49, 1946, 1943-44, 1937-40, 1934, 1927-29, 1907, 1791, 1765, 1685, 1613, <u>1529</u>, 1527?

344070 Negro, Cerro .... *12.51N   86.70W   728*   Pyroclastic cones.. 1999, 1995, 1992, 1971, 1968-69, 1960-63, 1957, 1954,
1947-50, 1929, 1923, 1919, 1914, 1899, 1867, 1850

344080 Pilas, Las .... *12.50N   86.69W   1088*   Complex............... 1954, 1952, @1528

344090 Momotombo .... *12.42N   86.54W   1297*   Stratovolcano .... 1905, 1902, 1886-87?, 1882, 1878, 1870, 1858-66, 1854,
1852, 1849, 1764, 1736, <u>1605-06</u>, 1578, 1524, C 1100±50, <u>C-0800±50</u>, C-2550±300

344091 **Apoyeque**................................. *12.24N   86.34W   518*   Pyroclastic shield . **T-0050±100**, <u>T-1050±1000</u>, **T-2550±1500**, **C-4160±30**

344092 Nejapa-Miraflores...................... *12.12N   86.32W   360*   Fissure vents .... T 1060±100, T-0550±500, T-3050±500, C-4390±100,
C-5230±200, C-5350±200, T-7300±3150, C-7430±300

344100 **Masaya..** .... *11.98N   86.16W   635*   Caldera................. 2008, 2005-06, 2003, 1996-2001, 1993-94, 1989, 1987,
1965-85, 1946-48, 1918-25, 1913, 1906, 1902-04, 1856-59, 1852-53, 1772, 1670, 1570, 1551, 1524-44?, **T 0150?**, **C-0170±100**, **T-4050?**

344120 **Concepción**............................. *11.54N   85.62W   1700*   Stratovolcano 2005-11, 1999, 1988, 1982-86, 1977-78, 1973-74, 1962-63,
1957, 1948-55, 1944-45, 1935, 1928-29, 1921-26, 1918-19, 1907-10, 1902, 1891, 1883-86, <u>C-0770±50</u>

Costa Rica

345020 Rincón de la Vieja .... *10.83N   85.32W   1916*   Complex............... 2014, 2011-12, 1998, 1995, 1991-92, 1983-87, 1969-70,
1966-67, 1922, 1912, 1853-63, @1849, @1844, 1765, G 0430±100, <u>G-1820±150</u>

345030 Miravalles................................. *10.75N   85.15W   2028*   Stratovolcano .... 1946, T-5050?

345033 **Arenal**... .... *10.46N   84.70W   1670*   Stratovolcano .... 1968-2010, 1922, T 1750±50, C 1440?, <u>C 1400?</u>, <u>T 1030?</u>,
T 1020?, <u>A 0750±50</u>, <u>T 0700?</u>, <u>G 0650±100</u>, <u>T 0550?</u>, <u>A 0400?</u>, <u>G-0170±200</u>, <u>T-0270?</u>, <u>G-0380±200</u>, T-0830±500, <u>G-1250±200</u>,
<u>T-1450?</u>, <u>T-1650?</u>, G-1770±100, <u>T-2250?</u>, <u>T-2800?</u>, C-3190±100, <u>T-3350?</u>, <u>T-3900?</u>, T-4450?, <u>G-5060±150</u>

345040 **Poás**....... .... *10.20N   84.23W   2708*   Stratovolcano 2008-14, 2006, 1996, 1987-94, 1972-81, 1967-70, 1963-65,
1948-61, 1941-46, 1932-34, 1929, 1925, 1914-16, 1910, 1898-1907, 1895, 1888-91, 1880, 1860, 1834, 1828, 1747, T 1280?,
T 0210?, G-0760±200, G-3950?, C-5590±100, C-7620±100, C-7920±75

345050 Barva .... *10.14N   84.10W   2906*   Complex............... <u>T-6050±2000</u>

345060 Irazú....... .... *9.98N   83.85W   3432*   Stratovolcano 1994, 1977, 1963-65, 1939-40, 1933, 1930, 1928, 1924,
1917-21, 1885-86, 1875±5, 1864, 1847, 1842, 1822-23, 1775?, 1726, 1723-24, G 1560±75, G 1110±100, G 0690±40, G 0430±500,
G-0640±500

345070 Turrialba .... *10.03N   83.77W   3340*   Stratovolcano .... 2010-13, 1864-66, 1855, 1853, C 1350?, G 0640±40,
<u>G 0040±50</u>, G-0830±150, T-1120±200, G-1420±300, G-7260±300

Panama

346010 Barú....... .... *8.81N   82.54W   3474*   Stratovolcano .... 1550±10, T 1340±75, G 1130±150, G 0710±30, G 0260±150,
G-1270±100, G-7420±75, G-9280±30

Colombia

351011 Romeral . ... *5.21N   75.36W   3858*   Stratovolcano .... <u>C-5950±500</u>

351012 Bravo, Cerro .... *5.09N   75.30W   4000*   Stratovolcano .... T 1720±150, C 1330±75, C 1050±75, C 0750±150,
C-0730±75, T-1050±200, C-1310±150, C-4280±150

351020 **Ruiz, Nevado del** .... *4.90N   75.32W   5321*   Stratovolcano 2012-13, 1984-91, 1916, 1845, 1831, 1828-29, 1805, 1623,
<u>1595</u>, 1570, <u>C 1350?</u>, C 0675±50, C 0350±300, <u>C-0200±100</u>, <u>T-0850?</u>, C-1245±150, T <-6660

351021 Santa Isabel.............................. *4.82N   75.37W   4950*   Shield.................. G-0850?, G-3550?, G-4800?, G-5500?

351030 Tolima, Nevado del .... *4.67N   75.33W   5200*   Stratovolcano .... 1943, 1825-26, 1822, G 0260±150, G-0200±200,
G-0610±200, **G-1990±200**, G-3500±300, G-5160±200, C-5310±100, <u>C-7800±300</u>

351040 Machín... .... *4.48N   75.39W   2650*   Stratovolcano .... G 1180±150, C 0750?, C-0650?, G-2100±200, G-2240±300,
C-2650?, G-3800±150

351050 **Huila, Nevado del**...................... *2.93N   76.03W   5364*   Stratovolcano .... 2007-12, 1555±5

351060 Puracé.... .... *2.32N   76.40W   4650*   Stratovolcanoes.... 1977, 1957, <1956, 1949, 1946-47, 1924-27, 1906, 1902?,
1899, 1885, 1881, 1878, 1869-70, 1860±9, 1847-52, 1840, 1835, 1827, 1816, C-0160±50

351070 Doña Juana .... *1.47N   76.92W   4150*   Stratovolcano .... <u>1897-1906</u>, <u>C-2550±150</u>

351080 **Galeras..** .... *1.22N   77.37W   4276*   Complex............... 2012-13, 2004-10, 2002, 2000, 1989-93, 1974-83, 1950,
1936, 1932, 1923-27, 1891, 1889, 1865-70, 1828-34, 1823, 1796-1801, 1754-56, 1670-1736, 1641-43, 1616, 1580, 1535,
G 0890±200, G-0490±100, G-1160±300, G-2580±500, G-3150±200, T-7050±1000

351090 Azufral... .... *1.08N   77.68W   4070*   Stratovolcano .... <u>C-0930?</u>, C-1650±150, C-1850?, C-2095±100

351100 Cumbal .. .. *0.95N   77.87W   4764*   Stratovolcano 1926, 1877

351110 Negro de Mayasquer, Cerro......... *0.83N   77.96W   4445*   Stratovolcano .... @1936

VNUM	VOLCANO NAME	LAT	LONG	ELEV	TYPE	ERUPTION YEARS (VEI≤3 <u>VEI=4</u> **VEI=5** <u>**VEI=6-7**</u>)

Ecuador

352001 Soche...... .... *0.55N* *77.58W* *3955* Stratovolcano .... **C-6650?**

352002 Chachimbiro................................ *0.47N* *78.29W* *4106* Stratovolcano .... C-3740?

352003 Cuicocha .... *0.31N* *78.36W* *3246* Caldera................. C 0650?, C-0950?, **C-1150±150**, C-2550?

352004 Imbabura .... *0.26N* *78.18W* *4609* Compound............ C-5550±500

352006 Cayambe .... *0.03N* *77.99W* *5790* Compound............ 1785-86, T 1700?, T 1590?, <u>T 1570?</u>, T 1440?, <u>T 1290?</u>, T 1270?, <u>T 1040?</u>, T 0880?, T 0260?, T 0200?, T 0170?, T 0010?, T-0180?, T-0230?, T-0260?, T-0460?, T-0510?, T-0560?, T-1300?, T-1650?, T-1800?

352010 **Reventador**...................................*0.08S* *77.66W* *3562* Stratovolcano ...... 2004-14, <u>2002-03</u>, 1976, 1972-74, 1960, 1958, 1955, 1944, 1936, 1929, 1926, 1912, 1898-1906, 1894, 1871, 1856, 1843-44, <1843, 1797, 1691, @1590, 1541

352011 Pululagua....................................*0.04N* *78.46W* *3356* Caldera................. C 0290?, <u>C-0450±150</u>, **C-0690±150**, C-4800?

352020 Guagua Pichincha......................... *0.17S* *78.60W* *4784* Stratovolcano ...... 1997-2002, 1993, 1990, 1985, 1981-82, 1881, 1868-69, 1830-31, <u>1660</u>, 1582-98, 1575, 1566, **G 0930±100**, G 0550±75, <u>G 0070±75</u>, G-1230±75, G-1860±100, G-2090±75, C-3500?, T-4850±1350, C-6200?, C-6300?, C-6400?, C-6650?, C-7000?

352021 **Atacazo**. ....*0.35S* *78.62W* *4463* Stratovolcano ...... **C-0320±16**, **C-2490±40**, <u>C-3490±100</u>, C <-6910

352022 **Chacana** ....*0.38S* *78.25W* *4643* Caldera................. 1773, 1760, T-0050?, C-1580±10, T-8050?

352030 Antisana. ....*0.48S* *78.14W* *5753* Stratovolcano ...... @1801-02, @<1748

352031 Aliso ....*0.53S* *78.00W* *4267* Stratovolcano ...... C-2450?

352040 Sumaco.. .... *0.54S* *77.63W* *3990* Stratovolcano ...... 1895±30

352050 **Cotopaxi** .... *0.68S* *78.44W* *5911* Stratovolcano ...... 1939-40, 1931, 1926, 1922, 1903-14, 1895, 1885-86, 1882-83, 1878-80, <u>1877</u>, 1866-76, 1850-63, 1844-45, 1803, <u>1768</u>, 1766, 1746-50, <u>1744</u>, 1740-43, 1738, 1698, <u>1534</u>, 1533, <u>1532</u>, T 1350?, T 1260±150, **C 1130±75**, T 0950?, <u>C 0770±75</u>, <u>C 0740±75</u>, T 0550±200, T 0370±200, <u>C 0180±100</u>, <u>T 0150?</u>, C 0070±150, T-0050?, <u>C-0230±200</u>, T-0400?, <u>T-1050?</u>, C-1510±150, **T-2050?**, C-2220±100, T-2250, **C-2640±200**, **T-3280±500**, **C-3880±75**, **C-4350±75**, **C-5820±75**, C-7690±75

352060 **Quilotoa** ....*0.85S* *78.90W* *3914* Caldera................. **G 1280?**

352071 Chimborazo .... *1.46S* *78.82W* *6310* Stratovolcano .... T 0550±150, G 0270±150, T-2500±1500, G-4130±150, G-5410±75, T-7500±2500

352080 **Tungurahua** .... *1.47S* *78.44W* *5023* Stratovolcano ...... 1999-2014, <u>1916-25</u>, <u>1886-88±1</u>, 1885, 1857, 1776, 1773, 1644-46?, 1640-41, 1557, G 1350±50, G 1250±50, C 1030±75, C 0800?, <u>C 0730±200</u>, C 0600?, C 0480±75, C 0350?, C 0200?, C 0100?, C-0050?, C-0100?, C-0270±100, C-0500?, **C-1010±100**, <u>C-7750?</u>

352090 **Sangay** .. ..*2.01S* *78.34W* *5286* Stratovolcano 1934-2013, 1728-1916, 1628

Galapagos Islands

353010 Fernandina.................................. *0.37S* *91.55W* *1476* Shield................. 2009, 2005, 1995, 1991, 1988, 1984, 1981, 1977-78, 1972-73, <u>1968</u>, 1961, 1958, 1937, 1926-27, 1888, 1846, 1825, 1819, @<1817, 1813-14, E >1550, E >1150, E 0950±500

353011 Ecuador.. .... *0.02S* *91.55W* *790* Shield................. E >1150

353020 Wolf...... ....*0.02N* *91.35W* *1710* Shield................. 1982, 1963, 1948, 1938, 1935, 1933, 1925-26, 1859, @1849, 1800, 1797, E >1450, E >0950, E 0150±800

353030 Darwin... .... *0.18S* *91.28W* *1330* Shield................. @1813, E 1150±300, E 0210±500

353040 Alcedo ... . *0.43S* *91.12W* *1130* Shield................. 1993, 1953±7

353050 Negra, Sierra................................*0.83S* *91.17W* *1124* Shield................. 2005, 1979-80, 1963, @1957, 1953-54, 1948-49, 1911, @1860, @1844, @1817, @1813, E 1350±500, E 1060±500, E 0370±1100, C-1250±100, E-8250±1600

353060 Azul, Cerro ....*0.92S* *91.41W* *1640* Shield................. 2008, 1998, 1979, 1959, 1951, 1948-49?, 1943, 1940, 1932, E >1850, E >1250, E-0550±1300, E-0950±900

353070 Pinta...... ....*0.58N* *90.75W* *780* Shield................. 1928

353080 Marchena....................................*0.33N* *90.47W* *343* Shield................. 1991

353090 Santiago. .... *0.22S* *90.77W* *920* Shield................. 1904-06, 1897, A 1759±75

Peru

354000 Quimsachata .... *14.20S* *71.33W* *3923* Lava dome .... C-4450?

354004 Andahua-Orcopampa................. *15.42S* *72.33W* *4713* Pyroclastic cones.. C 1490±40, C-0940±100, C-2110±50

354005 Huambo. .... *15.83S* *72.13W* *4550* Volcanic Field...... C-0700±50

354006 Sabancaya.................................. *15.78S* *71.85W* *5967* Stratovolcanoes.... 2014, 2003, 2000, 1990-98, 1988, 1986, 1784, 1750, T 1350±150, C-3490±40, T-6600?

354010 Misti, El. .... *16.29S* *71.41W* *5822* Stratovolcano .... 1985, 1787, 1784, 1677, 1454±16, G 1350±50, G 0760±100, G 0090±300, <u>G-0080±75</u>, G-0310±100, G-2230±200, G-3510±150, G-4020±200, G-5390±75, G-7190±150

354020 Ubinas ... . *16.36S* *70.90W* *5672* Stratovolcano .... 2013-14, 2006-09, 1969, 1956, 1951, 1937, 1906-07, 1869, 1867, 1865, 1862, 1830, 1784, 1677, 1667, 1662, 1550±50, **G 1080±75**, G-6850±150, G-8560±300

354030 **Huaynaputina** .... *16.61S* *70.85W* *4850* Stratovolcano .... <u>**1600**</u>, C-7750±200

354031 Ticsani................................... *16.76S* *70.60W* *5408* Lava domes .... T 1800±200

354050 Yucamane................................. *17.18S* *70.20W* *5550* Stratovolcanoes.... @1902, @1862, @1802, 1787, @1780, **C-1320?**

VNUM	VOLCANO NAME	LAT	LONG	ELEV	TYPE	ERUPTION YEARS (VEI≤3 VEI=4 **VEI=5** **VEI=6-7**)

Northern Chile, Bolivia and Argentina

355011 Taapaca.. .... *18.10S 69.50W 5860* Complex............... C-0320±50, C-1580±75, C-1860±100, C-2400±75, C-2950±75, C-4620±75, C-5490±50, C-7900±75

355012 Parinacota................................. *18.17S 69.15W 6348* Stratovolcano....... E 0290±300, A 0090±50, E-1100±500, E-4320±1200, C-5840±50

355020 Guallatiri .... *18.42S 69.09W 6071* Stratovolcano....... 1959-60, 1913, 1825±25

355030 Isluga..... .... *19.15S 68.83W 5550* Stratovolcano....... 1913, 1885, 1877-78, 1868-69, 1863

355040 Irruputuncu .... *20.73S 68.55W 5163* Stratovolcano....... 1995

355050 Olca-Paruma.............................. *20.93S 68.48W 5407* Stratovolcanoes.... 1865-67

355070 San Pedro................................. *21.88S 68.40W 6145* Stratovolcanoes.... 1960, 1938, 1911, 1901, 1891?, 1877?

355090 Putana.... .... *22.55S 67.85W 5890* Stratovolcano....... 1810±10

355100 **Láscar** ... . *23.37S 67.73W 5592* Stratovolcanoes.... 2013, 2005-07, 2000-02, 1994-96, <u>1993</u>, 1984-92, 1959-69, 1954, 1951-52, 1940, 1933, 1902, 1898-1900?, 1883-85, 1875, 1858, 1854, 1848, E-5150±1250, C-7250?

355109 Socompa .... *24.40S 68.25W 6031* Stratovolcano....... C-5250?

355110 Llullaillaco................................. *24.72S 68.53W 6739* Stratovolcano....... 1877, 1868, 1854

355130 Ojos del Salado, Nevados .... *27.11S 68.54W 6879* Stratovolcano....... T 0750±250

355210 **Blanco, Cerro** .... *26.79S 67.77W 4670* Caldera................. **G-2300±160**

Pacific Ocean (Chilean Islands)

356020 Robinson Crusoe .... *33.66S 78.85W 922* Shields 1835

Central Chile and Argentina

357010 Tupungatito .... *33.43S 69.80W 5660* Stratovolcano....... 1986-87, 1980, 1968, 1964, 1958-61, 1946-47, 1925, 1907, 1901, 1897, 1889-90, 1861, 1829

357020 San José. .... *33.79S 69.90W 6070* Stratovolcano....... 1959-60, 1895-97, 1889-90, 1881, 1822-38

357021 Maipo .... *34.16S 69.83W 5323* Caldera................. 1912, 1905, 1829, 1826

357030 Tinguiririca .... *34.81S 70.35W 4280* Stratovolcano....... 1917

357040 Planchón-Peteroa....................... *35.22S 70.57W 3977* Stratovolcanoes.... 2010-11, 1998, 1991, 1962, 1959-60, 1937-38, 1889-94?, 1878, 1860, 1837, 1835, <u>1762</u>, 1751, 1660, C 0900±100, C-5080±75

357041 Infiernillo................................. *35.14S 69.83W* Volcanic Field...... C-6890=40

357050 Descabezado Grande .... *35.58S 70.75W 3953* Stratovolcanoes.... 1932-33

357060 **Azul, Cerro [Quizapu]**............. *35.65S 70.76W 3788* Stratovolcano....... 1967, 1949, 1933-38, **1916-32**, 1914, 1912, 1906-07, 1846-53?

357063 Longaví, Nevado de................... *36.19S 71.16W 3242* Stratovolcano....... C-4890±75

357070 **Chillán, Nevados de** .... *36.86S 71.38W 3212* Stratovolcano....... 2003, 1973-86, 1946-47, 1934-35, 1927-29, 1914, 1906-07, 1898, 1893, 1891, 1877, 1872, 1860-65, 1749?-52, 1650±50, C-0320±75, C-1510±50, C-3660±500, C-6890±500

357072 Tromen .. .. *37.14S 70.03W 4114* Stratovolcanoes.... 1822, 1751

357080 Antuco... .... *37.41S 71.35W 2979* Stratovolcano....... 1869, 1863, 1861, 1852-53, 1845, 1828, 1820-21?, 1806?, 1752, 1750±10, C-7750?

357090 Copahue. .... *37.86S 71.18W 2953* Stratovolcano....... 2012-14, 2000, 1992-95, 1960-61, 1944, 1937, 1867?, 1750?, C-0250?, C-6820?

357091 Callaqui . ... *37.92S 71.45W 3164* Stratovolcano....... 1980, 1751

357100 Lonquimay................................. *38.38S 71.58W 2865* Stratovolcano....... 1988-90, 1933, 1887-90, 1853

357110 **Llaima**... .... *38.69S 71.73W 3125* Stratovolcano....... 2007-09, 2002-03, 1997-98, 1994-95, 1992, 1990, 1984, 1979, 1971-72, 1964, 1955-57, 1949, 1944-46, 1941-42, 1937-38, 1932-33, 1929-30, 1927, 1922, 1917, 1914, 1912, 1907-08, 1903, 1892-96, 1889, 1887, 1883, 1875-77, 1872, 1869, 1866, 1864, 1862, 1852-53, 1822, 1759, 1751-52, <u>1640</u>, C-5290±180, **C-6880±75**, C-7410±300

357111 Sollipulli .... *38.97S 71.52W 2282* Caldera................. C 1240±50, **C-0920±75**

357112 Caburgua-Huelemolle................. *39.25S 71.70W 1496* Pyroclastic cones.. T-5050±1000

357120 **Villarrica** .... *39.42S 71.93W 2847* Stratovolcano....... 1994-2013, 1991-92, 1983-85, 1980, 1977, 1971-72, 1963-64, 1961, 1958-59, 1956, 1947-49, 1938-39, 1935-36, 1933, 1927-29, 1922, 1920, 1915-18, 1906-10, 1904, 1897-98, 1893-94, 1883, 1879-80, 1874-77, 1869, 1864, 1859-60, 1853, 1837, 1832, 1822, 1815-18, 1806, 1790-1801, 1787, 1777-80, 1759, 1751, 1745, 1742, 1737, 1730, 1716, 1688?, 1657, 1647, 1594, 1562, 1558, C 0330?, C 0110?, <u>C-0670?</u>, C-1080?, **C-1810±200**, T-1980±150, C-2140?, C-2240?, T-2990±500, C-3730?, <u>C-6690?</u>, T-7520±900

357121 Quetrupillan.............................. *39.50S 71.70W 2360* Stratovolcano....... 1872

357122 Lanín .... *39.64S 71.50W 3776* Stratovolcano....... G 0560±150, G 0400±150, T 0090±300, G-0080±200, G-0220±200, G-0590±200, G-6340±200, G-9240±500

357123 Huanquihue Group .... *39.89S 71.58W 2189* Stratovolcanoes.... C 1750±100

357130 Mocho-Choshuenco..................... *39.93S 72.03W 2422* Stratovolcanoes.... 1937, 1864

357140 Carrán-Los Venados................... *40.35S 72.07W 1114* Pyroclastic cones.. 1979, <u>1955</u>, 1907-08

VNUM	VOLCANO NAME	LAT	LONG	ELEV	TYPE	ERUPTION YEARS (VEI≤3 VEI=4 **VEI=5 VEI=6-7**)
357150	**Puyehue-Cordón Caulle**	40.59S	72.12W	2236	Stratovolcano	**2011-12**, 1990, 1960, 1934, 1929, 1919-22, 1914, 1905?, 1893?, 1759?, G 1220±150, G 1140±100, **G 0860±75**, G 0500±100, T 0140±300, G 0110±200, G-0490±300, T-0990±500, G-1490±150, R-3250±2400, G-4230±200, R-4450±900, T-4460?, G-4690±200, **G-5080±150**
357153	**Antillanca Group**	40.77S	72.15W	1990	Stratovolcanoes	**G-0230±200, G-0960±150**
357160	Puntiagudo-Cordón Cenizos	40.97S	72.26W	2493	Stratovolcano	1850

Southern Chile and Argentina

VNUM	VOLCANO NAME	LAT	LONG	ELEV	TYPE	ERUPTION YEARS
358010	Osorno	41.10S	72.49W	2652	Stratovolcano	1869, 1855, 1851, 1837, 1834-35, 1790-91, 1765±14, 1719, 1644, 1640, 1575, C 1310±75, C 1220±100, C 0910±100, C 0420±100, C-0210±75, C-1710±75
358012	Cayutué-La Viguería	41.25S	72.27W	506	Pyroclastic cones..	G-0190±190?
358020	**Calbuco**.	41.33S	72.61W	2003	Stratovolcano	1972, 1961, 1945, 1932, 1929, 1917, 1911-12, 1909, 1906-07, 1894-95?, 1893-94, 1792?, C 1600±75, C 1380±50, G 0710±60, C 0520±200, C 0220±75, T 0160±135, C 0040±75, C-0100±100, C-0330±200, C-1920±50, C-4300±150, G-5030±180, T-5820±880, T-6300±1035, **T-6760±825**, G-7550±45, T-7930±275, T-7990±290, T-8100±1300, T-8210±290, T-8320±250, **G-8460±155**
358022	Yate	41.76S	72.40W	2187	Stratovolcano	G 1090±60
358023	Hornopirén................................	41.87S	72.43W	1572	Stratovolcano	G 0340±200, G-3720±175
358024	Apagado	41.88S	72.58W	1210	Pyroclastic cone ...	G-0590±175
358030	Huequi	42.38S	72.58W	1318	Lava domes..........	1920?, 1906-07, 1900, 1896, 1893, 1890
358040	**Minchinmávida**....................... G-8400±150	42.79S	72.44W	2404	Stratovolcano	1834-35, 1742, G 1550±100, G 0700±100, G-5500±150,
358041	**Chaitén** . ... G-7750±200	42.83S	72.65W	1122	Caldera................	2008-11, G 1640±18, **G-3100±220**, T-6650±1300,
358049	Yanteles.	43.50S	72.81W	2049	Stratovolcanoes	T-6650?, C-7240±150
358050	Corcovado	43.19S	72.79W	1826	Stratovolcano	C-4920±100, C-6030±100, T-6640±770
358052	Melimoyu	44.08S	72.88W	2400	Stratovolcano	C 0200±75, C-0820±100
358054	Mentolat	44.70S	73.08W	1660	Stratovolcano	1710±5, C-5010±50
358056	Maca......	45.10S	73.17W	2960	Stratovolcano	G 1560±110, C 0410±50
358057	**Hudson, Cerro**.........................	45.90S	72.97W	1905	Stratovolcano	2011, **1991**, 1971, 1891, C 1740±150, C 0860±100, G 0390±150, C-0120±200, G-0790±75, **G-1890?**, C <-2250, C-3890±500, C-4750?, C-4960±150, C-8010?
358059	Arenales.	47.20S	73.48W	3437	Stratovolcano	1979
358060	Lautaro	49.02S	73.55W	3607	Stratovolcano	1978-79, 1972, @1961, 1959-60, 1945, 1933, @1879, 1876
358061	Viedma, Volcan..........................	49.36S	73.28W	1500	Subglacial	1988
358062	Aguilera.	50.33S	73.75W	2546	Stratovolcano	**G-1250±150**, C-2610?
358063	Reclus....	50.96S	73.59W	1000	Pyroclastic cone ...	1908±1, 1879, 1869, T >-1830
358070	**Burney, Monte** .... T-7390±200, **G-7450±500**	52.33S	73.40W	1758	Stratovolcano	1910, G-0090±100, T-0800±500, **G-2320±100**, G-3740±10,
358080	Pali-Aike Volcanic Field	52.08S	69.70W	282	Pyroclastic cones..	A-5550±2500
358090	Fueguino	54.95S	70.25W	150	Lava domes..........	1820

West Indies

VNUM	VOLCANO NAME	LAT	LONG	ELEV	TYPE	ERUPTION YEARS
360010	Saba......	17.63N	63.23W	887	Stratovolcano	<1640
360020	Quill, The	17.48N	62.96W	601	Stratovolcano	C 0250±150, C-0550?, C-6140±200
360030	Liamuiga	17.37N	62.80W	1156	Stratovolcano	C 0160±200, C 0060±100, C-2010±150
360050	Soufrière Hills .... T-4050?, G-8050±2000	16.72N	62.18W	915	Stratovolcano	1995-2013, G 1550±50, G 1480±50, C 1180?, G-2460±70,
360060	Soufrière Guadeloupe................. G 1600±50, G 1370±150, G 1340±50, G 0370±75, C >-0580, G-0820±100, G-0980±200, G-1310±150, G-1810±150, G-3310±150, G-6450±150, G-7490±150	16.04N	61.66W	1467	Stratovolcano	1976-77, 1956, 1836-37, 1812, 1797-98, 1696, 1690,
360100	Trois Pitons, Morne	15.37N	61.33W	1387	Complex...............	C 0920±50, C 0790±50
360101	Watt, Morne..............................	15.31N	61.31W	1224	Stratovolcanoes	1997, 1880, C 0640±150, C-0950±300, C-1800±100
360110	Plat Pays, Morne.......................	15.26N	61.34W	940	Stratovolcano	C 1270±50, C 0390±40, C-0430?, C-4740?
360120	**Pelée**...... .... C 1370?, G 1340±50, C 1260±20, C 1190?, C 0910?, G 0890±100, C 0720?, C 0650?, C 0450?, G 0350±75, C 0300?, C 0220±75, C 0130?, C 0050?, G 0010±50, C-0200?, G-0300±100, C-0440?, G-0590±200, C-0600?, C-0620?, C-0730?, G-0890±50, G-1390±150, G-2100±200, C-2280?, C-2360?, C-2430?, G-2460±100, G-2660±200, C-3020?, G-3120±200, C-3250?, C-3290?, G-3430±75, T-3500±200, C-3820?, G-3930±100, G-4510±500, G-5500±200, C-5800?, G-6220±200, C-6450?, G-6610±150, U-7050±1000, G-7320±1730, U-7750±500, G-8210±200	14.81N	61.17W	1394	Stratovolcano	1929-32, 1902-05, 1851-52, 1792, <1635, C 1460±20,
360140	Qualibou	13.83N	61.05W	777	Caldera................	1766
360150	**Soufrière St. Vincent** .... C 1550±50, C 1480±150, C 1395±75, C 1325±75, C 0905±75, C-0530±75, C-0750±100, C-1600±75, C-2020±75, C-2135±50, C-2200±150, C-2310±100, C-2380±100	13.33N	61.18W	1220	Stratovolcano	1979, 1971-72, 1902-03, 1814, 1812, 1784, 1718, C 1640±50,

VNUM	VOLCANO NAME	LAT	LONG	ELEV	TYPE	ERUPTION YEARS (VEI≤3 <u>VEI=4</u> **VEI=5** <u>**VEI=6-7**</u>)
360160	Kick 'em Jenny	12.30N	61.64W	-185	Submarine	S 2001, S 1990, S 1988, S 1977, S 1977, 1974, S 1972, S 1966, S 1965, S 1953, S 1943, 1939, A 1000±200

Iceland (western)

370010	Snaefellsjökull	64.80N	23.78W	1448	Stratovolcano	C 0200±150, C-1000±500, C-2010±100, C-2270±300, C-2400±200, C-2970±300, T-4050?, T-4550±1500, T-6050±1000, G-8460±200
370030	Ljósufjöll	64.87N	22.23W	1063	Fissure vents	A 0960±10, C-0665±100, C-1750±150, T-2050?, T-7050?

Iceland (southwestern)

371020	Reykjanes	63.88N	22.50W	230	Crater rows...........	1926, 1879, 1830-31, 1783, 1583, 1422, 1340?, 1240, 1238, 1231, <u>1226-27?</u>, 1223, <u>1211</u>, 1210, <1179, T 0920?, C-0200?, C-0400±100, T-1800±300, T-3800±300, T-4000?, C-5040±100, T-8000?
371030	Krísuvik.	63.93N	22.10W	379	Crater rows...........	T 1340?, T 1325?, 1188, 1151, C 1075±75, T 0900?, C-0190±75, C-1060±75, C-5290±150, T-6000?, T-8500?
371040	Brennisteinsfjöll	63.92N	21.83W	621	Crater rows...........	1341±1, T 1200?, 1000, C 0950?, C 0910±75, C 0875±50, C-1040±75, C-2660±75, T-9000?
371050	Hengill...	64.08N	21.32W	803	Crater rows...........	G 0150±75, C-0080±75, G-1730±50, G-3250?, G-3750?, T-5000?, T-5550±500, T-7100?, T-7300?, T-7550?, T-8200?, G-8250?, G-8350?
371060	Grímsnes	64.03N	20.87W	214	Crater rows...........	T-3500?, T-3650?, T-3750?, T-3900?, T-4000?, T-4050?, C-4270±150, T-4450?, T-4500?, G-6250?, G-7750?
371070	Prestahnukur............................	64.60N	20.58W	1400	Subglacial	T-3350?, G-6950?, C-7550±500
371080	**Langjökull [Hveravellir]**	64.75N	19.98W	1360	Subglacial	T 0950±50, G-2050?, T-2550?, T-3550?, T-5850?, T-8600?

Iceland (southern)

372010	Vestmannaeyjar.........................	63.43N	20.28W	279	Submarine............	1973, 1963-67, 1896, 1637-38, U-3950±300, G-4270±200, T-4550±500, T-6050?, T-7550?, <u>T-8050?</u>
372020	Eyjafjallajökull	63.63N	19.62W	1666	Stratovolcano	<u>2010</u>, 1821-23, 1612, T 0920?, T 0550?
372030	**Katla**	63.63N	19.05W	1512	Subglacial	<u>1918</u>, <u>1860</u>, 1823, **1755-56**, **1721**, <u>1660-61</u>, **1625**, <u>1612</u>, <u>1580</u>, T 1550?, <u>T 1500?</u>, T 1450±50, <u>1440</u>, <u>1416</u>, <u>1357±3</u>, 1311, **1262**, **1245**, <u>T 1210?</u>, 1177±2, T 1150±50, T 0960?, T 0950?, <u>I 0934±2-40?</u>, <u>0920</u>, T 0820?, T 0780?, T 0680?, T 0610?, T 0590?, T 0540?, T 0500?, T 0400?, T 0290?, C 0270±12, T 0260?, T 0200?, T 0130?, T 0030?, T-0080?, T-0250?, T-0370?, T-0430?, T-0530?, T-0550?, T-0560?, T-0600?, T-0650?, T-0700?, T-0740?, T-0780?, <u>G-0850±50</u>, T-0860?, T-0920?, T-0990?, T-1160?, T-1190?, G-1220±12, T-1280?, T-1290?, <u>G-1440±40</u>, T-1540?, T-1640?, T-1670?, T-1700?, T-1850?, T-1910?, <u>G-1920?</u>, T-1950?, T-2000?, T-2020?, T-2050?, T-2110?, T-2160?, T-2190?, T-2220?, T-2250?, T-2420?, T-2480?, T-2540?, T-2680?, T-2850?, T-2920?, T-3180?, T-3280?, T-3370?, T-3390?, T-3480?, T-3510?, T-3640?, T-3670?, T-3720?, T-3790?, T-3810?, T-3930?, T-4060?, T-4210?, T-4240?, T-4280?, T-4370?, T-4430?, T-4610?, T-4660?, T-4750?, T-4810?, T-4880?, T-5020?, T-5040?, T-5070?, T-5180?, T-5230?, T-5360?, T-5460?, T-5470?, T-5550?, T-5560?, T-5630?, T-5710?, T-5720?, T-5730?, T-5850?, T-5890?, T-5960?, T-6050?, T-6170?, T-6200?, T-6230?, T-6380?
372050	Torfajökull...............................	63.92N	19.17W	1259	Stratovolcano	1477, T 1170?, T 0870?, T 0150±100, T-1150±100, T-1550±500, T-4550±500, T-4850?, T-5050?, T-6050?
372070	**Hekla**.....	63.98N	19.70W	1491	Stratovolcano	2000, 1991, 1980-81, 1970, <u>1947-48</u>, 1913, 1878, <u>1845-46</u>, <u>1766-68</u>, 1725, <u>1693</u>, 1636-37, <u>1597</u>, 1554, <u>1510</u>, T 1440?, 1389-90, 1341, <u>1300-01</u>, 1222, 1206, <u>1158</u>, **1104**, T 1050±500, T 0800±50, T 0750?, T 0650±500, T 0550±1500, T 0350±1500, T 0250±2500, T-0150±2500, T-0250±500, T-0650±2500, T-0750±500, T-0850±1500, **G-1100±50**, T-1150±1500, T-1250±1500, T-1350±2500, <u>T-1550?</u>, T-1650±2500, T-1750±500, T-1850±2500, **G-2310±20**, T-2450±1500, T-2750±2500, T-2950±500, T-3350±2500, T-3450±1500, T-3750±1500, T-3950±500, T-4050±500, **G-4110±100**, T-4150±2500, T-4250±500, T-4350±1500, T-4550±500, T-4650±500, <u>T-4700?</u>, T-4750±2500, T-4950±2500, T-5050?, **G-5150?**, T-5850±2500

Iceland (northeastern)

373010	**Grímsvötn**..............................	64.42N	17.33W	1725	Caldera................	<u>2011</u>, 2004, 1998, 1996, 1983, 1954, 1938, 1933-34, 1922, @1919, 1910, <u>1902-04</u>, @1897, 1891-92, 1887-89, 1883, <u>1873</u>, 1867, 1854, 1838, 1823, 1816, <u>1783-85</u>, 1774, @1768, 1753, 1730, 1725, 1716, 1706, 1697, 1684-85, @1681, 1665, 1659, 1638, I 1632?, 1629, I 1622?, 1619, I 1610?, @1603, 1598, I 1530±10, I 1521?, I 1510±10, I 1509?, I 1500?, I 1490±10, I 1471?, I 1470±10, I 1469?, I 1450±10, I 1430±10, I 1390±10, I 1370±10, I 1369?, 1354?, I 1350?, 1341, 1332, I 1310±10, I 1290±10, I 1270±10, I 1190?, I 1150?, I 1090?, T 1060?, I 1010?, I 0960?, T 0910?, T-0050?, T-1950±100, T-3550±500, T-4550±500, **G-8230±50**
373030	**Bárdarbunga**...........................	64.63N	17.53W	2009	Stratovolcano	2014, 1910, @1902-03, @1872?, 1862-64, @1797, 1794, @1769, 1766, I 1750±10, I 1739, @1729, 1726, I 1720, 1717, 1716, 1712, 1707, 1706, @1702, I 1697, <u>1477</u>, T 1410?, I 1350±10, I 1290±10, I 1270±10, T 1250±50, I 1210±10, 1159?, T 1080?, T 0940?, T 0880?, <u>T 0870?</u>, T 0150±100, T-1200?, T-4200?, T-4400?, T-4550?, T-4600?, T-4800?, T-5000?, G-6650±50, T-7050±1000, T-7100?
373050	**Kverkfjöll**	64.65N	16.72W	1929	Stratovolcano	1968, 1959, 1929, 1729, 1655, T-5000±1000, T-7050?
373060	**Askja**.....	65.03N	16.75W	1516	Stratovolcano	1961, 1938, 1926, 1924?, 1921-23, 1919, **1875**, 1797?, T 1300?, T-1250±300, T-2050±500, **G-8910±200**
373070	Fremrinamur............................	65.43N	16.65W	939	Stratovolcano	T-1200?, T-2300?, T-4000?, T-4050±1050

VNUM	VOLCANO NAME	LAT	LONG	ELEV	TYPE	ERUPTION YEARS (VEI≤3 VEI=4 **VEI=5** VEI=6-7)
373080	**Krafla**	65.73N	16.78W	818	Caldera	1984, 1980-81, 1977, 1975, 1746, 1727-29, 1724, T 1300±200, T 0850?, T 0250±300, T-0050?, G-0300?, T-0500±300, T-0650?, C-3050?, T-4050?, T-5750?, T-6150?, T-6500?, T-6800?, T-6850?, T-6950?, T-7400±300, T-7850?, T-8500?
373090	Theistareykjarbunga	65.88N	16.83W	564	Shield	T-0900±100, T-6800?, T-9500?
373100	Tjörnes Fracture Zone	66.30N	17.10W		Submarine	1867-68

Iceland (southeastern)

| 374010 | **Öraefajökull** | 64.00N | 16.65W | 2119 | Stratovolcano | 1727-28, **1362** |

North of Iceland

| 375010 | Kolbeinsey Ridge | 66.67N | 18.50W | 5 | Submarine | @1755, 1372 |

Atlantic Ocean (Jan Mayen)

| 376010 | Jan Mayen | 71.08N | 8.17W | 2277 | Stratovolcano | 1985, 1973, 1970-72?, 1851±30, 1818, 1732, T 1350±100 |

Atlantic Ocean (northern)

| 381040 | Pico Fracture Zone | 38.75N | 38.08W | -4200 | Submarine | 1865 |

Azores

382001	Flores	39.46N	31.22W	914	Stratovolcanoes	C-0950±100, C-1200±100
382010	Fayal	38.60N	28.73W	1043	Stratovolcano	1957-58, 1672-73
382020	Pico	38.47N	28.40W	2351	Stratovolcano	1720, 1718, 1562-64?
382030	**San Jorge**	38.65N	28.08W	1053	Fissure vent	1907, 1902, 1808, 1800, 1757, 1580
382040	Graciosa	39.02N	27.97W	402	Stratovolcano	R-1950±1400
382050	Terceira	38.73N	27.32W	1023	Stratovolcanoes	1998-2000, 1867, 1761, T 1400±50, T 1200±300, C 0920±50, C 0820±40, C 0190±40, C 0070±40, C-0060?, C-0090±100, C-0670±300, C-0940±50, C-2530±40, C-6720±50
382070	Don Joao de Castro Bank	38.23N	26.63W	-13	Submarine	1720
382080	Sete Cidades	37.87N	25.78W	856	Stratovolcano	@1880, @1861, 1811, 1713, 1682, 1638, 1444?, C 1110±50, C 0950±100, C 0670±150, T 0380±300, C 0090±100, C-0750±300, T-2050±1000, T-3050?
382081	Picos Volcanic System	37.78N	25.67W	350	Pyroclastic cones	1652, C 0940±100, C 0850±150, C 0600±100, C-0510±200, C-0850±100, C-4040?
382090	**Agua de Pau**	37.77N	25.47W	947	Stratovolcano	1564, **1563**, C 0700±150, C 0160±150, C-1290?, C-1850±500, C-2210±150, **C-2990?**, C-4550±100, C-6750±200
382100	**Furnas**	37.77N	25.32W	805	Stratovolcano	**1630**, 1441±2, C 1430±100, C 1170±100, C 0840±100, **C 0080±100**, C-0360±150, T-1670±1460, C <-4570
382110	Monaco Bank	37.60N	25.88W	-197	Submarine	1911, 1907
382120	Madeira	32.73N	16.97W	1862	Shield	G-4500±50

Canary Islands

383010	La Palma	28.57N	17.83W	2426	Stratovolcanoes	1971, 1949, 1712, 1677-78, 1646, 1585, 1480±10, C 0900±100, C-0360±50, C-1320±100, K-4050±3000, C-4900±50, K-6050±1500
383020	Hierro	27.73N	18.03W	1500	Shield	2011-12, C-0550±75, C-0950±150, C-4790?
383030	**Tenerife**	28.27N	16.64W	3715	Stratovolcano	1909, 1798, 1704-06, 1492, G 1060±100, G 0800±150, T 0700?, G 0240±150, T 0190?, G 0090±75, C 0040?, G 0030±150, G-0080±40, C-0520?, G-0580±200, G-0670±200, T-1050?, T-1150?, T-1400?, T-1650?, T-1700?, G-1980±200, T-2250?, T-2300?, T-2650?, T-2850?, T-3050?, T-3450?, G-3540±150, T-3750?, G-3960±300, G-4200±100, T-4650?, T-5250?, T-5550±1500, T-5750?, G-6200±75, T-6550?, T-6850?, G-7260±200, T-7550?
383040	Gran Canaria	28.00N	15.58W	1950	Fissure vents	G 0040±75, G 0010±75, G-0580±200, G-0590±200, G-0620±200, G-0920±200, G-1010±100, G-1180±50, G-1250±200, G-4630±75, G-4670±75
383060	**Lanzarote**	29.03N	13.63W	670	Fissure vents	1824, 1730-36, M 0700±50, M 0500±50

Cape Verde Islands

| 384010 | Fogo | 14.95N | 24.35W | 2829 | Stratovolcano | 1995, 1951, 1909, 1857, 1852, 1847, 1816, 1799, 1785, 1769, 1500-1761? |

Atlantic Ocean (central)

| 385052 | Unnamed | 32.96S | 5.22W | | Submarine | 2001-02 |

Atlantic Ocean (southern)

386010	Tristan da Cunha	37.09S	12.28W	2060	Shield	1961-62, T 1700±50
386011	Nightingale Island	37.42S	12.48W	365	Stratovolcano	2004
386020	Bouvet	54.42S	3.35E	780	Shield	M-0050?

VNUM	VOLCANO NAME	LAT	LONG	ELEV	TYPE	ERUPTION YEARS (VEI≤3 VEI=4 **VEI=5** VEI=6-7)

Antarctica and South Sandwich Islands

390010	Buckle Island	66.78S	163.25E	1239	Stratovolcano	1899, 1839
390013	Pleiades, The	72.67S	165.50E	3040	Stratovolcano	K-1050±14000
390015	Melbourne	74.35S	164.70E	2732	Stratovolcano	T 1892±30
390020	**Erebus**	77.53S	167.17E	3794	Stratovolcano	1972-2014, 1963, 1955, 1947, 1915, 1911-12, 1908, 1903, 1841, R 0950±4000, R-2050±3000, E-2950±300, E-4050±500, E-4550±500, R-7050±2000, R-8050±5000
390022	Berlin	76.05S	136.00W	3478	Shields	R-8350±5300
390027	Takahe	76.28S	112.08W	3460	Shield	I-5550?, R-6250±5400, I-7050?
390028	Hudson Mountains	74.33S	99.42W	749	Stratovolcanoes	I-0210±200
390030	Deception Island	63.00S	60.65W	602	Caldera	1969-70, 1967, 1912±5, T 1871±40, 1842, 1827±2, <1800, I 1641±3, C 1500?, C 1200?, C 0900?, C 0600?, C 0100?, C-0100?, C-0250?, C-0550?, C-0700?, C-0750?, C-0800?, C-1550?, C-2750?, C-3250?, C-6750?
390031	Penguin Island	62.10S	57.93W	180	Stratovolcano	L 1905?, 1850?, L 1683?
390070	Thule Islands	59.46S	27.19W	1075	Stratovolcanoes	1975±12
390080	Bristol Island	59.06S	26.59W	1100	Stratovolcano	1956, 1950, 1935-37, 1823
390081	Montagu Island	58.42S	26.33W	1370	Shield	2001-07
390090	Michael	57.79S	26.46W	990	Stratovolcano	2012, 2010, 2008, 2005-06, 2000-03, 1995-98, T 1900±10, 1819
390100	Candlemas Island	57.08S	26.67W	550	Stratovolcano	1911, 1823, I-1250?
390130	Zavodovski	56.30S	27.57W	551	Stratovolcano	1819
390140	Protector Shoal	56.01S	28.25W	-27	Submarine	1962

Index

Halogens, 167—168
Harker diagram. *See* Variation diagram
Harmonic infrasonic tremor, 1080—1082
Harmonic tremor, 1077—1079
Harzburgite, 37—38
Hawaii, 42—44
Hawaiian and Strombolian eruption styles, 485—490
    acoustic pressure, 496f
    conduit fluid dynamics, 496—498
    deposit geometry, 490—491
    explosive eruptions, 488f
    fountaining eruptions at Kīlauea Volcano, 488t
    high-speed video, 487f
    insights from infrasonic measurements, 494—496
    multiparametric studies of eruptive activity, 496b
    new directions, 498
    olivine-hosted melt inclusion volatile data, 494f
    pulsations, 489b
    pulsatory behavior, 489f
    pyroclast textures, 491
    style transitions, 491—492
    terminology, 485—486
    volatile role, 492—494
Hawaiian eruptions, 247, 1336—1337
Hawaiian volcano, 285f
Hazard(s)
    analysis, 436—438
    curves and maps, 904—906
    mitigation, 982—984
    modeling and mapping, 951—953
        Vesuvius hazard mapping case study, 954—955
    from volcanic debris avalanches, 681
    Vulcanian eruptions, 506—507
Heat
    capacity, 120—121
    flow, 750
    flux, 1144f
    hazards, 950—951
    pump, 1236—1237, 1241—1242, 1242f
Heavy Rare Earth Elements (HREE), 107
Hecates Tholus, 721
Helium (He), 785—787, 1127
Hellas highland volcanoes, 722—723. *See also* Large central volcanoes; Medusae Fossae Formation (MFF)
Helmholtz free energy, 150
Herculaneum, 1038
Hesperian, 720
Heterogeneous nucleation, 461
HF. *See* Hydrogen fluoride
HF earthquakes. *See* High-frequency earthquakes
HFA. *See* Hyogo Framework for Action
HFSE. *See* High field strength elements
High field strength elements (HFSE), 107b, 109
High nutrient low chlorophyll zones (HNLC zones), 885

High-frequency earthquakes (HF earthquakes), 1012, 1015, 1153
High-frequency events, 1077—1079
High-grade ignimbrites, 645
High-temperature fumarolic discharges, 1128
High-temperature systems, exploration and appraisal of, 821
Highland paterae, 722
HNLC zones. *See* High nutrient low chlorophyll zones
Hokusai and spell of Fuji, 1332—1334
Holocene record, 268
Homo habilis, 1315
Homo sapiens (*H. sapiens*), 1315
Homogeneous nucleation, 227
Hooke's law, 217
Hornblende, 204—205
Horst and Graben structure, 675—676
Hot spots, 245, 266, 749
    motion, 413—415
    volcanoes, 285, 444
"Hot Spring Museum", 1302
Hot-dry rock system, 1236
HREE. *See* Heavy Rare Earth Elements
Human health, gas and ash interactions, 1045—1046
Human impacts of eruptions, 1036
Hummocks, 675—676
Huygens probe, 772
Hyaloclastite, 363—364, 379—380, 415
Hybrid
    earthquakes, 1077—1079
    events, 1015
    vapor-cap systems, 812
Hydration "skin", 482—483
Hydraulic routing, 978—979
Hydrochloric acid. *See* Hydrogen chloride (HCl)
Hydrodynamic mingling, 478
Hydrodynamic suspension, 652—653
Hydrogen (H$_2$), 853, 1250
Hydrogen chloride (HCl), 853, 1054, 1127
    hazards, 990
Hydrogen fluoride (HF), 853, 1054, 1127
    hazards, 990
Hydrogen sulfide (H$_2$S), 853, 990, 1054, 1126, 1250, 1293
    hazards, 990
Hydrometeorologic monitoring, 1193
Hydrometeors, 912—913
Hydrosaline melt, 167—168
Hydrostatic pressure, 554, 556
Hydrothermal
    activity, 312
    alteration, 808—809
    circulation, 800—802, 804
    eruption, 538
    field, 1248
    processes, 417—418
    reservoirs, 1242—1243
    systems, 780, 853, 1236
        and roots, 1242—1243
Hydrothermal fluids, 866
    chemicals extraction, 1237

direct and multiple uses, 1239—1241
direct use of geothermal heat, 1238t—1239t
drilling for, 1244—1246
energy usage, 1237
    affecting factors, 1237—1239
heat pumps, 1241—1242
Líndal diagram, 1240f
power generation, 1239—1241
Hydrothermal gases. *See also* Volcanic gases
thermodynamics, 792—795
Hyogo Framework for Action (HFA), 1188
Hypabyssal intrusive complexes. *See* Shallow seafloor intrusions
Hyperacidic lakes, 853—854
Hyperconcentrated flows, 656, 972—974
    deposits, 662

I

I-DOAS. *See* Imaging-DOAS
IAB. *See* Island-arc basalt
IAVCEI. *See* International Association of Volcanology and Chemistry of the Earth's Interior
ICAO. *See* International Civil Aviation Organization
Ice cauldron, 379—380
Icefoots, 764—765
Icelandite, 101
Iconography, 1326—1327
Icy satellites, 764
IFOV. *See* Instantaneous Field of View
Igneous rocks, 94—96, 95t
Ignimbrites, 268, 306, 632—633
    associated with Phreatoplinian eruptions, 638—639
    eruptions, 522
    flow unit, 633
    high-grade, 645
    plateaus, 289, 294—295
    sheets, 633
        time-space sketches, 634f
    shields, 286—287
    welding and rheomorphism in, 642—645
    Zaragoza, 638f
IGS. *See* International GNSS Service
IHO. *See* International Hydrographic Organization
Imaging-DOAS (I-DOAS), 1132
Immiscible phase, 166—167
Imogolite, 1255
Impact basin, 702
Imperfections, 217—218
Impulse waves, 994
IMS. *See* International Monitoring System
Inclination, 765
Incompatible elements, 107b
Index of fractionation, 98b
Individual risk per annum (IRPA), 1225
Inflation, 331—333
    lava-rise pits, 333
    sheet lobes and lava rises, 332—333
    tumuli, 332

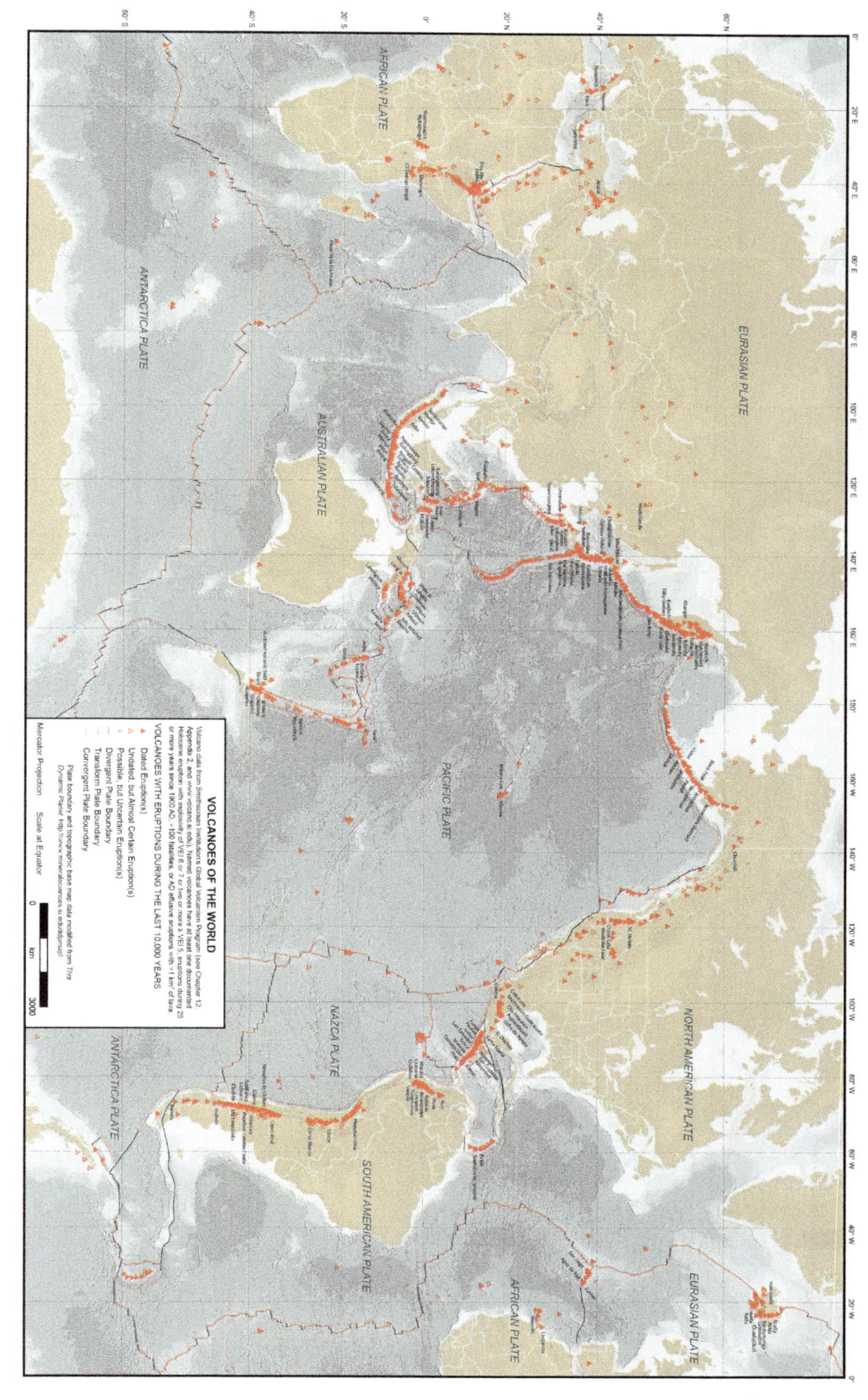

VOLCANOES OF THE WORLD

Volcano data from Smithsonian Institution's Global Volcanism Program (see Chapter 12, Appendix 2, and www.volcano.si.edu), named volcanoes have at least one documented Holocene eruption with explosivity of VEI 6 or 7 or two or more a VEI 5, eruptions during 25 or more years since 1900 AD, >100 fatalities, or AD effusive eruptions with >1 km³ of lava

VOLCANOES WITH ERUPTIONS DURING THE LAST 10,000 YEARS

▲ Dated Eruption(s)
▲ Undated Eruption(s)
▲ Possible, but Almost Certain Eruption(s)
▲ Possible, but Uncertain Eruption(s)

— Divergent Plate Boundary
— Transform Plate Boundary
— Convergent Plate Boundary

Plate boundary and topographic base map data modified from The
Dynamic Planet (http://www.mineralsciences.si.edu/tdpmap/)

Mercator Projection   Scale at Equator

0 ___ 3000 km